CW00729473

PRINCIPIA MATHEMATICA

BY

ALFRED NORTH WHITEHEAD, Sc.D., F.R.S.
Fellow and late Lecturer of Trinity College, Cambridge

AND

BERTRAND RUSSELL, M.A., F.R.S.
Lecturer and late Fellow of Trinity College, Cambridge

VOLUME II

1912

ISBN 1-60386-183-1

CONTENTS OF VOLUME II

ADDITIONAL ERRATA TO VOLUME I

p. 5, line 20, *delete* "π."
p. 34, line 20, *for* "yRx" *read* "xRy."
p. 36, line 7 and line 10, *for* "$Q \mid R$" *read* "$R \mid P$."
p. 44, line 17, *for* "$(p) \,.\, p \,.$ is false" *read* "$(p) \,.\, p$ is false."
p. 112, in $*2{\cdot}52$, *in place of* "$p \supset \sim q$" *read* "$\sim p \supset \sim q$."
p. 129, in $*5{\cdot}11$, *in place of reference to* "$*2{\cdot}51$" *read reference to* "$*2{\cdot}5$."
p. 129, in $*5{\cdot}12$, *in place of reference to* "$*2{\cdot}52$" *read reference to* "$*2{\cdot}51$."
p. 144, $*10{\cdot}23$ *should be* "$\vdash :. (x) \,.\, \phi x \supset p \,.\equiv: (\exists x) \,.\, \phi x \,.\supset.\, p$."
p. 157, line 11, *for* "$*10$" *read* "$*9$."
p. 184, last line of *Dem.* of $*14{\cdot}111$, *for second* "$x = c$" *read* "$x = b$."
p. 228, in $*23{\cdot}81$, *for* "$\dot{-} R \Subset \dot{-} S$" *read* "$\dot{-} S \Subset \dot{-} R$."
p. 242, in $*25{\cdot}37$, *for* "zRw" *read* "zSw."
p. 242, in $*25{\cdot}412$, *for* "R" *read* "S."
p. 253, 2nd and 4th lines of *Dem.* of $*31{\cdot}16$, *for* "$*21{\cdot}35$" *read* "$*23{\cdot}35$."
p. 259, in note to $*32{\cdot}35$, *for* "$*32{\cdot}2$" *read* "$*32{\cdot}3$."
p. 263, in $*33{\cdot}16$, 4th line of *Dem.*, *for* "$*20{\cdot}34$" *read* "$*22{\cdot}34$."
p. 265, in $*33{\cdot}26$, 2nd line of *Dem.*, *for* "$*21{\cdot}34$" *read* "$*23{\cdot}34$."
p. 275, in $*34{\cdot}6$, 4th line of *Dem.*, *for first* "S" *read* "R."
p. 289, 1st line, *for* "$= \beta \uparrow \gamma$" *read* "$= \alpha \uparrow \gamma$."
p. 322, in $*40{\cdot}18$, enunciation, *for* "$\equiv$" *read* "$=$".
p. 329, in $*40{\cdot}69$, *Dem.*, *for* "$\overrightarrow{P}$" *read* "$\overleftarrow{P}$" (*3 times*).
p. 387, in $*55{\cdot}224$, 1st line of *Dem.*, *for* "$\uparrow$" *read* "$\downarrow$" (*twice*).
p. 388, in $*55{\cdot}281$, *for third* "$=$" *read* "$\equiv .$".
p. 410, in $*60{\cdot}53$, last line of *Dem.*, *for* "γ" *read* "β."
p. 453, in $*71{\cdot}25$, *Dem.*, 1st line, *for* "$xRy \,.\, xRz$" *read* "$ySx \,.\, zSx$."
" " 2nd line, *for* "$xRy \,.\, ySu \,.\, xRz \,.\, zSv \,.\supset.\, y = z \,.\, ySu \,.\, zSv$" *read* "$uRy \,.\, ySx \,.\, vRz \,.\, zSx \,.\supset.\, y = z \,.\, uRy \,.\, vRz$."
" " 3rd line, *for* "$ySu \,.\, ySv$" *read* "$uRy \,.\, vRy$."
" " 6th line, *for* "$xRy \,.\, ySu$" *read* "$uRy \,.\, ySx$" *and for* "$xRz \,.\, zSv$" *read* "$vRz \,.\, zSx$."
" " 7th line, *for* "$x (R \mid S) u \,.\, x (R \mid S) v$" *read* "$u (R \mid S) x \,.\, v (R \mid S) x$."
p. 465, in $*72{\cdot}16$, *Dem.*, 1st line, *for last* "κ" *read* "x."
p. 483, in $*73{\cdot}44$, *Dem.*, 1st line, *for second* "y" *read* "x."
p. 485, in $*73{\cdot}511$, *for* "β" *read* "α."
p. 522, in $*81{\cdot}23$, enunciation and 2nd line of *Dem.*, *for* "$\overrightarrow{R}$" *read* "R."
p. 592, in $*91{\cdot}33$, *Dem.*, 1st line, *for* "P" *read* "R."
p. 614, in $*93{\cdot}36$, *Dem.*, *for* "R" *read* "P" *throughout*.
p. 628, in $*95{\cdot}21$, *Dem.*, line 6, *for* "Q'" *read* "T'."

ERRATA TO VOLUME II

p. 82, last line but one, *for* "Λ_α" *and* "Λ_β" *read* "$\Lambda \cap \alpha$" *and* "$\Lambda \cap \beta$."
p. 101, $*112{\cdot}23$, enunciation, the second time two dots occur, *read* one dot.
p. 573, $*205{\cdot}7$, enunciation, *for* "$\max_P$" *read* "$\overrightarrow{\max_P}$."

PREFATORY STATEMENT OF SYMBOLIC CONVENTIONS

THE purpose of the following observations is to bring together in one discussion various explanations which are required in applying the theory of types to cardinal arithmetic. It is convenient to collect these observations, since otherwise their dispersion throughout the several numbers of Part III makes it difficult to see what is their total effect. But although we have placed these observations at the beginning, they are to be read concurrently with the text of Part III, at least with so much of the text as consists of explanations of definitions. The earlier portion of what follows is merely a *résumé* of previous explanations; it is only in the later portions that the application to cardinal arithmetic is made.

I. *General Observations on Types.*

Three different kinds of typical ambiguity are involved in our propositions, concerning:

(1) the functional hierarchy,
(2) the propositional hierarchy,
(3) the extensional hierarchy.

The relevance of these must be separately considered.

We often speak as though the type represented by small Latin letters were not composed of functions. It is, however, compatible with all we have to say that it should be composed of functions. It is to be observed, further, that, given the number of individuals, there is nothing in our axioms to show how many predicative functions of individuals there are, *i.e.* their number is not a function of the number of individuals: we only know that their number $\geqslant 2^{\mathrm{Nc}\text{‘}\mathrm{Indiv}}$, where "Indiv" stands for the class of individuals.

In practice, we proceed along the extensional hierarchy after the early numbers of the book. If we have started from individuals, the result of this is to exclude functions wholly from our hierarchy; if we have started with functions of a given type, all functions of other types are excluded. Thus a fresh extensional hierarchy, wholly excluding every other, starts from each type of function. When we speak simply of "*the* extensional hierarchy," we mean the one which starts from individuals.

It is to be observed that when we have the assertion of a propositional function, say "$\vdash . \phi x$," the x must be of some definite type, *i.e.* we only assert that ϕx is true whatever x may be within some one type. Thus *e.g.* "$\vdash . x = x$" does not assert more than that this assertion holds for any x of a given type. It is true that symbolically the same assertion holds in other types, but other types cannot be included under one assertion-sign, because no variable can travel beyond its type.

The process of rendering the types of variables ambiguous is begun in ∗9, where we take the first step in regard to the *propositional* hierarchy. Before ∗9, our variables are *elementary propositions*. These are such as contain no apparent variables. Hence the only functions that occur are matrices, and these only occur through their values. The assumption involved in the transition from Section A to Section B (Part I) is that, given "$\vdash . fp$," where p is an elementary proposition, we may substitute for p "$\phi ! (x, y, z, \ldots)$," where ϕ is any matrix. Thus instead of "$\vdash . fp$," which contained one variable p of a given type, we have "$\vdash . f \{\phi ! (x, y, z, \ldots)\}$," which contains several variables of several types (any finite number of variables and types is possible). This assumption involves some rather difficult points. It is to be remembered that no *value* of ϕ contains ϕ as a constituent, and therefore ϕ is not a constituent of fp even if p is a value of ϕ. Thus we pass, above, from an assertion containing no function as a constituent to one containing one or more functions as constituents. The assertion "$\vdash . fp$" concerns *any* elementary proposition, whereas "$\vdash . f \{\phi ! (x, y, z, \ldots)\}$" concerns *any of a certain set* of elementary propositions, namely any of those that are values of ϕ. Different types of functions give different sorts of ways of picking out elementary propositions.

Having assumed or proved "$\vdash . fp$," where p is elementary and therefore involves no ambiguity of type, we thus assert

$$\vdash . f \{\phi ! (x, y, z, \ldots)\},$$

where the types of the arguments and the number of them are wholly arbitrary, except that they must belong to the functional hierarchy including individuals. (The assumption that propositions are incomplete symbols excludes the possibility that the arguments to ϕ are propositions.) The noteworthy point is that we thus obtain an assertion in which there may be any finite number of variables and the variables have unlimited typical ambiguity, from an assertion containing one variable of a perfectly definite type. All this is presupposed before we embark on the propositional hierarchy.

It should be observed that all elementary propositions are values of predicative functions of one individual, *i.e.* of $\phi ! \hat{x}$, where $\hat{x}$ is individual.

Thus we need not *assume* that elementary propositions form a type; we may replace p by "$\phi ! x$" in "$\vdash . fp$." In this way, propositions as variables wholly disappear.

In extending statements concerning elementary propositions so as formally to apply to first-order propositions, we have to assume afresh the primitive proposition ∗1·11 (∗1·1 is never used), *i.e.* given "$\vdash . \phi x$" and "$\vdash . \phi x \supset \psi x$," we have "$\vdash . \psi x$," which is practically ∗9·12. This was asserted in ∗1·11 for any case in which ϕx and ψx are elementary propositions. There was here already an ambiguity of type, owing to the fact that x need not be an individual, but might be a function of any order. *E.g.* we might use ∗1·11 to pass from

$$\text{“}\vdash . \phi ! a\text{” and “}\vdash . \phi ! a \supset \phi ! b\text{” to “}\vdash . \phi ! b\text{,”}$$

where ϕ replaces the x of ∗1·11, and $\hat{\phi} ! a$, $\hat{\phi} ! b$ replace $\phi \hat{x}$ and $\psi \hat{x}$. Thus ∗1·11, even before its extension in ∗9, already states a fresh primitive proposition for each fresh type of functions considered. The novelty in ∗9 is that we allow ϕ and ψ to contain one apparent variable. This may be of any functional type (including Indiv); thus we get another set of symbolically identical primitive propositions. In passing, as indicated at the end of ∗9, to more than one apparent variable, we introduce a new batch of primitive propositions with each additional apparent variable.

Similar remarks apply to the other primitive propositions of ∗9.

What makes the above process legitimate is that nothing in the treatment of functions of order n *presupposes* functions of higher order. We can deal with each new type of functions as it arises, without having to take account of the fact that there are later types. From symbolic analogy we "see" that the process can be repeated indefinitely. This possibility rests upon two things:

(1) A fresh interpretation of our constants—$\vee$, $\sim$, !, (x)., $(\exists x)$.—at each fresh stage;

(2) A fresh assumption, symbolically unchanged, of the primitive propositions which we found sufficient at an earlier stage—the possibility of avoiding symbolic change being due to the fresh interpretation of our constants.

The above remarks apply to the axiom of reducibility as well as to our other primitive propositions. If, at any stage, we wish to deal with a class defined by a function of the 30,000th type, we shall have to repeat our arguments and assumptions 30,000 times. But there is still no necessity to speak of the hierarchy as a whole, or to suppose that statements can be made about "all types."

We come now to the extensional hierarchy. This starts from some one point in the functional hierarchy. We usually suppose it to start from individuals, but any other starting-point is equally legitimate. Whatever type of functions (including Indiv) we start from, all higher types of functions are excluded from the extensional hierarchy, and also all lower types (if any). Some complications arise here. Suppose we start from Indiv. Then if $\phi ! \hat{z}$ is any predicative function of individuals, $\hat{z}(\phi ! z) = \phi ! \hat{z}$. But identity between a function and a class does not have the usual properties of identity; in fact, though every function is identical with some class, and vice versa, the number of functions is likely to be greater than the number of classes. This is due to the fact that we may have $\hat{z}(\phi ! z) = \psi ! \hat{z} \,.\, \hat{z}(\phi ! z) = \chi ! \hat{z}$ without having $\psi ! \hat{z} = \chi ! \hat{z}$.

In the extensional hierarchy, we prove the extension from classes to classes of classes, and so on, without fresh primitive propositions (*20, *21). The primitive propositions involved are those concerning the *functional* hierarchy.

From all these various modes of extension we "see" that whatever *can be proved* for lower types, whether functional or extensional, can also be proved for higher types*. Hence we assume that it is unnecessary to know the types of our variables, though they must always be confined within some one definite type.

Now although everything that can be proved for lower types can be proved for higher types, the converse does not hold. In Vol. I. only two propositions occur which can be proved for higher but not for lower types. These are $\exists ! 2$ and $\exists ! 2_r$. These can be proved for any type except that of individuals. It is to be observed that we do *not* state that whatever is *true* for lower types is *true* for higher types, but only that whatever *can be proved* for lower types *can be proved* for higher types. If, for example, Nc'Indiv $= \nu$, then this proposition is false for any higher type; but this proposition, Nc'Indiv $= \nu$, is one which cannot be proved logically; in fact, it is only ascertainable by a census, not by logic. Thus among the propositions which can be proved by logic, there are some which can only be proved for higher types, but none which can only be proved for lower types.

The propositions which can be proved in some types but not in others all are or depend upon existence-theorems for cardinals. We can prove

$\exists ! 0$, $\exists ! 1$, universally,

$\exists ! 2$, except for Indiv,

$\exists ! 3$, $\exists ! 4$, except for Indiv, Cl'Indiv, Rl'Indiv; and so on.

Exactly similar remarks would apply to the functional hierarchy. In both cases, the possibility of proving these propositions depends upon the axiom of reducibility and the definition of identity. Suppose there is only one individual, x. Then $\hat{y} = x$, $\hat{y} \neq x$ are two different functions, which, by the

* But cf. next page for a more exact statement of this principle.

axiom of reducibility, are equivalent to two different predicative functions. Hence there are at least two predicative functions of x, and at least two classes ι'x, Λ_x. This argument fails both for classes and functions if either we deny the axiom of reducibility or we suppose that there may be two different individuals which agree in all their predicates, *i.e.* that the definition of identity is misleading.

The statement that what can be proved for lower types can be proved for higher types requires certain limitations, or rather, a more exact formulation. Taking Indiv as a primitive idea, put

$$\text{Kl} = \text{Cl}‘\text{Indiv} \quad \text{Df}, \qquad \text{Kl}^2 = \text{Cl}‘\text{Kl} \quad \text{Df}, \ \text{ etc.}$$

Then consider the proposition Nc‘Kl = Λ. We can prove

$$\text{Nc}‘\text{Kl} \cap t‘\text{Indiv} = \Lambda \,.\, \exists ! \,\text{Nc}‘\text{Kl} \cap t‘\text{Kl} \,.\, \exists ! \,\text{Nc}‘\text{Kl} \cap t‘\text{Kl}^2 \,.\, \text{etc.}$$

Thus Nc‘Kl = Λ can be proved in the lowest type in which it is significant, and disproved in any other. The difficulty, however, is avoided if Indiv is replaced by a variable α, and Kl by $\text{Cl}‘t_0‘\alpha$. Then we have

$$\text{Nc}‘\text{Cl}‘t_0‘\alpha \cap t‘\alpha = \Lambda,$$

and this holds whatever the type of α may be. Thus in order that our principle about lower and higher types may be true, it is necessary that any relation there may be between two types occurring in a proposition should be preserved; in other words, when one constant type is *defined* in terms of another (as Kl and Indiv), the definition must be restored before the type is varied, so that when one type is varied, so is the other. With this proviso, our principle about higher and lower types holds.

With the above proviso, the truth of our statement is manifest. For we have shown that the same primitive propositions, symbolically, which hold for the lowest type concerned in our reasoning, hold also for subsequent types; and therefore all our proofs can be repeated symbolically unchanged.

The importance of this lies in the fact that, when we have proved a proposition for the lowest significant type, we "see" that it holds in any other assigned significant type. Hence every proposition which is proved without the mention of any type is to be regarded as proved for the lowest significant type, and extended by analogy to any other significant type.

By exactly similar considerations we "see" that a proposition which can be proved for some type other than the lowest significant type must hold for any type in the direct descent from this. *E.g.* suppose we can prove a proposition (such as $\exists ! 2$) for the type Kl (where Kl = Cl‘Indiv); then merely writing Cl‘Indiv for Kl, we have a proposition which is proved concerning Indiv, namely $\exists ! 2 \cap t‘\text{Cl}‘\text{Indiv}$, and here, by what was said before, Indiv may be replaced by any higher type.

Thus given a typically ambiguous relation R, such that, if τ is a type, $R‘\tau$ is a type (Cl or Rl is such a relation), we "see" that, if we can prove

ϕ (R'Indiv), we can also prove $\phi(R'\tau)$, where τ is any type, and ϕ is composed of typically ambiguous symbols. Similarly if we can prove ϕ (Indiv, R'Indiv), we can prove $\phi(\tau, R'\tau)$, where τ is any type. But we cannot in general prove ϕ (Indiv, $R'\tau$) or ϕ (τ, R'Indiv), and these may be in fact untrue. *E.g.* we have $\exists$! Nc (Kl)'Indiv . $\sim\exists$! Nc (Kl)'Kl^2.

Thus more generally, when a proposition containing several ambiguities can be proved for the types R'Indiv, S'Indiv, ..., but not for lower types, it is to be regarded as a function of Indiv, and then it becomes true for *any* type; that is, given

$$\phi\,(R\text{'Indiv}, S\text{'Indiv}, \ldots),$$

we shall also have

$$\phi\,(R'\tau, S'\tau, \ldots),$$

where τ is any type. In this way, *all* demonstrable propositions are in the first instance about Indiv, and when so expressed remain true if any other type is substituted for Indiv.

When a proposition containing typically ambiguous symbols *can be proved* to be true in the lowest significant type, and we can "see" that symbolically the same proof holds in any other assigned type, we say that the proposition has "permanent truth." (We may also say, loosely, that it is "true in all types.") When a proposition containing typically ambiguous symbols *can be proved* to be false in the lowest significant type, and we can "see" that it is false in any other assigned type, we say that it has "permanent falsehood." Any other proposition containing typically ambiguous symbols is said to be "fluctuating," or to have "fluctuating truth-value," as opposed to "permanent truth-value," which belongs to propositions that have either permanent truth or permanent falsehood.

In what follows, ambiguities concerned with the propositional hierarchy will be ignored, since they never lead to fluctuating propositions. Thus disjunction and negation and their derivatives will not receive explicit typical determination, but only such typical determination as results from assigning the types of the other typically ambiguous symbols involved.

It is convenient to call the symbolic form of a propositional function simply a "*symbolic form.*" Thus, if a symbolic form contains symbols of ambiguous type it represents different propositional functions according as the types of its ambiguous symbols are differently adjusted. The adjustment is of course always limited by the necessity for the preservation of meaning. It is evident that the ideas of "permanent truth-value" and "fluctuating truth-value" apply in reality to symbolic forms and not to propositions or propositional functions. Ambiguity of type can only exist in the process of determination of meaning. When the meaning has been assigned to a symbolic form and a propositional function thereby obtained, all ambiguity of type has vanished.

To "assert a symbolic form" is to assert each of the propositional functions arising for the set of possible typical determinations which are somewhere enumerated. We have in fact enumerated a very limited number of types starting from that of individuals, and we "see" that this process can be indefinitely continued by analogy. The form is always asserted so far as the enumeration has arrived; and this is sufficient for all purposes, since it is essentially impossible to use a type which has not been arrived at by successive enumeration from the lower types.

The only difficulties which arise in Cardinal Arithmetic in connection with the ambiguities of type of the symbols are those which enter through the use of the symbol sm, or of the symbol Nc, which is $\overrightarrow{\text{sm}}$. For it may happen that a class in one type has no class similar to it in some lower type (cf. ∗102·72·73). All fallacious reasoning in cardinal or ordinal arithmetic in connection with types, apart from that due to the mere absence of meaning in symbols, is due to this fact—in other words to the fact that in some types $\exists ! \text{Nc}'\alpha$ is true, and in other types $\exists ! \text{Nc}'\alpha$ may not be true. The fallacy consists in neglecting this latter possibility of the failure of $\exists ! \text{Nc}'\alpha$ for a limited number of types, that is, in taking the "fluctuating" form $\exists ! \text{Nc}'\alpha$ as though it possessed a "permanent" truth-value.

A fluctuating form however often possesses what is here termed a "stable" truth-value, which is as important as the permanent truth-value of other forms. For example, anticipating our definitions of elementary arithmetic, consider $2 +_c 3 = 5$. There is no abstract logical proof that there are two individuals; so suppose 2 and 3 refer to classes of individuals, but 5 refers to classes of a high enough type, then with these determinations $2 +_c 3 = 5$ cannot be proved. But $2 +_c 3 = 5$ has a *stable* truth-value, since it can always be proved when all the types are high enough. In this case the fact that our empirical census of individuals (at least of the "relative" individuals of ordinary life) has outrun the capacity of logical proof, makes the fluctuation in the truth-value of the form to be entirely unimportant.

In order to make this idea precise, it is necessary to have a convention as to the order in which the types of symbols in a symbolic form are assigned. The rule we adopt is that the types of the *real variables* are to be first assigned, and then those of the *constant symbols*. The types of the apparent variables, if any, will then be completely determinate.

A symbolic form has a *stable* truth-value if, after any assignment of types to the real variables, types can be assigned to the constant symbols so that the truth-value of the proposition thus obtained is the same as the truth-value of any proposition obtained by modifying it by the assignment of higher types to some or all of the constant symbols. This truth-value is the *stable* truth-value.

II. *Formal Numbers.*

The conventions, which we shall give below as to the assignment of types, practically restrict our interpretation of fluctuating symbolic forms to types in which the forms possess their stable truth-value. The assumption that these truth-values are stable never enters into the reasoning. But we judge a truth-value to be stable when any method of raising the types of the constant symbols by one step leaves it unaltered.

In practice the fluctuation of truth-values only enters into our consideration through a limited number of symbols called "formal numbers."

Formal numbers may be "constant" or "functional."

A *constant* formal number is any constant symbol for which there is a constant α such that, in whatever type the constant symbol is determined, it is, in that type, identical with $\text{Nc}'\alpha$. In other words if σ be a constant symbol, then σ is a formal number provided that "truth" is the permanent truth-value of $\sigma = \text{Nc}'\alpha$, for some constant α.

The *functional* formal numbers are defined by enumeration; they are

$$\text{Nc}'\alpha,\quad \Sigma\text{Nc}'\kappa,\quad \Pi\text{Nc}'\kappa,\quad \text{sm}''\mu,\quad \mu +_c \nu,\quad \mu -_c \nu,\quad \mu \times_c \nu,\quad \mu^\nu,$$

where in each formal number the symbols α, κ, μ, ν occurring in it are called the arguments of the functional form even when they are complex symbols. The argument of $\text{Nc}'(\alpha+\beta)$ is $\alpha+\beta$, and those of $\mu +_c (\nu +_c \varpi)$ are μ and $\nu +_c \varpi$, and those of $1 +_c 2$ are 1 and 2.

Thus among the constant formal numbers are

$$0,\ 1,\ 2,\ \ldots,\ \aleph_0,\ 1 +_c 2,\ 2 \times_c \aleph_0,\ 2^2.$$

The references which support this statement are

*101·11·21·32 . *123·36 . *110·42 . *113·23 . *116·23.

Among the functional formal numbers are

$$\text{Nc}'(\alpha+\beta),\quad \mu +_c (\nu +_c \varpi),\quad (\mu +_c \nu) \times_c \varpi,\quad (\mu +_c \nu)^\varpi.$$

It will be observed that *e.g.* $1 +_c 2$ is both a constant and a functional formal number, so that the two classes are not mutually exclusive. In fact they possess an indefinite number of members in common.

All the formal numbers, with the exception of $\text{sm}''\mu$ and $\mu -_c \nu$, are members of NC without any hypothesis [cf. *100·41·01·52.*110·42.*112·101. *113·23 . *114·1 . *116·23, note to *119·12, and *120·411].

A functional formal number consists of two parts, namely, its argument or arguments, and the constant "form." An argument of a functional formal number may be a complex symbol, and may be constant or variable. Thus $\mu +_c \nu$ is an argument of $(\mu +_c \nu) +_c \rho$, and of $(\mu +_c \nu) \times_c 1$ and of

$(\mu +_c \nu)^\rho$; also $2 +_c 3$ is an argument of $(2 +_c 3) \times_c 1$. The constant form is constituted by the other symbols which are constants. Two occurrences of functional formal numbers are only occurrences of the same formal number if the arguments and also the constant forms are identical in symbolism. Thus two occurrences of Nc'α are occurrences of the same formal number, even if they are determined to be in different types; but Nc'α and Nc'β are different formal numbers. Also μ^1 and $\mu \times_c 1$ are different formal numbers because their "forms" are different, though the arguments μ and 1 are the same and (in the same type) the entity denoted is the same. Thus the distinction between formal numbers depends on the symbolism and not on the entity denoted, and in considering them it is symbolic analogy and not denotation which is to be taken into account. For example two different occurrences of the same formal number will not denote the same entity, if in the two occurrences the ambiguity of type is determined differently.

The functional formal numbers are divided into three sets: (i) the *primary* set consisting of the forms Nc'α, ΣNc'κ, ΠNc'κ, (ii) the argumental set consisting only of sm''μ, (iii) the arithmetical set consisting of $\mu +_c \nu$, $\mu \times_c \nu$, μ^ν, and $\mu -_c \nu$.

A functional formal number has at most two arguments. But an argument of a functional formal number may itself be a functional formal number, and will accordingly possess either one or two arguments, which in their turn may be functional formal numbers, and so on. The whole set of arguments and of arguments of arguments, thus obtained, is called the set of *components* of the original formal number. Thus μ, ν, ρ, and $\mu +_c \nu$ are components of $(\mu +_c \nu) +_c \rho$; and μ, ν and sm''μ are components of $\nu +_c$ sm''μ; and μ, α and Nc'α are components of $\mu +_c$ Nc'α. The two arguments of $(\mu +_c \nu) +_c \rho$ are $\mu +_c \nu$ and ρ, and those of $\nu +_c$ sm''μ are ν and sm''μ, and those of $\mu +_c$ Nc'α are μ and Nc'α.

Addition, multiplication, exponentiation, and subtraction will be called the arithmetical operations; and in $\mu +_c \nu$, $\mu \times_c \nu$, μ^ν, $\mu -_c \nu$, μ and ν will each be said to be subjected to these respective operations. The *arithmetical components* of an arithmetical formal number (*i.e.* one belonging to the arithmetical set) consist of those of its components which do not appear in the capacity of components of a component which does not belong to the arithmetical set. Thus μ, ν, ρ, $\mu +_c \nu$ are arithmetical components of $(\mu +_c \nu) +_c \rho$; and ν and sm''μ are arithmetical components of $\nu +_c$ sm''μ, but μ is not one; and μ and Nc'α are arithmetical components of $\mu +_c$ Nc'α, but α is not one; and μ and sm''$(\nu +_c \rho)$ are arithmetical components of $\mu +_c$ sm''$(\nu +_c \rho)$, but $\nu +_c \rho$ and ν and ρ are components of sm''$(\nu +_c \rho)$ and are therefore not *arithmetical* components of $\mu +_c$ sm''$(\nu +_c \rho)$. Only arithmetical formal numbers possess arithmetical components.

A formal number of the arithmetical set having no components which are formal numbers of the argumental set is called a *pure* arithmetical formal number. For example $\mu +_c (\nu +_c \rho)$ and $\mu +_c \text{Nc}\text{‘}\alpha$ are pure, but $\mu +_c \text{sm}\text{“}(\nu +_c \rho)$ and $\mu +_c \text{sm}\text{“}\text{Nc}\text{‘}\alpha$ are not pure.

There are many types involved in the consideration of a formal number For example, in $\text{Nc}\text{‘}\alpha$ there is the type of $\text{Nc}\text{‘}\alpha$ and of α; in $\mu +_c \nu$ there is the type of $\mu +_c \nu$, the type of μ, and the type of ν; and so on for more complex formal numbers. The type of a formal number as a whole in any occurrence is called its *actual type.* This is the type of the entity which it then represents.

The other types involved in a formal number in any occurrence are called its subordinate types.

The actual types are not indicated in the symbolism for the various formal numbers as stated above. They can be indicated relatively to the type of the variable ξ by writing $\text{Nc}(\xi)\text{‘}\alpha$, $\text{sm}_\xi\text{“}\mu$, $(\mu +_c \nu)_\xi$, $(\mu \times_c \nu)_\xi$, $(\mu^\nu)_\xi$, $(\mu -_c \nu)_\xi$, by the notation of ∗65. Even when the actual type of a complex formal number, such as $\mu +_c (\nu +_c \varpi)$, is settled—so for instance that we have $\{\mu +_c (\nu +_c \varpi)\}_\xi$—the meaning of the symbol is not completely determined, for the type of $\nu +_c \varpi$ remains ambiguous. It follows, however, from

∗100·511 . ∗110·23 . ∗113·26 . ∗119·61·62,

that the subordinate types make no difference to the value of a formal number, so long as the components are not null.

We can therefore make a formal number definite as soon as its actual type is definite by securing that its components are not null. This is done by the convention IIT (below) combined with the definitions

∗110·03·04 . ∗113·04·05 . ∗116·03·04.

When the subordinate types are adjusted in accordance with these definitions and conventions, they will be said to be *normally adjusted.*

But in order to state this convention II T we require a definition of what is here called the *adequacy* of the actual type of a formal number. The general idea of adequacy is simple enough, namely that, given the subordinate types of σ, the actual type of σ should be high enough to enable us logically to prove $\exists ! \sigma$ when such a proof is possible for types which are not too low. For example, all types except the lowest for which it has meaning are adequate for the constant formal number 2. It is rather difficult however to state the meaning of adequacy with precision in a manner adapted to all formal numbers. Fortunately the definition of the lowest type which corresponds to this general idea of adequacy is not important for our purposes. It will be sufficient to define as adequate some types which certainly do have the property in question.

The method of definition which we adopt is to replace the formal number σ by another one σ' so related to σ that with the same actual type for both we can prove $\exists ! \sigma' . \supset . \exists ! \sigma$, whenever σ is not equal to Λ in all types. If σ be functional, we need only consider its argument, or its two arguments, and can dismiss from consideration the other components; then we replace these arguments by others so that the σ' has the required property. Thus:

(i) The actual types of $\mathrm{Nc}`\alpha$, $\Sigma\mathrm{Nc}`\kappa$, $\Pi\mathrm{Nc}`\kappa$, and $\mathrm{sm}``\mu$ are adequate when we can logically prove

$$\exists ! \mathrm{Nc}`t_0`\alpha,\ \exists ! \Sigma\mathrm{Nc}`t_0`\kappa,\ \exists ! \Pi\mathrm{Nc}`t_0`\kappa, \text{ and } \exists ! \mathrm{sm}``t_0`\mu;$$

(ii) The actual types of $\mu +_c \nu$, $\mu -_c \nu$, $\mu \times_c \nu$, and μ^ν are adequate when we can logically prove

$$\exists ! \mathrm{N_0c}`t_1`\mu +_c \mathrm{N_0c}`t_1`\nu,\quad \exists ! \mathrm{N_0c}`t_1`\mu -_c 0 \cap t_0`\nu,$$

$$\exists ! \mathrm{N_0c}`t_1`\mu \times_c \mathrm{N_0c}`t_1`\nu, \text{ and } \exists ! \mathrm{N_0c}`t_1`\mu^{\mathrm{N_0c}`t_1`\nu}.$$

It will be noticed that $t_0`\alpha$, $t_0`\kappa$, and $t_0`\mu$ are the greatest classes of the same type as $\alpha$, $\kappa$, and $\mu$ respectively, and that $\mathrm{N_0c}`t_1`\mu$ and $\mathrm{N_0c}`t_1`\nu$ are the greatest cardinal numbers of the same type as μ and ν respectively. These definitions hold even when any of α, κ, μ, ν are complex symbols.

The remaining formal numbers which are not functional must certainly be constant. The difficulty which arises here is that if σ be such a formal number and $\aleph_0$ occurs in its symbolism, we have no logical method of deciding as to the truth or falsehood of $\exists ! \aleph_0$ in any type. But we replace $\aleph_0$ by $\mathrm{N_0c}`t_1`\aleph_0$ which is the greatest existent cardinal of the same type as $\aleph_0$ in that occurrence. Thus:

(iii) If σ be a formal number which is not functional, an adequate actual type of σ is one for which we can logically prove $\exists ! \sigma'$, where σ' is derived from σ by replacing any occurrence of $\aleph_0$ in σ by $\mathrm{N_0c}`t_1`\aleph_0$. Accordingly if $\aleph_0$ does not occur in σ, an adequate type is any actual type for which we can logically prove $\exists ! \sigma$.

In the case of members of the primary and argumental groups we have substituted the V of the appropriate type in the place of each variable. When the actual type is adequate we have

$$(\alpha) . \exists ! \mathrm{Nc}`\alpha,\quad (\kappa) . \exists ! \Sigma\mathrm{Nc}`\kappa,\quad (\kappa) . \exists ! \Pi\mathrm{Nc}`\kappa,\quad (\mu) . \exists ! \mathrm{sm}``\mu.$$

In the case of members of the arithmetical group (except in the case of $\mu -_c \nu$), we have substituted for each argument the largest cardinal number which can be obtained in the type of that argument, namely the $\mathrm{N_0c}`\mathrm{V}$ for the V of the appropriate type. Accordingly we are sure (except in the case of $\mu -_c \nu$) that for all other values of the arguments which are existent cardinal numbers the formal number is not null.

It will be noticed that normal adjustment only concerns the subordinate types. For example *110·03 secures that in $\mathrm{Nc}`\alpha +_c \mu$ the actual type of

Nc'α is adequate, and ∗110·23 shows that any adequate actual type of Nc'α will do. But nothing is said about the actual type of Nc'$\alpha +_c \mu$. We make the following definition: When the subordinate types of a formal number are normally adjusted, and the actual type is adequate, the types of the formal number are said to be *arithmetically* adjusted.

We notice that for the primary set, the arithmetical adjustment of types means the same thing as the adequate adjustment of the actual type. Also if the arguments of a formal number of the arithmetical set are simple symbols, the two ideas come to the same thing.

In the case of variable formal numbers of the primary set, it follows from ∗117·22·32 that when their types are arithmetically adjusted they are not equal to Λ for any values of their variables.

Also in the case of those variable formal numbers which are of the pure arithmetical set (excluding $\mu -_c \nu$) it follows from ∗100·4·52·42 . ∗113·23 . ∗116·23 that, working from the ultimate components reached by successive analysis upwards, for all values of such ultimate components which are members of NC $-$ ι'Λ they can be reduced to the case of the formal numbers of the primary group; and that therefore they are not equal to Λ when their types are arithmetically adjusted. For example in $\mu +_c \{\nu +_c (\rho +_c \sigma)\}$, μ, ν, ρ, σ are these ultimate components; let them be existent cardinal numbers. Hence when the types are arithmetically adjusted, the actual type of $\rho +_c \sigma$ is adequate and $\rho +_c \sigma$ is an existent cardinal; we can therefore substitute $\mathrm{N_0c}$'α for it. By the same reasoning we can substitute $\mathrm{N_0c}$'β for $\nu +_c \mathrm{N_0c}$'α, and again $\mathrm{N_0c}$'γ for $\mu +_c \mathrm{N_0c}$'β.

A definite standard arithmetical adjustment of types for any formal number can always be found by making every use of sm, whether explicit or concealed in Nc or in some other symbol, to be homogeneous. Proofs which apply to any arithmetical adjustment of types start by dealing with this standard type, and then by the use of ∗104·21 . ∗106·21·211·212·213 the extension is made to the adjacent higher classical and relational types. We then "see" that by the analogy of symbolism this extension can always be formally proved at each stage, so that we are dealing with the stable truth-value. For some constant formal numbers a lower existential type can be found than that indicated by this method.

III. *Classification of Occurrences of Formal Numbers.*

A symbolic form of any of the kinds [cf. ∗117·01·04·05·06]

$$\mu > \nu, \quad \mu < \nu, \quad \mu \geqslant \nu, \quad \mu \leqslant \nu,$$

is called an *arithmetical inequality*.

These forms only arise when we are comparing cardinal numbers in respect to the relation of being "greater than" or "less than." It might seem natural to include equations among these arithmetical inequalities. Their use however, even as between cardinal numbers, is not so exclusively arithmetical, and it is convenient to consider them separately under another heading during our preliminary investigations.

In the arithmetical inequalities as above written, μ and ν, or any symbols replacing μ and ν, are called the *opposed sides* of the inequality, and either of μ or ν is called a *side* of the inequality.

Symbolic forms of the kinds $\sigma = \kappa$ and $\sigma \neq \kappa$, where either σ or κ is a formal number, will be called *equations* and *inequations* respectively; and σ and κ are called the *opposed sides* of the equation or inequation, and either of them is simply a *side* of the equation or inequation.

When we reach the exclusively arithmetical point of view, it will be convenient to put together equations, inequations and arithmetical inequalities as one sort of symbolic form. Their separation here is for the sake of investigations into the exceptions due to the failure of existence theorems in low types. It is unnecessary to consider arithmetical inequalities in this connection.

The ways in which a symbol σ can occur in a symbolic form are named as follows:

The occurrence of σ in sm"σ is called an *argumental* occurrence,

The occurrence of σ as an argument of an arithmetical formal number (which may be a component of another formal number) or as one side of an arithmetical inequality is called an *arithmetical* occurrence,

The occurrence of σ as one side of an equation is called an *equational* occurrence,

The occurrence of σ in "$\xi \epsilon \sigma$" is called an *attributive* occurrence,

Any other occurrence of σ is called a *logical* occurrence, so also is $\sigma = \Lambda$.

It is obvious that a pair of opposed sides of an equation or inequation must be of the same type. Furthermore, if σ be a formal number, and ∗20·18 is applied so as to give

$$\vdash :. \sigma = \kappa . \supset : f(\sigma) . \equiv . f(\kappa),$$

the equational occurrence of σ must be of the same type as its occurrence in $f(\sigma)$, otherwise the inference is fallacious. Accordingly substitution in arithmetical formulae can only be undertaken when the conventions as to the relations of ambiguous types secure this identity. This question is considered later in this prefatory statement, and the result appears in the text as ∗118·01.

At this point some examples will be useful; they will also be referred to subsequently in connection with the conventions limiting ambiguities of type.

***100·35.** $\vdash :. \exists ! \mathrm{Nc}'\alpha . \vee . \exists ! \mathrm{Nc}'\beta : \supset :$

$$\mathrm{Nc}'\alpha = \mathrm{Nc}'\beta . \equiv . \alpha \epsilon \mathrm{Nc}'\beta . \equiv . \beta \epsilon \mathrm{Nc}'\alpha . \equiv . \alpha \,\mathrm{sm}\, \beta$$

Here the formal numbers are $\mathrm{Nc}'\alpha$ and $\mathrm{Nc}'\beta$, each of which has three occurrences. The first occurrence of $\mathrm{Nc}'\alpha$ is logical, its second is equational, and its third is attributive.

***100·42** (in the demonstration).

$$\vdash : \mu, \nu \epsilon \mathrm{NC} . \exists ! \mu \cap \nu . \supset . (\exists \alpha, \beta) . \mu = \mathrm{Nc}'\alpha . \nu = \mathrm{Nc}'\beta . \mathrm{Nc}'\alpha = \mathrm{Nc}'\beta$$

Here $\mathrm{Nc}'\alpha$ and $\mathrm{Nc}'\beta$ are the only formal numbers, and all their occurrences are equational.

***100·44** (in the demonstration).

$$\vdash : \mu \epsilon \mathrm{NC} . \exists ! \mathrm{Nc}'\alpha . \alpha \epsilon \mu . \supset . (\exists \beta) . \mu = \mathrm{Nc}'\beta . \mathrm{Nc}'\alpha = \mathrm{Nc}'\beta$$

Here $\mathrm{Nc}'\alpha$ and $\mathrm{Nc}'\beta$ are the only formal numbers; the first occurrence of $\mathrm{Nc}'\alpha$ is logical, its second is equational; both the occurrences of $\mathrm{Nc}'\beta$ are equational.

***100·511.** $\vdash : \exists ! \mathrm{Nc}'\beta . \supset . \mathrm{sm}''\mathrm{Nc}'\beta = \mathrm{Nc}'\beta$

Here the formal numbers are $\mathrm{Nc}'\beta$ and $\mathrm{sm}''\mathrm{Nc}'\beta$. The first occurrence of $\mathrm{Nc}'\beta$ is logical, the second is argumental, the third is equational; the only occurrence of $\mathrm{sm}''\mathrm{Nc}'\beta$ is equational.

***100·521.** $\vdash : \mu \epsilon \mathrm{NC} . \exists ! \mathrm{sm}''\mu . \supset . \mathrm{sm}''\mathrm{sm}''\mu = \mu$

Here $\mathrm{sm}''\mu$ and $\mathrm{sm}''\mathrm{sm}''\mu$ are the only formal numbers; $\mathrm{sm}''\mu$ has two occurrences, the first logical, the second argumental; $\mathrm{sm}''\mathrm{sm}''\mu$ has one occurrence, which is equational.

***101·28** (in the demonstration).

$$\vdash : \gamma \epsilon \mathrm{sm}''1 . \equiv . (\exists \alpha) . \alpha \epsilon 1 . \gamma \,\mathrm{sm}\, \alpha$$

Here the formal numbers are 1 and $\mathrm{sm}''1$. The first occurrence of 1 is argumental, the second is attributive; the occurrence of $\mathrm{sm}''1$ is attributive.

***101·38.** $\vdash : \exists ! 2 . \supset . s'\mathrm{Cl}''2 = 0 \cup 1 \cup 2$

Here the formal numbers are 0, 1, and 2, and their occurrences are all logical.

***110·54.** $\vdash . (\mathrm{Nc}'\alpha +_{\mathrm{c}} \mathrm{Nc}'\beta) +_{\mathrm{c}} \mathrm{Nc}'\gamma = \mathrm{Nc}'(\alpha + \beta + \gamma)$

Here the formal numbers are

$$\mathrm{Nc}'\alpha,\ \mathrm{Nc}'\beta,\ \mathrm{Nc}'\gamma,\ \mathrm{Nc}'(\alpha+\beta+\gamma),\ \mathrm{Nc}'\alpha +_{\mathrm{c}} \mathrm{Nc}'\beta,\ (\mathrm{Nc}'\alpha +_{\mathrm{c}} \mathrm{Nc}'\beta) +_{\mathrm{c}} \mathrm{Nc}'\gamma.$$

The occurrence of $\mathrm{Nc}'(\alpha+\beta+\gamma)$ and that of $(\mathrm{Nc}'\alpha +_{\mathrm{c}} \mathrm{Nc}'\beta) +_{\mathrm{c}} \mathrm{Nc}'\gamma$ are both equational, and they must be of the same type since they are opposed sides of the same equation. The occurrences of the other formal numbers are as arithmetical components of a more complex arithmetical formal number and are therefore arithmetical.

∗116·63. $\vdash . \mu^{\nu \times_0 \varpi} = (\mu^\nu)^\varpi$

The formal numbers are $\nu \times_0 \varpi$, μ^ν, $\mu^{\nu \times_0 \varpi}$, and $(\mu^\nu)^\varpi$. Each formal number occurs once only. The occurrences of $\nu \times_0 \varpi$ and μ^ν are arithmetical, and those of the other two are equational.

∗117·108. $\vdash :. \text{Nc}`\alpha \geqslant \text{Nc}`\beta . \equiv : \text{Nc}`\alpha > \text{Nc}`\beta . \text{v} . \text{Nc}`\alpha = \text{Nc}`\beta$

The formal numbers are $\text{Nc}`\alpha$ and $\text{Nc}`\beta$, each with three occurrences. The first two occurrences of each formal number are arithmetical, the last occurrence of each is equational.

∗120·53 (in the demonstration).

$$\vdash : \beta = \gamma +_0 \delta . \exists ! \beta . \supset . \alpha^\beta = \alpha^\gamma \times_0 \alpha^\delta$$

Here the formal numbers are $\gamma +_0 \delta$, α^β, α^γ, α^δ, $\alpha^\gamma \times_0 \alpha^\delta$. Each formal number has one occurrence. Those of $\gamma +_0 \delta$, α^β and $\alpha^\gamma \times_0 \alpha^\delta$ are equational, and those of α^γ and α^δ are arithmetical.

∗120·53 (in the demonstration).

$$\vdash : \alpha^\beta = \alpha^\gamma . \beta = \gamma +_0 \delta . \exists ! \alpha^\beta . \supset . \alpha^\gamma = \alpha^\gamma \times_0 \alpha^\delta$$

Here the formal numbers are α^β, α^γ, α^δ, $\alpha^\gamma \times_0 \alpha^\delta$, $\gamma +_0 \delta$. The first occurrence of α^β is equational, its second occurrence is logical; the first two occurrences of α^γ are equational, its third occurrence is arithmetical; the only occurrence of α^δ is arithmetical; the only occurrences of $\alpha^\gamma \times_0 \alpha^\delta$ and of $\gamma +_c \delta$ are equational.

IV. *The Conventions* IT *and* IIT.

Two occurrences of a formal number with the same actual type are said to be *bound* to each other.

The choice of types for formal numbers, when they are not made definite in terms of variables by the notation of ∗65, is limited by the following conventions, which enable us to dispense largely with the elaboration produced by the definition of types.

IT. *All logical occurrences of the same formal number are in the same type; argumental occurrences are bound to logical and attributive occurrences; and, if there are no argumental occurrences, equational occurrences are bound to logical occurrences.*

This rule only applies, so far as meaning permits, to those types which remain ambiguous after the assignment of types to the real variables.

It will be noticed that if there are no argumental or logical occurrences of a formal number, IT does not in any way apply to the assignment of types to the occurrences in the form of that formal number.

The identification of types in argumental and attributive occurrences by IT is rendered necessary to secure the use of the equivalence

$$\gamma \epsilon \text{sm}``\sigma . \equiv . (\exists \alpha) . \alpha \epsilon \sigma . \gamma \text{ sm } \alpha,$$

where σ is a formal number. Without the convention, this application of ∗37·1 would be fallacious. The only one of our examples to which this part of the convention applies is ∗101·28 (demonstration), where it secures that the two occurrences of 1 are in the same type. It is relevant however to the symbolism in the demonstration of ∗100·521.

It will be found in practice that this convention relates the types of occurrences in the same way as would naturally be done by anyone who was not thinking of the convention at all. To see how the convention works, we will run through the examples which have already been given above.

In ∗100·35, IT directs the logical and equational occurrences of $\text{Nc}`\alpha$ to be in the same type, and similarly for $\text{Nc}`\beta$. Also "meaning" secures that the equational types of $\text{Nc}`\alpha$ and $\text{Nc}`\beta$ are the same. Thus these four occurrences are all in one type, which has no necessary relation to the types of the attributive occurrences of $\text{Nc}`\alpha$ and $\text{Nc}`\beta$. Thus, using the notation of ∗65·04 to secure typical definiteness, ∗100·35 is to mean

$$\vdash :. \exists ! \text{Nc}(\xi)`\alpha . \lor . \exists ! \text{Nc}(\xi)`\beta : \supset :$$
$$\text{Nc}(\xi)`\alpha = \text{Nc}(\xi)`\beta . \equiv . \alpha \epsilon \text{Nc}(\alpha)`\beta . \equiv . \beta \epsilon \text{Nc}(\beta)`\alpha . \equiv . \alpha \text{ sm } \beta.$$

The types of these attributive occurrences are settled by the necessity of "meaning."

In ∗100·42 (demonstration), since all the occurrences of formal numbers are equational, IT produces no limitation of types.

In ∗100·44 (demonstration), IT secures that the two occurrences of $\text{Nc}`\alpha$ are in the same type. Also we notice that the first occurrence of $\text{Nc}`\beta$ is really (cf. ∗65·04) $\text{Nc}(\alpha)`\beta$, since "$\alpha \epsilon \mu$" occurs, and thus "meaning" requires this relation of types, and the second occurrence of $\text{Nc}`\beta$ is in the type of the occurrences of $\text{Nc}`\alpha$.

In ∗100·511, IT directs that the logical and argumental occurrences are to have the same type. In ∗100·521, IT directs that the two occurrences of $\text{sm}``\mu$ are to have the same type. In ∗101·28 both occurrences of 1 are to be in the same type. In ∗101·38, IT directs that all the occurrences of 2 are to have the same type.

The convention IT in no way limits the types in ∗110·54, nor in ∗116·63, nor in ∗117·108.

In the first example from ∗120·53 (in the demonstration) convention IT has no application.

In the second example from ∗120·53 (in the demonstration) convention IT directs that the two occurrences of α^β shall be in the same type; and the necessity of "meaning" secures that the first occurrence of α^γ shall also be in this type. The same necessity secures that $\gamma +_c \delta$ shall be in the same type as β; and it also secures that in "$\alpha^\gamma = \alpha^\gamma \times_c \alpha^\delta$" the first occurrence of α^γ and that of $\alpha^\gamma \times_c \alpha^\delta$ shall have a common type, which is otherwise unfettered; also nothing has been decided as to the types of α^γ and α^δ in $\alpha^\gamma \times_c \alpha^\delta$.

We now come to conventions embodying the outcome of arithmetical ideas. The term "arithmetical" is here used to denote investigations in which the interest lies in the comparison of formal numbers in respect to equality or inequality, excluding the exceptional cases—whenever the cases are exceptional—due to the failure of existence in low types. The thoroughgoing arithmetical point of view, which we adopt later in the investigation on Ratio and Quantity and also in this volume in *117 and *126 and some earlier propositions, would sweep aside as uninteresting all investigation of the exact ways in which the failure of existence theorems is relevant to the truth of propositions, thus concentrating attention exclusively on stable truth-values. But the logical investigation has its own intrinsic interest among the principles of the subject. It is obvious however that it should be restrained to a consideration of the theorems of purely logical interest. In practice this extrusion of uninteresting cases of the failure of arithmetical theorems, even amid the logical investigations of the first part of this volume, is effected by securing that *all arithmetical occurrences of formal numbers have their actual types adequate.*

As far as formal numbers of the primary group, *i.e.* $\text{Nc}\text{‘}\alpha$, $\Sigma\text{Nc}\text{‘}\kappa$, $\Pi\text{Nc}\text{‘}\kappa$, are concerned, the arithmetical adjustment of types is secured formally in the symbolism by the definitions *110·03·04 for addition, and *113·04·05 for multiplication, and *116·03·04 for exponentiation, and *117·02·03 for arithmetical inequalities, and *119·02·03 for subtraction.

We save the symbolic elaboration which would arise from the extension of similar definitions to other formal numbers by the following convention:

IIT. *Whenever a formal number σ occurs, so that, if it were replaced by* $\text{Nc}\text{‘}\alpha$, *the actual type of* $\text{Nc}\text{‘}\alpha$ *would by definition have to be adequate, then the actual type of σ is also to be adequate.*

For example in $\mu +_c (\nu +_c \varpi)$, if $\nu +_c \varpi$ were replaced by $\text{Nc}\text{‘}\alpha$, then by *110·04 the actual type of $\text{Nc}\text{‘}\alpha$ is adequate. Hence by IIT the actual type of $\nu +_c \varpi$ is to be adequate: accordingly so long as ν and ϖ are simple variables and members of $\text{NC} - \iota\text{‘}\Lambda$, we can always assume $\exists ! (\nu +_c \varpi)$ for the type of the occurrence of $\nu +_c \varpi$ in $\mu +_c (\nu +_c \varpi)$.

It is essential to notice that so long as the argument of an argumental formal number, or the arguments of an arithmetical formal number, are adjusted arithmetically, the exact types chosen make no difference. This follows for argumental formal numbers from *102·862·87·88, for addition from

$*110\cdot25$, for multiplication from $*113\cdot26$, for exponentiation from $*116\cdot26$, for subtraction from $*119\cdot61\cdot62$. Thus (remembering also $*100\cdot511$) in any definite type a formal number has one definite meaning provided that any subordinate formal number which occurs in its symbolism is determined existentially. The convention IIT directs us always to take this definite meaning for any pure arithmetical formal number.

The convention does not determine completely the meaning of an arithmetical formal number which is not pure. For example, $\mu +_c (\nu +_c \rho)$ is a pure arithmetical formal number when μ, ν, ρ are determined in type; and convention IIT directs that the type of $(\nu +_c \rho)$ is to be adequate. But $\mu +_c \text{sm}``(\nu +_c \rho)$ is an arithmetical formal number which is not pure, and convention IIT directs that the type of the domain of sm is to be adequate, but does not affect the type of $\nu +_c \rho$. Thus it is easy to see that IIT secures the adequacy of the actual types of all *arithmetical* components of any arithmetical formal numbers which occur, but does not affect the actual type of a formal number which occurs as the argument of an argumental formal number. But in this case convention IT will bind the actual type of this occurrence of the argument to any logical or attributive occurrence of the same formal number. For example, if $\exists ! \, \nu +_c \rho$ and $\mu +_c \text{sm}``(\nu +_c \rho)$ occur in the same form, then these two occurrences of $\nu +_c \rho$ must have the same actual type. In practice argumental formal numbers are useful as components of arithmetical formal numbers for the very purpose of avoiding the automatic adjustment of types directed by IIT.

The meaning of IIT is best explained by examples. Among our previous examples we need only consider those in which arithmetical formal numbers occur.

In $*110\cdot54$ the convention or definitions direct us to determine the types of $\text{Nc}`\alpha$ and $\text{Nc}`\beta$ adequately when forming $\text{Nc}`\alpha +_c \text{Nc}`\beta$, also to determine $\text{Nc}`\alpha +_c \text{Nc}`\beta$ and $\text{Nc}`\gamma$ adequately when forming $(\text{Nc}`\alpha +_c \text{Nc}`\beta) +_c \text{Nc}`\gamma$. The convention does not apply to the types of $(\text{Nc}`\alpha +_c \text{Nc}`\beta) +_c \text{Nc}`\gamma$ and $\text{Nc}`(\alpha + \beta + \gamma)$. These types must be identical in order to secure meaning.

In $*116\cdot63$ the convention directs us to adjust the types of $\nu \times_c \varpi$ and μ^ν adequately; it does not affect the types of $\mu^{\nu \times_c \varpi}$ and $(\mu^\nu)^\varpi$, which must be identical to secure meaning. If we replace μ, ν, ϖ by formal numbers, by 2, $\aleph_0$, and 1 for example, we get "$\vdash . \, 2^{\aleph_0 \times_c 1} = (2^{\aleph_0})^1$." The convention now directs that 1 is to be determined adequately. It so happens that any type is adequate for it, since $\exists ! \, 1$ can be proved in any type. Then adequate types for $\aleph_0 \times_c 1$ and $2^{\aleph_0}$ are types for which we can prove $\exists ! \, (\text{N}_0\text{c}`t_1`\aleph_0) \times_c 1$ and $\exists ! \, 2^{\text{N}_0\text{c}`t_1`\aleph_0}$. Thus if τ is the type of $\aleph_0$ in both cases, an adequate type for $\aleph_0 \times_c 1$ is τ, and for $2^{\aleph_0}$ is $\text{Cl}`\tau$.

In $*117\cdot108$ we find arithmetical occurrences in arithmetical inequalities. Thus IIT directs us to take the first two occurrences of $\text{Nc}`\alpha$ and the first two of Nc'β with adequate actual types. The type of Nc'α and Nc'β in Nc'α = Nc'β is not affected by it. It is evident that the conventions IT, IIT are not sufficient to secure the truth of this proposition as thus symbolized. It is essential that in the equation the type be adjusted adequately for both formal numbers. In fact the general arithmetical convention, that types of equational as well as of arithmetical occurrences are adjusted arithmetically, is here used.

V. *Some Important Principles.*

Principle of Arithmetical Substitution. In ∗120·53, the application of IIT needs a consideration of the whole question of arithmetical substitution. Consider the first of the two examples. We have

$$\vdash : \beta = \gamma +_c \delta \,.\, \exists ! \beta \,.\, \supset \,.\, \alpha^\beta = \alpha^\gamma \times_c \alpha^\delta.$$

It is obvious that unless we can pass with practical immediateness from "$\beta = \gamma +_c \delta \,.\, \alpha^\beta = \alpha^\beta$" to "$\alpha^\beta = \alpha^{\gamma +_c \delta}$" by ∗20·18, arithmetic is made practically impossible by the theory of types. But a difficulty arises from the application of IIT. Suppose we assign the types of our real variables first. Then the types of α, β, γ, δ can be arbitrarily assigned, and there is no necessary connection between them which arises from the preservation of meaning. Thus β may be in a type which is not an adequate type for $\gamma +_c \delta$. Assume that this is the case. But the equational use of $\gamma +_c \delta$ is in the same type as β, and by IIT the arithmetical use of $\gamma +_c \delta$ in $\alpha^{\gamma +_c \delta}$ is in an adequate type. Thus, on the face of it, the reasoning, appealing to ∗20·18, by which the substitution was justified, is fallacious; for the two occurrences of $\gamma +_c \delta$ in fact mean different things.

In order to generalize our solution of this difficulty it is convenient to define the term "arithmetical equation." An *arithmetical equation* is an equation between purely arithmetical formal numbers whose actual types are both determined adequately. Then it is evident that from "$\sigma = \tau \,.\, f(\tau)$," where σ and τ are formal numbers and τ occurs arithmetically in $f(\tau)$, we cannot infer $f(\sigma)$ unless the equation $\sigma = \tau$ is arithmetical. For otherwise the τ in the equation cannot be identified with the τ in $f(\tau)$.

When we have "$\beta = \tau \,.\, f(\tau)$," where τ is a formal number and β is a number in a definite type, and wish to pass to "$f(\beta)$," or "$\beta = \tau \,.\, f(\beta)$" and wish to pass to "$f(\tau)$," the occurrence of τ in $f(\tau)$ being arithmetical, the type of β may not be an adequate type for τ. Accordingly the τ in "$\beta = \tau$" cannot be identified with the τ in $f(\tau)$. The type of the τ in the equation ought to be freed from dependence on that of β. Accordingly the transition is only legitimate when we can write instead

$$\text{"}\beta +_c 0 = \tau \,.\, f(\tau)\text{" or "}\beta +_c 0 = \tau \,.\, f(\beta)\text{,"}$$

where in both cases the equation is arithmetical. For now all the symbols are subject to the same rules.

If this modification can be made without altering the truth-value of the asserted propositions, the substitution is legitimate, otherwise it is not.

It is obvious that in the above our immediate passage is to or from $f(\beta +_c 0)$. But it is easy to see that, the occurrence of $\beta +_c 0$ being arithmetical, we always have

$$f(\beta) . \equiv . f(\beta +_c 0).$$

In order to prove this, we have only to prove

$$\alpha +_c (\beta +_c 0) = \alpha +_c \beta,$$
$$\alpha \times_c (\beta +_c 0) = \alpha \times_c \beta,$$
$$(\alpha +_c 0)^\beta = \alpha^\beta,$$
$$\alpha^{\beta +_c 0} = \alpha^\beta,$$

and $$\alpha > \beta +_c 0 . \equiv . \alpha > \beta . \equiv . \alpha +_c 0 > \beta.$$

The demonstration of the first of these propositions runs as follows:

$\vdash . *110{\cdot}4 . \supset \vdash :. \beta \sim\epsilon \text{NC} . \text{v} . \beta = \Lambda : \supset . \beta +_c 0 = \Lambda . \alpha +_c \beta = \Lambda .$

[$*110{\cdot}4$] $\supset . \alpha +_c (\beta +_c 0) = \Lambda = \alpha +_c \beta$ (1)

$\vdash . *110{\cdot}4 . \supset \vdash :. \alpha \sim\epsilon \text{NC} . \text{v} . \alpha = \Lambda : \supset . \alpha +_c (\beta +_c 0) = \Lambda = \alpha +_c \beta$ (2)

$\vdash . *110{\cdot}6 . \supset \vdash : \alpha, \beta \,\epsilon\, \text{NC} - \iota\text{'}\Lambda . \supset . \alpha +_c (\beta +_c 0) = \alpha +_c \text{sm}\text{''}\beta$

$= \alpha +_c \beta$ (3)

$\vdash . (1) . (2) . (3) . \supset \vdash : \alpha +_c (\beta +_c 0) = \alpha +_c \beta$

In the above demonstration the step to (3) is legitimate since by the hypothesis β is a determination of sm''β in an adequate type.

Similar proofs hold for the other propositions, using $*113{\cdot}204$ and $*116{\cdot}204$ and $*117{\cdot}12$ and $*103{\cdot}13$.

We must also consider the circumstances under which we can pass from "$\beta = \tau$" to "$\beta +_c 0 = \tau$," where the latter equation is arithmetical. In other words, using $*65{\cdot}01$ we require the hypothesis necessary for

$$\exists ! \tau_\eta . \beta = \tau_\xi . \supset . \beta +_c 0 = \tau_\eta .$$

We have

$\vdash . *20{\cdot}18 . \supset \vdash : \beta = \tau_\xi . \supset . \beta +_c 0 = \tau_\xi +_c 0$ (1)

$\vdash . *110{\cdot}35 . \supset \vdash : \exists ! \tau_\xi . \exists ! \tau_\eta . \supset . \tau_\xi +_c 0 = \tau_\eta +_c 0$ (2)

$\vdash . (1) . (2) . \supset \vdash :. \exists ! \tau_\xi . \exists ! \tau_\eta . \supset : \beta = \tau_\xi . \supset . \beta +_c 0 = \tau_\eta +_c 0$ (3)

$\vdash . (3) . \supset \vdash :. \exists ! \beta . \exists ! \tau_\eta . \supset : \beta = \tau_\xi . \supset . \beta +_c 0 = \tau_\eta +_c 0$ (4)

Now in (4) the occurrences of $\beta +_c 0$ and $\tau_\eta +_c 0$, which are in the same type, may be chosen to be in any type we like. Hence we deduce

$\vdash . (4) . *110{\cdot}6 . \supset \vdash :. \exists ! \beta . \exists ! \tau_\eta . \supset : \beta = \tau_\xi . \supset . (\beta +_c 0)_\zeta = \text{sm}_\zeta\text{''}\tau_\eta .$

[$*100{\cdot}511$] $\supset . (\beta +_c 0)_\zeta = \tau_\zeta$

Hence $\exists ! \beta$ is the requisite condition. Now since ζ can be in any type, we can also choose it in any existential type for τ. Thus with IIT applying to the arithmetical occurrence of τ in $f(\tau)$, we have, where τ is a formal number and β is a number in a definite type,

$$\vdash : \exists ! \beta . \beta = \tau . f(\tau) . \supset . f(\beta),$$
$$\vdash : \exists ! \beta . \beta = \tau . f(\beta) . \supset . f(\tau),$$
$$\vdash : \exists ! \sigma . \sigma = \tau . f(\tau) . \supset . f(\sigma).$$

In the last proposition by IT the equation $\sigma = \tau$ is arithmetical. These equations are summed up in $*118 \cdot 01$.

These three fundamental theorems embody the principle of arithmetical substitution. The hypothesis $\exists ! \beta$ is really less than is assumed in ordinary life, the usual tacit assumption being $\beta \epsilon \mathrm{NC} - \iota ` \Lambda$. In fact unless $\beta \epsilon \mathrm{NC}$, $\beta = \tau$ is necessarily false.

Principle of Identification of Types. Suppose we have proved "$\vdash : \mathrm{Hp} . \supset . \phi\sigma$" and "$\vdash : \phi(\sigma_\xi) . \supset . p$," where σ is a formal number whose occurrence in "$\vdash : \mathrm{Hp} . \supset . \phi\sigma$" is in an entirely ambiguous type, and σ_ξ is the same formal number σ with its type related to that of ξ by $*65 \cdot 01$. Then since the type of the σ in "$\vdash : \mathrm{Hp} . \supset . \phi\sigma$" is ambiguous, we can write "$\vdash : \mathrm{Hp} . \supset . \phi(\sigma_\xi)$," and thence infer "$\vdash . p$."

The principle is: An entirely undetermined type in an asserted symbolic form can be identified with any type ambiguous or otherwise in any other asserted symbolic form or in the same symbolic form.

For example in $*100 \cdot 42$ (demonstration) considered above, since $\exists ! \mu \cap \nu$ occurs, the first occurrences of $\mathrm{Nc} ` \alpha$ and $\mathrm{Nc} ` \beta$ are of the same type, and so are their second occurrences in $\mathrm{Nc} ` \alpha = \mathrm{Nc} ` \beta$. But the two types are not determined by our conventions to have any necessary connection. In fact the type in $\mathrm{Nc} ` \alpha = \mathrm{Nc} ` \beta$ is entirely arbitrary. Accordingly it can be identified with the other type, and thus the inference to the next line, *viz.* to "$\vdash : \mathrm{Hp} . \supset . \mu = \nu$," is justified.

In the case of arithmetical equations, it is important to notice that we have

$$\vdash . *100 \cdot 321 \cdot 33 . \supset \vdash :. \exists ! \mathrm{Nc}(\xi) ` \alpha . \supset : \mathrm{Nc}(\xi) ` \alpha = \mathrm{Nc}(\xi) ` \beta . \supset . \mathrm{Nc} ` \alpha = \mathrm{Nc} ` \beta.$$

Hence if σ and τ are formal numbers,

$$\vdash :. \exists ! \sigma_\xi . \supset : \sigma_\xi = \tau_\xi . \supset . \sigma = \tau.$$

Thus if we have "$\vdash : \mathrm{Hp} . \exists ! \sigma . \supset . \sigma = \tau$" and "$\vdash : \mathrm{Hp}' . \sigma_\eta = \tau_\eta . \supset . p$," we can infer from the former proposition "$\vdash : \mathrm{Hp} . \exists ! \sigma . \supset . \sigma_\eta = \tau_\eta$," and from this and the latter proposition, we infer "$\vdash : \mathrm{Hp}' . \mathrm{Hp} . \exists ! \sigma . \supset . p$," so the general principle of identification can be employed when the $\phi(\sigma)$ in the first proposition is an arithmetical equation.

For example, in an example given above, $*100 \cdot 44$ (demonstration), *viz.*

$$\vdash : \mu \epsilon \mathrm{NC} . \exists ! \mathrm{Nc} ` \alpha . \alpha \epsilon \mu . \supset . (\exists \beta) . \mu = \mathrm{Nc} ` \beta . \mathrm{Nc} ` \alpha = \mathrm{Nc} ` \beta,$$

the equation $\text{Nc}`\alpha = \text{Nc}`\beta$ is arithmetical. Accordingly we are justified in asserting the propositional function

$$\vdash : \mu \in \text{NC} \,.\, \exists ! \,\text{Nc}`\alpha \,.\, \alpha \in \mu \,.\, \supset \,.\, (\exists \beta) \,.\, \mu = \text{Nc}\,(\alpha)`\beta \,.\, \text{Nc}\,(\alpha)`\alpha = \text{Nc}\,(\alpha)`\beta,$$

where $\text{Nc}\,(\alpha)`\beta$ in "$\mu = \text{Nc}\,(\alpha)`\beta$" has all along been presupposed by the necessity of meaning.

Thus the inference follows,

$$\vdash : \mu \in \text{NC} \,.\, \exists ! \,\text{Nc}`\alpha \,.\, \alpha \in \mu \,.\, \supset \,.\, \text{Nc}\,(\alpha)`\alpha = \mu \,.$$
$$\supset \,.\, \text{Nc}`\alpha = \mu.$$

This proof loses its point when μ is looked on as a variable with necessarily the same type throughout. For then the proposition collapses into

$$\vdash :. \mu \in \text{NC} \,.\, \supset : \alpha \in \mu \,.\, \equiv \,.\, \text{Nc}\,(\alpha)`\alpha = \mu.$$

But if μ be a formal number necessarily a member of NC, the proposition is really

$$\vdash :. \exists ! \,\text{Nc}`\alpha \,.\, \supset : \alpha \in \mu \,.\, \equiv \,.\, \text{Nc}`\alpha = \mu.$$

With this presupposition we should have in the first line of the demonstration

$$\text{“} \vdash : \exists ! \,\text{Nc}`\alpha \,.\, \text{Nc}`\alpha = \mu \,.\, \supset \,.\, \alpha \in \mu, \text{”}$$

though with "μ" a single variable, the line is formally correct as it stands in the text.

Recognition of Particular Cases. It is important to notice the conditions under which $\phi\sigma$ can be recognized as a particular case of $\phi\xi$, where ξ is a real variable and σ is a formal number. In the first place obviously we must substitute $\sigma \cap t_0`\xi$ for $\sigma$, wherever it occurs in $\phi\sigma$, and thus obtain $\phi\,(\sigma \cap t_0`\xi)$. Then we may find that by the application of our conventions, we can replace this by $\phi\sigma$. For example we have

∗100·42. $\vdash : \mu, \nu \in \text{NC} \,.\, \exists ! \,\mu \cap \nu \,.\, \supset \,.\, \mu = \nu$

Now put $\text{Nc}`\alpha \cap t_0`\mu$ for μ, we obtain

$$\vdash : \text{Nc}`\alpha \cap t_0`\mu, \nu \in \text{NC} \,.\, \exists ! \,(\text{Nc}`\alpha \cap t_0`\mu) \cap \nu \,.\, \supset \,.\, \text{Nc}`\alpha \cap t_0`\mu = \nu \qquad (1)$$

$$\vdash \,.\, (1) \,.\, {*}100{\cdot}41 \,.\, \supset \vdash : \nu \in \text{NC} \,.\, \exists ! \,\text{Nc}`\alpha \cap t_0`\mu \cap \nu \,.\, \supset \,.\, \text{Nc}`\alpha \cap t_0`\mu = \nu \qquad (2)$$

Now by IT, even when ν is a formal number, the identity of types of the two occurrences of $\text{Nc}`\alpha$ is equally secured in

$$\vdash : \nu \in \text{NC} \,.\, \exists ! \,\text{Nc}`\alpha \cap \nu \,.\, \supset \,.\, \text{Nc}`\alpha = \nu.$$

Thus this is a particular case of ∗100·42. Such deductions can be made in general without any explicit formal statement.

Ambiguity of NC. It follows (cf. ∗100·02 and ∗103·02) from the typical ambiguity of Nc that NC is also typically ambiguous. Hence "$\mu \in \text{NC} \,.\, \nu \in \text{NC}$" according to our methods of interpretation would not necessitate that μ and ν should be of the same type. We shall always interpret "$\mu, \nu \in \text{NC}$" as standing for "$\mu \in \text{NC} \,.\, \nu \in \text{NC}$" and therefore as not necessarily identifying the types of μ and ν. Similarly for N_0C, NC induct, and NC ind. For example

∗110·402. $\vdash : \mu, \nu \epsilon N_0C . \supset . \exists ! (\mu +_c \nu) \cap t't'(\mu \uparrow \nu)$

Here the μ and ν need not be of the same type. Again

∗110·41. $\vdash : \mu, \nu \epsilon N_0C . t'\mu = t'\nu . \supset . \exists ! (\mu +_c \nu) \cap t'\mu$

Here the identification of the types of μ and ν requires the hypothesis "$t'\mu = t'\nu$."

VI. *Conventions* AT *and* Infin T.

General Arithmetical Convention. Conventions IT and IIT are always applied, but the following convention is not used at first. This convention limits the remaining ambiguity of type by sweeping away the exceptional cases in low types, due to the failure of existence theorems. The convention will be cited as AT.

AT. *All equations involving pure arithmetical formal numbers are to be arithmetical.*

We have seen that from an arithmetical equation the analogous equation in any other type can be deduced. Thus with AT all equations between formal numbers are so determined in type that their truth in "any type" is deducible. Thus in the few early propositions where AT is introduced, the fact is noted by stating that the equations hold "in *any* type." These propositions are ∗103·16, ∗110·71·72.

The effect of applying AT to other propositions in ∗100 is to render some of the hypotheses (usually logical forms affirming existence) unnecessary, but also materially to limit the scope of the propositions. Take for example

∗100·35. $\vdash : \exists ! Nc'\alpha . \vee . \exists ! Nc'\beta : \supset :$

$$Nc'\alpha = Nc'\beta . \equiv . \alpha \epsilon Nc'\beta . \equiv . \beta \epsilon Nc'\alpha . \equiv . \alpha \, \mathrm{sm} \, \beta$$

If we apply AT to this, we can write

$$\vdash : Nc'\alpha = Nc'\beta . \equiv . \alpha \epsilon Nc'\beta . \equiv . \beta \epsilon Nc'\alpha . \equiv . \alpha \, \mathrm{sm} \, \beta .$$

For the equational occurrences of $Nc'\alpha$ and $Nc'\beta$ are by AT and IIT to be with adequate actual types. But if α is a small class in a high type, an adequate actual type for $Nc'\alpha$ will be a high type, whereas $\exists ! Nc'\alpha$ may hold in a low type. Thus with AT, for the sake of simplicity we abandon the statement of the minimum of hypothesis necessary for our propositions. The enunciation of no other proposition in ∗100 is affected.

The enunciation of no proposition in ∗101 is affected by AT, though it would unduly limit the scope of ∗101·34. In ∗110, AT would unduly limit the scope of such propositions as

∗110·22·23·24·25·251·252·3·31·32·331·34·35·351·44·51·54

and of many others, without altering their enunciations. There is no proposition in ∗110 whose enunciation it would alter. AT is already applied to ∗110·71·72; if AT is removed from these propositions, then $\exists$! Nc'α must be added as an hypothesis to both of them. The effect of AT on ∗113 and ∗116 is entirely analogous to that on ∗110; in neither of these two numbers is there any proposition to which AT is applied in the text.

As regards ∗117, AT is applied throughout, so that the propositions are all in the form suitable for subsequent investigations in which the interest is purely arithmetical. It is important however to analyse the effect of AT on the enunciations for the sake of logical investigations, especially in connection with ∗120. First, AT can only affect propositions in which equations or inequations occur, and among such propositions it does not affect the enunciations of those in which both sides of the equations are not formal numbers, so that the equations are not arithmetical after the application of AT. These propositions are ∗117·104·14·24·241·243·31·551. These propositions, which are characterized by the presence of a single letter on one side of any equation involved, can be recognized at a glance. The propositions involving arithmetical equations whose enunciations are unaltered by the removal of AT are ∗117·21·54·592. Propositions involving inequations whose enunciations are unaltered by the removal of AT are ∗117·26·27. Finally the only propositions of ∗117 whose enunciations are altered by the removal of AT are ∗117·108·211·23·25·3.

In ∗118 and ∗119 AT is not used.

In ∗120, which is devoted to those properties of inductive cardinals which are of logical interest, AT is never used. None of the propositions ∗117·108·211·23·25·3 are cited in it, except ∗117·25 in the demonstration of ∗120·435 for a use where AT is not relevant. The application of AT to ∗120 would simplify the hypotheses of ∗120·31·41·451·53·55, and limit the scopes of the propositions.

One other convention, which we will call "Infin T," is required in certain propositions where the hypothesis implies that there are types in which every inductive cardinal exists, *i.e.* in which V is not an inductive class. Among such hypotheses are Infin ax, $\exists$! Prog, $\exists ! \aleph_0$ (or typically definite forms of these hypotheses), or $R \epsilon$ Prog or $\alpha \epsilon \aleph_0$. When such hypotheses occur, we shall assume that NC induct is, whenever significance permits, to be determined in a type in which every inductive cardinal exists, *i.e.* in which the axiom of infinity holds (cf. ∗120·03·04). The statement of this convention is as follows:

Infin T. *When the hypothesis of a proposition implies that there is a type in which every inductive cardinal exists, every occurrence of* "NC induct" *in this proposition is to be taken (if conditions of significance permit) in a sufficiently high type to insure the existence of every inductive cardinal.*

It is to be observed that this convention would be unnecessary if we confined ourselves to one extensional hierarchy, for in any one such hierarchy all types are inductive or all are non-inductive, so that if every inductive cardinal exists in one type in the hierarchy, the same holds for any other type in the hierarchy. But when we no longer confine ourselves to one extensional hierarchy, this result may not follow. For example, it may be the case that the number of individuals is inductive, but the number of predicative functions of individuals is not inductive; at any rate, no *logical* reason can be given against this possibility, which can only be rejected on empirical grounds, if at all.

The way in which this convention is used may be illustrated by the demonstration of ∗122·33. In the second line of this demonstration, we show that the hypothesis implies

$$E\,!\,\nu_R\,.\supset.\,E\,!\,(\nu +_c 1)_R \qquad (1)$$

where by ∗121·04 $\nu_R = \breve{R}_{\nu -_c 1}\text{‘}B\text{‘}R$ Df,

and by ∗121·02 $R_\nu = \hat{x}\hat{y}\,\{N_0c\text{‘}R\,(x \mapsto y) = \nu +_c 1\}$ Df.

It will be seen that these definitions do not suffice to determine the type of ν. Hence in (1), the ν on the left may not be of the same type as the $\nu +_c 1$ on the right. Now the use of ∗120·473, which occurs in the next line of the demonstration of ∗122·33, requires that the ν on the left and the $\nu +_c 1$ on the right should be of the same type. This requires that the ν should not be taken in a type in which we have $\exists\,!\,\nu\,.\,\nu +_c 1 = \Lambda$. Hence in order to apply ∗120·473, we must choose a type in which all inductive cardinals exist. Since "$R \epsilon$ Prog" occurs in the hypothesis, we know that all inductive cardinals exist in the type of $C\text{‘}R$. But it is unnecessary to restrict ourselves to the type of $C\text{‘}R$, since any other type in which all inductive cardinals exist will equally secure the validity of the demonstration. Thus the convention Infin T secures the restriction required, and no more.

The convention Infin T is often relevant when "Infin ax" without any typical determination occurs in the hypothesis. Whenever this is the case, if "NC induct" occurs in the proposition in a way which leaves its type undetermined so far as conditions of significance are concerned, it is to be taken in a type in which all its members exist.

VII. *Final Working Rule in Arithmetic.*

It is now (whenever AT is used, together with Infin T when necessary) possible finally to sweep aside all consideration of types in connection with inductive numbers. For by combining ∗126·121·122 and ∗120·4232·4622, we see that it is always possible to take the type high enough so that no definitely determined inductive number shall be null (Λ), and that all the inductive reasoning can take place within this type. Furthermore we have already seen that the arithmetical operations are independent of the types of the components, so long as they are existential. Thus, as far as the ordinary arithmetic of finite numbers is concerned, all the conventions (including AT), and the necessity for hypotheses as to the existence of inductive numbers, are finally superseded by the following single rule:

RULE OF INDEFINITE NUMBERS. *The type assigned to any symbol which represents an inductive number is such that the symbol is not equal to* Λ.

We make the definition

∗126·01. $\mathrm{Nc\,ind} = \mathrm{Nc\,induct} - \iota`\Lambda$ Df

Wherever this symbol "Nc ind" for the class of "indefinite inductive cardinal numbers" is used, the above rule is adhered to. In other words, "$\mu \in \mathrm{NC\,ind}$" can always be replaced by "$\mu = \mathrm{Nc}`\alpha \,.\, \alpha \in \mathrm{Cls\,induct}$," where $\mathrm{Nc}`\alpha$ is a homogeneous or ascending cardinal, and α is the appropriate constant, or is a variable, as the case may be. In the latter case, a symbolic form such as

$$(\mu) \,.\, f(\mu \in \mathrm{NC\,ind}, \mu)$$

can be replaced by

$$(\mu, \alpha) \,.\, f(\mu = \mathrm{Nc}`\alpha \,.\, \alpha \in \mathrm{Cls\,induct}, \mu).$$

Furthermore by ∗120·4622 it follows that with this rule the result of proceeding by induction in one type and then transforming to another type is the same as that of proceeding by induction in the latter type. Thus for example there is no advantage to be gained by discriminating between 2_ξ and 2_η; for $\mathrm{sm}_\eta``2_\xi = 2_\eta$, $\mathrm{sm}_\xi``2_\eta = 2_\xi$, $\mu +_c 2_\xi = \mu +_c 2_\eta$, $\mu \times_c 2_\xi = \mu \times_c 2_\eta$, $\mu^{2_\xi} = \mu^{2_\eta}$, $2_\xi{}^\mu = 2_\eta{}^\mu$, and $\mu \geqslant 2_\xi \,.\equiv .\, \mu \geqslant 2_\eta$, and so on.

Hence all discrimination of the types of indefinite inductive numbers may be dropped; and the types are entirely indefinite and irrelevant.

PART III.

CARDINAL ARITHMETIC.

SUMMARY OF PART III.

In this Part, we shall be concerned, first, with the definition and general logical properties of cardinal numbers (Section A); then with the operations of addition, multiplication and exponentiation, of which the definitions and formal laws do not require any restriction to finite numbers (Section B); then with the theory of finite and infinite, which is rendered somewhat complicated by the fact that there are two different senses of "finite," which cannot (so far as is known) be identified without assuming the multiplicative axiom. The theory of finite and infinite will be resumed, in connection with series, in Part V, Section E.

It is in this Part that the theory of types first becomes practically relevant. It will be found that contradictions concerning the maximum cardinal are solved by this theory. We have therefore devoted our first section in this Part (with the exception of two numbers giving the most elementary properties of cardinals in general, and of 0 and 1 and 2, respectively) to the application of types to cardinals. Every cardinal is typically ambiguous, and we confer typical definiteness by the notations of ∗63, ∗64, and ∗65. It is especially where existence-theorems are concerned that the theory of types is essential. The chief importance of the propositions of the present part lies, not only, as throughout the book, in the hypotheses necessary to secure the conclusions, but also in the typical ambiguity which can be allowed to the symbols consistently with the truth of the propositions in all the cases thereby included.

SECTION A.

DEFINITION AND LOGICAL PROPERTIES OF CARDINAL NUMBERS.

Summary of Section A.

The Cardinal Number of a class α, which we will denote by "$\text{Nc}'\alpha$," is defined as the class of all classes similar to α, *i.e.* as $\hat{\beta}(\beta \text{ sm } \alpha)$. This definition is due to Frege, and was first published in his *Grundlagen der Arithmetik**; its symbolic expression and use are to be found in his *Grundgesetze der Arithmetik*†. The chief merits of this definition are (1) that the formal properties which we expect cardinal numbers to have result from it; (2) that unless we adopt this definition or some more complicated and practically equivalent definition, it is necessary to regard the cardinal number of a class as an indefinable. Hence the above definition avoids a useless indefinable with its attendant primitive propositions.

It will be observed that, if x is any object, 1 is not the cardinal number of x, but that of $\iota'x$. This obviates a confusion which otherwise is liable to arise in dealing with classes. Suppose we have a class α consisting of many terms; we say, nevertheless, that it is *one* class. Thus it seems to be at once one and many. But in fact it is α that is many, and $\iota'\alpha$ that is one. In regard to zero, the analogous point is still clearer. Suppose we say "there are no Kings of France." This is equivalent to "the class of Kings of France has no members," or, in our language, "the class of Kings of France is a member of the class 0." It is obvious that we cannot say "the King of France is a member of the class 0," because there is no King of France. Thus in the case of 0 and 1, as more evidently in all other cases, a cardinal number appertains to a class, not to the members of the class.

For the purposes of formal definition, we subject the formula

$$\underset{\cdot}{\text{N}}\text{c}'\alpha = \hat{\beta}(\beta \text{ sm } \alpha)$$

to some simplification. It will be seen that, according to this formula, "Nc" is a relation, namely the relation of a cardinal number to any class of which it is the number. Thus for example 1 has to $\iota'x$ the relation Nc; so has

* Breslau, 1884. Cf. especially pp. 79, 80.

† Jena, Vol. I. 1893, Vol. II. 1903. Cf. Vol. I. §§ 40—42, pp. 57, 58. The grounds in favour of this definition will be found at length in *Principles of Mathematics*, Part II.

2 to $\iota{}'x \cup \iota{}'y$, provided $x \neq y$. The relation Nc is, in fact, the relation $\overrightarrow{\mathrm{sm}}$; for $\overrightarrow{\mathrm{sm}}{}'\alpha = \hat{\beta}(\beta \,\mathrm{sm}\, \alpha)$. Hence for formal purposes of definition we put

$$\mathrm{Nc} = \overrightarrow{\mathrm{sm}} \quad \mathrm{Df.}$$

The class of cardinal numbers is the class of objects which are the cardinal numbers of something or other, *i.e.* of objects which, for some α, are equal to $\mathrm{Nc}{}'\alpha$. We call the class of cardinal numbers NC; thus we have

$$\mathrm{NC} = \hat{\mu}\{(\exists\alpha) \,.\, \mu = \mathrm{Nc}{}'\alpha\}.$$

For purposes of formal definition, we replace this by the simpler formula

$$\mathrm{NC} = \mathrm{D}{}'\mathrm{Nc} \quad \mathrm{Df.}$$

In the present section, we shall be concerned with what we may call the purely logical properties of cardinal numbers, namely those which do not depend upon the arithmetical operations of addition, multiplication and exponentiation, nor upon the distinction of finite and infinite*. The chief point to be dealt with, as regards both importance and difficulty, is the relation of a cardinal number in one type to the same or an associated cardinal number in another type. When a symbol is ambiguous as to type, we will call it *typically ambiguous*; when, either always or in a given context, it is unambiguous as to type, we will call it *typically definite*. Now the symbol "sm" is typically ambiguous; the only limitation on its type is that its domain and converse domain must both consist of classes. When we have $\alpha \,\mathrm{sm}\, \beta$, α and β need not be of the same type, in fact, in any type of classes, there are classes similar to some of the classes of any other type of classes. For example, we have $\iota{}'x \,\mathrm{sm}\, \iota{}'y$, whatever types x and y may belong to. This ambiguity of "sm" is derived from that of $1 \to 1$, which in turn is derived from that of 1. We denote (cf. ∗65·01) by "1_α" all the unit classes which are of the same type as α. Then (according to the definition ∗70·01) $1_\alpha \to 1_\beta$ will be the class of those one-one relations whose domain is of the same type as α and whose converse domain is of the same type as β. Thus "$1_\alpha \to 1_\beta$" is typically definite as soon as α and β are given. Suppose now, instead of having merely $\gamma \,\mathrm{sm}\, \delta$, we have

$$(\exists R) \,.\, R \in 1_\alpha \to 1_\beta \,.\, \mathrm{D}{}'R = \gamma \,.\, \mathrm{Cl}{}'R = \delta;$$

then we know not only that $\gamma \,\mathrm{sm}\, \delta$, but also that γ belongs to the same type as α, and δ belongs to the same type as β. When the ambiguous symbol "sm" is rendered typically definite by having its domain defined as being of the same type as α, and its converse domain defined as being of the same type as β, we write it "$\mathrm{sm}_{(\alpha,\beta)}$," because generally, in accordance with ∗65·1, if R is a typically ambiguous relation, we write $R_{(\alpha,\beta)}$ for the typically

* The definitions of the arithmetical operations, and of finite and infinite, are really just as purely logical as what precedes them; but if we are to draw a line between logic and arithmetic somewhere, the arithmetical operations seem the natural point at which to place the beginning of arithmetic.

definite relation that results when the domain of R is to consist of terms of the same type as α, and the converse domain is to consist of terms of the same type as β. Thus we have

$$\gamma \, \mathrm{sm}_{(\alpha,\beta)} \, \delta \, . \equiv . \, (\exists R) \, . \, R \in 1_\alpha \rightarrow 1_\beta \, . \, \gamma = \mathrm{D}‘R \, . \, \delta = \mathrm{Cl}‘R.$$

Here everything is typically definite if α and β (or their types) are given.

Passing now to the relation "Nc," it will be seen that it shares the typical ambiguity of "sm." In order to render it typically definite, we must derive it from a typically definite "sm." So long as nothing is added to give typical definiteness, "Nc‘γ" will mean all the classes belonging to some one (unspecified) type and similar to γ. If α is a member of the type to which these classes are to belong, then Nc‘γ is contained in the type of α. For this case, it is convenient to introduce the following two notations, already defined in ∗65. When a typically ambiguous relation R is to be rendered typically definite as to its domain only, by deciding that every member of the domain is to be *contained in* the type of α, we write "$R(\alpha)$" in place of R. When we further wish to determine R as having members of the converse domain *contained in* the type of β, we write "$R(\alpha, \beta)$" in place of R; and when we wish members of the converse domain to be *members of* the type of β, we write "$R(\alpha_\beta)$" in place of R. Thus

$$\mathrm{sg}‘\{R_{(\alpha,\beta)}\} = \{\mathrm{sg}‘R\}(\alpha_\beta)$$

(cf. ∗65·2), and in particular, since $\mathrm{Nc} = \overrightarrow{\mathrm{sm}}$,

$$\mathrm{Nc}(\alpha_\beta) = \mathrm{sg}‘\mathrm{sm}_{(\alpha,\beta)}.$$

Thus "$\mathrm{Nc}(\alpha_\beta)‘\gamma$" is only significant when γ is of the same type as β, and then it means "classes of the same type as α and similar to γ (which is of the same type as β)."

"$\mathrm{Nc}(\alpha)‘\gamma$" will mean "classes of the same type as α and similar to γ." As soon as the types of α and γ are known, this is a typically definite symbol, being in fact equal to $\mathrm{Nc}(\alpha_\gamma)‘\gamma$. Hence so long as we only wish to consider "Nc‘γ," typical definiteness is secured by writing "$\mathrm{Nc}(\alpha)$" in place of "Nc."

When we come to the consideration of NC, "$\mathrm{Nc}(\alpha)$" is no longer a sufficient determination, although it suffices to determine the type. Suppose we put

$$\mathrm{NC}^\beta(\alpha) = \mathrm{D}‘\mathrm{Nc}(\alpha_\beta) \quad \mathrm{Df};$$

we have also, in virtue of the definitions in ∗65,

$$\mathrm{NC}(\alpha) = \mathrm{NC} \cap t^2‘\alpha = \mathrm{D}‘\mathrm{Nc}(\alpha).$$

Thus $\mathrm{NC}(\alpha)$ is definite as to type, but is the domain of a relation whose converse domain is ambiguous as to type; and it will appear that there are some propositions about $\mathrm{NC}(\alpha)$ whose truth or falsehood depends upon the determination chosen for the converse domain of $\mathrm{Nc}(\alpha)$. Hence if we wish to have a symbol which is completely definite, we must write "$\mathrm{NC}^\beta(\alpha)$."

This point is important in connection with the contradictions as to the maximum cardinal. The following remarks will illustrate it further.

Cantor has shown that, if β is any class, no class contained in β is similar to $\mathrm{Cl}'\beta$. Hence in particular if β is a type, no class contained in β is similar to $\mathrm{Cl}'\beta$, which is the next type above β. Consequently, if $\beta = \alpha \cup -\alpha$, where α is any class, we have

$$\sim (\exists\gamma) \,.\, \gamma \subset \alpha \cup -\alpha \,.\, \gamma \,\mathrm{sm}\, \mathrm{Cl}'(\alpha \cup -\alpha).$$

Now (cf. *63) we put

$$t_0{}'\alpha = \alpha \cup -\alpha \quad \mathrm{Df},$$

and we have $t'\alpha = \mathrm{Cl}'(\alpha \cup -\alpha)$. Thus we find

$$\sim (\exists\gamma) \,.\, \gamma \subset t_0{}'\alpha \,.\, \gamma \,\mathrm{sm}\, t'\alpha.$$

Hence $$\mathrm{Nc}\,(\alpha_{t'\alpha})'t'\alpha = \Lambda.$$

That is to say, no class of the same type as α has as many members as $t'\alpha$ has. Hence also

$$\Lambda \in \mathrm{NC}^{t'\alpha}(\alpha).$$

But $$\gamma \subset t_0{}'\alpha \,.\supset.\, \gamma \in \mathrm{Nc}\,(\alpha_\alpha)'\gamma \,.\supset.\, \exists \,!\, \mathrm{Nc}\,(\alpha_\alpha)'\gamma,$$

and "$\mathrm{Nc}\,(\alpha_\alpha)'\gamma$" is only significant when $\gamma \subset t_0{}'\alpha$; hence

$$\mu \in \mathrm{NC}^\alpha(\alpha) \,.\supset_\mu.\, \exists \,!\, \mu$$

and $$\Lambda \sim\in \mathrm{NC}^\alpha(\alpha).$$

Now the notation "$\mathrm{NC}(\alpha)$" will apply with equal justice to $\mathrm{NC}^\alpha(\alpha)$ or to $\mathrm{NC}^{t'\alpha}(\alpha)$; but we have just seen that in the first case we shall have $\Lambda \sim\in \mathrm{NC}(\alpha)$, and in the second we shall have $\Lambda \in \mathrm{NC}(\alpha)$. Consequently "$\mathrm{NC}(\alpha)$" has not sufficient definiteness to prevent practically important differences between the various determinations of which it is capable.

A converse procedure to the above yields similar results. Let α be a class of classes; then $s'\alpha$ is of lower type than α. Let us consider $\mathrm{NC}^{s'\alpha}(\alpha)$. In accordance with *63, we write $t_1{}'\alpha$ for the type containing $s'\alpha$, *i.e.* for $s'\alpha \cup -s'\alpha$. Then the greatest number in the class $\mathrm{NC}^{s'\alpha}(\alpha)$ will be $\mathrm{Nc}\,(\alpha)'t_1{}'\alpha$; but neither this nor any lesser member of the class will be equal to $\mathrm{Nc}\,(\alpha)'t_0{}'\alpha$, because, as before,

$$\sim (\exists\gamma) \,.\, \gamma \subset t_1{}'\alpha \,.\, \gamma \,\mathrm{sm}\, t_0{}'\alpha.$$

Hence $\mathrm{Nc}\,(\alpha)'t_0{}'\alpha$, which is a member of $\mathrm{NC}^\alpha(\alpha)$, is not a member of $\mathrm{NC}^{s'\alpha}(\alpha)$; but $\mathrm{NC}^\alpha(\alpha)$ and $\mathrm{NC}^{s'\alpha}(\alpha)$ have an equal right to be called $\mathrm{NC}(\alpha)$. Hence again "$\mathrm{NC}(\alpha)$" is a symbol not sufficiently definite for many of our purposes.

The solution of the paradox concerning the maximum cardinal is evident in view of what has been said. This paradox is as follows: It results from a theorem of Cantor's that there is no maximum cardinal, since, for all values of α,

$$\mathrm{Nc}'\mathrm{Cl}'\alpha > \mathrm{Nc}'\alpha.$$

But at first sight it would seem that the class which contains everything must be the greatest possible class, and must therefore contain the greatest possible number of terms. We have seen, however, that a class α must always be contained within some one type; hence all that is proved is that there are greater classes in the next type, which is that of $\text{Cl}`\alpha$. Since there is always a next higher type, we thus have a maximum cardinal in each type, without having any absolutely maximum cardinal. The maximum cardinal in the type of α is

$$\text{Nc}\,(\alpha)`(\alpha \cup -\alpha).$$

But if we take the corresponding cardinal in the next type, *i.e.*

$$\text{Nc}\,(\text{Cl}`\alpha)`(\alpha \cup -\alpha),$$

this is not as great as $\text{Nc}\,(\text{Cl}`\alpha)`\text{Cl}`(\alpha \cup -\alpha)$, and is therefore not the maximum cardinal of its type. This gives the complete solution of the paradox.

For most purposes, what we wish to know in order to have a sufficient amount of typical definiteness is not the absolute types of α and β, as above, but merely what we may call their *relative* types. Thus, for example, α and β may be of the same type; in that case, $\text{Nc}\,(\alpha_\beta)$ and $\text{NC}^\beta\,(\alpha)$ are respectively equal to $\text{Nc}\,(\alpha_\alpha)$ and $\text{NC}^\alpha\,(\alpha)$. We will call cardinals which, for some α, are members of the class $\text{NC}^\alpha\,(\alpha)$, *homogeneous* cardinals, because the "sm" from which they are derived is a homogeneous relation. We shall denote the homogeneous cardinal of α by "$\text{N}_0\text{c}`\alpha$," and we shall denote the class of homogeneous cardinals (in an unspecified type) by "N_0C"; thus we put

$$\text{N}_0\text{c}`\alpha = \text{Nc}`\alpha \cap t`\alpha \quad \text{Df},$$

$$\text{N}_0\text{C} = \text{D}`\text{N}_0\text{c} \qquad \text{Df}.$$

Almost all the properties of N_0C are the same in different types. When further typical definiteness is required, it can be secured by writing $\text{N}_0\text{c}\,(\alpha)$, $\text{N}_0\text{C}\,(\alpha)$ in place of N_0c, N_0C. For although $\text{Nc}\,(\alpha)$ and $\text{NC}\,(\alpha)$ were not wholly definite, $\text{N}_0\text{c}\,(\alpha)$ and $\text{N}_0\text{C}\,(\alpha)$ are wholly definite. Apart from the fact of being of different types, the only property in which $\text{N}_0\text{C}\,(\alpha)$ and $\text{N}_0\text{C}\,(\beta)$ differ when α and β are of different types is in regard to the magnitude of the cardinals belonging to them. Thus suppose the whole universe consisted (as monists aver) of a single individual. Let us call the type of this individual "Indiv." Then $\text{N}_0\text{C}\,(\text{Indiv})$ will consist of 0 and 1, *i.e.*

$$\text{N}_0\text{C}\,(\text{Indiv}) = \iota`0 \cup \iota`1.$$

But in the next higher type, there will be two members, namely Λ and Indiv. Thus

$$\text{N}_0\text{C}\,(t`\text{Indiv}) = \iota`0 \cup \iota`1 \cup \iota`2.$$

Similarly $\quad \text{N}_0\text{C}\,(t`t`\text{Indiv}) = \iota`0 \cup \iota`1 \cup \iota`2 \cup \iota`3 \cup \iota`4,$

the members of $t't'\text{Indiv}$ being $\Lambda \cap t'\text{Indiv}$, $\iota'\Lambda$, $\iota'\text{Indiv}$, $\iota'\Lambda \cup \iota'\text{Indiv}$; and so on. (The greatest cardinal in any except the lowest type is always a power of 2.)

The maximum of $N_0C(\alpha)$ is $N_0c't_0'\alpha$; but apart from this difference of maximum and its consequences, $N_0C(\alpha)$ and $N_0C(\beta)$ do not differ in any important properties. Hence for most purposes N_0C and N_0c have as much typical definiteness as is necessary.

Among cardinals which are not homogeneous we shall consider three kinds. The first of these we shall call *ascending* cardinals. A cardinal $NC^\beta(\alpha)$ is called an *ascending* cardinal if the type of β is $t'\alpha$ or $t't'\alpha$ or $t't't'\alpha$ or etc. We write $t^{2}{}'\alpha$ for $t't'\alpha$, $t^{3}{}'\alpha$ for $t't't'\alpha$, and so on. We put

$$N^1c'\alpha = Nc'\alpha \cap t't'\alpha \quad \text{Df}$$

$$N^2c'\alpha = Nc'\alpha \cap t't^{2}{}'\alpha \quad \text{Df}$$

$$N^3c'\alpha = Nc'\alpha \cap t't^{3}{}'\alpha \quad \text{Df} \quad \text{and so on,}$$

and

$$N^1C = D'N^1c \quad \text{Df}$$

$$N^2C = D'N^2c \quad \text{Df}$$

$$N^3C = D'N^3c \quad \text{Df} \quad \text{and so on.}$$

We then have obviously

$$N^1C(t'\alpha) \subset N_0C(t'\alpha).$$

We also have (by what was said earlier)

$$N_0c't'\alpha \sim\epsilon N^1C(t'\alpha).$$

Hence

$$\exists ! N_0C(t'\alpha) - N^1C(t'\alpha).$$

The members of $N_0C(t'\alpha) - N^1C(t'\alpha)$ will be all cardinals which exceed $Nc't_0'\alpha$ but do not exceed $Nc't'\alpha$.

Let us recur in illustration to our previous hypothesis of the universe consisting of a single individual. Then $N^1c'\text{Indiv}$ will consist of those classes which are similar to "Indiv" but of the next higher type. These are $\iota'\Lambda$ and $\iota'\text{Indiv}$. In our case we had $N_0c'\text{Indiv} = 1$. This leads to

$$N^1c'\text{Indiv} = 1 \,.\, N^2c'\text{Indiv} = 1 \text{ etc.}$$

or, introducing typical definiteness,

$$N^1c'\text{Indiv} = 1(t'\text{Indiv}) \,.\, N^2c'\text{Indiv} = 1(t^{2}{}'\text{Indiv}) \text{ etc.}$$

We have then $1(t'\text{Indiv}) \in N^1C(t't'\text{Indiv})$. Also

$$1(t'\text{Indiv}) \in N_0C(t't'\text{Indiv}).$$

And in the case supposed, $1(t'\text{Indiv})$ is the maximum of $N^1C(t't'\text{Indiv})$, but $2(t'\text{Indiv}) \in N_0C(t't'\text{Indiv})$. Hence

$$N_0C(t't'\text{Indiv}) - N^1C(t't'\text{Indiv}) = \iota'2.$$

Generalizing, we see that $N^1C(t'\alpha)$ consists of the same numbers as $N_0C(\alpha)$ each raised one degree in type. Similar propositions hold of $N^2C(t^{2}{}'\alpha)$, $N^3C(t^{3}{}'\alpha)$ etc.

It is often useful to have a notation for what we may call "the same cardinal in another type." Suppose μ is a typically definite cardinal; then we will denote by $\mu^{(1)}$ the same cardinal in the next type, *i.e.*

$$\text{sm}``\mu \cap t`\mu.$$

Note that, if μ is a cardinal, $\text{sm}``\mu \cap \mu = \mu$; and whether μ is a typically definite cardinal or not,

$$\text{sm}``\mu \cap t`\alpha$$

is a cardinal in a definite type. If μ is typically definite, then $\text{sm}``\mu \cap t`\alpha$ is wholly definite; if $\mu$ is typically ambiguous, $\text{sm}``\mu \cap t`\alpha$ has the same kind of indefiniteness as belongs to $\text{NC}(\alpha)$. The most important case is when μ is typically definite and α has an assigned relation of type to μ. We then put, as observed above,

$$\mu^{(1)} = \text{sm}``\mu \cap t`\mu \quad \text{Df}$$

$$\mu^{(2)} = \text{sm}``\mu \cap t^{2}`\mu \quad \text{Df etc.}$$

If μ is an N_0C, $\mu^{(1)}$ is an N^1C and $\mu^{(2)}$ is an N^2C and so on. $\text{N}^1\text{C}(t`\alpha)$ will consist of all numbers which are of the form $\mu^{(1)}$ for some μ which is a member of $\text{N}_0\text{C}(\alpha)$; *i.e.*

$$\text{N}^1\text{C}(t`\alpha) = \hat{\nu}\{(\exists\mu) \,.\, \mu \in \text{N}_0\text{C}(\alpha) \,.\, \nu = \mu^{(1)}\}.$$

The second kind of non-homogeneous cardinals to be considered is called the class of "descending cardinals." These are such as go into a lower type; *i.e.* $\text{Nc}(\alpha)`\beta$ is a descending cardinal if α is of a lower type than β. We put

$$\text{N}_1\text{c}`\alpha = \text{Nc}`\alpha \cap t`t_1`\alpha \quad \text{Df}$$

$$\text{N}_2\text{c}`\alpha = \text{Nc}`\alpha \cap t`t_2`\alpha \quad \text{Df etc.}$$

$$\text{N}_1\text{C} = \text{D}`\text{N}_1\text{c} \quad \text{Df}$$

$$\text{N}_2\text{C} = \text{D}`\text{N}_2\text{c} \quad \text{Df etc.}$$

$$\mu_{(1)} = \text{sm}``\mu \cap t_1`\mu \quad \text{Df}$$

$$\mu_{(2)} = \text{sm}``\mu \cap t_2`\mu \quad \text{Df etc.}$$

We have obviously $\quad \text{N}_0\text{c}`\alpha = \text{N}_1\text{c}`\iota``\alpha.$

Hence $\quad \text{N}_0\text{C}(\alpha) \subset \text{N}_1\text{C}(\alpha).$

Also $\quad \gamma \in \text{N}_1\text{c}`\delta \,.\, \supset \,.\, \text{N}_1\text{c}`\delta = \text{N}_0\text{c}`\gamma,$

whence $\quad \exists \,!\, \text{N}_1\text{c}`\delta \,.\, \supset \,.\, \text{N}_1\text{c}`\delta \in \text{N}_0\text{C},$

whence $\quad \text{N}_1\text{C} - \iota`\Lambda \subset \text{N}_0\text{C}.$

Since also $\Lambda \sim\in \text{N}_0\text{C}(\alpha)$, we find

$$\text{N}_0\text{C} = \text{N}_1\text{C} - \iota`\Lambda,$$

this proposition not requiring any further typical definiteness, since it holds however such definiteness may be introduced, remembering that such definiteness is necessarily so introduced as to secure significance. Further, in virtue of the fact that no class contained in $t_0`\alpha$ is similar to $t`\alpha$, we have

$$\Lambda \in \text{N}_1\text{C}(\alpha).$$

Consequently $$\mathrm{N}_1\mathrm{C} = \mathrm{N}_0\mathrm{C} \cup \iota\text{‘}\Lambda.$$

We can prove in just the same way

$$\mathrm{N}_2\mathrm{C} = \mathrm{N}_0\mathrm{C} \cup \iota\text{‘}\Lambda.$$

Hence $$\mathrm{N}_1\mathrm{C} = \mathrm{N}_2\mathrm{C},$$

and this result can obviously be extended to all descending cardinals.

The third kind of non-homogeneous cardinals to be considered may be called "relational cardinals." They are those applicable to classes of relations having a given relation of type to a given class. Consider for example $\mathrm{Nc}\text{‘}\epsilon_\Delta\text{‘}\kappa$. (We shall take this as the definition of the product of the numbers of the members of κ.) Suppose now that κ consists of a single term: we want to be able to say

$$\mathrm{Nc}\text{‘}\epsilon_\Delta\text{‘}\kappa = \mathrm{Nc}\text{‘}\breve{\iota}\text{‘}\kappa.$$

We have in this case, if $\kappa = \iota\text{‘}\alpha$,

$$\epsilon_\Delta\text{‘}\kappa = \downarrow \alpha\text{‘‘}\alpha,$$

and we know that $\downarrow \alpha\text{‘‘}\alpha \,\mathrm{sm}\, \alpha$. But if we put simply

$$\mathrm{Nc}\text{‘} \downarrow \alpha\text{‘‘}\alpha = \mathrm{Nc}\text{‘}\alpha,$$

our proposition, though not mistaken, requires care in interpretation. Just as we put $\iota\text{‘‘}\alpha \in \mathrm{N}^1\mathrm{c}\text{‘}\alpha$, so we want a notation giving typical definiteness to the proposition $\downarrow \alpha\text{‘‘}\alpha \in \mathrm{Nc}\text{‘}\alpha$. This is provided as follows.

Using the notation of ∗64, put

$$\mathrm{N}_{00}\mathrm{c}\text{‘}\alpha = \mathrm{Nc}\text{‘}\alpha \cap t\text{‘}t_{00}\text{‘}\alpha \quad \mathrm{Df}$$
$$\mathrm{N}_0{}^1\mathrm{c}\text{‘}\alpha = \mathrm{Nc}\text{‘}\alpha \cap t\text{‘}t_0{}^1\text{‘}\alpha \quad \mathrm{Df\ etc.}$$
$$\mathrm{N}_{00}\mathrm{C} = \mathrm{D}\text{‘}\mathrm{N}_{00}\mathrm{c} \quad \mathrm{Df}$$
$$\mathrm{N}_0{}^1\mathrm{C} = \mathrm{D}\text{‘}\mathrm{N}_0{}^1\mathrm{c} \quad \mathrm{Df\ etc.}$$
$$\mu_{(00)} = \mathrm{sm}\text{‘‘}\mu \cap t\text{‘}t_{00}\text{‘}t_1\text{‘}\mu \quad \mathrm{Df\ etc.}$$

Then we have, for example,

$$\downarrow \alpha\text{‘‘}\alpha \subset t_0{}^1\text{‘}\alpha, \text{ i.e. } \downarrow \alpha\text{‘‘}\alpha \in t\text{‘}t_0{}^1\text{‘}\alpha.$$

Hence $\downarrow \alpha\text{‘‘}\alpha \in \mathrm{N}_0{}^1\mathrm{c}\text{‘}\alpha$, where $\mathrm{N}_0{}^1\mathrm{c}\text{‘}\alpha = \mathrm{Nc}\text{‘}\alpha \cap t_0{}^1\text{‘}\alpha$.

Similarly $$x \in t\text{‘}\alpha \,.\supset.\, \downarrow x\text{‘‘}\alpha \in \mathrm{N}_{00}\mathrm{c}\text{‘}\alpha.$$

Thus the above definitions give us what is required.

In order to complete our notation for types, we should need to be able to express the type of the domain or converse domain of R, or of any relation whose domain and converse domain have respectively given relations of type to the domain and converse domain of R. Thus we might put

$$\mathrm{d}_0\text{‘}R = t_0\text{‘}\mathrm{D}\text{‘}R \quad \mathrm{Df}$$
$$\mathrm{b}_0\text{‘}R = t_0\text{‘}\mathrm{G}\text{‘}R \quad \mathrm{Df}$$

("b" appears here as "d" written backwards)

$$\mathrm{d}_{00}\text{‘}R = t\text{‘}(\mathrm{d}_0\text{‘}R \uparrow \mathrm{b}_0\text{‘}R) \quad \mathrm{Df}$$
$$= t\text{‘}R$$
$$\mathrm{d}^{mn}\text{‘}R = t\text{‘}(t^m\text{‘}\mathrm{d}_0\text{‘}R \uparrow t^n\text{‘}\mathrm{b}_0\text{‘}R) \quad \mathrm{Df\ and\ so\ on.}$$

This notation would enable us to deal with descending relational cardinals. But it is not required in the present work, and is therefore not introduced among the numbered propositions.

When a typically ambiguous symbol, such as "sm" or "Nc," occurs more than once in a given context, it must not be assumed, unless required by the conditions of significance, that it is to receive the same typical determination in each case. Thus *e.g.* we shall write "$\alpha \operatorname{sm} \beta . \supset . \beta \operatorname{sm} \alpha$," although, if α and β are of different types, the two symbols "sm" must receive different typical determinations.

Formulae which are typically ambiguous, or only partially definite as to type, must not be admitted unless every significant interpretation is true. Thus for example we may admit

$$\text{“} \vdash . \alpha \in \mathrm{Nc}\text{‘}\alpha \text{”}$$

because here "Nc" must mean "$\mathrm{Nc}(\alpha_a)$," so that the only ambiguity remaining is as to the type of α, and the formula holds whatever type α may belong to, provided "$\mathrm{Nc}\text{‘}\alpha$" is significant, *i.e.* provided α is a class. But we must not, from "$\alpha \in \mathrm{Nc}\text{‘}\alpha$," allow ourselves to infer

$$\text{“} \exists ! \mathrm{Nc}\text{‘}\alpha . \text{”}$$

For here the conditions of significance no longer demand that "Nc" should mean "$\mathrm{Nc}(\alpha_a)$": it might just as well mean "$\mathrm{Nc}(\beta_a)$." And as we saw, if β is a lower type than α, and α is sufficiently large of its type, we may have

$$\mathrm{Nc}(\beta_a)\text{‘}\alpha = \Lambda,$$

so that "$\exists ! \mathrm{Nc}\text{‘}\alpha$" is not admissible without qualification. Nevertheless, as we shall see in *100, there are a certain number of propositions to be made about a wholly ambiguous Nc or NC.

$*100$. DEFINITION AND ELEMENTARY PROPERTIES OF CARDINAL NUMBERS.

Summary of $*100$.

In this number we shall be concerned only with such immediate consequences of the definition of cardinal numbers as do not require typical definiteness, beyond what the inherent conditions of significance may bestow. We introduce here the fundamental definitions:

$*100\cdot01$. $\mathrm{Nc} = \overrightarrow{\mathrm{sm}}$ Df

$*100\cdot02$. $\mathrm{NC} = \mathrm{D}\text{‘}\mathrm{Nc}$ Df

The definition "Nc" is required chiefly for the sake of the descriptive function $\mathrm{Nc}\text{‘}\alpha$. We have

$*100\cdot1$. $\vdash . \mathrm{Nc}\text{‘}\alpha = \hat{\beta}(\beta \,\mathrm{sm}\, \alpha) = \hat{\beta}(\alpha \,\mathrm{sm}\, \beta)$

This may be stated in various equivalent forms, which are given at the beginning of this number ($*100\cdot1$—$\cdot16$). After a few propositions on Nc as a relation, we proceed to the elementary properties of $\mathrm{Nc}\text{‘}\alpha$. We have

$*100\cdot3$. $\vdash . \alpha \in \mathrm{Nc}\text{‘}\alpha$

$*100\cdot31$. $\vdash : \alpha \in \mathrm{Nc}\text{‘}\beta . \equiv . \beta \in \mathrm{Nc}\text{‘}\alpha . \equiv . \alpha \,\mathrm{sm}\, \beta$

$*100\cdot321$. $\vdash : \alpha \,\mathrm{sm}\, \beta . \supset . \mathrm{Nc}\text{‘}\alpha = \mathrm{Nc}\text{‘}\beta$

$*100\cdot33$. $\vdash : \exists ! \mathrm{Nc}\text{‘}\alpha \cap \mathrm{Nc}\text{‘}\beta . \supset . \alpha \,\mathrm{sm}\, \beta$

We proceed next to the elementary properties of NC. We have

$*100\cdot4$. $\vdash : \mu \in \mathrm{NC} . \equiv . (\exists \alpha) . \mu = \mathrm{Nc}\text{‘}\alpha$

$*100\cdot42$. $\vdash : \mu, \nu \in \mathrm{NC} . \exists ! \mu \cap \nu . \supset . \mu = \nu$

$*100\cdot45$. $\vdash : \mu \in \mathrm{NC} . \alpha \in \mu . \supset . \mathrm{Nc}\text{‘}\alpha = \mu$

$*100\cdot51$. $\vdash : \mu \in \mathrm{NC} . \alpha \in \mu . \supset . \mathrm{sm}\text{‘‘}\mu = \mathrm{Nc}\text{‘}\alpha$

Observe that when we have such a hypothesis as "$\mu \in \mathrm{NC}$," the μ, though it may be of any type, must be of *some* type; hence the μ cannot have the typical ambiguity which belongs to $\mathrm{Nc}\text{‘}\alpha$. If we put $\mu = \mathrm{Nc}\text{‘}\alpha$, this will hold only in the type of μ; but "$\mathrm{sm}\text{‘‘}\mu$" is a typically ambiguous symbol, which will represent in any type the "same" number as μ. Thus "$\text{sm}``\mu = \text{Nc}`\alpha$" is an equation which is applicable to all possible typical determinations of "sm" and "Nc."

***100·52.** $\vdash : \mu \in \text{NC} \,.\, \exists ! \mu \,.\, \supset . \, \text{sm}``\mu \in \text{NC}$

The hypothesis $\exists ! \mu$ is unnecessary, but we cannot prove this till later (*102).

We end the number with some propositions (*100·6—·64) stating that various classes (such as $\iota``\alpha$), which have already been proved to be similar to α, have $\text{Nc}`\alpha$ members.

***100·01.** $\text{Nc} = \overrightarrow{\text{sm}}$ Df

***100·02.** $\text{NC} = \text{D}`\text{Nc}$ Df

***100·1.** $\vdash . \, \text{Nc}`\alpha = \hat{\beta}(\beta \,\text{sm}\, \alpha) = \hat{\beta}(\alpha \,\text{sm}\, \beta)$ [*32·13 . *73·31 . (*100·01)]

***100·11.** $\vdash . \, \text{Nc}`\alpha = \hat{\beta}\{(\exists R) . R \in 1 \to 1 . \text{D}`R = \alpha . \text{Cl}`R = \beta\}$ [*100·1 . *73·1]

***100·12.** $\vdash . \, \text{Nc}`\alpha = \hat{\beta}\{(\exists R) . R \in 1 \to 1 . \alpha \subset \text{D}`R . \beta = \breve{R}``\alpha\}$
[*100·1 . *73·11]

***100·13.** $\vdash . \, \text{Nc}`\alpha = \text{Cl}``(1 \to 1 \cap \overleftarrow{\text{D}}`\alpha) = \text{D}``(1 \to 1 \cap \overleftarrow{\text{Cl}}`\alpha)$

Dem.

$\vdash . \, *100·11 . *33·6 . \quad \supset \vdash . \, \text{Nc}`\alpha = \hat{\beta}\{(\exists R) . R \in 1 \to 1 . R \in \overleftarrow{\text{D}}`\alpha . \text{Cl}`R = \beta\}$

[*22·33.*37·6] $\quad = \text{Cl}``(1 \to 1 \cap \overleftarrow{\text{D}}`\alpha)$ (1)

$\vdash . \, *100·1 . *73·1 . *33·61 . \supset \vdash . \, \text{Nc}`\alpha = \hat{\beta}\{(\exists R) . R \in 1 \to 1 . R \in \overleftarrow{\text{Cl}}`\alpha . \text{D}`R = \beta\}$

[*22·33.*37·6] $\quad = \text{D}``(1 \to 1 \cap \overleftarrow{\text{Cl}}`\alpha)$ (2)

$\vdash . (1) . (2) . \supset \vdash . \text{Prop}$

***100·14.** $\vdash . \, \text{Nc}`\alpha = \hat{\beta}\{(\exists R) . \alpha \subset \text{Cl}`R . R \restriction \alpha \in 1 \to 1 . \beta = R``\alpha\}$
[*73·15 . *100·1]

***100·15.** $\vdash . \, \text{Nc}`\alpha = \hat{\beta}\{(\exists R) : \text{E}!! \, R``\alpha :$
$x, y \in \alpha . R`x = R`y . \supset_{x,y} . x = y : \beta = R``\alpha\}$

Dem.

$\vdash . \, *74·1·11 . \supset$

$\vdash :. \text{E}!! \, R``\alpha : x, y \in \alpha . R`x = R`y . \supset_{x,y} . x = y : \beta = R``\alpha : \equiv :$
$R \restriction \alpha \in 1 \to \text{Cls} . \alpha \subset \text{Cl}`R . R \restriction \alpha \in 1 \to 1 . \beta = R``\alpha$ (1)

$\vdash . (1) . *4·71 . *100·14 . \supset \vdash . \text{Prop}$

***100·16.** $\vdash . \, \text{Nc}`\alpha = \hat{\beta}\{(\exists R) :. x, y \in \alpha . \supset_{x,y} : R`x = R`y . \equiv . x = y :. \beta = R``\alpha\}$

Dem.

$\vdash . \, *71·59 . \supset$

$\vdash :: x, y \in \alpha . \supset_{x,y} : R`x = R`y . \equiv . x = y :. \equiv . R \restriction \alpha \in 1 \to 1 . \alpha \subset \text{Cl}`R$ (1)

$\vdash . (1) . *100·14 . \supset \vdash . \text{Prop}$

∗100·2. $\vdash . E ! Nc`\alpha$ [∗32·12 . (∗100·01)]

∗100·21. $\vdash . \text{Ɑ}`Nc = Cls$

Dem.

$\vdash . ∗37·76 . (∗100·01) . \supset \vdash . \text{Ɑ}`Nc \subset Cls$ (1)

$\vdash . ∗33·431 . ∗100·2 . \supset \vdash . Cls \subset \text{Ɑ}`Nc$ (2)

$\vdash . (1) . (2) . \supset \vdash . Prop$

∗100·22. $\vdash . Nc \epsilon 1 \rightarrow Cls$ [∗72·12 . (∗100·01)]

∗100·3. $\vdash . \alpha \epsilon Nc`\alpha$ [∗73·3 . ∗100·1]

Note that it is fallacious to infer $\exists ! Nc`\alpha$, for reasons explained in the introduction to the present section.

∗100·31. $\vdash : \alpha \epsilon Nc`\beta . \equiv . \beta \epsilon Nc`\alpha . \equiv . \alpha \text{ sm } \beta$ [∗32·18 . ∗73·31 . (∗100·01)]

∗100·32. $\vdash : \alpha \epsilon Nc`\beta . \beta \epsilon Nc`\gamma . \supset . \alpha \epsilon Nc`\gamma$ [∗100·31 . ∗73·32]

∗100·321. $\vdash : \alpha \text{ sm } \beta . \supset . Nc`\alpha = Nc`\beta$

Dem.

$\vdash . ∗73·37 . \supset \vdash :. Hp . \supset : \gamma \text{ sm } \alpha . \equiv_\gamma . \gamma \text{ sm } \beta :$

[∗100·1] $\supset : Nc`\alpha = Nc`\beta :. \supset \vdash . Prop$

Note that $Nc`\alpha = Nc`\beta . \supset . \alpha \text{ sm } \beta$ is not always true. We might be tempted to prove it as follows:

$\vdash . ∗100·1 . \supset \vdash :. Nc`\alpha = Nc`\beta . \equiv : \gamma \text{ sm } \alpha . \equiv_\gamma . \gamma \text{ sm } \beta :$

[∗10·1] $\supset : \alpha \text{ sm } \alpha . \equiv . \alpha \text{ sm } \beta :$

[∗73·3] $\supset : \alpha \text{ sm } \beta$

But the use of ∗10·1 here is only legitimate when the "sm" concerned is a homogeneous relation. If $Nc`\alpha$, $Nc`\beta$ are descending cardinals, we may have $Nc`\alpha = \Lambda = Nc`\beta$ without having $\alpha \text{ sm } \beta$.

∗100·33. $\vdash : \exists ! Nc`\alpha \cap Nc`\beta . \supset . \alpha \text{ sm } \beta$

Dem.

$\vdash . ∗100·1 . \supset \vdash : Hp . \supset . (\exists\gamma) . \gamma \text{ sm } \alpha . \gamma \text{ sm } \beta .$

[∗73·31] $\supset . (\exists\gamma) . \alpha \text{ sm } \gamma . \gamma \text{ sm } \beta .$

[∗73·32] $\supset . \alpha \text{ sm } \beta : \supset \vdash . Prop$

Note that we do not always have

$$\alpha \text{ sm } \beta . \supset . \exists ! Nc`\alpha \cap Nc`\beta.$$

For if the Nc concerned is a descending Nc, and α and β are sufficiently great, $Nc`\alpha$ and $Nc`\beta$ may both be Λ. For example, we have

$$Cl`(\alpha \cup -\alpha) \text{ sm } Cl`(\alpha \cup -\alpha).$$

But $Nc(\alpha)`Cl`(\alpha \cup -\alpha) = \Lambda$, so that

$$\sim \exists ! Nc(\alpha)`Cl`(\alpha \cup -\alpha) \cap Nc(\alpha)`Cl`(\alpha \cup -\alpha).$$

Thus "$\alpha \,\text{sm}\, \beta \,.\supset.\, \exists \,!\, \text{Nc}‘\alpha \cap \text{Nc}‘\beta$" is not always true when it is significant.

$*100{\cdot}34$. $\vdash : \exists \,!\, \text{Nc}‘\alpha \cap \text{Nc}‘\beta \,.\supset.\, \text{Nc}‘\alpha = \text{Nc}‘\beta$ $[*100{\cdot}33{\cdot}321]$

$*100{\cdot}35$. $\vdash :. \exists \,!\, \text{Nc}‘\alpha \,.\vee.\, \exists \,!\, \text{Nc}‘\beta : \supset :$

$$\text{Nc}‘\alpha = \text{Nc}‘\beta \,.\equiv.\, \alpha \in \text{Nc}‘\beta \,.\equiv.\, \beta \in \text{Nc}‘\alpha \,.\equiv.\, \alpha \,\text{sm}\, \beta$$

Dem.

$\vdash . *22{\cdot}5 . \quad \supset \vdash :. \text{Hp} . \supset : \text{Nc}‘\alpha = \text{Nc}‘\beta \,.\supset.\, \exists \,!\, \text{Nc}‘\alpha \cap \text{Nc}‘\beta \,.$

$[*100{\cdot}33] \quad \supset . \alpha \,\text{sm}\, \beta \quad (1)$

$\vdash . (1) . *100{\cdot}321 . \supset \vdash :. \text{Hp} . \supset : \text{Nc}‘\alpha = \text{Nc}‘\beta \,.\equiv.\, \alpha \,\text{sm}\, \beta \quad (2)$

$\vdash . (2) . *100{\cdot}31 . \quad \supset \vdash . \text{Prop}$

Thus the only case in which the implications in $*100{\cdot}321{\cdot}33{\cdot}34$ cannot be turned into equivalences is the case in which $\text{Nc}‘\alpha$ and $\text{Nc}‘\beta$ are both Λ.

$*100{\cdot}36$. $\vdash :. \beta \in \text{Nc}‘\alpha . \supset : \exists \,!\, \alpha \,.\equiv.\, \exists \,!\, \beta$ $[*100{\cdot}31 . *73{\cdot}36]$

$*100{\cdot}4$. $\vdash : \mu \in \text{NC} \,.\equiv.\, (\exists \alpha) . \mu = \text{Nc}‘\alpha$ $[*37{\cdot}78{\cdot}79 . (*100{\cdot}02{\cdot}01)]$

$*100{\cdot}41$. $\vdash . \text{Nc}‘\alpha \in \text{NC}$ $[*100{\cdot}4{\cdot}2 . *14{\cdot}204]$

$*100{\cdot}42$. $\vdash : \mu, \nu \in \text{NC} . \exists \,!\, \mu \cap \nu \,.\supset.\, \mu = \nu$

Dem.

$\vdash . *100{\cdot}4 . \supset \vdash : \text{Hp} . \supset . (\exists \alpha, \beta) . \mu = \text{Nc}‘\alpha . \nu = \text{Nc}‘\beta . \exists \,!\, \text{Nc}‘\alpha \cap \text{Nc}‘\beta .$

$[*100{\cdot}34] \quad \supset . (\exists \alpha, \beta) . \mu = \text{Nc}‘\alpha . \nu = \text{Nc}‘\beta . \text{Nc}‘\alpha = \text{Nc}‘\beta .$

$[*14{\cdot}15] \quad \supset . \mu = \nu : \supset \vdash . \text{Prop}$

$*100{\cdot}43$. $\vdash . \text{NC} \in \text{Cls}^2\,\text{excl}$ $[*100{\cdot}42 . *84{\cdot}11]$

$*100{\cdot}44$. $\vdash :. \mu \in \text{NC} . \exists \,!\, \text{Nc}‘\alpha . \supset : \alpha \in \mu \,.\equiv.\, \text{Nc}‘\alpha = \mu$

Dem.

$\vdash . *100{\cdot}3 . \supset \vdash : \text{Nc}‘\alpha = \mu \,.\supset.\, \alpha \in \mu \quad (1)$

$\vdash . *10{\cdot}24 . \supset \vdash : \mu \in \text{NC} . \exists \,!\, \text{Nc}‘\alpha . \alpha \in \mu . \supset .$

$\mu \in \text{NC} . \exists \,!\, \mu . \exists \,!\, \text{Nc}‘\alpha . \alpha \in \mu .$

$[*100{\cdot}4] \quad \supset . (\exists \beta) . \mu = \text{Nc}‘\beta . \exists \,!\, \text{Nc}‘\beta . \exists \,!\, \text{Nc}‘\alpha . \alpha \in \text{Nc}‘\beta .$

$[*100{\cdot}35] \quad \supset . (\exists \beta) . \mu = \text{Nc}‘\beta . \text{Nc}‘\alpha = \text{Nc}‘\beta .$

$[*14{\cdot}15] \quad \supset . \text{Nc}‘\alpha = \mu \quad (2)$

$\vdash . (1) . (2) . \supset \vdash . \text{Prop}$

$*100{\cdot}45$. $\vdash : \mu \in \text{NC} . \alpha \in \mu \,.\supset.\, \text{Nc}‘\alpha = \mu$ $[*100{\cdot}4{\cdot}31{\cdot}321]$

$*100{\cdot}5$. $\vdash : \mu \in \text{NC} . \alpha, \beta \in \mu \,.\supset.\, \alpha \,\text{sm}\, \beta$

Dem.

$\vdash . *100{\cdot}4 . \supset \vdash : \text{Hp} . \supset . (\exists \gamma) . \mu = \text{Nc}‘\gamma . \alpha, \beta \in \text{Nc}‘\gamma .$

$[*100{\cdot}31] \quad \supset . (\exists \gamma) . \alpha \,\text{sm}\, \gamma . \beta \,\text{sm}\, \gamma .$

$[*73{\cdot}31{\cdot}32] \quad \supset . \alpha \,\text{sm}\, \beta : \supset \vdash . \text{Prop}$

***100·51.** $\vdash : \mu \in \text{NC} . \alpha \in \mu . \supset . \text{sm}``\mu = \text{Nc}`\alpha$

Dem.

$\vdash . *100·5 . \text{Fact} . \supset \vdash :. \text{Hp} . \supset : \beta \in \mu . \gamma \,\text{sm}\, \beta . \supset . \alpha \,\text{sm}\, \beta . \gamma \,\text{sm}\, \beta .$

[*73·31·32] $\supset . \alpha \,\text{sm}\, \gamma .$

[*100·31] $\supset . \gamma \in \text{Nc}`\alpha$ (1)

$\vdash . (1) . *10·11·21·23 . *37·1 . \supset \vdash : \text{Hp} . \supset . \text{sm}``\mu \subset \text{Nc}`\alpha$ (2)

$\vdash . *100·31 . \supset \vdash :. \text{Hp} . \supset : \gamma \in \text{Nc}`\alpha . \supset . \gamma \,\text{sm}\, \alpha . \alpha \in \mu .$

[*37·1] $\supset . \gamma \in \text{sm}``\mu$ (3)

$\vdash . (2) . (3) . \supset \vdash . \text{Prop}$

***100·511.** $\vdash : \exists ! \,\text{Nc}`\beta . \supset . \text{sm}``\text{Nc}`\beta = \text{Nc}`\beta$

Here the last "Nc‘β" may be of a different type from the others: the proposition holds however its type is determined.

Dem.

$\vdash . *100·51·41 . \supset \vdash : \alpha \in \text{Nc}`\beta . \supset . \text{sm}``\text{Nc}`\beta = \text{Nc}`\alpha$

[*100·31·321] $= \text{Nc}`\beta$ (1)

$\vdash . (1) . *10·11·23 . \supset \vdash . \text{Prop}$

***100·52.** $\vdash : \mu \in \text{NC} . \exists ! \mu . \supset . \text{sm}``\mu \in \text{NC}$ [*100·51·4]

This proposition still holds when $\mu = \Lambda$, but the proof is more difficult, since it depends upon the proof that every null-class of classes is an NC, which in turn depends upon the proof that Cl‘α is not similar to α or to any class contained in α.

***100·521.** $\vdash : \mu \in \text{NC} . \exists ! \,\text{sm}``\mu . \supset . \text{sm}``\text{sm}``\mu = \mu$

Dem.

$\vdash . *37·29 . \text{Transp} . \supset \vdash :. \text{Hp} . \supset : \exists ! \mu :$

[*100·52] $\supset : \text{sm}``\mu \in \text{NC} :$

[*100·51.Hp] $\supset : \gamma \in \text{sm}``\mu . \supset . \text{sm}``\text{sm}``\mu = \text{Nc}`\gamma$ (1)

$\vdash . *37·1 . \text{Fact} . \supset \vdash : \text{Hp} . \gamma \in \text{sm}``\mu . \supset . (\exists \alpha) . \alpha \in \mu . \mu \in \text{NC} . \gamma \,\text{sm}\, \alpha .$

[*100·45·321] $\supset . (\exists \alpha) . \text{Nc}`\alpha = \mu . \text{Nc}`\gamma = \text{Nc}`\alpha .$

[*13·17] $\supset . \text{Nc}`\gamma = \mu$ (2)

$\vdash . (1) . (2) . \supset \vdash :. \text{Hp} . \gamma \in \text{sm}``\mu . \supset . \text{sm}``\text{sm}``\mu = \mu$ (3)

$\vdash . (3) . *10·11·23·35 . \supset \vdash . \text{Prop}$

***100·53.** $\vdash :. \exists ! \mu . \exists ! \nu . \supset : \mu \in \text{NC} . \nu = \text{sm}``\mu . \equiv . \nu \in \text{NC} . \mu = \text{sm}``\nu$

Dem.

$\vdash . *100·52 . \supset \vdash :. \text{Hp} . \supset : \mu \in \text{NC} . \nu = \text{sm}``\mu . \supset . \nu \in \text{NC}$ (1)

$\vdash . *100·521 . \supset \vdash :. \text{Hp} . \supset : \mu \in \text{NC} . \nu = \text{sm}``\mu . \supset . \mu = \text{sm}``\nu$ (2)

$\vdash . (1) . (2) . \supset \vdash :. \text{Hp} . \supset : \mu \in \text{NC} . \nu = \text{sm}``\mu . \supset . \nu \in \text{NC} . \mu = \text{sm}``\nu$ (3)

$\vdash . (3) . (3)\dfrac{\nu, \mu}{\mu, \nu} . \supset \vdash . \text{Prop}$

∗100·6. $\vdash . \iota``\alpha \epsilon \text{Nc}`\alpha$ [∗73·41 . ∗100·31]

∗100·61. $\vdash . \hat{\beta}\{(\exists y) . y \epsilon \alpha . \beta = \iota`x \cup \iota`y\} \epsilon \text{Nc}`\alpha$ [∗73·27 . ∗54·21 . ∗100·31]

∗100·62. $\vdash . x \downarrow ``\alpha \epsilon \text{Nc}`\alpha$ [∗73·61 . ∗100·31]

∗100·621. $\vdash . \downarrow x``\alpha \epsilon \text{Nc}`\alpha$ [∗73·611 . ∗100·31]

∗100·63. $\vdash . \epsilon_\Delta`\iota`\alpha \epsilon \text{Nc}`\alpha$ [∗83·41 . ∗100·31]

∗100·631. $\vdash . \text{D}``\epsilon_\Delta`\iota`\alpha \epsilon \text{Nc}`\alpha$ [∗83·7 . ∗100·6]

∗100·64. $\vdash : \kappa \epsilon \text{Cls}^2 \text{excl} . \supset . \text{D}``\epsilon_\Delta`\kappa \subset \text{Nc}`\kappa$

Dem.

$\vdash$. ∗84·3 . ∗80·14 . $\supset \vdash : \text{Hp} . R \epsilon \epsilon_\Delta`\kappa . \supset . R \epsilon 1 \rightarrow 1 . \kappa = \text{G}`R$.

[∗73·2.∗100·31] $\supset . \text{D}`R \epsilon \text{Nc}`\kappa : \supset \vdash . \text{Prop}$

$*101$. ON 0 AND 1 AND 2.

Summary of $*101$.

In the present number, we have to show that 0 and 1 and 2 as previously defined are cardinal numbers in the sense defined in $*100$, and to add a few elementary propositions to those already given concerning them. We prove ($*101\cdot12\cdot241$) that 0 and 1 are not null, which cannot be proved, with our axioms, for any other cardinal, except (in the case of finite cardinals) when the type is specified as a sufficiently high one. Thus we prove ($*101\cdot42\cdot43$) that 2_{Cls} and 2_{Rel} exist; this follows from $\Lambda \neq V$ and $\dot{\Lambda} \neq \dot{V}$. We prove ($*101\cdot22\cdot34$) that 0 and 1 and 2 are all different from each other. We prove ($*101\cdot15\cdot28$) that $\text{sm}``0 = 0$ and $\text{sm}``1 = 1$, but we cannot prove $\text{sm}``2 = 2$ unless we assume the existence of at least two individuals, or define the first 2 in "$\text{sm}``2 = 2$" as a 2 of some type other than 2_{Indiv}, where "Indiv" stands for the type of individuals.

It should be observed that, since 0 and 1 and 2 are typically ambiguous, their properties are analogous to those of "$\text{Nc}`\alpha$" rather than to those of μ, where $\mu \in \text{NC}$. For example, we have

$*100\cdot511$. $\vdash : \exists ! \, \text{Nc}`\beta \, . \supset . \, \text{sm}``\text{Nc}`\beta = \text{Nc}`\beta$

but we shall not have $\mu \in \text{NC} \, . \, \exists ! \, \mu \, . \supset . \, \text{sm}``\mu = \mu$ unless the "sm" concerned is homogeneous, since in other cases the symbols do not express a significant proposition. But in $*100\cdot511$ we may substitute 0 or 1 or 2, and the proposition remains significant and true. In fact we have ($*101\cdot1\cdot2\cdot31$)

$$\vdash . \, 0 = \text{Nc}`\Lambda \, . \, 1 = \text{Nc}`\iota`x \, . \, 2 = \text{Nc}`(\iota`\iota`x \cup \iota`\Lambda),$$

where 0 and 1 and 2 have an ambiguity corresponding to that of "Nc."

$*101\cdot1$. $\vdash . \, 0 = \text{Nc}`\Lambda$ $\quad$ [$*73\cdot48 \, . \, *100\cdot1$]

$*101\cdot11$. $\vdash . \, 0 \in \text{NC}$ $\quad$ [$*101\cdot1 \, . \, *100\cdot4$]

$*101\cdot12$. $\vdash . \, \exists ! \, 0$ $\quad$ [$*51\cdot161 \, . \, (*54\cdot01)$]

$*101\cdot13$. $\vdash . \, \exists ! \, 0 \cap \text{Cl}`\alpha \, . \, \Lambda \in 0 \cap \text{Cl}`\alpha$ $\quad$ [$*51\cdot16 \, . \, *60\cdot3$]

***101·14.** $\vdash : \mathrm{Nc}`\gamma = 0 . \equiv . \gamma = \Lambda$

Dem.

$\vdash . *101\cdot1\cdot12 . \supset \vdash : \mathrm{Nc}`\gamma = 0 . \equiv . \mathrm{Nc}`\gamma = \mathrm{Nc}`\Lambda . \exists ! \mathrm{Nc}`\Lambda .$

$[*13\cdot194] \quad \equiv . \mathrm{Nc}`\gamma = \mathrm{Nc}`\Lambda . \exists ! \mathrm{Nc}`\Lambda . \exists ! \mathrm{Nc}`\gamma .$

$[*100\cdot35] \quad \equiv . \gamma \in \mathrm{Nc}`\Lambda . \exists ! \mathrm{Nc}`\Lambda . \exists ! \mathrm{Nc}`\gamma .$

$[*101\cdot1 . *54\cdot102] \quad \equiv . \gamma = \Lambda . \exists ! \mathrm{Nc}`\Lambda . \exists ! \mathrm{Nc}`\gamma .$

$[*101\cdot1\cdot12 . *13\cdot194] \quad \equiv . \gamma = \Lambda : \supset \vdash . \mathrm{Prop}$

***101·15.** $\vdash . \mathrm{sm}``0 = 0$

Dem.

$\vdash . *37\cdot1 . \supset \vdash : \gamma \in \mathrm{sm}``0 . \equiv . (\exists \alpha) . \alpha \in 0 . \gamma \,\mathrm{sm}\, \alpha .$

$[*54\cdot102] \quad \equiv . \gamma \,\mathrm{sm}\, \Lambda .$

$[*73\cdot48] \quad \equiv . \gamma \in 0 : \supset \vdash . \mathrm{Prop}$

***101·16.** $\vdash :. \mu \in \mathrm{NC} - \iota`0 . \supset : \alpha \in \mu . \supset_\alpha . \exists ! \alpha$

Dem.

$\vdash . *100\cdot45 . \quad \supset \vdash : \mu \in \mathrm{NC} . \Lambda \in \mu . \supset . \mu = \mathrm{Nc}`\Lambda$

$[*101\cdot1] \quad = 0 \qquad (1)$

$\vdash . (1) . \mathrm{Transp} . \supset \vdash :. \mu \in \mathrm{NC} - \iota`0 . \supset : \Lambda \sim\in \mu :$

$[*24\cdot63] \quad \supset : \alpha \in \mu . \supset_\alpha . \exists ! \alpha :. \supset \vdash . \mathrm{Prop}$

***101·17.** $\vdash : \Lambda \in \mathrm{Nc}`\alpha . \equiv . \mathrm{Nc}`\alpha = 0 . \equiv . \mathrm{Nc}`\alpha = \mathrm{Nc}`\Lambda . \equiv . \alpha = \Lambda$

Dem.

$\vdash . *100\cdot31\cdot321 . \supset \vdash : \Lambda \in \mathrm{Nc}`\alpha . \supset . \mathrm{Nc}`\alpha = \mathrm{Nc}`\Lambda .$

$[*101\cdot1] \quad \supset . \mathrm{Nc}`\alpha = 0 \qquad (1)$

$\vdash . *101\cdot13 . \quad \supset \vdash : \mathrm{Nc}`\alpha = 0 . \supset . \Lambda \in \mathrm{Nc}`\alpha \qquad (2)$

$\vdash . (1) . (2) . \quad \supset \vdash : \Lambda \in \mathrm{Nc}`\alpha . \equiv . \mathrm{Nc}`\alpha = 0 . \qquad (3)$

$[*101\cdot1] \quad \equiv . \mathrm{Nc}`\alpha = \mathrm{Nc}`\Lambda . \qquad (4)$

$[*101\cdot14] \quad \equiv . \alpha = \Lambda \qquad (5)$

$\vdash . (3) . (4) . (5) . \supset \vdash . \mathrm{Prop}$

***101·2.** $\vdash . 1 = \mathrm{Nc}`\iota`x \quad [*73\cdot45 . *100\cdot1]$

***101·21.** $\vdash . 1 \in \mathrm{NC} \quad [*101\cdot2 . *100\cdot4]$

***101·22.** $\vdash . 1 \neq 0$

Dem.

$\vdash . *52\cdot21 . *101\cdot13 . \supset \vdash . \Lambda \sim\in 1 . \Lambda \in 0 .$

$[*13\cdot14] \quad \supset \vdash . 1 \neq 0$

***101·23.** $\vdash . 1 \cap 0 = \Lambda$

Dem.

$\vdash . *52\cdot21 . \quad \supset \vdash : \alpha \in 1 . \supset . \alpha \neq \Lambda .$

$[*54\cdot102] \quad \supset . \alpha \sim\in 0 \qquad (1)$

$\vdash . (1) . *24\cdot39 . \supset \vdash . \mathrm{Prop}$

***101·24.** $\vdash : \exists ! \alpha . \supset . \exists ! 1 \cap \text{Cl}`\alpha$

Dem.

$\vdash . *52·22 . *60·6 . \supset \vdash : x \epsilon \alpha . \supset . \iota`x \epsilon 1 \cap \text{Cl}`\alpha$ (1)

$\vdash . (1) . *10·11·28 . \supset \vdash . \text{Prop}$

***101·241.** $\vdash . \exists ! 1$ [*52·23]

***101·25.** $\vdash : \alpha \epsilon 1 . \beta \subset \alpha . \beta \neq \alpha . \supset . \beta \epsilon 0$

Dem.

$\vdash . *52·64 . *22·621 . \supset \vdash : \alpha \epsilon 1 . \beta \subset \alpha . \supset . \beta \epsilon 1 \cup 0$ (1)

$\vdash . *52·46 . \quad \supset \vdash : \alpha, \beta \epsilon 1 . \beta \subset \alpha . \supset . \beta = \alpha :$

[Transp] $\quad \supset \vdash : \alpha \epsilon 1 . \beta \subset \alpha . \beta \neq \alpha . \supset . \beta \sim \epsilon 1$ (2)

$\vdash . (1) . (2) . \supset \vdash . \text{Prop}$

***101·26.** $\vdash . s`\text{Cl}``1 = 0 \cup 1$

Dem.

$\vdash . *60·371 . *40·43 . \supset \vdash . s`\text{Cl}``1 \subset 0 \cup 1$ (1)

$\vdash . *60·3·34 . \quad \supset \vdash . \Lambda \epsilon \text{Cl}`\iota`x . \iota`x \epsilon \text{Cl}`\iota`x .$

[*52·22.*40·4] $\quad \supset \vdash . \Lambda \epsilon s`\text{Cl}``1 , \iota`x \epsilon s`\text{Cl}``1 .$

[*51·2.*52·1] $\quad \supset \vdash . 0 \subset s`\text{Cl}``1 . 1 \subset s`\text{Cl}``1$ (2)

$\vdash . (1) . (2) . \supset \vdash . \text{Prop}$

***101·27.** $\vdash . 1 = \hat{\alpha} \{(\exists x) . x \epsilon \alpha . \alpha - \iota`x \epsilon 0\}$

Dem.

$\vdash . *54·102 . \supset \vdash : (\exists x) . x \epsilon \alpha . \alpha - \iota`x \epsilon 0 . \equiv . (\exists x) . x \epsilon \alpha . \alpha - \iota`x = \Lambda .$

[*24·3] $\quad \equiv . (\exists x) . x \epsilon \alpha . \alpha \subset \iota`x .$

[*51·2] $\quad \equiv . (\exists x) . \alpha = \iota`x .$

[*52·1] $\quad \equiv . \alpha \epsilon 1 : \supset \vdash . \text{Prop}$

***101·28.** $\vdash . \text{sm}``1 = 1$

Dem.

$\vdash . *37·1 . \supset \vdash : \gamma \epsilon \text{sm}``1 . \equiv . (\exists \alpha) . \alpha \epsilon 1 . \gamma \,\text{sm}\, \alpha .$

[*52·1] $\quad \equiv . (\exists x) . \gamma \,\text{sm}\, \iota`x .$

[*73·45] $\quad \equiv . \gamma \epsilon 1 : \supset \vdash . \text{Prop}$

***101·29.** $\vdash : \iota`x \epsilon \text{Nc}`\alpha . \equiv . \text{Nc}`\alpha = 1 . \equiv . \text{Nc}`\alpha = \text{Nc}`\iota`x . \equiv . \alpha \epsilon 1$

Dem.

$\vdash . *100·31·321 . \quad \supset \vdash : \iota`x \epsilon \text{Nc}`\alpha . \supset . \text{Nc}`\alpha = \text{Nc}`\iota`x .$

[*101·2] $\quad \supset . \text{Nc}`\alpha = 1$ (1)

$\vdash . *52·22 . \quad \supset \vdash : \text{Nc}`\alpha = 1 . \supset . \iota`x \epsilon \text{Nc}`\alpha$ (2)

$\vdash . (1) . (2) . \quad \supset \vdash : \iota`x \epsilon \text{Nc}`\alpha . \equiv . \text{Nc}`\alpha = 1 .$ (3)

[*101·2] $\quad \equiv . \text{Nc}`\alpha = \text{Nc}`\iota`x$ (4)

$\vdash . *101·2 . *52·1 . \supset \vdash : \alpha \epsilon 1 . \supset . \text{Nc}`\alpha = 1$ (5)

$\vdash . *100·3 . \quad \supset \vdash : \text{Nc}`\alpha = 1 . \supset . \alpha \epsilon 1$ (6)

$\vdash . (3) . (4) . (5) . (6) . \supset \vdash . \text{Prop}$

***101·3.** $\vdash : x \neq y . \supset . 2 = \text{Nc}`(\iota`x \cup \iota`y)$

Dem.

$\vdash . *73·71·43 . *51·231 . \supset \vdash :. \text{Hp} . \supset : z \neq w . \supset . (\iota`z \cup \iota`w) \text{ sm } (\iota`x \cup \iota`y) :$

$[*54·101] \quad \supset : \beta \in 2 . \supset . \beta \text{ sm } (\iota`x \cup \iota`y) :$

$[*100·1] \quad \supset : 2 \subset \text{Nc}`(\iota`x \cup \iota`y) \quad (1)$

$\vdash . *53·32 . *71·163 . \supset \vdash : R \in 1 \rightarrow 1 . x, y \in \text{C}`R . \supset .$

$R``(\iota`x \cup \iota`y) = \iota`R`x \cup \iota`R`y \quad (2)$

$\vdash . *71·56 . \text{Transp} . \supset \vdash : \text{Hp} . R \in 1 \rightarrow 1 . x, y \in \text{C}`R . \supset . R`x \neq R`y \quad (3)$

$\vdash . (2) . (3) . *54·26 . \supset$

$\vdash :. \text{Hp} . \supset : R \in 1 \rightarrow 1 . x, y \in \text{C}`R . \beta = R``(\iota`x \cup \iota`y) . \supset . \beta \in 2 :$

$[*10·11·21·23 . *51·234] \supset : (\exists R) . R \in 1 \rightarrow 1 . \iota`x \cup \iota`y \subset \text{C}`R . \beta = R``(\iota`x \cup \iota`y) .$

$\supset . \beta \in 2 :$

$[*73·12 . *100·1] \quad \supset : \text{Nc}`(\iota`x \cup \iota`y) \subset 2 \quad (4)$

$\vdash . (1) . (4) . \supset \vdash . \text{Prop}$

***101·301.** $\vdash . 2 = \hat{\alpha} \{(\exists x) . x \in \alpha . \alpha - \iota`x \in 1\}$ [*54·3]

In comparing *101·31 with *101·1·2·3, it should be observed that $\iota`x$ and Λ are both *classes*, whereas in *101·1·2·3 there was no typical limitation beyond what was imposed by the conditions of significance.

***101·31.** $\vdash . 2 = \text{Nc}`(\iota`\iota`x \cup \iota`\Lambda)$

Dem.

$\vdash . *51·161 . \quad \supset \vdash . \iota`x \neq \Lambda \quad (1)$

$\vdash . (1) . *101·3 . \supset \vdash . \text{Prop}$

***101·32.** $\vdash . 2 \in \text{NC}$ [*101·31 . *100·4]

***101·33.** $\vdash : \alpha, \beta \in 1 . \alpha \cap \beta = \Lambda . \supset . \alpha \cup \beta \in 2$ [*54·43]

***101·34.** $\vdash . 2 \neq 0 . 2 \neq 1$

Dem.

$\vdash . *101·13 . \quad \supset \vdash . \Lambda \in 0 \quad (1)$

$\vdash . *101·301 . \quad \supset \vdash : \alpha \in 2 . \supset . \exists ! \alpha :$

$[*24·63] \quad \supset \vdash . \Lambda \sim \in 2 \quad (2)$

$\vdash . (1) . (2) . *13·14 . \quad \supset \vdash . 2 \neq 0 \quad (3)$

$\vdash . *52·22 . *54·26 . *22·56 . \supset \vdash . \iota`y \in 1 . \iota`y \sim \in 2 .$

$[*13·14] \quad \supset \vdash . 1 \neq 2 \quad (4)$

$\vdash . (3) . (4) . \supset \vdash . \text{Prop}$

***101·35.** $\vdash . 2 \cap 0 = \Lambda . 2 \cap 1 = \Lambda$ [*100·42 . Transp . *101·11·21·32·34]

***101·36.** $\vdash : \alpha \in 2 . \beta \subset \alpha . \beta \neq \alpha . \supset . \beta \in 0 \cup 1$

Dem.

$\vdash . *54·42 . \supset \vdash : \alpha \in 2 . \beta \subset \alpha . \exists ! \beta . \beta \neq \alpha . \supset . \beta \in 1 \quad (1)$

$\vdash . *54·102 . \supset \vdash : \sim \exists ! \beta . \supset . \beta \in 0 \quad (2)$

$\vdash . (1) . (2) . \supset \vdash . \text{Prop}$

***101·37.** $\vdash . s{}^{\backprime}\text{Cl}{}^{\backprime\backprime}2 \subset 0 \cup 1 \cup 2$ [*54·411]

***101·38.** $\vdash : \exists ! 2 . \supset . s{}^{\backprime}\text{Cl}{}^{\backprime\backprime}2 = 0 \cup 1 \cup 2$

Dem.

$\vdash . *60{\cdot}3 . \quad \supset \vdash : \text{Hp} . \supset . (\exists \alpha) . \alpha \in 2 . \Lambda \in \text{Cl}{}^{\backprime}\alpha .$

[*40·4] $\quad \supset . \Lambda \in s{}^{\backprime}\text{Cl}{}^{\backprime\backprime}2 .$

[*51·2] $\quad \supset . 0 \subset s{}^{\backprime}\text{Cl}{}^{\backprime\backprime}2$ (1)

$\vdash . *60{\cdot}34 . \quad \supset \vdash . 2 \subset s{}^{\backprime}\text{Cl}{}^{\backprime\backprime}2$ (2)

$\vdash . *54{\cdot}101 . \supset \vdash :: \text{Hp} . \supset :. (\exists x, y) . x \neq y :.$

[*13·171.Transp] $\quad \supset :. (\exists x, y) :. (z) : z \neq x . \vee . z \neq y :.$

[*54·26] $\quad \supset :. (\exists x, y) :. (z) : \iota{}^{\backprime}z \cup \iota{}^{\backprime}x \in 2 . \vee . \iota{}^{\backprime}z \cup \iota{}^{\backprime}y \in 2 :.$

[*11·26.*22·58] $\quad \supset :. (z) :. (\exists \alpha, \beta) : \alpha \in 2 . \iota{}^{\backprime}z \in \text{Cl}{}^{\backprime}\alpha . \vee . \beta \in 2 . \iota{}^{\backprime}z \in \text{Cl}{}^{\backprime}\beta :.$

[*40·4] $\quad \supset :. (z) . \iota{}^{\backprime}z \in s{}^{\backprime}\text{Cl}{}^{\backprime\backprime}2 :.$

[*52·1] $\quad \supset :. 1 \subset s{}^{\backprime}\text{Cl}{}^{\backprime\backprime}2$ (3)

$\vdash . (1) . (2) . (3) . *101{\cdot}37 . \supset \vdash . \text{Prop}$

***101·4.** $\vdash : (\exists x, y) . x \neq y . \equiv . \exists ! 2$

Dem.

$\vdash . *54{\cdot}26 . \quad \supset \vdash : x \neq y . \supset . \exists ! 2 :$

[*11·11·35] $\quad \supset \vdash : (\exists x, y) . x \neq y . \supset . \exists ! 2$ (1)

$\vdash . *54{\cdot}101 . \supset \vdash : \alpha \in 2 . \supset . (\exists x, y) . x \neq y :$

[*10·11·23] $\quad \supset \vdash : \exists ! 2 . \supset . (\exists x, y) . x \neq y$ (2)

$\vdash . (1) . (2) . \supset \vdash . \text{Prop}$

When we are considering the lowest type occurring in a context, our premisses do not suffice to prove $(\exists x, y) . x \neq y$. For every other type, this can be proved. Thus $\Lambda \neq V$ and $\dot{\Lambda} \neq \dot{V}$ give the required result for classes and relations respectively.

***101·41.** $\vdash : (\exists x) . \iota{}^{\backprime}x \neq V . \equiv . \exists ! 2$

Dem.

$\vdash . *24{\cdot}14 . \text{Transp} . \supset$

$\vdash :. (\exists x) . \iota{}^{\backprime}x \neq V . \equiv : (\exists x) : (\exists y) . y \sim\in \iota{}^{\backprime}x :$

[*51·15] $\quad \equiv : (\exists x, y) . x \neq y :$

[*101·4] $\quad \equiv : \exists ! 2 :. \supset \vdash . \text{Prop}$

***101·42.** $\vdash . \exists ! 2_{\text{Cls}} . \iota{}^{\backprime}\Lambda \cup \iota{}^{\backprime}V \in 2_{\text{Cls}}$

Dem.

$\vdash . *20{\cdot}41 . *24{\cdot}1 . \supset \vdash . \Lambda, V \in \text{Cls} . \Lambda \neq V$ (1)

$\vdash . (1) . *54{\cdot}26 . \quad \supset \vdash . \iota{}^{\backprime}\Lambda \cup \iota{}^{\backprime}V \in 2 . \iota{}^{\backprime}\Lambda \cup \iota{}^{\backprime}V \subset \text{Cls} .$

[*63·371·105] $\quad \supset \vdash . \iota{}^{\backprime}\Lambda \cup \iota{}^{\backprime}V \in 2 \cap t{}^{\backprime}\text{Cls} .$

[(*65·01)] $\quad \supset \vdash . \iota{}^{\backprime}\Lambda \cup \iota{}^{\backprime}V \in 2_{\text{Cls}} . \supset \vdash . \text{Prop}$

***101·43.** $\vdash . \exists ! 2_{\text{Rel}}$ [Proof as in *101·42]

$*102$. ON CARDINAL NUMBERS OF ASSIGNED TYPES.

Summary of $*102$.

In this number, we shall consider a typically definite relation "Nc," *i.e.* we shall consider the relation, to a class δ which is given as of the same type as β, of the class μ of those classes γ which are similar to δ and of the same type as α. We shall then put

$$\mu = \mathrm{Nc}\,(\alpha_\beta)\text{‘}\delta,$$
$$\gamma \in \mathrm{Nc}\,(\alpha_\beta)\text{‘}\delta,$$
$$\gamma \,\mathrm{sm}_{(\alpha,\,\beta)}\, \delta,$$

and the class of all such numbers as μ for a given α and β we shall call $\mathrm{NC}^\beta(\alpha)$, so that

$$\mathrm{NC}^\beta(\alpha) = \mathrm{D}\text{‘}\mathrm{Nc}\,(\alpha_\beta).$$

The notations here introduced for giving typical definiteness to "sm" and "Nc" are those defined in $*65$ for any typically ambiguous relation.

By $*63{\cdot}01{\cdot}02$ we have, if α is a typically ambiguous symbol,

$$\vdash .\, \alpha_x = \alpha \cap t\text{‘}x,$$
$$\vdash .\, \alpha(x) = \alpha \cap t\text{‘}\iota\text{‘}x.$$

Thus $\vdash .\, \alpha(x) = \alpha_{\iota\text{‘}x}$. If we apply the definitions to 1, "1_x" is meaningless unless x is a class; we therefore write a Greek letter in place of x, and we have

$$\vdash .\, 1_\beta = 1 \cap t\text{‘}\beta = 1 \cap (\iota\text{‘}\beta \cup -\iota\text{‘}\beta).$$

If $x \in \beta$, we shall have $\iota\text{‘}x = \beta \,.\vee.\, \iota\text{‘}x \neq \beta$. Hence

$$\vdash : x \in \beta \,.\supset.\, \iota\text{‘}x \in 1_\beta.$$

Similarly $\quad \vdash : x \sim\in \beta \,.\supset.\, \iota\text{‘}x \in 1_\beta.$

Thus $\quad \vdash : x \in t_0\text{‘}\beta \,.\supset.\, \iota\text{‘}x \in 1_\beta.$

The converse implication also holds, so that

$$\vdash : x \in t_0\text{‘}\beta \,.\equiv.\, \iota\text{‘}x \in 1_\beta.$$

Thus 1_β consists of all unit classes whose sole members x either are or are not members of β, *i.e.* for which "$x \in \beta$" is significant.

In "$x \in t_0\text{‘}\beta \,.\supset.\, \iota\text{‘}x \in 1_\beta$," the hypothesis renders explicit the condition of significance; thus "$\iota\text{‘}x \in 1_\beta$" is always true when significant, and always significant when $x \in t_0\text{‘}\beta$. On the interpretation of negative statements concerning types, see the note at the end of this number.

It should be noted that all the constant relations introduced in this work are typically ambiguous. Consider *e.g.* $\dot\Lambda$, sg, D, s, $\dot s$, I, ι, ϵ, Cl, Rl. These all have more or less typical ambiguity, though all of them have what we will call *relative* typical definiteness, *i.e.* when the type of the relatum is given, that of the referent is given also. (In regard to D, it is not true that, conversely, when the type of the referent is given, that of the relatum is also given.) But "sm" and "Nc" have not even relative definiteness. When the type of the relatum is given, that of the referent becomes no more definite than before; the only restrictions are that the relatum for "sm" or "Nc" must be a class, that the referent for "sm" must be a class, and that the referent for "Nc" must be a class of classes. When a relation R has relative definiteness, it is enough to fix the type of the relatum; and if further $R \epsilon 1 \rightarrow \text{Cls}$, so that R leads to a descriptive function, "R‘y" has complete typical definiteness as soon as the type of y is given. Now the constant relations hitherto introduced, with the exception of "sm" and "$\dot{\text{V}}$," have all been one-many relations, and have been used almost exclusively in the form of descriptive functions. Hence no special notation has been required to give typical definiteness, since "R‘y," in these circumstances, has typical definiteness as soon as y is assigned. But with the consideration of "sm" and "Nc," which do not have even relative definiteness, an explicit means of giving typical definiteness becomes necessary. It should be observed, however, that "Nc‘δ" has typical definiteness, when δ is known, as soon as the *domain* of "Nc" has typical definiteness, since δ must belong to the converse domain. It is for the sake of this and similar cases that we introduced the two definitions in *65, which only give typical definiteness to the *domain*.

In virtue of the definitions in *65, if R is a typically ambiguous relation, and x is a referent, R becomes R_x; if, further, y is a relatum, R becomes $R_{(x,y)}$. If x is a referent for R, we have $(\exists y) . x \epsilon \overrightarrow{R}‘y$, and $\overrightarrow{R}‘y \epsilon \text{D}‘\overrightarrow{R}$. Thus $\text{D}‘\overrightarrow{R}$ has a member of the type next above that of x, *i.e.* of the type of $\iota‘x$. Thus

$$\vdash . \text{sg}‘(R_x) = (\overrightarrow{R})(x)$$

and

$$\vdash . \text{sg}‘\{R_{(x,y)}\} = (\overrightarrow{R})(x_y)$$

as was proved in *65. Hence in particular

$$\vdash . \text{sg}‘\{\text{sm}_{(\alpha,\beta)}\} = \text{Nc}(\alpha_\beta).$$

It is chiefly for this reason that it is worth while to introduce the definition of $R(x_y)$.

We have, in virtue of the above, as will be proved in *102·46,

$$\vdash : \gamma \epsilon t‘\alpha . \delta \epsilon t‘\beta . \gamma \,\text{sm}\, \delta . \equiv . \gamma \epsilon \text{Nc}(\alpha_\beta)‘\delta.$$

With regard to "Nc(α)," which is to be interpreted by *65·04, some caution is necessary. This will mean *some one* of those typically different relations called "Nc" which have their domains composed of terms of the same type as α. But it will not mean the logical sum of all such relations, because these relations are of different types according as their converse domains differ in type, and therefore their logical sum is meaningless. Thus for example if the type of β is lower than or equal to that of α, we shall have

$$\vdash . \exists ! \text{Nc}(\alpha)`\beta,$$

whence, if "Nc (α)" has its converse domain composed of terms of the same type as β,

$$\vdash . \Lambda \sim \epsilon \text{D}`\text{Nc}(\alpha).$$

But if β is of higher type than α, we shall find

$$\vdash . \Lambda \epsilon \text{D}`\text{Nc}(\alpha).$$

Thus "Nc (α)" is indeterminate in a way that makes a practical difference.

Exactly similar remarks apply to NC (α). We have

$$\vdash . \text{NC}(\alpha) = \text{D}`\text{Nc}(\alpha);$$

thus "NC (α)" shares the ambiguity of "Nc (α)." The question whether $\Lambda \epsilon \text{NC}(\alpha)$ depends upon the decision of this ambiguity. The difficulty is that "NC (α)" stands for the domain of any one determination of "Nc" which has its domain composed of objects of the type of $\iota`\alpha$; but it is the domain of *only one* such determination of "Nc," because different determinations are of different types, and therefore cannot be taken together, even when their domains are all of the same type. In consequence of this ambiguity, "NC (α)" is a symbol which is as a rule better avoided, and "Nc (α)" is not often useful except as a descriptive function, in which case the relatum supplies the requisite typical definiteness.

The peculiarity of "NC (α)" is that it is *typically* definite, and yet is capable of different meanings: it is not *wholly* definite, being defined as the domain of a relation whose converse domain is typically ambiguous. It results that we cannot profitably make "NC" half-definite, as "NC (α)" does, but must make it completely definite, as we do by taking $\text{D}`\text{Nc}(\alpha_\beta)$. For this we adopt the notation $\text{NC}^\beta(\alpha)$. We cannot adopt the notation $\text{NC}(\alpha_\beta)$, because that would conflict with $*65 \cdot 11$, nor $\text{NC}(\alpha)_\beta$, because that would conflict with $*65 \cdot 01$, nor $\text{NC}_\beta(\alpha)$, for the same reason. But $\text{NC}^\beta(\alpha)$ has no previously defined meaning. We may if we like regard "NC^β" as $\text{D}`(\text{Nc} \upharpoonright t`\beta)$. Then the required meaning of "$\text{NC}^\beta(\alpha)$" would result from $*65 \cdot 04$. But as "NC^β" so defined is not required, it is simpler to regard "$\text{NC}^\beta(\alpha)$" as a single symbol. We therefore put

$*102 \cdot 01$. $\text{NC}^\beta(\alpha) = \text{D}`\text{Nc}(\alpha_\beta)$ Df

The present number begins with various propositions ($*102 \cdot 2$—$\cdot 27$) on a typically definite relation of similarity, *i.e.* $\text{sm}_{(\alpha,\beta)}$. We then have a set of propositions ($*102 \cdot 3$—$\cdot 46$) on "$\text{Nc}(\alpha_\beta)`\delta$." This is only significant if β and δ are of the same type; it then denotes the class of those classes which are similar to δ and of the same type as α. We then have a set of propositions ($*102 \cdot 5$—$\cdot 64$) on $\text{NC}^\beta(\alpha)$, *i.e.* on cardinals consisting of classes of the same type as α which are similar to classes of the same type as β. We next prove

($*102\cdot71$—$\cdot75$) that no sub-class of α is similar to $\text{Cl}`\alpha$, and therefore (substituting $t_0`\alpha$ for α) no class of the same type as α is similar to $t`\alpha$, and therefore

$*102\cdot74$. $\vdash . \Lambda \in \text{NC}^{t`\alpha}(\alpha)$

This proves that Λ is a cardinal, which is a proposition constantly required. The remaining propositions of $*102$ are concerned with $\text{sm}``\mu$ where μ is a typically definite cardinal.

The most useful propositions in this number (apart from $*102\cdot74$) are

$*102\cdot3$. $\vdash : \gamma \,\text{sm}_{(\alpha,\beta)}\, \delta . \equiv . \gamma \in \text{Nc}(\alpha_\beta)`\delta$

$*102\cdot46$. $\vdash : \gamma \in \text{Nc}(\alpha_\beta)`\delta . \equiv . \delta \in \text{Nc}(\beta_\alpha)`\gamma . \equiv . \gamma \,\text{sm}\, \delta . \gamma \in t`\alpha . \delta \in t`\beta$

$*102\cdot5$. $\vdash : \mu \in \text{NC}^\beta(\alpha) . \equiv . (\exists\delta) . \mu = \text{Nc}(\alpha_\beta)`\delta$

$*102\cdot6$. $\vdash . \text{Nc}(\alpha)`\beta = \text{Nc}(\alpha_\beta)`\beta = \hat{\gamma}(\gamma \,\text{sm}\, \beta . \gamma \in t`\alpha) = \text{Nc}`\beta \cap t`\alpha$

$*102\cdot72$. $\vdash : \beta \subset \alpha . \supset . \sim(\beta \,\text{sm}\, \text{Cl}`\alpha)$

This is used in proving $\mu \in \text{NC} . \supset . 2^\mu > \mu$, which is the proposition from which Cantor deduced that there is no greatest cardinal. (If $\mu = \text{Nc}`\alpha$, $2^\mu = \text{Nc}`\text{Cl}`\alpha$, and thus there is a rise of type.)

$*102\cdot84$. $\vdash : (\exists\gamma) . \gamma \in t`\alpha . \gamma \,\text{sm}\, \alpha . \delta \,\text{sm}\, \gamma . \equiv . \delta \,\text{sm}\, \alpha$

$*102\cdot85$. $\vdash . \text{sm}``\mu \cap t`\beta = \text{sm}_\beta``\mu$

$*102\cdot01$. $\text{NC}^\beta(\alpha) = \text{D}`\text{Nc}(\alpha_\beta)$ Df

$*102\cdot11$. $\vdash : R \in 1 \rightarrow 1 . \supset . R_{(x,y)} \in 1(x) \rightarrow 1(y)$

Here, if R is a real variable, the conditions of significance require $R = R_{(x,y)}$. But if R is a typically ambiguous constant, such as I or $\dot{\Lambda}$ or sg, $R_{(x,y)}$ is a typically definite constant. It is chiefly for such cases that propositions such as the above are useful.

Dem.

$\vdash . *37\cdot402 . (*65\cdot1) . \supset \vdash . \text{D}`R_{(x,y)} \subset t`x .$

$[*33\cdot15] \quad \supset \vdash . \{\text{sg}`R_{(x,y)}\}`z \subset t`x .$

$[*63\cdot5] \quad \supset \vdash . \{\text{sg}`R_{(x,y)}\}`z \in t`t`x \quad (1)$

$\vdash . (1) . *71\cdot102 . \quad \supset \vdash : \text{Hp} . z \in \text{C}\!\text{I}`R_{(x,y)} . \supset . \{\text{sg}`R_{(x,y)}\}`z \in 1 \cap t`t`x .$

$[(*65\cdot02)] \quad \supset . \{\text{sg}`R_{(x,y)}\}`z \in 1(x) \quad (2)$

Similarly $\quad \vdash : \text{Hp} . w \in \text{D}`R_{(x,y)} . \supset . \{\text{gs}`R_{(x,y)}\}`w \in 1(y) \quad (3)$

$\vdash . (2) . (3) . *70\cdot1 . \supset \vdash . \text{Prop}$

$*102\cdot13$. $\vdash : R \in 1 \rightarrow 1 . \supset . R_x \in 1(x) \rightarrow 1$ [Proof as in $*102\cdot11$]

$*102\cdot2$. $\vdash : \gamma \,\text{sm}_{(\alpha,\beta)}\, \delta . \equiv . \gamma \,\text{sm}\, \delta . \gamma \in t`\alpha . \delta \in t`\beta$ [$*35\cdot102 . (*65\cdot1)$]

$*102\cdot21$. $\vdash : \gamma \,\text{sm}_{(\alpha,\beta)}\, \delta . \equiv . (\exists R) . R \in 1 \rightarrow 1 . \text{D}`R \in t`\alpha .$
$\text{C}\!\text{I}`R \in t`\beta . \text{D}`R = \gamma . \text{C}\!\text{I}`R = \delta$ [$*102\cdot2 . *73\cdot1$]

***102·22.** $\vdash : \gamma \,\mathrm{sm}\, (x, y)\, \delta \,.\equiv .\, \gamma \,\mathrm{sm}\, \delta \,.\, \gamma \subset t`x \,.\, \delta \subset t`y$ [*63·5 . (*65·12)]

***102·23.** $\vdash : \gamma \,\mathrm{sm}\, (x, y)\, \delta \,.\equiv .\, (\exists R) \,.\, R \epsilon 1 \rightarrow 1 \,.\, \mathrm{D}`R \subset t`x \,.$
$\text{Ɑ}`R \subset t`y \,.\, \mathrm{D}`R = \gamma \,.\, \text{Ɑ}`R = \delta$ [*102·22 . *73·1]

***102·24.** $\vdash : \gamma \,\mathrm{sm}\, (x, y)\, \delta \,.\equiv .\, (\exists R) \,.\, R \epsilon 1\,(x) \rightarrow 1\,(y) \,.\, \mathrm{D}`R = \gamma \,.\, \text{Ɑ}`R = \delta$

Dem.

$\vdash .$ *102·23 . *40·5·52·43 . *37·25 . $\supset$

$\vdash :. \gamma \,\mathrm{sm}\, (x, y)\, \delta \,.\equiv : (\exists R) : R \epsilon 1 \rightarrow 1 : w \epsilon \text{Ɑ}`R \,.\supset_w .\, \overrightarrow{R}`w \subset t`x :$
$z \epsilon \mathrm{D}`R \,.\supset_z .\, \overleftarrow{R}`z \subset t`y : \mathrm{D}`R = \gamma \,.\, \text{Ɑ}`R = \delta :$

[*63·5] $\equiv : (\exists R) : R \epsilon 1 \rightarrow 1 \,.\, \overrightarrow{R}``\text{Ɑ}`R \subset t`t`x \,.\, \overleftarrow{R}``\mathrm{D}`R \subset t`t`y \,.$
$\mathrm{D}`R = \gamma \,.\, \text{Ɑ}`R = \delta :$

[*71·102.(*65·02)] $\equiv : (\exists R) \,.\, \overrightarrow{R}``\text{Ɑ}`R \subset 1\,(x) \,.\, \overleftarrow{R}``\mathrm{D}`R \subset 1\,(y) \,.\, \mathrm{D}`R = \gamma \,.\, \text{Ɑ}`R = \delta :$

[*70·1] $\equiv : (\exists R) \,.\, R \epsilon 1\,(x) \rightarrow 1\,(y) \,.\, \mathrm{D}`R = \gamma \,.\, \text{Ɑ}`R = \delta :. \supset \vdash .$ Prop

***102·25.** $\vdash : \gamma \,\mathrm{sm}_{(\alpha, \beta)}\, \delta \,.\equiv .\, (\exists R) \,.\, R \epsilon 1_\alpha \rightarrow 1_\beta \,.\, \mathrm{D}`R = \gamma \,.\, \text{Ɑ}`R = \delta$
[Proof as in *102·24]

***102·26.** $\vdash : \gamma \,\mathrm{sm}_{(\alpha, \beta)}\, \delta \,.\, \gamma' \,\mathrm{sm}_{(\alpha, \beta)}\, \delta \,.\supset .\, \gamma \,\mathrm{sm}_{(\alpha, \alpha)}\, \gamma'$

Dem.

$\vdash .$ *102·2 . $\supset \vdash : \mathrm{Hp} \,.\supset .\, \gamma \,\mathrm{sm}\, \delta \,.\, \gamma' \,\mathrm{sm}\, \delta \,.\, \gamma, \gamma' \epsilon t`\alpha \,.$

[*73·32] $\supset .\, \gamma \,\mathrm{sm}\, \gamma' \,.\, \gamma, \gamma' \epsilon t`\alpha \,.$

[*102·2] $\supset .\, \gamma \,\mathrm{sm}_{(\alpha, \alpha)}\, \gamma' : \supset \vdash .$ Prop

***102·27.** $\vdash : \gamma \,\mathrm{sm}_{(\alpha, \beta)}\, \delta \,.\, \gamma' \,\mathrm{sm}_{(\alpha', \beta)}\, \delta \,.\supset .\, \gamma \,\mathrm{sm}_{(\alpha, \alpha')}\, \gamma'$ [Proof as in *102·26]

***102·3.** $\vdash : \gamma \,\mathrm{sm}_{(\alpha, \beta)}\, \delta \,.\equiv .\, \gamma \epsilon \mathrm{Nc}\,(\alpha_\beta)`\delta$

Dem.

$\vdash .$ *32·18 . $\supset$

$\vdash : \gamma \,\mathrm{sm}_{(\alpha, \beta)}\, \delta \,.\equiv .\, \gamma \epsilon \{\mathrm{sg}`\mathrm{sm}_{(\alpha, \beta)}\}`\delta \,.$

[*65·2] $\equiv .\, \gamma \epsilon \{(\mathrm{sg}`\mathrm{sm})\,(\alpha_\beta)\}`\delta \,.$

[(*100·01)] $\equiv \gamma \epsilon \mathrm{Nc}\,(\alpha_\beta)`\delta : \supset \vdash .$ Prop

***102·31.** $\vdash .\, \mathrm{Nc}\,(\alpha_\beta)`\delta = \mathrm{D}``\{1 \rightarrow 1 \cap \hat{R}\,(\mathrm{D}`R \epsilon t`\alpha \,.\, \text{Ɑ}`R \epsilon t`\beta \,.\, \text{Ɑ}`R = \delta)\}$

Dem.

$\vdash .$ *102·3·21 . $\supset$

$\vdash : \gamma \epsilon \mathrm{Nc}\,(\alpha_\beta)`\delta \,.\equiv .\, (\exists R) \,.\, R \epsilon 1 \rightarrow 1 \,.\, \mathrm{D}`R \epsilon t`\alpha \,.\, \text{Ɑ}`R \epsilon t`\beta \,.\, \mathrm{D}`R = \gamma \,.\, \text{Ɑ}`R = \delta \,.$

[*33·123.*37·1] $\equiv .\, \gamma \epsilon \mathrm{D}``\{1 \rightarrow 1 \cap \hat{R}\,(\mathrm{D}`R \epsilon t`\alpha \,.\, \text{Ɑ}`R \epsilon t`\beta \,.\, \text{Ɑ}`R = \delta)\} :$
$\supset \vdash .$ Prop

***102·32.** $\vdash . \mathrm{Nc}\,(\alpha_\beta)`\delta = \mathrm{D}``\{(1_\alpha \rightarrow 1_\beta) \cap \overleftarrow{\text{Ɑ}}`\delta\}$

Dem.

$\vdash . *102·3·25 . \supset$

$\vdash : \gamma \in \mathrm{Nc}\,(\alpha_\beta)`\delta . \equiv . (\exists R) . R \in 1_\alpha \rightarrow 1_\beta . \mathrm{D}`R = \gamma . \text{Ɑ}`R = \delta .$

$[*33·61] \quad \equiv . (\exists R) . R \in 1_\alpha \rightarrow 1_\beta . R \in \overleftarrow{\text{Ɑ}}`\delta . \mathrm{D}`R = \gamma .$

$[*33·123 . *37·1] \quad \equiv . \gamma \in \mathrm{D}``\{(1_\alpha \rightarrow 1_\beta) \cap \overleftarrow{\text{Ɑ}}`\delta\} : \supset \vdash . \mathrm{Prop}$

***102·34.** $\vdash . \mathrm{Nc}\,(\alpha, \beta)`\delta = \mathrm{D}``\{1 \rightarrow 1 \cap \hat{R}\,(\mathrm{D}`R \in t`\alpha . \text{Ɑ}`R \subset t`\beta . \text{Ɑ}`R = \delta)\}$

[Proof as in *102·31]

***102·35.** $\vdash . \mathrm{Nc}\,(\alpha, \beta)`\delta = \mathrm{D}``[\{1_\alpha \rightarrow 1\,(\beta)\} \cap \overleftarrow{\text{Ɑ}}`\delta]$ [Proof as in *102·32]

***102·36.** $\vdash . \mathrm{E}\,!\,\mathrm{Nc}\,(\alpha_\beta)`\delta$ [*102·31 . *14·21]

This proposition is true whenever it is significant, and is significant whenever $\delta \in t`\beta$. When δ belongs to some other type, the above proposition is not significant.

***102·361.** $\vdash . \mathrm{E}\,!\,\mathrm{Nc}\,(\alpha, \beta)`\delta$ [*102·34 . *14·21]

***102·37.** $\vdash . \text{Ɑ}`\mathrm{Nc}\,(\alpha_\beta) = t`\beta$

Dem.

$\vdash . *37·402 . (*65·11) . \supset \vdash . \text{Ɑ}`\mathrm{Nc}\,(\alpha_\beta) \subset t`\beta \quad (1)$

$\vdash . *102·36 . *33·43 . \quad \supset \vdash . (\delta) . \delta \in \text{Ɑ}`\mathrm{Nc}\,(\alpha_\beta) .$

$[*63·14] \quad \supset \vdash . t_0`\text{Ɑ}`\mathrm{Nc}\,(\alpha_\beta) = \text{Ɑ}`\mathrm{Nc}\,(\alpha_\beta) \quad (2)$

$\vdash . (1) . *63·21 . \quad \supset \vdash . t_0`\text{Ɑ}`\mathrm{Nc}\,(\alpha_\beta) = t`\beta \quad (3)$

$\vdash . (2) . (3) . \supset \vdash . \mathrm{Prop}$

***102·4.** $\vdash : \gamma \in \mathrm{Nc}\,(\alpha_\beta)`\delta . \gamma' \in \mathrm{Nc}\,(\alpha_\beta)`\delta . \supset . \gamma \in \mathrm{Nc}\,(\alpha_\alpha)`\gamma'$ [*102·3·26]

***102·41.** $\vdash : \gamma \in \mathrm{Nc}\,(\alpha_\beta)`\delta . \gamma' \in \mathrm{Nc}\,(\alpha'_\beta)`\delta . \supset . \gamma \in \mathrm{Nc}\,(\alpha_{\alpha'})`\gamma'$ [*102·3·27]

***102·42.** $\vdash . \alpha \in \mathrm{Nc}\,(\alpha_\alpha)`\alpha$ [*102·3·2 . *73·3 . *63·103]

***102·43.** $\vdash . \exists\,!\,\mathrm{Nc}\,(\alpha_\alpha)`\alpha$ [*102·42]

This inference is legitimate because, when α is given, "$\mathrm{Nc}\,(\alpha_\alpha)`\alpha$" is typically definite. The inference from "$\alpha \in \mathrm{Nc}`\alpha$" (which is true) to "$\exists\,!\,\mathrm{Nc}`\alpha$" is not valid, because "$\exists\,!\,\mathrm{Nc}`\alpha$" may hold only for *some* of the possible determinations of the ambiguity of "Nc."

***102·44.** $\vdash : \alpha \,\mathrm{sm}\, \beta . \equiv . \alpha \in \mathrm{Nc}\,(\alpha_\beta)`\beta . \equiv . \beta \in \mathrm{Nc}\,(\beta_\alpha)`\alpha$

Dem.

$\vdash . *63·102 . \supset$

$\vdash : \alpha \,\mathrm{sm}\, \beta . \equiv . \alpha \,\mathrm{sm}\, \beta . \alpha \in t`\alpha . \beta \in t`\beta \quad (1)$

$\vdash . (1) . *102·2·3 . \supset \vdash . \mathrm{Prop}$

***102·45.** $\vdash : \gamma \epsilon \mathrm{Nc}\,(\alpha_\beta)\text{‘}\delta \,.\supset .\, \gamma \epsilon \mathrm{Nc}\,(\alpha_\alpha)\text{‘}\gamma$

Dem.

$\vdash .\, *102\cdot 3\cdot 2 \,.\supset \vdash : \mathrm{Hp} \,.\supset .\, \gamma \epsilon t\text{‘}\alpha$ (1)

$\vdash .\, *73\cdot 3 \,.\quad \supset \vdash .\, \gamma \,\mathrm{sm}\, \gamma$ (2)

$\vdash .\, (1) \,.\, (2) \,.\, *102\cdot 3\cdot 2 \,.\supset \vdash .\, \mathrm{Prop}$

***102·46.** $\vdash : \gamma \epsilon \mathrm{Nc}\,(\alpha_\beta)\text{‘}\delta \,.\equiv .\, \delta \epsilon \mathrm{Nc}\,(\beta_\alpha)\text{‘}\gamma \,.\equiv .\, \gamma \,\mathrm{sm}\, \delta \,.\, \gamma \epsilon t\text{‘}\alpha \,.\, \delta \epsilon t\text{‘}\beta$
[*102·2·3 . *73·31]

***102·5.** $\vdash : \mu \epsilon \mathrm{NC}^\beta\,(\alpha) \,.\equiv .\, (\exists \delta) \,.\, \mu = \mathrm{Nc}\,(\alpha_\beta)\text{‘}\delta$ [*100·22 . *71·41 . (*102·01)]

In using propositions, such as those of *100, in which we have a typically ambiguous "Nc" or "NC," any significant typical definiteness may be added, since, when a typically ambiguous proposition is asserted, that includes the assertion of every possible proposition resulting from determining the ambiguity.

***102·501.** $\vdash .\, \mathrm{Nc}\,(\alpha_\beta)\text{‘}\delta \epsilon \mathrm{NC}^\beta\,(\alpha)$ [*102·5·36]

***102·51.** $\vdash : \gamma \epsilon \mathrm{Nc}\,(\alpha_\beta)\text{‘}\delta \,.\supset .\, \mathrm{Nc}\,(\alpha_\beta)\text{‘}\delta = \mathrm{Nc}\,(\alpha_\alpha)\text{‘}\gamma \,.$
$\mathrm{Nc}\,(\alpha_\beta)\text{‘}\delta \epsilon \mathrm{NC}^\beta\,(\alpha) \,.\, \mathrm{Nc}\,(\alpha_\alpha)\text{‘}\gamma \epsilon \mathrm{NC}^\alpha\,(\alpha)$

Dem.

$\vdash .\, *102\cdot 3\cdot 2 \,.\supset$

$\vdash :.\, \mathrm{Hp} \,.\supset : \gamma \,\mathrm{sm}\, \delta \,.\, \gamma \epsilon t\text{‘}\alpha \,.\, \delta \epsilon t\text{‘}\beta :$

[*73·37 . *4·73] $\supset : \xi \,\mathrm{sm}\, \delta \,.\equiv .\, \xi \,\mathrm{sm}\, \gamma : \xi \,\mathrm{sm}\, \delta \,.\equiv .\, \xi \,\mathrm{sm}\, \delta \,.\, \delta \epsilon t\text{‘}\beta :$
$\xi \,\mathrm{sm}\, \gamma \,.\equiv .\, \xi \,\mathrm{sm}\, \gamma \,.\, \gamma \epsilon t\text{‘}\alpha :$

[*4·22] $\supset : \xi \,\mathrm{sm}\, \delta \,.\, \delta \epsilon t\text{‘}\beta \,.\equiv .\, \xi \,\mathrm{sm}\, \gamma \,.\, \gamma \epsilon t\text{‘}\alpha :$

[Fact] $\supset : \xi \,\mathrm{sm}\, \delta \,.\, \xi \epsilon t\text{‘}\alpha \,.\, \delta \epsilon t\text{‘}\beta \,.\equiv .\, \xi \,\mathrm{sm}\, \gamma \,.\, \xi \epsilon t\text{‘}\alpha \,.\, \gamma \epsilon t\text{‘}\alpha :$

[*102·2·3] $\supset : \mathrm{Nc}\,(\alpha_\beta)\text{‘}\delta = \mathrm{Nc}\,(\alpha_\alpha)\text{‘}\gamma$ (1)

$\vdash .\, (1) \,.\, *102\cdot 501 \,.\supset \vdash .\, \mathrm{Prop}$

***102·52.** $\vdash : \exists ! \,\mathrm{Nc}\,(\alpha_\beta)\text{‘}\delta \,.\supset .\, \mathrm{Nc}\,(\alpha_\beta)\text{‘}\delta \epsilon \mathrm{NC}^\alpha\,(\alpha)$ [*102·51]

***102·53.** $\vdash .\, \mathrm{NC}^\beta\,(\alpha) - \iota\text{‘}\Lambda \subset \mathrm{NC}^\alpha\,(\alpha)$

Dem.

$\vdash .\, *102\cdot 52 \,.\supset \vdash : \mu = \mathrm{Nc}\,(\alpha_\beta)\text{‘}\delta \,.\, \exists ! \mu \,.\supset .\, \mu \epsilon \mathrm{NC}^\alpha\,(\alpha)$ (1)

$\vdash .\, (1) \,.\, *102\cdot 5 \,.\supset \vdash .\, \mathrm{Prop}$

***102·54.** $\vdash : \delta \epsilon \mathrm{Nc}\,(\beta_\alpha)\text{‘}\gamma \,.\supset .\, \mathrm{Nc}\,(\alpha_\beta)\text{‘}\delta = \mathrm{Nc}\,(\alpha_\alpha)\text{‘}\gamma$ [*102·51·46]

***102·541.** $\vdash : \exists ! \,\mathrm{Nc}\,(\beta_\alpha)\text{‘}\gamma \,.\supset .\, \mathrm{Nc}\,(\alpha_\alpha)\text{‘}\gamma \epsilon \mathrm{NC}^\beta\,(\alpha) - \iota\text{‘}\Lambda$

Dem.

$\vdash .\, *102\cdot 54\cdot 501 \,.\supset \vdash : \delta \epsilon \mathrm{Nc}\,(\beta_\alpha)\text{‘}\gamma \,.\supset .\, \mathrm{Nc}\,(\alpha_\alpha)\text{‘}\gamma \epsilon \mathrm{NC}^\beta\,(\alpha)$ (1)

$\vdash .\, *102\cdot 46\cdot 45 \,.\supset \vdash : \delta \epsilon \mathrm{Nc}\,(\beta_\alpha)\text{‘}\gamma \,.\supset .\, \gamma \epsilon \mathrm{Nc}\,(\alpha_\alpha)\text{‘}\gamma \,.$

[*10·24] $\supset .\, \exists ! \,\mathrm{Nc}\,(\alpha_\alpha)\text{‘}\gamma$ (2)

$\vdash .\, (1) \,.\, (2) \,.\supset$

$\vdash : \delta \epsilon \mathrm{Nc}\,(\beta_\alpha)\text{‘}\gamma \,.\supset .\, \mathrm{Nc}\,(\alpha_\alpha)\text{‘}\gamma \epsilon \mathrm{NC}^\beta\,(\alpha) - \iota\text{‘}\Lambda : \supset \vdash .\, \mathrm{Prop}$

$*102\cdot55$. $\vdash : \Lambda \sim\in \text{NC}^{\alpha}(\beta) . \supset . \text{NC}^{\beta}(\alpha) - \iota`\Lambda = \text{NC}^{\alpha}(\alpha)$

Dem.

$\vdash . *102\cdot5 . \supset$

$\vdash :. \text{Hp} . \supset : \mu = \text{Nc}(\beta_\alpha)`\gamma . \supset_{\mu,\gamma} . \exists ! \mu .$

$[*102\cdot541] \qquad \supset_{\mu,\gamma} . \text{Nc}(\alpha_\alpha)`\gamma \in \text{NC}^{\beta}(\alpha) - \iota`\Lambda :$

$[*10\cdot23] \quad \supset : (\exists\mu) . \mu = \text{Nc}(\beta_\alpha)`\gamma . \supset_\gamma . \text{Nc}(\alpha_\alpha)`\gamma \in \text{NC}^{\beta}(\alpha) - \iota`\Lambda :$

$[*102\cdot36] \supset : (\gamma) . \text{Nc}(\alpha_\alpha)`\gamma \in \text{NC}^{\beta}(\alpha) - \iota`\Lambda :$

$[*13\cdot191] \supset : \nu = \text{Nc}(\alpha_\alpha)`\gamma . \supset_{\nu,\gamma} . \nu \in \text{NC}^{\beta}(\alpha) - \iota`\Lambda :$

$[*102\cdot5] \quad \supset : \nu \in \text{NC}^{\alpha}(\alpha) . \supset_\nu . \nu \in \text{NC}^{\beta}(\alpha) - \iota`\Lambda \qquad (1)$

$\vdash . (1) . *102\cdot53 . \supset \vdash . \text{Prop}$

The above proposition shows that, if every class of the same type as β is similar to some class of the same type as α, then, given a class γ of the same type as α, there is a class δ, of the same type as β, such that the classes similar to δ and of the same type as α are the same as the classes similar to γ and of the same type as α; and conversely, given any class δ, of the same type as β, and similar to some class of the same type as α, then there is a class γ, of the same type as α, such that the classes similar to γ and of the same type as α are the same as the classes similar to δ and of the same type as α. We may express this by saying that, if the cardinals which go from the type of α to the type of β are never null, then those that go from the type of β to the type of α, with the exception of Λ (if Λ is one of them), are the same as those that begin and end within the type of α. The latter are what we call "homogeneous" cardinals. Thus our proposition is a step towards reducing the general study of cardinals to that of homogeneous cardinals.

$*102\cdot6$. $\vdash . \text{Nc}(\alpha)`\beta = \text{Nc}(\alpha_\beta)`\beta = \hat{\gamma}(\gamma \,\text{sm}\, \beta . \gamma \in t`\alpha) = \text{Nc}`\beta \cap t`\alpha$

Dem.

$\vdash . *35\cdot1 . (*65\cdot04) . \supset$

$\vdash : \mu = \text{Nc}(\alpha)`\beta . \equiv . \mu = \text{Nc}`\beta . \mu \in t^2`\alpha .$

$[*63\cdot5] \qquad \equiv . \mu = \text{Nc}`\beta . \mu \subset t`\alpha .$

$[*65\cdot13] \qquad \equiv . \mu = \text{Nc}`\beta \cap t`\alpha . \qquad (1)$

$[*100\cdot1] \qquad \equiv . \mu = \hat{\gamma}(\gamma \,\text{sm}\, \beta . \gamma \in t`\alpha) . \qquad (2)$

$[*63\cdot103] \qquad \equiv . \mu = \hat{\gamma}(\gamma \,\text{sm}\, \beta . \gamma \in t`\alpha . \beta \in t`\beta) .$

$[*102\cdot46] \qquad \equiv . \mu = \text{Nc}(\alpha_\beta)`\beta \qquad (3)$

$\vdash . (1) . (2) . (3) . *20\cdot2 . *100\cdot1 . \supset \vdash . \text{Prop}$

$*102\cdot61$. $\vdash : \delta \in t`\beta . \supset . \text{Nc}(\alpha)`\delta = \text{Nc}(\alpha_\beta)`\delta$

Dem.

$\vdash . *4\cdot73 . \supset \vdash : \text{Hp} . \supset . \hat{\gamma}(\gamma \,\text{sm}\, \delta . \gamma \in t`\alpha) = \hat{\gamma}(\gamma \,\text{sm}\, \delta . \gamma \in t`\alpha . \delta \in t`\beta)$

$[*102\cdot46] \qquad = \text{Nc}(\alpha_\beta)`\delta \qquad (1)$

$\vdash . (1) . *102\cdot6 . \supset \vdash . \text{Prop}$

∗102·62. $\vdash . \mathrm{NC}^{\beta}(\alpha) = \mathrm{Nc}(\alpha)``t`\beta$

Dem.

$\vdash . *37\cdot 7 . (*100\cdot 01) . \supset$

$\vdash . \mathrm{Nc}(\alpha)``t`\beta = \hat{\mu}\{(\exists\delta) . \delta \in t`\beta . \mu = \mathrm{Nc}(\alpha)`\delta\}$

$[*102\cdot 61] \quad = \hat{\mu}\{(\exists\delta) . \delta \in t`\beta . \mu = \mathrm{Nc}(\alpha_\beta)`\delta\}$

$[*102\cdot 37] \quad = \mathrm{D}`\mathrm{Nc}(\alpha_\beta)$

$[(*102\cdot 01)] \quad = \mathrm{NC}^{\beta}(\alpha) . \supset \vdash . \mathrm{Prop}$

∗102·63. $\vdash : \mu = \mathrm{Nc}`\gamma . \alpha \in \mu . \supset . \mu = \mathrm{Nc}(\alpha)`\gamma$

Dem.

$\vdash . *63\cdot 5 . \supset \vdash : \mathrm{Hp} . \supset . \mu = \mathrm{Nc}`\gamma . \mu \subset t`\alpha .$

$[*65\cdot 13] \quad \supset . \mu = \mathrm{Nc}`\gamma \cap t`\alpha .$

$[*102\cdot 6] \quad \supset . \mu = \mathrm{Nc}(\alpha)`\gamma : \supset \vdash . \mathrm{Prop}$

∗102·64. $\vdash : \mu \in \mathrm{NC} . \exists ! \mu . \supset . (\exists\alpha, \gamma) . \mu = \mathrm{Nc}(\alpha)`\gamma \quad [*102\cdot 63 . *100\cdot 4]$

The following propositions are part of Cantor's proof that there is no greatest cardinal. They are inserted here in order to enable us to prove that Λ is a cardinal, namely what we call a "descending" cardinal, *i.e.* one whose corresponding "sm" goes from a higher to a lower type.

∗102·71. $\vdash : R \in \mathrm{Cls} \to 1 . \mathrm{D}`R \subset \alpha . \mathrm{G}`R \subset \mathrm{Cl}`\alpha . \supset . \exists ! \mathrm{Cl}`\alpha - \mathrm{G}`R$

Dem.

$\vdash . *20\cdot 33 . *4\cdot 73 . \supset$

$\vdash :: \mathrm{Hp} . \varpi = \hat{x}(x \in \mathrm{D}`R . x \sim\in \breve{R}`x) . \supset :.$

$x \in \mathrm{D}`R . \supset_x : x \in \varpi . \equiv . x \sim\in \breve{R}`x :$

$[*5\cdot 18] \quad \supset_x : \sim\{x \in \varpi . \equiv . x \in \breve{R}`x\} :$

$[*20\cdot 43 . \mathrm{Transp} . *71\cdot 164] \supset_x : \varpi \neq \breve{R}`x :.$

$[*71\cdot 411 . \mathrm{Transp}] \supset :. \varpi \sim\in \mathrm{G}`R \quad (1)$

$\vdash . *20\cdot 33 . *3\cdot 26 . \supset \vdash : \mathrm{Hp}(1) . \supset . \varpi \subset \mathrm{D}`R .$

$[\mathrm{Hp}] \quad \supset . \varpi \subset \alpha \quad (2)$

$\vdash . (1) . (2) . *13\cdot 191 . \supset$

$\vdash : \mathrm{Hp} . \supset . \hat{x}(x \in \mathrm{D}`R . x \sim\in \breve{R}`x) \in \mathrm{Cl}`\alpha - \mathrm{G}`R : \supset \vdash . \mathrm{Prop}$

∗102·72. $\vdash : \beta \subset \alpha . \supset . \sim(\beta \,\mathrm{sm}\, \mathrm{Cl}`\alpha)$

Dem.

$\vdash . *102\cdot 71 . \supset \vdash :. \mathrm{Hp} . \supset : R \in 1 \to 1 . \mathrm{D}`R = \beta . \mathrm{G}`R \subset \mathrm{Cl}`\alpha . \supset_R . \exists ! \mathrm{Cl}`\alpha - \mathrm{G}`R :$

$[*24\cdot 55 . *22\cdot 41] \quad \supset : R \in 1 \to 1 . \mathrm{D}`R = \beta . \supset_R . \mathrm{G}`R \neq \mathrm{Cl}`\alpha :$

$[*10\cdot 51] \quad \supset : \sim(\exists R) . R \in 1 \to 1 . \mathrm{D}`R = \beta . \mathrm{G}`R = \mathrm{Cl}`\alpha :$

$[*73\cdot 1] \quad \supset : \sim(\beta \,\mathrm{sm}\, \mathrm{Cl}`\alpha) :. \supset \vdash . \mathrm{Prop}$

∗102·73. $\vdash . \mathrm{Nc}\,(\alpha)`t`\alpha = \Lambda$

Dem.

$\vdash . *102·6 . \supset \vdash . \mathrm{Nc}\,(\alpha)`t`\alpha = \hat{\gamma}\,(\gamma \,\mathrm{sm}\, t`\alpha . \gamma \epsilon t`\alpha)$

[∗63·65] $= \hat{\gamma}\,(\gamma \,\mathrm{sm}\, \mathrm{Cl}`t_0`\alpha . \gamma \subset t_0`\alpha)$

[∗102·72] $= \Lambda . \supset \vdash .$ Prop

This proposition proves that no class of the same type as α is similar to $t`\alpha$. Now $t`\alpha$ is the greatest class of its type; thus there are classes of the type next above that of α which are too great to be similar to any class of the type of α. Thus (as will be explicitly proved later) the maximum cardinal in one type is less than that in the next higher type. Cantor's proposition that there is no maximum cardinal only holds when we are allowed to rise to continually higher types: in each type, there is a maximum for that type, namely the number of members of the type.

∗102·74. $\vdash . \Lambda \epsilon \mathrm{NC}^{t`\alpha}(\alpha)$

Dem.

$\vdash . *102·6·501 . \supset \vdash . \mathrm{Nc}\,(\alpha)`t`\alpha \epsilon \mathrm{NC}^{t`\alpha}(\alpha)$ (1)

$\vdash . (1) . *102·73 . \supset \vdash .$ Prop

∗102·75. $\vdash . \mathrm{NC}^{t`\alpha}(\alpha) = \mathrm{NC}^{\alpha}(\alpha) \cup \iota`\Lambda$

Dem.

$\vdash . *100·6 . \supset \vdash : \gamma \epsilon t`\alpha . \supset . \gamma \epsilon t`\alpha . \iota``\gamma \epsilon \mathrm{Nc}`\gamma$ (1)

$\vdash . *63·64·5 . \supset \vdash : \gamma \epsilon t`\alpha . \supset . \iota``\gamma \epsilon t^2`\alpha$ (2)

$\vdash . (1) . (2) . *102·46 . \supset \vdash : \gamma \epsilon t`\alpha . \supset . \iota``\alpha \epsilon \mathrm{Nc}\,\{(t`\alpha)_\alpha\}`\gamma .$

[∗10·24] $\supset . \exists ! \mathrm{Nc}\,\{(t`\alpha)_\alpha\}`\gamma$ (3)

$\vdash . (3) . *102·55 . \supset \vdash . \mathrm{NC}^{t`\alpha}(\alpha) - \iota`\Lambda = \mathrm{NC}^{\alpha}(\alpha)$ (4)

$\vdash . (4) . *102·74 . \supset \vdash .$ Prop

∗102·8. $\vdash : \gamma \epsilon \mathrm{Nc}\,(\alpha_\beta)`\delta . \gamma \,\mathrm{sm}\, \zeta . \zeta \epsilon t`\xi . \supset . \zeta \epsilon \mathrm{Nc}\,(\xi_\beta)`\delta . \zeta \epsilon \mathrm{Nc}\,(\xi_\alpha)`\gamma$

Dem.

$\vdash . *102·46 . \supset$

$\vdash : \mathrm{Hp} . \supset . \gamma \,\mathrm{sm}\, \delta . \gamma \epsilon t`\alpha . \delta \epsilon t`\beta . \gamma \,\mathrm{sm}\, \zeta . \zeta \epsilon t`\xi .$

[∗73·31·32] $\supset . \zeta \,\mathrm{sm}\, \delta . \zeta \epsilon t`\xi . \delta \epsilon t`\beta . \zeta \,\mathrm{sm}\, \gamma . \zeta \epsilon t`\xi . \gamma \epsilon t`\alpha .$

[∗102·46] $\supset . \zeta \epsilon \mathrm{Nc}\,(\xi_\beta)`\delta . \zeta \epsilon \mathrm{Nc}\,(\xi_\alpha)`\gamma : \supset \vdash .$ Prop

∗102·81. $\vdash : \gamma \epsilon \mathrm{Nc}\,(\alpha_\beta)`\delta . \supset . \mathrm{sm}``\mathrm{Nc}\,(\alpha_\beta)`\delta \cap t`\xi = \mathrm{Nc}\,(\xi_\beta)`\delta = \mathrm{Nc}\,(\xi_\alpha)`\gamma$

Dem.

$\vdash . *102·8 . *73·31·32 . \supset$

$\vdash : \gamma, \gamma' \epsilon \mathrm{Nc}\,(\alpha_\beta)`\delta . \zeta \,\mathrm{sm}\, \gamma' . \zeta \epsilon t`\xi . \supset . \zeta \epsilon \mathrm{Nc}\,(\xi_\beta)`\delta . \zeta \epsilon \mathrm{Nc}\,(\xi_\alpha)`\gamma$ (1)

$\vdash . (1) . *37·1 . \supset \vdash : \mathrm{Hp} . \supset . \mathrm{sm}``\mathrm{Nc}\,(\alpha_\beta)`\delta \cap t`\xi \subset \mathrm{Nc}\,(\xi_\beta)`\delta .$

$\mathrm{sm}``\mathrm{Nc}\,(\alpha_\beta)`\delta \cap t`\xi \subset \mathrm{Nc}\,(\xi_\alpha)`\gamma$ (2)

$\vdash . *102·46 . *73·31·32 . \supset$

$\vdash : \mathrm{Hp} . \zeta \epsilon \mathrm{Nc}\,(\xi_\beta)`\delta . \supset . \zeta \epsilon t`\xi . \zeta \,\mathrm{sm}\, \gamma . \gamma \epsilon \mathrm{Nc}\,(\alpha_\beta)`\delta .$

[∗37·1] $\supset . \zeta \epsilon \mathrm{sm}``\mathrm{Nc}\,(\alpha_\beta)`\delta \cap t`\xi$ (3)

Similarly

$\vdash : \mathrm{Hp} . \zeta \epsilon \mathrm{Nc}\,(\xi_\alpha)`\gamma . \supset . \zeta \epsilon \mathrm{sm}``\mathrm{Nc}\,(\alpha_\beta)`\delta \cap t`\xi$ (4)

$\vdash . (2) . (3) . (4) . \supset \vdash .$ Prop

***102·82.** $\vdash : \mu \epsilon \mathrm{NC}^\beta (\alpha) \,.\, \exists ! \mu \,.\, \supset \,.\, \mathrm{sm}``\mu \cap t`\xi \epsilon \mathrm{NC}^\beta (\xi)$ [*102·81·5]

***102·83.** $\vdash : \mu \epsilon \mathrm{NC}^\beta (\alpha) \,.\, \exists ! \mu \,.\, \nu = \mathrm{sm}``\mu \cap t`\xi \,.\, \exists ! \nu \,.\, \supset \,.$
$\mathrm{sm}``\mu \cap t`\zeta = \mathrm{sm}``\nu \cap t`\zeta \,.\, \mu = \mathrm{sm}``\nu \cap t`\alpha$

Dem.

$\vdash \,.\, *102·81 \,.\, \supset$
$\vdash : \mathrm{Hp} \,.\, \gamma \epsilon \mu \,.\, \gamma' \epsilon \nu \,.\, \mu = \mathrm{Nc}\,(\alpha_\beta)`\delta \,.\, \supset \,.$
$\nu = \mathrm{Nc}\,(\xi_\beta)`\delta = \mathrm{Nc}\,(\xi_\alpha)`\gamma \,.\, \mathrm{sm}``\mu \cap t`\zeta = \mathrm{Nc}\,(\zeta_\alpha)`\gamma \,.$
[*102·81] $\supset \,.\, \mathrm{sm}``\nu \cap t`\zeta = \mathrm{Nc}\,(\zeta_\alpha)`\gamma = \mathrm{sm}``\mu \cap t`\zeta \,.$
$\mathrm{sm}``\nu \cap t`\alpha = \mathrm{Nc}\,(\alpha_\beta)`\delta = \mu$ (1)
$\vdash \,.\, (1) \,.\, *102·5 \,.\, \supset \vdash \,.\, \mathrm{Prop}$

***102·84.** $\vdash : (\exists \gamma) \,.\, \gamma \,\mathrm{sm}\, \alpha \,.\, \gamma \epsilon t`\alpha \,.\, \delta \,\mathrm{sm}\, \gamma \,.\, \equiv \,.\, \delta \,\mathrm{sm}\, \alpha$

Dem.

$\vdash \,.\, *73·32 \,.\, \supset \vdash : (\exists \gamma) \,.\, \gamma \,\mathrm{sm}\, \alpha \,.\, \gamma \epsilon t`\alpha \,.\, \delta \,\mathrm{sm}\, \gamma \,.\, \supset \,.\, \delta \,\mathrm{sm}\, \alpha$ (1)
$\vdash \,.\, *73·3 \,.\, *63·103 \,.\, \supset$
$\vdash : \delta \,\mathrm{sm}\, \alpha \,.\, \supset \,.\, \alpha \,\mathrm{sm}\, \alpha \,.\, \alpha \epsilon t`\alpha \,.\, \delta \,\mathrm{sm}\, \alpha \,.$
[*10·24] $\supset \,.\, (\exists \gamma) \,.\, \gamma \,\mathrm{sm}\, \alpha \,.\, \gamma \epsilon t`\alpha \,.\, \delta \,\mathrm{sm}\, \gamma$ (2)
$\vdash \,.\, (1) \,.\, (2) \,.\, \supset \vdash \,.\, \mathrm{Prop}$

***102·85.** $\vdash \,.\, \mathrm{sm}``\mu \cap t`\beta = \mathrm{sm}_\beta``\mu$ [*65·3]

***102·86.** $\vdash : \mu = \mathrm{Nc}\,(\alpha)`\delta \,.\, \exists ! \mu \,.\, \supset \,.\, \mathrm{sm}_\xi``\mu = \mathrm{Nc}\,(\xi)`\delta$

Dem.

$\vdash \,.\, *102·6·81 \,.\, \supset$
$\vdash : \gamma \epsilon \mathrm{Nc}\,(\alpha)`\delta \,.\, \supset \,.\, \mathrm{sm}``\mathrm{Nc}\,(\alpha)`\delta \cap t`\xi = \mathrm{Nc}\,(\xi)`\delta \,.$
[*102·85] $\supset \,.\, \mathrm{sm}_\xi``\mathrm{Nc}\,(\alpha)`\delta = \mathrm{Nc}\,(\xi)`\delta$ (1)
$\vdash \,.\, (1) \,.\, *13·12 \,.\, \supset$
$\vdash : \mu = \mathrm{Nc}\,(\alpha)`\delta \,.\, \gamma \epsilon \mu \,.\, \supset \,.\, \mathrm{sm}_\xi``\mu = \mathrm{Nc}\,(\xi)`\delta : \supset \vdash \,.\, \mathrm{Prop}$

***102·861.** $\vdash \,.\, \mathrm{sm}_\alpha``\mathrm{sm}_\xi``\mu \subset \mathrm{sm}_\alpha``\mu$

Dem.

$\vdash \,.\, *37·1 \,.\, \supset \vdash : \gamma \epsilon \mathrm{sm}_\alpha``\mathrm{sm}_\xi``\mu \,.\, \supset \,.\, (\exists \zeta, \eta) \,.\, \eta \epsilon \mu \,.\, \zeta \,\mathrm{sm}\, \eta \,.\, \zeta \epsilon t`\xi \,.\, \gamma \,\mathrm{sm}\, \zeta \,.\, \gamma \epsilon t`\alpha \,.$
[*73·32.*10·5] $\supset \,.\, (\exists \eta) \,.\, \eta \epsilon \mu \,.\, \gamma \,\mathrm{sm}\, \eta \,.\, \gamma \epsilon t`\alpha \,.$
[*37·1] $\supset \,.\, \gamma \epsilon \mathrm{sm}_\alpha``\mu : \supset \vdash \,.\, \mathrm{Prop}$

***102·862.** $\vdash :. \eta \epsilon \mu \,.\, \supset_\eta \,.\, \exists ! \mathrm{Nc}\,(\xi)`\eta : \supset \,.\, \mathrm{sm}_\alpha``\mu = \mathrm{sm}_\alpha``\mathrm{sm}_\xi``\mu$

Dem.

$\vdash \,.\, *102·6 \,.\, \supset \vdash :. \mathrm{Hp} \,.\, \supset : \eta \epsilon \mu \,.\, \supset \,.\, (\exists \zeta) \,.\, \zeta \,\mathrm{sm}\, \eta \,.\, \zeta \epsilon t`\xi :$
[Fact.*10·35] $\supset : \eta \epsilon \mu \,.\, \gamma \,\mathrm{sm}\, \eta \,.\, \gamma \epsilon t`\alpha \,.\, \supset \,.$
$(\exists \zeta) \,.\, \zeta \,\mathrm{sm}\, \eta \,.\, \zeta \epsilon t`\xi \,.\, \gamma \,\mathrm{sm}\, \eta \,.\, \gamma \epsilon t`\alpha \,.$
[*73·37] $\supset \,.\, (\exists \zeta) \,.\, \zeta \,\mathrm{sm}\, \eta \,.\, \zeta \epsilon t`\xi \,.\, \gamma \,\mathrm{sm}\, \zeta \,.\, \zeta \epsilon t`\alpha \,.$
[*37·1] $\supset \,.\, \gamma \epsilon \mathrm{sm}_\alpha``\mathrm{sm}_\xi``\mu$ (1)
$\vdash \,.\, (1) \,.\, *10·11·23 \,.\, *37·1 \,.\, \supset \vdash : \mathrm{Hp} \,.\, \supset \,.\, \mathrm{sm}_\alpha``\mu \subset \mathrm{sm}_\alpha``\mathrm{sm}_\xi``\mu$ (2)
$\vdash \,.\, (2) \,.\, *102·861 \,.\, \supset \vdash \,.\, \mathrm{Prop}$

∗102·863. $\vdash :. \mu = \mathrm{Nc}\,(\beta)`\delta \,.\, \exists\,!\, \mathrm{Nc}\,(\xi)`\delta \,.\, \supset : \eta \in \mu \,.\, \supset_\eta \,.\, \exists\,!\, \mathrm{Nc}\,(\xi)`\eta$

Dem.

$\vdash \,.\, ∗100·31·321 \,.\, \supset \vdash : \mathrm{Hp} \,.\, \eta \in \mu \,.\, \supset \,.\, \mathrm{Nc}\,(\xi)`\eta = \mathrm{Nc}\,(\xi)`\delta \,.$

[Hp] $\supset \,.\, \exists\,!\, \mathrm{Nc}\,(\xi)`\eta : \supset \vdash \,.\, \mathrm{Prop}$

∗102·87. $\vdash : \mu = \mathrm{Nc}\,(\beta)`\delta \,.\, \exists\,!\, \mathrm{Nc}\,(\xi)`\delta \,.\, \supset \,.\, \mathrm{sm}_\alpha``\mu = \mathrm{sm}_\alpha``\mathrm{sm}_\xi``\mu$

[∗102·862·863]

∗102·88. $\vdash : \mu = \mathrm{Nc}\,(\beta)`\delta \,.\, \exists\,!\, \mathrm{sm}_\xi``\mu \,.\, \supset . \mathrm{sm}_\xi``\mu = \mathrm{Nc}\,(\xi)`\delta \,.\, \mathrm{sm}_\alpha``\mu = \mathrm{Nc}(\alpha)`\delta \,.$

$\mathrm{sm}_\alpha``\mu = \mathrm{sm}_\alpha``\mathrm{sm}_\xi``\mu = \mathrm{sm}_\alpha``\mathrm{Nc}\,(\xi)`\delta$

Dem.

$\vdash \,.\, ∗37·29 \,.\, \mathrm{Transp} \,.\, \supset \vdash : \mathrm{Hp} \,.\, \supset \,.\, \exists\,!\, \mu \,.$

[∗102·86] $\supset \,.\, \mathrm{sm}_\xi``\mu = \mathrm{Nc}\,(\xi)`\delta \,.\, \mathrm{sm}_\alpha``\mu = \mathrm{Nc}\,(\alpha)`\delta \,.$ (1)

[Hp] $\supset \,.\, \exists\,!\, \mathrm{Nc}\,(\xi)`\delta \,.$

[∗102·87] $\supset \,.\, \mathrm{sm}_\alpha``\mu = \mathrm{sm}_\alpha``\mathrm{sm}_\xi``\mu$ (2)

[(1)] $= \mathrm{sm}_\alpha``\mathrm{Nc}\,(\xi)`\delta$ (3)

$\vdash \,.\, (1) \,.\, (2) \,.\, (3) \,.\, \supset \vdash \,.\, \mathrm{Prop}$

Note on negative statements concerning types. Statements such as "$x \sim\in t`y$" or "$x \sim\in t_0`\alpha$" are always false when they are significant. Hence when an object belongs to one type, there is no significant way of expressing what we mean when we say that it does not belong to some other type. The reason is that, when, for example, $t`\alpha$ and $t_0`\alpha$ are said to be different, the statement is only significant if interpreted as applying to the symbols, *i.e.* as meaning to deny that the two symbols denote the same class. We cannot assert that they denote *different* classes, since "$t`\alpha \neq t_0`\alpha$" is not significant, but we can deny that they denote the same class. Owing to this peculiarity, propositions dealing with types acquire their importance largely from the fact that they can be interpreted as dealing with the symbols rather than directly with the objects denoted by the symbols. Another reason for the importance of typically definite propositions is that, when they are implications of which the hypothesis can be asserted, they can be used for *inference, i.e.* for the assertion of the conclusion. Where typically ambiguous symbols occur in implications, on the contrary, the conditions of significance may be different for the hypothesis and the conclusion, so that fallacies may arise from the use of such implications in inference. *E.g.* it is a fallacy to infer "$\vdash \,.\, \exists\,!\, \mathrm{Nc}`\alpha$" from the (true) propositions "$\vdash : \alpha \in \mathrm{Nc}`\alpha \,.\, \supset \,.\, \exists\,!\, \mathrm{Nc}`\alpha$" and "$\vdash \,.\, \alpha \in \mathrm{Nc}`\alpha$." (The truth of the first of these two requires that "$\mathrm{Nc}`\alpha$" should receive the same typical determination in both its occurrences.) For these two reasons hypotheticals concerning types are often useful, in spite of the fact that their hypotheses are always true when they are significant.

$*$103. HOMOGENEOUS CARDINALS.

Summary of $*$103.

In this number, we shall consider cardinals generated by a homogeneous relation of similarity. A "homogeneous" cardinal is to mean all the classes similar to some class α and of the same type as α. The "homogeneous cardinal of α" will be defined as $\text{Nc}'\alpha \cap t'\alpha$; we shall denote it by "$\text{N}_0\text{c}'\alpha$." Then the class of homogeneous cardinals is the class of all such cardinals as "$\text{N}_0\text{c}'\alpha$," *i.e.* it is $\text{D}'\text{N}_0\text{c}$; this we shall denote by "N_0C." The symbol "$\text{N}_0\text{c}'\alpha$" is typically definite as soon as α is assigned; "N_0C," on the contrary, is typically ambiguous: it must be a Cls^3, but otherwise its type may vary indefinitely. Homogeneous cardinals have, however, many properties which do not require that the ambiguity of "N_0C" should be determined, and few which do require this. They are important also as being the simplest kind of cardinals, and as being a kind to which other kinds can usually be reduced.

The chief advantage of homogeneous cardinals is that they are never null ($*$103·13·22). This enables us to avoid by their means the explicit exclusion of exceptional cases; thus throughout Section B we shall use homogeneous cardinals in defining the arithmetical operations: the arithmetical sum of $\text{Nc}'\alpha$ and $\text{Nc}'\beta$, for example, will be defined by means of $\text{N}_0\text{c}'\alpha$ and $\text{N}_0\text{c}'\beta$, in order to exclude such a determination of the typical ambiguity of $\text{Nc}'\alpha$ and $\text{Nc}'\beta$ as would make either of them null. It is true that not only homogeneous cardinals, but also ascending cardinals (cf. $*$104), are never null. But homogeneous cardinals are much the simplest kind of cardinals that are never null, and are therefore the most convenient.

The fact that no homogeneous cardinal is null is derived from

$*$103·12. $\vdash . \alpha \in \text{N}_0\text{c}'\alpha$

Other important propositions in this number are the following:

$*$103·2. $\vdash : \mu \in \text{N}_0\text{C} . \equiv . (\exists\alpha) . \mu = \text{Nc}'\alpha \cap t'\alpha . \equiv . (\exists\alpha) . \mu = \text{N}_0\text{c}'\alpha$

$*$103·26. $\vdash :. \mu \in \text{NC} . \supset : \alpha \in \mu . \equiv . \text{N}_0\text{c}'\alpha = \mu$

The above proposition is used constantly.

***103·27.** $\vdash : \mu = \mathrm{N_0c}`\alpha \,.\equiv.\, \mu \in \mathrm{NC} \,.\, \alpha \in \mu$

Thus to say that μ is the homogeneous cardinal of α is equivalent to saying that μ is a cardinal of which α is a member.

***103·301.** $\vdash . \mathrm{NC}^{\alpha}(\alpha) = \mathrm{N_0C}(\alpha)$

***103·34.** $\vdash . \mathrm{NC} - \iota`\Lambda \subset \mathrm{N_0C}$

***103·4.** $\vdash . \mathrm{sm}``\mathrm{N_0c}`\alpha = \mathrm{Nc}`\alpha$

***103·41.** $\vdash . \mathrm{sm}``\mathrm{N_0c}`\alpha \cap t`\beta = \mathrm{Nc}(\beta)`\alpha$

***103·01.** $\mathrm{N_0c}`\alpha = \mathrm{Nc}`\alpha \cap t`\alpha$ Df

***103·02.** $\mathrm{N_0C} = \mathrm{D}`\mathrm{N_0c}$ Df

***103·1.** $\vdash . \mathrm{N_0c}`\alpha = (\mathrm{Nc}`\alpha)_{\alpha} = \mathrm{Nc}(\alpha)`\alpha = \mathrm{Nc}(\alpha_{\alpha})`\alpha$ [*102·6 . (*103·01)]

***103·11.** $\vdash : \beta \in \mathrm{N_0c}`\alpha \,.\equiv.\, \beta \,\mathrm{sm}\, \alpha \,.\, \beta \in t`\alpha \,.\equiv.\, \beta \in \mathrm{Nc}`\alpha \,.\, \beta \in t`\alpha$
[*103·1 . *102·6]

***103·12.** $\vdash . \alpha \in \mathrm{N_0c}`\alpha$ [*103·11 . *73·3 . *63·103]

***103·13.** $\vdash . \exists ! \mathrm{N_0c}`\alpha$ [*103·12 . *10·24]

This is a legitimate inference from *103·12 because, when α is given, $\mathrm{N_0c}`\alpha$ is typically definite.

***103·14.** $\vdash : \mathrm{N_0c}`\alpha = \mathrm{N_0c}`\beta \,.\equiv.\, \alpha \in \mathrm{N_0c}`\beta \,.\equiv.\, \beta \in \mathrm{N_0c}`\alpha \,.\equiv.\, \alpha \,\mathrm{sm}\, \beta \,.\, \alpha \in t`\beta$

Dem.

$\vdash . \text{*103·11} . \supset$

$\vdash :. \mathrm{N_0c}`\alpha = \mathrm{N_0c}`\beta \,.\equiv: \gamma \,\mathrm{sm}\, \alpha \,.\, \gamma \in t`\alpha \,.\equiv_{\gamma}.\, \gamma \,\mathrm{sm}\, \beta \,.\, \gamma \in t`\beta :$ (1)

[*10·1] $\supset : \alpha \,\mathrm{sm}\, \alpha \,.\, \alpha \in t`\alpha \,.\equiv.\, \alpha \,\mathrm{sm}\, \beta \,.\, \alpha \in t`\beta :$

[*73·3.*63·103] $\supset : \alpha \,\mathrm{sm}\, \beta \,.\, \alpha \in t`\beta$ (2)

$\vdash . \text{*73·32} . \text{*63·17}$

$\vdash : \alpha \,\mathrm{sm}\, \beta \,.\, \alpha \in t`\beta \,.\, \gamma \,\mathrm{sm}\, \alpha \,.\, \gamma \in t`\alpha \,.\supset.\, \gamma \,\mathrm{sm}\, \beta \,.\, \gamma \in t`\beta$ (3)

$\vdash . (3) \dfrac{\beta, \alpha}{\alpha, \beta} . \text{*73·31} . \text{*63·16} . \supset$

$\vdash : \alpha \,\mathrm{sm}\, \beta \,.\, \alpha \in t`\beta \,.\, \gamma \,\mathrm{sm}\, \beta \,.\, \gamma \in t`\beta \,.\supset.\, \gamma \,\mathrm{sm}\, \alpha \,.\, \gamma \in t`\alpha$ (4)

$\vdash . (3) . (4) . (1) . \supset$

$\vdash : \alpha \,\mathrm{sm}\, \beta \,.\, \alpha \in t`\beta \,.\supset.\, \mathrm{N_0c}`\alpha = \mathrm{N_0c}`\beta$ (5)

$\vdash . (2) . (5) . \text{*103·11} . \text{*73·31} . \text{*63·16} . \supset \vdash . \text{Prop}$

***103·15.** $\vdash : \exists ! \mathrm{N_0c}`\alpha \cap \mathrm{N_0c}`\beta \,.\equiv.\, \mathrm{N_0c}`\alpha = \mathrm{N_0c}`\beta$

Dem.

$\vdash . \text{*103·13} . \supset \vdash : \mathrm{N_0c}`\alpha = \mathrm{N_0c}`\beta \,.\supset.\, \exists ! \mathrm{N_0c}`\alpha \cap \mathrm{N_0c}`\beta$ (1)

$\vdash . \text{*103·14} . \supset \vdash : \gamma \in \mathrm{N_0c}`\alpha \,.\, \gamma \in \mathrm{N_0c}`\beta \,.\supset.\, \mathrm{N_0c}`\alpha = \mathrm{N_0c}`\gamma \,.\, \mathrm{N_0c}`\beta = \mathrm{N_0c}`\gamma \,.$

[*14·131·144] $\supset . \mathrm{N_0c}`\alpha = \mathrm{N_0c}`\beta :$

[*10·11·23] $\supset \vdash : \exists ! \mathrm{N_0c}`\alpha \cap \mathrm{N_0c}`\beta \,.\supset.\, \mathrm{N_0c}`\alpha = \mathrm{N_0c}`\beta$ (2)

$\vdash . (1) . (2) . \supset \vdash . \text{Prop}$

∗103·16. $\vdash : \mathrm{N_0c}`\alpha = \mathrm{Nc}`\beta . \equiv . \mathrm{Nc}`\alpha = \mathrm{Nc}`\beta$

In this proposition, the equation "$\mathrm{Nc}`\alpha = \mathrm{Nc}`\beta$" must be supposed to hold in *any* type for which it is significant. Otherwise, we might find a type for which $\mathrm{Nc}`\alpha = \Lambda = \mathrm{Nc}`\beta$, without having $\mathrm{N_0c}`\alpha = \mathrm{Nc}`\beta$.

Dem.

$\vdash . *103·12 . \supset \vdash : \mathrm{N_0c}`\alpha = \mathrm{Nc}`\beta . \supset . \alpha \epsilon \mathrm{Nc}`\beta .$
$[*100·31·321] \quad \supset . \mathrm{Nc}`\alpha = \mathrm{Nc}`\beta \quad (1)$
$\vdash . *22·481 . \supset \vdash : \mathrm{Nc}`\alpha = \mathrm{Nc}`\beta . \supset . \mathrm{Nc}`\alpha \cap t`\alpha = \mathrm{Nc}`\beta \cap t`\alpha .$
$[*65·13.(*103·01)] \quad \supset . \mathrm{N_0c}`\alpha = \mathrm{Nc}`\beta \quad (2)$
$\vdash . (1) . (2) . \supset \vdash . \mathrm{Prop}$

∗103·2. $\vdash : \mu \epsilon \mathrm{N_0C} . \equiv . (\exists \alpha) . \mu = \mathrm{Nc}`\alpha \cap t`\alpha . \equiv . (\exists \alpha) . \mu = \mathrm{N_0c}`\alpha$
$[*71·41 . *100·22 . (*103·01·02)]$

∗103·21. $\vdash . \mathrm{N_0c}`\alpha \epsilon \mathrm{N_0C} . \mathrm{N_0c}`\alpha \epsilon \mathrm{NC} \quad [*103·2 . *100·2·4 . *14·28 . *65·13]$

In adducing a proposition, such as ∗100·2, which is concerned with an "Nc" entirely undetermined in type, any degree of typical determination may be added to our "Nc," since an asserted proposition containing an ambiguous "Nc" is only legitimate if it is true for every possible determination of the ambiguity.

∗103·22. $\vdash : \mu \epsilon \mathrm{N_0C} . \supset . \exists ! \mu \quad [*103·13·2]$

∗103·23. $\vdash . \Lambda \sim \epsilon \mathrm{N_0C} \quad [*103·22]$

∗103·24. $\vdash . \mathrm{N_0C} \epsilon \mathrm{Cls\,ex^2\,excl} \quad [*100·43 . *103·23 . *84·13]$

∗103·25. $\vdash :. \mu, \nu \epsilon \mathrm{N_0C} . \supset : \exists ! \mu \cap \nu . \equiv . \mu = \nu \quad [*103·24 . *84·135]$

∗103·26. $\vdash :. \mu \epsilon \mathrm{NC} . \supset : \alpha \epsilon \mu . \equiv . \mathrm{N_0c}`\alpha = \mu$

Dem.

$\vdash . *100·45 . \quad \supset \vdash :. \mathrm{Hp} . \supset : \alpha \epsilon \mu . \supset . \mathrm{Nc}`\alpha = \mu \quad (1)$
$\vdash . *63·22 . \quad \supset \vdash : \alpha \epsilon \mu . \supset . \mu \subset t`\alpha \quad (2)$
$\vdash . (1) . (2) . *22·621 . \supset \vdash :. \mathrm{Hp} . \supset : \alpha \epsilon \mu . \supset . \mathrm{Nc}`\alpha \cap t`\alpha = \mu .$
$[(*103·01)] \quad \supset . \mathrm{N_0c}`\alpha = \mu \quad (3)$
$\vdash . *103·12 . \quad \supset \vdash : \mathrm{N_0c}`\alpha = \mu . \supset . \alpha \epsilon \mu \quad (4)$
$\vdash . (3) . (4) . \supset \vdash . \mathrm{Prop}$

∗103·27. $\vdash : \mu = \mathrm{N_0c}`\alpha . \equiv . \mu \epsilon \mathrm{NC} . \alpha \epsilon \mu$

Dem.

$\vdash . *103·26 . \supset \vdash : \mu \epsilon \mathrm{NC} . \mu = \mathrm{N_0c}`\alpha . \equiv . \mu \epsilon \mathrm{NC} . \alpha \epsilon \mu \quad (1)$
$\vdash . (1) . *103·21 . \supset \vdash . \mathrm{Prop}$

∗103·28. $\vdash : (\exists \alpha) . \gamma \,\mathrm{sm}\, \alpha . \mu = \mathrm{N_0c}`\alpha . \equiv . \exists ! \mu . \mu = \mathrm{Nc}`\gamma$

Dem.

$\vdash . *103·27 . \supset$
$\vdash : (\exists \alpha) . \gamma \,\mathrm{sm}\, \alpha . \mu = \mathrm{N_0c}`\alpha . \equiv . (\exists \alpha) . \gamma \,\mathrm{sm}\, \alpha . \mu \epsilon \mathrm{NC} . \alpha \epsilon \mu .$
$[*100·31] \quad \equiv . \mu \epsilon \mathrm{NC} . \exists ! \mu \cap \mathrm{Nc}`\gamma .$
$[*100·42·41] \quad \equiv . \mu \epsilon \mathrm{NC} . \exists ! \mu \cap \mathrm{Nc}`\gamma . \mu = \mathrm{Nc}`\gamma .$
$[*100·41] \quad \equiv . \exists ! \mu . \mu = \mathrm{Nc}`\gamma : \supset \vdash . \mathrm{Prop}$

∗103·3. $\vdash : \beta \in t\text{‘}\alpha \,.\supset.\, \mathrm{N_0c}\text{‘}\beta = \mathrm{Nc}\,(\alpha)\text{‘}\beta = \mathrm{Nc}\,(\alpha_\alpha)\text{‘}\beta = \mathrm{Nc}\text{‘}\beta \cap t\text{‘}\alpha$

Dem.

$$\vdash .\, {*}63\cdot16 \,.\supset \vdash : \mathrm{Hp} \,.\supset.\, t\text{‘}\beta = t\text{‘}\alpha \,.$$

$$[{*}22\cdot481.({*}103\cdot01)] \quad \supset .\, \mathrm{N_0c}\text{‘}\beta = \mathrm{Nc}\text{‘}\beta \cap t\text{‘}\alpha \qquad (1)$$

$$[{*}102\cdot6] \qquad = \mathrm{Nc}\,(\alpha)\text{‘}\beta \qquad (2)$$

$$[{*}102\cdot61] \qquad = \mathrm{Nc}\,(\alpha_\alpha)\text{‘}\beta \qquad (3)$$

$$\vdash .\, (1) \,.\, (2) \,.\, (3) \,.\supset \vdash .\, \mathrm{Prop}$$

∗103·301. $\vdash .\, \mathrm{NC}^\alpha\,(\alpha) = \mathrm{N_0C}\,(\alpha)$

Note that although "$\mathrm{NC}\,(\alpha)$" is not definite, "$\mathrm{N_0C}\,(\alpha)$" is absolutely definite as soon as α is assigned.

Dem.

$$\vdash .\, {*}103\cdot3 \,.\supset \vdash : \beta \in t\text{‘}\alpha \,.\, \mu = \mathrm{N_0c}\text{‘}\beta \,.\equiv.\, \beta \in t\text{‘}\alpha \,.\, \mu = \mathrm{Nc}\,(\alpha_\alpha)\text{‘}\beta \,.$$

$$[{*}102\cdot37] \qquad \equiv .\, \mu = \mathrm{Nc}\,(\alpha_\alpha)\text{‘}\beta \qquad (1)$$

$$\vdash .\, {*}63\cdot5 \,.\, ({*}103\cdot01) \,.\supset$$

$$\vdash :.\, \mu = \mathrm{N_0c}\text{‘}\beta \,.\supset : \beta \in t\text{‘}\alpha \,.\equiv.\, \mu \in t^2\text{‘}\alpha \qquad (2)$$

$$\vdash .\, (1) \,.\, (2) \,.\supset \vdash : \mu \in t^2\text{‘}\alpha \,.\, \mu = \mathrm{N_0c}\text{‘}\beta \,.\equiv.\, \mu = \mathrm{Nc}\,(\alpha_\alpha)\text{‘}\beta \qquad (3)$$

$$\vdash .\, (3) \,.\, {*}10\cdot11\cdot281\cdot35 \,.\supset$$

$$\vdash :.\, \mu \in t^2\text{‘}\alpha : (\exists\beta) \,.\, \mu = \mathrm{N_0c}\text{‘}\beta : \equiv .\, (\exists\beta) \,.\, \mu = \mathrm{Nc}\,(\alpha_\alpha)\text{‘}\beta \,.$$

$$[{*}102\cdot5] \qquad \equiv .\, \mu \in \mathrm{NC}^\alpha\,(\alpha) \qquad (4)$$

$$\vdash .\, (4) \,.\, {*}103\cdot2 \,.\supset \vdash : \mu \in t^2\text{‘}\alpha \cap \mathrm{N_0C} \,.\equiv.\, \mu \in \mathrm{NC}^\alpha\,(\alpha) \qquad (5)$$

$$\vdash .\, (5) \,.\, ({*}65\cdot02) \,.\supset \vdash .\, \mathrm{Prop}$$

∗103·31. $\vdash : \exists\,!\, \mathrm{Nc}\,(\alpha_\beta)\text{‘}\delta \,.\supset.\, \mathrm{Nc}\,(\alpha_\beta)\text{‘}\delta \in \mathrm{N_0C}\,(\alpha)$

Dem.

$$\vdash .\, {*}102\cdot52 \,.\supset \vdash : \mathrm{Hp} \,.\supset.\, \mathrm{Nc}\,(\alpha_\beta)\text{‘}\delta \in \mathrm{NC}^\alpha\,(\alpha) \,.$$

$$[{*}103\cdot301] \qquad \supset .\, \mathrm{Nc}\,(\alpha_\beta)\text{‘}\delta \in \mathrm{N_0C}\,(\alpha) : \supset \vdash .\, \mathrm{Prop}$$

∗103·32. $\vdash .\, \mathrm{NC}^\beta\,(\alpha) - \iota\text{‘}\Lambda \subset \mathrm{N_0C}\,(\alpha)$

Dem.

$$\vdash .\, {*}103\cdot31 \,.\supset \vdash : \mu = \mathrm{Nc}\,(\alpha_\beta)\text{‘}\delta \,.\, \exists\,!\,\mu \,.\supset.\, \mu \in \mathrm{N_0C}\,(\alpha) \qquad (1)$$

$$\vdash .\, (1) \,.\, {*}102\cdot5 \,.\supset \vdash .\, \mathrm{Prop}$$

In the above proposition, the "β" may be omitted, and we may write (cf. ∗103·33, below)

$$\vdash .\, \mathrm{NC}\,(\alpha) - \iota\text{‘}\Lambda \subset \mathrm{N_0C}\,(\alpha).$$

For the β is wholly arbitrary, so that any possible determination of $\mathrm{NC}\,(\alpha)$ makes the above proposition true. We may proceed a step further, and write (∗103·34, below)

$$\vdash .\, \mathrm{NC} - \iota\text{‘}\Lambda \subset \mathrm{N_0C}.$$

But although we also have $\mathrm{N_0C} \subset \mathrm{NC} - \iota\text{‘}\Lambda$, provided the "NC" on the right is suitably determined, we do not have this always. For example, if "NC" is determined as $\mathrm{NC}^\alpha\,(t\text{‘}\alpha)$, and "$\mathrm{N_0C}$" as $\mathrm{N_0C}\,(t\text{‘}\alpha)$, then $\mathrm{N_0c}\text{‘}t\text{‘}\alpha \in \mathrm{N_0C} - \mathrm{NC}$.

***103·33.** $\vdash . \mathrm{NC}(\alpha) - \iota`\Lambda \subset \mathrm{N_0C}(\alpha)$

Dem.

$\vdash . *4·2 . (*65·02) . \supset$

$\vdash :. \mu \in \mathrm{NC}(\alpha) - \iota`\Lambda . \equiv : \mu \in \mathrm{NC} . \mu \in t^2`\alpha . \exists ! \mu :$

[*100·4.*63·5] $\equiv : (\exists \beta) . \mu = \mathrm{Nc}`\beta : \mu \subset t`\alpha . \exists ! \mu :$

[*65·13] $\equiv : (\exists \beta) . \mu = \mathrm{Nc}`\beta \cap t`\alpha : \exists ! \mu :$

[*102·6] $\equiv : (\exists \beta) . \mu = \mathrm{Nc}(\alpha_\beta)`\beta . \exists ! \mu :$

[*103·31] $\supset : \mu \in \mathrm{N_0C}(\alpha) :. \supset \vdash . \mathrm{Prop}$

***103·34.** $\vdash . \mathrm{NC} - \iota`\Lambda \subset \mathrm{N_0C}$

Dem.

$\vdash . *100·31·321 . *63·5 . \supset$

$\vdash : \mu = \mathrm{Nc}`\alpha . \beta \in \mu . \supset . \mu = \mathrm{Nc}`\beta \cap t`\beta$

[(*103·01)] $= \mathrm{N_0c}`\beta .$

[*103·2] $\supset . \mu \in \mathrm{N_0C}$ (1)

$\vdash . (1) . *100·4 . *11·11·35·54 . \supset \vdash . \mathrm{Prop}$

Thus every cardinal except Λ is a homogeneous cardinal in the appropriate type. Note that although of course every homogeneous cardinal is a cardinal, yet "$\mathrm{N_0C} \subset \mathrm{NC}$" must not be asserted, because it is possible to determine the ambiguity of "NC" in such a way as to make this false. Hence we do not get $\mathrm{NC} - \iota`\Lambda = \mathrm{N_0C}$.

***103·35.** $\vdash : \Lambda \sim\in \mathrm{NC}^\alpha(\beta) . \supset . \mathrm{NC}^\beta(\alpha) - \iota`\Lambda = \mathrm{N_0C}(\alpha)$ [*102·55 . *103·301]

The hypothesis of this proposition is satisfied, as will appear later, if the type of β is in what we may call the direct ascent from that of α, *i.e.* if it can be reached from α by a finite number of steps each of which takes us from a type τ to either $\mathrm{Cl}`\tau$ or $\mathrm{Rl}`(\tau \uparrow \tau)$. Thus in such a case the cardinals (other than Λ) which go from $t`\beta$ to $t`\alpha$ are the same as those which begin and end within $t`\alpha$. It will also appear that in such a case $\Lambda$ always is a member of $\mathrm{NC}^\beta(\alpha)$. *If* two cardinals which are not equal must always be one greater and the other less, then $\Lambda \in \mathrm{NC}^\beta(\alpha)$ is the condition for $\mathrm{N_0c}`t`\beta > \mathrm{Nc}(\beta)`t`\alpha$. In that case, we shall have $\Lambda \in \mathrm{NC}^\beta(\alpha) . \supset . \Lambda \sim\in \mathrm{NC}^\alpha(\beta)$. But there is no known proof that of two different cardinals one must be the greater, except by assuming the multiplicative axiom and proving thence (by Zermelo's theorem) that every class can be well-ordered (cf. *258).

***103·4.** $\vdash . \mathrm{sm}``\mathrm{N_0c}`\alpha = \mathrm{Nc}`\alpha$

Dem.

$\vdash . *37·1 . \supset$

$\vdash : \delta \in \mathrm{sm}``\mathrm{N_0c}`\alpha . \equiv . (\exists \gamma) . \gamma \,\mathrm{sm}\, \alpha . \gamma \in t`\alpha . \delta \,\mathrm{sm}\, \gamma .$

[*102·84] $\equiv . \delta \,\mathrm{sm}\, \alpha : \supset \vdash . \mathrm{Prop}$

***103·41.** $\vdash . \mathrm{sm}``\mathrm{N_0c}`\alpha \cap t`\beta = \mathrm{Nc}(\beta)`\alpha$

Dem.

$\vdash . *103·4 . \supset \vdash . \mathrm{sm}``\mathrm{N_0c}`\alpha \cap t`\beta = \mathrm{Nc}`\alpha \cap t`\beta$

[*102·6] $= \mathrm{Nc}(\beta)`\alpha . \supset \vdash . \mathrm{Prop}$

$*103{\cdot}42$. $\vdash : \beta \operatorname{sm} \alpha . \equiv . \operatorname{Nc}(\beta)`\alpha = \operatorname{N_0c}`\beta$

Dem.

$\vdash . *100{\cdot}321 . \supset \vdash : \beta \operatorname{sm} \alpha . \supset . \operatorname{Nc}`\alpha = \operatorname{Nc}`\beta .$

$[*22{\cdot}481] \qquad \supset . \operatorname{Nc}`\alpha \cap t`\beta = \operatorname{Nc}`\beta \cap t`\beta .$

$[*102{\cdot}6 . (*103{\cdot}01)] \qquad \supset . \operatorname{Nc}(\beta)`\alpha = \operatorname{N_0c}`\beta \qquad (1)$

$\vdash . *103{\cdot}12 . \supset \vdash : \operatorname{Nc}(\beta)`\alpha = \operatorname{N_0c}`\beta . \supset . \beta \in \operatorname{Nc}(\beta)`\alpha .$

$[*100{\cdot}31] \qquad \supset . \beta \operatorname{sm} \alpha \qquad (2)$

$\vdash . (1) . (2) . \supset \vdash . \operatorname{Prop}$

$*103{\cdot}43$. $\vdash : \mu \in \operatorname{NC} . \supset . \operatorname{sm}``\mu \cap t_0`\mu = \mu$

Dem.

$\vdash . *37{\cdot}29 . \quad \supset \vdash : \mu = \Lambda . \supset . \operatorname{sm}``\mu \cap t_0`\mu = \Lambda \qquad (1)$

$\vdash . *103{\cdot}27 . \supset \vdash : \mu \in \operatorname{NC} . \alpha \in \mu . \supset . \mu = \operatorname{N_0c}`\alpha . t_0`\mu = t`\alpha .$

$[*103{\cdot}41] \qquad \supset . \operatorname{sm}``\mu \cap t_0`\mu = \operatorname{Nc}(\alpha)`\alpha$

$[*103{\cdot}3{\cdot}27] \qquad = \mu \qquad (2)$

$\vdash . (1) . (2) . \supset \vdash . \operatorname{Prop}$

$*103{\cdot}44$. $\vdash :. \mu, \nu \in \operatorname{N_0C} . \supset : \mu = \operatorname{sm}``\nu . \equiv . \nu = \operatorname{sm}``\mu$

Dem.

$\vdash . *100{\cdot}53 . \quad \supset \vdash :. \exists ! \mu . \exists ! \nu . \mu, \nu \in \operatorname{NC} . \supset : \mu = \operatorname{sm}``\nu . \equiv . \nu = \operatorname{sm}``\mu \quad (1)$

$\vdash . *103{\cdot}27{\cdot}2 . \supset \vdash : \operatorname{Hp} . \supset . \exists ! \mu . \exists ! \nu . \mu, \nu \in \operatorname{NC} \qquad (2)$

$\vdash . (1) . (2) . \quad \supset \vdash . \operatorname{Prop}$

$*103{\cdot}5$. $\vdash . 0 \in \operatorname{N_0C}$

Dem.

$\vdash . *101{\cdot}11{\cdot}12 . \supset \vdash . 0 \in \operatorname{NC} . \exists ! 0 .$

$[*103{\cdot}34] \qquad \supset \vdash . 0 \in \operatorname{N_0C} . \supset \vdash . \operatorname{Prop}$

$*103{\cdot}51$. $\vdash . 1 \in \operatorname{N_0C}$

Dem.

$\vdash . *101{\cdot}21{\cdot}241 . \supset \vdash . 1 \in \operatorname{NC} . \exists ! 1 .$

$[*103{\cdot}34] \qquad \supset \vdash . 1 \in \operatorname{N_0C} . \supset \vdash . \operatorname{Prop}$

0 and 1 are the only cardinals of which the above property can be proved universally with our assumptions. If (as is possible so far as our assumptions go) the lowest type is a unit class, we shall have *in that type* (though in no other) $2 = \Lambda$, so that in that type $2 \sim\in \operatorname{N_0C}$.

$*$104. ASCENDING CARDINALS.

Summary of $*$104.

In this number we have to consider cardinals derived from a relation of similarity which goes from the type of α to that of $t‘\alpha$, or to that of $t^2‘\alpha$. The propositions to be proved can be extended, by a mere repetition of the proofs, to $t^3‘\alpha$, $t^4‘\alpha$, etc. This extension must, however, be made afresh in each instance; we cannot prove that it can be made generally, because mathematical induction cannot be applied to the series

$$t_0‘\alpha,\ t‘\alpha,\ t^2‘\alpha,\ t^3‘\alpha,\ \ldots.$$

Ascending cardinals, though less important than homogeneous cardinals, yet have considerable importance in arithmetic, because $\mathrm{Nc}‘\alpha \times \mathrm{Nc}‘\beta$ and $(\mathrm{Nc}‘\alpha)^{\mathrm{Nc}‘\beta}$ are defined as the cardinals of classes of higher types than those of α and β, and the same applies to the product of the cardinals of members of a class of classes. In these cases, however, we also need cardinals of relational types, which will be dealt with in $*$106.

We have to deal, in this number, with three different sets of notions, namely

$*$**104·01.** $\quad \mathrm{N^1c}‘\alpha = \mathrm{Nc}‘\alpha \cap t‘t‘\alpha \quad$ Df

$*$**104·02.** $\quad \mathrm{N^1C} = \mathrm{D}‘\mathrm{N^1c} \quad$ Df

$*$**104·03.** $\quad \mu^{(1)} = \mathrm{sm}“\mu \cap t‘\mu \quad$ Df

with similar definitions of $\mathrm{N^2c}‘\alpha$, etc. Thus $\mathrm{N^1c}‘\alpha$ consists of all classes similar to α but of the next higher type, *i.e.* it is the cardinal number of α in the type next above that of $\mathrm{N_0c}‘\alpha$; $\mathrm{N^1C}$ is the class of all such cardinals as $\mathrm{N^1c}‘\alpha$, and is a typically ambiguous symbol, though $\mathrm{N^1c}‘\alpha$ is typically definite when α is given; $\mu^{(1)}$ (if μ is a cardinal which is not null) is the "same" cardinal in the next higher type, so that, *e.g.*, if μ is 1 determined as consisting of unit classes of *individuals*, $\mu^{(1)}$ will be 1 determined as consisting of unit classes of classes of individuals. (When μ is not an existent cardinal, $\mu^{(1)}$ is unimportant.)

The following are the most useful propositions in the present number:

$*$**104·12.** $\quad \vdash : \beta \in \mathrm{N^1c}‘\alpha \,.\, \gamma \in \mathrm{N^1c}‘\beta \,.\, \supset .\, \gamma \in \mathrm{N^2c}‘\alpha$

$*$**104·2.** $\quad \vdash .\, \iota“\alpha \in \mathrm{N^1c}‘\alpha$

$*104{\cdot}21.\quad \vdash . \exists ! \, N^1c`\alpha$

$*104{\cdot}24.\quad \vdash : \mu = N^1c`\alpha . \supset . \mu = N_0c`\iota``\alpha = N_0c`\hat{\beta}\{(\exists y) . y \in \alpha . \beta = \iota`x \cup \iota`y\}$

$*104{\cdot}25.\quad \vdash . N^1C \subset N_0C$

$*104{\cdot}26.\quad \vdash : \mu = N_0c`\alpha . \supset . \mu^{(1)} = N_0c`\iota``\alpha = N^1c`\alpha$

$*104{\cdot}265.\ \vdash . \mu^{(1)} = \mathrm{sm}_\mu``\mu$

$*104{\cdot}27.\quad \vdash :. \mu \in NC . \supset : \mu = N_0c`\alpha . \equiv . \mu^{(1)} = N^1c`\alpha$

$*104{\cdot}35.\quad \vdash . N^2C \subset N^1C . N^2C \subset N_0C$

$*104{\cdot}43.\quad \vdash : t`\alpha = t`\beta . \supset . (\exists \gamma, \delta) . \gamma \in N^1c`\alpha . \delta \in N^1c`\beta . \gamma \cap \delta = \Lambda$

$*104{\cdot}01.\quad N^1c`\alpha = Nc`\alpha \cap t`t`\alpha \quad$ Df

This defines the cardinal number of α in the next type above that of $N_0c`\alpha$; thus $N^1c`\alpha$ consists of all classes similar to α and of the next type above that of α.

$*104{\cdot}011.\ N^2c`\alpha = Nc`\alpha \cap t`t^{2}`\alpha \quad$ Df

Similar definitions are to be assumed for $N^3c`\alpha$, etc.

$*104{\cdot}02.\quad N^1C = D`N^1c \qquad$ Df

N^1C, like N_0C, is typically ambiguous; but $N^1C(\alpha)$ is typically definite.

$*104{\cdot}021.\ N^2C = D`N^2c \qquad$ Df

Similar definitions are to be assumed for N^3C, etc.

$*104{\cdot}03.\quad \mu^{(1)} = \mathrm{sm}``\mu \cap t`\mu \qquad$ Df

Here, if μ is a cardinal, $\mu^{(1)}$ is the same cardinal in the next higher type. For example, if μ is couples of individuals, $\mu^{(1)}$ is couples of classes of individuals.

$*104{\cdot}031.\ \mu^{(2)} = \mathrm{sm}``\mu \cap t^{2}`\mu \qquad$ Df

Similar definitions are to be assumed for $\mu^{(3)}$, etc.

$*104{\cdot}1.\quad \vdash : \beta \in N^1c`\alpha . \equiv . \beta \in Nc`\alpha . \beta \in t`t`\alpha . \equiv . \beta \in Nc`\alpha . \beta \subset t`\alpha$
[*63·5 . (*104·01)]

$*104{\cdot}101.\ \vdash : \beta \in N^1c`\alpha . \equiv . \beta \,\mathrm{sm}\, \alpha . \beta \subset t`\alpha \qquad$ [*100·31 . *104·1]

$*104{\cdot}102.\ \vdash . N^1c`\alpha = Nc(t`\alpha)`\alpha = Nc\{(t`\alpha)_\alpha\}`\alpha \qquad$ [*102·6 . (*104·01)]

$*104{\cdot}11.\quad \vdash : \beta \in N^2c`\alpha . \equiv . \beta \in Nc`\alpha . \beta \in t`t^{2}`\alpha . \equiv . \beta \in Nc`\alpha . \beta \subset t^{2}`\alpha$
[*63·5 . (*104·011)]

$*104{\cdot}111.\ \vdash : \beta \in N^2c`\alpha . \equiv . \beta \,\mathrm{sm}\, \alpha . \beta \subset t^{2}`\alpha \qquad$ [*100·31 . *104·11]

$*104{\cdot}112.\ \vdash . N^2c`\alpha = Nc(t^{2}`\alpha)`\alpha = Nc\{(t^{2}`\alpha)_\alpha\}`\alpha \qquad$ [*102·6 . (*104·011)]

***104·12.** $\vdash : \beta \,\epsilon\, \mathrm{N}^1\mathrm{c}\text{‘}\alpha \,.\, \gamma \,\epsilon\, \mathrm{N}^1\mathrm{c}\text{‘}\beta \,.\supset .\, \gamma \,\epsilon\, \mathrm{N}^2\mathrm{c}\text{‘}\alpha$

Dem.

$\vdash . *104\cdot1 . \supset \vdash : \mathrm{Hp} . \supset . \beta \,\epsilon\, \mathrm{Nc}\text{‘}\alpha . \beta \,\epsilon\, t\text{‘}t\text{‘}\alpha . \gamma \,\epsilon\, \mathrm{Nc}\text{‘}\beta . \gamma \,\epsilon\, t\text{‘}t\text{‘}\beta .$

$[*100\cdot32] \qquad \supset . \gamma \,\epsilon\, \mathrm{Nc}\text{‘}\alpha . \beta \,\epsilon\, t\text{‘}t\text{‘}\alpha . \gamma \,\epsilon\, t\text{‘}t\text{‘}\beta .$

$[*63\cdot16] \qquad \supset . \gamma \,\epsilon\, \mathrm{Nc}\text{‘}\alpha . t\text{‘}\beta = t\text{‘}t\text{‘}\alpha . \gamma \,\epsilon\, t\text{‘}t\text{‘}\beta .$

$[*13\cdot12] \qquad \supset . \gamma \,\epsilon\, \mathrm{Nc}\text{‘}\alpha . \gamma \,\epsilon\, t\text{‘}t\text{‘}t\text{‘}\alpha .$

$[*104\cdot11] \qquad \supset . \gamma \,\epsilon\, \mathrm{N}^2\mathrm{c}\text{‘}\alpha : \supset \vdash . \mathrm{Prop}$

***104·121.** $\vdash : \beta \,\epsilon\, \mathrm{N}^1\mathrm{c}\text{‘}\alpha . \gamma \,\epsilon\, \mathrm{N}^2\mathrm{c}\text{‘}\alpha . \supset . \gamma \,\epsilon\, \mathrm{N}^1\mathrm{c}\text{‘}\beta$

Dem.

$\vdash . *104\cdot102\cdot112 . \supset \vdash : \mathrm{Hp} . \supset . \beta \,\epsilon\, \mathrm{Nc}\,\{(t\text{‘}\alpha)_\alpha\}\text{‘}\alpha . \gamma \,\epsilon\, \mathrm{Nc}\,\{(t^2\text{‘}\alpha)_\alpha\}\text{‘}\alpha .$

$[*102\cdot41] \qquad \supset . \gamma \,\epsilon\, \mathrm{Nc}\,\{(t^2\text{‘}\alpha)_{t\text{‘}\alpha}\}\text{‘}\beta \qquad (1)$

$\vdash . *104\cdot1 . \qquad \supset \vdash : \mathrm{Hp} . \supset . \beta \,\epsilon\, t\text{‘}t\text{‘}\alpha .$

$[*63\cdot16] \qquad \supset . t\text{‘}\beta = t\text{‘}t\text{‘}\alpha .$

$[(*65\cdot11)] \qquad \supset . \mathrm{Nc}\,\{(t^2\text{‘}\alpha)_{t\text{‘}\alpha}\} = \mathrm{Nc}\,\{(t\text{‘}\beta)_\beta\} \qquad (2)$

$\vdash . (1) . (2) . *104\cdot102 . \supset \vdash . \mathrm{Prop}$

***104·122.** $\vdash : \beta \,\epsilon\, \mathrm{N}^1\mathrm{c}\text{‘}\alpha . \supset . \mathrm{N}^1\mathrm{c}\text{‘}\beta = \mathrm{N}^2\mathrm{c}\text{‘}\alpha \qquad [*104\cdot12\cdot121]$

***104·123.** $\vdash : \mathrm{N}_0\mathrm{c}\text{‘}\beta = \mathrm{N}^1\mathrm{c}\text{‘}\alpha . \supset . \mathrm{N}^1\mathrm{c}\text{‘}\beta = \mathrm{N}^2\mathrm{c}\text{‘}\alpha \qquad [*104\cdot122 . *103\cdot26]$

***104·13.** $\vdash : \mu \,\epsilon\, \mathrm{N}^1\mathrm{C} . \equiv . (\exists\alpha) . \mu = \mathrm{N}^1\mathrm{c}\text{‘}\alpha \qquad [*100\cdot22 . *71\cdot41 . (*104\cdot02)]$

***104·14.** $\vdash : \delta \,\epsilon\, \mu^{(1)} . \equiv . (\exists\gamma) . \gamma \,\epsilon\, \mu . \delta \,\mathrm{sm}\, \gamma . \delta \,\epsilon\, t\text{‘}\mu . \equiv . (\exists\gamma) . \gamma \,\epsilon\, \mu . \delta \,\mathrm{sm}\, \gamma . \delta \subset t\text{‘}\gamma$
$[*37\cdot1 . *63\cdot22 . (*104\cdot03)]$

***104·141.** $\vdash : \mu \,\epsilon\, \mathrm{NC} . \exists ! \mu . \supset . \mu^{(1)} \,\epsilon\, \mathrm{NC} \qquad [*100\cdot52]$

When the hypothesis "$\exists ! \mu$" is omitted, this proposition is still true, but with a difference. *E.g.* let us put

$$\mu = \mathrm{Nc}\,(\alpha)\text{‘}t\text{‘}\alpha.$$

Then $\mu = \Lambda . \mu^{(1)} = \Lambda$. Thus $\mu^{(1)} \neq \mathrm{Nc}\,(t\text{‘}\alpha)\text{‘}t\text{‘}\alpha$. But we still have

$$\mu^{(1)} = \mathrm{Nc}\,(t\text{‘}\alpha)\text{‘}t^2\text{‘}\alpha.$$

Thus $\mu^{(1)} \,\epsilon\, \mathrm{NC}$, but $\mu^{(1)}$ is not the same cardinal as μ in a higher type, *i.e.* there are classes whose cardinal in one type is μ, but whose cardinal in the next higher type is not $\mu^{(1)}$.

***104·142.** $\vdash : \mu \,\epsilon\, \mathrm{NC} . \exists ! \mu . \supset . \mu^{(2)} \,\epsilon\, \mathrm{NC} \qquad [*100\cdot52]$

***104·15.** $\vdash : \mu \,\epsilon\, \mathrm{N}^2\mathrm{C} . \equiv . (\exists\alpha) . \mu = \mathrm{N}^2\mathrm{c}\text{‘}\alpha \qquad [*100\cdot22 . *71\cdot41 . (*104\cdot021)]$

***104·2.** $\vdash . \iota\text{‘‘}\alpha \,\epsilon\, \mathrm{N}^1\mathrm{c}\text{‘}\alpha$

Dem.

$\vdash . *63\cdot621 . \supset \vdash : x \,\epsilon\, \alpha . \supset_x . \iota\text{‘}x \,\epsilon\, t\text{‘}\alpha :$

$[*37\cdot61] \qquad \supset \vdash . \iota\text{‘‘}\alpha \subset t\text{‘}\alpha \qquad (1)$

$\vdash . (1) . *100\cdot6 . *104\cdot1 . \supset \vdash . \mathrm{Prop}$

***104·201.** $\vdash : \beta \in \mathrm{N_0c}`\alpha \,.\supset .\, \iota``\beta \in \mathrm{N^1c}`\alpha \,.\, \mathrm{N^1c}`\alpha = \mathrm{N^1c}`\beta$

Dem.

$\vdash .\, *100{\cdot}31{\cdot}321 \,.\supset \vdash : \mathrm{Hp} \,.\supset .\, \mathrm{Nc}`\alpha = \mathrm{Nc}`\beta \quad (1)$

$\vdash .\, *103{\cdot}11 \,.\quad \supset \vdash : \mathrm{Hp} \,.\supset .\, \beta \in t`\alpha \,.$

$[*63{\cdot}16] \quad \supset .\, t`\alpha = t`\beta \,.$

$[*30{\cdot}37] \quad \supset .\, t`t`\alpha = t`t`\beta \quad (2)$

$\vdash .\, (1) \,.\, (2) \,.\, (*104{\cdot}01) \,.\supset \vdash .\, \mathrm{N^1c}`\alpha = \mathrm{N^1c}`\beta \quad (3)$

$\vdash .\, (3) \,.\, *104{\cdot}2 \,.\supset \vdash .\, \mathrm{Prop}$

***104·21.** $\vdash .\, \exists !\, \mathrm{N^1c}`\alpha \quad [*104{\cdot}2]$

It follows from this proposition that *ascending* cardinals are never null. The proof has to be made separately for each kind of ascending cardinal, *i.e.* $\mathrm{N^1C}$, $\mathrm{N^2C}$, etc.

***104·211.** $\vdash .\, \exists !\, \mathrm{N^1c}`\alpha \cap \mathrm{Cl}`1 \quad [*104{\cdot}2 \,.\, *52{\cdot}3]$

***104·23.** $\vdash .\, \hat{\beta}\{(\exists y) \,.\, y \in \alpha \,.\, \beta = \iota`x \cup \iota`y\} \in \mathrm{N^1c}`\alpha$

Dem.

$\vdash .\, *51{\cdot}16 \,.\quad \supset \vdash : y \in \alpha \,.\supset .\, y \in \alpha \cap (\iota`x \cup \iota`y) \,.$

$[*63{\cdot}16] \quad \supset .\, \iota`x \cup \iota`y \in t`\alpha \quad (1)$

$\vdash .\, (1) \,.\, *10{\cdot}11{\cdot}23 \,.\supset \vdash .\, \hat{\beta}\{(\exists y) \,.\, y \in \alpha \,.\, \beta = \iota`x \cup \iota`y\} \subset t`\alpha \quad (2)$

$\vdash .\, (2) \,.\, *100{\cdot}61 \,.\, *104{\cdot}1 \,.\supset \vdash .\, \mathrm{Prop}$

***104·231.** $\vdash : \mathrm{N^1c}`\alpha = \mathrm{N^1c}`\beta \,.\supset .\, \mathrm{N_0c}`\alpha = \mathrm{N_0c}`\beta$

Dem.

$\vdash .\, *104{\cdot}2 \,.\supset \vdash : \mathrm{Hp} \,.\supset .\, \iota``\beta \in \mathrm{N_1c}`\alpha \,.$

$[*104{\cdot}101] \quad \supset .\, \iota``\beta \,\mathrm{sm}\, \alpha \,.\, \iota``\beta \subset t`\alpha \,.$

$[*73{\cdot}41 . *63{\cdot}21{\cdot}64] \quad \supset .\, \beta \,\mathrm{sm}\, \alpha \,.\, t`\beta = t`\alpha \,.$

$[*103{\cdot}11 . *63{\cdot}16] \quad \supset .\, \beta \in \mathrm{N_0c}`\alpha .$

$[*103{\cdot}14] \quad \supset .\, \mathrm{N_0c}`\alpha = \mathrm{N_0c}`\beta : \supset \vdash .\, \mathrm{Prop}$

***104·232.** $\vdash : \mathrm{N^1c}`\alpha = \mathrm{N^1c}`\beta \,.\equiv .\, \mathrm{N_0c}`\alpha = \mathrm{N_0c}`\beta \,.\equiv .\, \beta \in \mathrm{N_0c}`\alpha$
$[*104{\cdot}231{\cdot}201 \,.\, *103{\cdot}14]$

***104·24.** $\vdash : \mu = \mathrm{N^1c}`\alpha \,.\supset .\, \mu = \mathrm{N_0c}`\iota``\alpha = \mathrm{N_0c}`\hat{\beta}\{(\exists y) \,.\, y \in \alpha \,.\, \beta = \iota`x \cup \iota`y\}$
$[*104{\cdot}2{\cdot}23 \,.\, *103{\cdot}26]$

***104·25.** $\vdash .\, \mathrm{N^1C} \subset \mathrm{N_0C} \quad [*104{\cdot}24{\cdot}13]$

This proposition holds for each possible determination of the typical ambiguities, *i.e.* for every α we have

$$\mathrm{N^1C}\,(t`\alpha) \subset \mathrm{N_0C}\,(t`\alpha).$$

We do not have $\quad \mathrm{N^1C}\,(t`\alpha) = \mathrm{N_0C}\,(t`\alpha),$

because $\quad \mathrm{N_0c}`t`\alpha \in \mathrm{N_0C}\,(t`\alpha) - \mathrm{N^1C}\,(t`\alpha).$

∗104·251. $\vdash . \Lambda \sim\epsilon \mathrm{N}^1\mathrm{C}$ [∗104·25 . ∗103·23]

∗104·252. $\vdash . \mathrm{N}^1\mathrm{C} \epsilon \mathrm{Cls\ ex^2\ excl}$ [∗104·25 . ∗103·24 . ∗84·26]

∗104·26. $\vdash : \mu = \mathrm{N_0c}`\alpha . \supset . \mu^{(1)} = \mathrm{N_0c}`\iota``\alpha = \mathrm{N^1c}`\alpha$

Dem.

$\vdash . ∗104·14 . ∗103·11 . \supset$

$\vdash :. \mathrm{Hp} . \supset : \delta \epsilon \mu^{(1)} . \equiv . (\exists\gamma) . \gamma \mathrm{\ sm\ } \alpha . \gamma \epsilon t`\alpha . \delta \mathrm{\ sm\ } \gamma . \delta \subset t`\gamma .$ (1)

[∗73·32.∗63·16] $\supset . \delta \mathrm{\ sm\ } \alpha . \delta \subset t`\alpha .$

[∗104·101] $\supset . \delta \epsilon \mathrm{N^1c}`\alpha$ (2)

$\vdash . ∗104·101 . \supset$

$\vdash : \delta \epsilon \mathrm{N^1c}`\alpha . \supset . \delta \mathrm{\ sm\ } \alpha . \delta \subset t`\alpha .$

[∗73·3.∗63·103] $\supset . \alpha \mathrm{\ sm\ } \alpha . \alpha \epsilon t`\alpha . \delta \mathrm{\ sm\ } \alpha . \delta \subset t`\alpha .$

[∗10·24] $\supset . (\exists\gamma) . \gamma \mathrm{\ sm\ } \alpha . \gamma \epsilon t`\alpha . \delta \mathrm{\ sm\ } \gamma . \delta \subset t`\gamma$ (3)

$\vdash . (3) . (1) . \supset \vdash :. \mathrm{Hp} . \supset : \delta \epsilon \mathrm{N^1c}`\alpha . \supset . \delta \epsilon \mu^{(1)}$ (4)

$\vdash . (2) . (4) . ∗104·24 . \supset \vdash . \mathrm{Prop}$

∗104·261. $\vdash : \mu^{(1)} = \mathrm{N^1c}`\alpha . \supset . \mu \subset \mathrm{N_0c}`\alpha$

Dem.

$\vdash . ∗104·14·101 . \supset$

$\vdash :. \mathrm{Hp} . \supset : (\exists\gamma) . \gamma \epsilon \mu . \delta \mathrm{\ sm\ } \gamma . \delta \subset t`\gamma . \equiv_\delta . \delta \mathrm{\ sm\ } \alpha . \delta \subset t`\alpha :$

[∗10·23] $\supset : \gamma \epsilon \mu . \delta \mathrm{\ sm\ } \gamma . \delta \subset t`\gamma . \supset_{\gamma,\delta} . \delta \mathrm{\ sm\ } \alpha . \delta \subset t`\alpha .$

[∗4·7] $\supset_{\gamma,\delta} . \delta \mathrm{\ sm\ } \alpha . \delta \mathrm{\ sm\ } \gamma . \delta \subset t`\alpha . \delta \subset t`\gamma .$

[∗73·32.∗63·13] $\supset_{\gamma,\delta} . \gamma \mathrm{\ sm\ } \alpha . \gamma \epsilon t`\alpha .$

[∗103·11] $\supset_{\gamma,\delta} . \gamma \epsilon \mathrm{N_0c}`\alpha$ (1)

$\vdash . (1) . ∗10·23·35 . ∗104·101 . \supset$

$\vdash :. \mathrm{Hp} . \supset : \gamma \epsilon \mu . \exists ! \mathrm{N^1c}`\gamma . \supset_\gamma . \gamma \epsilon \mathrm{N_0c}`\alpha :$

[∗104·21] $\supset : \gamma \epsilon \mu . \supset_\gamma . \gamma \epsilon \mathrm{N_0c}`\alpha :. \supset \vdash . \mathrm{Prop}$

∗104·262. $\vdash : \mu \epsilon \mathrm{NC} . \mu^{(1)} = \mathrm{N^1c}`\alpha . \supset . \mu = \mathrm{N_0c}`\alpha$

Dem.

$\vdash . ∗104·21 . \supset \vdash : \mathrm{Hp} . \supset . \exists ! \mu^{(1)} .$

[∗37·29.Transp] $\supset . \exists ! \mu$ (1)

$\vdash . ∗103·26 . \supset \vdash : \mathrm{Hp} . \gamma \epsilon \mu . \supset . \mu = \mathrm{N_0c}`\gamma$ (2)

$\vdash . (1) . (2) . \supset \vdash : \mathrm{Hp} . \supset . (\exists\gamma) . \mu = \mathrm{N_0c}`\gamma .$

[∗104·26.Hp] $\supset . (\exists\gamma) . \mu = \mathrm{N_0c}`\gamma . \mathrm{N^1c}`\alpha = \mathrm{N^1c}`\gamma .$

[∗104·231] $\supset . (\exists\gamma) . \mu = \mathrm{N_0c}`\gamma . \mathrm{N_0c}`\alpha = \mathrm{N_0c}`\gamma .$

[∗13·172] $\supset . \mu = \mathrm{N_0c}`\alpha : \supset \vdash . \mathrm{Prop}$

∗104·263. $\vdash : \alpha \epsilon \mu . \supset . \iota``\alpha \epsilon \mu^{(1)}$

Dem.

$\vdash . ∗73·41 . ∗37·1 . \supset \vdash : \mathrm{Hp} . \supset . \iota``\alpha \epsilon \mathrm{sm}``\mu$ (1)

$\vdash . ∗63·64 . \supset \vdash : \mathrm{Hp} . \supset . \iota``\alpha \epsilon t`\mu$ (2)

$\vdash . (1) . (2) . (∗104·03) . \supset \vdash . \mathrm{Prop}$

***104·264.** $\vdash : \exists ! \mu . \equiv . \exists ! \mu^{(1)}$

Dem.

$\vdash . *104·263 . \quad \supset \vdash : \exists ! \mu . \supset . \exists ! \mu^{(1)}$ (1)

$\vdash . *37·29 . \text{Transp} . (*104·03) . \supset \vdash : \exists ! \mu^{(1)} . \supset . \exists ! \mu$ (2)

$\vdash . (1) . (2) . \supset \vdash . \text{Prop}$

***104·265.** $\vdash . \mu^{(1)} = \text{sm}_\mu{}^{\text{``}}\mu$ [*102·85 . (*104·03)]

***104·27.** $\vdash :. \mu \in \text{NC} . \supset : \mu = \text{N}_0\text{c}^{\text{`}}\alpha . \equiv . \mu^{(1)} = \text{N}^1\text{c}^{\text{`}}\alpha$ [*104·26·262]

***104·28.** $\vdash : \mu \in \text{NC} - \iota^{\text{`}}\Lambda . \supset . \mu^{(1)} \in \text{N}^1\text{C}$ [*104·26 . *103·34]

***104·29.** $\vdash : \nu \in \text{N}^1\text{C} . \equiv . (\exists \mu) . \mu \in \text{N}_0\text{C} . \nu = \mu^{(1)}$

Dem.

$\vdash . *104·26 . \quad \supset \vdash : \mu = \text{N}_0\text{c}^{\text{`}}\alpha . \nu = \mu^{(1)} . \supset . \nu = \text{N}^1\text{c}^{\text{`}}\alpha :$

$[*10·11·28] \quad \supset \vdash : (\exists \alpha) . \mu = \text{N}_0\text{c}^{\text{`}}\alpha . \nu = \mu^{(1)} . \supset . (\exists \alpha) . \nu = \text{N}^1\text{c}^{\text{`}}\alpha :$

$[*103·2.*104·13] \quad \supset \vdash : \mu \in \text{N}_0\text{C} . \nu = \mu^{(1)} . \supset . \nu \in \text{N}^1\text{C}$ (1)

$\vdash . *104·26 . *103·2 . \supset$

$\vdash : \nu = \text{N}^1\text{c}^{\text{`}}\alpha . \mu = \text{N}_0\text{c}^{\text{`}}\alpha . \supset . \nu = \mu^{(1)} . \mu \in \text{N}_0\text{C}$ (2)

$\vdash . (2) . *10·11·28·35 . \supset$

$\vdash :. \nu = \text{N}^1\text{c}^{\text{`}}\alpha : (\exists \mu) . \mu = \text{N}_0\text{c}^{\text{`}}\alpha : \supset . (\exists \mu) . \mu \in \text{N}_0\text{C} . \nu = \mu^{(1)}$ (3)

$\vdash . (3) . *100·2 . *14·204 . \supset$

$\vdash : \nu = \text{N}^1\text{c}^{\text{`}}\alpha . \supset . (\exists \mu) . \mu \in \text{N}_0\text{C} . \nu = \mu^{(1)}$ (4)

$\vdash . (4) . *10·11·23 . *104·13 . \supset$

$\vdash : \nu \in \text{N}^1\text{C} . \supset . (\exists \mu) . \mu \in \text{N}_0\text{C} . \nu = \mu^{(1)}$ (5)

$\vdash . (1) . (5) . \supset \vdash . \text{Prop}$

***104·3.** $\vdash . \iota^{\text{``}}\iota^{\text{``}}\alpha \in \text{N}^2\text{c}^{\text{`}}\alpha$

Dem.

$\vdash . *104·2 . \supset \vdash . \iota^{\text{``}}\alpha \in \text{N}^1\text{c}^{\text{`}}\alpha . \iota^{\text{``}}\iota^{\text{``}}\alpha \in \text{N}^1\text{c}^{\text{`}}\iota^{\text{``}}\alpha .$

$[*104·12] \quad \supset \vdash . \iota^{\text{``}}\iota^{\text{``}}\alpha \in \text{N}^2\text{c}^{\text{`}}\alpha$

***104·31.** $\vdash . \exists ! \text{N}^2\text{c}^{\text{`}}\alpha$ [*104·3]

***104·311.** $\vdash . \text{N}^2\text{c}^{\text{`}}\alpha = \text{N}_0\text{c}^{\text{`}}\iota^{\text{``}}\iota^{\text{``}}\alpha = \text{N}^1\text{c}^{\text{`}}\iota^{\text{``}}\alpha$ [*104·3·2 . *103·26]

***104·32.** $\vdash : \mu = \text{N}_0\text{c}^{\text{`}}\alpha . \supset . \mu^{(2)} = \text{N}_0\text{c}^{\text{`}}\iota^{\text{``}}\iota^{\text{``}}\alpha = \text{N}^1\text{c}^{\text{`}}\iota^{\text{``}}\alpha = \text{N}^2\text{c}^{\text{`}}\alpha = \{\mu^{(1)}\}^{(1)}$

Dem.

$\vdash . *104·26 . \supset \vdash : \text{Hp} . \supset . \{\mu^{(1)}\}^{(1)} = \text{N}_0\text{c}^{\text{`}}\iota^{\text{``}}\iota^{\text{``}}\alpha$ (1)

$[*104·311] \quad = \text{N}^2\text{c}^{\text{`}}\alpha$ (2)

$\vdash . *103·11 . (*104·031) . \supset$

$\vdash :. \text{Hp} . \supset : \delta \in \mu^{(2)} . \equiv . (\exists \gamma) . \gamma \,\text{sm}\, \alpha . \gamma \in t^{\text{`}}\alpha . \delta \,\text{sm}\, \gamma . \delta \in t^{\text{`}}t^{2\text{`}}\gamma .$

$[*102·84.*63·16] \quad \equiv . \delta \,\text{sm}\, \alpha . \delta \in t^{\text{`}}t^{2\text{`}}\alpha .$

$[*104·11] \quad \equiv . \delta \in \text{N}^2\text{c}^{\text{`}}\alpha$ (3)

$\vdash . (1) . (2) . (3) . *104·24 . \supset \vdash . \text{Prop}$

***104·33.** $\vdash :. \mu \epsilon \mathrm{NC} . \supset : \mu = \mathrm{N_0c}`\alpha . \equiv . \mu^{(2)} = \mathrm{N^2c}`\alpha$

Dem.

$\vdash . *104{\cdot}27 . \supset \vdash :. \mathrm{Hp} . \supset : \mu = \mathrm{N_0c}`\alpha . \equiv . \mu^{(1)} = \mathrm{N^1c}`\alpha .$

$[*104{\cdot}24] \qquad \equiv . \mu^{(1)} = \mathrm{N_0c}`\iota``\alpha .$

$[*104{\cdot}27{\cdot}141 . *103{\cdot}13] \qquad \equiv . \{\mu^{(1)}\}^{(1)} = \mathrm{N^1c}`\iota``\alpha .$

$[*104{\cdot}32{\cdot}24] \qquad \equiv . \mu^{(2)} = \mathrm{N^2c}`\alpha :. \supset \vdash . \mathrm{Prop}$

***104·34.** $\vdash : \varpi \epsilon \mathrm{N^2C} . \equiv . (\exists \nu) . \nu \epsilon \mathrm{N^1C} . \varpi = \nu^{(1)} . \equiv . (\exists \mu) . \mu \epsilon \mathrm{N_0C} . \varpi = \mu^{(2)}$

Dem.

$\vdash . *104{\cdot}32 . \supset$

$\vdash : \varpi = \mathrm{N^2c}`\alpha . \mu = \mathrm{N_0c}`\alpha . \supset . \varpi = \mu^{(2)} . \mu \epsilon \mathrm{N_0C} \qquad (1)$

$\vdash . (1) . *100{\cdot}2 . *10{\cdot}11{\cdot}28{\cdot}35 . \supset$

$\vdash : (\exists \alpha) . \varpi = \mathrm{N^2c}`\alpha . \supset . (\exists \mu) . \mu \epsilon \mathrm{N_0C} . \varpi = \mu^{(2)} \qquad (2)$

$\vdash . *104{\cdot}32 . \qquad \supset \vdash : \mu = \mathrm{N_0c}`\alpha . \varpi = \mu^{(2)} . \supset . \varpi = \mathrm{N^2c}`\alpha .$

$[*104{\cdot}15 . *103{\cdot}2] \supset \vdash : \mu \epsilon \mathrm{N_0C} . \varpi = \mu^{(2)} . \supset . \varpi \epsilon \mathrm{N^2C} \qquad (3)$

$\vdash . (2) . (3) . \qquad \supset \vdash : \varpi \epsilon \mathrm{N^2C} . \equiv . (\exists \mu) . \mu \epsilon \mathrm{N_0C} . \varpi = \mu^{(2)} . \qquad (4)$

$[*104{\cdot}32] \qquad \equiv . (\exists \mu) . \mu \epsilon \mathrm{N_0C} . \varpi = \{\mu^{(1)}\}^{(1)} .$

$[*13{\cdot}195] \qquad \equiv . (\exists \mu, \nu) . \mu \epsilon \mathrm{N_0C} . \nu = \mu^{(1)} . \varpi = \nu^{(1)} .$

$[*104{\cdot}29] \qquad \equiv . (\exists \nu) . \nu \epsilon \mathrm{N^1C} . \varpi = \nu^{(1)} \qquad (5)$

$\vdash . (4) . (5) . \supset \vdash . \mathrm{Prop}$

***104·35.** $\vdash . \mathrm{N^2C} \subset \mathrm{N^1C} . \mathrm{N^2C} \subset \mathrm{N_0C} \quad [*104{\cdot}311{\cdot}13{\cdot}15]$

***104·36.** $\vdash : \gamma \epsilon \mathrm{N^2c}`\alpha . \gamma \epsilon \mathrm{N^1c}`\beta . \supset . \beta \epsilon \mathrm{N^1c}`\alpha . \mathrm{N^1c}`\alpha = \mathrm{N_0c}`\beta$

Dem.

$\vdash . *104{\cdot}1{\cdot}11 . \supset \vdash : \mathrm{Hp} . \supset . \gamma \epsilon \mathrm{Nc}`\alpha . \gamma \epsilon t`t^2`\alpha . \gamma \epsilon \mathrm{Nc}`\beta . \gamma \epsilon t`t`\beta .$

$[*100{\cdot}34 . *63{\cdot}16] \qquad \supset . \mathrm{Nc}`\alpha = \mathrm{Nc}`\beta . t`t^2`\alpha = t`t`\beta .$

$[*63{\cdot}35{\cdot}15] \qquad \supset . \mathrm{Nc}`\alpha = \mathrm{Nc}`\beta . t^2`\alpha = t`\beta .$

$[(*104{\cdot}01 . *103{\cdot}01)] \qquad \supset . \mathrm{N^1c}`\alpha = \mathrm{N_0c}`\beta \qquad (1)$

$\vdash . (1) . *103{\cdot}12 . \supset \vdash . \mathrm{Prop}$

***104·37.** $\vdash : \mathrm{N^2c}`\alpha = \mathrm{N^1c}`\beta . \equiv . \mathrm{N^1c}`\alpha = \mathrm{N_0c}`\beta$

Dem.

$\vdash . *104{\cdot}21 . \supset \vdash : \mathrm{N^2c}`\alpha = \mathrm{N^1c}`\beta . \supset . \exists ! \, \mathrm{N^2c}`\alpha \cap \mathrm{N^1c}`\beta .$

$[*104{\cdot}36] \qquad \supset . \mathrm{N^1c}`\alpha = \mathrm{N_0c}`\beta \qquad (1)$

$\vdash . (1) . *104{\cdot}123 . \supset \vdash . \mathrm{Prop}$

The following propositions are concerned with the proof that, given any two cardinals μ and ν, of the same type, we can find two mutually exclusive classes one of which has μ terms while the other has ν terms. The proof requires that we should raise the types of both μ and ν one degree above that in which they were originally given, *i.e.* that we should turn μ and ν into $\mu^{(1)}$ and $\nu^{(1)}$. Thus, for example, suppose the total number of individuals in the universe were finite (a supposition which is consistent with our primitive propositions), and suppose μ were this number. Then unless $\nu = 0$, a class of ν individuals will be an existent sub-class of the only class which consists of μ individuals, and therefore we shall have

$$\alpha \epsilon \mu . \beta \epsilon \nu . \supset_{\alpha, \beta} . \exists ! \alpha \cap \beta.$$

But if we consider classes of μ *classes* and ν *classes*, we shall always be able to find a γ and a δ such that

$$\gamma \epsilon \mu^{(1)} . \delta \epsilon \nu^{(1)} . \gamma \cap \delta = \Lambda.$$

The existence of such a γ and δ is important in connection with the arithmetical operations, and is therefore proved here.

***104·4.** $\vdash :. x \epsilon \alpha . x \neq y . x \neq z . y \neq z : (w) . w_\iota = \hat{\alpha}\hat{u} (\alpha = \iota \text{‘} w \cup \iota \text{‘} u) : \supset .$
$x_\iota \text{‘‘} (\alpha - \iota \text{‘} x) \cup \iota \text{‘} y_\iota \text{‘} z \, \epsilon \, \text{N}^1\text{c‘}\alpha \cap \text{Cl‘}2$

Dem.

$\vdash . *100{\cdot}61 . \quad \supset \vdash : \text{Hp} . \supset . x_\iota \text{‘‘} (\alpha - \iota \text{‘} x) \, \text{sm} \, (\alpha - \iota \text{‘} x)$ (1)
$\vdash . *73{\cdot}43 . \quad \supset \vdash : \text{Hp} . \supset . \iota \text{‘} y_\iota \text{‘} z \, \text{sm} \, \iota \text{‘} x$ (2)
$\vdash . *51{\cdot}232 . \text{Transp} . \supset \vdash : \text{Hp} . \supset . x \sim \epsilon \, y_\iota \text{‘} z$ (3)
$\vdash . *51{\cdot}232 . \quad \supset \vdash : \text{Hp} . \gamma \epsilon x_\iota \text{‘‘} (\alpha - \iota \text{‘} x) . \supset . x \epsilon \gamma$ (4)
$\vdash . (3) . (4) . \quad \supset \vdash : \text{Hp} . \supset . y_\iota \text{‘} z \sim \epsilon \, x_\iota \text{‘‘} (\alpha - \iota \text{‘} x) .$
$[*51{\cdot}211] \quad \supset . x_\iota \text{‘‘} (\alpha - \iota \text{‘} x) \cap \iota \text{‘} y_\iota \text{‘} z = \Lambda$ (5)
$\vdash . *51{\cdot}21{\cdot}211 . \quad \supset \vdash . (\alpha - \iota \text{‘} x) \cap \iota \text{‘} x = \Lambda$ (6)
$\vdash . (1) . (2) . (5) . (6) . *73{\cdot}71 . *51{\cdot}221 . \supset$
$\vdash : \text{Hp} . \supset . x_\iota \text{‘‘} (\alpha - \iota \text{‘} x) \cup \iota \text{‘} y_\iota \text{‘} z \, \text{sm} \, \alpha$ (7)
$\vdash . *63{\cdot}101{\cdot}16 . *51{\cdot}232{\cdot}16 . \supset$
$\vdash : \text{Hp} . \quad \supset . t \text{‘} x = t \text{‘} y . x \epsilon \alpha . y \epsilon y_\iota \text{‘} z . y_\iota \text{‘} z \epsilon x_\iota \text{‘‘} (\alpha - \iota \text{‘} x) \cup \iota \text{‘} y_\iota \text{‘} z .$
$[*63{\cdot}53{\cdot}2] \supset . t^2 \text{‘} x = t \text{‘} \alpha . t^2 \text{‘} y = t_0 \text{‘} \{x_\iota \text{‘‘} (\alpha - \iota \text{‘} x) \cup \iota \text{‘} y_\iota \text{‘} z\} . t^2 \text{‘} x = t^2 \text{‘} y .$
$[*13{\cdot}17] \quad \supset . t \text{‘} \alpha = t_0 \text{‘} \{x_\iota \text{‘‘} (\alpha - \iota \text{‘} x) \cup \iota \text{‘} y_\iota \text{‘} z\} .$
$[*63{\cdot}105] \supset . x_\iota \text{‘‘} (\alpha - \iota \text{‘} x) \cup \iota \text{‘} y_\iota \text{‘} z \subset t \text{‘} \alpha$ (8)
$\vdash . *54{\cdot}26 . \supset \vdash : \text{Hp} . \supset . x_\iota \text{‘‘} (\alpha - \iota \text{‘} x) \cup \iota \text{‘} y_\iota \text{‘} z \subset 2$ (9)
$\vdash . (7) . (8) . (9) . *104{\cdot}101 . \supset \vdash . \text{Prop}$

***104·41.** $\vdash :. t \text{‘} \alpha = t \text{‘} \beta : (\exists x, y, z) . x \epsilon \alpha . x \neq y . x \neq z . y \neq z : \supset .$
$(\exists \gamma, \delta) . \gamma \epsilon \text{N}^1\text{c‘}\alpha . \delta \epsilon \text{N}^1\text{c‘}\beta . \gamma \cap \delta = \Lambda$

Dem.

$\vdash . *104{\cdot}4{\cdot}2 . *52{\cdot}3 . \supset$
$\vdash : \text{Hp} . \text{Hp} \, *104{\cdot}4 . \supset . (\exists x, y, z) . x_\iota \text{‘‘} (\alpha - \iota \text{‘} x) \cup \iota \text{‘} y_\iota \text{‘} z \, \epsilon \, \text{N}^1\text{c‘}\alpha \cap \text{Cl‘}2 .$
$\iota \text{‘‘} \beta \, \epsilon \, \text{N}^1\text{c‘}\beta \cap \text{Cl‘}1 .$
$[*13{\cdot}22] \quad \supset . (\exists x, y, z, \gamma, \delta) . \gamma = x_\iota \text{‘‘} (\alpha - \iota \text{‘} x) \cup \iota \text{‘} y_\iota \text{‘} z . \delta = \iota \text{‘‘} \beta .$
$\gamma \epsilon \text{N}^1\text{c‘}\alpha \cap \text{Cl‘}2 . \delta \epsilon \text{N}^1\text{c‘}\beta \cap \text{Cl‘}1 .$
$[*11{\cdot}55] \quad \supset . (\exists \gamma, \delta) . \gamma \epsilon \text{N}^1\text{c‘}\alpha \cap \text{Cl‘}2 . \delta \epsilon \text{N}^1\text{c‘}\beta \cap \text{Cl‘}1$ (1)
$\vdash . (1) . *101{\cdot}35 . \supset \vdash . \text{Prop}$

This proposition proves the desired conclusions provided $\exists ! \alpha$, and $t_0`\alpha$ consists of at least three terms. The following propositions deal with the cases in which this hypothesis is not verified.

***104·411.** $\vdash : t`\alpha = t`\beta \,.\, \alpha \epsilon 0 \,.\, \gamma = \Lambda_\alpha \,.\, \delta = \iota``\beta \,.\supset .\, \gamma \epsilon \mathrm{N}^1\mathrm{c}`\alpha \,.\, \delta \epsilon \mathrm{N}^1\mathrm{c}`\beta \,.\, \gamma \cap \delta = \Lambda$

Dem.

$\vdash . *73{\cdot}47 . \quad \supset \vdash : \mathrm{Hp} . \supset . \gamma \,\mathrm{sm}\, \alpha \qquad (1)$

$*22{\cdot}43 . (*65{\cdot}01) . \quad \supset \vdash : \mathrm{Hp} . \supset . \gamma \subset t`\alpha \qquad (2)$

$\vdash . (1) . (2) . *104{\cdot}101 . \supset \vdash : \mathrm{Hp} . \supset . \gamma \epsilon \mathrm{N}^1\mathrm{c}`\alpha \qquad (3)$

$\vdash . (3) . *104{\cdot}2 . *24{\cdot}23 . \supset \vdash . \mathrm{Prop}$

***104·412.** $\vdash : t`\alpha = t`\beta \,.\, \alpha = \iota`x \,.\, \gamma = \iota`\Lambda_x \,.\, \delta = \iota``\beta \,.\supset .$
$\gamma \epsilon \mathrm{N}^1\mathrm{c}`\alpha \,.\, \delta \epsilon \mathrm{N}^1\mathrm{c}`\beta \,.\, \gamma \cap \delta = \Lambda$

Dem.

$\vdash . *73{\cdot}43 . \quad \supset \vdash : \mathrm{Hp} . \supset . \gamma \,\mathrm{sm}\, \alpha \qquad (1)$

$\vdash . *63{\cdot}61{\cdot}103 . \quad \supset \vdash : \mathrm{Hp} . \supset . \alpha \epsilon t^2`x \qquad (2)$

$\vdash . *22{\cdot}43 . (*65{\cdot}01) . \quad \supset \vdash : \mathrm{Hp} . \xi \epsilon \gamma . \supset_\xi . \xi \subset t`x .$

$[*63{\cdot}5] \quad \supset_\xi . \xi \epsilon t^2`x .$

$[(2) . *63{\cdot}13] \quad \supset_\xi . \xi \epsilon t`\alpha \qquad (3)$

$\vdash . (1) . (3) . *104{\cdot}101 . \supset \vdash : \mathrm{Hp} . \supset . \gamma \epsilon \mathrm{N}^1\mathrm{c}`\alpha \qquad (4)$

$\vdash . *101{\cdot}23 . \quad \supset \vdash : \mathrm{Hp} . \supset . \gamma \cap \delta = \Lambda \qquad (5)$

$\vdash . (4) . (5) . *104{\cdot}2 . \supset \vdash . \mathrm{Prop}$

***104·413.** $\vdash : t`\alpha = t`\beta \,.\, \alpha = \iota`x \cup \iota`y \,.\, x \neq y \,.\, \gamma = \iota`\Lambda \cup \iota`(\iota`x \cup \iota`y) \,.\, \delta = \iota``\beta \,.\supset .$
$\gamma \epsilon \mathrm{N}^1\mathrm{c}`\alpha \,.\, \delta \epsilon \mathrm{N}^1\mathrm{c}`\beta \,.\, \gamma \cap \delta = \Lambda$

Dem.

$\vdash . *54{\cdot}26 . \quad \supset \vdash : \mathrm{Hp} . \supset . \iota`x \cup \iota`y \epsilon 2 . \qquad (1)$

$[*101{\cdot}35] \quad \supset . \Lambda \neq \iota`x \cup \iota`y .$

$[*54{\cdot}26] \quad \supset . \iota`\Lambda \cup \iota`(\iota`x \cup \iota`y) \epsilon 2 .$

$[*101{\cdot}3] \quad \supset . \iota`\Lambda \cup \iota`(\iota`x \cup \iota`y) \epsilon \mathrm{Nc}`(\iota`x \cup \iota`y) \qquad (2)$

$\vdash . *51{\cdot}16 . \quad \supset \vdash : \mathrm{Hp} . \supset . \alpha \epsilon \gamma .$

$[*63{\cdot}5] \quad \supset . \gamma \subset t`\alpha \qquad (3)$

$\vdash . (2) . (3) . *104{\cdot}1 . \quad \supset \vdash : \mathrm{Hp} . \supset . \gamma \epsilon \mathrm{N}^1\mathrm{c}`\alpha \qquad (4)$

$\vdash . *52{\cdot}21{\cdot}3 . \quad \supset \vdash . \Lambda \sim\epsilon \iota``\beta \qquad (5)$

$\vdash . (1) . *52{\cdot}3 . *54{\cdot}25 . \supset \vdash : \mathrm{Hp} . \supset . \iota`x \cup \iota`y \sim\epsilon \iota``\beta \qquad (6)$

$\vdash . (5) . (6) . \quad \supset \vdash : \mathrm{Hp} . \supset . \gamma \cap \delta = \Lambda \qquad (7)$

$\vdash . (4) . (7) . *104{\cdot}2 . \supset \vdash . \mathrm{Prop}$

***104·42.** $\vdash : t`\alpha = t`\beta \,.\, \alpha \epsilon 0 \cup 1 \cup 2 \,.\supset .\, (\exists \gamma, \delta) \,.\, \gamma \epsilon \mathrm{N}^1\mathrm{c}`\alpha \,.\, \delta \epsilon \mathrm{N}^1\mathrm{c}`\beta \,.\, \gamma \cap \delta = \Lambda$
[*104·411·412·413 . *52·1 . *54·101]

∗104·43. $\vdash : t`\alpha = t`\beta . \supset . (\exists \gamma, \delta) . \gamma \in N^1c`\alpha . \delta \in N^1c`\beta . \gamma \cap \delta = \Lambda$

Dem.

$\vdash . ∗54·56 . \supset$

$\vdash : \mathrm{Hp} . \alpha \sim\in 0 \cup 1 \cup 2 . \supset . (\exists x, y, z) . x, y, z \in \alpha . x \neq y . x \neq z . y \neq z .$

[∗104·41] $\supset . (\exists \gamma, \delta) . \gamma \in N^1c`\alpha . \delta \in N^1c`\beta . \gamma \cap \delta = \Lambda$ (1)

$\vdash . (1) . ∗104·42 . \supset \vdash . \mathrm{Prop}$

The above proposition gives the desired result. The following propositions re-state this result in other forms.

∗104·44. $\vdash : \mu, \nu \in N^1C . t`\mu = t`\nu . \supset . (\exists \gamma, \delta) . \gamma \in \mu . \delta \in \nu . \gamma \cap \delta = \Lambda$
[∗104·13·43]

∗104·45. $\vdash : \mu, \nu \in N_0C . t`\mu = t`\nu . \supset . (\exists \gamma, \delta) . \gamma \in \mu^{(1)} . \delta \in \nu^{(1)} . \gamma \cap \delta = \Lambda$
[∗104·29·44]

∗104·46. $\vdash : \mu, \nu \in NC - \iota`\Lambda . t`\mu = t`\nu . \supset . (\exists \gamma, \delta) . \gamma \in \mu^{(1)} . \delta \in \nu^{(1)} . \gamma \cap \delta = \Lambda$
[∗104·28·44]

$*105$. DESCENDING CARDINALS.

Summary of $*105$.

In this number, we consider cardinals generated by a relation of similarity which goes from a higher to a lower type, *i.e.* given any class of classes κ, we consider $\mathrm{Nc}‘\kappa$ in the type of members of κ (which we shall call $\mathrm{N_1c}‘\kappa$) or in some lower type. Thus *e.g.* we shall have

$$\kappa = \iota``\alpha \,.\supset.\, \alpha \in \mathrm{N_1c}‘\kappa,$$

where "$\mathrm{N_1c}‘\kappa$" means "classes similar to κ but of the next lower type." Similarly

$$\kappa = \iota``\iota``\alpha \,.\supset.\, \alpha \in \mathrm{N_2c}‘\kappa,$$

and so on. We shall have generally

$$\beta \in \mathrm{N_1c}‘\alpha \,.\equiv.\, \alpha \in \mathrm{N^1c}‘\beta,$$

$$\beta \in \mathrm{N_2c}‘\alpha \,.\equiv.\, \alpha \in \mathrm{N^2c}‘\beta,$$

and so on. The chief difference between ascending and descending cardinals is that Λ is one of the latter, but not one of the former. Otherwise the propositions of the present number are mostly analogous to corresponding propositions of $*104$.

On the analogy of the definitions in $*104$, we put

$$\mathrm{N_1C} = \mathrm{D}‘\mathrm{N_1c} \quad \mathrm{Df},$$

$$\mu_{(1)} = \mathrm{sm}``\mu \cap t_1‘\mu \quad \mathrm{Df},$$

with similar definitions for $\mathrm{N_2C}$ and $\mu_{(2)}$.

No proposition of the present number is ever referred to in the sequel, and the reader who is not interested in the subject may therefore omit it without detriment to what follows. The principal propositions proved are the following:

$*105\cdot25$. $\vdash . \mathrm{N_0C} = \mathrm{N_1C} - \iota‘\Lambda$

$*105\cdot251$. $\vdash . \mathrm{N_0C} = \mathrm{N_2C} - \iota‘\Lambda$

$*105\cdot26$. $\vdash . \mathrm{N_1c}‘t‘\alpha = \Lambda$

Thus $\mathrm{N_1C}$ or $\mathrm{N_2C}$, in any given type, only differs from $\mathrm{N_0C}$ in that type by the addition of Λ.

***105·3.** $\vdash : \mu = \text{N}_0\text{c}'\alpha . \supset . \mu_{(1)} = \text{N}_1\text{c}'\alpha$

***105·322.** $\vdash :. \exists ! \, \text{N}_1\text{c}'\alpha . \supset : \text{N}_1\text{c}'\alpha = \text{N}_1\text{c}'\beta . \equiv . \text{N}_0\text{c}'\alpha = \text{N}_0\text{c}'\beta$

***105·34.** $\vdash :. \mu \in \text{NC} . \exists ! \, \mu_{(1)} . \supset : \mu_{(1)} = \text{N}_1\text{c}'\alpha . \equiv . \mu = \text{N}_0\text{c}'\alpha$

***105·35.** $\vdash :. \mu \in \text{NC} . \nu \in \text{N}_0\text{C} . \supset : \mu = \nu^{(1)} . \equiv . \mu_{(1)} = \nu$

***105·38.** $\vdash . \{\mu_{(1)}\}_{(1)} = \mu_{(2)}$

***105·01.** $\text{N}_1\text{c}'\alpha = \text{Nc}'\alpha \cap t't_1'\alpha$ Df

We might write

$$\text{N}_1\text{c}'\alpha = \text{Nc}'\alpha \cap t_0'\alpha \quad \text{Df},$$

which would be equivalent to the above. But we choose the above form for the sake of uniformity. If s is any suffix, we put, provided $t_s'\alpha$ has been defined,

$$\text{N}_s\text{c}'\alpha = \text{Nc}'\alpha \cap t't_s'\alpha \quad \text{Df},$$

and if i is any index for which $t^i{}'\alpha$ has been defined, we put

$$\text{N}^i\text{c}'\alpha = \text{Nc}'\alpha \cap t't^i{}'\alpha \quad \text{Df}.$$

Thus for the sake of uniformity it is better, in the above definition *105·01, to write "$t't_1'\alpha$" rather than "$t_0'\alpha$."

***105·011.** $\text{N}_2\text{c}'\alpha = \text{Nc}'\alpha \cap t't_2'\alpha$ Df

***105·02.** $\text{N}_1\text{C} = \text{D}'\text{N}_1\text{c}$ Df

***105·021.** $\text{N}_2\text{C} = \text{D}'\text{N}_2\text{c}$ Df

***105·03.** $\mu_{(1)} = \text{sm}''\mu \cap t_1'\mu$ Df

***105·031.** $\mu_{(2)} = \text{sm}''\mu \cap t_2'\mu$ Df

***105·1.** $\vdash . \text{N}_1\text{c}'\alpha = \text{Nc}'\alpha \cap t_0'\alpha$ [*63·383 . (*105·01)]

***105·101.** $\vdash . \text{N}_2\text{c}'\alpha = \text{Nc}'\alpha \cap t_1'\alpha$ [*63·41 . (*105·011)]

***105·11.** $\vdash : \beta \in \text{N}_1\text{c}'\alpha . \equiv . \beta \in \text{Nc}'\alpha . \beta \in t_0'\alpha . \equiv . \beta \,\text{sm}\, \alpha . \beta \in t_0'\alpha . \equiv . \beta \,\text{sm}\, \alpha . \beta \subset t_1'\alpha$
[*105·1 . *100·31 . *63·51]

***105·111.** $\vdash : \beta \in \text{N}_2\text{c}'\alpha . \equiv . \beta \in \text{Nc}'\alpha . \beta \in t_1'\alpha . \equiv . \beta \,\text{sm}\, \alpha . \beta \in t_1'\alpha . \equiv . \beta \,\text{sm}\, \alpha . \beta \subset t_2'\alpha$
[*105·101 . *100·31 . *63·52]

***105·12.** $\vdash : \beta \in \text{N}_1\text{c}'\alpha . \equiv . \beta \in \text{Nc}'\alpha . \alpha \subset t'\beta . \equiv . \beta \,\text{sm}\, \alpha . \alpha \subset t'\beta . \equiv . \alpha \in \text{N}^1\text{c}'\beta$
[*105·11 . *63·51 . *104·1]

***105·121.** $\vdash : \beta \in \text{N}_2\text{c}'\alpha . \equiv . \beta \in \text{Nc}'\alpha . \alpha \subset t^2{}'\beta . \equiv . \beta \,\text{sm}\, \alpha . \alpha \subset t^2{}'\beta . \equiv . \alpha \in \text{N}^2\text{c}'\beta$
[*105·111 . *63·52 . *104·11]

***105·13.** $\vdash . \text{N}_1\text{c}'\alpha = \text{Nc}\,(t_1'\alpha)'\alpha = \text{Nc}\,\{(t_1'\alpha)_\alpha\}'\alpha$ [*102·6 . (*105·01)]

***105·131.** $\vdash . \text{N}_2\text{c}'\alpha = \text{Nc}\,(t_2'\alpha)'\alpha = \text{Nc}\,\{(t_2'\alpha)_\alpha\}'\alpha$ [*102·6 . (*105·011)]

***105·14.** $\vdash : \alpha \epsilon t_0{}^\backprime\beta . \supset . N_1c{}^\backprime\beta = Nc(\alpha){}^\backprime\beta = Nc(\alpha_\beta){}^\backprime\beta$

Dem.

$\vdash . *63{\cdot}22 . \supset \vdash : Hp . \supset . t{}^\backprime\alpha = t_0{}^\backprime\beta .$

$[*105{\cdot}1] \qquad \supset . N_1c{}^\backprime\beta = Nc{}^\backprime\beta \cap t{}^\backprime\alpha \qquad (1)$

$\vdash . (1) . *102{\cdot}6 . \supset \vdash . Prop$

***105·141.** $\vdash : \alpha \epsilon t_1{}^\backprime\beta . \supset . N_2c{}^\backprime\beta = Nc(\alpha){}^\backprime\beta = Nc(\alpha_\beta){}^\backprime\beta$ [Proof as in *105·14]

***105·142.** $\vdash : \beta \subset t{}^\backprime\alpha . \supset . N_1c{}^\backprime\beta = Nc(\alpha){}^\backprime\beta = Nc(\alpha_\beta){}^\backprime\beta$ [*105·14 . *63·51]

***105·143.** $\vdash : \beta \subset t^2{}^\backprime\alpha . \supset . N_2c{}^\backprime\beta = Nc(\alpha){}^\backprime\beta = Nc(\alpha_\beta){}^\backprime\beta$ [*105·141 . *63·52]

***105·15.** $\vdash : \mu \epsilon N_1C . \equiv . (\exists\alpha) . \mu = N_1c{}^\backprime\alpha$ [*100·22 . *71·41 . (*105·02)]

***105·151.** $\vdash : \mu \epsilon N_2C . \equiv . (\exists\alpha) . \mu = N_2c{}^\backprime\alpha$

***105·16.** $\vdash : \delta \epsilon \mu_{(1)} . \equiv . (\exists\gamma) . \gamma \epsilon \mu . \delta \,\mathrm{sm}\, \gamma . \delta \epsilon t_1{}^\backprime\mu .$

$\equiv . (\exists\gamma) . \gamma \epsilon \mu . \delta \,\mathrm{sm}\, \gamma . \delta \epsilon t_0{}^\backprime\gamma .$

$\equiv . (\exists\gamma) . \gamma \epsilon \mu . \delta \,\mathrm{sm}\, \gamma . \gamma \subset t{}^\backprime\delta$ [*37·1 . *63·51·54]

***105·161.** $\vdash : \delta \epsilon \mu_{(2)} . \equiv . (\exists\gamma) . \gamma \epsilon \mu . \delta \,\mathrm{sm}\, \gamma . \delta \epsilon t_2{}^\backprime\mu .$

$\equiv . (\exists\gamma) . \gamma \epsilon \mu . \delta \,\mathrm{sm}\, \gamma . \delta \epsilon t_1{}^\backprime\gamma .$

$\equiv . (\exists\gamma) . \gamma \epsilon \mu . \delta \,\mathrm{sm}\, \gamma . \gamma \subset t^2{}^\backprime\delta$ [*37·1 . *63·52·55]

In what follows, propositions concerning N_2c or N_2C have proofs exactly analogous to those of the corresponding propositions concerning N_1c or N_1C.

***105·2.** $\vdash . N_0c{}^\backprime\alpha = N_1c{}^\backprime\iota{}^{\backprime\backprime}\alpha$

Dem.

$\vdash . *105{\cdot}12 . *104{\cdot}2 . \supset \vdash . \alpha \epsilon N_1c{}^\backprime\iota{}^{\backprime\backprime}\alpha .$

$[*103{\cdot}26] \qquad \supset \vdash . N_0c{}^\backprime\alpha = N_1c{}^\backprime\iota{}^{\backprime\backprime}\alpha$

***105·201.** $\vdash . N_0c{}^\backprime\alpha = N_2c{}^\backprime\iota{}^{\backprime\backprime}\iota{}^{\backprime\backprime}\alpha$

***105·21.** $\vdash . N_0C \subset N_1C$ [*105·2·15]

***105·211.** $\vdash . N_0C \subset N_2C$

***105·22.** $\vdash : \gamma \epsilon N_1c{}^\backprime\delta . \supset . N_1c{}^\backprime\delta = N_0c{}^\backprime\gamma$ [*103·26]

***105·221.** $\vdash : \gamma \epsilon N_2c{}^\backprime\delta . \supset . N_2c{}^\backprime\delta = N_0c{}^\backprime\gamma$

***105·23.** $\vdash : \exists ! N_1c{}^\backprime\delta . \supset . N_1c{}^\backprime\delta \epsilon N_0C$ [*105·22]

***105·231.** $\vdash : \exists ! N_2c{}^\backprime\delta . \supset . N_2c{}^\backprime\delta \epsilon N_0C$

***105·24.** $\vdash . N_1C - \iota{}^\backprime\Lambda \subset N_0C$ [*105·23]

***105·241.** $\vdash . N_2C - \iota{}^\backprime\Lambda \subset N_0C$

***105·25.** $\vdash . N_0C = N_1C - \iota{}^\backprime\Lambda$ [*105·21·24 . *103·23]

***105·251.** $\vdash . N_0C = N_2C - \iota{}^\backprime\Lambda$

***105·252.** $\vdash . N_1c`\beta = N_2c`\iota``\beta$

Dem.

$\vdash . *105·111 . \supset \vdash : \alpha \in N_2c`\iota``\beta . \equiv . \alpha \text{ sm } \iota``\beta . \alpha \in t_1`\iota``\beta .$
[*73·41.*63·64·54] $\equiv . \alpha \text{ sm } \beta . \alpha \in t_0`\beta .$
[*105·11] $\equiv . \alpha \in N_1c`\beta : \supset \vdash . \text{Prop}$

***105·26.** $\vdash . N_1c`t`\alpha = \Lambda$

Dem.

$\vdash . *105·142 . \supset \vdash . N_1c`t`\alpha = Nc(\alpha)`t`\alpha$ (1)
$\vdash . (1) . *102·73 . \supset \vdash . \text{Prop}$

***105·261.** $\vdash . N_2c`\iota``t`\alpha = \Lambda$ [*105·26·252]

***105·27.** $\vdash . \Lambda \in N_1C$ [*105·26]

***105·271.** $\vdash . \Lambda \in N_2C$

***105·28.** $\vdash . N_1C = N_0C \cup \iota`\Lambda$ [*105·25·27]

***105·281.** $\vdash . N_2C = N_1C = N_0C \cup \iota`\Lambda$

***105·29.** $\vdash . NC \subset N_1C . NC \subset N_2C$ [*105·281 . *103·34]

***105·3.** $\vdash : \mu = N_0c`\alpha . \supset . \mu_{(1)} = N_1c`\alpha$

Dem.

$\vdash . *103·4 . (*105·03) . \supset \vdash : \mu = N_0c`\alpha . \supset . \mu_{(1)} = Nc`\alpha \cap t_1`\mu$ (1)
$\vdash . *103·12 . \supset \vdash : \mu = N_0c`\alpha . \supset . \alpha \in \mu .$
[*63·105] $\supset . \alpha \in t_0`\mu .$
[*63·54] $\supset . t_0`\alpha = t_1`\mu$ (2)
$\vdash . (1) . (2) . \supset \vdash : \mu = N_0c`\alpha . \supset . \mu_{(1)} = Nc`\alpha \cap t_0`\alpha .$
[*105·1] $\supset . \mu_{(1)} = N_1c`\alpha : \supset \vdash . \text{Prop}$

***105·301.** $\vdash : \mu = N_0c`\alpha . \supset . \mu_{(2)} = N_2c`\alpha$

***105·31.** $\vdash : \mu \in N_0C . \supset . \mu_{(1)} \in N_1C$ [*105·3·15 . *103·2]

***105·311.** $\vdash : \mu \in N_0C . \supset . \mu_{(2)} \in N_2C$

***105·312.** $\vdash : \gamma \in N_1c`\alpha . \supset . \alpha \in N^1c`\gamma . N^1c`\gamma = N_0c`\alpha$ [*105·12 . *103·26]

***105·313.** $\vdash : \gamma \in N_2c`\alpha . \supset . \alpha \in N^2c`\gamma . N^2c`\gamma = N_0c`\alpha$

***105·314.** $\vdash : N_1c`\alpha = N_0c`\gamma . \supset . N_0c`\alpha = N^1c`\gamma$ [*105·312 . *103·12]

***105·315.** $\vdash : N_2c`\alpha = N_0c`\gamma . \supset . N_0c`\alpha = N^2c`\gamma$

***105·316.** $\vdash : \exists ! N_1c`\alpha . N_1c`\alpha = N_1c`\beta . \supset . N_0c`\alpha = N_0c`\beta$

Dem.

$\vdash . *105·312 . \supset \vdash : \gamma \in N_1c`\alpha . N_1c`\alpha = N_1c`\beta . \supset . N^1c`\gamma = N_0c`\alpha . N^1c`\gamma = N_0c`\beta .$
[*13·171] $\supset . N_0c`\alpha = N_0c`\beta$ (1)
$\vdash . (1) . *10·11·23·35 . \supset \vdash . \text{Prop}$

***105·317.** $\vdash : \exists ! N_2c`\alpha . N_2c`\alpha = N_2c`\beta . \supset . N_0c`\alpha = N_0c`\beta$

***105·32.** $\vdash : N_0c‘\alpha = N_0c‘\beta . \supset . N_1c‘\alpha = N_1c‘\beta$

Dem.

$\vdash . *103{\cdot}41 . \supset \vdash : Hp . \supset . Nc\,(t_1‘\alpha)‘\alpha = Nc\,(t_1‘\alpha)‘\beta$ (1)

$\vdash . *103{\cdot}14 . \supset \vdash : Hp . \supset . \beta \in t‘\alpha .$

$[*63{\cdot}16{\cdot}36] \qquad \supset . t_1‘\alpha = t_1‘\beta$ (2)

$\vdash . (1) . (2) . \supset \vdash : Hp . \supset . Nc\,(t_1‘\alpha)‘\alpha = Nc\,(t_1‘\beta)‘\beta .$

$[*105{\cdot}13] \qquad \supset . N_1c‘\alpha = N_1c‘\beta : \supset \vdash . Prop$

***105·321.** $\vdash : N_0c‘\alpha = N_0c‘\beta . \supset . N_2c‘\alpha = N_2c‘\beta$

***105·322.** $\vdash :. \exists ! N_1c‘\alpha . \supset : N_1c‘\alpha = N_1c‘\beta . \equiv . N_0c‘\alpha = N_0c‘\beta$ [*105·316·32]

***105·323.** $\vdash :. \exists ! N_2c‘\alpha . \supset : N_2c‘\alpha = N_2c‘\beta . \equiv . N_0c‘\alpha = N_0c‘\beta$

***105·324.** $\vdash : \exists ! \mu_{(1)} . \supset . \exists ! \mu$ [*37·29 . (*105·03)]

***105·325.** $\vdash : \exists ! \mu_{(2)} . \supset . \exists ! \mu$

***105·326.** $\vdash : \mu \in NC . \mu_{(1)} = N_0c‘\gamma . \supset . \mu = N^1c‘\gamma$

Dem.

$\vdash . *103{\cdot}26 . \qquad \supset \vdash : Hp . \alpha \in \mu . \supset . \mu = N_0c‘\alpha .$ (1)

$[*105{\cdot}3] \qquad \supset . \mu_{(1)} = N_1c‘\alpha .$

$[Hp] \qquad \supset . N_1c‘\alpha = N_0c‘\gamma .$

$[*105{\cdot}314] \qquad \supset . N_0c‘\alpha = N^1c‘\gamma .$

$[(1)] \qquad \supset . \mu = N^1c‘\gamma$ (2)

$\vdash . (2) . *10{\cdot}11{\cdot}23{\cdot}35 . \supset \vdash : Hp . \exists ! \mu . \supset . \mu = N^1c‘\gamma$ (3)

$\vdash . (3) . *105{\cdot}324 . *103{\cdot}13 . \supset \vdash . Prop$

***105·327.** $\vdash : \mu \in NC . \mu_{(2)} = N_0c‘\gamma . \supset . \mu = N^2c‘\gamma$

***105·33.** $\vdash : \mu \in NC . \exists ! \mu_{(1)} . \mu_{(1)} = N_1c‘\alpha . \supset . \mu = N_0c‘\alpha$

Dem.

$\vdash . *103{\cdot}26 . \supset \vdash : \gamma \in \mu_{(1)} . \mu_{(1)} = N_1c‘\alpha . \supset . N_1c‘\alpha = N_0c‘\gamma .$

$[*105{\cdot}314] \qquad \supset . N_0c‘\alpha = N^1c‘\gamma$ (1)

$\vdash . (1) . *105{\cdot}326 . \supset$

$\vdash : \gamma \in \mu_{(1)} . \mu_{(1)} = N_1c‘\alpha . \mu \in NC . \supset . \mu = N_0c‘\alpha$ (2)

$\vdash . (2) . *10{\cdot}11{\cdot}23{\cdot}35 . \supset \vdash . Prop$

***105·331.** $\vdash : \mu \in NC . \exists ! \mu_{(2)} . \mu_{(2)} = N_2c‘\alpha . \supset . \mu = N_0c‘\alpha$

***105·34.** $\vdash :. \mu \in NC . \exists ! \mu_{(1)} . \supset : \mu_{(1)} = N_1c‘\alpha . \equiv . \mu = N_0c‘\alpha$ [*105·33·3]

***105·341.** $\vdash :. \mu \in NC . \exists ! \mu_{(2)} . \supset : \mu_{(2)} = N_2c‘\alpha . \equiv . \mu = N_0c‘\alpha$

***105·342.** $\vdash . \mu \in NC . \supset . \mu_{(1)} \in N_1C$

Dem.

$\vdash . *103{\cdot}34 . \supset \vdash : Hp . \exists ! \mu . \supset . \mu \in N_0C .$

$[*105{\cdot}31] \qquad \supset . \mu_{(1)} \in N_1C$ (1)

$\vdash . *105{\cdot}324 . \supset \vdash : Hp . \sim\exists ! \mu . \supset . \sim\exists ! \mu_{(1)} .$

$[*105{\cdot}27] \qquad \supset . \mu_{(1)} \in N_1C$ (2)

$\vdash . (1) . (2) . \supset \vdash . Prop$

∗105·343. $\vdash : \mu \in \text{NC} \,.\, \supset \,.\, \mu_{(2)} \in \text{N}_2\text{C}$

∗105·344. $\vdash : \mu = \text{N}^1\text{c}'\gamma \,.\, \supset \,.\, \mu_{(1)} = \text{N}_0\text{c}'\gamma$

Dem.

$\vdash \,.\, ∗104{\cdot}24 \,.\, \supset \vdash : \text{Hp} \,.\, \supset \,.\, \mu = \text{N}_0\text{c}'\iota``\gamma \,.$

$[∗105{\cdot}3] \quad \supset \,.\, \mu_{(1)} = \text{N}_1\text{c}'\iota``\gamma \,.$

$[∗105{\cdot}2] \quad \supset \,.\, \mu_{(1)} = \text{N}_0\text{c}'\gamma : \supset \vdash \,.\, \text{Prop}$

∗105·345. $\vdash : \mu = \text{N}^2\text{c}'\gamma \,.\, \supset \,.\, \mu_{(2)} = \text{N}_0\text{c}'\gamma$

∗105·35. $\vdash :. \mu \in \text{NC} \,.\, \nu \in \text{N}_0\text{C} \,.\, \supset : \mu = \nu^{(1)} \,.\, \equiv \,.\, \mu_{(1)} = \nu$

Dem.

$\vdash \,.\, ∗105{\cdot}326 \,.\, ∗104{\cdot}26 \,.\, \supset$

$\vdash : \mu \in \text{NC} \,.\, \nu = \text{N}_0\text{c}'\gamma \,.\, \mu_{(1)} = \nu \,.\, \supset \,.\, \mu = \text{N}^1\text{c}'\gamma \,.\, \nu^{(1)} = \text{N}^1\text{c}'\gamma \,.$

$[∗13{\cdot}172] \quad \supset \,.\, \mu = \nu^{(1)} \quad (1)$

$\vdash \,.\, ∗104{\cdot}26 \,.\, \text{Fact} \,.\, \supset$

$\vdash : \mu \in \text{NC} \,.\, \nu = \text{N}_0\text{c}'\gamma \,.\, \mu = \nu^{(1)} \,.\, \supset \,.\, \mu = \text{N}^1\text{c}'\gamma \,.\, \nu = \text{N}_0\text{c}'\gamma \,.$

$[∗105{\cdot}344] \quad \supset \,.\, \mu_{(1)} = \text{N}_0\text{c}'\gamma \,.\, \nu = \text{N}_0\text{c}'\gamma \,.$

$[∗13{\cdot}172] \quad \supset \,.\, \mu_{(1)} = \nu \quad (2)$

$\vdash \,.\, (1) \,.\, (2) \,.\, \supset \vdash :. \mu \in \text{NC} \,.\, \nu = \text{N}_0\text{c}'\gamma \,.\, \supset : \mu = \nu^{(1)} \,.\, \equiv \,.\, \mu_{(1)} = \nu \quad (3)$

$\vdash \,.\, (3) \,.\, ∗103{\cdot}2 \,.\, \supset \vdash \,.\, \text{Prop}$

∗105·351. $\vdash :. \mu \in \text{NC} \,.\, \nu \in \text{N}_0\text{C} \,.\, \supset : \mu = \nu^{(2)} \,.\, \equiv \,.\, \mu_{(2)} = \nu$

∗105·352. $\vdash :. \mu, \nu \in \text{NC} \,.\, \exists ! \nu \,.\, \supset : \mu = \nu^{(1)} \,.\, \equiv \,.\, \mu_{(1)} = \nu$ [∗105·35 . ∗103·34]

∗105·353. $\vdash :. \mu, \nu \in \text{NC} \,.\, \exists ! \nu \,.\, \supset : \mu = \nu^{(2)} \,.\, \equiv \,.\, \mu_{(2)} = \nu$

∗105·354. $\vdash : \nu \in \text{NC} \,.\, \exists ! \nu \,.\, \supset \,.\, \{\nu^{(1)}\}_{(1)} = \nu$ [∗105·352]

∗105·355. $\vdash : \nu \in \text{NC} \,.\, \exists ! \nu \,.\, \supset \,.\, \{\nu^{(2)}\}_{(2)} = \nu$

∗105·356. $\vdash : \mu \in \text{NC} \,.\, \exists ! \mu_{(1)} \,.\, \supset \,.\, \{\mu_{(1)}\}^{(1)} = \mu$ [∗105·352]

∗105·357. $\vdash : \mu \in \text{NC} \,.\, \exists ! \mu_{(2)} \,.\, \supset \,.\, \{\mu_{(2)}\}^{(2)} = \mu$

∗105·36. $\vdash : \beta \in \text{N}_1\text{c}'\alpha \,.\, \gamma \in \text{N}_1\text{c}'\beta \,.\, \supset \,.\, \gamma \in \text{N}_2\text{c}'\alpha$

Dem.

$\vdash \,.\, ∗105{\cdot}11 \,.\, \supset \vdash : \text{Hp} \,.\, \supset \,.\, \beta \,\text{sm}\, \alpha \,.\, \beta \in t_0{}'\alpha \,.\, \gamma \,\text{sm}\, \beta \,.\, \gamma \in t_0{}'\beta \,.$

$[∗73{\cdot}32 . ∗63{\cdot}38] \quad \supset \,.\, \gamma \,\text{sm}\, \alpha \,.\, \gamma \in t_1{}'\alpha \,.$

$[∗105{\cdot}111] \quad \supset \,.\, \gamma \in \text{N}_2\text{c}'\alpha : \supset \vdash \,.\, \text{Prop}$

∗105·361. $\vdash : \beta \in \text{N}_1\text{c}'\alpha \,.\, \gamma \in \text{N}_2\text{c}'\alpha \,.\, \supset \,.\, \gamma \in \text{N}_1\text{c}'\beta$

Dem.

$\vdash \,.\, ∗105{\cdot}11{\cdot}111 \,.\, \supset \vdash : \text{Hp} \,.\, \supset \,.\, \beta \,\text{sm}\, \alpha \,.\, \beta \in t_0{}'\alpha \,.\, \gamma \,\text{sm}\, \alpha \,.\, \gamma \in t_1{}'\alpha \,.$

$[∗73{\cdot}31{\cdot}32] \quad \supset \,.\, \gamma \,\text{sm}\, \beta \,.\, \beta \in t_0{}'\alpha \,.\, \gamma \in t_1{}'\alpha \quad (1)$

$\vdash \,.\, ∗63{\cdot}54 \,.\, \quad \supset \vdash : \beta \in t_0{}'\alpha \,.\, \supset \,.\, t_0{}'\beta = t_1{}'\alpha \quad (2)$

$\vdash \,.\, (1) \,.\, (2) \,.\, \quad \supset \vdash : \text{Hp} \,.\, \supset \,.\, \gamma \,\text{sm}\, \beta \,.\, \gamma \in t_0{}'\beta \,.$

$[∗105{\cdot}11] \quad \supset \,.\, \gamma \in \text{N}_1\text{c}'\beta : \supset \vdash \,.\, \text{Prop}$

***105·362.** $\vdash : \beta \in \mathrm{N_1c}`\alpha . \supset . \mathrm{N_1c}`\beta = \mathrm{N_2c}`\alpha$ [*105·36·361]

***105·37.** $\vdash : \mathrm{N_0c}`\beta = \mathrm{N_1c}`\alpha . \supset . \mathrm{N_1c}`\beta = \mathrm{N_2c}`\alpha$ [*105·362 . *103·12]

***105·371.** $\vdash : \exists ! \mu_{(2)} . \supset . \exists ! \mu_{(1)}$

Dem.

$\vdash . *63 \cdot 381 . (*63 \cdot 05) . \supset$

$\vdash : \gamma \,\mathrm{sm}\, \alpha . \alpha \in \mu . \gamma \in t_2`\mu . \supset . \gamma \,\mathrm{sm}\, \alpha . \alpha \in \mu . t`\gamma = t_2`\mu .$

[*73·41.*63·64] $\supset . \iota``\gamma \,\mathrm{sm}\, \alpha . \alpha \in \mu . t_0`\iota``\gamma = t_2`\mu .$

[*63·57] $\supset . \iota``\gamma \,\mathrm{sm}\, \alpha . \alpha \in \mu . t`\iota``\gamma = t_1`\mu .$

[*63·103] $\supset . \iota``\gamma \,\mathrm{sm}\, \alpha . \alpha \in \mu . \iota``\gamma \in t_1`\mu .$

[*105·16] $\supset . \iota``\gamma \in \mu_{(1)} .$

[*10·24] $\supset . \exists ! \mu_{(1)}$ (1)

$\vdash . (1) . *10 \cdot 11 \cdot 23 . \supset$

$\vdash : (\exists \alpha) . \gamma \,\mathrm{sm}\, \alpha . \alpha \in \mu . \gamma \in t_2`\mu . \supset . \exists ! \mu_{(1)}$ (2)

$\vdash . (2) . *105 \cdot 161 . \supset \vdash : \gamma \in \mu_{(2)} . \supset . \exists ! \mu_{(1)}$ (3)

$\vdash . (3) . *10 \cdot 11 \cdot 23 . \supset \vdash . \mathrm{Prop}$

***105·372.** $\vdash : \mu_{(1)} = \Lambda . \supset . \mu_{(2)} = \Lambda$ [*105·371 . Transp]

***105·38.** $\vdash . \{\mu_{(1)}\}_{(1)} = \mu_{(2)}$

Dem.

$\vdash . *105 \cdot 16 . \supset \vdash : \gamma \in \{\mu_{(1)}\}_{(1)} . \equiv . (\exists \beta) . \beta \in \mu_{(1)} . \gamma \,\mathrm{sm}\, \beta . \gamma \in t_0`\beta .$

[*105·16] $\equiv . (\exists \alpha, \beta) . \alpha \in \mu . \beta \,\mathrm{sm}\, \alpha . \beta \in t_0`\alpha . \gamma \,\mathrm{sm}\, \beta . \gamma \in t_0`\beta .$ (1)

[*73·32.*63·38] $\supset . (\exists \alpha) . \alpha \in \mu . \gamma \,\mathrm{sm}\, \alpha . \gamma \in t_1`\alpha$ (2)

$\vdash . *73 \cdot 41 . *63 \cdot 64 \cdot 53 \cdot 57 . \supset$

$\vdash : \alpha \in \mu . \gamma \,\mathrm{sm}\, \alpha . \gamma \in t_1`\alpha . \supset . \alpha \in \mu . \iota``\gamma \,\mathrm{sm}\, \alpha . \gamma \,\mathrm{sm}\, \iota``\gamma . \gamma \in t_0`\iota``\gamma . \iota``\gamma \in t_0`\alpha .$

[(1)] $\supset . \gamma \in \{\mu_{(1)}\}_{(1)}$ (3)

$\vdash . (2) . (3) . \supset \vdash : \gamma \in \{\mu_{(1)}\}_{(1)} . \equiv . (\exists \alpha) . \alpha \in \mu . \gamma \,\mathrm{sm}\, \alpha . \gamma \in t_1`\alpha .$

[*105·161] $\equiv . \gamma \in \mu_{(2)} : \supset \vdash . \mathrm{Prop}$

***105·4.** $\vdash : \gamma \in \mathrm{N_2c}`\alpha . \supset . \iota``\gamma \in \mathrm{N_1c}`\alpha$

Dem.

$\vdash . *105 \cdot 111 . *73 \cdot 41 . *63 \cdot 64 . \supset \vdash : \mathrm{Hp} . \supset . \iota``\gamma \,\mathrm{sm}\, \alpha . \gamma \in t_1`\alpha . \gamma \in t_0`\iota``\gamma .$

[*63·41·383·16·55] $\supset . \iota``\gamma \,\mathrm{sm}\, \alpha . t_1`\alpha = t_0`\iota``\gamma .$

[*63·54] $\supset . \iota``\gamma \,\mathrm{sm}\, \alpha . \iota``\gamma \in t_0`\alpha .$

[*105·11] $\supset . \iota``\gamma \in \mathrm{N_1c}`\alpha : \supset \vdash . \mathrm{Prop}$

***105·41.** $\vdash : \exists ! \mathrm{N_2c}`\alpha . \supset . \exists ! \mathrm{N_1c}`\alpha$ [*105·4]

***105·42.** $\vdash : \mathrm{N_1c}`\alpha = \Lambda . \supset . \mathrm{N_2c}`\alpha = \Lambda$ [*105·41]

***105·43.** $\vdash : \mu_{(1)} = N_1c\text{‘}\alpha \,.\supset .\, \mu_{(2)} = N_2c\text{‘}\alpha$

Dem.

$\vdash . *105\cdot11 . \quad \supset \vdash : \text{Hp} . \beta \in \mu_{(1)} . \supset . \beta \in Nc\text{‘}\alpha \cap t_0\text{‘}\alpha .$

$[*63\cdot54 . *100\cdot31\cdot321] \quad \supset . Nc\text{‘}\beta = Nc\text{‘}\alpha . t_0\text{‘}\beta = t_1\text{‘}\alpha .$

$[*105\cdot1\cdot101] \quad \supset . N_1c\text{‘}\beta = N_2c\text{‘}\alpha \quad (1)$

$\vdash . *105\cdot3 . *103\cdot26 . \supset \vdash : \text{Hp} . \beta \in \mu_{(1)} . \supset . N_1c\text{‘}\beta = \{\mu_{(1)}\}_{(1)}$

$[*105\cdot38] \quad = \mu_{(2)} \quad (2)$

$\vdash . (1) . (2) . \quad \supset \vdash : \text{Hp} . \exists ! \mu_{(1)} . \supset . \mu_{(2)} = N_2c\text{‘}\alpha \quad (3)$

$\vdash . *105\cdot372\cdot42 . \quad \supset \vdash : \text{Hp} . \mu_{(1)} = \Lambda . \supset . \mu_{(2)} = \Lambda . N_2c\text{‘}\alpha = \Lambda \quad (4)$

$\vdash . (3) . (4) . \supset \vdash . \text{Prop}$

***105·44.** $\vdash . N_2c\text{‘}t^2\text{‘}\alpha = \Lambda$

Dem.

$\vdash . *105\cdot26 . \supset \vdash . N_1c\text{‘}t\text{‘}t\text{‘}\alpha = \Lambda .$

$[*105\cdot42] \quad \supset \vdash . N_2c\text{‘}t\text{‘}t\text{‘}\alpha = \Lambda . \supset \vdash . \text{Prop}$

$*106$. CARDINALS OF RELATIONAL TYPES.

Summary of $*106$.

In this number we have to consider the cardinals whose members are classes of relations which have a given relation of type to some given class. For example, we have $\downarrow x``\alpha \,\mathrm{sm}\, \alpha$, and $\downarrow x``\alpha$ has a given relation of type to α when x is given. Thus we want a notation for

$$\mathrm{Nc}`\alpha \cap t`\downarrow x``\alpha$$

and all the associated ideas. In this number, we shall deal only with relations in which the referent and relatum have a relation, as to type, which can be expressed by the notations of $*63$, *i.e.* roughly speaking, when, for suitable values of α, m, n, our relations are contained in

$$t^m`\alpha \uparrow t^n`\alpha \text{ or } t_m`\alpha \uparrow t_n`\alpha \text{ or } t^m`\alpha \uparrow t_n`\alpha \text{ or } t_m`\alpha \uparrow t^n`\alpha.$$

Thus if $t_{\mu\nu}`\alpha$ has been defined, we shall put

$$\mathrm{N}_{\mu\nu}\mathrm{c}`\alpha = \mathrm{Nc}`\alpha \cap t`t_{\mu\nu}`\alpha \quad \mathrm{Df},$$

$$\mathrm{N}_{\mu\nu}\mathrm{C} = \mathrm{D}`\mathrm{N}_{\mu\nu}\mathrm{c} \quad \mathrm{Df},$$

$$\xi_{(\mu\nu)} = \mathrm{sm}``\xi \cap t`t_{\mu\nu}`t_1`\xi \quad \mathrm{Df},$$

with analogous definitions for $t^{\mu\nu}`\alpha$, $t^{\mu}{}_{\nu}`\alpha$ and ${}^{\mu}t_{\nu}`\alpha$.

Much the most important case is that of $t_{00}`\alpha$. For this case we have

$*106\cdot1$. $\vdash : \beta \in \mathrm{N}_{00}\mathrm{c}`\alpha \,.\equiv .\, \beta \in \mathrm{Nc}`\alpha \,.\, \beta \in t`t_{00}`\alpha \,.\equiv .\, \beta \,\mathrm{sm}\, \alpha \,.\, \beta \in t`t`(t_0`\alpha \uparrow t_0`\alpha) \,.$
$\equiv .\, \beta \,\mathrm{sm}\, \alpha \,.\, \beta \subset t`(\alpha \uparrow \alpha)$

Thus $\mathrm{N}_{00}\mathrm{c}`\alpha$ will be the number of a class of relations whose fields are of the same type as $\alpha$, provided this class of relations is similar to $\alpha$. *E.g.* the number of terms such as $x \downarrow x$, where $x \in \alpha$, will be $\mathrm{N}_{00}\mathrm{c}`\alpha$.

We have

$*106\cdot21$. $\vdash .\, \exists !\, \mathrm{N}_{00}\mathrm{c}`\alpha \,.\, \mathrm{N}_{00}\mathrm{c}`\alpha \in \mathrm{N}_0\mathrm{C}$

$*106\cdot22$. $\vdash : \lambda \in \mathrm{N}_0{}^1\mathrm{c}`\alpha \,.\equiv .\, \mathrm{Cnv}``\lambda \in {}^1\mathrm{N}_0\mathrm{c}`\alpha$

$*106\cdot23$. $\vdash : \beta \in \mathrm{N}^1\mathrm{c}`\alpha \,.\supset .\, \mathrm{N}^{11}\mathrm{c}`\alpha = \mathrm{N}_{00}\mathrm{c}`\beta$

$*106\cdot32$. $\vdash : t_0`\alpha = t_0`\beta \,.\supset .\, (\exists \gamma, \delta) \,.\, \gamma \in \mathrm{N}_{00}\mathrm{c}`\alpha \,.\, \delta \in \mathrm{N}_{00}\mathrm{c}`\beta \,.\, \gamma \cap \delta = \Lambda$

$*106\cdot4\cdot41\cdot411$. $\vdash : \mu = \mathrm{N}_0\mathrm{c}`\alpha \,.\supset .\, \mu_{(00)} = \mathrm{N}_{00}\mathrm{c}`\alpha \,.\, \mu^{(11)} = \mathrm{N}^{11}\mathrm{c}`\alpha \,.\, \mu_{(11)} = \mathrm{N}_{11}\mathrm{c}`\alpha$

***106·53.** $\vdash . \mathrm{Nc}\,(\alpha)‘t_{00}‘\alpha = \Lambda$

whence it follows that

***106·54.** $\vdash . \mathrm{N_0c}‘t_{00}‘\alpha \sim\in \mathrm{N_{00}C}$

The propositions of this number, except *106·21, are never referred to again (except in *154·25·251·262, which are themselves never used again), but they have a somewhat greater importance than the propositions of *105, owing to the fact that the arithmetical operations are defined by means of classes of relations, *i.e.* the sum of two cardinals (for instance) is defined as the cardinal number of a certain class of relations (cf. *110).

***106·01.** $\mathrm{N_{00}c}‘\alpha = \mathrm{Nc}‘\alpha \cap t‘t_{00}‘\alpha$ Df

***106·011.** $\mathrm{N^{11}c}‘\alpha = \mathrm{Nc}‘\alpha \cap t‘t^{11}‘\alpha$ Df

***106·012.** $\mathrm{N_{01}c}‘\alpha = \mathrm{Nc}‘\alpha \cap t‘t_{01}‘\alpha$ Df etc.

***106·02.** $\mathrm{N_0{}^1c}‘\alpha = \mathrm{Nc}‘\alpha \cap t‘t_0{}^1‘\alpha$ Df etc.

***106·021.** $\mathrm{{}^1N_0c}‘\alpha = \mathrm{Nc}‘\alpha \cap t‘{}^1t_0‘\alpha$ Df etc.

***106·03.** $\mathrm{N_{00}C} = \mathrm{D}‘\mathrm{N_{00}c}$ Df etc.

***106·04.** $\mu_{(00)} = \mathrm{sm}‘‘\mu \cap t‘t_{00}‘t_1‘\mu$ Df

***106·041.** $\mu^{(11)} = \mathrm{sm}‘‘\mu \cap t‘t^{11}‘t_1‘\mu$ Df etc.

***106·1.** $\vdash : \beta \in \mathrm{N_{00}c}‘\alpha . \equiv . \beta \in \mathrm{Nc}‘\alpha . \beta \in t‘t_{00}‘\alpha .$

$\equiv . \beta \,\mathrm{sm}\, \alpha . \beta \in t‘t‘(t_0‘\alpha \uparrow t_0‘\alpha) .$

$\equiv . \beta \,\mathrm{sm}\, \alpha . \beta \subset t‘(\alpha \uparrow \alpha)$

[*100·1 . (*106·01 . *64·01) . *64·11]

***106·101.** $\vdash : \beta \in \mathrm{N^{11}c}‘\alpha . \equiv . \beta \in \mathrm{Nc}‘\alpha . \beta \in t‘t^{11}‘\alpha .$

$\equiv . \beta \,\mathrm{sm}\, \alpha . \beta \in t‘t‘(t‘\alpha \uparrow t‘\alpha) .$

$\equiv . \beta \,\mathrm{sm}\, \alpha . \beta \subset t‘(t‘\alpha \uparrow t‘\alpha)$

Similar propositions hold for any other double index mn for which $t^{mn}‘\alpha$ has been defined.

***106·11.** $\vdash : \beta \in \mathrm{N_{01}c}‘\alpha . \equiv . \beta \in \mathrm{Nc}‘\alpha . \beta \in t‘t_{01}‘\alpha .$

$\equiv . \beta \,\mathrm{sm}\, \alpha . \beta \in t‘t‘(t_0‘\alpha \uparrow t_1‘\alpha) .$

$\equiv . \beta \,\mathrm{sm}\, \alpha . \beta \subset t‘(t_0‘\alpha \uparrow t_1‘\alpha)$

Similar propositions hold for any other double suffix mn for which $t_{mn}‘\alpha$ has been defined.

***106·12.** $\vdash : \beta \in \mathrm{N_0{}^1c}‘\alpha . \equiv . \beta \in \mathrm{Nc}‘\alpha . \beta \in t‘t_0{}^1‘\alpha .$

$\equiv . \beta \,\mathrm{sm}\, \alpha . \beta \in t‘t‘(t_0‘\alpha \uparrow t‘\alpha) .$

$\equiv . \beta \,\mathrm{sm}\, \alpha . \beta \subset t‘(t_0‘\alpha \uparrow t‘\alpha)$

***106·121.** $\vdash : \beta \in \mathrm{{}^1N_0c}‘\alpha . \equiv . \beta \in \mathrm{Nc}‘\alpha . \beta \in t‘{}^1t_0‘\alpha .$

$\equiv . \beta \,\mathrm{sm}\, \alpha . \beta \in t‘t‘(t‘\alpha \uparrow t_0‘\alpha) .$

$\equiv . \beta \,\mathrm{sm}\, \alpha . \beta \subset t‘(t‘\alpha \uparrow t_0‘\alpha)$

Similar propositions hold for any other index and suffix for which $t_m{}^n{}'\alpha$ or ${}^n t_m{}'\alpha$ has been defined.

***106·13.** $\vdash : \mu \in \mathrm{N}_{00}\mathrm{C} . \equiv . (\exists \alpha) . \mu = \mathrm{N}_{00}\mathrm{c}'\alpha$ [*100·22 . *71·41]

Similar propositions hold for $\mathrm{N}^{11}\mathrm{C}'\alpha$ etc.

***106·14.** $\vdash : \beta \in \mu_{(00)} . \equiv . (\exists \alpha) . \alpha \in \mu . \beta \operatorname{sm} \alpha . \beta \in t' t'(t_1{}'\mu \uparrow t_1{}'\mu) .$
$\equiv . (\exists \alpha) . \alpha \in \mu . \beta \operatorname{sm} \alpha . \beta \in t' t_{00}{}'\alpha .$
$\equiv . (\exists \alpha) . \alpha \in \mu . \beta \operatorname{sm} \alpha . \beta \subset t'(\alpha \uparrow \alpha)$ [*64·33·11]

***106·141.** $\vdash : \beta \in \mu_0{}^1 . \equiv . (\exists \alpha) . \alpha \in \mu . \beta \operatorname{sm} \alpha . \beta \in t' t'(t_1{}'\mu \uparrow t_0{}'\mu) .$
$\equiv . (\exists \alpha) . \alpha \in \mu . \beta \operatorname{sm} \alpha . \beta \in t' t_0{}^1{}'\alpha .$
$\equiv . (\exists \alpha) . \alpha \in \mu . \beta \operatorname{sm} \alpha . \beta \subset t'(\alpha \uparrow t'\alpha)$

Similar propositions hold for ${}^1\mu_0$, μ^{11}, μ_{11} etc.

***106·2** $\vdash : x \in t_0{}'\alpha . \supset . \downarrow x''\alpha \in \mathrm{N}_{00}\mathrm{c}'\alpha . \downarrow x''\alpha \in \mathrm{N}_0\mathrm{c}' \downarrow x''\alpha$

Dem.

$\vdash . *55·15 . \supset \vdash : R \in \downarrow x''\alpha . \supset . \mathrm{D}'R \subset \alpha . \mathrm{Q}'R = \iota' x :$
[*63·105] $\supset \vdash :. x \in t_0{}'\alpha . \supset : R \in \downarrow x''\alpha . \supset_R . \mathrm{D}'R \subset t_0{}'\alpha . \mathrm{Q}'R \subset t_0{}'\alpha .$
[*35·83] $\supset_R . R \subseteq t_0{}'\alpha \uparrow t_0{}'\alpha .$
[*64·16·13] $\supset_R . R \in t'(\alpha \uparrow \alpha) :$
[*22·1] $\supset : \downarrow x''\alpha \subset t'(\alpha \uparrow \alpha)$ (1)
$\vdash . (1) . *73·611 . *106·1 . *103·12 . \supset \vdash .$ Prop

***106·201.** $\vdash : \beta \in t'\alpha . \supset . \downarrow \beta''\alpha \in \mathrm{N}_0{}^1\mathrm{c}'\alpha$

***106·202.** $\vdash : \beta \in t^2{}'\alpha . \supset . \downarrow \beta''\alpha \in \mathrm{N}_0{}^2\mathrm{c}'\alpha$

***106·203.** $\vdash . \downarrow \alpha''\alpha \in \mathrm{N}_0{}^1\mathrm{c}'\alpha$ [*106·201]

***106·204.** $\vdash . \downarrow (\iota''\alpha)''\alpha \in \mathrm{N}_0{}^2\mathrm{c}'\alpha$ [*106·202]

***106·21.** $\vdash . \exists ! \mathrm{N}_{00}\mathrm{c}'\alpha . \mathrm{N}_{00}\mathrm{c}'\alpha \in \mathrm{N}_0\mathrm{C}$ [*106·2 . *63·18]

***106·211.** $\vdash . \Lambda \sim\in \mathrm{N}_{00}\mathrm{C} . \mathrm{N}_{00}\mathrm{C} \subset \mathrm{N}_0\mathrm{C} . \mathrm{N}_{00}\mathrm{C} \in \mathrm{Cls\,ex^2\,excl}$ [*106·21 . *103·24]

***106·212.** $\vdash . \Lambda \sim\in \mathrm{N}_0{}^1\mathrm{C} . \mathrm{N}_0{}^1\mathrm{C} \subset \mathrm{N}_0\mathrm{C} . \mathrm{N}_0{}^1\mathrm{C} \in \mathrm{Cls\,ex^2\,excl}$ [*106·203]

***106·213.** $\vdash . \Lambda \sim\in \mathrm{N}_0{}^2\mathrm{C} . \mathrm{N}_0{}^2\mathrm{C} \subset \mathrm{N}_0\mathrm{C} . \mathrm{N}_0{}^2\mathrm{C} \in \mathrm{Cls\,ex^2\,excl}$ [*106·204]

***106·22.** $\vdash : \lambda \in \mathrm{N}_0{}^1\mathrm{c}'\alpha . \equiv . \mathrm{Cnv}''\lambda \in {}^1\mathrm{N}_0\mathrm{c}'\alpha$

Dem.

$\vdash . *73·4 . \supset \vdash : \lambda \operatorname{sm} \alpha . \equiv . \mathrm{Cnv}''\lambda \operatorname{sm} \alpha$ (1)
$\vdash . *64·16 . \supset \vdash :. \lambda \subset t'(t_0{}'\alpha \uparrow t'\alpha) . \equiv : R \in \lambda . \supset_R . R \subseteq t_0{}'\alpha \uparrow t'\alpha :$
[*35·84] $\equiv : R \in \lambda . \supset_R . \breve{R} \subseteq t'\alpha \uparrow t_0{}'\alpha :$
[*37·63] $\equiv : S \in \mathrm{Cnv}''\lambda . \supset_S . S \subseteq t'\alpha \uparrow t_0{}'\alpha :$
[*64·16] $\equiv : \mathrm{Cnv}''\lambda \subset t'(t'\alpha \uparrow t_0{}'\alpha)$ (2)
$\vdash . (1) . (2) . *106·12 . \supset \vdash .$ Prop

The proof requires, in addition to *106·12, its analogue for ${}^1\mathrm{N}_0\mathrm{c}'\alpha$. Such analogues will be assumed as required.

***106·221.** $\vdash : \lambda \in \mathrm{N}_0{}^2\mathrm{c}`\alpha \,.\equiv .\, \mathrm{Cnv}``\lambda \in {}^2\mathrm{N}_0\mathrm{c}`\alpha$

***106·222.** $\vdash .\, \Lambda \sim\in {}^1\mathrm{N}_0\mathrm{C} \,.\, {}^1\mathrm{N}_0\mathrm{C} \subset \mathrm{N}_0\mathrm{C} \,.\, {}^1\mathrm{N}_0\mathrm{C} \in \mathrm{Cls\,ex^2\,excl}$ [*106·22·212]

***106·223.** $\vdash .\, \Lambda \sim\in {}^2\mathrm{N}_0\mathrm{C} \,.\, {}^2\mathrm{N}_0\mathrm{C} \subset \mathrm{N}_0\mathrm{C} \,.\, {}^2\mathrm{N}_0\mathrm{C} \in \mathrm{Cls\,ex^2\,excl}$

Other propositions of the same kind as the above may be proved by observing that, if m and n are indices for which $t^m`\alpha$ and $t^n`\alpha$ have been defined, we have

$$\gamma \subset t^n`\alpha \,.\, \beta \in \mathrm{N}^m\mathrm{c}`\alpha \,.\supset .\, \downarrow \beta``\gamma \in \mathrm{N}^{mn}\mathrm{c}`\alpha,$$

of which the proof is direct and simple. Hence, since we always have $\exists ! \,\mathrm{N}^m\mathrm{c}`\alpha$, we also always have

$$\exists ! \,\mathrm{N}^{mn}\mathrm{c}`\alpha,$$

whence $$\mathrm{N}^{mn}\mathrm{C} \subset \mathrm{N}_0\mathrm{C} \,.\, \mathrm{N}^{mn}\mathrm{C} \in \mathrm{Cls\,ex^2\,excl}.$$

We have in like manner

$$\exists ! \,\mathrm{N}_0{}^m\mathrm{c}`\alpha \,.\, \exists ! \,{}^m\mathrm{N}_0\mathrm{c}`\alpha$$

But we do not always have

$$\exists ! \,\mathrm{N}_{mn}\mathrm{c}`\alpha \text{ or } \exists ! \,\mathrm{N}_n{}^m\mathrm{c}`\alpha \text{ or } \exists ! \,{}^m\mathrm{N}_n\mathrm{c}`\alpha.$$

***106·23.** $\vdash : \beta \in \mathrm{N}^1\mathrm{c}`\alpha \,.\supset .\, \mathrm{N}^{11}\mathrm{c}`\alpha = \mathrm{N}_{00}\mathrm{c}`\beta$

Dem.

$\vdash .\, *64{\cdot}33 \,.\!\cdot\, *104{\cdot}1 \,.\, *63{\cdot}5 \,.\supset \vdash : \mathrm{Hp} \,.\supset .\, t^{11}`\alpha = t_{00}`\beta$

$\vdash .\, (1) \,.\, (*106{\cdot}01{\cdot}011) \,.\, *100{\cdot}321 \,.\supset \vdash .\, \mathrm{Prop}$

***106·231.** $\vdash : \beta \in \mathrm{N}_1\mathrm{c}`\alpha \,.\supset .\, \mathrm{N}_{11}\mathrm{c}`\alpha = \mathrm{N}_{00}\mathrm{c}`\beta$ [Proof as in *106·23]

***106·24.** $\vdash : \mathrm{N}^1\mathrm{c}`\alpha = \mathrm{N}_0\mathrm{c}`\beta \,.\supset .\, \mathrm{N}^{11}\mathrm{c}`\alpha = \mathrm{N}_{00}\mathrm{c}`\beta$ [*106·23]

***106·241.** $\vdash : \mathrm{N}_1\mathrm{c}`\alpha = \mathrm{N}_0\mathrm{c}`\beta \,.\supset .\, \mathrm{N}_{11}\mathrm{c}`\alpha = \mathrm{N}_{00}\mathrm{c}`\beta$

The analogues of the above propositions for other indices or suffixes are similarly proved.

***106·25.** $\vdash .\, \mathrm{N}^{11}\mathrm{c}`\alpha = \mathrm{N}_{00}\mathrm{c}`\iota``\alpha$ [*106·23 . *104·2]

***106·251.** $\vdash .\, \mathrm{N}_{00}\mathrm{c}`\alpha = \mathrm{N}_{11}\mathrm{c}`\iota``\alpha$

***106·31.** $\vdash : x, y \in t_0`\alpha \,.\, t_0`\alpha = t_0`\beta \,.\, x \neq y \,.\supset .$

$$\downarrow x``\alpha \in \mathrm{N}_{00}\mathrm{c}`\alpha \,.\, \downarrow y``\beta \in \mathrm{N}_{00}\mathrm{c}`\beta \,.\, \downarrow x``\alpha \cap \downarrow x``\beta = \Lambda$$

[*106·2 . *55·233]

***106·311.** $\vdash :.\, x \in t_0`\alpha \,.\, t_0`\alpha = t_0`\beta : \alpha = \Lambda \,.\vee .\, \beta = \Lambda : \supset .$

$$\downarrow x``\alpha \in \mathrm{N}_{00}\mathrm{c}`\alpha \,.\, \downarrow x``\beta \in \mathrm{N}_{00}\mathrm{c}`\beta \,.\, \downarrow x``\alpha \cap \downarrow x``\beta = \Lambda$$

[*106·2 . *55·232 . Transp]

***106·312.** $\vdash : t_0`\alpha = \iota`x \,.\, \alpha = \beta = \iota`x \,.\supset .$

$$\iota`(\iota`x \uparrow \iota`x) \in \mathrm{N}_{00}\mathrm{c}`\alpha \,.\, \iota`(\Lambda \uparrow \iota`x) \in \mathrm{N}_{00}\mathrm{c}`\beta \,.\, \iota`(\iota`x \uparrow \iota`x) \cap \iota`(\Lambda \uparrow \iota`x) = \Lambda$$

Dem.

$\vdash .\, *73{\cdot}43 \,.\supset \vdash .\, \iota`(\iota`x \uparrow \iota`x) \,\mathrm{sm}\, \iota`x \,.\, \iota`(\Lambda \uparrow \iota`x) \,\mathrm{sm}\, \iota`x \,.$

$[*13{\cdot}12] \quad \supset \vdash : \mathrm{Hp} \,.\supset .\, \iota`(\iota`x \uparrow \iota`x) \,\mathrm{sm}\, \alpha \,.\, \iota`(\Lambda \uparrow \iota`x) \,\mathrm{sm}\, \beta \qquad (1)$

$\vdash .\, *64{\cdot}16 \,.\supset \vdash : \mathrm{Hp} \,.\supset .\, \iota`x \uparrow \iota`x \in t_{00}`\alpha \,.\, \Lambda \uparrow \iota`x \in t_{00}`\alpha \qquad (2)$

$\vdash .\, (1) \,.\, (2) \,.\, *106{\cdot}1 \,.\, *51{\cdot}161 \,.\, *24{\cdot}54 \,.\, *55{\cdot}202 \,.\supset \vdash .\, \mathrm{Prop}$

∗106·32. $\vdash : t_0\text{‘}\alpha = t_0\text{‘}\beta \,.\supset .\, (\exists\gamma, \delta) \,.\, \gamma \epsilon \mathrm{N}_{00}\mathrm{c}\text{‘}\alpha \,.\, \delta \epsilon \mathrm{N}_{00}\mathrm{c}\text{‘}\beta \,.\, \gamma \cap \delta = \Lambda$

Dem.

$\vdash \,.\, *106\cdot31 \,.\supset \vdash :. \mathrm{Hp} : (\exists x, y) \,.\, x, y \epsilon t_0\text{‘}\alpha \,.\, x \neq y : \supset .$

$\qquad\qquad (\exists\gamma, \delta) \,.\, \gamma \epsilon \mathrm{N}_{00}\mathrm{c}\text{‘}\alpha \,.\, \delta \epsilon \mathrm{N}_{00}\mathrm{c}\text{‘}\beta \,.\, \gamma \cap \delta = \Lambda \quad (1)$

$\vdash \,.\, *52\cdot4 \,.\quad \supset \vdash : \sim(\exists x, y) \,.\, x, y \epsilon t_0\text{‘}\alpha \,.\, x \neq y \,.\supset .\, t_0\text{‘}\alpha \epsilon 1 \cup \iota\text{‘}\Lambda \,.$

$[*63\cdot18] \qquad \supset .\, t_0\text{‘}\alpha \epsilon 1 \quad (2)$

$\vdash \,.\, (2) \,.\, *60\cdot38 \,.\, *63\cdot105 \,.\, *52\cdot46 \,.\supset \vdash : \sim(\exists x, y) \,.\, x, y \epsilon t_0\text{‘}\alpha \,.\, x \neq y \,.\, \exists ! \alpha \,.\, \exists ! \beta \,.\supset .$

$\qquad\qquad \alpha = \beta = t_0\text{‘}\alpha \,.\, t_0\text{‘}\alpha \epsilon 1 \,.$

$[*106\cdot312] \qquad \supset .\, (\exists\gamma, \delta) \,.\, \gamma \epsilon \mathrm{N}_{00}\mathrm{c}\text{‘}\alpha \,.\, \delta \epsilon \mathrm{N}_{00}\mathrm{c}\text{‘}\beta \,.\, \gamma \cap \delta = \Lambda \quad (3)$

$\vdash \,.\, *106\cdot311 \,.\, *63\cdot18 \,.\supset$

$\vdash :. \mathrm{Hp} : \sim(\exists ! \alpha \,.\, \exists ! \beta) : \supset .\, (\exists\gamma, \delta) \,.\, \gamma \epsilon \mathrm{N}_{00}\mathrm{c}\text{‘}\alpha \,.\, \delta \epsilon \mathrm{N}_{00}\mathrm{c}\text{‘}\beta \,.\, \gamma \cap \delta = \Lambda \quad (4)$

$\vdash \,.\, (1) \,.\, (3) \,.\, (4) \,.\supset \vdash \,.\, \mathrm{Prop}$

∗106·4. $\vdash : \mu = \mathrm{N}_0\mathrm{c}\text{‘}\alpha \,.\supset .\, \mu_{(00)} = \mathrm{N}_{00}\mathrm{c}\text{‘}\alpha$

Dem.

$\vdash \,.\, *106\cdot14 \,.\supset \vdash :: \mathrm{Hp} \,.\supset :. \beta \epsilon \mu_{00} \,.\equiv : (\exists\gamma) \,.\, \gamma \epsilon \mathrm{N}_0\mathrm{c}\text{‘}\alpha \,.\, \beta \,\mathrm{sm}\, \gamma \,.\, \beta \epsilon t\text{‘}t_{00}\text{‘}\gamma :$

$[*64\cdot3] \qquad \equiv : (\exists\gamma) \,.\, \gamma \epsilon \mathrm{N}_0\mathrm{c}\text{‘}\alpha \,.\, \beta \,\mathrm{sm}\, \gamma : \beta \epsilon t\text{‘}t_{00}\text{‘}\alpha :$

$[*102\cdot84] \qquad \equiv : \beta \,\mathrm{sm}\, \alpha \,.\, \beta \epsilon t\text{‘}t_{00}\text{‘}\alpha :$

$[*106\cdot1] \qquad \equiv : \beta \epsilon \mathrm{N}_{00}\mathrm{c}\text{‘}\alpha :: \supset \vdash \,.\, \mathrm{Prop}$

∗106·401. $\vdash : \mu = \mathrm{N}^1\mathrm{c}\text{‘}\alpha \,.\supset .\, \mu_{(00)} = \mathrm{N}^{11}\mathrm{c}\text{‘}\alpha$

Dem.

$\vdash \,.\, *104\cdot24 \,.\, *106\cdot4 \,.\supset \vdash : \mathrm{Hp} \,.\supset .\, \mu_{(00)} = \mathrm{N}_{00}\mathrm{c}\text{‘}\iota\text{‘‘}\alpha$

$[*106\cdot25] \qquad = \mathrm{N}^{11}\mathrm{c}\text{‘}\alpha : \supset \vdash \,.\, \mathrm{Prop}$

∗106·402. $\vdash : \mu = \mathrm{N}_1\mathrm{c}\text{‘}\alpha \,.\, \exists ! \mu \,.\supset .\, \mu_{(00)} = \mathrm{N}_{11}\mathrm{c}\text{‘}\alpha$

Dem.

$\vdash \,.\, *106\cdot231 \,.\qquad \supset \vdash : \mathrm{Hp} \,.\, \beta \epsilon \mu \,.\supset .\, \mathrm{N}_{11}\mathrm{c}\text{‘}\alpha = \mathrm{N}_{00}\mathrm{c}\text{‘}\beta$

$[*106\cdot4 \,.\, *103\cdot26] \qquad = \mu_{(00)} \quad (1)$

$\vdash \,.\, (1) \,.\, *10\cdot11\cdot23\cdot35 \,.\supset \vdash : \mathrm{Hp} \,.\, \exists ! \mu \,.\supset .\, \mu_{(00)} = \mathrm{N}_{11}\mathrm{c}\text{‘}\alpha : \supset \vdash \,.\, \mathrm{Prop}$

∗106·41. $\vdash : \mu = \mathrm{N}_0\mathrm{c}\text{‘}\alpha \,.\supset .\, \mu^{(11)} = \mathrm{N}^{11}\mathrm{c}\text{‘}\alpha$

Dem.

$\vdash \,.\, *63\cdot54 \,.\, (*106\cdot041) \,.\, *103\cdot27 \,.\supset$

$\vdash :: \mathrm{Hp} \,.\supset :. \beta \epsilon \mu^{(11)} \,.\equiv : (\exists\gamma) \,.\, \gamma \epsilon \mathrm{N}_0\mathrm{c}\text{‘}\alpha \,.\, \beta \,\mathrm{sm}\, \gamma : \beta \epsilon t\text{‘}t^{11}\text{‘}t_0\text{‘}\alpha :$

$[*102\cdot84 \,.\, *64\cdot32] \qquad \equiv : \beta \,\mathrm{sm}\, \alpha \,.\, \beta \epsilon t\text{‘}t^{11}\text{‘}\alpha :$

$[(*106\cdot011)] \qquad \equiv : \beta \epsilon \mathrm{N}^{11}\mathrm{c}\text{‘}\alpha :: \supset \vdash \,.\, \mathrm{Prop}$

∗106·411. $\vdash : \mu = \mathrm{N}_0\mathrm{c}\text{‘}\alpha \,.\supset .\, \mu_{(11)} = \mathrm{N}_{11}\mathrm{c}\text{‘}\alpha$ [Proof as in ∗106·41]

∗106·43. $\vdash : \mu, \nu \epsilon \mathrm{N}_0\mathrm{C} \,.\, t\text{‘}\mu = t\text{‘}\nu \,.\supset .\, (\exists\gamma, \delta) \,.\, \gamma \epsilon \mu_{(00)} \,.\, \delta \epsilon \nu_{(00)} \,.\, \gamma \cap \delta = \Lambda$

Dem.

$\vdash \,.\, *103\cdot2 \,.\supset \vdash : \mathrm{Hp} \,.\supset .\, (\exists\alpha, \beta) \,.\, \mu = \mathrm{N}_0\mathrm{c}\text{‘}\alpha \,.\, \nu = \mathrm{N}_0\mathrm{c}\text{‘}\beta \,.$

$[*106\cdot4] \qquad \supset .\, (\exists\alpha, \beta) \,.\, \mu_{(00)} = \mathrm{N}_{00}\mathrm{c}\text{‘}\alpha \,.\, \nu_{(00)} = \mathrm{N}_{00}\mathrm{c}\text{‘}\beta \,.$

$[*106\cdot32] \qquad \supset .\, (\exists\gamma, \delta) \,.\, \gamma \epsilon \mu_{(00)} \,.\, \delta \epsilon \nu_{(00)} \,.\, \gamma \cap \delta = \Lambda : \supset \vdash \,.\, \mathrm{Prop}$

∗106·44. $\vdash : \mu, \nu \epsilon \mathrm{N}_{00}\mathrm{C} \,.\, t\text{‘}\mu = t\text{‘}\nu \,.\supset .\, (\exists\gamma, \delta) \,.\, \gamma \epsilon \mu \,.\, \delta \epsilon \nu \,.\, \gamma \cap \delta = \Lambda$ [∗106·32]

The following propositions are analogous to $*102{\cdot}71$ ff., and similar remarks apply to them.

$*106{\cdot}5$. $\vdash : R \epsilon \mathrm{Cls} \to 1 \,.\, \mathrm{D}\text{‘}R \subset \alpha \,.\, \text{Ɑ}\text{‘}R \subset \mathrm{Rl}\text{‘}(\alpha \uparrow \alpha) \,.$
$$W = \hat{x}\hat{y}\,\{x, y \,\epsilon\, \alpha \,.\, \sim x\,(\breve{R}\text{‘}x)\,y\} \,.\supset .\, W \sim\epsilon\, \text{Ɑ}\text{‘}R \,.\, W \Subset \alpha \uparrow \alpha$$

Dem.

$\vdash . *4{\cdot}73 \,.\supset \vdash :: \mathrm{Hp} \,.\supset :.\, x, y \,\epsilon\, \alpha \,.\supset_{x,y} : x W y \,.\equiv .\, \sim x\,(\breve{R}\text{‘}x)\,y :$

$[*5{\cdot}18] \qquad \supset_{x,y} : \sim \{x W y \,.\equiv .\, x\,(\breve{R}\text{‘}x)\,y\} :.$

$[*10{\cdot}1] \qquad \supset :.\, x \,\epsilon\, \alpha \,.\supset_x .\sim \{x W x \,.\equiv .\, x\,(\breve{R}\text{‘}x)\,x\} \,.$

$[*21{\cdot}43 . \mathrm{Transp}] \qquad \supset_x .\, W \neq \breve{R}\text{‘}x :.$

$[\mathrm{Hp}] \qquad \supset :.\, x \,\epsilon\, \mathrm{D}\text{‘}R \,.\supset_x .\, W \neq \breve{R}\text{‘}x :.$

$[*71{\cdot}411 . \mathrm{Transp}] \qquad \supset :.\, W \sim\epsilon\, \text{Ɑ}\text{‘}R \qquad (1)$

$\vdash . *21{\cdot}33 \,.\, (*35{\cdot}04) \,.\supset \vdash : \mathrm{Hp} \,.\supset .\, W \Subset \alpha \uparrow \alpha \qquad (2)$

$\vdash . (1) . (2) \,.\supset \vdash .\, \mathrm{Prop}$

$*106{\cdot}51$. $\vdash : \beta \subset \alpha \,.\supset .\sim \{\beta \,\mathrm{sm}\, \mathrm{Rl}\text{‘}(\alpha \uparrow \alpha)\}$

Dem.

$\vdash . *106{\cdot}5 \,. \qquad \supset \vdash : \mathrm{Hp} \,.\, R \,\epsilon\, 1 \to 1 \,.\, \mathrm{D}\text{‘}R = \beta \,.\, \text{Ɑ}\text{‘}R \subset \mathrm{Rl}\text{‘}(\alpha \uparrow \alpha) \,.\supset .$
$$(\exists W) \,.\, W \,\epsilon\, \mathrm{Rl}\text{‘}(\alpha \uparrow \alpha) \,.\, W \sim\epsilon\, \text{Ɑ}\text{‘}R \,.$$

$[*13{\cdot}14] \qquad \supset .\, \text{Ɑ}\text{‘}R \neq \mathrm{Rl}\text{‘}(\alpha \uparrow \alpha) \qquad (1)$

$\vdash . (1) . *22{\cdot}41 \,.\supset \vdash :.\, \mathrm{Hp} \,.\supset : R \,\epsilon\, 1 \to 1 \,.\, \mathrm{D}\text{‘}R = \beta \,.\supset_R .\, \text{Ɑ}\text{‘}R \neq \mathrm{Rl}\text{‘}(\alpha \uparrow \alpha) :$

$[*10{\cdot}51 . *73{\cdot}1] \qquad \supset : \sim \{\beta \,\mathrm{sm}\, \mathrm{Rl}\text{‘}(\alpha \uparrow \alpha)\} :.\supset \vdash .\, \mathrm{Prop}$

$*106{\cdot}52$. $\vdash : \beta \subset t_0\text{‘}\alpha \,.\supset .\, \beta \sim\epsilon\, \mathrm{Nc}\text{‘}t_{00}\text{‘}\alpha$

Dem.

$\vdash . *106{\cdot}51 \,.\supset \vdash : \mathrm{Hp} \,.\supset .\sim \{\beta \,\mathrm{sm}\, \mathrm{Rl}\text{‘}(t_0\text{‘}\alpha \uparrow t_0\text{‘}\alpha)\} \,.$

$[*64{\cdot}54] \qquad \supset .\sim \{\beta \,\mathrm{sm}\, t_{00}\text{‘}\alpha\} \,.$

$[*100{\cdot}1] \qquad \supset .\, \beta \sim\epsilon\, \mathrm{Nc}\text{‘}t_{00}\text{‘}\alpha : \supset \vdash .\, \mathrm{Prop}$

$*106{\cdot}53$. $\vdash .\, \mathrm{Nc}\,(\alpha)\text{‘}t_{00}\text{‘}\alpha = \Lambda \quad [*106{\cdot}52 \,.\, *102{\cdot}6 \,.\, *63{\cdot}371]$

$*106{\cdot}54$. $\vdash .\, \mathrm{N_0c}\text{‘}t_{00}\text{‘}\alpha \sim\epsilon\, \mathrm{N_{00}C}$

Dem.

$\vdash . *100{\cdot}33 \,.\, *103{\cdot}15 \,.\supset$

$\vdash : \mathrm{N_{00}c}\text{‘}\beta = \mathrm{N_0c}\text{‘}t_{00}\text{‘}\alpha \,.\supset .\, \beta \,\mathrm{sm}\, t_{00}\text{‘}\alpha \qquad (1)$

$\vdash . *103{\cdot}12 \,.\, (*106{\cdot}01) \,.\supset$

$\vdash : \mathrm{N_{00}c}\text{‘}\beta = \mathrm{N_0c}\text{‘}t_{00}\text{‘}\alpha \,.\supset .\, t_{00}\text{‘}\alpha \,\epsilon\, t\text{‘}t_{00}\text{‘}\beta \,.$

$[*63{\cdot}16 . (*64{\cdot}01)] \qquad \supset .\, t\text{‘}t\text{‘}(t_0\text{‘}\alpha \uparrow t_0\text{‘}\alpha) = t\text{‘}t\text{‘}(t_0\text{‘}\beta \uparrow t_0\text{‘}\beta) \,.$

$[*63{\cdot}391] \qquad \supset .\, t\text{‘}(t_0\text{‘}\alpha \uparrow t_0\text{‘}\alpha) = t\text{‘}(t_0\text{‘}\beta \uparrow t_0\text{‘}\beta) \,.$

$[*64{\cdot}3 . (*64{\cdot}01)] \qquad \supset .\, t_0\text{‘}\alpha = t_0\text{‘}\beta \,.$

$[*63{\cdot}105] \qquad \supset .\, \beta \subset t_0\text{‘}\alpha \qquad (2)$

$\vdash . (1) . (2) \,.\supset \vdash : \mathrm{N_{00}c}\text{‘}\beta = \mathrm{N_0c}\text{‘}t_{00}\text{‘}\alpha \,.\supset .\, \beta \,\epsilon\, \mathrm{Nc}\text{‘}t_{00}\text{‘}\alpha \,.\, \beta \subset t_0\text{‘}\alpha \qquad (3)$

$\vdash . (3) \,.\, \mathrm{Transp} \,.\, *106{\cdot}52 \,.\supset \vdash .\, (\beta) \,.\, \mathrm{N_{00}c}\text{‘}\beta \neq \mathrm{N_0c}\text{‘}t_{00}\text{‘}\alpha \,.$

$[*106{\cdot}13 . \mathrm{Transp}] \qquad \supset \vdash .\, \mathrm{N_0c}\text{‘}t_{00}\text{‘}\alpha \sim\epsilon\, \mathrm{N_{00}C} \,.\supset \vdash .\, \mathrm{Prop}$

$*106{\cdot}55$. $\vdash .\, \exists !\, \mathrm{N_0C} - \mathrm{N_{00}C} \quad [*106{\cdot}54]$

SECTION B.

ADDITION, MULTIPLICATION AND EXPONENTIATION.

Summary of Section B.

In the present section, we have to consider the arithmetical operations as applied to cardinals, as well as the relation of greater and less between cardinals. Thus the topics to be dealt with in this section are the first that can properly be said to belong to Arithmetic.

The treatment of addition, multiplication and exponentiation to be given in what follows is guided by the desire to secure the greatest possible generality. In the first place, everything to be said generally about the arithmetical operations must apply equally to finite and infinite classes or cardinals. In the second place, we desire such definitions as shall allow the number of summands in a sum or of factors in a product to be infinite. In the third place, we wish to be able to add or multiply two numbers which are not necessarily of the same type. In the fourth place, we wish our definitions to be such that the sum of the cardinal numbers of two or more classes shall depend only upon the cardinal numbers of those classes, and shall be the same when the classes overlap as when they are mutually exclusive; with similar conditions for the product. The desire to obtain definitions fulfilling all these conditions leads to somewhat more complicated definitions than would otherwise be required; but in the outcome, the result is simpler than if we started with simpler definitions, since we avoid vexatious exceptions.

The above observations will become clearer through their applications. Let us begin with the case of arithmetical addition of two classes.

If α and β are mutually exclusive classes, the sum of their cardinal numbers will be the cardinal number of $\alpha \cup \beta$. But in order that α and β may be mutually exclusive, they must have no common members, and this is only significant when they are of the same type. Hence, given two perfectly general classes α and β, we require to find two classes which are mutually exclusive and are respectively similar to α and β; if these two classes are called α' and β', then $\mathrm{Nc}‘(\alpha' \cup \beta')$ will be the sum of the cardinal numbers of α and β. We note that $\Lambda \cap \alpha$ and $\Lambda \cap \beta$ indicate respectively the Λ's of the same types as α and β, and accordingly we take as α' and β' the two classes

$$\downarrow (\Lambda \cap \beta)“\iota“\alpha \text{ and } (\Lambda \cap \alpha) \downarrow “\iota“\beta;$$

these two classes are always of the same type, always mutually exclusive, and always similar to α and β respectively. Hence we define

$$\alpha + \beta = \downarrow (\Lambda \cap \beta)``\iota``\alpha \cup (\Lambda \cap \alpha) \downarrow ``\iota``\beta \quad \text{Df.}$$

The sum of the cardinal numbers of α and β will then be the cardinal number of $\alpha + \beta$; hence we may call $\alpha + \beta$ the *arithmetical* class-sum of two classes, in contradistinction to $\alpha \cup \beta$, which is the *logical* sum. It will be noted that $\alpha + \beta$, unlike $\alpha \cup \beta$, does not require that α and β should be of the same type. Also $\alpha + \alpha$ is not identical with α, but when $\alpha = \Lambda$, $\alpha + \alpha$ is also Λ, though in a different type. Thus the law of tautology does not hold of the arithmetical class-sum of two classes.

If μ and ν are two cardinals of assigned types, we denote their arithmetical sum by $\mu +_c \nu$. (As many kinds of arithmetical addition occur in our work, and as it is essential to our purpose to distinguish them, we effect the distinction by suffixes to the sign of addition. It is, of course, only in dealing with principles that these different symbols are needed: we do not wish to suggest that they should be adopted in ordinary mathematics.) Now if $\mu +_c \nu$ is to have the properties which we commonly associate with the sum of two cardinals, it must be typically ambiguous, and must be the cardinal number of any class which can be divided into two mutually exclusive parts having μ terms and ν terms respectively. Hence we are led to the following definition:

$$\mu +_c \nu = \hat{\xi} \{(\exists \alpha, \beta) \,.\, \mu = \text{N}_0\text{c}`\alpha \,.\, \nu = \text{N}_0\text{c}`\beta \,.\, \xi \,\text{sm}\, (\alpha + \beta)\} \quad \text{Df.}$$

In this definition, various points should be noted. In the first place, it does not require that μ and ν should be of the same type; $\mu +_c \nu$ is *significant* whenever μ and ν are classes of classes. Thus it is not necessary for significance that μ and ν should be cardinals, though if they are not both cardinals, $\mu +_c \nu = \Lambda$. If they are both cardinals, we find

$$\mu +_c \nu = \hat{\xi} \{(\exists \alpha, \beta) \,.\, \alpha \in \mu \,.\, \beta \in \nu \,.\, \xi \,\text{sm}\, (\alpha + \beta)\}.$$

Thus in this case $\quad \alpha \in \mu \,.\, \beta \in \nu \,.\, \supset \,.\, \alpha + \beta \in \mu +_c \nu.$

Hence if neither μ nor ν is null, and if α has μ terms and β has ν terms, $\alpha + \beta$ is a member of $\mu +_c \nu$. It easily follows that

$$\vdash : \mu = \text{N}_0\text{c}`\alpha \,.\, \nu = \text{N}_0\text{c}`\beta \,.\, \supset \,.\, \mu +_c \nu = \text{Nc}`(\alpha + \beta).$$

Hence when μ and ν are homogeneous cardinals (*i.e.* when they are cardinals other than Λ), their sum is the number of the arithmetical class-sum of any two classes having μ terms and ν terms respectively.

A few words are necessary to explain why, in the definition, we put $\mu = \text{N}_0\text{c}`\alpha \,.\, \nu = \text{N}_0\text{c}`\beta$ rather than $\mu = \text{Nc}`\alpha \,.\, \nu = \text{Nc}`\beta$. The reason is this. Suppose either μ or ν, say μ, is Λ. Then, by $*102\cdot73$, $\mu = \text{Nc}\,(\zeta)`t`\zeta$, if ζ is of the appropriate type. Hence if we had put

$$\mu +_c \nu = \hat{\xi} \{(\exists \alpha, \beta) \,.\, \mu = \text{Nc}`\alpha \,.\, \nu = \text{Nc}`\beta \,.\, \xi \,\text{sm}\, (\alpha + \beta)\} \quad \text{Df,}$$

where the ambiguities of type involved in $\text{Nc}'\alpha$ and $\text{Nc}'\beta$ may be determined as we please, we should have

$$\nu = \text{Nc}'\beta \,.\, \supset \,.\, t'\zeta + \beta \,\epsilon\, \mu +_c \nu,$$

i.e.

$$\nu = \text{Nc}'\beta \,.\, \supset \,.\, t'\zeta + \beta \,\epsilon\, \Lambda +_c \nu.$$

We should also have $t't'\zeta + \beta \,\epsilon\, \Lambda +_c \nu$ and so on. Thus $\Lambda +_c \nu$ would not have a definite value, *i.e.* it would not merely have typical ambiguity, which it ought to have, but it would not have a definite value even when its type was assigned. Thus such a definition would be unsuitable. For the above reasons, we put $\mu = \text{N}_0\text{c}'\alpha \,.\, \nu = \text{N}_0\text{c}'\beta$ in the definition, and obtain the typical ambiguity which we desire by means of the typical ambiguity of the "sm" in "$\xi \,\text{sm}\, (\alpha + \beta)$." It is always essential to right symbolism that the values of typically ambiguous symbols should be unique as soon as their type is assigned. The scope of these definitions and of the corresponding definitions for multiplication and exponentiation (*113·04·05 . *116·03·04) is extended by convention II*T* of the prefatory statement.

The above definition of $\mu +_c \nu$ is designed for the case in which μ and ν are typically definite. But we must be able to speak of "$\text{Nc}'\gamma +_c \text{Nc}'\delta$," and this must be a definite cardinal, namely $\text{Nc}'(\gamma + \delta)$. If we simply write $\text{Nc}'\gamma$, $\text{Nc}'\delta$ in place of μ, ν in the definition of $\mu +_c \nu$, we find

$$\text{Nc}'\gamma +_c \text{Nc}'\delta = \hat{\xi}\{(\exists \alpha, \beta) \,.\, \text{Nc}'\gamma = \text{N}_0\text{c}'\alpha \,.\, \text{Nc}'\delta = \text{N}_0\text{c}'\beta \,.\, \xi \,\text{sm}\, (\alpha + \beta)\}.$$

But this will not always have a definite value when the type of $\text{Nc}'\gamma +_c \text{Nc}'\delta$ is assigned. To take a simple case, write $t'\zeta$ for γ and $\iota'y$ for δ. Then

$$\text{Nc}'t'\zeta +_c \text{Nc}'\iota'y = \hat{\xi}\{(\exists \alpha, \beta) \,.\, \text{Nc}'t'\zeta = \text{N}_0\text{c}'\alpha \,.\, \text{Nc}'\iota'y = \text{N}_0\text{c}'\beta \,.\, \xi \,\text{sm}\, (\alpha + \beta)\},$$

whence we easily obtain

$$\text{Nc}'t'\zeta +_c \text{Nc}'\iota'y = \hat{\xi}\{(\exists \alpha) \,.\, \text{Nc}'t'\zeta = \text{N}_0\text{c}'\alpha \,.\, \xi \,\text{sm}\, (\alpha + \iota'y)\}.$$

If we determine the ambiguity of $\text{Nc}'t'\zeta$ to be $\text{N}_1\text{c}'t'\zeta$, we find

$$\text{Nc}'t'\zeta +_c \text{Nc}'\iota'y = \Lambda$$

in all types; but if we determine the ambiguity to be $\text{N}_0\text{c}'t'\zeta$, we have

$$\text{Nc}'t'\zeta +_c \text{Nc}'\iota'y = \text{Nc}'(t'\zeta + \iota'y),$$

and this exists in the type of $t'\zeta + \iota'y$, if not in lower types. Hence the value of $\text{Nc}'t'\zeta +_c \text{Nc}'\iota'y$ depends upon the determination of the ambiguity of $\text{Nc}'t'\zeta$. It is obvious that we want our definition to yield

$$\text{Nc}'\gamma +_c \text{Nc}'\delta = \text{Nc}'(\gamma + \delta)$$

in all types; but in order to insure that this shall hold even when, for some values of ζ, $\text{Nc}\,(\zeta)'\gamma = \Lambda$, we must introduce two new definitions, namely

$$\text{Nc}'\alpha +_c \mu = \text{N}_0\text{c}'\alpha +_c \mu \quad \text{Df},$$

$$\mu +_c \text{Nc}'\alpha = \mu +_c \text{N}_0\text{c}'\alpha \quad \text{Df},$$

whence $\vdash : \text{Nc}'\alpha +_c \text{Nc}'\beta = \text{N}_0\text{c}'\alpha +_c \text{N}_0\text{c}'\beta = \text{Nc}'(\alpha + \beta).$

This definition is to be applied when "$\text{Nc}'\gamma$" and "$\text{Nc}'\delta$" occur without any determination of type. On the other hand, if we have $\mathrm{Nc}\,(\zeta)\text{‘}\gamma$ and $\mathrm{Nc}\,(\eta)\text{‘}\delta$, we apply the definition of $\mu +_c \nu$. We shall find that whenever $\mathrm{Nc}\,(\zeta)\text{‘}\gamma$ and $\mathrm{Nc}\,(\eta)\text{‘}\delta$ both exist,

$$\mathrm{Nc}\,(\zeta)\text{‘}\gamma +_c \mathrm{Nc}\,(\eta)\text{‘}\delta = \mathrm{N_0c}\text{‘}\gamma +_c \mathrm{N_0c}\text{‘}\delta.$$

Thus the above definition is only required in order to exclude values of ζ or η for which either $\mathrm{Nc}\,(\zeta)\text{‘}\gamma$ or $\mathrm{Nc}\,(\eta)\text{‘}\delta$ is Λ.

The commutative and associative laws of arithmetical addition are easily deduced from the definition of $\alpha + \beta$. We shall have

$$\vdash . \alpha + \beta = \mathrm{Cnv}\text{“}(\beta + \alpha),$$

whence $\vdash . \mathrm{Nc}\text{‘}\alpha +_c \mathrm{Nc}\text{‘}\beta = \mathrm{Nc}\text{‘}\beta +_c \mathrm{Nc}\text{‘}\alpha,$

because each $= \mathrm{Nc}\text{‘}(\alpha + \beta)$. A similar though slightly longer proof shows that

$$\vdash . (\alpha + \beta) + \gamma \;\mathrm{sm}\; \alpha + (\beta + \gamma),$$

whence $\vdash . (\mathrm{Nc}\text{‘}\alpha +_c \mathrm{Nc}\text{‘}\beta) +_c \mathrm{Nc}\text{‘}\gamma = \mathrm{Nc}\text{‘}\alpha +_c (\mathrm{Nc}\text{‘}\beta +_c \mathrm{Nc}\text{‘}\gamma).$

The above definition of $\alpha + \beta$ enables us to proceed to the sum of any finite number of classes, and allows any one class to recur in the summation. But it does not enable us to define the sum of an infinite number of classes. For this we need a new definition. Since an infinite number of classes cannot be given by enumeration, but only by intension, we shall have to take a class of classes κ, and define the arithmetical sum of the members of κ. Thus now the classes which are the summands must all be of the same type (since they are all members of κ), and no one class can occur more than once, since each member of κ only counts once. (In order to deal with repetition, we must advance to multiplication, which will be explained shortly.) Thus in removing the limitation to a finite number of summands, we introduce certain other limitations. This is the reason which makes it worth while to introduce the above definition of $\alpha + \beta$ in addition to the definition now to be given.

If κ is a class of classes, the sum of the cardinal numbers of the members of κ will evidently be obtained by constructing a class of mutually exclusive classes whose members have a one-one relation to the members of corresponding members of κ. Suppose α, β are two different members of κ, and suppose x is a member both of α and of β. Then we wish to count x twice over, once as a member of α and once as a member of β. The simplest way to do this is to form the ordinal couples $x \downarrow \alpha$ and $x \downarrow \beta$, which are not identical except when α and β are identical. Thus if we take all such ordinal couples, *i.e.* if we take the class

$$\hat{R}\,\{(\exists x) . x \in \alpha . R = x \downarrow \alpha\},$$

for every α which is a member of κ, we get a class of mutually exclusive classes, namely the classes of the form $\downarrow \alpha\text{“}\alpha$, where $\alpha \in \kappa$, and each of these is similar to the corresponding member of κ. Hence the logical sum of this class of classes, *i.e.*

$$\hat{R}\,\{(\exists \alpha, x) . \alpha \in \kappa . x \in \alpha . R = x \downarrow \alpha\},$$

has the required number of terms. Now, by $*85{\cdot}601$,

$$\downarrow \alpha\text{‘‘}\alpha = \epsilon\,\overline{\downarrow}\text{‘}\alpha.$$

Hence the class whose logical sum we are taking is $\epsilon\,\overline{\downarrow}\text{‘‘}\kappa$. Hence we put

$$\Sigma\text{‘}\kappa = s\text{‘}\epsilon\,\overline{\downarrow}\text{‘‘}\kappa \quad \text{Df.}$$

$\Sigma\text{‘}\kappa$ may be called the *arithmetical* sum of κ, in contradistinction to $s\text{‘}\kappa$, which is the logical sum. Thus $\Sigma\text{‘}\kappa$ bears to $s\text{‘}\kappa$ a relation analogous to that which $\alpha + \beta$ bears to $\alpha \cup \beta$.

We put further $\quad \Sigma\text{Nc}\text{‘}\kappa = \text{Nc}\text{‘}s\text{‘}\epsilon\,\overline{\downarrow}\text{‘‘}\kappa \quad$ Df.

Thus $\Sigma\text{Nc}\text{‘}\kappa$ is the sum of the numbers of members of κ.

It is to be observed that $\Sigma\text{Nc}\text{‘}\kappa$ is not in general a function of $\text{Nc}\text{‘‘}\kappa$. For, if two members of κ have the same cardinal number, this will only count once in $\text{Nc}\text{‘‘}\kappa$, whereas it counts twice in $\Sigma\text{Nc}\text{‘}\kappa$.

We shall find that, provided $\alpha \neq \beta$,

$$\Sigma\text{Nc}\text{‘}(\iota\text{‘}\alpha \cup \iota\text{‘}\beta) = \text{Nc}\text{‘}\alpha +_c \text{Nc}\text{‘}\beta.$$

Thus where a finite number of summands are concerned, the two definitions of addition agree, except that the first allows one class to count several times over, while the second does not.

In dealing with multiplication, our procedure is closely analogous to the procedure for addition. We first define the *arithmetical class-product* of two classes α and β, which is a certain class whose cardinal number is the product of the cardinal numbers of α and β. We write $\beta \times \alpha$ for the arithmetical class-product of β and α, and define it as the class of all ordinal couples of which the referent is a member of α and the relatum a member of β, *i.e.* as

$$\hat{R}\{(\exists x, y) \,.\, x \,\epsilon\, \alpha \,.\, y \,\epsilon\, \beta \,.\, R = x \downarrow y\}.$$

By $*40{\cdot}7$, this class is $s\text{‘}\alpha \underset{,,}{\downarrow} \text{‘‘}\beta$. Hence we put

$$\beta \times \alpha = s\text{‘}\alpha \underset{,,}{\downarrow} \text{‘‘}\beta \quad \text{Df.}$$

The class $\alpha \underset{,,}{\downarrow} \text{‘‘}\beta$ is similar to β, and each member of it is similar to α; hence if $\text{N}_0\text{c}\text{‘}\alpha = \mu$ and $\text{N}_0\text{c}\text{‘}\beta = \nu$, $s\text{‘}\alpha \underset{,,}{\downarrow} \text{‘‘}\beta$ consists of ν classes having μ members each. The class $\alpha \underset{,,}{\downarrow} \text{‘‘}\beta$ is important also in connection with exponentiation. The product of two cardinals is defined as follows:

$$\mu \times_c \nu = \hat{\xi}\{(\exists \alpha, \beta) \,.\, \mu = \text{N}_0\text{c}\text{‘}\alpha \,.\, \nu = \text{N}_0\text{c}\text{‘}\beta \,.\, \xi \text{ sm } (\alpha \times \beta)\} \quad \text{Df.}$$

In regard to types, this definition calls for analogous remarks to those which were made on $\mu +_c \nu$. Also, as before, we need definitions of $\mu \times_c \text{Nc}\text{‘}\alpha$ and $\text{Nc}\text{‘}\alpha \times_c \mu$, whence we obtain

$$\text{Nc}\text{‘}\alpha \times_c \text{Nc}\text{‘}\beta = \text{N}_0\text{c}\text{‘}\alpha \times_c \text{N}_0\text{c}\text{‘}\beta \quad \text{Df.}$$

By means of these definitions, we can define the product of any finite number of cardinals; but in order to define products which have an infinite number of factors, we need a new definition.

If κ is a class of classes, we take $\epsilon_\Delta{}^{\backprime}\kappa$ as its arithmetical product. In simple cases, it is easy to see the justification of this decision. *E.g.* let κ consist of the three classes α_1, α_2, α_3, and let the members of α_1 be x_1, x_2; those of α_2, y_1, y_2; those of α_3, z_1, z_2. Then the members of $\epsilon_\Delta{}^{\backprime}\kappa$ are

$$x_1 \downarrow \alpha_1 \cup y_1 \downarrow \alpha_2 \cup z_1 \downarrow \alpha_3,$$
$$x_2 \downarrow \alpha_1 \cup y_1 \downarrow \alpha_2 \cup z_1 \downarrow \alpha_3,$$
$$x_1 \downarrow \alpha_1 \cup y_2 \downarrow \alpha_2 \cup z_1 \downarrow \alpha_3,$$
$$x_2 \downarrow \alpha_1 \cup y_2 \downarrow \alpha_2 \cup z_1 \downarrow \alpha_3,$$

with four more obtained by substituting z_2 for z_1 in the above. Thus $\text{Nc}{}^{\backprime}\epsilon_\Delta{}^{\backprime}\kappa = 8 = \text{Nc}{}^{\backprime}\alpha_1 \times_c \text{Nc}{}^{\backprime}\alpha_2 \times_c \text{Nc}{}^{\backprime}\alpha_3$. In general, however, the existence of $\epsilon_\Delta{}^{\backprime}\kappa$ is doubtful, owing to the doubt as to the validity of the multiplicative axiom. (We shall return to this point shortly.) Hence there is no proof that the product of an infinite number of factors cannot be zero unless one of the factors is zero.

When κ is a class of mutually exclusive classes, $\epsilon_\Delta{}^{\backprime}\kappa$ is similar to $\text{D}{}^{\backprime\backprime}\epsilon_\Delta{}^{\backprime}\kappa$. On account of its lower type, $\text{D}{}^{\backprime\backprime}\epsilon_\Delta{}^{\backprime}\kappa$ is often more convenient than $\epsilon_\Delta{}^{\backprime}\kappa$. Hence we put

$$\text{Prod}{}^{\backprime}\kappa = \text{D}{}^{\backprime\backprime}\epsilon_\Delta{}^{\backprime}\kappa \quad \text{Df},$$

or (what comes to the same thing)

$$\text{Prod} = \text{D}_\epsilon \,|\, \epsilon_\Delta \quad \text{Df}.$$

For the product of the cardinal numbers of the members of κ, we put

$$\Pi\text{Nc}{}^{\backprime}\kappa = \text{Nc}{}^{\backprime}\epsilon_\Delta{}^{\backprime}\kappa \quad \text{Df}.$$

As in the case of $\Sigma\text{Nc}{}^{\backprime}\kappa$, $\Pi\text{Nc}{}^{\backprime}\kappa$ is not in general a function of $\text{Nc}{}^{\backprime\backprime}\kappa$. We shall have

$$\vdash : \alpha \neq \beta \,.\supset.\, \Pi\text{Nc}{}^{\backprime}(\iota{}^{\backprime}\alpha \cup \iota{}^{\backprime}\beta) = \text{Nc}{}^{\backprime}\alpha \times_c \text{Nc}{}^{\backprime}\beta.$$

Thus for products of a finite number of different factors, the two definitions of multiplication agree.

It remains to define exponentiation. Since this is not a commutative operation, it essentially involves an order as between the base and the exponent; hence we do not obtain a definition of the exponentiation of a class κ, analogous to $\Sigma\text{Nc}{}^{\backprime}\kappa$ or $\Pi\text{Nc}{}^{\backprime}\kappa$, but only a definition of μ^ν, which may be extended to any finite number of exponentiations. We put

$$\alpha \exp \beta = \text{Prod}{}^{\backprime}\alpha \underset{,,}{\downarrow} {}^{\backprime\backprime}\beta \quad \text{Df},$$

where $\alpha \underset{,,}{\downarrow} {}^{\backprime\backprime}\beta$ has the meaning explained above, resulting from $*38{\cdot}03$. It will be observed that, if $\text{N}_0\text{c}{}^{\backprime}\alpha = \mu$ and $\text{N}_0\text{c}{}^{\backprime}\beta = \nu$, $\alpha \underset{,,}{\downarrow} {}^{\backprime\backprime}\beta$ is a class of ν mutually exclusive classes each of which has μ terms; hence $\alpha \exp \beta$ may suitably be used to define μ^ν. Hence we put

$$\mu^\nu = \hat{\xi}\{(\exists \alpha, \beta) \,.\, \mu = \text{N}_0\text{c}{}^{\backprime}\alpha \,.\, \nu = \text{N}_0\text{c}{}^{\backprime}\beta \,.\, \xi \,\text{sm}\, (\alpha \exp \beta)\} \quad \text{Df},$$

and for the same reasons as before, we put

$$\text{Nc}{}^{\backprime}\alpha^{\text{Nc}{}^{\backprime}\beta} = \text{N}_0\text{c}{}^{\backprime}\alpha^{\text{N}_0\text{c}{}^{\backprime}\beta} \quad \text{Df}.$$

The above definition of exponentiation gives the same value of μ^ν as results from Cantor's definition by means of "Belegungen." The class of Cantor's "Belegungen" is

$$\hat{R}\{R \in 1 \to \mathrm{Cls} \,.\, \mathrm{D}{}^{\backprime}R \subset \alpha \,.\, \mathrm{G}{}^{\backprime}R = \beta\},$$

i.e.

$$(\alpha \uparrow \beta)_{\Delta}{}^{\backprime}\beta,$$

and it is easily proved that this is similar to $\alpha \exp \beta$.

The usual formal properties of exponentiation result without much difficulty from the above definitions.

The above definition of exponentiation is so framed as to make propositions on exponentiation independent of the multiplicative axiom, except when exponentiation is to be connected with multiplication, *i.e.* when it is to be shown that the product of ν factors, each of which is μ, is μ^ν. This proposition cannot be proved generally without the multiplicative axiom. Similarly, in the theory of multiplication, the proposition that the sum of ν μ's is $\mu \times_c \nu$ requires the multiplicative axiom (as does also the proposition that a product is zero when and only when one of its factors is zero). Otherwise, the theory of multiplication proceeds without the need for employing the multiplicative axiom.

To take first the connection of addition and multiplication: this connection, in the form in which we naturally suppose it to hold, is affirmed in the proposition:

$$\mu, \nu \in \mathrm{NC} \,.\, \kappa \in \nu \cap \mathrm{Cls\,excl}{}^{\backprime}\mu \,.\, \supset \,.\, s{}^{\backprime}\kappa \in \mu \times_c \nu \qquad \text{(A)}$$

or

$$\mu, \nu \in \mathrm{NC} \,.\, \kappa \in \nu \cap \mathrm{Cl}{}^{\backprime}\mu \,.\, \supset \,.\, \Sigma{}^{\backprime}\kappa \in \mu \times_c \nu.$$

We will take the first of these as being simpler. It affirms that the sum of ν μ's is $\mu \times_c \nu$. This can be proved when ν is finite, whether μ is finite or not; but when ν is infinite, it cannot be proved without the multiplicative axiom. This may be seen as follows. We know that

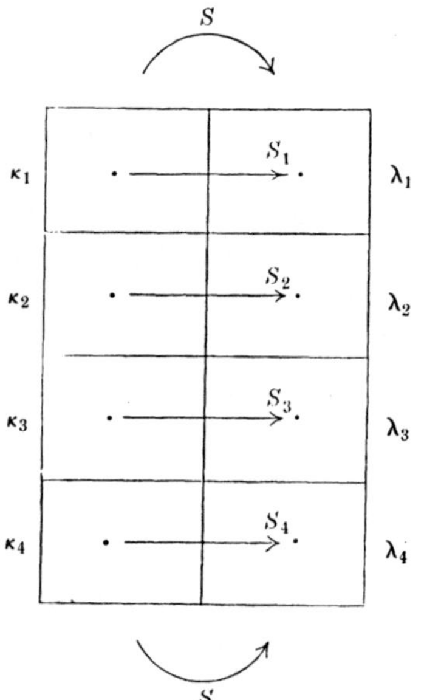

$$\vdash : \mu, \nu \in \mathrm{NC} \,.\, \alpha \in \mu \,.\, \beta \in \nu \,.\, \supset \,.$$
$$\alpha \downarrow {}_{\prime\prime}{}^{\prime\prime}\beta \in \nu \cap \mathrm{Cls\,excl}{}^{\backprime}\mu \,.\, s{}^{\backprime}\alpha \downarrow {}_{\prime\prime}{}^{\prime\prime}\beta \in \mu \times_c \nu \qquad \text{(B)}.$$

Thus (A) above will result if we can prove

$$\kappa, \lambda \in \nu \cap \mathrm{Cls\,excl}{}^{\backprime}\mu \,.\, \supset \,.\, s{}^{\backprime}\kappa \mathrel{\mathrm{sm}} s{}^{\backprime}\lambda,$$

since we shall put $\alpha \downarrow {}_{\prime\prime}{}^{\prime\prime}\beta$ for λ and use (B).

Since $\kappa, \lambda \in \nu$, we have $\kappa \mathrel{\mathrm{sm}} \lambda$. Assume

$$S \in 1 \to 1 \,.\, \mathrm{D}{}^{\backprime}S = \kappa \,.\, \mathrm{G}{}^{\backprime}S = \lambda.$$

Let $\kappa_1, \kappa_2, \ldots$ be members of κ, and let $\lambda_1, \lambda_2, \ldots$ be the members of λ which are correlated with $\kappa_1, \kappa_2, \ldots$ by S, *i.e.* $\lambda_1 = \breve{S}{}^{\backprime}\kappa_1 \,.\, \lambda_2 = \breve{S}{}^{\backprime}\kappa_2 \,.$ etc. We have, since $\kappa, \lambda \in \mathrm{Cl}{}^{\backprime}\mu$, $\kappa_1 \mathrel{\mathrm{sm}} \lambda_1 \,.\, \kappa_2 \mathrel{\mathrm{sm}} \lambda_2 \,.$ etc.

Thus $\alpha S \beta \,.\, \supset_{\alpha,\beta} .\, \alpha \text{ sm } \beta$, *i.e.* $S \subset \text{sm}$. If κ and λ are finite, we can pick out arbitrarily a correlation S_1 for κ_1 and λ_1, another S_2 for κ_2 and λ_2, and so on; then $S_1 \cup S_2 \cup \ldots$ correlates $s‘\kappa$ and $s‘\lambda$, and therefore $s‘\kappa \text{ sm } s‘\lambda$. But when κ and λ are infinite, this method is impracticable. In this case, we proceed as follows.

By $*73{\cdot}01$, $\quad \alpha \,\overline{\text{sm}}\, \beta = (1 \to 1) \cap \overleftarrow{\text{D}}‘\alpha \cap \overleftarrow{\text{Cl}}‘\beta \quad$ Df.

Thus "$\alpha \,\overline{\text{sm}}\, \alpha$" will stand for all the permutations of a class into itself; "$\alpha \,\overline{\text{sm}}\, \beta$" stands for all the permutations of α into β, *i.e.* all the $1 \to 1$'s whose domain is α and whose converse domain is β. It is obvious that

$$\vdash : \exists ! \, \alpha \,\overline{\text{sm}}\, \beta \cap \gamma \,\overline{\text{sm}}\, \delta \,.\, \supset .\, \alpha = \gamma \,.\, \beta = \delta.$$

In the case of the κ and λ above, we know that $\alpha \text{ sm } \beta$ when $\alpha S \beta$; thus

$$\alpha \in \kappa \,.\, \supset_\alpha .\, \exists ! \, \alpha \,\overline{\text{sm}}\, (\breve{S}‘\alpha)$$

or $$\beta \in \lambda \,.\, \supset_\beta .\, \exists ! \, (S‘\beta) \,\overline{\text{sm}}\, \beta.$$

Put $$\text{Crp}\,(S)‘\beta = (S‘\beta) \,\overline{\text{sm}}\, \beta \quad \text{Df},$$

where "Crp" stands for "correspondence." Thus $\text{Crp}\,(S)‘\beta$ is the class of all correspondences of $S‘\beta$ and β; $\text{Crp}\,(S)‘‘\lambda$ is the class of all such classes of correspondences. If we extract one member out of each of these classes of correspondences, we get a class of relations whose sum is a correlator of $s‘\kappa$ and $s‘\lambda$; *i.e.*

$$\varpi \in \text{D}‘‘\epsilon_\Delta‘\text{Crp}\,(S)‘‘\lambda \,.\, \supset .\, \dot{s}‘\varpi \in (s‘\kappa) \,\overline{\text{sm}}\, (s‘\lambda).$$

Thus the desired result follows whenever

$$\exists ! \, \epsilon_\Delta‘\text{Crp}\,(S)‘‘\lambda.$$

Now we have $S \in 1 \to 1 \,.\, S \subset \text{sm} \,.\, \supset .\, \text{Crp}\,(S)‘‘\lambda \in \text{Cls ex}^2 \text{ excl}.$

Consequently

$$\text{Mult ax} \,.\, \supset : S \in 1 \to 1 \,.\, S \subset \text{sm} \,.\, \text{D}‘S = \kappa \,.\, \text{Cl}‘S = \lambda \,.\, \kappa, \lambda \in \text{Cls}^2 \text{ excl} \,.$$
$$\supset .\, s‘\kappa \text{ sm } s‘\lambda,$$

whence, by what was said previously,

$$\text{Mult ax} \,.\, \supset : \kappa \in \nu \cap \text{Cls excl}‘\mu \,.\, \supset .\, s‘\kappa \in \mu \times_c \nu \,.\, \Sigma \text{Nc}‘\kappa = \mu \times_c \nu.$$

The consideration of $\epsilon_\Delta‘\text{Crp}\,(S)‘‘\lambda$ leads similarly to the proposition

$$\vdash :. \text{Mult ax} \,.\, \supset : \mu, \nu \in \text{NC} \,.\, \kappa \in \nu \cap \text{Cl}‘\mu \,.\, \supset .\, \epsilon_\Delta‘\kappa \in \mu^\nu \,.\, \Pi \text{Nc}‘\kappa = \mu^\nu.$$

The proof is closely analogous to that for the connection of addition and multiplication.

It will be seen that, in the above use of the multiplicative axiom, we have two classes of classes κ and λ concerning which we assume

$$(\exists S) \,.\, S \in 1 \to 1 \,.\, S \subset \text{sm} \,.\, \text{D}‘S = \kappa \,.\, \text{Cl}‘S = \lambda,$$

i.e. we assume that κ and λ are similar classes of similar classes. A slightly modified hypothesis concerning κ and λ will enable us to obtain many results, without the multiplicative axiom, which otherwise might be expected to require this axiom. This is effected as follows.

Put $\kappa \operatorname{sm} \operatorname{sm} \lambda . \equiv . (\exists T) . T \epsilon 1 \rightarrow 1 . \text{ɑ}`T = s`\lambda . \kappa = T_\epsilon``\lambda$,
where " sm sm " is a single symbol representing a relation.

When this relation holds between κ and λ, we shall say that κ and λ have " double similarity." In this case, T correlates $s`\kappa$ and $s`\lambda$, while T_ϵ correlates κ and λ, so that if β is a member of λ, $T_\epsilon`\beta$, *i.e.* $T``\beta$, is its correlate in κ. We shall then have

$$\vdash : \kappa \operatorname{sm} \operatorname{sm} \lambda . \supset . s`\kappa \operatorname{sm} s`\lambda,$$

$$\vdash : \kappa \operatorname{sm} \operatorname{sm} \lambda . \supset . \Sigma \mathrm{Nc}`\kappa = \Sigma \mathrm{Nc}`\lambda,$$

$$\vdash : \kappa \operatorname{sm} \operatorname{sm} \lambda . \supset . \Pi \mathrm{Nc}`\kappa = \Pi \mathrm{Nc}`\lambda.$$

Also we have

$$\vdash : \kappa \operatorname{sm} \operatorname{sm} \lambda . \supset . (\exists S) . S \epsilon 1 \rightarrow 1 . S \subseteq \operatorname{sm} . \mathrm{D}`S = \kappa . \text{ɑ}`S = \lambda.$$

Conversely,

$$\vdash : \kappa, \lambda \epsilon \mathrm{Cls}^2 \mathrm{excl} . S \epsilon 1 \rightarrow 1 . S \subseteq \operatorname{sm} . \mathrm{D}`S = \kappa . \text{ɑ}`S = \lambda .$$
$$\varpi \epsilon \mathrm{D}``\epsilon_\Delta`\mathrm{Crp}(S)``\lambda . T = \dot{s}`\varpi . \supset . T \epsilon 1 \rightarrow 1 . \text{ɑ}`T = s`\lambda . \kappa = T_\epsilon``\lambda,$$

whence

$$\vdash :: \mathrm{Mult\,ax} . \supset :. \kappa, \lambda \epsilon \mathrm{Cls}^2 \mathrm{excl} : (\exists S) . S \epsilon 1 \rightarrow 1 . S \subseteq \operatorname{sm} . \mathrm{D}`S = \kappa . \text{ɑ}`S = \lambda :$$
$$\supset . \kappa \operatorname{sm} \operatorname{sm} \lambda.$$

Hence the multiplicative axiom is only required in order to pass from

$$(\exists S) . S \epsilon 1 \rightarrow 1 . S \subseteq \operatorname{sm} . \mathrm{D}`S = \kappa . \text{ɑ}`S = \lambda$$

to $\kappa \operatorname{sm} \operatorname{sm} \lambda$. It is this fact, and the consequent possibility of diminishing the use of the multiplicative axiom, which has led us to the employment of " sm sm " in the present section.

We treat also, in this section, the relation of greater and less between cardinals. We say that $\mathrm{Nc}`\alpha > \mathrm{Nc}`\beta$ when there is a part of α which is similar to β, but no part of β is similar to α. The principal proposition in this subject is the Schröder-Bernstein theorem, *i.e.*

$$\vdash : \mu \geqslant \nu . \nu \geqslant \mu . \supset . \mu = \nu.$$

This is an immediate consequence of *73·88. It cannot be shown, without assuming the multiplicative axiom, that of any two cardinals one must be the greater, *i.e.*

$$\mu, \nu \epsilon \mathrm{NC} . \mu \neq \nu . \supset : \mu > \nu . \vee . \nu > \mu.$$

If we assume the multiplicative axiom, this results from Zermelo's proof that on that assumption, every class can be well-ordered, together with Cantor's proof that of any two well-ordered series which are not similar, one must be similar to a part of the other. But these propositions cannot be proved till a much later stage (*258).

*110. THE ARITHMETICAL SUM OF TWO CLASSES AND OF TWO CARDINALS.

Summary of *110.

In this number, we start from the definition:

*110·01. $\alpha + \beta = \downarrow (\Lambda \cap \beta)``\iota``\alpha \cup (\Lambda \cap \alpha) \downarrow ``\iota``\beta$ Df

$\alpha + \beta$ is called the "arithmetical class-sum" of α and β. The definition is framed so as to give two mutually exclusive classes respectively similar to α and β, so that the number of terms in the logical sum of these two classes is the arithmetical sum of the numbers of terms in α and β respectively. $\alpha + \beta$ is significant whenever α and β are classes, whatever their types may be.

By means of $\alpha + \beta$, we define the arithmetical sum of two cardinals as follows:

*110·02. $\mu +_c \nu = \hat{\xi}\{(\exists \alpha, \beta) . \mu = \mathrm{N_0c}`\alpha . \nu = \mathrm{N_0c}`\beta . \xi \,\mathrm{sm}\, (\alpha + \beta)\}$ Df

This defines the "arithmetical sum of two cardinals." (It is not necessary to *significance* that μ and ν should be cardinals, but only that they should be classes of classes. If, however, either is not a cardinal, $\mu +_c \nu = \Lambda$.) It will be observed that, when μ and ν are typically definite, so are α and β in the above definition; but ξ is typically ambiguous, on account of the ambiguity of "sm." Hence $\mu +_c \nu$ is also typically ambiguous.

It will be shown that $\mu +_c \nu$ is always a cardinal, and that, if

$$\mu = \mathrm{N_0c}`\alpha . \nu = \mathrm{N_0c}`\beta, \text{ then } \mu +_c \nu = \mathrm{Nc}`(\alpha + \beta).$$

Hence whenever μ and ν are cardinals other than Λ, $\mu +_c \nu$ is an existent cardinal in some types, though it may be Λ in others.

Two more definitions are required in this number, namely:

*110·03. $\mathrm{Nc}`\alpha +_c \mu = \mathrm{N_0c}`\alpha +_c \mu$ Df

*110·04. $\mu +_c \mathrm{Nc}`\alpha = \mu +_c \mathrm{N_0c}`\alpha$ Df

These definitions are needed in order to apply the definition of $\mu +_c \nu$ to the case in which μ and ν are replaced by typically ambiguous symbols $\mathrm{Nc}`\alpha$ and $\mathrm{Nc}`\beta$. It does not make any difference to the value of $\mathrm{Nc}`\alpha +_c \mathrm{Nc}`\beta$ how the ambiguities of $\mathrm{Nc}`\alpha$ and $\mathrm{Nc}`\beta$ are determined, so long as they are determined in a way that insures $\exists ! \mathrm{Nc}`\alpha . \exists ! \mathrm{Nc}`\beta$; but if there are types in which either $\mathrm{Nc}`\alpha$ or $\mathrm{Nc}`\beta$ is Λ, we get $\mathrm{Nc}`\alpha +_c \mathrm{Nc}`\beta = \Lambda$ in all types if we determine the ambiguities so that $\mathrm{Nc}`\alpha = \Lambda$ or $\mathrm{Nc}`\beta = \Lambda$. It is in order to exclude such determinations of the ambiguity that the above definitions are required. Also in connection with these definitions and the corresponding definitions $*113\cdot04\cdot05$ and $*116\cdot03\cdot04$ and $*117\cdot02\cdot03$, the convention IIT of the prefatory statement must be noted.

The propositions of the present number begin with the properties of $\alpha+\beta$. We show ($*110\cdot11\cdot12$) that $\alpha+\beta$ consists of two mutually exclusive parts, which are respectively similar to α and β; we show ($*110\cdot14$) that if α and β are mutually exclusive, $\alpha \cup \beta$ is similar to $\alpha+\beta$, and ($*110\cdot15$) that if γ and δ are respectively similar to α and β, then $\gamma+\delta$ is similar to $\alpha+\beta$. We show ($*110\cdot16$) that $\text{Nc}`(\alpha+\beta)$ consists of all classes which can be divided into two mutually exclusive parts which are respectively similar to α and β.

We then proceed ($*110\cdot2$—$\cdot252$) to the consideration of $\mu +_c \nu$. Here μ and ν are typically definite, and the definition $*110\cdot02$ applies to any typically definite symbols, such as $\text{N}_0\text{c}`\alpha$ or $\text{Nc}(\eta)`\alpha$. We prove ($*110\cdot21$) that if μ and ν are cardinals, their sum consists of all classes similar to some class of the form $\alpha+\beta$, where $\alpha \epsilon \mu \,.\, \beta \epsilon \nu$; we prove ($*110\cdot22$) that the sum of $\text{N}_0\text{c}`\alpha$ and $\text{N}_0\text{c}`\beta$ is $\text{Nc}`(\alpha+\beta)$, and ($*110\cdot25$) that if μ and ν are cardinals, their sum is equal to the sum of the "same" cardinals in any other types in which they are not null, *i.e.*

$*110\cdot25$. $\vdash : \mu, \nu \epsilon \text{NC} \,.\, \exists ! \,\text{sm}_\eta``\mu \,.\, \exists ! \,\text{sm}_\zeta``\nu \,.\, \supset .\, \mu +_c \nu = \text{sm}_\eta``\mu +_c \text{sm}_\zeta``\nu$

We then ($*110\cdot3$—$\cdot351$) consider $\text{Nc}`\alpha +_c \text{Nc}`\beta$, to which we apply the definitions $*110\cdot03\cdot04$. We have

$*110\cdot3$. $\vdash .\, \text{Nc}`\alpha +_c \text{Nc}`\beta = \text{N}_0\text{c}`\alpha +_c \text{N}_0\text{c}`\beta = \text{Nc}`(\alpha+\beta)$

whence the other properties of $\text{Nc}`\alpha +_c \text{Nc}`\beta$ follow from previous propositions.

We then have ($*110\cdot4$—$\cdot44$) various propositions on the type of $\mu +_c \nu$ and its existence and kindred matters. The chief of these are

$*110\cdot4$. $\vdash : \exists ! \,\mu +_c \nu \,.\, \supset .\, \mu, \nu \epsilon \text{NC} - \iota`\Lambda \,.\, \mu, \nu \epsilon \text{N}_0\text{C}$

$*110\cdot42$. $\vdash .\, \mu +_c \nu \epsilon \text{NC}$

This proposition requires no hypothesis, because, if μ and ν are not both cardinals, $\mu +_c \nu = \Lambda$, and Λ is a cardinal, by $*102\cdot74$.

Our next set of propositions ($*110\cdot5$—$\cdot57$) are concerned with the permutative and associative laws, which are $*110\cdot51$ and $*110\cdot56$ respectively.

We then ($*110\cdot6$—$\cdot643$) consider the addition of 0 or 1, proving ($*110\cdot61$) that a cardinal is unchanged by the addition of 0, and ($*110\cdot643$) that $1 +_c 1 = 2$.

$*110\cdot01$. $\alpha+\beta = \downarrow(\Lambda \cap \beta)``\iota``\alpha \cup (\Lambda \cap \alpha)\downarrow``\iota``\beta$ Df

$*110\cdot02$. $\mu +_c \nu = \hat{\xi}\{(\exists \alpha, \beta) \,.\, \mu = \text{N}_0\text{c}`\alpha \,.\, \nu = \text{N}_0\text{c}`\beta \,.\, \xi \,\text{sm}\,(\alpha+\beta)\}$ Df

$*110\cdot03$. $\text{Nc}`\alpha +_c \mu = \text{N}_0\text{c}`\alpha +_c \mu$ Df

$*110\cdot04$. $\mu +_c \text{Nc}`\alpha = \mu +_c \text{N}_0\text{c}`\alpha$ Df

These definitions are extended by IIT of the prefatory statement.

***110·1.** $\vdash :. R \in \alpha + \beta . \equiv : (\exists x) . x \in \alpha . R = (\iota\text{‘}x) \downarrow (\Lambda \cap \beta) . \vee .$
$(\exists y) . y \in \beta . R = (\Lambda \cap \alpha) \downarrow (\iota\text{‘}y)$
[*38·13·131 . (*110·01)]

***110·101.** $\vdash . (\iota\text{‘}x) \downarrow (\Lambda \cap \beta) \neq (\Lambda \cap \alpha) \downarrow (\iota\text{‘}y)$

Dem.

$\vdash . *55\cdot15 . \supset \vdash . D\text{‘}(\iota\text{‘}x) \downarrow (\Lambda \cap \beta) = \iota\text{‘}\iota\text{‘}x . D\text{‘}(\Lambda \cap \alpha) \downarrow (\iota\text{‘}y) = \iota\text{‘}(\Lambda \cap \alpha)$ (1)
$\vdash . *51\cdot161 . \supset \vdash . \iota\text{‘}x \neq (\Lambda \cap \alpha) .$
[*51·23] $\supset \vdash . \iota\text{‘}\iota\text{‘}x \neq \iota\text{‘}(\Lambda \cap \alpha)$ (2)
$\vdash . (1) . (2) . \supset \vdash . D\text{‘}(\iota\text{‘}x) \downarrow (\Lambda \cap \beta) \neq D\text{‘}(\Lambda \cap \alpha) \downarrow (\iota\text{‘}y) . \supset \vdash .$ Prop

***110·11.** $\vdash . \downarrow (\Lambda \cap \beta)\text{“}\iota\text{“}\alpha \cap (\Lambda \cap \alpha) \downarrow \text{“}\iota\text{“}\beta = \Lambda$

Dem.

$\vdash . *110\cdot101 . \supset \vdash : x \in \alpha . R = \downarrow (\Lambda \cap \beta)\text{‘}\iota\text{‘}x . y \in \beta . S = (\Lambda \cap \alpha) \downarrow \text{‘}\iota\text{‘}y . \supset . R \neq S :$
[*37·67] $\supset \vdash : R \in \downarrow (\Lambda \cap \beta)\text{“}\iota\text{“}\alpha . S \in (\Lambda \cap \alpha) \downarrow \text{“}\iota\text{“}\beta . \supset . R \neq S$ (1)
$\vdash . (1) . *24\cdot37 . \supset \vdash .$ Prop

***110·12.** $\vdash . \downarrow (\Lambda \cap \beta)\text{“}\iota\text{“}\alpha \operatorname{sm} \alpha . (\Lambda \cap \alpha) \downarrow \text{“}\iota\text{“}\beta \operatorname{sm} \beta$ [*73·41·61·611]

*110·11·12 give the justification for the use of $\alpha + \beta$ in defining arithmetical addition, since they show that $\alpha + \beta$ consists of two mutually exclusive parts which are respectively similar to α and β.

***110·13.** $\vdash : \gamma \operatorname{sm} \alpha . \delta \operatorname{sm} \beta . \gamma \cap \delta = \Lambda . \supset . \gamma \cup \delta \operatorname{sm} (\alpha + \beta)$

Dem.

$\vdash . *110\cdot12 . \supset \vdash : \text{Hp} . \supset . \gamma \operatorname{sm} \downarrow (\Lambda \cap \beta)\text{“}\iota\text{“}\alpha . \delta \operatorname{sm} (\Lambda \cap \alpha) \downarrow \text{“}\iota\text{“}\beta$ (1)
$\vdash . (1) . *110\cdot11 . *73\cdot71 . \supset \vdash .$ Prop

***110·14.** $\vdash : \alpha \cap \beta = \Lambda . \supset . \alpha \cup \beta \operatorname{sm} (\alpha + \beta)$ [*110·13 . *73·3]

Thus whenever α and β are mutually exclusive, their logical sum may replace their arithmetical sum in defining the sum of their cardinal numbers.

***110·15.** $\vdash : \gamma \operatorname{sm} \alpha . \delta \operatorname{sm} \beta . \supset . \gamma + \delta \operatorname{sm} \alpha + \beta$

Dem.

$\vdash . *110\cdot12 . \supset \vdash : \text{Hp} . \supset . \downarrow (\Lambda \cap \delta)\text{“}\iota\text{“}\gamma \operatorname{sm} \alpha . (\Lambda \cap \gamma) \downarrow \text{“}\iota\text{“}\delta \operatorname{sm} \beta$ (1)
$\vdash . *110\cdot11 . \supset \vdash . \downarrow (\Lambda \cap \delta)\text{“}\iota\text{“}\gamma \cap (\Lambda \cap \gamma) \downarrow \text{“}\iota\text{“}\delta = \Lambda$ (2)
$\vdash . (1) . (2) . *110\cdot13 . \supset$
$\vdash : \text{Hp} . \supset . \downarrow (\Lambda \cap \delta)\text{“}\iota\text{“}\gamma \cup (\Lambda \cap \gamma) \downarrow \text{“}\iota\text{“}\delta \operatorname{sm} \alpha + \beta : \supset \vdash .$ Prop

***110·151.** $\vdash :. \alpha \cap \beta = \Lambda . \supset : \xi \operatorname{sm} (\alpha \cup \beta) . \equiv . (\exists \gamma, \delta) . \gamma \operatorname{sm} \alpha . \delta \operatorname{sm} \beta . \gamma \cap \delta = \Lambda . \xi = \gamma \cup \delta$

Dem.

$\vdash . *73\cdot71 . \supset \vdash :. \text{Hp} . \supset :$
$(\exists \gamma, \delta) . \gamma \operatorname{sm} \alpha . \delta \operatorname{sm} \beta . \gamma \cap \delta = \Lambda . \xi = \gamma \cup \delta . \supset . \xi \operatorname{sm} (\alpha \cup \beta)$ (1)
$\vdash . *72\cdot411 . *37\cdot25\cdot22 . *73\cdot22 . \supset$
$\vdash : S \in 1 \rightarrow 1 . D\text{‘}S = \xi . \text{Ͻ}\text{‘}S = \alpha \cup \beta . \alpha \cap \beta = \Lambda . \supset .$
$S\text{“}\alpha \cap S\text{“}\beta = \Lambda . \xi = S\text{“}\alpha \cup S\text{“}\beta . S\text{“}\alpha \operatorname{sm} \alpha . S\text{“}\beta \operatorname{sm} \beta .$
[*11·36] $\supset . (\exists \gamma, \delta) . \gamma \operatorname{sm} \alpha . \delta \operatorname{sm} \beta . \gamma \cap \delta = \Lambda . \xi = \gamma \cup \delta$ (2)
$\vdash . (2) . *10\cdot11\cdot23\cdot35 . *73\cdot1 . \supset$
$\vdash :. \text{Hp} . \supset : \xi \operatorname{sm} (\alpha \cup \beta) . \supset . (\exists \gamma, \delta) . \gamma \operatorname{sm} \alpha . \delta \operatorname{sm} \beta . \gamma \cap \delta = \Lambda . \xi = \gamma \cup \delta$ (3)
$\vdash . (1) . (3) . \supset \vdash .$ Prop

***110·152.** $\vdash : \xi \,\mathrm{sm}\, (\alpha + \beta) . \equiv . (\exists \gamma, \delta) . \gamma \,\mathrm{sm}\, \alpha . \delta \,\mathrm{sm}\, \beta . \gamma \cap \delta = \Lambda . \xi = \gamma \cup \delta$

Dem.

$\vdash . *110·151·11 . \supset$

$\vdash : \xi \,\mathrm{sm}\, (\alpha + \beta) . \equiv . (\exists \gamma, \delta) . \gamma \,\mathrm{sm} \downarrow (\Lambda \cap \beta)``\iota``\alpha . \delta \,\mathrm{sm}\, (\Lambda \cap \alpha) \downarrow ``\iota``\beta .$

$\gamma \cap \delta = \Lambda . \xi = \gamma \cup \delta .$

[*73·37.*110·12] $\equiv . (\exists \gamma, \delta) . \gamma \,\mathrm{sm}\, \alpha . \delta \,\mathrm{sm}\, \beta . \gamma \cap \delta = \Lambda . \xi = \gamma \cup \delta : \supset \vdash .$ Prop

***110·16.** $\vdash . \mathrm{Nc}`(\alpha + \beta) = \hat{\xi} \{(\exists \gamma, \delta) . \gamma \,\mathrm{sm}\, \alpha . \delta \,\mathrm{sm}\, \beta . \gamma \cap \delta = \Lambda . \xi = \gamma \cup \delta\}$
[*110·152 . *100·1]

***110·17.** $\vdash : \alpha \in t`\beta . \supset . \exists ! \mathrm{Nc}\,(t`\alpha)`(\alpha + \beta)$

Dem.

$\vdash . *104·43 . \supset$

$\vdash : \mathrm{Hp} . \supset . (\exists \gamma, \delta) . \gamma \,\mathrm{sm}\, \alpha . \gamma \subset t`\alpha . \delta \,\mathrm{sm}\, \beta . \delta \subset t`\alpha . \gamma \cap \delta = \Lambda .$

[*22·59] $\supset . (\exists \gamma, \delta) . \gamma \,\mathrm{sm}\, \alpha . \delta \,\mathrm{sm}\, \beta . \gamma \cap \delta = \Lambda . \gamma \cup \delta \subset t`\alpha .$

[*110·16] $\supset . (\exists \xi) . \xi \subset t`\alpha . \xi \in \mathrm{Nc}`(\alpha + \beta) .$

[*102·6.*63·5] $\supset . \exists ! \mathrm{Nc}\,(t`\alpha)`(\alpha + \beta) : \supset \vdash .$ Prop

Thus when α and β are of the same type, $\mathrm{Nc}`(\alpha + \beta)$ exists at least in the type next above that of $\alpha$ and $\beta$. We cannot prove that it exists in the type of $\alpha$ and $\beta$. *E.g.* suppose the lowest type contained only one member; then if $x$ were that one member, $\mathrm{Nc}`(\iota`x + \iota`x)$ would not exist in the type to which $\iota`x$ belongs, but would exist in the next type, *i.e.* there would not be two individuals, but there would be two classes, namely $\Lambda$ and $\iota`x$, so that $\iota`\Lambda \cup \iota`\iota`x \in \mathrm{Nc}`(\iota`x + \iota`x)$.

***110·18.** $\vdash . \alpha + \beta \in t`t`(t`\alpha \uparrow t`\beta)$

Dem.

$\vdash . *64·53 . \quad \supset \vdash : x \in \alpha . \supset . \downarrow (\Lambda \cap \beta)`\iota`x \in t`(t`\alpha \uparrow t`\beta) \quad (1)$

$\vdash . (1) . *37·61 . \supset \vdash . \downarrow (\Lambda \cap \beta)``\iota``\alpha \subset t`(t`\alpha \uparrow t`\beta) \quad (2)$

Similarly $\vdash . (\Lambda \cap \alpha) \downarrow ``\iota``\beta \subset t`(t`\alpha \uparrow t`\beta) \quad (3)$

$\vdash . (2) . (3) . \quad \supset \vdash . \alpha + \beta \subset t`(t`\alpha \uparrow t`\beta) .$

[*63·5] $\supset \vdash . \alpha + \beta \in t`t`(t`\alpha \uparrow t`\beta) . \supset \vdash .$ Prop

***110·2.** $\vdash : \xi \in \mu +_c \nu . \equiv . (\exists \alpha, \beta) . \mu = \mathrm{N_0c}`\alpha . \nu = \mathrm{N_0c}`\alpha . \xi \,\mathrm{sm}\, (\alpha + \beta)$
[(*110·02)]

***110·201.** $\vdash :. \xi \in \mu +_c \nu . \equiv : \mu, \nu \in \mathrm{NC} : (\exists \alpha, \beta) . \alpha \in \mu . \beta \in \nu . \xi \,\mathrm{sm}\, (\alpha + \beta)$
[*103·27 . *110·2]

***110·202.** $\vdash :. \xi \in \mu +_c \nu . \equiv :$

$\exists ! \mu . \exists ! \nu : (\exists \gamma, \delta) . \mu = \mathrm{Nc}`\gamma . \nu = \mathrm{Nc}`\delta . \gamma \cap \delta = \Lambda . \xi = \gamma \cup \delta$

Dem.

$\vdash . *110·2·152 . \supset \vdash :. \xi \in \mu +_c \nu . \equiv :$

$(\exists \alpha, \beta, \gamma, \delta) . \mu = \mathrm{N_0c}`\alpha . \nu = \mathrm{N_0c}`\beta . \gamma \,\mathrm{sm}\, \alpha . \delta \,\mathrm{sm}\, \beta . \gamma \cap \delta = \Lambda . \xi = \gamma \cup \delta :$

[*103·28] $\equiv : (\exists \gamma, \delta) . \exists ! \mu . \exists ! \nu . \mu = \mathrm{Nc}`\gamma . \nu = \mathrm{Nc}`\delta . \gamma \cap \delta = \Lambda . \xi = \gamma \cup \delta :.$

$\supset \vdash .$ Prop

∗110·21. $\vdash :. \mu, \nu \in \mathrm{NC} . \supset : \xi \in \mu +_c \nu . \equiv . (\exists \alpha, \beta) . \alpha \in \mu . \beta \in \nu . \xi \,\mathrm{sm}\, (\alpha + \beta)$
[∗110·201]

∗110·211. $\vdash :. \mu, \nu \in \mathrm{NC} . \supset : \xi \in \mu +_c \nu . \equiv .$
$(\exists \gamma, \delta) . \gamma \in \mathrm{sm}``\mu . \delta \in \mathrm{sm}``\nu . \gamma \cap \delta = \Lambda . \xi = \gamma \cup \delta$

Dem.

$\vdash . ∗110·21·152 . \supset \vdash :. \mathrm{Hp} . \supset : \xi \in \mu +_c \nu . \equiv .$
$(\exists \alpha, \beta, \gamma, \delta) . \alpha \in \mu . \beta \in \nu . \gamma \,\mathrm{sm}\, \alpha . \delta \,\mathrm{sm}\, \beta . \gamma \cap \delta = \Lambda . \xi = \gamma \cup \delta .$
[∗37·1] $\equiv . (\exists \gamma, \delta) . \gamma \in \mathrm{sm}``\mu . \delta \in \mathrm{sm}``\nu . \gamma \cap \delta = \Lambda . \xi = \gamma \cup \delta :. \supset \vdash . \mathrm{Prop}$

∗110·212. $\vdash :. \mu, \nu \in \mathrm{NC} . \supset : \xi \in \mu +_c \nu . \equiv . (\exists \gamma) . \gamma \in \mathrm{sm}``\mu . \gamma \subset \xi . \xi - \gamma \in \mathrm{sm}``\nu$

Dem.

$\vdash . ∗110·211 . ∗24·47 . \supset$
$\vdash :. \mathrm{Hp} . \supset : \xi \in \mu +_c \nu . \equiv . (\exists \gamma, \delta) . \gamma \in \mathrm{sm}``\mu . \delta \in \mathrm{sm}``\nu . \gamma \subset \xi . \delta = \xi - \gamma .$
[∗13·195] $\equiv . (\exists \gamma) . \gamma \in \mathrm{sm}``\mu . \gamma \subset \xi . \xi - \gamma \in \mathrm{sm}``\nu :. \supset \vdash . \mathrm{Prop}$

∗110·22. $\vdash . \mathrm{N_0c}`\alpha +_c \mathrm{N_0c}`\beta = \mathrm{Nc}`(\alpha + \beta)$

Dem.

$\vdash . ∗103·4 . ∗110·211 . \supset$
$\vdash : \xi \in \mathrm{N_0c}`\alpha +_c \mathrm{N_0c}`\beta . \equiv . (\exists \gamma, \delta) . \gamma \in \mathrm{Nc}`\alpha . \delta \in \mathrm{Nc}`\beta . \gamma \cap \delta = \Lambda . \xi = \gamma \cup \delta .$
[∗100·31] $\equiv . (\exists \gamma, \delta) . \gamma \,\mathrm{sm}\, \alpha . \delta \,\mathrm{sm}\, \beta . \gamma \cap \delta = \Lambda . \xi = \gamma \cup \delta .$
[∗110·16] $\equiv . \xi \in \mathrm{Nc}`(\alpha + \beta) : \supset \vdash . \mathrm{Prop}$

∗110·221. $\vdash : \xi \in \mathrm{Nc}\,(\eta)`\alpha +_c \mathrm{Nc}\,(\zeta)`\beta . \equiv . \exists ! \mathrm{Nc}\,(\eta)`\alpha . \exists ! \mathrm{Nc}\,(\zeta)`\beta . \xi \in \mathrm{Nc}`(\alpha + \beta)$

Dem.

$\vdash . ∗110·202 . \supset \vdash :. \xi \in \mathrm{Nc}\,(\eta)`\alpha +_c \mathrm{Nc}\,(\zeta)`\beta .$
$\equiv : \exists ! \mathrm{Nc}\,(\eta)`\alpha . \exists ! \mathrm{Nc}\,(\zeta)`\beta : (\exists \gamma, \delta) . \mathrm{Nc}\,(\eta)`\alpha = \mathrm{Nc}`\gamma .$
$\mathrm{Nc}\,(\zeta)`\beta = \mathrm{Nc}`\delta . \gamma \cap \delta = \Lambda . \xi = \gamma \cup \delta :$
[∗100·35] $\equiv : \exists ! \mathrm{Nc}\,(\eta)`\alpha . \exists ! \mathrm{Nc}(\zeta)`\beta : (\exists \gamma, \delta) . \gamma \,\mathrm{sm}\, \alpha . \delta \,\mathrm{sm}\, \beta . \gamma \cap \delta = \Lambda . \xi = \gamma \cup \delta :$
[∗110·16] $\equiv : \exists ! \mathrm{Nc}\,(\eta)`\alpha . \exists ! \mathrm{Nc}\,(\zeta)`\beta . \xi \in \mathrm{Nc}`(\alpha + \beta) :. \supset \vdash . \mathrm{Prop}$

∗110·23. $\vdash : \exists ! \mathrm{Nc}\,(\eta)`\alpha . \exists ! \mathrm{Nc}\,(\zeta)`\beta . \supset .$
$\mathrm{Nc}\,(\eta)`\alpha +_c \mathrm{Nc}\,(\zeta)`\beta = \mathrm{Nc}`(\alpha + \beta) = \mathrm{N_0c}`\alpha +_c \mathrm{N_0c}`\beta$ [∗110·221·22]

Thus $\mathrm{Nc}\,(\eta)`\alpha +_c \mathrm{Nc}\,(\zeta)`\beta$ is independent of η and ζ so long as $\mathrm{Nc}`\alpha$ and $\mathrm{Nc}`\beta$ exist in the types of η and ζ respectively.

∗110·231. $\vdash :. \mathrm{Nc}\,(\eta)`\alpha = \Lambda . \lor . \mathrm{Nc}\,(\zeta)`\beta = \Lambda : \supset . \mathrm{Nc}\,(\eta)`\alpha +_c \mathrm{Nc}\,(\zeta)`\beta = \Lambda$
[∗110·221]

∗110·24. $\vdash : \eta \,\mathrm{sm}\, \alpha . \zeta \,\mathrm{sm}\, \beta . \supset . \mathrm{N_0c}`\eta +_c \mathrm{N_0c}`\zeta = \mathrm{N_0c}`\alpha +_c \mathrm{N_0c}`\beta$

Dem.

$\vdash . ∗103·42 . \supset \vdash : \mathrm{Hp} . \supset . \mathrm{N_0c}`\eta = \mathrm{Nc}\,(\eta)`\alpha . \mathrm{N_0c}`\zeta = \mathrm{Nc}\,(\zeta)`\beta$ (1)
$\vdash . (1) . ∗103·13 . \supset \vdash : \mathrm{Hp} . \supset . \exists ! \mathrm{Nc}\,(\eta)`\alpha . \exists ! \mathrm{Nc}\,(\zeta)`\beta .$
[∗110·23] $\supset . \mathrm{Nc}\,(\eta)`\alpha +_c \mathrm{Nc}\,(\zeta)`\beta = \mathrm{N_0c}`\alpha +_c \mathrm{N_0c}`\beta$ (2)
$\vdash . (1) . (2) . \supset \vdash . \mathrm{Prop}$

$*110\cdot25$. $\vdash : \mu, \nu \in \text{NC} \,.\, \exists ! \,\text{sm}_\eta``\mu \,.\, \exists ! \,\text{sm}_\zeta``\nu \,.\supset .\, \mu +_c \nu = \text{sm}_\eta``\mu +_c \text{sm}_\zeta``\nu$

Dem.

$\vdash . *103\cdot27 . \supset \vdash : \mu, \nu \in \text{NC} \,.\, \alpha \in \mu \,.\, \beta \in \nu \,.\, \exists ! \,\text{sm}_\eta``\mu \,.\, \exists ! \,\text{sm}_\zeta``\nu \,.$

$\supset .\, \mu = \text{N}_0\text{c}`\alpha \,.\, \nu = \text{N}_0\text{c}`\beta \,.\, \exists ! \,\text{sm}_\eta``\mu \,.\, \exists ! \,\text{sm}_\zeta``\nu \,.$

$[*103\cdot41 . *102\cdot85] \supset .\, \mu = \text{N}_0\text{c}`\alpha \,.\, \nu = \text{N}_0\text{c}`\beta \,.\, \text{sm}_\eta``\mu = \text{Nc}\,(\eta)`\alpha \,.$

$\text{sm}_\zeta``\nu = \text{Nc}\,(\zeta)`\beta \,.\, \exists ! \,\text{Nc}\,(\eta)`\alpha \,.\, \exists ! \,\text{Nc}\,(\zeta)`\beta \,.$

$[*110\cdot23] \quad \supset .\, \mu +_c \nu = \text{N}_0\text{c}`\alpha +_c \text{N}_0\text{c}`\beta = \text{sm}_\eta``\mu +_c \text{sm}_\zeta``\nu \quad (1)$

$\vdash . (1) . *10\cdot11\cdot23\cdot35 . \supset$

$\vdash : \mu, \nu \in \text{NC} \,.\, \exists ! \,\mu \,.\, \exists ! \,\nu \,.\, \exists ! \,\text{sm}_\eta``\mu \,.\, \exists ! \,\text{sm}_\zeta``\nu \,.\supset .\, \mu +_c \nu = \text{sm}_\eta``\mu +_c \text{sm}_\zeta``\nu \quad (2)$

$\vdash . *37\cdot29 . \text{Transp} . \supset \vdash : \exists ! \,\text{sm}_\eta``\mu \,.\, \exists ! \,\text{sm}_\zeta``\nu \,.\supset .\, \exists ! \,\mu \,.\, \exists ! \,\nu \quad (3)$

$\vdash . (2) . (3) . \supset \vdash . \text{Prop}$

$*110\cdot251$. $\vdash : \mu, \nu \in \text{NC} \,.\supset .\, \mu^{(1)} +_c \nu^{(1)} = \mu +_c \nu$

Dem.

$\vdash . *110\cdot25 . *104\cdot265 . \supset$

$\vdash : \text{Hp} \,.\, \exists ! \,\mu^{(1)} \,.\, \exists ! \,\nu^{(1)} \,.\supset .\, \mu^{(1)} +_c \nu^{(1)} = \mu +_c \nu \quad (1)$

$\vdash . *110\cdot202 . \supset \vdash : \sim(\exists ! \,\mu^{(1)} \,.\, \exists ! \,\nu^{(1)}) \,.\supset .\, \mu^{(1)} +_c \nu^{(1)} = \Lambda \quad (2)$

$\vdash . *104\cdot264 . \supset \vdash : \text{Hp}\,(2) \,.\supset .\, \sim(\exists ! \,\mu \,.\, \exists ! \,\nu) \,.$

$[*110\cdot202] \quad \supset .\, \mu +_c \nu = \Lambda \,.$

$[(2)] \quad \supset .\, \mu^{(1)} +_c \nu^{(1)} = \mu +_c \nu \quad (3)$

$\vdash . (1) . (3) . \supset \vdash . \text{Prop}$

$*110\cdot252$. $\vdash : \mu, \nu \in \text{NC} \,.\supset .\, \mu_{(00)} +_c \nu_{(00)} = \mu +_c \nu$ [Proof as in $*110\cdot251$]

A similar proof applies to $\mu^{(2)}$, $\nu^{(2)}$, etc., and to any such derived cardinals whose existence follows from that of μ and ν. The proposition does not hold generally for $\mu_{(1)}$, $\nu_{(1)}$ and other descending derived cardinals, because they may be null when μ and ν exist.

The following proposition ($*110\cdot3$) is more often used than any other in this number except $*110\cdot4$.

$*110\cdot3$. $\vdash . \text{Nc}`\alpha +_c \text{Nc}`\beta = \text{N}_0\text{c}`\alpha +_c \text{N}_0\text{c}`\beta = \text{Nc}`(\alpha + \beta)$ [$*110\cdot22 . (*110\cdot03\cdot04)$]

$*110\cdot31$. $\vdash : \gamma \,\text{sm}\, \alpha \,.\, \delta \,\text{sm}\, \beta \,.\supset .\, \text{Nc}`\gamma +_c \text{Nc}`\delta = \text{Nc}`\alpha +_c \text{Nc}`\beta$ [$*110\cdot24\cdot3$]

The following proposition is frequently used.

$*110\cdot32$. $\vdash : \alpha \cap \beta = \Lambda \,.\supset .\, \text{Nc}`\alpha +_c \text{Nc}`\beta = \text{Nc}`(\alpha \cup \beta)$ [$*110\cdot3\cdot14$]

$*110\cdot33$. $\vdash : \xi \in \text{Nc}`\alpha +_c \text{Nc}`\beta \,.\equiv .\, (\exists \gamma, \delta) \,.\, \gamma \,\text{sm}\, \alpha \,.\, \delta \,\text{sm}\, \beta \,.\, \gamma \cap \delta = \Lambda \,.\, \xi = \gamma \cup \delta$
[$*110\cdot3\cdot16$]

The above proposition is used in $*110\cdot63$. We might have used the above to define arithmetical addition, but this method would have been less convenient than the method adopted in this number, both because there would have been more difficulty in dealing with types, and because the existence of $\text{Nc}'\alpha +_c \text{Nc}'\beta$ (in the types in which it does exist) is less evident with the above definition than with the definitions given in ∗110·01·02·03·04.

∗110·331. $\vdash . \text{Nc}'\alpha +_c \text{Nc}'\beta = \hat{\xi} \{(\exists\gamma) . \gamma \text{ sm } \alpha . \xi - \gamma \text{ sm } \beta . \gamma \subset \xi\}$

Dem.

$\vdash . *110{\cdot}33 . *24{\cdot}47 . \supset$

$\vdash : \xi \in \text{Nc}'\alpha +_c \text{Nc}'\beta . \equiv . (\exists\gamma, \delta) . \gamma \text{ sm } \alpha . \delta \text{ sm } \beta . \gamma \subset \xi . \delta = \xi - \gamma .$

[∗13·195] $\equiv . (\exists\gamma) . \gamma \text{ sm } \alpha . \xi - \gamma \text{ sm } \beta . \gamma \subset \xi : \supset \vdash . \text{Prop}$

∗110·34. $\vdash : \exists ! \text{Nc}(\eta)'\alpha . \exists ! \text{Nc}(\zeta)'\beta . \supset . \text{Nc}(\eta)'\alpha +_c \text{Nc}(\zeta)'\beta = \text{Nc}'\alpha +_c \text{Nc}'\beta$ [∗110·23·3]

∗110·35. $\vdash . \text{N}^1\text{c}'\alpha +_c \text{N}^1\text{c}'\beta = \text{Nc}'\alpha +_c \text{Nc}'\beta$ [∗104·102·21 . ∗110·34]

∗110·351. $\vdash . \text{N}_{00}\text{c}'\alpha +_c \text{N}_{00}\text{c}'\beta = \text{Nc}'\alpha +_c \text{Nc}'\beta$ [∗106·21 . ∗110·34]

Similar propositions will hold generally for *ascending* cardinals.

The following proposition (∗110·4) is the most used of the propositions in this number. It is useful both in the form given, and in the form resulting from transposition, in which it shows that $\mu +_c \nu = \Lambda$ unless both μ and ν are existent cardinals. It is chiefly useful in avoiding the necessity of the hypothesis $\mu, \nu \in \text{NC}$ in such propositions as the commutative and associative laws.

∗110·4. $\vdash : \exists ! \mu +_c \nu . \supset . \mu, \nu \in \text{NC} - \iota'\Lambda . \mu, \nu \in \text{N}_0\text{C}$ [∗110·201·202·2]

The following propositions, down to ∗110·411 inclusive, are concerned with types. They are not referred to in the sequel.

∗110·401. $\vdash : \mu = \text{N}_0\text{c}'\alpha . \nu = \text{N}_0\text{c}'\beta . \supset . \alpha + \beta \in t't'(\mu \uparrow \nu)$

Dem.

$\vdash . *110{\cdot}18 . *103{\cdot}12 . \supset \vdash : \text{Hp} . \supset . \alpha + \beta \in t't'(t'\alpha \uparrow t'\beta) . \alpha \in \mu . \beta \in \nu .$

[∗63·11] $\supset . \alpha + \beta \in t't'(t'\alpha \uparrow t'\beta) . t'\alpha = t_0'\mu . t'\beta = t_0'\nu .$

[∗13·12] $\supset . \alpha + \beta \in t't'(t_0'\mu \uparrow t_0'\nu) .$

[∗64·13] $\supset . \alpha + \beta \in t't'(\mu \uparrow \nu) : \supset \vdash . \text{Prop}$

∗110·402. $\vdash : \mu, \nu \in \text{N}_0\text{C} . \supset . \exists ! (\mu +_c \nu) \cap t't'(\mu \uparrow \nu)$

Dem.

$\vdash . *110{\cdot}22 . *100{\cdot}3 . \supset$

$\vdash : \mu = \text{N}_0\text{c}'\alpha . \nu = \text{N}_0\text{c}'\beta . \supset . \alpha + \beta \in \mu +_c \nu .$

[∗110·401] $\supset . \alpha + \beta \in (\mu +_c \nu) \cap t't'(\mu \uparrow \nu) .$

[∗10·24] $\supset . \exists ! (\mu +_c \nu) \cap t't'(\mu \uparrow \nu)$ (1)

$\vdash . (1) . *103{\cdot}2 . \supset \vdash . \text{Prop}$

∗110·403. $\vdash : \mu, \nu \in \text{N}_0\text{C} . \equiv . \exists ! (\mu +_c \nu) \cap t't'(\mu \uparrow \nu)$ [∗110·402·4]

∗110·404. $\vdash . \exists ! (\text{Nc}'\alpha +_c \text{Nc}'\beta) \cap t't'(t'\alpha \uparrow t'\beta)$ [∗110·18·3 . ∗100·3]

***110·41.** $\vdash : \mu, \nu \in N_0C \,.\, t`\mu = t`\nu \,.\, \supset \,.\, \exists ! (\mu +_c \nu) \cap t`\mu$

Dem.

$\vdash \,.\, *103·11 \,.\, \supset \vdash : \mu = N_0c`\alpha \,.\, \nu = N_0c`\beta \,.\, t`\mu = t`\nu \,.\, \supset \,.$
$\mu \subset t`\alpha \,.\, \nu \subset t`\beta \,.\, t`\mu = t`\nu \,.$

[*63·21·35] $\supset \,.\, t_0`\mu = t`\alpha \,.\, t_0`\nu = t`\beta \,.\, t_0`\mu = t_0`\nu \,.$

[*13·16·17] $\supset \,.\, t`\alpha = t`\beta = t_0`\mu \,.$

[*110·17] $\supset \,.\, \exists ! \, Nc`(\alpha + \beta) \cap t`t_0`\mu \,.$

[*110·22.*63·19] $\supset \,.\, \exists ! (\mu +_c \nu) \cap t`\mu : \supset \vdash \,.\,$ Prop

***110·411.** $\vdash : t`\alpha = t`\beta \,.\, \supset \,.\, \exists ! (Nc`\alpha +_c Nc`\beta) \cap t`t`\alpha \,.\, \exists ! \, Nc\,(t`\alpha)`(\alpha + \beta)$
[*110·17·3]

It will be observed that the following proposition (*110·42) requires no hypothesis. This is owing to *110·4 and *102·74.

***110·42.** $\vdash \,.\, \mu +_c \nu \in NC$

Dem.

$\vdash \,.\, *110·22 \,.\, \supset \vdash : \mu = N_0c`\alpha \,.\, \nu = N_0c`\beta \,.\, \supset \,.\, \mu +_c \nu = Nc`(\alpha + \beta) \,.$

[*100·41] $\supset \,.\, \mu +_c \nu \in NC$ (1)

$\vdash \,.\, (1) \,.\, *103·2 \,.\, \supset \vdash : \mu, \nu \in N_0C \,.\, \supset \,.\, \mu +_c \nu \in NC$ (2)

$\vdash \,.\, *110·4 \,.\, \text{Transp} \,.\, \supset \vdash : \sim (\mu, \nu \in N_0C) \,.\, \supset \,.\, \mu +_c \nu = \Lambda \,.$

[*102·74] $\supset \,.\, \mu +_c \nu \in NC$ (3)

$\vdash \,.\, (2) \,.\, (3) \,.\, \supset \vdash \,.\,$ Prop

***110·43.** $\vdash : \mu +_c \nu = N_0c`\eta \,.\, \equiv \,.\, \eta \in \mu +_c \nu$ [*110·42 . *103·26]

***110·44.** $\vdash \,.\, \text{sm}``(\mu +_c \nu) = \mu +_c \nu$

Dem.

$\vdash \,.\, *37·1 \,.\, *110·2 \,.\, \supset$

$\vdash : \xi \in \text{sm}``(\mu +_c \nu) \,.\, \equiv \,.\, (\exists \eta, \alpha, \beta) \,.\, \mu = N_0c`\alpha \,.\, \nu = N_0c`\beta \,.\, \eta \,\text{sm}\, (\alpha + \beta) \,.\, \xi \,\text{sm}\, \eta \,.$

[*73·3·32] $\equiv \,.\, (\exists \alpha, \beta) \,.\, \mu = N_0c`\alpha \,.\, \nu = N_0c`\beta \,.\, \xi \,\text{sm}\, (\alpha + \beta) \,.$

[*110·2] $\equiv \,.\, \xi \in \mu +_c \nu : \supset \vdash \,.\,$ Prop

The above proposition depends upon the fact that $\mu +_c \nu$ is typically ambiguous, even when μ and ν are typically definite. It is used in the theory of inductive cardinals (*120·32·41·424).

The following propositions are concerned with the commutative and associative laws for arithmetical addition of cardinals.

***110·5.** $\vdash \,.\, \beta + \alpha = \text{Cnv}``(\alpha + \beta)$

Dem.

$\vdash \,.\, *55·14 \,.\, \supset \vdash \,.\, \text{Cnv}``(\alpha + \beta) = \Lambda_\beta \downarrow ``\iota``\alpha \cup \downarrow \Lambda_\alpha``\iota``\beta$

[(*110·01)] $= \beta + \alpha \,.\, \supset \vdash \,.\,$ Prop

***110·501.** $\vdash . \beta + \alpha \,\mathrm{sm}\, \alpha + \beta$ [*110·5 . *73·4]

***110·51.** $\vdash . \mu +_c \nu = \nu +_c \mu$ [*110·2·501 . *73·37]

It is not necessary to the truth of the above proposition that μ and ν should be cardinals. If either is not a cardinal, $\mu +_c \nu$ and $\nu +_c \mu$ are both Λ.

The following propositions lead to the associative law (*110·56).

***110·52.** $\vdash : \xi \,\mathrm{sm}\, (\alpha + \beta) + \gamma . \equiv . (\exists \pi, \rho, \sigma) . \pi \,\mathrm{sm}\, \alpha . \rho \,\mathrm{sm}\, \beta . \sigma \,\mathrm{sm}\, \gamma .$
$\pi \cap \rho = \Lambda . \pi \cap \sigma = \Lambda . \rho \cap \sigma = \Lambda . \xi = \pi \cup \rho \cup \sigma$

Dem.

$\vdash . $ *110·152 $. \supset \vdash :. \xi \,\mathrm{sm}\, (\alpha + \beta) + \gamma . \equiv : (\exists \eta, \sigma) . \eta \,\mathrm{sm}\, (\alpha + \beta) .$
$\sigma \,\mathrm{sm}\, \gamma . \eta \cap \sigma = \Lambda . \xi = \eta \cup \sigma :$

[*110·152] $\equiv : (\exists \pi, \rho, \eta, \sigma) . \pi \,\mathrm{sm}\, \alpha . \rho \,\mathrm{sm}\, \beta . \pi \cap \rho = \Lambda . \eta = \pi \cup \rho .$
$\sigma \,\mathrm{sm}\, \gamma . \eta \cap \sigma = \Lambda . \xi = \eta \cup \sigma :$

[*13·195.*22·68.*24·32] $\equiv : (\exists \pi, \rho, \sigma) . \pi \,\mathrm{sm}\, \alpha . \rho \,\mathrm{sm}\, \beta . \sigma \,\mathrm{sm}\, \gamma . \pi \cap \rho = \Lambda .$
$\pi \cap \sigma = \Lambda . \rho \cap \sigma = \Lambda . \xi = \pi \cup \rho \cup \sigma :. \supset \vdash . \mathrm{Prop}$

***110·521.** $\vdash : \xi \,\mathrm{sm}\, \alpha + (\beta + \gamma) . \equiv . (\exists \pi, \rho, \sigma) . \pi \,\mathrm{sm}\, \alpha . \rho \,\mathrm{sm}\, \beta . \sigma \,\mathrm{sm}\, \gamma .$
$\pi \cap \rho = \Lambda . \pi \cap \sigma = \Lambda . \rho \cap \sigma = \Lambda . \xi = \pi \cup \rho \cup \sigma$ [*110·501·52]

***110·53.** $\vdash . (\alpha + \beta) + \gamma \,\mathrm{sm}\, \alpha + (\beta + \gamma)$ [*110·52·521]

***110·531.** $\alpha + \beta + \gamma = (\alpha + \beta) + \gamma$ Df

***110·54.** $\vdash . (\mathrm{Nc}\text{‘}\alpha +_c \mathrm{Nc}\text{‘}\beta) +_c \mathrm{Nc}\text{‘}\gamma = \mathrm{Nc}\text{‘}(\alpha + \beta + \gamma)$

Dem

$\vdash . $ *110·3 $. \supset \vdash . (\mathrm{Nc}\text{‘}\alpha +_c \mathrm{Nc}\text{‘}\beta) +_c \mathrm{Nc}\text{‘}\gamma = \mathrm{Nc}\text{‘}(\alpha + \beta) +_c \mathrm{Nc}\text{‘}\gamma$

[*110·3.(*110·531)] $= \mathrm{Nc}\text{‘}(\alpha + \beta + \gamma) . \supset \vdash . \mathrm{Prop}$

***110·541.** $\vdash . \mathrm{Nc}\text{‘}\alpha +_c (\mathrm{Nc}\text{‘}\beta +_c \mathrm{Nc}\text{‘}\gamma) = \mathrm{Nc}\text{‘}(\alpha + \beta + \gamma)$

Dem.

$\vdash . $ *110·3 $. \supset \vdash . \mathrm{Nc}\text{‘}\alpha +_c (\mathrm{Nc}\text{‘}\beta +_c \mathrm{Nc}\text{‘}\gamma) = \mathrm{Nc}\text{‘}\{\alpha + (\beta + \gamma)\}$

[*110·53.(*110·531)] $= \mathrm{Nc}\text{‘}(\alpha + \beta + \gamma) . \supset \vdash . \mathrm{Prop}$

***110·55.** $\vdash . (\mathrm{Nc}\text{‘}\alpha +_c \mathrm{Nc}\text{‘}\beta) +_c \mathrm{Nc}\text{‘}\gamma = \mathrm{Nc}\text{‘}\alpha +_c (\mathrm{Nc}\text{‘}\beta +_c \mathrm{Nc}\text{‘}\gamma)$ [*110·54·541]

***110·551.** $\vdash . (\mathrm{N_0c}\text{‘}\alpha +_c \mathrm{N_0c}\text{‘}\beta) +_c \mathrm{N_0c}\text{‘}\gamma = \mathrm{N_0c}\text{‘}\alpha +_c (\mathrm{N_0c}\text{‘}\beta +_c \mathrm{N_0c}\text{‘}\gamma)$
[*110·55 . (*110·03·04)]

***110·56.** $\vdash . (\mu +_c \nu) +_c \varpi = \mu +_c (\nu +_c \varpi)$

Dem.

$\vdash . $ *110·551 . *103·2 $. \supset$

$\vdash : \mu, \nu, \varpi \in \mathrm{N_0C} . \supset . (\mu +_c \nu) +_c \varpi = \mu +_c (\nu +_c \varpi)$ (1)

$\vdash . $ *110·4 . Transp $. \supset$

$\vdash : \sim(\mu, \nu, \varpi \in \mathrm{N_0C}) . \supset . (\mu +_c \nu) +_c \varpi = \Lambda . \mu +_c (\nu +_c \varpi) = \Lambda .$

[*13·171] $\supset . (\mu +_c \nu) +_c \varpi = \mu +_c (\nu +_c \varpi)$ (2)

$\vdash . (1) . (2) . \supset \vdash . \mathrm{Prop}$

This is the associative law for arithmetical addition. It will be seen that, like the commutative law, it does not require that μ, ν, ϖ should be cardinals.

***110·561.** $\mu +_c \nu +_c \varpi = (\mu +_c \nu) +_c \varpi$ Df

***110·57.** $\vdash . (\mu +_c \nu) +_c (\varpi +_c \rho) = \mu +_c \nu +_c \varpi +_c \rho$ [*110·56 . (*110·561)]

The following propositions, concerning the addition of 0 or 1, are used frequently in dealing with inductive cardinals (*120).

***110·6.** $\vdash : \mu \epsilon \mathrm{NC} . \supset . \mu +_c 0 = \mathrm{sm}``\mu$

Dem.

$\vdash . *101{\cdot}11 . *110{\cdot}21 . \supset$

$\vdash :. \mathrm{Hp} . \supset : \xi \epsilon \mu +_c 0 . \equiv . (\exists \alpha, \beta) . \alpha \epsilon \mu . \beta \epsilon 0 . \xi \,\mathrm{sm}\, (\alpha + \beta) .$

[*54·102] $\equiv . (\exists \alpha) . \alpha \epsilon \mu . \xi \,\mathrm{sm}\, (\alpha + \Lambda) .$

[*110·152] $\equiv . (\exists \alpha, \gamma, \delta) . \alpha \epsilon \mu . \gamma \,\mathrm{sm}\, \alpha . \delta \,\mathrm{sm}\, \Lambda . \gamma \cap \delta = \Lambda . \xi = \gamma \cup \delta .$

[*73·47] $\equiv . (\exists \alpha, \gamma) . \alpha \epsilon \mu . \gamma \,\mathrm{sm}\, \alpha . \xi = \gamma .$

[*13·195] $\equiv . (\exists \alpha) . \alpha \epsilon \mu . \xi \,\mathrm{sm}\, \alpha .$

[*37·1] $\equiv . \xi \epsilon \mathrm{sm}``\mu :. \supset \vdash . \mathrm{Prop}$

When μ is a typically definite cardinal, $\mathrm{sm}``\mu$ is the same cardinal rendered typically ambiguous; when $\mu$ is a typically ambiguous cardinal, $\mathrm{sm}``\mu$ is μ. In place of the above proposition, we *might* write $\mu \epsilon \mathrm{NC} . \supset . \mu +_c 0 = \mu$; this would be true whenever the ambiguity of $\mu +_c 0$ was so determined as to make it significant. But the above form gives more information.

***110·61.** $\vdash . \mathrm{Nc}`\alpha +_c 0 = \mathrm{Nc}`\alpha$

Dem.

$\vdash . *101{\cdot}1 . \supset \vdash . \mathrm{Nc}`\alpha +_c 0 = \mathrm{Nc}`\alpha +_c \mathrm{Nc}`\Lambda$

[*110·32] $= \mathrm{Nc}`(\alpha \cup \Lambda)$

[*24·24] $= \mathrm{Nc}`\alpha . \supset \vdash . \mathrm{Prop}$

In this proposition, $\mathrm{Nc}`\alpha$ is typically ambiguous; hence we escape the necessity of putting $\mathrm{sm}``\mathrm{Nc}`\alpha$ on the right, as we should have to do if $\mathrm{Nc}`\alpha$ were typically definite. We can deduce *110·61 from *110·6 as follows:

$\vdash . *110{\cdot}3 . \supset \vdash . \mathrm{Nc}`\alpha +_c 0 = \mathrm{N_0c}`\alpha +_c 0$

[*110·6] $= \mathrm{sm}``\mathrm{N_0c}`\alpha$

[*103·4] $= \mathrm{Nc}`\alpha$

We have to travel via $\mathrm{N_0c}`\alpha$ in this proof, in order to avoid the possibility of a typical determination of $\mathrm{Nc}`\alpha$ which would make $\mathrm{Nc}`\alpha = \Lambda$. It is for the same reason that we cannot put "$\mathrm{sm}``\mathrm{Nc}`\alpha = \mathrm{Nc}`\alpha$"; for if the first $\mathrm{Nc}`\alpha$ is determined to a type in which $\mathrm{Nc}`\alpha = \Lambda$, while the second is not, this equation becomes false.

∗110·62. $\vdash : \mu +_{c} \nu = 0 \,.\equiv.\, \mu = 0 \,.\, \nu = 0$

Dem.

$\vdash . *103{\cdot}27 . *101{\cdot}11{\cdot}13 . \supset \vdash . 0 = N_0c`\Lambda$ (1)

$\vdash . (1) . *110{\cdot}43 . \supset$

$\vdash :. \mu +_c \nu = 0 .\equiv: \Lambda \in \mu +_c \nu :$

[∗110·202] $\equiv : \exists ! \mu . \exists ! \nu : (\exists \gamma, \delta) . \mu = Nc`\gamma . \nu = Nc`\delta . \gamma \cap \delta = \Lambda . \gamma \cup \delta = \Lambda :$

[∗24·32.∗13·22] $\equiv : \exists ! \mu . \exists ! \nu : \mu = Nc`\Lambda . \nu = Nc`\Lambda :$

[∗101·1·12] $\equiv : \mu = 0 . \nu = 0 :. \supset \vdash .$ Prop

∗110·63. $\vdash . Nc`\alpha +_c 1 = \hat{\xi}\{(\exists \gamma, y) . \gamma \text{ sm } \alpha . y \sim\in \gamma . \xi = \gamma \cup \iota`y\}$

Dem.

$\vdash . *101{\cdot}2 . \supset$

$\vdash . Nc`\alpha +_c 1 = Nc`\alpha +_c Nc`\iota`x$

[∗110·33] $= \hat{\xi}\{(\exists \gamma, \delta) . \gamma \text{ sm } \alpha . \delta \text{ sm } \iota`x . \gamma \cap \delta = \Lambda . \xi = \gamma \cup \delta\}$

[∗73·45] $= \hat{\xi}\{(\exists \gamma, \delta) . \gamma \text{ sm } \alpha . \delta \in 1 . \gamma \cap \delta = \Lambda . \xi = \gamma \cup \delta\}$

[∗52·1] $= \hat{\xi}\{(\exists \gamma, \delta, y) . \gamma \text{ sm } \alpha . \delta = \iota`y . \gamma \cap \delta = \Lambda . \xi = \gamma \cup \delta\}$

[∗13·195.∗51·211] $= \hat{\xi}\{(\exists \gamma, y) . \gamma \text{ sm } \alpha . y \sim\in \gamma . \xi = \gamma \cup \iota`y\} . \supset \vdash .$ Prop

The above proposition is much used in the theory of finite and infinite, both cardinal and ordinal. It connects mathematical induction for inductive cardinals with mathematical induction for inductive classes (cf. ∗120).

∗110·631. $\vdash : \mu \in NC . \supset . \mu +_c 1 = \hat{\xi}\{(\exists \gamma, y) . \gamma \in \text{sm}``\mu . y \sim\in \gamma . \xi = \gamma \cup \iota`y\}$

Dem.

$\vdash . *110{\cdot}211 . *101{\cdot}21 . \supset$

$\vdash : Hp . \supset . \mu +_c 1 = \hat{\xi}\{(\exists \gamma, \delta) . \gamma \in \text{sm}``\mu . \delta \in \text{sm}``1 . \gamma \cap \delta = \Lambda . \xi = \gamma \cup \delta\}$

[∗101·28] $= \hat{\xi}\{(\exists \gamma, \delta) . \gamma \in \text{sm}``\mu . \delta \in 1 . \gamma \cap \delta = \Lambda . \xi = \gamma \cup \delta\}$

[∗52·1.∗51·211] $= \hat{\xi}\{(\exists \gamma, y) . \gamma \in \text{sm}``\mu . y \sim\in \gamma . \xi = \gamma \cup \iota`y\} : \supset \vdash .$ Prop

The proposition

$$\mu \in NC . \supset . \mu +_c 1 = \hat{\xi}\{(\exists \gamma, y) . \gamma \in \mu . y \sim\in \gamma . \xi \text{ sm } \gamma \cup \iota`y\}$$

which might at first sight seem demonstrable, will only be true universally if the total number of objects in any one type is not finite. For suppose α is a type, and $\mu = N_0c`\alpha$. Then if $\alpha$ is a finite class, $\mu = \iota`\alpha$. Hence $\gamma \in \mu . \supset_{\gamma, y} . y \in \gamma$. Hence $\hat{\xi}\{(\exists \gamma, y) . \gamma \in \mu . y \sim\in \gamma . \xi \text{ sm } (\gamma \cup \iota`y)\} = \Lambda$ in all types. But $\mu +_c 1$ will exist in all types higher than that of γ. If on the other hand the number of entities in α is infinite, we shall have

$$y \in \alpha . \supset . \alpha - \iota`y \in Nc`\alpha . y \sim\in \alpha - \iota`y.$$

Hence in this case the above proposition will be true universally.

***110·632.** $\vdash : \mu \epsilon \text{NC} . \supset . \mu +_c 1 = \hat{\xi} \{(\exists y) . y \epsilon \xi . \xi - \iota ' y \epsilon \text{sm} `` \mu\}$

Dem.

$\vdash . *110·631 . *51·211·22 . \supset$

$\vdash : \text{Hp} . \supset . \mu +_c 1 = \hat{\xi} \{(\exists \gamma, y) . \gamma \epsilon \text{sm} `` \mu . y \epsilon \xi . \gamma = \xi - \iota ' y\}$

[*13·195] $= \hat{\xi} \{(\exists y) . y \epsilon \xi . \xi - \iota ' y \epsilon \text{sm} `` \mu\} : \supset \vdash . \text{Prop}$

***110·64.** $\vdash . 0 +_c 0 = 0$ [*110·62]

***110·641.** $\vdash . 1 +_c 0 = 0 +_c 1 = 1$ [*110·51·61 . *101·2]

***110·642.** $\vdash . 2 +_c 0 = 0 +_c 2 = 2$ [*110·51·61 . *101·31]

***110·643.** $\vdash . 1 +_c 1 = 2$

Dem.

$\vdash . *110·632 . *101·21·28 . \supset$

$\vdash . 1 +_c 1 = \hat{\xi}\{(\exists y) . y \epsilon \xi . \xi - \iota ' y \epsilon 1\}$

[*54·3] $= 2 . \supset \vdash . \text{Prop}$

The above proposition is occasionally useful. It is used at least three times, in *113·66 and *120·123·472.

*110·7·71 are required for proving *110·72, and *110·72 is used in *117·3, which is a fundamental proposition in the theory of greater and less.

***110·7.** $\vdash : \beta \subset \alpha . \supset . (\exists \mu) . \mu \epsilon \text{NC} . \text{Nc}'\alpha = \text{Nc}'\beta +_c \mu$

Dem.

$\vdash . *24·411·21 . \supset \vdash : \text{Hp} . \supset . \alpha = \beta \cup (\alpha - \beta) . \beta \cap (\alpha - \beta) = \Lambda .$

[*110·32] $\supset . \text{Nc}'\alpha = \text{Nc}'\beta +_c \text{Nc}'(\alpha - \beta) : \supset \vdash . \text{Prop}$

***110·71.** $\vdash : (\exists \mu) . \text{Nc}'\alpha = \text{Nc}'\beta +_c \mu . \supset . (\exists \delta) . \delta \,\text{sm}\, \beta . \delta \subset \alpha$

Dem.

$\vdash . *100·3 . *110·4 . \supset$

$\vdash : \text{Nc}'\alpha = \text{Nc}'\beta +_c \mu . \supset . \mu \epsilon \text{NC} - \iota ' \Lambda$ (1)

$\vdash . *110·3 . \supset \vdash : \text{Nc}'\alpha = \text{Nc}'\beta +_c \text{Nc}'\gamma . \equiv . \text{Nc}'\alpha = \text{Nc}'(\beta + \gamma) .$

[*100·3·31] $\supset . \alpha \,\text{sm}\, (\beta + \gamma) .$

[*73·1] $\supset . (\exists R) . R \epsilon 1 \rightarrow 1 . \text{D}'R = \alpha . \text{G}'R = \downarrow \Lambda_\gamma `` \iota `` \beta \cup \Lambda_\beta \downarrow `` \iota `` \gamma .$

[*37·15] $\supset . (\exists R) . R \epsilon 1 \rightarrow 1 . \downarrow \Lambda_\gamma `` \iota `` \beta \subset \text{G}'R . R `` \downarrow \Lambda_\gamma `` \iota `` \beta \subset \alpha .$

[*110·12.*73·22] $\supset . (\exists \delta) . \delta \subset \alpha . \delta \,\text{sm}\, \beta$ (2)

$\vdash . (1) . (2) . \supset \vdash . \text{Prop}$

The above proof depends upon the fact that "$\text{Nc}‘\alpha$" and "$\text{Nc}‘\beta +_c \mu$" are typically ambiguous, and therefore, when they are asserted to be equal, this must hold in *any* type, and therefore, in particular, in that type for which we have $\alpha \epsilon \text{Nc}‘\alpha$, *i.e.* for $\text{N}_0\text{c}‘\alpha$. This is why the use of ∗100·3 is legitimate.

∗110·72. $\vdash : (\exists \delta) . \delta \text{ sm } \beta . \delta \subset \alpha . \equiv . (\exists \mu) . \mu \epsilon \text{NC} . \text{Nc}‘\alpha = \text{Nc}‘\beta +_c \mu$

Dem.

$\vdash . *100·321 . *110·7 . \supset$

$\vdash :. \delta \text{ sm } \beta . \delta \subset \alpha . \supset : \text{Nc}‘\delta = \text{Nc}‘\beta : (\exists \mu) . \mu \epsilon \text{NC} . \text{Nc}‘\alpha = \text{Nc}‘\delta +_c \mu :$

[∗13·12] $\supset : (\exists \mu) . \mu \epsilon \text{NC} . \text{Nc}‘\alpha = \text{Nc}‘\beta +_c \mu$ (1)

$\vdash . (1) . *110·71 . \supset \vdash . \text{Prop}$

∗111. DOUBLE SIMILARITY.

Summary of ∗111.

The arithmetical properties of a class, so far as these do not require or assume that it is a class of classes, are the same for any similar class. But a class of classes has many arithmetical properties which it does not share with all similar classes of classes. For example, if κ is a class of classes, the number of members of $s'\kappa$ is an arithmetical property of κ, but it is obvious that this is not determined by the number of members of κ, but requires also a knowledge of the numbers of members of members of κ. For example, let κ consist of the two members α and β, and let λ consist of γ and δ. Then $\kappa \operatorname{sm} \lambda$; but in order to be able to infer $s'\kappa \operatorname{sm} s'\lambda$, we require $\kappa, \lambda \in \mathrm{Cls}^2\,\mathrm{excl}$ and $\alpha \operatorname{sm} \gamma \,.\, \beta \operatorname{sm} \delta$ or $\alpha \operatorname{sm} \delta \,.\, \beta \operatorname{sm} \gamma$ or some such further datum. The relation of "double similarity," to be defined in the present number, is a relation between classes of classes, which, when it holds between κ and λ, insures that all the arithmetical properties of κ and λ are the same, *e.g.* we have (in particular) $\mathrm{Nc}'s'\kappa = \mathrm{Nc}'s'\lambda$ and $\mathrm{Nc}'\epsilon_\Delta{}'\kappa = \mathrm{Nc}'\epsilon_\Delta{}'\lambda$. This relation we denote by "sm sm," which is to be read as one symbol. It is defined as follows: We define first the class of "double correlators" of κ and λ, which we denote by "$\kappa \,\overline{\mathrm{sm}}\,\overline{\mathrm{sm}}\, \lambda$," and of which the definition is

∗111·01. $\kappa \,\overline{\mathrm{sm}}\,\overline{\mathrm{sm}}\, \lambda = (1 \rightarrow 1) \cap \overleftarrow{\text{Ɑ}}'s'\lambda \cap \hat{T}(\kappa = T_\epsilon``\lambda)$ Df

so that

$$\vdash : T \in \kappa \,\overline{\mathrm{sm}}\,\overline{\mathrm{sm}}\, \lambda \,.\equiv .\, T \in 1 \rightarrow 1 \,.\, \text{Ɑ}'T = s'\lambda \,.\, \kappa = T_\epsilon``\lambda .$$

We then define "$\kappa \operatorname{sm} \operatorname{sm} \lambda$" as meaning that $\kappa \,\overline{\mathrm{sm}}\,\overline{\mathrm{sm}}\, \lambda$ is not null, *i.e.* that there is at least one double correlator of κ and λ.

To illustrate the nature of a double correlator, let us suppose that κ consists of the two classes α_1 and α_2, and that α_1 consists of x_{11}, x_{12}, while α_2 consists of x_{21}, x_{22}, x_{23}. Similarly let λ consist of β_1 and β_2, while β_1 consists of y_{11}, y_{12} and β_2 consists of y_{21}, y_{22}, y_{23}. Now let T correlate each x with the y having the same two suffixes. Then T is a one-one, and its converse domain is $s'\lambda$. Moreover $T_\epsilon{}'\beta_1$ (which is $T``\beta_1) = \alpha_1$, and $T_\epsilon{}'\beta_2 = \alpha_2$, so that $T_\epsilon``\lambda = \kappa$. Thus T is a double correlator according to the definition.

The essential characteristic of a double correlator T is that (1) T is a correlator of $s'\kappa$ and $s'\lambda$, (2) $T_\epsilon \upharpoonright \lambda$ is a correlator of κ and λ. If we write S

in place of $T_\epsilon \restriction \lambda$, then if $\beta \epsilon \lambda$, we have $S`\beta \epsilon \kappa$; moreover $T \restriction \beta$ is a correlator of $S`\beta$ and β. Thus κ and λ are similar classes of similar classes. They are not merely this, however, for we not only know that $S`\beta$ is similar to $\beta$, but we know a particular correlator of $S`\beta$ and β, namely $T \restriction \beta$. This is essential to the use of double similarity, as will appear shortly.

Let us consider the relation between κ and λ which consists in their being similar classes of similar classes. This means that there is a correlator S of κ and λ, such that, if $\beta \epsilon \lambda$, $S`\beta$ is similar to β. That is to say, we are to consider the hypothesis

$$(\exists S) . S \epsilon 1 \rightarrow 1 . \mathrm{D}`S = \kappa . \mathrm{Cl}`S = \lambda . S \subset \mathrm{sm}$$

or, as it may be more briefly expressed,

$$\exists ! \kappa \,\overline{\mathrm{sm}}\, \lambda \cap \mathrm{Rl}`\mathrm{sm}.$$

Let us assume $S \epsilon \kappa \,\overline{\mathrm{sm}}\, \lambda \cap \mathrm{Rl}`\mathrm{sm}$. If we attempt to prove (say) that $s`\kappa$ is similar to $s`\lambda$, we find that we are forced to assume the multiplicative axiom, unless κ and λ are finite. This necessity arises as follows. Let us put

$$\mathrm{Crp}\,(S)`\beta = (S`\beta) \,\overline{\mathrm{sm}}\, \beta,$$

where "Crp" stands for "correspondence." Then we know that whenever $\beta \epsilon \lambda$, $\mathrm{Crp}\,(S)`\beta$ is not null. Further it is easy to prove that, if $\kappa$ and $\lambda$ are classes of mutually exclusive classes, and if we can pick out one representative member of $\mathrm{Crp}\,(S)`\beta$ for each value of β which is a member of λ, then the relational sum of all these representative correlations gives us a correlator of $s`\kappa$ and $s`\lambda$. That is, we have

$$\vdash : \kappa, \lambda \epsilon \mathrm{Cls}^2 \,\mathrm{excl} . S \epsilon \kappa \,\overline{\mathrm{sm}}\, \lambda \cap \mathrm{Rl}`\mathrm{sm} . R \epsilon \epsilon_\Delta`\mathrm{Crp}\,(S)``\lambda . \supset . \dot{s}`\mathrm{D}`R \epsilon (s`\kappa) \,\overline{\mathrm{sm}}\, (s`\lambda).$$

But in order to infer hence $s`\kappa \;\mathrm{sm}\; s`\lambda$, we need $\exists ! \epsilon_\Delta`\mathrm{Crp}\,(S)``\lambda$, *i.e.* we need to be able to pick out a particular correlator for each pair of similar classes $S`\beta$ and $\beta$. This, however, cannot be done in general without assuming the multiplicative axiom. It follows that we must not define two classes as having double similarity when $\exists ! \kappa \,\overline{\mathrm{sm}}\, \lambda \cap \mathrm{Rl}`\mathrm{sm}$, but must give a definition which enables us to specify a particular correlator for each pair of similar classes. This is what is effected by the above definition of double correlators, where our S is given as of the form $T_\epsilon \restriction \lambda$, where $T \epsilon 1 \rightarrow 1 . \mathrm{Cl}`T = s`\lambda$. If the multiplicative axiom is assumed, but in general not otherwise, we have (∗111·5)

$$\kappa, \lambda \epsilon \mathrm{Cls}^2 \,\mathrm{excl} . \supset : \kappa \;\mathrm{sm}\;\mathrm{sm}\; \lambda . \equiv . \exists ! \kappa \,\overline{\mathrm{sm}}\, \lambda \cap \mathrm{Rl}`\mathrm{sm}.$$

In the present number, we shall begin with various properties of double correlators. We prove (∗111·11) that T is a double correlator of κ and λ when, and only when, T is a correlator of $s`\kappa$ and $s`\lambda$, and $T_\epsilon \restriction \lambda$ is a correlator of κ and λ. We prove (∗111·112) that in the same hypothesis, $T_\epsilon \restriction \lambda \epsilon \kappa \,\overline{\mathrm{sm}}\, \lambda \cap \mathrm{Rl}`\mathrm{sm}$. We prove (∗111·13) that $I \restriction s`\lambda$ is a double correlator of λ with itself; that (∗111·131) if T is a double correlator of

κ and λ, $\breve{T}$ is a double correlator of λ and κ; that (*111·132) if S, T are double correlators of κ with λ and of λ with μ respectively, $S|T$ is a double correlator of κ with μ. Hence it follows (*111·45·451·452) that double similarity is reflexive, symmetrical, and transitive.

We then proceed (*111·2—·34) to consider $\mathrm{Crp}(S)``\lambda$, where it is to be supposed that $S$ is a correlator of $S``\lambda$ and λ, and that $S`\beta$ is similar to β if $\beta \in \lambda$. We prove

***111·32.** $\vdash : \lambda, S``\lambda \in \mathrm{Cls}^2\,\mathrm{excl} \,.\, S \in 1 \to 1 \,.\, R \in \epsilon_\Delta`\mathrm{Crp}(S)``\lambda \,.\, M = \dot{s}`\mathrm{D}`R \,.\, \supset .$
$M \in 1 \to 1 \,.\, \mathrm{G}`M = s`\lambda \,.\, S``\lambda = M_\epsilon``\lambda \,.\, S \restriction \lambda = M_\epsilon \restriction \lambda$

Thus in the case supposed, M is a double correlator of $S``\lambda$ and λ. Thus

***111·322.** $\vdash : \kappa, \lambda \in \mathrm{Cls}^2\,\mathrm{excl} \,.\, S \in \kappa\, \overline{\mathrm{sm}}\, \lambda \,.\, R \in \epsilon_\Delta`\mathrm{Crp}(S)``\lambda \,.\, M = \dot{s}`\mathrm{D}`R \,.\, \supset .$
$M \in \kappa\, \overline{\mathrm{sm}}\, \overline{\mathrm{sm}}\, \lambda \,.\, S = M_\epsilon \restriction \lambda$

We then proceed (*111·4—·47) to various propositions on "sm sm," and finally (*111·5·51·53) state three propositions which assume the multiplicative axiom, namely

***111·5.** If $\kappa, \lambda \in \mathrm{Cls}^2\,\mathrm{excl}$, then $\kappa\, \mathrm{sm}\, \mathrm{sm}\, \lambda \,.\, \equiv .\, \exists ! \,\kappa\, \overline{\mathrm{sm}}\, \lambda \cap \mathrm{Rl}`\mathrm{sm}$.

***111·51.** In the same case, $\exists ! \,\kappa\, \overline{\mathrm{sm}}\, \lambda \cap \mathrm{Rl}`\mathrm{sm} \,.\, \supset .\, s`\kappa\, \mathrm{sm}\, s`\lambda$, *i.e.* if κ and λ are similar classes of mutually exclusive similar classes, their sums are similar.

***111·53.** In the same case, if $\kappa, \lambda \in \mathrm{Cls}^2\,\mathrm{excl}$, $\kappa\, \mathrm{sm}\, \mathrm{sm}\, \lambda$. Hence the multiplicative axiom implies that two mutually exclusive classes of μ classes each of which has ν terms, have the same number of terms in their sum.

***111·01.** $\kappa\, \overline{\mathrm{sm}}\, \overline{\mathrm{sm}}\, \lambda = (1 \to 1) \cap \overleftarrow{\mathrm{G}}`s`\lambda \cap \hat{T}(\kappa = T_\epsilon``\lambda)$ Df

***111·02.** $\mathrm{Crp}(S)`\beta = (S`\beta)\, \overline{\mathrm{sm}}\, \beta$ Df

***111·03.** $\mathrm{sm}\, \mathrm{sm} = \hat{\kappa}\hat{\lambda}(\exists ! \,\kappa\, \overline{\mathrm{sm}}\, \overline{\mathrm{sm}}\, \lambda)$ Df

***111·1.** $\vdash : T \in \kappa\, \overline{\mathrm{sm}}\, \overline{\mathrm{sm}}\, \lambda \,.\, \equiv .\, T \in 1 \to 1 \,.\, \mathrm{G}`T = s`\lambda \,.\, \kappa = T_\epsilon``\lambda$ [(*111·01)]

***111·11.** $\vdash : T \in \kappa\, \overline{\mathrm{sm}}\, \overline{\mathrm{sm}}\, \lambda \,.\, \equiv .\, T \in (s`\kappa)\, \overline{\mathrm{sm}}\, (s`\lambda) \,.\, T_\epsilon \restriction \lambda \in \kappa\, \overline{\mathrm{sm}}\, \lambda$

Dem.

$\vdash .\, *37·25 \,.\, \mathrm{Fact} \,.\, \supset \vdash : \mathrm{G}`T = s`\lambda \,.\, \kappa = T_\epsilon``\lambda \,.\, \supset .\, \mathrm{D}`T = T``s`\lambda \,.\, \kappa = T_\epsilon``\lambda \,.$

[*40·38] $\supset .\, \mathrm{D}`T = s`T```\lambda \,.\, \kappa = T_\epsilon``\lambda \,.$

[(*37·04)] $\supset .\, \mathrm{D}`T = s`\kappa$ (1)

$\vdash .\, *72·451 \,.\, *60·57 \,.\, *35·65 \,.\, \supset$

$\vdash : T \in 1 \to 1 \,.\, \mathrm{G}`T = s`\lambda \,.\, \supset .\, T_\epsilon \restriction \lambda \in 1 \to 1 \,.\, \lambda = \mathrm{G}`(T_\epsilon \restriction \lambda)$ (2)

$\vdash .\, *37·401 \,.\quad \supset \vdash : \kappa = T_\epsilon``\lambda \,.\, \equiv .\, \kappa = \mathrm{D}`(T_\epsilon \restriction \lambda)$ (3)

$\vdash .\, (1) \,.\, (2) \,.\, (3) \,.\, *4·71 \,.\, \supset \vdash : T \in 1 \to 1 \,.\, \mathrm{G}`T = s`\lambda \,.\, \kappa = T_\epsilon``\lambda \,.\, \equiv .$

$T \in 1 \to 1 \,.\, \mathrm{D}`T = s`\kappa \,.\, \mathrm{G}`T = s`\lambda \,.\, T_\epsilon \restriction \lambda \in 1 \to 1 \,.\, \mathrm{D}`(T_\epsilon \restriction \lambda) = \kappa \,.\, \mathrm{G}`(T_\epsilon \restriction \lambda) = \lambda$ (4)

$\vdash .\, (4) \,.\, *111·1 \,.\, *73·03 \,.\, \supset \vdash .$ Prop

***111·111.** $\vdash : T \in \kappa \,\overline{\mathrm{sm}}\,\overline{\mathrm{sm}}\, \lambda \,.\supset .\, T_\epsilon \restriction \lambda \in \mathrm{sm}$

Dem.

$\vdash . *111·1 . *60·57 . \supset \vdash : \mathrm{Hp} . \supset . T \in 1 \rightarrow 1 . \lambda \subset \mathrm{Cl}`\text{Ɑ}`T .$

$[*73·5] \quad \supset . T_\epsilon \restriction \lambda \in \mathrm{sm} : \supset \vdash . \mathrm{Prop}$

***111·112.** $\vdash : T \in \kappa \,\overline{\mathrm{sm}}\,\overline{\mathrm{sm}}\, \lambda \,.\supset .\, T_\epsilon \restriction \lambda \in \kappa \,\overline{\mathrm{sm}}\, \lambda \cap \mathrm{Rl}`\mathrm{sm} \quad [*111·11·111]$

The two following propositions are useful lemmas for the case when T is replaced (as it often is) by $T \restriction \alpha$.

***111·12.** $\vdash : s`\lambda \subset \alpha . \supset . (T \restriction \alpha)_\epsilon``\lambda = T_\epsilon``\lambda . (T \restriction \alpha)_\epsilon \restriction \lambda = T_\epsilon \restriction \lambda$

Dem.

$\vdash . *37·101·421 . \supset \vdash : \beta \subset \alpha . \supset . (T \restriction \alpha)_\epsilon`\beta = T_\epsilon`\beta \quad (1)$

$\vdash . *40·13 . \quad \supset \vdash :. \mathrm{Hp} . \supset : \beta \in \lambda . \supset . \beta \subset \alpha \quad (2)$

$\vdash . (1) . (2) . \quad \supset \vdash :. \mathrm{Hp} . \supset : \beta \in \lambda . \supset . (T \restriction \alpha)_\epsilon`\beta = T_\epsilon`\beta :$

$[*37·69 . *35·71] \quad \supset : (T \restriction \alpha)_\epsilon``\lambda = T_\epsilon``\lambda . (T \restriction \alpha)_\epsilon \restriction \lambda = T_\epsilon \restriction \lambda :. \supset \vdash . \mathrm{Prop}$

***111·121.** $\vdash . (T \restriction s`\lambda)_\epsilon``\lambda = T_\epsilon``\lambda = (T_\epsilon \restriction \lambda)``\lambda . (T \restriction s`\lambda)_\epsilon \restriction \lambda = T_\epsilon \restriction \lambda$

Dem.

$\vdash . *37·421 . \supset \vdash . T_\epsilon``\lambda = (T_\epsilon \restriction \lambda)``\lambda \quad (1)$

$\vdash . (1) . *111·12 \frac{s`\lambda}{\alpha} . \supset \vdash . \mathrm{Prop}$

***111·13.** $\vdash . I \restriction s`\lambda \in \lambda \,\overline{\mathrm{sm}}\,\overline{\mathrm{sm}}\, \lambda$

Dem.

$\vdash . *72·17 . *50·5·52 . \supset \vdash . I \restriction s`\lambda \in 1 \rightarrow 1 . \text{Ɑ}`(I \restriction s`\lambda) = s`\lambda \quad (1)$

$\vdash . *111·121 . \quad \supset \vdash . (I \restriction s`\lambda)_\epsilon``\lambda = I_\epsilon``\lambda$

$[*50·16·17] \quad = \lambda \quad (2)$

$\vdash . (1) . (2) . *111·1 . \supset \vdash . \mathrm{Prop}$

***111·131.** $\vdash : T \in \kappa \,\overline{\mathrm{sm}}\,\overline{\mathrm{sm}}\, \lambda . \equiv . \breve{T} \in \lambda \,\overline{\mathrm{sm}}\,\overline{\mathrm{sm}}\, \kappa$

Dem.

$\vdash . *71·212 . \supset \vdash : T \in 1 \rightarrow 1 . \equiv . \breve{T} \in 1 \rightarrow 1 \quad (1)$

$\vdash . *111·11 . \supset \vdash : T \in \kappa \,\overline{\mathrm{sm}}\,\overline{\mathrm{sm}}\, \lambda . \supset . \mathrm{D}`T = s`\kappa \quad (2)$

$\vdash . *111·1 . (2) . *60·57 . \supset$

$\vdash : T \in \kappa \,\overline{\mathrm{sm}}\,\overline{\mathrm{sm}}\, \lambda . \supset . T \in 1 \rightarrow 1 . \kappa \subset \mathrm{Cl}`\mathrm{D}`T . \lambda \subset \mathrm{Cl}`\text{Ɑ}`T . \kappa = T_\epsilon``\lambda .$

$[*74·6] \quad \supset . \lambda = (\breve{T})_\epsilon``\kappa \quad (3)$

$\vdash . (1) . (2) . (3) . *111·1 . \supset \vdash : T \in \kappa \,\overline{\mathrm{sm}}\,\overline{\mathrm{sm}}\, \lambda . \supset . \breve{T} \in \lambda \,\overline{\mathrm{sm}}\,\overline{\mathrm{sm}}\, \kappa \quad (4)$

$\vdash . (4) \frac{\breve{T}}{T} . \quad \supset \vdash : \breve{T} \in \lambda \,\overline{\mathrm{sm}}\,\overline{\mathrm{sm}}\, \kappa . \supset . T \in \kappa \,\overline{\mathrm{sm}}\,\overline{\mathrm{sm}}\, \lambda \quad (5)$

$\vdash . (4) . (5) . \supset \vdash . \mathrm{Prop}$

***111·132.** $\vdash : S \epsilon \kappa \,\overline{\text{sm}}\, \overline{\text{sm}}\, \lambda \,.\, T \epsilon \lambda \,\overline{\text{sm}}\, \overline{\text{sm}}\, \mu \,.\, \supset .\, S \,|\, T \epsilon \kappa \,\overline{\text{sm}}\, \overline{\text{sm}}\, \mu$

Dem.

$\vdash . *111{\cdot}11 . *73{\cdot}311 . \supset$

$\vdash : \text{Hp} . \supset . S \,|\, T \epsilon (s{`}\kappa) \,\overline{\text{sm}}\, (s{`}\mu) . (S_\epsilon \upharpoonright \lambda) \,|\, (T_\epsilon \upharpoonright \mu) \epsilon \kappa \,\overline{\text{sm}}\, \mu \quad (1)$

$\vdash . *35{\cdot}354 . \supset \vdash . (S_\epsilon \upharpoonright \lambda) \,|\, (T_\epsilon \upharpoonright \mu) = S_\epsilon \,|\, (\lambda \rceil T_\epsilon \upharpoonright \mu) \quad (2)$

$\vdash . *74{\cdot}251 . *111{\cdot}1 . \supset \vdash : \text{Hp} . \supset . S_\epsilon \,|\, (\lambda \rceil T_\epsilon \upharpoonright \mu) = S_\epsilon \,|\, (T_\epsilon \upharpoonright \mu)$

$[*35{\cdot}23] \qquad = (S_\epsilon \,|\, T_\epsilon) \upharpoonright \mu$

$[*37{\cdot}34] \qquad = (S \,|\, T)_\epsilon \upharpoonright \mu \quad (3)$

$\vdash . (1) . (2) . (3) . \supset \vdash : \text{Hp} . \supset . S \,|\, T \epsilon (s{`}\kappa) \,\overline{\text{sm}}\, (s{`}\mu) . (S \,|\, T)_\epsilon \upharpoonright \mu \epsilon \kappa \,\overline{\text{sm}}\, \mu .$

$[*111{\cdot}11] \qquad \supset . S \,|\, T \epsilon \kappa \,\overline{\text{sm}}\, \overline{\text{sm}}\, \mu : \supset \vdash . \text{Prop}$

***111·14.** $\vdash : T \upharpoonright s{`}\lambda \epsilon \kappa \,\overline{\text{sm}}\, \overline{\text{sm}}\, \lambda . \equiv . T \upharpoonright s{`}\lambda \epsilon 1 \to 1 . s{`}\lambda \subset \text{Cl}{`}T . \kappa = T_\epsilon{``}\lambda$

Dem.

$\vdash . *111{\cdot}1{\cdot}121 . \supset$

$\vdash : T \upharpoonright s{`}\lambda \epsilon \kappa \,\overline{\text{sm}}\, \overline{\text{sm}}\, \lambda . \equiv . T \upharpoonright s{`}\lambda \epsilon 1 \to 1 . \text{Cl}{`}(T \upharpoonright s{`}\lambda) = s{`}\lambda . \kappa = T_\epsilon{``}\lambda .$

$[*35{\cdot}65] \qquad \equiv . T \upharpoonright s{`}\lambda \epsilon 1 \to 1 . s{`}\lambda \subset \text{Cl}{`}T . \kappa = T_\epsilon{``}\lambda : \supset \vdash . \text{Prop}$

***111·15.** $\vdash : T \upharpoonright s{`}\lambda \epsilon \kappa \,\overline{\text{sm}}\, \overline{\text{sm}}\, \lambda . \equiv . T \upharpoonright s{`}\lambda \epsilon (s{`}\kappa) \,\overline{\text{sm}}\, (s{`}\lambda) . T_\epsilon \upharpoonright \lambda \epsilon \kappa \,\overline{\text{sm}}\, \lambda$

Dem.

$\vdash . *111{\cdot}11 . \supset$

$\vdash : T \upharpoonright s{`}\lambda \epsilon \kappa \,\overline{\text{sm}}\, \overline{\text{sm}}\, \lambda . \equiv . T \upharpoonright s{`}\lambda \epsilon (s{`}\kappa) \,\overline{\text{sm}}\, (s{`}\lambda) . (T \upharpoonright s{`}\lambda)_\epsilon \upharpoonright \lambda \epsilon \kappa \,\overline{\text{sm}}\, \lambda \quad (1)$

$\vdash . (1) . *111{\cdot}121 . \supset \vdash . \text{Prop}$

***111·16.** $\vdash : \exists ! \alpha \,\overline{\text{sm}}\, \beta \cap \gamma \,\overline{\text{sm}}\, \delta . \supset . \alpha = \gamma . \beta = \delta$

Dem.

$\vdash . *73{\cdot}03 . \supset \vdash : \text{Hp} . \supset . (\exists R) . \text{D}{`}R = \alpha . \text{Cl}{`}R = \beta . \text{D}{`}R = \gamma . \text{Cl}{`}R = \delta .$

$[*13{\cdot}171] \qquad \supset . \alpha = \gamma . \beta = \delta : \supset \vdash . \text{Prop}$

***111·18.** $\vdash . \alpha \,\overline{\text{sm}}\, \beta \subset (\alpha \uparrow \beta)_\Delta{`}\beta$

Dem.

$\vdash . *35{\cdot}83 . *73{\cdot}03 . \supset \vdash : R \epsilon \alpha \,\overline{\text{sm}}\, \beta . \supset . R \mathrel{\overline{\subset}} \alpha \uparrow \beta \quad (1)$

$\vdash . *73{\cdot}03 . \qquad \supset \vdash : R \epsilon \alpha \,\overline{\text{sm}}\, \beta . \supset . R \epsilon 1 \to \text{Cls} . \text{Cl}{`}R = \beta \quad (2)$

$\vdash . (1) . (2) . *80{\cdot}14 . \supset \vdash . \text{Prop}$

The class $(\alpha \uparrow \beta)_\Delta{`}\beta$ is important, being the class of Cantor's "Belegungen," used by him to define exponentiation; we have in fact

$$\text{Nc}{`}(\alpha \uparrow \beta)_\Delta{`}\beta = (\text{Nc}{`}\alpha)^{\text{Nc}{`}\beta}.$$

Thus the above proposition shows that $\text{Nc}{`}(\alpha \,\overline{\text{sm}}\, \beta)$ is less than or equal to $(\text{Nc}{`}\alpha)^{\text{Nc}{`}\beta}$; and since, whenever it is not zero, $\text{Nc}{`}\alpha = \text{Nc}{`}\beta$, it is less than or equal to

$$(\text{Nc}{`}\alpha)^{\text{Nc}{`}\alpha}.$$

The following propositions lead up to ∗111·32·33·34:

∗111·2. $\vdash : \text{E}\,!\,S`\beta \,.\supset .\, \text{Crp}\,(S)`\beta = (S`\beta)\,\overline{\text{sm}}\,\beta$ [∗14·28 . (∗111·02)]

∗111·201. $\vdash : f\{\text{Crp}\,(S)`\beta\} \,.\equiv .\, f\{(S`\beta)\,\overline{\text{sm}}\,\beta\}$ [∗4·2 . (∗111·02)]

∗111·202. $\vdash : R \,\epsilon\, \text{Crp}\,(S)`\beta \,.\equiv .\, R \,\epsilon\, 1 \rightarrow 1 \,.\, \text{D}`R = S`\beta \,.\, \text{Cl}`R = \beta$
[∗111·201 . ∗73·03]

∗111·21. $\vdash : \exists\,!\,\text{Crp}\,(S)`\beta \,.\equiv .\, S`\beta \,\text{sm}\, \beta$ [∗111·201 . ∗73·04]

∗111·211. $\vdash : \exists\,!\,\text{Crp}\,(S)`\beta \,.\supset .\, \text{E}\,!\,S`\beta \,.\, \beta \,\epsilon\, \text{Cl}`S$ [∗111·21 . ∗14·21 . ∗33·43]

∗111·22. $\vdash :.\, \beta \,\epsilon\, \text{Cl}`S \,.\supset_\beta .\, \exists\,!\,\text{Crp}\,(S)`\beta : \equiv .\, S \,\epsilon\, 1 \rightarrow \text{Cls} \,.\, S \subset \text{sm}$

Dem.

$\vdash . ∗111·21 . \supset \vdash :.\, \beta \,\epsilon\, \text{Cl}`S \,.\supset_\beta .\, \exists\,!\,\text{Crp}\,(S)`\beta : \equiv : \beta \,\epsilon\, \text{Cl}`S \,.\supset_\beta .\, S`\beta \,\text{sm}\, \beta :$

[∗72·93] $\equiv : S \,\epsilon\, 1 \rightarrow \text{Cls} \,.\, S \subset \text{sm} :. \supset \vdash . \text{Prop}$

∗111·221. $\vdash :.\, S \,\epsilon\, 1 \rightarrow \text{Cls} \,.\, S \subset \text{sm} \,.\supset : \exists\,!\,\text{Crp}\,(S)`\beta \,.\equiv .\, \beta \,\epsilon\, \text{Cl}`S$

Dem.

$\vdash . ∗111·22 . \supset \vdash :.\, \text{Hp} \,.\supset : \beta \,\epsilon\, \text{Cl}`S \,.\supset .\, \exists\,!\,\text{Crp}\,(S)`\beta$ (1)

$\vdash . (1) . ∗111·211 . \supset \vdash . \text{Prop}$

∗111·23. $\vdash : S \,\epsilon\, 1 \rightarrow 1 \,.\, \beta \,\epsilon\, \text{Cl}`S \,.\supset .\, \text{Crp}\,(S)`\beta = \text{Cnv}``\text{Crp}\,(\breve{S})`S`\beta$

Dem.

$\vdash . ∗111·2 . ∗71·163 . \supset$

$\vdash :.\, \text{Hp} \,.\supset : \text{Crp}\,(S)`\beta = (S`\beta)\,\overline{\text{sm}}\,\beta$

[∗73·301] $= \text{Cnv}``(\beta\,\overline{\text{sm}}\,S`\beta)$

[∗72·241] $= \text{Cnv}``(\breve{S}`S`\beta\,\overline{\text{sm}}\,S`\beta)$ (1)

$\vdash . (1) . ∗111·201 \dfrac{\breve{S}, S`\beta}{S, \beta} . \supset \vdash . \text{Prop}$

∗111·24. $\vdash : S \,\epsilon\, 1 \rightarrow \text{Cls} \,.\, \lambda \subset \text{Cl}`S \,.\supset .\, \text{Crp}\,(S)``\lambda \,\epsilon\, \text{Cls}^2\,\text{excl}$

Dem.

$\vdash . ∗111·2 . ∗71·163 . \supset$

$\vdash :.\, \text{Hp} \,.\supset : \beta, \gamma \,\epsilon\, \lambda \,.\supset_{\beta,\gamma} .\, \text{Crp}\,(S)`\beta = (S`\beta)\,\overline{\text{sm}}\,\beta \,.\, \text{Crp}\,(S)`\gamma = (S`\gamma)\,\overline{\text{sm}}\,\gamma \,.$ (1)

[∗111·16] $\supset_{\beta,\gamma} .\, \exists\,!\,\text{Crp}(S)`\beta \cap \text{Crp}(S)`\gamma \,.\supset .\, \beta = \gamma .$

[(1).∗30·37] $\supset . \text{Crp}(S)`\beta = \text{Crp}(S)`\gamma$ (2)

$\vdash . (2) . ∗37·63 . \supset \vdash :.\, \text{Hp} \,.\supset : \rho, \sigma \,\epsilon\, \text{Crp}\,(S)``\lambda \,.\, \exists\,!\,\rho \cap \sigma \,.\supset_{\rho,\sigma} .\, \rho = \sigma :. \supset \vdash . \text{Prop}$

∗111·25. $\vdash : S \,\epsilon\, 1 \rightarrow \text{Cls} \,.\, S \subset \text{sm} \,.\, \lambda \subset \text{Cl}`S \,.\supset .\, \text{Crp}\,(S)``\lambda \,\epsilon\, \text{Cls ex}^2\,\text{excl}$
[∗111·24·22]

***111·3.** $\vdash : \lambda \in \text{Cls}^2 \text{ excl} . \supset . \dot{s}''\text{D}''\epsilon_\Delta{}'\alpha\,\overline{\text{sm}}''\lambda \subset (\alpha \uparrow s'\lambda)_\Delta{}'s'\lambda$

Dem.

$\vdash . *37{\cdot}29 . *24{\cdot}12 . \supset$

$\vdash : \epsilon_\Delta{}'\alpha\,\overline{\text{sm}}''\lambda = \Lambda . \supset . \dot{s}''\text{D}''\epsilon_\Delta{}'\alpha\,\overline{\text{sm}}''\lambda \subset (\alpha \uparrow s'\lambda)_\Delta{}'s'\lambda \quad (1)$

$\vdash . *83{\cdot}1 . \supset$

$\vdash :. \text{Hp} . \exists ! \epsilon_\Delta{}'\alpha\,\overline{\text{sm}}''\lambda . \supset : \beta \in \lambda . \supset_\beta . \exists ! \alpha\,\overline{\text{sm}}'\beta .$

[*111·18] $\supset_\beta . \exists ! (\alpha \uparrow \beta)_\Delta{}'\beta .$

[*80·15] $\supset_\beta . \exists ! (\alpha \uparrow s'\lambda)_\Delta{}'\beta :$

[*80·83] $\supset : \{(\alpha \uparrow s'\lambda)_\Delta{}''\lambda\} \upharpoonleft (\alpha \uparrow s'\lambda)_\Delta \in 1 \rightarrow 1 \quad (2)$

$\vdash . (2) . *111{\cdot}18 . *85{\cdot}72 \dfrac{\alpha\,\overline{\text{sm}},\ (\alpha \uparrow s'\lambda)_\Delta}{R,\qquad S} . \supset$

$\vdash : \text{Hp} . \exists ! \epsilon_\Delta{}'\alpha\,\overline{\text{sm}}''\lambda . \supset . \text{D}''\epsilon_\Delta{}'\alpha\,\overline{\text{sm}}''\lambda \subset \text{D}''\epsilon_\Delta{}'(\alpha \uparrow s'\lambda)_\Delta{}''\lambda .$

[*37·2] $\supset . \dot{s}''\text{D}''\epsilon_\Delta{}'\alpha\,\overline{\text{sm}}''\lambda \subset \dot{s}''\text{D}''\epsilon_\Delta{}'(\alpha \uparrow s'\lambda)_\Delta{}''\lambda$

[*85·27] $\subset (\alpha \uparrow s'\lambda)_\Delta{}'s'\lambda \quad (3)$

$\vdash . (1) . (3) . \supset \vdash . \text{Prop}$

***111·31.** $\vdash : \lambda, S''\lambda \in \text{Cls}^2 \text{ excl} . S \in 1 \rightarrow 1 . R \in \epsilon_\Delta{}'\text{Crp}\,(S)''\lambda . \supset .$
$\dot{s}'\text{D}'R \in (s'S''\lambda)\,\overline{\text{sm}}\,(s'\lambda)$

Dem.

$\vdash . *83{\cdot}2 . \supset$

$\vdash :. \text{Hp} . \supset : \beta \in \lambda . \equiv . R'\text{Crp}\,(S)'\beta \in \text{Crp}\,(S)'\beta .$

[*111·202] $\equiv . R'\text{Crp}\,(S)'\beta \in 1 \rightarrow 1 . \text{D}'R'\text{Crp}\,(S)'\beta = S'\beta .$
$\text{Q}'R'\text{Crp}\,(S)'\beta = \beta \quad (1)$

$\vdash . (1) . *72{\cdot}322 . \quad \supset \vdash : \text{Hp} . \supset . \dot{s}'R''\text{Crp}\,(S)''\lambda \in 1 \rightarrow 1 .$

[*80·34] $\supset . \dot{s}'\text{D}'R \in 1 \rightarrow 1 \quad (2)$

$\vdash . (1) . *37{\cdot}68 . *50{\cdot}17 . \supset \vdash : \text{Hp} . \supset . \text{D}''R''\text{Crp}\,(S)''\lambda = S''\lambda .$
$\text{Q}''R''\text{Crp}\,(S)''\lambda = \lambda .$

[*80·34] $\supset . \text{D}''\text{D}'R = S''\lambda . \text{Q}''\text{D}'R = \lambda .$

[*41·43·44] $\supset . \text{D}'\dot{s}'\text{D}'R = s'S''\lambda . \text{Q}'\dot{s}'\text{D}'R = s'\lambda \quad (3)$

$\vdash . (2) . (3) . *73{\cdot}03 . \supset \vdash . \text{Prop}$

***111·311.** $\vdash : \lambda, S''\lambda \in \text{Cls}^2 \text{ excl} . S \in 1 \rightarrow 1 . \exists ! \epsilon_\Delta{}'\text{Crp}\,(S)''\lambda . \supset . s'S''\lambda \text{ sm } s'\lambda$
[*111·31 . *73·04]

***111·313.** $\vdash : \lambda \in \text{Cls}^2 \text{ excl} . R \in \epsilon_\Delta{}'\text{Crp}\,(S)''\lambda . \beta \in \lambda . M = \dot{s}'\text{D}'R . \supset .$
$M \upharpoonright \beta = R'\text{Crp}\,(S)'\beta . M \upharpoonright \beta \in \text{Crp}\,(S)'\beta$

Dem.

$\vdash . *83{\cdot}2 . \supset \vdash :: \text{Hp} . \supset :. \alpha \in \lambda . \supset_\alpha : R'\text{Crp}\,(S)'\alpha \in \text{Crp}\,(S)'\alpha : \quad (1)$

[*111·202] $\supset_\alpha : \text{Q}'R'\text{Crp}\,(S)'\alpha = \alpha :$

[*33·14 . *4·71] $\supset_\alpha : x\{R'\text{Crp}\,(S)'\alpha\}y . \equiv . x\{R'\text{Crp}\,(S)'\alpha\}y . y \in \alpha \quad (2)$

$\vdash . *35\cdot101 . *83\cdot23 . *41\cdot11 . \supset$

$\vdash :. \mathrm{Hp} . \supset : x(M \upharpoonright \beta)\, y . \equiv . (\exists\alpha) . \alpha \epsilon \lambda . x\, \{R\text{‘}\mathrm{Crp}\,(S)\text{‘}\alpha\}\, y . y \epsilon \beta .$

$[(2)] \qquad \equiv . (\exists\alpha) . \alpha \epsilon \lambda . x\, \{R\text{‘}\mathrm{Crp}\,(S)\text{‘}\alpha\}\, y . y \epsilon \alpha \cap \beta .$

$[*84\cdot11 . *22\cdot5] \qquad \equiv . (\exists\alpha) . \alpha \epsilon \lambda . x\, \{R\text{‘}\mathrm{Crp}\,(S)\text{‘}\alpha\}\, y . y \epsilon \beta . \alpha = \beta .$

$[*13\cdot195] \qquad \equiv . \beta \epsilon \lambda . x\, \{R\text{‘}\mathrm{Crp}\,(S)\text{‘}\beta\}\, y . y \epsilon \beta .$

$[\mathrm{Hp} . *4\cdot73 . (2)] \qquad \equiv . x\, \{R\text{‘}\mathrm{Crp}\,(S)\text{‘}\beta\}\, y \qquad (3)$

$\vdash . (1) . (3) . \supset \vdash . \mathrm{Prop}$

$*111\cdot32$. $\vdash : \lambda, S\text{‘‘}\lambda \epsilon \mathrm{Cls}^2\,\mathrm{excl} . S \epsilon 1 \rightarrow 1 . R \epsilon \epsilon_\Delta\text{‘}\mathrm{Crp}\,(S)\text{‘‘}\lambda . M = \dot{s}\text{‘}\mathrm{D}\text{‘}R . \supset .$
$M \epsilon 1 \rightarrow 1 . \text{Ɑ}\text{‘}M = s\text{‘}\lambda . S\text{‘‘}\lambda = M_\epsilon\text{‘‘}\lambda . S \upharpoonright \lambda = M_\epsilon \upharpoonright \lambda$

Dem.

$\vdash . *111\cdot31 . *73\cdot03 . \supset \vdash : \mathrm{Hp} . \supset . M \epsilon 1 \rightarrow 1 . \text{Ɑ}\text{‘}M = s\text{‘}\lambda \qquad (1)$

$\vdash . *111\cdot313\cdot202 . \quad \supset \vdash :. \mathrm{Hp} . \supset : \beta \epsilon \lambda . \supset . \mathrm{D}\text{‘}(M \upharpoonright \beta) = S\text{‘}\beta . \text{Ɑ}\text{‘}(M \upharpoonright \beta) = \beta .$

$[*37\cdot25] \qquad \supset . (M \upharpoonright \beta)\text{‘‘}\beta = S\text{‘}\beta .$

$[*37\cdot421\cdot11] \qquad \supset . M_\epsilon\text{‘}\beta = S\text{‘}\beta :$

$[*35\cdot71 . *37\cdot69] \qquad \supset : M_\epsilon \upharpoonright \lambda = S \upharpoonright \lambda . M_\epsilon\text{‘‘}\lambda = S\text{‘‘}\lambda \qquad (2)$

$\vdash . (1) . (2) . \supset \vdash . \mathrm{Prop}$

$*111\cdot321$. $\vdash : \lambda, S\text{‘‘}\lambda \epsilon \mathrm{Cls}^2\,\mathrm{excl} . S \epsilon 1 \rightarrow 1 . \exists ! \epsilon_\Delta\text{‘}\mathrm{Crp}\,(S)\text{‘‘}\lambda . \supset .$
$(\exists M) . M \epsilon 1 \rightarrow 1 . \text{Ɑ}\text{‘}M = s\text{‘}\lambda . S\text{‘‘}\lambda = M_\epsilon\text{‘‘}\lambda . S \upharpoonright \lambda = M_\epsilon \upharpoonright \lambda$
$[*111\cdot32]$

$*111\cdot322$. $\vdash : \kappa, \lambda \epsilon \mathrm{Cls}^2\,\mathrm{excl} . S \epsilon \kappa\, \overline{\mathrm{sm}}\, \lambda . R \epsilon \epsilon_\Delta\text{‘}\mathrm{Crp}\,(S)\text{‘‘}\lambda . M = \dot{s}\text{‘}\mathrm{D}\text{‘}R . \supset .$
$M \epsilon \kappa\, \overline{\mathrm{sm}}\, \overline{\mathrm{sm}}\, \lambda . S = M_\epsilon \upharpoonright \lambda \quad [*111\cdot32\cdot1 . *35\cdot66 . *73\cdot03]$

$*111\cdot33$. $\vdash :. \mathrm{Mult\,ax} . \supset : S \epsilon 1 \rightarrow 1 . S \subset \mathrm{sm} . \kappa, \lambda \epsilon \mathrm{Cls}^2\mathrm{excl} . \kappa = S\text{‘‘}\lambda . \lambda \subset \text{Ɑ}\text{‘}S . \supset .$
$s\text{‘}\kappa \;\mathrm{sm}\; s\text{‘}\lambda$

Dem.

$\vdash . *111\cdot221 . \supset$

$\vdash :. S \epsilon 1 \rightarrow 1 . S \subset \mathrm{sm} . \kappa, \lambda \epsilon \mathrm{Cls}^2\,\mathrm{excl} . \kappa = S\text{‘‘}\lambda . \lambda \subset \text{Ɑ}\text{‘}S . \supset :$
$\beta \epsilon \lambda . \supset_\beta . \exists ! \mathrm{Crp}\,(S)\text{‘}\beta :$

$[*88\cdot37] \quad \supset : \mathrm{Mult\,ax} . \supset . \exists ! \epsilon_\Delta\text{‘}\mathrm{Crp}\,(S)\text{‘‘}\lambda .$

$[*111\cdot311] \qquad \supset . s\text{‘}\kappa \;\mathrm{sm}\; s\text{‘}\lambda :. \supset \vdash . \mathrm{Prop}$

$*111\cdot34$. $\vdash :. \mathrm{Mult\,ax} . \supset :$
$(\exists S) . S \epsilon 1 \rightarrow 1 . S \subset \mathrm{sm} . \mathrm{D}\text{‘}S = \kappa . \text{Ɑ}\text{‘}S = \lambda . \kappa, \lambda \epsilon \mathrm{Cls}^2\,\mathrm{excl} . \supset .$
$(\exists M) . M \epsilon 1 \rightarrow 1 . \text{Ɑ}\text{‘}M = s\text{‘}\lambda . \kappa = M_\epsilon\text{‘‘}\lambda$

Dem.

$\vdash . *111\cdot25 . \supset$

$\vdash :. S \epsilon 1 \rightarrow 1 . S \subset \mathrm{sm} . \mathrm{D}\text{‘}S = \kappa . \text{Ɑ}\text{‘}S = \lambda . \kappa, \lambda \epsilon \mathrm{Cls}^2\,\mathrm{excl} . \supset :$
$\mathrm{Crp}\,(S)\text{‘‘}\lambda \epsilon \mathrm{Cls\,ex}^2\,\mathrm{excl} :$

$[*88\cdot32] \supset : \mathrm{Mult\,ax} . \supset . \exists ! \epsilon_\Delta\text{‘}\mathrm{Crp}\,(S)\text{‘‘}\lambda .$

$[*111\cdot321] \qquad \supset . (\exists M) . M \epsilon 1 \rightarrow 1 . \text{Ɑ}\text{‘}M = s\text{‘}\lambda . \kappa = M_\epsilon\text{‘‘}\lambda \quad (1)$

$\vdash . (1) . *10\cdot11\cdot23 . \mathrm{Comm} . \supset \vdash . \mathrm{Prop}$

The following propositions are concerned with the elementary properties of "sm sm." It will be seen that they are closely analogous to those of "sm."

✱111·4. $\vdash : \kappa \text{ sm sm } \lambda . \equiv . (\exists T) . T \epsilon 1 \rightarrow 1 . \text{Ɑ}`T = s`\lambda . \kappa = T_\epsilon``\lambda . \equiv . \exists ! \kappa \, \overline{\text{sm}} \, \overline{\text{sm}} \, \lambda$
[✱111·1 . (✱111·03)]

✱111·401. $\vdash : \kappa \text{ sm sm } \lambda . \equiv . (\exists T) . T \epsilon 1 \rightarrow 1 . s`\lambda \subset \text{Ɑ}`T . \kappa = T_\epsilon``\lambda$

Dem.

$\vdash . ✱22·42 . ✱111·4 . \supset \vdash : \kappa \text{ sm sm } \lambda . \supset . (\exists T) . T \epsilon 1 \rightarrow 1 . s`\lambda \subset \text{Ɑ}`T . \kappa = T_\epsilon``\lambda$ (1)

$\vdash . (1) . ✱111·14 . \supset \vdash . \text{Prop}$

✱111·402. $\vdash : \kappa \text{ sm sm } \lambda . \equiv . (\exists T) . T \upharpoonright s`\lambda \, \epsilon \, 1 \rightarrow 1 . s`\lambda \subset \text{Ɑ}`T . \kappa = T_\epsilon``\lambda$
[✱111·14·1·121]

✱111·43. $\vdash : \kappa \text{ sm sm } \lambda . \supset . (\exists S) . S \epsilon 1 \rightarrow 1 . S \mathrel{\Subset} \text{sm} . \text{D}`S = \kappa . \text{Ɑ}`S = \lambda$
[✱111·11·111]

✱111·44. $\vdash : \kappa \text{ sm sm } \lambda . \supset . \kappa \text{ sm } \lambda . s`\kappa \text{ sm } s`\lambda$ [✱111·11·4 . ✱73·03]

✱111·45. $\vdash . \lambda \text{ sm sm } \lambda$ [✱111·13·4]

✱111·451. $\vdash : \kappa \text{ sm sm } \lambda . \equiv . \lambda \text{ sm sm } \kappa$ [✱111·131·4]

✱111·452. $\vdash : \kappa \text{ sm sm } \lambda . \lambda \text{ sm sm } \mu . \supset . \kappa \text{ sm sm } \mu$ [✱111·132·4]

✱111·46. $\vdash : \lambda, S``\lambda \, \epsilon \, \text{Cls}^2 \text{ excl} . S \epsilon 1 \rightarrow 1 . \exists ! \epsilon_\Delta`\text{Crp}(S)``\lambda . \supset . S``\lambda \text{ sm sm } \lambda$
[✱111·32·4]

✱111·47. $\vdash :. \kappa \text{ sm sm } \lambda . \supset : \kappa \, \epsilon \, \text{Cls}^2 \text{ excl} . \equiv . \lambda \, \epsilon \, \text{Cls}^2 \text{ excl}$

Dem.

$\vdash . ✱111·4 . \qquad \supset \vdash :. \text{Hp} . \supset : (\exists T) . T \epsilon 1 \rightarrow 1 . \text{Ɑ}`T = s`\lambda . \kappa = T```\lambda :$

[✱84·53] $\qquad \supset : \lambda \, \epsilon \, \text{Cls}^2 \text{ excl} . \supset . \kappa \, \epsilon \, \text{Cls}^2 \text{ excl}$ (1)

$\vdash . (1) . ✱111·451 . \supset \vdash :. \text{Hp} . \supset : \kappa \, \epsilon \, \text{Cls}^2 \text{ excl} . \supset . \lambda \, \epsilon \, \text{Cls}^2 \text{ excl}$ (2)

$\vdash . (1) . (2) . \supset \vdash . \text{Prop}$

✱111·5. $\vdash :: \text{Mult ax} . \supset :. \kappa, \lambda \, \epsilon \, \text{Cls}^2 \text{ excl} . \supset :$
$\kappa \text{ sm sm } \lambda . \equiv . (\exists S) . S \epsilon 1 \rightarrow 1 . S \mathrel{\Subset} \text{sm} . \text{D}`S = \kappa . \text{Ɑ}`S = \lambda .$
$\equiv . \exists ! \kappa \, \overline{\text{sm}} \, \lambda \cap \text{Rl}`\text{sm}$ [✱111·34·43·4]

✱111·51. $\vdash :. \text{Mult ax} . \supset : \kappa, \lambda \, \epsilon \, \text{Cls}^2 \text{ excl} . \exists ! \kappa \, \overline{\text{sm}} \, \lambda \cap \text{Rl}`\text{sm} . \supset . s`\kappa \text{ sm } s`\lambda$
[✱111·5·44]

✱111·52. $\vdash : \mu, \nu \, \epsilon \, \text{NC} . \kappa, \lambda \, \epsilon \, \mu \cap \text{Cl}`\nu . \supset . \exists ! \kappa \, \overline{\text{sm}} \, \lambda \cap \text{Rl}`\text{sm}$

Dem.

$\vdash . ✱100·5 . ✱73·1 . \supset \vdash : \text{Hp} . \supset . (\exists S) . S \epsilon 1 \rightarrow 1 . \text{D}`S = \kappa . \text{Ɑ}`S = \lambda$ (1)

$\vdash . ✱100·5 . \qquad \supset \vdash :. \text{Hp} . \supset : \alpha \epsilon \kappa . \beta \epsilon \lambda . \supset . \alpha \text{ sm } \beta$ (2)

$\vdash . (1) . (2) . \supset \vdash . \text{Prop}$

✱111·53. $\vdash :. \text{Mult ax} . \supset : \mu, \nu \, \epsilon \, \text{NC} . \kappa, \lambda \, \epsilon \, \mu \cap \text{Cl excl}`\nu . \supset . \kappa \text{ sm sm } \lambda$
[✱111·52·5]

*112. THE ARITHMETICAL SUM OF A CLASS OF CLASSES.

Summary of *112.

In this number, we return to the arithmetical operations. The definition of addition in *110 was only applicable to a finite number of summands, because the summands had to be enumerated. In the present number, we define the arithmetical sum of a class of classes, so that the summands are given as the members of a class, and do not require to be enumerated. Hence the definition in this number is as applicable to an infinite number of summands as to a finite number.

If κ is a class of mutually exclusive classes, the number of $s\text{‘}\kappa$ will be the sum of the numbers of members of κ; *i.e.* if we write "$\Sigma\mathrm{Nc}\text{‘}\kappa$" for the sum of the numbers of members of κ,

$$\kappa \in \mathrm{Cls}^2\,\mathrm{excl} \,.\supset.\, \mathrm{Nc}\text{‘}s\text{‘}\kappa = \Sigma\mathrm{Nc}\text{‘}\kappa.$$

But when the members of κ are not mutually exclusive, a term x which is a member of two members (say α and β) of κ has to be counted twice over in obtaining the arithmetical sum of κ, whereas in the logical sum x is only counted once. Thus we need a construction which shall duplicate x, taking it first as a member of α, and then as a member of β. This is effected if we replace x first by $x \downarrow \alpha$, and then by $x \downarrow \beta$. In fact, $x \downarrow \alpha$ has the kind of arithmetical properties which we mean to secure when we speak of "x considered as a member of α"—a phrase which, as it stands, does not serve our purpose, for x is simply x however we may choose to consider it. Thus we replace α by $\downarrow \alpha\text{“}\alpha$ and β by $\downarrow \beta\text{“}\beta$ and so on; *i.e.* (using *85·5), we replace α by $\epsilon \,\underline{\downarrow}\, \alpha$ and β by $\epsilon \,\underline{\downarrow}\, \beta$ and so on. These new classes are similar to α and β and so on, and are mutually exclusive. Hence their *logical* sum has the number of terms which is wanted for the *arithmetical* sum of the members of κ. Thus we put

$$\Sigma\text{‘}\kappa = s\text{‘}\epsilon \,\underline{\downarrow}\text{“}\kappa \quad \mathrm{Df},$$

$$\Sigma\mathrm{Nc}\text{‘}\kappa = \mathrm{Nc}\text{‘}\Sigma\text{‘}\kappa \quad \mathrm{Df}.$$

With regard to the second of these definitions, it is to be observed that $\Sigma\mathrm{Nc}\text{‘}\kappa$ is not a function of $\mathrm{Nc}\text{“}\kappa$, unless no two members of κ are similar; for $\mathrm{Nc}\text{“}\kappa$ cannot contain the same number twice over. For the same reason, if λ is a class of cardinals, and we define "$\mathrm{Sum}\text{‘}\lambda$," we do not get what is wanted for arithmetical addition, because our definition will not enable us to deal with summations in which there are numbers that are repeated. We could, if it were worth while, define "Sum'λ" as follows: Take a class of classes κ, consisting of one class having each number which is a member of λ, *i.e.* let κ be a selection from λ; then Σ'κ will have the required number of terms. *I.e.* we might put

$$\text{Sum}`\lambda = \hat{\xi}\{(\exists\kappa) . \kappa \in \text{D}``\epsilon_\Delta`\lambda . \xi \,\text{sm}\, \Sigma`\kappa\} \quad \text{Df.}$$

But since this definition is only available for sums in which no number is repeated, it is not worth while to introduce it.

In this number we prove the following propositions among others.

***112·15.** $\vdash : \kappa \in \text{Cls}^2 \,\text{excl} . \supset . s`\kappa \in \Sigma\text{Nc}`\kappa$

This is an extension of *110·32.

***112·17.** $\vdash : \kappa \,\text{sm sm}\, \lambda . \supset . \Sigma\text{Nc}`\kappa = \Sigma\text{Nc}`\lambda . \Sigma`\kappa \,\text{sm}\, \Sigma`\lambda$

The chief point in the above proposition is that it does not require $\kappa, \lambda \in \text{Cls}^2$ excl.

*112·2—·24 are concerned with the use of the multiplicative axiom and the propositions of *111 in which it appears as hypothesis. We have

***112·22.** $\vdash :. \text{Mult ax} . \supset : \exists ! (\epsilon \downarrow ``\kappa) \,\overline{\text{sm}}\, (\epsilon \downarrow ``\lambda) \cap \text{Rl}`\text{sm} . \supset . \Sigma\text{Nc}`\kappa = \Sigma\text{Nc}`\lambda$

whence we derive the proposition

***112·24.** $\vdash :. \text{Mult ax} . \supset : \mu, \nu \in \text{NC} . \kappa, \lambda \in \mu \cap \text{Cl}`\nu . \supset . \Sigma\text{Nc}`\kappa = \Sigma\text{Nc}`\lambda$

I.e. assuming the multiplicative axiom, two classes which each consist of μ classes of ν terms each have the same number of terms in their sum. This number would naturally be defined as μ multiplied by ν, but owing to the necessity of the multiplicative axiom in this proposition, we have selected a different definition of multiplication (*113) which does not depend upon the multiplicative axiom. The reader should observe that the similarity of two classes, each of which consists of μ mutually exclusive sets of ν terms, cannot be proved in general without the multiplicative axiom.

The remaining propositions of this number give properties of Σ in special cases. We prove that $\Sigma`\Lambda = \Lambda$ (*112·3), that $\Sigma\text{Nc}`\iota`\alpha = \text{Nc}`\alpha$ (*112·321), that $\alpha \neq \beta . \supset . \Sigma\text{Nc}`(\iota`\alpha \cup \iota`\beta) = \text{Nc}`\alpha +_c \text{Nc}`\beta$ (*112·34), which connects the definition of addition in this number with that in *110. Finally we prove the general associative law for addition, in the following two forms:

***112·41.** $\vdash . s`\Sigma``\lambda = \Sigma`s`\lambda$

***112·43.** $\vdash : \lambda \in \text{Cls}^2 \,\text{excl} . \supset . \text{Nc}`\Sigma`\Sigma``\lambda = \text{Nc}`\Sigma`s`\lambda$

***112·01.** $\Sigma`\kappa = s`\epsilon \overline{\downarrow}``\kappa$ Df

***112·02.** $\Sigma\mathrm{Nc}`\kappa = \mathrm{Nc}`\Sigma`\kappa$ Df

***112·1.** $\vdash . \Sigma`\kappa = s`\epsilon \overline{\downarrow}``\kappa$ [*20·2 . (*112·01)]

***112·101.** $\vdash . \Sigma\mathrm{Nc}`\kappa = \mathrm{Nc}`\Sigma`\kappa = \mathrm{Nc}`s`\epsilon \overline{\downarrow}``\kappa$ [*20·2 . *112·1 . (*112·02)]

***112·102.** $\vdash . \Sigma`\kappa = \hat{R}\{(\exists \alpha, x) . \alpha \epsilon \kappa . x \epsilon \alpha . R = x \downarrow \alpha\}$

Dem.

$\vdash . *85·6 . *40·11 . *112·1 . \supset$

$\vdash . \Sigma`\kappa = \hat{R}\{(\exists \mu, \alpha) . \alpha \epsilon \kappa . \mu = \downarrow \alpha``\alpha . R \epsilon \mu\}$

[*13·195] $= \hat{R}\{(\exists \alpha) . \alpha \epsilon \kappa . R \epsilon \downarrow \alpha``\alpha\}$

[*55·231] $= \hat{R}\{(\exists \alpha, x) . \alpha \epsilon \kappa . x \epsilon \alpha . R = x \downarrow \alpha\} . \supset \vdash . \mathrm{Prop}$

***112·103.** $\vdash . \Sigma`\kappa = s`\hat{\mu}\{(\exists \alpha) . \alpha \epsilon \kappa . \mu = \downarrow \alpha``\alpha\}$ [*112·1 . *85·6]

***112·11.** $\vdash : \beta \epsilon \Sigma\mathrm{Nc}`\kappa . \equiv . \beta \,\mathrm{sm}\, s`\epsilon \overline{\downarrow}``\kappa$ [*112·101]

***112·12.** $\vdash . s`\epsilon \overline{\downarrow}``\kappa \epsilon \Sigma\mathrm{Nc}`\kappa$ [*112·11]

***112·13.** $\vdash : \lambda \,\mathrm{sm}\,\mathrm{sm}\, \epsilon \overline{\downarrow}``\kappa . \supset . s`\lambda \epsilon \Sigma\mathrm{Nc}`\kappa$ [*111·44 . *112·11]

***112·14.** $\vdash : \kappa \epsilon \mathrm{Cls}^2 \,\mathrm{excl} . \supset . \epsilon \overline{\downarrow}``\kappa \,\mathrm{sm}\,\mathrm{sm}\, \kappa$

Dem.

$\vdash . *21·33 . \supset \vdash :. \mathrm{Hp} . T = \hat{R}\hat{x}\{(\exists \alpha) . \alpha \epsilon \kappa . x \epsilon \alpha . R = x \downarrow \alpha\} . \supset :$

$xTR . yTR . \supset . (\exists \alpha, \beta) . R = x \downarrow \alpha . R = y \downarrow \beta .$

[*55·31] $\supset . x = y :$

[*71·17] $\supset : T \epsilon 1 \rightarrow \mathrm{Cls}$ (1)

$\vdash . *21·33 . \supset$

$\vdash : \mathrm{Hp}(1) . xTR . xTS . \supset . (\exists \alpha, \beta) . \alpha, \beta \epsilon \kappa . x \epsilon \alpha \cap \beta . R = x \downarrow \alpha . S = x \downarrow \beta .$

[*84·11.Hp] $\supset . (\exists \alpha, \beta) . \alpha = \beta . R = x \downarrow \alpha . S = x \downarrow \beta .$

[*13·195] $\supset . R = S :$

[*71·171] $\supset : T \epsilon \mathrm{Cls} \rightarrow 1$ (2)

$\vdash . *33·131 . \supset \vdash :. \mathrm{Hp}(1) . \supset : x \epsilon \mathrm{G}`T . \equiv . (\exists R, \alpha) . \alpha \epsilon \kappa . x \epsilon \alpha . R = x \downarrow \alpha .$

[*55·12] $\equiv . x \epsilon s`\kappa$ (3)

$\vdash . *37·1·11 . \supset$

$\vdash :: \mathrm{Hp} . \supset :. \alpha \epsilon \kappa . \supset : R \epsilon T_\epsilon`\alpha . \equiv . (\exists x, \beta) . x \epsilon \alpha \cap \beta . \beta \epsilon \kappa . R = x \downarrow \beta .$

[*84·11.Hp] $\equiv . (\exists x, \beta) . x \epsilon \alpha \cap \beta . \beta \epsilon \kappa . \alpha = \beta . R = x \downarrow \beta .$

[*13·195] $\equiv . (\exists x) . x \epsilon \alpha . R = x \downarrow \beta .$

[*85·601] $\equiv . R \epsilon \epsilon \overline{\downarrow}`\alpha :.$

[*37·69] $\supset :. T_\epsilon``\kappa = \epsilon \overline{\downarrow}``\kappa$ (4)

$\vdash . (1) . (2) . (3) . (4) . *111·4 . \supset \vdash . \mathrm{Prop}$

***112·15.** $\vdash : \kappa \epsilon \mathrm{Cls}^2 \,\mathrm{excl} . \supset . s`\kappa \epsilon \Sigma\mathrm{Nc}`\kappa$ [*112·14·11 . *111·44]

∗112·151. $s`\epsilon\,\overline{\downarrow}``\lambda = \hat{R}\{(\exists\alpha, x)\,.\,\alpha\,\epsilon\,\lambda\,.\,x\,\epsilon\,\alpha\,.\,R = x \downarrow \alpha\}\,.\,\dot{s}`s`\epsilon\,\overline{\downarrow}``\lambda = \epsilon \upharpoonright \lambda$

Dem.

$\vdash .\, *40{\cdot}11\,.\,(*85{\cdot}5)\,.\,\supset$

$\vdash .\, s`\epsilon\,\overline{\downarrow}``\lambda = \hat{R}\{(\exists\alpha)\,.\,\alpha\,\epsilon\,\lambda\,.\,R\,\epsilon \downarrow \alpha``\alpha\}$

$[*38{\cdot}131] \quad = \hat{R}\{(\exists\alpha, x)\,.\,\alpha\,\epsilon\,\lambda\,.\,x\,\epsilon\,\alpha\,.\,R = x \downarrow \alpha\} \qquad (1)$

$\vdash .\,(1)\,.\,*41{\cdot}11\,.\,\supset$

$\vdash .\, \dot{s}`s`\epsilon\,\overline{\downarrow}``\lambda = \hat{y}\hat{\beta}\{(\exists R, \alpha, x)\,.\,\alpha\,\epsilon\,\lambda\,.\,x\,\epsilon\,\alpha\,.\,R = x \downarrow \alpha\,.\,yR\beta\}$

$[*13{\cdot}195\,.\,*55{\cdot}13] = \hat{y}\hat{\beta}\{(\exists\alpha, x)\,.\,\alpha\,\epsilon\,\lambda\,.\,x\,\epsilon\,\alpha\,.\,y = x\,.\,\beta = \alpha\}$

$[*13{\cdot}22] \quad = \hat{y}\hat{\beta}\{\beta\,\epsilon\,\lambda\,.\,y\,\epsilon\,\beta\}$

$[*35{\cdot}101] \quad = \epsilon \upharpoonright \lambda \qquad (2)$

$\vdash .\,(1)\,.\,(2)\,.\,\supset \vdash .\,\text{Prop}$

The following proposition is a lemma for ∗112·153, which is required for ∗112·16. ∗112·16 in turn is used in ∗112·17, which is a fundamental proposition in the theory of addition.

∗112·152. $\vdash : T\,\epsilon\,1 \rightarrow \text{Cls}\,.\,\beta \subset \text{ᗡ}`T\,.\,\supset .\,(T \,\|\, \breve{T}_\epsilon)``\epsilon\,\overline{\downarrow}\,\beta = \epsilon\,\overline{\downarrow}\,(T``\beta)$

Dem.

$\vdash .\,*37{\cdot}6\,.\,*85{\cdot}601\,.\,\supset \vdash .\,(T \,\|\, \breve{T}_\epsilon)``\epsilon\,\overline{\downarrow}\,\beta = \hat{R}\{(\exists y)\,.\,y\,\epsilon\,\beta\,.\,R = (T \,\|\, \breve{T}_\epsilon)`(y \downarrow \beta)\} \qquad (1)$

$\vdash .\,(1)\,.\,*55{\cdot}61\,.\,\supset$

$\vdash : \text{Hp}\,.\,\supset .\,(T \,\|\, \breve{T}_\epsilon)``\epsilon\,\overline{\downarrow}\,\beta = \hat{R}\{(\exists y)\,.\,y\,\epsilon\,\beta\,.\,R = (T`y) \downarrow (T_\epsilon`\beta)\}$

$[*37{\cdot}11] \quad = \hat{R}\{(\exists y)\,.\,y\,\epsilon\,\beta\,.\,R = (T`y) \downarrow (T``\beta)\}$

$[*38{\cdot}131] \quad = \downarrow (T``\beta)``(T``\beta)$

$[*85{\cdot}601] \quad = \epsilon\,\overline{\downarrow}\,(T``\beta) : \supset \vdash .\,\text{Prop}$

In the following proposition, we have a double correlator of a sort which will frequently occur in cardinal arithmetic, namely $T \,\|\, \breve{T}_\epsilon$ with its converse domain limited, where T is a given double correlator (or single correlator, on other occasions). As appears from the propositions used in the above proof of ∗112·152, if T is a correlator whose converse domain includes β and has y as a member, $(T \,\|\, \breve{T}_\epsilon)`(y \downarrow \beta) = (T`y) \downarrow (T``\beta)$. Thus $T \,\|\, \breve{T}_\epsilon$ is an operation which, when operating on suitable relations of individuals to classes (including selectors), turns the individuals into their correlates and the classes into the classes of their members' correlates. This is why it is a useful relation.

∗112·153. $T\,\epsilon\,\kappa\,\overline{\text{sm}}\,\overline{\text{sm}}\,\lambda\,.\,\supset .\,(T \,\|\, \breve{T}_\epsilon) \upharpoonright s`\epsilon\,\overline{\downarrow}``\lambda\,\epsilon\,(\epsilon\,\overline{\downarrow}``\kappa)\,\overline{\text{sm}}\,\overline{\text{sm}}\,(\epsilon\,\overline{\downarrow}``\lambda)$

Dem.

$\vdash .\,*112{\cdot}151\,.\,*41{\cdot}43{\cdot}44\,.\,\supset \vdash .\,s`\text{D}``s`\epsilon\,\overline{\downarrow}``\lambda = \text{D}`(\epsilon \upharpoonright \lambda)\,.\,s`\text{ᗡ}``s`\epsilon\,\overline{\downarrow}``\lambda = \text{ᗡ}`(\epsilon \upharpoonright \lambda)\,.$

$[*62{\cdot}41{\cdot}43] \quad \supset \vdash .\,s`\text{D}``s`\epsilon\,\overline{\downarrow}``\lambda = s`\lambda\,.\,s`\text{ᗡ}``s`\epsilon\,\overline{\downarrow}``\lambda = \lambda - \iota`\Lambda \qquad (1)$

$\vdash.(1).*111\cdot1.*37\cdot231.\supset\vdash:\mathrm{Hp}.\supset.s`\mathrm{D}``s`\epsilon\downarrow``\lambda\subset\text{Ɑ}`T.s`\text{Ɑ}``s`\epsilon\downarrow``\lambda\subset\text{Ɑ}`T_\epsilon$ (2)

$\vdash.*111\cdot1.*71\cdot29. \quad \supset\vdash:\mathrm{Hp}.\supset.T\restriction s`\mathrm{D}``s`\epsilon\downarrow``\lambda\,\epsilon\,1\to1$ (3)

$\vdash.*111\cdot11.(1). \quad \supset\vdash:\mathrm{Hp}.\supset.T_\epsilon\restriction s`\text{Ɑ}``s`\epsilon\downarrow``\lambda\,\epsilon\,1\to1$ (4)

$\vdash.(2).(3).(4).*74\cdot775\dfrac{s`\epsilon\downarrow``\lambda,\,T,\,\breve{T}_\epsilon}{\lambda,\quad Q,\,R}.\supset\vdash:\mathrm{Hp}.\supset.(T\parallel\breve{T}_\epsilon)\restriction s`\epsilon\downarrow``\lambda\,\epsilon\,1\to1$ (5)

$\vdash.*43\cdot302.\ \supset\vdash.s`\epsilon\downarrow``\lambda\subset\text{Ɑ}`(T\parallel\breve{T}_\epsilon)$ (6)

$\vdash.*112\cdot152.\supset\vdash:\mathrm{Hp}.\supset.(T\parallel\breve{T}_\epsilon)```\epsilon\downarrow``\lambda=\epsilon\downarrow``T```\lambda.$

[*37·11] $\supset.(T\parallel\breve{T}_\epsilon)_\epsilon``\epsilon\downarrow``\lambda=\epsilon\downarrow``T_\epsilon``\lambda$

[*111·1.Hp] $=\epsilon\downarrow``\kappa$ (7)

$\vdash.(5).(6).(7).*111\cdot14.\supset\vdash.\mathrm{Prop}$

***112·16.** $\vdash:\kappa\,\mathrm{sm\,sm}\,\lambda.\supset.\epsilon\downarrow``\kappa\,\mathrm{sm\,sm}\,\epsilon\downarrow``\lambda$ [*112·153.*111·4]

***112·17.** $\vdash:\kappa\,\mathrm{sm\,sm}\,\lambda.\supset.\Sigma\mathrm{Nc}`\kappa=\Sigma\mathrm{Nc}`\lambda.\Sigma`\kappa\,\mathrm{sm}\,\Sigma`\lambda$

Dem.

$\vdash.*112\cdot16.*111\cdot44.\supset\vdash:\mathrm{Hp}.\supset.s`\epsilon\downarrow``\kappa\,\mathrm{sm}\,s`\epsilon\downarrow``\lambda$ (1)

$\vdash.(1).*112\cdot1\cdot101.\supset\vdash.\mathrm{Prop}$

***112·18.** $\vdash.\Sigma\mathrm{Nc}`\kappa=\Sigma\mathrm{Nc}`\epsilon\downarrow``\kappa$

Dem.

$\vdash.*85\cdot61.*112\cdot15.\supset\vdash.s`\epsilon\downarrow``\kappa\,\epsilon\,\Sigma\mathrm{Nc}`\epsilon\downarrow``\kappa$ (1)

$\vdash.(1).*112\cdot12.*100\cdot34.\supset\vdash.\mathrm{Prop}$

***112·2.** $\vdash:S\,\epsilon\,1\to1.\mathrm{D}`S=\epsilon\downarrow``\kappa.\text{Ɑ}`S=\epsilon\downarrow``\lambda.\exists!\,\epsilon_\Delta`\mathrm{Crp}(S)``\lambda.$
$\supset.\Sigma\mathrm{Nc}`\kappa=\Sigma\mathrm{Nc}`\lambda.\Sigma`\kappa\,\mathrm{sm}\,\Sigma`\lambda$

Dem.

$\vdash.*111\cdot311.*85\cdot61.\supset\vdash:\mathrm{Hp}.\supset.s`\epsilon\downarrow``\kappa\,\mathrm{sm}\,s`\epsilon\downarrow``\lambda$ (1)

$\vdash.(1).*112\cdot1\cdot101.\supset\vdash.\mathrm{Prop}$

***112·21.** $\vdash:.\mathrm{Mult\,ax}.\supset:(\exists S).S\,\epsilon\,1\to1.S\,\overline{\subset}\,\mathrm{sm}.\mathrm{D}`S=\epsilon\downarrow``\kappa.\text{Ɑ}`S=\epsilon\downarrow``\lambda.$
$\equiv.\epsilon\downarrow``\kappa\,\mathrm{sm\,sm}\,\epsilon\downarrow``\lambda$ [*111·5.*85·61]

***112·22.** $\vdash:.\mathrm{Mult\,ax}.\supset:\exists!(\epsilon\downarrow``\kappa)\,\overline{\mathrm{sm}}\,(\epsilon\downarrow``\lambda)\cap\mathrm{Rl}`\mathrm{sm}.\supset.$
$\Sigma\mathrm{Nc}`\kappa=\Sigma\mathrm{Nc}`\lambda$ [*112·17·18·21]

***112·23.** $\vdash:.\mathrm{Mult\,ax}.\supset:\kappa,\lambda\,\epsilon\,\mathrm{Cls^2excl}:\exists!\,\kappa\,\overline{\mathrm{sm}}\,\lambda\cap\mathrm{Rl}`\mathrm{sm}.\supset.$
$s`\kappa,s`\lambda\,\epsilon\,\Sigma\mathrm{Nc}`\kappa.\Sigma\mathrm{Nc}`\kappa=\Sigma\mathrm{Nc}`\lambda$

Dem.

$\vdash.*112\cdot15.\supset\vdash:\mathrm{Hp}.\kappa,\lambda\,\epsilon\,\mathrm{Cls^2\,excl}.\supset.s`\kappa\,\epsilon\,\Sigma\mathrm{Nc}`\kappa.s`\lambda\,\epsilon\,\Sigma\mathrm{Nc}`\lambda$ (1)

$\vdash.*111\cdot51.\supset\vdash:\mathrm{Hp}(1).\exists!\,\kappa\,\overline{\mathrm{sm}}\,\lambda\cap\mathrm{Rl}`\mathrm{sm}.\supset.s`\kappa\,\mathrm{sm}\,s`\lambda$ (2)

$\vdash.(1).(2).\supset\vdash.\mathrm{Prop}$

***112·231.** $\vdash:S\,\epsilon\,\kappa\,\overline{\mathrm{sm}}\,\lambda\cap\mathrm{Rl}`\mathrm{sm}.\supset.\epsilon\downarrow|S|\mathrm{Cnv}`\epsilon\downarrow\,\epsilon\,(\epsilon\downarrow``\kappa)\,\overline{\mathrm{sm}}\,(\epsilon\downarrow``\lambda)\cap\mathrm{Rl}`\mathrm{sm}$

Dem.

$\vdash.*73\cdot63.*85\cdot601.\supset\vdash:S\,\epsilon\,\kappa\,\overline{\mathrm{sm}}\,\lambda.\supset.\epsilon\downarrow|S|\mathrm{Cnv}`\epsilon\downarrow\,\epsilon\,(\epsilon\downarrow``\kappa)\,\overline{\mathrm{sm}}\,(\epsilon\downarrow``\lambda)$ (1)

$\vdash.*85\cdot601.*73\cdot33\cdot34.\supset\vdash:S\,\overline{\subset}\,\mathrm{sm}.\supset.\epsilon\downarrow|S|\mathrm{Cnv}`\epsilon\downarrow\,\overline{\subset}\,\mathrm{sm}$ (2)

$\vdash.(1).(2).\supset\vdash:S\,\epsilon\,\kappa\,\overline{\mathrm{sm}}\,\lambda\cap\mathrm{Rl}`\mathrm{sm}.\supset.\epsilon\downarrow|S|\mathrm{Cnv}`\epsilon\downarrow\,\epsilon\,(\epsilon\downarrow``\kappa)\,\overline{\mathrm{sm}}(\epsilon\downarrow``\lambda)\cap\mathrm{Rl}`\mathrm{sm}:$
$\supset\vdash.\mathrm{Prop}$

***112·24.** $\vdash :. \text{Mult ax} . \supset : \mu, \nu \in \text{NC} . \kappa, \lambda \in \mu \cap \text{Cl}\text{‘}\nu . \supset . \Sigma\text{Nc}\text{‘}\kappa = \Sigma\text{Nc}\text{‘}\lambda$

Dem.

$\vdash . *111·52 . \supset \vdash : \mu, \nu \in \text{NC} . \kappa, \lambda \in \mu \cap \text{Cl}\text{‘}\nu . \supset . \exists ! \kappa \,\overline{\text{sm}}\, \lambda \cap \text{Rl}\text{‘sm} .$

[*112·231] $\supset . \exists ! (\epsilon \downarrow\text{“}\kappa) \,\overline{\text{sm}}\, (\epsilon \downarrow\text{“}\lambda) \cap \text{Rl}\text{‘sm}$ (1)

$\vdash . (1) . *111·51 . *85·61 . \supset$

$\vdash :. \text{Mult ax} . \supset : \mu, \nu \in \text{NC} . \kappa, \lambda \in \mu \cap \text{Cl}\text{‘}\nu . \supset . s\text{‘}\epsilon \downarrow\text{“}\kappa \text{ sm } s\text{‘}\epsilon \downarrow\text{“}\lambda .$

[*112·101] $\supset . \Sigma\text{Nc}\text{‘}\kappa = \Sigma\text{Nc}\text{‘}\lambda :. \supset \vdash . \text{Prop}$

***112·3.** $\vdash . \Sigma\text{‘}\Lambda = \Lambda$ [*37·29 . *40·21 . *112·1]

***112·301.** $\vdash . \Sigma\text{‘}\iota\text{‘}\Lambda = \Lambda$

Dem.

$\vdash . *112·102 . \supset \vdash . \Sigma\text{‘}\iota\text{‘}\Lambda = \hat{R} \{(\exists \alpha, x) . \alpha \in \iota\text{‘}\Lambda . x \in \alpha . R = x \downarrow \alpha\}$

[*51·15] $= \hat{R} \{(\exists x) . x \in \Lambda . R = x \downarrow \Lambda\}$

[*24·15] $= \Lambda . \supset \vdash . \text{Prop}$

***112·302.** $\vdash . \Sigma\text{‘}\kappa = \Sigma\text{‘}(\kappa - \iota\text{‘}\Lambda)$

Dem.

$\vdash . *112·102 . \supset \vdash . \Sigma\text{‘}\kappa = \hat{R} \{(\exists \alpha, x) . \alpha \in \kappa . x \in \alpha . R = x \downarrow \alpha\}$

[*10·24] $= \hat{R} \{(\exists \alpha, x) . \alpha \in \kappa . \exists ! \alpha . x \in \alpha . R = x \downarrow \alpha\}$

[*53·52] $= \hat{R} \{(\exists \alpha, x) . \alpha \in \kappa - \iota\text{‘}\Lambda . x \in \alpha . R = x \downarrow \alpha\}$

[*112·102] $= \Sigma\text{‘}(\kappa - \iota\text{‘}\Lambda) . \supset \vdash . \text{Prop}$

Thus if Λ is a member of a class of classes, it does not affect the value of their arithmetical sum.

***112·303.** $\vdash : \kappa \cap \lambda = \Lambda . \supset . \Sigma\text{‘}\kappa \cap \Sigma\text{‘}\lambda = \Lambda$

Dem.

$\vdash . *112·102 . \supset$

$\vdash : R \in \Sigma\text{‘}\kappa \cap \Sigma\text{‘}\lambda . \equiv . (\exists \alpha, \beta, x, y) . \alpha \in \kappa . \beta \in \lambda . x \in \alpha . y \in \beta . R = x \downarrow \alpha = y \downarrow \beta .$

[*55·202] $\supset . (\exists \alpha, x) . \alpha \in \kappa \cap \lambda . x \in \alpha .$

[*24·5] $\supset . \exists ! \kappa \cap \lambda$ (1)

$\vdash . (1) . \text{Transp} . \supset \vdash . \text{Prop}$

***112·304.** $\vdash : \Sigma\text{‘}\kappa = \Lambda . \equiv . s\text{‘}\kappa = \Lambda$

Dem.

$\vdash . *112·3·301 . *53·24 . \supset \vdash : s\text{‘}\kappa = \Lambda . \supset . \Sigma\text{‘}\kappa = \Lambda$ (1)

$\vdash . *112·102 . \supset \vdash : \alpha \in \kappa . x \in \alpha . \supset . x \downarrow \alpha \in \Sigma\text{‘}\kappa :$

[*10·24.*40·11] $\supset \vdash : \exists ! s\text{‘}\kappa . \supset . \exists ! \Sigma\text{‘}\kappa$ (2)

$\vdash . (1) . (2) . \supset \vdash . \text{Prop}$

***112·31.** $\vdash . \Sigma\text{‘}(\kappa \cup \lambda) = \Sigma\text{‘}\kappa \cup \Sigma\text{‘}\lambda$

Dem.

$\vdash . *112·1 . \supset \vdash . \Sigma\text{‘}(\kappa \cup \lambda) = s\text{‘}\epsilon \downarrow\text{“}(\kappa \cup \lambda)$

[*40·31] $= s\text{‘}\epsilon \downarrow\text{“}\kappa \cup s\text{‘}\epsilon \downarrow\text{“}\lambda$

[*112·1] $= \Sigma\text{‘}\kappa \cup \Sigma\text{‘}\lambda . \supset \vdash . \text{Prop}$

∗112·311. $\vdash : \kappa \cap \lambda = \Lambda \,.\, \supset \,.\, \Sigma Nc`(\kappa \cup \lambda) = \Sigma Nc`\kappa +_c \Sigma Nc`\lambda$

Dem.

$\vdash \,.\, *112\cdot303 \,.\, *110\cdot32 \,.\, \supset$

$\vdash : Hp \,.\, \supset \,.\, Nc`(\Sigma`\kappa \cup \Sigma`\lambda) = Nc`\Sigma`\kappa +_c Nc`\Sigma`\lambda$

$[*112\cdot101] \qquad = \Sigma Nc`\kappa +_c \Sigma Nc`\lambda \qquad (1)$

$\vdash \,.\, (1) \,.\, *112\cdot31 \,.\, \supset \vdash \,.\, Prop$

∗112·32. $\vdash \,.\, \Sigma`\iota`\alpha = \epsilon \downarrow \alpha$

Dem.

$\vdash \,.\, *53\cdot31 \,.\, *112\cdot1 \,.\, \supset \vdash \,.\, \Sigma`\iota`\alpha = s`\iota`\epsilon \downarrow \alpha$

$[*53\cdot02] \qquad = \epsilon \downarrow \alpha \,.\, \supset \vdash \,.\, Prop$

∗112·321. $\vdash \,.\, \Sigma Nc`\iota`\alpha = Nc`\alpha$ [∗112·32·101 . ∗85·601]

∗112·33. $\vdash \,.\, \Sigma`(\iota`\alpha \cup \iota`\beta) = \epsilon \downarrow \alpha \cup \epsilon \downarrow \beta$ [∗112·32·31]

∗112·331. $\vdash \,.\, \Sigma`(\kappa \cup \iota`\beta) = \Sigma`\kappa \cup \epsilon \downarrow \beta$ [∗112·31·32]

∗112·34. $\vdash : \alpha \neq \beta \,.\, \supset \,.\, \Sigma Nc`(\iota`\alpha \cup \iota`\beta) = Nc`\alpha +_c Nc`\beta$

Dem.

$\vdash \,.\, *51\cdot231 \,.\, *112\cdot311 \,.\, \supset$

$\vdash : Hp \,.\, \supset \,.\, \Sigma Nc`(\iota`\alpha \cup \iota`\beta) = \Sigma Nc`\iota`\alpha +_c \Sigma Nc`\iota`\beta$

$[*112\cdot321] \qquad = Nc`\alpha +_c Nc`\beta : \supset \vdash \,.\, Prop$

This proposition establishes the agreement of the two definitions of addition, namely that in ∗110 and that in ∗112. It will be seen that the definition of ∗112 is inapplicable to the addition of a class to itself, if this is to give the double of the class, instead of (like logical addition) simply reproducing the class. Hence the need of the condition $\alpha \neq \beta$ in the above proposition.

∗112·341. $\vdash : \beta \sim \epsilon\, \kappa \,.\, \supset \,.\, \Sigma Nc`(\kappa \cup \iota`\beta) = \Sigma Nc`\kappa +_c Nc`\beta$

Dem.

$\vdash \,.\, *51\cdot211 \,.\, \supset \vdash : Hp \,.\, \supset \,.\, \kappa \cap \iota`\beta = \Lambda \,.$

$[*112\cdot311] \qquad \supset \,.\, \Sigma Nc`(\kappa \cup \iota`\beta) = \Sigma Nc`\kappa +_c \Sigma Nc`\iota`\beta$

$[*112\cdot321] \qquad = \Sigma Nc`\kappa +_c Nc`\beta : \supset \vdash \,.\, Prop$

∗112·35. $\vdash : \alpha \neq \beta \,.\, \alpha \neq \gamma \,.\, \beta \neq \gamma \,.\, \supset \,.\, \Sigma Nc`(\iota`\alpha \cup \iota`\beta \cup \iota`\gamma) = Nc`\alpha +_c Nc`\beta +_c Nc`\gamma$

Dem.

$\vdash \,.\, *51\cdot231 \,.\, *112\cdot311 \,.\, \supset$

$\vdash : Hp \,.\, \supset \,.\, \Sigma Nc`(\iota`\alpha \cup \iota`\beta \cup \iota`\gamma) = \Sigma Nc`(\iota`\alpha \cup \iota`\beta) +_c \Sigma Nc`\iota`\gamma$

$[*112\cdot34\cdot321] \qquad = Nc`\alpha +_c Nc`\beta +_c Nc`\gamma : \supset \vdash \,.\, Prop$

Similar propositions can obviously be proved for any finite number of summands.

$*112 \cdot 4$. $\vdash : s'\kappa, s''\kappa \in \mathrm{Cls}^2 \mathrm{excl} \,.\, \supset . \, \Sigma \mathrm{Nc}'s'\kappa = \Sigma \mathrm{Nc}'s''\kappa$

Dem.

$$\begin{aligned} \vdash . *112 \cdot 15 . \supset \vdash : \mathrm{Hp} . \supset . \, \Sigma \mathrm{Nc}'s'\kappa &= \mathrm{Nc}'s's'\kappa \\ [*42 \cdot 1] \quad &= \mathrm{Nc}'s's''\kappa \\ [*112 \cdot 15] \quad &= \Sigma \mathrm{Nc}'s''\kappa : \supset \vdash . \mathrm{Prop} \end{aligned}$$

$*112 \cdot 41$. $\vdash . \, s'\Sigma''\lambda = \Sigma' s'\lambda$

Dem.

$$\begin{aligned} \vdash . *112 \cdot 1 . \supset \vdash . \, s'\Sigma''\lambda &= s's''\epsilon \downarrow'''\lambda \\ [*42 \cdot 1] \quad &= s's'\epsilon \downarrow'''\lambda \\ [*40 \cdot 38] \quad &= s'\epsilon \downarrow''s'\lambda \\ [*112 \cdot 1] \quad &= \Sigma's'\lambda . \supset \vdash . \mathrm{Prop} \end{aligned}$$

$*112 \cdot 42$. $\vdash : \lambda \in \mathrm{Cls}^2 \mathrm{excl} \,.\, \supset . \, \Sigma''\lambda \in \mathrm{Cls}^2 \mathrm{excl}$

Dem.

$$\begin{aligned} \vdash . *112 \cdot 303 . \supset \vdash :. \lambda \in \mathrm{Cls}^2 \mathrm{excl} . \supset &: \beta, \gamma \in \lambda . \beta \neq \gamma . \supset_{\beta, \gamma} . \Sigma'\beta \cap \Sigma'\gamma = \Lambda : \\ [*30 \cdot 37 . \mathrm{Transp} . *37 \cdot 63] \quad \supset &: \mu, \nu \in \Sigma''\lambda . \mu \neq \nu . \supset_{\mu, \nu} . \mu \cap \nu = \Lambda : \\ [*84 \cdot 1] \quad \supset &: \Sigma''\lambda \in \mathrm{Cls}^2 \mathrm{excl} :. \supset \vdash . \mathrm{Prop} \end{aligned}$$

$*112 \cdot 43$. $\vdash : \lambda \in \mathrm{Cls}^2 \mathrm{excl} \,.\, \supset . \, \mathrm{Nc}'\Sigma'\Sigma''\lambda = \mathrm{Nc}'\Sigma's'\lambda$

Dem.

$$\begin{aligned} \vdash . *112 \cdot 15 \cdot 42 . \supset \vdash : \mathrm{Hp} . \supset . \, \mathrm{Nc}'\Sigma'\Sigma''\lambda &= \mathrm{Nc}'s'\Sigma''\lambda \\ [*112 \cdot 41] \quad &= \mathrm{Nc}'\Sigma's'\lambda : \supset \vdash . \mathrm{Prop} \end{aligned}$$

The above is the associative law for arithmetical addition.

*113. ON THE ARITHMETICAL PRODUCT OF TWO CLASSES OR OF TWO CARDINALS.

Summary of *113.

In this number, we give a definition of multiplication which can be extended to any finite number of factors, but not to an infinite number of factors. We define first the arithmetical class-product of two classes α and β, and thence the product of two cardinals μ and ν as the number of terms in the product of α and β when α has μ terms and β has ν terms. In *114, we shall give a definition of multiplication which is not restricted to a finite number of factors. The advantages of the definition to be given in this number are, that it does not require the factors to be of the same type, and that it enables us to multiply a class by itself without (as in logical addition and multiplication) simply reproducing the class in question. The disadvantage of the definition in this number is the impossibility of extending it to an infinite number of factors.

The arithmetical class-product of two classes α and β, which we denote by $\beta \times \alpha$[*], is the class of all ordinal couples which take their referent from α and their relatum from β, *i.e.* it is the class of all such relations as $x \downarrow y$, where $x \in \alpha$ and $y \in \beta$. For a given y, the class of couples we obtain is $\downarrow y``\alpha$, which is similar to α; and the number of such classes, for varying y, is $\mathrm{Nc}`\beta$. Thus we have $\mathrm{Nc}`\beta$ classes of $\mathrm{Nc}`\alpha$ couples, and $\beta \times \alpha$ is the logical sum of these classes of couples. The class of such classes as $\downarrow y``\alpha$, where $y \in \beta$, is important again in connection with exponentiation; we have $\downarrow y``\alpha = \alpha \underset{''}{\downarrow} y$, whence the class of such classes, when y is varied among the β's, is $\alpha \underset{''}{\downarrow} ``\beta$, and

$$\beta \times \alpha = s`\alpha \underset{''}{\downarrow} ``\beta \quad (\text{cf. } *40{\cdot}7),$$

which we take as the definition of $\beta \times \alpha$.

We represent the arithmetical product of μ and ν by $\mu \times_c \nu$. This, as well as $\mathrm{Nc}`\alpha \times_c \mathrm{Nc}`\beta$, is defined in terms of $\alpha \times \beta$ exactly as, in *110, the sum was defined in terms of $\alpha + \beta$.

[*] We define this as $\beta \times \alpha$, rather than $\alpha \times \beta$, for the sake of certain analogies with products in relation-arithmetic. Cf. *166.

The present number contains many propositions which belong to the theory of $\alpha \underset{\prime\prime}{\downarrow} {}^{\prime\prime}\beta$ rather than (specially) of $\beta \times \alpha$; and many propositions are rather logical than arithmetical in their nature, *i.e.* they might have been given in $*55$. The line is, however, so hard to draw that it has seemed better to deal simultaneously with all propositions on $\alpha \underset{\prime\prime}{\downarrow} {}^{\prime\prime}\beta$ or on its sum, which is $\beta \times \alpha$. Thus in the present number, the early propositions, down to $*113 \cdot 118$, deal mainly with logical properties of $\alpha \underset{\prime\prime}{\downarrow} {}^{\prime\prime}\beta$ and $\beta \times \alpha$; the following propositions, down to $*113 \cdot 13$, deal mainly with arithmetical properties of $\alpha \underset{\prime\prime}{\downarrow} {}^{\prime\prime}\beta$; the propositions $*113 \cdot 14$—$\cdot 191$ are concerned mainly with arithmetical properties of $\beta \times \alpha$; $*113 \cdot 2$—$\cdot 27$ deal with the simpler properties of $\mu \times_c \nu$; $*113 \cdot 3$—$\cdot 34$ give propositions involving the multiplicative axiom, and exhibiting the connection (assuming this axiom) of addition and multiplication; $*113 \cdot 4$—$\cdot 491$ are concerned with various forms of the distributive law; $*113 \cdot 5$—$\cdot 541$ deal with the associative law of multiplication, and the remaining propositions deal with multiplication by 0 or 1 or 2.

The most important propositions in the present number are the following:

$*113 \cdot 101$. $\vdash : R \in \beta \times \alpha . \equiv . (\exists x, y) . x \in \alpha . y \in \beta . R = x \downarrow y$

This merely embodies the definition of $\beta \times \alpha$.

$*113 \cdot 105$. $\vdash : \exists ! \alpha . \supset . \alpha \underset{\prime\prime}{\downarrow} \in 1 \rightarrow 1$

This proposition is especially useful in dealing with exponentiation ($*116$).

$*113 \cdot 114$. $\vdash :. \alpha = \Lambda . \vee . \beta = \Lambda : \equiv . \beta \times \alpha = \Lambda$

It is in virtue of this proposition that a product of a finite number of factors only vanishes when one of its factors vanishes.

$*113 \cdot 118$. $\vdash . s^{\prime} \mathrm{D}^{\prime\prime}(\beta \times \alpha) \subset \alpha . s^{\prime} \mathrm{G}^{\prime\prime}(\beta \times \alpha) \subset \beta$

This proposition is chiefly useful in the analogous theory of ordinal products ($*165$, $*166$), where it enables us to apply $*74 \cdot 773$. Unless $\beta = \Lambda$, we have $s^{\prime} \mathrm{D}^{\prime\prime}(\beta \times \alpha) = \alpha$, and unless $\alpha = \Lambda$, $s^{\prime} \mathrm{G}^{\prime\prime}(\beta \times \alpha) = \beta$ ($*113 \cdot 116$).

$*113 \cdot 12$. $\vdash : \exists ! \alpha . \supset . \alpha \underset{\prime\prime}{\downarrow} {}^{\prime\prime}\beta \in \mathrm{Nc}^{\prime}\beta \cap \mathrm{Cl\,excl}^{\prime}\mathrm{Nc}^{\prime}\alpha$

I.e. unless α is null, $\alpha \underset{\prime\prime}{\downarrow} {}^{\prime\prime}\beta$ consists of $\mathrm{Nc}^{\prime}\beta$ mutually exclusive classes each having $\mathrm{Nc}^{\prime}\alpha$ members.

$*113 \cdot 127$. $\vdash : R \upharpoonright \gamma \in \alpha \,\overline{\mathrm{sm}}\, \gamma . S \upharpoonright \delta \in \beta \,\overline{\mathrm{sm}}\, \delta . \supset .$

$$(R \parallel \breve{S}) \upharpoonright (\delta \times \gamma) \in (\alpha \underset{\prime\prime}{\downarrow} {}^{\prime\prime}\beta) \,\overline{\mathrm{sm}}\, \overline{\mathrm{sm}}\, (\gamma \underset{\prime\prime}{\downarrow} {}^{\prime\prime}\delta)$$

This is an important proposition, since it gives a double correlator of $\alpha \underset{\prime\prime}{\downarrow} {}^{\prime\prime}\beta$ with $\gamma \underset{\prime\prime}{\downarrow} {}^{\prime\prime}\delta$ whenever simple correlators of α with γ and of β with δ are given. It leads at once to

***113·13.** $\vdash : \alpha \operatorname{sm} \gamma . \beta \operatorname{sm} \delta . \supset . \alpha \underset{,,}{\downarrow} {}``\beta \operatorname{sm} \operatorname{sm} \gamma \underset{,,}{\downarrow} {}``\delta . (\beta \times \alpha) \operatorname{sm} (\delta \times \gamma)$

This proposition is fundamental in the theory of multiplication, since it shows that the number of members of $\beta \times \alpha$ depends only upon the numbers of members of α and β. It is also fundamental in the theory of exponentiation, as will appear in *116.

***113·141.** $\vdash . \operatorname{Nc}`(\alpha \times \beta) = \operatorname{Nc}`(\beta \times \alpha)$

This is the source of the commutative law of multiplication (*113·27).

***113·146.** $\vdash : \alpha \neq \beta . \supset . \alpha \times \beta \operatorname{sm} \epsilon_\Delta`(\iota`\alpha \cup \iota`\beta)$

This connects our present theory of multiplication with the theory of selections.

We come next to propositions concerning $\mu \times_c \nu$. We have

***113·204.** $\vdash :. \mu = \Lambda . \lor . \nu = \Lambda . \lor . \sim(\mu, \nu \in \operatorname{NC}) : \supset . \mu \times_c \nu = \Lambda$

The use of this proposition, like that of *110·4, is for avoiding trivial exceptions.

***113·23.** $\vdash . \mu \times_c \nu \in \operatorname{NC}$

***113·25.** $\vdash . \operatorname{Nc}`\gamma \times_c \operatorname{Nc}`\delta = \operatorname{Nc}`(\gamma \times \delta)$

This proposition enables us to infer propositions on products of cardinals from propositions on products of classes, and is therefore constantly used.

***113·27.** $\vdash . \mu \times_c \nu = \nu \times_c \mu$

This is the commutative law of cardinal multiplication.

The chief proposition using the multiplicative axiom is

***113·31.** $\vdash :. \operatorname{Mult ax} . \supset : \mu, \nu \in \operatorname{NC} . \kappa \in \nu \cap \operatorname{Cl}`\mu . \supset . \Sigma`\kappa \in \mu \times_c \nu$

I.e. assuming the multiplicative axiom, the sum of the numbers of members in ν classes of μ terms is $\mu \times_c \nu$. If we had taken this sum as *defining* $\mu \times_c \nu$, almost all propositions on multiplication would have required the multiplicative axiom. The advantage of $\alpha \underset{,,}{\downarrow} {}``\beta$ is that, given $\alpha \operatorname{sm} \gamma$ and $\beta \operatorname{sm} \delta$, we can construct a double correlator of $\alpha \underset{,,}{\downarrow} {}``\beta$ with $\gamma \underset{,,}{\downarrow} {}``\delta$, without using the multiplicative axiom. This is proved in *113·127 (mentioned above).

The distributive law, which is next considered, has various forms. We have, to begin with,

***113·4.** $\vdash . (\beta \cup \gamma) \times \alpha = (\beta \times \alpha) \cup (\gamma \times \alpha)$

whence, using also the commutative law, we easily deduce

***113·43.** $\vdash . (\nu +_c \varpi) \times_c \mu = \mu \times_c (\nu +_c \varpi) = (\mu \times_c \nu) +_c (\mu \times_c \varpi)$

But the distributive law also holds when, instead of enumerated summands β, γ or ν, ϖ, the summands are given as the members of a class κ, which may be infinite. We have

***113·48.** $\vdash . s`\alpha \times `` \kappa = \alpha \times s`\kappa = \text{Cnv}``(s`\kappa \times \alpha)$

whence, using the definitions of *112, we find

***113·491.** $\vdash : \kappa \in \text{Cls}^2 \text{excl} . \supset . \Sigma\text{Nc}`\alpha \times `` \kappa = \text{Nc}`(\alpha \times \Sigma`\kappa) = \text{Nc}`\alpha \times_c \Sigma\text{Nc}`\kappa$

This is an extension of the distributive law to the case where the number of summands may be infinite.

The associative law

***113·54.** $\vdash . (\mu \times_c \nu) \times_c \varpi = \mu \times_c (\nu \times_c \varpi)$

is proved without any difficulty.

We prove next that $\mu \times_c \nu = 0$ when, and only when, $\mu = 0$ or $\nu = 0$, μ, ν being existent cardinals (*113·602); that a cardinal is unchanged when it is multiplied by 1 (*113·62·621); that $\mu \times_c 2 = \mu +_c \mu$ (*113·66) and that $\mu \times_c (\nu +1) = (\mu \times_c \nu) +_c \mu$ (*113·671).

***113·02.** $\beta \times \alpha = s`\alpha \downarrow_{,,} `` \beta$ Df

***113·03.** $\mu \times_c \nu = \hat{\xi} \{(\exists \alpha, \beta) . \mu = \text{N}_0\text{c}`\alpha . \nu = \text{N}_0\text{c}`\beta . \xi \,\text{sm}\, (\alpha \times \beta)\}$ Df

***113·04.** $\text{Nc}`\beta \times_c \mu = \text{N}_0\text{c}`\beta \times_c \mu$ Df

***113·05.** $\mu \times_c \text{Nc}`\alpha = \mu \times_c \text{N}_0\text{c}`\alpha$ Df

In relation to types, *113·03·04·05 call for similar remarks to those made in *110 for addition.

***113·1.** $\vdash . \beta \times \alpha = s`\alpha \downarrow_{,,} `` \beta$ [(*113·02)]

***113·101.** $\vdash : R \in \beta \times \alpha . \equiv . (\exists x, y) . x \in \alpha . y \in \beta . R = x \downarrow y$ [*40·7 . *113·1]

***113·102.** $\vdash : y \in \beta . \supset . \alpha \downarrow_{,,} y = (\alpha \uparrow \beta)_\Delta ` \iota ` y$

Dem.

$\vdash . *35·103 . \supset$

$\vdash :. \text{Hp} . \supset : x (\alpha \uparrow \beta) y . \equiv . x \in \alpha :$

$[*85·51] \supset : (\alpha \uparrow \beta)_\Delta ` \iota ` y = \downarrow y `` \alpha$

$[(*38·03)] \qquad = \alpha \downarrow_{,,} y : \supset \vdash . \text{Prop}$

***113·103.** $\vdash . \alpha \downarrow_{,,} `` \beta = (\alpha \uparrow \beta)_\Delta `` \iota `` \beta = (\alpha \uparrow \beta) \,\underline{\downarrow}\, `` \beta$ [*113·102 . *85·52]

***113·104.** $\vdash . \text{E} ! \alpha \downarrow_{,,} ` y$ [*38·12]

✱113·105. $\vdash : \exists ! \alpha \,.\, \supset \,.\, \alpha \underset{,,}{\downarrow} \epsilon 1 \rightarrow 1$

Dem.

$\vdash \,.\, *113\cdot104 \,.\, *71\cdot166 \,.\, \supset \vdash \,.\, \alpha \underset{,,}{\downarrow} \epsilon 1 \rightarrow \text{Cls}$ (1)

$\vdash \,.\, *38\cdot131 \,.\, \supset \vdash : \alpha \underset{,,}{\downarrow} {}^{\backprime} y = \alpha \underset{,,}{\downarrow} {}^{\backprime} z \,.\, x \epsilon \alpha \,.\, \supset \,.\, x \downarrow y \epsilon \alpha \underset{,,}{\downarrow} {}^{\backprime} z \,.$

$[*38\cdot131] \qquad \supset \,.\, (\exists x') \,.\, x' \epsilon \alpha \,.\, x \downarrow y = x' \downarrow z \,.$

$[*55\cdot202] \qquad \supset \,.\, y = z$ (2)

$\vdash \,.\, (2) \,.\, *10\cdot11\cdot23\cdot35 \,.\, \supset \vdash : \exists ! \alpha \,.\, \alpha \underset{,,}{\downarrow} {}^{\backprime} y = \alpha \underset{,,}{\downarrow} {}^{\backprime} z \,.\, \supset \,.\, y = z$ (3)

$\vdash \,.\, (1) \,.\, (3) \,.\, *71\cdot54 \,.\, \supset \vdash \,.\, \text{Prop}$

✱113·106. $\vdash : x \epsilon \alpha \,.\, y \epsilon \beta \,.\, \supset \,.\, x \downarrow y \epsilon \beta \times \alpha$ [✱113·101]

✱113·107. $\vdash : \exists ! \alpha \,.\, \exists ! \beta \,.\, \supset \,.\, \exists ! \beta \times \alpha$ [✱113·106]

✱113·11. $\vdash : \exists ! \alpha \,.\, \supset \,.\, \alpha \underset{,,}{\downarrow} {}^{\backprime\backprime} \beta \epsilon \text{Nc}^{\backprime} \beta : (y) \,.\, \alpha \underset{,,}{\downarrow} y \epsilon \text{Nc}^{\backprime} \alpha$

Dem.

$\vdash \,.\, *113\cdot105\cdot104 \,.\, *73\cdot26 \,.\, \supset \vdash : \exists ! \alpha \,.\, \supset \,.\, \alpha \underset{,,}{\downarrow} {}^{\backprime\backprime} \beta \,\text{sm}\, \beta$ (1)

$\vdash \,.\, *38\cdot2 \,.\, *73\cdot611 \,.\, \qquad \supset \vdash \,.\, \alpha \underset{,,}{\downarrow} y \,\text{sm}\, \alpha$ (2)

$\vdash \,.\, (1) \,.\, (2) \,.\, \qquad \supset \vdash \,.\, \text{Prop}$

✱113·111. $\vdash \,.\, \alpha \underset{,,}{\downarrow} {}^{\backprime\backprime} \beta \epsilon \text{Cls}^2 \,\text{excl}$ [✱113·103 . ✱85·55]

✱113·112. $\vdash : \alpha = \Lambda \,.\, \exists ! \beta \,.\, \supset \,.\, \alpha \underset{,,}{\downarrow} {}^{\backprime\backprime} \beta = \iota^{\backprime} \Lambda$

Dem.

$\vdash \,.\, *38\cdot3 \,.\, \supset \vdash : \text{Hp} \,.\, \supset \,.\, \alpha \underset{,,}{\downarrow} {}^{\backprime\backprime} \beta = \hat{\mu} \{(\exists y) \,.\, y \epsilon \beta \,.\, \mu = \downarrow y^{\backprime\backprime} \Lambda\}$

$[*37\cdot29] \qquad = \hat{\mu} \{(\exists y) \,.\, y \epsilon \beta \,.\, \mu = \Lambda\}$

$[\text{Hp}] \qquad = \iota^{\backprime} \Lambda$

✱113·113. $\vdash : \beta = \Lambda \,.\, \supset \,.\, \alpha \underset{,,}{\downarrow} {}^{\backprime\backprime} \beta = \Lambda$ [✱37·29]

✱113·114. $\vdash :. \alpha = \Lambda \,.\, \mathsf{v} \,.\, \beta = \Lambda : \equiv \,.\, \beta \times \alpha = \Lambda$ [✱113·1·112·113·107 . ✱53·24]

✱113·115. $\vdash \,.\, \dot{s}^{\backprime}(\beta \times \alpha) = \alpha \uparrow \beta$

Dem.

$\vdash \,.\, *113\cdot101 \,.\, *41\cdot11 \,.\, \supset$

$\vdash : u \{\dot{s}^{\backprime}(\beta \times \alpha)\} v \,.\, \equiv \,.\, (\exists R, x, y) \,.\, x \epsilon \alpha \,.\, y \epsilon \beta \,.\, R = x \downarrow y \,.\, uRv \,.$

$[*13\cdot195 \,.\, *55\cdot13] \quad \equiv \,.\, (\exists x, y) \,.\, x \epsilon \alpha \,.\, y \epsilon \beta \,.\, u = x \,.\, v = y \,.$

$[*13\cdot22] \quad \equiv \,.\, u \epsilon \alpha \,.\, v \epsilon \beta \,.$

$[*35\cdot103] \quad \equiv \,.\, u \,(\alpha \uparrow \beta)\, v : \supset \vdash \,.\, \text{Prop}$

✱113·116. $\vdash : \exists ! \beta \,.\, \supset \,.\, s^{\backprime}\text{D}^{\backprime\backprime}(\beta \times \alpha) = \alpha : \exists ! \alpha \,.\, \supset \,.\, s^{\backprime}\text{Ꮖ}^{\backprime\backprime}(\beta \times \alpha) = \beta$
[✱113·115 . ✱41·43·44 . ✱35·85·86]

✱113·117. $\vdash :. \alpha = \Lambda \,.\, \mathsf{v} \,.\, \beta = \Lambda : \supset \,.\, s^{\backprime}\text{D}^{\backprime\backprime}(\beta \times \alpha) = \Lambda \,.\, s^{\backprime}\text{Ꮖ}^{\backprime\backprime}(\beta \times \alpha) = \Lambda$
[✱113·115 . ✱41·43·44 . ✱35·88]

✱113·118. $\vdash \,.\, s^{\backprime}\text{D}^{\backprime\backprime}(\beta \times \alpha) \subset \alpha \,.\, s^{\backprime}\text{Ꮖ}^{\backprime\backprime}(\beta \times \alpha) \subset \beta$ [✱113·116·117]

***113·12.** $\vdash : \exists ! \alpha . \supset . \alpha \underset{\prime\prime}{\downarrow} {}^{\prime\prime}\beta \in \mathrm{Nc}^{\prime}\beta \cap \mathrm{Cl\,excl}^{\prime}\mathrm{Nc}^{\prime}\alpha$ [*113·11·111]

***113·121.** $\vdash . \Sigma^{\prime}\alpha \underset{\prime\prime}{\downarrow} {}^{\prime\prime}\beta \,\mathrm{sm}\, \beta \times \alpha$ [*112·15 . *113·111·1]

***113·122.** $\vdash : R \upharpoonright \gamma, S \upharpoonright \delta \in \mathrm{Cls} \rightarrow 1 . \gamma \subset \mathrm{Cl}^{\prime}R . \delta \subset \mathrm{Cl}^{\prime}S . \supset . (R \parallel \breve{S}) \upharpoonright (\delta \times \gamma) \in 1 \rightarrow 1$ [*74·773 . *113·118]

***113·123.** $\vdash : R \upharpoonright \gamma, S \upharpoonright \delta \in 1 \rightarrow \mathrm{Cls} . \gamma \subset \mathrm{Cl}^{\prime}R . \delta \subset \mathrm{Cl}^{\prime}S . z \in \gamma . w \in \delta . \supset .$
$(R \parallel \breve{S})^{\prime}(z \downarrow w) = (R^{\prime}z) \downarrow (S^{\prime}w)$ [*55·61]

***113·124.** $\vdash : R \upharpoonright \gamma, S \upharpoonright \delta \in 1 \rightarrow \mathrm{Cls} . \gamma \subset \mathrm{Cl}^{\prime}R . \delta \subset \mathrm{Cl}^{\prime}S . w \in \delta . \supset .$
$(R \parallel \breve{S})^{\prime\prime}\gamma \underset{\prime\prime}{\downarrow} w = (R^{\prime\prime}\gamma) \underset{\prime\prime}{\downarrow} (S^{\prime}w)$

Dem.

$\vdash . *113·123 . *38·131 . \supset \vdash : \mathrm{Hp} . \supset . (R \parallel \breve{S})^{\prime\prime} \downarrow w^{\prime\prime}\gamma = \downarrow (S^{\prime}w)^{\prime\prime}R^{\prime\prime}\gamma .$
[*38·2] $\supset . (R \parallel \breve{S})^{\prime\prime}\gamma \underset{\prime\prime}{\downarrow} w = (R^{\prime\prime}\gamma) \underset{\prime\prime}{\downarrow} (S^{\prime}w) : \supset \vdash . \mathrm{Prop}$

***113·125.** $\vdash : R \upharpoonright \gamma, S \upharpoonright \delta \in 1 \rightarrow \mathrm{Cls} . \gamma \subset \mathrm{Cl}^{\prime}R . \delta \subset \mathrm{Cl}^{\prime}S . \supset .$
$(R \parallel \breve{S})_{\epsilon}{}^{\prime\prime}\gamma \underset{\prime\prime}{\downarrow} {}^{\prime\prime}\delta = (R^{\prime\prime}\gamma) \underset{\prime\prime}{\downarrow} {}^{\prime\prime}(S^{\prime\prime}\delta)$ [*113·124]

***113·126.** $\vdash : \mathrm{Hp}\, *113·125 . \supset . (R \parallel \breve{S})^{\prime\prime}(\delta \times \gamma) = (S^{\prime\prime}\delta) \times (R^{\prime\prime}\gamma)$

Dem.

$\vdash . *113·1 . *40·38 . \supset \vdash . (R \parallel \breve{S})^{\prime\prime}(\delta \times \gamma) = s^{\prime}(R \parallel \breve{S})^{\prime\prime\prime}\gamma \underset{\prime\prime}{\downarrow} {}^{\prime\prime}\delta$ (1)

$\vdash . (1) . *113·125 . \supset \vdash : \mathrm{Hp} . \supset . (R \parallel \breve{S})^{\prime\prime}(\delta \times \gamma) = s^{\prime}(R^{\prime\prime}\gamma) \underset{\prime\prime}{\downarrow} {}^{\prime\prime}(S^{\prime\prime}\delta)$
[*113·1] $= (S^{\prime\prime}\delta) \times (R^{\prime\prime}\gamma) : \supset \vdash . \mathrm{Prop}$

***113·127.** $\vdash : R \upharpoonright \gamma \in \alpha \,\overline{\mathrm{sm}}\, \gamma . S \upharpoonright \delta \in \beta \,\overline{\mathrm{sm}}\, \delta . \supset .$
$(R \parallel \breve{S}) \upharpoonright (\delta \times \gamma) \in (\alpha \underset{\prime\prime}{\downarrow} {}^{\prime\prime}\beta) \,\overline{\mathrm{sm}}\,\overline{\mathrm{sm}}\, (\gamma \underset{\prime\prime}{\downarrow} {}^{\prime\prime}\delta)$
[*113·122·125 . *43·302 . *73·142 . *111·14]

***113·128.** $\vdash : \mathrm{Hp}\, *113·127 . \supset . (R \parallel \breve{S}) \upharpoonright (\delta \times \gamma) \in (\beta \times \alpha) \,\overline{\mathrm{sm}}\, (\delta \times \gamma) .$
$(R \parallel \breve{S})_{\epsilon} \upharpoonright (\gamma \underset{\prime\prime}{\downarrow} {}^{\prime\prime}\delta) \in (\alpha \underset{\prime\prime}{\downarrow} {}^{\prime\prime}\beta) \,\overline{\mathrm{sm}}\, (\gamma \underset{\prime\prime}{\downarrow} {}^{\prime\prime}\delta)$ [*113·127 . *111·15]

***113·13.** $\vdash : \alpha \,\mathrm{sm}\, \gamma . \beta \,\mathrm{sm}\, \delta . \supset . \alpha \underset{\prime\prime}{\downarrow} {}^{\prime\prime}\beta \,\mathrm{sm\,sm}\, \gamma \underset{\prime\prime}{\downarrow} {}^{\prime\prime}\delta . (\beta \times \alpha) \,\mathrm{sm}\, (\delta \times \gamma)$
[*113·127 . *111·4·44 . *113·1]

***113·14.** $\vdash . \alpha \times \beta = \mathrm{Cnv}^{\prime\prime}(\beta \times \alpha)$ [*113·101 . *55·14]

***113·141.** $\vdash . \mathrm{Nc}^{\prime}(\alpha \times \beta) = \mathrm{Nc}^{\prime}(\beta \times \alpha)$ [*113·14 . *73·4]

***113·142.** $\vdash : \exists ! \beta . \supset . \mathrm{D}^{\prime\prime}(\beta \times \alpha) = \iota^{\prime\prime}\alpha : \exists ! \alpha . \supset . \mathrm{Cl}^{\prime\prime}(\beta \times \alpha) = \iota^{\prime\prime}\beta$

Dem.

$\vdash . *55·261 . *2·02 . \supset \vdash : y \in \beta . \supset . \mathrm{D}^{\prime\prime}\alpha \underset{\prime\prime}{\downarrow} y = \iota^{\prime\prime}\alpha$
[*37·63] $\supset \vdash : \gamma \in \mathrm{D}^{\prime\prime\prime}\alpha \underset{\prime\prime}{\downarrow} {}^{\prime\prime}\beta . \supset . \gamma = \iota^{\prime\prime}\alpha$ (1)

$\vdash . *37\cdot45 . \quad \supset\vdash : \exists ! \beta . \supset . \exists ! \text{D}``\text{`}\alpha \underset{,,}{\downarrow} ``\beta$ (2)

$\vdash . (1) . (2) . *51\cdot141 . \supset\vdash : \exists ! \beta . \supset . \text{D}``\text{`}\alpha \underset{,,}{\downarrow} ``\beta = \iota\text{`}\iota``\alpha .$

$[*40\cdot38 . *53\cdot02] \quad \supset . \text{D}``s\text{`}\alpha \underset{,,}{\downarrow} ``\beta = \iota``\alpha$ (3)

$\vdash . *55\cdot251 . \quad \supset\vdash : \exists ! \alpha . \supset . \text{Ɑ}``\alpha \underset{,,}{\downarrow} y = \iota\text{`}\iota\text{`}y .$

$[*37\cdot355] \quad \supset . \text{Ɑ}``\text{`}\alpha \underset{,,}{\downarrow} ``\beta = \iota``\iota``\beta .$

$[*40\cdot38 . *53\cdot22] \quad \supset . \text{Ɑ}``s\text{`}\alpha \underset{,,}{\downarrow} ``\beta = \iota``\beta$ (4)

$\vdash . (3) . (4) . *113\cdot1 . \quad \supset\vdash . \text{Prop}$

$*113\cdot143$. $\vdash : \alpha \neq \beta . P = x \downarrow y . R = x \downarrow \alpha \cup y \downarrow \beta . \supset .$

$P = (R\text{`}\alpha) \downarrow (R\text{`}\beta) . R = \text{D}\text{`}P \uparrow \iota\text{`}\alpha \cup \text{Ɑ}\text{`}P \uparrow \iota\text{`}\beta$

Dem.

$\vdash . *55\cdot62 . \supset\vdash : \text{Hp} . \supset . R\text{`}\alpha = x . R\text{`}\beta = y .$

$[*30\cdot19 . *13\cdot15] \quad \supset . P = (R\text{`}\alpha) \downarrow (R\text{`}\beta)$ (1)

$\vdash . *55\cdot15 . \supset\vdash : \text{Hp} . \supset . \text{D}\text{`}P = \iota\text{`}x . \text{Ɑ}\text{`}P = \iota\text{`}y .$

$[*55\cdot1] \quad \supset . R = \text{D}\text{`}P \uparrow \iota\text{`}\alpha \cup \text{Ɑ}\text{`}P \uparrow \iota\text{`}\beta$ (2)

$\vdash . (1) . (2) . \supset\vdash . \text{Prop}$

$*113\cdot144$. $\vdash : \alpha \neq \beta . T = \hat{P}\hat{R} \{(\exists x, y) . x \in \alpha . y \in \beta . P = x \downarrow y . R = x \downarrow \alpha \cup y \downarrow \beta\} .$

$\supset . T \in 1 \to 1 . \text{D}\text{`}T = \beta \times \alpha . \text{Ɑ}\text{`}T = \epsilon_\Delta\text{`}(\iota\text{`}\alpha \cup \iota\text{`}\beta)$

Dem.

$\vdash . *21\cdot33 . \supset\vdash :. \text{Hp} . \supset :$

$PTR . QTR . \supset . (\exists x, y, z, w) . x, z \in \alpha . y, w \in \beta . P = x \downarrow y . Q = z \downarrow w .$

$R = x \downarrow \alpha \cup y \downarrow \beta = z \downarrow \alpha \cup w \downarrow \beta .$

$[*113\cdot143] \supset . P = (R\text{`}\alpha) \downarrow (R\text{`}\beta) . Q = (R\text{`}\alpha) \downarrow (R\text{`}\beta) .$

$[*13\cdot172] \quad \supset . P = Q$ (1)

$\vdash . *21\cdot33 . \supset\vdash :. \text{Hp} . \supset : PTQ . PTR . \supset .$

$(\exists x, y, z, w) . x, z \in \alpha . y, w \in \beta . P = x \downarrow y = w \downarrow z . Q = x \downarrow \alpha \cup y \downarrow \beta . R = z \downarrow \alpha \cup w \downarrow \beta .$

$[*113\cdot143] \supset . Q = \text{D}\text{`}P \uparrow \iota\text{`}\alpha \cup \text{Ɑ}\text{`}P \uparrow \iota\text{`}\beta . R = \text{D}\text{`}P \uparrow \iota\text{`}\alpha \cup \text{Ɑ}\text{`}P \uparrow \iota\text{`}\beta .$

$[*13\cdot172] \quad \supset . Q = R$ (2)

$\vdash . *33\cdot13 . \quad \supset\vdash : \text{Hp} . \supset .$

$\text{D}\text{`}T = \hat{P} \{(\exists R, x, y) . x \in \alpha . y \in \beta . P = x \downarrow y . R = x \downarrow \alpha \cup y \downarrow \beta\}$

$[*11\cdot55 . *13\cdot19] \quad = \hat{P} \{(\exists x, y) . x \in \alpha . y \in \beta . P = x \downarrow y\}$

$[*113\cdot101] \quad = \beta \times \alpha$ (3)

$\vdash . *33\cdot131 . \supset\vdash : \text{Hp} . \supset .$

$\text{Ɑ}\text{`}T = \hat{R} \{(\exists P, x, y) . x \in \alpha . y \in \beta . P = x \downarrow y . R = x \downarrow \alpha \cup y \downarrow \beta\}$

$[*11\cdot55 . *13\cdot19] \quad = \hat{R} \{(\exists x, y) . x \in \alpha . y \in \beta . R = x \downarrow \alpha \cup y \downarrow \beta\}$

$[*80\cdot9] \quad = \epsilon_\Delta\text{`}(\iota\text{`}\alpha \cup \iota\text{`}\beta)$ (4)

$\vdash . (1) . (2) . (3) . (4) . \supset\vdash . \text{Prop}$

Note to ∗113·144. In virtue of ∗113·143 and ∗55·61 we have

$\vdash :. \mathrm{Hp} *113{\cdot}144 . \supset : PTR . \equiv . R \in \epsilon_\Delta{}^\text{‘}(\iota{}^\text{‘}\alpha \cup \iota{}^\text{‘}\beta) . P = (R \parallel \breve{R}){}^\text{‘}(\alpha \downarrow \beta).$

At a later stage (in ∗150) we shall put

$$R \dagger S = (R \parallel \breve{R}){}^\text{‘}S \quad \text{Df.}$$

Thus we shall have, anticipating this notation,

$$\vdash : \mathrm{Hp} *113{\cdot}144 . \supset . T = \{\dagger(\alpha \downarrow \beta)\} \upharpoonright \epsilon_\Delta{}^\text{‘}(\iota{}^\text{‘}\alpha \cup \iota{}^\text{‘}\beta).$$

Hence we have

$$\vdash : \alpha \neq \beta . \supset . \{\dagger(\alpha \downarrow \beta)\} \upharpoonright \epsilon_\Delta{}^\text{‘}(\iota{}^\text{‘}\alpha \cup \iota{}^\text{‘}\beta) \in (\beta \times \alpha) \,\overline{\mathrm{sm}}\, \epsilon_\Delta{}^\text{‘}(\iota{}^\text{‘}\alpha \cup \iota{}^\text{‘}\beta).$$

∗113·145. $\vdash : \alpha \neq \beta . \supset . \beta \times \alpha \,\mathrm{sm}\, \epsilon_\Delta{}^\text{‘}(\iota{}^\text{‘}\alpha \cup \iota{}^\text{‘}\beta)$ [∗113·144]

∗113·146. $\vdash : \alpha \neq \beta . \supset . \alpha \times \beta \,\mathrm{sm}\, \epsilon_\Delta{}^\text{‘}(\iota{}^\text{‘}\alpha \cup \iota{}^\text{‘}\beta)$ [∗113·141·145]

∗113·147. $\vdash : \mathrm{Hp} *113{\cdot}144 . \beta \times \alpha = \mu . \supset .$

$$T = \hat{P}\hat{R}\{P \in \mu . R = \mathrm{D}{}^\text{‘}P \uparrow \iota{}^\text{‘}s{}^\text{‘}\mathrm{D}{}^\text{‘‘}\mu \cup \mathrm{G}{}^\text{‘}P \uparrow \iota{}^\text{‘}s{}^\text{‘}\mathrm{G}{}^\text{‘‘}\mu\}$$

Dem.

$\vdash . *113{\cdot}114 . \mathrm{Transp} . \supset \vdash : \mathrm{Hp} . P \in \mu . \supset . \exists ! \alpha . \exists ! \beta .$

[∗113·142.∗53·22] $\supset . \alpha = s{}^\text{‘}\mathrm{D}{}^\text{‘‘}\mu . \beta = s{}^\text{‘}\mathrm{G}{}^\text{‘‘}\mu$ (1)

$\vdash . *113{\cdot}101{\cdot}143 . \supset \vdash :. \mathrm{Hp} . P \in \mu . \supset : PTR . \equiv . R = \mathrm{D}{}^\text{‘}P \uparrow \iota{}^\text{‘}\alpha \cup \mathrm{G}{}^\text{‘}P \uparrow \iota{}^\text{‘}\beta$ (2)

$\vdash . *113{\cdot}144 . \quad \supset \vdash : \mathrm{Hp} . PTR . \supset . P \in \mu$ (3)

$\vdash . (1) . (2) . (3) . *113{\cdot}101 . \supset \vdash . \mathrm{Prop}$

The advantage of this proposition is that it exhibits the correlator of $\beta \times \alpha$ and $\epsilon_\Delta{}^\text{‘}(\iota{}^\text{‘}\alpha \cup \iota{}^\text{‘}\beta)$ as a function of $\beta \times \alpha$.

∗113·148. $\vdash : \alpha \cap \beta = \Lambda . \supset . C \upharpoonright (\alpha \times \beta) \in 1 \to 1$

Dem.

$\vdash . *113{\cdot}101 . *55{\cdot}15 . \supset$

$\vdash :. \mathrm{Hp} . \supset : R, S \in \alpha \times \beta . C{}^\text{‘}R = C{}^\text{‘}S . \equiv .$

$(\exists x, x', y, y') . x, x' \in \alpha . y, y' \in \beta . R = y \downarrow x , S = y' \downarrow x' . \iota{}^\text{‘}x \cup \iota{}^\text{‘}y = \iota{}^\text{‘}x' \cup \iota{}^\text{‘}y' .$

[∗54·6] $\supset . (\exists x, x', y, y') . x, x' \in \alpha . y, y' \in \beta . R = y \downarrow x . S = y' \downarrow x' . x = x' . y = y' .$

[∗13·22·172] $\supset . R = S$ (1)

$\vdash . (1) . *71{\cdot}55 . \supset \vdash . \mathrm{Prop}$

∗113·15. $\vdash . C{}^\text{‘‘}(\alpha \times \beta) = C{}^\text{‘‘}(\beta \times \alpha) = \hat{\xi}\{(\exists x, y) . x \in \alpha . y \in \beta . \xi = \iota{}^\text{‘}x \cup \iota{}^\text{‘}y\}$

Dem.

$\vdash . *113{\cdot}1 . *40{\cdot}38 . \supset \vdash . C{}^\text{‘‘}(\beta \times \alpha) = s{}^\text{‘}C{}^\text{‘‘‘}\alpha \downarrow_{,,} {}^\text{‘‘}\beta$

[∗40·4] $= \hat{\xi}\{(\exists y) . y \in \beta . \xi \in C{}^\text{‘‘}\alpha \downarrow_{,,} y\}$

[∗55·27.∗38·2] $= \hat{\xi}\{(\exists x, y) . x \in \alpha . y \in \beta . \xi = \iota{}^\text{‘}x \cup \iota{}^\text{‘}y\}$ (1)

$\vdash . (1) \dfrac{\beta, \alpha}{\alpha, \beta} . \quad \supset \vdash . C{}^\text{‘‘}(\alpha \times \beta) = \hat{\xi}\{(\exists x, y) . x \in \alpha . y \in \beta . \xi = \iota{}^\text{‘}x \cup \iota{}^\text{‘}y\}$ (2)

$\vdash . (1) . (2) . \supset \vdash . \mathrm{Prop}$

***113·151.** $\vdash : \alpha \neq \beta . \supset . C``(\alpha \times \beta) = \mathrm{D}``\epsilon_\Delta`(\iota`\alpha \cup \iota`\beta)$ [*113·15 . *80·92]

***113·152.** $\vdash : \alpha \cap \beta = \Lambda . \supset . C``(\alpha \times \beta) \,\mathrm{sm}\, (\alpha \times \beta) . \mathrm{D}``\epsilon_\Delta`(\iota`\alpha \cup \iota`\beta) \,\mathrm{sm}\, (\alpha \times \beta)$

Dem.

$\vdash . *84·41·62 . \supset \vdash : \mathrm{Hp} . \alpha \neq \beta . \supset . \mathrm{D}``\epsilon_\Delta`(\iota`\alpha \cup \iota`\beta) \,\mathrm{sm}\, \epsilon_\Delta`(\iota`\alpha \cup \iota`\beta)$ (1)

$\vdash . (1) . *113·146·151 . \supset$

$\vdash : \mathrm{Hp} . \alpha \neq \beta . \supset . C``(\alpha \times \beta) \,\mathrm{sm}\, (\alpha \times \beta) . \mathrm{D}``\epsilon_\Delta`(\iota`\alpha \cup \iota`\beta) \,\mathrm{sm}\, \alpha \times \beta$ (2)

$\vdash . *24·38 . \supset \vdash : \mathrm{Hp} . \alpha = \beta . \supset . \alpha = \Lambda . \beta = \Lambda .$

[*113·114.*83·11.*37·29] $\supset . \alpha \times \beta = \Lambda . \mathrm{D}``\epsilon_\Delta`(\iota`\alpha \cup \iota`\beta) = \Lambda . C``(\alpha \times \beta) = \Lambda .$

[*73·47] $\supset . C``(\alpha \times \beta) \mathrm{sm} (\alpha \times \beta) . \mathrm{D}``\epsilon_\Delta`(\iota`\alpha \cup \iota`\beta) \mathrm{sm} (\alpha \times \beta)$ (3)

$\vdash . (2) . (3) . \supset \vdash .$ Prop

The following proposition is only significant when λ and μ are classes of relations. It is used in relation-arithmetic (*172·34).

***113·153.** $\vdash : \dot{s}`\lambda \dot{\cap} \dot{s}`\mu = \dot{\Lambda} . \supset . \dot{s} \,|\, C \upharpoonright (\lambda \times \mu) \epsilon (s`\lambda \underset{''}{\cup} ``\mu) \overline{\mathrm{sm}} (\lambda \times \mu) . s`\lambda \underset{''}{\cup} ``\mu \,\mathrm{sm}\, \lambda \times \mu$

Dem.

$\vdash . *55·15 . *53·13 . \supset \vdash : R = T \downarrow S . \supset . \dot{s}`C`R = S \cup T$ (1)

$\vdash . (1) . *113·101 . \supset$

$\vdash : R, R' \epsilon \lambda \times \mu . \dot{s}`C`R = \dot{s}`C`R' . \supset .$

$(\exists S, S', T, T') . S, S' \epsilon \lambda . T, T' \epsilon \mu . R = T \downarrow S . R' = T' \downarrow S' . S \cup T = S' \cup T'$ (2)

$\vdash . (2) . *25·48 . *41·13 . \supset$

$\vdash :. \mathrm{Hp} . \supset : R, R' \epsilon \lambda \times \mu . \dot{s}`C`R = \dot{s}`C`R' . \supset . R = R'$ (3)

$\vdash . (1) . *113·101 . \supset \vdash . \dot{s}``C``(\lambda \times \mu) = \hat{M} \{(\exists S, T) . S \epsilon \lambda . T \epsilon \mu . M = S \cup T\}$

[*40·7] $= s`\lambda \underset{''}{\cup} ``\mu$ (4)

$\vdash . (3) . (4) . *73·25 . \supset \vdash .$ Prop

***113·16.** $\vdash : t`\alpha = t`\beta . \supset . \mathrm{Nc}`(\alpha \times \beta) =$

$\hat{\xi} \{(\exists \gamma, \delta) . \gamma \epsilon \mathrm{N^1c}`\alpha . \delta \epsilon \mathrm{N^1c}`\beta . \gamma \cap \delta = \Lambda . \xi \,\mathrm{sm}\, \mathrm{D}``\epsilon_\Delta`(\iota`\gamma \cup \iota`\delta)\}$

Dem.

$\vdash . *113·152 . \supset \vdash :. \gamma \epsilon \mathrm{N^1c}`\alpha . \delta \epsilon \mathrm{N^1c}`\beta . \gamma \cap \delta = \Lambda . \supset :$

$\xi \,\mathrm{sm}\, \mathrm{D}``\epsilon_\Delta`(\iota`\gamma \cup \iota`\delta) . \equiv . \xi \,\mathrm{sm}\, (\gamma \times \delta) .$

[*113·13.*104·101] $\equiv . \xi \,\mathrm{sm}\, (\alpha \times \beta) .$

[*100·31] $\equiv . \xi \epsilon \mathrm{Nc}`(\alpha \times \beta)$ (1)

$\vdash . (1) . *5·32 . *11·11·341 . \supset$

$\vdash :. (\exists \gamma, \delta) . \gamma \epsilon \mathrm{N^1c}`\alpha . \delta \epsilon \mathrm{N^1c}`\beta . \gamma \cap \delta = \Lambda . \xi \,\mathrm{sm}\, \mathrm{D}``\epsilon_\Delta`(\iota`\gamma \cup \iota`\delta) . \equiv :$

$(\exists \gamma, \delta) . \gamma \epsilon \mathrm{N^1c}`\alpha . \delta \epsilon \mathrm{N^1c}`\beta . \gamma \cap \delta = \Lambda . \xi \epsilon \mathrm{Nc}`(\alpha \times \beta) :$

[*11·45] $\equiv : (\exists \gamma, \delta) . \gamma \epsilon \mathrm{N^1c}`\alpha . \delta \epsilon \mathrm{N^1c}`\beta . \gamma \cap \delta = \Lambda : \xi \epsilon \mathrm{Nc}`(\alpha \times \beta)$ (2)

$\vdash . (2) . *104·43 . \supset \vdash .$ Prop

***113·17.** $\vdash . \beta \times \alpha \in t' t' (\alpha \uparrow \beta)$

Dem.

$\vdash . *113·115 . *41·13 . \supset \vdash : R \in \beta \times \alpha . \supset . R \subset \alpha \uparrow \beta .$

[*64·201] $\supset . R \in t'(\alpha \uparrow \beta)$ (1)

$\vdash . (1) . *63·5 . \supset \vdash .$ Prop

***113·171.** $\vdash : \alpha \cap \beta = \Lambda . \supset . \exists ! \text{Nc}\,(t'\alpha)'(\alpha \times \beta)$

Dem.

$\vdash . *113·152·15 . \supset \vdash : \text{Hp} . \supset . \hat{\xi}\{(\exists x, y) . x \in \alpha . y \in \beta . \xi = \iota' x \cup \iota' y\} \in \text{Nc}'(\alpha \times \beta)$ (1)

$\vdash . *51·16 . \quad \supset \vdash : x \in \alpha . y \in \beta . \xi = \iota' x \cup \iota' y . \supset . x \in \alpha . x \in \xi .$

[*63·13] $\supset . \xi \in t'\alpha$ (2)

$\vdash . (2) . *11·11·35 . \supset$

$\vdash . \hat{\xi}\{(\exists x, y) . x \in \alpha . y \in \beta . \xi = \iota' x \cup \iota' y\} \subset t'\alpha .$

[*63·5] $\supset \vdash . \hat{\xi}\{(\exists x, y) . x \in \alpha . y \in \beta . \xi = \iota' x \cup \iota' y\} \in t' t' \alpha$ (3)

$\vdash . (1) . (3) . \supset \vdash : \text{Hp} . \supset . \exists ! \text{Nc}'(\alpha \times \beta) \cap t' t' \alpha$ (4)

$\vdash . (4) . *102·6 . \supset \vdash .$ Prop

Note that the hypothesis $\alpha \cap \beta = \Lambda$ is only significant when α and β are of the same type.

***113·172.** $\vdash : \alpha \in t'\beta . \supset . \exists ! \text{Nc}\,(t^{2}{}'\alpha)'(\alpha \times \beta)$

Dem.

$\vdash . *113·16 . \quad \supset \vdash :. \text{Hp} . \supset : \gamma \in \text{N}^1\text{c}'\alpha . \delta \in \text{N}^1\text{c}'\beta . \gamma \cap \delta = \Lambda . \supset .$

$D''\epsilon_\Delta'(\iota'\gamma \cup \iota'\delta) \in \text{Nc}'(\alpha \times \beta)$ (1)

$\vdash . (1) . *104·43 . \supset \vdash : \text{Hp} . \supset .$

$(\exists \gamma, \delta) . \gamma \in \text{N}^1\text{c}'\alpha . \delta \in \text{N}^1\text{c}'\beta . D''\epsilon_\Delta'(\iota'\gamma \cup \iota'\delta) \in \text{Nc}'(\alpha \times \beta)$ (2)

$\vdash . *104·1 . \supset \vdash : \gamma \in \text{N}^1\text{c}'\alpha . \supset . \gamma \in t^{2}{}'\alpha .$

[*63·61·621] $\supset . \iota'\gamma \cup \iota'\delta \in t' t^{2}{}'\alpha .$

[*83·81] $\supset . D''\epsilon_\Delta'(\iota'\gamma \cup \iota'\delta) \in t' t^{2}{}'\alpha$ (3)

$\vdash . (2) . (3) . \supset \vdash : \text{Hp} . \supset . \exists ! \text{Nc}'(\alpha \times \beta) \cap t' t^{2}{}'\alpha$ (4)

$\vdash . (4) . *102·6 . \supset \vdash .$ Prop

***113·18.** $\vdash : \exists ! \alpha . \exists ! \beta . \alpha \times \beta = \alpha' \times \beta' . \supset . \alpha = \alpha' . \beta = \beta'$

Dem.

$\vdash . *113·114 . \supset \vdash : \text{Hp} . \supset . \exists ! \alpha' \times \beta' .$

[*113·114] $\supset . \exists ! \alpha' . \exists ! \beta'$ (1)

$\vdash . *30·37 . \quad \supset \vdash : \text{Hp} . \supset . s'\text{G}''(\alpha \times \beta) = s'\text{G}''(\alpha' \times \beta') .$

[*113·142.(1)] $\supset . s'\iota''\alpha = s'\iota''\alpha' .$

[*53·22] $\supset . \alpha = \alpha'$ (2)

Similarly $\vdash : \text{Hp} . \supset . \beta = \beta'$ (3)

$\vdash . (2) . (3) . \supset \vdash .$ Prop

***113·181.** $\vdash : \exists ! \alpha . \exists ! \alpha' . \alpha \times \beta = \alpha' \times \beta' . \supset . \beta = \beta'$

Dem.

$\vdash . *13·172 . \supset \vdash : \beta = \Lambda . \beta' = \Lambda . \supset . \beta = \beta'$ (1)

$\vdash . *113·18 . \supset \vdash : \text{Hp} . \sim(\beta = \Lambda . \beta' = \Lambda) . \supset . \beta = \beta'$ (2)

$\vdash . (1) . (2) . \supset \vdash . \text{Prop}$

***113·182.** $\vdash : \exists ! \beta . \exists ! \beta' . \alpha \times \beta = \alpha' \times \beta' . \supset . \alpha = \alpha'$

[Proof as in *113·181]

***113·183.** $\vdash : \exists ! \alpha . \exists ! \beta . \supset . F``(\alpha \times \beta) = s`C``(\alpha \times \beta) = \alpha \cup \beta$

Dem.

$\vdash . *40·57 . \supset \vdash . s`C``(\alpha \times \beta) = s`\text{D}``(\alpha \times \beta) \cup s`\text{Ɑ}``(\alpha \times \beta)$ (1)

$\vdash . *40·56 . \supset \vdash . F``(\alpha \times \beta) = s`C``(\alpha \times \beta)$ (2)

$\vdash . *113·142 . \supset \vdash : \text{Hp} . \supset . s`\text{Ɑ}``(\alpha \times \beta) = s`\iota``\alpha$

$[*53·22] \qquad = \alpha$ (3)

$\vdash . *113·142 . \supset \vdash : \text{Hp} . \supset . s`\text{D}``(\alpha \times \beta) = s`\iota``\beta$

$[*53·22] \qquad = \beta$ (4)

$\vdash . (3) . (4) . \supset \vdash : \text{Hp} . \supset . s`\text{D}``(\alpha \times \beta) \cup s`\text{Ɑ}``(\alpha \times \beta) = \alpha \cup \beta$ (5)

$\vdash . (1) . (2) . (5) . \supset \vdash . \text{Prop}$

***113·19.** $\vdash : \exists ! (\alpha \times \beta) \cap (\gamma \times \delta) . \equiv . \exists ! \alpha \cap \gamma . \exists ! \beta \cap \delta$

Dem.

$\vdash . *113·101 . \supset \vdash :. \exists ! (\alpha \times \beta) \cap (\gamma \times \delta) . \equiv :$

$(\exists x, y, z, w) . x \epsilon \alpha . y \epsilon \beta . z \epsilon \gamma . w \epsilon \delta . x \downarrow y = w \downarrow z :$

$[*55·202] \equiv : (\exists x, y, z, w) . x \epsilon \alpha . y \epsilon \beta . z \epsilon \gamma . w \epsilon \delta . x = z . y = w :$

$[*13·22] \quad \equiv : (\exists x, y) . x \epsilon \alpha \cap \gamma . y \epsilon \beta \cap \delta :. \supset \vdash . \text{Prop}$

***113·191.** $\vdash :. \exists ! \alpha . \supset : \exists ! \alpha \underset{,,}{\downarrow}``\beta \cap \alpha \underset{,,}{\downarrow}``\gamma . \equiv . \exists ! \beta \cap \gamma$

Dem.

$\vdash . *37·6 . \supset \vdash : \exists ! \alpha \underset{,,}{\downarrow}``\beta \cap \alpha \underset{,,}{\downarrow}``\gamma . \equiv . (\exists y, z) . y \epsilon \beta . z \epsilon \gamma . \alpha \underset{,,}{\downarrow} y = \alpha \underset{,,}{\downarrow} z$ (1)

$\vdash . *113·105 . *71·57 . \supset \vdash :. \text{Hp} . \supset : \alpha \underset{,,}{\downarrow} y = \alpha \underset{,,}{\downarrow} z . \equiv . y = z :$

$[(1)] \qquad \supset : \exists ! \alpha \underset{,,}{\downarrow}``\beta \cap \alpha \underset{,,}{\downarrow}``\gamma . \equiv . (\exists y, z) . y \epsilon \beta . z \epsilon \gamma . y = z .$

$[*13·195] \qquad \equiv . \exists ! \alpha \cap \beta :. \supset \vdash . \text{Prop}$

***113·2.** $\vdash : \xi \epsilon \mu \times_c \nu . \equiv . (\exists \alpha, \beta) . \mu = \text{N}_0\text{c}`\alpha . \nu = \text{N}_0\text{c}`\beta . \xi \text{ sm } (\alpha \times \beta)$ [(*113·03)]

***113·201.** $\vdash :. \xi \epsilon \mu \times_c \nu . \equiv : \mu, \nu \epsilon \text{NC} : (\exists \alpha, \beta) . \alpha \epsilon \mu . \beta \epsilon \nu . \xi \text{ sm } (\alpha \times \beta)$

[*113·2 . *103·27]

∗113·202. $\vdash :. \xi \epsilon \mu \times_c \nu . \equiv : \exists ! \mu . \exists ! \nu : (\exists \gamma, \delta) . \mu = Nc'\gamma . \nu = Nc'\delta . \xi \text{ sm } (\gamma \times \delta)$

Dem.

$\vdash . *113·201 . *100·4 . \supset$

$\vdash :. \xi \epsilon \mu \times_c \nu . \equiv : (\exists \alpha, \beta, \gamma, \delta) . \mu = Nc'\gamma . \nu = Nc'\delta . \alpha \epsilon \mu . \beta \epsilon \nu . \xi \text{ sm } (\alpha \times \beta) .$

[∗100·31] $\equiv : (\exists \alpha, \beta, \gamma, \delta) . \mu = Nc'\gamma . \nu = Nc'\delta . \alpha \text{ sm } \gamma . \beta \text{ sm } \delta . \xi \text{ sm } (\alpha \times \beta) .$

[∗113·13.∗73·37] $\equiv : (\exists \alpha, \beta, \gamma, \delta) . \mu = Nc'\gamma . \nu = Nc'\delta . \alpha \text{ sm } \gamma . \beta \text{ sm } \delta . \xi \text{ sm } (\gamma \times \delta) .$

[∗100·31] $\equiv : (\exists \alpha, \beta, \gamma, \delta) . \mu = Nc'\gamma . \nu = Nc'\delta . \alpha \epsilon \mu . \beta \epsilon \nu . \xi \text{ sm } (\gamma \times \delta) .$

[∗10·35] $\equiv : \exists ! \mu . \exists ! \nu : (\exists \gamma, \delta) . \mu = Nc'\gamma . \nu = Nc'\delta . \xi \text{ sm } (\gamma \times \delta) :.$

$\supset \vdash . \text{Prop}$

∗113·203. $\vdash : \exists ! \mu \times_c \nu . \supset . \mu, \nu \epsilon NC - \iota'\Lambda . \mu, \nu \epsilon N_0C$ [∗113·201·202·2]

∗113·204. $\vdash :. \mu = \Lambda . \vee . \nu = \Lambda . \vee . \sim(\mu, \nu \epsilon NC) : \supset . \mu \times_c \nu = \Lambda$ [∗113·203]

∗113·205. $\vdash : \sim(\mu, \nu \epsilon N_0C) . \supset . \mu \times_c \nu = \Lambda$ [∗113·203]

∗113·21. $\vdash :. \mu, \nu \epsilon NC . \supset : \xi \epsilon \mu \times_c \nu . \equiv . (\exists \alpha, \beta) . \alpha \epsilon \mu . \beta \epsilon \nu . \xi \text{ sm } (\alpha \times \beta)$
[∗113·201]

∗113·22. $\vdash : \xi \epsilon Nc(\eta)'\gamma \times_c Nc(\zeta)'\delta . \equiv . \exists ! Nc(\eta)'\gamma . \exists ! Nc(\zeta)'\delta . \xi \text{ sm } (\gamma \times \delta)$

Dem.

$\vdash . *113·21 . *100·41 . \supset \vdash : \xi \epsilon Nc(\eta)'\gamma \times_c Nc(\zeta)'\delta . \equiv .$

$(\exists \alpha, \beta) . \alpha \epsilon Nc(\eta)'\gamma . \beta \epsilon Nc(\zeta)'\delta . \xi \text{ sm } (\alpha \times \beta) .$

[∗102·6] $\equiv . (\exists \alpha, \beta) . \alpha \epsilon Nc(\eta)'\gamma . \beta \epsilon Nc(\zeta)'\delta . \alpha \text{ sm } \gamma . \beta \text{ sm } \delta . \xi \text{ sm } (\alpha \times \beta) .$

[∗113·13.∗73·37] $\equiv . (\exists \alpha, \beta) . \alpha \epsilon Nc(\eta)'\gamma . \beta \epsilon Nc(\zeta)'\delta . \alpha \text{ sm } \gamma . \beta \text{ sm } \delta . \xi \text{ sm } (\gamma \times \delta) .$

[∗102·6] $\equiv . (\exists \alpha, \beta) . \alpha \epsilon Nc(\eta)'\gamma . \beta \epsilon Nc(\zeta)'\delta . \xi \text{ sm } (\gamma \times \delta) .$

[∗10·35] $\equiv . \exists ! Nc(\eta)'\gamma . \exists ! Nc(\zeta)'\delta . \xi \text{ sm } (\gamma \times \delta) : \supset \vdash . \text{Prop}$

∗113·221. $\vdash : \exists ! Nc(\eta)'\gamma . \exists ! Nc(\zeta)'\delta . \supset . Nc(\eta)'\gamma \times_c Nc(\zeta)'\delta = Nc'(\gamma \times \delta)$
[∗113·22]

∗113·222. $\vdash . N_0c'\gamma \times_c N_0c'\delta = Nc'(\gamma \times \delta)$

Dem.

$\vdash . *103·1·13 . \supset \vdash . N_0c'\gamma = Nc(\gamma)'\gamma . N_0c'\delta = Nc(\delta)'\delta . \exists ! N_0c'\gamma . \exists ! N_0c'\delta .$

[∗113·221] $\supset \vdash . N_0c'\gamma \times_c N_0c'\delta = Nc'(\gamma \times \delta) . \supset \vdash . \text{Prop}$

∗113·23. $\vdash . \mu \times_c \nu \epsilon NC$

Dem.

$\vdash . *113·222 . *100·41 . \supset \vdash : \mu, \nu \epsilon N_0C . \supset . \mu \times_c \nu \epsilon NC$ (1)

$\vdash . *113·205 . *102·74 . \supset \vdash : \sim(\mu, \nu \epsilon N_0C) . \supset . \mu \times_c \nu \epsilon NC$ (2)

$\vdash . (1) . (2) . \supset \vdash . \text{Prop}$

∗113·24. $\vdash . Nc'\gamma \times_c Nc'\delta = N_0c'\gamma \times_c N_0c'\delta$ [(∗113·04·05)]

∗113·25. $\vdash . Nc'\gamma \times_c Nc'\delta = Nc'(\gamma \times \delta)$ [∗113·24·222]

This proposition constitutes part of the reason for our definitions. It is obvious that such definitions ought, if possible, to be chosen as will yield this proposition.

***113·251.** $\vdash . \gamma \times \delta \in \mathrm{Nc}`\gamma \times_c \mathrm{Nc}`\delta$ [*113·25 . *100·3]

***113·26.** $\vdash : \mu, \nu \in \mathrm{NC} . \exists ! \mathrm{sm}_\eta``\mu . \exists ! \mathrm{sm}_\zeta``\nu . \supset . \mu \times_c \nu = \mathrm{sm}_\eta``\mu \times_c \mathrm{sm}_\zeta``\nu$

Dem.

$\vdash . *37·29 . \text{Transp} . \supset \vdash : \text{Hp} . \supset . \exists ! \mu . \exists ! \nu .$

[*102·64] $\supset . (\exists \alpha, \beta, \gamma, \delta) . \mu = \mathrm{Nc}(\alpha)`\gamma . \nu = \mathrm{Nc}(\beta)`\delta$ (1)

$\vdash . *102·88 . \supset \vdash : \mu = \mathrm{Nc}(\alpha)`\gamma . \nu = \mathrm{Nc}(\beta)`\delta . \exists ! \mathrm{sm}_\eta``\mu . \exists ! \mathrm{sm}_\zeta``\nu . \supset .$

$\mathrm{sm}_\eta``\mu = \mathrm{Nc}(\eta)`\gamma . \mathrm{sm}_\zeta``\nu = \mathrm{Nc}(\zeta)`\delta . \exists ! \mathrm{Nc}(\eta)`\gamma . \exists ! \mathrm{Nc}(\zeta)`\delta .$

[*113·221] $\supset . \mathrm{sm}_\eta``\mu \times_c \mathrm{sm}_\zeta``\nu = \mathrm{Nc}`(\gamma \times \delta)$ (2)

$\vdash . *37·29 . \text{Transp} . *113·221 . \supset$

$\vdash : \mu = \mathrm{Nc}(\alpha)`\gamma . \nu = \mathrm{Nc}(\beta)`\delta . \exists ! \mathrm{sm}_\eta``\mu . \exists ! \mathrm{sm}_\zeta``\nu . \supset . \mu \times_c \nu = \mathrm{Nc}`(\gamma \times \delta)$ (3)

$\vdash . (2) . (3) . \supset \vdash : \mu = \mathrm{Nc}(\alpha)`\gamma . \nu = \mathrm{Nc}(\beta)`\delta . \exists ! \mathrm{sm}_\eta``\mu . \exists ! \mathrm{sm}_\zeta``\nu . \supset .$

$\mu \times_c \nu = \mathrm{sm}_\eta``\mu \times_c \mathrm{sm}_\zeta``\nu$ (4)

$\vdash . (4) . *11·11·35·45 . (1) . \supset \vdash . \text{Prop}$

***113·261.** $\vdash : \mu, \nu \in \mathrm{NC} . \supset . \mu \times_c \nu = \mu^{(1)} \times_c \nu^{(1)} = \mu_{(00)} \times_c \nu_{(00)} = \text{etc.}$

Here "etc." includes all *ascending* derivatives of μ. We shall only prove the result for $\mu^{(1)}$ and $\nu^{(1)}$, since it is proved in just the same way for the other cases. $\mu^{(1)} \times_c \nu^{(2)}$ or $\mu^{(1)} \times_c \nu_{(00)}$ or etc. will serve equally well; *i.e.* it is not necessary to take the same derivative of μ as of ν.

Dem.

$\vdash . *104·264·265 . \supset$

$\vdash : \text{Hp} . \exists ! \mu . \exists ! \nu . \supset . \mu^{(1)} = \mathrm{sm}_\mu``\mu . \nu^{(1)} = \mathrm{sm}_\nu``\mu . \exists ! \mu^{(1)} . \exists ! \nu^{(1)} .$

[*113·26] $\supset : \mu \times_c \nu = \mu^{(1)} \times_c \nu^{(1)}$ (1)

$\vdash . *104·264 . *113·204 . \supset$

$\vdash : \sim (\exists ! \mu . \exists ! \nu) . \supset . \mu \times_c \nu = \Lambda . \mu^{(1)} \times_c \nu^{(1)} = \Lambda$ (2)

$\vdash . (1) . (2) . \supset \vdash . \text{Prop}$

As appears in the above proof, if μ^i and ν^j are any derivatives of μ and ν, the above proposition holds provided we have

$$\exists ! \mu . \exists ! \nu . \supset . \exists ! \mu^i . \exists ! \nu^j .$$

Thus it holds for all *ascending* derivatives, but not always for descending derivatives.

***113·27.** $\vdash . \mu \times_c \nu = \nu \times_c \mu$

Dem.

$\vdash . *113·2·141 . \supset$

$\vdash : \xi \in \mu \times_c \nu . \equiv . (\exists \alpha, \beta) . \mu = \mathrm{N_0c}`\alpha . \nu = \mathrm{N_0c}`\beta . \xi \,\mathrm{sm}\, (\beta \times \alpha) .$

[*113·2] $\equiv . \xi \in \nu \times_c \mu : \supset \vdash . \text{Prop}$

Note that this proposition is not confined to the case in which μ and ν are cardinals. When either or both are not cardinals,

$$\mu \times_c \nu = \Lambda = \nu \times_c \mu .$$

***113·3.** $\vdash :. \text{Mult ax} . \supset : \kappa \in \text{Nc}‘\beta \cap \text{Cl}‘\text{Nc}‘\alpha . \supset . \Sigma‘\kappa \in \text{Nc}‘\alpha \times_c \text{Nc}‘\beta$

Dem.

$\vdash . *112·24 . *113·12 . \supset$

$\vdash :. \text{Mult ax} . \exists ! \alpha . \supset : \kappa \in \text{Nc}‘\beta \cap \text{Cl}‘\text{Nc}‘\alpha . \supset . \Sigma‘\kappa \,\text{sm}\, \Sigma‘\alpha \underset{,,}{\downarrow} ‘‘\beta .$

[*113·121] $\supset . \Sigma‘\kappa \,\text{sm}\, \beta \times \alpha .$

[*113·141·25] $\supset . \Sigma‘\kappa \in \text{Nc}‘\alpha \times_c \text{Nc}‘\beta$ (1)

$\vdash . *113·114·25 . \supset \vdash : \alpha = \Lambda . \supset . \text{Nc}‘\alpha \times_c \text{Nc}‘\beta = 0$ (2)

$\vdash . *101·14 . \supset \vdash :. \alpha = \Lambda . \kappa \in \text{Nc}‘\beta \cap \text{Cl}‘\text{Nc}‘\alpha . \supset : \kappa \in \text{Cl}‘\iota‘\Lambda :$

[*60·362] $\supset : \kappa = \iota‘\Lambda . \vee . \kappa = \Lambda :$

[*112·3·301] $\supset : \Sigma‘\kappa = \Lambda$ (3)

$\vdash . (2) . (3) . *54·102 . \supset \vdash : \alpha = \Lambda . \kappa \in \text{Nc}‘\beta \cap \text{Cl}‘\text{Nc}‘\alpha . \supset . \Sigma‘\kappa \in \text{Nc}‘\alpha \times_c \text{Nc}‘\beta$ (4)

$\vdash . (1) . (4) . \supset \vdash . \text{Prop}$

***113·31.** $\vdash :. \text{Mult ax} . \supset : \mu, \nu \in \text{NC} . \kappa \in \nu \cap \text{Cl}‘\mu . \supset . \Sigma‘\kappa \in \mu \times_c \nu$ [*113·3]

***113·32.** $\vdash :. \text{Mult ax} . \supset : \mu, \nu \in \text{NC} . \kappa \in \nu \cap \text{Cl excl}‘\mu . \supset . s‘\kappa \in \mu \times_c \nu$
[*112·15 . *113·31·23]

***113·33.** $\vdash :. \text{Mult ax} . \supset : \mu, \nu \in \text{NC} . \kappa \in \nu \cap \text{Cl}‘\mu . \lambda \in \mu \cap \text{Cl}‘\nu . \supset .$
$\Sigma\text{Nc}‘\kappa = \Sigma\text{Nc}‘\lambda = \mu \times_c \nu$ [*113·31·27·23]

***113·34.** $\vdash :. \text{Mult ax} . \supset : \mu, \nu \in \text{NC} . \kappa \in \nu \cap \text{Cl excl}‘\mu . \lambda \in \mu \cap \text{Cl excl}‘\nu . \supset .$
$\text{Nc}‘s‘\kappa = \text{Nc}‘s‘\lambda = \mu \times_c \nu$ [*113·32·27]

The above propositions give the connection of addition and multiplication.

The following propositions are concerned with various forms of the distributive law.

***113·4.** $\vdash . (\beta \cup \gamma) \times \alpha = (\beta \times \alpha) \cup (\gamma \times \alpha)$

Dem.

$\vdash . *113·1 . \supset \vdash . (\beta \cup \gamma) \times \alpha = s‘\alpha \underset{,,}{\downarrow} ‘‘(\beta \cup \gamma)$

[*40·31] $= s‘\alpha \underset{,,}{\downarrow} ‘‘\beta \cup s‘\alpha \underset{,,}{\downarrow} ‘‘\gamma$

[*113·1] $= (\beta \times \alpha) \cup (\gamma \times \alpha) . \supset \vdash . \text{Prop}$

***113·401.** $\vdash : \beta \cap \gamma = \Lambda . \supset . (\beta \times \alpha) \cap (\gamma \times \alpha) = \Lambda$ [*113·19 . Transp]

***113·41.** $\vdash . \text{Nc}‘(\beta + \gamma) \times_c \text{Nc}‘\alpha = \text{Nc}‘\{(\beta + \gamma) \times \alpha\} = \text{Nc}‘\{(\beta \times \alpha) + (\gamma \times \alpha)\}$
$= \text{Nc}‘(\beta \times \alpha) +_c \text{Nc}‘(\gamma \times \alpha)$

Dem.

$\vdash . *113·25 . *110·3 . \quad \supset \vdash . \text{Nc}‘(\beta + \gamma) \times_c \text{Nc}‘\alpha = \text{Nc}‘\{(\beta + \gamma) \times \alpha\} .$
$\text{Nc}‘\{(\beta \times \alpha) + (\gamma \times \alpha)\} = \text{Nc}‘(\beta \times \alpha) +_c \text{Nc}‘(\gamma \times \alpha)$ (1)

$\vdash . *113·4 . (*110·01) . \quad \supset \vdash . (\beta + \gamma) \times \alpha = (\downarrow \Lambda_\gamma ‘‘\iota‘‘\beta \times \alpha) \cup (\Lambda_\beta \downarrow ‘‘\iota‘‘\gamma \times \alpha)$ (2)

$\vdash . *113·13 . *110·12 . \quad \supset \vdash . \downarrow \Lambda_\gamma ‘‘\iota‘‘\beta \times \alpha \,\text{sm}\, \beta \times \alpha . \Lambda_\beta \downarrow ‘‘\iota‘‘\gamma \times \alpha \,\text{sm}\, \gamma \times \alpha$ (3)

$\vdash . *113·401 . *110·11 . \quad \supset \vdash . (\downarrow \Lambda_\gamma ‘‘\iota‘‘\beta \times \alpha) \cap (\Lambda_\beta \downarrow ‘‘\iota‘‘\gamma \times \alpha) = \Lambda$ (4)

$\vdash . *110·152 . (2) . (3) . (4) . \supset \vdash . (\beta + \gamma) \times \alpha \,\text{sm}\, \{(\beta \times \alpha) + (\gamma \times \alpha)\}$ (5)

$\vdash . (1) . (5) . \supset \vdash . \text{Prop}$

***113·42.** $\vdash . (\text{Nc}'\beta +_c \text{Nc}'\gamma) \times_c \text{Nc}'\alpha = \text{Nc}'(\beta + \gamma) \times_c \text{Nc}'\alpha$
$= (\text{Nc}'\beta \times_c \text{Nc}'\alpha) +_c (\text{Nc}'\gamma \times_c \text{Nc}'\alpha)$
[*110·3 . *113·25 . *113·41]

***113·421.** $\vdash . \text{Nc}'\alpha \times_c (\text{Nc}'\beta +_c \text{Nc}'\gamma) = \text{Nc}'\alpha \times_c \text{Nc}'(\beta + \gamma)$
$= (\text{Nc}'\alpha \times_c \text{Nc}'\beta) +_c (\text{Nc}'\alpha \times_c \text{Nc}'\gamma)$ [*113·42·27]

***113·43.** $\vdash . (\nu +_c \varpi) \times_c \mu = \mu \times_c (\nu +_c \varpi) = (\mu \times_c \nu) +_c (\mu \times_c \varpi)$

Dem.

$\vdash . *113{\cdot}27{\cdot}421 . \supset \vdash : \mu, \nu, \varpi \in \text{NC} . \supset . (\nu +_c \varpi) \times_c \mu = \mu \times_c (\nu +_c \varpi)$
$= (\mu \times_c \nu) +_c (\mu \times_c \varpi)$ (1)

$\vdash . *113{\cdot}204 . *110{\cdot}4 . \supset$

$\vdash : \sim(\mu, \nu, \varpi \in \text{NC}) . \supset . (\nu +_c \varpi) \times_c \mu = \Lambda . \mu \times_c (\nu +_c \varpi) = \Lambda .$
$(\mu \times_c \nu) +_c (\mu \times_c \varpi) = \Lambda$ (2)

$\vdash . (1) . (2) . \supset \vdash . \text{Prop}$

The following propositions are concerned with various forms of the distributive law, when the summands are not enumerated, but given as the members of a class.

The first of them (*113·44) gives the distributive law with regard to arithmetical class-multiplication and logical addition of classes.

***113·44.** $\vdash . (s'\kappa) \times \alpha = s'(\times \alpha)``\kappa$

Dem.

$\vdash . *113{\cdot}1 . \supset \vdash . s'(\times \alpha)``\kappa = s's``\alpha \downarrow ```\kappa$
[*42·1] $= s's'\alpha \downarrow ```\kappa$
[*40·38] $= s'\alpha \downarrow ``s'\kappa$
[*113·1] $= (s'\kappa) \times \alpha . \supset \vdash . \text{Prop}$

***113·45.** $\vdash : \kappa \in \text{Cls}^2 \text{ excl} . \supset . \times \alpha``\kappa \in \text{Cls}^2 \text{ excl}$

Dem.

$\vdash . *113{\cdot}19 . \quad \supset \vdash : \exists ! \times \alpha'\beta \cap \times \alpha'\gamma . \supset . \exists ! \beta \cap \gamma$ (1)

$\vdash . (1) . *84{\cdot}11 . \supset \vdash :. \text{Hp} . \supset : \beta, \gamma \in \kappa . \exists ! \times \alpha'\beta \cap \times \alpha'\gamma . \supset_{\beta,\gamma} . \beta = \gamma .$
[*30·37] $\supset_{\beta,\gamma} . \times \alpha'\beta = \times \alpha'\gamma :$
[*37·63] $\supset : \rho, \sigma \in \times \alpha``\kappa . \exists ! \rho \cap \sigma . \supset_{\rho,\sigma} . \rho = \sigma$ (2)

$\vdash . (2) . *84{\cdot}11 . \supset \vdash . \text{Prop}$

***113·46.** $\vdash : \kappa \in \text{Cls}^2 \text{excl} . \supset . \Sigma'\times \alpha``\kappa \text{ sm } (\Sigma'\kappa) \times \alpha$

Dem.

$\vdash . *112{\cdot}15 . \quad \supset \vdash : \text{Hp} . \supset . \Sigma'\kappa \text{ sm } s'\kappa .$
[*113·13] $\supset . (\Sigma'\kappa) \times \alpha \text{ sm } (s'\kappa) \times \alpha$ (1)

$\vdash . *112{\cdot}15 . *113{\cdot}45 . \supset \vdash : \text{Hp} . \supset . \Sigma'\times \alpha``\kappa \text{ sm } s'\times \alpha``\kappa$ (2)

$\vdash . (1) . (2) . *113{\cdot}44 . \supset \vdash . \text{Prop}$

***113·47.** $\vdash : \kappa \in \text{Cls}^2\,\text{excl} \,.\supset .\, \Sigma\text{Nc}`\times\alpha``\kappa = \text{Nc}`\{(\Sigma`\kappa)\times\alpha\} = \Sigma\text{Nc}`\kappa \times_c \text{Nc}`\alpha$
[*113·46]

This is the distributive law for arithmetical multiplication and arithmetical addition of the kind defined in *112.

***113·48.** $\vdash .\, s`\alpha\times``\kappa = \alpha\times(s`\kappa) = \text{Cnv}``\{(s`\kappa)\times\alpha\}$

Dem.

$\vdash .\, *113{\cdot}14 \,.\supset \vdash .\, s`\alpha\times``\kappa = s`\text{Cnv}```\times\alpha``\kappa$

$[*40{\cdot}38] \qquad = \text{Cnv}``s`\times\alpha``\kappa$

$[*113{\cdot}44] \qquad = \text{Cnv}``\{(s`\kappa)\times\alpha\}$ (1)

$[*113{\cdot}14] \qquad = \alpha\times s`\kappa$ (2)

$\vdash .\,(1)\,.\,(2)\,.\supset\vdash .$ Prop

***113·49.** $\vdash : \kappa\in\text{Cls}^2\,\text{excl} \,.\supset .\, \Sigma`\alpha\times``\kappa \;\text{sm}\; \alpha\times(\Sigma`\kappa)$

Dem.

$\vdash .\, *113{\cdot}14 \,.\supset\vdash .\, \alpha\times``\kappa = \text{Cnv}```\times\alpha``\kappa$ (1)

$\vdash .\,(1)\,.\,*113{\cdot}45\,.\,*72{\cdot}11\,.\,*84{\cdot}53\,.\supset$

$\vdash : \text{Hp} \,.\supset .\, \alpha\times``\kappa\in\text{Cls}^2\,\text{excl}\,.$

$[*112{\cdot}15] \qquad \supset .\, \Sigma`\alpha\times``\kappa \;\text{sm}\; s`\alpha\times``\kappa\,.$

$[*113{\cdot}48] \qquad \supset .\, \Sigma`\alpha\times``\kappa \;\text{sm}\; \alpha\times(s`\kappa)\,.$

$[*112{\cdot}15 . *113{\cdot}13] \qquad \supset .\, \Sigma`\alpha\times``\kappa \;\text{sm}\; \alpha\times(\Sigma`\kappa) :\supset\vdash .$ Prop

***113·491.** $\vdash : \kappa\in\text{Cls}^2\,\text{excl}\,.\supset .\, \Sigma\text{Nc}`\alpha\times``\kappa = \text{Nc}`\{\alpha\times(\Sigma`\kappa)\} = \text{Nc}`\alpha\times_c\Sigma\text{Nc}`\kappa$
[*113·49·25]

The following propositions are concerned with the associative law for arithmetical multiplication.

***113·5.** $\vdash .\,(\gamma\times\beta)\times\alpha = \hat{R}\,\{(\exists x,y,z)\,.\,x\in\alpha\,.\,y\in\beta\,.\,z\in\gamma\,.\,R = x\downarrow(y\downarrow z)\}$

Dem.

$\vdash .\, *113{\cdot}101 \,.\supset$

$\vdash .\,(\gamma\times\beta)\times\alpha = \hat{R}\,\{(\exists x,P)\,.\,x\in\alpha\,.\,P\in(\gamma\times\beta)\,.\,R = x\downarrow P\}$

$[*113{\cdot}101] \qquad = \hat{R}\,\{(\exists x,y,z)\,.\,x\in\alpha\,.\,y\in\beta\,.\,z\in\gamma\,.\,R = x\downarrow(y\downarrow z)\}\,.\supset\vdash .$ Prop

***113·51.** $\vdash .\,(\alpha\times\beta)\times\gamma \;\text{sm}\; \alpha\times(\beta\times\gamma)$

Dem.

$\vdash .\, *113{\cdot}141 \,.\supset\vdash .\, \alpha\times(\beta\times\gamma) \;\text{sm}\; (\beta\times\gamma)\times\alpha$ (1)

$\vdash .\, *113{\cdot}5 \,.\quad \supset\vdash .\,(\alpha\times\beta)\times\gamma = \hat{R}\,\{(\exists x,y,z).x\in\alpha.y\in\beta.z\in\gamma.R = z\downarrow(y\downarrow x)\}\,.$

$(\beta\times\gamma)\times\alpha = \hat{P}\,\{(\exists x,y,z).x\in\alpha.y\in\beta.z\in\gamma.P = x\downarrow(z\downarrow y)\}$ (2)

$\vdash .\,(2)\,.\supset\vdash : T = \hat{R}\hat{P}\,\{(\exists x,y,z).x\in\alpha.y\in\beta\,.\,z\in\gamma.R = z\downarrow(y\downarrow x).P = x\downarrow(y\downarrow z)\}\,.\supset .$

$\text{D}`T = (\alpha\times\beta)\times\gamma\,.\,\text{G}`T = (\beta\times\gamma)\times\alpha$ (3)

$\vdash . *21\cdot33 . \supset \vdash : \text{Hp}(3) . RTP . RTQ . \supset$

$(\exists x, x', y, y', z, z') . x, x' \epsilon \alpha . y, y' \epsilon \beta . z, z' \epsilon \gamma . R = z \downarrow (y \downarrow x) = z' \downarrow (y' \downarrow x') .$

$P = x \downarrow (z \downarrow y) . Q = x' \downarrow (z' \downarrow y') .$

$[*55\cdot202] \supset . P = Q$ (4)

Similarly $\vdash : \text{Hp}(3) . RTP . QTP . \supset . R = Q$ (5)

$\vdash . (3) . (4) . (5) . \supset \vdash . (\alpha \times \beta) \times \gamma \text{ sm } (\beta \times \gamma) \times \alpha$ (6)

$\vdash . (1) . (6) . \supset \vdash . \text{Prop}$

***113·511.** $\alpha \times \beta \times \gamma = (\alpha \times \beta) \times \gamma$ Df

***113·52.** $\vdash . (\text{Nc}`\alpha \times_c \text{Nc}`\beta) \times_c \text{Nc}`\gamma = \text{Nc}`(\alpha \times \beta \times \gamma)$ [*113·25]

***113·53.** $\vdash . (\text{Nc}`\alpha \times_c \text{Nc}`\beta) \times_c \text{Nc}`\gamma = \text{Nc}`\alpha \times_c (\text{Nc}`\beta \times_c \text{Nc}`\gamma)$

Dem.

$\vdash . *113\cdot52\cdot51 . \supset$

$\vdash . (\text{Nc}`\alpha \times_c \text{Nc}`\beta) \times_c \text{Nc}`\gamma = \text{Nc}`\{\alpha \times (\beta \times \gamma)\}$

$[*113\cdot25] \quad = \text{Nc}`\alpha \times_c (\text{Nc}`\beta \times_c \text{Nc}`\gamma) . \supset \vdash . \text{Prop}$

***113·531.** $\vdash . (\text{N}_0\text{c}`\alpha \times_c \text{N}_0\text{c}`\beta) \times_c \text{N}_0\text{c}`\gamma = \text{N}_0\text{c}`\alpha \times_c (\text{N}_0\text{c}`\beta \times_c \text{N}_0\text{c}`\gamma)$
[*113·53 . (*113·04·05)]

***113·54.** $\vdash . (\mu \times_c \nu) \times_c \varpi = \mu \times_c (\nu \times_c \varpi)$

Dem.

$\vdash . *113\cdot531 . *103\cdot2 . \supset$

$\vdash : \mu, \nu, \varpi \epsilon \text{N}_0\text{C} . \supset . (\mu \times_c \nu) \times_c \varpi = \mu \times_c (\nu \times_c \varpi)$ (1)

$\vdash . *113\cdot204 . \supset$

$\vdash : \sim(\mu, \nu, \varpi \epsilon \text{N}_0\text{C}) . \supset . (\mu \times_c \nu) \times_c \varpi = \Lambda . \mu \times_c (\nu \times \varpi) = \Lambda$ (2)

$\vdash . (1) . (2) . \supset \vdash . \text{Prop}$

***113·541.** $\mu \times_c \nu \times_c \varpi = (\mu \times_c \nu) \times_c \varpi$ Df

***113·6.** $\vdash . \text{Nc}`\alpha \times_c 0 = 0$

Dem.

$\vdash . *113\cdot25 . *101\cdot1 . \supset \vdash . \text{Nc}`\alpha \times_c 0 = \text{Nc}`(\alpha \times \Lambda)$

$[*113\cdot114 . *101\cdot1] \quad = 0 . \supset \vdash . \text{Prop}$

***113·601.** $\vdash : \mu \epsilon \text{NC} - \iota`\Lambda . \supset . \mu \times_c 0 = 0$

Dem.

$\vdash . *103\cdot26 . \supset \vdash : \text{Hp} . \supset . (\exists \alpha) . \mu = \text{N}_0\text{c}`\alpha$ (1)

$\vdash . *101\cdot11\cdot13 . *103\cdot27 . \supset \vdash . 0 = \text{N}_0\text{c}`\Lambda$ (2)

$\vdash . (1) . (2) . \supset \vdash : \text{Hp} . \supset . (\exists \alpha) . \mu \times_c 0 = \text{N}_0\text{c}`\alpha \times_c \text{N}_0\text{c}`\Lambda$

$[*113\cdot222] \quad = \text{Nc}`(\alpha \times \Lambda)$

$[*113\cdot114 . *101\cdot1] \quad = 0 : \supset \vdash . \text{Prop}$

***113·602.** $\vdash :. \mu \times_c \nu = 0 . \equiv : \mu, \nu \in \text{NC} - \iota`\Lambda : \mu = 0 . \vee . \nu = 0$

Dem.

$\vdash . *113·203 . *101·12 . \supset$

$\vdash : \mu \times_c \nu = 0 . \supset . \mu, \nu \in \text{NC} - \iota`\Lambda \qquad (1)$

$\vdash . (1) . *113·201 . \supset$

$\vdash :: \mu \times_c \nu = 0 . \supset :. \xi \in 0 . \equiv_\xi : (\exists \alpha, \beta) . \alpha \in \mu . \beta \in \nu . \xi \,\text{sm}\, (\alpha \times \beta) :.$

[*54·102] $\supset :. \xi = \Lambda . \equiv_\xi : (\exists \alpha, \beta) . \alpha \in \mu . \beta \in \nu . \xi \,\text{sm}\, (\alpha \times \beta) :.$

[*10·1.*13·15] $\supset :. (\exists \alpha, \beta) . \alpha \in \mu . \beta \in \nu . \Lambda \,\text{sm}\, (\alpha \times \beta) :.$

[*73·47] $\supset :. (\exists \alpha, \beta) . \alpha \in \mu . \beta \in \nu . \alpha \times \beta = \Lambda :.$

[*113·114] $\supset :. (\exists \alpha, \beta) : \alpha \in \mu . \beta \in \nu : \alpha = \Lambda . \vee . \beta = \Lambda :.$

[*13·195] $\supset :. \Lambda \in \mu . \vee . \Lambda \in \nu :.$

[(1).*100·45] $\supset :. \mu = \text{Nc}`\Lambda . \vee . \nu = \text{Nc}`\Lambda :.$

[*101·1] $\supset :. \mu = 0 . \vee . \nu = 0 \qquad (2)$

$\vdash . *113·601·27 . \supset \vdash :. \mu, \nu \in \text{NC} - \iota`\Lambda : \mu = 0 . \vee . \nu = 0 : \supset . \mu \times_c \nu = 0 \qquad (3)$

$\vdash . (2) . (3) . \supset \vdash . \text{Prop}$

The following propositions are concerned with multiplication by a unit class or by 1 or 2.

***113·61.** $\vdash . \iota`z \times \alpha = \downarrow z``\alpha$

Dem.

$\vdash . *113·1 . \supset \vdash . \iota`z \times \alpha = s`\alpha \underset{,,}{\downarrow} ``\iota`z$

[*53·31·02] $= \alpha \underset{,,}{\downarrow} `z$

[*38·2] $= \downarrow z``\alpha . \supset \vdash . \text{Prop}$

***113·611.** $\vdash . \iota`z \times \alpha \,\text{sm}\, \alpha$ [*113·61 . *73·611]

***113·612.** $\vdash . \alpha \times \iota`z \,\text{sm}\, \alpha$ [*113·611·141]

***113·62.** $\vdash . \text{Nc}`\alpha \times_c 1 = \text{Nc}`\alpha$

Dem.

$\vdash . *101·2 . \supset \vdash . \text{Nc}`\alpha \times_c 1 = \text{Nc}`\alpha \times_c \text{Nc}`\iota`z$

[*113·25] $= \text{Nc}`(\alpha \times \iota`z)$

[*113·612] $= \text{Nc}`\alpha . \supset \vdash . \text{Prop}$

***113·621.** $\vdash : \mu \in \text{NC} . \supset . \mu \times_c 1 = \text{sm}``\mu$

Dem.

$\vdash . *113·204 . \supset \vdash : \mu = \Lambda . \supset . \mu \times_c 1 = \Lambda$

[*37·29] $= \text{sm}``\mu \qquad (1)$

$\vdash . *103·26 . \supset \vdash : \text{Hp} . \alpha \in \mu . \supset . \mu = \text{N}_0\text{c}`\alpha . \qquad (2)$

[(*113·04)] $\supset . \mu \times_c 1 = \text{Nc}`\alpha \times_c 1$

[*113·62] $= \text{Nc}`\alpha$

[*103·4.(2)] $= \text{sm}``\mu \qquad (3)$

$\vdash . (2) . *10·11·23·35 . \supset \vdash : \text{Hp} . \exists ! \mu . \supset . \mu \times_c 1 = \text{sm}``\mu \qquad (4)$

$\vdash . (1) . (4) . \supset \vdash . \text{Prop}$

Observe that if μ is a typically definite cardinal, sm$``\mu$ is the "same" cardinal rendered typically ambiguous; while if $\mu$ is typically ambiguous, $\mu = \text{sm}``\mu$ in every type.

***113·63.** $\vdash : z \sim\epsilon\, \alpha \,.\supset.\, \downarrow z``\alpha \,\text{sm}\, \text{D}``\epsilon_\Delta`(\iota`\alpha \cup \iota`\iota`z)$

Dem.

$\vdash . *113·152 . \supset \vdash : \text{Hp} . \supset . \text{D}``\epsilon_\Delta`(\iota`\alpha \cup \iota`\iota`z) \,\text{sm}\, \alpha \times \iota`z$ (1)

$\vdash . (1) . *113·61·141 . \supset \vdash . \text{Prop}$

***113·64.** $\vdash . \downarrow z``\alpha \times \downarrow z``\beta \,\text{sm}\, \alpha \times \beta \,.\, \downarrow z``\alpha \times \downarrow z``\beta \,\text{sm}\, \downarrow z``(\alpha \times \beta)$

Dem.

$\vdash . *73·611 . *113·13 . \supset \vdash . \downarrow z``\alpha \times \downarrow z``\beta \,\text{sm}\, \alpha \times \beta$ (1)

$\vdash . (1) . *73·611 . \quad \supset \vdash . \downarrow z``\alpha \times \downarrow z``\beta \,\text{sm}\, \downarrow z``(\alpha \times \beta)$ (2)

$\vdash . (1) . (2) . \supset \vdash . \text{Prop}$

***113·65.** $\vdash . \downarrow z``\alpha \times \downarrow z``\beta = (\downarrow z \,||\, \text{Cnv}`\downarrow z)``(\alpha \times \beta)$

Dem.

$\vdash . *72·184 . *55·21 . \supset \vdash . \downarrow z \,\epsilon\, 1 \rightarrow 1 \,.\, \alpha \subset \text{C}`\downarrow z \,.\, \beta \subset \text{C}`\downarrow z \,.$

[*113·126] $\quad \supset \vdash . \downarrow z``\alpha \times \downarrow z``\beta = (\downarrow z \,||\, \text{Cnv}`\downarrow z)``(\alpha \times \beta) \,.$

$\supset \vdash . \text{Prop}$

***113·66.** $\vdash . \mu \times_c 2 = \mu +_c \mu$

Dem.

$\vdash . *110·643 . \supset \vdash . \mu \times_c 2 = \mu \times_c (1 +_c 1)$

[*113·43] $\quad = (\mu \times_c 1) +_c (\mu \times_c 1)$ (1)

$\vdash . (1) . \quad \supset \vdash :. \mu = \text{N}_0\text{c}`\alpha . \supset . \mu \times_c 2 = (\text{N}_0\text{c}`\alpha \times_c 1) +_c (\text{N}_0\text{c}`\alpha \times_c 1)$

[*113·62.(*113·04)] $\quad = \text{Nc}`\alpha +_c \text{Nc}`\alpha$

[*110·3] $\quad = \mu +_c \mu$ (2)

$\vdash . (2) . *103·2 . \quad \supset \vdash : \mu \,\epsilon\, \text{N}_0\text{C} . \supset . \mu \times_c 2 = \mu +_c \mu$ (3)

$\vdash . *113·205 . *110·4 . \supset \vdash : \mu \sim\epsilon\, \text{N}_0\text{C} . \supset . \mu \times_c 2 = \Lambda \,.\, \mu +_c \mu = \Lambda$ (4)

$\vdash . (3) . (4) . \supset \vdash . \text{Prop}$

***113·67.** $\vdash . \text{Nc}`\alpha \times_c \text{Nc}`(\beta + \iota`y) = (\text{Nc}`\alpha \times_c \text{Nc}`\beta) +_c \text{Nc}`\alpha$

Dem.

$\vdash . *113·421 . *101·2 . \supset$

$\vdash . \text{Nc}`\alpha \times_c \text{Nc}`(\beta + \iota`y) = (\text{Nc}`\alpha \times_c \text{Nc}`\beta) +_c (\text{Nc}`\alpha \times_c 1)$

[*113·62] $\quad = (\text{Nc}`\alpha \times_c \text{Nc}`\beta) +_c \text{Nc}`\alpha . \supset \vdash . \text{Prop}$

***113·671.** $\vdash . \mu \times_c (\nu +_c 1) = (\mu \times_c \nu) +_c \mu$ [*113·67·205 . *110·4]

$*114$. THE ARITHMETICAL PRODUCT OF A CLASS OF CLASSES.

Summary of $*114$.

The kind of multiplication defined in $*113$ cannot be extended beyond a finite number of factors. We therefore, as in the case of addition, introduce another definition, defining the product of the numbers of a class of classes, and capable of being applied to an infinite number of factors. We define the product of the numbers of members of κ as $\text{Nc}`\epsilon_\Delta`\kappa$; thus we put

$$\Pi\text{Nc}`\kappa = \text{Nc}`\epsilon_\Delta`\kappa \quad \text{Df.}$$

It is to be observed that $\Pi\text{Nc}`\kappa$ is not a function of $\text{Nc}``\kappa$, because, if two members of $\kappa$ have the same number, this will count only once in $\text{Nc}``\kappa$, but will count twice in $\Pi\text{Nc}`\kappa$.

It is very easy to see that, in case κ is finite, $\text{Nc}`\epsilon_\Delta`\kappa$ will be what we should ordinarily regard as the product of the numbers of members of κ. For suppose (*e.g.*)

$$\kappa = \iota`\alpha \cup \iota`\beta \cup \iota`\gamma,$$

where $\alpha \neq \beta \,.\, \alpha \neq \gamma \,.\, \beta \neq \gamma$. Then

$$\epsilon_\Delta`\kappa = \hat{R}\{(\exists x, y, z) \,.\, R = x \downarrow \alpha \cup y \downarrow \beta \cup z \downarrow \gamma \,.\, x \,\epsilon\, \alpha \,.\, y \,\epsilon\, \beta \,.\, z \,\epsilon\, \gamma\}.$$

Thus if R is a member of $\epsilon_\Delta`\kappa$, R is determinate when x, y, z are given, x, y, z being the referents to α, β, γ. Whether α, β, γ overlap or not, the choice of any one of x, y, z is entirely independent of the choice of the other two, and therefore the total number of choices possible is obviously the product of the numbers of α, β, γ. Thus our definition will not conflict with what is commonly understood by a product.

The propositions of this number are less numerous and less important than those of $*113$. We shall deal first with products of a single factor, and products in which one factor is null ($*114{\cdot}2$—$\cdot27$). We shall then deal ($*114{\cdot}3$—$\cdot36$) with the relations between the sort of multiplication here defined and the sort defined in $*113$. Then we have a few propositions ($*114{\cdot}4$—$\cdot43$) showing that unit factors make no difference to the value of a product. Then we prove ($*114{\cdot}5$—$\cdot52$) that the value of the product is the same for two classes having double similarity, and then ($*114{\cdot}53$—$\cdot571$) we give extensions of this result which depend upon the multiplicative axiom. Finally, we give some new forms of the associative law of multiplication.

Among the more important propositions in this number are the following:

∗114·21. $\vdash . \Pi\text{Nc}`\iota`\alpha = \text{Nc}`\alpha$

I.e. a product of one factor is equal to that factor.

∗114·23. $\vdash : \Lambda \epsilon \kappa . \supset . \Pi\text{Nc}`\kappa = 0$

I.e. a product vanishes if one of its factors is zero. The converse requires the multiplicative axiom, as appears from the proposition

∗114·26. $\vdash :. \text{Mult ax} . \equiv : \Pi\text{Nc}`\kappa = 0 . \equiv_\kappa . \Lambda \epsilon \kappa$

I.e. the multiplicative axiom is equivalent to the assumption that a product vanishes when, and only when, one of its factors is zero.

∗114·301. $\vdash : \kappa \cap \lambda = \Lambda . \supset . \epsilon_\Delta`(\kappa \cup \lambda) \text{ sm } \epsilon_\Delta`\kappa \times \epsilon_\Delta`\lambda$

whence

∗114·31. $\vdash : \kappa \cap \lambda = \Lambda . \supset . \Pi\text{Nc}`\kappa \times_c \Pi\text{Nc}`\lambda = \Pi\text{Nc}`(\kappa \cup \lambda)$

which is a form of the associative law, and

∗114·35. $\vdash : \alpha \neq \beta . \supset . \Pi\text{Nc}`(\iota`\alpha \cup \iota`\beta) = \text{Nc}`\alpha \times_c \text{Nc}`\beta$

which connects the two sorts of multiplication.

∗114·41. $\vdash : \lambda \subset 1 . \supset . \Pi\text{Nc}`(\kappa \cup \lambda) = \Pi\text{Nc}`\kappa$

I.e. unit factors make no difference to the value of a product.

∗114·51. $\vdash : T \upharpoonright s`\lambda \epsilon \kappa \,\overline{\text{sm}}\, \overline{\text{sm}}\, \lambda . \supset . (T \,\|\, \breve{T}_\epsilon) \upharpoonright \epsilon_\Delta`\lambda \epsilon (\epsilon_\Delta`\kappa) \,\overline{\text{sm}}\, (\epsilon_\Delta`\lambda)$

This proposition gives a correlator of $\epsilon_\Delta`\kappa$ and $\epsilon_\Delta`\lambda$ as a function of a double correlator of κ and λ, and thus leads to

∗114·52. $\vdash : \kappa \text{ sm sm } \lambda . \supset . \Pi\text{Nc}`\kappa = \Pi\text{Nc}`\lambda . \epsilon_\Delta`\kappa \text{ sm } \epsilon_\Delta`\lambda$

Hence, by the propositions of ∗111, we infer

∗114·571. $\vdash :. \text{Mult ax} . \supset : \mu, \nu \epsilon \text{NC} . \kappa, \lambda \epsilon \mu \cap \text{Cl}`\nu . \supset . \Pi\text{Nc}`\kappa = \Pi\text{Nc}`\lambda$

I.e. assuming the multiplicative axiom, if κ and λ each consist of μ classes of ν terms each, their products are equal.

We have next various forms of the associative law, beginning with

∗114·6. $\vdash : \kappa \epsilon \text{Cls}^2 \text{ excl} . \supset . \Pi\text{Nc}`\epsilon_\Delta``\kappa = \Pi\text{Nc}`s`\kappa$

which is an immediate consequence of ∗85·44. The other form is

∗114·632. $\vdash : S \upharpoonright \gamma \epsilon 1 \rightarrow 1 . \gamma \subset \text{Cl}`S . \gamma \cap S``\gamma = \Lambda . \supset .$

$$\epsilon_\Delta`\hat{\mu}\{(\exists\alpha) . \alpha \epsilon \gamma . \mu = \alpha \times S`\alpha\} \text{ sm } \epsilon_\Delta`(\gamma \cup S``\gamma)$$

As to the sense in which this is a form of the associative law, see the observations following ∗114·6.

∗114·01. $\Pi\text{Nc}`\kappa = \text{Nc}`\epsilon_\Delta`\kappa$ Df

∗114·1. $\vdash . \Pi\text{Nc}`\kappa = \text{Nc}`\epsilon_\Delta`\kappa$ [(∗114·01)]

∗114·11. $\vdash : \beta \epsilon \Pi\text{Nc}`\kappa . \equiv . \beta \text{ sm } \epsilon_\Delta`\kappa . \equiv . \beta \epsilon \text{Nc}`\epsilon_\Delta`\kappa$ [∗114·1 . ∗100·31]

∗114·12. $\vdash . \epsilon_\Delta`\kappa \epsilon \Pi\text{Nc}`\kappa$ [∗100·3 . ∗114·1]

∗114·2. $\vdash . \Pi\text{Nc}`\Lambda = 1$ [∗83·15 . ∗101·2]

Thus a product of no factors is 1. This is the source of $\mu^0 = 1$, as we shall see later.

***114·21.** $\vdash . \Pi \mathrm{Nc}`\iota`\alpha = \mathrm{Nc}`\alpha$ [*83·41]

***114·22.** $\vdash . \Pi \mathrm{Nc}`\iota`\Lambda = 0$ [*114·21 . *101·1]

***114·23.** $\vdash : \Lambda \epsilon \kappa . \supset . \Pi \mathrm{Nc}`\kappa = 0$ [*83·11 . *101·1]

Thus an arithmetical product is zero if any of its factors is zero. To prove the converse, we have to assume the multiplicative axiom, which, in fact, is equivalent to the proposition that an arithmetical product is only zero when at least one of its factors is zero.

***114·24.** $\vdash : \Pi \mathrm{Nc}`\lambda \neq 0 . \kappa \subset \lambda . \supset . \Pi \mathrm{Nc}`\kappa \neq 0$

Dem.

$\vdash . *114·1 . *101·1 . \supset \vdash : \Pi \mathrm{Nc}`\lambda \neq 0 . \supset . \exists ! \epsilon_\Delta`\lambda$ (1)

$\vdash . (1) . *80·6 . \quad \supset \vdash : \Pi \mathrm{Nc}`\lambda \neq 0 . \kappa \subset \lambda . \supset . \exists ! \epsilon_\Delta`\kappa .$

[*114·1.*101·1] $\supset . \Pi \mathrm{Nc}`\kappa \neq 0 : \supset \vdash . \mathrm{Prop}$

***114·25.** $\vdash :. \mathrm{Mult\ ax} . \equiv : \Pi \mathrm{Nc}`\kappa = 0 . \supset_\kappa . \Lambda \epsilon \kappa$

Dem.

$\vdash . *88·37 . \mathrm{Transp} . \supset$

$\vdash :. \mathrm{Mult\ ax} . \quad \equiv : \epsilon_\Delta`\kappa = \Lambda . \supset_\kappa . \Lambda \epsilon \kappa :$

[*114·1.*101·1] $\equiv : \Pi \mathrm{Nc}`\kappa = 0 . \supset_\kappa . \Lambda \epsilon \kappa :. \supset \vdash . \mathrm{Prop}$

Note that $\Lambda \epsilon \kappa . \equiv . 0 \epsilon \mathrm{Nc}``\kappa$.

***114·26.** $\vdash :. \mathrm{Mult\ ax} . \equiv : \Pi \mathrm{Nc}`\kappa = 0 . \equiv_\kappa . \Lambda \epsilon \kappa$ [*88·372 . *101·1]

***114·261.** $\vdash :. \mathrm{Mult\ ax} . \equiv : \Pi \mathrm{Nc}`\kappa = 0 . \equiv_\kappa . 0 \epsilon \mathrm{Nc}``\kappa$ [*114·26 . *101·1]

***114·27.** $\vdash :: \mathrm{Mult\ ax} . \equiv :. \alpha \epsilon \kappa . \supset_\alpha . \exists ! \alpha : \equiv_\kappa . \Pi \mathrm{Nc}`\kappa \neq 0$
[*114·26 . Transp . *24·63]

***114·3.** $\vdash : \kappa \neq \lambda . \supset . \epsilon_\Delta`(\iota`\epsilon_\Delta`\kappa \cup \iota`\epsilon_\Delta`\lambda) \,\mathrm{sm}\, \epsilon_\Delta`\kappa \times \epsilon_\Delta`\lambda$

Dem.

$\vdash . *113·146 . \supset \vdash : \epsilon_\Delta`\kappa \neq \epsilon_\Delta`\lambda . \supset . \epsilon_\Delta`(\iota`\epsilon_\Delta`\kappa \cup \iota`\epsilon_\Delta`\lambda) \,\mathrm{sm}\, \epsilon_\Delta`\kappa \times \epsilon_\Delta`\lambda$ (1)

$\vdash . *80·81 . \quad \supset \vdash :. \exists ! \epsilon_\Delta`\kappa . \vee . \exists ! \epsilon_\Delta`\lambda : \kappa \neq \lambda : \supset . \epsilon_\Delta`\kappa \neq \epsilon_\Delta`\lambda$ (2)

$\vdash . *83·903 . *113·114 . \supset$

$\vdash : \epsilon_\Delta`\kappa = \Lambda . \epsilon_\Delta`\lambda = \Lambda . \supset . \epsilon_\Delta`(\iota`\epsilon_\Delta`\kappa \cup \iota`\epsilon_\Delta`\lambda) = \Lambda . \epsilon_\Delta`\kappa \times \epsilon_\Delta`\lambda = \Lambda$ (3)

$\vdash . (1) . (2) . (3) . \supset \vdash . \mathrm{Prop}$

***114·301.** $\vdash : \kappa \cap \lambda = \Lambda . \supset . \epsilon_\Delta`(\kappa \cup \lambda) \,\mathrm{sm}\, \epsilon_\Delta`\kappa \times \epsilon_\Delta`\lambda$

Dem.

$\vdash . *85·45 . *114·3 . \supset$

$\vdash : \kappa \cap \lambda = \Lambda . \kappa \neq \lambda . \supset . \epsilon_\Delta`(\kappa \cup \lambda) \,\mathrm{sm}\, \epsilon_\Delta`\kappa \times \epsilon_\Delta`\lambda$ (1)

$\vdash . *22·5 . \supset \vdash : \kappa \cap \lambda = \Lambda . \kappa = \lambda . \supset . \kappa = \Lambda . \lambda = \Lambda .$

[*83·15] $\supset . \epsilon_\Delta`(\kappa \cup \lambda) = \iota`\dot{\Lambda} . \epsilon_\Delta`\kappa = \iota`\dot{\Lambda} . \epsilon_\Delta`\lambda = \iota`\dot{\Lambda} .$

[*113·611] $\supset . \epsilon_\Delta`(\kappa \cup \lambda) \,\mathrm{sm}\, \epsilon_\Delta`\kappa \times \epsilon_\Delta`\lambda$ (2)

$\vdash . (1) . (2) . \supset \vdash . \mathrm{Prop}$

∗114·31. $\vdash : \kappa \cap \lambda = \Lambda \,.\supset .\, \Pi\text{Nc}‘\kappa \times_c \Pi\text{Nc}‘\lambda = \Pi\text{Nc}‘(\kappa \cup \lambda)$
[∗114·301·1 . ∗113·25]

The above is one form of the associative law of multiplication.

∗114·311. $\vdash .\, \Pi\text{Nc}‘(\kappa \cup \lambda) = \Pi\text{Nc}‘\kappa \times_c \Pi\text{Nc}‘(\lambda - \kappa)$ [∗114·31 . ∗22·91]

∗114·32. $\vdash : \Pi\text{Nc}‘(\kappa \cup \lambda) \neq 0 \,.\equiv .\, \Pi\text{Nc}‘\kappa \neq 0 \,.\, \Pi\text{Nc}‘\lambda \neq 0$

Dem.

$\vdash .\, *114\cdot311 \,.\, *113\cdot602 \,.\supset$

$\vdash : \Pi\text{Nc}‘(\kappa \cup \lambda) \neq 0 \,.\supset .\, \Pi\text{Nc}‘\kappa \neq 0 \qquad (1)$

$\vdash .\, (1)\dfrac{\lambda, \kappa}{\kappa, \lambda} \,.\supset \vdash : \Pi\text{Nc}‘(\kappa \cup \lambda) \neq 0 \,.\supset .\, \Pi\text{Nc}‘\lambda \neq 0 \qquad (2)$

$\vdash .\, *114\cdot24 \,.\supset \vdash : \Pi\text{Nc}‘\lambda \neq 0 \,.\supset .\, \Pi\text{Nc}‘(\lambda - \kappa) \neq 0 :$

[Fact] $\supset \vdash : \Pi\text{Nc}‘\kappa \neq 0 \,.\, \Pi\text{Nc}‘\lambda \neq 0 \,.\supset .\, \Pi\text{Nc}‘\kappa \neq 0 \,.\, \Pi\text{Nc}‘(\lambda - \kappa) \neq 0 \,.$

[∗113·602.∗114·311] $\supset .\, \Pi\text{Nc}‘(\kappa \cup \lambda) \neq 0 \qquad (3)$

$\vdash .\, (1) \,.\, (2) \,.\, (3) \,.\supset \vdash .\, \text{Prop}$

∗114·33. $\vdash : \alpha \sim\in \kappa \,.\supset .\, \Pi\text{Nc}‘(\kappa \cup \iota‘\alpha) = \Pi\text{Nc}‘\kappa \times_c \text{Nc}‘\alpha$ [∗114·31·21]

∗114·34. $\vdash : \Pi\text{Nc}‘\kappa \neq 0 \,.\, \exists ! \alpha \,.\equiv .\, \Pi\text{Nc}‘(\kappa \cup \iota‘\alpha) \neq 0$
[∗114·32·21 . ∗101·14]

∗114·35. $\vdash : \alpha \neq \beta \,.\supset .\, \Pi\text{Nc}‘(\iota‘\alpha \cup \iota‘\beta) = \text{Nc}‘\alpha \times_c \text{Nc}‘\beta$ [∗114·33·21]

∗114·36. $\vdash : \alpha \neq \beta \,.\, \alpha \neq \gamma \,.\, \beta \neq \gamma \,.\supset .\, \Pi\text{Nc}‘(\iota‘\alpha \cup \iota‘\beta \cup \iota‘\gamma) = \text{Nc}‘\alpha \times_c \text{Nc}‘\beta \times_c \text{Nc}‘\gamma$
[∗114·33·35]

∗114·4. $\vdash : \lambda \subset 1 \,.\supset .\, \Pi\text{Nc}‘\lambda = 1$ [∗83·44]

∗114·41. $\vdash : \lambda \subset 1 \,.\supset .\, \Pi\text{Nc}‘(\kappa \cup \lambda) = \Pi\text{Nc}‘\kappa$ [∗83·57]

∗114·42. $\vdash .\, \Pi\text{Nc}‘\kappa = \Pi\text{Nc}‘(\kappa - 1)$

Dem.

$\vdash .\, *24\cdot41 \,.\supset \vdash .\, \kappa = (\kappa - 1) \cup (\kappa \cap 1) \qquad (1)$

$\vdash .\, (1) \,.\, *114\cdot41 \,.\supset \vdash .\, \text{Prop}$

∗114·43. $\vdash .\, \Pi\text{Nc}‘(\kappa \cup \iota“\alpha) = \Pi\text{Nc}‘\kappa$ [∗114·41 . ∗52·3]

∗114·5. $\vdash : T \in \kappa \,\overline{\text{sm}}\,\overline{\text{sm}}\, \lambda \,.\supset .\, (T \,\|\, \breve{T}_\epsilon) \upharpoonright \epsilon_\Delta‘\lambda \in (\epsilon_\Delta‘\kappa) \,\overline{\text{sm}}\, (\epsilon_\Delta‘\lambda)$

Dem.

$\vdash .\, *111\cdot1\cdot11 \,.\supset \vdash : \text{Hp} \,.\supset .\, T, T_\epsilon \upharpoonright \lambda \in 1 \to 1 \qquad (1)$

$\vdash .\, *80\cdot14 \,.\, *83\cdot21 \,.\supset \vdash s‘\text{D}“\epsilon_\Delta‘\lambda \subset s‘\lambda \,.\, s‘\text{C}“\epsilon_\Delta‘\lambda \subset \lambda \qquad (2)$

$\vdash .\, (1) \,.\, (2) \,.\, *74\cdot773 \,.\supset$

$\vdash : \text{Hp} \,.\supset .\, (T \,\|\, \breve{T}_\epsilon) \upharpoonright \epsilon_\Delta‘\lambda \in \{(T \,\|\, \breve{T}_\epsilon)“\epsilon_\Delta‘\lambda\} \,\overline{\text{sm}}\, (\epsilon_\Delta‘\lambda) \qquad (3)$

$\vdash .\, *82\cdot43 \dfrac{\epsilon, T_\epsilon}{P, Q} \,.\, *62\cdot3 \,.\supset$

$\vdash : T, T_\epsilon \upharpoonright \lambda \in 1 \to 1 \,.\, s‘\lambda \subset \text{C}‘T \,.\, \lambda \subset \text{C}‘T_\epsilon \,.\, \kappa = T_\epsilon“\lambda \,.\supset .$

$(T \,|\, \epsilon \upharpoonright \lambda \,|\, \breve{T}_\epsilon)_\Delta‘\kappa = (T \,\|\, \breve{T}_\epsilon)“\epsilon_\Delta‘\lambda \qquad (4)$

$\vdash .\, (4) \,.\, (1) \,.\, *111\cdot1 \,.\, *37\cdot111 \,.\supset \vdash : \text{Hp} \,.\supset .\, (T \,|\, \epsilon \upharpoonright \lambda \,|\, \breve{T}_\epsilon)_\Delta‘\kappa = (T \,\|\, \breve{T}_\epsilon)“\epsilon_\Delta‘\lambda \qquad (5)$

$\vdash . *34\cdot1 . *37\cdot101 . \supset$

$\vdash : x(T \mid \epsilon \upharpoonright \lambda \mid \breve{T}_\epsilon)\alpha . \equiv . (\exists y, \beta) . xTy . y \epsilon \beta . \beta \epsilon \lambda . \alpha = T\text{‘‘}\beta$ (6)

$\vdash . (6) . *72\cdot52 . *111\cdot1 . \supset$

$\vdash :. \text{Hp} . \supset : x(T \mid \epsilon \upharpoonright \lambda \mid \breve{T}_\epsilon)\alpha . \equiv . (\exists y, \beta) . xTy . y \epsilon \beta . \beta \epsilon \lambda . \beta = \breve{T}\text{‘‘}\alpha . \alpha \subset \text{D‘}T .$

$[*111\cdot1\cdot131 . *13\cdot195]$ $\equiv . (\exists y) . xTy . y \epsilon \breve{T}\text{‘‘}\alpha . \alpha \epsilon \kappa .$

$[*37\cdot1]$ $\equiv . x \epsilon T\text{‘‘}\breve{T}\text{‘‘}\alpha . \alpha \epsilon \kappa .$

$[*72\cdot502 . *111\cdot1]$ $\equiv . x(\epsilon \upharpoonright \kappa)\alpha$ (7)

$\vdash . (5) . (7) . \supset \vdash : \text{Hp} . \supset . (\epsilon \upharpoonright \kappa)_\Delta\text{‘}\kappa = (T \parallel \breve{T}_\epsilon)\text{‘‘}\epsilon_\Delta\text{‘}\lambda .$

$[*83\cdot12]$ $\supset . \epsilon_\Delta\text{‘}\kappa = (T \parallel \breve{T}_\epsilon)\text{‘‘}\epsilon_\Delta\text{‘}\lambda$ (8)

$\vdash . (3) . (8) . \supset \vdash . \text{Prop}$

$*114\cdot501$. $\vdash : S = T \upharpoonright s\text{‘}\lambda . \supset . (S \parallel \breve{S}_\epsilon) \upharpoonright \epsilon_\Delta\text{‘}\lambda = (T \parallel \breve{T}_\epsilon) \upharpoonright \epsilon_\Delta\text{‘}\lambda$

Dem.

$\vdash . *80\cdot14 . *83\cdot21 . \supset$

$\vdash :. R \epsilon \epsilon_\Delta\text{‘}\lambda . \supset : yR\beta . \supset . y \epsilon s\text{‘}\lambda . \beta \epsilon \lambda .$

$[*40\cdot13]$ $\supset . y \epsilon s\text{‘}\lambda . \beta \subset s\text{‘}\lambda :$ (1)

$[*4\cdot71 . \text{Fact}] \supset : xTy . yR\beta . \beta\breve{T}_\epsilon\alpha . \equiv . xTy . y \epsilon s\text{‘}\lambda . yR\beta . \beta\breve{T}_\epsilon\alpha . \beta \subset s\text{‘}\lambda .$

$[*37\cdot101 . *22\cdot621]$ $\equiv . x(T \upharpoonright s\text{‘}\lambda)y . yR\beta . \alpha = T\text{‘‘}\beta . \beta = \beta \cap s\text{‘}\lambda .$

$[(1) . *37\cdot412]$ $\equiv . x(T \upharpoonright s\text{‘}\lambda)y . yR\beta . \alpha = (T \upharpoonright s\text{‘}\lambda)\text{‘‘}\beta$ (2)

$\vdash . (2) . \supset \vdash :. \text{Hp} . \supset : R \epsilon \epsilon_\Delta\text{‘}\kappa . \supset . T \mid R \mid \breve{T}_\epsilon = S \mid R \mid \breve{S}_\epsilon :$

$[*35\cdot71]$ $\supset : (T \parallel \breve{T}_\epsilon) \upharpoonright \epsilon_\Delta\text{‘}\kappa = (S \parallel \breve{S}_\epsilon) \upharpoonright \epsilon_\Delta\text{‘}\kappa :. \supset \vdash . \text{Prop}$

$*114\cdot51$. $\vdash : T \upharpoonright s\text{‘}\lambda \epsilon \kappa \,\overline{\text{sm}}\, \overline{\text{sm}}\, \lambda . \supset . (T \parallel \breve{T}_\epsilon) \upharpoonright \epsilon_\Delta\text{‘}\lambda \epsilon (\epsilon_\Delta\text{‘}\kappa) \,\overline{\text{sm}}\, (\epsilon_\Delta\text{‘}\lambda)$
$[*114\cdot5\cdot501]$

$*114\cdot52$. $\vdash : \kappa \text{ sm sm } \lambda . \supset . \Pi\text{Nc‘}\kappa = \Pi\text{Nc‘}\lambda . \epsilon_\Delta\text{‘}\kappa \text{ sm } \epsilon_\Delta\text{‘}\lambda$ $[*114\cdot51 . *111\cdot4]$

$*114\cdot53$. $\vdash :: \text{Mult ax} . \supset :. \kappa, \lambda \epsilon \text{Cls}^2 \text{ excl} :$
$(\exists S) . S \epsilon 1 \to 1 . S \Subset \text{sm} . \text{D‘}S = \kappa . \text{Ɑ‘}S = \lambda : \supset . \Pi\text{Nc‘}\kappa = \Pi\text{Nc‘}\lambda$
$[*114\cdot52 . *111\cdot5]$

$*114\cdot54$. $\vdash :. \text{Mult ax} . \supset : \mu, \nu \epsilon \text{NC} . \kappa, \lambda \epsilon \mu \cap \text{Cl excl‘}\nu . \supset . \Pi\text{Nc‘}\kappa = \Pi\text{Nc‘}\lambda$
$[*114\cdot52 . *111\cdot53]$

The condition $\kappa, \lambda \epsilon \text{Cls}^2 \text{ excl}$, which is involved in the hypothesis of $*114\cdot54$ (through $\kappa, \lambda \epsilon \text{Cl excl‘}\nu$), is not necessary. The following propositions enable us to remove it. We first prove

$$\epsilon_\Delta\text{‘}\kappa \text{ sm } \epsilon_\Delta\text{‘}\epsilon\downarrow\text{‘‘}\kappa$$

and then we use $*114\cdot54$ to take us from $\epsilon_\Delta\text{‘}\epsilon\downarrow\text{‘‘}\kappa$ to $\epsilon_\Delta\text{‘}\epsilon\downarrow\text{‘‘}\lambda$. Thence we arrive at $\epsilon_\Delta\text{‘}\kappa \text{ sm } \epsilon_\Delta\text{‘}\lambda$.

$*114\cdot56$. $\vdash . \epsilon_\Delta\text{‘}\kappa \text{ sm } \epsilon_\Delta\text{‘}\epsilon\downarrow\text{‘‘}\kappa . \Pi\text{Nc‘}\kappa = \Pi\text{Nc‘}\epsilon\downarrow\text{‘‘}\kappa$ $[*85\cdot54]$

$*114\cdot561$. $\vdash : S \epsilon \kappa \,\overline{\text{sm}}\, \lambda \cap \text{Rl‘sm} . \supset . \epsilon\downarrow \mid S \mid \text{Cnv‘}(\epsilon\downarrow) \epsilon (\epsilon\downarrow\text{‘‘}\kappa) \,\overline{\text{sm}}\, (\epsilon\downarrow\text{‘‘}\lambda) \cap \text{Rl‘sm}$
$[*73\cdot63 . *85\cdot601 . *38\cdot12 . *33\cdot432]$

∗114·562. $\vdash :. \text{Mult ax} . \supset :$

$$(\exists S) . S \in 1 \to 1 . S \subset \text{sm} . \text{D}'S = \kappa . \text{Cl}'S = \lambda . \supset . \epsilon\downarrow``\kappa \,\text{sm}\,\text{sm}\, \epsilon\downarrow``\lambda$$

Dem.

$\vdash . *114{\cdot}561 . *85{\cdot}61 . \supset$

$\vdash :. (\exists S) . S \in 1 \to 1 . S \subset \text{sm} . \text{D}'S = \kappa . \text{Cl}'S = \lambda . \supset :$

$\epsilon\downarrow``\kappa, \epsilon\downarrow``\lambda \in \text{Cls}^2 \text{excl} : (\exists T) . T \in 1 \to 1 . T \subset \text{sm} . \text{D}'T = \epsilon\downarrow``\kappa . \text{Cl}'T = \epsilon\downarrow``\lambda :$

[∗111·5] $\supset : \text{Mult ax} . \supset . \epsilon\downarrow``\kappa \,\text{sm}\,\text{sm}\, \epsilon\downarrow``\lambda :. \supset \vdash . \text{Prop}$

∗114·57. $\vdash :. \text{Mult ax} . \supset :$

$$(\exists S) . S \in 1 \to 1 . S \subset \text{sm} . \text{D}'S = \kappa . \text{Cl}'S = \lambda . \supset . \Pi\text{Nc}'\kappa = \Pi\text{Nc}'\lambda$$

Dem.

$\vdash . *114{\cdot}562{\cdot}52 . \supset$

$\vdash :. \text{Mult ax} . \supset : (\exists S) . S \in 1 \to 1 . S \subset \text{sm} . \text{D}'S = \kappa . \text{Cl}'S = \lambda . \supset .$

$\Pi\text{Nc}'\epsilon\downarrow``\kappa = \Pi\text{Nc}'\epsilon\downarrow``\lambda .$

[∗114·56] $\supset . \Pi\text{Nc}'\kappa = \Pi\text{Nc}'\lambda :. \supset \vdash . \text{Prop}$

∗114·571. $\vdash :. \text{Mult ax} . \supset : \mu, \nu \in \text{NC} . \kappa, \lambda \in \mu \cap \text{Cl}'\nu . \supset . \Pi\text{Nc}'\kappa = \Pi\text{Nc}'\lambda$
[∗111·52 . ∗114·57]

∗114·6. $\vdash : \kappa \in \text{Cls}^2 \text{excl} . \supset . \Pi\text{Nc}'\epsilon_\Delta``\kappa = \Pi\text{Nc}'s'\kappa$ [∗85·44]

This is the most general form of the associative law for arithmetical multiplication.

Owing to the fact that we have two kinds of multiplication, namely $\alpha \times \beta$ and $\epsilon_\Delta{}'\kappa$, we have four forms of the associative law of multiplication, namely:

(1) ∗114·6, above,

(2) ∗113·54, *i.e.* $\vdash . (\mu \times_c \nu) \times_c \varpi = \mu \times_c (\nu \times_c \varpi)$,

(3) ∗114·31, *i.e.* $\vdash : \kappa \cap \lambda = \Lambda . \supset . \Pi\text{Nc}'\kappa \times_c \Pi\text{Nc}'\lambda = \Pi\text{Nc}'(\kappa \cup \lambda)$,

(4) a form of the associative law which has not yet been proved, which may be explained as follows.

Suppose we have a number of pairs of classes, *e.g.* (α_1, β_1), (α_2, β_2), (α_3, β_3), ... Suppose we form the products $\alpha_1 \times \beta_1$, $\alpha_2 \times \beta_2$, $\alpha_3 \times \beta_3$, ... and multiply all these products together. We wish to prove that (with a suitable hypothesis) the result is similar to the product of all the α's and all the β's taken together as one class; *i.e.* if we call λ the class of products $\alpha_1 \times \beta_1$, $\alpha_2 \times \beta_2$, $\alpha_3 \times \beta_3$, ..., and μ the class whose members are $\alpha_1, \alpha_2, \alpha_3, \ldots, \beta_1, \beta_2, \beta_3, \ldots$, we wish to prove

$$\Pi\text{Nc}'\lambda = \Pi\text{Nc}'\mu.$$

In order to express this proposition in symbols, let S be the correlator of the α's and β's, so that $\beta_\nu = S'\alpha_\nu$. (The suffix ν will not be used further, since it implies that the number of α's and of β's is finite or denumerable.) Then our class of products of the form $\alpha \times \beta$ is

$$\hat{\mu}\{(\exists\alpha) . \alpha \in \gamma . \mu = \alpha \times S'\alpha\},$$

where γ is the class of all the α's; and the product of this class of products is

$$\epsilon_\Delta{}^{\backprime}\hat{\mu}\,\{(\exists\alpha)\,.\,\alpha\,\epsilon\,\gamma\,.\,\mu = \alpha \times S^{\backprime}\alpha\}.$$

On the other hand, the class of all the α's and β's is $\gamma \cup S^{\backprime\backprime}\gamma$, and the product of this class is

$$\epsilon_\Delta{}^{\backprime}(\gamma \cup S^{\backprime\backprime}\gamma).$$

Thus what we have to prove (with a suitable hypothesis) is

$$\epsilon_\Delta{}^{\backprime}\hat{\mu}\,\{(\exists\alpha)\,.\,\alpha\,\epsilon\,\gamma\,.\,\mu = \alpha \times S^{\backprime}\alpha\}\ \mathrm{sm}\ \epsilon_\Delta{}^{\backprime}(\gamma \cup S^{\backprime\backprime}\gamma).$$

The hypothesis required is

$$S \upharpoonright \gamma\,\epsilon\,1 \to 1\,.\,\gamma \subset \mathrm{C\!\!\!I}^{\backprime}S\,.\,\gamma \cap S^{\backprime\backprime}\gamma = \Lambda.$$

A smaller hypothesis suffices, however, for a proposition which, in virtue of ∗114·301, is closely allied to the above, namely

$$\epsilon_\Delta{}^{\backprime}\gamma \times \epsilon_\Delta{}^{\backprime}S^{\backprime\backprime}\gamma\ \mathrm{sm}\ \epsilon_\Delta{}^{\backprime}\hat{\mu}\,\{(\exists\alpha)\,.\,\alpha\,\epsilon\,\gamma\,.\,\mu = \alpha \times S^{\backprime}\alpha\}.$$

For this, a sufficient hypothesis is

$$S \upharpoonright \gamma\,\epsilon\,1 \to 1\,.\,\gamma \subset \mathrm{C\!\!\!I}^{\backprime}S.$$

Thus *e.g.* we may write I for S, and we find

$$\vdash .\,\epsilon_\Delta{}^{\backprime}\gamma \times \epsilon_\Delta{}^{\backprime}\gamma\ \mathrm{sm}\ \epsilon_\Delta{}^{\backprime}\hat{\mu}\,\{(\exists\alpha)\,.\,\alpha\,\epsilon\,\gamma\,.\,\mu = \alpha \times \alpha\}.$$

We shall now prove the above propositions. What follows, down to ∗114·621, consists of lemmas.

For convenience, we write $S_\times{}^{\backprime}\alpha$ for $\alpha \times S^{\backprime}\alpha$ in the course of these lemmas; this notation is introduced in the hypotheses of the lemmas.

∗114·601. $\vdash :. S \upharpoonright \gamma\,\epsilon\,1 \to 1\,.\,\gamma \subset \mathrm{C\!\!\!I}^{\backprime}S\,.\,\Lambda \sim \epsilon\,\gamma\,.\,S_\times = \hat{\mu}\hat{\alpha}(\alpha\,\epsilon\,\gamma\,.\,\mu = \alpha \times S^{\backprime}\alpha)\,.\supset:$
$S_\times\,\epsilon\,1 \to 1\,.\,\mathrm{C\!\!\!I}^{\backprime}S_\times = \gamma\,.\,\mathrm{D}^{\backprime}S_\times = \hat{\mu}\,\{(\exists\alpha)\,.\,\alpha\,\epsilon\,\gamma\,.\,\mu = \alpha \times S^{\backprime}\alpha\}:$
$\alpha\,\epsilon\,\gamma\,.\supset_\alpha.\,S_\times{}^{\backprime}\alpha = \alpha \times S^{\backprime}\alpha$

Dem.

$\vdash .\,∗33·11\,.\qquad \supset \vdash : \mathrm{Hp}\,.\supset.\,\mathrm{D}^{\backprime}S_\times = \hat{\mu}\,\{(\exists\alpha)\,.\,\alpha\,\epsilon\,\gamma\,.\,\mu = \alpha \times S^{\backprime}\alpha\} \quad (1)$

$\vdash .\,∗21·33\,.\qquad \supset \vdash :. \mathrm{Hp}\,.\,\alpha\,\epsilon\,\gamma\,.\supset: \mu\,(S_\times)\,\alpha\,.\equiv_\mu.\,\mu = \alpha \times S^{\backprime}\alpha:$

$[∗30·3]\qquad \supset: S_\times{}^{\backprime}\alpha = \alpha \times S^{\backprime}\alpha \quad (2)$

$\vdash .\,(2)\,.\,∗14·204\,.\qquad \supset \vdash :. \mathrm{Hp}\,.\supset: \alpha\,\epsilon\,\gamma\,.\supset_\alpha.\,\mathrm{E}\,!\,S_\times{}^{\backprime}\alpha\,. \quad (3)$

$[∗33·43]\qquad \supset_\alpha.\,\alpha\,\epsilon\,\mathrm{C\!\!\!I}^{\backprime}S_\times \quad (4)$

$\vdash .\,∗21·33\,.\,∗33·131\,.\supset \vdash :. \mathrm{Hp}\,.\supset: \alpha\,\epsilon\,\mathrm{C\!\!\!I}^{\backprime}S_\times\,.\supset_\alpha.\,\alpha\,\epsilon\,\gamma \quad (5)$

$\vdash .\,(4)\,.\,(5)\,.\qquad \supset \vdash : \mathrm{Hp}\,.\supset.\,\mathrm{C\!\!\!I}^{\backprime}S_\times = \gamma\,. \quad (6)$

$[(3).∗71·16]\qquad \supset.\,S_\times\,\epsilon\,1 \to \mathrm{Cls} \quad (7)$

$\vdash .\,∗113·181\,.\qquad \supset \vdash :. \mathrm{Hp}\,.\supset: \alpha, \alpha'\,\epsilon\,\gamma\,.\,\alpha \times S^{\backprime}\alpha = \alpha' \times S^{\backprime}\alpha'\,.\supset.\,S^{\backprime}\alpha = S^{\backprime}\alpha'\,.$

$[∗71·59]\qquad \supset.\,\alpha = \alpha' \quad (8)$

$\vdash .\,(8)\,.\,∗71·55\,.\,(2)\,.\,(6)\,.\,(7)\,.\supset \vdash : \mathrm{Hp}\,.\supset.\,S_\times\,\epsilon\,1 \to 1 \quad (9)$

$\vdash .\,(1)\,.\,(2)\,.\,(6)\,.\,(9)\,.\supset \vdash .\,\mathrm{Prop}$

∗114·602. $\vdash : \mathrm{Hp}\,{*}114{\cdot}601 \,.\, A = \hat{R}\hat{\alpha}\,\{\alpha \epsilon \gamma \,.\, R = (S`\alpha) \downarrow \alpha\} \,.\supset .\, A \epsilon 1 \to 1 \,.\, \text{Ɑ}`A = \gamma$

Dem.

As in ∗114·601, we prove

$\vdash : \mathrm{Hp} \,.\supset .\, A \epsilon 1 \to \mathrm{Cls} \,.\, \text{Ɑ}`A = \gamma$ (1)

$\vdash \,.\, {*}21{\cdot}33 \,.\, {*}13{\cdot}171 \,.\supset \vdash :.\, \mathrm{Hp} \,.\supset : RA\alpha \,.\, RA\beta \,.\supset .\, (S`\alpha) \downarrow \alpha = (S`\beta) \downarrow \beta \,.$

[∗55·202] $\supset .\, \alpha = \beta$ (2)

$\vdash \,.\, (1) \,.\, (2) \,.\supset \vdash \,.\, \mathrm{Prop}$

∗114·603. $\vdash : \mathrm{Hp}\,{*}114{\cdot}602 \,.\, X \epsilon \epsilon_\Delta`\gamma \,.\, Y \epsilon \epsilon_\Delta`S``\gamma \,.\, P = (Y \| \breve{X}) | A | \breve{S}_\times \,.\supset .\, P \epsilon \epsilon_\Delta`\mathrm{D}`S_\times$

Dem.

$\vdash \,.\, {*}43{\cdot}122 \,.\, {*}71{\cdot}166 \,.\, {*}114{\cdot}601{\cdot}602 \,.\supset \vdash : \mathrm{Hp} \,.\supset .\, P \epsilon 1 \to \mathrm{Cls}$ (1)

$\vdash \,.\, {*}43{\cdot}122 \,.\, {*}37{\cdot}32{\cdot}322 \,.\, {*}33{\cdot}431 \,.\supset \vdash : \mathrm{Hp} \,.\supset .\, \text{Ɑ}`P = S_\times``\text{Ɑ}`A$

[∗114·601·602] $= \mathrm{D}`S_\times$ (2)

$\vdash \,.\, {*}34{\cdot}1 \,.\supset :.\, \mathrm{Hp} \,.\supset :$

$MP\mu \,.\equiv .\, (\exists R, \alpha) \,.\, M = Y | R | \breve{X} \,.\, R = (S`\alpha) \downarrow \alpha \,.\, \alpha \epsilon \gamma \,.\, \mu = S_\times`\alpha \,.$

[∗113·123.∗80·14] $\equiv .\, (\exists \alpha) \,.\, M = (Y`S`\alpha) \downarrow (X`\alpha) \,.\, \mu = S_\times`\alpha \,.\, \alpha \epsilon \gamma \,.$

[∗13·195.∗114·601] $\equiv .\, (\exists \alpha, \beta) \,.\, \beta = S`\alpha \,.\, \alpha \epsilon \gamma \,.\, M = (Y`\beta) \downarrow (X`\alpha) \,.\, \mu = \alpha \times \beta \,.$

[∗83·2] $\supset .\, (\exists \alpha, \beta, u, v) \,.\, \beta = S`\alpha \,.\, \alpha \epsilon \gamma \,.\, u \epsilon \alpha \,.\, v \epsilon \beta \,.$

$M = (v \downarrow u) \,.\, \mu = \alpha \times \beta \,.$

[∗113·101] $\supset .\, M \epsilon \mu$ (3)

$\vdash \,.\, (1) \,.\, (2) \,.\, (3) \,.\, {*}80{\cdot}14 \,.\supset \vdash \,.\, \mathrm{Prop}$

∗114·604. $\vdash : \mathrm{Hp}\,{*}114{\cdot}602 \,.\, T = \hat{P}\hat{Q}\,\{(\exists X, Y) \,.\, X \epsilon \epsilon_\Delta`\gamma \,.\, Y \epsilon \epsilon_\Delta`S``\gamma \,.$

$Q = Y \downarrow X \,.\, P = (Y \| \breve{X}) | A | \breve{S}_\times\} \,.$

$\supset .\, T \epsilon 1 \to \mathrm{Cls} \,.\, \text{Ɑ}`T = \epsilon_\Delta`\gamma \times \epsilon_\Delta`S``\gamma \,.\, \mathrm{D}`T \subset \epsilon_\Delta`\mathrm{D}`S_\times$

The relation T here defined is the correlator required for proving

$$\epsilon_\Delta`\hat{\mu}\,\{(\exists \alpha) \,.\, \alpha \epsilon \gamma \,.\, \mu = \alpha \times S`\alpha\} \;\mathrm{sm}\; \epsilon_\Delta`\gamma \times \epsilon_\Delta`S``\gamma.$$

Besides what is proved in the present proposition, we shall have to prove

$$T \epsilon \mathrm{Cls} \to 1 \,.\, \epsilon_\Delta`\mathrm{D}`S_\times \subset \mathrm{D}`T.$$

The proof of the present proposition is as follows.

Dem.

$\vdash \,.\, {*}21{\cdot}33 \,.\, {*}13{\cdot}171 \,.\supset \vdash :.\, \mathrm{Hp} \,.\supset :$

$PTQ \,.\, P'TQ \,.\supset .\, (\exists X, Y, X', Y') \,.\, Y \downarrow X = Y' \downarrow X' \,.\, P = (Y \| \breve{X}) | A | \breve{S}_\times \,.$

$P' = (Y' \| \breve{X}') | A | \breve{S}_\times \,.$

[∗55·202] $\supset .\, P = P'$ (1)

$\vdash \,.\, {*}21{\cdot}33 \,.\, {*}114{\cdot}603 \,.\supset \vdash :.\, \mathrm{Hp} \,.\supset : PTQ \,.\supset .\, P \epsilon \epsilon_\Delta`\mathrm{D}`S_\times$ (2)

$\vdash \,.\, (1) \,.\, (2) \,.\, {*}113{\cdot}101 \,.\supset \vdash \,.\, \mathrm{Prop}$

$*114{\cdot}605$. $\vdash : \text{Hp}\, *114{\cdot}604 \,.\supset .\, T \epsilon \text{Cls} \to 1$

Dem.

$\vdash . *114{\cdot}601 . \supset \vdash : \text{Hp} . \supset . S_\times \epsilon 1 \to 1$ (1)

$\vdash . (1) . *74{\cdot}71 . *114{\cdot}601{\cdot}602 . \supset$

$\vdash :. \text{Hp} . X, X' \epsilon \epsilon_\Delta‘\gamma . Y, Y' \epsilon \epsilon_\Delta‘S‘‘\gamma . (Y \parallel \breve{X}) | A | \breve{S}_\times = (Y' \parallel \breve{X}') | A | \breve{S}_\times . \supset :$

$(Y \parallel \breve{X}) | A = (Y' \parallel \breve{X}') | A :$

$[*74{\cdot}7]$ $\supset : (Y \parallel \breve{X}) \upharpoonright \text{D}‘A = (Y' \parallel \breve{X}') \upharpoonright \text{D}‘A :$

$[*114{\cdot}602]$ $\supset : \alpha \epsilon \gamma . \supset_\alpha . (Y \parallel \breve{X})‘(S‘\alpha) \downarrow \alpha = (Y' \parallel \breve{X}')‘(S‘\alpha) \downarrow \alpha .$

$[*113{\cdot}123]$ $\supset_\alpha . (Y‘S‘\alpha) \downarrow (X‘\alpha) = (Y'‘S‘\alpha) \downarrow (X'‘\alpha) .$

$[*55{\cdot}202]$ $\supset_\alpha . X‘\alpha = X'‘\alpha . Y‘S‘\alpha = Y'‘S‘\alpha :$

$[*80{\cdot}14 . *33{\cdot}45]$ $\supset : X = X' . Y = Y' :$

$[*55{\cdot}202]$ $\supset : Y \downarrow X = Y' \downarrow X'$ (2)

$\vdash . (2) . *13{\cdot}22 . *21{\cdot}33 . \supset \vdash :. \text{Hp} . \supset : PTQ . PTQ' . \supset . Q = Q' :. \supset \vdash . \text{Prop}$

The following propositions are required for proving that, with the same hypothesis, $\epsilon_\Delta‘\text{D}‘S_\times \subset \text{D}‘T$.

$*114{\cdot}61$. $\vdash : \text{Hp}\, *114{\cdot}602 . P \epsilon \epsilon_\Delta‘\text{D}‘S_\times . X = \breve{\iota} | \text{ᗡ} | P | S_\times . Y = \breve{\iota} | \text{D} | P | S_\times | \breve{S} . \supset .$

$X \epsilon \epsilon_\Delta‘\gamma . Y \epsilon \epsilon_\Delta‘S‘‘\gamma$

Dem.

$\vdash . *72{\cdot}181{\cdot}13{\cdot}131 . *80{\cdot}14 . *114{\cdot}601 . \supset \vdash : \text{Hp} . \supset . X, Y \epsilon 1 \to \text{Cls}$ (1)

$\vdash . *72{\cdot}2{\cdot}181{\cdot}13{\cdot}131 . *80{\cdot}14 . *114{\cdot}601 . \supset$

$\vdash :. \text{Hp} . \supset : x X \alpha . \equiv . x = \breve{\iota}‘\text{ᗡ}‘P‘S_\times‘\alpha .$ (2)

$[*51{\cdot}53]$ $\supset . x \epsilon \text{ᗡ}‘P‘S_\times‘\alpha .$

$[*83{\cdot}2 . *114{\cdot}601]$ $\supset . (\exists R) . R \epsilon \alpha \times S‘\alpha . x \epsilon \text{ᗡ}‘R .$

$[*113{\cdot}142]$ $\supset . x \epsilon \alpha$ (3)

$\vdash . *114{\cdot}601 . \supset \vdash :. \text{Hp} . \supset : \alpha \epsilon \gamma . \equiv . S_\times‘\alpha \epsilon \text{D}‘S_\times .$

$[*83{\cdot}2]$ $\equiv . \text{E}\,!\, P‘S_\times‘\alpha$ (4)

$\vdash . *83{\cdot}2 . \supset \vdash :. \text{Hp} . \supset : \text{E}\,!\, P‘S_\times‘\alpha . \equiv . P‘S_\times‘\alpha \epsilon S_\times‘\alpha .$

$[*113{\cdot}142]$ $\supset . \text{ᗡ}‘P‘S_\times‘\alpha \epsilon 1 .$

$[*52{\cdot}15]$ $\supset . \text{E}\,!\, \breve{\iota}‘\text{ᗡ}‘P‘S_\times‘\alpha$ (5)

$\vdash . (2) . (4) . (5) . \supset \vdash : \text{Hp} . \supset . \gamma \subset \text{ᗡ}‘X$ (6)

$\vdash . *34{\cdot}36 . *114{\cdot}601 . \supset \vdash : \text{Hp} . \supset . \text{ᗡ}‘X \subset \gamma$ (7)

$\vdash . (1) . (3) . (6) . (7) . \supset \vdash : \text{Hp} . \supset . X \epsilon \epsilon_\Delta‘\gamma$ (8)

Similarly $\vdash : \text{Hp} . \supset . Y \epsilon \epsilon_\Delta‘S‘‘\gamma$ (9)

$\vdash . (8) . (9) . \supset \vdash . \text{Prop}$

$*114{\cdot}611$. $\vdash :. \text{Hp}\, *114{\cdot}61 . \supset : \alpha \epsilon \gamma . \supset . (Y‘S‘\alpha) \downarrow (X‘\alpha) = P‘S_\times‘\alpha$

Dem.

$\vdash . *72{\cdot}2 . \supset \vdash : \text{Hp} . \alpha \epsilon \gamma . \supset . X‘\alpha = \breve{\iota}‘\text{ᗡ}‘P‘S_\times‘\alpha . Y‘S‘\alpha = \breve{\iota}‘\text{D}‘P‘S_\times‘\alpha .$

$[*55{\cdot}16 . *51{\cdot}51]$ $\supset . (Y‘S‘\alpha) \downarrow (X‘\alpha) = P‘S_\times‘\alpha : \supset \vdash . \text{Prop}$

$*114\cdot612$. $\vdash : \text{Hp} *114\cdot61 . \supset . (Y \| \breve{X}) | A | \breve{S}_\times = P$

Dem.

$\vdash . *83\cdot15 . \supset \vdash : \text{Hp} . \dot{\exists} ! P . \supset . \exists ! \text{D}`S_\times .$

$[*114\cdot601] \qquad \supset . \exists ! \gamma \qquad (1)$

$\vdash . *34\cdot1 . \supset \vdash :. \text{Hp} . \supset : M \{(Y \| \breve{X}) | A | \breve{S}_\times\} \mu . \equiv .$

$(\exists Q, \alpha) . M = (Y \| \breve{X})`Q . QA\alpha . \mu = S_\times`\alpha .$

$[*114\cdot601\cdot602] \qquad \equiv . (\exists \alpha) . M = (Y \| \breve{X})`(S`\alpha) \downarrow \alpha . \mu = S_\times`\alpha . \alpha \epsilon \gamma .$

$[*113\cdot123] \qquad \equiv . (\exists \alpha) . M = (Y`S`\alpha) \downarrow (X`\alpha) . \mu = S_\times`\alpha . \alpha \epsilon \gamma .$

$[*114\cdot611] \qquad \equiv . (\exists \alpha) . M = P`S_\times`\alpha . \mu = S_\times`\alpha . \alpha \epsilon \gamma .$

$[*13\cdot193 . *114\cdot601 . *71\cdot16] \equiv . M = P`\mu . \exists ! \gamma .$

$[*71\cdot36 . *80\cdot14 . (1)] \qquad \equiv . MP\mu :. \supset \vdash . \text{Prop}$

$*114\cdot613$. $\vdash : \text{Hp} *114\cdot61 . \text{Hp} *114\cdot604 . \supset .$

$P = T`(Y \downarrow X) . (Y \downarrow X) \epsilon \epsilon_\Delta`\gamma \times \epsilon_\Delta`S``\gamma$

Dem.

$\vdash . *21\cdot33 . *114\cdot604 . \supset \vdash :. \text{Hp} *114\cdot604 . \supset :$

$X \epsilon \epsilon_\Delta`\gamma . Y \epsilon \epsilon_\Delta`S``\gamma . \supset . T`(Y \downarrow X) = (Y \| \breve{X}) | A | \breve{S}_\times \qquad (1)$

$\vdash . (1) . *114\cdot61\cdot612 . *113\cdot106 . \supset \vdash . \text{Prop}$

$*114\cdot614$. $\vdash : \text{Hp} *114\cdot604 . \supset . \epsilon_\Delta`\text{D}`S_\times \subset \text{D}`T$

Dem.

$\vdash . *114\cdot613 . \supset \vdash :. \text{Hp} . \supset : P \epsilon \epsilon_\Delta`\text{D}`S_\times . \supset . (\exists Q) . P = T`Q .$

$[*33\cdot43] \qquad \supset . P \epsilon \text{D}`T :. \supset \vdash . \text{Prop}$

$*114\cdot62$. $\vdash : \text{Hp} *114\cdot604 . \supset . T \epsilon 1 \to 1 . \text{D}`T = \epsilon_\Delta`\text{D}`S_\times . \text{Cl}`T = \epsilon_\Delta`\gamma \times \epsilon_\Delta`S``\gamma$

$[*114\cdot604\cdot605\cdot614]$

$*114\cdot621$. $\vdash : S \upharpoonright \gamma \epsilon 1 \to 1 . \gamma \subset \text{Cl}`S . \Lambda \sim\epsilon \gamma . \supset .$

$\epsilon_\Delta`\hat{\mu}\{(\exists \alpha) . \alpha \epsilon \gamma . \mu = \alpha \times S`\alpha\} \text{ sm } \epsilon_\Delta`\gamma \times \epsilon_\Delta`S``\gamma$

$[*114\cdot62\cdot601]$

The hypothesis $\Lambda \sim\epsilon \gamma$ is not necessary, since, when $\Lambda \epsilon \gamma$,

$$\epsilon_\Delta`\hat{\mu}\{(\exists \alpha) . \alpha \epsilon \gamma . \mu = \alpha \times S`\alpha\} \text{ and } \epsilon_\Delta`\gamma \times \epsilon_\Delta`S``\gamma$$

are both Λ. This is proved in $*114\cdot63$.

$*114\cdot63$. $\vdash : S \upharpoonright \gamma \epsilon 1 \to 1 . \gamma \subset \text{Cl}`S . \supset .$

$\epsilon_\Delta`\hat{\mu}\{(\exists \alpha) . \alpha \epsilon \gamma . \mu = \alpha \times S`\alpha\} \text{ sm } \epsilon_\Delta`\gamma \times \epsilon_\Delta`S``\gamma$

Dem.

$\vdash . *10\cdot24 . *83\cdot11 . \supset$

$\vdash : \text{Hp} . \Lambda \epsilon \gamma . \supset . \Lambda \times S`\Lambda \epsilon \hat{\mu}\{(\exists \alpha) . \alpha \epsilon \gamma . \mu = \alpha \times S`\alpha\} . \epsilon_\Delta`\gamma = \Lambda .$

$[*113\cdot114] \qquad \supset . \Lambda \epsilon \hat{\mu}\{(\exists \alpha) . \alpha \epsilon \gamma . \mu = \alpha \times S`\alpha\} . \epsilon_\Delta`\gamma \times \epsilon_\Delta`S``\gamma = \Lambda .$

$[*83\cdot11] \qquad \supset . \epsilon_\Delta`\hat{\mu}\{(\exists \alpha) . \alpha \epsilon \gamma . \mu = \alpha \times S`\alpha\} = \Lambda . \epsilon_\Delta`\gamma \times \epsilon_\Delta`S``\gamma = \Lambda \quad (1)$

$\vdash . (1) . *73\cdot47 . *114\cdot621 . \supset \vdash . \text{Prop}$

The above is one of the two variants of the associative law for ϵ_Δ and $\times$.

***114·631.** $\vdash . \epsilon_\Delta{}^{\backprime}\hat{\mu}\{(\exists\alpha) . \alpha \epsilon \gamma . \mu = \alpha \times \alpha\} \text{ sm } \epsilon_\Delta{}^{\backprime}\alpha \times \epsilon_\Delta{}^{\backprime}\alpha$ $\left[*114\cdot63 \frac{I}{S}\right]$

***114·632.** $\vdash : S \upharpoonright \gamma \epsilon 1 \rightarrow 1 . \gamma \subset \text{Ɑ}{}^{\backprime}S . \gamma \cap S^{\backprime\backprime}\gamma = \Lambda . \supset .$

$\epsilon_\Delta{}^{\backprime}\hat{\mu}\{(\exists\alpha) . \alpha \epsilon \gamma . \mu = \alpha \times S{}^{\backprime}\alpha\} \text{ sm } \epsilon_\Delta{}^{\backprime}(\gamma \cup S^{\backprime\backprime}\gamma)$ [*114·63·301]

This is the second variant of the associative law for ϵ_Δ and $\times$.

***114·64.** $\vdash : (R^{\backprime\backprime}\gamma) \daleth R, S \upharpoonright \gamma \epsilon 1 \rightarrow 1 . \gamma \subset \text{Ɑ}{}^{\backprime}R . \gamma \subset \text{Ɑ}{}^{\backprime}S . \supset .$

$\epsilon_\Delta{}^{\backprime}R^{\backprime\backprime}\gamma \times \epsilon_\Delta{}^{\backprime}S^{\backprime\backprime}\gamma \text{ sm } \epsilon_\Delta{}^{\backprime}\hat{\mu}\{(\exists z) . z \epsilon \gamma . \mu = R{}^{\backprime}z \times S{}^{\backprime}z\}$

Dem.

$\vdash . *114\cdot63 \frac{S | \breve{R}, R^{\backprime\backprime}\gamma}{S, \quad \gamma} . \supset$

$\vdash : S | \breve{R} \upharpoonright R^{\backprime\backprime}\gamma \epsilon 1 \rightarrow 1 . R^{\backprime\backprime}\gamma \subset \text{Ɑ}{}^{\backprime}(S | \breve{R}) . \supset .$

$\epsilon_\Delta{}^{\backprime}R^{\backprime\backprime}\gamma \times \epsilon_\Delta{}^{\backprime}S^{\backprime\backprime}\breve{R}^{\backprime\backprime}R^{\backprime\backprime}\gamma \text{ sm } \epsilon_\Delta{}^{\backprime}\hat{\mu}\{(\exists\alpha) . \alpha \epsilon R^{\backprime\backprime}\gamma . \mu = \alpha \times (S | \breve{R}){}^{\backprime}\alpha\}$ (1)

$\vdash . *74\cdot14 . *35\cdot354 . \supset \vdash : \text{Hp} . \supset . S | \breve{R} \upharpoonright R^{\backprime\backprime}\gamma = S \upharpoonright \gamma | \gamma \daleth \breve{R} . \breve{R} \upharpoonright R^{\backprime\backprime}\gamma = \gamma \daleth \breve{R} .$

[*71·252] $\supset . S | \breve{R} \upharpoonright R^{\backprime\backprime}\gamma \epsilon 1 \rightarrow 1$ (2)

$\vdash . *37\cdot2 . \quad \supset \vdash : \text{Hp} . \supset . R^{\backprime\backprime}\gamma \subset R^{\backprime\backprime}\text{Ɑ}{}^{\backprime}S .$

[*37·32] $\supset . R^{\backprime\backprime}\gamma \subset \text{Ɑ}{}^{\backprime}(S | \breve{R})$ (3)

$\vdash . *74\cdot171 . \quad \supset \vdash : \text{Hp} . \supset . \breve{R}^{\backprime\backprime}R^{\backprime\backprime}\gamma = \gamma$ (4)

$\vdash . (4) . *74\cdot14 . \supset \vdash : \text{Hp} . \supset . (R^{\backprime\backprime}\gamma) \daleth R = R \upharpoonright \gamma .$

[*35·7.*71·4] $\supset . \hat{\mu}\{(\exists\alpha) . \alpha \epsilon R^{\backprime\backprime}\gamma . \mu = \alpha \times (S | \breve{R}){}^{\backprime}\alpha\}$

$= \hat{\mu}\{(\exists z) . z \epsilon \gamma . \mu = R{}^{\backprime}z \times S{}^{\backprime}\breve{R}{}^{\backprime}R{}^{\backprime}z\}$

[*74·53] $= \hat{\mu}\{(\exists z) . z \epsilon \gamma . \mu = R{}^{\backprime}z \times S{}^{\backprime}z\}$ (5)

$\vdash . (1) . (2) . (3) . (4) . (5) . \supset \vdash . \text{Prop}$

In the above proposition, the hypothesis has to be such as to yield $\breve{R}^{\backprime\backprime}R^{\backprime\backprime}\gamma = \gamma$. Various other forms of hypothesis will secure this result, and will give other forms of the above proposition. This subject is treated in *74, above.

***114·65.** $\vdash :. (R^{\backprime\backprime}\gamma) \daleth R, S \upharpoonright \gamma \epsilon 1 \rightarrow 1 . \gamma \subset \text{Ɑ}{}^{\backprime}R . \gamma \subset \text{Ɑ}{}^{\backprime}S . R^{\backprime\backprime}\gamma \cap S^{\backprime\backprime}\gamma = \Lambda . \supset .$

$\epsilon_\Delta{}^{\backprime}(R^{\backprime\backprime}\gamma \cup S^{\backprime\backprime}\gamma) \text{ sm } \epsilon_\Delta{}^{\backprime}\hat{\mu}\{(\exists z) . z \epsilon \gamma . \mu = R{}^{\backprime}z \times S{}^{\backprime}z\}$

[*114·64·301]

$*115$. MULTIPLICATIVE CLASSES AND ARITHMETICAL CLASSES.

Summary of $*115$.

Whenever κ is a class of mutually exclusive classes, $\epsilon_\Delta{}^\backprime\kappa$ is similar to $D^{\backprime\backprime}\epsilon_\Delta{}^\backprime\kappa$; hence

$$\Pi Nc^\backprime\kappa = Nc^\backprime D^{\backprime\backprime}\epsilon_\Delta{}^\backprime\kappa.$$

Now $D^{\backprime\backprime}\epsilon_\Delta{}^\backprime\kappa$ is of the same type as κ; and when κ is a class of mutually exclusive classes, $D^{\backprime\backprime}\epsilon_\Delta{}^\backprime\kappa$ consists of all classes formed by selecting one representative from each member of κ. It often happens that $D^{\backprime\backprime}\epsilon_\Delta{}^\backprime\kappa$ is easier to deal with than $\epsilon_\Delta{}^\backprime\kappa$; hence when possible (*i.e.* when $\kappa \epsilon \text{Cls}^2 \text{excl}$), it is convenient to use $D^{\backprime\backprime}\epsilon_\Delta{}^\backprime\kappa$, rather than $\epsilon_\Delta{}^\backprime\kappa$, as the standard member of $\Pi Nc^\backprime\kappa$. We therefore put

$$\text{Prod}^\backprime\kappa = D^{\backprime\backprime}\epsilon_\Delta{}^\backprime\kappa \quad \text{Df.}$$

We shall call $\text{Prod}^\backprime\kappa$ the "multiplicative class" of κ.

The associative law,

$$\text{Prod}^\backprime s^\backprime\kappa \,\text{sm}\, \text{Prod}^\backprime\text{Prod}^{\backprime\backprime}\kappa,$$

requires not merely $\kappa \epsilon \text{Cls}^2 \text{excl}$, but also $s^\backprime\kappa \epsilon \text{Cls}^2 \text{excl}$. The combination of these two hypotheses gives a completely disjointed class of classes of classes, *i.e.* a class of classes of classes κ which can be obtained by dividing a given class ($s^\backprime s^\backprime\kappa$) into mutually exclusive portions, and then dividing each of those portions into mutually exclusive portions. For example, take a square (a class of points) and divide it by horizontal lines, and then divide each of the resulting rectangles by vertical lines; then the resulting rows of little rectangles form such a class, each row of rectangles being one member of the class. Such a class we call an "arithmetical" class, and denote by "Cls^3 arithm."

The present number is concerned with the properties of multiplicative classes and arithmetical classes. Some of these properties will be useful in dealing with exponentiation.

The present number begins with various propositions concerning $\text{Prod}^\backprime\kappa$ which are merely repetitions of previous propositions of $*83$, $*84$, $*85$ or $*113$. Thus we have

$*115\cdot141$. $\vdash : \exists ! \text{Prod}^\backprime\kappa . \supset . s^\backprime\text{Prod}^\backprime\kappa = s^\backprime\kappa$ by $*83\cdot66$,

$*115\cdot142$. $\vdash . \text{Prod}^\backprime\iota^\backprime\alpha = \iota^{\backprime\backprime}\alpha$ by $*83\cdot7$,

$*115\cdot143$. $\vdash . \text{Prod}^\backprime\iota^{\backprime\backprime}\alpha = \iota^\backprime\alpha$ by $*83\cdot71$,

$*115\cdot16$. $\vdash . \kappa \epsilon \text{Cls}^2 \text{excl} . \supset . \text{Prod}^\backprime\kappa \subset Nc^\backprime\kappa$ by $*100\cdot64$,

and various other properties.

We then proceed to consider $\text{Cls}^3\text{arithm}$. We prove

*115·22. $\vdash :. \kappa \epsilon \text{Cls}^3\text{arithm} . \supset : s``\kappa \epsilon \text{Cls}^2\text{excl} : \alpha, \beta \epsilon \kappa . \exists ! s`\alpha \cap s`\beta . \supset_{\alpha,\beta} . \alpha = \beta$

and *115·23 gives a similar proposition substituting "Prod" for s.

After a few more propositions on $\text{Cls}^3\text{arithm}$, we proceed to the associative law for Prod (*115·34), *i.e.*

$$\vdash : \kappa \epsilon \text{Cls}^3\text{arithm} . \supset . \text{Prod}`\text{Prod}``\kappa \text{ sm Prod}`s`\kappa.$$

(This proposition, *115·34, also states that, with the same hypothesis, $\text{Prod}`s`\kappa \text{ sm } \epsilon_\Delta`s`\kappa$.) Hence we have

*115·35. $\vdash : \kappa \epsilon \text{Cls}^3\text{arithm} . \supset . \text{Nc}`\text{Prod}`\text{Prod}``\kappa = \text{Nc}`\text{Prod}`s`\kappa = \Pi\text{Nc}`\text{Prod}``\kappa$
$= \Pi\text{Nc}`\epsilon_\Delta``\kappa = \Pi\text{Nc}`s`\kappa$

We have also

*115·42. $\vdash : \kappa \epsilon \text{Cls}^3\text{arithm} . \supset . \text{Prod}`\text{Prod}``\kappa = \text{D}```\text{Prod}`\epsilon_\Delta``\kappa = \text{D}```\text{D}``\epsilon_\Delta`\epsilon_\Delta``\kappa$

*115·44. $\vdash : \kappa \epsilon \text{Cls}^3\text{arithm} . \supset . \text{Prod}`s`\kappa = s``\text{Prod}`\text{Prod}``\kappa$

We have next to prove that if two classes of classes have double similarity, so have their multiplicative classes. The proof is simple, since the double correlator is the same as for the original classes, *i.e.*

*115·502. $\vdash : T \upharpoonright s`\lambda \epsilon \kappa \,\overline{\text{sm}}\,\overline{\text{sm}}\, \lambda . \supset . T \upharpoonright s`\text{Prod}`\lambda \epsilon (\text{Prod}`\kappa) \,\overline{\text{sm}}\,\overline{\text{sm}}\, (\text{Prod}`\lambda)$

whence

*115·51. $\vdash : \kappa \text{ sm sm } \lambda . \supset . \text{Prod}`\kappa \text{ sm sm Prod}`\lambda$

The number ends with some propositions which result from *114·64·65 and are analogous to them. One of these is used in the following number, in proving $\mu^\varpi \times_c \nu^\varpi = (\mu \times_c \nu)^\varpi$, namely,

*115·6. $\vdash : (R``\gamma) \daleth R, S \upharpoonright \gamma \epsilon 1 \rightarrow 1 . \gamma \subset \text{Ꮷ}`R . \gamma \subset \text{Ꮷ}`S . R``\gamma, S``\gamma \epsilon \text{Cls}^2\text{excl} . \supset .$
$\text{Prod}`R``\gamma \times \text{Prod}`S``\gamma \text{ sm } \epsilon_\Delta`\hat{\mu}\{(\exists z) . z \epsilon \gamma . \mu = R`z \times S`z\}$

The subject of this number will be useful in dealing with exponentiation, since we shall define μ^ν by means of $\text{Prod}`\alpha \downarrow ``\beta$, where $\mu = \text{N}_0\text{c}`\alpha$ and $\nu = \text{N}_0\text{c}`\beta$.

---

*115·01. $\text{Prod}`\kappa = \text{D}``\epsilon_\Delta`\kappa$ Df

*115·02. $\text{Cls}^3\text{arithm} = \hat{\kappa}(\kappa, s`\kappa \epsilon \text{Cls}^2\text{ excl})$ Df

*115·1. $\vdash . \text{Prod}`\kappa = \text{D}``\epsilon_\Delta`\kappa$ [(*115·01)]

*115·101. $\vdash :. \alpha \epsilon \kappa . \supset_\alpha . \varpi \cap \alpha \epsilon 1 : \varpi \subset s`\kappa : \supset . \varpi \epsilon \text{Prod}`\kappa$ [*84·411]

*115·11. $\vdash :: \kappa \epsilon \text{Cls}^2\text{ excl} . \supset :. \varpi \epsilon \text{Prod}`\kappa . \equiv : \alpha \epsilon \kappa . \supset_\alpha . \varpi \cap \alpha \epsilon 1 : \varpi \subset s`\kappa$
[*84·412]

Owing to this proposition, $\text{Prod}`\kappa$ can be treated without any reference to $\epsilon_\Delta`\kappa$ whenever $\kappa \epsilon \text{Cls}^2\text{ excl}$.

∗115·12. $\vdash : \kappa \in \text{Cls}^2 \text{ excl} . \supset . \text{Prod}`\kappa \in \Pi\text{Nc}`\kappa . \text{Prod}`\kappa \text{ sm } \epsilon_\Delta`\kappa$ [∗84·41]

It is this proposition that makes the notation Prod'κ appropriate for the multiplicative class.

∗115·13. $\vdash : \alpha \cap \beta = \Lambda . \supset . \text{Prod}`(\iota`\alpha \cup \iota`\beta) \text{ sm } (\alpha \times \beta)$ [∗113·152]

∗115·131. $\vdash : \alpha \neq \beta . \supset . \text{Prod}`(\iota`\alpha \cup \iota`\beta) = C``(\alpha \times \beta)$ [∗113·151]

∗115·14. $\vdash :. \kappa \cap \lambda = \Lambda . \vee . s`\kappa \cap s`\lambda = \Lambda : \supset :$
$\varpi \in \text{Prod}`(\kappa \cup \lambda) . \equiv . (\exists \rho, \sigma) . \rho \in \text{Prod}`\kappa . \sigma \in \text{Prod}`\lambda . \varpi = \rho \cup \sigma$
[∗83·64·641]

∗115·141. $\vdash : \exists ! \text{Prod}`\kappa . \supset . s`\text{Prod}`\kappa = s`\kappa$ [∗83·66]

∗115·142. $\vdash . \text{Prod}`\iota`\alpha = \iota``\alpha$ [∗83·7]

∗115·143. $\vdash . \text{Prod}`\iota``\alpha = \iota`\alpha$ [∗83·71]

∗115·144. $\vdash : \kappa \subset 1 . \supset . \text{Prod}`\kappa = \iota`s`\kappa$ [∗83·72]

∗115·145. $\vdash :. \kappa \in \text{Cls}^2 \text{ excl} . \alpha \in \kappa . \mu \cap \alpha \in 1 . \supset : \mu - \alpha \in \text{Prod}`(\kappa - \iota`\alpha) . \equiv . \mu \in \text{Prod}`\kappa$
[∗84·422]

∗115·15. $\vdash :. \kappa, \lambda \in \text{Cls}^2 \text{ excl} . s`\kappa = s`\lambda . \supset : \kappa \subset \text{Prod}`\lambda . \equiv . \lambda \subset \text{Prod}`\kappa$
[∗84·43]

∗115·151. $\vdash : \kappa \in \text{Cls}^2 \text{ excl} . \supset . \epsilon_\Delta`s`\kappa = \dot{s}``\text{Prod}`\epsilon_\Delta``\kappa$ [∗85·28]

∗115·152. $\vdash . P_\Delta`\alpha \text{ sm Prod}`P \downarrow``\alpha$ [∗85·55]

∗115·153. $\vdash . \epsilon_\Delta`\kappa \text{ sm Prod}`\epsilon \downarrow``\kappa$ [∗115·152]

∗115·154. $\vdash . \text{Prod}`\epsilon \downarrow``\kappa \in \Pi\text{Nc}`\kappa$ [∗115·153]

∗115·16. $\vdash : \kappa \in \text{Cls}^2 \text{ excl} . \supset . \text{Prod}`\kappa \subset \text{Nc}`\kappa$ [∗100·64]

The following proposition is used in the theory of well-ordered series (∗250·5).

∗115·17. $\vdash : \exists ! \epsilon_\Delta`\text{Cl ex}`\alpha . \supset . \text{Prod}`\text{Cl ex}`\alpha = \iota`\alpha$

Dem.

$\vdash . ∗80·14 . ∗115·1 . ∗37·45 . \supset \vdash : \text{Hp} . \supset . \exists ! \text{Prod}`\text{Cl ex}`\alpha$ (1)

$\vdash . ∗60·61 . \text{Fact} . \supset$

$\vdash :. R \in 1 \to \text{Cls} . R \subseteq \epsilon . \text{Ǝ}`R = \text{Cl ex}`\alpha . \supset : R \in 1 \to \text{Cls} . R \subseteq \epsilon . \iota``\alpha \subset \text{Ǝ}`R :$

[∗51·15] $\supset : x \in \alpha . \supset_x . xR(\iota`x) :$

[∗33·14] $\supset : \alpha \subset \text{D}`R$ (2)

$\vdash . ∗83·21 . \supset \vdash : \text{Hp}(2) . \supset . \text{D}`R \subset s`\text{Cl ex}`\alpha .$

[∗60·501] $\supset . \text{D}`R \subset \alpha$ (3)

$\vdash . (2) . (3) . \supset \vdash : R \in 1 \to \text{Cls} . R \subseteq \epsilon . \text{Ǝ}`R = \text{Cl ex}`\alpha . \supset . \text{D}`R = \alpha$ (4)

$\vdash . (4) . ∗115·1 . ∗80·14 . \supset \vdash . \text{Prod}`\text{Cl ex}`\alpha \subset \iota`\alpha$ (5)

$\vdash . (1) . (5) . ∗51·4 . \supset \vdash . \text{Prop}$

∗115·18. $\vdash . t`\text{Prod}`\kappa = t`\kappa$ [∗83·81]

∗115·2. $\vdash : \kappa \in \text{Cls}^3 \text{arithm} . \equiv . \kappa, s`\kappa \in \text{Cls}^2 \text{excl}$ [(∗115·02)]

∗115·21. $\vdash :. \kappa \in \text{Cls}^3 \text{arithm} . \equiv : \alpha, \beta \in \kappa . \exists ! \alpha \cap \beta . \supset_{\alpha,\beta} . \alpha = \beta :$

$\alpha, \beta \in \kappa . \rho \in \alpha . \sigma \in \beta . \exists ! \rho \cap \sigma . \supset_{\alpha,\beta,\rho,\sigma} . \rho = \sigma$

[∗115·2 . ∗84·11]

∗115·211. $\vdash : \kappa \in \text{Cls}^3 \text{arithm} . \alpha, \beta \in \kappa . \rho \in \alpha . \sigma \in \beta . \exists ! \rho \cap \sigma . \supset . \alpha = \beta$

Dem.

$\vdash . *115·21 . \supset \vdash : \text{Hp} . \supset . \rho = \sigma . \rho \in \alpha . \sigma \in \beta .$

[∗13·13] $\supset . \rho \in \alpha \cap \beta .$

[∗115·21] $\supset . \alpha = \beta : \supset \vdash . \text{Prop}$

∗115·22. $\vdash :. \kappa \in \text{Cls}^3 \text{arithm} . \supset : s``\kappa \in \text{Cls}^2 \text{excl} : \alpha, \beta \in \kappa . \exists ! s`\alpha \cap s`\beta . \supset_{\alpha,\beta} . \alpha = \beta$

Dem.

$\vdash . *40·11 . \supset \vdash : \exists ! s`\alpha \cap s`\beta . \equiv . (\exists x, \rho, \sigma) . \rho \in \alpha . \sigma \in \beta . x \in \rho . x \in \sigma .$

[∗10·35] $\equiv . (\exists \rho, \sigma) . \rho \in \alpha . \sigma \in \beta . \exists ! \rho \cap \sigma$ (1)

$\vdash . (1) . *115·211 . \supset$

$\vdash :. \text{Hp} . \supset : \alpha, \beta \in \kappa . \exists ! s`\alpha \cap s`\beta . \supset . \alpha = \beta .$ (2)

[∗30·37] $\supset . s`\alpha = s`\beta$ (3)

$\vdash . (2) . (3) . *84·11 . \supset \vdash . \text{Prop}$

Observe that, although "$s``\kappa \in \text{Cls}^2 \text{excl}$" follows from

"$\alpha, \beta \in \kappa . \exists ! s`\alpha \cap s`\beta . \supset_{\alpha,\beta} . \alpha = \beta$,"

the converse implication does not hold. If there were two different classes α and β having the same sum, we might have $\exists ! s`\alpha \cap s`\beta$, *i.e.* $\exists ! s`\alpha$, without having $\alpha = \beta$, in spite of "$s``\kappa \in \text{Cls}^2 \text{excl}$." In proofs, less use can be made of "$s``\kappa \in \text{Cls}^2 \text{excl}$" than of "$\alpha, \beta \in \kappa . \exists ! s`\alpha \cap s`\beta . \supset_{\alpha,\beta} . \alpha = \beta$." If $\Lambda \sim\in \kappa$ or $\iota`\Lambda \sim\in \kappa$, the latter implies $s \upharpoonright \kappa \in 1 \rightarrow 1$.

∗115·23. $\vdash :. \kappa \in \text{Cls}^3 \text{arithm} . \supset :$

$\text{Prod}``\kappa \in \text{Cls}^2 \text{excl} : \alpha, \beta \in \kappa . \exists ! \text{Prod}`\alpha \cap \text{Prod}`\beta . \supset_{\alpha,\beta} . \alpha = \beta$

Dem.

$\vdash . *83·62 .$ $\supset \vdash : \varpi \in \text{Prod}`\alpha \cap \text{Prod}`\beta . \supset . \varpi \subset s`\alpha \cap s`\beta$ (1)

$\vdash . (1) . *24·58 .$ $\supset \vdash : \varpi \in \text{Prod}`\alpha \cap \text{Prod}`\beta . \exists ! \varpi . \supset . \exists ! s`\alpha \cap s`\beta$ (2)

$\vdash . (2) . *115·22 .$ $\supset \vdash : \text{Hp} . \alpha, \beta \in \kappa . \varpi \in \text{Prod}`\alpha \cap \text{Prod}`\beta . \exists ! \varpi . \supset . \alpha = \beta$ (3)

$\vdash . *83·16 . \text{Transp} . \supset \vdash : \Lambda \in \text{Prod}`\alpha \cap \text{Prod}`\beta . \supset . \alpha = \Lambda . \beta = \Lambda$ (4)

$\vdash . (3) . (4) . \supset \vdash :. \text{Hp} . \supset : \alpha, \beta \in \kappa . \exists ! \text{Prod}`\alpha \cap \text{Prod}`\beta . \supset . \alpha = \beta .$ (5)

[∗30·37] $\supset . \text{Prod}`\alpha = \text{Prod}`\beta$ (6)

$\vdash . (5) . (6) . *84·11 . \supset \vdash . \text{Prop}$

∗115·24. $\vdash : \kappa \in \text{Cls}^3 \text{arithm} . \equiv . \epsilon \upharpoonright \kappa, \epsilon \upharpoonright s`\kappa \in \text{Cls} \rightarrow 1$ [∗115·2 . ∗84·14]

∗115·25. $\vdash : \kappa \in \text{Cls}^3 \text{arithm} . \supset . \epsilon_\Delta`\kappa \subset 1 \rightarrow 1 . \epsilon_\Delta`s`\kappa \subset 1 \rightarrow 1$ [∗84·3 . ∗115·2]

***115·26.** $\vdash : \kappa \epsilon \text{Cls}^3 \text{arithm} . \supset .$

$$\epsilon_\Delta{}^\text{‘}s^{\text{‘‘}}\kappa \subset 1 \to 1 . \epsilon_\Delta{}^\text{‘}\epsilon_\Delta{}^{\text{‘‘}}\kappa \subset 1 \to 1 . \epsilon_\Delta{}^\text{‘}\text{Prod}^{\text{‘‘}}\kappa \subset 1 \to 1$$

[*84·3 . *115·22 . *84·55 . *115·23]

In the above proposition, $\epsilon_\Delta{}^\text{‘}\epsilon_\Delta{}^{\text{‘‘}}\kappa \subset 1 \to 1$ does not require the hypothesis $\kappa \epsilon \text{Cls}^3$ arithm, being true always. It is merely included here for convenience of reference.

***115·27.** $\vdash : \kappa \epsilon \text{Cls}^3 \text{arithm} . \supset . \kappa \subset \text{Cls}^2 \text{excl}$ [*115·2 . *84·25 . *40·13]

We have now to prove the associative law for "Prod," *i.e.*

$$\kappa \epsilon \text{Cls}^3 \text{arithm} . \supset . \text{Prod}^\text{‘}s^\text{‘}\kappa \text{ sm } \text{Prod}^\text{‘}\text{Prod}^{\text{‘‘}}\kappa.$$

In virtue of *115·12, we have only to prove (under the hypothesis)

$$\epsilon_\Delta{}^\text{‘}s^\text{‘}\kappa \text{ sm } \epsilon_\Delta{}^\text{‘}\text{Prod}^{\text{‘‘}}\kappa$$

which, by *85·44, will follow from

$$\epsilon_\Delta{}^\text{‘}\epsilon_\Delta{}^{\text{‘‘}}\kappa \text{ sm } \epsilon_\Delta{}^\text{‘}\text{Prod}^{\text{‘‘}}\kappa$$

which, by *114·52, will follow from

$$\epsilon_\Delta{}^{\text{‘‘}}\kappa \text{ sm sm } \text{Prod}^{\text{‘‘}}\kappa.$$

Now $\qquad \text{Prod}^{\text{‘‘}}\kappa = \text{D}_\epsilon{}^{\text{‘‘}}\epsilon_\Delta{}^{\text{‘‘}}\kappa.$

Thus the correlator which will give our proposition will be $\text{D} \upharpoonright s^\text{‘}\epsilon_\Delta{}^{\text{‘‘}}\kappa$. We have only to prove that this is a $1 \to 1$, and the rest follows.

***115·3.** $\vdash : \kappa \epsilon \text{Cls}^3 \text{arithm} . R, S \epsilon s^\text{‘}\epsilon_\Delta{}^{\text{‘‘}}\kappa . \text{D}^\text{‘}R = \text{D}^\text{‘}S . \supset . R = S$

Dem.

$\vdash . *115·23 . \supset \vdash : \kappa \epsilon \text{Cls}^3\text{arithm} . \alpha, \beta \epsilon \kappa . R \epsilon \epsilon_\Delta{}^\text{‘}\alpha . S \epsilon \epsilon_\Delta{}^\text{‘}\beta . \text{D}^\text{‘}R = \text{D}^\text{‘}S . \supset . \alpha = \beta$ (1)

$\vdash . *115·27 . *84·4 . \supset \vdash : \kappa \epsilon \text{Cls}^3\text{arithm} . \alpha \epsilon \kappa . R, S \epsilon \epsilon_\Delta{}^\text{‘}\alpha . \text{D}^\text{‘}R = \text{D}^\text{‘}S . \supset . R = S$ (2)

$\vdash . (1) . (2) . \supset \vdash : \kappa \epsilon \text{Cls}^3\text{arithm} . \alpha, \beta \epsilon \kappa . R \epsilon \epsilon_\Delta{}^\text{‘}\alpha . S \epsilon \epsilon_\Delta{}^\text{‘}\beta . \text{D}^\text{‘}R = \text{D}^\text{‘}S . \supset . R = S$ (3)

$\vdash . (3) . *10·11·23·35 . *40·11 . \supset \vdash . \text{Prop}$

***115·31.** $\vdash : \kappa \epsilon \text{Cls}^3 \text{arithm} . \supset . \text{Prod}^{\text{‘‘}}\kappa \text{ sm sm } \epsilon_\Delta{}^{\text{‘‘}}\kappa$

Dem.

$\vdash . *115·3 . *71·55 . *72·13 . \supset \vdash : \text{Hp} . \supset . \text{D} \upharpoonright s^\text{‘}\epsilon_\Delta{}^{\text{‘‘}}\kappa \epsilon 1 \to 1$ (1)

$\vdash . *33·431 . \qquad \supset \vdash . s^\text{‘}\epsilon_\Delta{}^{\text{‘‘}}\kappa \subset \text{G}^\text{‘}\text{D}$ (2)

$\vdash . *37·11 . *115·1 . \qquad \supset \vdash . \text{Prod}^{\text{‘‘}}\kappa = \text{D}_\epsilon{}^{\text{‘‘}}\epsilon_\Delta{}^{\text{‘‘}}\kappa$ (3)

$\vdash . (1) . (2) . (3) . *111·402 . \supset \vdash . \text{Prop}$

***115·32.** $\vdash : \kappa \epsilon \text{Cls}^3 \text{arithm} . \supset . \epsilon_\Delta{}^\text{‘}\text{Prod}^{\text{‘‘}}\kappa \text{ sm } \epsilon_\Delta{}^\text{‘}\epsilon_\Delta{}^{\text{‘‘}}\kappa$ [*115·31 . *114·52]

***115·33.** $\vdash : \kappa \epsilon \text{Cls}^3 \text{arithm} . \supset . \epsilon_\Delta{}^\text{‘}\text{Prod}^{\text{‘‘}}\kappa \text{ sm } \epsilon_\Delta{}^\text{‘}s^\text{‘}\kappa$ [*115·32 . *85·44]

***115·34.** $\vdash : \kappa \epsilon \text{Cls}^3 \text{arithm} . \supset . \text{Prod}^\text{‘}\text{Prod}^{\text{‘‘}}\kappa \text{ sm } \text{Prod}^\text{‘}s^\text{‘}\kappa . \text{Prod}^\text{‘}s^\text{‘}\kappa \text{ sm } \epsilon_\Delta{}^\text{‘}s^\text{‘}\kappa$

[*115·33·12·23]

This proposition gives the associative law for "Prod."

The following proposition embodies the last three propositions.

***115·35.** $\vdash : \kappa \in \text{Cls}^3\,\text{arithm} \,.\, \supset .$

$$\text{Nc}`\text{Prod}`\text{Prod}``\kappa = \text{Nc}`\text{Prod}`s`\kappa = \Pi\,\text{Nc}`\text{Prod}``\kappa = \Pi\,\text{Nc}`\epsilon_\Delta``\kappa = \Pi\,\text{Nc}`s`\kappa$$

[*115·34·33·32]

In connection with $\text{Prod}`s`\kappa$ and $\text{Prod}`\text{Prod}``\kappa$, there remain two propositions of sufficient interest to deserve proof, namely

$$\kappa \in \text{Cls}^3\,\text{arithm} \,.\, \supset .\, \text{Prod}`s`\kappa = s``\text{Prod}`\text{Prod}``\kappa$$

and

$$\kappa \in \text{Cls}^3\,\text{arithm} \,.\, \supset .\, \text{Prod}`\text{Prod}``\kappa = \text{D}```\text{D}``\epsilon_\Delta`\epsilon_\Delta``\kappa.$$

Of these, the first is deduced from the second, while the second is proved by means of *114·51, putting D for the T which appears in that proposition, and $\epsilon_\Delta``\kappa$ for the λ of that proposition.

***115·4.** $\vdash : T \upharpoonright s`\lambda \in 1 \to 1 \,.\, s`\lambda \subset \text{Cl}`T \,.\, \supset .\, \text{Prod}`T```\lambda = T```\text{Prod}`\lambda$

Dem.

$\vdash .\, *111·14 \,.\, *37·103 \,.\, \supset \vdash : \text{Hp} \,.\, \kappa = T```\lambda \,.\, \supset .\, T \upharpoonright s`\lambda \in \kappa\, \overline{\text{sm}}\, \overline{\text{sm}}\, \lambda \,.$

$[*114·51.*73·142] \quad \supset .\, \epsilon_\Delta`\kappa = (T \,\|\, \breve{T}_\epsilon)``\epsilon_\Delta`\lambda \quad (1)$

$\vdash .\, (1) \,.\, *115·1 \,. \quad \supset \vdash : \text{Hp} \,.\, \supset .\, \text{Prod}`T```\lambda = \text{D}``(T \,\|\, \breve{T}_\epsilon)``\epsilon_\Delta`\lambda \quad (2)$

$\vdash .\, *37·321·231 \,. \quad \supset \vdash .\, \text{D}`(T \,|\, R \,|\, \breve{T}_\epsilon) = \text{D}`(T \,|\, R)$

$[*37·32] \quad = T``\text{D}`R \quad (3)$

$\vdash .\, (3) \,.\, *43·112 \,. \quad \supset \vdash .\, \text{D}``(T \,\|\, \breve{T}_\epsilon)``\epsilon_\Delta`\lambda = T```\text{D}``\epsilon_\Delta`\lambda$

$[*115·1] \quad = T```\text{Prod}`\lambda \quad (4)$

$\vdash .\, (2) \,.\, (4) \,.\, \supset \vdash .\, \text{Prop}$

***115·41.** $\vdash :.\, R, S \in s`\lambda \,.\, \text{D}`R = \text{D}`S \,.\, \supset_{R,S} .\, R = S : \supset .\, \text{Prod}`\text{D}```\lambda = \text{D}```\text{Prod}`\lambda$

$\left[*115·4 \dfrac{\text{D}}{T} \,.\, *71·55 \,.\, *72·13\right]$

***115·42.** $\vdash : \kappa \in \text{Cls}^3\,\text{arithm} \,.\, \supset .\, \text{Prod}`\text{Prod}``\kappa = \text{D}```\text{Prod}`\epsilon_\Delta``\kappa$

$$= \text{D}```\text{D}``\epsilon_\Delta`\epsilon_\Delta``\kappa$$

Dem.

$\vdash .\, *115·1 \,. \quad \supset \vdash .\, \text{Prod}`\text{Prod}``\kappa = \text{Prod}`\text{D}```\epsilon_\Delta``\kappa \quad (1)$

$\vdash .\, *115·3·41 \,.\, \supset \vdash : \text{Hp} \,.\, \supset .\, \text{Prod}`\text{D}```\epsilon_\Delta``\kappa = \text{D}```\text{Prod}`\epsilon_\Delta``\kappa \quad (2)$

$[*115·1] \quad = \text{D}```\text{D}``\epsilon_\Delta`\epsilon_\Delta``\kappa \quad (3)$

$\vdash .\, (1) \,.\, (2) \,.\, (3) \,.\, \supset \vdash .\, \text{Prop}$

***115·43.** $\vdash : \kappa \in \text{Cls}^2\,\text{excl} \,.\, \supset .\, \text{Prod}`s`\kappa = s``\text{D}```\text{D}``\epsilon_\Delta`\epsilon_\Delta``\kappa$

Dem.

$\vdash .\, *115·1 \,.\, *85·28 \,.\, \supset$

$\vdash : \text{Hp} \,.\, \supset .\, \text{Prod}`s`\kappa = \text{D}``\dot{s}``\text{D}``\epsilon_\Delta`\epsilon_\Delta``\kappa$

$[*41·43] \quad = s``\text{D}```\text{D}``\epsilon_\Delta`\epsilon_\Delta``\kappa : \supset \vdash .\, \text{Prop}$

***115·44.** $\vdash : \kappa \in \text{Cls}^3\,\text{arithm} \,.\, \supset .\, \text{Prod}`s`\kappa = s``\text{Prod}`\text{Prod}``\kappa$ [*115·43·42]

The following proposition is a lemma for *115·46.

***115·45.** $\vdash :. \alpha, \beta \epsilon \kappa \,.\, \exists ! \, s\text{‘}\alpha \cap s\text{‘}\beta \,.\, \supset_{\alpha,\beta} .\, \alpha = \beta : \supset .$

$$(s \,|\, \mathrm{D}) \upharpoonright \epsilon_\Delta\text{‘}\kappa \,\epsilon\, 1 \to 1 \,.\, s \upharpoonright \mathrm{Prod}\text{‘}\kappa \,\epsilon\, 1 \to 1$$

Dem.

$$\vdash .\, *83\cdot2 \,.\, *40\cdot13 \,.\, \supset \vdash : R \,\epsilon\, \epsilon_\Delta\text{‘}\kappa \,.\, \alpha \,\epsilon\, \kappa \,.\, \supset .\, R\text{‘}\alpha \subset s\text{‘}\alpha \tag{1}$$

$$\vdash .\, *83\cdot2 \,.\, *33\cdot43 \,.\, \supset \vdash : R \,\epsilon\, \epsilon_\Delta\text{‘}\kappa \,.\, \alpha \,\epsilon\, \kappa \,.\, \supset .\, R\text{‘}\alpha \subset s\text{‘}\mathrm{D}\text{‘}R \tag{2}$$

$\vdash .\, *83\cdot23 \,.\, \supset$

$\vdash : R \,\epsilon\, \epsilon_\Delta\text{‘}\kappa \,.\, \alpha \,\epsilon\, \kappa \,.\, x \,\epsilon\, (s\text{‘}\mathrm{D}\text{‘}R \cap s\text{‘}\alpha) \,.\, \supset .\, (\exists\beta) \,.\, \beta \,\epsilon\, \kappa \,.\, x \,\epsilon\, R\text{‘}\beta \,.\, x \,\epsilon\, s\text{‘}\alpha \,.$

$$[(1)] \qquad \supset .\, (\exists\beta) \,.\, \beta \,\epsilon\, \kappa \,.\, x \,\epsilon\, R\text{‘}\beta \,.\, x \,\epsilon\, s\text{‘}\beta \,.\, x \,\epsilon\, s\text{‘}\alpha \tag{3}$$

$\vdash .\, (3) \,.\, \supset \vdash :.\, \mathrm{Hp} \,.\, R \,\epsilon\, \epsilon_\Delta\text{‘}\kappa \,.\, \alpha \,\epsilon\, \kappa \,.\, \supset : x \,\epsilon\, (s\text{‘}\mathrm{D}\text{‘}R \cap s\text{‘}\alpha) \,.\, \supset .\, (\exists\beta) \,.\, x \,\epsilon\, R\text{‘}\beta \,.\, \beta = \alpha \,.$

$$[*13\cdot195] \qquad \supset .\, x \,\epsilon\, R\text{‘}\alpha \tag{4}$$

$$\vdash .\, (1) \,.\, (2) \,.\, (4) \,.\, \supset \vdash :.\, \mathrm{Hp} \,.\, \supset : R \,\epsilon\, \epsilon_\Delta\text{‘}\kappa \,.\, \alpha \,\epsilon\, \kappa \,.\, \supset .\, R\text{‘}\alpha = s\text{‘}\mathrm{D}\text{‘}R \cap s\text{‘}\alpha \tag{5}$$

$\vdash .\, (5) \,.\, \supset \vdash :: \mathrm{Hp} \,.\, \supset :.\, R, S \,\epsilon\, \epsilon_\Delta\text{‘}\kappa \,.\, s\text{‘}\mathrm{D}\text{‘}R = s\text{‘}\mathrm{D}\text{‘}S \,.\, \supset : \alpha \,\epsilon\, \kappa \,.\, \supset_\alpha .\, R\text{‘}\alpha = S\text{‘}\alpha :$

$$[*33\cdot45 \,.\, *80\cdot14] \qquad \supset : R = S :. \tag{6}$$

$$[*71\cdot55 \,.\, *72\cdot13\cdot161] \supset :.\, (s \,|\, \mathrm{D}) \upharpoonright \epsilon_\Delta\text{‘}\kappa \,\epsilon\, 1 \to 1 \tag{7}$$

$\vdash .\, (6) \,.\, *37\cdot63 \,.\, *115\cdot1 \,.\, *30\cdot37 \,.\, \supset \vdash :.\, \mathrm{Hp} \,.\, \supset : \mu, \nu \,\epsilon\, \mathrm{Prod}\text{‘}\kappa \,.\, s\text{‘}\mu = s\text{‘}\nu \,.\, \supset .\, \mu = \nu :$

$$[*71\cdot55 \,.\, *72\cdot161] \qquad \supset : s \upharpoonright \mathrm{Prod}\text{‘}\kappa \,\epsilon\, 1 \to 1 \tag{8}$$

$\vdash .\, (7) \,.\, (8) \,.\, \supset \vdash .\, \mathrm{Prop}$

***115·46.** $\vdash : \kappa \,\epsilon\, \mathrm{Cls}^3 \,\mathrm{arithm} \,.\, \supset .\, s \upharpoonright \mathrm{Prod}\text{‘}\mathrm{Prod}\text{“}\kappa \,\epsilon\, 1 \to 1$

Dem.

$\vdash .\, *115\cdot141 \,.\, \supset$

$$\vdash : \alpha, \beta \,\epsilon\, \kappa \,.\, \exists ! \, s\text{‘}\mathrm{Prod}\text{‘}\alpha \cap s\text{‘}\mathrm{Prod}\text{‘}\beta \,.\, \supset .\, \exists ! \, s\text{‘}\alpha \cap s\text{‘}\beta \tag{1}$$

$\vdash .\, (1) \,.\, *115\cdot22 \,.\, \supset$

$\vdash :.\, \kappa \,\epsilon\, \mathrm{Cls}^3 \,\mathrm{arithm} \,.\, \supset : \alpha, \beta \,\epsilon\, \kappa \,.\, \exists ! \, s\text{‘}\mathrm{Prod}\text{‘}\alpha \cap s\text{‘}\mathrm{Prod}\text{‘}\beta \,.\, \supset .\, \alpha = \beta \,.$

$[*30\cdot37] \qquad \supset .\, \mathrm{Prod}\text{‘}\alpha = \mathrm{Prod}\text{‘}\beta :$

$[*37\cdot63] \qquad \supset : \mu, \nu \,\epsilon\, \mathrm{Prod}\text{“}\kappa \,.\, \exists ! \, s\text{‘}\mu \cap s\text{‘}\nu \,.\, \supset .\, \mu = \nu :$

$[*115\cdot45] \qquad \supset : s \upharpoonright \mathrm{Prod}\text{‘}\mathrm{Prod}\text{“}\kappa \,\epsilon\, 1 \to 1 :.\, \supset \vdash .\, \mathrm{Prop}$

The above proposition is used in dealing with products in relation-arithmetic (*174·42).

***115·5.** $\vdash : T \upharpoonright s\text{‘}\lambda \,\epsilon\, \kappa \,\overline{\mathrm{sm}}\,\overline{\mathrm{sm}}\, \lambda \,.\, \supset .\, \mathrm{Prod}\text{‘}\kappa = T_\epsilon\text{“}\mathrm{Prod}\text{‘}\lambda \quad [*115\cdot4 \,.\, *111\cdot14]$

***115·501.** $\vdash : T \upharpoonright s\text{‘}\lambda \,\epsilon\, \kappa \,\overline{\mathrm{sm}}\,\overline{\mathrm{sm}}\, \lambda \,.\, \exists ! \, \mathrm{Prod}\text{‘}\lambda \,.\, \supset .\, T \upharpoonright s\text{‘}\lambda \,\epsilon\, (\mathrm{Prod}\text{‘}\kappa) \,\overline{\mathrm{sm}}\,\overline{\mathrm{sm}}\, (\mathrm{Prod}\text{‘}\lambda)$
$[*115\cdot5\cdot141 \,.\, *111\cdot14]$

***115·502.** $\vdash : T \upharpoonright s\text{‘}\lambda \,\epsilon\, \kappa \,\overline{\mathrm{sm}}\,\overline{\mathrm{sm}}\, \lambda \,.\, \supset .\, T \upharpoonright s\text{‘}\mathrm{Prod}\text{‘}\lambda \,\epsilon\, (\mathrm{Prod}\text{‘}\kappa) \,\overline{\mathrm{sm}}\,\overline{\mathrm{sm}}\, (\mathrm{Prod}\text{‘}\lambda)$

Dem.

$$\vdash .\, *35\cdot75 \,.\, \supset \vdash : \sim\exists ! \, \mathrm{Prod}\text{‘}\lambda \,.\, \supset .\, T \upharpoonright s\text{‘}\mathrm{Prod}\text{‘}\lambda = \dot{\Lambda} \tag{1}$$

$\vdash .\, *115\cdot5 \,.\, *37\cdot29 \,.\, \supset \vdash : \mathrm{Hp} \,.\, \sim\exists ! \, \mathrm{Prod}\text{‘}\lambda \,.\, \supset .\, \mathrm{Prod}\text{‘}\kappa = \Lambda \,.$

$$[*37\cdot29 \,.\, *40\cdot21] \qquad \supset .\, s\text{‘}\mathrm{Prod}\text{‘}\kappa = (T \upharpoonright s\text{‘}\mathrm{Prod}\text{‘}\lambda)\text{“}s\text{‘}\mathrm{Prod}\text{‘}\lambda \tag{2}$$

$\vdash .\, (1) \,.\, *72\cdot1 \,.\, (2) \,.\, *115\cdot5 \,.\, *111\cdot1 \,.\, \supset$

$$\vdash : \mathrm{Hp} \,.\, \sim\exists ! \, \mathrm{Prod}\text{‘}\lambda \,.\, \supset .\, T \upharpoonright s\text{‘}\mathrm{Prod}\text{‘}\lambda \,\epsilon\, (\mathrm{Prod}\text{‘}\kappa) \,\overline{\mathrm{sm}}\,\overline{\mathrm{sm}}\, (\mathrm{Prod}\text{‘}\lambda) \tag{3}$$

$\vdash .\, (3) \,.\, *115\cdot501\cdot141 \,.\, \supset \vdash .\, \mathrm{Prop}$

***115·51.** $\vdash : \kappa \operatorname{sm} \operatorname{sm} \lambda . \supset . \operatorname{Prod}`\kappa \operatorname{sm} \operatorname{sm} \operatorname{Prod}`\lambda$ [*115·502]

The above propositions show how, in certain respects, $\operatorname{Prod}`\kappa$ is more convenient than $\epsilon_\Delta`\kappa$. We cannot have $\epsilon_\Delta`\kappa \operatorname{sm} \operatorname{sm} \epsilon_\Delta`\lambda$, because $\epsilon_\Delta`\kappa$ is a class of *relations*, not a class of classes; and the correlator of $\epsilon_\Delta`\kappa$ and $\epsilon_\Delta`\lambda$ is by no means so simple a function of the correlator of $\kappa$ and $\lambda$ as $T_\epsilon \upharpoonright \operatorname{Prod}`\lambda$, which correlates $\operatorname{Prod}`\kappa$ and $\operatorname{Prod}`\lambda$, in virtue of *115·502.

The following propositions are a continuation of those given in *114·601 ff.

***115·6.** $\vdash :(R``\gamma) \rceil R, S \upharpoonright \gamma \epsilon 1 \to 1 . \gamma \subset \text{Cl}`R . \gamma \subset \text{Cl}`S . R``\gamma, S``\gamma \epsilon \text{Cls}^2 \text{excl} . \supset .$
$\operatorname{Prod}`R``\gamma \times \operatorname{Prod}`S``\gamma \operatorname{sm} \epsilon_\Delta`\hat{\mu}\{(\exists z) . z \epsilon \gamma . \mu = R`z \times S`z\}$

Dem.

$\vdash . *115·12 . *113·13 . \supset$

$\vdash : \text{Hp} . \supset . \operatorname{Prod}`R``\gamma \times \operatorname{Prod}`S``\gamma \operatorname{sm} \epsilon_\Delta`R``\gamma \times \epsilon_\Delta`S``\gamma \quad (1)$

$\vdash . (1) . *114·64 . \supset \vdash . \text{Prop}$

***115·601.** $\vdash : (R``\gamma) \rceil R, S \upharpoonright \gamma \epsilon 1 \to 1 . \gamma \subset \text{Cl}`R . \gamma \subset \text{Cl}`S . R``\gamma \epsilon \text{Cls}^2 \text{excl} . \supset .$
$\hat{\mu}\{(\exists z) . z \epsilon \gamma . \mu = R`z \times S`z\} \epsilon \text{Cls}^2 \text{excl}$

Dem.

$\vdash . *113·19 . \supset \vdash :. \text{Hp} . \supset :$

$z, w \epsilon \gamma . \exists ! (R`z \times S`z) \cap (R`w \times S`w) . \supset . \exists ! R`z \cap R`w .$

[*84·11] $\supset . R`z = R`w .$

[*74·53.*30·37] $\supset . z = w .$

[*30·37] $\supset . R`z \times S`z = R`w \times S`w \quad (1)$

$\vdash . (1) . *84·11 . \supset \vdash . \text{Prop}$

***115·602.** $\vdash : (R``\gamma) \rceil R, S \upharpoonright \gamma \epsilon 1 \to 1 . \gamma \subset \text{Cl}`R . \gamma \subset \text{Cl}`S . S``\gamma \epsilon \text{Cls}^2 \text{excl} . \supset .$
$\hat{\mu}\{(\exists z) . z \epsilon \gamma . \mu = R`z \times S`z\} \epsilon \text{Cls}^2 \text{excl}$

[Proof as in *115·601]

***115·61.** $\vdash :. (R``\gamma) \rceil R, S \upharpoonright \gamma \epsilon 1 \to 1 . \gamma \subset \text{Cl}`R . \gamma \subset \text{Cl}`S . R``\gamma \cap S``\gamma = \Lambda :$
$R``\gamma \epsilon \text{Cls}^2 \text{excl} . \vee . S``\gamma \epsilon \text{Cls}^2 \text{excl} : \supset .$
$\epsilon_\Delta`(R``\gamma \cup S``\gamma) \operatorname{sm} \operatorname{Prod}`\hat{\mu}\{(\exists z) . z \epsilon \gamma . \mu = R`z \times S`z\}$

[*115·601·602·12 . *114·65]

***115·62.** $\vdash : (R``\gamma) \rceil R, S \upharpoonright \gamma \epsilon 1 \to 1 . \gamma \subset \text{Cl}`R . \gamma \subset \text{Cl}`S . R``\gamma \cap S``\gamma = \Lambda .$
$(R``\gamma \cup S``\gamma) \epsilon \text{Cls}^2 \text{excl} . \supset .$
$\operatorname{Prod}`(R``\gamma \cup S``\gamma) \operatorname{sm} \operatorname{Prod}`\hat{\mu}\{(\exists z) . z \epsilon \gamma . \mu = R`z \times S`z\}$

[*115·61·12 . *84·25]

***115·63.** $\vdash :(R``\gamma) \rceil R, S \upharpoonright \gamma \epsilon 1 \to 1 . \gamma \subset \text{Cl}`R . \gamma \subset \text{Cl}`S . R``\gamma, S``\gamma \epsilon \text{Cls}^2 \text{excl} . \supset .$
$\operatorname{Prod}`R``\gamma \times \operatorname{Prod}`S``\gamma \operatorname{sm} \operatorname{Prod}`\hat{\mu}\{(\exists z) . z \epsilon \gamma . \mu = R`z \times S`z\}$

[*115·6·601·12]

$*116$. EXPONENTIATION.

Summary of $*116$.

In this number, we define "α exp β," meaning "α to the exponent β," where α and β are classes, as

$$\text{Prod}`\alpha \underset{;}{\downarrow} ``\beta.$$

Now $\text{Prod}`\alpha \underset{;}{\downarrow} ``\beta$ consists of all ways of selecting one each from the members of $\alpha \underset{;}{\downarrow} ``\beta$, *i.e.* from the classes $\downarrow y``\alpha$, where $y \in \beta$. Thus to get a member of $\text{Prod}`\alpha \underset{;}{\downarrow} ``\beta$, take a set of couples $x \downarrow y$, where x is always an α, and there is only one x for a given y, and y is each member of β in succession. Thus for each member of β, we have $\text{Nc}`\alpha$ possible referents; hence it is plain that the number of possible sets of couples consists of $\text{Nc}`\beta$ factors each equal to $\text{Nc}`\alpha$, and is therefore fit to be taken as defining $(\text{Nc}`\alpha)^{\text{Nc}`\beta}$.

The definitions of μ^ν and $(\text{Nc}`\alpha)^{\text{Nc}`\beta}$ are derived from the definition of $\alpha \exp \beta$ exactly as the definitions of $\mu +_c \nu$ and $\text{Nc}`\alpha +_c \text{Nc}`\beta$, or of $\mu \times_c \nu$ and $\text{Nc}`\alpha \times_c \text{Nc}`\beta$, were derived respectively from $\alpha + \beta$ and $\alpha \times \beta$.

The chief difficulty in this number lies in the proof of the three formal laws of exponentiation, namely

$$\mu^\nu \times_c \mu^\varpi = \mu^{\nu +_c \varpi},$$

$$\mu^\varpi \times_c \nu^\varpi = (\mu \times_c \nu)^\varpi,$$

and

$$(\mu^\nu)^\varpi = \mu^{\nu \times_c \varpi}.$$

The proofs of the second and third of these, in particular, require various lemmas; but there is no difficulty involved except the complexity of the classes and relations concerned.

The definition of μ^ν is so framed as to minimize the necessity for the multiplicative axiom (see the note on $*113\cdot31$ in the introduction to $*113$). We have

$*116\cdot36$. $\vdash :. \text{Mult ax} . \supset : \mu, \nu \in \text{NC} - \iota`\Lambda . \kappa \in \nu \cap \text{Cl}`\mu . \supset . \Pi\text{Nc}`\kappa = \mu^\nu$

that is, assuming the multiplicative axiom, the product of ν factors each equal to μ is μ^ν (assuming μ and ν to be cardinals which are not null). If we had *defined* μ^ν as the product of ν factors each equal to μ, we should have required the multiplicative axiom for almost all propositions on μ^ν; but by taking the particular class $\alpha \underset{\text{''}}{\downarrow} \text{``}\beta$, we avoid the multiplicative axiom except in a few propositions. Among these few is the above proposition connecting exponentiation with multiplication.

Cantor has defined μ^ν by means of the class of "Belegungen," *i.e.* the class

$$\hat{R}\,(R \in 1 \rightarrow \text{Cls} \,.\, \text{D}\text{`}R \subset \alpha \,.\, \text{G}\text{`}R = \beta)$$

which (*116·12) $= (\alpha \uparrow \beta)_\Delta\text{`}\beta$. By *85·53 and *113·103, this class is equal to $\dot{s}\text{``}(\alpha \exp \beta)$ (as is proved in *116·13), whence, since $\dot{s} \upharpoonright \alpha \exp \beta \in 1 \rightarrow 1$, it follows (*116·15) that the class of "Belegungen" is similar to $\alpha \exp \beta$. Hence our definition gives the same value of μ^ν as Cantor's.

The propositions of the present number begin with various simple properties of $\alpha \exp \beta$. Its existence follows from

***116·152.** $\vdash : x \in \alpha \,.\, \supset \,.\, x \downarrow \text{``}\beta \in (\alpha \exp \beta)$

whence (*116·16) $\vdash \,.\, \text{Cnv}\text{``}\text{`}\beta \underset{\text{''}}{\downarrow} \text{``}\alpha \subset \alpha \exp \beta$, and

***116·18.** $\vdash :.\, \exists \,!\, \alpha \,.\, \vee \,.\, \beta = \Lambda : \equiv \,.\, \exists \,!\, \alpha \exp \beta$

We have

***116·19.** $\vdash : \alpha \,\text{sm}\, \gamma \,.\, \beta \,\text{sm}\, \delta \,.\, \supset \,.\, (\alpha \exp \beta) \,\text{sm sm}\, (\gamma \exp \delta)$

in virtue of *113·13 and *115·51. *116·192 shows that, if $R \upharpoonright \gamma$ correlates α with γ, and $S \upharpoonright \delta$ correlates β with δ, then $(R \,\|\, \breve{S}) \upharpoonright (\delta \times \gamma)$ is a double correlator of $(\alpha \exp \beta)$ with $(\gamma \exp \delta)$.

We then proceed to a set of propositions on μ^ν, which are analogous to *113·2 ff. on $\mu \times_c \nu$. We have

***116·203.** $\vdash : \exists \,!\, \mu^\nu \,.\, \supset \,.\, \mu, \nu \in \text{NC} - \iota\text{`}\Lambda \,.\, \mu, \nu \in \text{N}_0\text{C}$

***116·25.** $\vdash \,.\, (\text{Nc}\text{`}\gamma)^{\text{Nc}\text{`}\delta} = \text{Nc}\text{`}(\gamma \exp \delta)$

and various other less useful propositions.

We then have various propositions on 0 and 1 and 2. We prove

***116·301.** $\vdash : \mu \in \text{NC} - \iota\text{`}\Lambda \,.\, \supset \,.\, \mu^0 = 1$

***116·311.** $\vdash : \nu \in \text{NC} - \iota\text{`}\Lambda - \iota\text{`}0 \,.\, \supset \,.\, 0^\nu = 0$

***116·321.** $\vdash : \mu \in \text{NC} - \iota\text{`}\Lambda \,.\, \supset \,.\, \mu^1 = \text{sm}\text{``}\mu$

(Observe that $\text{sm}\text{``}\mu$ is the same cardinal as μ, but rendered typically ambiguous.)

***116·331.** $\vdash : \mu \in \text{NC} - \iota\text{`}\Lambda \,.\, \supset \,.\, 1^\mu = 1$

***116·34.** $\vdash \,.\, \mu^2 = \mu \times_c \mu$

(This proposition does not require that μ should be a cardinal.)

After the proposition (∗116·36) already quoted, on the connection of exponentiation and multiplication, we proceed to a set of propositions on the case where a number of classes are all given as similar (by assignable correlations) to a given class. In ∗116·411, we prove that if κ is a class of mutually exclusive classes, each of which is similar to a given class γ, and if, when $\alpha \epsilon \kappa$, M'α is a correlator of α and γ, and T is the sum of M''κ, then

$$\text{Nc}`\epsilon_\Delta`\overrightarrow{T}``\gamma = \text{Nc}`T_\Delta`\gamma = \text{Nc}`(\kappa \exp \gamma) = (\text{Nc}`\kappa)^{\text{Nc}`\gamma}.$$

This is a further connection of multiplication and exponentiation. (On the purport of this and following propositions, see the explanation preceding ∗116·4.) In ∗116·43, the hypothesis is somewhat modified. We still have a set κ of classes which are all similar to γ, but the correlator for a given class α is not given as M'α, but is given as M'w, where w is a member of a class δ which is similar to κ. Then $\kappa = \text{D}``M``\delta$. We assume that $M \upharpoonright \delta$ is a one-one, and that if M'w and M'v have domains which overlap, then $w = v$. Thus κ is a class of mutually exclusive classes, each of which has Nc'γ terms, while κ has Nc'δ terms. Then it is proved in ∗116·43 that

$$\text{Prod}`\text{D}``M``\delta \text{ sm sm } (\gamma \exp \delta) \,.\, \Pi\text{Nc}`\text{D}``M``\delta = (\text{Nc}`\gamma)^{\text{Nc}`\delta}.$$

This proposition and another (∗116·45) which follows from it are useful in proving the formal laws of exponentiation. The proof of these occupies the following propositions from ∗116·5 to ∗116·68. We have

∗116·52. $\vdash . \mu^\nu \times_c \mu^\varpi = \mu^{\nu +_c \varpi}$

∗116·55. $\vdash . \mu^\varpi \times_c \nu^\varpi = (\mu \times_c \nu)^\varpi$

∗116·63. $\vdash . \mu^{\nu \times_c \varpi} = (\mu^\nu)^\varpi$

An extension of the first of these is

∗116·661. $\vdash . \Pi\text{Nc}`(\alpha \exp)``\kappa = (\text{Nc}`\alpha)^{\Sigma\text{Nc}`\kappa}$

Here the number of members of κ need not be finite. The purport of the proposition is as follows: Let $\beta, \gamma, \delta, \ldots$ be the members of κ; form $\alpha \exp \beta$, $\alpha \exp \gamma$, $\alpha \exp \delta$, ..., and take the product of the numbers of all these; then the resulting number is the same as if we first took the sum of the numbers of all the members of κ, thus obtaining (say) a number μ, and raised Nc'α to the μth power.

An extension of ∗116·55 is given by ∗116·68, where we prove

$$\vdash : \kappa \epsilon \text{Cls}^2 \text{ excl} \,.\, \supset \,.\, \Pi\text{Nc}`\exp \gamma``\kappa = (\Pi\text{Nc}`\kappa)^{\text{Nc}`\gamma}.$$

There is no analogous extension of ∗116·63.

We prove next Cantor's proposition (which is very useful)

∗116·72. $\vdash . \text{Nc}`\text{Cl}`\alpha = 2^{\text{Nc}`\alpha}$

I.e. the number of combinations of μ things any number at a time is 2^μ. (Observe that μ need not be finite.) The remainder of the number is concerned with consequences of this proposition.

***116·01.** $\alpha \exp \beta = \mathrm{Prod}`\alpha \underset{,,}{\downarrow} ``\beta$ Df

***116·02.** $\mu^\nu = \hat{\gamma} \{(\exists \alpha, \beta) . \mu = \mathrm{N_0c}`\alpha . \nu = \mathrm{N_0c}`\beta . \gamma \,\mathrm{sm}\, (\alpha \exp \beta)\}$ Df

***116·03.** $(\mathrm{Nc}`\alpha)^\nu = (\mathrm{N_0c}`\alpha)^\nu$ Df

***116·04.** $\mu^{\mathrm{Nc}`\beta} = \mu^{\mathrm{N_0c}`\beta}$ Df

***116·1.** $\vdash : \xi \epsilon (\alpha \exp \beta) . \equiv . (\exists R) . R \epsilon \epsilon_\Delta`\alpha \underset{,,}{\downarrow} ``\beta . \xi = \mathrm{D}`R$

[*115·1 . (*116·01)]

***116·11.** $\vdash :. \xi \epsilon (\alpha \exp \beta) . \equiv : y \epsilon \beta . \supset_y . \alpha \cap \hat{x} (x \downarrow y \epsilon \xi) \epsilon 1 : \xi \subset \beta \times \alpha$

Dem.

$\vdash . *113·111 . *115·11 . \supset$

$\vdash :. \xi \epsilon (\alpha \exp \beta) . \equiv : \rho \epsilon \alpha \underset{,,}{\downarrow} ``\beta . \supset_\rho . \rho \cap \xi \epsilon 1 : \xi \subset s`\alpha \underset{,,}{\downarrow} ``\beta :$

[*38·2.*113·1] $\equiv : y \epsilon \beta . \supset_y . \downarrow y``\alpha \cap \xi \epsilon 1 : \xi \subset \beta \times \alpha$ (1)

$\vdash . *37·6 . \supset$

$\vdash : \downarrow y``\alpha \cap \xi \epsilon 1 . \equiv . \hat{R} \{(\exists x) . x \epsilon \alpha . R = x \downarrow y . R \epsilon \xi\} \epsilon 1 .$

[*13·193] $\equiv . \hat{R} \{(\exists x) . x \epsilon \alpha . x \downarrow y \epsilon \xi . R = x \downarrow y\} \epsilon 1 .$

[*37·6] $\equiv . \downarrow y``\hat{x} (x \epsilon \alpha . x \downarrow y \epsilon \xi) \epsilon 1 .$

[*73·611·44] $\equiv . \hat{x} (x \epsilon \alpha . x \downarrow y \epsilon \xi) \epsilon 1$ (2)

$\vdash . (1) . (2) . \supset \vdash .$ Prop

***116·12.** $\vdash . (\alpha \uparrow \beta)_\Delta`\beta = \hat{R} \{R \epsilon 1 \to \mathrm{Cls} . \mathrm{D}`R \subset \alpha . \mathrm{C\!I}`R = \beta\}$

Dem.

$\vdash . *80·14 . \supset \vdash : R \epsilon (\alpha \uparrow \beta)_\Delta`\beta . \equiv . R \epsilon 1 \to \mathrm{Cls} . R \Subset \alpha \uparrow \beta . \mathrm{C\!I}`R = \beta .$

[*35·83] $\equiv . R \epsilon 1 \to \mathrm{Cls} . \mathrm{D}`R \subset \alpha . \mathrm{C\!I}`R \subset \beta . \mathrm{C\!I}`R = \beta .$

[*22·42] $\equiv . R \epsilon 1 \to \mathrm{Cls} . \mathrm{D}`R \subset \alpha . \mathrm{C\!I}`R = \beta : \supset \vdash .$ Prop

***116·13.** $\vdash . \dot{s}``(\alpha \exp \beta) = (\alpha \uparrow \beta)_\Delta`\beta$

Dem.

$\vdash . *85·53 . \supset \vdash . (\alpha \uparrow \beta)_\Delta`\beta = \dot{s}``\mathrm{D}``\epsilon_\Delta`(\alpha \uparrow \beta) \,\overline{\downarrow}\, ``\beta$

[*113·103] $= \dot{s}``\mathrm{D}``\epsilon_\Delta`\alpha \underset{,,}{\downarrow} ``\beta$

[*115·1.(*116·01)] $= \dot{s}``(\alpha \exp \beta) . \supset \vdash .$ Prop

$(\alpha \uparrow \beta)_\Delta`\beta$ is the class of one-many relations whose converse domain is β and whose domain is contained in α. This is what Cantor calls the "Belegungsmenge," and is used by him as the definition of exponentiation. In virtue of *116·15, his definition gives the same results as ours.

***116·131.** $\vdash . \dot{s} \upharpoonright (\alpha \exp \beta) \epsilon \{(\alpha \uparrow \beta)_\Delta`\beta\} \,\overline{\mathrm{sm}}\, (\alpha \exp \beta)$

Dem.

$\vdash . *84·241 . *113·103 . \supset \vdash . \iota``\beta \epsilon \mathrm{Cls^2 excl} . \alpha \underset{,,}{\downarrow} ``\beta = (\alpha \uparrow \beta)_\Delta``\iota``\beta$ (1)

$\vdash . (1) . *85·42 . \supset \vdash : M, N \epsilon \epsilon_\Delta`\alpha \underset{,,}{\downarrow} ``\beta . \dot{s}`\mathrm{D}`M = \dot{s}`\mathrm{D}`N . \supset . M = N .$

[*30·37] $\supset . \mathrm{D}`M = \mathrm{D}`N$ (2)

$\vdash . (2) . *37·63 . *115·1 . (*116·01) . \supset \vdash : \mu, \nu \epsilon (\alpha \exp \beta) . \dot{s}`\mu = \dot{s}`\nu . \supset . \mu = \nu :$

[*71·55.*72·163] $\supset \vdash . \dot{s} \upharpoonright (\alpha \exp \beta) \epsilon 1 \to 1$ (3)

$\vdash . (3) . *116·13 . \supset \vdash .$ Prop

***116·14.** $\vdash . (\alpha \exp \beta) \,\mathrm{sm}\, \epsilon_\Delta{}^{\backprime}\alpha \downarrow_{,,} {}^{\backprime\backprime}\beta$ [*115·12 . *113·111]

***116·15.** $\vdash . (\alpha \exp \beta) \,\mathrm{sm}\, (\alpha \uparrow \beta)_\Delta{}^{\backprime}\beta$ [*116·131]

*116·151 is a lemma for *116·152.

***116·151.** $\vdash : x \epsilon \alpha . \supset . x\downarrow \,|\, \mathrm{Cnv}{}^{\backprime}(\alpha \downarrow_{,,} \upharpoonright \beta) \,\epsilon\, \epsilon_\Delta{}^{\backprime}\alpha \downarrow_{,,} {}^{\backprime\backprime}\beta$

Dem.

$\vdash . *113{\cdot}105 . *72{\cdot}184 . \supset \vdash : \mathrm{Hp} . \supset . x\downarrow \,|\, \mathrm{Cnv}{}^{\backprime}(\alpha \downarrow_{,,} \upharpoonright \beta) \,\epsilon\, 1 \rightarrow \mathrm{Cls}$ (1)

$\vdash . *34{\cdot}1 . *38{\cdot}1 . \quad \supset \vdash :. \mathrm{Hp} . \supset : R \{x\downarrow \,|\, \mathrm{Cnv}{}^{\backprime}(\alpha \downarrow_{,,} \upharpoonright \beta)\} \lambda . \equiv .$

$(\exists y) . R = x \downarrow y . y \epsilon \beta . \lambda = \alpha \downarrow_{,,} y . x \epsilon \alpha .$

[*38·21] $\supset . R \epsilon \lambda$ (2)

$\vdash . *37{\cdot}322{\cdot}401 . \quad \supset \vdash . \text{ᗡ}{}^{\backprime}\{x\downarrow \,|\, \mathrm{Cnv}{}^{\backprime}(\alpha \downarrow_{,,} \upharpoonright \beta)\} = \alpha \downarrow_{,,} {}^{\backprime\backprime}\beta$ (3)

$\vdash . (1) . (2) . (3) . *80{\cdot}14 . \supset \vdash . \mathrm{Prop}$

***116·152.** $\vdash : x \epsilon \alpha . \supset . x \downarrow {}^{\backprime\backprime}\beta \,\epsilon\, (\alpha \exp \beta)$

Dem.

$\vdash . *37{\cdot}32 . *35{\cdot}65 . \supset \vdash . \mathrm{D}{}^{\backprime}\{x\downarrow \,|\, \mathrm{Cnv}{}^{\backprime}(\alpha \downarrow_{,,} \upharpoonright \beta)\} = x \downarrow {}^{\backprime\backprime}\beta$ (1)

$\vdash . (1) . *116{\cdot}151{\cdot}1 . \supset \vdash . \mathrm{Prop}$

***116·16.** $\vdash . \mathrm{Cnv}{}^{\backprime\backprime\backprime}\beta \downarrow_{,,} {}^{\backprime\backprime}\alpha \subset \alpha \exp \beta$

Dem.

$\vdash . *116{\cdot}152 . *55{\cdot}14 . \supset \vdash : x \epsilon \alpha . \supset . \mathrm{Cnv}{}^{\backprime\backprime} \downarrow x {}^{\backprime\backprime}\beta \,\epsilon\, (\alpha \exp \beta) .$

[*38·2] $\supset . \mathrm{Cnv}{}^{\backprime\backprime}\beta \downarrow_{,,} x \,\epsilon\, (\alpha \exp \beta) : \supset \vdash . \mathrm{Prop}$

The above propositions are useful in establishing existence-theorems, as appears in the following propositions.

***116·17.** $\vdash : \exists ! \beta \downarrow_{,,} {}^{\backprime\backprime}\alpha . \supset . \exists ! \alpha \exp \beta$ [*116·16 . *37·47]

***116·171.** $\vdash :. \exists ! \alpha . \mathsf{v} . \beta = \Lambda : \supset . \exists ! \alpha \exp \beta$

Dem.

$\vdash . *113{\cdot}113 . *83{\cdot}15 . *51{\cdot}161 . \supset \vdash : \beta = \Lambda . \supset . \exists ! \alpha \exp \beta$ (1)

$\vdash . *116{\cdot}152 . \quad \supset \vdash : \exists ! \alpha . \supset . \exists ! \alpha \exp \beta$ (2)

$\vdash . (1) . (2) . \supset \vdash . \mathrm{Prop}$

***116·172.** $\vdash :. \exists ! \alpha \exp \beta . \supset : \exists ! \alpha . \mathsf{v} . \beta = \Lambda$

Dem.

$\vdash . *83{\cdot}11 . \supset \vdash :. \mathrm{Hp} . \supset : \Lambda \sim\epsilon\, \alpha \downarrow_{,,} {}^{\backprime\backprime}\beta :$

[*113·112] $\supset : \sim(\alpha = \Lambda . \exists ! \beta) :$

[*24·51] $\supset : \exists ! \alpha . \mathsf{v} . \beta = \Lambda :. \supset \vdash . \mathrm{Prop}$

***116·18.** $\vdash :. \exists ! \alpha . \mathsf{v} . \beta = \Lambda : \equiv . \exists ! \alpha \exp \beta$ [*116·171·172]

***116·181.** $\vdash . \alpha \exp \Lambda = \iota{}^{\backprime}\Lambda$

Dem.

$\vdash . *113{\cdot}113 . \supset \vdash . \alpha \exp \Lambda = \mathrm{Prod}{}^{\backprime}\Lambda$

[*83·15.*33·241] $= \iota{}^{\backprime}\Lambda . \supset \vdash . \mathrm{Prop}$

∗116·182. $\vdash : \exists ! \beta . \supset . \Lambda \exp \beta = \Lambda$ [∗113·112 . ∗83·11]

∗116·183. $\vdash . s`(\alpha \exp \beta) = \beta \times \alpha$

Dem.

$\vdash . ∗115·141 . ∗116·18 . \supset \vdash :. \exists ! \alpha . \vee . \beta = \Lambda : \supset . s`(\alpha \exp \beta) = s`\alpha \downarrow_{,,} ``\beta$

[∗113·1] $= \beta \times \alpha$ (1)

$\vdash . ∗116·182 . \supset \vdash :. \alpha = \Lambda . \exists ! \beta . \supset . s`(\alpha \exp \beta) = \Lambda$

[∗113·114] $= \beta \times \alpha$ (2)

$\vdash . (1) . (2) . \supset \vdash .$ Prop

∗116·19. $\vdash : \alpha \operatorname{sm} \gamma . \beta \operatorname{sm} \delta . \supset . (\alpha \exp \beta) \operatorname{sm} \operatorname{sm} (\gamma \exp \delta)$

Dem.

$\vdash . ∗113·13 . \supset \vdash : \text{Hp} . \supset . \alpha \downarrow_{,,} ``\beta \operatorname{sm} \operatorname{sm} \gamma \downarrow_{,,} ``\delta .$

[∗115·51] $\supset . (\alpha \exp \beta) \operatorname{sm} \operatorname{sm} (\gamma \exp \delta) : \supset \vdash .$ Prop

∗116·191. $\vdash : R \epsilon \alpha \overline{\operatorname{sm}} \gamma . S \epsilon \beta \overline{\operatorname{sm}} \delta . \supset . (R \| \breve{S}) \upharpoonright (\delta \times \gamma) \epsilon (\alpha \exp \beta) \overline{\operatorname{sm}}\, \overline{\operatorname{sm}} (\gamma \exp \delta) .$
$(R \| \breve{S})_\epsilon ``(\gamma \exp \delta) = \alpha \exp \beta$

[∗113·127 . ∗115·502 . ∗116·183]

∗116·192. $\vdash : R \upharpoonright \gamma \epsilon \alpha \overline{\operatorname{sm}} \gamma . S \upharpoonright \delta \epsilon \beta \overline{\operatorname{sm}} \delta . \supset .$
$(R \| \breve{S}) \upharpoonright (\delta \times \gamma) \epsilon (\alpha \exp \beta) \overline{\operatorname{sm}}\, \overline{\operatorname{sm}} (\gamma \exp \delta) .$
$(R \| \breve{S})_\epsilon \upharpoonright (\gamma \exp \delta) \epsilon (\alpha \exp \beta) \overline{\operatorname{sm}} (\gamma \exp \delta)$

[∗113·127 . ∗115·502 . ∗116·183 . ∗111·15]

∗116·194. $\vdash : R \upharpoonright \gamma \epsilon \alpha \overline{\operatorname{sm}} \gamma . S \upharpoonright \delta \epsilon \beta \overline{\operatorname{sm}} \delta . \supset .$
$(R \| \breve{S}) \upharpoonright \{(\gamma \uparrow \delta)_\Delta `\delta\} \epsilon \{(\alpha \uparrow \beta)_\Delta `\beta\} \overline{\operatorname{sm}} \{(\gamma \uparrow \delta)_\Delta `\delta\}$

Dem.

$\vdash . ∗116·12 . \supset \vdash : \text{Hp} . \supset . s`\text{D}``(\gamma \uparrow \delta)_\Delta `\delta \subset \gamma . s`\text{G}``(\gamma \uparrow \delta)_\Delta `\delta \subset \delta .$

[∗74·773.∗73·142] $\supset . (R \| \breve{S}) \upharpoonright \{(\gamma \uparrow \delta)_\Delta `\delta\} \epsilon$
$\{(R \| \breve{S}) ``(\gamma \uparrow \delta)_\Delta `\delta\} \overline{\operatorname{sm}} \{(\gamma \uparrow \delta)_\Delta `\delta\}$ (1)

$\vdash . ∗116·192 . ∗111·14 . \supset \vdash : \text{Hp} . \supset . \alpha \exp \beta = (R \| \breve{S})_\epsilon ``(\gamma \exp \delta) .$

[∗116·13] $\supset . (\alpha \uparrow \beta)_\Delta `\beta = \dot{s}``(R \| \breve{S})_\epsilon ``(\gamma \exp \delta)$

[∗43·43] $= (R \| \breve{S}) `` \dot{s} `` (\gamma \exp \delta)$

[∗116·13] $= (R \| \breve{S}) `` (\gamma \uparrow \delta)_\Delta `\delta$ (2)

$\vdash . (1) . (2) . \supset \vdash .$ Prop

The following propositions (down to ∗116·27 exclusive) are the analogues of propositions with the same decimal part in ∗113.

∗116·2. $\vdash : \xi \epsilon \mu^\nu . \equiv . (\exists \alpha, \beta) . \mu = \text{N}_0\text{c}`\alpha . \nu = \text{N}_0\text{c}`\beta . \xi \operatorname{sm} (\alpha \exp \beta)$ [(∗116·02)]

∗116·201. $\vdash :. \xi \epsilon \mu^\nu . \equiv : \mu, \nu \epsilon \text{NC} : (\exists \alpha, \beta) . \alpha \epsilon \mu . \beta \epsilon \nu . \xi \operatorname{sm} (\alpha \exp \beta)$
[∗116·2 . ∗103·27]

$*116{\cdot}202$. $\vdash :. \xi \epsilon \mu^\nu . \equiv : \exists ! \mu . \exists ! \nu : (\exists \alpha, \beta) . \mu = Nc{`}\alpha . \nu = Nc{`}\beta . \xi sm (\alpha \exp \beta)$
[Proof as in $*113{\cdot}202$]

$*116{\cdot}203$. $\vdash : \exists ! \mu^\nu . \supset . \mu, \nu \epsilon NC - \iota{`}\Lambda . \mu, \nu \epsilon N_0C$ [$*116{\cdot}201{\cdot}202{\cdot}2$]

$*116{\cdot}204$. $\vdash :. \mu = \Lambda . v . \nu = \Lambda . v . \sim(\mu, \nu \epsilon NC) : \supset . \mu^\nu = \Lambda$ [$*116{\cdot}203$]

$*116{\cdot}205$. $\vdash : \sim (\mu, \nu \epsilon N_0C) . \supset . \mu^\nu = \Lambda$ [$*116{\cdot}203$]

$*116{\cdot}21$. $\vdash :. \mu, \nu \epsilon NC . \supset : \xi \epsilon \mu^\nu . \equiv . (\exists \alpha, \beta) . \alpha \epsilon \mu . \beta \epsilon \nu . \xi sm (\alpha \exp \beta)$ [$*116{\cdot}201$]

$*116{\cdot}22$. $\vdash : \xi \epsilon \{Nc(\eta){`}\gamma\}^{Nc(\zeta){`}\delta} . \equiv . \exists ! Nc(\eta){`}\gamma . \exists ! Nc(\zeta){`}\delta . \xi sm (\gamma \exp \delta)$
[Proof as in $*113{\cdot}22$, using $*116{\cdot}19$ in place of $*113{\cdot}13$]

$*116{\cdot}221$. $\vdash : \exists ! Nc(\eta){`}\gamma . \exists ! Nc(\zeta){`}\delta . \supset . \{Nc(\eta){`}\gamma\}^{Nc(\zeta){`}\delta} = Nc{`}(\gamma \exp \delta)$
[$*116{\cdot}22$]

$*116{\cdot}222$. $\vdash . (N_0c{`}\gamma)^{N_0c{`}\delta} = Nc{`}(\gamma \exp \delta)$ [Proof as in $*113{\cdot}222$]

$*116{\cdot}23$. $\vdash . \mu^\nu \epsilon NC$ [Proof as in $*113{\cdot}23$]

$*116{\cdot}24$. $\vdash . (Nc{`}\gamma)^{Nc{`}\delta} = (N_0c{`}\gamma)^{N_0c{`}\delta}$ [($*116{\cdot}03{\cdot}04$)]

$*116{\cdot}25$. $\vdash . (Nc{`}\gamma)^{Nc{`}\delta} = Nc{`}(\gamma \exp \delta)$ [$*116{\cdot}24{\cdot}222$]

$*116{\cdot}251$. $\vdash . (\gamma \exp \delta) \epsilon (Nc{`}\gamma)^{Nc{`}\delta}$ [$*116{\cdot}25 . *100{\cdot}3$]

$*116{\cdot}26$. $\vdash : \mu, \nu \epsilon NC . \exists ! sm_\eta{``}\mu . \exists ! sm_\zeta{``}\nu . \supset . \mu^\nu = (sm_\eta{``}\mu)^{sm_\zeta{``}\nu}$
[Proof as in $*113{\cdot}26$]

This proposition shows that we may raise or lower the types of μ and ν as we please, without affecting the value of μ^ν, provided μ and ν, or rather $sm{``}\mu$ and $sm{``}\nu$, exist in the new types.

$*116{\cdot}261$. $\vdash : \mu, \nu \epsilon NC . \supset . \mu^\nu = \{\mu^{(1)}\}^{\nu^{(1)}} = \{\mu_{(00)}\}^{\nu_{(00)}} =$ etc. [Proof as in $*113{\cdot}261$]

Here "etc." covers any derivative of μ or ν whose existence follows from that of μ or ν.

$*116{\cdot}27$. $\vdash . \mu^\nu = \hat{\xi} \{(\exists \alpha, \beta) . \mu = N_0c{`}\alpha . \nu = N_0c{`}\beta . \xi sm (\alpha \uparrow \beta)_\Delta{`}\beta\}$
[$*116{\cdot}15 . *73{\cdot}37 . (*116{\cdot}02)$]

$*116{\cdot}271$. $\vdash : \mu, \nu \epsilon NC . \alpha \epsilon \mu . \beta \epsilon \nu . \supset . (\alpha \exp \beta) \epsilon \mu^\nu$ [$*116{\cdot}21$]

$*116{\cdot}3$. $\vdash . (Nc{`}\alpha)^0 = 1$

Dem.

$\vdash . *101{\cdot}1 . *116{\cdot}25 . \supset \vdash . (Nc{`}\alpha)^0 = Nc{`}(\alpha \exp \Lambda)$
[$*116{\cdot}181$] $= Nc{`}\iota{`}\Lambda$
[$*101{\cdot}2$] $= 1 . \supset \vdash .$ Prop

$*116{\cdot}301$. $\vdash : \mu \epsilon NC - \iota{`}\Lambda . \supset . \mu^0 = 1$ [Proof as in $*113{\cdot}601$]

$*116{\cdot}31$. $\vdash : \beta \neq \Lambda . \supset . 0^{Nc{`}\beta} = 0$

Dem.

$\vdash . *101{\cdot}1 . *116{\cdot}25 . \supset \vdash . 0^{Nc{`}\beta} = Nc{`}(\Lambda \exp \beta)$
[$*116{\cdot}182$] $\supset \vdash : Hp . \supset . 0^{Nc{`}\beta} = Nc{`}\Lambda$
[$*101{\cdot}1$] $= 0 : \supset \vdash .$ Prop

***116·311.** $\vdash : \nu \in \text{NC} - \iota`\Lambda - \iota`0 . \supset . 0^\nu = 0$

Dem.

$\vdash . *103·34 . *101·1 . \supset \vdash : \text{Hp} . \supset . (\exists\beta) . \beta \neq \Lambda . \nu = \text{N}_0\text{c}`\beta .$

[*13·12·15] $\supset . (\exists\beta) . \beta \neq \Lambda . 0^\nu = 0^{\text{N}_0\text{c}`\beta}$

[*116·31.(*116·04)] $= 0 : \supset \vdash . \text{Prop}$

***116·32.** $\vdash . (\text{Nc}`\alpha)^1 = \text{Nc}`\alpha$

Dem.

$\vdash . *116·25 . *101·2 . \supset \vdash . (\text{Nc}`\alpha)^1 = \text{Nc}`\{\alpha \exp (\iota`x)\}$

[(*116·01)] $= \text{Nc}`\text{Prod}`\alpha \downarrow ``\iota`x$

[*115·142.*53·31] $= \text{Nc}`\iota``\alpha \downarrow x$

[*113·11.*100·6] $= \text{Nc}`\alpha . \supset \vdash . \text{Prop}$

***116·321.** $\vdash . \mu \in \text{NC} - \iota`\Lambda . \supset . \mu^1 = \text{sm}``\mu$ [*116·32]

It would not be an error to write "$\mu^1 = \mu$" instead of "$\mu^1 = \text{sm}``\mu$" in the above proposition. For if the "sm" is typically determined so that $\text{sm}``\mu \in t`\mu$, then $\text{sm}``\mu = \mu$. Thus in virtue of *116·321, $\mu^1 = \mu$ is true whenever it is significant. But the above form gives more information, since it preserves the typical ambiguity of $\mu^1$ and $\text{sm}``\mu$.

***116·33.** $\vdash . 1^{\text{Nc}`\beta} = 1$

Dem.

$\vdash . *113·11 . \quad \supset \vdash : \alpha \in 1 . \supset . \alpha \downarrow ``\beta \subset 1 .$

[*115·144.*101·2] $\supset . \text{Nc}`\text{Prod}`\alpha \downarrow ``\beta = 1$ (1)

$\vdash . (1) . *101·2 . \quad \supset \vdash . \text{Nc}`\{(\iota`x) \exp \beta\} = 1$ (2)

$\vdash . *101·2 . *116·25 . \supset \vdash . 1^{\text{Nc}`\beta} = \text{Nc}`\{(\iota`x) \exp \beta\}$ (3)

$\vdash . (2) . (3) . \supset \vdash . \text{Prop}$

***116·331.** $\vdash : \mu \in \text{NC} - \iota`\Lambda . \supset . 1^\mu = 1$

Dem.

$\vdash . *103·34 . \supset \vdash : \text{Hp} . \supset . (\exists\beta) . \mu = \text{N}_0\text{c}`\beta .$

[*13·12·15] $\supset . (\exists\beta) . 1^\mu = 1^{\text{N}_0\text{c}`\beta} .$

[(*116·04)] $\supset . (\exists\beta) . 1^\mu = 1^{\text{Nc}`\beta} .$

[*116·33] $\supset . 1^\mu = 1 : \supset \vdash . \text{Prop}$

***116·34.** $\vdash . \mu^2 = \mu \times_c \mu$

Dem.

$\vdash . *24·1 . *101·3 . \supset \vdash . \iota`\Lambda \cup \iota`\text{V} \in 2 .$

[*116·222] $\supset \vdash : \mu = \text{N}_0\text{c}`\alpha . \supset . \mu^2 = \text{Nc}`\text{Prod}`\alpha \downarrow ``(\iota`\Lambda \cup \iota`\text{V})$

[*53·32] $= \text{Nc}`\text{Prod}`(\iota`\alpha \downarrow \Lambda \cup \iota`\alpha \downarrow \text{V})$

[*115·13.*55·233.*38·2] $= \text{Nc}`(\alpha \downarrow \Lambda \times \alpha \downarrow \text{V})$

[*113·11·25·13] $= \text{Nc}`\alpha \times_c \text{Nc}`\alpha$

[*113·24] $= \mu \times_c \mu$ (1)

$\vdash . (1) . *103\cdot2 . \supset \vdash : \mu \epsilon N_0C . \supset . \mu^2 = \mu \times_c \mu$ (2)

$\vdash . *116\cdot205 . \quad \supset \vdash : \mu \sim\epsilon N_0C . \supset . \mu^2 = \Lambda$

[*113·205] $= \mu \times_c \mu$ (3)

$\vdash . (2) . (3) . \supset \vdash .$ Prop

***116·35.** $\vdash : \mu^\nu = 0 . \equiv . \mu = 0 . \nu \epsilon NC - \iota`0 - \iota`\Lambda$

Dem.

$\vdash . *116\cdot311 . \supset \vdash : \mu = 0 . \nu \epsilon NC - \iota`0 - \iota`\Lambda . \supset . \mu^\nu = 0$ (1)

$\vdash . *101\cdot12 . \quad \supset \vdash : \mu^\nu = 0 . \supset . \exists ! \mu^\nu .$

[*116·203] $\supset . \mu, \nu \epsilon NC - \iota`\Lambda$ (2)

$\vdash . (2) . *116\cdot21 . *54\cdot102 . \supset$

$\vdash :. \mu^\nu = 0 . \supset : \xi = \Lambda . \equiv . (\exists \alpha, \beta) . \alpha \epsilon \mu . \beta \epsilon \nu . \xi \text{ sm } (\alpha \exp \beta) :$

[*73·47] $\supset : (\exists \alpha, \beta) . \alpha \epsilon \mu . \beta \epsilon \nu . \alpha \exp \beta = \Lambda :$

[*116·18] $\supset : (\exists \alpha, \beta) . \alpha \epsilon \mu . \beta \epsilon \nu . \alpha = \Lambda . \beta \neq \Lambda :$

[*13·195] $\supset : \Lambda \epsilon \mu . \nu \neq \iota`\Lambda . \exists ! \nu :$

[*101·1.*100·45.(2)] $\supset : \mu = 0 . \nu \epsilon NC - \iota`\Lambda - \iota`0$ (3)

$\vdash . (1) . (3) . \supset \vdash .$ Prop

***116·351.** $\vdash : \mu \epsilon NC - \iota`\Lambda . \kappa = \Lambda . \nu = 0 . \supset . \mu^\nu = \Pi Nc`\kappa = 1$

[*116·301 . *114·2]

***116·352.** $\vdash : \mu = 0 . \nu \epsilon NC - \iota`\Lambda . \kappa \epsilon \nu . \Lambda \epsilon \kappa . \supset . \mu^\nu = \Pi Nc`\kappa = 0$

[*116·311 . *114·23]

***116·353.** $\vdash : \mu = 0 . \nu \epsilon NC - \iota`\Lambda . \kappa \epsilon \nu \cap Cl`\mu . \supset . \mu^\nu = \Pi Nc`\kappa$

Dem.

$\vdash . *60\cdot362 . *54\cdot1 . \quad \supset \vdash :. \text{Hp} . \supset : \kappa = \Lambda . \vee . \kappa = \iota`\Lambda$ (1)

$\vdash . *100\cdot45 . *101\cdot1 . \supset \vdash : \text{Hp} . \kappa = \Lambda . \supset . \nu = 0 .$

[*116·351] $\supset . \mu^\nu = \Pi Nc`\kappa$ (2)

$\vdash . *51\cdot16 . \quad \supset \vdash : \text{Hp} . \kappa = \iota`\Lambda . \supset . \Lambda \epsilon \kappa .$

[*116·352] $\supset . \mu^\nu = \Pi Nc`\kappa$ (3)

$\vdash . (1) . (2) . (3) . \supset \vdash .$ Prop

***116·36.** $\vdash :. \text{Mult ax} . \supset : \mu, \nu \epsilon NC - \iota`\Lambda . \kappa \epsilon \nu \cap Cl`\mu . \supset . \Pi Nc`\kappa = \mu^\nu$

Dem.

$\vdash . *113\cdot12 . *100\cdot45 . \supset \vdash : \mu, \nu \epsilon NC . \alpha \epsilon \mu . \beta \epsilon \nu . \exists ! \alpha . \supset . \alpha \downarrow_{,,} `` \beta \epsilon \nu \cap Cl`\mu$ (1)

$\vdash . (1) . *114\cdot571 . \quad \supset \vdash :. \text{Mult ax} . \supset :$

$\mu, \nu \epsilon NC . \alpha \epsilon \mu . \beta \epsilon \nu . \exists ! \alpha . \kappa \epsilon \nu \cap Cl`\mu . \supset . \Pi Nc`\kappa = \Pi Nc`\alpha \downarrow_{,,} `` \beta$

[*116·14.*114·1] $= Nc`(\alpha \exp \beta)$

[*116·271] $= \mu^\nu$ (2)

$\vdash . (2) . \supset \vdash :. \text{Mult ax} . \supset : \mu, \nu \epsilon NC - \iota`\Lambda . \exists ! \mu - \iota`\Lambda . \kappa \epsilon \nu \cap Cl`\mu . \supset . \Pi Nc`\kappa = \mu^\nu$ (3)

$\vdash . *51\cdot4 . *54\cdot1 . \supset \vdash : \mu \epsilon NC - \iota`\Lambda . \sim\exists ! \mu - \iota`\Lambda . \supset . \mu = 0$ (4)

$\vdash . (4) . *116\cdot353 . \supset$

$\vdash : \mu, \nu \epsilon NC - \iota`\Lambda . \sim\exists ! \mu - \iota`\Lambda . \kappa \epsilon \nu \cap Cl`\mu . \supset . \Pi Nc`\kappa = \mu^\nu$ (5)

$\vdash . (3) . (5) . \supset \vdash .$ Prop

In the above proposition, "$\nu \epsilon \mathrm{NC}$" is sufficient hypothesis as to ν, since "$\nu \neq \Lambda$" is implied by $\kappa \epsilon \nu \cap \mathrm{Cl}'\mu$. But $\mu \neq \Lambda$ is essential, since if $\mu = \Lambda$, $\mu^\nu = \Lambda$ and $\kappa = \Lambda$ (provided $\nu = 0$), whence $\Pi \mathrm{Nc}'\kappa = 1$.

The above proposition connects exponentiation with multiplication.

∗116·361. $\vdash :. \text{Mult ax} . \supset : \mu, \nu \epsilon \mathrm{NC} - \iota'\Lambda . \kappa \epsilon \nu \cap \mathrm{Cl\,excl}'\mu . \supset . \mathrm{Prod}'\kappa \epsilon \mu^\nu$

Dem.

$$\vdash . *115{\cdot}12 . \supset \vdash : \kappa \epsilon \nu \cap \mathrm{Cl\,excl}'\mu . \supset . \mathrm{Prod}'\kappa \epsilon \Pi \mathrm{Nc}'\kappa \quad (1)$$

$$\vdash . (1) . *116{\cdot}36 . \supset \vdash . \text{Prop}$$

The following propositions, which illustrate certain generalizations of the relations of rows and columns, may be made clearer by the accompanying

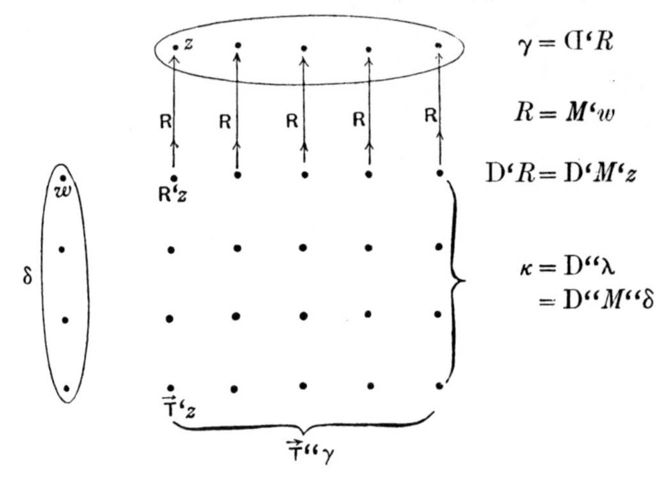

figure, in which, for the sake of simplicity, all the classes concerned are taken to be finite.

Let κ be a set of classes, constituted by four rows of five dots in the figure, which are each given as similar to a given class γ, represented by the top row of five dots in the figure, namely the row enclosed in an oval. We assume that an actual correlating relation is given correlating each member of κ with γ. Let λ be the class of these relations, and assume that λ consists of one correlator for each member of κ, and that $\kappa \epsilon \mathrm{Cls}^2\,\mathrm{excl}$. Thus $\mathrm{D}``\lambda = \kappa$, and $R \epsilon \lambda . \supset . \text{Ɑ}'R = \gamma$. Put $T = \dot{s}'\lambda$. Then, if $z \epsilon \gamma$, T relates to z every member of the column below z, *i.e.* $\overrightarrow{T}'z$ consists of the four dots which are vertically below z; assuming, what in the circumstances is possible, that each dot is placed below its correlate in γ. Thus $\overrightarrow{T}``\gamma$ represents the columns, while $\mathrm{D}``\lambda$ represents the rows.

We prove, in ∗116·41, that $\overrightarrow{T}``\gamma$, the class of rows, has double similarity with $\lambda \downarrow_{,,} ``\gamma$, or, what comes to the same thing, with $\kappa \downarrow_{,,} ``\gamma$. Hence it follows that $T``\gamma$, which is the whole class of dots, is similar to $\gamma \times \lambda$ or $\gamma \times \kappa$, and that $\mathrm{Nc}'\epsilon_\Delta{}'\overrightarrow{T}``\gamma$, which is the product of the numbers of the columns, is equal to

$(\text{Nc}'\lambda)^{\text{Nc}'\gamma}$ or $(\text{Nc}'\kappa)^{\text{Nc}'\gamma}$. The correlator which is used for proving these propositions is W, where, if R is a member of λ and z is a member of γ, W correlates $R'z$ with $R \downarrow z$.

Similarly, by correlating $R'z$ with $z \downarrow R$, calling the correlator U, we have $U``\downarrow R``\gamma = R``\gamma$, *i.e.* $U_\epsilon`\gamma \underset{,,}{\downarrow} R = \text{D}'R$, whence $U_\epsilon``\gamma \underset{,,}{\downarrow} ``\lambda = \text{D}``\lambda$. Hence $\text{D}``\lambda$, *i.e.* the class of rows, has double similarity with $\gamma \underset{,,}{\downarrow} ``\lambda$ or $\gamma \underset{,,}{\downarrow} ``\kappa$, whence the product of the numbers of the rows is $(\text{Nc}'\gamma)^{\text{Nc}'\lambda}$ or $(\text{Nc}'\gamma)^{\text{Nc}'\kappa}$.

Finally, we take a class δ similar to κ or λ (illustrated in the figure by the column of dots enclosed in an oval), and calling M a correlator of λ and δ, we replace λ by $M``\delta$ and $\kappa$ by $\text{D}``M``\delta$. We thus find that, if $M \upharpoonright \delta$ correlates with $\delta$ a class of relations whose domains are mutually exclusive, and which each correlate their domains with a given class $\gamma$, then $\text{D}``M``\delta$ has double similarity with $\gamma \underset{,,}{\downarrow} ``\delta$, whence the same results as before with δ in place of κ or λ.

The following propositions are useful in connecting multiplication with exponentiation, and in proving the formal laws of exponentiation.

*116·4·401 are lemmas for *116·41.

***116·4.** $\vdash :. \lambda \subset 1 \to 1 : R, S \in \lambda . \exists ! \text{D}'R \cap \text{D}'S . \supset_{R,S} . R = S :$

$$\text{G}``\lambda \subset \iota`\gamma . W = \hat{x}\hat{P}\{(\exists R, z) . R \in \lambda . x = R'z . P = R \downarrow z\} :$$
$$\supset . W \in 1 \to 1 . \text{G}'W = \gamma \times \lambda . \text{D}'W = \text{D}'\dot{s}'\lambda$$

Dem.

$\vdash . *21{\cdot}33 . \supset \vdash :. \text{Hp} . \supset : xWP . xWQ . \equiv .$

$\quad (\exists R, S, z, w) . R, S \in \lambda . x = R'z = S'w . P = R \downarrow z . Q = S \downarrow w .$

[*33·43] $\quad \equiv . (\exists R, S, z, w) . R, S \in \lambda . x = R'z = S'w . x \in \text{D}'R \cap \text{D}'S .$

$\quad P = R \downarrow z . Q = S \downarrow w .$

[Hp.*13·195] $\quad \supset . (\exists R, z, w) . R \in \lambda . x = R'z = R'w . P = R \downarrow z . Q = R \downarrow w .$

[*71·532.*13·195] $\supset . (\exists R, z) . R \in \lambda . x = R'z . P = R \downarrow z . Q = R \downarrow z .$

[*13·172] $\quad \supset . P = Q \quad (1)$

$\vdash . *21{\cdot}33 . \supset \vdash :. \text{Hp} . \supset : xWP . yWP . \equiv .$

$\quad (\exists R, S, z, w) . R, S \in \lambda . x = R'z . y = S'w . P = R \downarrow z = Q \downarrow w .$

[*55·202] $\quad \supset . (\exists R, S, z, w) . R, S \in \lambda . x = R'z . y = S'w . R = S . z = w .$

[*13·22·172] $\supset . x = y \quad (2)$

$\vdash . *33{\cdot}131 . \supset \vdash :. \text{Hp} . \supset : P \in \text{G}'W . \equiv . (\exists x, R, z) . R \in \lambda . x = R'z . P = R \downarrow z .$

[*71·411] $\quad \equiv . (\exists R, z) . R \in \lambda . z \in \text{G}'R . P = R \downarrow z .$

[Hp] $\quad \equiv . (\exists R, z) . R \in \lambda . z \in \gamma . P = R \downarrow z .$

[*113·101] $\quad \equiv . P \in \gamma \times \lambda \quad (3)$

$\vdash . *33{\cdot}13 . \supset \vdash :. \text{Hp} . \supset : x \in \text{D}'W . \equiv . (\exists P, R, z) . R \in \lambda . x = R'z . P = R \downarrow z .$

[*55·12.*71·36] $\quad \equiv . (\exists R, z) . R \in \lambda . xRz .$

[*41·11.*33·13] $\quad \equiv . x \in \text{D}'\dot{s}'\lambda \quad (4)$

$\vdash . (1) . (2) . (3) . (4) . \supset \vdash . \text{Prop}$

***116·401.** $\vdash : \text{Hp} *116\cdot4 . T = \dot{s}`\lambda . \supset . \overrightarrow{T}``\gamma = W_\epsilon``\lambda \downarrow_{,,} ``\gamma$

Dem.

$\vdash . *37\cdot11\cdot1 . *38\cdot2 . \supset \vdash :. \text{Hp} . z \epsilon \gamma . \supset :$

$x \epsilon W_\epsilon`\lambda \downarrow_{,,} z . \equiv . (\exists R) . R \epsilon \lambda . x W (R \downarrow z) .$

[*21·33] $\equiv . (\exists R, S, w) . R, S \epsilon \lambda . x = S`w . R \downarrow z = S \downarrow w .$

[*55·202.*13·22] $\equiv . (\exists R) . R \epsilon \lambda . x = R`z .$

[*71·36] $\equiv . (\exists R) . R \epsilon \lambda . xRz .$

[*41·11] $\equiv . x (\dot{s}`\lambda) z .$

[Hp.*32·18] $\equiv . x \epsilon \overrightarrow{T}`z$ (1)

$\vdash . (1) . *37\cdot68 . \supset \vdash . \text{Prop}$

***116·41.** $\vdash :. \lambda \subset 1 \to 1 . \text{G}``\lambda \subset \iota`\gamma : R, S \epsilon \lambda . \exists ! \text{D}`R \cap \text{D}`S . \supset_{R,S} . R = S : T = \dot{s}`\lambda :$
$\supset . \overrightarrow{T}``\gamma \text{ sm sm } \lambda \downarrow_{,,} ``\gamma . \overrightarrow{T}``\gamma \text{ sm } \gamma \times \lambda . T \epsilon \text{Cls} \to 1 . \overrightarrow{T}``\gamma \epsilon \text{Cls}^2 \text{ excl} .$

$$\text{Nc}`\epsilon_\Delta`\overrightarrow{T}``\gamma = \text{Nc}`T_\Delta`\gamma = \text{Nc}`\text{Prod}`\overrightarrow{T}``\gamma = \text{Nc}`(\lambda \exp \gamma) = (\text{Nc}`\lambda)^{\text{Nc}`\gamma}$$

Dem.

$\vdash . *116\cdot4\cdot401 . *111\cdot4 . *113\cdot1 . \supset \vdash : \text{Hp} . \supset . \overrightarrow{T}``\gamma \text{ sm sm } \lambda \downarrow_{,,} ``\gamma .$ (1)

[*111·44.*40·5] $\supset . \overrightarrow{T}``\gamma \text{ sm } \gamma \times \lambda$ (2)

$\vdash . *72\cdot321 . *85\cdot14 . \supset \vdash : \text{Hp} . \supset . T \epsilon \text{Cls} \to 1 . \text{Nc}`\epsilon_\Delta`\overrightarrow{T}``\gamma = \text{Nc}`T_\Delta`\gamma$ (3)

$\vdash . (3) . *84\cdot51 . \quad \supset \vdash : \text{Hp} . \supset . \overrightarrow{T}``\gamma \epsilon \text{Cls}^2 \text{ excl} .$ (4)

[*115·12] $\supset . \text{Nc}`\epsilon_\Delta`\overrightarrow{T}``\gamma = \text{Nc}`\text{Prod}`\overrightarrow{T}``\gamma$ (5)

$\vdash . (1) . *114\cdot52 . \quad \supset \vdash : \text{Hp} . \supset . \text{Nc}`\epsilon_\Delta`\overrightarrow{T}``\gamma = \text{Nc}`\epsilon_\Delta`\lambda \downarrow_{,,} ``\gamma$

[*116·14] $= \text{Nc}`(\lambda \exp \gamma)$ (6)

[*116·25] $= (\text{Nc}`\lambda)^{\text{Nc}`\gamma}$ (7)

$\vdash . (1) . (2) . (3) . (4) . (5) . (6) . (7) . \supset \vdash . \text{Prop}$

The following proposition is merely another form of *116·41.

***116·411.** $\vdash :. \kappa \epsilon \text{Cls}^2 \text{ excl} : \alpha \epsilon \kappa . \supset_\alpha . M`\alpha \epsilon \alpha \overline{\text{sm}} \gamma : T = \dot{s}`M``\kappa : \supset .$
$\overrightarrow{T}``\gamma \text{ sm sm } \kappa \downarrow_{,,} ``\gamma . \overrightarrow{T}``\gamma \text{ sm } \gamma \times \kappa . T \epsilon \text{Cls} \to 1 . \overrightarrow{T}``\gamma \epsilon \text{Cls}^2 \text{ excl} .$

$$\text{Nc}`\epsilon_\Delta`\overrightarrow{T}``\gamma = \text{Nc}`T_\Delta`\gamma = \text{Nc}`\text{Prod}`\overrightarrow{T}``\gamma = \text{Nc}`(\kappa \exp \gamma) = (\text{Nc}`\kappa)^{\text{Nc}`\gamma}$$

Dem.

$\vdash . *73\cdot03 . \supset \vdash : \text{Hp} . \supset . M``\kappa \subset 1 \to 1 . \text{G}``M``\kappa \subset \iota`\gamma$ (1)

$\vdash . *111\cdot16 . \supset \vdash :. \text{Hp} . \supset : \alpha, \beta \epsilon \kappa . M`\alpha = M`\beta . \supset . \alpha = \beta$ (2)

$\vdash . *14\cdot21 . \supset \vdash :. \text{Hp} . \supset : \alpha \epsilon \kappa . \supset . \text{E} ! M`\alpha$ (3)

$\vdash . (2) . (3) . *73\cdot24 . \supset \vdash : \text{Hp} . \supset . M``\kappa \text{ sm } \kappa$ (4)

$\vdash . *73\cdot03 . \supset \vdash :. \text{Hp} . \supset : \alpha \epsilon \kappa . \supset . \text{D}`M`\alpha = \alpha :$

[*13·12] $\supset : \alpha, \beta \epsilon \kappa . \exists ! \text{D}`M`\alpha \cap \text{D}`M`\beta . \supset . \exists ! \alpha \cap \beta .$

[*84·11] $\supset . \alpha = \beta .$

[*30·37.(3)] $\supset . M`\alpha = M`\beta :$

[*37·63] $\supset : R, S \epsilon M``\kappa . \exists ! \text{D}`R \cap \text{D}`S . \supset . R = S$ (5)

$\vdash . (1) . (4) . (5) . *116\cdot41 \frac{M``\kappa}{\lambda} . *113\cdot13 . *116\cdot19 . \supset \vdash . \text{Prop}$

*116·412·413 are lemmas for *116·414.

***116·412.** $\vdash :. \lambda \subset 1 \to 1 : R, S \epsilon \lambda . \exists ! \text{D}`R \cap \text{D}`S . \supset_{R,S} . R = S : \text{Ɑ}``\lambda \subset \iota`\gamma :$

$U = \hat{x}\hat{P} \{(\exists R, z) . R \epsilon \lambda . x = R`z . P = z \downarrow R\} :$

$\supset . U \epsilon (s`\text{D}``\lambda) \overline{\text{sm}} (\lambda \times \gamma)$ [Proof as in *116·4]

***116·413.** $\vdash : \text{Hp} *116\cdot412 . \supset . \text{D}``\lambda = U_\epsilon``\gamma \underset{,,}{\downarrow} ``\lambda$ [Proof as in *116·401]

***116·414.** $\vdash : \text{Hp} *116\cdot412 . \supset . U \epsilon (\text{D}``\lambda) \overline{\text{sm}}\, \overline{\text{sm}} (\gamma \underset{,,}{\downarrow} ``\lambda) . (\text{D}``\lambda) \text{ sm sm } (\gamma \underset{,,}{\downarrow} ``\lambda)$

[*116·412·413]

***116·42.** $\vdash :. \lambda \subset 1 \to 1 : R, S \epsilon \lambda . \exists ! \text{D}`R \cap \text{D}`S . \supset_{R,S} . R = S : \text{Ɑ}``\lambda \subset \iota`\gamma :$

$\supset . \text{D}``\lambda \text{ sm sm } (\gamma \underset{,,}{\downarrow} ``\lambda) . (\text{D}`\dot{s}`\lambda) \text{ sm } (\lambda \times \gamma) . (\epsilon_\Delta`\text{D}``\lambda) \text{ sm } (\gamma \exp \lambda) .$

$\text{Nc}`\text{Prod}`\text{D}``\lambda = \Pi \text{Nc}`\text{D}``\lambda = (\text{Nc}`\gamma)^{\text{Nc}`\lambda}$

[*116·414·25 . *115·51 . *111·44 . *41·43]

***116·422.** $\vdash :. M \restriction \delta \epsilon 1 \to 1 : w, v \epsilon \delta . \exists ! \text{D}`M`w \cap \text{D}`M`v . \supset_{w,v} . w = v :$

$w \epsilon \delta . \supset_w . M`w \epsilon 1 \to 1 . \text{Ɑ}`M`w = \gamma : \supset . \text{D}``M``\delta \text{ sm sm } \gamma \underset{,,}{\downarrow} ``\delta$

Dem.

$\vdash . *116\cdot42 \frac{M``\delta}{\lambda} . \supset$

$\vdash :. M``\delta \subset 1 \to 1 : R, S \epsilon M``\delta . \exists ! \text{D}`R \cap \text{D}`S . \supset_{R,S} . R = S : \text{Ɑ}``M``\delta \subset \iota`\gamma :$

$\supset . \text{D}``M``\delta \text{ sm sm } \gamma \underset{,,}{\downarrow} ``M``\delta$ (1)

$\vdash . *14\cdot21 . \supset \vdash :. \text{Hp} . \supset : w \epsilon \delta . \supset . \text{E} ! M`w :$ (2)

[*33·43] $\supset : \delta \subset \text{Ɑ}`M :$

[*73·15] $\supset : (M``\delta) \text{ sm } \delta$ (3)

$\vdash . *51\cdot15 . \supset \vdash :. \text{Hp} . \supset : w \epsilon \delta . \supset . \text{Ɑ}`M`w \epsilon \iota`\gamma :$

[*37·61] $\supset : \text{Ɑ}``M``\delta \subset \iota`\gamma$ (4)

$\vdash . (2) . *30\cdot37 . \supset \vdash :. \text{Hp} . \supset : w, v \epsilon \delta . \exists ! \text{D}`M`w \cap \text{D}`M`v . \supset_{w,v} . M`w = M`v :$

[*37·63] $\supset : R, S \epsilon M``\delta . \exists ! \text{D}`R \cap \text{D}`S . \supset_{R,S} . R = S$ (5)

$\vdash . (1) . (4) . (5) . \supset \vdash : \text{Hp} . \supset . \text{D}``M``\delta \text{ sm sm } \gamma \underset{,,}{\downarrow} ``M``\delta .$

[(3).*113·13] $\supset . \text{D}``M``\delta \text{ sm sm } \gamma \underset{,,}{\downarrow} ``\delta : \supset \vdash . \text{Prop}$

∗116·43. $\vdash :. M \upharpoonright \delta \in 1 \to 1 : w, v \in \delta . \exists ! \text{D}`M`w \cap \text{D}`M`v . \supset_{w,v} . w = v :$
$w \in \delta . \supset_w . M`w \in 1 \to 1 . \text{Cl}`M`w = \gamma :$
$\supset . \text{Prod}`\text{D}``M``\delta \text{ sm sm } (\gamma \exp \delta) . \Pi\text{Nc}`\text{D}``M``\delta = (\text{Nc}`\gamma)^{\text{Nc}`\delta}$

Dem.

$\vdash . ∗115·51 . ∗116·422 . \supset \vdash : \text{Hp} . \supset . \text{Prod}`\text{D}``M``\delta \text{ sm sm } (\gamma \exp \delta)$ (1)

$\vdash . ∗116·422 . ∗114·52 . \supset \vdash : \text{Hp} . \supset . \Pi\text{Nc}`\text{D}``M``\delta = \Pi\text{Nc}`\gamma \downarrow_{,,} ``\delta$ (2)

$\vdash . ∗116·14·25 . \quad \supset \vdash . \Pi\text{Nc}`\gamma \downarrow_{,,} ``\delta = (\text{Nc}`\gamma)^{\text{Nc}`\delta}$ (3)

$\vdash . (1) . (2) . (3) . \supset \vdash . \text{Prop}$

The above proposition is used in ∗116·534·61.

∗116·44. $\vdash :. \exists ! \gamma : (z) . M`z \in 1 \to 1 . \text{Cl}`M`z = \text{V} :$
$w, v \in \delta . \exists ! (M`w)``\gamma \cap (M`v)``\gamma . \supset_{w,v} . w = v :$
$\supset . \text{D}`` \upharpoonright \gamma``M``\delta \text{ sm sm } \gamma \downarrow_{,,} ``\delta . \text{Prod}`\text{D}`` \upharpoonright \gamma``M``\delta \text{ sm sm } (\gamma \exp \delta)$

Dem.

$\vdash . ∗71·29 . ∗35·65 . \supset$

$\vdash :. \text{Hp} : (z) . N`z = (M`z) \upharpoonright \gamma : \supset . (z) . N`z \in 1 \to 1 . \text{Cl}`N`z = \gamma$ (1)

$\vdash . ∗37·401 . \supset \vdash :. \text{Hp} . \text{Hp}(1) . \supset : w, v \in \delta . \exists ! \text{D}`N`w \cap \text{D}`N`v . \supset_{w,v} . w = v$ (2)

$\vdash . ∗35·7 . \quad \supset \vdash :. \text{Hp} . \text{Hp}(1) . \supset : x \in \gamma . w, v \in \delta . N`w = N`v . \supset . (N`w)`x = (N`v)`x .$

[(2)] $\supset . w = v$ (3)

$\vdash . (3) . ∗10·11·23·35 . \supset \vdash :. \text{Hp} . \text{Hp}(1) . \supset : w, v \in \delta . N`w = N`v . \supset . w = v :$

[∗71·55·166] $\supset : N \upharpoonright \delta \in 1 \to 1$ (4)

$\vdash . (1) . (2) . (4) . ∗116·422 . ∗115·51 . \supset$

$\vdash : \text{Hp} . \text{Hp}(1) . \supset . \text{D}``N``\delta \text{ sm sm } \gamma \downarrow_{,,} ``\delta . \text{Prod}`\text{D}``N``\delta \text{ sm sm } (\gamma \exp \delta)$ (5)

$\vdash . ∗38·11 . \supset \vdash : \text{Hp} . \text{Hp}(1) . \supset . \text{D}`N`z = \text{D}` \upharpoonright \gamma`M`z .$

[∗37·353] $\supset . \text{D}``N``\delta = \text{D}`` \upharpoonright \gamma``M``\delta$ (6)

$\vdash . (5) . (6) . \supset \vdash . \text{Prop}$

∗116·45. $\vdash :. (z) . M`z \in 1 \to 1 . \text{Cl}`M`z = \text{V} :$
$w, v \in \delta . \exists ! (M`w)``\gamma \cap (M`v)``\gamma . \supset_{w,v} . w = v : \supset . \text{Prod}`\text{D}`` \upharpoonright \gamma``M``\delta \text{ sm} (\gamma \exp \delta)$

Dem.

$\vdash . ∗116·182 . ∗115·142 . ∗37·29 . \supset$

$\vdash : \text{Hp} . \gamma = \Lambda . \exists ! \delta . \supset . \text{Prod}`\text{D}`` \upharpoonright \gamma``M``\delta = \Lambda . \gamma \exp \delta = \Lambda$ (1)

$\vdash . ∗115·1 . ∗83·15 . ∗116·181 . \supset$

$\vdash : \text{Hp} . \delta = \Lambda . \supset . \text{Prod}`\text{D}`` \upharpoonright \gamma``M``\delta = \iota`\Lambda . \gamma \exp \delta = \iota`\Lambda$ (2)

$\vdash . (1) . (2) . ∗116·44 . \supset \vdash . \text{Prop}$

The above proposition is used in ∗116·676.

We have now to prove the three formal laws of exponentiation, namely

$$\mu^\nu \times_c \mu^\varpi = \mu^{\nu +_c \varpi},$$

$$\mu^\varpi \times_c \nu^\varpi = (\mu \times_c \nu)^\varpi,$$

and

$$(\mu^\nu)^\varpi = \mu^{\nu \times_c \varpi}.$$

Of these the first is an immediate consequence of the distributive law, while the second and third result from forms of the associative law of multiplication.

***116·5.** $\vdash : \beta \cap \gamma = \Lambda . \supset . (\alpha \exp \beta) \times (\alpha \exp \gamma) \operatorname{sm} \alpha \exp (\beta \cup \gamma)$

Dem.

$\vdash . *113{\cdot}191 . \supset$

$\vdash : \mathrm{Hp} . \exists ! \alpha . \supset . \alpha \underset{,,}{\downarrow} {}^{\text{“}}\beta \cap \alpha \underset{,,}{\downarrow} {}^{\text{“}}\gamma = \Lambda .$

[*114·301] $\supset . \epsilon_\Delta{}^{\text{‘}}\alpha \underset{,,}{\downarrow} {}^{\text{“}}\beta \times \epsilon_\Delta{}^{\text{‘}}\alpha \underset{,,}{\downarrow} {}^{\text{“}}\gamma \operatorname{sm} \epsilon_\Delta{}^{\text{‘}}(\alpha \underset{,,}{\downarrow} {}^{\text{“}}\beta \cup \alpha \underset{,,}{\downarrow} {}^{\text{“}}\gamma) .$

[*116·14.*113·13] $\supset . (\alpha \exp \beta) \times (\alpha \exp \gamma) \operatorname{sm} \epsilon_\Delta{}^{\text{‘}}(\alpha \underset{,,}{\downarrow} {}^{\text{“}}\beta \cup \alpha \underset{,,}{\downarrow} {}^{\text{“}}\gamma) .$

[*37·22] $\supset . (\alpha \exp \beta) \times (\alpha \exp \gamma) \operatorname{sm} \epsilon_\Delta{}^{\text{‘}}\alpha \underset{,,}{\downarrow} {}^{\text{“}}(\beta \cup \gamma) .$

[*116·14] $\supset . (\alpha \exp \beta) \times (\alpha \exp \gamma) \operatorname{sm} \alpha \exp (\beta \cup \gamma)$ (1)

$\vdash . *116{\cdot}182 . \supset \vdash : \alpha = \Lambda . \exists ! \beta . \supset . \alpha \exp \beta = \Lambda .$

[*113·114] $\supset . (\alpha \exp \beta) \times (\alpha \exp \gamma) = \Lambda$ (2)

$\vdash . *116{\cdot}182 . *24{\cdot}56 . \supset \vdash : \alpha = \Lambda . \exists ! \beta . \supset . \alpha \exp (\beta \cup \gamma) = \Lambda$ (3)

$\vdash . (2) . (3) . \supset \vdash : \alpha = \Lambda . \exists ! \beta . \supset . (\alpha \exp \beta) \times (\alpha \exp \gamma) \operatorname{sm} \alpha \exp (\beta \cup \gamma)$ (4)

Similarly $\vdash : \alpha = \Lambda . \exists ! \gamma . \supset . (\alpha \exp \beta) \times (\alpha \exp \gamma) \operatorname{sm} \alpha \exp (\beta \cup \gamma)$ (5)

$\vdash . *116{\cdot}181 . \supset \vdash : \alpha = \Lambda . \beta = \Lambda . \gamma = \Lambda . \supset . (\alpha \exp \beta) \times (\alpha \exp \gamma) = \iota{}^{\text{‘}}\Lambda \times \iota{}^{\text{‘}}\Lambda$ (6)

$\vdash . *116{\cdot}181 . \supset \vdash : \alpha = \Lambda . \beta = \Lambda . \gamma = \Lambda . \supset . \alpha \exp (\beta \cup \gamma) = \iota{}^{\text{‘}}\Lambda$ (7)

$\vdash . (6) . (7) . *113{\cdot}611 . *73{\cdot}43 . \supset$

$\vdash : \alpha = \Lambda . \beta = \Lambda . \gamma = \Lambda . \supset . (\alpha \exp \beta) \times (\alpha \exp \gamma) \operatorname{sm} \alpha \exp (\beta \cup \gamma)$ (8)

$\vdash . (1) . (4) . (5) . (8) . \supset \vdash .$ Prop

In the last line of the above proof, *73·43 is required because the two Λ's involved have not been proved to be of the same type. They are in fact of the same type, but it is unnecessary to prove this.

***116·51.** $\vdash . (\alpha \exp \beta) \times (\alpha \exp \gamma) \operatorname{sm} \alpha \exp (\beta + \gamma)$

Dem.

$\vdash . *116{\cdot}19 . *110{\cdot}12 . \supset \vdash . (\alpha \exp \beta) \operatorname{sm} (\alpha \exp \downarrow \Lambda_\gamma{}^{\text{“}}\iota{}^{\text{“}}\beta) .$

$(\alpha \exp \gamma) \operatorname{sm} (\alpha \exp \Lambda_\beta \downarrow {}^{\text{“}}\iota{}^{\text{“}}\gamma) .$

[*113·13] $\supset \vdash . (\alpha \exp \beta) \times (\alpha \exp \gamma) \operatorname{sm}$

$(\alpha \exp \downarrow \Lambda_\gamma{}^{\text{“}}\iota{}^{\text{“}}\beta) \times (\alpha \exp \Lambda_\beta \downarrow {}^{\text{“}}\iota{}^{\text{“}}\gamma) .$

[*110·11.*116·5] $\supset \vdash . (\alpha \exp \beta) \times (\alpha \exp \gamma) \operatorname{sm}$

$\alpha \exp (\downarrow \Lambda_\gamma{}^{\text{“}}\iota{}^{\text{“}}\beta \cup \Lambda_\beta \downarrow {}^{\text{“}}\iota{}^{\text{“}}\gamma)$ (1)

$\vdash . (1) . (*110{\cdot}01) . \supset \vdash .$ Prop

***116·52.** $\vdash . \mu^{\nu} \times_c \mu^{\varpi} = \mu^{\nu +_c \varpi}$

Dem.

$\vdash . *116·51 . *110·22 . \supset$

$\vdash . (N_0c`\alpha)^{N_0c`\beta} \times_c (N_0c`\alpha)^{N_0c`\gamma} = (N_0c`\alpha)^{N_0c`\beta +_c N_0c`\gamma}$ (1)

$\vdash . (1) . *103·2 . \supset \vdash : \mu, \nu, \varpi \in N_0C . \supset . \mu^{\nu} \times_c \mu^{\varpi} = \mu^{\nu +_c \varpi}$ (2)

$\vdash . *116·205 . *113·204 . \supset$

$\vdash : \mu \sim\in N_0C . \supset . \mu^{\nu} \times_c \mu^{\varpi} = \Lambda = \mu^{\nu +_c \varpi}$ (3)

$\vdash . *116·205 . *113·204 . \supset$

$\vdash : \sim(\nu, \varpi \in N_0C) . \supset . \mu^{\nu} \times_c \mu^{\varpi} = \Lambda$ (4)

$\vdash . *110·4 . *116·204 . \supset \vdash : \sim(\nu, \varpi \in N_0C) . \supset . \mu^{\nu +_c \varpi} = \Lambda$ (5)

$\vdash . (4) . (5) . \supset \vdash : \sim(\nu, \varpi \in N_0C) . \supset . \mu^{\nu} \times_c \mu^{\varpi} = \mu^{\nu +_c \varpi}$ (6)

$\vdash . (2) . (3) . (6) . \supset \vdash .$ Prop

The following propositions are lemmas for

$$\mu^{\varpi} \times_c \nu^{\varpi} = (\mu \times_c \nu)^{\varpi}.$$

The principal previous propositions used in the proof are *115·6 and *116·43. The proof proceeds as follows.

$(\alpha \exp \gamma) \times (\beta \exp \gamma)$ is $\text{Prod}`\alpha \underset{,,}{\downarrow} ``\gamma \times \text{Prod}`\beta \underset{,,}{\downarrow} ``\gamma$. This, using *115·6, and putting $\alpha \underset{,,}{\downarrow}, \beta \underset{,,}{\downarrow}$ in place of R and S of that proposition, is similar to $\epsilon_\Delta`\hat{\mu}\{(\exists z) . z \in \gamma . \mu = \alpha \underset{,,}{\downarrow} z \times \beta \underset{,,}{\downarrow} z\}$, *i.e.* to $\epsilon_\Delta`\hat{\mu}\{(\exists z) . z \in \gamma . \mu = \downarrow z``\alpha \times \downarrow z``\beta\}$.

Now by *113·65, putting $R\dagger = R \| \breve{R}$ Dft, $\downarrow z``\alpha \times \downarrow z``\beta = (\downarrow z)\dagger``(\alpha \times \beta)$. We now apply *116·43, taking $(\downarrow z)\dagger$ as the $M`z$ of that proposition, or rather, taking $(\downarrow z)\dagger \upharpoonright (\alpha \times \beta)$. Thus we find

$$\epsilon_\Delta`\hat{\mu}\{(\exists z) . z \in \gamma . \mu = \downarrow z``\alpha \times \downarrow z``\beta\} \text{ sm } (\alpha \times \beta) \exp \gamma.$$

Hence our proposition follows.

***116·529.** $R\dagger = R \| \breve{R}$ Dft [*116]

In *150, this notation will be introduced as a permanent definition. For the present, we only introduce it to avoid $(\downarrow z \| \text{Cnv}`\downarrow z)$, which is awkward.

***116·53.** $\vdash : \exists ! \alpha . \exists ! \beta . \supset .$

$$(\alpha \exp \gamma) \times (\beta \exp \gamma) \text{ sm } \epsilon_\Delta`\hat{\mu}\{(\exists z) . z \in \gamma . \mu = \downarrow z``\alpha \times \downarrow z``\beta\}$$

Dem.

$\vdash . *113·104·111 . \supset \vdash . \gamma \subset \text{Cl}`\alpha \underset{,,}{\downarrow} . \gamma \subset \text{Cl}`\beta \underset{,,}{\downarrow} . \alpha \underset{,,}{\downarrow} ``\gamma, \beta \underset{,,}{\downarrow} ``\gamma \in \text{Cls}^2 \text{ excl}$ (1)

$\vdash . *113·105 . \quad \supset \vdash : \text{Hp} . \supset . (\alpha \underset{,,}{\downarrow} ``\gamma) \upharpoonright \alpha \underset{,,}{\downarrow}, \beta \underset{,,}{\downarrow} \upharpoonright \gamma \in 1 \to 1$ (2)

$\vdash . (1) . (2) . *115·6 \dfrac{\alpha \underset{,,}{\downarrow}, \beta \underset{,,}{\downarrow}}{R, S} . \supset$

$\vdash : \text{Hp} . \supset . (\alpha \exp \gamma) \times (\beta \exp \gamma) \text{ sm } \epsilon_\Delta`\hat{\mu}\{(\exists z) . z \in \gamma . \mu = \alpha \underset{,,}{\downarrow} z \times \beta \underset{,,}{\downarrow} z\}$ (3)

$\vdash . (3) . *38·2 . \supset \vdash .$ Prop

The hypothesis $\exists ! \alpha . \exists ! \beta$ is not necessary in the above proposition; but the proof is simpler with the hypothesis, and we do not need the proposition without the hypothesis.

***116·531.** $\vdash :. M = \hat{R}\hat{z}\{z \epsilon \gamma . R = (\downarrow z) \dagger \upharpoonright (\alpha \times \beta)\} . \supset :$

$$z \epsilon \gamma . \supset_z . M \text{‘} z = (\downarrow z) \dagger \upharpoonright (\alpha \times \beta) . M \text{‘} z \epsilon 1 \rightarrow 1 . \text{Ɑ} \text{‘} M \text{‘} z = \alpha \times \beta$$

Dem.

$\vdash . *74·772 . *55·12 . *72·184 . \supset \vdash . (\downarrow z) \dagger \epsilon 1 \rightarrow 1$ (1)

$\vdash . *21·33 . \supset \vdash :. \text{Hp} . z \epsilon \gamma . \supset : RMz . \equiv . R = (\downarrow z) \dagger \upharpoonright (\alpha \times \beta) :$

[*30·3] $\supset : M \text{‘} z = (\downarrow z) \dagger \upharpoonright (\alpha \times \beta) :$ (2)

[(1).*43·122] $\supset : M \text{‘} z \epsilon 1 \rightarrow 1 . \text{Ɑ} \text{‘} M \text{‘} z = \alpha \times \beta$ (3)

$\vdash . (2) . (3) . \supset \vdash . \text{Prop}$

***116·532.** $\vdash : \text{Hp} *116·531 . \exists ! \alpha . \exists ! \beta . \supset . M \epsilon 1 \rightarrow 1 . \text{Ɑ} \text{‘} M = \gamma$

Dem.

$\vdash . *116·531 . *14·21 . *71·16 . \supset \vdash : \text{Hp} . \supset . M \epsilon 1 \rightarrow \text{Cls}$ (1)

$\vdash . *116·531 . \supset \vdash :. \text{Hp} . z, w \epsilon \gamma . M \text{‘} z = M \text{‘} w . \supset :$

$$(\downarrow z) \dagger \upharpoonright (\alpha \times \beta) = (\downarrow w) \dagger \upharpoonright (\alpha \times \beta) :$$

[*71·35] $\supset : R \epsilon (\alpha \times \beta) . \supset . (\downarrow z) \dagger \text{‘} R = (\downarrow w) \dagger \text{‘} R :$

[*113·101] $\supset : x \epsilon \alpha . y \epsilon \beta . \supset . (\downarrow z) \dagger \text{‘} (y \downarrow x) = (\downarrow w) \dagger \text{‘} (y \downarrow x) .$

[*113·123] $\supset . (y \downarrow z) \downarrow (x \downarrow z) = (y \downarrow w) \downarrow (x \downarrow w) .$

[*55·202] $\supset . z = w$ (2)

$\vdash . (2) . \quad \supset \vdash :. \text{Hp} . \supset : z, w \epsilon \gamma . M \text{‘} z = M \text{‘} w . \supset . z = w$ (3)

$\vdash . *116·531 . *14·21 . *33·43 . \supset \vdash : \text{Hp} . z \epsilon \gamma . \supset . z \epsilon \text{Ɑ} \text{‘} R$ (4)

$\vdash . *21·33 . \supset \vdash :. \text{Hp} . \supset : RMz . \supset_{R, z} . z \epsilon \gamma :$

[*33·351] $\supset : \text{Ɑ} \text{‘} R \subset \gamma$ (5)

$\vdash . (1) . (3) . (4) . (5) . *71·55 . \supset \vdash . \text{Prop}$

***116·533.** $\vdash :. \text{Hp} *116·531 . \supset : \text{D} \text{‘‘} M \text{‘‘} \gamma = \hat{\mu}\{(\exists z) . z \epsilon \gamma . \mu = \downarrow z \text{‘‘} \alpha \times \downarrow z \text{‘‘} \beta\} :$

$$z, w \epsilon \gamma . \exists ! \text{D} \text{‘} M \text{‘} z \cap \text{D} \text{‘} M \text{‘} w . \supset_{z, w} . z = w$$

Dem.

$\vdash . *116·531 . \supset \vdash : \text{Hp} . z \epsilon \gamma . \supset . \text{D} \text{‘} M \text{‘} z = \text{D} \text{‘} \{(\downarrow z) \dagger \upharpoonright (\alpha \times \beta)\}$

[*37·401] $= (\downarrow z) \dagger \text{‘‘} (\alpha \times \beta)$

[*113·65] $= \downarrow z \text{‘‘} \alpha \times \downarrow z \text{‘‘} \beta$ (1)

$\vdash . (1) . *37·6 . \supset \vdash : \text{Hp} . \supset . \text{D} \text{‘‘} M \text{‘‘} \gamma = \hat{\mu}\{(\exists z) . z \epsilon \gamma . \mu = \downarrow z \text{‘‘} \alpha \times \downarrow z \text{‘‘} \beta\}$ (2)

$\vdash . *113·19 . \quad \supset \vdash : \exists ! (\downarrow z \text{‘‘} \alpha \times \downarrow z \text{‘‘} \beta) \cap (\downarrow w \text{‘‘} \alpha \times \downarrow w \text{‘‘} \beta) . \supset .$

$$\exists ! \downarrow z \text{‘‘} \alpha \cap \downarrow w \text{‘‘} \alpha .$$

[*55·232] $\supset . z = w$ (3)

$\vdash . (1) . (2) . (3) . \supset \vdash . \text{Prop}$

***116·534.** $\vdash : \text{Hp} *116{\cdot}532 . \supset . \epsilon_\Delta{}^{\text{`}}\text{D}^{\text{``}}M^{\text{``}}\gamma \operatorname{sm} (\alpha \times \beta) \exp \gamma$

Dem.

$\vdash . *116{\cdot}531{\cdot}532{\cdot}533 . \supset$

$\vdash :. \text{Hp} . \supset : M \epsilon 1 \to 1 : z, w \epsilon \gamma . \exists ! \text{D}^{\text{`}}M^{\text{`}}z \cap \text{D}^{\text{`}}M^{\text{`}}w . \supset_{z,w} . z = w :$

$z \epsilon \gamma . \supset_z . M^{\text{`}}z \epsilon 1 \to 1 . \text{Q}^{\text{`}}M^{\text{`}}z = \alpha \times \beta :$

$[*116{\cdot}43] \quad \supset : \text{Prod}^{\text{`}}\text{D}^{\text{``}}M^{\text{``}}\gamma \operatorname{sm} (\alpha \times \beta) \exp \gamma :$

$[*115{\cdot}12 . *30{\cdot}37 . *84{\cdot}11] \supset : \epsilon_\Delta{}^{\text{`}}\text{D}^{\text{``}}M^{\text{``}}\gamma \operatorname{sm} (\alpha \times \beta) \exp \gamma :. \supset \vdash . \text{Prop}$

***116·535.** $\vdash : \exists ! \alpha . \exists ! \beta . \supset . (\alpha \exp \gamma) \times (\beta \exp \gamma) \operatorname{sm} (\alpha \times \beta) \exp \gamma$

$[*116{\cdot}53{\cdot}533{\cdot}534]$

The hypothesis $\exists ! \alpha . \exists ! \beta$ is not necessary, as we shall now prove.

***116·54.** $\vdash . (\alpha \exp \gamma) \times (\beta \exp \gamma) \operatorname{sm} (\alpha \times \beta) \exp \gamma$

Dem.

$\vdash . *116{\cdot}182 . \supset \vdash : \alpha = \Lambda . \exists ! \gamma . \supset . \alpha \exp \gamma = \Lambda .$

$[*113{\cdot}114] \quad \supset . (\alpha \exp \gamma) \times (\beta \exp \gamma) = \Lambda \quad (1)$

$\vdash . *113{\cdot}114 . *116{\cdot}182 . \supset \vdash : \alpha = \Lambda . \exists ! \gamma . \supset . (\alpha \times \beta) \exp \gamma = \Lambda \quad (2)$

$\vdash . (1) . (2) . \quad \supset \vdash : \alpha = \Lambda . \exists ! \gamma . \supset . (\alpha \exp \gamma) \times (\beta \exp \gamma) \operatorname{sm} (\alpha \times \beta) \exp \gamma \quad (3)$

Similarly $\quad \vdash : \beta = \Lambda . \exists ! \gamma . \supset . (\alpha \exp \gamma) \times (\beta \exp \gamma) \operatorname{sm} (\alpha \times \beta) \exp \gamma \quad (4)$

$\vdash . *116{\cdot}181 . \supset \vdash : \gamma = \Lambda . \supset . (\alpha \exp \gamma) \times (\beta \exp \gamma) = \iota^{\text{`}}\Lambda \times \iota^{\text{`}}\Lambda .$

$[*113{\cdot}611 . *73{\cdot}43] \quad \supset . (\alpha \exp \gamma) \times (\beta \exp \gamma) \operatorname{sm} \iota^{\text{`}}\Lambda \quad (5)$

$\vdash . *116{\cdot}181 . \supset \vdash : \gamma = \Lambda . \supset . (\alpha \times \beta) \exp \gamma = \iota^{\text{`}}\Lambda .$

$[(5)] \quad \supset . (\alpha \exp \gamma) \times (\beta \exp \gamma) \operatorname{sm} (\alpha \times \beta) \exp \gamma \quad (6)$

$\vdash . (3) . (4) . (6) . *116{\cdot}535 . \supset \vdash . \text{Prop}$

In obtaining (5), we use *73·43 as well as *113·611, because Λ's of different types are involved.

***116·55.** $\vdash . \mu^\varpi \times_c \nu^\varpi = (\mu \times_c \nu)^\varpi$

Dem.

$\vdash . *116{\cdot}54{\cdot}222 . *113{\cdot}222 . \supset \vdash . (\text{N}_0\text{c}^{\text{`}}\alpha)^{\text{N}_0\text{c}^{\text{`}}\gamma} \times_c (\text{N}_0\text{c}^{\text{`}}\beta)^{\text{N}_0\text{c}^{\text{`}}\gamma}$

$= (\text{N}_0\text{c}^{\text{`}}\alpha \times_c \text{N}_0\text{c}^{\text{`}}\beta)^{\text{N}_0\text{c}^{\text{`}}\gamma} \quad (1)$

$\vdash . (1) . *103{\cdot}2 . \quad \supset \vdash : \mu, \nu, \varpi \epsilon \text{N}_0\text{C} . \supset . \mu^\varpi \times_c \nu^\varpi = (\mu \times_c \nu)^\varpi \quad (2)$

$\vdash . *116{\cdot}205 . *113{\cdot}204 . \supset \vdash : \varpi \sim\epsilon \text{N}_0\text{C} . \supset . \mu^\varpi \times_c \nu^\varpi = \Lambda = (\mu \times_c \nu)^\varpi \quad (3)$

$\vdash . *116{\cdot}205 . *113{\cdot}204 . \supset \vdash : \sim(\mu, \nu \epsilon \text{N}_0\text{C}) . \supset . \mu^\varpi \times_c \nu^\varpi = \Lambda \quad (4)$

$\vdash . *113{\cdot}204 . *116{\cdot}204 . \supset \vdash : \sim(\mu, \nu \epsilon \text{N}_0\text{C}) . \supset . (\mu \times_c \nu)^\varpi = \Lambda \quad (5)$

$\vdash . (4) . (5) \quad \supset \vdash : \sim(\mu, \nu \epsilon \text{N}_0\text{C}) . \supset . \mu^\varpi \times_c \nu^\varpi = (\mu \times_c \nu)^\varpi \quad (6)$

$\vdash . (2) . (3) . (6) . \quad \supset \vdash . \text{Prop}$

This completes the proof of the second of the formal laws of exponentiation. The following propositions are lemmas for the third of these laws, namely

$$(\mu^\nu)^\varpi = \mu^{\nu \times_c \varpi}.$$

∗116·6. $\vdash : \exists ! \alpha . \supset . \alpha \exp (\beta \times \gamma) \text{ sm Prod'Prod}``\alpha \downarrow_{,,} ```\beta \downarrow_{,,} ``\gamma .$
$\alpha \downarrow_{,,} ```\beta \downarrow_{,,} ``\gamma \epsilon \text{Cls}^3 \text{ arithm}$

Dem.

$\vdash . *113\cdot105 . *84\cdot53 \frac{\alpha \downarrow_{,,}}{R} . *113\cdot111 . \supset \vdash : \text{Hp} . \supset . \alpha \downarrow_{,,} ```\beta \downarrow_{,,} ``\gamma \epsilon \text{Cls}^2 \text{ excl}$ (1)

$\vdash . *40\cdot38 . \quad \supset \vdash . s`\alpha \downarrow_{,,} ```\beta \downarrow_{,,} ``\gamma = \alpha \downarrow_{,,} ``s`\beta \downarrow_{,,} ``\gamma$ (2)

$[*113\cdot111] \quad \supset \vdash . s`\alpha \downarrow_{,,} ```\beta \downarrow_{,,} ``\gamma \epsilon \text{Cls}^2 \text{ excl}$ (3)

$\vdash . (1) . (3) . *115\cdot2 . \quad \supset \vdash : \text{Hp} . \supset . \alpha \downarrow_{,,} ```\beta \downarrow_{,,} ``\gamma \epsilon \text{Cls}^3 \text{ arithm}$ (4)

$\vdash . *113\cdot141 . *116\cdot19 . \supset \vdash . \text{Nc}`\{\alpha \exp (\beta \times \gamma)\} = \text{Nc}`\{\alpha \exp (\gamma \times \beta)\}$

$[(*116\cdot01 . *113\cdot02)] \quad = \text{Nc}`\text{Prod}`\alpha \downarrow_{,,} ``s`\beta \downarrow_{,,} ``\gamma$

$[(2)] \quad = \text{Nc}`\text{Prod}`s`\alpha \downarrow_{,,} ```\beta \downarrow_{,,} ``\gamma$

$[*115\cdot35 . (4)] \quad = \text{Nc}`\text{Prod}`\text{Prod}``\alpha \downarrow_{,,} ```\beta \downarrow_{,,} ``\gamma$ (5)

$\vdash . (4) . (5) . \supset \vdash . \text{Prop}$

∗116·601. $\vdash . | (\text{Cnv}` \downarrow z) \epsilon 1 \rightarrow 1 \quad [*74\cdot774 . *72\cdot184]$

∗116·602. $\vdash :. M = \hat{R}\hat{z} [z \epsilon \gamma . R = \{|(\text{Cnv}` \downarrow z)\}_\epsilon \upharpoonright (\alpha \exp \beta)] . \supset :$
$z \epsilon \gamma . \supset . M`z = \{|(\text{Cnv}` \downarrow z)\}_\epsilon \upharpoonright (\alpha \exp \beta) : \text{Ɑ}`M = \gamma$

Dem.

$\vdash . *21\cdot33 . \supset$

$\vdash :: \text{Hp} . \supset :. z \epsilon \gamma . \supset : RMz . \equiv . R = \{|(\text{Cnv}` \downarrow z)\}_\epsilon \upharpoonright (\alpha \exp \beta)$ (1)

$\vdash . (1) . *30\cdot3 . \supset \vdash :. \text{Hp} . \supset : z \epsilon \gamma . \supset . M`z = \{|(\text{Cnv}` \downarrow z)\}_\epsilon \upharpoonright (\alpha \exp \beta)$ (2)

$\vdash . *21\cdot33 . *33\cdot131 . \supset \vdash : \text{Hp} . \supset . \text{Ɑ}`M = \gamma$ (3)

$\vdash . (2) . (3) . \supset \vdash . \text{Prop}$

∗116·603. $\vdash :. \text{Hp} *116\cdot602 . \supset : z \epsilon \gamma . \supset . \text{Ɑ}`M`z = \alpha \exp \beta$
$[*116\cdot602 . *37\cdot231 . *35\cdot65]$

∗116·604. $\vdash :. \text{Hp} *116\cdot602 . \supset : z \epsilon \gamma . \supset . \text{D}`M`z = \text{Prod}`\alpha \downarrow_{,,} ``\beta \downarrow_{,,} z$

Dem.

$\vdash . *37\cdot401 . *116\cdot602 . \supset$

$\vdash : \text{Hp} . z \epsilon \gamma . \supset . \text{D}`M`z = \{|(\text{Cnv}` \downarrow z)\}_\epsilon ``(\alpha \exp \beta)$

$[*115\cdot4 . *116\cdot601 . *43\cdot301] = \text{Prod}`\{|(\text{Cnv}` \downarrow z)\}_\epsilon ``\alpha \downarrow_{,,} ``\beta$

$[*113\cdot125 \frac{I}{R} . *50\cdot75\cdot16] \quad = \text{Prod}`\alpha \downarrow_{,,} `` \downarrow z``\beta$

$[*38\cdot2] \quad = \text{Prod}`\alpha \downarrow_{,,} ``\beta \downarrow_{,,} z : \supset \vdash . \text{Prop}$

∗116·605. $\vdash :. \text{Hp} *116\cdot602 . \supset : z \epsilon \gamma . \supset . M`z \epsilon 1 \rightarrow 1$

Dem.

$\vdash . *116\cdot601 . *72\cdot451 . \supset$

$\vdash . \{|(\text{Cnv}` \downarrow z)\}_\epsilon \upharpoonright \text{Cl}`\text{Ɑ}`|(\text{Cnv}` \downarrow z) \epsilon 1 \rightarrow 1 .$

$[*43\cdot301] \supset \vdash . \{|(\text{Cnv}` \downarrow z)\}_\epsilon \upharpoonright (\alpha \exp \beta) \epsilon 1 \rightarrow 1$ (1)

$\vdash . (1) . *116\cdot602 . \supset \vdash . \text{Prop}$

***116·606.** $\vdash :. \mathrm{Hp} *116\cdot602 \,.\, \exists ! \alpha \,.\, \exists ! \beta \,.\, \supset :$

$$M \epsilon 1 \rightarrow 1 : z, w \epsilon \gamma \,.\, \mathrm{D}\text{‘}M\text{‘}z = \mathrm{D}\text{‘}M\text{‘}w \,.\, \supset_{z,w} .\, z = w$$

Dem.

$\vdash .\, *116\cdot602 \,.\, *14\cdot21 \,.\, \supset \vdash :. \mathrm{Hp} \,.\, \supset : z \epsilon \text{ᗡ}\text{‘}M \,.\, \supset_z .\, \mathrm{E}! M\text{‘}z :$

$[*71\cdot16]$ $\supset : M \epsilon 1 \rightarrow \mathrm{Cls}$ (1)

$\vdash .\, *30\cdot37 \,.\, \supset \vdash :. \mathrm{Hp} \,.\, \supset : z, w \epsilon \gamma \,.\, M\text{‘}z = M\text{‘}w \,.\, \supset .\, \mathrm{D}\text{‘}M\text{‘}z = \mathrm{D}\text{‘}M\text{‘}w$ (2)

$\vdash .\, *116\cdot604 \,.\, \supset \vdash :. \mathrm{Hp} \,.\, \supset : z, w \epsilon \gamma \,.\, \mathrm{D}\text{‘}M\text{‘}z = \mathrm{D}\text{‘}M\text{‘}w \,.\, \supset .$

$$\mathrm{Prod}\text{‘}\alpha \underset{,,}{\downarrow} \text{“}\beta \underset{,,}{\downarrow} z = \mathrm{Prod}\text{‘}\alpha \underset{,,}{\downarrow} \text{“}\beta \underset{,,}{\downarrow} w \,.$$

$[*30\cdot37]$ $\supset .\, s\text{‘}\mathrm{Prod}\text{‘}\alpha \underset{,,}{\downarrow} \text{“}\beta \underset{,,}{\downarrow} z = s\text{‘}\mathrm{Prod}\text{‘}\alpha \underset{,,}{\downarrow} \text{“}\beta \underset{,,}{\downarrow} w \,.$

$[*116\cdot171 \,.\, *115\cdot141 \,.\, (*116\cdot01)]$ $\supset .\, s\text{‘}\alpha \underset{,,}{\downarrow} \text{“}\beta \underset{,,}{\downarrow} z = s\text{‘}\alpha \underset{,,}{\downarrow} \text{“}\beta \underset{,,}{\downarrow} w \,.$

$[*113\cdot1]$ $\supset .\, \beta \underset{,,}{\downarrow} z \times \alpha = \beta \underset{,,}{\downarrow} w \times \alpha \,.$

$[*113\cdot182]$ $\supset .\, \beta \underset{,,}{\downarrow} z = \beta \underset{,,}{\downarrow} w \,.$

$[*113\cdot105 \,.\, \mathrm{Hp}]$ $\supset .\, z = w$ (3)

$\vdash .\, (1) \,.\, (2) \,.\, (3) \,.\, *71\cdot55 \,.\, *116\cdot602 \,.\, \supset \vdash .\, \mathrm{Prop}$

***116·607.** $\vdash :. \mathrm{Hp} *116\cdot602 \,.\, \exists ! \alpha \,.\, \exists ! \beta \,.\, \supset :$

$M \epsilon 1 \rightarrow 1 \,.\, \mathrm{D}\text{“}M\text{“}\gamma = \mathrm{Prod}\text{“}\alpha \underset{,,}{\downarrow} \text{‘“}\beta \underset{,,}{\downarrow} \text{“}\gamma :$

$z, w \epsilon \gamma \,.\, \mathrm{D}\text{‘}M\text{‘}z = \mathrm{D}\text{‘}M\text{‘}w \,.\, \supset_{z,w} .\, z = w :$

$z \epsilon \gamma \,.\, \supset_z .\, M\text{‘}z \epsilon 1 \rightarrow 1 \,.\, \text{ᗡ}\text{‘}M\text{‘}z = \alpha \exp \beta$ $[*116\cdot606\cdot604\cdot605\cdot603]$

***116·61.** $\vdash : \exists ! \alpha \,.\, \exists ! \beta \,.\, \supset .\, \mathrm{Prod}\text{‘}\mathrm{Prod}\text{“}\alpha \underset{,,}{\downarrow} \text{‘“}\beta \underset{,,}{\downarrow} \text{“}\gamma \ \mathrm{sm}\ (\alpha \exp \beta) \exp \gamma$

$[*116\cdot607\cdot43]$

***116·611.** $\vdash : \exists ! \alpha \,.\, \exists ! \beta \,.\, \supset .\, \alpha \exp (\beta \times \gamma) \ \mathrm{sm}\ (\alpha \exp \beta) \exp \gamma$ $[*116\cdot6\cdot61]$

***116·62.** $\vdash .\, \alpha \exp (\beta \times \gamma) \ \mathrm{sm}\ (\alpha \exp \beta) \exp \gamma$

Dem.

$\vdash .\, *116\cdot181 \,.\, *113\cdot114 \,.\, \supset \vdash : \beta = \Lambda \,.\, \supset .\, \alpha \exp (\beta \times \gamma) = \iota\text{‘}\Lambda$ (1)

$\vdash .\, *116\cdot181 \,.\, \supset \vdash : \beta = \Lambda \,.\, \supset .\, (\alpha \exp \beta) \exp \gamma = (\iota\text{‘}\Lambda) \exp \gamma$ (2)

$\vdash .\, *116\cdot33\cdot25 \,.\, \supset \vdash .\, \mathrm{Nc}\text{‘}\{(\iota\text{‘}\Lambda) \exp \gamma\} = 1$ (3)

$\vdash .\, (1) . (2) . (3) .\, *52\cdot22 \,.\, *100\cdot31 \,.\, \supset \vdash : \beta = \Lambda \,.\, \supset .\, \alpha \exp (\beta \times \gamma) \ \mathrm{sm}\ (\alpha \exp \beta) \exp \gamma$ (4)

$\vdash .\, *113\cdot107 \,.\, *116\cdot182 \,.\, \supset \vdash : \alpha = \Lambda \,.\, \exists ! \beta \,.\, \exists ! \gamma \,.\, \supset .\, \alpha \exp (\beta \times \gamma) = \Lambda$ (5)

$\vdash .\, *116\cdot182 \,.\, \supset \vdash : \alpha = \Lambda \,.\, \exists ! \beta \,.\, \exists ! \gamma \,.\, \supset .\, (\alpha \exp \beta) \exp \gamma = \Lambda$ (6)

$\vdash .\, (5) \,.\, (6) \,.\, \supset \vdash : \alpha = \Lambda \,.\, \exists ! \beta \,.\, \exists ! \gamma \,.\, \supset .\, \alpha \exp (\beta \times \gamma) \ \mathrm{sm}\ (\alpha \exp \beta) \exp \gamma$ (7)

$\vdash .\, *113\cdot114 \,.\, *116\cdot181 \,.\, \supset \vdash : \gamma = \Lambda \,.\, \supset .\, \alpha \exp (\beta \times \gamma) = \iota\text{‘}\Lambda \,.\, (\alpha \exp \beta) \exp \gamma = \iota\text{‘}\Lambda \,.$

$[*73\cdot43]$ $\supset .\, \alpha \exp (\beta \times \gamma) \ \mathrm{sm}\ (\alpha \exp \beta) \exp \gamma$ (8)

$\vdash .\, (4) \,.\, (7) \,.\, (8) \,.\, \supset \vdash :. \alpha = \Lambda \,.\, \mathsf{v} \,.\, \beta = \Lambda \,.\, \mathsf{v} \,.\, \gamma = \Lambda : \supset .$

$$\alpha \exp (\beta \times \gamma) \ \mathrm{sm}\ (\alpha \exp \beta) \exp \gamma \quad (9)$$

$\vdash .\, (9) \,.\, *116\cdot611 \,.\, \supset \vdash .\, \mathrm{Prop}$

***116·63.** $\vdash . \mu^{\nu \times_c \varpi} = (\mu^\nu)^\varpi$

Dem.

$\vdash . *113\cdot222 . \supset \vdash . (N_0c`\alpha)^{N_0c`\beta \times_c N_0c`\gamma} = (N_0c`\alpha)^{Nc`(\beta\times\gamma)}$

$[*116\cdot222.(*116\cdot04)] = Nc`\{\alpha \exp (\beta \times \gamma)\}$

$[*116\cdot62] \quad = Nc`\{(\alpha \exp \beta) \exp \gamma\}$

$[*116\cdot222] \quad = \{N_0c`(\alpha \exp \beta)\}^{N_0c`\gamma}$

$[*116\cdot222.(*116\cdot03)] = \{(N_0c`\alpha)^{N_0c`\beta}\}^{N_0c`\gamma}$ (1)

$\vdash . (1) . *103\cdot2 . \quad \supset \vdash : \mu, \nu, \varpi \in N_0C . \supset . \mu^{\nu\times_c\varpi} = (\mu^\nu)^\varpi$ (2)

$\vdash . *116\cdot204\cdot205 . \quad \supset \vdash : \sim(\mu, \nu \in N_0C) . \supset . (\mu^\nu)^\varpi = \Lambda$ (3)

$\vdash . *113\cdot205 . *116\cdot204\cdot205 . \supset \vdash : \sim(\mu, \nu \in N_0C) . \supset . \mu^{\nu\times_c\varpi} = \Lambda$ (4)

$\vdash . *116\cdot205 . \quad \supset \vdash : \varpi \sim\in N_0C . \supset . (\mu^\nu)^\varpi = \Lambda$ (5)

$\vdash . *113\cdot205 . *116\cdot204 . \supset \vdash : \varpi \sim\in N_0C . \supset . \mu^{\nu\times_c\varpi} = \Lambda$ (6)

$\vdash . (3) . (4) . (5) . (6) . \quad \supset \vdash : \sim(\mu, \nu, \varpi \in N_0C) . \supset . \mu^{\nu\times_c\varpi} = (\mu^\nu)^\varpi$ (7)

$\vdash . (2) . (7) . \quad \supset \vdash .$ Prop

This completes the proof of the third of the formal laws of exponentiation.

***116·64.** $\vdash . (\mu^\nu)^\varpi = (\mu^\varpi)^\nu$ $[*116\cdot63 . *113\cdot27]$

***116·651.** $\vdash : Q \in \text{Cls} \to 1 . \kappa \in \text{Cls}^2 \text{excl} . \supset . \epsilon_\Delta`P_\Delta``Q```\kappa \text{ sm } P_\Delta`Q``s`\kappa$

Dem.

$\vdash . *84\cdot53 . \supset \vdash : \text{Hp} . \supset . Q```\kappa \in \text{Cls}^2 \text{excl} .$

$[*85\cdot43] \quad \supset . \epsilon_\Delta`P_\Delta``Q```\kappa \text{ sm } P_\Delta`s`Q```\kappa .$

$[*40\cdot38] \quad \supset . \epsilon_\Delta`P_\Delta``Q```\kappa \text{ sm } P_\Delta`Q``s`\kappa : \supset \vdash .$ Prop

***116·652.** $\vdash : Q \in \text{Cls} \to 1 . \kappa \in \text{Cls}^2 \text{excl} . \supset . \epsilon_\Delta`\epsilon_\Delta``Q```\kappa \text{ sm } \epsilon_\Delta`Q``s`\kappa$

$\left[*116\cdot651 \dfrac{\epsilon}{P}\right]$

The following propositions are lemmas for *116·661, which is an extension of *116·52.

***116·653.** $\vdash : \kappa \in \text{Cls}^2 \text{excl} . \supset . \alpha \downarrow_{,,} ```\kappa \in \text{Cls}^2 \text{arithm}$

Dem.

$\vdash . *113\cdot105 . *84\cdot53 . \supset \vdash : \text{Hp} . \exists ! \alpha . \supset . \alpha \downarrow_{,,} ```\kappa \in \text{Cls}^2 \text{excl}$ (1)

$\vdash . *113\cdot111 . \quad \supset \vdash . \alpha \downarrow_{,,} ``s`\kappa \in \text{Cls}^2 \text{excl} .$

$[*40\cdot38] \quad \supset \vdash . s`\alpha \downarrow_{,,} ```\kappa \in \text{Cls}^2 \text{excl} .$ (2)

$\vdash . *113\cdot112\cdot113 . \supset \vdash :. \alpha = \Lambda . \supset : \beta \in \kappa . \exists ! \beta . \supset . \alpha \downarrow_{,,} ``\beta = \iota`\Lambda :$

$\beta \in \kappa . \beta = \Lambda . \supset . \alpha \downarrow_{,,} ``\beta = \Lambda$ (3)

$\vdash . (3) . \supset \vdash :. \alpha = \Lambda . \supset : \alpha \downarrow_{,,} ```\kappa \subset \iota`\iota`\Lambda \cup \iota`\Lambda :$

$[*24\cdot43\cdot561] \quad \supset : \rho, \sigma \in \alpha \downarrow_{,,} ```\kappa . \exists ! \rho \cap \sigma . \supset . \rho, \sigma \in \iota`\iota`\Lambda .$

$[*51\cdot15] \quad \supset . \rho = \sigma :$

$[*84\cdot11] \quad \supset : \alpha \downarrow_{,,} ```\kappa \in \text{Cls}^2 \text{excl}$ (4)

$\vdash . (1) . (4) . \supset \vdash : \text{Hp} . \supset . \alpha \downarrow_{,,} ```\kappa \in \text{Cls}^2 \text{excl}$ (5)

$\vdash . (2) . (5) . \supset \vdash .$ Prop

∗116·654. $\vdash : \kappa \epsilon \text{Cls}^2 \text{excl} . \supset . \{\text{Prod}`(\alpha \exp)``\kappa\} \text{sm} \{\alpha \exp (s`\kappa)\}$

Dem.

$\vdash . ∗38·13 . (∗116·01) . \supset \vdash . \text{Prod}`(\alpha \exp)``\kappa = \text{Prod}`\text{Prod}``\alpha \downarrow ```\kappa$ (1)

$\vdash . (1) . ∗116·653 . ∗115·34 . \supset$

$\vdash . \{\text{Prod}`(\alpha \exp)``\kappa\} \text{sm} \{\text{Prod}`s`\alpha \downarrow ```\kappa\}$ (2)

$\vdash . (2) . ∗40·38 . (∗116·01) . \supset \vdash . \text{Prop}$

∗116·655. $\vdash : \kappa \epsilon \text{Cls}^2 \text{excl} . \supset . \Pi \text{Nc}`(\alpha \exp)``\kappa = (\text{Nc}`\alpha)^{\Sigma \text{Nc}`\kappa}$ [∗116·654]

This proposition is an extension of ∗116·5.

The hypothesis $\kappa \epsilon \text{Cls}^2 \text{excl}$ is unnecessary in the above proposition, as we shall now prove.

∗116·656. $\vdash : \exists ! \alpha \exp \beta \cap \alpha \exp \gamma . \supset . \beta = \gamma$

Dem.

$\vdash . ∗116·11 . ∗52·16 . \supset$

$\vdash :. \mu \epsilon (\alpha \exp \beta) \cap (\alpha \exp \gamma) . \supset : y \epsilon \beta . \supset . (\exists x) . x \epsilon \alpha . x \downarrow y \epsilon \mu : \mu \subset \gamma \times \alpha :$

[∗113·101] $\supset : y \epsilon \beta . \supset . (\exists x) . x \epsilon \alpha . x \downarrow y \epsilon \mu : x \downarrow y \epsilon \mu . \supset . y \epsilon \gamma :$

[Syll] $\supset : y \epsilon \beta . \supset . (\exists x) . x \epsilon \alpha . x \downarrow y \epsilon \mu . y \epsilon \gamma .$

[∗10·35] $\supset . y \epsilon \gamma$ (1)

Similarly $\vdash :. \mu \epsilon \alpha \exp \beta \cap \alpha \exp \gamma . \supset : y \epsilon \gamma . \supset . y \epsilon \beta$ (2)

$\vdash . (1) . (2) . \supset \vdash . \text{Prop}$

∗116·657. $\vdash . (\alpha \exp)``\kappa \epsilon \text{Cls}^2 \text{excl}$ [∗116·656]

∗116·658. $\vdash . \alpha \exp (\epsilon \downarrow \beta) = \{|(\text{Cnv}` \downarrow \beta)\}_\epsilon``(\alpha \exp \beta)$

Dem.

$\vdash . ∗116·602·604 . ∗37·401 . \supset \vdash . \{|(\text{Cnv}` \downarrow \beta)\}_\epsilon``(\alpha \exp \beta) = \alpha \exp (\beta \downarrow \beta)$

[∗85·601] $= \alpha \exp (\epsilon \downarrow \beta) . \supset \vdash . \text{Prop}$

∗116·659. $T = \hat{\nu} \hat{\mu} \{(\exists \beta) . \beta \epsilon \kappa . \mu \epsilon \alpha \exp \beta . \nu = |(\text{Cnv}` \downarrow \beta)``\mu\} . \supset .$

$T \epsilon (\alpha \exp)``\epsilon \downarrow ``\kappa \, \overline{\text{sm}} \, \overline{\text{sm}} \, (\alpha \exp)``\kappa$

Dem.

$\vdash . ∗40·4 . \supset \vdash : \text{Hp} . \supset . \text{G}`T = s`(\alpha \exp)``\kappa$ (1)

$\vdash . ∗21·33 . \supset \vdash :. \text{Hp} . \supset : \nu T \mu . \varpi T \mu . \supset .$

$(\exists \beta, \gamma) . \beta, \gamma \epsilon \kappa . \mu \epsilon \alpha \exp \beta . \mu \epsilon \alpha \exp \gamma .$

$\nu = |(\text{Cnv}` \downarrow \beta)``\mu . \varpi = |(\text{Cnv}` \downarrow \gamma)``\mu .$

[∗116·656] $\supset . (\exists \beta, \gamma) . \beta = \gamma . \nu = |(\text{Cnv}` \downarrow \beta)``\mu . \varpi = |(\text{Cnv}` \downarrow \gamma)``\mu .$

[∗13·195] $\supset . \nu = \varpi$ (2)

$\vdash . ∗21·33 . \supset \vdash :. \text{Hp} . \supset : \varpi T \mu . \varpi T \nu . \supset .$

$(\exists \beta, \gamma) . \beta, \gamma \epsilon \kappa . \mu \epsilon \alpha \exp \beta . \nu \epsilon \alpha \exp \gamma .$

$\varpi = |(\text{Cnv}` \downarrow \beta)``\mu = |(\text{Cnv}` \downarrow \gamma)``\nu .$

[*116·658] $\supset . (\exists \beta, \gamma) . \beta, \gamma \epsilon \kappa . \mu \epsilon \alpha \exp \beta . \nu \epsilon \alpha \exp \gamma .$
$\varpi = |(\text{Cnv}' \downarrow \beta)``\mu = |(\text{Cnv}' \downarrow \gamma)``\nu .$
$\varpi \epsilon \alpha \exp (\epsilon \downarrow \beta) \cap \alpha \exp (\epsilon \downarrow \gamma) .$

[*116·656] $\supset . (\exists \beta, \gamma) . \beta, \gamma \epsilon \kappa . |(\text{Cnv}' \downarrow \beta)``\mu = |(\text{Cnv}' \downarrow \gamma)``\nu .$
$\epsilon \downarrow \beta = \epsilon \downarrow \gamma .$

[*85·601] $\supset . (\exists \beta) . \beta \epsilon \kappa . |(\text{Cnv}' \downarrow \beta)``\mu = |(\text{Cnv}' \downarrow \beta)``\nu .$

[*116·601.*72·441] $\supset . \mu = \nu$ (3)

$\vdash . (2) . (3) . \supset \vdash : \text{Hp} . \supset . T \epsilon 1 \to 1$ (4)

$\vdash . *116·658 . \supset \vdash :. \text{Hp} . \supset : \beta \epsilon \kappa . \supset . T``(\alpha \exp \beta) = \alpha \exp (\epsilon \downarrow \beta) :$

[*37·69] $\supset : T_\epsilon``(\alpha \exp)``\kappa = (\alpha \exp)``\epsilon \downarrow``\kappa$ (5)

$\vdash . (1) . (4) . (5) . *111·1 . \supset \vdash . \text{Prop}$

***116·66.** $\vdash . \text{Prod}`(\alpha \exp)``\kappa \text{ sm } \{\alpha \exp (\Sigma`\kappa)\}$

Dem.

$\vdash . *116·659 . *115·51 . \supset \vdash . \text{Prod}`(\alpha \exp)``\kappa \text{ sm Prod}`(\alpha \exp)``\epsilon \downarrow``\kappa$ (1)

$\vdash . *85·61 . *116·654 . \supset \vdash . \text{Prod}`(\alpha \exp)``\epsilon \downarrow``\kappa \text{ sm } \{\alpha \exp (s`\epsilon \downarrow``\kappa)\}$ (2)

$\vdash . (1) . (2) . *112·1 . \supset \vdash . \text{Prop}$

***116·661.** $\vdash . \Pi \text{Nc}`(\alpha \exp)``\kappa = (\text{Nc}`\alpha)^{\Sigma \text{Nc}`\kappa}$ [*116·66·657 . *115·12 . *112·101]

This proposition is an extension of *116·52.

The following propositions are concerned in proving *116·68, which is an extension of *116·54, where the α and β of that proposition are replaced by the members of a class κ.

***116·67.** $\vdash :. \rho = \hat{\lambda} \{(\exists \alpha) . \alpha \epsilon \kappa . \lambda = \alpha \downarrow_{,,} ``\gamma\} . \supset : \kappa \epsilon \text{Cls}^2 \text{ excl} . \supset . \rho \epsilon \text{Cls}^3 \text{ arithm}$

Dem.

$\vdash . *20·3 . \supset \vdash : \text{Hp} . \lambda, \mu \epsilon \rho . \exists ! \lambda \cap \mu . \supset .$
$(\exists \alpha, \beta) . \alpha, \beta \epsilon \kappa . \lambda = \alpha \downarrow_{,,} ``\gamma . \mu = \beta \downarrow_{,,} ``\gamma . \exists ! \lambda \cap \mu .$

[*37·6] $\supset . (\exists \alpha, \beta, z, w) . \alpha, \beta \epsilon \kappa . z, w \epsilon \gamma . \alpha \downarrow_{,,} z = \beta \downarrow_{,,} w . \lambda = \alpha \downarrow_{,,} ``\gamma .$
$\mu = \beta \downarrow_{,,} ``\gamma .$

[*55·262.*38·2] $\supset . (\exists \alpha, z, w) . \alpha \epsilon \kappa . z, w \epsilon \gamma . \lambda = \alpha \downarrow_{,,} ``\gamma . \mu = \alpha \downarrow_{,,} ``\gamma .$

[*13·172] $\supset . \lambda = \mu$ (1)

$\vdash . *37·6 . *40·11 . \supset$

$\vdash : \text{Hp} . \xi, \eta \epsilon s`\rho . \exists ! \xi \cap \eta . \supset .$
$(\exists \alpha, \beta, z, w) . \alpha, \beta \epsilon \kappa . z, w \epsilon \gamma . \xi = \alpha \downarrow_{,,} z . \eta = \beta \downarrow_{,,} w . \exists ! \xi \cap \eta .$

[*55·232.*38·2] $\supset . (\exists \alpha, \beta, z) . \alpha, \beta \epsilon \kappa . z \epsilon \gamma . \xi = \alpha \downarrow_{,,} z . \eta = \beta \downarrow_{,,} z . \exists ! \alpha \cap \beta$ (2)

$\vdash . (2) . *84·11 . \supset$

$\vdash : \text{Hp} . \kappa \epsilon \text{Cls}^2 \text{ excl} . \xi, \eta \epsilon s`\rho . \exists ! \xi \cap \eta . \supset . (\exists \alpha, \beta, z) . \xi = \alpha \downarrow_{,,} z . \eta = \beta \downarrow_{,,} z . \alpha = \beta .$

[*13·195·172] $\supset . \xi = \eta$ (3)

$\vdash . (1) . (3) . \supset \vdash . \text{Prop}$

***116·671.** $\vdash :. \sigma = \hat{\mu}\{(\exists z) . z \epsilon \gamma . \mu = \downarrow z ``` \kappa\} . \supset : \text{Hp} *116·67 . \supset . s`\rho = s`\sigma$

Dem.

$\vdash . *40·11 . \supset \vdash :. \text{Hp} *116·67 . \supset : \xi \epsilon s`\rho . \equiv . (\exists \alpha) . \alpha \epsilon \kappa . \xi \epsilon \alpha \downarrow_{,,} ``\gamma .$

[*38·3] $\equiv . (\exists \alpha, z) . \alpha \epsilon \kappa . z \epsilon \gamma . \xi = \downarrow z``\alpha .$

[*37·103] $\equiv . (\exists z) . z \epsilon \gamma . \xi \epsilon \downarrow z```\kappa .$

[*40·11] $\equiv . \xi \epsilon s`\hat{\mu}\{(\exists z) . z \epsilon \gamma . \mu = \downarrow z```\kappa\} :.$

$\supset \vdash . \text{Prop}$

***116·672.** $\vdash : \text{Hp} *116·671 . \kappa \epsilon \text{Cls}^2 \text{excl} . \Lambda \sim \epsilon \kappa . \supset . \sigma \epsilon \text{Cls}^2 \text{excl}$

Dem.

$\vdash . *37·103 . \supset \vdash : \text{Hp} . \mu, \nu \epsilon \sigma . \exists ! \mu \cap \nu . \supset .$

$(\exists z, w, \alpha, \beta) . z, w \epsilon \gamma . \alpha, \beta \epsilon \kappa . \downarrow z``\alpha = \downarrow w``\beta .$

$\mu = \downarrow z```\alpha . \nu = \downarrow w```\beta .$

[*55·262] $\supset . (\exists z, w, \alpha) . z, w \epsilon \gamma . \alpha \epsilon \kappa . \downarrow z``\alpha = \downarrow w``\alpha .$

$\mu = \downarrow z```\alpha . \nu = \downarrow w```\alpha .$

[*113·105.*38·2.Hp] $\supset . (\exists z, \alpha) . \mu = \downarrow z```\alpha . \nu = \downarrow z```\alpha .$

[*13·172] $\supset . \mu = \nu : \supset \vdash . \text{Prop}$

***116·673.** $\vdash : \text{Hp} *116·672 . \supset . \epsilon_\Delta`(\exp \gamma)``\kappa \text{ sm } \epsilon_\Delta`\epsilon_\Delta``\sigma$

Dem.

$\vdash . *38·131 . (*116·01) . \supset \vdash . \epsilon_\Delta`(\exp \gamma)``\kappa = \epsilon_\Delta`\hat{\xi}\{(\exists \alpha) . \alpha \epsilon \kappa . \xi = \text{Prod}`\alpha \downarrow_{,,} ``\gamma\}$

[*37·6] $= \epsilon_\Delta`\text{Prod}``\hat{\lambda}\{(\exists \alpha) . \alpha \epsilon \kappa . \lambda = \alpha \downarrow_{,,} ``\gamma\}$ (1)

$\vdash . (1) . *115·33 . *116·67 . \supset$

$\vdash : \text{Hp} . \supset . \epsilon_\Delta`(\exp \gamma)``\kappa \text{ sm } \epsilon_\Delta`s`\hat{\lambda}\{(\exists \alpha) . \alpha \epsilon \kappa . \lambda = \alpha \downarrow_{,,} ``\gamma\} .$

[*116·671] $\supset . \epsilon_\Delta`(\exp \gamma)``\kappa \text{ sm } \epsilon_\Delta`s`\sigma .$

[*85·44.*116·672] $\supset . \epsilon_\Delta`(\exp \gamma)``\kappa \text{ sm } \epsilon_\Delta`\epsilon_\Delta``\sigma : \supset \vdash . \text{Prop}$

***116·674.** $\vdash :. M = \hat{R}\hat{z}\{R = (\downarrow z) \| \text{Cnv}`(\downarrow z)_\epsilon\} . \supset :$

$(z) . M`z \epsilon 1 \to 1 . \text{D}`(M`z) \upharpoonright \epsilon_\Delta`\kappa = \epsilon_\Delta` \downarrow z```\kappa$

Dem.

$\vdash . *30·3 . \quad \supset \vdash : \text{Hp} . \supset . M`z = (\downarrow z) \| \text{Cnv}`(\downarrow z)_\epsilon$ (1)

$\vdash . *72·184 . *111·14 . \supset \vdash . \downarrow z \upharpoonright \kappa \epsilon (\downarrow z```\kappa) \overline{\text{sm}}\, \overline{\text{sm}}\, \kappa .$

[*114·51] $\supset \vdash . \{\downarrow z \| \text{Cnv}`(\downarrow z)_\epsilon\} \upharpoonright \epsilon_\Delta`\kappa \epsilon (\epsilon_\Delta` \downarrow z```\kappa) \overline{\text{sm}} (\epsilon_\Delta`\kappa)$ (2)

$\vdash . (1) . (2) . *73·03 . \supset \vdash . \text{Prop}$

***116·675.** $\vdash :. \text{Hp} *116·674 . \exists ! s`\kappa . \supset : \exists ! (M`w)``\epsilon_\Delta`\kappa \cap (M`v)``\epsilon_\Delta`\kappa . \supset . w = v$

Dem.

$\vdash . *116·674 . \supset \vdash :. \text{Hp} . \supset : \exists ! (M`w)``\epsilon_\Delta`\kappa \cap (M`v)``\epsilon_\Delta`\kappa . \supset .$

$\exists ! \epsilon_\Delta` \downarrow w```\kappa \cap \epsilon_\Delta` \downarrow v```\kappa .$

[*80·32] $\supset . \downarrow w```\kappa = \downarrow v```\kappa .$

[*40·38] $\supset . \downarrow w``s`\kappa = \downarrow v``s`\kappa .$

[*113·105.*38·2] $\supset . w = v :. \supset \vdash . \text{Prop}$

***116·676.** $\vdash : \mathrm{Hp} *116\cdot672\cdot675 \,.\supset .\, \mathrm{Prod}‘\mathrm{D}“\upharpoonright(\epsilon_\Delta‘\kappa)“M“\gamma \;\mathrm{sm}\; (\epsilon_\Delta‘\kappa) \exp \gamma \,.$
$\mathrm{D}“\upharpoonright(\epsilon_\Delta‘\kappa)“M“\gamma = \epsilon_\Delta“\sigma$

Dem.

$\vdash . *116\cdot674\cdot675 \,.\, *116\cdot45 \dfrac{\epsilon_\Delta‘\kappa, \gamma}{\gamma, \quad \delta} . \supset$

$\vdash : \mathrm{Hp} \,.\supset .\, \mathrm{Prod}‘\mathrm{D}“\upharpoonright(\epsilon_\Delta‘\kappa)“M“\gamma \;\mathrm{sm}\; (\epsilon_\Delta‘\kappa) \exp \gamma$ (1)

$\vdash . *116\cdot674 \,.\supset \vdash : \mathrm{Hp} \,.\supset .\, \mathrm{D}“\upharpoonright(\epsilon_\Delta‘\kappa)“M“\gamma = \hat{\mu}\{(\exists z) \,.\, z \in \gamma \,.\, \mu = \epsilon_\Delta‘\downarrow z“‘\kappa\}$

$[*37\cdot6.\mathrm{Hp}] \qquad = \epsilon_\Delta“\sigma$ (2)

$\vdash . (1) . (2) . \supset \vdash . \mathrm{Prop}$

***116·68.** $\vdash : \kappa \in \mathrm{Cls}^2 \,\mathrm{excl} \,.\supset .\, \epsilon_\Delta‘(\exp \gamma)“\kappa \;\mathrm{sm}\; (\epsilon_\Delta‘\kappa) \exp \gamma \,.$
$\Pi \mathrm{Nc}‘(\exp \gamma)“\kappa = (\Pi \mathrm{Nc}‘\kappa)^{\mathrm{Nc}‘\gamma}$

Dem.

$\vdash . *115\cdot12 \,.\, *84\cdot55 \,.\supset \vdash .\, \mathrm{Prod}‘\epsilon_\Delta“\sigma \;\mathrm{sm}\; \epsilon_\Delta‘\epsilon_\Delta“\sigma$ (1)

$\vdash . (1) \,.\, *116\cdot673\cdot676 \,.\supset$

$\vdash : \mathrm{Hp} \,.\, \Lambda \sim\in \kappa \,.\, \exists ! s‘\kappa \,.\supset .\, \epsilon_\Delta‘(\exp \gamma)“\kappa \;\mathrm{sm}\; (\epsilon_\Delta‘\kappa) \exp \gamma$ (2)

$\vdash . *53\cdot24 \,.\supset \vdash : \Lambda \sim\in \kappa \,.\, \sim\exists ! s‘\kappa \,.\supset .\, \kappa = \Lambda \,.$

$[*83\cdot15.*116\cdot33] \qquad \supset .\, \epsilon_\Delta‘(\exp \gamma)“\kappa = \iota‘\dot{\Lambda} \,.\, (\epsilon_\Delta‘\kappa) \exp \gamma \in 1 \,.$

$[*73\cdot45] \qquad \supset .\, \epsilon_\Delta‘(\exp \gamma)“\kappa \;\mathrm{sm}\; (\epsilon_\Delta‘\kappa) \exp \gamma$ (3)

$\vdash . *83\cdot11 \,.\, *116\cdot182 \,.\supset \vdash : \Lambda \in \kappa \,.\, \exists ! \gamma \,.\supset .\, \epsilon_\Delta‘\kappa = \Lambda \,.\, \Lambda \in (\exp \gamma)“\kappa \,.$

$[*116\cdot182.*83\cdot11] \qquad \supset . (\epsilon_\Delta‘\kappa) \exp \gamma = \Lambda \,.\, \epsilon_\Delta‘(\exp \gamma)“\kappa = \Lambda$ (4)

$\vdash . *116\cdot181 \,.\supset \vdash : \Lambda \in \kappa \,.\, \gamma = \Lambda \,.\supset .\, (\epsilon_\Delta‘\kappa) \exp \gamma = \iota‘\Lambda$ (5)

$\vdash . *116\cdot181 \,.\supset \vdash : \Lambda \in \kappa \,.\, \gamma = \Lambda \,.\supset .\, (\exp \gamma)“\kappa = \iota‘\iota‘\Lambda \,.$

$[*83\cdot41] \qquad \supset .\, \epsilon_\Delta‘(\exp \gamma)“\kappa \;\mathrm{sm}\; \iota‘\Lambda$ (6)

$\vdash . (4) . (5) . (6) \,.\supset \vdash : \Lambda \in \kappa \,.\supset .\, \epsilon_\Delta‘(\exp \gamma)“\kappa \;\mathrm{sm}\; (\epsilon_\Delta‘\kappa) \exp \gamma$ (7)

$\vdash . (2) . (3) . (7) \,.\, *114\cdot1 \,.\, *116\cdot25 \,.\supset \vdash . \mathrm{Prop}$

The above proposition is an extension of *116·54·55.

The following propositions are lemmas for

$$\mathrm{Nc}‘\mathrm{Cl}‘\alpha = 2^{\mathrm{Nc}‘\alpha}.$$

The proposition and its proof are due to Cantor.

***116·7.** $\vdash . \mathrm{Nc}‘\{(\iota‘\Lambda \cup \iota‘V) \uparrow \alpha\}_\Delta‘\alpha = 2^{\mathrm{Nc}‘\alpha}$

Dem.

$\vdash . *24\cdot1 \,.\, *101\cdot3 \,.\supset \vdash . \mathrm{Nc}‘(\iota‘\Lambda \cup \iota‘V) = 2$ (1)

$\vdash . *116\cdot15 \,.\supset \vdash .\, \mathrm{Nc}‘\{(\iota‘\Lambda \cup \iota‘V) \uparrow \alpha\}_\Delta‘\alpha = \mathrm{Nc}‘\{(\iota‘\Lambda \cup \iota‘V) \exp \alpha\}$

$[*116\cdot25] \qquad = \{\mathrm{Nc}‘(\iota‘\Lambda \cup \iota‘V)\}^{\mathrm{Nc}‘\alpha}$

$[(1)] \qquad = 2^{\mathrm{Nc}‘\alpha} \,.\supset \vdash . \mathrm{Prop}$

In this and following propositions, the class $\iota‘\Lambda \cup \iota‘V$ is introduced solely as a known class consisting of two terms. Any other class of two terms will serve equally well.

***116·71.** $\vdash : R \epsilon \{(\iota`\Lambda \cup \iota`V) \uparrow \alpha\}_\Delta`\alpha . \supset . \overleftarrow{R}`V = \alpha - \overleftarrow{R}`\Lambda$

Dem.

$\vdash . *116·12 . \supset \vdash : Hp . \supset . R \epsilon 1 \to Cls . D`R \subset \iota`\Lambda \cup \iota`V . \text{ᗡ}`R = \alpha \quad (1)$

$[*37·271] \quad \supset . \alpha = \breve{R}``(\iota`\Lambda \cup \iota`V)$

$[*53·302] \quad = \overleftarrow{R}`\Lambda \cup \overleftarrow{R}`V \quad (2)$

$\vdash . (1) . *71·18 . \supset \vdash : Hp . \supset . \overleftarrow{R}`\Lambda \cap \overleftarrow{R}`V = \Lambda \quad (3)$

$\vdash . (2) . (3) . *24·47 . \supset \vdash . Prop$

***116·711.** $\vdash : R, S \epsilon \{(\iota`\Lambda \cup \iota`V) \uparrow \alpha\}_\Delta`\alpha . \overleftarrow{R}`\Lambda = \overleftarrow{S}`\Lambda . \supset . R = S$

Dem.

$\vdash . *116·71 . \quad \supset \vdash : Hp . \supset . \overleftarrow{R}`V = \overleftarrow{S}`V \quad (1)$

$\vdash . (1) . *116·12 . \supset \vdash :. Hp . \supset : \gamma \epsilon D`R \cup D`S . \supset_\gamma . \overleftarrow{R}`\gamma = \overleftarrow{S}`\gamma :$

$[*33·48] \quad \supset : R = S :. \supset \vdash . Prop$

***116·712.** $\vdash :. T = \hat{\mu}\hat{R}[R \epsilon \{(\iota`\Lambda \cup \iota`V) \uparrow \alpha\}_\Delta`\alpha . \mu = \overleftarrow{R}`\Lambda] . \supset :$

$R \epsilon \{(\iota`\Lambda \cup \iota`V) \uparrow \alpha\}_\Delta`\alpha . \supset . T`R = \overleftarrow{R}`\Lambda : \text{ᗡ}`T = \{(\iota`\Lambda \cup \iota`V) \uparrow \alpha\}_\Delta`\alpha$

Dem.

$\vdash . *21·33 . \supset \vdash :: Hp . \supset :. R \epsilon \{(\iota`\Lambda \cup \iota`V) \uparrow \alpha\}_\Delta`\alpha . \supset : \mu T R . \equiv_\mu . \mu = \overleftarrow{R}`\Lambda :$

$[*30·3] \quad \supset : T`R = \overleftarrow{R}`\Lambda \quad (1)$

$\vdash . (1) . *14·21 . \supset \vdash :. Hp . \supset : R \epsilon \{(\iota`\Lambda \cup \iota`V) \uparrow \alpha\}_\Delta`\alpha . \supset . E ! T`R .$

$[*33·43] \quad \supset . R \epsilon \text{ᗡ}`T \quad (2)$

$\vdash . *21·33 . *33·131 . \supset \vdash : Hp . \supset . \text{ᗡ}`T \subset \{(\iota`\Lambda \cup \iota`V) \uparrow \alpha\}_\Delta`\alpha \quad (3)$

$\vdash . (1) . (2) . (3) . \supset \vdash . Prop$

***116·713.** $\vdash : Hp *116·712 . \supset . T \epsilon 1 \to 1$

Dem.

$\vdash . *116·712 . *14·21 . \supset \vdash :. Hp . \supset : R \epsilon \text{ᗡ}`T . \supset . E ! T`R :$

$[*71·16] \quad \supset : R \epsilon 1 \to Cls \quad (1)$

$\vdash . *116·712·711 . \quad \supset \vdash :. Hp . \supset : R, S \epsilon \text{ᗡ}`T . T`R = T`S . \supset . R = S \quad (2)$

$\vdash . (1) . (2) . *71·55 . \quad \supset \vdash . Prop$

***116·714.** $\vdash : Hp *116·712 . \mu \epsilon Cl`\alpha . R = \hat{\gamma}\hat{x}\{\gamma = \Lambda . x \epsilon \mu . \vee . \gamma = V . x \epsilon \alpha - \mu\} . \supset .$

$R \epsilon \{(\iota`\Lambda \cup \iota`V) \uparrow \alpha\}_\Delta`\alpha . \mu = T`R$

Dem.

$\vdash . *21·33 . *33·13 . \quad \supset \vdash :. Hp . \supset : \gamma \epsilon D`R . \supset . \gamma \epsilon \iota`\Lambda \cup \iota`V \quad (1)$

$\vdash . *21·33 . *33·131 . \supset \vdash :: Hp . \supset :. x \epsilon \text{ᗡ}`R . \equiv :$

$(\exists \gamma) : \gamma = \Lambda . x \epsilon \mu . \vee . \gamma = V . x \epsilon \alpha - \mu :$

$[*10·42 . *13·19] \equiv : x \epsilon \mu . \vee . x \epsilon \alpha - \mu : \quad (2)$

$[*24·411 . Hp] \quad \equiv : x \epsilon \alpha \quad (3)$

$\vdash . *21\cdot33 . *30\cdot3 . \supset \vdash :. \mathrm{Hp} . \supset : x \in \mu . \supset_x . R`x = \Lambda : x \in \alpha - \mu . \supset_x . R`x = \mathrm{V} :$

[(2).*14·21] $\supset : x \in \text{Cl}`R . \supset_x . \mathrm{E}\,!\,R`x :$

[*71·16] $\supset : R \in 1 \rightarrow \mathrm{Cls}$ (4)

$\vdash . *21\cdot33 . \quad \supset \vdash :: \mathrm{Hp} . \supset :. \gamma = \Lambda . \supset : \gamma R x . \equiv_x . x \in \mu :$

[*32·181] $\supset : \overleftarrow{R}`\gamma = \mu$ (5)

$\vdash . (1) . (3) . (4) . *116\cdot12 . \supset \vdash : \mathrm{Hp} . \supset . R \in \{(\iota`\Lambda \cup \iota`\mathrm{V}) \uparrow \alpha\}_\Delta`\alpha$ (6)

$\vdash . (5) . (6) . *116\cdot712 . \quad \supset \vdash : \mathrm{Hp} . \supset . \mu = T`R$ (7)

$\vdash . (6) . (7) . \supset \vdash . \mathrm{Prop}$

***116·715.** $\vdash : \mathrm{Hp} *116\cdot712 . \supset . \mathrm{D}`T = \mathrm{Cl}`\alpha$

Dem.

$\vdash . *116\cdot714 . *33\cdot43 . \supset \vdash : \mathrm{Hp} . \supset . \mathrm{Cl}`\alpha \subset \mathrm{D}`T$ (1)

$\vdash . *21\cdot33 . *33\cdot13 . \supset$

$\vdash :. \mathrm{Hp} . \supset : \mu \in \mathrm{D}`T . \supset . (\exists R) . R \in \{(\iota`\Lambda \cup \iota`\mathrm{V}) \uparrow \alpha\}_\Delta`\alpha . \mu = \overleftarrow{R}`\Lambda .$

[*33·151] $\supset . (\exists R) . R \in \{(\iota`\Lambda \cup \iota`\mathrm{V}) \uparrow \alpha\}_\Delta`\alpha . \mu \subset \text{Cl}`R .$

[*80·14] $\supset . \mu \subset \alpha$ (2)

$\vdash . (1) . (2) . \supset \vdash . \mathrm{Prop}$

***116·72.** $\vdash . \mathrm{Nc}`\mathrm{Cl}`\alpha = 2^{\mathrm{Nc}`\alpha}$

Dem.

$\vdash . *116\cdot712\cdot713\cdot715 . \supset \vdash . \mathrm{Cl}`\alpha \,\mathrm{sm}\, \{(\iota`\Lambda \cup \iota`\mathrm{V}) \uparrow \alpha\}_\Delta`\alpha$ (1)

$\vdash . (1) . *116\cdot7 . \supset \vdash . \mathrm{Prop}$

***116·8.** $\vdash . \mathrm{Rl}`(\rho \uparrow \sigma) = \dot{s}``\mathrm{Cl}`(\sigma \times \rho)$

Dem.

$\vdash . *60\cdot2 . \supset \vdash :. R \in \dot{s}``\mathrm{Cl}`(\sigma \times \rho) . \equiv : (\exists \lambda) . \lambda \subset \sigma \times \rho . R = \dot{s}`\lambda :$

[*113·101] $\equiv : (\exists \lambda) : P \in \lambda . \supset_P . (\exists x, y) . x \in \rho . y \in \sigma . P = x \downarrow y : R = \dot{s}`\lambda :$

[*41·11] $\equiv : (\exists \lambda) : P \in \lambda . \supset_P . (\exists x, y) . x \in \rho . y \in \sigma . P = x \downarrow y :$

$uRv . \equiv_{u,v} . (\exists P) . P \in \lambda . uPv :$

[*10·56] $\supset : uRv . \supset_{u,v} . (\exists x, y) . x \in \rho . y \in \sigma . u (x \downarrow y) v :$

[*55·13] $\supset : uRv . \supset_{u,v} . u \in \rho . v \in \sigma :$

[*35·103] $\supset : R \mathrel{\dot{\subset}} \rho \uparrow \sigma$ (1)

$\vdash . *35\cdot103 . *113\cdot101 . \supset$

$\vdash : R \mathrel{\dot{\subset}} \rho \uparrow \sigma . \lambda = \hat{P} \{(\exists x, y) . xRy . P = x \downarrow y\} . \supset . \lambda \subset \sigma \times \rho$ (2)

$\vdash . *41\cdot11 . *13\cdot195 . \supset$

$\vdash :. \mathrm{Hp}\,(2) . \supset : u (\dot{s}`\lambda) v . \equiv . (\exists x, y) . xRy . u (x \downarrow y) v .$

[*55·13] $\equiv . uRv$ (3)

$\vdash . (2) . (3) . \supset \vdash : R \mathrel{\dot{\subset}} \rho \uparrow \sigma . \supset . (\exists \lambda) . \lambda \subset \sigma \times \rho . R = \dot{s}`\lambda$ (4)

$\vdash . (1) . (4) . \supset \vdash . \mathrm{Prop}$

$*116{\cdot}81$. $\vdash . \dot{s} \upharpoonright \mathrm{Cl}\text{‘}(\sigma \times \rho) \in 1 \to 1$

Dem.

$\vdash . *41{\cdot}13 . \supset \vdash :. \alpha, \beta \in \mathrm{Cl}\text{‘}(\sigma \times \rho) . \dot{s}\text{‘}\alpha = \dot{s}\text{‘}\beta . x \downarrow y \in \alpha . \supset : x \downarrow y \subseteq \dot{s}\text{‘}\beta :$

$[*41{\cdot}11] \quad \supset : (\exists P) . P \in \beta . x \downarrow y \subseteq P :$

$[*113{\cdot}101 . \mathrm{Hp}] \supset : (\exists P, u, v) . P \in \beta . P = u \downarrow v . x \downarrow y \subseteq u \downarrow v :$

$[*55{\cdot}134{\cdot}34] \quad \supset : (\exists P, u, v) . P \in \beta . P = u \downarrow v . x \downarrow y = u \downarrow v :$

$[*13{\cdot}172{\cdot}13] \quad \supset : x \downarrow y \in \beta \qquad (1)$

$\vdash . (1) \dfrac{\beta, \alpha}{\alpha, \beta} . \supset \vdash : \alpha, \beta \in \mathrm{Cl}\text{‘}(\sigma \times \rho) . \dot{s}\text{‘}\alpha = \dot{s}\text{‘}\beta . x \downarrow y \in \beta . \supset . x \downarrow y \in \alpha \qquad (2)$

$\vdash . (1) . (2) . *113{\cdot}101 . \supset \vdash : \alpha, \beta \in \mathrm{Cl}\text{‘}(\sigma \times \rho) . \dot{s}\text{‘}\alpha = \dot{s}\text{‘}\beta . \supset . \alpha = \beta \qquad (3)$

$\vdash . (3) . *71{\cdot}55 . *72{\cdot}163 . \supset \vdash . \mathrm{Prop}$

$*116{\cdot}82$. $\vdash . \mathrm{Rl}\text{‘}(\rho \uparrow \sigma) \,\mathrm{sm}\, \mathrm{Cl}\text{‘}(\sigma \times \rho) \qquad [*116{\cdot}8{\cdot}81]$

$*116{\cdot}83$. $\vdash . \mathrm{Nc}\text{‘}\mathrm{Rl}\text{‘}(\rho \uparrow \sigma) = 2^{\mathrm{Nc}\text{‘}\rho \times_c \mathrm{Nc}\text{‘}\sigma} \qquad [*116{\cdot}82{\cdot}72 . *113{\cdot}25]$

$*116{\cdot}9$. $\vdash : \mathrm{Nc}\text{‘}t\text{‘}x = \mu . \supset . \mathrm{Nc}\text{‘}t^2\text{‘}x = 2^\mu \qquad [*116{\cdot}72 . *63{\cdot}66]$

$*116{\cdot}901$. $\vdash : \mathrm{Nc}\text{‘}t_0\text{‘}\alpha = \mu . \supset . \mathrm{Nc}\text{‘}t\text{‘}\alpha = 2^\mu \qquad [*116{\cdot}72 . *63{\cdot}65]$

$*116{\cdot}91$. $\vdash : \mathrm{Nc}\text{‘}t_0\text{‘}\alpha = \mu . \supset . \mathrm{Nc}\text{‘}t_{00}\text{‘}\alpha = 2^{\mu^2} \qquad [*116{\cdot}83 . *64{\cdot}5{\cdot}11]$

$*116{\cdot}92$. $\vdash : \mathrm{Nc}\text{‘}t_0\text{‘}\alpha = \mu . \supset . \mathrm{Nc}\text{‘}t_0{}^1\text{‘}\alpha = 2^{\mu \times_c 2^\mu} . \mathrm{Nc}\text{‘}t^{11}\text{‘}\alpha = 2^{2^\mu \times_c 2^\mu} .$ etc.
$[*116{\cdot}83 . *64{\cdot}16 . *116{\cdot}901]$

$*117$. GREATER AND LESS.

Summary of $*117$.

A cardinal μ is said to be greater than another cardinal ν when there is a class α which has μ terms and has a part which has ν terms, while there is no class β which has ν terms and has a part which has μ terms. The relation "greater than" is transitive and asymmetrical; and by the Schröder-Bernstein theorem, if μ is greater than or equal to ν, and ν is greater than or equal to μ, then $\mu = \nu$. But we cannot prove that of any two cardinals one must be the greater, unless we assume the multiplicative axiom. The proof then follows from Zermelo's theorem that on that assumption every class can be well-ordered. This subject will be dealt with at a later stage.

The form of the definitions is so arranged as to allow of the inequality of two cardinals in different types. The relevant considerations are the same as for the definitions of addition, multiplication and exponentiation.

Our definition of "$\mu > \nu$" is

$*117{\cdot}01$. $\mu > \nu \,.=.\, (\exists \alpha, \beta) \,.\, \mu = \mathrm{N_0c}`\alpha \,.\, \nu = \mathrm{N_0c}`\beta \,.$
$\exists ! \,\mathrm{Cl}`\alpha \cap \mathrm{Nc}`\beta \,.\, \sim \exists ! \,\mathrm{Cl}`\beta \cap \mathrm{Nc}`\alpha$ Df

We also define "$\mu > \mathrm{Nc}`\alpha$" as meaning "$\mu > \mathrm{N_0c}`\alpha$," and "$\mathrm{Nc}`\alpha > \nu$" as meaning "$\mathrm{N_0c}`\alpha > \nu$," for the reasons explained in $*110$. It then easily follows that if $\mu > \nu$, μ and ν must be homogeneous cardinals (this is part of $*117{\cdot}15$); that if μ and ν are homogeneous cardinals, and $\mu > \nu$, the same holds if we substitute $\mathrm{sm}``\mu$ and $\mathrm{sm}``\nu$ for one or both of μ and ν ($*117{\cdot}16$); that

$*117{\cdot}13$. $\vdash : \mathrm{Nc}`\alpha > \mathrm{Nc}`\beta \,.\equiv.\, \exists ! \,\mathrm{Cl}`\alpha \cap \mathrm{Nc}`\beta \,.\, \sim \exists ! \,\mathrm{Cl}`\beta \cap \mathrm{Nc}`\alpha$

and that

$*117{\cdot}14$. $\vdash : \mu > \nu \,.\equiv.\, (\exists \alpha, \beta) \,.\, \mu = \mathrm{N_0c}`\alpha \,.\, \nu = \mathrm{N_0c}`\beta \,.\, \mathrm{Nc}`\alpha > \mathrm{Nc}`\beta$

We cannot define "$\mu \geqslant \nu$" as "$\mu > \nu \,.\vee.\, \mu = \nu$," because "$\mu = \nu$" restricts μ and ν too much by requiring that they should be of the same type, and restricts them too little by not requiring that they should both be existent cardinals. To avoid both these inconveniences, we put

$*117{\cdot}05$. $\mu \geqslant \nu \,.=:\, \mu > \nu \,.\vee.\, \mu, \nu \in \mathrm{N_0C} \,.\, \mu = \mathrm{sm}``\nu$ Df

The use of this definition is chiefly through the propositions

$*117{\cdot}108$. $\vdash :. \mathrm{Nc}`\alpha \geqslant \mathrm{Nc}`\beta \,.\equiv:\, \mathrm{Nc}`\alpha > \mathrm{Nc}`\beta \,.\vee.\, \mathrm{Nc}`\alpha = \mathrm{Nc}`\beta$

$*117{\cdot}24$. $\vdash : \mu \geqslant \nu \,.\equiv.\, (\exists \alpha, \beta) \,.\, \mu = \mathrm{N_0c}`\alpha \,.\, \nu = \mathrm{N_0c}`\beta \,.\, \mathrm{Nc}`\alpha \geqslant \mathrm{Nc}`\beta$

In $*117\cdot2$, we repeat the Schröder-Bernstein theorem ($*73\cdot88$), which is required in most of the remaining propositions of this number. It leads at once to the propositions

$*117\cdot22$. $\vdash : \exists ! \text{Cl}`\alpha \cap \text{Nc}`\beta \,.\equiv .\, \text{Nc}`\alpha \geqslant \text{Nc}`\beta$

(which practically supersedes the definition of "$\geqslant$")

$*117\cdot221$. $\vdash : \text{Nc}`\alpha \geqslant \text{Nc}`\beta \,.\equiv .\, (\exists\rho) \,.\, \rho \subset \alpha \,.\, \rho \,\text{sm}\, \beta$

$*117\cdot222$. $\vdash : \beta \subset \alpha \,.\supset .\, \text{Nc}`\alpha \geqslant \text{Nc}`\beta$

$*117\cdot23$. $\vdash : \text{Nc}`\alpha \geqslant \text{Nc}`\beta \,.\, \text{Nc}`\beta \geqslant \text{Nc}`\alpha \,.\equiv .\, \text{Nc}`\alpha = \text{Nc}`\beta$

This last proposition may be called the Schröder-Bernstein theorem with as much propriety as $*73\cdot88$; the two are scarcely different.

If we now revert to the definition of $\mu > \nu$, or to $*117\cdot13$, and apply $*117\cdot22$, we see ($*117\cdot26$) that "$\text{Nc}`\alpha > \text{Nc}`\beta$" may be conveniently regarded as asserting $\text{Nc}`\alpha \geqslant \text{Nc}`\beta \,.\, \sim(\text{Nc}`\beta \geqslant \text{Nc}`\alpha)$; in fact, the best ideas to work with are $\geqslant$ and its converse $\leqslant$, which for practical purposes we regard as defined by $*117\cdot22$, and from which we derive $>$ and $<$. The relation $>$ will be the product of $\geqslant$ into the negation of its converse; this holds for μ and ν ($*117\cdot281$) as well as for $\text{Nc}`\alpha$ and $\text{Nc}`\beta$.

$*117\cdot3\cdot31$ constitute an important use of $*110\cdot72$, namely to prove that one existent cardinal is greater than another or equal to it when the first can be obtained by adding to the second (where what is added must be a cardinal). That is to say, we have

$*117\cdot3$. $\vdash : \text{Nc}`\alpha \geqslant \text{Nc}`\beta \,.\equiv .\, (\exists\varpi) \,.\, \varpi \in \text{NC} \,.\, \text{Nc}`\alpha = \text{Nc}`\beta +_c \varpi$

$*117\cdot31$. $\vdash :. \mu \geqslant \nu \,.\equiv : \mu, \nu \in \text{N}_0\text{C} : (\exists\varpi) \,.\, \varpi \in \text{NC} \,.\, \mu = \nu +_c \varpi$

$*117\cdot4$—$\cdot471$ are concerned in proving that $>$ and $\geqslant$ are transitive, that $>$ is asymmetrical ($*117\cdot42$), and allied propositions.

Our next set of propositions is concerned with 0 and 1 and 2. We prove that a homogeneous cardinal is whatever is greater than or equal to 0 ($*117\cdot501$); that a homogeneous cardinal other than 0 is whatever is greater than 0 ($*117\cdot511$); that a homogeneous cardinal other than 0 is whatever is greater than or equal to 1 ($*117\cdot531$); and that a homogeneous cardinal other than 0 and 1 is whatever is greater than 1 ($*117\cdot55$), and is whatever is greater than or equal to 2 ($*117\cdot551$).

We next prove a set of propositions concerning $\geqslant$ which have no analogues for $>$, except when the cardinals concerned are finite. Thus *e.g.* we prove

$*117\cdot561$. $\vdash : \mu \geqslant \nu \,.\, \varpi \in \text{N}_0\text{C} \,.\supset .\, \mu +_c \varpi \geqslant \nu +_c \varpi$

If we substitute $>$ for $\geqslant$, this no longer holds. Thus *e.g.* put $\mu = 2$, $\nu = 1$, $\varpi = \aleph_0$ (cf. $*123$); then $\mu > \nu$, but $\mu +_c \varpi = \nu +_c \varpi = \varpi$. Similar remarks apply to the analogous propositions ($*117\cdot571\cdot581\cdot591$) on multiplication and exponentiation.

We prove next that a sum is greater than or equal to either of its summands (*117·6); that a product neither of whose factors vanishes is greater than or equal to either of its factors (*117·62); that, assuming μ and ν are existent cardinals, then if they are neither 0 nor 1, their product is greater than or equal to their sum (*117·631), and if μ is neither 0 nor 1, then $\mu^\nu \geqslant \mu \times_c \nu$ (*117·652).

The last important proposition in this number is Cantor's theorem

***117·661.** $\vdash : \mu \in N_0C . \supset . 2^\mu > \mu$

which follows immediately from *102·72 and *116·72.

The propositions of this number are much used in the following section, on finite and infinite.

***117·01.** $\mu > \nu . = . (\exists \alpha, \beta) . \mu = N_0c`\alpha . \nu = N_0c`\beta .$
$\exists ! Cl`\alpha \cap Nc`\beta . \sim \exists ! Cl`\beta \cap Nc`\alpha$ Df

***117·02.** $\mu > Nc`\alpha . = . \mu > N_0c`\alpha$ Df

***117·03.** $Nc`\alpha > \nu . = . N_0c`\alpha > \nu$ Df

***117·04.** $\mu < \nu . = . \nu > \mu$ Df

***117·05.** $\mu \geqslant \nu . = : \mu > \nu . \vee . \mu, \nu \in N_0C . \mu = sm``\nu$ Df

***117·06.** $\mu \leqslant \nu . = . \nu \geqslant \mu$ Df

The analogues of *117·02·03 are to be applied also to *117·04·05·06.

***117·1.** $\vdash : \mu > \nu . \equiv . (\exists \alpha, \beta) . \mu = N_0c`\alpha . \nu = N_0c`\beta .$
$\exists ! Cl`\alpha \cap Nc`\beta . \sim \exists ! Cl`\beta \cap Nc`\alpha$ [(*117·01)]

***117·101.** $\vdash : \mu > Nc`\beta . \equiv . \mu > N_0c`\beta$ [(*117·02)]

***117·102.** $\vdash : Nc`\alpha > \nu . \equiv . N_0c`\alpha > \nu$ [(*117·03)]

***117·103.** $\vdash : \mu < \nu . \equiv . \nu > \mu$ [(*117·04)]

***117·104.** $\vdash :. \mu \geqslant \nu . \equiv : \mu > \nu . \vee . \mu, \nu \in N_0C . \mu = sm``\nu$ [(*117·05)]

***117·105.** $\vdash : \mu \leqslant \nu . \equiv . \nu \geqslant \mu$ [(*117·06)]

***117·106.** $\vdash : Nc`\alpha > Nc`\beta . \equiv . N_0c`\alpha > N_0c`\beta$ [*117·101·102]

***117·107.** $\vdash : Nc`\alpha \geqslant Nc`\beta . \equiv . N_0c`\alpha \geqslant N_0c`\beta$

Dem.

$\vdash . *117·104·106 . \supset$

$\vdash :. Nc`\alpha \geqslant Nc`\beta . \equiv : N_0c`\alpha > N_0c`\beta . \vee . Nc`\alpha, Nc`\beta \in N_0C . Nc`\alpha = sm``Nc`\beta :$

[*100·511.*103·22] $\equiv : N_0c`\alpha > N_0c`\beta . \vee . Nc`\alpha, Nc`\beta \in N_0C . Nc`\alpha = Nc`\beta :$

[*103·16] $\equiv : N_0c`\alpha > N_0c`\beta . \vee . Nc`\alpha, Nc`\beta \in N_0C . Nc`\alpha = N_0c`\beta :$

[*103·21] $\equiv : N_0c`\alpha > N_0c`\beta . \vee . Nc`\beta \in N_0C . Nc`\alpha = N_0c`\beta :$

[*103·16] $\equiv : N_0c`\alpha > N_0c`\beta . \vee . Nc`\beta \in N_0C . N_0c`\alpha = Nc`\beta :$

[*103·2] $\equiv : N_0c`\alpha > N_0c`\beta . \vee . N_0c`\alpha = Nc`\beta :$

[*103·4] $\equiv : N_0c`\alpha > N_0c`\beta . \vee . N_0c`\alpha = sm``N_0c`\beta :$

[*103·21.*117·104] $\equiv : N_0c`\alpha \geqslant N_0c`\beta :. \supset \vdash . Prop$

***117·108.** $\vdash :. \text{Nc}`\alpha \geqslant \text{Nc}`\beta . \equiv : \text{Nc}`\alpha > \text{Nc}`\beta . \vee . \text{Nc}`\alpha = \text{Nc}`\beta$
[*117·107·106·104 . *103·16·4]

***117·11.** $\vdash :. \alpha \,\text{sm}\, \alpha' . \beta \,\text{sm}\, \beta' . \supset : \exists ! \text{Cl}`\alpha \cap \text{Nc}`\beta . \equiv . \exists ! \text{Cl}`\alpha' \cap \text{Nc}`\beta'$

Dem.

$\vdash . *100·321 . \supset \vdash :. \text{Hp} . \supset : \exists ! \text{Cl}`\alpha' \cap \text{Nc}`\beta . \equiv . \exists ! \text{Cl}`\alpha' \cap \text{Nc}`\beta'$ (1)

$\vdash . *73·21 . \supset$

$\vdash : R \in 1 \rightarrow 1 . \text{D}`R = \alpha . \text{Ꟁ}`R = \alpha' . \gamma \subset \alpha . \gamma \in \text{Nc}`\beta . \supset .$
$\breve{R}``\gamma \subset \alpha' . \breve{R}``\gamma \in \text{Nc}`\beta .$

[*60·2] $\supset . \exists ! \text{Cl}`\alpha' \cap \text{Nc}`\beta$ (2)

$\vdash . (2) . *10·11·23·35 . *73·1 . \supset$

$\vdash : \alpha \,\text{sm}\, \alpha' . \exists ! \text{Cl}`\alpha \cap \text{Nc}`\beta . \supset . \exists ! \text{Cl}`\alpha' \cap \text{Nc}`\beta$ (3)

$\vdash . (3) \frac{\alpha', \alpha}{\alpha, \alpha'} . \supset \vdash : \alpha \,\text{sm}\, \alpha' . \exists ! \text{Cl}`\alpha' \cap \text{Nc}`\beta . \supset . \exists ! \text{Cl}`\alpha \cap \text{Nc}`\beta$ (4)

$\vdash . (3) . (4) . \supset \vdash :. \alpha \,\text{sm}\, \alpha' . \supset : \exists ! \text{Cl}`\alpha \cap \text{Nc}`\beta . \equiv . \exists ! \text{Cl}`\alpha' \cap \text{Nc}`\beta$ (5)

$\vdash . (1) . (5) . \supset \vdash . \text{Prop}$

***117·12.** $\vdash :. \mu > \nu . \equiv : \mu, \nu \in \text{N}_0\text{C} :$
$\gamma \in \mu . \delta \in \nu . \supset_{\gamma,\delta} . \exists ! \text{Cl}`\gamma \cap \text{Nc}`\delta . \sim \exists ! \text{Cl}`\delta \cap \text{Nc}`\gamma$

Dem.

$\vdash . *117·1·11 . \supset$

$\vdash :. \mu > \nu . \equiv : (\exists \alpha, \beta) : \mu = \text{N}_0\text{c}`\alpha . \nu = \text{N}_0\text{c}`\beta . \exists ! \text{Cl}`\alpha \cap \text{Nc}`\beta . \sim \exists ! \text{Cl}`\beta \cap \text{Nc}`\alpha :$
$\gamma \in \mu . \delta \in \nu . \supset_{\gamma,\delta} . \exists ! \text{Cl}`\gamma \cap \text{Nc}`\delta . \sim \exists ! \text{Cl}`\delta \cap \text{Nc}`\gamma :$

[*103·12] $\equiv : (\exists \alpha, \beta) : \mu = \text{N}_0\text{c}`\alpha . \nu = \text{N}_0\text{c}`\beta . \alpha \in \mu . \beta \in \nu . \exists ! \text{Cl}`\alpha \cap \text{Nc}`\beta .$
$\sim \exists ! \text{Cl}`\beta \cap \text{Nc}`\alpha :$
$\gamma \in \mu . \delta \in \nu . \supset_{\gamma,\delta} . \exists ! \text{Cl}`\gamma \cap \text{Nc}`\delta . \sim \exists ! \text{Cl}`\delta \cap \text{Nc}`\gamma :$

[*10·55] $\equiv : (\exists \alpha, \beta) : \mu = \text{N}_0\text{c}`\alpha . \nu = \text{N}_0\text{c}`\beta . \alpha \in \mu . \beta \in \nu :$
$\gamma \in \mu . \delta \in \nu . \supset_{\gamma,\delta} . \exists ! \text{Cl}`\gamma \cap \text{Nc}`\delta . \sim \exists ! \text{Cl}`\delta \cap \text{Nc}`\gamma :$

[*103·12·2] $\equiv : \mu, \nu \in \text{N}_0\text{C} : \gamma \in \mu . \delta \in \nu . \supset_{\gamma,\delta} . \exists ! \text{Cl}`\gamma \cap \text{Nc}`\delta . \sim \exists ! \text{Cl}`\delta \cap \text{Nc}`\gamma :.$
$\supset \vdash . \text{Prop}$

***117·121.** $\vdash :. \mu > \nu . \equiv : \mu, \nu \in \text{N}_0\text{C} :$
$\alpha \in \mu . \supset_\alpha . (\exists \beta) . \beta \in \nu . \exists ! \text{Cl}`\alpha \cap \text{Nc}`\beta . \sim \exists ! \text{Cl}`\beta \cap \text{Nc}`\alpha$

Dem.

$\vdash . *117·1·11 . \supset$

$\vdash :. \mu > \nu . \equiv : (\exists \alpha, \beta) : \mu = \text{N}_0\text{c}`\alpha . \nu = \text{N}_0\text{c}`\beta . \exists ! \text{Cl}`\alpha \cap \text{Nc}`\beta .$
$\sim \exists ! \text{Cl}`\beta \cap \text{Nc}`\alpha :$
$\gamma \in \mu . \supset_\gamma . (\exists \delta) . \delta \in \nu . \exists ! \text{Cl}`\gamma \cap \text{Nc}`\delta . \sim \exists ! \text{Cl}`\delta \cap \text{Nc}`\gamma$

[*103·12.*10·55] $\equiv : (\exists \alpha, \beta) : \mu = \text{N}_0\text{c}`\alpha . \nu = \text{N}_0\text{c}`\beta . \alpha \in \mu :$
$\gamma \in \mu . \supset_\gamma . (\exists \delta) . \delta \in \nu . \exists ! \text{Cl}`\gamma \cap \text{Nc}`\delta . \sim \exists ! \text{Cl}`\delta \cap \text{Nc}`\gamma$

[*103·12·2] $\equiv : \mu, \nu \in \text{N}_0\text{C} : \gamma \in \mu . \supset_\gamma . (\exists \delta) . \delta \in \nu . \exists ! \text{Cl}`\gamma \cap \text{Nc}`\delta .$
$\sim \exists ! \text{Cl}`\delta \cap \text{Nc}`\gamma :. \supset \vdash . \text{Prop}$

The above proof is given shortly because it proceeds on the same lines as $*117\cdot12$. In applying $*10\cdot55$, the ϕx of that proposition is replaced by $\alpha \in \mu$, and the ψx is replaced by

$$(\exists\beta) . \beta \in \nu . \exists ! \text{Cl}`\alpha \cap \text{Nc}`\beta . \sim\exists ! \text{Cl}`\beta \cap \text{Nc}`\alpha.$$

$*117\cdot13$. $\vdash : \text{Nc}`\alpha > \text{Nc}`\beta . \equiv . \exists ! \text{Cl}`\alpha \cap \text{Nc}`\beta . \sim\exists ! \text{Cl}`\beta \cap \text{Nc}`\alpha$

Dem.

$\vdash . *117\cdot106 . \supset$

$\vdash :. \text{Nc}`\alpha > \text{Nc}`\beta . \equiv : \text{N}_0\text{c}`\alpha > \text{N}_0\text{c}`\beta :$

$[*103\cdot2 . *117\cdot12] \quad \equiv : \gamma \in \text{N}_0\text{c}`\alpha . \delta \in \text{N}_0\text{c}`\beta . \supset_{\gamma,\delta} .$
$\exists ! \text{Cl}`\gamma \cap \text{Nc}`\delta . \sim\exists ! \text{Cl}`\delta \cap \text{Nc}`\gamma :$

$[*100\cdot31 . *117\cdot11] \equiv : \gamma \in \text{N}_0\text{c}`\alpha . \delta \in \text{N}_0\text{c}`\beta . \supset_{\gamma,\delta} .$
$\exists ! \text{Cl}`\alpha \cap \text{Nc}`\beta . \sim\exists ! \text{Cl}`\beta \cap \text{Nc}`\alpha :$

$[*10\cdot23] \quad \equiv : \exists ! \text{N}_0\text{c}`\alpha . \exists ! \text{N}_0\text{c}`\beta . \supset .$
$\exists ! \text{Cl}`\alpha \cap \text{Nc}`\beta . \sim\exists ! \text{Cl}`\beta \cap \text{Nc}`\alpha :$

$[*103\cdot13] \quad \equiv : \exists ! \text{Cl}`\alpha \cap \text{Nc}`\beta . \sim\exists ! \text{Cl}`\beta \cap \text{Nc}`\alpha :. \supset \vdash . \text{Prop}$

$*117\cdot14$. $\vdash : \mu > \nu . \equiv . (\exists\alpha, \beta) . \mu = \text{N}_0\text{c}`\alpha . \nu = \text{N}_0\text{c}`\beta . \text{Nc}`\alpha > \text{Nc}`\beta$
$[*117\cdot1\cdot13]$

$*117\cdot15$. $\vdash : \mu > \nu . \equiv . \mu, \nu \in \text{N}_0\text{C} . \exists ! s`\text{Cl}``\mu \cap \text{sm}``\nu . \sim\exists ! s`\text{Cl}``\nu \cap \text{sm}``\mu$

Dem.

$\vdash . *103\cdot4 . *117\cdot1 . \supset$

$\vdash :. \mu > \nu . \equiv : (\exists\alpha, \beta) . \mu = \text{N}_0\text{c}`\alpha . \nu = \text{N}_0\text{c}`\beta . \exists ! \text{Cl}`\alpha \cap \text{sm}``\nu .$
$\sim\exists ! \text{Cl}`\beta \cap \text{sm}``\mu :$

$[*103\cdot2\cdot26] \quad \equiv : \mu, \nu \in \text{N}_0\text{C} : (\exists\alpha, \beta) . \alpha \in \mu . \beta \in \nu . \exists ! \text{Cl}`\alpha \cap \text{sm}``\nu .$
$\sim\exists ! \text{Cl}`\beta \cap \text{sm}``\mu :$

$[*117\cdot11] \quad \equiv : \mu, \nu \in \text{N}_0\text{C} : (\exists\alpha, \beta) . \alpha \in \mu . \beta \in \nu . \exists ! \text{Cl}`\alpha \cap \text{sm}``\nu :$
$\delta \in \nu . \supset_\delta . \sim\exists ! \text{Cl}`\delta \cap \text{sm}``\mu :$

$[*103\cdot13 . *10\cdot51] \equiv : \mu, \nu \in \text{N}_0\text{C} : (\exists\alpha) . \alpha \in \mu . \exists ! \text{Cl}`\alpha \cap \text{sm}``\nu :$
$\sim (\exists\delta) . \delta \in \nu . \exists ! \text{Cl}`\delta \cap \text{sm}``\mu :$

$[*40\cdot4 . *60\cdot2] \quad \equiv : \mu, \nu \in \text{N}_0\text{C} . \exists ! s`\text{Cl}``\mu \cap \text{sm}``\nu .$
$\sim\exists ! s`\text{Cl}``\nu \cap \text{sm}``\mu :. \supset \vdash . \text{Prop}$

The advantage of this proposition is that it expresses "$\mu > \nu$" in terms of μ and ν alone, without the auxiliary α and β of the definition.

$*117\cdot16$. $\vdash :. \mu, \nu \in \text{N}_0\text{C} . \supset : \mu > \nu . \equiv . \text{sm}``\mu > \nu . \equiv . \mu > \text{sm}``\nu . \equiv . \text{sm}``\mu > \text{sm}``\nu$
$[*117\cdot14 . *103\cdot4]$

$*117\cdot2$. $\vdash : \alpha \text{ sm } \alpha' . \beta \text{ sm } \beta' . \beta' \subset \alpha . \alpha' \subset \beta . \supset . \alpha \text{ sm } \beta \quad [*73\cdot88]$

This proposition (which is the Schröder-Bernstein theorem) is fundamental in the theory of greater and less.

***117·21.** $\vdash : \exists ! \mathrm{Cl}'\alpha \cap \mathrm{Nc}'\beta \,.\, \exists ! \mathrm{Cl}'\beta \cap \mathrm{Nc}'\alpha \,.\supset .\, \mathrm{Nc}'\alpha = \mathrm{Nc}'\beta$
[*117·2 . *100·321]

***117·211.** $\vdash : \exists ! \mathrm{Cl}'\alpha \cap \mathrm{Nc}'\beta \,.\, \exists ! \mathrm{Cl}'\beta \cap \mathrm{Nc}'\alpha \,.\equiv .\, \mathrm{Nc}'\alpha = \mathrm{Nc}'\beta$

Dem.

$\vdash . *100{\cdot}3 . *60{\cdot}34 . \supset \vdash : \mathrm{Nc}'\alpha = \mathrm{Nc}'\beta \,.\supset .\, \alpha \in \mathrm{Cl}'\alpha \cap \mathrm{Nc}'\beta \,.\, \beta \in \mathrm{Cl}'\beta \cap \mathrm{Nc}'\alpha$ (1)

$\vdash . (1) . *117{\cdot}21 . \supset \vdash . \mathrm{Prop}$

***117·22.** $\vdash : \exists ! \mathrm{Cl}'\alpha \cap \mathrm{Nc}'\beta \,.\equiv .\, \mathrm{Nc}'\alpha \geqslant \mathrm{Nc}'\beta$

Dem.

$\vdash . *117{\cdot}13 . \supset \vdash : \mathrm{Hp} . \sim \exists ! \mathrm{Cl}'\beta \cap \mathrm{Nc}'\alpha \,.\equiv .\, \mathrm{Nc}'\alpha > \mathrm{Nc}'\beta$ (1)

$\vdash . *117{\cdot}211 . \supset \vdash : \mathrm{Hp} . \exists ! \mathrm{Cl}'\beta \cap \mathrm{Nc}'\alpha \,.\equiv .\, \mathrm{Nc}'\alpha = \mathrm{Nc}'\beta$ (2)

$\vdash . (1) . (2) . *117{\cdot}108 . \supset \vdash . \mathrm{Prop}$

***117·221.** $\vdash : \mathrm{Nc}'\alpha \geqslant \mathrm{Nc}'\beta \,.\equiv .\, (\exists \rho) . \rho \subset \alpha . \rho \,\mathrm{sm}\, \beta$ [*117·22 . *60·2 . *100·1]

***117·222.** $\vdash : \beta \subset \alpha \,.\supset .\, \mathrm{Nc}'\alpha \geqslant \mathrm{Nc}'\beta$ [*117·221]

***117·23.** $\vdash : \mathrm{Nc}'\alpha \geqslant \mathrm{Nc}'\beta \,.\, \mathrm{Nc}'\beta \geqslant \mathrm{Nc}'\alpha \,.\equiv .\, \mathrm{Nc}'\alpha = \mathrm{Nc}'\beta$ [*117·211·22]

***117·24.** $\vdash : \mu \geqslant \nu \,.\equiv .\, (\exists \alpha, \beta) . \mu = \mathrm{N_0c}'\alpha . \nu = \mathrm{N_0c}'\beta . \mathrm{Nc}'\alpha \geqslant \mathrm{Nc}'\beta$

Dem.

$\vdash . *117{\cdot}104{\cdot}14 . \supset \vdash :. \mu \geqslant \nu . \equiv : (\exists \alpha, \beta) . \mu = \mathrm{N_0c}'\alpha . \nu = \mathrm{N_0c}'\beta . \mathrm{Nc}'\alpha > \mathrm{Nc}'\beta . \mathsf{v} .$
$(\exists \alpha, \beta) . \mu = \mathrm{N_0c}'\alpha . \nu = \mathrm{N_0c}'\beta . \mu = \mathrm{sm}''\nu :$

[*103·4.*13·193] $\equiv : (\exists \alpha, \beta) . \mu = \mathrm{N_0c}'\alpha . \nu = \mathrm{N_0c}'\beta . \mathrm{Nc}'\alpha > \mathrm{Nc}'\beta . \mathsf{v} .$
$(\exists \alpha, \beta) . \mu = \mathrm{N_0c}'\alpha . \nu = \mathrm{N_0c}'\beta . \mathrm{N_0c}'\alpha = \mathrm{Nc}'\beta :$

[*103·16] $\equiv : (\exists \alpha, \beta) . \mu = \mathrm{N_0c}'\alpha . \nu = \mathrm{N_0c}'\beta . \mathrm{Nc}'\alpha > \mathrm{Nc}'\beta . \mathsf{v} .$
$(\exists \alpha, \beta) . \mu = \mathrm{N_0c}'\alpha . \nu = \mathrm{N_0c}'\beta . \mathrm{Nc}'\alpha = \mathrm{Nc}'\beta :$

[*11·41.*117·108] $\equiv : (\exists \alpha, \beta) . \mu = \mathrm{N_0c}'\alpha . \nu = \mathrm{N_0c}'\beta . \mathrm{Nc}'\alpha \geqslant \mathrm{Nc}'\beta :.$
$\supset \vdash . \mathrm{Prop}$

***117·241.** $\vdash : \mu \geqslant \nu \,.\equiv .\, (\exists \alpha, \beta) . \mu = \mathrm{N_0c}'\alpha . \nu = \mathrm{N_0c}'\beta . \exists ! \mathrm{Cl}'\alpha \cap \mathrm{Nc}'\beta$
[*117·24·22]

***117·242.** $\vdash :. \mu, \nu \in \mathrm{NC} . \supset : \mu \geqslant \nu \,.\equiv .\, (\exists \alpha, \beta) . \alpha \in \mu . \beta \in \nu . \exists ! \mathrm{Cl}'\alpha \cap \mathrm{Nc}'\beta$
[*117·241 . *103·26]

***117·243.** $\vdash :. \mu \geqslant \nu . \equiv : (\exists \alpha, \beta) : \mu = \mathrm{N_0c}'\alpha . \nu = \mathrm{N_0c}'\beta : (\exists \rho) . \rho \subset \alpha . \rho \,\mathrm{sm}\, \beta$
[*117·24·221]

***117·244.** $\vdash :. \mu, \nu \in \mathrm{N_0C} . \supset : \mu \geqslant \nu \,.\equiv .\, \mathrm{sm}''\mu \geqslant \nu \,.\equiv .\, \mu \geqslant \mathrm{sm}''\nu \,.\equiv .$
$\mathrm{sm}''\mu \geqslant \mathrm{sm}''\nu$ [*117·24 . *103·4]

***117·25.** $\vdash : \mu \geqslant \nu . \nu \geqslant \mu . \equiv . \mu, \nu \in N_0C . sm``\mu = sm``\nu$

Dem.

$\vdash . *117·24 . \supset$

$\vdash : \mu \geqslant \nu . \nu \geqslant \mu . \equiv . (\exists \alpha, \beta, \gamma, \delta) . \mu = N_0c`\alpha = N_0c`\gamma . \nu = N_0c`\beta = N_0c`\delta .$
$Nc`\alpha \geqslant Nc`\beta . Nc`\delta \geqslant Nc`\gamma .$

[*117·107] $\equiv . (\exists \alpha, \beta, \gamma \delta) . \mu = N_0c`\alpha = N_0c`\gamma . \nu = N_0c`\beta = N_0c`\delta .$
$N_0c`\alpha \geqslant N_0c`\beta . N_0c`\delta \geqslant N_0c`\gamma .$

[*13·193] $\equiv . (\exists \alpha, \beta, \gamma, \delta) . \mu = N_0c`\alpha = N_0c`\gamma . \nu = N_0c`\beta = N_0c`\delta .$
$N_0c`\alpha \geqslant N_0c`\beta . N_0c`\beta \geqslant N_0c`\alpha .$

[*117·107·23] $\equiv . (\exists \alpha, \beta, \gamma, \delta) . \mu = N_0c`\alpha = N_0c`\gamma . \nu = N_0c`\beta = N_0c`\delta .$
$Nc`\alpha = Nc`\beta .$

[*11·45.*103·2] $\equiv . (\exists \alpha, \beta) . \mu = N_0c`\alpha . \nu = N_0c`\beta . \mu, \nu \in N_0C . Nc`\alpha = Nc`\beta .$

[*103·4] $\equiv . (\exists \alpha, \beta) . \mu = N_0c`\alpha . \nu = N_0c`\beta . \mu, \nu \in N_0C . sm``\mu = sm``\nu .$

[*11·45.*103·2] $\equiv . \mu, \nu \in N_0C . sm``\mu = sm``\nu : \supset \vdash .$ Prop

***117·26.** $\vdash : Nc`\alpha > Nc`\beta . \equiv . Nc`\alpha \geqslant Nc`\beta . Nc`\alpha \neq Nc`\beta$

Dem.

$\vdash . *117·13 . *13·12 .$ Transp $. \supset \vdash : Nc`\alpha > Nc`\beta . \supset . Nc`\alpha \neq Nc`\beta :$

[*117·108] $\supset \vdash : Nc`\alpha > Nc`\beta . \supset . Nc`\alpha \geqslant Nc`\beta . Nc`\alpha \neq Nc`\beta$ (1)

$\vdash . *117·108 . *5·6 . \supset \vdash : Nc`\alpha \geqslant Nc`\beta . Nc`\alpha \neq Nc`\beta . \supset . Nc`\alpha > Nc`\beta$ (2)

$\vdash . (1) . (2) . \supset \vdash .$ Prop

***117·27.** $\vdash : Nc`\alpha < Nc`\beta . \equiv . Nc`\alpha \leqslant Nc`\beta . Nc`\alpha \neq Nc`\beta$
[*117·26·103·105]

***117·28.** $\vdash : Nc`\alpha > Nc`\beta . \equiv . Nc`\alpha \geqslant Nc`\beta . \sim (Nc`\beta \geqslant Nc`\alpha)$
[*117·22·13]

***117·281.** $\vdash : \mu > \nu . \equiv . \mu \geqslant \nu . \sim (\nu \geqslant \mu)$ [*117·14·28·24]

***117·29.** $\vdash : Nc`\alpha < Nc`\beta . \equiv . Nc`\alpha \leqslant Nc`\beta . \sim (Nc`\beta \leqslant Nc`\alpha)$ [*117·28]

***117·291.** $\vdash : \mu < \nu . \equiv . \mu \leqslant \nu . \sim (\nu \leqslant \mu)$ [*117·281]

***117·3.** $\vdash : Nc`\alpha \geqslant Nc`\beta . \equiv . (\exists \varpi) . \varpi \in NC . Nc`\alpha = Nc`\beta +_c \varpi$

Dem.

$\vdash . *117·221 . \supset \vdash : Nc`\alpha \geqslant Nc`\beta . \equiv . (\exists \delta) . \delta \, sm \, \beta . \delta \subset \alpha .$

[*110·72] $\equiv . (\exists \varpi) . \varpi \in NC . Nc`\alpha = Nc`\beta +_c \varpi :$
$\supset \vdash .$ Prop

***117·31.** $\vdash :. \mu \geqslant \nu . \equiv : \mu, \nu \in \text{N}_0\text{C} : (\exists \varpi) . \varpi \in \text{NC} . \mu = \nu +_c \varpi$

Dem.

$\vdash . *117·24·3 . \supset$

$\vdash :. \mu \geqslant \nu . \equiv : (\exists \alpha, \beta, \varpi) . \mu = \text{N}_0\text{c}‘\alpha . \nu = \text{N}_0\text{c}‘\beta . \text{Nc}‘\alpha = \text{Nc}‘\beta +_c \varpi :$

[(*110·03)] $\equiv : (\exists \alpha, \beta, \varpi) . \mu = \text{N}_0\text{c}‘\alpha . \nu = \text{N}_0\text{c}‘\beta . \text{Nc}‘\alpha = \nu +_c \varpi :$

[*103·16.*110·42] $\equiv : (\exists \alpha, \beta, \varpi) . \mu = \text{N}_0\text{c}‘\alpha . \nu = \text{N}_0\text{c}‘\beta . \mu = \nu +_c \varpi :$

[*103·2] $\equiv : \mu, \nu \in \text{N}_0\text{C} : (\exists \varpi) . \mu = \nu +_c \varpi :. \supset \vdash . \text{Prop}$

***117·32.** $\vdash : \mu \geqslant \nu . \exists ! \, \text{sm}‘‘\mu \cap t‘\alpha . \supset . \exists ! \, \text{sm}‘‘\nu \cap t‘\alpha$

Dem.

$\vdash . *117·241 . *103·4 . \supset$

$\vdash : \text{Hp} . \supset . (\exists \beta, \gamma) . \mu = \text{N}_0\text{c}‘\beta . \nu = \text{N}_0\text{c}‘\gamma . \exists ! \, \text{Cl}‘\beta \cap \text{Nc}‘\gamma . \text{sm}‘‘\mu = \text{Nc}‘\beta .$
$\text{sm}‘‘\nu = \text{Nc}‘\gamma \quad (1)$

$\vdash . *63·105·371 . *73·12 . \supset$

$\vdash : R \in \rho \, \overline{\text{sm}} \, \beta . \rho \in t‘\alpha . \sigma \subset \beta . \sigma \, \text{sm} \, \gamma . \supset . R‘‘\sigma \in t‘\alpha . R‘‘\sigma \, \text{sm} \, \gamma \quad (2)$

$\vdash . (2) . *73·04 . \supset \vdash : \rho \in \text{Nc}‘\beta \cap t‘\alpha . \sigma \in \text{Cl}‘\beta \cap \text{Nc}‘\gamma . \supset . \exists ! \, \text{Nc}‘\gamma \cap t‘\alpha \quad (3)$

$\vdash . (1) . (3) . \supset \vdash . \text{Prop}$

The above proposition shows that if a cardinal μ exists in a given type, so do all smaller cardinals.

***117·4.** $\vdash : \mu \geqslant \nu . \nu \geqslant \varpi . \supset . \mu \geqslant \varpi$

Dem.

$\vdash . *117·243 . \supset \vdash :. \text{Hp} . \supset : (\exists \alpha, \beta, \gamma) : \mu = \text{N}_0\text{c}‘\alpha . \nu = \text{N}_0\text{c}‘\beta . \varpi = \text{N}_0\text{c}‘\gamma :$
$(\exists \rho) . \rho \subset \alpha . \rho \, \text{sm} \, \beta : (\exists \sigma) . \sigma \subset \beta . \sigma \, \text{sm} \, \gamma :$

[*117·11.*60·2.*100·1] $\supset : (\exists \alpha, \beta, \gamma) : \mu = \text{N}_0\text{c}‘\alpha . \nu = \text{N}_0\text{c}‘\beta . \varpi = \text{N}_0\text{c}‘\gamma :$
$(\exists \rho, \tau) . \rho \subset \alpha . \rho \, \text{sm} \, \beta . \tau \subset \rho . \tau \, \text{sm} \, \gamma :$

[*22·44] $\supset : (\exists \alpha, \gamma) : \mu = \text{N}_0\text{c}‘\alpha . \varpi = \text{N}_0\text{c}‘\gamma : (\exists \tau) . \tau \subset \alpha . \tau \, \text{sm} \gamma :$

[*117·243] $\supset : \mu \geqslant \varpi :. \supset \vdash . \text{Prop}$

***117·41.** $\vdash : \mu \leqslant \nu . \nu \leqslant \varpi . \supset . \mu \leqslant \varpi$ [*117·4]

***117·42.** $\vdash : \sim (\mu > \mu) . \sim (\mu < \mu)$

Dem.

$\vdash . *117·15 . *13·12 . \text{Transp} . \supset \vdash : \mu > \nu . \supset . \mu \neq \nu : \supset \vdash . \text{Prop}$

***117·43.** $\vdash : \mu \geqslant \nu . \sim (\mu \geqslant \varpi) . \supset . \sim (\nu \geqslant \varpi)$ [*117·4 . Transp]

***117·44.** $\vdash : \nu \geqslant \varpi . \sim (\mu \geqslant \varpi) . \supset . \sim (\mu \geqslant \nu)$ [*117·4 . Transp]

***117·45.** $\vdash : \mu \geqslant \nu . \nu > \varpi . \supset . \mu > \varpi$

Dem.

$\vdash . *117·281 . \supset \vdash : \text{Hp} . \supset . \mu \geqslant \nu . \nu \geqslant \varpi . \sim (\varpi \geqslant \nu) .$

[*117·4·44] $\supset . \mu \geqslant \varpi . \sim (\varpi \geqslant \mu) .$

[*117·281] $\supset . \mu > \varpi : \supset \vdash . \text{Prop}$

***117·46.** $\vdash : \mu > \nu . \nu \geqslant \varpi . \supset . \mu > \varpi$ [Proof as in *117·45]

***117·47.** $\vdash : \mu > \nu . \nu > \varpi . \supset . \mu > \varpi$ [*117·45·104]

∗117·471. $\vdash : \mu < \nu . \nu < \varpi . \supset . \mu < \varpi$ [∗117·47·103]

∗117·5. $\vdash : \mu \in N_0C . \supset . \mu \geqslant 0$

Dem.

$\vdash . {*}60{\cdot}3 . {*}100{\cdot}3 . \supset \vdash . \exists ! \mathrm{Cl}`\alpha \cap \mathrm{Nc}`\Lambda .$

$[{*}117{\cdot}22] \quad \supset \vdash . \mathrm{Nc}`\alpha \geqslant \mathrm{Nc}`\Lambda .$

$[{*}117{\cdot}107 . {*}101{\cdot}1] \supset \vdash . N_0c`\alpha \geqslant 0 \quad (1)$

$\vdash . (1) . {*}103{\cdot}2 . \quad \supset \vdash . \mathrm{Prop}$

∗117·501. $\vdash : \mu \in N_0C . \equiv . \mu \geqslant 0$ [∗117·5·104]

∗117·51. $\vdash : \mu \in N_0C - \iota`0 . \supset . \mu > 0$

Dem.

$\vdash . {*}101{\cdot}15 . \supset \vdash : \mathrm{Hp} . \supset . \mu \neq \mathrm{sm}``0 \quad (1)$

$\vdash . (1) . {*}117{\cdot}5{\cdot}104 . \supset \vdash . \mathrm{Prop}$

∗117·511. $\vdash : \mu \in N_0C - \iota`0 . \equiv . \mu > 0$ [∗117·51·15·42]

∗117·52. $\vdash : \exists ! \xi . \supset . \mathrm{Nc}`\xi \geqslant 1$

Dem.

$\vdash . {*}51{\cdot}2 . \supset \vdash : \mathrm{Hp} . \supset . (\exists x) . \iota`x \subset \xi .$

$[{*}117{\cdot}222] \quad \supset . (\exists x) . \mathrm{Nc}`\xi \geqslant \mathrm{Nc}`\iota`x .$

$[{*}101{\cdot}2] \quad \supset . \mathrm{Nc}`\xi \geqslant 1 : \supset \vdash . \mathrm{Prop}$

∗117·53. $\vdash : \mu \in N_0C - \iota`0 . \supset . \mu \geqslant 1$

Dem.

$\vdash . {*}101{\cdot}16 . {*}103{\cdot}2 . \supset \vdash : \mathrm{Hp} . \supset . (\exists \alpha) . N_0c`\alpha = \mu . \exists ! \alpha .$

$[{*}117{\cdot}52] \quad \supset . (\exists \alpha) . N_0c`\alpha = \mu . \mathrm{Nc}`\alpha \geqslant 1 .$

$[{*}117{\cdot}107] \quad \supset . \mu \geqslant 1 : \supset \vdash . \mathrm{Prop}$

∗117·531. $\vdash : \mu \in N_0C - \iota`0 . \equiv . \mu \geqslant 1$

Dem.

$\vdash . {*}117{\cdot}104 . \quad \supset \vdash : \mu \geqslant 1 . \supset . \mu \in N_0C \quad (1)$

$\vdash . {*}117{\cdot}51 . {*}101{\cdot}22 . \supset \vdash . 1 > 0 .$

$[{*}117{\cdot}45] \quad \supset \vdash : \mu \geqslant 1 . \supset . \mu > 0 .$

$[{*}117{\cdot}42] \quad \supset . \mu \neq 0 \quad (2)$

$\vdash . (1) . (2) . {*}117{\cdot}53 . \supset \vdash . \mathrm{Prop}$

∗117·54. $\vdash :. 1 \geqslant \mu . \equiv : \mu = 0 . \vee . \mu = 1$

Dem.

$\vdash . {*}117{\cdot}241 . {*}101{\cdot}2 . {*}52{\cdot}22 . \supset$

$\vdash :. 1 \geqslant \mu . \equiv : (\exists \alpha, x) . \mu = N_0c`\alpha . \exists ! \mathrm{Nc}`\alpha \cap \mathrm{Cl}`\iota`x :$

$[{*}60{\cdot}362] \quad \equiv : (\exists \alpha, x) : \mu = N_0c`\alpha : \exists ! \mathrm{Nc}`\alpha \cap \iota`\Lambda . \vee . \exists ! \mathrm{Nc}`\alpha \cap \iota`\iota`x :$

$[{*}51{\cdot}31] \quad \equiv : (\exists \alpha, x) : \mu = N_0c`\alpha : \Lambda \in \mathrm{Nc}`\alpha . \vee . \iota`x \in \mathrm{Nc}`\alpha :$

$[{*}101{\cdot}17{\cdot}29] \quad \equiv : (\exists \alpha, x) : \mu = N_0c`\alpha : \mathrm{Nc}`\alpha = \mathrm{Nc}`\Lambda . \vee . \mathrm{Nc}`\alpha = \mathrm{Nc}`\iota`x :$

$[{*}103{\cdot}16] \quad \equiv : (\exists \alpha, x) . \mu = N_0c`\alpha : \mu = \mathrm{Nc}`\Lambda . \vee . \mu = \mathrm{Nc}`\iota`x :$

$[{*}101{\cdot}1{\cdot}2] \quad \equiv : (\exists \alpha) . \mu = N_0c`\alpha : \mu = 0 . \vee . \mu = 1 ;$

$[{*}103{\cdot}2{\cdot}5{\cdot}51] \quad \equiv : \mu = 0 . \vee . \mu = 1 :. \supset \vdash . \mathrm{Prop}$

***117·55.** $\vdash : \mu > 1 . \equiv . \mu \epsilon N_0C - \iota`0 - \iota`1$

Dem.

$\vdash . *117·281 . \supset \vdash : \mu > 1 . \equiv . \mu \geqslant 1 . \sim(1 \geqslant \mu) .$

[*117·531·54] $\equiv . \mu \epsilon N_0C - \iota`0 . \mu \neq 0 . \mu \neq 1 .$

[*51·15] $\equiv . \mu \epsilon N_0C - \iota`0 - \iota`1 : \supset \vdash . \text{Prop}$

***117·551.** $\vdash :. \mu \epsilon N_0C - \iota`0 - \iota`1 . \equiv :$

$(\exists \alpha) : \mu = N_0c`\alpha : (\exists x, y) . x, y \epsilon \alpha . x \neq y : \equiv . \mu \geqslant 2$

Dem.

$\vdash . *103·2 . \supset \vdash :. \mu \epsilon N_0C - \iota`0 - \iota`1 . \equiv :$

$(\exists \alpha) . \mu = N_0c`\alpha . N_0c`\alpha \neq 0 . N_0c`\alpha \neq 1 :$

[*101·14] $\equiv : (\exists \alpha) . \mu = N_0c`\alpha . \exists ! \alpha . N_0c`\alpha \neq 1 :$

[*103·26] $\equiv : (\exists \alpha) . \mu = N_0c`\alpha . \exists ! \alpha . \alpha \sim \epsilon 1 :$

[*52·41] $\equiv : (\exists \alpha) : \mu = N_0c`\alpha : (\exists x, y) . x, y \epsilon \alpha . x \neq y :$ (1)

[*54·26.*51·2] $\equiv : (\exists \alpha) : \mu = N_0c`\alpha : (\exists x, y) . \iota`x \cup \iota`y \subset \alpha . \iota`x \cup \iota`y \epsilon 2 :$

[*13·195] $\equiv : (\exists \alpha) : \mu = N_0c`\alpha : (\exists x, y, \beta) . \beta = \iota`x \cup \iota`y . \beta \subset \alpha . \beta \epsilon 2 :$

[*54·101] $\equiv : (\exists \alpha) : \mu = N_0c`\alpha : (\exists \beta) . \beta \subset \alpha . \beta \epsilon 2 :$

[*117·241] $\equiv : \mu \geqslant 2$ (2)

$\vdash . (1) . (2) . \supset \vdash . \text{Prop}$

***117·56.** $\vdash : Nc`\alpha \geqslant Nc`\beta . \supset . Nc`\alpha +_c Nc`\gamma \geqslant Nc`\beta +_c Nc`\gamma$

Dem.

$\vdash . *110·12 . *117·221 . \supset$

$\vdash : \text{Hp} . \supset . (\exists \delta) . \delta \subset \downarrow \Lambda_\gamma``\iota``\alpha . \delta \text{ sm } \downarrow \Lambda_\gamma``\iota``\beta .$

[*110·11.*73·71.(*110·01)] $\supset . (\exists \delta) . \delta \cup \Lambda_\alpha \downarrow ``\iota``\gamma \subset \alpha + \gamma .$

$\delta \cup \Lambda_\alpha \downarrow ``\iota``\gamma \text{ sm } (\beta + \gamma) .$

[*117·221] $\supset . Nc`(\alpha + \gamma) \geqslant Nc`(\beta + \gamma) .$

[*110·3] $\supset . Nc`\alpha +_c Nc`\gamma \geqslant Nc`\beta +_c Nc`\gamma : \supset \vdash . \text{Prop}$

***117·561.** $\vdash : \mu \geqslant \nu . \varpi \epsilon N_0C . \supset . \mu +_c \varpi \geqslant \nu +_c \varpi$ [*117·56]

The proof of *117·561 follows from *117·56 in the same way as the proof of *117·31 follows from *117·3. In the remainder of this number we shall omit proofs of this kind.

***117·57.** $\vdash : Nc`\alpha \geqslant Nc`\beta . \supset . Nc`\alpha \times_c Nc`\gamma \geqslant Nc`\beta \times_c Nc`\gamma$

Dem.

$\vdash . *37·2 . \supset \vdash : \rho \subset \alpha . \supset . \gamma \underset{,,}{\downarrow} ``\rho \subset \gamma \underset{,,}{\downarrow} ``\alpha .$

[*40·161.*113·1] $\supset . \rho \times \gamma \subset \alpha \times \gamma$ (1)

$\vdash . *113·13 . \supset \vdash \rho \text{ sm } \beta . \supset . \rho \times \gamma \text{ sm } \beta \times \gamma$ (2)

$\vdash . (1) . (2) . \supset \vdash : \rho \subset \alpha . \rho \text{ sm } \beta . \supset . \rho \times \gamma \subset \alpha \times \gamma . \rho \times \gamma \text{ sm } \beta \times \gamma .$

[*117·221] $\supset . Nc`(\alpha \times \gamma) \geqslant Nc`(\beta \times \gamma)$ (3)

$\vdash . (3) . *117·221 . \supset \vdash . \text{Prop}$

***117·571.** $\vdash . \mu \geqslant \nu . \varpi \in \mathrm{N_0C} . \supset . \mu \times_c \varpi \geqslant \nu \times_c \varpi$ [*117·57]

***117·58.** $\vdash : \mathrm{Nc}‘\alpha \geqslant \mathrm{Nc}‘\beta . \supset . (\mathrm{Nc}‘\alpha)^{\mathrm{Nc}‘\gamma} \geqslant (\mathrm{Nc}‘\beta)^{\mathrm{Nc}‘\gamma}$

Dem.

$\vdash . *35{\cdot}432{\cdot}82 . \supset \vdash : \rho \subset \alpha . \supset . \rho \uparrow \gamma \mathrel{\dot\subset} \alpha \uparrow \gamma .$

[*80·15] $\supset . (\rho \uparrow \gamma)_\Delta‘\gamma \subset (\alpha \uparrow \gamma)_\Delta‘\gamma$ (1)

$\vdash . *116{\cdot}15{\cdot}19 . \supset \vdash : \rho \,\mathrm{sm}\, \beta . \supset . (\rho \uparrow \gamma)_\Delta‘\gamma \,\mathrm{sm}\, (\beta \uparrow \gamma)_\Delta‘\gamma$ (2)

$\vdash . (1) . (2) . *117{\cdot}221 . \supset$

$\vdash : \rho \subset \alpha . \rho \,\mathrm{sm}\, \beta . \supset . \mathrm{Nc}‘(\alpha \uparrow \gamma)_\Delta‘\gamma \geqslant \mathrm{Nc}‘(\beta \uparrow \gamma)_\Delta‘\gamma .$

[*116·15·25] $\supset . (\mathrm{Nc}‘\alpha)^{\mathrm{Nc}‘\gamma} \geqslant (\mathrm{Nc}‘\beta)^{\mathrm{Nc}‘\gamma}$ (3)

$\vdash . (3) . *117{\cdot}221 . \supset \vdash .$ Prop

***117·581.** $\vdash : \mu \geqslant \nu . \varpi \in \mathrm{N_0C} . \supset . \mu^\varpi \geqslant \nu^\varpi$ [*117·58]

The two following propositions are lemmas for *117·59.

***117·582.** $\vdash : \exists ! \gamma . \beta \subset \alpha . \sigma \in \gamma \exp (\alpha - \beta) . \supset . (\cup \sigma) \upharpoonright (\gamma \exp \beta) \in 1 \rightarrow 1 .$
$(\cup \sigma)‘‘(\gamma \exp \beta) \subset \gamma \exp \alpha$

Dem.

$\vdash . *116{\cdot}183 . \supset \vdash : \rho \in (\gamma \exp \beta) . \sigma \in \gamma \exp (\alpha - \beta) . \supset . \rho \subset \beta \times \gamma . \sigma \subset (\alpha - \beta) \times \gamma .$

[*113·19.*24·21] $\supset . \rho \cap \sigma = \Lambda$ (1)

$\vdash . (1) . *24{\cdot}481 . \supset \vdash :: \mathrm{Hp} . \supset :. \rho, \rho' \in (\gamma \exp \beta) . \supset : \rho \cup \sigma = \rho' \cup \sigma . \equiv . \rho = \rho' :.$

[*71·58] $\supset :. (\cup \sigma) \upharpoonright (\gamma \exp \beta) \in 1 \rightarrow 1$ (2)

$\vdash . *113{\cdot}191 . \supset \vdash :. \mathrm{Hp} . \supset : \gamma \downarrow ‘‘\beta \cap \gamma \downarrow ‘‘(\alpha - \beta) = \Lambda :$

[*115·14.(*116·01)] $\supset : \rho \in (\gamma \exp \beta) . \supset . \rho \cup \sigma \in \mathrm{Prod}‘\{\gamma \downarrow ‘‘\beta \cup \gamma \downarrow ‘‘(\alpha - \beta)\} .$

[*37·22.*24·411] $\supset . \rho \cup \sigma \in (\gamma \exp \alpha) :$

[*37·61] $\supset : (\cup \sigma)‘‘(\gamma \exp \beta) \subset \gamma \exp \alpha$ (3)

$\vdash . (2) . (3) . \supset \vdash .$ Prop

***117·583.** $\vdash : \beta \subset \alpha . \exists ! \gamma . \supset . (\exists \tau) . \tau \subset \gamma \exp \alpha . \tau \,\mathrm{sm}\, (\gamma \exp \beta)$

Dem.

$\vdash . *116{\cdot}171 . \supset \vdash : \mathrm{Hp} . \supset . \exists ! \gamma \exp (\alpha - \beta)$ (1)

$\vdash . (1) . *117{\cdot}582 . *73{\cdot}15 . \supset \vdash .$ Prop

***117·59.** $\vdash : \mathrm{Nc}‘\alpha \geqslant \mathrm{Nc}‘\beta . \exists ! \gamma . \supset . (\mathrm{Nc}‘\gamma)^{\mathrm{Nc}‘\alpha} \geqslant (\mathrm{Nc}‘\gamma)^{\mathrm{Nc}‘\beta}$

Dem.

$\vdash . *117{\cdot}221 . \supset \vdash :. \mathrm{Hp} . \supset : (\exists \rho) . \rho \subset \alpha . \rho \,\mathrm{sm}\, \beta : \exists ! \gamma :$

[*117·583] $\supset : (\exists \rho, \tau) . \rho \subset \alpha . \rho \,\mathrm{sm}\, \beta . \tau \subset \gamma \exp \alpha . \tau \,\mathrm{sm}\, (\gamma \exp \rho) :$

[*116·19] $\supset : (\exists \tau) . \tau \subset \gamma \exp \alpha . \tau \,\mathrm{sm}\, (\gamma \exp \beta) :$

[*117·221] $\supset : \mathrm{Nc}‘(\gamma \exp \alpha) \geqslant \mathrm{Nc}‘(\gamma \exp \beta)$ (1)

$\vdash . (1) . *116{\cdot}25 . \supset \vdash .$ Prop

The hypothesis is essential in the above proposition, for $0^0 = 1$ while $0^1 = 0$, so that $0^0 > 0^1$.

*117·591. $\vdash : \mu \geqslant \nu \,.\, \varpi \in \mathrm{N_0C} - \iota`0 \,.\, \supset .\, \varpi^\mu \geqslant \varpi^\nu$ [*117·59]

*117·592. $\vdash : \alpha^\delta = 1 \,.\, \alpha \neq 0 \,.\, \alpha \neq 1 \,.\, \supset .\, \delta = 0$

Dem.

$\vdash$. *116·203 . $\supset \vdash :.\ \mathrm{Hp} \,.\, \supset : \alpha, \delta \in \mathrm{N_0C} :$

[*117·551·53] $\supset : \alpha \geqslant 2 : \delta \neq 0 \,.\, \supset .\, \delta \geqslant 1 :$

[*117·581·591] $\supset : \delta \neq 0 \,.\, \supset .\, \alpha^\delta \geqslant 2^1 \,.$

[*116·321.*117·244] $\supset .\, \alpha^\delta \geqslant 2 \,.$

[*117·551] $\supset .\, \alpha^\delta \neq 1$ (1)

$\vdash$. (1) . Transp . $\supset \vdash$. Prop

The above proposition is used in *120·53.

*117·6. $\vdash : \mu, \nu \in \mathrm{N_0C} \,.\, \supset .\, \mu +_c \nu \geqslant \mu \,.\, \mu +_c \nu \geqslant \nu$

Dem.

$\vdash$. *117·561·5 . $\supset \vdash : \mathrm{Hp} \,.\, \supset .\, \mu +_c \nu \geqslant \mu +_c 0 \,.\, \mu +_c \nu \geqslant 0 +_c \nu$ (1)

$\vdash$. (1) . *110·6 . *117·244 . $\supset \vdash$. Prop

*117·61. $\vdash : \nu > \mu \,.\, \supset .\, \mu +_c \nu > \mu$ [*117·6·45]

*117·62. $\vdash : \mu, \nu \in \mathrm{N_0C} - \iota`0 \,.\, \supset .\, \mu \times_c \nu \geqslant \mu \,.\, \mu \times_c \nu \geqslant \nu$

Dem.

$\vdash$. *117·571·53 . $\supset \vdash : \mathrm{Hp} \,.\, \supset .\, \mu \times_c \nu \geqslant \mu \times_c 1 \,.\, \mu \times_c \nu \geqslant 1 \times_c \nu$ (1)

$\vdash$. (1) . *113·621 . *117·244 . $\supset \vdash$. Prop

*117·63. $\vdash : \alpha, \beta \sim\in 0 \cup 1 \,.\, \supset .\, \mathrm{Nc}`\alpha \times_c \mathrm{Nc}`\beta \geqslant \mathrm{Nc}`\alpha +_c \mathrm{Nc}`\beta$

Dem.

$\vdash$. *52·4 . Transp . $\supset \vdash : \mathrm{Hp} \,.\, \supset .\, (\exists x, x', y, y') \,.\, x, x' \in \alpha \,.\, y, y' \in \beta \,.\, x \neq x' \,.\, y \neq y'$ (1)

$\vdash$. *113·101 . $\supset \vdash : \mathrm{Hp} \,.\, x, x' \in \alpha \,.\, y, y' \in \beta \,.\, x \neq x' \,.\, y \neq y' \,.\, \rho = \downarrow y``\alpha \,.$

$\sigma = x \downarrow ``(\beta - \iota`y) \cup \iota`x' \downarrow y' \,.\, \supset .\, \rho \cup \sigma \subset \beta \times \alpha$ (2)

$\vdash$. *55·15 . $\supset \vdash :.\ \mathrm{Hp}(2) \,.\, \supset : R \in \rho \,.\, \supset_R .\, \mathrm{G}`R = \iota`y :$

$S \in x \downarrow ``(\beta - \iota`y) \,.\, \supset_S .\, \mathrm{G}`S \in \beta - \iota`y : \mathrm{G}`x' \downarrow y' = \iota`y' :$

[*51·23] $\supset : R \in \rho \,.\, S \in \sigma \,.\, \supset_{R,S} .\, \mathrm{G}`R \neq \mathrm{G}`S :$

[*24·37.*30·37] $\supset : \rho \cap \sigma = \Lambda$ (3)

$\vdash$. *73·61·611 . $\supset \vdash : \mathrm{Hp}(2) \,.\, \supset .\, \rho \,\mathrm{sm}\, \alpha \,.\, x \downarrow ``(\beta - \iota`y) \,\mathrm{sm}\, (\beta - \iota`y)$ (4)

$\vdash$. *55·202 . $\supset \vdash : \mathrm{Hp}(2) \,.\, \supset .\, x' \downarrow y' \sim\in x \downarrow ``(\beta - \iota`y)$ (5)

$\vdash$. (4) . (5) . *73·71 . $\supset \vdash : \mathrm{Hp}(2) \,.\, \supset .\, \rho \,\mathrm{sm}\, \alpha \,.\, \sigma \,\mathrm{sm}\, \beta$ (6)

$\vdash$. (3) . (6) . *110·13 . $\supset \vdash : \mathrm{Hp}(2) \,.\, \supset .\, \rho \cup \sigma \in \mathrm{Nc}`(\alpha + \beta)$ (7)

$\vdash$. (2) . (7) . *117·221 . $\supset \vdash : \mathrm{Hp}(2) \,.\, \supset .\, \mathrm{Nc}`(\beta \times \alpha) \geqslant \mathrm{Nc}`(\alpha + \beta)$ (8)

$\vdash$. (1) . (8) . *113·141·25 . *110·3 . $\supset \vdash$. Prop

$*117 \cdot 631$. $\vdash : \mu, \nu \in N_0C - \iota`0 - \iota`1 . \supset . \mu \times_c \nu \geqslant \mu +_c \nu$ [$*117 \cdot 63$]

The two following propositions are lemmas for $*117 \cdot 64$.

$*117 \cdot 632$. $\vdash : \kappa \in \text{Cls}^2 \text{excl} . \kappa \sim \in 0 \cup 1 . \rho, \sigma \in \text{Prod}`\kappa . \rho \cap \sigma = \Lambda .$
$T = \hat{\mu}\hat{x}\{(\exists \alpha, \beta) . \alpha, \beta \in \kappa . \alpha \neq \beta . x \in \beta . \mu = (\rho - \alpha - \beta) \cup (\sigma \cap \alpha) \cup \iota`x\} .$
$\supset . T \in 1 \to 1 . \text{D}`T \subset \text{Prod}`\kappa . \text{Ɑ}`T = s`\kappa$

Dem.

$\vdash . *115 \cdot 11 \cdot 145 . \supset \vdash :. \text{Hp} . \alpha, \beta \in \kappa . \alpha \neq \beta . \supset : \rho - \alpha - \beta \in \text{Prod}`(\kappa - \iota`\alpha - \iota`\beta) :$
$[*115 \cdot 11 \cdot 145] \supset : (\rho - \alpha - \beta) \cup (\sigma \cap \alpha) \in \text{Prod}`(\kappa - \iota`\beta) :$
$[*115 \cdot 145] \quad \supset : x \in \beta . \supset . (\rho - \alpha - \beta) \cup (\sigma \cap \alpha) \cup \iota`x \in \text{Prod}`\kappa$ (1)
$\vdash . (1) . *21 \cdot 33 . \quad \supset \vdash : \text{Hp} . \mu T x . \supset . \mu \in \text{Prod}`\kappa$ (2)
$\vdash . *52 \cdot 4 . \text{Transp} . \quad \supset \vdash :. \text{Hp} . \supset : \beta \in \kappa . x \in \beta . \supset . (\exists \alpha) . \alpha \in \kappa . \alpha \neq \beta .$
$[*21 \cdot 33 . *33 \cdot 131] \quad \supset . x \in \text{Ɑ}`T$ (3)
$\vdash . *21 \cdot 33 . *33 \cdot 131 . \supset \vdash :. \text{Hp} . \supset : x \in \text{Ɑ}`T . \supset . (\exists \beta) . \beta \in \kappa . x \in \beta$ (4)
$\vdash . (3) . (4) . \quad \supset \vdash : \text{Hp} . \supset . \text{Ɑ}`T = s`\kappa$ (5)
$\vdash : *21 \cdot 33 . *13 \cdot 172 . \supset \vdash :. \text{Hp} . \supset : \mu T x . \nu T x . \supset . \mu = \nu$ (6)
$\vdash . *21 \cdot 33 . *13 \cdot 171 . \supset \vdash :. \text{Hp} . \supset : \mu T x . \mu T x' . \supset .$
$(\exists \alpha, \alpha', \beta, \beta') . \alpha, \alpha' \in \kappa . \beta, \beta' \in \kappa . \alpha \neq \beta . \alpha' \neq \beta' .$
$(\rho - \alpha - \beta) \cup (\sigma \cap \alpha) \cup \iota`x = (\rho - \alpha' - \beta') \cap (\sigma \cap \alpha') \cup \iota`x' .$
$[*24 \cdot 48 . \text{Hp}] \supset . \iota`x = \iota`x'$ (7)
$\vdash . (2) . (5) . (6) . (7) . \supset \vdash . \text{Prop}$

$*117 \cdot 633$. $\vdash :. \kappa \in \text{Cls}^2 \text{excl} . \kappa \sim \in 0 \cup 1 : (\exists \rho, \sigma) . \rho, \sigma \in \text{Prod}`\kappa . \rho \cap \sigma = \Lambda : \supset .$
$\Pi \text{Nc}`\kappa \geqslant \Sigma \text{Nc}`\kappa$

Dem.

$\vdash . *117 \cdot 632 . \supset \vdash : \text{Hp} . \supset . (\exists \gamma) . \gamma \subset \text{Prod}`\kappa . \gamma \text{ sm } s`\kappa .$
$[*117 \cdot 221] \quad \supset . \text{Nc}`\text{Prod}`\kappa \geqslant \text{Nc}`s`\kappa$ (1)
$\vdash . (1) . *115 \cdot 12 . *112 \cdot 15 . \supset \vdash . \text{Prop}$

$*117 \cdot 64$. $\vdash :. \kappa \in \text{Cls}^2 \text{excl} : (\exists \rho, \sigma) . \rho, \sigma \in \text{Prod}`\kappa . \rho \cap \sigma = \Lambda : \supset .$
$\Pi \text{Nc}`\kappa \geqslant \Sigma \text{Nc}`\kappa$

Dem.

$\vdash . *112 \cdot 321 . *114 \cdot 21 . \supset \vdash : \kappa \in 1 . \supset . \Pi \text{Nc}`\kappa = \Sigma \text{Nc}`\kappa$ (1)
$\vdash . *114 \cdot 2 . *112 \cdot 3 . \quad \supset \vdash : \kappa \in 0 . \supset . \Pi \text{Nc}`\kappa = 1 . \Sigma \text{Nc}`\kappa = 0 .$
$[*117 \cdot 51] \quad \supset . \Pi \text{Nc}`\kappa > \Sigma \text{Nc}`\kappa$ (2)
$\vdash . (1) . (2) . *117 \cdot 633 . \supset \vdash . \text{Prop}$

$*117 \cdot 651$. $\vdash : \alpha \sim \in 0 \cup 1 . \supset . (\text{Nc}`\alpha)^{\text{Nc}`\beta} \geqslant \text{Nc}`\alpha \times_c \text{Nc}`\beta$

Dem.

$\vdash . *52 \cdot 4 . \text{Transp} . \quad \supset \vdash : \text{Hp} . \supset . (\exists x, y) . x, y \in \alpha . x \neq y$ (1)
$\vdash . *116 \cdot 152 . *55 \cdot 23 \cdot 202 . \supset \vdash : x, y \in \alpha . x \neq y . \supset . x \downarrow ``\beta, y \downarrow ``\beta \in (\alpha \exp \beta) .$
$x \downarrow ``\beta \cap y \downarrow ``\beta = \Lambda$ (2)
$\vdash . *113 \cdot 111 . \quad \supset \vdash . \alpha \underset{,,}{\downarrow} ``\beta \in \text{Cls}^2 \text{excl}$ (3)
$\vdash . (1) . (2) . (3) . *117 \cdot 64 . *113 \cdot 1 \cdot 141 \cdot 25 . *116 \cdot 25 . (*116 \cdot 01) . \supset \vdash . \text{Prop}$

***117·652.** $\vdash : \mu \epsilon \mathrm{N_0C} - \iota\text{‘}0 - \iota\text{‘}1 \,.\, \nu \epsilon \mathrm{N_0C} \,.\supset.\, \mu^\nu \geqslant \mu \times_c \nu$ [*117·651]

***117·66.** $\vdash . \mathrm{Nc}\text{‘}\mathrm{Cl}\text{‘}\alpha > \mathrm{Nc}\text{‘}\alpha$

Dem.

$\vdash . *102{\cdot}72 . \quad \supset \vdash . \sim (\exists\beta) . \beta \subset \alpha . \beta \,\mathrm{sm}\, \mathrm{Cl}\text{‘}\alpha$ (1)

$\vdash . *100{\cdot}6 . *60{\cdot}61 . \quad \supset \vdash . \iota\text{“}\alpha \subset \mathrm{Cl}\text{‘}\alpha . \iota\text{“}\alpha \,\mathrm{sm}\, \alpha$ (2)

$\vdash . (1) . (2) . *117{\cdot}13 . \supset \vdash . \mathrm{Prop}$

***117·661.** $\vdash : \mu \epsilon \mathrm{N_0C} \,.\supset.\, 2^\mu > \mu$ [*117·66 . *116·72]

The above proposition is important.

***117·67.** $\vdash : \kappa \epsilon \mathrm{Cls^2\,excl} \,.\, \exists ! \,\mathrm{Prod}\text{‘}\kappa \,.\supset.\, \mathrm{Nc}\text{‘}s\text{‘}\kappa \geqslant \mathrm{Nc}\text{‘}\kappa$

Dem.

$\vdash . *115{\cdot}16{\cdot}11 . \supset \vdash : \kappa \epsilon \mathrm{Cls^2\,excl} . \mu \epsilon \mathrm{Prod}\text{‘}\kappa . \supset . \mu \,\mathrm{sm}\, \kappa . \mu \subset s\text{‘}\kappa .$

[*117·22] $\supset . \mathrm{Nc}\text{‘}s\text{‘}\kappa \geqslant \mathrm{Nc}\text{‘}\kappa : \supset \vdash . \mathrm{Prop}$

***117·68.** $\vdash : R, S \epsilon \epsilon_\Delta\text{‘}\kappa \,.\, R \dot{\cap} S = \dot{\Lambda} \,.\, T = \hat{P}\hat{\rho}\,\{\rho \epsilon \kappa \,.\, P = R \upharpoonright -\iota\text{‘}\rho \cup S \upharpoonright \iota\text{‘}\rho\} \,.$

$\supset . T \epsilon 1 \rightarrow 1 \,.\, \mathrm{D}\text{‘}T \subset \epsilon_\Delta\text{‘}\kappa \,.\, \mathrm{Cl}\text{‘}T = \kappa$

Dem.

$\vdash . *21{\cdot}33 . *13{\cdot}172 . \supset \vdash :. \mathrm{Hp} . \supset : PT\rho . QT\rho . \supset . P = Q$ (1)

$\vdash . *23{\cdot}631 . \quad \supset \vdash : \mathrm{Hp} . \rho \epsilon \kappa . \supset . (T\text{‘}\rho) \dot{\cap} S = S \upharpoonright \iota\text{‘}\rho :$

[*13·17] $\supset \vdash : \mathrm{Hp} . \rho, \sigma \epsilon \kappa . T\text{‘}\rho = T\text{‘}\sigma . \supset . S \upharpoonright \iota\text{‘}\rho = S \upharpoonright \iota\text{‘}\sigma .$

[*35·65] $\supset . \iota\text{‘}\rho = \iota\text{‘}\sigma .$

[*51·23] $\supset . \rho = \sigma$ (2)

$\vdash . (1) . (2) . \quad \supset \vdash : \mathrm{Hp} . \supset . T \epsilon 1 \rightarrow 1$ (3)

$\vdash . *21{\cdot}33 . *33{\cdot}131 . \supset \vdash : \mathrm{Hp} . \supset . \mathrm{Cl}\text{‘}T = \kappa$ (4)

$\vdash . *80{\cdot}36 . \quad \supset \vdash : \mathrm{Hp} . \supset . \mathrm{D}\text{‘}T \subset \epsilon_\Delta\text{‘}\kappa$ (5)

$\vdash . (3) . (4) . (5) . \quad \supset \vdash . \mathrm{Prop}$

***117·681.** $\vdash : (\exists R, S) \,.\, R, S \epsilon \epsilon_\Delta\text{‘}\kappa \,.\, R \dot{\cap} S = \dot{\Lambda} \,.\supset.\, \mathrm{Nc}\text{‘}\epsilon_\Delta\text{‘}\kappa \geqslant \mathrm{Nc}\text{‘}\kappa$ [*117·68·22]

***117·682.** $\vdash : \kappa \subset \lambda \,.\, \exists ! \,\epsilon_\Delta\text{‘}(\lambda - \kappa) \,.\supset.\, \mathrm{Nc}\text{‘}\epsilon_\Delta\text{‘}\lambda \geqslant \mathrm{Nc}\text{‘}\epsilon_\Delta\text{‘}\kappa$

Dem.

$\vdash . *80{\cdot}65 . \supset \vdash :. \mathrm{Hp} . \supset : R \epsilon \epsilon_\Delta\text{‘}\kappa . S \epsilon \epsilon_\Delta\text{‘}(\lambda - \kappa) . \supset . R \cup S \epsilon \epsilon_\Delta\text{‘}\lambda$ (1)

$\vdash . *80{\cdot}14 . \supset \vdash : R \epsilon \epsilon_\Delta\text{‘}\lambda . S \epsilon \epsilon_\Delta\text{‘}(\lambda - \kappa) . \supset . \mathrm{Cl}\text{‘}R \cap \mathrm{Cl}\text{‘}S = \Lambda .$

[*33·33] $\supset . R \dot{\cap} S = \dot{\Lambda} .$

[*25·4] $\supset . (R \cup S) \dot{-} S = R$ (2)

$\vdash . (2) . *13{\cdot}171 . \supset \vdash : Q, R \epsilon \epsilon_\Delta\text{‘}\lambda . S \epsilon \epsilon_\Delta\text{‘}(\lambda - \kappa) . Q \cup S = R \cup S . \supset . Q = R$ (3)

$\vdash . (1) . (3) . \supset \vdash : \mathrm{Hp} . S \epsilon \epsilon_\Delta\text{‘}(\lambda - \kappa) . \supset . (\cup S) \upharpoonright \epsilon_\Delta\text{‘}\kappa \epsilon 1 \rightarrow 1 . (\cup S)\text{“}\epsilon_\Delta\text{‘}\kappa \subset \epsilon_\Delta\text{‘}\lambda .$

[*117·22] $\supset . \mathrm{Nc}\text{‘}\epsilon_\Delta\text{‘}\lambda \geqslant \mathrm{Nc}\text{‘}\epsilon_\Delta\text{‘}\kappa : \supset \vdash . \mathrm{Prop}$

***117·683.** $\vdash :. \kappa \subset \lambda \,.\, \exists ! \,\epsilon_\Delta\text{‘}(\lambda - \kappa) : (\exists R, S) \,.\, R, S \epsilon \epsilon_\Delta\text{‘}\kappa \,.\, R \dot{\cap} S = \dot{\Lambda} : \supset .$

$\mathrm{Nc}\text{‘}\epsilon_\Delta\text{‘}\lambda \geqslant \mathrm{Nc}\text{‘}\kappa$ [*117·681·682]

***117·684.** $\vdash : \kappa \subset \lambda \,.\, \exists ! \,\epsilon_\Delta\text{‘}\lambda : (\exists R, S) \,.\, R, S \epsilon \epsilon_\Delta\text{‘}\kappa \,.\, R \dot{\cap} S = \dot{\Lambda} : \supset .$

$\mathrm{Nc}\text{‘}\epsilon_\Delta\text{‘}\lambda \geqslant \mathrm{Nc}\text{‘}\kappa$ [*117·683 . *88·22]

The above proposition is used in *120·765.

GENERAL NOTE ON CARDINAL CORRELATORS.

The correlators established at various stages throughout Section B present certain analogies to each other, and they or others closely resembling them will be found to be the correlators required in relation-arithmetic (Part IV). We shall therefore here collect together the most important propositions hitherto proved on correlators.

When we have to deal with correlators of two different functions of a single class, as *e.g.* $\epsilon_\Delta`\kappa$ and Prod$`\kappa$, the correlator is usually D or $\dot{s}$ or $\dot{s} | D$, with a suitable limitation on the converse domain. Sometimes it is $\breve{\iota} | D$ or $\epsilon | D$. Thus for example the class $\epsilon \downarrow``\kappa$, by means of which $\Sigma`\kappa$ is defined (*112), has double similarity with κ if $\kappa \epsilon \text{Cls}^2 \text{excl}$ (*112·14); in this case, the double correlator is $\breve{\iota} | D$ with its converse domain limited, *i.e.*

$$\vdash : \kappa \epsilon \text{Cls}^2 \text{excl} . \supset . \breve{\iota} | D \upharpoonright \Sigma`\kappa \epsilon \kappa \, \overline{\text{sm}} \, \overline{\text{sm}} \, (\epsilon \downarrow``\kappa).$$

In the case of Prod$`\kappa$ and $\epsilon_\Delta`\kappa$, the correlator is D, *i.e.*

$$\vdash : \kappa \epsilon \text{Cls}^2 \text{excl} . \supset . D \upharpoonright \epsilon_\Delta`\kappa \epsilon (\text{Prod}`\kappa) \, \overline{\text{sm}} \, (\epsilon_\Delta`\kappa).$$

In the case of $\epsilon_\Delta`s`\kappa$ and $\epsilon_\Delta`\epsilon_\Delta``\kappa$, the correlator is $\dot{s} | D$, *i.e.*

$$\vdash : \kappa \epsilon \text{Cls}^2 \text{excl} . \supset . \dot{s} | D \upharpoonright \epsilon_\Delta`\epsilon_\Delta``\kappa \epsilon (\epsilon_\Delta`s`\kappa) \, \overline{\text{sm}} \, (\epsilon_\Delta`\epsilon_\Delta``\kappa).$$

$\dot{s} | D$ also correlates $\epsilon_\Delta`\kappa$ with $\epsilon_\Delta`\epsilon \downarrow``\kappa$ (*85·61) and $P_\Delta`\alpha$ with $\epsilon_\Delta`P \downarrow``\alpha$ (*85·53), and $P_\Delta`s`\kappa$ with $\epsilon_\Delta`P_\Delta``\kappa$ (*85·27·42) if $\kappa \epsilon \text{Cls}^2 \text{excl}$.

The correlator of $(\alpha \uparrow \beta)_\Delta`\beta$ with $(\alpha \exp \beta)$ is $\dot{s}$ (*116·131).

Another kind of correlators arises where we are given a correlator of κ and λ, and we wish to construct a correlator for some associated classes $W`\kappa$ and $W`\lambda$, or where we are given correlators of α with γ and of β with δ, and we wish to construct a correlator of $\alpha \,♀\, \beta$ with $\gamma \,♀\, \delta$, where ♀ is some double descriptive function in the sense of *38. In this case, the correlator will usually be of the form $R \| \breve{S}$ (with a limited converse domain). Sometimes R and S will be identical; sometimes S will be R_ϵ. Such correlators always depend upon

***55·61.** $\vdash : \text{E} ! R`x . \text{E} ! S`y . \supset . (R \| \breve{S})`(x \downarrow y) = (R`x) \downarrow (S`y)$

together with the propositions *74·77 *seq.* giving cases in which $(R \| \breve{S}) \upharpoonright \lambda$ is a one-one relation. It follows from *55·61 that if R and S are correlators whose converse domains include the domain and converse domain respectively of a relation P, then $(R \| \breve{S})`P$ will be a relation holding between $R`x$ and $S`y$ whenever P holds between x and y. Examples of such correlators as $R \| \breve{S}$ are

$*112{\cdot}153.\ \vdash : T \in \kappa \,\overline{\text{sm}}\,\overline{\text{sm}}\, \lambda \,.\supset.\, (T \| \breve{T}_\epsilon) \upharpoonright s`\epsilon \downarrow``\lambda \in (\epsilon \downarrow``\kappa) \,\overline{\text{sm}}\,\overline{\text{sm}}\, (\epsilon \downarrow``\lambda)$

$*113{\cdot}127.\ \vdash : R \upharpoonright \gamma \in \alpha \,\overline{\text{sm}}\, \gamma \,.\, S \upharpoonright \delta \in \beta \,\overline{\text{sm}}\, \delta \,.\supset.$

$$(R \| \breve{S}) \upharpoonright (\delta \times \gamma) \in (\alpha \downarrow_{,,} ``\beta) \,\overline{\text{sm}}\,\overline{\text{sm}}\, (\gamma \downarrow_{,,} ``\delta)$$

$*113{\cdot}65.\ \vdash . \downarrow z``\alpha \times \downarrow z``\beta = (\downarrow z \| \text{Cnv}`\downarrow z)``(\alpha \times \beta)$

$*114{\cdot}51.\ \vdash : T \upharpoonright s`\lambda \in \kappa \,\overline{\text{sm}}\,\overline{\text{sm}}\, \lambda \,.\supset.\, (T \| \breve{T}_\epsilon) \upharpoonright \epsilon_\Delta`\lambda \in (\epsilon_\Delta`\kappa) \,\overline{\text{sm}}\, (\epsilon_\Delta`\lambda)$

$*116{\cdot}192.\ \vdash : R \upharpoonright \gamma \in \alpha \,\overline{\text{sm}}\, \gamma \,.\, S \upharpoonright \delta \in \beta \,\overline{\text{sm}}\, \delta \,.\supset.$

$$(R \| \breve{S}) \upharpoonright (\delta \times \gamma) \in (\alpha \exp \beta) \,\overline{\text{sm}}\,\overline{\text{sm}}\, (\gamma \exp \delta) \,.$$
$$(R \| \breve{S})_\epsilon \upharpoonright (\gamma \exp \delta) \in (\alpha \exp \beta) \,\overline{\text{sm}}\, (\gamma \exp \delta)$$

An exceptionally simple correlator is given by

$*115{\cdot}502.\ \vdash : T \upharpoonright s`\lambda \in \kappa \,\overline{\text{sm}}\,\overline{\text{sm}}\, \lambda \,.\supset.\, T \upharpoonright s`\text{Prod}`\lambda \in (\text{Prod}`\kappa) \,\overline{\text{sm}}\,\overline{\text{sm}}\, (\text{Prod}`\lambda)$

Another exceptionally simple case is

$*73{\cdot}63.\ \vdash : S \in \alpha \,\overline{\text{sm}}\, \beta \,.\, T \upharpoonright \alpha, T \upharpoonright \beta \in 1 \to 1 \,.\, \alpha \cup \beta \subset \text{Ɑ}`T \,.\supset.$

$$T \,|\, S \,|\, \breve{T} \in (T``\alpha) \,\overline{\text{sm}}\, (T``\beta)$$

By means of the above correlators, most correlators that are required can be calculated. Thus it will be seen that $*116{\cdot}192$ in the above list is an immediate consequence of $*113{\cdot}127$ and $*115{\cdot}502$, since

$$\alpha \exp \beta = \text{Prod}`\alpha \downarrow_{,,} ``\beta \text{ and } s`\text{Prod}`\gamma \downarrow_{,,} ``\delta = \delta \times \gamma.$$

In order to develop the subject, it is almost always necessary, not merely to prove that two classes are similar, but actually to construct a correlator of the two classes. This applies equally to relation-arithmetic, in which analogous correlators are used to prove ordinal similarity.

SECTION C.

FINITE AND INFINITE.

Summary of Section C.

The distinction of finite and infinite is not required, as appears from Section B, for the definition of the arithmetical operations or for the proof of their formal laws. There are, however, many important respects in which finite cardinals and classes differ respectively from infinite cardinals and classes, and these differences must now be investigated.

There are two different ways in which we may define the finite and the infinite, and these two ways cannot (so far as is known at present) be shown to be equivalent except by assuming the multiplicative axiom. As there seems no good reason for regarding one of these ways as giving more exactly than the other what is usually meant by the words "finite" and "infinite," we shall, to avoid confusion, give other names than these to each of the two ways of dividing classes and cardinals. The division effected by the first method of definition we shall call the division into *inductive* and *non-inductive*; that effected by the second method we shall call the division into *non-reflexive* and *reflexive.*

The division into inductive and non-inductive, which is treated in *120, is defined as follows. An inductive cardinal is one which can be reached from 0 by successive additions of 1; that is, an inductive cardinal is one which has to 0 the relation $(+_c 1)_*$, where (by *38·02) $+_c 1$ is the relation of $\alpha +_c 1$ to α, and the subscript asterisk has the meaning defined in *90. Hence we put

$$\text{NC induct} = \hat{\alpha}\{\alpha(+_c 1)_* 0\} \quad \text{Df.}$$

By applying the definition of *90, this gives

$$\vdash :: \alpha \epsilon \text{NC induct} . \equiv :. \xi \epsilon \mu . \supset_\xi . \xi +_c 1 \epsilon \mu : 0 \epsilon \mu : \supset_\mu . \alpha \epsilon \mu.$$

This proposition may be regarded as stating that an inductive cardinal is one which obeys mathematical induction starting from 0, *i.e.* it is one which possesses every property possessed by 0 and by the numbers obtained by adding 1 to numbers possessing the property. In elementary mathematics, it is customary to regard mathematical induction, as applied to the series of natural numbers, as a principle rather than a definition, but according to the above procedure it becomes a definition rather than a principle. This procedure is unavoidable as soon as it is perceived that there are cardinals which do not obey mathematical induction starting from 0. (This only holds on the assumption that the total number of objects in any one type is not one of the inductive cardinals. This assumption, in a slightly different form, is introduced below as the "axiom of infinity.") Thus for example $0 \neq 1$, and $\xi \neq \xi +_c 1 . \supset . \xi +_c 1 \neq \xi +_c 2$. Hence if α is any inductive cardinal, $\alpha \neq \alpha +_c 1$. But we know that $\aleph_0$, the first of Cantor's transfinite cardinals*, satisfies $\aleph_0 = \aleph_0 +_c 1$. Thus mathematical induction starting from 0 cannot be validly applied to prove properties of $\aleph_0$. It follows that the inductive cardinals as above defined are only some among cardinals; nor does it appear that there is any way of defining them except as those that obey mathematical induction starting from 0. It follows that mathematical induction is not a principle, to be either proved or assumed as an axiom, but is merely a characteristic defining a certain class of cardinals, namely the class of inductive cardinals.

By a syllogism in Barbara, it is evident that 0 is an inductive cardinal; hence by the definition 1 is an inductive cardinal, and hence 2, 3, ... are inductive cardinals. Thus any given cardinal in the series of natural numbers can be shown to be an inductive cardinal. The usual elementary properties of inductive cardinals, such as the uniqueness of subtraction and division, are easily proved by mathematical induction.

We define an inductive class as a class the number of whose terms is an inductive cardinal. More simply, we put

$$\text{Cls induct} = s\text{`NC induct} \quad \text{Df.}$$

It is then easily shown that an inductive class is one which can be reached from Λ by successive additions of single members. That is, if we put

$$M = \hat{\eta}\hat{\zeta}\{(\exists y) . \zeta = \eta \cup \iota\text{`}y\},$$

then

$$\text{Cls induct} = \overleftarrow{M_*}\text{`}\Lambda.$$

Thus we have

$$\vdash :: \rho \in \text{Cls induct} . \equiv :. \eta \in \mu . \supset_{\eta, y} . \eta \cup \iota\text{`}y \in \mu : \Lambda \in \mu : \supset_\mu . \rho \in \mu.$$

We might equally well have begun by defining inductive classes, and proceeded to define inductive cardinals as the cardinals of inductive classes; in that case, we should have used the above relation M to define inductive classes.

Some of the properties which we expect inductive cardinals to possess, such for example as $\alpha \neq \alpha +_c 1$, can only be proved by assuming that no inductive cardinal is null, *i.e.* that

$$\alpha \in \text{NC induct} . \supset_\alpha . \exists ! \alpha.$$

This amounts to the assumption that, in any fixed type, a class can be found

* For the definition of $\aleph_0$, cf. *123·01 and p. 192 of this summary.

having any assigned inductive number of terms. If this were false, there would have to be some definite member of the series of natural numbers which gave the total number of objects of the type in question. Thus suppose there were exactly n individuals in the universe, and no more, where n is an inductive cardinal. We should then have 2^n classes, 2^{2^n} classes of classes, and so on. In that case, in the type of individuals we should have $n +_c 1 = \Lambda$, $n +_c 2 = \Lambda$, etc. Hence we should have

$$n +_c 1 = (n +_c 1) +_c 1, \text{ etc.}$$

In the type of classes, we should get similar results for 2^n, and so on. It is plain (though not demonstrable except in each particular case) that if the assumption $\alpha \epsilon \text{NC induct} . \supset_\alpha . \exists ! \alpha$ fails in any one type, it fails in any other type in the same hierarchy, and if it holds in any one, it holds in any other; for if n be the total number of individuals, then if n is an inductive cardinal, the total number of any other type is an inductive cardinal, while if n is not an inductive cardinal, no more is the total number of any other type. Hence the assumption $\alpha \epsilon \text{NC induct} . \supset_\alpha . \exists ! \alpha$ is either true in any type or false in any type in one hierarchy. We shall call it the "axiom of infinity," putting

$$\text{Infin ax} . = : \alpha \epsilon \text{NC induct} . \supset_\alpha . \exists ! \alpha \quad \text{Df.}$$

This assumption, like the multiplicative axiom, will be adduced as a hypothesis whenever it is relevant. It seems plain that there is nothing in logic to necessitate its truth or falsehood, and that it can only be legitimately believed or disbelieved on empirical grounds. When we wish to use a typically definite form of the axiom, we shall employ the definition

$$\text{Infin ax}(x) . = : \alpha \epsilon \text{NC induct} . \supset_\alpha . \exists ! \alpha(x) \quad \text{Df}$$

which asserts that, if α is any inductive cardinal, there are at least α terms of the same type as x.

It is important to observe that, although the axiom of infinity cannot (so far as appears) be proved a priori, we can prove that any given inductive cardinal exists in a sufficiently high type. For if the total number of individuals be n, the numbers of objects in succeeding types are 2^n, 2^{2^n}, etc., and these numbers grow beyond any assigned inductive cardinal. Owing, however, to the fact that we cannot add together an infinite number of classes whose types increase without limit, we cannot hence show that there is a type in which every inductive cardinal exists, though we can show of every inductive cardinal that there is a type in which it exists. *I.e.* if α is any inductive cardinal, there must be a type for x such that $\exists ! \alpha(x)$ is true; but there need not be a type for x such that if α is any inductive cardinal, $\exists ! \alpha(x)$ is true.

The axiom of infinity suffices to prove the existence, in appropriate types, of $\aleph_0$, $2^{\aleph_0}$, $2^{2^{\aleph_0}}$, ... $\aleph_1$, $\aleph_2$, ...*. It does not suffice, so far as we know, to

* For the definitions of $\aleph_1$, $\aleph_2$, etc., see *265.

prove the existence of $\aleph_\omega$ or any Aleph with a greater suffix than ω, because the existences of $\aleph_1, \aleph_2, \ldots$ are proved in successively rising types, and no meaning can be found for a type whose order is infinite.

The other definition of finite and infinite is of less importance in practice than the definition by induction. It is dealt with in *124. According to this definition, we call a class *reflexive* when it contains a proper part similar to itself, *i.e.* we put

$$\text{Cls refl} = \hat{\alpha} \{(\exists R) . R \epsilon 1 \rightarrow 1 . \text{D}'R = \alpha . \text{G}'R \subset \alpha . \text{G}'R \neq \alpha\} \quad \text{Df},$$

or, what comes to the same thing,

$$\text{Cls refl} = \hat{\alpha} \{(\exists R) . R \epsilon 1 \rightarrow 1 . \text{G}'R \subset \text{D}'R . \exists ! \overrightarrow{B}'R . \alpha = \text{D}'R\} \quad \text{Df}.$$

We call a cardinal reflexive when it is the homogeneous cardinal of a reflexive class, *i.e.* we put

$$\text{NC refl} = \text{N}_0\text{c}``\text{Cls refl} \quad \text{Df}.$$

It is easy to show that

$$\text{NC refl} = \hat{\alpha} \{\exists ! \alpha . \alpha = \alpha +_c 1\}.$$

We find that inductive classes and cardinals are non-reflexive, and reflexive classes and cardinals are non-inductive. We find also that reflexive cardinals are those that are equal to or greater than $\aleph_0$, while inductive cardinals are those that are less than $\aleph_0$. By assuming the multiplicative axiom, we can show that every cardinal is equal to, greater than, or less than $\aleph_0$, whence it follows that every cardinal is either reflexive or inductive, thus identifying the two definitions of finite and infinite. But so long as we refrain from assuming either the multiplicative axiom or some special axiom *ad hoc*, it remains possible (so far as is known at present) that there may be cardinals neither greater than, nor equal to, nor less than $\aleph_0$. Such cardinals, if they exist, are neither inductive nor reflexive: they are infinite if we define infinity by the negation of induction, but finite if we define infinity by reflexiveness. It is possible that further investigation may either prove or disprove the existence of such cardinals; for the present, their existence must remain an open question, except for those who regard the multiplicative axiom as a self-evident truth.

In *121 we shall consider *intervals* in a discrete series; *i.e.* in a series generated by a one-one relation between consecutive terms. If P be the generating relation of such a series, and x and y be two members of the series, of which y is the later, the terms which lie between x and y are the terms z for which we have

$$xP_{\text{po}}z . zP_{\text{po}}y,$$

where P_{po} has the meaning defined in *91. Hence we put

$$P(x - y) = \overleftarrow{P}_{\text{po}}`x \cap \overrightarrow{P}_{\text{po}}`y \quad \text{Df},$$

where "$P(x-y)$" means "the P-interval between x and y." We want also symbols for the interval together with one or both of its end-points. For these we put

$$P(x \dashv y) = \overleftarrow{P_{po}}`x \cap \overrightarrow{P_*}`y \quad \text{Df},$$
$$P(x \vdash y) = \overleftarrow{P_*}`x \cap \overrightarrow{P_{po}}`y \quad \text{Df},$$
$$P(x \vdash\dashv y) = \overleftarrow{P_*}`x \cap \overrightarrow{P_*}`y \quad \text{Df}^*.$$

Thus, for example, if x and y be inductive cardinals, and P be the relation of n to $n +_c 1$, and $x < y$, $P(x - y)$ will be the numbers greater than x and less than y, while $P(x \dashv y)$ will be these numbers together with y, $P(x \vdash y)$ will be these numbers together with x, and $P(x \vdash\dashv y)$ will be these numbers together with both x and y. By means of intervals, we define a class of relations P_ν (where ν is any inductive cardinal), where "$xP_\nu z$" means that we can pass from x to z in ν steps. In order to fit the case in which x and z are identical, and to insure that no relation such as P_ν shall hold between terms which do not both belong to the field of P, we put

$$P_\nu = \hat{x}\hat{y}\,\{\text{Nc}`P(x \vdash\dashv y) = \nu +_c 1\} \quad \text{Df}.$$

Then, provided $P_{po} \subset J$, $P_0 = I \upharpoonright C`P$, and if further $P \in 1 \to 1$, then $P_1 = P$, $P_2 = P^2$, etc. If P is a transitive serial relation, P_1 is the relation "immediately preceding," which has great importance in well-ordered series. In this case, $P_1 = P \dot{-} P^2$. If P is a transitive serial relation generating a finite series or a progression or a series of the type of the negative and positive integers in order of magnitude, we have

$$P = (P_1)_{po}.$$

In $*121$ we shall only consider P_ν in the case where

$$P \in (1 \to \text{Cls}) \cup (\text{Cls} \to 1),$$

and generally we shall have the further hypothesis $P_{po} \subset J$. We can then prove that the interval between x and y is always an inductive class (it will be null unless $xP_* y$); this proposition is useful in its application to the number-series and to progressions generally.

When $P \in (1 \to \text{Cls}) \cup (\text{Cls} \to 1) \,.\, P_{po} \subset J$, the class of such relations as P_ν (where ν is an inductive cardinal) is identical with Potid$`P$, the class of powers of P (cf. $*91$ *seq.*). This identification (which does not hold in general without the above hypothesis) leads to many useful propositions. In $*91$ *seq.*, we treated powers of a relation without the use of numbers, *i.e.* without defining the νth power of P. When the powers of P are the class of such relations as P_ν, we can of course take P_ν as the νth power of P. The general definition of the νth power of P (where ν is an inductive cardinal) will be given later, in $*301$; we shall denote it by P^ν, thereby including the notation P^2 already defined.

* These symbols are suggested by those given in Peano's *Formulaire*, Vol. IV. p. 116. (*Algèbre*, § 46.)

In $*122$ we shall deal with progressions, *i.e.* with series of the type of the series of natural numbers. In this number, we shall deal with such series as generated by one-one relations; they will be dealt with at a later stage ($*263$) as generated by transitive relations. We define a progression as a one-one relation whose domain is the posterity of its first term, *i.e.*

$$\text{Prog} = (1 \rightarrow 1) \cap \hat{R}(\text{D}\text{'}R = \overleftarrow{R_*}\text{'}B\text{'}R) \quad \text{Df.}$$

According to this definition, there must be a first term $B\text{'}R$; $\text{Q}\text{'}R$ will be $\breve{R}\text{''}\overleftarrow{R_*}\text{'}B\text{'}R$, *i.e.* $\overleftarrow{R_{po}}\text{'}B\text{'}R$, which is contained in $\overleftarrow{R_*}\text{'}B\text{'}R$, *i.e.* in $\text{D}\text{'}R$; since $\text{Q}\text{'}R \subset \text{D}\text{'}R$, every term of the field of R has a successor, so that there is no end to the series; since $C\text{'}R = \text{D}\text{'}R = \overleftarrow{R_*}\text{'}B\text{'}R$, every term of the series can be reached from the beginning by successive steps. These characteristics suffice to define progressions.

In $*123$ we proceed to the definition and discussion of $\aleph_0$, the smallest of reflexive cardinals. This is the cardinal number of any class whose terms can be arranged in a progression; hence it is the class of domains of progressions, *i.e.* we may put

$$\aleph_0 = \text{D}\text{''}\text{Prog} \quad \text{Df.}$$

With this definition, remembering that Λ is a cardinal, we can prove that $\aleph_0$ is a cardinal; but to prove that $\aleph_0$ is an *existent* cardinal, we need the axiom of infinity. The existence-theorem for $\aleph_0$ is then derived from the inductive cardinals, which, if no one of them is null, form a progression when arranged in order of magnitude. It should be observed that this existence-theorem is for a higher type than that for which the axiom of infinity is assumed. In order to get an existence-theorem for the same type, we need the multiplicative axiom as well.

After a number on reflexive classes and cardinals ($*124$) and a number on the axiom of infinity ($*125$), the Section ends with a number ($*126$) on "typically indefinite inductive cardinals." The constant inductive cardinals are the typically ambiguous symbols 0, 1, 2, ... ; thus we want to define the class of inductive cardinals in such a way that a variable member of the class shall be typically ambiguous. This is not possible without a sacrifice of rigour, but in $*126$ it is shown how to minimize the sacrifice of rigour, and how to obviate the resulting logical dangers. A variable whose values are typically ambiguous is said to be "typically indefinite."

A proof that all inductive cardinals exist has often been derived from $*120{\cdot}57$ (below). But according to the doctrine of types, this proof is invalid, since "$\mu +_0 1$" in $*120{\cdot}57$ is necessarily of higher type than "μ."

$*118$. ARITHMETICAL SUBSTITUTION AND UNIFORM FORMAL NUMBERS.

Summary of $*118$.

A difficulty arises respecting substitution in arithmetic. For if μ is a formal number and its occurrence in $f\mu$ is arithmetical, then by IIT μ is always to be taken in an existential type. Hence we can only substitute a real variable ξ for μ under the hypothesis $\exists ! \xi$, and we can only substitute another formal number σ for μ provided that the equation $\mu = \sigma$, which justifies the substitution, is arithmetical, *i.e.* provided that in this equation the type of μ is such that $\exists ! \mu$.

The result is that the application of $*20{\cdot}18$ is apt to lead to fallacies owing to the different meanings which a formal number may possess in different occurrences. Hitherto we have considered each case in detail, *e.g.* note on $*110{\cdot}61$, and proof of $*110{\cdot}56$.

The condition for the safe application of $*20{\cdot}18$ is given in $*118{\cdot}01$, namely

$*118{\cdot}01$. $\vdash :. \exists ! \mu . \mu = \sigma . \supset : f\mu . \equiv . f\sigma \quad [*20{\cdot}18]$

This question is more fully discussed in the prefatory statement of this volume. The first reference to $*118{\cdot}01$ is in $*120{\cdot}222$. Another way of evading the difficulty is to work with formal numbers which, together with all their components, are of the same type. This leads to the consideration of Uniform Formal Numbers, which with the exception of $*118{\cdot}01$ occupies the rest of the number.

The *dominant type* of a formal number as used in any context is the type of the formal number itself in that context, and the *subordinate types* of the formal number are the dominant types of its component formal numbers.

When the dominant types of some of the formal numbers are not expressly indicated by an explicit notation (cf. $*65$), the rules according to which the dominant types thus left ambiguous are to be related, so far as they are related, including the rules governing the relation of subordinate types, if left ambiguous, to dominant types, are given by conventions IT, IIT, and AT of the prefatory statement in this volume.

We have now to consider an important special case which arises when types are explicitly indicated by the use of $*65{\cdot}01{\cdot}03$. A formal number, whose subordinate types are the same as its dominant type, is called *uniform*; and if some of its subordinate types are the same as its dominant type, it is called *partially uniform*. A formal number can only be partially uniform, or at least so designated as to be necessarily partially uniform, when the dominant type and those subordinate types identical with it are expressly indicated by ∗65·01·03. For otherwise the conventions IT, IIT, and perhaps also AT, apply; and these do not secure uniformity, and may perhaps in some contexts be inconsistent with it.

Common sense in its consideration of arithmetic habitually disregards the possibility of a formal number representing Λ. In other words, it always applies conventions IIT and AT. But also, owing to its disregard of types, it assumes that the formal numbers are all uniform. The assumption which is really essential to this common sense reasoning, so far as the form of its arithmetical conclusions are concerned, is the assumption that none of the numerical symbols represent Λ. This assumption is secured here, when no types are expressly indicated, by IIT and AT. We have now to consider the effect on arithmetical operations of the other assumption, that the formal numbers are uniform, or partially uniform. There is no difficulty arising from any change of convention for symbolism, since, as stated above, partial or complete uniformity is secured by express indication of type. Accordingly conventions IT, IIT continue, as always, to apply when the types of formal numbers are left ambiguous.

Convention AT will not be applied either in ∗118 or ∗119 or ∗120: in ∗118 the fact is entirely unimportant since the dominant types of equational occurrences are always indicated, so that no case arises when it could apply.

Apart from its intrinsic interest and its bearing on substitution, the arithmetic of uniform formal numbers is necessary for ∗120, where the fundamental arithmetical properties of inductive numbers are investigated.

The propositions of this number are proved by the use of the results of ∗117. The basis of the reasoning is

∗118·13. $\vdash :. \mu \leqslant \nu . \supset : \exists ! \, \mathrm{sm}_\xi{}^{\iota\iota}\nu . \supset . \exists ! \, \mathrm{sm}_\xi{}^{\iota\iota}\mu$

In ∗118·2·3·4 the meaning of the symbolism for dominant types is stated, namely

∗118·2. $\vdash . (\mu +_c \nu)_\xi = \hat{\eta} \{(\exists \alpha, \beta) . \mu = \mathrm{N_0c}{}^{\iota}\alpha . \nu = \mathrm{N_0c}{}^{\iota}\beta . \eta \, \mathrm{sm}_\xi (\alpha + \beta)\}$

∗118·3. $\vdash . (\mu \times_c \nu)_\xi = \hat{\eta} \{(\exists \alpha, \beta) . \mu = \mathrm{N_0c}{}^{\iota}\alpha . \nu = \mathrm{N_0c}{}^{\iota}\beta . \eta \, \mathrm{sm}_\xi (\alpha \times \beta)\}$

∗118·4. $\vdash . (\mu^\nu)_\xi = \hat{\eta} \{(\exists \alpha, \beta) . \mu = \mathrm{N_0c}{}^{\iota}\alpha . \nu = \mathrm{N_0c}{}^{\iota}\beta . \eta \, \mathrm{sm}_\xi (\alpha \exp \beta)\}$

The important propositions which are finally reached for addition are

∗118·23. $\vdash : \mu, \nu \in \mathrm{NC} . \supset . (\mu +_c \nu)_\xi = (\mathrm{sm}_\xi{}^{\iota\iota}\mu +_c \mathrm{sm}_\xi{}^{\iota\iota}\nu)_\xi$

∗118·24. $\vdash : \nu \in \mathrm{NC} . \supset . (\mu +_c \nu)_\xi = (\mu +_c \mathrm{sm}_\xi{}^{\iota\iota}\nu)_\xi$

***118·241.** $\vdash : \mu \in \mathrm{NC} \,.\, \supset \,.\, (\mu +_c \nu)_\xi = (\mathrm{sm}_\xi{}^{\prime\prime}\mu +_c \nu)_\xi$

***118·25.** $\vdash \,.\, (\mu +_c \nu +_c \varpi)_\xi = \{(\mu +_c \nu)_\xi +_c \varpi\}_\xi = \{\mu +_c (\nu +_c \varpi)_\xi\}_\xi$

The important propositions for multiplication are

***118·33.** $\vdash : \mu, \nu \in \mathrm{NC} - \iota\text{‘}0 \,.\, \supset \,.\, (\mu \times_c \nu)_\xi = (\mathrm{sm}_\xi{}^{\prime\prime}\mu \times_c \mathrm{sm}_\xi{}^{\prime\prime}\nu)_\xi$

***118·34.** $\vdash : \nu \in \mathrm{NC} \,.\, \mu \neq 0 \,.\, \supset \,.\, (\mu \times_c \nu)_\xi = (\mu \times_c \mathrm{sm}_\xi{}^{\prime\prime}\nu)_\xi$

***118·341.** $\vdash : \mu \in \mathrm{NC} \,.\, \nu \neq 0 \,.\, \supset \,.\, (\mu \times_c \nu)_\xi = (\mathrm{sm}_\xi{}^{\prime\prime}\mu \times_c \nu)_\xi$

***118·35.** $\vdash : \varpi \neq 0 \,.\, \supset \,.\, (\mu \times_c \nu \times_c \varpi)_\xi = \{(\mu \times_c \nu)_\xi \times_c \varpi\}_\xi$

***118·351.** $\vdash : \mu \neq 0 \,.\, \supset \,.\, (\mu \times_c \nu \times_c \varpi)_\xi = \{\mu \times_c (\nu \times_c \varpi)_\xi\}_\xi$

The important propositions for exponentiation are

***118·43.** $\vdash : \mu, \nu \in \mathrm{NC} - \iota\text{‘}0 \,.\, \mu \neq 1 \,.\, \supset \,.\, (\mu^\nu)_\xi = \{(\mathrm{sm}_\xi{}^{\prime\prime}\mu)^{\mathrm{sm}_\xi{}^{\prime\prime}\nu}\}_\xi$

***118·44.** $\vdash : \nu \in \mathrm{NC} \,.\, \mu \neq 0 \,.\, \mu \neq 1 \,.\, \supset \,.\, (\mu^\nu)_\xi = (\mu^{\mathrm{sm}_\xi{}^{\prime\prime}\nu})_\xi$

***118·441.** $\vdash : \mu \in \mathrm{NC} \,.\, \nu \neq 0 \,.\, \supset \,.\, (\mu^\nu)_\xi = \{(\mathrm{sm}_\xi{}^{\prime\prime}\mu)^\nu\}_\xi$

***118·45.** $\vdash : \mu \neq 0 \,.\, \mu \neq 1 \,.\, \supset \,.\, (\mu^{\nu \times_c \varpi})_\xi = \{\mu^{(\nu \times_c \varpi)_\xi}\}_\xi$

***118·451.** $\vdash : \varpi \neq 0 \,.\, \supset \,.\, (\mu^{\nu \times_c \varpi})_\xi = [\{(\mu^\nu)_\xi\}^\varpi]_\xi$

***118·46.** $\vdash : \mu \neq 0 \,.\, \mu \neq 1 \,.\, \supset \,.\, (\mu^{\nu +_c \varpi})_\xi = \{\mu^{(\nu +_c \varpi)_\xi}\}_\xi$

***118·461.** $\vdash \,.\, (\mu^{\nu +_c \varpi})_\xi = \{(\mu^\nu)_\xi \times_c (\mu^\varpi)_\xi\}_\xi$

with two analogous propositions *118·462·463,

***118·47.** $\vdash : \varpi \neq 0 \,.\, \supset \,.\, \{(\mu \times_c \nu)^\varpi\}_\xi = [\{(\mu \times_c \nu)_\xi\}^\varpi]_\xi$

***118·471.** $\vdash :. \mu \neq 0 \,.\, \nu \neq 0 \,.\, \mathbf{v} \,.\, \varpi = 0 \,.\, \mathbf{v} \,.\, \sim(\mu, \nu, \varpi \in \mathrm{N}_0\mathrm{C}) : \supset \,.$

$$\{(\mu \times_c \nu)^\varpi\}_\xi = \{(\mu^\varpi)_\xi \times_c (\nu^\varpi)_\xi\}_\xi$$

with two analogous propositions *118·472·473.

It is thus seen that, apart from some exceptional cases connected with 0 and 1, in all arithmetical operations uniform, or partially uniform, formal numbers can replace those constructed in obedience to convention IIT.

***118·01.** $\vdash :. \exists ! \mu \,.\, \mu = \sigma \,.\, \supset : f\mu \,.\, \equiv \,.\, f\sigma$ [*20·18]

As far as the symbolism is concerned, this proposition with the omission of $\exists ! \mu$ from the hypothesis is a transcript of *20·18. But if μ or σ (not excluding both) is a formal number, $\exists ! \mu$ is required in case the occurrence of μ in $f\mu$ is arithmetical. In fact this proposition embodies the three fundamental propositions of the Principle of Arithmetical Substitution arrived at in the Prefatory Explanations on Types. Its necessity arises from the convention IIT which is explained there.

✳118·11. $\vdash : \exists ! \mathrm{Nc}(\xi)`\beta \,.\, \alpha \subset \beta \,.\, \supset .\, \exists ! \mathrm{Nc}(\xi)`\alpha$

Dem.

$\vdash .\, ✳100{\cdot}31 \,.\, \supset \vdash :. \mathrm{Hp} \,.\, \supset :$

$\gamma \in \mathrm{Nc}(\xi)`\beta \,.\, \supset .\, \gamma \,\mathrm{sm}_\xi\, \beta \,.$

$[✳73{\cdot}1] \quad \supset .\, (\exists R) \,.\, R \in 1(\xi) \to 1 \,.\, \gamma = \mathrm{D}`R \,.\, \beta = \mathrm{G}`R \,.$

$[✳22{\cdot}55] \quad \supset .\, (\exists R) \,.\, R \in 1(\xi) \to 1 \,.\, \alpha \subset \mathrm{G}`R \,.\, R``\alpha = R``\alpha \,.$

$[✳73{\cdot}12] \quad \supset .\, (\exists R) \,.\, R``\alpha \,\mathrm{sm}_\xi\, \alpha \,.$

$[✳100{\cdot}31] \quad \supset .\, \exists ! \mathrm{Nc}(\xi)`\alpha : \supset \vdash .\, \mathrm{Prop}$

✳118·12. $\vdash :. \mathrm{Nc}`\alpha \leqslant \mathrm{Nc}`\beta \,.\, \supset : \exists ! \mathrm{Nc}(\xi)`\beta \,.\, \supset .\, \exists ! \mathrm{Nc}(\xi)`\alpha$
[✳117·32·107 . ✳100·511]

✳118·13. $\vdash :. \mu \leqslant \nu \,.\, \supset : \exists ! \mathrm{sm}_\xi``\nu \,.\, \supset .\, \exists ! \mathrm{sm}_\xi``\mu$ [✳117·32]

✳118·2. $\vdash .\, (\mu +_c \nu)_\xi = \hat{\eta} \{(\exists \alpha, \beta) \,.\, \mu = \mathrm{N_0c}`\alpha \,.\, \nu = \mathrm{N_0c}`\beta \,.\, \eta \,\mathrm{sm}_\xi\, (\alpha + \beta)\}$
[(✳65·01·03) . ✳110·2]

✳118·201. $\vdash : \exists ! (\mu +_c \nu) \,.\, \supset .\, \mathrm{sm}_\xi``(\mu +_c \nu) = (\mu +_c \nu)_\xi$
[✳110·44. Note Erratum in enunciation]

✳118·21. $\vdash : \exists ! (\mu +_c \nu)_\xi \,.\, \supset .\, \exists ! \mathrm{sm}_\xi``\mu \,.\, \exists ! \mathrm{sm}_\xi``\nu$

Dem.

$\vdash .\, ✳110{\cdot}4 \,.\, ✳118{\cdot}2 \,.\, \supset \vdash : \mathrm{Hp} \,.\, \supset .\, \mu, \nu \in \mathrm{N_0C} \,.$

$[✳117{\cdot}6] \quad \supset .\, \mu +_c \nu \geqslant \mu \,.\, \mu +_c \nu \geqslant \nu \,.$

$[✳118{\cdot}13{\cdot}201.(\mathrm{IIT})] \quad \supset .\, \exists ! \mathrm{sm}_\xi``\mu \,.\, \exists ! \mathrm{sm}_\xi``\nu : \supset \vdash .\, \mathrm{Prop}$

Here the reference (IIT) is to the convention IIT explained in the prefatory statement.

✳118·22. $\vdash :. \mu, \nu \in \mathrm{NC} \,.\, \supset : \exists ! (\mu +_c \nu)_\xi \,.\, \equiv .\, \exists ! (\mathrm{sm}_\xi``\mu +_c \mathrm{sm}_\xi``\nu)_\xi \,.\, \equiv .$
$\exists ! (\mu +_c \mathrm{sm}_\xi``\nu)_\xi \,.\, \equiv .\, \exists ! (\mathrm{sm}_\xi``\mu +_c \nu)_\xi$

Dem.

$\vdash .\, ✳118{\cdot}21 \,.\, \supset \vdash :. \mathrm{Hp} \,.\, \supset : \exists ! (\mu +_c \nu)_\xi \,.\, \equiv .\, \exists ! (\mu +_c \nu)_\xi \,.\, \exists ! \mathrm{sm}_\xi``\mu \,.\, \exists ! \mathrm{sm}_\xi``\nu \,.$

$[✳110{\cdot}25{\cdot}4] \quad \equiv .\, \exists ! (\mathrm{sm}_\xi``\mu +_c \mathrm{sm}_\xi``\nu)_\xi \qquad (1)$

$\vdash .\, ✳118{\cdot}21 \,.\, ✳103{\cdot}43 \,.\, ✳110{\cdot}4 \,.\, \supset$

$\vdash :. \mathrm{Hp} \,.\, \supset : \exists ! (\mu +_c \nu)_\xi \,.\, \equiv .\, \exists ! (\mu +_c \nu)_\xi \,.\, \exists ! \mathrm{sm}``\mu \cap t_0`\mu \,.\, \exists ! \mathrm{sm}_\xi``\nu \,.$

$[✳103{\cdot}43 . ✳110{\cdot}25{\cdot}4] \quad \equiv .\, \exists ! (\mu +_c \mathrm{sm}_\xi``\nu)_\xi \qquad (2)$

Similarly $\quad \vdash :. \mathrm{Hp} \,.\, \supset : \exists ! (\mu +_c \nu)_\xi \,.\, \equiv .\, \exists ! (\mathrm{sm}_\xi``\mu +_c \nu)_\xi \qquad (3)$

$\vdash .\, (1) \,.\, (2) \,.\, (3) \,.\, \supset \vdash .\, \mathrm{Prop}$

✳118·23. $\vdash : \mu, \nu \in \mathrm{NC} \,.\, \supset .\, (\mu +_c \nu)_\xi = (\mathrm{sm}_\xi``\mu +_c \mathrm{sm}_\xi``\nu)_\xi$

Dem.

$\vdash .\, ✳118{\cdot}21 \,.\, ✳110{\cdot}4{\cdot}25 \,.\, \supset \vdash : \exists ! (\mu +_c \nu)_\xi \,.\, \supset .\, (\mu +_c \nu)_\xi = (\mathrm{sm}_\xi``\mu +_c \mathrm{sm}_\xi``\nu)_\xi \qquad (1)$

$\vdash .\, ✳118{\cdot}22 \,.\, \supset \vdash : \mathrm{Hp} \,.\, \sim \exists ! (\mu +_c \nu)_\xi \,.\, \supset .\, (\mu +_c \nu)_\xi = (\mathrm{sm}_\xi``\mu +_c \mathrm{sm}_\xi``\nu)_\xi \qquad (2)$

$\vdash .\, (1) \,.\, (2) \,.\, \supset \vdash .\, \mathrm{Prop}$

***118·24.** $\vdash : \nu \in \text{NC} . \supset . (\mu +_c \nu)_\xi = (\mu +_c \text{sm}_\xi``\nu)_\xi$

Dem.

$\vdash . *118·21 . *110·4·25 . *103·43 . \supset$

$\vdash : \exists ! (\mu +_c \nu)_\xi . \supset . (\mu +_c \nu)_\xi = (\mu +_c \text{sm}_\xi``\nu)_\xi$ (1)

$\vdash . *110·4 . \supset \vdash : \mu \sim \in \text{NC} . \supset . (\mu +_c \nu)_\xi = (\mu +_c \text{sm}_\xi``\nu)_\xi$ (2)

$\vdash . *118·22 . \supset \vdash : \text{Hp} . \mu \in \text{NC} . \sim \exists ! (\mu +_c \nu)_\xi . \supset . (\mu +_c \nu)_\xi = (\mu +_c \text{sm}_\xi``\nu)_\xi$ (3)

$\vdash . (1) . (2) . (3) . \supset \vdash .$ Prop

***118·241.** $\vdash : \mu \in \text{NC} . \supset . (\mu +_c \nu)_\xi = (\text{sm}_\xi``\mu +_c \nu)_\xi$ [*118·24 . *110·51]

***118·25.** $\vdash . (\mu +_c \nu +_c \varpi)_\xi = \{(\mu +_c \nu)_\xi +_c \varpi\}_\xi = \{\mu +_c (\nu +_c \varpi)_\xi\}_\xi$

Dem.

$\vdash . *110·42 . *118·241·201 . (\text{IIT}) . \supset$

$\vdash : \mu, \nu \in \text{N}_0\text{C} . \supset . (\mu +_c \nu +_c \varpi)_\xi = \{(\mu +_c \nu)_\xi +_c \varpi\}_\xi$ (1)

$\vdash . *110·4 . \supset \vdash : \sim (\mu, \nu \in \text{N}_0\text{C}) . \supset . \mu +_c \nu = \Lambda . (\mu +_c \nu)_\xi = \Lambda .$

[*110·4] $\supset . (\mu +_c \nu +_c \varpi)_\xi = \{(\mu +_c \nu)_\xi +_c \varpi\}_\xi$ (2)

$\vdash . (1) . (2) . \supset \vdash . (\mu +_c \nu +_c \varpi)_\xi = \{(\mu +_c \nu)_\xi +_c \varpi\}_\xi$ (3)

Similarly $\vdash . (\mu +_c \nu +_c \varpi)_\xi = \{\mu +_c (\nu +_c \varpi)_\xi\}_\xi$ (4)

$\vdash . (3) . (4) . \supset \vdash .$ Prop

***118·3.** $\vdash . (\mu \times_c \nu)_\xi = \hat{\eta} \{(\exists \alpha, \beta) . \mu = \text{N}_0\text{c}`\alpha . \nu = \text{N}_0\text{c}`\beta . \eta \,\text{sm}_\xi\, (\alpha \times \beta)\}$ [(*65·01·03) . *113·2]

***118·301.** $\vdash : \exists ! (\mu \times_c \nu) . \supset . \text{sm}_\xi``(\mu \times_c \nu) = (\mu \times_c \nu)_\xi$ [Proof as in *118·201]

***118·31.** $\vdash : \exists ! (\mu \times_c \nu)_\xi . \nu \neq 0 . \supset . \exists ! \text{sm}_\xi``\mu$

Dem.

$\vdash . *101·15·12 . \supset \vdash : \mu = 0 . \supset . \exists ! \text{sm}_\xi``\mu$ (1)

$\vdash . *113·203 . *118·3 . \supset \vdash : \text{Hp} . \mu \neq 0 . \supset . \mu, \nu \in \text{N}_0\text{C} - \iota`0 .$

[*117·62] $\supset . \mu \times_c \nu \geqslant \mu .$

[*118·13·301.(IIT)] $\supset . \exists ! \text{sm}_\xi``\mu$ (2)

$\vdash . (1) . (2) . \supset \vdash .$ Prop

***118·311.** $\vdash : \exists ! (\mu \times_c \nu)_\xi . \mu \neq 0 . \supset . \exists ! \text{sm}_\xi``\nu$ [*118·31 . *113·27]

***118·32.** $\vdash :. \nu \in \text{NC} . \mu \neq 0 . \supset : \exists ! (\mu \times_c \nu)_\xi . \equiv . \exists ! (\mu \times_c \text{sm}_\xi``\nu)_\xi$

Dem.

$\vdash . *113·203 . \supset \vdash : \exists ! (\mu \times_c \nu)_\xi . \supset . \mu \in \text{NC}$ (1)

$\vdash . *113·203 . \supset \vdash : \exists ! (\mu \times_c \text{sm}_\xi``\nu)_\xi . \supset . \mu \in \text{NC}$ (2)

$\vdash . *113·203 . *118·311 . \supset$

$\vdash :. \text{Hp} . \supset : \exists ! (\mu \times_c \nu)_\xi . \supset . \exists ! \mu . \exists ! \text{sm}_\xi``\nu .$

[*103·43] $\supset . \exists ! \text{sm}``\mu \cap t_0`\mu . \exists ! \text{sm}_\xi``\nu .$

[(1).*113·26.*103·43] $\supset . \exists ! (\mu \times_c \text{sm}_\xi``\nu)_\xi$ (3)

$\vdash . *113·203 . *103·43 . \supset$

$\vdash :. \text{Hp} . \supset : \exists ! (\mu \times_c \text{sm}_\xi``\nu)_\xi . \supset . \exists ! \text{sm}``\mu \cap t_0`\mu . \exists ! \text{sm}_\xi``\nu .$

[(2).*113·26.*103·43] $\supset . \exists ! (\mu \times_c \nu)_\xi$ (4)

$\vdash . (3) . (4) . \supset \vdash .$ Prop

***118·33.** $\vdash : \mu, \nu \in \text{NC} - \iota`0 . \supset . (\mu \times_c \nu)_\xi = (\text{sm}_\xi``\mu \times_c \text{sm}_\xi``\nu)_\xi$
[Proof as in *118·23, using *118·31·311 . *113·203·26]

***118·34.** $\vdash : \nu \in \text{NC} . \mu \neq 0 . \supset . (\mu \times_c \nu)_\xi = (\mu \times_c \text{sm}_\xi``\nu)_\xi$

Dem.

$\vdash . *118·311 . *113·203·26 . *103·43 . \supset$
$\vdash : \exists ! (\mu \times_c \nu)_\xi . \mu \neq 0 . \supset . (\mu \times_c \nu)_\xi = (\mu \times_c \text{sm}_\xi``\nu)_\xi$ (1)
$\vdash . *118·32 . \supset \vdash : \text{Hp} . \sim \exists ! (\mu \times_c \nu)_\xi . \supset . \sim \exists ! (\mu \times_c \text{sm}_\xi``\nu)_\xi .$
[*24·51] $\supset . (\mu \times_c \nu)_\xi = (\mu \times_c \text{sm}_\xi``\nu)_\xi$ (2)
$\vdash . (1) . (2) . \supset \vdash .$ Prop

***118·341.** $\vdash : \mu \in \text{NC} . \nu \neq 0 . \supset . (\mu \times_c \nu)_\xi = (\text{sm}_\xi``\mu \times_c \nu)_\xi$ [*118·34 . *113·27]

***118·35.** $\vdash : \varpi \neq 0 . \supset . (\mu \times_c \nu \times_c \varpi)_\xi = \{(\mu \times_c \nu)_\xi \times_c \varpi\}_\xi$
[Proof similar to *118·25, using *118·341·301 . *113·203·23]

***118·351.** $\vdash : \mu \neq 0 . \supset . (\mu \times_c \nu \times_c \varpi)_\xi = \{\mu \times_c (\nu \times_c \varpi)_\xi\}_\xi$ [*118·35 . *113·27]

***118·352.** $\vdash : \mu \neq 0 . \varpi \neq 0 . \supset . \{\mu \times_c (\nu \times_c \varpi)_\xi\}_\xi = \{(\mu \times_c \nu)_\xi \times_c \varpi\}_\xi$
[*118·35·351]

***118·4.** $\vdash . (\mu^\nu)_\xi = \hat{\eta} \{(\exists \alpha, \beta) . \mu = \text{N}_0\text{c}`\alpha . \nu = \text{N}_0\text{c}`\beta . \eta \, \text{sm}_\xi \, (\alpha \exp \beta)\}$
[(*65·01·03) . *116·2]

***118·401.** $\vdash : \exists ! \mu^\nu . \supset . \text{sm}_\xi``\mu^\nu = (\mu^\nu)_\xi$ [Proof as in *118·201]

***118·402.** $\vdash :. \mu, \nu \in \text{N}_0\text{C} . \mu \neq 0 . \mu \neq 1 . \supset : \exists ! (\mu^\nu)_\xi . \supset . \exists ! (\mu \times_c \nu)_\xi$

Dem.

$\vdash . *103·2 . \supset \vdash :. \text{Hp} . \supset : (\exists \alpha, \beta) . \mu = \text{N}_0\text{c}`\alpha . \nu = \text{N}_0\text{c}`\beta . \alpha \sim \in 0 \cup 1 :$
[*117·651] $\supset : (\exists \alpha, \beta) . \mu = \text{N}_0\text{c}`\alpha . \nu = \text{N}_0\text{c}`\beta .$
$(\text{N}_0\text{c}`\alpha)^{\text{N}_0\text{c}`\beta} \geqslant \text{N}_0\text{c}`\alpha \times_c \text{N}_0\text{c}`\beta :$
[*118·13·301·401.(IIT)] $\supset : \exists ! (\mu^\nu)_\xi . \supset . \exists ! (\mu \times_c \nu)_\xi :. \supset \vdash .$ Prop

***118·41.** $\vdash : \exists ! (\mu^\nu)_\xi . \nu \neq 0 . \supset . \exists ! \text{sm}_\xi``\mu$

Dem.

$\vdash . *118·402·31 . \supset \vdash : \text{Hp} . \mu \neq 1 . \mu \neq 0 . \supset . \exists ! \text{sm}_\xi``\mu$ (1)
$\vdash . *101·12·15·241·28 . \supset \vdash :. \mu = 0 . \vee . \mu = 1 : \supset . \exists ! \text{sm}_\xi``\mu$ (2)
$\vdash . (1) . (2) . \supset \vdash .$ Prop

***118·411.** $\vdash : \exists ! (\mu^\nu)_\xi . \mu \neq 0 . \mu \neq 1 . \supset . \exists ! \text{sm}_\xi``\nu$ [*118·402·311]

***118·42.** $\vdash :. \nu \in \text{NC} . \mu \neq 0 . \mu \neq 1 . \supset : \exists ! (\mu^\nu)_\xi . \equiv . \exists ! (\mu^{\text{sm}_\xi``\nu})_\xi$
[Proof as in *118·32, using *116·203·26 . *118·411]

***118·421.** $\vdash :. \mu \in \text{NC} . \nu \neq 0 . \supset : \exists ! (\mu^\nu)_\xi . \equiv . \exists ! \{(\text{sm}_\xi``\mu)^\nu\}_\xi$
[Proof as in *118·32, using *116·203·26 . *118·41]

***118·43.** $\vdash : \mu, \nu \in \text{NC} - \iota`0 . \mu \neq 1 . \supset . (\mu^\nu)_\xi = \{(\text{sm}_\xi``\mu)^{\text{sm}_\xi``\nu}\}_\xi$
[Proof as in *118·23, using *118·41·411 . *116·203·26]

***118·44.** $\vdash : \nu \epsilon \mathrm{NC} \,.\, \mu \neq 0 \,.\, \mu \neq 1 \,.\, \supset \,.\, (\mu^\nu)_\xi = (\mu^{\mathrm{sm}_\xi{}^{\prime\prime}\nu})_\xi$

[Proof as in *118·34, using *116·203·26 . *118·411·42]

***118·441.** $\vdash : \mu \epsilon \mathrm{NC} \,.\, \nu \neq 0 \,.\, \supset \,.\, (\mu^\nu)_\xi = \{(\mathrm{sm}_\xi{}^{\prime\prime}\mu)^\nu\}_\xi$

[Proof as in *118·34, using *116·203·26 . *118·41·421]

***118·45.** $\vdash : \mu \neq 0 \,.\, \mu \neq 1 \,.\, \supset \,.\, (\mu^{\nu \times_c \varpi})_\xi = \{\mu^{(\nu \times_c \varpi)_\xi}\}_\xi$

Dem.

$\vdash \,.\, *113·23 \,.\, *118·44·301 \,.\, (\mathrm{IIT}) \,.\, \supset$

$\vdash : \mathrm{Hp} \,.\, \nu, \varpi \epsilon \mathrm{N_0C} \,.\, \supset \,.\, (\mu^{\nu \times_c \varpi})_\xi = \{\mu^{(\nu \times_c \varpi)_\xi}\}_\xi$ (1)

$\vdash \,.\, *113·203 \,.\, \supset \vdash : \sim(\nu, \varpi \epsilon \mathrm{N_0C}) \,.\, \supset \,.\, \nu \times_c \varpi = \Lambda \,.$

[*116·203] $\supset \,.\, (\mu^{\nu \times_c \varpi})_\xi = \{\mu^{(\nu \times_c \varpi)_\xi}\}_\xi$ (2)

$\vdash \,.\, (1) \,.\, (2) \,.\, \supset \vdash \,.$ Prop

***118·451.** $\vdash : \varpi \neq 0 \,.\, \supset \,.\, (\mu^{\nu \times_c \varpi})_\xi = [\{(\mu^\nu)_\xi\}^\varpi]_\xi$

Dem.

$\vdash \,.\, *116·63 \,.\, \supset \vdash : \mathrm{Hp} \,.\, \mu \epsilon \mathrm{NC} \,.\, \supset \,.\, (\mu^{\nu \times_c \varpi})_\xi = \{(\mu^\nu)^\varpi\}_\xi$

[*116·23.*118·441·401.(IIT)] $= [\{(\mu^\nu)_\xi\}^\varpi]_\xi$ (1)

$\vdash \,.\, *116·204 \,.\, \supset \vdash : \mu \sim\epsilon \mathrm{NC} \,.\, \supset \,.\, (\mu^{\nu \times_c \varpi})_\xi = [\{(\mu^\nu)_\xi\}^\varpi]_\xi$ (2)

$\vdash \,.\, (1) \,.\, (2) \,.\, \supset \vdash \,.$ Prop

***118·46.** $\vdash : \mu \neq 0 \,.\, \mu \neq 1 \,.\, \supset \,.\, (\mu^{\nu +_c \varpi})_\xi = \{\mu^{(\nu +_c \varpi)_\xi}\}_\xi$

[Proof as in *118·45, using *118·44·201 . *116·203 . *110·4·42]

***118·461.** $\vdash \,.\, (\mu^{\nu +_c \varpi})_\xi = \{(\mu^\nu)_\xi \times_c (\mu^\varpi)_\xi\}_\xi$

Dem.

$\vdash \,.\, *116·52 \,.\, \supset \vdash : \mu \neq 0 \,.\, \supset \,.\, (\mu^{\nu +_c \varpi})_\xi = (\mu^\nu \times_c \mu^\varpi)_\xi$

[*116·35·23.*118·33·401.(IIT)] $= \{(\mu^\nu)_\xi \times_c (\mu^\varpi)_\xi\}_\xi$ (1)

$\vdash \,.\, *110·4 \,.\, *113·203 \,.\, *116·203 \,.\, \supset$

$\vdash : \sim(\nu, \varpi \epsilon \mathrm{N_0C}) \,.\, \supset \,.\, (\mu^{\nu +_c \varpi})_\xi = \{(\mu^\nu)_\xi \times_c (\mu^\varpi)_\xi\}_\xi$ (2)

$\vdash \,.\, *116·311 \,.\, *113·601 \,.\, *110·62 \,.\, \supset$

$\vdash : \nu, \varpi \epsilon \mathrm{N_0C} - \iota`0 \,.\, \mu = 0 \,.\, \supset \,.\, (\mu^{\nu +_c \varpi})_\xi = \{(\mu^\nu)_\xi \times_c (\mu^\varpi)_\xi\}_\xi$ (3)

$\vdash \,.\, *116·311·301 \,.\, *110·6 \,.\, *113·601 \,.\, \supset$

$\vdash : \nu \epsilon \mathrm{N_0C} - \iota`0 \,.\, \varpi = 0 \,.\, \mu = 0 \,.\, \supset \,.\, (\mu^{\nu +_c \varpi})_\xi = \{(\mu^\nu)_\xi \times_c (\mu^\varpi)_\xi\}_\xi$ (4)

Similarly $\vdash : \varpi \epsilon \mathrm{N_0C} - \iota`0 \,.\, \nu = 0 \,.\, \mu = 0 \,.\, \supset \,.\, (\mu^{\nu \times_c \varpi})_\xi = \{(\mu^\nu)_\xi \times_c (\mu^\varpi)_\xi\}_\xi$ (5)

$\vdash \,.\, *116·301 \,.\, *113·621 \,.\, \supset$

$\vdash : \nu = 0 \,.\, \varpi = 0 \,.\, \mu = 0 \,.\, \supset \,.\, (\mu^{\nu +_c \varpi})_\xi = \{(\mu^\nu)_\xi \times_c (\mu^\varpi)_\xi\}_\xi$ (6)

$\vdash \,.\, (1) \,.\, (2) \,.\, (3) \,.\, (4) \,.\, (5) \,.\, (6) \,.\, \supset \vdash \,.$ Prop

***118·462.** $\vdash \,.\, (\mu^{\nu +_c \varpi})_\xi = \{\mu^\nu \times_c (\mu^\varpi)_\xi\}_\xi$ [Proof as in *118·461, using *118·34]

***118·463.** $\vdash \,.\, (\mu^{\nu +_c \varpi})_\xi = \{(\mu^\nu)_\xi \times_c \mu^\varpi\}_\xi$ [Proof as in *118·461, using *118·341]

***118·47.** $\vdash : \varpi \neq 0 . \supset . \{(\mu \times_c \nu)^\varpi\}_\xi = [\{(\mu \times_c \nu)_\xi\}^\varpi]_\xi$

[Proof as in *118·45, using *118·441]

***118·471.** $\vdash :. \mu \neq 0 . \nu \neq 0 . \vee . \varpi = 0 . \vee . \sim(\mu, \nu, \varpi \in N_0C) : \supset .$

$$\{(\mu \times_c \nu)^\varpi\}_\xi = \{(\mu^\varpi)_\xi \times_c (\nu^\varpi)_\xi\}_\xi$$

Dem.

$\vdash . *116·55 . \supset \vdash : \mu \neq 0 . \nu \neq 0 . \supset . \{(\mu \times_c \nu)^\varpi\}_\xi = \{\mu^\varpi \times_c \nu^\varpi\}_\xi$

[*116·35·23.*118·33·401.(IIT)] $= \{(\mu^\varpi)_\xi \times_c (\nu^\varpi)_\xi\}_\xi$ (1)

$\vdash . *110·4 . *113·203 . *116·203 . \supset$

$\vdash : \sim(\mu, \nu, \varpi \in N_0C) . \supset . \{(\mu \times_c \nu)^\varpi\}_\xi = \{(\mu^\varpi)_\xi \times_c (\nu^\varpi)_\xi\}_\xi$ (2)

$\vdash . *116·301 . *113·621 . \supset$

$\vdash : \mu, \nu \in N_0C . \varpi = 0 . \supset . \{(\mu \times_c \nu)^\varpi\}_\xi = \{(\mu^\varpi)_\xi \times_c (\nu^\varpi)_\xi\}_\xi$ (3)

$\vdash . (1) . (2) . (3) . \supset \vdash .$ Prop

***118·472.** $\vdash :. \mu \neq 0 . \vee . \varpi = 0 . \vee . \sim(\mu, \nu, \varpi \in N_0C) : \supset . \{(\mu \times_c \nu)^\varpi\}_\xi = \{\mu^\varpi \times_c (\nu^\varpi)_\xi\}_\xi$

[Proof as in *118·471, using *118·34]

***118·473.** $\vdash :. \nu \neq 0 . \vee . \varpi = 0 . \vee . \sim(\mu, \nu, \varpi \in N_0C) : \supset . \{(\mu \times_c \nu)^\varpi\}_\xi = \{(\mu^\varpi)_\xi \times_c \nu^\varpi\}_\xi$

[Proof as in *118·471, using *118·341]

$*119$. SUBTRACTION.

Summary of $*119$.

The treatment of subtraction follows the same general lines as that of addition, and is simplified by the results in $*110$. A difficulty arises from the fact that subtraction (in any ordinary sense of the term) is not always possible; and also from the fact that the result, when possible, is not always a cardinal number.

We put

$*119 \cdot 01$. $\gamma -_c \nu = \hat{\xi} \{ \text{Nc}`\xi +_c \nu = \gamma \,.\, \exists ! \, \text{Nc}`\xi +_c \nu \}$ Df

Thus when subtraction (in the ordinary sense of the term) is not possible,

$$\gamma -_c \nu = \Lambda.$$

The question of existential adjustment of types is dealt with by IIT of the prefatory statement combined with the following definitions:

$*119 \cdot 02$. $\text{Nc}`\alpha -_c \nu = \text{N}_0\text{c}`\alpha -_c \nu$ Df

$*119 \cdot 03$. $\gamma -_c \text{Nc}`\beta = \gamma -_c \text{N}_0\text{c}`\beta$ Df

We then proceed to deduce the elementary properties derivable from these definitions.

$*119 \cdot 11$. $\vdash : \exists ! \, \gamma -_c \nu \,.\, \supset \,.\, \gamma, \nu \in \text{N}_0\text{C}$

$*119 \cdot 12$. $\vdash : \xi \in \text{Nc}`\alpha -_c \text{Nc}`\beta \,.\, \equiv \,.\, \alpha \, \text{sm} \, \xi + \beta$

$*119 \cdot 14$. $\vdash : \xi \in \gamma -_c \nu \,.\, \supset \,.\, \text{N}_0\text{c}`\xi \subset \gamma -_c \nu$

$*119 \cdot 25$. $\vdash : \gamma \geqslant \nu \,.\, \supset \,.\, \exists ! \, (\gamma -_c \nu) \cap t_0`\gamma$

$*119 \cdot 26$. $\vdash : \exists ! \, \gamma -_c \nu \,.\, \supset \,.\, \gamma \geqslant \nu$

The next group of propositions is concerned with some simple results of subtraction.

$*119 \cdot 32$. $\vdash : (\gamma +_c \nu) -_c \nu \in \text{N}_0\text{C} \,.\, \supset \,.\, \text{sm}``\gamma = (\gamma +_c \nu) -_c \nu$

$*119 \cdot 34$. $\vdash : \gamma -_c \nu \in \text{N}_0\text{C} \,.\, \supset \,.\, (\gamma -_c \nu) +_c \nu = \text{sm}``\gamma$

$*119 \cdot 35$. $\vdash : \gamma -_c \nu \in \text{N}_0\text{C} \,.\, \supset \,.\, \alpha +_c \gamma = (\alpha +_c \nu) +_c (\gamma -_c \nu)$

Associative laws are then considered.

$*119 \cdot 44$. $\vdash : \mu +_c (\nu -_c \varpi) \subset (\mu +_c \nu) -_c \varpi$

$*119 \cdot 45$. $\vdash : (\mu +_c \nu) -_c \varpi \in \text{NC} \,.\, \exists ! \, \{\mu +_c (\nu -_c \varpi)\} \,.\, \supset \,.\, \mu +_c (\nu -_c \varpi) = (\mu +_c \nu) -_c \varpi$

The question of types is then dealt with:

$*119 \cdot 52$. $\vdash : \text{sm}_{\delta,\gamma}``(\mu -_c \nu)_\gamma = (\mu -_c \nu)_\delta \cap \text{D}`\text{sm}_{\delta,\gamma}$

A difficulty arises from the fact that if τ_1 and τ_2 are two complete types whose members are classes, we cannot prove that, either $\tau_1 = \text{sm}``\tau_2$ or $\tau_2 = \text{sm}``\tau_1$. We put

***119·54.** $\text{SM}(\delta, \gamma) . = : t`\delta = \text{D}`\text{sm}_{\delta,\gamma} . \vee . t`\gamma = \text{D}`\text{sm}_{\gamma,\delta}$ Df

Then we obtain

***119·541.** $\vdash : \text{SM}(\delta, \gamma) . (\mu -_c \nu)_\gamma \in \text{N}_0\text{C} . (\mu -_c \nu)_\delta \in \text{NC} . \supset .$
$\text{sm}_{\delta,\gamma}``(\mu -_c \nu)_\gamma = (\mu -_c \nu)_\delta$

Finally we show that any existential adjustment of types will suffice for the components:

***119·61.** $\vdash : \mu \in \text{N}_0\text{C} . \exists ! \text{sm}_\xi``\mu . \supset . \mu -_c \nu = \text{sm}_\xi``\mu -_c \nu$

***119·62.** $\vdash : \nu \in \text{N}_0\text{C} . \exists ! \text{sm}_\xi``\nu . \supset . \mu -_c \nu = \mu -_c \text{sm}_\xi``\nu$

Also *119·25·26 are now extended to

***119·64.** $\vdash :. \exists ! \text{sm}_\xi``\mu . \supset : \mu \geqslant \nu . \equiv . \exists ! (\mu -_c \nu)_\xi$

The only applications of the propositions of this number are in connection with Inductive Cardinals (cf. *120).

***119·01.** $\gamma -_c \nu = \hat{\xi}\{\text{Nc}`\xi +_c \nu = \gamma . \exists ! \text{Nc}`\xi +_c \nu\}$ Df

Here the suffix to the sign of subtraction is introduced to show that we are concerned with *cardinal* subtraction. It will be found that $\gamma -_c \nu$ is not an NC except under hypotheses for γ and ν.

***119·02.** $\text{Nc}`\alpha -_c \nu = \text{N}_0\text{c}`\alpha -_c \nu$ Df

***119·03.** $\gamma -_c \text{Nc}`\beta = \gamma -_c \text{N}_0\text{c}`\beta$ Df

***119·04.** $\vdash . \text{Nc}`\alpha -_c \text{Nc}`\beta = \text{N}_0\text{c}`\alpha -_c \text{N}_0\text{c}`\beta$ [*119·02·03]

Note that the occurrence of a formal number in the place of γ or ν in $\gamma -_c \nu$ is an arithmetic occurrence, and accordingly IIT applies to it.

***119·1.** $\vdash : \xi \in \gamma -_c \nu . \equiv . \text{Nc}`\xi +_c \nu = \gamma . \exists ! \text{Nc}`\xi +_c \nu$ [(*119·01)]

***119·101.** $\vdash : \xi \in \text{Nc}`\alpha -_c \nu . \equiv . \text{Nc}`\xi +_c \nu = \text{N}_0\text{c}`\alpha$ [(*119·02) . *103·13]

***119·102.** $\vdash : \xi \in \gamma -_c \text{Nc}`\beta . \equiv . \text{Nc}`\xi +_c \text{Nc}`\beta = \gamma . \exists ! \text{Nc}`\xi +_c \text{Nc}`\beta$
[(*119·03) . *110·3]

***119·103.** $\vdash : \xi \in \text{Nc}`\alpha -_c \text{Nc}`\beta . \equiv . \text{Nc}`\xi +_c \text{Nc}`\beta = \text{N}_0\text{c}`\alpha$
[*119·04 . *110·3 . *103·13]

***119·11.** $\vdash : \exists ! \gamma -_c \nu . \supset . \gamma, \nu \in \text{N}_0\text{C}$ [*110·4·42 . *103·34]

***119·12.** $\vdash : \xi \in \text{Nc}`\alpha -_c \text{Nc}`\beta . \equiv . \alpha \,\text{sm}\, \xi + \beta$

Dem.

$\vdash . \text{*}119·103 . \supset \vdash : \xi \in \text{Nc}`\alpha -_c \text{Nc}`\beta . \equiv . \text{Nc}`\xi +_c \text{Nc}`\beta = \text{N}_0\text{c}`\alpha .$
[*110·3] $\equiv . \text{Nc}`(\xi + \beta) = \text{N}_0\text{c}`\alpha .$
[*100·35.*103·13] $\equiv . \alpha \,\text{sm}\, \xi + \beta : \supset \vdash . \text{Prop}$

Thus $\text{Nc}`\alpha -_c \text{Nc}`\beta$ is an NC when $\hat{\xi}(\alpha \,\text{sm}\, \xi + \beta)$ is an NC.

∗119·13. $\vdash : N_0c`\gamma \subset Nc`\alpha -_c Nc`\beta . \equiv . \alpha \,\mathrm{sm}\, (\gamma + \beta)$

Dem.

$\vdash . *22·1 . \supset \vdash :. N_0c`\gamma \subset Nc`\alpha -_c Nc`\beta . \equiv : \xi \in N_0c`\gamma . \supset_\xi . \xi \in Nc`\alpha -_c Nc`\beta :$

[∗103·12.∗119·12] $\supset : \alpha \,\mathrm{sm}\, (\gamma + \beta)$ (1)

$\vdash . *110·15 . *100·31 . \supset \vdash :. \alpha \,\mathrm{sm}\, (\gamma + \beta) . \supset : \xi \in N_0c`\gamma . \supset . (\xi + \beta) \,\mathrm{sm}\, (\gamma + \beta) .$

[∗73·32] $\supset . \alpha \,\mathrm{sm}\, (\xi + \beta) .$

[∗119·12] $\supset . \xi \in Nc`\alpha -_c Nc`\beta$ (2)

$\vdash . (1) . (2) . \supset \vdash .$ Prop

∗119·14. $\vdash : \xi \in \gamma -_c \nu . \supset . N_0c`\xi \subset \gamma -_c \nu$ [∗119·1 . ∗100·31·321]

∗119·21. $\vdash : \beta \subset \alpha . \supset . \exists ! (Nc`\alpha -_c Nc`\beta)_\alpha$

The notation is defined in ∗65·01.

Dem.

$\vdash . *24·411·21 . \supset \vdash : \mathrm{Hp} . \supset . \alpha = \beta \cup (\alpha - \beta) . \beta \cap (\alpha - \beta) = \Lambda .$

[∗110·32] $\supset . Nc`\alpha = Nc`\beta +_c Nc`(\alpha - \beta) .$

[∗10·24] $\supset . (\exists \xi) . \xi \in t`\alpha . N_0c`\alpha = Nc`\beta +_c Nc`\xi .$

[∗119·103] $\supset . \exists ! (Nc`\alpha -_c Nc`\beta)_\alpha : \supset \vdash .$ Prop

∗119·22. $\vdash : Nc`\alpha \geqslant Nc`\beta . \supset . \exists ! (Nc`\alpha -_c Nc`\beta)_\alpha$

Dem.

$\vdash . *117·221 . \supset \vdash : \mathrm{Hp} . \supset . (\exists \rho) . \rho \subset \alpha . \rho \,\mathrm{sm}\, \beta .$

[∗119·21] $\supset . (\exists \rho) . \exists ! (Nc`\alpha -_c Nc`\rho)_\alpha . \rho \,\mathrm{sm}\, \beta .$

[∗100·35.∗119·04] $\supset . \exists ! (Nc`\alpha -_c Nc`\beta)_\alpha : \supset \vdash .$ Prop

∗119·23. $\vdash : \exists ! (Nc`\alpha -_c Nc`\beta) . \supset . (\exists \delta) . \delta \,\mathrm{sm}\, \beta . \delta \subset \alpha$

Dem.

$\vdash . *119·103 . \supset \vdash : \mathrm{Hp} . \supset . (\exists \xi) . N_0c`\alpha = Nc`\beta +_c Nc`\xi .$

[∗110·71] $\supset . (\exists \delta) . \delta \,\mathrm{sm}\, \beta . \delta \subset \alpha : \supset \vdash .$ Prop

∗119·24. $\vdash : \exists ! (Nc`\alpha -_c Nc`\beta) . \supset . Nc`\alpha \geqslant Nc`\beta$ [∗119·23 . ∗117·221]

∗119·25. $\vdash : \gamma \geqslant \nu . \supset . \exists ! (\gamma -_c \nu) \cap t_0`\gamma$

Dem.

$\vdash . *117·24 . \supset \vdash : \mathrm{Hp} . \supset . (\exists \alpha, \beta) . \gamma = N_0c`\alpha . \nu = N_0c`\beta . N_0c`\alpha \geqslant N_0c`\beta .$

[∗117·107] $\supset . (\exists \alpha, \beta) . \gamma = N_0c`\alpha . \nu = N_0c`\beta . Nc`\alpha \geqslant Nc`\beta .$

[∗119·22·04] $\supset . (\exists \alpha, \beta) . \gamma = N_0c`\alpha . \nu = N_0c`\beta . \exists ! (N_0c`\alpha -_c N_0c`\beta)_\alpha .$

[(∗63·02).∗13·193] $\supset . \exists ! (\gamma -_c \nu) \cap t_0`\gamma : \supset \vdash .$ Prop

∗119·26. $\vdash : \exists ! \gamma -_c \nu . \supset . \gamma \geqslant \nu$

Dem.

$\vdash . *119·11 . \supset \vdash : \mathrm{Hp} . \supset . (\exists \alpha, \beta) . \gamma = N_0c`\alpha . \nu = N_0c`\beta . \exists ! (N_0c`\alpha -_c N_0c`\beta) .$

[∗119·04·24] $\supset . (\exists \alpha, \beta) . \gamma = N_0c`\alpha . \nu = N_0c`\beta . Nc`\alpha \geqslant Nc`\beta .$

[∗117·107.∗13·193] $\supset . \gamma \geqslant \nu : \supset \vdash .$ Prop

∗119·27. $\vdash : \gamma \geqslant \nu . \equiv . \exists ! (\gamma -_c \nu) \cap t_0`\gamma$ [∗119·25·26]

For the extension of this theorem cf. ∗119·64.

∗119·31. $\vdash : \gamma, \nu \in \mathrm{N_0C} . \supset . \mathrm{sm}``\gamma \subset (\gamma +_c \nu) -_c \nu$

Dem.

$\vdash . *119\cdot1 . (\mathrm{IIT}) . \supset \vdash : \xi \in (\gamma +_c \nu) -_c \nu . \equiv . \mathrm{Nc}`\xi +_c \nu = \gamma +_c \nu . \exists ! \gamma +_c \nu$ (1)

$\vdash . *100\cdot51\cdot521 . \quad \supset \vdash :. \mathrm{Hp} . \supset : \xi \in \mathrm{sm}``\gamma . \supset . \mathrm{Nc}`\xi = \gamma .$

[∗103·22.∗118·01] $\supset . \mathrm{Nc}`\xi +_c \nu = \gamma +_c \nu .$

[∗110·22·03.∗103·13] $\supset . \mathrm{Nc}`\xi +_c \nu = \gamma +_c \nu . \exists ! \gamma +_c \nu .$

[(1)] $\supset . \xi \in (\gamma +_c \nu) -_c \nu : \supset \vdash . \mathrm{Prop}$

The penultimate step in the proof employs the principle, explained in the prefatory statement, that, since in the previous line the equation

$$\mathrm{Nc}`\xi +_c \nu = \gamma +_c \nu$$

has its sides undetermined in type by the conventions IT and IIT, any convenient type can be chosen for them. The type chosen in this line is such that $\exists ! \gamma +_c \nu$, and the references indicate the existence of at least one such type.

∗119·32. $\vdash : (\gamma +_c \nu) -_c \nu \in \mathrm{N_0C} . \supset . \mathrm{sm}``\gamma = (\gamma +_c \nu) -_c \nu$
[∗119·11·31 . ∗103·22 . ∗100·52·42]

∗119·33. $\vdash : \mathrm{Nc}`\alpha -_c \mathrm{Nc}`\beta \in \mathrm{N_0C} . \supset . (\mathrm{Nc}`\alpha -_c \mathrm{Nc}`\beta) +_c \mathrm{Nc}`\beta = \mathrm{Nc}`\alpha$

Dem.

$\vdash . *119\cdot13 . \supset \vdash : \mathrm{N_0c}`\gamma = \mathrm{Nc}`\alpha -_c \mathrm{Nc}`\beta . \supset . \alpha \,\mathrm{sm}\, (\gamma + \beta)$ (1)

$\vdash . *20\cdot18 . *118\cdot01 . \supset \vdash :. \mathrm{Hp}(1) . \supset :$

$(\mathrm{Nc}`\alpha -_c \mathrm{Nc}`\beta) +_c \mathrm{Nc}`\beta = \mathrm{N_0c}`\xi . \equiv_\xi . \mathrm{Nc}`\gamma +_c \mathrm{Nc}`\beta = \mathrm{N_0c}`\xi .$

[∗110·3.∗100·35] $\equiv_\xi . \xi \,\mathrm{sm}\, (\gamma + \beta) .$

[(1).∗103·42] $\equiv_\xi . \mathrm{N_0c}`\xi = \mathrm{Nc}`\alpha$ (2)

$\vdash . *103\cdot2\cdot34 . \supset \vdash :. \mathrm{Hp} . \supset : \exists ! \mathrm{Nc}`\alpha . \supset . (\exists \xi) . \mathrm{N_0c}`\xi = \mathrm{Nc}`\alpha .$

[(2).∗10·1] $\supset . (\mathrm{Nc}`\alpha -_c \mathrm{Nc}`\beta) +_c \mathrm{Nc}`\beta = \mathrm{Nc}`\alpha$ (3)

$\vdash . *110\cdot42 . *103\cdot34\cdot2 . \supset \vdash :. \mathrm{Hp} . \supset :$

$\exists ! \{(\mathrm{Nc}`\alpha -_c \mathrm{Nc}`\beta) +_c \mathrm{Nc}`\beta\} . \supset . (\exists \xi) . \mathrm{N_0c}`\xi = (\mathrm{Nc}`\alpha -_c \mathrm{Nc}`\beta) +_c \mathrm{Nc}`\beta .$

[(2).∗10·1] $\supset . (\mathrm{Nc}`\alpha -_c \mathrm{Nc}`\beta) +_c \mathrm{Nc}`\beta = \mathrm{Nc}`\alpha$ (4)

$\vdash . (3) . (4) . \supset \vdash . \mathrm{Prop}$

∗119·34. $\vdash : \gamma -_c \nu \in \mathrm{N_0C} . \supset . (\gamma -_c \nu) +_c \nu = \mathrm{sm}``\gamma$
[∗119·11·33 . ∗103·2 . ∗100·51 . ∗118·01]

∗119·35. $\vdash : \gamma -_c \nu \in \mathrm{N_0C} . \supset . \alpha +_c \gamma = (\alpha +_c \nu) +_c (\gamma -_c \nu)$

Dem.

$\vdash . *110\cdot51\cdot56 . \supset \vdash : \mathrm{Hp} . \supset . (\alpha +_c \nu) +_c (\gamma -_c \nu) = \alpha +_c \{(\gamma -_c \nu) +_c \nu\}$

[∗119·34] $= \alpha +_c \mathrm{sm}``\gamma$

[∗118·24.∗119·11] $= \alpha +_c \gamma : \supset \vdash . \mathrm{Prop}$

∗119·41. $\vdash :. \delta \in \mathrm{Nc}`\beta -_c \mathrm{Nc}`\gamma . \supset :$

$\xi \in (\mathrm{Nc}`\alpha +_c \mathrm{Nc}`\beta) -_c \mathrm{Nc}`\gamma . \equiv . \{(\alpha + \delta) + \gamma\} \,\mathrm{sm}\, (\xi + \gamma)$

Dem.

$\vdash . *119\cdot12 . *110\cdot3 . \supset \vdash : \xi \in (\mathrm{Nc}`\alpha +_c \mathrm{Nc}`\beta) -_c \mathrm{Nc}`\gamma . \equiv . (\alpha + \beta) \,\mathrm{sm}\, (\xi + \gamma)$ (1)

$\vdash . *119\cdot12 . \quad \supset \vdash : \mathrm{Hp} . \equiv . \beta \,\mathrm{sm}\, (\delta + \gamma)$ (2)

$\vdash . (1) . (2) . *110\cdot15\cdot53 . \supset \vdash . \mathrm{Prop}$

∗119·42. $\vdash :. \text{Nc}'\beta -_c \text{Nc}'\gamma \in \text{N}_0\text{C} . \eta \in \text{Nc}'\alpha +_c (\text{Nc}'\beta -_c \text{Nc}'\gamma) . \supset :$
$\xi \in (\text{Nc}'\alpha +_c \text{Nc}'\beta) -_c \text{Nc}'\gamma . \equiv . (\eta + \gamma) \text{sm} (\xi + \gamma)$

Dem.

$\vdash . *118{\cdot}01 . *110{\cdot}3 . *103{\cdot}2 . *100{\cdot}31 . \supset \vdash :. \text{N}_0\text{c}'\delta = \text{Nc}'\beta -_c \text{Nc}'\gamma . \supset :$
$\eta \in \text{Nc}'\alpha +_c (\text{Nc}'\beta -_c \text{Nc}'\gamma) . \equiv . \eta \text{ sm} (\alpha + \delta)$ (1)

$\vdash . *119{\cdot}41 . (1) . *103{\cdot}12 . *110{\cdot}15 . \supset \vdash . \text{Prop}$

Note that if γ be an infinite class, it does not follow from $(\eta + \gamma) \text{sm} (\xi + \gamma)$ that $\eta \text{ sm } \xi$. This will be proved, however, when γ is an inductive class (cf. ∗120·41).

∗119·43. $\vdash : \text{Nc}'\beta -_c \text{Nc}'\gamma \in \text{N}_0\text{C} . \supset .$
$\text{Nc}'\alpha +_c (\text{Nc}'\beta -_c \text{Nc}'\gamma) \subset (\text{Nc}'\alpha +_c \text{Nc}'\beta) -_c \text{Nc}'\gamma$

Dem.

$\vdash . *119{\cdot}42 . \supset \vdash :. \text{Hp} . \eta \in \text{Nc}'\alpha +_c (\text{Nc}'\beta -_c \text{Nc}'\gamma) . \supset :$
$\eta \in (\text{Nc}'\alpha +_c \text{Nc}'\beta) -_c \text{Nc}'\gamma . \equiv . (\eta + \gamma) \text{sm} (\eta + \gamma) :$

[∗73·3] $\supset : \eta \in (\text{Nc}'\alpha +_c \text{Nc}'\beta) -_c \text{Nc}'\gamma$ (1)

$\vdash . (1) . *22{\cdot}1 . \supset \vdash . \text{Prop}$

∗119·44. $\vdash : \mu +_c (\nu -_c \varpi) \subset (\mu +_c \nu) -_c \varpi$

Dem.

$\vdash . *119{\cdot}11{\cdot}43 . *103{\cdot}2 . \supset$

$\vdash : \nu -_c \varpi \in \text{N}_0\text{C} . \mu \in \text{N}_0\text{C} . \supset . \mu +_c (\nu -_c \varpi) \subset (\mu +_c \nu) -_c \varpi$ (1)

$\vdash . *110{\cdot}4{\cdot}42 . *119{\cdot}11 . \supset$

$\vdash : \sim \{\nu -_c \varpi \in \text{N}_0\text{C} . \mu \in \text{N}_0\text{C}\} . \supset . \mu +_c (\nu -_c \varpi) = \Lambda .$

[∗24·12] $\supset . \mu +_c (\nu -_c \varpi) \subset (\mu +_c \nu) -_c \varpi$ (2)

$\vdash . (1) . (2) . \supset \vdash . \text{Prop}$

∗119·45. $\vdash : (\mu +_c \nu) -_c \varpi \in \text{NC} . \exists ! \{\mu +_c (\nu -_c \varpi)\} . \supset . \mu +_c (\nu -_c \varpi) = (\mu +_c \nu) -_c \varpi$
[∗119·44 . ∗100·33·321 . ∗110·42]

∗119·51. $\vdash : \text{sm}_{\delta,\gamma}``(\text{Nc}'\alpha -_c \text{Nc}'\beta)_\gamma = (\text{Nc}'\alpha -_c \text{Nc}'\beta)_\delta \cap \text{D}'\text{sm}_{\delta,\gamma}$

Dem.

$\vdash . *119{\cdot}12 . \supset \vdash : \eta \in (\text{Nc}'\alpha -_c \text{Nc}'\beta)_\gamma . \zeta \text{sm}_{\delta,\gamma} \eta . \equiv . \alpha \text{ sm } \eta + \beta . \zeta \text{sm}_{\delta,\gamma} \eta .$

[∗110·15] $\equiv . \alpha \text{ sm } \zeta + \beta . \zeta \text{sm}_{\delta,\gamma} \eta .$

[∗119·12] $\equiv . \zeta \in (\text{Nc}'\alpha -_c \text{Nc}'\beta)_\delta . \zeta \text{sm}_{\delta,\gamma} \eta :$

[∗37·1.∗33·13] $\supset \vdash . \text{sm}_{\delta,\gamma}``(\text{Nc}'\alpha -_c \text{Nc}'\beta)_\gamma = (\text{Nc}'\alpha -_c \text{Nc}'\beta)_\delta \cap \text{D}'\text{sm}_{\delta,\gamma} : \supset \vdash . \text{Prop}$

∗119·52. $\vdash : \text{sm}_{\delta,\gamma}``(\mu -_c \nu)_\gamma = (\mu -_c \nu)_\delta \cap \text{D}'\text{sm}_{\delta,\gamma}$ [∗119·51·11]

The difficulty in respect to types, which arises from the fact that $\text{sm}_{\delta,\gamma}``(\mu -_c \nu)_\gamma$ and $(\mu -_c \nu)_\delta$ have not been proved to be identical, does not exist when ν is an "inductive number"; cf. ∗120·413.

∗119·53. $\vdash :. t'\delta = \text{D}'\text{sm}_{\delta,\gamma} . \supset : \text{sm}_{\delta,\gamma}``(\mu -_c \nu)_\gamma = (\mu -_c \nu)_\delta$ [∗119·52 . (∗65·01)]

∗119·531. $\vdash : t'\delta = \text{D}'\text{sm}_{\delta,\gamma} . (\mu -_c \nu)_\delta \in \text{N}_0\text{C} . \supset . \text{sm}_{\gamma,\delta}``(\mu -_c \nu)_\delta \in \text{N}_0\text{C}$

Dem.

$\vdash . *65{\cdot}13 . \quad \supset \vdash : \text{Hp} . \supset . (\mu -_c \nu)_\delta \subset \text{D}'\text{sm}_{\delta,\gamma} .$

[∗37·43.∗103·22.(∗65·1)] $\supset . \exists ! \text{sm}_{\gamma,\delta}``(\mu -_c \nu)_\delta .$

[∗100·52.∗103·34] $\supset . \text{sm}_{\gamma,\delta}``(\mu -_c \nu)_\delta \in \text{N}_0\text{C} : \supset \vdash . \text{Prop}$

***119·532.** $\vdash : t'\delta = \text{D}'\text{sm}_{\delta,\gamma} \,.\, (\mu -_c \nu)_\delta \in \text{N}_0\text{C} \,.\, (\mu -_c \nu)_\gamma \in \text{NC} \,.\, \supset .$

$$\text{sm}_{\gamma,\delta}``(\mu -_c \nu)_\delta = (\mu -_c \nu)_\gamma$$

Dem.

$\vdash . *119\cdot52\cdot531 . \supset \vdash : \text{Hp} . \supset . \exists ! (\mu -_c \nu)_\gamma .$

$[*119\cdot52\cdot531 . *100\cdot34] \qquad \supset . \text{sm}_{\gamma,\delta}``(\mu -_c \nu)_\delta = (\mu -_c \nu)_\gamma : \supset \vdash . \text{Prop}$

***119·54.** $\text{SM}(\delta, \gamma) . = : t'\delta = \text{D}'\text{sm}_{\delta,\gamma} . \vee . t'\gamma = \text{D}'\text{sm}_{\gamma,\delta}$ Df

***119·541.** $\vdash : \text{SM}(\delta, \gamma) . (\mu -_c \nu)_\gamma \in \text{N}_0\text{C} . (\mu -_c \nu)_\delta \in \text{NC} . \supset .$

$$\text{sm}_{\delta,\gamma}``(\mu -_c \nu)_\gamma = (\mu -_c \nu)_\delta \quad [*119\cdot53\cdot532]$$

***119·61.** $\vdash : \mu \in \text{N}_0\text{C} . \exists ! \text{sm}_\xi``\mu . \supset . \mu -_c \nu = \text{sm}_\xi``\mu -_c \nu$

Dem.

$\vdash . *119\cdot1 . \supset \vdash :. \text{Hp} . \supset : \eta \in \mu -_c \nu . \equiv . \text{Nc}'\eta +_c \nu = \mu . \exists ! \mu .$

$[*103\cdot16 . *118\cdot201 . *37\cdot29] \qquad \equiv . (\text{Nc}'\eta +_c \nu)_\xi = \text{sm}_\xi``\mu .$

$[*119\cdot1] \qquad \equiv . \eta \in \text{sm}_\xi``\mu -_c \nu :. \supset \vdash . \text{Prop}$

***119·62.** $\vdash : \nu \in \text{N}_0\text{C} . \exists ! \text{sm}_\xi``\nu . \supset . \mu -_c \nu = \mu -_c \text{sm}_\xi``\nu$

Dem.

$\vdash . *119\cdot1 . \supset \vdash :. \text{Hp} . \supset : \eta \in \mu -_c \nu . \equiv . \text{Nc}'\eta +_c \nu = \mu . \exists ! \mu .$

$[*110\cdot25] \qquad \equiv . \text{Nc}'\eta +_c \text{sm}_\xi``\nu = \mu . \exists ! \mu .$

$[*119\cdot1] \qquad \equiv . \eta \in \mu -_c \text{sm}_\xi``\nu :. \supset \vdash . \text{Prop}$

***119·63.** $\vdash : \mu, \nu \in \text{N}_0\text{C} . \exists ! \text{sm}_\xi``\mu . \supset . \mu -_c \nu = \text{sm}_\xi``\mu -_c \text{sm}_\xi``\nu$

Dem.

$\vdash . *119\cdot26 . \supset \vdash : \text{Hp} . \exists ! \mu -_c \nu . \supset . \mu \geqslant \nu .$

$[*118\cdot13] \qquad \supset . \exists ! \text{sm}_\xi``\nu .$

$[*119\cdot61\cdot62] \qquad \supset . \mu -_c \nu = \text{sm}_\xi``\mu -_c \text{sm}_\xi``\nu \quad (1)$

$\vdash . *119\cdot11 . *103\cdot13 . \supset$

$\vdash : \text{Hp} . \exists ! \text{sm}_\xi``\mu -_c \text{sm}_\xi``\nu . \supset . \exists ! \text{sm}_\xi``\nu .$

$[*119\cdot61\cdot62] \qquad \supset . \mu -_c \nu = \text{sm}_\xi``\mu -_c \text{sm}_\xi``\nu \quad (2)$

$\vdash . (1) . (2) . \supset \vdash . \text{Prop}$

***119·64.** $\vdash :. \exists ! \text{sm}_\xi``\mu . \supset : \mu \geqslant \nu . \equiv . \exists ! (\mu -_c \nu)_\xi$

Dem.

$\vdash . *117\cdot24 . \qquad \supset \vdash :. \text{Hp} . \supset : \mu \geqslant \nu . \supset . \mu, \nu \in \text{N}_0\text{C} . \exists ! \text{sm}_\xi``\mu .$

$[*119\cdot61] \qquad \supset . (\mu -_c \nu)_\xi = (\text{sm}_\xi``\mu -_c \nu)_\xi \quad (1)$

$\vdash . *117\cdot24\cdot244 . \supset \vdash :. \text{Hp} . \supset : \mu \geqslant \nu . \supset . \text{sm}_\xi``\mu \geqslant \nu .$

$[*119\cdot27] \qquad \supset . \exists ! (\text{sm}_\xi``\mu -_c \nu)_\xi .$

$[(1)] \qquad \supset . \exists ! (\mu -_c \nu)_\xi \quad (2)$

$\vdash . (2) . *119\cdot26 . \supset \vdash . \text{Prop}$

$*120$. INDUCTIVE CARDINALS.

Summary of $*120$.

Inductive Cardinals are those that obey mathematical induction starting from 0, *i.e.* in the language of Part II, Section E, they are the posterity of 0 with respect to the relation of ν to $\nu +_c 1$, or, in more popular language, they are those that can be reached from 0 by successive additions of 1. In former days, these were supposed to be all the cardinals, and mathematical induction was treated as a kind of self-evident axiom. We now know that only certain cardinals obey mathematical induction starting from 0. It is these cardinals which are to be considered in this number. They embrace 0, 1, 2, ... and generally all those cardinals which would be commonly called finite, all those which can be expressed in the usual Arabic system of numeration, and no others. The propositions to be proved concerning them in this number are elementary and familiar; the interest lies entirely in the definition and method of proof, not in the propositions themselves.

Put $\qquad\qquad \text{NC induct} = \hat{\alpha} \{\alpha (+_c 1)_* 0\} \quad \text{Df.}$

Since $(+_c 1)_*$ has necessarily its domain and converse domain of the same type, it is important to be careful in noting the relations of type. Accordingly we also put

$$\text{N}_\xi\text{C induct} = \hat{\alpha} \{\alpha (+_c 1)_* 0_\xi\} \quad \text{Df.}$$

We begin by applying the propositions of $*90$. Thus we have

$*120{\cdot}11$. $\vdash :. \alpha \in \text{N}_\eta\text{C induct} : \phi\xi . \supset_\xi . \phi(\xi +_c 1) : \phi 0_\eta : \supset . \phi\alpha$

$*120{\cdot}12$. $\vdash . 0 \in \text{NC induct}$

$*120{\cdot}121$. $\vdash : \alpha \in \text{N}_\xi\text{C induct} . \supset . (\alpha +_c 1)_\xi \in \text{N}_\xi\text{C induct}$

$*120{\cdot}13$. $\vdash :. \alpha \in \text{N}_\eta\text{C induct} : \xi \in \text{N}_\eta\text{C induct} . \phi\xi . \supset_\xi . \phi(\xi +_c 1) : \phi 0_\eta : \supset . \phi\alpha$

$*120{\cdot}15$. $\vdash : \alpha \in \text{NC induct} . \exists ! \alpha . \supset . \text{sm}``\alpha \in \text{NC induct}$

$*120{\cdot}151$. $\vdash : \alpha \in \text{NC induct} . \exists ! \alpha . \supset . \alpha +_c 1 \in \text{NC induct}$

$*120{\cdot}152$. $\vdash : \alpha \in \text{NC} . \text{sm}``\alpha \in \text{NC induct} - \iota`\Lambda . \supset . \alpha \in \text{NC induct} - \iota`\Lambda$

We then proceed to deduce the elementary properties of inductive *classes*, putting

$$\text{Cls induct} = s`\text{NC induct}.$$

We have

$*120{\cdot}21$. $\vdash : \rho \in \text{Cls induct} . \equiv . \text{N}_0\text{c}`\rho \in \text{NC induct}$

∗120·211. $\vdash : \text{Nc}'\rho \in \text{NC induct} - \iota'\Lambda \,.\supset .\, \rho \in \text{Cls induct}$

(We do not have an equivalence here, because, for aught we know, it might be possible to determine the ambiguity of $\text{Nc}'\rho$ so that $\text{Nc}'\rho = \Lambda$, even when $\rho \in \text{Cls induct}$. This will not be possible, however, if the axiom of infinity is assumed.)

∗120·212·213. $\vdash . \Lambda, \iota'x \in \text{Cls induct}$

∗120·214. $\vdash :. \rho \,\text{sm}\, \sigma \,.\supset : \rho \in \text{Cls induct} \,.\equiv .\, \sigma \in \text{Cls induct}$

We have a set of propositions applying induction to classes *directly*, and not through the intermediary of cardinals. Thus we have

∗120·251. $\vdash : \eta \in \text{Cls induct} \,.\supset .\, \eta \cup \iota'y \in \text{Cls induct}$

∗120·26. $\vdash :. \rho \in \text{Cls induct} : \phi\eta \,.\supset_{\eta, x} .\, \phi(\eta \cup \iota'x) : \phi\Lambda : \supset .\, \phi\rho$

We then state the axiom of infinity, and prove (∗120·33) that it is equivalent to the assumption that if α is an inductive cardinal, $\alpha \neq \alpha +_c 1$. To prove this, we first prove various propositions about $\alpha +_c 1$, among others the following:

∗120·311. $\vdash : \exists ! \alpha +_c 1 \,.\, \alpha +_c 1 = \beta +_c 1 \,.\supset .\, \alpha = \text{sm}``\beta \,.\, \exists ! \alpha$

∗120·322. $\vdash :. \alpha \in \text{NC induct} \,.\supset : \exists ! \alpha \,.\equiv .\, \alpha \neq \alpha +_c 1$

We then proceed to consider subtraction (∗120·41—·418), which only gives a cardinal number when the subtrahend is an inductive cardinal. We have

∗120·41. $\vdash :. \nu \in \text{NC induct} \,.\, \exists ! \alpha +_c \nu \,.\supset : \alpha +_c \nu = \beta +_c \nu \,.\supset .\, \alpha = \text{sm}``\beta$

We might validly put $\alpha = \beta$ instead of $\alpha = \text{sm}``\beta$, since $\alpha = \beta$ will be true whenever it is significant.

We have

∗120·411. $\vdash :. \nu \in \text{NC induct} \,.\supset :$

$$\exists ! \gamma -_c \nu \,.\supset .\, \gamma -_c \nu \in \text{N}_0\text{C} : \gamma \geqslant \nu \,.\equiv .\, (\gamma -_c \nu) \cap t_0{}'\gamma \in \text{N}_0\text{C}$$

∗120·4111. $\vdash :. \nu \in \text{NC induct} \,.\, \exists ! \text{sm}_\xi``\gamma \,.\supset : \gamma \geqslant \nu \,.\equiv .\, (\gamma -_c \nu)_\xi \in \text{N}_0\text{C}$

Hence we arrive at the conditions requisite for the usual point of view of subtraction; namely,

∗120·412. $\vdash : \nu \in \text{NC induct} \,.\, \gamma \geqslant \nu \,.\, \exists ! \text{sm}_\xi``\gamma \,.\supset .\, (\gamma -_c \nu)_\xi = \{(\imath\alpha)(\alpha +_c \nu = \gamma)\}_\xi$

Also from ∗120·4111 we deduce

∗120·414. $\vdash : \mu \in \text{N}_0\text{C} - \iota'0 \,.\, \exists ! \text{sm}_\xi``\mu \,.\supset .\, (\mu -_c 1)_\xi \in \text{N}_0\text{C}$

And from ∗120·411 . ∗119·34, we find

∗120·416. $\vdash : \nu \in \text{NC induct} \,.\, \exists ! \gamma -_c \nu \,.\supset .\, (\gamma -_c \nu) +_c \nu = \text{sm}``\gamma$

We prove next that no proper part of an inductive class is similar to the whole (∗120·426), *i.e.* that inductive classes are non-reflexive, and various connected propositions, *e.g.*

∗120·423. $\vdash : \alpha \in \text{N}_\eta\text{C induct} - \iota'0 \,.\equiv .\, (\exists\beta) \,.\, \beta \in \text{N}_\eta\text{C induct} \,.\, \alpha = (\beta +_c 1)_\eta$

$*120 \cdot 4232$. $\vdash : \alpha \in \mathrm{N}_\eta\mathrm{C}\ \mathrm{induct} - \iota`0 . \equiv . (\exists\beta) . \beta \in \mathrm{N}_\eta\mathrm{C}\ \mathrm{induct} - \iota`\Lambda . \alpha = (\beta +_c 1)_\eta$

$*120 \cdot 428$. $\vdash : \nu \in \mathrm{NC}\ \mathrm{induct} . \exists ! \alpha +_c \nu . \alpha \neq 0 . \supset . \alpha +_c \nu > \nu$

$*120 \cdot 429$. $\vdash :. \nu \in \mathrm{NC}\ \mathrm{induct} . \supset : \mu > \nu . \equiv . \mu \geqslant \nu +_c 1$

The last two of the above propositions do not hold in general when ν is a cardinal which is not inductive.

We prove next that if α is an existent inductive cardinal, then any existent cardinal is greater than, equal to, or less than α ($*120 \cdot 441$); that if α, β are inductive cardinals, so is $\alpha +_c \beta$ ($*120 \cdot 45 \cdot 450$), and if $\alpha +_c \beta$ is an inductive cardinal other than Λ, so are α and β ($*120 \cdot 452$). We then have some propositions dealing with mathematical induction starting from 1 or 2, *e.g.*

$*120 \cdot 4622$. $\vdash :. \alpha \in \mathrm{NC} . \beta \in \mathrm{NC}(\eta) . \exists ! \mathrm{sm}_\xi``\beta . \supset :$

$$\beta\,(+_c 1)_* \,\mathrm{sm}_\eta``\alpha . \equiv . \mathrm{sm}_\xi``\beta\,(+_c 1)_* \,\mathrm{sm}_\xi``\alpha$$

$*120 \cdot 47$. $\vdash :: \beta \in \mathrm{N}_\eta\mathrm{C}\ \mathrm{induct} - \iota`0 . \equiv :. \xi \in \mu . \supset_\xi . (\xi +_c 1)_\eta \in \mu : 1_\eta \in \mu : \supset_\mu . \beta \in \mu$

From $*120 \cdot 452$ we deduce

$*120 \cdot 48$. $\vdash : \beta \in \mathrm{NC}\ \mathrm{induct} . \beta \geqslant \alpha . \supset . \alpha \in \mathrm{NC}\ \mathrm{induct} - \iota`\Lambda$

so that any number less than an inductive number is inductive. Hence

$*120 \cdot 481$. $\vdash : \eta \in \mathrm{Cls}\ \mathrm{induct} . \xi \subset \eta . \supset . \xi \in \mathrm{Cls}\ \mathrm{induct}$

which is a proposition constantly used, and

$*120 \cdot 491$. $\vdash :. \xi \sim \in \mathrm{Cls}\ \mathrm{induct} . \equiv : \beta \in \mathrm{NC}\ \mathrm{induct} . \supset_\beta . \exists ! \beta \cap \mathrm{Cl}`\xi$

We then prove that if α, β are inductive cardinals, $\alpha \times_c \beta$ and α^β are either inductive cardinals or Λ ($*120 \cdot 5 \cdot 52$), while conversely if $\alpha \times_c \beta$ or α^β is an existent inductive cardinal, α and β are so also, with exceptions for 0 and 1 ($*120 \cdot 512 \cdot 56 \cdot 561$). Hence we infer the uniqueness of division and the taking of roots ($*120 \cdot 51 \cdot 53 \cdot 55$) so long as inductive numbers are concerned.

We have next a set of propositions on the axiom of infinity and the multiplicative axiom. We prove ($*120 \cdot 61$) that if there is any existent cardinal which is not inductive, the axiom of infinity is true. From $*83 \cdot 9 \cdot 904$, we infer by induction that if κ is an inductive class of which Λ is not a number, $\epsilon_\Delta`\kappa$ exists ($*120 \cdot 62$), whence it follows that either the multiplicative axiom or the axiom of infinity must be true ($*120 \cdot 64$).

Finally, we have a set of propositions on inductive *classes*. We prove

$*120 \cdot 71$. $\vdash : \rho, \sigma \in \mathrm{Cls}\ \mathrm{induct} . \equiv . \rho \cup \sigma \in \mathrm{Cls}\ \mathrm{induct} . \equiv . \rho + \sigma \in \mathrm{Cls}\ \mathrm{induct}$

$*120 \cdot 74$. $\vdash : \rho \in \mathrm{Cls}\ \mathrm{induct} . \equiv . \mathrm{Cl}`\rho \in \mathrm{Cls}\ \mathrm{induct}$

$*120 \cdot 75$. $\vdash : s`\kappa \in \mathrm{Cls}\ \mathrm{induct} . \equiv . \kappa \in \mathrm{Cls}\ \mathrm{induct} . \kappa \subset \mathrm{Cls}\ \mathrm{induct}$

with analogous propositions (involving however a hypothesis as to κ) on the subject of $\epsilon_\Delta`\kappa$.

The propositions of the present number are essential to the ordinary arithmetic of finite numbers. In the present work, however, they are not much used after the present section until we reach Part V, Section E, where we deal with the ordinal theory of finite and infinite.

***120·01.** $\text{NC induct} = \hat{\alpha} \{\alpha (+_c 1)_* 0\}$ Df

Note that in virtue of our general conventions for descriptive functions of two arguments (*38),

$$+_c 1 = \hat{\alpha} \hat{\beta} (\alpha = \beta +_c 1).$$

That is, $+_c 1$ is the relation of a cardinal to its immediate predecessor. It is the number written in the usual mathematical notation as $+1$ in the series of positive and negative integers, just as its converse is the number -1. (It should be observed that if ν is any cardinal, $+\nu$ is not identical with ν, since $+\nu$ is a relation, while ν is a class of classes.)

***120·011.** $\text{N}_\xi\text{C induct} = \hat{\alpha} \{\alpha (+_c 1)_* 0_\xi\}$ Df

All members of N_ξC induct belong to the same type as 0_ξ, so that, if α is any member of N_ξC induct, "$\xi \in \alpha$" is significant.

***120·02.** $\text{Cls induct} = s`\text{NC induct}$ Df

***120·021.** $\text{Cls}_\xi \text{ induct} = s`\text{N}_\xi\text{C induct}$ Df

In virtue of these definitions an inductive class is one whose cardinal is an inductive cardinal.

***120·03.** $\text{Infin ax} . = : \alpha \in \text{NC induct} . \supset_\alpha . \exists ! \alpha$ Df

"Infin ax," like "Mult ax," is an arithmetical hypothesis which some will consider self-evident, but which we prefer to keep as a hypothesis, and to adduce in that form whenever it is relevant. Like "Mult ax," it states an existence-theorem. In the above form, it states that, if α is any inductive cardinal, there is at least one class (of the type in question) which has α terms. An equivalent assumption would be that, if ρ is any inductive class, there are objects which are not members of ρ. For in that case, if x be such an object, $\text{Nc}`(\rho \cup \iota`x) = \text{Nc}`\rho +_c 1$. Hence by induction, every inductive cardinal must exist. Another equivalent assumption would be that V (the class of all objects of the type in question) is not an inductive class. The assumption that $\aleph_0$ exists in the type in question is, as we shall see, a stronger assumption than the above, unless we assume the multiplicative axiom.

If the axiom of infinity is true, the inductive cardinals are all different one from another, *i.e.* $\alpha +_c \beta$, where α and β are inductive cardinals, is not equal to α unless $\beta = 0$. But if the axiom of infinity is false, then, in any assigned type, all the cardinals after a certain one are Λ. (Except in the lowest type, the last existent cardinal must be a power of 2.) That is, if (say) 8 were the largest existent cardinal in the type in question, we should have, in that type, $9 = \Lambda$, and the same would hold of 10, 11, ... This possibility has to be taken account of in what follows.

In order to give typical definiteness to the axiom of infinity, we write

***120·04.** Infin ax $(x) . = : \alpha \epsilon \text{NC induct} . \supset_\alpha . \exists ! \alpha(x)$ Df

Then "Infin ax (x)" states that, if α is any inductive cardinal, there are at least α objects of the same type as x.

***120·1.** $\vdash : \alpha \epsilon \text{NC induct} . \equiv . \alpha (+_c 1)_* 0$ [(*120·01)]

***120·101.** $\vdash :: \alpha \epsilon \text{NC induct} . \equiv :. \xi \epsilon \mu . \supset_\xi . \xi +_c 1 \epsilon \mu : 0 \epsilon \mu : \supset_\mu . \alpha \epsilon \mu$
[*120·1 . *90·131 . *38·12]

The right-hand side of the above equivalence gives the usual formula for mathematical induction. Observe that the conditions of significance require that $\xi +_c 1$ should be taken in the same type as ξ. This fact is specially relevant in the proof of *120·15.

The symbol "NC induct" is of ambiguous type not necessarily the same in different occurrences; also, according to the convention explained in the prefatory statement as holding for NC and NC induct, "$\alpha, \beta \epsilon$ NC induct" will not imply that α and β are of the same type. Accordingly to avoid error in connection with *120·1·101 typical definiteness is required as in the three following propositions.

***120·102.** $\vdash : \alpha \epsilon N_\eta\text{C induct} . \equiv . \alpha (+_c 1)_* 0_\eta$ [(*120·011)]

***120·103.** $\vdash :: \alpha \epsilon N_\eta\text{C induct} . \equiv :. \xi \epsilon \mu . \supset_\xi . (\xi +_c 1)_\eta \epsilon \mu : 0_\eta \epsilon \mu : \supset_\mu . \alpha \epsilon \mu$
[*120·101]

***120·11.** $\vdash :. \alpha \epsilon N_\eta\text{C induct} : \phi\xi . \supset_\xi . \phi(\xi +_c 1) : \phi 0_\eta : \supset . \phi\alpha$
[*120·102 . *90·112]

***120·12.** $\vdash . 0 \epsilon \text{NC induct}$ $\left[*120\cdot101 \frac{0}{\alpha} \right]$

***120·121.** $\vdash : \alpha \epsilon N_\xi\text{C induct} . \supset . (\alpha +_c 1)_\xi \epsilon N_\xi\text{C induct}$ [*90·172 . *120·102]

By means of this proposition and *120·12, any assigned cardinal in the series of natural numbers can be shown to be an inductive cardinal; thus *e.g.* to show that 27 is an inductive cardinal, we shall only have to use *120·121 twenty-seven times in succession.

***120·122.** $\vdash . 1 \epsilon \text{NC induct}$ [*120·12·121 . *110·641]

***120·123.** $\vdash . 2 \epsilon \text{NC induct} . \text{etc.}$ [*120·122·121 . *110·643]

***120·124.** $\vdash . \alpha +_c 1 \neq 0$

Dem.

$\vdash . *110\cdot4 . \text{Transp} . \supset \vdash : \alpha \sim\epsilon \text{NC} . \supset . \alpha +_c 1 = \Lambda .$
[*101·12] $\supset . \alpha +_c 1 \neq 0$ (1)
$\vdash . *110\cdot632 . \supset \vdash :. \alpha \epsilon \text{NC} . \supset : \xi \epsilon \alpha +_c 1 . \supset . \exists ! \xi :$
[*24·63] $\supset : \Lambda \sim\epsilon \alpha +_c 1 :$
[*54·102] $\supset : \alpha +_c 1 \neq 0$ (2)
$\vdash . (1) . (2) . \supset \vdash . \text{Prop}$

***120·13.** $\vdash :. \alpha \in N_\eta C \text{ induct} : \xi \in N_\eta C \text{ induct} . \phi\xi . \supset_\xi . \phi(\xi +_c 1) : \phi 0_\eta : \supset . \phi\alpha$

Dem.

$\vdash . *120·121 . \supset \vdash :. \xi \in N_\eta C \text{ induct} . \phi\xi . \supset_\xi . \phi(\xi +_c 1) : \supset :$

$\xi \in N_\eta C \text{ induct} . \phi\xi . \supset_\xi . (\xi +_c 1)_\eta \in N_\eta C \text{ induct} . \phi(\xi +_c 1)$ (1)

$\vdash . *120·12 . \supset \vdash : \phi 0_\eta . \supset . 0_\eta \in N_\eta C \text{ induct} . \phi 0_\eta$ (2)

$\vdash . (1) . (2) . \supset \vdash :. \text{Hp} . \supset :$

$\xi \in N_\eta C \text{ induct} . \phi\xi . \supset_\xi . (\xi +_c 1)_\eta \in N_\eta C \text{ induct} . \phi(\xi +_c 1) : 0_\eta \in N_\eta C \text{induct} . \phi 0_\eta :$

$\left[*120·11 \frac{\xi \in N_\eta C \text{ induct} . \phi\xi}{\phi\xi} \right] \supset : \alpha \in N_\eta C \text{ induct} . \phi\alpha :. \supset \vdash . \text{Prop}$

The above proposition is often convenient for inductive proofs.

***120·14.** $\vdash . \text{NC induct} \subset \text{NC}$

Dem.

$\vdash . *110·42 . \text{Simp} . \supset \vdash : \alpha \in \text{NC} . \supset . \alpha +_c 1 \in \text{NC}$ (1)

$\vdash . (1) . *101·11 . *120·11 \frac{\alpha \in \text{NC}}{\phi\alpha} . \supset \vdash . \text{Prop}$

This proposition does not show that every inductive cardinal is an *existent* cardinal; to obtain this, we require the axiom of infinity.

***120·15.** $\vdash : \alpha \in \text{NC induct} . \exists ! \alpha . \supset . \text{sm}``\alpha \in \text{NC induct}$

I.e. a cardinal which is not null and is inductive in any one type is also inductive in any other type.

Dem.

$\vdash . *101·15 . *120·12 . \supset \vdash . \text{sm}_\eta``0_\xi \in N_\eta C \text{ induct}$ (1)

$\vdash . *110·4 . \supset \vdash . \alpha = \Lambda_\xi . \supset . (\alpha +_c 1)_\xi = \Lambda_\xi$ (2)

$\vdash . *118·201 . \supset \vdash : \exists ! (\alpha +_c 1)_\xi . \supset . \text{sm}_\eta``(\alpha +_c 1)_\xi = (\alpha +_c 1)_\eta$

$[*118·241 . *110·4] \quad = (\text{sm}_\eta``\alpha +_c 1)_\eta$ (3)

$\vdash . *120·121 . \supset \vdash : \exists ! (\alpha +_c 1)_\xi . \text{sm}_\eta``\alpha \in N_\eta C \text{induct} . \supset . (\text{sm}_\eta``\alpha +_c 1)_\eta \in N_\eta C \text{ induct} .$

$[(3)] \quad \supset . \text{sm}_\eta``(\alpha +_c 1)_\xi \in N_\eta C \text{ induct}$ (4)

$\vdash . (4) . *2·2 . \supset \vdash :. \text{sm}_\eta``\alpha \in N_\eta C \text{ induct} . \supset :$

$(\alpha +_c 1)_\xi = \Lambda_\xi . \lor . \text{sm}_\eta``(\alpha +_c 1)_\xi \in N_\eta C \text{ induct}$ (5)

$\vdash . (2) . (5) . *3·48 . \supset \vdash :. \alpha = \Lambda_\xi . \lor . \text{sm}_\eta``\alpha \in N_\eta C \text{ induct} : \supset :$

$(\alpha +_c 1)_\xi = \Lambda_\xi . \lor . \text{sm}_\eta``(\alpha +_c 1)_\xi \in N_\eta C \text{ induct}$ (6)

$\vdash . (1) . (6) . *120·11 . *4·6 . \supset \vdash . \text{Prop}$

***120·151.** $\vdash : \alpha \in \text{NC induct} . \exists ! \alpha . \supset . \alpha +_c 1 \in \text{NC induct}$

Dem.

$\vdash . *120·15 . \supset \vdash : \alpha \in N_\xi C \text{ induct} . \exists ! \alpha . \supset . \text{sm}_\eta``\alpha \in N_\eta C \text{ induct} .$

$[*120·121] \quad \supset . (\text{sm}_\eta``\alpha +_c 1)_\eta \in N_\eta C \text{ induct} .$

$[*118·241 . *120·14] \quad \supset . (\alpha +_c 1)_\eta \in N_\eta C \text{ induct} : \supset \vdash . \text{Prop}$

***120·152.** $\vdash : \alpha \in \text{NC} . \text{sm}``\alpha \in \text{NC induct} - \iota`\Lambda . \supset . \alpha \in \text{NC induct} - \iota`\Lambda$

Dem.

$\vdash . *100·521 . \supset \vdash : \text{Hp} . \supset . \text{sm}``\text{sm}``\alpha = \alpha .$

$[*120·15] \quad \supset . \alpha \in \text{NC induct}$ (1)

$\vdash . *37·29 . \supset \vdash : \text{Hp} . \supset . \exists ! \alpha$ (2)

$\vdash . (1) . (2) . \supset \vdash . \text{Prop}$

The following propositions, giving alternative forms for the definition of inductive *classes*, are inserted in order to show that the theory of inductive classes might be treated in a less arithmetical manner than we have adopted.

∗120·2. $\vdash : \rho \,\epsilon\, \text{Cls induct} \,.\equiv .\, (\exists \alpha) \,.\, \alpha \,\epsilon\, \text{NC induct} \,.\, \rho \,\epsilon\, \alpha$ [(∗120·02)]

∗120·201. $\vdash :. \rho \,\text{sm}\, \sigma \,.\supset : \text{N}_0\text{c}`\rho \,\epsilon\, \text{NC induct} \,.\equiv .\, \text{N}_0\text{c}`\sigma \,\epsilon\, \text{NC induct}$

Dem.

$\vdash .\, *100{\cdot}35 \,.\, *103{\cdot}13 \,.\, *100{\cdot}511 \,.\supset$

$\vdash : \text{Hp} \,.\supset .\, \text{N}_0\text{c}`\rho = \text{sm}``\text{N}_0\text{c}`\sigma \,.\, \text{N}_0\text{c}`\sigma = \text{sm}``\text{N}_0\text{c}`\rho :$

[∗120·152.∗103·13] $\supset \vdash .$ Prop

∗120·21. $\vdash : \rho \,\epsilon\, \text{Cls induct} \,.\equiv .\, \text{N}_0\text{c}`\rho \,\epsilon\, \text{NC induct}$

Dem.

$\vdash .\, *120{\cdot}14{\cdot}2 \,.\supset \vdash : \rho \,\epsilon\, \text{Cls induct} \,.\equiv .\, (\exists \alpha) \,.\, \alpha \,\epsilon\, \text{NC induct} \,.\, \alpha \,\epsilon\, \text{NC} \,.\, \rho \,\epsilon\, \alpha \,.$

[∗103·27] $\equiv .\, (\exists \alpha) \,.\, \alpha \,\epsilon\, \text{NC induct} \,.\, \text{N}_0\text{c}`\rho = \alpha \,.$

[∗13·195] $\equiv .\, \text{N}_0\text{c}`\rho \,\epsilon\, \text{NC induct} : \supset \vdash .$ Prop

Note that "$\rho \,\epsilon\, \text{Cls induct} \,.\equiv .\, \text{Nc}`\rho \,\epsilon\, \text{NC induct}$" is not proved above. The proof encounters the difficulty that we may have $\text{Nc}`\rho = \Lambda$; in order to establish our proposition in this case, we have to show that if $\Lambda \,\epsilon\, \text{NC induct}$, then *every* class is an inductive class. We can however prove the following implication.

∗120·211. $\vdash : \text{Nc}`\rho \,\epsilon\, \text{NC induct} - \iota`\Lambda \,.\supset .\, \rho \,\epsilon\, \text{Cls induct}$

Dem.

$\vdash .\, *100{\cdot}511 \,.\supset \vdash : \text{Hp} \,.\supset .\, \text{sm}``\text{Nc}`\rho = \text{N}_0\text{c}`\rho \,.$

[∗120·15] $\supset .\, \text{N}_0\text{c}`\rho \,\epsilon\, \text{NC induct} \,.$

[∗120·21] $\supset .\, \rho \,\epsilon\, \text{Cls induct} : \supset \vdash .$ Prop

∗120·212. $\vdash .\, \Lambda \,\epsilon\, \text{Cls induct}$ [∗120·211·12]

∗120·213. $\vdash .\, \iota`x \,\epsilon\, \text{Cls induct}$ [∗120·211·122]

∗120·214. $\vdash :. \rho \,\text{sm}\, \sigma \,.\supset : \rho \,\epsilon\, \text{Cls induct} \,.\equiv .\, \sigma \,\epsilon\, \text{Cls induct}$ [∗120·201·21]

The following propositions are lemmas for ∗120·24.

∗120·22. $\vdash :: \eta \,\epsilon\, \mu \,.\supset_{\eta, y} .\, \eta \cup \iota`y \,\epsilon\, \mu : \Lambda \,\epsilon\, \mu : \supset_\mu .\, \rho \,\epsilon\, \mu :. \supset .\, \rho \,\epsilon\, \text{Cls induct}$

Dem.

$\vdash .\, *120{\cdot}212 \,.\supset \vdash .\, \Lambda \,\epsilon\, \text{Cls induct}$ (1)

$\vdash .\, *51{\cdot}2 \,.\quad \supset \vdash :. y \,\epsilon\, \eta \,.\supset : \eta \cup \iota`y = \eta :$

[∗13·12] $\supset : \eta \,\epsilon\, \text{Cls induct} \,.\supset .\, \eta \cup \iota`y \,\epsilon\, \text{Cls induct}$ (2)

$\vdash .\, *110{\cdot}63 \,.\quad \supset \vdash :. y \sim\epsilon\, \eta \,.\supset : \text{Nc}`(\eta \cup \iota`y) = \text{Nc}`\eta +_c 1$

[(∗110·03)] $= \text{N}_0\text{c}`\eta +_c 1 :$

[∗120·121] $\supset : \text{N}_0\text{c}`\eta \,\epsilon\, \text{NC induct} \,.\supset .\, \text{N}_0\text{c}`(\eta \cup \iota`y) \,\epsilon\, \text{NC induct} :$

[∗120·21·211] $\supset : \eta \,\epsilon\, \text{Cls induct} \,.\supset .\, \eta \cup \iota`y \,\epsilon\, \text{Cls induct}$ (3)

$\vdash .\, (2) \,.\, (3) \,.\quad \supset \vdash : \eta \,\epsilon\, \text{Cls induct} \,.\supset .\, \eta \cup \iota`y \,\epsilon\, \text{Cls induct}$ (4)

$\vdash .\, *10{\cdot}1 \,.\, (1) \,.\, (4) \,.\supset \vdash .$ Prop

***120·221.** $\vdash :. \eta \epsilon \mu . \supset_{\eta, y} . \eta \cup \iota`y \epsilon \mu : \mathrm{Nc}`\rho \subset \mu : \supset . \mathrm{Nc}`\rho +_{\mathrm{c}} 1 \subset \mu$

Dem.

$\vdash . *110·63 . *100·31 . \supset$

$\vdash : \zeta \epsilon \mathrm{Nc}`\rho +_{\mathrm{c}} 1 . \equiv . (\exists \eta, y) . \eta \epsilon \mathrm{Nc}`\rho . y \sim\epsilon \eta . \zeta = \eta \cup \iota`y$ (1)

$\vdash . *22·1 . \quad \supset \vdash :. \mathrm{Hp} . \supset : \eta \epsilon \mathrm{Nc}`\rho . \supset . \eta \epsilon \mu .$

[*10·1] $\qquad \supset . \eta \cup \iota`y \epsilon \mu :$

[*3·41] $\qquad \supset : \eta \epsilon \mathrm{Nc}`\rho . y \sim\epsilon \eta . \supset . \eta \cup \iota`y \epsilon \mu :$

[*13·12] $\qquad \supset : \eta \epsilon \mathrm{Nc}`\rho . y \sim\epsilon \eta . \zeta = \eta \cup \iota`y . \supset . \zeta \epsilon \mu$ (2)

$\vdash . (1) . (2) . \supset \vdash :. \mathrm{Hp} . \supset : \zeta \epsilon \mathrm{Nc}`\rho +_{\mathrm{c}} 1 . \supset . \zeta \epsilon \mu :. \supset \vdash . \mathrm{Prop}$

***120·222.** $\vdash :. \eta \epsilon \mu . \supset_{\eta, y} . \eta \cup \iota`y \epsilon \mu : \xi \epsilon \mathrm{NC} . \xi \subset \mu : \supset . \xi +_{\mathrm{c}} 1 \subset \mu$

Dem.

$\vdash . *100·4 . \supset \vdash : \mathrm{Hp} . \exists ! \xi . \supset . (\exists \alpha) . \xi = \mathrm{Nc}\,(\zeta)`\alpha . \mathrm{Nc}\,(\zeta)`\alpha \subset \mu .$

[*120·221] $\qquad \supset . (\exists \alpha) . \xi = \mathrm{Nc}\,(\zeta)`\alpha . \mathrm{Nc}`\alpha +_{\mathrm{c}} 1 \subset \mu .$

[*118·01] $\qquad \supset . \xi +_{\mathrm{c}} 1 \subset \mu$ (1)

$\vdash . *110·4 . \supset \vdash : \sim\exists ! \xi . \supset . \xi +_{\mathrm{c}} 1 \subset \mu$ (2)

$\vdash . (1) . (2) . \supset \vdash . \mathrm{Prop}$

The proof of this proposition might also proceed by the use of uniform formal numbers, employing *118·241.

***120·23.** $\vdash :. \eta \epsilon \mu . \supset_{\eta, y} . \eta \cup \iota`y \epsilon \mu : \Lambda \epsilon \mu : \supset . \mathrm{Cls\ induct} \subset \mu$

Dem.

$\vdash . *51·2 . *54·1 . \quad \supset \vdash : \mathrm{Hp} . \supset . 0 \subset \mu$ (1)

$\vdash . *120·222·14 . \quad \supset \vdash :. \mathrm{Hp} . \supset : \xi \epsilon \mathrm{NC\ induct} . \xi \subset \mu . \supset_{\xi} . \xi +_{\mathrm{c}} 1 \subset \mu$ (2)

$\vdash . (1) . (2) . *120·13 . \supset \vdash :. \mathrm{Hp} . \supset : \xi \epsilon \mathrm{NC\ induct} . \supset . \xi \subset \mu :$

[*40·151.(*120·02)] $\qquad \supset : \mathrm{Cls\ induct} \subset \mu :. \supset \vdash . \mathrm{Prop}$

***120·24.** $\vdash :: \rho \epsilon \mathrm{Cls\ induct} . \equiv :. \eta \epsilon \mu . \supset_{\eta, y} . \eta \cup \iota`y \epsilon \mu : \Lambda \epsilon \mu : \supset_{\mu} . \rho \epsilon \mu$

Dem.

$\vdash . *120·23 . \supset \vdash :: \rho \epsilon \mathrm{Cls\ induct} . \supset :. \eta \epsilon \mu . \supset_{\eta, y} . \eta \cup \iota`y \epsilon \mu : \Lambda \epsilon \mu : \supset . \rho \epsilon \mu$ (1)

$\vdash . (1) . *120·22 . \supset \vdash . \mathrm{Prop}$

This proposition might be used to define inductive classes. It gives a form of mathematical induction applicable to classes instead of to numbers. Virtually it states that an inductive class is one which can be formed by adding members one at a time, starting from Λ. This is made more explicit in *120·25. Instead of $\eta \epsilon \mu . \supset_{\eta, y} . \eta \cup \iota`y \epsilon \mu$, in the above propositions, as well as in those that follow, we may plainly substitute

$$\eta \epsilon \mu . y \sim\epsilon \eta . \supset_{\eta, y} . \eta \cup \iota`y \epsilon \mu .$$

***120·25.** $\vdash : M = \hat{\eta}\hat{\zeta}\,\{(\exists y) . \zeta = \eta \cup \iota`y\} . \supset . \mathrm{Cls\ induct} = \overleftarrow{M_*}`\Lambda$
[*120·24 . *90·131]

***120·251.** $\vdash : \eta \epsilon \mathrm{Cls\ induct} . \supset . \eta \cup \iota`y \epsilon \mathrm{Cls\ induct}$ [*90·172 . *120·25]

***120·26.** $\vdash :. \rho \epsilon \mathrm{Cls\ induct} : \phi\eta . \supset_{\eta, x} . \phi\,(\eta \cup \iota`x) : \phi\Lambda : \supset . \phi\rho$
[*120·25 . *90·112]

***120·261.** $\vdash :. \rho \in \text{Cls induct} : \eta \in \text{Cls induct} . \phi\eta . \supset_{\eta, x} . \phi(\eta \cup \iota`x) : \phi\Lambda : \supset . \phi\rho$
[*120·26·251·212]

***120·27.** $\vdash : \rho \in \text{Cls induct} . \supset . \text{Nc}`\rho \cap t`\gamma \in \text{NC induct}$

Dem.

$\vdash . *120 \cdot 12 . \supset \vdash . \text{Nc}`\Lambda \cap t`\gamma \in \text{NC induct}$ (1)

$\vdash . *13 \cdot 12 . \supset \vdash : \text{Nc}`\eta \cap t`\gamma \in \text{NC induct} . y \in \eta . \supset .$

$\text{Nc}`(\eta \cup \iota`y) \cap t`\gamma \in \text{NC induct}$ (2)

$\vdash . *110 \cdot 63 . *120 \cdot 121 . \supset$

$\vdash : \text{Nc}`\eta \cap t`\gamma \in \text{NC induct} . y \sim\in \eta . \supset . \text{Nc}`(\eta \cup \iota`y) \cap t`\gamma \in \text{NC induct}$ (3)

$\vdash . (1) . (2) . (3) . *120 \cdot 26 . \supset \vdash . \text{Prop}$

This proposition also follows immediately from *120·21·15.

***120·3.** $\vdash :. \text{Infin ax} . \equiv : \alpha \in \text{NC induct} . \supset_\alpha . \exists ! \alpha$ [(*120·03)]

***120·301.** $\vdash :. \text{Infin ax}\,(x) . \equiv : \alpha \in \text{NC induct} . \supset_\alpha . \exists ! \alpha(x)$ [(*120·04)]

***120·31.** $\vdash : \exists ! \text{Nc}`\alpha +_c 1 . \text{Nc}`\alpha +_c 1 = \text{Nc}`\beta +_c 1 . \supset . \text{Nc}`\alpha = \text{Nc}`\beta . \alpha \,\text{sm}\, \beta$

Dem.

$\vdash . *110 \cdot 63 . \supset \vdash :. \text{Nc}`\alpha +_c 1 = \text{Nc}`\beta +_c 1 . \equiv :$

$(\exists \gamma, y) . \gamma \,\text{sm}\, \alpha . y \sim\in \gamma . \xi = \gamma \cup \iota`y . \equiv_\xi . (\exists \delta, z) . \delta \,\text{sm}\, \beta . z \sim\in \delta . \xi = \delta \cup \iota`z :$

[*10·1] $\supset : \gamma \,\text{sm}\, \alpha . y \sim\in \gamma . \supset . (\exists \delta, z) . \delta \,\text{sm}\, \beta . z \sim\in \delta . \gamma \cup \iota`y = \delta \cup \iota`z .$

[*73·72·3] $\supset . (\exists \delta) . \delta \,\text{sm}\, \beta . \gamma \,\text{sm}\, \delta .$

[*73·32] $\supset . \gamma \,\text{sm}\, \beta .$

[*73·32] $\supset . \alpha \,\text{sm}\, \beta$ (1)

$\vdash . *110 \cdot 63 . \supset \vdash : \text{Hp} . \supset . (\exists \gamma, y) . \gamma \,\text{sm}\, \alpha . y \sim\in \gamma$ (2)

$\vdash . (1) . (2) . *100 \cdot 321 . \supset \vdash . \text{Prop}$

***120·311.** $\vdash : \exists ! \alpha +_c 1 . \alpha +_c 1 = \beta +_c 1 . \supset . \alpha = \text{sm}``\beta . \exists ! \alpha$
[*120·31 . *110·4 . *103·16·4·2]

***120·32.** $\vdash : \alpha \in \text{NC induct} . \exists ! \alpha . \supset . \alpha \neq \alpha +_c 1$

Dem.

$\vdash . *101 \cdot 22 . *110 \cdot 641 . \supset \vdash . 0_\xi \neq 0_\xi +_c 1$ (1)

$\vdash . *120 \cdot 311 . *110 \cdot 44 . \supset \vdash : \alpha \in \text{NC} . \exists ! \alpha +_c 1 . \alpha +_c 1 = \alpha +_c 1 +_c 1 . \supset . \alpha = \alpha +_c 1 :$

[Transp] $\supset \vdash : \alpha \in \text{NC} . \exists ! \alpha +_c 1 . \alpha \neq \alpha +_c 1 . \supset . \alpha +_c 1 \neq \alpha +_c 1 +_c 1 :$

[*118·2·25] $\supset \vdash : \alpha \in \text{NC}(\xi) . \exists ! (\alpha +_c 1)_\xi . \alpha \neq (\alpha +_c 1)_\xi . \supset . (\alpha +_c 1)_\xi \neq \{(\alpha +_c 1)_\xi +_c 1\}_\xi$ (2)

$\vdash . (2) . \supset \vdash :. \alpha \in \text{NC}(\xi) . \alpha \neq (\alpha +_c 1)_\xi . \supset : (\alpha +_c 1)_\xi = \Lambda . \vee . (\alpha +_c 1)_\xi \neq \{(\alpha +_c 1)_\xi +_c 1\}_\xi$ (3)

$\vdash . *110 \cdot 4 . \text{Transp} . \supset \vdash :. \alpha \sim\in \text{NC}(\xi) . \vee . \alpha = \Lambda_\xi : \supset . (\alpha +_c 1)_\xi = \Lambda_\xi$ (4)

$\vdash . (3) . (4) . \supset \vdash :. \alpha = \Lambda_\xi . \vee . \alpha \neq (\alpha +_c 1)_\xi : \supset :$

$(\alpha +_c 1)_\xi = \Lambda_\xi . \vee . (\alpha +_c 1)_\xi \neq \{(\alpha +_c 1)_\xi +_c 1\}_\xi$ (5)

$\vdash . (1) . (5) . *120 \cdot 11 . \supset \vdash :. \alpha \in \text{N}_\xi\text{C induct} . \supset : \alpha = \Lambda_\xi . \vee . \alpha \neq (\alpha +_c 1)_\xi :. \supset \vdash . \text{Prop}$

***120·321.** $\vdash : \alpha \neq \alpha +_c 1 . \supset . \exists ! \alpha$

Dem.

$\vdash . *110\cdot4 . \text{Transp} . \supset \vdash : \alpha = \Lambda . \supset . \alpha +_c 1 = \Lambda$ (1)

$\vdash . (1) . \text{Transp} . \quad \supset \vdash . \text{Prop}$

***120·322.** $\vdash :. \alpha \in \text{NC induct} . \supset : \exists ! \alpha . \equiv . \alpha \neq \alpha +_c 1$ [*120·32·321]

***120·33.** $\vdash :. \text{Infin ax} . \equiv : \alpha \in \text{NC induct} . \supset_\alpha . \alpha \neq \alpha +_c 1$ [*120·3·322]

***120·41.** $\vdash :. \nu \in \text{NC induct} . \exists ! \alpha +_c \nu . \supset : \alpha +_c \nu = \beta +_c \nu . \supset . \alpha = \text{sm}``\beta$

Dem.

$\vdash . *110\cdot4 . \text{Transp} . *118\cdot25 . \supset \vdash : (\alpha +_c \nu)_\xi = \Lambda . \supset . \{\alpha +_c (\nu +_c 1)_\xi\}_\xi = \Lambda$ (1)

$\vdash . *118\cdot25 . \supset \vdash :: \exists ! \{\alpha +_c (\nu +_c 1)_\xi\}_\xi . \supset :. \exists ! \{(\alpha +_c \nu)_\xi +_c 1\}_\xi :.$

[*120·311.*110·4.*118·201]

$\supset :. \{(\alpha +_c \nu)_\xi +_c 1\}_\xi = \{(\beta +_c \nu)_\xi +_c 1\}_\xi . \supset . (\alpha +_c \nu)_\xi = (\beta +_c \nu)_\xi :.$

[Syll.*118·25] $\supset :. (\alpha +_c \nu)_\xi = (\beta +_c \nu)_\xi . \supset . \alpha = \text{sm}``\beta : \supset :$

$\{\alpha +_c (\nu +_c 1)_\xi\}_\xi = \{\beta +_c (\nu +_c 1)_\xi\}_\xi . \supset . \alpha = \text{sm}``\beta$ (2)

$\vdash . (2) . \text{Comm} . \supset \vdash :: (\alpha +_c \nu)_\xi = (\beta +_c \nu)_\xi . \supset . \alpha = \text{sm}``\beta : \supset :.$

$\{\alpha +_c (\nu +_c 1)_\xi\}_\xi = \Lambda : \vee : \{\alpha +_c (\nu +_c 1)_\xi\}_\xi = \{\beta +_c (\nu +_c 1)_\xi\}_\xi . \supset . \alpha = \text{sm}``\beta$ (3)

$\vdash . (1) . (3) . \supset \vdash :: (\alpha +_c \nu)_\xi = \Lambda : \vee : (\alpha +_c \nu)_\xi = (\beta +_c \nu)_\xi . \supset . \alpha = \text{sm}``\beta :. \supset :.$

$\{\alpha +_c (\nu +_c 1)_\xi\}_\xi = \Lambda : \vee : \{\alpha +_c (\nu +_c 1)_\xi\}_\xi = \{\beta +_c (\nu +_c 1)_\xi\}_\xi . \supset . \alpha = \text{sm}``\beta$ (4)

$\vdash . *110\cdot4 . *118\cdot21 . \supset \vdash :. \exists ! (\beta +_c 0)_\xi . \supset : \beta \in \text{NC} . \exists ! \text{sm}_\xi``\beta :$

[*102·87.*100·51] $\supset : \text{sm}_\xi``\alpha = \text{sm}_\xi``\beta . \supset . \alpha = \text{sm}``\beta$ (5)

$\vdash . *110\cdot6\cdot4 . \supset \vdash : \exists ! (\alpha +_c 0)_\xi . (\alpha +_c 0)_\xi = (\beta +_c 0)_\xi . \supset . \text{sm}_\xi``\alpha = \text{sm}_\xi``\beta .$

[(5)] $\supset . \alpha = \text{sm}``\beta$ (6)

$\vdash . (6) . \text{Exp} . *4\cdot6 . \supset \vdash :. (\alpha +_c 0)_\xi = \Lambda : \vee : (\alpha +_c 0)_\xi = (\beta +_c 0)_\xi . \supset . \alpha = \text{sm}``\beta$ (7)

$\vdash . (4) . (7) . *120\cdot11 . \supset$

$\vdash :: \nu \in \text{N}_\xi\text{C induct} . \supset :. (\alpha +_c \nu)_\xi = \Lambda : \vee : (\alpha +_c \nu)_\xi = (\beta +_c \nu)_\xi . \supset . \alpha = \text{sm}``\beta$ (8)

$\vdash . *110\cdot4 . \quad \supset \vdash : \nu = \Lambda_\eta . \supset . (\alpha +_c \nu)_\xi = \Lambda$ (9)

$\vdash . *120\cdot15 . \supset \vdash :: \nu \in \text{N}_\eta\text{C induct} - \iota`\Lambda . \supset :. \text{sm}_\xi``\nu = \text{N}_\xi\text{C induct} :.$

[(8)] $\supset :. (\alpha +_c \text{sm}_\xi``\nu)_\xi = \Lambda : \vee : (\alpha +_c \text{sm}_\xi``\nu)_\xi = (\beta +_c \text{sm}_\xi``\nu)_\xi . \supset . \alpha = \text{sm}``\beta :.$

[*118·24] $\supset :. (\alpha +_c \nu)_\xi = \Lambda : \vee : (\alpha +_c \nu)_\xi = (\beta +_c \nu)_\xi . \supset . \alpha = \text{sm}``\beta$ (10)

$\vdash . (9) . (10) . \supset \vdash . \text{Prop}$

The above proposition establishes (with the natural limitations) the uniqueness (within each type) of subtraction (conceived as in *120·412) when the subtrahend is an inductive cardinal. (When the subtrahend is a non-inductive cardinal, subtraction ceases to give a unique result.) Hence we are led to the following extensions of *118 for the case of inductive cardinals:

***120·411.** $\vdash :. \nu \in \text{NC induct} . \supset :$

$$\exists ! \gamma -_c \nu . \supset . \gamma -_c \nu \in \text{N}_0\text{C} : \gamma \geqslant \nu . \equiv . (\gamma -_c \nu) \cap t_0{}^{\backprime}\gamma \in \text{N}_0\text{C}$$

Dem.

$\vdash . *119·1 . \supset \vdash :. \nu \in \text{NC induct} . \supset :$

$\xi, \eta \in \gamma -_c \nu . \supset . \text{Nc}{}^{\backprime}\xi +_c \nu = \gamma . \text{Nc}{}^{\backprime}\eta +_c \nu = \gamma . \exists ! \text{Nc}{}^{\backprime}\xi +_c \nu .$

[*20·22] $\supset . \text{Nc}{}^{\backprime}\xi +_c \nu = \text{Nc}{}^{\backprime}\eta +_c \nu . \exists ! \text{Nc}{}^{\backprime}\xi +_c \nu .$

[*120·41.*100·511.(*110·03)] $\supset . \text{Nc}{}^{\backprime}\xi = \text{Nc}{}^{\backprime}\eta$ (1)

$\vdash . (1) . *119·14 . \supset \vdash :. \text{Hp} . \supset : \exists ! \gamma -_c \nu . \supset . \gamma -_c \nu \in \text{N}_0\text{C}$ (2)

$\vdash . *119·27 . (2) . \supset \vdash :. \text{Hp} . \supset : \gamma \geqslant \nu . \supset . (\gamma -_c \nu) \cap t_0{}^{\backprime}\gamma \in \text{N}_0\text{C}$ (3)

$\vdash . *103·22 . *119·27 . \supset \vdash :. \text{Hp} . \supset : (\gamma -_c \nu) \cap t_0{}^{\backprime}\gamma \in \text{N}_0\text{C} . \supset . \gamma \geqslant \nu$ (4)

$\vdash . (2) . (3) . (4) . \supset \vdash . \text{Prop}$

***120·4111.** $\vdash :. \nu \in \text{NC induct} . \exists ! \text{sm}_\xi{}^{\backprime\backprime}\gamma . \supset : \gamma \geqslant \nu . \equiv . (\gamma -_c \nu)_\xi \in \text{N}_0\text{C}$

Dem.

$\vdash . *119·64 . \supset \vdash :. \text{Hp} . \supset : \gamma \geqslant \nu . \supset . \exists ! (\gamma -_c \nu)_\xi .$

[*120·411] $\supset . (\gamma -_c \nu)_\xi \in \text{N}_0\text{C}$ (1)

$\vdash . (1) . *119·26 . *103·13 . \supset \vdash . \text{Prop}$

***120·412.** $\vdash : \nu \in \text{NC induct} . \gamma \geqslant \nu . \exists ! \text{sm}_\xi{}^{\backprime\backprime}\gamma . \supset . (\gamma -_c \nu)_\xi = \{(\imath\alpha)(\alpha +_c \nu = \gamma)\}_\xi$

Dem.

$\vdash . *120·4111 . \supset \vdash :. \text{Hp} . \supset . (\gamma -_c \nu)_\xi \in \text{N}_0\text{C} .$

[*119·34] $\supset . (\gamma -_c \nu)_\xi +_c \nu = \gamma$ (1)

$\vdash . *120·41 . *103·43 . *37·29 . \supset \vdash :. \text{Hp} . \supset : \alpha +_c \nu = \gamma . \beta +_c \nu = \gamma . \supset_{\alpha, \beta} . \alpha = \beta$ (2)

$\vdash . (1) . (2) . \supset \vdash . \text{Prop}$

***120·413.** $\vdash : \mu \in \text{N}_0\text{C} . \supset . \mu -_c 0 = \text{sm}{}^{\backprime\backprime}\mu$

Dem.

$\vdash . *119·1 . \supset \vdash :. \text{Hp} . \supset : \xi \in \mu -_c 0 . \equiv . \text{N}_0\text{c}{}^{\backprime}\xi +_c 0 = \mu . \exists ! \mu .$

[*110·61.*103·13] $\equiv . \text{Nc}{}^{\backprime}\xi = \mu .$

[*103·44·4] $\equiv . \text{N}_0\text{c}{}^{\backprime}\xi = \text{sm}{}^{\backprime\backprime}\mu .$

[*103·26] $\equiv . \xi \in \text{sm}{}^{\backprime\backprime}\mu :. \supset \vdash . \text{Prop}$

***120·414.** $\vdash : \mu \in \text{N}_0\text{C} - \iota{}^{\backprime}0 . \exists ! \text{sm}_\xi{}^{\backprime\backprime}\mu . \supset . (\mu -_c 1)_\xi \in \text{N}_0\text{C}$

[*120·4111 . *117·53]

***120·415.** $\vdash : \mu \in \text{N}_0\text{C} - \iota{}^{\backprime}0 - \iota{}^{\backprime}1 . \exists ! \text{sm}_\xi{}^{\backprime\backprime}\mu . \supset . (\mu -_c 2)_\xi \in \text{N}_0\text{C}$

[*120·4111 . *117·551]

***120·416.** $\vdash : \nu \in \text{NC induct} . \exists ! \gamma -_c \nu . \supset . (\gamma -_c \nu) +_c \nu = \text{sm}{}^{\backprime\backprime}\gamma$

[*120·411 . *119·34]

***120·417.** $\vdash : \mu \in \text{N}_0\text{C} - \iota{}^{\backprime}0 . \exists ! \text{sm}_\xi{}^{\backprime\backprime}\gamma . \supset . \alpha +_c \gamma = (\alpha +_c 1) +_c (\gamma -_c 1)_\xi$

[*120·414 . *119·35]

***120·418.** $\vdash : \nu \in \text{NC induct} . \exists ! \text{sm}_\xi{}^{\backprime\backprime}\gamma . \gamma \geqslant \nu . \supset . \alpha +_c \gamma = (\alpha +_c \nu) +_c (\gamma -_c \nu)_\xi$

[*120·4111 . *119·35]

***120·42.** $\vdash : \nu \in \text{NC induct} . \exists ! \nu . \alpha \neq 0 . \supset . \nu \neq \alpha +_c \nu$

Dem.

$\vdash . *110·61 . *120·14 . \supset \vdash : \nu \in \text{NC induct} . \supset . \nu = 0 +_c \nu$ (1)

$\vdash . *120·41 . \supset \vdash : \nu \in \text{NC induct} . \exists ! 0 +_c \nu . 0 +_c \nu = \alpha +_c \nu . \supset . 0 = \alpha$ (2)

$\vdash . (1) . (2) . \supset \vdash : \nu \in \text{NC induct} . \exists ! \nu . \nu = \alpha +_c \nu . \supset . \alpha = 0 : \supset \vdash . \text{Prop}$

***120·422.** $\vdash : \alpha +_c 1 \in \text{NC induct} - \iota'\Lambda . \supset . \alpha \in \text{NC induct} - \iota'\Lambda$

Dem.

$\vdash . *120·1·124 . *91·542 . \supset \vdash : \alpha +_c 1 \in \text{NC induct} . \supset . (\alpha +_c 1)(+_c 1)_{po} 0 .$

[*91·52] $\supset . (\exists \beta) . (\alpha +_c 1)(+_c 1) \beta . \beta (+_c 1)_* 0 .$

[*120·1] $\supset . (\exists \beta) . \alpha +_c 1 = \beta +_c 1 . \beta \in \text{NC induct}$ (1)

$\vdash . *120·311 . \supset \vdash :. \text{Hp} . \supset : \alpha +_c 1 = \beta +_c 1 . \supset . \alpha = \text{sm}``\beta . \exists ! \alpha$ (2)

$\vdash . (1) . (2) . *120·15 . \supset \vdash : \text{Hp} . \supset . \alpha \in \text{NC induct}$ (3)

$\vdash . (3) . *110·4 . \supset \vdash . \text{Prop}$

***120·423.** $\vdash : \alpha \in N_\eta C \text{ induct} - \iota'0 . \equiv . (\exists \beta) . \beta \in N_\eta C \text{ induct} . \alpha = (\beta +_c 1)_\eta$

Dem.

$\vdash . *120·121·124 . \supset \vdash : \beta \in N_\eta C \text{ induct} . \alpha = (\beta +_c 1)_\eta . \supset . \alpha \in N_\eta C \text{ induct} - \iota'0$ (1)

$\vdash . *120·102 . *91·542 . \supset \vdash : \alpha \in N_\eta C \text{ induct} - \iota'0 . \supset . \alpha (+_c 1)_{po} 0_\eta .$

[*91·52] $\supset . (\exists \beta) . \alpha (+_c 1) \beta . \beta (+_c 1)_* 0_\eta .$

[*120·102] $\supset . (\exists \beta) . \beta \in N_\eta C \text{ induct} . \alpha = (\beta +_c 1)_\eta$ (2)

$\vdash . (1) . (2) . \supset \vdash . \text{Prop}$

***120·4231.** $\vdash : \alpha \in N_\eta C \text{ induct} . \supset . (\exists \beta) . \beta \in N_\eta C \text{ induct} - \iota'\Lambda . (\alpha +_c 1)_\eta = (\beta +_c 1)_\eta$

Dem.

$\vdash . *10·24 . *101·12 . *120·12 . \supset$

$\vdash . (\exists \beta) . \beta \in N_\eta C \text{ induct} - \iota'\Lambda . (0 +_c 1)_\eta = (\beta +_c 1)_\eta$ (1)

$\vdash . *120·121 . \supset \vdash :. \exists ! \xi . \supset :$

$\beta \in N_\eta C \text{ induct} - \iota'\Lambda . \xi = (\beta +_c 1)_\eta . \supset . \xi \in N_\eta C \text{ induct} - \iota'\Lambda . (\xi +_c 1)_\eta = (\xi +_c 1)_\eta :$

[*10·23·24] $\supset : (\exists \beta) . \beta \in N_\eta C \text{ induct} - \iota'\Lambda . \xi = (\beta +_c 1)_\eta . \supset .$

$(\exists \gamma) . \gamma \in N_\eta C \text{ induct} - \iota'\Lambda . (\xi +_c 1)_\eta = (\gamma +_c 1)_\eta$ (2)

$\vdash . *110·4 . *13·17 . \supset$

$\vdash :. \sim \exists ! \xi . \supset : \beta \in N_\eta C \text{ induct} - \iota'\Lambda . \xi = (\beta +_c 1)_\eta . \supset . (\xi +_c 1)_\eta = (\beta +_c 1)_\eta :$

[*10·28] $\supset : (\exists \beta) . \beta \in N_\eta C \text{ induct} - \iota'\Lambda . \xi = (\beta +_c 1)_\eta . \supset .$

$(\exists \beta) . \beta \in N_\eta C \text{ induct} - \iota'\Lambda . (\xi +_c 1)_\eta = (\beta +_c 1)_\eta$ (3)

$\vdash . (2) . (3) . \supset \vdash : (\exists \beta) . \beta \in N_\eta C \text{ induct} - \iota'\Lambda . \xi = (\beta +_c 1)_\eta . \supset .$

$(\exists \beta) . \beta \in N_\eta C \text{ induct} - \iota'\Lambda . (\xi +_c 1)_\eta = (\beta +_c 1)_\eta$ (4)

$\vdash . (1) . (4) \dfrac{(\xi +_c 1)_\eta}{\xi} . *120·11 . \supset$

$\vdash : \alpha \in N_\eta C \text{ induct} . \supset . (\exists \beta) . \beta \in N_\eta C \text{ induct} - \iota'\Lambda . (\alpha +_c 1)_\eta = (\beta +_c 1)_\eta :$

$\supset \vdash . \text{Prop}$

***120·4232.** $\vdash : \alpha \in N_\eta C \text{ induct} - \iota'0 . \equiv . (\exists \beta) . \beta \in N_\eta C \text{ induct} - \iota'\Lambda . \alpha = (\beta +_c 1)_\eta$
[*120·423·4231]

∗120·424. $\vdash : \beta \neq 0 \,.\, \exists ! (\alpha +_c \beta)_\xi \,.\supset .\, (\alpha +_c \beta)_\xi -_c 1 = \alpha +_c (\beta -_c 1)_\xi$

Dem.

$\vdash . *110{\cdot}42{\cdot}62 . \supset \vdash : \mathrm{Hp} . \supset . (\alpha +_c \beta)_\xi \,\epsilon\, \mathrm{NC} - \iota`\Lambda - \iota`0 .$

$[*120{\cdot}414 . *103{\cdot}13] \quad \supset . \exists ! (\alpha +_c \beta)_\xi -_c 1 \qquad (1)$

$\vdash . *110{\cdot}4 . *118{\cdot}21 . *120{\cdot}414 . *103{\cdot}13 . \supset \vdash : \mathrm{Hp} . \supset . \exists ! (\beta -_c 1)_\xi \qquad (2)$

$\vdash . (1) . (2) . *120{\cdot}416 . \supset$

$\vdash : \mathrm{Hp} . \supset . \{(\alpha +_c \beta)_\xi -_c 1\} +_c 1 = \alpha +_c \beta \,.\, (\beta -_c 1)_\xi +_c 1 = \beta . \qquad (3)$

$[*110{\cdot}56] \supset . \{(\alpha +_c \beta)_\xi -_c 1\} +_c 1 = \{\alpha +_c (\beta -_c 1)_\xi\} +_c 1 \qquad (4)$

$\vdash . (3) . \supset \vdash : \mathrm{Hp} . \supset . \exists ! [\{(\alpha +_c \beta)_\xi -_c 1\} +_c 1]_\xi \qquad (5)$

$\vdash . (4) . (5) . *120{\cdot}311 . *110{\cdot}44 . \supset$

$\vdash : \mathrm{Hp} . \supset . (\alpha +_c \beta)_\xi -_c 1 = \alpha +_c (\beta -_c 1)_\xi : \supset \vdash . \mathrm{Prop}$

∗120·425. $\vdash :. (\alpha +_c \beta)_\xi \,\epsilon\, \mathrm{N_0C} - \iota`0 . \supset :$

$(\alpha +_c \beta)_\xi -_c 1 = \alpha +_c (\beta -_c 1)_\xi \,.\vee.\, (\alpha +_c \beta)_\xi -_c 1 = (\alpha -_c 1)_\xi +_c \beta$

Dem.

$\vdash . *110{\cdot}62 . *103{\cdot}22 . \supset \vdash :. \mathrm{Hp} . \supset : \alpha \neq 0 \,.\vee.\, \beta \neq 0 : \exists ! (\alpha +_c \beta)_\xi \qquad (1)$

$\vdash . (1) . *120{\cdot}424 . \supset \vdash . \mathrm{Prop}$

∗120·426. $\vdash : \rho \,\epsilon\, \mathrm{Cls\ induct} \,.\, \rho \subset \sigma \,.\, \exists ! \sigma - \rho . \supset . \sim(\rho \,\mathrm{sm}\, \sigma) \,.\, \mathrm{Nc}`\rho < \mathrm{Nc}`\sigma$

Dem.

$\vdash . *110{\cdot}32 . \supset \vdash : \mathrm{Hp} . \supset . \mathrm{Nc}`\sigma = \mathrm{Nc}`\rho +_c \mathrm{Nc}`(\sigma - \rho) \qquad (1)$

$\vdash . *101{\cdot}14 . \supset \vdash : \mathrm{Hp} . \supset . \mathrm{Nc}`(\sigma - \rho) \neq 0 \qquad (2)$

$\vdash . (1) . (2) . *120{\cdot}42 . *117{\cdot}222{\cdot}26 . \supset \vdash . \mathrm{Prop}$

∗120·427. $\vdash : R \,\epsilon\, 1 \rightarrow 1 \,.\, \mathrm{Cl}`R \subset \mathrm{D}`R \,.\, \exists ! \mathrm{D}`R - \mathrm{Cl}`R . \supset . \mathrm{D}`R \sim\epsilon\, \mathrm{Cls\ induct}$

[∗120·426 . Transp]

The above proposition shows that no reflexive class is inductive.

∗120·428. $\vdash : \nu \,\epsilon\, \mathrm{NC\ induct} \,.\, \exists ! \alpha +_c \nu \,.\, \alpha \neq 0 . \supset . \alpha +_c \nu > \nu$

Dem.

$\vdash . *117{\cdot}511 . *110{\cdot}4 . \supset \vdash : \mathrm{Hp} . \supset . \alpha > 0 \,.\, \nu \,\epsilon\, \mathrm{N_0C} .$

$[*117{\cdot}561 . *110{\cdot}6] \quad \supset . \alpha +_c \nu \geqslant \nu \qquad (1)$

$\vdash . *120{\cdot}42 . *110{\cdot}4 . \quad \supset \vdash : \mathrm{Hp} . \supset . \alpha +_c \nu \neq \nu \qquad (2)$

$\vdash . (1) . (2) . *117{\cdot}26 . \supset \vdash . \mathrm{Prop}$

∗120·429. $\vdash :. \nu \,\epsilon\, \mathrm{NC\ induct} . \supset : \mu > \nu \,.\equiv.\, \mu \geqslant \nu +_c 1$

Dem.

$\vdash . *120{\cdot}428 . \quad \supset \vdash :. \mathrm{Hp} . \supset : \mu \,\epsilon\, \mathrm{N_0C} \,.\, \mu = \nu +_c 1 . \supset . \mu > \nu : \qquad (1)$

$[*117{\cdot}47{\cdot}12] \quad \supset : \mu > \nu +_c 1 . \supset . \mu > \nu \qquad (2)$

$\vdash . *117{\cdot}31 . \quad \supset \vdash : \mu > \nu . \supset . (\exists \varpi) . \varpi \,\epsilon\, \mathrm{N_0C} \,.\, \mu = \nu +_c \varpi \qquad (3)$

$\vdash . *117{\cdot}26{\cdot}12 . \supset \vdash : \mu > \nu . \supset . \mu \neq \nu +_c 0 \qquad (4)$

$\vdash . (3) . (4) . \quad \supset \vdash : \mu > \nu . \supset . (\exists \varpi) . \varpi \,\epsilon\, \mathrm{N_0C} - \iota`0 \,.\, \mu = \nu +_c \varpi .$

$[*117{\cdot}531] \quad \supset . (\exists \varpi) . \varpi \geqslant 1 \,.\, \mu = \nu +_c \varpi .$

$[*117{\cdot}31] \quad \supset . (\exists \varpi, \rho) . \rho \,\epsilon\, \mathrm{N_0C} \,.\, \varpi = \rho +_c 1 \,.\, \mu = \nu +_c \varpi .$

$[*13{\cdot}195] \quad \supset . (\exists \rho) . \rho \,\epsilon\, \mathrm{N_0C} \,.\, \mu = \nu +_c \rho +_c 1 .$

$[*117{\cdot}31] \quad \supset . \mu \geqslant \nu +_c 1 \qquad (5)$

$\vdash . (1) . (2) . (5) . \supset \vdash . \mathrm{Prop}$

The following definition, in which "spec" stands for "species," defines the "species" of a cardinal β as all cardinals which are less than, equal to, or greater than β. We cannot prove, unless by assuming the multiplicative axiom, that all cardinals belong to the species of β, except in the case where β is an inductive cardinal. In all other cases there may, so far as is known at present, be other cardinals which are neither greater nor less than β.

***120·43.** $\text{spec}`\beta = \hat{\alpha}\{\alpha < \beta . \vee . \alpha \geqslant \beta\}$ Df

***120·431.** $\vdash :. \alpha \in \text{spec}`\beta . \equiv : \alpha < \beta . \vee . \alpha \geqslant \beta$ [(*120·43)]

***120·432.** $\vdash :. \alpha \in \text{spec}`\beta . \equiv : \alpha \leqslant \beta . \vee . \alpha \geqslant \beta$ [*117·281 . *120·431]

***120·433.** $\vdash :. \text{Nc}`\rho \in \text{spec}`\text{Nc}`\sigma . \equiv : \exists ! \text{Cl}`\rho \cap \text{Nc}`\sigma . \vee . \exists ! \text{Cl}`\sigma \cap \text{Nc}`\rho$
[*117·22 . *120·432]

***120·434.** $\vdash . \text{spec}`\beta \subset \text{N}_0\text{C}$ [*117·105·104·12 . *120·432]

***120·435.** $\vdash : \beta \in \text{N}_0\text{C} . \equiv . \beta \in \text{spec}`\beta . \equiv . \exists ! \text{spec}`\beta$ [*117·104 . *120·434]

***120·436.** $\vdash :. \alpha \in \text{spec}`\beta . \equiv : \alpha, \beta \in \text{N}_0\text{C} : (\exists \gamma) : \alpha +_c \gamma = \beta . \vee . \beta +_c \gamma = \alpha$
[*120·432 . *117·31]

***120·437.** $\vdash : \beta \in \text{N}_0\text{C} . \supset . 0 \in \text{spec}`\beta$ [*117·5 . *120·432]

***120·438.** $\vdash : \alpha \in \text{spec}`\beta . \exists ! \alpha +_c 1 . \supset . \alpha +_c 1 \in \text{spec}`\beta$

Dem.

$\vdash . \text{*120·436} . \text{*110·4} . \supset \vdash :. \text{Hp} . \equiv : \alpha, \beta \in \text{N}_0\text{C} . \exists ! \alpha +_c 1 :$
$(\exists \gamma) : \gamma \in \text{N}_0\text{C} : \alpha +_c \gamma = \beta . \vee . \beta +_c \gamma = \alpha$ (1)

$\vdash . \text{*110·61} . \supset \vdash : \alpha, \beta \in \text{N}_0\text{C} . \alpha +_c 0 = \beta . \supset . \alpha = \beta .$

[*13·12·15] $\supset . \alpha +_c 1 = \beta +_c 1 .$

[*120·436] $\supset . \alpha +_c 1 \in \text{spec}`\beta$ (2)

$\vdash . \text{*120·417} . \supset \vdash . \alpha, \beta, \gamma \in \text{N}_0\text{C} . \gamma \neq 0 . \alpha +_c \gamma = \beta . \supset . \alpha +_c 1 +_c (\gamma -_c 1) = \beta .$

[*120·436] $\supset . \alpha +_c 1 \in \text{spec}`\beta$ (3)

$\vdash . \text{*13·12·15} . \supset \vdash : \alpha, \beta, \gamma \in \text{N}_0\text{C} . \beta +_c \gamma = \alpha . \exists ! \alpha +_c 1 . \supset . \beta +_c \gamma +_c 1 = \alpha +_c 1 .$

[*120·436] $\supset . \alpha +_c 1 \in \text{spec}`\beta$ (4)

$\vdash . (1) . (2) . (3) . (4) . \supset \vdash . \text{Prop}$

***120·44.** $\vdash : \beta \in \text{N}_0\text{C} . \supset . \text{NC induct} - \iota`\Lambda \subset \text{spec}`\beta$

Dem.

$\vdash . \text{*120·437} . \supset \vdash : \text{Hp} . \supset . 0 \in \text{spec}`\beta$ (1)

$\vdash . \text{*120·438} . \text{*110·4} . \supset \vdash :: \text{Hp} . \supset :. \alpha = \Lambda . \vee . \alpha \in \text{spec}`\beta : \supset :$
$\alpha +_c 1 = \Lambda . \vee . \alpha +_c 1 \in \text{spec}`\beta$ (2)

$\vdash . (1) . (2) . \text{*120·11} . \supset \vdash :: \text{Hp} . \supset :. \alpha \in \text{NC induct} . \supset :$
$\alpha = \Lambda . \vee . \alpha \in \text{spec}`\beta :: \supset \vdash . \text{Prop}$

***120·441.** $\vdash :. \alpha \in \text{NC induct} - \iota`\Lambda . \beta \in \text{NC} - \iota`\Lambda . \supset : \alpha < \beta . \vee . \alpha = \text{sm}``\beta . \vee . \alpha > \beta$
[*120·44 . *103·34]

***120·442.** $\vdash :. \alpha \in \text{NC induct} - \iota\text{‘}\Lambda \,.\, \beta \in \text{NC} - \iota\text{‘}\Lambda \,.\supset :$

$$\alpha < \beta \,.\equiv .\, \sim(\alpha \geqslant \beta) : \alpha > \beta \,.\equiv .\, \sim(\alpha \leqslant \beta)$$

Dem.

$\vdash . *117\cdot104 . *120\cdot441 . \supset \vdash :. \text{Hp} . \supset : \alpha < \beta \,.\mathbf{v}.\, \alpha \geqslant \beta$ (1)

$\vdash . *117\cdot291 . \qquad \supset \vdash : \alpha < \beta \,.\supset .\, \sim(\alpha \geqslant \beta)$ (2)

$\vdash . (1) . (2) . *5\cdot17 . \supset \vdash :. \text{Hp} . \supset : \alpha < \beta \,.\equiv .\, \sim(\alpha \geqslant \beta)$ (3)

Similarly $\qquad \vdash :. \text{Hp} . \supset : \alpha > \beta \,.\equiv .\, \sim(\alpha \leqslant \beta)$ (4)

$\vdash . (3) . (4) . \supset \vdash . \text{Prop}$

***120·45.** $\vdash : \alpha, \beta \in \text{N}_\xi\text{C induct} \,.\supset .\, (\alpha +_c \beta)_\xi \in \text{N}_\xi\text{C induct}$

Dem.

$\vdash . *110\cdot6 . \supset \vdash : \alpha \in \text{N}_\xi\text{C induct} \,.\supset .\, (\alpha +_c 0_\xi)_\xi \in \text{N}_\xi\text{C induct}$ (1)

$\vdash . *120\cdot121 . *118\cdot25 . \supset$

$\vdash : (\alpha +_c \beta)_\xi \in \text{N}_\xi\text{C induct} \,.\supset .\, \{\alpha +_c (\beta +_c 1)_\xi\}_\xi \in \text{N}_\xi\text{C induct}$ (2)

$\vdash . (1) . (2) . *120\cdot11 . \supset \vdash . \text{Prop}$

***120·4501.** $\vdash : \alpha, \beta \in \text{NC induct} - \iota\text{‘}\Lambda \,.\supset .\, \alpha +_c \beta \in \text{NC induct}$

Dem.

$\vdash . *120\cdot15 . \supset \vdash : \text{Hp} . \supset . \text{sm}_\xi\text{“}\alpha, \text{sm}_\xi\text{“}\beta \in \text{N}_\xi\text{C induct} .$

$[*120\cdot45] \qquad \supset . (\text{sm}_\xi\text{“}\alpha +_c \text{sm}_\xi\text{“}\beta)_\xi \in \text{N}_\xi\text{C induct} .$

$[*118\cdot23] \qquad \supset . (\alpha +_c \beta)_\xi \in \text{N}_\xi\text{C induct} : \supset \vdash . \text{Prop}$

The following proposition is a lemma in the proof of *120·452.

***120·451.** $\vdash :. \gamma = (\alpha +_c \beta)_\xi \,.\supset_{\alpha,\beta} .\, \alpha, \beta \in \text{NC induct} - \iota\text{‘}\Lambda :$

$$\exists ! (\gamma +_c 1)_\xi \,.\, (\gamma +_c 1)_\xi = (\alpha' +_c \beta')_\xi : \supset . \alpha', \beta' \in \text{NC induct} - \iota\text{‘}\Lambda$$

Dem.

$\vdash . *120\cdot414\cdot124 . *110\cdot42 . \supset \vdash : \exists ! (\gamma +_c 1)_\xi \,.\supset .\, \{(\gamma +_c 1)_\xi -_c 1\}_\xi \in \text{N}_0\text{C} .$

$[*119\cdot32] \qquad \supset . \gamma = \{(\gamma +_c 1)_\xi -_c 1\}_\xi$ (1)

$\vdash . (1) . *120\cdot124 . \supset \vdash :. \text{Hp} . \supset : \gamma = \{(\alpha' + \beta')_\xi -_c 1\}_\xi . (\alpha' + \beta')_\xi \neq 0 . \exists ! (\alpha' + \beta')_\xi :$

$[*120\cdot425] \supset : \gamma = \{\alpha' +_c (\beta' -_c 1)_\xi\}_\xi \,.\mathbf{v}.\, \gamma = \{(\alpha' -_c 1)_\xi +_c \beta'\}_\xi :$

$[\text{Hp}] \qquad \supset : \alpha', (\beta' -_c 1)_\xi \in \text{NC induct} - \iota\text{‘}\Lambda \,.\mathbf{v}.\, (\alpha' -_c 1)_\xi, \beta' \in \text{NC induct} - \iota\text{‘}\Lambda :$

$[*119\cdot11] \quad \supset : \alpha', \beta' \in \text{NC induct} - \iota\text{‘}\Lambda :. \supset \vdash . \text{Prop}$

This proposition could be extended to greater generality as regards types; but its sole use is as a lemma.

***120·452.** $\vdash : \alpha +_c \beta \in \text{NC induct} - \iota\text{‘}\Lambda \,.\supset .\, \alpha, \beta \in \text{NC induct} - \iota\text{‘}\Lambda$

Dem.

$\vdash . *110\cdot4 . \text{Transp} . \supset \vdash : \gamma = \Lambda \,.\supset .\, (\gamma +_c 1)_\eta = \Lambda$ (1)

$\vdash . *120\cdot451 . \supset \vdash :: \gamma = (\alpha +_c \beta)_\eta \,.\supset_{\alpha,\beta} .\, \alpha, \beta \in \text{NC induct} - \iota\text{‘}\Lambda : \supset :.$

$\quad (\gamma +_c 1)_\eta = \Lambda : \mathbf{v} : (\gamma +_c 1)_\eta = (\alpha' +_c \beta')_\eta \,.\supset_{\alpha',\beta'} .\, \alpha', \beta' \in \text{NC induct} - \iota\text{‘}\Lambda$ (2)

$\vdash . (1) . (2) . \quad \supset \vdash :: \gamma = \Lambda : \mathbf{v} : \gamma = (\alpha +_c \beta)_\eta \,.\supset_{\alpha,\beta} .\, \alpha, \beta \in \text{NC induct} - \iota\text{‘}\Lambda :. \supset :.$

$\quad (\gamma +_c 1)_\eta = \Lambda : \mathbf{v} : (\gamma +_c 1)_\eta = (\alpha' +_c \beta')_\eta \,.\supset_{\alpha',\beta'} .\, \alpha', \beta' \in \text{NC induct} - \iota\text{‘}\Lambda$ (3)

$\vdash . *110\cdot62 . *120\cdot12 . \supset \vdash : 0 = (\alpha +_c \beta)_\eta \,.\supset_{\alpha,\beta} .\, \alpha, \beta \in \text{NC induct} - \iota\text{‘}\Lambda$ (4)

$\vdash . (3) . (4) . *120\cdot11 . \supset \vdash :: \gamma \in \text{N}_\eta\text{C induct} . \supset :.$

$$\gamma = \Lambda : \mathbf{v} : \gamma = (\alpha +_c \beta)_\eta \,.\supset_{\alpha,\beta} .\, \alpha, \beta \in \text{NC induct} - \iota\text{‘}\Lambda ::$$

$[*13\cdot15] \qquad \supset \vdash :. (\alpha +_c \beta)_\eta \in \text{N}_\eta\text{C induct} . \supset :$

$$(\alpha +_c \beta)_\eta = \Lambda \,.\mathbf{v}.\, \alpha, \beta \in \text{NC induct} - \iota\text{‘}\Lambda :. \supset \vdash . \text{Prop}$$

In the last line but one of the above proof, we substitute for the $\phi\xi$ of $*120\cdot11$ the function

$$\xi = \Lambda : \mathsf{v} : \xi = (\alpha +_c \beta)_\eta \,.\, \supset_{\alpha,\beta} .\, \alpha, \beta \in \text{NC induct} - \iota`\Lambda.$$

The following propositions are chiefly required as leading to $*120\cdot4621$ $\cdot4622\cdot47$, which are useful in proving propositions concerning all inductive cardinals other than zero.

$*120\cdot46$. $\vdash : \alpha \in \text{NC} \,.\, \gamma \in \text{N}_\eta\text{C induct} \,.\, \supset .\, (\alpha +_c \gamma)_\eta \, (+_c 1)_* \, \text{sm}_\eta``\alpha$

Dem.

$\vdash . *110\cdot6 . *118\cdot241 . \supset \vdash : \alpha \in \text{NC} . \supset . (\alpha +_c 0)_\eta \, (+_c 1)_* \, \text{sm}_\eta``\alpha$ (1)

$\vdash . *90\cdot172 . *118\cdot25 . \supset \vdash : (\alpha +_c \gamma)_\eta (+_c 1)_* \, \text{sm}_\eta``\alpha . \supset .$

$\{\alpha +_c (\gamma +_c 1)_\eta\}_\eta (+_c 1)_* \text{sm}_\eta``\alpha$ (2)

$\vdash . (1) . (2) . *120\cdot11 . \supset \vdash . \text{Prop}$

$*120\cdot461$. $\vdash : \alpha \in \text{NC} \,.\, \beta \, (+_c 1)_* \, \text{sm}_\eta``\alpha \,.\, \supset .\, (\exists\gamma) \,.\, \gamma \in \text{N}_\eta\text{C induct} \,.\, \beta = (\alpha +_c \gamma)_\eta$

Dem.

$\vdash . *110\cdot6 . *118\cdot23 . \quad \supset \vdash : \alpha \in \text{NC} . \beta = \text{sm}_\eta``\alpha . \supset . \beta = (\alpha +_c 0)_\eta$ (1)

$\vdash . *120\cdot121 . *118\cdot25 . \supset \vdash : \beta = (\alpha +_c \gamma)_\eta . \gamma \in \text{N}_\eta\text{C induct} . \supset .$

$(\beta +_c 1)_\eta = \{\alpha +_c (\gamma +_c 1)_\eta\}_\eta . (\gamma +_c 1)_\eta \in \text{N}_\eta\text{C induct}$ (2)

$\vdash . (1) . (2) . *90\cdot112 . \supset \vdash . \text{Prop}$

$*120\cdot462$. $\vdash :. \alpha \in \text{NC} \,.\, \supset : (\exists\gamma) \,.\, \gamma \in \text{N}_\eta\text{C induct} \,.\, \beta = (\alpha +_c \gamma)_\eta \,.\, \equiv .\, \beta (+_c 1)_* \text{sm}_\eta``\alpha$ $[*120\cdot46\cdot461]$

$*120\cdot4621$. $\vdash :. \alpha \in \text{NC} \,.\, \exists ! \beta \,.\, \supset : \beta \, (+_c 1)_* \, \text{sm}_\eta``\alpha \,.\, \supset .\, \text{sm}_\xi``\beta \, (+_c 1)_* \, \text{sm}_\xi``\alpha$

Dem.

$\vdash . *120\cdot461 . \supset \vdash :. \text{Hp} . \supset :$

$\beta \, (+_c 1)_* \, \text{sm}_\eta``\alpha . \supset . (\exists\gamma) . \gamma \in \text{N}_\eta\text{C induct} . \beta = (\alpha +_c \gamma)_\eta .$

$[*110\cdot4]$ $\quad \supset . (\exists\gamma) . \gamma \in \text{N}_\eta\text{C induct} - \iota`\Lambda . \beta = (\alpha +_c \gamma)_\eta .$

$[*120\cdot15 . *118\cdot201]$ $\quad \supset . (\exists\gamma) . \text{sm}_\xi``\gamma \in \text{N}_\xi\text{C induct} . \text{sm}_\xi``\beta = (\alpha +_c \gamma)_\xi .$

$[*118\cdot24 . *120\cdot14]$ $\quad \supset . (\exists\gamma') . \gamma' \in \text{N}_\xi\text{C induct} . \text{sm}_\xi``\beta = (\alpha +_c \gamma')_\xi .$

$[*120\cdot462]$ $\quad \supset . \text{sm}_\xi``\beta \, (+_c 1)_* \, \text{sm}_\xi``\alpha :. \supset \vdash . \text{Prop}$

$*120\cdot4622$. $\vdash :. \alpha \in \text{NC} \,.\, \beta \in \text{NC}(\eta) \,.\, \exists ! \text{sm}_\xi``\beta \,.\, \supset :$

$\beta \, (+_c 1)_* \, \text{sm}_\eta``\alpha \,.\, \equiv .\, \text{sm}_\xi``\beta \, (+_c 1)_* \, \text{sm}_\xi``\alpha$

Dem.

$\vdash . *110\cdot4 . *37\cdot29 . *120\cdot461 . \supset$

$\vdash :. \text{Hp} . \supset : \text{sm}_\xi``\beta \, (+_c 1)_* \, \text{sm}_\xi``\alpha . \supset . \exists ! \text{sm}_\xi``\alpha . \exists ! \alpha .$ (1)

$[*100\cdot52]$ $\quad \supset . \text{sm}_\xi``\alpha \in \text{NC}$ (2)

$\vdash . *120\cdot4621 . (2) . \supset \vdash :. \text{Hp} . \supset :$

$\text{sm}_\xi``\beta \, (+_c 1)_* \, \text{sm}_\xi``\alpha . \supset . \text{sm}_\eta``\text{sm}_\xi``\beta \, (+_c 1)_* \, \text{sm}_\eta``\text{sm}_\xi``\alpha .$

$[*102\cdot87 . \text{Hp} . (1)]$ $\quad \supset . \text{sm}_\eta``\beta \, (+_c 1)_* \, \text{sm}_\eta``\alpha .$

$[*103\cdot34]$ $\quad \supset . \beta \, (+_c 1)_* \, \text{sm}_\eta``\alpha$ (3)

$\vdash . *37\cdot29 . *120\cdot4621 . \supset \vdash :. \text{Hp} . \supset : \beta(+_c 1)_* \text{sm}_\eta``\alpha . \supset . \text{sm}_\xi``\beta(+_c 1)_* \text{sm}_\xi``\alpha$ (4)

$\vdash . (3) . (4) . \supset \vdash . \text{Prop}$

It is on this proposition that the irrelevance of types in the consideration of inductive cardinals depends.

$*120\cdot463$. $\vdash ::. \alpha \epsilon \text{NC} . \supset :: (\exists\gamma) . \gamma \epsilon \text{N}_\eta\text{C induct} . \beta = (\alpha +_c \gamma)_\eta . \equiv :.$
$\xi \epsilon \mu . \supset_\xi . (\xi +_c 1)_\eta \epsilon \mu : \text{sm}_\eta``\alpha \epsilon \mu : \supset_\mu . \beta \epsilon \mu$

[$*120\cdot462 . *90\cdot11$]

$*120\cdot47$. $\vdash :: \beta \epsilon \text{N}_\eta\text{C induct} - \iota`0 . \equiv :. \xi \epsilon \mu . \supset_\xi . (\xi +_c 1)_\eta \epsilon \mu : 1_\eta \epsilon \mu : \supset_\mu . \beta \epsilon \mu$
[$*120\cdot423\cdot463$]

Thus mathematical induction starting from 1 will apply to all inductive cardinals except 0. Similar propositions can be similarly proved for 2, 3, ...

$*120\cdot471$. $\vdash : (\exists\alpha) . \alpha \epsilon \text{NC induct} - \iota`0 . f\alpha . \equiv . (\exists\beta) . \beta \epsilon \text{NC induct} . f(\beta +_c 1)$

Dem.

$\vdash . *120\cdot423 . \supset$

$\vdash : (\exists\alpha) . \alpha \epsilon \text{NC induct} - \iota`0 . f\alpha . \equiv . (\exists\beta) . \beta \epsilon \text{NC induct} . \alpha = \beta +_c 1 . f\alpha .$

[$*13\cdot195$] $\equiv . (\exists\beta) . \beta \epsilon \text{NC induct} . f(\beta +_c 1) : \supset \vdash . \text{Prop}$

$*120\cdot472$. $\vdash : (\exists\alpha) . \alpha \epsilon \text{NC induct} - \iota`0 - \iota`1 . f\alpha . \equiv .$
$(\exists\beta) . \beta \epsilon \text{NC induct} - \iota`0 . f(\beta +_c 1) . \equiv . (\exists\gamma) . \gamma \epsilon \text{NC induct} . f(\gamma +_c 2)$

Dem.

$\vdash . *120\cdot471 . \supset$

$\vdash : (\exists\alpha) . \alpha \epsilon \text{NC induct} - \iota`0 - \iota`1 . f\alpha . \equiv .$
$(\exists\beta) . \beta \epsilon \text{NC induct} . \beta +_c 1 \neq 1 . f(\beta +_c 1) .$

[$*120\cdot42 . *110\cdot641$] $\equiv . (\exists\beta) . \beta \epsilon \text{NC induct} - \iota`0 . f(\beta +_c 1) .$ (1)

[$*120\cdot471$] $\equiv . (\exists\gamma) . \gamma \epsilon \text{NC induct} . f(\gamma +_c 1 +_c 1) .$

[$*110\cdot643$] $\equiv . (\exists\gamma) . \gamma \epsilon \text{NC induct} . f(\gamma +_c 2)$ (2)

$\vdash . (1) . (2) . \supset \vdash . \text{Prop}$

$*120\cdot473$. $\vdash :. \phi 1 : \xi \epsilon \text{N}_\eta\text{C induct} - \iota`0 . \phi\xi . \supset_\xi . \phi(\xi +_c 1) : \supset :$
$\xi \epsilon \text{N}_\eta\text{C induct} - \iota`0 . \supset . \phi\xi$

Dem.

$\vdash . *120\cdot122 . *101\cdot22 . \supset \vdash : \phi 1 . \supset . 1 \epsilon \text{N}_\eta\text{C induct} - \iota`0 . \phi 1$ (1)

$\vdash . *120\cdot121\cdot124 . \supset \vdash : \xi \epsilon \text{N}_\eta\text{C induct} - \iota`0 . \supset . \xi +_c 1 \epsilon \text{N}_\eta\text{C induct} - \iota`0$ (2)

$\vdash . (1) . (2) . \supset$

$\vdash :. \text{Hp} . \supset : 1 \epsilon \text{N}_\eta\text{C induct} - \iota`0 . \phi 1 : \xi \epsilon \text{N}_\eta\text{C induct} - \iota`0 . \phi\xi . \supset_\xi .$
$\xi +_c 1 \epsilon \text{N}_\eta\text{C induct} - \iota`0 . \phi(\xi +_c 1)$ (3)

$\vdash . (3) . *120\cdot47 \dfrac{\hat{\xi}(\xi \epsilon \text{N}_\eta\text{C induct} - \iota`0 . \phi\xi)}{\mu} . \supset \vdash . \text{Prop}$

***120·48.** $\vdash : \beta \in \text{NC induct} . \beta \geqslant \alpha . \supset . \alpha \in \text{NC induct} - \iota`\Lambda$
[*120·452 . *117·31]

Thus every cardinal which is not greater than every inductive cardinal is an inductive cardinal.

***120·481.** $\vdash : \eta \in \text{Cls induct} . \xi \subset \eta . \supset . \xi \in \text{Cls induct}$ [*117·222 . *120·21·48]

Thus if any inductive class can be found which contains a given class, the given class is also inductive.

***120·49.** $\vdash : \alpha \in \text{NC} - \text{NC induct} - \iota`\Lambda . \beta \in \text{NC induct} - \iota`\Lambda . \supset . \alpha > \beta$

Dem.

$\vdash . *120·48 . \text{Transp} . \supset \vdash : \text{Hp} . \supset . \sim(\beta \geqslant \alpha)$ (1)

$\vdash . *120·441 . \quad \supset \vdash :. \text{Hp} . \supset : \alpha > \beta . \text{v} . \beta \geqslant \alpha$ (2)

$\vdash . (1) . (2) . \supset \vdash . \text{Prop}$

Thus every non-inductive cardinal (except Λ) is greater than every inductive cardinal (except Λ).

***120·491.** $\vdash :. \xi \sim \in \text{Cls induct} . \equiv : \beta \in \text{NC induct} . \supset_\beta . \exists ! \beta \cap \text{Cl}`\xi$

Dem.

$\vdash . *120·49 . \supset \vdash : \xi \sim \in \text{Cls induct} . \beta \in \text{NC induct} - \iota`\Lambda . \supset . \text{N}_0\text{c}`\xi > \beta .$

$[*120·429.*117·12] \supset . \text{N}_0\text{c}`\xi \geqslant \beta +_c 1 . \exists ! \beta \cap \text{Cl}`\xi$ (1)

$\vdash . (1) . *117·104·12 . *103·13 . \supset$

$\vdash : \xi \sim \in \text{Cls induct} . \beta \in \text{NC induct} - \iota`\Lambda . \supset . \beta +_c 1 \neq \Lambda$ (2)

$\vdash . (2) . *101·12 . *120·13 . \supset$

$\vdash :. \xi \sim \in \text{Cls induct} . \supset : \beta \in \text{NC induct} . \supset . \beta \neq \Lambda$ (3)

$\vdash . (1) . (3) . \supset \vdash : \xi \sim \in \text{Cls induct} . \beta \in \text{NC induct} . \supset . \exists ! \beta \cap \text{Cl}`\xi$ (4)

$\vdash . *120·121 . \supset$

$\vdash :. \beta \in \text{NC induct} . \supset_\beta . \exists ! \beta \cap \text{Cl}`\xi : \supset : \beta \in \text{NC induct} . \supset_\beta . \exists ! (\beta +_c 1) \cap \text{Cl}`\xi .$

$[*117·242.*120·429] \quad \supset_\beta . \text{Nc}`\xi > \beta .$

$[*117·42.(*117·03)] \quad \supset_\beta . \text{N}_0\text{c}`\xi \neq \beta :$

$[*13·196] \quad \supset : \text{N}_0\text{c}`\xi \sim \in \text{NC induct} :$

$[*120·21] \quad \supset : \xi \sim \in \text{Cls induct}$ (5)

$\vdash . (4) . (5) . \supset \vdash . \text{Prop}$

***120·492.** $\vdash : \alpha \in \text{NC} - \text{NC induct} . \beta \geqslant \alpha . \supset . \beta \in \text{NC} - \text{NC induct}$
[*120·48 . Transp]

In virtue of *120·491, a class ξ which is not inductive contains sub-classes having 0, 1, 2, 3, ... terms. If we take the successive classes of sub-classes

$$0 \cap \text{Cl}`\xi,\ 1 \cap \text{Cl}`\xi,\ 2 \cap \text{Cl}`\xi,\ \ldots,$$

these are mutually exclusive, and all exist provided Λ is not an inductive cardinal, *i.e.* provided the axiom of infinity holds. Thus if the axiom of infinity holds, we get $\aleph_0$ classes of sub-classes contained in any non-inductive class. It follows, as we shall see later, that if ξ is a non-inductive class, $\text{Cl}'\text{Cl}'\xi$ is a reflexive class. This seems to be the nearest approach possible to identifying the two definitions of finite and infinite when the multiplicative axiom is not assumed. When the multiplicative axiom is assumed as well as the axiom of infinity, we pick out one class from $1 \cap \text{Cl}'\xi$, one from $2 \cap \text{Cl}'\xi$, and so on; then, forming the logical sum of all these classes, we get $\aleph_0$ terms which are members of ξ. Hence it follows that ξ is a reflexive class; for, as we shall see later, a reflexive class is one which contains sub-classes of $\aleph_0$ terms. Thus with the help of the multiplicative axiom, the two definitions of finite and infinite can be identified.

***120·493.** $\vdash :. \sigma \in \text{Cls induct} . \supset :$

$$\text{Nc}'\xi < \text{Nc}'\sigma . \equiv . (\exists \rho) . \rho \text{ sm } \xi . \rho \subset \sigma . \exists ! \sigma - \rho . \equiv . \exists ! \text{Nc}'\xi \cap \text{Cl}'\sigma - \iota'\sigma$$

Dem.

$\vdash . *117·26·221 . \supset \vdash :. \text{Nc}'\xi < \text{Nc}'\sigma . \supset : \sim(\xi \text{ sm } \sigma) : (\exists \rho) . \rho \text{ sm } \xi . \rho \subset \sigma :$

$[*73·3·37] \quad \supset : (\exists \rho) . \rho \text{ sm } \xi . \rho \subset \sigma . \rho \neq \sigma \quad (1)$

$\vdash . *120·481 . \supset \vdash :. \text{Hp} . \supset : \rho \subset \sigma . \exists ! \sigma - \rho . \supset . \rho \in \text{Cls induct} . \rho \subset \sigma . \exists ! \sigma - \rho .$

$[*120·426] \quad \supset . \text{Nc}'\rho < \text{Nc}'\sigma :$

$[*100·321] \quad \supset : \rho \text{ sm } \xi . \rho \subset \sigma . \exists ! \sigma - \rho . \supset . \text{Nc}'\xi < \text{Nc}'\sigma \quad (2)$

$\vdash . (1) . (2) . *24·6 . \supset \vdash . \text{Prop}$

***120·5.** $\vdash : \alpha, \beta \in \text{NC induct} . \exists ! \alpha \times_c \beta . \supset . \alpha \times_c \beta \in \text{NC induct}$

Dem.

$\vdash . *113·203 . \supset \vdash : \alpha \in \text{NC induct} . \exists ! \alpha \times_c 0 . \supset . \alpha \in \text{NC} - \iota'\Lambda .$

$[*113·601] \quad \supset . \alpha \times_c 0 = 0 .$

$[*120·12] \quad \supset . \alpha \times_c 0 \in \text{NC induct} \quad (1)$

$\vdash . *113·671 . \quad \supset \vdash . \alpha \times_c (\beta +_c 1) = (\alpha \times_c \beta) +_c \alpha .$

$[*120·4501 . *113·203] \supset \vdash : \alpha \in \text{NC induct} . \alpha \times_c \beta \in \text{NC induct} - \iota'\Lambda . \supset .$

$\alpha \times_c (\beta +_c 1) \in \text{NC induct} \quad (2)$

$\vdash . (1) . (2) . *120·13 . \supset \vdash . \text{Prop}$

The restriction involved in $\exists ! \alpha \times_c \beta$ in the hypothesis of the above proposition is not necessary if we assume that the axiom of infinity must fail in any one type if it fails in any other, *i.e.*

$$\Lambda \cap t'\alpha \in \text{NC induct} . \supset . \Lambda \cap t'\beta \in \text{NC induct},$$

where α and β are any two objects of any two types. To prove this proposition would require assumptions, as to the interrelation of various types, which have not been made in our previous proofs.

***120·51.** $\vdash : \alpha, \beta, \gamma \in \text{NC induct} . \alpha \neq 0 . \exists ! \alpha \times_c \beta . \alpha \times_c \beta = \alpha \times_c \gamma . \supset . \beta = \text{sm}``\gamma$

This proposition establishes the uniqueness of division among inductive cardinals.

Dem.

$\vdash . *120·44·436 . \supset \vdash :. \text{Hp} . \supset : (\exists \delta) : \beta = \gamma +_c \delta . \vee . \gamma = \beta +_c \delta$ (1)

$\vdash . *113·43 . \quad \supset \vdash : \text{Hp} . \beta = \gamma +_c \delta . \supset . \alpha \times_c \gamma = (\alpha \times_c \gamma) +_c (\alpha \times_c \delta) .$

[*120·42.Transp] $\supset . \alpha \times_c \delta = 0 .$

[*113·602] $\supset . \delta = 0 .$

[*110·6] $\supset . \beta = \text{sm}``\gamma$ (2)

Similarly $\vdash : \text{Hp} . \gamma = \beta +_c \delta . \supset . \gamma = \text{sm}``\beta .$

[*100·53.*113·203] $\supset . \beta = \text{sm}``\gamma$ (3)

$\vdash . (1) . (2) . (3) . \supset \vdash . \text{Prop}$

If β, γ in the above are typically ambiguous symbols, such as

$$0, 1, 2, \ldots \text{Nc}`\rho, \text{Nc}`\sigma, \ldots,$$

we have $\beta = \gamma$; for in this case, $\beta = \text{sm}``\beta . \gamma = \text{sm}``\gamma$. Also if β and γ are of the same type, we have $\beta = \gamma$, in virtue of *103·43. Hence "$\beta = \gamma$" may, with truth, be substituted for "$\beta = \text{sm}``\gamma$" in the above proposition, since the result is true whenever significant. But in this form the proposition gives less information, since it tells us nothing as to what happens when β and γ are not of the same type.

***120·511.** $\vdash : \alpha, \beta \in \text{NC induct} . \alpha \neq 0 . \exists ! \alpha . \alpha \times_c \beta = \alpha . \supset . \beta = 1$

Dem.

$\vdash . *113·621 . \supset \vdash : \text{Hp} . \supset . \alpha \times_c \beta = \alpha \times_c 1$ (1)

$\vdash . (1) . *120·51 . *101·28 . \supset \vdash . \text{Prop}$

***120·512.** $\vdash : \alpha \times_c \beta \in \text{NC induct} - \iota`0 - \iota`\Lambda . \supset . \alpha, \beta \in \text{NC induct} - \iota`0 - \iota`\Lambda$

Dem.

$\vdash . *113·602·203 . \supset \vdash : \text{Hp} . \supset . \alpha, \beta \in \text{NC} - \iota`0 - \iota`\Lambda$ (1)

$\vdash . (1) . *117·62 . \supset \vdash : \text{Hp} . \supset . \alpha \times_c \beta \geqslant \alpha . \alpha \times_c \beta \geqslant \beta .$

[*120·48] $\supset . \alpha, \beta \in \text{NC induct}$ (2)

$\vdash . (1) . (2) . \supset \vdash . \text{Prop}$

***120·513.** $\vdash : \alpha \in \text{NC induct} - \iota`0 - \iota`\Lambda . \alpha \times_c \beta = \alpha . \supset . \beta = 1$ [*120·511·512]

This proposition does not hold when α is a non-inductive cardinal.

***120·52.** $\vdash : \alpha, \beta \in \text{NC induct} . \exists ! \alpha^\beta . \supset . \alpha^\beta \in \text{NC induct}$

Dem.

$\vdash . *116·203·301 . \supset \vdash : \alpha \in \text{NC induct} . \exists ! \alpha^0 . \supset . \alpha^0 = 1 .$

[*120·122] $\supset . \alpha^0 \in \text{NC induct}$ (1)

$\vdash . *116·321·52 . \supset \vdash : \exists ! \alpha^{\beta +_c 1} . \supset . \alpha^{\beta +_c 1} = \alpha^\beta \times_c \alpha .$

[*120·5] $\supset \vdash : \alpha \in \text{NC induct} . \alpha^\beta \in \text{NC induct} . \exists ! \alpha^{\beta +_c 1} . \supset .$

$\alpha^{\beta +_c 1} \in \text{NC induct}$ (2)

$\vdash . *116·52 . *113·204 . \supset \vdash : \alpha^\beta = \Lambda . \supset . \alpha^{\beta +_c 1} = \Lambda$ (3)

$\vdash . (1) . (2) . (3) . *120·11 . \supset \vdash . \text{Prop}$

***120·53.** $\vdash : \alpha, \beta, \gamma \in \text{NC induct} . \alpha \neq 0 . \alpha \neq 1 . \exists ! \alpha^\beta . \alpha^\beta = \alpha^\gamma . \supset . \beta = \text{sm}``\gamma$

Dem.

$\vdash . *116·203 . \quad \supset \vdash : \exists ! \alpha^\beta . \supset . \exists ! \beta \quad (1)$

$\vdash . *120·44·436 . \quad \supset \vdash :. \text{Hp} . \supset : (\exists \delta) : \beta = \gamma +_c \delta . \vee . \gamma = \beta +_c \delta \quad (2)$

$\vdash . *118·01 . *116·52 . \quad \supset \vdash : \beta = \gamma +_c \delta . \exists ! \beta . \supset . \alpha^\beta = \alpha^\gamma \times_c \alpha^\delta :$

$[*13·171 . *118·01 . (1)] \quad \supset \vdash : \alpha^\beta = \alpha^\gamma . \beta = \gamma +_c \delta . \exists ! \alpha^\beta . \supset . \alpha^\gamma = \alpha^\gamma \times_c \alpha^\delta \quad (3)$

$\vdash . *120·52 . *116·35 . (1) . \supset \vdash : \text{Hp} . \beta = \gamma +_c \delta . \supset . \alpha^\gamma \in \text{NC induct} - \iota`\Lambda - \iota`0 . \exists ! \beta .$

$[(3) . *120·513] \quad \supset . \alpha^\delta = 1 .$

$[*117·592] \quad \supset . \delta = 0 .$

$[*110·6] \quad \supset . \beta = \text{sm}``\gamma \quad (4)$

Similarly $\quad \vdash : \text{Hp} . \gamma = \beta +_c \delta . \supset . \gamma = \text{sm}``\beta .$

$[*100·53 . (1)] \quad \supset . \beta = \text{sm}``\gamma \quad (5)$

$\vdash . (2) . (4) . (5) . \supset \vdash . \text{Prop}$

If α, β, γ are typically ambiguous symbols, we have $\beta = \gamma$ in the conclusion of the above proposition, instead of $\beta = \text{sm}``\gamma$. Also if β and γ are of the same type, $\beta = \gamma$; thus $\beta = \gamma$ whenever "$\beta = \gamma$" is significant.

***120·54.** $\vdash : \xi, \rho \in \text{Cls induct} . \exists ! \xi . \rho \subset \sigma . \exists ! \sigma - \rho . \supset . (\text{Nc}`\rho)^{\text{Nc}`\xi} < (\text{Nc}`\sigma)^{\text{Nc}`\xi}$

For the proof, which is here given shortly, compare *117·58.

Dem.

$\vdash . *35·432·82 . *80·15 . *116·12 . \supset \vdash : \text{Hp} . \supset . (\rho \uparrow \xi)_\Delta`\xi \subset (\sigma \uparrow \xi)_\Delta`\xi .$

$\exists ! (\sigma \uparrow \xi)_\Delta`\xi - (\rho \uparrow \xi)_\Delta`\xi \quad (1)$

$\vdash . *120·52 . *116·15·251 . *120·2 . \supset \vdash : \text{Hp} . \supset . (\rho \uparrow \xi)_\Delta`\xi \in \text{Cls induct} \quad (2)$

$\vdash . (1) . (2) . *120·426 . \supset \vdash : \text{Hp} . \supset . \text{Nc}`(\rho \uparrow \xi)_\Delta`\xi < \text{Nc}`(\sigma \uparrow \xi)_\Delta`\xi : \supset \vdash . \text{Prop}$

***120·541.** $\vdash : \alpha, \beta \in \text{NC induct} - \iota`\Lambda . \alpha \neq 0 . \beta < \gamma . \supset . \beta^\alpha < \gamma^\alpha$ [*120·54·493]

***120·542.** $\vdash : \alpha, \gamma \in \text{NC induct} - \iota`\Lambda . \alpha \neq 0 . \beta > \gamma . \supset . \beta^\alpha > \gamma^\alpha$ [*120·541]

***120·55.** $\vdash : \alpha, \beta, \gamma \in \text{NC induct} . \alpha \neq 0 . \exists ! \beta^\alpha . \beta^\alpha = \gamma^\alpha . \supset . \beta = \text{sm}``\gamma$

Dem.

$\vdash . *120·541·542 . \supset \vdash : \text{Hp} . \supset . \sim(\beta < \gamma) . \sim(\beta > \gamma) .$

$[*120·441] \quad \supset . \beta = \text{sm}``\gamma : \supset \vdash . \text{Prop}$

***120·56.** $\vdash : \alpha \geqslant 2 . \alpha^\beta \in \text{NC induct} - \iota`\Lambda . \supset . \beta \in \text{NC induct}$

Dem.

$\vdash . *117·581 . \supset \vdash : \text{Hp} . \supset . \alpha^\beta \geqslant 2^\beta .$

$[*117·661] \quad \supset . \alpha^\beta > \beta \quad (1)$

$\vdash . (1) . *120·48 . \supset \vdash . \text{Prop}$

***120·561.** $\vdash : \beta \geqslant 1 . \alpha^\beta \in \text{NC induct} - \iota`\Lambda . \supset . \alpha \in \text{NC induct}$

Dem.

$\vdash . *117·591 . *116·321 . \supset \vdash : \text{Hp} . \supset . \alpha^\beta \geqslant \alpha \quad (1)$

$\vdash . (1) . *120·48 . \supset \vdash . \text{Prop}$

***120·57.** $\vdash : \mu \in \mathrm{NC\ induct} - \iota`\Lambda . \supset . \mathrm{Nc}`\hat{\nu}(\nu \leqslant \mu) = \mu +_c 1$

Here "$\mu +_c 1$" is necessarily in a higher type than "μ," because it applies to a class of which μ is a member.

Dem.

$\vdash . *117·511 . \supset \vdash . \mathrm{Nc}`\hat{\nu}(\nu \leqslant 0) \in 1$ (1)

$\vdash . *110·4 . \quad \supset \vdash : \mu = \Lambda . \supset . \mu +_c 1 = \Lambda$ (2)

$\vdash . *120·429·442 . \supset$

$\vdash : \mu \in \mathrm{NC\ induct} . \exists ! \mu +_c 1 . \supset . \hat{\nu}(\nu \leqslant \mu) = \hat{\nu}(\nu < \mu +_c 1) .$

[*117·104·105] $\supset . \hat{\nu}(\nu \leqslant \mu +_c 1) = \hat{\nu}(\nu \leqslant \mu) \cup \iota`(\mu +_c 1)$ (3)

$\vdash . *120·428 . \supset \vdash : \mathrm{Hp}(3) . \supset . \mu +_c 1 \sim \in \hat{\nu}(\nu \leqslant \mu)$ (4)

$\vdash . (3) . (4) . *110·631 . \supset$

$\vdash : \mathrm{Hp}(3) . \mathrm{Nc}`\hat{\nu}(\nu \leqslant \mu) = \mu +_c 1 . \supset . \mathrm{Nc}`\hat{\nu}(\nu \leqslant \mu +_c 1) = \mu +_c 2$ (5)

$\vdash . (2) . (5) . \supset \vdash :. \mu \in \mathrm{NC\ induct} : \mu = \Lambda . \vee . \mathrm{Nc}`\hat{\nu}(\nu \leqslant \mu) = \mu +_c 1 : \supset :$

$\mu +_c 1 = \Lambda . \vee . \mathrm{Nc}`\hat{\nu}(\nu \leqslant \mu +_c 1) = \mu +_c 2$ (6)

$\vdash . (1) . (6) . *120·13 . \supset \vdash . \mathrm{Prop}$

***120·6.** $\vdash : (\exists \gamma) . \gamma > \alpha . \gamma \subset t`\eta . \supset . \exists ! (\alpha +_c 1) \cap t`\eta$

Dem.

$\vdash . *117·1 . \supset$

$\vdash :. \mathrm{Hp} . \supset : (\exists \gamma, \rho, \sigma) . \mathrm{N_0c}`\rho = \alpha . \mathrm{N_0c}`\sigma = \gamma . \exists ! \mathrm{Nc}`\rho \cap \mathrm{Cl}`\sigma . \sim \exists ! \mathrm{Nc}`\sigma \cap \mathrm{Cl}`\rho :$

[*100·1] $\supset : (\exists \gamma, \rho, \sigma, \xi) . \mathrm{N_0c}`\rho = \alpha . \mathrm{N_0c}`\sigma = \gamma . \xi \,\mathrm{sm}\, \rho . \xi \subset \sigma . \xi \neq \sigma :$

[*24·6] $\supset : (\exists \gamma, \rho, \sigma, \xi, x) . \mathrm{N_0c}`\rho = \alpha . \mathrm{N_0c}`\sigma = \gamma . \xi \,\mathrm{sm}\, \rho . x \in \sigma - \xi :$

[*110·631] $\supset : (\exists \xi, x) . \xi \cup \iota`x \in \alpha +_c 1 \cap t`\eta :. \supset \vdash . \mathrm{Prop}$

***120·61.** $\vdash : \exists ! \mathrm{N_0C} \cap t^3`x - \mathrm{NC\ induct} . \supset . \mathrm{Infin\ ax}(x)$

Dem.

$\vdash . *120·49 . \supset \vdash :. \gamma \in \mathrm{N_0C} \cap t^3`x - \mathrm{NC\ induct} . \supset :$

$\alpha \in \mathrm{NC\ induct} . \exists ! \alpha . \quad \supset_\alpha . \gamma > \alpha . \gamma \subset t^2`x .$

[*120·6] $\supset_\alpha . \exists ! \alpha +_c 1 \cap t^2`x$ (1)

$\vdash . (1) . *101·12 . *120·13 . \supset$

$\vdash :. \gamma \in \mathrm{N_0C} - \mathrm{NC\ induct} . \supset : \alpha \in \mathrm{NC\ induct} . \quad \supset_\alpha . \exists ! \alpha(x) :. \supset \vdash . \mathrm{Prop}$

***120·611.** $\vdash : \beta \in \mathrm{Cls\ induct} . \beta \subset \mathrm{C}`P . \supset . \exists ! P_\Delta`\beta$

Dem.

$\vdash . *80·26 . \supset \vdash . \exists ! P_\Delta`\Lambda .$

[Simp] $\supset \vdash : \Lambda \subset \mathrm{C}`P . \supset . \exists ! P_\Delta`\Lambda$ (1)

$\vdash . *80·94 . \supset \vdash : \exists ! P_\Delta`\beta . z \in \mathrm{C}`P . \supset . \exists ! P_\Delta`(\beta \cup \iota`z) :$

[Syll] $\supset \vdash :. \beta \subset \mathrm{C}`P . \supset . \exists ! P_\Delta`\beta : \supset : \beta \subset \mathrm{C}`P . z \in \mathrm{C}`P . \supset . \exists ! P_\Delta`(\beta \cup \iota`z) :$

[*51·238] $\supset : \beta \cup \iota`z \subset \mathrm{C}`P . \supset . \exists ! P_\Delta`(\beta \cup \iota`z)$ (2)

$\vdash . (1) . (2) . *120·26 . \supset \vdash . \mathrm{Prop}$

$*120\cdot62$. $\vdash : \kappa \epsilon \text{Cls induct} . \Lambda \sim \epsilon \kappa . \supset . \exists ! \epsilon_\Delta{}^\backprime\kappa$

Dem.

$\vdash . *83\cdot9 . \quad \supset \vdash . \exists ! \epsilon_\Delta{}^\backprime\Lambda \qquad (1)$

$\vdash . *83\cdot904 . \supset \vdash : \exists ! \epsilon_\Delta{}^\backprime\kappa . \exists ! \alpha . \supset . \exists ! \epsilon_\Delta{}^\backprime(\kappa \cup \iota^\backprime\alpha) :$

[Syll] $\quad \supset \vdash :. \Lambda \sim \epsilon \kappa . \supset . \exists ! \epsilon_\Delta{}^\backprime\kappa : \supset : \Lambda \sim \epsilon \kappa . \exists ! \alpha . \supset . \exists ! \epsilon_\Delta{}^\backprime(\kappa \cup \iota^\backprime\alpha) :$

[$*24\cdot54$] $\quad \supset : \Lambda \sim \epsilon (\kappa \cup \iota^\backprime\alpha) . \supset . \exists ! \epsilon_\Delta{}^\backprime(\kappa \cup \iota^\backprime\alpha) \qquad (2)$

$\vdash . (1) . (2) . *120\cdot26 . \supset \vdash . \text{Prop}$

The above proposition may also be deduced from $*120\cdot611$, by $*62\cdot231$.

$*120\cdot63$. $\vdash . \text{Cls induct} - \overleftarrow{\epsilon}{}^\backprime\Lambda \subset \text{Cls}^2 \text{mult} \quad [*120\cdot62 . *88\cdot2]$

In virtue of this proposition the multiplicative axiom is not required in dealing with a finite number of factors, even when some or all of the factors are themselves infinite.

$*120\cdot64$. $\vdash : \text{Infin ax} . \lor . \text{Mult ax}$

Dem.

$\vdash . *120\cdot61 . \text{Transp} . \supset \vdash :. \sim \text{Infin ax} . \supset : N_0C \subset \text{NC induct} :$

[$*120\cdot21$] $\quad \supset : (\kappa) . \kappa \epsilon \text{Cls induct} :$

[$*120\cdot62$] $\quad \supset : (\kappa) : \Lambda \sim \epsilon \kappa . \supset . \exists ! \epsilon_\Delta{}^\backprime\kappa :$

[$*88\cdot37$] $\quad \supset : \text{Mult ax} :. \supset \vdash . \text{Prop}$

Thus of our two arithmetical axioms, the multiplicative axiom and the axiom of infinity, at least one must be true.

$*120\cdot7$. $\vdash : \alpha \epsilon \text{Cls induct} . \alpha \subset \beta . \alpha \neq \beta . \supset . \text{Nc}^\backprime\alpha < \text{Nc}^\backprime\beta \quad [*120\cdot426 . *24\cdot6]$

$*120\cdot71$. $\vdash : \rho, \sigma \epsilon \text{Cls induct} . \equiv . \rho \cup \sigma \epsilon \text{Cls induct} . \equiv . \rho + \sigma \epsilon \text{Cls induct}$

Dem.

$\vdash . *120\cdot481 . \supset \vdash : \rho \cup \sigma \epsilon \text{Cls induct} . \supset . \rho, \sigma \epsilon \text{Cls induct} \qquad (1)$

$\vdash . *120\cdot481 . \supset \vdash : \rho, \sigma \epsilon \text{Cls induct} . \supset . \rho, \sigma - \rho \epsilon \text{Cls induct} .$

[$*120\cdot21$] $\quad \supset . N_0c^\backprime\rho, N_0c^\backprime(\sigma - \rho) \epsilon \text{NC induct} .$

[$*120\cdot45$] $\quad \supset . N_0c^\backprime\rho +_c N_0c^\backprime(\sigma - \rho) \epsilon \text{NC induct} .$

[$*110\cdot32$] $\quad \supset . \text{Nc}^\backprime(\rho \cup \sigma) \epsilon \text{NC induct} .$

[$*120\cdot211$] $\quad \supset . \rho \cup \sigma \epsilon \text{Cls induct} \qquad (2)$

$\vdash . (1) . (2) . \quad \supset \vdash : \rho, \sigma \epsilon \text{Cls induct} . \equiv . \rho \cup \sigma \epsilon \text{Cls induct} \qquad (3)$

$\vdash . *110\cdot12 . *120\cdot214 . \supset \vdash : \rho, \sigma \epsilon \text{Cls induct} . \equiv .$

$\quad \downarrow (\Lambda \cap \sigma)^{\backprime\backprime}\iota^{\backprime\backprime}\rho, (\Lambda \cap \rho) \downarrow {}^{\backprime\backprime}\iota^{\backprime\backprime}\sigma \epsilon \text{Cls induct} .$

[$(3).(*110\cdot01)$] $\quad \equiv . \rho + \sigma \epsilon \text{Cls induct} \qquad (4)$

$\vdash . (3) . (4) . \supset \vdash . \text{Prop}$

The above proposition is frequently used.

***120·72.** $\vdash : \rho, \sigma \in \text{Cls induct} . \supset . \rho \times \sigma \in \text{Cls induct}$

Dem.

$\vdash . *120\cdot21 . \supset \vdash : \text{Hp} . \supset . \text{N}_0\text{c}`\rho, \text{N}_0\text{c}`\sigma \in \text{NC induct} .$

$[*120\cdot5] \quad \supset . \text{Nc}`(\rho \times \sigma) \in \text{NC induct} .$

$[*120\cdot211] \quad \supset . \rho \times \sigma \in \text{Cls induct} : \supset \vdash . \text{Prop}$

***120·721.** $\vdash :. \exists ! \rho . \exists ! \sigma . \supset : \rho, \sigma \in \text{Cls induct} . \equiv . \rho \times \sigma \in \text{Cls induct}$

Dem.

$\vdash . *120\cdot512 . *113\cdot107 . \supset$

$\vdash :. \text{Hp} . \supset : \rho \times \sigma \in \text{Cls induct} . \supset . \text{Nc}`\rho, \text{Nc}`\sigma \in \text{NC induct} .$

$[*120\cdot211] \quad \supset . \rho, \sigma \in \text{Cls induct} \quad (1)$

$\vdash . (1) . *120\cdot72 . \supset \vdash . \text{Prop}$

***120·73.** $\vdash : \rho, \sigma \in \text{Cls induct} . \supset . (\rho \exp \sigma) \in \text{Cls induct} \quad [*120\cdot52 . *116\cdot251]$

***120·731.** $\vdash :. \exists ! \rho . \exists ! \sigma . \rho \sim\in 1 . \supset : \rho, \sigma \in \text{Cls induct} . \equiv . (\rho \exp \sigma) \in \text{Cls induct}$ $[*120\cdot56\cdot561\cdot73]$

***120·74.** $\vdash : \rho \in \text{Cls induct} . \equiv . \text{Cl}`\rho \in \text{Cls induct}$

Dem.

$\vdash . *116\cdot72 . *120\cdot21 . \supset \vdash : \text{Cl}`\rho \in \text{Cls induct} . \equiv . 2^{\text{Nc}`\rho} \cap t`\text{Cl}`\rho \in \text{NC induct} .$

$[*120\cdot123\cdot52\cdot56 . *116\cdot72 . (*116\cdot04)] \quad \equiv . \text{N}_0\text{c}`\rho \in \text{NC induct} .$

$[*120\cdot21] \quad \equiv . \rho \in \text{Cls induct} : \supset \vdash . \text{Prop}$

***120·741.** $\vdash : s`\kappa \in \text{Cls induct} . \supset . \kappa \in \text{Cls induct} . \kappa \subset \text{Cls induct}$

Dem.

$\vdash . *120\cdot74 . \quad \supset \vdash : \text{Hp} . \supset . \text{Cl}`s`\kappa \in \text{Cls induct} .$

$[*60\cdot57 . *120\cdot481] \quad \supset . \kappa \in \text{Cls induct} \quad (1)$

$\vdash . *40\cdot13 . *120\cdot481 . \supset \vdash :. \text{Hp} . \supset : \rho \in \kappa . \supset . \rho \in \text{Cls induct} \quad (2)$

$\vdash . (1) . (2) . \supset \vdash . \text{Prop}$

***120·75.** $\vdash : s`\kappa \in \text{Cls induct} . \equiv . \kappa \in \text{Cls induct} . \kappa \subset \text{Cls induct}$

Dem.

$\vdash . *22\cdot58 . \quad \supset \vdash : \exists ! \kappa - \text{Cls induct} . \supset . \exists ! (\kappa \cup \iota`\alpha) - \text{Cls induct} \quad (1)$

$\vdash . *120\cdot71 . *53\cdot15 . \supset \vdash : s`\kappa \in \text{Cls induct} . \alpha \in \text{Cls induct} . \supset .$
$s`(\kappa \cup \iota`\alpha) \in \text{Cls induct} :$

$[*5\cdot6] \quad \supset \vdash :. s`\kappa \in \text{Cls induct} . \supset :$
$\alpha \sim\in \text{Cls induct} . \vee . s`(\kappa \cup \iota`\alpha) \in \text{Cls induct} :$

$[*51\cdot16] \supset : \exists ! (\kappa \cup \iota`\alpha) - \text{Cls induct} . \vee . s`(\kappa \cup \iota`\alpha) \in \text{Cls induct} \quad (2)$

$\vdash . (1) . (2) . \supset \vdash :. \exists ! \kappa - \text{Cls induct} . \vee . s`\kappa \in \text{Cls induct} : \supset :$
$\exists ! (\kappa \cup \iota`\alpha) - \text{Cls induct} . \vee . s`(\kappa \cup \iota`\alpha) \in \text{Cls induct} \quad (3)$

$\vdash . *40\cdot21 . *120\cdot212 . \supset \vdash . s`\Lambda \in \text{Cls induct} \quad (4)$

$\vdash . (3) . (4) . *120\cdot26 . \supset \vdash :. \kappa \in \text{Cls induct} . \supset :$
$\exists ! \kappa - \text{Cls induct} . \vee . s`\kappa \in \text{Cls induct} :.$

$[*5\cdot6] \supset \vdash : \kappa \in \text{Cls induct} . \kappa \subset \text{Cls induct} . \supset . s`\kappa \in \text{Cls induct} \quad (5)$

$\vdash . (5) . *120\cdot741 . \supset \vdash . \text{Prop}$

***120·76.** $\vdash : \kappa \epsilon \text{Cls induct} . \kappa \subset \text{Cls induct} . \supset . \epsilon_\Delta{}^\backprime\kappa \epsilon \text{Cls induct}$

Dem.

$\vdash . *51{\cdot}2 . \supset \vdash :. \alpha \epsilon \kappa . \supset : \kappa = \kappa \cup \iota{}^\backprime\alpha :$

$[*13{\cdot}12] \qquad \supset : \epsilon_\Delta{}^\backprime\kappa \epsilon \text{Cls induct} . \supset . \epsilon_\Delta{}^\backprime(\kappa \cup \iota{}^\backprime\alpha) \epsilon \text{Cls induct} \quad (1)$

$\vdash . *83{\cdot}41 . *114{\cdot}301 . \supset \vdash : \alpha \sim\epsilon \kappa . \supset . \epsilon_\Delta{}^\backprime(\kappa \cup \iota{}^\backprime\alpha) \text{ sm } \epsilon_\Delta{}^\backprime\kappa \times \alpha \quad (2)$

$\vdash . (2) . *120{\cdot}214 . \supset \vdash :. \alpha \sim\epsilon \kappa . \supset :$

$\qquad \epsilon_\Delta{}^\backprime(\kappa \cup \iota{}^\backprime\alpha) \epsilon \text{Cls induct} . \equiv . \epsilon_\Delta{}^\backprime\kappa \times \alpha \epsilon \text{Cls induct} \quad (3)$

$\vdash . (3) . *120{\cdot}72 . \quad \supset \vdash :. \alpha \sim\epsilon \kappa . \supset :$

$\qquad \epsilon_\Delta{}^\backprime\kappa, \alpha \epsilon \text{Cls induct} . \supset . \epsilon_\Delta{}^\backprime(\kappa \cup \iota{}^\backprime\alpha) \epsilon \text{Cls induct} \quad (4)$

$\vdash . (1) . (4) . \qquad \supset \vdash : \epsilon_\Delta{}^\backprime\kappa, \alpha \epsilon \text{Cls induct} . \supset . \epsilon_\Delta{}^\backprime(\kappa \cup \iota{}^\backprime\alpha) \epsilon \text{Cls induct} \quad (5)$

$\vdash . (5) . *51{\cdot}2 . \text{Syll} . \quad \supset \vdash :. \kappa \subset \text{Cls induct} . \supset . \epsilon_\Delta{}^\backprime\kappa \epsilon \text{Cls induct} : \supset :$

$\qquad \kappa \cup \iota{}^\backprime\alpha \subset \text{Cls induct} . \supset . \epsilon_\Delta{}^\backprime(\kappa \cup \iota{}^\backprime\alpha) \epsilon \text{Cls induct} \quad (6)$

$\vdash . *83{\cdot}15 . *120{\cdot}213 . \supset \vdash . \epsilon_\Delta{}^\backprime\Lambda \epsilon \text{Cls induct} .$

$[\text{Simp}] \qquad \supset \vdash : \Lambda \subset \text{Cls induct} . \supset . \epsilon_\Delta{}^\backprime\Lambda \epsilon \text{Cls induct} \quad (7)$

$\vdash . (6) . (7) . *120{\cdot}26 . \quad \supset \vdash :. \kappa \epsilon \text{Cls induct} . \supset :$

$\qquad \kappa \subset \text{Cls induct} . \supset . \epsilon_\Delta{}^\backprime\kappa \epsilon \text{Cls induct} :. \supset \vdash . \text{Prop}$

The following propositions are concerned in establishing the converse of *120·76 subject to a suitable hypothesis. The final outcome is given in *120·77.

***120·761.** $\vdash : \exists ! \epsilon_\Delta{}^\backprime\kappa . \epsilon_\Delta{}^\backprime\kappa \epsilon \text{Cls induct} . \supset . \kappa \subset \text{Cls induct}$

Dem.

$\vdash . *83{\cdot}41 . *114{\cdot}301 . \supset \vdash :. \alpha \epsilon \kappa . \supset : \epsilon_\Delta{}^\backprime\kappa \text{ sm } \alpha \times \epsilon_\Delta{}^\backprime(\kappa - \iota{}^\backprime\alpha) : \quad (1)$

$[*120{\cdot}214] \supset : \epsilon_\Delta{}^\backprime\kappa \epsilon \text{Cls induct} . \equiv . \alpha \times \epsilon_\Delta{}^\backprime(\kappa - \iota{}^\backprime\alpha) \epsilon \text{Cls induct} \quad (2)$

$\vdash . (1) . *113{\cdot}114 . \qquad \supset \vdash : \exists ! \epsilon_\Delta{}^\backprime\kappa . \alpha \epsilon \kappa . \supset . \exists ! \alpha . \exists ! \epsilon_\Delta{}^\backprime(\kappa - \iota{}^\backprime\alpha) \quad (3)$

$\vdash . (2) . (3) . *120{\cdot}721 . \supset \vdash :. \exists ! \epsilon_\Delta{}^\backprime\kappa . \alpha \epsilon \kappa . \supset :$

$\qquad \epsilon_\Delta{}^\backprime\kappa \epsilon \text{Cls induct} . \supset . \alpha \epsilon \text{Cls induct} \quad (4)$

$\vdash . (4) . \text{Comm} . \supset \vdash . \text{Prop}$

***120·762.** $\vdash : \kappa \epsilon \text{Cls induct} . \Lambda \sim\epsilon \kappa . \sim\exists ! 1 \cap \kappa . \supset . (\exists R, S) . R, S \epsilon \epsilon_\Delta{}^\backprime\kappa . R \dot{\cap} S = \dot{\Lambda}$

Dem.

$\vdash . *51{\cdot}2 . \supset \vdash : R, S \epsilon \epsilon_\Delta{}^\backprime\kappa . R \dot{\cap} S = \dot{\Lambda} . \alpha \epsilon \kappa . \supset . R, S \epsilon \epsilon_\Delta{}^\backprime(\kappa \cup \iota{}^\backprime\alpha) . R \dot{\cap} S = \dot{\Lambda} \quad (1)$

$\vdash . *83{\cdot}5 . *55{\cdot}201 . \supset$

$\vdash : R, S \epsilon \epsilon_\Delta{}^\backprime\kappa . R \dot{\cap} S = \dot{\Lambda} . x, y \epsilon \alpha . x \neq y . \alpha \sim\epsilon \kappa . \supset .$

$\qquad R \dot{\cup} x \downarrow \alpha, S \dot{\cup} y \downarrow \alpha \epsilon \epsilon_\Delta{}^\backprime(\kappa \cup \iota{}^\backprime\alpha) . (R \dot{\cup} x \downarrow \alpha) \dot{\cap} (S \dot{\cup} y \downarrow \alpha) = \dot{\Lambda} \quad (2)$

$\vdash . (1) . (2) . *52{\cdot}41 . \supset \vdash : R, S \epsilon \epsilon_\Delta{}^\backprime\kappa . R \dot{\cap} S = \dot{\Lambda} . \alpha \neq \Lambda . \alpha \sim\epsilon 1 . \supset .$

$\qquad (\exists P, Q) . P, Q \epsilon \epsilon_\Delta{}^\backprime(\kappa \cup \iota{}^\backprime\alpha) . P \dot{\cap} Q = \dot{\Lambda} \quad (3)$

$\vdash . *51{\cdot}16 . \supset \vdash :. \alpha = \Lambda . \mathsf{v} . \alpha \epsilon 1 : \supset : \Lambda \epsilon (\kappa \cup \iota{}^\backprime\alpha) . \mathsf{v} . \exists ! 1 \cap (\kappa \cup \iota{}^\backprime\alpha) \quad (4)$

$\vdash . *22{\cdot}58 . \supset \vdash :. \Lambda \epsilon \kappa . \mathsf{v} . \exists ! 1 \cap \kappa : \supset : \Lambda \epsilon (\kappa \cup \iota{}^\backprime\alpha) . \mathsf{v} . \exists ! 1 \cap (\kappa \cup \iota{}^\backprime\alpha) \quad (5)$

$\vdash . (3) . (4) . (5) . \supset \vdash :. \Lambda \epsilon \kappa . \mathsf{v} . \exists ! 1 \cap \kappa . \mathsf{v} . (\exists R, S) . R, S \epsilon \epsilon_\Delta{}^\backprime\kappa . R \dot{\cap} S = \dot{\Lambda} : \supset :$

$\qquad \Lambda \epsilon (\kappa \cup \iota{}^\backprime\alpha) . \mathsf{v} . \exists ! 1 \cap (\kappa \cup \iota{}^\backprime\alpha) . \mathsf{v} . (\exists R, S) . R, S \epsilon \epsilon_\Delta{}^\backprime(\kappa \cup \iota{}^\backprime\alpha) . R \dot{\cap} S = \dot{\Lambda} \quad (6)$

$\vdash . *83{\cdot}15 . \supset \vdash . (\exists R, S) . R, S \epsilon \epsilon_\Delta{}^\backprime\Lambda . R \dot{\cap} S = \dot{\Lambda} \quad (7)$

$\vdash . (6) . (7) . *120{\cdot}26 . \supset \vdash . \text{Prop}$

$*120\cdot764$. $\vdash : \kappa \epsilon \text{Cls induct} . \Lambda \sim \epsilon \kappa . \sim \exists ! (1 \cap \kappa) . \supset . \text{Nc}`\epsilon_\Delta`\kappa \geqslant \text{Nc}`\kappa$
[$*120\cdot762 . *117\cdot681$]

$*120\cdot765$. $\vdash : \kappa \epsilon \text{Cls induct} . \Lambda \sim \epsilon \kappa . \sim \exists ! (1 \cap \kappa) . \kappa \subset \lambda . \exists ! \epsilon_\Delta`\lambda . \supset .$
$\text{Nc}`\epsilon_\Delta`\lambda \geqslant \text{Nc}`\kappa$ [$*120\cdot762 . *117\cdot684$]

$*120\cdot766$. $\vdash : \lambda \sim \epsilon \text{Cls induct} . \Lambda \sim \epsilon \lambda . \sim \exists ! (1 \cap \lambda) . \exists ! \epsilon_\Delta`\lambda . \supset .$
$\text{Nc}`\epsilon_\Delta`\lambda \sim \epsilon \text{NC induct}$

Dem.

$\vdash . *120\cdot491 . \supset \vdash :. \text{Hp} . \supset : \nu \epsilon \text{NC induct} . \supset .$
$(\exists \kappa) . \kappa \subset \lambda . \text{Nc}`\kappa = \nu . \Lambda \sim \epsilon \kappa . \sim \exists ! (1 \cap \kappa) .$
[$*120\cdot765$] $\supset . \text{Nc}`\epsilon_\Delta`\lambda \geqslant \nu :$
[$*120\cdot121$] $\supset : \nu \epsilon \text{NC induct} . \supset . \text{Nc}`\epsilon_\Delta`\lambda \geqslant \nu +_c 1 .$
[$*120\cdot429$] $\supset . \text{Nc}`\epsilon_\Delta`\lambda > \nu :$
[$*117\cdot42$] $\supset : \text{Nc}`\epsilon_\Delta`\lambda \sim \epsilon \text{NC induct} :. \supset \vdash . \text{Prop}$

$*120\cdot767$. $\vdash : \epsilon_\Delta`\lambda \epsilon \text{Cls induct} . \Lambda \sim \epsilon \lambda . \sim \exists ! (1 \cap \lambda) . \exists ! \epsilon_\Delta`\lambda . \supset . \lambda \epsilon \text{Cls induct}$
[$*120\cdot766$. Transp]

$*120\cdot77$. $\vdash :. \Lambda \sim \epsilon \kappa . \sim \exists ! (1 \cap \kappa) . \exists ! \epsilon_\Delta`\kappa . \supset :$
$\epsilon_\Delta`\kappa \epsilon \text{Cls induct} . \equiv . \kappa \epsilon \text{Cls induct} . \kappa \subset \text{Cls induct}$
[$*120\cdot76\cdot761\cdot767$]

$*121$. INTERVALS.

Summary of $*121$.

The present number is concerned with the class of terms between x and y with respect to some relation P, *i.e.* those terms which lie on a road from x to y on which any two consecutive terms have the relation P. Such a road may be called a P-road, and if zPw, the step from z to w may be called a P-step. In order that a P-road from x to y should exist, it is necessary and sufficient that we should have $xP_{po}y$. When this condition is fulfilled, there will in general be many P-roads from x to y. But if $P \epsilon \text{Cls} \rightarrow 1 . \sim (yP_{po}y)$, or if $P \epsilon 1 \rightarrow \text{Cls} . \sim (xP_{po}x)$, then at most one road leads from x to y. This follows from the propositions of $*96$. In virtue of those propositions, if $P \epsilon \text{Cls} \rightarrow 1 . \sim (yP_{po}y) . xP_{po}y$, P is $1 \rightarrow 1$ throughout the road from x to y, and this road forms an open series. The two other possibilities with a $\text{Cls} \rightarrow 1$ are (assuming $xP_{po}y$)

$$(1) \quad xP_{po}x,$$

$$(2) \quad yP_{po}y . \sim (xP_{po}x).$$

In the first case, there is a cyclic road from x to x, and there are two roads from x to y, one consisting of that part of the cycle which is required to reach y, the other consisting of this part together with the whole cycle required to travel from y back to y. Thus the class of terms which can be reached in some journey from x to y is the whole class of descendants of x, *i.e.* the class $\overleftarrow{R_*}`x$, which is the cycle composing the road from x to x.

In the second case, the descendants of x form a Q, and y is in the circular part of the Q. Here, as before, there are two roads from x to y, of which the first stops as soon as it reaches y, while the second proceeds to travel round the circle until it comes to y again. Thus here again, all the descendants of x lie on some road between x and y.

The *interval* between x and y is defined as the class of terms lying on some road from x to y. There will be four kinds of interval, according as we do or do not include the end-points as such. We denote the kind including both end-points by

$$P(x \mapsto y),$$

that excluding both by

$$P(x - y),$$

and the other two respectively by

$$P(x \dashv y), \; P(x \vdash y).$$

The definitions are

$$\overleftarrow{P_*}\text{‘}x \cap \overrightarrow{P_*}\text{‘}y,\ \overleftarrow{P_{\text{po}}}\text{‘}x \cap \overrightarrow{P_{\text{po}}}\text{‘}y,\ \overleftarrow{P_{\text{po}}}\text{‘}x \cap \overrightarrow{P_*}\text{‘}y,\ \overleftarrow{P_*}\text{‘}x \cap \overrightarrow{P_{\text{po}}}\text{‘}y.$$

If P is either one-many or many-one, it will be one-one throughout the interval $P(x \mapsto y)$, except at most at one exceptional point, namely the junction of the tail and circle of the Q. If $xP_{\text{po}}x$ or $\sim(yP_{\text{po}}y)$, the interval between x and y cannot be Q-shaped, but must be either open or cyclic; in either case, P is $1 \rightarrow 1$ throughout $P(x \mapsto y)$, with no exceptions; for if $P \in \text{Cls} \rightarrow 1$, P is $1 \rightarrow 1$ throughout the interval because the interval is contained in $\overleftarrow{P_*}\text{‘}x$, and if $P \in 1 \rightarrow \text{Cls}$, because the interval is contained in $\overrightarrow{P_*}\text{‘}y$. Thus throughout this number we shall constantly have the hypothesis $P \in (\text{Cls} \rightarrow 1) \cup (1 \rightarrow \text{Cls})$; if $P \in \text{Cls} \rightarrow 1$, the interval is to be supposed traversed from x to y, while if $P \in 1 \rightarrow \text{Cls}$, it is to be supposed traversed from y to x. In either case the interval between x and y must be an *inductive* class. This is proved in $*121\cdot47$. If, however, P is serial (cf. $*204$), and thus neither many-one nor one-many, the interval between x and y is the stretch of the series between x and y, with or without end-points according to the definition chosen, and need not be an inductive class.

If the interval between x and y (both included) has $\nu +_c 1$ members, we say that $xP_\nu y$. Thus if there is only one road from x to y, "$xP_\nu y$" means that it requires ν steps to get from x to y. Assuming $P \in \text{Cls} \rightarrow 1$, if we also have $P_{\text{po}} \subseteq J$ (*i.e.* if none of the families of P are cyclic), then if $xP_\nu y$ and yPz, we shall have $xP_{\nu +_c 1}z$. On this basis an inductive theory of P_ν is built up, and it is shown that the class of such relations as P_ν for different inductive values of ν is the same as $\text{Potid}\text{‘}P$, the class of powers of P including $I \upharpoonright C\text{‘}P$ ($*121\cdot5$). The definition of P_ν is

$$P_\nu = \hat{x}\hat{y}\,\{\text{N}_0\text{c}\text{‘}P(x \mapsto y) = \nu +_c 1\} \quad \text{Df.}$$

The whole class of such relations as P_ν for different inductive values of ν is called $\text{finid}\text{‘}P$, *i.e.* we put

$$\text{finid}\text{‘}P = \hat{R}\,\{(\exists \nu) \,.\, \nu \in \text{NC induct} - \iota\text{‘}\Lambda \,.\, R = P_\nu\} \quad \text{Df.}$$

If $B\text{‘}P$ exists, and if $P \in \text{Cls} \rightarrow 1$, then the descendants of $B\text{‘}P$, so long as we do not reach a term y for which $yP_{\text{po}}y$, may be unambiguously described as the 2nd, 3rd, ... νth, ... terms of the posterity of $B\text{‘}P$, $B\text{‘}P$ itself being the 1st term. The correlation thus effected with the inductive cardinals is the logical essence of the process of counting; the last cardinal used in the correlation is the cardinal number of terms counted. We will call these terms 1_P, 2_P, ... ν_P, ..., defining ν_P as follows:

$$\nu_P = \breve{P}_{\nu -_c 1}\text{‘}B\text{‘}P \quad \text{Df,}$$

This notation does not conflict with ν_ξ as defined in $*65\cdot01$. There ξ must be a class if ν is a cardinal, here ν must be a cardinal and P a relation.

Hence whenever ν_P exists, the number of terms from the beginning to ν_P (both included) is ν. This is the fact upon which counting relies. If P is a many-one and P_{po} is contained in diversity, and ν is any inductive cardinal other than 0, then ν_P exists when and only when $\overleftarrow{P_*}$'B'P has at least ν members; *i.e.* roughly speaking, ν_P exists whenever it could possibly be expected to exist. In this case the whole posterity of B'P is contained in the series $1_P, 2_P, \ldots \nu_P, \ldots$ (∗121·62). If the posterity is an inductive class, this series stops; if not, it forms a *progression* (cf. ∗122).

The propositions of the present number are very useful, not only in this section, but in the ordinal theory of finite and infinite and in parts of the book subsequent to that theory.

After some propositions which merely repeat definitions and give immediate consequences, we proceed (∗121·3 ff.) to the theory of P_ν. We have

∗121·302. $\vdash : P_{po} \subseteq J . \supset . P_0 = I \upharpoonright C\text{'}P$

∗121·305. $\vdash : P_{po} \subseteq J . \supset . P_1 \subseteq P$

∗121·31. $\vdash : P \epsilon (1 \to \text{Cls}) \cup (\text{Cls} \to 1) . P_{po} \subseteq J . \supset . P_1 = P$

When P is a transitive serial relation, we shall have $P_1 = P \dot{-} P^2$.

∗121·321. $\vdash : \nu > 0 . \supset . P_\nu \subseteq P_{po}$

∗121·333. $\vdash : P \epsilon \text{Cls} \to 1 . P_{po} \subseteq J . \supset . P_{\nu +_c 1} = P | P_\nu$

∗121·35·351·352. $\vdash : P \epsilon (1 \to \text{Cls}) \cup (\text{Cls} \to 1) . P_{po} \subseteq J . \mu, \nu \epsilon \text{NC induct} . \supset .$
$$P_\mu | P_\nu = P_\nu | P_\mu = P_{\mu +_c \nu}$$

A similar result holds for $(P_\mu)_\nu$, which $= P_{\mu \times_c \nu}$ in the same circumstances.

We next proceed to the proof that an interval (under a similar hypothesis) is always an inductive class. This occupies ∗121·4—·47, being summed up in the proposition

∗121·47. $\vdash : R \epsilon (\text{Cls} \to 1) \cup (1 \to \text{Cls}) . \supset . R(x \mapsto z) \epsilon \text{Cls induct}$

This is an important proposition. It leads to

∗121·481. $\vdash :. R \epsilon \text{Cls} \to 1 . \supset : \text{Nc}\text{'}R(x \mapsto y) \leqslant \text{Nc}\text{'}R(x \mapsto z) . \equiv .$
$$R(x \mapsto y) \subset R(x \mapsto z)$$
with a similar proposition if $R \epsilon 1 \to \text{Cls}$.

The next set of propositions (∗121·5—·52) is concerned with finid'P. Assuming $P \epsilon (\text{Cls} \to 1) \cup (1 \to \text{Cls}) . P_{po} \subseteq J$, we prove that finid'$P$ = Potid'P and finid'$P - \iota$'$P_0 \subset$ Pot'P (∗121·5); that if P is not null, finid'$P - \iota$'P_0 = Pot'P (∗121·501); that $\dot{s}$'finid'$P = P_*$ (∗121·52) and $\dot{s}$'(finid'$P - \iota$'$P_0) = P_{po}$ (∗121·502); and that $P_2 = P^2 . P_3 = P^3$ etc. (∗121·51).

Our next set of propositions is concerned with ν_P (∗121·6—·638). We have

$*121\cdot601.\ \vdash : E!\, B`P\,.\supset .\, B`P = 1_P\,.\sim\{(B`P)\, P_{po}\,(B`P)\}$

$*121\cdot602.\ \vdash : E!\, B`P\,.\, P\in 1\to 1\,.\supset .\, \breve{P}`B`P = 2_P$

$*121\cdot634.\ \vdash :.\, P\in \mathrm{Cls}\to 1\,.\, P_{po}\subseteq J\,.\,\nu\in \mathrm{NC\,induct} - \iota`0\,.\supset :$
$\nu_P \in \mathrm{D}`P\,.\equiv .\, E!\,(\nu +_c 1)_P$

Finally we have three propositions ($*121\cdot7$—$\cdot72$) on $\overrightarrow{R_*}`x$, of which the most useful is

$*121\cdot7.\quad \vdash : R\in 1\to 1\,.\, aBR\,.\, aR_*x\,.\supset .\, \overrightarrow{R_*}`x = R\,(a \mapsto x)\,.\, \overrightarrow{R_*}`x \in \mathrm{Cls\,induct}$

$*121\cdot01.\quad P\,(x - y) = \overleftarrow{P_{po}}`x \cap \overrightarrow{P_{po}}`y \quad \mathrm{Df}$

$*121\cdot011.\ P\,(x \dashv y) = \overleftarrow{P_{po}}`x \cap \overrightarrow{P_*}`y \quad \mathrm{Df}$

$*121\cdot012.\ P\,(x \vdash y) = \overleftarrow{P_*}`x \cap \overrightarrow{P_{po}}`y \quad \mathrm{Df}$

$*121\cdot013.\ P\,(x \mapsto y) = \overleftarrow{P_*}`x \cap \overrightarrow{P_*}`y \quad \mathrm{Df}$

$*121\cdot02.\quad P_\nu = \hat{x}\hat{y}\,\{\mathrm{N_0c}`P\,(x\mapsto y) = \nu +_c 1\} \quad \mathrm{Df}$

$*121\cdot03.\quad \mathrm{finid}`P = \hat{R}\,\{(\exists\nu)\,.\,\nu\in \mathrm{NC\,induct} - \iota`\Lambda\,.\, R = P_\nu\} \quad \mathrm{Df}$

$*121\cdot031.\ \mathrm{fin}`P = \hat{R}\,\{(\exists\nu)\,.\,\nu\in \mathrm{NC\,induct} - \iota`\Lambda - \iota`0\,.\, R = P_\nu\} \quad \mathrm{Df}$

$*121\cdot04.\quad \nu_P = \breve{P}_{\nu -_c 1}`B`P \quad \mathrm{Df}$

$*121\cdot1.\quad \vdash : z\in P\,(x - y)\,.\equiv .\, xP_{po}z\,.\, zP_{po}y \quad [(*121\cdot01)]$

$*121\cdot101.\ \vdash : z\in P\,(x \dashv y)\,.\equiv .\, xP_{po}z\,.\, zP_*y$

$*121\cdot102.\ \vdash : z\in P\,(x \vdash y)\,.\equiv .\, xP_*z\,.\, zP_{po}y$

$*121\cdot103.\ \vdash : z\in P\,(x \mapsto y)\,.\equiv .\, xP_*z\,.\, zP_*y$

$*121\cdot11.\quad \vdash : xP_\nu y\,.\equiv .\, \mathrm{N_0c}`P\,(x\mapsto y) = \nu +_c 1 \quad [(*121\cdot02)]$

$*121\cdot12.\quad \vdash : R\in \mathrm{finid}`P\,.\equiv .\,(\exists\nu)\,.\,\nu\in \mathrm{NC\,induct} - \iota`\Lambda\,.\, R = P_\nu \quad [(*121\cdot03)]$

$*121\cdot121.\ \vdash : R\in \mathrm{fin}`P\,.\equiv .\,(\exists\nu)\,.\,\nu\in \mathrm{NC\,induct} - \iota`\Lambda - \iota`0\,.\, R = P_\nu \quad [(*121\cdot031)]$

$*121\cdot13.\quad \vdash : f(\nu_P)\,.\equiv .\, f(\breve{P}_{\nu -_c 1}`B`P) \quad [(*121\cdot04)]$

$*121\cdot131.\ \vdash : E!\,\breve{P}_{\nu -_c 1}`B`P\,.\supset .\,\nu_P = \breve{P}_{\nu -_c 1}`B`P \quad [*121\cdot13\,.\,*14\cdot28]$

$*121\cdot14.\quad \vdash .\, P\,(x - y) = \breve{P}\,(y - x) \quad [*121\cdot1\,.\,*91\cdot53]$

$*121\cdot141.\ \vdash .\, P\,(x \dashv y) = \breve{P}\,(y \vdash x)$

$*121\cdot142.\ \vdash .\, P\,(x \vdash y) = \breve{P}\,(y \dashv x)$

$*121\cdot143.\ \vdash .\, P\,(x \mapsto y) = \breve{P}\,(y \mapsto x)$

$*121\cdot2.\quad \vdash : \sim(xP_{po}x)\,.\supset .\, x \sim\in P\,(x - y) \quad [*121\cdot1]$

$*121\cdot201.\ \vdash : \sim(yP_{po}y)\,.\supset .\, y \sim\in P\,(x - y)$

$*121{\cdot}202$. $\vdash : P_{po} \subseteq J . \supset . x, y \sim \epsilon P(x - y)$ [$*121{\cdot}2{\cdot}201$]

$*121{\cdot}21$. $\vdash : x P_{po} y . \equiv . y \epsilon P(x \dashv y) . \equiv . \exists ! P(x \dashv y)$

Dem.

$$\vdash . *90{\cdot}12 . *91{\cdot}54 . \supset \vdash : x P_{po} y . \equiv . x P_{po} y . y P_* y .$$
$$[*121{\cdot}101] \qquad \equiv . y \epsilon P(x \dashv y) \quad (1)$$
$$\vdash . *121{\cdot}101 . \supset \vdash : \exists ! P(x \dashv y) . \equiv . x P_{po} | P_* y .$$
$$[*91{\cdot}574] \qquad \equiv . x P_{po} y \quad (2)$$
$$\vdash . (1) . (2) . \supset \vdash . \text{Prop}$$

$*121{\cdot}22$. $\vdash : x P_{po} y . \equiv . x \epsilon P(x \vdash y) . \equiv . \exists ! P(x \vdash y)$

$*121{\cdot}23$. $\vdash : x P_* y . \equiv . x, y \epsilon P(x \mapsto y) . \equiv . \exists ! P(x \mapsto y)$

$*121{\cdot}231$. $\vdash : x \epsilon C{}^{\backprime}P . \equiv . x \epsilon P(x \mapsto x) . \equiv . \exists ! P(x \mapsto x)$ [$*121{\cdot}23 . *90{\cdot}12$]

$*121{\cdot}24$. $\vdash : x P_{po} y . \supset . P(x \dashv y) = P(x - y) \cup \iota{}^{\backprime}y$

Dem.

$$\vdash . *91{\cdot}54 . *121{\cdot}101 . \supset$$
$$\vdash :. z \epsilon P(x \dashv y) . \equiv : x P_{po} z : z P_{po} y . \vee . z = y . y \epsilon C{}^{\backprime}P :$$
$$[*13{\cdot}193 . *91{\cdot}504] \equiv : x P_{po} z . z P_{po} y . \vee . x P_{po} y . z = y \quad (1)$$
$$\vdash . (1) . *4{\cdot}73 . \supset \vdash :: \text{Hp} . \supset :. z \epsilon P(x \dashv y) . \equiv : x P_{po} z . z P_{po} y . \vee . z = y :: \supset \vdash . \text{Prop}$$

$*121{\cdot}241$. $\vdash : x P_{po} y . \supset . P(x \vdash y) = P(x - y) \cup \iota{}^{\backprime}x$

$*121{\cdot}242$. $\vdash : x P_* y . \supset . P(x \mapsto y) = P(x \dashv y) \cup \iota{}^{\backprime}x = P(x \vdash y) \cup \iota{}^{\backprime}y$
$$= P(x - y) \cup \iota{}^{\backprime}x \cup \iota{}^{\backprime}y$$

$*121{\cdot}25$. $\vdash . P_{po}(x - y) = P(x - y)$ [$*91{\cdot}601 . *121{\cdot}1$]

$*121{\cdot}251$. $\vdash . P_{po}(x \dashv y) = P(x \dashv y)$

$*121{\cdot}252$. $\vdash . P_{po}(x \vdash y) = P(x \vdash y)$

$*121{\cdot}253$. $\vdash . P_{po}(x \mapsto y) = P(x \mapsto y)$

$*121{\cdot}254$. $\vdash . P_\nu = (P_{po})_\nu$ [$*121{\cdot}253{\cdot}11$]

$*121{\cdot}254$ is frequently used in the theory of series.

$*121{\cdot}26$. $\vdash . \breve{P}_\nu = (\breve{P})_\nu$

Dem.

$$\vdash . *121{\cdot}11{\cdot}143 . \supset \vdash : x \breve{P}_\nu y . \equiv . \text{N}_0\text{c}{}^{\backprime}\breve{P}(x \mapsto y) = \nu +_c 1 .$$
$$[*90{\cdot}132 . *121{\cdot}11] \qquad \equiv . x (\breve{P})_\nu y : \supset \vdash . \text{Prop}$$

$*121{\cdot}27$. $\vdash : x P_\nu y . \supset . \nu, \nu +_c 1 \epsilon \text{NC} - \iota{}^{\backprime}\Lambda$

Dem.

$$\vdash . *121{\cdot}11 . *103{\cdot}12 . \supset \vdash : \text{Hp} . \supset . P(x \mapsto y) \epsilon \nu +_c 1 \quad (1)$$
$$\vdash . (1) . *110{\cdot}4{\cdot}42 . \supset \vdash . \text{Prop}$$

$*121{\cdot}271$. $\vdash : \sim(\nu, \nu +_c 1 \epsilon \text{NC} - \iota{}^{\backprime}\Lambda) . \supset . P_\nu = \dot{\Lambda}$ [$*121{\cdot}27 . \text{Transp}$]

***121·272.** $\vdash : \dot{\exists} ! P_\nu . \supset . \nu \geqslant 0 . \nu +_c 1 > 0 . \nu +_c 1 \geqslant 1$

Dem.

$\vdash . *117·5 . *121·27 . \supset \vdash : \text{Hp} . \supset . \nu \geqslant 0 .$ (1)

[*117·561.*110·641] $\supset . \nu +_c 1 \geqslant 1 .$ (2)

[*117·511·531] $\supset . \nu +_c 1 > 0$ (3)

$\vdash . (1) . (2) . (3) . \supset \vdash . \text{Prop}$

***121·273.** $\vdash : \dot{\exists} ! P_{\nu +_c 1} . \supset . \nu +_c 1 > 0$

Dem.

$\vdash . *121·27 . *110·4 . \supset \vdash : \text{Hp} . \supset . \nu \in \text{NC} - \iota`\Lambda .$

[*117·6] $\supset . \nu +_c 1 \geqslant 1 .$

[*117·511·531] $\supset . \nu +_c 1 > 0 : \supset \vdash . \text{Prop}$

***121·3.** $\vdash . P_0 \subset I \upharpoonright C`P$

Dem.

$\vdash . *121·11 . \supset \vdash : x P_0 y . \equiv . P(x \mapsto y) \in 1 .$

[*121·23] $\supset . x P_* y . x = y .$

[*90·12] $\supset . x (I \upharpoonright C`P) y : \supset \vdash . \text{Prop}$

***121·301.** $\vdash : \sim (x P_{\text{po}} x) . \supset : x P_0 y . \equiv . x \in C`P . x = y$

Dem.

$\vdash . *91·542·56 . \supset \vdash : x P_* z . z P_* x . x \neq z . \supset . x P_{\text{po}} x$ (1)

$\vdash . (1) . \text{Transp} . \supset \vdash :. \text{Hp} . \supset : x P_* z . z P_* x . \supset_{x,z} . x = z :$

[*121·231] $\supset : x \in C`P . \supset . P(x \mapsto x) = \iota`x :$

[*13·12.*52·22] $\supset : x \in C`P . x = y . \supset . P(x \mapsto y) \in 1 .$

[*121·11] $\supset . x P_0 y$ (2)

$\vdash . (2) . *121·3 . \supset \vdash . \text{Prop}$

***121·302.** $\vdash : P_{\text{po}} \subset J . \supset . P_0 = I \upharpoonright C`P$ [*121·301]

***121·303.** $\vdash : \text{Nc}`P(x \mapsto y) > 1 . \supset . x P_{\text{po}} y$

Dem.

$\vdash . *121·23 . *52·22 . *117·42 . \supset \vdash :. \text{Hp} . \supset : x \in P(x \mapsto y) . P(x \mapsto y) \neq \iota`x :$

[*51·4.Transp] $\supset : (\exists z) . z \neq x . z \in P(x \mapsto y) :$

[*121·103.*91·542] $\supset : (\exists z) . x P_{\text{po}} z . z P_* y :$

[*91·574] $\supset : x P_{\text{po}} y :. \supset \vdash . \text{Prop}$

***121·304.** $\vdash :. P_{\text{po}} \subset J . \supset : x P_1 y . \equiv . P(x \mapsto y) = \iota`x \cup \iota`y . x \neq y$

Dem.

$\vdash . *121·303·11 . \supset \vdash : \text{Hp} . x P_1 y . \supset . x P_{\text{po}} y .$

[Hp] $\supset . x \neq y$ (1)

$\vdash . (1) . *54·53·101 . *121·23·11 . \supset \vdash . \text{Prop}$

∗121·305. $\vdash : P_{po} \in J . \supset . P_1 \in P$

Dem.

$\vdash . *121·303 . \supset \vdash : \text{Hp} . xP_1y . \supset . xP_{po}y .$

$[*91·52] \qquad \supset . (\exists z) . xPz . zP_*y \qquad (1)$

$\vdash . *121·304 . *91·542 . \supset$

$\vdash :. \text{Hp} . xP_1y . \supset : xP_{po}z . zP_*y . \supset . z = y :$

$[*91·502] \qquad \supset : xPz . zP_*y . \supset . z = y \qquad (2)$

$\vdash . (1) . (2) . \supset \vdash . \text{Prop}$

∗121·306. $\vdash : P \epsilon 1 \to \text{Cls} . \sim (xP_{po}x) . xPy . \supset . P(x \mapsto y) = \iota`x \cup \iota`y . x \neq y$

Dem.

$\vdash . *91·542 . \supset \vdash : xP_*z . zP_*y . z \neq x . z \neq y . xPy . \supset : xP_{po}z . zP_{po}y . xPy :$

$[*34·1] \qquad \supset : xP_{po}z . zP_{po} | \breve{P}x :$

$[*92·11] \qquad \supset : P \epsilon 1 \to \text{Cls} . \supset . xP_{po}z . zP_*x :$

$[*91·574] \qquad \supset : P \epsilon 1 \to \text{Cls} . \supset . xP_{po}x \qquad (1)$

$\vdash . (1) . \text{Transp} . \supset \vdash :: \text{Hp} . \supset :. xP_*z . zP_*y . \supset_z : z = x . \vee . z = y \qquad (2)$

$\vdash . *121·23 . \quad \supset \vdash : \text{Hp} . \supset . x, y \epsilon P(x \mapsto y) \qquad (3)$

$\vdash . *91·502 . \quad \supset \vdash : \text{Hp} . \supset . x \neq y \qquad (4)$

$\vdash . (2) . (3) . (4) . *121·103 . \supset \vdash . \text{Prop}$

∗121·307. $\vdash : P \epsilon \text{Cls} \to 1 . \sim (yP_{po}y) . xPy . \supset . P(x \mapsto y) = \iota`x \cup \iota`y . x \neq y$
[∗121·306·143]

∗121·308. $\vdash : P \epsilon (1 \to \text{Cls}) \cup (\text{Cls} \to 1) . P_{po} \in J . \supset . P \in P_1$
[∗121·306·307·11 . ∗54·101]

∗121·31. $\vdash : P \epsilon (1 \to \text{Cls}) \cup (\text{Cls} \to 1) . P_{po} \in J . \supset . P_1 = P$ [∗121·305·308]

∗121·32. $\vdash . P_\nu \in P_*$

Dem.

$\vdash . *121·11 . *120·421 . *101·14 . \text{Transp} . \supset \vdash : xP_\nu y . \supset . \exists ! P(x \mapsto y) .$

$[*121·23] \qquad \supset . xP_*y : \supset \vdash . \text{Prop}$

If ν is not a cardinal, or if $\nu +_c 1 = \Lambda$, $P_\nu = \dot{\Lambda}$.

∗121·321. $\vdash : \nu > 0 . \supset . P_\nu \in P_{po}$

Dem.

$\vdash . *120·428 . *121·11 . \supset \vdash : \text{Hp} . xP_\nu y . \supset . \text{Nc}`P(x \mapsto y) > 1 .$

$[*117·55.*52·181.*121·23] \qquad \supset . (\exists z) . z \epsilon P(x \mapsto y) . z \neq x .$

$[*121·103.*91·542] \qquad \supset . (\exists z) . xP_{po}z . zR_*y .$

$[*91·574] \qquad \supset . xP_{po}y : \supset \vdash . \text{Prop}$

∗121·322. $\vdash . C`P_\nu \subset C`P$ [∗121·32 . ∗90·14]

∗121·323. $\vdash : \nu > 0 . \supset . \text{D}`P_\nu \subset \text{D}`P . \text{G}`P_\nu \subset \text{G}`P$ [∗121·321 . ∗91·504]

∗121·324. $\vdash . D`P_{\nu +_c 1} \subset D`P . ⅁`P_{\nu +_c 1} \subset ⅁`P$

Dem.

$\vdash . ∗121·273·323 . \supset \vdash : \dot{\exists} ! P_{\nu +_c 1} . \supset . D`P_{\nu +_c 1} \subset D`P . ⅁`P_{\nu +_c 1} \subset ⅁`P$ (1)

$\vdash . (1) . ∗33·241 . \supset \vdash .$ Prop

∗121·325. $\vdash : \dot{\exists} ! P_\mu \dot{\cap} P_\nu . \supset . \mu = \nu$

Dem.

$\vdash . ∗121·11 . \supset \vdash : \text{Hp} . \supset . \exists ! (\mu +_c 1) \cap (\nu +_c 1) \cap t_0`\mu .$

[∗100·42.∗110·4] $\supset . \exists ! (\mu +_c 1) \cap t_0`\mu . (\mu +_c 1) \cap t_0`\mu = \nu +_c 1 .$

[∗120·311] $\supset . \mu = \nu : \supset \vdash .$ Prop

∗121·326. $\vdash . \text{fin}`P \subset \text{finid}`P . \text{finid}`P - \iota`P_0 \subset \text{fin}`P$ [∗121·12·121]

∗121·327. $\vdash : \dot{\exists} ! P_0 . \supset . \text{fin}`P = \text{finid}`P - \iota`P_0$

Dem.

$\vdash . ∗121·325 . \text{Transp} . ∗121·121 . \supset \vdash :. \text{Hp} . \supset : R \epsilon \text{fin}`P . \supset . R \neq P_0$ (1)

$\vdash . (1) . ∗121·326 . \supset \vdash .$ Prop

∗121·33·331 are lemmas for ∗121·332, which is a very useful proposition.

∗121·33. $\vdash :. P \epsilon 1 \to \text{Cls} . \supset : z \epsilon P(x - y) . \equiv . z \epsilon P(x \dashv P`y) :$
$z \epsilon P(x \vdash y) . \equiv . z \epsilon P(x \vdash\!\dashv P`y)$

Dem.

$\vdash . ∗71·7 . \supset \vdash :. \text{Hp} . \supset : zP_*(P`y) . \equiv . zP_* | Py .$

[∗91·52] $\equiv . zP_{po} y$ (1)

$\vdash . (1) . ∗121·1·101·102·103 . \supset \vdash .$ Prop

From the above proposition it follows that

$$P \epsilon 1 \to \text{Cls} . y \epsilon ⅁`P . \supset . P(x - y) = P(x \dashv P`y) . P(x \vdash y) = P(x \vdash\!\dashv P`y).$$

This does not follow unless $y \epsilon ⅁`P$, because

$$P(x - y) = P(x \dashv P`y) . \supset . E ! P`y,$$

whereas $$z \epsilon P(x - y) . \equiv_z . z \epsilon P(x \dashv P`y)$$

will always be true if $y \sim\epsilon ⅁`P$, and therefore (when $P \epsilon 1 \to \text{Cls}$) if $\sim E ! P`y$.

∗121·331. $\vdash :. P \epsilon 1 \to \text{Cls} . P_{po} \in J . \supset : xP_\nu(P`y) . \equiv . xP_{\nu +_c 1} y$

Dem.

$\vdash . ∗121·324 . ∗71·16 . \supset \vdash :. \text{Hp} . \supset : xP_{\nu +_c 1} y . \supset . E ! P`y$ (1)

$\vdash . ∗121·33 . \qquad \supset \vdash : \text{Hp} . E ! P`y . \supset . P(x \vdash y) = P(x \vdash\!\dashv P`y)$ (2)

$\vdash . ∗121·242·32 . (2) . \supset \vdash : \text{Hp}(2) . xP_* y . \supset . P(x \vdash\!\dashv y) = P(x \vdash\!\dashv P`y) \cup \iota`y$ (3)

$\vdash . ∗91·52 . \qquad \supset \vdash : \text{Hp} . \supset . \sim(yP_* | Py) .$

[∗71·7] $\supset . \sim \{yP_*(P`y)\} .$

[∗121·103] $\supset . \sim \{y \epsilon P(x \vdash\!\dashv P`y)\}$ (4)

$\vdash . (3) . (4) . ∗110·63 . \supset \vdash : \text{Hp}(3) . \supset . \text{Nc}`P(x \vdash\!\dashv y) = \text{Nc}`P(x \vdash\!\dashv P`y) +_c 1$ (5)

$\vdash . (1) . (5) . *121{\cdot}11{\cdot}32 . \supset$

$\vdash : \mathrm{Hp} . xP_{\nu +_c 1}y . \supset . (\nu +_c 1) +_c 1 = \mathrm{Nc}{`}P(x \vdash\!\dashv P{`}y) +_c 1 .$

$[*120{\cdot}311 . *121{\cdot}27] \supset . \nu +_c 1 = \mathrm{Nc}{`}P(x \vdash\!\dashv P{`}y) .$

$[*121{\cdot}11] \qquad \supset . xP_\nu(P{`}y) \qquad (6)$

$\vdash . (5) . *14{\cdot}21 . *121{\cdot}11{\cdot}32 . \supset \vdash : \mathrm{Hp} . xP_\nu(P{`}y) . \supset . \mathrm{Nc}{`}P(x \vdash\!\dashv y) = (\nu +_c 1) +_c 1 .$

$[*121{\cdot}11] \qquad \supset . xP_{\nu +_c 1}y \qquad (7)$

$\vdash . (6) . (7) . \supset \vdash . \mathrm{Prop}$

***121·332.** $\vdash : P \in 1 \rightarrow \mathrm{Cls} . P_{\mathrm{po}} \mathrel{\dot\subset} J . \supset . P_{\nu +_c 1} = P_\nu \,|\, P \quad [*121{\cdot}331]$

***121·333.** $\vdash : P \in \mathrm{Cls} \rightarrow 1 . P_{\mathrm{po}} \mathrel{\dot\subset} J . \supset . P_{\nu +_c 1} = P \,|\, P_\nu$

***121·34.** $\vdash : P \in 1 \rightarrow \mathrm{Cls} . P_{\mathrm{po}} \mathrel{\dot\subset} J . \nu \in \mathrm{NC\,induct} . \supset . P_\nu \in 1 \rightarrow \mathrm{Cls}$

Dem.

$\vdash . *121{\cdot}3 . \quad \supset \vdash . P_0 \in 1 \rightarrow \mathrm{Cls} \qquad (1)$

$\vdash . *121{\cdot}332 . \supset \vdash :. \mathrm{Hp} . \supset : P_\nu \in 1 \rightarrow \mathrm{Cls} . \supset . P_{\nu +_c 1} \in 1 \rightarrow \mathrm{Cls} \qquad (2)$

$\vdash . (1) . (2) . *120{\cdot}11 . \supset \vdash . \mathrm{Prop}$

***121·341.** $\vdash : P \in \mathrm{Cls} \rightarrow 1 . P_{\mathrm{po}} \mathrel{\dot\subset} J . \nu \in \mathrm{NC\,induct} . \supset . P_\nu \in \mathrm{Cls} \rightarrow 1$

***121·342.** $\vdash : P \in 1 \rightarrow 1 . P_{\mathrm{po}} \mathrel{\dot\subset} J . \nu \in \mathrm{NC\,induct} . \supset . P_\nu \in 1 \rightarrow 1 \quad [*121{\cdot}34{\cdot}341]$

***121·35.** $\vdash : P \in 1 \rightarrow \mathrm{Cls} . P_{\mathrm{po}} \mathrel{\dot\subset} J . \mu, \nu \in \mathrm{NC\,induct} . \supset . P_\mu \,|\, P_\nu = P_{\mu +_c \nu}$

Dem.

$\vdash . *50{\cdot}62 . *121{\cdot}302{\cdot}322 . \supset \vdash : \mathrm{Hp} . \supset . P_\mu \,|\, P_0 = P_{\mu +_c 0} \qquad (1)$

$\vdash . *121{\cdot}332 . \qquad \supset \vdash :. \mathrm{Hp} . \supset : \mu, \nu \in \mathrm{NC\,induct} . P_\mu \,|\, P_\nu = P_{\mu +_c \nu} . \supset .$

$P_\mu \,|\, P_{\nu +_c 1} = P_{\mu +_c \nu} \,|\, P$

$[*121{\cdot}332] \qquad = P_{\mu +_c \nu +_c 1} \qquad (2)$

$\vdash . (1) . (2) . *120{\cdot}13 . \supset \vdash . \mathrm{Prop}$

***121·351.** $\vdash : P \in \mathrm{Cls} \rightarrow 1 . P_{\mathrm{po}} \mathrel{\dot\subset} J . \mu, \nu \in \mathrm{NC\,induct} . \supset . P_\mu \,|\, P_\nu = P_{\mu +_c \nu}$

***121·352.** $\vdash : P \in (1 \rightarrow \mathrm{Cls}) \cup (\mathrm{Cls} \rightarrow 1) . P_{\mathrm{po}} \mathrel{\dot\subset} J . \mu, \nu \in \mathrm{NC\,induct} . \supset .$

$P_\mu \,|\, P_\nu = P_\nu \,|\, P_\mu \quad [*121{\cdot}35{\cdot}351 . *110{\cdot}51]$

***121·36.** $\vdash : P \in (1 \rightarrow \mathrm{Cls}) \cup (\mathrm{Cls} \rightarrow 1) . P_{\mathrm{po}} \mathrel{\dot\subset} J . \mu, \nu \in \mathrm{NC\,induct} - \iota{`}0 . \supset .$

$(P_\mu)_\nu = P_{\mu \times_c \nu}$

Dem.

$\vdash . *121{\cdot}321 . \supset \vdash : \mathrm{Hp} . \supset . P_\mu \mathrel{\dot\subset} P_{\mathrm{po}} .$

$[*91{\cdot}59{\cdot}601] \qquad \supset . (P_\mu)_{\mathrm{po}} \mathrel{\dot\subset} J . \qquad (1)$

$[*121{\cdot}31{\cdot}34{\cdot}341] \qquad \supset . (P_\mu)_1 = P_\mu \qquad (2)$

$\vdash . *121{\cdot}332{\cdot}333{\cdot}352 . (1) . \supset$

$\vdash :. \mathrm{Hp} . \supset : (P_\mu)_{\nu +_c 1} = (P_\mu)_\nu \,|\, P_\mu :$

$[*34{\cdot}27] \supset : (P_\mu)_\nu = P_{\mu \times_c \nu} . \supset . (P_\mu)_{\nu +_c 1} = P_{\mu \times_c \nu} \,|\, P_\mu$

$[*121{\cdot}35{\cdot}351] \qquad = P_{(\mu \times_c \nu) +_c \mu}$

$[*113{\cdot}671] \qquad = P_{\mu \times_c (\nu +_c 1)} \qquad (3)$

$\vdash . (2) . (3) . *120{\cdot}47 . \supset \vdash . \mathrm{Prop}$

***121·361.** $\vdash : P \epsilon (1 \rightarrow \text{Cls}) \cup (\text{Cls} \rightarrow 1) . P_{\text{po}} \subseteq J . \mu, \nu \epsilon \text{NC induct} - \iota`0 . \supset .$
$(P_\mu)_\nu = (P_\nu)_\mu$ [*121·36 . *113·27]

***121·37.** $\vdash : P \epsilon \text{Cls} \rightarrow 1 . y \epsilon P(x \vdash\!\dashv z) . \supset . P(x \vdash\!\dashv z) = P(x \vdash\!\dashv y) \cup P(y \vdash\!\dashv z)$

Dem.

$\vdash . *121·103 . \supset \vdash : \text{Hp} . \supset . xP_*y . yP_*z$ (1)

$\vdash . (1) . *121·103 . \supset$

$\vdash :. \text{Hp} . \supset : w \epsilon P(x \vdash\!\dashv z) . \equiv . xP_*w . wP_*z . xP_*y . yP_*z$ (2)

$\vdash . *96·302 . \supset \vdash :: \text{Hp} . \supset :. xP_*w . xP_*y . \supset : wP_*y . \vee . yP_*w$ (3)

$\vdash . (2) . (3) . *4·73 . \supset$

$\vdash :: \text{Hp} . \supset :. w \epsilon P(x \vdash\!\dashv z) . \equiv : xP_*w . wP_*z . xP_*y . yP_*z . wP_*y . \vee .$
$xP_*w . wP_*z . xP_*y . yP_*z . yP_*w$ (4)

$\vdash . *90·17 . *4·73 . \supset \vdash : wP_*y . yP_*z . \equiv . wP_*z . wP_*y . yP_*z :$
$yP_*w . wP_*z . \equiv . yP_*z . wP_*z . yP_*w$ (5)

$\vdash . (4) . (5) . \supset \vdash :: \text{Hp} . \supset :. w \epsilon P(x \vdash\!\dashv z) . \equiv : xP_*w . xP_*y . yP_*z . wP_*y . \vee .$
$xP_*w . wP_*z . xP_*y . yP_*w :$

[*90·17.*4·73] $\equiv : xP_*w . wP_*y . yP_*z . \vee . xP_*y . yP_*w . wP_*z :$

[(1).*4·73] $\equiv : xP_*w . wP_*y . \vee . yP_*w . wP_*z :$

[*121·103] $\equiv : w \epsilon P(x \vdash\!\dashv y) \cup P(y \vdash\!\dashv z) :: \supset \vdash . \text{Prop}$

***121·371.** $\vdash : P \epsilon (\text{Cls} \rightarrow 1) \cup (1 \rightarrow \text{Cls}) . y \epsilon P(x \vdash\!\dashv z) . \supset .$
$P(x \vdash\!\dashv z) = P(x \vdash\!\dashv y) \cup P(y \vdash\!\dashv z) = P(x \vdash y) \cup P(y \vdash\!\dashv z)$
$= P(x \vdash\!\dashv y) \cup P(y \dashv z)$ [Proof as in *121·37]

***121·372.** $\vdash : P \epsilon (\text{Cls} \rightarrow 1) \cup (1 \rightarrow \text{Cls}) . y \epsilon P(x \dashv z) . \supset .$
$P(x \dashv z) = P(x \dashv y) \cup P(y \dashv z) = P(x \dashv y) \cup P(y \vdash\!\dashv z)$

***121·373.** $\vdash : P \epsilon (\text{Cls} \rightarrow 1) \cup (1 \rightarrow \text{Cls}) . y \epsilon P(x \vdash z) . \supset .$
$P(x \vdash z) = P(x \vdash y) \cup P(y \vdash z) = P(x \vdash\!\dashv y) \cup P(y \vdash z)$

***121·374.** $\vdash : P \epsilon (\text{Cls} \rightarrow 1) \cup (1 \rightarrow \text{Cls}) . y \epsilon P(x - z) . \supset .$
$P(x - z) = P(x \dashv y) \cup P(y - z) = P(x - y) \cup P(y \vdash z)$
$= P(x \dashv y) \cup P(y \vdash z)$

The proofs of these propositions are analogous to the proof of *121·37.

***121·38.** $\vdash : R \epsilon \text{Cls} \rightarrow 1 . xR_{\text{po}}x . \supset . R(x \vdash\!\dashv x) = \overleftarrow{R_*}`x$ [*97·5]

***121·381.** $\vdash : R \epsilon 1 \rightarrow \text{Cls} . xR_{\text{po}}x . \supset . R(x \vdash\!\dashv x) = \overrightarrow{R_*}`x$ [*97·501]

***121·382.** $\vdash : R \epsilon \text{Cls} \rightarrow 1 . xR_{\text{po}}x . xR_{\text{po}}y . \supset .$
$R(x \vdash\!\dashv x) = R(x \vdash\!\dashv y) = \overleftarrow{R_*}`x = R(y \vdash\!\dashv y)$ [*97·5 . *91·56]

***121·383.** $\vdash : R \epsilon 1 \rightarrow \text{Cls} . xR_{\text{po}}x . yR_{\text{po}}x . \supset .$
$R(x \vdash\!\dashv x) = R(y \vdash\!\dashv x) = \overrightarrow{R_*}`x = R(y \vdash\!\dashv y)$

***121·384.** $\vdash : R \epsilon (\text{Cls} \rightarrow 1) \cup (1 \rightarrow \text{Cls}) . xR_{\text{po}}x . y \epsilon R(x \vdash\!\dashv x) . \supset .$
$R(x \vdash\!\dashv x) = R(x \vdash\!\dashv y) = R(y \vdash\!\dashv x) = R(y \vdash\!\dashv y)$ [*121·382·383]

***121·39.** $\vdash :. R \epsilon \text{Cls} \rightarrow 1 . \supset : R(x \mapsto y) \subset R(x \mapsto z) . \vee . R(x \mapsto z) \subset R(x \mapsto y)$

Dem.

$\vdash . *96·302 . \quad \supset \vdash :. \text{Hp} . xR_* y . xR_* z . \supset : yR_* z . \vee . zR_* y \quad (1)$

$\vdash . *121·37 . \quad \supset \vdash : \text{Hp} . xR_* y . yR_* z . \supset . R(x \mapsto y) \subset R(x \mapsto z) \quad (2)$

$\vdash . *121·37 . \quad \supset \vdash : \text{Hp} . xR_* z . zR_* y . \supset . R(x \mapsto z) \subset R(x \mapsto y) \quad (3)$

$\vdash . (1) . (2) . (3) . \supset \vdash :. \text{Hp} . xR_* y . xR_* z . \supset :$

$R(x \mapsto y) \subset R(x \mapsto z) . \vee . R(x \mapsto z) \subset R(x \mapsto y) \quad (4)$

$\vdash . *121·23 . \quad \supset \vdash : \sim(xR_* y) . \supset . R(x \mapsto y) = \Lambda .$

$[*24·12] \quad \supset . R(x \mapsto y) \subset R(x \mapsto z) \quad (5)$

$\vdash . (5) \frac{z, y}{y, z} . \quad \supset \vdash : \sim(xR_* z) . \supset . R(x \mapsto z) \subset R(x \mapsto y) \quad (6)$

$\vdash . (4) . (5) . (6) . \supset \vdash . \text{Prop}$

The following series of propositions are concerned with proving *121·47, *i.e.*

$$R \epsilon (\text{Cls} \rightarrow 1) \cup (1 \rightarrow \text{Cls}) . \supset . R(x \mapsto z) \epsilon \text{Cls induct.}$$

The proof for $R \epsilon 1 \rightarrow \text{Cls}$ follows from that for $R \epsilon \text{Cls} \rightarrow 1$ by *121·143. Confining ourselves, therefore, to $R \epsilon \text{Cls} \rightarrow 1$, we proceed as follows.

We prove first that, starting from z and going backwards, each new step adds only one term (which may not be distinct from all its predecessors); *i.e.* we have

$$R \epsilon \text{Cls} \rightarrow 1 . xRy . yR_* z . \supset . R(x \mapsto z) = \iota`x \cup R(y \mapsto z).$$

From this it follows by induction that if $R(z \mapsto z)$ is an inductive class, so is $R(x \mapsto z)$. Thus we only have to prove that $R(z \mapsto z)$ is an inductive class. Here we must distinguish two cases, according as $\sim(zR_{\text{po}} z)$ or $zR_{\text{po}} z$. In the former case, we have

$$\exists ! R(z \mapsto z) . \supset . R(z \mapsto z) = \iota`z,$$

whence $R(z \mapsto z)$ is an inductive class, and therefore so is $R(x \mapsto z)$.

But in the latter case, when $zR_{\text{po}} z$, the matter is more difficult. In this case, z is a member of a cycle, the cycle being $R(z \mapsto z)$. We have to prove that this cycle must be an inductive class. Given $xR_* z$, x will be a member of this cycle if $xR_{\text{po}} x$, and may be at the end of the tail of a Q, if $\sim(xR_{\text{po}} x)$. (Cf. *96.)

By *96·453, we know that R is $1 \rightarrow 1$ when confined to $R(z \mapsto z)$. Hence

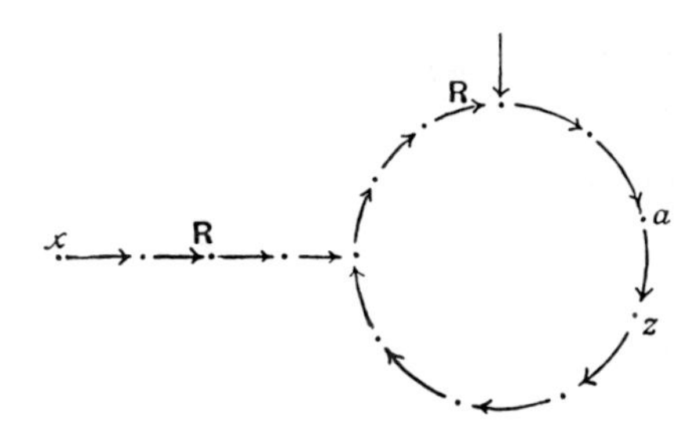

in $R(z \vdash\!\dashv z)$, z has a unique predecessor, say a. Assume $a \neq z$. We then imagine a barrier placed between a and z, *i.e.* we construct a relation S which is to hold between any two consecutive members of $R(z \vdash\!\dashv z)$ except a and z. Putting $\alpha = R(z \vdash\!\dashv z) - \iota`a$, we have $S = \alpha \upharpoonleft R$. Then the relation S generates an *open* series consisting of all the terms of $R(z \vdash\!\dashv z)$; *i.e.* we have

$$\sim(aS_{po}a) \,.\, S(z \vdash\!\dashv a) = R(z \vdash\!\dashv z).$$

Hence, by our previous case, since $S(z \vdash\!\dashv a)$ is an inductive class, so is $R(z \vdash\!\dashv z)$.

If $a = z$, then by $*96{\cdot}33$ the cycle reduces to the single term z, and therefore $R(z \vdash\!\dashv z)$ is still an inductive class.

Hence $R(z \vdash\!\dashv z)$, and therefore $R(x \vdash\!\dashv z)$, is always an inductive class when $R \,\epsilon\, \text{Cls} \to 1$, which was to be proved.

$*121{\cdot}4$. $\vdash : R \,\epsilon\, \text{Cls} \to 1 \,.\, xRy \,.\, yR_* z \,.\, \supset .\, R(x \vdash\!\dashv z) = \iota`x \cup R(y \vdash\!\dashv z)$

Dem.

$\vdash .\, *90{\cdot}311 \,.\, \supset \vdash :: \text{Hp} \,.\, \supset :.\, xR_* w \,.\, \equiv : x = w \,.\, \mathsf{v} \,.\, xR \,|\, R_* w :$

$[*71{\cdot}701 . \text{Hp}] \qquad \equiv : x = w \,.\, \mathsf{v} \,.\, yR_* w \qquad (1)$

$\vdash .\, *90{\cdot}172 \,.\, \supset \vdash :.\, \text{Hp} \,.\, \supset : x = w \,.\, \supset .\, wR_* z \qquad (2)$

$\vdash .\, (1) \,.\, (2) \,.\, \supset \vdash :: \text{Hp} \,.\, \supset :.\, xR_* w \,.\, wR_* z \,.\, \equiv : x = w \,.\, \mathsf{v} \,.\, yR_* w \,.\, wR_* z \qquad (3)$

$\vdash .\, (3) \,.\, *121{\cdot}103 \,.\, \supset \vdash .\, \text{Prop}$

$*121{\cdot}41$. $\vdash : R \,\epsilon\, \text{Cls} \to 1 \,.\, R(z \vdash\!\dashv z) \,\epsilon\, \text{Cls induct} \,.\, \supset .\, R(x \vdash\!\dashv z) \,\epsilon\, \text{Cls induct}$

Dem.

$\vdash .\, *121{\cdot}4 \,.\, *120{\cdot}251 \,.\, *90{\cdot}172 \,.\, \supset \vdash :.\, \text{Hp} \,.\, \supset :$

$\quad yR_* z \,.\, R(y \vdash\!\dashv z) \,\epsilon\, \text{Cls induct} \,.\, xRy \,.\, \supset .\, xR_* z \,.\, R(x \vdash\!\dashv z) \,\epsilon\, \text{Cls induct} \qquad (1)$

$\vdash .\, (1) \,.\, *90{\cdot}112 \dfrac{\breve{R}}{R} . \qquad \supset \vdash : \text{Hp} \,.\, xR_* z \,.\, \supset .\, R(x \vdash\!\dashv z) \,\epsilon\, \text{Cls induct} \qquad (2)$

$\vdash .\, *121{\cdot}23 \,.\, *120{\cdot}212 \,.\, \supset \vdash : \sim(xR_* z) \,.\, \supset .\, R(x \vdash\!\dashv z) \,\epsilon\, \text{Cls induct} \qquad (3)$

$\vdash .\, (2) \,.\, (3) \,.\, \supset \vdash .\, \text{Prop}$

In virtue of this proposition, we have only to prove $R(z \vdash\!\dashv z) \,\epsilon\, \text{Cls induct}$. This is obvious when $\sim(zR_{po}z)$, for then either $R(z \vdash\!\dashv z) = \iota`z$ or $R(z \vdash\!\dashv z) = \Lambda$. But when $zR_{po}z$, it is more difficult.

$*121{\cdot}42$. $\vdash : R \,\epsilon\, \text{Cls} \to 1 \,.\, \sim(zR_{po}z) \,.\, \supset .\, R(x \vdash\!\dashv z) \,\epsilon\, \text{Cls induct}$

Dem.

$\vdash .\, *121{\cdot}303 \,.\, \text{Transp} \,.\, *120{\cdot}441 \,.\, \supset \vdash : \text{Hp} \,.\, \supset .\, \text{Nc}`R(z \vdash\!\dashv z) \leqslant 1 \,.$

$[*120{\cdot}48] \qquad \supset .\, \text{Nc}`R(z \vdash\!\dashv z) \,\epsilon\, \text{NC induct} \,.$

$[*120{\cdot}211] \qquad \supset .\, R(z \vdash\!\dashv z) \,\epsilon\, \text{Cls induct} \qquad (1)$

$\vdash .\, (1) \,.\, *121{\cdot}41 \,.\, \supset \vdash .\, \text{Prop}$

$*121{\cdot}43$. $\vdash : R \in \text{Cls} \to 1 \,.\, zR_{po}z \,.\, \supset \,.\, E\,!\,\breve{\iota}`(\overrightarrow{R}`z \cap \overleftarrow{R_*}`z)$

Dem.

$\vdash \,.\, *91{\cdot}52 \,.\, \supset \vdash : \text{Hp} \,.\, \supset \,.\, (\exists a) \,.\, zR_*a \,.\, aRz$ (1)

$\vdash \,.\, *96{\cdot}453 \,.\, \supset \vdash : \text{Hp} \,.\, \supset \,.\, (\overleftarrow{R_*}`z) \upharpoonright R \in 1 \to 1 \,.$

$[*71{\cdot}122] \quad \supset \,.\, \hat{a}(zR_*a \,.\, aRz) \in 1 \cup \iota`\Lambda$ (2)

$\vdash \,.\, (1) \,.\, (2) \,.\, \supset \vdash : \text{Hp} \,.\, \supset \,.\, \hat{a}(zR_*a \,.\, aRz) \in 1 \,.$

$[*52{\cdot}15] \quad \supset \,.\, E\,!\,\breve{\iota}`(\overrightarrow{R}`z \cap \overleftarrow{R_*}`z) : \supset \vdash \,.\, \text{Prop}$

$*121{\cdot}431$. $\vdash : R \in \text{Cls} \to 1 \,.\, zR_{po}z \,.\, a = \breve{\iota}`(\overrightarrow{R}`z \cap \overleftarrow{R_*}`z) \,.\, \alpha = \overleftarrow{R_*}`z - \iota`a \,.$
$S = \alpha \upharpoonright R \,.\, \supset \,.\, \sim(aS_{po}a)$

Dem.

$\vdash \,.\, *35{\cdot}61 \,.\, \supset \vdash : \text{Hp} \,.\, \supset \,.\, a \sim\in \text{D}`S \,.$

$[*91{\cdot}504] \quad \supset \,.\, a \sim\in \text{D}`S_{po} \,.$

$[*33{\cdot}14] \quad \supset \,.\, \sim(aS_{po}a) : \supset \vdash \,.\, \text{Prop}$

$*121{\cdot}432$. $\vdash : \text{Hp}\, *121{\cdot}431 \,.\, \supset \,.\, S(z \mapsto a) \in \text{Cls induct}$

Dem.

$\vdash \,.\, *71{\cdot}261 \,.\, \supset \vdash : \text{Hp} \,.\, \supset \,.\, S \in \text{Cls} \to 1$ (1)

$\vdash \,.\, (1) \,.\, *121{\cdot}431{\cdot}42 \,.\, \supset \vdash \,.\, \text{Prop}$

$*121{\cdot}433$. $\vdash : \text{Hp}\, *121{\cdot}431 \,.\, z \neq a \,.\, \supset \,.\, S(z \mapsto a) = \overleftarrow{R_*}`z = R(z \mapsto z)$

Dem.

$\vdash \,.\, *96{\cdot}11 \,.\, \supset \vdash :. \text{Hp} \,.\, \supset : zS_*w \,.\, \supset \,.\, zR_*w$ (1)

$\vdash \,.\, *51{\cdot}3 \,.\, *91{\cdot}504 \,.\, \supset \vdash :. \text{Hp} \,.\, \supset : z \in \alpha \,.\, z \in \text{D}`R :$

$[*35{\cdot}61] \quad \supset : z \in \text{D}`S :$

$[*90{\cdot}12] \quad \supset : zS_*z$ (2)

$\vdash \,.\, (1) \,.\, *90{\cdot}16 \,.\, \supset \vdash :: \text{Hp} \,.\, \supset :. zS_*w \,.\, wRy \,.\, \supset : w \in \alpha \cup \iota`a \,.\, wRy :$

$[*35{\cdot}1] \quad \supset : wSy \,.\, \vee \,.\, w = a \,.\, wRy :$

$[\text{Hp}.*71{\cdot}171] \quad \supset : wSy \,.\, \vee \,.\, y = z :$

$[*90{\cdot}16{\cdot}17.(2)] \quad \supset : zS_*y$ (3)

$\vdash \,.\, (2) \,.\, (3) \,.\, *90{\cdot}112 \,.\, \supset \vdash :. \text{Hp} \,.\, \supset : zR_*w \,.\, \supset \,.\, zS_*w :$ (4)

$[\text{Hp}] \quad \supset : zS_*a$ (5)

$\vdash \,.\, *71{\cdot}171 \,.\, \supset \vdash : \text{Hp} \,.\, aRy \,.\, \supset \,.\, y = z$ (6)

$\vdash \,.\, *91{\cdot}542{\cdot}504 \,.\, *35{\cdot}61 \,.\, \supset \vdash :. \text{Hp} \,.\, \supset : wS_*a \,.\, w \neq a \,.\, wRy \,.\, \supset \,.\, wS_{po}a \,.\, wSy \,.$

$[*92{\cdot}111] \quad \supset \,.\, yS_*a$ (7)

$\vdash \,.\, (5) \,.\, (6) \,.\, (7) \,.\, \supset \vdash :. \text{Hp} \,.\, \supset : wS_*a \,.\, wRy \,.\, \supset \,.\, yS_*a$ (8)

$\vdash \,.\, (5) \,.\, (8) \,.\, *90{\cdot}112 \,.\, \supset \vdash :. \text{Hp} \,.\, \supset : zR_*y \,.\, \supset \,.\, yS_*a$ (9)

$\vdash \,.\, (4) \,.\, (9) \,.\, \supset \vdash :. \text{Hp} \,.\, \supset : zR_*y \,.\, \supset \,.\, zS_*y \,.\, yS_*a$ (10)

$\vdash \,.\, (1) \frac{y}{w} \,.\, (10) \,.\, \supset \vdash :. \text{Hp} \,.\, \supset : zS_*y \,.\, yS_*a \,.\, \equiv \,.\, zR_*y :$

$[*121{\cdot}103] \quad \supset : S(z \mapsto a) = \overleftarrow{R_*}`z$

$[*121{\cdot}38] \quad = R(z \mapsto z) :. \supset \vdash \,.\, \text{Prop}$

∗121·434. $\vdash : \text{Hp} *121\cdot431 \,.\, z = a \,.\, \supset .\, \overleftarrow{R_*}`z = R(z \mapsto z) = \iota`z$

Dem.

$\vdash . *32\cdot18 \,.\, \supset \vdash : \text{Hp} \,.\, \supset .\, zRz \,.$

$[*96\cdot33] \quad \supset .\, \overleftarrow{R_*}`z = \iota`z \,. \quad (1)$

$[*121\cdot38] \quad \supset .\, R(z \mapsto z) = \iota`z \quad (2)$

$\vdash . (1) . (2) . \supset \vdash . \text{Prop}$

∗121·44. $\vdash : R \in \text{Cls} \to 1 \,.\, zR_{\text{po}}z \,.\, \supset .\, R(z \mapsto z) \in \text{Cls induct}$

Dem.

$\vdash . *121\cdot43\cdot432\cdot433 \,.\, \supset$

$\vdash : \text{Hp} \,.\, z \neq \iota`(\breve{\overrightarrow{R}}`z \cap \overleftarrow{R_*}`z) \,.\, \supset .\, R(z \mapsto z) \in \text{Cls induct} \quad (1)$

$\vdash . *121\cdot434 \,.\, *120\cdot213 \,.\, \supset$

$\vdash : \text{Hp} \,.\, z = \iota`(\breve{\overrightarrow{R}}`z \cap \overleftarrow{R_*}`z) \,.\, \supset .\, R(z \mapsto z) \in \text{Cls induct} \quad (2)$

$\vdash . (1) . (2) . \supset \vdash . \text{Prop}$

∗121·441. $\vdash : R \in \text{Cls} \to 1 \,.\, zR_{\text{po}}z \,.\, \supset .\, R(x \mapsto z) \in \text{Cls induct}$ [∗121·44·41]

∗121·45. $\vdash : R \in \text{Cls} \to 1 \,.\, \supset .\, R(x \mapsto z) \in \text{Cls induct}$ [∗121·42·441]

∗121·46. $\vdash : R \in 1 \to \text{Cls} \,.\, \supset .\, R(x \mapsto z) \in \text{Cls induct}$ [∗121·45·143]

∗121·47. $\vdash : R \in (\text{Cls} \to 1) \cup (1 \to \text{Cls}) \,.\, \supset .\, R(x \mapsto z) \in \text{Cls induct}$ [∗121·45·46]

∗121·48. $\vdash :. R \in \text{Cls} \to 1 \,.\, \supset :$

$$\text{Nc}`R(x \mapsto y) < \text{Nc}`R(x \mapsto z) \,.\, \equiv .\, \exists ! R(x \mapsto z) - R(x \mapsto y)$$

Dem.

$\vdash . *121\cdot39 \,.\, \supset \vdash :. \text{Hp} \,.\, \supset : \exists ! R(x \mapsto z) - R(x \mapsto y) \,.\, \equiv .$

$\quad R(x \mapsto y) \subset R(x \mapsto z) \,.\, R(x \mapsto y) \neq R(x \mapsto z) \,.$

$[*120\cdot7 . *121\cdot45] \quad \supset .\, \text{Nc}`R(x \mapsto y) < \text{Nc}`R(x \mapsto z) \quad (1)$

$\vdash . *117\cdot222\cdot29 \,.\, \supset \vdash : \text{Nc}`R(x \mapsto y) < \text{Nc}`R(x \mapsto z) \,.\, \supset .$

$\quad \sim \{R(x \mapsto z) \subset R(x \mapsto y)\} \,.$

$[*24\cdot55] \quad \supset .\, \exists ! R(x \mapsto z) - R(x \mapsto y) \quad (2)$

$\vdash . (1) . (2) . \supset \vdash . \text{Prop}$

∗121·481. $\vdash :. R \in \text{Cls} \to 1 \,.\, \supset : \text{Nc}`R(x \mapsto y) \leqslant \text{Nc}`R(x \mapsto z) \,.\, \equiv .$

$$R(x \mapsto y) \subset R(x \mapsto z)$$

Dem.

$\vdash . *121\cdot45 \,.\, *120\cdot441 \,.\, \supset$

$\vdash :. \text{Hp} \,.\, \supset : \text{Nc}`R(x \mapsto y) \leqslant \text{Nc}`R(x \mapsto z) \,.\, \equiv .\, \sim \{\text{Nc}`R(x \mapsto z) < \text{Nc}`R(x \mapsto y)\} \,.$

$[*121\cdot48] \quad \equiv .\, \sim \exists ! R(x \mapsto y) - R(x \mapsto z) \,.$

$[*24\cdot55] \quad \equiv .\, R(x \mapsto y) \subset R(x \mapsto z) :. \supset \vdash . \text{Prop}$

The above proposition is used in the proof of ∗122·35, which is an important proposition in the theory of progressions.

The following propositions are concerned with the identification of such relations as P_ν with powers of P in the sense of ∗91.

***121·5.** $\vdash : P \epsilon (\text{Cls} \to 1) \cup (1 \to \text{Cls}) \,.\, P_{\text{po}} \subseteq J \,.\, \supset .$
$\text{finid}'P = \text{Potid}'P \,.\, \text{fin}'P = \text{Pot}'P$

Dem.

$\vdash . *121{\cdot}302{\cdot}31 . \supset \vdash : \text{Hp} . \supset . P_0 = I \upharpoonright C'P \,.\, P_1 = P$ (1)

$\vdash . (1) . *121{\cdot}332{\cdot}333{\cdot}352 . \supset \vdash :. \text{Hp} . \nu \epsilon \text{NC induct} . \supset : P_{\nu +_c 1} = P_\nu | P :$ (2)

[*91·341] $\supset : P_\nu \epsilon \text{Potid}'P . \supset . P_{\nu +_c 1} \epsilon \text{Potid}'P : P_\nu \epsilon \text{Pot}'P . \supset . P_{\nu +_c 1} \epsilon \text{Pot}'P$ (3)

$\vdash . (1) . *91{\cdot}35 . \supset \vdash : \text{Hp} . \supset . P_0 \epsilon \text{Potid}'P \,.\, P_1 \epsilon \text{Pot}'P$ (4)

$\vdash . (3) . (4) . *120{\cdot}13{\cdot}47 . \supset \vdash :. \text{Hp} . \supset : \nu \epsilon \text{NC induct} . \supset . P_\nu \epsilon \text{Potid}'P :$
$\nu \epsilon \text{NC induct} - \iota'0 . \supset . P_\nu \epsilon \text{Pot}'P :$

[*121·12·121] $\supset : \text{finid}'P \subset \text{Potid}'P \,.\, \text{fin}'P \subset \text{Pot}'P$ (5)

$\vdash . (2) . *121{\cdot}121 . \supset \vdash :. \text{Hp} . \supset : \nu \epsilon \text{NC induct} . \supset . P_\nu | P \epsilon \text{fin}'P :$

[*121·12] $\supset : Q \epsilon \text{finid}'P . \supset . Q | P \epsilon \text{fin}'P :$

[(1).*91·17·171] $\supset : \text{Potid}'P \subset \text{finid}'P \,.\, \text{Pot}'P \subset \text{fin}'P$ (6)

$\vdash . (5) . (6) . \supset \vdash . \text{Prop}$

***121·501.** $\vdash : P \epsilon (\text{Cls} \to 1) \cup (1 \to \text{Cls}) \,.\, P_{\text{po}} \subseteq J \,.\, \dot{\exists} ! P . \supset .$
$\text{Pot}'P = \text{finid}'P - \iota'P_0 = \text{fin}'P$

Dem.

$\vdash . *121{\cdot}302 . \supset \vdash : \text{Hp} . \supset . \dot{\exists} ! P_0$ (1)

$\vdash . (1) . *121{\cdot}5{\cdot}327 . \supset \vdash . \text{Prop}$

***121·502.** $\vdash : P \epsilon (\text{Cls} \to 1) \cup (1 \to \text{Cls}) \,.\, P_{\text{po}} \subseteq J \,.\, \supset .$
$\dot{s}'(\text{finid}'P - \iota'P_0) = P_{\text{po}} = \dot{s}'\text{fin}'P$

Dem.

$\vdash . *91{\cdot}504 . *33{\cdot}24 . *121{\cdot}5 . \supset \vdash : P = \dot{\Lambda} . \supset . \dot{s}'(\text{finid}'P - \iota'P_0) = \dot{\Lambda} = P_{\text{po}}$ (1)

$\vdash . (1) . *121{\cdot}501{\cdot}5 . \supset \vdash . \text{Prop}$

***121·51.** $\vdash : P \epsilon (\text{Cls} \to 1) \cup (1 \to \text{Cls}) \,.\, P_{\text{po}} \subseteq J \,.\, \supset . P_2 = P^2 \,.\, P_3 = P^3 \,.\, \text{etc.}$

Dem.

$\vdash . *121{\cdot}31 . \quad \supset \vdash : \text{Hp} . \supset . P_1 = P$ (1)

$\vdash . *121{\cdot}332{\cdot}333 . \quad \supset \vdash : \text{Hp} . \supset . P_2 = P_1 | P_1$

[(1)] $= P^2$ (2)

$\vdash . *121{\cdot}332{\cdot}333{\cdot}352 . \supset \vdash : \text{Hp} . \supset . P_3 = P_2 | P_1$

[(1).(2)] $= P^3$ (3)

$\vdash . (2) . (3) . \text{etc.} . \supset \vdash . \text{Prop}$

***121·52.** $\vdash : P \epsilon (\text{Cls} \to 1) \cup (1 \to \text{Cls}) \,.\, P_{\text{po}} \subseteq J \,.\, \supset . \dot{s}'\text{finid}'P = P_*$
[*121·5 . *91·55]

We shall at a later stage (*301) give a general definition of P^ν. When this definition has been introduced, we shall be able to prove, with the hypothesis of *121·51,

$$\nu \epsilon \text{NC induct} . \supset . P_\nu = P^\nu.$$

The definition of P^ν is postponed on account of various complications which render a general definition of P^ν difficult. The chief difficulty arises when

$\dot{\exists} ! P \dot{\wedge} I$. Thus suppose we have yPy; we shall also have yP^2y, yP^3y, etc. Hence if we have xPy, we have

$$\nu \in \text{NC induct} - \iota`0 . \supset_\nu . xP^\nu y.$$

Again, suppose this case excluded, but suppose

$$(\exists \mu, y) . \mu \in \text{NC induct} . y \in P(x \mapsto z) . yP^\mu y.$$

Then we shall have

$$\nu \in \text{NC induct} - \iota`0 - \iota`\Lambda . \supset . yP^{\mu \times_c \nu} y.$$

Thus the general definition of P^ν has to be complicated, except when $P_{po} \subseteq J$.

The following propositions are concerned with the series of relations P_ν and the series of terms ν_P. The relation P_ν holds between two terms (roughly speaking) when it requires ν steps to get from the first to the second; the term ν_P is the νth term starting from $B`P$, which, when it exists, is $1_P$. In order that $\nu_P$ should exist, it is necessary that $B`P$ should exist, and that there should be just one term x in the field of P such that the interval from $B`P$ to $x$ (both included) consists of $\nu$ terms. When this is the case for all inductive cardinals from 1 to $\nu$, we can say that $P$ generates a series starting from $B`P$ and having at least ν terms, each correlated with one of the cardinals in the interval from 1 to ν, both included; *i.e.* the series has a μth term, whenever $1 \leqslant \mu \leqslant \nu$. If this holds for all inductive values of ν, the family of $B`P$ is a progression*. (It will be observed that all such terms as $\nu_P$ belong to the family of $B`P$, which need not form the whole field of P.)

***121·6.** $\vdash :. \nu \neq 0 . \supset . f(\nu_P) . \equiv . f[\iota`\breve{y} \{\text{N}_0\text{c}`P(B`P \mapsto y) = \nu\}]$

Dem.

$\vdash . *121·11 . *120·414·416 . \supset \vdash :. \text{Hp} . \supset :$

$$f[\iota`\breve{y} \{\text{Nc}`P(B`P \mapsto y) = \nu\}] . \equiv . f[\iota`\breve{y} \{(B`P) P_{\nu -_c 1} y\}] .$$

$[*121·13]$ $\equiv . f(\nu_P) : \supset \vdash . \text{Prop}$

***121·601.** $\vdash : \text{E} ! B`P . \supset . B`P = 1_P . \sim \{(B`P) P_{po} (B`P)\}$

Dem.

$\vdash . *91·504 . *93·1 . \supset \vdash . \sim \{(B`P) P_{po} (B`P)\} .$ (1)

$[*121·301]$ $\supset \vdash :. \text{E} ! B`P . \supset : (B`P) P_0 y . \equiv_y . B`P = y :$

$[*31·17]$ $\supset : B`P = \breve{P}_0`B`P :$

$[*121·13]$ $\supset : B`P = 1_P$ (2)

$\vdash . (1) . (2) . \supset \vdash . \text{Prop}$

***121·602.** $\vdash : \text{E} ! B`P . P \in 1 \to 1 . \supset . \breve{P}`B`P = 2_P$

Dem.

$\vdash . *121·306·601 . \supset \vdash : \text{Hp} . \supset . P(B`P \mapsto \breve{P}`B`P) \in 2$ (1)

* Cf. *122, below.

$\vdash . *121{\cdot}23{\cdot}601 . \supset \vdash :: \text{Hp} . \supset :. (B\text{‘}P) P_{\text{po}} y . \supset . B\text{‘}P, y \in P(B\text{‘}P \mapsto y) . B\text{‘}P \neq y :.$

$[*54{\cdot}53 . *121{\cdot}303] \supset :. P(B\text{‘}P \mapsto y) \in 2 . \supset : P(B\text{‘}P \mapsto y) = \iota\text{‘}B\text{‘}P \cup \iota\text{‘}y . (B\text{‘}P) P_{\text{po}} y :$

$[*92{\cdot}111] \qquad \supset : (\breve{P}\text{‘}B\text{‘}P) P_* y . P(B\text{‘}P \mapsto y) = \iota\text{‘}B\text{‘}P \cup \iota\text{‘}y :$

$[*121{\cdot}103{\cdot}601] \qquad \supset : \breve{P}\text{‘}B\text{‘}P \in \iota\text{‘}B\text{‘}P \cup \iota\text{‘}y . \breve{P}\text{‘}B\text{‘}P \neq B\text{‘}P :$

$[*51{\cdot}232] \qquad \supset : y = \breve{P}\text{‘}B\text{‘}P \qquad (2)$

$\vdash . (1) . (2) . *121{\cdot}6 . \supset \vdash . \text{Prop}$

$*121{\cdot}61$. $\vdash : P \in 1 \to \text{Cls} . P_{\text{po}} \subseteq J . x \in s\text{‘gen‘}P . \supset .$
$$(\exists a, \nu) . aBP . \nu \in \text{NC induct} . aP_\nu x$$

Dem.

$\vdash . *93{\cdot}36 . \supset \vdash :. P \in 1 \to \text{Cls} . x \in s\text{‘gen‘}P . \supset . (\exists a) . aBP . aP_* x \qquad (1)$

$\vdash . *121{\cdot}52 . \supset \vdash :. P \in 1 \to \text{Cls} . P_{\text{po}} \subseteq J . \supset : aP_* x . \equiv . a(\dot{s}\text{‘finid‘}P) x \qquad (2)$

$\vdash . (1) . (2) . \supset \vdash : \text{Hp} . \supset . (\exists a) . aBP . a(\dot{s}\text{‘finid‘}P) x .$

$[*121{\cdot}12] \qquad \supset . (\exists a, \nu) . aBP . \nu \in \text{NC induct} . aP_\nu x : \supset \vdash . \text{Prop}$

$*121{\cdot}62$. $\vdash : P \in \text{Cls} \to 1 . P_{\text{po}} \subseteq J . (B\text{‘}P) P_* x . \supset .$
$$(\exists \nu) . \nu \in \text{NC induct} - \iota\text{‘}0 . x = \nu_P$$

Dem.

$\vdash . *121{\cdot}52 . \supset \vdash : \text{Hp} . \supset . (B\text{‘}P)(\dot{s}\text{‘finid‘}P) x .$

$[*121{\cdot}12] \qquad \supset . (\exists \nu) . \nu \in \text{NC induct} . (B\text{‘}P) P_\nu x \qquad (1)$

$\vdash . *121{\cdot}341 . \supset \vdash : \text{Hp} . \nu \in \text{NC induct} . \supset . P_\nu \in \text{Cls} \to 1 \qquad (2)$

$\vdash . (1) . (2) . \supset \vdash : \text{Hp} . \supset . (\exists \nu) . \nu \in \text{NC induct} . x = \breve{P}_\nu\text{‘}B\text{‘}P .$

$[*121{\cdot}13] \qquad \supset . (\exists \nu) . \nu \in \text{NC induct} . x = (\nu +_c 1)_P .$

$[*120{\cdot}471] \qquad \supset . (\exists \mu) . \mu \in \text{NC induct} - \iota\text{‘}0 . x = \mu_P : \supset \vdash . \text{Prop}$

$*121{\cdot}63$. $\vdash : \text{E} ! \nu_P . \supset . \text{N}_0\text{c‘}P(B\text{‘}P \mapsto \nu_P) = \nu$

Dem.

$\vdash . *121{\cdot}13{\cdot}131 . \supset \vdash : \text{Hp} . \supset . (B\text{‘}P) P_{\nu -_c 1} \nu_P .$

$[*121{\cdot}11] \qquad \supset . \text{N}_0\text{c‘}P(B\text{‘}P \mapsto \nu_P) = \nu : \supset \vdash . \text{Prop}$

$*121{\cdot}631$. $\vdash :. P \in \text{Cls} \to 1 . P_{\text{po}} \subseteq J . \nu \in \text{NC induct} - \iota\text{‘}0 . \supset :$
$$\text{N}_0\text{c‘}P(B\text{‘}P \mapsto y) = \nu . \equiv . y = \nu_P . \equiv . (B\text{‘}P) P_{\nu -_c 1} y$$

Dem.

$\vdash . *120{\cdot}414{\cdot}416 . *121{\cdot}11 . \supset$

$\vdash :. \text{Hp} . \supset : \text{N}_0\text{c‘}P(B\text{‘}P \mapsto y) = \nu . \equiv . (B\text{‘}P) P_{\nu -_c 1} y . \qquad (1)$

$[*121{\cdot}341] \qquad \equiv . y = \breve{P}_{\nu -_c 1}\text{‘}B\text{‘}P .$

$[*121{\cdot}13] \qquad \equiv . y = \nu_P \qquad (2)$

$\vdash . (1) . (2) . \supset \vdash . \text{Prop}$

$*121{\cdot}632{\cdot}633$ are required for proving $*121{\cdot}634$.

***121·632.** $\vdash : P \in \text{Cls} \to 1 \,.\, P_{\text{po}} \mathrel{\dot\subset} J \,.\, \nu \in \text{NC induct} - \iota\text{`}0 \,.\, y = \nu_P \,.\, yPz \,.\supset .\, z = (\nu +_c 1)_P$

Dem.

$\vdash . *121\cdot13 . \supset \vdash : \text{Hp} . \supset . (B\text{`}P)\, P_{\nu -_c 1}\, y \,.\, yPz \,.$

$[*121\cdot333\cdot352] \qquad \supset . (B\text{`}P)\, P_\nu z \,.$

$[*121\cdot631] \qquad \supset . z = (\nu +_c 1)_P : \supset \vdash . \text{Prop}$

***121·633.** $\vdash : P \in \text{Cls} \to 1 \,.\, P_{\text{po}} \mathrel{\dot\subset} J \,.\, \nu \in \text{NC induct} - \iota\text{`}0 \,.\, \nu_P \in \text{D`}P \,.\supset .$
$\qquad E\,!\,(\nu +_c 1)_P \,.\, (\nu +_c 1)_P = \breve{P}\text{`}\nu_P$

$[*121\cdot632]$

***121·634.** $\vdash :. P \in \text{Cls} \to 1 \,.\, P_{\text{po}} \mathrel{\dot\subset} J \,.\, \nu \in \text{NC induct} - \iota\text{`}0 \,.\supset : \nu_P \in \text{D`}P \,.\equiv .\, E\,!\,(\nu +_c 1)_P$

$[*121\cdot633\cdot631\cdot333\cdot352]$

***121·635.** $\vdash : P \in \text{Cls} \to 1 \,.\, P_{\text{po}} \mathrel{\dot\subset} J \,.\, E\,!\,\nu_P \,.\supset .\, \nu \in \text{NC induct} - \iota\text{`}0$

Dem.

$\vdash . *121\cdot63\cdot45 . \supset \vdash : \text{Hp} . \supset . \nu \in \text{NC induct} \qquad (1)$

$\vdash . *121\cdot13 . \quad \supset \vdash : E\,!\,\nu_P . \supset . \dot{\exists}\,!\, P_{(\nu -_c 1)} \,.$

$[*121\cdot272] \qquad \supset . (\nu -_c 1) +_c 1 > 0 \,.$

$[*120\cdot416] \qquad \supset . \nu > 0 \qquad (2)$

$\vdash . (1) . (2) . \supset \vdash . \text{Prop}$

***121·636.** $\vdash : P \in \text{Cls} \to 1 \,.\, P_{\text{po}} \mathrel{\dot\subset} J \,.\, E\,!\,\nu_P \,.\, \sim E\,!\,(\nu +_c 1)_P \,.\supset .$
$\qquad \overleftarrow{P_*}\text{`}B\text{`}P = P\,(B\text{`}P \mapsto \nu_P) \,.\, \text{N}_0\text{c`}\overleftarrow{P_*}\text{`}B\text{`}P = \nu$

Dem.

$\vdash . *121\cdot635 . \qquad \supset \vdash : \text{Hp} . \supset . \nu \in \text{NC induct} - \iota\text{`}0 \,. \qquad (1)$

$[*121\cdot634 . \text{Hp}] \qquad \supset . \nu_P \sim\in \text{D`}P \qquad (2)$

$\vdash . (1) . *121\cdot63 . \quad \supset \vdash :: \text{Hp} . \supset :. \dot{\exists}\,!\, P\,(B\text{`}P \mapsto \nu_P) :.$

$[*121\cdot23] \qquad \supset :. (B\text{`}P)\, P_* \nu_P :.$

$[*96\cdot302 . *91\cdot542] \qquad \supset :. (B\text{`}P)\, P_* z . \supset : z P_* \nu_P . \vee . \nu_P P_{\text{po}} z \qquad (3)$

$\vdash . (2) . (3) . *91\cdot504 . \supset \vdash :. \text{Hp} . \supset : (B\text{`}P)\, P_* z . \supset . z P_* \nu_P :$

$[*4\cdot71] \qquad \supset : (B\text{`}P)\, P_* z . \equiv . (B\text{`}P)\, P_* z \,.\, z P_* \nu_P :$

$[*121\cdot103] \qquad \supset : \overleftarrow{P_*}\text{`}B\text{`}P = P\,(B\text{`}P \mapsto \nu_P) : \qquad (4)$

$[*121\cdot63] \qquad \supset : \text{N}_0\text{c`}\overleftarrow{P_*}\text{`}B\text{`}P = \nu \qquad (5)$

$\vdash . (4) . (5) . \supset \vdash . \text{Prop}$

***121·637.** $\vdash : E\,!\,\nu_P . \supset . \nu_P \in C\text{`}P$

Dem.

$\vdash . *121\cdot13 . *14\cdot28 . \supset \vdash : E\,!\,\nu_P . \equiv . \nu_P = \breve{P}_{\nu -_c 1}\text{`}B\text{`}P \,.$

$[*121\cdot322] \qquad \supset . \nu_P \in C\text{`}P : \supset \vdash . \text{Prop}$

***121·638.** $\vdash :. E!(\nu +_c 1)_P . \supset : (B'P) P_\nu x . \equiv . x = (\nu +_c 1)_P : (\nu +_c 1) -_c 1 = \nu$

Dem.

$\vdash . *121\cdot13 . \supset \vdash : E!(\nu +_c 1)_P . \equiv . E! \breve{P}_{(\nu +_c 1) -_c 1} ` B'P .$ (1)

[*121·272] $\supset . (\nu +_c 1) -_c 1 \geqslant 0 .$

[*14·21] $\supset . E!(\nu +_c 1) -_c 1 .$

[*14·22.(*120·411)] $\supset . (\nu +_c 1) -_c 1 = \nu$ (2)

$\vdash . (2) . \supset \vdash :. Hp . \supset : (B'P) P_\nu x . \equiv . (B'P) P_{(\nu +_c 1) -_c 1} x .$

[(1).*30·4] $\equiv . x = \breve{P}_{(\nu +_c 1) -_c 1} ` B'P .$

[*121·13] $\equiv . x = (\nu +_c 1)_P$ (3)

$\vdash . (3) . (2) . \supset \vdash .$ Prop

***121·64.** $\vdash : P \in Cls \to 1 . P_{po} \subseteq J . \nu \in NC\,induct - \iota`0 . Nc`\overleftarrow{P_*}`B'P \geqslant \nu . \supset . E!\nu_P$

Dem.

$\vdash . *121\cdot636 . \supset \vdash :. Hp . E!\nu_P . \supset : \sim E!(\nu +_c 1)_P . \supset . N_0c`\overleftarrow{P_*}`B'P = \nu$ (1)

$\vdash . *120\cdot428 . \supset \vdash : \nu \in NC\,induct . \exists ! \nu +_c 1 . \supset . \nu +_c 1 > \nu .$

[*117·281] $\supset . \sim(\nu \geqslant \nu +_c 1)$ (2)

$\vdash . *117\cdot15 . \supset \vdash : \sim \exists ! \nu +_c 1 . \supset . \sim(\nu \geqslant \nu +_c 1)$ (3)

$\vdash . (2) . (3) . \supset \vdash : \nu \in NC\,induct . \supset . \sim(\nu \geqslant \nu +_c 1)$ (4)

$\vdash . (1) . (4) . \supset \vdash :. Hp . E!\nu_P . \supset :$

$\sim E!(\nu +_c 1)_P . \supset . \sim(Nc`\overleftarrow{P_*}`B'P \geqslant \nu +_c 1) :$

[Transp] $\supset : Nc`\overleftarrow{P_*}`B'P \geqslant \nu +_c 1 . \supset . E!(\nu +_c 1)_P$ (5)

$\vdash . (5) . Syll . *117\cdot6 . \supset \vdash :. Hp : Nc`\overleftarrow{P_*}`B'P \geqslant \nu . \supset . E!\nu_P : \supset :$

$Nc`\overleftarrow{P_*}`B'P \geqslant \nu +_c 1 . \supset . E!(\nu +_c 1)_P$ (6)

$\vdash . *14\cdot21 . *121\cdot601 . \supset \vdash : Nc`\overleftarrow{P_*}`B'P \geqslant 1 . \supset . E!1_P$ (7)

$\vdash . (6) . (7) . *120\cdot473 . \supset$

$\vdash :. Hp . \supset : Nc`\overleftarrow{P_*}`B'P \geqslant \nu . \supset . E!\nu_P :. \supset \vdash .$ Prop

***121·641.** $\vdash :. P \in Cls \to 1 . P_{po} \subseteq J . \nu \in NC\,induct - \iota`0 . \supset :$

$Nc`\overleftarrow{P_*}`B'P \geqslant \nu . \equiv . E!\nu_P$

[*121·64·63·32]

***121·65.** $\vdash : P \in Cls \to 1 . P_{po} \subseteq J . \mu \neq 0 . E!(\mu +_c \nu)_P . \supset . \mu_P P_\nu (\mu +_c \nu)_P$

Dem.

$\vdash . *121\cdot631\cdot635\cdot64 . *120\cdot452 . \supset$

$\vdash : Hp . \supset . (B'P) P_{\mu -_c 1} \mu_P . (B'P) P_{\mu +_c \nu -_c 1} (\mu +_c \nu)_P .$

[*121·351.*120·424] $\supset . (B'P) P_{\mu -_c 1} \mu_P . (B'P)(P_{\mu -_c 1} | P_\nu)(\mu +_c \nu)_P .$

[*121·341.*72·591] $\supset . \mu_P P_\nu (\mu +_c \nu)_P : \supset \vdash .$ Prop

***121·66.** $\vdash : P \epsilon \text{Cls} \to 1 \,.\, P_{\text{po}} \subseteq J \,.\, \text{Nc}`P(B`P \mapsto x) > \nu \,.\, \supset \,.\, x \epsilon \text{C}`P_\nu$

Dem.

$\vdash \,.\, *121·45 \,.\, *120·48 \,.\, \supset \vdash : \text{Hp} \,.\, \supset \,.\, \nu \epsilon \text{NC induct} \,.$

[*120·429] $\supset \,.\, \text{Nc}`P(B`P \mapsto x) \geqslant \nu +_c 1 \,.$

[*117·31] $\supset \,.\, (\exists \mu) \,.\, \text{Nc}`P(B`P \mapsto x) = \nu +_c 1 +_c \mu \,.$

[*121·11] $\supset \,.\, (\exists \mu) \,.\, (B`P) P_{\nu +_c \mu} x \,.$

[*121·351·352] $\supset \,.\, (\exists \mu) \,.\, (B`P)(P_\mu \,|\, P_\nu) x \,.$

[*34·36] $\supset \,.\, x \epsilon \text{C}`P_\nu : \supset \vdash \,.\, \text{Prop}$

The following proposition is used in *122·38·381.

***121·7.** $\vdash : R \epsilon 1 \to 1 \,.\, aBR \,.\, aR_* x \,.\, \supset \,.\, \overrightarrow{R_*}`x = R(a \mapsto x) \,.\, \overrightarrow{R_*}`x \epsilon \text{Cls induct}$

Dem.

$\vdash \,.\, *96·25 \,.\, \supset \vdash :. \text{Hp} \,.\, \supset : yR_* x \,.\, \supset \,.\, aR_* y :$

[*4·71] $\supset : yR_* x \,.\, \equiv \,.\, aR_* y \,.\, yR_* x :$

[*121·103] $\supset : \overrightarrow{R_*}`x = R(a \mapsto x)$ (1)

$\vdash \,.\, (1) \,.\, *121·45 \,.\, \supset \vdash \,.\, \text{Prop}$

***121·71.** $\vdash :. R \epsilon 1 \to 1 : x \epsilon s`\text{gen}`R \,.\, \lor \,.\, (\exists y) \,.\, y \epsilon \overleftrightarrow{R_*}`x \,.\, yR_{\text{po}} y : \supset \,.$
$\overrightarrow{R_*}`x \epsilon \text{Cls induct}$

Dem.

$\vdash \,.\, *121·7 \,.\, *93·36 \,.\, \supset \vdash : R \epsilon 1 \to 1 \,.\, x \epsilon s`\text{gen}`R \,.\, \supset \,.\, \overrightarrow{R_*}`x \epsilon \text{Cls induct}$ (1)

$\vdash \,.\, *97·55·111 \,.\, \supset \vdash :. R \epsilon 1 \to 1 : (\exists y) \,.\, y \epsilon \overleftrightarrow{R_*}`x \,.\, yR_{\text{po}} y : \supset :$
$y \epsilon \overleftrightarrow{R_*}`x \,.\, \supset_y \,.\, yR_{\text{po}} y : x \epsilon \overleftrightarrow{R_*}`x :$

[*10·26] $\supset : xR_{\text{po}} x :$

[*121·381] $\supset : \overrightarrow{R_*}`x = R(x \mapsto x) :$

[*121·45] $\supset : \overrightarrow{R_*}`x \epsilon \text{Cls induct}$ (2)

$\vdash \,.\, (1) \,.\, (2) \,.\, \supset \vdash \,.\, \text{Prop}$

***121·72.** $\vdash : R \epsilon 1 \to 1 \,.\, \overrightarrow{R_*}`x \sim \epsilon \text{Cls induct} \,.\, \supset \,.\, x \epsilon p`\text{C}``\text{Pot}`R \,.\, R_{\text{po}} \upharpoonright \overleftrightarrow{R_*}`x \subseteq J$

[*121·71 . Transp . *93·271 . *120·212 . *50·24]

$*$122. PROGRESSIONS.

Summary of $*$122.

By a "progression" we mean a series which is like the series of the inductive cardinals in order of magnitude (assuming that all inductive cardinals exist), *i.e.* a series whose terms can be called

$$1_R,\ 2_R,\ 3_R, \ldots\ \nu_R, \ldots,$$

where every term of the series is correlated with some inductive cardinal, and every inductive cardinal is correlated with some term of the series. Such series belong to the relation-number (cf. $*$152 and $*$263) which Cantor calls ω. Their generating relation may be taken to be the transitive relation of earlier and later, or the one-one relation of immediate predecessor to immediate successor. We shall reserve the notation $\boldsymbol{\omega}$ for the *transitive* generating relations of progressions; for the present, we are concerned with the one-one relations which generate progressions. The class of these relations we shall call "Prog."

It is not convenient to *define* a progression as a series which is ordinally similar to that of the inductive cardinals, both because this definition only applies if we assume the axiom of infinity, and because we have in any case to show that (assuming the axiom of infinity) the series of inductive cardinals has certain properties, which can be used to afford a direct definition of progressions. The existence of progressions, however, is only obtainable by means of the axiom of infinity, and is then most easily obtained from the fact that the inductive cardinals form a progression. We shall not consider the existence-theorem until the next number ($*$123).

From this number onwards convention Infin T of the Prefatory Statement is used when relevant.

The characteristics of the generating relation R of a progression, which we employ in the definition, are the following:

(1) R is a one-one relation;

(2) there is a first term, *i.e.* $E\,!\,B{}^{\backprime}R$;

(3) the whole field is contained in the posterity of the first term, *i.e.* $C{}^{\backprime}R = \overleftarrow{R_*}{}^{\backprime}B{}^{\backprime}R$. (If this failed, $C{}^{\backprime}R$ would consist of two or more distinct families, of which, since we have $E\,!\,B{}^{\backprime}R$, all but one would have to be cyclic families.)

(4) every term of the field has a successor, *i.e.* the series is endless. This is secured by $\text{Ꟈ}{}^{\backprime}R \subset \text{D}{}^{\backprime}R$, or (what is equivalent) $C{}^{\backprime}R = \text{D}{}^{\backprime}R$.

These four properties suffice to define the one-one generating relations of progressions. It will be observed that (2), (3) and (4) are all secured by

$$\text{D}`R = \overleftarrow{R_*}`B`R.$$

This secures $\text{E} ! B`R$, by $*14{\cdot}21$; it secures $\text{Œ}`R \subset \text{D}`R$, by $*37{\cdot}25$ and $*90{\cdot}163$; hence, by $*33{\cdot}181$, $\text{D}`R = C`R$, and therefore

$$C`R = \overleftarrow{R_*}`B`R.$$

Hence our definition of progressions is

$$\text{Prog} = (1 \rightarrow 1) \cap \hat{R}\,(\text{D}`R = \overleftarrow{R_*}`B`R) \quad \text{Df.}$$

Instead of stating in the definition that R is to be a one-one relation, it is sufficient to put $R \epsilon \text{Cls} \rightarrow 1 \,.\, R_{\text{po}} \Subset J$, which, with $\text{D}`R = \overleftarrow{R_*}`B`R$, implies $R \epsilon 1 \rightarrow 1$, and may be substituted for $R \epsilon 1 \rightarrow 1$ without altering the force of the definition ($*122{\cdot}17$).

In the present number we shall prove, among other propositions, that every existent class contained in a progression has a first term ($*122{\cdot}23$), *i.e.* that progressions are well-ordered series; that in a progression $R_{\text{po}} \Subset J$ ($*122{\cdot}16$), which makes the propositions of $*121$ available; that if ν is any inductive cardinal other than 0, ν_R exists ($*122{\cdot}33$), *i.e.* the series has a νth term; that any class contained in $\text{D}`R$ and having a last term is an inductive class ($*122{\cdot}43$), and that any class contained in $\text{D}`R$ and not having a last term is itself the domain of a progression ($*122{\cdot}45$), so that every class contained in $\text{D}`R$ is either inductive or the domain of a progression ($*122{\cdot}46$); that if P is a many-one, and x a member of its domain, and if the descendants of x have no last term and are none of them descendants of themselves, then P arranges these descendants in a progression ($*122{\cdot}51$); and that the same holds if P is a one-one and $\sim(xPx)$ ($*122{\cdot}52$); and that if $P \epsilon 1 \rightarrow 1$ and x belongs to one of the generations of P, but not to one of the generations of $\breve{P}$, then P arranges the whole family of x in a progression ($*122{\cdot}54$).

The following general observations on the families of one-one relations may serve to elucidate the bearing of the propositions of this section.

Given any relation P, we call $\overleftrightarrow{P_*}`x$, *i.e.* $\overrightarrow{P_*}`x \cup \overleftarrow{P_*}`x$, the *family* of x. If P is a one-one, this family may be of four different kinds. (1) It may be a closed series, like the angles of a polygon. This occurs if $xP_{\text{po}}x$. In this case the family forms an inductive class. (2) It may be an open series with a beginning and an end; this occurs if

$$\sim(xP_{\text{po}}x) \,.\, \text{E} ! \min_P`\overleftrightarrow{P_*}`x \,.\, \text{E} ! \max_P`\overleftrightarrow{P_*}`x.$$

In this case also the family forms an inductive class. (3) It may be an open series with a beginning and no end, or an end and no beginning. This occurs if

$$\sim(xP_{\text{po}}x) \,.\, \text{E}\,!\, \text{min}_P{}^{\backprime}\overleftrightarrow{P}_{*}{}^{\backprime}x \,.\, \sim \text{E}\,!\, \text{max}_P{}^{\backprime}\overleftrightarrow{P}_{*}{}^{\backprime}x,$$

or if

$$\sim(xP_{\text{po}}x) \,.\, \sim \text{E}\,!\, \text{min}_P{}^{\backprime}\overleftrightarrow{P}_{*}{}^{\backprime}x \,.\, \text{E}\,!\, \text{max}_P{}^{\backprime}\overleftrightarrow{P}_{*}{}^{\backprime}x.$$

In this case, the series is of the type ω or $\text{Cnv}``\omega$, and is non-inductive and reflexive. (4) The series may be open and have neither beginning nor end. This occurs if

$$\sim(xP_{\text{po}}x) \,.\, \sim \text{E}\,!\, \text{min}_P{}^{\backprime}\overleftrightarrow{P}_{*}{}^{\backprime}x \,.\, \sim \text{E}\,!\, \text{max}_P{}^{\backprime}\overleftrightarrow{P}_{*}{}^{\backprime}x.$$

In this case we get a series whose relation-number is the sum (in the sense of $*180$) of $\text{Cnv}``\omega$ and ω, which again is non-inductive and reflexive. In all four cases, if y and z be any two members of the family of x, the interval between y and z is an inductive class.

If x is a member of $\overrightarrow{B}{}^{\backprime}P$, or if the family of x contains a member of $\overrightarrow{B}{}^{\backprime}P$, cases (1) and (4) are excluded, since the series has a beginning. In this case the number of predecessors of any term is an inductive number. It will be observed that every family is either wholly contained in $s{}^{\backprime}\text{gen}{}^{\backprime}P$ or wholly contained in $p{}^{\backprime}\text{Cl}``\text{Pot}{}^{\backprime}P$; families of kinds (2) and (3) (excluding, in (2), those which have an end but no beginning) are contained in $s{}^{\backprime}\text{gen}{}^{\backprime}P$, while families of kinds (1) and (4), and those of (2) which have an end but no beginning, are contained in $p{}^{\backprime}\text{Cl}``\text{Pot}{}^{\backprime}P$; families containing a member of $\overrightarrow{B}{}^{\backprime}P$ are contained in $s{}^{\backprime}\text{gen}{}^{\backprime}P$, while all others are contained in $p{}^{\backprime}\text{Cl}``\text{Pot}{}^{\backprime}P$.

Thus a one-one relation in general gives rise to a number of wholly disconnected series, some closed, others open and with or without a beginning or an end. The condition that all the series should be open is $P_{\text{po}} \subset J$.

The case of a Q-shaped family, considered in $*96$, cannot arise when $P \epsilon 1 \rightarrow 1$, for in a Q-shaped family the term at the junction of the tail and the circle has two predecessors, one in the tail and one in the circle, so that the relation in question is not $1 \rightarrow 1$. It follows that, when $P \epsilon 1 \rightarrow 1$, if α is a family containing a member of $\overrightarrow{B}{}^{\backprime}P$, $\alpha \upharpoonleft P_{\text{po}} \subset J$ (cf. $*96{\cdot}23$).

When $B{}^{\backprime}P$ exists, there is only one family which has a beginning. In this case, ignoring the other families (if any), we call the members of the family of $B{}^{\backprime}P$ respectively 1_P, 2_P, 3_P, ... If the family has ν members, where ν is an inductive cardinal, its last member will be ν_P. If on the other hand the number of members of the family is not an inductive cardinal, it must be $\aleph_0$; in this case, the family forms a progression, whose members are 1_P, 2_P, 3_P, ..., ν_P, ..., where ν_P always exists when ν is an inductive cardinal.

In addition to the propositions already mentioned, the following are important:

***122·21.** $\vdash :. R \in \text{Prog} . x, y \in C\text{‘}R . \supset : x R_{po} y . \vee . x = y . \vee . y R_{po} x$

(Cf. note to *122·21, below.)

***122·34.** $\vdash :. R \in \text{Prog} . \supset : \nu \in \text{NC induct} - \iota\text{‘}0 . \equiv . \text{E} ! \nu_R$

***122·341.** $\vdash : R \in \text{Prog} . \supset . \text{D}\text{‘}R = \hat{x} \{(\exists \nu) . \nu \in \text{NC induct} - \iota\text{‘}0 . x = \nu_R\}$

In virtue of these two propositions, the terms of a progression are

$$1_R,\ 2_R,\ 3_R,\ \dots\ \nu_R,\ \dots,$$

where every inductive cardinal occurs. This is the same fact as is usually assumed when the terms are represented as

$$x_1,\ x_2,\ x_3,\ \dots\ x_\nu,\ \dots.$$

***122·35.** $\vdash : R \in \text{Prog} . \nu \in \text{NC induct} - \iota\text{‘}0 . \supset . \overrightarrow{B\text{‘}R_\nu} = R (1_R \mapsto \nu_R) . \overrightarrow{B\text{‘}R_\nu} \in \nu$

***122·36.** $\vdash : \exists ! \text{Prog} \cap t^{11}\text{‘}x . \supset . \text{Infin ax} (x)$

***122·37.** $\vdash : R \in \text{Prog} . \supset . \text{D}\text{‘}R \sim \in \text{Cls induct} . \text{N}_0\text{c}\text{‘D}\text{‘}R \sim \in \text{NC induct}$

***122·38.** $\vdash : R \in \text{Prog} . \supset . \overrightarrow{R_*}\text{‘}x \in \text{Cls induct}$

I.e. the number of terms up to any given point of a progression is inductive.

***122·01.** $\text{Prog} = (1 \rightarrow 1) \cap \hat{R} (\text{D}\text{‘}R = \overleftarrow{R_*}\text{‘}B\text{‘}R)$ Df

***122·1.** $\vdash : R \in \text{Prog} . \equiv . R \in 1 \rightarrow 1 . \text{D}\text{‘}R = \overleftarrow{R_*}\text{‘}B\text{‘}R$ [(*122·01)]

***122·11.** $\vdash :. R \in \text{Prog} . \equiv : R \in 1 \rightarrow 1 . \text{E} ! B\text{‘}R : x \in \text{D}\text{‘}R . \equiv_x . x \in \overleftarrow{R_*}\text{‘}B\text{‘}R$

Dem.

$\vdash . *122·1 . *14·205 . \supset$

$\vdash :: R \in \text{Prog} . \equiv :. R \in 1 \rightarrow 1 : (\exists a) . a = B\text{‘}R . \text{D}\text{‘}R = \overleftarrow{R_*}\text{‘}a :.$

[*20·43] $\equiv :. R \in 1 \rightarrow 1 :. (\exists a) : a = B\text{‘}R : x \in \text{D}\text{‘}R . \equiv_x . x \in \overleftarrow{R_*}\text{‘}a :.$

[*14·15] $\equiv :. R \in 1 \rightarrow 1 :. (\exists a) : a = B\text{‘}R : x \in \text{D}\text{‘}R . \equiv_x . x \in \overleftarrow{R_*}\text{‘}B\text{‘}R :.$

[*14·204] $\equiv :. R \in 1 \rightarrow 1 . \text{E} ! B\text{‘}R : x \in \text{D}\text{‘}R . \equiv_x . x \in \overleftarrow{R_*}\text{‘}B\text{‘}R :: \supset \vdash . \text{Prop}$

Observe that, by the conventions as to descriptive symbols, $\text{D}\text{‘}R = \overleftarrow{R_*}\text{‘}B\text{‘}R$ involves the existence of $B\text{‘}R$, whereas $x \in \text{D}\text{‘}R . \equiv_x . x \in \overleftarrow{R_*}\text{‘}B\text{‘}R$ does not, since, if $B\text{‘}R$ does not exist, we have $(x) . x \sim \in \overleftarrow{R_*}\text{‘}B\text{‘}R$, and therefore $(x) . x \sim \in \text{D}\text{‘}R$ will satisfy the equivalence, *i.e.* $\dot{\Lambda}$ will satisfy the equivalence although it has no first term. This is the reason why $\text{E} ! B\text{‘}R$ appears explicitly in *122·11, though it was only implicit in *122·1.

***122·12.** $\vdash :: R \in \text{Prog} . \equiv :. R \in 1 \rightarrow 1 . \text{E} ! B\text{‘}R :. x \in \text{D}\text{‘}R . \equiv_x :$

$$B\text{‘}R \in \alpha . \breve{R}\text{‘‘}\alpha \subset \alpha . \supset_\alpha . x \in \alpha \qquad [*122·11 . *90·1]$$

∗122·14. $\vdash : R \in \text{Prog} \,.\supset .\, \overleftarrow{R_{po}}\text{‘}B\text{‘}R = \text{Ɑ}\text{‘}R$

Dem.

$$\vdash .\, {*}122{\cdot}1 \,.\, {*}37{\cdot}25 \,.\supset \vdash : \text{Hp} \,.\supset .\, \text{Ɑ}\text{‘}R = \breve{R}\text{“}\overleftarrow{R_{*}}\text{‘}B\text{‘}R$$

$$[{*}91{\cdot}52] \qquad = \overleftarrow{R_{po}}\text{‘}B\text{‘}R : \supset \vdash .\, \text{Prop}$$

∗122·141. $\vdash : R \in \text{Prog} \,.\supset .\, \text{Ɑ}\text{‘}R \subset \text{D}\text{‘}R \,.\, C\text{‘}R = \text{D}\text{‘}R$

Dem.

$$\vdash .\, {*}122{\cdot}1 \,.\, {*}37{\cdot}25 \,.\supset \vdash : \text{Hp} \,.\supset .\, \text{Ɑ}\text{‘}R = \breve{R}\text{“}\overleftarrow{R_{*}}\text{‘}B\text{‘}R \,.$$

$$[{*}90{\cdot}163] \qquad \supset .\, \text{Ɑ}\text{‘}R \subset \overleftarrow{R_{*}}\text{‘}B\text{‘}R \,.$$

$$[{*}122{\cdot}1 . {*}33{\cdot}181] \qquad \supset .\, \text{Ɑ}\text{‘}R \subset \text{D}\text{‘}R \,.\, C\text{‘}R = \text{D}\text{‘}R : \supset \vdash .\, \text{Prop}$$

∗122·142. $\vdash : R \in \text{Prog} \,.\, P \in \text{Pot}\text{‘}R \,.\supset .\, \text{D}\text{‘}P = \text{D}\text{‘}R$ $\quad [{*}122{\cdot}141 \,.\, {*}92{\cdot}14]$

∗122·143. $\vdash : R \in \text{Prog} \,.\, P \in \text{Pot}\text{‘}R \,.\supset .\, \text{Ɑ}\text{‘}P \subset \text{D}\text{‘}P$ $\quad [{*}122{\cdot}142{\cdot}141 \,.\, {*}91{\cdot}271]$

∗122·15. $\vdash : R \in \text{Prog} \,.\supset .\, R = (\overleftarrow{R_{*}}\text{‘}B\text{‘}R) \rceil R = R \upharpoonright (\overleftarrow{R_{po}}\text{‘}B\text{‘}R) = R \upharpoonright (\overleftarrow{R_{*}}\text{‘}B\text{‘}R)$

Dem.

$$\vdash .\, {*}122{\cdot}1 \,.\, {*}35{\cdot}63 \,.\supset \vdash : \text{Hp} \,.\supset .\, R = (\overleftarrow{R_{*}}\text{‘}B\text{‘}R) \rceil R$$

$$[{*}96{\cdot}2] \qquad = R \upharpoonright (\overleftarrow{R_{po}}\text{‘}B\text{‘}R)$$

$$[{*}96{\cdot}21] \qquad = R \upharpoonright (\overleftarrow{R_{*}}\text{‘}B\text{‘}R) : \supset \vdash .\, \text{Prop}$$

∗122·151. $\vdash : R \in \text{Prog} \,.\supset .\, R_{*} = (\overleftarrow{R_{*}}\text{‘}B\text{‘}R) \rceil R_{*} = R_{*} \upharpoonright (\overleftarrow{R_{*}}\text{‘}B\text{‘}R)$

$[35{\cdot}63{\cdot}66 \,.\, {*}90{\cdot}14 \,.\, {*}122{\cdot}141{\cdot}1]$

∗122·152. $\vdash : R \in \text{Prog} \,.\supset .\, R_{po} = (\overleftarrow{R_{*}}\text{‘}B\text{‘}R) \rceil R_{po} = R_{po} \upharpoonright (\overleftarrow{R_{po}}\text{‘}B\text{‘}R)$

$$= R_{po} \upharpoonright (\overleftarrow{R_{*}}\text{‘}B\text{‘}R)$$

$[{*}35{\cdot}63{\cdot}66 \,.\, {*}91{\cdot}504 \,.\, {*}121{\cdot}1{\cdot}14]$

∗122·16. $\vdash : R \in \text{Prog} \,.\supset .\, R_{po} \in J$ $\quad [{*}96{\cdot}23 \,.\, {*}122{\cdot}152]$

This proposition enables us to apply to progressions all the propositions of ∗121 in which we have as hypothesis

$$R \in \text{Cls} \to 1 \,.\, R_{po} \in J, \quad \text{or} \quad R \in 1 \to \text{Cls} \,.\, R_{po} \in J.$$

∗122·17. $\vdash : R \in \text{Prog} \,.\equiv .\, R \in \text{Cls} \to 1 \,.\, R_{po} \in J \,.\, \text{D}\text{‘}R = \overleftarrow{R_{*}}\text{‘}B\text{‘}R$

Dem.

$$\vdash .\, {*}35{\cdot}63 \,.\supset \vdash : \text{D}\text{‘}R = \overleftarrow{R_{*}}\text{‘}B\text{‘}R \,.\supset .\, R = (\overleftarrow{R_{*}}\text{‘}B\text{‘}R) \rceil R \qquad (1)$$

$$\vdash .\, {*}96{\cdot}453 \,.\supset \vdash : R \in \text{Cls} \to 1 \,.\, (\overleftarrow{R_{*}}\text{‘}B\text{‘}R) \rceil R_{po} \in J \,.\supset .\, (\overleftarrow{R_{*}}\text{‘}B\text{‘}R) \rceil R \in 1 \to 1 \qquad (2)$$

$$\vdash .\, (1) \,.\, (2) \,.\, {*}122{\cdot}1 \,.\supset \vdash : \text{D}\text{‘}R = \overleftarrow{R_{*}}\text{‘}B\text{‘}R \,.\, R \in \text{Cls} \to 1 \,.\, R_{po} \in J \,.\supset .\, R \in \text{Prog} \qquad (3)$$

$$\vdash .\, (3) \,.\, {*}122{\cdot}1{\cdot}16 \,.\supset \vdash .\, \text{Prop}$$

To illustrate this proposition, consider its application to the inductive cardinals arranged in order of magnitude; *i.e.* take as a value of R the relation

$$\hat{\mu}\hat{\nu}(\mu \in \text{NC induct} \,.\, \nu = \mu +_c 1).$$

We then have $R \in \text{Cls} \rightarrow 1 \,.\, 0 = B`R$; also

$$\text{NC induct} = \text{D}`R = \overleftarrow{R_*}`B`R.$$

We have also

$$\exists ! \mu +_c 1 \,.\, \mu +_c 1 = \nu +_c 1 \,.\supset.\, \mu = \nu,$$

so that $$R \upharpoonright (-\iota`\Lambda) \in 1 \rightarrow \text{Cls}.$$

Again

$$\mu R_{po} \nu \,.\equiv.\, (\exists \varpi) \,.\, \varpi \in \text{NC induct} - \iota`0 \,.\, \nu = \mu +_c \varpi,$$

whence $$\mu R_{po} \nu \,.\, \exists ! \mu \,.\supset.\, \mu \neq \nu,$$

i.e. $$(-\iota`\Lambda) \upharpoonleft R_{po} \in J.$$

But we do not get $R \in 1 \rightarrow \text{Cls}$ or $R_{po} \in J$ unless we have

$$\Lambda \sim\in \text{NC induct},$$

which is the axiom of infinity. If this condition fails, we reach at last an inductive cardinal which $= \Lambda$, and we have

$$\Lambda = \Lambda +_c 1,$$

so that Λ has two immediate predecessors, namely itself and the last existent cardinal. The posterity of 0, in this case, is a Q in which the circle has narrowed to a single term, namely Λ.

Thus we need the axiom of infinity in order to prove

$$\hat{\mu}\hat{\nu}(\mu \in \text{NC induct} \,.\, \nu = \mu +_c 1) \in \text{Prog}.$$

***122·2.** $\vdash :. R \in \text{Prog} \,.\, x, y \in C`R \,.\supset:\, xR_*y \,.\vee.\, yR_*x$ [*96·302 . *122·1·141]

***122·21.** $\vdash :. R \in \text{Prog} \,.\, x, y \in C`R \,.\supset:\, xR_{po}y \,.\vee.\, x = y \,.\vee.\, yR_{po}x$ [*96·303 . *122·1·141]

This proposition, together with *122·16 and *91·56, shows that if $R \in \text{Prog}$, R_{po} has the three properties by which transitive serial relations are defined (cf. *204), namely it is (1) transitive, (2) contained in diversity, (3) connected, *i.e.* such that it relates any two distinct members of its field. We shall at a later stage define the ordinal number ω as the class of such relations as R_{po}, where $R \in \text{Prog}$.

***122·22.** $\vdash : R \in \text{Prog} \,.\, \alpha \subset \text{D}`R \,.\, x, y \in \alpha - \breve{R}_{po}``\alpha \,.\supset.\, x = y$

Dem.

$\vdash . *122·21 . \supset \vdash :. \text{Hp} . \supset : xR_{po}y \,.\vee.\, x = y \,.\vee.\, yR_{po}x$ (1)

$\vdash . *37·105 . \supset \vdash : x \in \alpha \,.\, xR_{po}y \,.\supset.\, y \in \breve{R}_{po}``\alpha :$

[Transp] $\supset \vdash : x \in \alpha \,.\, y \sim\in \breve{R}_{po}``\alpha \,.\supset.\, \sim(xR_{po}y)$ (2)

$\vdash . (2) .$ $\supset \vdash : \text{Hp} \,.\supset.\, \sim(xR_{po}y) \,.\, \sim(yR_{po}x)$ (3)

$\vdash . (1) . (3) . \supset \vdash . \text{Prop}$

∗122·23. $\vdash : R \epsilon \text{Prog} . \alpha \subset \text{D}'R . \exists ! \alpha . \supset .$

$$\text{E} ! \min(R_{\text{po}})'\alpha . \alpha - \breve{R}_{\text{po}}``\alpha = \iota'\min(R_{\text{po}})'\alpha$$

Dem.

$\vdash . *96{\cdot}52 . \quad \supset \vdash : \text{Hp} . \supset . \exists ! \overrightarrow{\min}(R_{\text{po}})'\alpha \quad (1)$

$\vdash . *93{\cdot}111 . *122{\cdot}22 . \supset \vdash :. \text{Hp} . \supset : x, y \epsilon \overrightarrow{\min}(R_{\text{po}})'\alpha . \supset_{x,y} . x = y \quad (2)$

$\vdash . (1) . (2) . *32{\cdot}4 . *93{\cdot}111 . \supset \vdash . \text{Prop}$

This proposition shows that every existent class contained in a progression has a first term, *i.e.* that a progression is a well-ordered series (cf. ∗250).

∗122·231. $\vdash : R \epsilon \text{Prog} . \alpha \subset \breve{R}_{\text{po}}``\alpha . \supset . \alpha = \Lambda$

Dem.

$\vdash . *91{\cdot}504 . \supset \vdash : \text{Hp} . \supset . \alpha \subset \text{Ꮖ}'R \quad (1)$

$\vdash . *93{\cdot}11 . \quad \supset \vdash : \text{Hp} . \supset . \sim \text{E} ! \overrightarrow{\min}(R_{\text{po}})'\alpha \quad (2)$

$\vdash . (1) . (2) . *122{\cdot}23{\cdot}141 . \text{Transp} . \supset \vdash . \text{Prop}$

∗122·24. $\vdash : R \epsilon \text{Prog} . P \epsilon \text{Pot}'R . \supset . \text{D}'P = \breve{P}_*``\overrightarrow{B}'P = s'\text{gen}'P$

Dem.

$\vdash . *122{\cdot}1 . *92{\cdot}102 . \supset \vdash : \text{Hp} . \supset . P \epsilon 1 \rightarrow 1 .$

$[*93{\cdot}42] \quad \supset . p'\text{Ꮖ}``\text{Pot}'P = \breve{P}``p'\text{Ꮖ}``\text{Pot}'P .$

$[*91{\cdot}581] \quad \supset . p'\text{Ꮖ}``\text{Pot}'P \subset \breve{R}_{\text{po}}``p'\text{Ꮖ}``\text{Pot}'P .$

$[*122{\cdot}231] \quad \supset . p'\text{Ꮖ}``\text{Pot}'P = \Lambda \quad (1)$

$\vdash . (1) . *93{\cdot}37{\cdot}36 . \quad \supset \vdash : \text{Hp} . \supset . C'P = \breve{P}_*``\overrightarrow{B}'P = s'\text{gen}'P \quad (2)$

$\vdash . (2) . *122{\cdot}143 . \supset \vdash . \text{Prop}$

Except when $P = R$, $\overrightarrow{B}'P$ will not reduce to a single term. In fact, if $P = R_\nu$, $\overrightarrow{B}'P = R(1_R \mapsto \nu_R)$, *i.e.* $\overrightarrow{B}'P$ consists of the first ν terms of the progression.

∗122·25. $\vdash : R \epsilon \text{Prog} . P \epsilon \text{Pot}'R . x \epsilon \text{D}'R . \supset .$

$$(\overleftarrow{P_*}'x) \upharpoonleft P \epsilon \text{Prog} . x = B'\{(\overleftarrow{P_*}'x) \upharpoonleft P\}$$

Dem.

$\vdash . *122{\cdot}1 . *92{\cdot}102 . \supset \vdash : \text{Hp} . \supset . (\overleftarrow{P_*}'x) \upharpoonleft P \epsilon 1 \rightarrow 1 \quad (1)$

$\vdash . *122{\cdot}143 . \quad \supset \vdash : \text{Hp} . \supset . \overleftarrow{P_*}'x \subset \text{D}'P .$

$[*35{\cdot}62] \quad \supset . \text{D}'\{(\overleftarrow{P_*}'x) \upharpoonleft P\} = \overleftarrow{P_*}'x \quad (2)$

$\vdash . *37{\cdot}4 . *91{\cdot}52 . \quad \supset \vdash . \text{Ꮖ}'\{(\overleftarrow{P_*}'x) \upharpoonleft P\} = \overleftarrow{P}_{\text{po}}'x \quad (3)$

$\vdash . *122{\cdot}16 . *91{\cdot}6 . \quad \supset \vdash : \text{Hp} . \supset . x \sim \epsilon \overleftarrow{P}_{\text{po}}'x \quad (4)$

$\vdash . *91{\cdot}542 . \quad \supset \vdash : y \epsilon \overleftarrow{P_*}'x . y \neq x . \supset . y \epsilon \overleftarrow{P}_{\text{po}}'x \quad (5)$

$\vdash . (2) . (3) . (4) . (5) . \supset \vdash : \text{Hp} . \supset . x = B'\{(\overleftarrow{P_*}'x) \upharpoonleft P\} \quad (6)$

$\vdash . (1) . (2) . (6) . *96{\cdot}131 . \supset \vdash . \text{Prop}$

The above proposition shows that what we may call an "arithmetical progression" in a progression is a progression, *i.e.* if, starting from any term of a progression, we take every other term, or every third term, or every νth term, we still have a progression.

***122·26.** $\vdash : R \epsilon \text{Prog} . \alpha \subset R_{\text{po}}``\alpha . \exists ! \alpha . \supset . \text{D}`R = R_{\text{po}}``\alpha$

Dem.

$\vdash . *22·1 . \quad \supset \vdash :. \text{Hp} . \supset : B`R \epsilon \alpha . \supset . B`R \epsilon R_{\text{po}}``\alpha \quad (1)$

$\vdash . *91·542 . *122·11 . \supset \vdash :. \text{Hp} . B`R \sim \epsilon \alpha . \supset : y \epsilon \alpha \cap \text{D}`R . \supset_y . (B`R) R_{\text{po}} y :$

[*91·504.*37·15] $\quad \supset : y \epsilon \alpha . \supset_y . (B`R) R_{\text{po}} y :$

[*10·55.Hp] $\quad \supset : (\exists y) . y \epsilon \alpha . (B`R) R_{\text{po}} y :$

[*37·1] $\quad \supset : B`R \epsilon R_{\text{po}}``\alpha \quad (2)$

$\vdash . (1) . (2) . \supset \vdash : \text{Hp} . \supset . B`R \epsilon R_{\text{po}}``\alpha \quad (3)$

$\vdash . *92·111 . \supset \vdash :. \text{Hp} . \supset : x \epsilon R_{\text{po}}``\alpha . xRy . \supset . y \epsilon R_*``\alpha .$

[*91·545] $\quad \supset . y \epsilon \alpha \cup R_{\text{po}}``\alpha .$

[Hp] $\quad \supset . y \epsilon R_{\text{po}}``\alpha \quad (4)$

$\vdash . (3) . (4) . *90·112 . \supset \vdash :. \text{Hp} . \supset : (B`R) R_* y . \supset . y \epsilon R_{\text{po}}``\alpha \quad (5)$

$\vdash . (5) . *122·1 . \supset \vdash . \text{Prop}$

The above proposition shows that if an existent class contained in a progression has no maximum, then any assigned member of the progression is succeeded by members of the class.

The following proposition states that if α has members belonging to a progression, and there are members of the progression which do not precede any member of α, then there is in the progression a last member of α.

***122·27.** $\vdash : R \epsilon \text{Prog} . \exists ! \text{D}`R - R_{\text{po}}``\alpha . \exists ! \alpha \cap \text{D}`R . \supset .$
$\text{E} ! \max (R_{\text{po}})`\alpha . \exists ! \alpha \cap \text{D}`R - R_{\text{po}}``\alpha$

Dem.

$\vdash . *122·26 . \text{Transp} . *37·265 . \quad \supset \vdash : \text{Hp} . \supset . \exists ! \alpha \cap C`R - R_{\text{po}}``\alpha \quad (1)$

$\vdash . *122·21 . \quad \supset \vdash : \text{Hp} . x, y \epsilon \alpha \cap C`R - R_{\text{po}}``\alpha . \supset . x = y \quad (2)$

$\vdash . (1) . (2) . *93·115 . *122·141 . \supset \vdash . \text{Prop}$

***122·28.** $\vdash : R \epsilon \text{Prog} . \alpha \subset \overrightarrow{R_*}`x . \exists ! \alpha . \supset . \text{E} ! \max (R_{\text{po}})`\alpha . \exists ! \alpha \cap \text{D}`R - R_{\text{po}}``\alpha$

Dem.

$\vdash . *90·13 . *122·141 . \supset \vdash : \text{Hp} . \supset . \alpha \subset \text{D}`R \quad (1)$

$\vdash . *90·14 . *122·141 . \supset \vdash : \text{Hp} . \supset . x \epsilon \text{D}`R .$

[*71·161.*122·16] $\quad \supset . \breve{R}`x \sim \epsilon R_{\text{po}}``\alpha .$

[*122·1] $\quad \supset . \exists ! \text{D}`R - R_{\text{po}}``\alpha \quad (2)$

$\vdash . (1) . (2) . *122·27 . \supset \vdash . \text{Prop}$

***122·3.** $\vdash : R \epsilon \mathrm{Prog} . \supset . \mathrm{D}'R = \hat{x}\{(\exists \nu) . \nu \epsilon \mathrm{NC\,induct} . (B'R) R_\nu x\}$
[*121·52 . *122·1·16]

***122·31.** $\vdash : R \epsilon \mathrm{Prog} . \nu \epsilon \mathrm{NC\,induct} - \iota'0 . \supset . \mathrm{G}'R_\nu = \hat{y}\{\mathrm{Nc}'R(B'R \mapsto y) > \nu\}$

Dem.

$\vdash . *120·429 . \supset \vdash :. \mathrm{Hp} . \supset : \mathrm{Nc}'R(B'R \mapsto y) > \nu . \equiv . \mathrm{Nc}'R(B'R \mapsto y) \geqslant \nu +_c 1 .$

[*117·31] $\equiv . (\exists \mu) . \mu \epsilon \mathrm{NC} . \mathrm{Nc}'R(B'R \mapsto y) = \mu +_c \nu +_c 1 .$
$\nu +_c 1, \mu +_c \nu +_c 1 \epsilon \mathrm{N_0C} .$

[*121·45.*120·452.*110·4] $\equiv . (\exists \mu) . \mu \epsilon \mathrm{NC\,induct} .$
$\mathrm{Nc}'R(B'R \mapsto y) = \mu +_c \nu +_c 1 . \mu +_c \nu +_c 1 \epsilon \mathrm{N_0C} .$

[*121·11·35.*110·43.*100·3] $\equiv . (\exists \mu) . \mu \epsilon \mathrm{NC\,induct} . (B'R) R_\mu | R_\nu y .$

[*34·1] $\equiv . (\exists \mu, x) . \mu \epsilon \mathrm{NC\,induct} . (B'R) R_\mu x . x R_\nu y .$

[*122·3] $\equiv . (\exists x) . x \epsilon \mathrm{D}'R . x R_\nu y .$

[*121·323] $\equiv . (\exists x) . x R_\nu y .$

[*33·131] $\equiv . y \epsilon \mathrm{G}'R_\nu :. \supset \vdash . \mathrm{Prop}$

***122·32.** $\vdash : R \epsilon \mathrm{Prog} . \nu \epsilon \mathrm{NC\,induct} - \iota'0 . \supset .$

$$\overrightarrow{B}'R_\nu = \mathrm{D}'R \cap \hat{x}\{\mathrm{Nc}'R(B'R \mapsto x) \leqslant \nu\}$$

Dem.

$\vdash . *122·142 . *121·501 . \supset \vdash : \mathrm{Hp} . \supset . \mathrm{D}'R_\nu = \mathrm{D}'R$ (1)

$\vdash . *122·31 . *120·442 . \supset \vdash : \mathrm{Hp} . \supset . -\mathrm{G}'R_\nu = \hat{x}\{\mathrm{Nc}'R(B'R \mapsto x) \leqslant \nu\}$ (2)

$\vdash . (1) . (2) . *93·101 . \supset \vdash . \mathrm{Prop}$

***122·33.** $\vdash : R \epsilon \mathrm{Prog} . \nu \epsilon \mathrm{NC\,induct} - \iota'0 . \supset . \mathrm{E}!\, \nu_R$

Dem.

$\vdash . *121·601 . *122·11 . \quad \supset \vdash : \mathrm{Hp} . \supset . \mathrm{E}!\, 1_R$ (1)

$\vdash . *121·634·637 . *122·141 . \supset \vdash :. \mathrm{Hp} . \supset : \mathrm{E}!\, \nu_R . \supset . \mathrm{E}!\, (\nu +_c 1)_R$ (2)

$\vdash . (1) . (2) . *120·473 . \supset \vdash . \mathrm{Prop}$

***122·34.** $\vdash :. R \epsilon \mathrm{Prog} . \supset : \nu \epsilon \mathrm{NC\,induct} - \iota'0 . \equiv . \mathrm{E}!\, \nu_R$ [*122·33.*121·635]

***122·341.** $\vdash : R \epsilon \mathrm{Prog} . \supset . \mathrm{D}'R = \hat{x}\{(\exists \nu) . \nu \epsilon \mathrm{NC\,induct} - \iota'0 . x = \nu_R\}$

Dem.

$\vdash . *122·3·34 . *121·638 . \supset$

$\vdash : \mathrm{Hp} . \supset . \mathrm{D}'R = \hat{x}\{(\exists \nu) . \nu \epsilon \mathrm{NC\,induct} . x = (\nu +_c 1)_R\}$

[*120·471] $= \hat{x}\{(\exists \nu) . \nu \epsilon \mathrm{NC\,induct} - \iota'0 . x = \nu_R\} : \supset \vdash . \mathrm{Prop}$

In virtue of *122·34·341, all the terms of a progression occur in the series $1_R, 2_R, \ldots \nu_R, \ldots$, and every inductive cardinal except 0 is used in forming this series.

***122·35.** $\vdash : R \epsilon \text{Prog} . \nu \epsilon \text{NC induct} - \iota`0 . \supset . \overrightarrow{B`R_\nu} = R(1_R \mapsto \nu_R) . \overrightarrow{B`R_\nu} \epsilon \nu$

Dem.

$\vdash . *121·63 . *122·33 . \supset \vdash : \text{Hp} . \supset . \text{Nc}`R(B`R \mapsto \nu_R) = \nu .$ (1)

[*122·32] $\supset . \overrightarrow{B`R_\nu} = \text{D}`R \cap \hat{x}\{\text{Nc}`R(B`R \mapsto x) \leqslant \text{Nc}`R(B`R \mapsto \nu_R)\}$

[*121·481] $= \text{D}`R \cap \hat{x}\{R(B`R \mapsto x) \subset R(B`R \mapsto \nu_R)\}$

[*122·1.*121·103] $= \hat{x}\{(B`R) R_* x : y R_* x . \supset_y . y R_* \nu_R\}$

[*90·17·13.*10·1] $= \hat{x}\{(B`R) R_* x . x R_* \nu_R\}$

[*121·103] $= R(B`R \mapsto \nu_R)$ (2)

[*121·601.*122·11] $= R(1_R \mapsto \nu_R)$ (3)

$\vdash . (1) . (2) . (3) . \supset \vdash . \text{Prop}$

***122·36.** $\vdash : \exists ! \text{Prog} \cap t^{11}`x . \supset . \text{Infin ax}(x)$

Dem.

$\vdash . *122·35 . \supset \vdash : R \epsilon \text{Prog} \cap t^{11}`x . \nu \epsilon \text{NC induct} - \iota`0 . \supset . \exists ! \nu(x)$ (1)

$\vdash . (1) . *101·12 . \supset \vdash :. R \epsilon \text{Prog} \cap t^{11}`x . \supset : \nu \epsilon \text{NC induct} . \supset_\nu . \exists ! \nu(x) :$

[*120·301] $\supset : \text{Infin ax}(x) :. \supset \vdash . \text{Prop}$

***122·37.** $\vdash : R \epsilon \text{Prog} . \supset . \text{D}`R \sim \epsilon \text{Cls induct} . \text{N}_0\text{c}`\text{D}`R \sim \epsilon \text{NC induct}$

Dem.

$\vdash . *122·35 . \supset \vdash :. R \epsilon \text{Prog} . \supset : \nu \epsilon \text{NC induct} . \supset_\nu . \exists ! \text{Cl}`\text{D}`R \cap (\nu +_c 1) .$

[*117·22·107] $\supset_\nu . \text{N}_0\text{c}`\text{D}`R \geqslant \nu +_c 1 .$

[*120·429] $\supset_\nu . \text{N}_0\text{c}`\text{D}`R > \nu .$

[*117·42] $\supset_\nu . \text{N}_0\text{c}`\text{D}`R \neq \nu :$

[*13·196] $\supset : \text{N}_0\text{c}`\text{D}`R \sim \epsilon \text{NC induct}$ (1)

$\vdash . (1) . *120·21 . \supset \vdash . \text{Prop}$

***122·38.** $\vdash : R \epsilon \text{Prog} . \supset . \overrightarrow{R_*}`x \epsilon \text{Cls induct}$ [*121·7 . *90·13 . *120·212]

***122·381.** $\vdash : R \epsilon \text{Prog} . \nu \epsilon \text{NC induct} - \iota`0 . \supset . \overrightarrow{R_*}`\nu_R = R(1_R \mapsto \nu_R) . \overrightarrow{R_*}`\nu_R \epsilon \nu$ [*121·7 . *122·35]

The following series of propositions are concerned in proving that any class contained in a progression is inductive if it has a last term, and is a progression if it has no last term. In the latter case, it is supposed arranged in the same order as it had in the original progression. A certain complication is necessary in order to define its one-one generating relation. If R is the generating relation of the original progression, we proceed first to R_{po}, then to $R_{po} \upharpoonright \alpha$, where α is the class in question; this gives us a transitive generating relation for α. Calling this relation P, we then proceed to $P \dot{-} P^2$, *i.e.* the relation of consecutive members of the series generated by P. This relation turns out to be one-one, and to arrange α in a progression; hence our proposition is proved. The reason for the necessity of this detour is that consecutive members of α may not be consecutive members of the original progression.

***122·41.** $\vdash : R \in \text{Prog} . \alpha \subset \text{D}`R . y \in \alpha - R_{\text{po}}``\alpha . \supset . \alpha \subset R(B`R \mapsto y)$

Dem.

$\vdash . *37·1 . *10·51 . \supset \vdash :. \text{Hp} . \supset : z \in \alpha . \supset_z . \sim(y R_{\text{po}} z) .$
$[*122·21] \quad \supset_z . z R_* y$ (1)
$\vdash . *122·1 . \quad \supset \vdash :. \text{Hp} . \supset : z \in \alpha . \supset_z . (B`R) R_* z$ (2)
$\vdash . (1) . (2) . *121·103 . \supset \vdash . \text{Prop}$

***122·42.** $\vdash : R \in \text{Prog} . \alpha \subset R(B`R \mapsto y) . y \in \alpha . \supset . y = \max_R`\alpha$

Dem.

$\vdash . *121·103 . \supset \vdash :. \text{Hp} . \supset : z \in \alpha . \supset_z . z R_* y .$
$[*91·574.*122·16] \quad \supset_z . \sim(y R_{\text{po}} z) :$
$[*37·1.*10·51] \quad \supset : y \sim\in R_{\text{po}}``\alpha :$ (1)
$[*96·303] \quad \supset : z \in \alpha - R_{\text{po}}``\alpha . \supset_z . z = y$ (2)
$\vdash . (1) . (2) . *93·115 . \supset \vdash . \text{Prop}$

***122·43.** $\vdash : R \in \text{Prog} . \alpha \subset \text{D}`R . \exists ! \alpha - R_{\text{po}}``\alpha . \supset . \alpha \in \text{Cls induct}$
[*122·41 . *121·45 . *120·481]

Thus every class which is contained in a progression and has a last term is inductive. We have next to prove

$$R \in \text{Prog} . \alpha \subset \text{D}`R . \exists ! \alpha . \sim \exists ! \alpha - R_{\text{po}}``\alpha . \supset . \alpha \in \text{D}``\text{Prog}.$$

This is effected in the following propositions.

***122·44.** $\vdash : R \in \text{Prog} . \alpha \subset R_{\text{po}}``\alpha . \exists ! \alpha . P = R_{\text{po}} \upharpoonright \alpha . Q = P \dot{-} P^2 . \supset .$
$Q \in 1 \rightarrow 1 . Q \subseteq R_{\text{po}}$

Note. The hypothesis here exceeds what is necessary for the conclusion, but is the hypothesis required for *122·45, for which the present and the following propositions are lemmas.

Dem.

$\vdash . *23·43 . *35·442 . \supset \vdash : \text{Hp} . \supset . Q \subseteq R_{\text{po}}$ (1)
$\vdash . *36·13 . \quad \supset \vdash :. \text{Hp} . \supset : x, y, z \in \alpha . x R_{\text{po}} y . y R_{\text{po}} z . \supset . x P^2 z :$
$[\text{Transp}] \quad \supset : x, y, z \in \alpha . x R_{\text{po}} y . \sim(x P^2 z) . \supset . \sim(y R_{\text{po}} z) :$
$[*36·13] \quad \supset : x P y . \sim(x P^2 z) . \supset . \sim(y R_{\text{po}} z) :$
$[*3·47] \quad \supset : x Q y . x Q z . \supset . \sim(y R_{\text{po}} z) . \sim(z R_{\text{po}} y) .$
$[*122·21.(1)] \quad \supset . y = z$ (2)
Similarly $\quad \vdash :. \text{Hp} . \supset : x Q z . y Q z . \supset . x = y$ (3)
$\vdash . (1) . (2) . (3) . \supset \vdash . \text{Prop}$

***122·441.** $\vdash : \text{Hp} *122·44 . \supset . \text{D}`Q = \alpha$

Dem.

$\vdash . *37·41 . \supset \vdash : \text{Hp} . \supset . \text{D}`Q \subset \alpha$ (1)
$\vdash . *37·1 . \quad \supset \vdash :. \text{Hp} . \supset : x \in \alpha . \supset . (\exists y) . y \in \alpha . x R_{\text{po}} y .$
$[*36·13] \quad \supset . \exists ! \overleftarrow{P}`x .$
$[*122·23.*93·11] \quad \supset . \exists ! \overleftarrow{P}`x - \breve{R}_{\text{po}}``\overleftarrow{P}`x .$
$[*35·442] \quad \supset . \exists ! \overleftarrow{P}`x - \breve{P}``\overleftarrow{P}`x .$
$[*37·311.*32·31·35] \quad \supset . \exists ! \overleftarrow{Q}`x$ (2)
$\vdash . (1) . (2) . *33·4 . \supset \vdash . \text{Prop}$

∗122·442. $\vdash : \text{Hp} \ast 122{\cdot}44 \,.\supset .\, P = Q_{po}$

In proving $P \subseteq Q_{po}$ below, we assume xPz and consider the maximum of $\overrightarrow{R_{po}}{}^{\backprime}z \cap \overleftarrow{Q_\ast}{}^{\backprime}x$, which is shown to exist and be $Q{}^{\backprime}z$, whence $xQ_{po}z$.

Dem.

$\vdash . \ast 23{\cdot}43 . \supset \vdash : \text{Hp} . \supset . Q \subseteq P$ (1)

$\vdash . \ast 91{\cdot}56 . \supset \vdash :. \text{Hp} . \supset : P^2 \subseteq P :$

[(1)] $\supset : S \subseteq P . \supset . S | Q \subseteq P$ (2)

$\vdash . (1) . (2) . \ast 91{\cdot}171 . \ast 41{\cdot}151 . \supset \vdash : \text{Hp} . \supset . Q_{po} \subseteq P$ (3)

$\vdash . \ast 36{\cdot}13 . \ast 121{\cdot}1 . \supset \vdash :. \text{Hp} . \supset : xP^2z . \equiv . x, z \in \alpha . \exists ! \alpha \cap R(x - z) :$

[Transp.Fact] $\supset : xQz . \equiv . x, z \in \alpha . xR_{po}z . \alpha \cap R(x - z) = \Lambda$ (4)

$\vdash . \ast 122{\cdot}441 . \supset \vdash :. \text{Hp} . xPz . \supset : x \in (\overrightarrow{R_{po}}{}^{\backprime}z \cap \overleftarrow{Q_\ast}{}^{\backprime}x) :$

[∗122·27] $\supset : \exists ! \overrightarrow{R_{po}}{}^{\backprime}z \cap \overleftarrow{Q_\ast}{}^{\backprime}x - R_{po}{}^{\backprime\backprime}(\overrightarrow{R_{po}}{}^{\backprime}z \cap \overleftarrow{Q_\ast}{}^{\backprime}x) :$

[∗37·461] $\supset : (\exists y) . y \in \overrightarrow{R_{po}}{}^{\backprime}z \cap \overleftarrow{Q_\ast}{}^{\backprime}x . \overleftarrow{R_{po}}{}^{\backprime}y \cap \overrightarrow{R_{po}}{}^{\backprime}z \cap \overleftarrow{Q_\ast}{}^{\backprime}x = \Lambda :$

[∗90·151] $\supset : (\exists y) . y \in \overrightarrow{R_{po}}{}^{\backprime}z \cap \overleftarrow{Q_\ast}{}^{\backprime}x . \overleftarrow{R_{po}}{}^{\backprime}y \cap \overrightarrow{R_{po}}{}^{\backprime}z \cap \overleftarrow{Q}{}^{\backprime}y = \Lambda :$

[(4)] $\supset : (\exists y) : y \in \overrightarrow{R_{po}}{}^{\backprime}z \cap \overleftarrow{Q_\ast}{}^{\backprime}x :$

$\sim (\exists w) . w \in \overleftarrow{R_{po}}{}^{\backprime}y \cap \overrightarrow{R_{po}}{}^{\backprime}z . \alpha \cap \overleftarrow{R_{po}}{}^{\backprime}y \cap \overrightarrow{R_{po}}{}^{\backprime}w = \Lambda :$

[∗22·43.∗91·56] $\supset : (\exists y) : y \in \overrightarrow{R_{po}}{}^{\backprime}z \cap \overleftarrow{Q_\ast}{}^{\backprime}x :$

$\sim (\exists w) . w \in \alpha \cap \overleftarrow{R_{po}}{}^{\backprime}y \cap \overrightarrow{R_{po}}{}^{\backprime}z . \alpha \cap \overleftarrow{R_{po}}{}^{\backprime}y \cap \overrightarrow{R_{po}}{}^{\backprime}z \cap \overrightarrow{R_{po}}{}^{\backprime}w = \Lambda :$

[∗37·461] $\supset : (\exists y) . y \in \overrightarrow{R_{po}}{}^{\backprime}z \cap \overleftarrow{Q_\ast}{}^{\backprime}x .$

$\sim \exists ! \alpha \cap \overleftarrow{R_{po}}{}^{\backprime}y \cap \overrightarrow{R_{po}}{}^{\backprime}z - \breve{R}_{po}{}^{\backprime\backprime}(\alpha \cap \overleftarrow{R_{po}}{}^{\backprime}y \cap \overrightarrow{R_{po}}{}^{\backprime}z) :$

[∗122·28.Transp] $\supset : (\exists y) . y \in \overrightarrow{R_{po}}{}^{\backprime}z \cap \overleftarrow{Q_\ast}{}^{\backprime}x . \alpha \cap \overleftarrow{R_{po}}{}^{\backprime}y \cap \overrightarrow{R_{po}}{}^{\backprime}z = \Lambda :$

[(4)] $\supset : (\exists y) . y \in \overleftarrow{Q_\ast}{}^{\backprime}x . yQz :$

[∗91·52] $\supset : xQ_{po}z$ (5)

$\vdash . (3) . (5) . \supset \vdash . \text{Prop}$

∗122·443. $\vdash : \text{Hp} \ast 122{\cdot}44 . \supset . \min(R_{po}){}^{\backprime}\alpha = B{}^{\backprime}Q . \text{Ꭰ}{}^{\backprime}Q = \alpha \cap \breve{R}_{po}{}^{\backprime\backprime}\alpha$

Dem.

$\vdash . \ast 91{\cdot}504 . \ast 122{\cdot}442 . \supset \vdash : \text{Hp} . \supset . \text{Ꭰ}{}^{\backprime}Q = \text{Ꭰ}{}^{\backprime}P$

[∗37·41] $= \alpha \cap \breve{R}_{po}{}^{\backprime\backprime}\alpha$ (1)

$\vdash . (1) . \ast 122{\cdot}441 . \supset \vdash : \text{Hp} . \supset . \overrightarrow{B}{}^{\backprime}Q = \alpha - \breve{R}_{po}{}^{\backprime\backprime}\alpha$ (2)

$\vdash . (1) . (2) . \ast 122{\cdot}23 . \supset \vdash . \text{Prop}$

∗122·444. $\vdash : \text{Hp} \ast 122{\cdot}44 . \supset . \text{D}{}^{\backprime}Q = \overleftarrow{Q_\ast}{}^{\backprime}B{}^{\backprime}Q$

Dem.

$\vdash . \ast 122{\cdot}443 . \ast 14{\cdot}21 . \supset \vdash : \text{Hp} . \supset . \text{E} ! B{}^{\backprime}Q .$

[∗90·13] $\supset . \overleftarrow{Q_\ast}{}^{\backprime}B{}^{\backprime}Q \subset C{}^{\backprime}Q .$

[∗122·441·443] $\supset . \overleftarrow{Q_\ast}{}^{\backprime}B{}^{\backprime}Q \subset \alpha$ (1)

$\vdash . \ast 122{\cdot}443 . \ast 96{\cdot}303 . \supset$

$\vdash : \text{Hp} . x \in \alpha . x \neq B{}^{\backprime}Q . \supset . (B{}^{\backprime}Q) R_{po} x . B{}^{\backprime}Q, x \in \alpha .$

$[\text{Hp}] \qquad \supset . (B\text{'}Q) Px .$

$[*122\cdot442] \qquad \supset . (B\text{'}Q) Q_{po} x \qquad (2)$

$\vdash . (2) . *91\cdot54 . \supset \vdash : \text{Hp} . x \in \alpha . \supset . (B\text{'}Q) Q_* x \qquad (3)$

$\vdash . (1) . (3) . \quad \supset \vdash : \text{Hp} . \supset . \overleftarrow{Q_*}\text{'}B\text{'}Q = \alpha$

$[*122\cdot441] \qquad = \text{D'}Q : \supset \vdash . \text{Prop}$

***122·45.** $\vdash : R \in \text{Prog} . \alpha \subset R_{po}\text{''}\alpha . \exists ! \alpha . P = R_{po} \upharpoonright \alpha . Q = P \dot{-} P^2 . \supset .$
$Q \in \text{Prog} . \text{D'}Q = \alpha \quad [*122\cdot44\cdot444\cdot441]$

This proposition shows that every series extracted from a progression and having no last term is a progression.

***122·46.** $\vdash : R \in \text{Prog} . \alpha \subset \text{D'}R . \supset . \alpha \in \text{Cls induct} \cup \text{D''Prog}$
$[*122\cdot43\cdot45 . *120\cdot212]$

This proposition shows that any number less than the number of terms in a progression is inductive. This result will be developed in the next number (*123).

***122·47.** $\vdash :. R \in \text{Prog} . \alpha \subset \text{D'}R . \supset : \alpha \in \text{Cls induct} - \iota\text{'}\Lambda . \equiv . \exists ! \alpha - R_{po}\text{''}\alpha$

Dem.

$\vdash . *122\cdot45 . \supset \vdash : \text{Hp} . \exists ! \alpha . \sim \exists ! \alpha - R_{po}\text{''}\alpha . \supset . \alpha \in \text{D''Prog} .$

$[*122\cdot37] \qquad \supset . \alpha \sim \in \text{Cls induct} \qquad (1)$

$\vdash . (1) . *122\cdot43 . \supset \vdash . \text{Prop}$

***122·48.** $\vdash : R \in \text{Prog} . \alpha \subset \text{D'}R . \alpha \in \text{Cls induct} . \supset . \text{D'}R - \alpha \sim \in \text{Cls induct}$

Dem.

$\vdash . *120\cdot71 . \supset \vdash : \alpha \subset \text{D'}R . \alpha, \text{D'}R - \alpha \in \text{Cls induct} . \supset . \text{D'}R \in \text{Cls induct} :$

$[\text{Transp}] \quad \supset \vdash : \alpha \subset \text{D'}R . \alpha \in \text{Cls induct} . \text{D'}R \sim \in \text{Cls induct} . \supset .$
$\text{D'}R - \alpha \sim \in \text{Cls induct} \qquad (1)$

$\vdash . (1) . *122\cdot37 . \supset \vdash . \text{Prop}$

***122·49.** $\vdash : R \in \text{Prog} . \alpha \subset \text{D'}R . \alpha \in \text{Cls induct} . \supset . \text{D'}R - \alpha \in \text{D''Prog}$
$[*122\cdot46\cdot48]$

The following propositions are concerned with circumstances under which the posterity or the family of a term forms a progression.

***122·51.** $\vdash : P \in \text{Cls} \to 1 . I_P\text{'}x = \Lambda . x \in \text{D'}P . \overleftarrow{P_*}\text{'}x \subset \text{D'}P . \supset . (\overleftarrow{P_*}\text{'}x) \upharpoonright P \in \text{Prog}$

Here $I_P\text{'}x$ has the meaning defined in *96.

Dem.

$\vdash . *71\cdot261 . *96\cdot13 . \supset \vdash : \text{Hp} . Q = (\overleftarrow{P_*}\text{'}x) \upharpoonright P . \supset .$
$Q \in \text{Cls} \to 1 . Q_{po} = (\overleftarrow{P_*}\text{'}x) \upharpoonright P_{po} . \quad (1)$

$[*96\cdot104] \qquad \supset . Q_{po} \Subset J \qquad (2)$

$\vdash . *35\cdot61 . *37\cdot4 . \quad \supset \vdash : \text{Hp}(1) . \supset . \text{D}'Q = \overleftarrow{P_*}'x . \text{Cl}'Q = \breve{P}``\overleftarrow{P_*}'x$ (3)

[$*91\cdot52$] $= \overleftarrow{P}_{\text{po}}'x$

[(1)] $= \overleftarrow{Q}_{\text{po}}'x$

[(2).$*91\cdot542$] $= \overleftarrow{Q_*}'x - \iota'x$ (4)

$\vdash . (1) . (3) . (4) . \quad \supset \vdash : \text{Hp}(1) . \supset . \text{D}'Q = \overleftarrow{Q_*}'x . \text{Cl}'Q = \overleftarrow{Q_*}'x - \iota'x .$

[$*93\cdot101$] $\supset . \overrightarrow{B}'Q = \iota'x . \text{D}'Q = \overleftarrow{Q_*}'x$ (5)

$\vdash . (1) . (2) . (5) . \supset$

$\vdash : \text{Hp} . Q = (\overleftarrow{P_*}'x) \upharpoonright P . \supset . Q \in \text{Cls} \rightarrow 1 . Q_{\text{po}} \subseteq J . \text{D}'Q = \overleftarrow{Q_*}'B'Q .$

[$*122\cdot17$] $\supset . Q \in \text{Prog} : \supset \vdash . \text{Prop}$

The following proposition ($*122\cdot52$) is used in $*123\cdot191$, $*261\cdot4$ and $*264\cdot22$.

$*122\cdot52$. $\vdash : P \in 1 \rightarrow 1 . x \in \text{D}'P . \sim(xP_{\text{po}}x) . \overleftarrow{P_*}'x \subset \text{D}'P . \supset . (\overleftarrow{P_*}'x) \upharpoonright P \in \text{Prog}$

Dem.

$\vdash . *96\cdot492 . \supset \vdash : \text{Hp} . \supset . I_P'x = \Lambda$ (1)

$\vdash . (1) . *122\cdot51 . \supset \vdash . \text{Prop}$

The remaining propositions ($*122\cdot53\cdot54\cdot55$) are not used in the sequel.

$*122\cdot53$. $\vdash : P \in 1 \rightarrow 1 . x \in s'\text{gen}'P . \overleftarrow{P_*}'x \subset \text{D}'P . \supset . (\overleftrightarrow{P_*}'x) \upharpoonright P \in \text{Prog}$

Dem.

$\vdash . *97\cdot21 . \supset \vdash : \text{Hp} . \supset . (\exists y) . yBP . \overleftrightarrow{P_*}'x = \overleftarrow{P_*}'y .$

[$*96\cdot23 . *93\cdot1$] $\supset . (\exists y) . y \in \text{D}'P . \overleftrightarrow{P_*}'x = \overleftarrow{P_*}'y . I_P'y = \Lambda .$

[$*97\cdot17 . *91\cdot504$.Hp] $\supset . (\exists y) . y \in \text{D}'P . \overleftarrow{P_*}'y \subset \text{D}'P . I_P'y = \Lambda . \overleftrightarrow{P_*}'x = \overleftarrow{P_*}'y .$

[$*122\cdot51$] $\supset . (\overleftrightarrow{P_*}'x) \upharpoonright P \in \text{Prog} : \supset \vdash . \text{Prop}$

$*122\cdot54$. $\vdash : P \in 1 \rightarrow 1 . x \in s'\text{gen}'P - s'\text{gen}'\breve{P} . \supset . (\overleftrightarrow{P_*}'x) \upharpoonright P \in \text{Prog}$

Dem.

$\vdash . *93\cdot27\cdot272 . \supset \vdash : \text{Hp} . \supset . x \in s'\text{gen}'P \cap p'\text{Cl}``\text{Pot}'\breve{P} .$

[$*93\cdot381$] $\supset . x \in s'\text{gen}'P . \overleftarrow{P_*}'x \subset \text{D}'P$ (1)

$\vdash . (1) . *122\cdot53 . \supset \vdash . \text{Prop}$

$*122\cdot55$. $\vdash :. P \in 1 \rightarrow 1 . \supset : x \in s'\text{gen}'P - s'\text{gen}'\breve{P} . \equiv . (\overleftrightarrow{P_*}'x) \upharpoonright P \in \text{Prog}$

Dem.

$\vdash . *35\cdot61 . \supset \vdash : Q = (\overleftrightarrow{P_*}'x) \upharpoonright P . \supset . \text{D}'Q = \overleftrightarrow{P_*}'x \cap \text{D}'P$ (1)

$\vdash . *37 \cdot 4 . \quad \supset \vdash :. Q = (\overleftrightarrow{P}_{*}{}^{\backprime}x) \upharpoonright P . \supset : \text{ꓷ}{}^{\backprime}Q = \breve{P}{}^{\backprime\backprime}\overleftrightarrow{P}_{*}{}^{\backprime}x :$

$[*97 \cdot 17 . *92 \cdot 111 . *91 \cdot 54 \cdot 52] \supset : Q \epsilon 1 \rightarrow 1 . \supset . \text{ꓷ}{}^{\backprime}Q = \overleftrightarrow{P}_{*}{}^{\backprime}x \cap \text{ꓷ}{}^{\backprime}P \quad (2)$

$\vdash . (1) . (2) . \supset \vdash :. \text{Hp} . \text{Hp}(1) . \supset : \exists ! \overrightarrow{B}{}^{\backprime}Q . \supset . \exists ! \overleftrightarrow{P}_{*}{}^{\backprime}x - \text{ꓷ}{}^{\backprime}P .$

$[*97 \cdot 17 . *91 \cdot 504] \qquad \supset . \exists ! \overrightarrow{P}_{*}{}^{\backprime}x - \text{ꓷ}{}^{\backprime}P .$

$[*93 \cdot 38 \cdot 27] \qquad \supset . x \epsilon s{}^{\backprime}\text{gen}{}^{\backprime}P \quad (3)$

$\vdash . (1) . (2) . \supset \vdash :. \text{Hp}(3) . \supset : \text{D}{}^{\backprime}Q = C{}^{\backprime}Q . \supset . \overleftrightarrow{P}_{*}{}^{\backprime}x \cap \text{D}{}^{\backprime}P = \overleftrightarrow{P}_{*}{}^{\backprime}x .$

$[*22 \cdot 621] \qquad \supset . \overleftrightarrow{P}_{*}{}^{\backprime}x \subset \text{D}{}^{\backprime}P .$

$[*97 \cdot 13] \qquad \supset . \overleftarrow{P}_{*}{}^{\backprime}x \subset \text{D}{}^{\backprime}P .$

$[*93 \cdot 381 \cdot 275] \qquad \supset . x \sim \epsilon s{}^{\backprime}\text{gen}{}^{\backprime}\breve{P} \quad (4)$

$\vdash . (3) . (4) . *122 \cdot 11 \cdot 141 \cdot 54 . \supset \vdash . \text{Prop}$

$*123.\ \aleph_0$.

Summary of $*123$.

In this number we are concerned with the arithmetical properties of $\aleph_0$, the smallest of Cantor's transfinite cardinals. Cantor defines $\aleph_0$ as the cardinal number of any class which can be put into one-one relation with the inductive cardinals. This definition assumes that $\nu \neq \nu +_c 1$, when ν is an inductive cardinal; in other words, it assumes the axiom of infinity; for without this, the inductive cardinals would form a finite series, with a last term, namely Λ. For this reason among others, we do not make similarity with the inductive cardinals our *definition*. We define $\aleph_0$ as the class of those classes which can be arranged in progressions, *i.e.* as D''Prog. We then have to prove that $\aleph_0$ so defined is a cardinal, and that if it is not null, it is the number of the inductive numbers.

For convenience we put for the moment N for the relation of μ to $\mu +_c 1$ when μ is an inductive cardinal. We then easily prove

$*123{\cdot}21{\cdot}23$. $\vdash . N \in \mathrm{Cls} \to 1 . \mathrm{D}'N = \mathrm{NC\,induct} . B'N = 0 . \overleftarrow{N_*}'0 = \mathrm{NC\,induct}$

The only thing further required to prove $N \in \mathrm{Prog}$ is $N \in 1 \to \mathrm{Cls}$, *i.e.*

$$\mu, \nu \in \mathrm{NC\,induct} . \mu +_c 1 = \nu +_c 1 . \supset . \mu = \nu .$$

By $*120{\cdot}311$, this holds if $\exists ! \mu +_c 1$, which holds if Infin ax holds. Hence

$*123{\cdot}25{\cdot}26$. $\vdash : \mathrm{Infin\,ax}\,(x) . \supset . N \upharpoonright t^3{}'x \in \mathrm{Prog} . \mathrm{NC\,induct} \cap t^3{}'x \in \aleph_0$

whence, by $*122{\cdot}36$,

$*123{\cdot}27$. $\vdash : \exists ! \aleph_0 (x) . \supset . \mathrm{NC\,induct} \cap t^3{}'x \in \aleph_0$

Again it is obvious from $*122{\cdot}34{\cdot}341$ that if R is a progression, D'R can always be put into a $1 \to 1$ relation to the inductive cardinals ($*123{\cdot}3$) since D'R consists of the terms $1_R, 2_R, \ldots \nu_R, \ldots$, and all the inductive cardinals are used in putting D'R into this form. Hence

$*123{\cdot}31$. $\vdash : \alpha \in \aleph_0 . \supset . \alpha \,\mathrm{sm}\, \mathrm{NC\,induct}$

whence also

$*123{\cdot}311$. $\vdash : \alpha, \beta \in \aleph_0 . \supset . \alpha \,\mathrm{sm}\, \beta$

It remains to prove that any class similar to the inductive cardinals is an $\aleph_0$; this can only be proved by assuming the axiom of infinity. We prove first (*123·32) that if R is a progression, and S is a one-one whose converse domain is D'R, then $S \mid R \mid \breve{S}$ is a progression whose domain is D'S. Hence

***123·321.** $\vdash : \alpha \epsilon \aleph_0 . \alpha \text{ sm } \beta . \supset . \beta \epsilon \aleph_0$

From this and $\alpha, \beta \epsilon \aleph_0 . \supset . \alpha \text{ sm } \beta$, we obtain

***123·322.** $\vdash : \alpha \epsilon \aleph_0 . \supset . \aleph_0 = \text{Nc}\text{‘}\alpha$

Hence by our previous results

***123·34.** $\vdash : \text{Infin ax}(x) . \supset . \aleph_0 = \text{Nc}\text{‘}(\text{NC induct} \cap t^3\text{‘}x)$

Also we have, by *123·322 above,

$$\exists ! \aleph_0 . \supset . \aleph_0 \epsilon \text{NC},$$

whence, since $\Lambda \epsilon \text{NC}$, we obtain at last

***123·36.** $\vdash . \aleph_0 \epsilon \text{NC}$

As to the existence of $\aleph_0$ in various types, if Infin ax (x) holds, *i.e.* if, given any inductive cardinal ν, there are classes having ν terms and composed of terms of the same type as x, then NC induct $(t\text{‘}x) \epsilon \aleph_0 (t^2\text{‘}x)$. Thus

***123·37.** $\vdash : \text{Infin ax}(x) . \supset . \exists ! \aleph_0 (t^2\text{‘}x) . \aleph_0 (t^2\text{‘}x) \epsilon \text{N}_0\text{C}$

The arithmetical properties of $\aleph_0$ in regard to addition, multiplication and exponentiation by an inductive cardinal are easily proved. We have

***123·41.** $\vdash : \nu \epsilon \text{NC induct} . \supset . \aleph_0 = \aleph_0 +_c \nu$

***123·421.** $\vdash . \aleph_0 = \aleph_0 +_c \aleph_0 = 2 \times_c \aleph_0$

***123·422.** $\vdash : \nu \epsilon \text{NC induct} - \iota\text{‘}0 . \supset . \nu \times_c \aleph_0 = \aleph_0$

***123·52.** $\vdash . \aleph_0 = \aleph_0 \times_c \aleph_0 = \aleph_0{}^2$

***123·53.** $\vdash : \nu \epsilon \text{NC induct} - \iota\text{‘}0 . \supset . \aleph_0{}^\nu = \aleph_0$

All these propositions are well known.

The early propositions of the present number are for the most part immediate consequences of propositions proved in *122.

***123·01.** $\aleph_0 = \text{D}\text{‘‘Prog}$ Df

***123·02.** $N = \hat{\mu}\hat{\nu} \{\mu \epsilon \text{NC induct} . \nu = (\mu +_c 1) \cap t_0\text{‘}\mu\}$ Dft [*123—4]

***123·1.** $\vdash : \alpha \epsilon \aleph_0 . \equiv . (\exists R) . R \epsilon \text{Prog} . \alpha = \text{D}\text{‘}R$ [*37·1 . (*123·01)]

***123·101.** $\vdash : R \epsilon \text{Prog} . \supset . \text{D}\text{‘}R \epsilon \aleph_0$ [*123·1]

***123·11.** $\vdash : R \epsilon 1 \rightarrow 1 . \text{D}\text{‘}R = \overleftarrow{R_*}\text{‘}B\text{‘}R . \supset . \text{D}\text{‘}R \epsilon \aleph_0$ [*123·101 . *122·1]

***123·12.** $\vdash : \alpha \epsilon \aleph_0 . \supset . (\exists R) . \text{D}\text{‘}R = \alpha . R \epsilon 1 \rightarrow 1 . \text{Ꮧ}\text{‘}R \subset \text{D}\text{‘}R . \overrightarrow{B}\text{‘}R \epsilon 1$
[*123·1 . *122·141·11]

***123·13.** $\vdash : \alpha \in \aleph_0 . \supset . \text{Nc}‘\alpha = \text{Nc}‘\alpha +_c 1$

Dem.

$\vdash . *123·12 . *110·32 . \supset$

$\vdash : \alpha \in \aleph_0 . \supset . (\exists R) . \text{D}‘R = \alpha . R \in 1 \to 1 . \text{Nc}‘\text{D}‘R = \text{Nc}‘\text{G}‘R +_c 1 .$

[*100·321] $\supset . (\exists R) . \text{D}‘R = \alpha . \text{Nc}‘\text{D}‘R = \text{Nc}‘\text{D}‘R +_c 1 .$

[*35·94.*13·195] $\supset . \text{Nc}‘\alpha = \text{Nc}‘\alpha +_c 1 : \supset \vdash . \text{Prop}$

***123·14.** $\vdash : \alpha \in \aleph_0 . \nu \in \text{NC induct} . \supset . \exists ! \nu \cap \text{Cl}‘\alpha$ [*122·35]

***123·15.** $\vdash : \alpha \in \aleph_0 . \supset . \alpha \sim \in \text{Cls induct}$ [*122·37]

***123·16.** $\vdash : \alpha \in \aleph_0 . \supset . \text{Cl}‘\alpha \subset \text{Cls induct} \cup \aleph_0$ [*122·46]

***123·17.** $\vdash : \alpha \in \aleph_0 . \beta \in \text{Cls induct} . \supset . \alpha - \beta \in \aleph_0$

Dem.

$\vdash . *120·481 . \supset \vdash : \text{Hp} . \supset . \alpha \cap \beta \in \text{Cls induct} .$

[*122·49] $\supset . \alpha - (\alpha \cap \beta) \in \aleph_0 : \supset \vdash . \text{Prop}$

***123·18.** $\vdash : \exists ! \aleph_0 (x) . \supset . \text{Infin ax} (x)$ [*122·36]

***123·19.** $\vdash : R \in \text{Prog} . \exists ! \alpha . \alpha \subset R_{po}“\alpha . \supset . \alpha \in \aleph_0$ [*122·45]

***123·191.** $\vdash : R \in 1 \to 1 . x \in \text{D}‘R . \sim (x R_{po} x) . \overleftarrow{R_*}‘x \subset \text{D}‘R . \supset . \overleftarrow{R_*}‘x \in \aleph_0$ [*122·52]

***123·192.** $\vdash : R \in 1 \to 1 . \text{G}‘R \subset \text{D}‘R . \supset . \overleftarrow{R_*}“\overrightarrow{B}‘R \subset \aleph_0$

Dem.

$\vdash . *93·101 . \quad \supset \vdash : x \in \overrightarrow{B}‘R . \supset . x \in \text{D}‘R \quad (1)$

$\vdash . *91·504 . *93·101 . \quad \supset \vdash : x \in \overrightarrow{B}‘R . \supset . \sim (x R_{po} x) \quad (2)$

$\vdash . *90·13 . \quad \supset \vdash : \text{G}‘R \subset \text{D}‘R . \supset . \overleftarrow{R_*}‘x \subset \text{D}‘R \quad (3)$

$\vdash . (1) . (2) . (3) . *123·191 . \supset \vdash : \text{Hp} . x \in \overrightarrow{B}‘R . \supset . \overleftarrow{R_*}‘x \in \aleph_0 : \supset \vdash . \text{Prop}$

***123·2.** $\vdash : \mu N \nu . \equiv . \mu \in \text{NC induct} . \nu = (\mu +_c 1) \cap t_0‘\mu$ [(*123·02)]

***123·21.** $\vdash . N \in \text{Cls} \to 1 . \text{D}‘N = \text{NC induct} . \text{G}‘N = \text{NC induct} - \iota‘0 . B‘N = 0$

Dem.

$\vdash . *123·2 . *13·172 . \supset \vdash : \mu N \nu . \mu N \varpi . \supset . \nu = \varpi :$

[*71·171] $\supset \vdash . N \in \text{Cls} \to 1 \quad (1)$

$\vdash . *123·2 . \quad \supset \vdash . \text{D}‘N = \text{NC induct} \quad (2)$

$\vdash . *123·2 . \quad \supset \vdash . \text{G}‘N = \hat{\nu} \{(\exists \mu) . \mu \in \text{NC induct} . \nu = \mu +_c 1\}$

[*120·423] $= \text{NC induct} - \iota‘0 \quad (3)$

$\vdash . (2) . (3) . *93·101 . \supset \vdash . B‘N = 0 \quad (4)$

$\vdash . (1) . (2) . (3) . (4) . \supset \vdash . \text{Prop}$

***123·22.** $\vdash . \breve{N} = (+_c 1) \upharpoonright \text{NC induct}$ [*123·2]

***123·23.** $\vdash . \overleftarrow{N_*}\text{‘}0 = \text{NC induct} = \text{D‘}N$

Dem.

$\vdash . *123{\cdot}22 . \supset \vdash . \overleftarrow{N_*}\text{‘}0 = \hat{\mu}[\mu \{(+_c 1) \upharpoonright \text{NC induct}\}_* 0]$

$[*120{\cdot}1 . *96{\cdot}21{\cdot}131] \quad = \hat{\mu}[\mu \{\text{NC induct} \upharpoonleft (+_c 1)_*\} 0]$

$[*120{\cdot}1] \quad = \text{NC induct} \quad (1)$

$\vdash . (1) . *123{\cdot}21 . \supset \vdash . \text{Prop}$

***123·24.** $\vdash : \text{Infin ax}(x) . \supset . N \upharpoonright t^3\text{‘}x \in 1 \to 1$

Dem.

$\vdash . *120{\cdot}301{\cdot}121 . \supset \vdash :: \text{Hp} . \supset :. \mu \in \text{NC induct} . \supset : \exists ! (\mu +_c 1) \cap t^2\text{‘}x :$

$[*120{\cdot}311] \quad \supset : (\mu +_c 1) \cap t^2\text{‘}x = \nu +_c 1 . \supset . \mu = \nu :$

$[*123{\cdot}2 . *71{\cdot}17] \quad \supset : N \upharpoonright t^3\text{‘}x \in 1 \to \text{Cls} \quad (1)$

$\vdash . (1) . *123{\cdot}21 . \supset \vdash . \text{Prop}$

***123·25.** $\vdash : \text{Infin ax}(x) . \supset . N \upharpoonright t^3\text{‘}x \in \text{Prog}$ $\quad [*123{\cdot}21{\cdot}23{\cdot}24 . *122{\cdot}1]$

***123·26.** $\vdash : \text{Infin ax}(x) . \supset . \text{NC induct} \cap t^3\text{‘}x \in \aleph_0$ $\quad [*123{\cdot}25{\cdot}21{\cdot}101]$

***123·27.** $\vdash : \exists ! \aleph_0(x) . \supset . \text{NC induct} \cap t^3\text{‘}x \in \aleph_0$ $\quad [*123{\cdot}26{\cdot}18]$

***123·3.** $\vdash : R \in \text{Prog} . S = \hat{x}\hat{\nu}\{\nu \in \text{NC induct} . x = (\nu +_c 1)_R\} . \supset .$

$S \in 1 \to 1 . \text{D‘}S = \text{D‘}R . \text{Ꙃ‘}S = \text{NC induct}$

Dem.

$\vdash . *120{\cdot}423 . \supset \vdash : \text{Hp} . \supset . \text{D‘}S = \hat{x}\{(\exists \mu) . \mu \in \text{NC induct} - \iota\text{‘}0 . x = \mu_R\}$

$[*122{\cdot}341] \quad = \text{D‘}R \quad (1)$

$\vdash . *14{\cdot}204 . *122{\cdot}34 . \supset \vdash : \text{Hp} . \supset . \text{Ꙃ‘}S = \hat{\nu}\{\text{E} ! (\nu +_c 1)_R\}$

$[*122{\cdot}34] \quad = \hat{\nu}\{\nu +_c 1 \in \text{NC induct} - \iota\text{‘}0\} \quad (2)$

$\vdash . *122{\cdot}36 . *120{\cdot}3 . \supset \vdash :. \text{Hp} . \supset : \nu +_c 1 \in \text{NC induct} . \supset . \exists ! \nu +_c 1 .$

$[*120{\cdot}422] \quad \supset . \nu \in \text{NC induct} \quad (3)$

$\vdash . (3) . *120{\cdot}421{\cdot}121 . \supset \vdash :. \text{Hp} . \supset : \nu +_c 1 \in \text{NC induct} - \iota\text{‘}0 . \equiv .$

$\nu \in \text{NC induct} \quad (4)$

$\vdash . (2) . (4) . \quad \supset \vdash : \text{Hp} . \supset . \text{Ꙃ‘}S = \text{NC induct} \quad (5)$

$\vdash . *13{\cdot}172 . *71{\cdot}17 . \quad \supset \vdash : \text{Hp} . \supset . S \in 1 \to \text{Cls} \quad (6)$

$\vdash . *121{\cdot}631 . \quad \supset \vdash :. \text{Hp} . \supset : xS\mu . xS\nu . \supset .$

$\text{Nc‘}R(B\text{‘}R \mapsto x) = \mu +_c 1 . \text{Nc‘}R(B\text{‘}R \mapsto x) = \nu +_c 1 .$

$[*13{\cdot}171] \quad \supset . \mu +_c 1 = \nu +_c 1 \quad (7)$

$\vdash . (5) . *122{\cdot}36 . *120{\cdot}3 . \supset \vdash :. \text{Hp} . \supset : xS\mu . \supset . \exists ! \mu +_c 1 :$

$[*120{\cdot}41] \quad \supset : xS\mu . \mu +_c 1 = \nu +_c 1 . \supset . \mu = \nu :$

$[(7)] \quad \supset : xS\mu . xS\nu . \supset . \mu = \nu :$

$[*71{\cdot}171] \quad \supset : S \in \text{Cls} \to 1 \quad (8)$

$\vdash . (1) . (5) . (6) . (8) . \supset \vdash . \text{Prop}$

***123·31.** $\vdash : \alpha \in \aleph_0 \,.\, \supset .\, \alpha \,\mathrm{sm}\, \mathrm{NC\,induct}$ [*123·3]

***123·311.** $\vdash : \alpha, \beta \in \aleph_0 \,.\, \supset .\, \alpha \,\mathrm{sm}\, \beta$ [*123·31 . *73·31·32]

It is not assumed here that α and β are of the same type.

***123·312.** $\vdash : R \in \mathrm{Prog} \,.\, S \in 1 \rightarrow 1 \,.\, \text{Ɑ}\text{‘}S = \mathrm{D}\text{‘}R \,.\, \supset .$

$$S \,|\, R \,|\, \breve{S} \in 1 \rightarrow 1 \,.\, \mathrm{D}\text{‘}S = \mathrm{D}\text{‘}(S \,|\, R \,|\, \breve{S}) \,.\, S\text{‘}B\text{‘}R = B\text{‘}(S \,|\, R \,|\, \breve{S})$$

Dem.

$\vdash . *71{\cdot}252 . *122{\cdot}1 . \supset \vdash : \mathrm{Hp} . \supset . S \,|\, R \,|\, \breve{S} \in 1 \rightarrow 1$ (1)

$\vdash . *122{\cdot}141 . *37{\cdot}321 . \supset \vdash : \mathrm{Hp} . \supset . \mathrm{D}\text{‘}(R \,|\, \breve{S}) = \mathrm{D}\text{‘}R = \text{Ɑ}\text{‘}S .$ (2)

[*37·323] $\supset . \mathrm{D}\text{‘}(S \,|\, R \,|\, \breve{S}) = \mathrm{D}\text{‘}S$ (3)

$\vdash . (2) . *37{\cdot}32 . \supset \vdash : \mathrm{Hp} . \supset . \text{Ɑ}\text{‘}(S \,|\, R \,|\, \breve{S}) = S\text{“}\text{Ɑ}\text{‘}R$ (4)

$\vdash . (3) . (4) . \supset \vdash : \mathrm{Hp} . \supset . \overrightarrow{B}\text{‘}(S \,|\, R \,|\, \breve{S}) = \mathrm{D}\text{‘}S - S\text{“}\text{Ɑ}\text{‘}R$

[*37·25.Hp] $= S\text{“}\mathrm{D}\text{‘}R - S\text{“}\text{Ɑ}\text{‘}R$

[*71·381] $= S\text{“}\overrightarrow{B}\text{‘}R$

[*122·11.*53·31] $= \iota\text{‘}S\text{‘}B\text{‘}R$ (5)

$\vdash . (1) . (3) . (5) . \supset \vdash . \mathrm{Prop}$

***123·313.** $\vdash : R \in \mathrm{Prog} \,.\, S \in 1 \rightarrow 1 \,.\, \text{Ɑ}\text{‘}S = \mathrm{D}\text{‘}R \,.\, P = S \,|\, R \,|\, \breve{S} \,.\, \supset .\, \mathrm{D}\text{‘}P = \overleftarrow{P}_*\text{‘}B\text{‘}P$

Dem.

$\vdash . *34{\cdot}36 . *123{\cdot}312 . \supset \vdash : \mathrm{Hp} . \supset . \text{Ɑ}\text{‘}P \subset \mathrm{D}\text{‘}P . \mathrm{E}\,!\, B\text{‘}P .$

[*90·13] $\supset . \overleftarrow{P}_*\text{‘}B\text{‘}P \subset \mathrm{D}\text{‘}P$ (1)

$\vdash . *123{\cdot}312 . \supset \vdash : \mathrm{Hp} . \supset . S\text{‘}B\text{‘}R \in \overleftarrow{P}_*\text{‘}B\text{‘}P$ (2)

$\vdash . *33{\cdot}14 . \supset \vdash : \mathrm{Hp} . S\text{‘}x \in \overleftarrow{P}_*\text{‘}B\text{‘}P . xRy . \supset . y \in \text{Ɑ}\text{‘}R .$

[*122·141.Hp] $\supset . y \in \text{Ɑ}\text{‘}S .$

[*71·16] $\supset . \mathrm{E}\,!\, S\text{‘}y .$

[*30·32.*34·1] $\supset . S\text{‘}x \,(S \,|\, R \,|\, \breve{S})\, S\text{‘}y .$

[Hp] $\supset . S\text{‘}x \, P \, S\text{‘}y .$

[*90·163] $\supset . S\text{‘}y \in \overleftarrow{P}_*\text{‘}B\text{‘}P$ (3)

$\vdash . (2) . (3) . *90{\cdot}112 . \supset \vdash :. \mathrm{Hp} . \supset : (B\text{‘}R) R_* x . \supset . S\text{‘}x \in \overleftarrow{P}_*\text{‘}B\text{‘}P :$

[*37·63] $\supset : S\text{“}\overleftarrow{R}_*\text{‘}B\text{‘}R \subset \overleftarrow{P}_*\text{‘}B\text{‘}P :$

[*122·1] $\supset : S\text{“}\mathrm{D}\text{‘}R \subset \overleftarrow{P}_*\text{‘}B\text{‘}P :$

[*37·25.Hp] $\supset : \mathrm{D}\text{‘}S \subset \overleftarrow{P}_*\text{‘}B\text{‘}P :$

[*123·312] $\supset : \mathrm{D}\text{‘}P \subset \overleftarrow{P}_*\text{‘}B\text{‘}P$ (4)

$\vdash . (1) . (4) . \supset \vdash . \mathrm{Prop}$

***123·32.** $\vdash : R \in \mathrm{Prog} \,.\, S \in 1 \rightarrow 1 \,.\, \text{Ɑ}\text{‘}S = \mathrm{D}\text{‘}R \,.\, \supset .$

$$S \,|\, R \,|\, \breve{S} \in \mathrm{Prog} \,.\, \mathrm{D}\text{‘}S = \mathrm{D}\text{‘}S \,|\, R \,|\, \breve{S} \,.\, S\text{‘}B\text{‘}R = B\text{‘}(S \,|\, R \,|\, \breve{S})$$ [*123·312·313]

***123·321.** $\vdash : \alpha \epsilon \aleph_0 . \alpha \operatorname{sm} \beta . \supset . \beta \epsilon \aleph_0$ [*123·32]

***123·322.** $\vdash : \alpha \epsilon \aleph_0 . \supset . \aleph_0 = \operatorname{Nc}‘\alpha$

Dem.

$\vdash . *123·311·321 . \supset \vdash :. \alpha \epsilon \aleph_0 . \supset : \beta \epsilon \aleph_0 . \equiv . \beta \operatorname{sm} \alpha$ (1)

$\vdash . (1) . *100·1 . \supset \vdash . \text{Prop}$

***123·323.** $\vdash : R \epsilon \text{Prog} . \supset . \aleph_0 = \operatorname{Nc}‘\text{D}‘R$ [*123·322]

***123·33.** $\vdash :. \text{Infin ax}(x) . \supset : \alpha \epsilon \aleph_0 . \equiv . \alpha \operatorname{sm} (\text{NC induct} \cap t^3‘x)$ [*123·26·321·31]

***123·34.** $\vdash : \text{Infin ax}(x) . \supset . \aleph_0 = \operatorname{Nc}‘(\text{NC induct} \cap t^3‘x)$ [*123·33]

***123·35.** $\vdash : \exists ! \aleph_0 (x) . \supset . \aleph_0 (x) = \operatorname{Nc}‘(\text{NC induct} \cap t^3‘x)$ [*123·34·18]

***123·36.** $\vdash . \aleph_0 \epsilon \text{NC}$ [*123·35 . *102·74]

***123·361.** $\vdash : \exists ! \aleph_0 . \supset . \aleph_0 \sim \epsilon \text{ NC induct}$ [*123·15·322 . *120·211]

***123·37.** $\vdash : \text{Infin ax}(x) . \supset . \exists ! \aleph_0 (t^2‘x) . \aleph_0 (t^2‘x) \epsilon \text{N}_0\text{C}$

Dem.

$\vdash . *120·301 . \supset \vdash :. \text{Hp} . \supset : \nu \epsilon \text{NC induct} . \supset_\nu . \exists ! \nu (x) :$

[*65·13] $\supset : \nu \epsilon \text{NC induct} . \supset_\nu . \exists ! \nu . \nu = \nu (x) :$

[(*65·02)] $\supset : \nu \epsilon \text{NC induct} . \supset_\nu . \exists ! \nu : \text{NC induct} \subset t^3‘x :$

[*123·34] $\supset : \text{NC induct} \epsilon \aleph_0 . \text{NC induct} \subset t^3‘x :$

[(*65·02)] $\supset : \text{NC induct} \epsilon \aleph_0 (t^2‘x)$ (1)

$\vdash . (1) . *103·34 . *123·36 . \supset \vdash . \text{Prop}$

***123·39.** $\vdash . (\aleph_0)_\eta = (\aleph_0 +_c 1)_\eta$

Dem.

$\vdash . *118·12 . *117·6 . *123·322 . \supset \vdash : (\aleph_0)_\eta = \Lambda . \supset . (\aleph_0 +_c 1)_\eta = \Lambda$ (1)

$\vdash . *123·13·322 . \quad \supset \vdash : \exists ! (\aleph_0)_\eta . \supset . (\aleph_0)_\eta = (\aleph_0 +_c 1)_\eta$ (2)

$\vdash . (1) . (2) . \supset \vdash . \text{Prop}$

***123·4.** $\vdash . \aleph_0 = \aleph_0 +_c 1$ [*123·39]

***123·401.** $\vdash : \exists ! \aleph_0 . \supset . \aleph_0 = \aleph_0 -_c 1$

Dem.

$\vdash . *120·124 . *123·36·4 . \supset \vdash : \exists ! \aleph_0 . \supset . \aleph_0 \epsilon \text{NC} - \iota‘0 .$

[*120·414·416] $\supset . (\aleph_0 -_c 1) +_c 1 = \aleph_0$

[*123·4] $= \aleph_0 +_c 1 .$

[*120·311] $\supset . \aleph_0 -_c 1 = \aleph_0$ (1)

$\vdash . *119·11 . \supset \vdash : (\aleph_0)_\eta = \Lambda . \supset . (\aleph_0)_\eta = (\aleph_0 -_c 1)_\eta$ (2)

$\vdash . (1) . (2) . \supset \vdash . \text{Prop}$

***123·41.** $\vdash : \nu \epsilon \text{NC induct} . \supset . \aleph_0 = \aleph_0 +_c \nu$ [*123·4 . *120·11]

***123·411.** $\vdash : \nu \epsilon \text{NC induct} . \supset . \aleph_0 = \aleph_0 -_c \nu$ [*123·401 . *120·11]

∗123·42. $\vdash : P \epsilon \text{Prog} \,.\, Q = P^2 \,.\, \supset \,.\, \overleftarrow{Q_*}\text{‘}1_P, \overleftarrow{Q_*}\text{‘}2_P \epsilon \aleph_0 \,.\, \overleftarrow{Q_*}\text{‘}1_P \cap \overleftarrow{Q_*}\text{‘}2_P = \Lambda$

Note that $\overleftarrow{Q_*}\text{‘}1_P$ is the odd terms and $\overleftarrow{Q_*}\text{‘}2_P$ the even terms of $\text{D‘}P$.

Dem.

$\vdash . ∗91·6 . \supset \vdash :. \text{Hp} . \supset : \overleftarrow{Q_*}\text{‘}1_P \subset \overleftarrow{P_*}\text{‘}1_P . \overleftarrow{Q_*}\text{‘}2_P \subset \overleftarrow{P_*}\text{‘}2_P :$

[∗122·1] $\supset : \overleftarrow{Q_*}\text{‘}1_P \subset \text{D‘}P :$

[∗33·13] $\supset : y \epsilon \overleftarrow{Q_*}\text{‘}1_P . \supset . (\exists z) . yPz .$

[∗122·141] $\supset . (\exists z, w) . yPz . zPw .$

[Hp.∗90·163.∗91·503] $\supset . (\exists w) . yQw . w \epsilon \overleftarrow{Q_*}\text{‘}1_P . yP_{\text{po}}w :$

[∗37·1] $\supset : \overleftarrow{Q_*}\text{‘}1_P \subset P_{\text{po}}\text{‘‘}\overleftarrow{Q_*}\text{‘}1_P :$

[∗123·19] $\supset : \overleftarrow{Q_*}\text{‘}1_P \epsilon \aleph_0$ (1)

Similarly $\vdash : \text{Hp} . \supset . \overleftarrow{Q_*}\text{‘}2_P \epsilon \aleph_0$ (2)

$\vdash . ∗121·601·602 . \supset \vdash : \text{Hp} . \supset . 1_P P 2_P .$

[∗122·16.∗91·52·6] $\supset . \sim(2_P Q_* 1_P)$ (3)

$\vdash . ∗121·602 . ∗53·31 . ∗93·1 . \supset \vdash :. \text{Hp} . \supset : \overrightarrow{Q}\text{‘}2_P = \overrightarrow{P}\text{‘}1_P = \Lambda :$

[∗13·14] $\supset : yQz . \supset . z \neq 2_P :$

[∗91·542] $\supset : 2_P Q_* z . yQz . \supset . 2_P Q_{\text{po}} z . yQz .$

[∗92·11] $\supset . 2_P Q_* y :$

[Transp] $\supset : \sim(2_P Q_* y) . yQz . \supset . \sim(2_P Q_* z)$ (4)

$\vdash . (3) . (4) . ∗90·112 . \supset \vdash :. \text{Hp} . \supset : 1_P Q_* z . \supset . \sim(2_P Q_* z)$ (5)

$\vdash . (1) . (2) . (5) . \supset \vdash . \text{Prop}$

∗123·421. $\vdash . \aleph_0 = \aleph_0 +_c \aleph_0 = 2 \times_c \aleph_0$

Dem.

$\vdash . ∗123·42 . \supset \vdash : \alpha \epsilon \aleph_0 . \supset . (\exists \beta, \gamma) . \beta, \gamma \epsilon \aleph_0 . \beta \cap \gamma = \Lambda . \beta \cup \gamma \subset \alpha .$

[∗110·32.∗117·22] $\supset . \text{Nc‘}\alpha \geqslant \aleph_0 +_c \aleph_0$ (1)

$\vdash . (1) . ∗117·6·23 . \supset \vdash : \exists ! \aleph_0 . \supset . \aleph_0 = \aleph_0 +_c \aleph_0$ (2)

$\vdash . (2) . ∗118·12 . ∗117·6 . \supset \vdash . \aleph_0 = \aleph_0 +_c \aleph_0$ (3)

$\vdash . (3) . ∗113·66 . \supset \vdash . \text{Prop}$

∗123·422. $\vdash : \nu \epsilon \text{NC induct} - \iota\text{‘}0 . \supset . \nu \times_c \aleph_0 = \aleph_0$

Dem.

$\vdash : ∗113·671 . \supset \vdash : \nu \times_c \aleph_0 = \aleph_0 . \supset . (\nu +_c 1) \times \aleph_0 = \aleph_0 +_c \aleph_0$

[∗123·421] $= \aleph_0$ (1)

$\vdash . (1) . ∗120·47 . \supset \vdash . \text{Prop}$

∗123·43. $\vdash :. \exists ! \aleph_0 . \supset : \nu \epsilon \text{NC induct} . \supset_\nu . \aleph_0 > \nu$

Dem.

$\vdash . ∗123·18·36·361 . \supset \vdash : \text{Hp} . \supset . \aleph_0 \epsilon \text{NC} - \text{NC induct} - \iota\text{‘}\Lambda .$
$\text{NC induct} \subset - \iota\text{‘}\Lambda$ (1)

$\vdash . (1) . ∗120·49 . \supset \vdash . \text{Prop}$

***123·44.** $\vdash :. \exists ! \aleph_0 . \supset : \nu \in \text{NC induct} \cup \iota`\aleph_0 . \equiv . \aleph_0 \geqslant \nu$

Dem.

$\vdash . *123\cdot322 . \supset \vdash :. \alpha \in \aleph_0 . \supset : \aleph_0 \geqslant \nu . \supset . \text{Nc}`\alpha \geqslant \nu .$

[*117·22·104·12] $\supset . \exists ! \nu \cap \text{Cl}`\alpha . \nu \in \text{N}_0\text{C} .$

[*123·16] $\supset . \exists ! \nu \cap (\text{Cls induct} \cup \aleph_0) . \nu \in \text{N}_0\text{C} .$

[*103·26] $\supset . (\exists \beta) . \nu = \text{N}_0\text{c}`\beta . \beta \in \text{Cls induct} \cup \aleph_0 .$

[*120·21.*103·26] $\supset . \nu \in \text{NC induct} \cup \iota`\aleph_0$ (1)

$\vdash . (1) . *123\cdot43 . \supset \vdash . \text{Prop}$

***123·45.** $\vdash :. \exists ! \aleph_0 . \supset : \nu \in \text{NC induct} . \equiv . \aleph_0 > \nu . \equiv . \nu < \aleph_0$ [*123·43·44]

***123·46.** $\vdash : \alpha \in \text{Cls induct} . \beta \in \aleph_0 . \supset . \alpha \cup \beta \in \aleph_0$

Dem.

$\vdash . *110\cdot32 . *22\cdot91 . \supset \vdash . \text{Nc}`(\alpha \cup \beta) = \text{Nc}`\beta +_c \text{Nc}`(\alpha - \beta)$ (1)

$\vdash . *120\cdot481\cdot21 . \supset \vdash : \text{Hp} . \supset . \text{N}_0\text{c}`(\alpha - \beta) \in \text{NC induct}$ (2)

$\vdash . *123\cdot322 . \supset \vdash : \text{Hp} . \supset . \aleph_0 = \text{Nc}`\beta$ (3)

$\vdash . (2) . (3) . (*110\cdot04) . *123\cdot41 . \supset \vdash : \text{Hp} . \supset . \text{Nc}`\beta +_c \text{Nc}`(\alpha - \beta) = \aleph_0$ (4)

$\vdash . (1) . (4) . *100\cdot44 . \supset \vdash . \text{Prop}$

***123·47.** $\vdash :. \exists ! \aleph_0 . \supset : \alpha \in \text{Cls induct} \cup \aleph_0 . \equiv . (\exists \gamma) . \gamma \in \aleph_0 . \alpha \subset \gamma .$

$\equiv . \text{Nc}`\alpha \leqslant \aleph_0$

Dem.

$\vdash . *123\cdot46 . \supset \vdash :. \text{Hp} . \supset : \alpha \in \text{Cls induct} . \supset . (\exists \gamma) . \gamma \in \aleph_0 . \alpha \subset \gamma$ (1)

$\vdash . *22\cdot42 . \supset \vdash : \alpha \in \aleph_0 . \supset . (\exists \gamma) . \gamma \in \aleph_0 . \alpha \subset \gamma$ (2)

$\vdash . *123\cdot16 . \supset \vdash : (\exists \gamma) . \gamma \in \aleph_0 . \alpha \subset \gamma . \supset . \alpha \in \text{Cls induct} \cup \aleph_0$ (3)

$\vdash . (1) . (2) . (3) . \supset \vdash :. \text{Hp} . \supset : \alpha \in \text{Cls induct} \cup \aleph_0 . \equiv . (\exists \gamma) . \gamma \in \aleph_0 . \alpha \subset \gamma$ (4)

$\vdash . *123\cdot44\cdot322 . \supset \vdash :. \beta \in \aleph_0 . \supset : \text{N}_0\text{c}`\alpha \in \text{NC induct} \cup \iota`\aleph_0 . \equiv . \text{N}_0\text{c}`\alpha \leqslant \text{N}_0\text{c}`\beta :$

[*103·26.*120·21.*117·107] $\supset : \alpha \in \text{Cls induct} \cup \aleph_0 . \equiv . \text{Nc}`\alpha \leqslant \text{Nc}`\beta .$

[*123·322] $\equiv . \text{Nc}`\alpha \leqslant \aleph_0$ (5)

$\vdash . (5) . *10\cdot11\cdot23 . \supset \vdash :. \exists ! \aleph_0 . \supset : \alpha \in \text{Cls induct} \cup \aleph_0 . \equiv . \text{Nc}`\alpha \leqslant \aleph_0$ (6)

$\vdash . (4) . (6) . \supset \vdash . \text{Prop}$

The following propositions are concerned in proving $\aleph_0^2 = \aleph_0$. The proof given is roughly Cantor's. It consists in showing that the relation R defined in the hypothesis of *123·5 is a progression.

***123·5.** $\vdash : P, Q \in \text{Prog} .$

$$R = \hat{X}\hat{Y}[(\exists \mu, \nu) : X = \mu_P \downarrow \nu_Q . Y = (\mu +_c 1)_P \downarrow (\nu -_c 1)_Q . \vee . X = \mu_P \downarrow 1_Q . Y = 1_P \downarrow (\mu +_c 1)_Q] . \supset . R \in 1 \to 1$$

Dem.

$\vdash . *122·34 . \supset \vdash :. \text{Hp} . \supset : X = \mu_P \downarrow \nu_Q . Y = (\mu +_c 1)_P \downarrow (\nu -_c 1)_Q . \supset . \mu, \nu \in \text{NC induct} - \iota`0 . \nu \neq 1$ (1)

$\vdash . (1) . \supset \vdash :. \text{Hp} . \supset : (\exists \mu, \nu) . X = \mu_P \downarrow \nu_Q . Y = (\mu +_c 1)_P \downarrow (\nu -_c 1)_Q . \supset . \sim(\exists \mu) . X = \mu_P \downarrow 1_Q . Y = 1_P \downarrow (\mu +_c 1)_Q$ (2)

$\vdash . (2) . *123·3 . \supset$

$\vdash :: \text{Hp} . \supset :. (\exists \mu, \nu) . X = \mu_P \downarrow \nu_Q . Y = (\mu +_c 1)_Q \downarrow (\nu -_c 1)_Q : XRY' . X'RY : \supset . X = X' . Y = Y'$ (3)

$\vdash . (2) . \text{Transp} . *123·3 . \supset$

$\vdash :: \text{Hp} . \supset :. (\exists \mu) . X = \mu_P \downarrow 1_Q . Y = 1_P \downarrow (\mu +_c 1)_Q : XRY' . X'RY : \supset . X = X' . Y = Y'$ (4)

$\vdash . (3) . (4) . \supset \vdash : \text{Hp} . \supset . R \in 1 \to 1 : \supset \vdash . \text{Prop}$

***123·501.** $\vdash : \text{Hp} *123·5 . \supset . \text{D}`R = \text{D}`P \times \text{D}`Q$

Dem.

$\vdash . *122·34 . \supset \vdash :. \text{Hp} . \supset : \mu, \nu \in \text{NC induct} - \iota`0 . \nu \neq 1 . \supset . (\mu_P \downarrow \nu_Q) R \{(\mu +_c 1)_P \downarrow (\nu -_c 1)_Q\}$ (1)

$\vdash . *122·34 . \supset \vdash :. \text{Hp} . \supset : \mu \in \text{NC induct} - \iota`0 . \supset . (\mu_P \downarrow 1_Q) R \{1_P \downarrow (\mu +_c 1)_Q\}$ (2)

$\vdash . (1) . (2) . \supset \vdash : \text{Hp} . \supset : \mu, \nu \in \text{NC induct} - \iota`0 . \supset . \mu_P \downarrow \nu_Q \in \text{D}`R :$

[*122·341] $\supset : x \in \text{D}`P . y \in \text{D}`Q . \supset . x \downarrow y \in \text{D}`R$ (3)

$\vdash . *21·33 . \supset \vdash :. \text{Hp} . \supset : X \in \text{D}`R . \supset . (\exists \mu, \nu) . X = \mu_P \downarrow \nu_Q .$

[*122·341] $\supset . (\exists x, y) . x \in \text{D}`P . y \in \text{D}`Q . X = x \downarrow y$ (4)

$\vdash . (3) . (4) . *113·101 . \supset \vdash . \text{Prop}$

***123·502.** $\vdash : \text{Hp} *123·5 . \supset . \text{G}`R \subset \text{D}`R . \overleftarrow{R}_*`(1_P \downarrow 1_Q) \subset \text{D}`R$

Dem.

$\vdash . *21·33 . \supset \vdash : \text{Hp} . Y = (\mu +_c 1)_P \downarrow (\nu -_c 1)_Q . \nu -_c 1 \neq 1 . \supset . YR \{(\mu +_c 2)_P \downarrow (\nu -_c 2)_Q\}$ (1)

$\vdash . *21·33 . \supset \vdash : \text{Hp} . Y = (\mu +_c 1)_P \downarrow 1_Q . \supset . YR \{1_P \downarrow (\mu +_c 2)_Q\}$ (2)

$\vdash . *21·33 . \supset \vdash : \text{Hp} . Y = 1_P \downarrow (\mu +_c 1)_Q . \supset . YR\, 2_P \downarrow \mu_Q$ (3)

$\vdash . (1) . (2) . (3) . \supset \vdash : \text{Hp} . \supset . \text{G}`R \subset \text{D}`R : \supset \vdash . \text{Prop}$

∗123·503. $\vdash : \text{Hp} * 123\cdot5 . \supset . \text{D}`R \subset \overleftarrow{R_*}`(1_P \downarrow 1_Q)$

Dem.

$\vdash . *123\cdot501 . *122\cdot11 . \supset \vdash : \text{Hp} . \supset . 1_P \downarrow 1_Q \in \overleftarrow{R_*}`(1_P \downarrow 1_Q)$ (1)

$\vdash . *90\cdot16 . \supset \vdash : \text{Hp} . (1_P \downarrow 1_Q) R_* (\mu_P \downarrow \nu_Q) . \nu \neq 1 . \supset .$

$(1_P \downarrow 1_Q) R_* \{(\mu +_c 1)_P \downarrow (\nu -_c 1)_Q\}$ (2)

$\vdash . (2) . *120\cdot47 . \supset$

$\vdash : \text{Hp} . (1_P \downarrow 1_Q) R_* (\mu_P \downarrow \nu_Q) . \supset . (1_P \downarrow 1_Q) R_* \{(\mu +_c \nu -_c 1)_P \downarrow 1_Q\} .$

[∗90·16] $\supset . (1_P \downarrow 1_Q) R_* \{1_P \downarrow (\mu +_c \nu)_Q\} .$

[(2).∗120·47] $\supset . (1_P \downarrow 1_Q) R_* \{\mu_P \downarrow (\nu +_c 1)_Q\} .$ (3)

[∗90·16] $\supset . (1_P \downarrow 1_Q) R_* \{(\mu +_c 1)_P \downarrow \nu_Q\}$ (4)

$\vdash . (1) . (3) . (4) . *120\cdot47 . \supset$

$\vdash :. \text{Hp} . \supset : \mu, \nu \in \text{NC induct} - \iota`0 . \supset . (1_P \downarrow 1_Q) R_* (\mu_P \downarrow \nu_Q)$ (5)

$\vdash . (5) . *122\cdot341 . \supset \vdash . \text{Prop}$

∗123·504. $\vdash : \text{Hp} * 123\cdot5 . \supset . B`R = 1_P \downarrow 1_Q$ [∗123·34 . ∗120·414]

∗123·51. $\vdash : \text{Hp} * 123\cdot5 . \supset . R \in \text{Prog} . \text{D}`R = \text{D}`P \times \text{D}`Q$

[∗123·5·501·502·503·504]

∗123·52. $\vdash . \aleph_0 = \aleph_0 \times_c \aleph_0 = \aleph_0{}^2$ [∗123·51 . ∗116·34 . ∗113·25·204]

∗123·53. $\vdash : \nu \in \text{NC induct} - \iota`0 . \supset . \aleph_0{}^\nu = \aleph_0$ [∗123·52 . ∗116·52]

∗123·7. $\vdash : \text{Infin ax}(x) . \text{Mult ax} . \supset . \exists ! \aleph_0(t`x)$

Dem.

$\vdash . *123\cdot34 . *120\cdot301 . \supset \vdash : \text{Hp} . \supset . \text{NC induct}(t`x) \in \aleph_0$ (1)

$\vdash . *100\cdot43 . *120\cdot301 . \supset \vdash : \text{Hp} . \supset . \text{NC induct}(t`x) \in \text{Cls ex}^2 \text{ excl}$ (2)

$\vdash . (1) . (2) . *88\cdot32 . \supset \vdash : \text{Hp} . \supset . \exists ! \text{Prod}`\text{NC induct}(t`x)$ (3)

$\vdash . (1) . (2) . *115\cdot16 . \supset \vdash : \text{Hp} . \supset . \text{Prod}`\text{NC induct}(t`x) \subset \aleph_0$ (4)

$\vdash . *115\cdot18 . (*65\cdot02) . \supset \vdash : \kappa \in \text{Prod}`\text{NC induct}(t`x) . \supset . \kappa \in t`t`t`x$ (5)

$\vdash . (3) . (4) . (5) . (*65\cdot02) . \supset \vdash . \text{Prop}$

$*124$. REFLEXIVE CLASSES AND CARDINALS.

Summary of $*124$.

In this number, we have to take up the second definition of infinity mentioned in the introduction to this Section. A class which is infinite according to this definition we propose to call a reflexive class, because a class which is of this kind is capable of *reflexion* into a part of itself. A class is called *reflexive* when there is a one-one relation which correlates the class with a proper part of itself. (A *proper part* is a part not the whole.) A reflexive cardinal is the homogeneous cardinal of a reflexive class.

We prove easily that reflexive classes are not inductive ($*124 \cdot 271$), that reflexive cardinals are such as are greater than or equal to $\aleph_0$ ($*124 \cdot 23$), and such as are unchanged by adding 1 (excepting Λ) ($*124 \cdot 25$). To prove that classes which are not inductive must be reflexive has not hitherto been found possible without assuming the multiplicative axiom. We do not need, however, to assume the axiom generally, but only as applied to products of $\aleph_0$ factors. With this assumption, the result follows by a series of propositions explained below. Thus if a product of $\aleph_0$ factors, no one of which is zero, is never zero, then the two definitions of the finite and the infinite coincide ($*124 \cdot 56$).

We will call a cardinal ν a "multiplicative cardinal" if a product of ν factors none of which are zero is never zero. Thus all inductive cardinals are multiplicative cardinals; and the assumption needed for identifying the two definitions of finite and infinite is that $\aleph_0$ should be a multiplicative cardinal.

For a reflexive class we use the notation "Cls refl," and for a reflexive cardinal we use "NC refl." We define a reflexive cardinal as the *homogeneous* cardinal of a reflexive class, *i.e.* we put

$$\text{NC refl} = \text{N}_0\text{c}``\text{Cls refl} \quad \text{Df.}$$

The only effect of this is to exclude Λ from reflexive cardinals, which is convenient. We then need (on the analogy of $*110 \cdot 03 \cdot 04$) a definition of what is meant when an ambiguous symbol such as $\text{Nc}`\alpha$ is said to be reflexive, and we therefore put

$$\text{Nc}`\rho \in \text{NC refl} \,.=.\, \text{N}_0\text{c}`\rho \in \text{NC refl} \quad \text{Df.}$$

For the class of multiplicative cardinals we use the notation "NC mult." Thus we put

$$\text{NC mult} = \text{NC} \cap \hat{\alpha}\{\kappa \epsilon \alpha \cap \text{Cls ex}^2 \text{excl} \,.\, \supset_\kappa \,.\, \exists ! \, \epsilon_\Delta{}'\kappa\} \quad \text{Df},$$

whence it follows that if $\alpha \epsilon$ NC mult, a product of α factors, none of which is zero, will never be zero.

We begin, in this number, with the more obvious properties of Cls refl, proving that a Cls refl is one which contains sub-classes of $\aleph_0$ terms (∗124·15), that it is one whose number is unchanged when a single term is taken away (∗124·17), and that it remains reflexive if any inductive class is taken away from it (∗124·182).

We then give corresponding propositions concerning NC refl (∗124·2·34), proving, in addition to propositions already mentioned, that a reflexive cardinal is greater than every inductive cardinal (∗124·26), and that a class which is neither inductive nor reflexive (if there be such) is one which neither contains nor is contained in any progression (∗124·34). On such classes, see the remarks at the end of this number.

We then (∗124·4·41) give a proposition merely embodying the definition of NC mult, and show that all inductive cardinals are multiplicative, which follows immediately from ∗120·62.

The following series of propositions (∗124·51 ff.) are concerned with the proof that, if $\aleph_0$ is a multiplicative cardinal, then the two definitions of finite and infinite coalesce. The proof, which is somewhat complicated, proceeds as follows.

To begin with, we know that if ρ is a class which is not inductive, it contains classes having ν terms, if ν is any inductive cardinal. Thus we have

$$\exists ! \, 0 \cap \text{Cl}'\rho, \; \exists ! \, 1 \cap \text{Cl}'\rho, \ldots \exists ! \, \nu \cap \text{Cl}'\rho, \ldots.$$

The classes of classes $0 \cap \text{Cl}'\rho$, $1 \cap \text{Cl}'\rho$, ... $\nu \cap \text{Cl}'\rho$, ... thus form a progression, which is contained in $\text{Cl}'\text{Cl}'\rho$. Hence (∗124·511)

$$\vdash : \rho \sim \epsilon \, \text{Cls induct} \,.\, \supset \,.\, \text{Cl}'\text{Cl}'\rho \,\epsilon\, \text{Cls refl}.$$

So far, the multiplicative axiom is not required.

The above progression of classes of classes is

$$(\cap \text{Cl}'\rho)``\text{NC induct}.$$

If P is a selective relation for this class of classes, $\text{D}'P$ is a progression contained in $\text{Cl}'\rho$. Hence

∗124·513. $\vdash : \exists ! \, \epsilon_\Delta{}'(\cap \text{Cl}'\rho)``\text{NC induct} \,.\, \supset \,.\, \text{Cl}'\rho \,\epsilon\, \text{Cls refl}$

whence

∗124·514. $\vdash :. \aleph_0 \,\epsilon\, \text{NC mult} \,.\, \supset : \rho \sim \epsilon \, \text{Cls induct} \,.\, \supset \,.\, \text{Cl}'\rho \,\epsilon\, \text{Cls refl}$

To prove the next step, namely

$$\aleph_0 \,\epsilon\, \text{NC mult} \,.\, \exists ! \, \aleph_0 \cap \text{Cl}'\text{Cl}'\rho \,.\, \supset \,.\, \exists ! \, \aleph_0 \cap \text{Cl}'\rho,$$

we make a fresh start. We have, by hypothesis, a progression R whose domain is contained in $\text{Cl}`\rho$; hence $s`\text{D}`R \subset \rho$. Thus it will suffice to prove

$$\aleph_0 \in \text{NC mult} . R \in \text{Prog} . \text{D}`R \subset \text{Cls induct} . \supset . \exists ! \aleph_0 \cap s`\text{D}`R,$$

where the conditions of significance require that $\text{D}`R$ should consist of classes.

For this purpose, we prove that no member of $\text{D}`R$ can be the last that has new members which have not occurred before. The proof proceeds by showing that if this were not so, $s`\text{D}`R$ would be an inductive class, and therefore, by *120·75, $\text{D}`R$ would be an inductive class. Hence (*124·534) the members of $\text{D}`R$ which introduce new terms form an $\aleph_0$, by *123·19; and so therefore do the classes of new terms which they introduce (*124·535). Hence (*124·536) a selection from these classes of new terms, which is a sub-class of $s`\text{D}`R$, is also an $\aleph_0$, and therefore (*124·54) there is a progression contained in $s`\text{D}`R$ if the selection in question exists. This completes the proof.

In virtue of *124·511 and *120·74, we have, without the multiplicative axiom,

***124·6.** $\vdash : \rho \sim\in \text{Cls induct} . \equiv . \text{Cl}`\text{Cl}`\rho \,\dot{\in}\, \text{Cls refl}$

Hence if it could be shown that $\text{Cl}`\rho$ cannot be reflexive unless ρ is reflexive, a double application of this would enable us, by means of *124·6, to identify the two definitions of the finite without the multiplicative axiom.

***124·01.** $\text{Cls refl} = \hat{\rho} \{(\exists R) . R \in 1 \to 1 . \text{C}`R \subset \text{D}`R . \exists ! \overrightarrow{B}`R . \rho = \text{D}`R\}$ Df

An equivalent definition would be

$$\text{Cls refl} = \text{D}``\{(1 \to 1) \cap \text{C}`B - \text{Cnv}``\text{C}`B\} \quad \text{Df.}$$

***124·02.** $\text{NC refl} = \text{N}_0\text{c}``\text{Cls refl}$ Df

***124·021.** $\text{Nc}`\rho \in \text{NC refl} . = . \text{N}_0\text{c}`\rho \in \text{NC refl}$ Df

***124·03.** $\text{NC mult} = \text{NC} \cap \hat{\alpha} \{\kappa \in \alpha \cap \text{Cls ex}^2 \text{excl} . \supset_\kappa . \exists ! \epsilon_\Delta`\kappa\}$ Df

***124·1.** $\vdash : \rho \in \text{Cls refl} . \equiv . (\exists R) . R \in 1 \to 1 . \text{C}`R \subset \text{D}`R . \exists ! \overrightarrow{B}`R . \rho = \text{D}`R$ [(*124·01)]

***124·11.** $\vdash : R \in 1 \to 1 . \text{C}`R \subset \text{D}`R . \exists ! \overrightarrow{B}`R . \supset . \text{D}`R \in \text{Cls refl}$ [*124·1]

***124·12.** $\vdash . \aleph_0 \subset \text{Cls refl}$ [*123·12 . *124·1]

***124·13.** $\vdash : \rho \in \text{Cls refl} . \supset . \exists ! \aleph_0 \cap \text{Cl}`\rho$ [*124·1 . *123·192]

***124·14.** $\vdash : \rho \in \text{Cls refl} . \supset . \rho \cup \sigma \in \text{Cls refl}$

Dem.

$\vdash . *71\cdot242 . *50\cdot5\cdot52 . \supset$

$\vdash : R \in 1 \to 1 . \text{G}'R \subset \text{D}'R . \exists ! \overrightarrow{B}'R . \text{D}'R = \rho . S = I \upharpoonright (\sigma - \rho) . \supset .$

$R \cup S \in 1 \to 1 . \text{D}'(R \cup S) = \text{D}'R \cup \sigma . \text{G}'(R \cup S) = \text{G}'R \cup (\sigma - \rho) .$

$[\text{Hp}.*93\cdot101] \quad \supset . R \cup S \in 1 \to 1 . \text{D}'(R \cup S) = \rho \cup \sigma . \overrightarrow{B}'(R \cup S) = \overrightarrow{B}'R .$

$[\text{Hp}.*13\cdot12] \quad \supset . R \cup S \in 1 \to 1 . \text{D}'(R \cup S) = \rho \cup \sigma . \exists ! \overrightarrow{B}'(R \cup S) .$

$[*124\cdot11] \quad \supset . \rho \cup \sigma \in \text{Cls refl} \quad (1)$

$\vdash . (1) . *124\cdot1 . \supset \vdash . \text{Prop}$

***124·141.** $\vdash : \exists ! \text{Cl}'\rho \cap \text{Cls refl} . \supset . \rho \in \text{Cls refl}$

Dem.

$\vdash . *124\cdot14 . \supset \vdash : \mu \in \text{Cls refl} . \supset . \mu \cup (\rho - \mu) \in \text{Cls refl} .$

$[*24\cdot411] \quad \supset \vdash : \mu \subset \rho . \mu \in \text{Cls refl} . \supset . \rho \in \text{Cls refl} : \supset \vdash . \text{Prop}$

***124·15.** $\vdash : \rho \in \text{Cls refl} . \equiv . \exists ! \aleph_0 \cap \text{Cl}'\rho$

Dem.

$\vdash . *124\cdot12 . \supset \vdash : \exists ! \aleph_0 \cap \text{Cl}'\rho . \supset . \exists ! \text{Cls refl} \cap \text{Cl}'\rho .$

$[*124\cdot141] \quad \supset . \rho \in \text{Cls refl} \quad (1)$

$\vdash . (1) . *124\cdot13 . \supset \vdash . \text{Prop}$

***124·151.** $\vdash : \rho \in \text{Cls refl} . \equiv . \text{Nc}'\rho \geqslant \aleph_0 \quad [*124\cdot15 . *117\cdot22]$

***124·16.** $\vdash : \rho \in \text{Cls refl} . \equiv . (\exists \sigma) . \sigma \subset \rho . \exists ! \rho - \sigma . \rho \text{ sm } \sigma .$

$\equiv . \exists ! \text{Nc}'\rho \cap \text{Cl}'\rho - \iota'\rho$

Dem.

$\vdash . *73\cdot1 . \supset \vdash : (\exists \sigma) . \sigma \subset \rho . \exists ! \rho - \sigma . \rho \text{ sm } \sigma . \equiv .$

$(\exists R, \sigma) . \sigma \subset \rho . \exists ! \rho - \sigma . R \in 1 \to 1 . \text{D}'R = \rho . \text{G}'R = \sigma .$

$[*13\cdot195] \quad \equiv . (\exists R) . \text{G}'R \subset \rho . \exists ! \rho - \text{G}'R . R \in 1 \to 1 . \text{D}'R = \rho .$

$[*13\cdot193] \quad \equiv . (\exists R) . \text{G}'R \subset \text{D}'R . \exists ! \text{D}'R - \text{G}'R . R \in 1 \to 1 . \text{D}'R = \rho .$

$[*93\cdot101.*124\cdot1] \quad \equiv . \rho \in \text{Cls refl} : \supset \vdash . \text{Prop}$

***124·17.** $\vdash : \rho \in \text{Cls refl} . \equiv . (\exists x) . x \in \rho . \rho - \iota'x \text{ sm } \rho$

Dem.

$\vdash . *124\cdot16 . \supset \vdash : (\exists x) . x \in \rho . \rho - \iota'x \text{ sm } \rho . \supset . \rho \in \text{Cls refl} \quad (1)$

$\vdash . *123\cdot17\cdot192\cdot311 . \supset$

$\vdash : R \in 1 \to 1 . \text{G}'R \subset \text{D}'R . x \in \overrightarrow{B}'R . \supset . \overleftarrow{R_*}'x \text{ sm } \overleftarrow{R_*}'x - \iota'x .$

$[*73\cdot7] \quad \supset . (\text{D}'R - \overleftarrow{R_*}'x) \cup \overleftarrow{R_*}'x \text{ sm } (\text{D}'R - \overleftarrow{R_*}'x) \cup (\overleftarrow{R_*}'x - \iota'x) .$

$[*24\cdot411\cdot412] \quad \supset . \text{D}'R \text{ sm } \text{D}'R - \iota'x \quad (2)$

$\vdash . (2) . *124\cdot1 . \supset \vdash : \rho \in \text{Cls refl} . \supset . (\exists x) . x \in \rho . \rho \text{ sm } \rho - \iota'x \quad (3)$

$\vdash . (1) . (3) . \supset \vdash . \text{Prop}$

***124·18.** $\vdash : \rho \in \text{Cls refl} \,.\, \rho \text{ sm } \sigma \,.\, \supset \,.\, \sigma \in \text{Cls refl}$ [*124·151 . *100·321]

***124·181.** $\vdash : \rho \in \text{Cls refl} \,.\, \supset \,.\, \rho - \iota \text{‘} x \in \text{Cls refl} \,.\, \rho - \iota \text{‘} x \text{ sm } \rho$

Dem.

$\vdash \,.\, *124{\cdot}17{\cdot}18 \,.\, *73{\cdot}72 \,.\, \supset$

$\vdash : \rho \in \text{Cls refl} \,.\, x \in \rho \,.\, \supset \,.\, \rho - \iota \text{‘} x \text{ sm } \rho \,.\, \rho - \iota \text{‘} x \in \text{Cls refl}$ (1)

$\vdash \,.\, (1) \,.\, *51{\cdot}222 \,.\, \supset \vdash \,.\, \text{Prop}$

***124·182.** $\vdash : \rho \in \text{Cls refl} \,.\, \sigma \in \text{Cls induct} \,.\, \supset \,.\, \rho - \sigma \in \text{Cls refl} \,.\, \rho - \sigma \text{ sm } \rho$
[*124·181 . *120·26]

***124·2.** $\vdash : \mu \in \text{NC refl} \,.\, \equiv \,.\, (\exists \rho) \,.\, \rho \in \text{Cls refl} \,.\, \mu = \text{N}_0\text{c‘}\rho$ [(*124·02)]

***124·21.** $\vdash : \mu \in \text{NC refl} \,.\, \equiv \,.$

$(\exists R) \,.\, R \in 1 \rightarrow 1 \,.\, \text{G‘}R \subset \text{D‘}R \,.\, \exists ! \overrightarrow{B}\text{‘}R \,.\, \mu = \text{N}_0\text{c‘D‘}R$ [*124·2·1]

***124·23.** $\vdash : \mu \in \text{NC refl} \,.\, \equiv \,.\, \mu \geqslant \aleph_0$

Dem.

$\vdash \,.\, *117{\cdot}241 \,.\, \supset \vdash : \mu \geqslant \aleph_0 \,.\, \equiv \,.\, (\exists \alpha, \beta) \,.\, \mu = \text{N}_0\text{c‘}\alpha \,.\, \aleph_0 = \text{N}_0\text{c‘}\beta \,.\, \exists ! \text{Cl‘}\alpha \cap \text{Nc‘}\beta \,.$

[*123·36·322.*103·26] $\equiv \,.\, (\exists \alpha, \beta) \,.\, \mu = \text{N}_0\text{c‘}\alpha \,.\, \beta \in \aleph_0 \,.\, \exists ! \text{Cl‘}\alpha \cap \aleph_0 \,.$

[*10·35] $\equiv \,.\, (\exists \alpha) \,.\, \mu = \text{N}_0\text{c‘}\alpha \,.\, \exists ! \text{Cl‘}\alpha \cap \aleph_0 \,.$

[*124·15] $\equiv \,.\, (\exists \alpha) \,.\, \mu = \text{N}_0\text{c‘}\alpha \,.\, \alpha \in \text{Cls refl} \,.$

[*124·2] $\equiv \,.\, \mu \in \text{NC refl} : \supset \vdash \,.\, \text{Prop}$

***124·231.** $\vdash : \exists ! \text{NC refl} \,.\, \equiv \,.\, \exists ! \text{Cls refl} \,.\, \equiv \,.\, \exists ! \aleph_0$ [*124·2·12·13]

***124·232.** $\vdash : \exists ! \text{NC refl} \,.\, \supset \,.\, \text{Infin ax}$ [*124·231 . *123·18]

***124·24.** $\vdash :. \mu \in \text{NC refl} \,.\, \equiv : \mu \in \text{N}_0\text{C} : (\exists \nu) \,.\, \mu = \aleph_0 +_c \nu \,.\, \nu \in \text{NC}$

Dem.

$\vdash \,.\, *124{\cdot}23 \,.\, *117{\cdot}31 \,.\, \supset$

$\vdash :. \mu \in \text{NC refl} \,.\, \equiv : \mu, \aleph_0 \in \text{N}_0\text{C} : (\exists \nu) \,.\, \nu \in \text{NC} \,.\, \mu = \aleph_0 +_c \nu$ (1)

$\vdash \,.\, *110{\cdot}4 \,.\, \supset \vdash : \mu = \aleph_0 +_c \nu \,.\, \mu \in \text{N}_0\text{C} \,.\, \supset \,.\, \aleph_0 \in \text{N}_0\text{C}$ (2)

$\vdash \,.\, (1) \,.\, (2) \,.\, \supset \vdash \,.\, \text{Prop}$

***124·25.** $\vdash : \mu \in \text{NC refl} \,.\, \equiv \,.\, \mu \in \text{N}_0\text{C} \,.\, \mu = \mu +_c 1 \,.\, \equiv \,.\, \exists ! \mu \,.\, \mu = \mu +_c 1$
[*124·17·2]

***124·251.** $\vdash : \mu \in \text{NC refl} \,.\, \supset \,.\, \mu = \mu +_c 1$ [*124·25]

***124·252.** $\vdash : \mu \in \text{NC refl} \,.\, \nu \in \text{NC induct} \,.\, \supset \,.\, \mu = \mu +_c \nu$

Dem.

$\vdash \,.\, *124{\cdot}251 \,.\, \supset \vdash : \mu \in \text{NC refl} \,.\, \mu = \mu +_c \nu \,.\, \supset \,.\, \mu = \mu +_c \nu +_c 1$ (1)

$\vdash \,.\, (1) \,.\, *120{\cdot}11 \,.\, \supset \vdash \,.\, \text{Prop}$

***124·253.** $\vdash : \mu \in \text{NC refl} \,.\, \supset \,.\, \mu = \mu +_c \aleph_0$

Dem.

$\vdash \,.\, *124{\cdot}24 \,.\, \supset \vdash : \text{Hp} \,.\, \supset \,.\, (\exists \nu) \,.\, \mu = \aleph_0 +_c \nu \,.$

[*123·421] $\supset \,.\, (\exists \nu) \,.\, \mu = \aleph_0 +_c \aleph_0 +_c \nu \,.\, \mu = \aleph_0 +_c \nu \,.$

[*13·13] $\supset \,.\, \mu = \aleph_0 +_c \mu : \supset \vdash \,.\, \text{Prop}$

***124·26.** $\vdash :. \mu \in \text{NC refl} . \supset : \nu \in \text{NC induct} . \supset_\nu . \mu > \nu$

Dem.

$\vdash . *124\cdot231 . \supset \vdash :. \text{Hp} . \supset : \exists ! \aleph_0 :$

$[*123\cdot43] \qquad \supset : \nu \in \text{NC induct} . \supset_\nu . \aleph_0 > \nu \qquad (1)$

$\vdash . (1) . *124\cdot23 . \supset \vdash . \text{Prop}$

***124·27.** $\vdash . \text{NC refl} \cap \text{NC induct} = \Lambda \quad [*124\cdot26 . *117\cdot42]$

***124·271.** $\vdash . \text{Cls refl} \cap \text{Cls induct} = \Lambda$

Dem.

$\vdash . *124\cdot2 . \supset \vdash : \rho \in \text{Cls refl} . \supset . \text{N}_0\text{c}`\rho \in \text{NC refl} .$

$[*124\cdot27] \qquad \supset . \text{N}_0\text{c}`\rho \sim \in \text{NC induct} .$

$[*120\cdot21] \qquad \supset . \rho \sim \in \text{Cls induct} : \supset \vdash . \text{Prop}$

***124·28.** $\vdash : \rho \in \text{Cls refl} . \equiv . \text{N}_0\text{c}`\rho \in \text{NC refl} . \equiv . \text{Nc}`\rho \in \text{NC refl}$

Dem.

$\vdash . *4\cdot2 . (*124\cdot021) . \supset \vdash : \text{Nc}`\rho \in \text{NC refl} . \equiv . \text{N}_0\text{c}`\rho \in \text{NC refl} .$

$[*124\cdot2] \qquad \equiv . (\exists \sigma) . \sigma \in \text{Cls refl} . \text{N}_0\text{c}`\rho = \text{N}_0\text{c}`\sigma .$

$[*103\cdot14] \qquad \equiv . (\exists \sigma) . \sigma \in \text{Cls refl} . \rho \,\text{sm}\, \sigma . \rho \in t`\sigma .$

$[*124\cdot18 . *73\cdot3 . *63\cdot103] \qquad \equiv . \rho \in \text{Cls refl} : \supset \vdash . \text{Prop}$

***124·29.** $\vdash . s`\text{NC refl} = \text{Cls refl}$

Dem.

$\vdash . *40\cdot11 . \supset \vdash : \rho \in s`\text{NC refl} . \equiv . (\exists \mu) . \mu \in \text{NC refl} . \rho \in \mu .$

$[*103\cdot26] \qquad \equiv . (\exists \mu) . \mu \in \text{NC refl} . \mu = \text{N}_0\text{c}`\rho .$

$[*13\cdot195] \qquad \equiv . \text{N}_0\text{c}`\rho \in \text{NC refl} .$

$[*124\cdot28] \qquad \equiv . \rho \in \text{Cls refl} : \supset \vdash . \text{Prop}$

***124·3.** $\vdash :: \exists ! \aleph_0 . \supset :. \mu < \aleph_0 . \vee . \mu \geqslant \aleph_0 : \equiv . \mu \in \text{NC induct} \cup \text{NC refl}$
$[*123\cdot45 . *124\cdot23]$

***124·31.** $\vdash : \exists ! \aleph_0 . \supset . \text{spec}`\aleph_0 = \text{NC induct} \cup \text{NC refl} \quad [*124\cdot3 . *120\cdot431]$

In virtue of the above proposition, if there are any numbers which are neither inductive nor reflexive, they are such as are neither greater than, less than, nor equal to $\aleph_0$. (The existence of $\aleph_0$ in a suitable type can be deduced from the existence of numbers which are neither inductive nor reflexive; cf. *124·6.) Two further propositions (*124·33·34) are given below on non-inductive non-reflexive classes and cardinals. The subject is resumed in the remarks at the end of the number.

***124·33.** $\vdash :. \exists ! \aleph_0 . \supset : \mu \in \text{NC} - \text{NC induct} - \text{NC refl} . \equiv .$
$\mu \in \text{NC} . \sim(\mu < \aleph_0) . \sim(\mu \geqslant \aleph_0) \quad [*124\cdot3 . \text{Transp}]$

∗124·34. $\vdash :: \exists ! \aleph_0 . \supset :. \alpha \sim \epsilon (\text{Cls induct} \cup \text{Cls refl}) . \equiv :$
$\sim (\exists \gamma) : \gamma \epsilon \aleph_0 : \alpha \subset \gamma . \vee . \gamma \subset \alpha$

Dem.

$\vdash . *120·21 . *124·28 . \supset \vdash : \alpha \sim \epsilon (\text{Cls induct} \cup \text{Cls refl}) . \equiv .$
$N_0c\text{‘}\alpha \sim \epsilon (\text{NC induct} \cup \text{NC refl})$ (1)

$\vdash . *123·36 . *103·26 . \supset \vdash : \beta \epsilon \aleph_0 . \supset . \aleph_0 = N_0c\text{‘}\beta$ (2)

$\vdash . (1) . (2) . *124·31 . \supset \vdash :: \beta \epsilon \aleph_0 . \supset :. \alpha \sim \epsilon (\text{Cls induct} \cup \text{Cls refl}) . \equiv :$
$N_0c\text{‘}\alpha \sim \epsilon \text{ spec‘} N_0c\text{‘}\beta :$

[∗120·432] $\equiv : \sim (N_0c\text{‘}\alpha \leqslant N_0c\text{‘}\beta) . \sim (N_0c\text{‘}\alpha \geqslant N_0c\text{‘}\beta) :$

[∗117·107·22] $\equiv : \sim (Nc\text{‘}\alpha \leqslant Nc\text{‘}\beta) : \sim (\exists \gamma) . \gamma \epsilon Nc\text{‘}\beta . \gamma \subset \alpha :$

[∗123·322] $\equiv : \sim (Nc\text{‘}\alpha \leqslant \aleph_0) : \sim (\exists \gamma) . \gamma \epsilon \aleph_0 . \gamma \subset \alpha :$

[∗123·47] $\equiv : \sim (\exists \gamma) . \gamma \epsilon \aleph_0 . \alpha \subset \gamma : \sim (\exists \gamma) . \gamma \epsilon \aleph_0 . \gamma \subset \alpha$ (3)

$\vdash . (3) . *10·11·21 . \supset \vdash .$ Prop

∗124·4. $\vdash :. \mu \epsilon \text{NC mult} . \equiv : \mu \epsilon \text{NC} : \kappa \epsilon \mu \cap \text{Cls ex}^2 \text{ excl} . \supset_\kappa . \exists ! \epsilon_\Delta\text{‘}\kappa$
[(∗124·03)]

∗124·41. $\vdash . \text{NC induct} \subset \text{NC mult}$ [∗120·62 . ∗124·4]

The following propositions give the proof of ∗124·56, which identifies the two definitions of the finite, on the assumption that $\aleph_0$ is a multiplicative cardinal. (∗124·513, however, is only used in proving ∗124·514, and ∗124·514 is not used in the proof. It is retained as marking a stage in the argument, although the actual propositions subsequently used are not it, but the lemmas which lead to it.)

∗124·51. $\vdash : \rho \sim \epsilon \text{Cls induct} . Q = (\cap Cl\text{‘}\rho) | N | \text{Cnv‘}(\cap Cl\text{‘}\rho) . \supset .$
$Q \epsilon \text{Prog} . D\text{‘}Q \subset Cl\text{‘}Cl\text{‘}\rho . D\text{‘}Q = (\cap Cl\text{‘}\rho)\text{“NC induct}$

N here has the meaning defined in ∗123·02.

Dem.

$\vdash . *120·61·21 . *123·25 . \supset \vdash : \text{Hp} . \supset . N \epsilon \text{Prog}$ (1)

$\vdash . *120·491 . \supset \vdash :. \text{Hp} . \supset : \mu, \nu \epsilon \text{NC induct} . \supset_{\mu,\nu} . \exists ! \mu \cap Cl\text{‘}\rho . \exists ! \nu \cap Cl\text{‘}\rho :$

[∗22·5] $\supset : \mu, \nu \epsilon \text{NC induct} . \mu \cap Cl\text{‘}\rho = \nu \cap Cl\text{‘}\rho . \supset_{\mu,\nu} .$
$\exists ! \mu \cap \nu \cap Cl\text{‘}\rho .$

[∗100·43] $\supset_{\mu,\nu} . \mu = \nu$

[∗71·55] $\supset : (\cap Cl\text{‘}\rho) \upharpoonright \text{NC induct} \epsilon 1 \rightarrow 1$ (2)

$\vdash . (1) . (2) . *123·32 . \supset \vdash : \text{Hp} . \supset . Q \epsilon \text{Prog}$ (3)

$\vdash . *22·43 . \quad \supset \vdash : \alpha \epsilon D\text{‘}Q . \supset . \alpha \subset Cl\text{‘}\rho$ (4)

$\vdash . (3) . (4) . *37·32·321 . \supset \vdash .$ Prop

∗124·511. $\vdash : \rho \sim \epsilon \text{Cls induct} . \supset .$
$Cl\text{‘}Cl\text{‘}\rho \epsilon \text{Cls refl} . (\cap Cl\text{‘}\rho)\text{“NC induct} \epsilon \aleph_0 \cap \text{Cls ex}^2 \text{ excl}$
[∗124·51·15 . ∗120·491 . ∗100·43]

***124·512.** $\vdash : P \in \epsilon_\Delta\text{‘}(\cap \text{Cl‘}\rho)\text{“NC induct} \,.\, \supset .$

$$\text{D‘}P \in \aleph_0 \cap \text{Cl‘Cl‘}\rho \,.\, \text{D‘}P \subset \text{Cls induct}$$

Dem.

$\vdash . *83·11 \,.\, \text{Transp} \,.\, \supset \vdash :. \text{Hp} \,.\, \supset : \nu \in \text{NC induct} \,.\, \supset_\nu . \exists ! \nu \cap \text{Cl‘}\rho$ (1)

$\vdash . *115·16 \,.\, (1) \,.\, *124·511 \,.\, *120·491 \,.\, \supset$

$\vdash : \text{Hp} \,.\, \supset . \text{D‘}P \in \text{Nc‘}(\cap \text{Cl‘}\rho)\text{“NC induct} \,.\, \rho \sim\in \text{Cls induct} \,.$

[*124·511] $\supset . \text{D‘}P \in \aleph_0$ (2)

$\vdash . *83·21 \,.\, \supset \vdash :. \text{Hp} \,.\, \supset : \alpha \in \text{D‘}P \,.\, \supset . (\exists \nu) . \nu \in \text{NC induct} \,.\, \alpha \in \nu \cap \text{Cl‘}\rho \,.$

[*10·5.*120·2] $\supset . \alpha \in \text{Cls induct} \,.\, \alpha \in \text{Cl‘}\rho$ (3)

$\vdash . (2) . (3) . \supset \vdash . \text{Prop}$

***124·513.** $\vdash : \exists ! \epsilon_\Delta\text{‘}(\cap \text{Cl‘}\rho)\text{“NC induct} \,.\, \supset . \text{Cl‘}\rho \in \text{Cls refl}$ [*124·512·15]

***124·514.** $\vdash :. \aleph_0 \in \text{NC mult} \,.\, \supset : \rho \sim\in \text{Cls induct} \,.\, \supset . \text{Cl‘}\rho \in \text{Cls refl}$

[*124·511·513·4]

The following propositions are concerned in proving that, if $\aleph_0$ is a multiplicative cardinal, then a class such as D‘P in *124·512 must be such that a progression is contained in s‘D‘P. The characteristics of D‘P which are used in the proof are $\text{D‘}P \in \aleph_0 \,.\, \text{D‘}P \subset \text{Cls induct}$. Since $\text{D‘}P \in \aleph_0$, we have $(\exists R) . R \in \text{Prog} \,.\, \text{D‘}P = \text{D‘}R$. Hence the hypothesis with which the following series of propositions is concerned is

$$R \in \text{Prog} \,.\, \text{D‘}R \subset \text{Cls induct},$$

but the earlier propositions do not need the full hypothesis.

In what follows, note that if $\gamma \in \text{D‘}R$, $\gamma - s\text{‘}\overrightarrow{R_{\text{po}}}\text{‘}\gamma$ is the class of those terms which occur in γ and have never occurred before in any earlier member of D‘R. We prove that, with our hypothesis, members of D‘R for which this class of new terms is not null form a class which has no last member, and therefore form a progression.

***124·52.** $\vdash :. R \in \text{Prog} \,.\, \sigma = \hat{\beta} \{(\exists \gamma) . \gamma \in \text{D‘}R \,.\, \beta = \gamma - s\text{‘}\overrightarrow{R_{\text{po}}}\text{‘}\gamma \,.\, \exists ! \beta\} \,.\, \supset :$

$$\sigma \in \text{Cls ex}^2 \text{ excl} : \gamma, \delta \in \text{D‘}R \,.\, \gamma \neq \delta \,.\, \supset . (\gamma - s\text{‘}\overrightarrow{R_{\text{po}}}\text{‘}\gamma) \cap (\delta - s\text{‘}\overrightarrow{R_{\text{po}}}\text{‘}\delta) = \Lambda$$

Dem.

$\vdash . *20·33 \,.\, \supset \vdash :. \text{Hp} \,.\, \supset : \beta \in \sigma \,.\, \supset_\beta . \exists ! \beta$ (1)

$\vdash . *122·21 \,.\, \supset \vdash :. \text{Hp} \,.\, \gamma, \delta \in \text{D‘}R \,.\, \gamma \neq \delta \,.\, \supset : \gamma R_{\text{po}} \delta \,.\, \vee \,.\, \delta R_{\text{po}} \gamma$ (2)

$\vdash . *40·13 \,.\, \supset \vdash :. \text{Hp} \,.\, \gamma R_{\text{po}} \delta \,.\, \supset : \gamma \subset s\text{‘}\overrightarrow{R_{\text{po}}}\text{‘}\delta :$

[*24·3] $\supset : (\gamma - s\text{‘}\overrightarrow{R_{\text{po}}}\text{‘}\gamma) \cap (\delta - s\text{‘}\overrightarrow{R_{\text{po}}}\text{‘}\delta) = \Lambda$ (3)

Similarly $\vdash :. \text{Hp} \,.\, \delta R_{\text{po}} \gamma \,.\, \supset : (\gamma - s\text{‘}\overrightarrow{R_{\text{po}}}\text{‘}\gamma) \cap (\delta - s\text{‘}\overrightarrow{R_{\text{po}}}\text{‘}\delta) = \Lambda$ (4)

$\vdash . (2) . (3) . (4) . \supset \vdash : \text{Hp} \,.\, \gamma, \delta \in \text{D‘}R \,.\, \gamma \neq \delta \,.\, \supset .$

$(\gamma - s\text{‘}\overrightarrow{R_{\text{po}}}\text{‘}\gamma) \cap (\delta - s\text{‘}\overrightarrow{R_{\text{po}}}\text{‘}\delta) = \Lambda$ (5)

$\vdash . (5) . *20·33 \,.\, \supset \vdash : \text{Hp} \,.\, \beta, \beta' \in \sigma \,.\, \beta \neq \beta' \,.\, \supset . \beta \cap \beta' = \Lambda$ (6)

$\vdash . (1) . (6) . (5) . \supset \vdash . \text{Prop}$

∗124·521. $\vdash : \text{Hp} ∗124·52 . \pi = \hat{\gamma}\{\gamma \in D‘R . \exists ! \gamma - s‘\overrightarrow{R_{po}}‘\gamma\} . \supset . \sigma \text{ sm } \pi$

Dem.

$\vdash . ∗124·52 . ∗24·57 . \supset$

$\vdash : \text{Hp} . \gamma, \delta \in \pi . \gamma \neq \delta . \supset . \gamma - s‘\overrightarrow{R_{po}}‘\gamma \neq \delta - s‘\overrightarrow{R_{po}}‘\delta$ (1)

$\vdash . (1) . \supset \vdash : \text{Hp} . S = \hat{\beta}\hat{\gamma}\{\gamma \in D‘R . \beta = \gamma - s‘\overrightarrow{R_{po}}‘\gamma . \exists ! \beta\} . \supset .$

$S \in 1 \to 1 . D‘S = \sigma . \text{Ɑ}‘S = \pi :. \supset \vdash . \text{Prop}$

∗124·53. $\vdash : R \in \text{Prog} . \supset . s‘D‘R \sim\in \text{Cls induct}$ [∗120·75 . ∗122·37]

∗124·531. $\vdash : R \in \text{Prog} . D‘R \subset \text{Cls induct} . \supset . s‘\overrightarrow{R_*}‘\gamma \in \text{Cls induct}$

Dem.

$\vdash . ∗122·38 . \supset \vdash : \text{Hp} . \supset . \overrightarrow{R_*}‘\gamma \in \text{Cls induct}$ (1)

$\vdash . (1) . ∗120·75 . \supset \vdash . \text{Prop}$

∗124·532. $\vdash : R \in \text{Prog} . D‘R \subset \text{Cls induct} . \supset . \exists ! s‘D‘R - s‘\overrightarrow{R_*}‘\gamma$

[∗124·53·531 . ∗120·481 . Transp]

∗124·533. $\vdash : R \in \text{Prog} . D‘R \subset \text{Cls induct} . \gamma \in D‘R . \supset .$

$(\exists\beta) . \gamma R_{po} \beta . \exists ! \beta - s‘\overrightarrow{R_{po}}‘\beta$

Dem.

$\vdash . ∗124·532 . \supset \vdash : \text{Hp} . \supset . (\exists\beta) . \beta \in D‘R . \exists ! \beta - s‘\overrightarrow{R_*}‘\gamma$ (1)

$\vdash . ∗40·13 . \quad \supset \vdash : \beta R_* \gamma . \supset . \beta \subset s‘\overrightarrow{R_*}‘\gamma :$

[Transp] $\quad \supset \vdash : \exists ! \beta - s‘\overrightarrow{R_*}‘\gamma . \supset . \sim(\beta R_* \gamma) :$

[∗122·21] $\quad \supset \vdash : \text{Hp} . \beta \in D‘R . \exists ! \beta - s‘\overrightarrow{R_*}‘\gamma . \supset . \gamma R_{po} \beta$ (2)

$\vdash . (1) . (2) . \supset$

$\vdash :. \text{Hp} . \quad \supset : (\exists\beta) . \gamma R_{po} \beta . \exists ! \beta - s‘\overrightarrow{R_*}‘\gamma :$

[∗122·23] $\supset : E ! \min (R_{po})‘\hat{\beta}\{\gamma R_{po} \beta . \exists ! \beta - s‘\overrightarrow{R_{po}}‘\beta\} :$

[∗93·111] $\supset : (\exists\beta) : \gamma R_{po} \beta . \exists ! \beta - s‘\overrightarrow{R_*}‘\gamma : \delta R_{po} \beta . \supset_\delta . \delta \subset s‘\overrightarrow{R_*}‘\gamma :$

[∗40·151] $\supset : (\exists\beta) . \gamma R_{po} \beta . \exists ! \beta - s‘\overrightarrow{R_*}‘\gamma . s‘\overrightarrow{R_{po}}‘\beta \subset s‘\overrightarrow{R_*}‘\gamma :$

[∗22·81] $\supset : (\exists\beta) . \gamma R_{po} \beta . \exists ! \beta - s‘\overrightarrow{R_{po}}‘\beta :. \supset \vdash . \text{Prop}$

∗124·534. $\vdash : R \in \text{Prog} . D‘R \subset \text{Cls induct} .$

$\pi = \hat{\gamma}\{\gamma \in D‘R . \exists ! \gamma - s‘\overrightarrow{R_{po}}‘\gamma\} . \supset . \pi \in \aleph_0$

Dem.

$\vdash . ∗124·533 . \supset \vdash : \text{Hp} . \supset . \exists ! \pi . \pi \subset R_{po}“\pi$ (1)

$\vdash . (1) . ∗123·19 . \supset \vdash . \text{Prop}$

∗124·535. $\vdash : R \in \text{Prog} . D‘R \subset \text{Cls induct} .$

$\sigma = \hat{\beta}\{(\exists\gamma) . \gamma \in D‘R . \beta = \gamma - s‘\overrightarrow{R_{po}}‘\gamma . \exists ! \beta\} . \supset . \sigma \in \aleph_0$

[∗124·534·521 . ∗123·321]

∗124·536. $\vdash : R \epsilon \text{Prog} . \text{D}`R \subset \text{Cls induct} .$

$\sigma = \hat{\beta}\{(\exists\gamma) . \gamma \epsilon \text{D}`R . \beta = \gamma - s`\overrightarrow{R_{\text{po}}}`\gamma . \exists ! \beta\} .$

$S \epsilon \epsilon_\Delta`\sigma . \supset . \text{D}`S \epsilon \aleph_0 . \text{D}`S \subset s`\text{D}`R$

Dem.

$\vdash . ∗115·16 . ∗124·52·535 . \supset \vdash : \text{Hp} . \supset . \text{D}`S \epsilon \aleph_0$ (1)

$\vdash . ∗83·21 . \supset \vdash :. \text{Hp} . \supset : \text{D}`S \subset s`\sigma :$

[∗40·11] $\supset : x \epsilon \text{D}`S . \supset . (\exists\beta, \gamma) . \gamma \epsilon \text{D}`R . \beta = \gamma - s`\overrightarrow{R_{\text{po}}}`\gamma . \exists ! \beta . x \epsilon \beta .$

[∗13·195] $\supset . (\exists\gamma) . \gamma \epsilon \text{D}`R . x \epsilon \gamma - s`\overrightarrow{R_{\text{po}}}`\gamma .$

[∗22·43] $\supset . (\exists\gamma) . \gamma \epsilon \text{D}`R . x \epsilon \gamma .$

[∗40·11] $\supset . x \epsilon s`\text{D}`R$ (2)

$\vdash . (1) . (2) . \supset \vdash . \text{Prop}$

∗124·54. $\vdash : \aleph_0 \epsilon \text{NC mult} . R \epsilon \text{Prog} . \text{D}`R \subset \text{Cls induct} . \supset . \exists ! \aleph_0 \cap \text{Cl}`s`\text{D}`R$

Dem.

$\vdash . ∗124·52·535·4 . \supset$

$\vdash :. \text{Hp} . \supset : \sigma = \hat{\beta}\{(\exists\gamma) . \gamma \epsilon \text{D}`R . \beta = \gamma - s`\overrightarrow{R_{\text{po}}}`\gamma . \exists ! \beta\} . \supset . \exists ! \epsilon_\Delta`\sigma .$

[∗124·536] $\supset . \exists ! \aleph_0 \cap \text{Cl}`s`\text{D}`R :. \supset \vdash . \text{Prop}$

∗124·541. $\vdash : \aleph_0 \epsilon \text{NC mult} . P \epsilon \epsilon_\Delta`(\cap \text{Cl}`\rho)``\text{NC induct} . \supset .$

$\exists ! \aleph_0 \cap \text{Cl}`s`\text{D}`P . s`\text{D}`P \subset \rho$

Dem.

$\vdash . ∗124·512 . \supset \vdash : \text{Hp} . \supset . \text{D}`P \epsilon \aleph_0 . \text{D}`P \subset \text{Cls induct} .$

[∗123·1] $\supset . (\exists R) . \text{D}`P = \text{D}`R . R \epsilon \text{Prog} . \text{D}`R \subset \text{Cls induct} .$

[∗124·54] $\supset . (\exists R) . \text{D}`P = \text{D}`R . \exists ! \aleph_0 \cap \text{Cl}`s`\text{D}`R .$

[∗13·193.∗10·35] $\supset . \exists ! \aleph_0 \cap \text{Cl}`s`\text{D}`P$ (1)

$\vdash . ∗124·512 . \supset \vdash : \text{Hp} . \supset . \text{D}`P \epsilon \text{Cl}`\text{Cl}`\rho .$

[∗60·2] $\supset . \text{D}`P \subset \text{Cl}`\rho .$

[∗60·52] $\supset . s`\text{D}`P \subset \rho$ (2)

$\vdash . (1) . (2) . \supset \vdash . \text{Prop}$

∗124·55. $\vdash : \aleph_0 \epsilon \text{NC mult} . \rho \sim\epsilon \text{Cls induct} . \supset . \exists ! \aleph_0 \cap \text{Cl}`\rho$

Dem.

$\vdash . ∗124·511·4 . \supset \vdash : \text{Hp} . \supset . \exists ! \epsilon_\Delta`(\cap \text{Cl}`\rho)``\text{NC induct} .$

[∗124·541.∗60·4] $\supset . \exists ! \aleph_0 \cap \text{Cl}`\rho : \supset \vdash . \text{Prop}$

∗124·56. $\vdash : \aleph_0 \epsilon \text{NC mult} . \supset . -\text{Cls induct} = \text{Cls refl} . \text{N}_0\text{C} - \text{NC induct} = \text{NC refl}$

Dem.

$\vdash . ∗124·55·15 . \supset \vdash : \text{Hp} . \supset . - \text{Cls induct} \subset \text{Cls refl}$ (1)

$\vdash . ∗124·271 . \supset \vdash : \text{Hp} . \supset . \text{Cls refl} \subset - \text{Cls induct}$ (2)

$\vdash . (1) . (2) . \supset \vdash :. \text{Hp} . \supset : - \text{Cls induct} = \text{Cls refl} :$ (3)

[∗120·21.∗124·28] $\supset : \text{N}_0\text{c}`\rho \sim\epsilon \text{NC induct} . \equiv . \text{N}_0\text{c}`\rho \epsilon \text{NC refl} :$

[∗103·2.∗124·2] $\supset : \alpha \epsilon \text{N}_0\text{C} - \text{NC induct} . \equiv . \alpha \epsilon \text{NC refl}$ (4)

$\vdash . (3) . (4) . \supset \vdash . \text{Prop}$

The above proposition identifies the two definitions of the finite, on the hypothesis $\aleph_0 \in \text{NC mult}$.

***124·57.** $\vdash : \mu \in \text{N}_0\text{C} - \text{NC induct} . \supset . 2^{2^\mu} \in \text{NC refl}$ [*124·511 . *116·72]

***124·58.** $\vdash :. 2^\mu \in \text{NC refl} . \supset_\mu . \mu \in \text{NC refl} : \supset . \text{N}_0\text{C} - \text{NC induct} = \text{NC refl}$

Dem.

$\vdash . *124·57 . \supset \vdash :. \text{Hp} . \supset : \mu \in \text{N}_0\text{C} - \text{NC induct} . \supset . 2^\mu \in \text{NC refl} .$

[Hp] $\supset . \mu \in \text{NC refl}$ (1)

$\vdash . (1) . *124·2·27 . \supset \vdash . \text{Prop}$

The above proposition gives another hypothesis which would enable us to identify the two definitions of the finite if it could be proved, namely

$$2^\mu \in \text{NC refl} . \supset_\mu . \mu \in \text{NC refl},$$

or, what comes to the same thing,

$$\text{Cl}‘\rho \in \text{Cls refl} . \supset . \rho \in \text{Cls refl}.$$

***124·6.** $\vdash : \rho \sim\in \text{Cls induct} . \equiv . \text{Cl}‘\text{Cl}‘\rho \in \text{Cls refl}$

Dem.

$\vdash . *124·511 . \supset \vdash : \rho \sim\in \text{Cls induct} . \supset . \text{Cl}‘\text{Cl}‘\rho \in \text{Cls refl}$ (1)

$\vdash . *120·74 . \supset \vdash : \rho \in \text{Cls induct} . \supset . \text{Cl}‘\text{Cl}‘\rho \in \text{Cls induct} .$

[*124·271] $\supset . \text{Cl}‘\text{Cl}‘\rho \sim\in \text{Cls refl}$ (2)

$\vdash . (1) . (2) . \supset \vdash . \text{Prop}$

***124·61.** $\vdash :. \aleph_0 \in \text{NC mult} . \supset : \rho \in \text{Cls refl} . \equiv . \text{Cl}‘\rho \in \text{Cls refl} . \equiv . \text{Cl}‘\text{Cl}‘\rho \in \text{Cls refl}$

Dem.

$\vdash . *124·6·271 . \supset \vdash : \rho \in \text{Cls refl} . \supset . \text{Cl}‘\rho \in \text{Cls refl} . \supset . \text{Cl}‘\text{Cl}‘\rho \in \text{Cls refl}$ (1)

$\vdash . *124·6·56 . \supset \vdash :. \aleph_0 \in \text{NC mult} . \supset : \text{Cl}‘\text{Cl}‘\rho \in \text{Cls refl} . \supset . \rho \in \text{Cls refl} .$ (2)

[(1)] $\supset . \text{Cl}‘\rho \in \text{Cls refl}$ (3)

$\vdash . (1) . (2) . (3) . \supset \vdash . \text{Prop}$

The following properties of cardinals which are neither inductive nor reflexive (supposing there are such) are easily proved. Let us put

$$\text{NC med} = \text{N}_0\text{C} - \text{NC induct} - \text{NC refl} \quad \text{Df},$$

$$\text{Cls med} = - \text{Cls induct} - \text{Cls refl} \quad \text{Df},$$

where "med" stands for "mediate." Then

$$\mu \in \text{NC med} . \supset . \mu +_c 1 \in \text{NC med} . \mu -_c 1 \in \text{NC med} . \mu \neq \mu +_c 1 . \mu \neq \mu -_c 1.$$

Hence mediate cardinals have no maximum or minimum.

$$\mu, \nu \in \text{NC med} . \supset . \mu +_c \nu \in \text{NC med},$$

$$\mu \in \text{NC med} . \nu \in \text{NC med} \cup \text{NC induct} - \iota‘0 . \supset . \mu \times_c \nu \in \text{NC med},$$

whence $\mu \in \text{NC med} . \supset . \mu^2, \mu^3, \ldots \in \text{NC med},$

$$\mu^\nu \in \text{NC med} . \supset : \mu \in \text{NC med} . \vee . \nu \in \text{NC med},$$

$$\mu \in \text{NC med} . \supset . 2^{2^\mu} \in \text{NC refl},$$

whence $\exists ! \text{NC med} . \supset . (\exists \nu) . \nu \in \text{NC med} . 2^\nu \in \text{NC refl},$

since we have either $\mu \in \text{NC med} . 2^\mu \in \text{NC refl}$ or $2^\mu \in \text{NC med} . 2^{2^\mu} \in \text{NC refl}$.

$*125$. THE AXIOM OF INFINITY.

Summary of $*125$.

The present number is merely concerned to give a few equivalent forms of the axiom of infinity, and of the kindred assumption of the existence of $\aleph_0$.

In virtue of $*125\cdot24\cdot25$ below, if the axiom of infinity holds in any one type, then it holds in any other type which can be derived from this one, or from any type from which this one can be derived. Hence if we assume, as it seems natural to do, that all extensional types are derived from a first type, namely that of individuals, then the axiom of infinity in any such type is equivalent to the assumption that the number of individuals is not inductive.

We deal, in this number, first with equivalent forms of Infin ax, then with equivalent forms of Infin ax (x), then with equivalent forms of $\exists ! \aleph_0$ or $\exists ! \aleph_0(x)$. When "Infin ax" or "$\exists ! \aleph_0$" occurs in this number without typical definition, it and all other typically ambiguous symbols are to be taken in the lowest logically possible types, or with the same *relative* types as if this had been done. The propositions of this number are often not referred to in the sequel, but are here collected together on account of their intrinsic interest.

$*125\cdot1$. $\vdash :. \text{Infin ax} . \equiv : \alpha \in \text{NC induct} . \supset_\alpha . \exists ! \alpha$ $[*120\cdot3]$

$*125\cdot11$. $\vdash :. \text{Infin ax} . \equiv : \alpha \in \text{NC induct} . \supset_\alpha . \alpha \neq \alpha +_c 1$ $[*120\cdot33]$

$*125\cdot12$. $\vdash :. \text{Infin ax} . \equiv : \alpha \in \text{NC induct} . \supset_\alpha . \exists ! \alpha +_c 1$

Dem.

$\vdash . *101\cdot12 . *125\cdot1 . \supset$

$\vdash :. \text{Infin ax} . \equiv : \alpha \in \text{NC induct} - \iota`0 . \supset_\alpha . \exists ! \alpha :$

$[*120\cdot423]$ $\equiv : \alpha \in \text{NC induct} . \supset_\alpha . \exists ! \alpha +_c 1 :. \supset \vdash . \text{Prop}$

$*125\cdot13$. $\vdash : \text{Infin ax} . \equiv . \Lambda \sim\in \text{NC induct}$ $[*125\cdot1 . *24\cdot63]$

$*125\cdot14$. $\vdash : \text{Infin ax} . \equiv . (+_c 1) \upharpoonright \text{NC induct} \in 1 \rightarrow 1$

Dem.

$\vdash . *123\cdot22\cdot24 . \supset \vdash : \text{Infin ax} . \supset . (+_c 1) \upharpoonright \text{NC induct} \in 1 \rightarrow 1$ (1)

$\vdash . *71\cdot55 .$ $\supset \vdash :. (+_c 1) \upharpoonright \text{NC induct} \in 1 \rightarrow 1 . \supset :$

$\alpha, \beta \in \text{NC induct} . \alpha +_c 1 = \beta +_c 1 . \supset_{\alpha,\beta} . \alpha = \beta :$

[Transp] $\supset : \alpha, \beta \in \text{NC induct} . \alpha \neq \beta . \supset_{\alpha,\beta} . \alpha +_c 1 \neq \beta +_c 1 :$

$[*10\cdot1]$ $\supset : \Lambda, \beta \in \text{NC induct} . \Lambda \neq \beta . \supset_\beta . \Lambda +_c 1 \neq \beta +_c 1 :$

[$*110\cdot4$.Transp] $\supset : \Lambda \epsilon \mathrm{NC\,induct} \,.\, \beta \epsilon \mathrm{NC\,induct} \,.\, \exists ! \beta \,.\, \supset_\beta \,.\, \exists ! (\beta +_c 1)$ (2)

$\vdash . (2) . *101\cdot12 . *120\cdot13 . \supset$

$\vdash :. (+_c 1) \upharpoonright \mathrm{NC\,induct} \,\epsilon\, 1 \to 1 \,.\, \Lambda \epsilon \mathrm{NC\,induct} \,.\, \supset : \beta \epsilon \mathrm{NC\,induct} \,.\, \supset_\beta \,.\, \exists ! \beta :$

[$*24\cdot63$] $\supset : \Lambda \sim\epsilon \mathrm{NC\,induct}$ (3)

$\vdash . (3) . *2\cdot01 . *125\cdot13 . \supset \vdash : (+_c 1) \upharpoonright \mathrm{NC\,induct} \,\epsilon\, 1 \to 1 \,.\, \supset .\, \mathrm{Infin\,ax}$ (4)

$\vdash . (1) . (4) . \supset \vdash . \mathrm{Prop}$

$*125\cdot15$. $\vdash :. \mathrm{Infin\,ax} \,.\, \equiv : \rho \epsilon \mathrm{Cls\,induct} \,.\, \supset_\rho \,.\, \exists ! - \rho$

Dem.

$\vdash . *110\cdot63 . \supset \vdash : x \sim\epsilon \rho \,.\, \supset_x \,.\, \rho \cup \iota`x \,\epsilon\, \mathrm{Nc}`\rho +_c 1 :$

[$*10\cdot28$] $\supset \vdash : \exists ! - \rho \,.\, \supset .\, \exists ! \mathrm{Nc}`\rho +_c 1 :$

[Syll] $\supset \vdash :. \rho \epsilon \mathrm{Cls\,induct} \,.\, \supset_\rho \,.\, \exists ! - \rho : \supset :$

$\rho \epsilon \mathrm{Cls\,induct} \,.\, \supset_\rho \,.\, \exists ! \mathrm{Nc}`\rho +_c 1 :$

[$*120\cdot2$] $\supset : \alpha \epsilon \mathrm{NC\,induct} \,.\, \rho \epsilon \alpha \,.\, \supset_{\alpha,\rho} \,.\, \exists ! \mathrm{Nc}`\rho +_c 1 :$

[$*100\cdot45$] $\supset : \alpha \epsilon \mathrm{NC\,induct} \,.\, \exists ! \alpha \,.\, \supset_\alpha \,.\, \exists ! \alpha +_c 1 :$

[$*120\cdot13.*101\cdot12$] $\supset : \alpha \epsilon \mathrm{NC\,induct} \,.\, \supset_\alpha \,.\, \exists ! \alpha$ (1)

$\vdash . *13\cdot12 . \supset \vdash :. \alpha \epsilon \mathrm{NC\,induct} \,.\, \supset_\alpha \,.\, \exists ! (\alpha +_c 1) : \supset :$

$\alpha \epsilon \mathrm{NC\,induct} \,.\, \mathrm{N_0c}`\rho = \alpha \,.\, \supset_{\alpha,\rho} \,.\, \exists ! (\mathrm{N_0c}`\rho +_c 1) :$

[$*120\cdot21$] $\supset : \rho \epsilon \mathrm{Cls\,induct} \,.\, \supset_\rho \,.\, \exists ! (\mathrm{N_0c}`\rho +_c 1) \,.$

[$*103\cdot11.*63\cdot101.*110\cdot63$] $\supset_\rho \,.\, (\exists \gamma, z) \,.\, \gamma \,\mathrm{sm}\, \rho \,.\, z \sim\epsilon \gamma \,.\, \gamma \epsilon \iota`\rho \cup - \iota`\rho$ (2)

$\vdash . *13\cdot12 . *10\cdot24 . \supset \vdash : \gamma = \rho \,.\, z \sim\epsilon \gamma \,.\, \supset .\, \exists ! - \rho$ (3)

$\vdash . *120\cdot426 . *24\cdot6 . \supset \vdash : \rho \epsilon \mathrm{Cls\,induct} \,.\, \gamma \neq \rho \,.\, \gamma \subset \rho \,.\, \supset .\, \sim(\gamma \,\mathrm{sm}\, \rho) :$

[Transp] $\supset \vdash : \rho \epsilon \mathrm{Cls\,induct} \,.\, \gamma \,\mathrm{sm}\, \rho \,.\, \gamma \neq \rho \,.\, \supset .\, \exists ! \gamma - \rho \,.$

[$*24\cdot561$] $\supset .\, \exists ! - \rho$ (4)

$\vdash . (3) . (4) . \supset \vdash :. \rho \epsilon \mathrm{Cls\,induct} : (\exists \gamma, z) \,.\, \gamma \,\mathrm{sm}\, \rho \,.\, z \sim\epsilon \gamma \,.\, \gamma \epsilon \iota`\rho \cup - \iota`\rho : \supset .$

$\exists ! - \rho$ (5)

$\vdash . (2) . (5) . \supset \vdash :. \alpha \epsilon \mathrm{NC\,induct} \,.\, \supset_\alpha \,.\, \exists ! (\alpha +_c 1) : \supset :$

$\rho \epsilon \mathrm{Cls\,induct} \,.\, \supset_\rho \,.\, \exists ! - \rho$ (6)

$\vdash . (1) . (6) . *125\cdot121 . \supset \vdash . \mathrm{Prop}$

$*125\cdot16$. $\vdash : \mathrm{Infin\,ax} \,.\, \equiv .\, \exists ! \mathrm{Cls} - \mathrm{Cls\,induct} \,.\, \equiv .\, \exists ! \mathrm{N_0C} - \mathrm{NC\,induct} \,.\, \equiv .$

$\mathrm{V} \sim\epsilon \mathrm{Cls\,induct}$

Dem.

$\vdash . *125\cdot15 . \quad \supset \vdash :. \mathrm{Infin\,ax} \,.\, \equiv : \rho \epsilon \mathrm{Cls\,induct} \,.\, \supset_\rho \,.\, \rho \neq \mathrm{V} :$

[$*13\cdot196$] $\equiv : \mathrm{V} \sim\epsilon \mathrm{Cls\,induct}$ (1)

$\vdash . *120\cdot481 . \mathrm{Transp} . \supset \vdash : \exists ! \mathrm{Cls} - \mathrm{Cls\,induct} \,.\, \supset .\, \mathrm{V} \sim\epsilon \mathrm{Cls\,induct}$ (2)

$\vdash . (1) . (2) . *120\cdot21 . \supset \vdash . \mathrm{Prop}$

$*125\cdot2$. $\vdash :. \mathrm{Infin\,ax}(x) \,.\, \equiv : \alpha \epsilon \mathrm{NC\,induct} \,.\, \supset_\alpha \,.\, \exists ! \alpha(x)$ [$*120\cdot301$]

∗125·21. $\vdash :$ Infin ax $(x) . \equiv . t`x \sim\epsilon$ Cls induct

Dem.

$\vdash . *125·15 . \supset \vdash :.$ Infin ax $(x) . \equiv : \rho \epsilon$ Cls induct $\cap$ Cl$`t`x . \supset_\rho . \exists ! - \rho :$

[∗63·102] $\equiv : \rho \epsilon$ Cls induct $\cap$ Cl$`t`x . \supset_\rho . \rho \neq t`x :$

[∗13·196] $\equiv : t`x \sim\epsilon$ Cls induct $:. \supset \vdash .$ Prop

∗125·22. $\vdash :$ Infin ax $(x) . \equiv . t^3`x \epsilon$ Cls refl [∗125·21 . ∗63·66 . ∗124·6]

∗125·23. $\vdash :$ Infin ax $(x) . \equiv . \exists ! \aleph_0 (t^2`x)$ [∗125·22 . ∗124·15]

∗125·24. $\vdash :$ Infin ax $(x) . \equiv .$ Infin ax $(t`x) . \equiv .$ Infin ax $(t^2`x) . \equiv .$ etc.

Dem.

$\vdash . *125·21 . \supset \vdash :$ Infin ax $(x) . \equiv . t`x \sim\epsilon$ Cls induct .

[∗120·74] $\equiv .$ Cl$`t`x \sim\epsilon$ Cls induct .

[∗63·66] $\equiv . t^2`x \sim\epsilon$ Cls induct .

[∗125·21] $\equiv .$ Infin ax $(t`x) : \supset \vdash .$ Prop

∗125·25. $\vdash :$ Infin ax $(\alpha) . \equiv .$ Infin ax $(t_{00}`\alpha) . \equiv .$ Infin ax $(t_0{}^1`\alpha) . \equiv .$
Infin ax $(t^{11}`\alpha) . \equiv .$ etc.
[∗116·91·92 . ∗120·56·52 . ∗125·21]

∗125·3. $\vdash : \exists ! \aleph_0 . \equiv . \exists ! (1 \rightarrow 1) \cap \hat{R} (\exists ! \overrightarrow{B}`R . \sim \exists ! \overrightarrow{B}`\breve{R})$

Dem.

$\vdash . *123·1 . \quad \supset \vdash : \exists ! \aleph_0 . \equiv . \exists !$ Prog .

[∗122·11·141] $\supset . \exists ! (1 \rightarrow 1) \cap \hat{R} (\exists ! \overrightarrow{B}`R . \sim \exists ! \overrightarrow{B}`\breve{R})$ (1)

$\vdash . *123·192 . \supset \vdash : \exists ! (1 \rightarrow 1) \cap \hat{R} (\exists ! \overrightarrow{B}`R . \sim \exists ! \overrightarrow{B}`\breve{R}) . \supset . \exists ! \aleph_0$ (2)

$\vdash . (1) . (2) . \supset \vdash .$ Prop

∗125·31. $\vdash : \exists ! \aleph_0 (x) . \equiv . t`x \epsilon$ Cls refl [∗124·15]

∗125·32. $\vdash : \exists ! \aleph_0 (x) . \equiv . \exists ! (1 \rightarrow 1) \cap \overleftarrow{D}`t`x - \overleftarrow{Cl}`t`x$

Dem.

$\vdash . *63·102 . \supset \vdash : \exists ! (1 \rightarrow 1) \cap \overleftarrow{D}`t`x - \overleftarrow{Cl}`t`x . \equiv .$
$(\exists R) . R \epsilon 1 \rightarrow 1 . D`R = t`x . Cl`R \subset t`x . \exists ! t`x - Cl`R .$

[∗124·1] $\equiv . t`x \epsilon$ Cls refl (1)

$\vdash . (1) . *125·31 . \supset \vdash .$ Prop

∗125·33. $\vdash :. \exists ! \aleph_0 (x) . \equiv : \alpha \subset t`x . \exists ! \alpha . \supset_\alpha . \exists ! (1 \rightarrow 1) \cap \overleftarrow{D}`\alpha - \overleftarrow{Cl}`\alpha$

Dem.

$\vdash . *73·7 . 51·222 . \supset \vdash : \alpha \subset t`x . y \epsilon \alpha . z \epsilon t`x - \alpha . \supset . \alpha \text{ sm } (\alpha - \iota`y) \cup \iota`z .$

[∗73·1] $\supset . (\exists R) . R \epsilon 1 \rightarrow 1 . D`R = \alpha . Cl`R = (\alpha - \iota`y) \cup \iota`z .$

[∗33·6·61] $\supset . \exists ! (1 \rightarrow 1) \cap \overleftarrow{D}`\alpha - \overleftarrow{Cl}`\alpha$ (1)

$\vdash . (1) . \quad \supset \vdash : \exists ! \alpha . \exists ! t`x - \alpha . \supset . \exists ! (1 \rightarrow 1) \cap \overleftarrow{D}`\alpha - \overleftarrow{Cl}`\alpha :$

[∗63·102] $\supset \vdash : \exists ! \alpha . \alpha \subset t`x . \alpha \neq t`x . \supset . \exists ! (1 \rightarrow 1) \cap \overleftarrow{D}`\alpha - \overleftarrow{Cl}`\alpha$ (2)

$\vdash . (2) . *125·32 . \supset \vdash .$ Prop

***125·34.** $\vdash : \exists ! \aleph_0(x) . \equiv . \iota`t`x \sim \epsilon \text{NC}$

Dem.

$\vdash . *125·32 . \supset \vdash :. \sim \exists ! \aleph_0(x) . \equiv : R \epsilon 1 \to 1 . \text{D}`R = t`x . \supset_R . \text{Cl}`R = t`x :$

[*100·13] $\equiv : \text{Nc}`t`x = \iota`t`x .$

[*100·41·45] $\equiv : \iota`t`x \epsilon \text{NC}$ (1)

$\vdash . (1) . \text{Transp} . \supset \vdash . \text{Prop}$

***125·35.** $\vdash :. \aleph_0 \epsilon \text{NC mult} . \supset : \exists ! \aleph_0(x) . \equiv . \text{Infin ax}(x)$

Dem.

$\vdash . *125·21 . *124·56 . \supset \vdash :. \text{Hp} . \supset : \text{Infin ax}(x) . \equiv . t`x \epsilon \text{Cls refl} .$

[*125·31] $\equiv . \exists ! \aleph_0(x) :. \supset \vdash . \text{Prop}$

***125·36.** $\vdash : \text{Infin ax}(\text{Cls}) . \equiv . \exists ! \aleph_0(\text{Cls})$

Dem.

$\vdash . *24·14 . *63·102·66 . \supset \vdash . t^2`x = \text{Cl}`\text{V}$

[*24·11] $= \text{Cls}$ (1)

$\vdash . (1) . *125·24 . \supset \vdash : \text{Infin ax}(\text{Cls}) . \equiv . \text{Infin ax}(x) .$

[*125·23.(1)] $\equiv . \exists ! \aleph_0(\text{Cls}) : \supset \vdash . \text{Prop}$

$*126$. ON TYPICALLY INDEFINITE INDUCTIVE CARDINALS.

Recapitulation of Conventions and Summary of $*126$.

We have now arrived at the stage where we can adopt the standpoint of ordinary arithmetic, and can for the future in arithmetical operations with cardinals ignore differences of type. In order to understand how this is so, it will be necessary briefly to recall the line of thought of some of the previous numbers and the conventions upon which the symbolism is based.

The symbolism of $*102$, though perfectly precise as to the typical relations of the various symbols, is in fact too complex for use, except in cases of absolute necessity. It is better to use the typically ambiguous symbols Nc and sm, combined with some simple rules of interpretation of the symbolism, so as to secure that the various occurrences of the same symbols are in their proper relationships of type. This is the course followed in $*100$, $*101$, and in every number from $*110$ onwards.

The important symbols which involve an explicit or implicit use of Nc or sm are called 'formal numbers,' and it is only necessary to make the rules of interpretation apply to them.

A constant formal number is any symbol representing a typically ambiguous constant such that there is a constant α such that, however the ambiguities of type may be determined, the former constant is identical with $\text{Nc}'\alpha$. The variable formal numbers are defined by enumeration. They are divided into three Sets, the Primary Set, the Argumental Set, and the Arithmetical Set.

The Primary Set consists of $\text{Nc}'\alpha$, $\Sigma\,\text{Nc}'\kappa$, $\Pi\,\text{Nc}'\kappa$, where α is a variable Cls of any type and κ is a variable Cls^2 of any type. Also α and κ may themselves be complex symbols which in some way involve variables.

The Argumental Set has only one member $\text{sm}``\mu$, where $\mu$ is a variable $\text{Cls}^2$ of any type. In its capacity of a formal number $\text{sm}``\mu$ is only interesting when μ is an NC; then $\text{sm}``\mu$ gives the corresponding NC in another type, provided that $\mu$ is not $\Lambda$. Also $\mu$ may be a complex symbol which in some way involves a variable, *e.g.* $\text{sm}``\text{Nc}'\alpha$ is a formal number of the Argumental Set: μ is called the *argument* of $\text{sm}``\mu$.

The Arithmetical Set consists of $\mu +_c \nu$, $\mu \times_c \nu$, μ^ν, $\mu -_c \nu$. These formal numbers are only interesting when μ and ν are also members of NC. Also μ and ν may be complex symbols, so long as one of them at least involves a variable. For example $2^{3 +_c \nu}$ is a formal number, and so is $\alpha +_c (3 +_c \nu)$.

The Primary and Argumental and Arithmetical Sets of Formal Numbers are derived from the corresponding sets of *variable* formal numbers, by adding to them the constant formal numbers obtained by substituting constants for the variables occurring in the expressions for the members of the variable set in question.

In the formal numbers of the arithmetical set as written above, μ and ν are called the *first components*. Thus every formal number of this set has two first components. The first components (if any) of the first components are also called *components* of the original formal number, and so on; so that components of components are components of the original symbol.

A formal number of the arithmetical set, whose components are all formal numbers, either constant or variable but *not* belonging to the argumental set, is called a *pure arithmetical* formal number. These are the formal numbers which it is important in arithmetic to secure from assuming the value Λ owing to lowness of type.

The logical investigation of ∗100 and ∗101, where typically ambiguous formal numbers are used, is directly concerned in investigating the premisses necessary to secure various propositions from fluctuating truth-values owing to the intrusion of null-values among the cardinals. The convention, necessary to avoid determinations of type which we never wish to consider, is as follows, where the terms used are explained fully in the prefatory statement:

IT. Argumental occurrences are bound to logical and attributive occurrences; and, if there are no argumental occurrences, equational occurrences are bound to logical occurrences. This rule only applies so far as meaning permits after the assignment of types to the real variables.

In ∗110, ∗113, ∗116, ∗119 we consider the arithmetical operations of addition, multiplication, exponentiation, and subtraction. Also in ∗117 we consider the comparison of cardinal numbers in respect to the relation of greater and less.

There is no interest in complicating our theorems by allowing for the cases when a pure arithmetical formal number, whose components are ambiguous as to type, becomes equal to Λ owing to the low type of one of its *components*. Also in the theory of greater and less the possibility of null-values in low types has no real interest. Accordingly these are excluded from any consideration by the definitions

∗110·03·04, ∗113·04·05, ∗116·03·04, ∗117·02·03,

as far as members of the primary set of formal numbers are concerned; and for other formal numbers by the following convention:

IIT. Whenever a formal number σ occurs, so that, if it were replaced by $\mathrm{Nc}\text{‘}\alpha$, the dominant type of $\mathrm{Nc}\text{‘}\alpha$ would by definition have to be adequate, then the dominant type of σ is also to be adequate.

When σ is a pure arithmetical formal number, this convention secures that the type of every component is adequate.

But in arithmetic we also wish to avoid the intrusion of null-values into the consideration of equations, so far as this avoidance can be attained by the use of high types. Accordingly when we are concerned with the purely arithmetical point of view, we add also the following definition and convention (AT).

Definition. An *arithmetical equation* is an equation between pure arithmetical formal numbers whose dominant types are both determined adequately.

AT. All equations involving pure arithmetical formal numbers are to be arithmetical.

This convention is used in $*117$ and in some earlier propositions which are noted in the prefatory statement.

Its effect is to render the statement of hypotheses often unnecessary. Examples of its application to the numbers where it is not used in the symbolism are also considered in the prefatory statement.

In the case of the inductive numbers we cannot logically prove, apart from Infin ax, that one type exists which is adequate for all the formal numbers 0, 1, 2, 3, etc. But we can prove that for any particular inductive number, say 521, a type exists for which 521 is not equal to Λ. Accordingly for a given symbolic form, in which the symbolism necessarily has only finite complexity, when the types of variables which by hypothesis represent inductive classes or inductive numbers, not Λ, have been settled, it is always possible to fix on a type which will be adequate for all the pure arithmetical formal numbers produced by the symbolism of the form, and also at the same time (and here the peculiar properties of inductive numbers come in) to have chosen the original types of the variables so that any of the variables can assume the value of any assigned constant inductive number, say 521, without being null.

The result is that we may assume that the symbols representing inductive numbers are never null, and thereby obtain the stable truth-values of propositions about them.

Accordingly we proceed as follows: we put

$*$126·01. $\mathrm{NC\,ind} = \mathrm{NC\,induct} - \iota\text{‘}\Lambda \quad \mathrm{Df.}$

We make the rule that when NC ind appears, convention AT is always applied. The result is that when a formal number is an NC ind we need never think about its type, and accordingly all the conventions vanish from the mind, as far as pure arithmetical indefinite inductive cardinals are concerned. We supersede all other conventions by the single one that, if it has been proved or assumed that a formal number represents an inductive cardinal, the types are so arranged that that formal number is not equal to Λ. The proofs of propositions in this number consist largely of the production of a definite type in which this result is attained.

The important propositions are

∗126·12. $\vdash : \nu \in \text{NC ind} \,.\supset.\, (\nu +_c 1) \cap t`\nu \in \text{NC ind}$

∗126·121. $\vdash . 1, 2, 3, \ldots \in \text{NC ind}$

∗126·13·14·15. $\vdash : \alpha, \beta \in \text{NC ind} \,.\supset.\, \alpha +_c \beta,\ \alpha \times_c \beta,\ \alpha^\beta \in \text{NC ind}$

∗126·141. $\vdash : \alpha, \beta \in \text{NC ind} - \iota`0 \,.\equiv.\, \alpha \times_c \beta \in \text{NC ind} - \iota`0$

∗126·151. $\vdash : \alpha, \beta \in \text{NC ind} - \iota`0 \,.\, \alpha \neq 1 \,.\equiv.\, \alpha^\beta \in \text{NC ind} - \iota`0 - \iota`1$

Also ∗126·4·42·43 give the fundamental propositions for subtraction, division, and "inverse exponentiation"; and ∗126·5·51·52·53 the fundamental propositions for the relations of greater and less.

∗126·01. $\text{NC ind} = \text{Nc induct} - \iota`\Lambda$ Df

Whenever the symbol NC ind is used the *Rule of Indefinite Numbers* is adhered to, so that all consideration of distinctions in type among inductive cardinals can be laid aside (cf. Prefatory Statement and also the Summary of this number).

∗126·011. $\vdash : \nu \in \text{NC ind} \,.\equiv.\, \nu \in \text{NC induct} - \iota`\Lambda$ [(∗126·01)]

∗126·1. $\vdash : \nu \in \text{NC ind} \,.\equiv.\, (\exists\alpha) \,.\, \alpha \in \text{Cls induct} \,.\, \nu = \text{Nc}`\alpha \,.\, \exists ! \nu$

Dem.

$\vdash . ∗120·14 . ∗100·4 . ∗126·011 . \supset$

$\vdash : \nu \in \text{NC ind} \,.\supset.\, (\exists\alpha) \,.\, \nu = \text{Nc}`\alpha \,.\, \nu \in \text{NC induct} - \iota`\Lambda \,.$

[∗118·01] $\supset . (\exists\alpha) \,.\, \nu = \text{Nc}`\alpha \,.\, \text{Nc}`\alpha \in \text{NC induct} - \iota`\Lambda \,.\, \exists ! \nu \,.$

[∗120·211] $\supset . (\exists\alpha) \,.\, \alpha \in \text{Cls induct} \,.\, \nu = \text{Nc}`\alpha \,.\, \exists ! \nu$ (1)

$\vdash . ∗120·21 . \supset \vdash : (\exists\alpha) \,.\, \alpha \in \text{Cls induct} \,.\, \nu = \text{Nc}`\alpha \,.\, \exists ! \nu \,.$

$\supset . (\exists\alpha) \,.\, \text{N}_0\text{c}`\alpha \in \text{NC induct} \,.\, \nu = \text{Nc}`\alpha \,.\, \exists ! \nu \,.$

[∗120·15.∗100·511] $\supset . \nu \in \text{NC induct} - \iota`\Lambda$ (2)

$\vdash . (1) . (2) . \supset \vdash . \text{Prop}$

∗126·101. $\vdash :. \mu, \nu \in \text{NC ind} \,.\, \exists ! \mu_\lambda \,.\supset:\, \mu_\lambda = \nu_\lambda \,.\equiv.\, \mu_\lambda = \nu \,.\equiv.\, \mu = \nu$
[∗126·1 . ∗103·16]

∗126·11. $\vdash . 0 \in \text{NC ind}$ [∗120·12 . ∗101·12]

∗126·12. $\vdash : \nu \in \mathrm{NC\,ind} . \supset . (\nu +_c 1) \cap t`\nu \in \mathrm{NC\,ind}$

Dem.

$\vdash . ∗120·151 . \supset \vdash : \nu \in \mathrm{NC\,ind} . \supset . \nu +_c 1 \in \mathrm{NC\,induct}$ (1)

$\vdash . ∗117·66 . ∗118·01 . \supset \vdash : \alpha \in \mathrm{Cls\,induct} . \nu = \mathrm{N_0c}`\alpha . \exists ! \nu . \supset . \mathrm{Nc}`\mathrm{Cl}`\alpha > \nu .$

[∗126·1.∗120·429] $\supset . \mathrm{Nc}`\mathrm{Cl}`\alpha \geqslant \nu +_c 1 .$

[∗103·13.∗117·32] $\supset . \exists ! (\nu +_c 1) \cap t`\mathrm{Cl}`\alpha .$

[∗103·12.∗60·34] $\supset . \exists ! (\nu +_c 1) \cap t`\nu$ (2)

$\vdash . (1) . (2) . ∗126·1 . \supset \vdash .$ Prop

∗126·121. $\vdash . 1, 2, 3, \ldots \in \mathrm{NC\,ind}$ [∗126·11·12]

This proposition, taken in connection with ∗120·4232, embodies the convention named the Rule of Indefinite Numbers and its justification. The convention is that 1, 2, 3, ... are always in future to be used in existential types. In other words whenever any particular inductive number is employed, it is determined in a type in which it is not Λ. The justification is that by ∗126·11·12 such a type can always be found for each particular inductive number.

The convention is also applied to arithmetical formal numbers in ∗126·13·14·15.

For all arithmetical and equational occurrences this convention is really the outcome of IT, IIT, and AT.

∗126·13. $\vdash : \alpha, \beta \in \mathrm{NC\,ind} . \equiv . \alpha +_c \beta \in \mathrm{NC\,ind}$
[∗120·71 . ∗126·1 . ∗110·3 . ∗103·13]

∗126·14. $\vdash : \alpha, \beta \in \mathrm{NC\,ind} . \supset . \alpha \times_c \beta \in \mathrm{NC\,ind}$
[∗120·72 . ∗126·1 . ∗113·25 . ∗103·13]

∗126·141. $\vdash : \alpha, \beta \in \mathrm{NC\,ind} - \iota`0 . \equiv . \alpha \times_c \beta \in \mathrm{NC\,ind} - \iota`0$
[∗120·721 . ∗113·114]

∗126·15. $\vdash : \alpha, \beta \in \mathrm{NC\,ind} . \supset . \alpha^\beta \in \mathrm{NC\,ind}$ [∗120·73 . ∗116·25 . ∗103·13]

∗126·151. $\vdash : \alpha, \beta \in \mathrm{NC\,ind} - \iota`0 . \alpha \neq 1 . \equiv . \alpha^\beta \in \mathrm{NC\,ind} - \iota`0 - \iota`1$
[∗120·731 . ∗116·35 . ∗117·592]

∗126·23. $\vdash : \mu \in \mathrm{NC} . \exists ! \mu \cap t`\alpha . \supset . \exists ! 2^\mu \cap t`t`\alpha . \exists ! (\mu +_c 1) \cap t`t`\alpha$

Dem.

$\vdash . ∗63·661 . ∗116·72 . \supset$

$\vdash : \mu \in \mathrm{NC} . \beta \in \mu \cap t`\alpha . \supset . \mathrm{Cl}`\beta \in 2^\mu \cap t`t`\alpha$ (1)

$\vdash . (1) . ∗117·32 . \supset$

$\vdash : \mathrm{Hp}\,(1) . 2^\mu \geqslant \nu . \supset . \exists ! \mathrm{sm}``\nu \cap t`t`\alpha$ (2)

$\vdash . ∗117·661·31 . \supset \vdash : \mathrm{Hp}\,(1) . \supset . 2^\mu \geqslant \mu +_c 1$ (3)

$\vdash . (1) . (2) . (3) . ∗100·511 . \supset \vdash .$ Prop

∗126·31. $\vdash : \alpha +_c 1 \in \text{NC ind} . \equiv . \alpha \in \text{NC ind}$ [∗126·12·13·121 . ∗120·452]

Note that the specification of the type of $\alpha +_c 1$ is omitted in accordance with the convention. The reference to ∗126·12 shows that it is always possible to apply the convention.

∗126·32. $\vdash : \alpha \in \text{NC} - \iota\text{‘}0 - \iota\text{‘}\Lambda . \nu \in \text{NC ind} . \supset . \alpha +_c \nu > \nu$ [∗120·428 . ∗110·3]

∗126·33. $\vdash :. \alpha \in \text{NC ind} . \beta \in \text{NC} - \iota\text{‘}\Lambda . \supset : \alpha < \beta . \vee . \alpha = \beta . \vee . \alpha > \beta$ [∗120·441]

∗126·4. $\vdash :. \mu, \nu, \varpi \in \text{NC ind} . \supset : \mu +_c \varpi = \nu +_c \varpi . \equiv . \mu = \nu$
[∗126·13 . ∗120·41]

∗126·41. $\vdash :. \mu, \nu, \varpi \in \text{NC ind} . \varpi \neq 0 . \supset : \mu \times_c \varpi = \nu \times_c \varpi . \equiv . \mu = \nu$
[∗120·51 . ∗126·14]

∗126·42. $\vdash :. \mu, \nu, \varpi \in \text{NC ind} . \varpi \neq 0 . \supset : \mu^\varpi = \nu^\varpi . \equiv . \mu = \nu$
[∗120·55 . ∗126·15]

∗126·43. $\vdash :. \mu, \nu, \varpi \in \text{NC ind} . \varpi \neq 0 . \varpi \neq 1 . \supset : \varpi^\mu = \varpi^\nu . \equiv . \mu = \nu$
[∗120·53 . ∗126·15]

∗126·5. $\vdash :. \mu, \nu, \varpi \in \text{NC ind} . \supset : \mu +_c \varpi > \nu +_c \varpi . \equiv . \mu > \nu$

Dem.

$\vdash . ∗117·561 . \quad \supset \vdash : \text{Hp} . \mu > \nu . \supset . \mu +_c \varpi \geqslant \nu +_c \varpi \quad (1)$

$\vdash . ∗126·4 . \quad \supset \vdash : \text{Hp} . \mu > \nu . \supset . \mu +_c \varpi \neq \nu +_c \varpi \quad (2)$

$\vdash . (1) . (2) . ∗117·26 . \supset \vdash : \text{Hp} . \mu > \nu . \supset . \mu +_c \varpi > \nu +_c \varpi \quad (3)$

$\vdash . ∗117·561 . \text{Transp} . ∗117·281 . \supset$

$\vdash : \text{Hp} . \mu +_c \varpi > \nu +_c \varpi . \supset . \sim(\nu \geqslant \mu) .$

[∗126·33] $\supset . \mu > \nu \quad (4)$

$\vdash . (3) . (4) . \supset \vdash . \text{Prop}$

∗126·51. $\vdash :. \mu, \nu, \varpi \in \text{NC ind} . \varpi \neq 0 . \supset : \mu \times_c \varpi > \nu \times_c \varpi . \equiv . \mu > \nu$
[∗117·571 . ∗126·41]

The proof proceeds as in ∗126·5.

∗126·52. $\vdash :. \mu, \nu, \varpi \in \text{NC ind} . \varpi \neq 0 . \supset : \mu^\varpi > \nu^\varpi . \equiv . \mu > \nu$
[∗117·581 . ∗126·42]

∗126·53. $\vdash :. \mu, \nu, \varpi \in \text{NC ind} . \varpi \neq 0 . \varpi \neq 1 . \supset : \varpi^\mu > \varpi^\nu . \equiv . \mu > \nu$
[∗117·591 . ∗126·43]

PART IV.

RELATION-ARITHMETIC.

SUMMARY OF PART IV.

THE subject to be treated in this Part is a general kind of arithmetic of which ordinal arithmetic is a particular application. The form of arithmetic to be treated in this Part is applicable to all relations, though its chief importance is in regard to such relations as generate series. The analogy with cardinal arithmetic is very close, and the reader will find that what follows is much facilitated by bearing the analogy in mind.

The outlines of relation-arithmetic are as follows. We first define a relation between relations, which we shall call *ordinal similarity* or *likeness*, and which plays the same part for relations as similarity plays for classes. Likeness between P and Q is constituted by the fact that the fields of P and Q can be so correlated by a one-one relation that if any two terms have the relation P, their correlates have the relation Q, and vice versa. If P and Q generate series, we may express this by saying that P and Q are like if their fields can be correlated without change of order. Having defined likeness, our next step is to define the *relation-number* of a relation P as the class of relations which are like P, just as the cardinal number of a class α is the class of classes which are similar to α. We then proceed to addition. The ordinal sum of two relations P and Q is defined as the relation which holds between x and y when x and y have the relation P or the relation Q, or when x is a member of C‘P and y is a member of C‘Q. If P and Q generate series, it will be seen that this defines the sum of P and Q as the series resulting from adding the Q-series after the end of the P-series. The sum is thus not commutative. The sum of the relation-numbers of P and Q is of course the relation-number of their sum, provided C‘P and C‘Q have no common terms.

The ordinal product of two relations P and Q is the relation between two couples $z \downarrow x$, $w \downarrow y$, when x, y belong to C‘P and z, w belong to C‘Q and either xPy or $x = y \,.\, zQw$. Thus, for example, if the field of P consists of 1_P, 2_P, 3_P, and the field of Q consists of 1_Q, 2_Q, the relation $P \times Q$ will hold from any earlier to any later term of the following series:

$$1_Q \downarrow 1_P,\ 2_Q \downarrow 1_P,\ 1_Q \downarrow 2_P,\ 2_Q \downarrow 2_P,\ 1_Q \downarrow 3_P,\ 2_Q \downarrow 3_P.$$

It is plain that, denoting the ordinal product of P and Q by $P \times Q$, we have

$$C‘(P \times Q) = C‘P \times C‘Q,$$

where the second "×" as standing between classes has the meaning defined in ∗113·01.

Infinite ordinal sums and products will also be defined, but the definitions are somewhat complicated.

The arithmetic which results from the above definitions satisfies all those of the formal laws which are satisfied in ordinal arithmetic, when this is not confined to finite ordinals; that is to say, relation-numbers satisfy the associative law for addition and for multiplication*, they satisfy the distributive law in the shape (where the + and × are those appropriate to relation-numbers)

$$(\beta + \gamma) \times \alpha = (\beta \times \alpha) + (\gamma \times \alpha),$$

and they satisfy the exponential laws

$$\alpha^\beta \times \alpha^\gamma = \alpha^{\beta+\gamma},$$

$$(\alpha^\beta)^\gamma = \alpha^{\beta\times\gamma}.$$

They do not in general satisfy the commutative law either in addition or in multiplication, nor do they satisfy the distributive law in the form

$$\alpha \times (\beta + \gamma) = (\alpha \times \beta) + (\alpha \times \gamma),$$

nor the exponential law

$$\alpha^\gamma \times \beta^\gamma = (\alpha \times \beta)^\gamma.$$

But in the particular case in which the relations concerned are finite serial relations, the corresponding relation-numbers do satisfy these additional formal laws; hence the arithmetic of *finite* ordinals is exactly analogous to that of inductive cardinals (cf. Part V, Section E).

If the relations concerned are limited to well-ordered relations, relation-arithmetic becomes ordinal arithmetic as developed by Cantor; but many of Cantor's propositions, as we shall see in this Part, do not require the limitation to well-ordered relations.

* For the associative law of multiplication, a hypothesis is required as to the kind of relation concerned. Cf. ∗174·241·25.

SECTION A.

ORDINAL SIMILARITY AND RELATION-NUMBERS.

Summary of Section A.

Two series generated by the relations P and Q respectively are said to be ordinally similar when their terms can be correlated as they stand, without

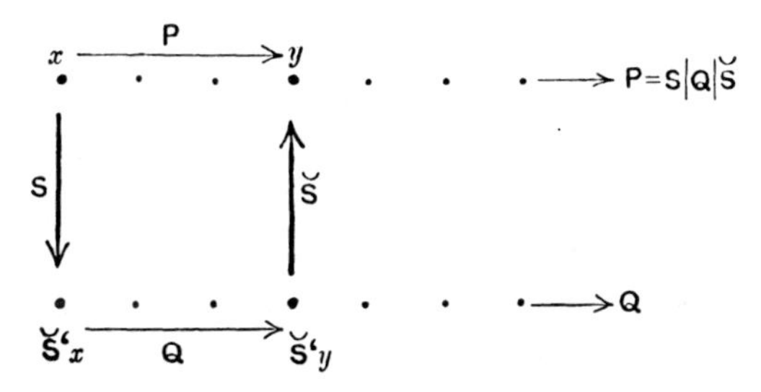

change of order. In the accompanying figure, the relation S correlates the members of $C‘P$ and $C‘Q$ in such a way that if xPy, then $(\breve{S}‘x)\,Q\,(\breve{S}‘y)$, and if zQw, then $(S‘z)\,P\,(S‘w)$. It is evident that the journey from x to y (where xPy) may, in such a case, be taken by going first to $\breve{S}‘x$, thence to $\breve{S}‘y$, and thence back to y, so that $xPy \,.\equiv.\, x\,(S \,|\, Q \,|\, \breve{S})\,y$, *i.e.* $P = S \,|\, Q \,|\, \breve{S}$. Hence to say that P and Q are ordinally similar is equivalent to saying that there is a one-one relation S which has $C‘Q$ for its converse domain and gives $P = S \,|\, Q \,|\, \breve{S}$. In this case we call S a *correlator* of Q and P.

We denote the relation of ordinal similarity by "smor," which is short for "similar ordinally." Thus

$$P \operatorname{smor} Q \,.\equiv.\, (\exists S) \,.\, S \in 1 \to 1 \,.\, C‘Q = \text{Cl}‘S \,.\, P = S \,|\, Q \,|\, \breve{S}.$$

It will be found that the relation $S \,|\, Q \,|\, \breve{S}$ plays the same part in relation to Q in relation-arithmetic as $S‘‘\beta$ plays in relation to β in cardinal arithmetic. It is therefore desirable to have a simpler notation for $S \,|\, Q \,|\, \breve{S}$. We put

$$S;Q = S \,|\, Q \,|\, \breve{S} \quad \text{Df.}$$

We shall find that the semi-colon so defined has the same kind of properties in relation-arithmetic as the two inverted commas have in cardinal arithmetic. Corresponding to the notation $S_\epsilon{}^\epsilon\beta$, we put

$$S\dagger Q = S \mid Q \; \breve{S} \quad \text{Df.}$$

We shall thus have $S\dagger = S \parallel \breve{S}$. It will appear that $S\dagger$ has ordinal properties analogous to the cardinal properties of S_ϵ. Thus *e.g.* where $S \parallel \breve{S}_\epsilon$ appears as a cardinal correlator, $S \parallel \text{Cnv}{}^\epsilon S\dagger$ will appear as an ordinal correlator (in each case with the converse domain suitably limited).

The elementary properties of $S;Q$ will be considered in $*150$. We shall then, in $*151$, be able to study ordinal similarity, taking as our definition of an ordinal correlator

$$P \,\overline{\text{smor}}\, Q = \hat{S}\{S \in 1 \rightarrow 1 \,.\, C{}^\epsilon Q = \text{Ɑ}{}^\epsilon S \,.\, P = S;Q\} \quad \text{Df,}$$

and defining two relations as ordinally similar when they have at least one ordinal correlator, *i.e.* putting (on the analogy of $*73$)

$$\text{smor} = \hat{P}\hat{Q}\{\exists !\, P \,\overline{\text{smor}}\, Q\} \quad \text{Df.}$$

There is no need to confine the notion of ordinal similarity (or likeness, as we shall also call it) to *serial* relations. When two relations have ordinal similarity, their internal structures are analogous, and they therefore have many common properties. Whenever similarity has been proved between two classes α and β, then if β is given as the field of some relation Q, and S is the correlating relation, $S;Q$ is like Q, and has α for its field. Hence similar classes are the fields of like relations. It must not be supposed, however, that like relations are coextensive with relations whose fields are similar. This does not hold even when we confine ourselves to serial relations, except in the special case of *finite* serial relations.

The definition of relation-numbers ($*152$) is as follows: The relation-number of P, which we call $\text{Nr}{}^\epsilon P$, is the class of relations which are ordinally similar to P; and the class of relation-numbers, which we denote by NR, is the class of all classes of the form $\text{Nr}{}^\epsilon P$. The elementary properties of relation-numbers, treated in $*152$, are closely analogous to those of cardinal numbers treated in $*100$.

After a few propositions about the ordinal 0 and the ordinal 2, which we call 0_r and 2_r ($*153$), we pass to the consideration of relation-numbers of various types. It will be observed that "smor," like "sm," is a relation which is ambiguous as to the type both of its domain and of its converse domain. Thus "P smor Q" only has an unambiguous meaning when the types of P and Q are determined. P and Q may or may not be of the same type; the only restriction upon the type of either is that both must be "homogeneous" relations, *i.e.* relations whose domain and converse domain are of the same type. This restriction results from the fact that $C\text{‘}Q$ occurs in the definition of "$P \operatorname{smor} Q$," and a relation does not have a field unless it is homogeneous; hence Q must be homogeneous, and therefore, whatever S may be, $S \mid Q \mid \breve{S}$ must be homogeneous, *i.e.* P must be homogeneous. Thus *e.g.* such relations as D, ι, or ϵ are not ordinally similar either to themselves or to anything else. Whenever "$P \operatorname{smor} Q$" is *significant* for a suitable Q, we have $P \operatorname{smor} P$; but if P is not homogeneous, "$P \operatorname{smor} Q$" is never significant. Hence throughout the theory of ordinal similarity, the relations of which ordinal similarity is affirmed or denied must be homogeneous. The correlators, on the contrary, need not be homogeneous.

Owing to the homogeneity of our relations, the types of relation-numbers are much more easily dealt with than they otherwise would be; for the type of a homogeneous relation is determined by that of a single class, namely its field, whereas the type of a relation in general depends upon the types of *two* classes, namely its domain and its converse domain. Since, where likeness is concerned, the type of the field determines the type of the relation, propositions concerning the relations between different typical determinations of a given relation-number are, for the most part, exactly analogous to and deducible from those for cardinals. In fact, a relation ordinally similar to Q exists in the type of P when, and only when, a class similar to $C\text{‘}Q$ exists in the type of $C\text{‘}P$, *i.e.*

$$\exists ! \operatorname{Nr}(P)\text{‘}Q \,.\equiv .\, \exists ! \operatorname{Nc}(C\text{‘}P)\text{‘}C\text{‘}Q.$$

The half of this proposition follows from the fact that, if P is like Q, $C\text{‘}P$ is similar to $C\text{‘}Q$. The other half follows from the fact, mentioned above, that if $\beta = C\text{‘}Q$ and $\alpha \operatorname{sm} \beta$, then there is a relation like Q and having α for its field. Now if α belongs to the type of $C\text{‘}P$, any relation having α for its field is contained in $t_0\text{‘}C\text{‘}P \uparrow t_0\text{‘}C\text{‘}P$. Hence in the case supposed there is a relation like Q and contained in $t_0\text{‘}C\text{‘}P \uparrow t_0\text{‘}C\text{‘}P$. But the relations contained in $t_0\text{‘}C\text{‘}P \uparrow t_0\text{‘}C\text{‘}P$ constitute $t\text{‘}P$. Hence there is a relation which is like Q and is a member of $t\text{‘}P$, whence our proposition results. By means of this proposition and those of *102—6, the properties of relation-numbers with respect to types follow easily. The conventions IT, IIT and AT apply to relation-numbers as to cardinals; they are to be applied in the same way as in the analogous propositions of Part III, Section A.

$*150$. INTERNAL TRANSFORMATION OF A RELATION.

Summary of $*150$.

In this number we introduce two notations which have uses in regard to relations closely analogous to the uses of $R``\alpha$ and R_ϵ in regard to classes. These two notations are defined as follows:

$$S;Q = S \mid Q \mid \breve{S} \quad \text{Df},$$

$$S\dagger Q = S \mid Q \mid \breve{S} \quad \text{Df}.$$

We then have $\qquad \vdash . S\dagger Q = S;Q = S \mid Q \mid \breve{S} = (S \parallel \breve{S})`Q$.

$S\dagger Q$ is merely an alternative to $S;Q$, just as $R_\epsilon`\alpha$ is an alternative to $R``\alpha$. Also $S\dagger = S \parallel \breve{S}$, in virtue of $*38{\cdot}01$ and $*43{\cdot}01$.

The uses of $S;Q$ occur chiefly when S is a one-one relation and $C`Q \subset \text{Ɑ}`S$. This case is illustrated in the figure in the introduction to this section. Here if Q relates x and y, $S;Q$ relates $S`x$ and $S`y$. Thus given a class α similar to $C`Q$, if S is the correlating relation, $S;Q$ has α for its field, and has, in very many respects, properties analogous to those of Q.

$S;Q$ is important for many special values of S. For example, let Q be a relation between relations; then $C;Q$ will be the corresponding relation of the fields of these relations. If Q be any relation, $\downarrow x;Q$ will be the corresponding relation between ordered couples of which x is the relatum; *i.e.* if yQz, the relation $\downarrow x;Q$ will hold between $y \downarrow x$ and $z \downarrow x$. If Q is a relation between classes, and we have $\beta Q \gamma$, then the relation $\alpha \cup ;Q$ will hold between $\alpha \cup \beta$ and $\alpha \cup \gamma$. In short, whenever S is a one-many relation, and therefore gives rise to a descriptive function, then $S;Q$ is the relation which holds between $S`y$ and $S`z$ whenever Q holds between y and z.

We introduce one other new notation in this number, corresponding to $\alpha \underset{,,}{\text{♀}} y$ in $*38$. This notation is thus defined:

$$Q \underset{;}{\text{♀}} y = \text{♀} y ; Q \quad \text{Df}.$$

The purpose of this notation is to enable us to proceed to $Q \underset{;}{\text{♀}} ;P$ and other similar notations; or, otherwise stated, to enable us to treat $\text{♀} y ; Q$ as a function of y rather than of Q. Take for example the case of $x \downarrow ;Q$. We may wish to consider various relations $x \downarrow {;}Q$, $y \downarrow {;}Q$, where we are to have (say) xPy. To express the relation of $x \downarrow {;}Q$ to $y \downarrow {;}Q$ resulting from xPy, we need the above notation. By its help, we have

$$x \downarrow {;}Q = Q \underset{;}{\downarrow} {}^{\backprime}x \,.\, y \downarrow {;}Q = Q \underset{;}{\downarrow} {}^{\backprime}y.$$

Hence $$xPy \,.\equiv.\, (Q \underset{;}{\downarrow} {}^{\backprime}x)(Q \underset{;}{\downarrow} {;}P)(Q \underset{;}{\downarrow} {}^{\backprime}y).$$

Thus $Q \underset{;}{\downarrow} {;}P$ is the relation between $x \downarrow {;}Q$ and $y \downarrow {;}Q$ corresponding to the relation P between x and y. $Q \underset{;}{\downarrow} {;}P$ plays the same part in relation-arithmetic as is played by $\alpha \underset{\prime\prime}{\downarrow} {}^{\backprime\backprime}\beta$ in cardinal arithmetic.

The notations of this number are capable of occasional uses in cardinal arithmetic*, but their chief utility is in relation-arithmetic, in which they are fundamental.

In order to minimize the use of brackets, we put

$$R^{\backprime}S{;}Q = R^{\backprime}(S{;}Q) \quad \text{Df},$$
$$R{;}S{;}Q = R{;}(S{;}Q) \quad \text{Df}.$$

As an immediate result of the definition of $S{;}Q$, we have

***150·11.** $\vdash : x\,(S{;}Q)\,y \,.\equiv.\, (\exists z, w) \,.\, xSz \,.\, ySw \,.\, zQw$

We have also

***150·12.** $\vdash .\, \text{Cnv}^{\backprime}S{;}Q = S{;}\breve{Q}$

***150·13.** $\vdash .\, R{;}S{;}Q = (R \,|\, S){;}Q$

This proposition, which is the analogue of $(P \,|\, Q)^{\backprime\backprime}\gamma = P^{\backprime\backprime}Q^{\backprime\backprime}\gamma$ (*37·33), is very often used. We have also

***150·3.** $\vdash .\, S{;}(Q \cup R) = S{;}Q \cup S{;}R$

***150·42.** $\vdash .\, S{;}\dot{\Lambda} = \dot{\Lambda}$

The remaining propositions of this number (with a few exceptions) may be thus classified:

(1) Propositions concerning the domain, converse domain, and field of $S{;}Q$ (*150·2—·23). Owing to the fact that the chief applications of this subject are to cases where Q and $S{;}Q$ are serial, the *field* of $S{;}Q$ is more important than its domain or converse domain. Thus the chief propositions here are

***150·22.** $\vdash : C^{\backprime}Q \subset \text{Ɑ}^{\backprime}S \,.\supset.\, C^{\backprime}S{;}Q = S^{\backprime\backprime}C^{\backprime}Q$

***150·23.** $\vdash : C^{\backprime}Q = \text{Ɑ}^{\backprime}S \,.\supset.\, C^{\backprime}S{;}Q = \text{D}^{\backprime}S$

The hypothesis $C^{\backprime}Q \subset \text{Ɑ}^{\backprime}S$ is verified in almost all applications of $S{;}Q$. When it is not verified, the part of $C^{\backprime}Q$ not contained in $\text{Ɑ}^{\backprime}S$ is irrelevant to the value of $S{;}Q$. The hypothesis $C^{\backprime}Q = \text{Ɑ}^{\backprime}S$ is very often verified in practice, since it is verified when S is a correlator of $S{;}Q$ and Q.

* *E.g.* in *116·53 and following propositions, where the notation $S\dagger$ was introduced by a temporary definition.

(2) Propositions concerning relations with limited domains, converse domains, or fields (∗150·32—·38). Broadly speaking, a limitation on the *field* of Q is equivalent to a limitation on the *converse domain* of S, and both are equivalent to a corresponding limitation on the field of $S;Q$ provided $S \epsilon \text{Cls} \rightarrow 1$. The limitations that occur in practice are limitations on the converse domain of S, with consequent limitations on the fields of Q and $S;Q$.

The chief propositions on this subject are

∗150·32. $\vdash . (S \upharpoonright C`Q);Q = S;Q$

∗150·35. $\vdash :. y \epsilon C`Q . \supset_y . R`y = S`y : \supset . R;Q = S;Q$

(This follows from ∗150·32 and ∗35·71.)

∗150·36. $\vdash . (S \upharpoonright \beta);Q = S;(Q \upharpoonright \beta)$

∗150·37. $\vdash : S \epsilon \text{Cls} \rightarrow 1 . \supset . S;(Q \upharpoonright \beta) = (S;Q) \upharpoonright S``\beta = (S \upharpoonright \beta);Q = \{(S``\beta) \urcorner S\};Q$

(3) Propositions on $S;Q$ when S is one-many or many-one (∗150·4—·56). We have

∗150·4. $\vdash :. S \epsilon 1 \rightarrow \text{Cls} . \supset : x (S;Q) y . \equiv . (\exists z, w) . x = S`z . y = S`w . zQw$

This proposition is used constantly. Only slightly less useful is

∗150·41. $\vdash :. S \epsilon \text{Cls} \rightarrow 1 . \supset : x (S;Q) y . \equiv . (\breve{S}`x) \, Q \, (\breve{S}`y)$

The remaining propositions of this set are chiefly applications of ∗150·4·41 to special cases.

(4) A few propositions on $Q \dot{\text{♀}} y$ (∗150·6—·62). These are immediate consequences of the definition.

(5) A set of propositions on couples and matters connected with them (∗150·7—·75). The chief of these is

∗150·71. $\vdash : S \epsilon 1 \rightarrow \text{Cls} . z, w \epsilon \text{Ꮖ}`S . \supset . S;(z \downarrow w) = (S`z) \downarrow (S`w)$

This proposition is very often used in relation-arithmetic. Useful also is

∗150·73. $\vdash . S;(\alpha \uparrow \beta) = S``\alpha \uparrow S``\beta$

(6) We next have four propositions (∗150·8—·83) on $S;P$ when P is a power of Q. These belong with the propositions of ∗92; they are useful in the ordinal theory of finite and infinite. We have

∗150·82·83. $\vdash :. S \epsilon \text{Cls} \rightarrow 1 : \text{D}`Q \subset \text{Ꮖ}`S . \vee . \text{Ꮖ}`Q \subset \text{Ꮖ}`S : \supset .$
$\text{Pot}`S;Q = S\dagger``\text{Pot}`Q . (S;Q)_{po} = S;Q_{po}$

It follows that, in the hypothesis supposed, if S is a correlator of P and Q, it is also a correlator of P_{po} and Q_{po}.

(7) Propositions concerning the relation $S\dagger$ (∗150·14—·171 and ∗150·9—·94). These have uses analogous to those of propositions concerning S_ϵ. The most important are

***150·14.** $\vdash . R\dagger | S\dagger = (R | S)\dagger$

(This follows immediately from *150·13, above.)

***150·141.** $\vdash . S\dagger = S \| \breve{S}$

(This follows immediately from the definition.)

***150·16.** $\vdash . \dot{s}`R\dagger``\lambda = R\dagger(\dot{s}`\lambda) = R;\dot{s}`\lambda$

This proposition is analogous to $s`R```\kappa = R``s`\kappa$ (*40·38), *i.e.* to

$$s`R_\epsilon``\kappa = R_\epsilon`s`\kappa = R``s`\kappa,$$

as appears on substituting $\dot{s}$ and $R\dagger$ for s and R_ϵ in this variant of *40·38.

The remaining propositions are mainly of the nature of lemmas, to be used once or twice each in relation-arithmetic.

***150·01.** $S;Q = S | Q | \breve{S}$ Df

***150·02.** $S\dagger Q = S | Q | \breve{S}$ Df

***150·03.** $Q \underset{;}{\female} y = \female y ; Q$ Df

Here, as in *38, "$\female$" stands for any sign which, when placed between two letters, defines a descriptive function of the arguments represented by those letters. Thus for example "$\female$" may represent any of the following:

$$\cap, \cup, \dot{\cap}, \dot{\cup}, |, \upharpoonright, \restriction, \sqsubset, \uparrow, \downarrow.$$

The two following definitions serve merely for the avoidance of brackets.

***150·04.** $R`S;Q = R`(S;Q)$ Df

***150·05.** $R;S;Q = R;(S;Q)$ Df

***150·1.** $\vdash . S;Q = S | Q | \breve{S} = (S \| \breve{S})`Q = S\dagger Q = S\dagger`Q$
[*43·112 . *38·11 . (*150·01·02)]

***150·11.** $\vdash : x (S;Q) y . \equiv . (\exists z, w) . xSz . ySw . zQw$ [*34·1 . *31·11]

***150·12.** $\vdash . \text{Cnv}`S;Q = S;\breve{Q}$ [*34·2 . *31·33]

***150·13.** $\vdash . R;S;Q = (R | S);Q$

Dem.

$\vdash . *150·1 . (*150·05) . \supset \vdash . R;S;Q = R;(S | Q | \breve{S})$

[*150·1] $= R | S | Q | \breve{S} | \breve{R}$

[*34·2] $= R | S | Q | \text{Cnv}`(R | S)$

[*150·1] $= (R | S);Q . \supset \vdash . \text{Prop}$

∗150·131. $\vdash . (R \mathbin{;} S) \mathbin{;} Q = R \mathbin{;} (S \mid \breve{R}) \mathbin{;} Q$

Dem.

$$\vdash . *150·13 . \supset \vdash . R \mathbin{;} (S \mid \breve{R}) \mathbin{;} Q = (R \mid S \mid \breve{R}) \mathbin{;} Q$$
$$[*150·1] \qquad = (R \mathbin{;} S) \mathbin{;} Q . \supset \vdash . \text{Prop}$$

Observe that we do not have $(R \mathbin{;} S) \mathbin{;} Q = R \mathbin{;} (S \mathbin{;} Q)$.

∗150·14. $\vdash . R\dagger \mid S\dagger = (R \mid S)\dagger$

Dem.

$$\vdash . *150·1·13 . \supset \vdash . R\dagger ‘ S\dagger ‘ Q = (R \mid S)\dagger ‘ Q \qquad (1)$$
$$\vdash . (1) . *34·42 . \supset \vdash . \text{Prop}$$

This proposition is the relational analogue of ∗37·34.

∗150·141. $\vdash . S\dagger = S \parallel \breve{S} \quad [*150·1 . *30·41]$

∗150·15. $\vdash . S\dagger \in 1 \rightarrow \text{Cls} \quad [*72·14]$

∗150·151. $\vdash :. (x) . \text{E} ! S‘x : S \in \text{Cls} \rightarrow 1 : \supset . S\dagger \in 1 \rightarrow 1 \quad [*74·772 . *150·141]$

The following proposition is used in the theory of double ordinal similarity (∗164·13).

∗150·152. $\vdash : S \upharpoonright s‘C‘‘\lambda \in \text{Cls} \rightarrow 1 . s‘C‘‘\lambda \subset \text{Ɑ}‘S . \supset . (S\dagger) \upharpoonright \lambda \in 1 \rightarrow 1$

Dem.

$$\vdash . *74·775 \frac{S, S}{Q, R} . \supset$$
$$\vdash : S \upharpoonright s‘C‘‘\lambda \in \text{Cls} \rightarrow 1 . s‘\text{D}‘‘\lambda \subset \text{Ɑ}‘S . s‘\text{Ɑ}‘‘\lambda \subset \text{Ɑ}‘S . \supset .$$
$$(S \parallel \breve{S}) \upharpoonright \lambda \in 1 \rightarrow 1 \qquad (1)$$
$$\vdash . (1) . *40·57 . *150·141 . \supset \vdash . \text{Prop}$$

∗150·153. $\vdash :. S \upharpoonright s‘C‘‘\lambda \in \text{Cls} \rightarrow 1 . s‘C‘‘\lambda \subset \text{Ɑ}‘S . Q, R \in \lambda . \supset :$
$$S \mathbin{;} Q = S \mathbin{;} R . \supset . Q = R$$

Dem.

$$\vdash . *150·152 . *71·55 . \supset \vdash :. \text{Hp} . \supset : S\dagger ‘ Q = S\dagger ‘ R . \supset . Q = R :$$
$$[*150·1] \qquad \supset : S \mathbin{;} Q = S \mathbin{;} R . \supset . Q = R :. \supset \vdash . \text{Prop}$$

The above proposition is used in dealing with relations of relations of couples (∗165·23).

∗150·16. $\vdash . \dot{s}‘R\dagger‘‘\lambda = R\dagger(\dot{s}‘\lambda) = R \mathbin{;} \dot{s}‘\lambda \quad \left[*43·43 \frac{\breve{R}}{S} . *150·141·1 \right]$

The following proposition is a lemma for ∗150·171.

∗150·17. $\vdash . (R \upharpoonright \lambda)\dagger = R\dagger \mid \lceil \lambda$

Dem.

$$\vdash . *150·1 . \supset \vdash . (R \upharpoonright \lambda)\dagger ‘ P = (R \upharpoonright \lambda) \mid P \mid (\lambda \upharpoonleft \breve{R})$$
$$[*35·354] \qquad = R \mid \lambda \upharpoonleft P \upharpoonright \lambda \mid \breve{R}$$
$$[*150·1 . *36·11] \qquad = R\dagger (P \lceil \lambda)$$
$$[*38·11] \qquad = R\dagger ‘ \lceil \lambda ‘ P \qquad (1)$$
$$\vdash . (1) . *34·42 . \supset \vdash . \text{Prop}$$

∗150·171. $\vdash : s^{\prime}C^{\prime\prime}C^{\prime}Q \subset \lambda . \supset . (R \upharpoonright \lambda)\dagger ; Q = R\dagger ; Q . \upharpoonright \lambda ; Q = Q$

Dem.

$\vdash . *150\cdot17\cdot13 . \supset \vdash . (R \upharpoonright \lambda)\dagger ; Q = R\dagger ; \upharpoonright \lambda ; Q$ (1)

$\vdash . *150\cdot11 . \quad \supset \vdash : M(\upharpoonright \lambda ; Q) N . \equiv . (\exists S, T) . M = S \upharpoonright \lambda . N = T \upharpoonright \lambda . SQT$ (2)

$\vdash . *33\cdot17 . \quad \supset \vdash : SQT . \supset . S, T \in C^{\prime}Q .$

[∗37·62] $\supset . C^{\prime}S, C^{\prime}T \in C^{\prime\prime}C^{\prime}Q .$

[∗40·13] $\supset . C^{\prime}S \subset s^{\prime}C^{\prime\prime}C^{\prime}Q . C^{\prime}T \subset s^{\prime}C^{\prime\prime}C^{\prime}Q$ (3)

$\vdash . (3) . \quad \supset \vdash :. \mathrm{Hp} . \supset : SQT . \supset . C^{\prime}S \subset \lambda . C^{\prime}T \subset \lambda .$

[∗36·25] $\supset . S \upharpoonright \lambda = S . T \upharpoonright \lambda = T$ (4)

$\vdash . (2) . (4) . \supset \vdash :. \mathrm{Hp} . \supset : M(\upharpoonright \lambda ; Q) N . \equiv . (\exists S, T) . M = S . N = T . SQT .$

[∗13·22] $\equiv . MQT :$

[∗21·43] $\supset : \upharpoonright \lambda ; Q = Q$ (5)

$\vdash . (1) . (5) . \supset \vdash . \mathrm{Prop}$

The above proposition is required in the theory of double ordinal similarity. It is used in proving ∗164·141, which is used in ∗164·18, which is a fundamental proposition in the theory of double ordinal similarity.

The following propositions, on the domain, converse domain and field of $S ; Q$, are much used, especially ∗150·202·22·23. ∗150·201 is hardly ever used, but is inserted in order that the general case may not remain unconsidered.

∗150·2. $\vdash . \mathrm{D}^{\prime}S ; Q = S^{\prime\prime}Q^{\prime\prime}\mathrm{Ɑ}^{\prime}S . \mathrm{Ɑ}^{\prime}S ; Q = S^{\prime\prime}\breve{Q}^{\prime\prime}\mathrm{Ɑ}^{\prime}S$ [∗37·32 . ∗150·1]

∗150·201. $\vdash . C^{\prime}S ; Q = S^{\prime\prime}(Q \cup \breve{Q})^{\prime\prime}\mathrm{Ɑ}^{\prime}S = \mathrm{D}^{\prime}S ; (Q \cup \breve{Q})$

Dem.

$\vdash . *150\cdot2 . *37\cdot22 . \supset \vdash . C^{\prime}S ; Q = S^{\prime\prime}(Q^{\prime\prime}\mathrm{Ɑ}^{\prime}S \cup \breve{Q}^{\prime\prime}\mathrm{Ɑ}^{\prime}S)$

[∗37·221] $= S^{\prime\prime}(Q \cup \breve{Q})^{\prime\prime}\mathrm{Ɑ}^{\prime}S$

[∗150·2] $= \mathrm{D}^{\prime}S ; (Q \cup \breve{Q}) . \supset \vdash . \mathrm{Prop}$

∗150·202. $\vdash . \mathrm{D}^{\prime}S ; Q \subset S^{\prime\prime}\mathrm{D}^{\prime}Q . \mathrm{Ɑ}^{\prime}S ; Q \subset S^{\prime\prime}\mathrm{Ɑ}^{\prime}Q . C^{\prime}S ; Q \subset S^{\prime\prime}C^{\prime}Q$

Dem.

$\vdash . *37\cdot15\cdot16 . \supset \vdash . Q^{\prime\prime}\mathrm{Ɑ}^{\prime}S \subset \mathrm{D}^{\prime}Q . \breve{Q}^{\prime\prime}\mathrm{Ɑ}^{\prime}S \subset \mathrm{Ɑ}^{\prime}Q .$

[∗37·2.∗150·2] $\supset \vdash . \mathrm{D}^{\prime}S ; Q \subset S^{\prime\prime}\mathrm{D}^{\prime}Q . \mathrm{Ɑ}^{\prime}S ; Q \subset S^{\prime\prime}\mathrm{Ɑ}^{\prime}Q$ (1)

[∗37·22] $\supset \vdash . C^{\prime}S ; Q \subset S^{\prime\prime}C^{\prime}Q$ (2)

$\vdash . (1) . (2) . \quad \supset \vdash . \mathrm{Prop}$

∗150·203. $\vdash . C^{\prime}S ; Q \subset \mathrm{D}^{\prime}S$ [∗150·202 . ∗37·15]

∗150·21. $\vdash : \mathrm{Ɑ}^{\prime}Q \subset \mathrm{Ɑ}^{\prime}S . \supset . \mathrm{D}^{\prime}S ; Q = S^{\prime\prime}\mathrm{D}^{\prime}Q = \mathrm{D}^{\prime}(S \mid Q)$ [∗150·2 . ∗37·27·32]

∗150·211. $\vdash : \mathrm{D}^{\prime}Q \subset \mathrm{Ɑ}^{\prime}S . \supset . \mathrm{Ɑ}^{\prime}S ; Q = S^{\prime\prime}\mathrm{Ɑ}^{\prime}Q = \mathrm{D}^{\prime}(S \mid \breve{Q})$
[∗150·2 . ∗37·271·32]

***150·22.** $\vdash : C'Q \subset \text{Ɑ}'S . \supset . C'S;Q = S``C'Q$ [*150·21·211 . *37·22]

In practice, when $S;Q$ is used, we almost always have $C'Q \subset \text{Ɑ}'S$. For the use of $S;Q$ is to obtain a relation analogous to Q and having a different field; now $S;Q$ is analogous to $Q \upharpoonleft \text{Ɑ}'S$, for the part of $C'Q$ which lies outside $\text{Ɑ}'S$ is unaffected by S. Hence if we have, to start with, a relation Q whose field is *not* contained in $\text{Ɑ}'S$, we shall usually find it profitable to limit the field to $\text{Ɑ}'S$, and consider the transformed relation rather as $S;(Q \upharpoonleft \text{Ɑ}'S)$ than as $S;Q$. Thus the hypothesis $C'Q \subset \text{Ɑ}'S$ will be verified in almost all useful applications of the notion of $S;Q$.

***150·23.** $\vdash : C'Q = \text{Ɑ}'S . \supset . C'S;Q = \text{D}'S$ [*150·22 . *37·25]

***150·24.** $\vdash :. C'Q \subset \text{Ɑ}'S . \supset : \dot{\exists} ! S;Q . \equiv . \dot{\exists} ! Q$

Dem.

$\vdash . *37·43 . *150·22 . \supset \vdash :. \text{Hp} . \supset : \exists ! C'S;Q . \equiv . \exists ! C'Q :$

[*33·24] $\supset : \dot{\exists} ! S;Q . \equiv . \dot{\exists} ! Q :. \supset \vdash . \text{Prop}$

***150·25.** $\vdash :. (y) . \text{E} ! S'y . \supset : \dot{\exists} ! S;Q . \equiv . \dot{\exists} ! Q$ [*150·24 . *33·431]

***150·3.** $\vdash . S;(Q \dot{\cup} R) = S;Q \dot{\cup} S;R$ [*34·25·26 . *150·1]

***150·301.** $\vdash . S;(Q \dot{\cap} R) \mathrel{\dot{\subset}} (S;Q) \dot{\cap} (S;R)$ [*34·23·24 . *150·1]

***150·31.** $\vdash : P \mathrel{\dot{\subset}} Q . R \mathrel{\dot{\subset}} S . \supset . R;P \mathrel{\dot{\subset}} S;Q$ [*34·34 . *150·1]

The following propositions are frequently useful when we have to deal with correlators of the form $S \upharpoonright C'Q$, which often happens.

***150·32.** $\vdash . (S \upharpoonright C'Q);Q = S;Q$ $\left[*43·5 \dfrac{S, \breve{S}, C'Q, C'\breve{Q}}{Q, R, \alpha, \beta} . *150·141\right]$

***150·33.** $\vdash : C'Q \subset \beta . \supset . (S \upharpoonright \beta);Q = S;Q$ $\left[*43·5 \dfrac{S, \breve{S}}{Q, R} . *150·141\right]$

***150·34.** $\vdash : \text{D}'Q \subset \alpha . \text{Ɑ}'Q \subset \beta . \supset . S \upharpoonright \alpha | Q | \beta \upharpoonleft \breve{S} = S;\alpha \upharpoonleft Q \upharpoonright \beta = S;Q$
[*43·5 . *35·354 . *150·141·1]

***150·35.** $\vdash :. y \in C'Q . \supset_y . R'y = S'y : \supset . R;Q = S;Q$

Dem.

$\vdash . *35·71 . \supset \vdash : \text{Hp} . \supset . R \upharpoonright C'Q = S \upharpoonright C'Q .$

[*34·27·28.*150·1] $\supset . (R \upharpoonright C'Q);Q = (S \upharpoonright C'Q);Q .$

[*150·32] $\supset . R;Q = S;Q : \supset \vdash . \text{Prop}$

The above proposition, which is the analogue of *37·69, is much used in relation-arithmetic.

The following proposition is much used after we reach the theory of well-ordered series, but not before (except in *150·37).

***150·36.** $\vdash . (S \upharpoonright \beta) ; Q = S ; (Q \sqsubset \beta)$

Dem.

$$\vdash . *150 \cdot 11 . *35 \cdot 101 . \supset$$
$$\vdash : x \{(S \upharpoonright \beta) ; Q\} w . \equiv . (\exists y, z) . xSy . y \in \beta . yQz . z \in \beta . wSz .$$
$$[*36 \cdot 13] \qquad \equiv . (\exists y, z) . xSy . y (Q \sqsubset \beta) z . wSz .$$
$$[*150 \cdot 11] \qquad \equiv . x \{S ; (Q \sqsubset \beta)\} w : \supset \vdash . \text{Prop}$$

***150·361.** $\vdash . (\alpha \upharpoonleft S) ; Q = (S ; Q) \sqsubset \alpha$ [Proof as in *150·36]

***150·37.** $\vdash : S \in \text{Cls} \to 1 . \supset . S ; (Q \sqsubset \beta) = (S ; Q) \sqsubset S``\beta$
$$= (S \upharpoonright \beta) ; Q = \{(S``\beta) \upharpoonleft S\} ; Q$$

Dem.

$$\vdash . *74 \cdot 141 . \supset \vdash : \text{Hp} . \supset . (S \upharpoonright \beta) ; Q = \{(S``\beta) \upharpoonleft S\} ; Q \qquad (1)$$
$$\vdash . (1) . *150 \cdot 36 \cdot 361 . \supset \vdash . \text{Prop}$$

The above proposition is not used until we reach the theory of series.

***150·38.** $\vdash : S \in 1 \to 1 . \supset . S ; \breve{S} ; Q = Q \sqsubset \text{D‘}S$

Dem.

$$\vdash . *150 \cdot 1 . \qquad \supset \vdash . S ; \breve{S} ; Q = S \mid \breve{S} \mid Q \mid S \mid \breve{S} \qquad (1)$$
$$\vdash . (1) . *72 \cdot 59 \cdot 591 . \supset \vdash : \text{Hp} . \supset . S ; \breve{S} ; Q = (\text{D‘}S) \upharpoonleft Q \upharpoonright \text{D‘}S$$
$$[*36 \cdot 11] \qquad = Q \sqsubset \text{D‘}S : \supset \vdash . \text{Prop}$$

The above proposition is used in dealing with the correlation of series (*208·2).

***150·4.** $\vdash :. S \in 1 \to \text{Cls} . \supset : x (S ; Q) y . \equiv . (\exists z, w) . x = S‘z . y = S‘w . zQw$
[*150·11 . *71·36]

This proposition is fundamental in the theory of $S;Q$, because in most of the uses of this notion S is one-many. The proposition states that when S is one-many, $S;Q$ is the relation between the S's of terms related by Q. Thus if S is the relation of wife to husband, and Q is the relation of brother to brother, $S;Q$ is the relation between wives of brothers. If Q is a relation between relations, $C;P$ will be the corresponding relation of their fields; and so on.

***150·41.** $\vdash :. S \in \text{Cls} \to 1 . \supset : x (S ; Q) y . \equiv . (\breve{S}‘x) \, Q \, (\breve{S}‘y)$ [*150·11 . *71·331]

***150·42.** $\vdash . S ; \dot{\Lambda} = \dot{\Lambda}$ [*150·1 . *34·32]

The following propositions, down to *150·56, are, with the exception of *150·52—·535, all illustrations of *150·4·41.

***150·5.** $\vdash : \alpha (\overrightarrow{R} ; P) \beta . \equiv . (\exists x, y) . \alpha = \overrightarrow{R}‘x . \beta = \overrightarrow{R}‘y . xPy$

***150·51.** $\vdash : \alpha (\text{D} ; R) \beta . \equiv . (\exists P, Q) . \alpha = \text{D‘}P . \beta = \text{D‘}Q . PRQ$

***150·511.** $\vdash : \alpha (\text{ᗡ} ; R) \beta . \equiv . (\exists P, Q) . \alpha = \text{ᗡ‘}P . \beta = \text{ᗡ‘}Q . PRQ$

***150·512.** $\vdash : \alpha (C ; R) \beta \,.\equiv .\, (\exists P, Q) \,.\, \alpha = C`R \,.\, \beta = C`Q \,.\, PRQ$

***150·52.** $\vdash : x (F ; R) y \,.\equiv .\, (\exists P, Q) \,.\, x \in C`P \,.\, y \in C`Q \,.\, PRQ$
[*150·11 . *33·51]

$F ; R$ is a relation which plays a great part in relation-arithmetic.

***150·53.** $\vdash . I ; P = P$ [*50·4]

***150·531.** $\vdash . P ; I = P \mid \breve{P}$ [*50·4]

***150·532.** $\vdash . P ; I ; Q = P ; Q$ [*150·13 . *50·4]

***150·534.** $\vdash . (I \upharpoonright C`P) ; P = P$ [*150·53·32]

***150·535.** $\vdash : C`P \subset \alpha \,.\supset .\, (I \upharpoonright \alpha) ; P = P$ [*150·53·33]

***150·54.** $\vdash : \alpha (\breve{\iota} ; R) \beta \,.\equiv .\, (\breve{\iota}`\alpha) R (\breve{\iota}`\beta)$

***150·541.** $\vdash : x (\breve{\iota} ; R) y \,.\equiv .\, (\iota`x) R (\iota`y)$

***150·55.** $\vdash : Q (\downarrow z ; P) R \,.\equiv .\, (\exists u, v) \,.\, Q = u \downarrow z \,.\, R = v \downarrow z \,.\, uPv$

***150·56.** $\vdash : M (S \dagger ; Q) N \,.\equiv .\, (\exists X, Y) \,.\, XQY \,.\, M = S ; X \,.\, N = S ; Y$
[*150·4·15·1]

***150·6.** $\vdash . P \underset{\cdot ;}{\text{♀}} y = \text{♀} y ; P$ [(*150·03)]

***150·601.** $\vdash . P \underset{\cdot ;}{\text{♀}} \in 1 \to \mathrm{Cls}$ [*150·6 . *14·21 . *71·166]

***150·61.** $\vdash : z (P \underset{\cdot ;}{\text{♀}} y) w \,.\equiv .\, (\exists u, v) \,.\, z = u \text{♀} y \,.\, w = v \text{♀} y \,.\, uPv$

[*150·11 . *38·101 . *150·6]

***150·62.** $\vdash : R (P \underset{\cdot ;}{\text{♀}} ; Q) S \,.\equiv .\, (\exists z, w) \,.\, R = \text{♀} z ; P \,.\, S = \text{♀} w ; P \,.\, zQw$

[*150·4·601·6]

Relations of the form $P \underset{\cdot ;}{\text{♀}} ; Q$ are frequently useful in relation-arithmetic, especially in the particular case of $P \underset{\cdot ;}{\downarrow} ; Q$, which takes the place taken by $\alpha \underset{;;}{\downarrow} ``\beta$ in cardinal arithmetic. Relations of the form $P \underset{\cdot ;}{\downarrow} ; Q$ will be considered in *165.

The following propositions are chiefly concerned with correlations of couples. They are of great utility in relation-arithmetic. *150·71, in particular, is fundamental.

***150·7.** $\vdash . S ; (z \downarrow w) = \overrightarrow{S}`z \uparrow \overrightarrow{S}`w$ [*55·6]

***150·71.** $\vdash : S \in 1 \to \mathrm{Cls} \,.\, z, w \in \text{ᗡ}`S \,.\supset .\, S ; (z \downarrow w) = (S`z) \downarrow (S`w)$ [*55·61]

***150·72.** $\vdash : z \neq w \,.\, S = x \downarrow z \cup y \downarrow w \,.\supset .\, S ; (z \downarrow w) = x \downarrow y$ [*55·62·61]

***150·73.** $\vdash . S ; (\alpha \uparrow \beta) = S``\alpha \uparrow S``\beta$ $\left[*37·82 \dfrac{S, \breve{S}}{R, S} \right]$

***150·74.** $\vdash . (S \cup T) ; Q = S ; Q \cup T ; Q \cup S \mid Q \mid \breve{T} \cup T \mid Q \mid \breve{S}$ [*150·1]

∗150·75. $\vdash : \sim(yQy) \,.\supset .\, (S \cup x \downarrow y){;}Q = S{;}Q \cup S``\overrightarrow{Q}`y \uparrow \iota`x \cup \iota`x \uparrow S``\overleftarrow{Q}`y$

Dem.

$\vdash . *150\cdot1 . \supset \vdash : \mathrm{Hp} . \supset . (x \downarrow y){;}Q = \dot{\Lambda} .$

$[*150\cdot74] \qquad \supset . (S \cup x \downarrow y){;}Q = S{;}Q \cup S | Q | y \downarrow x \cup x \downarrow y | Q | \breve{S}$

$[*55\cdot57\cdot571] \qquad = S{;}Q \cup S``\overrightarrow{Q}`y \uparrow \iota`x \cup \iota`x \uparrow S``\overleftarrow{Q}`y : \supset \vdash . \mathrm{Prop}$

The four following propositions belong to the subject of ∗92, but could not be given in that number owing to the fact that they involve the notations of ∗150. They are required for proving that, if S is a correlator of P and Q, it is also a correlator of P_{po} and Q_{po} (∗151·45), and for one of the fundamental propositions in the ordinal theory of progressions (∗263·17).

∗150·8. $\vdash :. S \epsilon \mathrm{Cls} \rightarrow 1 : \mathrm{D}`Q \subset \text{Ɑ}`S \,.\vee .\, \text{Ɑ}`Q \subset \text{Ɑ}`S : P \epsilon \mathrm{Pot}`Q : \supset .$
$\qquad S{;}P \epsilon \mathrm{Pot}`(S{;}Q) . (S{;}P) | (S{;}Q) = S{;}(P | Q)$

Dem.

$\vdash . *91\cdot351 . \qquad \supset \vdash . S{;}Q \epsilon \mathrm{Pot}`(S{;}Q) \qquad (1)$

$\vdash . *150\cdot1 . \qquad \supset \vdash . (S{;}P) | (S{;}Q) = S | P | \breve{S} | S | Q | \breve{S} \qquad (2)$

$\vdash . (2) . *71\cdot191 . \qquad \supset \vdash : \mathrm{Hp} . \supset . (S{;}P) | (S{;}Q) = S | P | I \upharpoonright \text{Ɑ}`S | Q | \breve{S} \qquad (3)$

$\vdash . *50\cdot63 . \qquad \supset \vdash : \mathrm{D}`Q \subset \text{Ɑ}`S . \supset . I \upharpoonright \text{Ɑ}`S | Q = Q \qquad (4)$

$\vdash . *50\cdot62 . *91\cdot271 . \supset \vdash : P \epsilon \mathrm{Pot}`Q . \text{Ɑ}`Q \subset \text{Ɑ}`S . \supset . P | I \upharpoonright \text{Ɑ}`S = P \qquad (5)$

$\vdash . (3) . (4) . (5) . \qquad \supset \vdash : \mathrm{Hp} . P \epsilon \mathrm{Pot}`Q . \supset . (S{;}P) | (S{;}Q) = S | P | Q | \breve{S}$

$[*150\cdot1] \qquad = S{;}(P | Q) \qquad (6)$

$\vdash . *91\cdot282 . \qquad \supset \vdash : S{;}P \epsilon \mathrm{Pot}`S{;}Q . \supset . (S{;}P) | (S{;}Q) \epsilon \mathrm{Pot}`S{;}Q \qquad (7)$

$\vdash . (6) . (7) . \qquad \supset \vdash :. \mathrm{Hp} . P \epsilon \mathrm{Pot}`Q . \supset : S{;}P \epsilon \mathrm{Pot}`S{;}Q . \supset .$
$\qquad S{;}(P | Q) \epsilon \mathrm{Pot}`S{;}Q \qquad (8)$

$\vdash . (1) . (8) . *91\cdot373 \dfrac{S{;}P \epsilon \mathrm{Pot}`S{;}Q}{\phi P} . \supset$

$\vdash :. \mathrm{Hp} . \supset : P \epsilon \mathrm{Pot}`Q . \supset_P . S{;}P \epsilon \mathrm{Pot}`(S{;}Q) \qquad (9)$

$\vdash . (6) . (9) . \supset \vdash . \mathrm{Prop}$

∗150·81. $\vdash :. S \epsilon \mathrm{Cls} \rightarrow 1 : \mathrm{D}`Q \subset \text{Ɑ}`S \,.\vee .\, \text{Ɑ}`Q \subset \text{Ɑ}`S : T \epsilon \mathrm{Pot}`S{;}Q : \supset .$
$\qquad (\exists P) . P \epsilon \mathrm{Pot}`Q . T = S{;}P$

Dem.

$\vdash . *91\cdot351 . \supset \vdash . (\exists P) . P \epsilon \mathrm{Pot}`Q . S{;}Q = S{;}P \qquad (1)$

$\vdash . *150\cdot8 . \quad \supset \vdash : \mathrm{Hp} . P \epsilon \mathrm{Pot}`Q . T = S{;}P . \supset . T | (S{;}Q) = S{;}(P | Q) .$

$[*91\cdot282] \qquad \supset . (\exists R) . R \epsilon \mathrm{Pot}`Q . T | (S{;}Q) = S{;}R \qquad (2)$

$\vdash . (2) . *10\cdot23 . \supset$

$\vdash :. \mathrm{Hp} . \supset : (\exists P) . P \epsilon \mathrm{Pot}`Q . T = S{;}P . \supset .$
$\qquad (\exists R) . R \epsilon \mathrm{Pot}`Q . T | (S{;}Q) = S{;}R \qquad (3)$

$\vdash . (1) . (3) . *91\cdot171 \dfrac{S{;}Q, T, (\exists P) . P \epsilon \mathrm{Pot}`Q . T = S{;}P}{R, \quad S, \qquad \phi T} . \supset$

$\vdash : \mathrm{Hp} . T \epsilon \mathrm{Pot}`S{;}Q . \supset . (\exists P) . P \epsilon \mathrm{Pot}`Q . T = S{;}P : \supset \vdash . \mathrm{Prop}$

***150·82.** $\vdash :. S \epsilon \text{Cls} \rightarrow 1 : \text{D}`Q \subset \text{Ɑ}`S . \text{v} . \text{Ɑ}`Q \subset \text{Ɑ}`S : \supset . \text{Pot}`S;Q = S\dagger``\text{Pot}`Q$

Dem.

$\vdash . *150·8·81 . \supset$

$\vdash : \text{Hp} . \supset . \text{Pot}`S;Q = \hat{T}\{(\exists P) . P \epsilon \text{Pot}`Q . T = S;P\}$

$[*150·1] \qquad = S\dagger``\text{Pot}`Q : \supset \vdash . \text{Prop}$

***150·83.** $\vdash :. S \epsilon \text{Cls} \rightarrow 1 : \text{D}`Q \subset \text{Ɑ}`S . \text{v} . \text{Ɑ}`Q \subset \text{Ɑ}`S : \supset . (S;Q)_{\text{po}} = S;Q_{\text{po}}$

Dem.

$\vdash . *150·82 . (*91·05) . \supset \vdash : \text{Hp} . \dot{\supset} . (S;Q)_{\text{po}} = \dot{s}`S\dagger``\text{Pot}`Q$

$[*150·16.(*91·05)] \qquad = S;Q_{\text{po}} : \supset \vdash . \text{Prop}$

The following propositions, down to *150·94 inclusive, resume the subject of the relation $S\dagger$, which has already been treated in *150·14—·171.

***150·9.** $\vdash . (I\dagger);Q = Q$

Dem.

$\vdash . *150·56 . \supset \vdash : M (I\dagger;Q) N . \equiv . (\exists X, Y) . XQY . M = I;X . N = I;Y .$

$[*150·53] \qquad \equiv . (\exists X, Y) . XQY . M = X . N = Y .$

$[*13·22] \qquad \equiv . MQY : \supset \vdash . \text{Prop}$

The following propositions lead up to *150·931·94, which are used in the theory of double ordinal similarity (*164·3·21).

***150·91.** $\vdash : s`C``C`Q \subset \alpha . \supset . (I \upharpoonright \alpha)\dagger;Q = Q$

Dem.

$\vdash . *150·535 . \quad \supset \vdash :. \text{Hp} . \supset : X \epsilon C`Q . \supset . (I \upharpoonright \alpha);X = X \qquad (1)$

$\vdash . (1) . *150·56 . \supset \vdash :. \text{Hp} . \supset :$

$M \{(I \upharpoonright \alpha)\dagger;Q\} N . \equiv . (\exists X, Y) . XQY . M = X . N = Y .$

$[*13·22] \qquad \equiv . MQN :. \supset \vdash . \text{Prop}$

***150·92.** $\vdash : S \epsilon \text{Cls} \rightarrow 1 . s`C``C`Q \subset \text{Ɑ}`S . \supset . \breve{S}\dagger;S\dagger;Q = Q$

Dem.

$\vdash . *150·13·14 . \quad \supset \vdash . \breve{S}\dagger;S\dagger;Q = (\breve{S} | S)\dagger;Q \qquad (1)$

$\vdash . (1) . *71·191 . \supset \vdash : \text{Hp} . \supset . \breve{S}\dagger;S\dagger;Q = (I \upharpoonright \text{Ɑ}`S)\dagger;Q$

$[*150·91] \qquad = Q : \supset \vdash . \text{Prop}$

***150·921.** $\vdash : S \epsilon 1 \rightarrow \text{Cls} . s`C``C`P \subset \text{D}`S . \supset . S\dagger;\breve{S}\dagger;P = P$

***150·93.** $\vdash :. S \epsilon 1 \rightarrow 1 . s`C``C`P \subset \text{D}`S . s`C``C`Q \subset \text{Ɑ}`S . \supset :$

$P = S\dagger;Q . \equiv . Q = \breve{S}\dagger;P \quad [*150·92·921]$

***150·931.** $\vdash : s`C``C`Q \subset \text{Ɑ}`S . \supset . C``C`S\dagger;Q = S_{\epsilon}``C``C`Q$

Dem.

$\vdash . *150·22 . \quad \supset \vdash . C`S\dagger;Q = S\dagger``C`Q \qquad (1)$

$\vdash . *150·22·1 . \supset \vdash :. \text{Hp} . \supset : M \epsilon C`Q . \supset . C`S\dagger`M = S``C`M :$

$[*37·68·11] \qquad \supset : C``S\dagger``C`Q = S_{\epsilon}``C``C`Q \qquad (2)$

$\vdash . (1) . (2) . \supset \vdash . \text{Prop}$

***150·932.** $\vdash : s`C``C`Q \subset \text{ɑ}`S . \supset . s`C``C`S\dagger;Q = S``s`C``C`Q$ [*150·931 . *37·11 . *40·38]

***150·933.** $\vdash : s`C``C`Q \subset \text{ɑ}`S . \supset . s`C``C`S\dagger;Q \subset \text{D}`S$ [*150·932 . *37·15]

***150·94.** $\vdash :. S \epsilon 1 \to 1 . \supset : s`C``C`Q \subset \text{ɑ}`S . P = S\dagger;Q . \equiv . s`C``C`P \subset \text{D}`S . Q = \breve{S}\dagger;P$

Dem.

$\vdash . *150{\cdot}933 . \supset \vdash : s`C``C`Q \subset \text{ɑ}`S . P = S\dagger;Q . \equiv .$
$s`C``C`P \subset \text{D}`S . s`C``C`Q \subset \text{ɑ}`S . P = S\dagger;Q$ (1)

$\vdash . *150{\cdot}933 \frac{\breve{S}}{S} . \supset \vdash : s`C``C`P \subset \text{D}`S . Q = \breve{S}\dagger;P . \equiv .$
$s`C``C`P \subset \text{D}`S . s`C``C`Q \subset \text{ɑ}`S . Q = \breve{S}\dagger;P$ (2)

$\vdash . *150{\cdot}93 . *5{\cdot}32 . \supset$
$\vdash :. S \epsilon 1 \to 1 . \supset : s`C``C`P \subset \text{D}`S . s`C``C`Q \subset \text{ɑ}`S . P = S\dagger;Q . \equiv .$
$s`C``C`P \subset \text{D}`S . s`C``C`Q \subset \text{ɑ}`S . Q = \breve{S}\dagger;P$ (3)

$\vdash . (1) . (2) . (3) . \supset \vdash .$ Prop

The above proposition is the analogue of *74·61, which (with a few trivial transformations) may be written

$$\vdash :. S \epsilon 1 \to 1 . \supset : s`\lambda \subset \text{ɑ}`S . \kappa = S_\epsilon``\lambda . \equiv . s`\kappa \subset \text{D}`S . \lambda = (\breve{S})_\epsilon``\kappa.$$

In obtaining ordinal analogues of such propositions, S_ϵ will be replaced by $S\dagger$, and the two inverted commas will be replaced by the semi-colon; a class of classes κ will be replaced, in most of its occurrences, by a relation of relations P, but will sometimes be replaced by $C``C`P$.

The above proposition (*150·94) is used in proving that the converse of a double correlator of P and Q is a double correlator of Q and P (*164·21). The corresponding cardinal proposition (*111·131) uses *74·6, which is practically the same proposition as *74·61, which is the analogue of *150·94.

***150·95.** $\vdash : C`R \subset \text{Cl}`\alpha . \supset . (S \upharpoonright \alpha)_\epsilon;R = S;R$

Dem.

$\vdash . *37{\cdot}421 . \supset \vdash :. \text{Hp} . \supset : \beta \epsilon C`R . \supset . (S \upharpoonright \alpha)``\beta = S``\beta .$
[*37·11] $\supset . (S \upharpoonright \alpha)_\epsilon`\beta = S_\epsilon`\beta$ (1)

$\vdash . (1) . *150{\cdot}35 . \supset \vdash .$ Prop

The above proposition is used in the theory of "first differences" (*170·41).

***150·96.** $\vdash : \text{C}`\dot{s}`\lambda \subset \text{C}`S \,.\supset .\, \text{D}{;}(T \parallel \breve{S}) \upharpoonright \lambda = T_\epsilon \upharpoonright \text{D}``\lambda$

Dem.

$\vdash . *150{\cdot}51 . \supset \vdash : \alpha \{\text{D}{;}(T \parallel \breve{S}) \upharpoonright \lambda\} \beta . \equiv .$

$(\exists M, N) . N \epsilon \lambda . M = T \,|\, N \,|\, \breve{S} . \alpha = \text{D}`M . \beta = \text{D}`N$ (1)

$\vdash . *41{\cdot}13 . \supset \vdash :. \text{Hp} . \supset : N \epsilon \lambda . \supset . \text{C}`N \subset \text{C}`S .$

[*37·321] $\supset . \text{D}`(N \,|\, \breve{S}) = \text{D}`N .$

[*37·32] $\supset . \text{D}`(T \,|\, N \,|\, \breve{S}) = T``\text{D}`N$ (2)

$\vdash . (1) . (2) . *37{\cdot}6 . \supset$

$\vdash :. \text{Hp} . \supset : \alpha \{\text{D}{;}(T \parallel \breve{S}) \upharpoonright \lambda\} \beta . \equiv . \beta \epsilon \text{D}``\lambda . \alpha = T``\beta .$

[*37·101] $\equiv . \alpha (T_\epsilon \upharpoonright \text{D}``\lambda) \beta :. \supset \vdash . \text{Prop}$

***150·961.** $\vdash . \dot{s}{;}(U \parallel \breve{W})_\epsilon \upharpoonright \lambda = (U \parallel \breve{W}) \upharpoonright \dot{s}``\lambda$

Dem.

$\vdash . *150{\cdot}4 . \supset \vdash : R \{\dot{s}{;}(U \parallel \breve{W})_\epsilon \upharpoonright \lambda\} S . \equiv . (\exists \beta) . \beta \epsilon \lambda . S = \dot{s}`\beta . R = \dot{s}`(U \parallel \breve{W})``\beta .$

[*43·43] $\equiv . (\exists \beta) . \beta \epsilon \lambda . S = \dot{s}`\beta . R = (U \parallel \breve{W})`\dot{s}`\beta .$

[*13·193.*37·6] $= . S \epsilon \dot{s}``\lambda . R = (U \parallel \breve{W})`S : \supset \vdash . \text{Prop}$

The above proposition is used in the theory of ordinal exponentiation (*176·21).

$*151$. ORDINAL SIMILARITY.

Summary of $*151$.

In this number, we give the definition of ordinal similarity, and various equivalent forms; we prove that ordinal similarity is reflexive ($*151{\cdot}13$), symmetrical ($*151{\cdot}14$) and transitive ($*151{\cdot}15$), and we give some particular cases of ordinal similarity ($*151{\cdot}6$ ff.). Propositions in this number should be compared with those in $*73$, to which they are analogous.

The class of ordinal correlators of P and Q is written $P \,\overline{\text{smor}}\, Q$, where "smor" stands for "similar ordinally." We put

$$P \,\overline{\text{smor}}\, Q = \hat{S}\{S \in 1 \to 1 \,.\, C`Q = \text{Ɑ}`S \,.\, P = S;Q\} \quad \text{Df.}$$

(We might equally well put

$$P \,\overline{\text{smor}}\, Q = (1 \to 1) \cap \overleftarrow{\text{Ɑ}}`C`Q \cap \overleftarrow{\dagger Q}`P \quad \text{Df,}$$

which is an equivalent but more condensed form of the definition.) We then define "P is ordinally similar to Q" as meaning that there is at least one ordinal correlator of P and Q, *i.e.*

$$\text{smor} = \hat{P}\hat{Q}(\exists \,!\, P \,\overline{\text{smor}}\, Q) \quad \text{Df.}$$

We shall find that if P and Q generate well-ordered series, they have at most one correlator ($*250{\cdot}6$), but this does not hold in general for other series.

After giving the elementary properties of ordinal similarity, we have three important propositions on its connection with cardinal similarity, namely: ($*151{\cdot}18$) if P is similar to Q, the field of P is similar to the field of Q (the converse does not hold in general, but holds if P and Q are finite serial relations); ($*151{\cdot}19$) if $C`P$ is similar to $C`Q$, there is a relation R similar to Q and having $C`P$ for its field, and vice versa; ($*151{\cdot}191$) $S$ is an ordinal correlator of $P$ and $Q$ when, and only when, it is a cardinal correlator of $C`P$ and $C`Q$ and $P = S;Q$.

We then have a set of propositions on correlators of the form $S \upharpoonright C`Q$ ($*151{\cdot}2$—${\cdot}243$). Most of the correlators with which we shall be concerned are of this form. The most useful proposition here is

***151·22.** $\vdash : S \upharpoonright C\text{'}Q \in 1 \rightarrow 1 \,.\, C\text{'}Q \subset \text{ᗡ}\text{'}S \,.\, P = S;Q \,.\equiv.\, S \upharpoonright C\text{'}Q \in P \,\overline{\text{smor}}\, Q$

A useful consequence of this proposition is

***151·231.** $\vdash :. (y) \,.\, \text{E}\,!\, S\text{'}y : S \upharpoonright C\text{'}Q \in 1 \rightarrow 1 \,.\, P = S;Q : \supset . \, S \upharpoonright C\text{'}Q \in P \,\overline{\text{smor}}\, Q$

This consequence is useful because the hypothesis $(y) \,.\, \text{E}\,!\, S\text{'}y$ is satisfied by most of the relations which occur as correlators.

We have next a number of propositions on the inferribility of $Q = \breve{S};P$ or $Q \subseteq \breve{S};P$ from $P = S;Q$ or $P \subseteq S;Q$, and connected matters (*151·25—·29). We have

***151·25.** $\vdash : S \in \text{Cls} \rightarrow 1 \,.\, C\text{'}Q \subset \text{ᗡ}\text{'}S \,.\, P = S;Q \,.\supset.\, Q = \breve{S};P$

***151·26.** $\vdash :. S \in \text{Cls} \rightarrow 1 \,.\, C\text{'}Q \subset \text{ᗡ}\text{'}S \,.\supset : P \subseteq S;Q \,.\supset.\, \breve{S};P \subseteq Q :$
$S;Q \subseteq P \,.\supset.\, Q \subseteq \breve{S};P$

***151·29.** $\vdash :. P \,\text{smor}\, Q \,.\equiv : (\exists S) : xPy \,.\supset_{x,y}.\, (\breve{S}\text{'}x)\, Q\, (\breve{S}\text{'}y) :$
$zQw \,.\supset_{z,w}.\, (S\text{'}z)\, P\, (S\text{'}w)$

*151·29 is never used, but is inserted in order to show that our definition of "ordinal similarity" agrees with what is commonly understood by that term. If P and Q are regarded as serial, so that "xPy" means "x precedes y in the P-series," and "zQw" means "z precedes w in the Q-series," then our proposition states that two series are ordinally similar when their terms can be so correlated that predecessors in either are correlated with predecessors in the other, and successors with successors, *i.e.* when the two series can be correlated without change of order.

We have next (*151·31—·52) a set of miscellaneous propositions, of which the most useful are

***151·401.** $\vdash : T \upharpoonright C\text{'}P \in X \,\overline{\text{smor}}\, P \,.\, T \upharpoonright C\text{'}Q \in Y \,\overline{\text{smor}}\, Q \,.\, S \in P \,\overline{\text{smor}}\, Q \,.\supset.$
$T;S \in X \,\overline{\text{smor}}\, Y$

***151·5.** $\vdash : S \upharpoonright C\text{'}Q \in P \,\overline{\text{smor}}\, Q \,.\supset.\, \text{D}\text{'}P = S\text{''}\text{D}\text{'}Q \,.\, \text{ᗡ}\text{'}P = S\text{''}\text{ᗡ}\text{'}Q \,.\, \overrightarrow{B}\text{'}P = S\text{''}\overrightarrow{B}\text{'}Q \,.$
$\overrightarrow{B}\text{'}\breve{P} = S\text{''}\overrightarrow{B}\text{'}\breve{Q}$

*151·401 will be useful in such cases as the following: Let P and Q be relations between relations, then $\text{D};P$ and $\text{D};Q$ will be the corresponding relations of their domains. Suppose $\text{D} \upharpoonright C\text{'}P$, $\text{D} \upharpoonright C\text{'}Q \in 1 \rightarrow 1$. Then, by *151·401, if S is a correlator of P and Q, $\text{D};S$ is a correlator of $\text{D};P$ and $\text{D};Q$.

*151·5 shows that if S is a correlator of P and Q, it correlates $\text{D}\text{'}P$ with $\text{D}\text{'}Q$, $\text{ᗡ}\text{'}P$ with $\text{ᗡ}\text{'}Q$, $\overrightarrow{B}\text{'}P$ with $\overrightarrow{B}\text{'}Q$, and $\overrightarrow{B}\text{'}\breve{P}$ with $\overrightarrow{B}\text{'}\breve{Q}$.

Our next set of propositions (*151·53—·59) is concerned with the correlation of powers of P and Q and kindred matters. We show (*151·55) that a correlator of P and Q is also a correlator of P_{po} and Q_{po}, and therefore if P and Q are similar, so are P_{po} and Q_{po} (*151·56); we show also (*151·59) that if P and Q are similar, so are P_ν and Q_ν. These propositions are used in the theory of progressions (*263·17).

The remaining propositions (*151·6 to the end) are concerned with applications to particular cases. The most useful of these are

***151·61.** $\vdash . \iota ; P \text{ smor } P$

which shows how to raise the type of a relation without changing its relation-number;

***151·64.** $\vdash . x \downarrow ; P \text{ smor } P . (x \downarrow) \upharpoonright C`P \in (x \downarrow ; P) \overline{\text{smor}} P$

***151·65.** $\vdash . \downarrow x ; P \text{ smor } P . (\downarrow x) \upharpoonright C`P \in (\downarrow x ; P) \overline{\text{smor}} P$

We prove also that all members of 2_r (*i.e.* all relations of the form $x \downarrow y$, where $x \neq y$) are similar (*151·63), and that all relations of the form $x \downarrow x$ are similar (*151·631).

***151·01.** $P \overline{\text{smor}} Q = \hat{S} \{S \in 1 \rightarrow 1 . C`Q = \text{Ǫ}`S . P = S ; Q\}$ Df

***151·02.** $\text{smor} = \hat{P}\hat{Q} \{\exists ! P \overline{\text{smor}} Q\}$ Df

***151·1.** $\vdash : P \text{ smor } Q . \equiv . (\exists S) . S \in 1 \rightarrow 1 . C`Q = \text{Ǫ}`S . P = S ; Q$ [(*151·02)]

***151·11.** $\vdash : S \in P \overline{\text{smor}} Q . \equiv . S \in 1 \rightarrow 1 . C`Q = \text{Ǫ}`S . P = S ; Q$ [(*151·01)]

***151·12.** $\vdash : P \text{ smor } Q . \equiv . \exists ! P \overline{\text{smor}} Q$ [(*151·02)]

***151·121.** $\vdash . I \upharpoonright C`Q \in (Q \overline{\text{smor}} Q)$ [*72·17 . *50·5·52 . *150·534 . *151·11]

***151·13.** $\vdash . Q \text{ smor } Q$ [*151·121·12]

***151·131.** $\vdash : S \in P \overline{\text{smor}} Q . \equiv . \breve{S} \in Q \overline{\text{smor}} P$

Dem.

$\vdash . *71·212 . \supset \vdash : S \in 1 \rightarrow 1 . \equiv . \breve{S} \in 1 \rightarrow 1$ (1)

$\vdash . *150·13 . \supset \vdash : P = S ; Q . \supset . \breve{S} ; P = (\breve{S} | S) ; Q :$

[*71·192] $\supset \vdash : S \in 1 \rightarrow 1 . P = S ; Q . \supset . \breve{S} ; P = (I \upharpoonright \text{Ǫ}`S) ; Q :$

[*150·534] $\supset \vdash : S \in 1 \rightarrow 1 . C`Q = \text{Ǫ}`S . P = S ; Q . \supset . \breve{S} ; P = Q$ (2)

$\vdash . *150·23 . \supset \vdash : C`Q = \text{Ǫ}`S . P = S ; Q . \supset . C`P = \text{D}`S$ (3)

$\vdash . (1) . (2) . (3) . *33·21 . \supset$

$\vdash : S \in 1 \rightarrow 1 . C`Q = \text{Ǫ}`S . P = S ; Q . \supset . \breve{S} \in 1 \rightarrow 1 . C`P = \text{Ǫ}`\breve{S} . Q = \breve{S} ; P$ (4)

$\vdash . (4) \dfrac{\breve{S}, Q, P}{S, P, Q} . *31·33 . \supset$

$\vdash : \breve{S} \in 1 \rightarrow 1 . C`P = \text{Ǫ}`\breve{S} . Q = \breve{S} ; P . \supset . S \in 1 \rightarrow 1 . C`Q = \text{Ǫ}`S . P = S ; Q$ (5)

$\vdash . (4) . (5) . *151·11 . \supset \vdash . \text{Prop}$

***151·14.** $\vdash : P \text{ smor } Q . \equiv . Q \text{ smor } P$ [*151·131·12 . *31·52]

***151·141.** $\vdash : S \epsilon P \,\overline{\text{smor}}\, Q \,.\, T \epsilon Q \,\overline{\text{smor}}\, R \,.\supset .\, S \mid T \epsilon P \,\overline{\text{smor}}\, R$

Dem.

$\vdash . *151·11 . *71·252 . \supset \vdash : \text{Hp} . \supset . S \mid T \epsilon 1 \rightarrow 1 \quad (1)$

$\vdash . *151·11 . *150·23 . \supset \vdash : \text{Hp} . \supset . \text{Ɑ}‘S = C‘Q \,.\, \text{D}‘T = C‘Q \,.\, \text{Ɑ}‘T = C‘R \,.$

$[*37·323] \quad \supset . \text{Ɑ}‘(S \mid T) = \text{Ɑ}‘T \,.\, \text{Ɑ}‘T = C‘R \,.$

$[*13·17] \quad \supset . \text{Ɑ}‘(S \mid T) = C‘R \quad (2)$

$\vdash . *151·11 . \quad \supset \vdash : \text{Hp} . \supset . P = S ;T ;R$

$[*150·13] \quad = (S \mid T) ;R \quad (3)$

$\vdash . (1) . (2) . (3) . *151·11 . \supset \vdash . \text{Prop}$

***151·15.** $\vdash : P \text{ smor } Q \,.\, Q \text{ smor } R \,.\supset .\, P \text{ smor } R \quad [*151·141]$

***151·16.** $\vdash . I \subseteq \text{smor} \quad [*151·13]$

***151·161.** $\vdash . \text{smor} = \text{Cnv}‘\text{smor} \quad [*151·14]$

***151·162.** $\vdash . (\text{smor})^2 = \text{smor} \quad [*151·15·161 . *34·81]$

***151·17.** $\vdash :. P \text{ smor } Q . \supset : R \text{ smor } P . \equiv . R \text{ smor } Q \quad [*151·14·15]$

***151·18.** $\vdash : P \text{ smor } Q . \supset . C‘P \text{ sm } C‘Q$

Dem.

$\vdash . *151·11 . *150·23 . \supset \vdash : \text{Hp} . \supset . (\exists S) . S \epsilon 1 \rightarrow 1 \,.\, \text{D}‘S = C‘P \,.\, \text{Ɑ}‘S = C‘Q \,.$

$[*73·1] \quad \supset . C‘P \text{ sm } C‘Q : \supset \vdash . \text{Prop}$

***151·19.** $\vdash : C‘P \text{ sm } C‘Q . \equiv . (\exists R) . C‘R = C‘P \,.\, R \text{ smor } Q$

Dem.

$\vdash . *73·1 . \supset \vdash : C‘P \text{ sm } C‘Q . \equiv . (\exists S) . S \epsilon 1 \rightarrow 1 \,.\, \text{D}‘S = C‘P \,.\, \text{Ɑ}‘S = C‘Q \,.$

$[*150·23] \quad \equiv . (\exists S) . S \epsilon 1 \rightarrow 1 \,.\, \text{Ɑ}‘S = C‘Q \,.\, C‘S ;Q = C‘P \,.$

$[*13·195] \quad \equiv . (\exists R, S) . S \epsilon 1 \rightarrow 1 \,.\, \text{Ɑ}‘S = C‘Q \,.\, R = S ;Q \,.\, C‘R = C‘P \,.$

$[*151·1] \quad \equiv . (\exists R) . C‘R = C‘P \,.\, R \text{ smor } Q : \supset \vdash . \text{Prop}$

***151·191.** $\vdash : S \epsilon P \,\overline{\text{smor}}\, Q . \equiv . S \epsilon (C‘P) \,\overline{\text{sm}}\, (C‘Q) \,.\, P = S ;Q$

Dem.

$\vdash . *151·131·11 . \supset \vdash : S \epsilon P \,\overline{\text{smor}}\, Q . \supset . C‘P = \text{D}‘S :$

$[*4·71 . *151·11] \supset \vdash : S \epsilon P \,\overline{\text{smor}}\, Q . \equiv . S \epsilon 1 \rightarrow 1 \,.\, C‘Q = \text{Ɑ}‘S \,.\, P = S ;Q \,.\, C‘P = \text{D}‘S \,.$

$[*73·03] \quad \equiv . S \epsilon (C‘P) \,\overline{\text{sm}}\, (C‘Q) \,.\, P = S ;Q : \supset \vdash . \text{Prop}$

***151·2.** $\vdash : S \epsilon 1 \rightarrow 1 \,.\, C‘Q \subset \text{Ɑ}‘S \,.\, P = S ;Q \,.\supset .\, S \upharpoonright C‘Q \epsilon P \,\overline{\text{smor}}\, Q$

Dem.

$\vdash . *71·29 . \supset \vdash : \text{Hp} . \supset . S \upharpoonright C‘Q \epsilon 1 \rightarrow 1 \quad (1)$

$\vdash . *35·65 . \supset \vdash : \text{Hp} . \supset . \text{Ɑ}‘S \upharpoonright C‘Q = C‘Q \quad (2)$

$\vdash . *150·32 . \supset \vdash : \text{Hp} . \supset . P = S \upharpoonright C‘Q ;Q \quad (3)$

$\vdash . (1) . (2) . (3) . *151·11 . \supset \vdash . \text{Prop}$

***151·21.** $\vdash : P \operatorname{smor} Q . \equiv . (\exists S) . S \in 1 \to 1 . C`Q \subset \text{Ɑ}`S . P = S;Q$ [*151·2]

***151·22.** $\vdash : S \upharpoonright C`Q \in 1 \to 1 . C`Q \subset \text{Ɑ}`S . P = S;Q . \equiv . S \upharpoonright C`Q \in P \,\overline{\operatorname{smor}}\, Q$

Dem.

$\vdash . *35·65 . *150·32 . \supset \vdash : S \upharpoonright C`Q \in 1 \to 1 . C`Q \subset \text{Ɑ}`S . P = S;Q . \supset .$
$S \upharpoonright C`Q \in P \,\overline{\operatorname{smor}}\, Q$ (1)

$\vdash . *151·11 . *150·32 . \supset \vdash : S \upharpoonright C`Q \in P \,\overline{\operatorname{smor}}\, Q . \supset . S \upharpoonright C`Q \in 1 \to 1 . P = S;Q$ (2)

$\vdash . *151·11 . \quad \supset \vdash : S \upharpoonright C`Q \in P \,\overline{\operatorname{smor}}\, Q . \supset . C`Q = \text{Ɑ}`(S \upharpoonright C`Q)$

[*35·64] $= C`Q \cap \text{Ɑ}`S .$

[*22·621] $\supset . C`Q \subset \text{Ɑ}`S$ (3)

$\vdash . (1) . (2) . (3) . \supset \vdash .$ Prop

***151·23.** $\vdash : P \operatorname{smor} Q . \equiv . (\exists S) . S \upharpoonright C`Q \in 1 \to 1 . C`Q \subset \text{Ɑ}`S . P = S;Q$ [*151·22]

The above proposition (*151·23) is very useful. It is the analogue of *73·15. (It should be observed that, in all propositions concerning likeness, $S;Q$ plays the same part as $S``\beta$ plays in propositions concerning similarity.) By means of *151·23, we can establish likeness in all those numerous cases in which a relation which is not usually one-one becomes one-one when confined to a certain converse domain, as for example if we have to deal with $D \upharpoonright \epsilon_\Delta`\kappa$, where $\kappa \in \text{Cls}^2 \text{excl}$, or with $D \upharpoonright P_\Delta`\kappa$, where $P \upharpoonright \kappa \in \text{Cls} \to 1$. Thus *e.g.* by the above proposition, if Q is any relation whose field is $P_\Delta`\kappa$, where $P \upharpoonright \kappa \in \text{Cls} \to 1$, $D;Q$ will be an ordinally similar relation whose field is $D``P_\Delta`\kappa$.

***151·231.** $\vdash :. (y) . E ! S`y : S \upharpoonright C`Q \in 1 \to 1 . P = S;Q : \supset . S \upharpoonright C`Q \in P \,\overline{\operatorname{smor}}\, Q$
[*151·22 . *33·431]

***151·232.** $\vdash :. (\exists S) : (y) . E ! S`y : S \upharpoonright C`Q \in 1 \to 1 . P = S;Q : \supset . P \operatorname{smor} Q$
[*151·231·12]

***151·24.** $\vdash :. (y) . E ! S`y : y, z \in C`Q . S`y = S`z . \supset_{y,z} . y = z : P = S;Q : \supset .$
$S \upharpoonright C`Q \in P \,\overline{\operatorname{smor}}\, Q . P \operatorname{smor} Q$ [*71·166·55 . *33·431 . *151·22·23]

***151·241.** $\vdash :. S \in 1 \to \text{Cls} . C`Q \subset \text{Ɑ}`S : y, z \in C`Q . S`y = S`z . \supset_{y,z} . y = z : P = S;Q : \supset .$
$S \upharpoonright C`Q \in P \,\overline{\operatorname{smor}}\, Q . P \operatorname{smor} Q$ [*71·55 . *151·22·23]

***151·242.** $\vdash :: y, z \in C`Q . \supset_{y,z} : S`y = S`z . \equiv . y = z :. P = S;Q :. \equiv . S \upharpoonright C`Q \in P \,\overline{\operatorname{smor}}\, Q$
[*71·59 . *151·22]

***151·243.** $\vdash :: y, z \in C`Q . \supset_{y,z} : S`y = S`z . \equiv . y = z :. P = S;Q :. \supset . P \operatorname{smor} Q$
[*151·242·12]

***151·25.** $\vdash : S \in \text{Cls} \to 1 . C`Q \subset \text{Ɑ}`S . P = S;Q . \supset . Q = \breve{S};P$

Dem.

$\vdash . *150·13 . \supset \vdash : \text{Hp} . \supset . \breve{S};P = (\breve{S} \mid S);Q$

[*71·191] $= (I \upharpoonright \text{Ɑ}`S);Q$

[*150·535] $= Q : \supset \vdash .$ Prop

$*151{\cdot}251$. $\vdash :. S \epsilon 1 \to 1 . \supset : C'Q \subset \text{ꓷ}'S . P = S ; Q . \equiv . C'P \subset \text{D}'S . Q = \breve{S} ; P$
$[*151{\cdot}25 . *150{\cdot}22 . *37{\cdot}15]$

$*151{\cdot}252$. $\vdash : S \epsilon \text{Cls} \to 1 . C'Q \subset \text{ꓷ}'S . \supset . Q = \breve{S} ; S ; Q \quad [*151{\cdot}25]$

$*151{\cdot}253$. $\vdash : S \epsilon 1 \to \text{Cls} . C'P \subset \text{D}'S . \supset . P = S ; \breve{S} ; P \quad \left[*151{\cdot}252 \frac{\breve{S}}{S}\right]$

$*151{\cdot}254$. $\vdash : S \epsilon 1 \to 1 . \supset . S\dagger \upharpoonright \breve{C}``\text{Cl}'\text{ꓷ}'S = \text{Cnv}'\{\breve{S}\dagger \upharpoonright \breve{C}``\text{Cl}'\text{D}'S\}$

Dem.

$\vdash . *151{\cdot}251 . \supset \vdash :. \text{Hp} . \supset : C'Q \epsilon \text{Cl}'\text{ꓷ}'S . P (S\dagger) Q . \equiv .$
$C'P \epsilon \text{Cl}'\text{D}'S . Q (\breve{S}\dagger) P :. \supset \vdash . \text{Prop}$

This proposition is the analogue of $*72{\cdot}54$. "$\breve{S}\dagger$" means "$(\text{Cnv}'S)\dagger$," not "$\text{Cnv}'(S\dagger)$."

$*151{\cdot}26$. $\vdash :. S \epsilon \text{Cls} \to 1 . C'Q \subset \text{ꓷ}'S . \supset :$
$P \mathrel{\dot\subset} S ; Q . \supset . \breve{S} ; P \mathrel{\dot\subset} Q : S ; Q \mathrel{\dot\subset} P . \supset . Q \mathrel{\dot\subset} \breve{S} ; P$

Dem.

$\vdash . *150{\cdot}31 . \supset \vdash : P \mathrel{\dot\subset} S ; Q . \supset . \breve{S} ; P \mathrel{\dot\subset} \breve{S} ; S ; Q :$
$[*151{\cdot}252] \quad \supset \vdash :. \text{Hp} . \supset : P \mathrel{\dot\subset} S ; Q . \supset . \breve{S} ; P \mathrel{\dot\subset} Q \quad (1)$
Similarly $\quad \vdash :. \text{Hp} . \supset : S ; Q \mathrel{\dot\subset} P . \supset . Q \mathrel{\dot\subset} \breve{S} ; P \quad (2)$
$\vdash . (1) . (2) . \supset \vdash . \text{Prop}$

$*151{\cdot}261$. $\vdash :. S \epsilon 1 \to \text{Cls} . C'P \subset \text{D}'S . \supset :$
$Q \mathrel{\dot\subset} \breve{S} ; P . \supset . S ; Q \mathrel{\dot\subset} P : \breve{S} ; P \mathrel{\dot\subset} Q . \supset . P \mathrel{\dot\subset} S ; Q \quad \left[*151{\cdot}26 \frac{\breve{S}, Q, P}{S, P, Q}\right]$

$*151{\cdot}262$. $\vdash :. S \epsilon 1 \to 1 . C'P \subset \text{D}'S . C'Q \subset \text{ꓷ}'S . \supset :$
$P \mathrel{\dot\subset} S ; Q . \equiv . \breve{S} ; P \mathrel{\dot\subset} Q : Q \mathrel{\dot\subset} \breve{S} ; P . \equiv . S ; Q \mathrel{\dot\subset} P \quad [*151{\cdot}26{\cdot}261]$

$*151{\cdot}263$. $\vdash :. S \epsilon 1 \to 1 . C'P \subset \text{D}'S . C'Q \subset \text{ꓷ}'S . \supset :$
$P \mathrel{\dot\subset} S ; Q . Q \mathrel{\dot\subset} \breve{S} ; P . \equiv . \breve{S} ; P \mathrel{\dot\subset} Q . S ; Q \mathrel{\dot\subset} P . \equiv . P = S ; Q . \equiv . Q = \breve{S} ; P$
$[*151{\cdot}262]$

$*151{\cdot}264$. $\vdash :. S \upharpoonright C'Q \epsilon 1 \to 1 . \supset : P \mathrel{\dot\subset} S ; Q . Q \mathrel{\dot\subset} \breve{S} ; P . \equiv . P = S ; Q$

Dem.

$\vdash . *150{\cdot}202 . *37{\cdot}401 . \quad \supset \vdash : P \mathrel{\dot\subset} S ; Q . \supset . C'P \subset \text{D}'S \upharpoonright C'Q \quad (1)$
$\vdash . (1) . *151{\cdot}262 \frac{S \upharpoonright C'Q}{Q} . \supset \vdash : \text{Hp} . P \mathrel{\dot\subset} S ; Q . \supset :$
$Q \mathrel{\dot\subset} (C'Q \upharpoonleft \breve{S}) ; P . \equiv . (S \upharpoonright C'Q) ; Q \mathrel{\dot\subset} P :$
$[*150{\cdot}361{\cdot}32] \supset : Q \mathrel{\dot\subset} (\breve{S} ; P) \upharpoonright C'Q . \equiv . S ; Q \mathrel{\dot\subset} P :$
$[*35{\cdot}9 . *36{\cdot}29] \supset : Q \mathrel{\dot\subset} \breve{S} ; P . \equiv . S ; Q \mathrel{\dot\subset} P \quad (2)$
$\vdash . (2) . *5{\cdot}32 . \supset \vdash . \text{Prop}$

***151·27.** $\vdash : S \epsilon 1 \rightarrow 1 . P \mathrel{\mathsf{G}} S ; Q . Q \mathrel{\mathsf{G}} \breve{S} ; P .$
$\equiv . S \epsilon 1 \rightarrow 1 . C`P \subset \mathrm{D}`S . C`Q \subset \mathrm{G}`S . \breve{S} ; P \mathrel{\mathsf{G}} Q . S ; Q \mathrel{\mathsf{G}} P .$
$\equiv . S \epsilon 1 \rightarrow 1 . C`Q \subset \mathrm{G}`S . P = S ; Q .$
$\equiv . S \epsilon 1 \rightarrow 1 . C`P \subset \mathrm{D}`S . Q = \breve{S} ; P$
[*151·263 . *5·32 . *150·203 . *4·73]

***151·271.** $\vdash : (\exists S) . S \epsilon 1 \rightarrow 1 . P \mathrel{\mathsf{G}} S ; Q . Q \mathrel{\mathsf{G}} \breve{S} ; P .$
$\equiv . (\exists S) . S \epsilon 1 \rightarrow 1 . C`P \subset \mathrm{D}`S . C`Q \subset \mathrm{G}`S . \breve{S} ; P \mathrel{\mathsf{G}} Q . S ; Q \mathrel{\mathsf{G}} P .$
$\equiv . P \text{ smor } Q$ [*151·27·21]

***151·28.** $\vdash :. P \text{ smor } Q . \equiv : (\exists S) : S \epsilon 1 \rightarrow 1 : xPy . \supset_{x,y} . (\breve{S}`x) \, Q \, (\breve{S}`y) :$
$zQw . \supset_{z,w} . (S`z) \, P \, (S`w)$

Dem.

$\vdash . *150·41 . \supset \vdash :: S \epsilon 1 \rightarrow 1 . \supset :. (\breve{S}`x) Q (\breve{S}`y) . \equiv . x S ; Q y : (S`z) \, P (S`w) . \equiv . z \breve{S} ; P w :.$
[*23·1] $\supset :. xPy . \supset_{x,y} . (\breve{S}`x) \, Q \, (\breve{S}`y) : \equiv . P \mathrel{\mathsf{G}} S ; Q :$
$zQw . \supset_{z,w} . (S`z) \, P \, (S`w) : \equiv . Q \mathrel{\mathsf{G}} \breve{S} ; P :.$
[*151·27] $\supset :. xPy . \supset_{x,y} . (\breve{S}`x) \, Q \, (\breve{S}`y) : zQw . \supset_{z,w} . (S`z) \, P \, (S`w) : \equiv .$
$C`Q \subset \mathrm{G}`S . P = S ; Q$ (1)

$\vdash . (1) . *5·32 . *151·21 . \supset \vdash .$ Prop

The above proposition shows that ordinal similarity as we have defined it has the properties which are commonly associated with the term "ordinal similarity," namely that P and Q are ordinally similar when their fields can be so correlated that two terms having the relation P are always correlated with two terms having the relation Q, and vice versa.

The hypothesis $S \epsilon 1 \rightarrow 1$ is redundant in *151·28; this is shown in the following proposition.

***151·281.** $\vdash :. xPy . \supset_{x,y} . (\breve{S}`x) \, Q \, (\breve{S}`y) : zQw . \supset_{z,w} . (S`z) \, P \, (S`w) :$
$\supset . C`P \upharpoonleft S = S \upharpoonright C`Q . S \upharpoonright C`Q \epsilon P \, \overline{\text{smor}} \, Q$

Dem.

$\vdash . *14·21 . \supset \vdash :. \text{Hp} . \supset : xPy . \supset . \text{E} ! \breve{S}`x . \text{E} ! \breve{S}`y :$
[*33·352] $\supset : x \epsilon C`P . \supset . \text{E} ! \breve{S}`x :$
[*71·571] $\supset : (C`P) \upharpoonleft S \epsilon \text{Cls} \rightarrow 1 . C`P \subset \mathrm{D}`S$ (1)
Similarly $\vdash : \text{Hp} . \supset . S \upharpoonright C`Q \epsilon 1 \rightarrow \text{Cls} . C`Q \subset \mathrm{G}`S$ (2)
$\vdash . *33·17 . \supset \vdash :. \text{Hp} . \supset : xPy . \supset . \breve{S}`x, \breve{S}`y \epsilon C`Q :$
[*33·352] $\supset : x \epsilon C`P . \supset . \breve{S}`x \epsilon C`Q :$
[*14·21·26] $\supset : x \epsilon C`P . xSz . \supset . z \epsilon C`Q :$

[*4·71] $\supset : x \in C‘P \,.\, xSz \,.\, \equiv \,.\, x \in C‘P \,.\, xSz \,.\, z \in C‘Q :$

[*35·1·102] $\supset : (C‘P) \upharpoonleft S = (C‘P) \upharpoonleft S \upharpoonright C‘Q$ (3)

Similarly $\vdash : \text{Hp} \,.\, \supset \,.\, S \upharpoonright C‘Q = (C‘P) \upharpoonleft S \upharpoonright C‘Q$ (4)

$\vdash \,.\, (3) \,.\, (4) \,.\, \supset \vdash : \text{Hp} \,.\, \supset \,.\, (C‘P) \upharpoonleft S = S \upharpoonright C‘Q \,.$ (5)

[(1).(2)] $\supset \,.\, S \upharpoonright C‘Q \in 1 \rightarrow 1 \,.\, C‘Q \subset \text{ꓷ}‘S$ (6)

$\vdash \,.\, (6) \,.\, *35·7 \,.\, (1) \,.\, (2) \,.\, *150·4·41 \,.\, \supset \vdash : \text{Hp} \,.\, \supset \,.\, P \Subset S \,;\, Q \,.\, Q \Subset \breve{S} \,;\, P .$

[*151·264.(6)] $\supset \,.\, P = S \,;\, Q$ (7)

$\vdash \,.\, (5) \,.\, (6) \,.\, (7) \,.\, *151·22 \,.\, \supset \vdash \,.\,$ Prop

***151·29.** $\vdash :. P \,\text{smor}\, Q \,.\, \equiv : (\exists S) : xPy \,.\, \supset_{x,y} \,.\, (\breve{S}‘x) Q (\breve{S}‘y) : zQw \,.\, \supset_{z,w} \,.\, (S‘z) P (S‘w)$
[*151·28·281]

***151·31.** $\vdash : S \in \text{Cls} \rightarrow 1 \,.\, S \,;\, Q = S \,;\, R \,.\, C‘Q \subset \text{ꓷ}‘S \,.\, C‘R \subset \text{ꓷ}‘S \,.\, \supset \,.\, Q = R$

Dem.

$\vdash \,.\, *151·252 \,.\, \supset \vdash : \text{Hp} \,.\, \supset \,.\, Q = \breve{S} \,;\, S \,;\, Q$

[Hp] $= \breve{S} \,;\, S \,;\, R$

[*151·252] $= R : \supset \vdash \,.\,$ Prop

***151·32.** $\vdash :. P \,\text{smor}\, Q \,.\, \supset : \dot{\exists} \,!\, P \,.\, \equiv \,.\, \dot{\exists} \,!\, Q$ [*151·18 . *73·36 . *33·24]

***151·33.** $\vdash : S \in P \,\overline{\text{smor}}\, Q \,.\, \supset \,.\, P \,|\, S = S \,|\, Q \,.\, \breve{S} \,|\, P = Q \,|\, \breve{S}$

Dem.

$\vdash \,.\, *151·11 \,.\, \supset \vdash : \text{Hp} \,.\, \supset \,.\, P \,|\, S = S \,|\, Q \,|\, \breve{S} \,|\, S \,.\, S \in 1 \rightarrow 1 \,.\, C‘Q = \text{ꓷ}‘S \,.$

[*72·601] $\supset \,.\, P \,|\, S = S \,|\, Q$ (1)

Similarly $\vdash : \text{Hp} \,.\, \supset \,.\, \breve{S} \,|\, P = Q \,|\, \breve{S}$ (2)

$\vdash \,.\, (1) \,.\, (2) \,.\, \supset \vdash \,.\,$ Prop

***151·4.** $\vdash : T \upharpoonright C‘Q \in 1 \rightarrow 1 \,.\, C‘P = T``C‘Q \,.\, Q = \breve{T} \,;\, P \,.\, \supset \,.\, T \upharpoonright C‘Q \in P \,\overline{\text{smor}}\, Q$

Dem.

$\vdash \,.\, *35·52 \,.\, *37·4 \,.\, \supset \vdash : \text{Hp} \,.\, \supset \,.\, (C‘Q) \upharpoonleft \breve{T} \in 1 \rightarrow 1 \,.\, \text{ꓷ}‘\{(C‘Q) \upharpoonleft \breve{T}\} = C‘P$ (1)

$\vdash \,.\, *36·33 \,.\, \supset \vdash : \text{Hp} \,.\, \supset \,.\, Q = (\breve{T} \,;\, P) \restriction C‘Q$

[*150·361] $= \{(C‘Q) \upharpoonleft \breve{T}\} \,;\, P$ (2)

$\vdash \,.\, (1) \,.\, (2) \,.\, *151·11 \,.\, \supset \vdash : \text{Hp} \,.\, \supset \,.\, (C‘Q) \upharpoonleft \breve{T} \in Q \,\overline{\text{smor}}\, P \,.$

[*151·131] $\supset \,.\, T \upharpoonright C‘Q \in P \,\overline{\text{smor}}\, Q : \supset \vdash \,.\,$ Prop

***151·401.** $\vdash : T \upharpoonright C‘P \in X \,\overline{\text{smor}}\, P \,.\, T \upharpoonright C‘Q \in Y \,\overline{\text{smor}}\, Q \,.\, S \in P \,\overline{\text{smor}}\, Q \,.\, \supset \,.$
$T \,;\, S \in X \,\overline{\text{smor}}\, Y$

Dem.

$\vdash \,.\, *151·131·141 \,.\, \supset \vdash : \text{Hp} \,.\, \supset \,.\, T \upharpoonright C‘P \,|\, S \,|\, (C‘Q) \upharpoonleft \breve{T} \in X \,\overline{\text{smor}}\, Y$ (1)

$\vdash \,.\, *151·11·131 \,.\, \supset \vdash : \text{Hp} \,.\, \supset \,.\, \text{D}‘S = C‘P \,.\, \text{ꓷ}‘S = C‘Q \,.$

[*150·34] $\supset \,.\, T \upharpoonright C‘P \,|\, S \,|\, (C‘Q) \upharpoonleft \breve{T} = T \,;\, S$ (2)

$\vdash \,.\, (1) \,.\, (2) \,.\, \supset \vdash \,.\,$ Prop

∗151·41. $\vdash : S \epsilon P \, \overline{\text{smor}} \, Q \,.\, T \upharpoonright C\text{‘}P, T \upharpoonright C\text{‘}Q \epsilon 1 \to 1 \,.\, C\text{‘}P \cup C\text{‘}Q \subset \text{Ɑ}\text{‘}T \,.\supset.$
$T \mathbin{;} S \epsilon (T \mathbin{;} P) \, \overline{\text{smor}} \, (T \mathbin{;} Q)$ [∗151·401·22]

This proposition is the analogue of ∗73·63.

The following proposition is used frequently both in relation-arithmetic and in the theory of series.

∗151·5. $\vdash : S \upharpoonright C\text{‘}Q \epsilon P \, \overline{\text{smor}} \, Q \,.\supset.$
$\text{D}\text{‘}P = S\text{“}\text{D}\text{‘}Q \,.\, \text{Ɑ}\text{‘}P = S\text{“}\text{Ɑ}\text{‘}Q \,.\, \overrightarrow{B}\text{‘}P = S\text{“}\overrightarrow{B}\text{‘}Q \,.\, \overrightarrow{B}\text{‘}\breve{P} = S\text{“}\overrightarrow{B}\text{‘}\breve{Q}$

Dem.

$\vdash . ∗151·22 . ∗150·21·211 . \supset \vdash : \text{Hp} . \supset . \text{D}\text{‘}P = S\text{“}\text{D}\text{‘}Q \,.\, \text{Ɑ}\text{‘}P = S\text{“}\text{Ɑ}\text{‘}Q \,.$ (1)

[∗93·101] $\supset . \overrightarrow{B}\text{‘}P = S\text{“}\text{D}\text{‘}Q - S\text{“}\text{Ɑ}\text{‘}Q \,.$

[∗37·421.∗151·22] $\supset . \overrightarrow{B}\text{‘}P = (S \upharpoonright C\text{‘}Q)\text{“}\text{D}\text{‘}Q - (S \upharpoonright C\text{‘}Q)\text{“}\text{Ɑ}\text{‘}Q$

[∗71·381.∗151·22] $= (S \upharpoonright C\text{‘}Q)\text{“}(\text{D}\text{‘}Q - \text{Ɑ}\text{‘}Q)$

[∗93·101.∗37·421] $= S\text{“}\overrightarrow{B}\text{‘}Q$ (2)

Similarly $\vdash : \text{Hp} . \supset . \overrightarrow{B}\text{‘}\breve{P} = S\text{“}\overrightarrow{B}\text{‘}\breve{Q}$ (3)

$\vdash . (1) . (2) . (3) . \supset \vdash .$ Prop

∗151·51. $\vdash : S \upharpoonright C\text{‘}Q \epsilon P \, \overline{\text{smor}} \, Q \,.\, R \Subset Q \,.\supset.\, S \upharpoonright C\text{‘}R \epsilon (S \mathbin{;} R) \, \overline{\text{smor}} \, R \,.\, S \mathbin{;} R \Subset P$

Dem.

$\vdash . ∗151·22 . ∗33·265 . \supset \vdash : \text{Hp} . \supset . C\text{‘}R \subset \text{Ɑ}\text{‘}S$ (1)

$\vdash . ∗151·22 . ∗71·222 . \supset \vdash : \text{Hp} . \supset . S \upharpoonright C\text{‘}R \epsilon 1 \to 1$ (2)

$\vdash . ∗150·31 . ∗151·22 . \supset \vdash : \text{Hp} . \supset . S \mathbin{;} R \Subset P$ (3)

$\vdash . (1) . (2) . (3) . ∗151·22 . \supset \vdash .$ Prop

∗151·52. $\vdash : P \, \text{smor} \, Q \,.\supset.\, \text{Rl}\text{‘}Q \subset \text{smor}\text{“}\text{Rl}\text{‘}P$ [∗151·51·12·14]

∗151·53. $\vdash : S \upharpoonright C\text{‘}Q \epsilon P \, \overline{\text{smor}} \, Q \,.\, T \epsilon \text{Pot}\text{‘}Q \,.\supset.$
$S \upharpoonright C\text{‘}T \epsilon (S \mathbin{;} T) \, \overline{\text{smor}} \, T \,.\, S \mathbin{;} T \epsilon \text{Pot}\text{‘}P$

Dem.

$\vdash . ∗150·8 . \supset \vdash : \text{Hp} . \supset . S \mathbin{;} T \epsilon \text{Pot}\text{‘}P$ (1)

$\vdash . ∗91·27 . \supset \vdash : \text{Hp} . \supset . C\text{‘}T \subset \text{Ɑ}\text{‘}S$ (2)

$\vdash . (1) . (2) . ∗151·22 . \supset \vdash .$ Prop

∗151·54. $\vdash : S \upharpoonright C\text{‘}Q \epsilon P \, \overline{\text{smor}} \, Q \,.\supset.\, S \upharpoonright C\text{‘}Q \epsilon P_{\text{po}} \, \overline{\text{smor}} \, Q_{\text{po}}$

Dem.

$\vdash . ∗91·504 . ∗151·22 . \supset \vdash : \text{Hp} . \supset . S \upharpoonright C\text{‘}Q = S \upharpoonright C\text{‘}Q_{\text{po}} \,.\, C\text{‘}Q_{\text{po}} \subset \text{Ɑ}\text{‘}S$ (1)

$\vdash . ∗150·83 . ∗151·22 . \supset \vdash : \text{Hp} . \supset . P_{\text{po}} = S \mathbin{;} Q_{\text{po}}$ (2)

$\vdash . (1) . (2) . ∗151·22 . \supset \vdash .$ Prop

∗151·55. $\vdash : S \epsilon P \, \overline{\text{smor}} \, Q \,.\supset.\, S \epsilon P_{\text{po}} \, \overline{\text{smor}} \, Q_{\text{po}}$ [∗151·54]

∗151·56. $\vdash : P \, \text{smor} \, Q \,.\supset.\, P_{\text{po}} \, \text{smor} \, Q_{\text{po}}$ [∗151·55]

∗151·56 is used in ∗263·17.

The two following propositions are lemmas for ∗151·59, which is used in ∗263·17.

∗151·57. $\vdash : S \in P \overline{\text{smor}}\, Q \,.\, z, w \in C'Q \,.\, \supset .\, P(S'z \mapsto S'w) = S``Q(z \mapsto w)$

Dem.

$\vdash .\, *151{\cdot}33{\cdot}55 \,.\, \supset \vdash : \text{Hp} \,.\, \supset .\, \overleftarrow{P}_{\text{po}}`S`z = S``\overleftarrow{Q}_{\text{po}}`z \,.\, \overrightarrow{P}_{\text{po}}`S`w = S``\overrightarrow{Q}_{\text{po}}`w \,.$

[∗91·54] $\supset .\, \overleftarrow{P}_{*}`S`z = S``\overleftarrow{Q}_{\text{po}}`z \cup \iota`S`z \,.$

$\overrightarrow{P}_{*}`S`w = S``\overrightarrow{Q}_{\text{po}}`w \cup \iota`S`w \,.$

[∗53·31.∗91·54] $\supset .\, \overleftarrow{P}_{*}`S`z = S``\overleftarrow{Q}_{*}`z \,.\, \overrightarrow{P}_{*}`S`w = S``\overrightarrow{Q}_{*}`w \,.$

[(∗121·103)] $\supset .\, P(S'z \mapsto S'w) = S``Q(z \mapsto w) : \supset \vdash .\, \text{Prop}$

∗151·58. $\vdash : S \in P \overline{\text{smor}}\, Q \,.\, \supset .\, S \upharpoonright C'Q_\nu \in P_\nu \overline{\text{smor}}\, Q_\nu$

Dem.

$\vdash .\, *151{\cdot}57 \,.\, *73{\cdot}22 \,.\, \supset \vdash :.\, \text{Hp} \,.\, \supset : z, w \in C'Q \,.\, \supset .$

$$\text{Nc}`P(S'z \mapsto S'w) = \text{Nc}`Q(z \mapsto w) \quad (1)$$

$\vdash .\, (1) \,.\, *121{\cdot}11 \,.\quad \supset \vdash :.\, \text{Hp} \,.\, z, w \in C'Q \,.\, \supset : zQ_\nu w \,.\, \equiv .\, (S'z)\, P_\nu\, (S'w)$ (2)

$\vdash .\, (2) \,.\, *150{\cdot}41 \,.\quad \supset \vdash : \text{Hp} \,.\, \supset .\, Q_\nu = \breve{S} ; P_\nu \,.$

[∗151·253.∗121·322] $\supset .\, S ; Q_\nu = P_\nu \,.\, C'Q_\nu \subset \text{ᗡ}`S$ (3)

$\vdash .\, (3) \,.\, *151{\cdot}22 \,.\, \supset \vdash .\, \text{Prop}$

∗151·59. $\vdash : P\, \text{smor}\, Q \,.\, \supset .\, P_\nu\, \text{smor}\, Q_\nu$ [∗151·58]

The remaining propositions of this number consist of applications to particular cases.

∗151·6. $\vdash .\, \text{Cnv} ; P\, \text{smor}\, P \,.\, \text{Cnv} \upharpoonright C'P \in (\text{Cnv} ; P)\, \overline{\text{smor}}\, P$
[∗151·231 . ∗31·13 . ∗72·11]

This proposition is only significant when P is a relation between relations.

∗151·61. $\vdash .\, \iota ; P\, \text{smor}\, P$ [∗151·232 . ∗51·12 . ∗72·18]

∗151·62. $\vdash : C'P \subset 1 \,.\, \supset .\, \breve{\iota} ; P\, \text{smor}\, P$ [∗52·62 . ∗151·243]

∗151·63. $\vdash : x \neq y \,.\, z \neq w \,.\, \supset .\, x \downarrow y\, \text{smor}\, z \downarrow w \,.\, x \downarrow z \cup y \downarrow w \in (x \downarrow y)\, \overline{\text{smor}}\, (z \downarrow w)$

Dem.

$\vdash .\, *150{\cdot}72 \,.\, \supset \vdash : S = x \downarrow z \cup y \downarrow w \,.\, z \neq w \,.\, \supset .\, S ; (z \downarrow w) = x \downarrow y$ (1)

$\vdash .\, *72{\cdot}182 \,.\, *71{\cdot}242 \,.\, \supset \vdash : \text{Hp} \,.\, \text{Hp}(1) \,.\, \supset .\, S \in 1 \to 1$ (2)

$\vdash .\, *55{\cdot}15 \,.\quad \supset \vdash : \text{Hp}(1) \,.\, \supset .\, \text{ᗡ}`S = C`(z \downarrow w)$ (3)

$\vdash .\, (1) \,.\, (2) \,.\, (3) \,.\, *151{\cdot}1{\cdot}11 \,.\, \supset \vdash .\, \text{Prop}$

The above proposition shows that all ordinal couples (*i.e.* all members of 2_r) are ordinally similar. The following proposition shows the same for couples whose referent and relatum are identical.

$*151{\cdot}631$. $\vdash . x \downarrow x \,\mathrm{smor}\, z \downarrow z$

Dem.

$\vdash . *72{\cdot}182 . *55{\cdot}15 . \supset \vdash . x \downarrow z \in 1 \to 1 . \mathrm{G}`(x \downarrow z) = C`(z \downarrow z)$ (1)

$\vdash . *55{\cdot}13 . \quad \supset \vdash : u \{x \downarrow z \mid z \downarrow z \mid \mathrm{Cnv}`(x \downarrow z)\} u' . \equiv .$

$u (x \downarrow z) z . u' (x \downarrow z) z .$

$[*55{\cdot}13] \quad \equiv . u = x . u' = x .$

$[*55{\cdot}13] \quad \equiv . u (x \downarrow x) u'$ (2)

$\vdash . (2) . *150{\cdot}1 . \quad \supset \vdash . (x \downarrow z) ; (z \downarrow z) = x \downarrow x$ (3)

$\vdash . (1) . (3) . *151{\cdot}1 . \supset \vdash . \mathrm{Prop}$

$*151{\cdot}64$. $\vdash . x \downarrow ; P \,\mathrm{smor}\, \dot{P} . (x \downarrow) \upharpoonright C`P \in (x \downarrow ; P) \,\overline{\mathrm{smor}}\, P$

$[*72{\cdot}184 . *55{\cdot}12 . *151{\cdot}231]$

The following proposition is frequently used in relation-arithmetic.

$*151{\cdot}65$. $\vdash . \downarrow x ; P \,\mathrm{smor}\, P . (\downarrow x) \upharpoonright C`P \in (\downarrow x ; P) \,\overline{\mathrm{smor}}\, P$

$[*72{\cdot}184 . *55{\cdot}121 . *151{\cdot}231]$

$*152$. DEFINITION AND ELEMENTARY PROPERTIES OF RELATION-NUMBERS.

Summary of $*152$.

The relation-number of P, which we denote by $\text{Nr}'P$, is defined as the class of relations which are ordinally similar to P, *i.e.*

$$\text{Nr}'P = \overrightarrow{\text{smor}}'P.$$

Hence our definition is

$$\text{Nr} = \overrightarrow{\text{smor}} \quad \text{Df.}$$

The class of relation-numbers consists of all such classes as $\text{Nr}'P$, *i.e.*

$$\text{NR} = \text{D}'\text{Nr} \quad \text{Df.}$$

These two definitions are analogous to those of $*100$, merely substituting "smor" for "sm." They are justified by similar considerations, and lead to similar results. With the exception of $*152{\cdot}7{\cdot}71{\cdot}72$, the propositions of this number are the analogues of those of $*100$, and call for no remarks other than those in the introduction to $*100$ (*mutatis mutandis*).

$*152{\cdot}7{\cdot}71{\cdot}72$ give relations between relation-numbers and cardinals. $*152{\cdot}7$, which is constantly used, states that the cardinal number of $C'Q$ consists of the fields of the relation-number of Q, *i.e.* the classes similar to $C'Q$ are the fields of the relations similar to Q; in symbols,

$*152{\cdot}7$. $\vdash . \text{Nc}'C'Q = C''\text{Nr}'Q$

Hence it follows that the fields of a relation-number form a cardinal number, *i.e.*

$*152{\cdot}71$. $\vdash : \mu \in \text{NR} . \supset . C''\mu \in \text{NC}$

Hence also it follows that cardinals other than Λ consist of classes of the form $C''\mu$, where μ is a relation-number other than Λ, *i.e.*

$*152{\cdot}72$. $\vdash . \text{NC} - \iota'\Lambda = C'''(\text{NR} - \iota'\Lambda)$

In $*154{\cdot}9$, we shall show how to remove the restriction to numbers other than Λ, thus arriving at

$$\vdash . \text{NC} = C'''\text{NR}.$$

$*152{\cdot}01$. $\text{Nr} = \overrightarrow{\text{smor}}$ Df

$*152{\cdot}02$. $\text{NR} = \text{D}'\text{Nr}$ Df

∗152·1. $\vdash . \mathrm{Nr}`P = \hat{Q}(Q \,\mathrm{smor}\, P) = \hat{Q}(P \,\mathrm{smor}\, Q)$
[∗32·11 . (∗152·01) . ∗151·14]

∗152·11. $\vdash : Q \in \mathrm{Nr}`P . \equiv . Q \,\mathrm{smor}\, P . \equiv . P \,\mathrm{smor}\, Q$ [∗152·1]

∗152·2. $\vdash . \mathrm{E}\,!\, \mathrm{Nr}`P$ [∗152·1 . ∗14·21]

∗152·21. $\vdash . \mathrm{C}\!\mathrm{l}`\mathrm{Nr} = \mathrm{Rel}$ [∗152·2 . ∗33·432]

∗152·22. $\vdash . \mathrm{Nr} \in 1 \rightarrow \mathrm{Cls}$ [∗152·2 . ∗71·166]

∗152·3. $\vdash . P \in \mathrm{Nr}`P$ [∗151·13 . ∗152·11]

∗152·31. $\vdash : P \in \mathrm{Nr}`Q . \equiv . Q \in \mathrm{Nr}`P$ [∗152·11]

∗152·32. $\vdash : P \in \mathrm{Nr}`Q . Q \in \mathrm{Nr}`R . \supset . P \in \mathrm{Nr}`R$ [∗151·15 . ∗152·11]

∗152·321. $\vdash : P \,\mathrm{smor}\, Q . \supset . \mathrm{Nr}`P = \mathrm{Nr}`Q$ [∗151·17 . ∗152·1]

∗152·33. $\vdash : \exists\,!\, \mathrm{Nr}`P \cap \mathrm{Nr}`Q . \supset . P \,\mathrm{smor}\, Q . \mathrm{Nr}`P = \mathrm{Nr}`Q$

Dem.

$\vdash .$ ∗152·11 . ∗151·14 . $\supset \vdash : \mathrm{Hp} . \supset . (\exists R) . P \,\mathrm{smor}\, R . R \,\mathrm{smor}\, Q .$
[∗151·15] $\supset . P \,\mathrm{smor}\, Q$ (1)
$\vdash .$ (1) . ∗152·321 . $\supset \vdash .$ Prop

∗152·35. $\vdash :. \exists\,!\, \mathrm{Nr}`P . \vee . \exists\,!\, \mathrm{Nr}`Q : \supset :$
$\mathrm{Nr}`P = \mathrm{Nr}`Q . \equiv . P \in \mathrm{Nr}`Q . \equiv . Q \in \mathrm{Nr}`P . \equiv . P \,\mathrm{smor}\, Q$

Dem.

$\vdash .$ ∗24·571 . $\supset \vdash :. \mathrm{Hp} . \supset : \mathrm{Nr}`P = \mathrm{Nr}`Q . \supset . \exists\,!\, \mathrm{Nr}`P \cap \mathrm{Nr}`Q .$
[∗152·33] $\supset . P \,\mathrm{smor}\, Q$ (1)
$\vdash .$ (1) . ∗152·321 . $\supset \vdash :. \mathrm{Hp} . \supset : \mathrm{Nr}`P = \mathrm{Nr}`Q . \equiv . P \,\mathrm{smor}\, Q$ (2)
$\vdash .$ (2) . ∗152·11 . $\supset \vdash .$ Prop

In the above proposition, the same remarks as to types are to be made as in the case of ∗100·35. If in a certain type Nr‘P and Nr‘Q are both null, we have in that type Nr‘P = Nr‘Q, but we need not have P smor Q. Thus for example we shall find that, in the type of $x \downarrow x$,

$$\mathrm{Nr}`(t^{2}`x \uparrow t^{2}`x) = \Lambda = \mathrm{Nr}`(t^{3}`x \uparrow t^{3}`x).$$

But we do not have

$$(t^{2}`x \uparrow t^{2}`x) \,\mathrm{smor}\, (t^{3}`x \uparrow t^{3}`x).$$

∗152·4. $\vdash : \mu \in \mathrm{NR} . \equiv . (\exists P) . \mu = \mathrm{Nr}`P$ [∗37·78·79 . (∗152·02·01)]

Note that "Nr‘P," like "Nc‘α," is a formal number, and may be subjected to the conventions IT, IIT, AT.

∗152·41. $\vdash . \mathrm{Nr}`P \in \mathrm{NR}$ [∗152·4·2]

∗152·42. $\vdash : \mu, \nu \in \mathrm{NR} . \exists\,!\, \mu \cap \nu . \supset . \mu = \nu$ [∗152·33·4]

∗152·43. $\vdash . \mathrm{NR} \in \mathrm{Cls}^{2}\,\mathrm{excl}$ [∗152·42]

***152·44.** $\vdash :. \mu \in \mathrm{NR} : \exists ! \mu . \lor . \exists ! \mathrm{Nr}'P . \supset : P \in \mu . \equiv . \mathrm{Nr}'P = \mu$
[*152·35·4]

***152·45.** $\vdash : \mu \in \mathrm{NR} . P \in \mu . \supset . \mathrm{Nr}'P = \mu$ [*152·44 . *10·24]

***152·5.** $\vdash : \mu \in \mathrm{NR} . P, Q \in \mu . \supset . P \,\mathrm{smor}\, Q$ [*152·31·32·4]

***152·51.** $\vdash : \mu \in \mathrm{NR} . P \in \mu . \supset . \mathrm{smor}``\mu = \mathrm{Nr}'P$

Dem.

$\vdash . *37·1 . \supset \vdash : R \in \mathrm{smor}``\mu . \equiv . (\exists Q) . Q \in \mu . R \,\mathrm{smor}\, Q$ (1)
$\vdash . *152·5 . \supset \vdash :. \mu \in \mathrm{NR} . P \in \mu . \supset : Q \in \mu . P \,\mathrm{smor}\, Q . \equiv . Q \in \mu$ (2)
$\vdash . (1) . (2) . \supset \vdash :. \mathrm{Hp} . \supset : R \in \mathrm{smor}``\mu . \equiv . (\exists Q) . Q \in \mu . P \,\mathrm{smor}\, Q . R \,\mathrm{smor}\, Q .$
[*151·17] $\equiv . (\exists Q) . Q \in \mu . P \,\mathrm{smor}\, Q . R \,\mathrm{smor}\, P .$
[(2)] $\equiv . (\exists Q) . Q \in \mu . R \,\mathrm{smor}\, P .$
[*10·35] $\equiv . \exists ! \mu . R \,\mathrm{smor}\, P$ (3)
$\vdash . *10·24 . \supset \vdash :. \mathrm{Hp} . \supset : \exists ! \mu :$
[*4·73] $\supset : R \,\mathrm{smor}\, P . \equiv . \exists ! \mu . R \,\mathrm{smor}\, P$ (4)
$\vdash . (3) . (4) . \supset \vdash :. \mathrm{Hp} . \supset : R \in \mathrm{smor}``\mu . \equiv . R \,\mathrm{smor}\, P .$
[*152·11] $\equiv . R \in \mathrm{Nr}'P :. \supset \vdash .$ Prop

***152·52.** $\vdash : \mu \in \mathrm{NR} . \exists ! \mu . \supset . \mathrm{smor}``\mu \in \mathrm{NR}$ [*152·51·4]

The restriction involved in $\exists ! \mu$ is, as we shall see later, not necessary, since $\Lambda \in \mathrm{NR}$ in any assigned type.

***152·53.** $\vdash : \exists ! \mathrm{Nr}'Q . \supset . \mathrm{smor}``\mathrm{Nr}'Q = \mathrm{Nr}'Q$

Dem.

$\vdash . *152·51 . \supset \vdash : P \in \mathrm{Nr}'Q . \supset . \mathrm{smor}``\mathrm{Nr}'Q = \mathrm{Nr}'P$ (1)
$\vdash . *152·321 . \supset \vdash : P \in \mathrm{Nr}'Q . \supset . \mathrm{Nr}'P = \mathrm{Nr}'Q$ (2)
$\vdash . (1) . (2) . \supset \vdash .$ Prop

***152·54.** $\vdash :. \exists ! \mu . \exists ! \nu . \supset : \mu \in \mathrm{NR} . \nu = \mathrm{smor}``\mu . \equiv . \nu \in \mathrm{NR} . \mu = \mathrm{smor}``\nu$
[Proof as in *100·53]

***152·6.** $\vdash . \iota ; P \in \mathrm{Nr}'P$ [*151·61]

***152·62.** $\vdash . x \downarrow ; P \in \mathrm{Nr}'P$ [*151·64]

***152·63.** $\vdash . \downarrow x ; P \in \mathrm{Nr}'P$ [*151·65]

The utility of *152·6·62·63 is that they enable us to raise the type of a relation-number to any required extent. Thus $\iota ; P$ gives a relation whose field is a class of the next type above that of $C'P$, *i.e.* of the type $t^2{}'C'P$; while $x \downarrow ; P$ gives a relation whose field is $x \downarrow ``C'P$, which is of the type $t't'(\iota'x \uparrow C'P)$. If $x \in C'P$, or, more generally, if $x \in t_0{}'C'P$, this is the type $t^2{}'P$. Thus if we put $Q = x \downarrow ; P$, we have

$$t'Q = t'(C'Q \uparrow C'Q) = t'(t'P \uparrow t'P) = t'(P \downarrow P).$$

Thus $x \downarrow ; P$ is a relation whose field consists of terms of the same type as P.

The following propositions on the relations of cardinals and relation-numbers are very important.

***152·7.** $\vdash . \text{Nc}`C`Q = C``\text{Nr}`Q$

Dem.

$\vdash . *151\cdot19 . *35\cdot942 . \supset \vdash : \alpha \in \text{Nc}`C`Q . \supset . (\exists R) . C`R = \alpha . R \in \text{Nr}`Q .$

[*37·6] $\supset . \alpha \in C``\text{Nr}`Q$ (1)

$\vdash . *151\cdot18 . \supset \vdash : P \in \text{Nr}`Q . \supset . C`P \in \text{Nc}`C`Q$

[*37·61] $\supset \vdash . C``\text{Nr}`Q \subset \text{Nc}`C`Q$ (2)

$\vdash . (1) . (2) . \supset \vdash . \text{Prop}$

***152·71.** $\vdash : \mu \in \text{NR} . \supset . C``\mu \in \text{NC}$ [*152·7]

***152·72.** $\vdash . \text{NC} - \iota`\Lambda = C```(\text{NR} - \iota`\Lambda)$

Dem.

$\vdash . *152\cdot71 . \quad \supset \vdash . C```\text{NR} \subset \text{NC}$ (1)

$\vdash . *152\cdot7 . *50\cdot5\cdot52 . \supset \vdash : \mu \in \text{NC} . \alpha \in \mu . \supset . C``\text{Nr}`(I \upharpoonright \alpha) = \text{NC}`\alpha .$

[*100·45] $\supset . C``\text{Nr}`(I \upharpoonright \alpha) = \mu .$

[*37·103] $\supset . \mu \in C```\text{NR}$ (2)

$\vdash . (2) . *10\cdot11\cdot23\cdot35 . \supset \vdash : \mu \in \text{NC} . \exists ! \mu . \supset . \mu \in C```\text{NR}$ (3)

$\vdash . *37\cdot45 . \quad \supset \vdash : \mu = C``\nu . \exists ! \mu . \supset . \exists ! \nu :$

[*37·103] $\supset \vdash : \mu \in C```\text{NR} . \exists ! \mu . \supset . \mu \in C```(\text{NR} - \iota`\Lambda)$ (4)

$\vdash . (1) . (3) . (4) . \supset \vdash . \text{Prop}$

We shall show in *154·9 that the exclusion of Λ in *152·72 is unnecessary.

$*153$. THE RELATION-NUMBERS 0_r, 2_r AND 1_s.

Summary of $*153$.

The relation-numbers 0_r and 2_r have already been defined (in $*56$), though it remains for the present number to show that they are relation-numbers. They are the ordinal 0 and 2 respectively, *i.e.* they are the ordinal numbers of series of no terms and series of two terms respectively. But there is no means of introducing an ordinal 1 which shall be analogous to the cardinal 1 as completely as 0_r and 2_r are analogous to 0 and 2. The only relations whose fields are unit classes are relations of the form $x \downarrow x$. We therefore put

$*153 \cdot 01$. $1_s = \hat{R}\{(\exists x) . R = x \downarrow x\}$ Df

The above definition gives the nearest possible approach to an ordinal 1. 1_s so defined is a relation-number, and is the relation-number corresponding to 1 in the sense that it is the relation-number of all such relations as have a field consisting of one term. But 1_s is not what is called an "ordinal number," because this term is confined by usage to the relation-numbers of well-ordered series, and $x \downarrow x$ is not a serial relation. It is essential to a serial relation to be contained in diversity; and if, by definition, we include $x \downarrow x$ among series, we introduce more exceptions than we avoid. Moreover 1_s does not have the kind of properties which we wish 1 to have; *e.g.* $1_s \dot{+} 1_s$ is not 2_r.

We do not use 1_r, because we shall at a later stage define ν_r as the class of those *series* whose fields have ν terms, so that $1_r = \Lambda$, while 0_r and 2_r have the values $\iota`\dot{\Lambda}$ and $\hat{R}\{(\exists x, y) . x \neq y . R = x \downarrow y\}$, as already defined. On account of this general definition of ν_r, we choose a different symbol for the relation-number 1, and 1_s has the merit of being as like 1_r as possible.

To illustrate, by anticipation, the way in which 1_s differs from proper ordinal numbers, we may point out that if 1_s is added to 2_r, we do not obtain 3_r. We shall define 3_r as the class of series which consist of three terms, *i.e.* the class of relations of the form

$$x \downarrow y \cup x \downarrow z \cup y \downarrow z,$$

where $x \neq y . x \neq z . y \neq z$. We shall define the sum of two ordinal numbers as the ordinal number of the sum of two relations having these ordinal numbers (cf. ∗180), and it will appear that if P and Q are relations whose fields have no members in common, then

$$P \cup Q \cup C`P \uparrow C`Q$$

has a relation-number which is the sum of those of P and Q. Suppose now $P = x \downarrow y$ and $Q = z \downarrow z$, where $x \neq y \,.\, x \neq z \,.\, y \neq z$. Then

$$P \cup Q \cup C`P \uparrow C`Q = x \downarrow y \cup x \downarrow z \cup y \downarrow z \cup z \downarrow z.$$

This is not a member of 3_r, because of the additional term $z \downarrow z$. Thus the addition of one term to a series P does not give the same number as results from the addition of 1_s to Nr‘P. Hence the addition of 1 to an ordinal number has to be separately treated*.

We prove in this number that $0_r = \text{Nr}`\dot{\Lambda}$ (∗153·11), that $2_r = \text{Nr}`(\Lambda \downarrow \iota`x)$ (∗153·24; observe that we have to take a couple of *classes* (or relations) in order to be sure of the existence of two different objects of the class in question), and that $1_s = \text{Nr}`(y \downarrow y)$ (∗153·32). We prove $C``0_r = 0$ (∗153·18), $C``2_r = 2$ (∗153·212), and $C``1_s = 1$ (∗153·36). We have also $\breve{C}``0 = 0_r$ (not proved) and $\breve{C}``1 = 1_s$ (∗153·301). But we do not have $\breve{C}``2 = 2_r$; *e.g.* $(x \downarrow y \cup y \downarrow x) \,\epsilon\, \breve{C}``2$ if $x \neq y$, but $(x \downarrow y \cup y \downarrow x) \sim\epsilon\, 2_r$. We have $\exists ! 0_r$ (∗153·12) and $\exists ! 1_s$ (∗153·34), but from our primitive propositions we cannot deduce $\exists ! 2_r$ unless we rise above the lowest type of relations. The case is exactly analogous to that of $\exists ! 2$ (cf. ∗101); we have

∗153·26·262. $\vdash . \exists ! 2_r \cap \text{Rl}`(\text{Cls} \uparrow \text{Cls}) \,.\, \exists ! 2_r \cap \text{Rel}^2$

But if, as monists aver, there is only one individual, we shall not have $\exists ! 2_r$ in the type of relations of individuals to individuals. Our primitive propositions do not suffice to disprove this supposition.

∗153·01. $1_s = \hat{R}\{(\exists x) \,.\, R = x \downarrow x\}$ Df

∗153·1. $\vdash : P \,\epsilon\, 0_r \,.\equiv .\, P = \dot{\Lambda}$ [∗56·104]

∗153·101. $\vdash : P \text{ smor } \dot{\Lambda} \,.\equiv .\, P = \dot{\Lambda}$

Dem.

$\vdash . ∗151·32 . \text{Transp} . \supset \vdash : P \text{ smor } \dot{\Lambda} \,.\supset .\, \sim \dot{\exists} ! P$ (1)

$\vdash . ∗151·13 . \supset \vdash : P = \dot{\Lambda} \,.\supset .\, P \text{ smor } \dot{\Lambda}$ (2)

$\vdash . (1) . (2) . \supset \vdash . \text{Prop}$

∗153·11. $\vdash . 0_r = \text{Nr}`\dot{\Lambda}$ [∗153·1·101 . ∗152·1]

∗153·111. $\vdash . 0_r \,\epsilon\, \text{NR}$ [∗152·41 . ∗153·11]

∗153·12. $\vdash . \exists ! 0_r$ [∗51·161]

∗153·13. $\vdash . \exists ! 0_r \cap \text{Rl}`R \,.\, \Lambda \,\epsilon\, 0_r \cap \text{Rl}`R$ [∗61·3]

* Cf. ∗161 and ∗181, where this point is more fully elucidated.

***153·14.** $\vdash : \text{Nr}`P = 0_r . \equiv . P = \dot{\Lambda}$

Dem.

$\vdash . *152·44 . *153·111·12 . \supset \vdash : \text{Nr}`P = 0_r . \equiv . P \epsilon 0_r .$

[*152·1] $\equiv . P = \dot{\Lambda} : \supset \vdash . \text{Prop}$

***153·15.** $\vdash . \text{smor}``0_r = 0_r$

Dem.

$\vdash . *152·51 . *153·111·13 . \supset \vdash . \text{smor}``0_r = \text{Nr}`\dot{\Lambda}$

[*153·11] $= 0_r . \supset \vdash . \text{Prop}$

***153·16.** $\vdash :. \mu \epsilon \text{NR} - \iota`0_r . \supset : P \epsilon \mu . \supset_P . \dot{\exists} ! P$

Dem.

$\vdash . *153·13 . *152·42 . \supset \vdash :. \mu \epsilon \text{NR} . \supset : \dot{\Lambda} \epsilon \mu . \supset . \mu = 0_r :$

[Transp] $\supset : \mu \neq 0_r . \supset . \dot{\Lambda} \sim \epsilon \mu$ (1)

$\vdash . (1) . *25·63 . \supset \vdash . \text{Prop}$

***153·17.** $\vdash : \dot{\Lambda} \epsilon \text{Nr}`P . \equiv . \text{Nr}`P = 0_r . \equiv . \text{Nr}`P = \text{Nr}`\dot{\Lambda} . \equiv . P = \dot{\Lambda}$
[*152·35 . *153·11·14]

***153·18.** $\vdash . C``0_r = 0$

Dem.

$\vdash . *53·31 . \supset \vdash . C``\iota`\dot{\Lambda} = \iota`C`\dot{\Lambda}$ (1)

$\vdash . (1) . *33·241 . (*56·03 . *54·01) . \supset \vdash . \text{Prop}$

***153·2.** $\vdash : P \epsilon 2_r . \equiv . (\exists x, y) . x \neq y . P = x \downarrow y$ [*56·11]

***153·201.** $\vdash : x \neq y . \equiv . x \downarrow y \epsilon 2_r$ [*56·17]

***153·202.** $\vdash : P, Q \epsilon 2_r . \supset . P \,\text{smor}\, Q$ [*151·63 . *153·2]

***153·203.** $\vdash : Q \epsilon 2_r . P \,\text{smor}\, Q . \supset . P \epsilon 2_r$

Dem.

$\vdash . *113·123 . \supset \vdash : S \epsilon 1 \to \text{Cls} . z, w \epsilon \text{ᗡ}`S . \supset . S;(z \downarrow w) = (S`z) \downarrow (S`w) :$

[*55·15] $\supset \vdash : S \epsilon 1 \to \text{Cls} . C`(z \downarrow w) = \text{ᗡ}`S . \supset .$

$S;(z \downarrow w) = (S`z) \downarrow (S`w)$ (1)

$\vdash . *71·56 . \supset \vdash :. S \epsilon 1 \to 1 . C`(z \downarrow w) = \text{ᗡ}`S . \supset : z = w . \equiv . S`z = S`w :$

[Transp] $\supset : z \neq w . \equiv . S`z \neq S`w$ (2)

$\vdash . (1) . (2) . *153·201 . \supset$

$\vdash : S \epsilon 1 \to 1 . z \neq w . C`(z \downarrow w) = \text{ᗡ}`S . P = S;(z \downarrow w) . \supset . P \epsilon 2_r :$

[*151·1] $\supset \vdash : z \neq w . P \,\text{smor}\, (z \downarrow w) . \supset . P \epsilon 2_r :$

[*153·2] $\supset \vdash : Q \epsilon 2_r . P \,\text{smor}\, Q . \supset . P \epsilon 2_r : \supset \vdash . \text{Prop}$

***153·21.** $\vdash : P \epsilon 2_r . \supset . 2_r = \text{Nr}`P$ [*153·202·203]

***153·211.** $\vdash : x \neq y . \supset . 2_r = \text{Nr}`(x \downarrow y)$ [*153·21·201]

***153·212.** $\vdash . C``2_r = 2$ [*55·15 . *56·11 . *54·101]

***153·22.** $\vdash : \exists ! 2_r \cap t''{}^{\backprime}z . \equiv . \exists ! 2(z) . \equiv . (\exists x, y) . x \neq y x \epsilon t^{\backprime}z$ [*153·211 . *101·4]

***153·23.** $\vdash : P \epsilon 2_r . \supset . \mathrm{Rl}^{\backprime}P \subset 0_r \cup 2_r$ [*56·261]

This proposition illustrates the reasons for not putting

$$1_r = \hat{P}\{(\exists x) . P = x \downarrow x\} \quad \text{Df.}$$

We want the inductive ordinals, like the inductive cardinals, to form a series in order of magnitude; but, as the above proposition illustrates, the relation-number of such relations as $x \downarrow x$ is not in the same series with 0_r and 2_r. The above proposition should be contrasted with *54·411.

***153·24.** $\vdash . 2_r = \mathrm{Nr}^{\backprime}(\Lambda \downarrow \iota^{\backprime}x)$ [*153·211 . *51·161]

***153·25.** $\vdash . 2_r \epsilon \mathrm{NR}$ [*153·24 . *152·41]

***153·251.** $\vdash . 2_r \neq 0_r . 2_r \cap 0_r = \Lambda$

Dem.

$\vdash . *153·212·18 . *101·34·35 . \supset \vdash . C``2_r \neq C``0_r . C``2_r \cap C``0_r = \Lambda .$

[*13·12.Transp.*37·21] $\supset \vdash . 2_r \neq 0_r . C``(2_r \cap 0_r) = \Lambda .$

[*37·45] $\supset \vdash . 2_r \neq 0_r . 2_r \cap 0_r = \Lambda$

***153·26.** $\vdash . \exists ! 2_r \cap \mathrm{Rl}^{\backprime}(\mathrm{Cls} \uparrow \mathrm{Cls})$ [*153·24 . *152·3]

***153·261.** $\vdash . \dot{\Lambda} \downarrow (x \downarrow x) \epsilon 2_r$ [*55·134 . *56·11]

***153·262.** $\vdash . \exists ! 2_r \cap \mathrm{Rel}^2$ [*153·261 . (*61·03)]

***153·27.** $\vdash . 2_r = \mathrm{smor}``(2_r \cap \mathrm{Rl}^{\backprime}\mathrm{Cls}) = \mathrm{smor}``(2_r \cap \mathrm{Rel}^2)$
[*152·53 . *153·26·262·24]

***153·28.** $\vdash : x \neq y . \supset . B^{\backprime}(x \downarrow y) = x . B^{\backprime}\mathrm{Cnv}^{\backprime}(x \downarrow y) = y$

Dem.

$\vdash . *93·101 . *55·15 . \supset \vdash : \mathrm{Hp} . \supset . \overrightarrow{B}^{\backprime}(x \downarrow y) = \iota^{\backprime}x . \overrightarrow{B}^{\backprime}\mathrm{Cnv}^{\backprime}(x \downarrow y) = \iota^{\backprime}y : \supset \vdash . \mathrm{Prop}$

***153·281.** $\vdash : P \epsilon 2_r . \supset . B^{\backprime}P = \iota^{\backprime}\mathrm{D}^{\backprime}P . B^{\backprime}\breve{P} = \iota^{\backprime}\mathrm{G}^{\backprime}P$ [*153·28 . *55·15]

The above proposition is used in the theory of series (*204·48).

***153·3.** $\vdash . 1_s = \dot{2} - 2_r = \hat{R}\{(\exists x) . R = x \downarrow x\}$ [*56·13 . (*153·01)]

***153·301.** $\vdash . 1_s = \breve{C}``1$ [*153·3 . *56·39]

***153·31.** $\vdash . x \downarrow y \epsilon (x \downarrow x) \overline{\mathrm{smor}} (y \downarrow y)$

Dem.

$\vdash . *72·182 . *55·15 . \supset \vdash . x \downarrow y \epsilon 1 \rightarrow 1 . \mathrm{G}^{\backprime}(x \downarrow y) = C^{\backprime}(y \downarrow y)$ (1)

$\vdash . *35·89 . *55·1 . \supset \vdash . x \downarrow y \,|\, y \downarrow y = x \downarrow y .$

[*150·1.*55·14] $\supset \vdash . (x \downarrow y) ; (y \downarrow y) = x \downarrow y \,|\, y \downarrow x$

[*35·89.*55·1] $= x \downarrow x$ (2)

$\vdash . (1) . (2) . *151·11 . \supset \vdash . \mathrm{Prop}$

$*153\cdot311$. $\vdash : Q \in 1_s \,.\, P \text{ smor } Q \,.\supset .\, P \in 1_s$

Dem.

$\vdash \,.\, *153\cdot3 \,.\, *151\cdot1 \,.\supset \vdash : \text{Hp} \,.\supset .$

$$(\exists S, y) \,.\, Q = y \downarrow y \,.\, S \in 1 \to 1 \,.\, \text{Cl}'S = \iota' y \,.\, P = S ; Q \,.$$

[$*150\cdot71$] $\supset .\, (\exists S, y) \,.\, P = (S'y) \downarrow (S'y) \,.$

[$*153\cdot3$] $\supset .\, P \in 1_s : \supset \vdash .\, \text{Prop}$

$*153\cdot32$. $\vdash .\, 1_s = \text{Nr}'(y \downarrow y)$ [$*153\cdot31\cdot311$]

$*153\cdot33$. $\vdash .\, 1_s \in \text{NR}$ [$*153\cdot32$]

$*153\cdot34$. $\vdash .\, \exists ! 1_s \,.\, 1_s \neq 0_r \,.\, 1_s \neq 2_r \,.\, 1_s \cap 0_r = \Lambda \,.\, 1_s \cap 2_r = \Lambda$

Dem.

$\vdash .\, *153\cdot3 \,.$ $\supset \vdash .\, x \downarrow x \in 1_s \,.$

[$*10\cdot24$] $\supset \vdash .\, \exists ! 1_s$ (1)

$\vdash .\, *56\cdot103\cdot104 \,.$ $\supset \vdash .\, 1_s \cap 0_r = \Lambda$ (2)

$\vdash .\, (1) \,.\, (2) \,.$ $\supset \vdash .\, 1_s \neq 0_r$ (3)

$\vdash .\, *153\cdot301 \,.\, *56\cdot113 \,.\supset \vdash .\, 1_s \cap 2_r \subset \breve{C}``1 \cap \breve{C}``2 \,.$

[$*72\cdot41 . *101\cdot35$] $\supset \vdash .\, 1_s \cap 2_r = \Lambda$ (4)

[(1)] $\supset \vdash .\, 1_s \neq 2_r$ (5)

$\vdash .\, (1) \,.\, (2) \,.\, (3) \,.\, (4) \,.\, (5) \,.\supset \vdash .\, \text{Prop}$

$*153\cdot341$. $\vdash : R \in 1_s \,.\equiv .\, \text{Nr}'R = 1_s$ [$*153\cdot33\cdot34 \,.\, *152\cdot44$]

$*153\cdot35$. $\vdash : R \in 1_s \,.\supset .\, \text{Nc}'C'R = C``\text{Nr}'R = 1$

Dem.

$\vdash .\, *55\cdot15 \,.\, *153\cdot3 \,.\supset \vdash : \text{Hp} \,.\supset .\, \text{Nc}'C'R = 1$ (1)

$\vdash .\, (1) \,.\, *152\cdot7 \,.\supset \vdash .\, \text{Prop}$

$*153\cdot36$. $\vdash .\, C``1_s = 1$

Dem.

$\vdash .\, *153\cdot301 \,.\supset \vdash .\, C``1_s = C``\breve{C}``1$

[$*72\cdot502$] $= 1 \,.\supset \vdash .\, \text{Prop}$

$*154$. RELATION-NUMBERS OF ASSIGNED TYPES.

Summary of $*154$.

This number gives propositions analogous to those of $*102$. In accordance with our general notations for typical definiteness, "$\text{Nr}(P)`Q$" means "the class of relations like Q and of the same type as P," "$\text{Nr}(P_Q)$" means "the relation to a relation of the type of Q of the class of relations like it and of the type of P." By a special definition, "$\text{NR}^Q(P)$" is to mean all typically definite relation-numbers of the form "$\text{Nr}(P_Q)`R$," *i.e.* all relation-numbers generated by the relation $\text{Nr}(P_Q)$, *i.e.* the domain of $\text{Nr}(P_Q)$.

Existence-theorems in this subject can be proved by means of $*154{\cdot}14$, which states that relations like Q exist in the type of P when, and only when, classes similar to $C`Q$ exist in the type of $C`P$. In virtue of this proposition, the existence-theorems of our present topic are deducible from those for cardinals. In symbols, this proposition is

$*154{\cdot}14$. $\vdash : \exists ! \, \text{Nr}(P)`Q \, . \equiv . \, \exists ! \, \text{Nc}(C`P)`C`Q$

Hence by $*102{\cdot}73$ we deduce

$*154{\cdot}242$. $\vdash . \, \Lambda \in \text{NR}^{t`P}(P)$

whence, by $*152{\cdot}72$,

$*154{\cdot}9$. $\vdash . \, \text{NC} = C``\text{NR}$

The remaining propositions are chiefly analogues of those in $*102$. Very few of them are subsequently referred to.

$*154{\cdot}01$. $\text{NR}^Y(X) = \text{D}`\text{Nr}(X_Y) \quad \text{Df}$

$*154{\cdot}1$. $\vdash : \exists ! \, \text{Rl}`P \cap \text{Nr}`Q \, . \supset . \, \exists ! \, \text{Cl}`C`P \cap \text{Nc}`C`Q$

Dem.

$\vdash . *152{\cdot}1 . \supset \vdash : \text{Hp} . \supset . (\exists R) . R \subseteq P . R \, \text{smor} \, Q .$

$[*151{\cdot}18] \qquad \supset . (\exists R) . R \subseteq P . C`R \, \text{sm} \, C`Q .$

$[*33{\cdot}265] \qquad \supset . (\exists R) . C`R \subset C`P . C`R \, \text{sm} \, C`Q .$

$[*100{\cdot}1] \qquad \supset . \exists ! \, \text{Cl}`C`P \cap \text{Nc}`C`Q : \supset \vdash . \text{Prop}$

***154·11.** $\vdash : \exists ! \,\text{Cl}‘C‘P \cap \text{Nc}‘C‘Q . \supset . (\exists R) . R \,\text{smor}\, Q . C‘R \subset C‘P$

Dem.

$\vdash . *100·1 . *73·1 . \supset \vdash : \text{Hp} . \supset . (\exists S) . S \in 1 \rightarrow 1 . \text{D}‘S \subset C‘P . \text{ᗡ}‘S = C‘Q .$

[*151·1] $\supset . (\exists S) . \text{D}‘S \subset C‘P . S ; Q \,\text{smor}\, Q .$

[*150·203] $\supset . (\exists S) . C‘S;Q \subset C‘P . S;Q \,\text{smor}\, Q : \supset \vdash . \text{Prop}$

***154·12.** $\vdash : \exists ! \,\text{Rl}‘(\alpha \uparrow \alpha) \cap \text{Nr}‘Q . \equiv . \exists ! \,\text{Cl}‘\alpha \cap \text{Nc}‘C‘Q . \equiv . \text{Nc}‘\alpha \geqslant \text{Nc}‘C‘Q$

Dem.

$\vdash . *154·1 . *35·9 . \quad \supset \vdash : \exists ! \,\text{Rl}‘(\alpha \uparrow \alpha) \cap \text{Nr}‘Q . \supset . \exists ! \,\text{Cl}‘\alpha \cap \text{Nc}‘C‘Q \quad (1)$

$\vdash . *154·11 . *35·92 . \supset \vdash : \exists ! \,\text{Cl}‘\alpha \cap \text{Nc}‘C‘Q . \supset . \exists ! \,\text{Rl}‘(\alpha \uparrow \alpha) \cap \text{Nr}‘Q \quad (2)$

$\vdash . (1) . (2) . \supset \vdash . \text{Prop}$

***154·121.** $\vdash . \text{Rl}‘(t_0‘C‘P \uparrow t_0‘C‘P) = t‘P = t_{00}‘C‘P$

Dem.

$\vdash . *64·5 . \quad \supset \vdash . \text{Rl}‘(t_0‘C‘P \uparrow t_0‘C‘P) = t‘(C‘P \uparrow C‘P) \quad (1)$

$\vdash . *64·201 . \supset \vdash . t‘(C‘P \uparrow C‘P) = t‘P \quad (2)$

$\vdash . (1) . (2) . *64·54 . \supset \vdash . \text{Prop}$

***154·13.** $\vdash : \exists ! \, t‘P \cap \text{Nr}‘Q . \equiv . \exists ! \, t‘C‘P \cap \text{Nc}‘C‘Q . \equiv . \text{Nc}‘t_0‘C‘P \geqslant \text{Nc}‘C‘Q$

Dem.

$\vdash . *154·12 \dfrac{t_0‘C‘P}{\alpha} . *154·121 . \supset$

$\vdash : \exists ! \, t‘P \cap \text{Nr}‘Q . \equiv . \exists ! \,\text{Cl}‘t_0‘C‘P \cap \text{Nc}‘C‘Q \quad (1)$

$\vdash . (1) . *63·65 . *117·22 . \supset \vdash . \text{Prop}$

***154·14.** $\vdash : \exists ! \,\text{Nr}(P)‘Q . \equiv . \exists ! \,\text{Nc}(C‘P)‘C‘Q$ [*154·13 . (*65·04)]

In virtue of *154·14 and the propositions of *102, *103, *104, *105, *106, we see that all homogeneous or ascending relation-numbers exist, while Λ is a member of every descending type of relation-numbers. Remembering that the relations concerned must be homogeneous, we see that there are two kinds of steps by which their types may be raised, namely (1) from P to relations of the type of $t‘C‘P \uparrow t‘C‘P$, *i.e.* from P to relations of the type of $C‘P \downarrow C‘P$, or of $\iota ; P$; (2) from P to relations of the type of $t‘P \uparrow t‘P$, *i.e.* from P to relations of the type of $P \downarrow P$, or of $\downarrow x ; P$ if $x \in t_0‘C‘P$. Thus repetitions of the two steps from P to $\iota ; P$, and from P to $\downarrow x ; P$, where $x \in t_0‘C‘P$, will enable us, without changing the relation-number, to raise its type indefinitely. It will be observed that, in accordance with our general definitions for relative types, the type of $\iota ; P$ is $t^{11}‘C‘P$, and the type of $\downarrow x ; P$ (where $x \in t_0‘C‘P$) is $t^{11}‘P$.

***154·2.** $\vdash . \text{Nr}(X_Y)‘Q = \hat{P}\{P \,\text{smor}_{(X,Y)}\, Q\}$ [*65·2 . (*152·01)]

***154·201.** $\vdash . \text{Nr}(X)‘Q = \text{Nr}‘Q \cap t‘X$ [Proof as in *102·6]

***154·202.** $\vdash : P \epsilon \mathrm{Nr}(X_Y)`Q . \equiv . P \epsilon \mathrm{Nr}(X)`Q . Q \epsilon t`Y . \equiv .$
$P \epsilon \mathrm{Nr}`Q . P \epsilon t`X . Q \epsilon t`Y$ [*152·2·201 . (*65·1)]

***154·203.** $\vdash : Q \epsilon t`Y . \supset . \mathrm{Nr}(X_Y)`Q = \mathrm{Nr}(X)`Q$ [*154·202]

When Q belongs to any other type than $t`Y$, $\mathrm{Nr}(X_Y)`Q$ is meaningless.

***154·21.** $\vdash . \mathrm{NR}^Y(X) = \hat{\lambda}\{(\exists Q) . \lambda = \mathrm{Nr}(X_Y)`Q\}$ [(*154·01)]

***154·22.** $\vdash . \mathrm{NR}^Y(X) = \mathrm{Nr}(X)``t`Y = (\cap t`X)``\mathrm{Nr}``t`Y$

Dem.

$\vdash . *154·21·202 . \supset$

$\vdash :. \lambda \epsilon \mathrm{NR}^Y(X) . \equiv : (\exists Q) : P \epsilon \lambda . \equiv_P . P \epsilon \mathrm{Nr}(X)`Q . Q \epsilon t`Y :$

[*63·108.*4·73] $\equiv : (\exists Q) : Q \epsilon t`Y : P \epsilon \lambda . \equiv_P . P \epsilon \mathrm{Nr}(X)`Q :$

[*20·43] $\equiv : (\exists Q) . Q \epsilon t`Y . \lambda = \mathrm{Nr}(X)`Q :$

[*37·6] $\equiv : \lambda \epsilon \mathrm{Nr}(X)``t`Y$ (1)

$\vdash . (1) . *154·201 . \supset \vdash .$ Prop

***154·23.** $\vdash : \Lambda \epsilon \mathrm{NR}^Q(P) . \equiv . \Lambda \epsilon \mathrm{NC}^{C`Q}(C`P) . \equiv . \Lambda \epsilon \mathrm{NC}(C`P)``t`C`Q$

Dem.

$\vdash . *154·22 . \supset \vdash : \Lambda \epsilon \mathrm{NR}^Q(P) . \equiv . \Lambda \epsilon \mathrm{Nr}(P)``t`Q .$

[*37·6] $\equiv . (\exists R) . R \epsilon t`Q . \Lambda = \mathrm{Nr}(P)`R .$

[*154·14.Transp] $\equiv . (\exists R) . R \epsilon t`Q . \Lambda = \mathrm{Nc}(C`P)`C`R .$

[*64·24] $\equiv . (\exists R) . C`R \epsilon t`C`Q . \Lambda = \mathrm{Nc}(C`P)`C`R .$

[*35·942] $\equiv . (\exists \alpha) . \alpha \epsilon t`C`Q . \Lambda = \mathrm{Nc}(C`P)`\alpha .$

[*37·6] $\equiv . \Lambda \epsilon \mathrm{Nc}(C`P)``t`C`Q .$ (1)

[*102·62] $\equiv . \Lambda \epsilon \mathrm{NC}^{C`Q}(C`P)$ (2)

$\vdash . (1) . (2) . \supset \vdash .$ Prop

***154·24.** $\vdash : C`Q = t`C`P . \supset . \mathrm{Nr}(P)`Q = \Lambda$ [*102·73 . *154·14]

***154·241.** $\vdash . \mathrm{Nr}(P)`I \upharpoonright t`C`P = \Lambda$ [*154·24]

***154·242.** $\vdash . \Lambda \epsilon \mathrm{NR}^{\iota;P}(P)$

Dem.

$\vdash . *35·91 . \supset \vdash . I \upharpoonright t`C`P \subset t`C`P \uparrow t`C`P$

[*63·64] $\subset t_0`\iota``C`P \uparrow t_0`\iota``C`P$

[*150·22] $\subset t_0`C`\iota;P \uparrow t_0`C`\iota;P$ (1)

$\vdash . (1) . *154·121 . \supset \vdash . I \upharpoonright t`C`P \epsilon t`\iota;P$ (2)

$\vdash . (2) . *154·22·241 . \supset \vdash .$ Prop

***154·25.** $\vdash : C`Q = t_{00}`C`P . \supset . \mathrm{Nr}(P)`Q = \Lambda$ [*106·53 . *154·14]

***154·251.** $\vdash . \Lambda \epsilon \mathrm{NR}^{P \downarrow P}(P)$

Dem.

$\vdash . *154·23 . \supset \vdash : \Lambda \epsilon \mathrm{NR}^{P \downarrow P}(P) . \equiv . \Lambda \epsilon \mathrm{Nc}(C`P)``t`C`(P \downarrow P) .$

[*55·15] $\equiv . \Lambda \epsilon \mathrm{Nc}(C`P)``t`\iota`P .$

[*63·61] $\equiv . \Lambda \epsilon \mathrm{Nc}(C`P)``t`t`P .$

[*154·121] $\equiv . \Lambda \epsilon \mathrm{Nc}(C`P)``t`t_{00}`C`P$ (1)

$\vdash . (1) . *106·53 . *104·264 . \supset \vdash .$ Prop

***154·26.** $\vdash : P \epsilon t^{\backprime}Q . \supset . \exists ! \mathrm{Nr}(P)^{\backprime}Q$ [*64·231 . *103·3·13 . *154·14]

***154·261.** $\vdash : C^{\backprime}P \epsilon t^{2\backprime}C^{\backprime}Q . \supset . \exists ! \mathrm{Nr}(P)^{\backprime}Q$ [*104·21·1 . *154·14]

***154·262.** $\vdash : C^{\backprime}P \epsilon t_{00}{}^{\backprime}C^{\backprime}Q . \supset . \exists ! \mathrm{Nr}(P)^{\backprime}Q$ [*106·21·1 . *154·14]

The following propositions are concerned with the two particular transformations from P to $\iota ; P$ and from P to $x \downarrow ; P$, which are useful in raising the type of a relation-number.

***154·31.** $\vdash . t^{\backprime}\iota ; P = t^{11\backprime}C^{\backprime}P$

Dem.

$\vdash . *154·121 . *150·22 . \supset \vdash . t^{\backprime}\iota ; P = \mathrm{Rl}^{\backprime}(t_0{}^{\backprime}\iota^{\backprime\backprime}C^{\backprime}P \uparrow t_0{}^{\backprime}\iota^{\backprime\backprime}C^{\backprime}P)$

[*63·64] $= \mathrm{Rl}^{\backprime}(t^{\backprime}C^{\backprime}P \uparrow t^{\backprime}C^{\backprime}P)$

[*64·56] $= t^{11\backprime}C^{\backprime}P . \supset \vdash .$ Prop

***154·311.** $\vdash . \exists ! \mathrm{Nr}(t^{11\backprime}C^{\backprime}P)^{\backprime}P$ [*154·31 . *152·6]

***154·32.** $\vdash : x \epsilon t_0{}^{\backprime}C^{\backprime}P . \supset . t^{\backprime}x \downarrow ; P = t^{11\backprime}P . t_0{}^{\backprime}x \downarrow {}^{\backprime\backprime}C^{\backprime}P = t^{\backprime}P$

Dem.

$\vdash . *154·121 . *150·22 . \supset \vdash . t^{\backprime}x \downarrow ; P = \mathrm{Rl}^{\backprime}\{(t_0{}^{\backprime}x \downarrow {}^{\backprime\backprime}C^{\backprime}P) \uparrow (t_0{}^{\backprime}x \downarrow {}^{\backprime\backprime}C^{\backprime}P)\}$ (1)

$\vdash . *64·52 . \quad \supset \vdash : x, y \epsilon t_0{}^{\backprime}C^{\backprime}P . \supset . x \downarrow y \epsilon t^{\backprime}(t_0{}^{\backprime}C^{\backprime}P \uparrow t_0{}^{\backprime}C^{\backprime}P) .$

[*154·121] $\supset . x \downarrow y \epsilon t^{\backprime}P$ (2)

$\vdash . (2) . \quad \supset \vdash : x \epsilon t_0{}^{\backprime}C^{\backprime}P . \supset . x \downarrow {}^{\backprime\backprime}C^{\backprime}P \subset t^{\backprime}P .$

[*63·21] $\supset . t_0{}^{\backprime}x \downarrow {}^{\backprime\backprime}C^{\backprime}P = t^{\backprime}P$ (3)

$\vdash . (1) . (3) . \quad \supset \vdash : \mathrm{Hp} . \supset . t^{\backprime}x \downarrow ; P = \mathrm{Rl}^{\backprime}(t^{\backprime}P \uparrow t^{\backprime}P)$

[*64·56] $= t^{11\backprime}P$ (4)

$\vdash . (3) . (4) . \supset \vdash .$ Prop

***154·321.** $\vdash . \exists ! \mathrm{Nr}(t^{11\backprime}P)^{\backprime}P$ [*154·32 . *152·62 . *63·18]

***154·322.** $\vdash : x \epsilon t_0{}^{\backprime}C^{\backprime}P . \supset . t^{\backprime} \downarrow x ; P = t^{11\backprime}P$ [Proof as in *154·32]

***154·33.** $\vdash : x \epsilon t_0{}^{\backprime}C^{\backprime}P . \supset . t^{\backprime}P \downarrow ; x \downarrow ; P = t^{11\backprime}s^{\backprime}t^{11\backprime}P$

Dem.

$\vdash . *154·32 . \supset \vdash : \mathrm{Hp} . \supset . P \epsilon t_0{}^{\backprime}x \downarrow {}^{\backprime\backprime}C^{\backprime}P .$

[*150·22] $\supset . P \epsilon t_0{}^{\backprime}C^{\backprime}x \downarrow ; P .$

[*154·32] $\supset . t^{\backprime}P \downarrow ; x \downarrow ; P = t^{11\backprime}x \downarrow ; P$

[*64·23] $= t^{11\backprime}s^{\backprime}t^{\backprime}x \downarrow ; P$

[*154·32] $= t^{11\backprime}s^{\backprime}t^{11\backprime}P : \supset \vdash .$ Prop

***154·331.** $\vdash . \exists ! \mathrm{Nr}(t^{11\backprime}s^{\backprime}t^{11\backprime}P)$ [*154·33 . *152·62 . *63·18]

***154·4.** $\vdash . \mathrm{Nr}(X_Y)^{\backprime}Q = \hat{P}\{(\exists S) . S \epsilon 1 \rightarrow 1 . \mathrm{Cl}^{\backprime}S = C^{\backprime}Q . P = S ; Q .$
$\mathrm{D}^{\backprime}S \epsilon t^{\backprime}C^{\backprime}X . \mathrm{Cl}^{\backprime}S \epsilon t^{\backprime}C^{\backprime}Y\}$

Dem.

$\vdash . *154·202 . *152·1 . \supset$

$\vdash :. P \epsilon \mathrm{Nr}(X_Y)^{\backprime}Q . \equiv : (\exists S) . S \epsilon 1 \rightarrow 1 . \mathrm{Cl}^{\backprime}S = C^{\backprime}Q . P = S ; Q : P \epsilon t^{\backprime}X . Q \epsilon t^{\backprime}Y :$

[*64·24] $\equiv : (\exists S) . S \epsilon 1 \rightarrow 1 . \mathrm{Cl}^{\backprime}S = C^{\backprime}Q . P = S ; Q . C^{\backprime}P \epsilon t^{\backprime}C^{\backprime}X .$
$C^{\backprime}Q \epsilon t^{\backprime}C^{\backprime}Y :$

[*13·193.*150·23] $\equiv : (\exists S) . S \epsilon 1 \rightarrow 1 . \mathrm{Cl}^{\backprime}S = C^{\backprime}Q . P = S ; Q .$
$\mathrm{D}^{\backprime}S \epsilon t^{\backprime}C^{\backprime}X . \mathrm{Cl}^{\backprime}S \epsilon t^{\backprime}C^{\backprime}Y :. \supset \vdash .$ Prop

∗154·401. $\vdash . \mathrm{Nr}(X_Y)`Q = \hat{P}\{\exists ! (P \,\overline{\mathrm{smor}}\, Q) \cap t`(C`X \uparrow C`Y)\}$
[∗154·4 . ∗151·11 . ∗64·63]

The remaining propositions of this number (except ∗154·9) are the analogues of those whose numbers have the same decimal part in ∗102. They are here given without proof, because the proofs are, step by step, analogous to the proofs of the corresponding propositions in ∗102.

∗154·41. $\vdash : P \in \mathrm{Nr}(X_Z)`R . Q \in \mathrm{Nr}(Y_Z)`R . \supset . P \in \mathrm{Nr}(X_Y)`Q . Q \in \mathrm{Nr}(Y_X)`P$

∗154·42. $\vdash . P \in \mathrm{Nr}(P_P)`P$

∗154·43. $\vdash . \exists ! \mathrm{Nr}(P_P)`P$

∗154·46. $\vdash : P \in \mathrm{Nr}(X_Y)`Q . \equiv . Q \in \mathrm{Nr}(Y_X)`P . \equiv . P \,\mathrm{smor}\, Q . P \in t`X . Q \in t`Y$

∗154·52. $\vdash : \exists ! \mathrm{Nr}(X_Y)`Q . \supset . \mathrm{Nr}(X_Y)`Q \in \mathrm{NR}^X(X)$

∗154·53. $\vdash : \mathrm{NR}^Y(X) - \iota`\Lambda \subset \mathrm{NR}^X(X)$

∗154·55. $\vdash : \Lambda \sim\in \mathrm{NR}^X(Y) . \supset . \mathrm{NR}^Y(X) - \iota`\Lambda = \mathrm{NR}^X(X)$

∗154·64. $\vdash : \mu \in \mathrm{NR} . \exists ! \mu . \supset . (\exists P, Q) . \mu = \mathrm{Nr}(P)`Q$

∗154·641. $\vdash : \mu \in \mathrm{NR} . \supset . (\exists P, Q) . \mu = \mathrm{Nr}(P)`Q$ [∗154·64·241]

∗154·8. $\vdash : P \in \mathrm{Nr}(X_Y)`Q . P \,\mathrm{smor}\, R . R \in t`S . \supset . R \in \mathrm{Nr}(S_Y)`Q . R \in \mathrm{Nr}(S_X)`P$

∗154·81. $\vdash : P \in \mathrm{Nr}(X_Y)`Q . \supset . \mathrm{smor}``\mathrm{Nr}(X_Y)`Q \cap t`S = \mathrm{Nr}(S_Y)`Q = \mathrm{Nr}(S_X)`P$

∗154·82. $\vdash : \mu \in \mathrm{NR}^Y(X) . \exists ! \mu . \supset . \mathrm{smor}``\mu \cap t`S \in \mathrm{NR}^Y(S)$

∗154·83. $\vdash : \mu \in \mathrm{NR}^Y(X) . \nu = \mathrm{smor}``\mu \cap t`S . \exists ! \nu . \supset .$
$\mathrm{smor}``\mu \cap t`S = \mathrm{smor}``\nu \cap t`S . \mu = \mathrm{smor}``\nu \cap t`X$

∗154·84. $\vdash : (\exists P) . P \,\mathrm{smor}\, X . P \in t`X . Q \,\mathrm{smor}\, P . \equiv . Q \,\mathrm{smor}\, X$

∗154·85. $\vdash . \mathrm{smor}``\mu \cap t`Y = \mathrm{smor}_Y``\mu$

∗154·86. $\vdash : \mu = \mathrm{Nr}(X)`Q . \exists ! \mu . \supset . \mathrm{smor}_Y``\mu = \mathrm{Nr}(Y)`Q$

∗154·861. $\vdash . \mathrm{smor}_X``\mathrm{smor}_Y``\mu \subset \mathrm{smor}_X``\mu$

∗154·87. $\vdash : \mu = \mathrm{Nr}(Y)`Q . \exists ! \mathrm{Nr}(X)`Q . \supset . \mathrm{smor}_P``\mu = \mathrm{smor}_P``\mathrm{smor}_X``\mu$

∗154·88. $\vdash : \mu = \mathrm{Nr}(Y)`Q . \exists ! \mathrm{smor}_P``\mu . \supset .$
$\mathrm{smor}_P``\mu = \mathrm{Nr}(P)`Q . \mathrm{smor}_X``\mu = \mathrm{Nr}(X)`Q .$
$\mathrm{smor}_X``\mu = \mathrm{smor}_X``\mathrm{smor}_P``\mu = \mathrm{smor}_X``\mathrm{Nr}(P)`Q$

∗154·9. $\vdash \mathrm{NC} = C```\mathrm{NR}$

Dem.

$\vdash . ∗37·29 . \supset \vdash : \mu = \Lambda . \supset . \mu = C``\Lambda .$
[∗154·241] $\supset . \mu \in C```\mathrm{NR}$ (1)
$\vdash . ∗37·29 . \supset \vdash : \nu = \Lambda . \supset . C``\nu = \Lambda .$
[∗102·73] $\supset . C``\nu \in \mathrm{NC}$ (2)
$\vdash . (1) . (2) . ∗152·72 . \supset \vdash . \mathrm{Prop}$

$*155$. HOMOGENEOUS RELATION-NUMBERS.

Summary of $*155$.

A relation-number is called *homogeneous* when it is generated by a homogeneous relation of likeness, *i.e.* when it consists of all relations which are like a given relation P and of the same type as P. For the homogeneous relation-number of P we write "$\mathrm{N_0r}`P$"; thus $\mathrm{N_0r}`P = \mathrm{Nr}`P \cap t`P$. When P is given, $\mathrm{N_0r}`P$ is typically definite. We have always $P \epsilon \mathrm{N_0r}`P$, hence $\exists ! \mathrm{N_0r}`P$. Conversely, if a typically definite relation-number is not null, it is a homogeneous relation-number; in fact, if $P$ is a member of it, it is $\mathrm{N_0r}`P$. Thus the homogeneous relation-numbers are all the relation-numbers except Λ.

Homogeneous relation-numbers play the same part in relation-arithmetic as homogeneous cardinals play in cardinal arithmetic. The propositions of this number (except $*155{\cdot}6{\cdot}61$) are the analogues of those with the same decimal part in $*103$. Their proofs are exactly analogous to the proofs of their analogues in $*103$, and are therefore omitted.

The following propositions are the most useful in this number.

$*155{\cdot}11$. $\vdash : Q \epsilon \mathrm{N_0r}`P \,.\equiv.\, Q \,\mathrm{smor}\, P \,.\, Q \epsilon t`P \,.\equiv.\, Q \epsilon \mathrm{Nr}`P \,.\, Q \epsilon t`P$

This merely embodies the definition.

$*155{\cdot}12$. $\vdash . P \epsilon \mathrm{N_0r}`P$

whence

$*155{\cdot}13$. $\vdash . \exists ! \mathrm{N_0r}`P$

$*155{\cdot}16$. $\vdash : \mathrm{N_0r}`P = \mathrm{Nr}`Q \,.\equiv.\, \mathrm{Nr}`P = \mathrm{Nr}`Q$

This proposition is used in the theory of well-ordered series ($*253$ and $*255$). It requires that the equation "$\mathrm{Nr}`P = \mathrm{Nr}`Q$" on the right-hand side should be subject to the convention AT. Otherwise, the typical ambiguities might be so determined as to give $\mathrm{Nr}`P = \mathrm{Nr}`Q = \Lambda$, which would not imply $\mathrm{N_0r}`P = \mathrm{Nr}`Q$.

$*155{\cdot}2$. $\vdash : \mu \epsilon \mathrm{N_0R} \,.\equiv.\, (\exists P) . \mu = \mathrm{Nr}`P \cap t`P \,.\equiv.\, (\exists P) . \mu = \mathrm{N_0r}`P$

This merely embodies the definition of $\mathrm{N_0R}$.

$*155{\cdot}22$. $\vdash : \mu \epsilon \mathrm{N_0R} \,.\supset.\, \exists ! \mu$

*155·26. $\vdash :. \mu \epsilon \mathrm{NR} . \supset : P \epsilon \mu . \equiv . \mathrm{N_0r}\text{‘}P = \mu$

*155·27. $\vdash : \mu = \mathrm{N_0r}\text{‘}P . \equiv . \mu \epsilon \mathrm{NR} . P \epsilon \mu$

*155·34. $\vdash . \mathrm{NR} - \iota\text{‘}\Lambda \subset \mathrm{N_0R}$

*155·4. $\vdash . \mathrm{smor}\text{“}\mathrm{N_0r}\text{‘}P = \mathrm{Nr}\text{‘}P$

*155·5. $\vdash . 0_r \epsilon \mathrm{N_0R}$

*155·6. $\vdash . C\text{“}\mathrm{N_0r}\text{‘}P = \mathrm{N_0c}\text{‘}C\text{‘}P$

This last proposition connects homogeneous relation-numbers with homogeneous cardinals.

---

*155·01. $\mathrm{N_0r}\text{‘}P = \mathrm{Nr}\text{‘}P \cap t\text{‘}P$ Df

*155·02. $\mathrm{N_0R} = \mathrm{D}\text{‘}\mathrm{N_0r}$ Df

*155·1. $\vdash . \mathrm{N_0r}\text{‘}P = (\mathrm{Nr}\text{‘}P)_P = \mathrm{Nr}\,(P)\text{‘}P = \mathrm{Nr}\,(P_P)\text{‘}P$

*155·11. $\vdash : Q \epsilon \mathrm{N_0r}\text{‘}P . \equiv . Q \,\mathrm{smor}\, P . Q \epsilon t\text{‘}P . \equiv . Q \epsilon \mathrm{Nr}\text{‘}P . Q \epsilon t\text{‘}P$

*155·12. $\vdash . P \epsilon \mathrm{N_0r}\text{‘}P$

*155·13. $\vdash . \exists ! \mathrm{N_0r}\text{‘}P$

*155·14. $\vdash : \mathrm{N_0r}\text{‘}P = \mathrm{N_0r}\text{‘}Q . \equiv . P \epsilon \mathrm{N_0r}\text{‘}Q . \equiv . Q \epsilon \mathrm{N_0r}\text{‘}P . \equiv . P \,\mathrm{smor}\, Q . Q \epsilon t\text{‘}P$

*155·15. $\vdash : \exists ! \mathrm{N_0r}\text{‘}P \cap \mathrm{N_0r}\text{‘}Q . \equiv . \mathrm{N_0r}\text{‘}P = \mathrm{N_0r}\text{‘}Q$

*155·16. $\vdash : \mathrm{N_0r}\text{‘}P = \mathrm{Nr}\text{‘}Q . \equiv . \mathrm{Nr}\text{‘}P = \mathrm{Nr}\text{‘}Q$

*155·2. $\vdash : \mu \epsilon \mathrm{N_0R} . \equiv . (\exists P) . \mu = \mathrm{Nr}\text{‘}P \cap t\text{‘}P . \equiv . (\exists P) . \mu = \mathrm{N_0r}\text{‘}P$

*155·21. $\vdash . \mathrm{N_0r}\text{‘}P \epsilon \mathrm{N_0R} . \mathrm{N_0r}\text{‘}P \epsilon \mathrm{NR}$

*155·22. $\vdash : \mu \epsilon \mathrm{N_0R} . \supset . \exists ! \mu$

*155·23. $\vdash . \Lambda \sim \epsilon \mathrm{N_0R}$

*155·24. $\vdash . \mathrm{N_0R} \epsilon \mathrm{Cls\,ex^2\,excl}$

*155·25. $\vdash :. \mu, \nu \epsilon \mathrm{N_0R} . \supset : \exists ! \mu \cap \nu . \equiv . \mu = \nu$

*155·26. $\vdash :. \mu \epsilon \mathrm{NR} . \supset : P \epsilon \mu . \equiv . \mathrm{N_0r}\text{‘}P = \mu$

*155·27. $\vdash : \mu = \mathrm{N_0r}\text{‘}P . \equiv . \mu \epsilon \mathrm{NR} . P \epsilon \mu$

*155·28. $\vdash : (\exists R) . R \,\mathrm{smor}\, P . \mu = \mathrm{N_0r}\text{‘}R . \equiv . \exists ! \mu . \mu = \mathrm{Nr}\text{‘}P$

*155·3. $\vdash : Q \epsilon t\text{‘}P . \supset . \mathrm{N_0r}\text{‘}Q = \mathrm{Nr}\,(P)\text{‘}Q = \mathrm{Nr}\,(P_P)\text{‘}Q = \mathrm{Nr}\text{‘}Q \cap t\text{‘}P$

*155·301. $\vdash . \mathrm{NR}^P(P) = \mathrm{N_0R}\,(P)$

*155·31. $\vdash : \exists ! \mathrm{Nr}\,(X_Y)\text{‘}Q . \supset . \mathrm{Nr}\,(X_Y)\text{‘}Q \epsilon \mathrm{N_0R}\,(X)$

*155·32. $\vdash . \mathrm{NR}^Y(X) - \iota\text{‘}\Lambda \subset \mathrm{N_0R}\,(X)$

*155·33. $\vdash . \mathrm{NR}\,(X) - \iota\text{‘}\Lambda \subset \mathrm{N_0R}\,(X)$

*155·34. $\vdash . \mathrm{NR} - \iota\text{‘}\Lambda \subset \mathrm{N_0R}$

*155·35. $\vdash : \Lambda \sim \epsilon \mathrm{NR}^X(Y) . \supset . \mathrm{NR}^Y(X) - \iota\text{‘}\Lambda = \mathrm{N_0R}\,(X)$

*155·4. $\vdash . \mathrm{smor}\text{“}\mathrm{N_0r}\text{‘}P = \mathrm{Nr}\text{‘}P$

*155·41. $\vdash . \mathrm{smor}\text{“}\mathrm{N_0r}\text{‘}P \cap t\text{‘}Q = \mathrm{Nr}\,(Q)\text{‘}P$

***155·42.** $\vdash : Q \text{ smor } P . \equiv . \text{Nr}(Q)`P = \text{N}_0\text{r}`Q$

***155·43.** $\vdash : \mu \in \text{NR} . \supset . \text{smor}``\mu \cap t_0`\mu = \mu$

***155·44.** $\vdash :. \mu, \nu \in \text{N}_0\text{R} . \supset : \mu = \text{smor}``\nu . \equiv . \nu = \text{smor}``\mu$

***155·5.** $\vdash . 0_r \in \text{N}_0\text{R}$

***155·51.** $\vdash . 2_r \cap \text{Rl}`\text{Cls} \in \text{N}_0\text{R}$

***155·52.** $\vdash . 2_r \cap \text{Rel}^2 \in \text{N}_0\text{R}$

The following propositions have no analogue in *103.

***155·6.** $\vdash . C``\text{N}_0\text{r}`P = \text{N}_0\text{c}`C`P$

Dem.

$\vdash . *100·11 . *103·11 . \supset \vdash :. \alpha \in \text{N}_0\text{c}`C`P . \equiv :$

$\alpha \in t`C`P : (\exists S) . S \in 1 \rightarrow 1 . \text{D}`S = \alpha . \text{Ͻ}`S = C`P :$

[*150·23] $\equiv : \alpha \in t`C`P : (\exists S) . S \in 1 \rightarrow 1 . C`S;P = \alpha . \text{Ͻ}`S = C`P :$

[*151·11] $\equiv : \alpha \in t`C`P : (\exists Q) . Q \text{ smor } P . \alpha = C`Q :$

[*64·24] $\equiv : (\exists Q) . Q \text{ smor } P . Q \in t`P . \alpha = C`Q :$

[*152·11.*155·11] $\equiv : \alpha \in C``\text{N}_0\text{r}`P :. \supset \vdash . \text{Prop}$

***155·61.** $\vdash . C```\text{N}_0\text{R} = \text{N}_0\text{C}$ [*155·6]

On ascending and descending relation-numbers, propositions analogous to those of *104, *105, and *106 might be proved by proofs analogous to those given in those numbers. It is, however, scarcely necessary to add anything to the propositions already proved, namely *154·24·241·242·25·251 on descending relation-numbers, *154·26·261·262·31·311·32·321·322·33·331 on ascending relation-numbers, and *155·23·34 giving the relations of non-homogeneous to homogeneous relation-numbers. Ascending relation-numbers all exist, and those that start from the type of P, wherever they end*, are the correspondents† of the homogeneous relation-numbers of the type of P, and are only some of the homogeneous relation-numbers of the type in which they end. Descending relation-numbers consist of Λ together with the homogeneous relation-numbers of the type in which they end: they are the correspondents of only some of the type in which they begin, or rather, Λ is the common correspondent of all those relation-numbers in the initial type which are not correspondents of any homogeneous relation-number in the end-type. These properties are exactly the same as in the case of cardinals, as might be foreseen by *154·14.

* We say that $\text{Nr}(P)`Q$ starts from the type of Q and ends in the type of P.

† We call two typically definite relation-numbers *correspondents* when they only differ as to the typical determination, *i.e.* $\text{Nr}(X)`P$ and $\text{Nr}(Y)`P$ are correspondents.

SECTION B.

ADDITION OF RELATIONS, AND THE PRODUCT OF TWO RELATIONS.

Summary of Section B.

In the present section, we have to consider the kind of addition of relations which is required in ordinal arithmetic. In cardinal arithmetic, if κ is a class of mutually exclusive classes, $s\text{‘}\kappa$ has the properties required of their sum, and thus we do not require a new kind of logical addition before dealing with arithmetical addition. But in ordinal arithmetic this is not so. Suppose P and Q are the generating relations of two series, and we wish to add the Q-series at the end of the P-series. Then we wish every term of the P-series to precede every term of the Q-series; thus $P \mathbin{\dot{\cup}} Q$ is not the generating relation of the new series, since $P \mathbin{\dot{\cup}} Q$ gives no relation between the terms of the P-series and the terms of the Q-series. The relation we want is

$$P \mathbin{\dot{\cup}} Q \mathbin{\dot{\cup}} C\text{‘}P \uparrow C\text{‘}Q,$$

since this makes every term of the P-series precede every term of the Q-series. Hence we put

$$P \mathbin{\text{⫩}} Q = P \mathbin{\dot{\cup}} Q \mathbin{\dot{\cup}} C\text{‘}P \uparrow C\text{‘}Q \quad \text{Df.}$$

It will be seen that $P \mathbin{\text{⫩}} Q$ is in general different from $Q \mathbin{\text{⫩}} P$.

If $C\text{‘}P$ and $C\text{‘}Q$ have no common terms, the sum of the relation-numbers of P and Q is the relation-number of $P \mathbin{\text{⫩}} Q$ (cf. $*180$).

The addition of a single term to a series requires a new definition, and cannot be dealt with as a particular case of the addition of two relations. It might be thought that, just as $\alpha \cup \iota\text{‘}x$ gives the result of adding the one term x to the series α, so $P \mathbin{\text{⫩}} (x \downarrow x)$ would give the result of adding the one term x to the series P. But this is not the case, since, when we add a term to a series, we do not want this term to precede itself, whereas $P \mathbin{\text{⫩}} (x \downarrow x)$ is a relation which x has to itself. What we want is a relation which every member of $C\text{‘}P$ has to x but which x does not have to itself; thus we take $P \mathbin{\dot{\cup}} C\text{‘}P \uparrow \iota\text{‘}x$ as our relation, and put

$$P \mathbin{+\!\!\!\rightarrow} x = P \mathbin{\dot{\cup}} C\text{‘}P \uparrow \iota\text{‘}x \quad \text{Df.}$$

This definition defines the generating relation of the series obtained by adding x at the *end* of the P-series; similarly for adding x at the *beginning* we put

$$x \leftarrow\!\!\!\!+\, P = \iota\text{‘}x \uparrow C\text{‘}P \cup P \quad \text{Df.}$$

If x is not a member of $C\text{‘}P$, the relation-number of $P +\!\!\!\!\rightarrow x$ is the sum of the relation-number of P and the ordinal 1, which we represent by $\dot{1}$. (The ordinal 1 has no meaning by itself, but only as a summand.)

The sum of a series of series is defined in the same way as the sum of two series was defined. Let P be a serial relation whose field consists of serial relations. Then the sum of all the series generated by members of $C\text{‘}P$, when these series are taken in the order generated by P, must be a relation which holds between x and y whenever either (1) x and y both belong to the field of one of the series, and x precedes y in this series, or (2) x belongs to the field of an earlier series than that to which y belongs. In the first case, we have $(\exists Q) . Q \in C\text{‘}P . xQy$, *i.e.* $x \,(\dot{s}\text{‘}C\text{‘}P)\, y$. In the second case, we have $(\exists Q, R) . QPR . x \in C\text{‘}Q . y \in C\text{‘}R$, *i.e.* $(\exists Q, R) . QPR . xFQ . yFR$, *i.e.* $x \,(F;P)\, y$. Hence the generating relation of the sum of all the series is $\dot{s}\text{‘}C\text{‘}P \cup F;P$. Hence we put

$$\Sigma\text{‘}P = \dot{s}\text{‘}C\text{‘}P \cup F;P \quad \text{Df.}$$

The relation $\Sigma\text{‘}P$ has all the properties which we should expect of the sum of a series of series.

If a series is to result from the addition of a series of series, it is necessary that no two of the series should have any common terms. For if we have

$$QPR . x \in C\text{‘}Q \cap C\text{‘}R,$$

we shall also have

$$x \,(\Sigma\text{‘}P)\, x.$$

Hence instead of a series, we shall have cycles; for it is essential to a series that no term should precede itself. (What seem to be series in which there is repetition are always the result of a one-many correlation with series in which there is no repetition, so that a term can be counted once as the correlate of one term, and again as the correlate of a later term.) For this reason, as well as for many others, it is important to consider relations between mutually exclusive relations, *i.e.* between relations whose fields have no common terms. We put

$$\text{Rel}^2\,\text{excl} = \hat{P}\,\{Q, R \in C\text{‘}P . Q \neq R . \supset_{Q,R} . C\text{‘}Q \cap C\text{‘}R = \Lambda\} \quad \text{Df.}$$

Then Rel^2 excl has much the same utility in relation-arithmetic as Cls^3 excl has in cardinal arithmetic. We have

$$\vdash : R \in \text{Rel}^2\,\text{excl} . \equiv . F \upharpoonright C\text{‘}P \in \text{Cls} \rightarrow 1,$$

which is analogous to the proposition (*84·14)

$$\vdash : \kappa \in \text{Cls}^2\,\text{excl} . \equiv . \epsilon \upharpoonright \kappa \in \text{Cls} \rightarrow 1.$$

It will be found that in relation-arithmetic the relation F often appears where ϵ appears in the analogous proposition of cardinal arithmetic.

Analogous to "sm sm" is the relation of double ordinal similarity. This holds between two relations P and Q when they are ordinally similar relations between ordinally similar relations with known correlators, *i.e.* when, if T is an ordinal correlator of P and Q, so that $P = T ; Q$, then if X is a member of C'P, and Y is the corresponding member of C'Q, so that XTY, we shall have X smor Y, and shall be able to specify a member of $X\ \overline{\text{smor}}\ Y$. But as in cardinals, so here, we have to frame our definition of double ordinal similarity in such a way as to minimize the use of the multiplicative axiom. We therefore take as our definition the following: P and Q are said to have double ordinal similarity when there is a one-one relation S which has C'Σ'Q for its converse domain, and is such that $P = S\dagger ; Q$. A relation S which has these properties is called a *double correlator* of P and Q, *i.e.* we put

$$P\ \overline{\text{smor}}\ \overline{\text{smor}}\ Q = (1 \rightarrow 1) \cap \overleftarrow{\text{C}}\text{'}C\text{'}\Sigma\text{'}Q \cap \hat{S}\,(P = S\dagger ; Q) \quad \text{Df},$$

a definition which, as will be perceived, is closely analogous to that of $\kappa\ \overline{\text{sm}}\ \overline{\text{sm}}\ \lambda$ in *111. Two relations have double similarity when they have a double correlator, *i.e.*

$$\text{smor smor} = \hat{P}\hat{Q}\,\{\exists\,!\, P\ \overline{\text{smor}}\ \overline{\text{smor}}\ Q\} \quad \text{Df}.$$

S is a double correlator of P and Q when S is a correlator of Σ'P and Σ'Q and $S\dagger \upharpoonright C$'$Q$ is a correlator of P and Q. This might be taken as the definition of a double correlator, since it is equivalent to the above definition.

If we assume the multiplicative axiom, we can prove that double similarity holds between similar relations of mutually exclusive similar relations, *i.e.* between two relations of mutually exclusive relations P and Q which have a correlator S such that, if $Y \epsilon C$'Q, then Y and S'Y are always similar. In this case, $S \subset$ smor. Thus if we assume the multiplicative axiom we have, if $P, Q \epsilon \text{Rel}^2\ \text{excl}$,

$$P \text{ smor smor } Q \,.\equiv.\, \exists\,!\, P\ \overline{\text{smor}}\ Q \cap \text{Rl'smor}.$$

In the particular case in which the fields of P and Q consist of *well-ordered* relations (*i.e.* relations generating well-ordered series), this equivalence can be proved without the use of the multiplicative axiom, because two similar well-ordered relations have only one correlator, so that the difficulty of selecting among correlators does not arise.

Double ordinal correlators have the same importance in proving the formal laws of relation-arithmetic that double cardinal correlators have in cardinal arithmetic. The construction of double correlators in various cases constitutes a large part of relation-arithmetic.

In defining the ordinal product of two relation-numbers, and in defining exponentiation, we use a relation which has properties analogous to those of

$\alpha \underset{,,}{\downarrow} \text{“}\beta$. This relation is $P \underset{\cdot\cdot}{\downarrow} ;Q$, of which the structure is as follows: Let z, w be two terms having the relation Q; then form the two relations $\downarrow z;P$, $\downarrow w;P$. The relation $\downarrow z;P$ holds between two couples $x \downarrow z$ and $y \downarrow z$ whenever xPy; thus it arranges couples whose referents are members of $C\text{‘}P$, and whose relata are z, in an order similar to P. The relations $\downarrow z;P$ and $\downarrow w;P$ are (by $*150{\cdot}03$) the same as $P \underset{\cdot\cdot}{\downarrow} z$ and $P \underset{\cdot\cdot}{\downarrow} w$. Thus $P \underset{\cdot\cdot}{\downarrow} ;Q$ arranges such relations as $\downarrow z;P$ in an order similar to Q. Thus $P \underset{\cdot\cdot}{\downarrow} ;Q$ is similar to Q, and every member of its field is similar to P. Thus the relation-number of $P \underset{\cdot\cdot}{\downarrow} ;Q$ is $\text{Nr}\text{‘}Q$, and every member of its field has the relation-number $\text{Nr}\text{‘}P$. Moreover $P \underset{\cdot\cdot}{\downarrow} ;Q$, as it is easy to see, is a relation of mutually exclusive relations. Hence it is suitable for defining the product of Q and P, and we put

$$Q \times P = \Sigma\text{‘}P \underset{\cdot\cdot}{\downarrow} ;Q \quad \text{Df.}$$

In the next section, after we have defined the product of a relation of relations, we shall use the same relation $P \underset{\cdot\cdot}{\downarrow} ;Q$ for the definition of exponentiation, putting

$$P \exp Q = \text{Prod}\text{‘}P \underset{\cdot\cdot}{\downarrow} ;Q \quad \text{Df.}$$

These two definitions should be compared with those in $*113$ and $*116$.

In virtue of the definition of Σ, the relation $\Sigma\text{‘}P \underset{\cdot\cdot}{\downarrow} ;Q$ holds between terms which either have one of the relations of the form $P \underset{\cdot\cdot}{\downarrow} z$, or belong respectively to the fields of two relations $P \underset{\cdot\cdot}{\downarrow} z$, $P \underset{\cdot\cdot}{\downarrow} w$, where zQw. Thus the relation $\Sigma\text{‘}P \underset{\cdot\cdot}{\downarrow} ;Q$ holds between $x \downarrow z$ and $y \downarrow z$ whenever xPy and $z \in C\text{‘}Q$, and also between $x \downarrow z$ and $y \downarrow w$ whenever $x, y \in C\text{‘}P \,.\, zQw$. Thus if, for the sake of illustration, P and Q generate finite series, so that their fields are

$$1_P,\ 2_P,\ \ldots,\ \mu_P,$$
$$1_Q,\ 2_Q,\ \ldots,\ \nu_Q,$$

then the field of $\Sigma\text{‘}P \underset{\cdot\cdot}{\downarrow} ;Q$ will consist of the couples

$$1_P \downarrow 1_Q,\ 2_P \downarrow 1_Q,\ \ldots,\ \mu_P \downarrow 1_Q;$$
$$1_P \downarrow 2_Q,\ 2_P \downarrow 2_Q,\ \ldots,\ \mu_P \downarrow 2_Q;$$
$$\ldots\ldots\ldots\ldots\ldots\ldots\ldots\ldots\ldots\ldots$$
$$1_P \downarrow \nu_Q,\ 2_P \downarrow \nu_Q,\ \ldots,\ \mu_P \downarrow \nu_Q;$$

and their order as arranged by $\Sigma\text{‘}P \underset{\cdot\cdot}{\downarrow} ;Q$ is that in which they are written above. Thus the above couples in the above order constitute the series $Q \times P$, and it is evident that this series has $\nu \times \mu$ terms.

When the factors of a product are not enumerated, but are given as the field of a relation, a new definition of multiplication is required. This definition, which has the advantage of being applicable to infinite products, will be dealt with in the following section.

$*160$. THE SUM OF TWO RELATIONS.

Summary of $*160$.

In this number, we introduce the definition

$$P \dotplus Q = P \cup Q \cup C\text{‘}P \uparrow C\text{‘}Q \quad \text{Df},$$

which was explained in the introduction to this section. Although the propositions of this and other numbers in this Part do not require that P and Q should be such as to generate series, yet the reader will find it convenient to imagine them to be such, since the important applications of the ideas of this Part are to series. Thus we may regard the sum of P and Q as a relation which holds between x and y when either x precedes y in the P-series, or x precedes y in the Q-series, or x belongs to the P-series and y belongs to the Q-series.

The most important propositions of this number are:

$*160\cdot14$. $\vdash . C\text{‘}(P \dotplus Q) = C\text{‘}P \cup C\text{‘}Q$

$*160\cdot21$. $\vdash . P \dotplus \dot{\Lambda} = P$

$*160\cdot22$. $\vdash . \dot{\Lambda} \dotplus Q = Q$

$*160\cdot31$. $\vdash . (P \dotplus Q) \dotplus R = P \dotplus (Q \dotplus R)$

which is the associative law, and

$*160\cdot4$. $\vdash . (P \cup Q) \dotplus R = (P \dotplus R) \cup (Q \dotplus R)$

which is the distributive law for logical and arithmetical addition;

$*160\cdot44$. $\vdash : C\text{‘}P \subset \text{G}\text{‘}S . C\text{‘}Q \subset \text{G}\text{‘}S . \supset . S ; (P \dotplus Q) = S ; P \dotplus S ; Q$

which is also a kind of distributive law;

$*160\cdot47$. $\vdash : C\text{‘}P \cap C\text{‘}Q = \Lambda . C\text{‘}P' \cap C\text{‘}Q' = \Lambda . S \in P \,\overline{\text{smor}}\, P' . T \in Q \,\overline{\text{smor}}\, Q' . \supset .$
$S \cup T \in (P \dotplus Q) \,\overline{\text{smor}}\, (P' \dotplus Q')$

whence

$*160\cdot48$. $\vdash : C\text{‘}P \cap C\text{‘}Q = \Lambda . C\text{‘}P' \cap C\text{‘}Q' = \Lambda . P \,\text{smor}\, P' . Q \,\text{smor}\, Q' . \supset .$
$P \dotplus Q \,\text{smor}\, P' \dotplus Q'$

whence it follows that if P and Q are mutually exclusive, the relation-number of their sum depends only upon the relation-numbers of P and Q;

$*160\cdot5$. $\vdash : C\text{‘}P \cap C\text{‘}Q = \Lambda . \supset . (P \dotplus Q) \upharpoonright C\text{‘}P = P . (P \dotplus Q) \upharpoonright C\text{‘}Q = Q$

$*160\cdot52$. $\vdash : C\text{‘}P \cap C\text{‘}Q = \Lambda . C\text{‘}P \cap C\text{‘}R = \Lambda . P \dotplus Q = P \dotplus R . \supset . Q = R$

***160·01.** $P \dotplus Q = P \mathbin{\dot\cup} Q \mathbin{\dot\cup} C\text{‘}P \uparrow C\text{‘}Q$ Df

***160·1.** $\vdash . P \dotplus Q = P \mathbin{\dot\cup} Q \mathbin{\dot\cup} C\text{‘}P \uparrow C\text{‘}Q$ [(*160·01)]

***160·11.** $\vdash :. x (P \dotplus Q) y . \equiv : xPy . \vee . xQy . \vee . x \in C\text{‘}P . y \in C\text{‘}Q$ [*160·1]

***160·111.** $\vdash :. x (P \dotplus Q) y . \equiv : xPy . \vee . xQy . \vee . xFP . yFQ$ [*160·11 . *33·51]

***160·12.** $\vdash : \exists ! Q . \supset . \mathrm{D}\text{‘}(P \dotplus Q) = C\text{‘}P \cup \mathrm{D}\text{‘}Q$ [*33·26 . *35·85 . *160·1]

***160·13.** $\vdash : \exists ! P . \supset . \text{ꓷ}\text{‘}(P \dotplus Q) = \text{ꓷ}\text{‘}P \cup C\text{‘}Q$

***160·14.** $\vdash . C\text{‘}(P \dotplus Q) = C\text{‘}P \cup C\text{‘}Q$

Dem.

$\vdash . *33\cdot262 . *160\cdot1 . \supset \vdash . C\text{‘}(P \dotplus Q) = C\text{‘}P \cup C\text{‘}Q \cup C\text{‘}(C\text{‘}P \uparrow C\text{‘}Q)$ (1)

$\vdash . *35\cdot85\cdot86\cdot88 . \supset \vdash . C\text{‘}(C\text{‘}P \uparrow C\text{‘}Q) \subset C\text{‘}P \cup C\text{‘}Q$ (2)

$\vdash . (1) . (2) . \supset \vdash . \text{Prop}$

The above proposition is constantly used. The following propositions (*160·15—·161) are not used, but are inserted to show that $P \dotplus Q$ has the kind of structure that we should expect of a sum.

***160·15.** $\vdash : \exists ! P . \supset . \overrightarrow{B}\text{‘}(P \dotplus Q) = \overrightarrow{B}\text{‘}P - C\text{‘}Q$

Dem.

$\vdash . *160\cdot12\cdot13 . \supset \vdash : \exists ! P . \exists ! Q . \supset . \overrightarrow{B}\text{‘}(P \dotplus Q) = (C\text{‘}P \cup \mathrm{D}\text{‘}Q) - (\text{ꓷ}\text{‘}P \cup C\text{‘}Q)$

[*93·101 . *33·161] $= \overrightarrow{B}\text{‘}P - C\text{‘}Q$ (1)

$\vdash . *160\cdot1 . \supset \vdash : Q = \dot{\Lambda} . \supset . P \dotplus Q = P .$

[*30·37] $\supset . \overrightarrow{B}\text{‘}(P \dotplus Q) = \overrightarrow{B}\text{‘}P$

[*33·241] $= \overrightarrow{B}\text{‘}P - C\text{‘}Q$ (2)

$\vdash . (1) . (2) . \supset \vdash . \text{Prop}$

***160·151.** $\vdash : \exists ! Q . \supset . \overrightarrow{B}\text{‘}\mathrm{Cnv}\text{‘}(P \dotplus Q) = \overrightarrow{B}\text{‘}\breve{Q} - C\text{‘}P$

***160·16.** $\vdash : \exists ! P . \overrightarrow{B}\text{‘}P \cap C\text{‘}Q = \Lambda . \supset . \overrightarrow{B}\text{‘}(P \dotplus Q) = \overrightarrow{B}\text{‘}P$ [*160·15]

***160·161.** $\vdash : \exists ! Q . \overrightarrow{B}\text{‘}\breve{Q} \cap C\text{‘}P = \Lambda . \supset . \overrightarrow{B}\text{‘}\mathrm{Cnv}\text{‘}(P \dotplus Q) = \overrightarrow{B}\text{‘}\breve{Q}$

***160·2.** $\vdash . \mathrm{Cnv}\text{‘}(P \dotplus Q) = \breve{Q} \dotplus \breve{P}$ [*31·15 . *35·84]

***160·21.** $\vdash . P \dotplus \dot{\Lambda} = P$ [*35·88 . *25·24]

***160·22.** $\vdash . \dot{\Lambda} \dotplus Q = Q$

***160·3.** $\vdash . (P \dotplus Q) \dotplus R = P \mathbin{\dot\cup} Q \mathbin{\dot\cup} R \mathbin{\dot\cup} C\text{‘}P \uparrow C\text{‘}Q \mathbin{\dot\cup} C\text{‘}P \uparrow C\text{‘}R \mathbin{\dot\cup} C\text{‘}Q \uparrow C\text{‘}R$

Dem.

$\vdash . *160\cdot14\cdot1 . \supset$

$\vdash . (P \dotplus Q) \dotplus R = (P \dotplus Q) \mathbin{\dot\cup} R \mathbin{\dot\cup} (C\text{‘}P \cup C\text{‘}Q) \uparrow C\text{‘}R$

[*160·1 . *35·41·82] $= P \mathbin{\dot\cup} Q \mathbin{\dot\cup} C\text{‘}P \uparrow C\text{‘}Q \mathbin{\dot\cup} R \mathbin{\dot\cup} C\text{‘}P \uparrow C\text{‘}R \mathbin{\dot\cup} C\text{‘}Q \uparrow C\text{‘}R . \supset \vdash . \text{Prop}$

***160·31.** $\vdash . (P \updownarrow Q) \updownarrow R = P \updownarrow (Q \updownarrow R)$

Dem.

$\vdash . *160·14·1 . \supset$

$\vdash . P \updownarrow (Q \updownarrow R) = P \cup Q \cup R \cup C`P \uparrow C`Q \cup C`P \uparrow C`R \cup C`Q \uparrow C`R$ (1)

$\vdash . (1) . *160·3 . \supset \vdash .$ Prop

***160·32.** $P \updownarrow Q \updownarrow R = (P \updownarrow Q) \updownarrow R$ Df

This definition serves merely for the avoidance of brackets.

***160·33.** $\vdash : P \subseteq Q . \supset . P \updownarrow R \subseteq Q \updownarrow R$ [*33·265 . *160·1]

***160·34.** $\vdash : R \subseteq S . \supset . Q \updownarrow R \subseteq Q \updownarrow S$ [*33·265 . *160·1]

***160·35.** $\vdash : P \subseteq Q . R \subseteq S . \supset . P \updownarrow Q \subseteq R \updownarrow S$ [*160·33·34]

***160·4.** $\vdash . (P \cup Q) \updownarrow R = (P \updownarrow R) \cup (Q \updownarrow R)$

Dem.

$\vdash . *160·1 . \supset \vdash . (P \cup Q) \updownarrow R = P \cup Q \cup R \cup C`(P \cup Q) \uparrow C`R$

[*33·262.*23·56] $= P \cup R \cup Q \cup R \cup (C`P \cup C`Q) \uparrow C`R$

[*35·41·82] $= P \cup R \cup Q \cup R \cup C`P \uparrow C`R \cup C`Q \uparrow C`R$

[*160·1] $= (P \updownarrow R) \cup (Q \updownarrow R) . \supset \vdash .$ Prop

***160·401.** $\vdash . P \updownarrow (Q \cup R) = (P \updownarrow Q) \cup (P \updownarrow R)$

The above two propositions state the distributive law for logical and arithmetical addition. The three following propositions give the generalized form of this law, when $\dot{s}`\lambda$ replaces $P \cup Q$; these propositions are not subsequently used but are inserted for the sake of their intrinsic interest.

***160·41.** $\vdash : \exists ! \lambda . \supset . \dot{s}`\lambda \updownarrow R = \dot{s}` \updownarrow R``\lambda = \dot{s}`(\lambda \underset{,,}{\updownarrow} R)$

Dem.

$\vdash . *41·11 . \supset \vdash :. x (\dot{s}` \updownarrow R``\lambda) y . \equiv : (\exists P) . P \epsilon \lambda . x (P \updownarrow R) y :$

[*160·11] $\equiv : (\exists P) : P \epsilon \lambda : xPy . \vee . xRy . \vee . x \epsilon C`P . y \epsilon C`R :$

[*10·42] $\equiv : (\exists P) . P \epsilon \lambda . xPy . \vee . (\exists P) . P \epsilon \lambda . xRy . \vee .$
$(\exists P) . P \epsilon \lambda . x \epsilon C`P . y \epsilon C`R :$

[*41·11.*10·35.*41·45] $\equiv : x (\dot{s}`\lambda) y . \vee . \exists ! \lambda . xRy . \vee . x \epsilon C`\dot{s}`\lambda . y \epsilon C`R$ (1)

$\vdash . (1) . \supset \vdash :: \text{Hp} . \supset :. x (\dot{s}` \updownarrow R``\lambda) . \equiv : x (\dot{s}`\lambda) y . \vee . xRy . \vee . x \epsilon C`\dot{s}`\lambda . y \epsilon C`R :$

[*160·11] $\equiv : x (\dot{s}`\lambda \updownarrow R) y :: \supset \vdash .$ Prop

***160·411.** $\vdash : \exists ! \lambda . \supset . P \updownarrow \dot{s}`\lambda = \dot{s}`P \updownarrow ``\lambda$ [Proof as in *160·41]

***160·412.** $\vdash : \exists ! \lambda . \exists ! \mu . \supset . \dot{s}`\lambda \updownarrow \dot{s}`\mu = \dot{s}`s`\lambda \underset{,,}{\updownarrow} ``\mu$

Dem.

$\vdash . *160·411 . \supset \vdash : \exists ! \mu . \supset . \dot{s}`\lambda \updownarrow \dot{s}`\mu = \dot{s}`(\dot{s}`\lambda) \updownarrow ``\mu$ (1)

$\vdash . *160·41 . \supset \vdash : \exists ! \lambda . \supset . (\dot{s}`\lambda) \updownarrow ``\mu = \dot{s}``\lambda \underset{,,}{\updownarrow} ``\mu$ (2)

$\vdash . (1) . (2) . \supset \vdash : \exists ! \lambda . \exists ! \mu . \supset . \dot{s}`\lambda \updownarrow \dot{s}`\mu = \dot{s}`\dot{s}``\lambda \underset{,,}{\updownarrow} ``\mu$

[*42·12] $= \dot{s}`s`\lambda \underset{,,}{\updownarrow} ``\mu : \supset \vdash .$ Prop

The following propositions lead up to $*160{\cdot}44$, which is frequently used.

$*160{\cdot}42$. $\vdash . (P \ddagger Q) \mid S = P \mid S \cup Q \mid S \cup C\text{‘}P \uparrow \breve{S}\text{“}C\text{‘}Q$

Dem.

$\vdash . *160{\cdot}1 . \supset \vdash . (P \ddagger Q) \mid S = P \mid S \cup Q \mid S \cup (C\text{‘}P \uparrow C\text{‘}Q) \mid S$

$[*37{\cdot}8] \qquad = P \mid S \cup Q \mid S \cup C\text{‘}P \uparrow \breve{S}\text{“}C\text{‘}Q . \supset \vdash . \text{Prop}$

$*160{\cdot}421$. $\vdash . S \mid (P \ddagger Q) = S \mid P \cup S \mid Q \cup S\text{“}C\text{‘}P \uparrow C\text{‘}Q$

$*160{\cdot}43$. $\vdash . S;(P \ddagger Q) = S;P \cup S;Q \cup S\text{“}C\text{‘}P \uparrow S\text{“}C\text{‘}Q$

Dem.

$\vdash . *150{\cdot}1 . *160{\cdot}421 . \supset$

$\vdash . S;(P \ddagger Q) \quad = (S \mid P \cup S \mid Q \cup S\text{“}C\text{‘}P \uparrow C\text{‘}Q) \mid \breve{S}$

$[*150{\cdot}1 . *37{\cdot}8] = S;P \cup S;Q \cup S\text{“}C\text{‘}P \uparrow S\text{“}C\text{‘}Q . \supset \vdash . \text{Prop}$

$*160{\cdot}44$. $\vdash : C\text{‘}P \subset \text{Ɑ}\text{‘}S . C\text{‘}Q \subset \text{Ɑ}\text{‘}S . \supset . S;(P \ddagger Q) = S;P \ddagger S;Q$

Dem.

$\vdash . *160{\cdot}43 . *150{\cdot}22 . \supset$

$\vdash : \text{Hp} . \supset . S;(P \ddagger Q) = S;P \cup S;Q \cup (C\text{‘}S;P) \uparrow (C\text{‘}S;Q)$

$[*160{\cdot}1] \qquad = S;P \ddagger S;Q : \supset \vdash . \text{Prop}$

$*160{\cdot}45$. $\vdash : S \upharpoonright (C\text{‘}P' \cup C\text{‘}Q') \in 1 \rightarrow 1 . S \upharpoonright C\text{‘}P' \in P \,\overline{\text{smor}}\, P' . S \upharpoonright C\text{‘}Q' \in Q \,\overline{\text{smor}}\, Q' . \supset .$
$S \upharpoonright C\text{‘}(P' \ddagger Q') \in (P \ddagger Q) \,\overline{\text{smor}}\, (P' \ddagger Q')$

Dem.

$\vdash . *151{\cdot}22 . \supset \vdash : \text{Hp} . \supset . C\text{‘}P' \subset \text{Ɑ}\text{‘}S . C\text{‘}Q' \subset \text{Ɑ}\text{‘}S . P = S;P' . Q = S;Q' . \quad (1)$

$[*160{\cdot}44] \qquad \supset . P \ddagger Q = S;(P' \ddagger Q') \quad (2)$

$\vdash . (1) . *160{\cdot}14 . \supset \vdash : \text{Hp} . \supset . C\text{‘}(P' \ddagger Q') \subset \text{Ɑ}\text{‘}S \quad (3)$

$\vdash . *160{\cdot}14 . \quad \supset \vdash : \text{Hp} . \supset . S \upharpoonright C\text{‘}(P' \ddagger Q') \in 1 \rightarrow 1 \quad (4)$

$\vdash . (2) . (3) . (4) . *151{\cdot}22 . \supset \vdash . \text{Prop}$

$*160{\cdot}451$. $\vdash : S \upharpoonright C\text{‘}P' \in P \,\overline{\text{smor}}\, P' . S \upharpoonright C\text{‘}Q' \in Q \,\overline{\text{smor}}\, Q' . S\text{“}(C\text{‘}P' - C\text{‘}Q') \cap C\text{‘}Q = \Lambda .$
$\supset . S \upharpoonright C\text{‘}(P' \ddagger Q') \in (P \ddagger Q) \,\overline{\text{smor}}\, (P' \ddagger Q')$

Dem.

$\vdash . *151{\cdot}22 . *150{\cdot}22 . \supset \vdash : \text{Hp} . \supset . C\text{‘}Q = S\text{“}C\text{‘}Q' .$

$[*71{\cdot}381 . *37{\cdot}421] \qquad \supset . S\text{“}(C\text{‘}P' - C\text{‘}Q') \cap S\text{“}C\text{‘}Q' = \Lambda .$

$[*74{\cdot}823] \qquad \supset . S \upharpoonright (C\text{‘}P' \cup C\text{‘}Q') \in 1 \rightarrow 1 \quad (1)$

$\vdash . (1) . *160{\cdot}45 . \supset \vdash . \text{Prop}$

$*160{\cdot}452$. $\vdash : S \upharpoonright C\text{‘}P' \in P \,\overline{\text{smor}}\, P' . S \upharpoonright C\text{‘}Q' \in Q \,\overline{\text{smor}}\, Q' . C\text{‘}P \cap C\text{‘}Q = \Lambda . \supset .$
$S \upharpoonright C\text{‘}(P' \ddagger Q') \in (P \ddagger Q) \,\overline{\text{smor}}\, (P' \ddagger Q')$

Dem.

$\vdash . *151{\cdot}22 . *150{\cdot}22 . \supset \vdash : \text{Hp} . \supset . C\text{‘}P = S\text{“}C\text{‘}P' . C\text{‘}Q = S\text{“}C\text{‘}Q' .$

$[\text{Hp}] \qquad \supset . S\text{“}C\text{‘}P' \cap S\text{“}C\text{‘}Q' = \Lambda .$

$[*74{\cdot}833] \qquad \supset . S \upharpoonright C\text{‘}(P' \ddagger Q') \in 1 \rightarrow 1 \quad (1)$

$\vdash . (1) . *160{\cdot}45 . \supset \vdash . \text{Prop}$

$*160 \cdot 46$. $\vdash : C`P = ᗡ`S \,.\, C`Q = ᗡ`T \,.\, C`P \cap C`Q = \Lambda \,.\supset .$
$$(S \cup T);(P ⤉ Q) = S;P ⤉ T;Q$$

Dem.

$\vdash . *160 \cdot 44 . \supset \vdash : \text{Hp} . \supset . (S \cup T);(P ⤉ Q) = (S \cup T);P ⤉ (S \cup T);Q$

$[*150 \cdot 32] \qquad = \{(S \cup T) \upharpoonright C`P\};P ⤉ \{(S \cup T) \upharpoonright C`Q\};Q$

$[*35 \cdot 644 . \text{Hp}] \qquad = (S \upharpoonright C`P);P ⤉ (T \upharpoonright C`Q);Q$

$[*150 \cdot 32] \qquad = S;P ⤉ T;Q : \supset \vdash . \text{Prop}$

$*160 \cdot 47$. $\vdash : C`P \cap C`Q = \Lambda \,.\, C`P' \cap C`Q' = \Lambda \,.\, S \in P \, \overline{\text{smor}} \, P' \,.\, T \in Q \, \overline{\text{smor}} \, Q' . \supset .$
$$S \cup T \in (P ⤉ Q) \, \overline{\text{smor}} \, (P' ⤉ Q')$$

Dem.

$\vdash . *151 \cdot 11 \cdot 131 . \supset \vdash : \text{Hp} . \supset . D`S = C`P \,.\, D`T = C`Q \,.\, ᗡ`S = C`P' \,.$
$ᗡ`T = C`Q' . \qquad (1)$

$[\text{Hp}] \qquad \supset . D`S \cap D`T = \Lambda \,.\, ᗡ`S \cap ᗡ`T = \Lambda \,.$

$[*151 \cdot 11 . *71 \cdot 242] \qquad \supset . S \cup T \in 1 \to 1 \qquad (2)$

$\vdash . (1) . *160 \cdot 14 . \supset \vdash : \text{Hp} . \supset . C`(P' ⤉ Q') = ᗡ`S \cup ᗡ`T$

$[*33 \cdot 261] \qquad = ᗡ`(S \cup T) \qquad (3)$

$\vdash . *160 \cdot 46 . *151 \cdot 11 . \supset \vdash : \text{Hp} . \supset . P ⤉ Q = (S \cup T);(P' ⤉ Q') \qquad (4)$

$\vdash . (2) . (3) . (4) . *151 \cdot 11 . \supset \vdash . \text{Prop}$

$*160 \cdot 48$. $\vdash : C`P \cap C`Q = \Lambda \,.\, C`P' \cap C`Q' = \Lambda \,.\, P \,\text{smor}\, P' \,.\, Q \,\text{smor}\, Q' . \supset .$
$$P ⤉ Q \,\text{smor}\, P' ⤉ Q' \quad [*160 \cdot 47 . *151 \cdot 12]$$

$*160 \cdot 5$. $\vdash : C`P \cap C`Q = \Lambda . \supset . (P ⤉ Q) \sqsubset C`P = P \,.\, (P ⤉ Q) \sqsubset C`Q = Q$

Dem.

$\vdash . *160 \cdot 1 . *36 \cdot 23 . \supset$

$\vdash . (P ⤉ Q) \sqsubset C`P = P \sqsubset C`P \cup (C`P \uparrow C`Q) \sqsubset C`P \cup Q \sqsubset C`P$

$[*36 \cdot 29 \cdot 33] \qquad = P \cup \{(C`P \uparrow C`Q) \dot{\cap} (C`P \uparrow C`P)\} \cup Q \sqsubset C`P \qquad (1)$

$\vdash . *36 \cdot 31 . \qquad \supset \vdash : \text{Hp} . \supset . Q \sqsubset C`P = \dot{\Lambda} \qquad (2)$

$\vdash . *35 \cdot 834 \cdot 88 . \supset \vdash : \text{Hp} . \supset . \{(C`P \uparrow C`Q) \dot{\cap} (C`P \uparrow C`P)\} = \dot{\Lambda} \qquad (3)$

$\vdash . (1) . (2) . (3) . \supset \vdash : \text{Hp} . \supset . (P ⤉ Q) \sqsubset C`P = P \qquad (4)$

Similarly $\qquad \vdash : \text{Hp} . \supset . (P ⤉ Q) \sqsubset C`Q = Q \qquad (5)$

$\vdash . (4) . (5) . \supset \vdash . \text{Prop}$

$*160 \cdot 51$. $\vdash : C`P \cap C`Q = \Lambda . \supset . (P ⤉ Q)^2 = P^2 \cup Q^2 \cup D`P \uparrow C`Q \cup C`P \uparrow ᗡ`Q$

Dem.

$\vdash . *34 \cdot 73 . \quad \supset \vdash : \text{Hp} . \supset . (P \cup Q)^2 = P^2 \cup Q^2 \qquad (1)$

$\vdash . *35 \cdot 895 . \supset \vdash : \text{Hp} . \supset . (C`P \uparrow C`Q)^2 = \dot{\Lambda} \qquad (2)$

$\vdash . *34 \cdot 62 . \quad \supset \vdash . (P ⤉ Q)^2 = (P \cup Q)^2 \cup (C`P \uparrow C`Q)^2$
$\cup (P \cup Q) \,|\, (C`P \uparrow C`Q) \cup (C`P \uparrow C`Q) \,|\, (P \cup Q)$

$[(1).(2)] \qquad = P^2 \cup Q^2 \cup (P \cup Q) \,|\, (C`P \uparrow C`Q) \cup (C`P \uparrow C`Q) \,|\, (P \cup Q) \qquad (3)$

$\vdash . *37 \cdot 81 . \quad \supset \vdash : \text{Hp} . \supset . (P \cup Q) \,|\, (C`P \uparrow C`Q) = D`P \uparrow C`Q \qquad (4)$

$\vdash . *37 \cdot 8 . \quad \supset \vdash : \text{Hp} . \supset . (C`P \uparrow C`Q) \,|\, (P \cup Q) = C`P \uparrow ᗡ`Q \qquad (5)$

$\vdash . (3) . (4) . (5) . \supset \vdash . \text{Prop}$

The above proposition is useful in proving that, if $C'P \cap C'Q = \Lambda$, $P \ddagger Q$ is transitive when P and Q are transitive (cf. *201·4).

***160·52.** $\vdash : C'P \cap C'Q = \Lambda \,.\, C'P \cap C'R = \Lambda \,.\, P \ddagger Q = P \ddagger R \,.\supset .\, Q = R$

Dem.

$\vdash . *160{\cdot}14 . \supset \vdash : \text{Hp} . \supset . (P \ddagger Q) \upharpoonright (-C'P) = (P \ddagger Q) \upharpoonright C'Q \,.$

$(P \ddagger Q) \upharpoonright (-C'P) = (P \ddagger R) \upharpoonright C'R \,.$

[*160·5] $\supset . (P \ddagger Q) \upharpoonright (-C'P) = Q \,.\, (P \ddagger R) \upharpoonright (-C'P) = R \,.$

[Hp] $\supset . Q = R : \supset \vdash . \text{Prop}$

The above proposition is used in dealing with the series of segments of a series (*213·561).

$*161$. ADDITION OF A TERM TO A RELATION.

Summary of $*161$.

The addition of a term has two forms, according as it occurs at the beginning or end of the field of the relation in question. If we add first x and then y at the end, the result is the same as if we added $x \downarrow y$ ($*161\cdot22$); if at the beginning, it is the same as if we added $y \downarrow x$ ($*161\cdot221$). The propositions of the present number are all obvious, and offer no difficulties of any kind. As explained in the introduction to this section, we put

$$P \nrightarrow x = P \cup C`P \uparrow \iota`x \quad \text{Df,}$$

$$x \nleftarrow P = \iota`x \uparrow C`P \cup P \quad \text{Df.}$$

Most of the propositions of this number require the hypothesis $\dot{\exists} ! P$, because if $P = \dot{\Lambda}$, $P \nrightarrow x = x \nleftarrow P = \dot{\Lambda}$ ($*161\cdot2\cdot201$). This is connected with the fact that there is no ordinal number 1. Apart from propositions already mentioned, the chief propositions of this number are the following (we omit propositions about $x \nleftarrow P$ when they are merely analogues of propositions about $P \nrightarrow x$):

$*161\cdot12$. $\vdash . x \nleftarrow P = \text{Cnv}`(\breve{P} \nrightarrow x)$

$*161\cdot14$. $\vdash : \dot{\exists} ! P . \supset . C`(P \nrightarrow x) = C`P \cup \iota`x = C`(x \nleftarrow P)$

$*161\cdot15$. $\vdash : \dot{\exists} ! P . x \sim\epsilon C`P . \supset .$

$$\overrightarrow{B}`(P \nrightarrow x) = \overrightarrow{B}`P . \overrightarrow{B}`\text{Cnv}`(P \nrightarrow x) = \iota`x . B`(x \nleftarrow \breve{P}) = x$$

$*161\cdot211$. $\vdash . x \nleftarrow (y \downarrow z) = x \downarrow y \cup x \downarrow z \cup y \downarrow z = (x \downarrow y) \nrightarrow z$

$*161\cdot31$. $\vdash : P \,\text{smor}\, Q . x \sim\epsilon C`P . y \sim\epsilon C`Q . \supset .$

$$P \nrightarrow x \,\text{smor}\, Q \nrightarrow y . x \nleftarrow P \,\text{smor}\, y \nleftarrow Q$$

$*161\cdot4$. $\vdash : C`Q \subset \text{G}`S . x \,\epsilon\, \text{G}`S . S \,\epsilon\, 1 \rightarrow \text{Cls} . \supset . S;(Q \nrightarrow x) = S;Q \nrightarrow S`x$

$*161\cdot01$. $P \nrightarrow x = P \cup C`P \uparrow \iota`x$ Df

$*161\cdot02$. $x \nleftarrow P = \iota`x \uparrow C`P \cup P$ Df

$*161\cdot1$. $\vdash . P \nrightarrow x = P \cup C`P \uparrow \iota`x$ [($*161\cdot01$)]

$*161\cdot101$. $\vdash . x \nleftarrow P = \iota`x \uparrow C`P \cup P$ [($*161\cdot02$)]

$*161\cdot11$. $\vdash :. y (P \nrightarrow x) z . \equiv : yPz . \vee . y \,\epsilon\, C`P . z = x$ [$*161\cdot1$]

***161·111.** $\vdash :.\, y\,(x \nleftarrow P)\,z \,.\equiv :\, y = x \,.\, z \,\epsilon\, C‘P \,.\vee .\, yPz$ [*161·101]

***161·12.** $\vdash .\, x \nleftarrow P = \text{Cnv}‘(\breve{P} \nrightarrow x)$ [*161·1·101 . *35·84 . *33·22]

***161·13.** $\vdash .\, \text{D}‘(P \nrightarrow x) = C‘P \,.\, \text{Ɑ}‘(x \nleftarrow P) = C‘P$

Dem.

$\vdash .\, *161·1 \,.\supset \vdash .\, \text{D}‘(P \nrightarrow x) = \text{D}‘P \cup \text{D}‘(C‘P \uparrow \iota‘x)$

[*35·85] $= \text{D}‘P \cup C‘P$

[*33·161] $= C‘P$ (1)

Similarly $\vdash .\, \text{Ɑ}‘(x \nleftarrow P) = C‘P$ (2)

$\vdash .\, (1) \,.\, (2) \,.\supset \vdash .$ Prop

***161·131.** $\vdash :\, \dot{\exists} \,!\, P \,.\supset .\, \text{Ɑ}‘(P \nrightarrow x) = \text{Ɑ}‘P \cup \iota‘x \,.\, \text{D}‘(x \nleftarrow P) = \text{D}‘P \cup \iota‘x$
[*35·86 . *161·1]

***161·14.** $\vdash :\, \dot{\exists} \,!\, P \,.\supset .\, C‘(P \nrightarrow x) = C‘P \cup \iota‘x = C‘(x \nleftarrow P)$ [*161·13·131]

The hypothesis $\dot{\exists} \,!\, P$ is necessary in this proposition, since without it we have $P \nrightarrow x = \Lambda$.

***161·141.** $\vdash :\, \dot{\exists} \,!\, P \,.\supset .\, \overrightarrow{B}‘(P \nrightarrow x) = \overrightarrow{B}‘P - \iota‘x \,.\, \overrightarrow{B}‘\text{Cnv}‘(P \nrightarrow x) = \iota‘x - C‘P$
[*161·13·131 . *93·101]

***161·15.** $\vdash :\, \dot{\exists} \,!\, P \,.\, x \sim\epsilon\, C‘P \,.\supset .$
$\overrightarrow{B}‘(P \nrightarrow x) = \overrightarrow{B}‘P \,.\, \overrightarrow{B}‘\text{Cnv}‘(P \nrightarrow x) = \iota‘x \,.\, B‘(x \nleftarrow \breve{P}) = x$ [*161·141]

***161·16.** $\vdash :\, x \sim\epsilon\, C‘P \,.\supset .\, (P \nrightarrow x) \upharpoonright C‘P = (P \nrightarrow x) \upharpoonright (-\iota‘x) = P$ [*161·1]

The above proposition is used in the theory of connected relations (*202·412).

***161·161.** $\vdash :\, x \sim\epsilon\, C‘P \,.\supset .\, (x \nleftarrow P) \upharpoonright C‘P = (x \nleftarrow P) \upharpoonright (-\iota‘x) = P$

The two following propositions are frequently used.

***161·2.** $\vdash .\, \dot{\Lambda} \nrightarrow x = \dot{\Lambda}$ [*35·75·82 . *161·1]

***161·201.** $\vdash .\, x \nleftarrow \dot{\Lambda} = \dot{\Lambda}$

***161·21.** $\vdash .\, (x \downarrow y) \nrightarrow z = x \downarrow y \,\dot{\cup}\, x \downarrow z \,\dot{\cup}\, y \downarrow z$

Dem.

$\vdash .\, *161·1 \,.\, *55·15 \,.\supset \vdash .\, (x \downarrow y) \nrightarrow z = x \downarrow y \,\dot{\cup}\, (\iota‘x \cup \iota‘y) \uparrow \iota‘z$

[*35·82·41.*55·1] $= x \downarrow y \,\dot{\cup}\, x \downarrow z \,\dot{\cup}\, y \downarrow z \,.\supset \vdash .$ Prop

Note that $x \downarrow y \,\dot{\cup}\, x \downarrow z \,\dot{\cup}\, y \downarrow z$ is the relation which orders x and y and z in the order x, y, z.

***161·211.** $\vdash .\, x \nleftarrow (y \downarrow z) = x \downarrow y \,\dot{\cup}\, x \downarrow z \,\dot{\cup}\, y \downarrow z = (x \downarrow y) \nrightarrow z$
[Proof as in *161·21]

***161·212.** $P \nrightarrow x \nrightarrow y = (P \nrightarrow x) \nrightarrow y$ Df

***161·213.** $x \nleftarrow y \nleftarrow P = x \nleftarrow (y \nleftarrow P)$ Df

These definitions serve merely for the avoidance of brackets.

$*161{\cdot}22.\quad \vdash : \dot{\exists} ! P . \supset . (P \rightarrowtail x) \rightarrowtail y = P \ddagger (x \downarrow y)$

Dem.

$\vdash . *161{\cdot}14{\cdot}1 . \supset \vdash : \text{Hp} . \supset . (P \rightarrowtail x) \rightarrowtail y = P \mathbin{\dot{\cup}} C`P \uparrow \iota`x \mathbin{\dot{\cup}} (C`P \cup \iota`x) \uparrow \iota`y$

$[*35{\cdot}82{\cdot}41] \qquad = P \mathbin{\dot{\cup}} C`P \uparrow \iota`x \mathbin{\dot{\cup}} C`P \uparrow \iota`y \mathbin{\dot{\cup}} \iota`x \uparrow \iota`y$

$[*35{\cdot}82{\cdot}412] \qquad = P \mathbin{\dot{\cup}} C`P \uparrow (\iota`x \cup \iota`y) \mathbin{\dot{\cup}} \iota`x \uparrow \iota`y$

$[*55{\cdot}1{\cdot}15] \qquad = P \mathbin{\dot{\cup}} C`P \uparrow C`(x \downarrow y) \mathbin{\dot{\cup}} x \downarrow y$

$[*160{\cdot}1] \qquad = P \ddagger (x \downarrow y) : \supset \vdash . \text{Prop}$

$*161{\cdot}221.\quad \vdash : \dot{\exists} ! P . \supset . x \leftarrowtail (y \leftarrowtail P) = (x \downarrow y) \ddagger P$

$*161{\cdot}23.\quad \vdash : \dot{\exists} ! Q . \supset . (P \ddagger Q) \rightarrowtail y = P \ddagger (Q \rightarrowtail y)$

Dem.

$\vdash . *161{\cdot}14{\cdot}1 . *160{\cdot}1 . \supset \vdash : \text{Hp} . \supset .$

$P \ddagger (Q \rightarrowtail y) = P \mathbin{\dot{\cup}} Q \mathbin{\dot{\cup}} C`Q \uparrow \iota`y \mathbin{\dot{\cup}} C`P \uparrow (C`Q \cup \iota`y)$

$[*35{\cdot}82{\cdot}412] \qquad = P \mathbin{\dot{\cup}} Q \mathbin{\dot{\cup}} C`P \uparrow C`Q \mathbin{\dot{\cup}} C`P \uparrow \iota`y \mathbin{\dot{\cup}} C`Q \uparrow \iota`y$

$[*160{\cdot}1] \qquad = P \ddagger Q \mathbin{\dot{\cup}} C`P \uparrow \iota`y \mathbin{\dot{\cup}} C`Q \uparrow \iota`y$

$[*35{\cdot}82{\cdot}41 . *160{\cdot}14] \qquad = P \ddagger Q \mathbin{\dot{\cup}} C`(P \ddagger Q) \uparrow \iota`y$

$[*161{\cdot}1] \qquad = (P \ddagger Q) \rightarrowtail y : \supset \vdash . \text{Prop}$

$*161{\cdot}231.\quad \vdash : \dot{\exists} ! P . \supset . x \leftarrowtail (P \ddagger Q) = (x \leftarrowtail P) \ddagger Q$

$*161{\cdot}232.\quad \vdash : \dot{\exists} ! P . \dot{\exists} ! Q . \supset . P \ddagger (x \leftarrowtail Q) = (P \rightarrowtail x) \ddagger Q$

Dem.

$\vdash . *161{\cdot}14{\cdot}101 . *160{\cdot}1 . \supset \vdash : \text{Hp} . \supset .$

$P \ddagger (x \leftarrowtail Q) = P \mathbin{\dot{\cup}} \iota`x \uparrow C`Q \mathbin{\dot{\cup}} Q \mathbin{\dot{\cup}} C`P \uparrow (\iota`x \cup C`Q)$

$[*35{\cdot}82{\cdot}412] \qquad = P \mathbin{\dot{\cup}} C`P \uparrow \iota`x \mathbin{\dot{\cup}} Q \mathbin{\dot{\cup}} C`P \uparrow C`Q \mathbin{\dot{\cup}} \iota`x \uparrow C`Q$

$[*161{\cdot}1{\cdot}14 . *35{\cdot}82{\cdot}41] \qquad = (P \rightarrowtail x) \mathbin{\dot{\cup}} Q \mathbin{\dot{\cup}} C`(P \rightarrowtail x) \uparrow C`Q$

$[*160{\cdot}1] \qquad = (P \rightarrowtail x) \ddagger Q : \supset \vdash . \text{Prop}$

$*161{\cdot}24.\quad \vdash . x \leftarrowtail (P \rightarrowtail y) = (x \leftarrowtail P) \rightarrowtail y$

Dem.

$\vdash . *161{\cdot}101{\cdot}14 . \supset \vdash : \dot{\exists} ! P . \supset .$

$x \leftarrowtail (P \rightarrowtail y) = \iota`x \uparrow (C`P \cup \iota`y) \mathbin{\dot{\cup}} P \mathbin{\dot{\cup}} C`P \uparrow \iota`y$

$[*35{\cdot}82{\cdot}412] \qquad = \iota`x \uparrow C`P \mathbin{\dot{\cup}} P \mathbin{\dot{\cup}} \iota`x \uparrow \iota`y \mathbin{\dot{\cup}} C`P \uparrow \iota`y$

$[*35{\cdot}82{\cdot}41 . *161{\cdot}101{\cdot}14] \qquad = (x \leftarrowtail P) \mathbin{\dot{\cup}} C`(x \leftarrowtail P) \uparrow \iota`y$

$[*161{\cdot}1] \qquad = (x \leftarrowtail P) \rightarrowtail y \qquad (1)$

$\vdash . *161{\cdot}2{\cdot}201 . \supset \vdash : P = \dot{\Lambda} . \supset . x \leftarrowtail (P \rightarrowtail y) = \dot{\Lambda} . (x \leftarrowtail P) \rightarrowtail y = \dot{\Lambda} \qquad (2)$

$\vdash . (1) . (2) . \supset \vdash . \text{Prop}$

$*161\cdot25$. $\vdash : \dot{\exists} ! P . \dot{\exists} ! Q . \supset . (P \mathbin{+\!\!\rightarrow} x) \ddagger (y \mathbin{\leftarrow\!\!+} Q) = P \ddagger (x \downarrow y) \ddagger Q$

Dem.

$\vdash . *161\cdot14 . *160\cdot1 . \supset$

$\vdash : \text{Hp} . \supset . (P \mathbin{+\!\!\rightarrow} x) \ddagger (y \mathbin{\leftarrow\!\!+} Q) = (P \mathbin{+\!\!\rightarrow} x) \mathbin{\dot\cup} (y \mathbin{\leftarrow\!\!+} Q) \mathbin{\dot\cup} (C\text{‘}P \cup \iota\text{‘}x) \uparrow (C\text{‘}Q \cup \iota\text{‘}y)$

$[*161\cdot1\cdot101] \quad = P \mathbin{\dot\cup} C\text{‘}P \uparrow \iota\text{‘}x \mathbin{\dot\cup} \iota\text{‘}y \uparrow C\text{‘}Q \mathbin{\dot\cup} Q$
$\qquad \mathbin{\dot\cup} (C\text{‘}P \cup \iota\text{‘}x) \uparrow (C\text{‘}Q \cup \iota\text{‘}y)$

$[*35\cdot82\cdot41\cdot412] \quad = P \mathbin{\dot\cup} C\text{‘}P \uparrow (\iota\text{‘}x \cup \iota\text{‘}y) \mathbin{\dot\cup} \iota\text{‘}x \uparrow \iota\text{‘}y \mathbin{\dot\cup} Q$
$\qquad \mathbin{\dot\cup} (C\text{‘}P \cup \iota\text{‘}x \cup \iota\text{‘}y) \uparrow C\text{‘}Q$

$[*55\cdot15\cdot1 . *160\cdot14\cdot1] \quad = \{P \ddagger (x \downarrow y)\} \mathbin{\dot\cup} Q \mathbin{\dot\cup} C\text{‘}\{P \ddagger (x \downarrow y)\} \uparrow C\text{‘}Q$

$[*160\cdot1 . (*160\cdot32)] \quad = P \ddagger (x \downarrow y) \ddagger Q : \supset \vdash . \text{Prop}$

$*161\cdot26$. $\vdash . x \mathbin{\leftarrow\!\!+} \{y \mathbin{\leftarrow\!\!+} (z \downarrow w)\} = (x \downarrow y) \ddagger (z \downarrow w) = \{(x \downarrow y) \mathbin{+\!\!\rightarrow} z\} \mathbin{+\!\!\rightarrow} w$
$\qquad = \{x \mathbin{\leftarrow\!\!+} (y \downarrow z)\} \mathbin{+\!\!\rightarrow} w$

Dem.

$\vdash . *161\cdot221 . *55\cdot134 . \supset \vdash . x \mathbin{\leftarrow\!\!+} \{y \mathbin{\leftarrow\!\!+} (z \downarrow w)\} = (x \downarrow y) \ddagger (z \downarrow w)$

$[*161\cdot22 . *55\cdot134] \quad = \{(x \downarrow y) \mathbin{+\!\!\rightarrow} z\} \mathbin{+\!\!\rightarrow} w$

$[*161\cdot211] \quad = \{x \mathbin{\leftarrow\!\!+} (y \downarrow z)\} \mathbin{+\!\!\rightarrow} w . \supset \vdash . \text{Prop}$

The following propositions lead up to $*161\cdot33$.

$*161\cdot3$. $\vdash : \dot{\exists} ! Q . S \in P \,\overline{\text{smor}}\, Q . x \sim\in C\text{‘}P . y \sim\in C\text{‘}Q . \supset .$
$\qquad S \mathbin{\dot\cup} x \downarrow y \in (P \mathbin{+\!\!\rightarrow} x) \,\overline{\text{smor}}\, (Q \mathbin{+\!\!\rightarrow} y)$

Dem.

$\vdash . *151\cdot11\cdot131 . \supset \vdash : \text{Hp} . \supset . S \in 1 \rightarrow 1 . C\text{‘}Q = \text{Ɑ}\text{‘}S . P = S ; Q . C\text{‘}P = \text{D}\text{‘}S \quad (1)$

$\vdash . (1) . *55\cdot15 . \supset \vdash : \text{Hp} . \supset . \text{D}\text{‘}S \cap \text{D}\text{‘}(x \downarrow y) = \Lambda . \text{Ɑ}\text{‘}S \cup \text{Ɑ}\text{‘}(x \downarrow y) = \Lambda . \quad (2)$

$[*72\cdot182 . *71\cdot242] \quad \supset . S \mathbin{\dot\cup} x \downarrow y \in 1 \rightarrow 1 \quad (3)$

$\vdash . *55\cdot15 . *151\cdot11 . \supset \vdash : \text{Hp} . \supset . \text{Ɑ}\text{‘}(S \mathbin{\dot\cup} x \downarrow y) = C\text{‘}Q \cup \iota\text{‘}y$

$[*161\cdot14] \quad = C\text{‘}(Q \mathbin{+\!\!\rightarrow} y) \quad (4)$

$\vdash . (1) . (2) . *34\cdot301 . \supset \vdash : \text{Hp} . \supset . (x \downarrow y) | Q = \dot\Lambda . Q | (y \downarrow x) = \dot\Lambda .$
$\qquad (x \downarrow y) | (C\text{‘}Q \uparrow \iota\text{‘}y) = \dot\Lambda . (C\text{‘}Q \uparrow \iota\text{‘}y) | \breve{S} = \dot\Lambda .$

$[*34\cdot25\cdot26] \quad \supset . (S \mathbin{\dot\cup} x \downarrow y) | (Q \mathbin{\dot\cup} C\text{‘}Q \uparrow \iota\text{‘}y) = S | (Q \mathbin{\dot\cup} C\text{‘}Q \uparrow \iota\text{‘}y) .$
$\qquad (Q \mathbin{\dot\cup} C\text{‘}Q \uparrow \iota\text{‘}y) | (y \downarrow x \mathbin{\dot\cup} \breve{S}) = Q | \breve{S} \mathbin{\dot\cup} (C\text{‘}Q \uparrow \iota\text{‘}y) | (y \downarrow x)$

$[*35\cdot89 . *55\cdot1] \quad = Q | \breve{S} \mathbin{\dot\cup} C\text{‘}Q \uparrow \iota\text{‘}x \quad (5)$

$\vdash . (5) . *150\cdot1 . \supset \vdash : \text{Hp} . \supset . (S \mathbin{\dot\cup} x \downarrow y) ; (Q \mathbin{\dot\cup} C\text{‘}Q \uparrow \iota\text{‘}y) = S | \{Q | \breve{S} \mathbin{\dot\cup} C\text{‘}Q \uparrow \iota\text{‘}x\}$

$[*150\cdot1] \quad = S ; Q \mathbin{\dot\cup} S | C\text{‘}Q \uparrow \iota\text{‘}x$

$[*37\cdot81 . (1) . *150\cdot23] \quad = P \mathbin{\dot\cup} C\text{‘}P \uparrow \iota\text{‘}x \quad (6)$

$\vdash . (6) . *161\cdot1 . \supset \vdash : \text{Hp} . \supset . (S \mathbin{\dot\cup} x \downarrow y) ; (Q \mathbin{+\!\!\rightarrow} y) = P \mathbin{+\!\!\rightarrow} x \quad (7)$

$\vdash . (3) . (4) . (7) . *151\cdot11 . \supset \vdash . \text{Prop}$

$*161\cdot301.\ \vdash : \dot{\exists}\,!\,Q\,.\,S\in P\,\overline{\mathrm{smor}}\,Q\,.\,x\sim\in C`P\,.\,y\sim\in C`Q\,.\supset.$
$$x\downarrow y\,\dot{\cup}\,S\in(x\mathbin{\leftarrow\!\!\!\!+}P)\,\overline{\mathrm{smor}}\,(y\mathbin{\leftarrow\!\!\!\!+}Q)$$

$*161\cdot31.\ \vdash : P\ \mathrm{smor}\ Q\,.\,x\sim\in C`P\,.\,y\sim\in C`Q\,.\supset.$
$$P\mathbin{+\!\!\!\!\rightarrow}x\ \mathrm{smor}\ Q\mathbin{+\!\!\!\!\rightarrow}y\,.\,x\mathbin{\leftarrow\!\!\!\!+}P\ \mathrm{smor}\ y\mathbin{\leftarrow\!\!\!\!+}Q$$

Dem.

$\vdash.*161\cdot3\cdot301\,.\,*151\cdot12\,.\supset$

$\vdash : \mathrm{Hp}\,.\,\dot{\exists}\,!\,Q\,.\supset.\,P\mathbin{+\!\!\!\!\rightarrow}x\ \mathrm{smor}\ Q\mathbin{+\!\!\!\!\rightarrow}y\,.\,x\mathbin{\leftarrow\!\!\!\!+}P\ \mathrm{smor}\ y\mathbin{\leftarrow\!\!\!\!+}Q \qquad (1)$

$\vdash.*151\cdot32\,.\,*161\cdot2\cdot201\,.\supset$

$\vdash : \mathrm{Hp}\,.\,Q=\dot{\Lambda}\,.\supset.\,P\mathbin{+\!\!\!\!\rightarrow}x=\dot{\Lambda}\,.\,Q\mathbin{+\!\!\!\!\rightarrow}y=\dot{\Lambda}\,.\,x\mathbin{\leftarrow\!\!\!\!+}P=\dot{\Lambda}\,.\,y\mathbin{\leftarrow\!\!\!\!+}Q=\dot{\Lambda}\,.$

$[*153\cdot101]\quad \supset.\,P\mathbin{+\!\!\!\!\rightarrow}x\ \mathrm{smor}\ Q\mathbin{+\!\!\!\!\rightarrow}y\,.\,x\mathbin{\leftarrow\!\!\!\!+}P\ \mathrm{smor}\ y\mathbin{\leftarrow\!\!\!\!+}Q \qquad (2)$

$\vdash.(1).(2).\supset\vdash.\mathrm{Prop}$

$*161\cdot32.\ \vdash : \dot{\exists}\,!\,Q\,.\,x\sim\in C`P\,.\,y\sim\in C`Q\,.\,S\in(P\mathbin{+\!\!\!\!\rightarrow}x)\,\overline{\mathrm{smor}}\,(Q\mathbin{+\!\!\!\!\rightarrow}y)\,.\supset.$
$$S\upharpoonright(-\iota`y)\in P\,\overline{\mathrm{smor}}\,Q\,.\,xSy$$

Dem.

$\vdash.*151\cdot5\,.\,*161\cdot15\,.\supset\vdash:\mathrm{Hp}\,.\supset.\,xSy \qquad (1)$

$\vdash.(1).*150\cdot1\,.\supset\vdash:.\,\mathrm{Hp}\,.\supset:u\,\{S\upharpoonright(-\iota`y)\,;\,Q\}\,v\,.$

$\equiv.(\exists z,w)\,.\,z\,(Q\mathbin{+\!\!\!\!\rightarrow}y)\,w\,.\,u\neq x\,.\,v\neq x\,.\,uSz\,.\,vSw\,.$

$[*151\cdot11]\quad \equiv.\,u\neq x\,.\,v\neq x\,.\,u\,(P\mathbin{+\!\!\!\!\rightarrow}x)\,v\,.$

$[*161\cdot11]\quad \equiv.\,uPv \qquad (2)$

$\vdash.*35\cdot64\,.\supset\vdash:\mathrm{Hp}\,.\supset.\,\text{Ɑ}`S\upharpoonright(-\iota`y)=C`(Q\mathbin{+\!\!\!\!\rightarrow}y)-\iota`y$

$[*161\cdot14\cdot2]\quad =C`Q \qquad (3)$

$\vdash.(1).(2).(3).\supset\vdash.\mathrm{Prop}$

$*161\cdot321.\ \vdash : \dot{\exists}\,!\,Q\,.\,x\sim\in C`P\,.\,y\sim\in C`Q\,.\,S\in(x\mathbin{\leftarrow\!\!\!\!+}P)\,\overline{\mathrm{smor}}\,(y\mathbin{\leftarrow\!\!\!\!+}Q)\,.\supset.$
$$S\upharpoonright(-\iota`y)\in P\,\overline{\mathrm{smor}}\,Q\,.\,xSy$$

$*161\cdot33.\ \vdash :.\,x\sim\in C`P\,.\,y\sim\in C`Q\,.\supset:$
$$P\ \mathrm{smor}\ Q\,.\equiv.\,(P\mathbin{+\!\!\!\!\rightarrow}x)\ \mathrm{smor}\ (Q\mathbin{+\!\!\!\!\rightarrow}y)\,.\equiv.\,(x\mathbin{\leftarrow\!\!\!\!+}P)\ \mathrm{smor}\ (y\mathbin{\leftarrow\!\!\!\!+}Q)$$
$[*161\cdot31\cdot32\cdot321\cdot2\cdot201\,.\,*153\cdot101]$

The above proposition justifies addition of 1 or subtraction of 1 in ordinal arithmetic.

The following proposition ($*161\cdot4$) is much used.

$*161\cdot4.\ \vdash : C`Q\subset\text{Ɑ}`S\,.\,x\in\text{Ɑ}`S\,.\,S\in 1\rightarrow\mathrm{Cls}\,.\supset.\,S\,;(Q\mathbin{+\!\!\!\!\rightarrow}x)=S\,;Q\mathbin{+\!\!\!\!\rightarrow}S`x$

Dem.

$\vdash.*161\cdot1\,.\,*150\cdot3\,.\quad\supset\vdash.\,S\,;(Q\mathbin{+\!\!\!\!\rightarrow}x)=S\,;Q\,\dot{\cup}\,S\,;(C`Q\uparrow\iota`x)$

$[*150\cdot73]\quad =S\,;Q\,\dot{\cup}\,(S``C`Q)\uparrow(S``\iota`x) \qquad (1)$

$\vdash.(1).*150\cdot22\,.\,*53\cdot31\,.\supset\vdash:\mathrm{Hp}\,.\supset.\,S\,;(Q\mathbin{+\!\!\!\!\rightarrow}x)=S\,;Q\,\dot{\cup}\,(C`S\,;Q)\uparrow(\iota`S`x)$

$[*161\cdot1]\quad =S\,;Q\mathbin{+\!\!\!\!\rightarrow}S`x:\supset\vdash.\mathrm{Prop}$

$*161\cdot41.\ \vdash : C`Q\subset\text{Ɑ}`S\,.\,x\in\text{Ɑ}`S\,.\,S\in 1\rightarrow\mathrm{Cls}\,.\supset.\,S\,;(x\mathbin{\leftarrow\!\!\!\!+}Q)=S`x\mathbin{\leftarrow\!\!\!\!+}S\,;Q$

$*161\cdot42.\ \vdash.\downarrow y\,;(Q\mathbin{+\!\!\!\!\rightarrow}x)=\downarrow y\,;Q\mathbin{+\!\!\!\!\rightarrow}(x\downarrow y)\quad[*161\cdot4\,.\,*55\cdot21\,.\,*72\cdot184]$

$*161\cdot43.\ \vdash.\downarrow y\,;(x\mathbin{\leftarrow\!\!\!\!+}Q)=(x\downarrow y)\mathbin{\leftarrow\!\!\!\!+}\downarrow y\,;Q$

$*162$. THE SUM OF THE RELATIONS OF A FIELD.

Summary of $*162$.

The form of summation defined in $*160$ cannot be extended beyond a finite number of summands, since it involves explicit mention of all the summands. In the present number, we shall be concerned with a form of summation which is not subject to this restriction. It will be observed that, since relational summation is not permutative, we cannot define the sum of a *class* of relations, for this would not determine the order in which the summation is to be effected. Our relations must be given as the field of some relation which orders them; thus the sum appears not as the sum of a class, but as the sum of a relation, namely of a relation whose field is the relations to be summed. In the case of two relations Q and R, the sum of $Q \downarrow R$, as defined in the present number, will be equal to $Q \Uparrow R$; similarly for three, the sum of $Q \downarrow R \cup Q \downarrow S \cup R \downarrow S$ will be equal to $Q \Uparrow R \Uparrow S$, and so on for any finite number of summands.

As explained in the introduction to this Section, if P is a relation between relations, we put

$$\Sigma\text{‘}P = \dot{s}\text{‘}C\text{‘}P \cup F;P \quad \text{Df.}$$

It is convenient to suppose that P is serial, and that every member of $C\text{‘}P$ is also serial. Then $\Sigma\text{‘}P$ holds between x and y if either (1) there is a series, in the field of P, in which x precedes y, or (2) x belongs to a series which is earlier, in the P-series, than the series to which y belongs. The following are the chief propositions of this number:

$*162{\cdot}22{\cdot}23$. $\vdash . C\text{‘}\Sigma\text{‘}P = s\text{‘}C\text{‘‘}C\text{‘}P = C\text{‘}\dot{s}\text{‘}C\text{‘}P = F\text{‘‘}C\text{‘}P = \overrightarrow{F^{2}}\text{‘}P$

$*162{\cdot}26$. $\vdash . \Sigma\text{‘}(P \cup Q) = \Sigma\text{‘}P \cup \Sigma\text{‘}Q$

$*162{\cdot}3$. $\vdash . \Sigma\text{‘}(Q \downarrow R) = Q \Uparrow R$

$*162{\cdot}31$. $\vdash . \Sigma\text{‘}Q \Uparrow \Sigma\text{‘}R = \Sigma\text{‘}(Q \Uparrow R)$

$*162{\cdot}34$. $\vdash . \Sigma\text{‘}\Sigma;P = \Sigma\text{‘}\Sigma\text{‘}P$ [Associative Law. Cf. $*42{\cdot}1$]

$*162{\cdot}35$. $\vdash : C\text{‘}\Sigma\text{‘}Q \subset \text{ᗡ}\text{‘}R . \supset . \Sigma\text{‘}R\dagger;Q = R;\Sigma\text{‘}Q$

This is the analogue of $*40{\cdot}38$. (Cf. note to $*162{\cdot}35$, below.)

$*162{\cdot}4$. $\vdash . \Sigma\text{‘}\dot{\Lambda} = \dot{\Lambda}$

$*162{\cdot}42$. $\vdash : \dot{\exists} ! \Sigma\text{‘}P . \equiv . \dot{\exists} ! \dot{s}\text{‘}C\text{‘}P . \equiv . \exists ! C\text{‘}P - \iota\text{‘}\dot{\Lambda}$

***162·43.** $\vdash : \dot{\exists} ! P . \supset . \Sigma\text{‘}(P \text{⇸} R) = \Sigma\text{‘}P \text{⤉} R$

It should be observed that the ordinal analogues of propositions about classes of classes often involve the substitution of Σ (not $\dot{s}$) for s. Examples are afforded by *162·34·35, quoted above.

***162·01.** $\Sigma\text{‘}P = \dot{s}\text{‘}C\text{‘}P \cup F \,;\, P$ Df

***162·1.** $\vdash . \Sigma\text{‘}P = \dot{s}\text{‘}C\text{‘}P \cup F \,;\, P$ [(*162·01)]

***162·11.** $\vdash :. x (\Sigma\text{‘}P) y . \equiv : x (\dot{s}\text{‘}C\text{‘}P) y . \vee . x (F \,;\, P) y$ [*162·1]

***162·12.** $\vdash :. x (\Sigma\text{‘}P) y . \equiv : (\exists Q) . Q \in C\text{‘}P . xQy . \vee . (\exists Q, R) . xFQ . yFR . QPR$
[*162·1 . *41·11 . *150·11]

***162·13.** $\vdash :. x (\Sigma\text{‘}P) y . \equiv : (\exists Q) . Q \in C\text{‘}P . xQy .$
$\vee . (\exists Q, R) . x \in C\text{‘}Q . y \in C\text{‘}R . QPR$ [*161·12 . *33·51]

***162·14.** $\vdash :. x (\Sigma\text{‘}P) y . \equiv : (\exists Q) . QFP . xQy . \vee . (\exists Q, R) . xFQ . yFR . QPR$
[*161·12 . *33·51]

***162·2.** $\vdash . \text{Cnv}\text{‘}\Sigma\text{‘}P = \Sigma\text{‘}\text{Cnv} \,;\, \breve{P}$

Dem.

$\vdash . *162\cdot13 . \supset \vdash :. x (\Sigma\text{‘}\text{Cnv} \,;\, \breve{P}) y . \equiv : (\exists Q) . Q \in C\text{‘}\text{Cnv} \,;\, \breve{P} . xQy .$
$\vee . (\exists Q, R) . Q (\text{Cnv} \,;\, \breve{P}) R . x \in C\text{‘}Q . y \in C\text{‘}R :$

[*150·22·41] $\equiv : (\exists Q) . Q \in \text{Cnv}\text{‘‘}C\text{‘}P . xQy .$
$\vee . (\exists Q, R) . \breve{Q}\breve{P}\breve{R} . x \in C\text{‘}Q . y \in C\text{‘}R :$

[*37·64 . *33·22] $\equiv : (\exists Q) . Q \in C\text{‘}P . yQx . \vee . (\exists Q, R) . RPQ . x \in C\text{‘}Q . y \in C\text{‘}R :$

[*162·13] $\equiv : y (\Sigma\text{‘}P) x :. \supset \vdash . \text{Prop}$

***162·21.** $\vdash . \text{D}\text{‘}\Sigma\text{‘}P = s\text{‘}\text{D}\text{‘‘}C\text{‘}P \cup s\text{‘}C\text{‘‘}\text{D}\text{‘}(P \upharpoonright - \iota\text{‘}\dot{\Lambda})$

Dem.

$\vdash . *162\cdot13 . \supset \vdash :. x \in \text{D}\text{‘}\Sigma\text{‘}P . \equiv : (\exists Q, y) . Q \in C\text{‘}P . xQy .$
$\vee . (\exists Q, R, y) . QPR . x \in C\text{‘}Q . y \in C\text{‘}R :$

[*33·13·24] $\equiv : (\exists Q) . Q \in C\text{‘}P . x \in \text{D}\text{‘}Q . \vee . (\exists Q, R) . QPR . x \in C\text{‘}Q . \dot{\exists} ! R :$

[*40·4 . *35·101] $\equiv : x \in s\text{‘}\text{D}\text{‘‘}C\text{‘}P . \vee . x \in s\text{‘}C\text{‘‘}\text{D}\text{‘}(P \upharpoonright - \iota\text{‘}\dot{\Lambda}) :. \supset \vdash . \text{Prop}$

***162·211.** $\vdash . \text{Ꭰ}\text{‘}\Sigma\text{‘}P = s\text{‘}\text{Ꭰ}\text{‘‘}C\text{‘}P \cup s\text{‘}C\text{‘‘}\text{Ꭰ}\text{‘}(- \iota\text{‘}\dot{\Lambda}) \upharpoonleft P$

***162·212.** $\vdash : \dot{\Lambda} \sim \in \text{Ꭰ}\text{‘}P . \supset . \text{D}\text{‘}\Sigma\text{‘}P = s\text{‘}C\text{‘‘}\text{D}\text{‘}P \cup s\text{‘}\text{D}\text{‘‘}\overrightarrow{B}\text{‘}\breve{P}$

Dem.

$\vdash . *162\cdot21 . \supset \vdash : \text{Hp} . \supset . \text{D}\text{‘}\Sigma\text{‘}P = s\text{‘}\text{D}\text{‘‘}C\text{‘}P \cup s\text{‘}C\text{‘‘}\text{D}\text{‘}P$

[*40·31 . *93·12] $= s\text{‘}\text{D}\text{‘‘}\text{D}\text{‘}P \cup s\text{‘}\text{D}\text{‘‘}\overrightarrow{B}\text{‘}\breve{P} \cup s\text{‘}C\text{‘‘}\text{D}\text{‘}P$

[*40·57] $= s\text{‘}\text{D}\text{‘‘}\overrightarrow{B}\text{‘}\breve{P} \cup s\text{‘}C\text{‘‘}\text{D}\text{‘}P : \supset \vdash . \text{Prop}$

***162·213.** $\vdash : \dot{\Lambda} \sim\epsilon \mathrm{D}‘P \,.\supset.\, \mathrm{G}‘\Sigma‘P = s‘C‘‘\mathrm{G}‘P \cup s‘\mathrm{G}‘‘\overrightarrow{B}‘P$

The above proposition is used in *163·22.

The two following propositions are used very often.

***162·22.** $\vdash . C‘\Sigma‘P = s‘C‘‘C‘P$

Dem.

$\vdash . *162·21·211 . *40·57 . \supset$

$\vdash . C‘\Sigma‘P = s‘C‘‘C‘P \cup s‘C‘‘\mathrm{D}‘(P \upharpoonright -\iota‘\dot{\Lambda}) \cup s‘C‘‘\mathrm{G}‘(-\iota‘\dot{\Lambda}) \upharpoonleft P$

$[*40·161] = s‘C‘‘C‘P . \supset \vdash . \mathrm{Prop}$

***162·23.** $\vdash . C‘\Sigma‘P = C‘\dot{s}‘C‘P = F‘‘C‘P = \overrightarrow{F^2}‘P$ [*162·22 . *42·2]

***162·26.** $\vdash . \Sigma‘(P \uplus Q) = \Sigma‘P \uplus \Sigma‘Q$

Dem.

$\vdash . *162·1 . \supset \vdash . \Sigma‘(P \uplus Q) = \dot{s}‘C‘(P \uplus Q) \uplus F;(P \uplus Q)$

$[*33·262 . *41·171 . *150·3] \quad = \dot{s}‘C‘P \uplus \dot{s}‘C‘Q \uplus F;P \uplus F;Q$

$[*162·1] \quad = \Sigma‘P \uplus \Sigma‘Q . \supset \vdash . \mathrm{Prop}$

***162·27.** $\vdash . \Sigma‘S;(P \uplus Q) = \Sigma‘S;P \uplus \Sigma‘S;Q$ [*162·26 . *150·3]

***162·3.** $\vdash . \Sigma‘(Q \downarrow R) = Q \ddagger R$

Dem.

$\vdash . *160·1 . \supset \vdash . \Sigma‘(Q \downarrow R) = \dot{s}‘C‘(Q \downarrow R) \uplus F;(Q \downarrow R)$

$[*55·15 . *150·7] \quad = \dot{s}‘(\iota‘Q \cup \iota‘R) \uplus \overrightarrow{F}‘Q \uparrow \overrightarrow{F}‘R$

$[*53·13 . *33·5] \quad = Q \uplus R \uplus C‘Q \uparrow C‘R$

$[*160·1] \quad = Q \ddagger R . \supset \vdash . \mathrm{Prop}$

This proposition establishes the connection between the two kinds of arithmetical addition of relations.

***162·31.** $\vdash . \Sigma‘Q \ddagger \Sigma‘R = \Sigma‘(Q \ddagger R)$

Dem.

$\vdash . *160·1 . \supset \vdash . \Sigma‘Q \ddagger \Sigma‘R = \Sigma‘Q \uplus \Sigma‘R \uplus C‘\Sigma‘Q \uparrow C‘\Sigma‘R$

$[*162·123] \quad = \dot{s}‘C‘Q \uplus F;Q \uplus \dot{s}‘C‘R \uplus F;R \uplus (F‘‘C‘Q) \uparrow (F‘‘C‘R)$

$[*150·73] \quad = \dot{s}‘C‘Q \uplus \dot{s}‘C‘R \uplus F;Q \uplus F;R \uplus F;(C‘Q \uparrow C‘R)$

$[*41·171 . *160·14 . *150·3 . *160·1] = \dot{s}‘C‘(Q \ddagger R) \uplus F;(Q \ddagger R)$

$[*162·1] \quad = \Sigma‘(Q \ddagger R) . \supset \vdash . \mathrm{Prop}$

The following propositions lead up to *162·34.

***162·32.** $\vdash . \Sigma‘\dot{s}‘\kappa = \dot{s}‘\Sigma‘‘\kappa$

Dem.

$\vdash . *41·6 . *162·1 . *150·1 . \supset \vdash . \dot{s}‘\Sigma‘‘\kappa = \dot{s}‘\dot{s}‘‘C‘‘\kappa \uplus \dot{s}‘F\dagger‘‘\kappa$

$[*42·12 . *150·16] \quad = \dot{s}‘s‘C‘‘\kappa \uplus F;\dot{s}‘\kappa$

$[*41·45] \quad = \dot{s}‘C‘\dot{s}‘\kappa \uplus F;\dot{s}‘\kappa$

$[*162·1] \quad = \Sigma‘\dot{s}‘\kappa . \supset \vdash . \mathrm{Prop}$

∗162·33. $\vdash . \Sigma`\Sigma`P = \dot{s}`C`\dot{s}`C`P \cup F;\dot{s}`C`P \cup F^2;P$

Dem.

$\vdash . *162{\cdot}1 . \supset \vdash . \Sigma`\Sigma`P = \dot{s}`C`\Sigma`P \cup F;\Sigma`P$

[∗162·23] $= \dot{s}`C`\dot{s}`C`P \cup F;(\dot{s}`C`P \cup F;P)$

[∗150·3·13] $= \dot{s}`C`\dot{s}`C`P \cup F;\dot{s}`C`P \cup F^2;P . \supset \vdash .$ Prop

∗162·331. $\vdash . F | \Sigma = F | \dot{s} | C = F^2$

Dem.

$\vdash . *71{\cdot}7 . \supset \vdash : x (F | \Sigma) P . \equiv . xF(\Sigma`P) .$

[∗33·51] $\equiv . x \in C`\Sigma`P .$

[∗162·23] $\equiv . xF^2P$ (1)

$\vdash . *71{\cdot}7 . \supset \vdash : x (F | \dot{s} | C) P . \equiv . xF(\dot{s}`C`P) .$

[∗33·51] $\equiv . x \in C`\dot{s}`C`P .$

[∗42·2] $\equiv . xF^2P$ (2)

$\vdash . (1) . (2) . \supset \vdash .$ Prop

∗162·332. $\vdash . \Sigma`\Sigma;P = \dot{s}`C`\dot{s}`C`P \cup F;\dot{s}`C`P \cup F^2;P$

Dem.

$\vdash . *162{\cdot}1 . \supset \vdash . \Sigma`\Sigma;P = \dot{s}`C`\Sigma;P \cup F;\Sigma;P$

[∗150·22·13] $= \dot{s}`\Sigma``C`P \cup (F | \Sigma);P$

[∗162·32·331] $= \Sigma`\dot{s}`C`P \cup F^2;P$

[∗162·1] $= \dot{s}`C`\dot{s}`C`P \cup F;\dot{s}`C`P \cup F^2;P . \supset \vdash .$ Prop

∗162·34. $\vdash . \Sigma`\Sigma;P = \Sigma`\Sigma`P$ [∗162·33·332]

This is the associative law for arithmetical sums of relations.

The following propositions lead up to ∗162·35.

∗162·341. $\vdash :. C`Q \subset \mathrm{Cl}`R . \supset : x (F | R\dagger) Q . \equiv . x (R | F) Q$

Dem.

$\vdash . *71{\cdot}7 . *150{\cdot}1 . \supset \vdash : x (F | R\dagger) Q . \equiv . xF(R;Q) .$

[∗33·51] $\equiv . x \in C`R;Q$ (1)

$\vdash . (1) . *150{\cdot}22 . \supset \vdash :. \mathrm{Hp} . \supset : x (F | R\dagger) Q . \equiv . x \in R``C`Q .$

[∗33·5] $\equiv . x \in R``\overrightarrow{F}`Q .$

[∗37·3.∗32·18] $\equiv . x (R | F) Q :. \supset \vdash .$ Prop

∗162·342. $\vdash : C`\dot{s}`\lambda \subset \mathrm{Cl}`R . \supset . (F | R\dagger) \upharpoonright \lambda = (R | F) \upharpoonright \lambda$

Dem.

$\vdash . *41{\cdot}13 . \supset \vdash :. \mathrm{Hp} . \supset : Q \in \lambda . \supset . C`Q \subset \mathrm{Cl}`R :$

[∗162·341] $\supset : Q \in \lambda . x (F | R\dagger) Q . \equiv . Q \in \lambda . x (R | F) Q :. \supset \vdash .$ Prop

∗162·343. $\vdash : C`\Sigma`P \subset \mathrm{Cl}`R . \supset . F;R\dagger;P = R;F;P$

Dem.

$\vdash . *162{\cdot}23 . \supset \vdash : \mathrm{Hp} . \supset . C`\dot{s}`C`P \subset \mathrm{Cl}`R .$

[∗162·342] $\supset . (F | R\dagger) \upharpoonright (C`P);P = (R | F) \upharpoonright (C`P);P .$

[∗150·32] $\supset . (F | R\dagger);P = (R | F);P .$

[∗150·13] $\supset . F;R\dagger;P = R;F;P : \supset \vdash .$ Prop

***162·35.** $\vdash : C\text{‘}\Sigma\text{‘}Q \subset \text{Ɑ}\text{‘}R \,.\supset .\, \Sigma\text{‘}R\dagger;Q = R;\Sigma\text{‘}Q$

Dem.

$\vdash . *162\cdot1 . *150\cdot22 . \supset \vdash . \Sigma\text{‘}R\dagger;Q = \dot{s}\text{‘}R\dagger\text{“}C\text{‘}Q \cup F;R\dagger;Q$

$[*150\cdot16] \qquad = R;\dot{s}\text{‘}C\text{‘}Q \cup F;R\dagger;Q \qquad (1)$

$\vdash . (1) . *162\cdot343 . \quad \supset \vdash : \mathrm{Hp} . \supset . \Sigma\text{‘}R\dagger;Q = R;\dot{s}\text{‘}C\text{‘}Q \cup R;F;Q$

$[*150\cdot3 . *162\cdot1] \qquad = R;\Sigma\text{‘}Q : \supset \vdash . \mathrm{Prop}$

This proposition is important, since it enables us to infer (with a suitable hypothesis) that if $R;M$ is always like M when $R \in C\text{‘}Q$, then the arithmetical sum of all such relations as M is like $\Sigma\text{‘}Q$, being in fact $R;\Sigma\text{‘}Q$. In other words, if, whenever $M \in C\text{‘}Q$, $R \upharpoonright C\text{‘}M$ is a correlator of $R;M$ and M, then $R \upharpoonright \Sigma\text{‘}Q$ is a correlator if $\Sigma\text{‘}R\dagger;Q$ and $\Sigma\text{‘}Q$. This proposition is analogous in its uses to the proposition

$$s\text{‘}R_\epsilon\text{“}\kappa = R\text{“}s\text{‘}\kappa,$$

which is *40·38. In general, in obtaining relational analogues of cardinal propositions, $R\text{“}\kappa$ is to be replaced by $R;Q$, R_ϵ by $R\dagger$, and s by Σ. When these substitutions are made in $s\text{‘}R_\epsilon\text{“}\kappa = R\text{“}s\text{‘}\kappa$, *162·35 results, except for its hypothesis.

If we regard $R;Q$ as a kind of product of R and Q, *162·35 becomes a distributive law. For it asserts that if we multiply each member of $C\text{‘}Q$ by R, and then sum the resulting products, we get the same relation as if we first sum $C\text{‘}Q$, and then multiply by R. The following application of *162·35 to the sum of two relations makes its distributive character more evident.

***162·36.** $\vdash : C\text{‘}P \cup C\text{‘}Q \subset \text{Ɑ}\text{‘}R \,.\supset .\, R;P \dotplus R;Q = R;(P \dotplus Q)$

Dem.

$\vdash . *162\cdot3 . \quad \supset \vdash . R;P \dotplus R;Q = \Sigma\text{‘}\{(R;P) \downarrow (R;Q)\}$

$[*150\cdot1\cdot71] \qquad = \Sigma\text{‘}R\dagger;(P \downarrow Q) \qquad (1)$

$\vdash . (1) . *162\cdot35 . \supset \vdash : \mathrm{Hp} . \supset . R;P \dotplus R;Q = R;\Sigma\text{‘}(P \downarrow Q)$

$[*162\cdot3] \qquad = R;(P \dotplus Q) : \supset \vdash . \mathrm{Prop}$

This proposition can be extended to any finite number of summands.

***162·37.** $\vdash : \exists ! \lambda . \exists ! \mu . \supset . \Sigma\text{‘}(\lambda \uparrow \mu) = \dot{s}\text{‘}\lambda \dotplus \dot{s}\text{‘}\mu$

Dem.

$\vdash . *35\cdot85\cdot86 . \supset \vdash : \mathrm{Hp} . \supset . C\text{‘}(\lambda \uparrow \mu) = \lambda \cup \mu .$

$[*162\cdot1] \qquad \supset . \Sigma\text{‘}(\lambda \uparrow \mu) = \dot{s}\text{‘}(\lambda \cup \mu) \cup F;(\lambda \uparrow \mu)$

$[*41\cdot171 . *150\cdot73] \qquad = \dot{s}\text{‘}\lambda \cup \dot{s}\text{‘}\mu \cup (F\text{“}\lambda) \uparrow (F\text{“}\mu)$

$[*41\cdot45 . *40\cdot56] \qquad = \dot{s}\text{‘}\lambda \cup \dot{s}\text{‘}\mu \cup (C\text{‘}\dot{s}\text{‘}\lambda) \uparrow (C\text{‘}\dot{s}\text{‘}\mu)$

$[*160\cdot1] \qquad = \dot{s}\text{‘}\lambda \dotplus \dot{s}\text{‘}\mu : \supset \vdash . \mathrm{Prop}$

***162·371.** $\vdash : \exists ! \alpha . \supset . \Sigma\text{‘}(\alpha \uparrow \iota\text{‘}Q) = \dot{s}\text{‘}\alpha \dotplus Q \quad [*162\cdot37 . *53\cdot04]$

***162·372.** $\vdash : \exists ! \beta . \supset . \Sigma\text{‘}(\iota\text{‘}P) \uparrow \beta = P \dotplus \dot{s}\text{‘}\beta$

***162·4.** $\vdash . \Sigma`\dot{\Lambda} = \dot{\Lambda}$

Dem.

$\vdash . *33·241 . *41·21 . \supset \vdash . \dot{s}`C`\dot{\Lambda} = \dot{\Lambda}$ (1)

$\vdash . *150·42 . \supset \vdash . F ; \dot{\Lambda} = \dot{\Lambda}$ (2)

$\vdash . (1) . (2) . *162·1 . \supset \vdash .$ Prop

***162·41.** $\vdash . \Sigma`(\dot{\Lambda} \downarrow \dot{\Lambda}) = \dot{\Lambda}$

Dem.

$\vdash . *162·3 . \supset \vdash . \Sigma`(\dot{\Lambda} \downarrow \dot{\Lambda}) = \dot{\Lambda} \ddagger \dot{\Lambda}$

[*160·21] $= \dot{\Lambda} . \supset \vdash .$ Prop

***162·42.** $\vdash : \dot{\exists} ! \Sigma`P . \equiv . \dot{\exists} ! \dot{s}`C`P . \equiv . \exists ! C`P - \iota`\dot{\Lambda}$

Dem.

$\vdash . *162·23 . *33·24 . \supset \vdash : \dot{\exists} ! \Sigma`P . \equiv . \dot{\exists} ! \dot{s}`C`P .$

[*41·26] $\equiv . \exists ! C`P - \iota`\dot{\Lambda}$

***162·43.** $\vdash : \dot{\exists} ! P . \supset . \Sigma`(P \dotplus\!\!\rightarrow R) = \Sigma`P \ddagger R$

Dem.

$\vdash . *162·26 . *161·1 . \supset \vdash . \Sigma`(P \dotplus\!\!\rightarrow R) = \Sigma`P \cup \Sigma`(C`P \uparrow \iota`R)$ (1)

$\vdash . *162·371 . *33·24 . \supset \vdash : \dot{\exists} ! P . \supset . \Sigma`(C`P \uparrow \iota`R) = \dot{s}`C`P \ddagger R$ (2)

$\vdash . (1) . (2) . *160·1 . \supset$

$\vdash : \text{Hp} . \supset . \Sigma`(P \dotplus\!\!\rightarrow R) = \Sigma`P \cup \dot{s}`C`P \cup R \cup (C`\dot{s}`C`P) \uparrow C`R$

[*162·1·23] $= \Sigma`P \cup R \cup (C`\Sigma`P) \uparrow C`R$

[*160·1] $= \Sigma`P \ddagger R : \supset \vdash .$ Prop

***162·431.** $\vdash : \dot{\exists} ! P . \supset . \Sigma`(R \leftarrow\!\!\dotplus P) = R \ddagger \Sigma`P$ [Proof as in *162·43]

Observe that in *162·43·431, P and R must be of different types, in fact R must be of the type to which members of $C`P$ belong. *162·43·431 are often useful.

***162·44.** $\vdash . \Sigma`(P \dotplus\!\!\rightarrow \dot{\Lambda}) = \Sigma`(\dot{\Lambda} \leftarrow\!\!\dotplus P) = \Sigma`P$

Dem.

$\vdash . *162·43 . \supset \vdash : \dot{\exists} ! P . \supset . \Sigma`(P \dotplus\!\!\rightarrow \dot{\Lambda}) = \Sigma`P \ddagger \dot{\Lambda}$

[*160·21] $= \Sigma`P$ (1)

$\vdash . *33·241 . *35·88 . \supset \vdash : P = \dot{\Lambda} . \supset . C`P \uparrow \iota`\dot{\Lambda} = \dot{\Lambda} .$

[*162·4] $\supset . \Sigma`(C`P \uparrow \iota`\Lambda) = \dot{\Lambda} .$

[*25·24] $\supset . \Sigma`P = \Sigma`P \cup \Sigma`(C`P \uparrow \iota`\dot{\Lambda})$

[*162·26] $= \Sigma`(P \cup C`P \uparrow \iota`\dot{\Lambda})$

[*161·1] $= \Sigma`(P \dotplus\!\!\rightarrow \dot{\Lambda})$ (2)

$\vdash . (1) . (2) . \supset \vdash . \Sigma`(P \dotplus\!\!\rightarrow \dot{\Lambda}) = \Sigma`P$ (3)

Similarly $\vdash . \Sigma`(\dot{\Lambda} \leftarrow\!\!\dotplus P) = \Sigma`P$ (4)

$\vdash . (3) . (4) . \supset \vdash .$ Prop

***162·45.** $\vdash : \dot{\exists} ! P . \Sigma`P = \dot{\Lambda} . \equiv . P = \dot{\Lambda} \downarrow \dot{\Lambda}$

Dem.

$\vdash . *162·42 . \supset \vdash : \Sigma`P = \dot{\Lambda} . \equiv . C`P \subset \iota`\dot{\Lambda} .$

$[*33·16] \quad \equiv . \mathrm{D}`P \subset \iota`\dot{\Lambda} . \mathrm{G}`P \subset \iota`\dot{\Lambda}$ (1)

$\vdash . *33·24 . \supset \vdash : \dot{\exists} ! P . \equiv . \exists ! \mathrm{D}`P . \exists ! \mathrm{G}`P$ (2)

$\vdash . (1) . (2) . *51·4 . \supset$

$\vdash : \dot{\exists} ! P . \Sigma`P = \dot{\Lambda} . \equiv . \mathrm{D}`P = \iota`\dot{\Lambda} . \mathrm{G}`P = \iota`\dot{\Lambda} .$

$[*55·16] \quad \equiv . P = \dot{\Lambda} \downarrow \dot{\Lambda} : \supset \vdash .$ Prop

The above proposition is used in *174·162.

*163. RELATIONS OF MUTUALLY EXCLUSIVE RELATIONS.

Summary of *163.

In the present number we have to define mutually exclusive relations, and to give a few of their properties. Mutually exclusive relations play much the same part in relation-arithmetic as mutually exclusive classes play in cardinal arithmetic. *Prima facie*, there are various ways in which we might define them. We might define P as a relation of mutually exclusive relations when

$$QPR \,.\, Q \neq R \,.\, \supset_{Q,R} \,.\, Q \dot{\cap} R = \dot{\Lambda},$$

or when $$Q, R \in C\text{`}P \,.\, Q \neq R \,.\, \supset_{Q,R} \,.\, Q \dot{\cap} R = \dot{\Lambda},$$

or when

$$Q, R \in C\text{`}P \,.\, Q \neq R \,.\, \supset_{Q,R} \,.\, \text{D`}Q \cap \text{D`}R = \Lambda \,.\, \text{Ꮖ`}Q \cap \text{Ꮖ`}R = \Lambda,$$

or in several other ways. But in fact the most useful property to choose is the property that any two members of the field have mutually exclusive fields, *i.e.*

$$Q, R \in C\text{`}P \,.\, Q \neq R \,.\, \supset_{Q,R} \,.\, C\text{`}Q \cap C\text{`}R = \Lambda.$$

The principal applications of the subjects studied in this Part are to series, and in series it is always the *fields* of the relations that are important. We want, for instance, to define relations of mutually exclusive relations in such a way that, if P is a serial relation, and every member of $C\text{`}P$ is a serial relation, then $\Sigma\text{`}P$ is a serial relation. For this purpose it is necessary that $\Sigma\text{`}P$ should be contained in diversity, which requires that $F;P$ should be contained in diversity, *i.e.* that

$$QPR \,.\, \supset_{Q,R} \,.\, C\text{`}Q \cap C\text{`}R = \Lambda.$$

If P is a serial relation, as we are supposing, this is equivalent to

$$Q, R \in C\text{`}P \,.\, Q \neq P \,.\, \supset_{Q,R} \,.\, C\text{`}Q \cap C\text{`}R = \Lambda.$$

Again we want to define relations of mutually exclusive relations in such a way that, if P and Q are two such relations, and P and Q have double likeness (cf. *164), then $\Sigma\text{`}P$ is like $\Sigma\text{`}Q$; *i.e.* if we are given a correlator S of P and Q, and for every M and N which S correlates, we are again given a correlator, then $\Sigma\text{`}P$ is to be like $\Sigma\text{`}Q$. That is, if λ is the class of relations which correlate pairs of relations M and N, where $N \in C\text{`}Q \,.\, MSN$, we want

$\dot{s}`\lambda$ to be a correlator of $P$ and $Q$. Now this requires that $\dot{s}`\lambda$ should be a one-one relation, which requires

$$M, M' \epsilon C`P . M \neq M' . \supset_{M, M'} . \mathrm{D}`M \cap \mathrm{D}`M' = \Lambda . \text{Ꟶ}`M \cap \text{Ꟶ}`M' = \Lambda.$$

This is secured by

$$M, M' \epsilon C`P . M \neq M' . \supset_{M, M'} . C`M \cap C`M' = \Lambda,$$

but except for special classes of relations it is not secured by

$$MPM' . \supset_{M, M'} . C`M \cap C`M' = \Lambda,$$

since there may be two relations M and M' which both belong to the field of P, but of which neither has the relation P to the other. Again, the analogy with cardinal arithmetic fails at many points unless, when P is a relation of mutually exclusive relations, $C``C`P$ is a class of mutually exclusive classes. But this is not secured by any of the other possible definitions we have been considering. There are further reasons, connected with the arithmetical product of a relation of relations, for choosing as the definition

$$Q, R \epsilon C`P . Q \neq R . \supset_{Q, R} . C`Q \cap C`R = \Lambda.$$

From a technical point of view, the properties of a Cls^2 excl depend mainly upon the fact that when κ is such a class, $\epsilon \upharpoonright \kappa \epsilon \mathrm{Cls} \to 1$ (*84·14); in like manner the properties of a Rel^2 excl depend upon

$$F \upharpoonright C`P \epsilon \mathrm{Cls} \to 1,$$

which requires our definition, and is equivalent to it (*163·12). We thus become able to use the propositions of *81 on selections from many-one relations, which would not otherwise be the case.

It should be observed that

$$Q, R \epsilon C`P . Q \neq R . \supset_{Q, R} . C`Q \cap C`R = \Lambda$$

is not equivalent to

$$C``C`P \epsilon \mathrm{Cls}^2 \text{ excl},$$

though it implies this. The converse implication will fail if $C`P$ contains two different relations with the same field. *E.g.* take a relation $P$ whose field consists of the four relations $S$, $\breve{S}$, $T$, $\breve{T}$, and suppose $C`S \cap C`T = \Lambda$. Then $C``C`P = \iota`C`S \cup \iota`C`T$, and $C``C`P \epsilon \mathrm{Cls}^2$ excl. But unless $S = \breve{S}$ and $T = \breve{T}$ we shall not have

$$Q, R \epsilon C`P . Q \neq R . \supset_{Q, R} . C`Q \cap C`R = \Lambda.$$

The property by which we define relations of mutually exclusive relations is a property which only depends on the field, so that we might equally well put

$$(\mathrm{Cl}`\mathrm{Rel}) \text{ excl} = \hat{\lambda} \{Q, R \epsilon \lambda . Q \neq R . \supset_{Q, R} . C`Q \cap C`R = \Lambda\} \quad \text{Df.}$$

But for our purposes this would be less convenient than the definition of Rel^2 excl.

We thus put

***163·01.** $\text{Rel}^2\,\text{excl} = \hat{P}\{Q, R \,\epsilon\, C\text{‘}P \,.\, Q \neq R \,.\, \supset_{Q,R} .\, C\text{‘}Q \cap C\text{‘}R = \Lambda\}$ Df

We have

***163·11.** $\vdash :.\, P \,\epsilon\, \text{Rel}^2\,\text{excl} \,.\equiv :\, Q, R \,\epsilon\, C\text{‘}P \,.\, \exists !\, C\text{‘}Q \cap C\text{‘}R \,.\, \supset_{Q,R} .\, Q = R$

***163·12.** $\vdash :\, P \,\epsilon\, \text{Rel}^2\,\text{excl} \,.\equiv .\, F \upharpoonright C\text{‘}P \,\epsilon\, \text{Cls} \rightarrow 1$

***163·17.** $\vdash :\, P \,\epsilon\, \text{Rel}^2\,\text{excl} \,.\equiv .\, C \upharpoonright C\text{‘}P \,\epsilon\, 1 \rightarrow 1 \,.\, C\text{“}C\text{‘}P \,\epsilon\, \text{Cls}^2\,\text{excl}$

Any of the above might have been used to define Rel² excl. The following propositions are important.

***163·3.** $\vdash :\, Q \,\epsilon\, \text{Rel}^2\,\text{excl} \,.\, S \,\epsilon\, \text{Cls} \rightarrow 1 \,.\supset .\, S \dagger ; Q \,\epsilon\, \text{Rel}^2\,\text{excl}$

This is the analogue of *84·53.

***163·4·41.** $\vdash .\, \dot{\Lambda}, P \downarrow P \,\epsilon\, \text{Rel}^2\,\text{excl}$

***163·441.** $\vdash :\, P, Q \,\epsilon\, \text{Rel}^2\,\text{excl} \,.\, C\text{‘}\Sigma\text{‘}P \cap C\text{‘}\Sigma\text{‘}Q = \Lambda \,.\supset .\, P \ddagger Q \,\epsilon\, \text{Rel}^2\,\text{excl}$

***163·451.** $\vdash :\, P \,\epsilon\, \text{Rel}^2\,\text{excl} \,.\, C\text{‘}\Sigma\text{‘}P \cap C\text{‘}R = \Lambda \,.\supset .\, P \nrightarrow R \,\epsilon\, \text{Rel}^2\,\text{excl}$

***163·01.** $\text{Rel}^2\,\text{excl} = \hat{P}\{Q, R \,\epsilon\, C\text{‘}P \,.\, Q \neq R \,.\, \supset_{Q,R} .\, C\text{‘}Q \cap C\text{‘}R = \Lambda\}$ Df

***163·1.** $\vdash :.\, P \,\epsilon\, \text{Rel}^2\,\text{excl} \,.\equiv :\, Q, R \,\epsilon\, C\text{‘}P \,.\, Q \neq R \,.\, \supset_{Q,R} .\, C\text{‘}Q \cap C\text{‘}R = \Lambda$ [(*163·01)]

***163·11.** $\vdash :.\, P \,\epsilon\, \text{Rel}^2\,\text{excl} \,.\equiv :\, Q, R \,\epsilon\, C\text{‘}P \,.\, \exists !\, C\text{‘}Q \cap C\text{‘}R \,.\, \supset_{Q,R} .\, Q = R$ [*163·1 . Transp]

***163·12.** $\vdash :\, P \,\epsilon\, \text{Rel}^2\,\text{excl} \,.\equiv .\, F \upharpoonright C\text{‘}P \,\epsilon\, \text{Cls} \rightarrow 1$ [*163·1 . *74·632]

For many purposes, this proposition gives the most useful equivalent of $P \,\epsilon\, \text{Rel}^2\,\text{excl}$.

Instead of the above proof, we may use *74·62, which gives us the result in virtue of *33·5.

***163·13.** $\vdash :.\, P \,\epsilon\, \text{Rel}^2\,\text{excl} \,.\supset :$

$$Q, R \,\epsilon\, C\text{‘}P \,.\, Q \neq R \,.\, \supset_{Q,R} .\, D\text{‘}Q \cap D\text{‘}R = \Lambda \,.\, Ɑ\text{‘}Q \cap Ɑ\text{‘}R = \Lambda$$

[*24·402 . *163·1]

***163·14.** $\vdash :\, P \,\epsilon\, \text{Rel}^2\,\text{excl} \,.\supset .\, C \upharpoonright C\text{‘}P \,\epsilon\, 1 \rightarrow 1$ [*163·12 . *74·32 . *33·5]

***163·15.** $\vdash :\, P \,\epsilon\, \text{Rel}^2\,\text{excl} \,.\supset .\, D \upharpoonright C\text{‘}P, Ɑ \upharpoonright C\text{‘}P \,\epsilon\, 1 \rightarrow 1$

Dem.

$\vdash .\, \text{*}74·63 \,.\, \text{*}163·13 \,.\supset \vdash :\, \text{Hp} \,.\supset .\, (\epsilon \mid D) \upharpoonright C\text{‘}P \,\epsilon\, \text{Cls} \rightarrow 1 \,.$

[*74·32] $\supset .\, \overrightarrow{\epsilon \mid D} \upharpoonright C\text{‘}P \,\epsilon\, 1 \rightarrow 1 \,.$

[*72·27] $\supset .\, D \upharpoonright C\text{‘}P \,\epsilon\, 1 \rightarrow 1$ (1)

Similarly $\vdash :\, \text{Hp} \,.\supset .\, Ɑ \upharpoonright C\text{‘}P \,\epsilon\, 1 \rightarrow 1$ (2)

$\vdash .\, (1) \,.\, (2) \,.\supset \vdash .\, \text{Prop}$

***163·16.** $\vdash :\, P \,\epsilon\, \text{Rel}^2\,\text{excl} \,.\supset .\, C\text{“}C\text{‘}P \,\epsilon\, \text{Cls}^2\,\text{excl}$ [*84·51 . *33·5 . *163·12]

***163·17.** $\vdash : P \in \text{Rel}^2\,\text{excl} . \equiv . C \upharpoonright C`P \in 1 \to 1 . C``C`P \in \text{Cls}^2\,\text{excl}$
[*163·12 . *84·522 . *33·5]

***163·2.** $\vdash : P \in \text{Rel}^2\,\text{excl} . \supset . \text{D} \upharpoonright F_\Delta`C`P \in 1 \to 1 . F_\Delta`C`P \subset 1 \to 1$
[*81·21·1 . *163·12]

***163·21.** $\vdash : P \in \text{Rel}^2\,\text{excl} . \supset . \text{D}``F_\Delta`C`P = \text{Prod}`C``C`P$

Dem.

$$\vdash . *85·1 \frac{F, C`P}{Q, \lambda} . *163·12 . \supset \vdash : \text{Hp} . \supset . \text{D}``F_\Delta`C`P = \text{D}``\epsilon_\Delta`\overrightarrow{F}``C`P$$

$$[*115·1 . *33·5] \qquad = \text{Prod}`C``C`P : \supset \vdash . \text{Prop}$$

This proposition is important in connection with the multiplication of relations, for we shall define as the product of a relation P (whose field consists of relations) a relation whose field is $\text{D}``F_\Delta`C`P$. Thus by the above proposition, whenever P is a Rel^2 excl, the field of its product is the product (in the cardinal sense) of the fields of its field, just as the field of its sum is (by *162·22) the sum of the fields of its field.

***163·22.** $\vdash : P \in \text{Rel}^2\,\text{excl} . \dot\Lambda \sim \in C`P . \supset .$
$$\overrightarrow{B}`\Sigma`P = B``\overrightarrow{B}`P . \overrightarrow{B}`\text{Cnv}`\Sigma`P = B``\text{Cnv}``\overrightarrow{B}`\breve{P}$$

Dem.

$$\vdash . *162·23·213 . *93·103 . \supset \vdash : \text{Hp} . \supset . \overrightarrow{B}`\Sigma`P = F``C`P - s`C``\text{Ꮖ}`P - s`\text{Ꮖ}``\overrightarrow{B}`P$$

$$[*40·56] \qquad = F``C`P - F``\text{Ꮖ}`P - s`\text{Ꮖ}``\overrightarrow{B}`P$$

$$[*71·381 . *37·421 . *163·12] \qquad = F``(C`P - \text{Ꮖ}`P) - s`\text{Ꮖ}``\overrightarrow{B}`P$$

$$[*40·56 . *93·103] \qquad = s`C``\overrightarrow{B}`P - s`\text{Ꮖ}``\overrightarrow{B}`P \quad (1)$$

$$\vdash . *163·11 . \supset \vdash :: \text{Hp} . \supset :. Q \in \overrightarrow{B}`P . x \in C`Q . \supset : R \in \overrightarrow{B}`P . x \in \text{Ꮖ}`R . \supset . R = Q .$$

$$[*13·12] \qquad \supset . x \in \text{Ꮖ}`Q :$$

$$[*40·4] \qquad \supset : x \in s`\text{Ꮖ}``\overrightarrow{B}`P . \supset . x \in \text{Ꮖ}`Q :$$

$$[*40·13] \qquad \supset : x \in s`\text{Ꮖ}``\overrightarrow{B}`P . \equiv . x \in \text{Ꮖ}`Q \quad (2)$$

$$\vdash . (2) . *5·32 . \supset \vdash :. \text{Hp} . \supset : Q \in \overrightarrow{B}`P . x \in C`Q . x \sim \in s`\text{Ꮖ}``\overrightarrow{B}`P . \equiv .$$
$$Q \in \overrightarrow{B}`P . x \in C`Q . x \sim \in \text{Ꮖ}`Q :$$

$$[*10·281 . *40·4 . *93·103] \supset : x \in s`C``\overrightarrow{B}`P - s`\text{Ꮖ}``\overrightarrow{B}`P . \equiv . (\exists Q) . Q \in \overrightarrow{B}`P . xBQ .$$

$$[*37·1] \qquad \equiv . x \in B``\overrightarrow{B}`P \quad (3)$$

$$\vdash . (1) . (3) . \supset \vdash : \text{Hp} . \supset . \overrightarrow{B}`\Sigma`P = B``\overrightarrow{B}`P \quad (4)$$

$$\vdash . (4) . *162·2 . *33·22 . *163·1 . \supset \vdash : \text{Hp} . \supset . \overrightarrow{B}`\text{Cnv}`\Sigma`P = B``\overrightarrow{B}`\text{Cnv} ; \breve{P}$$

$$[*151·6·5] \qquad = B``\text{Cnv}``\overrightarrow{B}`\breve{P} \quad (5)$$

$$\vdash . (4) . (5) . \supset \vdash . \text{Prop}$$

***163·3.** $\vdash : Q \in \text{Rel}^2\text{excl} . S \in \text{Cls} \to 1 . \supset . S \dagger {}^{;} Q \in \text{Rel}^2\text{excl}$

Dem.

$\vdash . *72\cdot421 . \supset \vdash :. \text{Hp} . \supset : M, N \in C`Q . \exists ! S``C`M \cap S``C`N . \supset . \exists ! C`M \cap C`N .$

[*163·11] $\supset . M = N .$

[*30·37] $\supset . S \dagger M = S \dagger N$ (1)

$\vdash . (1) . *150\cdot202 . \supset \vdash :. \text{Hp} . \supset :$

$M, N \in C`Q . \exists ! C`(S \dagger M) \cap C`(S \dagger N) . \supset . S \dagger M = S \dagger N$ (2)

$\vdash . (2) . *163\cdot11 . \supset \vdash . \text{Prop}$

***163·31.** $\vdash :. C`P = C`Q . \supset : P \in \text{Rel}^2\text{excl} . \equiv . Q \in \text{Rel}^2\text{excl}$ [*163·1 . *13·12]

***163·311.** $\vdash :. C`Q = \text{Cnv}``C`P . \supset : P \in \text{Rel}^2\text{excl} . \equiv . Q \in \text{Rel}^2\text{excl}$

Dem.

$\vdash . *72\cdot513 . \supset \vdash :: \text{Hp} . \supset :. M, N \in C`P . \equiv . \breve{M}, \breve{N} \in C`Q :.$

[*31·32] $\supset :. M, N \in C`P . M \neq N . \equiv . \breve{M}, \breve{N} \in C`Q . \breve{M} \neq \breve{N} :.$

[*33·22] $\supset :. M, N \in C`P . M \neq N . \supset . C`M \cap C`N = \Lambda : \equiv :$

$\breve{M}, \breve{N} \in C`Q . \breve{M} \neq \breve{N} . \supset . C`\breve{M} \cap C`\breve{N} = \Lambda :.$

[*11·33.*163·1] $\supset :. P \in \text{Rel}^2\text{excl} . \equiv :$

$\breve{M}, \breve{N} \in C`Q . \breve{M} \neq \breve{N} . \supset_{M,N} . C`\breve{M} \cap C`\breve{N} = \Lambda :$

[*31·51] $\equiv : M, N \in C`Q . M \neq N . \supset_{M,N} . C`M \cap C`N = \Lambda :$

[*163·1] $\equiv : Q \in \text{Rel}^2\text{excl} :: \supset \vdash . \text{Prop}$

***163·32.** $\vdash : P \in \text{Rel}^2\text{excl} . \equiv . \breve{P} \in \text{Rel}^2\text{excl} . \equiv . \text{Cnv}{}^{;}P \in \text{Rel}^2\text{excl} . \equiv .$

$\text{Cnv}{}^{;}\breve{P} \in \text{Rel}^2\text{excl}$ [*163·31·311 . *33·22 . *150·22·12]

***163·33.** $\vdash : P \Uparrow Q \in \text{Rel}^2\text{excl} . \equiv . Q \Uparrow P \in \text{Rel}^2\text{excl}$ [*163·31 . *160·14]

***163·331.** $\vdash : P \nrightarrow R \in \text{Rel}^2\text{excl} . \equiv . R \nleftarrow P \in \text{Rel}^2\text{excl}$

[*163·31 . *161·14·2·201]

***163·4.** $\vdash . \dot{\Lambda} \in \text{Rel}^2\text{excl}$

Dem.

$\vdash . *33\cdot241 . *24\cdot105 . \supset \vdash . (Q) . Q \sim\in C`\dot{\Lambda} .$

[*11·57] $\supset \vdash . (Q, R) . Q, R \sim\in C`\dot{\Lambda} .$

[*11·63] $\supset \vdash : Q, R \in C`\dot{\Lambda} . Q \neq R . \supset_{Q,R} . C`Q \cap C`R = \Lambda$ (1)

$\vdash . (1) . *163\cdot1 . \supset \vdash . \text{Prop}$

***163·41.** $\vdash . P \downarrow P \in \text{Rel}^2\text{excl}$

Dem.

$\vdash . *54\cdot25 . *55\cdot15 . \supset \vdash . C`(P \downarrow P) \in 1 .$

[*52·41.Transp] $\supset \vdash . \sim(\exists Q, R) . Q, R \in C`(P \downarrow P) . Q \neq R .$

[*11·63] $\supset \vdash : Q, R \in C`(P \downarrow P) . Q \neq R . \supset_{Q,R} . C`Q \cap C`R = \Lambda$ (1)

$\vdash . (1) . *163\cdot1 . \supset \vdash . \text{Prop}$

∗163·42. $\vdash :. P \downarrow Q \in \text{Rel}^2\,\text{excl} . \equiv : P = Q . \vee . C^{\text{‘}}P \cap C^{\text{‘}}Q = \Lambda$

Dem.

$\vdash . *163·1 . *55·15 . \supset$

$\vdash :. P \downarrow Q \in \text{Rel}^2\,\text{excl} . \equiv : M, N \in \iota^{\text{‘}}P \cup \iota^{\text{‘}}Q . M \neq N . \supset_{M,N} . C^{\text{‘}}M \cap C^{\text{‘}}N = \Lambda :$

[∗54·441] $\equiv : P = Q . \vee . C^{\text{‘}}P \cap C^{\text{‘}}Q = \Lambda . C^{\text{‘}}Q \cap C^{\text{‘}}P = \Lambda :$

[∗22·51] $\equiv : P = Q . \vee . C^{\text{‘}}P \cap C^{\text{‘}}Q = \Lambda :. \supset \vdash . \text{Prop}$

The above proposition is used in ∗251·22.

∗163·43. $\vdash : P \in \text{Rel}^2\,\text{excl} . Q \Subset P . \supset . Q \in \text{Rel}^2\,\text{excl}$

Dem.

$\vdash . *33·265 . \supset \vdash :. \text{Hp} . \supset : M, N \in C^{\text{‘}}Q . \supset . M, N \in C^{\text{‘}}P :$

[Fact] $\supset : M, N \in C^{\text{‘}}Q . M \neq N . \supset . M, N \in C^{\text{‘}}P . M \neq N :$

[∗163·1.Hp] $\supset . C^{\text{‘}}M \cap C^{\text{‘}}N = \Lambda$ (1)

$\vdash . (1) . *163·1 . \supset \vdash . \text{Prop}$

∗163·431. $\vdash : P \in \text{Rel}^2\,\text{excl} . \supset . \text{Rl}^{\text{‘}}P \subset \text{Rel}^2\,\text{excl}$ [∗163·43]

∗163·44. $\vdash : P \ddagger Q \in \text{Rel}^2\,\text{excl} . \equiv .$

$P, Q \in \text{Rel}^2\,\text{excl} . s^{\text{‘}}C^{\text{‘‘}}C^{\text{‘}}P \cap s^{\text{‘}}C^{\text{‘‘}}(C^{\text{‘}}Q - C^{\text{‘}}P) = \Lambda$

Dem.

$\vdash . *163·12 . *160·14 . \supset \vdash : P \ddagger Q \in \text{Rel}^2\,\text{excl} . \equiv . F \upharpoonright (C^{\text{‘}}P \cup C^{\text{‘}}Q) \in \text{Cls} \rightarrow 1 .$

[∗74·821] $\equiv . F \upharpoonright C^{\text{‘}}P, F \upharpoonright C^{\text{‘}}Q \in \text{Cls} \rightarrow 1 . F^{\text{‘‘}}C^{\text{‘}}P \cap F^{\text{‘‘}}(C^{\text{‘}}Q - C^{\text{‘}}P) = \Lambda .$

[∗163·12.∗40·56] $\equiv . P, Q \in \text{Rel}^2\,\text{excl} . s^{\text{‘}}C^{\text{‘‘}}C^{\text{‘}}P \cap s^{\text{‘}}C^{\text{‘‘}}(C^{\text{‘}}Q - C^{\text{‘}}P) = \Lambda : \supset \vdash . \text{Prop}$

∗163·441. $\vdash : P, Q \in \text{Rel}^2\,\text{excl} . C^{\text{‘}}\Sigma^{\text{‘}}P \cap C^{\text{‘}}\Sigma^{\text{‘}}Q = \Lambda . \supset . P \ddagger Q \in \text{Rel}^2\,\text{excl}$

[∗163·44 . ∗162·22]

The above proposition is used in ∗173·26.

∗163·442. $\vdash :. C^{\text{‘}}P \cap C^{\text{‘}}Q = \Lambda . \supset :$

$P \ddagger Q \in \text{Rel}^2\,\text{excl} . \equiv . P, Q \in \text{Rel}^2\,\text{excl} . C^{\text{‘}}\Sigma^{\text{‘}}P \cap C^{\text{‘}}\Sigma^{\text{‘}}Q = \Lambda$

Dem.

$\vdash . *24·313 . \supset \vdash : \text{Hp} . \supset . C^{\text{‘}}Q - C^{\text{‘}}P = C^{\text{‘}}Q$ (1)

$\vdash . (1) . *163·44 . *162·22 . \supset \vdash . \text{Prop}$

∗163·45. $\vdash : P \rightarrowtail R \in \text{Rel}^2\,\text{excl} . \equiv . P \in \text{Rel}^2\,\text{excl} . s^{\text{‘}}C^{\text{‘‘}}(C^{\text{‘}}P - \iota^{\text{‘}}R) \cap C^{\text{‘}}R = \Lambda$

Dem.

$\vdash . *161·14 . *163·12 . \supset$

$\vdash :. \dot{\exists} ! P . \supset : P \rightarrowtail R \in \text{Rel}^2\,\text{excl} . \equiv . F \upharpoonright (C^{\text{‘}}P \cup \iota^{\text{‘}}R) \in \text{Cls} \rightarrow 1 .$

[∗74·821.∗53·301.∗33·5]

$\equiv . F \upharpoonright C^{\text{‘}}P, F \upharpoonright \iota^{\text{‘}}R \in \text{Cls} \rightarrow 1 . F^{\text{‘‘}}(C^{\text{‘}}P - \iota^{\text{‘}}R) \cap C^{\text{‘}}R = \Lambda .$

[∗35·101.∗71·171] $\equiv . F \upharpoonright C^{\text{‘}}P \in \text{Cls} \rightarrow 1 . F^{\text{‘‘}}(C^{\text{‘}}P - \iota^{\text{‘}}R) \cap C^{\text{‘}}R = \Lambda .$

[∗163·12.∗40·56] $\equiv : P \in \text{Rel}^2\,\text{excl} . s^{\text{‘}}C^{\text{‘‘}}(C^{\text{‘}}P - \iota^{\text{‘}}R) \cap C^{\text{‘}}R = \Lambda$ (1)

$\vdash . *161 \cdot 2 . *163 \cdot 4 . \supset \vdash : P = \dot{\Lambda} . \supset . P \not\mapsto R \in \text{Rel}^2 \text{excl} . P \in \text{Rel}^2 \text{excl}$ (2)

$\vdash . *33 \cdot 241 . *37 \cdot 29 . *40 \cdot 21 . \supset \vdash : P = \dot{\Lambda} . \supset . s`C``(C`P - \iota`R) \cap C`R = \Lambda$ (3)

$\vdash . (2) . (3) . \text{Comp} . *5 \cdot 1 . \supset \vdash :. P = \dot{\Lambda} . \supset :$

$P \not\mapsto R \in \text{Rel}^2 \text{excl} . \equiv . P \in \text{Rel}^2 \text{excl} . s`C``(C`P - \iota`R) \cap C`R = \Lambda$ (4)

$\vdash . (1) . (4) . \supset \vdash . \text{Prop}$

$*163 \cdot 451$. $\vdash : P \in \text{Rel}^2 \text{excl} . C`\Sigma`P \cap C`R = \Lambda . \supset . P \not\mapsto R \in \text{Rel}^2 \text{excl}$
[$*163 \cdot 45 . *162 \cdot 22$]

The above proposition is used in $*173 \cdot 25$.

$*163 \cdot 452$. $\vdash :. R \sim \in C`P . \supset : P \not\mapsto R \in \text{Rel}^2 \text{excl} . \equiv . P \in \text{Rel}^2 \text{excl} . C`\Sigma`P \cap C`R = \Lambda$
[$*51 \cdot 222 . *163 \cdot 45 . *162 \cdot 22$]

$*163 \cdot 46$. $\vdash : R \not\mapsfrom P \in \text{Rel}^2 \text{excl} . \equiv . P \in \text{Rel}^2 \text{excl} . s`C``(C`P - \iota`R) \cap C`R = \Lambda$
[$*163 \cdot 45 \cdot 331$]

$*163 \cdot 461$. $\vdash : P \in \text{Rel}^2 \text{excl} . C`\Sigma`P \cap C`R = \Lambda . \supset . R \not\mapsfrom P \in \text{Rel}^2 \text{excl}$
[$*163 \cdot 451 \cdot 331$]

$*163 \cdot 462$. $\vdash :. R \sim \in C`P . \supset : R \not\mapsfrom P \in \text{Rel}^2 \text{excl} . \equiv . P \in \text{Rel}^2 \text{excl} . C`\Sigma`P \cap C`R = \Lambda$
[$*163 \cdot 452 \cdot 331$]

*164. DOUBLE LIKENESS.

Summary of *164.

The subject of this number is of great importance throughout relation-arithmetic and its applications. Double likeness, or double ordinal similarity, is a relation which is to hold between P and Q when (1) P and Q are like, (2) correlated members of the fields of P and Q are like, with a specific given correlator in each case. (It is necessary, in general, to have a given correlator in each case, to avoid the necessity of the multiplicative axiom for selecting among correlators.) This definition can be somewhat simplified by starting from a relation correlating $\Sigma\text{‘}P$ and $\Sigma\text{‘}Q$. If S is such a correlator, so that

$$S \epsilon 1 \rightarrow 1 \,.\, \text{Ɑ‘}S = C\text{‘}\Sigma\text{‘}Q \,.\, \Sigma\text{‘}P = S{;}\Sigma\text{‘}Q,$$

we want S to be such that it not only correlates the whole of $\Sigma\text{‘}P$ with the whole of $\Sigma\text{‘}Q$, but also correlates each member of $C\text{‘}P$ with the corresponding member of $C\text{‘}Q$, *i.e.* such that, if N is any member of $C\text{‘}Q$, $S{;}N$ is the corresponding member of $C\text{‘}P$. This requires

$$NQN' \,.\equiv.\, (S{;}N)\, P\, (S{;}N'),$$

i.e. writing $S\dagger\text{‘}N$, $S\dagger\text{‘}N'$ in place of $S{;}N$, $S{;}N'$, it requires

$$P = S\dagger{;}Q.$$

When $P = S\dagger{;}Q$ and $\text{Ɑ‘}S = C\text{‘}\Sigma\text{‘}Q$, we have $\Sigma\text{‘}P = S{;}\Sigma\text{‘}Q$ by *162·35. Hence double likeness will subsist if there is a relation S such that

$$S \epsilon 1 \rightarrow 1 \,.\, \text{Ɑ‘}S = C\text{‘}\Sigma\text{‘}Q \,.\, P = S\dagger{;}Q.$$

A relation S fulfilling this condition will be called a *double correlator* of P and Q. Thus two relations P and Q have double likeness when there exists a double correlator of P and Q, *i.e.* when

$$(\exists S) \,.\, S \epsilon 1 \rightarrow 1 \,.\, \text{Ɑ‘}S = C\text{‘}\Sigma\text{‘}Q \,.\, P = S\dagger{;}Q.$$

A double correlator of P and Q is a relation S which is a correlator of $\Sigma\text{‘}P$ and $\Sigma\text{‘}Q$ and is such that $S\dagger \upharpoonright C\text{‘}Q$ is a correlator of P and Q.

It will be seen that this definition has the usual analogy to the corresponding definition in cardinals (*111·01). The two inverted commas of the cardinal definition are replaced by the semi-colon, and S_ϵ is replaced by $S\dagger$, and $s\text{‘}\lambda$ is replaced by $\Sigma\text{‘}Q$ or $C\text{‘}\Sigma\text{‘}Q$. The propositions of the present number consist largely of analogues of the propositions of *111, in accordance with the above substitutions.

If it were not for the difficulty of choice among correlators, we could define two relations as having double likeness when they are like relations of like relations, *i.e.* when, if P and Q are the two relations, they have a correlator S such that, if MSN, then M smor N. In this case, $S \in P \,\overline{\text{smor}}\, Q \cap \text{Rl}\text{'smor}$. Thus we have to consider the relations of the class $P \,\overline{\text{smor}}\, Q \cap \text{Rl}\text{'smor}$ to the class of double correlators, and we have to consider the relation of the relation "$\exists ! P \,\overline{\text{smor}}\, Q \cap \text{Rl}\text{'smor}$" to the relation of double likeness. The propositions to be proved on this subject in the present number are analogous to the propositions of $*111$. But at a later stage ($*251{\cdot}61$) we shall show that if the field of P consists entirely of relations which generate *well-ordered* series, then the use of the multiplicative axiom ceases to be necessary in identifying double likeness with the relation $\exists ! P \,\overline{\text{smor}}\, Q \cap \text{Rl}\text{'smor}$, the reason being that two well-ordered series can never be correlated in more than one way.

Our definitions are

$*164{\cdot}01$. $P \,\overline{\text{smor}}\,\overline{\text{smor}}\, Q = (1 \to 1) \cap \overleftarrow{\text{ᗡ}}\text{'}C\text{'}\Sigma\text{'}Q \cap \hat{S}(P = S\dagger{;}Q)$ Df

$*164{\cdot}02$. smor smor $= \hat{P}\hat{Q}(\exists ! P \,\overline{\text{smor}}\,\overline{\text{smor}}\, Q)$ Df

The principal propositions of this number are

$*164{\cdot}15$. $\vdash : S \in P \,\overline{\text{smor}}\,\overline{\text{smor}}\, Q \,.\equiv.\, S \in \Sigma\text{'}P \,\overline{\text{smor}}\, \Sigma\text{'}Q \,.\, (S\dagger) \restriction C\text{'}Q \in P \,\overline{\text{smor}}\, Q$

whence

$*164{\cdot}151$. $\vdash : P \text{ smor smor } Q \,.\supset.\, \Sigma\text{'}P \text{ smor } \Sigma\text{'}Q \,.\, P \text{ smor } Q$

$*164{\cdot}18$. $\vdash : S \restriction C\text{'}\Sigma\text{'}Q \in P \,\overline{\text{smor}}\,\overline{\text{smor}}\, Q \,.\equiv.$

$$S \restriction C\text{'}\Sigma\text{'}Q \in 1 \to 1 \,.\, C\text{'}\Sigma\text{'}Q \subset \text{ᗡ}\text{'}S \,.\, P = S\dagger{;}Q$$

This is usually the most convenient proposition when a double correlation has to be proved.

$*164{\cdot}201{\cdot}211{\cdot}221$. Double likeness is reflexive, symmetrical and transitive.

$*164{\cdot}31$. $\vdash : S \in P \,\overline{\text{smor}}\,\overline{\text{smor}}\, Q \,.\equiv.\, S \in (C\text{''}C\text{'}P)\,\overline{\text{sm}}\,\overline{\text{sm}}\,(C\text{''}C\text{'}Q) \,.\, P = S\dagger{;}Q$

(Cf. note to $*164{\cdot}31$, below.)

We then have a set of propositions ($*164{\cdot}4$ to the end) on the identification of $\exists ! P \,\overline{\text{smor}}\, Q \cap \text{Rl}\text{'smor}$ with double likeness by means of the multiplicative axiom. We have

$*164{\cdot}43$. $\vdash :. P, Q \in \text{Rel}^2\,\text{excl} \,.\, S \in P \,\overline{\text{smor}}\, Q \,.$

$$\mu = \hat{\lambda}\{(\exists N) \,.\, N \in C\text{'}Q \,.\, \lambda = (S\text{'}N) \,\overline{\text{smor}}\, N\} \,.\supset :$$

$$R \in \epsilon_\Delta\text{'}\mu \,.\supset.\, \dot{s}\text{'D'}R \in P \,\overline{\text{smor}}\,\overline{\text{smor}}\, Q \,.\, S = (\dot{s}\text{'D'}R)\dagger \restriction C\text{'}Q$$

That is to say, given that P and Q are like relations of like mutually exclusive relations, if we can pick out one correlator for each pair of correlated members of $C\text{'}P$ and $C\text{'}Q$, then the sum ($\dot{s}$) of such selected correlators is a double correlator of P and Q. Hence, observing that if S is a double correlator of P and Q, $(S\dagger) \restriction C\text{'}Q \in P \,\overline{\text{smor}}\, Q \cap \text{Rl}\text{'smor}$ ($*164{\cdot}15{\cdot}16$), we arrive at

$*164\cdot45.\ \vdash :: \text{Mult ax} . \supset :.$

$P, Q \in \text{Rel}^2 \text{excl} . \supset : \exists ! P \overline{\text{smor}} Q \cap \text{Rl}`\text{smor} . \equiv . P \text{ smor smor } Q$

From $*164\cdot43$ we deduce also

$*164\cdot46.\ \vdash :. \text{Mult ax} . \supset :$

$P, Q \in \text{Rel}^2 \text{excl} . \exists ! P \overline{\text{smor}} Q \cap \text{Rl}`\text{smor} . \supset . \Sigma`P \text{ smor } \Sigma`Q$

$*164\cdot48.\ \vdash :. \text{Mult ax} . \supset : R, S \in \text{Rel}^2 \text{excl} \cap \text{Nr}`Q . C`R, C`S \in \text{Cl}`\text{Nr}`P . \supset .$

$R \text{ smor smor } S . \Sigma`R \text{ smor } \Sigma`S$

I.e. in effect, assuming the multiplicative axiom, if two series ($\Sigma`R$ and $\Sigma`S$) can each be divided into β sets of α terms (α, β being relation-numbers), then the two series are ordinally similar, and the β sets in the one case have double similarity with the β sets in the other. (Here we have written α, β in place of the $\text{Nr}`P$ and $\text{Nr}`Q$ of the enunciation.)

It is by means of the above propositions that ordinal addition and multiplication are connected, as will appear in $*166$.

$*164\cdot01.\quad P \overline{\text{smor}}\,\overline{\text{smor}} Q = (1 \rightarrow 1) \cap \overleftarrow{\text{G}}`C`\Sigma`Q \cap \hat{S}(P = S\dagger;Q) \quad \text{Df}$

$*164\cdot02.\quad \text{smor smor} = \hat{P}\hat{Q}(\exists ! P \overline{\text{smor}}\,\overline{\text{smor}} Q) \quad \text{Df}$

$*164\cdot1.\quad \vdash : S \in P \overline{\text{smor}}\,\overline{\text{smor}} Q . \equiv . S \in 1 \rightarrow 1 . \text{G}`S = C`\Sigma`Q . P = S\dagger;Q$
$[(*164\cdot01)]$

$*164\cdot11.\quad \vdash : P \text{ smor smor } Q . \equiv . \exists ! P \overline{\text{smor}}\,\overline{\text{smor}} Q \quad [(*164\cdot02)]$

$*164\cdot12.\quad \vdash : P \text{ smor smor } Q . \equiv . (\exists S) . S \in 1 \rightarrow 1 . \text{G}`S = C`\Sigma`Q . P = S\dagger;Q$
$[*164\cdot1\cdot11]$

$*164\cdot13.\quad \vdash : S \upharpoonright C`\Sigma`Q \in 1 \rightarrow 1 . C`\Sigma`Q \subset \text{G}`S . \supset . (S\dagger) \upharpoonright C`Q \in 1 \rightarrow 1$
$[*150\cdot152 . *162\cdot22]$

$*164\cdot131.\ \vdash : \text{G}`S = C`\Sigma`Q . P = S\dagger;Q . \supset . \text{D}`S = C`\Sigma`P . \Sigma`P = S;\Sigma`Q$

Dem.

$\vdash . *162\cdot35 . \supset \vdash : \text{Hp} . \supset . \Sigma`P = S;\Sigma`Q \quad (1)$

$[*150\cdot23.\text{Hp}] \quad \supset . C`\Sigma`P = \text{D}`S \quad (2)$

$\vdash . (1) . (2) . \supset \vdash . \text{Prop}$

$*164\cdot14.\quad \vdash : S \in P \overline{\text{smor}}\,\overline{\text{smor}} Q . \supset . S \in \Sigma`P \overline{\text{smor}} \Sigma`Q \quad [*164\cdot1\cdot131 . *151\cdot11]$

The two following propositions are required for proving $*164\cdot18$.

$*164\cdot141.\ \vdash : C`\Sigma`Q \subset \alpha . \supset . (T \upharpoonright \alpha)\dagger;Q = T\dagger;Q \quad [*150\cdot171 . *162\cdot22]$

$*164\cdot142.\ \vdash . (T \upharpoonright C`\Sigma`Q)\dagger;Q = T\dagger;Q = \{(T\dagger) \upharpoonright C`Q\};Q \quad [*164\cdot141 . *150\cdot32]$

$*164\cdot143.\ \vdash : S \in P \overline{\text{smor}}\,\overline{\text{smor}} Q . \supset . (S\dagger) \upharpoonright C`Q \in P \overline{\text{smor}} Q$

Dem.

$\vdash . *164\cdot1\cdot13 . \quad \supset \vdash : \text{Hp} . \supset . (S\dagger) \upharpoonright C`Q \in 1 \rightarrow 1 \quad (1)$

$\vdash . *35\cdot65 . \quad \supset \vdash . \text{G}`(S\dagger) \upharpoonright C`Q = C`Q \quad (2)$

$\vdash . *164\cdot1 . *150\cdot32 . \supset \vdash : \text{Hp} . \supset . P = \{(S\dagger) \upharpoonright C`Q\};Q \quad (3)$

$\vdash . (1) . (2) . (3) . *151\cdot11 . \supset \vdash . \text{Prop}$

***164·15.** $\vdash : S \in P \,\overline{\text{smor}}\, \overline{\text{smor}}\, Q . \equiv . S \in \Sigma\text{‘}P \,\overline{\text{smor}}\, \Sigma\text{‘}Q . (S\dagger) \upharpoonright C\text{‘}Q \in P \,\overline{\textbf{smor}}\, Q$

Dem.

$\vdash . *164·14·143 . \supset$

$\vdash : S \in P \,\overline{\text{smor}}\, \overline{\text{smor}}\, Q . \supset . S \in \Sigma\text{‘}P \,\overline{\text{smor}}\, \Sigma\text{‘}Q . (S\dagger) \upharpoonright C\text{‘}Q \in P \,\overline{\text{smor}}\, Q \quad (1)$

$\vdash . *151·11 . \supset$

$\vdash : S \in \Sigma\text{‘}P \,\overline{\text{smor}}\, \Sigma\text{‘}Q . (S\dagger) \upharpoonright C\text{‘}Q \in P \,\overline{\text{smor}}\, Q . \supset .$

$S \in 1 \rightarrow 1 . \text{Ɑ}\text{‘}S = C\text{‘}\Sigma\text{‘}Q . P = \{(S\dagger) \upharpoonright C\text{‘}Q\} ; Q \quad (2)$

$\vdash . (2) . *150·32 . *164·1 . \supset$

$\vdash : S \in \Sigma\text{‘}P \,\overline{\text{smor}}\, \Sigma\text{‘}Q . (S\dagger) \upharpoonright C\text{‘}Q \in P \,\overline{\text{smor}}\, Q . \supset . S \in P \,\overline{\text{smor}}\, \overline{\text{smor}}\, Q \quad (3)$

$\vdash . (1) . (3) . \supset \vdash . \text{Prop}$

***164·151.** $\vdash : P \text{ smor smor } Q . \supset . \Sigma\text{‘}P \text{ smor } \Sigma\text{‘}Q . P \text{ smor } Q$ [*164·15·11]

***164·16.** $\vdash : S \in P \,\overline{\text{smor}}\, \overline{\text{smor}}\, Q . \supset . (S\dagger) \upharpoonright C\text{‘}Q \in \text{smor}$

Dem.

$\vdash . *35·101 . *150·1 . \supset \vdash : M \{(S\dagger) \upharpoonright C\text{‘}Q\} N . \equiv . N \in C\text{‘}Q . M = S ; N \quad (1)$

$\vdash . *164·1 . *162·22 . \supset \vdash :. \text{Hp} . \supset : S \in 1 \rightarrow 1 : N \in C\text{‘}Q . \supset_N . C\text{‘}N \subset \text{Ɑ}\text{‘}S :$

[*151·23] $\supset : N \in C\text{‘}Q . M = S ; N . \supset_{M, N} . M \text{ smor } N$

[(1)] $\supset : M \{(S\dagger) \upharpoonright C\text{‘}Q\} N . \supset_{M, N} . M \text{ smor } N :. \supset \vdash . \text{Prop}$

***164·17.** $\vdash : P \text{ smor smor } Q . \supset . \exists ! P \,\overline{\text{smor}}\, Q \cap \text{Rl}\text{‘smor}$ [*164·143·16]

This proposition states that when P and Q have double likeness, there is a correlator of P and Q which couples like with like relations; *i.e.* if S is the correlator, then, if MSN, M and N are ordinally similar. The converse of this proposition, namely, that if P and Q have a correlator which couples ordinally similar relations, then P and Q have double likeness, can be proved if the multiplicative axiom is assumed, but not otherwise, except in special cases, such as that of well-ordered series.

The following proposition is used frequently, owing to the fact that, in the cases we are concerned with, double correlators generally have the form $S \upharpoonright C\text{‘}\Sigma\text{‘}Q$, where S is some relation for which we have $(y) . \text{E} ! S\text{‘}y$.

***164·18.** $\vdash : S \upharpoonright C\text{‘}\Sigma\text{‘}Q \in P \,\overline{\text{smor}}\, \overline{\text{smor}}\, Q . \equiv .$

$S \upharpoonright C\text{‘}\Sigma\text{‘}Q \in 1 \rightarrow 1 . C\text{‘}\Sigma\text{‘}Q \subset \text{Ɑ}\text{‘}S . P = \dot{S}\dagger ; Q$

Dem.

$\vdash . *35·64 . *22·621 . \supset \vdash : \text{Ɑ}\text{‘}(S \upharpoonright C\text{‘}\Sigma\text{‘}Q) = C\text{‘}\Sigma\text{‘}Q . \equiv . C\text{‘}\Sigma\text{‘}Q \subset \text{Ɑ}\text{‘}S \quad (1)$

$\vdash . *164·142 . \quad \supset \vdash : P = (S \upharpoonright C\text{‘}\Sigma\text{‘}Q)\dagger ; Q . \equiv . P = S\dagger ; Q \quad (2)$

$\vdash . *164·1 . \quad \supset \vdash : S \upharpoonright C\text{‘}\Sigma\text{‘}Q \in P \,\overline{\text{smor}}\, \overline{\text{smor}}\, Q . \equiv .$

$S \upharpoonright C\text{‘}\Sigma\text{‘}Q \in 1 \rightarrow 1 . \text{Ɑ}\text{‘}(S \upharpoonright C\text{‘}\Sigma\text{‘}Q) = C\text{‘}\Sigma\text{‘}Q . P = (S \upharpoonright C\text{‘}\Sigma\text{‘}Q)\dagger ; Q \quad (3)$

$\vdash . (1) . (2) . (3) . \supset \vdash . \text{Prop}$

∗164·181. $\vdash : P \,\text{smor smor}\, Q . \equiv . (\exists S) . S \upharpoonright C`\Sigma`Q \in 1 \to 1 . C`\Sigma`Q \subset \text{Ɑ}`S . P = S \dagger ; Q$

Dem.

$\vdash . *35{\cdot}66 . *164{\cdot}1 . \supset$

$\vdash : S \in P \,\overline{\text{smor}}\,\overline{\text{smor}}\, Q . \supset . S \upharpoonright C`\Sigma`Q \in 1 \to 1 . C`\Sigma`Q \subset \text{Ɑ}`S . P = S \dagger ; Q \quad (1)$

$\vdash . (1) . *164{\cdot}11 . \supset$

$\vdash : P \,\text{smor smor}\, Q . \supset . (\exists S) . S \upharpoonright C`\Sigma`Q \in 1 \to 1 . C`\Sigma`Q \subset \text{Ɑ}`S . P = S \dagger ; Q \quad (2)$

$\vdash . *164{\cdot}18{\cdot}11 . \supset$

$\vdash : (\exists S) . S \upharpoonright C`\Sigma`Q \in 1 \to 1 . C`\Sigma`Q \subset \text{Ɑ}`S . P = S \dagger ; Q . \supset . P \,\text{smor smor}\, Q \quad (3)$

$\vdash . (2) . (3) . \supset \vdash . \text{Prop}$

The following propositions are concerned in proving that double likeness is reflexive, symmetrical, and transitive.

∗164·2. $\vdash . I \upharpoonright C`\Sigma`P \in P \,\overline{\text{smor}}\,\overline{\text{smor}}\, P$

Dem.

$\vdash . *151{\cdot}121 . \supset \vdash . I \upharpoonright C`\Sigma`P \in \Sigma`P \,\overline{\text{smor}}\, \Sigma`P . I \upharpoonright C`P \in P \,\overline{\text{smor}}\, P \quad (1)$

$\vdash . *35{\cdot}101 . *150{\cdot}1 . \supset$

$\vdash : M \{(I \upharpoonright C`\Sigma`P) \dagger \upharpoonright C`P\} N . \equiv . N \in C`P . M = (I \upharpoonright C`\Sigma`P) ; N .$

$[*150{\cdot}33 . *162{\cdot}22] \qquad \equiv . N \in C`P . M = I ; N .$

$[*150{\cdot}53] \qquad \equiv . M (I \upharpoonright C`P) N \quad (2)$

$\vdash . (1) . (2) . \supset \vdash . I \upharpoonright C`\Sigma`P \in \Sigma`P \,\overline{\text{smor}}\, \Sigma`P . (I \upharpoonright C`\Sigma`P) \dagger \upharpoonright C`P \in P \,\overline{\text{smor}}\, P .$

$[*164{\cdot}15] \quad \supset \vdash . \text{Prop}$

∗164·201. $\vdash . P \,\text{smor smor}\, P \quad [*164{\cdot}2{\cdot}11]$

∗164·21. $\vdash : S \in P \,\overline{\text{smor}}\,\overline{\text{smor}}\, Q . \equiv . \breve{S} \in Q \,\overline{\text{smor}}\,\overline{\text{smor}}\, P$

Dem.

$\vdash . *164{\cdot}1 . *71{\cdot}212 . \supset \vdash : S \in P \,\overline{\text{smor}}\,\overline{\text{smor}}\, Q . \supset . \breve{S} \in 1 \to 1 \quad (1)$

$\vdash . *164{\cdot}131{\cdot}1 . \qquad \supset \vdash : S \in P \,\overline{\text{smor}}\,\overline{\text{smor}}\, Q . \supset . \text{Ɑ}`\breve{S} = C`\Sigma`P \quad (2)$

$\vdash . *150{\cdot}94 . *164{\cdot}1 . *162{\cdot}22 . \supset \vdash : S \in P \,\overline{\text{smor}}\,\overline{\text{smor}}\, Q . \supset . Q = \breve{S} \dagger ; P \quad (3)$

$\vdash . (1) . (2) . (3) . *164{\cdot}1 . \supset \vdash : S \in P \,\overline{\text{smor}}\,\overline{\text{smor}}\, Q . \supset . \breve{S} \in Q \,\overline{\text{smor}}\,\overline{\text{smor}}\, P \quad (4)$

$\vdash . (4) \dfrac{\breve{S}, Q, P}{S, P, Q} . \qquad \supset \vdash : \breve{S} \in Q \,\overline{\text{smor}}\,\overline{\text{smor}}\, P . \supset . S \in P \,\overline{\text{smor}}\,\overline{\text{smor}}\, Q \quad (5)$

$\vdash . (4) . (5) . \supset \vdash . \text{Prop}$

∗164·211. $\vdash : P \,\text{smor smor}\, Q . \equiv . Q \,\text{smor smor}\, P \quad [*164{\cdot}21{\cdot}11]$

∗164·22. $\vdash : S \in P \,\overline{\text{smor}}\,\overline{\text{smor}}\, Q . T \in Q \,\overline{\text{smor}}\,\overline{\text{smor}}\, R . \supset . S | T \in P \,\overline{\text{smor}}\,\overline{\text{smor}}\, R$

Dem.

$\vdash . *164{\cdot}1 . \qquad \supset \vdash : \text{Hp} . \supset . S, T \in 1 \to 1 .$

$[*71{\cdot}252] \qquad \supset . S | T \in 1 \to 1 \quad (1)$

$\vdash . *164{\cdot}1{\cdot}131 . \supset \vdash : \text{Hp} . \supset . \text{Ɑ}`S = C`\Sigma`Q . \text{D}`T = C`\Sigma`Q .$

$[*37{\cdot}323] \qquad \supset . \text{Ɑ}`(S | T) = \text{Ɑ}`T .$

$[*164{\cdot}1] \qquad \supset . \text{Ɑ}`(S | T) = C`\Sigma`R \quad (2)$

$\vdash . *150\cdot13\cdot14 . \supset \vdash . (S \mid T)\dagger;R = S\dagger;T\dagger;R$ (3)

$\vdash . *164\cdot1 . \quad \supset \vdash : \text{Hp} . \supset . T\dagger;R = Q . S\dagger;Q = P$ (4)

$\vdash . (3) . (4) . \quad \supset \vdash : \text{Hp} . \supset . (S \mid T)\dagger;R = P$ (5)

$\vdash . (1) . (2) . (5) . *164\cdot1 . \supset \vdash . \text{Prop}$

***164·221.** $\vdash : P \text{ smor smor } Q . Q \text{ smor smor } R . \supset . P \text{ smor smor } R$ [*164·22·11]

***164·23.** $\vdash :. P \text{ smor smor } Q . \supset : P \epsilon \text{Rel}^2 \text{excl} . \equiv . Q \epsilon \text{Rel}^2 \text{excl}$

Dem.

$\vdash . *164\cdot12 . \quad \supset \vdash :. \text{Hp} . \supset : (\exists T) . T \epsilon 1 \to 1 . \text{D}\text{'}T = C\text{'}\Sigma\text{'}Q . P = T\dagger;Q :$

$[*163\cdot3] \quad \supset : Q \epsilon \text{Rel}^2 \text{excl} . \supset . P \epsilon \text{Rel}^2 \text{excl}$ (1)

$\vdash . (1) . *164\cdot211 . \supset \vdash :. \text{Hp} . \supset : P \epsilon \text{Rel}^2 \text{excl} . \supset . Q \epsilon \text{Rel}^2 \text{excl}$ (2)

$\vdash . (1) . (2) . \supset \vdash . \text{Prop}$

***164·3.** $\vdash : S \epsilon P \,\overline{\text{smor}}\, \overline{\text{smor}}\, Q . \supset . S \epsilon (C\text{''}C\text{'}P) \,\overline{\text{sm}}\, \overline{\text{sm}}\, (C\text{''}C\text{'}Q)$

Dem.

$\vdash . *164\cdot1 . *162\cdot22 . \supset \vdash : \text{Hp} . \supset . S \epsilon 1 \to 1 . \text{D}\text{'}S = s\text{'}C\text{''}C\text{'}Q . P = S\dagger;Q .$ (1)

$[*150\cdot931] \quad \supset . C\text{''}C\text{'}P = S_\epsilon\text{''}C\text{''}C\text{'}Q$ (2)

$\vdash . (1) . (2) . *111\cdot1 . \supset \vdash . \text{Prop}$

***164·301.** $\vdash : P \text{ smor smor } Q . \supset . C\text{''}C\text{'}P \text{ sm sm } C\text{''}C\text{'}Q$ [*164·3·11 . *111·4]

***164·31.** $\vdash : S \epsilon P \,\overline{\text{smor}}\, \overline{\text{smor}}\, Q . \equiv . S \epsilon (C\text{''}C\text{'}P) \,\overline{\text{sm}}\, \overline{\text{sm}}\, (C\text{''}C\text{'}Q) . P = S\dagger;Q$

Dem.

$\vdash . *164\cdot3\cdot1 . \supset$

$\vdash : S \epsilon P \,\overline{\text{smor}}\, \overline{\text{smor}}\, Q . \supset . S \epsilon (C\text{''}C\text{'}P) \,\overline{\text{sm}}\, \overline{\text{sm}}\, (C\text{''}C\text{'}Q) . P = S\dagger;Q$ (1)

$\vdash . *111\cdot1 . *162\cdot22 . \supset$

$\vdash : S \epsilon (C\text{''}C\text{'}P) \,\overline{\text{sm}}\, \overline{\text{sm}}\, (C\text{''}C\text{'}Q) . \supset . S \epsilon 1 \to 1 . \text{D}\text{'}S = C\text{'}\Sigma\text{'}Q$ (2)

$\vdash . (2) . \text{Fact} . *164\cdot1 . \supset$

$\vdash : S \epsilon (C\text{''}C\text{'}P) \,\overline{\text{sm}}\, \overline{\text{sm}}\, (C\text{''}C\text{'}Q) . P = S\dagger;Q . \supset . S \epsilon P \,\overline{\text{smor}}\, \overline{\text{smor}}\, Q$ (3)

$\vdash . (1) . (3) . \supset \vdash . \text{Prop}$

This proposition has the merit of reducing the ordinal element in double likeness to a minimum. The proof of

$$S \epsilon (C\text{''}C\text{'}P) \,\overline{\text{sm}}\, \overline{\text{sm}}\, (C\text{''}C\text{'}Q)$$

is a cardinal problem, and what has to be added for ordinal purposes is merely $P = S\dagger;Q$.

***164·32.** $\vdash . \dot\Lambda \epsilon (\dot\Lambda \,\overline{\text{smor}}\, \overline{\text{smor}}\, \dot\Lambda) . \dot\Lambda \text{ smor smor } \dot\Lambda$

In this proposition, the various $\dot\Lambda$'s need not be of the same type. Hence "$\dot\Lambda$ smor smor $\dot\Lambda$" is not an immediate consequence of *164·201.

Dem.

$\vdash . *72\cdot1 . *162\cdot4 . \quad \supset \vdash . \dot\Lambda \epsilon 1 \to 1 . \text{D}\text{'}\dot\Lambda = C\text{'}\Sigma\text{'}\dot\Lambda$ (1)

$\vdash . *150\cdot42 . \quad \supset \vdash . \dot\Lambda = \dot\Lambda ; \dot\Lambda$ (2)

$\vdash . (1) . (2) . *164\cdot1 . \supset \vdash . \dot\Lambda \epsilon (\dot\Lambda \,\overline{\text{smor}}\, \overline{\text{smor}}\, \dot\Lambda) .$ (3)

$[*164\cdot11] \quad \supset . \dot\Lambda \text{ smor smor } \dot\Lambda$ (4)

$\vdash . (3) . (4) . \supset \vdash . \text{Prop}$

***164·33.** $\vdash : M \in P \, \overline{\text{smor}} \, R \, . \, N \in Q \, \overline{\text{smor}} \, S \, . \, C\text{‘}P \cap C\text{‘}Q = \Lambda \, . \, C\text{‘}R \cap C\text{‘}S = \Lambda \, . \supset .$
$M \cup N \in (P \downarrow Q) \, \overline{\text{smor}} \, \overline{\text{smor}} \, (R \downarrow S)$

Dem.

$\vdash . *160·47 . \supset \vdash : \text{Hp} . \supset . M \cup N \in (P \ddagger Q) \, \overline{\text{smor}} \, (R \ddagger S) .$

$[*162·3 . *151·11] \quad \supset . M \cup N \in 1 \to 1 . \text{D}\text{‘}(M \cup N) = C\text{‘}\Sigma\text{‘}(R \downarrow S) \quad (1)$

$\vdash . *150·32 . \supset \vdash : \text{Hp} . \supset . (M \cup N) ; R = \{(M \cup N) \upharpoonright C\text{‘}R\} ; R$

$[*35·644 . *150·32] \quad = M ; R$

$[*151·11] \quad = P \quad (2)$

Similarly $\quad \vdash : \text{Hp} . \supset . (M \cup N) ; S = Q \quad (3)$

$\vdash . *150·71·1 . \supset \vdash : \text{Hp} . \supset . (M \cup N)\dagger ; (R \downarrow S) = \{(M \cup N) ; R\} \downarrow \{(M \cup N) ; S\}$

$[(2).(3)] \quad = P \downarrow Q \quad (4)$

$\vdash . (1) . (4) . *164·1 . \supset \vdash . \text{Prop}$

***164·34.** $\vdash : P \text{ smor } R \, . \, Q \text{ smor } S \, . \, C\text{‘}P \cap C\text{‘}Q = \Lambda \, . \, C\text{‘}R \cap C\text{‘}S = \Lambda \, . \supset .$
$P \downarrow Q \text{ smor smor } R \downarrow S$

$[*164·33·11 . *151·12]$

The following propositions are concerned in showing that, if P and Q are like relations, and the correlator of P and Q is contained in likeness (*i.e.* correlates relations which have the relation of likeness), a correlator being given for each pair of relations coupled by the correlator of P and Q, then the logical sum of such correlators is a double correlator of P and Q, provided P and Q are relations of mutually exclusive relations. That is, assuming S to be the correlator of P and Q, and assuming that $S\text{‘}N$ smor N whenever $N \in C\text{‘}Q$, let it be possible to choose one correlator out of the class of correlators $(S\text{‘}N) \, \overline{\text{smor}} \, N$, for every N which belongs to $C\text{‘}Q$. That is, assume that it is possible to make a selection from the class of classes of correlators. If μ is such a selection, then $\dot{s}\text{‘}\mu$ will be a double correlator of P and Q, if $P, Q \in \text{Rel}^2 \text{ excl}$.

The following propositions, down to *164·421, are lemmas for *164·43.

***164·4.** $\vdash :. N \in C\text{‘}Q . \supset_N . R\text{‘}N \in (S\text{‘}N) \, \overline{\text{smor}} \, N : \supset . \text{D}\text{‘}\dot{s}\text{‘}R\text{“}C\text{‘}Q = C\text{‘}\Sigma\text{‘}Q$

Dem.

$\vdash . *41·44 . \supset \vdash . \text{D}\text{‘}\dot{s}\text{‘}R\text{“}C\text{‘}Q = s\text{‘}\text{D}\text{“}R\text{“}C\text{‘}Q \quad (1)$

$\vdash . *151·11 . \supset \vdash :. \text{Hp} . \supset : N \in C\text{‘}Q . \supset . \text{D}\text{‘}R\text{‘}N = C\text{‘}N :$

$[*37·68] \quad \supset : \text{D}\text{“}R\text{“}C\text{‘}Q = C\text{“}C\text{‘}Q \quad (2)$

$\vdash . (1) . (2) . \supset \vdash :. \text{Hp} . \supset : \text{D}\text{‘}\dot{s}\text{‘}R\text{“}C\text{‘}Q = s\text{‘}C\text{“}C\text{‘}Q$

$[*162·22] \quad = C\text{‘}\Sigma\text{‘}Q : \supset \vdash . \text{Prop}$

***164·41.** $\vdash :. Q \in \text{Rel}^2 \text{ excl} : N \in C\text{‘}Q . \supset_N . R\text{‘}N \in (S\text{‘}N) \, \overline{\text{smor}} \, N : \supset .$
$\dot{s}\text{‘}R\text{“}C\text{‘}Q \in 1 \to \text{Cls}$

Dem.

$\vdash . *151·11 . \supset \vdash :. \text{Hp} . \supset : M, N \in C\text{‘}Q . \exists ! \text{D}\text{‘}R\text{‘}M \cap \text{D}\text{‘}R\text{‘}N . \supset .$
$\exists ! C\text{‘}M \cap C\text{‘}N .$

[*163·11] $\supset . M = N .$

[*30·37] $\supset . R^{\prime}M = R^{\prime}N$ (1)

$\vdash . *151·11 . \supset \vdash :. \text{Hp} . \supset : M \in C^{\prime}Q . \supset . R^{\prime}M \in 1 \rightarrow 1$ (2)

$\vdash . (1) . (2) . *72·32 . \supset \vdash . \text{Prop}$

***164·411.** $\vdash : S ; Q \in \text{Rel}^2 \text{excl} . S \upharpoonright C^{\prime}Q \in 1 \rightarrow 1 . \text{Hp} *164·4 . \supset . \dot{s}^{\prime}R^{\prime\prime}C^{\prime}Q \in \text{Cls} \rightarrow 1$

Dem.

$\vdash . *151·11 . \supset \vdash :. \text{Hp} . \supset : M, N \in C^{\prime}Q . \exists ! D^{\prime}R^{\prime}M \cap D^{\prime}R^{\prime}N . \supset .$

$\exists ! C^{\prime}S^{\prime}M \cap C^{\prime}S^{\prime}N .$

[*163·11.*150·22] $\supset . S^{\prime}M = S^{\prime}N .$

[*71·532] $\supset . M = N .$

[*30·37] $\supset . R^{\prime}M = R^{\prime}N$ (1)

$\vdash . *151·11 . \supset \vdash :. \text{Hp} . \supset : M \in C^{\prime}Q . \supset . R^{\prime}M \in 1 \rightarrow 1$ (2)

$\vdash . (1) . (2) . *72·321 . \supset \vdash . \text{Prop}$

***164·412.** $\vdash :. S ; Q, Q \in \text{Rel}^2 \text{excl} . S \upharpoonright C^{\prime}Q \in 1 \rightarrow 1 :$

$N \in C^{\prime}Q . \supset_N . R^{\prime}N \in (S^{\prime}N) \overline{\text{smor}} N : \supset . \dot{s}^{\prime}R^{\prime\prime}C^{\prime}Q \in 1 \rightarrow 1$

[*164·41·411]

***164·413.** $\vdash :. \text{Hp} *164·41 . \supset :$

$N \in C^{\prime}Q . \supset . R^{\prime}N = (\dot{s}^{\prime}R^{\prime\prime}C^{\prime}Q) \upharpoonright C^{\prime}N . S^{\prime}N = (\dot{s}^{\prime}R^{\prime\prime}C^{\prime}Q) ; N$

Dem.

$\vdash . *41·13 . \supset \vdash : \text{Hp} . N \in C^{\prime}Q . \supset . R^{\prime}N \mathrel{\dot{\subset}} \dot{s}^{\prime}R^{\prime\prime}C^{\prime}Q .$

[*72·92.*164·41] $\supset . R^{\prime}N = (\dot{s}^{\prime}R^{\prime\prime}C^{\prime}Q) \upharpoonright \text{Cl}^{\prime}R^{\prime}N$

[*151·11.Hp] $= (\dot{s}^{\prime}R^{\prime\prime}C^{\prime}Q) \upharpoonright C^{\prime}N$ (1)

$\vdash . *151·11 . \supset \vdash : \text{Hp} . N \in C^{\prime}Q . \supset . S^{\prime}N = (R^{\prime}N) ; N$

[(1).*150·32] $= (\dot{s}^{\prime}R^{\prime\prime}C^{\prime}Q) ; N$ (2)

$\vdash . (1) . (2) . \supset \vdash . \text{Prop}$

***164·414.** $\vdash : \text{Hp} *164·41 . \supset . S ; Q = (\dot{s}^{\prime}R^{\prime\prime}C^{\prime}Q) \dagger ; Q$ [*164·413 . *150·1·35]

***164·42.** $\vdash :. Q, S ; Q \in \text{Rel}^2 \text{excl} . S \upharpoonright C^{\prime}Q \in 1 \rightarrow 1 :$

$N \in C^{\prime}Q . \supset_N . R^{\prime}N \in (S^{\prime}N) \overline{\text{smor}} N : \supset .$

$\dot{s}^{\prime}R^{\prime\prime}C^{\prime}Q \in (S ; Q) \overline{\text{smor}} \overline{\text{smor}} Q$ [*164·4·412·414·1]

***164·421.** $\vdash :. P, Q \in \text{Rel}^2 \text{excl} . S \upharpoonright C^{\prime}Q \in P \overline{\text{smor}} Q :$

$N \in C^{\prime}Q . \supset_N . R^{\prime}N \in (S^{\prime}N) \overline{\text{smor}} N : \supset .$

$\dot{s}^{\prime}R^{\prime\prime}C^{\prime}Q \in P \overline{\text{smor}} \overline{\text{smor}} Q$ [*164·42]

The following proposition, besides being used in proving all subsequent propositions of this number (except *164·432·433, which are mere lemmas for *164·44), is used in *251·6, in the theory of ordinal numbers.

***164·43.** $\vdash :. P, Q \in \text{Rel}^2\,\text{excl} . S \in P \,\overline{\text{smor}}\, Q .$
$$\mu = \hat{\lambda}\{(\exists N) . N \in C\text{‘}Q . \lambda = (S\text{‘}N) \,\overline{\text{smor}}\, N\} . \supset :$$
$$R \in \epsilon_\Delta\text{‘}\mu . \supset . \dot{s}\text{‘D‘}R \in P \,\overline{\text{smor}}\,\overline{\text{smor}}\, Q . S = (\dot{s}\text{‘D‘}R)\dagger \upharpoonright C\text{‘}Q$$

Dem.

$\vdash . *83{\cdot}2{\cdot}22 . \supset \vdash :. \text{Hp} . R \in \epsilon_\Delta\text{‘}\mu . \supset :$
$$N \in C\text{‘}Q . \supset . R\text{‘}\{(S\text{‘}N) \,\overline{\text{smor}}\, N\} \in (S\text{‘}N) \,\overline{\text{smor}}\, N : \dot{s}\text{‘D‘}R = R\text{‘‘}\mu \quad (1)$$
$\vdash . (1) . \supset \vdash :. \text{Hp}(1) . T = \hat{\lambda}\hat{N}\{N \in C\text{‘}Q . \lambda = (S\text{‘}N) \,\overline{\text{smor}}\, N\} . \supset :$
$$N \in C\text{‘}Q . \supset . R\text{‘}T\text{‘}N \in (S\text{‘}N) \,\overline{\text{smor}}\, N : \dot{s}\text{‘D‘}R = R\text{‘‘}T\text{‘‘}C\text{‘}Q : \quad (2)$$
$\left[*164{\cdot}42 \frac{R \mid T}{R}\right] \supset : \dot{s}\text{‘D‘}R \in P \,\overline{\text{smor}}\,\overline{\text{smor}}\, Q \quad (3)$

$\vdash . (2) . *164{\cdot}413 \frac{R \mid T}{R} . *151{\cdot}11 . *35{\cdot}71 . \supset \vdash : \text{Hp}(2) . \supset . S = (\dot{s}\text{‘D‘}R)\dagger \upharpoonright C\text{‘}Q \quad (4)$

$\vdash . (3) . (4) . \supset \vdash . \text{Prop}$

***164·431.** $\vdash :. P, Q \in \text{Rel}^2\,\text{excl} : (\exists S) . S \in P \,\overline{\text{smor}}\, Q .$
$$\exists ! \epsilon_\Delta\text{‘}\hat{\lambda}\{(\exists N) . N \in C\text{‘}Q . \lambda = (S\text{‘}N) \,\overline{\text{smor}}\, N\} : \supset . P \text{ smor smor } Q$$
[*163·43·11]

***164·432.** $\vdash : S \in P \,\overline{\text{smor}}\, Q \cap \text{Rl‘smor} . \supset .$
$$\Lambda \sim \in \hat{\lambda}\{(\exists N) . N \in C\text{‘}Q . \lambda = (S\text{‘}N) \,\overline{\text{smor}}\, N\}$$

Dem.

$\vdash . *151{\cdot}11 . \supset \vdash :. \text{Hp} . \supset : N \in C\text{‘}Q . \supset . N \in \text{Ɑ‘}S .$
[*71·31] $\supset . (S\text{‘}N) S N .$
[Hp] $\supset . (S\text{‘}N) \text{ smor } N .$
[*151·12] $\supset . \exists ! (S\text{‘}N) \,\overline{\text{smor}}\, N :. \supset \vdash . \text{Prop}$

***164·433.** $\vdash :. \text{Mult ax} . \supset : S \in P \,\overline{\text{smor}}\, Q \cap \text{Rl‘smor} . \supset .$
$$\exists ! \epsilon_\Delta\text{‘}\hat{\lambda}\{(\exists N) . N \in C\text{‘}Q . \lambda = (S\text{‘}N) \,\overline{\text{smor}}\, N\}$$
[*164·432 . *88·37]

All the remaining propositions of the number are important.

***164·44.** $\vdash :. \text{Mult ax} . \supset : P, Q \in \text{Rel}^2\,\text{excl} . \exists ! P \,\overline{\text{smor}}\, Q \cap \text{Rl‘smor} . \supset .$
$$P \text{ smor smor } Q \quad [*164{\cdot}433{\cdot}431]$$

***164·45.** $\vdash :: \text{Mult ax} . \supset :. P, Q \in \text{Rel}^2\,\text{excl} . \supset :$
$$\exists ! P \,\overline{\text{smor}}\, Q \cap \text{Rl‘smor} . \equiv . P \text{ smor smor } Q \quad [*164{\cdot}44{\cdot}17]$$

***164·46.** $\vdash :. \text{Mult ax} . \supset : P, Q \in \text{Rel}^2\,\text{excl} . \exists ! P \,\overline{\text{smor}}\, Q \cap \text{Rl‘ smor} . \supset .$
$$\Sigma\text{‘}P \text{ smor } \Sigma\text{‘}Q \quad [*164{\cdot}44{\cdot}151]$$

∗164·47. $\vdash : R, S \epsilon \text{Nr}`Q . C`R, C`S \epsilon \text{Cl}`\text{Nr}`P . \supset . \exists ! R \overline{\text{smor}} S \cap \text{Rl}`\text{smor}$

Dem.

$\vdash . *152 \cdot 5 \cdot 4 . \quad \supset \vdash : \text{Hp} . \supset . R \,\text{smor}\, S .$

$[*151 \cdot 12] \quad \supset . \exists ! R \overline{\text{smor}} S \quad (1)$

$\vdash . *60 \cdot 2 . \quad \supset \vdash :. \text{Hp} . \supset : M \epsilon C`R . N \epsilon C`S . \supset . M, N \epsilon \text{Nr}`P .$

$[*152 \cdot 5 \cdot 4] \quad \supset . M \,\text{smor}\, N \quad (2)$

$\vdash . *151 \cdot 1 \cdot 131 . \supset \vdash :. T \epsilon R \overline{\text{smor}} S . \supset : MTN . \supset . M \epsilon C`R . N \epsilon C`S \quad (3)$

$\vdash . (2) . (3) . \quad \supset \vdash :. \text{Hp} . \supset : T \epsilon R \overline{\text{smor}} S . \supset . T \subseteq \text{smor} \quad (4)$

$\vdash . (1) . (4) . \supset \vdash . \text{Prop}$

∗164·48. $\vdash :. \text{Mult ax} . \supset : R, S \epsilon \text{Rel}^2 \text{excl} \cap \text{Nr}`Q . C`R, C`S \epsilon \text{Cl}`\text{Nr}`P . \supset .$
$R \,\text{smor smor}\, S . \Sigma`R \,\text{smor}\, \Sigma`S \quad [*164 \cdot 47 \cdot 44 \cdot 46]$

$*$165. RELATIONS OF RELATIONS OF COUPLES.

Summary of $*$165.

In the present number, we shall give various propositions concerning the relation $P \underset{\cdot,}{\downarrow} {}^{;}Q$, which has the same uses in relation-arithmetic as $\alpha \underset{,,}{\downarrow} {}^{\text{“}}\beta$ has in cardinal arithmetic. The propositions of this number will be used in the next number to establish the properties of the arithmetical product of two relations Q and P, which is defined as $\Sigma\text{‘}P \underset{\cdot,}{\downarrow} {}^{;}Q$. Again in connection with exponentiation the propositions of the present number will be useful, since, after the product of a relation of relations has been defined ($*$172), we shall define exponentiation by means of the definition

$$P \exp Q = \text{Prod}\text{‘}P \underset{\cdot,}{\downarrow} {}^{;}Q \quad \text{Df.} \quad (\text{Cf. } *176.)$$

There will also be occasional uses of the propositions of this number throughout the theory of series. The relation $P \underset{\cdot,}{\downarrow} {}^{;}Q$ is important because its structure is thoroughly known. It is a Rel^2excl which consists of $\text{Nr}\text{‘}Q$ relations, each like P ($*$165·27); and if $P \,\text{smor}\, P' \,.\, Q \,\text{smor}\, Q'$, we can construct a double correlator of $P \underset{\cdot,}{\downarrow} {}^{;}Q$ and $P' \underset{\cdot,}{\downarrow} {}^{;}Q'$ without invoking the multiplicative axiom. In fact we have

$*$165·362. $\vdash : R \upharpoonright C\text{‘}P' \,\epsilon\, P \,\overline{\text{smor}}\, P' \,.\, S \upharpoonright C\text{‘}Q' \,\epsilon\, Q \,\overline{\text{smor}}\, Q' \,.\, \supset .$

$$(R \,\|\, \breve{S}) \upharpoonright C\text{‘}\Sigma\text{‘}P' \underset{\cdot,}{\downarrow} {}^{;}Q' \,\epsilon\, (P \underset{\cdot,}{\downarrow} {}^{;}Q) \,\overline{\text{smor}}\,\overline{\text{smor}}\, (P' \underset{\cdot,}{\downarrow} {}^{;}Q')$$

This proposition should be compared with $*$113·127. In virtue of $*$164·31, together with various propositions of $*$165 and $*$166, it will appear that $*$165·362 includes $*$113·127 as part of what it asserts.

In the present number, we begin with a set of propositions on fields. We have

$*$165·12. $\vdash . C\text{‘}P \underset{\cdot,}{\downarrow} {}^{;}Q = P \underset{\cdot,}{\downarrow} {}^{\text{“}}C\text{‘}Q$

$*$165·13. $\vdash . C\text{‘}P \underset{\cdot,}{\downarrow} z = \downarrow z^{\text{“}}C\text{‘}P = (C\text{‘}P) \underset{,,}{\downarrow} z$

whence

***165·14.** $\vdash . C\text{‘‘}C\text{‘}P\underset{;}{\downarrow}\,;Q=(C\text{‘}P)\underset{,,}{\downarrow}\text{‘‘}C\text{‘}Q$

which connects the theory of $P\underset{;}{\downarrow}\,;Q$ with that of $\alpha\underset{,,}{\downarrow}\text{‘‘}\beta$ (*113 and *116). Hence

***165·16.** $\vdash . C\text{‘}\Sigma\text{‘}P\underset{;}{\downarrow}\,;Q=C\text{‘}Q\times C\text{‘}P$

In *166, we shall define $Q\times P$ as $\Sigma\text{‘}P\underset{;}{\downarrow}\,;Q$; thus the above will become

$$\vdash . C\text{‘}(Q\times P)=C\text{‘}Q\times C\text{‘}P.$$

We next have a set of propositions concerned with $P\underset{;}{\downarrow}$ as a relation, and with the circumstances under which we can infer $x=y$ or $P=Q$ from data as to $P\underset{;}{\downarrow}x$ and $Q\underset{;}{\downarrow}y$. We have

***165·21.** $\vdash . P\underset{;}{\downarrow}\,;Q\in\text{Rel}^2\,\text{excl}$

***165·211.** $\vdash : \exists ! C\text{‘}P\underset{;}{\downarrow}x\cap C\text{‘}P\underset{;}{\downarrow}y . \supset . x=y$

***165·22.** $\vdash : \dot{\exists} ! P . \supset . P\underset{;}{\downarrow}\in 1\rightarrow 1$

We then have various propositions concerning $\dot{\Lambda}$, of which the chief are

***165·241.** $\vdash : Q=\dot{\Lambda} . \supset . P\underset{;}{\downarrow}\,;Q=\dot{\Lambda}$

***165·242.** $\vdash : P=\dot{\Lambda} . \dot{\exists} ! Q . \supset . P\underset{;}{\downarrow}\,;Q=\dot{\Lambda}\downarrow\dot{\Lambda}$

We have next four propositions which are constantly used, proving that $P\underset{;}{\downarrow}\,;Q$ consists of $\text{Nr}\text{‘}Q$ relations each like P. These propositions are

***165·25.** $\vdash : \dot{\exists} ! P . \supset . P\underset{;}{\downarrow}\,;Q \;\text{smor}\; Q . (P\underset{;}{\downarrow})\upharpoonright C\text{‘}Q\in(P\underset{;}{\downarrow}\,;Q)\,\overline{\text{smor}}\,Q$

***165·251.** $\vdash . P\underset{;}{\downarrow}x \;\text{smor}\; P . (\downarrow x)\upharpoonright C\text{‘}P\in(P\underset{;}{\downarrow}x)\,\overline{\text{smor}}\,P$

***165·26.** $\vdash . C\text{‘}P\underset{;}{\downarrow}\,;Q\subset\text{Nr}\text{‘}P$

***165·27.** $\vdash : \dot{\exists} ! P . \supset . P\underset{;}{\downarrow}\,;Q\in\text{Rel}^2\,\text{excl}\cap\text{Nr}\text{‘}Q . C\text{‘}P\underset{;}{\downarrow}\,;Q\in\text{Cl}\text{‘}\text{Nr}\text{‘}P$

From *165·3 to *165·372, we are concerned with constructing a double correlator of $P\underset{;}{\downarrow}\,;Q$ and $P'\underset{;}{\downarrow}\,;Q'$ when we are given simple correlators of P with P' and of Q with Q'. The result (*165·362) has already been given. Hence we have

***165·37.** $\vdash : P\;\text{smor}\;P' . Q\;\text{smor}\;Q' . \supset . P\underset{;}{\downarrow}\,;Q\;\text{smor}\;\text{smor}\;P'\underset{;}{\downarrow}\,;Q'$

and by *164·48 and *165·27 we have

***165·38.** $\vdash :. \text{Mult ax} . \supset :$

$$R\in\text{Rel}^2\,\text{excl}\cap\text{Nr}\text{‘}Q . C\text{‘}R\subset\text{Nr}\text{‘}P . \supset . R\;\text{smor}\;\text{smor}\;P\underset{;}{\downarrow}\,;Q$$

Hence propositions concerning a series of β series, each containing α terms (where α and β are relation-numbers), which in general require the multiplicative axiom, can be deduced, assuming that axiom, from propositions (not requiring the axiom) concerning $P \underset{;}{\downarrow} ; Q$, where $\mathrm{Nr}\text{‘}P = \alpha$ and $\mathrm{Nr}\text{‘}Q = \beta$. Thus the use of $P \underset{;}{\downarrow} ; Q$ enables us to minimize the use of the multiplicative axiom.

***165·01.** $\vdash . P \underset{;}{\downarrow} z = \downarrow z ; P$ [*150·6]

***165·1.** $\vdash : R (P \underset{;}{\downarrow} ; Q) S . \equiv . (\exists z, w) . zQw . R = \downarrow z ; P . S = \downarrow w ; P$ [*150·62]

***165·11.** $\vdash : X (\downarrow z ; P) Y . \equiv . (\exists x, y) . xPy . X = x \downarrow z . Y = y \downarrow z$ [*150·55]

***165·12.** $\vdash . C\text{‘}P \underset{;}{\downarrow} ; Q = P \underset{;}{\downarrow} \text{“}C\text{‘}Q$ [*150·22]

***165·13.** $\vdash . C\text{‘}P \underset{;}{\downarrow} z = \downarrow z \text{“}C\text{‘}P = (C\text{‘}P) \underset{,,}{\downarrow} z$ [*165·01 . *150·22 . *38·2]

***165·131.** $\vdash . C\text{“}P \underset{;}{\downarrow} \text{“}\beta = (C\text{‘}P) \underset{,,}{\downarrow} \text{“}\beta$ [*165·13 . *38·11 . *37·68]

***165·14.** $\vdash . C\text{“}C\text{‘}P \underset{;}{\downarrow} ; Q = (C\text{‘}P) \underset{,,}{\downarrow} \text{“}C\text{‘}Q$ [*165·12·131]

***165·15.** $\vdash . s\text{‘}C\text{“}C\text{‘}P \underset{;}{\downarrow} ; Q = C\text{‘}Q \times C\text{‘}P$ [*165·14 . *113·1]

***165·16.** $\vdash . C\text{‘}\Sigma\text{‘}P \underset{;}{\downarrow} ; Q = C\text{‘}Q \times C\text{‘}P$ [*165·15 . *162·22]

***165·161.** $\vdash : M (F ; P \underset{;}{\downarrow} ; Q) N . \equiv .$

$$(\exists x, y, z, w) . x, y \in C\text{‘}P . zQw . M = x \downarrow z . N = y \downarrow w$$

Dem.

$\vdash . *150·52 . \supset$

$\vdash :. M (F ; P \underset{;}{\downarrow} ; Q) N . \equiv : (\exists R, S) . R (P \underset{;}{\downarrow} ; Q) S . M \in C\text{‘}R . N \in C\text{‘}S .$

[*165·1] $\equiv : (\exists R, S, z, w) . zQw . R = \downarrow z ; P . S = \downarrow w ; P . M \in C\text{‘}R . N \in C\text{‘}S .$

[*165·01·13] $\equiv : (\exists R, S, z, w) . zQw . R = \downarrow z ; P . S = \downarrow w ; P .$

$M \in \downarrow z \text{“}C\text{‘}P . N \in \downarrow w \text{“}C\text{‘}P .$

[*21·151] $\equiv : (\exists z, w) . zQw . M \in \downarrow z \text{“}C\text{‘}P . N \in \downarrow w \text{“}C\text{‘}P .$

[*38·131] $\equiv : (\exists x, y, z, w) . zQw . x, y \in C\text{‘}P . M = x \downarrow z . N = y \downarrow w :. \supset \vdash .$ Prop

***165·162.** $\vdash : M (\dot{s}\text{‘}C\text{‘}P \underset{;}{\downarrow} ; Q) N . \equiv . (\exists x, y, z) . xPy . z \in C\text{‘}Q . M = x \downarrow z . N = y \downarrow z$

Dem.

$\vdash . *165·12 . *41·11 . \supset$

$\vdash : M (\dot{s}\text{‘}C\text{‘}P \underset{;}{\downarrow} ; Q) N . \equiv . (\exists R) . R \in P \underset{;}{\downarrow} \text{“}C\text{‘}Q . MRN .$

[*38·13] $\equiv . (\exists z) . z \in C\text{‘}Q . M (P \underset{;}{\downarrow} z) N .$

[*165·01·11] $\equiv . (\exists x, y, z) . xPy . z \in C\text{‘}Q . M = x \downarrow z . N = y \downarrow z : \supset \vdash .$ Prop

***165·17.** $\vdash :. M (\Sigma`P \underset{;}{\downarrow} ;Q) N . \equiv : (\exists x, y, z, w) :$

$$x, y \in C`P . z, w \in C`Q : zQw . \vee . z = w . xPy : M = x \downarrow z . N = y \downarrow w$$

Dem.

$\vdash . *165·161·162 . *162·11 . \supset$

$\vdash :. M (\Sigma`P \underset{;}{\downarrow} ;Q) N . \equiv : (\exists x, y, z, w) . x, y \in C`P . zQw . M = x \downarrow z . N = y \downarrow w . \vee .$

$(\exists x, y, z) . xPy . z \in C`Q . M = x \downarrow z . N = y \downarrow w :$

[*13·195] $\equiv : (\exists x, y, z, w) . x, y \in C`P . zQw . M = x \downarrow z . N = y \downarrow w . \vee .$

$(\exists x, y, z, w) . xPy . z, w \in C`Q . z = w . M = x \downarrow z . N = y \downarrow w :$

[*33·17.*4·71] $\equiv : (\exists x, y, z, w) . x, y \in C`P . z, w \in C`Q . zQw . M = x \downarrow z . N = y \downarrow w . \vee .$

$(\exists x, y, z, w) . x, y \in C`P . z, w \in C`Q . xPy . z = w . M = x \downarrow z . N = y \downarrow w :$

[*11·41.*4·4] $\equiv : (\exists x, y, z, w) : x, y \in C`P . z, w \in C`Q : zQw . \vee . z = w . xPy :$

$M = x \downarrow z . N = y \downarrow w :. \supset \vdash .$ Prop

***165·18.** $\vdash . \text{Cnv}`P \underset{;}{\downarrow} ;Q = P \underset{;}{\downarrow} ;\breve{Q}$ [*150·12]

***165·181.** $\vdash . \text{Cnv}`P \underset{;}{\downarrow} z = \breve{P} \underset{;}{\downarrow} z$ [*165·01 . *150·12]

***165·182.** $\vdash . \text{Cnv};P \underset{;}{\downarrow} ;Q = \breve{P} \underset{;}{\downarrow} ;Q$ [*165·181 . *150·35]

***165·19.** $\vdash . \text{Cnv}`\text{Cnv};P \underset{;}{\downarrow} ;Q = \breve{P} \underset{;}{\downarrow} ;\breve{Q} = \text{Cnv};\text{Cnv}`P \underset{;}{\downarrow} ;Q$ [*165·18·182]

***165·2.** $\vdash . P \underset{;}{\downarrow} \in 1 \rightarrow \text{Cls}$ [*72·14]

***165·201.** $\vdash . C`(P \underset{;}{\downarrow} z) = (C`P \uparrow \iota`z)_{\Delta}`\iota`z$

Dem.

$\vdash . *35·103 . \supset \vdash : y (C`P \uparrow \iota`z) z . \equiv . y \in C`P :$

[*85·51] $\supset \vdash . (C`P \uparrow \iota`z)_{\Delta}`\iota`z = \downarrow z``C`P$

[*165·13] $= C`(P \underset{;}{\downarrow} z) . \supset \vdash .$ Prop

***165·202.** $\vdash . C``C`P \underset{;}{\downarrow} ;Q = (C`P \uparrow C`Q)_{\Delta}``\iota``C`Q$ [*165·14 . *113·103]

***165·203.** $\vdash . C``C`P \underset{;}{\downarrow} ;Q \in \text{Cls}^2 \text{ excl}$ [*84·55 . *165·202]

***165·204.** $\vdash : C`P \underset{;}{\downarrow} x = C`P \underset{;}{\downarrow} y . \equiv . P \underset{;}{\downarrow} x = P \underset{;}{\downarrow} y$

Dem.

$\vdash . *165·13 . *55·232 . \supset$

$\vdash : C`P \underset{;}{\downarrow} x = C`P \underset{;}{\downarrow} y . \exists ! C`P \underset{;}{\downarrow} x . \supset . x = y .$

[*30·37] $\supset . P \underset{;}{\downarrow} x = P \underset{;}{\downarrow} y$ (1)

$\vdash . *33·241 . \supset \vdash : C`P \underset{;}{\downarrow} x = C`P \underset{;}{\downarrow} y . C`P \underset{;}{\downarrow} x = \Lambda . \supset . P \underset{;}{\downarrow} x = \dot{\Lambda} . P \underset{;}{\downarrow} y = \dot{\Lambda}$ (2)

$\vdash . (1) . (2) . \supset \vdash : C`P \underset{;}{\downarrow} x = C`P \underset{;}{\downarrow} y . \supset . P \underset{;}{\downarrow} x = P \underset{;}{\downarrow} y$ (3)

$\vdash . (3) . *30·37 . \supset \vdash .$ Prop

***165·205.** $\vdash . C \upharpoonright \text{D}`P_{\downarrow ;} \in 1 \to 1$ [*165·204 . *71·58]

***165·206.** $\vdash : (x) . \text{E} ! P_{\downarrow ;} `x : (\alpha) . \alpha \subset \text{G}`P_{\downarrow ;}$ [*38·12 . *33·431]

***165·21.** $\vdash . P_{\downarrow ;} \, {}^{;}Q \in \text{Rel}^2 \text{ excl}$

Dem.

$\vdash . *165·205 . *150·203 . \supset \vdash . C \upharpoonright C`P_{\downarrow ;} \, {}^{;}Q \in 1 \to 1$ (1)

$\vdash . (1) . *165·203 . *163·17 . \supset \vdash . \text{Prop}$

***165·211.** $\vdash : \exists ! C`P_{\downarrow ;} x \cap C`P_{\downarrow ;} y . \supset . x = y$ [*165·13 . *55·232]

***165·212.** $\vdash : \dot{\exists} ! P . \equiv . \dot{\exists} ! P_{\downarrow ;} x$

Dem.

$\vdash . *165·11·01 . \supset \vdash : \dot{\exists} ! P_{\downarrow ;} x . \equiv . (\exists X, Y, x, y) . xPy . X = x \downarrow z . Y = y \downarrow z .$

[*13·19] $\equiv . (\exists x, y) . xPy : \supset \vdash . \text{Prop}$

***165·22.** $\vdash : \dot{\exists} ! P . \supset . P_{\downarrow ;} \in 1 \to 1$

Dem.

$\vdash . *165·212 . \supset \vdash :. \text{Hp} . \supset : \dot{\exists} ! P_{\downarrow ;} x :$

[*30·37 . *24·571 . *33·24] $\supset : P_{\downarrow ;} x = P_{\downarrow ;} y . \supset . \exists ! C`P_{\downarrow ;} x \cap C`P_{\downarrow ;} y .$

[*165·211] $\supset . x = y$ (1)

$\vdash . (1) . *71·54 . *165·2 . \supset \vdash . \text{Prop}$

***165·221.** $\vdash :. \dot{\exists} ! P . \supset : \dot{\exists} ! P_{\downarrow ;} x \,\dot{\cap}\, P_{\downarrow ;} y . \equiv . P_{\downarrow ;} x = P_{\downarrow ;} y . \equiv . x = y$

Dem.

$\vdash . *33·252 . \supset \vdash : \dot{\exists} ! P_{\downarrow ;} x \,\dot{\cap}\, P_{\downarrow ;} y . \supset . \exists ! C`P_{\downarrow ;} x \cap C`P_{\downarrow ;} y .$

[*165·211] $\supset . x = y$ (1)

$\vdash . *165·212 . *25·571 . \supset \vdash :. \dot{\exists} ! P . \supset : x = y . \supset . \dot{\exists} ! P_{\downarrow ;} x \,\dot{\cap}\, P_{\downarrow ;} y$ (2)

$\vdash . (1) . (2) . *165·212 . *30·37 . \supset \vdash . \text{Prop}$

***165·222.** $\vdash :. \dot{\exists} ! P . \supset : \exists ! C`P_{\downarrow ;} x \cap C`P_{\downarrow ;} y . \equiv . C`P_{\downarrow ;} x = C`P_{\downarrow ;} y . \equiv . x = y$

[Proof as in *165·221]

***165·223.** $\vdash :. \dot{\exists} ! P . \supset : P_{\downarrow ;} \, {}^{;}Q = P_{\downarrow ;} \, {}^{;}R . \equiv . Q = R$

Dem.

$\vdash . *151·31 . *165·22 . \supset \vdash :. \text{Hp} . \supset : P_{\downarrow ;} \, {}^{;}Q = P_{\downarrow ;} \, {}^{;}R . \supset . Q = R$ (1)

$\vdash . *34·29 . *150·1 . \quad \supset \vdash : Q = R . \supset . P_{\downarrow ;} \, {}^{;}Q = P_{\downarrow ;} \, {}^{;}R$ (2)

$\vdash . (1) . (2) . \supset \vdash . \text{Prop}$

***165·23.** $\vdash : P_{\downarrow ;} x = Q_{\downarrow ;} y . \supset . P = Q$

Dem.

$\vdash . *72·184 . *150·153 . \supset \vdash : \downarrow x \, {}^{;}P = \downarrow x \, {}^{;}Q . \supset . P = Q$ (1)

$\vdash . (1) . *165·01 . \supset \vdash . \text{Prop}$

***165·231.** $\vdash : P \underset{;}{\downarrow} x = Q \underset{;}{\downarrow} x \,.\equiv .\, P = Q$ [*165·23 . *30·37]

***165·232.** $\vdash :.\, \dot{\exists} ! P \,.\mathsf{v}.\, \dot{\exists} ! Q : \supset : P \underset{;}{\downarrow} x = Q \underset{;}{\downarrow} y \,.\equiv .\, P = Q \,.\, x = y$

Dem.

$\vdash .\, *165 \cdot 23 \,.\supset \vdash :.\, P \underset{;}{\downarrow} x = Q \underset{;}{\downarrow} y \,.\supset : P = Q :$ (1)

[*13·12.Hp (1)] $\supset : P \underset{;}{\downarrow} x = P \underset{;}{\downarrow} y \,.\, Q \underset{;}{\downarrow} x = Q \underset{;}{\downarrow} y :$

[*165·221] $\supset : \dot{\exists} ! P \,.\supset .\, x = y : \dot{\exists} ! Q \,.\supset .\, x = y$ (2)

$\vdash .\, (1) \,.\, (2) \,.\supset \vdash :.\, \dot{\exists} ! P \,.\mathsf{v}.\, \dot{\exists} ! Q : \supset : P \underset{;}{\downarrow} x = Q \underset{;}{\downarrow} y \,.\supset .\, P = Q \,.\, x = y$ (3)

$\vdash .\, (3) \,.\, *13 \cdot 12 \cdot 15 \,.\supset \vdash .$ Prop

***165·233.** $\vdash : \exists ! C\text{‘}P \underset{;}{\downarrow} x \cap C\text{‘}Q \underset{;}{\downarrow} y \,.\equiv .\, x = y \,.\, \exists ! C\text{‘}P \cap C\text{‘}Q$

[*55·232 . *165·13]

***165·24.** $\vdash : P = \dot{\Lambda} \,.\supset .\, P \underset{;}{\downarrow} x = \dot{\Lambda} \,.\, P \underset{;}{\downarrow} = \iota\text{‘}\dot{\Lambda} \uparrow \mathrm{V}$

Dem.

$\vdash .\, *165 \cdot 212 \,.\,$Transp$\,.\supset \vdash : P = \dot{\Lambda} \,.\supset .\, P \underset{;}{\downarrow} x = \dot{\Lambda}$ (1)

$\vdash .\, (1) \,.\, *38 \cdot 1 \,.\quad \supset \vdash :.\, P = \dot{\Lambda} \,.\supset : R\, (P \underset{;}{\downarrow})\, x \,.\equiv .\, R = \dot{\Lambda} \,.$

[*51·15.*24·104] $\equiv .\, R \in \iota\text{‘}\dot{\Lambda} \,.\, x \in \mathrm{V} \,.$

[*35·103] $\equiv .\, R\, (\iota\text{‘}\dot{\Lambda} \uparrow \mathrm{V})\, x$ (2)

$\vdash .\, (1) \,.\, (2) \,.\supset \vdash .$ Prop

***165·241.** $\vdash : Q = \dot{\Lambda} \,.\supset .\, P \underset{;}{\downarrow} {}^{;}Q = \dot{\Lambda}$ [*150·42]

***165·242.** $\vdash : P = \dot{\Lambda} \,.\, \dot{\exists} ! Q \,.\supset .\, P \underset{;}{\downarrow} {}^{;}Q = \dot{\Lambda} \downarrow \dot{\Lambda}$

Dem.

$\vdash .\, *165 \cdot 1 \cdot 24 \,.\supset \vdash :.\, P = \dot{\Lambda} \,.\supset : R\, (P \underset{;}{\downarrow} {}^{;}Q)\, S \,.\equiv .\, (\exists z, w) \,.\, zQw \,.\, R = \dot{\Lambda} \,.\, S = \dot{\Lambda} \,.$

[*10·35] $\equiv .\, \dot{\exists} ! Q \,.\, R = \dot{\Lambda} \,.\, S = \dot{\Lambda}$ (1)

$\vdash .\, (1) \,.\, *55 \cdot 13 \,.\supset \vdash .$ Prop

***165·243.** $\vdash : \dot{\exists} ! Q \,.\equiv .\, \dot{\exists} ! P \underset{;}{\downarrow} {}^{;}Q$

Dem.

$\vdash .\, *165 \cdot 1 \,.\supset \vdash : \dot{\exists} ! P \underset{;}{\downarrow} {}^{;}Q \,.\equiv .\, (\exists x, y, R, S) \,.\, xQy \,.\, R = P \underset{;}{\downarrow} x \,.\, S = P \underset{;}{\downarrow} y \,.$

[*13·19] $\equiv .\, (\exists x, y) \,.\, xQy : \supset \vdash .$ Prop

***165·244.** $\vdash : \dot{\Lambda} \in C\text{‘}P \underset{;}{\downarrow} {}^{;}Q \,.\equiv .\, P = \dot{\Lambda} \,.\, \dot{\exists} ! Q \,.\equiv .\, P \underset{;}{\downarrow} {}^{;}Q = \dot{\Lambda} \downarrow \dot{\Lambda}$

Dem.

$\vdash .\, *165 \cdot 212 \cdot 12 \,.\quad \supset \vdash : \dot{\Lambda} \in C\text{‘}P \underset{;}{\downarrow} {}^{;}Q \,.\supset .\, P = \dot{\Lambda}$ (1)

$\vdash .\, *10 \cdot 24 \,.\, *33 \cdot 24 \,.\supset \vdash : \dot{\Lambda} \in C\text{‘}P \underset{;}{\downarrow} {}^{;}Q \,.\supset .\, \dot{\exists} ! P \underset{;}{\downarrow} {}^{;}Q \,.$

[*165·243] $\supset .\, \dot{\exists} ! Q$ (2)

$\vdash . *165\cdot242 . *55\cdot15 . \supset \vdash : P = \dot{\Lambda} . \dot{\exists} ! Q . \supset . \dot{\Lambda} \in C\text{‘}P \downarrow_{;} ;Q$ (3)

$\vdash . (1) . (2) . (3) . \quad \supset \vdash : \dot{\Lambda} \in C\text{‘}P \downarrow_{;} ;Q . \equiv . P = \dot{\Lambda} . \dot{\exists} ! Q$ (4)

$\vdash . *55\cdot15 . \quad \supset \vdash : P \downarrow_{;} ;Q = \dot{\Lambda} \downarrow \dot{\Lambda} . \supset . \dot{\Lambda} \in C\text{‘}P \downarrow_{;} ;Q .$

[(4)] $\supset . P = \dot{\Lambda} . \dot{\exists} ! Q$ (5)

$\vdash . (5) . *165\cdot242 . \quad \supset \vdash : P = \dot{\Lambda} . \dot{\exists} ! Q . \equiv . P \downarrow_{;} ;Q = \dot{\Lambda} \downarrow \dot{\Lambda}$ (6)

$\vdash . (4) . (6) . \supset \vdash . \text{Prop}$

***165·245.** $\vdash :. \dot{\exists} ! P . \mathsf{v} . Q = \dot{\Lambda} : \equiv . \dot{\Lambda} \sim\in C\text{‘}P \downarrow_{;} ;Q . \equiv . \dot{\Lambda} \sim\in (C\text{‘}P) \downarrow_{"} \text{“}C\text{‘}Q$

[*165·244 . Transp . *33·241 . *165·14]

***165·25.** $\vdash : \dot{\exists} ! P . \supset . P \downarrow_{;} ;Q \,\text{smor}\, Q . (P \downarrow_{;}) \upharpoonright C\text{‘}Q \in (P \downarrow_{;} ;Q) \,\overline{\text{smor}}\, Q$

[*165·22·206 . *151·231]

***165·251.** $\vdash . P \downarrow_{;} x \,\text{smor}\, P . (\downarrow x) \upharpoonright C\text{‘}P \in (P \downarrow_{;} x) \,\overline{\text{smor}}\, P$

[*72·184 . *55·21 . *151·22]

***165·26.** $\vdash . C\text{‘}P \downarrow_{;} ;Q \subset \text{Nr}\text{‘}P$ [*165·251·12 . *152·11]

***165·27.** $\vdash : \dot{\exists} ! P . \supset . P \downarrow_{;} ;Q \in \text{Rel}^2 \text{excl} \cap \text{Nr}\text{‘}Q . C\text{‘}P \downarrow_{;} ;Q \in \text{Cl}\text{‘}\text{Nr}\text{‘}P$

[*165·21·25 . *152·11 . *165·26]

The following propositions are concerned in proving that, if R is a correlator of P and P', and S is a correlator of Q and Q', then $R \parallel \breve{S}$ (with its converse domain limited) is a double correlator of $P \downarrow_{;} ;Q$ and $P' \downarrow_{;} ;Q'$. This proposition is required subsequently in establishing likenesses.

***165·3.** $\vdash : \text{E} ! R\text{‘}y . \supset . \downarrow z\text{‘}R\text{‘}y = R \,|\text{‘} \downarrow z\text{‘}y$

Dem.

$\vdash . *34\cdot1 . *38\cdot11 . \supset \vdash : u \{R \,|\text{‘} \downarrow z\text{‘}y\} w . \equiv . (\exists v) . uRv . v (y \downarrow z) w .$

[*55·13] $\equiv . uRy . w = z$ (1)

$\vdash . (1) . *30\cdot4 . \quad \supset \vdash :. \text{Hp} . \supset : u \{R \,|\text{‘} \downarrow z\text{‘}y\} w . \equiv . u = R\text{‘}y . w = z .$

[*55·13.*38·11] $\equiv . u (\downarrow z\text{‘}R\text{‘}y) w :. \supset \vdash . \text{Prop}$

***165·301.** $\vdash : R \in 1 \rightarrow \text{Cls} . \supset . \downarrow z \,|\, R = (R \,|) \,|\, (\downarrow z) \upharpoonright \text{Ɑ}\text{‘}R$

Dem.

$\vdash . *165\cdot3 . \quad \supset \vdash :. \text{E} ! R\text{‘}y . \supset : M \{(\downarrow z) \,|\, R\} y . \equiv . M \{(R \,|) \,|\, \downarrow z\} y :.$

[*71·16.*34·36] $\supset \vdash :. \text{Hp} . \supset : M \{(\downarrow z) \,|\, R\} y . \equiv .$

$M \{(R \,|) \,|\, \downarrow z\} y . y \in \text{Ɑ}\text{‘}R :. \supset \vdash . \text{Prop}$

***165·302.** $\vdash : \text{E} !! R\text{“}C\text{‘}P . \supset . \downarrow z ; R ; P = R \,|; \downarrow z ; P$

Dem.

$\vdash . *165\cdot3 . \supset \vdash :. \text{Hp} . \supset : y \in C\text{‘}P . \supset . \downarrow z\text{‘}R\text{‘}y = R \,|\text{‘} \downarrow z\text{‘}y$ (1)

$\vdash . (1) . *150\cdot35\cdot13 . \supset \vdash . \text{Prop}$

✱165·31. $\vdash : E!! R``C`P . \supset . (R;P) \underset{\cdot,}{\downarrow} z = R|;P \underset{\cdot,}{\downarrow} z . (R;P) \underset{\cdot,}{\downarrow} ;Q = (R|)\dagger;P \underset{\cdot,}{\downarrow} ;Q$

Dem.

$\vdash . *165\cdot302\cdot01 . \supset \vdash : \mathrm{Hp} . \supset . (R;P) \underset{\cdot,}{\downarrow} z = R|;P \underset{\cdot,}{\downarrow} z$ (1)

[✱150·1] $= (R|)\dagger`P \underset{\cdot,}{\downarrow} z$ (2)

$\vdash . (1) . (2) . *150\cdot35 . \supset \vdash . \mathrm{Prop}$

✱165·311. $\vdash : R \upharpoonright C`P \in 1 \to \mathrm{Cls} . C`P \subset \text{ꓷ}`R . \supset .$
$(R;P) \underset{\cdot,}{\downarrow} z = R|;P \underset{\cdot,}{\downarrow} z . (R;P) \underset{\cdot,}{\downarrow} ;Q = (R|)\dagger;P \underset{\cdot,}{\downarrow} ;Q$

[✱165·31 . ✱71·571]

✱165·32. $\vdash : E! S`z . \supset . \downarrow (S`z) = (|\breve{S})|\downarrow z . \downarrow (S`z);P = |\breve{S};\downarrow z;P$

Dem.

$\vdash . *34\cdot1 . *43\cdot101 . *38\cdot101 . \supset$

$\vdash : M\{(|\breve{S})|\downarrow z\} x . \equiv . (\exists N) . M = N|\breve{S} . N = x \downarrow z .$

[✱13·195] $\equiv . M = (x \downarrow z)|\breve{S}$ (1)

$\vdash . (1) . *55\cdot581 . \supset$

$\vdash :. \mathrm{Hp} . \supset : M\{(|\breve{S})|\downarrow z\} x . \equiv . M = x \downarrow (S`z) .$

[✱38·101] $\equiv . M\{\downarrow (S`z)\} x$ (2)

$\vdash . (2) . *21\cdot43 . \supset \vdash : \mathrm{Hp} . \supset . \downarrow (S`z) = (|\breve{S})|\downarrow z .$

[✱150·13] $\supset . \downarrow (S`z);P = |\breve{S};\downarrow z;P : \supset \vdash . \mathrm{Prop}$

✱165·321. $\vdash : E! S`z . \supset . P \underset{\cdot,}{\downarrow} (S`z) = |\breve{S};P \underset{\cdot,}{\downarrow} z$ [✱165·32·01]

✱165·33. $\vdash : E!! S``C`Q . \supset . P \underset{\cdot,}{\downarrow} ;S;Q = (|\breve{S})\dagger;P \underset{\cdot,}{\downarrow} ;Q$

Dem.

$\vdash . *165\cdot321 . *38\cdot11 . *150\cdot1 . \supset$

$\vdash :. \mathrm{Hp} . \supset : z \in C`Q . \supset . P \underset{\cdot,}{\downarrow} `S`z = (|\breve{S})\dagger`P \underset{\cdot,}{\downarrow} `z$ (1)

$\vdash . (1) . *150\cdot35 . \supset \vdash . \mathrm{Prop}$

✱165·331. $\vdash : S \upharpoonright C`Q \in 1 \to \mathrm{Cls} . C`Q \subset \text{ꓷ}`S . \supset . P \underset{\cdot,}{\downarrow} ;S;Q = (|\breve{S})\dagger;P \underset{\cdot,}{\downarrow} ;Q$

[✱165·33 . ✱71·571]

✱165·34. $\vdash : E!! R``C`P . E!! S``C`Q . \supset . (R;P) \underset{\cdot,}{\downarrow} ;(S;Q) = (R \| \breve{S})\dagger;(P \underset{\cdot,}{\downarrow} ;Q)$

Dem.

$\vdash . *165\cdot31 . \supset \vdash : \mathrm{Hp} . \supset . (R;P) \underset{\cdot,}{\downarrow} ;(S;Q) = (R|)\dagger;P \underset{\cdot,}{\downarrow} ;(S;Q)$

[✱165·33] $= (R|)\dagger;(|\breve{S})\dagger;P \underset{\cdot,}{\downarrow} ;Q$

[✱150·13·14.(✱43·01)] $= (R \| \breve{S})\dagger;(P \underset{\cdot,}{\downarrow} ;Q) : \supset \vdash . \mathrm{Prop}$

***165·341.** $\vdash : R \upharpoonright C\text{‘}P, S \upharpoonright C\text{‘}Q \,\epsilon\, 1 \rightarrow \mathrm{Cls} \,.\, C\text{‘}P \subset \text{Ɑ}\text{‘}R \,.\, C\text{‘}Q \subset \text{Ɑ}\text{‘}S \,.\, \supset .$

$$(R{;}P) \underset{\cdot,}{\downarrow} {;}(S{;}Q) = (R \,\|\, \breve{S})\dagger{;}P \underset{\cdot,}{\downarrow} {;}Q \quad [*165·34 \,.\, *71·571]$$

***165·35.** $\vdash : R \upharpoonright C\text{‘}P \,\epsilon\, \mathrm{Cls} \rightarrow 1 \,.\, C\text{‘}P \subset \text{Ɑ}\text{‘}R \,.\, \supset . (R\,|) \upharpoonright C\text{‘}\Sigma\text{‘}P \underset{\cdot,}{\downarrow} {;}Q \,\epsilon\, 1 \rightarrow 1$

Dem.

$\vdash . *113·118 . *165·16 . \supset \vdash . s\text{‘}\mathrm{D}\text{“}C\text{‘}\Sigma\text{‘}P \underset{\cdot,}{\downarrow} {;}Q \subset C\text{‘}P \quad (1)$

$\vdash . (1) . *74·751 \dfrac{C\text{‘}\Sigma\text{‘}P \underset{\cdot,}{\downarrow} {;}Q,\; C\text{‘}P}{\lambda, \qquad \alpha} . \supset \vdash . \mathrm{Prop}$

***165·351.** $\vdash : S \upharpoonright C\text{‘}Q \,\epsilon\, \mathrm{Cls} \rightarrow 1 \,.\, C\text{‘}Q \subset \text{Ɑ}\text{‘}S \,.\, \supset . (\,|\breve{S}) \upharpoonright C\text{‘}\Sigma\text{‘}P \underset{\cdot,}{\downarrow} {;}Q \,\epsilon\, 1 \rightarrow 1$

Dem.

$\vdash . *113·118 . *165·16 . \supset \vdash . s\text{‘}\text{Ɑ}\text{“}C\text{‘}\Sigma\text{‘}P \underset{\cdot,}{\downarrow} {;}Q \subset C\text{‘}Q .$

$\left[*74·75 \dfrac{\breve{S}}{Q}\right] \quad \supset \vdash : \mathrm{Hp} . \supset . (\,|\breve{S}) \upharpoonright C\text{‘}\Sigma\text{‘}P \underset{\cdot,}{\downarrow} {;}Q \,\epsilon\, 1 \rightarrow 1 : \supset \vdash . \mathrm{Prop}$

***165·352.** $\vdash : R \upharpoonright C\text{‘}P, S \upharpoonright C\text{‘}Q \,\epsilon\, \mathrm{Cls} \rightarrow 1 \,.\, C\text{‘}P \subset \text{Ɑ}\text{‘}R \,.\, C\text{‘}Q \subset \text{Ɑ}\text{‘}S \,.\, \supset .$

$$(R \,\|\, \breve{S}) \upharpoonright C\text{‘}\Sigma\text{‘}P \underset{\cdot,}{\downarrow} {;}Q \,\epsilon\, 1 \rightarrow 1$$

Dem.

$\vdash . *113·118 . *165·16 . \supset$

$\vdash : \mathrm{Hp} . \supset . s\text{‘}\mathrm{D}\text{“}C\text{‘}\Sigma\text{‘}P \underset{\cdot,}{\downarrow} {;}Q \subset C\text{‘}P \,.\, s\text{‘}\text{Ɑ}\text{“}C\text{‘}\Sigma\text{‘}P \underset{\cdot,}{\downarrow} {;}Q \subset C\text{‘}Q .$

$[*74·773] \supset . (R \,\|\, \breve{S}) \upharpoonright C\text{‘}\Sigma\text{‘}P \underset{\cdot,}{\downarrow} {;}Q \,\epsilon\, 1 \rightarrow 1 : \supset \vdash . \mathrm{Prop}$

***165·36.** $\vdash : R \upharpoonright C\text{‘}P' \,\epsilon\, P \,\overline{\mathrm{smor}}\, P' . \supset .$

$$(R\,|) \upharpoonright C\text{‘}\Sigma\text{‘}P' \underset{\cdot,}{\downarrow} {;}Q \,\epsilon\, (P \underset{\cdot,}{\downarrow} {;}Q) \,\overline{\mathrm{smor}}\, \overline{\mathrm{smor}}\, (P' \underset{\cdot,}{\downarrow} {;}Q)$$

Dem.

$\vdash . *151·22 . *165·35 . \quad \supset \vdash : \mathrm{Hp} . \supset . (R\,|) \upharpoonright C\text{‘}\Sigma\text{‘}P' \underset{\cdot,}{\downarrow} {;}Q \,\epsilon\, 1 \rightarrow 1 \quad (1)$

$\vdash . *43·3 . \quad \supset \vdash . C\text{‘}\Sigma\text{‘}P' \underset{\cdot,}{\downarrow} {;}Q \subset \text{Ɑ}\text{‘}R\,| \quad (2)$

$\vdash . *151·22 . *165·311 . \supset \vdash : \mathrm{Hp} . \supset . P \underset{\cdot,}{\downarrow} {;}Q = (\,|\,R)\dagger{;}P' \underset{\cdot,}{\downarrow} {;}Q \quad (3)$

$\vdash . (1) . (2) . (3) . *164·18 . \supset \vdash . \mathrm{Prop}$

***165·361.** $\vdash : S \upharpoonright C\text{‘}Q' \,\epsilon\, Q \,\overline{\mathrm{smor}}\, Q' . \supset .$

$$(\,|\breve{S}) \upharpoonright C\text{‘}\Sigma\text{‘}P \underset{\cdot,}{\downarrow} {;}Q' \,\epsilon\, (P \underset{\cdot,}{\downarrow} {;}Q) \,\overline{\mathrm{smor}}\, \overline{\mathrm{smor}}\, (P \underset{\cdot,}{\downarrow} {;}Q')$$

[*165·351·331]

The proof proceeds as in *165·36.

***165·362.** $\vdash : R \upharpoonright C\text{‘}P' \,\epsilon\, P \,\overline{\mathrm{smor}}\, P' \,.\, S \upharpoonright C\text{‘}Q' \,\epsilon\, Q \,\overline{\mathrm{smor}}\, Q' . \supset .$

$$(R \,\|\, \breve{S}) \upharpoonright C\text{‘}\Sigma\text{‘}P' \underset{\cdot,}{\downarrow} {;}Q' \,\epsilon\, (P \underset{\cdot,}{\downarrow} {;}Q) \,\overline{\mathrm{smor}}\, \overline{\mathrm{smor}}\, (P' \underset{\cdot,}{\downarrow} {;}Q')$$

[*165·352·341]

The above three propositions are of great utility in relation-arithmetic.

***165·37.** $\vdash : P \operatorname{smor} P' \,.\, Q \operatorname{smor} Q' \,.\, \supset .\, P \downarrow ; Q \operatorname{smor} \operatorname{smor} P' \downarrow ; Q'$

[*165·362 . *164·11 . *151·12]

***165·38.** $\vdash :. \operatorname{Mult\,ax} . \supset : R \epsilon \operatorname{Rel}^2 \operatorname{excl} \cap \operatorname{Nr}`Q \,.\, C`R \subset \operatorname{Nr}`P \,.\, \supset .$

$R \operatorname{smor} \operatorname{smor} P \downarrow ; Q$

Dem.

$\vdash . *164·48 . *165·27 . \supset \vdash : \operatorname{Hp} . \dot{\exists} ! P . \supset . R \operatorname{smor} \operatorname{smor} P \downarrow ; Q \quad (1)$

$\vdash . *153·17 . *165·241 . \supset$

$\vdash : Q = \dot{\Lambda} \,.\, R \epsilon \operatorname{Rel}^2 \operatorname{excl} \cap \operatorname{Nr}`Q \,.\, \supset .\, R = \dot{\Lambda} \,.\, P \downarrow ; Q = \dot{\Lambda} \,.$

[*164·32] $\supset . R \operatorname{smor} \operatorname{smor} P \downarrow ; Q \quad (2)$

$\vdash . *165·242 . \supset \vdash : P = \dot{\Lambda} \,.\, \dot{\exists} ! Q \,.\, \supset .\, P \downarrow ; Q = \dot{\Lambda} \downarrow \dot{\Lambda} \quad (3)$

$\vdash . *153·17 . *51·4 . *151·32 . \supset$

$\vdash : R \epsilon \operatorname{Nr}`Q \,.\, C`R \subset \operatorname{Nr}`P \,.\, P = \dot{\Lambda} \,.\, \dot{\exists} ! Q \,.\, \supset .\, C`R = \iota`\dot{\Lambda} \,.$

[*56·381] $\supset . R = \dot{\Lambda} \downarrow \dot{\Lambda} \quad (4)$

$\vdash . (3) . (4) . *153·101 . *164·34 . \supset$

$\vdash : R \epsilon \operatorname{Nr}`Q \,.\, C`R \subset \operatorname{Nr}`P \,.\, P = \dot{\Lambda} \,.\, \dot{\exists} ! Q \,.\, \supset .\, R \operatorname{smor} \operatorname{smor} P \downarrow ; Q \quad (5)$

$\vdash . (1) . (2) . (5) . \supset \vdash . \operatorname{Prop}$

$*166$. THE PRODUCT OF TWO RELATIONS.

Summary of $*166$.

The product $Q \times P$ is defined as $\Sigma`P \downarrow ;Q$. This is a relation which has for its field all the couples that can be formed by choosing the referent in $C`P$ and the relatum in $C`Q$. These couples are arranged by $Q \times P$ on the following principle: If the relatum of the one couple has the relation Q to the relatum of the other, we put the one before the other, and if the relata of the two couples are equal while the referent of the one has the relation P to the referent of the other, we put the one before the other. Thus in advancing from any term $x \downarrow y$ in the field of $Q \times P$, we first keep y fixed and alter x into later terms as long as possible; then we alter y into a later term, move x back to the beginning, and so on. Thus with a given y, we get a series which is like P, and this series is wholly followed or wholly preceded by the series with the referent y', where y' follows or precedes y.

The propositions of this number are for the most part immediate consequences of those of $*165$. The most important of them are:

$*166 \cdot 12$. $\vdash . C`(P \times Q) = C`P \times C`Q$

$*166 \cdot 13$. $\vdash :. P \times Q = \dot{\Lambda} . \equiv : P = \dot{\Lambda} . \vee . Q = \dot{\Lambda}$

Hence it follows that an ordinal product of a finite number of factors vanishes when, and only when, one of its factors vanishes.

$*166 \cdot 16$. $\vdash . \overrightarrow{B}`(P \times Q) = \overrightarrow{B}`P \times \overrightarrow{B}`Q . \overrightarrow{B}`\text{Cnv}`(P \times Q) = \overrightarrow{B}`\breve{P} \times \overrightarrow{B}`\breve{Q}$

$*166 \cdot 23$. $\vdash : P \text{ smor } P' . Q \text{ smor } Q' . \supset . Q \times P \text{ smor } Q' \times P'$

This proposition shows that the relation-number of a product $Q \times P$ depends only upon the relation-numbers of its factors.

$*166 \cdot 24$. $\vdash :. \text{Mult ax} . \supset : R \in \text{Rel}^2 \text{excl} \cap \text{Nr}`Q . C`R \subset \text{Nr}`P . \supset .$

$$\Sigma`R \text{ smor } Q \times P$$

This proposition connects addition and multiplication (cf. note to $*166 \cdot 24$, below).

$*166 \cdot 42$. $\vdash . (P \times Q) \times R \text{ smor } P \times (Q \times R)$

This is the associative law. The distributive law has two forms:

***166·44.** $\vdash . \Sigma` \times P;Q = (\Sigma`Q) \times P$

***166·45.** $\vdash . (Q \mathbin{\updownarrow} R) \times P = (Q \times P) \mathbin{\updownarrow} (R \times P)$

We do not have in general (cf. note before *166·44, below)

$$P \times (Q \mathbin{\updownarrow} R) = (P \times Q) \mathbin{\updownarrow} (P \times R).$$

We have also a distributive law for the addition of a single term, *i.e.*

***166·53.** $\vdash : \dot{\exists} ! Q . \supset . (Q \mathbin{\dashrightarrow} y) \times P = (Q \times P) \mathbin{\updownarrow} (P \downarrow ; y)$

***166·531.** $\vdash : \dot{\exists} ! Q . \supset . (y \mathbin{\dashleftarrow} Q) \times P = (P \downarrow ; y) \mathbin{\updownarrow} (Q \times P)$

Here again the law does not hold in general for $P \times (Q \mathbin{\dashrightarrow} y)$ or $P \times (y \mathbin{\dashleftarrow} Q)$.

***166·01.** $Q \times P = \Sigma`P \downarrow ; ;Q$ Df

***166·1.** $\vdash . Q \times P = \Sigma`P \downarrow ; ;Q$ [(*166·01)]

***166·11.** $\vdash :. M (Q \times P) N . \equiv : (\exists x, y, z, w) : x, y \in C`P . z, w \in C`Q : zQw . \lor . z = w . xPy : M = x \downarrow z . N = y \downarrow w$ [*165·17 . *166·1]

***166·111.** $\vdash :. M (P \times Q) N . \equiv : (\exists x, y, z, w) : x, y \in C`P . z, w \in C`Q : xPy . \lor . x = y . zQw : M = z \downarrow x . N = w \downarrow y$ [*165·17 . *166·1]

***166·112.** $\vdash :. (x \downarrow z) (Q \times P) (y \downarrow w) . \equiv : x, y \in C`P . z, w \in C`Q : zQw . \lor . z = w . xPy$ [*166·11 . *55·202 . *13·22]

***166·113.** $\vdash :: x, y \in C`P . z, w \in C`Q . \supset :. (x \downarrow z) (Q \times P) (y \downarrow w) . \equiv : zQw . \lor . z = w . xPy$ [*166·112]

***166·12.** $\vdash . C`(P \times Q) = C`P \times C`Q$ [*165·16 . *166·1]

***166·13.** $\vdash :. P \times Q = \dot{\Lambda} . \equiv : P = \dot{\Lambda} . \lor . Q = \dot{\Lambda}$ [*166·12 . *113·114 . *33·241]

***166·14.** $\vdash : \dot{\exists} ! P \times Q . \equiv . \dot{\exists} ! P . \dot{\exists} ! Q$ [*166·13]

***166·15.** $\vdash . \text{Cnv}`(P \times Q) = \breve{P} \times \breve{Q}$ [*165·19 . *162·2]

***166·16.** $\vdash . \overrightarrow{B}`(P \times Q) = \overrightarrow{B}`P \times \overrightarrow{B}`Q . \overrightarrow{B}`\text{Cnv}`(P \times Q) = \overrightarrow{B}`\breve{P} \times \overrightarrow{B}`\breve{Q}$

Dem.

$\vdash . *166·111 . *93·103 . \supset$

$\vdash :. M \in \overrightarrow{B}`(P \times Q) . \equiv : (\exists x, z) : x \in C`P . z \in C`Q . M = z \downarrow x : \sim (\exists y) . yPx : \sim (\exists w) . wQz :$

[*93·103] $\equiv : (\exists x, z) . x \in \overrightarrow{B}`P . y \in \overrightarrow{B}`Q . M = z \downarrow x :$

[*113·101] $\equiv : M \in \overrightarrow{B}`P \times \overrightarrow{B}`Q$ (1)

$\vdash . (1) . *166·15 . \supset \vdash . \overrightarrow{B}`\text{Cnv}`(P \times Q) = \overrightarrow{B}`\breve{P} \times \overrightarrow{B}`\breve{Q}$ (2)

$\vdash . (1) . (2) . \supset \vdash .$ Prop

The above proposition is used in the ordinal theory of progressions (*263·62·65).

***166·2.** $\vdash : R \upharpoonright C`P' \in P \,\overline{\text{smor}}\, P' . \supset . (R \| ) \upharpoonright C`(Q \times P') \in (Q \times P) \,\overline{\text{smor}}\, (Q \times P')$
[*165·36 . *166·1 . *164·14]

***166·21.** $\vdash : S \upharpoonright C`Q' \in Q \,\overline{\text{smor}}\, Q' . \supset . (\| \breve{S}) \upharpoonright C`(Q' \times P) \in (Q \times P) \,\overline{\text{smor}}\, (Q' \times P)$
[*165·361 . *166·1 . *164·14]

***166·22.** $\vdash : R \upharpoonright C`P' \in P \,\overline{\text{smor}}\, P' . S \upharpoonright C`Q' \in Q \,\overline{\text{smor}}\, Q' . \supset .$
$(R \| \breve{S}) \upharpoonright C`(Q' \times P') \in (Q \times P) \,\overline{\text{smor}}\, (Q' \times P')$
[*165·362 . *166·1 . *164·14]

This proposition gives the correlator for the product when correlators are given for the factors.

***166·23.** $\vdash : P \text{ smor } P' . Q \text{ smor } Q' . \supset . Q \times P \text{ smor } Q' \times P'$
[*166·22 . *151·12]

This proposition enables us to use $Q \times P$ to define the product of the relation-numbers of Q and P, for it shows that the relation-number of $Q \times P$ is determinate when the relation-numbers of Q and P are given. We shall therefore (in Section D of this part) define the product of two relation-numbers ν and μ as the relation-number of $Q \times P$ when $\mathrm{N_0r}`Q = \nu$ and $\mathrm{N_0r}`P = \mu$.

***166·24.** $\vdash :. \text{Mult ax} . \supset : R \in \text{Rel}^2 \text{ excl} \cap \mathrm{Nr}`Q . C``R \subset \mathrm{Nr}`P . \supset .$
$\Sigma`R \text{ smor } Q \times P$ [*165·38 . *164·151 . *166·1]

This proposition exhibits the connection of addition and multiplication. If we put $\mathrm{Nr}`P = \mu$ and $\mathrm{Nr}`Q = \nu$, then $\Sigma`R$ in the above proposition is the sum of $\nu$ relations of which each is a $\mu$. In virtue of the above proposition, it follows that (if the multiplicative axiom is assumed) $\mathrm{Nr}`\Sigma`R = \nu \times \mu$. In other words, assuming the multiplicative axiom, the sum of ν series (or other relations), each of which has μ terms, has $\nu \times \mu$ terms.

***166·3.** $\vdash : \exists ! C`(P \times Q) \cap C`(P' \times Q') . \equiv . \exists ! C`P \cap C`P' . \exists ! C`Q \cap C`Q'$
[*166·12 . *113·19]

The analogous proposition

$\dot{\exists} ! (P \times Q) \dot{\cap} (P' \times Q') . \equiv :$
$\dot{\exists} ! (P \dot{\cap} P') . \exists ! C`Q \cap C`Q' . \vee . \dot{\exists} ! (Q \dot{\cap} Q') . \exists ! C`P \cap C`P'$

is only true in general if $P \Subset J . P' \Subset J$

***166·31.** $\vdash . s`C``(Q \times P) = C`P \uparrow C`Q$ [*113·115 . *166·12]

***166·311.** $\vdash : \dot{\exists} ! Q . \supset . s`\mathrm{D}``C``(Q \times P) = C`P : \dot{\exists} ! P . \supset . s`\text{Ɑ}``C``(Q \times P) = C`Q$
[*113·116 . *166·12 . *33·24]

***166·312.** $\vdash . s`\mathrm{D}``C``(Q \times P) \subset C`P . s`\text{Ɑ}``C``(Q \times P) \subset C`Q$
[*113·118 . *166·12]

The following propositions are lemmas for the associative law (*166·42).

***166·4.** $\vdash :. M\{(P \times Q) \times R\} M' . \equiv : (\exists x, y, z, x', y', z') :$
$x, x' \in C\text{‘}P . y, y' \in C\text{‘}Q . z, z' \in C\text{‘}R :$
$xPx' . \vee . x = x' . yQy' . \vee . x = x' . y = y' . zRz' :$
$M = z \downarrow (y \downarrow x) . M' = z' \downarrow (y' \downarrow x')$

Dem.

$\vdash . *116·111 . \supset \vdash :. M\{(P \times Q) \times R\} M' . \equiv : (\exists N, N', z, z') :$
$N, N' \in C\text{‘}(P \times Q) . z, z' \in C\text{‘}R : N(P \times Q) N' . \vee . N = N' . zRz' :$
$M = z \downarrow N . M' = z' \downarrow N' :$

$[*116·12.*113·101] \equiv : (\exists N, N', x, x', y, y', z, z') :$
$x, x' \in C\text{‘}P . y, y' \in C\text{‘}Q . z, z' \in C\text{‘}R . N = y \downarrow x . N' = y' \downarrow x' :$
$N(P \times Q)N' . \vee . N = N' . zRz' : M = z \downarrow N . M' = z' \downarrow N' :$

$[*13·22.*116·113] \equiv : (\exists x, x', y, y', z, z') : x, x' \in C\text{‘}P . y, y' \in C\text{‘}Q . z, z' \in C\text{‘}R :$
$xPx' . \vee . x = x' . yQy' . \vee . y \downarrow x = y' \downarrow x' . zRz' :$
$M = z \downarrow (y \downarrow x) . M' = z' \downarrow (y' \downarrow x') :$

$[*55·202] \equiv : (\exists x, x', y, y', z, z') : x, x' \in C\text{‘}P . y, y' \in C\text{‘}Q . z, z' \in C\text{‘}R :$
$xPx' . \vee . x = x' . yQy' . \vee . x = x' . y = y' . zRz' :$
$M = z \downarrow (y \downarrow x) . M' = z' \downarrow (y' \downarrow x') :. \supset \vdash .$ Prop

***166·401.** $\vdash :. N\{P \times (Q \times R)\} N' . \equiv : (\exists x, x', y, y', z, z') :$
$x, x' \in C\text{‘}P . y, y' \in C\text{‘}Q . z, z' \in C\text{‘}R :$
$xPx' . \vee . x = x' . yQy' . \vee . x = x' . y = y' . zRz' :$
$N = (z \downarrow y) \downarrow x . N' = (z' \downarrow y') \downarrow x'$

[Proof as in *166·4]

***166·41.** $\vdash : T = \hat{M}\hat{N}\{(\exists x, y, z) . x \in C\text{‘}P . y \in C\text{‘}Q . z \in C\text{‘}R . M = z \downarrow (y \downarrow x) .$
$N = (z \downarrow y) \downarrow x\} . \supset . T \in \{(P \times Q) \times R\} \overline{\text{smor}} \{P \times (Q \times R)\}$

Dem.

$\vdash . *21·33 . \supset \vdash :: \text{Hp} . \supset :. MTN . M'TN . \supset :$
$(\exists x, x', y, y', z, z') : x, x' \in C\text{‘}P . y, y' \in C\text{‘}Q . z, z' \in C\text{‘}R :$
$M = z \downarrow (y \downarrow x) . M' = z' \downarrow (y' \downarrow x') :$
$N = (z \downarrow y) \downarrow x . N = (z' \downarrow y') \downarrow x' :$

$[*55·202] \supset : (\exists x, x', y, y', z, z') . M = z \downarrow (y \downarrow x) . M' = z' \downarrow (y' \downarrow x') .$
$x = x' . y = y' . z = z' :$

$[*13·22] \supset : M = M' \quad (1)$

Similarly $\vdash :. \text{Hp} . \supset : MTN . MTN' . \supset . N = N' \quad (2)$

$\vdash . (1) . (2) . \supset \vdash : \text{Hp} . \supset . T \in 1 \rightarrow 1 \quad (3)$

$\vdash . *21·33 . *13·19 . \supset \vdash : \text{Hp} . \supset .$

$\text{Ɑ}\text{‘}T = \hat{N}\{(\exists x, y, z) . x \in C\text{‘}P . y \in C\text{‘}Q . z \in C\text{‘}R . N = (z \downarrow y) \downarrow x\}$

$[*113·101] = C\text{‘}P \times (C\text{‘}Q \times C\text{‘}R)$

$[*166·12] = C\text{‘}\{P \times (Q \times R)\} \quad (4)$

$\vdash . *166{\cdot}401 . \supset \vdash :: \text{Hp} . \supset :. M\{T{\text{;}}(P \times (Q \times R))\} M' . \equiv : (\exists x, x', y, y', z, z', N, N') :$
$x, x' \in C{\text{‘}}P . y, y' \in C{\text{‘}}Q . z, z' \in C{\text{‘}}R . N = (z \downarrow y) \downarrow x . N' = (z' \downarrow y') \downarrow x' .$
$M = z \downarrow (y \downarrow x) . M' = z' \downarrow (y' \downarrow x') :$
$xPx' . \mathbf{v} . x = x' . yQy' . \mathbf{v} . x = x' . y = y' . zRz' :$

$[*13{\cdot}19 . *166{\cdot}4] \quad \equiv : M \{(P \times Q) \times R\} M' \qquad (5)$

$\vdash . (3) . (4) . (5) . *151{\cdot}11 . \supset \vdash . \text{Prop}$

***166·42.** $\vdash . (P \times Q) \times R \text{ smor } P \times (Q \times R) \quad [*166{\cdot}41]$

This is the associative law for the kind of multiplication concerned in this number.

***166·421.** $P \times Q \times R = (P \times Q) \times R \quad \text{Df}$

This definition serves merely for the avoidance of brackets.

The two following propositions give the distributive law. In relation-arithmetic, this is in general only true in one of its two forms, *i.e.* we have

$$(Q \ddagger R) \times P = (Q \times P) \ddagger (R \times P),$$

but not $\quad P \times (Q \ddagger R) = (P \times Q) \ddagger (P \times R).$

The latter is true for finite series, but not for infinite series or (except in exceptional cases) for relations which are not serial.

***166·44.** $\vdash . \Sigma{\text{‘}} \times P{\text{;}}Q = (\Sigma{\text{‘}}Q) \times P$

Dem.

$\vdash . *166{\cdot}1 . *38{\cdot}11 . *150{\cdot}1 . \supset \vdash . \Sigma{\text{‘}} \times P{\text{;}}Q = \Sigma{\text{‘}}\Sigma{\text{;}}(P \underset{\cdot\cdot}{\downarrow}){\dagger}{\text{;}}Q$

$[*162{\cdot}34{\cdot}35] \qquad = \Sigma{\text{‘}}P \underset{\cdot\cdot}{\downarrow} {\text{;}}\Sigma{\text{‘}}Q$

$[*166{\cdot}1] \qquad = (\Sigma{\text{‘}}Q) \times P . \supset \vdash . \text{Prop}$

***166·45.** $\vdash . (Q \ddagger R) \times P = (Q \times P) \ddagger (R \times P)$

Dem.

$\vdash . *166{\cdot}1 . \supset \vdash . (Q \times P) \ddagger (R \times P) = \Sigma{\text{‘}}P \underset{\cdot\cdot}{\downarrow} {\text{;}}Q \ddagger \Sigma{\text{‘}}P \underset{\cdot\cdot}{\downarrow} {\text{;}}R$

$[*162{\cdot}31] \qquad = \Sigma{\text{‘}}(P \underset{\cdot\cdot}{\downarrow} {\text{;}}Q \ddagger P \underset{\cdot\cdot}{\downarrow} {\text{;}}R)$

$[*162{\cdot}36] \qquad = \Sigma{\text{‘}}P \underset{\cdot\cdot}{\downarrow} {\text{;}}(Q \ddagger R)$

$[*166{\cdot}1] \qquad = (Q \ddagger R) \times P . \supset \vdash . \text{Prop}$

The following propositions (*166·46—·472) exhibit the failure of the distributive law in the form $P \times (Q \ddagger R) = (P \times Q) \ddagger (P \times R)$, and give certain results for special cases. They are not referred to except in this number.

***166·46.** $\vdash . (P \cup Q) \underset{\cdot\cdot}{\downarrow} z = P \underset{\cdot\cdot}{\downarrow} z \cup Q \underset{\cdot\cdot}{\downarrow} z \quad [*165{\cdot}01 . *150{\cdot}3]$

***166·461.** $\vdash . \dot{s}{\text{‘}}C{\text{‘‘}}(P \cup Q) \underset{\cdot\cdot}{\downarrow} {\text{;}}R = \dot{s}{\text{‘}}C{\text{‘‘}}P \underset{\cdot\cdot}{\downarrow} {\text{;}}R \cup \dot{s}{\text{‘}}C{\text{‘‘}}Q \underset{\cdot\cdot}{\downarrow} {\text{;}}R$
$[*41{\cdot}6 . *165{\cdot}12 . *166{\cdot}46]$

∗166·462. $\vdash . F;(P \cup Q) \underset{\cdot\cdot}{\downarrow} ;R = F;P \underset{\cdot\cdot}{\downarrow} ;R \cup F;Q \underset{\cdot\cdot}{\downarrow} ;R \cup \hat{M}\hat{N} \{(\exists x, y, z, w) :$
$zRw : x \in C`P . y \in C`Q . \vee . x \in C`Q . y \in C`P : M = x \downarrow z . N = y \downarrow w\}$

Dem.

$\vdash . *165{\cdot}161 . \supset \vdash . F;(P \cup Q) \underset{\cdot\cdot}{\downarrow} ;R$

$= \hat{M}\hat{N} \{(\exists x, y, z, w) . x, y \in C`P \cup C`Q . zRw . M = x \downarrow z . N = y \downarrow w\}$

[∗22·34] $= \hat{M}\hat{N} \{(\exists x, y, z, w) : x, y \in C`P . \vee . x, y \in C`Q . \vee .$
$x \in C`P . y \in C`Q . \vee . x \in C`Q . y \in C`P : zRw . M = x \downarrow z . N = y \downarrow w\}$

[∗11·41.∗165·161] $= F;P \underset{\cdot\cdot}{\downarrow} ;R \cup F;Q \underset{\cdot\cdot}{\downarrow} ;R \cup \hat{M}\hat{N} \{(\exists x, y, z, w) :$
$x \in C`P . y \in C`Q . \vee . x \in C`Q . y \in C`P :$
$zRw . M = x \downarrow z . N = y \downarrow w\} . \supset \vdash .$ Prop

∗166·463. $\vdash : C`P \subset C`Q . \supset . F;P \underset{\cdot\cdot}{\downarrow} ;R \Subset F;Q \underset{\cdot\cdot}{\downarrow} ;R$ [∗165·161]

∗166·464. $\vdash : C`P \subset C`Q . \supset . F;(P \cup Q) \underset{\cdot\cdot}{\downarrow} ;R = F;Q \underset{\cdot\cdot}{\downarrow} ;R = F;P \underset{\cdot\cdot}{\downarrow} ;R \cup F;Q \underset{\cdot\cdot}{\downarrow} ;R$

Dem.

$\vdash . *166{\cdot}463 . \supset \vdash : \text{Hp} . \supset . F;P \underset{\cdot\cdot}{\downarrow} ;R \Subset F;Q \underset{\cdot\cdot}{\downarrow} ;R$ (1)

$\vdash . *33{\cdot}262 . \supset \vdash : \text{Hp} . \supset . C`(P \cup Q) = C`Q .$

[∗166·463] $\supset . F;(P \cup Q) \underset{\cdot\cdot}{\downarrow} ;R = F;Q \underset{\cdot\cdot}{\downarrow} ;R$ (2)

$\vdash . (1) . (2) . \supset \vdash .$ Prop

∗166·47. $\vdash . R \times (P \cup Q) = (R \times P) \cup (R \times Q) \cup \hat{M}\hat{N} \{(\exists x, y, z, w) :$
$x \in C`P . y \in C`Q . \vee . x \in C`Q . y \in C`P : zRw . M = x \downarrow z . N = y \downarrow w\}$
[∗166·461·462·1 . ∗162·1]

∗166·471. $\vdash : C`P \subset C`Q . \supset . R \times (P \cup Q) = (R \times P) \cup (R \times Q)$
[∗166·461·464]

∗166·472. $\vdash . R \times (P \mathbin{\updownarrow} Q) = (R \times P) \cup (R \times Q) \cup R \times (C`P \downarrow C`Q)$

Dem.

$\vdash . *166{\cdot}471 . *35{\cdot}85 . \supset$

$\vdash :. \dot{\exists} ! Q . \supset : R \times (P \mathbin{\updownarrow} Q) = (R \times P) \cup R \times \{Q \cup (C`P \uparrow C`Q)\} :$

[∗166·471.∗35·86]

$\supset : \dot{\exists} ! P . \supset . R \times (P \mathbin{\updownarrow} Q) = (R \times P) \cup (R \times Q) \cup R \times (C`P \uparrow C`Q)$ (1)

$\vdash . *160{\cdot}21 . *166{\cdot}13 . \supset$

$\vdash : Q = \dot{\Lambda} . \supset . P \mathbin{\updownarrow} Q = P . R \times Q = \dot{\Lambda} . R \times (C`P \uparrow C`Q) = \dot{\Lambda} .$

[∗25·24] $\supset . R \times (P \mathbin{\updownarrow} Q) = (R \times P) \cup (R \times Q) \cup R \times (C`P \downarrow C`Q)$ (2)

Similarly

$\vdash : P = \dot{\Lambda} . \supset . R \times (P \mathbin{\updownarrow} Q) = (R \times P) \cup (R \times Q) \cup R \times (C`P \uparrow C`Q)$ (3)

$\vdash . (1) . (2) . (3) . \supset \vdash .$ Prop

The following propositions are concerned with the distributive law for the addition of a single term to a relation. This law, in the form in which it holds, is given in ∗166·53·531 (remembering $\text{Nr}\text{‘}P \underset{\cdot\cdot}{\downarrow} y = \text{Nr}\text{‘}P$). ∗166·54·541 exhibit the failure of the other form.

∗166·5. $\vdash . (Q \mathbin{\dot{\cup}} R) \times P = (Q \times P) \mathbin{\dot{\cup}} (R \times P)$

Dem.

$$\begin{aligned} \vdash . *166{\cdot}1 . \supset \vdash . (Q \mathbin{\dot{\cup}} R) \times P &= \Sigma\text{‘}P \underset{\cdot\cdot}{\downarrow} ;(Q \mathbin{\dot{\cup}} R) \\ [*162{\cdot}27] \quad &= \Sigma\text{‘}P \underset{\cdot\cdot}{\downarrow} ;Q \mathbin{\dot{\cup}} \Sigma\text{‘}P \underset{\cdot\cdot}{\downarrow} ;R \\ [*166{\cdot}1] \quad &= (Q \times P) \mathbin{\dot{\cup}} (R \times P) . \supset \vdash . \text{Prop} \end{aligned}$$

∗166·51. $\vdash . (Q \mathbin{+\!\!\!\rightarrow} y) \times P = (Q \times P) \mathbin{\dot{\cup}} (C\text{‘}Q \uparrow \iota\text{‘}y) \times P$ [∗166·5 . ∗161·1]

∗166·511. $\vdash . (y \mathbin{\leftarrow\!\!\!+} Q) \times P = (\iota\text{‘}y \uparrow C\text{‘}P) \times P \mathbin{\dot{\cup}} (Q \times P)$

∗166·52. $\vdash . P \underset{\cdot\cdot}{\downarrow} ;(Q \mathbin{+\!\!\!\rightarrow} y) = P \underset{\cdot\cdot}{\downarrow} ;Q \mathbin{+\!\!\!\rightarrow} P \underset{\cdot\cdot}{\downarrow} y$ [∗161·4 . ∗165·2]

∗166·521. $\vdash . P \underset{\cdot\cdot}{\downarrow} ;(y \mathbin{\leftarrow\!\!\!+} Q) = P \underset{\cdot\cdot}{\downarrow} y \mathbin{\leftarrow\!\!\!+} P \underset{\cdot\cdot}{\downarrow} ;Q$

∗166·53. $\vdash : \dot{\exists} ! \, Q . \supset . (Q \mathbin{+\!\!\!\rightarrow} y) \times P = (Q \times P) \ddagger (P \underset{\cdot\cdot}{\downarrow} y)$

Dem.

$$\begin{aligned} &\vdash . *162{\cdot}43 . *165{\cdot}243 . \supset \vdash : \text{Hp} . \supset . \Sigma\text{‘}(P \underset{\cdot\cdot}{\downarrow} ;Q \mathbin{+\!\!\!\rightarrow} P \underset{\cdot\cdot}{\downarrow} y) = \Sigma\text{‘}P \underset{\cdot\cdot}{\downarrow} ;Q \ddagger P \underset{\cdot\cdot}{\downarrow} y . \\ &[*166{\cdot}52] \qquad \supset . \Sigma\text{‘}P \underset{\cdot\cdot}{\downarrow} ;(Q \mathbin{+\!\!\!\rightarrow} y) = \Sigma\text{‘}P \underset{\cdot\cdot}{\downarrow} ;Q \ddagger P \underset{\cdot\cdot}{\downarrow} y \qquad (1) \\ &\vdash . (1) . *166{\cdot}1 . \supset \vdash . \text{Prop} \end{aligned}$$

∗166·531. $\vdash : \dot{\exists} ! \, Q . \supset . (y \mathbin{\leftarrow\!\!\!+} Q) \times P = (P \underset{\cdot\cdot}{\downarrow} y) \ddagger (Q \times P)$

∗166·54. $\vdash . Q \times (P \mathbin{+\!\!\!\rightarrow} x) = (Q \times P) \mathbin{\dot{\cup}} Q \times (C\text{‘}P \uparrow \iota\text{‘}x)$

Dem.

$$\begin{aligned} \vdash . *161{\cdot}1 . \supset \vdash . Q \times (P \mathbin{+\!\!\!\rightarrow} x) &= Q \times \{P \mathbin{\dot{\cup}} (C\text{‘}P \uparrow \iota\text{‘}x)\} \\ [*35{\cdot}85 . *166{\cdot}471] \quad &= (Q \times P) \mathbin{\dot{\cup}} Q \times (C\text{‘}P \uparrow \iota\text{‘}x) . \supset \vdash . \text{Prop} \end{aligned}$$

∗166·541. $\vdash . Q \times (x \mathbin{\leftarrow\!\!\!+} P) = Q \times (\iota\text{‘}x \uparrow C\text{‘}P) \mathbin{\dot{\cup}} (Q \times P)$

SECTION C.

THE PRINCIPLE OF FIRST DIFFERENCES, AND THE MULTIPLICATION AND EXPONENTIATION OF RELATIONS.

Summary of Section C.

In the present section, we have to consider various forms of a principle which is of the utmost utility in relation-arithmetic. This principle may be called "the principle of first differences." It has been explained and used by Hausdorff in brilliant articles*. The results there obtained by its use give some measure of its importance in relation-arithmetic. It has, however, other uses besides those that are concerned with the multiplication and exponentiation of relation-numbers, as, for example, in the ordering of segments and stretches in a series, or of any other set of classes which are contained in the field of a given relation. In the present section, after the first two numbers, we shall be concerned with its arithmetical uses, but other uses will occur later.

The principle of first differences has various forms which, though analogous, cannot, in the general case, be reduced to one common genus. The simplest of these is the relation P_{cl}, by which the sub-classes of $C{\text{‘}}P$ are ordered. This is defined as follows. If α and β are both contained in $C{\text{‘}}P$, we say that $\alpha P_{\text{cl}} \beta$ if there are terms belonging to α but not to β such that no terms belonging to β and not to α precede them; *i.e.* if, after taking away the terms (if any) which are common to α and β, there are terms left in α which do not come after any of the terms left in β, *i.e.* if $\exists ! \alpha - \beta - \breve{P}{\text{“}}(\beta - \alpha)$. Thus the definition is

$$P_{\text{cl}} = \hat{\alpha}\hat{\beta} \{\alpha, \beta \in \text{Cl}{\text{‘}}C{\text{‘}}P \,.\, \exists ! \alpha - \beta - \breve{P}{\text{“}}(\beta - \alpha)\} \quad \text{Df.}$$

It will be seen that this relation holds if $\beta \subset \alpha \,.\, \beta \neq \alpha$. Thus it holds between any existent member of $\text{Cl}{\text{‘}}C{\text{‘}}P$ and Λ, and between $C{\text{‘}}P$ and any member of $\text{Cl}{\text{‘}}C{\text{‘}}P$ other than $C{\text{‘}}P$ itself. When P is a serial relation (which is the important case for all the relations in this section), P_{cl} is transitive ($P^2_{\text{cl}} \subset P_{\text{cl}}$) and asymmetrical ($P_{\text{cl}} \dot{\cap} \breve{P}_{\text{cl}} = \dot{\Lambda}$), but not necessarily *connected*, *i.e.* there may

* "Untersuchungen über Ordnungstypen," *Berichte der mathematischphysischen Klasse der Königlich Sächsischen Gesellschaft der Wissenschaften zu Leipzig*, Feb. 1906 and Feb. 1907. Cf. also his "Grundzüge einer Theorie der geordneten Mengen," *Math. Annalen*, 65 (1908).

be two members of its field of which neither has the relation P_{cl} to the other. This happens whenever P is not well-ordered; but when P is well-ordered, P_{cl} is connected, and therefore generates a series.

To illustrate the order generated by P_{cl} in a simple case, consider a series of three terms, x, y, z. Let us for the moment write $(x \downarrow y \downarrow z)$ for the relation

$$x \downarrow y \cup x \downarrow z \cup y \downarrow z, \quad i.e.\ (x \downarrow y) \mathbin{+\!\!\!\rightarrow} z,$$

and similarly we will write $(x \downarrow y \downarrow z \downarrow w)$ for $x \downarrow y \downarrow z \mathbin{+\!\!\!\rightarrow} w$, and so on. Then assuming $x \neq y \,.\, x \neq z \,.\, y \neq z$,

$$\begin{aligned}(x \downarrow y \downarrow z)_{cl} = (\iota{}^\backprime x \cup \iota{}^\backprime y \cup \iota{}^\backprime z) \downarrow (\iota{}^\backprime x \cup \iota{}^\backprime y) \\ \downarrow (\iota{}^\backprime x \cup \iota{}^\backprime z) \downarrow \iota{}^\backprime x \downarrow (\iota{}^\backprime y \cup \iota{}^\backprime z) \downarrow \iota{}^\backprime y \downarrow \iota{}^\backprime z \downarrow \Lambda.\end{aligned}$$

In this series, a class containing x is always earlier than one not containing x; and of two classes of which both or neither contain x, one containing y is earlier than one not containing y; and of two classes of which both or neither contain x, and both or neither contain y, one containing z is earlier than one not containing z. Thus our relation may be generated as follows: Begin with $(\iota{}^\backprime z) \downarrow \Lambda$, which is $(z \downarrow z)_{cl}$. Add before these terms what results from adding $\iota{}^\backprime y$ to each; then we have $(y \downarrow z)_{cl}$, which is

$$(\iota{}^\backprime y \cup \iota{}^\backprime z) \downarrow \iota{}^\backprime y \downarrow \iota{}^\backprime z \downarrow \Lambda.$$

Now add at the beginning what results from adding $\iota{}^\backprime x$ to each of the above four classes, and we have $(x \downarrow y \downarrow z)_{cl}$. Thus generally, if $x \sim \epsilon\, C{}^\backprime P$,

$$(x \mathbin{\leftarrow\!\!\!+} P)_{cl} = (\iota{}^\backprime x \cup)^{;} P_{cl} \mathbin{\uparrow\!\!\!\!+} P_{cl}.$$

Thus by adding one term to P, we double the number of terms in P_{cl}.

Again, if P and Q are two relations which have no common terms in their fields, we shall have

$$\alpha P_{cl} \beta \,.\, \gamma, \delta \,\epsilon\, \mathrm{Cl}{}^\backprime C{}^\backprime Q \,.\supset.\, (\alpha \cup \gamma)(P \mathbin{\uparrow\!\!\!\!+} Q)_{cl}(\beta \cup \delta)$$

and $$\alpha \,\epsilon\, \mathrm{Cl}{}^\backprime C{}^\backprime P \,.\, \gamma Q_{cl} \delta \,.\supset.\, (\alpha \cup \gamma)(P \mathbin{\uparrow\!\!\!\!+} Q)_{cl}(\alpha \cup \delta),$$

while conversely

$$\begin{aligned}\alpha, \beta \,\epsilon\, \mathrm{Cl}{}^\backprime C{}^\backprime P \,.\, \gamma, \delta \,\epsilon\, \mathrm{Cl}{}^\backprime C{}^\backprime Q \,.\, (\alpha \cup \gamma)(P \mathbin{\uparrow\!\!\!\!+} Q)_{cl}(\beta \cup \delta) \,.\supset: \\ \alpha P_{cl} \beta \,.\vee.\, \alpha = \beta \,.\, \gamma Q_{cl} \delta.\end{aligned}$$

Hence $$\begin{aligned}(\alpha \cup \gamma)(P \mathbin{\uparrow\!\!\!\!+} Q)_{cl}(\beta \cup \delta) \,.\equiv.\, (\gamma \downarrow \alpha)(P_{cl} \times Q_{cl})(\delta \downarrow \beta) \\ \equiv.\, (\alpha \cup \gamma)\{s^{;} C^{;}(P_{cl} \times Q_{cl})\}(\beta \cup \delta),\end{aligned}$$

so that $$\mathrm{Nr}{}^\backprime(P \mathbin{\uparrow\!\!\!\!+} Q)_{cl} = \mathrm{Nr}{}^\backprime P_{cl} \times \mathrm{Nr}{}^\backprime Q_{cl}.$$

These propositions illustrate the connection of P_{cl} with multiplication.

Besides P_{cl}, we often require (though not in this Part) the relation which is the converse of $(\breve{P})_{cl}$. This relation we call P_{lc}, so that

$$P_{lc} = \mathrm{Cnv}{}^\backprime(\breve{P})_{cl} \quad \mathrm{Df.}$$

This begins with Λ, and ends with $C{}^\backprime P$.

Thus we shall have, for example,

$$(x \downarrow y \downarrow z)_{lc} = \Lambda \downarrow \iota`x \downarrow \iota`y \downarrow (\iota`x \cup \iota`y)$$
$$\downarrow \iota`z \downarrow (\iota`x \cup \iota`z) \downarrow (\iota`y \cup \iota`z) \downarrow (\iota`x \cup \iota`y \cup \iota`z).$$

Here, if we start from $\Lambda \downarrow \iota`x$, which is $(x \downarrow x)_{lc}$, the series grows by adding terms at the end: we add $\iota`y$ to each member of $\Lambda \downarrow \iota`x$, and put the resulting terms $\iota`y$, $\iota`x \cup \iota`y$ after Λ and $\iota`x$; we then add $\iota`z$ to each of the four terms we already have, and add the resulting terms at the end; and so we can proceed indefinitely.

The relation P_{lc} with its field limited arranges the *segments* of P in ascending order of magnitude; if the class of segments is σ, $P_{lc} \upharpoonright \sigma$ generates what may be called the natural order among the segments (cf. *212).

A variant of P_{cl} is afforded by the relation P_{df} (*171), which is to hold between two members α, β of $\text{Cl}`C`P$ when the first term of either which does not belong to both belongs to α, *i.e.* the "first difference" belongs to α. This relation implies P_{cl}, and coincides with it if P is well-ordered; but when P is not well-ordered, P_{cl} may hold between two classes which have no *first* point of difference, *e.g.* (if P is "less than" among rationals) if α consists of rationals between 0 and 1 (both excluded) and β of rationals between 1 and 2 (both excluded). The definition of P_{df} is

$$P_{df} = \hat{\alpha}\hat{\beta} \{\alpha, \beta \in \text{Cl}`C`P : (\exists z) . z \in \alpha - \beta . \overrightarrow{P}`z \cap \alpha - \iota`z = \overrightarrow{P}`z \cap \beta\} \quad \text{Df.}$$

The relation P_{df} has the interesting property that its relation-number is found by raising 2_r to the power $\text{Nr}`P$ (cf. *177). As the field of $P_{df}$ is $\text{Cl}`C`P$, this theorem is the ordinal analogue of $\text{Nc}`\text{Cl}`\alpha = 2^{\text{Nc}`\alpha}$ (*116·72).

A somewhat more complicated form of the relation of first differences arises when we have a series of series. Let us suppose, to begin with, that P is a serial relation whose field consists of mutually exclusive serial relations. Thus in the accompanying figure, each row represents a series, the generating relations of these series being $Q, \ldots R, \ldots$. But the series themselves form a series, which may be regarded as generated by a relation P whose field consists of the relations $Q, \ldots R, \ldots$. (It might be thought more natural to take $C`Q$, $C`R$, ... as the field of P; but this would lead to confusion in the case when two or more of the series have the same field.)

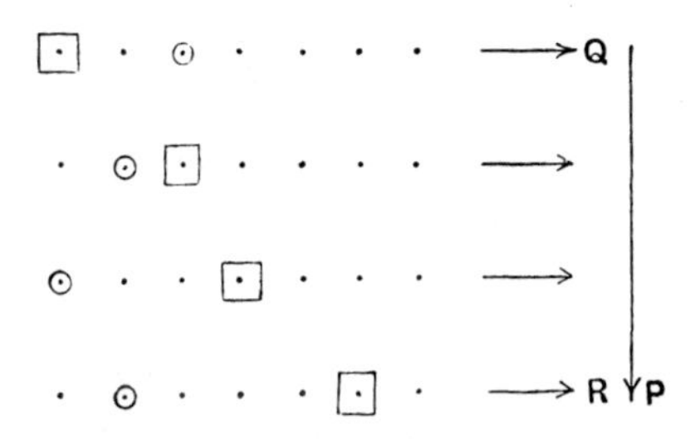

Suppose we now wish to find a relation which will order the multiplicative class of the fields of $Q, R, \ldots$, *i.e.* the class $\text{Prod}`C``C`P$. In the case illustrated in the figure, in which $P$ generates a well-ordered series, and all the members of $C`P$ are serial, and $P \in \text{Rel}^2\text{excl}$, we might use $(\Sigma`P)_{cl}$; this relation, with its field limited to $\text{Prod}`C``C`P$, will then give us what we want. This relation will, in the case supposed, put a selected class μ before another selected class ν if, where they first differ, μ chooses an earlier term than ν. But if the series P is not well-ordered—if it is (say) of the type Cnv$``\omega$ (cf. *263)—there may be no *first* member of the field of P where μ and ν differ. This will happen, for example, if μ consists of all the first terms, and ν of all the second terms. Our ordering relation can be so defined as to put μ before ν in this case also, but if it is so defined, the associative law of multiplication only holds if P is well-ordered. For this reason, we define our ordering relation so that, in such a case, μ comes neither before nor after ν. Again, if P is not a Rel^2 excl, a member of a selected class may occur twice, once as the representative of $C`Q$, and once as that of $C`R$, if $C`Q$ and $C`R$ have terms in common. We wish to distinguish these two occurrences. Hence we proceed as follows: If μ and ν are two selected classes of $C``C`P$, let there be one or more members of $C`P$ in which the μ-representative precedes the ν-representative, and which are such that, among all earlier* members of $C`P$, the μ-representative is identical with the ν-representative.

But a further modification is desirable in order to meet the case in which two or more of the members of $C`P$ have the same field. Suppose, for example, we had to deal with a series consisting of all the series that can be formed out of a given set of terms: in this case, we should have to distinguish occurrences of any given term not by the field, but by the generating relation. This requires that we should make an F-selection from $C`P$, not an ϵ-selection from $C``C`P$. Hence we take two members of $F_\Delta`C`P$, say $M$ and $N$, and we arrange them or their domains on the following principle: We put $M$ before $N$ (or $\mathrm{D}`M$ before $\mathrm{D}`N$) if there is a relation $Q$ in the field of $P$ such that the $M$-representative of $Q$, *i.e.* $M`Q$, has the relation Q to the N-representative of Q, and such that, if R is any earlier member of $C`P$, then $M`R$ is identical with $N`R$. That is, M precedes N if

$$(\exists Q) : (M`Q)\, Q\, (N`Q) : RPQ \,.\, R \neq Q \,.\, \supset_R \,.\, M`R = N`R.$$

The relation between M and N so defined has the properties required of an arithmetical product; hence we put

$$\Pi`P = \hat{M}\hat{N}\{M, N \,\epsilon\, F_\Delta`C`P :. (\exists Q) : (M`Q)\, Q\, (N`Q) : RPQ \,.\, R \neq Q \,.\, \supset_R \,.\, M`R = N`R\} \quad \text{Df.}$$

This relation is the ordinal analogue of $\epsilon_\Delta`\kappa$. The ordinal analogue of Prod$`\kappa$ is the corresponding relation of the domains of M and N, *i.e.* $\mathrm{D};\Pi`P$; hence we put

$$\text{Prod}`P = \mathrm{D};\Pi`P \quad \text{Df.}$$

In case P is a Rel^2 excl, we have Nr$`$Prod$`P$ = Nr$`\Pi`P$. But when P is not a Rel^2 excl, Prod$`P$ and $\Pi`P$ are in general not ordinally similar. We can, however, always make a Rel^2 excl by replacing the members x, y, etc. of

* Here Q is said to be *earlier* than R if Q has the relation P to R and is not identical with R.

$C'Q$ (where $Q \epsilon C'P$) by $x \downarrow Q, y \downarrow Q$, etc. In this way, if x occurs twice in $C'\Sigma'P$, once as a member of $C'Q$, and once as a member of $C'R$, the two occurrences are made to correspond to $x \downarrow Q$ and $x \downarrow R$ respectively, and thus we get a new relation which *is* a Rel^2 excl.

If every member of $C'P$ has a first term, $B \upharpoonright C'P$ will be the first term of $\Pi'P$, and $B``C'P$ will be the first term of Prod$'P$. If further there is a last member of $C'P$, *i.e.* if $\text{E}\,!\,B'\breve{P}$, and if this last member has a second term, the second member of $\Pi'P$ is obtained by taking this second term as the representative of $B'\breve{P}$, and leaving all the other representatives unchanged. In any case, if $B'\breve{P}$ exists, the earliest successors of any member of $\Pi'P$ are those obtained by only varying the representative in $B'\breve{P}$. Thus, if $B'\breve{P}$ exists, those members of $\Pi'P$ which have a given set of representatives in all members of $\text{D}'P$ form a consecutive stretch of the series, and this stretch is like $B'\breve{P}$. If $B'\breve{P}$ has an immediate predecessor, the stretches obtained by varying only the representative in this predecessor are again consecutive, and form a series like the said predecessor; and so on. This makes it plain why $\Pi'P$ has the properties of a product.

As in the case of cardinals, the definition of exponentiation is derived from that of multiplication. We put

$$P \exp Q = \text{Prod}'P \underset{\because}{\downarrow} ;Q \quad \text{Df.}$$

We put also $$P^Q = \dot{s};(P \exp Q) \quad \text{Df.}$$

This is an important relation, which deserves consideration apart from the fact that it is useful in connection with exponentiation. It will be found that

$$P^Q = \hat{M}\hat{N}\{M, N \epsilon (C'P \uparrow C'Q)_\Delta{}'C'Q :. (\exists y) : y \epsilon C'Q \,.\, (M'y)\, P\, (N'y) : xQy \,.\, x \neq y \,.\supset_x .\, M'x = N'x\}.$$

This is a form of the principle of first differences which is appropriate when *two* relations are concerned, instead of only one as in P_{cl}. The principle, in this case, is as follows: Let M, N be any two one-many relations which relate part (or the whole) of $C'P$ to the whole of $C'Q$. That is, each of the two relations assigns a representative in $C'P$ to every term of $C'Q$, but different terms of $C'Q$ may have the same representative. Then in travelling along the series Q, there is to be, sooner or later, a term y whose M-representative is earlier than its N-representative, and terms which come earlier than y in Q are all to have their M-representatives identical with their N-representatives.

The relation P^Q may be subjected to various restrictions which give important results. This subject has been treated by Hausdorff. For example, if $P = x \downarrow y$ (where $x \neq y$), and Q is of the ordinal type which Cantor calls ω, *i.e.* the type of progressions (generated by transitive relations), then if z is any member of $C`Q$, $M`z$ is always either x or y. If we impose the condition that $M`z$ is to be $x$ except for a finite number of values of $z$, the resulting series is of the type of the rationals in order of magnitude, *i.e.* the type called $\eta$. If we impose the condition that there are to be an infinite number of values of $z$ for which $M`z = y$, the resulting series is a continuum, *i.e.* it is of the ordinal type called θ; in this case, the contained "rational" series consists of those M's for which there are only a finite number of z's having $M`z = x$. If we impose no limitation, P^Q is of the type presented by the real numbers when decimals ending in 9 recurring are counted separately from the terminating decimals having the same value.

We may generalize P^Q, instead of restricting it. To begin with, we may allow our M and N to have only part of $C`Q$ for their converse domain, and remove the assumption that there is a *first* member of $C`Q$ for which $M`y$ and $N`y$ differ; this leads to the relation

$$\hat{M}\hat{N}\{M, N \in (1 \rightarrow \text{Cls}) \cap \text{Rl}`(C`P \uparrow C`Q) :. \\ (\exists y) : (M`y)\, P\, (N`y) : xQy \,.\, x \in \text{Cl}`N \,.\, \supset_x .\, (M`x)\,(P \cup I)\,(N`x)\}.$$

Further, we may drop the restriction to one-many relations. It will be observed that if $(M`y)\,P\,(N`y)$, we have $y\,(\breve{M} \mid P \mid N)\,y$. Thus we may consider the relation

$$MN[\hat{M}, \hat{N} \in \text{Rl}`(C`P \uparrow C`Q) :. \\ (\exists y) : y\,(\breve{M} \mid P \mid N)\,y : xQy \,.\, \supset_x .\, x\,\{\breve{M} \mid (P \cup I) \mid N\}\,x].$$

This relation has for its field all relations contained in $C`P \uparrow C`Q$. We may, if we like, drop even this restriction, and consider

$$\hat{M}\hat{N}[(\exists y) : y \in C`Q \,.\, y\,(\breve{M} \mid P \mid N)\,y : xQy \,.\, \supset_x .\, x\,\{\breve{M} \mid (P \cup I \upharpoonright C`P) \mid N\}\,x].$$

This represents the most general form of the principle of first differences as applied to a couple of relations P and Q. In ordinal arithmetic, however, P^Q is sufficiently general for the uses we wish to make of it.

The formal laws, as far as they are true, can be proved without excessive difficulty. We have

$$\vdash : P \neq Q \,.\, \supset .\, \text{Nr}`\Pi`(P \downarrow Q) = \text{Nr}`(P \times Q),$$

which connects the two kinds of multiplication;

$$\vdash : P \,\text{smor smor}\, Q \,.\, \supset .\, \text{Nr}`\Pi`P = \text{Nr}`\Pi`Q,$$

$$\vdash : P \in \text{Rel}^2\,\text{excl} \,.\, P \subset J \,.\, \supset .\, \text{Nr}`\Pi`\Pi;P = \text{Nr}`\Pi`\Sigma`P,$$

which is one form of the associative law, of which another form is

$$\vdash : P \neq Q \,.\, \supset .\, \text{Nr}`(\Pi`P \times \Pi`Q) = \text{Nr}`\Pi`(P \,\updownarrow\, Q).$$

Also

$\vdash : P, \Sigma\text{`}P \in \text{Rel}^2\,\text{excl} \,.\, P \mathrel{G} J \,.\, \supset .\, \text{Nr}\text{`}\text{Prod}\text{`}\text{Prod}\,;P = \text{Nr}\text{`}\text{Prod}\text{`}\Sigma\text{`}P = \text{Nr}\text{`}\Pi\text{`}\Sigma\text{`}P$,

which is the associative law for "Prod." We have

$$\vdash : C\text{`}Q \cap C\text{`}R = \Lambda \,.\, \supset .\, \text{Nr}\text{`}(P^Q \times P^R) = \text{Nr}\text{`}P^{Q \dagger R},$$

$$\vdash .\, \text{Nr}\text{`}(P^Q)^R = \text{Nr}\text{`}(P^{R \times Q}).$$

But we do not have in general

$$\text{Nr}\text{`}(P^R \times Q^R) = \text{Nr}\text{`}(P \times Q)^R,$$

which obviously would require the commutative law for multiplication, and therefore does not hold in general in spite of the fact that its cardinal analogue does always hold.

As regards the connection with cardinals, we have

$$\vdash : P \in \text{Rel}^2\,\text{excl} \,.\, \supset .\, C\text{`}\text{Prod}\text{`}P = \text{Prod}\text{`}C\text{``}C\text{`}P,$$

$$\vdash : \dot{\exists}\,!\, Q \,.\, \supset .\, C\text{`}(P \exp Q) = (C\text{`}P) \exp (C\text{`}Q),$$

and we have already had

$$\vdash .\, C\text{`}(P \times Q) = C\text{`}P \times C\text{`}Q.$$

Moreover the correlators by which similarity is established in cardinals generally suffice to establish likeness in the analogous cases in relation-arithmetic. Thus we have

$$\vdash : S \in P \,\overline{\text{smor}}\, Q \,.\, \supset .\, S_\epsilon \upharpoonright \text{Cl}\text{`}C\text{`}Q \in P_{\text{cl}} \,\overline{\text{smor}}\, Q_{\text{cl}},$$

$$\vdash : P, Q \in \text{Rel}^2\,\text{excl} \,.\, S \upharpoonright C\text{`}\Sigma\text{`}Q \in P \,\overline{\text{smor}}\,\overline{\text{smor}}\, Q \,.\, \supset .$$
$$S_\epsilon \upharpoonright C\text{`}\text{Prod}\text{`}Q \in (\text{Prod}\text{`}P) \,\overline{\text{smor}}\, (\text{Prod}\text{`}Q),$$

$$\vdash : S \upharpoonright C\text{`}P' \in P \,\overline{\text{smor}}\, P' \,.\, T \upharpoonright C\text{`}Q' \in Q \,\overline{\text{smor}}\, Q' \,.\, \supset .$$
$$(S \,\|\, \breve{T}) \upharpoonright C\text{`}(P' \exp Q') \in (P \exp Q) \,\overline{\text{smor}}\, (P' \exp Q'),$$

which are all closely analogous to propositions which were proved in cardinals.

The applications of the propositions of this section are almost wholly to series, and it is convenient to imagine our relations to be serial. But the hypothesis that they are serial is not necessary to the truth of any of the propositions of the present section, and it is a remarkable fact that so many of the formal laws of ordinal arithmetic hold for relations in general.

It should be observed that $\Pi\text{`}P$ is not always a series when P is a series and all the relations in the field of P are series. A series (cf. *204) is a relation P which is (1) contained in diversity, (2) transitive, (3) connected, *i.e.* such that every term of the field of P has the relation P or the relation $\breve{P}$ to every other term of the field. It is the third condition which may fail for $\Pi\text{`}P$, and which in fact does fail whenever P is not well-ordered. Thus suppose, for the sake of simplicity, that P is of the type $\text{Cnv}\text{``}\omega$, which we will call a *regression*, *i.e.* the converse of a progression (cf. *263); and suppose that the field of P consists entirely of couples. Take a selection M

which chooses the first term of every odd couple, and the second term of every even couple; and take another selection N which chooses the second term of every odd couple, and the first term of every even couple. Neither of these two selections has the relation $\Pi\text{‘}P$ to the other, for whatever term Q of $C\text{‘}P$ we choose, if M is the selection which chooses the first term of Q, there is an earlier term of $C\text{‘}P$ (namely the immediate predecessor of Q) in which N chooses the first term while M chooses the second. Hence there is no such Q as is required for $M(\Pi\text{‘}P)N$; and a similar argument holds against $N(\Pi\text{‘}P)M$. In such a case, $\Pi\text{‘}P$ generates a number of different series, and by suitable restrictions of the field, one of these series can be extracted. Exactly similar remarks apply to P^Q.

*170. ON THE RELATION OF FIRST DIFFERENCES AMONG THE SUB-CLASSES OF A GIVEN CLASS.

Summary of *170.

The definition to be given in this number of the relation of first differences among the sub-classes of a given class is by no means the only one possible, in fact a different definition will be considered in *171. In the present number, the definition we choose is this: α is said to precede β according to this definition when α has at least one member which neither belongs to β nor follows any term belonging to β and not to α (α and β being both sub-classes of C‘P). In other words, if we consider the two classes $\alpha - \beta$ and $\beta - \alpha$, there are members of $\alpha - \beta$ which are not preceded by any members of $\beta - \alpha$. Pictorially, we may conceive the relation as follows (P being supposed serial): α and β each pick out terms from C‘P, and these terms have an order conferred by P; we suppose that the earlier terms selected by α and β are perhaps the same, but sooner or later, if $\alpha \neq \beta$, we must come to terms which belong to one but not to the other. We assume that the earliest terms of this sort belong to α, not to β; in this case, α has to β the relation P_{cl}. That is, where α and β begin to differ, it is terms of α that we come to, not terms of β. We do not assume that there is a *first* term which belongs to α and not to β, since this would introduce undesirable restrictions in case P is not well-ordered.

A few of the propositions of the present number will be used in the next number, which deals with a slightly different form of the relation of first differences, but with this exception the propositions of this number will not be referred to again until we come to series. Their chief use occurs in the section on compact series, rational series, and continuous series (Part V, Section F), especially in *274 and *276, which respectively establish the existence of rational series (assuming the axiom of infinity) and the fact that the cardinal number of terms in a continuous series is the same as the number of classes contained in the field of a progression, *i.e.* $2^{\aleph_0}$. The definitions and a few of the simpler propositions are also used in connection with the series of segments of a series, since, as explained above, the segments of a series P are arranged in the series generated by P_{lc}.

The propositions of this number which will be used in dealing with series are the following:

***170·1.** $\vdash : \alpha P_{cl} \beta . \equiv . \alpha, \beta \in \mathrm{Cl}`C`P . \exists ! \alpha - \beta - \breve{P}``(\beta - \alpha)$

***170·101.** $\vdash . P_{1c} = \mathrm{Cnv}`(\breve{P})_{cl}$

***170·102.** $\vdash : \alpha P_{1c} \beta . \equiv . \alpha, \beta \in \mathrm{Cl}`C`P . \exists ! \beta - \alpha - P``(\alpha - \beta)$

(These propositions merely embody the definitions.)

***170·11.** $\vdash :. \alpha P_{cl} \beta . \equiv : \alpha, \beta \in \mathrm{Cl}`C`P : (\exists y) . y \in \alpha - \beta . \overrightarrow{P}`y \cap \beta \subset \alpha$

This form is often more convenient than *170·1.

***170·16.** $\vdash : \alpha \subset C`P . \beta \subset \alpha . \beta \neq \alpha . \supset . \alpha P_{cl} \beta$

I.e. every sub-class of $C`P$ has the relation P_{cl} to every proper part of itself.

***170·17.** $\vdash . P_{cl} \in J . P_{1c} \in J$

***170·2.** $\vdash :. \alpha, \beta \in \mathrm{Cl}`C`P : (\exists y) . y \in \alpha - \beta . \overrightarrow{P}`y \cap \alpha = \overrightarrow{P}`y \cap \beta : \supset . \alpha P_{cl} \beta$

This proposition deals with the case where there is a definite first term y which belongs to α and not to β, and whose predecessors all belong to both or neither.

***170·23.** $\vdash :. \alpha \subset C`P . y \in \alpha - \beta - \breve{P}``(\beta - \alpha) . \supset :$

$$y \min_P (\alpha - \beta) . \equiv . \overrightarrow{P}`y \cap \alpha = \overrightarrow{P}`y \cap \beta$$

This proposition is useful in case P is well-ordered, since then $\alpha - \beta$ must have a minimum if it exists (α and β being supposed sub-classes of $C`P$).

***170·31.** $\vdash : \beta \subset C`P . \beta \neq C`P . \equiv . (C`P) P_{cl} \beta$

This follows from *170·16, as does the following proposition:

***170·32.** $\vdash : \alpha \subset C`P . \exists ! \alpha . \equiv . \alpha P_{cl} \Lambda$

***170·35.** $\vdash . \dot{\Lambda}_{cl} = \dot{\Lambda}$

***170·38.** $\vdash : \dot{\exists} ! P . \supset . B`P_{cl} = C`P . B`\mathrm{Cnv}`P_{cl} = \Lambda$

***170·6.** $\vdash : \Lambda P_{1c} \beta . \equiv . \beta \subset C`P . \exists ! \beta$

Besides the above, the following propositions should be noted:

***170·36.** $\vdash . \mathrm{D}`P_{cl} = \mathrm{Cl\,ex}`C`P . \mathrm{Cl}`P_{cl} = \mathrm{Cl}`C`P - \iota`C`P$

***170·37.** $\vdash : \dot{\exists} ! P . \supset . C`P_{cl} = \mathrm{Cl}`C`P$

***170·44.** $\vdash : P \,\mathrm{smor}\, Q . \supset . P_{cl} \,\mathrm{smor}\, Q_{cl}$

***170·64.** $\vdash : x \sim \in C`P . \supset . (x \leftarrow\!\!\!+ P)_{cl} = (\iota`x \cup)^{;} P_{cl} \,\text{\textpm}\, P_{cl}$

This proposition shows that every term added to P doubles the number of terms in P_{cl}; hence it is not surprising that P_{cl} (when P is well-ordered) has a power of 2_r for its relation-number (cf. *177).

***170·67.** $\vdash : \dot{\exists} ! P . \dot{\exists} ! Q . C`P \cap C`Q = \Lambda . \supset . (P \,\text{\textpm}\, Q)_{cl} = s^{;}C^{;}(P_{cl} \times Q_{cl})$

whence

***170·69.** $\vdash : \dot{\exists} ! P . \dot{\exists} ! Q . C`P \cap C`Q = \Lambda . \supset . (P \,\text{\textpm}\, Q)_{cl} \,\mathrm{smor}\, (P_{cl} \times Q_{cl})$

***170·01.** $P_{\text{cl}} = \hat{\alpha}\hat{\beta}\,\{\alpha, \beta \in \text{Cl}'C'P \,.\, \exists\,!\, \alpha - \beta - \breve{P}``(\beta - \alpha)\}$ Df

***170·02.** $P_{\text{lc}} = \text{Cnv}'(\breve{P})_{\text{cl}}$ Df

***170·1.** $\vdash : \alpha P_{\text{cl}} \beta \,.\equiv .\, \alpha, \beta \in \text{Cl}'C'P \,.\, \exists\,!\, \alpha - \beta - \breve{P}``(\beta - \alpha)$ [(*170·01)]

***170·101.** $\vdash .\, P_{\text{lc}} = \text{Cnv}'(\breve{P})_{\text{cl}}$ [(*170·02)]

***170·102.** $\vdash : \alpha P_{\text{lc}} \beta \,.\equiv .\, \alpha, \beta \in \text{Cl}'C'P \,.\, \exists\,!\, \beta - \alpha - P``(\alpha - \beta)$ [*170·1·101]

Thus $\alpha P_{\text{lc}} \beta$ means, roughly speaking, that $\beta - \alpha$ goes on longer than $\alpha - \beta$, just as $\alpha P_{\text{cl}} \beta$ means that $\alpha - \beta$ begins sooner. Thus if P is the relation of earlier and later in time, and α and β are the times when A and B respectively are out of bed, "$\alpha P_{\text{cl}} \beta$" will mean that A gets up earlier than B, and "$\alpha P_{\text{lc}} \beta$" will mean that B goes to bed later than A.

***170·103.** $\vdash : y \sim\in \breve{P}``(\beta - \alpha) \,.\equiv .\, \overrightarrow{P}'y \cap \beta \subset \alpha$

Dem.

$\vdash .\, *37{\cdot}105 \,.\supset \vdash :.\, y \sim\in \breve{P}``(\beta - \alpha) \,.\equiv : \sim(\exists x) \,.\, x \in \beta - \alpha \,.\, xPy :$

[*10·51] $\equiv : x \in \beta \,.\, xPy \,.\supset_x .\, x \in \alpha :$

[*32·18] $\equiv : \overrightarrow{P}'y \cap \beta \subset \alpha :.\supset \vdash .$ Prop

***170·11.** $\vdash :.\, \alpha P_{\text{cl}} \beta \,.\equiv : \alpha, \beta \in \text{Cl}'C'P : (\exists y) \,.\, y \in \alpha - \beta \,.\, \overrightarrow{P}'y \cap \beta \subset \alpha$
[*170·1·103]

***170·12.** $\vdash : \alpha P_{\text{cl}} \beta \,.\equiv .\, \alpha, \beta \in \text{Cl}'C'P \,.\, \exists\,!\, \alpha - (\alpha \cap \beta) - \breve{P}``\{\beta - (\alpha \cap \beta)\}$
[*170·1 . *22·93]

***170·121.** $\vdash :.\, \alpha P_{\text{cl}} \beta \,.\equiv .\, \alpha, \beta \in \text{Cl}'C'P \,.\, \exists\,!\, (\alpha \cup \beta) - \beta - \breve{P}``\{(\alpha \cup \beta) - \alpha\}$
[*170·1 . *22·9]

***170·13.** $\vdash :.\, \alpha P_{\text{cl}} \beta \,.\equiv : (\exists \rho, \sigma, \gamma) \,.\, \rho, \sigma, \gamma \in \text{Cl}'C'P \,.$

$\rho \cap \gamma = \Lambda \,.\, \sigma \cap \gamma = \Lambda \,.\, \rho \cap \sigma = \Lambda \,.\, \alpha = \gamma \cup \rho \,.\, \beta = \gamma \cup \sigma \,.\, \exists\,!\, \rho - \breve{P}``\sigma$

Dem.

$\vdash .\, *24{\cdot}24 \,.\, *22{\cdot}69 \,.\supset \vdash : \rho \cap \sigma = \Lambda \,.\, \alpha = \gamma \cup \rho \,.\, \beta = \gamma \cup \sigma \,.\supset .\, \alpha \cap \beta = \gamma$ (1)

$\vdash .\, *24{\cdot}4 \,.\quad \supset \vdash :.\, \alpha = \gamma \cup \rho \,.\supset : \rho \cap \gamma = \Lambda \,.\equiv .\, \alpha - \gamma = \rho$ (2)

$\vdash .\, *24{\cdot}4 \,.\quad \supset \vdash :.\, \beta = \gamma \cup \sigma \,.\supset : \sigma \cap \gamma = \Lambda \,.\equiv .\, \beta - \gamma = \sigma$ (3)

$\vdash .\, (1) \,.\, (2) \,.\, (3) \,.\quad \supset \vdash :.\, \rho \cap \sigma = \Lambda \,.\, \alpha = \gamma \cup \rho \,.\, \beta = \gamma \cup \sigma \,.\supset :$

$\rho \cap \gamma = \Lambda \,.\, \sigma \cap \gamma = \Lambda \,.\equiv .\, \alpha - (\alpha \cap \beta) = \rho \,.\, \beta - (\alpha \cap \beta) = \sigma \,.$

[*22·93] $\equiv .\, \alpha - \beta = \rho \,.\, \beta - \alpha = \sigma$ (4)

$\vdash .\, (1) \,.\, (4) \,.\supset \vdash : (\exists \rho, \sigma, \gamma) \,.\, \rho, \sigma, \gamma \in \text{Cl}'C'P \,.\, \rho \cap \gamma = \Lambda \,.\, \sigma \cap \gamma = \Lambda \,.\, \rho \cap \sigma = \Lambda \,.$

$\alpha = \gamma \cup \rho \,.\, \beta = \gamma \cup \sigma \,.\, \exists\,!\, \rho - \breve{P}``\sigma \,.\equiv .$

$(\exists \rho, \sigma, \gamma) \,.\, \rho, \sigma, \gamma \in \text{Cl}'C'P \,.\, \rho \cap \sigma = \Lambda \,.\, \alpha = \gamma \cup \rho \,.\, \beta = \gamma \cup \sigma \,.$

$\alpha \cap \beta = \gamma \,.\, \alpha - \beta = \rho \,.\, \beta - \alpha = \sigma \,.\, \exists\,!\, \rho - \breve{P}``\sigma \,.$

[*13·22] $\equiv .\, \alpha - \beta, \beta - \alpha, \alpha \cap \beta \in \text{Cl}'C'P \,.\, \exists\,!\, \alpha - \beta - \breve{P}``(\beta - \alpha) \,.$

[*60·43.*24·41] $\equiv .\, \alpha, \beta \in \text{Cl}'C'P \,.\, \exists\,!\, \alpha - \beta - \breve{P}``(\beta - \alpha) \,.$

[*170·1] $\equiv .\, \alpha P_{\text{cl}} \beta : \supset \vdash .$ Prop

***170·14.** $\vdash :. \alpha, \beta \epsilon \text{Cl}`C`P . \supset : \alpha \mathbin{\dot{-}} P_{cl} \beta . \equiv . \alpha - \beta \subset \breve{P}``(\beta - \alpha)$
[*170·1 . *24·55]

***170·141.** $\vdash :. \alpha, \beta \epsilon \text{Cl}`C`P . \supset : \alpha \mathbin{\dot{-}} P_{lc} \beta . \equiv . \beta - \alpha \subset P``(\alpha - \beta)$
[*170·14·101]

***170·15.** $\vdash : \alpha P_{cl} \beta . \supset . \beta \cap p`\overrightarrow{P}``(\alpha - \beta) \subset \alpha$

Dem.

$\vdash . *40·12 . \supset \vdash : y \epsilon \alpha - \beta . \supset . p`\overrightarrow{P}``(\alpha - \beta) \subset \overrightarrow{P}`y .$

[*22·48] $\supset . \beta \cap p`\overrightarrow{P}``(\alpha - \beta) \subset \beta \cap \overrightarrow{P}`y :$

[*22·44] $\supset \vdash : y \epsilon \alpha - \beta . \beta \cap \overrightarrow{P}`y \subset \alpha . \supset . \beta \cap p`\overrightarrow{P}``(\alpha - \beta) \subset \alpha :$

[*10·11·23] $\supset \vdash : (\exists y) . y \epsilon \alpha - \beta . \beta \cap \overrightarrow{P}`y \subset \alpha . \supset . \beta \cap p`\overrightarrow{P}``(\alpha - \beta) \subset \alpha$ (1)

$\vdash . (1) . *170·11 . \supset \vdash .$ Prop

***170·16.** $\vdash : \alpha \subset C`P . \beta \subset \alpha . \beta \neq \alpha . \supset . \alpha P_{cl} \beta$

Dem.

$\vdash . *24·6 . \supset \vdash : \text{Hp} . \supset . \exists ! \alpha - \beta$ (1)

$\vdash . *24·3 . \supset \vdash : \text{Hp} . \supset . \beta - \alpha = \Lambda .$

[*37·29] $\supset . \breve{P}``(\beta - \alpha) = \Lambda .$

[*24·101·26] $\supset . \alpha - \beta - \breve{P}``(\beta - \alpha) = \alpha - \beta$ (2)

$\vdash . (1) . (2) . *170·1 . \supset \vdash .$ Prop

***170·161.** $\vdash : \alpha \subset C`P . \beta \subset \alpha . \beta \neq \alpha . \supset . \beta P_{lc} \alpha$

Dem.

$\vdash . *170·16 . \supset \vdash : \text{Hp} . \supset . \alpha (\breve{P})_{cl} \beta .$

[*170·101] $\supset . \beta P_{lc} \alpha : \supset \vdash .$ Prop

***170·17.** $\vdash . P_{cl} \in J . P_{lc} \in J$

Dem.

$\vdash . *170·1 . \supset \vdash : \alpha P_{cl} \beta . \supset . \exists ! \alpha - \beta .$

[*24·55.*22·42] $\supset . \alpha \neq \beta .$

[*50·11] $\supset . \alpha J \beta : \supset \vdash .$ Prop

In order that P_{cl} should be serial, we need further that it should be transitive and connected. P_{cl} is transitive if P is transitive and connected. But P_{cl} may still not be connected: there may be many distinct families in its field, though all of them must begin with $C`P$ and end with Λ. For example, if P is a regression, the class which takes every odd member does not have either of the relations P_{cl}, $\breve{P}_{cl}$ to the class which takes every even member. In order that P_{cl} should be serial, we require that P should be not only serial, but well-ordered, *i.e.* that every existent sub-class of $C`P$ should have a first term. When P is serial but not well-ordered, P_{cl} will, however, generate various series contained in it by imposing suitable limitations on the field.

$*170\cdot2$. $\vdash :. \alpha, \beta \epsilon \mathrm{Cl}'C'P : (\exists y) . y \epsilon \alpha - \beta . \overrightarrow{P}'y \cap \alpha = \overrightarrow{P}'y \cap \beta : \supset . \alpha P_{\mathrm{cl}} \beta$
[$*170\cdot11 . *22\cdot43$]

$*170\cdot21$. $\vdash :. \alpha \subset C'P . \supset : y \min_P (\alpha - \beta) . \equiv . y \epsilon \alpha - \beta . \overrightarrow{P}'y \cap \alpha \subset \beta$

Dem.

$\vdash . *93\cdot11 . \supset \vdash :. \mathrm{Hp} . \supset : y \min_P (\alpha - \beta) . \equiv . y \epsilon \alpha - \beta - \breve{P}``(\alpha - \beta) .$

[$*170\cdot103$] $\equiv . y \epsilon \alpha - \beta . \overrightarrow{P}'y \cap \alpha \subset \beta :. \supset \vdash . \mathrm{Prop}$

$*170\cdot22$. $\vdash :. \alpha \subset C'P . y \min_P (\alpha - \beta) . \supset : \overrightarrow{P}'y \cap \beta \subset \alpha . \equiv . \overrightarrow{P}'y \cap \alpha = \overrightarrow{P}'y \cap \beta$

Dem.

$\vdash . *170\cdot21 . *4\cdot73 . \supset \vdash :. \mathrm{Hp} . \supset : \overrightarrow{P}'y \cap \beta \subset \alpha . \equiv . \overrightarrow{P}'y \cap \alpha \subset \beta . \overrightarrow{P}'y \cap \beta \subset \alpha .$

[$*22\cdot74$] $\equiv . \overrightarrow{P}'y \cap \alpha = \overrightarrow{P}'y \cap \beta :. \supset \vdash . \mathrm{Prop}$

$*170\cdot23$. $\vdash :. \alpha \subset C'P . y \epsilon \alpha - \beta - \breve{P}``(\beta - \alpha) . \supset :$
$y \min_P (\alpha - \beta) . \equiv . \overrightarrow{P}'y \cap \alpha = \overrightarrow{P}'y \cap \beta$

Dem.

$\vdash . *170\cdot103\cdot21 . \supset \vdash :. \mathrm{Hp} \supset :$

$y \min_P (\alpha - \beta) . \equiv . y \epsilon \alpha - \beta . \overrightarrow{P}'y \cap \beta \subset \alpha . \overrightarrow{P}'y \cap \alpha \subset \beta .$

[$*22\cdot74 . *4\cdot73$] $\equiv . y \epsilon \alpha - \beta . \overrightarrow{P}'y \cap \beta \subset \alpha . \overrightarrow{P}'y \cap \alpha = \overrightarrow{P}'y \cap \beta .$

[$*170\cdot103$] $\equiv . y \epsilon \alpha - \beta - \breve{P}``(\beta - \alpha) . \overrightarrow{P}'y \cap \alpha = \overrightarrow{P}'y \cap \beta$ (1)

$\vdash . (1) . *5\cdot32 . \supset \vdash . \mathrm{Prop}$

$*170\cdot3$. $\vdash : \alpha \epsilon \mathrm{Cl}'C'P . \beta \subset \alpha . \exists ! \alpha - \beta . \supset . \alpha P_{\mathrm{cl}} \beta$ [$*170\cdot16$]

$*170\cdot31$. $\vdash : \beta \subset C'P . \beta \neq C'P . \equiv . (C'P) P_{\mathrm{cl}} \beta$ [$*170\cdot16$]

$*170\cdot32$. $\vdash : \alpha \subset C'P . \exists ! \alpha . \equiv . \alpha P_{\mathrm{cl}} \Lambda$ [$*170\cdot3$]

$*170\cdot33$. $\vdash : \dot{\exists} ! P . \equiv . (C'P) P_{\mathrm{cl}} \Lambda$

Dem.

$\vdash . *33\cdot24 . *170\cdot32 . \supset \vdash : \dot{\exists} ! P . \supset . (C'P) P_{\mathrm{cl}} \Lambda$ (1)

$\vdash . *170\cdot1 . \supset \vdash : (C'P) P_{\mathrm{cl}} \Lambda . \supset . \exists ! (C'P) - \Lambda .$

[$*33\cdot24$] $\supset . \dot{\exists} ! P$ (2)

$\vdash . (1) . (2) . \supset \vdash . \mathrm{Prop}$

$*170\cdot34$. $\vdash : \dot{\exists} ! P . \equiv . \dot{\exists} ! P_{\mathrm{cl}}$

Dem.

$\vdash . *170\cdot33 . \supset \vdash : \dot{\exists} ! P . \supset . \dot{\exists} ! P_{\mathrm{cl}}$ (1)

$\vdash . *170\cdot1 . \supset \vdash : \dot{\exists} ! P_{\mathrm{cl}} . \supset . (\exists \alpha, \beta) . \alpha, \beta \epsilon \mathrm{Cl}'C'P . \exists ! \alpha - \beta .$

[$*24\cdot561$] $\supset . (\exists \alpha) . \alpha \epsilon \mathrm{Cl}'C'P . \exists ! \alpha .$

[$*60\cdot361$] $\supset . \exists ! C'P .$

[$*33\cdot24$] $\supset . \dot{\exists} ! P$ (2)

$\vdash . (1) . (2) . \supset \vdash . \mathrm{Prop}$

***170·35.** $\vdash . \dot{\Lambda}_{cl} = \dot{\Lambda}$ [*170·34 . Transp]

***170·36.** $\vdash . \text{D}‘P_{cl} = \text{Cl ex}‘C‘P . \text{Cl}‘P_{cl} = \text{Cl}‘C‘P - \iota‘C‘P$

Dem.

$\vdash . *170·32 . \supset \vdash . \text{Cl ex}‘C‘P \subset \text{D}‘P_{cl}$ (1)

$\vdash . *170·31 . \supset \vdash . \text{Cl}‘C‘P - \iota‘C‘P \subset \text{Cl}‘P_{cl}$ (2)

$\vdash . *170·1 . \supset \vdash : \alpha \in \text{D}‘P_{cl} . \supset . (\exists \beta) . \alpha, \beta \in \text{Cl}‘C‘P . \exists ! \alpha - \beta .$

[*24·561] $\supset . \alpha \in \text{Cl}‘C‘P . \exists ! \alpha$ (3)

$\vdash . *170·1 . \supset \vdash : \alpha \in \text{Cl}‘P_{cl} . \supset . (\exists \beta) . \alpha, \beta \in \text{Cl}‘C‘P . \exists ! \beta - \alpha .$

[*60·2] $\supset . \alpha \in \text{Cl}‘C‘P . \exists ! C‘P - \alpha .$

[*24·6] $\supset . \alpha \in \text{Cl}‘C‘P - \iota‘C‘P$ (4)

$\vdash . (1) . (2) . (3) . (4) . \supset \vdash .$ Prop

***170·37.** $\vdash : \dot{\exists} ! P . \supset . C‘P_{cl} = \text{Cl}‘C‘P$ [*170·36]

***170·371.** $\vdash . C‘P_{cl} \subset \text{Cl}‘C‘P$ [*170·37·35 . *33·241]

***170·38.** $\vdash : \dot{\exists} ! P . \supset . B‘P_{cl} = C‘P . B‘\text{Cnv}‘P_{cl} = \Lambda$ [*170·36]

The following propositions lead up to *170·44.

***170·4.** $\vdash : S \in 1 \to 1 . C‘Q = \text{Cl}‘S . \supset . (S;Q)_{cl} = S_\epsilon;Q_{cl}$

Dem.

$\vdash . *170·1 . *150·4 . *37·11 . \supset \vdash : \alpha (S_\epsilon;Q_{cl}) \beta . \equiv .$

$(\exists \gamma, \delta) . \gamma, \delta \in \text{Cl}‘C‘Q . \alpha = S“\gamma . \beta = S“\delta . \exists ! \gamma - \delta - \breve{Q}“(\delta - \gamma)$ (1)

$\vdash . (1) . \supset \vdash :. \text{Hp} . \supset :$

$\alpha (S_\epsilon;Q_{cl}) \beta . \equiv . (\exists \gamma, \delta) . \gamma, \delta \in \text{Cl}‘\text{Cl}‘S . \alpha = S“\gamma . \beta = S“\delta .$
$\exists ! \gamma - \delta - \breve{Q}“(\delta - \gamma) .$

[*37·43] $\equiv . (\exists \gamma, \delta) . \gamma, \delta \in \text{Cl}‘\text{Cl}‘S . \alpha = S“\gamma . \beta = S“\delta .$
$\exists ! S“\{\gamma - \delta - \breve{Q}“(\delta - \gamma)\} .$

[*71·381] $\equiv . (\exists \gamma, \delta) . \gamma, \delta \in \text{Cl}‘\text{Cl}‘S . \alpha = S“\gamma . \beta = S“\delta .$
$\exists ! S“\gamma - S“\delta - S“\breve{Q}“(\delta - \gamma) .$

[*72·511.*71·38] $\equiv . (\exists \gamma, \delta) . \gamma, \delta \in \text{Cl}‘\text{Cl}‘S . \alpha = S“\gamma . \beta = S“\delta .$
$\exists ! S“\gamma - S“\delta - S“\breve{Q}“\breve{S}“(\beta - \alpha) .$

[*13·193.*37·33] $\equiv . (\exists \gamma, \delta) . \gamma, \delta \in \text{Cl}‘\text{Cl}‘S . \alpha = S“\gamma . \beta = S“\delta .$
$\exists ! \alpha - \beta - (S;\breve{Q})“(\beta - \alpha) .$

[*71·48.*37·23] $\equiv . \alpha, \beta \in \text{Cl}‘\text{D}‘S . \exists ! \alpha - \beta - (S;\breve{Q})“(\beta - \alpha) .$

[*150·23] $\equiv . \alpha, \beta \in \text{Cl}‘C‘(S;Q) . \exists ! \alpha - \beta - (S;\breve{Q})“(\beta - \alpha) .$

[*150·12.*170·1] $\equiv . \alpha (S;Q)_{cl} \beta :. \supset \vdash .$ Prop

***170·41.** $\vdash . (S \upharpoonright C\text{‘}Q)_\epsilon \mathbin{;} Q_{\text{cl}} = S_\epsilon \mathbin{;} Q_{\text{cl}}$ [*150·95 . *170·371]

***170·42.** $\vdash : S \upharpoonright C\text{‘}Q \in 1 \rightarrow 1 . C\text{‘}Q \subset \text{Ɑ}\text{‘}S . \supset . (S \mathbin{;} Q)_{\text{cl}} = S_\epsilon \mathbin{;} Q_{\text{cl}}$

Dem.

$\vdash . *150\cdot32 . \quad \supset \vdash . (S \mathbin{;} Q)_{\text{cl}} = \{(S \upharpoonright C\text{‘}Q) \mathbin{;} Q\}_{\text{cl}}$ (1)

$\vdash . (1) . *170\cdot4 . \supset \vdash : \text{Hp} . \supset . (S \mathbin{;} Q)_{\text{cl}} = (S \upharpoonright C\text{‘}Q)_\epsilon \mathbin{;} Q_{\text{cl}}$

[*170·41] $\qquad = S_\epsilon \mathbin{;} Q_{\text{cl}} : \supset \vdash . \text{Prop}$

***170·43.** $\vdash : S \upharpoonright C\text{‘}Q \in P \, \overline{\text{smor}} \, Q . \supset . S_\epsilon \upharpoonright C\text{‘}Q_{\text{cl}} \in P_{\text{cl}} \, \overline{\text{smor}} \, Q_{\text{cl}}$

Dem.

$\vdash . *151\cdot22 . *170\cdot42 . \supset \vdash : \text{Hp} . \supset . P_{\text{cl}} = S_\epsilon \mathbin{;} Q_{\text{cl}}$ (1)

$\vdash . *74\cdot131 . *170\cdot371 . \supset \vdash : \text{Hp} . \supset . S_\epsilon \upharpoonright C\text{‘}Q_{\text{cl}} \in 1 \rightarrow 1$ (2)

$\vdash . *37\cdot231 . \qquad \supset \vdash . C\text{‘}Q_{\text{cl}} \subset \text{Ɑ}\text{‘}S_\epsilon$ (3)

$\vdash . (1) . (2) . (3) . *151\cdot22 . \supset \vdash . \text{Prop}$

***170·44.** $\vdash : P \, \text{smor} \, Q . \supset . P_{\text{cl}} \, \text{smor} \, Q_{\text{cl}}$ [*170·43 . *151·23·12]

***170·5.** $\vdash . (x \downarrow x)_{\text{cl}} = (\iota\text{‘}x) \downarrow \Lambda$

Dem.

$\vdash . *170\cdot36 . *55\cdot15 . \supset \vdash . \text{D}\text{‘}(x \downarrow x)_{\text{cl}} = \text{Cl ex}\text{‘}\iota\text{‘}x$

[*60·37] $\qquad = \iota\text{‘}\iota\text{‘}x$ (1)

$\vdash . *170\cdot36 . *55\cdot15 . \supset \vdash . \text{Ɑ}\text{‘}(x \downarrow x)_{\text{cl}} = \text{Cl}\text{‘}\iota\text{‘}x - \iota\text{‘}\iota\text{‘}x$

[*60·362] $\qquad = \iota\text{‘}\Lambda$ (2)

$\vdash . (1) . (2) . *55\cdot16 . \supset \vdash . \text{Prop}$

***170·51.** $\vdash : x \neq y . \supset . (x \downarrow y)_{\text{cl}} = (\iota\text{‘}x \cup \iota\text{‘}y) \downarrow \iota\text{‘}x \cup (\iota\text{‘}x \cup \iota\text{‘}y) \downarrow \iota\text{‘}y \cup (\iota\text{‘}x \cup \iota\text{‘}y) \downarrow \Lambda$
$\qquad \cup \, \iota\text{‘}x \downarrow \iota\text{‘}y \cup \iota\text{‘}x \downarrow \Lambda \cup \iota\text{‘}y \downarrow \Lambda$

Dem.

$\vdash . *55\cdot13 . \supset \vdash : \text{Hp} . \supset . \overrightarrow{x \downarrow y}\text{‘}x = \Lambda . \overrightarrow{x \downarrow y}\text{‘}y = \iota\text{‘}x$ (1)

$\vdash . *170\cdot11 . *55\cdot15 . \supset \vdash :: \text{Hp} . \supset :. \alpha (x \downarrow y)_{\text{cl}} \beta . \equiv :$

$\qquad \alpha, \beta \in \text{Cl}\text{‘}(\iota\text{‘}x \cup \iota\text{‘}y) : (\exists z) . z \in \alpha - \beta . \overrightarrow{x \downarrow y}\text{‘}z \cap \beta \subset \alpha$

[*60·39] $\qquad \equiv : \alpha = \iota\text{‘}x \cup \iota\text{‘}y . \vee . \alpha = \iota\text{‘}x . \vee . \alpha = \iota\text{‘}y : \beta \subset \iota\text{‘}x \cup \iota\text{‘}y :$

$\qquad (\exists z) . z \in \alpha - \beta . \overrightarrow{x \downarrow y}\text{‘}z \cap \beta \subset \alpha$ (2)

$\vdash . *51\cdot235 . \supset \vdash :: \alpha = \iota\text{‘}x \cup \iota\text{‘}y . \supset :. (\exists z) . z \in \alpha - \beta . \overrightarrow{x \downarrow y}\text{‘}z \cap \beta \subset \alpha . \equiv :$

$\qquad x \in \alpha - \beta . \overrightarrow{x \downarrow y}\text{‘}x \cap \beta \subset \alpha . \vee . y \in \alpha - \beta . \overrightarrow{x \downarrow y}\text{‘}y \cap \beta \subset \alpha :$

[(1)] $\qquad \equiv : x \in \alpha - \beta . \vee . y \in \alpha - \beta . \iota\text{‘}x \cap \beta \subset \alpha :$

[Hp.*22·43·58] $\equiv : x \in \alpha - \beta . \vee . y \in \alpha - \beta :$

[*51·232.*4·73] $\equiv : x \sim\in \beta . \vee . y \sim\in \beta$ (3)

$\vdash . *54\cdot4 . \supset \vdash :: \text{Hp} . \supset :. \beta \subset \iota\text{‘}x \cup \iota\text{‘}y . x \sim\in \beta . \equiv : \beta = \iota\text{‘}y . \vee . \beta = \Lambda$ (4)

$\vdash . *54\cdot4 . \supset \vdash :: \text{Hp} . \supset :. \beta \subset \iota\text{‘}x \cup \iota\text{‘}y . y \sim\in \beta . \equiv : \beta = \iota\text{‘}x . \vee . \beta = \Lambda$ (5)

$\vdash.(2).(3).(4).(5).\supset$

$\vdash :: \alpha=\iota'x\cup\iota'y.\supset:.\alpha(x\downarrow y)_{cl}\beta.\equiv:\beta=\iota'x.\vee.\beta=\iota'y.\vee.\beta=\Lambda$ (6)

$\vdash.(1).(2).\supset\vdash ::.\ \mathrm{Hp}.\supset::\alpha=\iota'x.\supset:.\alpha(x\downarrow y)_{cl}\beta.\equiv:\beta\subset\iota'x\cup\iota'y.x\sim\epsilon\beta.$

[(4)] $\equiv:\beta=\iota'y.\vee.\beta=\Lambda$ (7)

$\vdash.(1).(2).\supset\vdash::\mathrm{Hp}.\supset:.\alpha=\iota'y.\supset:\alpha(x\downarrow y)_{cl}\beta.\equiv.$

$\beta\subset\iota'x\cup\iota'y.y\sim\epsilon\beta.\iota'x\cap\beta\subset\alpha.$

[*51·211] $\equiv.\beta\subset\iota'x\cup\iota'y.y\sim\epsilon\beta.x\sim\epsilon\beta.$

[*54·4] $\equiv.\beta=\Lambda$ (8)

$\vdash.(2).(6).(7).(8).\supset\vdash.\mathrm{Prop}$

***170·52.** $\vdash:x\neq y.\supset.(x\downarrow y)_{cl}=(\iota'x\cup\iota'y)\downarrow\iota'x \mathbin{\dot\uparrow} \iota'y\downarrow\Lambda$

Dem.

$\vdash.*55\cdot15.\supset\vdash.C'\{(\iota'x\cup\iota'y)\downarrow\iota'x\}\uparrow C'\{\iota'y\downarrow\Lambda\}=$

$\{\iota'(\iota'x\cup\iota'y)\cup\iota'\iota'x\}\uparrow\{\iota'\iota'y\cup\iota'\Lambda\}$

[*55·52] $=(\iota'x\cup\iota'y)\downarrow\iota'y\mathbin{\dot\cup}(\iota'x\cup\iota'y)\downarrow\iota'\Lambda\mathbin{\dot\cup}\iota'x\downarrow\iota'y\mathbin{\dot\cup}\iota'x\downarrow\Lambda$ (1)

$\vdash.(1).*170\cdot51.*160\cdot1.\supset\vdash.\mathrm{Prop}$

***170·6.** $\vdash:\Lambda P_{lc}\beta.\equiv.\beta\subset C'P.\exists!\beta$ [*170·32·101]

***170·601.** $\vdash:\alpha P_{lc}(C'P).\equiv.\alpha\subset C'P.\alpha\neq C'P$ [*170·31·101]

***170·61.** $\vdash:.x\sim\epsilon C'P.\dot\exists!P.x\epsilon\alpha\cap\beta.\supset:$

$\alpha(x\leftarrow\!\!\!+ P)_{cl}\beta.\equiv.\alpha\{(\iota'x\cup)^{;}P_{cl}\}\beta.\equiv.(\alpha-\iota'x)P_{cl}(\beta-\iota'x)$

This and the following propositions are lemmas for

$x\sim\epsilon C'P.\supset.(x\leftarrow\!\!\!+ P)_{cl}=(\iota'x\cup)^{;}P_{cl}\mathbin{\dot\uparrow}P_{cl}$ (*170·64).

Dem.

$\vdash.*161\cdot111.\supset\vdash::\mathrm{Hp}.\supset:.y\epsilon\beta.y(x\leftarrow\!\!\!+ P)z.\supset_y.y\epsilon\alpha:\equiv:$

$y\epsilon\beta.yPz.\supset_y.y\epsilon\alpha:y\epsilon\beta.y=x.z\epsilon C'P.\supset_y.y\epsilon\alpha:$

[*13·191.*33·17] $\equiv:y\epsilon\beta-\iota'x.yPz.\supset_y.y\epsilon\alpha-\iota'x:x\epsilon\beta.z\epsilon C'P.\supset.x\epsilon\alpha:$

[Hp] $\equiv:y\epsilon\beta-\iota'x.yPz.\supset_y.y\epsilon\alpha-\iota'x$ (1)

$\vdash.*51\cdot34.\supset\vdash:\mathrm{Hp}.\supset.-\beta=-\iota'x\cap-\beta.$

[*22·481] $\supset.\alpha-\beta=\alpha-\iota'x-\beta$

[*24·21] $=\alpha-\iota'x\cap(\iota'x\cup-\beta)$

[*22·86] $=\alpha-\iota'x-(\beta-\iota'x)$ (2)

$\vdash.*170\cdot11.*161\cdot101\cdot14.\supset\vdash::\mathrm{Hp}.\supset:.\alpha(x\leftarrow\!\!\!+ P)_{cl}\beta.\equiv:$

$\alpha,\beta\epsilon\mathrm{Cl}'(C'P\cup\iota'x):(\exists z):z\epsilon\alpha-\beta:y\epsilon\beta.y(x\leftarrow\!\!\!+ P)z.\supset_z.y\epsilon\alpha:$

[(1).(2)] $\equiv:\alpha,\beta\epsilon\mathrm{Cl}'(C'P\cup\iota'x):(\exists z):z\epsilon\alpha-\iota'x-(\beta-\iota'x):y\epsilon\beta-\iota'x.yPz.\supset_y.$

$y\epsilon\alpha-\iota'x:$

[*24·43] $\equiv:\alpha-\iota'x,\beta-\iota'x\epsilon\mathrm{Cl}'C'P:(\exists z).z\epsilon\alpha-\iota'x-(\beta-\iota'x).$

$\overrightarrow{P}'z\cap(\beta-\iota'x)\subset\alpha-\iota'x:$

[*170·11] $\equiv:(\alpha-\iota'x)P_{cl}(\beta-\iota'x):$

[*51·221] $\equiv:\alpha\{(\iota'x\cup)^{;}P_{cl}\}\beta::\supset\vdash.\mathrm{Prop}$

***170·62.** $\vdash :. x \sim\epsilon C`P . \dot{\exists} ! P . x \epsilon \alpha - \beta . \supset :$

$$\alpha (x \leftarrow\!\!\!\!+ P)_{cl} \beta . \equiv . \alpha \subset \iota`x \cup C`P . \beta \subset C`P$$

Dem.

$\vdash . *161·13 . \supset \vdash :. \text{Hp} . \supset : x \sim\epsilon \text{Cl}`(x \leftarrow\!\!\!\!+ P) :$

$[*10·53] \quad \supset : y \epsilon \beta . y (x \leftarrow\!\!\!\!+ P) x . \supset_y . y \epsilon \alpha :$

$[\text{Hp}] \quad \supset : x \epsilon \alpha - \beta : y \epsilon \beta . y (x \leftarrow\!\!\!\!+ P) x . \supset_y . y \epsilon \alpha :$

$[*170·11] \quad \supset : \alpha, \beta \epsilon \text{Cl}`C`(x \leftarrow\!\!\!\!+ P) . \supset . \alpha (x \leftarrow\!\!\!\!+ P)_{cl} \beta :$

$[*161·14.\text{Hp}.*24·49] \quad \supset : \alpha \subset \iota`x \cup C`P . \beta \subset C`P . \supset . \alpha (x \leftarrow\!\!\!\!+ P)_{cl} \beta \quad (1)$

$\vdash . *170·11 . *161·14 . \supset$

$\vdash :. \text{Hp} . \supset : \alpha (x \leftarrow\!\!\!\!+ P)_{cl} \beta . \supset . \alpha, \beta \epsilon \text{Cl}`(\iota`x \cup C`P) .$

$[*24·49.\text{Hp}] \quad \supset . \alpha \subset \iota`x \cup C`P . \beta \subset C`P \quad (2)$

$\vdash . (1) . (2) . \supset \vdash . \text{Prop}$

***170·63.** $\vdash :. x \sim\epsilon (\alpha \cup \beta) . \supset : \alpha (x \leftarrow\!\!\!\!+ P)_{cl} \beta . \equiv . \alpha P_{cl} \beta$

Dem.

$\vdash . *24·49 . *161·14 . \supset \vdash :. \text{Hp} . \supset : \alpha, \beta \epsilon \text{Cl}`C`(x \leftarrow\!\!\!\!+ P) . \equiv . \alpha, \beta \epsilon \text{Cl}`C`P \quad (1)$

$\vdash . *13·14 . \supset \vdash :: \text{Hp} . \supset :. y \epsilon \beta . \supset : y \neq x :$

$[*161·111] \quad \supset : y (x \leftarrow\!\!\!\!+ P) z . \equiv . y P z \quad (2)$

$\vdash . *170·11 . \supset \vdash :: \text{Hp} . \supset :. \alpha (x \leftarrow\!\!\!\!+ P)_{cl} \beta . \equiv :$

$$\alpha, \beta \epsilon \text{Cl}`(x \leftarrow\!\!\!\!+ P) : (\exists z) : z \epsilon \alpha - \beta : y \epsilon \beta . y (x \leftarrow\!\!\!\!+ P) z . \supset_y . y \epsilon \alpha :$$

$[(1).(2)] \quad \equiv : \alpha, \beta \epsilon \text{Cl}`C`P : (\exists z) : z \epsilon \alpha - \beta : y \epsilon \beta . y P z . \supset_y . y \epsilon \alpha :$

$[*170·11] \quad \equiv : \alpha P_{cl} \beta :: \supset \vdash . \text{Prop}$

***170·64.** $\vdash : x \sim\epsilon C`P . \supset . (x \leftarrow\!\!\!\!+ P)_{cl} = (\iota`x \cup)^{;} P_{cl} \uparrow\!\!\!\!+ P_{cl}$

Dem.

$\vdash . *170·61·62·63·37 . \supset$

$\vdash :: \text{Hp} . \dot{\exists} ! P . \supset :. \alpha (x \leftarrow\!\!\!\!+ P)_{cl} \beta . \equiv :$

$$x \epsilon \alpha \cap \beta . \alpha \{(\iota`x \cup)^{;} P_{cl}\} \beta . \vee . x \epsilon \alpha - \beta . \alpha \epsilon C`(\iota`x \cup)^{;} P_{cl} . \beta \epsilon C`P_{cl} . \vee .$$

$$x \sim\epsilon (\alpha \cup \beta) . \alpha P_{cl} \beta \quad (1)$$

$\vdash . *150·4 . \quad \supset \vdash : \alpha \{(\iota`x \cup)^{;} P\} \beta . \supset . x \epsilon \alpha \cap \beta \quad (2)$

$\vdash . *150·22 . *170·37 . \supset \vdash :. \text{Hp} . \supset : \alpha \epsilon C`(\iota`x \cup)^{;} P_{cl} . \beta \epsilon C`P_{cl} . \supset . x \epsilon \alpha - \beta \quad (3)$

$\vdash . *170·1 . \quad \supset \vdash :. \text{Hp} . \supset : \alpha P_{cl} \beta . \supset . x \sim\epsilon (\alpha \cap \beta) \quad (4)$

$\vdash . (1) . (2) . (3) . (4) . \supset \vdash :: \text{Hp} . \dot{\exists} ! P . \supset :.$

$\alpha (x \leftarrow\!\!\!\!+ P)_{cl} \beta . \equiv : \alpha \{(\iota`x \cup)^{;} P_{cl}\} \beta . \vee . \alpha \epsilon C`(\iota`x \cup)^{;} P_{cl} . \beta \epsilon C`P_{cl} . \vee . \alpha P_{cl} \beta :$

$[*160·11] \quad \equiv : \alpha \{(\iota`x \cup)^{;} P_{cl} \uparrow\!\!\!\!+ P_{cl}\} \beta \quad (5)$

$\vdash . *161·201 . \supset \vdash : P = \dot{\Lambda} . \supset . x \leftarrow\!\!\!\!+ P = \dot{\Lambda} .$

$[*170·35] \quad \supset . (x \leftarrow\!\!\!\!+ P)_{cl} = \dot{\Lambda} \quad (6)$

$\vdash . *150·42 . *160·22 . *170·35 . \supset \vdash : P = \dot{\Lambda} . \supset . (\iota`x \cup)^{;} P_{cl} \uparrow\!\!\!\!+ P_{cl} = \dot{\Lambda} \quad (7)$

$\vdash . (6) . (7) . \supset \vdash : P = \dot{\Lambda} . \supset . (x \leftarrow\!\!\!\!+ P)_{cl} = (\iota`x \cup)^{;} P_{cl} \uparrow\!\!\!\!+ P_{cl} \quad (8)$

$\vdash . (5) . (8) . \supset \vdash . \text{Prop}$

The following propositions are lemmas for $*170{\cdot}67$, *i.e.*

$$\dot{\exists} ! P . \dot{\exists} ! Q . C`P \cap C`Q = \Lambda . \supset . (P \ddagger Q)_{cl} = s`C``(P_{cl} \times Q_{cl}),$$

which itself leads to $*170{\cdot}69$, *i.e.*

$$\dot{\exists} ! P . \dot{\exists} ! Q . C`P \cap C`Q = \Lambda . \supset . (P \ddagger Q)_{cl} \text{ smor } (P_{cl} \times Q_{cl}).$$

$*170{\cdot}65$. $\vdash :. \rho (P \ddagger Q)_{cl} \sigma . \equiv : (\exists \alpha, \beta, \gamma, \delta) : \alpha, \beta \in \text{Cl}`C`P . \gamma, \delta \in \text{Cl}`C`Q .$
$\rho = \alpha \cup \gamma . \sigma = \beta \cup \delta : (\exists y) . y \in (\alpha \cup \gamma) - (\beta \cup \delta) . \overrightarrow{P \ddagger Q}`y \cap (\beta \cup \delta) \subset \alpha \cup \gamma$

Dem.

$\vdash . *13{\cdot}193 . \supset \vdash :. (\exists \alpha, \beta, \gamma, \delta) : \alpha, \beta \in \text{Cl}`C`P . \gamma, \delta \in \text{Cl}`C`Q . \rho = \alpha \cup \gamma . \sigma = \beta \cup \delta :$
$(\exists y) . y \in (\alpha \cup \gamma) - (\beta \cup \delta) . \overrightarrow{P \ddagger Q}`y \cap (\beta \cup \delta) \subset \alpha \cup \gamma :$
$\equiv : (\exists \alpha, \beta, \gamma, \delta) : \alpha, \beta \in \text{Cl}`C`P . \gamma, \delta \in \text{Cl}`C`Q . \rho = \alpha \cup \gamma . \sigma = \beta \cup \delta :$
$(\exists y) . y \in \rho - \sigma . \overrightarrow{P \ddagger Q}`y \cap \sigma \subset \rho :$
[$*60{\cdot}45$] $\equiv : \rho, \sigma \in \text{Cl}`(C`P \cup C`Q) : (\exists y) . y \in \rho - \sigma . \overrightarrow{P \ddagger Q}`y \cap \sigma \subset \rho :$
[$*160{\cdot}14$] $\equiv : \rho, \sigma \in \text{Cl}`C`(P \ddagger Q) : (\exists y) . y \in \rho - \sigma . \overrightarrow{P \ddagger Q}`y \cap \sigma \subset \rho :$
[$*170{\cdot}11$] $\equiv : \rho (P \ddagger Q)_{cl} \sigma :. \supset \vdash . \text{Prop}$

$*170{\cdot}651$. $\vdash :. C`P \cap C`Q = \Lambda . \alpha, \beta \in \text{Cl}`C`P . \gamma, \delta \in \text{Cl}`C`Q . y \in \alpha . \supset :$
$y \in (\alpha \cup \gamma) - (\beta \cup \delta) . \overrightarrow{P \ddagger Q}`y \cap (\beta \cup \delta) \subset \alpha \cup \gamma . \equiv . y \in \alpha - \beta . \overrightarrow{P}`y \cap \beta \subset \alpha$

Dem.

$\vdash . *24{\cdot}402{\cdot}313 . \supset \vdash : \text{Hp} . \supset . (\alpha \cup \gamma) - (\beta \cup \delta) = (\alpha - \beta) \cup (\gamma - \delta)$ (1)
$\vdash . *160{\cdot}11 . \quad \supset \vdash : \text{Hp} . \supset . \overrightarrow{P \ddagger Q}`y = \overrightarrow{P}`y$ (2)
$\vdash . *24{\cdot}402 . \quad \supset \vdash :. \text{Hp} . \supset : y \sim\in \gamma :$
[(1)] $\quad \supset : y \in (\alpha \cup \gamma) - (\beta \cup \delta) . \equiv . y \in \alpha - \beta$ (3)
$\vdash . *33{\cdot}15{\cdot}161 . \supset \vdash . \overrightarrow{P}`y \subset C`P .$
[$*24{\cdot}402$] $\quad \supset \vdash : \text{Hp} . \supset . \overrightarrow{P}`y \cap \delta = \Lambda .$
[(2)] $\quad \supset . \overrightarrow{P \ddagger Q}`y \cap (\beta \cup \delta) = \overrightarrow{P}`y \cap \beta .$ (4)
[$*24{\cdot}402$] $\quad \supset . \overrightarrow{P \ddagger Q}`y \cap (\beta \cup \delta) \cap \gamma = \Lambda$ (5)
$\vdash . (4) . (5) . *24{\cdot}49 . \supset \vdash :. \text{Hp} . \supset : \overrightarrow{P \ddagger Q}`y \cap (\beta \cup \delta) \subset \alpha \cup \gamma . \equiv .$
$\overrightarrow{P}`y \cap \beta \subset \alpha$ (6)
$\vdash . (3) . (6) . \supset \vdash . \text{Prop}$

$*170{\cdot}652$. $\vdash :. C`P \cap C`Q = \Lambda . \alpha, \beta \in \text{Cl}`C`P . \gamma, \delta \in \text{Cl}`C`Q . y \in \gamma . \supset :$
$y \in (\alpha \cup \gamma) - (\beta \cup \delta) . \overrightarrow{P \ddagger Q}`y \cap (\beta \cup \delta) \subset \alpha \cup \gamma . \equiv .$
$\beta \subset \alpha . y \in \gamma - \delta . \overrightarrow{Q}`y \cap \delta \subset \gamma$

Dem.

$\vdash . *24{\cdot}402{\cdot}313 . \supset \vdash : \text{Hp} . \supset . (\alpha \cup \gamma) - (\beta \cup \delta) = (\alpha - \beta) \cup (\gamma - \delta)$ (1)
$\vdash . *24{\cdot}402 . \quad \supset \vdash : \text{Hp} . \supset . y \sim\in \alpha$ (2)

$\vdash . (1) . (2) . \supset \vdash :. \text{Hp} . \supset : y \in (\alpha \cup \gamma) - (\beta \cup \delta) . \equiv . y \in \gamma - \delta$ (3)

$\vdash . *160 \cdot 11 . \supset \vdash : \text{Hp} . \supset . \overrightarrow{P \ddagger Q}\text{'}y = C\text{'}P \cup \overrightarrow{Q}\text{'}y .$

[*22·621.*24·402] $\supset . \overrightarrow{P \ddagger Q}\text{'}y \cap (\beta \cup \delta) = \beta \cup (\overrightarrow{Q}\text{'}y \cap \delta)$ (4)

$\vdash . *24 \cdot 49 . \supset \vdash :. \text{Hp} . \supset : \beta \subset \alpha \cup \gamma . \equiv . \beta \subset \alpha :$

$\overrightarrow{Q}\text{'}y \cap \delta \subset \alpha \cup \gamma . \equiv . \overrightarrow{Q}\text{'}y \cap \delta \subset \gamma$ (5)

$\vdash . (4) . (5) . \supset \vdash :. \text{Hp} . \supset :$

$\overrightarrow{P \ddagger Q}\text{'}y \cap (\beta \cup \delta) \subset \alpha \cup \gamma . \equiv . \beta \subset \alpha . \overrightarrow{Q}\text{'}y \cap \delta \subset \gamma$ (6)

$\vdash . (3) . (6) . \supset \vdash . \text{Prop}$

***170·653.** $\vdash :: C\text{'}P \cap C\text{'}Q = \Lambda . \alpha, \beta \in \text{Cl}\text{'}C\text{'}P . \gamma, \delta \in \text{Cl}\text{'}C\text{'}Q . \supset :.$

$(\alpha \cup \gamma)(P \ddagger Q)_{\text{cl}}(\beta \cup \delta) . \equiv : \alpha P_{\text{cl}} \beta . \text{v} . \alpha = \beta . \gamma Q_{\text{cl}} \delta$

Dem.

$\vdash . *170 \cdot 11 . \supset \vdash :: \text{Hp} . \supset :. (\alpha \cup \gamma)(P \ddagger Q)_{\text{cl}}(\beta \cup \delta) . \equiv :$

$(\exists y) . y \in (\alpha \cup \gamma) - (\beta \cup \delta) . \overrightarrow{P \ddagger Q}\text{'}y \cap (\beta \cup \delta) \subset \alpha \cup \gamma :$

[*170·651·652] $\equiv : (\exists y) . y \in \alpha - \beta . \overrightarrow{P}\text{'}y \cap \beta \subset \alpha : \text{v} :$

$\beta \subset \alpha : (\exists y) . y \in \gamma - \delta . \overrightarrow{Q}\text{'}y \cap \delta \subset \beta :$

[*170·11] $\equiv : \alpha P_{\text{cl}} \beta . \text{v} . \beta \subset \alpha . \gamma Q_{\text{cl}} \delta :$

[*170·16] $\equiv : \alpha P_{\text{cl}} \beta . \text{v} . \alpha P_{\text{cl}} \beta . \gamma Q_{\text{cl}} \delta . \text{v} . \alpha = \beta . \gamma Q_{\text{cl}} \delta :$

[*4·44] $\equiv : \alpha P_{\text{cl}} \beta . \text{v} . \alpha = \beta . \gamma Q_{\text{cl}} \delta :: \supset \vdash . \text{Prop}$

***170·66.** $\vdash :. \dot{\exists} ! P . \dot{\exists} ! Q . C\text{'}P \cap C\text{'}Q = \Lambda . \supset :$

$\rho (P \ddagger Q) \sigma . \equiv . (\exists \alpha, \beta, \gamma, \delta) . (\gamma \downarrow \alpha)(P_{\text{cl}} \times Q_{\text{cl}})(\delta \downarrow \beta) . \rho = \alpha \cup \gamma . \sigma = \beta \cup \delta$

Dem.

$\vdash . *170 \cdot 65 \cdot 11 . \supset$

$\vdash : \rho (P \ddagger Q)_{\text{cl}} \sigma . \equiv . (\exists \alpha, \beta, \gamma, \delta) . \alpha, \beta \in \text{Cl}\text{'}C\text{'}P . \gamma, \delta \in \text{Cl}\text{'}C\text{'}Q .$

$\rho = \alpha \cup \gamma . \sigma = \beta \cup \delta . (\alpha \cup \gamma)(P \ddagger Q)_{\text{cl}}(\beta \cup \delta)$ (1)

$\vdash . (1) . *170 \cdot 653 . \supset \vdash :: \text{Hp} . \supset :.$

$\rho (P \ddagger Q)_{\text{cl}} \sigma . \equiv : (\exists \alpha, \beta, \gamma, \delta) : \alpha, \beta \in \text{Cl}\text{'}C\text{'}P . \gamma, \delta \in \text{Cl}\text{'}C\text{'}Q . \rho = \alpha \cup \gamma . \sigma = \beta \cup \delta :$

$\alpha P_{\text{cl}} \beta . \text{v} . \alpha = \beta . \gamma Q_{\text{cl}} \delta :$

[*170·37] $\equiv : (\exists \alpha, \beta, \gamma, \delta) : \alpha, \beta \in C\text{'}P_{\text{cl}} . \gamma, \delta \in C\text{'}Q_{\text{cl}} . \rho = \alpha \cup \gamma . \sigma = \beta \cup \delta :$

$\alpha P_{\text{cl}} \beta . \text{v} . \alpha = \beta . \gamma Q_{\text{cl}} \delta :$

[*166·112] $\equiv : (\exists \alpha, \beta, \gamma, \delta) . (\gamma \downarrow \alpha)(P_{\text{cl}} \times Q_{\text{cl}})(\delta \downarrow \beta) . \rho = \alpha \cup \gamma . \sigma = \beta \cup \delta ::$

$\supset \vdash . \text{Prop}$

***170·67.** $\vdash : \dot{\exists} ! P . \dot{\exists} ! Q . C\text{'}P \cap C\text{'}Q = \Lambda . \supset . (P \ddagger Q)_{\text{cl}} = s ; C ; (P_{\text{cl}} \times Q_{\text{cl}})$

Dem.

$\vdash . *170 \cdot 66 . *13 \cdot 22 . \supset \vdash :: \text{Hp} . \supset :. \rho (P \ddagger Q) \sigma . \equiv :$

$(\exists \alpha, \beta, \gamma, \delta, R, S) . R = \gamma \downarrow \alpha . S = \delta \downarrow \beta . \rho = \alpha \cup \gamma . \sigma = \beta \cup \delta .$

$R (P_{\text{cl}} \times Q_{\text{cl}}) S :$

[*55·15.*53·11] $\equiv : (\exists \alpha, \beta, \gamma, \delta, R, S) . R = \gamma \downarrow \alpha . S = \delta \downarrow \beta .$

$\rho = s\text{‘}C\text{‘}R . \sigma = s\text{‘}C\text{‘}S . R\,(P_{cl} \times Q_{cl})\,S :$

[*166·111] $\equiv : (\exists R, S) . \rho = s\text{‘}C\text{‘}R . \sigma = s\text{‘}C\text{‘}S . R\,(P_{cl} \times Q_{cl})\,S :$

[*150·4] $\equiv : \rho\,\{s;C;(P_{cl} \times Q_{cl})\}\,\sigma :: \supset \vdash .$ Prop

***170·68.** $\vdash : \dot{\exists} ! P . \dot{\exists} ! Q . C\text{‘}P \cap C\text{‘}Q = \Lambda . \supset .$

$(s \mid C) \upharpoonright C\text{‘}(P_{cl} \times Q_{cl}) \in (P \ddagger Q)_{cl}\,\overline{\text{smor}}\,(P_{cl} \times Q_{cl})$

Dem.

$\vdash . *55{\cdot}15 . *53{\cdot}11 . \supset$

$\vdash :. R = \gamma \downarrow \alpha . S = \delta \downarrow \beta . s\text{‘}C\text{‘}R = s\text{‘}C\text{‘}S . \supset . \alpha \cup \gamma = \beta \cup \delta$ (1)

$\vdash . (1) . *24{\cdot}48 . \supset$

$\vdash :: \text{Hp} . \supset :. \alpha, \beta \in \text{Cl}\text{‘}C\text{‘}P . \gamma, \delta \in \text{Cl}\text{‘}C\text{‘}Q . R = \gamma \downarrow \alpha . S = \delta \downarrow \beta . s\text{‘}C\text{‘}R = s\text{‘}C\text{‘}S . \supset .$

$\alpha = \beta . \gamma = \delta .$

[*55·202] $\supset . R = S$ (2)

$\vdash . (2) . *166{\cdot}12 . *170{\cdot}37 . \supset$

$\vdash :. \text{Hp} . \supset : R, S \in C\text{‘}(P_{cl} \times Q_{cl}) . s\text{‘}C\text{‘}R = s\text{‘}C\text{‘}S . \supset . R = S$ (3)

$\vdash . (3) . *151{\cdot}24 . *170{\cdot}67 . \supset \vdash .$ Prop

***170·69.** $\vdash : \dot{\exists} ! P . \dot{\exists} ! Q . C\text{‘}P \cap C\text{‘}Q = \Lambda . \supset . (P \ddagger Q)_{cl}\,\text{smor}\,(P_{cl} \times Q_{cl})$
[*170·68]

$*171$. THE PRINCIPLE OF FIRST DIFFERENCES (*continued*).

Summary of $*171$.

In this number, we shall consider a more restricted form of the principle of first differences, which is applicable when there is a definite first member of one class not belonging to the other class. In this case, if z is the first differing member, the part of α which precedes z is to be the same as the part of β which precedes z. If z belongs to α and not to β, we put α before β; in the converse case, we put β before α. In case zPz, z itself is not to be counted among its own predecessors; thus the predecessors of z are to be $\overrightarrow{P}\text{'}z - \iota\text{'}z$. Hence the relation in question will hold between two sub-classes (α and β) of $C\text{'}P$ when there is a z such that

$$z \in \alpha - \beta \,.\, \overrightarrow{P}\text{'}z - \iota\text{'}z \cap \alpha = \overrightarrow{P}\text{'}z - \iota\text{'}z \cap \beta,$$

or, what comes to the same thing (owing to $z \sim\in \beta$),

$$z \in \alpha - \beta \,.\, \overrightarrow{P}\text{'}z \cap \alpha - \iota\text{'}z = \overrightarrow{P}\text{'}z \cap \beta.$$

This relation between α and β we denote by "P_{df}," where "df" stands for "difference."

Thus our definition is

$$P_{df} = \hat{\alpha}\hat{\beta}\,\{\alpha, \beta \in \text{Cl}\text{'}C\text{'}P : (\exists z) \,.\, z \in \alpha - \beta \,.\, \overrightarrow{P}\text{'}z \cap \alpha - \iota\text{'}z = \overrightarrow{P}\text{'}z \cap \beta\} \quad \text{Df.}$$

On the analogy of P_{lc}, we put also

$$P_{fd} = \text{Cnv}\text{'}(\breve{P})_{df}.$$

When P is well-ordered, P_{df} and P_{fd} coincide respectively with P_{cl} and P_{lc}. Their properties are closely analogous to those of P_{cl} and P_{lc}. Thus *e.g.* the following propositions remain true when P_{df} is substituted for P_{cl}:

$*170\cdot17\cdot35\cdot36\cdot37\cdot38\cdot44\cdot5\cdot51\cdot52\cdot64\cdot67\cdot68\cdot69$.

The only new propositions to be noted in this number are

$*171\cdot2$. $\vdash : P \in J \,.\, \supset \,.$

$$P_{df} = \hat{\alpha}\hat{\beta}\,\{\alpha, \beta \in \text{Cl}\text{'}C\text{'}P : (\exists z) \,.\, z \in \alpha - \beta \,.\, \overrightarrow{P}\text{'}z \cap \alpha = \overrightarrow{P}\text{'}z \cap \beta\}$$

$*171\cdot21$. $\vdash \,.\, P_{df} \in P_{cl}$

and the following formulae suggesting an inductive identification of P_{cl} and P_{df} in cases to which such induction is applicable:

$*171\cdot7.\quad \vdash : P_{df} = P_{cl} . x \sim\in C`P . \supset . (x \leftarrow\!\!\!+ P)_{df} = (x \leftarrow\!\!\!+ P)_{cl}$

$*171\cdot71.\quad \vdash : C`P \cap C`Q = \Lambda . P_{df} = P_{cl} . Q_{df} = Q_{cl} . \supset . (P \updownarrow Q)_{df} = (P \updownarrow Q)_{cl}$

These propositions are however superseded (at a later stage) by the proof that P_{cl} and P_{df} coincide if P is well-ordered (*251·37).

The chief property of P_{df} is that its relation-number is 2_r to the power Nr‘P. This will be proved in *177 and *186·4.

$*171\cdot01.\quad P_{df} = \hat{\alpha}\hat{\beta}\{\alpha, \beta \in \text{Cl}`C`P : (\exists z) . z \in \alpha - \beta . \overrightarrow{P}`z \cap \alpha - \iota`z = \overrightarrow{P}`z \cap \beta\}$ Df

$*171\cdot02.\quad P_{fd} = \text{Cnv}`(\breve{P})_{df}$ Df

$*171\cdot1.\quad \vdash :. \alpha P_{df} \beta . \equiv : \alpha, \beta \in \text{Cl}`C`P : (\exists z) . z \in \alpha - \beta . \overrightarrow{P}`z \cap \alpha - \iota`z = \overrightarrow{P}`z \cap \beta$
[(*171·01)]

$*171\cdot101.\quad \vdash . P_{fd} = \text{Cnv}`(\breve{P})_{df}$ [(*171·02)]

$*171\cdot102.\quad \vdash :. \alpha P_{fd} \beta . \equiv : \alpha, \beta \in \text{Cl}`C`P : (\exists z) . z \in \beta - \alpha . \overleftarrow{P}`z \cap \beta - \iota`z = \overleftarrow{P}`z \cap \alpha$
[*171·1·101]

$*171\cdot11.\quad \vdash ::. \alpha P_{df} \beta . \equiv :: \alpha, \beta \in \text{Cl}`C`P ::$
$(\exists z) :. z \in \alpha - \beta :. yPz . y \neq z . \supset_y : y \in \alpha . \equiv . y \in \beta$ [*171·1]

$*171\cdot12.\quad \vdash :. \alpha P_{df} \beta . \equiv : \alpha, \beta \in \text{Cl}`C`P :$
$(\exists z) . z \in \alpha - \beta . \overrightarrow{P}`z \cap \alpha - \iota`z = \overrightarrow{P}`z \cap \beta - \iota`z$ [*171·1 . *51·222]

$*171\cdot13.\quad \vdash . C`P_{df} \subset \text{Cl}`C`P$ [*171·1]

$*171\cdot14.\quad \vdash : \alpha \subset C`P . z \in \alpha . \supset . \alpha P_{df} (\alpha - \iota`z)$

Dem.

$\vdash . *51\cdot21 . \supset \vdash : \text{Hp} . \supset . z \in \alpha - (\alpha - \iota`z) .$

[*13·15] $\supset . z \in \alpha - (\alpha - \iota`z) . \overrightarrow{P}`z \cap \alpha - \iota`z = \overrightarrow{P}`z \cap (\alpha - \iota`z) .$

[*171·12] $\supset . \alpha P_{df} (\alpha - \iota`z) : \supset \vdash . \text{Prop}$

$*171\cdot15.\quad \vdash : \beta \subset C`P . z \in C`P - \beta . \supset . (\beta \cup \iota`z) P_{df} \beta$

Dem.

$\vdash . *51\cdot16 . \supset \vdash : \text{Hp} . \supset . z \in (\beta \cup \iota`z) - \beta$ (1)

$\vdash . *51\cdot211\cdot22 . \supset \vdash : \text{Hp} . \supset . (\beta \cup \iota`z) - \iota`z = \beta .$

[*22·481] $\supset . \overrightarrow{P}`z \cap (\beta \cup \iota`z) - \iota`z = \overrightarrow{P}`z \cap \beta$ (2)

$\vdash . (1) . (2) . *171\cdot1 . \supset \vdash . \text{Prop}$

$*171\cdot16.\quad \vdash . \text{D}`P_{df} = \text{Cl ex}`C`P . \text{Ɑ}`P_{df} = \text{Cl}`C`P - \iota`C`P$

Dem.

$\vdash . *171\cdot14 . \supset \vdash : \alpha \in \text{Cl ex}`C`P . \supset . \alpha \in \text{D}`P_{df}$ (1)

$\vdash . *171\cdot1 . \supset \vdash : \alpha \in \text{D}`P_{df} . \supset . \alpha \in \text{Cl ex}`C`P$ (2)

$\vdash . (1) . (2) . \supset \vdash . \text{D}\text{‘}P_{\text{df}} = \text{Cl ex}\text{‘}C\text{‘}P$ (3)

$\vdash . *171{\cdot}15 . \supset \vdash : \beta \in \text{Cl}\text{‘}C\text{‘}P . \exists ! C\text{‘}P - \beta . \supset . \beta \in \text{Ɑ}\text{‘}P_{\text{df}} :$

$[*24{\cdot}6] \quad \supset \vdash : \beta \in \text{Cl}\text{‘}C\text{‘}P - \iota\text{‘}C\text{‘}P . \supset . \beta \in \text{Ɑ}\text{‘}P_{\text{df}}$ (4)

$\vdash . *171{\cdot}1 . \supset \vdash : \beta \in \text{Ɑ}\text{‘}P_{\text{df}} . \supset . \beta \in \text{Cl}\text{‘}C\text{‘}P . \exists ! C\text{‘}P - \beta .$

$[*24{\cdot}6] \quad \supset . \beta \in \text{Cl}\text{‘}C\text{‘}P - \iota\text{‘}C\text{‘}P$ (5)

$\vdash . (4) . (5) . \supset \vdash . \text{Ɑ}\text{‘}P_{\text{df}} = \text{Cl}\text{‘}C\text{‘}P - \iota\text{‘}C\text{‘}P$ (6)

$\vdash . (3) . (6) . \supset \vdash . \text{Prop}$

$*171{\cdot}17$. $\vdash : \dot{\exists} ! P . \supset . C\text{‘}P_{\text{df}} = \text{Cl}\text{‘}C\text{‘}P$

Dem.

$\vdash . *171{\cdot}16 . \supset \vdash : \alpha \in \text{Cl}\text{‘}C\text{‘}P . \alpha \neq \Lambda . \supset . \alpha \in \text{D}\text{‘}P_{\text{df}}$ (1)

$\vdash . *171{\cdot}16 . \supset \vdash : \alpha \in \text{Cl}\text{‘}C\text{‘}P . \alpha \neq C\text{‘}P . \supset . \alpha \in \text{Ɑ}\text{‘}P_{\text{df}}$ (2)

$\vdash . (1) . (2) . \supset \vdash : \alpha \in \text{Cl}\text{‘}C\text{‘}P . \sim(\alpha = \Lambda . \alpha = C\text{‘}P) . \supset . \alpha \in C\text{‘}P_{\text{df}} :$

$[*13{\cdot}171] \quad \supset \vdash : \alpha \in \text{Cl}\text{‘}C\text{‘}P . C\text{‘}P \neq \Lambda . \supset . \alpha \in C\text{‘}P_{\text{df}}$ (3)

$\vdash . (3) . *33{\cdot}24 . \supset \vdash . \text{Prop}$

$*171{\cdot}18$. $\vdash : \dot{\exists} ! P . \supset . B\text{‘}P_{\text{df}} = C\text{‘}P . B\text{‘}\text{Cnv}\text{‘}P_{\text{df}} = \Lambda$

Dem.

$\vdash . *171{\cdot}16 . \quad \supset \vdash . \overrightarrow{B}\text{‘}P_{\text{df}} = \text{Cl ex}\text{‘}C\text{‘}P - (\text{Cl}\text{‘}C\text{‘}P - \iota\text{‘}C\text{‘}P)$

$[*24{\cdot}3] \quad = \text{Cl ex}\text{‘}C\text{‘}P \cap \iota\text{‘}C\text{‘}P$ (1)

$\vdash . (1) . *60{\cdot}35 . \supset \vdash : \dot{\exists} ! P . \supset . \overrightarrow{B}\text{‘}P_{\text{df}} = \iota\text{‘}C\text{‘}P$ (2)

$\vdash . *171{\cdot}16 . \quad \supset \vdash . \overrightarrow{B}\text{‘}\text{Cnv}\text{‘}P_{\text{df}} = \text{Cl}\text{‘}C\text{‘}P - \iota\text{‘}C\text{‘}P - \text{Cl ex}\text{‘}C\text{‘}P$

$[*60{\cdot}24] \quad = \iota\text{‘}\Lambda - \iota\text{‘}C\text{‘}P$ (3)

$\vdash . (3) . *33{\cdot}24 . \supset \vdash : \dot{\exists} ! P . \supset . \overrightarrow{B}\text{‘}\text{Cnv}\text{‘}P_{\text{df}} = \iota\text{‘}\Lambda$ (4)

$\vdash . (2) . (4) . \supset \vdash . \text{Prop}$

$*171{\cdot}19$. $\vdash : P = \dot{\Lambda} . \supset . P_{\text{df}} = \dot{\Lambda}$

Dem.

$\vdash . *60{\cdot}33 . *171{\cdot}16 . \supset \vdash : \text{Hp} . \supset . \text{D}\text{‘}P_{\text{df}} = \Lambda .$

$[*33{\cdot}241] \quad \supset . P_{\text{df}} = \dot{\Lambda} : \supset \vdash . \text{Prop}$

$*171{\cdot}2$. $\vdash : P \mathrel{\dot{\subset}} J . \supset . P_{\text{df}} = \hat{\alpha}\hat{\beta}\{\alpha, \beta \in \text{Cl}\text{‘}C\text{‘}P : (\exists z) . z \in \alpha - \beta . \overrightarrow{P}\text{‘}z \cap \alpha = \overrightarrow{P}\text{‘}z \cap \beta\}$

Dem.

$\vdash . *50{\cdot}11 . *32{\cdot}19 . \supset \vdash : \text{Hp} . \supset . \overrightarrow{P}\text{‘}z \subset -\iota\text{‘}z .$

$[*22{\cdot}621] \quad \supset . \overrightarrow{P}\text{‘}z \cap \alpha - \iota\text{‘}z = \overrightarrow{P}\text{‘}z \cap \alpha$ (1)

$\vdash . (1) . *171{\cdot}1 . \supset \vdash . \text{Prop}$

***171·21.** $\vdash . P_{\mathrm{df}} \subseteq P_{\mathrm{cl}}$

Dem.

$\vdash . *171·1 . *22·43 . \supset$

$\vdash :. \alpha P_{\mathrm{df}} \beta . \supset : \alpha, \beta \in \mathrm{Cl}`C`P : (\exists z) . z \in \alpha - \beta . \overrightarrow{P}`z \cap \beta \subset \alpha :$

[*170·11] $\supset : \alpha P_{\mathrm{cl}} \beta :. \supset \vdash . \mathrm{Prop}$

***171·22.** $\vdash . P_{\mathrm{df}} \subseteq J$ [*170·17 . *171·21]

***171·4.** $\vdash : S \in 1 \to 1 . C`Q = \mathrm{G}`S . \supset . (S;Q)_{\mathrm{df}} = S_\epsilon ; Q_{\mathrm{df}}$ [Proof as in *170·4]

***171·41.** $\vdash : (S \upharpoonright C`Q)_\epsilon ; Q_{\mathrm{df}} = S_\epsilon ; Q_{\mathrm{df}}$ [Proof as in *170·41]

***171·42.** $\vdash : S \upharpoonright C`Q \in 1 \to 1 . C`Q \subset \mathrm{G}`S . \supset . (S;Q)_{\mathrm{df}} = S_\epsilon ; Q_{\mathrm{df}}$ [*171·4·41]

***171·43.** $\vdash : S \upharpoonright C`Q \in P \,\overline{\mathrm{smor}}\, Q . \supset . S_\epsilon \upharpoonright C`Q_{\mathrm{df}} \in P_{\mathrm{df}} \,\overline{\mathrm{smor}}\, Q_{\mathrm{df}}$

[Proof as in *170·43]

***171·44.** $\vdash : P \,\mathrm{smor}\, Q . \supset . P_{\mathrm{df}} \,\mathrm{smor}\, Q_{\mathrm{df}}$ [*171·43]

***171·5.** $\vdash . (x \downarrow x)_{\mathrm{df}} = (\iota`x) \downarrow \Lambda = (x \downarrow x)_{\mathrm{cl}}$

Dem.

$\vdash . *171·1 . *55·15 . \supset$

$\vdash :. \alpha (x \downarrow x)_{\mathrm{df}} \beta . \equiv : \alpha, \beta \in \mathrm{Cl}`\iota`x : (\exists z) . z \in \alpha - \beta . \overrightarrow{x \downarrow x}`z \cap \alpha - \iota`z = \overrightarrow{x \downarrow x}`z \cap \beta :$

[*171·16] $\equiv : \alpha \in \mathrm{Cl\,ex}`\iota`x . \beta \in \mathrm{Cl}`\iota`x - \iota`\iota`x :$

$(\exists z) . z \in \alpha - \beta . \overrightarrow{x \downarrow x}`z \cap \alpha - \iota`z = \overrightarrow{x \downarrow x}`z \cap \beta :$

[*60·362·37] $\equiv : \alpha = \iota`x . \beta = \Lambda : (\exists z) . z \in \iota`x . \overrightarrow{x \downarrow x}`z \cap \alpha - \iota`z = \overrightarrow{x \downarrow x}`z \cap \beta :$

[*13·195] $\equiv : \alpha = \iota`x . \beta = \Lambda . \iota`x \cap \alpha - \iota`x = \iota`x \cap \beta :$

[*24·21·23] $\equiv : \alpha = \iota`x . \beta = \Lambda . \Lambda = \Lambda :$

[*13·15.*55·13] $\equiv : \alpha \{(\iota`x) \downarrow \Lambda\} \beta$ (1)

$\vdash . (1) . *170·5 . \supset \vdash . \mathrm{Prop}$

***171·51.** $\vdash . (x \downarrow y)_{\mathrm{df}} = (x \downarrow y)_{\mathrm{cl}}$

Dem.

$\vdash . *171·1 . \supset \vdash :. \alpha (x \downarrow y)_{\mathrm{df}} \beta . \equiv : \alpha, \beta \in \mathrm{Cl}`(\iota`x \cup \iota`y) :$

$(\exists z) . z \in \alpha - \beta . \overrightarrow{x \downarrow y}`z \cap \alpha - \iota`z = \overrightarrow{x \downarrow y}`z \cap \beta :$

[*171·16] $\equiv : \alpha \in \mathrm{Cl\,ex}`(\iota`x \cup \iota`y) . \beta \in \mathrm{Cl}`(\iota`x \cup \iota`y) - \iota`(\iota`x \cup \iota`y) :$

$(\exists z) . z \in \alpha - \beta . \overrightarrow{x \downarrow y}`z \cap \alpha - \iota`z = \overrightarrow{x \downarrow y}`z \cap \beta :$

[*60·39] $\equiv : \alpha = \iota`x \cup \iota`y . \mathrm{v} . \alpha = \iota`x . \mathrm{v} . \alpha = \iota`y : \beta = \iota`x . \mathrm{v} . \beta = \iota`y . \mathrm{v} . \beta = \Lambda :$

$(\exists z) . z \in \alpha - \beta . \overrightarrow{x \downarrow y}`z \cap \alpha - \iota`z = \overrightarrow{x \downarrow y}`z \cap \beta$ (1)

$\vdash . *55·13 . \supset \vdash :. x \neq y . \supset : \overrightarrow{x \downarrow y}`y = \iota`x . \overrightarrow{x \downarrow y}`x = \Lambda :$ (2)

[*51·222] $\supset : \alpha = \iota`x \cup \iota`y . \beta = \iota`x . \supset .$

$y \in \alpha - \beta . \overrightarrow{x \downarrow y}`y \cap \alpha - \iota`y = \iota`x = \overrightarrow{x \downarrow y}`y \cap \beta .$

[(1)] $\supset . \alpha P_{\mathrm{df}} \beta$ (3)

$\vdash . (2) . \supset \vdash : x \neq y . \alpha = \iota`x \cup \iota`y . \beta = \iota`y . \supset .$

$x \in \alpha - \beta . \overrightarrow{x \downarrow y}`x \cap \alpha - \iota`x = \Lambda = \overrightarrow{x \downarrow y}`x \cap \beta$

[(1)] $\supset . \alpha P_{df} \beta$ (4)

$\vdash . (2) . \supset \vdash : x \neq y . \alpha = \iota`x \cup \iota`y . \beta = \Lambda . \supset .$

$x \in \alpha - \beta . \overrightarrow{x \downarrow y}`x \cap \alpha - \iota`x = \Lambda = \overrightarrow{x \downarrow y}`x \cap \beta .$

[(1)] $\supset . \alpha P_{df} \beta$ (5)

$\vdash . (2) . \supset \vdash :. x \neq y . \alpha = \iota`x : \beta = \Lambda . \vee . \beta = \iota`y : \supset .$

$x \in \alpha - \beta . \overrightarrow{x \downarrow y}`x \cap \alpha - \iota`x = \Lambda = \overrightarrow{x \downarrow y}`x \cap \beta .$

$\supset . \alpha P_{df} \beta$ (6)

$\vdash . (2) . *24 \cdot 23 . \supset \vdash : x \neq y . \alpha = \iota`y . \beta = \Lambda . \supset .$

$y \in \alpha - \beta . \overrightarrow{x \downarrow y}`x \cap \alpha - \iota`y = \Lambda = \overrightarrow{x \downarrow y}`y \cap \beta$ (7)

$\vdash . (3) . (4) . (5) . (6) . (7) . *170 \cdot 51 . \supset \vdash : x \neq y . \supset . (x \downarrow y)_{cl} \subset (x \downarrow y)_{df} .$

[*171·21] $\supset . (x \downarrow y)_{df} = (x \downarrow y)_{cl}$ (8)

$\vdash . (8) . *171 \cdot 5 . \supset \vdash .$ Prop

***171·52.** $\vdash : x \neq y . \supset . (x \downarrow y)_{df} = (\iota`x \cup \iota`y) \downarrow (\iota`x) \mathbin{\dot{+}} (\iota`y) \downarrow \Lambda$
[*171·51 . *170·52]

***171·64.** $\vdash : x \sim \in C`P . \supset . (x \leftarrow\!\!\!+ P)_{df} = (\iota`x \cup)^{;} P_{df} \mathbin{\dot{+}} P_{df}$

The proof proceeds by the same stages as the proof of *170·64.

***171·67.** $\vdash : \dot{\exists} ! P . \dot{\exists} ! Q . C`P \cap C`Q = \Lambda . \supset . (P \mathbin{\dot{+}} Q)_{df} = s^{;} C^{;} (P_{df} \times Q_{df})$
[Proof as in *170·67]

***171·68.** $\vdash : \dot{\exists} ! P . \dot{\exists} ! Q . C`P \cap C`Q = \Lambda . \supset .$

$s \mid C \upharpoonright (P_{df} \times Q_{df}) \in (P \mathbin{\dot{+}} Q)_{df} \, \overline{\text{smor}} \, (P_{df} \times Q_{df})$

[Proof as in *170·68]

***171·69.** $\vdash : \dot{\exists} ! P . \dot{\exists} ! Q . C`P \cap C`Q = \Lambda . \supset . (P \mathbin{\dot{+}} Q)_{df} \, \text{smor} \, (P_{df} \times Q_{df})$
[*171·68]

***171·7.** $\vdash : P_{df} = P_{cl} . x \sim \in C`P . \supset . (x \leftarrow\!\!\!+ P)_{df} = (x \leftarrow\!\!\!+ P)_{cl}$
[*171·64 . *170·64]

***171·71.** $\vdash : C`P \cap C`Q = \Lambda . P_{df} = P_{cl} . Q_{df} = Q_{cl} . \supset . (P \mathbin{\dot{+}} Q)_{df} = (P \mathbin{\dot{+}} Q)_{cl}$
[*170·67 . *171·67 . *160·21·22]

$*172$. THE PRODUCT OF THE RELATIONS OF A FIELD.

Summary of $*172$.

In this number we have to consider the form of product which is applicable to any relation of relations, whether mutually exclusive or not. If our relation were a Rel^2excl, we could take $C``C`P$, and order selected classes from $C``C`P$ by first differences. This would give us a relation whose field would be $\text{Prod}`C``C`P$. But if any two fields overlap, this method fails. We might substitute $\epsilon_\Delta`C``C`P$ for $\text{Prod}`C``C`P$, and order the members of $\epsilon_\Delta`C``C`P$ by first differences; but this method will not give what we want if two or more members of $C`P$ have the same field. In order to avoid any confusion due to repetition, we must, if $Q \epsilon C`P$ and $x \epsilon C`Q$, consider x in connection with Q, not merely with $C`Q$. That is, the relations in the field of the product of $P$ must be such as concern themselves with the ordered couple $x \downarrow Q$, not merely with $x$. The simplest way of effecting this is to consider $F_\Delta`C`P$. A member of $F_\Delta`C`P$, say $M$, is a relation which picks out a representative of $Q$ from the field of every $Q$ which is a member of $C`P$; that is, whenever $Q \epsilon C`P$, $M`Q \epsilon C`Q$. Since we have $M`Q$, not $M`C`Q$, two relations may have the same field and yet we can distinguish the occurrence of a given term as the representative of the one from its occurrence as the representative of the other. Thus no degree of overlapping will cause confusion.

The relations which compose $F_\Delta`C`P$ are to be ordered by first differences, but in order to distinguish different occurrences of a given term, we must give a slightly different form to the principle of first differences from that employed in $*170$ or $*171$. The new form of the principle is as follows: Consider two relations M and N which are members of $F_\Delta`C`P$. Let Q be a member of $C`P$ in which $M$ chooses a representative which precedes that of $N$, *i.e.* in which $(M`Q)\, Q\, (N`Q)$; and let all earlier relations than $Q$, *i.e.* all relations $R$ such that $RPQ$ and $R \neq Q$, have $M`R = N`R$. Then we say that $M$ precedes $N$. This principle may also be stated as follows: We may divide the members of $C`P$ into four classes, not in general mutually exclusive, namely:

(1) those in which $(M`Q) Q (N`Q)$, *i.e.* in which the M-representative precedes the N-representative;

(2) those in which $(N`Q) Q (M`Q)$,

(3) those in which $M`Q = N`Q$,

(4) those in which no one of the above three relations of $M`Q$ and $N`Q$ occurs.

Then we shall say that M precedes N if there is a member of class (1) whose predecessors all belong to class (3).

In case all the members of $C`P$ are serial, the fourth of the above classes is null, and the other three are mutually exclusive. If, further, $P$ is well-ordered, any two different members of $F_\Delta`C`P$ must be such that one precedes the other in the above-defined order. Thus in this case the product of a series of series is a series (cf. $*251$).

The definition of the product $\Pi`P$ is

$$\Pi`P = \hat{M}\hat{N}\{M, N \in F_\Delta`C`P :. (\exists Q) : (M`Q) Q (N`Q) : RPQ . R \neq Q . \supset_R . M`R = N`R\} \quad \text{Df.}$$

Owing to the complication of this definition, the proofs of propositions of the present number are apt to be long.

Various other definitions might be adopted for $\Pi`P$, but we have found the above definition on the whole the best.

We might, for example, drop the condition $R \neq Q$ in the definition; we could then write our definition in the simpler form:

$$\Pi`P = \hat{M}\hat{N}\{M, N \in F_\Delta`C`P : (\exists Q) . (M`Q) Q (N`Q) . M \upharpoonright \overrightarrow{P}`Q = N \upharpoonright \overrightarrow{P}`Q\},$$

which, with our definition, is only available when $P \subset J$. But if we adopt this simplification, we no longer have

$$\Pi`(P \downarrow P) = P \underset{\bullet\bullet}{\downarrow} P \quad (*172 \cdot 2),$$

which is a very useful proposition, required in the proofs of $*183 \cdot 13$, $*185 \cdot 21$ and other important propositions.

On the other hand, we might frame our definition on the analogy of P_{cl} rather than, as above, on the analogy of P_{df}. The definition would then be:

$$\Pi`P = \hat{M}\hat{N}\{M, N \in F_\Delta`C`P :. (\exists Q) : (M`Q) Q (N`Q) : RPQ . \supset_R . (M`R)(R \cup I)(N`R)\}.$$

This definition does not assume that there is a *first* relation Q for which the M-representative precedes the N-representative. Thus it might be thought that it would give better results in cases where P is not well-ordered. But in fact this is not the case. If P is not well-ordered, it may happen that every Q for which $(M`Q) Q (N`Q)$ is preceded by one for which $(N`Q) Q (M`Q)$, *and vice versa*; in this case, we shall have neither $M (\Pi`P) N$ nor $N (\Pi`P) M$.

Thus our suggested new definition does not secure that $\Pi\text{‘}P$ shall be a series whenever P and all the members of $C\text{‘}P$ are series, and therefore has no substantial advantage over the simpler definition which we have adopted, and has the disadvantage of greater complication.

In the present number, we first prove that $\Pi\text{‘}\dot{\Lambda} = \dot{\Lambda}$ (*172·13) and that $\dot{\Lambda} \in C\text{‘}P . \supset . \Pi\text{‘}P = \dot{\Lambda}$ (*172·14), so that a product is null if any one of its factors is null. We then proceed to propositions about $C\text{‘}\Pi\text{‘}P$, $\overrightarrow{B}\text{‘}\Pi\text{‘}P$, etc. We have

*172·162. $\vdash : \dot{\exists} ! P . \supset . \overrightarrow{B}\text{‘}\Pi\text{‘}P = B_\Delta\text{‘}C\text{‘}P . \overrightarrow{B}\text{‘}\mathrm{Cnv}\text{‘}\Pi\text{‘}P = B_\Delta\text{‘}\mathrm{Cnv}\text{“}C\text{‘}P$

*172·17. $\vdash : \dot{\exists} ! P . \supset . C\text{‘}\Pi\text{‘}P = F_\Delta\text{‘}C\text{‘}P$

Hence we derive propositions as to the existence of $\Pi\text{‘}P$. We have

*172·181. $\vdash :. \text{Mult ax} . \supset : \Lambda \sim\in C\text{‘}P . \dot{\exists} ! P . \equiv . \dot{\exists} ! \Pi\text{‘}P$

Thus assuming the multiplicative axiom, a product which has factors none of which are null is not null.

We then consider $\Pi\text{‘}(P \downarrow P)$, and $\Pi\text{‘}(P \downarrow Q)$ where $P \neq Q$. We have

*172·2. $\vdash . \Pi\text{‘}(P \downarrow P) = P \underset{\cdot\cdot}{\downarrow} P$

which is a useful proposition, and

*172·23. $\vdash : P \neq Q . \supset . \Pi\text{‘}(P \downarrow Q) \text{ smor } P \times Q$

which connects the two definitions of multiplication, showing that they lead to equivalent results for any finite number of factors, *i.e.* whenever the definition of *166 is applicable.

We next consider $\Pi\text{‘}(P \nrightarrow Z)$ and $\Pi\text{‘}(P \not\uparrow Q)$, proving

*172·32. $\vdash : Z \sim\in C\text{‘}P . \supset . \Pi\text{‘}(P \nrightarrow Z) \text{ smor } \Pi\text{‘}P \times Z$

with a similar proposition for $Z \nleftarrow P$ (*172·321), and

*172·35. $\vdash : \dot{\exists} ! P . \dot{\exists} ! Q . C\text{‘}P \cap C\text{‘}Q = \Lambda . \supset . \Pi\text{‘}(P \not\uparrow Q) \text{ smor } \Pi\text{‘}P \times \Pi\text{‘}Q$

which is a form of the associative law using both kinds of multiplication. The kind which uses only Π will be proved in *174.

We have next the proof (with its immediate consequences) that if P and Q have double likeness, $\Pi\text{‘}P \text{ smor } \Pi\text{‘}Q$. We prove

*172·43. $\vdash : T \upharpoonright C\text{‘}\Sigma\text{‘}Q \in P \,\overline{\text{smor}}\, \overline{\text{smor}}\, Q . \supset .$

$$(T \,\|\, \mathrm{Cnv}\text{‘}T\dagger) \upharpoonright C\text{‘}\Pi\text{‘}Q \in (\Pi\text{‘}P) \,\overline{\text{smor}}\, (\Pi\text{‘}Q)$$

This proposition should be compared with *114·51, which is its cardinal analogue. It will be seen that the correlator only differs by the substitution of $T\dagger$ for T_ϵ. From *172·43 we obtain

*172·44. $\vdash : P \text{ smor smor } Q . \supset . \Pi\text{‘}P \text{ smor } \Pi\text{‘}Q$

whence

∗172·45. $\vdash :. \text{Mult ax} . \supset : P, Q \in \text{Rel}^2 \text{excl} . \exists ! P \overline{\text{smor}} Q \cap \text{Rl}\text{‘smor} . \supset . \Pi\text{‘}P \text{ smor } \Pi\text{‘}Q$

Other propositions about $\Pi\text{‘}P$ will be given in ∗174.

∗172·01. $\Pi\text{‘}P = \hat{M}\hat{N} \{M, N \in F_\Delta\text{‘}C\text{‘}P :. (\exists Q) : (M\text{‘}Q) Q (N\text{‘}Q) : RPQ . R \neq Q . \supset_R . M\text{‘}R = N\text{‘}R\}$ Df

∗172·1. $\vdash :: M (\Pi\text{‘}P) N . \equiv :. M, N \in F_\Delta\text{‘}C\text{‘}P :. (\exists Q) : (M\text{‘}Q) Q (N\text{‘}Q) : RPQ . R \neq Q . \supset_R . M\text{‘}R = N\text{‘}R$ [(∗172·01)]

∗172·11. $\vdash :: M (\Pi\text{‘}P) N . \equiv :. M, N \in F_\Delta\text{‘}C\text{‘}P :. (\exists Q) : Q \in C\text{‘}P . (M\text{‘}Q) Q (N\text{‘}Q) : RPQ . R \neq Q . \supset_R . M\text{‘}R = N\text{‘}R$

Dem.

$\vdash . ∗14·21 . \supset \vdash : (M\text{‘}Q) Q (N\text{‘}Q) . \supset . E ! M\text{‘}Q .$

[∗33·43] $\supset . Q \in \text{Cl}\text{‘}M$ (1)

$\vdash . (1) . ∗80·14 . \supset \vdash :. M \in F_\Delta\text{‘}C\text{‘}P . \supset : (M\text{‘}Q) Q (N\text{‘}Q) . \supset . Q \in C\text{‘}P :$

[∗4·73] $\supset : (M\text{‘}Q) Q (N\text{‘}Q) . \equiv . Q \in C\text{‘}P . (M\text{‘}Q) Q (N\text{‘}Q)$ (2)

$\vdash . (2) . ∗172·1 . \supset \vdash . \text{Prop}$

∗172·12. $\vdash . C\text{‘}\Pi\text{‘}P \subset F_\Delta\text{‘}C\text{‘}P$

Dem.

$\vdash . ∗172·1 . \supset \vdash : M (\Pi\text{‘}P) N . \supset . M, N \in F_\Delta\text{‘}C\text{‘}P$ (1)

$\vdash . (1) . ∗33·352 . \supset \vdash . \text{Prop}$

∗172·13. $\vdash . \Pi\text{‘}\dot{\Lambda} = \dot{\Lambda}$

Dem.

$\vdash . ∗172·11 . \supset \vdash : M (\Pi\text{‘}P) N . \supset . \exists ! C\text{‘}P .$

[∗33·24] $\supset . \dot{\exists} ! P$ (1)

$\vdash . (1) . \text{Transp} . \supset \vdash : P = \dot{\Lambda} . \supset . (M, N) . \sim \{M (\Pi\text{‘}P) N\} : \supset \vdash . \text{Prop}$

∗172·14. $\vdash : \dot{\Lambda} \in C\text{‘}P . \supset . \Pi\text{‘}P = \dot{\Lambda}$

Dem.

$\vdash . ∗33·24·5 . \supset \vdash . \overrightarrow{F}\text{‘}\dot{\Lambda} = \Lambda .$

[∗33·41] $\supset \vdash . \dot{\Lambda} \sim\in \text{Cl}\text{‘}F .$

[∗80·21] $\supset \vdash : \dot{\Lambda} \in C\text{‘}P . \supset . F_\Delta\text{‘}C\text{‘}P = \Lambda .$

[∗172·12.∗33·24] $\supset . \Pi\text{‘}P = \dot{\Lambda} : \supset \vdash . \text{Prop}$

∗172·141. $\vdash :. \dot{\exists} ! \Pi\text{‘}P . \supset : Q \in C\text{‘}P . \supset_Q . \dot{\exists} ! Q$ [∗172·14 . Transp]

The following propositions are concerned with $C\text{‘}\Pi\text{‘}P$, $\overrightarrow{B}\text{‘}\Pi\text{‘}P$, etc. ∗172·15·151·16·161 are lemmas for ∗172·162·17.

***172·15.** $\vdash : M \epsilon F_\Delta\text{‘}C\text{‘}P \,.\, Q \epsilon C\text{‘}P \,.\, (M\text{‘}Q)\, Qy \,.\supset.\, M\,(\Pi\text{‘}P)\,\{M \upharpoonright -\iota\text{‘}Q \,\dot{\cup}\, y \downarrow Q\}$

Dem.

$\vdash . *80{\cdot}41 . \supset \vdash : \mathrm{Hp} . \supset . M \upharpoonright -\iota\text{‘}Q \,\dot{\cup}\, y \downarrow Q \,\epsilon\, F_\Delta\text{‘}C\text{‘}P \qquad (1)$

$\vdash . *35{\cdot}101 . *55{\cdot}13 . \supset$

$\vdash :. z\,\{M \upharpoonright -\iota\text{‘}Q \,\dot{\cup}\, y \downarrow Q\}\,R \,.\equiv:\, R \neq Q \,.\, zMR \,.\vee.\, R = Q \,.\, z = y \qquad (2)$

$\vdash . (2) . *80{\cdot}3 . \supset$

$\vdash :. \mathrm{Hp} . \supset : R = Q . \supset . \{M \upharpoonright -\iota\text{‘}Q \,\dot{\cup}\, y \downarrow Q\}\text{‘}R = y :$

$R \epsilon C\text{‘}P \,.\, R \neq Q . \supset . \{M \upharpoonright -\iota\text{‘}Q \,\dot{\cup}\, y \downarrow Q\}\text{‘}R = M\text{‘}R :$

[Hp] $\supset : (M\text{‘}Q)\, Q\, \{M \upharpoonright -\iota\text{‘}Q \,\dot{\cup}\, y \downarrow Q\}\text{‘}Q :$

$R \epsilon C\text{‘}P \,.\, R \neq Q . \supset . \{M \upharpoonright -\iota\text{‘}Q \,\dot{\cup}\, y \downarrow Q\}\text{‘}R = M\text{‘}R :$

[*33·17] $\supset : (M\text{‘}Q)\, Q\, \{M \upharpoonright -\iota\text{‘}Q \,\dot{\cup}\, y \downarrow Q\}\text{‘}Q :$

$RPQ \,.\, R \neq Q . \supset_R . M\text{‘}R = \{M \upharpoonright -\iota\text{‘}Q \,\dot{\cup}\, y \downarrow Q\}\text{‘}R :$

[*172·1.(1)] $\supset : M\,(\Pi\text{‘}P)\,\{M \upharpoonright -\iota\text{‘}Q \,\dot{\cup}\, y \downarrow Q\} :. \supset \vdash . \mathrm{Prop}$

***172·151.** $\vdash : N \epsilon F_\Delta\text{‘}C\text{‘}P \,.\, Q \epsilon C\text{‘}P \,.\, yQ\,(N\text{‘}Q) \,.\supset.$

$\{N \upharpoonright -\iota\text{‘}Q \,\dot{\cup}\, y \downarrow Q\}\,(\Pi\text{‘}P)\,N$ [Proof as in *172·15]

***172·16.** $\vdash : M \epsilon F_\Delta\text{‘}C\text{‘}P \,.\, \dot{\exists} !\, M \dot{-} B \,.\supset.\, M \epsilon \text{ᗡ}\text{‘}\Pi\text{‘}P$

Dem.

$\vdash . *72{\cdot}93 . *80{\cdot}14 . \supset \vdash :: \mathrm{Hp} . \supset :. M \Subset B \,.\equiv:\, Q \epsilon C\text{‘}P . \supset_Q . (M\text{‘}Q)\, BQ :.$

[Transp] $\supset :. \dot{\exists} !\, M \dot{-} B \,.\equiv:\, (\exists Q) . Q \epsilon C\text{‘}P . \sim\{(M\text{‘}Q) BQ\} :$

[*93·1.*80·3] $\supset : (\exists Q) . Q \epsilon C\text{‘}P . M\text{‘}Q \epsilon \text{ᗡ}\text{‘}Q :$

[*33·131] $\supset : (\exists Q, y) . Q \epsilon C\text{‘}P . yQ\,(M\text{‘}Q) :$

[*172·151] $\supset : M \epsilon \text{ᗡ}\text{‘}\Pi\text{‘}P :: \supset \vdash . \mathrm{Prop}$

***172·161.** $\vdash : M \epsilon F_\Delta\text{‘}C\text{‘}P \,.\, \dot{\exists} !\, M \dot{-} B \,|\, \mathrm{Cnv} \,.\supset.\, M \epsilon \mathrm{D}\text{‘}\Pi\text{‘}P$

Dem.

$\vdash . *72{\cdot}93 . *80{\cdot}14 . \supset \vdash :: \mathrm{Hp} . \supset :.$

$M \Subset B \,|\, \mathrm{Cnv} \,.\equiv:\, Q \epsilon C\text{‘}P . \supset_Q . (M\text{‘}Q)\,(B \,|\, \mathrm{Cnv})\, Q :$

[*71·7] $\equiv : Q \epsilon C\text{‘}P . \supset_Q . (M\text{‘}Q)\, B\breve{Q} :.$

[Transp] $\supset :. \dot{\exists} !\, M \dot{-} B \,|\, \mathrm{Cnv} \,.\equiv:\, (\exists Q) . Q \epsilon C\text{‘}P . \sim\{(M\text{‘}Q)\, B\breve{Q}\} :$

[*93·1.*80·3] $\supset : (\exists Q) . Q \epsilon C\text{‘}P . M\text{‘}Q \epsilon \mathrm{D}\text{‘}Q :$

[*33·13] $\supset : (\exists Q, y) . Q \epsilon C\text{‘}P . (M\text{‘}Q)\, Qy :$

[*172·15] $\supset : M \epsilon \mathrm{D}\text{‘}\Pi\text{‘}P :: \supset \vdash . \mathrm{Prop}$

The following proposition is important. It shows that, if $C\text{‘}P$ consists of series, if any member of $C\text{‘}P$ has no first term, $\Pi\text{‘}P$ has no first term, but if every member of $C\text{‘}P$ has a first term, the selection of all these first terms is the first term of $\Pi\text{‘}P$.

***172·162.** $\vdash : \dot{\exists} ! P . \supset . \overrightarrow{B}`\Pi`P = B_\Delta`C`P . \overrightarrow{B}`\mathrm{Cnv}`\Pi`P = B_\Delta`\mathrm{Cnv}``C`P$

Dem.

$\vdash . *93 \cdot 103 . *172 \cdot 12 \cdot 16 . \mathrm{Transp} . \supset \vdash . \overrightarrow{B}`\Pi`P \subset F_\Delta`C`P \cap \mathrm{Rl}`B$ (1)

$\vdash . *72 \cdot 93 . \supset$

$\vdash :. M \in F_\Delta`C`P . M \mathrel{G} B . Q \in C`P . \supset : (M`Q) B Q :$

[*93·1] $\supset : (M`Q) \in \mathrm{D}`Q :$

[*33·13] $\supset : (\exists y) . (M`Q) Q y :$

[*172·15] $\supset : M \in \mathrm{D}`\Pi`P$ (2)

$\vdash . (2) . *10 \cdot 11 \cdot 23 \cdot 35 . \supset \vdash :. \dot{\exists} ! P . \supset : M \in F_\Delta`C`P \cap \mathrm{Rl}`B . \supset . M \in \mathrm{D}`\Pi`P$ (3)

$\vdash . *172 \cdot 11 .$ $\supset \vdash : N \in \mathrm{Cl}`\Pi`P . \supset . (\exists Q, M) . Q \in C`P . (M`Q) Q (N`Q) .$

[*93·1] $\supset . (\exists Q) . Q \in C`P . \sim \{(N`Q) B Q\} .$

[*72·93] $\supset . \sim (N \mathrel{G} B) :$

$\left[\mathrm{Transp} . \frac{M}{N}\right]$ $\supset \vdash : M \mathrel{G} B . \supset . M \sim \in \mathrm{Cl}`\Pi`P$ (4)

$\vdash . (1) . (3) . (4) .$ $\supset \vdash : \mathrm{Hp} . \supset . \overrightarrow{B}`\Pi`P = F_\Delta`C`P \cap \mathrm{Rl}`B$

[*80·17] $= B_\Delta`C`P$ (5)

Similarly $\vdash : \mathrm{Hp} . \supset . \overrightarrow{B}`\mathrm{Cnv}`\Pi`P = B_\Delta`\mathrm{Cnv}``C`P$ (6)

$\vdash . (5) . (6) . \supset \vdash . \mathrm{Prop}$

The following proposition is much used.

***172·17.** $\vdash : \dot{\exists} ! P . \supset . C`\Pi`P = F_\Delta`C`P$

Dem.

$\vdash . *172 \cdot 16 \cdot 162 . \supset$

$\vdash :. \mathrm{Hp} . M \in F_\Delta`C`P . \supset : \dot{\exists} ! M \dot{-} B . \supset . M \in \mathrm{Cl}`\Pi`P : M \mathrel{G} B . \supset . M \in \overrightarrow{B}`\Pi`P :$

[*93·11.*25·55] $\supset : M \in C`\Pi`P$ (1)

$\vdash . (1) . *172 \cdot 12 . \supset \vdash . \mathrm{Prop}$

***172·171.** $\vdash : \dot{\exists} ! P . \supset . \mathrm{D}`\Pi`P = F_\Delta`C`P - B_\Delta`\mathrm{Cnv}``C`P .$
$\mathrm{Cl}`\Pi`P = F_\Delta`C`P - B_\Delta`C`P$ [*172·162·17]

***172·18.** $\vdash :. \dot{\exists} ! P . \supset : \dot{\exists} ! \Pi`P . \equiv . \exists ! F_\Delta`C`P$ [*172·17]

***172·181.** $\vdash :. \text{Mult ax} . \supset : \dot{\Lambda} \sim \in C`P . \dot{\exists} ! P . \equiv . \dot{\exists} ! \Pi`P$

Dem.

$\vdash . *88 \cdot 361 . *172 \cdot 18 . \supset \vdash :: \mathrm{Hp} . \supset :. \dot{\exists} ! P . \supset : \dot{\exists} ! \Pi`P . \equiv . C`P \subset \mathrm{Cl}`F .$

[*33·41·5] $\equiv . C`P \subset \hat{Q}(\exists ! C`Q) .$

[*33·241] $\equiv . \dot{\Lambda} \sim \in C`P$ (1)

$\vdash . *172 \cdot 13 .$ $\supset \vdash : \dot{\exists} ! \Pi`P . \supset . \dot{\exists} ! P$ (2)

$\vdash . (1) . (2) . \supset \vdash . \mathrm{Prop}$

$*172\cdot182.$ $\vdash :: \text{Mult ax} \,.\, \supset :. \dot{\Lambda} \in C`P \,.\, \vee \,.\, P = \dot{\Lambda} : \equiv . \Pi`P = \dot{\Lambda}$

[$*172\cdot181$. Transp]

$*172\cdot19.$ $\vdash : \dot{\exists} ! \Pi`P \,.\, \supset . \dot{s}`C`\Pi`P = F \upharpoonright C`P$ [$*172\cdot17 \,.\, *80\cdot42$]

Note that we cannot proceed to $\Sigma`\Pi`P$, because $F ; \Pi`P$ is meaningless, owing to the fact that the field of $\Pi`P$ consists of non-homogeneous relations.

$*172\cdot191.$ $\vdash . \dot{s}`C`\Pi`P \subset F \upharpoonright C`P$

Dem.

$\vdash . *172\cdot19 \,.\, *23\cdot42 \,.\, \supset \vdash : \dot{\exists} ! \Pi`P \,.\, \supset . \dot{s}`C`\Pi`P \subset F \upharpoonright C`P$ (1)

$\vdash . *41\cdot21 \,.\, \supset \vdash : \Pi`P = \dot{\Lambda} \,.\, \supset . \dot{s}`C`\Pi`P = \dot{\Lambda} \,.$

[$*25\cdot12$] $\supset . \dot{s}`C`\Pi`P \subset F \upharpoonright C`P$ (2)

$\vdash . (1) . (2) . \supset \vdash . \text{Prop}$

$*172\cdot192.$ $\vdash . \text{Cl}`(F \upharpoonright \beta) = \beta - \iota`\dot{\Lambda}$

Dem.

$\vdash . *35\cdot101 \,.\, \supset \vdash : Q \in \text{Cl}`(F \upharpoonright \beta) \,.\, \equiv . (\exists x) . xFQ \,.\, Q \in \beta \,.$

[$*33\cdot5$] $\equiv . \exists ! C`Q \,.\, Q \in \beta \,.$

[$*33\cdot24$] $\equiv . \dot{\exists} ! Q \,.\, Q \in \beta : \supset \vdash . \text{Prop}$

The following proposition is sometimes useful. (It is used in $*173\cdot22$. $*182\cdot2$. $*185\cdot21$.)

$*172\cdot2.$ $\vdash . \Pi`(P \downarrow P) = P \underset{;}{\downarrow} P$

Dem.

$\vdash . *172\cdot11 \,.\, *55\cdot15 \,.\, \supset$

$\vdash :: M \{\Pi`(P \downarrow P)\} N \,.\, \equiv :. M, N \in F_\Delta`\iota`P :.$

$(\exists Q) : Q \in \iota`P \,.\, (M`Q) \, Q \, (N`Q) : R = P \,.\, R \neq Q \,.\, \supset_R . M`R = N`R :.$

[$*13\cdot195\cdot191$] $\equiv :. M, N \in F_\Delta`\iota`P :. (M`P) \, P \, (N`P) :.$

[$*85\cdot51 . *33\cdot5$] $\equiv :. M, N \in \downarrow P``C`P \,.\, (M`P) \, P \, (N`P) :.$

[$*38\cdot131$] $\equiv :. (\exists x, y) . x, y \in C`P \,.\, M = x \downarrow P \,.\, N = y \downarrow P \,.\, (M`P) \, P \, (N`P) :.$

[$*55\cdot13$] $\equiv :. (\exists x, y) . x, y \in C`P \,.\, M = x \downarrow P \,.\, N = y \downarrow P \,.\, xPy :.$

[$*150\cdot11$] $\equiv :. M (\downarrow P ; P) N :.$

[$*150\cdot6$] $\equiv :. M (P \underset{;}{\downarrow} P) N :: \supset \vdash . \text{Prop}$

The following propositions are concerned with the nature of the connection between $\Pi`(P \downarrow Q)$ and $P \times Q$. The connection is such as might be desired, except when $P = Q$, in which case, as shown above, $\Pi`(P \downarrow P)$ is like P, and is therefore not like $P \times P$.

∗172·21. $\vdash : P \neq Q . \supset . P \times Q = \dagger(Q \downarrow P) ; \Pi`(P \downarrow Q)$

Dem.

$\vdash . *172·11 . *55·15 . \supset$

$\vdash :: . M \{\Pi`(P \downarrow Q)\} N . \equiv :: M, N \in F_\Delta`(\iota`P \cup \iota`Q) :.$

$(\exists R) : R \in \iota`P \cup \iota`Q . (M`R) R (N`R) : S (P \downarrow Q) R . S \neq R . \supset_S . M`S = N`S ::$

$[*51·235] \equiv :: M, N \in F_\Delta`(\iota`P \cup \iota`Q) ::$

$(M`P) P (N`P) : S (P \downarrow Q) P . S \neq P . \supset_S . M`S = N`S :. \vee :.$

$(M`Q) Q (N`Q) : S (P \downarrow Q) Q . S \neq Q . \supset_S . M`S = N`S$ (1)

$\vdash . (1) . *55·13 . \supset \vdash :: \mathrm{Hp} . \supset :.$

$M \{\Pi`(P \downarrow Q)\} N . \equiv : M, N \in F_\Delta`(\iota`P \cup \iota`Q) :$

$(M`P) P (N`P) . \vee . (M`Q) Q (N`Q) . M`P = N`P :$

$[*80·9·91] \equiv : (\exists x, x', y, y') : x, x' \in C`P . y, y' \in C`Q .$

$M = x \downarrow P \cup y \downarrow Q . N = x' \downarrow P \cup y' \downarrow Q :$

$xPx' . \vee . x = x' . yQy'$ (2)

$\vdash . *150·72 . \supset \vdash : M = x \downarrow P \cup y \downarrow Q . \supset . M ; (Q \downarrow P) = y \downarrow x .$

$[*150·1]$ $\supset . \dagger(Q \downarrow P)`M = y \downarrow x$ (3)

$\vdash . *150·4 . \supset \vdash : R \{\dagger(Q \downarrow P) ; \Pi`(P \downarrow Q)\} S . \equiv .$

$(\exists M, N) . M \{\Pi`(P \downarrow Q)\} N . R = \dagger(Q \downarrow P)`M . S = \dagger(Q \downarrow P)`N$ (4)

$\vdash . (2) . (3) . (4) . \supset \vdash :: \mathrm{Hp} . \supset :.$

$R \{\dagger(Q \downarrow P) ; \Pi`(P \downarrow Q)\} S . \equiv : (\exists M, N, x, x', y, y') :$

$x, x' \in C`P . y, y' \in C`Q . M = x \downarrow P \cup y \downarrow Q . N = x' \downarrow P \cup y' \downarrow Q :$

$R = y \downarrow x . S = y' \downarrow x' : xPx' . \vee . x = x' . yQy' :$

$[*13·19] \equiv : (\exists x, x', y, y') : x, x' \in C`P . y, y' \in C`Q . R = y \downarrow x . S = y' \downarrow x' :$

$xPx' . \vee . x = x' . yQy' :$

$[*166·111] \equiv : R (P \times Q) S :: \supset \vdash . \mathrm{Prop}$

∗172·22. $\vdash : P \neq Q . \supset . \{\dagger(Q \downarrow P)\} \upharpoonright F_\Delta`(\iota`P \cup \iota`Q) \in (P \times Q) \overline{\mathrm{smor}}\, \Pi`(P \downarrow Q)$

Dem.

$\vdash . *80·9 . *150·71 . \supset \vdash :. \mathrm{Hp} . \supset :$

$M \in F_\Delta`(\iota`P \cup \iota`Q) . \supset . M ; (Q \downarrow P) = (M`Q) \downarrow (M`P)$ (1)

$\vdash . (1) . *150·1 . \supset \vdash :. \mathrm{Hp} . \supset :$

$M, N \in F_\Delta`(\iota`P \cup \iota`Q) . \dagger(Q \downarrow P)`M = \dagger(Q \downarrow P)`N . \supset .$

$(M`Q) \downarrow (M`P) = (N`Q) \downarrow (N`P) .$

$[*55·202]$ $\supset . M`P = N`P . M`Q = N`Q .$

$[*80·91]$ $\supset . M = N$ (2)

$\vdash . (2) . *151·241 . *172·21·17 . \supset \vdash . \mathrm{Prop}$

∗172·23. $\vdash : P \neq Q . \supset . \Pi`(P \downarrow Q) \,\mathrm{smor}\, P \times Q$ $[*172·22]$

The following propositions are lemmas for ∗172·32.

***172·3.** $\vdash :. \dot{\exists} ! P . Z \sim \epsilon C`P . \supset : M \{\Pi`(P \nrightarrow Z)\} N . \equiv .$

$(\exists S, T, u, v) . (u \downarrow S)(\Pi`P \times Z)(v \downarrow T) . M = S \cup u \downarrow Z . N = T \cup v \downarrow Z$

Dem.

$\vdash . *80·66·44 . *161·14 . \supset \vdash :. \text{Hp} . \supset : M \epsilon F_\Delta`C`(P \nrightarrow Z) . \equiv .$

$(\exists S, u) . S \epsilon F_\Delta`C`P . u \epsilon C`Z . M = S \cup u \downarrow Z \quad (1)$

$\vdash . *55·13 . *33·14 . *4·73 . \supset \vdash :: M = S \cup u \downarrow Z . \supset :.$

$xMQ . \equiv : xSQ . Q \epsilon \text{Cl}`S . \text{v} . x = u . Q = Z \quad (2)$

$\vdash . (2) . *80·14 . \supset \vdash :: \text{Hp} . \text{Hp}(2) . S \epsilon F_\Delta`C`P . u \epsilon C`Z . \supset :.$

$xMQ . \equiv : xSQ . Q \epsilon C`P . \text{v} . x = u . Q = Z :.$

[*24·37] $\supset :. Q \epsilon C`P . \supset : xMQ . \equiv . xSQ :. Q = Z . \supset : xMQ . \equiv . x = u :.$

[*30·341.*80·3.*30·3] $\supset :. Q \epsilon C`P . \supset . M`Q = S`Q : Q = Z . \supset . M`Q = u \quad (3)$

$\vdash . (1) . *172·11 . *161·14 . *172·17 . \supset$

$\vdash ::. \text{Hp} . \supset :: M \{\Pi`(P \nrightarrow Z)\} N . \equiv :.$

$(\exists S, T, u, v) :. S, T \epsilon F_\Delta`C`P . u, v \epsilon C`Z . M = S \cup u \downarrow Z . N = T \cup v \downarrow Z :.$

$(\exists Q) : Q \epsilon C`P \cup \iota`Z . (M`Q) Q (N`Q) : R (P \nrightarrow Z) Q . R \neq Q . \supset_R . M`R = N`R :.$

[*51·239.(3).*161·11]

$\equiv :. (\exists S, T, u, v) :. S, T \epsilon F_\Delta`C`P . u, v \epsilon C`Z . M = S \cup u \downarrow Z . N = T \cup v \downarrow Z :.$

$(\exists Q) : Q \epsilon C`P . (S`Q) Q (T`Q) : RPQ . R \neq Q . \supset_R . S`R = T`R : \text{v} :$

$uZv : R \epsilon C`P . \supset_R . S`R = T`R :.$

[*172·11·17.*71·35.*80·14]

$\equiv :. (\exists S, T, u, v) :. S, T \epsilon C`\Pi`P . u, v \epsilon C`Z . M = S \cup u \downarrow Z . N = T \cup v \downarrow Z :$

$S (\Pi`P) T . \text{v} . S = T . uZv :.$

[*166·112] $\equiv :. (\exists S, T, u, v) : (u \downarrow S)(\Pi`P \times Z)(v \downarrow T) .$

$M = S \cup u \downarrow Z . N = T \cup v \downarrow Z ::. \supset \vdash . \text{Prop}$

***172·31.** $\vdash : \dot{\exists} ! P . Z \sim \epsilon C`P .$

$W = \hat{M} \hat{R} \{(\exists S, u) . S \epsilon F_\Delta`C`P . u \epsilon C`Z . R = u \downarrow S . M = S \cup u \downarrow Z\} . \supset .$

$W \epsilon \Pi`(P \nrightarrow Z) \,\overline{\text{smor}}\, (\Pi`P \times Z)$

Dem.

$\vdash . *172·3 . \supset \vdash : \text{Hp} . \supset . \Pi`(P \nrightarrow Z) = W ; (\Pi`P \times Z) \quad (1)$

$\vdash . *21·33 . \supset \vdash :. \text{Hp} . \supset : MWR . M'WR . \equiv .$

$(\exists S, S', u, u') . S, S' \epsilon F_\Delta`C`P . u, u' \epsilon C`Z . R = u \downarrow S = u' \downarrow S' .$

$M = S \cup u \downarrow Z . M' = S' \cup u' \downarrow Z .$

[*55·202] $\supset . (\exists S, S', u, u') . S, S' \epsilon F_\Delta`C`P . u, u' \epsilon C`Z . S = S' . u = u' .$

$M = S \cup u \downarrow Z . M' = S' \cup u' \downarrow Z .$

[*13·22·172] $\supset . M = M' \quad (2)$

$\vdash . *21\cdot33 . \supset \vdash :. \text{Hp} . \supset : MWR . MWR' . \equiv .$

$(\exists S, S', u, u') . S, S' \in F_\Delta{}^{\backprime}C^{\backprime}P . u, u' \in C^{\backprime}Z . R = u \downarrow S . R' = u' \downarrow S' .$

$M = S \cup u \downarrow Z . M = S' \cup u' \downarrow Z .$

$[*80\cdot45\cdot661] \supset . (\exists S, S', u, u') . R = u \downarrow S . R' = u' \downarrow S' .$

$S = M \upharpoonright C^{\backprime}P . u \downarrow Z = M \upharpoonright \iota^{\backprime}Z . S' = M \upharpoonright C^{\backprime}P . u' \downarrow Z = M \upharpoonright C^{\backprime}P .$

$[*13\cdot172] \quad \supset . (\exists S, S', u, u') . R = u \downarrow S . R' = u' \downarrow S' . S = S' . u \downarrow Z = u' \downarrow Z .$

$[*55\cdot202] \quad \supset . R = R' \qquad (3)$

$\vdash . (2) . (3) . \supset \vdash : \text{Hp} . \supset . W \in 1 \rightarrow 1 \qquad (4)$

$\vdash . *166\cdot12 . *113\cdot101 . *172\cdot17 . \supset \vdash : \text{Hp} . \supset . \text{Ꮧ}^{\backprime}W = C^{\backprime}(\Pi^{\backprime}P \times Z) \qquad (5)$

$\vdash . (1) . (4) . (5) . *151\cdot11 . \supset \vdash . \text{Prop}$

$*172\cdot32$. $\vdash : Z \sim\in C^{\backprime}P . \supset . \Pi^{\backprime}(P \nrightarrow Z) \text{ smor } \Pi^{\backprime}P \times Z$

Dem.

$\vdash . *172\cdot31 . \quad \supset \vdash : \text{Hp} . \dot{\exists} ! P . \supset . \Pi^{\backprime}(P \nrightarrow Z) \text{ smor } \Pi^{\backprime}P \times Z \qquad (1)$

$\vdash . *172\cdot13 . *161\cdot2 . \quad \supset \vdash : \sim\dot{\exists} ! P . \supset . \Pi^{\backprime}(P \nrightarrow Z) = \dot{\Lambda} \qquad (2)$

$\vdash . *172\cdot13 . *166\cdot13 . \quad \supset \vdash : \sim\dot{\exists} ! P . \supset . \Pi^{\backprime}P \times Z = \dot{\Lambda} \qquad (3)$

$\vdash . (2) . (3) . *153\cdot101 . \supset \vdash : \sim\dot{\exists} ! P . \supset . \Pi^{\backprime}(P \nrightarrow Z) \text{ smor } \Pi^{\backprime}P \times Z \qquad (4)$

$\vdash . (1) . (4) . \supset \vdash . \text{Prop}$

$*172\cdot321$. $\vdash : Z \sim\in C^{\backprime}P . \supset . \Pi^{\backprime}(Z \nleftarrow C^{\backprime}P) \text{ smor } Z \times \Pi^{\backprime}P$

[Proof by similar stages to those in proof of *172·32]

The following proposition is a lemma for *172·34, which is required in proving *172·35 (as well as *176·4).

$*172\cdot33$. $\vdash :: \dot{\exists} ! P . \dot{\exists} ! Q . C^{\backprime}P \cap C^{\backprime}Q = \Lambda . \supset :.$

$M \{\Pi^{\backprime}(P \ddagger Q)\} N . \equiv : (\exists S, T, S', T') : S, S' \in F_\Delta{}^{\backprime}C^{\backprime}P . T, T' \in F_\Delta{}^{\backprime}C^{\backprime}Q :$

$S(\Pi^{\backprime}P) S' . \vee . S = S' . T (\Pi^{\backprime}Q) T' : M = S \cup T . M' = S' \cup T'$

Dem.

$\vdash . *80\cdot66 . \supset \vdash :. \text{Hp} . \supset : M \in F_\Delta{}^{\backprime}(C^{\backprime}P \cup C^{\backprime}Q) . \equiv .$

$(\exists S, T) . S \in F_\Delta{}^{\backprime}C^{\backprime}P . T \in F_\Delta{}^{\backprime}C^{\backprime}Q . M = S \cup T \qquad (1)$

$\vdash . *80\cdot661 . *35\cdot7 . \supset \vdash :. \text{Hp} . S \in F_\Delta{}^{\backprime}C^{\backprime}P . T \in F_\Delta{}^{\backprime}C^{\backprime}Q . M = S \cup T . \supset :$

$R \in C^{\backprime}P . \supset . M^{\backprime}R = S^{\backprime}R : R \in C^{\backprime}Q . \supset . M^{\backprime}R = T^{\backprime}R \qquad (2)$

$\vdash . (1) . *172\cdot11\cdot17 . *160\cdot14 . \supset \vdash ::. \text{Hp} . \supset :: M \{\Pi^{\backprime}(P \ddagger Q)\} N . \equiv :.$

$(\exists S, T, S', T') :. S, S' \in F_\Delta{}^{\backprime}C^{\backprime}P . T, T' \in F_\Delta{}^{\backprime}C^{\backprime}Q . M = S \cup T . N = S' \cup T' :.$

$(\exists R) : R \in C^{\backprime}P \cup C^{\backprime}Q . (M^{\backprime}R) R (N^{\backprime}R) : R' (P \ddagger Q) R . R \neq R' . \supset_{R'} . M^{\backprime}R' = N^{\backprime}R' :.$

$[(2).*160\cdot11] \equiv :. (\exists S, T, S', T') :. S, S' \in F_\Delta{}^{\backprime}C^{\backprime}P . T, T' \in F_\Delta{}^{\backprime}C^{\backprime}Q .$

$M = S \cup T . N = S' \cup T' :.$

$(\exists R) : R \in C^{\backprime}P . (S^{\backprime}R) R (S'^{\backprime}R) : R'PR . R' \neq R . \supset_{R'} . S^{\backprime}R' = S'^{\backprime}R' : \vee :$

$(\exists R) : R \epsilon C`Q . (T`R) R (T'`R) : R'QR . R' \neq R . \supset_{R'} . T`R' = T'`R' :$
$R' \epsilon C`P . \supset_{R'} . S`R' = S'`R' :.$

$[*10\cdot35] \equiv :. (\exists S, T, S', T') :. S, S' \epsilon F_\Delta`C`P . T, T' \epsilon F_\Delta`C`Q . M = S \cup T . N = S' \cup T' :.$
$(\exists R) : R \epsilon C`P . (S`R) R (S'`R) : R'PR . R' \neq R . \supset_{R'} . S`R' = S'`R' : \vee :$
$R' \epsilon C`P . \supset_{R'} . S`R' = S'`R' :$
$(\exists R) : R \epsilon C`Q . (T`R) R (T'`R) : R'QR . R' \neq R . \supset_{R'} . T`R' = T'`R' :.$

$[*172\cdot11 . *71\cdot35 . *80\cdot14]$
$\equiv :. (\exists S, T, S', T') :. S, S' \epsilon F_\Delta`C`P . T, T' \epsilon F_\Delta`C`Q . M = S \cup T, M' = S' \cup T' :.$
$S (\Pi`P) S' . \vee . S = S' . T (\Pi`Q) T' ::. \supset \vdash .$ Prop

***172·34.** $\vdash : \dot{\exists} ! P . \dot{\exists} ! Q . C`P \cap C`Q = \Lambda . \supset .$
$(\dot{s} | C) \epsilon \{\Pi`(P \ddagger Q)\} \overline{\text{smor}} \{\Pi`P \times \Pi`Q\}$

Dem.

$\vdash . *172\cdot33 . *55\cdot15 . *53\cdot13 . \supset$
$\vdash :: \text{Hp} . \supset :. M \{\Pi`(P \ddagger Q)\} N . \equiv : (\exists S, T, S', T', R, R') :$
$S, S' \epsilon F_\Delta`C`P . T, T' \epsilon F_\Delta`C`Q . R = T \downarrow S . R' = T' \downarrow S' . M = \dot{s}`C`R . N = \dot{s}`C`R' :$
$S (\Pi`P) S' . \vee . S = S' . T (\Pi`Q) T' :$
$[*166\cdot11 . *172\cdot17] \equiv : (\exists R, R') . R \{\Pi`P \times \Pi`Q\} R' . M = \dot{s}`C`R . N = \dot{s}`C`R' :$
$[*150\cdot4] \qquad \equiv : M \{\dot{s};C;(\Pi`P \times \Pi`Q)\} N \qquad (1)$
$\vdash . *113\cdot153 . *172\cdot19 . *166\cdot12 . \supset \vdash : \text{Hp} . \supset . (\dot{s} | C) \upharpoonright C`(\Pi`P \times \Pi`Q) \epsilon 1 \to 1 \quad (2)$
$\vdash . (1) . (2) . *151\cdot231 . \supset \vdash .$ Prop

***172·35.** $\vdash : \dot{\exists} ! P . \dot{\exists} ! Q . C`P \cap C`Q = \Lambda . \supset . \Pi`(P \ddagger Q) \text{ smor } \Pi`P \times \Pi`Q$
$[*172\cdot34]$

The above proposition is important, being a form of the associative law.

The following propositions are extensions of *172·23. It is obvious that they may be extended to any finite number of factors.

***172·36.** $\vdash : X \neq Y . X \neq Z . Y \neq Z . \supset . \Pi`\{(X \downarrow Y) \rightarrow Z\} \text{ smor } X \times Y \times Z$

Dem.

$\vdash . *172\cdot32 . \supset \vdash : \text{Hp} . \supset . \Pi`\{(X \downarrow Y) \rightarrow Z\} \text{ smor } \Pi`(X \downarrow Y) \times Z \quad (1)$
$\vdash . *172\cdot23 . *166\cdot23 . \supset \vdash : \text{Hp} . \supset . \Pi`(X \downarrow Y) \times Z \text{ smor } X \times Y \times Z \quad (2)$
$\vdash . (1) . (2) . \supset \vdash .$ Prop

***172·361.** $\vdash : X \neq Y . X \neq Z . Y \neq Z . \supset . \Pi`\{X \leftarrow (Y \downarrow Z)\} \text{ smor } X \times Y \times Z$
[Proof as in *172·36]

***172·37.** $\vdash : X \neq Y . X \neq Z . X \neq W . Y \neq Z . Y \neq W . Z \neq W . \supset .$
$\Pi`\{(X \downarrow Y) \ddagger (Z \downarrow W)\} \text{ smor } X \times Y \times Z \times W$

Dem.

$\vdash . *172\cdot35 . \supset$
$\vdash : \text{Hp} . \supset . \Pi`\{(X \downarrow Y) \ddagger (Z \downarrow W)\} \text{ smor } \Pi`(X \downarrow Y) \times \Pi`(Z \downarrow W) \quad (1)$
$\vdash . *172\cdot23 . *166\cdot23 . \supset$
$\vdash : \text{Hp} . \supset . \Pi`(X \downarrow Y) \times \Pi`(Z \downarrow W) \text{ smor } (X \times Y) \times (Z \times W) \quad (2)$
$\vdash . (1) . (2) . *166\cdot42 . \supset \vdash .$ Prop

The following propositions are concerned with the construction of a correlator of $\Pi`P$ with $\Pi`Q$ when we are given a double correlator of P with Q. If the double correlator is T or $T \upharpoonright C`\Sigma`Q$, the correlator of $\Pi`P$ with $\Pi`Q$ is

$$\{T \| \text{Cnv}`T\dagger\} \upharpoonright C`\Pi`Q.$$

***172·4.** $\vdash : T \epsilon P \overline{\text{smor}}\, \overline{\text{smor}}\, Q . \supset . \{T \| \text{Cnv}`T\dagger\} \upharpoonright C`\Pi`Q \epsilon 1 \to 1$

Dem.

$\vdash . *164{\cdot}15 . \supset$

$\vdash : \text{Hp} . \supset . T \upharpoonright C`\Sigma`Q, T\dagger \upharpoonright C`Q \epsilon 1 \to 1 . C`\Sigma`Q = \text{Ꭰ}`T . C`Q \subset \text{Ꭰ}`T\dagger$ (1)

$\vdash . *41{\cdot}43 . \supset \vdash . s`\text{D}``C`\Pi`Q = \text{D}`\dot{s}`C`\Pi`Q .$

$[*172{\cdot}191]$ $\supset \vdash . s`\text{D}``C`\Pi`Q \subset \text{D}`(F \upharpoonright C`Q)$

$[*37{\cdot}401 . *162{\cdot}23]$ $\subset C`\Sigma`Q$ (2)

$\vdash . *41{\cdot}44 . \supset \vdash . s`\text{Ꭰ}``C`\Pi`Q = \text{Ꭰ}`\dot{s}`C`\Pi`Q .$

$[*172{\cdot}191]$ $\supset \vdash . s`\text{Ꭰ}``C`\Pi`Q \subset \text{Ꭰ}`(F \upharpoonright C`Q)$

$[*35{\cdot}64]$ $\subset C`Q$ (3)

$\vdash . (1) . (2) . (3) . *74{\cdot}773 \dfrac{T, T\dagger, C`\Sigma`Q, C`Q, C`\Pi`Q}{Q, R, \alpha, \beta, \lambda} . \supset \vdash . \text{Prop}$

***172·401.** $\vdash : T \epsilon P \overline{\text{smor}}\, \overline{\text{smor}}\, Q . N \epsilon F_\Delta`C`Q . S \epsilon C`Q . \supset .$

$\{(T \| \text{Cnv}`T\dagger)`N\}`T;S = T`N`S$

Dem.

$\vdash . *43{\cdot}112 . *150{\cdot}1 . \supset \vdash . \{(T \| \text{Cnv}`T\dagger)`N\}`T;S = (T | N | \text{Cnv}`T\dagger)`T\dagger`S$ (1)

$\vdash . (1) . *35{\cdot}7{\cdot}48 . *80{\cdot}14 . \supset$

$\vdash : \text{Hp} . \supset . \{(T \| \text{Cnv}`T\dagger)`N\}`T;S = \{T | N | \text{Cnv}`(T\dagger \upharpoonright C`Q)\}`(T\dagger \upharpoonright C`Q)`S$

$[*34{\cdot}41 . *72{\cdot}601 . *164{\cdot}13]$ $= (T | N)`S$

$[*34{\cdot}41]$ $= T`N`S : \supset \vdash . \text{Prop}$

***172·402.** $\vdash : T \epsilon P \overline{\text{smor}}\, \overline{\text{smor}}\, Q . N, N' \epsilon F_\Delta`C`Q . S \epsilon C`Q . M = (T \| \text{Cnv}`T\dagger)`N .$

$M' = (T \| \text{Cnv}`T\dagger)`N' . R = T;S . \supset :$

$N`S = N'`S . \equiv . M`R = M'`R : (N`S) \, S \, (N'`S) . \equiv . (M`R) \, R \, (M'`R)$

Dem.

$\vdash . *172{\cdot}401 .$ $\supset \vdash : \text{Hp} . \supset . M`R = T`N`S . M'`R = T`N'`S$ (1)

$\vdash . *162{\cdot}22 . *40{\cdot}13 . \supset \vdash : \text{Hp} . \supset . C`S \subset C`\Sigma`Q .$

$[*164{\cdot}1]$ $\supset . C`S \subset \text{Ꭰ}`T$ (2)

$\vdash . *80{\cdot}31 . *33{\cdot}5 .$ $\supset \vdash :. \text{Hp} . \supset : N`S, N'`S \epsilon C`S :$

$[(2)]$ $\supset : N`S, N'`S \epsilon \text{Ꭰ}`T :$

$[*71{\cdot}56]$ $\supset : N`S = N'`S . \equiv . T`N`S = T`N'`S .$

$[(1)]$ $\equiv . M`R = M'`R$ (3)

$\vdash . (1) . \supset \vdash :. \text{Hp} . \supset : (M`R) \, R \, (M'`R) . \equiv . (T`N`S) (T;S) (T`N'`S) .$

$[*150{\cdot}41]$ $\equiv . (N`S) (\breve{T};T;S) (N'`S) .$

$[*151{\cdot}252 . (2)]$ $\equiv . (N`S) \, S \, (N'`S)$ (4)

$\vdash . (3) . (4) . \supset \vdash . \text{Prop}$

***172·403.** $\vdash : T \epsilon P \,\overline{\text{smor}}\,\overline{\text{smor}}\, Q . \supset . (T \parallel \text{Cnv}\text{‘}T\dagger)\text{“}F_\Delta\text{‘}C\text{‘}Q \subset F_\Delta\text{‘}C\text{‘}P$

Dem.

$\vdash . *80·14 . *35·48 . \supset$

$\vdash :. \text{Hp} . \supset : N \epsilon F_\Delta\text{‘}C\text{‘}Q . \supset . T | N | \text{Cnv}\text{‘}T\dagger = T | N | \text{Cnv}\text{‘}(T\dagger \upharpoonright C\text{‘}Q) .$

[*80·14.*164·13] $\supset . (T | N | \text{Cnv}\text{‘}T\dagger) \epsilon 1 \rightarrow \text{Cls}$ (1)

$\vdash . *37·32 . \supset \vdash :. \text{Hp} . \supset : N \epsilon F_\Delta\text{‘}C\text{‘}Q . \supset . \text{ᗡ}\text{‘}(T | N | \text{Cnv}\text{‘}T\dagger) = T\dagger\text{“}\breve{N}\text{“}\text{ᗡ}\text{‘}T$

[*37·271.*164·1.*80·33.*162·23] $= T\dagger\text{“}\text{ᗡ}\text{‘}N$

[*80·14] $= T\dagger\text{“}C\text{‘}Q$

[*164·1.*150·22] $= C\text{‘}P$ (2)

$\vdash . *80·14 . \supset \vdash :. \text{Hp} . \supset : N \epsilon F_\Delta\text{‘}C\text{‘}Q . x (T | N | \text{Cnv}\text{‘}T\dagger) R . \supset .$

$(\exists y, S) . xTy . yFS . R = T \,;\, S . S \epsilon C\text{‘}Q .$

[*33·51.*37·1] $\supset . (\exists S) . x \epsilon T\text{“}C\text{‘}S . R = T \,;\, S . S \epsilon C\text{‘}Q .$

[*150·22.*164·1] $\supset . x \epsilon C\text{‘}R :$

[*33·51] $\supset : N \epsilon F_\Delta\text{‘}C\text{‘}Q . \supset . T | N | \text{Cnv}\text{‘}T\dagger \mathrel{\text{Ⓔ}} F$ (3)

$\vdash . (1) . (2) . (3) . *80·14 . \supset$

$\vdash :. \text{Hp} . \supset : N \epsilon F_\Delta\text{‘}C\text{‘}Q . \supset . (T | N | \text{Cnv}\text{‘}T\dagger) \epsilon F_\Delta\text{‘}C\text{‘}P$ (4)

$\vdash . (4) . *43·112 . \supset \vdash . \text{Prop}$

***172·404.** $\vdash :. T \epsilon P \,\overline{\text{smor}}\,\overline{\text{smor}}\, Q . \supset : N \epsilon F_\Delta\text{‘}C\text{‘}Q . M = T | N | \text{Cnv}\text{‘}T\dagger . \equiv .$

$M \epsilon F_\Delta\text{‘}C\text{‘}P . N = \breve{T} | M | \text{Cnv}\text{‘}\breve{T}\dagger$

Dem.

$\vdash . *164·1 . *162·23 . *80·33 . \supset \vdash :. \text{Hp} . \supset : N \epsilon F_\Delta\text{‘}C\text{‘}Q . \supset . \text{D}\text{‘}N \subset \text{ᗡ}\text{‘}T .$

[*71·191.*50·63] $\supset . \breve{T} | T | N = N :$

[*34·28] $\supset : N \epsilon F_\Delta\text{‘}C\text{‘}Q . M = T | N | \text{Cnv}\text{‘}T\dagger . \supset . \breve{T} | M = N | \text{Cnv}\text{‘}T\dagger .$

[*34·27] $\supset . \breve{T} | M | \text{Cnv}\text{‘}\breve{T}\dagger = N | \text{Cnv}\text{‘}T\dagger | \text{Cnv}\text{‘}\breve{T}\dagger$ (1)

$\vdash . *80·14 . \quad \supset \vdash : N \epsilon F_\Delta\text{‘}C\text{‘}Q . S \epsilon \text{ᗡ}\text{‘}N . \supset . S \epsilon C\text{‘}Q .$

[*40·13] $\supset . C\text{‘}S \subset s\text{‘}C\text{“}C\text{‘}Q$ (2)

$\vdash . (2) . *164·1 . \supset \vdash :. \text{Hp} . \supset : N \epsilon F_\Delta\text{‘}C\text{‘}Q . S \epsilon \text{ᗡ}\text{‘}N . \supset . C\text{‘}S \subset \text{ᗡ}\text{‘}T$ (3)

$\vdash . (1) . *150·1 . \supset \vdash :. \text{Hp} . N \epsilon F_\Delta\text{‘}C\text{‘}Q . M = T | N | \text{Cnv}\text{‘}T\dagger . \supset :$

$y (\breve{T} | M | \text{Cnv}\text{‘}\breve{T}\dagger) Y . \equiv . (\exists S, R) . yNS . R = T \,;\, S . Y = \breve{T} \,;\, R .$

[(3).*151·25] $\equiv . (\exists S, R) . yNS . R = T \,;\, S . Y = S .$

[*13·19·195] $\equiv . yNS$ (4)

$\vdash . (4) . *172·403 . \supset \vdash :. \text{Hp} . \supset : N \epsilon F_\Delta\text{‘}C\text{‘}Q . M = T | N | \text{Cnv}\text{‘}T\dagger . \supset .$

$M \epsilon F_\Delta\text{‘}C\text{‘}P . N = \breve{T} | M | \text{Cnv}\text{‘}\breve{T}\dagger$ (5)

$\vdash . (5) \dfrac{\breve{T}, Q, P}{T, P, Q} . *164·21 . \supset \vdash :. \text{Hp} . \supset : M \epsilon F_\Delta\text{‘}C\text{‘}P . N = \breve{T} | M | \text{Cnv}\text{‘}\breve{T}\dagger . \supset .$

$N \epsilon F_\Delta\text{‘}C\text{‘}Q . M = T | N | \text{Cnv}\text{‘}T\dagger$ (6)

$\vdash . (5) . (6) . \supset \vdash . \text{Prop}$

***172·41.** $\vdash : T \in P \,\overline{\text{smor}}\,\overline{\text{smor}}\, Q . \supset . F_\Delta{}^{\boldsymbol{\cdot}}C^{\boldsymbol{\cdot}}P = (T \| \text{Cnv}^{\boldsymbol{\cdot}}T\dagger)^{\boldsymbol{\cdot}\boldsymbol{\cdot}}F_\Delta{}^{\boldsymbol{\cdot}}C^{\boldsymbol{\cdot}}Q$

Dem.

$\vdash . *172{\cdot}404 . *43{\cdot}112 . \supset \vdash :. \text{Hp} . \supset : M \in F_\Delta{}^{\boldsymbol{\cdot}}C^{\boldsymbol{\cdot}}P . \supset .$

$\breve{T} | M | \text{Cnv}^{\boldsymbol{\cdot}}\breve{T}\dagger \in F_\Delta{}^{\boldsymbol{\cdot}}C^{\boldsymbol{\cdot}}Q . M = (T \| \text{Cnv}^{\boldsymbol{\cdot}}T\dagger)^{\boldsymbol{\cdot}}(\breve{T} | M | \text{Cnv}^{\boldsymbol{\cdot}}\breve{T}\dagger) .$

$[*37{\cdot}6] \quad \supset . M \in (T \| \text{Cnv}^{\boldsymbol{\cdot}}T\dagger)^{\boldsymbol{\cdot}\boldsymbol{\cdot}}F_\Delta{}^{\boldsymbol{\cdot}}C^{\boldsymbol{\cdot}}Q \qquad (1)$

$\vdash . (1) . *172{\cdot}403 . \supset \vdash . \text{Prop}$

The following proposition is important, since it gives the required correlator of $\Pi^{\boldsymbol{\cdot}}P$ with $\Pi^{\boldsymbol{\cdot}}Q$.

***172·42.** $\vdash : T \in P \,\overline{\text{smor}}\,\overline{\text{smor}}\, Q . \supset . (T \| \text{Cnv}^{\boldsymbol{\cdot}}T\dagger) \upharpoonright C^{\boldsymbol{\cdot}}\Pi^{\boldsymbol{\cdot}}Q \in (\Pi^{\boldsymbol{\cdot}}P) \,\overline{\text{smor}}\, (\Pi^{\boldsymbol{\cdot}}Q)$

Dem.

$\vdash . *164{\cdot}1 . *150{\cdot}22 . \supset \vdash :. \text{Hp} . \supset : C^{\boldsymbol{\cdot}}P = T\dagger^{\boldsymbol{\cdot}\boldsymbol{\cdot}}C^{\boldsymbol{\cdot}}Q :$

$[*37{\cdot}6 . *150{\cdot}1] \quad \supset : R \in C^{\boldsymbol{\cdot}}P . \equiv . (\exists S) . S \in C^{\boldsymbol{\cdot}}Q . R = T ; S \qquad (1)$

$\vdash . *164{\cdot}1 . \supset \vdash :. \text{Hp} . \supset : R'PR . \equiv . (\exists S', Y) . R' = T ; S' . R = T ; Y . S'QY \qquad (2)$

$\vdash . *151{\cdot}31 . *164{\cdot}1 . \supset \vdash :: \text{Hp} . \supset :. S, Y \in C^{\boldsymbol{\cdot}}Q . R = T ; S . R = T ; Y . \supset . S = Y :. \qquad (3)$

$[*13{\cdot}13] \supset :. S \in C^{\boldsymbol{\cdot}}Q . R = T ; S . \supset : Y \in C^{\boldsymbol{\cdot}}Q . R = T ; Y . \equiv . S = Y \qquad (4)$

$\vdash . (2) . (4) . *33{\cdot}17 . \supset$

$\vdash :. \text{Hp} . S \in C^{\boldsymbol{\cdot}}Q . R = T ; S . \supset : R'PR . \equiv . (\exists S', Y) . R' = T ; S' . S = Y . S'QY .$

$[*13{\cdot}195] \quad \equiv . (\exists S') . R' = T ; S' . S'QS \qquad (5)$

$\vdash . *150{\cdot}4 . *172{\cdot}11 . \supset \vdash ::. \text{Hp} . \supset :: M \{(T \| \text{Cnv}^{\boldsymbol{\cdot}}T\dagger) ; \Pi^{\boldsymbol{\cdot}}Q\} M' . \equiv :.$

$(\exists N, N') :. M = (T \| \text{Cnv}^{\boldsymbol{\cdot}}T\dagger)^{\boldsymbol{\cdot}}N . M' = (T \| \text{Cnv}^{\boldsymbol{\cdot}}T\dagger)^{\boldsymbol{\cdot}}N' . N, N' \in F_\Delta{}^{\boldsymbol{\cdot}}C^{\boldsymbol{\cdot}}Q :.$

$(\exists S) : S \in C^{\boldsymbol{\cdot}}Q . (N^{\boldsymbol{\cdot}}S) S (N'^{\boldsymbol{\cdot}}S) : S'QS . S' \neq S . \supset_{S'} . N^{\boldsymbol{\cdot}}S' = N'^{\boldsymbol{\cdot}}S' :.$

$[*172{\cdot}41{\cdot}402] \equiv :. M, M' \in F_\Delta{}^{\boldsymbol{\cdot}}C^{\boldsymbol{\cdot}}P :. (\exists S, R) : S \in C^{\boldsymbol{\cdot}}Q . R = T ; S . (M^{\boldsymbol{\cdot}}R) R (M'^{\boldsymbol{\cdot}}R) :$

$S'QS . R' = T ; S . S' \neq S . \supset_{S', R'} . M^{\boldsymbol{\cdot}}R' = M'^{\boldsymbol{\cdot}}R' :.$

$[*10{\cdot}23 . (3) . (5)] \equiv :. M, M' \in F_\Delta{}^{\boldsymbol{\cdot}}C^{\boldsymbol{\cdot}}P :. (\exists S, R) : S \in C^{\boldsymbol{\cdot}}Q . R = T ; S . (M^{\boldsymbol{\cdot}}R) R (M'^{\boldsymbol{\cdot}}R) :$

$R' \neq R . R'PR . \supset_{R'} . M^{\boldsymbol{\cdot}}R' = M'^{\boldsymbol{\cdot}}R' :.$

$[(1) . *172{\cdot}11] \quad \equiv :. M (\Pi^{\boldsymbol{\cdot}}P) M' \qquad (6)$

$\vdash . (6) . *43{\cdot}302 . *172{\cdot}4 . *151{\cdot}22 . \supset \vdash . \text{Prop}$

The following proposition is a lemma for *172·43.

***172·421.** $\vdash : S = T \upharpoonright C^{\boldsymbol{\cdot}}\Sigma^{\boldsymbol{\cdot}}Q . S \in P \,\overline{\text{smor}}\,\overline{\text{smor}}\, Q . \supset .$

$(S \| \text{Cnv}^{\boldsymbol{\cdot}}S\dagger) \upharpoonright F_\Delta{}^{\boldsymbol{\cdot}}C^{\boldsymbol{\cdot}}Q = (T \| \text{Cnv}^{\boldsymbol{\cdot}}T\dagger) \upharpoonright F_\Delta{}^{\boldsymbol{\cdot}}C^{\boldsymbol{\cdot}}Q$

Dem.

$\vdash . *80{\cdot}33 . *164{\cdot}18 . *162{\cdot}23 . \supset \vdash :. \text{Hp} . N \in F_\Delta{}^{\boldsymbol{\cdot}}C^{\boldsymbol{\cdot}}Q . \supset . \text{D}^{\boldsymbol{\cdot}}N \subset C^{\boldsymbol{\cdot}}\Sigma^{\boldsymbol{\cdot}}Q .$

$[*35{\cdot}481] \quad \supset . T | N = S | N \qquad (1)$

$\vdash . *80{\cdot}14 . \supset \vdash :. \text{Hp} . N \in F_\Delta{}^{\boldsymbol{\cdot}}C^{\boldsymbol{\cdot}}Q . Y \in \text{Ɑ}^{\boldsymbol{\cdot}}N . \supset : Y \in C^{\boldsymbol{\cdot}}Q :$

$[*150{\cdot}33 . *162{\cdot}22] \supset : X = T ; Y . \equiv . X = S ; Y .$

$[*150{\cdot}1] \quad \supset : Y (\text{Cnv}^{\boldsymbol{\cdot}}T\dagger) X . \equiv . Y (\text{Cnv}^{\boldsymbol{\cdot}}S\dagger) X \qquad (2)$

$\vdash . (2) . *33{\cdot}14 . \supset \vdash :. \text{Hp} . N \in F_\Delta{}^{\boldsymbol{\cdot}}C^{\boldsymbol{\cdot}}Q . \supset :$

$(\exists Y) . yNY . Y (\text{Cnv}^{\boldsymbol{\cdot}}T\dagger) X . \equiv . (\exists Y) . yNY . Y (\text{Cnv}^{\boldsymbol{\cdot}}S\dagger) X :$

$[*34{\cdot}1] \quad \supset : N | \text{Cnv}^{\boldsymbol{\cdot}}T\dagger = N | \text{Cnv}^{\boldsymbol{\cdot}}S\dagger \qquad (3)$

$\vdash . (1) . (3) . \supset \vdash :. \text{Hp} . \supset : N \in F_\Delta{}^{\boldsymbol{\cdot}}C^{\boldsymbol{\cdot}}Q . \supset . T | N | \text{Cnv}^{\boldsymbol{\cdot}}T\dagger = S | N | \text{Cnv}^{\boldsymbol{\cdot}}S\dagger :$

$[*43{\cdot}112 . *35{\cdot}71] \supset : (T \| \text{Cnv}^{\boldsymbol{\cdot}}T\dagger) \upharpoonright F_\Delta{}^{\boldsymbol{\cdot}}C^{\boldsymbol{\cdot}}Q = (S \| \text{Cnv}^{\boldsymbol{\cdot}}S\dagger) \upharpoonright F_\Delta{}^{\boldsymbol{\cdot}}C^{\boldsymbol{\cdot}}Q :. \supset \vdash . \text{Prop}$

***172·43.** $\vdash : T \upharpoonright C\text{'}\Sigma\text{'}Q \in P \,\overline{\text{smor}}\, \overline{\text{smor}}\, Q \,.\supset.$

$$(T \parallel \text{Cnv}\text{'}T\dagger) \upharpoonright C\text{'}\Pi\text{'}Q \in (\Pi\text{'}P) \,\overline{\text{smor}}\, (\Pi\text{'}Q)$$

[*172·42·421]

***172·44.** $\vdash : P \text{ smor smor } Q \,.\supset.\, \Pi\text{'}P \text{ smor } \Pi\text{'}Q$ [*172·42]

***172·45.** $\vdash :. \text{Mult ax} \,.\supset: P, Q \in \text{Rel}^2 \text{ excl} \,.\, \exists ! P \,\overline{\text{smor}}\, Q \cap \text{Rl}\text{'smor} \,.\supset.$

$$\Pi\text{'}P \text{ smor } \Pi\text{'}Q$$

[*164·44 . *172·44]

The following proposition shows that if two relations have the same field, and if the parts of them that are contained in diversity are the same, they have the same product. Thus *e.g.* $\Pi\text{'}P_{po} = \Pi\text{'}P_{*}$, in virtue of *91·541.

***172·5.** $\vdash : C\text{'}P = C\text{'}Q \,.\, P \dot{\cap} J = Q \dot{\cap} J \,.\supset.\, \Pi\text{'}P = \Pi\text{'}Q$

Dem.

$$\vdash . *50\cdot11 \,.\supset \vdash :. \text{Hp} \,.\supset: RPS \,.\, R \neq S \,.\equiv.\, RQS \,.\, R \neq S \quad (1)$$

$$\vdash . (1) . *172\cdot11 \,.\supset \vdash . \text{Prop}$$

The following proposition is used in *182·42.

***172·51.** $\vdash . \Pi\text{'}P = \Pi\text{'}(P \cup I \upharpoonright C\text{'}P)$ [*172·5]

***172·52.** $\vdash :. Q \in \text{Cl}\text{'}P \,.\supset_Q.\, (\exists R) \,.\, RPQ \,.\, R \neq Q : \supset .\, \Pi\text{'}P = \Pi\text{'}(P \dot{\cap} J)$

Dem.

$$\vdash . *50\cdot11 . \qquad \supset \vdash : \text{Hp} \,.\supset.\, \text{Cl}\text{'}P \subset \text{Cl}\text{'}(P \dot{\cap} J) \quad (1)$$

$$\vdash . *33\cdot14 . *93\cdot101 . \text{Transp} . \supset \vdash : QPQ \,.\supset.\, Q \sim\in \overrightarrow{B}\text{'}P :$$

$$[\text{Transp}.*33\cdot13] \qquad \supset \vdash : Q \in \overrightarrow{B}\text{'}P \,.\supset.\, (\exists R) \,.\, QPR \,.\, R \neq Q \,.$$

$$[*50\cdot11] \qquad \supset .\, Q \in C\text{'}(P \dot{\cap} J) \quad (2)$$

$$\vdash . (1) . (2) . *93\cdot103 . \qquad \supset \vdash : \text{Hp} \,.\supset.\, C\text{'}P \subset C\text{'}(P \dot{\cap} J) \,.$$

$$[*33\cdot265] \qquad \supset .\, C\text{'}P = C\text{'}(P \dot{\cap} J) \quad (3)$$

$$\vdash . (3) . *172\cdot5 \,.\supset \vdash . \text{Prop}$$

Thus we shall always have $\Pi\text{'}P = \Pi\text{'}(P \dot{\cap} J)$ unless there are members of $\text{Cl}\text{'}P$ which have no referent except themselves.

$*173$. THE PRODUCT OF THE RELATIONS OF A FIELD (*continued*).

Summary of $*173$.

In this number, we shall consider the relation between the domains of relations related by $\Pi`P$, *i.e.* we shall consider $\mathrm{D};\Pi`P$. This relation bears to $\Pi`P$ a relation analogous to that which $\mathrm{Prod}`\kappa$ bears to $\epsilon_\Delta`\kappa$. We shall denote it by "$\mathrm{Prod}`P$." When $P \in \mathrm{Rel}^2\,\mathrm{excl}$, $\mathrm{Prod}`P$ is like $\Pi`P$, and is often more convenient than $\Pi`P$. When $P \in \mathrm{Rel}^2\,\mathrm{excl}$, $\mathrm{Prod}`P$ arranges the multiplicative class of $C``C`P$ by first differences, taking first differences to mean that the earliest member $Q$ of $C`P$ for which $\mu \cap C`Q \neq \nu \cap C`Q$ has the μ-member earlier than the ν-member in the Q-series.

The properties of $\mathrm{Prod}`P$ all result immediately from those of $\Pi`P$, and offer no difficulty of any kind. The most important of them are:

$*173\cdot14.$ $\vdash : \dot{\exists} ! P \,.\, C \upharpoonright C`P \in 1 \to 1 \,.\, \supset . \, C`\mathrm{Prod}`P = \mathrm{Prod}`C``C`P$

I.e. if P is not null, and no two members of $C`P$ have the same field, then the field of $\mathrm{Prod}`P$ is the product of the fields of $C`P$. Observe that $C \upharpoonright C`P \in 1 \to 1$ if $P \in \mathrm{Rel}^2\,\mathrm{excl}$.

$*173\cdot16.$ $\vdash : P \in \mathrm{Rel}^2\,\mathrm{excl} \,.\, \supset .$
$\qquad \mathrm{Prod}`P \,\mathrm{smor}\, \Pi`P \,.\, \mathrm{D} \upharpoonright C`\Pi`P \in (\mathrm{Prod}`P)\, \overline{\mathrm{smor}}\, (\Pi`P)$

$*173\cdot2.$ $\vdash . \, \mathrm{Prod}`\dot{\Lambda} = \dot{\Lambda}$

$*173\cdot22.$ $\vdash . \, \mathrm{Prod}`(P \downarrow P) = \iota ; P$

$*173\cdot23.$ $\vdash : P \neq Q \,.\, \supset . \, \mathrm{Prod}`(P \downarrow Q) = C;(P \times Q)$

$*173\cdot3.$ $\vdash : T \upharpoonright C`\Sigma`Q \in P \,\overline{\mathrm{smor}}\, \overline{\mathrm{smor}}\, Q \,.\, \supset .$
$\qquad T_\epsilon \upharpoonright C`\mathrm{Prod}`Q \in (\mathrm{Prod}`P)\, \overline{\mathrm{smor}}\, (\mathrm{Prod}`Q)$

$*173\cdot31.$ $\vdash : P \,\mathrm{smor}\,\mathrm{smor}\, Q \,.\, \supset . \, \mathrm{Prod}`P \,\mathrm{smor}\, \mathrm{Prod}`Q$

$*173\cdot01.$ $\mathrm{Prod}`P = \mathrm{D};\Pi`P$ Df

$*173\cdot1.$ $\vdash . \, \mathrm{Prod}`P = \mathrm{D};\Pi`P$ [$(*173\cdot01)$]

$*173\cdot11.$ $\vdash : \mu\, (\mathrm{Prod}`P)\, \nu \,.\, \equiv . \, (\exists M, N) \,.\, M\, (\Pi`P)\, N \,.\, \mu = \mathrm{D}`M \,.\, \nu = \mathrm{D}`N$
[$*173\cdot1 \,.\, *150\cdot51$]

$*173\cdot12.$ $\vdash . \, C`\mathrm{Prod}`P \subset \mathrm{D}``F_\Delta`C`P$ [$*172\cdot12 \,.\, *150\cdot202$]

***173·121.** $\vdash . C\text{‘}\mathrm{Prod}\text{‘}P = \mathrm{D}\text{“}C\text{‘}\Pi\text{‘}P$ [*173·1 . *150·22]

***173·13.** $\vdash : \dot{\exists} ! P . \supset . C\text{‘}\mathrm{Prod}\text{‘}P = \mathrm{D}\text{“}F_\Delta\text{‘}C\text{‘}P$ [*172·17 . *173·121]

***173·14.** $\vdash : \dot{\exists} ! P . C \upharpoonright C\text{‘}P \in 1 \rightarrow 1 . \supset . C\text{‘}\mathrm{Prod}\text{‘}P = \mathrm{Prod}\text{‘}C\text{“}C\text{‘}P$

Dem.

$\vdash . *85{\cdot}12 . *33{\cdot}5 . \supset \vdash : \mathrm{Hp} . \supset . \mathrm{D}\text{“}F_\Delta\text{‘}C\text{‘}P = \mathrm{D}\text{“}\epsilon_\Delta\text{‘}C\text{“}C\text{‘}P .$

[*173·13.*115·1] $\supset . C\text{‘}\mathrm{Prod}\text{‘}P = \mathrm{Prod}\text{‘}C\text{“}C\text{‘}P : \supset \vdash . \mathrm{Prop}$

***173·15.** $\vdash : \mathrm{D} \upharpoonright F_\Delta\text{‘}C\text{‘}P \in 1 \rightarrow 1 . \supset . \mathrm{D} \upharpoonright C\text{‘}\Pi\text{‘}P \in (\mathrm{Prod}\text{‘}P) \overline{\mathrm{smor}} (\Pi\text{‘}P)$

[*173·1 . *172·12 . *151·231]

***173·151.** $\vdash : \mathrm{D} \upharpoonright F_\Delta\text{‘}C\text{‘}P \in 1 \rightarrow 1 . \supset . \mathrm{Prod}\text{‘}P \mathrm{\ smor\ } \Pi\text{‘}P$ [*173·15]

***173·16.** $\vdash : P \in \mathrm{Rel}^2 \mathrm{\ excl} . \supset .$

$\mathrm{Prod}\text{‘}P \mathrm{\ smor\ } \Pi\text{‘}P . \mathrm{D} \upharpoonright C\text{‘}\Pi\text{‘}P \in (\mathrm{Prod}\text{‘}P) \overline{\mathrm{smor}} (\Pi\text{‘}P)$

Dem.

$\vdash . *163{\cdot}12 . \supset \vdash : \mathrm{Hp} . \supset . F \upharpoonright C\text{‘}P \in \mathrm{Cls} \rightarrow 1 .$

[*81·21] $\supset . \mathrm{D} \upharpoonright F_\Delta\text{‘}C\text{‘}P \in 1 \rightarrow 1$ (1)

$\vdash . (1) . *173{\cdot}151{\cdot}15 . \supset \vdash . \mathrm{Prop}$

***173·161.** $\vdash : P \in \mathrm{Rel}^2 \mathrm{\ excl} . \dot{\exists} ! P . \supset . C\text{‘}\mathrm{Prod}\text{‘}P = \mathrm{Prod}\text{‘}C\text{“}C\text{‘}P$

[*173·14 . *163·14]

***173·17.** $\vdash : \dot{\exists} ! \mathrm{Prod}\text{‘}P . \supset . s\text{‘}C\text{‘}\mathrm{Prod}\text{‘}P = C\text{‘}\Sigma\text{‘}P$

Dem.

$\vdash . *173{\cdot}13 . \supset \vdash : \mathrm{Hp} . \supset . s\text{‘}C\text{‘}\mathrm{Prod}\text{‘}P = s\text{‘}\mathrm{D}\text{“}F_\Delta\text{‘}C\text{‘}P$

[*41·43.*80·42] $= \mathrm{D}\text{‘}F \upharpoonright C\text{‘}P$

[*37·401.*162·23] $= C\text{‘}\Sigma\text{‘}P : \supset \vdash . \mathrm{Prop}$

***173·2.** $\vdash . \mathrm{Prod}\text{‘}\dot{\Lambda} = \dot{\Lambda}$ [*172·13 . *150·42]

***173·21.** $\vdash : \dot{\exists} ! \mathrm{Prod}\text{‘}P . \equiv . \dot{\exists} ! \Pi\text{‘}P$ [*173·1 . *150·24 . *33·12]

***173·22.** $\vdash . \mathrm{Prod}\text{‘}(P \downarrow P) = \iota ; P$

Dem.

$\vdash . *172{\cdot}2 . \supset \vdash . \mathrm{Prod}\text{‘}(P \downarrow P) = \mathrm{D} ; \downarrow P ; P$

[*150·4] $= \hat{\mu}\hat{\nu} \{(\exists x, y) . xPy . \mu = \mathrm{D}\text{‘}(x \downarrow P) . \nu = \mathrm{D}\text{‘}(y \downarrow P)\}$

[*55·16] $= \hat{\mu}\hat{\nu} \{(\exists x, y) . xPy . \mu = \iota\text{‘}x . \nu = \iota\text{‘}y\}$

[*150·4] $= \iota ; P . \supset \vdash . \mathrm{Prop}$

***173·23.** $\vdash : P \neq Q . \supset . \mathrm{Prod}\text{‘}(P \downarrow Q) = C ; (P \times Q)$

Dem.

$\vdash . *172{\cdot}21 . \supset \vdash : \mathrm{Hp} . \supset . C ; (P \times Q) = C ; \dagger(Q \downarrow P) ; \Pi\text{‘}(P \downarrow Q)$ (1)

$\vdash . *80{\cdot}14 . *150{\cdot}23 . \supset \vdash : M \in F_\Delta\text{‘}C\text{‘}(Q \downarrow P) . \supset . C\text{‘}M ; (Q \downarrow P) = \mathrm{D}\text{‘}M :$

[*55·15.*150·1] $\supset \vdash : M \in F_\Delta\text{‘}C\text{‘}(P \downarrow Q) . \supset . C\text{‘}\dagger(Q \downarrow P)\text{‘}M = \mathrm{D}\text{‘}M :$

[*172·12] $\supset \vdash : M \in C\text{‘}\Pi\text{‘}(P \downarrow Q) . \supset . C\text{‘}\dagger(Q \downarrow P)\text{‘}M = \mathrm{D}\text{‘}M :$

[*150·35] $\supset \vdash . C ; \dagger(Q \downarrow P) ; \Pi\text{‘}(P \downarrow Q) = \mathrm{D} ; \Pi\text{‘}(P \downarrow Q)$ (2)

$\vdash . (1) . (2) . *173{\cdot}1 . \supset \vdash . \mathrm{Prop}$

∗173·24. $\vdash : C\text{‘}P \cap C\text{‘}Q = \Lambda . \supset . C \upharpoonright C\text{‘}(P \times Q) \in \{\text{Prod}\text{‘}(P \downarrow Q)\} \,\overline{\text{smor}}\, (P \times Q) .$
$\text{Prod}\text{‘}(P \downarrow Q) \text{ smor } P \times Q$

Dem.

$\vdash . ∗166·12 . \quad \supset \vdash . C \upharpoonright C\text{‘}(P \times Q) = C \upharpoonright (C\text{‘}P \times C\text{‘}Q) \quad (1)$
$\vdash . (1) . ∗113·148 . \supset \vdash : \text{Hp} . \supset . C \upharpoonright C\text{‘}(P \times Q) \in 1 \to 1 \quad (2)$
$\vdash . (2) . ∗173·23 . \supset \vdash . \text{Prop}$

∗173·25. $\vdash : P \in \text{Rel}^2 \text{ excl} . Z \sim \in C\text{‘}P . C\text{‘}Z \cap C\text{‘}\Sigma\text{‘}P = \Lambda . \supset .$
$\text{Prod}\text{‘}(P \leftrightarrow Z) \text{ smor } (\text{Prod}\text{‘}P \times Z) . \text{Prod}\text{‘}(Z \leftarrow P) \text{ smor } (Z \times \text{Prod}\text{‘}P)$

Dem.

$\vdash . ∗163·451 . \supset \vdash : \text{Hp} . \supset . P \leftrightarrow Z \in \text{Rel}^2 \text{ excl} .$
$[∗173·16] \quad \supset . \text{Prod}\text{‘}(P \leftrightarrow Z) \text{ smor } \Pi\text{‘}(P \leftrightarrow Z) .$
$[∗172·32] \quad \supset . \text{Prod}\text{‘}(P \leftrightarrow Z) \text{ smor } \Pi\text{‘}P \times Z .$
$[∗173·16 . ∗166·23] \quad \supset . \text{Prod}\text{‘}(P \leftrightarrow Z) \text{ smor Prod}\text{‘}P \times Z \quad (1)$
Similarly $\quad \vdash : \text{Hp} . \supset . \text{Prod}\text{‘}(Z \leftarrow P) \text{ smor } Z \times \text{Prod}\text{‘}P \quad (2)$
$\vdash . (1) . (2) . \supset \vdash . \text{Prop}$

∗173·26. $\vdash : P, Q \in \text{Rel}^2 \text{excl} . \dot{\exists} ! P . \dot{\exists} ! Q . C\text{‘}P \cap C\text{‘}Q = \Lambda . C\text{‘}\Sigma\text{‘}P \cap C\text{‘}\Sigma\text{‘}Q = \Lambda . \supset .$
$\text{Prod}\text{‘}(P \ddagger Q) \text{ smor Prod}\text{‘}P \times \text{Prod}\text{‘}Q$

Dem.

$\vdash . ∗163·441 . ∗173·16 . \supset \vdash : \text{Hp} . \supset . \text{Prod}\text{‘}(P \ddagger Q) \text{ smor } \Pi\text{‘}(P \ddagger Q) .$
$[∗172·35] \quad \supset . \text{Prod}\text{‘}(P \ddagger Q) \text{ smor } \Pi\text{‘}P \times \Pi\text{‘}Q .$
$[∗173·16 . ∗166·23] \supset . \text{Prod}\text{‘}(P \ddagger Q) \text{ smor Prod}\text{‘}P \times \text{Prod}\text{‘}Q : \supset \vdash . \text{Prop}$

∗173·27. $\vdash : C\text{‘}P \cap C\text{‘}Q = \Lambda . C\text{‘}P \cap C\text{‘}R = \Lambda . C\text{‘}Q \cap C\text{‘}R = \Lambda . \supset .$
$\text{Prod}\text{‘}\{(P \downarrow Q) \leftrightarrow R\} \text{ smor } P \times Q \times R$

Dem.

$\vdash . ∗173·25 . \supset \vdash : \text{Hp} . R \neq P . R \neq Q . \supset .$
$\text{Prod}\text{‘}\{(P \downarrow Q) \leftrightarrow R\} \text{ smor } \{\text{Prod}\text{‘}(P \downarrow Q)\} \times R .$
$[∗173·24] \quad \supset . \text{Prod}\text{‘}\{(P \downarrow Q) \leftrightarrow R\} \text{ smor } P \times Q \times R \quad (1)$
$\vdash . ∗33·241 . \supset \vdash : \text{Hp} . R = P . \supset . R = \dot{\Lambda} . P = \dot{\Lambda} .$
$[∗172·14 . ∗166·13] \quad \supset . \Pi\text{‘}\{(P \downarrow Q) \leftrightarrow R\} = \dot{\Lambda} . P \times Q \times R = \dot{\Lambda} .$
$[∗173·1 . ∗150·42] \quad \supset . \text{Prod}\text{‘}\{(P \downarrow Q) \leftrightarrow R\} = \dot{\Lambda} . P \times Q \times R = \dot{\Lambda} .$
$[∗153·101] \quad \supset . \text{Prod}\text{‘}\{(P \downarrow Q) \leftrightarrow R\} \text{ smor } (P \times Q \times R) \quad (2)$
Similarly $\quad \vdash : \text{Hp} . R = Q . \supset . \text{Prod}\text{‘}\{(P \downarrow Q) \leftrightarrow R\} \text{ smor } (P \times Q \times R) \quad (3)$
$\vdash . (1) . (2) . (3) . \supset \vdash . \text{Prop}$

The following proposition gives a correlator of Prod‘P and Prod‘Q when we are given a double correlator of P and Q.

***173·3.** $\vdash : T \upharpoonright C\text{‘}\Sigma\text{‘}Q \in P \,\overline{\text{smor}}\,\overline{\text{smor}}\, Q \,.\supset.$

$$T_\epsilon \upharpoonright C\text{‘}\text{Prod}\text{‘}Q \in (\text{Prod}\text{‘}P) \,\overline{\text{smor}}\, (\text{Prod}\text{‘}Q)$$

Dem.

$\vdash . *173{\cdot}11 . *172{\cdot}43 . \supset$

$\vdash :. \text{Hp} . \supset : \mu \,(\text{Prod}\text{‘}P)\, \mu' . \equiv . (\exists N, N') . N \,(\Pi\text{‘}Q)\, N' . \mu = \text{D}\text{‘}(T \,|\, N \,|\, \text{Cnv}\text{‘}T\dagger) .$
$\mu' = \text{D}\text{‘}(T \,|\, N' \,|\, \text{Cnv}\text{‘}T\dagger) .$

$[*37{\cdot}32{\cdot}321] \equiv . (\exists N, N') . N \,(\Pi\text{‘}Q)\, N' . \mu = T\text{‘‘}\text{D}\text{‘}N . \mu' = T\text{‘‘}\text{D}\text{‘}N' .$

$[*173{\cdot}11] \quad \equiv . (\exists \nu, \nu') . \nu \,(\text{Prod}\text{‘}Q)\, \nu' . \mu = T\text{‘‘}\nu . \mu' = T\text{‘‘}\nu' .$

$[*37{\cdot}101] \quad \equiv . \mu \,(T_\epsilon ; \text{Prod}\text{‘}Q)\, \mu'$ (1)

$\vdash . *173{\cdot}17 . \supset \vdash . s\text{‘}C\text{‘}\text{Prod}\text{‘}Q \subset C\text{‘}\Sigma\text{‘}Q$ (2)

$[*111{\cdot}12] \quad \supset \vdash . (T \upharpoonright C\text{‘}\Sigma\text{‘}Q)_\epsilon \upharpoonright C\text{‘}\text{Prod}\text{‘}Q = T_\epsilon \upharpoonright C\text{‘}\text{Prod}\text{‘}Q$ (3)

$\vdash . (2) . (3) . *72{\cdot}451 . \supset \vdash : \text{Hp} . \supset . T_\epsilon \upharpoonright C\text{‘}\text{Prod}\text{‘}Q \in 1 \to 1$ (4)

$\vdash . (1) . (4) . *151{\cdot}231 . \supset \vdash . \text{Prop}$

***173·31.** $\vdash : P \text{ smor smor } Q . \supset . \text{Prod}\text{‘}P \text{ smor Prod}\text{‘}Q \quad [*173{\cdot}3]$

***173·32.** $\vdash : R \upharpoonright C\text{‘}\Sigma\text{‘}Q \in 1 \to 1 . C\text{‘}\Sigma\text{‘}Q \subset \text{G}\text{‘}R . \supset . \text{Prod}\text{‘}R\dagger ; Q = R_\epsilon ; \text{Prod}\text{‘}Q$

Dem.

$\vdash . *164{\cdot}18 . \supset \vdash : \text{Hp} . \supset . R \upharpoonright C\text{‘}\Sigma\text{‘}Q \in (R\dagger ; Q) \,\overline{\text{smor}}\,\overline{\text{smor}}\, Q .$

$[*173{\cdot}3] \quad \supset . R_\epsilon \upharpoonright C\text{‘}\text{Prod}\text{‘}Q \in (\text{Prod}\text{‘}R\dagger ; Q) \,\overline{\text{smor}}\, (\text{Prod}\text{‘}Q) .$

$[*151{\cdot}22] \quad \supset . \text{Prod}\text{‘}R\dagger ; Q = R_\epsilon ; \text{Prod}\text{‘}Q : \supset \vdash . \text{Prop}$

***173·33.** $\vdash : \text{D} \upharpoonright C\text{‘}\Sigma\text{‘}Q \in 1 \to 1 . \supset . \text{Prod}\text{‘}\text{D}\dagger ; Q = \text{D}_\epsilon ; \text{Prod}\text{‘}Q \left[*173{\cdot}32 \frac{\text{D}}{R} \right] .$

The above proposition is used in proving the associative law for "Prod" (*174·401).

*174. THE ASSOCIATIVE LAW OF RELATIONAL MULTIPLICATION.

Summary of *174.

In the present number, we have to prove the associative law for Π and for Prod, *i.e.* we have to prove (with a suitable hypothesis)

$$\Pi‘\Pi;P \text{ smor } \Pi‘\Sigma‘P$$

and

$$\text{Prod}‘\text{Prod};P \text{ smor } \text{Prod}‘\Sigma‘P.$$

The first of these requires $P \in \text{Rel}^2 \text{ excl}$ and either $P \subseteq J$ or

$$QPQ \,.\supset_Q .\, C‘Q \in 0 \cup 1;$$

the second requires not only this, but also $\Sigma‘P \in \text{Rel}^2 \text{excl}$. When both P and $\Sigma‘P$ are relations of mutually exclusive relations, we call P an *arithmetical* relation, which we denote by "Rel^3 arithm." Arithmetical relations serve exactly analogous purposes to those served by arithmetical classes in cardinal arithmetic.

The proof of the associative law for Π consists in showing that, under a suitable hypothesis, $\dot{s}\,|\,\text{D}$ (with its converse domain limited) is a correlator of $\Pi‘\Sigma‘P$ and $\Pi‘\Pi;P$ (*174·221·23). To prove this, we first prove

***174·17.** $\vdash : P \in \text{Rel}^2 \text{ excl} \,.\supset .\, \dot{s}“\text{D}“C‘\Pi‘\Pi;P = C‘\Pi‘\Sigma‘P$

and

***174·19.** $\vdash : P \in \text{Rel}^2 \text{ excl} \,.\supset .\, (\dot{s}\,|\,\text{D}) \upharpoonright C‘\Pi‘\Pi;P \in 1 \rightarrow 1$

This gives what we may call the cardinal part of the proof, *i.e.* it shows that $(\dot{s}\,|\,\text{D}) \upharpoonright C‘\Pi‘\Pi;P$ is a cardinal correlator of the fields of $\Pi‘\Sigma‘P$ and $\Pi‘\Pi;P$. We then prove that if M and N belong to the field of $\Pi‘\Pi;P$, they have the relation $\Pi‘\Pi;P$ when the relational sums of their domains have the relation $\Pi‘\Sigma‘P$. Here, in addition to the hypothesis $P \in \text{Rel}^2 \text{ excl}$, we require that if any relation Q has the relation P to itself, then $C‘Q$ is not to have more than one term. Thus we have

***174·215.** $\vdash :. P \in \text{Rel}^2 \text{ excl} : QPQ \,.\supset_Q .\, C‘Q \in 0 \cup 1 : \supset :$

$$M(\Pi‘\Pi;P)N \,.\equiv .\, M, N \in F_\Delta‘\Pi“C‘P \,.\, (\dot{s}‘\text{D}‘M)(\Pi‘\Sigma‘P)(\dot{s}‘\text{D}‘N)$$

The hypothesis $QPQ \,.\supset_Q .\, C‘Q \in 0 \cup 1$ is verified if $P \subseteq J$ (*174·216); thus for most purposes it is more convenient to substitute the simpler hypothesis $P \subseteq J$ for $QPQ \,.\supset_Q .\, C‘Q \in 0 \cup 1$. We shall, however, have occasion to use the hypothesis $QPQ \,.\supset_Q .\, C‘Q \in 0 \cup 1$ in *182·42·43·431, where our P is a relation whose field consists entirely of relations of the form $Q \downarrow Q$, whose fields are always unit classes, so that our P satisfies the above hypothesis even if P is not contained in J.

The proof of $*174\cdot215$ (above) is effected by first proving

$*174\cdot2$. $\vdash : P \in \text{Rel}^2 \text{ excl} . Q \in C'P . M \in C'\Pi'\Pi;P . \supset . M'\Pi'Q = (\dot{s}'D'M) \upharpoonright C'Q$

From $*174\cdot17\cdot19\cdot215$ we deduce

$*174\cdot221$. $\vdash :. P \in \text{Rel}^2 \text{ excl} : QPQ . \supset_Q . C'Q \in 0 \cup 1 : \supset .$
$\Pi'\Sigma'P = \dot{s};D;\Pi'\Pi;P . (\dot{s} \mid D) \upharpoonright C'\Pi'\Pi;P \in (\Pi'\Sigma'P) \overline{\text{smor}} (\Pi'\Pi;P)$

whence we obtain the more convenient proposition

$*174\cdot23$. $\vdash : P \in \text{Rel}^2 \text{ excl} . P \subset J . \supset .$
$\Pi'\Sigma'P = \dot{s};D;\Pi'\Pi;P . (\dot{s} \mid D) \upharpoonright C'\Pi'\Pi;P \in (\Pi'\Sigma'P) \overline{\text{smor}} (\Pi'\Pi;P)$

Thus if the hypothesis of $*174\cdot221$ or of $*174\cdot23$ holds, the associative law holds for Π ($*174\cdot241\cdot25$).

To prove the associative law for Prod, *i.e.*

$$P \in \text{Rel}^3 \text{ arithm} . P \subset J . \supset . \text{Prod}'\Sigma'P \text{ smor } \text{Prod}'\text{Prod};P,$$

we observe that, since $\Pi'\Sigma'P = \dot{s};D;\Pi'\Pi;P$ ($*174\cdot23$)
$= \dot{s};\text{Prod}'\Pi;P$, by the definition of Prod,

we have ($*174\cdot41$) $\text{Prod}'\Sigma'P = D;\dot{s};\text{Prod}'\Pi;P$
$= s;D_\epsilon;\text{Prod}'\Pi;P$, by $*41\cdot33$,
$= s;\text{Prod}'D\dagger;\Pi;P$, by $*173\cdot33$,
$= s;\text{Prod}'\text{Prod};P$, by the definition of Prod.

Also $s \upharpoonright C'\text{Prod}'\text{Prod};P \in 1 \to 1$, by $*115\cdot46$. Hence the associative law follows ($*174\cdot43$). It will be observed that in this case the correlator is simply s with its converse domain limited ($*174\cdot42$).

As in the case of Π, "$P \subset J$" is a stronger hypothesis than we really need: what we need is $QPQ . \supset_Q . C'Q \in 0 \cup 1$.

$*174\cdot01$. $\text{Rel}^3 \text{ arithm} = \hat{P}(P, \Sigma'P \in \text{Rel}^2 \text{ excl})$ Df

$*174\cdot12$. $\vdash : C \upharpoonright C'P \in 1 \to 1 . \supset . \Pi;P \in \text{Rel}^2 \text{ excl}$

Dem.

$\vdash . *150\cdot202 . \supset$
$\vdash : M, N \in C'\Pi;P . \exists ! C'M \cap C'N . \supset . M, N \in \Pi``C'P . \exists ! C'M \cap C'N .$
$[*37\cdot6] \quad \supset . (\exists Q, R) . Q, R \in C'P . M = \Pi'Q . N = \Pi'R . \exists ! C'M \cap C'N .$
$[*172\cdot12] \supset . (\exists Q, R) . Q, R \in C'P . M = \Pi'Q . N = \Pi'R . \exists ! F_\Delta'C'Q \cap F_\Delta'C'R .$
$[*80\cdot82.\text{Transp}] \supset . (\exists Q, R) . Q, R \in C'P . M = \Pi'Q . N = \Pi'R . C'Q = C'R$ (1)
$\vdash . (1) . *71\cdot59 . \supset \vdash :. \text{Hp} . \supset :$
$M, N \in C'\Pi;P . \exists ! C'M \cap C'N . \supset . (\exists Q, R) . Q = R . M = \Pi'Q . N = \Pi'R .$
$[*13\cdot195\cdot172] \quad \supset . M = N$ (2)
$\vdash . (2) . *163\cdot11 . \supset \vdash . \text{Prop}$

$*174\cdot13$. $\vdash : P \in \text{Rel}^2 \text{ excl} . \supset . \Pi;P \in \text{Rel}^2 \text{ excl}$ [$*174\cdot12 . *163\cdot14$]

***174·16.** $\vdash : \dot{\exists} ! P . \supset . C\text{‘}\Pi\text{‘}\Pi\,;P = F_\Delta\text{‘}\Pi\text{“}C\text{‘}P$

Dem.

$\vdash . *150\cdot25 . \supset \vdash : \mathrm{Hp} . \supset . \dot{\exists} ! \Pi\,;P .$

$[*172\cdot17] \quad \supset . C\text{‘}\Pi\text{‘}\Pi\,;P = F_\Delta\text{‘}C\text{‘}\Pi\,;P$

$[*150\cdot22] \quad = F_\Delta\text{‘}\Pi\text{“}C\text{‘}P : \supset \vdash . \mathrm{Prop}$

***174·161.** $\vdash : \dot{\exists} ! P . P \epsilon \mathrm{Rel}^2 \mathrm{excl} . \supset .$

$C\text{‘}\mathrm{Prod}\text{‘}\Pi\,;P = \mathrm{D}\text{“}C\text{‘}\Pi\text{‘}\Pi\,;P = \mathrm{Prod}\text{‘}F_\Delta\text{“}C\text{“}C\text{‘}P$

Dem.

$\vdash . *173\cdot121 . \quad \supset \vdash . C\text{‘}\mathrm{Prod}\text{‘}\Pi\,;P = \mathrm{D}\text{“}C\text{‘}\Pi\text{‘}\Pi\,;P \quad (1)$

$\vdash . *173\cdot161 . \quad \supset \vdash : \mathrm{Hp} . \supset . C\text{‘}\mathrm{Prod}\text{‘}\Pi\,;P = \mathrm{Prod}\text{‘}C\text{“}C\text{‘}\Pi\,;P$

$[*150\cdot22] \quad = \mathrm{Prod}\text{‘}C\text{“}\Pi\text{“}C\text{‘}P \quad (2)$

$\vdash . *172\cdot17 . \quad \supset \vdash : \dot\Lambda \sim\epsilon C\text{‘}P . \supset . C\text{“}\Pi\text{“}C\text{‘}P = F_\Delta\text{“}C\text{“}C\text{‘}P \quad (3)$

$\vdash . *172\cdot14 . *173\cdot21 . \supset \vdash : \dot\Lambda \epsilon C\text{‘}P . \supset . C\text{‘}\mathrm{Prod}\text{‘}\Pi\,;P = \Lambda \quad (4)$

$\vdash . *80\cdot26 . *83\cdot11 . \quad \supset \vdash : \dot\Lambda \epsilon C\text{‘}P . \supset . \mathrm{Prod}\text{‘}F_\Delta\text{“}C\text{“}C\text{‘}P = \Lambda \quad (5)$

$\vdash . (2) . (3) . \supset \vdash : \mathrm{Hp} . \dot\Lambda \sim\epsilon C\text{‘}P . \supset . C\text{‘}\mathrm{Prod}\text{‘}\Pi\,;P = \mathrm{Prod}\text{‘}F_\Delta\text{“}C\text{“}C\text{‘}P \quad (6)$

$\vdash . (4) . (5) . \supset \vdash : \mathrm{Hp} . \dot\Lambda \epsilon C\text{‘}P . \supset . C\text{‘}\mathrm{Prod}\text{‘}\Pi\,;P = \mathrm{Prod}\text{‘}F_\Delta\text{“}C\text{“}C\text{‘}P \quad (7)$

$\vdash . (1) . (6) . (7) . \supset \vdash . \mathrm{Prop}$

***174·162.** $\vdash : \dot{\exists} ! P . P \epsilon \mathrm{Rel}^2 \mathrm{excl} . \supset . \dot{s}\text{“}\mathrm{D}\text{“}C\text{‘}\Pi\text{‘}\Pi\,;P = C\text{‘}\Pi\text{‘}\Sigma\text{‘}P = F_\Delta\text{‘}C\text{‘}\Sigma\text{‘}P$

Dem.

$\vdash . *174\cdot161 . *115\cdot1 . \supset \vdash : \mathrm{Hp} . \supset . \dot{s}\text{“}\mathrm{D}\text{“}C\text{‘}\Pi\text{‘}\Pi\,;P = \dot{s}\text{“}\mathrm{D}\text{“}\epsilon_\Delta\text{‘}F_\Delta\text{“}C\text{“}C\text{‘}P$

$[*85\cdot27 . *163\cdot16] \quad = F_\Delta\text{‘}s\text{‘}C\text{“}C\text{‘}P$

$[*162\cdot22] \quad = F_\Delta\text{‘}C\text{‘}\Sigma\text{‘}P \quad (1)$

$\vdash . (1) . *172\cdot17 . \supset \vdash : \mathrm{Hp} . \dot{\exists} ! \Sigma\text{‘}P . \supset . \dot{s}\text{“}\mathrm{D}\text{“}C\text{‘}\Pi\text{‘}\Pi\,;P = C\text{‘}\Pi\text{‘}\Sigma\text{‘}P \quad (2)$

$\vdash . *162\cdot45 . \quad \supset \vdash : \mathrm{Hp} . \Sigma\text{‘}P = \dot\Lambda . \supset . P = \dot\Lambda \downarrow \dot\Lambda .$

$[*172\cdot13 . *150\cdot71] \quad \supset . \Pi\,;P = \dot\Lambda \downarrow \Lambda .$

$[*172\cdot14] \quad \supset . \Pi\text{‘}\Pi\,;P = \dot\Lambda .$

$[*33\cdot241] \quad \supset . \dot{s}\text{“}\mathrm{D}\text{“}C\text{‘}\Pi\text{‘}\Pi\,;P = \Lambda \quad (3)$

$\vdash . *172\cdot13 . *33\cdot241 . \supset \vdash : \Sigma\text{‘}P = \dot\Lambda . \supset . C\text{‘}\Pi\text{‘}\Sigma\text{‘}P = \Lambda \quad (4)$

$\vdash . (3) . (4) . \supset \vdash : \mathrm{Hp} . \Sigma\text{‘}P = \dot\Lambda . \supset . \dot{s}\text{“}\mathrm{D}\text{“}C\text{‘}\Pi\text{‘}\Pi\,;P = C\text{‘}\Pi\text{‘}\Sigma\text{‘}P \quad (5)$

$\vdash . (1) . (2) . (5) . \supset \vdash . \mathrm{Prop}$

***174·17.** $\vdash : P \epsilon \mathrm{Rel}^2 \mathrm{excl} . \supset . \dot{s}\text{“}\mathrm{D}\text{“}C\text{‘}\Pi\text{‘}\Pi\,;P = C\text{‘}\Pi\text{‘}\Sigma\text{‘}P$

Dem.

$\vdash . *150\cdot42 . *172\cdot13 . \supset \vdash : P = \dot\Lambda . \supset . \dot{s}\text{“}\mathrm{D}\text{“}C\text{‘}\Pi\text{‘}\Pi\,;P = \Lambda \quad (1)$

$\vdash . *162\cdot4 . *172\cdot13 . \supset \vdash : P = \dot\Lambda . \supset . C\text{‘}\Pi\text{‘}\Sigma\text{‘}P = \Lambda \quad (2)$

$\vdash . (1) . (2) . *174\cdot162 . \supset \vdash . \mathrm{Prop}$

***174·18.** $\vdash : P \epsilon \mathrm{Rel}^2\,\mathrm{excl} \,.\supset .\, \mathrm{D} \upharpoonright C\text{‘}\Pi\text{‘}\Pi\,;P \epsilon 1 \to 1$

Dem.

$\vdash .\, *174\cdot12 \,.\, *163\cdot14\cdot12 \,.\supset \vdash : \mathrm{Hp} \,.\supset .\, F \upharpoonright C\text{‘}\Pi\,;P \epsilon \mathrm{Cls} \to 1 \,.$

$[*81\cdot21] \qquad \supset .\, \mathrm{D} \upharpoonright F_\Delta\text{‘}C\text{‘}\Pi\,;P \epsilon 1 \to 1 \,.$

$[*172\cdot12] \qquad \supset .\, \mathrm{D} \upharpoonright C\text{‘}\Pi\text{‘}\Pi\,;P \epsilon 1 \to 1 : \supset \vdash .\, \mathrm{Prop}$

***174·19.** $\vdash : P \epsilon \mathrm{Rel}^2\,\mathrm{excl} \,.\supset .\, (\dot{s} \,|\, \mathrm{D}) \upharpoonright C\text{‘}\Pi\text{‘}\Pi\,;P \epsilon 1 \to 1$

Dem.

$\vdash .\, *163\cdot1 \,.\, *35\cdot14 \,.\supset \vdash :.\, \mathrm{Hp} \,.\supset :$

$Q, R \epsilon C\text{‘}P \,.\, Q \neq R \,.\supset_{Q,R} .\, F \upharpoonright C\text{‘}Q \,\dot{\cap}\, F \upharpoonright C\text{‘}R = \dot{\Lambda} \,.$

$[*172\cdot191] \qquad \supset_{Q,R} .\, \dot{s}\text{‘}C\text{‘}\Pi\text{‘}Q \,\dot{\cap}\, \dot{s}\text{‘}C\text{‘}\Pi\text{‘}R = \dot{\Lambda} \qquad (1)$

$\vdash .\, (1) \,.\, *33\cdot5 \,.\, *85\cdot31 \dfrac{F, \Pi\text{“}C\text{‘}P}{P, \quad \alpha} \,.\supset$

$\vdash :.\, \mathrm{Hp} \,.\supset : M, N \epsilon F_\Delta\text{‘}\Pi\text{“}C\text{‘}P \,.\, \dot{s}\text{‘}\mathrm{D}\text{‘}M = \dot{s}\text{‘}\mathrm{D}\text{‘}N \,.\supset .\, M = N :$

$[*172\cdot12.*150\cdot22] \supset : M, N \epsilon C\text{‘}\Pi\text{‘}\Pi\,;P \,.\, \dot{s}\text{‘}\mathrm{D}\text{‘}M = \dot{s}\text{‘}\mathrm{D}\text{‘}N \,.\supset .\, M = N :.\supset \vdash .\, \mathrm{Prop}$

***174·191.** $\vdash : P \epsilon \mathrm{Rel}^2\,\mathrm{excl} \,.\supset .\, \dot{s} \upharpoonright C\text{‘}\mathrm{Prod}\text{‘}\Pi\,;P \epsilon 1 \to 1$

Dem.

$\vdash .\, *174\cdot19 \,.\supset \vdash :.\, \mathrm{Hp} \,.\supset : M, N \epsilon C\text{‘}\Pi\text{‘}\Pi\,;P \,.\, \dot{s}\text{‘}\mathrm{D}\text{‘}M = \dot{s}\text{‘}\mathrm{D}\text{‘}N \,.\supset .\, M = N \,.$

$[*30\cdot37] \qquad \supset .\, \mathrm{D}\text{‘}M = \mathrm{D}\text{‘}N :$

$[*37\cdot63] \qquad \supset : \mu, \nu \epsilon \mathrm{D}\text{“}C\text{‘}\Pi\text{‘}\Pi\,;P \,.\, \dot{s}\text{‘}\mu = \dot{s}\text{‘}\nu \,.\supset .\, \mu = \nu :$

$[*173\cdot121] \qquad \supset : \mu, \nu \epsilon C\text{‘}\mathrm{Prod}\text{‘}\Pi\,;P \,.\, \dot{s}\text{‘}\mu = \dot{s}\text{‘}\nu \,.\supset .\, \mu = \nu :.\supset \vdash .\, \mathrm{Prop}$

***174·2.** $\vdash : P \epsilon \mathrm{Rel}^2\,\mathrm{excl} \,.\, Q \epsilon C\text{‘}P \,.\, M \epsilon C\text{‘}\Pi\text{‘}\Pi\,;P \,.\supset .\, M\text{‘}\Pi\text{‘}Q = (\dot{s}\text{‘}\mathrm{D}\text{‘}M) \upharpoonright C\text{‘}Q$

Dem.

$\vdash .\, *172\cdot12 \,.\, *150\cdot22 \,.\supset \vdash : \mathrm{Hp} \,.\supset .\, M \epsilon F_\Delta\text{‘}\Pi\text{“}C\text{‘}P \,. \qquad (1)$

$[*80\cdot31.*33\cdot5] \qquad \supset .\, M\text{‘}\Pi\text{‘}Q \epsilon C\text{‘}\Pi\text{‘}Q \,.$

$[*172\cdot12] \qquad \supset .\, M\text{‘}\Pi\text{‘}Q \epsilon F_\Delta\text{‘}C\text{‘}Q \,.$

$[*80\cdot14] \qquad \supset .\, \text{Ɑ}\text{‘}M\text{‘}\Pi\text{‘}Q = C\text{‘}Q \qquad (2)$

$\vdash .\, (1) \,.\, *80\cdot3 \,.\, *41\cdot13 \,.\supset \vdash : \mathrm{Hp} \,.\supset .\, M\text{‘}\Pi\text{‘}Q \mathrel{\dot{\subset}} \dot{s}\text{‘}\mathrm{D}\text{‘}M \qquad (3)$

$\vdash .\, *174\cdot17 \,. \qquad \supset \vdash : \mathrm{Hp} \,.\supset .\, \dot{s}\text{‘}\mathrm{D}\text{‘}M \epsilon C\text{‘}\Pi\text{‘}\Sigma\text{‘}P \,.$

$[*172\cdot12.*80\cdot14] \qquad \supset .\, \dot{s}\text{‘}\mathrm{D}\text{‘}M \epsilon 1 \to \mathrm{Cls} \qquad (4)$

$\vdash .\, (3) \,.\, (4) \,.\, *72\cdot92 \,. \qquad \supset \vdash : \mathrm{Hp} \,.\supset .\, M\text{‘}\Pi\text{‘}Q = (\dot{s}\text{‘}\mathrm{D}\text{‘}M) \upharpoonright \text{Ɑ}\text{‘}M\text{‘}\Pi\text{‘}Q$

$[(2)] \qquad = (\dot{s}\text{‘}\mathrm{D}\text{‘}M) \upharpoonright C\text{‘}Q : \supset \vdash .\, \mathrm{Prop}$

***174·21.** $\vdash :: P \epsilon \mathrm{Rel}^2\,\mathrm{excl} \,.\, Q \epsilon C\text{‘}P \,.\, M, N \epsilon C\text{‘}\Pi\text{‘}\Pi\,;P \,.\supset :.$

$M\text{‘}\Pi\text{‘}Q = N\text{‘}\Pi\text{‘}Q \,.\equiv : R \epsilon C\text{‘}Q \,.\supset_R .\, (\dot{s}\text{‘}\mathrm{D}\text{‘}M)\text{‘}R = (\dot{s}\text{‘}\mathrm{D}\text{‘}N)\text{‘}R$

Dem.

$\vdash .\, *71\cdot35 \,.\, *80\cdot14 \,.\, *172\cdot12 \,.\supset \vdash :: \mathrm{Hp} \,.\supset :.$

$M\text{‘}\Pi\text{‘}Q = N\text{‘}\Pi\text{‘}Q \,.\equiv : R \epsilon C\text{‘}Q \,.\supset_R .\, (M\text{‘}\Pi\text{‘}Q)\text{‘}R = (N\text{‘}\Pi\text{‘}Q)\text{‘}R :$

$[*174\cdot2] \qquad \equiv : R \epsilon C\text{‘}Q \,.\supset_R .\, \{(\dot{s}\text{‘}\mathrm{D}\text{‘}M) \upharpoonright C\text{‘}Q\}\text{‘}R = \{(\dot{s}\text{‘}\mathrm{D}\text{‘}N) \upharpoonright C\text{‘}Q\}\text{‘}R :$

$[*35\cdot7] \qquad \equiv : R \epsilon C\text{‘}Q \,.\supset_R .\, (\dot{s}\text{‘}\mathrm{D}\text{‘}M)\text{‘}R = (\dot{s}\text{‘}\mathrm{D}\text{‘}N)\text{‘}R :: \supset \vdash .\, \mathrm{Prop}$

∗174·211. $\vdash ::. P \in \text{Rel}^2\text{excl} . \supset :: M(\Pi`\Pi;P)N . \equiv :.$
$M, N \in F_\Delta`\Pi``C`P :. (\exists Q, S) : Q \in C`P . S \in C`Q . \{(M`\Pi`Q)`S\} S \{(N`\Pi`Q)`S\} :$
$TQS . T \neq S . \supset_T . (M`\Pi`Q)`T = (N`\Pi`Q)`T :$
$RPQ . R \neq Q . T \in C`R . \supset_{R,T} . (M`\Pi`R)`T = (N`\Pi`R)`T$

Dem.

$\vdash . ∗172·11 . ∗150·22 . \supset \vdash :: M(\Pi`\Pi;P)N . \equiv :.$
$M, N \in F_\Delta`\Pi``C`P :. (\exists Q) : Q \in C`P . (M`\Pi`Q)(\Pi`Q)(N`\Pi`Q) :$
$RPQ . \Pi`R \neq \Pi`Q . \supset_R . M`\Pi`R = N`\Pi`R$ (1)
$\vdash . ∗80·31 . \supset \vdash :. M \in F_\Delta`\Pi``C`P . \supset : Q \in C`P . \supset_Q . M`\Pi`Q \in C`\Pi`Q .$ (2)
[∗33·24] $\supset_Q . \dot{\exists} ! \Pi`Q .$ (3)
[∗172·19] $\supset_Q . \dot{s}`C`\Pi`Q = F \upharpoonright C`Q$ (4)
$\vdash . (3) . (4) . \supset \vdash : M \in F_\Delta`\Pi``C`P . Q, R \in C`P . \Pi`Q = \Pi`R . \supset .$
$F \upharpoonright C`Q = F \upharpoonright C`R . \dot{\exists} ! \Pi`Q . \dot{\exists} ! \Pi`R .$
[∗172·141·192] $\supset . C`Q = C`R$ (5)
$\vdash . (5) . ∗163·14 . \supset \vdash :: \text{Hp} . \supset :. M \in F_\Delta`\Pi``C`P . Q, R \in C`P . \supset :$
$\Pi`Q = \Pi`R . \supset . Q = R :$
[∗30·37.Transp] $\supset : \Pi`Q \neq \Pi`R . \equiv . Q \neq R$ (6)
$\vdash . ∗71·35 . (2) . ∗172·12 . ∗80·14 . \supset$
$\vdash :: M, N \in F_\Delta`\Pi``C`P . R \in C`P . \supset :.$
$M`\Pi`R = N`\Pi`R . \equiv : T \in C`R . \supset_T . (M`\Pi`R)`T = (N`\Pi`R)`T$ (7)
$\vdash . ∗172·11 . \supset \vdash :: (M`\Pi`Q)(\Pi`Q)(N`\Pi`Q) . \equiv :. M`\Pi`Q, N`\Pi`Q \in F_\Delta`C`Q :.$
$(\exists S) : S \in C`Q . \{(M`\Pi`Q)`S\} S \{(N`\Pi`Q)`S\} :$
$TQS . T \neq S . \supset_T . (M`\Pi`Q)`T = (N`\Pi`Q)`T$ (8)
$\vdash . (1) . (2) . (6) . (7) . (8) . \supset \vdash .$ Prop

∗174·212. $\vdash ::. P \in \text{Rel}^2\text{excl} . \supset :: M(\Pi`\Pi;P)N . \equiv :.$
$M, N \in F_\Delta`\Pi``C`P :. (\exists Q, S) : Q \in C`P . S \in C`Q . \{(\dot{s}`D`M)`S\} S \{(\dot{s}`D`N)`S\} :$
$TQS . T \neq S . \supset_T . (\dot{s}`D`M)`T = (\dot{s}`D`N)`T :$
$RPQ . R \neq Q . T \in C`R . \supset_{R,T} . (\dot{s}`D`M)`T = (\dot{s}`D`N)`T$
[∗174·2·211 . ∗35·7]

∗174·213. $\vdash :. RPQ . S \in C`Q . T \in C`R . S \neq T . \supset_{Q,R,S,T} . R \neq Q : P \in \text{Rel}^2\text{excl} : \supset :$
$RPQ . S \in C`Q . T \in C`R . R \neq Q . \equiv . RPQ . S \in C`Q . T \in C`R . S \neq T$

Dem.

$\vdash . ∗163·1 . \supset \vdash :. \text{Hp} . \supset : RPQ . R \neq Q . S \in C`Q . T \in C`R . \supset . S \neq T$ (1)
$\vdash . ∗11·1 . \supset \vdash :. \text{Hp} . \supset : RPQ . S \in C`Q . T \in C`R . S \neq T . \supset . R \neq Q$ (2)
$\vdash . (1) . (2) . \supset \vdash .$ Prop

***174·214.** $\vdash :: P \in \text{Rel}^2\,\text{excl} : QPQ \,.\, \supset_Q \,.\, C\text{‘}Q \in 0 \cup 1 : \supset :.$
$(\exists Q) : Q \in C\text{‘}P : TQS \,.\, T \neq S \,.\, \vee \,.\, (\exists R) \,.\, RPQ \,.\, R \neq Q \,.\, S \in C\text{‘}Q \,.\, T \in C\text{‘}R : \equiv \,.$
$T(\Sigma\text{‘}P)S \,.\, T \neq S$

Dem.

$\vdash \,.\, *52\cdot41 \,.\, \supset \vdash :. \text{Hp} \,.\, \supset : S, T \in C\text{‘}Q \,.\, S \neq T \,.\, \supset \,.\, \sim(QPQ) :$
$[*13\cdot12.\text{Transp}] \quad \supset : RPQ \,.\, S \in C\text{‘}Q \,.\, T \in C\text{‘}R \,.\, S \neq T \,.\, \supset \,.\, Q \neq R \quad (1)$
$\vdash \,.\, (1) \,.\, *174\cdot213 \,.\, \supset$
$\vdash :: \text{Hp} \,.\, \supset :. RPQ \,.\, S \in C\text{‘}Q \,.\, T \in C\text{‘}R \,.\, R \neq Q \,.\, \equiv \,.\, RPQ \,.\, S \in C\text{‘}Q \,.\, T \in C\text{‘}R \,.\, S \neq T :.$
$[*4\cdot37.*11\cdot341] \supset :.$
$(\exists Q) \,.\, Q \in C\text{‘}P \,.\, TQS \,.\, T \neq S \,.\, \vee \,.\, (\exists Q, R) \,.\, RPQ \,.\, S \in C\text{‘}Q \,.\, T \in C\text{‘}R \,.\, R \neq Q : \equiv :$
$(\exists Q) \,.\, Q \in C\text{‘}P \,.\, TQS \,.\, T \neq S \,.\, \vee \,.\, (\exists Q, R) \,.\, RPQ \,.\, S \in C\text{‘}Q \,.\, T \in C\text{‘}R \,.\, T \neq S :$
$[*162\cdot13] \quad \equiv : T(\Sigma\text{‘}P)S \,.\, T \neq S \quad (2)$
$\vdash \,.\, (2) \,.\, *33\cdot17 \,.\, \supset \vdash \,.\, \text{Prop}$

***174·215.** $\vdash :. P \in \text{Rel}^2\,\text{excl} : QPQ \,.\, \supset_Q \,.\, C\text{‘}Q \in 0 \cup 1 : \supset :$
$M(\Pi\text{‘}\Pi\text{;}P)N \,.\, \equiv \,.\, M, N \in F_\Delta\text{‘}\Pi\text{‘‘}C\text{‘}P \,.\, (\dot{s}\text{‘}\text{D}\text{‘}M)(\Pi\text{‘}\Sigma\text{‘}P)(\dot{s}\text{‘}\text{D}\text{‘}N)$

Dem.

$\vdash \,.\, *174\cdot212\cdot214 \,.\, \supset \vdash ::. \text{Hp} \,.\, \supset :: M(\Pi\text{‘}\Pi\text{;}P)N \,.\, \equiv :.$
$M, N \in F_\Delta\text{‘}\Pi\text{‘‘}C\text{‘}P :. (\exists Q, S) : Q \in C\text{‘}P \,.\, S \in C\text{‘}Q \,.\, \{(\dot{s}\text{‘}\text{D}\text{‘}M)\text{‘}S\}\, S\, \{(\dot{s}\text{‘}\text{D}\text{‘}N)\text{‘}S\} :$
$T(\Sigma\text{‘}P)S \,.\, T \neq S \,.\, \supset_T \,.\, (\dot{s}\text{‘}\text{D}\text{‘}M)\text{‘}T = (\dot{s}\text{‘}\text{D}\text{‘}N)\text{‘}T \quad (1)$
$\vdash \,.\, *172\cdot13 \,.\, *152\cdot42 \,.\, \supset \vdash : M(\Pi\text{‘}\Pi\text{;}P)N \,.\, \supset \,.\, \dot{\exists}\,!\,P \,.$
$[*172\cdot162] \quad \supset \,.\, \dot{s}\text{‘}\text{D}\text{‘}M, \dot{s}\text{‘}\text{D}\text{‘}N \in F_\Delta\text{‘}C\text{‘}\Sigma\text{‘}P \quad (2)$
$\vdash \,.\, *162\cdot22 \,.\, \supset \vdash : (\exists Q) \,.\, Q \in C\text{‘}P \,.\, S \in C\text{‘}Q \,.\, \equiv \,.\, S \in C\text{‘}\Sigma\text{‘}P \quad (3)$
$\vdash \,.\, (1) \,.\, (2) \,.\, (3) \,.\, *172\cdot11 \,.\, \supset \vdash \,.\, \text{Prop}$

***174·216.** $\vdash :. P \mathrel{\dot{\subset}} J \,.\, \supset : QPQ \,.\, \supset_Q \,.\, C\text{‘}Q \in 0 \cup 1$

Dem.

$\vdash \,.\, *50\cdot24 \,.\, \supset \vdash :. \text{Hp} \,.\, \supset : (Q) \,.\, \sim(QPQ) :$
$[*10\cdot53] \quad \supset : QPQ \,.\, \supset_Q \,.\, C\text{‘}Q \in 0 \cup 1 :. \supset \vdash \,.\, \text{Prop}$

***174·22.** $\vdash :. P \in \text{Rel}^2\,\text{excl} \,.\, P \mathrel{\dot{\subset}} J \,.\, \supset :$
$M(\Pi\text{‘}\Pi\text{;}P)N \,.\, \equiv \,.\, M, N \in F_\Delta\text{‘}\Pi\text{‘‘}C\text{‘}P \,.\, (\dot{s}\text{‘}\text{D}\text{‘}M)(\Pi\text{‘}\Sigma\text{‘}P)(\dot{s}\text{‘}\text{D}\text{‘}N)$
$[*174\cdot215\cdot216]$

***174·221.** $\vdash :. P \in \text{Rel}^2\,\text{excl} : QPQ \,.\, \supset_Q \,.\, C\text{‘}Q \in 0 \cup 1 : \supset \,.$
$\Pi\text{‘}\Sigma\text{‘}P = \dot{s}\text{;}\text{D}\text{;}\Pi\text{‘}\Pi\text{;}P \,.\, (\dot{s}\,|\,\text{D}) \upharpoonright C\text{‘}\Pi\text{‘}\Pi\text{;}P \in (\Pi\text{‘}\Sigma\text{‘}P)\,\overline{\text{smor}}\,(\Pi\text{‘}\Pi\text{;}P)$

Dem.

$\vdash \,.\, *174\cdot215 \,.\, *150\cdot41 \,.\, \supset$
$\vdash : \text{Hp} \,.\, T = (\dot{s}\,|\,\text{D}) \upharpoonright C\text{‘}\Pi\text{‘}\Pi\text{;}P \,.\, \supset \,.\, \Pi\text{‘}\Pi\text{;}P = \breve{T}\text{;}\Pi\text{‘}\Sigma\text{‘}P \quad (1)$
$\vdash \,.\, *174\cdot19 \,.\, \quad \supset \vdash : \text{Hp}(1) \,.\, \supset \,.\, T \in 1 \to 1 \quad (2)$
$\vdash \,.\, *174\cdot17 \,.\, \quad \supset \vdash : \text{Hp}(1) \,.\, \supset \,.\, \text{D}\text{‘}T = C\text{‘}\Pi\text{‘}\Sigma\text{‘}P \quad (3)$
$\vdash \,.\, (1) \,.\, (2) \,.\, (3) \,.\, *151\cdot11 \,.\, \supset \vdash : \text{Hp}(1) \,.\, \supset \,.\, \breve{T} \in (\Pi\text{‘}\Pi\text{;}P)\,\overline{\text{smor}}\,(\Pi\text{‘}\Sigma\text{‘}P) \,.$
$[*151\cdot131] \quad \supset \,.\, T \in (\Pi\text{‘}\Sigma\text{‘}P)\,\overline{\text{smor}}\,(\Pi\text{‘}\Pi\text{;}P) \quad (4)$
$\vdash \,.\, (4) \,.\, *151\cdot22 \,.\, \supset \vdash \,.\, \text{Prop}$

***174·23.** $\vdash : P \in \text{Rel}^2\,\text{excl} \,.\, P \Subset J \,.\, \supset \,.\, \Pi`\Sigma`P = \dot{s};\text{D};\Pi`\Pi;P \,.$
$(\dot{s} \,|\, \text{D}) \upharpoonright C`\Pi`\Pi;P \in (\Pi`\Sigma`P)\,\overline{\text{smor}}\,(\Pi`\Pi;P)$ [*174·221·216]

***174·231.** $\vdash :. P \in \text{Rel}^2\,\text{excl} : QPQ \,.\, \supset_Q \,.\, C`Q \in 0 \cup 1 : \supset \,.$
$\dot{s} \upharpoonright C`\text{Prod}`\Pi;P \in (\Pi`\Sigma`P)\,\overline{\text{smor}}\,(\text{Prod}`\Pi;P)$

Dem.

$\vdash \,.\, *174·221 \,.\, *173·1 \,.\, \supset \vdash : \text{Hp} \,.\, \supset \,.\, \Pi`\Sigma`P = \dot{s};\text{Prod}`\Pi;P$ (1)

$\vdash \,.\, (1) \,.\, *174·191 \,.\, *151·231 \,.\, \supset \vdash \,.\, \text{Prop}$

***174·24.** $\vdash : P \in \text{Rel}^2\,\text{excl} \,.\, P \Subset J \,.\, \supset \,.$
$\dot{s} \upharpoonright C`\text{Prod}`\Pi;P \in (\Pi`\Sigma`P)\,\overline{\text{smor}}\,(\text{Prod}`\Pi;P)$ [*174·231·216]

***174·241.** $\vdash :. P \in \text{Rel}^2\,\text{excl} : QPQ \,.\, \supset_Q \,.\, C`Q \in 0 \cup 1 : \supset \,.$
$\Pi`\Sigma`P\ \text{smor}\ \Pi`\Pi;P \,.\, \Pi`\Sigma`P\ \text{smor}\ \text{Prod}`\Pi;P$ [*174·221·231]

***174·25.** $\vdash : P \in \text{Rel}^2\,\text{excl} \,.\, P \Subset J \,.\, \supset \,.$
$\Pi`\Sigma`P\ \text{smor}\ \Pi`\Pi;P \,.\, \Pi`\Sigma`P\ \text{smor}\ \text{Prod}`\Pi;P$ [*174·23·24]

This proposition gives the associative law for Π. It remains to prove the associative law for Prod.

The following propositions are concerned with various properties of "arithmetical" relations, down to *174·4, where the proof of the associative law for Prod begins.

***174·3.** $\vdash : P \in \text{Rel}^3\,\text{arithm} \,.\, \equiv \,.\, P, \Sigma`P \in \text{Rel}^2\,\text{excl}$ [(*174·01)]

***174·31.** $\vdash :. P \in \text{Rel}^3\,\text{arithm} \,.\, \equiv : Q, Q' \in C`P \,.\, Q \neq Q' \,.\, \supset_{Q,Q'} \,.\, C`Q \cap C`Q' = \Lambda :$
$R, R' \in C`\Sigma`P \,.\, R \neq R' \,.\, \supset_{R,R'} \,.\, C`R \cap C`R' = \Lambda$ [*174·3 . *163·1]

***174·311.** $\vdash :. P \in \text{Rel}^3\,\text{arithm} \,.\, \equiv : Q, Q' \in C`P \,.\, \exists ! C`Q \cap C`Q' \,.\, \supset_{Q,Q'} \,.\, Q = Q' :$
$R, R' \in C`\Sigma`P \,.\, \exists ! C`R \cap C`R' \,.\, \supset_{R,R'} \,.\, R = R'$ [*174·3 . *163·11]

***174·32.** $\vdash : P \in \text{Rel}^3\,\text{arithm} \,.\, \equiv \,.\, F \upharpoonright C`P, F \upharpoonright C`\Sigma`P \in \text{Cls} \rightarrow 1$
[*174·3 . *163·12]

***174·321.** $\vdash : P \in \text{Rel}^3\,\text{arithm} \,.\, \supset \,.\, C \upharpoonright C`P, C \upharpoonright C`\Sigma`P \in 1 \rightarrow 1$ [*174·3 . *163·14]

***174·322.** $\vdash : P \in \text{Rel}^3\,\text{arithm} \,.\, Q, Q' \in C`P \,.\, \exists ! C``C`Q \cap C``C`Q' \,.\, \supset \,.\, Q = Q'$

Dem.

$\vdash \,.\, *37·6 \,.\, \supset \vdash : \text{Hp} \,.\, \supset \,.\, (\exists R, R') \,.\, R \in C`Q \,.\, R' \in C`Q' \,.\, C`R = C`R' \,.$

[*174·321] $\supset \,.\, (\exists R, R') \,.\, R \in C`Q \,.\, R' \in C`Q' \,.\, R = R' \,.$

[*13·195] $\supset \,.\, \exists ! C`Q \cap C`Q' \,.$

[*174·311] $\supset \,.\, Q = Q' : \supset \vdash \,.\, \text{Prop}$

***174·33.** $\vdash : P \in \text{Rel}^3\,\text{arithm} \,.\, \supset \,.\, C```C``C`P \in \text{Cls}^3\,\text{arithm}$

Dem.

$\vdash \,.\, *174·322 \,.\, \supset$

$\vdash :. \text{Hp} \,.\, \supset : Q, Q' \in C`P \,.\, \exists ! C``C`Q \cap C``C`Q' \,.\, \supset_{Q,Q'} \,.\, C``C`Q = C``C`Q' :$

[*37·63] $\supset : \gamma, \delta \in C```C``C`P \,.\, \exists ! \gamma \cap \delta \,.\, \supset_{\gamma,\delta} \,.\, \gamma = \delta :$

[*84·11] $\supset : C```C``C`P \in \text{Cls}^2\,\text{excl}$ (1)

$\vdash \,.\, *174·3 \,.\, *163·16 \,.\, *162·22 \,.\, \supset \vdash : \text{Hp} \,.\, \supset \,.\, C``s`C``C`P \in \text{Cls}^2\,\text{excl} \,.$

[*40·38] $\supset \,.\, s`C```C``C`P \in \text{Cls}^2\,\text{excl}$ (2)

$\vdash \,.\, (1) \,.\, (2) \,.\, *115·2 \,.\, \supset \vdash \,.\, \text{Prop}$

***174·34.** $\vdash : P \in \text{Rel}^3\,\text{arithm} . \equiv .$

$$C\text{‘‘‘}C\text{‘‘}C\text{‘}P \in \text{Cls}^3\,\text{arithm} . C \upharpoonright C\text{‘}P, C \upharpoonright C\text{‘}\Sigma\text{‘}P \in 1 \to 1$$

Dem.

$\vdash . *174{\cdot}321{\cdot}33 . \supset$

$\vdash : P \in \text{Rel}^3\,\text{arithm} . \supset . C\text{‘‘‘}C\text{‘‘}C\text{‘}P \in \text{Cls}^3\,\text{arithm} . C \upharpoonright C\text{‘}P, C \upharpoonright C\text{‘}\Sigma\text{‘}P \in 1 \to 1 \quad (1)$

$\vdash . *115{\cdot}2 . \supset \vdash : C\text{‘‘‘}C\text{‘‘}C\text{‘}P \in \text{Cls}^3\,\text{arithm} . C \upharpoonright C\text{‘}\Sigma\text{‘}P \in 1 \to 1 . \supset .$

$s\text{‘}C\text{‘‘‘}C\text{‘‘}C\text{‘}P \in \text{Cls}^2\,\text{excl} . C \upharpoonright C\text{‘}\Sigma\text{‘}P \in 1 \to 1 .$

[*40·38.*162·22] $\supset . C\text{‘‘}C\text{‘}\Sigma\text{‘}P \in \text{Cls}^2\,\text{excl} . C \upharpoonright C\text{‘}\Sigma\text{‘}P \in 1 \to 1 .$

[*163·17] $\supset . \Sigma\text{‘}P \in \text{Rel}^2\,\text{excl} \quad (2)$

$\vdash . *37{\cdot}62 . \supset \vdash : Q, Q' \in C\text{‘}P . R \in C\text{‘}Q \cap C\text{‘}Q' . \supset . C\text{‘}R \in C\text{‘‘}C\text{‘}Q \cap C\text{‘‘}C\text{‘}Q' \quad (3)$

$\vdash . (3) . *115{\cdot}2 . *84{\cdot}11 . \supset$

$\vdash : C\text{‘‘‘}C\text{‘‘}C\text{‘}P \in \text{Cls}^3\,\text{arithm} . Q, Q' \in C\text{‘}P . \exists ! C\text{‘}Q \cap C\text{‘}Q' . \supset .$

$C\text{‘‘}C\text{‘}Q = C\text{‘‘}C\text{‘}Q' \quad (4)$

$\vdash . (4) . *72{\cdot}481 . *37{\cdot}421 . \supset$

$\vdash : C\text{‘‘‘}C\text{‘‘}C\text{‘}P \in \text{Cls}^3\,\text{arithm} . C \upharpoonright C\text{‘}\Sigma\text{‘}P \in 1 \to 1 .$

$Q, Q' \in C\text{‘}P . \exists ! C\text{‘}Q \cap C\text{‘}Q' . \supset . C\text{‘}Q = C\text{‘}Q' \quad (5)$

$\vdash . (5) . *71{\cdot}59 . \supset$

$\vdash :. C\text{‘‘‘}C\text{‘‘}C\text{‘}P \in \text{Cls}^3\,\text{arithm} . C \upharpoonright C\text{‘}\Sigma\text{‘}P \in 1 \to 1 . C \upharpoonright C\text{‘}P \in 1 \to 1 . \supset :$

$Q, Q' \in C\text{‘}P . \exists ! C\text{‘}Q \cap C\text{‘}Q' . \supset_{Q, Q'} . Q = Q' :$

[*163·11] $\supset : P \in \text{Rel}^2\,\text{excl} \quad (6)$

$\vdash . (2) . (6) . *174{\cdot}3 . \supset$

$\vdash : C\text{‘‘‘}C\text{‘‘}C\text{‘}P \in \text{Cls}^3\,\text{arithm} . C \upharpoonright C\text{‘}\Sigma\text{‘}P \in 1 \to 1 . C \upharpoonright C\text{‘}P \in 1 \to 1 . \supset .$

$P \in \text{Rel}^3\,\text{arithm} \quad (7)$

$\vdash . (1) . (7) . \supset \vdash . \text{Prop}$

***174·35.** $\vdash : P \in \text{Rel}^3\,\text{arithm} . Q, Q' \in C\text{‘}P . Q \neq Q' . \supset . C\text{‘}\Sigma\text{‘}Q \cap C\text{‘}\Sigma\text{‘}Q' = \Lambda$

Dem.

$\vdash . *174{\cdot}3 . *163{\cdot}1 . \quad \supset \vdash :. \text{Hp} . \supset : R \in C\text{‘}Q . R' \in C\text{‘}Q' . \supset_{R, R'} . R \neq R' \quad (1)$

$\vdash . *162{\cdot}22 . \quad \supset \vdash :. \text{Hp} . \supset : R \in C\text{‘}Q . R' \in C\text{‘}Q' . \supset_{R, R'} . R, R' \in C\text{‘}\Sigma\text{‘}P \quad (2)$

$\vdash . (1) . (2) . *174{\cdot}31 . \supset \vdash :. \text{Hp} . \supset : R \in C\text{‘}Q . R' \in C\text{‘}Q' . \supset_{R, R'} . C\text{‘}R \cap C\text{‘}R' = \Lambda :$

[*40·27] $\supset : s\text{‘}C\text{‘‘}C\text{‘}Q \cap s\text{‘}C\text{‘‘}C\text{‘}Q' = \Lambda :$

[*162·22] $\supset : C\text{‘}\Sigma\text{‘}Q \cap C\text{‘}\Sigma\text{‘}Q' = \Lambda :. \supset \vdash . \text{Prop}$

***174·36.** $\vdash : P \in \text{Rel}^3\,\text{arithm} . \supset . \Sigma ; P \in \text{Rel}^2\,\text{excl}$

Dem.

$\vdash . *174{\cdot}35 . *37{\cdot}63 . *150{\cdot}22 . \supset$

$\vdash :. \text{Hp} . \supset : R, R' \in C\text{‘}\Sigma ; P . R \neq R' . \supset . C\text{‘}R \cap C\text{‘}R' = \Lambda \quad (1)$

$\vdash . (1) . *163{\cdot}1 . \supset \vdash . \text{Prop}$

$*174{\cdot}361$. $\vdash : P \epsilon \text{Rel}^3 \text{arithm} . \supset . C\text{‘}P \subset \text{Rel}^2 \text{excl}$

Dem.

$\vdash . *162{\cdot}1 . \supset \vdash : Q \epsilon C\text{‘}P . \supset . Q \mathrel{\dot\subset} \Sigma\text{‘}P \quad (1)$

$\vdash . *174{\cdot}3 . \supset \vdash : \text{Hp} . \supset . \Sigma\text{‘}P \epsilon \text{Rel}^2 \text{excl} \quad (2)$

$\vdash . (1) . (2) . *163{\cdot}43 . \supset \vdash . \text{Prop}$

$*174{\cdot}362$. $\vdash : P \epsilon \text{Rel}^3 \text{arithm} . Q, Q' \epsilon C\text{‘}P . C\text{“}C\text{‘}Q = C\text{“}C\text{‘}Q' . \supset . Q = Q'$

Dem.

$\vdash . *174{\cdot}322 . \quad \supset \vdash : \text{Hp} . \exists ! C\text{“}C\text{‘}Q . \supset . Q = Q' \quad (1)$

$\vdash . *37{\cdot}45 . \quad \supset \vdash : C\text{“}C\text{‘}Q = \Lambda . \supset . C\text{‘}Q = \Lambda .$

$[*33{\cdot}241] \quad \supset . Q = \dot\Lambda \quad (2)$

$\vdash . (2) . *13{\cdot}172 . \supset \vdash : \text{Hp} . C\text{“}C\text{‘}Q = \Lambda . \supset . Q = Q' \quad (3)$

$\vdash . (1) . (3) . \supset \vdash . \text{Prop}$

$*174{\cdot}363$. $\vdash : P \epsilon \text{Rel}^3 \text{arithm} . \supset . \text{Prod} ; P \epsilon \text{Rel}^2 \text{excl}$

Dem.

$\vdash . *173{\cdot}161 . *174{\cdot}361 . *173{\cdot}2 . \text{Transp} . \supset$

$\vdash :. \text{Hp} . \supset : Q, Q' \epsilon C\text{‘}P . \exists ! C\text{‘}\text{Prod}\text{‘}Q \cap C\text{‘}\text{Prod}\text{‘}Q' . \supset .$

$\exists ! \text{Prod}\text{‘}C\text{“}C\text{‘}Q \cap \text{Prod}\text{‘}C\text{“}C\text{‘}Q' .$

$[*115{\cdot}23 . *174{\cdot}33] \quad \supset . C\text{“}C\text{‘}Q = C\text{“}C\text{‘}Q' .$

$[*174{\cdot}362] \quad \supset . Q = Q' .$

$[*30{\cdot}37] \quad \supset . \text{Prod}\text{‘}Q = \text{Prod}\text{‘}Q' \quad (1)$

$\vdash . (1) . *163{\cdot}11 . *150{\cdot}22 . \supset \vdash . \text{Prop}$

$*174{\cdot}4$. $\vdash : P \epsilon \text{Rel}^2 \text{excl} . P \mathrel{\dot\subset} J . \supset .$

$\text{Prod}\text{‘}\Sigma\text{‘}P = \text{D} ; \dot{s} ; \text{Prod}\text{‘}\Pi ; P = s ; \text{D}_\epsilon ; \text{Prod}\text{‘}\Pi ; P$

Dem.

$\vdash . *173{\cdot}1 . \quad \supset \vdash . \text{Prod}\text{‘}\Sigma\text{‘}P = \text{D} ; \Pi\text{‘}\Sigma\text{‘}P \quad (1)$

$\vdash . (1) . *174{\cdot}24 . \supset \vdash : P \epsilon \text{Rel}^2 \text{excl} . P \mathrel{\dot\subset} J . \supset . \text{Prod}\text{‘}\Sigma\text{‘}P = \text{D} ; \dot{s} ; \text{Prod}\text{‘}\Pi ; P$

$[*41{\cdot}43] \quad = s ; \text{D}_\epsilon ; \text{Prod}\text{‘}\Pi ; P : \supset \vdash . \text{Prop}$

$*174{\cdot}401$. $\vdash : P \epsilon \text{Rel}^3 \text{arithm} . \supset . \text{Prod}\text{‘}\text{Prod} ; P = \text{D}_\epsilon ; \text{Prod}\text{‘}\Pi ; P$

Dem.

$\vdash . *80{\cdot}33 . *162{\cdot}23 . \supset \vdash : R \epsilon F_\Delta\text{‘}C\text{‘}Q . \supset . \text{D}\text{‘}R \subset C\text{‘}\Sigma\text{‘}Q \quad (1)$

$\vdash . (1) . \supset \vdash : R \epsilon F_\Delta\text{‘}C\text{‘}Q . R' \epsilon F_\Delta\text{‘}C\text{‘}Q' . \exists ! \text{D}\text{‘}R \cap \text{D}\text{‘}R' . \supset .$

$\exists ! C\text{‘}\Sigma\text{‘}Q \cap C\text{‘}\Sigma\text{‘}Q' \quad (2)$

$\vdash . (2) . *174{\cdot}35 . \supset \vdash :. \text{Hp} . \supset :$

$Q, Q' \epsilon C\text{‘}P . R \epsilon F_\Delta\text{‘}C\text{‘}Q . R' \epsilon F_\Delta\text{‘}C\text{‘}Q' . \text{D}\text{‘}R = \text{D}\text{‘}R' . \exists ! \text{D}\text{‘}R . \supset .$

$Q = Q' .$

$[*81{\cdot}21 . *174{\cdot}361 . *163{\cdot}12] \quad \supset . R = R' \quad (3)$

$\vdash . (3) . *33{\cdot}241 . \supset$

$\vdash :. \text{Hp} . \supset : Q, Q' \epsilon C\text{‘}P . R \epsilon F_\Delta\text{‘}C\text{‘}Q . R' \epsilon F_\Delta\text{‘}C\text{‘}Q' . \text{D}\text{‘}R = \text{D}\text{‘}R' . \supset . R = R' :$

$[*172{\cdot}12 . *150{\cdot}22] \supset : \text{D} \upharpoonright s\text{‘}C\text{“}\Pi\text{“}C\text{‘}P \epsilon 1 \rightarrow 1 :$

$[*162{\cdot}22] \quad \supset : \text{D} \upharpoonright C\text{‘}\Sigma\text{‘}\Pi ; P \epsilon 1 \rightarrow 1 :$

$[*173{\cdot}33] \quad \supset : \text{D}_\epsilon ; \text{Prod}\text{‘}\Pi ; P = \text{Prod}\text{‘}\text{D} \dagger ; \Pi ; P$

$[*173{\cdot}1] \quad = \text{Prod}\text{‘}\text{Prod} ; P :. \supset \vdash . \text{Prop}$

$*174{\cdot}41$. $\vdash : P \in \text{Rel}^3\text{arithm} \,.\, P \Subset J \,.\, \supset .\, \text{Prod}‘\Sigma‘P = s;\text{Prod}‘\text{Prod};P$ $\quad$ $[*174{\cdot}4{\cdot}401]$

$*174{\cdot}42$. $\vdash : P \in \text{Rel}^3\text{arithm} \,.\, P \Subset J \,.\, \supset .$

$$s \upharpoonright (C‘\text{Prod}‘\text{Prod};P) \in (\text{Prod}‘\Sigma‘P)\,\overline{\text{smor}}\,(\text{Prod}‘\text{Prod};P)$$

Dem.

$\vdash .\, *173{\cdot}161{\cdot}2 \,.\, *174{\cdot}363 \,.\, \supset$

$\vdash : \text{Hp} \,.\, \supset .\, C‘\text{Prod}‘\text{Prod};P \subset \text{Prod}‘C‘‘C‘\text{Prod};P$

$[*150{\cdot}22] \qquad \subset \text{Prod}‘C‘‘\text{Prod}‘‘C‘P$

$[*173{\cdot}161 . *174{\cdot}361] \qquad \subset \text{Prod}‘\text{Prod}‘‘C‘‘‘C‘‘C‘P \qquad (1)$

$\vdash .\, (1) \,.\, *174{\cdot}33 \,.\, *115{\cdot}46 \,.\, \supset \vdash : \text{Hp} \,.\, \supset .\, s \upharpoonright C‘\text{Prod}‘\text{Prod};P \in 1 \to 1 \qquad (2)$

$\vdash .\, (2) \,.\, *174{\cdot}41 \,.\, *151{\cdot}231 \,.\, \supset \vdash .\, \text{Prop}$

$*174{\cdot}43$. $\vdash : P \in \text{Rel}^3\text{arithm} \,.\, P \Subset J \,.\, \supset .\, \text{Prod}‘\Sigma‘P \;\text{smor}\; \text{Prod}‘\text{Prod};P$ $\quad$ $[*174{\cdot}42]$

This is the associative law for Prod.

$*174{\cdot}44$. $\vdash : P \in \text{Rel}^3\text{arithm} \,.\, \supset .\, \text{Prod}‘\text{Prod};P = \text{D}_\epsilon;\text{D};\Pi‘\Pi;P$ $\quad$ $[*174{\cdot}401 \,.\, *173{\cdot}1]$

$*174{\cdot}45$. $\vdash : P \in \text{Rel}^3\text{arithm} \,.\, \supset .$

$$(\text{D}_\epsilon \,|\, \text{D}) \upharpoonright C‘\Pi‘\Pi;P \in (\text{Prod}‘\text{Prod};P)\,\overline{\text{smor}}\,(\Pi‘\Pi;P)$$

Dem.

$\vdash .\, *174{\cdot}18 \,.\, \supset \vdash : \text{Hp} \,.\, \supset .\, \text{D} \upharpoonright C‘\Pi‘\Pi;P \in 1 \to 1 \qquad (1)$

$\vdash .\, *80{\cdot}33 \,.\, *162{\cdot}23 \,.\, \supset$

$\vdash : R \in F_\Delta‘C‘Q \,.\, R' \in F_\Delta‘C‘Q' \,.\, \exists !\, \text{D}‘R \cap \text{D}‘R' \,.\, \supset .\, \exists !\, C‘\Sigma‘Q \cap C‘\Sigma‘Q' \qquad (2)$

$\vdash .\, (2) \,.\, *174{\cdot}35 \,.\, \supset$

$\vdash :.\, \text{Hp} \,.\, \supset : Q, Q' \in C‘P \,.\, R \in F_\Delta‘C‘Q \,.\, R' \in F_\Delta‘C‘Q' \,.\, \exists !\, \text{D}‘R \,.\, \text{D}‘R = \text{D}‘R' \,.\, \supset .$

$Q = Q' \,.$

$[*81{\cdot}21 . *174{\cdot}361 . *163{\cdot}12] \qquad \supset .\, R = R' \qquad (3)$

$\vdash .\, (3) \,.\, *33{\cdot}241 \,.\, \supset$

$\vdash :.\, \text{Hp} \,.\, \supset : Q, Q' \in C‘P \,.\, R \in F_\Delta‘C‘Q \,.\, R' \in F_\Delta‘C‘Q' \,.\, \text{D}‘R \doteq \text{D}‘R' \,.\, \supset .\, R = R' :$

$[*172{\cdot}12] \supset : Q, Q' \in C‘P \,.\, R \in C‘\Pi‘Q \,.\, R' \in C‘\Pi‘Q' \,.\, \text{D}‘R = \text{D}‘R' \,.\, \supset .\, R = R' \quad (4)$

$\vdash .\, *173{\cdot}161 \,.\, *37{\cdot}6 \,.\, *173{\cdot}2 \,.\, \text{Transp} \,.\, \supset$

$\vdash ::\, \text{Hp} \,.\, \mu, \nu \in C‘\text{Prod}‘\Pi;P \,.\, \text{D}‘‘\mu = \text{D}‘‘\nu \,.\, \supset :.$

$R \in \mu \,.\, \supset : (\exists Q) : Q \in C‘P \,.\, R \in C‘\Pi‘Q :$

$(\exists Q', R') \,.\, Q' \in C‘P \,.\, R' \in C‘\Pi‘Q' \,.\, R' \in \nu \,.\, \text{D}‘R = \text{D}‘R' :$

$[(4)] \qquad \supset : (\exists R') \,.\, R' \in \nu \,.\, R = R' :$

$[*13{\cdot}195] \supset : R \in \nu \qquad (5)$

Similarly $\quad \vdash :.\, \text{Hp}\,(5) \,.\, \supset : R \in \nu \,.\, \supset .\, R \in \mu \qquad (6)$

$\vdash .\, (5) \,.\, (6) \,.\, \supset \vdash :.\, \text{Hp} \,.\, \supset : \mu, \nu \in C‘\text{Prod}‘\Pi;P \,.\, \text{D}‘‘\mu = \text{D}‘‘\nu \,.\, \supset .\, \mu = \nu :$

$[*71{\cdot}55] \qquad \supset : \text{D}_\epsilon \upharpoonright C‘\text{Prod}‘\Pi;P \in 1 \to 1 :$

$[*150{\cdot}22 . *173{\cdot}1] \qquad \supset : \text{D}_\epsilon \upharpoonright \text{D}‘‘C‘\Pi‘\Pi;P \in 1 \to 1 \qquad (7)$

$\vdash .\, (1) \,.\, (7) \,.\, *35{\cdot}481 \,.\, \supset \vdash : \text{Hp} \,.\, \supset .\, (\text{D}_\epsilon \,|\, \text{D}) \upharpoonright C‘\Pi‘\Pi;P \in 1 \to 1 \qquad (8)$

$\vdash .\, (8) \,.\, *174{\cdot}44 \,.\, \supset \vdash .\, \text{Prop}$

***174·46.** $\vdash : P \epsilon \mathrm{Rel}^3 \mathrm{arithm} . \supset . \mathrm{Prod}‘\mathrm{Prod}“P \mathrm{smor} \Pi‘\Pi“P$ [*174·45]

***174·461.** $\vdash : P \epsilon \mathrm{Rel}^3 \mathrm{arithm} . P \subseteq J . \supset . \mathrm{Prod}‘\mathrm{Prod}“P \mathrm{smor} \Pi‘\Sigma‘P$
[*174·46·25]

***174·462.** $\vdash : P \epsilon \mathrm{Rel}^3 \mathrm{arithm} . \supset . \Pi‘\mathrm{Prod}“P \mathrm{smor} \mathrm{Prod}‘\mathrm{Prod}“P$
[*174·363 . *173·16]

The two following propositions merely sum up previous results.

***174·47.** $\vdash : P \epsilon \mathrm{Rel}^3 \mathrm{arithm} . P \subseteq J . \supset .$
$\mathrm{Prod}‘\Sigma‘P = s“\mathrm{Prod}‘\mathrm{Prod}“P = s“\mathrm{D}_\epsilon“\mathrm{D}“\Pi‘\Pi“P = \mathrm{D}“\dot{s}“\mathrm{Prod}‘\Pi“P .$
$s \upharpoonright C“\mathrm{Prod}‘\mathrm{Prod}“P,\ s \,|\, \mathrm{D}_\epsilon \,|\, \mathrm{D} \upharpoonright C“\Pi‘\Pi“P,\ \mathrm{D} \,|\, \dot{s} \upharpoonright C“\mathrm{Prod}‘\Pi“P \epsilon 1 \rightarrow 1$
[*174·42·45·24 . *41·43]

***174·48.** $\vdash : P \epsilon \mathrm{Rel}^3 \mathrm{arithm} . P \subseteq J . \supset .$
$\mathrm{Nr}‘\mathrm{Prod}‘\mathrm{Prod}“P = \mathrm{Nr}‘\mathrm{Prod}‘\Sigma‘P = \mathrm{Nr}‘\Pi‘\Sigma‘P = \mathrm{Nr}‘\Pi‘\Pi“P$
$= \mathrm{Nr}‘\mathrm{Prod}‘\Pi“P = \mathrm{Nr}‘\Pi‘\mathrm{Prod}“P$
[*174·43·46·25·462 . *152·321]

$*176$. EXPONENTIATION.

Summary of $*176$.

The definition of exponentiation is framed on the analogy of the definition in cardinals, *i.e.* we put

$$P \exp Q = \text{Prod}‘P \underset{;}{\downarrow} {}^{;}Q \quad \text{Df.}$$

We put also, what is often a more convenient form,

$$P^Q = \dot{s}{}^{;}(P \exp Q) \quad \text{Df.}$$

The relation P^Q has for its field (unless $Q = \dot{\Lambda}$) the class of Cantor's "Belegungen," *i.e.* the class $(C‘P \uparrow C‘Q)_{\Delta}‘C‘Q$. It arranges these by a form of the principle of first differences, namely as follows: Suppose M and N are two members of $(C‘P \uparrow C‘Q)_{\Delta}‘C‘Q$, and suppose there is in $C‘Q$ a term y for which the M-representative $(M‘y)$ precedes the N-representative $(N‘y)$, *i.e.* for which $(M‘y)\, P\, (N‘y)$, and suppose further that all terms in $C‘Q$ which are earlier than y, *i.e.* for which $zQy \,.\, z \neq y$, have their M-representative and their N-representative identical; in this case we say that M has to N the relation P^Q. This may be stated as follows, provided we assume that P and Q are series: Let M and N be two one-valued functions whose possible arguments are all the members of $C‘Q$, while their values are some or all of the members of $C‘P$. Then we say that M has to N the relation P^Q if the first argument for which the two functions do not have the same value gives an earlier value to M than to N. Thus for example let P be the series a_1, a_2, a_3, a_4, a_5, and let Q be the series b_1, b_2, b_3, b_4. Then M and N are to be such that $M‘b$ or $N‘b$ is defined when, and only when, b is b_1 or b_2 or b_3 or b_4, and the value of $M‘b$ or $N‘b$ is a_1 or a_2 or a_3 or a_4 or a_5. Then if $M‘b_1 = a_1$ and $N‘b_1 \neq a_1$, M precedes N; if $M‘b_1 = N‘b_1 = a_1$, and $M‘b_2 = a_1 \,.\, N‘b_2 \neq a_1$, M precedes N; and so on. Thus in this case the first term of the series generated by P^Q is the one for which $M‘b = a_1$ when b has any of the values b_1, b_2, b_3, b_4. Thus the first term of the series is $\iota‘a_1 \uparrow C‘Q$, *i.e.* $\iota‘B‘P \uparrow C‘Q$. The next term will be

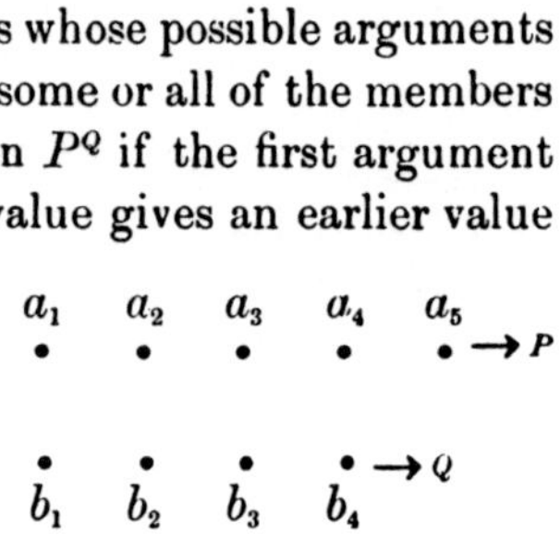

$a_1 \quad a_2 \quad a_3 \quad a_4 \quad a_5$
$\bullet \quad \bullet \quad \bullet \quad \bullet \quad \bullet \rightarrow P$

$\bullet \quad \bullet \quad \bullet \quad \bullet \rightarrow Q$
$b_1 \quad b_2 \quad b_3 \quad b_4$

$$\iota‘a_1 \uparrow (\iota‘b_1 \cup \iota‘b_2 \cup \iota‘b_3) \mathbin{\dot{\cup}} \iota‘a_2 \uparrow \iota‘b_4,$$

i.e.

$$\iota‘B‘P \uparrow \text{D}‘Q \cup 2_P \downarrow B‘Q.$$

The next is $$\iota`B`P \uparrow D`Q \cup 3_P \downarrow B`\breve{Q},$$

and so on. This makes it evident that our series has the structure required of a series which is to represent the Qth power of P.

The two relations $P \exp Q$ and P^Q are ordinally similar, since $\dot{s}$ is one-one when its field is limited to $C``(P \exp Q)$. This follows from $*116{\cdot}131$, together with

$$\dot{\exists} ! Q . \supset . C``(P \exp Q) = (C`P) \exp (C`Q).$$

If S is a correlator of P and P', and T is a correlator of Q and Q', then $(S \| \breve{T})_\epsilon$ and $(S \| \breve{T})$, with their converse domains limited, are respectively correlators of $(P \exp Q)$ with $(P' \exp Q')$ and of P^Q with $P'^{Q'}$. This shows that the relation-number of $(P \exp Q)$ depends only upon those of P and Q, which is of course essential if $(P \exp Q)$ is to afford a definition of exponentiation.

If the multiplicative axiom is assumed, then if R is a relation which is like Q, and whose field consists of relations which are like P, and $R \epsilon \text{Rel}^2 \text{excl}$, the product of R is like $(P \exp Q)$. That is, if we put $\mu = \text{Nr}`P . \nu = \text{Nr}`Q$, so that R consists of ν terms each of which has μ terms, the product of R has μ^ν terms. This gives the connection of multiplication with exponentiation.

There are two formal laws of exponentiation which hold for relation-numbers, namely

$$P^Q \times P^R \text{ smor } P^{Q \updownarrow R}$$

and $$(P^Q)^R \text{ smor } P^{R \times Q}.$$

They both need a hypothesis: the first needs

$$\dot{\exists} ! Q . \dot{\exists} ! R . C`Q \cap C`R = \Lambda,$$

while the second needs $$R \in J$$

because it is proved by means of the associative law ($*174{\cdot}43$).

The first of the above formal laws can be generalized, by putting $\Sigma`S$ in place of $Q \updownarrow R$, and taking the product of the various powers

$$P \exp Q,\ P \exp Q', \ldots,$$

where $Q, Q', \ldots \epsilon C`S$, and the products are taken in the order determined by S. The resulting generalization is

$$S \epsilon \text{Rel}^2 \text{excl} . S \in J . \supset . \{\text{Prod}`(P \exp);S\} \text{ smor } \{P \exp (\Sigma`S)\}.$$

The proof of this proposition results immediately from $*174{\cdot}43$ and $*162{\cdot}35$.

The proof of the second of the formal laws is more difficult. We observe, to begin with, that

$$P \exp (R \times Q) = \text{Prod}`P \downarrow_{;} ;\Sigma`Q \downarrow_{;} ;R.$$

Assuming suitable hypotheses, this, by $*162{\cdot}35$,

$$= \text{Prod}‘\Sigma‘(P \underset{;}{\downarrow})\dagger{}^{;} Q \underset{;}{\downarrow}{}^{;} R,$$

which is like $\quad \text{Prod}‘\text{Prod}^{;}(P \underset{;}{\downarrow})\dagger{}^{;} Q \underset{;}{\downarrow}{}^{;} R$, by $*174{\cdot}43$.

But $\quad (P \exp Q) \exp R = \text{Prod}‘\{\text{Prod}‘P \underset{;}{\downarrow}{}^{;} Q\} \underset{;}{\downarrow}{}^{;} R.$

Thus our result will follow if we can prove

$$\{\text{Prod}^{;}(P \underset{;}{\downarrow})\dagger{}^{;} Q \underset{;}{\downarrow}{}^{;} R\} \text{ smor smor } \{(\text{Prod}‘P \underset{;}{\downarrow}{}^{;} Q) \underset{;}{\downarrow}{}^{;} R\}.$$

Now one member of the field of $\text{Prod}^{;}(P \underset{;}{\downarrow})\dagger{}^{;} Q \underset{;}{\downarrow}{}^{;} R$ will be

$$\text{Prod}‘P \underset{;}{\downarrow}{}^{;} Q \underset{;}{\downarrow} z, \text{ where } z \in C‘R.$$

This is like $\text{Prod}‘P \underset{;}{\downarrow}{}^{;} Q$, because $Q \underset{;}{\downarrow} z \text{ smor } Q$. Hence $\text{Prod}^{;}(P \underset{;}{\downarrow})\dagger{}^{;} Q \underset{;}{\downarrow}{}^{;} R$ is a series of terms each of which is like $\text{Prod}‘P \underset{;}{\downarrow}{}^{;} Q$, and the whole series of such terms is like R. If we assumed the multiplicative axiom, this would suffice to prove the result. But it is possible to obtain our result without assuming the multiplicative axiom.

For this purpose, we proceed as follows. The correlator of

$$\text{Prod}‘P \underset{;}{\downarrow}{}^{;} Q \underset{;}{\downarrow} z \text{ and } \text{Prod}‘P \underset{;}{\downarrow}{}^{;} Q$$

is $\{|(\text{Cnv}‘\downarrow z)\}_\epsilon$, by $*165{\cdot}361$ and $*172{\cdot}3$. Call this $M‘z$. Then

$$M \in 1 \rightarrow 1 : z \in C‘R \,.\supset_z.\, (M‘z) \in (\text{Prod}‘P \underset{;}{\downarrow}{}^{;} Q \underset{;}{\downarrow} z) \,\overline{\text{smor}}\, (\text{Prod}‘P \underset{;}{\downarrow}{}^{;} Q) :$$

$$z, w \in C‘R \,.\, \exists ! \, D‘M‘z \cap D‘M‘w \,.\supset_{z,w}.\, z = w.$$

This, by the help of two or three lemmas, suffices to prove that

$$\{\text{Prod}^{;}(P \underset{;}{\downarrow})\dagger{}^{;} Q \underset{;}{\downarrow}{}^{;} R\} \text{ smor smor } \{(\text{Prod}‘P \underset{;}{\downarrow}{}^{;} Q) \underset{;}{\downarrow}{}^{;} R\},$$

whence the result follows.

The principal propositions of the present number are the following:

$*176{\cdot}1$. $\vdash .\, P \exp Q = \text{Prod}‘P \underset{;}{\downarrow}{}^{;} Q = D^{;}\Pi‘P \underset{;}{\downarrow}{}^{;} Q$

$*176{\cdot}11$. $\vdash .\, P^Q = \dot{s}^{;}(P \exp Q) = \dot{s}^{;}\, \text{Prod}‘P \underset{;}{\downarrow}{}^{;} Q = \dot{s}^{;} D^{;} \Pi‘P \underset{;}{\downarrow}{}^{;} Q$

These propositions merely embody the definitions.

$*176{\cdot}14$. $\vdash : \dot{\exists} ! \, Q \,.\supset.\, C‘(P \exp Q) = (C‘P) \exp (C‘Q) \,.\, C‘P^Q = (C‘P \uparrow C‘Q)_\Delta‘C‘Q$

$*176{\cdot}151$. $\vdash :.\, P = \dot{\Lambda} \,.\vee.\, Q = \dot{\Lambda} : \equiv .\, P \exp Q = \dot{\Lambda} \,.\equiv.\, P^Q = \dot{\Lambda}$

It will be observed that in relation-arithmetic, $\mu^0 = 0$, whereas in cardinal arithmetic $\mu^0 = 1$. The difference is due to the fact that there is no ordinal number 1 (cf. $*153$).

$*176{\cdot}181$. $\vdash .\, P^Q \text{ smor } (P \exp Q)$

***176·182.** $\vdash . (P \exp Q) \text{ smor } (\Pi\text{‘}P \underset{;}{\downarrow} {}^{;}Q)$

***176·19.** $\vdash :: S (P^Q) T . \equiv :. S, T \in (C\text{‘}P \uparrow C\text{‘}Q)_{\Delta}\text{‘}C\text{‘}Q :.$
$$(\exists y) : y \in C\text{‘}Q . (S\text{‘}y) P (T\text{‘}y) : y' Q y . y' \neq y . \supset_{y'} . S\text{‘}y' = T\text{‘}y'$$

***176·2.** $\vdash : U \upharpoonright C\text{‘}R \in P \,\overline{\text{smor}}\, R . W \upharpoonright C\text{‘}S \in Q \,\overline{\text{smor}}\, S . \supset .$
$$(U \parallel \breve{W})_{\epsilon} \upharpoonright C\text{‘}(R \exp S) \in (P \exp Q) \,\overline{\text{smor}}\, (R \exp S)$$

***176·21.** With the same hypothesis, $(U \parallel \breve{W}) \upharpoonright C\text{‘}(R^S)$ correlates P^Q and R^S

***176·22.** $\vdash : P \text{ smor } R . Q \text{ smor } S . \supset . (P \exp Q) \text{ smor } (R \exp S) . P^Q \text{ smor } R^S$

***176·24.** $\vdash :. \text{Mult ax} . \supset :$
$$R \in \text{Rel}^2 \text{ excl} \cap \text{Nr}\text{‘}Q . C\text{‘}R \subset \text{Nr}\text{‘}P . \supset . \Pi\text{‘}R \text{ smor } (P \exp Q)$$

This proposition connects multiplication and exponentiation.

***176·31.** $\vdash : \dot{\exists} ! Q . \supset . \overrightarrow{B}\text{‘}(P \exp Q) = (\overrightarrow{B}\text{‘}P) \exp (C\text{‘}Q)$

***176·311·32·321.** Similar propositions for $\overrightarrow{B}\text{‘}\text{Cnv}\text{‘}(P \exp Q), \overrightarrow{B}\text{‘}(P^Q), \overrightarrow{B}\text{‘}(\text{Cnv}\text{‘}P^Q)$

***176·34.** $\vdash : \dot{\exists} ! Q . \text{E} ! B\text{‘}P . \supset .$
$$B\text{‘}(P \exp Q) = (B\text{‘}P) \downarrow \text{“}C\text{‘}Q . B\text{‘}(P^Q) = (\iota\text{‘}B\text{‘}P) \uparrow C\text{‘}Q$$

We come next to the formal laws. We have

***176·42.** $\vdash : \dot{\exists} ! Q . \dot{\exists} ! R . C\text{‘}Q \cap C\text{‘}R = \Lambda . \supset . P^Q \times P^R \text{ smor } P^{Q \ddagger R} .$
$$(P \exp Q) \times (P \exp R) \text{ smor } P \exp (Q \ddagger R)$$

***176·44.** $\vdash : S \in \text{Rel}^2 \text{excl} . S \in J . \supset . \{\text{Prod}\text{‘}(P \exp){}^{;}S\} \text{ smor } \{P \exp (\Sigma\text{‘}S)\}$

This is an extension of *176·42.

***176·57.** $\vdash : R \in J . \supset . \{(P \exp Q) \exp R\} \text{ smor } \{P \exp (R \times Q)\} .$
$$(P^Q)^R \text{ smor } P^{R \times Q}$$

***176·01.** $P \exp Q = \text{Prod}\text{‘}P \underset{;}{\downarrow} {}^{;}Q$ Df

***176·02.** $P^Q = \dot{s}{}^{;}(P \exp Q)$ Df

***176·1.** $\vdash . P \exp Q = \text{Prod}\text{‘}P \underset{;}{\downarrow} {}^{;}Q = \text{D}{}^{;}\Pi\text{‘}P \underset{;}{\downarrow} {}^{;}Q$ [(*176·01)]

***176·11.** $\vdash . P^Q = \dot{s}{}^{;}(P \exp Q) = \dot{s}{}^{;}\text{Prod}\text{‘}P \underset{;}{\downarrow} {}^{;}Q = \dot{s}{}^{;}\text{D}{}^{;}\Pi\text{‘}P \underset{;}{\downarrow} {}^{;}Q$ [(*176·02)]

***176·12.** $\vdash :: \mu (P \exp Q) \nu . \equiv :. \mu, \nu \in (C\text{‘}P) \exp (C\text{‘}Q) :.$
$$(\exists y, x, x') : x \downarrow y \in \mu . x' \downarrow y \in \nu . x P x' : z Q y . z \neq y . w \downarrow z \in \mu . \supset_{w,z} . w \downarrow z \in \nu$$

Dem.

$\vdash . *165·21·12 . *163·12 . \supset \vdash . F \upharpoonright P \underset{;}{\downarrow} \text{“}C\text{‘}Q \in \text{Cls} \to 1 .$ (1)

[*85·1.*115·1.*33·5] $\supset \vdash . \text{D}\text{“}F_{\Delta}\text{‘}P \underset{;}{\downarrow} \text{“}C\text{‘}Q = \text{Prod}\text{‘}C\text{“}P \underset{;}{\downarrow} \text{“}C\text{‘}Q$

[*165·12·14.(*116·01)] $= (C\text{‘}P) \exp (C\text{‘}Q)$ (2)

$\vdash . *176·1 . *173·11 . *172·11 . *165·12 . \supset$

$\vdash ::. \mu (P \exp Q) \nu . \equiv :: (\exists M, N, y) :. M, N \in F_{\Delta}\text{‘}P \underset{;}{\downarrow} \text{“}C\text{‘}Q :$
$$y \in C\text{‘}Q . (M\text{‘}P \underset{;}{\downarrow} y)(P \underset{;}{\downarrow} y)(N\text{‘}P \underset{;}{\downarrow} y) :$$
$$z Q y . z \neq y . \supset_z . M\text{‘}P \underset{;}{\downarrow} z = N\text{‘}P \underset{;}{\downarrow} z : \mu = \text{D}\text{‘}M . \nu = \text{D}\text{‘}N ::$$

$$[*81\cdot15.(1).*150\cdot6] \equiv :: (\exists M, N, y) :. M, N \in F_{\Delta}\text{‘}P \underset{\cdot\cdot}{\downarrow}\text{‘‘}C\text{‘}Q \,.\, \mu = \text{D}\text{‘}M \,.\, \nu = \text{D}\text{‘}N :$$

$$y \in C\text{‘}Q \,.\, \breve{\iota}\text{‘}(\mu \cap \downarrow y\text{‘‘}C\text{‘}P)(\downarrow y ; P)\, \breve{\iota}\text{‘}(\nu \cap \downarrow y\text{‘‘}C\text{‘}P) :$$

$$zQy \,.\, z \neq y \,.\, \supset_z \,.\, \breve{\iota}\text{‘}(\mu \cap \downarrow y\text{‘‘}C\text{‘}P) = \breve{\iota}\text{‘}(\nu \cap \downarrow y\text{‘‘}C\text{‘}P) ::$$

$$[(2).*150\cdot55] \equiv :: (\exists y) :: \mu, \nu \in (C\text{‘}P) \exp (C\text{‘}Q) \,.\, y \in C\text{‘}Q :. (\exists x, x') :.$$

$$x \downarrow y = \breve{\iota}\text{‘}(\mu \cap \downarrow y\text{‘‘}C\text{‘}P) \,.\, x' \downarrow y = \breve{\iota}\text{‘}(\nu \cap \downarrow y\text{‘‘}C\text{‘}P) \,.\, xPx' :$$

$$zQy \,.\, z \neq y \,.\, w \downarrow z = \breve{\iota}\text{‘}(\mu \cap \downarrow z\text{‘‘}C\text{‘}P) \,.$$

$$w' \downarrow z = \breve{\iota}\text{‘}(\nu \cap \downarrow z\text{‘‘}C\text{‘}P) \,.\, \supset_{z,w,w'} \,.\, w = w' ::$$

$$[*116\cdot11] \equiv :: (\exists y) :: \mu, \nu \in (C\text{‘}P) \exp (C\text{‘}Q) :.$$

$$(\exists x, x') : x \downarrow y \in \mu \,.\, x' \downarrow y \in \nu \,.\, xPx' :$$

$$zQy \,.\, z \neq y \,.\, w \downarrow z \in \mu \,.\, \supset_{w,z} \,.\, w \downarrow z \in \nu ::. \supset \vdash . \text{Prop}$$

The above proposition is used in $*176\cdot19$. It has the merit of giving a direct formula for $P \exp Q$, instead of one which proceeds by way of $\Pi\text{‘}P \underset{\cdot\cdot}{\downarrow} ; Q$.

$*176\cdot13$. $\vdash : \dot{\exists} ! (P \exp Q) \,.\equiv.\, \dot{\exists} ! P^Q \,.\equiv.\, \dot{\exists} ! \Pi\text{‘}P \underset{\cdot\cdot}{\downarrow} ; Q$ $[*150\cdot25 \,.\, *176\cdot1\cdot11]$

$*176\cdot131$. $\vdash : Q = \dot{\Lambda} \,.\supset.\, P \exp Q = \dot{\Lambda} \,.\, P^Q = \dot{\Lambda}$ $[*165\cdot241 \,.\, *173\cdot2 \,.\, *150\cdot42]$

Owing to this proposition, propositions stating analogies between ordinal and cardinal powers mostly require the hypothesis $\dot{\exists} ! Q$ or its equivalent, because an ordinal power whose index is zero is itself zero, whereas a cardinal power whose index is zero is 1.

$*176\cdot132$. $\vdash : P = \dot{\Lambda} \,.\, \dot{\exists} ! Q \,.\supset.\, P \exp Q = \dot{\Lambda} \,.\, P^Q = \dot{\Lambda}$
$[*165\cdot244 \,.\, *172\cdot14 \,.\, *176\cdot13 \,.\, *150\cdot42]$

$*176\cdot133$. $\vdash \,.\, C\text{‘}P^Q = \dot{s}\text{‘‘}C\text{‘}(P \exp Q)$ $[*176\cdot11 \,.\, *150\cdot22]$

$*176\cdot14$. $\vdash : \dot{\exists} ! Q \,.\supset.\, C\text{‘}(P \exp Q) = (C\text{‘}P) \exp (C\text{‘}Q) \,.\, C\text{‘}P^Q = (C\text{‘}P \uparrow C\text{‘}Q)_{\Delta}\text{‘}C\text{‘}Q$

Dem.

$$\vdash . *165\cdot243 \,.\supset \vdash : \text{Hp} \,.\supset.\, \dot{\exists} ! P \underset{\cdot\cdot}{\downarrow} ; Q \,.$$

$$[*173\cdot161.*165\cdot21] \quad \supset . \, C\text{‘}\text{Prod}\text{‘}P \underset{\cdot\cdot}{\downarrow} ; Q = \text{Prod}\text{‘}C\text{‘‘}C\text{‘}P \underset{\cdot\cdot}{\downarrow} ; Q \,.$$

$$[*176\cdot1.*165\cdot14] \quad \supset . \, C\text{‘}(P \exp Q) = \text{Prod}\text{‘}(C\text{‘}P) \underset{\cdot\cdot}{\downarrow} \text{‘‘}(C\text{‘}Q)$$

$$[(*116\cdot01)] \quad = (C\text{‘}P) \exp (C\text{‘}Q) \qquad (1)$$

$$\vdash . (1) . *176\cdot133 \,.\supset \vdash : \text{Hp} \,.\supset.\, C\text{‘}P^Q = \dot{s}\text{‘‘}\{(C\text{‘}P) \exp (C\text{‘}Q)\}$$

$$[*116\cdot13] \quad = (C\text{‘}P \uparrow C\text{‘}Q)_{\Delta}\text{‘}C\text{‘}Q \qquad (2)$$

$$\vdash . (1) . (2) \,.\supset \vdash . \text{Prop}$$

$*176\cdot15$. $\vdash : \dot{\exists} ! P \,.\, \dot{\exists} ! Q \,.\equiv.\, \dot{\exists} ! (P \exp Q) \,.\equiv.\, \dot{\exists} ! P^Q$

Dem.

$$\vdash . *176\cdot131\cdot132 \,. \quad \supset \vdash : \dot{\exists} ! (P \exp Q) \,.\supset.\, \dot{\exists} ! P \,.\, \dot{\exists} ! Q \qquad (1)$$

$$\vdash . *116\cdot18 \,.\, *176\cdot14 \,.\supset \vdash : \dot{\exists} ! P \,.\, \dot{\exists} ! Q \,.\supset.\, \dot{\exists} ! C\text{‘}(P \exp Q) \,.$$

$$[*33\cdot24] \quad \supset . \, \dot{\exists} ! (P \exp Q) \qquad (2)$$

$$\vdash . (1) . (2) . *176\cdot13 \,.\supset \vdash . \text{Prop}$$

***176·151.** $\vdash :. P = \dot{\Lambda} \,.\, \mathsf{v} \,.\, Q = \dot{\Lambda} : \equiv . P \exp Q = \dot{\Lambda} . \equiv . P^Q = \dot{\Lambda}$ [*176·15]

***176·16.** $\vdash . C\text{‘}(P \exp Q) \subset (C\text{‘}P) \exp (C\text{‘}Q) . C\text{‘}P^Q \subset (C\text{‘}P \uparrow C\text{‘}Q)_\Delta\text{‘}C\text{‘}Q$ [*176·14·151]

***176·18.** $\vdash . \dot{s} \upharpoonright C\text{‘}(P \exp Q) \in (P^Q) \overline{\text{smor}} (P \exp Q)$

Dem.

$\vdash . *116·131 . *176·14 . \supset$

$\vdash : \dot{\exists} ! Q . \supset . \dot{s} \upharpoonright C\text{‘}(P \exp Q) \in (C\text{‘}P^Q) \overline{\text{sm}}\, C\text{‘}(P \exp Q)$ (1)

$\vdash . (1) . *176·11 . *151·191 . \supset$

$\vdash : \dot{\exists} ! Q . \supset . \dot{s} \upharpoonright C\text{‘}(P \exp Q) \in (P^Q) \overline{\text{smor}} (P \exp Q)$ (2)

$\vdash . *176·151 . *150·42 . *72·1 . \supset$

$\vdash : Q = \dot{\Lambda} . \supset . \dot{s} \upharpoonright C\text{‘}(P \exp Q) \in (P^Q) \overline{\text{smor}} (P \exp Q)$ (3)

$\vdash . (2) . (3) . \supset \vdash$. Prop

***176·181.** $\vdash . P^Q \text{ smor } (P \exp Q)$ [*176·18]

***176·182.** $\vdash . (P \exp Q) \text{ smor } (\Pi\text{‘}P \downarrow ; Q)$ [*176·1 . *173·16 . *165·21]

***176·19.** $\vdash :: S (P^Q) T . \equiv :. S, T \in (C\text{‘}P \uparrow C\text{‘}Q)_\Delta\text{‘}C\text{‘}Q :.$
$(\exists y) : y \in C\text{‘}Q . (S\text{‘}y) P (T\text{‘}y) : y'Qy . y' \neq y . \supset_{y'} . S\text{‘}y' = T\text{‘}y'$

Dem.

$\vdash . *176·11·12 . \supset$

$\vdash :: S (P^Q) T . \equiv :. (\exists \mu, \nu) :. \mu, \nu \in (C\text{‘}P) \exp (C\text{‘}Q) . S = \dot{s}\text{‘}\mu . T = \dot{s}\text{‘}\nu :.$
$(\exists y, x, x') : y \in C\text{‘}Q . x \downarrow y \in \mu . x' \downarrow y \in \nu . xPx' :$
$y'Qy . y' \neq y . w \downarrow y' \in \mu . \supset_{y', w} . w \downarrow y' \in \nu :.$

[*56·4] $\equiv :. (\exists \mu, \nu) :. \mu, \nu \in (C\text{‘}P) \exp (C\text{‘}Q) . S = \dot{s}\text{‘}\mu . T = \dot{s}\text{‘}\nu :.$
$(\exists y, x, x') : y \in C\text{‘}Q . xSy . x'Ty . xPx' : y'Qy . y' \neq y . wSy' . \supset_{y', w} . wTy' :.$

[*116·13.*80·3] $\equiv :. S, T \in (C\text{‘}P \uparrow C\text{‘}Q)_\Delta\text{‘}C\text{‘}Q :. (\exists y) : y \in C\text{‘}Q . (S\text{‘}y) P (T\text{‘}y) :$
$y'Qy . y' \neq y . \supset_{y'} . S\text{‘}y' = T\text{‘}y' :: \supset \vdash$. Prop

The above proposition is often useful, since it gives a direct formula for P^Q, not one which passes by way of $P \exp Q$ or $\Pi\text{‘}P \downarrow ; Q$.

***176·2.** $\vdash : U \upharpoonright C\text{‘}R \in P \overline{\text{smor}} R . W \upharpoonright C\text{‘}S \in Q \overline{\text{smor}} S . \supset .$
$(U \| \breve{W})_\epsilon \upharpoonright C\text{‘}(R \exp S) \in (P \exp Q) \overline{\text{smor}} (R \exp S)$

Dem.

$\vdash . *165·362 . \supset \vdash : \text{Hp} . \supset . (U \| \breve{W}) \upharpoonright C\text{‘}\Sigma\text{‘}R \downarrow ; S \in (P \downarrow ; Q) \overline{\text{smor}}\, \overline{\text{smor}} (R \downarrow ; S) .$

[*173·3] $\supset . (U \| \breve{W})_\epsilon \upharpoonright C\text{‘}\text{Prod}\text{‘}R \downarrow ; S \in (\text{Prod}\text{‘}P \downarrow ; Q) \overline{\text{smor}} (\text{Prod}\text{‘}R \downarrow ; S)$ (1)

$\vdash . (1) . *176·1 . \supset \vdash$. Prop

***176·21.** $\vdash : U \upharpoonright C`R \in P \,\overline{\text{smor}}\, R \,.\, W \upharpoonright C`S \in Q \,\overline{\text{smor}}\, S \,.\, \supset .$
$$(U \,\|\, \breve{W}) \upharpoonright C`(R^S) \in (P^Q) \,\overline{\text{smor}}\, (R^S)$$

Dem.

$\vdash . *176·2·18 . *151·401 . \supset \vdash : \text{Hp} . \supset . \dot{s}`(U \,\|\, \breve{W})_\epsilon \upharpoonright C`(R \exp S) \in (P^Q) \,\overline{\text{smor}}\, (R^S)$

[*150·961] $\supset . (U \,\|\, \breve{W}) \upharpoonright \dot{s}``C`(R \exp S) \in (P^Q) \,\overline{\text{smor}}\, (R^S)$ (1)

$\vdash . (1) . *176·11 . *150·22 . \supset \vdash . \text{Prop}$

***176·22.** $\vdash : P \text{ smor } R \,.\, Q \text{ smor } S \,.\, \supset . (P \exp Q) \text{ smor } (R \exp S) \,.\, P^Q \text{ smor } R^S$
[*176·2·21]

***176·23.** $\vdash : R \text{ smor smor } P \downarrow ; Q \,.\, \supset . \Pi`R \text{ smor } (P \exp Q)$

Dem.

$\vdash . *172·44 . \supset \vdash : \text{Hp} . \supset . \Pi`R \text{ smor } \Pi`P \downarrow ; Q$ (1)

$\vdash . (1) . *176·182 . \supset \vdash . \text{Prop}$

***176·24.** $\vdash :. \text{Mult ax} . \supset :$
$$R \in \text{Rel}^2 \text{excl} \cap \text{Nr}`Q \,.\, C`R \subset \text{Nr}`P \,.\, \supset . \Pi`R \text{ smor } (P \exp Q)$$
[*165·38 . *176·23]

***176·3.** $\vdash . \text{Cnv}`(P^Q) = (\breve{P})^Q$

Dem.

$\vdash . *176·19 . \supset$

$\vdash :: T (\breve{P})^Q S . \equiv :. S, T \in (C`P \uparrow C`Q)_\Delta`C`Q :.$
$$(\exists y) : y \in C`Q \,.\, (T`y) \breve{P} (S`y) : y'Qy \,.\, y' \neq y \,.\, \supset_{y'} . S`y' = T`y' :.$$
[*176·19] $\equiv :. S (P^Q) T :: \supset \vdash . \text{Prop}$

***176·31.** $\vdash : \dot{\exists} ! Q \,.\, \supset . \overrightarrow{B}`(P \exp Q) = (\overrightarrow{B}`P) \exp (C`Q)$

Dem.

$\vdash . *165·21 . *163·12 . *71·221 . *93·1 . \supset \vdash . B \upharpoonright C`P \downarrow ; Q \in \text{Cls} \to 1$ (1)

$\vdash . *165·12·01 . *37·67 . \supset \vdash . \overrightarrow{B}``C`P \downarrow ; Q = \hat{\alpha} \{(\exists z) . z \in C`Q . \alpha = \overrightarrow{B}` \downarrow z ; P\}$

[*165·251.*151·5.*38·3] $= (\overrightarrow{B}`P) \downarrow ``C`Q$ (2)

$\vdash . *172·162 . *165·243 . \supset \vdash : \dot{\exists} ! Q \,.\, \supset . \overrightarrow{B}`\Pi`P \downarrow ; Q = B_\Delta`C`P \downarrow ; Q \,.$

[*173·16.*165·21.*151·5] $\supset . \overrightarrow{B}`(P \exp Q) = \text{D}``B_\Delta`C`P \downarrow ; Q$

[*85·1.(1).*115·1] $= \text{Prod}`\overrightarrow{B}``C`P \downarrow ; Q$

[(2)] $= \text{Prod}`(\overrightarrow{B}`P) \downarrow ``C`Q$

[(*116·01)] $= (\overrightarrow{B}`P) \exp (C`Q) : \supset \vdash . \text{Prop}$

***176·311.** $\vdash : \dot{\exists} ! Q \,.\, \supset . \overrightarrow{B}`\text{Cnv}`(P \exp Q) = (\overrightarrow{B}`\breve{P}) \exp (C`Q)$
[Proof as in *176·31]

***176·32.** $\vdash : \dot{\exists} ! Q . \supset . \overrightarrow{B}\text{‘}(P^Q) = (\overrightarrow{B}\text{‘}P \uparrow C\text{‘}Q)_\Delta\text{‘}C\text{‘}Q$

Dem.

$\vdash . *176\cdot31\cdot18 . *151\cdot5 . \supset$

$\vdash : \text{Hp} . \supset . \overrightarrow{B}\text{‘}(P^Q) = \dot{s}\text{‘‘}(\overrightarrow{B}\text{‘}P) \exp (C\text{‘}Q)$

[*116·13] $\quad = (\overrightarrow{B}\text{‘}P \uparrow C\text{‘}Q)_\Delta\text{‘}C\text{‘}Q : \supset \vdash . \text{Prop}$

***176·321.** $\vdash : \dot{\exists} ! Q . \supset . \overrightarrow{B}\text{‘}\text{Cnv}\text{‘}(P^Q) = (\overrightarrow{B}\text{‘}\breve{P} \uparrow C\text{‘}Q)_\Delta\text{‘}C\text{‘}Q$ [*176·32·3]

***176·33.** $\vdash :. \dot{\exists} ! Q . \supset : \exists ! \overrightarrow{B}\text{‘}(P \exp Q) . \equiv . \exists ! \overrightarrow{B}\text{‘}(P^Q) . \equiv . \exists ! \overrightarrow{B}\text{‘}P :$

$\exists ! \overrightarrow{B}\text{‘}\text{Cnv}\text{‘}(P \exp Q) . \equiv . \exists ! \overrightarrow{B}\text{‘}\text{Cnv}\text{‘}(P^Q) . \equiv . \exists ! \overrightarrow{B}\text{‘}\breve{P}$

[*176·31·311·32·321 . *116·18·15]

***176·34.** $\vdash : \dot{\exists} ! Q . \text{E} ! B\text{‘}P . \supset .$

$B\text{‘}(P \exp Q) = (B\text{‘}P) \downarrow \text{‘‘}C\text{‘}Q . B\text{‘}(P^Q) = (\iota\text{‘}B\text{‘}P) \uparrow C\text{‘}Q$

Dem.

$\vdash . *176\cdot31 . \supset \vdash : \text{Hp} . \supset . \overrightarrow{B}\text{‘}(P \exp Q) = (\iota\text{‘}B\text{‘}P) \exp (C\text{‘}Q)$

[(*116·01)] $\quad = \text{Prod}\text{‘}(\iota\text{‘}B\text{‘}P) \underset{,,}{\downarrow} \text{‘‘}C\text{‘}Q$

[*38·3.*53·31] $\quad = \text{Prod}\text{‘}\iota\text{‘‘}(B\text{‘}P) \downarrow \text{‘‘}C\text{‘}Q$

[*115·143] $\quad = \iota\text{‘}\{(B\text{‘}P) \downarrow \text{‘‘}C\text{‘}Q\}$ (1)

$\vdash . *176\cdot32 . \supset \vdash : \text{Hp} . \supset . \overrightarrow{B}\text{‘}(P^Q) = (\iota\text{‘}B\text{‘}P \uparrow C\text{‘}Q)_\Delta\text{‘}C\text{‘}Q$

[*116·12.*51·4] $\quad = \iota\text{‘}\{(\iota\text{‘}B\text{‘}P) \uparrow C\text{‘}Q\}$ (2)

$\vdash . (1) . (2) . \supset \vdash . \text{Prop}$

***176·341.** $\vdash : \dot{\exists} ! Q . \text{E} ! B\text{‘}\breve{P} . \supset .$

$B\text{‘}\text{Cnv}\text{‘}(P \exp Q) = (B\text{‘}\breve{P}) \downarrow \text{‘‘}C\text{‘}Q . B\text{‘}\text{Cnv}\text{‘}(P^Q) = (\iota\text{‘}B\text{‘}\breve{P}) \uparrow C\text{‘}Q$

[Proof as in *176·34]

***176·35.** $\vdash : P \Subset Q . \supset . P^R \Subset Q^R$

Dem.

$\vdash . *116\cdot12 . \supset \vdash : \text{Hp} . \supset . (C\text{‘}P \uparrow C\text{‘}R)_\Delta\text{‘}C\text{‘}R \subset (C\text{‘}Q \uparrow C\text{‘}R)_\Delta\text{‘}C\text{‘}R$ (1)

$\vdash . (1) . *176\cdot19 . \supset \vdash . \text{Prop}$

The above proposition is used in the theory of finite ordinals (*261·64).

The following propositions are concerned in proving (with a suitable hypothesis)

$$P^Q \times P^R \text{ smor } P^{Q \ddagger R}$$

and its extension

$$\{\text{Prod}\text{‘}(P \exp)\text{;}S\} \text{ smor } \{P \exp (\Sigma\text{‘}S)\}.$$

∗176·4. $\vdash : \dot{\exists} ! Q . \dot{\exists} ! R . P``C`Q \cap P``C`R = \Lambda . C`Q \cup C`R \subset \text{Ɑ}`P . \supset .$
$\dot{s} \mid C \upharpoonright C`\{(\Pi`P;Q) \times (\Pi`P;R)\} \in \{\Pi`P;(Q \ddagger R)\} \,\overline{\text{smor}}\, \{(\Pi`P;Q) \times (\Pi`P;R)\}$

Dem.

$\vdash . ∗172·34 . ∗150·22·24 . \supset \vdash : \text{Hp} . \supset .$

$\dot{s} \mid C \upharpoonright C`\{(\Pi`P;Q) \times (\Pi`P;R)\} \in \{\Pi`(P;Q \ddagger P;R)\} \,\overline{\text{smor}}\, \{(\Pi`P;Q) \times (\Pi`P;R)\}$ (1)

$\vdash . ∗162·36 . \supset \vdash : \text{Hp} . \supset . P;Q \ddagger P;R = P;(Q \ddagger R)$ (2)

$\vdash . (1) . (2) . \supset \vdash . \text{Prop}$

∗176·41. $\vdash : \dot{\exists} ! Q . \dot{\exists} ! R . P``C`Q \cap P``C`R = \Lambda . C`Q \cup C`R \subset \text{Ɑ}`P . \supset .$
$\Pi`P;(Q \ddagger R) \,\text{smor}\, (\Pi`P;Q) \times (\Pi`P;R)$ [∗176·4]

∗176·42. $\vdash : \dot{\exists} ! Q . \dot{\exists} ! R . C`Q \cap C`R = \Lambda . \supset . P^Q \times P^R \,\text{smor}\, P^{Q \ddagger R} .$
$(P \exp Q) \times (P \exp R) \,\text{smor}\, P \exp (Q \ddagger R)$

Dem.

$\vdash . ∗72·411 . ∗165·22 . \supset$

$\vdash : \dot{\exists} ! P . C`Q \cap C`R = \Lambda . \supset . P \underset{;}{\downarrow} ``C`Q \cap P \underset{;}{\downarrow} ``C`R = \Lambda$ (1)

$\vdash . (1) . ∗176·41 \dfrac{P \underset{;}{\downarrow}}{P} . ∗38·12 . ∗33·431 . \supset$

$\vdash : \text{Hp} . \dot{\exists} ! P . \supset . \Pi`P \underset{;}{\downarrow} ;(Q \ddagger R) \,\text{smor}\, (\Pi`P \underset{;}{\downarrow} ;Q) \times (\Pi`P \underset{;}{\downarrow} ;R) .$

[∗176·182.∗166·23] $\supset . P \exp (Q \ddagger R) \,\text{smor}\, (P \exp Q) \times (P \exp R) .$ (2)

[∗176·181.∗166·23] $\supset . P^{Q \ddagger R} \,\text{smor}\, P^Q \times P^R$ (3)

$\vdash . ∗176·151 . ∗166·13 . ∗153·101 . \supset$

$\vdash : P = \dot{\Lambda} . \supset . P \exp (Q \ddagger R) \,\text{smor}\, (P \exp Q) \times (P \exp R) .$
$P^{Q \ddagger R} \,\text{smor}\, P^Q \times P^R$ (4)

$\vdash . (2) . (3) . (4) . \supset \vdash . \text{Prop}$

∗176·43. $\vdash : S \in \text{Rel}^2 \text{excl} . S \Subset J . \supset .$
$s \upharpoonright C`\text{Prod}`(P \exp);S \in \{P \exp (\Sigma`S)\} \,\overline{\text{smor}}\, \text{Prod}`(P \exp);S$

Dem.

$\vdash . ∗165·22 . ∗163·3 . \quad \supset \vdash : \text{Hp} . \dot{\exists} ! P . \supset . (P \underset{;}{\downarrow})\dagger;S \in \text{Rel}^2 \text{excl}$ (1)

$\vdash . ∗162·35 . ∗38·12 . ∗33·431 . \supset \vdash . \Sigma`(P \underset{;}{\downarrow})\dagger;S = P \underset{;}{\downarrow} ;\Sigma`S$ (2)

$\vdash . ∗165·21 . (2) . \quad \supset \vdash . \Sigma`(P \underset{;}{\downarrow})\dagger;S \in \text{Rel}^2 \text{excl}$ (3)

$\vdash . (1) . (3) . ∗174·3 . \quad \supset \vdash : \text{Hp} . \dot{\exists} ! P . \supset . (P \underset{;}{\downarrow})\dagger;S \in \text{Rel}^3 \text{arithm}$ (4)

$\vdash . ∗165·223 . \text{Transp} . \quad \supset \vdash :. \text{Hp} . \dot{\exists} ! P . \supset : Q \neq R . \supset . P \underset{;}{\downarrow} ;Q \neq P \underset{;}{\downarrow} ;R :$

[∗150·4.∗72·14] $\supset : (P \underset{;}{\downarrow})\dagger;S \Subset J$ (5)

$\vdash . ∗176·1 . \quad \supset \vdash . \text{Prod}`(P \exp);S = \text{Prod}`\text{Prod};(P \underset{;}{\downarrow})\dagger;S$ (6)

$\vdash . (4) . (5) . (6) . ∗174·42 . \supset$

$\vdash : \text{Hp} \,.\, \dot{\exists} \,!\, P \,.\, \supset .$

$s \upharpoonright C\text{'Prod'}(P\,\text{exp}) ; S \in \{\text{Prod'}\Sigma\text{'}(P \underset{\centerdot}{\downarrow}) \dagger ; S\} \,\overline{\text{smor}}\, \{\text{Prod'}(P\,\text{exp}) ; S\} \,.$

$[(2)] \quad \supset . \, s \upharpoonright C\text{'Prod'}(P\,\text{exp}) ; S \in \{\text{Prod'}P \underset{\centerdot}{\downarrow} ; \Sigma\text{'}S\} \,\overline{\text{smor}}\, \{\text{Prod'}(P\,\text{exp}) ; S\} \,.$

$[*176\cdot1] \supset . \, s \upharpoonright C\text{'}(P\,\text{exp}) ; S \in \{P\,\text{exp}\,(\Sigma\text{'}S)\} \,\overline{\text{smor}}\, \{\text{Prod'}(P\,\text{exp}) ; S\} \quad (7)$

$\vdash . *176\cdot151 . *173\cdot21 . *172\cdot13\cdot14 . \supset$

$\vdash : P = \dot{\Lambda} . \supset . \text{Prod'}(P\,\text{exp}) ; S = \dot{\Lambda} \,.\, P\,\text{exp}\,(\Sigma\text{'}S) = \dot{\Lambda} \quad (8)$

$\vdash . (7) . (8) . *173\cdot2 . *164\cdot32 . \supset \vdash . \text{Prop}$

***176·44.** $\vdash : S \in \text{Rel}^2\,\text{excl} \,.\, S \subseteq J \,.\, \supset . \{\text{Prod'}(P\,\text{exp}) ; S\} \,\text{smor}\, \{P\,\text{exp}\,(\Sigma\text{'}S)\}$
[*176·43]

The following propositions are lemmas for

$$R \subseteq J \,.\, \supset . \, (P^Q)^R \,\text{smor}\, P^{R \times Q}.$$

***176·5.** $\vdash :. \, M \upharpoonright C\text{'}R \in 1 \to 1 \,.\, C\text{'}R \subset \text{Cl'}M \,.\, C\text{'}Q \subset p\text{'Cl''}M\text{''}C\text{'}R \,.$

$M\text{''}C\text{'}R \subset 1 \to 1 : z, z' \in C\text{'}R \,.\, \exists \,!\, \text{D'}M\text{'}z \cap \text{D'}M\text{'}z' \,.\, \supset_{z, z'} . \, z = z' :$

$T = \hat{x}\hat{X}\,\{(\exists u, z) \,.\, u \in C\text{'}Q \,.\, z \in C\text{'}R \,.\, x = (M\text{'}z)\text{'}u \,.\, X = u \downarrow (M\text{'}z)\} :$

$\supset . \, T \in 1 \to 1$

Dem.

$\vdash . *21\cdot33 . \supset \vdash :. \text{Hp} . \supset : xTX \,.\, x'TX \,.\, \supset .$

$(\exists u, u', z, z') \,.\, u, u' \in C\text{'}Q \,.\, z, z' \in C\text{'}R \,.\, x = (M\text{'}z)\text{'}u \,.\, x' = (M\text{'}z')\text{'}u' \,.$

$X = u \downarrow (M\text{'}z) \,.\, X = u' \downarrow (M\text{'}z') \,.$

$[*55\cdot202] \quad \supset . (\exists u, u' z, z') . \, x = (M\text{'}z)\text{'}u \,.\, x' = (M\text{'}z')\text{'}u' \,.\, u = u' \,.\, M\text{'}z = M\text{'}z' \,.$

$[*13\cdot22] \quad \supset . \, x = x' \quad (1)$

$\vdash . *21\cdot33 . \supset \vdash :. \text{Hp} . \supset : xTX \,.\, xTX' \,.\, \supset .$

$(\exists u, u', z, z') \,.\, u, u' \in C\text{'}Q \,.\, z, z' \in C\text{'}R \,.\, x = (M\text{'}z)\text{'}u \,.\, x = (M\text{'}z')\text{'}u' \,.$

$X = u \downarrow (M\text{'}z) \,.\, X' = u' \downarrow (M\text{'}z') \,.$

$[*33\cdot43.\text{Hp}] \supset . (\exists u, u', z, z') \,.\, u, u' \in C\text{'}Q \,.\, z \in C\text{'}R \,.\, z = z' \,.\, x = (M\text{'}z)\text{'}u = (M\text{'}z')\text{'}u' \,.$

$X = u \downarrow (M\text{'}z) \,.\, X' = u' \downarrow (M\text{'}z') \,.$

$[*13\cdot195] \quad \supset . (\exists u, u', z) \,.\, u, u' \in C\text{'}Q \,.\, z \in C\text{'}R \,.\, x = (M\text{'}z)\text{'}u = (M\text{'}z)\text{'}u' \,.$

$X = u \downarrow (M\text{'}z) \,.\, X' = u' \downarrow (M\text{'}z) \,.$

$[*71\cdot59.\text{Hp}] \supset . (\exists u, u', z) \,.\, u = u' \,.\, X = u \downarrow (M\text{'}z) \,.\, X' = u' \downarrow (M\text{'}z) \,.$

$[*13\cdot195] \quad \supset . \, X = X' \quad (2)$

$\vdash . (1) . (2) . \supset \vdash . \text{Prop}$

***176·501.** $\vdash : \text{Hp}\,*176\cdot5 \,.\, \supset . \, \text{Cl'}T = C\text{'}\Sigma\text{'}Q \underset{\centerdot}{\downarrow} ; M ; R$

Dem.

$\vdash . *71\cdot16 . \supset \vdash :. \text{Hp} . \supset : z \in C\text{'}R \,.\, u \in C\text{'}Q \,.\, \supset . \, \text{E} \,!\, (M\text{'}z)\text{'}u \,.$

$[*21\cdot33] \quad \supset . \, u \downarrow (M\text{'}z) \in \text{Cl'}T \quad (1)$

$\vdash . *21\cdot33 . \supset$

$\vdash :. \text{Hp} . \supset : X \in \text{Ɑ}\text{‘}T . \supset . (\exists z, u) . z \in C\text{‘}R . u \in C\text{‘}Q . X = u \downarrow (M\text{‘}z)$ (2)

$\vdash . (1) . (2) . \supset \vdash :. \text{Hp} . \supset : X \in \text{Ɑ}\text{‘}T . \equiv . (\exists z, u) . z \in C\text{‘}R . u \in C\text{‘}Q . X = u \downarrow (M\text{‘}z) .$

$[*71\cdot4] \quad \equiv . (\exists Z, u) . Z \in M\text{“}C\text{‘}R . u \in C\text{‘}Q . X = u \downarrow Z .$

$[*150\cdot22] \quad \equiv . (\exists Z, u) . Z \in C\text{‘}M ; R . u \in C\text{‘}Q . X = u \downarrow Z .$

$[*165\cdot16 . *113\cdot101] \quad \equiv . X \in C\text{‘}\Sigma\text{‘}Q \underset{\cdot\,,}{\downarrow} ; M ; R :. \supset \vdash . \text{Prop}$

***176·502.** $\vdash : \text{Hp} *176\cdot5 . z \in C\text{‘}R . \supset . T ; Q \underset{\cdot\,,}{\downarrow} \text{‘}M\text{‘}z = \dagger Q\text{‘}M\text{‘}z$

Dem.

$\vdash . *150\cdot4 . *165\cdot01 . *176\cdot5 . \supset$

$\vdash : \text{Hp} . \supset . T ; Q \underset{\cdot\,,}{\downarrow} \text{‘}M\text{‘}z = \hat{x}\hat{y} \{(\exists u, v) . uQv . x = T\text{‘} \downarrow (M\text{‘}z)\text{‘}u . y = T\text{‘} \downarrow (M\text{‘}z)\text{‘}v\}$

$[\text{Hp} . *176\cdot5\cdot501] \quad = \hat{x}\hat{y} \{(\exists u, v) . uQv . x = (M\text{‘}z)\text{‘}u . y = (M\text{‘}z)\text{‘}v\}$

$[*150\cdot4] \quad = (M\text{‘}z) ; Q$

$[*150\cdot1] \quad = \dagger Q\text{‘}M\text{‘}z : \supset \vdash . \text{Prop}$

***176·503.** $\vdash : \text{Hp} *176\cdot5 . \supset . T \in (\dagger Q ; M ; R) \, \overline{\text{smor}} \, \overline{\text{smor}} \, (Q \underset{\cdot\,,}{\downarrow} ; M ; R)$

Dem.

$\vdash . *176\cdot502 . *150\cdot1\cdot35 . \supset \vdash : \text{Hp} . \supset . T \dagger ; Q \underset{\cdot\,,}{\downarrow} ; M ; R = \dagger Q ; M ; R$ (1)

$\vdash . (1) . *176\cdot5\cdot501 . *164\cdot1 . \supset \vdash . \text{Prop}$

***176·51.** $\vdash :. M \upharpoonright C\text{‘}R \in 1 \to 1 . M\text{“}C\text{‘}R \subset 1 \to 1 .$

$C\text{‘}R \subset \text{Ɑ}\text{‘}M . C\text{‘}Q \subset p\text{‘}\text{Ɑ}\text{“}M\text{“}C\text{‘}R :$

$z, z' \in C\text{‘}R . \exists ! \text{D}\text{‘}M\text{‘}z \cap \text{D}\text{‘}M\text{‘}z' . \supset_{z, z'} . z = z' : \supset . \dagger Q ; M ; R \text{ smor smor } Q \underset{\cdot\,,}{\downarrow} ; R$

Dem.

$\vdash . *165\cdot361 . \supset \vdash : \text{Hp} . \supset . Q \underset{\cdot\,,}{\downarrow} ; M ; R \text{ smor smor } Q \underset{\cdot\,,}{\downarrow} ; R$ (1)

$\vdash . (1) . *176\cdot503 . *164\cdot221 . \supset \vdash . \text{Prop}$

***176·52.** $\vdash :. z \in C\text{‘}R . \supset_z . M\text{‘}z \in (P\text{‘}z) \, \overline{\text{smor}} \, Q : \supset . P ; R = \dagger Q ; M ; R$

Dem.

$\vdash . *151\cdot11 . \supset \vdash :. \text{Hp} . \supset : z \in C\text{‘}R . \supset_z . P\text{‘}z = (M\text{‘}z) ; Q$

$[*150\cdot1] \quad = \dagger Q\text{‘}M\text{‘}z$ (1)

$\vdash . (1) . *150\cdot35 . \supset \vdash . \text{Prop}$

***176·53.** $\vdash :. M \upharpoonright C\text{‘}R \in 1 \to 1 : z \in C\text{‘}R . \supset_z . M\text{‘}z \in (P\text{‘}z) \, \overline{\text{smor}} \, Q :$

$z, z' \in C\text{‘}R . \exists ! C\text{‘}P\text{‘}z \cap C\text{‘}P\text{‘}z' . \supset_{z, z'} . z = z' : \supset . P ; R \text{ smor smor } Q \underset{\cdot\,,}{\downarrow} ; R$

Dem.

$\vdash . *14\cdot21 . \supset \vdash :. \text{Hp} . \supset : z \in C\text{‘}R . \supset_z . \text{E} ! M\text{‘}z :$ (1)

$[*33\cdot43] \quad \supset : C\text{‘}R \subset \text{Ɑ}\text{‘}M$ (2)

$\vdash . *151\cdot11 . \supset \vdash :. \text{Hp} . \supset : z \in C\text{‘}R . \supset_z . M\text{‘}z \in 1 \to 1 :$

$[*37\cdot61 . (1)] \quad \supset : M\text{“}C\text{‘}R \subset 1 \to 1$ (3)

$\vdash . *151 \cdot 11 \cdot 131 . \supset \vdash :. \text{Hp} . \supset : z \in C\text{`}R . \supset_z . \text{D`}M\text{`}z = C\text{`}P\text{`}z :$

$[\text{Hp}] \qquad \supset : z, z' \in C\text{`}R . \exists ! \text{D`}M\text{`}z \cap \text{D`}M\text{`}z' . \supset_{z, z'} . z = z' \quad (4)$

$\vdash . *151 \cdot 11 . \quad \supset \vdash :. \text{Hp} . \supset : z \in C\text{`}R . \supset . \text{G`}M\text{`}z = C\text{`}Q :$

$[*37 \cdot 63] \qquad \supset : Z \in M\text{``}C\text{`}R . \supset . \text{G`}Z = C\text{`}Q :$

$[*40 \cdot 15] \qquad \supset : C\text{`}Q \subset p\text{`G``}M\text{``}C\text{`}R \quad (5)$

$\vdash . (2) . (3) . (4) . (5) . *176 \cdot 51 . \supset \vdash : \text{Hp} . \supset . \dagger Q ; M ; R \text{ smor smor } Q \underset{\cdot\cdot}{\downarrow} ; R \quad (6)$

$\vdash . (6) . *176 \cdot 52 . \supset \vdash . \text{Prop}$

$*176 \cdot 54$. $\vdash :. \dot{\exists} ! P . \dot{\exists} ! Q . M = \hat{Z}\hat{z} [z \in C\text{`}R . Z = \{ | (\text{Cnv`} \downarrow z)\}_\epsilon \upharpoonright C\text{`}(P \exp Q)] . \supset :$
$M \in 1 \to 1 : z \in C\text{`}R . \supset_z . M\text{`}z \in (\text{Prod`}P \underset{\cdot\cdot}{\downarrow} ; Q \underset{\cdot\cdot}{\downarrow} z) \overline{\text{smor}} (\text{Prod`}P \underset{\cdot\cdot}{\downarrow} ; Q)$

Dem.

$\vdash . *116 \cdot 606 . *176 \cdot 14 . \supset \vdash : \text{Hp} . \supset . M \in 1 \to 1 \quad (1)$

$\vdash . *21 \cdot 33 . *30 \cdot 3 . \supset \vdash : \text{Hp} . z \in C\text{`}R . \supset . M\text{`}z = \{ | (\text{Cnv`} \downarrow z)\}_\epsilon \upharpoonright C\text{`}(P \exp Q) \quad (2)$

$\vdash . *151 \cdot 65 . *165 \cdot 361 . *166 \cdot 1 . *165 \cdot 01 . \supset$
$\vdash . \{ | (\text{Cnv`} \downarrow z)\} \upharpoonright C\text{`}(Q \times P) \in (P \underset{\cdot\cdot}{\downarrow} ; Q \underset{\cdot\cdot}{\downarrow} z) \overline{\text{smor}}\, \overline{\text{smor}} (P \underset{\cdot\cdot}{\downarrow} ; Q) .$

$[(2) . *173 \cdot 3] \supset \vdash : \text{Hp} . z \in C\text{`}R . \supset .$
$M\text{`}z \in (\text{Prod`}P \underset{\cdot\cdot}{\downarrow} ; Q \underset{\cdot\cdot}{\downarrow} z) \overline{\text{smor}} (\text{Prod`}P \underset{\cdot\cdot}{\downarrow} ; Q) \quad (3)$

$\vdash . (1) . (3) . \supset \vdash . \text{Prop}$

$*176 \cdot 541$. $\vdash . (P \underset{\cdot\cdot}{\downarrow})\dagger ; Q \underset{\cdot\cdot}{\downarrow} ; R \in \text{Rel}^3 \text{ arithm} . \Sigma\text{`}(P \underset{\cdot\cdot}{\downarrow})\dagger ; Q \underset{\cdot\cdot}{\downarrow} ; R = P \underset{\cdot\cdot}{\downarrow} ; \Sigma\text{`}Q \underset{\cdot\cdot}{\downarrow} ; R$

Dem.

$\vdash . *163 \cdot 3 . *165 \cdot 21 \cdot 22 . \supset \vdash : \dot{\exists} ! P . \supset . (P \underset{\cdot\cdot}{\downarrow})\dagger ; Q \underset{\cdot\cdot}{\downarrow} ; R \in \text{Rel}^2 \text{ excl} \quad (1)$

$\vdash . *165 \cdot 242 . \supset \vdash : P = \dot{\Lambda} . \dot{\exists} ! S . \dot{\exists} ! S' . \supset . P \underset{\cdot\cdot}{\downarrow} ; S = \dot{\Lambda} \downarrow \dot{\Lambda} . P \underset{\cdot\cdot}{\downarrow} ; S' = \dot{\Lambda} \downarrow \dot{\Lambda} :$

$[\text{Transp}] \qquad \supset \vdash :. P = \dot{\Lambda} . P \underset{\cdot\cdot}{\downarrow} ; S \neq P \underset{\cdot\cdot}{\downarrow} ; S' . \supset : S = \Lambda . \vee . S' = \Lambda :$

$[*165 \cdot 241] \qquad \supset : P \underset{\cdot\cdot}{\downarrow} ; S = \dot{\Lambda} . \vee . P \underset{\cdot\cdot}{\downarrow} ; S' = \dot{\Lambda} :$

$[*33 \cdot 241] \qquad \supset : C\text{`}P \underset{\cdot\cdot}{\downarrow} ; S \cap C\text{`}P \underset{\cdot\cdot}{\downarrow} ; S' = \Lambda \quad (2)$

$\vdash . *150 \cdot 22 \cdot 1 . \supset \vdash :. P = \dot{\Lambda} . \supset :$
$T, T' \in C\text{`}(P \underset{\cdot\cdot}{\downarrow})\dagger ; Q \underset{\cdot\cdot}{\downarrow} ; R . T \neq T' . \supset .$
$(\exists z, z') . z \neq z' . z, z' \in C\text{`}R . T = (P \underset{\cdot\cdot}{\downarrow}) ; Q \underset{\cdot\cdot}{\downarrow} z . T' = (P \underset{\cdot\cdot}{\downarrow}) ; Q \underset{\cdot\cdot}{\downarrow} z' .$

$[(2)] \qquad \supset . C\text{`}T \cap C\text{`}T' = \Lambda \quad (3)$

$\vdash . (1) . (3) . *163 \cdot 1 . \supset \vdash . (P \underset{\cdot\cdot}{\downarrow})\dagger ; Q \underset{\cdot\cdot}{\downarrow} ; R \in \text{Rel}^2 \text{ excl} \quad (4)$

$\vdash . *162 \cdot 35 . \qquad \supset \vdash . \Sigma\text{`}(P \underset{\cdot\cdot}{\downarrow})\dagger ; Q \underset{\cdot\cdot}{\downarrow} ; R = P \underset{\cdot\cdot}{\downarrow} ; \Sigma\text{`}Q \underset{\cdot\cdot}{\downarrow} ; R . \quad (5)$

$[*165 \cdot 21] \qquad \supset \vdash . \Sigma\text{`}(P \underset{\cdot\cdot}{\downarrow})\dagger ; Q \underset{\cdot\cdot}{\downarrow} ; R \in \text{Rel}^2 \text{ excl} \quad (6)$

$\vdash . (4) . (5) . (6) . *174 \cdot 3 . \supset \vdash . \text{Prop}$

***176·55.** $\vdash : \dot{\exists} ! P . \dot{\exists} ! Q . \supset . \mathrm{Prod} ; (P \downarrow_{;}) \dagger ; Q \downarrow_{;} ; R \, \mathrm{smor} \, \mathrm{smor} (\mathrm{Prod} ‘ P \downarrow_{;} ; Q) \downarrow_{;} ; R$

Dem.

$\vdash . *176·133·15 . *37·44·21 . \supset$

$\vdash : \exists ! C ‘ \mathrm{Prod} ‘ P \downarrow_{;} ; Q \downarrow_{;} z \cap C ‘ \mathrm{Prod} ‘ P \downarrow_{;} ; Q \downarrow_{;} w . \supset .$

$\exists ! C ‘ P^{Q \dagger_{;} z} \cap C ‘ P^{Q \dagger_{;} w} . \dot{\exists} ! Q \downarrow_{;} z .$

[*176·16.*80·14.*165·212] $\supset . (\exists R) . \mathrm{G} ‘ R = C ‘ Q \downarrow_{;} z . \mathrm{G} ‘ R = C ‘ Q \downarrow_{;} w . \dot{\exists} ! Q .$

[*13·171.*150·22] $\supset . \downarrow z ‘‘ C ‘ Q = \downarrow w ‘‘ C ‘ Q . \dot{\exists} ! Q .$

[*55·232] $\supset . z = w$ (1)

$\vdash . (1) . *176·54 . *176·53 \dfrac{\mathrm{Prod} ‘ P \downarrow_{;} ; Q \downarrow_{;} z, \; \mathrm{Prod} ‘ P \downarrow_{;} ; Q}{P ‘ z, \qquad Q} . \supset \vdash . \mathrm{Prop}$

***176·56.** $\vdash : \dot{\exists} ! P . \dot{\exists} ! Q . R \mathrel{G} J . \supset .$

$\mathrm{Prod} ‘ \Sigma ‘ (P \downarrow_{;}) \dagger ; Q \downarrow_{;} ; R \; \mathrm{smor} \; \mathrm{Prod} ‘ (\mathrm{Prod} ‘ P \downarrow_{;} ; Q) \downarrow_{;} ; R$

Dem.

$\vdash . *165·223 . \supset \vdash :. \dot{\exists} ! P . P \downarrow_{;} ; Q \downarrow_{;} z = P \downarrow_{;} ; Q \downarrow_{;} z' . \supset : Q \downarrow_{;} z = Q \downarrow_{;} z' :$

[*165·22] $\supset : \dot{\exists} ! Q . \supset . z = z'$ (1)

$\vdash . (1) . \mathrm{Transp} . \supset \vdash :. \mathrm{Hp} . z R z' . \supset . P \downarrow_{;} ; Q \downarrow_{;} z \neq P \downarrow_{;} ; Q \downarrow_{;} z'$ (2)

$\vdash . (2) . *150·4 . \supset \vdash : \mathrm{Hp} . \supset . (P \downarrow_{;}) \dagger ; Q \downarrow_{;} ; R \mathrel{G} J$ (3)

$\vdash . (3) . *176·541 . *174·43 . \supset$

$\vdash : \mathrm{Hp} . \supset . \mathrm{Prod} ‘ \Sigma ‘ (P \downarrow_{;}) \dagger ; Q \downarrow_{;} ; R \, \mathrm{smor} \, \mathrm{Prod} ‘ \mathrm{Prod} ; (P \downarrow_{;}) \dagger ; Q \downarrow_{;} ; R$ (4)

$\vdash . *176·55 . *173·31 . \supset$

$\vdash : \mathrm{Hp} . \supset . \mathrm{Prod} ‘ \mathrm{Prod} ; (P \downarrow_{;}) \dagger ; Q \downarrow_{;} ; R \, \mathrm{smor} \, \mathrm{Prod} ‘ (\mathrm{Prod} ‘ P \downarrow_{;} ; Q) \downarrow_{;} ; R$ (5)

$\vdash . (4) . (5) . \supset \vdash . \mathrm{Prop}$

***176·57.** $\vdash : R \mathrel{G} J . \supset . \{(P \exp Q) \exp R\} \, \mathrm{smor} \, \{P \exp (R \times Q)\} . (P^Q)^R \, \mathrm{smor} \, P^{R \times Q}$

Dem.

$\vdash . *176·151 . \qquad \supset \vdash :. P = \dot{\Lambda} . \vee . Q = \dot{\Lambda} : \supset . (P \exp Q) \exp R = \dot{\Lambda}$ (1)

$\vdash . *176·151 . *166·13 . \supset \vdash :. P = \dot{\Lambda} . \vee . Q = \dot{\Lambda} : \supset . P \exp (R \times Q) = \dot{\Lambda}$ (2)

$\vdash . (1) . (2) . *153·101 . \supset \vdash :. P = \dot{\Lambda} . \vee . Q = \dot{\Lambda} : \supset .$

$\{(P \exp Q) \exp R\} \, \mathrm{smor} \, \{P \exp (R \times Q)\}$ (3)

$\vdash . *176·56·541·1 . *166·1 . \supset \vdash : \dot{\exists} ! P . \dot{\exists} ! Q . R \mathrel{G} J . \supset .$

$\{(P \exp Q) \exp R\} \, \mathrm{smor} \, \{P \exp (R \times Q)\}$ (4)

$\vdash . (3) . (4) . \supset \vdash : \mathrm{Hp} . \supset . \{(P \exp Q) \exp R\} \, \mathrm{smor} \, \{P \exp (R \times Q)\}$ (5)

[*176·181·22] $\supset . (P^Q)^R \, \mathrm{smor} \, P^{R \times Q}$ (6)

$\vdash . (5) . (6) . \supset \vdash . \mathrm{Prop}$

This completes the proof of the second formal law of exponentiation.

*177. PROPOSITIONS CONNECTING P_{df} WITH PRODUCTS AND POWERS.

Summary of *177.

The principal proposition on this subject is

***177·13.** $\vdash : x \neq y . \supset . P_{\text{df}} \text{ smor } \{(x \downarrow y)^P\}$

which is the analogue of *116·72, or rather leads to the analogue of *116·72 as soon as powers of relation-numbers have been defined; for then it becomes

$$P_{\text{df}} \in 2_r^{\text{Nr}'P}.$$

Another proposition is an extension of *171·69, namely

***177·22.** $\vdash : P \in \text{Rel}^2 \text{ excl} . P \mathrel{G} J . \supset . \text{Prod}'\text{df}``P \text{ smor } (\Sigma'P)_{\text{df}}$

where we put $\text{df}'Q = Q_{\text{df}}$.

The remaining propositions of this number are lemmas for the above two.

*177·13 shows, for example, that all classes of finite integers can be arranged in a series of which the relation-number is 2_r^ω, where ω is the relation-number of the series of finite integers. 2_r^ω is not the relation-number of the continuum, but is closely allied to it.

***177·1.** $\vdash : x \neq y . T = \hat{\mu}\hat{R}[R \in \{(\iota'x \cup \iota'y) \uparrow \alpha\}_\Delta{}'\alpha . \mu = \overleftarrow{R}'x] . \supset .$

$T \in (\text{Cl}'\alpha) \,\overline{\text{sm}}\, \{(\iota'x \cup \iota'y) \uparrow \alpha\}_\Delta{}'\alpha$ [*116·712·713·715]

In the propositions of *116 referred to, Λ and V appear in place of x and y, but no property of Λ and V is used in the proof except $\Lambda \neq \text{V}$.

***177·11.** $\vdash : \text{Hp} *177·1 . \alpha = C'P . \supset . T``(x \downarrow y)^P = P_{\text{df}}$

Dem.

$\vdash . *176·19 . \supset$

$\vdash ::. \text{Hp} . \supset :: \mu \{T``(x \downarrow y)^P\} \nu . \equiv :. (\exists R, S) : R, S \in \{(\iota'x \cup \iota'y) \uparrow C'P\}_\Delta{}'C'P :$

$(\exists z) : z \in C'P . R'z \, (x \downarrow y) \, S'z :$

$wPz . w \neq z . \supset_w . R'w = S'w : \mu = \overleftarrow{R}'x . \nu = \overleftarrow{S}'x :.$

[*55·13] $\equiv :. (\exists R, S) : R, S \in \{(\iota'x \cup \iota'y) \uparrow C'P\}_\Delta{}'C'P :.$

$(\exists z) :. z \in C'P . R'z = x . S'z = y :. \mu = \overleftarrow{R}'x . \nu = \overleftarrow{S}'x :.$

$wPz . w \neq z . \supset_w : xRw . \equiv . xSw : yRw . \equiv . ySw :.$

[*71·36] $\equiv :. (\exists R, S) : R, S \epsilon \{(\iota`x \cup \iota`y) \uparrow C`P\}_\Delta`C`P :.$

$(\exists z) :. z \epsilon C`P . z \epsilon \mu - \nu . \mu = \overleftarrow{R}`x . \nu = \overleftarrow{S}`x :.$

$wPz . w \neq z . \supset_w : w \epsilon \mu . \equiv . w \epsilon \nu :.$

[*177·1] $\equiv :. \mu, \nu \epsilon \text{Cl}`C`P :. (\exists z) :. z \epsilon C`P . z \epsilon \mu - \nu :.$

$wPz . w \neq z . \supset_w : w \epsilon \mu . \equiv . w \epsilon \nu :.$

[*171·11] $\equiv :. \mu (P_{\text{df}}) \nu ::. \supset \vdash .$ Prop

***177·12.** $\vdash : \text{Hp} *177·11 . \supset . T \epsilon P_{\text{df}} \overline{\text{smor}} \{(x \downarrow y)^P\}$ [*177·1·11 . *151·191]

***177·13.** $\vdash : x \neq y . \supset . P_{\text{df}} \text{ smor } \{(x \downarrow y)^P\}$ [*177·12]

***177·2.** $\text{df}`Q = Q_{\text{df}}$ Dft [*177]

***177·21.** $\vdash : P \epsilon \text{Rel}^2 \text{excl} . P \Subset J . \supset . s \upharpoonright C`\text{Prod}`\text{df};P \epsilon (\Sigma`P)_{\text{df}} \overline{\text{smor}} (\text{Prod}`\text{df};P)$

The proof proceeds as the proof of *174·24 proceeds. If $Q \epsilon C`P$, we shall have, if $M \epsilon F_\Delta`\text{df}``C`P$,

$$M`Q_{\text{df}} = (s`\text{D}`M) \cap C`Q.$$

Hence we easily obtain

$$M (\Pi`\text{df};P) N . \equiv . M, N \epsilon F_\Delta`\text{df}``C`P . (s`\text{D}`M) (\Sigma`P)_{\text{df}} (s`\text{D}`N),$$

whence

$$\mu (\text{Prod}`\text{df};P) \nu . \equiv . \mu, \nu \epsilon \text{Prod}`\text{Cl}``C``C`P . (s`\mu) (\Sigma`P)_{\text{df}} (s`\nu),$$

whence the result follows easily.

***177·22.** $\vdash : P \epsilon \text{Rel}^2 \text{excl} . P \Subset J . \supset . \text{Prod}`\text{df};P \text{ smor } (\Sigma`P)_{\text{df}}$ [*177·21]

SECTION D.

ARITHMETIC OF RELATION-NUMBERS.

Summary of Section D.

In the present section, we shall be concerned with the arithmetical operations on relation-numbers. Their purely logical properties have been dealt with in Section A; in the present section, it is their arithmetical properties that are to be established. These properties result immediately from the arithmetical properties of relations which have been established in Sections B and C. The subjects treated of in the present section are analogous to those treated of in Section B of Part III, with the exception of such as have already had their analogues discussed in Sections B and C of Part IV. The analogy is sufficiently close to render it often unnecessary to give proofs, since these are often step by step analogous to the proofs of corresponding propositions in Part III, Section B.

The two chief requisites in defining the arithmetical operations with relation-numbers are (1) to take due account of types, (2) to construct what may be called *separated* relations, *i.e.* relations of mutually exclusive relations derived from and ordinally similar to given relations. Each of these points calls for some preliminary explanations.

The sum of two relation-numbers μ, ν will be denoted by "$\mu \dot{+} \nu$," in order to distinguish this kind of addition from $\mu + \nu$ (the arithmetical addition of classes) and $\mu +_c \nu$ (the addition of cardinals). In defining $\mu \dot{+} \nu$, we have to take account of the following considerations.

Suppose P and Q are two relations which are of the same type, and have mutually exclusive fields. Then obviously we shall want to frame our definition of the sum of two relation-numbers in such a way that the sum of $\text{Nr}`P$ and $\text{Nr}`Q$ shall be $\text{Nr}`(P \updownarrow Q)$. But if $P$ and $Q$ are not of the same type, $P \updownarrow Q$ is meaningless; and if $C`P$ and $C`Q$ overlap, $P \updownarrow Q$ may be too small to have as its relation-number the sum of the relation-numbers of $P$ and $Q$. Both these difficulties can be met by observing that, if $\text{Nr}`P = \text{Nr}`R$ and $\text{Nr}`Q = \text{Nr}`S$, we must make such definitions as to have

$$\text{Nr}`P \dot{+} \text{Nr}`Q = \text{Nr}`R \dot{+} \text{Nr}`S.$$

Hence, in defining the sum of the relation-numbers of P and Q, we may replace P and Q by any two relations R and S which are respectively like P and Q. Therefore what we require for our definition is to find two relations R and S which (1) are respectively like P and Q, (2) are of the same type, (3) have mutually exclusive fields. All these three requisites are satisfied if we put

$$R = \downarrow(\Lambda \cap C\text{‘}Q) ; \iota ; P \,.\, S = (\Lambda \cap C\text{‘}P) \downarrow ; \iota ; Q.$$

We then define $P + Q$ as meaning $R \updownarrow S$, and we define the sum of the relation-numbers of P and Q as the relation-number of $P + Q$. This procedure is exactly analogous to that of $*110$; in fact, we have

$$C\text{‘}(P + Q) = C\text{‘}P + C\text{‘}Q.$$

In defining the sum of the relation-numbers of a field, we do not have to consider types, because the members of a field are necessarily all of the same type. But we do have to consider the question of overlapping. If a term x occurs both in $C\text{‘}Q$ and in $C\text{‘}R$, where $Q, R \in C\text{‘}P$, we want a method of counting x twice over in forming the arithmetical sum. Thus $\mathrm{Nr}\text{‘}\Sigma\text{‘}P$ cannot be taken as the sum of the relation-numbers of members of $C\text{‘}P$, unless $P \in \mathrm{Rel}^2\,\mathrm{excl}$. Suppose, for instance, we have three series

$$(a, b, c), \quad (b, c, a), \quad (c, a, b).$$

These each have three terms; and we want the sum of their relation-numbers to be the relation-number of a series of nine terms. But if we put

$$Q = a \downarrow b \downarrow c \text{ (where } a \downarrow b \downarrow c \text{ is written for } a \downarrow b \cup a \downarrow c \cup b \downarrow c\text{)},$$

$$R = b \downarrow c \downarrow a,$$

$$S = c \downarrow a \downarrow b,$$

and if we further put

$$P = Q \downarrow R \downarrow S,$$

so that P places the above three series in the above order, we have

$$\Sigma\text{‘}P = (\iota\text{‘}a \cup \iota\text{‘}b \cup \iota\text{‘}c) \uparrow (\iota\text{‘}a \cup \iota\text{‘}b \cup \iota\text{‘}c),$$

which is not a series, and does not have the relation-number which we require as the sum of the relation-numbers of Q, R, S.

What is wanted is a method of distinguishing the various occurrences of a and b and c. For this reason, when a occurs as a member of the field of Q, we replace it by $a \downarrow Q$; when as a member of the field of R, by $a \downarrow R$; and when as a member of the field of S, by $a \downarrow S$. Thus the series (a, b, c) is replaced by $(a \downarrow Q, b \downarrow Q, c \downarrow Q)$; (b, c, a) is replaced by $(b \downarrow R, c \downarrow R, a \downarrow R)$; and (c, a, b) is replaced by $(c \downarrow S, a \downarrow S, b \downarrow S)$. The sum of these three series then has the relation-number which is required as the sum of the relation-numbers of Q, R, S.

The above process is symbolized as follows. The generating relation of the series $(a \downarrow Q, b \downarrow Q, c \downarrow Q)$ is $\downarrow Q ; Q$; thus the three relations whose sum is to be taken are $\downarrow Q ; Q$, $\downarrow R ; R$, $\downarrow S ; S$, *i.e.* using the notation of $*182$, according to which we put $\overset{\frown}{\text{♀}} \text{‘} x = x \, \text{♀} \, x$, our three relations are $\overset{\frown}{\downarrow}_{;} \text{‘} Q$, $\overset{\frown}{\downarrow}_{;} \text{‘} R$, $\overset{\frown}{\downarrow}_{;} \text{‘} S$.

But the generating relation of the series $(\overset{\frown}{\downarrow}_{;} \text{‘} Q, \overset{\frown}{\downarrow}_{;} \text{‘} R, \overset{\frown}{\downarrow}_{;} \text{‘} S)$ is $\overset{\frown}{\downarrow}_{;} ; P$, since $P = (Q \downarrow R \downarrow S)$. Thus $\overset{\frown}{\downarrow}_{;} ; P$ is the relation required for defining the sum of the relation-numbers of members of the field of P; *i.e.* we put

$$\Sigma \text{Nr‘} P = \text{Nr‘} \Sigma \text{‘} \overset{\frown}{\downarrow}_{;} ; P \quad \text{Df.}$$

We will call $\overset{\frown}{\downarrow}_{;} ; P$ the *separated* relation corresponding to P. $\overset{\frown}{\downarrow}_{;} ; P$ is constructed, as above, by replacing every member x of $C \text{‘} Q$, where $Q \in C \text{‘} R$, by $x \downarrow Q$; so that if x belongs both to $C \text{‘} Q$ and to $C \text{‘} R$, it is duplicated by being transformed once into $x \downarrow Q$, and once again into $x \downarrow R$.

For the treatment of products, we do not require $\overset{\frown}{\downarrow}_{;} ; P$, because $\Pi \text{‘} P$ has been so defined as to effect the requisite separation. We might, however, by the use of $\overset{\frown}{\downarrow}_{;} ; P$, have dispensed with $\Pi \text{‘} P$ as a fundamental notion, and contented ourselves with $\text{Prod‘} P$; for we have

$$\Pi \text{‘} P = \dot{s} ; \text{Prod‘} \overset{\frown}{\downarrow}_{;} ; P.$$

Thus we might have taken Prod as the fundamental notion, and defined Π by means of it.

The addition of unity to a relation-number has to be treated separately from the addition of two relation-numbers, for the same reasons which necessitate the treatment of $P \rightarrow\!\!\!\!+ \, x$ and $x \leftarrow\!\!\!\!+ \, P$ separately from $P \updownarrow Q$. There is no ordinal number 1, but we can define the *addition* of one to a relation-number. If $\text{Nr‘} P = \mu$ and $x \sim \in C \text{‘} P$, we must have

$$\text{Nr‘}(P \rightarrow\!\!\!\!+ \, x) = \mu \dot{+} \dot{1},$$

where we write "$\dot{1}$" for unity as an addendum. We do not write "1_r," because we shall, at a later stage, give a general definition of μ_r, in virtue of which, if μ is an inductive cardinal, μ_r will be the corresponding ordinal. This definition entails $1_r = \Lambda$, and therefore we use a different symbol "$\dot{1}$" for 1 as addendum. The symbol $\dot{1}$ is only defined in its uses, and has no significance except in a use which has been specially defined.

We define the product $\mu \dot{\times} \nu$ as the relation-number of $P \times Q$, when $\mu = \text{N}_0\text{r‘} P$ and $\nu = \text{N}_0\text{r‘} Q$. The product so defined obeys the associative law, and obeys the distributive law in the form

$$(\nu \dot{+} \varpi) \dot{\times} \mu = (\nu \dot{\times} \mu) \dot{+} (\varpi \dot{\times} \mu)$$

but not, in general, in the form

$$\mu \dot{\times} (\nu \dot{+} \varpi) = (\mu \dot{\times} \nu) \dot{+} (\mu \dot{\times} \varpi).$$

The latter form holds when μ, ν, ϖ are finite ordinals, as we shall prove at a later stage (*262). The commutative law also does not hold in general for ordinal addition and multiplication, but holds where finite ordinals are concerned.

The product of the numbers of the members of $C^{\prime}P$, in the order generated by P, is defined as being $\text{Nr}^{\prime}\Pi^{\prime}P$, and is denoted by $\Pi\text{Nr}^{\prime}P$. It will be seen that $\dot{\Pi}\text{Nr}^{\prime}P$ is not a function of $C^{\prime}P$, since the value of a product depends upon the order of the factors; it is also not a function of $\text{Nr}{;}P$, unless no two members of $C^{\prime}P$ have the same relation-number. The properties of $\Pi\text{Nr}^{\prime}P$ result from *172 and *174.

"μ to the νth power" is denoted by "$\mu \exp_r \nu$" and is defined as the relation-number of $P \exp Q$, where $\mu = \text{N}_0\text{r}^{\prime}P$ and $\nu = \text{N}_0\text{r}^{\prime}Q$. Its properties result from the propositions of *176 and *177.

$*180$. THE SUM OF TWO RELATION-NUMBERS.

Summary of $*180$.

In order to define the sum of two relation-numbers, we proceed (as in $*110$) to construct a relation whose relation-number shall be the required sum. For this purpose, we put

$$P + Q = \{\downarrow(\Lambda \cap C\text{‘}Q) ; \iota ; P\} \Uparrow \{(\Lambda \cap C\text{‘}P) \downarrow ; \iota ; Q\} \quad \text{Df.}$$

This definition has the following merits: (1) whatever may be the types of P and Q, $\downarrow(\Lambda \cap C\text{‘}Q) ; \iota ; P$ is of the same type as $(\Lambda \cap C\text{‘}P) \downarrow ; \iota ; Q$; (2) however the fields of P and Q may overlap, and even if $P = Q$, the fields of $\downarrow(\Lambda \cap C\text{‘}Q) ; \iota ; P$ and $(\Lambda \cap C\text{‘}P) \downarrow ; \iota ; Q$ are mutually exclusive; (3) these two relations are respectively similar to P and Q. Hence it is evident that, without placing any restriction upon P and Q, we may take the relation-number of $P + Q$ as defining the sum of the relation-numbers of P and Q. Hence we put

$$\mu \dotplus \nu = \hat{R}\{(\exists P, Q) . \mu = \mathrm{N_0r}\text{‘}P . \nu = \mathrm{N_0r}\text{‘}Q . R \,\mathrm{smor}\, (P + Q)\} \quad \text{Df.}$$

From this definition it follows that $\mu \dotplus \nu$ is null unless μ and ν are homogeneous relation-numbers, but that if they are the homogeneous relation-numbers of P and Q, then $\mu \dotplus \nu$ is the relation-number of $P + Q$.

In order to be able to deal with typically ambiguous relation-numbers, we put, as in $*110$,

$$\mathrm{Nr}\text{‘}P \dotplus \nu = \mathrm{N_0r}\text{‘}P \dotplus \nu \quad \text{Df,}$$

$$\mu \dotplus \mathrm{Nr}\text{‘}Q = \mu \dotplus \mathrm{N_0r}\text{‘}Q \quad \text{Df.}$$

The principal propositions of the present number are

$*180{\cdot}111$. $\vdash . C\text{‘}(P + Q) = C\text{‘}P + C\text{‘}Q$

$*180{\cdot}3$. $\vdash . \mathrm{Nr}\text{‘}P \dotplus \mathrm{Nr}\text{‘}Q = \mathrm{N_0r}\text{‘}P \dotplus \mathrm{Nr}\text{‘}Q = \mathrm{Nr}\text{‘}P \dotplus \mathrm{N_0r}\text{‘}Q$
$= \mathrm{N_0r}\text{‘}P \dotplus \mathrm{N_0r}\text{‘}Q = \mathrm{Nr}\text{‘}(P + Q)$

$*180{\cdot}31$. $\vdash : P \,\mathrm{smor}\, R . Q \,\mathrm{smor}\, S . \supset . \mathrm{Nr}\text{‘}P \dotplus \mathrm{Nr}\text{‘}Q = \mathrm{Nr}\text{‘}R \dotplus \mathrm{Nr}\text{‘}S$

This proposition is essential, since otherwise $\mathrm{Nr}\text{‘}P \dotplus \mathrm{Nr}\text{‘}Q$ would not be a function of $\mathrm{Nr}\text{‘}P$ and $\mathrm{Nr}\text{‘}Q$, but would depend upon the particular P and Q.

$*180{\cdot}32$. $\vdash : C\text{‘}P \cap C\text{‘}Q = \Lambda . \supset . \mathrm{Nr}\text{‘}P \dotplus \mathrm{Nr}\text{‘}Q = \mathrm{Nr}\text{‘}(P \Uparrow Q)$

$*180{\cdot}4$. $\vdash : \exists ! \mu \dotplus \nu . \supset . \mu, \nu \in \mathrm{NR} - \iota\text{‘}\Lambda . \mu, \nu \in \mathrm{N_0R}$

***180·42.** $\vdash . \mu \dotplus \nu \in \mathrm{NR}$

***180·56.** $\vdash . (\mu \dotplus \nu) \dotplus \varpi = \mu \dotplus (\nu \dotplus \varpi)$

which is the associative law.

***180·61.** $\vdash . \mathrm{Nr}`P \dotplus 0_r = \mathrm{Nr}`P = 0_r \dotplus \mathrm{Nr}`P$

***180·71.** $\vdash : \mu, \nu \in \mathrm{NR} . \supset . C``(\mu \dotplus \nu) = C``\mu +_c C``\nu$

This proposition gives the connection of ordinal and cardinal addition. It should be observed that, in virtue of *154·9, $C``\mu$ and $C``\nu$ are cardinals when μ and ν are relation-numbers.

***180·01.** $P + Q = \{\downarrow (\Lambda \cap C`Q) ; \iota ; P\} \updownarrow \{(\Lambda \cap C`P) \downarrow ; \iota ; Q\}$ Df

***180·02.** $\mu \dotplus \nu = \hat{R}\{(\exists P, Q) . \mu = \mathrm{N_0r}`P . \nu = \mathrm{N_0r}`Q . R \,\mathrm{smor}\, (P + Q)\}$ Df

***180·03.** $\mathrm{Nr}`P \dotplus \nu = \mathrm{N_0r}`P \dotplus \nu$ Df

***180·031.** $\mu \dotplus \mathrm{Nr}`Q = \mu \dotplus \mathrm{N_0r}`Q$ Df

On the purpose of the definitions *180·03·031, see the remarks on the corresponding definitions in *110 and IIT of the Prefatory Statement.

***180·1.** $\vdash . P + Q = \{\downarrow (\Lambda \cap C`Q) ; \iota ; P\} \updownarrow \{(\Lambda \cap C`P) \downarrow ; \iota ; Q\}$ [(*180·01)]

***180·101.** $\vdash . C` \downarrow (\Lambda \cap C`Q) ; \iota ; P = \downarrow (\Lambda \cap C`Q)``\iota``C`P .$
$C`(\Lambda \cap C`P) \downarrow ; \iota ; Q = (\Lambda \cap C`P) \downarrow ``\iota``C`Q$ [*150·22]

***180·11.** $\vdash . C` \downarrow (\Lambda \cap C`Q) ; \iota ; P \cap C`(\Lambda \cap C`P) \downarrow ; \iota ; Q = \Lambda$ [*180·101 . *110·11]

***180·111.** $\vdash . C`(P + Q) = C`P + C`Q$

Dem.

$\vdash . *180{\cdot}101 . *160{\cdot}14 . \supset$
$\vdash . C`(P + Q) = \downarrow (\Lambda \cap C`Q)``\iota``C`P \cup (\Lambda \cap C`P) \downarrow ``\iota``C`Q$
[(*110·01)] $= C`P + C`Q . \supset \vdash . \mathrm{Prop}$

***180·12.** $\vdash . \downarrow (\Lambda \cap C`Q) ; \iota ; P \,\mathrm{smor}\, P . (\Lambda \cap C`P) \downarrow ; \iota ; Q \,\mathrm{smor}\, Q$ [*151·61·64·65]

***180·13.** $\vdash : R \,\mathrm{smor}\, P . S \,\mathrm{smor}\, Q . C`R \cap C`S = \Lambda . \supset . R \updownarrow S \,\mathrm{smor}\, P + Q$

Dem.

$\vdash . *180{\cdot}12 . \supset \vdash : \mathrm{Hp} . \supset . R \,\mathrm{smor} \downarrow (\Lambda \cap C`Q) ; \iota ; P . S \,\mathrm{smor}\, (\Lambda \cap C`P) \downarrow ; \iota ; Q$ (1)
$\vdash . (1) . *180{\cdot}11 . *160{\cdot}48 . \supset$
$\vdash : \mathrm{Hp} . \supset . R \updownarrow S \,\mathrm{smor}\, \{\downarrow (\Lambda \cap C`Q) ; \iota ; P \updownarrow (\Lambda \cap C`P) \downarrow ; \iota ; Q\} .$
[(*180·01)] $\supset . R \updownarrow S \,\mathrm{smor}\, (P + Q) : \supset \vdash . \mathrm{Prop}$

***180·14.** $\vdash : C`P \cap C`Q = \Lambda . \supset . P \updownarrow Q \,\mathrm{smor}\, P + Q$ [*180·13 . *151·13]

***180·15.** $\vdash : R \,\mathrm{smor}\, P . S \,\mathrm{smor}\, Q . \supset . R + S \,\mathrm{smor}\, P + Q$

Dem.

$\vdash . *180{\cdot}12 . \supset \vdash : \mathrm{Hp} . \supset . \downarrow (\Lambda \cap C`S) ; \iota ; R \,\mathrm{smor}\, P . (\Lambda \cap C`R) \downarrow ; \iota ; S \,\mathrm{smor}\, Q .$
[*180·13] $\supset . \{\downarrow (\Lambda \cap C`S) ; \iota ; R \updownarrow (\Lambda \cap C`R) \downarrow ; \iota ; S\} \,\mathrm{smor}\, P + Q .$
[(*180·01)] $\supset . R + S \,\mathrm{smor}\, P + Q : \supset \vdash . \mathrm{Prop}$

∗180·151. $\vdash :. C`P \cap C`Q = \Lambda . \supset : Z \text{ smor } (P \ddagger Q) . \equiv .$
$(\exists R, S) . R \text{ smor } P . S \text{ smor } Q . C`R \cap C`S = \Lambda . Z = R \ddagger S$

Dem.

$\vdash . *160·48 . \supset \vdash :. \text{Hp} . \supset : (\exists R, S) . R \text{ smor } P . S \text{ smor } Q . C`R \cap C`S = \Lambda .$
$Z = R \ddagger S . \supset . Z \text{ smor } (P \ddagger Q)$ (1)

$\vdash . *160·44 . \supset \vdash : T \epsilon Z \overline{\text{smor}} (P \ddagger Q) . \supset . Z = T;P \ddagger T;Q$ (2)

$\vdash . *160·14 . *151·11 . \supset \vdash : T \epsilon Z \overline{\text{smor}} (P \ddagger Q) . \supset . C`P \subset \text{Ɑ}`T . C`Q \subset \text{Ɑ}`T$ (3)

$\vdash . (3) . *151·21 . \quad \supset \vdash : T \epsilon Z \overline{\text{smor}} (P \ddagger Q) . \supset . T;P \text{ smor } P . T;Q \text{ smor } Q$ (4)

$\vdash . *72·411 . *150·22 . \supset \vdash :. \text{Hp} . \supset :$
$T \epsilon Z \overline{\text{smor}} (P \ddagger Q) . \supset . C`T;P \cap C`T;Q = \Lambda$ (5)

$\vdash . (2) . (4) . (5) . \supset \vdash :. \text{Hp} . \supset : T \epsilon Z \overline{\text{smor}} (P \ddagger Q) . \supset .$
$T;P \text{ smor } P . T;Q \text{ smor } Q . C`T;P \cap C`T;Q = \Lambda . Z = T;P \ddagger T;Q :$

[∗151·12] $\supset : Z \text{ smor } (P \ddagger Q) . \supset .$
$(\exists R, S) . R \text{ smor } P . S \text{ smor } Q . C`R \cap C`S = \Lambda . Z = R \ddagger S$ (6)

$\vdash . (1) . (6) . \supset \vdash . \text{Prop}$

∗180·152. $\vdash : Z \text{ smor } (P + Q) . \equiv .$
$(\exists R, S) . R \text{ smor } P . S \text{ smor } Q . C`R \cap C`S = \Lambda . Z = R \ddagger S$
[∗180·151·11·12]

∗180·16. $\vdash . \text{Nr}`(P + Q) =$
$\hat{Z} \{(\exists R, S) . R \epsilon \text{Nr}`P . S \epsilon \text{Nr}`Q . C`R \cap C`S = \Lambda . Z = R \ddagger S\}$
[∗180·152 . ∗152·11]

∗180·2. $\vdash : Z \epsilon \mu \dot{+} \nu . \equiv . (\exists P, Q) . \mu = \text{N}_0\text{r}`P . \nu = \text{N}_0\text{r}`Q . Z \text{ smor } (P + Q)$
[(∗180·02)]

∗180·201. $\vdash :. Z \epsilon \mu \dot{+} \nu . \equiv : \mu, \nu \epsilon \text{N}_0\text{R} : (\exists P, Q) . P \epsilon \mu . Q \epsilon \nu . Z \text{ smor } (P + Q)$
[∗155·27 . ∗180·2]

∗180·202. $\vdash :. Z \epsilon \mu \dot{+} \nu . \equiv :$
$\exists ! \mu . \exists ! \nu : (\exists P, Q) . \mu = \text{Nr}`P . \nu = \text{Nr}`Q . Z \text{ smor } (P + Q)$

Dem.

$\vdash . *155·34·22 . *180·201 . \supset$

$\vdash :. Z \epsilon \mu \dot{+} \nu . \equiv : \exists ! \mu . \exists ! \nu . \mu, \nu \epsilon \text{NR} : (\exists P, Q) . P \epsilon \mu . Q \epsilon \nu . Z \text{ smor } (P + Q) :$

[∗152·44] $\equiv : \exists ! \mu . \exists ! \nu . \mu, \nu \epsilon \text{NR} : (\exists P, Q) . \mu = \text{Nr}`P . \nu = \text{Nr}`Q . Z \text{ smor } (P + Q) :$

[∗152·41] $\equiv : \exists ! \mu . \exists ! \nu : (\exists P, Q) . \mu = \text{Nr}`P . \nu = \text{Nr}`Q . Z \text{ smor } (P + Q) :.$
$\supset \vdash . \text{Prop}$

In the following propositions proofs are omitted, since they are exactly analogous to proofs of propositions in ∗110 whose numbers have the same decimal part.

∗180·21. $\vdash :. \mu, \nu \epsilon \text{NR} . \supset : Z \epsilon \mu \dot{+} \nu . \equiv . (\exists P, Q) . P \epsilon \mu . Q \epsilon \nu . Z \text{ smor } (P + Q)$

***180·211.** $\vdash :. \mu, \nu \in \mathrm{NR} . \supset : Z \in \mu \dotplus \nu . \equiv .$
$(\exists R, S) . R \in \mathrm{smor}``\mu . S \in \mathrm{smor}``\nu . C`R \cap C`S = \Lambda . Z = R \Uparrow S$

***180·212.** $\vdash :. \mu, \nu \in \mathrm{NR} . \supset : Z \in \mu \dotplus \nu . \equiv .$
$(\exists R) . R \in \mathrm{smor}``\mu . R \subseteq Z . Z \upharpoonright (- C`R) \in \mathrm{smor}``\nu$

***180·22.** $\vdash . \mathrm{N_0r}`P \dotplus \mathrm{N_0r}`Q = \mathrm{Nr}`(P + Q)$

***180·24.** $\vdash : R \;\mathrm{smor}\; P . S \;\mathrm{smor}\; Q . \supset . \mathrm{N_0r}`R \dotplus \mathrm{N_0r}`S = \mathrm{N_0r}`P \dotplus \mathrm{N_0r}`Q$
[*180·15·22]

***180·3.** $\vdash . \mathrm{Nr}`P \dotplus \mathrm{Nr}`Q = \mathrm{N_0r}`P \dotplus \mathrm{Nr}`Q = \mathrm{Nr}`P \dotplus \mathrm{N_0r}`Q$
$= \mathrm{N_0r}`P \dotplus \mathrm{N_0r}`Q = \mathrm{Nr}`(P + Q)$ [*180·22 . (*180·03·031)]

***180·31.** $\vdash : P \;\mathrm{smor}\; R . Q \;\mathrm{smor}\; S . \supset . \mathrm{Nr}`P \dotplus \mathrm{Nr}`Q = \mathrm{Nr}`R \dotplus \mathrm{Nr}`S$

***180·32.** $\vdash : C`P \cap C`Q = \Lambda . \supset . \mathrm{Nr}`P \dotplus \mathrm{Nr}`Q = \mathrm{Nr}`(P \Uparrow Q)$ [*180·14·3]

***180·4.** $\vdash : \exists ! \mu \dotplus \nu . \supset . \mu, \nu \in \mathrm{NR} - \iota`\Lambda . \mu, \nu \in \mathrm{N_0R}$

***180·42.** $\vdash . \mu \dotplus \nu \in \mathrm{NR}$

***180·43.** $\vdash : \mu \dotplus \nu = \mathrm{N_0r}`Z . \equiv . Z \in \mu \dotplus \nu$

***180·53.** $\vdash . (P + Q) + R \;\mathrm{smor}\; P + (Q + R)$

Dem.

$\vdash . *160·44 . (*180·01) . \supset$
$\vdash : P' = \downarrow (\Lambda \cap C`R) ; \iota ; \downarrow (\Lambda \cap C`Q) ; \iota ; P . Q' = \downarrow (\Lambda \cap C`R) ; \iota ; (\Lambda \cap C`P) \downarrow ; \iota ; Q .$
$R' = \{\Lambda \cap C`(P + Q)\} \downarrow ; \iota ; R . \supset . P' \Uparrow Q' = \downarrow (\Lambda \cap C`R) ; \iota ; (P + Q) .$ (1)
[(*180·01)] $\supset . (P' \Uparrow Q') \Uparrow R' = (P + Q) + R .$
[*160·31] $\supset . P' \Uparrow (Q' \Uparrow R') = (P + Q) + R$ (2)
$\vdash . (1) . *180·11 \dfrac{P + Q, R}{P, \quad Q} . *160·14 . \supset \vdash : \mathrm{Hp}(1) . \supset .$
$C`P' \cap C`R' = \Lambda . C`Q' \cap C`R' = \Lambda$ (3)
$\vdash . *180·11 . *72·411 . *150·22 . \supset \vdash : \mathrm{Hp}(1) . \supset . C`P' \cap C`Q' = \Lambda$ (4)
$\vdash . (3) . (4) . *160·14 . \supset \vdash : \mathrm{Hp}(1) . \supset . C`P' \cap C`(Q' \Uparrow R') = \Lambda$ (5)
$\vdash . *180·12 . \quad \supset \vdash : \mathrm{Hp}(1) . \supset . P' \;\mathrm{smor}\; P . Q' \;\mathrm{smor}\; Q . R' \;\mathrm{smor}\; R$ (6)
$\vdash . (3) . (6) . *180·13 . \supset \vdash : \mathrm{Hp}(1) . \supset . Q' \Uparrow R' \;\mathrm{smor}\; Q + R .$
[(5).(6).*180·13] $\supset . P' \Uparrow (Q' \Uparrow R') \;\mathrm{smor}\; P + (Q + R) .$
[(2)] $\supset . (P + Q) + R \;\mathrm{smor}\; P + (Q + R)$ (7)
$\vdash . (7) . *13·19 . \supset \vdash . \mathrm{Prop}$

***180·531.** $P + Q + R = (P + Q) + R$ Df

***180·54.** $\vdash . (\mathrm{Nr}`P \dotplus \mathrm{Nr}`Q) \dotplus \mathrm{Nr}`R = \mathrm{Nr}`(P + Q + R)$

***180·541.** $\vdash . \mathrm{Nr}`P \dotplus (\mathrm{Nr}`Q \dotplus \mathrm{Nr}`R) = \mathrm{Nr}`(P + Q + R)$

***180·55.** $\vdash . (\mathrm{Nr}`P \dotplus \mathrm{Nr}`Q) \dotplus \mathrm{Nr}`R = \mathrm{Nr}`P \dotplus (\mathrm{Nr}`Q \dotplus \mathrm{Nr}`R)$

***180·551.** $\vdash . (\mathrm{N_0r}`P \dotplus \mathrm{N_0r}`Q) \dotplus \mathrm{N_0r}`R = \mathrm{N_0r}`P \dotplus (\mathrm{N_0r}`Q \dotplus \mathrm{N_0r}`R)$

***180·56.** $\vdash . (\mu \dotplus \nu) \dotplus \varpi = \mu \dotplus (\nu \dotplus \varpi)$

***180·561.** $\mu \dotplus \nu \dotplus \varpi = (\mu \dotplus \nu) \dotplus \varpi$ Df

***180·57.** $\vdash . (\mu \dotplus \nu) \dotplus (\varpi \dotplus \rho) = \mu \dotplus \nu \dotplus \varpi \dotplus \rho$

***180·6.** $\vdash : \mu \epsilon \mathrm{NR} . \supset . \mu \dotplus 0_r = \mathrm{smor}``\mu = 0_r \dotplus \mu$

Observe that $\mu \dotplus 0_r = 0_r \dotplus \mu$ is an equation depending upon the peculiar properties of 0_r. We do not in general have $\mu \dotplus \nu = \nu \dotplus \mu$ unless μ and ν are *finite* ordinals.

***180·61.** $\vdash . \mathrm{Nr}`P \dotplus 0_r = \mathrm{Nr}`P = 0_r \dotplus \mathrm{Nr}`P$

***180·62.** $\vdash : \mu \dotplus \nu = 0_r . \equiv . \mu = 0_r . \nu = 0_r$

***180·64.** $\vdash . 0_r \dotplus 0_r = 0_r$

***180·642.** $\vdash . 2_r \dotplus 0_r = 0_r \dotplus 2_r = 2_r$

Note that $\dot{1} \dotplus 0_r$, which will be defined in *181, is 0_r, not $\dot{1}$.

The following propositions, being concerned with the relations of relation-numbers and cardinal numbers, have no analogues in *110.

***180·7.** $\vdash . C``\mathrm{Nr}`(P + Q) = C``\mathrm{Nr}`P +_c C``\mathrm{Nr}`Q = \mathrm{Nc}`C`P +_c \mathrm{Nc}`C`Q$

Dem.

$\vdash . *152·7 . \supset \vdash . C``\mathrm{Nr}`(P + Q) = \mathrm{Nc}`C`(P + Q)$

[*180·111] $= \mathrm{Nc}`(C`P + C`Q)$

[*110·3] $= \mathrm{Nc}`C`P +_c \mathrm{Nc}`C`Q$ (1)

[*152·7] $= C``\mathrm{Nr}`P +_c C``\mathrm{Nr}`Q$ (2)

$\vdash . (1) . (2) . \supset \vdash .$ Prop

***180·71.** $\vdash : \mu, \nu \epsilon \mathrm{NR} . \supset . C``(\mu \dotplus \nu) = C``\mu +_c C``\nu$

Dem.

$\vdash . *152·4 . \supset \vdash : \mathrm{Hp} . \supset . (\exists P, Q) . \mu = \mathrm{Nr}`P . \nu = \mathrm{Nr}`Q .$

[*180·3] $\supset . (\exists P, Q) . \mu = \mathrm{Nr}`P . \nu = \mathrm{Nr}`Q . \mu \dotplus \nu = \mathrm{Nr}`(P + Q) .$

[*180·7] $\supset . (\exists P, Q) . \mu = \mathrm{Nr}`P . \nu = \mathrm{Nr}`Q .$
$C``(\mu \dotplus \nu) = C``\mathrm{Nr}`P +_c C``\mathrm{Nr}`Q .$

[*13·193] $\supset . C``(\mu \dotplus \nu) = C``\mu +_c C``\nu : \supset \vdash .$ Prop

*181. ON THE ADDITION OF UNITY TO A RELATION-NUMBER.

Summary of *181.

The relation-number $\dot{1}$ has, according to our definitions, no meaning in isolation, because our definitions are framed with a view to series, and a series cannot consist of one term. But we can *add* one term to a series; hence $\dot{1}$ is required as an addendum. In order to get our definitions in the most manageable form, we first construct a relation, which we call $P \dot{+\!\!\!\rightarrow} x$, which is such that, whenever P exists, $P \dot{+\!\!\!\rightarrow} x$ has one more term in its field than P; the relation-number of this relation is then defined as $\mathrm{Nr}`P \dot{+} \dot{1}$. We add also a definition

$$\dot{1} \dot{+} \dot{1} = 2_r \quad \mathrm{Df}$$

which is purely formal, and serves to minimize exceptions to the associative law of addition.

The definitions are closely analogous to those of *180. We put

$$P \dot{+\!\!\!\rightarrow} x = \downarrow \Lambda_x ; \iota ; P +\!\!\!\rightarrow (\Lambda \cap C`P) \downarrow \iota`x \quad \mathrm{Df}$$

with a similar definition for $x \leftarrow\!\!\!\!\dot{+} P$. x and P may be of any relative types, and we have always

$$\downarrow \Lambda_x ; \iota ; P \text{ smor } P \,.\, (\Lambda \cap C`P) \downarrow \iota`x \sim\in C` \downarrow \Lambda_x ; \iota ; P \quad (*181{\cdot}11{\cdot}12).$$

We put

$$\mu \dot{+} \dot{1} = \hat{R}\{(\exists P, x) \,.\, \mathrm{N_0r}`P = \mu \,.\, R \text{ smor } (P \dot{+\!\!\!\rightarrow} x)\} \quad \mathrm{Df}$$

with a similar definition for $\dot{1} \dot{+} \mu$. We also introduce definitions analogous to *180·03·031.

The principal propositions of this number are

***181·3.** $\vdash . \mathrm{Nr}`P \dot{+} \dot{1} = \mathrm{N_0r}`P \dot{+} \dot{1} = \mathrm{Nr}`(P \dot{+\!\!\!\rightarrow} x)$

***181·31.** $\vdash : P \text{ smor } Q \,.\supset .\, \mathrm{Nr}`P \dot{+} \dot{1} = \mathrm{Nr}`Q \dot{+} \dot{1}$

***181·32.** $\vdash : x \sim\in C`P \,.\supset .\, \mathrm{Nr}`P \dot{+} \dot{1} = \mathrm{Nr}`(P +\!\!\!\rightarrow x)$

***181·33.** $\vdash :. \mu, \nu \in \mathrm{NR} \,.\, \exists ! \mu \dot{+} \dot{1} \,.\supset : \mu = \nu \,.\equiv .\, \mu \dot{+} \dot{1} = \nu \dot{+} \dot{1} \,.\equiv .\, \dot{1} \dot{+} \mu = \dot{1} \dot{+} \nu$

***181·4.** $\vdash : \exists ! \mu \dot{+} \dot{1} \,.\supset .\, \mu \in \mathrm{NR} - \iota`\Lambda \,.\, \mu \in \mathrm{N_0R}$

***181·42.** $\vdash . \mu \dot{+} \dot{1} \in \mathrm{NR}$

The following propositions are formally forms of the associative law, but they need separate proof on account of the peculiarity of $\dot{1}$.

***181·54.** $\vdash : \nu \neq 0_r . \supset . (\mu \dotplus \nu) \dotplus \dot{1} = \mu \dotplus (\nu \dotplus \dot{1})$

***181·56.** $\vdash : \mu \neq 0_r . \supset . (\mu \dotplus \dot{1}) \dotplus \dot{1} = \mu \dotplus (\dot{1} \dotplus \dot{1}) = \mu \dotplus 2_r$

***181·58.** $\vdash : \mu \neq 0_r . \nu \neq 0_r . \supset . (\mu \dotplus \dot{1}) \dotplus \nu = \mu \dotplus (\dot{1} \dotplus \nu)$

***181·59.** $\vdash : \mu \neq 0_r . \nu \neq 0_r . \supset . (\mu \dotplus \dot{1}) \dotplus (\dot{1} \dotplus \nu) = \mu \dotplus 2_r \dotplus \nu$

The hypotheses in the above propositions are essential.

***181·6.** $\vdash : \dot{\exists} ! P . \supset . C``\text{Nr}`(P \dot{\rightarrow} x) = \text{Nc}`C`P +_c 1$

***181·62.** $\vdash : \mu \in \text{NR} - \iota`0_r . \supset . C``(\mu \dotplus \dot{1}) = C``(\dot{1} \dotplus \mu) = C``\mu +_c 1$

These propositions give the connection with cardinals.

***181·01.** $P \dot{\rightarrow} x = \downarrow \Lambda_x ;\iota ;P \rightarrow (\Lambda \cap C`P) \downarrow \iota`x$ Df

***181·011.** $x \dot{\leftarrow} P = (\iota`x) \downarrow (\Lambda \cap C`P) \leftarrow \Lambda_x \downarrow ;\iota ;P$ Df

***181·02.** $\mu \dotplus \dot{1} = \hat{R} \{(\exists P, x) . \text{N}_0\text{r}`P = \mu . R \text{ smor } (P \dot{\rightarrow} x)\}$ Df

***181·021.** $\dot{1} \dotplus \mu = \hat{R} \{(\exists P, x) . \text{N}_0\text{r}`P = \mu . R \text{ smor } (x \dot{\leftarrow} P)\}$ Df

***181·03.** $\text{Nr}`P \dotplus \dot{1} = \text{N}_0\text{r}`P \dotplus \dot{1}$ Df

***181·031.** $\dot{1} \dotplus \text{Nr}`P = \dot{1} \dotplus \text{N}_0\text{r}`P$ Df

***181·04.** $\dot{1} \dotplus \dot{1} = 2_r$ Df

Propositions concerning $x \dot{\leftarrow} P$ are omitted in what follows, since they are proved exactly as the analogous propositions concerning $P \dot{\rightarrow} x$ are proved.

***181·1.** $\vdash :. R (P \dot{\rightarrow} x) S . \equiv : (\exists y, z) . yPz . R = (\iota`y) \downarrow \Lambda_x . S = (\iota`z) \downarrow \Lambda_x . \vee .$
$(\exists y) . y \in C`P . R = (\iota`y) \downarrow \Lambda_x . S = (\Lambda \cap C`P) \downarrow \iota`x$ [(*181·01)]

***181·11.** $\vdash . (\Lambda \cap C`P) \downarrow \iota`x \sim\in C` \downarrow \Lambda_x ;\iota ;P$

Dem.

$\vdash . *150·22 . \quad \supset \vdash . C` \downarrow \Lambda_x ;\iota ;P = \downarrow \Lambda_x``\iota``C`P .$

[*55·15] $\quad \supset \vdash : Q \in C` \downarrow \Lambda_x ;\iota ;P . \supset_Q . \text{Cl}`Q = \iota`\Lambda_x$ (1)

$\vdash . *55·15 . \quad \supset \vdash . \text{Cl}`(\Lambda \cap C`P) \downarrow \iota`x = \iota`\iota`x$ (2)

$\vdash . (1) . (2) . *51·161 . \supset \vdash : Q \in C` \downarrow \Lambda_x ;\iota ;P . \supset_Q . \text{Cl}`Q \neq \text{Cl}`(\Lambda \cap C`P) \downarrow \iota`x .$

[*30·37.Transp] $\quad \supset_Q . Q \neq (\Lambda \cap C`P) \downarrow \iota`x :$

[*13·196] $\quad \supset \vdash . (\Lambda \cap C`P) \downarrow \iota`x \sim\in C` \downarrow \Lambda_x ;\iota ;P . \supset \vdash .$ Prop

***181·12.** $\vdash . \downarrow \Lambda_x ;\iota ;P \text{ smor } P$ [*151·61·65]

***181·13.** $\vdash : Q \text{ smor } P . y \sim\in C`Q . \supset . Q \rightarrow y \text{ smor } P \dot{\rightarrow} x$

Dem.

$\vdash . *181·12 . \supset \vdash : \text{Hp} . \supset . Q \text{ smor} \downarrow \Lambda_x ;\iota ;P$ (1)

$\vdash . (1) . *161·31 . *181·11 . \supset \vdash .$ Prop

***181·2.** $\vdash : Z \epsilon \mu \dot{+} \dot{1} . \equiv . (\exists P, x) . \mu = \mathrm{N_0r}‘P . Z \,\mathrm{smor}\, (P \mathbin{\dot{+}\!\!\rightarrow} x)$ [(*181·02)]

***181·21.** $\vdash :. \mu \epsilon \mathrm{NR} . \supset : Z \epsilon \mu \dot{+} \dot{1} . \equiv . (\exists P, x) . P \epsilon \mu . Z \,\mathrm{smor}\, (P \mathbin{\dot{+}\!\!\rightarrow} x)$
[*181·2 . *155·26]

***181·22.** $\vdash . \mathrm{N_0r}‘P \dot{+} \dot{1} = \mathrm{Nr}‘(P \mathbin{\dot{+}\!\!\rightarrow} x)$

Dem.

$\vdash . *181·21 . \supset \vdash : Z \epsilon \mathrm{N_0r}‘P \dot{+} \dot{1} . \equiv . (\exists Q, y) . Q \epsilon \mathrm{N_0r}‘P . Z \,\mathrm{smor}\, (Q \mathbin{\dot{+}\!\!\rightarrow} y)$ (1)

$\vdash . (1) . *155·12 . *152·11 . \supset \vdash . \mathrm{Nr}‘(P \mathbin{\dot{+}\!\!\rightarrow} x) \subset \mathrm{N_0r}‘P \dot{+} \dot{1}$ (2)

$\vdash . *181·12·11 . *161·31 . \supset \vdash : Q \epsilon \mathrm{N_0r}‘P . Z \mathrm{smor} (Q \mathbin{\dot{+}\!\!\rightarrow} y) . \supset . Z \,\mathrm{smor}\, (P \mathbin{\dot{+}\!\!\rightarrow} x)$ (3)

$\vdash . (1) . (3) . *152·11 . \quad \supset \vdash : Z \epsilon \mathrm{N_0r}‘P \dot{+} \dot{1} . \supset . Z \epsilon \mathrm{Nr}‘(P \mathbin{\dot{+}\!\!\rightarrow} x)$ (4)

$\vdash . (2) . (4) . \supset \vdash .$ Prop

***181·24.** $\vdash : P \,\mathrm{smor}\, Q . \supset . \mathrm{N_0r}‘P \dot{+} \dot{1} = \mathrm{N_0r}‘Q \dot{+} \dot{1}$ [*181·22·12·11 . *161·31]

***181·3.** $\vdash . \mathrm{Nr}‘P \dot{+} \dot{1} = \mathrm{N_0r}‘P \dot{+} \dot{1} = \mathrm{Nr}‘(P \mathbin{\dot{+}\!\!\rightarrow} x)$ [*181·22 . (*181·03)]

***181·31.** $\vdash : P \,\mathrm{smor}\, Q . \supset . \mathrm{Nr}‘P \dot{+} \dot{1} = \mathrm{Nr}‘Q \dot{+} \dot{1}$ [*181·3·24]

***181·32.** $\vdash : x \sim \epsilon C‘P . \supset . \mathrm{Nr}‘P \dot{+} \dot{1} = \mathrm{Nr}‘(P \mathbin{+\!\!\rightarrow} x)$ [*181·3·13]

***181·33.** $\vdash :. \mu, \nu \epsilon \mathrm{NR} . \exists ! \mu \dot{+} \dot{1} . \supset : \mu = \nu . \equiv . \mu \dot{+} \dot{1} = \nu \dot{+} \dot{1} . \equiv . \dot{1} \dot{+} \mu = \dot{1} \dot{+} \nu$
[*161·33 . *181·3·11·12]

The above proposition is used in *253·23·571.

***181·4.** $\vdash : \exists ! \mu \dot{+} \dot{1} . \supset . \mu \epsilon \mathrm{NR} - \iota‘\Lambda . \mu \epsilon \mathrm{N_0R}$ [*181·2 . *155·22]

***181·42.** $\vdash . \mu \dot{+} \dot{1} \epsilon \mathrm{NR}$

Dem.

$\vdash . *181·3 . \supset \vdash : \mu \epsilon \mathrm{N_0R} . \supset . (\exists P, x) . \mu \dot{+} \dot{1} = \mathrm{Nr}‘(P \mathbin{\dot{+}\!\!\rightarrow} x) .$

[*152·4] $\supset . \mu \dot{+} \dot{1} \epsilon \mathrm{NR}$ (1)

$\vdash . *181·4 . \supset \vdash : \mu \sim \epsilon \mathrm{N_0R} . \supset . \mu \dot{+} \dot{1} = \Lambda .$

[*154·242] $\supset . \mu \dot{+} \dot{1} \epsilon \mathrm{NR}$ (2)

$\vdash . (1) . (2) . \supset \vdash .$ Prop

***181·43.** $\vdash : \mu \dot{+} \dot{1} = \mathrm{N_0r}‘Z . \equiv . Z \epsilon \mu \dot{+} \dot{1}$ [*155·26 . *181·42]

The following propositions are concerned with the associative law when $\dot{1}$ is one of the addenda.

***181·53.** $\vdash : \dot{\exists} ! P . x \neq y . \supset . (P \mathbin{\dot{+}\!\!\rightarrow} x) \mathbin{\dot{+}\!\!\rightarrow} y \,\mathrm{smor}\, P + (x \downarrow y)$

Dem.

$\vdash . *13·15 . (*181·01) . \supset$

$\vdash . (P \mathbin{\dot{+}\!\!\rightarrow} x) \mathbin{\dot{+}\!\!\rightarrow} y = \downarrow \Lambda_y ; \iota ; \{ \downarrow \Lambda_x ; \iota ; P \mathbin{+\!\!\rightarrow} (\Lambda \cap C‘P) \downarrow \iota‘x \} \mathbin{+\!\!\rightarrow} \{\Lambda \cap C‘(P \mathbin{\dot{+}\!\!\rightarrow} x)\} \downarrow \iota‘y$

[*161·4] $= \downarrow \Lambda_y ; \iota ; \downarrow \Lambda_x ; \iota ; P \mathbin{+\!\!\rightarrow} \downarrow \Lambda_y ‘\iota‘(\Lambda \cap C‘P) \downarrow \iota‘x \mathbin{+\!\!\rightarrow} \{\Lambda \cap C‘(P \mathbin{\dot{+}\!\!\rightarrow} x)\} \downarrow \iota‘y$ (1)

$\vdash . (1) . *161·22 . \supset \vdash : \mathrm{Hp} . \supset . (P \mathbin{\dot{+}\!\!\rightarrow} x) \mathbin{\dot{+}\!\!\rightarrow} y$

$= \downarrow \Lambda_y ; \iota ; \downarrow \Lambda_x ; \iota ; P \mathbin{\ddagger} \{ \downarrow \Lambda_y ‘\iota‘(\Lambda \cap C‘P) \downarrow \iota‘x \} \downarrow [\{\Lambda \cap C‘(P \mathbin{\dot{+}\!\!\rightarrow} x)\} \downarrow \iota‘y]$ (2)

$\vdash . *180·13 . *181·11 . \supset$

$\vdash : \mathrm{Hp} . \supset . \downarrow \Lambda_y ; \iota ; \downarrow \Lambda_x ; \iota ; P \mathbin{\ddagger} \{ \downarrow \Lambda_y ‘\iota‘(\Lambda \cap C‘P) \downarrow \iota‘x \} \downarrow [\{\Lambda \cap C‘(P \mathbin{\dot{+}\!\!\rightarrow} x)\} \downarrow \iota‘y]$

$\mathrm{smor}\, P + (x \downarrow y)$ (3)

$\vdash . (2) . (3) . \supset \vdash .$ Prop

***181·54.** $\vdash : \nu \neq 0_r . \supset . (\mu \dot{+} \nu) \dot{+} \dot{1} = \mu \dot{+} (\nu \dot{+} \dot{1})$

Dem.

$\vdash . *181·2 . *180·2 . \supset \vdash : Y \epsilon (\mu \dot{+} \nu) \dot{+\!\!\rightarrow} \dot{1} . \equiv .$

$(\exists P, Q, R, x) . N_0r`P = \mu . N_0r`Q = \nu . R \text{ smor } P + Q . Y \text{ smor } R \dot{+\!\!\rightarrow} x .$

$[*181·22·31] \equiv . (\exists P, Q, x) . N_0r`P = \mu . N_0r`Q = \nu . Y \text{ smor } (P + Q) \dot{+\!\!\rightarrow} x$ (1)

$\vdash . *153·14 . \supset \vdash :. \text{Hp} . \supset : N_0r`Q = \nu . \supset . \dot{\exists} ! Q$ (2)

$\vdash . *160·44 . (*180·01 . *181·01) . \supset$

$\vdash : P' = \downarrow \Lambda_x ; \iota ; \downarrow \Lambda_{C`Q} ; \iota ; P . Q' = \downarrow \Lambda_x ; \iota ; \Lambda_{C`P} \downarrow ; \iota ; Q . X = \{\Lambda \cap C`(P + Q)\} \downarrow \iota`x .$

$\supset . P' \ddagger Q' = \downarrow \Lambda_x ; \iota ; (P + Q) . (P' \ddagger Q') +\!\!\rightarrow X = (P + Q) \dot{+\!\!\rightarrow} x$ (3)

$\vdash . *180·12 . *181·12 . \supset \vdash : \text{Hp}(3) . \supset . P' \text{ smor } P . Q' \text{ smor } Q$ (4)

$\vdash . *180·11 . *72·411 . *181·11 . (3) . \supset$

$\vdash : \text{Hp}(3) . \supset . C`P' \cap C`Q' = \Lambda . X \sim \epsilon C`P' . X \sim \epsilon C`Q'$ (5)

$\vdash . *161·23 . (4) . \supset \vdash : \text{Hp}(3) . \dot{\exists} ! Q . \supset . (P' \ddagger Q') +\!\!\rightarrow X = P' \ddagger (Q' +\!\!\rightarrow X)$ (6)

$\vdash . *181·13 . (4) . (5) . \supset \vdash : \text{Hp}(3) . \supset . Q' +\!\!\rightarrow X \text{ smor } (Q \dot{+\!\!\rightarrow} x) .$

$[*180·13.(4).(5)] \qquad \supset . P' \ddagger (Q' +\!\!\rightarrow X) \text{ smor } P + (Q \dot{+\!\!\rightarrow} x)$ (7)

$\vdash . (1) . (2) . (6) . (7) . \supset \vdash :. \text{Hp} . \text{Hp}(3) . \supset : Y \epsilon (\mu \dot{+} \nu) \dot{+} \dot{1} . \equiv .$

$(\exists P, Q, x) . N_0r`P = \mu . N_0r`Q = \nu . Y \text{ smor } P + (Q \dot{+\!\!\rightarrow} x) .$

$[*180·3.*181·3] \equiv . (\exists P, Q, x) . N_0r`P = \mu . N_0r`Q = \nu . Y \epsilon N_0r`P \dot{+} (N_0r`Q \dot{+} \dot{1}) .$

$[*13·193.*155·2] \equiv . \mu, \nu \epsilon N_0R . Y \epsilon \mu \dot{+} (\nu \dot{+} \dot{1}) .$

$[*181·4.*180·4] \equiv . Y \epsilon \mu \dot{+} (\nu \dot{+} \dot{1})$ (8)

$\vdash . (8) . *13·19 . \supset \vdash . \text{Prop}$

***181·55.** $\vdash : \mu \neq 0_r . \supset . \dot{1} \dot{+} (\mu \dot{+} \nu) = (\dot{1} \dot{+} \mu) \dot{+} \nu$ [Proof as in *181·54]

***181·56.** $\vdash : \mu \neq 0_r . \supset . (\mu \dot{+} \dot{1}) \dot{+} \dot{1} = \mu \dot{+} (\dot{1} \dot{+} \dot{1}) = \mu \dot{+} 2_r$

Dem.

$\vdash . *153·2 . *180·2 . \supset$

$\vdash : Z \epsilon \mu \dot{+} 2_r . \equiv . (\exists P, x, y) . \mu = N_0r`P . x \neq y . Z \text{ smor } P + (x \downarrow y)$ (1)

$\vdash . (1) . *181·53 . \supset \vdash :. \text{Hp} . \supset :$

$Z \epsilon \mu \dot{+} 2_r . \equiv . (\exists P, x, y) . \mu = N_0r`P . x \neq y . Z \text{ smor } (P \dot{+\!\!\rightarrow} x) \dot{+\!\!\rightarrow} y .$

$[*181·22] \equiv . (\exists P, x, y) . \mu = N_0r`P . x \neq y . Z \epsilon (N_0r`P \dot{+} \dot{1}) \dot{+} \dot{1} .$

$[*181·4] \equiv . (\exists x, y) . x \neq y . Z \epsilon (\mu \dot{+} \dot{1}) \dot{+} \dot{1} .$

$[*24·1] \equiv . Z \epsilon (\mu \dot{+} \dot{1}) \dot{+} \dot{1}$ (2)

$\vdash . (2) . (*181·04) . \supset \vdash . \text{Prop}$

The last line in the above proof, in which *24·1 is used, is legitimate because x and y may be of any type whatever, and therefore the fact that $\Lambda \neq V$ is sufficient to establish $(\exists x, y) . x \neq y$ in the sense wanted.

***181·561.** $\mu \dotplus \dot{1} \dotplus \dot{1} = \mu \dotplus (\dot{1} \dotplus \dot{1})$ Df

This definition adopts the opposite convention to that usually adopted. But it is convenient to have $0_r \dotplus \dot{1} \dotplus \dot{1} = 2_r$, and also to have as much similarity as possible between the results of adding $\dot{1}$ at the beginning and end of a relation. Both reasons lead to the adoption of the above convention. (Cf. *181·57·571, below.)

***181·57.** $\vdash : \mu \neq 0_r . \supset . \dot{1} \dotplus (\dot{1} \dotplus \mu) = (\dot{1} \dotplus \dot{1}) \dotplus \mu = 2_r \dotplus \mu$

[Proof as in *181·56]

***181·571.** $\dot{1} \dotplus \dot{1} \dotplus \mu = (\dot{1} \dotplus \dot{1}) \dotplus \mu$ Df

***181·58.** $\vdash : \mu \neq 0_r . \nu \neq 0_r . \supset . (\mu \dotplus \dot{1}) \dotplus \nu = \mu \dotplus (\dot{1} \dotplus \nu)$ [*161·232]

The proof proceeds in the same way as that of *181·54.

***181·59.** $\vdash : \mu \neq 0_r . \nu \neq 0_r . \supset . (\mu \dotplus \dot{1}) \dotplus (\dot{1} \dotplus \nu) = \mu \dotplus 2_r \dotplus \nu$ [*161·25]

The above propositions show that, except when one of the summands is zero, the associative law holds for $\dot{1}$ just as if it were a relation-number.

The following propositions are concerned with relations to cardinal addition.

***181·6.** $\vdash : \dot{\exists} ! P . \supset . C``\text{Nr}`(P \dot{\rightarrow} x) = \text{Nc}`C`P +_c 1$

Dem.

$\vdash . *152·7 . \supset \vdash . C``\text{Nr}`(P \dot{\rightarrow} x) = \text{Nc}`C`(P \dot{\rightarrow} x) .$

$\vdash . (1) . *161·14 . \supset \vdash : \text{Hp} . \supset . C``\text{Nr}`(P \dot{\rightarrow} x)$

$= \text{Nc}`[\downarrow \Lambda_x``\iota``C`P \cup \iota`\{(\Lambda \cap C`P) \downarrow \iota`x\}]$

$[*110·13·3 . *181·11 . *110·12] = \text{Nc}`C`P +_c 1 : \supset \vdash . \text{Prop}$

***181·61.** $\vdash : \dot{\exists} ! P . \supset . C``\text{Nr}`(x \dot{\leftarrow} P) = 1 +_c \text{Nc}`C`P = \text{Nc}`C`P +_c 1$

[Proof as in *181·6]

***181·62.** $\vdash : \mu \in \text{NR} - \iota`0_r . \supset . C``(\mu \dotplus \dot{1}) = C``(\dot{1} \dotplus \mu) = C``\mu +_c 1$

Dem.

$\vdash . *153·16 . *152·4 . \supset \vdash : \text{Hp} . \supset . (\exists P) . \mu = \text{Nr}`P . \dot{\exists} ! P .$

$[*181·3·6] \quad \supset . (\exists P) . \mu = \text{Nr}`P . C``(\text{Nr}`P \dotplus \dot{1}) = \text{Nc}`C`P +_c 1$

$[*152·7] \quad = C``\text{Nr}`P +_c 1 .$

$[*13·193] \quad \supset . C``(\mu \dotplus \dot{1}) = C``\mu +_c 1 \quad (1)$

Similarly $\vdash : \text{Hp} . \supset . C``(1 \dotplus \mu) = 1 +_c C``\mu \quad (2)$

$\vdash . (1) . (2) . *110·51 . \supset \vdash . \text{Prop}$

*182. ON SEPARATED RELATIONS.

Summary of *182.

In this number, we have to consider, as a preliminary to the addition of the relation-numbers of a field, the properties of the relation $\underset{;}{\overset{\frown}{\downarrow}}{}^{\,;}P$, which is defined as follows. If $x \,\text{♀}\, y$ is any function of two arguments in the sense of *38, we put $\overset{\frown}{\text{♀}}\text{‘}x = x \,\text{♀}\, x$ Df. Thus $\underset{;}{\overset{\frown}{\downarrow}}\text{‘}Q = Q \underset{;}{\downarrow} Q$, *i.e.* $\underset{;}{\overset{\frown}{\downarrow}}\text{‘}Q = \downarrow Q{}^{;}Q$. Hence $\underset{;}{\overset{\frown}{\downarrow}}{}^{;}P$ is the relation of $\downarrow Q{}^{;}Q$ to $\downarrow R{}^{;}R$ when QPR. Thus the symbol $\underset{;}{\overset{\frown}{\downarrow}}{}^{;}P$ is only significant when P is a relation of relations; when this is the case, $\underset{;}{\overset{\frown}{\downarrow}}{}^{;}P$ is the relation which results when, for every Q which is a member of $C\text{“}P$, every member x of $C\text{‘}Q$ is replaced by $x \downarrow Q$. The result is a Rel^2 excl, whose arithmetical properties serve to define the arithmetical properties of the sum of the relation-numbers of members of $C\text{‘}P$. In the next number, we shall put

$$\Sigma\,\text{Nr}\text{‘}P = \text{Nr}\text{‘}\Sigma\text{‘}\underset{;}{\overset{\frown}{\downarrow}}{}^{;}P \quad \text{Df.}$$

We shall put later

$$\Pi\text{Nr}\text{‘}P = \text{Nr}\text{‘}\Pi\text{‘}P$$

and we shall find

$$\Pi\text{‘}P = \dot{s}{}^{;}\text{Prod}\text{‘}\underset{;}{\overset{\frown}{\downarrow}}{}^{;}P \,.\, \text{Nr}\text{‘}\Pi\text{‘}P = \text{Nr}\text{‘}\text{Prod}\text{‘}\underset{;}{\overset{\frown}{\downarrow}}{}^{;}P.$$

Thus we might have dispensed with $\Pi\text{‘}P$ as a fundamental notion, using Prod instead, and putting

$$\Pi\text{‘}P = \dot{s}{}^{;}\text{Prod}\text{‘}\underset{;}{\overset{\frown}{\downarrow}}{}^{;}P \quad \text{Df.}$$

But this course is on the whole less convenient than that adopted in *172 and *173.

The notation $\overset{\frown}{\text{♀}}$ is thus required in connection with ordinal addition, where it is almost indispensable. It has besides certain minor uses. The object of the notation is to enable us to exhibit as a function of x an expression of the form $x \,\text{♀}\, x$, where ♀ is any descriptive double function which exists for all possible pairs of arguments. Thus for example $x \downarrow x$ is a function of x, but the notations hitherto introduced do not enable us to exhibit it in the form $R\text{‘}x$. Hence if we wish (say) to deal with the class

$$\hat{P}\,\{(\exists x) \,.\, x \in \alpha \,.\, P = x \downarrow x\}$$

we cannot write it in the form $R\text{“}\alpha$ unless we introduce a new notation. We put

$$\overset{\frown}{\downarrow}\text{‘}x = x \downarrow x$$

whence $$\hat{P}\{(\exists x) . x \in \alpha . P = x \downarrow x\} = \overset{\frown}{\downarrow}\text{“}\alpha.$$

We introduce the notation generally for all descriptive double functions which exist for all possible pairs of arguments. Thus " $\overset{\frown}{♀}$ " in this number corresponds to " ♀ " in ∗38.

In the present number, we shall begin by a few propositions illustrating possible uses of the notation $\overset{\frown}{♀}$. Thus for example if λ is a class of relations, we have hitherto had no simple notation for expressing the class of their squares. But since $R^2 = \overset{\frown}{|}\text{‘}R$, the class of the squares of λ's is $\overset{\frown}{|}\text{“}\lambda$. The notation is, however, introduced chiefly in order to be applied to $\downarrow$ and $\underset{;}{\downarrow}$. We therefore proceed almost at once to propositions on $\overset{\frown}{\underset{;}{\downarrow}}$, and especially on $\overset{\frown}{\underset{;}{\downarrow}};P$. We have

∗182·16·162. $\vdash . \overset{\frown}{\underset{;}{\downarrow}};P \in \text{Rel}^2\,\text{excl} . \overset{\frown}{\underset{;}{\downarrow}} \in 1 \rightarrow 1 . \overset{\frown}{\underset{;}{\downarrow}};P \;\text{smor}\; P$

∗182·2. $\vdash . \overset{\frown}{\underset{;}{\downarrow}}\text{‘}Q = \Pi\text{‘}(Q \downarrow Q) = \Pi\text{‘}\overset{\frown}{\downarrow}\text{‘}Q$

∗182·21. $\vdash . \overset{\frown}{\underset{;}{\downarrow}};P = \Pi;\overset{\frown}{\downarrow};P$

We next prove (∗182·27) that if $P \in \text{Rel}^2\,\text{excl}$, then P has double likeness to $\overset{\frown}{\underset{;}{\downarrow}};P$, the double correlator being $\breve{\iota} \,|\, \text{D}$ with its converse domain limited to $C\text{‘}\Sigma\text{‘}\overset{\frown}{\underset{;}{\downarrow}};P$ (∗182·26). We then prove (∗182·33) that if $T \upharpoonright C\text{‘}\Sigma\text{‘}P$ is a double correlator of P with Q, then $T \,||\, \text{Cnv‘}T\dagger$ (with its converse domain limited) is a double correlator of $\overset{\frown}{\underset{;}{\downarrow}};P$ and $\overset{\frown}{\underset{;}{\downarrow}};Q$, whence we deduce

∗182·34. $\vdash : P \;\text{smor smor}\; Q . \supset . \overset{\frown}{\underset{;}{\downarrow}};P \;\text{smor smor}\; \overset{\frown}{\underset{;}{\downarrow}};Q$

We next proceed to prove

∗182·42. $\vdash . \Pi\text{‘}P = \dot{s};\text{Prod‘}\overset{\frown}{\underset{;}{\downarrow}};P = \dot{s};\text{D};\Pi\text{‘}\overset{\frown}{\underset{;}{\downarrow}};P = \Pi\text{‘}\Sigma\text{‘}\overset{\frown}{\downarrow};P$

The proof of this is as follows: In virtue of ∗182·21 and the associative law for Π, we have

$$\dot{s};\text{D};\Pi\text{‘}\overset{\frown}{\underset{;}{\downarrow}};P = \Pi\text{‘}\Sigma\text{‘}\overset{\frown}{\downarrow};P.$$

Now $$\Sigma\text{‘}\overset{\frown}{\downarrow};P = P \cup I \upharpoonright C\text{‘}P \quad (∗182·413),$$

and $$\Pi\text{‘}(P \cup I \upharpoonright C\text{‘}P) = \Pi\text{‘}P \quad (∗172·51).$$

Hence our proposition results. Hence we arrive at

***182·44.** $\vdash . \text{Nr}\text{‘}\Pi\text{‘}P = \text{Nr}\text{‘}\text{Prod}\text{‘}\underset{;}{\overset{\frown}{\downarrow}};P = \text{Nr}\text{‘}\Pi\text{‘}\underset{;}{\overset{\frown}{\downarrow}};P$

Finally we have some propositions showing how the notation $\overset{\frown}{\text{♀}}$ can be applied in cardinals. It is then applied to $\underset{,,}{\downarrow}$, instead of, as above, to $\underset{;}{\downarrow}$. We have (*182·5·51·52) $\epsilon \overline{\downarrow} \alpha = \underset{,,}{\overset{\frown}{\downarrow}}\text{‘}\alpha . \epsilon \overline{\downarrow}\text{“}\kappa = \underset{,,}{\overset{\frown}{\downarrow}}\text{“}\kappa . \Sigma\text{‘}\kappa = s\text{‘}\underset{,,}{\overset{\frown}{\downarrow}}\text{“}\kappa$. Thus the notation of the present number might have been employed in dealing with cardinal addition (*112) instead of the notation $\epsilon \overline{\downarrow} \alpha$. The general notation $P \overline{\downarrow} x$ was, however, required for other purposes (cf. *85) and could not have been dispensed with.

In *183 we shall put

$$\Sigma\text{Nr}\text{‘}P = \text{Nr}\text{‘}\Sigma\text{‘}\underset{;}{\overset{\frown}{\downarrow}};P,$$

and by *182·52 we have

$$\Sigma\text{Nc}\text{‘}\kappa = \text{Nc}\text{‘}s\text{‘}\underset{,,}{\overset{\frown}{\downarrow}}\text{“}\kappa.$$

It will be seen that these formulae have the usual kind of analogy.

***182·01.** $\overset{\frown}{\text{♀}} = \hat{y}\hat{x}(y = x \text{♀} x)$ Df

***182·02.** $\vdash : y \overset{\frown}{\text{♀}} x . \equiv . y = x \text{♀} x$ [(*182·01)]

***182·021.** $\vdash . \overset{\frown}{\text{♀}}\text{‘}x = x \text{♀} x$ [*182·02 . *30·3]

***182·022.** $\vdash . \text{E} ! \overset{\frown}{\text{♀}}\text{‘}x$ [*182·021 . *14·21]

***182·023.** $\vdash : \overset{\frown}{\text{♀}} \epsilon 1 \rightarrow \text{Cls} : (\alpha) . \alpha \subset \text{Ɑ}\text{‘}\overset{\frown}{\text{♀}}$ [*182·022 . *71·166 . *33·431]

***182·03.** $\vdash . \overset{\frown}{|}\text{‘}R = R^2$ [*182·021 . (*34·02)]

Thus if λ is a class of relations, the class of their squares is $\overset{\frown}{|}\text{“}\lambda$.

***182·031.** $\vdash . \overset{\frown}{\uparrow}\text{‘}\alpha = \alpha \uparrow \alpha$ [*182·021]

***182·032.** $\vdash . \overset{\frown}{\downarrow}\text{‘}x = x \downarrow x$ [*182·021]

***182·033.** $\vdash . \dot{2} - 2_r = \text{D}\text{‘}\overset{\frown}{\downarrow} = 1_s$ [*56·13 . *182·032 . *153·3]

***182·04.** $\vdash . \underset{,,}{\overset{\frown}{\downarrow}}\text{‘}\alpha = \downarrow \alpha\text{“}\alpha$ [*182·021 . *38·2]

Observe that in $\underset{,,}{\overset{\frown}{\downarrow}}$, we first take $\underset{,,}{\downarrow}$, and then put a circumflex over it. If we first took $\overset{\frown}{\downarrow}$, we could not then place two commas under it, because $\overset{\frown}{\downarrow}$ is a relation, not a double descriptive function, and two commas can only significantly be placed under a double descriptive function.

***182·05.** $\vdash . \underset{;}{\overset{\frown}{\downarrow}}\text{‘}Q = \downarrow Q;Q = Q \underset{;}{\downarrow} Q$ [*182·021 . *150·6]

The relation for the sake of which the above notation is chiefly introduced is $\underset{;}{\overset{\frown}{\downarrow}};P$, where P is a relation of relations. If P relates Q and R, then $\underset{;}{\overset{\frown}{\downarrow}};P$ relates $\downarrow Q;Q$ and $\downarrow R;R$. This is stated in the following proposition:

***182·1.** $\vdash . \underset{;}{\overset{\frown}{\downarrow}};P = \hat{X}\hat{Y}\{(\exists Q, R) . QPR . X = Q \underset{;}{\downarrow} Q . Y = R \underset{;}{\downarrow} R\}$
[*182·023·05 . *150·4]

***182·11.** $\vdash . C`\underset{;}{\overset{\frown}{\downarrow}};P = \underset{;}{\overset{\frown}{\downarrow}}``C`P$ [*150·22]

***182·12.** $\vdash . C`\underset{;}{\overset{\frown}{\downarrow}}`Q = \downarrow Q``C`Q = F\bar{\downarrow} Q$ [*182·05 . *150·22 . *33·5 . (*85·5)]

***182·13.** $\vdash . C``C`\underset{;}{\overset{\frown}{\downarrow}};P = F\bar{\downarrow}``C`P$ [*182·11·12]

***182·14.** $\vdash . F\bar{\downarrow} \epsilon 1 \rightarrow 1$

Dem.

$\vdash . *182{\cdot}12 . \quad \supset \vdash : F\bar{\downarrow} Q = F\bar{\downarrow} R . \supset . \downarrow Q``C`Q = \downarrow R``C`R$ (1)

$\vdash . (1) . *55{\cdot}232 . *37{\cdot}45 . \supset \vdash : F\bar{\downarrow} Q = F\bar{\downarrow} R . \dot{\exists} ! Q . \supset . Q = R$ (2)

$\vdash . *37{\cdot}45 . \quad \supset \vdash : \downarrow Q``C`Q = \downarrow R``C`R . Q = \dot{\Lambda} . \supset . \downarrow R``C`R = \dot{\Lambda} .$

$[*37{\cdot}45 . *33{\cdot}241] \quad \supset . R = \dot{\Lambda}$ (3)

$\vdash . (1) . (2) . (3) . \quad \supset \vdash : F\bar{\downarrow} Q = F\bar{\downarrow} R . \supset . Q = R : \supset \vdash . \mathrm{Prop}$

***182·15.** $\vdash : \exists ! F\bar{\downarrow} Q \cap F\bar{\downarrow} R . \supset . Q = R$

Dem.

$\vdash . *182{\cdot}12 . \supset \vdash : \mathrm{Hp} . \supset . \exists ! \downarrow Q``C`Q \cap \downarrow R``C`R .$

$[*55{\cdot}232] \quad \supset . Q = R : \supset \vdash . \mathrm{Prop}$

***182·16.** $\vdash . \underset{;}{\overset{\frown}{\downarrow}};P \epsilon \mathrm{Rel}^2 \mathrm{excl}$ [*182·12·15]

***182·161.** $\vdash : \underset{;}{\overset{\frown}{\downarrow}}`Q = \underset{;}{\overset{\frown}{\downarrow}}`R . \equiv . Q = R$

Dem.

$\vdash . *182{\cdot}05 . \supset \vdash : \underset{;}{\overset{\frown}{\downarrow}}`Q = \underset{;}{\overset{\frown}{\downarrow}}`R . \equiv . Q \underset{;}{\downarrow} Q = R \underset{;}{\downarrow} R .$

$[*165{\cdot}23] \quad \supset . Q = R$ (1)

$\vdash . (1) . *30{\cdot}37 . \supset \vdash . \mathrm{Prop}$

***182·162.** $\vdash . \underset{;}{\overset{\frown}{\downarrow}} \epsilon 1 \rightarrow 1 . \underset{;}{\overset{\frown}{\downarrow}};P \ \mathrm{smor}\ P$ [*182·161 . *71·57 . *151·243]

***182·17.** $\vdash . C`\Sigma`\underset{;}{\overset{\frown}{\downarrow}};P = \hat{S}\{(\exists Q, x) . Q \epsilon C`P . x \epsilon C`Q . S = x \downarrow Q\}$

Dem.

$\vdash . *182{\cdot}12 . *162{\cdot}22 . *40{\cdot}4 . \supset$

$\vdash . C`\Sigma`\underset{;}{\overset{\frown}{\downarrow}};P = \hat{S}\{(\exists Q) . Q \epsilon C`P . S \epsilon \downarrow Q``C`Q\}$

$[*55{\cdot}231] \quad = \hat{S}\{(\exists Q, x) . Q \epsilon C`P . x \epsilon C`Q . S = x \downarrow Q\} . \supset \vdash . \mathrm{Prop}$

∗182·18. $\vdash . \dot{s}`C`\Sigma`\overset{\frown}{\underset{\cdot\cdot}{\downarrow}} ; P = F \upharpoonright C`P$

Dem.

$\vdash . *182·17 . \supset$

$\vdash . \dot{s}`C`\Sigma`\overset{\frown}{\underset{\cdot\cdot}{\downarrow}} ; P = \hat{y}\hat{R}\{(\exists Q, x) . Q \epsilon C`P . x \epsilon C`Q . y (x \downarrow Q) R\}$

[∗55·13] $= \hat{y}\hat{R}\{(\exists Q, x) . Q \epsilon C`P . x \epsilon C`Q . y = x . R = Q\}$

[∗13·22] $= \hat{y}\hat{R}\{R \epsilon C`P . y \epsilon C`R\}$

[∗33·51] $= \hat{y}\hat{R}\{R \epsilon C`P . yFR\}$

[∗35·101] $= F \upharpoonright C`P . \supset \vdash .$ Prop

∗182·19. $\vdash . s`D``C`\Sigma`\overset{\frown}{\underset{\cdot\cdot}{\downarrow}} ; P = C`\Sigma`P . s`\text{Ɑ}``C`\Sigma`\overset{\frown}{\underset{\cdot\cdot}{\downarrow}} ; P = C`P - \iota`\dot{\Lambda}$

Dem.

$\vdash . *41·43 . *182·18 . \supset \vdash . s`D``C`\Sigma`\overset{\frown}{\underset{\cdot\cdot}{\downarrow}} ; P = D`(F \upharpoonright C`P)$

[∗162·23] $= C`\Sigma`P$ (1)

$\vdash . *41·44 . *182·18 . \supset \vdash . s`\text{Ɑ}``C`\Sigma`\overset{\frown}{\underset{\cdot\cdot}{\downarrow}} ; P = \text{Ɑ}`(F \upharpoonright C`P)$

[∗172·192] $= C`P - \iota`\dot{\Lambda}$ (2)

$\vdash . (1) . (2) . \supset \vdash .$ Prop

∗182·2. $\vdash . \overset{\frown}{\underset{\cdot\cdot}{\downarrow}} `Q = \Pi`(Q \downarrow Q) = \Pi` \overset{\frown}{\downarrow} `Q$ [∗182·05·032 . ∗172·2]

∗182·21. $\vdash . \overset{\frown}{\underset{\cdot\cdot}{\downarrow}} ; P = \Pi ; \overset{\frown}{\downarrow} ; P$ [∗182·2]

The following propositions lead up to ∗182·26·27.

∗182·22. $\vdash . D ; \overset{\frown}{\underset{\cdot\cdot}{\downarrow}} `Q = \text{Prod}` \overset{\frown}{\downarrow} `Q = \iota ; Q$ [∗182·2 . ∗173·1·22]

∗182·23. $\vdash . \breve{\iota} ; D ; \overset{\frown}{\underset{\cdot\cdot}{\downarrow}} `Q = Q$ [∗182·22 . ∗151·252]

∗182·24. $\vdash . \breve{\iota}\dagger ; D\dagger ; \overset{\frown}{\underset{\cdot\cdot}{\downarrow}} ; P = P$ [∗182·23]

∗182·25. $\vdash . \breve{\iota} ; D ; \Sigma` \overset{\frown}{\underset{\cdot\cdot}{\downarrow}} ; P = \Sigma`P . C`\Sigma`\overset{\frown}{\underset{\cdot\cdot}{\downarrow}} ; P \subset \text{Ɑ}`(\breve{\iota} \mid D)$

Dem.

$\vdash . *55·15 . \supset \vdash . \breve{\iota}`D` \downarrow Q`y = y .$

[∗33·43] $\supset \vdash . \downarrow Q`y \epsilon \text{Ɑ}`(\breve{\iota} \mid D) .$

[∗182·12] $\supset \vdash . C` \overset{\frown}{\underset{\cdot\cdot}{\downarrow}} `Q \subset \text{Ɑ}`(\breve{\iota} \mid D) .$

[∗162·22] $\supset \vdash . C`\Sigma` \overset{\frown}{\underset{\cdot\cdot}{\downarrow}} ; P \subset \text{Ɑ}`(\breve{\iota} \mid D) .$

[∗162·35] $\supset \vdash . \breve{\iota} ; D ; \Sigma` \overset{\frown}{\underset{\cdot\cdot}{\downarrow}} ; P = \Sigma`\breve{\iota}\dagger ; D\dagger ; \overset{\frown}{\underset{\cdot\cdot}{\downarrow}} ; P$

[∗182·24] $= \Sigma`P . \supset \vdash .$ Prop

***182·26.** $\vdash : P \in \text{Rel}^2\,\text{excl} \,.\supset .\, \breve{\iota}\,|\,\text{D} \restriction C\text{‘}\Sigma\text{‘}\underset{\cdot,}{\overset{\frown}{\downarrow}}\,;P \in P \,\overline{\text{smor}}\,\overline{\text{smor}}\, \underset{\cdot,}{\overset{\frown}{\downarrow}}\,;P$

Dem.

$\vdash . *182\cdot24\cdot25 \,.\supset \vdash .\, P = (\breve{\iota}\,|\,\text{D})\dagger ; \underset{\cdot,}{\overset{\frown}{\downarrow}}\,;P \,.\, C\text{‘}\Sigma\text{‘}\underset{\cdot,}{\overset{\frown}{\downarrow}}\,;P \subset \text{Ɑ}\text{‘}(\breve{\iota}\,|\,\text{D})$ (1)

$\vdash . *182\cdot17 \,.\, *55\cdot15 \,.\supset$

$\vdash : S, T \in C\text{‘}\Sigma\text{‘}\underset{\cdot,}{\overset{\frown}{\downarrow}}\,;P \,.\, \breve{\iota}\text{‘D‘}S = \breve{\iota}\text{‘D‘}T \,.\supset .$

$(\exists Q, R, x, y) \,.\, Q, R \in C\text{‘}P \,.\, x \in C\text{‘}Q \,.\, y \in C\text{‘}R \,.\, x = y \,.\, S = x \downarrow Q \,.\, T = y \downarrow R \,.$

$[*13\cdot195] \supset .\, (\exists Q, R, x) \,.\, Q, R \in C\text{‘}P \,.\, x \in C\text{‘}Q \cap C\text{‘}R \,.\, S = x \downarrow Q \,.\, T = x \downarrow R$ (2)

$\vdash . (2) \,.\, *163\cdot11 \,.\supset$

$\vdash : \text{Hp} \,.\, S, T \in C\text{‘}\Sigma\text{‘}\underset{\cdot,}{\overset{\frown}{\downarrow}}\,;P \,.\, \breve{\iota}\text{‘D‘}S = \breve{\iota}\text{‘D‘}T \,.\supset .$

$(\exists Q, R, x) \,.\, Q = R \,.\, S = x \downarrow Q \,.\, T = x \downarrow R \,.$

$[*13\cdot195\cdot172] \supset .\, S = T$ (3)

$\vdash . (3) \,.\, *71\cdot55 \,.\supset \vdash : \text{Hp} \,.\supset .\, \breve{\iota}\,|\,\text{D} \restriction C\text{‘}\Sigma\text{‘}\underset{\cdot,}{\overset{\frown}{\downarrow}}\,;P \in 1 \rightarrow 1$ (4)

$\vdash . (1) \,.\, (4) \,.\, *164\cdot18 \,.\supset \vdash .$ Prop

***182·27.** $\vdash : P \in \text{Rel}^2\,\text{excl} \,.\supset .\, P \,\text{smor}\,\text{smor}\, \underset{\cdot,}{\overset{\frown}{\downarrow}}\,;P$ [*182·26]

The following propositions lead up to *182·33·34.

***182·3.** $\vdash : T \restriction C\text{‘}\Sigma\text{‘}Q \in P \,\overline{\text{smor}}\,\overline{\text{smor}}\, Q \,.\supset .\, (T \,\|\, \text{Cnv‘}T\dagger) \restriction C\text{‘}\Sigma\text{‘}\underset{\cdot,}{\overset{\frown}{\downarrow}}\,;Q \in 1 \rightarrow 1$

Dem.

$\vdash . *74\cdot775 \dfrac{T, T\dagger, C\text{‘}\Sigma\text{‘}\underset{\cdot,}{\overset{\frown}{\downarrow}}\,;Q}{Q, \quad R, \quad \lambda} \,.\supset$

$\vdash : T \restriction s\text{‘D“}C\text{‘}\Sigma\text{‘}\underset{\cdot,}{\overset{\frown}{\downarrow}}\,;Q, T\dagger \restriction s\text{‘Ɑ“}C\text{‘}\Sigma\text{‘}\underset{\cdot,}{\overset{\frown}{\downarrow}}\,;Q \in \text{Cls} \rightarrow 1 \,.$

$s\text{‘D“}C\text{‘}\Sigma\text{‘}\underset{\cdot,}{\overset{\frown}{\downarrow}}\,;Q \subset \text{Ɑ‘}T \,.\, s\text{‘Ɑ“}C\text{‘}\Sigma\text{‘}\underset{\cdot,}{\overset{\frown}{\downarrow}}\,;Q \subset \text{Ɑ‘}T\dagger \,.\supset .$

$(T \,\|\, \text{Cnv‘}\breve{T}) \restriction C\text{‘}\Sigma\text{‘}\underset{\cdot,}{\overset{\frown}{\downarrow}}\,;Q \in 1 \rightarrow 1$ (1)

$\vdash . *182\cdot19 \,.\, *164\cdot18 \,.\supset \vdash : \text{Hp} \,.\supset .\, T \restriction s\text{‘D“}C\text{‘}\Sigma\text{‘}\underset{\cdot,}{\overset{\frown}{\downarrow}}\,;Q \in 1 \rightarrow 1$ (2)

$\vdash . *164\cdot18\cdot13 \,. \qquad \supset \vdash : \text{Hp} \,.\supset .\, T\dagger \restriction C\text{‘}Q \in 1 \rightarrow 1 \,.$

$[*182\cdot19] \qquad \supset .\, T\dagger \restriction s\text{‘Ɑ“}C\text{‘}\Sigma\text{‘}\underset{\cdot,}{\overset{\frown}{\downarrow}}\,;Q \in 1 \rightarrow 1$ (3)

$\vdash . *164\cdot18 \,.\, *182\cdot19 \,.\supset \vdash : \text{Hp} \,.\supset .\, s\text{‘D“}C\text{‘}\Sigma\text{‘}\underset{\cdot,}{\overset{\frown}{\downarrow}}\,;Q \subset \text{Ɑ‘}T$ (4)

$\vdash . *150\cdot1 \,.\, *33\cdot431 \,. \quad \supset \vdash .\, s\text{‘Ɑ“}C\text{‘}\Sigma\text{‘}\underset{\cdot,}{\overset{\frown}{\downarrow}}\,;Q \subset \text{Ɑ‘}T\dagger$ (5)

$\vdash . (1) \,.\, (2) \,.\, (3) \,.\, (4) \,.\, (5) \,.\supset \vdash .$ Prop

∗182·31. $\vdash : E!!\, T``C`S . \supset . \overset{\frown}{\underset{\cdot,}{\downarrow}}`T;S = (T \| \mathrm{Cnv}`T\dagger); \overset{\frown}{\underset{\cdot,}{\downarrow}}`S$

Dem.

$\vdash . ∗182·05 . \supset \vdash . \overset{\frown}{\underset{\cdot,}{\downarrow}}`T;S = (T;S) \underset{\cdot,}{\downarrow} (T;S)$ (1)

$\vdash . (1) . ∗165·31 . \supset$

$\vdash : \mathrm{Hp} . \supset . \overset{\frown}{\underset{\cdot,}{\downarrow}}`T;S = T|;S \underset{\cdot,}{\downarrow} (T;S)$

$[∗150·1 . ∗165·321] = T|;(|\mathrm{Cnv}`T\dagger);S \underset{\cdot,}{\downarrow} S$

$[∗150·13 . ∗182·05] = (T \| \mathrm{Cnv}`T\dagger); \overset{\frown}{\underset{\cdot,}{\downarrow}}`S : \supset \vdash . \mathrm{Prop}$

∗182·32. $\vdash : E!!\, T``C`\Sigma`Q . \supset . \overset{\frown}{\underset{\cdot,}{\downarrow}};T\dagger;Q = (T \| \mathrm{Cnv}`T\dagger)\dagger; \overset{\frown}{\underset{\cdot,}{\downarrow}};Q$

Dem.

$\vdash . ∗162·22 . \supset \vdash :. \mathrm{Hp} . \supset : S \epsilon C`Q . \supset_S . E!!\, T``C`S .$

$[∗182·31] \quad \supset_S . \overset{\frown}{\underset{\cdot,}{\downarrow}}`T;S = (T \| \mathrm{Cnv}`T\dagger); \overset{\frown}{\underset{\cdot,}{\downarrow}}`S :$

$[∗150·35·1] \quad \supset : \overset{\frown}{\underset{\cdot,}{\downarrow}};T\dagger;S = (T \| \mathrm{Cnv}`T\dagger)\dagger; \overset{\frown}{\underset{\cdot,}{\downarrow}};Q :. \supset \vdash . \mathrm{Prop}$

∗182·33. $\vdash : T \upharpoonright C`\Sigma`P \epsilon P \,\overline{\mathrm{smor}}\, \overline{\mathrm{smor}}\, Q . \supset .$

$(T \| \mathrm{Cnv}`T\dagger) \upharpoonright C`\Sigma` \overset{\frown}{\underset{\cdot,}{\downarrow}};Q \epsilon (\overset{\frown}{\underset{\cdot,}{\downarrow}};P) \,\overline{\mathrm{smor}}\, \overline{\mathrm{smor}}\, (\overset{\frown}{\underset{\cdot,}{\downarrow}};Q)$

Dem.

$\vdash . ∗164·18 . \supset \vdash : \mathrm{Hp} . \supset . \mathrm{C\!\!I}`T \subset C`\Sigma`Q . T \upharpoonright C`\Sigma`P \epsilon 1 \to 1 . P = T\dagger;Q .$

$[∗74·11] \quad \supset . E!!\, T``C`\Sigma`Q . P = T\dagger;Q .$

$[∗182·32] \quad \supset . \overset{\frown}{\underset{\cdot,}{\downarrow}};P = (T \| \mathrm{Cnv}`T\dagger)\dagger; \overset{\frown}{\underset{\cdot,}{\downarrow}};Q$ (1)

$\vdash . (1) . ∗182·3 . ∗164·18 . \supset \vdash . \mathrm{Prop}$

∗182·34. $\vdash : P \,\mathrm{smor}\, \mathrm{smor}\, Q . \supset . \overset{\frown}{\underset{\cdot,}{\downarrow}};P \,\mathrm{smor}\, \mathrm{smor}\, \overset{\frown}{\underset{\cdot,}{\downarrow}};Q$ [∗182·33]

The converse of the above proposition is false. For example, if $Q = \overset{\frown}{\underset{\cdot,}{\downarrow}};P$, we shall have $\overset{\frown}{\underset{\cdot,}{\downarrow}};P \,\mathrm{smor}\, \mathrm{smor}\, \overset{\frown}{\underset{\cdot,}{\downarrow}};Q$, by ∗182·16·27, but we shall not have $P \,\mathrm{smor}\, \mathrm{smor}\, Q$ unless $P \epsilon \mathrm{Rel}^2 \mathrm{excl}$, as appears from ∗182·16 and ∗164·23.

∗182·411·412 are lemmas for ∗182·413. All the following propositions lead up to ∗182·42, which leads to ∗182·44.

∗182·411. $\vdash . \dot{s}`C`\overset{\frown}{\downarrow};P = I \upharpoonright C`P$

Dem.

$\vdash . ∗150·22 . \supset \vdash . \dot{s}`C`\overset{\frown}{\downarrow};P = \dot{s}`\overset{\frown}{\downarrow}``C`P$

$[∗182·032] \quad = \dot{s}`\hat{R}\{(\exists y) . y \epsilon C`P . R = y \downarrow y\}$

$[∗41·11] \quad = \hat{x}\hat{z}\{(\exists y) . y \epsilon C`P . x(y \downarrow y)z\}$

$[∗55·13 . ∗13·195] \quad = \hat{x}\hat{z}\{z \epsilon C`P . x = z\}$

$[∗50·1 . ∗35·101] \quad = I \upharpoonright C`P . \supset \vdash . \mathrm{Prop}$

∗182·412. $\vdash . F ; \breve{\downarrow} ; P = P$

Dem.

$\vdash . *150\cdot11 . *182\cdot032 . \supset \vdash . F ; \breve{\downarrow} ; P = \hat{x}\hat{y} \{(\exists z, w) . zPw . xF(z \downarrow z) . yF(w \downarrow w)\}$

$[*33\cdot51 . *55\cdot15] \qquad = \hat{x}\hat{y} \{(\exists z, w) . zPw . x = z . y = w\}$

$[*13\cdot22] \qquad = P . \supset \vdash . \text{Prop}$

∗182·413. $\vdash . \Sigma ‘ \breve{\downarrow} ; P = P \cup I \upharpoonright C‘P \quad [*182\cdot411\cdot412 . *162\cdot1]$

∗182·414. $\vdash . \breve{\downarrow} ; P \in \text{Rel}^2 \text{ excl}$

Dem.

$\vdash . *150\cdot22 . \supset \vdash . C‘ \breve{\downarrow} ; P = \breve{\downarrow} “C‘P$

$[*182\cdot032] \qquad = \hat{Q} \{(\exists x) . x \in C‘P . Q = x \downarrow x\} \qquad (1)$

$\vdash . (1) . *55\cdot15 . \supset \vdash : Q, R \in C‘ \breve{\downarrow} ; P . \exists ! C‘Q \cap C‘R . \supset .$

$(\exists x, y) . x, y \in C‘P . Q = x \downarrow x . R = y \downarrow y . \exists ! \iota‘x \cap \iota‘y .$

$[*51\cdot231 . \text{Transp}] \supset . (\exists x, y) . Q = x \downarrow x . R = y \downarrow y . x = y .$

$[*13\cdot195\cdot172] \qquad \supset . Q = R : \supset \vdash . \text{Prop}$

∗182·415. $\vdash : Q \in C‘ \breve{\downarrow} ; P . \supset . C‘Q \in 1$

Dem.

$\vdash . *150\cdot22 . \supset \vdash : \text{Hp} . \supset . (\exists x) . x \in C‘P . Q = x \downarrow x .$

$[*55\cdot15] \qquad \supset . (\exists x) . x \in C‘P . C‘Q = \iota‘x .$

$[*52\cdot1] \qquad \supset . C‘Q \in 1 : \supset \vdash . \text{Prop}$

The purpose of the above proposition is to enable us to apply ∗174·221·231 to $\Pi‘\Pi ; \breve{\downarrow} ; P$, as is done in ∗182·42·43·431 below.

∗182·42. $\vdash . \Pi‘P = \dot{s} ; \text{Prod}‘ \underset{\cdot,}{\breve{\downarrow}} ; P = \dot{s} ; \text{D} ; \Pi‘ \underset{\cdot,}{\breve{\downarrow}} ; P = \Pi‘\Sigma‘ \breve{\downarrow} ; P$

Dem.

$\vdash . *182\cdot21 . \supset \vdash . \dot{s} ; \text{D} ; \Pi‘ \underset{\cdot,}{\breve{\downarrow}} ; P = \dot{s} ; \text{D} ; \Pi‘\Pi ; \breve{\downarrow} ; P$

$[*174\cdot221 . *182\cdot414\cdot415] \qquad = \Pi‘\Sigma‘ \breve{\downarrow} ; P \qquad (1)$

$[*182\cdot413] \qquad = \Pi‘(P \cup I \upharpoonright C‘P)$

$[*172\cdot51] \qquad = \Pi‘P \qquad (2)$

$\vdash . (1) . (2) . *173\cdot1 . \supset \vdash . \text{Prop}$

∗182·43. $\vdash . \dot{s} \upharpoonright (C‘\text{Prod}‘ \underset{\cdot,}{\breve{\downarrow}} ; P) \in (\Pi‘P) \,\overline{\text{smor}}\, (\text{Prod}‘ \underset{\cdot,}{\breve{\downarrow}} ; P)$

Dem.

$\vdash . *174\cdot231 . *182\cdot414\cdot415 . \supset$

$\vdash . \dot{s} \upharpoonright (C‘\text{Prod}‘\Pi ; \breve{\downarrow} ; P) \in (\Pi‘\Sigma‘ \breve{\downarrow} ; P) \,\overline{\text{smor}}\, (\text{Prod}‘\Pi ; \breve{\downarrow} ; P) \qquad (1)$

$\vdash . (1) . *182\cdot21\cdot42 . \supset \vdash . \text{Prop}$

***182·431.** $\vdash . \dot{s} \,|\, \text{D} \upharpoonright (C``\Pi`\overset{\frown}{\underset{;}{\downarrow}} ; P) \, \epsilon \, (\Pi`P) \, \text{smor} \, (\Pi`\overset{\frown}{\underset{;}{\downarrow}} ; P)$

[*174·221 . *182·414·415·21·42]

***182·44.** $\vdash . \text{Nr}`\Pi`P = \text{Nr}`\text{Prod}`\overset{\frown}{\underset{;}{\downarrow}} ; P = \text{Nr}`\Pi`\overset{\frown}{\underset{;}{\downarrow}} ; P$ [*182·43·431 . *152·321]

***182·45.** $\vdash : P \, \epsilon \, \text{Rel}^2\text{excl} \, . \supset . \, \text{Nr}`\text{Prod}`P = \text{Nr}`\text{Prod}`\overset{\frown}{\underset{;}{\downarrow}} ; P$ [*182·44 . *173·16]

The following propositions are concerned with cardinals. They show how to express the propositions and definitions of *112 in the notation of this number, and they thereby illustrate the analogy of cardinal and ordinal addition.

***182·5.** $\vdash . \, \epsilon \, \underline{\downarrow} \, \alpha = \overset{\frown}{\underset{,,}{\downarrow}} `\alpha$ [*182·04 . *85·601]

***182·51.** $\vdash . \, \epsilon \, \underline{\downarrow} \, ``\kappa = \overset{\frown}{\underset{,,}{\downarrow}} ``\kappa$ [*182·5]

***182·52.** $\vdash . \, \Sigma`\kappa = s`\overset{\frown}{\underset{,,}{\downarrow}} ``\kappa \, . \, \Sigma\text{Nc}`\kappa = \text{Nc}`s`\overset{\frown}{\underset{,,}{\downarrow}} ``\kappa$ [*182·51 . *112·1·101]

***182·53.** $\vdash : C \upharpoonright C`P \, \epsilon \, 1 \to 1 \, . \supset .$

$$(|\breve{C}) \upharpoonright (C`\Sigma`\overset{\frown}{\underset{;}{\downarrow}} ; P) \, \epsilon \, (\overset{\frown}{\underset{,,}{\downarrow}} ``C``C`P) \, \overline{\text{sm}} \, \overline{\text{sm}} \, (C``C`\overset{\frown}{\underset{;}{\downarrow}} ; P)$$

Dem.

$\vdash . \, *182{\cdot}18 \, . \, *41{\cdot}44 \, . \supset \vdash . \, s`\text{Cl}``C`\Sigma`\overset{\frown}{\underset{;}{\downarrow}} ; P = \text{Cl}`(F \upharpoonright C`P)$

$[*35{\cdot}64] \qquad \subset C`P \qquad (1)$

$\vdash . (1) . \, *74{\cdot}75 \, . \supset \vdash : \text{Hp} \, . \supset . \, (|\breve{C}) \upharpoonright C`\Sigma`\overset{\frown}{\underset{;}{\downarrow}} ; P \, \epsilon \, 1 \to 1 \qquad (2)$

$\vdash . \, *55{\cdot}581 \dfrac{\breve{C}}{S} . \supset \vdash . \, (x \downarrow Q) \,|\, \breve{C} = x \downarrow C`Q \, .$

$[*38{\cdot}11] \quad \supset \vdash . \, |\breve{C}` \downarrow Q`x = \downarrow (C`Q)`x \, .$

$[*182{\cdot}12{\cdot}04] \quad \supset \vdash . \, |\breve{C}``C`\overset{\frown}{\underset{;}{\downarrow}} `Q = \overset{\frown}{\underset{,,}{\downarrow}} `C`Q \, .$

$[*150{\cdot}22] \quad \supset \vdash . \, |\breve{C}```C``C`\overset{\frown}{\underset{;}{\downarrow}} ; P = \overset{\frown}{\underset{,,}{\downarrow}} ``C``C`P \qquad (3)$

$\vdash . (2) . (3) . \, *111{\cdot}14 \, . \, *162{\cdot}22 \, . \supset \vdash . \, \text{Prop}$

***182·54.** $\vdash : C \upharpoonright C`P \, \epsilon \, 1 \to 1 \, . \supset . \, \text{Nc}`C`\Sigma`\overset{\frown}{\underset{;}{\downarrow}} ; P = \Sigma\text{Nc}`C``C`P$

[*182·52·53 . *111·44]

$*183$. THE SUM OF THE RELATION-NUMBERS OF A FIELD.

Summary of $*183$.

In this number we have to define and consider the sum of the relation-numbers of the members of $C\text{‘}P$, where P is a relation of relations. Since relational sums are not commutative, we cannot define the sum of the relation-numbers of members of a class of relations λ: it is necessary that λ should be given as the field of a relation P, where P determines the order in which the summation is to be effected.

In order to avoid repetition, we replace P by $\underset{\cdot\,,}{\overset{\frown}{\downarrow}}{}^{;}P$, so that if Q is a member of $C\text{‘}P$, Q is replaced by $\underset{\cdot\,,}{\overset{\frown}{\downarrow}}\text{‘}Q$, *i.e.* by $\downarrow Q{}^{;}Q$. This relation is like Q, and its field has no members in common with the field of $\downarrow R{}^{;}R$, unless $Q = R$. Hence we are led to the following definition:

$*183 \cdot 01$. $\Sigma\text{Nr}\text{‘}P = \text{Nr}\text{‘}\Sigma\text{‘}\underset{\cdot\,,}{\overset{\frown}{\downarrow}}{}^{;}P$ Df

This definition is analogous to $*112 \cdot 01$, as appears from $*182 \cdot 52$, and the propositions of the present number are analogous to some of the propositions of $*112$.

We have not merely

$*183 \cdot 11$. $\vdash : P \operatorname{smor} \operatorname{smor} Q . \supset . \Sigma\text{Nr}\text{‘}P = \Sigma\text{Nr}\text{‘}Q$

but also

$*183 \cdot 15$. $\vdash : \underset{\cdot\,,}{\overset{\frown}{\downarrow}}{}^{;}P \operatorname{smor} \operatorname{smor} \underset{\cdot\,,}{\overset{\frown}{\downarrow}}{}^{;}Q . \supset . \Sigma\text{Nr}\text{‘}P = \Sigma\text{Nr}\text{‘}Q$

which is a proposition with a weaker hypothesis than that of $*183 \cdot 11$ (cf. note to $*182 \cdot 34$).

Important propositions in this number are

$*183 \cdot 13$. $\vdash : P \in \text{Rel}^2 \operatorname{excl} . \supset . \text{Nr}\text{‘}\Sigma\text{‘}P = \Sigma\text{Nr}\text{‘}P$

$*183 \cdot 2$. $\vdash : \Sigma\text{Nr}\text{‘}P = 0_r . \equiv . \Sigma\text{‘}P = \dot{\Lambda}$

I.e. a sum is only zero when there is no summand except (at most) zero. (Cf. $*162 \cdot 4 \cdot 45$.)

$*183 \cdot 25$. $\vdash . \Sigma\text{Nr}\text{‘}P \underset{\cdot\,,}{\downarrow}{}^{;}Q = \text{Nr}\text{‘}(Q \times P)$

***183·26.** $\vdash :. \text{Mult ax} . \supset : P \in \text{Nr}`R . C`P \subset \text{Nr}`S . \supset . \Sigma\text{Nr}`P = \text{Nr}`(R \times S)$

This proposition connects addition and multiplication.

***183·31.** $\vdash : P \mp Q . \supset . \Sigma\text{Nr}`(P \downarrow Q) = \text{Nr}`P \dotplus \text{Nr}`Q$

This proposition connects the two kinds of addition. We have also

***183·33.** $\vdash : \dot{\exists} ! P . Z \sim \in C`P . \supset . \Sigma\text{Nr}`(P \leftrightarrow Z) = \Sigma\text{Nr}`P \dotplus \text{Nr}`Z$

The associative law of addition in a very general form is

***183·43.** $\vdash : P \in \text{Rel}^2 \text{excl} . \supset . \Sigma\text{Nr}`\Sigma; \overset{\frown}{\downarrow} \dagger ; P = \Sigma\text{Nr}`\Sigma`P$

Finally the connection of ordinal and cardinal addition is given by

***183·5.** $\vdash : C \upharpoonright C`P \in 1 \to 1 . \supset . C``\Sigma\text{Nr}`P = \Sigma\text{Nc}`C``C`P$

***183·01.** $\Sigma\text{Nr}`P = \text{Nr}`\Sigma` \overset{\frown}{\downarrow} ; P$ Df

***183·1.** $\vdash . \Sigma\text{Nr}`P = \text{Nr}`\Sigma` \overset{\frown}{\downarrow} ; P$ [(*183·01)]

***183·11.** $\vdash : P \text{ smor smor } Q . \supset . \Sigma\text{Nr}`P = \Sigma\text{Nr}`Q$

Dem.

$\vdash . *182·34 . \supset \vdash : \text{Hp} . \supset . (\overset{\frown}{\downarrow} ; P) \text{ smor smor } (\overset{\frown}{\downarrow} ; Q) .$

[*164·151] $\supset . (\Sigma` \overset{\frown}{\downarrow} ; P) \text{ smor } (\Sigma` \overset{\frown}{\downarrow} ; Q) .$

[*183·1.*152·321] $\supset . \Sigma\text{Nr}`P = \Sigma\text{Nr}`Q : \supset \vdash . \text{Prop}$

***183·12.** $\vdash : P \text{ smor smor } \overset{\frown}{\downarrow} ; Q . \supset . \text{Nr}`\Sigma`P = \Sigma\text{Nr}`Q$ [*164·151 . *183·1]

***183·13.** $\vdash : P \in \text{Rel}^2 \text{excl} . \supset . \text{Nr}`\Sigma`P = \Sigma\text{Nr}`P$ [*182·27 . *183·12]

***183·14.** $\vdash . \Sigma\text{Nr}`P = \Sigma\text{Nr}` \overset{\frown}{\downarrow} ; P$

Dem.

$\vdash . *182·16 . *183·13 . \supset \vdash . \text{Nr}`\Sigma` \overset{\frown}{\downarrow} ; P = \Sigma\text{Nr}` \overset{\frown}{\downarrow} ; P$ (1)

$\vdash . (1) . *183·1 . \supset \vdash . \text{Prop}$

***183·15.** $\vdash : \overset{\frown}{\downarrow} ; P \text{ smor smor } \overset{\frown}{\downarrow} ; Q . \supset . \Sigma\text{Nr}`P = \Sigma\text{Nr}`Q$ [*183·11·14]

***183·2.** $\vdash : \Sigma\text{Nr}`P = 0_r . \equiv . \Sigma`P = \dot{\Lambda}$

Dem.

$\vdash . *183·1 . *153·17 . \supset$

$\vdash :. \Sigma\text{Nr}`P = 0_r . \equiv : \Sigma` \overset{\frown}{\downarrow} ; P = \dot{\Lambda} :$

[*162·42] $\equiv : C` \overset{\frown}{\downarrow} ; P \subset \iota`\dot{\Lambda} :$

[*182·05] $\equiv : Q \in C`P . \supset_Q . \downarrow Q ; Q = \dot{\Lambda} :$

[*151·65.*153·101] $\equiv : Q \in C`P . \supset_Q . Q = \dot{\Lambda} :$

[*162·42] $\equiv : \Sigma`P = \dot{\Lambda} :. \supset \vdash . \text{Prop}$

$*183\cdot22$. $\vdash :. \text{Mult ax} . \supset : \exists ! (\overset{\frown}{\underset{\cdot,}{\downarrow}} ; P) \overline{\text{smor}} (\overset{\frown}{\underset{\cdot,}{\downarrow}} ; Q) \cap \text{Rl}`\text{smor} . \supset . \Sigma \text{Nr}`P = \Sigma \text{Nr}`Q$

Dem.

$\vdash . *164\cdot46 . *182\cdot16 . \supset$

$\vdash :. \text{Mult ax} . \supset : \exists ! (\overset{\frown}{\underset{\cdot,}{\downarrow}} ; P) \overline{\text{smor}} (\overset{\frown}{\underset{\cdot,}{\downarrow}} ; Q) \cap \text{Rl}`\text{smor} . \supset . \Sigma`\overset{\frown}{\underset{\cdot,}{\downarrow}} ; P \,\text{smor}\, \Sigma`\overset{\frown}{\underset{\cdot,}{\downarrow}} ; Q .$

$[*183\cdot1 . *152\cdot321]$ $\supset . \Sigma \text{Nr}`P = \Sigma \text{Nr}`Q : \supset \vdash . \text{Prop}$

$*183\cdot23$. $\vdash :. \text{Mult ax} . \supset : P, Q \in \text{Rel}^2 \text{excl} . \exists ! P \,\overline{\text{smor}}\, Q \cap \text{Rl}`\text{smor} . \supset .$
$\Sigma \text{Nr}`P = \Sigma \text{Nr}`Q$ $[*164\cdot46 . *183\cdot13]$

$*183\cdot231$. $\vdash : P \in \text{Nr}`R . C`P \subset \text{Nr}`S . \equiv . \overset{\frown}{\underset{\cdot,}{\downarrow}} ; P \in \text{Nr}`R . C`\overset{\frown}{\underset{\cdot,}{\downarrow}} ; P \subset \text{Nr}`S$

Dem.

$\vdash . *182\cdot162 . *152\cdot31\cdot321 . \supset \vdash : P \in \text{Nr}`R . \equiv . \overset{\frown}{\underset{\cdot,}{\downarrow}} ; P \in \text{Nr}`R$ (1)

$\vdash . *182\cdot05\cdot11 . \supset \vdash : C`\overset{\frown}{\underset{\cdot,}{\downarrow}} ; P \subset \text{Nr}`S . \equiv : Q \in C`P . \supset_Q . \downarrow Q ; Q \in \text{Nr}`S :$

$[*151\cdot65]$ $\equiv : Q \in C`P . \supset_Q . Q \in \text{Nr}`S :$

$[*22\cdot1]$ $\equiv : C`P \subset \text{Nr}`S$ (2)

$\vdash . (1) . (2) . \supset \vdash . \text{Prop}$

$*183\cdot24$. $\vdash :. \text{Mult ax} . \supset : P, Q \in \text{Nr}`R . C`P, C`Q \in \text{Cl}`\text{Nr}`S . \supset . \Sigma \text{Nr}`P = \Sigma \text{Nr}`Q$

Dem.

$\vdash . *183\cdot231 . \supset \vdash :. P, Q \in \text{Nr}`R . C`P, C`Q \in \text{Cl}`\text{Nr}`S . \supset :$
$\overset{\frown}{\underset{\cdot,}{\downarrow}} ; P, \overset{\frown}{\underset{\cdot,}{\downarrow}} ; Q \in \text{Nr}`R . C`\overset{\frown}{\underset{\cdot,}{\downarrow}} ; P, C`\overset{\frown}{\underset{\cdot,}{\downarrow}} ; Q \in \text{Cl}`\text{Nr}`S :$

$[*164\cdot48 . *182\cdot16] \supset : \text{Mult ax} . \supset . \overset{\frown}{\underset{\cdot,}{\downarrow}} ; P \,\text{smor smor}\, \overset{\frown}{\underset{\cdot,}{\downarrow}} ; Q .$

$[*183\cdot15]$ $\supset . \Sigma \text{Nr}`P = \Sigma \text{Nr}`Q$ (1)

$\vdash . (1) . \text{Comm} . \supset \vdash . \text{Prop}$

$*183\cdot25$. $\vdash . \Sigma \text{Nr}`P \underset{\cdot,}{\downarrow} ; Q = \text{Nr}`(Q \times P)$

Dem.

$\vdash . *165\cdot21 . *183\cdot13 . \supset \vdash . \Sigma \text{Nr}`P \underset{\cdot,}{\downarrow} ; Q = \text{Nr}`\Sigma`P \underset{\cdot,}{\downarrow} ; Q$

$[*166\cdot1]$ $= \text{Nr}`(Q \times P) . \supset \vdash . \text{Prop}$

$*183\cdot26$. $\vdash :. \text{Mult ax} . \supset : P \in \text{Nr}`R . C`P \subset \text{Nr}`S . \supset . \Sigma \text{Nr}`P = \text{Nr}`(R \times S)$

Dem.

$\vdash . *165\cdot27 . *183\cdot24 . \supset$

$\vdash :. \text{Mult ax} . \supset : \dot{\exists} ! S . P \in \text{Nr}`R . C`P \subset \text{Nr}`S . \supset . \Sigma \text{Nr}`P = \Sigma \text{Nr}`S \underset{\cdot,}{\downarrow} ; R$

$[*183\cdot13 . *166\cdot1]$ $= \text{Nr}`(R \times S)$ (1)

$\vdash . *153\cdot11\cdot101 . \supset$

$\vdash : S = \dot{\Lambda} . P \in \text{Nr}`R . C`P \subset \text{Nr}`S . \supset . C`P \subset \iota`\dot{\Lambda} .$

$[*162\cdot42]$ $\supset . \Sigma`P = \dot{\Lambda} .$

$[*183\cdot2]$ $\supset . \Sigma \text{Nr}`P = 0_r$ (2)

$\vdash . *166\cdot13 . \supset \vdash : S = \dot{\Lambda} . \supset . R \times S = \dot{\Lambda}$ (3)

$\vdash . (2) . (3) . *153\cdot17 . \supset$

$\vdash : S = \dot{\Lambda} . P \in \text{Nr}`R . C`P \subset \text{Nr}`S . \supset . \Sigma \text{Nr}`P = \text{Nr}`(R \times S)$ (4)

$\vdash . (1) . (4) . \supset \vdash . \text{Prop}$

***183·3.** $\vdash . \Sigma\text{Nr}‘\dot{\Lambda} = 0_r$ $[*183·2 . *162·4]$

***183·301.** $\vdash . \Sigma\text{Nr}‘(\dot{\Lambda} \downarrow \dot{\Lambda}) = 0_r$ $[*183·2 . *162·41]$

***183·302.** $\vdash . \Sigma\text{Nr}‘(P \downarrow P) = \text{Nr}‘(C‘P \uparrow C‘P)$

Dem.

$\vdash . *183·13 . *163·41 . \supset \vdash . \Sigma\text{Nr}‘(P \downarrow P) = \text{Nr}‘\Sigma‘(P \downarrow P)$

$[*162·3.*160·1]$ $= \text{Nr}‘(C‘P \uparrow C‘P) . \supset \vdash . \text{Prop}$

***183·31.** $\vdash : P \ddagger Q . \supset . \Sigma\text{Nr}‘(P \downarrow Q) = \text{Nr}‘P \dotplus \text{Nr}‘Q$

Dem.

$\vdash . *183·1 . \supset \vdash . \Sigma\text{Nr}‘(P \downarrow Q) = \text{Nr}‘\Sigma‘\underset{;}{\overset{\frown}{\downarrow}}{}^{;}(P \downarrow Q)$

$[*150·71]$ $= \text{Nr}‘\Sigma‘\{(\underset{;}{\overset{\frown}{\downarrow}}‘P) \downarrow (\underset{;}{\overset{\frown}{\downarrow}}‘Q)\}$

$[*162·3]$ $= \text{Nr}‘(\underset{;}{\overset{\frown}{\downarrow}}‘P \ddagger \underset{;}{\overset{\frown}{\downarrow}}‘Q)$ (1)

$\vdash . (1) . *180·32 . *182·12·15 . \supset \vdash : \text{Hp} . \supset .$

$\Sigma\,\text{Nr}‘(P \downarrow Q) = \text{Nr}‘\underset{;}{\overset{\frown}{\downarrow}}‘P \dotplus \text{Nr}‘\underset{;}{\overset{\frown}{\downarrow}}‘Q$

$[*182·05.*151·65.*180·31]$ $= \text{Nr}‘P \dotplus \text{Nr}‘Q : \supset \vdash . \text{Prop}$

***183·32.** $\vdash : C‘P \cap C‘Q = \Lambda . \supset . \Sigma\text{Nr}‘(P \ddagger Q) = \Sigma\text{Nr}‘P \dotplus \Sigma\text{Nr}‘Q$

Dem.

$\vdash . *183·1 . \supset \vdash . \Sigma\text{Nr}‘(P \ddagger Q) = \text{Nr}‘\Sigma‘\underset{;}{\overset{\frown}{\downarrow}}{}^{;}(P \ddagger Q)$

$[*162·31.*160·44]$ $= \text{Nr}‘(\Sigma‘\underset{;}{\overset{\frown}{\downarrow}}{}^{;}P \ddagger \Sigma‘\underset{;}{\overset{\frown}{\downarrow}}{}^{;}Q)$ (1)

$\vdash . *182·17 . *55·202 . \supset$

$\vdash : \exists ! C‘\Sigma‘\underset{;}{\overset{\frown}{\downarrow}}{}^{;}P \cap C‘\Sigma‘\underset{;}{\overset{\frown}{\downarrow}}{}^{;}Q . \equiv . (\exists S, R, x) . R \in C‘P \cap C‘Q . x \in C‘R . S = x \downarrow R .$

$[*10·5]$ $\supset . \exists ! C‘P \cap C‘Q$ (2)

$\vdash . (2) . \text{Transp} . \supset \vdash : \text{Hp} . \supset . C‘\Sigma‘\underset{;}{\overset{\frown}{\downarrow}}{}^{;}P \cap C‘\Sigma‘\underset{;}{\overset{\frown}{\downarrow}}{}^{;}Q = \Lambda .$

$[*180·32]$ $\supset . \text{Nr}‘(\Sigma‘\underset{;}{\overset{\frown}{\downarrow}}{}^{;}P \ddagger \Sigma‘\underset{;}{\overset{\frown}{\downarrow}}{}^{;}Q) = \text{Nr}‘\Sigma‘\underset{;}{\overset{\frown}{\downarrow}}{}^{;}P \dotplus \text{Nr}‘\Sigma‘\underset{;}{\overset{\frown}{\downarrow}}{}^{;}Q$

$[*183·1]$ $= \Sigma\text{Nr}‘P \dotplus \Sigma\text{Nr}‘Q$ (3)

$\vdash . (1) . (3) . \supset \vdash . \text{Prop}$

***183·33.** $\vdash : \dot{\exists} ! P . Z \sim\in C‘P . \supset . \Sigma\text{Nr}‘(P \mathbin{+\!\!\!\rightarrow} Z) = \Sigma\text{Nr}‘P \dotplus \text{Nr}‘Z$

Dem.

$\vdash . *183·1 . \supset \vdash . \Sigma\text{Nr}‘(P \mathbin{+\!\!\!\rightarrow} Z) = \text{Nr}‘\Sigma‘\underset{;}{\overset{\frown}{\downarrow}}{}^{;}(P \mathbin{+\!\!\!\rightarrow} Z)$

$[*161·4]$ $= \text{Nr}‘\Sigma‘(\underset{;}{\overset{\frown}{\downarrow}}{}^{;}P \mathbin{+\!\!\!\rightarrow} \underset{;}{\overset{\frown}{\downarrow}}‘Z)$ (1)

$\vdash . (1) . *162·43 . \supset \vdash : \text{Hp} . \supset . \Sigma\text{Nr}‘(P \mathbin{+\!\!\!\rightarrow} Z) = \text{Nr}‘(\Sigma‘\underset{;}{\overset{\frown}{\downarrow}}{}^{;}P \ddagger \underset{;}{\overset{\frown}{\downarrow}}‘Z)$

$[*182·12·15.*180·32]$ $= \text{Nr}‘\Sigma‘\underset{;}{\overset{\frown}{\downarrow}}{}^{;}P \dotplus \text{Nr}‘\underset{;}{\overset{\frown}{\downarrow}}‘Z$

$[*183·1.*182·05.*151·65]$ $= \Sigma\text{Nr}‘P \dotplus \text{Nr}‘Z : \supset \vdash . \text{Prop}$

∗183·331. $\vdash : \dot{\exists} ! P \,.\, Z \sim \epsilon\, C\text{‘}P \,.\supset .\, \Sigma\mathrm{Nr}\text{‘}(Z \leftarrow\!\!\!+ P) = \mathrm{Nr}\text{‘}Z \dotplus \Sigma\mathrm{Nr}\text{‘}P$

[Proof as in ∗183·33]

∗183·42. $\vdash : P \,\epsilon\, \mathrm{Rel}^2\,\mathrm{excl} \,.\supset .\, \overset{\frown}{\downarrow}\!\dagger ; P \,\epsilon\, \mathrm{Rel}^3\,\mathrm{arithm}$

Dem.

$\vdash . *163{\cdot}3 . *182{\cdot}162 . \supset \vdash : \mathrm{Hp} . \supset . \overset{\frown}{\downarrow}\!\dagger ; P \,\epsilon\, \mathrm{Rel}^2\,\mathrm{excl}$ (1)

$\vdash . *162{\cdot}35 . \quad \supset \vdash . \Sigma\text{‘}\overset{\frown}{\downarrow}\!\dagger ; P = \overset{\frown}{\downarrow} ; \Sigma\text{‘}P .$

$[*182{\cdot}16] \quad \supset \vdash . \Sigma\text{‘}\overset{\frown}{\downarrow}\!\dagger ; P \,\epsilon\, \mathrm{Rel}^2\,\mathrm{excl}$ (2)

$\vdash . (1) . (2) . *174{\cdot}3 . \supset \vdash . \mathrm{Prop}$

∗183·43. $\vdash : P \,\epsilon\, \mathrm{Rel}^2\,\mathrm{excl} \,.\supset .\, \Sigma\mathrm{Nr}\text{‘}\Sigma ; \overset{\frown}{\downarrow}\!\dagger ; P = \Sigma\mathrm{Nr}\text{‘}\Sigma\text{‘}P$

This is a form of the associative law of addition.

Dem.

$\vdash . *183{\cdot}42 . *174{\cdot}36 . \supset \vdash : \mathrm{Hp} . \supset . \Sigma ; \overset{\frown}{\downarrow}\!\dagger ; P \,\epsilon\, \mathrm{Rel}^2\,\mathrm{excl} .$

$[*183{\cdot}13] \quad \supset . \Sigma\mathrm{Nr}\text{‘}\Sigma ; \overset{\frown}{\downarrow}\!\dagger ; P = \mathrm{Nr}\text{‘}\Sigma\text{‘}\Sigma ; \overset{\frown}{\downarrow}\!\dagger ; P$

$[*162{\cdot}34] \quad = \mathrm{Nr}\text{‘}\Sigma\text{‘}\Sigma\text{‘}\overset{\frown}{\downarrow}\!\dagger ; P$

$[*162{\cdot}35] \quad = \mathrm{Nr}\text{‘}\Sigma\text{‘}\overset{\frown}{\downarrow} ; \Sigma\text{‘}P$

$[*183{\cdot}1] \quad = \Sigma\mathrm{Nr}\text{‘}\Sigma\text{‘}P : \supset \vdash . \mathrm{Prop}$

∗183·5. $\vdash : C \upharpoonright C\text{‘}P \,\epsilon\, 1 \to 1 \,.\supset .\, C\text{“}\Sigma\mathrm{Nr}\text{‘}P = \Sigma\mathrm{Nc}\text{‘}C\text{“}C\text{‘}P$

Dem.

$\vdash . *152{\cdot}7 . *183{\cdot}1 . \supset \vdash : \mathrm{Hp} . \supset . C\text{“}\Sigma\mathrm{Nr}\text{‘}P = \mathrm{Nc}\text{‘}C\text{‘}\Sigma\text{‘}\overset{\frown}{\downarrow} ; P$

$[*182{\cdot}54] \quad = \Sigma\mathrm{Nc}\text{‘}C\text{“}C\text{‘}P . \supset \vdash . \mathrm{Prop}$

*184. THE PRODUCT OF TWO RELATION-NUMBERS.

Summary of *184.

The propositions of this number are for the most part analogous to those of the propositions of *113 which are concerned with $\mu \times_c \nu$. Those of *113 which are concerned with $\alpha \times \beta$ have their analogues in *166. We put

***184·01.** $\mu \dot{\times} \nu = \hat{R}\{(\exists P, Q) \,.\, \mu = \mathrm{N_0r}`P \,.\, \nu = \mathrm{N_0r}`Q \,.\, R \,\mathrm{smor}\, (P \times Q)\}$ Df

***184·02.** $\mathrm{Nr}`P \dot{\times} \nu = \mathrm{N_0r}`P \dot{\times} \nu$ Df

***184·03.** $\mu \dot{\times} \mathrm{Nr}`Q = \mu \dot{\times} \mathrm{N_0r}`Q$ Df

We prove that $\mu \dot{\times} \nu$ is only zero when one of its factors is zero (*184·16); we prove the associative law (*184·31), and the distributive law in the forms

***184·33.** $\vdash : P \in \mathrm{Rel}^2 \,\mathrm{excl} \,.\, \supset \,.\, \Sigma\mathrm{Nr}`P \dot{\times} \mathrm{Nr}`R = \Sigma\mathrm{Nr}`(\times R);P$

***184·35.** $\vdash \,.\, (\nu \dotplus \varpi) \dot{\times} \mu = (\nu \dot{\times} \mu) \dotplus (\varpi \dot{\times} \mu)$

and we prove $2_r \dot{\times} \mu = \mu \dotplus \mu$ (*184·4). Also we extend the distributive law to the case where one of the summands is $\dot{1}$, *i.e.* we prove

***184·41.** $\vdash : \nu \neq 0_r \,.\, \supset \,.\, (\nu \dotplus \dot{1}) \dot{\times} \mu = (\nu \dot{\times} \mu) \dotplus \mu$

***184·42.** $\vdash : \nu \neq 0_r \,.\, \supset \,.\, (\dot{1} \dotplus \nu) \dot{\times} \mu = \mu \dotplus (\nu \dot{\times} \mu)$

and the connection of cardinal and ordinal multiplication is given by

***184·5.** $\vdash : \mu, \nu \in \mathrm{NR} \,.\, \supset \,.\, C``(\mu \dot{\times} \nu) = C``\mu \times_c C``\nu$

***184·01.** $\mu \dot{\times} \nu = \hat{R}\{(\exists P, Q) \,.\, \mu = \mathrm{N_0r}`P \,.\, \nu = \mathrm{N_0r}`Q \,.\, R \,\mathrm{smor}\, (P \times Q)\}$ Df

***184·02.** $\mathrm{Nr}`P \dot{\times} \nu = \mathrm{N_0r}`P \dot{\times} \nu$ Df

***184·03.** $\mu \dot{\times} \mathrm{Nr}`Q = \mu \dot{\times} \mathrm{N_0r}`Q$ Df

***184·1.** $\vdash : R \in \mu \dot{\times} \nu \,.\, \equiv \,.\, (\exists P, Q) \,.\, \mu = \mathrm{N_0r}`P \,.\, \nu = \mathrm{N_0r}`Q \,.\, R \,\mathrm{smor}\, (P \times Q)$ [(*184·01)]

The proofs of the following propositions are omitted, since they are analogous to those of the corresponding propositions of *113.

***184·11.** $\vdash : \exists ! \mu \dot\times \nu . \supset . \mu, \nu \in \mathrm{N_0R} . \exists ! \mu . \exists ! \nu$

***184·111.** $\vdash : \sim (\mu, \nu \in \mathrm{N_0R}) . \supset . \mu \dot\times \nu = \Lambda$

***184·12.** $\vdash :. \mu, \nu \in \mathrm{NR} . \supset : R \in \mu \dot\times \nu . \equiv . (\exists P, Q) . P \in \mu . Q \in \nu . R \,\mathrm{smor}\, (P \times Q)$

***184·13.** $\vdash . \mathrm{Nr}\text{‘}P \dot\times \mathrm{Nr}\text{‘}Q = \mathrm{N_0r}\text{‘}P \dot\times \mathrm{Nr}\text{‘}Q = \mathrm{Nr}\text{‘}P \dot\times \mathrm{N_0r}\text{‘}Q$

$$= \mathrm{N_0r}\text{‘}P \dot\times \mathrm{N_0r}\text{‘}Q = \mathrm{Nr}\text{‘}(P \times Q)$$

***184·14.** $\vdash : P \,\mathrm{smor}\, R . Q \,\mathrm{smor}\, S . \supset . \mathrm{Nr}\text{‘}P \dot\times \mathrm{Nr}\text{‘}Q = \mathrm{Nr}\text{‘}R \dot\times \mathrm{Nr}\text{‘}S$

***184·15.** $\vdash . \mu \dot\times \nu \in \mathrm{NR}$

***184·16.** $\vdash :. \mu \dot\times \nu = 0_r . \equiv : \mu, \nu \in \mathrm{NR} - \iota\text{‘}\Lambda : \mu = 0_r . \vee . \nu = 0_r$

***184·2.** $\vdash :. \text{Mult ax} . \supset : P \in \mathrm{Nr}\text{‘}R . C\text{‘}P \subset \mathrm{Nr}\text{‘}S . \supset . \Sigma \mathrm{Nr}\text{‘}P = \mathrm{Nr}\text{‘}R \dot\times \mathrm{Nr}\text{‘}S$
[*183·26 . *184·13]

***184·21.** $\vdash :. \text{Mult ax} . \supset : \mu, \nu \in \mathrm{NR} . \nu \neq \Lambda . P \in \mu . C\text{‘}P \subset \nu . \supset . \Sigma \mathrm{Nr}\text{‘}P = \mu \dot\times \nu$

Dem.

$\vdash . *152·45 . \supset \vdash : \mu \in \mathrm{NR} . P \in \mu . \supset . \mu = \mathrm{Nr}\text{‘}P$ (1)

$\vdash . *152·45 . \supset \vdash : \nu \in \mathrm{NR} . C\text{‘}P \subset \nu . S \in C\text{‘}P . \supset . \nu = \mathrm{Nr}\text{‘}S . C\text{‘}P \subset \mathrm{Nr}\text{‘}S$ (2)

$\vdash . (1) . (2) . *184·2 . \supset$

$\vdash :. \text{Mult ax} . \supset : \mu, \nu \in \mathrm{NR} . P \in \mu . C\text{‘}P \subset \nu . S \in C\text{‘}P . \supset .$

$\Sigma \mathrm{Nr}\text{‘}P = \mathrm{Nr}\text{‘}P \dot\times \mathrm{Nr}\text{‘}S . \mu = \mathrm{Nr}\text{‘}P . \nu = \mathrm{Nr}\text{‘}S .$

[*13·13] $\supset . \Sigma \mathrm{Nr}\text{‘}P = \mu \dot\times \nu$ (3)

$\vdash . (3) . *10·11·21·23 . \supset$

$\vdash :. \text{Mult ax} . \supset : \mu, \nu \in \mathrm{NR} . P \in \mu . C\text{‘}P \subset \nu . \exists ! C\text{‘}P . \supset . \Sigma \mathrm{Nr}\text{‘}P = \mu \dot\times \nu$ (4)

$\vdash . *183·2 . *162·4 . \supset \vdash : P = \dot\Lambda . \supset . \Sigma \mathrm{Nr}\text{‘}P = 0_r$ (5)

$\vdash . *153·16 . \text{Transp} . \supset \vdash :. \mu \in \mathrm{NR} . P \in \mu . P = \dot\Lambda . \supset : \mu = 0_r :$

[*184·16] $\supset : \nu \in \mathrm{NR} - \iota\text{‘}\Lambda . \supset . \mu \dot\times \nu = 0_r$ (6)

$\vdash . (5) . (6) . \supset \vdash : \mu, \nu \in \mathrm{NR} . \nu \neq \Lambda . P \in \mu . C\text{‘}P \subset \nu . P = \dot\Lambda . \supset .$

$\Sigma \mathrm{Nr}\text{‘}P = \mu \dot\times \nu$ (7)

$\vdash . (4) . (7) . \supset \vdash . \text{Prop}$

***184·3.** $\vdash . (\mathrm{Nr}\text{‘}P \dot\times \mathrm{Nr}\text{‘}Q) \dot\times \mathrm{Nr}\text{‘}R = \mathrm{Nr}\text{‘}P \dot\times (\mathrm{Nr}\text{‘}Q \dot\times \mathrm{Nr}\text{‘}R) = \mathrm{Nr}\text{‘}(P \times Q \times R)$

Dem.

$\vdash . *184·13 . \supset \vdash . (\mathrm{Nr}\text{‘}P \dot\times \mathrm{Nr}\text{‘}Q) \dot\times \mathrm{Nr}\text{‘}R = \mathrm{Nr}\text{‘}(P \times Q) \dot\times \mathrm{Nr}\text{‘}R$

[*184·13] $= \mathrm{Nr}\text{‘}(P \times Q \times R)$

[*166·42] $= \mathrm{Nr}\text{‘}\{P \times (Q \times R)\}$

[*184·13] $= \mathrm{Nr}\text{‘}P \dot\times (\mathrm{Nr}\text{‘}Q \dot\times \mathrm{Nr}\text{‘}R) . \supset \vdash . \text{Prop}$

***184·31.** $\vdash . (\mu \dot\times \nu) \dot\times \varpi = \mu \dot\times (\nu \dot\times \varpi)$

Dem.

$\vdash . *184·111 . \supset \vdash : \sim (\mu, \nu, \varpi \in \mathrm{N_0R}) . \supset . (\mu \dot\times \nu) \dot\times \varpi = \Lambda . \mu \dot\times (\nu \dot\times \varpi) = \Lambda$ (1)

$\vdash . *155·2 . \supset \vdash : \mu, \nu, \varpi \in \mathrm{N_0R} . \supset .$

$(\exists P, Q, R) . \mu = \mathrm{N_0r}\text{‘}P . \nu = \mathrm{N_0r}\text{‘}Q . \varpi = \mathrm{N_0r}\text{‘}R$ (2)

$\vdash . *184\cdot13 . \supset$

$\vdash : \mu = \mathrm{N_c r}`P . \nu = \mathrm{N_0 r}`Q . \varpi = \mathrm{N_0 r}`R . \supset . (\mu \dot{\times} \nu) \dot{\times} \varpi = \mathrm{Nr}`\{(P \times Q) \times R\}$

[$*184\cdot3$] $= \mathrm{Nr}`\{P \times (Q \times R)\}$

[$*184\cdot13$] $= \mu \dot{\times} (\nu \dot{\times} \varpi)$ (3)

$\vdash . (2) . (3) . \supset \vdash : \mu, \nu, \varpi \in \mathrm{N_0 R} . \supset . (\mu \dot{\times} \nu) \dot{\times} \varpi = \mu \dot{\times} (\nu \dot{\times} \varpi)$ (4)

$\vdash . (1) . (4) . \supset \vdash .$ Prop

$*184\cdot32$. $\mu \dot{\times} \nu \dot{\times} \varpi = (\mu \dot{\times} \nu) \dot{\times} \varpi$ Df

$*184\cdot33$. $\vdash : P \in \mathrm{Rel}^2 \mathrm{excl} . \supset . \Sigma \mathrm{Nr}`P \dot{\times} \mathrm{Nr}`R = \Sigma \mathrm{Nr}`(\times R)$;$P$

Dem.

$\vdash . *183\cdot13 . \supset \vdash : \mathrm{Hp} . \supset . \Sigma \mathrm{Nr}`P \dot{\times} \mathrm{Nr}`R = \mathrm{Nr}`\Sigma`P \dot{\times} \mathrm{Nr}`R$

[$*184\cdot13$] $= \mathrm{Nr}`(\Sigma`P \times R)$

[$*166\cdot44$] $= \mathrm{Nr}`\Sigma`(\times R)$;$P$ (1)

$\vdash . *166\cdot3 . \supset \vdash : \mathrm{Hp} . \supset . (\times R)$;$P \in \mathrm{Rel}^2 \mathrm{excl} .$

[$*183\cdot13$] $\supset . \mathrm{Nr}`\Sigma`(\times R)$;$P = \Sigma \mathrm{Nr}`(\times R)$;$P$ (2)

$\vdash . (1) . (2) . \supset \vdash .$ Prop

$*184\cdot34$. $\vdash . (\mathrm{Nr}`P \dotplus \mathrm{Nr}`Q) \dot{\times} \mathrm{Nr}`R = (\mathrm{Nr}`P \dot{\times} \mathrm{Nr}`R) \dotplus (\mathrm{Nr}`Q \dot{\times} \mathrm{Nr}`R)$

Dem.

$\vdash . *180\cdot3 . *184\cdot13 . \supset$

$\vdash . (\mathrm{Nr}`P \dotplus \mathrm{Nr}`Q) \dot{\times} \mathrm{Nr}`R = \mathrm{Nr}`\{(P + Q) \times R\}$

[$*166\cdot45 . *180\cdot1$] $= \mathrm{Nr}`[\{\downarrow (\Lambda \cap C`Q)$;ι;$P\} \times R \ \ddagger \ \{(\Lambda \cap C`P) \downarrow$;$\iota$;$Q\} \times R]$

[$*166\cdot3 . *180\cdot11\cdot32$] $= \mathrm{Nr}`[\{\downarrow (\Lambda \cap C`Q)$;ι;$P\} \times R] \dotplus \mathrm{Nr}`[\{(\Lambda \cap C`P) \downarrow$;$\iota$;$Q\} \times R]$

[$*184\cdot14 . *180\cdot12$] $= \mathrm{Nr}`(P \times R) \dotplus \mathrm{Nr}`(Q \times R)$

[$*184\cdot13$] $= (\mathrm{Nr}`P \dot{\times} \mathrm{Nr}`Q) \dotplus (\mathrm{Nr}`Q \dot{\times} \mathrm{Nr}`R) . \supset \vdash .$ Prop

$*184\cdot35$. $\vdash . (\nu \dotplus \varpi) \dot{\times} \mu = (\nu \dot{\times} \mu) \dotplus (\varpi \dot{\times} \mu)$ [$*184\cdot34$]

The proof proceeds as in $*184\cdot31$.

$*184\cdot4$. $\vdash . 2_r \dot{\times} \mu = \mu \dotplus \mu$

Dem.

$\vdash . *184\cdot111 . *180\cdot4 . \supset \vdash : \mu \sim \in \mathrm{N_0 R} . \supset . 2_r \dot{\times} \mu = \Lambda . \mu \dotplus \mu = \Lambda$ (1)

$\vdash . *153\cdot24 . *184\cdot13 . \supset$

$\vdash : \mu = \mathrm{N_0 r}`P . \supset . 2_r \dot{\times} \mu = \mathrm{Nr}`\{\Lambda \downarrow (\iota`x) \times P\}$ (2)

$\vdash . *166\cdot1 . \supset \vdash . \Lambda \downarrow (\iota`x) \times P = \Sigma`P \underset{;}{\downarrow}$;$(\Lambda \downarrow \iota`x)$

[$*150\cdot71$] $= \Sigma`\{(P \underset{;}{\downarrow} \Lambda) \downarrow (P \underset{;}{\downarrow} \iota`x)\}$

[$*162\cdot3$] $= (P \underset{;}{\downarrow} \Lambda) \ddagger (P \underset{;}{\downarrow} \iota`x)$ (3)

$\vdash . *180\cdot31\cdot32 . *165\cdot251\cdot211 .$ Transp $. \supset$

$\vdash . \mathrm{Nr}`\{(P \underset{;}{\downarrow} \Lambda) \ddagger (P \underset{;}{\downarrow} \iota`x)\} = \mathrm{Nr}`P \dotplus \mathrm{Nr}`P$ (4)

$\vdash . (2) . (3) . (4) . \supset \vdash : \mu = \mathrm{N_0 r}`P . \supset . 2_r \dot{\times} \mu = \mathrm{Nr}`P \dotplus \mathrm{Nr}`P$

[$*180\cdot3$] $= \mu \dotplus \mu$ (5)

$\vdash . (1) . (5) . \supset \vdash .$ Prop

***184·41.** $\vdash : \nu \neq 0_r . \supset . (\nu \dot{+} \dot{1}) \dot{\times} \mu = (\nu \dot{\times} \mu) \dot{+} \mu$

Dem.

$\vdash . *166 \cdot 53 . *180 \cdot 32 . *165 \cdot 251 . \supset$

$\vdash : \dot{\exists} ! Q . y \sim \epsilon C`Q . \supset . \text{Nr}`\{(Q \rightarrow\!\!\!\!\!+ y) \times P\} = \text{Nr}`(Q \times P) \dot{+} \text{Nr}`P \quad (1)$

$\vdash . (1) . *181 \cdot 32 . *184 \cdot 13 . \supset$

$\vdash : \mu = \text{Nr}`P . \nu = \text{Nr}`Q . \nu \neq 0_r . y \sim \epsilon C`Q . \supset . (\nu \dot{+} \dot{1}) \dot{\times} \mu = (\nu \dot{\times} \mu) \dot{+} \mu \quad (2)$

$\vdash . *181 \cdot 11 \cdot 12 . \supset \vdash : \nu \epsilon \text{NR} - \iota`\Lambda . \supset . (\exists Q, y) . \nu = \text{Nr}`Q . y \sim \epsilon C`Q \quad (3)$

$\vdash . (2) . (3) . \supset$

$\vdash : \mu \epsilon \text{NR} . \nu \epsilon \text{NR} - \iota`\Lambda . \nu \neq 0_r . \supset . (\nu \dot{+} \dot{1}) \dot{\times} \mu = (\nu \dot{\times} \mu) \dot{+} \mu \quad (4)$

$\vdash . *184 \cdot 111 . *181 \cdot 4 . \supset$

$\vdash : \sim (\mu \epsilon \text{NR} . \nu \epsilon \text{NR} - \iota`\Lambda) . \supset . (\nu \dot{+} 1) \dot{\times} \mu = \Lambda . (\nu \dot{\times} \mu) \dot{+} \mu = \Lambda \quad (5)$

$\vdash . (4) . (5) . \supset \vdash .$ Prop

***184·42.** $\vdash : \nu \neq 0_r . \supset . (\dot{1} \dot{+} \nu) \dot{\times} \mu = \mu \dot{+} (\nu \dot{\times} \mu)$ [Proof as in *184·41]

***184·5.** $\vdash : \mu, \nu \epsilon \text{NR} . \supset . C``(\mu \dot{\times} \nu) = C``\mu \times_c C``\nu$

Dem.

$\vdash . *184 \cdot 13 . \supset \vdash : \text{Hp} . P \epsilon \mu . Q \epsilon \nu . \supset . C``(\mu \dot{\times} \nu) = C``\text{Nr}`(P \times Q)$

[*152·7.*166·12] $= \text{Nc}`(C`P \times C`Q)$

[*152·7.*113·25] $= C``\text{Nr}`P \times_c C``\text{Nr}`Q$

[*152·45] $= C``\mu \times_c C``\nu \quad (1)$

$\vdash . *184 \cdot 11 . *113 \cdot 204 . \supset$

$\vdash : \sim (\exists ! \mu . \exists ! \nu) . \supset . C``(\mu \dot{\times} \nu) = \Lambda . C``\mu \times_c C``\nu = \Lambda \quad (2)$

$\vdash . (1) . (2) . \supset \vdash .$ Prop

$*185$. THE PRODUCT OF THE RELATION-NUMBERS OF A FIELD.

Summary of $*185$.

The subject of this number is analogous to part of the subject of $*114$. The propositions concerned are immediate consequences of previously proved properties of $\Pi‘P$, and offer no difficulty of any kind.

$*185\cdot01$. $\Pi \mathrm{Nr}‘P = \mathrm{Nr}‘\Pi‘P \quad \mathrm{Df}$

$*185\cdot1$. $\vdash . \Pi \mathrm{Nr}‘P = \mathrm{Nr}‘\Pi‘P$ $\quad [(*185\cdot01)]$

$*185\cdot11$. $\vdash : P \,\mathrm{smor}\, \mathrm{smor}\, Q . \supset . \Pi \mathrm{Nr}‘P = \Pi \mathrm{Nr}‘Q$ $\quad [*172\cdot44]$

$*185\cdot12$. $\vdash . \Pi \mathrm{Nr}‘P = \mathrm{Nr}‘\mathrm{Prod}‘\overset{\frown}{\underset{;}{\downarrow}} ;P = \mathrm{Nr}‘\Pi‘\overset{\frown}{\underset{;}{\downarrow}} ;P$ $\quad [*182\cdot44]$

$*185\cdot2$. $\vdash . \Pi \mathrm{Nr}‘\dot{\Lambda} = 0_r$ $\quad [*172\cdot13]$

$*185\cdot21$. $\vdash . \Pi \mathrm{Nr}‘(P \downarrow P) = \mathrm{Nr}‘P$ $\quad [*172\cdot2 . *165\cdot251]$

$*185\cdot22$. $\vdash . \Pi \mathrm{Nr}‘(\dot{\Lambda} \downarrow \dot{\Lambda}) = 0_r$ $\quad [*185\cdot21]$

$*185\cdot23$. $\vdash : \dot{\Lambda} \in C‘P . \supset . \Pi \mathrm{Nr}‘P = 0_r$ $\quad [*172\cdot14]$

$*185\cdot25$. $\vdash :: \mathrm{Mult\,ax} . \supset :. \Pi \mathrm{Nr}‘P = 0_r . \equiv : \dot{\Lambda} \in C‘P . \vee . P = \dot{\Lambda}$ $\quad [*172\cdot182]$

$*185\cdot27$. $\vdash :. \mathrm{Mult\,ax} . \supset : P, Q \in \mathrm{Rel}^2 \mathrm{excl} . \exists ! P \,\overline{\mathrm{smor}}\, Q \cap \mathrm{Rl}‘\mathrm{smor} . \supset .$
$\Pi \mathrm{Nr}‘P = \Pi \mathrm{Nr}‘Q$ $\quad [*172\cdot45]$

$*185\cdot28$. $\vdash :. \mathrm{Mult\,ax} . \supset : P, Q \in \mathrm{Rel}^2 \mathrm{excl} . P, Q \in \mathrm{Nr}‘R . C‘P, C‘Q \in \mathrm{Cl}‘\mathrm{Nr}‘S . \supset .$
$\Pi \mathrm{Nr}‘P = \Pi \mathrm{Nr}‘Q$ $\quad [*164\cdot48 . *185\cdot11]$

$*185\cdot29$. $\vdash :. \mathrm{Mult\,ax} . \supset : P \in \mathrm{Rel}^2 \mathrm{excl} . P \in \mathrm{Nr}‘R . C‘P \subset \mathrm{Nr}‘S . \supset .$
$\Pi \mathrm{Nr}‘P = \mathrm{Nr}‘(S \exp R)$ $\quad [*176\cdot24]$

$*185\cdot31$. $\vdash : \dot{\exists} ! P . \dot{\exists} ! Q . C‘P \cap C‘Q = \Lambda . \supset . \Pi \mathrm{Nr}‘(P \Lsh Q) = \Pi \mathrm{Nr}‘P \dot{\times} \Pi \mathrm{Nr}‘Q$
$[*172\cdot35]$

$*185\cdot32$. $\vdash : Z \sim\in C‘P . \supset . \Pi \mathrm{Nr}‘(P \rightarrow Z) = \Pi \mathrm{Nr}‘P \dot{\times} \mathrm{Nr}‘Z$ $\quad [*172\cdot32]$

$*185\cdot321$. $\vdash : Z \sim\in C‘P . \supset . \Pi \mathrm{Nr}‘(Z \leftarrow P) = \mathrm{Nr}‘Z \dot{\times} \Pi \mathrm{Nr}‘P$ $\quad [*172\cdot321]$

$*185\cdot35$. $\vdash : P \ddagger Q . \supset . \Pi \mathrm{Nr}‘(P \downarrow Q) = \mathrm{Nr}‘P \dot{\times} \mathrm{Nr}‘Q$ $\quad [*172\cdot23]$

∗185·4. $\vdash :. P \epsilon \text{Rel}^2 \text{excl} : QPQ . \supset_Q . C\text{‘}Q \epsilon 0 \cup 1 : \supset . \Pi\text{Nr‘}\Pi\text{;}P = \Pi\text{Nr‘}\Sigma\text{‘}P$ [∗174·241]

∗185·41. $\vdash : P \epsilon \text{Rel}^2 \text{excl} . P \Subset J . \supset . \Pi\text{Nr‘}\Pi\text{;}P = \Pi\text{Nr‘}\Sigma\text{‘}P$ [∗174·25]

The following proposition gives the connection between ordinal and cardinal multiplication.

∗185·5. $\vdash : P \epsilon \text{Rel}^2 \text{excl} . \dot{\exists} ! P . \supset . C\text{“}\Pi\text{Nr‘}P = \Pi\text{Nc‘}C\text{“}C\text{‘}P$

Dem.

$\vdash . ∗173·16 . \supset \vdash : \text{Hp} . \supset . C\text{“}\Pi\text{Nr‘}P = C\text{“}\text{Nr‘Prod‘}P$

[∗152·7] $= \text{Nc‘}C\text{‘Prod‘}P$

[∗173·161] $= \text{Nc‘Prod‘}C\text{“}C\text{‘}P$

[∗163·16.∗115·12] $= \Pi\text{Nc‘}C\text{“}C\text{‘}P : \supset \vdash . \text{Prop}$

$*186$. POWERS OF RELATION-NUMBERS.

Summary of $*186$.

For "μ to the νth power," where ordinal powers are concerned, we use the notation "$\mu \exp_r \nu$." We cannot use "μ^ν" or "$\mu \exp \nu$" because these have been already used for cardinals and classes ($*116$). We therefore put a suffix r to "exp" to show that it is *relational* powers that we are dealing with. We put

$$\mu \exp_r \nu = \hat{R}\{(\exists P, Q) . \mu = N_0 r`P . \nu = N_0 r`Q . R \operatorname{smor} (P \exp Q)\} \quad \text{Df.}$$

The following are the principal propositions of this number:

$*186 \cdot 2$. $\vdash : \mu \in N_0 R . \supset . 0_r \exp_r \mu = 0_r . \mu \exp_r 0_r = 0_r$

We do not have $\mu \exp_r 0_r = 1$, because there is no ordinal 1.

$*186 \cdot 21$. $\vdash . \mu \exp_r 2_r = \mu \dot{\times} \mu$

$*186 \cdot 22$. $\vdash . \alpha \exp_r (\beta \dot{+} \dot{1}) = (\alpha \exp_r \beta) \dot{\times} \alpha$

$*186 \cdot 23$. $\vdash . \alpha \exp_r (\dot{1} \dot{+} \beta) = \alpha \dot{\times} (\alpha \exp_r \beta)$

$*186 \cdot 14$. $\vdash : \nu \neq 0_r . \varpi \neq 0_r . \supset . \mu \exp_r (\nu \dot{+} \varpi) = (\mu \exp_r \nu) \dot{\times} (\mu \exp_r \varpi)$

$*186 \cdot 15$. $\vdash : \varpi \subset \text{Rl}`J . \supset . \mu \exp_r (\varpi \dot{\times} \nu) = (\mu \exp_r \nu) \exp_r \varpi$

$*186 \cdot 31$. $\vdash :. \text{Mult ax} . \supset : \mu, \nu \in \text{NR} - \iota`\Lambda . P \in \text{Rel}^2 \text{excl} \cap \mu . C`P \subset \nu . \supset .$
$\Pi \text{Nr}`P = \mu \exp_r \nu$

which connects exponentiation with multiplication.

$*186 \cdot 4$. $\vdash . \text{Nr}`P_{\text{df}} = 2_r \exp_r (\text{Nr}`P)$ (cf. $*177$)

$*186 \cdot 5$. $\vdash : \mu, \nu \in N_0 R . \nu \neq 0_r . \supset . C``(\mu \exp_r \nu) = (C``\mu)^{C``\nu}$

which connects ordinal and cardinal exponentiation.

$*186 \cdot 01$. $\mu \exp_r \nu = \hat{R}\{(\exists P, Q) . \mu = N_0 r`P . \nu = N_0 r`Q . R \operatorname{smor} (P \exp Q)\}$ Df

$*186 \cdot 02$. $(\text{Nr}`P) \exp_r \nu = (N_0 r`P) \exp_r \nu$ Df

$*186 \cdot 03$. $\mu \exp_r (\text{Nr}`Q) = \mu \exp_r (N_0 r`Q)$ Df

$*186 \cdot 1$. $\vdash : R \in \mu \exp_r \nu . \equiv . (\exists P, Q) . \mu = N_0 r`P . \nu = N_0 r`Q . R \operatorname{smor} (P \exp Q)$
[($*186 \cdot 01$)]

$*186 \cdot 11$. $\vdash . \exists ! \mu \exp_r \nu . \supset . \mu, \nu \in N_0 R . \mu, \nu \in \text{NR} - \iota`\Lambda$

***186·111.** $\vdash : \sim(\mu, \nu \in \mathrm{N_0R}) . \supset . \mu \exp_r \nu = \Lambda$

***186·12.** $\vdash : R \in \mu \exp_r \nu . \equiv . (\exists P, Q) . \mu = \mathrm{N_0r}`P . \nu = \mathrm{N_0r}`Q . R \,\mathrm{smor}\, P^Q$
[*176·181 . *186·1]

***186·13.** $\vdash . (\mathrm{Nr}`P) \exp_r (\mathrm{Nr}`Q) = (\mathrm{N_0r}`P) \exp_r (\mathrm{Nr}`Q) = (\mathrm{Nr}`P) \exp_r (\mathrm{N_0r}`Q)$
$= (\mathrm{N_0r}`P) \exp_r (\mathrm{N_0r}`Q) = \mathrm{Nr}`(P \exp Q) = \mathrm{Nr}`(P^Q)$
[Proof as in *180·3]

***186·14.** $\vdash : \nu \neq 0_r . \varpi \neq 0_r . \supset . \mu \exp_r (\nu \dot{+} \varpi) = (\mu \exp_r \nu) \dot{\times} (\mu \exp_r \varpi)$

Dem.

$\vdash . *180·4 . *186·111 . \supset$

$\vdash : \sim(\mu, \nu, \varpi \in \mathrm{N_0R}) . \supset . \mu \exp_r (\nu \dot{+} \varpi) = \Lambda . (\mu \exp_r \nu) \dot{\times} (\mu \exp_r \varpi) = \Lambda \quad (1)$

$\vdash . *186·13 . *180·3 . \supset$

$\vdash : \mu = \mathrm{N_0r}`P . \nu = \mathrm{N_0r}`Q . \varpi = \mathrm{N_0r}`R . \supset . \mu \exp_r (\nu \dot{+} \varpi) = \mathrm{Nr}`P^{Q+R} \quad (2)$

$\vdash . *176·42 . *180·11 . \supset$

$\vdash : \mathrm{Hp} . \mathrm{Hp}(2) . \supset . \mathrm{Nr}`P^{Q+R} = \mathrm{Nr}`(P\downarrow^{(\Lambda \cap C`P);\iota;Q} \times P^{(\Lambda \cap C`P)\downarrow;\iota;R})$

[*180·12.*176·22.*166·23] $= \mathrm{Nr}`(P^Q \times P^R)$

[*186·13.*184·13] $= (\mu \exp_r \nu) \dot{\times} (\mu \exp_r \varpi) \quad (3)$

$\vdash . (2) . (3) . *155·2 . \supset \vdash : \mu, \nu, \varpi \in \mathrm{N_0R} . \nu \neq 0_r . \varpi \neq 0_r . \supset .$

$\mu \exp_r (\nu \dot{+} \varpi) = (\mu \exp_r \nu) \dot{\times} (\mu \exp_r \varpi) \quad (4)$

$\vdash . (1) . (4) . \supset \vdash . \mathrm{Prop}$

***186·15.** $\vdash : \varpi \subset \mathrm{Rl}`J . \supset . \mu \exp_r (\varpi \dot{\times} \nu) = (\mu \exp_r \nu) \exp_r \varpi$

Dem.

$\vdash . *186·111 . *184·111 . \supset$

$\vdash : \sim(\mu, \nu, \varpi \in \mathrm{N_0R}) . \supset . \mu \exp_r (\varpi \dot{\times} \nu) = \Lambda . (\mu \exp_r \nu) \exp_r \varpi = \Lambda \quad (1)$

$\vdash . *186·13 . *184·13 . \supset$

$\vdash : \mu = \mathrm{N_0r}`P . \nu = \mathrm{N_0r}`Q . \varpi = \mathrm{N_0r}`R . \supset . \mu \exp_r (\varpi \dot{\times} \nu) = \mathrm{Nr}`(P^{R \times Q}) \quad (2)$

$\vdash . *176·57 . \supset \vdash : \mathrm{Hp} . \mathrm{Hp}(2) . \supset . \mathrm{Nr}`(P^{R \times Q}) = \mathrm{Nr}`(P^Q)^R$

[*186·13] $= \{(\mathrm{N_0r}`P) \exp_r (\mathrm{N_0r}`Q)\} \exp_r (\mathrm{N_0r}`R)$

[Hp] $= (\mu \exp_r \nu) \exp_r \varpi \quad (3)$

$\vdash . (2) . (3) . *155·2 . \supset$

$\vdash : \mu, \nu, \varpi \in \mathrm{N_0R} . \varpi \subset \mathrm{Rl}`J . \supset . \mu \exp_r (\varpi \dot{\times} \nu) = (\mu \exp_r \nu) \exp_r \varpi \quad (4)$

$\vdash . (1) . (4) . \supset \vdash . \mathrm{Prop}$

***186·2.** $\vdash : \mu \in \mathrm{N_0R} . \supset . 0_r \exp_r \mu = 0_r . \mu \exp_r 0_r = 0_r$ [*176·151]

***186·21.** $\vdash . \mu \exp_r 2_r = \mu \dot{\times} \mu$

Dem.

$\vdash . *186·111 . *184·111 . \supset \vdash : \mu \sim\in \mathrm{N_0R} . \supset . \mu \exp_r 2_r = \Lambda . \mu \dot{\times} \mu = \Lambda \quad (1)$

$\vdash . *186\cdot13 . *176\cdot1 . \supset \vdash : \mu = \mathrm{N_0r}`P . x \neq y . \supset .$

$\mu \exp_r 2_r = \mathrm{Nr}`\mathrm{Prod}`P \downarrow ; `(x \downarrow y)$

[*150·71] $= \mathrm{Nr}`\mathrm{Prod}`\{(P \downarrow x) \downarrow (P \downarrow y)\}$

[*173·24.*165·211.Transp] $= \mathrm{Nr}`\{(P \downarrow x) \times (P \downarrow y)\}$

[*165·251.*166·23] $= \mathrm{Nr}`(P \times P)$

[*184·13] $= \mu \dot{\times} \mu$ (2)

$\vdash . (2) . *155\cdot2 . *24\cdot1 . \supset \vdash : \mu \in \mathrm{N_0R} . \supset . \mu \exp_r 2_r = \mu \dot{\times} \mu$ (3)

$\vdash . (1) . (3) . \supset \vdash .$ Prop

***186·22.** $\vdash . \alpha \exp_r (\beta \dot{+} \dot{1}) = (\alpha \exp_r \beta) \dot{\times} \alpha$

Dem.

$\vdash . *186\cdot111 . *181\cdot4 . \supset$

$\vdash : \sim(\alpha, \beta \in \mathrm{N_0R}) . \supset . \alpha \exp_r (\beta \dot{+} \dot{1}) = \Lambda . (\alpha \exp_r \beta) \dot{\times} \alpha = \Lambda$ (1)

$\vdash . *186\cdot13 . *181\cdot22 . \supset$

$\vdash : \alpha = \mathrm{N_0r}`P . \beta = \mathrm{N_0r}`Q . \supset . \alpha \exp_r (\beta \dot{+} \dot{1}) = \mathrm{Nr}`\{P \exp (Q \dot{+\!\!\!\rightarrow} z)\} .$

$(\alpha \exp_r \beta) \dot{\times} \alpha = \mathrm{Nr}`(P \exp Q) \dot{\times} \mathrm{Nr}`P$ (2)

$\vdash . (2) . *176\cdot151 . *166\cdot13 . \supset$

$\vdash : \mathrm{Hp}(2) . P = \dot{\Lambda} . \supset . \alpha \exp_r (\beta \dot{+} \dot{1}) = 0_r . (\alpha \exp_r \beta) \dot{\times} \alpha = 0_r$ (3)

$\vdash . *165\cdot2 . *161\cdot4 . *176\cdot1 . (*181\cdot01) . \supset$

$\vdash . \mathrm{Nr}`\{P \exp (Q \dot{+\!\!\!\rightarrow} z)\} = \mathrm{Nr}`\mathrm{Prod}`[P \downarrow ; `\downarrow \Lambda_x ; `\iota ; `Q \dot{+\!\!\!\rightarrow} P \downarrow \{(\Lambda \cap C`P) \downarrow \iota`x\}]$ (4)

$\vdash . *165\cdot221\cdot222 . *181\cdot11 . *162\cdot22 . \supset$

$\vdash : \dot{\exists} ! P . \supset . P \downarrow \{(\Lambda \cap C`P) \downarrow \iota`x\} \sim\in C`P \downarrow ; `\downarrow \Lambda_x ; `\iota ; `Q .$

$C`P \downarrow \{(\Lambda \cap C`P) \downarrow \iota`x\} \cap C`\Sigma`P \downarrow ; `\downarrow \Lambda_x ; `\iota ; `Q = \Lambda$ (5)

$\vdash . (4) . (5) . *165\cdot21 . *173\cdot25 . \supset \vdash : \dot{\exists} ! P . \supset .$

$\mathrm{Nr}`\{P \exp (Q \dot{+\!\!\!\rightarrow} z)\} = \mathrm{Nr}`[(\mathrm{Prod}`P \downarrow ; `\downarrow \Lambda_x ; `\iota ; `Q) \times P \downarrow \{(\Lambda \cap C`P) \downarrow \iota`x\}]$

[*181·12.*165·251.*176·1·22.*184·13] $= \mathrm{Nr}`(P \exp Q) \dot{\times} \mathrm{Nr}`P$ (6)

$\vdash . (2) . (6) . \supset \vdash : \mathrm{Hp}(2) . \dot{\exists} ! P . \supset . \alpha \exp_r (\beta \dot{+} \dot{1}) = (\alpha \exp_r \beta) \dot{\times} \alpha$ (7)

$\vdash . (1) . (3) . (7) . \supset \vdash .$ Prop

***186·23.** $\vdash . \alpha \exp_r (\dot{1} \dot{+} \beta) = \alpha \dot{\times} (\alpha \exp_r \beta)$ [Proof as in *186·22]

***186·3.** $\vdash :. \text{Mult ax} . \supset : P \in \mathrm{Rel}^2 \mathrm{excl} \cap \mathrm{Nr}`R . C`P \subset \mathrm{Nr}`S . \supset .$

$\Pi \mathrm{Nr}`P = (\mathrm{Nr}`P) \exp_r (\mathrm{Nr}`S)$ [*185·29]

***186·31.** $\vdash :. \text{Mult ax} . \supset : \mu, \nu \in \mathrm{NR} - \iota`\Lambda . P \in \mathrm{Rel}^2 \mathrm{excl} \cap \mu . C`P \subset \nu . \supset .$

$\Pi \mathrm{Nr}`P = \mu \exp_r \nu$ [*186·3]

***186·4.** $\vdash . \text{Nr}`P_{\text{df}} = 2_r \exp_r (\text{Nr}`P)$ [*177·13]

***186·5.** $\vdash : \mu, \nu \in \text{N}_0\text{R} . \nu \neq 0_r . \supset . C``(\mu \exp_r \nu) = (C``\mu)^{C``\nu}$

Dem.

$\vdash . *152·7 . *186·13 . \supset \vdash : \mu = \text{N}_0\text{r}`P . \nu = \text{N}_0\text{r}`Q . \supset .$

$$C``(\mu \exp_r \nu) = \text{Nc}`C`(P \exp Q) \quad (1)$$

$\vdash . (1) . *176·14 . \supset \vdash : \text{Hp}(1) . \nu \neq 0_r . \supset . C``(\mu \exp_r \nu) = \text{Nc}`\{(C`P) \exp (C`Q)\}$

[*116·222] $\qquad = (\text{N}_0\text{c}`C`P)^{\text{N}_0\text{c}`C`Q}$

[*155·6] $\qquad = (C``\text{N}_0\text{r}`P)^{C``\text{N}_0\text{r}`Q}$

[Hp] $\qquad = (C``\mu)^{C``\nu} : \supset \vdash . \text{Prop}$

PART V.

SERIES.

SUMMARY OF PART V.

A RELATION is said to be *serial*, or to generate a series, when it possesses three different properties, namely (1) being contained in diversity, (2) transitiveness, (3) connexity, *i.e.* the property that the relation or its converse holds between any two different members of its field. Thus P is a serial relation if (1) $P \subset J$, (2) $P^2 \subset P$, (3) $x, y \in C\text{‘}P \,.\, x \neq y \,.\, \supset_{x,y} : xPy \,.\, \vee \,.\, yPx$. The third characteristic, that of connexity, may be written more shortly

$$x \in C\text{‘}P \,.\, \supset_x \,.\, \overrightarrow{P}\text{‘}x \cup \iota\text{‘}x \cup \overleftarrow{P}\text{‘}x = C\text{‘}P,$$

i.e.

$$x \in C\text{‘}P \,.\, \supset_x \,.\, \overleftrightarrow{P}\text{‘}x = C\text{‘}P,$$

using the notation of ∗97; and this, in virtue of ∗97·23, is equivalent to

$$\overleftrightarrow{P}\text{‘‘}C\text{‘}P \in 0 \cup 1.$$

In virtue of ∗50·47, the first two characteristics are equivalent to

$$P \dot{\cap} \breve{P} = \dot{\Lambda} \,.\, P^2 \subset P.$$

When $P \dot{\cap} \breve{P} = \dot{\Lambda}$, we say that P is "asymmetrical." Thus serial relations are such as are asymmetrical, transitive, and connected.

It might be thought that a serial relation need not be contained in diversity, since we commonly speak of series in which there are repetitions, *i.e.* in which an earlier term is identical with a later term. Thus, *e.g.*

$$a, \ b, \ c, \ a, \ e, \ f, \ b, \ g, \ h$$

would be called a series of letters, although the letters a and b recur. But in all such cases, there is some means (in the above case, position in space) by which one *occurrence* of a given term is distinguished from another occurrence, and this will be found to mean that there is some other series (in the above case, the series of positions in a line) free from repetitions, with which our pseudo-series has a one-many correlation. Thus, in the above instance, we have a series of nine positions, which we may call

$$1, \ 2, \ 3, \ 4, \ 5, \ 6, \ 7, \ 8, \ 9,$$

which form a true series without repetitions; we have a one-many relation, that of *occupying* these positions, by means of which we distinguish occurrences of a, the first occurrence being a as the correlate of 1, the second being

a as the correlate of 4. All series in which there are repetitions (which we may call pseudo-series) are thus obtained by correlation with true series, *i.e.* with series in which there is no repetition. That is to say, a pseudo-series has as its generating relation a relation of the form $S;P$, where P is a serial relation, and S is a one-many relation whose converse domain contains the field of P. Thus what we may call self-subsistent series must be series without repetitions, *i.e.* series whose generating relations are contained in diversity.

For our purposes, there is no use in distinguishing a series from its generating relation. A series is not a class, since it has a definite order, while a class has no order, but is capable of many orders (unless it contains only one term or none). The generating relation determines the order, and also the class of terms ordered, since this class is the field of the generating relation. Hence the generating relation completely determines the series, and may, for all mathematical purposes, be taken to *be* the series.

When P is transitive, we have

$$P_{po} = P \,.\, P_* = P \cup I \upharpoonright C\text{`}P.$$

Hence all the propositions of Part II, Section E become greatly simplified when applied to series.

Also, since the field of a connected relation consists of a single family, a series has one first term or none, and one last term or none.

In the case of a serial relation P, the relation P_1 (defined in ∗121·02) becomes $P \dot{-} P^2$, *i.e.* the relation "immediately preceding." In a *discrete* series, the terms in general immediately precede other terms. A *compact* series, on the contrary, is defined as one in which there are terms between any two: in such a series, $P_1 = \dot{\Lambda}$.

It very frequently occurs that we wish to consider the relations of various series which are all contained in some one series; for example, we may wish to consider various series of real numbers, all arranged in order of magnitude. In such a case, if P is the series in which all the others are contained, and $\alpha, \beta, \gamma, \ldots$ are the fields of the contained series, the contained series themselves are $P \upharpoonright \alpha, P \upharpoonright \beta, P \upharpoonright \gamma, \ldots$. Thus when series are given as contained in a given series, they are completely determined by their fields.

In what follows, Section A deals with the elementary properties of series, including maximum and minimum points, sequent points and limits.

Section B will deal with the theory of segments and kindred topics; in this section we shall define "Dedekindian" series, and shall prove the important proposition that the series of segments of a series is always Dedekindian, *i.e.* that every class of segments has either a maximum or a limit.

Section C, which stands outside the main developments of the book, is concerned with convergence and the limits of functions and the definition of a continuous function. Its purpose is to show how these notions can be expressed, and many of their properties established, in a much more general way than is usually done, and without assuming that the arguments or values of the functions concerned are either numerical or numerically measurable.

Section D will deal with "well-ordered" series, *i.e.* series in which every class containing members of the field has a first term. The properties of well-ordered series are many and important; most of them depend upon the fact that an extended variety of mathematical induction is possible in dealing with well-ordered series. The term "ordinal number" is confined by usage to the relation-number of a well-ordered series; ordinal numbers will also be considered in our fourth section.

Section E will deal with finite and infinite. We shall show that the distinction between "inductive" and "non-reflexive" does not arise in well-ordered series.

Section F will deal with "compact" series, *i.e.* series in which there is a term between any two, *i.e.* in which $P^2 = P$. In particular we shall consider "rational" series (*i.e.* series like the series of rationals in order of magnitude) and continuous series (*i.e.* series like the series of real numbers in order of magnitude). Our treatment of this subject will follow Cantor closely.

SECTION A.

GENERAL THEORY OF SERIES.

Summary of Section A.

In the present section, we shall be concerned with the properties common to all series. Such properties, for the most part, are very simple, and present no difficulties of any kind. Many of the properties of series do not require all the three characteristics by which serial relations are defined, but only one or two of these properties: we therefore begin with numbers in which, though the properties proved derive their chief importance from their applicability to series, the hypotheses are only that the relations in question have one or two of the properties of serial relations. Thence we proceed to the most elementary properties peculiar to series, and thence to the theory of minimum and maximum members of classes contained in a series, and of the successors and limits of classes. We then proceed to the correlation of a series with part of itself. The ground covered is familiar, and the difficulties encountered are less than in most previous sections.

It will be observed that where series are concerned, if α is an existent class contained in $C\text{'}P$, $p\text{'}\overleftarrow{P}\text{``}\alpha$ is correlative to $P\text{``}\alpha$ (which is $s\text{'}\overrightarrow{P}\text{``}\alpha$): $P\text{``}\alpha$ is "predecessors of some α," and $p\text{'}\overleftarrow{P}\text{``}\alpha$ is "successors of all $\alpha$'s." If $\alpha$ is an existent class contained in $C\text{'}P$, the whole of $C\text{'}P$, with the exception of the last term of $\alpha$ (if there is such a term), belongs to one or other of the classes $P\text{``}\alpha$, $p\text{'}\overleftarrow{P}\text{``}\alpha$, of which the first wholly precedes the second. The division of $C\text{'}P$ into these two classes is the Dedekind "cut" defined by $\alpha$. But when only part of $\alpha$ is contained in $C\text{'}P$, we must replace $p\text{'}\overleftarrow{P}\text{``}\alpha$ by $p\text{'}\overleftarrow{P}\text{``}(\alpha \cap C\text{'}P)$, since $p\text{'}\overleftarrow{P}\text{``}\alpha = \Lambda$ if α has any member not belonging to $C\text{'}P$. Again, if $\alpha \cap C\text{'}P = \Lambda$, we have $p\text{'}\overleftarrow{P}\text{``}(\alpha \cap C\text{'}P) = \mathrm{V}$. But what we want is the complement to $P\text{``}\alpha$, which in this case is null. Hence we must replace $p\text{'}\overleftarrow{P}\text{``}(\alpha \cap C\text{'}P)$ by $C\text{'}P \cap p\text{'}\overleftarrow{P}\text{``}(\alpha \cap C\text{'}P)$: this is $C\text{'}P$ when $P\text{``}\alpha = \Lambda$, *i.e.* when $\alpha \cap C\text{'}P = \Lambda$. In any other event it is equal to $p\text{'}\overleftarrow{P}\text{``}(\alpha \cap C\text{'}P)$. If α

is contained in $C'P$ and is not null, $C'P \cap p'\overleftarrow{P}``(\alpha \cap C'P) = p'\overleftarrow{P}``\alpha$. Thus the Dedekind "cut" defined by a class α, whether or not this class is contained in whole or part in $C'P$, is always the two classes

$$P``\alpha, \quad C'P \cap p'\overleftarrow{P}``(\alpha \cap C'P).$$

Throughout the elementary propositions of this section, we have been careful to avoid stronger hypotheses than are required: we have not assumed P to be serial, if our conclusion would follow (*e.g.*) from the hypothesis that P is transitive and connected. It will be found that many properties of series depend upon the fact that, if x, y are two different terms of a series P, then $xPy \,.\equiv.\, \sim(yPx)$ (*204·3). Here the implication $xPy \,.\supset.\, \sim(yPx)$ requires that P should be asymmetrical, *i.e.* that we should have $P \dot{\cap} \breve{P} = \dot{\Lambda}$, or $P^2 \subset J$. The implication $\sim(yPx) \,.\supset.\, xPy$ requires that P should be connected. Thus the hypothesis required is not that P should be serial, but that P should be connected and asymmetrical (*202·5).

Again, consider the proposition that if P is a series, $P_1 = P \dot{-} P^2$. This relation P_1 is the very useful relation "immediately preceding"; thus the above proposition is important, as is the further proposition that if P is a series, P_1 is a one-one relation. It will be remembered that (by *121) "xP_1y" means that $P(x \mapsto y)$ consists of two terms. It was shown in *121·304·305 that if P_{po} is contained in diversity, "xP_1y" implies "xPy," and is equivalent to the statement that x and y constitute the whole interval $P(x \mapsto y)$ and are not identical. Also by *121·254, $P_1 = (P_{po})_1$. It is evident that, if P_{po} is contained in diversity, and xP_1y, we cannot have xP^2y, because there is no term other than x and y in the interval $P(x \mapsto y)$, and we cannot have xPx or yPy. Hence if $P_{po} \subset J$, we have $P_1 \subset \dot{-} P^2$. Hence by what was said above (*121·305), if $P_{po} \subset J$, we shall have $P_1 \subset P \dot{-} P^2$. On the other hand, if P is transitive, we have $P \dot{-} P^2 \subset P_1$ (*201·61). Combining these two facts, and remembering that if P is transitive, $P = P_{po}$ (*201·18), we find that $P_1 = P \dot{-} P^2$ if P is transitive and contained in diversity. We find further (*202·7) that if P is connected, $P \dot{-} P^2$ is one-one. Hence we need the full hypothesis that P is a series in order to prove that P_1 is a one-one (*204·7). This is a good example of the way in which the various separate characteristics that make up the definition of series are relevant in proving the properties of series.

*200. RELATIONS CONTAINED IN DIVERSITY.

Summary of *200.

Some of the propositions of this number are repetitions or immediate consequences of previous propositions, especially those of the propositions of *50 which deal with diversity. But we are chiefly concerned here with propositions which will be useful in the theory of series; this leads us to introduce propositions on $p`\overleftarrow{P}``\alpha$ and on matters connected with relation-arithmetic and other topics. It will be seen that "$P^2 \subseteq J$" (*i.e.* "P is asymmetrical") is an important hypothesis, as is also $P_{po} \subseteq J$, of the use of which we have already had examples in *96 and *121.

The following are among the most useful propositions in this number:

***200·12.** $\vdash : P \in \mathrm{Rl}`J \,.\supset.\, C`P \sim\in 1$

This is the proposition which makes it impossible to define an ordinal number 1 which shall take its place among relation-numbers applicable to series.

***200·35.** $\vdash : P \subseteq J \,.\, \alpha \in 1 \,.\supset.\, P \upharpoonright \alpha = \dot{\Lambda}$

This is a consequence of *200·12.

***200·36.** $\vdash : P^2 \subseteq J \,.\supset.\, P \subseteq J$

***200·361.** $\vdash : P^2 \subseteq J \,.\supset.\, \overrightarrow{P}`x \cap (\iota`x \cup \overleftarrow{P}`x) = \Lambda \,.\, \overleftarrow{P}`x \cap (\overrightarrow{P}`x \cup \iota`x) = \Lambda$

I.e. if $P^2 \subseteq J$, no term precedes itself or any of its predecessors, and no term succeeds itself or any of its successors.

***200·38.** $\vdash : P_{po} \subseteq J \,.\supset.\, P_{po} = P_* \mathbin{\dot\cap} J$

***200·39.** $\vdash : P_{po} \subseteq J \,.\, x \in C`P \,.\supset.\, \overrightarrow{P_*}`x \cap \overleftarrow{P_*}`x = \iota`x$

We then have a collection of propositions concerned with relation-arithmetic.

***200·211.** $\vdash : P \subseteq J \,.\, P \mathbin{\mathrm{smor}} Q \,.\supset.\, Q \subseteq J$

I.e. the property of being contained in diversity is invariant for likeness-transformations;

***200·4.** $\vdash : P \mathbin{\ddagger} Q \in \mathrm{Rl}`J \,.\equiv.\, P, Q \in \mathrm{Rl}`J \,.\, C`P \cap C`Q = \Lambda$

***200·41.** $\vdash : P \nrightarrow x \mathrel{G} J \,.\equiv.\, x \nleftarrow P \mathrel{G} J \,.\equiv.\, P \mathrel{G} J \,.\, x \sim\in C`P$

and other such propositions.

We then have a set of propositions concerned with $p`\overrightarrow{P}``\alpha$ and $p`\overleftarrow{P}``\alpha$. The most important are

***200·5.** $\vdash : P \mathrel{G} J \,.\supset.\, \alpha \cap p`\overrightarrow{P}``\alpha = \Lambda \,.\, \alpha \cap p`\overleftarrow{P}``\alpha = \Lambda$

***200·52.** $\vdash : P \mathrel{G} J \,.\supset.\, C`P \sim\in \overrightarrow{P}``C`P$

***200·53.** $\vdash : P^2 \mathrel{G} J \,.\supset.\, P``\alpha \cap p`\overleftarrow{P}``\alpha = \Lambda \,.\, \breve{P}``\alpha \cap p`\overrightarrow{P}``\alpha = \Lambda$

I.e. if P is asymmetrical, the terms which precede part of α do not succeed the whole of α, and vice versa.

***200·11.** $\vdash : P \in \text{Rl}`J \,.\equiv.\, \breve{P} \in \text{Rl}`J$ [*50·23]

***200·12.** $\vdash : P \in \text{Rl}`J \,.\supset.\, C`P \sim\in 1$

Dem.

$\vdash .\, *50{\cdot}11 \,.\, *33{\cdot}17 \,.\supset \vdash :.\, \text{Hp} : xPy \,.\vee.\, yPx : \supset .\, y \neq x \,.\, y \in C`P$ (1)

$\vdash .\, (1) \,.\, *33{\cdot}132 \,.\quad \supset \vdash :.\, \text{Hp} \,.\supset : x \in C`P \,.\supset .\, (\exists y) \,.\, y \neq x \,.\, y \in C`P :$

[*52·181] $\supset : C`P \sim\in 1 :. \supset \vdash .\, \text{Prop}$

***200·2.** $\vdash : T \in 1 \rightarrow 1 \,.\supset.\, T{;}(P \dot{\cap} J) = T{;}P \dot{\cap} J$

Dem.

$\vdash .\, *150{\cdot}4 \,.\supset \vdash :.\, \text{Hp} \,.\supset :$

$x \{T{;}(P \dot{\cap} J)\} y \,.\equiv.\, (\exists z, w) \,.\, x = T`z \,.\, y = T`w \,.\, zPw \,.\, z \neq w \,.$

[*71·56] $\equiv .\, (\exists z, w) \,.\, x \neq y \,.\, x = T`z \,.\, y = T`w \,.\, zPw \,.$

[*150·4] $\equiv .\, x \{T{;}P \dot{\cap} J\} y :. \supset \vdash .\, \text{Prop}$

***200·21.** $\vdash : T \in \text{Cls} \rightarrow 1 \,.\supset.\, T{;}P \mathrel{G} J$

Dem.

$\vdash .\, *150{\cdot}1 \,.\, *50{\cdot}24 \,.\supset \vdash :.\, \text{Hp} \,.\supset : x (T{;}P) y \,.\supset .\, (\exists z, w) \,.\, xTz \,.\, yTw \,.\, z \neq w \,.$

[*71·171.Transp] $\supset .\, x \neq y :. \supset \vdash .\, \text{Prop}$

***200·211.** $\vdash : P \mathrel{G} J \,.\, P \text{ smor } Q \,.\supset.\, Q \mathrel{G} J$ [*200·21. *151·1]

The properties of relations are very frequently common to all relations which are like a given relation, and this applies specially to the kinds of properties with which we are most concerned. The above proposition is an illustration of this fact: it shows that the property of being contained in diversity is invariant for likeness-transformations.

***200·22.** $\vdash : P \mathrel{G} J \,.\equiv.\, \text{N}_0\text{r}`P \subset \text{Rl}`J \,.\equiv.\, \exists !\, \text{N}_0\text{r}`P \cap \text{Rl}`J$

Dem.

$\vdash .\, *155{\cdot}11 \,.\, *200{\cdot}211 \,.\supset \vdash : P \mathrel{G} J \,.\supset.\, \text{N}_0\text{r}`P \subset \text{Rl}`J$ (1)

$\vdash .\, *155{\cdot}12 \,.\quad \supset \vdash : \text{N}_0\text{r}`P \subset \text{Rl}`J \,.\supset.\, P \mathrel{G} J$ (2)

$\vdash .\, *155{\cdot}12 \,.\quad \supset \vdash : P \mathrel{G} J \,.\supset.\, \exists !\, \text{N}_0\text{r}`P \cap \text{Rl}`J$ (3)

$\vdash .\, *155{\cdot}11 \,.\, *200{\cdot}211 \,.\supset \vdash : \exists !\, \text{N}_0\text{r}`P \cap \text{Rl}`J \,.\supset.\, P \mathrel{G} J$ (4)

$\vdash .\, (1) \,.\, (2) \,.\, (3) \,.\, (4) \,.\supset \vdash .\, \text{Prop}$

We have, without the need of typical definiteness,

$$\vdash : P \mathrel{\mathsf{G}} J . \supset . \mathrm{Nr}`P \subset \mathrm{Rl}`J$$

and $$\vdash : \exists ! \mathrm{Nr}`P \cap \mathrm{Rl}`J . \supset . P \mathrel{\mathsf{G}} J,$$

both of which are immediate consequences of $*200{\cdot}211$. The converse implications, however, fail if $\mathrm{Nr}`P$ is taken in a type in which $\mathrm{Nr}`P = \Lambda$.

***200·3.** $\vdash . \dot{\Lambda} \in \mathrm{Rl}`J$ [*25·12]

***200·31.** $\vdash : x \neq y . \equiv . x \downarrow y \in \mathrm{Rl}`J$ [*55·3]

***200·32.** $\vdash : \alpha \uparrow \beta \mathrel{\mathsf{G}} J . \equiv . \alpha \cap \beta = \Lambda$ [*50·55]

***200·33.** $\vdash : P \mathrel{\mathsf{G}} J . \supset . P \restriction \alpha \mathrel{\mathsf{G}} J$ [*35·442]

***200·34.** $\vdash : P \restriction \alpha \mathrel{\mathsf{G}} J . \equiv . P \upharpoonright \alpha \mathrel{\mathsf{G}} J . \equiv . \alpha \upharpoonleft P \mathrel{\mathsf{G}} J$ [*50·58]

***200·35.** $\vdash : P \mathrel{\mathsf{G}} J . \alpha \in 1 . \supset . P \restriction \alpha = \dot{\Lambda}$

Dem.

$\vdash . *52{\cdot}16 . \supset \vdash :. \mathrm{Hp} . \supset : x, y \in \alpha . \supset_{x,y} . \sim (xJy) .$

[*23·81] $\supset_{x,y} . \sim (xPy) :$

[*11·521] $\supset : (x, y) . \sim \{x, y \in \alpha . xPy\} :. \supset \vdash . \mathrm{Prop}$

***200·36.** $\vdash : P^2 \mathrel{\mathsf{G}} J . \supset . P \mathrel{\mathsf{G}} J$ [*50·45]

***200·361.** $\vdash : P^2 \mathrel{\mathsf{G}} J . \supset . \overrightarrow{P}`x \cap (\iota`x \cup \overleftarrow{P}`x) = \Lambda . \overleftarrow{P}`x \cap (\overrightarrow{P}`x \cup \iota`x) = \Lambda$

Dem.

$\vdash . *51{\cdot}15 . \quad \supset \vdash : y \in \overrightarrow{P}`x \cap \iota`x . \supset . xPx$ (1)

$\vdash . *200{\cdot}36 . \quad \supset \vdash : \mathrm{Hp} . \supset . \sim (xPx) .$

[(1).Transp] $\supset . \overrightarrow{P}`x \cap \iota`x = \Lambda$ (2)

$\vdash . *34{\cdot}11 . \quad \supset \vdash : \exists ! \overrightarrow{P}`x \cap \overleftarrow{P}`x . \equiv . xP^2x$ (3)

$\vdash . (3) . \mathrm{Transp} . \supset \vdash : \mathrm{Hp} . \supset . \overrightarrow{P}`x \cap \overleftarrow{P}`x = \Lambda$ (4)

$\vdash . (2) . (4) . \quad \supset \vdash : \mathrm{Hp} . \supset . \overrightarrow{P}`x \cap (\iota`x \cup \overleftarrow{P}`x) = \Lambda$ (5)

Similarly $\vdash : \mathrm{Hp} . \supset . \overleftarrow{P}`x \cap (\overrightarrow{P}`x \cup \iota`x) = \Lambda$ (6)

$\vdash . (5) . (6) . \supset \vdash . \mathrm{Prop}$

***200·37.** $\vdash : \exists ! \mathrm{Pot}`P \cap \mathrm{Rl}`J . \supset . P \mathrel{\mathsf{G}} J$

Dem.

$\vdash . *91{\cdot}373 \dfrac{xSx}{\phi S} . \supset$

$\vdash :: xPx : S \in \mathrm{Pot}`P . xSx . \supset_S . x (S \mid P) x : \supset : Q \in \mathrm{Pot}`P . \supset_Q . xQx$ (1)

$\vdash . *3{\cdot}2 . \quad \supset \vdash :. xPx . \supset : xSx . \supset . xSx . xPx .$

[*34·1] $\supset . x (S \mid P) x$ (2)

$\vdash . (1) . (2) . \quad \supset \vdash :. xPx . \supset : Q \in \mathrm{Pot}`P . \supset_Q . xQx .$

[*50·24] $\supset_Q . \sim (Q \mathrel{\mathsf{G}} J)$ (3)

$\vdash . (3) . \mathrm{Transp} . \supset \vdash : (\exists Q) . Q \in \mathrm{Pot}`P . Q \mathrel{\mathsf{G}} J . \supset . \sim (xPx) .$

[*50·24] $\supset . P \mathrel{\mathsf{G}} J : \supset \vdash . \mathrm{Prop}$

***200·38.** $\vdash : P_{po} \subseteq J . \supset . P_{po} = P_* \dot{\cap} J$ [*91·541]

***200·381.** $\vdash : P_{po} \subseteq J . \supset . \overrightarrow{P_{po}}‘x \cap \overleftarrow{P_*}‘x = \Lambda . \overleftarrow{P_{po}}‘x \cap \overrightarrow{P_*}‘x = \Lambda$

Dem.

$\vdash . *91·56 . \supset \vdash : \text{Hp} . \supset . P_{po}{}^2 \subseteq J .$

[*200·361] $\supset . \overrightarrow{P_{po}}‘x \cap (\iota‘x \cup \overleftarrow{P_{po}}‘x) = \Lambda . \overleftarrow{P_{po}}‘x \cap (\overrightarrow{P_{po}}‘x \cup \iota‘x) = \Lambda .$

[*91·54] $\supset . \overrightarrow{P_{po}}‘x \cap \overleftarrow{P_*}‘x = \Lambda . \overleftarrow{P_{po}}‘x \cap \overrightarrow{P_*}‘x = \Lambda : \supset \vdash . \text{Prop}$

***200·39.** $\vdash : P_{po} \subseteq J . x \in C‘P . \supset . \overrightarrow{P_*}‘x \cap \overleftarrow{P_*}‘x = \iota‘x$

Dem.

$\vdash . *91·54 . \quad \supset \vdash : \text{Hp} . \supset . \overrightarrow{P_*}‘x \cap \overleftarrow{P_*}‘x = (\overrightarrow{P_{po}}‘x \cup \iota‘x) \cap (\overleftarrow{P_{po}}‘x \cup \iota‘x)$

[*22·69] $= (\overrightarrow{P_{po}}‘x \cap \overleftarrow{P_{po}}‘x) \cup \iota‘x \quad (1)$

$\vdash . *91·56 . \quad \supset \vdash : y \in \overrightarrow{P_{po}}‘x \cap \overleftarrow{P_{po}}‘x . \supset . y P_{po} y \quad (2)$

$\vdash . (2) . \text{Transp} . \supset \vdash : \text{Hp} . \supset . \overrightarrow{P_{po}}‘x \cap \overleftarrow{P_{po}}‘x = \Lambda \quad (3)$

$\vdash . (1) . (3) . \supset \vdash . \text{Prop}$

***200·391.** $\vdash : P_{po} \subseteq J . \supset . \overrightarrow{P_*};P \text{ smor } P . \overrightarrow{P_*} \upharpoonright C‘P \in (\overrightarrow{P_*};P) \overline{\text{smor}}\, P$

Dem.

$\vdash . *90·12 . \supset \vdash : \text{Hp} . x, y \in C‘P . \overrightarrow{P_*}‘x = \overrightarrow{P_*}‘y . \supset . x P_* y . y P_* x .$

[*200·39] $\supset . x = y \quad (1)$

$\vdash . (1) . *151·24 . \supset \vdash . \text{Prop}$

The above proposition is useful in the theory of segments.

The following propositions are concerned with the ideas of relation-arithmetic. Analogous propositions will be proved for transitiveness and connection in *201 and *202, whence analogous propositions concerning series will be deduced in *204.

***200·4.** $\vdash : P \updownarrow Q \in \text{Rl}‘J . \equiv . P, Q \in \text{Rl}‘J . C‘P \cap C‘Q = \Lambda$

Dem.

$\vdash . *23·59 . *160·1 . \supset$

$\vdash : P \updownarrow Q \in \text{Rl}‘J . \equiv . P, Q \in \text{Rl}‘J . C‘P \uparrow C‘Q \subseteq J .$

[*200·32] $\equiv . P, Q \in \text{Rl}‘J . C‘P \cap C‘Q = \Lambda : \supset \vdash . \text{Prop}$

This proposition is part of the proof that the sum of two mutually exclusive series is a series.

***200·41.** $\vdash : P \rightarrow\!\!\!\!+ x \subseteq J . \equiv . x \leftarrow\!\!\!\!+ P \subseteq J . \equiv . P \subseteq J . x \sim\in C‘P$ [*23·59 . *200·32]

***200·42.** $\vdash : \Sigma‘P \subseteq J . \equiv . C‘P \subset \text{Rl}‘J . F;P \subseteq J$

Dem.

$\vdash . *23·59 . *162·1 . \supset \vdash : \Sigma‘P \subseteq J . \equiv . \dot{s}‘C‘P \subseteq J . F;P \subseteq J .$

[*61·52] $\equiv . C‘P \subset \text{Rl}‘J . F;P \subseteq J : \supset \vdash . \text{Prop}$

The following propositions (*200·421·422·423) are lemmas for *204·53.

***200·421.** $\vdash : P \in \text{Rel}^2\,\text{excl} \,.\, P \mathrel{G} J \,.\, Q \in C`P \,.\supset .\, Q = (\Sigma`P) \upharpoonright C`Q$

Dem.

$\vdash . *163·11 . *162·13 . \supset$

$\vdash :: \text{Hp} . \supset :. x \{(\Sigma`P) \upharpoonright C`Q\} y . \equiv :$

$(\exists R) . R \in C`P . x, y \in C`Q . xRy . R = Q . \mathbf{v} .$

$(\exists R, S) . RPS . x, y \in C`Q . x \in C`R . y \in C`S . R = Q . S = Q :$

[*13·195·22] $\equiv : xQy . \mathbf{v} . QPQ . x, y \in C`Q :$

[*50·24.Hp] $\equiv : xQy :: \supset \vdash . \text{Prop}$

***200·422.** $\vdash : \Sigma`P \mathrel{G} J . \supset . P \upharpoonright (-\iota`\dot{\Lambda}) \mathrel{G} J$

Dem.

$\vdash . *162·13 . *50·24 . \supset \vdash :: \text{Hp} . \supset :. QPR . \supset : x \in C`Q . y \in C`R . \supset . x \neq y :$

[*24·37] $\supset : C`Q \cap C`R = \Lambda :$

[*24·57] $\supset : \dot{\exists} ! Q . \supset . C`Q \neq C`R .$

[*30·37] $\supset . Q \neq R :: \supset \vdash . \text{Prop}$

***200·423.** $\vdash :. P \in \text{Rel}^2\,\text{excl} . \Lambda \sim \in C`P . \supset : \Sigma`P \mathrel{G} J . \equiv . P \mathrel{G} J . C`P \subset \text{Rl}`J$

Dem.

$\vdash . *200·422·42 . \quad \supset \vdash : \text{Hp} . \Sigma`P \mathrel{G} J . \supset . P \mathrel{G} J . C`P \subset \text{Rl}`J \quad (1)$

$\vdash . *61·52 . \quad \supset \vdash : C`P \subset \text{Rl}`J . \supset . \dot{s}`C`P \mathrel{G} J \quad (2)$

$\vdash . *163·12 . *200·21 . \supset \vdash : \text{Hp} . P \mathrel{G} J . \supset . F ; P \mathrel{G} J \quad (3)$

$\vdash . (2) . (3) . *162·1 . \quad \supset \vdash : \text{Hp} . P \mathrel{G} J . C`P \subset \text{Rl}`J . \supset . \Sigma`P \mathrel{G} J \quad (4)$

$\vdash . (1) . (4) . \supset \vdash . \text{Prop}$

***200·43.** $\vdash : P \mathrel{G} J . \supset . \Pi`P =$

$$\hat{M}\hat{N} \{M, N \in F_\Delta`C`P : (\exists Q) . (M`Q)\, Q\, (N`Q) . M \upharpoonright \overrightarrow{P}`Q = N \upharpoonright \overrightarrow{P}`Q\}$$

Dem.

$\vdash . *4·71 . *172·1 . \supset \vdash : \text{Hp} . \supset .$

$\Pi`P = \hat{M}\hat{N} \{M, N \in F_\Delta`C`P :. (\exists Q) : (M`Q)\, Q\, (N`Q) : RPQ . \supset_R . M`R = N`R\}$

[*35·71.*71·35]

$= \hat{M}\hat{N} \{M, N \in F_\Delta`C`\dot{P} : (\exists Q) . (M`Q)\, Q\, (N`Q) . M \upharpoonright \overrightarrow{P}`Q = N \upharpoonright \overrightarrow{P}`Q\} . \supset \vdash . \text{Prop}$

The following propositions, with the exception of *200·52, are concerned with $p`\overrightarrow{P}``\alpha$ and $p`\overleftarrow{P}``\alpha$, *i.e.* the class of terms preceding (or succeeding) the whole of α.

∗200·5. $\vdash : P \mathbin{\text{G}} J \,.\, \supset .\, \alpha \cap p\text{‘}\overrightarrow{P}\text{“}\alpha = \Lambda \,.\, \alpha \cap p\text{‘}\overleftarrow{P}\text{“}\alpha = \Lambda$

Dem.

$\vdash .\, *40\cdot51 \,.\, \supset \vdash :.\, x \in \alpha \cap p\text{‘}\overrightarrow{P}\text{“}\alpha \,.\, \supset : x \in \alpha : y \in \alpha \,.\, \supset_y .\, xPy :$

[∗10·26] $\supset : xPx :$

[∗50·24] $\supset : \sim(P \mathbin{\text{G}} J)$ (1)

$\vdash .\, (1) \,.\, \text{Transp} \,.\, \supset \vdash : \text{Hp} \,.\, \supset .\, \alpha \cap p\text{‘}\overrightarrow{P}\text{“}\alpha = \Lambda$ (2)

Similarly $\vdash : \text{Hp} \,.\, \supset .\, \alpha \cap p\text{‘}\overleftarrow{P}\text{“}\alpha = \Lambda$ (3)

$\vdash .\, (2) \,.\, (3) \,.\, \supset \vdash .\, \text{Prop}$

∗200·51. $\vdash : P \mathbin{\text{G}} J \,.\, \dot{\exists} ! P \,.\, \supset .\, p\text{‘}\overrightarrow{P}\text{“}C\text{‘}P = \Lambda \,.\, p\text{‘}\overleftarrow{P}\text{“}C\text{‘}P = \Lambda$

Dem.

$\vdash .\, *40\cdot62 \,.\, \supset \vdash : \text{Hp} \,.\, \supset .\, p\text{‘}\overrightarrow{P}\text{“}C\text{‘}P \subset C\text{‘}P \,.$

[∗22·621] $\supset .\, p\text{‘}\overrightarrow{P}\text{“}C\text{‘}P = C\text{‘}P \cap p\text{‘}\overrightarrow{P}\text{“}C\text{‘}P$

[∗200·5] $= \Lambda$ (1)

Similarly $\vdash : \text{Hp} \,.\, \supset .\, p\text{‘}\overleftarrow{P}\text{“}C\text{‘}P = \Lambda$ (2)

$\vdash .\, (1) \,.\, (2) \,.\, \supset \vdash .\, \text{Prop}$

∗200·52. $\vdash : P \mathbin{\text{G}} J \,.\, \supset .\, C\text{‘}P \sim\in \overrightarrow{P}\text{“}C\text{‘}P$

Dem.

$\vdash .\, *50\cdot24 \,.\, \supset \vdash :.\, \text{Hp} \,.\, \supset : x \in C\text{‘}P \,.\, \supset_x .\, x \sim\in \overrightarrow{P}\text{‘}x \,.$

[∗13·14] $\supset_x .\, C\text{‘}P \neq \overrightarrow{P}\text{‘}x :$

[∗37·7.Transp] $\supset : C\text{‘}P \sim\in \overrightarrow{P}\text{“}C\text{‘}P :.\, \supset \vdash .\, \text{Prop}$

This proposition is often used in the theory of well-ordered series.

∗200·53. $\vdash : P^2 \mathbin{\text{G}} J \,.\, \supset .\, P\text{“}\alpha \cap p\text{‘}\overleftarrow{P}\text{“}\alpha = \Lambda \,.\, \breve{P}\text{“}\alpha \cap p\text{‘}\overrightarrow{P}\text{“}\alpha = \Lambda$

Dem.

$\vdash .\, *37\cdot1 \,.\, *40\cdot53 \,.\, \supset \vdash :.\, x \in P\text{“}\alpha \cap p\text{‘}\overleftarrow{P}\text{“}\alpha \,.\, \supset : (\exists y) \,.\, y \in \alpha \,.\, xPy : y \in \alpha \,.\, \supset_y .\, yPx :$

[∗10·56] $\supset : (\exists y) \,.\, xPy \,.\, yPx :$

[∗34·5] $\supset : xP^2x :$

[∗50·24] $\supset : \sim(P^2 \mathbin{\text{G}} J)$ (1)

$\vdash .\, (1) \,.\, \text{Transp} \,.\, \supset \vdash : \text{Hp} \,.\, \supset .\, (x) \,.\, x \sim\in P\text{“}\alpha \cap p\text{‘}\overleftarrow{P}\text{“}\alpha$ (2)

Similarly $\vdash : \text{Hp} \,.\, \supset .\, (x) \,.\, x \sim\in \breve{P}\text{“}\alpha \cap p\text{‘}\overrightarrow{P}\text{“}\alpha$ (3)

$\vdash .\, (2) \,.\, (3) \,.\, \supset \vdash .\, \text{Prop}$

The above proposition is frequently used. If α is an existent class contained in $C\text{‘}P$, $P\text{“}\alpha$ and $p\text{‘}\overleftarrow{P}\text{“}\alpha$ are the two parts of the Dedekind "cut" determined by α (excluding the maximum of α, if any). The above proposition shows that these two parts are mutually exclusive.

***200·54.** $\vdash : P \in J . \dot{\exists} ! P . \supset . p`\overrightarrow{P}``\{C`P \cap p`\overleftarrow{P}``\alpha\} = p`\overrightarrow{P}``p`\overleftarrow{P}``\alpha$

Dem.

$\vdash . *40{\cdot}62 . \supset \vdash : \exists ! \alpha . \supset . C`P \cap p`\overleftarrow{P}``\alpha = p`\overleftarrow{P}``\alpha$ (1)

$\vdash . *40{\cdot}2 . \supset \vdash : \alpha = \Lambda . \supset . p`\overleftarrow{P}``\alpha = V .$ (2)

$[*40{\cdot}16]$ $\supset . p`\overrightarrow{P}``p`\overleftarrow{P}``\alpha \subset p`\overrightarrow{P}``C`P$ (3)

$\vdash . (3) . *200{\cdot}51 . \supset \vdash : \mathrm{Hp} . \alpha = \Lambda . \supset . p`\overrightarrow{P}``p`\overleftarrow{P}``\alpha = \Lambda$ (4)

$\vdash . (2) . *24{\cdot}26 . \supset \vdash : \alpha = \Lambda . \supset . C`P \cap p`\overleftarrow{P}``\alpha = C`P$ (5)

$\vdash . (5) . *200{\cdot}51 . \supset \vdash : \mathrm{Hp} . \alpha = \Lambda . \supset . p`\overrightarrow{P}``(C`P \cap p`\overleftarrow{P}``\alpha) = \Lambda$ (6)

$\vdash . (1) . (4) . (6) . \supset \vdash . \mathrm{Prop}$

This proposition is a lemma whose purpose is to avoid the necessity of introducing the hypothesis $\exists ! \alpha$ in proofs in which it is not really necessary. The first use of this proposition occurs in *206·551.

*201. TRANSITIVE RELATIONS.

Summary of *201.

There are two main varieties of transitive relations, namely those that are symmetrical ($P = \breve{P}$), and those that are asymmetrical ($P \dot{\cap} \breve{P} = \dot{\Lambda}$). Transitive *symmetrical* relations have the formal properties of equality; examples of such relations have occurred above, *e.g.* identity, similarity, and likeness. The propositions of the present number, however, are rather such as will be useful in connection with transitive *asymmetrical* relations, since they are intended to be applied to series.

We denote the class of transitive relations by "trans"; thus

$$\text{trans} = \hat{P}(P^2 \in P) \quad \text{Df.}$$

Many propositions of this number are analogous to propositions whose numbers have the same decimal part in *200. Such are: If P is transitive, so is its converse (*201·11), and so is any relation which is like P (*201·211); $\dot{\Lambda}$ and $x \downarrow y$ are transitive (*201·3·31); if P is transitive, so is $P \upharpoonleft \alpha$ (*201·33). The propositions *201·4—·42, which deal with the ideas of relation-arithmetic, are also analogous to *200·4—·42.

Most of the other propositions of this number, however, have no analogues in *200. Among the most important of these are the following:

***201·14.** $\vdash : P \epsilon \text{trans} . xPy . \supset . \overrightarrow{P}\text{`}x \subset \overrightarrow{P}\text{`}y$

***201·15.** $\vdash . R_* \epsilon \text{trans}$

***201·18.** $\vdash : P^2 \in P . \supset . P_{po} = P . P_* = P \cup I \upharpoonright C\text{`}P$

This proposition is very important, since it effects an immense simplification in the use of all propositions involving P_{po} or P_*, when these propositions are to be applied to transitive relations. Owing to the above proposition, P_{po} drops out where transitive relations are concerned. P_*, on the other hand, remains useful: If $y \epsilon C\text{`}P$, "xP_*y" will mean "x precedes or is y," which, if P generates a series of which x and y are members, is equivalent to "x does not follow y."

We have a series of propositions (*201·5—·56) on $P\text{``}\alpha$ and $p\text{`}\overrightarrow{P}\text{``}\alpha$. The chief of these are

*201·5. $\vdash : P \epsilon \text{trans} . \supset . P``P``\alpha \subset P``\alpha$

*201·501. $\vdash : P \epsilon \text{trans} . \supset . P``\overrightarrow{P}`x \subset \overrightarrow{P}`x$

These two propositions express the fact that a predecessor of a predecessor is a predecessor.

*201·52. $\vdash : P \epsilon \text{trans} . \supset . P_*``\alpha = P``\alpha \cup (\alpha \cap C`P)$

Thus if $\alpha \subset C`P$, $P_*``\alpha$ consists of α together with the predecessors of its members.

*201·521. $\vdash : P \epsilon \text{trans} . x \epsilon C`P . \supset . \overrightarrow{P_*}`x = \overrightarrow{P}`x \cup \iota`x$

*201·55. $\vdash : P \epsilon \text{trans} . \supset . P``(\alpha \cup P``\alpha) = P``\alpha$

We have next a set of important propositions on $P \dot{-} P^2$ and P_1. The chief are

*201·63. $\vdash : P \epsilon \text{trans} \cap \text{Rl}`J . \supset . P_1 = P \dot{-} P^2$

*201·65. $\vdash :. P \epsilon \text{trans} \cap \text{Rl}`J . \supset : P_1 = \dot{\Lambda} . \equiv . P^2 = P$

On these two propositions, see the notes appended to them below.

---

*201·01. $\text{trans} = \hat{P}(P^2 \subseteq P)$ Df

*201·1. $\vdash : P \epsilon \text{trans} . \equiv . P^2 \subseteq P$ [(*201·01)]

*201·11. $\vdash : P \epsilon \text{trans} . \equiv . \breve{P} \epsilon \text{trans}$

Dem.

$\vdash . *201·1 . *31·4 . \supset \vdash : P \epsilon \text{trans} . \equiv . \text{Cnv}`P^2 \subseteq \breve{P} .$

[*34·63.*201·1] $\equiv . \breve{P} \epsilon \text{trans} : \supset \vdash . \text{Prop}$

*201·12. $\vdash :. P \epsilon \text{trans} . \supset : P \subseteq J . \equiv . P^2 \subseteq J . \equiv . P \dot{\cap} \breve{P} = \dot{\Lambda}$ [*50·47]

In virtue of this proposition, being contained in diversity is equivalent (where transitive relations are concerned) to asymmetry. This is not in general the case with relations which are not transitive; thus *e.g.* diversity itself is contained in diversity, but is symmetrical.

*201·13. $\vdash . \text{Rl}`I \subset \text{trans}$

Dem.

$\vdash . *34·34 . \supset \vdash : R \subseteq I . \supset . R^2 \subseteq R | I .$

[*50·4] $\supset . R^2 \subseteq R : \supset \vdash . \text{Prop}$

*201·14. $\vdash : P \epsilon \text{trans} . xPy . \supset . \overrightarrow{P}`x \subset \overrightarrow{P}`y$

Dem.

$\vdash . *201·1 . \supset \vdash : \text{Hp} . zPx . \supset . zPy$ (1)

$\vdash . (1) . *32·18 . \supset \vdash . \text{Prop}$

The following propositions (*201·15—·19) are concerned with R_* and R_{po}.

***201·15.** $\vdash . R_* \epsilon \text{trans}$ [*90·17]

***201·16.** $\vdash . R_{po} \epsilon \text{trans}$ [*91·56]

This proposition is important, since it often happens that a series is given as defined by a one-one relation R, as in *122 for example, and in such cases R_{po} is a serial relation in our present sense. By the above proposition, R_{po} is always transitive; by *96·421, R_{po} is connected when confined to the posterity of a given term, provided $R \epsilon \text{Cls} \rightarrow 1$; by *96·23, if $R \epsilon 1 \rightarrow \text{Cls}$ and xBR, R_{po} is contained in diversity throughout the posterity of x. Thus if R is a one-one, R_{po} confined to any family which has a beginning will be a serial relation.

***201·17.** $\vdash : P^2 \subseteq P . Q \epsilon \text{Pot}'P . \supset . Q \subseteq P$

Dem.

$$\vdash . *34·34 . \supset \vdash :. \text{Hp} . \supset : S \subseteq P . \supset_S . S | P \subseteq P \quad (1)$$

$$\vdash . *91·171 \frac{S \subseteq P}{\phi S} . \supset$$

$$\vdash :. Q \epsilon \text{Pot}'P : S \subseteq P . \supset_S . S | P \subseteq P : P \subseteq P : \supset . Q \subseteq P \quad (2)$$

$$\vdash . (1) . (2) . *23·42 . \supset \vdash . \text{Prop}$$

***201·18.** $\vdash : P^2 \subseteq P . \supset . P_{po} = P . P_* = P \cup I \upharpoonright C'P$

Dem.

$$\vdash . *201·17 . *41·151 . (*91·05) . \supset \vdash : \text{Hp} . \supset . P_{po} \subseteq P \quad (1)$$

$$\vdash . (1) . *91·502 . \quad \supset \vdash : \text{Hp} . \supset . P_{po} = P \quad (2)$$

$$\vdash . (2) . *91·54 . \supset \vdash . \text{Prop}$$

This proposition is important, since it simplifies all propositions concerning P_{po} and P_* in case P is transitive. The following proposition is an instance of this simplification.

***201·19.** $\vdash : P \epsilon \text{trans} . \supset . P(x - y) = \overleftarrow{P}'x \cap \overrightarrow{P}'y$ [*201·18 . (*121·01)]

The following propositions (*201·2—·22) are concerned in proving that transitiveness is unaffected by likeness-transformations, and therefore belongs to every member of a relation-number or to none.

***201·2.** $\vdash : S \epsilon \text{Cls} \rightarrow 1 . \text{C}'Q \subset \text{C}'S . \supset . (S ; Q)^2 = S ; Q^2$

Dem.

$$\vdash . *150·1 . \supset \vdash . (S ; Q)^2 = S | Q | \breve{S} | S | Q | \breve{S} \quad (1)$$

$$\vdash . *72·601 . \supset \vdash : \text{Hp} . \supset . Q | \breve{S} | S = Q \quad (2)$$

$$\vdash . (1) . (2) . \supset \vdash : \text{Hp} . \supset . (S ; Q)^2 = S | Q^2 | \breve{S} : \supset \vdash . \text{Prop}$$

***201·201.** $\vdash : S \epsilon \text{Cls} \rightarrow 1 . \text{D}'Q \subset \text{C}'S . \supset . (S ; Q)^2 = S ; Q^2$

[Proof as in *201·2]

***201·21.** $\vdash : S \,\epsilon\, \mathrm{Cls} \to 1 \,.\, Q \,\epsilon\, \mathrm{trans} \,.\supset .\, S;Q \,\epsilon\, \mathrm{trans}$

Dem.

$$\vdash . *150{\cdot}36 . *35{\cdot}452 . \supset \vdash . S;Q = S;Q \upharpoonright \text{Ɑ}`S \qquad (1)$$

$$\vdash . (1) . *201{\cdot}2 . \supset \vdash : \mathrm{Hp} . \supset . (S;Q)^2 = S;(Q \upharpoonright \text{Ɑ}`S)^2 .$$

$$[*150{\cdot}31 . *201{\cdot}1] \quad \supset . (S;Q)^2 \mathrel{\dot{\subset}} S;Q : \supset \vdash . \mathrm{Prop}$$

***201·211.** $\vdash : P \,\epsilon\, \mathrm{trans} \,.\, Q \,\mathrm{smor}\, P \,.\supset .\, Q \,\epsilon\, \mathrm{trans}$ [*201·21 . *151·1]

This shows that transitiveness is a property which is unchanged by likeness-transformations. Hence

***201·212.** $\vdash : P \,\epsilon\, \mathrm{trans} \,.\supset .\, \mathrm{Nr}`P \subset \mathrm{trans}$ [*201·211]

***201·22.** $\vdash : P \,\epsilon\, \mathrm{trans} \,.\equiv .\, \mathrm{N_0r}`P \subset \mathrm{trans} \,.\equiv .\, \exists ! \,\mathrm{N_0r}`P \cap \mathrm{trans}$

[Proof as in *200·22]

***201·3.** $\vdash . \dot{\Lambda} \,\epsilon\, \mathrm{trans}$

Dem.

$$\vdash . *34{\cdot}32 . \supset \vdash . \dot{\Lambda}^2 = \dot{\Lambda} \qquad (1)$$

$$\vdash . (1) . *23{\cdot}42 . \supset \vdash . \dot{\Lambda}^2 \mathrel{\dot{\subset}} \dot{\Lambda} . \supset \vdash . \mathrm{Prop}$$

***201·31.** $\vdash . x \downarrow y \,\epsilon\, \mathrm{trans}$

Dem.

$$\vdash . *55{\cdot}13 . \supset \vdash : z (x \downarrow y)^2 w . \equiv . (\exists u) . z = x . u = y . u = x . w = y .$$

$$[*10{\cdot}35] \quad \supset . z = x . w = y .$$

$$[*55{\cdot}13] \quad \supset . z (x \downarrow y) w : \supset \vdash . \mathrm{Prop}$$

Unless $x = y$, $(x \downarrow y)^2 = \dot{\Lambda}$. A relation whose square is $\dot{\Lambda}$ is transitive, because $\dot{\Lambda}$ is contained in every relation.

***201·32.** $\vdash . \alpha \uparrow \beta \,\epsilon\, \mathrm{trans}$

Dem.

$$\vdash . *35{\cdot}103 . \supset \vdash : x (\alpha \uparrow \beta)^2 z . \equiv . (\exists y) . x \,\epsilon\, \alpha . y \,\epsilon\, \beta . y \,\epsilon\, \alpha . z \,\epsilon\, \beta .$$

$$[*10{\cdot}35] \quad \supset . x \,\epsilon\, \alpha . z \,\epsilon\, \beta .$$

$$[*35{\cdot}103] \quad \supset . x (\alpha \uparrow \beta) z : \supset \vdash . \mathrm{Prop}$$

***201·33.** $\vdash : P \,\epsilon\, \mathrm{trans} \,.\supset .\, P \upharpoonright \alpha \,\epsilon\, \mathrm{trans}$

Dem.

$$\vdash . *36{\cdot}13 . \supset \vdash : x (P \upharpoonright \alpha)^2 z . \equiv . (\exists y) . x, y, z \,\epsilon\, \alpha . xPy . yPz \qquad (1)$$

$$\vdash . (1) . \supset \vdash :. \mathrm{Hp} . \supset : x (P \upharpoonright \alpha)^2 z . \supset . (\exists y) . x, y, z \,\epsilon\, \alpha . xPz .$$

$$[*10{\cdot}35 . *36{\cdot}13] \quad \supset . x (P \upharpoonright \alpha) z :. \supset \vdash . \mathrm{Prop}$$

The following propositions (*201·4—·42) are concerned with the ideas of relation-arithmetic.

***201·4.** $\vdash : P, Q \,\epsilon\, \mathrm{trans} \,.\, C`P \cap C`Q = \Lambda \,.\supset .\, P \mathbin{\hat{+}} Q \,\epsilon\, \mathrm{trans}$

Dem.

$$\vdash . *160{\cdot}51 . \supset \vdash : \mathrm{Hp} . \supset . (P \mathbin{\hat{+}} Q)^2 = P^2 \cup Q^2 \cup \mathrm{D}`P \uparrow C`Q \cup C`P \uparrow \text{Ɑ}`Q \qquad (1)$$

$$\vdash . *201{\cdot}1 . \supset \vdash : \mathrm{Hp} . \supset . P^2 \mathrel{\dot{\subset}} P . Q^2 \mathrel{\dot{\subset}} Q \qquad (2)$$

$$\vdash . *35{\cdot}432{\cdot}82 . \supset \vdash . \mathrm{D}`P \uparrow C`Q \mathrel{\dot{\subset}} C`P \uparrow C`Q . C`P \uparrow \text{Ɑ}`Q \mathrel{\dot{\subset}} C`P \uparrow C`Q \qquad (3)$$

$$\vdash . (1) . (2) . (3) . \supset \vdash : \mathrm{Hp} . \supset . (P \mathbin{\hat{+}} Q)^2 \mathrel{\dot{\subset}} P \cup Q \cup C`P \uparrow C`Q : \supset \vdash . \mathrm{Prop}$$

***201·401.** $\vdash :. C`P \dot\cap C`Q = \Lambda . \supset : P \ddagger Q \epsilon \text{trans} . \equiv . P, Q \epsilon \text{trans}$

Dem.

$\vdash . *160·51 . \supset$

$\vdash :. \text{Hp} . \supset : P \ddagger Q \epsilon \text{trans} . \equiv . P^2 \dot\cup Q^2 \dot\cup D`P \uparrow C`Q \dot\cup C`P \uparrow \text{G}`Q \subseteq P \ddagger Q .$

[*160·1] $\equiv . P^2 \dot\cup Q^2 \subseteq P \ddagger Q .$

[*160·5] $\supset . (P^2 \dot\cup Q^2) \upharpoonright C`P \subseteq P . (P^2 \dot\cup Q^2) \upharpoonright C`Q \subseteq Q .$

[*36·4.*34·56] $\supset . P^2 \subseteq P . Q^2 \subseteq Q$ (1)

$\vdash . (1) . *201·4 . \supset \vdash . \text{Prop}$

***201·41.** $\vdash :. x \sim \epsilon C`P . \supset : P \epsilon \text{trans} . \equiv . P \nrightarrow x \epsilon \text{trans} . \equiv . x \nleftarrow P \epsilon \text{trans}$

Dem.

$\vdash . *34·301 . \supset \vdash : \text{Hp} . \supset . (C`P \uparrow \iota`x) | P = \dot\Lambda .$

[*161·1] $\supset . (P \nrightarrow x)^2 = P^2 \dot\cup (C`P \uparrow \iota`x)^2 \dot\cup P | (C`P \uparrow \iota`x)$

[*35·881] $= P^2 \dot\cup (C`P \uparrow \iota`x)^2 \dot\cup (D`P \uparrow \iota`x)$

[*35·895] $= P^2 \dot\cup (D`P \uparrow \iota`x)$ (1)

$\vdash . (1) . *201·1 . \supset$

$\vdash :. \text{Hp} . \supset : (P \nrightarrow x) \epsilon \text{trans} . \equiv . P^2 \dot\cup (D`P \uparrow \iota`x) \subseteq P \dot\cup (C`P \uparrow \iota`x) .$

[*35·432·82] $\equiv . P^2 \subseteq P \dot\cup (C`P \uparrow \iota`x)$ (2)

$\vdash . *33·33 . *34·56 . *35·86 . \supset \vdash : \text{Hp} . \supset . P^2 \dot\cap (C`P \uparrow \iota`x) = \dot\Lambda$ (3)

$\vdash . (2) . (3) . *25·49 . \supset \vdash :. \text{Hp} . \supset : P \nrightarrow x \epsilon \text{trans} . \equiv . P^2 \subseteq P .$

[*201·1] $\equiv . P \epsilon \text{trans}$ (4)

Similarly $\vdash :. \text{Hp} . \supset : x \nleftarrow P \epsilon \text{trans} . \equiv . P \epsilon \text{trans}$ (5)

$\vdash . (4) . (5) . \supset \vdash . \text{Prop}$

***201·411.** $\vdash : z \neq x . z \neq y . \supset . x \downarrow y \nrightarrow z \epsilon \text{trans}$ [*201·41·31]

***201·42.** $\vdash : P \epsilon \text{trans} \cap \text{Rel}^2 \text{excl} . C`P \subset \text{trans} . \supset . \Sigma`P \epsilon \text{trans}$

Dem.

$\vdash . *162·1 . \supset$

$\vdash . (\Sigma`P)^2 = (\dot{s}`C`P)^2 \dot\cup (F;P)^2 \dot\cup (\dot{s}`C`P) | (F;P) \dot\cup (F;P) | (\dot{s}`C`P)$ (1)

$\vdash . *41·11 . \supset \vdash : x (\dot{s}`C`P)^2 z . \equiv . (\exists Q, R, y) . Q, R \epsilon C`P . xQy . yRz .$

[*33·17] $\equiv . (\exists Q, R, y) . Q, R \epsilon C`P . xQy . yRz . \exists ! C`Q \dot\cap C`R$ (2)

$\vdash . (2) . *163·11 . \supset$

$\vdash :. \text{Hp} . \supset : x (\dot{s}`C`P)^2 z . \supset . (\exists Q, R, y) . Q, R \epsilon C`P . xQy . yRz . Q = R .$

[*13·195] $\supset . (\exists Q) . Q \epsilon C`P . xQ^2 z .$

[*201·1.Hp] $\supset . (\exists Q) . Q \epsilon C`P . xQz .$

[*41·11] $\supset . x (\dot{s}`C`P) z$ (3)

$\vdash . *201·21 . *163·12 . \supset \vdash : \text{Hp} . \supset . (F;P)^2 \subseteq F;P$ (4)

$\vdash . *34 \cdot 1 . *41 \cdot 11 . *150 \cdot 52 . \supset$

$\vdash : x (\dot{s}`C`P) | (F ; P) z . \supset . (\exists Q, R, S, y) . Q \epsilon C`P . xQy . RPS . y \epsilon C`R . z \epsilon C`S \quad (5)$

$\vdash . (5) . *163 \cdot 11 . *13 \cdot 195 . \supset$

$\vdash :. \text{Hp} . \supset : x (\dot{s}`C`P) | (F ; P) z . \supset . (\exists Q, S, y) . Q \epsilon C`P . xQy . QPS . z \epsilon C`S .$

[*33·17.*150·52] $\supset . x (F ; P) z \quad (6)$

Similarly $\vdash :. \text{Hp} . \supset : x (F ; P) | (\dot{s}`C`P) z . \supset . x (F ; P) z \quad (7)$

$\vdash . (1) . (3) . (4) . (6) . (7) . \supset$

$\vdash : \text{Hp} . \supset . (\Sigma`P)^2 \in \dot{s}`C`P \cup F ; P : \supset \vdash . \text{Prop}$

The following propositions (*201·5—·56) are concerned with $P``\alpha$ and $p`\overrightarrow{P}``\alpha$, *i.e.* with the predecessors of some part of a class and the predecessors of the whole of a class.

***201·5.** $\vdash : P \epsilon \text{trans} . \supset . P``P``\alpha \subset P``\alpha$ [*37·33·201]

***201·501.** $\vdash : P \epsilon \text{trans} . \supset . P``\overrightarrow{P}`x \subset \overrightarrow{P}`x$ [*53·301 . *201·5]

***201·51.** $\vdash : P \epsilon \text{trans} . \supset . P``p`\overrightarrow{P}``\alpha \subset p`\overrightarrow{P}``\alpha$

Dem.

$\vdash . *37 \cdot 1 . *40 \cdot 51 . \supset \vdash :. x \epsilon P``p`\overrightarrow{P}``\alpha . \equiv : (\exists y) : z \epsilon \alpha . \supset_z . yPz : xPy :$

[*5·31] $\supset : z \epsilon \alpha . \supset_z . xP^2 z \quad (1)$

$\vdash . (1) . *201 \cdot 1 . \supset \vdash :. \text{Hp} . \supset : x \epsilon P``p`\overrightarrow{P}``\alpha . \supset : z \epsilon \alpha . \supset_z . xPz :$

[*40·51] $\supset : x \epsilon p`\overrightarrow{P}``\alpha :. \supset \vdash . \text{Prop}$

***201·52.** $\vdash : P \epsilon \text{trans} . \supset . P_*``\alpha = P``\alpha \cup (\alpha \cap C`P)$ [*91·543 . *201·18]

***201·521.** $\vdash : P \epsilon \text{trans} . x \epsilon C`P . \supset . \overrightarrow{P_*}`x = \overrightarrow{P}`x \cup \iota`x$ [*201·52 . *53·301]

***201·53.** $\vdash : P \epsilon \text{trans} . \supset . P_*``P``\alpha = P``\alpha$ [*201·5·52 . *37·265]

***201·54.** $\vdash : P \epsilon \text{trans} . \supset . P_*``p`\overrightarrow{P}``\alpha \subset p`\overrightarrow{P}``\alpha$ [*201·51·52]

***201·55.** $\vdash : P \epsilon \text{trans} . \supset . P``(\alpha \cup P``\alpha) = P``\alpha$

Dem.

$\vdash . *201 \cdot 5 . \supset \vdash : \text{Hp} . \supset . P``\alpha = P``\alpha \cup P``P``\alpha$

[*37·22] $= P``(\alpha \cup P``\alpha) : \supset \vdash . \text{Prop}$

The following proposition is a lemma which is used in *205·192 and *206·24.

***201·56.** $\vdash : P \epsilon \text{trans} . \beta \subset P``\alpha . \supset .$

$$P``(\alpha \cup \beta) = P``\alpha . p`\overleftarrow{P}``\{(\alpha \cup \beta) \cap C`P\} = p`\overleftarrow{P}``(\alpha \cap C`P)$$

Dem.

$\vdash . *37 \cdot 22 . \supset \vdash . P``(\alpha \cup \beta) = P``\alpha \cup P``\beta \quad (1)$

$\vdash . *37 \cdot 2 . \supset \vdash : \text{Hp} . \supset . P``\beta \subset P``P``\alpha .$

[*201·5] $\supset . P``\beta \subset P``\alpha \quad (2)$

$\vdash . (1) . (2) . \supset \vdash : \text{Hp} . \supset . P``(\alpha \cup \beta) = P``\alpha \quad (3)$

$\vdash . *40{\cdot}51 . *37{\cdot}265 . \supset$

$\vdash :: \text{Hp} . \supset :. z \epsilon p'\overleftarrow{P}``(\alpha \cap C'P) . x \epsilon \beta \cap C'P . \supset :$

$y \epsilon \alpha \cap C'P . \supset_y . yPz : (\exists y) . y \epsilon \alpha \cap C'P . xPy :$

[*10·56] $\supset : (\exists y) . xPy . yPz :$

[*34·5.Hp] $\supset : xPz$ (4)

$\vdash . (4) . *40{\cdot}51 . \supset \vdash : \text{Hp} . \supset . p'\overleftarrow{P}``(\alpha \cap C'P) \subset p'\overleftarrow{P}``(\beta \cap C'P) .$

[*22·621] $\supset . p'\overleftarrow{P}``(\alpha \cap C'P) = p'\overleftarrow{P}``(\alpha \cap C'P) \cap p'\overleftarrow{P}``(\beta \cap C'P)$

[*40·18.*37·22] $= p'\overleftarrow{P}``\{(\alpha \cup \beta) \cap C'P\}$ (5)

$\vdash . (3) . (5) . \supset \vdash . \text{Prop}$

The following propositions, to the end of the number, are concerned with the relation P_1 defined in *121. We may regard P_1 as meaning "immediately precedes." *201·6·61·62 are lemmas for *201·63.

***201·6.** $\vdash : P \epsilon \text{trans} . \sim(xPx) . \sim(yPy) . xP_1y . \supset . x(P \dot{-} P^2)y$

Dem.

$\vdash . *121{\cdot}32{\cdot}242 . \supset \vdash : \text{Hp} . \supset . P(x \vdash\!\dashv y) = \iota'x \cup \iota'y \cup P(x - y)$

[*201·19] $= \iota'x \cup \iota'y \cup \overleftarrow{P}'x \cap \overrightarrow{P}'y$ (1)

$\vdash . *121{\cdot}321 . *201{\cdot}18 . \quad \supset \vdash : \text{Hp} . \supset . xPy$ (2)

$\vdash . (2) . *13{\cdot}14 . \quad \supset \vdash : \text{Hp} . \supset . x \neq y$ (3)

$\vdash . (1) . (3) . *54{\cdot}53 . *121{\cdot}11 . \supset \vdash : \text{Hp} . \supset . \overleftarrow{P}'x \cap \overrightarrow{P}'y \subset \iota'x \cup \iota'y$ (4)

$\vdash . *32{\cdot}18{\cdot}181 . \quad \supset \vdash : \text{Hp} . \supset . x \sim\epsilon \overleftarrow{P}'x . y \sim\epsilon \overrightarrow{P}'y$ (5)

$\vdash . (4) . (5) . \quad \supset \vdash : \text{Hp} . \supset . \overleftarrow{P}'x \cap \overrightarrow{P}'y = \Lambda .$

[*34·11] $\supset . \sim(xP^2y)$ (6)

$\vdash . (2) . (6) . \supset \vdash . \text{Prop}$

***201·61.** $\vdash : P \epsilon \text{trans} . \supset . P \dot{-} P^2 \Subset P_1$

Dem.

$\vdash . *121{\cdot}242 . *90{\cdot}151 . \supset \vdash : xPy . \supset . P(x \vdash\!\dashv y) = \iota'x \cup \iota'y \cup P(x - y)$ (1)

$\vdash . (1) . *201{\cdot}19 . \supset \vdash :. \text{Hp} . \supset : xPy . \supset . P(x \vdash\!\dashv y) = \iota'x \cup \iota'y \cup (\overleftarrow{P}'x \cap \overrightarrow{P}'y)$ (2)

$\vdash . *34{\cdot}11 . \quad \supset \vdash : \sim(xP^2y) . \supset . \overleftarrow{P}'x \cap \overrightarrow{P}'y = \Lambda$ (3)

$\vdash . *34{\cdot}54 . \quad \supset \vdash : xPy . \sim(xP^2y) . \supset . x \neq y$ (4)

$\vdash . (2) . (3) . (4) . \supset \vdash :. \text{Hp} . \supset : xPy . \sim(xP^2y) . \supset . P(x \vdash\!\dashv y) = \iota'x \cup \iota'y . x \neq y .$

[*54·101] $\supset . P(x \vdash\!\dashv y) \epsilon 2 .$

[*121·11] $\supset . xP_1y :. \supset \vdash . \text{Prop}$

***201·62.** $\vdash :. P \epsilon \text{trans} . \sim(xPx) . \sim(yPy) . \supset : xP_1y . \equiv . x(P \dot{-} P^2)y$
[*201·6·61]

***201·63.** $\vdash : P \epsilon \text{trans} \cap \text{Rl}'J . \supset . P_1 = P \dot{-} P^2$ [*201·62]

The above proposition is of fundamental importance. The relation P_1 (defined in *121) plays a great part in the theory of series. It is the relation "immediately preceding." Its domain consists of those terms which have immediate successors; its converse domain, of those that have immediate predecessors. In well-ordered series, $\text{D}'P_1 = \text{D}'P$, while $\text{Ɑ}'P_1$ consists of all terms (except the first) which do not belong to the first derivative (cf. *216). In any series, $\text{Ɑ}'P - \text{Ɑ}'P_1$ consists of all the terms which are limits of ascending series, and $\text{D}'P - \text{D}'P_1$ consists of all the terms which are limits of descending series.

***201·64.** $\vdash :. P \epsilon \text{trans} . \supset : P \dot{-} P^2 = \dot{\Lambda} . \equiv . P^2 = P$

Dem.

$$\vdash . *23\cdot41 . \supset \vdash :. \text{Hp} . \supset : P^2 = P . \equiv . P \Subset P^2 .$$

$$[*25\cdot3] \qquad \equiv . P \dot{-} P^2 = \dot{\Lambda} :. \supset \vdash . \text{Prop}$$

***201·65.** $\vdash :. P \epsilon \text{trans} \cap \text{Rl}'J . \supset : P_1 = \dot{\Lambda} . \equiv . P^2 = P \quad [*201\cdot64\cdot63]$

When P is a series, $P^2 = P$ is the condition for its being a *compact* series, *i.e.* one in which there are terms between any two. In virtue of *201·65, this condition is equivalent to $P_1 = \dot{\Lambda}$, which states that no term has an immediate predecessor.

The following proposition is first used in *253·521.

***201·66.** $\vdash : P \epsilon \text{trans} . \text{E} ! P'x . P'x \neq x . \supset . (P'x) P_1 x$

Dem.

$$\vdash . *201\cdot521 . *121\cdot11 . \supset$$

$$\vdash :. \text{Hp} . \supset : (P'x) P_1 x . \equiv . (\iota'P'x \cup \overleftarrow{P}'P'x) \cap (\iota'x \cup \overrightarrow{P}'x) \epsilon 2 \quad (1)$$

$$\vdash . *53\cdot31 . \supset \vdash : \text{Hp} . \supset .$$

$$(\iota'P'x \cup \overleftarrow{P}'P'x) \cap (\iota'x \cup \overrightarrow{P}'x) = (\iota'P'x \cup \overleftarrow{P}'P'x) \cap (\iota'x \cup \iota'P'x)$$

$$[*30\cdot32 . *22\cdot68] \qquad = \iota'x \cup \iota'P'x \quad (2)$$

$$\vdash . *54\cdot26 . \supset \vdash : \text{Hp} . \supset . (\iota'x \cup \iota'P'x) \epsilon 2 \quad (3)$$

$$\vdash . (1) . (2) . (3) . \supset \vdash . \text{Prop}$$

***201·661.** $\vdash : P \epsilon \text{trans} . \text{Ɑ}'P \epsilon 1 . \exists ! \text{D}'P - \text{Ɑ}'P . \supset . \text{Ɑ}'P \subset \text{Ɑ}'P_1$

Dem.

$$\vdash . *33\cdot151\cdot4 . *60\cdot38 . \supset$$

$$\vdash : \text{Hp} . y \epsilon \text{D}'P - \text{Ɑ}'P . \supset . \overleftarrow{P}'y \epsilon 1 . y \sim \epsilon \overleftarrow{P}'y . \overleftarrow{P}'y = \text{Ɑ}'P .$$

$$[*53\cdot3] \qquad \supset . \text{E} ! \breve{P}'y . y \neq \breve{P}'y . \iota'\breve{P}'y = \text{Ɑ}'P .$$

$$[*201\cdot66\cdot11 . *121\cdot26] \qquad \supset . y P_1 (\breve{P}'y) . \iota'\breve{P}'y = \text{Ɑ}'P : \supset \vdash . \text{Prop}$$

The above proposition is a lemma for the following.

***201·662.** $\vdash : P \epsilon \text{trans} . \exists ! \overrightarrow{B}'P . \exists ! \text{Ɑ}'P - \text{Ɑ}'P_1 . \supset . \text{Ɑ}'P \sim \epsilon 1$

$$[*201\cdot661 . \text{Transp}]$$

This proposition is first used in *253·521.

$*202$. CONNECTED RELATIONS.

Summary of $*202$.

A relation is said to be *connected* when either it or its converse holds between any two different members of its field, *i.e.* when, if $x, y \in C`P \,.\, x \neq y$, we have $xPy \,.\vee.\, yPx$. Thus the field of a connected relation consists of a single family, unless the relation is null, in which case it has no families. Conversely, a relation which has one family or none is connected. Connection is necessary, in addition to transitiveness and asymmetry, in order that a relation may generate a single series. If λ is a class of transitive or asymmetrical relations, $\dot{s}`\lambda$ is transitive or asymmetrical; but if $\lambda$ is a class of connected relations, $\dot{s}`\lambda$ is not in general connected. Hence if λ is a class of series, $\dot{s}`\lambda$ is not one series, but many detached series. This is one reason why the arithmetical sum of a relation of relations is not defined as $\dot{s}`C`P$, but as $\dot{s}`C`P \mathbin{\dot{\cup}} F \dot{;} P$ (cf. $*162$), because the latter, but not in general the former, is connected when $P$ and all the members of $C`P$ are connected ($*202\cdot42$).

When P is connected, if α is any class contained in $C`P$, we have

$$C`P = P``\alpha \cup \alpha \cup (C`P \cap p`\overleftarrow{P}``\alpha),$$

and there is at most one member of α belonging neither to $P``\alpha$ nor to $C`P \cap p`\overleftarrow{P}``\alpha$. This member of α, if it exists, is the maximum of α. If, further, $P^2 \mathbin{\dot{\subset}} J$ (*i.e.* if P is asymmetrical), $(P``\alpha \cup \alpha) \cap (C`P \cap p`\overleftarrow{P}``\alpha) = \Lambda$. Thus when P is both connected and asymmetrical, $P``\alpha \cup \alpha$ and $C`P \cap p`\overleftarrow{P}``\alpha$ are each other's complements, and the two together constitute the Dedekind cut defined by α, $P``\alpha \cup \alpha$ being all the terms that do not follow the whole of $\alpha$, and $C`P \cap p`\overleftarrow{P}``\alpha$ being all the terms that do follow the whole of α.

More generally, if α is any class, not necessarily contained in $C`P$, then when P is connected, we have

$$C`P - p`\overleftarrow{P}``(\alpha \cap C`P) \subset P``\alpha \cup (\alpha \cap C`P),$$

and when P is asymmetrical, we have

$$P``\alpha \cup (\alpha \cap C`P) \subset C`P - p`\overleftarrow{P}``\alpha.$$

Thus when both conditions are fulfilled, we have ($*202\cdot503$)

$$C`P - p`\overleftarrow{P}``(\alpha \cap C`P) = P``\alpha \cup (\alpha \cap C`P).$$

The above inclusions and the consequent equality will be constantly required throughout what follows. The division of $C'P$ into the two mutually exclusive parts

$$P``\alpha \cup (\alpha \cap C'P) \text{ and } C'P \cap p'\overleftarrow{P}``(\alpha \cap C'P)$$

is the Dedekind "cut" defined by the class α. If $\alpha \subset C'P$, the two parts become, as above mentioned,

$$P``\alpha \cup \alpha \text{ and } C'P \cap p'\overleftarrow{P}``\alpha.$$

If, further, α is not null, they become

$$P``\alpha \cup \alpha \text{ and } p'\overleftarrow{P}``\alpha.$$

If α is contained in $C'P$ and contains all its own predecessors, they become

$$\alpha \text{ and } C'P \cap p'\overleftarrow{P}``\alpha.$$

In this simplified form, Dedekind "cuts" will be considered later (*211).

We take as our definition

$$\text{connex} = \hat{P} \{x \in C'P . \supset_x . \overleftrightarrow{P}'x = C'P\} \quad \text{Df.}$$

Some of the propositions of the present number are analogues of propositions in *200 and *201. Such are: If P is connected, so is $\breve{P}$ (*202·11); if P is connected, so is any similar relation (*202·211); $\dot{\Lambda}$ and $x \downarrow y$ are connected (*202·3·31); if P is connected, so is $P \upharpoonright \alpha$ (*202·33); and various propositions connected with relation-arithmetic (*202·4—·42). The majority of the propositions of this number, however, deal with properties peculiar to connexity. Among the most important of these are:

***202·101.** $\vdash :. P \in \text{connex} . \equiv : x \in C'P . \supset_x . \overrightarrow{P}'x \cup \iota'x \cup \overleftarrow{P}'x = C'P$

***202·103.** $\vdash :: P \in \text{connex} . \equiv :. x, y \in C'P . \supset_{x,y} : xPy . \vee . x = y . \vee . yPx$

These are merely alternative forms of the definition.

***202·13.** $\vdash : R_* \in \text{connex} . \equiv . R_{po} \in \text{connex}$

***202·5.** $\vdash :. P \in \text{connex} . P^2 \in J . x, y \in C'P . \supset : x \neq y . \sim(xPy) . \equiv . yPx$

***202·501.** $\vdash : P \in \text{connex} . \supset . C'P - \alpha - P``\alpha \subset p'\overleftarrow{P}``(\alpha \cap C'P)$

***202·503.** $\vdash : P \in \text{connex} . P^2 \in J . \supset . C'P - p'\overleftarrow{P}``(\alpha \cap C'P) = (\alpha \cap C'P) \cup P``\alpha$

***202·505.** $\vdash : P \in \text{connex} . \supset . C'P = P``\alpha \cup (\alpha \cap C'P) \cup \{C'P \cap p'\overleftarrow{P}``(\alpha \cap C'P)\}$

***202·52.** $\vdash : P \in \text{connex} . \supset . \overrightarrow{B}'P, \overrightarrow{B}'\breve{P} \in 0 \cup 1$

***202·524.** $\vdash : P \in \text{connex} . \exists ! \overrightarrow{B}'P . \supset . \text{Cl}'P = \overleftarrow{P}'B'P$

***202·55.** $\vdash : P \upharpoonright \alpha \in \text{connex} . \alpha \subset C'P . \alpha \sim\in 1 . \supset . C'P \upharpoonright \alpha = \alpha$

In virtue of this proposition (and others) if P is a series and α is a class (not a unit class) contained in $C‘P$, $P \upharpoonright \alpha$ is the generating relation of the series consisting of the class α in the order which it has in the series P.

***202·7.** $\vdash : P \in \text{connex} . \supset . P \dot{-} P^2 \in 1 \rightarrow 1$

This proposition is to be taken in connection with *201·63. The two together show that when P is a series, P_1 is one-one.

***202·01.** $\text{connex} = \hat{P}\{x \in C‘P . \supset_x . \overleftrightarrow{P}‘x = C‘P\}$ Df

For the definition of $\overleftrightarrow{P}‘x$, see *97·01.

***202·1.** $\vdash :. P \in \text{connex} . \equiv : x \in C‘P . \supset_x . \overleftrightarrow{P}‘x = C‘P$ [(*202·01)]

***202·101.** $\vdash :. P \in \text{connex} . \equiv : x \in C‘P . \supset_x . \overrightarrow{P}‘x \cup \iota‘x \cup \overleftarrow{P}‘x = C‘P$ [*202·1 . *97·1]

***202·102.** $\vdash : P \in \text{connex} . \equiv . \overleftrightarrow{P}‘‘C‘P \in 0 \cup 1$ [*97·231 . *202·101]

***202·103.** $\vdash :: P \in \text{connex} . \equiv :. x, y \in C‘P . \supset_{x,y} : xPy . \vee . x = y . \vee . yPx$ [*97·23 . *202·102]

***202·104.** $\vdash :: P \in \text{connex} . \equiv :. x, y \in C‘P . x \neq y . \supset_{x,y} : xPy . \vee . yPx$ [*202·103 . *5·6]

***202·11.** $\vdash : P \in \text{connex} . \equiv . \breve{P} \in \text{connex}$ [*202·104 . *33·22]

***202·12.** $\vdash :. \dot{\exists} ! P . \supset : P \in \text{connex} . \equiv . \overleftrightarrow{P}‘‘C‘P \in 1 . \equiv . \overleftrightarrow{P}‘‘C‘P = \iota‘C‘P$

Dem.

$\vdash . \text{*}202·1 . \supset \vdash : P \in \text{connex} . \equiv . \overleftrightarrow{P}‘‘C‘P \subset \iota‘C‘P$ (1)

$\vdash . \text{*}37·45 . \supset \vdash :. \text{Hp} . \supset : \exists ! \overleftrightarrow{P}‘‘C‘P :$

[*54·102] $\supset : \overleftrightarrow{P}‘‘C‘P \sim \in 0 :$

[*202·102] $\supset : P \in \text{connex} . \supset . \overleftrightarrow{P}‘‘C‘P \in 1$ (2)

$\vdash . \text{*}202·102 . \supset \vdash : \overleftrightarrow{P}‘‘C‘P \in 1 . \supset . P \in \text{connex}$ (3)

$\vdash . (2) . (3) . \supset \vdash :. \text{Hp} . \supset : P \in \text{connex} . \equiv . \overleftrightarrow{P}‘‘C‘P \in 1$ (4)

$\vdash . (1) . (4) . \text{*}52·46 . \supset \vdash :. \text{Hp} . \supset : P \in \text{connex} . \supset . \overleftrightarrow{P}‘‘C‘P = \iota‘C‘P$ (5)

$\vdash . (1) . \text{*}22·42 . \supset \vdash : \overleftrightarrow{P}‘‘C‘P = \iota‘C‘P . \supset . P \in \text{connex}$ (6)

$\vdash . (4) . (5) . (6) . \supset \vdash . \text{Prop}$

The following propositions, down to *202·181 inclusive (excepting *202·16·161) are concerned with R_* and R_{po}. It often happens that these are connected when R is not so, *e.g.* if R is the relation $+_c 1$ among inductive cardinals.

***202·13.** $\vdash : R_* \in \text{connex} . \equiv . R_{po} \in \text{connex}$

Dem.

$\vdash . *202·104 . \supset$

$\vdash :: R_* \in \text{connex} . \equiv :. x, y \in C`R_* . x \neq y . \supset_{x,y} : x R_* y . \vee . y R_* x :.$

[*91·542] $\equiv :. x, y \in C`R_* . x \neq y . \supset_{x,y} : x R_{po} y . \vee . y R_{po} x :.$

[*90·14.*91·504] $\equiv :. x, y \in C`R_{po} . x \neq y . \supset_{x,y} : x R_{po} y . \vee . y R_{po} x :.$

[*202·104] $\equiv :. R_{po} \in \text{connex} :: \supset \vdash . \text{Prop}$

***202·131.** $\vdash : P \in \text{connex} . C`P = C`Q . P \subseteq Q . \supset . Q \in \text{connex}$ [*202·103]

***202·132.** $\vdash : P \in \text{connex} . \supset . P_{po}, P_* \in \text{connex}$
[*202·131 . *90·14·151 . *91·502·504]

***202·133.** $\vdash :: I \upharpoonright C`P \subseteq P . \supset :. P \in \text{connex} . \equiv : x \in C`P . \supset_x . C`P = \overrightarrow{P}`x \cup \overleftarrow{P}`x$

Dem.

$\vdash . *35·101 . \supset \vdash :. \text{Hp} . \supset : x \in C`P . \supset . \iota`x \subset \overrightarrow{P}`x$ (1)

$\vdash . (1) . *202·101 . \supset \vdash . \text{Prop}$

***202·134.** $\vdash ::. I \upharpoonright C`P \subseteq P . \supset :: P \in \text{connex} . \equiv :. x, y \in C`P . \supset_{x,y} : x P y . \vee . y P x$
[*202·103]

***202·135.** $\vdash : P \in \text{connex} . \equiv . P \cup I \upharpoonright C`P \in \text{connex}$

Dem.

$\vdash . *202·134 . \supset \vdash :: P \cup I \upharpoonright C`P \in \text{connex} . \equiv :.$

$x, y \in C`P . \supset_{x,y} : x (P \cup I \upharpoonright C`P) y . \vee . y (P \cup I \upharpoonright C`P) x :.$

[*202·103] $\equiv :. P \in \text{connex} :: \supset \vdash . \text{Prop}$

***202·136.** $\vdash :. P_* \in \text{connex} . \equiv : x \in C`P . \supset_x . C`P = \overrightarrow{P_*}`x \cup \overleftarrow{P_*}`x$
[*202·133 . *90·14·15]

***202·137.** $\vdash :: P_* \in \text{connex} . \equiv :. x, y \in C`P . \supset_{x,y} : x P_* y . \vee . y P_* x$
[*202·134 . *90·15]

***202·138.** $\vdash :. P \in \text{trans} . \supset : P \in \text{connex} . \equiv . P_* \in \text{connex}$ [*202·13 . *201·18]

***202·14.** $\vdash : R \in \text{Cls} \rightarrow 1 . \supset . R_{po} \upharpoonright \overleftarrow{R_*}`x \in \text{connex}$ [*96·303 . *202·104]

***202·141.** $\vdash : R \in 1 \rightarrow \text{Cls} . \supset . R_{po} \upharpoonright \overrightarrow{R_*}`x \in \text{connex}$ [$*202·14 \frac{\breve{R}}{R} . *202·11$]

***202·15.** $\vdash : R \in 1 \rightarrow 1 . \supset . R_{po} \upharpoonright \overleftrightarrow{R_*}`x \in \text{connex}$

Dem.

$\vdash . *97·13 . \supset \vdash :. y, z \in \overleftrightarrow{R_*}`x . \supset_{y,z} : y, z \in \overrightarrow{R_*}`x . \vee . y, z \in \overleftarrow{R_*}`x . \vee .$

$y \in \overrightarrow{R_*}`x . z \in \overleftarrow{R_*}`x . \vee . y \in \overleftarrow{R_*}`x . z \in \overrightarrow{R_*}`x$ (1)

$\vdash . *202\cdot141\cdot104 . \supset \vdash :: \text{Hp} . \supset :. y, z \in \overrightarrow{R_*}\text{'}x . y \neq z . \supset : yR_{\text{po}}x . \vee . xR_{\text{po}}y$ (2)

$\vdash . *202\cdot14\cdot104 . \supset \vdash :: \text{Hp} . \supset :. y, z \in \overleftarrow{R_*}\text{'}x . y \neq z . \supset : yR_{\text{po}}x . \vee . xR_{\text{po}}y$ (3)

$\vdash . *90\cdot17 . \supset \vdash : y \in \overrightarrow{R_*}\text{'}x . z \in \overleftarrow{R_*}\text{'}x . y \neq z . \supset . yR_*z . y \neq z .$

[*91·542] $\supset . yR_{\text{po}}z$ (4)

Similarly $\vdash : y \in \overleftarrow{R_*}\text{'}x . z \in \overrightarrow{R_*}\text{'}x . y \neq z . \supset . zR_{\text{po}}y$ (5)

$\vdash . (1) . (2) . (3) . (4) . (5) . \supset$

$\vdash :: \text{Hp} . \supset :. y, z \in \overleftrightarrow{R_*}\text{'}x . y \neq z . \supset_{y,z} : yR_{\text{po}}z . \vee . zR_{\text{po}}y$ (6)

$\vdash . (6) . *202\cdot104 . \supset \vdash . \text{Prop}$

The above proposition is used in the ordinal theory of finite and infinite (*260·4).

***202·16.** $\vdash : P \in \text{connex} . x, y \in C\text{'}P . \sim(xPx) . \sim(yPy) . \overrightarrow{P}\text{'}x = \overrightarrow{P}\text{'}y . \supset . x = y$

Dem.

$\vdash . *32\cdot18\cdot181 . \supset \vdash : \text{Hp} . \supset . \sim(xPy) . \sim(yPx) .$

[*202·103] $\supset . x = y : \supset \vdash . \text{Prop}$

***202·161.** $\vdash : P \in \text{connex} \cap \text{Rl}\text{'}J . \supset . \overrightarrow{P} \upharpoonright C\text{'}P \in 1 \rightarrow 1 . \overrightarrow{P} \upharpoonright C\text{'}P \in (\overrightarrow{P} ; P) \overline{\text{smor}}\, P$

Dem.

$\vdash . *202\cdot16 . \supset \vdash :. \text{Hp} . \supset : x, y \in C\text{'}P . \overrightarrow{P}\text{'}x = \overrightarrow{P}\text{'}y . \supset . x = y$ (1)

$\vdash . (1) . *71\cdot55 . *151\cdot24 . \supset \vdash . \text{Prop}$

***202·162.** $\vdash : P \in \text{connex} . P_{\text{po}} \subseteq J . \supset . P \mathrel{\text{⸦}} ; \overrightarrow{P_*} ; P \text{ smor } P . P \mathrel{\text{⸦}} | \overrightarrow{P_*} \upharpoonright C\text{'}P \in 1 \rightarrow 1$

Dem.

$\vdash . *36\cdot13 . \supset \vdash :. P \mathrel{\text{⸦}} \overrightarrow{P_*}\text{'}x = P \mathrel{\text{⸦}} \overrightarrow{P_*}\text{'}y . \equiv :$

$uPv . u, v \in \overrightarrow{P_*}\text{'}x . \equiv_{u,v} . uPv . u, v \in \overrightarrow{P_*}\text{'}y$ (1)

$\vdash . (1) . *11\cdot1 . *90\cdot12 . \supset$

$\vdash :. x, y \in C\text{'}P . P \mathrel{\text{⸦}} \overrightarrow{P_*}\text{'}x = P \mathrel{\text{⸦}} \overrightarrow{P_*}\text{'}y . \supset : xPy . yP_*x . \equiv . xPy . xP_*y :$

$yPx . yP_*x . \equiv . yPx . xP_*y :$

[*90·151.*91·52] $\supset : xPy . \supset . xP_{\text{po}}x : yPx . \supset . yP_{\text{po}}y$ (2)

$\vdash . (2) . \supset \vdash : \text{Hp} . x, y \in C\text{'}P . P \mathrel{\text{⸦}} \overrightarrow{P_*}\text{'}x = P \mathrel{\text{⸦}} \overrightarrow{P_*}\text{'}y . \supset . \sim(xPy) . \sim(yPx) .$

[*202·103] $\supset . x = y : \supset \vdash . \text{Prop}$

***202·17.** $\vdash : P_{\text{po}} \in \text{connex} . y \in P(x \mapsto z) . \supset . P(x \mapsto y) \cup P(y \mapsto z) = P(x \mapsto z)$

Dem.

$\vdash . *201\cdot14\cdot15 . *121\cdot103 . \supset$

$\vdash : \text{Hp} . \supset . P(x \mapsto y) \subset P(x \mapsto z) . P(y \mapsto z) \subset P(x \mapsto z)$ (1)

$\vdash . *202\cdot13\cdot137 . *121\cdot103 . \supset$

$\vdash :. \text{Hp} . w \in P(x \mapsto z) . \supset : wP_*y . \vee . yP_*w : xP_*w . wP_*z :$

[*121·103] $\supset : w \in P(x \mapsto y) \cup P(y \mapsto z)$ (2)

$\vdash . (1) . (2) . \supset \vdash . \text{Prop}$

***202·171.** $\vdash : P_{\text{po}} \epsilon \text{ connex} \,.\, y \epsilon P(x \vdash\!\dashv z) \,.\, \supset .$

$$P(x \dashv z) = P(x \dashv y) \cup P(y \dashv z) \,.\, P(x \vdash z) = P(x \vdash y) \cup P(y \vdash z)$$

[Proof as in *202·17]

***202·172.** $\vdash : P_{\text{po}} \epsilon \text{ connex} \,.\, y \epsilon P(x - z) \,.\, \supset .$

$$P(x - y) = P(x \dashv y) \cup P(y - z) = P(x - y) \cup P(y \vdash z)$$

[Proof as in *202·17]

***202·18.** $\vdash : P_{\text{po}} \epsilon \text{ connex} \,.\, E\,!\,B\text{‘}P \,.\, \supset .\, C\text{‘}P = \overleftarrow{P_*}\text{‘}B\text{‘}P$

Dem.

$$\vdash .\, *202\cdot1 \,.\, \supset \vdash : \text{Hp} \,.\, \supset .\, C\text{‘}P = \overleftrightarrow{P_{\text{po}}}\text{‘}B\text{‘}P$$

$$[*97\cdot2 . *91\cdot504] \qquad = \overleftarrow{P_*}\text{‘}B\text{‘}P : \supset \vdash .\, \text{Prop}$$

***202·181.** $\vdash : P_{\text{po}} \epsilon \text{ connex} \,.\, E\,!\,B\text{‘}P \,.\, E\,!\,B\text{‘}\breve{P} \,.\, \supset .\, C\text{‘}P = P(B\text{‘}P \vdash\!\dashv B\text{‘}\breve{P})$

Dem.

$$\vdash .\, *202\cdot18 \,.\, \supset \vdash : \text{Hp} \,.\, \supset .\, C\text{‘}P = \overleftarrow{P_*}\text{‘}B\text{‘}P \cap \overrightarrow{P_*}\text{‘}B\text{‘}\breve{P}$$

$$[*121\cdot103] \qquad = P(B\text{‘}P \vdash\!\dashv B\text{‘}\breve{P}) : \supset \vdash .\, \text{Prop}$$

The above proposition is used in the ordinal theory of finite and infinite (*261·2).

The following proposition is a lemma for *202·211, which shows that if a relation is connected, so are all similar relations.

***202·21.** $\vdash : P \epsilon \text{ connex} \,.\, S \epsilon 1 \rightarrow \text{Cls} \,.\, \supset .\, S ; P \epsilon \text{ connex}$

Dem.

$$\vdash .\, *150\cdot202 \,.\, \supset \vdash :: \text{Hp} \,.\, \supset :.\, x, y \epsilon C\text{‘}S;P \,.\, x \neq y \,.\, \supset : x, y \epsilon S\text{“}C\text{‘}P \,.\, x \neq y :$$

$$[*71\cdot4 . *30\cdot37] \supset : (\exists z, w) \,.\, z, w \epsilon C\text{‘}P \,.\, x = S\text{‘}z \,.\, y = S\text{‘}w \,.\, z \neq w :$$

$$[*202\cdot104] \quad \supset : (\exists z, w) : x = S\text{‘}z \,.\, y = S\text{‘}w : zPw \,.\, \text{v} \,.\, wPz :$$

$$[*150\cdot4] \quad \supset : x\,(S;P)\,y \,.\, \text{v} \,.\, y\,(S;P)\,x \qquad (1)$$

$$\vdash .\, (1) \,.\, *202\cdot104 \,.\, \supset \vdash .\, \text{Prop}$$

The proofs of the three following propositions proceed like the proofs of the analogous propositions in *200 and *201.

***202·211.** $\vdash : P \epsilon \text{ connex} \,.\, Q \text{ smor } P \,.\, \supset .\, Q \epsilon \text{ connex}$

***202·212.** $\vdash : P \epsilon \text{ connex} \,.\, \supset .\, \text{Nr‘}P \subset \text{connex}$

***202·22.** $\vdash : P \epsilon \text{ connex} \,.\, \equiv .\, N_0\text{r‘}P \subset \text{connex} \,.\, \equiv .\, \exists\,!\, N_0\text{r‘}P \cap \text{connex}$

***202·3.** $\vdash .\, \dot{\Lambda} \epsilon \text{ connex}$

Dem.

$$\vdash .\, *37\cdot29 \,.\, \supset \vdash : P = \dot{\Lambda} \,.\, \supset .\, \overleftrightarrow{P}\text{“}C\text{‘}P = \Lambda \,.$$

$$[*202\cdot102] \qquad \supset .\, P \epsilon \text{ connex} : \supset \vdash .\, \text{Prop}$$

***202·31.** $\vdash . x \downarrow y \in \text{connex}$

Dem.

$\vdash . *55·15 . \supset \vdash :. z, w \in C‘(x \downarrow y) . \supset :$

$z, w \in \iota‘x . \vee . z, w \in \iota‘y . \vee . z \in \iota‘x . w \in \iota‘y . \vee . z \in \iota‘y . w \in \iota‘x :$

[*51·15.*13·172] $\supset : z = w . \vee . z = x . w = y . \vee . z = y . w = x :$

[*55·15] $\supset : z = w . \vee . z (x \downarrow y) w . \vee . w (x \downarrow y) z$ (1)

$\vdash . (1) . *202·103 . \supset \vdash . \text{Prop}$

***202·33.** $\vdash : P \in \text{connex} . \supset . P \upharpoonright \alpha \in \text{connex}$

Dem.

$\vdash . *37·41 . \supset \vdash : x, y \in C‘P \upharpoonright \alpha . \supset . x, y \in \alpha . x, y \in C‘P$ (1)

$\vdash . (1) . *202·103 . \supset$

$\vdash :: \text{Hp} . \supset :. x, y \in C‘P \upharpoonright \alpha . \supset : x, y \in \alpha : xPy . \vee . x = y . \vee . yPx :$

[*36·13] $\supset : x (P \upharpoonright \alpha) y . \vee . x = y . \vee . y (P \upharpoonright \alpha) x$ (2)

$\vdash . (2) . *202·103 . \supset \vdash . \text{Prop}$

The following propositions (*202·4—·42) are concerned with applications of relation-arithmetic.

***202·4.** $\vdash : P, Q \in \text{connex} . \supset . P \dagger Q \in \text{connex}$

Dem.

$\vdash . *160·14 . \supset \vdash :. x, y \in C‘(P \dagger Q) . \equiv :$

$x, y \in C‘P . \vee . x, y \in C‘Q . \vee . x \in C‘P . y \in C‘Q . \vee . x \in C‘Q . y \in C‘P$ (1)

$\vdash . *202·103 . \supset \vdash :: \text{Hp} . \supset :. x, y \in C‘P . \supset : xPy . \vee . x = y . \vee . yPx :$

[*160·1] $\supset : x (P \dagger Q) y . \vee . x = y . \vee . y (P \dagger Q) x$ (2)

Similarly $\vdash :: \text{Hp} . \supset :. x, y \in C‘Q . \supset : x (P \dagger Q) y . \vee . x = y . \vee . y (P \dagger Q) x$ (3)

$\vdash . *160·1 . *35·103 . \supset \vdash : x \in C‘P . y \in C‘Q . \supset . x (P \dagger Q) y$ (4)

$\vdash . *160·1 . *35·103 . \supset \vdash : x \in C‘Q . y \in C‘P . \supset . y (P \dagger Q) x$ (5)

$\vdash . (1) . (2) . (3) . (4) . (5) . \supset$

$\vdash :: \text{Hp} . \supset :. x, y \in C‘(P \dagger Q) . \supset : x (P \dagger Q) y . \vee . x = y . \vee . y (P \dagger Q) x$ (6)

$\vdash . (6) . *202·103 . \supset \vdash . \text{Prop}$

The above proposition illustrates the reasons for defining $P \dagger Q$ as was done in *160. When P and Q are connected, $P \cup Q$ is in general not connected: it is the additional term $C‘P \uparrow C‘Q$ which insures connection.

***202·401.** $\vdash :. C‘P \cap C‘Q = \Lambda . \supset : P \dagger Q \in \text{connex} . \equiv . P, Q \in \text{connex}$

Dem.

$\vdash . *202·33 . \supset \vdash : P \dagger Q \in \text{connex} . \supset . (P \dagger Q) \upharpoonright C‘P, (P \dagger Q) \upharpoonright C‘Q \in \text{connex}$ (1)

$\vdash . (1) . *160·5 . \supset \vdash :. \text{Hp} . \supset : P \dagger Q \in \text{connex} . \supset . P, Q \in \text{connex}$ (2)

$\vdash . (2) . *202·4 . \supset \vdash . \text{Prop}$

$*202{\cdot}41$. $\vdash : P \epsilon \text{connex} . \supset . P \mathbin{+\!\!\!\rightarrow} z \,\epsilon\, \text{connex} . z \mathbin{\leftarrow\!\!\!+} P \epsilon\, \text{connex}$

Dem.

$\vdash . *161{\cdot}14{\cdot}2 . \supset \vdash :. x, y \,\epsilon\, C\text{`}(P \mathbin{+\!\!\!\rightarrow} z) . x \neq y . \supset : x, y \,\epsilon\, (C\text{`}P \cup \iota\text{`}z) . x \neq y :$

$[*51{\cdot}236] \quad \supset : x, y \,\epsilon\, C\text{`}P . x \neq y . \vee . x \,\epsilon\, C\text{`}P . y = z . \vee . y \,\epsilon\, C\text{`}P . x = z \quad (1)$

$\vdash . (1) . *202{\cdot}104 . \supset$

$\vdash :: P \,\epsilon\, \text{connex} . \supset :. x, y \,\epsilon\, C\text{`}(P \mathbin{+\!\!\!\rightarrow} z) . x \neq y . \supset :$

$\qquad xPy . \vee . yPx . \vee . x \,\epsilon\, C\text{`}P . y = z . \vee . y \,\epsilon\, C\text{`}P . x = z :$

$[*161{\cdot}11] \quad \supset : x \,(P \mathbin{+\!\!\!\rightarrow} z)\, y . \vee . y \,(P \mathbin{+\!\!\!\rightarrow} z)\, x :.$

$[*202{\cdot}104] \quad \supset :. P \mathbin{+\!\!\!\rightarrow} z \,\epsilon\, \text{connex} \quad (2)$

Similarly $\vdash : P \,\epsilon\, \text{connex} . \supset . z \mathbin{\leftarrow\!\!\!+} P \,\epsilon\, \text{connex} \quad (3)$

$\vdash . (2) . (3) . \supset \vdash . \text{Prop}$

$*202{\cdot}411$. $\vdash . x \downarrow y \mathbin{+\!\!\!\rightarrow} z \,\epsilon\, \text{connex} \quad [*202{\cdot}41{\cdot}31]$

$*202{\cdot}412$. $\vdash :. z \sim\epsilon\, C\text{`}P . \supset : P \,\epsilon\, \text{connex} . \equiv . P \mathbin{+\!\!\!\rightarrow} z \,\epsilon\, \text{connex} . \equiv . z \mathbin{\leftarrow\!\!\!+} P \,\epsilon\, \text{connex}$

Dem.

$\vdash . *161{\cdot}16 . \supset \vdash :. \text{Hp} . \supset : P = (P \mathbin{+\!\!\!\rightarrow} z) \upharpoonright C\text{`}P :$

$[*202{\cdot}33] \quad \supset : P \mathbin{+\!\!\!\rightarrow} z \,\epsilon\, \text{connex} . \supset . P \,\epsilon\, \text{connex} \quad (1)$

Similarly $\vdash :. \text{Hp} . \supset . z \mathbin{\leftarrow\!\!\!+} P \,\epsilon\, \text{connex} . \supset . P \,\epsilon\, \text{connex} \quad (2)$

$\vdash . (1) . (2) . *202{\cdot}41 . \supset \vdash . \text{Prop}$

$*202{\cdot}42$. $\vdash : P \,\epsilon\, \text{connex} . C\text{`}P \subset \text{connex} . \supset . \Sigma\text{`}P \,\epsilon\, \text{connex}$

Dem.

$\vdash . *162{\cdot}22 . \supset \vdash : x, y \,\epsilon\, C\text{`}\Sigma\text{`}P . \equiv . (\exists Q, R) . Q, R \,\epsilon\, C\text{`}P . x \,\epsilon\, C\text{`}Q . y \,\epsilon\, C\text{`}R \quad (1)$

$\vdash . (1) . *202{\cdot}103 . \supset$

$\vdash :: P \,\epsilon\, \text{connex} . \supset :. x, y \,\epsilon\, C\text{`}\Sigma\text{`}P . \supset :$

$\qquad (\exists Q, R) : QPR . \vee . Q = R . Q, R \,\epsilon\, C\text{`}P . \vee . RPQ : x \,\epsilon\, C\text{`}Q . y \,\epsilon\, C\text{`}R \quad (2)$

$\vdash . *162{\cdot}13 . \supset \vdash :. QPR . \vee . RPQ : x \,\epsilon\, C\text{`}Q . y \,\epsilon\, C\text{`}R : \supset :$

$\qquad x \,(\Sigma\text{`}P)\, y . \vee . y \,(\Sigma\text{`}P)\, x \quad (3)$

$\vdash . *13{\cdot}195 . \supset \vdash : (\exists Q, R) . Q = R . Q, R \,\epsilon\, C\text{`}P . x \,\epsilon\, C\text{`}Q . y \,\epsilon\, C\text{`}R . \supset .$

$\qquad (\exists Q) . Q \,\epsilon\, C\text{`}P . x, y \,\epsilon\, C\text{`}Q \quad (4)$

$\vdash . *202{\cdot}103 . \supset \vdash :: C\text{`}P \subset \text{connex} . \supset :. (\exists Q) . Q \,\epsilon\, C\text{`}P . x, y \,\epsilon\, C\text{`}Q . \supset :$

$\qquad (\exists Q) : Q \,\epsilon\, C\text{`}P : xQy . \vee . x = y . \vee . yQx :$

$[*162{\cdot}13] \quad \supset : x \,(\Sigma\text{`}P)\, y . \vee . x = y . \vee . y \,(\Sigma\text{`}P)\, x \quad (5)$

$\vdash . (4) . (5) . \supset$

$\vdash :: C\text{`}P \subset \text{connex} . \supset :. (\exists Q, R) . Q = R . Q, R \,\epsilon\, C\text{`}P . x \,\epsilon\, C\text{`}Q . y \,\epsilon\, C\text{`}R . \supset :$

$\qquad x \,(\Sigma\text{`}P)\, y . \vee . x = y . \vee . y \,(\Sigma\text{`}P)\, x \quad (6)$

$\vdash . (2) . (3) . (6) . \supset \vdash :: \text{Hp} . \supset :.$

$\qquad x, y \,\epsilon\, C\text{`}\Sigma\text{`}P . \supset : x \,(\Sigma\text{`}P)\, y . \vee . x = y . \vee . y \,(\Sigma\text{`}P)\, x \quad (7)$

$\vdash . (7) . *202{\cdot}103 . \supset \vdash . \text{Prop}$

$*202 \cdot 5$. $\vdash :. P \epsilon \text{connex} . P^2 \Subset J . x, y \epsilon C\text{'}P . \supset : x \neq y . \sim(xPy) . \equiv . yPx$

Dem.

$\vdash . *50 \cdot 43 . \quad \supset \vdash :. P^2 \Subset J . \supset : yPx . \supset . \sim(xPy)$ (1)

$\vdash . *200 \cdot 36 . \quad \supset \vdash :. P^2 \Subset J . \supset : yPx . \supset . x \neq y$ (2)

$\vdash . *202 \cdot 104 . \supset \vdash :. P \epsilon \text{connex} . x, y \epsilon C\text{'}P . \supset : x \neq y . \sim(xPy) . \supset . yPx$ (3)

$\vdash . (1) . (2) . (3) . \supset \vdash . \text{Prop}$

The following propositions ($*202 \cdot 501$—$\cdot 51$) are concerned with the relations of $P\text{''}\alpha$ and $p\text{'}\overleftarrow{P}\text{''}(\alpha \cap C\text{'}P)$. They are important, and $*202 \cdot 501 \cdot 503 \cdot 505$ will be often used.

$*202 \cdot 501$. $\vdash : P \epsilon \text{connex} . \supset . C\text{'}P - \alpha - P\text{''}\alpha \subset p\text{'}\overleftarrow{P}\text{''}(\alpha \cap C\text{'}P)$

Dem.

$\vdash . *13 \cdot 14 . *37 \cdot 1 . \supset \vdash :. y \epsilon C\text{'}P - \alpha - P\text{''}\alpha . x \epsilon \alpha . \supset . x \neq y . \sim(yPx)$ (1)

$\vdash . (1) . *202 \cdot 103 . \supset \vdash :. \text{Hp} . \supset : y \epsilon C\text{'}P - \alpha - P\text{''}\alpha . x \epsilon \alpha \cap C\text{'}P . \supset . xPy :$

$[*40 \cdot 53] \supset : y \epsilon C\text{'}P - \alpha - P\text{''}\alpha . \supset . y \epsilon p\text{'}\overleftarrow{P}\text{''}(\alpha \cap C\text{'}P) :. \supset \vdash . \text{Prop}$

$*202 \cdot 502$. $\vdash : P \epsilon \text{connex} . P^2 \Subset J . \exists ! \alpha \cap C\text{'}P . \supset . C\text{'}P - \alpha - P\text{''}\alpha = p\text{'}\overleftarrow{P}\text{''}(\alpha \cap C\text{'}P)$

Dem.

$\vdash . *40 \cdot 62 . \quad \supset \vdash : \text{Hp} . \supset . p\text{'}\overleftarrow{P}\text{''}(\alpha \cap C\text{'}P) \subset C\text{'}P$ (1)

$\vdash . *200 \cdot 5 . \quad \supset \vdash : \text{Hp} . \supset . p\text{'}\overleftarrow{P}\text{''}(\alpha \cap C\text{'}P) \subset -\alpha$ (2)

$\vdash . *200 \cdot 53 . \supset \vdash : \text{Hp} . \supset . p\text{'}\overleftarrow{P}\text{''}(\alpha \cap C\text{'}P) \subset - P\text{''}\alpha$ (3)

$\vdash . (1) . (2) . (3) . *202 \cdot 501 . \supset \vdash . \text{Prop}$

$*202 \cdot 503$. $\vdash : P \epsilon \text{connex} . P^2 \Subset J . \supset . C\text{'}P - p\text{'}\overleftarrow{P}\text{''}(\alpha \cap C\text{'}P) = (\alpha \cap C\text{'}P) \cup P\text{''}\alpha$

Dem.

$\vdash . *202 \cdot 501 . *24 \cdot 43 . \supset \vdash : \text{Hp} . \supset . C\text{'}P - p\text{'}\overleftarrow{P}\text{''}(\alpha \cap C\text{'}P) \subset \alpha \cup P\text{''}\alpha$ (1)

$\vdash . (1) . *22 \cdot 43 . \quad \supset \vdash : \text{Hp} . \supset . C\text{'}P - p\text{'}\overleftarrow{P}\text{''}(\alpha \cap C\text{'}P) \subset (\alpha \cup P\text{''}\alpha) \cap C\text{'}P$

$[*22 \cdot 68 . *37 \cdot 15] \quad \subset (\alpha \cap C\text{'}P) \cup P\text{''}\alpha$ (2)

$\vdash . *200 \cdot 5 \cdot 36 . \quad \supset \vdash : \text{Hp} . \supset . \alpha \cap C\text{'}P \subset - p\text{'}\overleftarrow{P}\text{''}(\alpha \cap C\text{'}P)$ (3)

$\vdash . *200 \cdot 53 . \quad \supset \vdash : \text{Hp} . \supset . P\text{''}\alpha \subset - p\text{'}\overleftarrow{P}\text{''}(\alpha \cap C\text{'}P)$ (4)

$\vdash . *22 \cdot 43 . *37 \cdot 15 . \quad \supset \vdash . \alpha \cap C\text{'}P \subset C\text{'}P . P\text{''}\alpha \subset C\text{'}P$ (5)

$\vdash . (3) . (4) . (5) . \supset \vdash : \text{Hp} . \supset . (\alpha \cap C\text{'}P) \cup P\text{''}\alpha \subset C\text{'}P - p\text{'}\overleftarrow{P}\text{''}(\alpha \cap C\text{'}P)$ (6)

$\vdash . (2) . (6) . \supset \vdash . \text{Prop}$

$*202 \cdot 504$. $\vdash : P \epsilon \text{connex} . P^2 \Subset J . \supset . C\text{'}P \cap p\text{'}\overleftarrow{P}\text{''}(\alpha \cap C\text{'}P) = C\text{'}P - \alpha - P\text{''}\alpha$

Dem.

$\vdash . *200 \cdot 5 \cdot 36 . \quad \supset \vdash : \text{Hp} . \supset . p\text{'}\overleftarrow{P}\text{''}(\alpha \cap C\text{'}P) \subset -\alpha$ (1)

$\vdash . *200 \cdot 53 . \quad \supset \vdash : \text{Hp} . \supset . p\text{'}\overleftarrow{P}\text{''}(\alpha \cap C\text{'}P) \subset - P\text{''}\alpha$ (2)

$\vdash . (1) . (2) . *22 \cdot 48 . \supset \vdash : \text{Hp} . \supset . C\text{'}P \cap p\text{'}\overleftarrow{P}\text{''}(\alpha \cap C\text{'}P) \subset C\text{'}P - \alpha - P\text{''}\alpha$ (3)

$\vdash . (3) . *202 \cdot 501 . \supset \vdash . \text{Prop}$

***202·505.** $\vdash : P \in \text{connex} . \supset . C'P = P''\alpha \cup (\alpha \cap C'P) \cup \{C'P \cap p'\overleftarrow{P}''(\alpha \cap C'P)\}$

Dem.

$\vdash . *202\cdot501 . \supset \vdash : \text{Hp} . \supset . C'P - \alpha - P''\alpha \subset p'\overleftarrow{P}''(\alpha \cap C'P) .$

$[*24\cdot43] \quad \supset . C'P \subset \alpha \cup P''\alpha \cup \{p'\overleftarrow{P}''(\alpha \cap C'P)\} .$

$[*22\cdot621 . *37\cdot15] \supset . C'P = (\alpha \cap C'P) \cup P''\alpha \cup \{C'P \cap p'\overleftarrow{P}''(\alpha \cap C'P)\} : \supset \vdash . \text{Prop}$

***202·51.** $\vdash : P \in \text{connex} . \alpha \subset C'P . \exists ! \alpha . \supset .$

$$C'P = P''\alpha \cup \alpha \cup p'\overleftarrow{P}''\alpha = \breve{P}''\alpha \cup \alpha \cup p'\overrightarrow{P}''\alpha$$

Dem.

$\vdash . *40\cdot62 . \quad \supset \vdash : \text{Hp} . \supset . p'\overleftarrow{P}''\alpha \subset C'P \quad (1)$

$\vdash . *22\cdot621 . \quad \supset \vdash : \text{Hp} . \supset . \alpha = \alpha \cap C'P \quad (2)$

$\vdash . (1) . (2) . *202\cdot505 . \supset \vdash : \text{Hp} . \supset . C'P = P''\alpha \cup \alpha \cup p'\overleftarrow{P}''\alpha \quad (3)$

$\vdash . (3) \dfrac{\breve{P}}{P} . *202\cdot11 . \quad \supset \vdash : \text{Hp} . \supset . C'P = \breve{P}''\alpha \cup \alpha \cup p'\overrightarrow{P}''\alpha \quad (4)$

$\vdash . (3) . (4) . \supset \vdash . \text{Prop}$

The following propositions (*202·511—·524) are concerned with $\overrightarrow{B}'P$. *202·52 shows that if $P \in$ connex, P cannot have more than one first term or more than one last term, and *202·523 shows that this still holds if only P_* is connected. *202·511 shows that if P is a connected relation which has a first term, then if α is any class, there are predecessors of the whole of $\alpha \cap C'P$ when and only when $B'P$ is such a predecessor, and when and only when $B'P \sim \in \alpha$. *202·524 shows that if P is connected and has a first term, $\text{Ɑ}'P$ consists of the successors of the first term. These propositions are much used.

***202·511.** $\vdash :. P \in \text{connex} . \text{E} ! B'P . \supset :$

$$\exists ! p'\overrightarrow{P}''(\alpha \cap C'P) . \equiv . B'P \sim \in \alpha . \equiv . B'P \in p'\overrightarrow{P}''(\alpha \cap C'P)$$

Dem.

$\vdash . *202\cdot104 . *93\cdot1 . \supset \vdash :. \text{Hp} . B'P \sim \in \alpha . \supset : x \in (\alpha \cap C'P) . \supset_x . (B'P) P x :$

$[*40\cdot51] \quad \supset : B'P \in p'\overrightarrow{P}''(\alpha \cap C'P) : \quad (1)$

$[*10\cdot24] \quad \supset : \exists ! p'\overrightarrow{P}''(\alpha \cap C'P) \quad (2)$

$\vdash . *93\cdot1 . \supset \vdash : \text{Hp} . B'P \in \alpha . \supset . (x) . \sim \{x P (B'P)\} . B'P \in \alpha \cap C'P .$

$[*40\cdot51] \quad \supset . p'\overrightarrow{P}''(\alpha \cap C'P) = \Lambda . \quad (3)$

$[*24\cdot105] \quad \supset . B'P \sim \in p'\overrightarrow{P}''(\alpha \cap C'P) \quad (4)$

$\vdash . (2) . (3) . \supset \vdash :. \text{Hp} . \supset : B'P \sim \in \alpha . \equiv . \exists ! p'\overrightarrow{P}''(\alpha \cap C'P) \quad (5)$

$\vdash . (1) . (4) . \supset \vdash :. \text{Hp} . \supset : B'P \sim \in \alpha . \equiv . B'P \in p'\overrightarrow{P}''(\alpha \cap C'P) \quad (6)$

$\vdash . (5) . (6) . \supset \vdash . \text{Prop}$

***202·52.** $\vdash : P \in \text{connex} . \supset . \overrightarrow{B}\text{‘}P, \overrightarrow{B}\text{‘}\breve{P} \in 0 \cup 1$

Dem.

$\vdash . *93\cdot103 . \quad \supset \vdash : x, y \in \overrightarrow{B}\text{‘}P . \supset . x, y \in C\text{‘}P . x \sim\in \text{Ɑ‘}P . y \sim\in \text{Ɑ‘}P .$

$[*33\cdot14] \quad \supset . x, y \in C\text{‘}P . \sim(xPy) . \sim(yPx) \quad (1)$

$\vdash . (1) . *202\cdot103 . \supset \vdash :. \text{Hp} . \supset : x, y \in \overrightarrow{B}\text{‘}P . \supset . x = y :$

$[*52\cdot4] \quad \supset : \overrightarrow{B}\text{‘}P \in 0 \cup 1 \quad (2)$

$\vdash . (2) . *202\cdot11 . \supset \vdash : \text{Hp} . \supset . \overrightarrow{B}\text{‘}\breve{P} \in 0 \cup 1 \quad (3)$

$\vdash . (2) . (3) . \supset \vdash . \text{Prop}$

***202·521.** $\vdash : P_* \in \text{connex} . \supset . \overrightarrow{B}\text{‘}P \subset p\text{‘}\overrightarrow{P_*}\text{“}C\text{‘}P$

Dem.

$\vdash . *202\cdot13\cdot103 . \supset$

$\vdash :: \text{Hp} . \supset :. x \in \overrightarrow{B}\text{‘}P . y \in C\text{‘}P . \supset : xP_{\text{po}}y . \vee . x = y . \vee . yP_{\text{po}}x \quad (1)$

$\vdash . *91\cdot504 . \supset \vdash : x \in \overrightarrow{B}\text{‘}P . \supset . \sim(yP_{\text{po}}x) \quad (2)$

$\vdash . (1) . (2) . \supset \vdash :: \text{Hp} . \supset :. x \in \overrightarrow{B}\text{‘}P . y \in C\text{‘}P . \supset : xP_{\text{po}}y . \vee . x = y :$

$[*91\cdot54] \quad \supset : xP_* y :: \supset \vdash . \text{Prop}$

***202·522.** $\vdash . \overrightarrow{B}\text{‘}P = \overrightarrow{B}\text{‘}P_{\text{po}} \quad [*91\cdot504]$

***202·523.** $\vdash : P_* \in \text{connex} . \supset . \overrightarrow{B}\text{‘}P \in 0 \cup 1 \quad [*202\cdot13\cdot52\cdot522]$

***202·524.** $\vdash : P \in \text{connex} . \exists ! \overrightarrow{B}\text{‘}P . \supset . \text{Ɑ‘}P = \overleftarrow{P}\text{‘}B\text{‘}P$

Dem.

$\vdash . *202\cdot52 . \supset \vdash :. \text{Hp} . \supset : E ! B\text{‘}P :$

$[*202\cdot104 . *93\cdot103] \quad \supset : x \in \text{Ɑ‘}P . \supset . (B\text{‘}P) Px \quad (1)$

$\vdash . (1) . *33\cdot151 . \supset \vdash . \text{Prop}$

The following propositions (*202·53—·55) are concerned with relations with limited fields. Such relations are constantly used in the theory of series.

***202·53.** $\vdash : Q \in \text{connex} . P^2 \subseteq J . Q \subseteq P . \supset . Q = P \upharpoonright C\text{‘}Q$

Dem.

$\vdash . *33\cdot17 . *36\cdot13 . \quad \supset \vdash :. \text{Hp} . \supset : xQy . \supset . x (P \upharpoonright C\text{‘}Q) y \quad (1)$

$\vdash . *50\cdot43 . \quad \supset \vdash :. \text{Hp} . \supset : xPy . \supset . \sim(yPx) .$

$[*23\cdot81] \quad \supset . \sim(yQx) . \quad (2)$

$\vdash . *200\cdot36 . \quad \supset \vdash :. \text{Hp} . \supset : xPy . \supset . x \neq y \quad (3)$

$\vdash . (2) . (3) . *202\cdot104 . \supset \vdash :. \text{Hp} . \supset : x, y \in C\text{‘}Q . xPy . \supset . xQy :$

$[*36\cdot13] \quad \supset : x (P \upharpoonright C\text{‘}Q) y . \supset . xQy \quad (4)$

$\vdash . (1) . (4) . \supset \vdash . \text{Prop}$

This proposition is important in series. If P and Q are serial relations, and $Q \subseteq P$, they verify the above hypothesis; hence if Q is a series contained in a given series P, Q is simply P with its field limited. Thus series contained in a given series are completely determined by their fields.

***202·54.** $\vdash : P \sqsubset \alpha \in \text{connex} . \alpha \cap C'P \sim\in 1 . \supset . C'P \sqsubset \alpha = \alpha \cap C'P$

Dem.

$\vdash . *52\cdot181 . \supset$

$\vdash :: \text{Hp} . \supset :. x \in \alpha \cap C'P . \supset_x : (\exists y) . y \in \alpha \cap C'P . y \neq x :$

$[*202\cdot104] \quad \supset_x : (\exists y) : y \in \alpha \cap C'P : xPy . \mathsf{v} . yPx :$

$[*36\cdot13] \quad \supset_x : (\exists y) : x (P \sqsubset \alpha) y . \mathsf{v} . y (P \sqsubset \alpha) x :$

$[*33\cdot132] \quad \supset_x : x \in C'P \sqsubset \alpha \quad (1)$

$\vdash . *37\cdot41\cdot15\cdot16 . \supset \vdash . C'P \sqsubset \alpha \subset \alpha \cap C'P \quad (2)$

$\vdash . (1) . (2) . \supset \vdash . \text{Prop}$

The above proposition is frequently used. *202·55, which is an immediate consequence of *202·54, is used incessantly.

The following proposition is used in *232·14.

***202·541.** $\vdash : P \in \text{trans} \cap \text{connex} . \alpha \cap C'P \sim\in 1 . \supset . (P \sqsubset \alpha)_* = P_* \sqsubset \alpha$

Dem.

$\vdash . *201\cdot18\cdot33 . \supset \vdash : \text{Hp} . \supset . (P \sqsubset \alpha)_* = P \sqsubset \alpha \cup I \upharpoonright (C'P \sqsubset \alpha)$

$[*202\cdot54] \quad = P \sqsubset \alpha \cup I \upharpoonright (C'P \cap \alpha)$

$[*201\cdot18 . *36\cdot23 . *50\cdot5] \quad = P_* \sqsubset \alpha$

***202·55.** $\vdash : P \sqsubset \alpha \in \text{connex} . \alpha \subset C'P . \alpha \sim\in 1 . \supset . C'P \sqsubset \alpha = \alpha \quad [*202\cdot54]$

***202·56.** $\vdash . P \in \text{connex} . P \Subset J . x \in C'P . \beta \subset C'P . P``\beta \subset \overrightarrow{P}'x . \supset . \beta \subset \overrightarrow{P}'x \cup \iota'x$

Dem.

$\vdash . *37\cdot1 . \quad \supset \vdash : P``\beta \subset \overrightarrow{P}'x . y \in \beta . xPy . \supset . xPx \quad (1)$

$\vdash . (1) . \text{Transp} . \supset \vdash : \text{Hp} . y \in \beta . \supset . \sim(xPy) \quad (2)$

$\vdash . (2) . *32\cdot18 . \supset \vdash : \text{Hp} . y \in \beta - \overrightarrow{P}'x . \supset . \sim(xPy) . \sim(yPx) .$

$[*202\cdot103] \quad \supset . y = x : \supset \vdash . \text{Prop}$

The above proposition is used in *212·652.

***202·6.** $\vdash :: P \in \text{connex} . P \Subset J . \supset :. x, y \in C'P . x \neq y . \equiv : xPy . \mathsf{v} . yPx$

Dem.

$\vdash . *202\cdot104 . \quad \supset \vdash :: \text{Hp} . \supset :. x, y \in C'P . x \neq y . \supset : xPy . \mathsf{v} . yPx \quad (1)$

$\vdash . *50\cdot11 . *33\cdot17 . \supset \vdash :: \text{Hp} . \supset :. xPy . \mathsf{v} . yPx : \supset . x, y \in C'P . x \neq y \quad (2)$

$\vdash . (1) . (2) . \supset \vdash . \text{Prop}$

The following proposition is a lemma for *202·62, which is itself a lemma for *204·52.

***202·61.** $\vdash :: P \in \text{connex} . P \Subset J : \phi(x, y) . \equiv_{x,y} . \phi(y, x) : \supset :.$
$xPy . \supset_{x,y} . \phi(x, y) : \equiv : x, y \in C'P . x \neq y . \supset_{x,y} . \phi(x, y)$

Dem.

$\vdash . *202\cdot6 . \supset \vdash ::. \text{Hp} . \supset :: x, y \in C'P . x \neq y . \supset_{x,y} . \phi(x, y) : \equiv :.$
$xPy . \mathsf{v} . yPx : \supset_{x,y} . \phi(x, y) :.$

$[*4\cdot77] \quad \equiv :. xPy . \supset_{x,y} . \phi(x, y) : yPx . \supset_{x,y} . \phi(x, y) :.$

$[*4\cdot85 . \text{Hp}] \quad \equiv :. xPy . \supset_{x,y} . \phi(x, y) : yPx . \supset_{x,y} . \phi(y, x) :.$

$[*4\cdot24] \quad \equiv :. xPy . \supset_{x,y} . \phi(x, y) ::. \supset \vdash . \text{Prop}$

***202·611.** $\vdash :. P \epsilon \text{connex} . P \subseteq J . R = \breve{R} . \supset : P \subseteq R . \equiv . J \upharpoonright C'P \subseteq R$

$\left[*202\cdot61 \dfrac{xRy}{\phi(x, y)}\right]$

***202·62.** $\vdash :. P \epsilon \text{connex} . P \subseteq J . \supset : P \epsilon \text{Rel}^2 \text{excl} . \equiv . F ; P \subseteq J$

Dem.

$\vdash . *202\cdot61 . *163\cdot1 . \supset \vdash :: \text{Hp} . \supset :.$

$P \epsilon \text{Rel}^2 \text{excl} . \equiv : QPR . \supset_{Q,R} . C'Q \cap C'R = \Lambda :$

$[*24\cdot37] \quad \equiv : QPR . x \epsilon C'Q . y \epsilon C'R . \supset_{Q,R,x,y} . x \neq y :$

$[*150\cdot52] \quad \equiv . x (F;P) y . \supset_{x,y} . x \neq y :: \supset \vdash . \text{Prop}$

The three following propositions (*202·7—·72) are concerned with $P \dot{-} P^2$. Of these, *202·7 is important: it shows that if P is connected, no term can have more than one immediate predecessor or successor. *202·72 is used in *204·71, which is an important proposition.

***202·7.** $\vdash : P \epsilon \text{connex} . \supset . P \dot{-} P^2 \epsilon 1 \rightarrow 1$

Dem.

$\vdash . *34\cdot5 . \text{Transp} . \supset \vdash : zPx . \sim(yP^2x) . \supset . \sim(yPz)$ (1)

Similarly $\vdash : yPx . \sim(zP^2x) . \supset . \sim(zPy)$ (2)

$\vdash . (1) . (2) . \supset \vdash : y (P \dot{-} P^2) x . z (P \dot{-} P^2) x . \supset . \sim(yPz) . \sim(zPy)$ (3)

$\vdash . (3) . *202\cdot103 . \supset \vdash :. \text{Hp} . \supset : y (P \dot{-} P^2) x . z (P \dot{-} P^2) x . \supset . y = z$ (4)

Similarly $\vdash :. \text{Hp} . \supset : x (P \dot{-} P^2) y . x (P \dot{-} P^2) z . \supset . y = z$ (5)

$\vdash . (4) . (5) . \supset \vdash . \text{Prop}$

***202·71.** $\vdash : P \epsilon \text{connex} . x (P \dot{-} P^2) y . \supset . \overrightarrow{P_{po}}'y = \overrightarrow{P_*}'x$

Dem.

$\vdash . *91\cdot52 . \supset \vdash : \text{Hp} . \supset . \overrightarrow{P_*}'x \subset \overrightarrow{P_{po}}'y$ (1)

$\vdash . *91\cdot57 . \supset \vdash :. zP_{po}y . \supset : zPy . \vee . zP_{po} | Py :$

$[*25\cdot41] \quad \supset : z (P \dot{-} P^2) y . \vee . z (P \dot{\wedge} P^2) y . \vee . z (P_{po} | P) y :$

$[*91\cdot502] \quad \supset : z (P \dot{-} P^2) y . \vee . z (P_{po} | P) y$ (2)

$\vdash . *202\cdot7 . \supset \vdash : \text{Hp} . z (P \dot{-} P^2) y . \supset . z = x$ (3)

$\vdash . (2) . (3) . \supset \vdash : \text{Hp} . zP_{po}y . z \neq x . \supset . z (P_{po} | P) y .$

$[*34\cdot1] \quad \supset . (\exists w) . zP_{po}w . wPy$ (4)

$\vdash . *34\cdot5 . \quad \supset \vdash : wPy . xPw . \supset . xP^2y$ (5)

$\vdash . (5) . \text{Transp} . \supset \vdash :. \text{Hp} . wPy . \supset : \sim(xPw) :$

$[*202\cdot103] \quad \supset : wPx . \vee . w = x$ (6)

$\vdash . (4) . (6) . \supset \vdash :. \text{Hp} . zP_{po}y . z \neq x . \supset : zP_{po}x . \vee . (\exists w) . zP_{po}w . wPx :$

$[*91\cdot511] \quad \supset : zP_{po}x$ (7)

$\vdash . (7) . *91\cdot54 . \supset \vdash : \text{Hp} . \supset . \overrightarrow{P_{po}}'y \subset \overrightarrow{P_*}'x$ (8)

$\vdash . (1) . (8) . \supset \vdash . \text{Prop}$

***202·72.** $\vdash : P \in \text{trans} \cap \text{connex} \,.\, x(P \dot{-} P^2)y \,.\supset .\, \overrightarrow{P}\text{'}y = \overrightarrow{P}\text{'}x \cup \iota\text{'}x$
[*202·71 . *201·18·521]

***202·8.** $\vdash : Q \in \text{connex} \,.\, S \in P \,\overline{\text{smor}}\, Q \,.\, C\text{'}Q \cap \beta \sim\in 1 \,.\supset .$
$S \upharpoonright \beta \in (P \upharpoonleft S\text{''}\beta)\,\overline{\text{smor}}\, Q \upharpoonleft \beta$

Dem.

$\vdash . *71\cdot29 . \quad \supset \vdash : \text{Hp} \,.\supset .\, S \upharpoonright \beta \in 1 \to 1$ (1)
$\vdash . *35\cdot64 . *151\cdot11 . \quad \supset \vdash : \text{Hp} \,.\supset .\, \text{Ɑ}\text{'}(S \upharpoonright \beta) = C\text{'}Q \cap \beta$
[*202·54] $\quad = C\text{'}(Q \upharpoonleft \beta)$ (2)
$\vdash . *150\cdot37 . \quad \supset \vdash : \text{Hp} \,.\supset .\, (S \upharpoonright \beta) ; Q = P \upharpoonleft S\text{''}\beta$ (3)
$\vdash . (1) . (2) . (3) . \supset \vdash . \text{Prop}$

***202·81.** $\vdash : Q \in \text{connex} \,.\, S \in P \,\overline{\text{smor}}\, Q \,.\supset .\, (P \upharpoonleft S\text{''}\beta) \text{ smor } Q \upharpoonleft \beta$

Dem.

$\vdash . *202\cdot8 . \quad \supset \vdash : \text{Hp} \,.\, C\text{'}Q \cap \beta \sim\in 1 \,.\supset .\, (P \upharpoonleft S\text{''}\beta) \text{ smor } Q \upharpoonleft \beta$ (1)
$\vdash . *36\cdot13 . *33\cdot17 . \quad \supset \vdash : C\text{'}Q \cap \beta = \iota\text{'}y \,.\supset .\, Q \upharpoonleft \beta \subseteq y \downarrow y$ (2)
$\vdash . *36\cdot13 . \quad \supset \vdash :. \text{Hp}(2) \,.\supset : y(Q \upharpoonleft \beta)y \,.\equiv .\, yQy$ (3)
$\vdash . (2) . (3) . *55\cdot341 . \supset \vdash : \text{Hp}(2) \,.\, yQy \,.\supset .\, Q \upharpoonleft \beta = y \downarrow y$ (4)
$\vdash . *35\cdot64 . *151\cdot11 . \supset \vdash : \text{Hp} \,.\supset .\, \text{Ɑ}\text{'}(S \upharpoonright \beta) = C\text{'}Q \cap \beta$ (5)
$\vdash . (4) . (5) . \quad \supset \vdash : \text{Hp}(4) \,.\supset .\, \text{Ɑ}\text{'}(S \upharpoonright \beta) = C\text{'}(Q \upharpoonleft \beta)$ (6)
$\vdash . *71\cdot29 . *150\cdot37 . \supset \vdash : \text{Hp} \,.\supset .\, S \upharpoonright \beta \in 1 \to 1 \,.\, P \upharpoonleft S\text{''}\beta = S ; (Q \upharpoonleft \beta)$ (7)
$\vdash . (6) . (7) . *151\cdot1 . \supset \vdash : \text{Hp}(4) \,.\supset .\, (P \upharpoonleft S\text{''}\beta) \text{ smor } Q \upharpoonleft \beta$ (8)
$\vdash . (2) . (3) . *55\cdot341 . \supset \vdash : \text{Hp}(2) \,.\, \sim(yQy) \,.\supset .\, Q \upharpoonleft \beta = \dot{\Lambda} \,.$ (9)
[(7).*150·42] $\quad \supset .\, P \upharpoonleft S\text{''}\beta = \dot{\Lambda}$ (10)
$\vdash . (9) . (10) . *153\cdot101 . \supset \vdash : \text{Hp}(9) \,.\supset .\, (P \upharpoonleft S\text{''}\beta) \text{ smor } Q \upharpoonleft \beta$ (11)
$\vdash . (8) . (11) . *52\cdot1 . \quad \supset \vdash : \text{Hp} \,.\, C\text{'}Q \cap \beta \in 1 \,.\supset .\, (P \upharpoonleft S\text{''}\beta) \text{ smor } Q \upharpoonleft \beta$ (12)
$\vdash . (1) . (12) . \supset \vdash . \text{Prop}$

The above proposition shows that if Q is connected, and any class β is picked out of $C\text{'}Q$, then Q arranges β in an order which is similar to that in which P arranges the correlates of β.

*204. ELEMENTARY PROPERTIES OF SERIES.

Summary of *204.

In this number we give the definition and a few of the simpler properties of series. Most of the propositions of this number result immediately from those of *200, *201, and *202. Our definition is

$$\text{Ser} = \text{Rl}‘J \cap \text{trans} \cap \text{connex} \quad \text{Df.}$$

We have

***204·16.** $\vdash : P \in \text{Ser} . \equiv . P \in \text{connex} . P^2 \subseteq J . P^3 \subseteq J . \equiv . P \in \text{connex} . P_{\text{po}} \subseteq J$

either of which might have been taken as the definition.

After a few propositions giving other possible forms of the definition of series, we proceed to a set of propositions which follow immediately from those of *200, *201, and *202. Such are

***204·2.** $\vdash : P \in \text{Ser} . \equiv . \breve{P} \in \text{Ser}$

***204·21.** $\vdash : P \in \text{Ser} . P \text{ smor } Q . \supset . Q \in \text{Ser}$

***204·24.** $\vdash . \dot{\Lambda} \in \text{Ser}$

***204·25.** $\vdash : x \neq y . \equiv . x \downarrow y \in \text{Ser}$

Another important proposition on couples is

***204·272.** $\vdash :. P \in \text{Ser} . \supset : \text{D}‘P \in 1 . \equiv . P \in 2_r . \equiv . \text{ᗡ}‘P \in 1$

so that couples are the only series having unit classes for their domains or converse domains.

We then proceed to a set of propositions on $\overrightarrow{P}‘x$. We have

***204·33.** $\vdash :. P \in \text{Ser} . x, y \in C‘P . \supset : x \neq y . \overrightarrow{P}‘y \subset \overrightarrow{P}‘x . \equiv . yPx$

Also, if $P \in \text{Ser}$, $\overrightarrow{P} \upharpoonright C‘P$ is a one-one and $\overrightarrow{P} ; P \text{ smor } P$ (*204·34·35).

We then have some propositions (*204·4—·44) on relations with limited fields. The most important of these are

***204·4.** $\vdash : P \in \text{Ser} . \supset . P \mathbin{\text{⦍}} \alpha \in \text{Ser}$

***204·41.** $\vdash : P, Q \in \text{Ser} . Q \subseteq P . \supset . Q = P \mathbin{\text{⦍}} C‘Q$

This proposition is important, since it shows that any series contained in a given series is wholly determined when its field is given.

We have next a number of propositions (∗204·45—·59) applying relation-arithmetic to series. The first set of these (∗204·45—·483) are concerned with the proof that if a "cut" is made in a series, the series is the sum of the two parts into which the cut divides it, where the sum is taken in the sense of ∗160 or ∗161, according as one part of the cut does not or does consist of a single term. Most of these propositions do not require the full hypothesis that P is a series, but only some part of it. Thus we have for instance

∗204·46. $\vdash : P \in \text{connex} \,.\, \text{E} ! \, B`P \,.\, \text{ᗡ}`P \sim\in 1 \,.\supset .$
$$P = B`P \twoheadleftarrow P \restriction \text{ᗡ}`P \,.\, \text{Nr}`P = \dot{1} \dotplus \text{Nr}`(P \restriction \text{ᗡ}`P)$$

with a similar proposition for $B`\breve{P}$ and $\text{D}`P$ (∗204·461).

We next prove that if P, Q are mutually exclusive series, their sum $(P \maltese Q)$ is a series, and vice versa (∗204·5); that if P is a series to which x does not belong, $P \twoheadrightarrow x$ and $x \twoheadleftarrow P$ are series, and vice versa (∗204·51); that if P is a series of mutually exclusive series, its sum $\Sigma`P$ is a series (∗204·52); that if $P$, $Q$ are series, so is $P \times Q$ (∗204·55); that if $P$ is a series of series, $\Pi`P$ is contained in diversity and is transitive (∗204·561), while if P is also well-ordered, *i.e.* such that every existent sub-class of $C`P$ has a first term, then $\Pi`P$ is a series (∗204·57); and that if P and Q are series, and Q is well-ordered, then P^Q and $P \exp Q$ are series (∗204·59). These propositions are essential to ordinal arithmetic, but they will not be referred to again until we reach that stage (Sections D and E of this Part).

We have next a collection of propositions (∗204·6—·65) on $p`\overrightarrow{P}``\alpha$ for various values of α, and finally three propositions on P_1. Two of these are much used, namely

∗204·7. $\vdash : P \in \text{Ser} \,.\supset .\, P_1 \in 1 \to 1$

∗204·71. $\vdash : P \in \text{Ser} \,.\, xP_1y \,.\supset .\, \overrightarrow{P}`y = \overrightarrow{P}`x \cup \iota`x$

∗204·01. $\text{Ser} = \text{Rl}`J \cap \text{trans} \cap \text{connex}$ Df

∗204·1. $\vdash : P \in \text{Ser} \,.\equiv .\, P \subseteq J \,.\, P^2 \subseteq P \,.\, P \in \text{connex} \,.$
$$\equiv .\, P \in \text{Rl}`J \,.\, P \in \text{trans} \,.\, P \in \text{connex} \quad [(∗204·01)]$$

∗204·11. $\vdash :. P \in \text{Ser} \,.\equiv : P \subseteq J \,.\, P^2 \subseteq P : x \in C`P \,.\supset_x .\, \overrightarrow{P}`x \cup \iota`x \cup \overleftarrow{P}`x = C`P$
[∗204·1 . ∗202·101]

∗204·12. $\vdash :: P \in \text{Ser} \,.\equiv :. P \subseteq J \,.\, P^2 \subseteq P :. x, y \in C`P \,.\supset_{x,y} :$
$$xPy \,.\vee .\, x = y \,.\vee .\, yPx \quad [∗204·1 \,.\, ∗202·103]$$

∗204·121. $\vdash :: P \in \text{Ser} \,.\equiv :. P \subseteq J \,.\, P^2 \subseteq P :. x, y \in C`P \,.\, x \neq y \,.\supset_{x,y} :$
$$xPy \,.\vee .\, yPx \quad [∗204·1 \,.\, ∗202·104]$$

***204·13.** $\vdash : P \epsilon \text{Ser} . \supset . P^2 \subseteq J . P \dot{\cap} \breve{P} = \dot{\Lambda}$

Dem.

$\vdash . *204·1 . *23·44 . \supset \vdash : P \epsilon \text{Ser} . \supset . P^2 \subseteq J . P \epsilon \text{trans}$ (1)

$\vdash . (1) . *201·12 . \supset \vdash . \text{Prop}$

***204·14.** $\vdash : P \epsilon \text{Ser} . \equiv . P \dot{\cap} \breve{P} = \dot{\Lambda} . P^2 \subseteq P . P \epsilon \text{connex}$
[*204·1 . *50·47]

***204·15.** $\vdash : P \epsilon \text{connex} . P^2 \subseteq J . P^3 \subseteq J . \supset . P \epsilon \text{trans}$

Dem.

$\vdash . *34·5 . \supset \vdash :. P^2 \subseteq J . \supset : xPy . yPz . \supset . x \neq z$ (1)

$\vdash . *50·41 . \supset \vdash :. P^3 \subseteq J . \supset : xPy . yPz . \supset . \sim (zPx)$ (2)

$\vdash . (1) . (2) . \supset \vdash :. \text{Hp} . \supset : xPy . yPz . \supset . x \neq z . \sim (zPx) .$

[*202·103] $\supset . xPz :. \supset \vdash . \text{Prop}$

***204·151.** $\vdash : P \epsilon \text{connex} . P_{po} \subseteq J . \supset . P \epsilon \text{trans}$
[*204·15 . *91·502·503·511]

***204·16.** $\vdash : P \epsilon \text{Ser} . \equiv . P \epsilon \text{connex} . P^2 \subseteq J . P^3 \subseteq J . \equiv . P \epsilon \text{connex} . P_{po} \subseteq J$
[*204·15·151 . *200·36 . *201·18]

We have also

$$\vdash : P \epsilon \text{Ser} . \equiv . P \epsilon \text{connex} . P^6 \subseteq J.$$

For, by *200·37, since $P^6 = (P^2)^3 = (P^3)^2$, it follows that

$$P^6 \subseteq J . \supset . P^2 \subseteq J . P^3 \subseteq J.$$

A relation such as $x \downarrow y \cup y \downarrow z \cup z \downarrow x$, where $x \neq y . y \neq z . z \neq x$, satisfies $P \epsilon \text{connex} . P^2 \subseteq J$, but not $P^3 \subseteq J$. On the other hand,

$$x \downarrow y \cup y \downarrow z \cup z \downarrow w \cup w \downarrow x$$

satisfies $P^2 \subseteq J . P^3 \subseteq J$, but not $P \epsilon \text{connex}$.

***204·2.** $\vdash : P \epsilon \text{Ser} . \equiv . \breve{P} \epsilon \text{Ser}$ [*200·11 . *201·11 . *202·11]

***204·21.** $\vdash : P \epsilon \text{Ser} . P \,\text{smor}\, Q . \supset . Q \epsilon \text{Ser}$
[*200·211 . *201·211 . *202·211]

***204·22.** $\vdash : P \epsilon \text{Ser} . \supset . \text{Nr}`P \subset \text{Ser}$ [*204·21]

***204·23.** $\vdash : P \epsilon \text{Ser} . \equiv . \text{N}_0\text{r}`P \subset \text{Ser} . \equiv . \exists ! \text{N}_0\text{r}`P \cap \text{Ser}$
[*200·22 . *201·22 . *202·22]

***204·24.** $\vdash . \dot{\Lambda} \epsilon \text{Ser}$ [*200·3 . *201·3 . *202·3]

***204·25.** $\vdash : x \neq y . \equiv . x \downarrow y \epsilon \text{Ser}$ [*200·31 . *201·31 . *202·31]

***204·26.** $\vdash : x \neq y . x \neq z . y \neq z . \supset . x \downarrow y \rightarrow z \epsilon \text{Ser}$
[*200·31·41 . *201·411 . *202·411]

The three following propositions deal with couples. Couples often require special treatment, owing to the fact that, if P is a couple, $P \upharpoonright \text{D}`P = \dot{\Lambda}$, so that $C`(P \upharpoonright \text{D}`P) \neq \text{D}`P$, whereas in any other case, if P is a series, $C'(P \upharpoonright D'P) = D'P$. Hence the following propositions are often required.

***204·27.** $\vdash : P \epsilon \text{Ser} . xPy . D'P = \iota'x . \supset . P = x \downarrow y$

Dem.

$\vdash . *33·14 . \quad \supset \vdash : \text{Hp} . zPw . \supset . z = x \quad (1)$

$\vdash . (1) . *50·24 . \supset \vdash : \text{Hp} . zPw . \supset . w \neq x .$

$\left[(1) . \text{Transp} \frac{w, y}{z, w}\right] \quad \supset . \sim(wPy) \quad (2)$

$\vdash . (1) . \text{Transp} . *50·24 . \supset \vdash : \text{Hp} . \supset . \sim(yPw) \quad (3)$

$\vdash . (2) . (3) . *204·12 . \quad \supset \vdash : \text{Hp} . zPw . \supset . y = w \quad (4)$

$\vdash . (1) . (4) . \supset \vdash :. \text{Hp} . \supset : zPw . \supset . z = x . y = w \quad (5)$

$\vdash . (5) . *55·34 . \supset \vdash . \text{Prop}$

***204·271.** $\vdash : P \epsilon \text{Ser} . D'P \epsilon 1 . \supset . P \epsilon 2_r$

Dem.

$\vdash . *204·27 . \supset \vdash : \text{Hp} . \supset . (\exists x, y) . P = x \downarrow y .$

$[*204·25] \quad \supset . (\exists x, y) . x \neq y . P = x \downarrow y .$

$[*56·11] \quad \supset . P \epsilon 2_r : \supset \vdash . \text{Prop}$

***204·272.** $\vdash :. P \epsilon \text{Ser} . \supset : D'P \epsilon 1 . \equiv . P \epsilon 2_r . \equiv . Ɑ'P \epsilon 1$

[*204·271·2 . *56·111]

***204·3.** $\vdash :. P \epsilon \text{Ser} . x, y \epsilon C'P . \supset : x \neq y . \sim(yPx) . \equiv . xPy$

[*202·5 . *204·13]

***204·32.** $\vdash :. P \epsilon \text{Ser} . x, y \epsilon C'P . \supset : \overrightarrow{P}'y \subset \overrightarrow{P}'x . \equiv . y \epsilon \overrightarrow{P}'x \cup \iota'x$

Dem.

$\vdash . *204·1 . \supset \vdash :. \text{Hp} . \supset : yPx . zPy . \supset . zPx :$

$[*32·18] \quad \supset : y \epsilon \overrightarrow{P}'x . \supset . \overrightarrow{P}'y \subset \overrightarrow{P}'x \quad (1)$

$\vdash . *22·42 . \supset \vdash : y = x . \supset . \overrightarrow{P}'y \subset \overrightarrow{P}'x \quad (2)$

$\vdash . (1) . (2) . \supset \vdash :. \text{Hp} . \supset : y \epsilon \overrightarrow{P}'x \cup \iota'x . \supset . \overrightarrow{P}'y \subset \overrightarrow{P}'x \quad (3)$

$\vdash . *204·11 . \supset \vdash :. \text{Hp} . \supset : y \sim\epsilon \overrightarrow{P}'x \cup \iota'x . \supset . y \epsilon \overleftarrow{P}'x .$

$[*32·18·181] \quad \supset . x \epsilon \overrightarrow{P}'y \quad (4)$

$\vdash . *50·24 . \supset \vdash : \text{Hp} . \supset . x \sim\epsilon \overrightarrow{P}'x \quad (5)$

$\vdash . (4) . (5) . \supset \vdash :. \text{Hp} . \supset : y \sim\epsilon \overrightarrow{P}'x \cup \iota'x . \supset . \sim(\overrightarrow{P}'y \subset \overrightarrow{P}'x) \quad (6)$

$\vdash . (3) . (6) . \supset \vdash . \text{Prop}$

***204·33.** $\vdash :. P \epsilon \text{Ser} . x, y \epsilon C'P . \supset : x \neq y . \overrightarrow{P}'y \subset \overrightarrow{P}'x . \equiv . yPx$

Dem.

$\vdash . *204·32 . \supset \vdash :. \text{Hp} . \supset : x \neq y . \overrightarrow{P}'y \subset \overrightarrow{P}'x . \equiv . x \neq y . y \epsilon \overrightarrow{P}'x \cup \iota'x .$

$[*51·15] \quad \equiv . x \neq y . y \epsilon \overrightarrow{P}'x .$

$[\text{Hp} . *4·71] \quad \equiv . yPx :. \supset \vdash . \text{Prop}$

The three following propositions only require $P \in \text{Rl}`J \cap \text{connex}$, but are required for application to series, and are therefore convenient in the form here given.

***204·331.** $\vdash :. P \in \text{Ser} . x, y \in C`P . \supset : \overrightarrow{P}`x = \overrightarrow{P}`y . \equiv . x = y$
[*202·161 . *71·55]

***204·34.** $\vdash : P \in \text{Ser} . \supset . \overrightarrow{P} \upharpoonright C`P \in 1 \rightarrow 1 . \overrightarrow{P} \upharpoonright C`P \in (\overrightarrow{P} ; P) \overline{\text{smor}}\, P$ [*202·161]

***204·35.** $\vdash : P \in \text{Ser} . \supset . \overrightarrow{P} ; P \text{ smor } P$ [*204·34]

This proposition shows that the series of segments which have upper limits is like the original series, for a segment whose upper limit is x is $\overrightarrow{P}`x$, and the series of such segments is $\overrightarrow{P} ; P$.

The following propositions (*204·4—·44) are concerned with relations with limited fields.

***204·4.** $\vdash : P \in \text{Ser} . \supset . P \upharpoonright\!\!\!\!\lfloor \alpha \in \text{Ser}$ [*200·33 . *201·33 . *202·33]

***204·41.** $\vdash : P, Q \in \text{Ser} . Q \subseteq P . \supset . Q = P \upharpoonright\!\!\!\!\lfloor C`Q$ [*202·53 . *204·13]

In virtue of the above two propositions, the series contained in a given series are the relations resulting from limitations of the field; the process of limiting the field is merely the process of selecting a part of the original series without changing the order.

***204·42.** $\vdash :. P \in \text{Ser} . \supset : Q \in \text{Ser} . Q \subseteq P . \equiv . (\exists \alpha) . Q = P \upharpoonright\!\!\!\!\lfloor \alpha . \equiv . Q \in \text{D}`P \upharpoonright\!\!\!\!\lfloor$
[*204·4·41]

***204·421.** $\vdash : P \in \text{Ser} . \supset . \text{Ser} \cap \text{Rl}`P = \text{D}`P \upharpoonright\!\!\!\!\lfloor$ [*204·42]

***204·43.** $\vdash : P^2 \subseteq P . P \subseteq J . Q \subseteq P . Q \in \text{connex} . \supset . Q \in \text{Ser}$

Dem.

$\vdash . *23·1 . *34·55 . \supset \vdash :. \text{Hp} . \supset : xQy . yQz . \supset . xPz .$
$[*50·43.\text{Hp}] \qquad \supset . \sim(zPx) . x \neq z .$
$[*23·81.\text{Hp}] \qquad \supset . \sim(zQx) . x \neq z .$
$[*202·103] \qquad \supset . xQz :$
$[*34·55] \qquad \supset : Q^2 \subseteq Q \qquad (1)$
$\vdash . *23·44 . \supset \vdash : \text{Hp} . \supset . Q \subseteq J \qquad (2)$
$\vdash . (1) . (2) . *204·1 . \supset \vdash . \text{Prop}$

***204·44.** $\vdash : P \in \text{Rl}`J \cap \text{trans} . \supset . \text{Rl}`P \cap \text{connex} \subset \text{Ser}$ [*204·43]

The following propositions (*204·45—·483) are concerned with the division of a series into two parts, one of which wholly precedes the other. The case where one of the parts consists of a single term requires special treatment, and so does the case where both parts consist of single terms, *i.e.* where the series is a couple.

***204·45.** $\vdash : P \epsilon \text{connex} . \alpha \epsilon \text{Cl}`C`P - 1 . P``\alpha \subset \alpha . \beta = C`P - \alpha . \beta \sim\epsilon 1 . \supset .$
$P = P \upharpoonright \alpha \mathbin{+\!\!\!\!\uparrow} P \upharpoonright \beta . \text{Nr}`P = \text{Nr}`P \upharpoonright \alpha \dotplus \text{Nr}`P \upharpoonright \beta$

Dem.

$\vdash . *24{\cdot}411 . *33{\cdot}17 . \supset \vdash :: \text{Hp} . \supset :.$
$xPy . \equiv : y \epsilon \alpha . xPy . \vee . x \epsilon \alpha . y \epsilon \beta . xPy . \vee . x, y \epsilon \beta . xPy$ (1)
$\vdash . *37{\cdot}17 . \qquad \supset \vdash :. \text{Hp} . \supset : y \epsilon \alpha . xPy . \supset . x \epsilon \alpha$ (2)
$\vdash . (2) . \text{Transp} . *202{\cdot}103 . \supset \vdash :. \text{Hp} . \supset : y \epsilon \alpha . x \epsilon \beta . \supset . yPx$ (3)
$\vdash . *202{\cdot}55 . \qquad \supset \vdash : \text{Hp} . \supset . \alpha = C`P \upharpoonright \alpha . \beta = C`P \upharpoonright \beta$ (4)
$\vdash . (1) . (2) . (3) . (4) . \supset \vdash :: \text{Hp} . \supset :.$
$xPy . \equiv : x (P \upharpoonright \alpha) y . \vee . x \epsilon C`P \upharpoonright \alpha . y \epsilon C`P \upharpoonright \beta . \vee . x (P \upharpoonright \beta) y :$
[*160·1] $\qquad \equiv : x \{P \upharpoonright \alpha \mathbin{+\!\!\!\!\uparrow} P \upharpoonright \beta\} y$ (5)
$\vdash . (5) . *180{\cdot}32 . \supset \vdash : \text{Hp} . \supset . \text{Nr}`P = \text{Nr}`P \upharpoonright \alpha \dotplus \text{Nr}`P \upharpoonright \beta$ (6)
$\vdash . (5) . (6) . \supset \vdash . \text{Prop}$

***204·46.** $\vdash : P \epsilon \text{connex} . \text{E} ! B`P . \text{ꓷ}`P \sim\epsilon 1 . \supset .$
$P = B`P \mathbin{\leftarrow\!\!\!\!+} P \upharpoonright \text{ꓷ}`P . \text{Nr}`P = \dot{1} \dotplus \text{Nr}`(P \upharpoonright \text{ꓷ}`P)$

Dem.

$\vdash . *202{\cdot}524 . \qquad \supset \vdash :. \text{Hp} . \supset : x = B`P . y \epsilon \text{ꓷ}`P . \supset . xPy$ (1)
$\vdash . (1) . *161{\cdot}111 . \supset \vdash :: \text{Hp} . \supset :. x (B`P \mathbin{\leftarrow\!\!\!\!+} P \upharpoonright \text{ꓷ}`P) y . \equiv :$
$x = B`P . y \epsilon \text{ꓷ}`P . xPy . \vee . x, y \epsilon \text{ꓷ}`P . xPy :$
[*93·103] $\qquad \equiv : x \epsilon C`P . y \epsilon \text{ꓷ}`P . xPy :$
[*33·14·17] $\qquad \equiv : xPy$ (2)
$\vdash . (2) . *181{\cdot}32 . \quad \supset \vdash : \text{Hp} . \supset . \text{Nr}`P = \dot{1} \dotplus \text{Nr}`(P \upharpoonright \text{ꓷ}`P)$ (3)
$\vdash . (2) . (3) . \supset \vdash . \text{Prop}$

***204·461.** $\vdash : P \epsilon \text{connex} . \text{E} ! B`\breve{P} . \text{D}`P \sim\epsilon 1 . \supset .$
$P = P \upharpoonright \text{D}`P \mathbin{+\!\!\!\!\rightarrow} B`\breve{P} . \text{Nr}`P = \text{Nr}`(P \upharpoonright \text{D}`P) \dotplus \dot{1}$
[Proof as in *204·46]

***204·462.** $\vdash :. P, Q \epsilon \text{connex} . \text{E} ! B`P . \text{ꓷ}`P \sim\epsilon 1 . \text{E} ! B`Q . \text{ꓷ}`Q \sim\epsilon 1 . \supset :$
$P \,\text{smor}\, Q . \equiv . P \upharpoonright \text{ꓷ}`P \,\text{smor}\, Q \upharpoonright \text{ꓷ}`Q$ [*161·33 . *204·46]

***204·463.** $\vdash : P, Q \epsilon \text{Rl}`J . \text{E} ! B`P . \text{ꓷ}`P \epsilon 1 . \text{E} ! B`Q . \text{ꓷ}`Q \epsilon 1 . \supset .$
$P \,\text{smor}\, Q . P \upharpoonright \text{ꓷ}`P \,\text{smor}\, Q \upharpoonright \text{ꓷ}`Q$

Dem.

$\vdash . *56{\cdot}37 . \quad \supset \vdash : \text{Hp} . \supset . P, Q \epsilon 2_r$ (1)
$\vdash . *200{\cdot}35 . \supset \vdash : \text{Hp} . \supset . P \upharpoonright \text{ꓷ}`P = \dot{\Lambda} . Q \upharpoonright \text{ꓷ}`Q = \dot{\Lambda}$ (2)
$\vdash . (1) . (2) . *153{\cdot}202{\cdot}101 . \supset$
$\vdash : \text{Hp} . \supset . P \,\text{smor}\, Q . P \upharpoonright \text{ꓷ}`P \,\text{smor}\, Q \upharpoonright \text{ꓷ}`Q : \supset \vdash . \text{Prop}$

***204·47.** $\vdash :. P, Q \epsilon \text{connex} \cap \text{Rl}`J . \text{E} ! B`P . \text{E} ! B`Q . \supset :$
$P \,\text{smor}\, Q . \equiv . P \upharpoonright \text{ꓷ}`P \,\text{smor}\, Q \upharpoonright \text{ꓷ}`Q$

Dem.

$\vdash . *151{\cdot}18 . *200{\cdot}35 . *202{\cdot}55 . *153{\cdot}102 . \supset$
$\vdash : \text{Hp} . \text{ꓷ}`P \epsilon 1 . \text{ꓷ}`Q \sim\epsilon 1 . \supset . \sim (P \,\text{smor}\, Q) . \sim (P \upharpoonright \text{ꓷ}`P \,\text{smor}\, Q \upharpoonright \text{ꓷ}`Q)$ (1)
$\vdash . (1) . *204{\cdot}462{\cdot}463 . \supset \vdash . \text{Prop}$

***204·48.** $\vdash :: P \epsilon \text{Ser} . \supset :.$

$$\text{E} ! B‘P . \equiv : (\exists Q) . \dot{\exists} ! Q . \text{Nr}‘P = \dot{1} \dotplus \text{Nr}‘Q . \vee . \text{Nr}‘P = 2_r$$

Dem.

$\vdash . *204·46 . \supset \vdash : \text{Hp} . \text{E} ! B‘P . \text{Ɑ}‘P \sim \epsilon 1 . \supset . (\exists Q) . \text{Nr}‘P = \dot{1} \dotplus \text{Nr}‘Q$ (1)

$\vdash . *161·2 . \supset \vdash : \dot{\exists} ! P . \text{Nr}‘P = \dot{1} \dotplus \text{Nr}‘Q . \supset . \dot{\exists} ! Q$ (2)

$\vdash . (1) . (2) . \supset \vdash : \text{Hp} . \text{E} ! B‘P . \text{Ɑ}‘P \sim \epsilon 1 . \supset .$

$(\exists Q) . \dot{\exists} ! Q . \text{Nr}‘P = \dot{1} \dotplus \text{Nr}‘Q$ (3)

$\vdash . *204·272 . \supset \vdash : \text{Hp} . \text{Ɑ}‘P \epsilon 1 . \supset . P \epsilon 2_r$ (4)

$\vdash . (3) . (4) . \supset \vdash : \text{Hp} . \text{E} ! B‘P . \supset :$

$(\exists Q) . \dot{\exists} ! Q . \text{Nr}‘P = \dot{1} \dotplus \text{Nr}‘Q . \vee . \text{Nr}‘P = 2_r$ (5)

$\vdash . *181·11·12·32 . \supset \vdash : \text{Nr}‘P = \dot{1} \dotplus \text{Nr}‘Q . \supset .$

$(\exists R, z) . R \text{ smor } Q . z \sim \epsilon C‘R . \text{Nr}‘P = \text{Nr}‘(z \leftarrow\!\!\!\!+ R)$ (6)

$\vdash . *161·15·12 . \supset \vdash :. \dot{\exists} ! R . z \sim \epsilon C‘R . \supset : \text{E} ! B‘(z \leftarrow\!\!\!\!+ R) :$

[*151·5] $\supset : \text{Nr}‘P = \text{Nr}‘(z \leftarrow\!\!\!\!+ R) . \supset . \text{E} ! B‘P$ (7)

$\vdash . (6) . (7) . \supset \vdash : \text{Nr}‘P = \dot{1} \dotplus \text{Nr}‘Q . \dot{\exists} ! Q . \supset . \text{E} ! B‘P$ (8)

$\vdash . *153·281 . \supset \vdash : P \epsilon 2_r . \supset . \text{E} ! B‘P$ (9)

$\vdash . (5) . (8) . (9) . \supset \vdash . \text{Prop}$

***204·481.** $\vdash :: P \epsilon \text{Ser} . \supset :.$

$$\text{E} ! B‘\breve{P} . \equiv : (\exists Q) . \dot{\exists} ! Q . \text{Nr}‘P = \text{Nr}‘Q \dotplus \dot{1} . \vee . \text{Nr}‘P = 2_r$$

[Proof as in *204·48]

***204·482.** $\vdash :: \alpha \epsilon \text{N}_0\text{r}‘‘\text{Ser} . \supset :. \alpha \subset \text{Ɑ}‘B : \equiv : \exists ! \alpha \cap \text{Ɑ}‘B :$

$$\equiv : (\exists \beta) . \beta \epsilon \text{NR} - \iota‘0_r . \alpha = \dot{1} \dotplus \beta . \vee . \alpha = 2_r$$

Dem.

$\vdash . *151·5 . *155·13 . \supset \vdash :. \text{Hp} . \supset : \alpha \subset \text{Ɑ}‘B . \equiv . \exists ! \alpha \cap \text{Ɑ}‘B$ (1)

$\vdash . *204·23·48 . \supset \vdash :: \text{Hp} . P \epsilon \alpha . \supset :.$

$\text{E} ! B‘P . \equiv : (\exists \beta) . \beta \epsilon \text{NR} - \iota‘0_r . \alpha = \dot{1} \dotplus \beta . \vee . \alpha = 2_r$ (2)

$\vdash . (1) . (2) . *202·52 . \supset \vdash . \text{Prop}$

***204·483.** $\vdash :: \alpha \epsilon \text{N}_0\text{r}‘‘\text{Ser} . \supset :. \alpha \subset \text{Ɑ}‘(B \,|\, \text{Cnv}) : \equiv : \exists ! \alpha \cap \text{Ɑ}‘(B \,|\, \text{Cnv}) :$

$$\equiv : (\exists \beta) . \beta \epsilon \text{NR} - \iota‘0_r . \alpha = \beta \dotplus \dot{1} . \vee . \alpha = 2_r$$

[Proof as in *204·482]

The following propositions are concerned with the application of relation-arithmetic to series.

***204·5.** $\vdash : P, Q \epsilon \text{Ser} . C‘P \cap C‘Q = \Lambda . \equiv . P \updownarrow Q \epsilon \text{Ser}$

[*200·4 . *201·401 . *202·401]

***204·51.** $\vdash : P \epsilon \text{Ser} . x \sim \epsilon C‘P . \equiv . P \rightarrow\!\!\!\!+ x \epsilon \text{Ser} . \equiv . x \leftarrow\!\!\!\!+ P \epsilon \text{Ser}$

[*200·41 . *201·41 . *202·412]

***204·52.** $\vdash : P \epsilon \text{Rel}^2 \text{ excl} \cap \text{Ser} . C`P \subset \text{Ser} . \supset . \Sigma`P \epsilon \text{Ser}$

Dem.

$\vdash . *200·42 . *202·62 . \supset \vdash : \text{Hp} . \supset . \Sigma`P \in J$ (1)

$\vdash . (1) . *201·42 . *202·42 . \supset \vdash . \text{Prop}$

***204·53.** $\vdash :. P \epsilon \text{Rel}^2 \text{ excl} . \dot{\Lambda} \sim \epsilon C`P . \supset : \Sigma`P \epsilon \text{Ser} . \equiv . P \epsilon \text{Ser} . C`P \subset \text{Ser}$

Dem.

$\vdash . *200·423 . \supset \vdash :. \text{Hp} . \Sigma`P \epsilon \text{Ser} . \supset : P \in J :$ (1)

[*200·421] $\supset : Q \epsilon C`P . \supset . Q = (\Sigma`P) \upharpoonright C`Q .$

[*204·4] $\supset . Q \epsilon \text{Ser}$ (2)

$\vdash . *162·13 . \supset$

$\vdash :. \text{Hp} . \Sigma`P \epsilon \text{Ser} . QPR . RPS . x \epsilon C`Q . y \epsilon C`R . z \epsilon C`S . \supset : x (\Sigma`P) z :$ (3)

[*162·13.*163·11] $\supset : (\exists M, N) . MPN . x \epsilon C`M . z \epsilon C`N . M = Q . N = S . \vee .$

$(\exists M) . M \epsilon C`P . xMz . M = Q . M = S :$

[*13·22·195] $\supset : QPS . \vee . Q = S$ (4)

$\vdash . (3) . *50·24 . *24·37 . \supset \vdash : \text{Hp} (3) . \supset . C`Q \cap C`S = \Lambda .$

[*24·57.*30·37] $\supset . Q \neq S$ (5)

$\vdash . (4) . (5) . \supset \vdash :. \text{Hp} . \Sigma`P \epsilon \text{Ser} . \supset : QPR . RPS . \supset . QPS$ (6)

$\vdash . *162·1 . \supset \vdash :. \text{Hp} . \Sigma`P \epsilon \text{Ser} . Q, R \epsilon C`P . x \epsilon C`P . y \epsilon C`Q . Q \neq R . \supset : x \neq y :$

[*202·104] $\supset : x (\Sigma`P) y . \vee . y (\Sigma`P) x :$

[*162·13.*163·11] $\supset : QPR . \vee . RPQ$ (7)

$\vdash . (6) . (7) . \supset \vdash : \text{Hp} . \Sigma`P \epsilon \text{Ser} . \supset . P \epsilon \text{trans} \cap \text{connex}$ (8)

$\vdash . (1) . (2) . (8) . *204·52 . \supset \vdash . \text{Prop}$

***204·54.** $\vdash : P \epsilon \text{Rel}^3 \text{ arithm} \cap \text{Ser} . C`P \subset \text{Ser} . C`\Sigma`P \subset \text{Ser} . \supset . \Sigma`\Sigma`P \epsilon \text{Ser}$

Dem.

$\vdash . *204·52 . \supset \vdash : \text{Hp} . \supset . \Sigma`P \epsilon \text{Ser}$ (1)

$\vdash . *174·3 . \supset \vdash : \text{Hp} . \supset . \Sigma`P \epsilon \text{Rel}^2 \text{ excl}$ (2)

$\vdash . (1) . (2) . *204·52 . \supset \vdash . \text{Prop}$

***204·55.** $\vdash : P, Q \epsilon \text{Ser} . \supset . Q \times P \epsilon \text{Ser}$

Dem.

$\vdash . *165·27 . *204·22 . \supset \vdash :. \text{Hp} . \supset : \dot{\exists} ! P . \supset . P \downarrow ; Q \epsilon \text{Ser}$ (1)

$\vdash . *165·26 . *204·22 . \supset \vdash : \text{Hp} . \supset . C`P \downarrow ; Q \subset \text{Ser}$ (2)

$\vdash . (1) . (2) . *165·21 . *204·52 . \supset \vdash : \text{Hp} . \dot{\exists} ! P . \supset . \Sigma`P \downarrow ; Q \epsilon \text{Ser} .$

[*166·1] $\supset . Q \times P \epsilon \text{Ser}$ (3)

$\vdash . *166·13 . *204·24 . \supset \vdash : P = \dot{\Lambda} . \supset . Q \times P \epsilon \text{Ser}$ (4)

$\vdash . (3) . (4) . \supset \vdash . \text{Prop}$

∗204·551. $\vdash :. \dot{\exists} ! P . \dot{\exists} ! Q . \supset : P \times Q \in \mathrm{Ser} . \equiv . P, Q \in \mathrm{Ser}$

Dem.

$\vdash . {*}165{\cdot}21{\cdot}212 . \supset \vdash :. \mathrm{Hp} . \supset : P \underset{\cdot\cdot}{\downarrow} {}^{;}Q \in \mathrm{Rel}^2\,\mathrm{excl} . \dot{\Lambda} \sim \in C{}^{\backprime}P \underset{\cdot\cdot}{\downarrow} {}^{;}Q :$

$[{*}204{\cdot}53 . {*}166{\cdot}1] \quad \supset : P \times Q \in \mathrm{Ser} . \equiv . P \underset{\cdot\cdot}{\downarrow} {}^{;}Q \in \mathrm{Ser} . C{}^{\backprime}P \underset{\cdot\cdot}{\downarrow} {}^{;}Q \subset \mathrm{Ser} .$

$[{*}165{\cdot}27 . {*}204{\cdot}22] \quad \equiv . P, Q \in \mathrm{Ser} :. \supset \vdash . \mathrm{Prop}$

∗204·56. $\vdash : C{}^{\backprime}P \subset \mathrm{Rl}{}^{\backprime}J . \supset . \Pi{}^{\backprime}P \in J$

Dem.

$\vdash . {*}172{\cdot}11 . \supset \vdash : M (\Pi{}^{\backprime}P) N . \supset . (\exists Q) . Q \in C{}^{\backprime}P . (M{}^{\backprime}Q) Q (N{}^{\backprime}Q) \quad (1)$

$\vdash . (1) . \supset \vdash :. \mathrm{Hp} . \supset : M (\Pi{}^{\backprime}P) N . \supset . (\exists Q) . M{}^{\backprime}Q \neq N{}^{\backprime}Q .$

$[{*}30{\cdot}37 . \mathrm{Transp}] \quad \supset . M \neq N :. \supset \vdash . \mathrm{Prop}$

∗204·561. $\vdash : P \in \mathrm{Ser} . C{}^{\backprime}P \subset \mathrm{Ser} . \supset . \Pi{}^{\backprime}P \in \mathrm{Rl}{}^{\backprime}J \cap \mathrm{trans}$

Dem.

$\vdash . {*}200{\cdot}43 . \supset \vdash :: \mathrm{Hp} . \supset :. L (\Pi{}^{\backprime}P) M . M (\Pi{}^{\backprime}P) N . \supset :$

$(\exists Q, R) . Q, R \in C{}^{\backprime}P . (L{}^{\backprime}Q) Q (M{}^{\backprime}Q) . (M{}^{\backprime}R) R (N{}^{\backprime}R) . L \upharpoonright \overrightarrow{P}{}^{\backprime}Q = M \upharpoonright \overrightarrow{P}{}^{\backprime}Q .$

$M \upharpoonright \overrightarrow{P}{}^{\backprime}R = N \upharpoonright \overrightarrow{P}{}^{\backprime}R :$

$[{*}204{\cdot}12]$

$\supset : (\exists Q, R) : Q = R . \vee . QPR . \vee . RPQ : (L{}^{\backprime}Q) Q (M{}^{\backprime}Q) . (M{}^{\backprime}R) R (N{}^{\backprime}R) .$

$L \upharpoonright \overrightarrow{P}{}^{\backprime}Q = M \upharpoonright \overrightarrow{P}{}^{\backprime}Q . M \upharpoonright \overrightarrow{P}{}^{\backprime}R = N \upharpoonright \overrightarrow{P}{}^{\backprime}R \quad (1)$

$\vdash . {*}204{\cdot}1 . \supset \vdash : \mathrm{Hp} . L (\Pi{}^{\backprime}P) M . M (\Pi{}^{\backprime}P) N .$

$Q = R . (L{}^{\backprime}Q) Q (M{}^{\backprime}Q) . (M{}^{\backprime}R) R (N{}^{\backprime}R) . L \upharpoonright \overrightarrow{P}{}^{\backprime}Q = M \upharpoonright \overrightarrow{P}{}^{\backprime}Q .$

$M \upharpoonright \overrightarrow{P}{}^{\backprime}R = N \upharpoonright \overrightarrow{P}{}^{\backprime}R . \supset . (L{}^{\backprime}Q) Q (N{}^{\backprime}Q) . L \upharpoonright \overrightarrow{P}{}^{\backprime}Q = N \upharpoonright \overrightarrow{P}{}^{\backprime}Q \quad (2)$

$\vdash . {*}204{\cdot}33 . \supset$

$\vdash : \mathrm{Hp} . L (\Pi{}^{\backprime}P) M . M (\Pi{}^{\backprime}P) N . QPR . (L{}^{\backprime}Q) Q (M{}^{\backprime}Q) . (M{}^{\backprime}R) R (N{}^{\backprime}R) .$

$L \upharpoonright \overrightarrow{P}{}^{\backprime}Q = M \upharpoonright \overrightarrow{P}{}^{\backprime}Q . M \upharpoonright \overrightarrow{P}{}^{\backprime}R = N \upharpoonright \overrightarrow{P}{}^{\backprime}R . \supset .$

$L \upharpoonright \overrightarrow{P}{}^{\backprime}Q = N \upharpoonright \overrightarrow{P}{}^{\backprime}Q . M{}^{\backprime}Q = N{}^{\backprime}Q .$

$[{*}13{\cdot}12] \quad \supset . L \upharpoonright \overrightarrow{P}{}^{\backprime}Q = N \upharpoonright \overrightarrow{P}{}^{\backprime}Q . (L{}^{\backprime}Q) Q (N{}^{\backprime}Q) \quad (3)$

$\vdash . {*}204{\cdot}33 . \supset$

$\vdash : \mathrm{Hp} . L (\Pi{}^{\backprime}P) M . M (\Pi{}^{\backprime}P) N . RPQ . (L{}^{\backprime}Q) Q (M{}^{\backprime}Q) . (M{}^{\backprime}R) R (N{}^{\backprime}R) .$

$L \upharpoonright \overrightarrow{P}{}^{\backprime}Q = M \upharpoonright \overrightarrow{P}{}^{\backprime}Q . M \upharpoonright \overrightarrow{P}{}^{\backprime}R = N \upharpoonright \overrightarrow{P}{}^{\backprime}R . \supset .$

$L \upharpoonright \overrightarrow{P}{}^{\backprime}R = N \upharpoonright \overrightarrow{P}{}^{\backprime}R . L{}^{\backprime}R = M{}^{\backprime}R .$

$[{*}13{\cdot}12] \quad \supset . L \upharpoonright \overrightarrow{P}{}^{\backprime}R = N \upharpoonright \overrightarrow{P}{}^{\backprime}R . (L{}^{\backprime}R) R (N{}^{\backprime}R) \quad (4)$

$\vdash . (1) . (2) . (3) . (4) . {*}200{\cdot}43 . \supset$

$\vdash :. \mathrm{Hp} . \supset : L (\Pi{}^{\backprime}P) M . M (\Pi{}^{\backprime}P) N . \supset . L (\Pi{}^{\backprime}P) N \quad (5)$

$\vdash . (5) . {*}204{\cdot}56 . \supset \vdash . \mathrm{Prop}$

In order to prove that $\Pi{}^{\backprime}P$ is connected, we require a further hypothesis, namely that P is *well-ordered*, *i.e.* that every class contained in $C{}^{\backprime}P$ and not null has a first term.

∗204·562. $\vdash :. C‘P \subset \text{Ser} : \alpha \subset C‘P \,.\, \exists ! \alpha \,.\, \supset_\alpha .\, \exists ! \alpha - \breve{P}“\alpha : \supset .\, \Pi‘P \in \text{connex}$

Dem.

$\vdash .\, *172{\cdot}11 \,.\, *33{\cdot}45 \,.\, \text{Transp} \,.\, \supset$

$\vdash :: \text{Hp} \,.\, \supset :. M, N \in C‘\Pi‘P \,.\, M \neq N \,.\, \supset : (\exists Q) \,.\, Q \in C‘P \,.\, M‘Q \neq N‘Q :$

[Hp] $\qquad \supset : (\exists Q) : Q \in C‘P \,.\, M‘Q \neq N‘Q : RPQ \,.\, \supset_R .\, M‘R = N‘R :$

[∗204·121.∗172·12] $\qquad \supset : (\exists Q) : Q \in C‘P : (M‘Q)\, Q\, (N‘Q) \,.\, \mathbf{v} \,.\, (N‘Q)\, Q\, (M‘Q) :$

$\qquad RPQ \,.\, \supset_R .\, M‘R = N‘R :$

[∗172·11] $\qquad \supset : M\, (\Pi‘P)\, N \,.\, \mathbf{v} \,.\, N\, (\Pi‘P)\, M \qquad (1)$

$\vdash .\, (1) \,.\, *202{\cdot}104 \,.\, \supset \vdash .\, \text{Prop}$

∗204·57. $\vdash :. P \in \text{Ser} \,.\, C‘P \subset \text{Ser} : \alpha \subset C‘P \,.\, \exists ! \alpha \,.\, \supset_\alpha .\, \exists ! \alpha - \breve{P}“\alpha : \supset .\, \Pi‘P \in \text{Ser}$
[∗204·561·562]

∗204·58. $\vdash :. P \in \text{Ser} \,.\, C‘P \subset \text{Ser} \,.\, C‘\Sigma‘P \subset \text{Ser} \,.\, P \in \text{Rel}^2 \,\text{excl} :$

$\alpha \subset C‘\Sigma‘P \,.\, \exists ! \alpha \,.\, \supset_\alpha .\, \exists ! \alpha - (\text{Cnv}‘\Sigma‘P)“\alpha : \supset .\, \Pi‘\Sigma‘P, \Pi‘\Pi ; P \in \text{Ser}$

Dem.

$\vdash .\, *204{\cdot}52 \,.\qquad \supset \vdash : \text{Hp} \,.\, \supset .\, \Sigma‘P \in \text{Ser} \qquad (1)$

$\vdash .\, (1) \,.\, *204{\cdot}57 \,.\qquad \supset \vdash : \text{Hp} \,.\, \supset .\, \Pi‘\Sigma‘P \in \text{Ser} \qquad (2)$

$\vdash .\, *174{\cdot}25 \,.\qquad \supset \vdash : \text{Hp} \,.\, \supset .\, \Pi‘\Sigma‘P \,\text{smor}\, \Pi‘\Pi ; P \qquad (3)$

$\vdash .\, (2) \,.\, (3) \,.\, *204{\cdot}21 \,.\, \supset \vdash : \text{Hp} \,.\, \supset .\, \Pi‘\Pi ; P \in \text{Ser} \qquad (4)$

$\vdash .\, (2) \,.\, (4) \,.\, \supset \vdash .\, \text{Prop}$

∗204·581. $\vdash : \text{Hp}\, *204{\cdot}58 \,.\, \Sigma‘P \in \text{Rel}^2 \,\text{excl} \,.\, \supset .\, \text{Prod}‘\text{Prod} ; P, \text{Prod}‘\Sigma‘P \in \text{Ser}$
[∗174·461·43 . ∗204·58·21]

∗204·59. $\vdash :. P, Q \in \text{Ser} : \alpha \subset C‘Q \,.\, \exists ! \alpha \,.\, \supset_\alpha .\, \exists ! \alpha - \breve{Q}“\alpha : \supset .$

$P^Q \in \text{Ser} \,.\, (P \exp Q) \in \text{Ser}$

Dem.

$\vdash .\, *165{\cdot}27{\cdot}241 \,.\, *204{\cdot}22{\cdot}24 \,.\, \supset \vdash : \text{Hp} \,.\, \supset .\, P \underset{;}{\downarrow} ; Q \in \text{Ser} \qquad (1)$

$\vdash .\, *165{\cdot}26 \,.\, *204{\cdot}22 \,.\qquad \supset \vdash : \text{Hp} \,.\, \supset .\, C‘P \underset{;}{\downarrow} ; Q \subset \text{Ser} \qquad (2)$

$\vdash .\, *150{\cdot}22 \,.\, *71{\cdot}47 \,.\, \supset \vdash : \beta \subset C‘P \underset{;}{\downarrow} ; Q \,.\, \exists ! \beta \,.\, \supset .\, (\exists \alpha) \,.\, \alpha \subset C‘Q \,.\, \exists ! \alpha \,.\, \beta = P \underset{;}{\downarrow} “\alpha :$

[Hp] $\qquad \supset \vdash : \text{Hp} \,.\, \beta \subset C‘P \underset{;}{\downarrow} ; Q \,.\, \exists ! \beta \,.\, \supset .\, (\exists \alpha) \,.\, \exists ! \alpha - \breve{Q}“\alpha \,.\, \beta = P \underset{;}{\downarrow} “\alpha \qquad (3)$

$\vdash .\, *37{\cdot}45 \,.\, \supset \vdash : \exists ! \alpha - \breve{Q}“\alpha \,.\, \equiv .\, \exists ! P \underset{;}{\downarrow} “(\alpha - \breve{Q}“\alpha) \qquad (4)$

$\vdash .\, (4) \,.\, *71{\cdot}381 \,.\, *165{\cdot}22 \,.\, \supset \vdash : \dot{\exists} ! P \,.\, \exists ! \alpha - \breve{Q}“\alpha \,.\, \supset .\, \exists ! P \underset{;}{\downarrow} “\alpha - P \underset{;}{\downarrow} “\breve{Q}“\alpha \qquad (5)$

$\vdash .\, *72{\cdot}503 \,.\, *165{\cdot}22 \,.\qquad \supset \vdash : \dot{\exists} ! P \,.\, \supset .\, \alpha = (\text{Cnv}‘P \underset{;}{\downarrow})“P \underset{;}{\downarrow} “\alpha \qquad (6)$

$\vdash .\, (5) \,.\, (6) \,.\, \supset \vdash : \dot{\exists} ! P \,.\, \exists ! \alpha - \breve{Q}“\alpha \,.\, \supset .\, \exists ! P \underset{;}{\downarrow} “\alpha - P \underset{;}{\downarrow} “\breve{Q}“(\text{Cnv}‘P \underset{;}{\downarrow})“P \underset{;}{\downarrow} “\alpha .$

[∗165·18] $\qquad \supset .\, \exists ! P \underset{;}{\downarrow} “\alpha - (\text{Cnv}‘P \underset{;}{\downarrow} ; Q)“P \underset{;}{\downarrow} “\alpha \qquad (7)$

$\vdash . (3) . (7) . \supset \vdash :. \mathrm{Hp} . \dot{\exists} ! P . \supset :$

$$\beta \subset C'P \underset{\cdot\cdot}{\downarrow} ; Q . \exists ! \beta . \supset . \exists ! \beta - (\mathrm{Cnv}'P \underset{\cdot\cdot}{\downarrow} ; Q)``\beta \quad (8)$$

$\vdash . (1) . (2) . (8) . *204{\cdot}57 . \quad \supset \vdash : \mathrm{Hp} . \dot{\exists} ! P . \supset . \Pi'P \underset{\cdot\cdot}{\downarrow} ; Q \, \epsilon \, \mathrm{Ser} \quad (9)$

$\vdash . (9) . *176{\cdot}182 . *204{\cdot}21 . \supset \vdash : \mathrm{Hp} . \dot{\exists} ! P . \supset . (P \exp Q) \, \epsilon \, \mathrm{Ser} \quad (10)$

$\vdash . *176{\cdot}151 . *204{\cdot}24 . \quad \supset \vdash : P = \dot{\Lambda} . \supset . (P \exp Q) \, \epsilon \, \mathrm{Ser} \quad (11)$

$\vdash . (10) . (11) . \quad \supset \vdash : \mathrm{Hp} . \supset . (P \exp Q) \, \epsilon \, \mathrm{Ser} \quad (12)$

$\vdash . (12) . *176{\cdot}181 . *204{\cdot}21 . \supset \vdash : \mathrm{Hp} . \supset . P^Q \, \epsilon \, \mathrm{Ser} \quad (13)$

$\vdash . (12) . (13) . \supset \vdash . \mathrm{Prop}$

The two following propositions are lemmas for $*204{\cdot}62$.

$*204{\cdot}6$. $\vdash : P \, \epsilon \, \mathrm{trans} . \supset . \alpha \cup \breve{P}``\alpha \subset p'\overleftarrow{P}``p'\overrightarrow{P}``\alpha$

Dem.

$\vdash . *40{\cdot}53 . \supset \vdash :: x \, \epsilon \, p'\overleftarrow{P}``p'\overrightarrow{P}``\alpha . \equiv :. y \, \epsilon \, p'\overrightarrow{P}``\alpha . \supset_y . yPx :.$

$[*40{\cdot}51] \quad \equiv :. z \, \epsilon \, \alpha . \supset_z . yPz : \supset_y . yPx \quad (1)$

$\vdash . *10{\cdot}26 . \supset \vdash :. x \, \epsilon \, \alpha : z \, \epsilon \, \alpha . \supset_z . yPz : \supset . yPx :$

$[\mathrm{Exp.}(1)] \quad \supset \vdash : x \, \epsilon \, \alpha . \supset . x \, \epsilon \, p'\overleftarrow{P}``p'\overrightarrow{P}``\alpha \quad (2)$

$\vdash . *10{\cdot}1 . \quad \supset \vdash :. u \, \epsilon \, \alpha . uPx : z \, \epsilon \, \alpha . \supset_z . yPz : \supset . yPu . uPx \quad (3)$

$\vdash . (3) . *201{\cdot}1 . \supset \vdash :: \mathrm{Hp} . \supset :. u \, \epsilon \, \alpha . u\dot{P}x : z \, \epsilon \, \alpha . \supset_z . yPz : \supset . yPx :.$

$[*37{\cdot}105] \quad \supset :. x \, \epsilon \, \breve{P}``\alpha : z \, \epsilon \, \alpha . \supset_z . yPz : \supset . yPx :.$

$[\mathrm{Exp.}(1)] \quad \supset :. x \, \epsilon \, \breve{P}``\alpha . \supset . x \, \epsilon \, p'\overleftarrow{P}``p'\overrightarrow{P}``\alpha \quad (4)$

$\vdash . (2) . (4) . \supset \vdash . \mathrm{Prop}$

$*204{\cdot}61$. $\vdash : P \, \epsilon \, \mathrm{Rl}'J \cap \mathrm{connex} . \supset . C'P \cap p'\overleftarrow{P}``p'\overrightarrow{P}``(\alpha \cap C'P) \subset \alpha \cup \breve{P}``\alpha$

Dem.

$\vdash . *200{\cdot}5 . \supset \vdash : \mathrm{Hp} . \supset . p'\overrightarrow{P}``(\alpha \cap C'P) \cap p'\overleftarrow{P}``p'\overrightarrow{P}``(\alpha \cap C'P) = \Lambda .$

$[*24{\cdot}311] \quad \supset . p'\overleftarrow{P}``p'\overrightarrow{P}``(\alpha \cap C'P) \subset - p'\overrightarrow{P}``(\alpha \cap C'P) .$

$[*22{\cdot}48] \quad \supset . C'P \cap p'\overleftarrow{P}``p'\overrightarrow{P}``(\alpha \cap C'P) \subset C'P - p'\overrightarrow{P}``(\alpha \cap C'P)$

$[*24{\cdot}43 . *202{\cdot}505] \quad \subset \alpha \cup \breve{P}``\alpha : \supset \vdash . \mathrm{Prop}$

$*204{\cdot}62$. $\vdash : P \, \epsilon \, \mathrm{Ser} . \supset . C'P \cap p'\overleftarrow{P}``p'\overrightarrow{P}``(\alpha \cap C'P) = (\alpha \cap C'P) \cup \breve{P}``\alpha$

Dem.

$\vdash . *204{\cdot}6 . *37{\cdot}265 . \supset \vdash : \mathrm{Hp} . \supset . (\alpha \cap C'P) \cup \breve{P}``\alpha \subset p'\overleftarrow{P}``p'\overrightarrow{P}``(\alpha \cap C'P) \quad (1)$

$\vdash . *37{\cdot}16 . *22{\cdot}43 . \quad \supset \vdash . (\alpha \cap C'P) \cup \breve{P}``\alpha \subset C'P \quad (2)$

$\vdash . *204{\cdot}61 . *22{\cdot}43 . \supset \vdash : \mathrm{Hp} . \supset . C'P \cap p'\overleftarrow{P}``p'\overrightarrow{P}``(\alpha \cap C'P) \subset (\alpha \cup \breve{P}``\alpha) \cap C'P .$

$[*37{\cdot}16] \quad \supset . C'P \cap p'\overleftarrow{P}``p'\overrightarrow{P}``(\alpha \cap C'P) \subset (\alpha \cap C'P) \cup \breve{P}``\alpha \quad (3)$

$\vdash . (1) . (2) . (3) . \supset \vdash . \mathrm{Prop}$

***204·63.** $\vdash : P \epsilon \mathrm{Ser} \,.\, \exists ! \, p{}^{\backprime}\overrightarrow{P}{}^{\backprime\backprime}\alpha \,.\, \supset . \, p{}^{\backprime}\overleftarrow{P}{}^{\backprime\backprime}p{}^{\backprime}\overrightarrow{P}{}^{\backprime\backprime}\alpha = \alpha \cup \breve{P}{}^{\backprime\backprime}\alpha$

Dem.

$\vdash . *40{\cdot}65 \,.\, \mathrm{Transp} \,.\, \supset \vdash : \mathrm{Hp} \,.\, \supset . \, \alpha \subset C{}^{\backprime}P$ (1)

$\vdash . *40{\cdot}62 \,.\, \supset \vdash : \mathrm{Hp} \,.\, \supset . \, p{}^{\backprime}\overleftarrow{P}{}^{\backprime\backprime}p{}^{\backprime}\overrightarrow{P}{}^{\backprime\backprime}\alpha \subset C{}^{\backprime}P$ (2)

$\vdash . (1) . (2) . *204{\cdot}62 \,.\, \supset \vdash . \mathrm{Prop}$

***204·64.** $\vdash : P \epsilon \mathrm{Ser} \,.\, x \epsilon \mathrm{D}{}^{\backprime}P \,.\, \supset . \, p{}^{\backprime}\overrightarrow{P}{}^{\backprime\backprime}\overleftarrow{P}{}^{\backprime}x = \overrightarrow{P_*}{}^{\backprime}x$

Dem.

$\vdash . *40{\cdot}62 \,.\, \supset \vdash : \mathrm{Hp} \,.\, \supset . \, p{}^{\backprime}\overrightarrow{P}{}^{\backprime\backprime}\overleftarrow{P}{}^{\backprime}x \subset C{}^{\backprime}P$ (1)

$\vdash . *40{\cdot}51 \,.\, \supset \vdash :. \, z \epsilon p{}^{\backprime}\overrightarrow{P}{}^{\backprime\backprime}\overleftarrow{P}{}^{\backprime}x \,.\, \equiv : xPy \,.\, \supset_y . \, zPy$ (2)

$\vdash . (2) . *50{\cdot}11 \,.\, \supset \vdash :: \mathrm{Hp} \,.\, \supset :. \, z \epsilon p{}^{\backprime}\overrightarrow{P}{}^{\backprime\backprime}\overleftarrow{P}{}^{\backprime}x \,.\, \supset : xPy \,.\, \supset_y . \, z \neq y :$

[(1).*202·103] $\supset : zPx \,.\, \vee . \, z = x$ (3)

$\vdash . *201{\cdot}521 \,.\, \supset \vdash :: \mathrm{Hp} \,.\, \supset :. \, z \epsilon \overrightarrow{P_*}{}^{\backprime}x \,.\, \equiv : zPx \,.\, \vee . \, z = x :$ (4)

[*201·1.*13·12] $\supset : xPy \,.\, \supset . \, zPy :$

[(2)] $\supset : z \epsilon p{}^{\backprime}\overrightarrow{P}{}^{\backprime\backprime}\overleftarrow{P}{}^{\backprime}x$ (5)

$\vdash . (3) . (4) . (5) \,.\, \supset \vdash . \mathrm{Prop}$

The following proposition is used in *234·101.

***204·65.** $\vdash : P \epsilon \mathrm{Ser} \,.\, x \epsilon C{}^{\backprime}P \,.\, \supset . \, p{}^{\backprime}\overrightarrow{P}{}^{\backprime\backprime}\overleftarrow{P}{}^{\backprime}x \cap C{}^{\backprime}P = \overrightarrow{P_*}{}^{\backprime}x$

Dem.

$\vdash . *40{\cdot}2 \,.\, \supset \vdash : \mathrm{Hp} \,.\, x \sim \epsilon \dot{\mathrm{D}}{}^{\backprime}P \,.\, \supset . \, p{}^{\backprime}\overrightarrow{P}{}^{\backprime\backprime}\overleftarrow{P}{}^{\backprime}x \cap C{}^{\backprime}P = C{}^{\backprime}P$

[*204·11] $= \overrightarrow{P}{}^{\backprime}x \cup \iota{}^{\backprime}x$

[*201·521] $= \overrightarrow{P_*}{}^{\backprime}x$ (1)

$\vdash . *40{\cdot}62 \,.\, *204{\cdot}64 \,.\, \supset \vdash : \mathrm{Hp} \,.\, x \epsilon \mathrm{D}{}^{\backprime}P \,.\, \supset . \, p{}^{\backprime}\overrightarrow{P}{}^{\backprime\backprime}\overleftarrow{P}{}^{\backprime}x \cap C{}^{\backprime}P = \overrightarrow{P_*}{}^{\backprime}x$ (2)

$\vdash . (1) . (2) \,.\, \supset \vdash . \mathrm{Prop}$

***204·7.** $\vdash : P \epsilon \mathrm{Ser} \,.\, \supset . \, P_1 \epsilon 1 \rightarrow 1$ [*201·63 . *202·7]

On this proposition, compare the remarks preceding *201·6.

***204·71.** $\vdash : P \epsilon \mathrm{Ser} \,.\, xP_1y \,.\, \supset . \, \overrightarrow{P}{}^{\backprime}y = \overrightarrow{P}{}^{\backprime}x \cup \iota{}^{\backprime}x$ [*202·72 . *201·63]

***204·72.** $\vdash :: P \epsilon \mathrm{Ser} \,.\, \supset :. \, xP_1y \,.\, \equiv : xPy : xPz \,.\, z \neq y \,.\, \supset_z . \, yPz$

Dem.

$\vdash . *201{\cdot}63 \,.\, \supset \vdash :. \mathrm{Hp} \,.\, \supset : xP_1y \,.\, \supset . \, xPy$ (1)

$\vdash . *204{\cdot}71 \,.\, *121{\cdot}26 \,.\, \supset \vdash :. \mathrm{Hp} \,.\, xP_1y \,.\, \supset : \overleftarrow{P}{}^{\backprime}x = \overleftarrow{P}{}^{\backprime}y \cup \iota{}^{\backprime}y :$

[*24·43.*32·181] $\supset : xPz \,.\, z \neq y \,.\, \supset_z . \, yPz$ (2)

$\vdash . *24{\cdot}43 \,.\, *32{\cdot}181 \,.\, \supset \vdash :. \, xPz \,.\, z \neq y \,.\, \supset_z . \, yPz : \supset . \, \overleftarrow{P}{}^{\backprime}x \subset \iota{}^{\backprime}y \cup \overleftarrow{P}{}^{\backprime}y$ (3)

$\vdash . (3) . *200{\cdot}361 \,.\, \supset \vdash :. \mathrm{Hp} : xPz \,.\, z \neq y \,.\, \supset_z . \, yPz : \supset . \, \overleftarrow{P}{}^{\backprime}x \cap \overrightarrow{P}{}^{\backprime}y = \Lambda \,.$

[*34·11] $\supset . \sim (xP^2y)$ (4)

$\vdash . (4) . \mathrm{Fact} \,.\, \supset \vdash :. \mathrm{Hp} : xPy : xPz \,.\, z \neq y \,.\, \supset_z . \, yPz : \supset . \, x(P \dot{-} P^2)y \,.$

[*201·63] $\supset . \, xP_1y$ (5)

$\vdash . (1) . (2) . (5) \,.\, \supset \vdash . \mathrm{Prop}$

The above proposition is used in *274·23.

*205. MAXIMUM AND MINIMUM POINTS.

Summary of *205.

The minimum points of a class α with respect to a relation P are those members of α which belong to the field of P but to which no members of α have the relation P; that is, they are those members of α which belong to $C‘P$ but have no predecessors in α. Similarly the maximum points of α are those members of α which belong to $C‘P$ but have no successors in α. Both these notions have been already defined in *93, but they were there only used for the special purpose of studying generations. Their chief utility is in connection with *series*, and it is in this connection that we shall now consider them. Many of the properties of maxima and minima in series do not demand the whole hypothesis "$P \in \text{Ser}$," but only "$P \in \text{connex}$." This is the case, in particular, with the fundamental property of maxima and minima in series, namely that each class has at most one maximum and one minimum. The minimum of a class, if it exists, is the first term of the class, and the maximum, if it exists, is the last term. The maxima with respect to P are the minima with respect to $\breve{P}$; hence properties of maxima result immediately from the corresponding properties of minima, and will be set down without proof in what follows.

It will be seen that the maxima and minima of α depend only upon $\alpha \cap C‘P$: the part of α (if any) which is not contained in $C‘P$ is irrelevant.

In accordance with the definitions of *93, the class of minima of α is denoted by $\overrightarrow{\text{min}}_P‘\alpha$, where

$$\overrightarrow{\text{min}}_P‘\alpha = (\alpha \cap C‘P) - \breve{P}‘‘\alpha,$$

the definition being

$$\text{min}_P = \hat{x}\hat{\alpha}\{x \in (\alpha \cap C‘P) - \breve{P}‘‘\alpha\}.$$

Thus min_P is a relation contained in ϵ. When P is connected, we have $\overrightarrow{\text{min}}_P‘\alpha \in 0 \cup 1$, *i.e.* (by *71·12)

$$\text{min}_P \in 1 \rightarrow \text{Cls}.$$

It follows that, if κ is a set of classes which all have minima, $\text{min}_P \upharpoonright \kappa$ is a selective relation for κ, *i.e.*

$$\text{min}_P \upharpoonright \kappa \in \epsilon_\Delta‘\kappa.$$

Owing to this fact, the existence of selections can sometimes be proved in dealing with series (especially with well-ordered series), in cases where such proof would be impossible if no serial arrangement were given.

The definition of min_P is so chosen as to exclude from $\overrightarrow{\text{min}}_P{}^{\backprime}\alpha$ whatever part of α is not contained in $C^{\backprime}P$, and to make $\overrightarrow{\text{min}}_P{}^{\backprime}\iota^{\backprime}x = \iota^{\backprime}x$, *i.e.* $\text{min}_P{}^{\backprime}\iota^{\backprime}x = x$, provided $x \in C^{\backprime}P \,.\, \sim(xPx)$. For these two reasons we have to reject two simpler definitions which might otherwise be thought preferable. One of these would give

$$\overrightarrow{\text{min}}_P{}^{\backprime}\alpha = \alpha - \breve{P}^{\backprime\backprime}\alpha,$$

which might be obtained by putting

$$\text{min}_P = \epsilon \mathbin{\dot{-}} \epsilon \,|\, \breve{P} \quad \text{Df.}$$

This agrees with our definition whenever $\alpha \subset C^{\backprime}P$, but not otherwise, since it includes in $\overrightarrow{\text{min}}_P{}^{\backprime}\alpha$ any part of α not contained in $C^{\backprime}P$. Hence it necessitates the hypothesis $\alpha \subset C^{\backprime}P$ in many propositions which, with our definition, do not require this hypothesis, and in particular in the proposition

$$P \in \text{connex} \,.\supset.\, \overrightarrow{\text{min}}_P{}^{\backprime}\alpha \in 0 \cup 1,$$

so that instead of having (as with our definition)

$$P \in \text{connex} \,.\supset.\, \text{min}_P \in 1 \rightarrow \text{Cls}$$

we should only have

$$P \in \text{connex} \,.\supset.\, \text{min}_P \upharpoonright \text{Cl}^{\backprime}C^{\backprime}P \in 1 \rightarrow \text{Cls}.$$

For these reasons, this definition is less convenient than the one we have adopted.

The other definition which suggests itself is one which will give

$$\overrightarrow{\text{min}}_P{}^{\backprime}\alpha = \overrightarrow{B}^{\backprime}P \upharpoonright \alpha.$$

If this definition were adopted, we might dispense with a special notation altogether, using $\overrightarrow{B}^{\backprime}P \upharpoonright \alpha$, $B^{\backprime}P \upharpoonright \alpha$ in place of $\overrightarrow{\text{min}}_P{}^{\backprime}\alpha$, $\text{min}_P{}^{\backprime}\alpha$. This definition, however, has the drawback that, if $\alpha \in 1$ and $P \subseteq J$,

$$P \upharpoonright \alpha = \dot{\Lambda},$$

so that we have

$$\overrightarrow{\text{min}}_P{}^{\backprime}\alpha = \dot{\Lambda} \quad \text{when} \quad \alpha \in 1 \,.\, \alpha \subset C^{\backprime}P.$$

This necessitates the addition of the hypothesis $\alpha \sim\in 1$ (as in $*204{\cdot}45$ above, for example) in cases where, with our definition, no such hypothesis is required. If we take $\overrightarrow{B}^{\backprime}\alpha \upharpoonleft P$, instead of $\overrightarrow{B}^{\backprime}P \upharpoonright \alpha$, as the class of minimum points, we secure $\text{min}_P{}^{\backprime}\iota^{\backprime}x = x$ when $P \subseteq J$ and $x \in \text{D}^{\backprime}P$, but not when $x \in \overrightarrow{B}^{\backprime}\breve{P}$. Thus we still have exceptions to provide against which do not arise with the definition we have adopted.

The first few propositions of this number have already been proved in *93, but are repeated here for convenience of reference.

The propositions of this number are numerous and much used. Among the elementary properties of $\max_P$ and $\min_P$ with which the number begins, the following should be noted:

***205·12.** $\vdash . \overrightarrow{B}\text{‘}P = \overrightarrow{\min}_P\text{‘D‘}P = \overrightarrow{\min}_P\text{‘}C\text{‘}P$

***205·123.** $\vdash : \overrightarrow{\max}_P\text{‘}\alpha = \Lambda . \equiv . \alpha \cap C\text{‘}P \subset P\text{“}\alpha$

***205·14.** $\vdash . \overrightarrow{\min}_P\text{‘}\alpha = \hat{x}\{x \in \alpha \cap C\text{‘}P . \alpha \cap \overrightarrow{P}\text{‘}x = \Lambda\}$

***205·15.** $\vdash . \overrightarrow{\min}_P\text{‘}(\alpha \cap C\text{‘}P) = \overrightarrow{\min}_P\text{‘}\alpha$

***205·16.** $\vdash . \overrightarrow{\min}_P\text{‘}\Lambda = \Lambda$

***205·18.** $\vdash : \sim (xPx) . x \in C\text{‘}P . \supset . \min_P\text{‘}\iota\text{‘}x = \max_P\text{‘}\iota\text{‘}x = x$

***205·19.** $\vdash : P \in \text{trans} . \supset . \overrightarrow{\min}_P\text{‘}\alpha = \overrightarrow{\min}_P\text{‘}(\alpha \cup \breve{P}\text{“}\alpha) = \overrightarrow{\min}_P\text{‘}\breve{P}_*\text{“}\alpha$

***205·194.** $\vdash : x \min_P \alpha . \supset . \sim (xPx)$

Owing to this proposition, we can sometimes dispense with the hypothesis $P \Subset J$ in propositions about minima which would otherwise require this hypothesis.

***205·197.** $\vdash :. P \in \text{Rl‘}J \cap \text{trans} . \supset : x \in C\text{‘}P . \equiv . x = \max_P\text{‘}(\overrightarrow{P}\text{‘}x \cup \iota\text{‘}x)$

Our next set of propositions (*205·2—·27) introduces the hypothesis that P is connected, or transitive and connected. The chief of them are

***205·21.** $\vdash : P \in \text{connex} . \text{E} ! \min_P\text{‘}\alpha . y \in \alpha \cap C\text{‘}P - \iota\text{‘}\min_P\text{‘}\alpha . \supset . \min_P\text{‘}\alpha P y$

I.e. if the minimum of α exists, it precedes every other member of $\alpha \cap C\text{‘}P$.

***205·22.** $\vdash : P \in \text{trans} \cap \text{connex} . \text{E} ! \min_P\text{‘}\alpha . \supset . \breve{P}\text{“}\alpha = \overleftarrow{P}\text{‘}\min_P\text{‘}\alpha$

I.e. the terms which come after some part of α are those that come after its minimum (when the minimum exists).

***205·25.** $\vdash . \overrightarrow{\min}_P\text{‘}\overleftarrow{P}\text{‘}x = (\overleftarrow{P \dot{-} P^2})\text{‘}x$

We have next the fundamental proposition:

***205·3.** $\vdash : P \in \text{connex} . \supset . \overrightarrow{\min}_P\text{‘}\alpha \in 0 \cup 1 . \overrightarrow{\max}_P\text{‘}\alpha \in 0 \cup 1$
whence

***205·31.** $\vdash : P \in \text{connex} . \supset . \min_P, \max_P \in 1 \rightarrow \text{Cls}$
which leads to

***205·33.** $\vdash : P \in \text{connex} . \kappa \subset \text{Ɑ‘}\min_P . \supset . \min_P \upharpoonright \kappa \in \epsilon_\Delta\text{‘}\kappa$

This proposition is useful in the theory of well-ordered series. Observe that "$\kappa \subset \text{D‘}\min_P$" means that κ consists of classes which have minima.

We have next a set of propositions (∗205·4—·44) dealing with the relations of $\min_P{}^{\backprime}\alpha$ to $B{}^{\backprime}P \upharpoonright \alpha$ and $B{}^{\backprime}\alpha \upharpoonleft P$; next we have propositions on the relations of the minima of two different classes, of which the most useful is

∗205·55. $\vdash : P \in \text{connex} \,.\, B{}^{\backprime}P \in \alpha \,.\, \supset \,.\, B{}^{\backprime}P = \min_P{}^{\backprime}\alpha$

We have next various propositions on $p{}^{\backprime}\overrightarrow{P}{}^{\backprime\backprime}(\alpha \cap C{}^{\backprime}P)$, of which the chief is

∗205·65. $\vdash : P \in \text{trans} \cap \text{connex} \,.\, \text{E}\,!\, \min_P{}^{\backprime}\alpha \,.\, \supset \,.\, p{}^{\backprime}\overrightarrow{P}{}^{\backprime\backprime}(\alpha \cap C{}^{\backprime}P) = \overrightarrow{P}{}^{\backprime}\min_P{}^{\backprime}\alpha$

I.e. the predecessors of the whole of a class contained in $C{}^{\backprime}P$ are the predecessors of its minimum (if it has one).

A useful proposition is

∗205·68. $\vdash : \breve{P}{}^{\backprime\backprime}\alpha \subset \alpha \,.\, \supset \,.\, \overrightarrow{\min_P}{}^{\backprime}\alpha = \overrightarrow{\min}\,(P_{po}){}^{\backprime}\alpha$

I.e. if α is a hereditary class, its minima with respect to P are the same as its minima with respect to P_{po}.

We prove next that if $P{}^{\backprime\backprime}\alpha$ has a maximum, so has α (∗205·7), and that if $P \in$ connex, only a unit class can have its maximum identical with its minimum (∗205·73).

∗205·8—·85 are concerned with relation-arithmetic. The chief proposition here is

∗205·8. $\vdash : S \in P \,\overline{\text{smor}}\, Q \,.\, \supset \,.\, \overrightarrow{\min_P}{}^{\backprime}\alpha = S{}^{\backprime\backprime}\overrightarrow{\min_Q}{}^{\backprime}\breve{S}{}^{\backprime\backprime}\alpha$

I.e. in any correlation, the minima of the correlates of a class are the correlates of the minima.

We end with two propositions on relations with limited fields. The more useful of these is

∗205·9. $\vdash : P \in \text{connex} \,.\, \kappa \subset C{}^{\backprime}P \,.\, \kappa \sim\in 1 \,.\, \supset \,.\, \overrightarrow{\min}\,(P \upharpoonright \kappa){}^{\backprime}\alpha = \overrightarrow{\min_P}{}^{\backprime}(\alpha \cap \kappa)$

∗205·1. $\vdash : x \min_P \alpha \,.\, \equiv \,.\, x \in \alpha \cap C{}^{\backprime}P - \breve{P}{}^{\backprime\backprime}\alpha$ [∗93·11]

∗205·101. $\vdash : x \max_P \alpha \,.\, \equiv \,.\, x \in \alpha \cap C{}^{\backprime}P - P{}^{\backprime\backprime}\alpha \,.\, \equiv \,.\, x \min\,(\breve{P})\, \alpha$ [∗93·115]

∗205·102. $\vdash \,.\, \max_P = \min\,(\breve{P})$ [∗93·114]

∗205·11. $\vdash \,.\, \overrightarrow{\min_P}{}^{\backprime}\alpha = \alpha \cap C{}^{\backprime}P - \breve{P}{}^{\backprime\backprime}\alpha$ [∗93·111]

∗205·111. $\vdash \,.\, \overrightarrow{\max_P}{}^{\backprime}\alpha = \alpha \cap C{}^{\backprime}P - P{}^{\backprime\backprime}\alpha$ [∗93·116]

∗205·12. $\vdash \,.\, \overrightarrow{B}{}^{\backprime}P = \overrightarrow{\min_P}{}^{\backprime}\text{D}{}^{\backprime}P = \overrightarrow{\min_P}{}^{\backprime}C{}^{\backprime}P$ [∗93·112]

∗205·121. $\vdash \,.\, \overrightarrow{B}{}^{\backprime}\breve{P} = \overrightarrow{\max_P}{}^{\backprime}\text{Ꭰ}{}^{\backprime}P = \overrightarrow{\max_P}{}^{\backprime}C{}^{\backprime}P$ [∗93·117]

∗205·122. $\vdash : \overrightarrow{\min_P}{}^{\backprime}\alpha = \Lambda \,.\, \equiv \,.\, \alpha \cap C{}^{\backprime}P \subset \breve{P}{}^{\backprime\backprime}\alpha$ [∗205·11 . ∗24·3]

∗205·123. $\vdash : \overrightarrow{\max_P}{}^{\backprime}\alpha = \Lambda \,.\, \equiv \,.\, \alpha \cap C{}^{\backprime}P \subset P{}^{\backprime\backprime}\alpha$

∗205·13. $\vdash . \overrightarrow{\min}_P{}^{\backprime}\alpha \cup \breve{P}{}^{\backprime\backprime}\alpha = (\alpha \cap C{}^{\backprime}P) \cup \breve{P}{}^{\backprime\backprime}\alpha$ [∗22·91 . ∗205·11]

∗205·131. $\vdash . \overrightarrow{\max}_P{}^{\backprime}\alpha \cup P{}^{\backprime\backprime}\alpha = (\alpha \cap C{}^{\backprime}P) \cup P{}^{\backprime\backprime}\alpha$

∗205·14. $\vdash . \overrightarrow{\min}_P{}^{\backprime}\alpha = \hat{x}\{x \in \alpha \cap C{}^{\backprime}P . \alpha \cap \overrightarrow{P}{}^{\backprime}x = \Lambda\}$ [∗37·462 . ∗205·11]

∗205·141. $\vdash . \overrightarrow{\max}_P{}^{\backprime}\alpha = \hat{x}\{x \in \alpha \cap C{}^{\backprime}P . \alpha \cap \overleftarrow{P}{}^{\backprime}x = \Lambda\}$

∗205·15. $\vdash . \overrightarrow{\min}_P{}^{\backprime}(\alpha \cap C{}^{\backprime}P) = \overrightarrow{\min}_P{}^{\backprime}\alpha$ [∗37·265 . ∗205·11]

∗205·151. $\vdash . \overrightarrow{\max}_P{}^{\backprime}(\alpha \cap C{}^{\backprime}P) = \overrightarrow{\max}_P{}^{\backprime}\alpha$

∗205·16. $\vdash . \overrightarrow{\min}_P{}^{\backprime}\Lambda = \Lambda$ [∗205·11 . ∗24·23]

∗205·161. $\vdash . \overrightarrow{\max}_P{}^{\backprime}\Lambda = \Lambda$

∗205·17. $\vdash :. x \in (\alpha \cap C{}^{\backprime}P) . \supset_x . \sim(xPx) : \alpha \cap C{}^{\backprime}P \in 1 : \supset .$

$$\overrightarrow{\min}_P{}^{\backprime}\alpha = \overrightarrow{\max}_P{}^{\backprime}\alpha = \alpha \cap C{}^{\backprime}P$$

Dem.

$\vdash . *13{\cdot}14 . \quad \supset \vdash :. \text{Hp} . \supset : x \in \alpha . xPy . \supset_{x,y} . x \neq y$ (1)

$\vdash . *52{\cdot}16 . \quad \supset \vdash :. \text{Hp} . \supset : x, y \in \alpha \cap C{}^{\backprime}P . \supset_{x,y} . x = y$ (2)

$\vdash . (1) . (2) . *33{\cdot}17 . \supset \vdash :. \text{Hp} . \supset : x \in \alpha . xPy . \supset_{x,y} . y \sim\in \alpha :$

[∗37·1] $\quad \supset : \breve{P}{}^{\backprime\backprime}\alpha \subset -\alpha :$

[∗22·811] $\quad \supset : \alpha \subset -\breve{P}{}^{\backprime\backprime}\alpha$ (3)

$\vdash . (3) . *205{\cdot}11 . \supset \vdash . \text{Prop}$

∗205·18. $\vdash : \sim(xPx) . x \in C{}^{\backprime}P . \supset . \min_P{}^{\backprime}\iota{}^{\backprime}x = \max_P{}^{\backprime}\iota{}^{\backprime}x = x$

Dem.

$\vdash . *205{\cdot}17 . \supset \vdash : \text{Hp} . \supset . \overrightarrow{\min}_P{}^{\backprime}\iota{}^{\backprime}x = \overrightarrow{\max}_P{}^{\backprime}\iota{}^{\backprime}x = \iota{}^{\backprime}x$ (1)

$\vdash . (1) . *53{\cdot}4 . \supset \vdash . \text{Prop}$

∗205·181. $\vdash : xPy . \sim(xPx) . \sim(yPx) . \supset . \min_P{}^{\backprime}(\iota{}^{\backprime}x \cup \iota{}^{\backprime}y) = x$

Dem.

$\vdash . *37{\cdot}105 . \supset \vdash : \text{Hp} . \supset . x \sim\in \breve{P}{}^{\backprime\backprime}(\iota{}^{\backprime}x \cup \iota{}^{\backprime}y) . y \in \breve{P}{}^{\backprime\backprime}(\iota{}^{\backprime}x \cup \iota{}^{\backprime}y)$ (1)

$\vdash . *33{\cdot}17 . \supset \vdash : \text{Hp} . \supset . \iota{}^{\backprime}x \cup \iota{}^{\backprime}y \subset C{}^{\backprime}P$ (2)

$\vdash . (1) . (2) . *205{\cdot}11 . \supset \vdash : \text{Hp} . \supset . \overrightarrow{\min}_P{}^{\backprime}(\iota{}^{\backprime}x \cup \iota{}^{\backprime}y) = \iota{}^{\backprime}x : \supset \vdash . \text{Prop}$

∗205·182. $\vdash : P^2 \mathrel{\subset\!\!\!\cdot} J . xPy . \supset . \min_P{}^{\backprime}(\iota{}^{\backprime}x \cup \iota{}^{\backprime}y) = x$

Dem.

$\vdash . *200{\cdot}36 . *50{\cdot}43 . \supset \vdash : \text{Hp} . \supset . \sim(xPx) . \sim(yPx)$ (1)

$\vdash . (1) . *205{\cdot}181 . \supset \vdash . \text{Prop}$

***205·183.** $\vdash :. P^2 \subseteq J . P \epsilon \text{connex} . x, y \epsilon C'P . \supset :$

$$\text{min}_P`(\iota`x \cup \iota`y) = x . \vee . \text{min}_P`(\iota`x \cup \iota`y) = y$$

Dem.

$\vdash . *202\cdot103 . \supset \vdash :. \text{Hp} . \supset : x = y . \vee . xPy . \vee . yPx$ (1)

$\vdash . *205\cdot18 . \supset \vdash : \text{Hp} . x = y . \supset . \text{min}_P`(\iota`x \cup \iota`y) = x$ (2)

$\vdash . *205\cdot182 . \supset \vdash : \text{Hp} . xPy . \supset . \text{min}_P`(\iota`x \cup \iota`y) = x$ (3)

$\vdash . *205\cdot182 . \supset \vdash : \text{Hp} . yPx . \supset . \text{min}_P(\iota`x \cup \iota`y) = y$ (4)

$\vdash . (1) . (2) . (3) . (4) . \supset \vdash . \text{Prop}$

***205·19.** $\vdash : P \epsilon \text{trans} . \supset . \overrightarrow{\text{min}}_P`\alpha = \overrightarrow{\text{min}}_P`(\alpha \cup \breve{P}``\alpha) = \overrightarrow{\text{min}}_P`\breve{P}_*``\alpha$

Dem.

$\vdash . *205\cdot11 . \quad \supset \vdash . \overrightarrow{\text{min}}_P`(\alpha \cup \breve{P}``\alpha) = (\alpha \cup \breve{P}``\alpha) \cap C`P - \breve{P}``(\alpha \cup \breve{P}``\alpha)$ (1)

$\vdash . (1) . *201\cdot55 . \supset \vdash : \text{Hp} . \supset . \overrightarrow{\text{min}}_P`(\alpha \cup \breve{P}``\alpha) = (\alpha \cup \breve{P}``\alpha) \cap C`P - \breve{P}``\alpha$

[*22·9] $= \alpha \cap C`P - \breve{P}``\alpha$

[*205·11] $= \overrightarrow{\text{min}}_P`\alpha$ (2)

$\vdash . *201\cdot52 . *37\cdot265 . \supset \vdash : \text{Hp} . \supset . \breve{P}_*``\alpha = (\alpha \cap C`P) \cup \breve{P}``(\alpha \cap C`P) .$

[(2)] $\supset . \overrightarrow{\text{min}}_P`\breve{P}_*``\alpha = \overrightarrow{\text{min}}_P`(\alpha \cap C`P)$

[*205·15] $= \overrightarrow{\text{min}}_P`\alpha$ (3)

$\vdash . (2) . (3) . \supset \vdash . \text{Prop}$

***205·191.** $\vdash : P \epsilon \text{trans} . \supset . \overrightarrow{\text{max}}_P`\alpha = \overrightarrow{\text{max}}_P`(\alpha \cup P``\alpha) = \overrightarrow{\text{max}}_P`P_*``\alpha$

***205·192.** $\vdash : P \epsilon \text{trans} . \beta \subset \breve{P}``\alpha . \supset . \overrightarrow{\text{min}}_P`(\alpha \cup \beta) = \overrightarrow{\text{min}}_P`\alpha$

Dem.

$\vdash . *205\cdot11 . *201\cdot56 . \supset$

$\vdash : \text{Hp} . \supset . \overrightarrow{\text{min}}_P`(\alpha \cup \beta) = (\alpha \cup \beta) \cap C`P - \breve{P}``\alpha$

[*22·68] $= (\alpha \cap C`P - \breve{P}``\alpha) \cup (\beta \cap C`P - \breve{P}``\alpha)$

[*24·3] $= \alpha \cap C`P - \breve{P}``\alpha$

[*205·11] $= \overrightarrow{\text{min}}_P`\alpha : \supset \vdash . \text{Prop}$

***205·193.** $\vdash : P \epsilon \text{trans} . \beta \subset P``\alpha . \supset . \overrightarrow{\text{max}}_P`(\alpha \cup \beta) = \overrightarrow{\text{max}}_P`\alpha$

***205·194.** $\vdash : x \text{ min}_P \alpha . \supset . \sim(xPx)$

Dem.

$\vdash . *37\cdot105 . \quad \supset \vdash . x \epsilon \alpha . xPx . \supset . x \epsilon \breve{P}``\alpha$ (1)

$\vdash . (1) . \text{Transp} . \supset \vdash : x \epsilon \alpha - \breve{P}``\alpha . \supset . \sim(xPx)$ (2)

$\vdash . (2) . *205\cdot1 . \supset \vdash . \text{Prop}$

***205·195.** $\vdash : x \text{ max}_P \alpha . \supset . \sim(xPx)$

***205·196.** $\vdash :. P \epsilon \text{Rl}`J \cap \text{trans} . \supset : x \epsilon C`P . \equiv . x = \text{min}_P`(\iota`x \cup \overleftarrow{P}`x)$

Dem.

$\vdash . *205·19 . \supset \vdash :. \text{Hp} . \supset : \overrightarrow{\text{min}_P}`(\iota`x \cup \overleftarrow{P}`x) = \overrightarrow{\text{min}_P}`\iota`x :$

$[*205·18] \quad \supset : x \epsilon C`P . \supset . \text{min}_P`(\iota`x \cup \overleftarrow{P}`x) = x \quad (1)$

$\vdash . *205·11 . \supset \vdash : \text{min}_P`(\iota`x \cup \overleftarrow{P}`x) = x . \supset . x \epsilon C`P \quad (2)$

$\vdash . (1) . (2) . \supset \vdash . \text{Prop}$

***205·197.** $\vdash :. P \epsilon \text{Rl}`J \cap \text{trans} . \supset : x \epsilon C`P . \equiv . x = \text{max}_P`(\overrightarrow{P}`x \cup \iota`x)$

***205·2.** $\vdash :. P \epsilon \text{connex} . E ! \text{min}_P`\alpha . y \epsilon \alpha \cap C`P . \supset : \text{min}_P`\alpha = y . \vee . \text{min}_P`\alpha P y$

Dem.

$\vdash . *202·103 . \supset \vdash :. \text{Hp} . \supset : y P \text{min}_P`\alpha . \vee . \text{min}_P`\alpha = y . \vee . \text{min}_P`\alpha P y \quad (1)$

$\vdash . *205·14 . \supset \vdash : \text{Hp} . \supset . \sim (y P \text{min}_P`\alpha) \quad (2)$

$\vdash . (1) . (2) . \supset \vdash . \text{Prop}$

In the remainder of the present number, when a proposition has been proved for min_P, we shall not state the corresponding proposition for max_P, unless it is specially important. When propositions concerning max_P are required for reference in the sequel, we shall refer to the corresponding propositions for min_P, in case no reference exists for max_P.

***205·21.** $\vdash : P \epsilon \text{connex} . E ! \text{min}_P`\alpha . y \epsilon \alpha \cap C`P - \iota`\text{min}_P`\alpha . \supset . \text{min}_P`\alpha P y$ [*205·2]

***205·211.** $\vdash : P \epsilon \text{trans} \cap \text{connex} . E ! \text{min}_P`\alpha . y \epsilon \breve{P}``\alpha . \supset . \text{min}_P`\alpha P y$

Dem.

$\vdash . *37·105 . \supset \vdash : \text{Hp} . \supset . (\exists x) . x \epsilon \alpha . x P y \quad (1)$

$\vdash . *13·13 . \supset \vdash : x \epsilon \alpha . x P y . x = \text{min}_P`\alpha . \supset . \text{min}_P`\alpha P y \quad (2)$

$\vdash . *205·21 . \supset \vdash : \text{Hp} . x \epsilon \alpha . x P y . x \neq \text{min}_P`\alpha . \supset . \text{min}_P`\alpha P x . x P y .$

$[\text{Hp}.*201·1] \quad \supset . \text{min}_P`\alpha P y \quad (3)$

$\vdash . (1) . (2) . (3) . \supset \vdash . \text{Prop}$

***205·22.** $\vdash : P \epsilon \text{trans} \cap \text{connex} . E ! \text{min}_P`\alpha . \supset . \breve{P}``\alpha = \overleftarrow{P}`\text{min}_P`\alpha$ [*205·211 . *37·181]

***205·23.** $\vdash : P \epsilon \text{connex} . x \epsilon D`P . y \epsilon \overrightarrow{B}`\breve{P} . \supset . x P y$

Dem.

$\vdash . *93·101 . \supset \vdash : \text{Hp} . \supset . x \neq y . \sim (y P x) .$

$[*202·103] \quad \supset . x P y : \supset \vdash . \text{Prop}$

***205·24.** $\vdash : P \epsilon \text{connex} . \supset . \overrightarrow{B}`\breve{P} \subset p`\overleftarrow{P}``D`P$ [*205·23]

***205·241.** $\vdash : P \epsilon \text{connex} . \supset . \overrightarrow{B}`P \subset p`\overrightarrow{P}``\text{Ɑ}`P$ [Proof as in *205·24]

∗205·25. $\vdash . \overrightarrow{\min}_P{}^{\backprime}\overleftarrow{P}{}^{\backprime}x = (\overleftarrow{P \dot{-} P^2}){}^{\backprime}x$

Dem.

$$\begin{aligned} \vdash . *205{\cdot}11 . \supset \vdash . \overrightarrow{\min}_P{}^{\backprime}\overleftarrow{P}{}^{\backprime}x &= \overleftarrow{P}{}^{\backprime}x - \breve{P}{}^{\backprime\backprime}\overleftarrow{P}{}^{\backprime}x \\ [*37{\cdot}301] \quad &= \overleftarrow{P}{}^{\backprime}x - \overleftarrow{P^2}{}^{\backprime}x \\ [*32{\cdot}31{\cdot}35] \quad &= (\overleftarrow{P \dot{-} P^2}){}^{\backprime}x . \supset \vdash . \text{Prop} \end{aligned}$$

The following proposition is used in the theory of well-ordered series (∗250·2).

∗205·251. $\vdash : \exists ! \overrightarrow{\min}_P{}^{\backprime}\overleftarrow{P}{}^{\backprime}x . \equiv . x \in \text{D}{}^{\backprime}(P \dot{-} P^2)$ [∗205·25]

∗205·252. $\vdash : \exists ! \overrightarrow{\max}_P{}^{\backprime}\overrightarrow{P}{}^{\backprime}x . \equiv . x \in \text{Ɑ}{}^{\backprime}(P \dot{-} P^2)$

∗205·253. $\vdash : P \in \text{connex} . \text{E} ! B{}^{\backprime}P . \supset . \text{Ɑ}{}^{\backprime}P = \overleftarrow{P}{}^{\backprime}B{}^{\backprime}P$ [∗202·524]

∗205·254. $\vdash : P \in \text{connex} . \text{E} ! B{}^{\backprime}P . \supset . \overleftarrow{\min}_P{}^{\backprime}\text{Ɑ}{}^{\backprime}P = \overleftarrow{P \dot{-} P^2}{}^{\backprime}B{}^{\backprime}P$ [∗205·253·25]

∗205·255. $\vdash : \exists ! \overrightarrow{\min}_P{}^{\backprime}\text{Ɑ}{}^{\backprime}P . \supset . \exists ! \overrightarrow{B}{}^{\backprime}P$

Dem.

$$\begin{aligned} \vdash . *93{\cdot}101 . \supset \vdash : \overrightarrow{B}{}^{\backprime}P = \Lambda . &\supset . \text{D}{}^{\backprime}P \subset \text{Ɑ}{}^{\backprime}P . \\ [*37{\cdot}271] \quad &\supset . \text{Ɑ}{}^{\backprime}P = \breve{P}{}^{\backprime\backprime}\text{Ɑ}{}^{\backprime}P . \\ [*205{\cdot}122] \quad &\supset . \overrightarrow{\min}_P{}^{\backprime}\text{Ɑ}{}^{\backprime}P = \Lambda \quad (1) \\ \vdash . (1) . \text{Transp} . \supset \vdash . \text{Prop} & \end{aligned}$$

∗205·256. $\vdash :. P \in \text{Ser} . \supset :$

$$\text{E} ! \min_P{}^{\backprime}\text{Ɑ}{}^{\backprime}P . \equiv . \text{E} ! \breve{P}_1{}^{\backprime}B{}^{\backprime}P . \equiv . \min_P{}^{\backprime}\text{Ɑ}{}^{\backprime}P = \breve{P}_1{}^{\backprime}B{}^{\backprime}P$$

[∗205·254·255 . ∗201·63 . ∗202·52·7]

∗205·26. $\vdash : Q \subseteq P . \supset . \min_P \restriction \text{Cl}{}^{\backprime}C{}^{\backprime}Q \subseteq \min_Q$

Dem.

$$\begin{aligned} \vdash . *37{\cdot}201 . \supset \vdash :. \text{Hp} . \alpha \subset C{}^{\backprime}Q . \supset &: \breve{Q}{}^{\backprime\backprime}\alpha \subset \breve{P}{}^{\backprime\backprime}\alpha . \alpha \subset C{}^{\backprime}Q . \alpha \subset C{}^{\backprime}P : \\ [\text{Transp}. *22{\cdot}621] \quad &\supset : \alpha - \breve{P}{}^{\backprime\backprime}\alpha \subset \alpha - \breve{Q}{}^{\backprime\backprime}\alpha . \alpha = \alpha \cap C{}^{\backprime}Q = \alpha \cap C{}^{\backprime}P : \\ [*205{\cdot}11] \quad &\supset : \overrightarrow{\min}_P{}^{\backprime}\alpha \subset \overrightarrow{\min}_Q{}^{\backprime}\alpha : \\ [*32{\cdot}18] \quad &\supset : x \min_P \alpha . \supset . x \min_Q \alpha :. \supset \vdash . \text{Prop} \end{aligned}$$

∗205·261. $\vdash : P \upharpoonleft \beta \in \text{connex} . \beta \cap C{}^{\backprime}P \sim\in 1 . \supset . \overrightarrow{\min}(P \upharpoonleft \beta){}^{\backprime}\alpha = \overrightarrow{\min}_P{}^{\backprime}(\alpha \cap \beta)$

Dem.

$$\begin{aligned} &\vdash . *205{\cdot}11 . *202{\cdot}54 . *37{\cdot}413 . *36{\cdot}34 . \supset \\ &\vdash : \text{Hp} . \supset . \overrightarrow{\min}(P \upharpoonleft \beta){}^{\backprime}\alpha = \alpha \cap \beta \cap C{}^{\backprime}P - \{\beta \cap \breve{P}{}^{\backprime\backprime}(\alpha \cap \beta)\} \\ &[*22{\cdot}93 . *205{\cdot}11] \quad = \overrightarrow{\min}_P{}^{\backprime}(\alpha \cap \beta) : \supset \vdash . \text{Prop} \end{aligned}$$

***205·262.** $\vdash : P \in \text{trans} \cap \text{connex} . x \in \alpha \cap C\text{'}P . \beta = \overrightarrow{P}\text{'}x \cup \iota\text{'}x . \supset .$
$\overrightarrow{\min}_P\text{'}\alpha = \overrightarrow{\min}_P\text{'}(\alpha \cap \beta)$

Dem.

$\vdash . *32\cdot18 . \supset \vdash :. \text{Hp} . y \in \alpha . yPx . \supset : y \in \alpha \cap \beta :$

$[*37\cdot105] \quad \supset : yPz . \supset . z \in \breve{P}\text{''}(\alpha \cap \beta) \quad (1)$

$\vdash . *51\cdot15 . \supset \vdash :. \text{Hp} . y \in \alpha . y = x . \supset : y \in \alpha \cap \beta :$

$[*37\cdot105] \quad \supset : yPz . \supset . z \in \breve{P}\text{''}(\alpha \cap \beta) \quad (2)$

$\vdash . *51\cdot15 . *201\cdot1 . \supset \vdash : \text{Hp} . y \in \alpha . xPy . yPz . \supset . x \in \alpha . xPz .$

$[*37\cdot105] \quad \supset . z \in \breve{P}\text{''}(\alpha \cap \beta) \quad (3)$

$\vdash . (1) . (2) . (3) . *202\cdot103 . \supset \vdash :. \text{Hp} . \supset : y \in \alpha . yPz . \supset . z \in \breve{P}\text{''}(\alpha \cap \beta) :$

$[*37\cdot105\cdot2] \quad \supset : \breve{P}\text{''}\alpha = \breve{P}\text{''}(\alpha \cap \beta) \quad (4)$

$\vdash . *37\cdot181 . *202\cdot101 . \supset \vdash : \text{Hp} . \supset . \overleftarrow{P}\text{'}x \subset \breve{P}\text{''}\alpha . \overleftarrow{P}\text{'}x = C\text{'}P - \beta .$

$[*22\cdot82] \quad \supset . C\text{'}P - \breve{P}\text{''}\alpha \subset \beta .$

$[*22\cdot621.(4)] \quad \supset . C\text{'}P \cap \alpha - \breve{P}\text{''}\alpha = C\text{'}P \cap \alpha \cap \beta - \breve{P}\text{''}(\alpha \cap \beta) .$

$[*205\cdot11] \quad \supset . \overrightarrow{\min}_P\text{'}\alpha = \overrightarrow{\min}_P\text{'}(\alpha \cap \beta) : \supset \vdash . \text{Prop}$

***205·27.** $\vdash : P \in \text{trans} \cap \text{connex} . x \in \alpha \cap C\text{'}P . \beta = \overrightarrow{P}\text{'}x \cup \iota\text{'}x . \supset .$
$\overrightarrow{\min}_P\text{'}\alpha = \overrightarrow{\min}\,(P \upharpoonright \beta)\text{'}\alpha = \overrightarrow{\min}_P\text{'}(\alpha \cap \beta)$

Dem.

$\vdash . *52\cdot41 . \quad \supset \vdash : \text{Hp} . \overrightarrow{P}\text{'}x \neq \iota\text{'}x . \supset . \beta \sim \in 1 .$

$[*205\cdot261] \quad \supset . \overrightarrow{\min}\,(P \upharpoonright \beta)\text{'}\alpha = \overrightarrow{\min}_P\text{'}(\alpha \cap \beta) \quad (1)$

$\vdash . *202\cdot101 . \supset \vdash : \text{Hp} . \overrightarrow{P}\text{'}x = \iota\text{'}x . \supset . C\text{'}P - \iota\text{'}x = \overleftarrow{P}\text{'}x . xPx .$

$[*37\cdot105] \quad \supset . C\text{'}P \subset \breve{P}\text{''}(\alpha \cap \beta) .$

$[*205\cdot122.*37\cdot413] \quad \supset . \overrightarrow{\min}_P\text{'}(\alpha \cap \beta) = \Lambda . \overrightarrow{\min}\,(P \upharpoonright \beta)\text{'}\alpha = \Lambda \quad (2)$

$\vdash . (1) . (2) . \supset \vdash : \text{Hp} . \supset . \overrightarrow{\min}\,(P \upharpoonright \beta)\text{'}\alpha = \overrightarrow{\min}_P\text{'}(\alpha \cap \beta) \quad (3)$

$\vdash . (3) . *205\cdot262 . \supset \vdash . \text{Prop}$

The above proposition is used in $*250\cdot7$.

***205·3.** $\vdash : P \in \text{connex} . \supset . \overrightarrow{\min}_P\text{'}\alpha \in 0 \cup 1 . \overrightarrow{\max}_P\text{'}\alpha \in 0 \cup 1$

Dem.

$\vdash . *205\cdot11 . \supset \vdash :. x, y \in \overrightarrow{\min}_P\text{'}\alpha . \supset : x, y \in \alpha \cap C\text{'}P : z \in \alpha . \supset_z . \sim(zPx) . \sim(zPy) :$

$[*10\cdot1] \quad \supset : x, y \in \alpha \cap C\text{'}P . \sim(yPx) . \sim(xPy) \quad (1)$

$\vdash . (1) . *202\cdot103 . \supset \vdash :. \text{Hp} . \supset : x, y \in \overrightarrow{\min}_P\text{'}\alpha . \supset . x = y :$

$[*52\cdot4] \quad \supset : \overrightarrow{\min}_P\text{'}\alpha \in 0 \cup 1 \quad (2)$

Similarly $\quad \vdash :. \text{Hp} . \supset : \overrightarrow{\max}_P\text{'}\alpha \in 0 \cup 1 \quad (3)$

$\vdash . (2) . (3) . \supset \vdash . \text{Prop}$

The above proposition is of great importance in the theory of maxima and minima.

***205·31.** $\vdash : P \epsilon \text{connex} . \supset . \min_P, \max_P \epsilon 1 \rightarrow \text{Cls}$ [*205·3 . *71·12]

***205·32.** $\vdash :. P \epsilon \text{connex} . \supset : \exists ! \overrightarrow{\min_P}\text{‘}\alpha . \equiv . E ! \min_P\text{‘}\alpha . \equiv . \alpha \epsilon \text{ꓷ‘}\min_P$
[*205·31 . *71·163 . *33·41]

***205·33.** $\vdash : P \epsilon \text{connex} . \kappa \subset \text{ꓷ‘}\min_P . \supset . \min_P \upharpoonright \kappa \epsilon \epsilon_\Delta\text{‘}\kappa$

Dem.

$$\vdash . *205{\cdot}31 . \supset \vdash : \text{Hp} . \supset . \min_P \upharpoonright \kappa \epsilon 1 \rightarrow \text{Cls} \quad (1)$$
$$\vdash . *205{\cdot}1 . \supset \vdash : \text{Hp} . \supset . \min_P \upharpoonright \kappa \subseteq \epsilon \quad (2)$$
$$\vdash . *35{\cdot}65 . \supset \vdash : \text{Hp} . \supset . \text{ꓷ‘}\min_P \upharpoonright \kappa = \kappa \quad (3)$$
$$\vdash . (1) . (2) . (3) . *80{\cdot}14 . \supset \vdash . \text{Prop}$$

***205·34.** $\vdash : P \epsilon \text{connex} . \kappa \subset \text{ꓷ‘}\min_P . \supset . \kappa \epsilon \text{Cls}^2 \text{ mult}$ [*205·33 . *88·2]

The following proposition is used in *260·17.

***205·35.** $\vdash :: P^2 \subseteq J . P \epsilon \text{connex} . \supset :.$
$$x = \min_P\text{‘}\alpha . \equiv : x \epsilon \alpha \cap C\text{‘}P : y \epsilon \alpha \cap C\text{‘}P - \iota\text{‘}x . \supset_y . xPy$$

Dem.

$\vdash . *205{\cdot}31 . *71{\cdot}36 . \supset \vdash :: \text{Hp} . \supset :. x = \min_P\text{‘}\alpha . \equiv : x \min_P \alpha :$

[*205·1.*37·265] $\equiv : x \epsilon \alpha \cap C\text{‘}P - \breve{P}\text{“}(\alpha \cap C\text{‘}P) :$

[*37·105] $\equiv : x \epsilon \alpha \cap C\text{‘}P : y \epsilon \alpha \cap C\text{‘}P . \supset_y . \sim(yPx) :$

[*51·221] $\equiv : x \epsilon \alpha \cap C\text{‘}P : y = x . \supset_y . \sim(yPx) : y \epsilon \alpha \cap C\text{‘}P - \iota\text{‘}x . \supset_x . \sim(yPx)$ (1)

$\vdash . *200{\cdot}36 . \supset \vdash :. \text{Hp} . \supset : yPz . \supset_{z,y} . y \neq z :$

[Transp.*10·1] $\supset : y = x . \supset_y . \sim(yPx)$ (2)

$\vdash . (1) . (2) . \supset \vdash :: \text{Hp} . \supset :.$

$x = \min_P\text{‘}\alpha . \equiv : x \epsilon \alpha \cap C\text{‘}P : y \epsilon \alpha \cap C\text{‘}P - \iota\text{‘}x . \supset_y . \sim(yPx) :$

[*202·5] $\equiv : x \epsilon \alpha \cap C\text{‘}P : y \epsilon \alpha \cap C\text{‘}P - \iota\text{‘}x . \supset_y . xPy :: \supset \vdash . \text{Prop}$

***205·36.** $\vdash : P \epsilon \text{trans} \cap \text{connex} . \supset . \overrightarrow{\min_P}\text{‘}\alpha \subset p\text{‘}P_*\text{“}(\alpha \cap C\text{‘}P)$

Dem.

$\vdash . *205{\cdot}2 . *201{\cdot}18 . \supset \vdash :. \text{Hp} . x = \min_P\text{‘}\alpha . \supset : y \epsilon (\alpha \cap C\text{‘}P) . \supset_y . xP_*y :. \supset \vdash . \text{Prop}$

The above proposition is used in *230·53.

***205·37.** $\vdash : P \epsilon \text{trans} . \overrightarrow{\max_P}\text{‘}\alpha = \Lambda . \supset . P_*\text{“}\alpha = P\text{“}\alpha$ [*201·52.*205·123]

The following proposition is used in *257·21.

***205·38.** $\vdash : P_{\text{po}} \subseteq J . \supset . \mu \cap p\text{‘}\overrightarrow{P_*}\text{“}\mu \subset \overrightarrow{\min}(P_{\text{po}})\text{‘}\mu$

Dem.

$\vdash . *200{\cdot}381 . \supset \vdash :: \text{Hp} . \supset :. x \epsilon \mu . \supset_x . yP_*x : \supset : x \epsilon \mu . \supset_x . \sim(xP_{\text{po}}y) :.$

[*40·51.*37·105] $\supset :. p\text{‘}\overrightarrow{P_*}\text{“}\mu \subset - \breve{P}_{\text{po}}\text{“}\mu$ (1)

$\vdash . *40{\cdot}62 . \supset \vdash : \exists ! \mu . \supset . p\text{‘}\overrightarrow{P_*}\text{“}\mu \subset C\text{‘}P$ (2)

$\vdash . *24{\cdot}12 . \supset \vdash : \sim \exists ! \mu . \supset . \mu \subset C\text{‘}P$ (3)

$\vdash . (1) . (2) . (3) . \supset \vdash : \text{Hp} . \supset . \mu \cap p\text{‘}\overrightarrow{P_*}\text{“}\mu \subset \mu \cap C\text{‘}P - \breve{P}_{\text{po}}\text{“}\mu$

[*205·11.*91·504] $\subset \overrightarrow{\min}(P_{\text{po}})\text{‘}\mu : \supset \vdash . \text{Prop}$

***205·381.** $\vdash : P_{po} \subseteq J \,.\, \overrightarrow{\max}_P{}^{\backprime}\mu = \Lambda \,.\, \supset \,.\, p^{\backprime}\overleftarrow{P_*}{}^{\backprime\backprime}\mu = p^{\backprime}\overleftarrow{P_{po}}{}^{\backprime\backprime}\mu$

Dem.

$\vdash \,.\, *205\cdot38 \frac{\breve{P}}{P} \,.\, \supset \vdash : \text{Hp} \,.\, \supset \,.\, \mu \cap p^{\backprime}\overleftarrow{P_*}{}^{\backprime\backprime}\mu = \Lambda$ (1)

$\vdash \,.\, (1) \,.\, *40\cdot53 \,.\, *24\cdot37 \,.\, \supset$

$\vdash :: \text{Hp} \,.\, \supset :. x \in p^{\backprime}\overleftarrow{P_*}{}^{\backprime\backprime}\mu \,.\, \equiv : y \in \mu \,.\, \supset_y \,.\, yP_*x \,.\, y \neq x :$

[*200·38] $\equiv : y \in \mu \,.\, \supset_y \,.\, yP_{po}x :$

[*40·53] $\equiv : x \in p^{\backprime}\overleftarrow{P_{po}}{}^{\backprime\backprime}\mu :: \supset \vdash \,.\, \text{Prop}$

The three following propositions lead up to *205·42, which is used in *261·26.

***205·4.** $\vdash : C^{\backprime}P \in 1 \,.\, \supset . \overrightarrow{B}{}^{\backprime}P = \Lambda \,.\, \overrightarrow{B}{}^{\backprime}\breve{P} = \Lambda$

Dem.

$\vdash \,.\, *56\cdot381 \,.\, *55\cdot15 \,.\, \supset \vdash : \text{Hp} \,.\, \supset \,.\, (\exists x) \,.\, \text{D}^{\backprime}P = \iota^{\backprime}x \,.\, \text{Ɑ}^{\backprime}P = \iota^{\backprime}x \,.$

[*93·101] $\supset \,.\, \overrightarrow{B}{}^{\backprime}P = \Lambda \,.\, \overrightarrow{B}{}^{\backprime}\breve{P} = \Lambda : \supset \vdash \,.\, \text{Prop}$

***205·401.** $\vdash : \exists ! \overrightarrow{B}{}^{\backprime}P \upharpoonright \alpha \,.\, \supset \,.\, \alpha \cap C^{\backprime}P \sim\in 0 \cup 1 \,.\, C^{\backprime}P \upharpoonright \alpha \sim\in 0 \cup 1$

Dem.

$\vdash \,.\, *205\cdot4 \,.\, \text{Transp} \,.\, \supset \vdash : \text{Hp} \,.\, \supset \,.\, C^{\backprime}P \upharpoonright \alpha \sim\in 1$ (1)

$\vdash \,.\, *93\cdot103 \,.\, \supset \vdash : \text{Hp} \,.\, \supset \,.\, \exists ! C^{\backprime}P \upharpoonright \alpha$ (2)

$\vdash \,.\, (1) \,.\, (2) \,.\, \supset \vdash : \text{Hp} \,.\, \supset \,.\, C^{\backprime}P \upharpoonright \alpha \sim\in 0 \cup 1$ (3)

$\vdash \,.\, *37\cdot41 \,.\, \supset \vdash \,.\, C^{\backprime}P \upharpoonright \alpha \subset \alpha \cap C^{\backprime}P$ (4)

$\vdash \,.\, (4) \,.\, *60\cdot32\cdot371 \,.\, \text{Transp} \,.\, \supset \vdash : C^{\backprime}P \upharpoonright \alpha \sim\in 0 \cup 1 \,.\, \supset \,.\, \alpha \cap C^{\backprime}P \sim\in 0 \cup 1$ (5)

$\vdash \,.\, (3) \,.\, (5) \,.\, \supset \vdash \,.\, \text{Prop}$

The following proposition, besides being required for *205·41, is used in *250·151.

***205·41.** $\vdash : P \in \text{connex} \,.\, \alpha \cap C^{\backprime}P \sim\in 1 \,.\, \supset \,.\, \overrightarrow{\min}_P{}^{\backprime}\alpha = \overrightarrow{B}{}^{\backprime}P \upharpoonright \alpha$

Dem.

$\vdash \,.\, *202\cdot54 \,.\, \supset \vdash : \text{Hp} \,.\, \supset \,.\, C^{\backprime}P \upharpoonright \alpha = \alpha \cap C^{\backprime}P$ (1)

$\vdash \,.\, *37\cdot41 \,.\, \supset \vdash \,.\, \text{Ɑ}^{\backprime}P \upharpoonright \alpha = \alpha \cap \breve{P}^{\backprime\backprime}\alpha$ (2)

$\vdash \,.\, (1) \,.\, (2) \,.\, *93\cdot103 \,.\, \supset \vdash : \text{Hp} \,.\, \supset \,.\, \overrightarrow{B}{}^{\backprime}P \upharpoonright \alpha = \alpha \cap C^{\backprime}P - (\alpha \cap \breve{P}^{\backprime\backprime}\alpha)$

[*22·93.*205·11] $= \overrightarrow{\min}_P{}^{\backprime}\alpha : \supset \vdash \,.\, \text{Prop}$

***205·42.** $\vdash : P \in \text{connex} \,.\, \text{E} ! B^{\backprime}P \upharpoonright \alpha \,.\, \supset \,.\, B^{\backprime}P \upharpoonright \alpha = \min_P{}^{\backprime}\alpha$

Dem.

$\vdash \,.\, *205\cdot401 \,.\, \supset \vdash : \text{Hp} \,.\, \supset \,.\, \alpha \cap C^{\backprime}P \sim\in 1 \,.$

[*205·41] $\supset \,.\, \overrightarrow{\min}_P{}^{\backprime}\alpha = \overrightarrow{B}{}^{\backprime}P \upharpoonright \alpha$ (1)

$\vdash \,.\, (1) \,.\, *32\cdot41 \,.\, \supset \vdash \,.\, \text{Prop}$

The following proposition leads up to ∗205·44, which is used in ∗263·11.

∗205·43. $\vdash : P \in \text{connex} \,.\, \exists ! \, \alpha \cap \text{D}\text{‘}P \,.\supset .\, \overrightarrow{\min_P}\text{‘}\alpha = \overrightarrow{B}\text{‘}\alpha \rceil P$

Dem.

$\vdash . ∗205·11 . \quad \supset \vdash . \overrightarrow{\min_P}\text{‘}\alpha = (\alpha \cap \text{D}\text{‘}P - \breve{P}\text{“}\alpha) \cup (\alpha \cap \overrightarrow{B}\text{‘}\breve{P} - \breve{P}\text{“}\alpha)$

$[∗35·61 . ∗37·4] \quad = \overrightarrow{B}\text{‘}\alpha \rceil P \cup (\alpha \cap \overrightarrow{B}\text{‘}\breve{P} - \breve{P}\text{“}\alpha)$ (1)

$\vdash . ∗205·23 . \quad \supset \vdash : P \in \text{connex} \,.\, x \in \alpha \cap \text{D}\text{‘}P \,.\, y \in \overrightarrow{B}\text{‘}\breve{P} \,.\supset .\, x \in \alpha \,.\, xPy \,.$

$[∗37·1] \quad \supset .\, y \in \breve{P}\text{“}\alpha$ (2)

$\vdash . (2) . ∗10·23 . \supset \vdash : \text{Hp} . \supset . \overrightarrow{B}\text{‘}\breve{P} \subset \breve{P}\text{“}\alpha .$

$[∗24·3] \quad \supset . \alpha \cap \overrightarrow{B}\text{‘}\breve{P} - \breve{P}\text{“}\alpha = \Lambda$ (3)

$\vdash . (1) . (3) . \supset \vdash . \text{Prop}$

∗205·44. $\vdash : P \in \text{connex} \,.\, \text{E} ! \, B\text{‘}\alpha \rceil P \,.\supset .\, \min_P\text{‘}\alpha = B\text{‘}\alpha \rceil P$ [∗205·43 . ∗32·41]

The following propositions deal with the circumstances under which the minimum of one class is identical with, or earlier than, that of another.

∗205·5. $\vdash : P \in \text{connex} \,.\, \alpha \subset \beta \,.\, \min_P\text{‘}\beta \in \alpha \,.\supset .\, \text{E} ! \min_P\text{‘}\alpha \,.\, \min_P\text{‘}\alpha = \min_P\text{‘}\beta$

Dem.

$\vdash . ∗37·2 . \text{Transp} . \supset \vdash : \text{Hp} . \supset . - P\text{“}\beta \subset - P\text{“}\alpha .$

$[∗205·11 . \text{Hp}] \quad \supset . \min_P\text{‘}\beta \in \alpha - P\text{“}\alpha .$

$[∗205·1] \quad \supset . \min_P\text{‘}\beta \in \overrightarrow{\min_P}\text{‘}\alpha$ (1)

$\vdash . (1) . ∗205·3 . \supset \vdash . \text{Prop}$

∗205·501. $\vdash : P \in \text{connex} \,.\, \min_P\text{‘}\alpha = \min_P\text{‘}\beta \,.\supset .\, \beta \subset - p\text{‘}\overrightarrow{P}\text{“}\alpha$

Dem.

$\vdash . ∗205·11 . \supset \vdash :. \text{Hp} . \supset : \min_P\text{‘}\alpha \sim\in \breve{P}\text{“}\beta :$

$[∗37·105] \quad \supset : y \in \beta . \supset_y . \sim (yP \min_P\text{‘}\alpha) :$

$[∗205·11] \quad \supset : y \in \beta . \supset_y . (\exists x) . x \in \alpha . \sim (yPx) :$

$[∗40·51] \quad \supset : \beta \subset - p\text{‘}\overrightarrow{P}\text{“}\alpha :. \supset \vdash . \text{Prop}$

∗205·51. $\vdash :. P \in \text{connex} \,.\, \alpha \subset \beta \,.\, \text{E} ! \min_P\text{‘}\alpha \,.\, \text{E} ! \min_P\text{‘}\beta \,.\supset :$

$\min_P\text{‘}\alpha = \min_P\text{‘}\beta \,.\vee .\, \min_P\text{‘}\beta \, P \min_P\text{‘}\alpha$

Dem.

$\vdash . ∗22·1 . ∗205·1 . \supset \vdash : \text{Hp} . \supset . \min_P\text{‘}\alpha \in \beta \cap C\text{‘}P$ (1)

$\vdash . (1) . ∗205·2 . \supset \vdash . \text{Prop}$

∗205·52. $\vdash : P \in \text{trans} \cap \text{connex} \,.\, \exists ! \, \alpha \cap p\text{‘}\overrightarrow{P}\text{“}\beta \,.$

$\text{E} ! \min_P\text{‘}\alpha \,.\, \text{E} ! \min_P\text{‘}\beta \,.\supset .\, \min_P\text{‘}\alpha \, P \min_P\text{‘}\beta$

Dem.

$\vdash . ∗40·51 . \supset \vdash :. \text{Hp} . \supset : (\exists x) : x \in \alpha : y \in \beta . \supset_y . xPy$ (1)

$\vdash . ∗205·2 . \supset \vdash :: \text{Hp} . \supset :. x \in \alpha \cap C\text{‘}P . \supset_x : \min_P\text{‘}\alpha = x . \vee . \min_P\text{‘}\alpha \, Px$ (2)

$\vdash . (1) . *205\cdot1 . \supset \vdash :. \mathrm{Hp} . \supset : (\exists x) . x \in \alpha . xP \, \mathrm{min}_P\text{‘}\beta :$

$[*33\cdot17] \qquad \supset : (\exists x) . x \in \alpha \cap C\text{‘}P . xP \, \mathrm{min}_P\text{‘}\beta \qquad (3)$

$\vdash . (2) . (3) . \quad \supset \vdash :. \mathrm{Hp} . \supset : (\exists x) : xP \, \mathrm{min}_P\text{‘}\beta : \mathrm{min}_P\text{‘}\alpha = x . \vee . \mathrm{min}_P\text{‘}\alpha \, Px :$

$[*201\cdot1 . *13\cdot195] \qquad \supset : \mathrm{min}_P\text{‘}\alpha \, P \, \mathrm{min}_P\text{‘}\beta :. \supset \vdash . \mathrm{Prop}$

***205·53.** $\vdash : P \in \mathrm{connex} \cap \mathrm{Rl}\text{‘}J . x \in \alpha \cap C\text{‘}P . \overrightarrow{P}\text{‘}x = P\text{‘‘}\alpha . \supset . x = \mathrm{max}_P\text{‘}\alpha$

Dem.

$\vdash . *50\cdot24 . \supset \vdash : \mathrm{Hp} . \supset . x \in \alpha \cap C\text{‘}P - \overrightarrow{P}\text{‘}x .$

$[\mathrm{Hp}] \qquad \supset . x \in \alpha \cap C\text{‘}P - P\text{‘‘}\alpha .$

$[*205\cdot111] \qquad \supset . x \in \overrightarrow{\mathrm{max}}_P\text{‘}\alpha \qquad (1)$

$\vdash . (1) . *205\cdot3 . \supset \vdash . \mathrm{Prop}$

***205·54.** $\vdash :. P \in \mathrm{Ser} . \supset : x \in \alpha \cap C\text{‘}P . \overrightarrow{P}\text{‘}x = P\text{‘‘}\alpha . \equiv . x = \mathrm{max}_P\text{‘}\alpha$
$[*205\cdot53\cdot22]$

***205·55.** $\vdash : P \in \mathrm{connex} . B\text{‘}P \in \alpha . \supset . B\text{‘}P = \mathrm{min}_P\text{‘}\alpha$

Dem.

$\vdash . *93\cdot101 . *37\cdot16 . \supset \vdash : \mathrm{E} ! B\text{‘}P . \supset . B\text{‘}P \in C\text{‘}P - \breve{P}\text{‘‘}\alpha \qquad (1)$

$\vdash . (1) . *205\cdot1 . \qquad \supset \vdash : B\text{‘}P \in \alpha . \supset . B\text{‘}P \, \mathrm{min}_P \alpha \qquad (2)$

$\vdash . (2) . *205\cdot31 . \supset \vdash . \mathrm{Prop}$

***205·56.** $\vdash . \overrightarrow{\mathrm{max}}_P\text{‘}s\text{‘}\kappa \subset \mathrm{max}_P\text{‘‘}\kappa$

Dem.

$\vdash . *205\cdot111 . *40\cdot38 . \supset \vdash . \overrightarrow{\mathrm{max}}_P\text{‘}s\text{‘}\kappa \subset s\text{‘}\kappa \cap C\text{‘}P - s\text{‘}P\text{‘‘‘}\kappa$

$[*40\cdot11] \qquad \subset \hat{y} \{(\exists \alpha) . \alpha \in \kappa . y \in \alpha \cap C\text{‘}P : \sim (\exists \alpha) . \alpha \in \kappa . y \in P\text{‘‘}\alpha\}$

$[*10\cdot56] \qquad \subset \hat{y} \{(\exists \alpha) . \alpha \in \kappa . y \in \alpha \cap C\text{‘}P - P\text{‘‘}\alpha\}$

$[*205\cdot111] \qquad \subset \hat{y} \{(\exists \alpha) . \alpha \in \kappa . y \in \overrightarrow{\mathrm{max}}_P\text{‘}\alpha\}$

$[*40\cdot5] \qquad \subset \mathrm{max}_P\text{‘‘}\kappa . \supset \vdash . \mathrm{Prop}$

***205·561.** $\vdash : \kappa \subset - \text{Ꭰ}\text{‘}\mathrm{max}_P . \supset . s\text{‘}\kappa \sim \in \text{Ꭰ}\text{‘}\mathrm{max}_P \quad [*205\cdot56 . *37\cdot26\cdot29]$

***205·6.** $\vdash :. P \in \mathrm{connex} . \supset : \sim \mathrm{E} ! \mathrm{min}_P\text{‘}\alpha . \equiv . \alpha \cap C\text{‘}P \subset \breve{P}\text{‘‘}\alpha \quad [*205\cdot32\cdot122]$

***205·601.** $\vdash : P \in \mathrm{connex} . \alpha \subset C\text{‘}P . \supset : \sim \mathrm{E} ! \mathrm{min}_P\text{‘}\alpha . \equiv . \alpha \subset \breve{P}\text{‘‘}\alpha \quad [*205\cdot6]$

***205·61.** $\vdash : P \in \mathrm{connex} . \supset . C\text{‘}P = \{C\text{‘}P \cap p\text{‘}\overrightarrow{P}\text{‘‘}(\alpha \cap C\text{‘}P)\} \cup \overrightarrow{\mathrm{min}}_P\text{‘}\alpha \cup \breve{P}\text{‘‘}\alpha$
$[*202\cdot505 . *205\cdot13]$

***205·62.** $\vdash : P \in \mathrm{connex} . \exists ! \alpha \cap C\text{‘}P . \supset . C\text{‘}P = p\text{‘}\overrightarrow{P}\text{‘‘}(\alpha \cap C\text{‘}P) \cup \overrightarrow{\mathrm{min}}_P\text{‘}\alpha \cup \breve{P}\text{‘‘}\alpha$
$[*40\cdot62 . *205\cdot61]$

***205·63.** $\vdash : P \in \mathrm{connex} . P^2 \in J . \exists ! (\alpha \cap C\text{‘}P) . \supset .$

$$p\text{‘}\overrightarrow{P}\text{‘‘}(\alpha \cap C\text{‘}P) = C\text{‘}P - \breve{P}\text{‘‘}\alpha - \overrightarrow{\mathrm{min}}_P\text{‘}\alpha$$

$[*202\cdot502 . *205\cdot13]$

***205·64.** $\vdash : P \in \text{connex} . \exists ! (\alpha \cap C\text{‘}P) . \supset .$

$$\overrightarrow{\min}_P\text{‘}\alpha = C\text{‘}P - \breve{P}\text{“}\alpha - p\text{‘}\overrightarrow{P}\text{“}(\alpha \cap C\text{‘}P)$$

Dem.

$\vdash . *205·62 . \supset \vdash : \text{Hp} . \supset .$

$$C\text{‘}P - \breve{P}\text{“}\alpha - p\text{‘}\overrightarrow{P}\text{“}(\alpha \cap C\text{‘}P) = \overrightarrow{\min}_P\text{‘}\alpha - \breve{P}\text{“}\alpha - p\text{‘}\overrightarrow{P}\text{“}(\alpha \cap C\text{‘}P) \quad (1)$$

$\vdash . *205·11 . \supset \vdash . \overrightarrow{\min}_P\text{‘}\alpha - \breve{P}\text{“}\alpha = \overrightarrow{\min}_P\text{‘}\alpha$ (2)

$\vdash . *205·14 . \supset \vdash :. x \in \overrightarrow{\min}_P\text{‘}\alpha . \supset : y \in \alpha . \supset_y . \sim (yPx) :$

[*205·11.*10·1] $\supset : \sim (xPx) . x \in \alpha \cap C\text{‘}P :$

[*40·51] $\supset : x \sim \in p\text{‘}\overrightarrow{P}\text{“}(\alpha \cap C\text{‘}P)$ (3)

$\vdash . (3) . \supset \vdash . \overrightarrow{\min}_P\text{‘}\alpha - p\text{‘}\overrightarrow{P}\text{“}(\alpha \cap C\text{‘}P) = \overrightarrow{\min}_P\text{‘}\alpha$ (4)

$\vdash . (1) . (2) . (4) . \supset \vdash . \text{Prop}$

***205·65.** $\vdash : P \in \text{trans} \cap \text{connex} . \text{E} ! \min_P\text{‘}\alpha . \supset . p\text{‘}\overrightarrow{P}\text{“}(\alpha \cap C\text{‘}P) = \overrightarrow{P}\text{‘}\min_P\text{‘}\alpha$

Dem.

$\vdash . *205·2 . \supset \vdash :: \text{Hp} . \supset :. xP \min_P\text{‘}\alpha . \supset : y \in \alpha \cap C\text{‘}P . \supset_y . xPy :$

[*40·51] $\supset : x \in p\text{‘}\overrightarrow{P}\text{“}(\alpha \cap C\text{‘}P)$ (1)

$\vdash . *205·1 . *40·12 . \supset \vdash : \text{Hp} . \supset . p\text{‘}\overrightarrow{P}\text{“}(\alpha \cap C\text{‘}P) \subset \overrightarrow{P}\text{‘}\min_P\text{‘}\alpha$ (2)

$\vdash . (1) . (2) . \supset \vdash . \text{Prop}$

***205·66.** $\vdash : P \in \text{trans} \cap \text{connex} . \text{E} ! \min_P\text{‘}\alpha . \supset .$

$$p\text{‘}\overrightarrow{P}\text{“}(\alpha \cap C\text{‘}P) = \overrightarrow{P}\text{‘}\min_P\text{‘}\alpha . \breve{P}\text{“}\alpha = \overleftarrow{P}\text{‘}\min_P\text{‘}\alpha .$$

$$C\text{‘}P = p\text{‘}\overrightarrow{P}\text{“}(\alpha \cap C\text{‘}P) \cup \iota\text{‘}\min_P\text{‘}\alpha \cup \breve{P}\text{“}\alpha$$

[*205·65·22 . *202·101]

***205·67.** $\vdash :. P \in \text{Ser} . \supset : x = \min_P\text{‘}\alpha . \equiv . \overrightarrow{P}\text{‘}x = p\text{‘}\overrightarrow{P}\text{“}(\alpha \cap C\text{‘}P) . x \in C\text{‘}P$

Dem.

$\vdash . *205·65·11 . \supset$

$\vdash :. \text{Hp} . \supset : x = \min_P\text{‘}\alpha . \supset . \overrightarrow{P}\text{‘}x = p\text{‘}\overrightarrow{P}\text{“}(\alpha \cap C\text{‘}P) . x \in C\text{‘}P$ (1)

$\vdash . *50·24 . \supset \vdash : \text{Hp} . \overrightarrow{P}\text{‘}x = p\text{‘}\overrightarrow{P}\text{“}(\alpha \cap C\text{‘}P) . \supset . x \sim \in p\text{‘}\overrightarrow{P}\text{“}(\alpha \cap C\text{‘}P)$ (2)

$\vdash . *200·5 . \supset \vdash : \text{Hp}(2) . \supset . \alpha \cap \overrightarrow{P}\text{‘}x = \Lambda .$

[*37·462] $\supset . x \sim \in \breve{P}\text{“}\alpha$ (3)

$\vdash . (2) . (3) . *202·505 . \supset \vdash : \text{Hp}(2) . x \in C\text{‘}P . \supset . x \in \alpha \cap C\text{‘}P - \breve{P}\text{“}\alpha .$

[*205·3·11] $\supset . x = \min_P\text{‘}\alpha$ (4)

$\vdash . (1) . (4) . \supset \vdash . \text{Prop}$

***205·68.** $\vdash : \breve{P}``\alpha \subset \alpha . \supset . \overrightarrow{\min}_P`\alpha = \overrightarrow{\min}(P_{po})`\alpha$

Dem.

$\vdash . *91·711 . \supset \vdash : \mathrm{Hp} . \supset . \breve{P}_{po}``\alpha = \breve{P}``\alpha .$

$[*205·11] \quad \supset . \overrightarrow{\min}(P_{po})`\alpha = \overrightarrow{\min}_P`\alpha : \supset \vdash . \mathrm{Prop}$

***205·681.** $\vdash : P_{po} \epsilon \mathrm{connex} . \breve{P}``\alpha \subset \alpha . \supset . \overrightarrow{\min}_P`\alpha \epsilon 0 \cup 1 \quad [*205·68·3]$

***205·7.** $\vdash : \exists ! \overrightarrow{\max}_P`P``\alpha . \supset . \exists ! \max_P`\alpha$

Dem.

$\vdash . *37·2·265 . \quad \supset \vdash : \alpha \cap C`P \subset P``\alpha . \supset . P``\alpha \subset P``P``\alpha \quad (1)$

$\vdash . (1) . \mathrm{Transp} . \supset \vdash : \exists ! P``\alpha - P``P``\alpha . \supset . \exists ! \alpha \cap C`P - P``\alpha \quad (2)$

$\vdash . (2) . *205·111 . \supset \vdash . \mathrm{Prop}$

***205·71.** $\vdash : P \epsilon \mathrm{connex} . \exists ! \overrightarrow{\max}_P`P``\alpha . \supset . \max_P`P``\alpha (P \dot{-} P^2) \max_P`\alpha$

Dem.

$\vdash . *205·7·3 . \supset \vdash : \mathrm{Hp} . \supset . E ! \max_P`P``\alpha . E ! \max_P`\alpha . \quad (1)$

$[*205·101] \quad \supset . \max_P`P``\alpha \epsilon P``\alpha \quad (2)$

$\vdash . (1) . *205·101 . \supset \vdash :. \mathrm{Hp} . \supset : \max_P`P``\alpha \sim\epsilon P``P``\alpha :$

$[*37·39] \quad \supset : y \epsilon \alpha . \supset_y . \sim(\max_P`P``\alpha \, P^2 y) :$

$[(1)] \quad \supset : \sim(\max_P`P``\alpha \, P^2 \max_P`\alpha) : \quad (3)$

$[*34·5.\mathrm{Transp}] \quad \supset : zP \max_P`\alpha . \supset . \sim(\max_P`P``\alpha \, Pz) :$

$[*205·21] \quad \supset : z \epsilon \alpha - \iota`\max_P`\alpha . \supset . \sim(\max_P`P``\alpha \, Pz) \quad (4)$

$\vdash . (2) . *37·1 . \supset \vdash : \mathrm{Hp} . \supset . (\exists z) . z \epsilon \alpha . \max_P`P``\alpha Pz \quad (5)$

$\vdash . (4) . (5) . \quad \supset \vdash : \mathrm{Hp} . \supset . \max_P`P``\alpha \, P \max_P`\alpha \quad (6)$

$\vdash . (3) . (6) . \supset \vdash . \mathrm{Prop}$

***205·72.** $\vdash : P \epsilon \mathrm{connex} . P \in P^2 . \supset . \sim \exists ! \overrightarrow{\max}_P`P``\alpha \quad [*205·71 . \mathrm{Transp}]$

***205·73.** $\vdash : P \epsilon \mathrm{connex} . \min_P`\gamma = \max_P`\gamma . \supset . \gamma \cap C`P \epsilon 1 . \gamma \cap C`P = \iota`\min_P`\gamma$

Dem.

$\vdash . *205·21 . \quad \supset \vdash :. \mathrm{Hp} . \supset : x \epsilon \gamma \cap C`P - \iota`\min_P`\gamma . \supset . \max_P`\gamma Px .$

$[*37·1] \quad \supset . \max_P`\gamma \epsilon P``\gamma \quad (1)$

$\vdash . *205·111 . \supset \vdash : \mathrm{Hp} . \supset . \max_P`\gamma \sim\epsilon P``\gamma \quad (2)$

$\vdash . (2) . (1) . \mathrm{Transp} . \supset \vdash : \mathrm{Hp} . \supset . \gamma \cap C`P - \iota`\min_P`\gamma = \Lambda .$

$[*205·11] \quad \supset . \gamma \cap C`P = \iota`\min_P`\gamma : \supset \vdash . \mathrm{Prop}$

***205·731.** $\vdash :. P \epsilon \mathrm{connex} \cap \mathrm{Rl}`J . \supset : \min_P`\gamma = \max_P`\gamma . \equiv . \gamma \cap C`P \epsilon 1$
$[*205·17·73]$

***205·732.** $\vdash : P \epsilon \mathrm{connex} . \gamma \cap C`P \sim\epsilon 1 . E ! \min_P`\gamma . E ! \max_P`\gamma . \supset .$
$\min_P`\gamma P \max_P`\gamma$

Dem.

$\vdash . *205·73 . \mathrm{Transp} . \supset \vdash : \mathrm{Hp} . \supset . \max_P`\alpha \neq \min_P`\alpha .$

$[*205·21] \quad \supset . \min_P`\alpha P \max_P`\alpha : \supset \vdash . \mathrm{Prop}$

The following propositions lead up to *205·75, which shows that the minimum of a class belongs to $D`P$ unless the part of the class contained in $C`P$ is $\iota`B`\breve{P}$.

***205·74.** $\vdash : \alpha \cap C`P \subset \overrightarrow{B}`\breve{P} \,.\, \supset \,.\, \overrightarrow{\min}_P`\alpha = \alpha \cap C`P$

Dem.

$\vdash \,.\, *93\cdot101 \,.\, \supset \vdash : \mathrm{Hp} \,.\, \supset \,.\, \alpha \cap \mathrm{D}`P = \Lambda \,.$

$[*37\cdot261\cdot29] \quad \supset \,.\, \breve{P}``\alpha = \Lambda \,.$

$[*205\cdot11] \quad \supset \,.\, \overrightarrow{\min}_P`\alpha = \alpha \cap C`P : \supset \vdash \,.\, \mathrm{Prop}$

***205·741.** $\vdash : P \in \mathrm{connex} \,.\, \alpha \cap C`P \sim\in 1 \,.\, \supset \,.\, \overrightarrow{\min}_P`\alpha \subset \mathrm{D}`P$

Dem.

$\vdash \,.\, *205\cdot21 \,.\, \supset \vdash : P \in \mathrm{connex} \,.\, y = \min_P`\alpha \,.\, z \in \alpha \cap C`P - \iota`y \,.\, \supset \,.\, yPz :$

$[*205\cdot3] \quad \supset \vdash : P \in \mathrm{connex} \,.\, y \in \overrightarrow{\min}_P`\alpha \,.\, z \in \alpha \cap C`P - \iota`y \,.\, \supset \,.\, yPz :$

$[*33\cdot13] \quad \supset \vdash : P \in \mathrm{connex} \,.\, y \in \overrightarrow{\min}_P`\alpha \,.\, \exists ! \alpha \cap C`P - \iota`y \,.\, \supset \,.\, y \in \mathrm{D}`P :$

$[*52\cdot181] \quad \supset \vdash : P \in \mathrm{connex} \,.\, \alpha \cap C`P \sim\in 1 \,.\, \supset \,.\, \overrightarrow{\min}_P`\alpha \subset \mathrm{D}`P : \supset \vdash \,.\, \mathrm{Prop}$

***205·742.** $\vdash :. P \in \mathrm{connex} \,.\, \supset : \exists ! \overrightarrow{\min}_P`\alpha - \mathrm{D}`P \,.\, \equiv \,.\, \alpha \cap C`P = \iota`B`\breve{P}$

Dem.

$\vdash \,.\, *205\cdot74 \,.\, \supset \vdash : \alpha \cap C`P = \iota`B`\breve{P} \,.\, \supset \,.\, \overrightarrow{\min}_P`\alpha = \iota`B`\breve{P} \,.$

$[*93\cdot101] \quad \supset \,.\, \exists ! \min_P`\alpha - \mathrm{D}`P \qquad (1)$

$\vdash \,.\, *205\cdot741 \,.\, \supset \vdash : \mathrm{Hp} \,.\, \exists ! \overrightarrow{\min}_P`\alpha - \mathrm{D}`P \,.\, \supset \,.\, \alpha \cap C`P \in 1 \qquad (2)$

$\vdash \,.\, *205\cdot11 \,.\, \supset \vdash : \exists ! \overrightarrow{\min}_P`\alpha - \mathrm{D}`P \,.\, \supset \,.\, \exists ! \alpha \cap C`P - \mathrm{D}`P \,.$

$[*93\cdot103] \quad \supset \,.\, \exists ! \alpha \cap \overrightarrow{B}`\breve{P} \qquad (3)$

$\vdash \,.\, (2) \,.\, (3) \,.\, *202\cdot52 \,.\, \supset \vdash : \mathrm{Hp} \,.\, \exists ! \overrightarrow{\min}_P`\alpha - \mathrm{D}`P \,.\, \supset \,.\, \alpha \cap C`P = \iota`B`\breve{P} \qquad (4)$

$\vdash \,.\, (1) \,.\, (4) \,.\, \supset \vdash \,.\, \mathrm{Prop}$

***205·75.** $\vdash :. P \in \mathrm{connex} \,.\, \supset : \sim(\alpha \cap C`P = \iota`B`\breve{P}) \,.\, \equiv \,.\, \overrightarrow{\min}_P`\alpha \subset \mathrm{D}`P$

$[*205\cdot742]$

Observe that $\sim(\alpha \cap C`P = \iota`B`\breve{P})$ is not in general equivalent to $\alpha \cap C`P \neq \iota`B`\breve{P}$, since the latter implies $\mathrm{E} ! B`\breve{P}$, while the former does not.

The following proposition is important.

***205·8.** $\vdash : S \in P \,\overline{\mathrm{smor}}\, Q \,.\, \supset \,.\, \overrightarrow{\min}_P`\alpha = S``\overrightarrow{\min}_Q`\breve{S}``\alpha$

Dem.

$\vdash \,.\, *205\cdot11 \,.\, \supset \vdash \,.\, S``\overrightarrow{\min}_Q`\breve{S}``\alpha = S``\{\breve{S}``\alpha \cap C`Q - \breve{Q}``\breve{S}``\alpha\} \qquad (1)$

$\vdash \,.\, *151\cdot11 \,.\, \supset \vdash : \mathrm{Hp} \,.\, \supset \,.\, \breve{S}``\alpha \subset C`Q \qquad (2)$

$\vdash \,.\, (1) \,.\, (2) \,.\, \supset \vdash : \mathrm{Hp} \,.\, \supset \,.\, S``\overrightarrow{\min}_Q`\breve{S}``\alpha = S``\{\breve{S}``\alpha - \breve{Q}``\breve{S}``\alpha\}$

$[*71\cdot381] \quad = S``\breve{S}``\alpha - S``\breve{Q}``\breve{S}``\alpha$

$[*72\cdot5 \,.\, *150\cdot23] \quad = \alpha \cap C`P - \breve{P}``\alpha$

$[*205\cdot11] \quad = \overrightarrow{\min}_P`\alpha : \supset \vdash \,.\, \mathrm{Prop}$

$*205\cdot81.\ \vdash :.\ S \epsilon P \,\overline{\text{smor}}\, Q\ .\ \supset :\ E\,!\,\overrightarrow{\min}_P{}^{\backprime}\alpha\ .\ \equiv\ .\ E\,!\,\overrightarrow{\min}_Q{}^{\backprime}\breve{S}{}^{\backprime\backprime}\alpha$

Dem.

$\vdash\ .\ *205\cdot8\ .\ *73\cdot22\ .\ \supset \vdash :.\ \text{Hp}\ .\ \supset :\ \overrightarrow{\min}_P{}^{\backprime}\alpha \ \text{sm}\ \overrightarrow{\min}_Q{}^{\backprime}\breve{S}{}^{\backprime\backprime}\alpha :$

$[*73\cdot44] \qquad \supset :\ \overrightarrow{\min}_P{}^{\backprime}\alpha \epsilon 1\ .\ \equiv\ .\ \overrightarrow{\min}_Q{}^{\backprime}\breve{S}{}^{\backprime\backprime}\alpha \epsilon 1 :$

$[*53\cdot3] \qquad \supset :\ E\,!\,\overrightarrow{\min}_P{}^{\backprime}\alpha\ .\ \equiv\ .\ E\,!\,\overrightarrow{\min}_Q{}^{\backprime}\breve{S}{}^{\backprime\backprime}\alpha :.\ \supset \vdash\ .\ \text{Prop}$

$*205\cdot82.\ \vdash :\ S \epsilon P\, \overline{\text{smor}}\, Q\ .\ E\,!\,\overrightarrow{\min}_P{}^{\backprime}\alpha\ .\ \supset\ .\ \overrightarrow{\min}_P{}^{\backprime}\alpha = S{}^{\backprime}\overrightarrow{\min}_Q{}^{\backprime}\breve{S}{}^{\backprime\backprime}\alpha$
$[*53\cdot31\ .\ *205\cdot8\cdot81]$

The two following propositions are used in $*251\cdot13$.

$*205\cdot83.\ \vdash :\ z \sim\epsilon C{}^{\backprime}P\ .\ \exists\,!\,C{}^{\backprime}P \cap \alpha\ .\ \supset\ .\ \overrightarrow{\min}_P{}^{\backprime}\alpha = \overrightarrow{\min}\,(P \nrightarrow z){}^{\backprime}\alpha$

Dem.

$\vdash\ .\ *161\cdot1\ .\ \supset \vdash :\ \text{Hp}\ .\ \supset\ .\ \{\text{Cnv}{}^{\backprime}(P \nrightarrow z)\}{}^{\backprime\backprime}\alpha = \breve{P}{}^{\backprime\backprime}\alpha \cup \iota{}^{\backprime}z\ .$

$[*161\cdot14\cdot2 . *24\cdot495] \quad \supset\ .\ \alpha \cap C{}^{\backprime}(P \nrightarrow z) - \{\text{Cnv}{}^{\backprime}(P \nrightarrow z)\}{}^{\backprime\backprime}\alpha = \alpha \cap C{}^{\backprime}P - \breve{P}{}^{\backprime\backprime}\alpha\ .$

$[*205\cdot11] \qquad \supset\ .\ \overrightarrow{\min}\,(P \nrightarrow z){}^{\backprime}\alpha = \overrightarrow{\min}_P{}^{\backprime}\alpha :\ \supset \vdash\ .\ \text{Prop}$

$*205\cdot831.\ \vdash :\ z \sim\epsilon C{}^{\backprime}P\ .\ C{}^{\backprime}(P \nrightarrow z) \cap \alpha = \iota{}^{\backprime}z\ .\ \supset\ .\ \overrightarrow{\min}\,(P \nrightarrow z){}^{\backprime}\alpha = \iota{}^{\backprime}z$

Dem.

$\vdash\ .\ *161\cdot11\ . \qquad \supset \vdash :.\ \text{Hp}\ .\ \supset :\ x \epsilon \alpha\ .\ \supset_x .\ \sim \{x\,(P \nrightarrow z)\,z\} :$

$[*37\cdot1 . \text{Transp}] \qquad \supset :\ z \sim\epsilon \{\text{Cnv}{}^{\backprime}(P \nrightarrow z)\}{}^{\backprime\backprime}\alpha \qquad (1)$

$\vdash\ .\ (1)\ .\ *22\cdot621\ .\ \supset \vdash :\ \text{Hp}\ .\ \supset\ .\ \iota{}^{\backprime}z = C{}^{\backprime}(P \nrightarrow z) \cap \alpha - \{\text{Cnv}{}^{\backprime}(P \nrightarrow z)\}{}^{\backprime\backprime}\alpha$

$[*205\cdot11] \qquad = \overrightarrow{\min}\,(P \nrightarrow z){}^{\backprime\backprime}\alpha :\ \supset \vdash\ .\ \text{Prop}$

The two following propositions are used in $*251\cdot14$.

$*205\cdot832.\ \vdash :\ z \sim\epsilon C{}^{\backprime}P\ .\ z \sim\epsilon \alpha\ .\ \supset\ .\ \overrightarrow{\max}_P{}^{\backprime}\alpha = \overrightarrow{\max}\,(P \nrightarrow z){}^{\backprime}\alpha$

Dem.

$\vdash\ .\ *205\cdot111\ .\ *161\cdot2\ .\ \supset \vdash :\ P = \dot{\Lambda}\ .\ \supset\ .\ \overrightarrow{\max}_P{}^{\backprime}\alpha = \Lambda\ .\ \overrightarrow{\max}\,(P \nrightarrow z){}^{\backprime}\alpha = \Lambda \qquad (1)$

$\vdash\ .\ *205\cdot111\ .\ *161\cdot11\cdot14\ .\ \supset$

$\vdash :\ \text{Hp}\ .\ \exists\,!\,C{}^{\backprime}P \cap \alpha\ .\ \supset\ .\ \overrightarrow{\max}\,(P \nrightarrow z){}^{\backprime}\alpha = \alpha \cap (C{}^{\backprime}P \cup \iota{}^{\backprime}z) - (\breve{P}{}^{\backprime\backprime}\alpha \cup \iota{}^{\backprime}z)$

$[*24\cdot495 . *205\cdot111] \qquad = \overrightarrow{\max}_P{}^{\backprime}\alpha \qquad (2)$

$\vdash\ .\ *161\cdot14\ .\ *205\cdot151\cdot161\ .\ \supset$

$\vdash :\ \text{Hp}\ .\ \dot{\exists}\,!\,P\ .\ C{}^{\backprime}P \cap \alpha = \Lambda\ .\ \supset\ .\ \overrightarrow{\max}_P{}^{\backprime}\alpha = \Lambda\ .\ \overrightarrow{\max}\,(P \nrightarrow z){}^{\backprime}\alpha = \Lambda \qquad (3)$

$\vdash\ .\ (1)\ .\ (2)\ .\ (3)\ .\ \supset \vdash\ .\ \text{Prop}$

$*205\cdot833.\ \vdash :\ \dot{\exists}\,!\,P\ .\ z \sim\epsilon C{}^{\backprime}P\ .\ z \epsilon \alpha\ .\ \supset\ .\ \overrightarrow{\max}\,(P \nrightarrow z){}^{\backprime}\alpha = \iota{}^{\backprime}z$

Dem.

$\vdash\ .\ *161\cdot11\ .\ \supset \vdash :\ \text{Hp}\ .\ \supset\ .\ (P \nrightarrow z){}^{\backprime\backprime}\alpha = C{}^{\backprime}P\ .$

$[*161\cdot14 . *205\cdot111] \quad \supset\ .\ \overrightarrow{\max}\,(P \nrightarrow z){}^{\backprime}\alpha = \alpha \cap (C{}^{\backprime}P \cup \iota{}^{\backprime}z) - C{}^{\backprime}P$

$[*22\cdot621 . \text{Hp}] \qquad = \iota{}^{\backprime}z :\ \supset \vdash\ .\ \text{Prop}$

The following proposition is used in *251·25.

***205·84.** $\vdash : C'P \cap C'Q = \Lambda \,.\, \exists ! \, C'P \cap \alpha \,.\, \supset \,.\, \overrightarrow{\min}\,(P \Updownarrow Q)'\alpha = \overrightarrow{\min}_P{}'\alpha$

Dem.

$\vdash \,.\, *160{\cdot}11 \,.\, \supset \vdash : \mathrm{Hp} \,.\, \supset \,.\, \{\mathrm{Cnv}'(P \Updownarrow Q)\}''\alpha = \breve{P}''\alpha \cup C'Q \,.$

$[*205{\cdot}11 \,.\, *160{\cdot}14] \quad \supset \,.\, \overrightarrow{\min}\,(P \Updownarrow Q)'\alpha = \alpha \cap (C'P \cup C'Q) - (\breve{P}''\alpha \cup C'Q)$

$[*24{\cdot}495] \quad = \alpha \cap C'P - \breve{P}''\alpha$

$[*205{\cdot}11] \quad = \overrightarrow{\min}_P{}'\alpha : \supset \vdash \,.\, \mathrm{Prop}$

***205·841.** $\vdash : C'P \cap \alpha = \Lambda \,.\, \supset \,.\, \overrightarrow{\min}\,(P \Updownarrow Q)'\alpha = \overrightarrow{\min}_Q{}'\alpha$

Dem.

$\vdash \,.\, *160{\cdot}11 \,.\, \supset \vdash : \mathrm{Hp} \,.\, \supset \,.\, \{\mathrm{Cnv}'(P \Updownarrow Q)\}''\alpha = \breve{Q}''\alpha \,.$

$[*205{\cdot}11 \,.\, *160{\cdot}14] \quad \supset \,.\, \overrightarrow{\min}\,(P \Updownarrow Q)'\alpha = \alpha \cap (C'P \cup C'Q) - \breve{Q}''\alpha$

$[\mathrm{Hp}] \quad = \alpha \cap C'Q - \breve{Q}''\alpha$

$[*205{\cdot}11] \quad = \overrightarrow{\min}_Q{}'\alpha : \supset \vdash \,.\, \mathrm{Prop}$

The following proposition is used in *251·2.

***205·85.** $\vdash :. \, P \in \mathrm{Rel}^2\mathrm{excl} \,.\, \supset : x \,\{\min(\Sigma'P)\}\, \alpha \,.\, \equiv \,.\, (\exists Q) \,.\, Q \min_P(\breve{F}''\alpha) \,.\, x \min_Q \alpha$

Dem.

$\vdash \,.\, *162{\cdot}12{\cdot}23 \,.\, *205{\cdot}1 \,.\, \supset \vdash :. \, x \,\{\min(\Sigma'P)\}\, \alpha \,.\, \equiv :$

$x \in \alpha : (\exists Q) \,.\, Q \in C'P \,.\, xFQ : \sim (\exists Q, y) \,.\, Q \in C'P \,.\, y \in \alpha \,.\, yQx :$

$\sim (\exists Q, R, y) \,.\, xFQ \,.\, RPQ \,.\, yFR \,.\, y \in \alpha :$

$[*37{\cdot}105] \equiv : x \in \alpha : (\exists Q) \,.\, Q \in C'P \,.\, xFQ : xFQ \,.\, Q \in C'P \,.\, \supset_Q \,.\, x \sim \in \breve{Q}''\alpha :$

$xFQ \,.\, Q \in C'P \,.\, \supset_Q \,.\, Q \sim \in \breve{P}''\breve{F}''\alpha \quad (1)$

$\vdash \,.\, (1) \,.\, *163{\cdot}12 \,.\, *14{\cdot}26 \,.\, \supset \vdash :: \mathrm{Hp} \,.\, \supset :. \, x \,\{\min(\Sigma'P)\}\, \alpha \,.\, \equiv :$

$(\exists Q) \,.\, xFQ \,.\, Q \in C'P \,.\, x \in \alpha - \breve{Q}''\alpha : xFQ \,.\, Q \in C'P \,.\, \supset_Q \,.\, Q \in \breve{F}''\alpha - \breve{P}''\breve{F}''\alpha :$

$[*163{\cdot}12 \,.\, *14{\cdot}26] \equiv : (\exists Q) \,.\, xFQ \,.\, Q \in C'P \,.\, x \in \alpha - \breve{Q}''\alpha \,.\, Q \in \breve{F}''\alpha - \breve{P}''\breve{F}''\alpha :$

$[*205{\cdot}1] \quad \equiv : (\exists Q) \,.\, Q \min_P (\breve{F}''\alpha) \,.\, x \min_Q \alpha :: \supset \vdash \,.\, \mathrm{Prop}$

***205·9.** $\vdash : P \in \mathrm{connex} \,.\, \kappa \subset C'P \,.\, \kappa \sim \in 1 \,.\, \supset \,.\, \overrightarrow{\min}\,(P \upharpoonright \kappa)'\alpha = \overrightarrow{\min}_P{}'(\alpha \cap \kappa)$
[*205·261]

***205·91.** $\vdash : \breve{P}''\alpha \subset \alpha \,.\, P_{\mathrm{po}} \upharpoonright \alpha \in \mathrm{connex} \,.\, \supset \,.\, \overrightarrow{\min}_P{}'\alpha \in 0 \cup 1$

Dem.

$\vdash \,.\, *205{\cdot}261 \,.\, \supset \vdash : \mathrm{Hp} \,.\, \alpha \cap C'P \sim \in 1 \,.\, \supset \,.\, \overrightarrow{\min}\,(P_{\mathrm{po}} \upharpoonright \alpha)'\alpha = \overrightarrow{\min}\,(P_{\mathrm{po}})'\alpha$

$[*205{\cdot}68] \quad = \overrightarrow{\min}_P{}'\alpha \,.$

$[*205{\cdot}3] \quad \supset \,.\, \overrightarrow{\min}_P{}'\alpha \in 0 \cup 1 \quad (1)$

$\vdash \,.\, *93{\cdot}113 \,.\, *60{\cdot}371 \,.\, \supset \vdash : \alpha \cap C'P \in 1 \,.\, \supset \,.\, \overrightarrow{\min}_P{}'\alpha \in 0 \cup 1 \quad (2)$

$\vdash \,.\, (1) \,.\, (2) \,.\, \supset \vdash \,.\, \mathrm{Prop}$

$*206$. SEQUENT POINTS.

Summary of $*206$.

A "sequent" of a class α is a minimum of the terms that come after the whole of $\alpha \cap C'P$; that is, we put

$$\overrightarrow{\mathrm{seq}}_P{}'\alpha = \overrightarrow{\min}_P{}'p'\overleftarrow{P}''(\alpha \cap C'P).$$

Thus the sequents of α are its immediate successors. If α has a maximum, the sequents are the immediate successors of the maximum; but if α has no maximum, there will be no one term of α which is immediately succeeded by a sequent of α; in this case, if α has a single sequent, the sequent is the "upper limit" of α. Whenever P is connected, and therefore whenever P is serial, every class has one sequent or none with respect to P, by $*205 \cdot 3$.

It will be seen that the sequents of α are the same as the sequents of $\alpha \cap C'P$, and therefore that $\overrightarrow{\mathrm{seq}}_P{}'\alpha$ depends only upon $\alpha \cap C'P$: if α has terms not belonging to $C'P$, they are irrelevant.

For the immediate predecessors of a class α, we put

$$\overrightarrow{\mathrm{prec}}_P{}'\alpha = \overrightarrow{\max}_P{}'p'\overrightarrow{P}''(\alpha \cap C'P).$$

We have $\mathrm{prec}_P = \mathrm{seq}\,(\breve{P})$, so that propositions about prec_P result from those about seq_P by merely writing $\breve{P}$ in place of P; they will therefore not be given in what follows.

Among the elementary properties of seq_P with which this number begins, the following are the most important:

$*206 \cdot 13$. $\vdash . \overrightarrow{\mathrm{seq}}_P{}'\alpha = \overrightarrow{\min}_P{}'p'\overleftarrow{P}''(\alpha \cap C'P)$

This merely embodies the definition.

$*206 \cdot 131$. $\vdash . \overrightarrow{\mathrm{seq}}_P{}'\alpha = \overrightarrow{\mathrm{seq}}_P{}'(\alpha \cap C'P)$

$*206 \cdot 134$. $\vdash . \overrightarrow{\mathrm{seq}}_P{}'\alpha = C'P \cap \hat{x}\{\alpha \cap C'P \subset \overrightarrow{P}'x \,.\, \overrightarrow{P}'x \subset -p'\overleftarrow{P}''(\alpha \cap C'P)\}$

$*206 \cdot 14$. $\vdash : \alpha \cap C'P = \Lambda \,.\, \supset . \overrightarrow{\mathrm{seq}}_P{}'\alpha = \overrightarrow{B}'P$

Thus if P has a first term, this is the sequent of the null class, or of any other class which has no members in common with $C\text{‘}P$.

***206·16.** $\vdash : P \in \text{connex} \,.\, \supset \,.\, \overrightarrow{\text{seq}}_P\text{‘}\alpha \in 0 \cup 1$

This follows at once from *205·3. It leads to

***206·161.** $\vdash : P \in \text{connex} \,.\, \supset \,.\, \text{seq}_P \in 1 \rightarrow \text{Cls}$

Thus if P is a connected relation, no class has more than one sequent. This is not in general the case with relations which are not connected, even where the idea of sequents is quite naturally applicable. Take, *e.g.*, the relation of descendent to ancestor, and let α be the class of monarchs of England. Then $\overrightarrow{\text{seq}}_P\text{‘}\alpha$ will be such parents of monarchs as were not themselves monarchs.

***206·171.** $\vdash : P \in \text{connex} \,.\, P^2 \in J \,.\, \supset \,.$

$$\overrightarrow{\text{seq}}_P\text{‘}\alpha = C\text{‘}P \cap \hat{x}\{\alpha \cap C\text{‘}P \subset \overrightarrow{P}\text{‘}x \,.\, \overrightarrow{P}\text{‘}x \subset (\alpha \cap C\text{‘}P) \cup P\text{“}\alpha\}$$

This proposition states that x is a sequent of α if the whole of $\alpha \cap C\text{‘}P$ precedes x, but every term that precedes x either belongs to α or precedes some term of α. When P is a series and α has no maximum, we have

$$\overrightarrow{\text{seq}}_P\text{‘}\alpha = C\text{‘}P \cap \hat{x}\,(\overrightarrow{P}\text{‘}x = P\text{“}\alpha) \quad (*206·174),$$

i.e. the sequent of α, if any, is a term whose predecessors are identical with the predecessors of members of α. This is the case of a *limit* (cf. *207).

We have next a set of propositions (*206·211·28) concerned with $\overrightarrow{P}\text{‘}\text{seq}_P\text{‘}\alpha$ and $\overleftarrow{P}\text{‘}\text{seq}_P\text{‘}\alpha$. When P is transitive and connected, and α is an existent class contained in $C\text{‘}P$ and having a sequent, we shall have

$$\overrightarrow{P}\text{‘}\text{seq}_P\text{‘}\alpha = \alpha \cup P\text{“}\alpha \,.\, \iota\text{‘}\text{seq}_P\text{‘}\alpha \cup \overleftarrow{P}\text{‘}\text{seq}_P\text{‘}\alpha = p\text{‘}\overleftarrow{P}\text{“}\alpha.$$

That is, the predecessors of the sequent are the members of α and the predecessors of members, while the sequent and its successors are the successors of the whole of α. The various parts of this statement require various parts of the hypothesis. Thus we have

***206·211.** $\vdash : \text{E}\,!\,\text{seq}_P\text{‘}\alpha \,.\, \supset \,.\, \alpha \cap C\text{‘}P \subset \overrightarrow{P}\text{‘}\text{seq}_P\text{‘}\alpha$

***206·213.** $\vdash : P \in \text{connex} \,.\, \text{E}\,!\,\text{seq}_P\text{‘}\alpha \,.\, \supset \,.\, \overrightarrow{P}\text{‘}\text{seq}_P\text{‘}\alpha \subset (\alpha \cap C\text{‘}P) \cup P\text{“}\alpha$

***206·22.** $\vdash : P \in \text{trans} \cap \text{connex} \,.\, \text{E}\,!\,\text{seq}_P\text{‘}\alpha \,.\, \supset \,.$

$$\overrightarrow{P}\text{‘}\text{seq}_P\text{‘}\alpha = (\alpha \cap C\text{‘}P) \cup P\text{“}\alpha = \overrightarrow{\text{max}}_P\text{‘}\alpha \cup P\text{“}\alpha$$

***206·23.** $\vdash : P \in \text{trans} \cap \text{connex} \,.\, \text{E}\,!\,\text{seq}_P\text{‘}\alpha \,.\, \supset \,.$

$$\iota\text{‘}\text{seq}_P\text{‘}\alpha \cup \overleftarrow{P}\text{‘}\text{seq}_P\text{‘}\alpha = p\text{‘}\overleftarrow{P}\text{“}(\alpha \cap C\text{‘}P) \cap C\text{‘}P$$

If P is transitive, the value of $\overrightarrow{\text{seq}}_P{}^{\backprime}\alpha$ is unchanged if we add to α any set of terms contained in $P^{\backprime\backprime}\alpha$ (*206·24); thus in particular, $\overrightarrow{\text{seq}}_P{}^{\backprime}(\alpha \cup P^{\backprime\backprime}\alpha) = \overrightarrow{\text{seq}}_P{}^{\backprime}\alpha$ (*206·25). Thus we can fill up any gaps in α, and take the whole series up to the end of α, without altering the sequent.

We have next a set of propositions (*206·3—·38) on the sequent of $P^{\backprime\backprime}\alpha$, *i.e.* of the segment defined by α. If P is a series, $\text{seq}_P{}^{\backprime}P^{\backprime\backprime}\alpha$ is the maximum of α if α has a maximum, the sequent of α if α has a sequent but no maximum, and non-existent if α has neither a maximum nor a sequent (*206·35·331·36).

Our next set of propositions (*206·4—·52) concerns the sequents of unit classes, especially of $\iota^{\backprime}\text{max}_P{}^{\backprime}\alpha$, and of classes of the form $\overrightarrow{P}{}^{\backprime}x$. We have

***206·4.** $\vdash : P \in J \,.\, x \in C^{\backprime}P \,.\, \supset \,.\, x \, \text{seq}_P \, \overrightarrow{P}{}^{\backprime}x$

***206·42.** $\vdash : x \in C^{\backprime}P \,.\, \supset \,.\, \overrightarrow{\text{seq}}_P{}^{\backprime}\iota^{\backprime}x = \overleftarrow{P \dot{-} P^2}{}^{\backprime}x = \overrightarrow{\text{min}}_P{}^{\backprime}\overleftarrow{P}{}^{\backprime}x$

whence the three following propositions:

***206·43.** $\vdash : P \in \text{trans} \cap \text{Rl}^{\backprime}J \,.\, x \in C^{\backprime}P \,.\, \supset \,.\, \overrightarrow{\text{seq}}_P{}^{\backprime}\iota^{\backprime}x = \overleftarrow{P_1}{}^{\backprime}x$

***206·45.** $\vdash :. P \in \text{Ser} \,.\, x \in C^{\backprime}P \,.\, \supset : \text{E} ! \, \text{seq}_P{}^{\backprime}\iota^{\backprime}x \,.\, \equiv \,.\, x \in \text{D}^{\backprime}P_1$

***206·46.** $\vdash : P \in \text{trans} \cap \text{connex} \,.\, \text{E} ! \, \text{max}_P{}^{\backprime}\alpha \,.\, \supset \,.\, \overrightarrow{\text{seq}}_P{}^{\backprime}\alpha = \overrightarrow{\text{seq}}_P{}^{\backprime}\overrightarrow{\text{max}}_P{}^{\backprime}\alpha$

From the above propositions it results that, when P is a series, any member of $C^{\backprime}P$ is the sequent of the class of its predecessors, $\breve{P}_1{}^{\backprime}x$ is the sequent of $\iota^{\backprime}x$ if either exists, and the sequent of a class which has a maximum is the immediate successor (if any) of the maximum, *i.e.*

***206·5.** $\vdash : P \in \text{trans} \cap \text{connex} \,.\, \text{E}! \, \text{max}_P{}^{\backprime}\alpha \,.\, \text{E}! \, \text{seq}_P{}^{\backprime}\alpha \,.\, \supset \,.\, \text{max}_P{}^{\backprime}\alpha (P \dot{-} P^2) \text{seq}_P{}^{\backprime}\alpha$

We then have a set of propositions (*206·53—·57) on the sequent of $p^{\backprime}\overrightarrow{P}{}^{\backprime\backprime}(\alpha \cap C^{\backprime}P)$, *i.e.* the sequent of the predecessors of the whole of $\alpha \cap C^{\backprime}P$. These propositions are specially useful in connection with "Dedekindian" series, *i.e.* series in which every class has either a maximum or a sequent (*214). These propositions all require the full hypothesis that P is a series. In this case, $\overrightarrow{\text{seq}}_P{}^{\backprime}p^{\backprime}\overrightarrow{P}{}^{\backprime\backprime}(\alpha \cap C^{\backprime}P) = \overrightarrow{\text{min}}_P{}^{\backprime}\alpha$, *i.e.* the sequent (if any) of the predecessors of the whole of $\alpha \cap C^{\backprime}P$ is the minimum (if any) of α. Moreover by definition the maximum of $p^{\backprime}\overrightarrow{P}{}^{\backprime\backprime}(\alpha \cap C^{\backprime}P)$, if any, is the precedent of α. Hence α has either a minimum or a precedent if $p^{\backprime}\overrightarrow{P}{}^{\backprime\backprime}(\alpha \cap C^{\backprime}P)$ has either a sequent or a maximum (*206·54). Moreover the sequent and maximum of α are respectively (if they exist) the sequent and maximum of the predecessors of all the successors of the whole of $\alpha \cap C^{\backprime}P$ (*206·551). Hence we arrive at the conclusion that the assumption that every class of the form $p^{\backprime}\overrightarrow{P}{}^{\backprime\backprime}(\alpha \cap C^{\backprime}P)$ has either a maximum or a sequent is equivalent both to the assumption that every class has either a maximum or a sequent (*206·56) and to the assumption that every class has either a minimum or a precedent (*206·55). It follows that these two latter assumptions are equivalent (*206·57), *i.e.* that a series is Dedekindian when, and only when, its converse is Dedekindian (*214·14).

We deal next (*206·6—·63) with correlations, showing that if two relations are correlated, the sequents of the correlates of any class are the correlates of the sequents, *i.e.*

***206·61.** $\vdash : S \epsilon P \overline{\text{smor}} Q . \supset . \overrightarrow{\text{seq}_P}`\alpha = S``\overrightarrow{\text{seq}_Q}`\breve{S}``\alpha$

We end with a set of propositions (*206·7—·732) showing that the sequent of a class is unchanged if we remove from the class any term other than its maximum (*206·72); that if a class has terms in $C`P$, and has both a precedent and a sequent, the precedent has the relation P^2 to the sequent (*206·73), and that the precedent is not identical with the sequent (*206·732). These propositions are in the nature of lemmas, whose use is chiefly in the theory of stretches (*215).

***206·01.** $\text{seq}_P = \hat{x}\hat{\alpha} \{x \min_P p`\overleftarrow{P}``(\alpha \cap C`P)\}$ Df

***206·02.** $\text{prec}_P = \hat{x}\hat{\alpha} \{x \max_P p`\overrightarrow{P}``(\alpha \cap C`P)\}$ Df

***206·1.** $\vdash : x \,\text{seq}_P\, \alpha . \equiv . x \min_P p`\overleftarrow{P}``(\alpha \cap C`P)$ [(*206·01)]

***206·101.** $\vdash . \text{prec}_P = \text{seq}(\breve{P})$ [*32·241 . *33·22 . *205·102]

We shall not enunciate any other propositions on prec_P (unless for some special reason), since the above proposition enables them to be immediately deduced from the corresponding propositions on seq_P.

***206·11.** $\vdash : x \,\text{seq}_P\, \alpha . \equiv . x \epsilon p`\overleftarrow{P}``(\alpha \cap C`P) \cap C`P - \breve{P}``p`\overleftarrow{P}``(\alpha \cap C`P)$
[*206·1 . *205·1]

Observe that when $\alpha \cap C`P$ is not null, $p`\overleftarrow{P}``(\alpha \cap C`P) \subset C`P$, so that the factor $C`P$ on the right is unnecessary; but when $\alpha \cap C`P = \Lambda$, we have $p`\overleftarrow{P}``(\alpha \cap C`P) = V$, so that the factor $C`P$ becomes relevant. Owing to this factor, the sequents of Λ are $\overrightarrow{B}`P$, so that if $B`P$ exists, $B`P$ is the sequent of Λ.

***206·12.** $\vdash :: x \,\text{seq}_P \alpha . \equiv :. y \epsilon \alpha \cap C`P . \supset_y . yPx : x \epsilon C`P :.$
$y \epsilon \alpha \cap C`P . \supset_y . yPz : \supset_z . \sim (zPx)$ [*206·11 . *40·53 . *37·105]

***206·13.** $\vdash . \overrightarrow{\text{seq}_P}`\alpha = \overrightarrow{\min_P}`p`\overleftarrow{P}``(\alpha \cap C`P)$ [*206·1]

***206·131.** $\vdash . \overrightarrow{\text{seq}_P}`\alpha = \overrightarrow{\text{seq}_P}`(\alpha \cap C`P)$ [*206·13 . *22·43·621]

$*206\cdot132$. $\vdash . \overrightarrow{\text{seq}}_P{}^{\backprime}\alpha = p{}^{\backprime}\overleftarrow{P}{}^{\backprime\backprime}(\alpha \cap C{}^{\backprime}P) \cap C{}^{\backprime}P - \breve{P}{}^{\backprime\backprime}p{}^{\backprime}\overleftarrow{P}{}^{\backprime\backprime}(\alpha \cap C{}^{\backprime}P)$ [$*206\cdot11$]

$*206\cdot133$. $\vdash : x \,\text{seq}_P\, \alpha . \supset . \sim(xPx)$ [$*205\cdot194 . *206\cdot13$]

$*206\cdot134$. $\vdash . \overrightarrow{\text{seq}}_P{}^{\backprime}\alpha = C{}^{\backprime}P \cap \hat{x}\{\alpha \cap C{}^{\backprime}P \subset \overrightarrow{P}{}^{\backprime}x . \overrightarrow{P}{}^{\backprime}x \subset - p{}^{\backprime}\overleftarrow{P}{}^{\backprime\backprime}(\alpha \cap C{}^{\backprime}P)\}$

Dem.

$\vdash . *206\cdot12 . *32\cdot18 . \supset$

$\vdash . \overrightarrow{\text{seq}}_P{}^{\backprime}\alpha = C{}^{\backprime}P \cap \hat{x}(\alpha \cap C{}^{\backprime}P \subset \overrightarrow{P}{}^{\backprime}x) \cap \hat{x}\{y \in \alpha \cap C{}^{\backprime}P . \supset_y . yPz : \supset_z . \sim(zPx)\}$

$[*40\cdot53] \quad = C{}^{\backprime}P \cap \hat{x}(\alpha \cap C{}^{\backprime}P \subset \overrightarrow{P}{}^{\backprime}x) \cap \hat{x}\{z \in p{}^{\backprime}\overleftarrow{P}{}^{\backprime\backprime}(\alpha \cap C{}^{\backprime}P) . \supset_z . \sim(zPx)\}$

[Transp.$*32\cdot18$]

$= C{}^{\backprime}P \cap \hat{x}(\alpha \cap C{}^{\backprime}P \subset \overrightarrow{P}{}^{\backprime}x) \cap \hat{x}\{\overrightarrow{P}{}^{\backprime}x \subset - p{}^{\backprime}\overleftarrow{P}{}^{\backprime\backprime}(\alpha \cap C{}^{\backprime}P)\} . \supset \vdash . \text{Prop}$

This formula for $\overrightarrow{\text{seq}}_P{}^{\backprime}\alpha$ is usually more convenient than $*206\cdot13\cdot132$.

$*206\cdot14$. $\vdash : \alpha \cap C{}^{\backprime}P = \Lambda . \supset . \overrightarrow{\text{seq}}_P{}^{\backprime}\alpha = \overrightarrow{B}{}^{\backprime}P$

Dem.

$\vdash . *206\cdot13 . *40\cdot2 . \supset \vdash : \text{Hp} . \supset . \overrightarrow{\text{seq}}_P{}^{\backprime}\alpha = \overrightarrow{\text{min}}_P{}^{\backprime}\text{V}$

$[*205\cdot15 . *24\cdot26] \quad = \overrightarrow{\text{min}}_P{}^{\backprime}C{}^{\backprime}P$

$[*205\cdot12] \quad = \overrightarrow{B}{}^{\backprime}P : \supset \vdash . \text{Prop}$

$*206\cdot141$. $\vdash : \exists ! \alpha \cap C{}^{\backprime}P . \supset . \overrightarrow{\text{seq}}_P{}^{\backprime}\alpha = p{}^{\backprime}\overleftarrow{P}{}^{\backprime\backprime}(\alpha \cap C{}^{\backprime}P) - \breve{P}{}^{\backprime\backprime}p{}^{\backprime}\overleftarrow{P}{}^{\backprime\backprime}(\alpha \cap C{}^{\backprime}P)$

Dem.

$\vdash . *40\cdot62 . \supset \vdash : \text{Hp} . \supset . p{}^{\backprime}\overleftarrow{P}{}^{\backprime\backprime}(\alpha \cap C{}^{\backprime}P) \subset C{}^{\backprime}P \quad (1)$

$\vdash . (1) . *206\cdot132 . \supset \vdash . \text{Prop}$

$*206\cdot142$. $\vdash : \exists ! \alpha \cap C{}^{\backprime}P . \supset . \overrightarrow{\text{seq}}_P{}^{\backprime}\alpha \subset \breve{P}{}^{\backprime\backprime}\alpha$ [$*40\cdot61 . *206\cdot141$]

$*206\cdot143$. $\vdash : \alpha \subset C{}^{\backprime}P . \supset . \overrightarrow{\text{seq}}_P{}^{\backprime}\alpha = p{}^{\backprime}\overleftarrow{P}{}^{\backprime\backprime}\alpha \cap C{}^{\backprime}P - \breve{P}{}^{\backprime\backprime}p{}^{\backprime}\overleftarrow{P}{}^{\backprime\backprime}\alpha$

[$*206\cdot132 . *22\cdot621$]

$*206\cdot144$. $\vdash : \exists ! \overrightarrow{\text{seq}}_P{}^{\backprime}\alpha . \supset . \exists ! p{}^{\backprime}\overleftarrow{P}{}^{\backprime\backprime}(\alpha \cap C{}^{\backprime}P)$ [$*206\cdot132$]

$*206\cdot15$. $\vdash : \alpha \subset C{}^{\backprime}P . \exists ! \alpha . \supset . \overrightarrow{\text{seq}}_P{}^{\backprime}\alpha = p{}^{\backprime}\overleftarrow{P}{}^{\backprime\backprime}\alpha - \breve{P}{}^{\backprime\backprime}p{}^{\backprime}\overleftarrow{P}{}^{\backprime\backprime}\alpha$

[$*206\cdot141 . *22\cdot621$]

$*206\cdot16$. $\vdash : P \in \text{connex} . \supset . \overrightarrow{\text{seq}}_P{}^{\backprime}\alpha \in 0 \cup 1$ [$*205\cdot3 . *206\cdot13$]

$*206\cdot161$. $\vdash : P \in \text{connex} . \supset . \text{seq}_P \in 1 \to \text{Cls}$ [$*206\cdot16 . *71\cdot12$]

Thus in a series, or in any connected relation, no class has more than one sequent.

***206·17.** $\vdash :. x \, \mathrm{seq}_P \, \alpha . \equiv : y \in \alpha \cap C'P . \supset_y . yPx : x \in C'P :$

$$yPx . \supset_y . (\exists z) . z \in \alpha \cap C'P . \sim (zPy)$$

Dem.

$\vdash . *37{\cdot}462 . *206{\cdot}11 . \supset$

$\vdash :. x \, \mathrm{seq}_P \, \alpha . \equiv : x \in p'\overleftarrow{P}``(\alpha \cap C'P) \cap C'P . \overrightarrow{P}'x \subset - p'\overleftarrow{P}``(\alpha \cap C'P) :$

[*40·53] $\equiv : y \in \alpha \cap C'P . \supset_y . yPx : x \in C'P :$

$$yPx . \supset_y . (\exists z) . z \in \alpha \cap C'P . \sim (zPy) :. \supset \vdash . \mathrm{Prop}$$

The following propositions give simplified formulae for $\overrightarrow{\mathrm{seq}}_P{}'\alpha$ in various special cases.

***206·171.** $\vdash : P \in \mathrm{connex} . P^2 \in J . \supset .$

$$\overrightarrow{\mathrm{seq}}_P{}'\alpha = C'P \cap \hat{x} \{\alpha \cap C'P \subset \overrightarrow{P}'x . \overrightarrow{P}'x \subset (\alpha \cap C'P) \cup P``\alpha\}$$

Dem.

$\vdash . *206{\cdot}134 . *33{\cdot}152 . \supset$

$$\vdash . \overrightarrow{\mathrm{seq}}_P{}'\alpha = C'P \cap \hat{x} \{\alpha \cap C'P \subset \overrightarrow{P}'x . \overrightarrow{P}'x \subset C'P - p'\overleftarrow{P}``(\alpha \cap C'P)\} \quad (1)$$

$\vdash . (1) . *202{\cdot}503 . \supset \vdash . \mathrm{Prop}$

***206·172.** $\vdash : P \in \mathrm{connex} . P^2 \in J . P``\alpha \subset \alpha . \supset .$

$$\overrightarrow{\mathrm{seq}}_P{}'\alpha = C'P \cap \hat{x} (\alpha \cap C'P = \overrightarrow{P}'x) \quad [*206{\cdot}171 . *22{\cdot}62]$$

***206·173.** $\vdash : P \in \mathrm{connex} . P^2 \in J . \alpha \cap C'P \subset P``\alpha . \supset .$

$$\overrightarrow{\mathrm{seq}}_P{}'\alpha = C'P \cap \hat{x} \{\alpha \cap C'P \subset \overrightarrow{P}'x . \overrightarrow{P}'x \subset P``\alpha\}$$

[*206·171 . *22·62]

***206·174.** $\vdash : P \in \mathrm{Ser} . \alpha \cap C'P \subset P``\alpha . \supset . \overrightarrow{\mathrm{seq}}_P{}'\alpha = C'P \cap \hat{x} (\overrightarrow{P}'x = P``\alpha)$

Dem.

$\vdash . *13{\cdot}12 . *22{\cdot}42 . \supset \vdash :. \mathrm{Hp} . \supset :$

$$\overrightarrow{P}'x = P``\alpha . \supset . \alpha \cap C'P \subset \overrightarrow{P}'x . \overrightarrow{P}'x \subset P``\alpha \quad (1)$$

$\vdash . *37{\cdot}265 .$ $\supset \vdash : \alpha \cap C'P \subset \overrightarrow{P}'x . \supset . P``\alpha \subset P``\overrightarrow{P}'x :$

[*201·501] $\supset \vdash :. \mathrm{Hp} . \supset : \alpha \cap C'P \subset \overrightarrow{P}'x . \supset . P``\alpha \subset \overrightarrow{P}'x :$

[Fact] $\supset : \alpha \cap C'P \subset \overrightarrow{P}'x . \overrightarrow{P}'x \subset P``\alpha . \supset . P``\alpha = \overrightarrow{P}'x \quad (2)$

$\vdash . (1) . (2) . *206{\cdot}173 . \supset \vdash . \mathrm{Prop}$

The propositions *206·173·174 deal with *limits*. When a class α has no maximum, *i.e.* when $\alpha \cap C'P \subset P``\alpha$, its sequent (if any) is called its *limit*. By the above propositions, the limit is a term x such that $\alpha \cap C'P$ precedes x, but every predecessor of x precedes some member of $\alpha \cap C'P$ (*206·173); it is also a term x whose predecessors are identical with the predecessors of α (*206·174). The subject of limits will be explicitly treated in *207.

***206·18.** $\vdash . \overrightarrow{\text{seq}}_P\text{‘}\alpha \subset C\text{‘}P$ [*206·132]

***206·181.** $\vdash : \exists ! \, \alpha \cap C\text{‘}P \,.\supset .\, \overrightarrow{\text{seq}}_P\text{‘}\alpha \subset \text{Cl}\text{‘}P$ [*206·142 . *37·16]

***206·2.** $\vdash . \overrightarrow{\text{seq}}_P\text{‘}\alpha \subset -\alpha$

Dem.

$\vdash . *40\cdot68 . \text{Transp} . \supset \vdash . p\text{‘}\overleftarrow{P}\text{‘‘}(\alpha \cap C\text{‘}P) - \breve{P}\text{‘‘}p\text{‘}\overleftarrow{P}\text{‘‘}(\alpha \cap C\text{‘}P) \subset -(\alpha \cap C\text{‘}P)$ (1)

$\vdash . (1) . *206\cdot132 . \supset \vdash . \text{Prop}$

***206·21.** $\vdash : P^2 \mathrel{\mathsf{G}} J \,.\supset .\, \overrightarrow{\text{seq}}_P\text{‘}\alpha \subset -P\text{‘‘}\alpha$ [*200·53 . *206·132]

***206·211.** $\vdash : \text{E} ! \, \text{seq}_P\text{‘}\alpha \,.\supset .\, \alpha \cap C\text{‘}P \subset \overrightarrow{P}\text{‘}\text{seq}_P\text{‘}\alpha$

Dem.

$\vdash . *206\cdot17 . \supset \vdash :. \text{Hp} . \supset : y \in \alpha \cap C\text{‘}P \,.\supset_y .\, yP \, \text{seq}_P\text{‘}\alpha :. \supset \vdash . \text{Prop}$

***206·212.** $\vdash : P \in \text{trans} \,.\, \text{E} ! \, \text{seq}_P\text{‘}\alpha \,.\supset .\, P\text{‘‘}\alpha \subset \overrightarrow{P}\text{‘}\text{seq}_P\text{‘}\alpha$

Dem.

$\vdash . *206\cdot211 . \supset \vdash : \text{Hp} . \supset . P\text{‘‘}\alpha \subset P\text{‘‘}\overrightarrow{P}\text{‘}\text{seq}_P\text{‘}\alpha$

[*201·501] $\subset \overrightarrow{P}\text{‘}\text{seq}_P\text{‘}\alpha : \supset \vdash . \text{Prop}$

***206·213.** $\vdash : P \in \text{connex} \,.\, \text{E} ! \, \text{seq}_P\text{‘}\alpha \,.\supset .\, \overrightarrow{P}\text{‘}\text{seq}_P\text{‘}\alpha \subset (\alpha \cap C\text{‘}P) \cup P\text{‘‘}\alpha$

Dem.

$\vdash . *206\cdot17 . \supset \vdash :: \text{Hp} . \supset :. yP \, \text{seq}_P\text{‘}\alpha \,.\supset_y : (\exists z) . z \in (\alpha \cap C\text{‘}P) . \sim(zPy) :$

[*202·103] $\supset_y : (\exists z) : z \in \alpha \cap C\text{‘}P : y = z \,.\vee.\, yPz :$

[*13·195.*37·1] $\supset_y : y \in \alpha \cap C\text{‘}P \,.\vee.\, y \in P\text{‘‘}(\alpha \cap C\text{‘}P) :$

[*37·265] $\supset_y : y \in (\alpha \cap C\text{‘}P) \cup P\text{‘‘}\alpha :: \supset \vdash . \text{Prop}$

***206·22.** $\vdash : P \in \text{trans} \cap \text{connex} \,.\, \text{E} ! \, \text{seq}_P\text{‘}\alpha \,.\supset .$

$\overrightarrow{P}\text{‘}\text{seq}_P\text{‘}\alpha = (\alpha \cap C\text{‘}P) \cup P\text{‘‘}\alpha = \overrightarrow{\text{max}}_P\text{‘}\alpha \cup P\text{‘‘}\alpha$

[*206·211·212·213 . *205·131]

***206·23.** $\vdash : P \in \text{trans} \cap \text{connex} \,.\, \text{E} ! \, \text{seq}_P\text{‘}\alpha \,.\supset .$

$\iota\text{‘}\text{seq}_P\text{‘}\alpha \cup \overleftarrow{P}\text{‘}\text{seq}_P\text{‘}\alpha = p\text{‘}\overleftarrow{P}\text{‘‘}(\alpha \cap C\text{‘}P) \cap C\text{‘}P$

Dem.

$\vdash . *205\cdot22 . *206\cdot13 . \supset$

$\vdash : \text{Hp} . \supset . \iota\text{‘}\text{seq}_P\text{‘}\alpha \cup \overleftarrow{P}\text{‘}\text{seq}_P\text{‘}\alpha = \iota\text{‘}\text{seq}_P\text{‘}\alpha \cup \breve{P}\text{‘‘}p\text{‘}\overleftarrow{P}\text{‘‘}(\alpha \cap C\text{‘}P)$

[*206·13.*53·31] $= \overrightarrow{\text{min}}_P\text{‘}p\text{‘}\overleftarrow{P}\text{‘‘}(\alpha \cap C\text{‘}P) \cup \breve{P}\text{‘‘}p\text{‘}\overleftarrow{P}\text{‘‘}(\alpha \cap C\text{‘}P)$

[*205·13] $= p\text{‘}\overleftarrow{P}\text{‘‘}(\alpha \cap C\text{‘}P) \cap C\text{‘}P \cup \breve{P}\text{‘‘}p\text{‘}\overleftarrow{P}\text{‘‘}(\alpha \cap C\text{‘}P)$

[*201·51.*37·16] $= p\text{‘}\overleftarrow{P}\text{‘‘}(\alpha \cap C\text{‘}P) \cap C\text{‘}P : \supset \vdash . \text{Prop}$

***206·24.** $\vdash : P \epsilon \text{trans} . \beta \subset P``\alpha . \supset . \overrightarrow{\text{seq}}_P`(\alpha \cup \beta) = \overrightarrow{\text{seq}}_P`\alpha$

Dem.

$\vdash . *201·56 . \supset \vdash : \text{Hp} . \supset . p`\overleftarrow{P}``\{(\alpha \cup \beta) \cap C`P\} = p`\overleftarrow{P}``(\alpha \cap C`P)$ (1)

$\vdash . (1) . *206·13 . \supset \vdash . \text{Prop}$

***206·25.** $\vdash : P \epsilon \text{trans} . \supset . \overrightarrow{\text{seq}}_P`(\alpha \cup P``\alpha) = \overrightarrow{\text{seq}}_P`\alpha$ [*206·24]

***206·26.** $\vdash : P \epsilon \text{trans} \cap \text{connex} . \exists ! \alpha \cap C`P . E ! \text{seq}_P`\alpha . \supset .$

$$p`\overleftarrow{P}``(\alpha \cap C`P) = \iota`\text{seq}_P`\alpha \cup \overleftarrow{P}`\text{seq}_P`\alpha$$

Dem.

$\vdash . *40·62 . \supset \vdash : \text{Hp} . \supset . p`\overleftarrow{P}``(\alpha \cap C`P) \subset C`P$ (1)

$\vdash . (1) . *206·23 . \supset \vdash . \text{Prop}$

***206·27.** $\vdash : P \epsilon \text{trans} \cap \text{connex} . E ! \text{seq}_P`\alpha . E ! \max_P`\alpha . \supset .$

$$\overrightarrow{P}`\text{seq}_P`\alpha = \overrightarrow{P}`\max_P`\alpha \cup \iota`\max_P`\alpha .$$
$$\overleftarrow{P}`\max_P`\alpha = \overleftarrow{P}`\text{seq}_P`\alpha \cup \iota`\text{seq}_P`\alpha$$

Dem.

$\vdash . *206·22 . \quad \supset \vdash : \text{Hp} . \supset . \overrightarrow{P}`\text{seq}_P`\alpha = \overrightarrow{\max}_P`\alpha \cup P``\alpha$

[*205·22] $\quad = \iota`\max_P`\alpha \cup \overrightarrow{P}`\max_P`\alpha$ (1)

$\vdash . *205·65 . \quad \supset \vdash : \text{Hp} . \supset . \overleftarrow{P}`\max_P`\alpha = p`\overleftarrow{P}``(\alpha \cap C`P)$ (2)

$\vdash . *205·151·161 . \supset \vdash : \text{Hp} . \supset . \exists ! (\alpha \cap C`P)$ (3)

$\vdash . (3) . *206·26 . \supset \vdash : \text{Hp} . \supset . p`\overleftarrow{P}``(\alpha \cap C`P) = \iota`\text{seq}_P`\alpha \cup \overleftarrow{P}`\text{seq}_P`\alpha$ (4)

$\vdash . (2) . (4) . \quad \supset \vdash : \text{Hp} . \supset . \overleftarrow{P}`\max_P`\alpha = \iota`\text{seq}_P`\alpha \cup \overleftarrow{P}`\text{seq}_P`\alpha$ (5)

$\vdash . (1) . (5) . \supset \vdash . \text{Prop}$

***206·28.** $\vdash :. P \epsilon \text{Ser} . \supset :$

$$x \epsilon C`P - \alpha . \overrightarrow{P}`x = P``\alpha . \equiv . x = \text{seq}_P`\alpha . \sim E ! \max_P`\alpha$$

Dem.

$\vdash . *206·174 . *205·6 . \supset$

$\vdash :. \text{Hp} . \supset : x = \text{seq}_P`\alpha . \sim E ! \max_P`\alpha . \supset . x \epsilon C`P . \overrightarrow{P}`x = P``\alpha .$

[*206·2] $\quad \supset . x \epsilon C`P - \alpha . \overrightarrow{P}`x = P``\alpha$ (1)

$\vdash . *37·1 . \quad \supset \vdash : xPy . y \epsilon \alpha . \overrightarrow{P}`x = P``\alpha . \supset . x \epsilon \overrightarrow{P}`x$ (2)

$\vdash . (2) . \text{Transp} . \supset \vdash : P \in J . y \epsilon \alpha . \overrightarrow{P}`x = P``\alpha . \supset . \sim(xPy)$ (3)

$\vdash . *13·14 . \quad \supset \vdash : x \epsilon C`P - \alpha . y \epsilon \alpha . \supset . x \neq y$ (4)

$\vdash . (3) . (4) . *202·103 . \supset \vdash :. \text{Hp} . \supset :$

$$x \epsilon C`P - \alpha . \overrightarrow{P}`x = P``\alpha . y \epsilon \alpha \cap C`P . \supset . yPx :$$

[*32·18] $\quad \supset : x \epsilon C`P - \alpha . \overrightarrow{P}`x = P``\alpha . \supset . \alpha \cap C`P \subset \overrightarrow{P}`x$ (5)

$\vdash . (5) . *206{\cdot}171{\cdot}16 . \supset \vdash :. \text{Hp} . \supset :$

$$x \in C`P - \alpha . \overrightarrow{P}`x = P``\alpha . \supset . x = \text{seq}_P`\alpha \quad (6)$$

$\vdash . (5) . *205{\cdot}123 . \quad \supset \vdash :. \text{Hp} . \supset :$

$$x \in C`P - \alpha . \overrightarrow{P}`x = P``\alpha . \supset . \sim \text{E} ! \max_P`\alpha \quad (7)$$

$\vdash . (1) . (6) . (7) . \supset \vdash . \text{Prop}$

***206·3.** $\vdash : P \in \text{trans} \cap \text{connex} . \alpha \subset C`P . P``\alpha \subset \alpha . \text{E} ! \text{seq}_P`\alpha . \supset .$

$$\overrightarrow{P}`\text{seq}_P`\alpha = \alpha \quad [*206{\cdot}22]$$

***206·31.** $\vdash : P \in \text{trans} \cap \text{connex} . \text{E} ! \text{seq}_P`P``\alpha . \supset . \overrightarrow{P}`\text{seq}_P`P``\alpha = P``\alpha$
$[*206{\cdot}3 . *201{\cdot}5]$

***206·32.** $\vdash : P \in \text{trans} \cap \text{connex} . \text{E} ! \max_P`\alpha . \text{E} ! \text{seq}_P`P``\alpha . \supset .$

$$\max_P`\alpha = \text{seq}_P`P``\alpha$$

Dem.

$\vdash . *206{\cdot}31 . *205{\cdot}22 . \supset \vdash :. \text{Hp} . \supset : \overrightarrow{P}`\max_P`\alpha = \overrightarrow{P}`\text{seq}_P`P``\alpha :$
$[*205{\cdot}194 . *206{\cdot}133] \supset : \sim (\text{seq}_P`P``\alpha \, P \max_P`\alpha) . \sim (\max_P`\alpha \, P \, \text{seq}_P`P``\alpha) :$
$[*202{\cdot}103] \qquad \supset : \max_P`\alpha = \text{seq}_P`P``\alpha :. \supset \vdash . \text{Prop}$

In the hypothesis of *206·32, we have both $\text{E} ! \max_P`\alpha$ and $\text{E} ! \text{seq}_P`P``\alpha$. So long as P is not contained in diversity, these are both necessary. For example, suppose we take

$$P = \alpha \uparrow (\alpha \cup \iota`x), \text{ where } x \sim \in \alpha . \exists ! \alpha .$$

Then P is transitive and connected, but not contained in diversity. We have

$$\alpha \cup \iota`x = C`P . P``(\alpha \cup \iota`x) = \alpha = \text{D}`P.$$

Also $\qquad \max_P`(\alpha \cup \iota`x) = x,$

$$\overrightarrow{\text{seq}}_P`P``(\alpha \cup \iota`x) = \overrightarrow{\min}_P`p`\overleftarrow{P}``\alpha = \overrightarrow{\min}_P`(\alpha \cup \iota`x) = \Lambda.$$

Thus in this case $\max_P`(\alpha \cup \iota`x)$ exists, but $\text{seq}_P`P``(\alpha \cup \iota`x)$ does not exist. When $P$ is serial, *i.e.* when $P$ is contained in diversity, in addition to being transitive and connected, the existence of $\max_P`\alpha$ involves that of $\text{seq}_P`P``\alpha$, and therefore the hypothesis $\text{E}! \text{seq}_P`P``\alpha$, which appears in *206·32, becomes unnecessary.

***206·33.** $\vdash : P \in \text{trans} \cap \text{connex} . \sim \text{E} ! \max_P`\alpha . \supset . \overrightarrow{\text{seq}}_P`P``\alpha = \overrightarrow{\text{seq}}_P`\alpha$

Dem.

$\vdash . *205{\cdot}6 . \supset \vdash : \text{Hp} . \supset . \alpha \cap C`P \subset P``\alpha .$
$[*22{\cdot}62 . *37{\cdot}15] \qquad \supset . (\alpha \cup P``\alpha) \cap C`P = P``\alpha \quad (1)$
$\vdash . *206{\cdot}25 . \supset \vdash : \text{Hp} . \supset . \overrightarrow{\text{seq}}_P`\alpha = \overrightarrow{\text{seq}}_P`(\alpha \cup P``\alpha)$
$[*206{\cdot}131] \qquad = \overrightarrow{\text{seq}}_P`\{(\alpha \cup P``\alpha) \cap C`P\}$
$[(1)] \qquad = \overrightarrow{\text{seq}}_P`P``\alpha : \supset \vdash . \text{Prop}$

***206·331.** $\vdash : P \epsilon \text{trans} \cap \text{connex} . \sim E ! \max_P `\alpha . E ! \text{seq}_P `\alpha . \supset . \text{seq}_P `P``\alpha = \text{seq}_P `\alpha$ [*206·33]

***206·34.** $\vdash : P \epsilon \text{Ser} . \supset . \overrightarrow{\max_P} `\alpha \subset \overrightarrow{\text{seq}_P} `P``\alpha$

Dem.

$\vdash . *205·101 . *37·265 . \supset$

$\vdash :. y \epsilon \overrightarrow{\max_P} `\alpha . \equiv : y \epsilon \alpha \cap C`P : z \epsilon \alpha \cap C`P . \supset_z . \sim (yPz)$ (1)

$\vdash . (1) . *202·103 . \supset \vdash ::. \text{Hp} . \supset ::$

$y \epsilon \overrightarrow{\max_P} `\alpha . \supset :. y \epsilon \alpha \cap C`P :. z \epsilon \alpha \cap C`P . \supset_z : z = y . \vee . zPy$ (2)

$\vdash . (2) . *13·195 . *201·1 . \supset \vdash ::. \text{Hp} . \supset ::$

$y \epsilon \overrightarrow{\max_P} `\alpha . \supset :. y \epsilon \alpha \cap C`P :. z \epsilon \alpha \cap C`P . uPz . \supset_{u,z} . uPy :.$

[*37·1·265] $\supset :. u \epsilon P``\alpha . \supset_u . uPy :.$

[*40·53] $\supset :. y \epsilon p`\overleftarrow{P}``P``\alpha$ (3)

$\vdash . (1) . *37·1 . \quad \supset \vdash : y \epsilon \overrightarrow{\max_P} `\alpha . vPy . \supset . v \epsilon P``\alpha$ (4)

$\vdash . *50·24 . \quad \supset \vdash : \text{Hp} . \supset . \sim (vPv)$ (5)

$\vdash . (4) . (5) . \quad \supset \vdash :. \text{Hp} . \supset : y \epsilon \overrightarrow{\max_P} `\alpha . vPy . \supset . (\exists w) . w \epsilon P``\alpha . \sim (wPv) .$

[*40·53] $\supset . v \sim \epsilon p`\overleftarrow{P}``P``\alpha :$

[*10·51] $\supset : y \epsilon \overrightarrow{\max_P} `\alpha . \supset . \sim (\exists v) . v \epsilon p`\overleftarrow{P}``P``\alpha . vPy .$

[*37·105] $\supset . y \sim \epsilon \breve{P}``p`\overleftarrow{P}``P``\alpha$ (6)

$\vdash . (3) . (6) . (1) . \supset \vdash :. \text{Hp} . \supset :$

$y \epsilon \overrightarrow{\max_P} `\alpha . \supset . y \epsilon p`\overleftarrow{P}``P``\alpha \cap C`P - \breve{P}``p`\overleftarrow{P}``P``\alpha .$

[*206·143] $\supset . y \epsilon \overrightarrow{\text{seq}_P} `P``\alpha :. \supset \vdash . \text{Prop}$

***206·35.** $\vdash : P \epsilon \text{Ser} . E ! \max_P `\alpha . \supset . \max_P `\alpha = \text{seq}_P `P``\alpha . E ! \text{seq}_P `P``\alpha$

Dem.

$\vdash . *206·34 . \quad \supset \vdash : \text{Hp} . \supset . \max_P `\alpha \epsilon \overrightarrow{\text{seq}_P} `P``\alpha$ (1)

$\vdash . (1) . *206·16 . \supset \vdash : \text{Hp} . \supset . \max_P `\alpha = \text{seq}_P `P``\alpha$ (2)

$\vdash . (2) . *14·21 . \supset \vdash : \text{Hp} . \supset . E ! \text{seq}_P `P``\alpha$ (3)

$\vdash . (2) . (3) . \supset \vdash . \text{Prop}$

***206·36.** $\vdash :: P \epsilon \text{Ser} . \supset :. E ! \text{seq}_P `P``\alpha . \equiv : E ! \max_P `\alpha . \vee . E ! \text{seq}_P `\alpha$

Dem.

$\vdash . *206·35·331 . \supset \vdash :. \text{Hp} : E ! \max_P `\alpha . \vee . E ! \text{seq}_P `\alpha : \supset . E ! \text{seq}_P `P``\alpha$ (1)

$\vdash . *206·34 . \quad \supset \vdash :. \text{Hp} . \supset : \sim E ! \text{seq}_P `P``\alpha . \supset . \sim E ! \max_P `\alpha .$ (2)

[*206·33] $\supset . \sim E ! \text{seq}_P `\alpha$ (3)

$\vdash . (1) . (2) . (3) . \supset \vdash . \text{Prop}$

The condition $(\alpha): \text{E}\,!\,\text{max}_P{}^{\backprime}\alpha \,.\vee.\, \text{E}\,!\,\text{seq}_P{}^{\backprime}\alpha$ is the definition of what may be called "Dedekindian" series, *i.e.* series in which, when any division of the field into two parts is made in such a way that the first part wholly precedes the second, then either the first part has a last term or the second part has a first term. (When these alternatives are also mutually exclusive, the series has "Dedekindian continuity.") If α is any class, $P^{\backprime\backprime}\alpha$ is the segment of $C^{\backprime}P$ defined by α. In virtue of the above proposition, every segment of a Dedekindian series has a sequent. The sequent of a class having no maximum is what is commonly called a *limit*. Thus in a series having Dedekindian continuity (in which segments never have maxima), every segment has a limit.

***206·37.** $\vdash : P \in \text{Ser} \,.\supset.\, \overrightarrow{\text{seq}}_P{}^{\backprime}P^{\backprime\backprime}\alpha = \overrightarrow{\text{min}}_P{}^{\backprime}(\overrightarrow{\text{max}}_P{}^{\backprime}\alpha \cup \overrightarrow{\text{seq}}_P{}^{\backprime}\alpha)$

Dem.

$\vdash .\, *205·16 \,.\supset \vdash : \overrightarrow{\text{max}}_P{}^{\backprime}\alpha = \Lambda \,.\, \overrightarrow{\text{seq}}_P{}^{\backprime}\alpha = \Lambda \,.\supset.$
$\overrightarrow{\text{min}}_P{}^{\backprime}(\overrightarrow{\text{max}}_P{}^{\backprime}\alpha \cup \overrightarrow{\text{seq}}_P{}^{\backprime}\alpha) = \Lambda \quad (1)$

$\vdash .\, *206·36 \,.\supset \vdash : \text{Hp} \,.\, \text{Hp}(1) \,.\supset.\, \sim \text{E}\,!\,\text{seq}_P{}^{\backprime}P^{\backprime\backprime}\alpha \,.$

[*206·16] $\supset.\, \overrightarrow{\text{seq}}_P{}^{\backprime}P^{\backprime\backprime}\alpha = \Lambda \quad (2)$

$\vdash .\, *24·24 \,.\supset \vdash : \text{Hp} \,.\, \overrightarrow{\text{max}}_P{}^{\backprime}\alpha = \Lambda \,.\, \exists\,!\,\overrightarrow{\text{seq}}_P{}^{\backprime}\alpha \,.\supset.$
$\overrightarrow{\text{min}}_P{}^{\backprime}(\overrightarrow{\text{max}}_P{}^{\backprime}\alpha \cup \overrightarrow{\text{seq}}_P{}^{\backprime}\alpha) = \overrightarrow{\text{min}}_P{}^{\backprime}\overrightarrow{\text{seq}}_P{}^{\backprime}\alpha$

[*205·17.*206·16] $= \overrightarrow{\text{seq}}_P{}^{\backprime}\alpha$

[*206·33] $= \overrightarrow{\text{seq}}_P{}^{\backprime}P^{\backprime\backprime}\alpha \quad (3)$

$\vdash .\, *205·17·3 \,.\supset \vdash : \text{Hp} \,.\, \exists\,!\,\overrightarrow{\text{max}}_P{}^{\backprime}\alpha \,.\, \overrightarrow{\text{seq}}_P{}^{\backprime}\alpha = \Lambda \,.\supset.$
$\overrightarrow{\text{min}}_P{}^{\backprime}(\overrightarrow{\text{max}}_P{}^{\backprime}\alpha \cup \overrightarrow{\text{seq}}_P{}^{\backprime}\alpha) = \overrightarrow{\text{max}}_P{}^{\backprime}\alpha$

[*206·35] $= \overrightarrow{\text{seq}}_P{}^{\backprime}P^{\backprime\backprime}\alpha \quad (4)$

$\vdash .\, *206·16 \,.\, *205·3 \,.\supset$

$\vdash : \text{Hp} \,.\, \exists\,!\,\overrightarrow{\text{max}}_P{}^{\backprime}\alpha \,.\, \exists\,!\,\overrightarrow{\text{seq}}_P{}^{\backprime}\alpha \,.\supset.$
$\overrightarrow{\text{min}}_P{}^{\backprime}(\overrightarrow{\text{max}}_P{}^{\backprime}\alpha \cup \overrightarrow{\text{seq}}_P{}^{\backprime}\alpha) = \overrightarrow{\text{min}}_P{}^{\backprime}(\iota^{\backprime}\text{max}_P{}^{\backprime}\alpha \cup \iota^{\backprime}\text{seq}_P{}^{\backprime}\alpha)$

[*206·27.*205·182] $= \iota^{\backprime}\text{max}_P{}^{\backprime}\alpha$

[*206·35] $= \overrightarrow{\text{seq}}_P{}^{\backprime}P^{\backprime\backprime}\alpha \quad (5)$

$\vdash .\, (1) \,.\, (2) \,.\, (3) \,.\, (4) \,.\, (5) \,.\supset \vdash .\,\text{Prop}$

***206·38.** $\vdash : P \in \text{Ser} \,.\supset.\, \overrightarrow{\text{max}}_P{}^{\backprime}\alpha = \alpha \cap \overrightarrow{\text{seq}}_P{}^{\backprime}P^{\backprime\backprime}\alpha$

Dem.

$\vdash .\, *206·35 \,.\, *205·111 \,.\supset$

$\vdash : \text{Hp} \,.\, \text{E}\,!\,\text{max}_P{}^{\backprime}\alpha \,.\supset.\, \overrightarrow{\text{max}}_P{}^{\backprime}\alpha = \overrightarrow{\text{seq}}_P{}^{\backprime}P^{\backprime\backprime}\alpha \,.\, \overrightarrow{\text{max}}_P{}^{\backprime}\alpha \subset \alpha \,.$

[*22·621] $\supset.\, \overrightarrow{\text{max}}_P{}^{\backprime}\alpha = \alpha \cap \overrightarrow{\text{seq}}_P{}^{\backprime}P^{\backprime\backprime}\alpha \quad (1)$

$\vdash . *205\cdot3 . \supset \vdash : \text{Hp} . \sim \text{E}!\, \text{max}_P\text{‘}\alpha . \supset . \overrightarrow{\text{max}}_P\text{‘}\alpha = \Lambda$ (2)

$\vdash . *206\cdot33 . \supset \vdash : \text{Hp} . \sim \text{E}!\, \text{max}_P\text{‘}\alpha . \supset . \overrightarrow{\text{seq}}_P\text{‘}P\text{“}\alpha = \overrightarrow{\text{seq}}_P\text{‘}\alpha .$

[*206·2] $\supset . \alpha \cap \overrightarrow{\text{seq}}_P\text{‘}P\text{“}\alpha = \Lambda .$

[(2)] $\supset . \overrightarrow{\text{max}}_P\text{‘}\alpha = \alpha \cap \overrightarrow{\text{seq}}_P\text{‘}P\text{“}\alpha$ (3)

$\vdash . (1) . (3) . \supset \vdash . \text{Prop}$

***206·4.** $\vdash : P \in J . x \in C\text{‘}P . \supset . x\, \text{seq}_P\, \overrightarrow{P}\text{‘}x$

Dem.

$\vdash . *206\cdot134 . *22\cdot43 . \supset$

$\vdash : x\, \text{seq}_P\, \overrightarrow{P}\text{‘}x . \equiv . x \in C\text{‘}P . \overrightarrow{P}\text{‘}x \subset -p\text{‘}\overleftarrow{P}\text{“}\overrightarrow{P}\text{‘}x$ (1)

$\vdash . *200\cdot5 . \supset \vdash : P \in J . \supset . \overrightarrow{P}\text{‘}x \subset -p\text{‘}\overleftarrow{P}\text{“}\overrightarrow{P}\text{‘}x$ (2)

$\vdash . (1) . (2) . \supset \vdash . \text{Prop}$

***206·401.** $\vdash : P \in \text{connex} \cap \text{Rl}\text{‘}J . x \in C\text{‘}P . \supset . x = \text{seq}_P\text{‘}\overrightarrow{P}\text{‘}x$ [*206·4·161]

***206·41.** $\vdash . \overrightarrow{\text{min}}_P\text{‘}\overleftarrow{P}\text{‘}x = \overleftarrow{P \dot{-} P^2}\text{‘}x$ [*205·25]

***206·42.** $\vdash : x \in C\text{‘}P . \supset . \overrightarrow{\text{seq}}_P\text{‘}\iota\text{‘}x = \overleftarrow{P \dot{-} P^2}\text{‘}x = \overrightarrow{\text{min}}_P\text{‘}\overleftarrow{P}\text{‘}x$

Dem.

$\vdash . *53\cdot01\cdot31 . \supset \vdash . p\text{‘}\overleftarrow{P}\text{“}\iota\text{‘}x = \overleftarrow{P}\text{‘}x$ (1)

$\vdash . (1) . *206\cdot41\cdot143 . \supset \vdash . \text{Prop}$

***206·43.** $\vdash : P \in \text{trans} \cap \text{Rl}\text{‘}J . x \in C\text{‘}P . \supset . \overrightarrow{\text{seq}}_P\text{‘}\iota\text{‘}x = \overleftarrow{P}_1\text{‘}x$
[*206·42 . *201·63]

***206·44.** $\vdash :. P \in \text{trans} \cap \text{Rl}\text{‘}J . x \in C\text{‘}P . \supset :$

$\text{E}!\, \text{seq}_P\text{‘}\iota\text{‘}x . \equiv . \text{E}!\, \breve{P}_1\text{‘}x : \text{E}!\, \text{seq}_P\text{‘}\iota\text{‘}x . \supset . \text{seq}_P\text{‘}\iota\text{‘}x = \breve{P}_1\text{‘}x$
[*206·43]

***206·45.** $\vdash :. P \in \text{Ser} . x \in C\text{‘}P . \supset : \text{E}!\, \text{seq}_P\text{‘}\iota\text{‘}x . \equiv . x \in \text{D}\text{‘}P_1$
[*206·44 . *204·7 . *71·165]

***206·451.** $\vdash : P \in \text{Ser} . \text{E}!\, \text{seq}_P\text{‘}\alpha . \supset . \overrightarrow{\text{max}}_P\text{‘}\alpha = \overrightarrow{P}_1\text{‘}\text{seq}_P\text{‘}\alpha$

Dem.

$\vdash . *206\cdot41 . \supset \vdash : \text{Hp} . \supset . \overrightarrow{P}_1\text{‘}\text{seq}_P\text{‘}\alpha = \overrightarrow{\text{max}}_P\text{‘}\overrightarrow{P}\text{‘}\text{seq}_P\text{‘}\alpha$

[*206·22] $= \overrightarrow{\text{max}}_P\text{‘}\{(\alpha \cap C\text{‘}P) \cup P\text{“}\alpha\}$

[*205·191] $= \overrightarrow{\text{max}}_P\text{‘}\alpha : \supset \vdash . \text{Prop}$

***206·46.** $\vdash : P \epsilon \text{trans} \cap \text{connex} . \text{E} ! \text{max}_P\text{‘}\alpha . \supset . \overrightarrow{\text{seq}}_P\text{‘}\alpha = \overrightarrow{\text{seq}}_P\text{‘}\overrightarrow{\text{max}}_P\text{‘}\alpha$

Dem.

$\vdash . *206\cdot42 . \supset \vdash : \text{Hp} . \supset . \overrightarrow{\text{seq}}_P\text{‘}\overrightarrow{\text{max}}_P\text{‘}\alpha = \overrightarrow{\text{min}}_P\text{‘}\overleftarrow{P}\text{‘max}_P\text{‘}\alpha$

[*205·65] $= \overrightarrow{\text{min}}_P\text{‘}p\text{‘}\overleftarrow{P}\text{‘‘}(\alpha \cap C\text{‘}P)$

[*206·13] $= \overrightarrow{\text{seq}}_P\text{‘}\alpha : \supset \vdash . \text{Prop}$

***206·47.** $\vdash : P \epsilon \text{trans} . \text{E} ! \text{seq}_P\text{‘}\alpha . \supset . \text{seq}_P\text{‘}\alpha = \text{max}_P\text{‘}(\alpha \cup \overrightarrow{\text{seq}}_P\text{‘}\alpha)$

Dem.

$\vdash . *206\cdot134 . \supset \vdash : \text{Hp} . \supset . \alpha \cap C\text{‘}P \subset \overrightarrow{P}\text{‘seq}_P\text{‘}\alpha .$

[*205·193·151] $\supset . \overrightarrow{\text{max}}_P\text{‘}(\alpha \cup \overrightarrow{\text{seq}}_P\text{‘}\alpha) = \overrightarrow{\text{max}}_P\text{‘}\overrightarrow{\text{seq}}_P\text{‘}\alpha$

[*206·133.*205·18] $= \iota\text{‘seq}_P\text{‘}\alpha : \supset \vdash . \text{Prop}$

***206·48.** $\vdash : P \epsilon \text{trans} \cap \text{connex} . \text{E} ! \text{seq}_P\text{‘}\alpha . \supset . \overrightarrow{\text{seq}}_P\text{‘}\overrightarrow{\text{seq}}_P\text{‘}\alpha = \overrightarrow{\text{seq}}_P\text{‘}(\alpha \cup \overrightarrow{\text{seq}}_P\text{‘}\alpha)$

Dem.

$\vdash . *206\cdot47 . \supset \vdash : \text{Hp} . \supset .$

$\overrightarrow{\text{seq}}_P\text{‘}\overrightarrow{\text{seq}}_P\text{‘}\alpha = \overrightarrow{\text{seq}}_P\text{‘}\overrightarrow{\text{max}}_P\text{‘}(\alpha \cup \overrightarrow{\text{seq}}_P\text{‘}\alpha) . \text{E} ! \text{max}_P\text{‘}(\alpha \cup \text{seq}_P\text{‘}\alpha) .$

[*206·46] $\supset . \overrightarrow{\text{seq}}_P\text{‘}\overrightarrow{\text{seq}}_P\text{‘}\alpha = \overrightarrow{\text{seq}}_P\text{‘}(\alpha \cup \overrightarrow{\text{seq}}_P\text{‘}\alpha) : \supset \vdash . \text{Prop}$

***206·5.** $\vdash : P \epsilon \text{trans} \cap \text{connex} . \text{E} ! \text{max}_P\text{‘}\alpha . \text{E} ! \text{seq}_P\text{‘}\alpha . \supset .$

$\text{max}_P\text{‘}\alpha \, (P \dot{-} P^2) \, \text{seq}_P\text{‘}\alpha$

Dem.

$\vdash . *206\cdot46 . \supset \vdash : \text{Hp} . \supset . \overrightarrow{\text{sep}}_P\text{‘}\alpha = \overrightarrow{\text{seq}}_P\text{‘}\iota\text{‘max}_P\text{‘}\alpha$

[*206·42] $= \overleftarrow{P \dot{-} P^2}\text{‘max}_P\text{‘}\alpha : \supset \vdash . \text{Prop}$

***206·51.** $\vdash : \exists ! \overrightarrow{\text{max}}_P\text{‘}\overrightarrow{P}\text{‘}x . \supset . x \, \text{seq}_P \, \overrightarrow{P}\text{‘}x$

Dem.

$\vdash . *205\cdot161 . \quad \supset \vdash : \text{Hp} . \supset . \exists ! \overrightarrow{P}\text{‘}x .$

[*33·42] $\supset . x \epsilon C\text{‘}P$ (1)

$\vdash . (1) . *206\cdot134 . \supset \vdash :. \text{Hp} . \supset : x \, \text{seq}_P \overrightarrow{P}\text{‘}x . \equiv . \overrightarrow{P}\text{‘}x \subset \overrightarrow{P}\text{‘}x . \overrightarrow{P}\text{‘}x \subset - p\text{‘}\overleftarrow{P}\text{‘‘}\overrightarrow{P}\text{‘}x .$

[*22·42] $\equiv . \overrightarrow{P}\text{‘}x \subset - p\text{‘}\overleftarrow{P}\text{‘‘}\overrightarrow{P}\text{‘}x$ (2)

$\vdash . *205\cdot101 . \supset \vdash :. y \epsilon \overrightarrow{\text{max}}_P\text{‘}\overrightarrow{P}\text{‘}x . \supset : yPx . y \sim\epsilon P\text{‘‘}\overrightarrow{P}\text{‘}x :$

[*37·1] $\supset : yPx : zPx . \supset_z . \sim(yPz) :$

[*32·18.*5·31] $\supset : zPx . \supset_z . y \epsilon \overrightarrow{P}\text{‘}x . \sim(yPz) .$

[*40·53] $\supset_z . z \sim\epsilon p\text{‘}\overleftarrow{P}\text{‘‘}\overrightarrow{P}\text{‘}x :$

[*32·18] $\supset : \overrightarrow{P}\text{‘}x \subset - p\text{‘}\overleftarrow{P}\text{‘‘}\overrightarrow{P}\text{‘}x$ (3)

$\vdash . (2) . (3) . \supset \vdash : \text{Hp} . y \epsilon \overrightarrow{\text{max}}_P\text{‘}\overrightarrow{P}\text{‘}x . \supset . x \, \text{seq}_P \, \overrightarrow{P}\text{‘}x : \supset \vdash . \text{Prop}$

∗206·52. $\vdash : P \epsilon \text{trans} \cap \text{connex} . \text{E} ! \max_P{}^{\backprime}P^{\backprime\backprime}\alpha . \supset .$
$\text{E} ! \text{seq}_P{}^{\backprime}P^{\backprime\backprime}\alpha . \text{seq}_P{}^{\backprime}P^{\backprime\backprime}\alpha = \max_P{}^{\backprime}\alpha$

Dem.

$\vdash . ∗205·7 . \supset \vdash : \text{Hp} . \supset . \text{E} ! \max_P{}^{\backprime}\alpha .$ (1)

$[∗205·22] \supset . P^{\backprime\backprime}\alpha = \overrightarrow{P}{}^{\backprime}\max_P{}^{\backprime}\alpha$ (2)

$\vdash . (2) . ∗206·51 . \supset \vdash : \text{Hp} . \supset . \max_P{}^{\backprime}\alpha \, \text{seq}_P \, P^{\backprime\backprime}\alpha .$

$[∗206·161] \supset . \max_P{}^{\backprime}\alpha = \text{seq}_P{}^{\backprime}P^{\backprime\backprime}\alpha$ (3)

$\vdash . (1) . (3) . \supset \vdash . \text{Prop}$

∗206·53. $\vdash : P \epsilon \text{Ser} . \supset . \overrightarrow{\text{seq}}_P{}^{\backprime}p{}^{\backprime}\overrightarrow{P}{}^{\backprime\backprime}(\alpha \cap C{}^{\backprime}P) = \overrightarrow{\min}_P{}^{\backprime}\alpha$

Dem.

$\vdash . ∗206·13 . \supset \vdash . \overrightarrow{\text{seq}}_P{}^{\backprime}p{}^{\backprime}\overrightarrow{P}{}^{\backprime\backprime}(\alpha \cap C{}^{\backprime}P) = \overrightarrow{\min}_P{}^{\backprime}p{}^{\backprime}\overleftarrow{P}{}^{\backprime\backprime}\{p{}^{\backprime}\overrightarrow{P}{}^{\backprime\backprime}(\alpha \cap C{}^{\backprime}P) \cap C{}^{\backprime}P\}$

$[∗205·15·16 . ∗206·18 . ∗200·54] = \overrightarrow{\min}_P{}^{\backprime}\{C{}^{\backprime}P \cap p{}^{\backprime}\overleftarrow{P}{}^{\backprime\backprime}p{}^{\backprime}\overrightarrow{P}{}^{\backprime\backprime}(\alpha \cap C{}^{\backprime}P)\}$ (1)

$\vdash . (1) . ∗204·62 . \supset \vdash : \text{Hp} . \supset . \overrightarrow{\text{seq}}_P{}^{\backprime}p{}^{\backprime}\overrightarrow{P}{}^{\backprime\backprime}(\alpha \cap C{}^{\backprime}P) = \overrightarrow{\min}_P{}^{\backprime}\{(\alpha \cap C{}^{\backprime}P) \cup \breve{P}{}^{\backprime\backprime}\alpha\}$

$[∗205·19 . ∗201·52] = \overrightarrow{\min}_P{}^{\backprime}\alpha : \supset \vdash . \text{Prop}$

∗206·531. $\vdash : P \epsilon \text{Ser} . \supset .$
$C{}^{\backprime}P \cap \hat{x}\{p{}^{\backprime}\overrightarrow{P}{}^{\backprime\backprime}(\alpha \cap C{}^{\backprime}P) = \overrightarrow{P}{}^{\backprime}x\} = \overrightarrow{\text{seq}}_P{}^{\backprime}p{}^{\backprime}\overrightarrow{P}{}^{\backprime\backprime}(\alpha \cap C{}^{\backprime}P) = \overrightarrow{\min}_P{}^{\backprime}\alpha$

Dem.

$\vdash . ∗206·172 . ∗201·51 . \supset$

$\vdash : \text{Hp} . \supset . \overrightarrow{\text{seq}}_P{}^{\backprime}p{}^{\backprime}\overleftarrow{P}{}^{\backprime\backprime}(\alpha \cap C{}^{\backprime}P) = C{}^{\backprime}P \cap \hat{x}\{p{}^{\backprime}\overrightarrow{P}{}^{\backprime\backprime}(\alpha \cap C{}^{\backprime}P) \cap C{}^{\backprime}P = \overrightarrow{P}{}^{\backprime}x\}$ (1)

$\vdash . (1) . ∗40·62 . \supset \vdash : \text{Hp} . \exists ! (\alpha \cap C{}^{\backprime}P) . \supset .$
$\overrightarrow{\text{seq}}_P{}^{\backprime}p{}^{\backprime}\overleftarrow{P}{}^{\backprime\backprime}(\alpha \cap C{}^{\backprime}P) = C{}^{\backprime}P \cap \hat{x}\{p{}^{\backprime}\overrightarrow{P}{}^{\backprime\backprime}(\alpha \cap C{}^{\backprime}P) = \overrightarrow{P}{}^{\backprime}x\}$ (2)

$\vdash . ∗205·16 . ∗206·53 . \supset \vdash : \text{Hp} . \alpha \cap C{}^{\backprime}P = \Lambda . \supset . \overrightarrow{\text{seq}}_P{}^{\backprime}p{}^{\backprime}\overleftarrow{P}{}^{\backprime\backprime}(\alpha \cap C{}^{\backprime}P) = \Lambda$ (3)

$\vdash . ∗40·2 . \supset \vdash : \alpha \cap C{}^{\backprime}P = \Lambda . \supset .$
$C{}^{\backprime}P \cap \hat{x}\{p{}^{\backprime}\overrightarrow{P}{}^{\backprime\backprime}(\alpha \cap C{}^{\backprime}P) = \overrightarrow{P}{}^{\backprime}x\} = C{}^{\backprime}P \cap \hat{x}(\text{V} = \overrightarrow{P}{}^{\backprime}x)$ (4)

$\vdash . ∗50·24 . \supset \vdash : \text{Hp} . \supset . (x) . x \sim \epsilon \overrightarrow{P}{}^{\backprime}x .$

$[∗24·104] \supset . (x) . \overrightarrow{P}{}^{\backprime}x \neq \text{V}$ (5)

$\vdash . (4) . (5) . \supset \vdash : \text{Hp} . \alpha \cap C{}^{\backprime}P = \Lambda . \supset . C{}^{\backprime}P \cap \hat{x}\{p{}^{\backprime}\overrightarrow{P}{}^{\backprime\backprime}(\alpha \cap C{}^{\backprime}P) = \overrightarrow{P}{}^{\backprime}x\} = \Lambda$

$[(3)] = \overrightarrow{\text{seq}}_P{}^{\backprime}p{}^{\backprime}\overrightarrow{P}{}^{\backprime\backprime}(\alpha \cap C{}^{\backprime}P)$ (6)

$\vdash . (2) . (6) . ∗206·53 . \supset \vdash . \text{Prop}$

∗206·54. $\vdash :. P \epsilon \text{Ser} . \supset : \text{E} ! \text{seq}_P{}^{\backprime}p{}^{\backprime}\overrightarrow{P}{}^{\backprime\backprime}(\alpha \cap C{}^{\backprime}P) . \equiv . \text{E} ! \min_P{}^{\backprime}\alpha :$
$\text{E} ! \max_P{}^{\backprime}p{}^{\backprime}\overrightarrow{P}{}^{\backprime\backprime}(\alpha \cap C{}^{\backprime}P) . \equiv . \text{E} ! \text{prec}_P{}^{\backprime}\alpha$

Dem.

$\vdash . ∗206·53 . \supset \vdash :. \text{Hp} . \supset : \text{E} ! \text{seq}_P{}^{\backprime}p{}^{\backprime}\overrightarrow{P}{}^{\backprime\backprime}(\alpha \cap C{}^{\backprime}P) . \equiv . \text{E} ! \min_P{}^{\backprime}\alpha$. (1)

$\vdash . ∗206·13·101 . \supset \vdash : \text{E} ! \max_P{}^{\backprime}p{}^{\backprime}\overrightarrow{P}{}^{\backprime\backprime}(\alpha \cap C{}^{\backprime}P) . \equiv . \text{E} ! \text{prec}_P{}^{\backprime}\alpha$ (2)

$\vdash . (1) . (2) . \supset \vdash . \text{Prop}$

∗206·55. $\vdash :. P \epsilon \text{Ser} . \supset : (\alpha) . \alpha \epsilon \text{Cl}{}^{\backprime}\min_P \cup \text{Cl}{}^{\backprime}\text{prec}_P . \equiv .$
$(\alpha) . p{}^{\backprime}\overrightarrow{P}{}^{\backprime\backprime}(\alpha \cap C{}^{\backprime}P) \epsilon \text{Cl}{}^{\backprime}\max_P \cup \text{Cl}{}^{\backprime}\text{seq}_P$ [∗206·54·161 . ∗205·32]

***206·551.** $\vdash : P \in \mathrm{Ser} \,.\supset .\, \overrightarrow{\mathrm{seq}}_P{}'\alpha = \overrightarrow{\mathrm{seq}}_P{}'p'\overrightarrow{P}{}''p'\overleftarrow{P}{}''(\alpha \cap C'P) \,.$

$$\overrightarrow{\mathrm{max}}_P{}'\alpha = \overrightarrow{\mathrm{max}}_P{}'p'\overrightarrow{P}{}''p'\overleftarrow{P}{}''(\alpha \cap C'P)$$

Dem.

$\vdash . *206·13 . \quad \supset \vdash . \overrightarrow{\mathrm{seq}}_P{}'\alpha = \overrightarrow{\mathrm{min}}_P{}'p'\overleftarrow{P}{}''(\alpha \cap C'P)$ (1)

$\vdash . (1) . *206·53 . \supset \vdash : \mathrm{Hp} . \supset . \overrightarrow{\mathrm{seq}}_P{}'\alpha = \overrightarrow{\mathrm{seq}}_P{}'p'\overrightarrow{P}{}''\{p'\overleftarrow{P}{}''(\alpha \cap C'P) \cap C'P\}$ (2)

$\vdash . (2) . *200·54 . \supset \vdash : \mathrm{Hp} . \dot{\exists} ! P . \supset . \overrightarrow{\mathrm{seq}}_P{}'\alpha = \overrightarrow{\mathrm{seq}}_P{}'p'\overrightarrow{P}{}''p'\overleftarrow{P}{}''(\alpha \cap C'P)$ (3)

$\vdash . *206·18 . \quad \supset \vdash : P = \dot{\Lambda} . \supset . \overrightarrow{\mathrm{seq}}_P{}'\alpha = \Lambda . \overrightarrow{\mathrm{seq}}_P{}'p'\overrightarrow{P}{}''p'\overleftarrow{P}{}''(\alpha \cap C'P) = \Lambda$ (4)

$\vdash . (3) . (4) . \quad \supset \vdash : \mathrm{Hp} . \supset . \overrightarrow{\mathrm{seq}}_P{}'\alpha = \overrightarrow{\mathrm{seq}}_P{}'p'\overrightarrow{P}{}''p'\overleftarrow{P}{}''(\alpha \cap C'P)$ (5)

$\vdash . *206·53 . \quad \supset \vdash : \mathrm{Hp} . \supset . \overrightarrow{\mathrm{max}}_P{}'\alpha = \overrightarrow{\mathrm{prec}}_P{}'p'\overleftarrow{P}{}''(\alpha \cap C'P)$

[*206·13·101.*200·54] $\quad = \overrightarrow{\mathrm{max}}_P{}'p'\overrightarrow{P}{}''p'\overleftarrow{P}{}''(\alpha \cap C'P)$ (6)

$\vdash . (5) . (6) . \supset \vdash . \mathrm{Prop}$

***206·56.** $\vdash :. P \in \mathrm{Ser} . \supset : (\alpha) . p'\overrightarrow{P}{}''(\alpha \cap C'P) \in \text{ᗡ}'\mathrm{max}_P \cup \text{ᗡ}'\mathrm{seq}_P . \equiv .$

$$(\alpha) . \alpha \in \text{ᗡ}'\mathrm{max}_P \cup \text{ᗡ}'\mathrm{seq}_P$$

Dem.

$\vdash . *10·1·11 . \supset \vdash : (\alpha) . \alpha \in \text{ᗡ}'\mathrm{max}_P \cup \text{ᗡ}'\mathrm{seq}_P . \supset .$

$\quad (\alpha) . p'\overrightarrow{P}{}''(\alpha \cap C'P) \in \text{ᗡ}'\mathrm{max}_P \cup \text{ᗡ}'\mathrm{seq}_P$ (1)

$\vdash . *10·1 . \quad \supset \vdash : (\alpha) . p'\overrightarrow{P}{}''(\alpha \cap C'P) \in \text{ᗡ}'\mathrm{max}_P \cup \text{ᗡ}'\mathrm{seq}_P . \supset .$

$\quad p'\overrightarrow{P}{}''p'\overleftarrow{P}{}''(\beta \cap C'P) \in \text{ᗡ}'\mathrm{max}_P \cup \text{ᗡ}'\mathrm{seq}_P .$

[*206·551] $\quad \supset . \beta \in \text{ᗡ}'\mathrm{max}_P \cup \text{ᗡ}'\mathrm{seq}_P$ (2)

$\vdash . (1) . (2) . \supset \vdash . \mathrm{Prop}$

***206·57.** $\vdash :. P \in \mathrm{Ser} . \supset : (\alpha) . \alpha \in \text{ᗡ}'\mathrm{min}_P \cup \text{ᗡ}'\mathrm{prec}_P . \equiv .$

$\quad (\alpha) . \alpha \in \text{ᗡ}'\mathrm{max}_P \cup \text{ᗡ}'\mathrm{seq}_P$ [*206·55·56]

This proposition is important, since it shows that when a serial relation satisfies Dedekind's axiom, so does its converse. Thus if all classes which have no maximum have an upper limit, then all classes which have no minimum have a lower limit, and vice versa.

***206·6.** $\vdash : S \in P \, \overline{\mathrm{smor}} \, Q . \supset . p'\overleftarrow{P}{}''(\alpha \cap C'P) = S''p'\overleftarrow{Q}{}''\breve{S}{}''\alpha$

Dem.

$\vdash . *151·11 . \supset \vdash : \mathrm{Hp} . \supset . p'\overleftarrow{P}{}''(\alpha \cap C'P) = p'S'''\breve{Q}{}'''\overleftarrow{S}{}''(\alpha \cap \mathrm{D}'S)$

[*72·341] $\quad = S''p'\breve{Q}{}'''\overleftarrow{S}{}''(\alpha \cap \mathrm{D}'S)$

[*71·613] $\quad = S''p'\overleftarrow{Q}{}''\breve{S}{}''\alpha : \supset \vdash . \mathrm{Prop}$

***206·61.** $\vdash : S \in P \,\overline{\text{smor}}\, Q \,.\supset .\, \overrightarrow{\text{seq}}_P‘\alpha = S“\overrightarrow{\text{seq}}_Q‘\breve{S}“\alpha$

Dem.

$\vdash . *205·8 . *206·6·13 . \supset \vdash : \text{Hp} . \supset . \overrightarrow{\text{seq}}_P‘\alpha = S“\overrightarrow{\text{min}}_Q‘\breve{S}“S“p‘\overleftarrow{Q}“\breve{S}“\alpha$

$[*72·501 . *151·11] \qquad = S“\overrightarrow{\text{min}}_Q‘(p‘\overleftarrow{Q}“\breve{S}“\alpha \cap C‘Q)$

$[*206·13 . *205·15] \qquad = S“\overrightarrow{\text{seq}}_Q‘\breve{S}“\alpha : \supset \vdash . \text{Prop}$

***206·62.** $\vdash :. S \in P \,\overline{\text{smor}}\, Q . \supset : \text{E}\,!\, \text{seq}_P‘\alpha . \equiv . \text{E}\,!\, \text{seq}_Q‘\breve{S}“\alpha$

$[*206·61 . *73·22·44 . *53·3]$

***206·63.** $\vdash : S \in P \,\overline{\text{smor}}\, Q . \text{E}\,!\, \text{seq}_P‘\alpha . \supset . \text{seq}_P‘\alpha = S‘\text{seq}_Q‘\breve{S}“\alpha$

$[*206·61·62 . *53·31]$

***206·7.** $\vdash : P \in \text{trans} . \beta \subset C‘P . \sim(yPy) . y \sim\in \overrightarrow{\text{max}}_P‘\beta . \supset .$

$$p‘\overleftarrow{P}“\beta = p‘\overleftarrow{P}“(\beta - \iota‘y)$$

Dem.

$\vdash . *51·222 . \supset \vdash : y \sim\in \beta . \supset . p‘\overleftarrow{P}“\beta = p‘\overleftarrow{P}“(\beta - \iota‘y) \qquad (1)$

$\vdash . *205·111 . \supset \vdash :. \text{Hp} . y \in \beta . \supset : y \in P“\beta . \sim(yPy) :$

$[*37·1] \qquad \supset : (\exists x) . x \in \beta - \iota‘y . yPx :$

$[*10·56 . \text{Hp}] \qquad \supset : z \in p‘\overleftarrow{P}“(\beta - \iota‘y) . \supset . yPz :$

$[*53·14 . *51·221] \qquad \supset : p‘\overleftarrow{P}“(\beta - \iota‘y) \subset p‘\overleftarrow{P}“\beta :$

$[*40·16] \qquad \supset : p‘\overleftarrow{P}“(\beta - \iota‘y) = p‘\overleftarrow{P}“\beta \qquad (2)$

$\vdash . (1) . (2) . \supset \vdash . \text{Prop}$

***206·71.** $\vdash : P \in \text{trans} . \beta \subset C‘P . \sim(yPy) . y \sim\in \overrightarrow{\text{max}}_P‘\beta . \supset . \overrightarrow{\text{seq}}_P‘\beta = \overrightarrow{\text{seq}}_P‘(\beta - \iota‘y)$

Dem.

$\vdash . *51·222 . \supset \vdash : y \sim\in \beta . \supset . \overrightarrow{\text{seq}}_P‘\beta = \overrightarrow{\text{seq}}_P‘(\beta - \iota‘y) \qquad (1)$

$\vdash . *205·111 . \supset \vdash : \text{Hp} . y \in \beta . \supset . y \in P“\beta . \sim(yPy) .$

$[*37·1] \qquad \supset . (\exists z) . z \in \beta - \iota‘y . yPz \qquad (2)$

$\vdash . (2) . *10·56 . *201·1 . \supset \vdash : \text{Hp} . y \in \beta . \beta - \iota‘y \subset \overrightarrow{P}‘x . \supset . yPx .$

$[*32·18] \qquad \supset . \beta \subset \overrightarrow{P}‘x \qquad (3)$

$\vdash . (3) . *206·7 . \supset \vdash :. \text{Hp}\,(2) . \supset :$

$$\beta \subset \overrightarrow{P}‘x . \overrightarrow{P}‘x \subset - p‘\overrightarrow{P}“\beta . \equiv . \beta - \iota‘y \subset \overrightarrow{P}‘x . \overrightarrow{P}‘x \subset - p‘\overrightarrow{P}“(\beta - \iota‘y) :$$

$[*206·134] \supset : x \,\text{seq}_P\, \beta . \equiv . x \,\text{seq}_P\, (\beta - \iota‘y) \qquad (4)$

$\vdash . (1) . (4) . \supset \vdash . \text{Prop}$

***206·72.** $\vdash : P \in \text{trans} . \sim(yPy) . y \sim\in \overrightarrow{\text{max}}_P‘\beta . \supset . \overrightarrow{\text{seq}}_P‘\beta = \overrightarrow{\text{seq}}_P‘(\beta - \iota‘y)$

Dem.

$\vdash . *206·71·131 . *205·151 . \supset \vdash : \text{Hp} . \supset . \overrightarrow{\text{seq}}_P‘\beta = \overrightarrow{\text{seq}}_P‘(\beta \cap C‘P - \iota‘y)$

$[*206·131] \qquad = \overrightarrow{\text{seq}}_P‘(\beta - \iota‘y) : \supset \vdash . \text{Prop}$

***206·73.** $\vdash : \exists ! \gamma \cap C'P . E ! \text{prec}_P{}'\gamma . E ! \text{seq}_P{}'\gamma . \supset . \text{prec}_P{}'\gamma P^2 \text{seq}_P{}'\gamma$

Dem.

$\vdash . *206·211 . \supset \vdash : \text{Hp} . \supset . \gamma \cap C'P \subset \overrightarrow{P}'\text{seq}_P{}'\gamma \cap \overleftarrow{P}'\text{prec}_P{}'\gamma . \exists ! \gamma \cap C'P .$

[*34·11] $\supset . \text{prec}_P{}'\gamma P^2 \text{seq}_P{}'\gamma : \supset \vdash . \text{Prop}$

***206·731.** $\vdash :. \exists ! \gamma \cap C'P : P \epsilon \text{trans} . \text{v} . P^2 \mathrel{\subseteq} J : \supset . \sim(\text{prec}_P{}'\gamma = \text{seq}_P{}'\gamma)$

Dem.

$\vdash . *206·73 . \supset$

$\vdash : \exists ! \gamma \cap C'P . E ! \text{prec}_P{}'\gamma . E ! \text{seq}_P{}'\gamma . P \epsilon \text{trans} . \supset . \text{prec}_P{}'\gamma P \text{seq}_P{}'\gamma .$

[*206·133] $\supset . \text{prec}_P{}'\gamma \neq \text{seq}_P{}'\gamma$ (1)

$\vdash . *206·73 . \supset$

$\vdash : \exists ! \gamma \cap C'P . E ! \text{prec}_P{}'\gamma . E ! \text{seq}_P{}'\gamma . P^2 \mathrel{\subseteq} J . \supset . \text{prec}_P{}'\gamma \neq \text{seq}_P{}'\gamma$ (2)

$\vdash . *14·21 . \supset \vdash : \sim(E ! \text{prec}_P{}'\gamma . E ! \text{seq}_P{}'\gamma) . \supset . \sim(\text{prec}_P{}'\gamma = \text{seq}_P{}'\gamma)$ (3)

$\vdash . (1) . (2) . (3) . \supset \vdash . \text{Prop}$

Note that "$\text{prec}_P{}'\gamma \neq \text{seq}_P{}'\gamma$" is not the same proposition as $\sim(\text{prec}_P{}'\gamma = \text{seq}_P{}'\gamma)$. The former involves $E ! \text{prec}_P{}'\gamma . E ! \text{seq}_P{}'\gamma$, while the latter does not, in virtue of the conventions as to descriptive symbols explained in *14.

***206·732.** $\vdash :. P \epsilon \text{trans} . \text{v} . P^2 \mathrel{\subseteq} J : \supset . \sim(\text{prec}_P{}'\gamma = \text{seq}_P{}'\gamma)$

Dem.

$\vdash . *206·14 . \supset \vdash : \gamma \cap C'P = \Lambda . \supset . \overrightarrow{\text{prec}_P}{}'\gamma = \overrightarrow{B}'\breve{P} . \text{seq}_P{}'\gamma = \overrightarrow{B}'P .$

[*93·101] $\supset . \overrightarrow{\text{prec}_P}{}'\gamma \cap \overrightarrow{\text{seq}_P}{}'\gamma = \Lambda .$

[*53·4] $\supset . \sim(\text{prec}_P{}'\gamma = \text{seq}_P{}'\gamma)$ (1)

$\vdash . (1) . *206·731 . \supset \vdash . \text{Prop}$

*207. LIMITS.

Summary of *207.

A term x is said to be the "upper limit" of α in P if α has no maximum and x is the sequent of α. In this case, x immediately follows the class α, though there is no one member of α which x immediately follows. Sequents which are limits have special importance, and it is convenient to have a special notation for them. We write "$\text{lt}_P\text{‘}\alpha$" for the upper limit of α; or, if it is more convenient, "$\text{lt}(P)\text{‘}\alpha$." (This is more convenient when P is replaced by an expression consisting of several letters, or by a letter with a suffix.) The *lower* limit of α will be the immediate predecessor of α when α has no minimum; this we denote by $\text{tl}_P\text{‘}\alpha$.

The following propositions on limits for the most part follow immediately from the propositions of *206 on sequents.

Our definition is so framed that the limit of the null-class is the first member of our series (if any). This departure from usage is convenient in order that, whenever our series contains any limiting point in the ordinary sense, the *series* of limiting points may exist, *i.e.* in order that $P \upharpoonleft \text{D‘lt}_P$ may exist whenever there are existent parts of $C\text{‘}P$ which have upper limits. The series $P \upharpoonleft \text{D‘lt}_P$ is the "first derivative" of P. The definition of a limit is

$$\text{lt}_P = \text{seq}_P \upharpoonright (-\text{Œ‘max}_P) \quad \text{Df.}$$

Besides the limit, we require, for many purposes, a single notation for the "limit or maximum." This we denote by "limax_P," putting

$$\text{limax}_P = \text{max}_P \cup \text{lt}_P \quad \text{Df.}$$

Similarly for the lower limit or minimum we use "limin_P," putting

$$\text{limin}_P = \text{min}_P \cup \text{tl}_P \quad \text{Df.}$$

We have $\text{tl}_P = \text{lt}(\breve{P})$ (*207·101) and $\text{limin}_P = \text{limax}(\breve{P})$ (*207·401). Hence it is unnecessary to prove propositions concerning lower limits, since they result immediately from propositions concerning upper limits.

In virtue of our definition of a limit, x limits α if x is a sequent of α and α has no maximum (*207·1). Thus if α has a maximum, it has no limit (*207·11), but if it has no maximum, the class of its limits is the class of its sequents (*207·12). Thus the existence of the class of limits is equivalent to the existence of the class of sequents combined with the non-existence of the class of maxima, *i.e.*

∗207·13. $\vdash : \exists ! \overrightarrow{\mathrm{lt}}_P{}^{\backprime}\alpha \,.\equiv .\sim \exists ! \overrightarrow{\mathrm{max}}_P{}^{\backprime}\alpha \,.\, \exists ! \overrightarrow{\mathrm{seq}}_P{}^{\backprime}\alpha$

∗207·2—·232 consist of various formulae for $\overrightarrow{\mathrm{lt}}_P{}^{\backprime}\alpha$. We have

∗207·2. $\vdash : P \in \mathrm{connex} \,.\, x \,\mathrm{lt}_P\, \alpha \,.\supset .\, \alpha \cap C{}^{\backprime}P \subset \overrightarrow{P}{}^{\backprime}x \,.\, \overrightarrow{P}{}^{\backprime}x \subset P{}^{\backprime\backprime}\alpha$

I.e. the whole of $\alpha \cap C{}^{\backprime}P$ precedes x, but any predecessor of x precedes some member of α.

∗207·231. $\vdash : P \in \mathrm{Ser} \,.\, \exists ! \overrightarrow{\mathrm{lt}}_P{}^{\backprime}\alpha \,.\supset .\, \overrightarrow{\mathrm{lt}}_P{}^{\backprime}\alpha = C{}^{\backprime}P \cap \hat{x}\,(\overrightarrow{P}{}^{\backprime}x = P{}^{\backprime\backprime}\alpha)$

I.e. the limit of α, if it exists, is the term whose predecessors are identical with the predecessors of some part of α.

We have also

∗207·232. $\vdash :. P \in \mathrm{Ser} \,.\supset : x = \mathrm{lt}_P{}^{\backprime}\alpha \,.\equiv .\, x \in C{}^{\backprime}P - \alpha \,.\, \overrightarrow{P}{}^{\backprime}x = P{}^{\backprime\backprime}\alpha$

This proposition should be compared with ∗205·54, which (slightly re-written) is

$$\vdash :. P \in \mathrm{Ser} \,.\supset : x = \mathrm{max}_P{}^{\backprime}\alpha \,.\equiv .\, x \in C{}^{\backprime}P \cap \alpha \,.\, \overrightarrow{P}{}^{\backprime}x = P{}^{\backprime\backprime}\alpha$$

From the two together we arrive at

∗207·51. $\vdash :. P \in \mathrm{Ser} \,.\supset : x = \mathrm{limax}_P{}^{\backprime}\alpha \,.\equiv .\, x \in C{}^{\backprime}P \,.\, \overrightarrow{P}{}^{\backprime}x = P{}^{\backprime\backprime}\alpha$

which serves to illustrate the utility of "limax_P."

We have

∗207·24. $\vdash : P \in \mathrm{connex} \,.\supset .\, \overrightarrow{\mathrm{lt}}_P{}^{\backprime}\alpha \in 0 \cup 1 \,.\, \mathrm{lt}_P \in 1 \rightarrow \mathrm{Cls}$

I.e. if P is connected, a class cannot have more than one limit; also

∗207·25. $\vdash : P \in \mathrm{trans} \,.\, \beta \subset P{}^{\backprime\backprime}\alpha \,.\supset .\, \overrightarrow{\mathrm{lt}}_P{}^{\backprime}(\alpha \cup \beta) = \overrightarrow{\mathrm{lt}}_P{}^{\backprime}\alpha$

I.e. any terms which have some α's beyond them may be added to α without altering the limit.

We next have a set of propositions (∗207·251—·27) proving that if a class has a limit, any single term of the class may be removed without altering the limit (∗207·261), and that in any case, provided the class is not a unit class, its minimum (if any) may be removed without altering the limit (∗207·27). We then prove (∗207·291) that if P is a series, and α is a class which has a limit, the predecessors of the limit are the class $P_*{}^{\backprime\backprime}\alpha$.

We then have a set of propositions (∗207·3—·36) on the limit of $\overrightarrow{P}{}^{\backprime}x$ and kindred matters. If x has no immediate predecessor, the limit of $\overrightarrow{P}{}^{\backprime}x$ is x, and vice versa (∗207·32·33). Hence

∗207·35. $\vdash : P \in \mathrm{Rl}{}^{\backprime}J \cap \mathrm{connex} \,.\supset .\, \mathrm{D}{}^{\backprime}\mathrm{lt}_P = C{}^{\backprime}P - \mathrm{G}{}^{\backprime}(P \mathbin{\dot{-}} P^2)$

I.e. the limit-points of P are those which have no immediate predecessors.

We next turn our attention to "limax_P." This again is one-many, provided P is connected (*207·41). We have by the definition

*207·42. $\vdash : \exists ! \overrightarrow{\text{max}}_P{}'\alpha \,.\, \supset .\, \overrightarrow{\text{limax}}_P{}'\alpha = \overrightarrow{\text{max}}_P{}'\alpha$

*207·43. $\vdash : \overrightarrow{\text{max}}_P{}'\alpha = \Lambda \,.\, \supset .\, \overrightarrow{\text{limax}}_P{}'\alpha = \overrightarrow{\text{seq}}_P{}'\alpha = \overrightarrow{\text{lt}}_P{}'\alpha$

*207·44. $\vdash .\, \text{Cl}'\text{limax}_P = \text{Cl}'\text{max}_P \cup \text{Cl}'\text{lt}_P = \text{Cl}'\text{max}_P \cup \text{Cl}'\text{seq}_P$

*207·45. $\vdash .\, \overrightarrow{\text{limax}}_P{}'\alpha = \overrightarrow{\text{max}}_P{}'\alpha \cup \overrightarrow{\text{lt}}_P{}'\alpha$

Also we have

*207·46. $\vdash :.\, x = \text{limax}_P{}'\alpha \,.\, \equiv : x = \text{max}_P{}'\alpha \,.\, \vee .\, x = \text{lt}_P{}'\alpha$

which is a very useful proposition, as is also *207·51 (given above).

A useful proposition in dealing with classes of classes contained in a series is

*207·54. $\vdash : P \epsilon \text{Ser} \,.\, \kappa \subset \text{Cl}'\text{lt}_P \,.\, \supset .\, \overrightarrow{\text{limax}}_P{}'\text{lt}_P{}''\kappa = \overrightarrow{\text{limax}}_P{}'s'\kappa = \overrightarrow{\text{lt}}_P{}'s'\kappa$

I.e. if every member of κ has a limit, the limit or maximum (if any) of the limits is the limit or maximum, and in fact the limit, of $s'\kappa$.

We have next a set of propositions (*207·6—·66) on correlations, proving that the limit, or the limax, of the correlates is the correlate of the limit or limax, *i.e.*

*207·6. $\vdash : S \epsilon P \,\overline{\text{smor}}\, Q \,.\, \supset .\, \overrightarrow{\text{lt}}_P{}'\alpha = S''\overrightarrow{\text{lt}}_Q{}'\breve{S}''\alpha$

*207·64. $\vdash : S \epsilon P \,\overline{\text{smor}}\, Q \,.\, \supset .\, \overrightarrow{\text{limax}}_P{}'\alpha = S''\overrightarrow{\text{limax}}_Q{}'\breve{S}''\alpha$

The last three propositions (*207·7—·72) are lemmas for use in the theory of stretches (*215·5·51).

---

*207·01. $\text{lt}_P = \text{lt}(P) = \text{seq}_P \upharpoonright (-\text{Cl}'\text{max}_P)$ Df

*207·02. $\text{tl}_P = \text{tl}(P) = \text{prec}_P \upharpoonright (-\text{Cl}'\text{min}_P)$ Df

*207·03. $\text{limax}_P = \text{max}_P \,\dot{\cup}\, \text{lt}_P$ Df

*207·04. $\text{limin}_P = \text{min}_P \,\dot{\cup}\, \text{tl}_P$ Df

*207·1. $\vdash : x \,\text{lt}_P\, \alpha \,.\, \equiv .\, x \,\text{seq}_P\, \alpha \,.\, \sim \exists ! \overrightarrow{\text{max}}_P{}'\alpha$ [(*207·01)]

*207·101. $\vdash .\, \text{tl}_P = \text{lt}(\breve{P})$ [*205·102 . *206·101 . (*207·02)]

We shall not give further propositions on lower limits, unless for some special reason, since all of them result from propositions on upper limits by means of *207·101.

*207·11. $\vdash : \exists ! \overrightarrow{\text{max}}_P{}'\alpha \,.\, \supset .\, \overrightarrow{\text{lt}}_P{}'\alpha = \Lambda$ [*207·1]

*207·12. $\vdash : \overrightarrow{\text{max}}_P{}'\alpha = \Lambda \,.\, \supset .\, \overrightarrow{\text{lt}}_P{}'\alpha = \overrightarrow{\text{seq}}_P{}'\alpha$ [*207·1]

$*207{\cdot}121.\ \vdash : \alpha \cap C`P \subset P``\alpha \,.\supset.\, \overrightarrow{\mathrm{lt}}_P`\alpha = \overrightarrow{\mathrm{seq}}_P`\alpha$ [$*207{\cdot}12 \,.\, *205{\cdot}123$]

$*207{\cdot}13.\ \vdash : \exists ! \overrightarrow{\mathrm{lt}}_P`\alpha \,.\equiv.\, \sim \exists ! \overrightarrow{\max}_P`\alpha \,.\, \exists ! \overrightarrow{\mathrm{seq}}_P`\alpha$ [$*207{\cdot}1$]

$*207{\cdot}14.\ \vdash :. \exists ! \overrightarrow{\max}_P`\alpha \,.\vee.\, \exists ! \overrightarrow{\mathrm{seq}}_P`\alpha : \equiv : \exists ! \overrightarrow{\max}_P`\alpha \,.\vee.\, \exists ! \overrightarrow{\mathrm{lt}}_P`\alpha$
[$*207{\cdot}13 \,.\, *5{\cdot}63$]

The above proposition is important because

$$(\alpha) : \exists ! \overrightarrow{\max}_P`\alpha \,.\vee.\, \exists ! \overrightarrow{\mathrm{lt}}_P`\alpha$$

is the characteristic of "Dedekindian" series, *i.e.* of such as fulfil Dedekind's axiom.

$*207{\cdot}15.\ \vdash : x \,\mathrm{lt}_P\, \alpha \,.\equiv.\, x \in C`P \,.\, \alpha \cap C`P \subset P``\alpha \cap \overrightarrow{P}`x \,.\, \overrightarrow{P}`x \subset - p`\overleftarrow{P}``(\alpha \cap C`P)$
[$*207{\cdot}1 \,.\, *205{\cdot}123 \,.\, *206{\cdot}134$]

$*207{\cdot}16.\ \vdash .\, \overrightarrow{\mathrm{lt}}_P`\alpha = \overrightarrow{\mathrm{lt}}_P`(\alpha \cap C`P)$ [$*207{\cdot}15 \,.\, *37{\cdot}265$]

$*207{\cdot}17.\ \vdash .\, \overrightarrow{\mathrm{lt}}_P`\Lambda = \overrightarrow{B}`P$ [$*207{\cdot}12 \,.\, *205{\cdot}161 \,.\, *206{\cdot}14$]

$*207{\cdot}18.\ \vdash : \mathrm{Cl}`P \subset \mathrm{D}`\mathrm{lt}_P \,.\equiv.\, C`P = \mathrm{D}`\mathrm{lt}_P$

Dem.

$\vdash .\, *207{\cdot}17 \,.\supset \vdash : \mathrm{Cl}`P \subset \mathrm{D}`\mathrm{lt}_P \,.\equiv.\, \mathrm{Cl}`P \cup \overrightarrow{B}`P \subset \mathrm{D}`\mathrm{lt}_P \,.$
[$*93{\cdot}103$] $\equiv .\, C`P \subset \mathrm{D}`\mathrm{lt}_P \,.$
[$*207{\cdot}15$] $\equiv .\, C`P = \mathrm{D}`\mathrm{lt}_P : \supset \vdash .\, \mathrm{Prop}$

$*207{\cdot}2.\ \vdash : P \in \mathrm{connex} \,.\, x \,\mathrm{lt}_P\, \alpha \,.\supset.\, \alpha \cap C`P \subset \overrightarrow{P}`x \,.\, \overrightarrow{P}`x \subset P``\alpha$
[$*207{\cdot}15 \,.\, *202{\cdot}503$]

$*207{\cdot}21.\ \vdash : P^2 \subseteq J \,.\, x \in C`P \,.\, \alpha \cap C`P \subset \overrightarrow{P}`x \,.\, \overrightarrow{P}`x \subset P``\alpha \,.\supset.\, x \,\mathrm{lt}_P\, \alpha$

Dem.

$\vdash .\, *200{\cdot}53 \,.\supset \vdash : P^2 \subseteq J \,.\supset.\, P``\alpha \subset - p`\overleftarrow{P}``(\alpha \cap C`P)$ (1)
$\vdash .\, (1) \,.\quad \supset \vdash : \mathrm{Hp} \,.\supset.\, x \in C`P \,.\, \alpha \cap C`P \subset \overrightarrow{P}`x \,.\, \overrightarrow{P}`x \subset - p`\overrightarrow{P}``(\alpha \cap C`P) \,.$
[$*206{\cdot}134$] $\supset .\, x \,\mathrm{seq}_P\, \alpha$ (2)
$\vdash .\, *22{\cdot}44 \,.\quad \supset \vdash : \mathrm{Hp} \,.\supset.\, \alpha \cap C`P \subset P``\alpha \,.$
[$*205{\cdot}123$] $\supset .\, \overrightarrow{\max}_P`\alpha = \Lambda$ (3)
$\vdash .\, (2) \,.\, (3) \,.\, *207{\cdot}1 \,.\supset \vdash .\, \mathrm{Prop}$

$*207{\cdot}22.\ \vdash : P \in \mathrm{connex} \,.\, P^2 \subseteq J \,.\supset.\, \overrightarrow{\mathrm{lt}}_P`\alpha = C`P \cap \hat{x}(\alpha \cap C`P \subset \overrightarrow{P}`x \,.\, \overrightarrow{P}`x \subset P``\alpha)$
[$*207{\cdot}2{\cdot}21$]

This is very often the most convenient form for $\overrightarrow{\mathrm{lt}}_P`\alpha$. It states that a limit of $\alpha$ is a member $x$ of $C`P$ such that $\alpha \cap C`P$ wholly precedes x, but every predecessor of x precedes some member of α.

$*207{\cdot}23$. $\vdash : P \in \mathrm{Ser} \,.\supset .\, \overrightarrow{\mathrm{lt}}_P`\alpha = C`P \cap \hat{x}(\overrightarrow{P}`x = P``\alpha \,.\, \alpha \cap C`P \subset P``\alpha)$

Dem.

$\vdash .\, *13{\cdot}12 \,.\, *22{\cdot}42 \,.\supset$

$\vdash : \overrightarrow{P}`x = P``\alpha \,.\, \alpha \cap C`P \subset P``\alpha \,.\supset .\, \alpha \cap C`P \subset \overrightarrow{P}`x \,.\, \overrightarrow{P}`x \subset P``\alpha$ (1)

$\vdash .\, *201{\cdot}501 \,.\, *37{\cdot}265 \,.\supset \vdash :.\, P \in \mathrm{trans} \,.\supset : \alpha \cap C`P \subset \overrightarrow{P}`x \,.\supset .\, P``\alpha \subset \overrightarrow{P}`x :$

[Fact] $\supset : \alpha \cap C`P \subset \overrightarrow{P}`x \,.\, \overrightarrow{P}`x \subset P``\alpha \,.\supset .\, \overrightarrow{P}`x = P``\alpha$ (2)

$\vdash .\, *22{\cdot}44 \,.\supset \vdash : \alpha \cap C`P \subset \overrightarrow{P}`x \,.\, \overrightarrow{P}`x \subset P``\alpha \,.\supset .\, \alpha \cap C`P \subset P``\alpha$ (3)

$\vdash .\, (1) \,.\, (2) \,.\, (3) \,.\supset$

$\vdash :.\, P \in \mathrm{trans} \,.\supset : \alpha \cap C`P \subset \overrightarrow{P}`x \,.\, \overrightarrow{P}`x \subset P``\alpha \,.\equiv .\, \overrightarrow{P}`x = P``\alpha \,.\, \alpha \cap C`P \subset P``\alpha$ (4)

$\vdash .\, (4) \,.\, *207{\cdot}22 \,.\supset \vdash .\,$ Prop

$*207{\cdot}231$. $\vdash : P \in \mathrm{Ser} \,.\, \exists ! \overrightarrow{\mathrm{lt}}_P`\alpha \,.\supset .\, \overrightarrow{\mathrm{lt}}_P`\alpha = C`P \cap \hat{x}(\overrightarrow{P}`x = P``\alpha)$ [$*207{\cdot}23$]

$*207{\cdot}232$. $\vdash :.\, P \in \mathrm{Ser} \,.\supset : x = \mathrm{lt}_P`\alpha \,.\equiv .\, x \in C`P - \alpha \,.\, \overrightarrow{P}`x = P``\alpha$
[$*206{\cdot}28 \,.\, *207{\cdot}1$]

$*207{\cdot}24$. $\vdash : P \in \mathrm{connex} \,.\supset .\, \overrightarrow{\mathrm{lt}}_P`\alpha \in 0 \cup 1 \,.\, \mathrm{lt}_P \in 1 \rightarrow \mathrm{Cls}$

Dem.

$\vdash .\, *206{\cdot}161 \,.\, *71{\cdot}26 \,.\, (*207{\cdot}01) \,.\supset \vdash : \mathrm{Hp} \,.\supset .\, \mathrm{lt}_P \in 1 \rightarrow \mathrm{Cls} \,.$ (1)

[$*71{\cdot}12$] $\supset .\, \overrightarrow{\mathrm{lt}}_P`\alpha \in 0 \cup 1$ (2)

$\vdash .\, (1) \,.\, (2) \,.\supset \vdash .\,$ Prop

$*207{\cdot}25$. $\vdash : P \in \mathrm{trans} \,.\, \beta \subset P``\alpha \,.\supset .\, \overrightarrow{\mathrm{lt}}_P`(\alpha \cup \beta) = \overrightarrow{\mathrm{lt}}_P`\alpha$

Dem.

$\vdash .\, *205{\cdot}193 \,.\quad \supset \vdash : \mathrm{Hp} \,.\, \exists ! \overrightarrow{\mathrm{max}}_P`\alpha \,.\supset .\, \exists ! \overrightarrow{\mathrm{max}}_P`(\alpha \cup \beta)$ (1)

$\vdash .\, (1) \,.\, *207{\cdot}11 \,.\supset \vdash : \mathrm{Hp} \,.\, \exists ! \overrightarrow{\mathrm{max}}_P`\alpha \,.\supset .\, \overrightarrow{\mathrm{lt}}_P`\alpha = \Lambda \,.\, \overrightarrow{\mathrm{lt}}_P`(\alpha \cup \beta) = \Lambda$ (2)

$\vdash .\, *205{\cdot}193 \,.\, *207{\cdot}12 \,.\supset$

$\vdash : \mathrm{Hp} \,.\, \overrightarrow{\mathrm{max}}_P`\alpha = \Lambda \,.\supset .\, \overrightarrow{\mathrm{lt}}_P`\alpha = \overrightarrow{\mathrm{seq}}_P`\alpha \,.\, \overrightarrow{\mathrm{lt}}_P`(\alpha \cup \beta) = \overrightarrow{\mathrm{seq}}_P`(\alpha \cup \beta) \,.$

[$*206{\cdot}24$] $\supset .\, \overrightarrow{\mathrm{lt}}_P`\alpha = \overrightarrow{\mathrm{lt}}_P`(\alpha \cup \beta)$ (3)

$\vdash .\, (2) \,.\, (3) \,.\supset \vdash .\,$ Prop

$*207{\cdot}251$. $\vdash : P \in \mathrm{trans} \,.\, y \in P``(\beta - \iota`y) \,.\supset .\, \overrightarrow{\mathrm{lt}}_P`\beta = \overrightarrow{\mathrm{lt}}_P`(\beta - \iota`y)$

Dem.

$\vdash .\, *51{\cdot}222 \,.\quad \supset \vdash : y \sim\in \beta \,.\supset .\, \overrightarrow{\mathrm{lt}}_P`\beta = \overrightarrow{\mathrm{lt}}_P`(\beta - \iota`y)$ (1)

$\vdash .\, *207{\cdot}25 \,.\quad \supset \vdash : \mathrm{Hp} \,.\supset .\, \overrightarrow{\mathrm{lt}}_P`\{(\beta - \iota`y) \cup \iota`y\} = \overrightarrow{\mathrm{lt}}_P`(\beta - \iota`y)$ (2)

$\vdash .\, (2) \,.\, *51{\cdot}221 \,.\supset \vdash : \mathrm{Hp} \,.\, y \in \beta \,.\supset .\, \overrightarrow{\mathrm{lt}}_P`\beta = \overrightarrow{\mathrm{lt}}_P`(\beta - \iota`y)$ (3)

$\vdash .\, (1) \,.\, (3) \,.\supset \vdash .\,$ Prop

***207·26.** $\vdash : P \epsilon \text{trans} . \sim(yPy) . \exists ! \overrightarrow{\text{lt}}_P{}^{\backprime}\beta . \supset . \overrightarrow{\text{lt}}_P{}^{\backprime}\beta = \overrightarrow{\text{lt}}_P{}^{\backprime}(\beta - \iota{}^{\backprime}y)$
[*207·13·12 . *206·72]

***207·261.** $\vdash : P \epsilon \text{trans} . y \epsilon \overrightarrow{\text{min}}_P{}^{\backprime}\beta . \exists ! \overrightarrow{\text{lt}}_P{}^{\backprime}\beta . \supset . \overrightarrow{\text{lt}}_P{}^{\backprime}\beta = \overrightarrow{\text{lt}}_P{}^{\backprime}(\beta - \iota{}^{\backprime}y)$
[*207·26 . *205·194]

***207·262.** $\vdash : P \epsilon \text{trans} \cap \text{connex} . \exists ! \overrightarrow{\text{lt}}_P{}^{\backprime}\beta . \supset . \overrightarrow{\text{lt}}_P{}^{\backprime}\beta = \overrightarrow{\text{lt}}_P{}^{\backprime}(\beta - \overrightarrow{\text{min}}_P{}^{\backprime}\beta)$
[*207·261 . *205·3]

***207·263.** $\vdash : P \epsilon \text{trans} \cap \text{connex} . \supset . \overrightarrow{\text{lt}}_P{}^{\backprime}\beta \subset \overrightarrow{\text{lt}}_P{}^{\backprime}(\beta - \overrightarrow{\text{min}}_P{}^{\backprime}\beta)$
[*207·262 . *24·12]

***207·27.** $\vdash : P \epsilon \text{trans} \cap \text{connex} . \beta \cap C{}^{\backprime}P \sim\epsilon 1 . \supset . \overrightarrow{\text{lt}}_P{}^{\backprime}\beta = \overrightarrow{\text{lt}}_P{}^{\backprime}(\beta - \overrightarrow{\text{min}}_P{}^{\backprime}\beta)$

Dem.

$\vdash . *24{\cdot}26{\cdot}101 . \supset \vdash : \overrightarrow{\text{min}}_P{}^{\backprime}\beta = \Lambda . \supset . \overrightarrow{\text{lt}}_P{}^{\backprime}\beta = \overrightarrow{\text{lt}}_P{}^{\backprime}(\beta - \overrightarrow{\text{min}}_P{}^{\backprime}\beta)$ (1)

$\vdash . *52{\cdot}181 . \supset$

$\vdash : \text{Hp} . \exists ! \overrightarrow{\text{min}}_P{}^{\backprime}\beta . \supset . (\exists y) . y \epsilon \beta \cap C{}^{\backprime}P . y \neq \text{min}_P{}^{\backprime}\beta .$

[*205·2] $\supset . (\exists y) . y \epsilon (\beta \cap C{}^{\backprime}P) - \iota{}^{\backprime}\text{min}_P{}^{\backprime}\beta . \text{min}_P{}^{\backprime}\beta P y .$

[*37·1] $\supset . \text{min}_P{}^{\backprime}\beta \epsilon P{}^{\backprime\backprime}(\beta - \iota{}^{\backprime}\text{min}_P{}^{\backprime}\beta) .$

[*207·251] $\supset . \overrightarrow{\text{lt}}_P{}^{\backprime}\beta = \overrightarrow{\text{lt}}_P{}^{\backprime}(\beta - \iota{}^{\backprime}\text{min}_P{}^{\backprime}\beta)$ (2)

$\vdash . (1) . (2) . \supset \vdash . \text{Prop}$

***207·28.** $\vdash : P \epsilon \text{trans} . \supset . \overrightarrow{\text{lt}}_P{}^{\backprime}(\alpha \cup P{}^{\backprime\backprime}\alpha) = \overrightarrow{\text{lt}}_P{}^{\backprime}\alpha$ [*207·25]

***207·281.** $\vdash : P \epsilon \text{trans} . \sim \exists ! \overrightarrow{\text{max}}_P{}^{\backprime}\alpha . \supset . \overrightarrow{\text{lt}}_P{}^{\backprime}P{}^{\backprime\backprime}\alpha = \overrightarrow{\text{lt}}_P{}^{\backprime}\alpha$
[*207·28·16 . *205·123]

***207·282.** $\vdash : P \epsilon \text{trans} . \sim \exists ! \overrightarrow{\text{max}}_P{}^{\backprime}\alpha . \sim \exists ! \overrightarrow{\text{max}}_P{}^{\backprime}\beta . P{}^{\backprime\backprime}\alpha = P{}^{\backprime\backprime}\beta . \supset . \overrightarrow{\text{lt}}_P{}^{\backprime}\alpha = \overrightarrow{\text{lt}}_P{}^{\backprime}\beta$
[*207·281]

***207·29.** $\vdash : P \epsilon \text{trans} . \supset . \overrightarrow{\text{lt}}_P{}^{\backprime}\alpha = \overrightarrow{\text{lt}}_P{}^{\backprime}P_*{}^{\backprime\backprime}\alpha$

Dem.

$\vdash . *207{\cdot}16{\cdot}28 . \supset \vdash : \text{Hp} . \supset . \overrightarrow{\text{lt}}_P{}^{\backprime}\alpha = \overrightarrow{\text{lt}}_P{}^{\backprime}\{(\alpha \cup P{}^{\backprime\backprime}\alpha) \cap C{}^{\backprime}P\}$

[*201·52] $= \text{lt}_P{}^{\backprime}P_*{}^{\backprime\backprime}\alpha : \supset \vdash . \text{Prop}$

***207·291.** $\vdash : P \epsilon \text{trans} \cap \text{connex} . \text{E} ! \text{lt}_P{}^{\backprime}\alpha . \supset . \overrightarrow{P}{}^{\backprime}\text{lt}_P{}^{\backprime}\alpha = P_*{}^{\backprime\backprime}\alpha$

Dem.

$\vdash . *207{\cdot}29 . \supset \vdash : \text{Hp} . \supset . \overrightarrow{P}{}^{\backprime}\text{lt}_P{}^{\backprime}\alpha = \overrightarrow{P}{}^{\backprime}\text{lt}_P{}^{\backprime}P_*{}^{\backprime\backprime}\alpha$ (1)

$\vdash . *90{\cdot}14{\cdot}172 . \supset \vdash . P_*{}^{\backprime\backprime}\alpha \subset C{}^{\backprime}P . P{}^{\backprime\backprime}P_*{}^{\backprime\backprime}\alpha \subset P_*{}^{\backprime\backprime}\alpha$ (2)

$\vdash . *207{\cdot}11{\cdot}12 . \supset \vdash : \text{Hp} . \supset . \text{seq}_P{}^{\backprime}P_*{}^{\backprime\backprime}\alpha = \text{lt}_P{}^{\backprime}P_*{}^{\backprime\backprime}\alpha$ (3)

$\vdash . (2) . (3) . *206{\cdot}3 . \supset \vdash : \text{Hp} . \supset . \overrightarrow{P}{}^{\backprime}\text{seq}_P{}^{\backprime}P_*{}^{\backprime\backprime}\alpha = P_*{}^{\backprime\backprime}\alpha$ (4)

$\vdash . (1) . (3) . (4) . \supset \vdash . \text{Prop}$

***207·3.** $\vdash : \alpha \cap C`P = \Lambda . \supset . \overrightarrow{\mathrm{lt}}_P`\alpha = \overrightarrow{B}`P$

Dem.

$\vdash . *205·151·161 . \supset \vdash : \mathrm{Hp} . \supset . \overrightarrow{\mathrm{max}}_P`\alpha = \Lambda$ (1)

$\vdash . *206·14 . \supset \vdash : \mathrm{Hp} . \supset . \overrightarrow{\mathrm{seq}}_P`\alpha = \overrightarrow{B}`P$ (2)

$\vdash . (1) . (2) . *207·12 . \supset \vdash . \mathrm{Prop}$

***207·31.** $\vdash : P \mathrel{\mathsf{G}} J . x \in C`P - \text{Π}`(P \dot{-} P^2) . \supset . x \, \mathrm{lt}_P \, \overrightarrow{P}`x$

Dem.

$\vdash . *206·41 . \supset \vdash : \mathrm{Hp} . \supset . \overrightarrow{\mathrm{max}}_P`\overrightarrow{P}`x = \Lambda$ (1)

$\vdash . *206·4 . \supset \vdash : \mathrm{Hp} . \supset . x \, \mathrm{seq}_P \, \overrightarrow{P}`x$ (2)

$\vdash . (1) . (2) . *207·1 . \supset \vdash . \mathrm{Prop}$

***207·32.** $\vdash : P \in \mathrm{Rl}`J \cap \mathrm{connex} . x \in C`P - \text{Π}`(P \dot{-} P^2) . \supset . x = \mathrm{lt}_P`\overrightarrow{P}`x$
[*207·31·24]

***207·33.** $\vdash : x \in \text{Π}`(P \dot{-} P^2) . \supset . \overrightarrow{\mathrm{lt}}_P`\overrightarrow{P}`x = \Lambda$ [*205·252 . *207·11]

***207·34.** $\vdash : P \in \mathrm{connex} . x \, \mathrm{lt}_P \, \alpha . \supset . x \, \mathrm{lt}_P \, \overrightarrow{P}`x . x \sim \in \text{Π}`(P \dot{-} P^2)$

Dem.

$\vdash . *207·15 . \supset \vdash : \mathrm{Hp} . \supset . x \in C`P . \alpha \cap C`P \subset P``\alpha . \alpha \cap C`P \subset \overrightarrow{P}`x .$
$\overrightarrow{P}`x \subset - p`\overleftarrow{P}``(\alpha \cap C`P)$ (1)

$\vdash . *40·16 . \supset \vdash : \alpha \cap C`P \subset \overrightarrow{P}`x . \supset . p`\overleftarrow{P}``\overrightarrow{P}`x \subset p`\overleftarrow{P}``(\alpha \cap C`P) .$

[*22·81] $\supset . - p`\overleftarrow{P}``(\alpha \cap C`P) \subset - p`\overleftarrow{P}``\overrightarrow{P}`x$ (2)

$\vdash . (1) . (2) . \supset \vdash : \mathrm{Hp} . \supset . x \in C`P . \overrightarrow{P}`x \subset - p`\overleftarrow{P}``\overrightarrow{P}`x .$

[*22·42] $\supset . x \in C`P . \overrightarrow{P}`x \subset \overrightarrow{P}`x . \overrightarrow{P}`x \subset - p`\overleftarrow{P}``\overrightarrow{P}`x$ (3)

$\vdash . (1) . *202·505 . \supset \vdash : \mathrm{Hp} . \supset . \overrightarrow{P}`x \subset (\alpha \cap C`P) \cup P``\alpha . \alpha \cap C`P \subset P``\alpha .$

[*22·62] $\supset . \overrightarrow{P}`x \subset P``\alpha$ (4)

$\vdash . (1) . *37·2·265 . \supset \vdash : \mathrm{Hp} . \supset . P``\alpha \subset P``\overrightarrow{P}`x$ (5)

$\vdash . (4) . (5) . \supset \vdash : \mathrm{Hp} . \supset . \overrightarrow{P}`x \subset P``\overrightarrow{P}`x$ (6)

$\vdash . (3) . (6) . *207·15 . \supset \vdash . \mathrm{Prop}$

***207·35.** $\vdash : P \in \mathrm{Rl}`J \cap \mathrm{connex} . \supset . \mathrm{D}`\mathrm{lt}_P = C`P - \text{Π}`(P \dot{-} P^2)$

Dem.

$\vdash . *207·34 . \supset \vdash : \mathrm{Hp} . \supset . \mathrm{D}`\mathrm{lt}_P \subset - \text{Π}`(P \dot{-} P^2)$ (1)

$\vdash . *207·15 . \supset \vdash . \mathrm{D}`\mathrm{lt}_P \subset C`P$ (2)

$\vdash . *207·32 . \supset \vdash : \mathrm{Hp} . \supset . C`P - \text{Π}`(P \dot{-} P^2) \subset \mathrm{D}`\mathrm{lt}_P$ (3)

$\vdash . (1) . (2) . (3) . \supset \vdash . \mathrm{Prop}$

***207·36.** $\vdash : P \epsilon \text{Rl}`J \cap \text{connex} . \supset .$

$$\text{D}`\text{lt}_P = \text{lt}_P``\overrightarrow{P}``\{C`P - \text{Ǝ}`(P \dot{-} P^2)\} = \text{lt}_P``\overrightarrow{P}``C`P$$

Dem.

$\vdash . *207·32 . \supset \vdash : \text{Hp} . \supset . C`P - \text{Ǝ}`(P \dot{-} P^2) = \text{lt}_P``\overrightarrow{P}``\{C`P - \text{Ǝ}`(P \dot{-} P^2)\}$ (1)

$\vdash . (1) . *207·35 . \supset \vdash : \text{Hp} . \supset . \text{D}`\text{lt}_P = \text{lt}_P``\overrightarrow{P}``\{C`P - \text{Ǝ}`(P \dot{-} P^2)\}$ (2)

$\vdash . *207·33 . \quad \supset \vdash . \text{lt}_P``\overrightarrow{P}``\{C`P \cap \text{Ǝ}`(P \dot{-} P^2)\} = \Lambda$ (3)

$\vdash . (2) . (3) . \quad \supset \vdash : \text{Hp} . \supset . \text{D}`\text{lt}_P = \text{lt}_P``\overrightarrow{P}``C`P$ (4)

$\vdash . (2) . (4) . \supset \vdash . \text{Prop}$

In virtue of this proposition, all limits are limits of classes of the form $\overrightarrow{P}`x$. In this respect, limits (in general) differ from segments. If we call $P``\alpha$ the segment defined by α, there will in general be segments not of the form $\overrightarrow{P}`x$. These, however, will be the segments which have no sequents, and therefore no limits; thus their existence does not introduce limits not derivable from classes of the form $\overrightarrow{P}`x$.

***207·4.** $\vdash :. x \,\text{limax}_P\, \alpha . \equiv : x \,\text{max}_P\, \alpha . \vee . x \,\text{lt}_P\, \alpha :$

$$\equiv : x \,\text{max}_P\, \alpha . \vee . \sim \exists ! \overrightarrow{\text{max}_P}`\alpha . x \,\text{seq}_P\, \alpha \quad [(*207·03)]$$

***207·401.** $\vdash . \text{limin}_P = \text{limax}\,(\breve{P}) \quad [(*207·04)]$

***207·41.** $\vdash : P \epsilon \text{connex} . \supset . \text{limax}_P, \text{limin}_P \epsilon 1 \rightarrow \text{Cls}$
[*71·24 . *205·31 . *207·24 . (*207·03·04)]

***207·42.** $\vdash : \exists ! \overrightarrow{\text{max}_P}`\alpha . \supset . \overrightarrow{\text{limax}_P}`\alpha = \overrightarrow{\text{max}_P}`\alpha$ [*207·4]

***207·43.** $\vdash : \overrightarrow{\text{max}_P}`\alpha = \Lambda . \supset . \overrightarrow{\text{limax}_P}`\alpha = \overrightarrow{\text{seq}_P}`\alpha = \overrightarrow{\text{lt}_P}`\alpha$ [*207·4]

***207·44.** $\vdash . \text{Ǝ}`\text{limax}_P = \text{Ǝ}`\text{max}_P \cup \text{Ǝ}`\text{lt}_P = \text{Ǝ}`\text{max}_P \cup \text{Ǝ}`\text{seq}_P$
[*207·14 . (*207·03)]

***207·45.** $\vdash . \overrightarrow{\text{limax}_P}`\alpha = \overrightarrow{\text{max}_P}`\alpha \cup \overrightarrow{\text{lt}_P}`\alpha \quad [(*207·03)]$

***207·46.** $\vdash :. x = \text{limax}_P`\alpha . \equiv : x = \text{max}_P`\alpha . \vee . x = \text{lt}_P`\alpha$

Dem.

$\vdash . *207·45·11 . \supset \vdash :. \exists ! \overrightarrow{\text{max}_P}`\alpha . \supset : x = \text{limax}_P`\alpha . \equiv . x = \text{max}_P`\alpha$ (1)

$\vdash . *207·45·12 . \supset \vdash :. \overrightarrow{\text{max}_P}`\alpha = \Lambda . \supset : x = \text{limax}_P`\alpha . \equiv . x = \text{lt}_P`\alpha$ (2)

$\vdash . (1) . (2) . *5·32 . \supset$

$\vdash :. \exists ! \overrightarrow{\text{max}_P}`\alpha . x = \text{limax}_P`\alpha . \vee . \overrightarrow{\text{max}_P}`\alpha = \Lambda . x = \text{limax}_P`\alpha : \equiv :$

$\exists ! \overrightarrow{\text{max}_P}`\alpha . x = \text{max}_P`\alpha . \vee . \overrightarrow{\text{max}_P}`\alpha = \Lambda . x = \text{lt}_P`\alpha$ (3)

$\vdash . (3) . *4·42 . \supset$

$\vdash :. x = \text{limax}_P`\alpha . \equiv : \exists ! \overrightarrow{\text{max}_P}`\alpha . x = \text{max}_P`\alpha . \vee . \overrightarrow{\text{max}_P}`\alpha = \Lambda . x = \text{lt}_P`\alpha :$

[*30·32] $\quad \equiv : x = \text{max}_P`\alpha . \vee . \overrightarrow{\text{max}_P}`\alpha = \Lambda . x = \text{lt}_P`\alpha :$

[*207·13] $\quad \equiv : x = \text{max}_P`\alpha . \vee . x = \text{lt}_P`\alpha :. \supset \vdash . \text{Prop}$

***207·47.** $\vdash : \exists ! \overrightarrow{\mathrm{lt}}_P\text{‘}\alpha . \equiv . \exists ! \overrightarrow{\mathrm{limax}}_P\text{‘}\alpha . \sim \exists ! \overrightarrow{\mathrm{max}}_P\text{‘}\alpha$

Dem.

$\vdash . *207\cdot45\cdot11 . \supset \vdash : \exists ! \overrightarrow{\mathrm{lt}}_P\text{‘}\alpha . \supset . \exists ! \overrightarrow{\mathrm{limax}}_P\text{‘}\alpha . \sim \exists ! \overrightarrow{\mathrm{max}}_P\text{‘}\alpha \quad (1)$

$\vdash . *207\cdot45 . \quad \supset \vdash : \exists ! \overrightarrow{\mathrm{limax}}_P\text{‘}\alpha . \sim \exists ! \overrightarrow{\mathrm{max}}_P\text{‘}\alpha . \supset . \exists ! \overrightarrow{\mathrm{lt}}_P\text{‘}\alpha \quad (2)$

$\vdash . (1) . (2) . \supset \vdash . \mathrm{Prop}$

***207·48.** $\vdash . \overrightarrow{\mathrm{limax}}_P\text{‘}\alpha = \overrightarrow{\mathrm{limax}}_P\text{‘}(\alpha \cap C\text{‘}P)$ [*207·45 . *205·151 . *207·16]

***207·481.** $\vdash : P \in \mathrm{trans} . \supset . \overrightarrow{\mathrm{limax}}_P\text{‘}\alpha = \overrightarrow{\mathrm{limax}}_P\text{‘}P_*\text{‘‘}\alpha$
[*207·45 . *205·191 . *207·29]

***207·482.** $\vdash : P \in \mathrm{Ser} . \alpha \subset C\text{‘}P . a = \mathrm{limax}_P\text{‘}\alpha . \supset . \alpha \subset \overrightarrow{P_*}\text{‘}a$

Dem.

$\vdash . *205\cdot22 . *90\cdot151 . \supset \vdash : \mathrm{Hp} . a = \mathrm{max}_P\text{‘}\alpha . \supset . \alpha \subset \overrightarrow{P_*}\text{‘}a \quad (1)$

$\vdash . *207\cdot291 . *90\cdot151 . \supset \vdash : \mathrm{Hp} . a = \mathrm{lt}_P\text{‘}\alpha . \supset . P_*\text{‘‘}\alpha \subset \overrightarrow{P_*}\text{‘}a .$

$[*90\cdot21] \quad \supset . \alpha \subset \overrightarrow{P_*}\text{‘}a \quad (2)$

$\vdash . (1) . (2) . *207\cdot46 . \supset \vdash . \mathrm{Prop}$

***207·5.** $\vdash : P \in \mathrm{Ser} . \supset . \overrightarrow{\mathrm{limax}}_P\text{‘}\alpha = \overrightarrow{\mathrm{seq}}_P\text{‘}P\text{‘‘}\alpha = \overrightarrow{\mathrm{min}}_P\text{‘}(\overrightarrow{\mathrm{max}}_P\text{‘}\alpha \cup \overrightarrow{\mathrm{seq}}_P\text{‘}\alpha)$
[*206·33·35·37]

***207·51.** $\vdash :. P \in \mathrm{Ser} . \supset : x = \mathrm{limax}_P\text{‘}\alpha . \equiv . x \in C\text{‘}P . \overrightarrow{P}\text{‘}x = P\text{‘‘}\alpha$
[*205·54 . *207·232·46]

***207·52.** $\vdash :. P \in \mathrm{Ser} . \exists ! P\text{‘‘}\alpha . \supset : x = \mathrm{limax}_P\text{‘}\alpha . \equiv . \overrightarrow{P}\text{‘}x = P\text{‘‘}\alpha$ [*207·51]

***207·521.** $\vdash :. P \in \mathrm{Ser} . \supset : x = \mathrm{lt}_P\text{‘}\alpha . \equiv . x \in C\text{‘}P . \overrightarrow{P}\text{‘}x = P\text{‘‘}\alpha . \sim \mathrm{E} ! \mathrm{max}_P\text{‘}\alpha$

Dem.

$\vdash . *207\cdot51 . \supset \vdash :. \mathrm{Hp} . \supset :$

$x \in C\text{‘}P . \overrightarrow{P}\text{‘}x = P\text{‘‘}\alpha . \sim \mathrm{E} ! \mathrm{max}_P\text{‘}\alpha . \equiv . x = \mathrm{limax}_P\text{‘}\alpha . \sim \mathrm{E} ! \mathrm{max}_P\text{‘}\alpha .$

$[*207\cdot46] \quad \equiv . x = \mathrm{lt}_P\text{‘}\alpha :. \supset \vdash . \mathrm{Prop}$

***207·53.** $\vdash : P \in \mathrm{Ser} . \kappa \subset \mathrm{Cl}\text{‘}\mathrm{limax}_P . \supset . \overrightarrow{\mathrm{limax}}_P\text{‘}\mathrm{limax}_P\text{‘‘}\kappa = \overrightarrow{\mathrm{limax}}_P\text{‘}s\text{‘}\kappa$

Dem.

$\vdash . *207\cdot51 . \supset \vdash :. \mathrm{Hp} . \supset : \alpha \in \kappa . \supset_\alpha . \overrightarrow{P}\text{‘}\mathrm{limax}_P\text{‘}\alpha = P\text{‘‘}\alpha :$

$[*37\cdot68] \quad \supset : \overrightarrow{P}\text{‘‘}\mathrm{limax}_P\text{‘‘}\kappa = P\text{‘‘‘}\kappa :$

$[*40\cdot5\cdot38] \quad \supset : P\text{‘‘}\mathrm{limax}_P\text{‘‘}\kappa = P\text{‘‘}s\text{‘}\kappa :$

$[*207\cdot51] \quad \supset : x = \mathrm{limax}_P\text{‘}\mathrm{limax}_P\text{‘‘}\kappa . \equiv . x = \mathrm{limax}_P\text{‘}s\text{‘}\kappa :. \supset \vdash . \mathrm{Prop}$

***207·54.** $\vdash : P \in \mathrm{Ser} \,.\, \kappa \subset \text{Ɑ}\text{‘}\mathrm{lt}_P \,.\, \supset \,.\, \overrightarrow{\mathrm{limax}}_P\text{‘}\mathrm{lt}_P\text{‘‘}\kappa = \overrightarrow{\mathrm{limax}}_P\text{‘}s\text{‘}\kappa = \overrightarrow{\mathrm{lt}}_P\text{‘}s\text{‘}\kappa$

Dem.

$\vdash \,.\, *205\cdot561 \,.\, *207\cdot13 \,.\, \supset \vdash : \mathrm{Hp} \,.\, \supset \,.\, s\text{‘}\kappa \sim\in \text{Ɑ}\text{‘}\mathrm{max}_P \,.$

$[*207\cdot43] \qquad \supset \,.\, \overrightarrow{\mathrm{limax}}_P\text{‘}s\text{‘}\kappa = \overrightarrow{\mathrm{lt}}_P\text{‘}s\text{‘}\kappa \qquad (1)$

$\vdash \,.\, *207\cdot13\cdot43 \,.\, \supset \vdash : \mathrm{Hp} \,.\, \supset \,.\, \mathrm{lt}_P\text{‘‘}\kappa = \mathrm{limax}_P\text{‘‘}\kappa \,.$

$[*207\cdot53] \qquad \supset \,.\, \overrightarrow{\mathrm{limax}}_P\text{‘}\mathrm{lt}_P\text{‘‘}\kappa = \overrightarrow{\mathrm{limax}}_P\text{‘}s\text{‘}\kappa \qquad (2)$

$\vdash \,.\, (1) \,.\, (2) \,.\, \supset \vdash \,.\, \mathrm{Prop}$

***207·55.** $\vdash : P \in \mathrm{Ser} \,.\, \kappa \subset \text{Ɑ}\text{‘}\mathrm{lt}_P \,.\, s\text{‘}\kappa \in \text{Ɑ}\text{‘}\mathrm{lt}_P \,.\, \supset \,.\, \mathrm{limax}_P\text{‘}\mathrm{lt}_P\text{‘‘}\kappa = \mathrm{lt}_P\text{‘}s\text{‘}\kappa$

$[*207\cdot54]$

***207·6.** $\vdash : S \in P \,\overline{\mathrm{smor}}\, Q \,.\, \supset \,.\, \overrightarrow{\mathrm{lt}}_P\text{‘}\alpha = S\text{‘‘}\overrightarrow{\mathrm{lt}}_Q\text{‘}\breve{S}\text{‘‘}\alpha$

Dem.

$\vdash \,.\, *205\cdot8 \,.\, *37\cdot43 \,.\, \supset \vdash :.\, \mathrm{Hp} \,.\, \supset : \exists \,!\, \overrightarrow{\mathrm{max}}_P\text{‘}\alpha \,.\, \equiv \,.\, \exists \,!\, \overrightarrow{\mathrm{max}}_Q\text{‘}\breve{S}\text{‘‘}\alpha : \qquad (1)$

$[*207\cdot11] \qquad \supset : \exists \,!\, \overrightarrow{\mathrm{max}}_P\text{‘}\alpha \,.\, \supset \,.\, \overrightarrow{\mathrm{lt}}_P\text{‘}\alpha = \Lambda \,.\, \overrightarrow{\mathrm{lt}}_Q\text{‘}\breve{S}\text{‘‘}\alpha = \Lambda \qquad (2)$

$\vdash \,.\, (1) \,.\, \mathrm{Transp} \,.\, *207\cdot12 \,.\, \supset$

$\vdash :.\, \mathrm{Hp} \,.\, \overrightarrow{\mathrm{max}}_P\text{‘}\alpha = \Lambda \,.\, \supset : \overrightarrow{\mathrm{lt}}_P\text{‘}\alpha = \overrightarrow{\mathrm{seq}}_P\text{‘}\alpha \,.\, \overrightarrow{\mathrm{lt}}_Q\text{‘}\breve{S}\text{‘‘}\alpha = \overrightarrow{\mathrm{seq}}_Q\text{‘}\breve{S}\text{‘‘}\alpha :$

$[*206\cdot61] \qquad \supset : \overrightarrow{\mathrm{lt}}_P\text{‘}\alpha = S\text{‘‘}\overrightarrow{\mathrm{lt}}_Q\text{‘}\breve{S}\text{‘‘}\alpha \qquad (3)$

$\vdash \,.\, (2) \,.\, (3) \,.\, *37\cdot29 \,.\, \supset \vdash \,.\, \mathrm{Prop}$

***207·61.** $\vdash :.\, S \in P \,\overline{\mathrm{smor}}\, Q \,.\, \supset : \mathrm{E}\,!\, \mathrm{lt}_P\text{‘}\alpha \,.\, \equiv \,.\, \mathrm{E}\,!\, \mathrm{lt}_Q\text{‘}\breve{S}\text{‘‘}\alpha \qquad [*207\cdot6 \,.\, *53\cdot3]$

***207·62.** $\vdash : S \in P \,\overline{\mathrm{smor}}\, Q \,.\, \mathrm{E}\,!\, \mathrm{lt}_P\text{‘}\alpha \,.\, \supset \,.\, \mathrm{lt}_P\text{‘}\alpha = S\text{‘}\mathrm{lt}_Q\text{‘}\breve{S}\text{‘‘}\alpha \qquad [*207\cdot6 \,.\, *53\cdot31]$

***207·63.** $\vdash : S \in P \,\overline{\mathrm{smor}}\, Q \,.\, \supset \,.\, \mathrm{lt}_P\text{‘‘}\kappa = S\text{‘‘}\mathrm{lt}_Q\text{‘‘}\breve{S}\text{‘‘‘}\kappa$

Dem.

$\vdash \,.\, *207\cdot6 \,.\, *40\cdot5 \,.\, \supset \vdash : \mathrm{Hp} \,.\, \supset \,.\, \mathrm{lt}_P\text{‘‘}\kappa = s\text{‘}S\text{‘‘‘}\overrightarrow{\mathrm{lt}}_Q\text{‘‘}\breve{S}\text{‘‘‘}\kappa$

$[*40\cdot38\cdot5] \qquad = S\text{‘‘}\mathrm{lt}_Q\text{‘‘}\breve{S}\text{‘‘‘}\kappa : \supset \vdash \,.\, \mathrm{Prop}$

***207·64.** $\vdash : S \in P \,\overline{\mathrm{smor}}\, Q \,.\, \supset \,.\, \overrightarrow{\mathrm{limax}}_P\text{‘}\alpha = S\text{‘‘}\overrightarrow{\mathrm{limax}}_Q\text{‘}\breve{S}\text{‘‘}\alpha$

$[*205\cdot8 \,.\, *207\cdot6\cdot45]$

***207·65.** $\vdash :.\, S \in P \,\overline{\mathrm{smor}}\, Q \,.\, \supset : \mathrm{E}\,!\, \mathrm{limax}_P\text{‘}\alpha \,.\, \equiv \,.\, \mathrm{E}\,!\, \mathrm{limax}_Q\text{‘}\breve{S}\text{‘‘}\alpha$

$[*207\cdot64]$

***207·66.** $\vdash : S \in P \,\overline{\mathrm{smor}}\, Q \,.\, \mathrm{E}\,!\, \mathrm{limax}_P\text{‘}\alpha \,.\, \supset \,.\, \mathrm{limax}_P\text{‘}\alpha = S\text{‘}\mathrm{limax}_Q\text{‘}\breve{S}\text{‘‘}\alpha$

$[*207\cdot64]$

∗207·7. $\vdash :. P \epsilon \text{trans} . \lor . P^2 \subset J : \supset :$

$\text{limin}_P`\gamma = \text{limax}_P`\gamma . \supset . \text{limin}_P`\gamma = \text{min}_P`\gamma = \text{max}_P`\gamma$

Dem.

$\vdash . ∗207·42·43 . \supset \vdash : E ! \text{min}_P`\gamma . E ! \text{limax}_P`\gamma . \sim E ! \text{max}_P`\gamma . \supset .$

$\text{limin}_P`\gamma = \text{min}_P`\gamma . \text{limax}_P`\gamma = \text{seq}_P`\gamma .$

[∗205·11.∗206·2] $\supset . \text{limin}_P`\gamma \epsilon \gamma . \text{limax}_P`\gamma \sim \epsilon \gamma .$

[∗13·14] $\supset . \text{limin}_P`\gamma \neq \text{limax}_P`\gamma$ (1)

Similarly

$\vdash : E ! \text{max}_P`\gamma . E ! \text{limin}_P`\gamma . \sim E ! \text{min}_P`\gamma . \supset . \text{limin}_P`\gamma \neq \text{limax}_P`\gamma$ (2)

$\vdash . ∗206·732 . ∗207·43·12 . \supset$

$\vdash : \text{Hp} . \sim E ! \text{min}_P`\gamma . \sim E ! \text{max}_P`\gamma . \supset . \sim \{\text{limin}_P`\gamma = \text{limax}_P`\gamma\}$ (3)

$\vdash . (1) . (2) . (3) . \supset \vdash : \text{Hp} . \text{limin}_P`\gamma = \text{limax}_P`\gamma . \supset . E ! \text{min}_P`\gamma . E ! \text{max}_P`\gamma .$

[∗207·42] $\supset . \text{limin}_P`\gamma = \text{min}_P`\gamma = \text{max}_P`\gamma : \supset \vdash . \text{Prop}$

∗207·71. $\vdash :. P \epsilon \text{connex} : P \epsilon \text{trans} . \lor . P^2 \subset J : \text{limin}_P`\gamma = \text{limax}_P`\gamma : \supset .$

$\gamma \cap C`P \epsilon 1 . \gamma \cap C`P = \iota`\text{limax}_P`\gamma$

[∗207·7 . ∗205·73]

∗207·72. $\vdash :. P \epsilon \text{connex} . P^2 \subset J . \supset : \text{limin}_P`\gamma = \text{limax}_P`\gamma . \equiv . \gamma \cap C`P \epsilon 1$

[∗207·71 . ∗205·731·17 . ∗207·42]

*208. THE CORRELATION OF SERIES.

Summary of *208.

The propositions of this number are chiefly important on account of their consequences in the theory of well-ordered series (*250 ff.) and in the theory of vector-families (*330 ff.). When two well-ordered series are ordinally similar, they have only one correlator; and a well-ordered series is not ordinally similar to any of its segments. Of these two propositions, the first is an immediate consequence of *208·41, and the second is an immediate consequence of *208·47.

Propositions concerning correlators of two relations P and Q are obtained from propositions concerning correlators of P with itself, by means of the fact that, if S, T are two correlators of P and Q, $S \mid \breve{T}$ is a correlator of P with itself. Again, correlators of P with itself are considered, in this number, as a special case of correlators of P with parts of itself. This latter is a notion which will prove important for other reasons than those for which it is used in our present context. If P is connected, and S correlates P with part of itself (so that $S ; P \subseteq P$), $C'P$ will contain terms of three kinds, (1) those for which $S'x = x$, (2) those for which $(S'x) P x$, (3) those for which $x P (S'x)$. Our propositions result from the non-existence (under certain circumstances) of maxima or minima of classes (2) and (3).

The following definition defines "correlations of P with parts (or the whole) of itself." The letters "cror" stand for "ordinal correlation." For a cardinal correlation, should occasion arise, we should use "cr," *i.e.* we should put

$$\mathrm{cr}'\alpha = s'\overline{\mathrm{sm}}\,\alpha``\mathrm{Cl}'\alpha \quad \mathrm{Df},$$

so that $$S \in \mathrm{cr}'\alpha \,.\equiv.\, S \in 1 \rightarrow 1 \,.\, \mathrm{G}'S = \alpha \,.\, \mathrm{D}'S \subset \alpha.$$

For the present, we are concerned with the corresponding ordinal notion; thus we require

$$S \in \mathrm{cror}'P \,.\equiv.\, S \in 1 \rightarrow 1 \,.\, \mathrm{G}'S = C'P \,.\, S ; P \subseteq P.$$

This is secured by putting

$$\mathrm{cror}'P = s'\overline{\mathrm{smor}}\,P``\mathrm{Rl}'P \quad \mathrm{Df}.$$

It will be observed that if α is what we called a "non-reflexive" class (cf. ∗124), $\text{cr}'\alpha = \iota' I \upharpoonright \alpha$, and $S \in \text{cr}'\alpha \,.\supset.\, \text{D}'S = \alpha$. When $C'P$ is non-reflexive, the same is true of P; and when $C'P$ is reflexive, P is also reflexive, in the sense that it contains proper parts similar to itself, though if P is well-ordered, such proper parts cannot be *segments* of P, but must extend to the end of $C'P$.

The class of correlators of P with the whole of itself, *i.e.* $P \,\overline{\text{smor}}\, P$, is a sub-class of $\text{cror}'P$, and is specially important. This class differs widely in its properties from the corresponding cardinal class. If α has more than one member, the class $\alpha \,\overline{\text{sm}}\, \alpha$ (which is the "permutations" of α in the usual elementary sense) always has more than one member. But the class $P \,\overline{\text{smor}}\, P$ (which consists of such permutations of $C'P$ as keep the order unchanged) will consist of the single term $I \upharpoonright C'P$, unless $C'P$ contains classes which have neither a minimum nor a maximum, in which case there will be many correlators of P with itself. As a simple illustration, take the series of negative and positive integers in their natural order. Then if ν is any one of these integers, $+\nu$ is a correlator of the whole series with itself. If we take only the positive integers, $+\nu$ is no longer a correlator of the *whole* series with itself, since all integers less than ν are omitted from the correlate.

The first important use of the propositions of this number is in the beginning of the theory of well-ordered series (∗250). The propositions there used are

∗208·4. $\vdash : P \in \text{connex} \,.\, P^2 \mathbin{\mathsf{G}} J \,.\, \text{Cl ex}'C'P \subset \text{Cl}'\text{min}_P \cup \text{Cl}'\text{max}_P \,.$
$S, T \in P \,\overline{\text{smor}}\, Q \,.\supset.\, S = T$

I.e. if P is connected and asymmetrical, and every existent sub-class of $C'P$ has either a minimum or a maximum, P and Q cannot have more than one correlator.

∗208·42. In the same circumstances, $P \,\overline{\text{smor}}\, P = \iota'(I \upharpoonright C'P)$

∗208·43. $\vdash : \text{Cl ex}'C'P \subset \text{Cl}'\text{min}_P \,.\, S \in \text{cror}'P \,.\supset.\, \sim(\exists x) \,.\, (S'x)\, P x$

I.e. if every existent sub-class of $C'P$ has a minimum, a correlator of P with part of itself can never move terms backwards. Thus for example, to take a simple instance, an infinite series consisting of some of the natural numbers in order of magnitude cannot have its μth term less than μ.

∗208·45. $\vdash : P \in \text{connex} \,.\, \text{Cl ex}'C'P \subset \text{Cl}'\text{min}_P \cap \text{Cl}'\text{max}_P \,.\supset.\, \text{Rl}'P \cap \text{Nr}'P = \iota'P$

I.e. if P is connected and every existent sub-class of $C'P$ has both a maximum and a minimum, no proper part of P is similar to P. This proposition is important in the theory of finite series and finite ordinals.

∗208·46. $\vdash : \text{Cl ex}'C'P \subset \text{Cl}'\text{min}_P \,.\, S \in \text{cror}'P \,.\supset.\, C'P \cap p'\overleftarrow{P}``\text{D}'S = \Lambda$

I.e. if every existent sub-class of $C'P$ has a minimum, a part of P which is similar to P must go up to the end of P, *i.e.* must not wholly precede any member of $C'P$.

∗208·47. $\vdash : \text{Cl ex}‘C‘P \subset \text{Cl}‘\text{min}_P \,.\, Q \subseteq P \,.\, \exists ! \, C‘P \cap p‘\overleftarrow{P}“C‘Q \,.\supset .\, \sim(Q \,\text{smor}\, P)$

This is an immediate consequence of ∗208·46.

The proof of the above propositions proceeds simply by showing that if $S \in \text{cror}‘P$ and $(S‘x)\,Px$, then $(S‘S‘x)\,P\,(S‘x)$, so that x is not the earliest term for which $(S‘x)\,Px$, since $S‘x$ is an earlier term for which the same thing holds. Hence $\hat{x}\,\{(S‘x)\,Px\}$ can have no minimum; and similarly $\hat{x}\,\{xP\,(S‘x)\}$ can have no maximum (∗208·14). So far we require no hypothesis as to P. Assuming now $P \in \text{connex} \,.\, P^2 \subseteq J$, we show similarly that if S correlates the whole of P with itself, $\hat{x}\,\{(S‘x)\,Px\}$ can have no maximum and $\hat{x}\,\{xP\,(S‘x)\}$ can have no minimum.

Propositions about correlators of P with Q follow from the above by taking two correlators S and T, and applying the above propositions to $S \,|\, \breve{T}$, which is a correlator of P with the whole of itself.

∗208·01. $\text{cror}‘P = s‘\overline{\text{smor}}\, P“\text{Rl}‘P$ Df

∗208·1. $\vdash : S \in \text{cror}‘P \,.\equiv.\, S \in 1 \to 1 \,.\, \text{Cl}‘S = C‘P \,.\, S;P \subseteq P$

Dem.

$\vdash .\, ∗40·4 \,.\, (∗208·01) \,.\, ∗151·11 \,.\supset$

$\vdash : S \in \text{cror}‘P \,.\equiv.\, (\exists Q) \,.\, Q \subseteq P \,.\, S \in 1 \to 1 \,.\, \text{Cl}‘S = C‘P \,.\, Q = S;P \,.$

$[∗13·195] \quad \equiv .\, S \in 1 \to 1 \,.\, \text{Cl}‘S = C‘P \,.\, S;P \subseteq P : \supset \vdash .\, \text{Prop}$

∗208·11. $\vdash : S \in \text{cror}‘P \,.\supset .\, S;P \subseteq P \upharpoonright \text{D}‘S$

Dem.

$\vdash .\, ∗150·203 \,.\supset \vdash :. \text{Hp} \,.\supset : x\,(S;P)\,y \,.\supset .\, x, y \in \text{D}‘S \quad (1)$

$\vdash .\, ∗208·1 \,. \quad \supset \vdash :. \text{Hp} \,.\supset : x\,(S;P)\,y \,.\supset .\, xPy \quad (2)$

$\vdash .\, (1) \,.\, (2) \,. \quad \supset \vdash .\, \text{Prop}$

∗208·111. $\vdash : S \in \text{cror}‘P \,.\supset .\, \text{D}‘S = C‘S;P = S“C‘P \,.\, \text{D}‘S \subset \text{Cl}‘S$
$[∗150·22·23 \,.\, ∗208·1 \,.\, ∗33·265]$

∗208·12. $\vdash : S \in \text{cror}‘P \,.\supset .\, \breve{S};S;P = P \,.\, P \subseteq \breve{S};P \quad [∗151·252·26 \,.\, ∗208·1]$

∗208·13. $\vdash : S \in \text{cror}‘P \,.\, (S‘x)\,Px \,.\supset .\, (S‘S‘x)\,P\,(S‘x)$

Dem.

$\vdash .\, ∗208·12 \,.\supset \vdash : \text{Hp} \,.\supset .\, (S‘x)\,(\breve{S};P)\,x \,.$

$[∗150·41] \quad \supset .\, (S‘S‘x)\,P\,(S‘x) : \supset \vdash .\, \text{Prop}$

∗208·131. $\vdash : S \in \text{cror}‘P \,.\, xP\,(S‘x) \,.\supset .\, (S‘x)\,P\,(S‘S‘x)$ [Proof as in ∗208·13]

***208·14.** $\vdash : S \in \text{cror}`P . \supset . \overrightarrow{\min}_P`\hat{x}\{(S`x)Px\} = \Lambda . \overrightarrow{\max}_P`\hat{x}\{xP(S`x)\} = \Lambda$

Dem.

$\vdash . *208·13 . *20·3 . \supset \vdash :. \text{Hp} . \supset : x \in \hat{x}\{(S`x)Px\} . \supset . S`x \in \hat{x}\{(S`x)Px\} . (S`x)Px .$

$[*37·105] \qquad \supset . x \in \breve{P}``\hat{x}\{(S`x)Px\} \quad (1)$

$\vdash . (1) . *24·3 . \quad \supset \vdash : \text{Hp} . \supset . \hat{x}\{(S`x)Px\} - \breve{P}``\hat{x}\{(S`x)Px\} = \Lambda .$

$[*205·11] \qquad \supset . \overrightarrow{\min}_P`\hat{x}\{(S`x)Px\} = \Lambda \quad (2)$

Similarly $\qquad \vdash : \text{Hp} . \supset . \overrightarrow{\max}_P`\hat{x}\{xP(S`x)\} = \Lambda \quad (3)$

$\vdash . (2) . (3) . \supset \vdash . \text{Prop}$

Thus the proof that $\hat{x}\{(S`x)Px\}$ has no minimum, and $\hat{x}\{xP(S`x)\}$ no maximum, requires no hypothesis as to P. The proof that $\hat{x}\{(S`x)Px\}$ has no maximum, and $\hat{x}\{xP(S`x)\}$ no minimum, requires the hypothesis $P \in \text{connex} . P^2 \subset J$. This proof results from the following propositions.

***208·2.** $\vdash : P \in \text{connex} . P^2 \subset J . S \in \text{cror}`P . \supset . P = \breve{S};P . S;P = P \upharpoonright D`S$

Dem.

$\vdash . *150·41 . \supset \vdash :. \text{Hp} . \supset : x(\breve{S};P)y . \equiv . (S`x)P(S`y) .$

$[*50·43·45] \qquad \supset . S`x \neq S`y . \sim\{(S`y)P(S`x)\} .$

$[*30·37 . *150·41] \qquad \supset . x \neq y . \sim\{y(\breve{S};P)x\} .$

$[*208·12 . \text{Transp}] \qquad \supset . x \neq y . \sim(yPx) \quad (1)$

$\vdash . *150·203 . \qquad \supset \vdash : x(\breve{S};P)y . \supset . x, y \in \text{G}`S \quad (2)$

$\vdash . (2) . *208·1 . \qquad \supset \vdash :. \text{Hp} . \supset : x(\breve{S};P)y . \supset . x, y \in C`P \quad (3)$

$\vdash . (1) . (3) . *202·103 . \supset \vdash :. \text{Hp} . \supset : x(\breve{S};P)y . \supset . xPy \quad (4)$

$\vdash . (4) . *208·12 . \qquad \supset \vdash : \text{Hp} . \supset . P = \breve{S};P \quad (5)$

$\vdash . (5) . \qquad \supset \vdash : \text{Hp} . \supset . S;P = S;\breve{S};P$

$[*150·38] \qquad = P \upharpoonright D`S \quad (6)$

$\vdash . (5) . (6) . \supset \vdash . \text{Prop}$

***208·21.** $\vdash : P \in \text{connex} . P^2 \subset J . S \in \text{cror}`P . (S`x)Px . x \in D`S . \supset . xP(\breve{S}`x)$

Dem.

$\vdash . *33·43 . \supset \vdash : \text{Hp} . \supset . (S`x)(P \upharpoonright D`S)x .$

$[*208·2] \qquad \supset . (S`x)(S;P)x .$

$[*150·41] \qquad \supset . (\breve{S}`S`x)P(\breve{S}`x) .$

$[*72·241 . *33·43] \qquad \supset . xP(\breve{S}`x) : \supset \vdash . \text{Prop}$

***208·211.** $\vdash : P \in \text{connex} . P^2 \subset J . S \in \text{cror}`P . xP(S`x) . x \in D`S . \supset . (\breve{S}`x)Px$

[Proof as in *208·21]

$*208\cdot22$. $\vdash : P \epsilon \text{connex} . P^2 \subseteq J . S \epsilon \text{cror}' P . \text{Cl}' S \subset \text{D}' S . \supset .$

$$\overrightarrow{\max}_P ` \hat{x} \{(\breve{S}`x) P x\} = \Lambda . \overrightarrow{\min}_P ` \hat{x} \{x P (\breve{S}`x)\} = \Lambda$$

Dem.

$\vdash . *33\cdot43 . \quad \supset \vdash :. \text{Hp} . \supset : (\breve{S}`x) P x . \supset . x \epsilon \text{D}`S . x \epsilon \text{Cl}`S .$

$[*208\cdot21] \quad \supset . x P (\breve{S}`x) . x \epsilon \text{Cl}`S .$

$[*72\cdot241] \quad \supset . x P (\breve{S}`x) . \breve{S}`x \epsilon \hat{x} \{(\breve{S}`x) P x\} .$

$[*37\cdot1] \quad \supset . x \epsilon P `` \hat{x} \{(\breve{S}`x) P x\} \quad (1)$

$\vdash . (1) . *205\cdot123 . \supset \vdash : \text{Hp} . \supset . \overrightarrow{\max}_P ` \hat{x} \{(\breve{S}`x) P x\} = \Lambda \quad (2)$

Similarly $\quad \vdash : \text{Hp} . \supset . \overrightarrow{\min}_P ` \hat{x} \{x P (\breve{S}`x)\} = \Lambda \quad (3)$

$\vdash . (2) . (3) . \supset \vdash . \text{Prop}$

Observe that, in virtue of $*208\cdot111$, the above hypothesis gives $\text{D}`S = \text{Cl}`S = C`P$, so that $S \epsilon P \,\overline{\text{smor}}\, P$. Hence we are led to $*208\cdot3$.

$*208\cdot3$. $\vdash : P \epsilon \text{connex} . P^2 \subseteq J . S \epsilon P \,\overline{\text{smor}}\, P . \supset .$

$$\sim \exists ! \min_P ` \hat{x} \{(\breve{S}`x) P x\} . \sim \exists ! \overrightarrow{\max}_P ` \hat{x} \{(\breve{S}`x) P x\} .$$
$$\sim \exists ! \min_P ` \hat{x} \{x P (\breve{S}`x)\} . \sim \exists ! \overrightarrow{\max}_P ` \hat{x} \{x P (\breve{S}`x)\}$$

Dem.

$\vdash . *151\cdot11 . *150\cdot23 . \supset \vdash : \text{Hp} . \supset . S \epsilon 1 \to 1 . \text{Cl}`S = C`P . S ; P = P . \text{D}`S = C`P .$

$[*208\cdot1] \quad \supset . S \epsilon \text{cror}`P . \text{Cl}`S = \text{D}`S \quad (1)$

$\vdash . (1) . *208\cdot14\cdot22 . \supset \vdash . \text{Prop}$

$*208\cdot31$. $\vdash : S, T \epsilon P \,\overline{\text{smor}}\, Q . \supset . S | \breve{T} \epsilon P \,\overline{\text{smor}}\, P \quad [*151\cdot131\cdot141]$

$*208\cdot32$. $\vdash : P \epsilon \text{connex} . P^2 \subseteq J . S, T \epsilon P \,\overline{\text{smor}}\, Q . \supset .$

$$\sim \exists ! \overrightarrow{\min}_P ` \hat{x} \{(S`\breve{T}`x) P x\} . \sim \exists ! \overrightarrow{\max}_P ` \hat{x} \{(S`\breve{T}`x) P x\} .$$
$$\sim \exists ! \overrightarrow{\min}_P ` \hat{x} \{x P (S`\breve{T}`x)\} . \sim \exists ! \overrightarrow{\max}_P ` \hat{x} \{x P (S`\breve{T}`x)\}$$

$[*208\cdot3\cdot31 . *34\cdot41]$

$*208\cdot4$. $\vdash : P \epsilon \text{connex} . P^2 \subseteq J . \text{Cl ex}`C`P \subset \text{Cl}`\min_P \cup \text{Cl}`\max_P .$

$$S, T \epsilon P \,\overline{\text{smor}}\, Q . \supset . S = T$$

Dem.

$\vdash . *208\cdot32 . \quad \supset \vdash : \text{Hp} . \supset . \hat{x} \{(S`\breve{T}`x) P x\} = \Lambda . \hat{x} \{x P (S`\breve{T}`x)\} = \Lambda \quad (1)$

$\vdash . *208\cdot31 . *34\cdot41 . \quad \supset \vdash :. \text{Hp} . \supset : x \epsilon C`P . \supset . S`\breve{T}`x \epsilon C`P \quad (2)$

$\vdash . (1) . (2) . *202\cdot103 . \supset \vdash :. \text{Hp} . \supset : x \epsilon C`P . \supset . S`\breve{T}`x = x .$

$[*72\cdot241] \quad \supset . \breve{T}`x = \breve{S}`x :$

$[*150\cdot23] \quad \supset : x \epsilon \text{D}`S \cup \text{D}`T . \supset . \breve{T}`x = \breve{S}`x :$

$[*33\cdot46] \quad \supset : S = T :. \supset \vdash . \text{Prop}$

$*208\cdot41$. $\vdash : P \epsilon \text{connex} . P^2 \subseteq J . \text{Cl ex}`C`P \subset \text{Cl}`\min_P \cup \text{Cl}`\max_P .$

$$P \text{ smor } Q . \supset . (P \,\overline{\text{smor}}\, Q) \epsilon 1$$

$[*208\cdot4 . *151\cdot12 . *52\cdot16]$

The above proposition is of great importance in the theory of well-ordered series.

***208·42.** $\vdash : P \epsilon \text{connex} . P^2 \in J . \text{Cl ex}`C`P \subset \text{D}`\text{min}_P \cup \text{D}`\text{max}_P . \supset .$
$P \,\overline{\text{smor}}\, P = \iota`(I \upharpoonright C`P)$

[*208·4 . *51·141 . *151·121]

***208·43.** $\vdash : \text{Cl ex}`C`P \subset \text{D}`\text{min}_P . S \epsilon \text{cror}`P . \supset . \sim(\exists x) . (S`x) P x$ [*208·14]

***208·431.** $\vdash : \text{Cl ex}`C`P \subset \text{D}`\text{max}_P . S \epsilon \text{cror}`P . \supset . \sim(\exists x) . x P (S`x)$ [*208·14]

***208·44.** $\vdash : P \epsilon \text{connex} . \text{Cl ex}`C`P \subset \text{D}`\text{min}_P \cap \text{D}`\text{max}_P . S \epsilon \text{cror}`P . \supset .$
$S = I \upharpoonright C`P$

Dem.

$\vdash . *208·43·431 . *202·103 . \supset \vdash :. \text{Hp} . \supset : x \epsilon C`P . \supset . S`x = x .$

[*50·14.*35·7] $\supset . S`x = (I \upharpoonright C`P)`x :$

[*208·1.*50·5·52] $\supset : x \epsilon \text{D}`S \cup \text{D}`(I \upharpoonright C`P) . \supset . S`x = (I \upharpoonright C`P)`x :$

[*33·45] $\supset : S = I \upharpoonright C`P :. \supset \vdash . \text{Prop}$

In virtue of this proposition, if P is a finite series, no proper part of P is ordinally similar to P. (It will be shown later that a finite series is one in which every existent contained class has both a maximum and a minimum.) The following proposition gives a more explicit form of the above result.

***208·45.** $\vdash : P \epsilon \text{connex} . \text{Cl ex}`C`P \subset \text{D}`\text{min}_P \cap \text{D}`\text{max}_P . \supset . \text{Rl}`P \cap \text{Nr}`P = \iota`P$

Dem.

$\vdash . *208·44·1 . \supset \vdash :. \text{Hp} . \supset : S \epsilon 1 \rightarrow 1 . \text{D}`S = C`P . S;P \in P . \supset . S = I \upharpoonright C`P .$

[*150·534] $\supset . S;P = P$ (1)

$\vdash . (1) . *13·12 . \supset \vdash :. \text{Hp} . \supset : Q \in P . S \epsilon 1 \rightarrow 1 . \text{D}`S = C`P . Q = S;P . \supset . Q = P :$

[*151·1] $\supset : Q \in P . Q \,\text{smor}\, P . \supset . Q = P :$

[*152·1] $\supset : \text{Rl}`P \cap \text{Nr}`P \subset \iota`P$ (2)

$\vdash . *61·34 . *152·3 . \supset \vdash . \iota`P \subset \text{Rl}`P \cap \text{Nr}`P$ (3)

$\vdash . (2) . (3) . \supset \vdash . \text{Prop}$

The following propositions are useful in the theory of segments of well-ordered series, since they show that a well-ordered series is never ordinally similar to any of its segments.

***208·46.** $\vdash : \text{Cl ex}`C`P \subset \text{D}`\text{min}_P . S \epsilon \text{cror}`P . \supset . C`P \cap p`\overleftarrow{P}``\text{D}`S = \Lambda$

Dem.

$\vdash . *208·1 . \supset \vdash :. S \epsilon \text{cror}`P . \supset : x \epsilon C`P \cap p`\overleftarrow{P}``\text{D}`S . \supset . (S`x) P x :$

[Transp] $\supset : \sim\{(S`x) P x\} . \supset . x \sim\epsilon C`P \cap p`\overleftarrow{P}``\text{D}`S$ (1)

$\vdash . (1) . *208·43 . \supset \vdash : \text{Hp} . \supset . (x) . x \sim\epsilon C`P \cap p`\overleftarrow{P}``\text{D}`S : \supset \vdash . \text{Prop}$

∗208·461. $\vdash : \mathrm{Cl\,ex}`C`P \subset \mathrm{G}`\mathrm{min}_P \,.\, S \in \mathrm{cror}`P \,.\, \dot{\exists} ! P \,.\, \supset \,.\, p`\overleftarrow{P}``\mathrm{D}`S = \Lambda$
[∗208·46·1 . ∗40·62]

∗208·47. $\vdash : \mathrm{Cl\,ex}`C`P \subset \mathrm{G}`\mathrm{min}_P \,.\, Q \mathrel{\mathsf{G}} P \,.\, \exists ! C`P \cap p`\overleftarrow{P}``C`Q \,.\, \supset \,.\, \sim(Q \,\mathrm{smor}\, P)$

Dem.

$\vdash \,.\, *208{\cdot}46 \,.\, (*208{\cdot}01) \,.\, \supset$

$\vdash :. \mathrm{Hp} \,.\, \supset : Q \mathrel{\mathsf{G}} P \,.\, S \in Q \,\overline{\mathrm{smor}}\, P \,.\, \supset \,.\, C`P \cap p`\overleftarrow{P}``\mathrm{D}`S = \Lambda$ (1)

$\vdash \,.\, (1) \,.\, \mathrm{Transp} \,.\, *151{\cdot}11 \,.\, *150{\cdot}23 \,.\, \supset$

$\vdash :. \mathrm{Hp} \,.\, \supset : Q \mathrel{\mathsf{G}} P \,.\, \exists ! C`P \cap p`\overleftarrow{P}``C`Q \,.\, \supset \,.\, (S) \,.\, S \sim\in Q \,\overline{\mathrm{smor}}\, P \,.$

[∗151·12] $\supset \,.\, \sim(Q \,\mathrm{smor}\, P) :. \supset \vdash$. Prop

SECTION B.

ON SECTIONS, SEGMENTS, STRETCHES, AND DERIVATIVES.

In this section, our chief topic will be *sections* and *segments.* This topic will occupy $*211$, $*212$ and $*213$, and $*210$ will consist of propositions whose chief utility lies in their application to segments. In $*214$, we shall consider Dedekindian series, which are intimately connected with segments, owing to the fact that one of the chief propositions in the subject is that the series of segments of a series is Dedekindian. In $*215$, we shall consider "stretches," which consist of any consecutive piece of a series, and are constituted by the product of an upper and lower section. Finally, in $*216$, we shall consider the derivative of a series, or of a class α contained in a series: the former is the series of limit-points of the series, *i.e.* $P \upharpoonright \text{D}\text{`}\text{lt}_P$, the latter is the class of limits of existent sub-classes of $\alpha \cap C\text{`}P$, *i.e.* $\text{lt}_P\text{``}\text{Cl ex}\text{`}(\alpha \cap C\text{`}P)$.

A class is called a *section* of P when it is contained in $C\text{`}P$, and contains all the predecessors of its members, *i.e.* $\alpha$ is a section of $P$ if $\alpha \subset C\text{`}P \,.\, P\text{``}\alpha \subset \alpha$. Thus a section consists of all the field up to a certain point. It may consist of all the predecessors of x, *i.e.* it may be of the form $\overrightarrow{P}\text{`}x$; or again, it may consist of these together with $x$, in which case it is of the form $\overrightarrow{P}\text{`}x \cup \iota\text{`}x$; or again, it may be not definable by means of a single sequent or maximum, but be of the form $P\text{``}\alpha$, where α is a class without a limit or maximum. The class of sections of P is denoted by $\text{sect}\text{`}P$. A section of $\breve{P}$ will be called an "upper section" of P.

The idea of a *segment* is slightly less general than that of a *section.* We define a segment of P as any class of the form $P\text{``}\alpha$, *i.e.* as any member of $\text{D}\text{`}P_\epsilon$. Provided $P$ is transitive, segments are contained among sections. But even in a series sections are not, in general, contained among segments: if $P$ is a series, and if $x$ is a member of $C\text{`}P$ which has no immediate successor, $\overrightarrow{P}\text{`}x \cup \iota\text{`}x$ will be a section but not a segment.

If a segment has a maximum, it must also have a sequent. Segments which have no maximum form a specially important class of segments: these are classes α such that $\alpha = P\text{``}\alpha$; they form the class $\text{D}\text{`}(P_\epsilon \dot{\cap} I)$.

The properties of sections and segments considered as classes of classes are many and various: they are considered in $*211$. In $*212$, we pass to the consideration of the *series* of sections and segments. These series are $P_{1c} \upharpoonright \text{sect}`P$ and $P_{1c} \upharpoonright D`P_\epsilon$ (cf. $*170$). The series of such segments as have no maximum is $P_{1c} \upharpoonright D`(P_\epsilon \dot{\cap} I)$. We put

$$\varsigma`P = P_{1c} \upharpoonright D`P_\epsilon \quad \text{Df},$$

$$\text{sgm}`P = P_{1c} \upharpoonright D`(P_\epsilon \dot{\cap} I) \quad \text{Df}.$$

It then appears that

$$\varsigma`P_* = \text{sgm}`P_* = P_{1c} \upharpoonright \text{sect}`P,$$

so that it is unnecessary to introduce a special notation for the series of sections.

Whenever P is connected and transitive, $P_{1c} \upharpoonright D`P_\epsilon$ turns out to be equivalent to logical inclusion combined with diversity (with the field limited to $D`P_\epsilon$). That is to say ($*212{\cdot}23$),

$$\vdash : P \in \text{trans} \cap \text{connex} . \supset . \varsigma`P = \hat{\alpha}\hat{\beta} \{\alpha, \beta \in D`P_\epsilon . \alpha \subset \beta . \alpha \neq \beta\}.$$

Hence it follows ($*212{\cdot}24$) that

$$\vdash : P_* \in \text{connex} . \supset . \varsigma`P_* = \hat{\alpha}\hat{\beta} \{\alpha, \beta \in \text{sect}`P . \alpha \subset \beta . \alpha \neq \beta\}.$$

We have also ($*211{\cdot}6{\cdot}17$)

$$\vdash :. P_* \in \text{connex} . \alpha, \beta \in \text{sect}`P . \supset : \alpha \subset \beta . \vee . \beta \subset \alpha.$$

Hence it easily follows that whenever P_* is connected, $\varsigma`P_*$ is a series. Similarly $\varsigma`P$ will be a series if P is transitive and connected.

The fact of connection, which is required in order that $\varsigma`P$ or $\varsigma`P_*$ may be a series, results from

$$\alpha, \beta \in \text{sect}`P . \supset : \alpha \subset \beta . \vee . \beta \subset \alpha$$

or

$$\alpha, \beta \in D`P_\epsilon . \supset : \alpha \subset \beta . \vee . \beta \subset \alpha.$$

In order to deal with such cases generally, we study, in a preliminary number ($*210$), the consequences to be deduced from the hypothesis

$$\alpha, \beta \in \kappa . \supset_{\alpha, \beta} : \alpha \subset \beta . \vee . \beta \subset \alpha.$$

We find that, with this hypothesis, putting

$$Q = \hat{\alpha}\hat{\beta} (\alpha, \beta \in \kappa . \alpha \subset \beta . \alpha \neq \beta),$$

$Q = P_{1c} \upharpoonright \kappa$ if $\kappa \subset \text{Cl}`C`P$ ($*210{\cdot}13$), and thus in the same circumstances $P_{1c} \upharpoonright \kappa$ is a series ($*210{\cdot}14$).

The interesting point about such series is their behaviour with regard to limits. Assuming that κ is not a unit class (so as to insure $\dot{\exists} ! Q$), if λ is any sub-class of κ, the logical product $p`\lambda$ is the minimum of $\lambda$ if it is a member of $\lambda$ ($*210{\cdot}21$), and the lower limit of $\lambda$ if it is a member of $\kappa$ but not of $\lambda$ ($*210{\cdot}23$). Similarly $s`\lambda$ is the maximum of λ if it is a member of λ ($*210{\cdot}211$), and the upper limit of λ if it is not a member of λ but is a member of κ ($*210{\cdot}231$). Thus if κ is such that, whenever $\lambda \subset \kappa$, we have

$s'\lambda \in \kappa$, it follows that every sub-class of κ has either a maximum or a limit, *i.e.* the series $P_{1c} \upharpoonright \kappa$ is Dedekindian. Now each of the three classes sect'P, D'P_ϵ, D'$(P_\epsilon \dot{\cap} I)$ verifies this condition, *i.e.* the sum of any sub-class of any one of these classes belongs to the class in question (*211·63·64·65). (This holds without any hypothesis as to P.) Hence we arrive at the result that $\varsigma'P_*$ (*i.e.* the series of sections) is a Dedekindian series whenever P_* is connected and P is not null (*214·31), while $\varsigma'P$ (*i.e.* the series of segments) is a Dedekindian series whenever P is transitive and connected and not null (*214·33), and sgm'P (the series of segments having no maximum) is a Dedekindian series whenever it exists and P is connected (*214·34). These propositions are important, and are the source of much of the utility of sections and segments.

For many purposes, especially in ordinal arithmetic, it is necessary to consider sections not as classes, but as series. That is to say, if α is a member of sect'P, we want to deal with $P \upharpoonright \alpha$ rather than with α. The series of all such terms as $P \upharpoonright \alpha$ might be supposed to be $P \upharpoonright ; \varsigma'P_*$. But here a limitation is necessary owing to the fact that, if $B'P$ exists, Λ and $\iota'B'P$ are both sections, and $P \upharpoonright \Lambda$ and $P \upharpoonright \iota'B'P$ are both $\dot{\Lambda}$, so that $P \upharpoonright ; \varsigma'P_*$ will be a relation which $\dot{\Lambda}$ will have to itself. In order to avoid this, we first exclude Λ from the sections to be considered, and thus put

$$P_\varsigma = P \upharpoonright ; (\varsigma'P_*) \upharpoonright (-\iota'\Lambda) \quad \text{Df.}$$

Then P_ς is the series of segments considered as series. Provided P_{po} is a series, the relation P_ς holds between any two members Q and R of its field when, and only when, $Q \subseteq R \,.\, Q \neq R$. The subject of P_ς is considered in *213; the utility of the propositions of this number will not appear until we come to ordinal arithmetic.

The subject of Dedekindian relations is next considered (*214). We define a Dedekindian relation as one such that every class has either a maximum or a sequent. A Dedekindian series must have a first and a last term, since the first term must be the sequent of Λ, and the last must be the maximum of the field. A Dedekindian series may be discrete, or compact (*i.e.* such that there is a term between any two, *i.e.* such that $P^2 = P$), or partly one and partly the other. A finite series must be Dedekindian: a well-ordered series is Dedekindian if it has a last term. But the chief importance of the Dedekindian property is in connection with compact series. A compact Dedekindian series is said to possess "Dedekindian continuity"; such series have many important properties. They are a wider class than series possessing Cantorian continuity; these latter will be considered in Section F of this Part.

*210. ON SERIES OF CLASSES GENERATED BY THE RELATION OF INCLUSION.

Summary of *210.

In the theory of series it frequently happens that we have to deal with a class of classes such that, of any two, one is contained in the other. *I.e.* if κ is the class of classes, we have

$$\alpha, \beta \in \kappa . \supset_{\alpha,\beta} : \alpha \subset \beta . \vee . \beta \subset \alpha.$$

Instances of this are afforded by the various classes of sections, to be considered in *211. When κ fulfils the above condition, the classes composing κ can be arranged in a series by the relation of inclusion (combined with inequality), *i.e.* by the relation

$$\hat{\alpha}\hat{\beta} (\alpha, \beta \in \kappa . \alpha \subset \beta . \alpha \neq \beta),$$

or, what comes to the same,

$$\hat{\alpha}\hat{\beta} (\alpha, \beta \in \kappa . \exists ! \beta - \alpha).$$

If P is any relation such that $\kappa \subset \mathrm{Cl}\text{‘}C\text{‘}P$, the above relation of inclusion is equal to

$$P_{1c} \upharpoonright \kappa.$$

(For the definition of P_{1c}, see *170.) Thus under the above circumstances, $P_{1c} \upharpoonright \kappa$ is a series, whatever P may be.

The importance of such relations of inclusion, as generators of series, is in connection with the existence of maxima and minima or limits. If we put

$$Q = \hat{\alpha}\hat{\beta} (\alpha, \beta \in \kappa . \exists ! \beta - \alpha),$$

where κ satisfies the above condition, then if $\lambda \subset \kappa$, and if $s\text{‘}\lambda \in \kappa$, $s\text{‘}\lambda$ is the maximum or the upper limit of λ with respect to Q, according as $s\text{‘}\lambda$ is a member of λ or not. Similarly if $p\text{‘}\lambda \in \kappa$, $p\text{‘}\lambda$ is the minimum or lower limit of λ, according as $p\text{‘}\lambda$ is a member of λ or not. Hence if κ is such that the sum of any sub-class of κ is a member of κ, every sub-class of κ has either a maximum or an upper limit; and if the product of every sub-class of κ is a member of κ, every sub-class of κ has either a minimum or a lower limit.

In order that every sub-class of κ should have a minimum or a lower limit, it is sufficient that the *sum* of every sub-class of κ should be a member of κ. For, if λ is any sub-class of κ, consider those members of κ which are contained in $p‘\lambda$, *i.e.*

$$\kappa \cap \mathrm{Cl}‘p‘\lambda.$$

If $p‘\lambda \in \kappa$, the sum of these classes $= p‘\lambda$, and is the lower limit or minimum of κ. But if $p‘\lambda \sim\in \kappa$, then every member of κ which is not contained in $s‘(\kappa \cap \mathrm{Cl}‘p‘\lambda)$ is also not contained in $p‘\lambda$, and is therefore not contained in some member of λ. Hence $s‘(\kappa \cap \mathrm{Cl}‘p‘\lambda)$ is the lower limit of λ.

It is owing to these propositions that segments of series are of such great importance in connection with limits.

The hypothesis that if $\lambda \subset \kappa$, $p‘\lambda$ is a member of κ, will usually fail to be verified in the case when $\lambda = \Lambda$, since in this case $p‘\lambda = V$. But all the results desired can be obtained from the hypothesis that, if $\lambda \subset \kappa$, $(p‘\lambda \cap s‘\kappa) \in \kappa$. This hypothesis is equivalent to the other except in the case of Λ, in which case it requires $s‘\kappa \in \kappa$, which is much more often verified than $V \in \kappa$, which was required by the other hypothesis.

The principal propositions of this number are the following:

***210·1.** $\vdash :: \alpha, \beta \in \kappa . \supset_{\alpha,\beta} : \alpha \subset \beta . \vee . \beta \subset \alpha :. \supset :. \alpha, \beta \in \kappa . \supset : \alpha \subset \beta . \alpha \neq \beta . \equiv . \exists ! \beta - \alpha$

***210·11.** $\vdash : Q = \hat{\alpha}\hat{\beta}(\alpha, \beta \in \kappa . \alpha \subset \beta . \alpha \neq \beta) . \supset . Q \in \mathrm{trans} \cap \mathrm{Rl}‘J$

***210·12.** $\vdash : \mathrm{Hp} *210·1·11 . \supset . Q \in \mathrm{Ser}$

***210·13.** $\vdash : \mathrm{Hp} *210·12 . \kappa \subset \mathrm{Cl}‘C‘P . \supset . Q = P_{\mathrm{lc}} \upharpoonright \kappa$

***210·2.** $\vdash : \mathrm{Hp} *210·12 . \kappa \sim\in 1 . \supset . \overrightarrow{\min}_Q‘\lambda = \lambda \cap \kappa \cap \iota‘p‘(\lambda \cap \kappa)$

***210·21.** $\vdash : \mathrm{Hp} *210·2 . \lambda \subset \kappa . p‘\lambda \in \lambda . \supset . \min_Q‘\lambda = p‘\lambda$

*210·211 gives an analogous proposition for $s‘\lambda$ and $\max_Q$. We shall not here mention such analogues, unless for some special reason.

***210·23.** $\vdash : \mathrm{Hp} *210·2 . \lambda \subset \kappa . p‘\lambda \in \kappa - \lambda . \supset . p‘\lambda = \mathrm{prec}_Q‘\lambda = \mathrm{tl}_Q‘\lambda$

***210·232.** $\vdash : \mathrm{Hp} *210·2 . \lambda \subset \kappa . p‘\lambda \in \kappa . \supset . p‘\lambda = \mathrm{limin}_P‘\lambda$

***210·251.** $\vdash :. \mathrm{Hp} *210·2 : \lambda \subset \kappa . \supset_\lambda . s‘\lambda \in \kappa : \supset : \lambda \subset \kappa . \supset . s‘\lambda \in (\overrightarrow{\max}_Q‘\lambda \cup \overrightarrow{\mathrm{seq}}_Q‘\lambda)$

***210·252.** $\vdash :. \mathrm{Hp} *210·2 : \lambda \subset \kappa . \supset_\lambda . p‘\lambda \cap s‘\kappa \in \kappa : \supset :$

$$\lambda \subset \kappa . \supset . p‘\lambda \cap s‘\kappa \in (\overrightarrow{\min}_Q‘\lambda \cup \overrightarrow{\mathrm{prec}}_Q‘\lambda) . p‘\lambda \cap s‘\kappa = \mathrm{limin}_Q‘\lambda$$

***210·254.** $\vdash : \mathrm{Hp} *210·251 . \supset . (\lambda) . \lambda \in \mathrm{G}‘\max_Q \cup \mathrm{G}‘\mathrm{seq}_Q$

***210·26.** $\vdash : \mathrm{Hp} *210·2 . \lambda \subset \kappa . p‘\lambda \sim\in \lambda . s‘(\kappa \cap \mathrm{Cl}‘p‘\lambda) \in \kappa . \supset .$

$$s‘(\kappa \cap \mathrm{Cl}‘p‘\lambda) = \mathrm{prec}_Q‘\lambda$$

***210·28.** $\vdash : \mathrm{Hp} *210·2 . s‘‘\mathrm{Cl}‘\kappa \subset \kappa . \supset .$

$$(\lambda) . \lambda \in (\mathrm{G}‘\max_Q \cup \mathrm{G}‘\mathrm{seq}_Q) \cap (\mathrm{G}‘\min_Q \cup \mathrm{G}‘\mathrm{prec}_Q)$$

Thus if κ is a class of not less than two classes such that, of any two of its members, one must be contained in the other, and if Q is the relation

$\alpha \subset \beta . \alpha \neq \beta$ confined to members of κ, then Q is a series (*210·12) in which, provided the sums of sub-classes of κ are always members of κ, every class has either a maximum or an upper limit, and every class has either a minimum or a lower limit (*210·28).

The reader will observe that, if $\alpha, \beta \in \kappa . \supset_{\alpha, \beta} : \alpha \subset \beta . \vee . \beta \subset \alpha$, any *finite* sub-class of κ must contain its own sum and product as members. For example, if we have two classes α and β, if $\alpha \subset \beta$, then $\alpha = p`(\iota`\alpha \cup \iota`\beta)$ and $\beta = s`(\iota`\alpha \cup \iota`\beta)$; if we have three classes α, β, γ, and $\alpha \subset \beta . \beta \subset \gamma$, then $\alpha = p`(\iota`\alpha \cup \iota`\beta \cup \iota`\gamma)$ and $\gamma = s`(\iota`\alpha \cup \iota`\beta \cup \iota`\gamma)$; and so on. Thus the hypothesis $s``\text{Cl}`\kappa \subset \kappa$ is only required in order to enable us to deal with infinite sub-classes of κ.

***210·1.** $\vdash :: \alpha, \beta \in \kappa . \supset_{\alpha,\beta} : \alpha \subset \beta . \vee . \beta \subset \alpha :. \supset :. \alpha, \beta \in \kappa . \supset : \alpha \subset \beta . \alpha \neq \beta . \equiv . \exists ! \beta - \alpha$

Dem.

$\vdash . *24·6 . \quad \supset \vdash : \alpha \subset \beta . \alpha \neq \beta . \supset . \exists ! \beta - \alpha \quad (1)$

$\vdash . *24·55 . \quad \supset \vdash : \exists ! \beta - \alpha . \supset . \sim(\beta \subset \alpha) .$ (2)

[*22·42] $\quad \supset . \alpha \neq \beta \quad (3)$

$\vdash . *2·53 . \quad \supset \vdash :. \text{Hp} . \alpha, \beta \in \kappa . \sim(\beta \subset \alpha) . \supset . \alpha \subset \beta \quad (4)$

$\vdash . (2) . (3) . (4) . \supset \vdash :. \text{Hp} . \alpha, \beta \in \kappa . \supset : \exists ! \beta - \alpha . \supset . \alpha \subset \beta . \alpha \neq \beta \quad (5)$

$\vdash . (1) . (5) . \supset \vdash . \text{Prop}$

***210·11.** $\vdash : Q = \hat{\alpha}\hat{\beta}(\alpha, \beta \in \kappa . \alpha \subset \beta . \alpha \neq \beta) . \supset . Q \in \text{trans} \cap \text{Rl}`J$

Dem.

$\vdash . *50·11 . \quad \supset \vdash : \text{Hp} . \supset . Q \in \text{Rl}`J \quad (1)$

$\vdash . *22·44 . \quad \supset \vdash :. \text{Hp} . \supset : \alpha Q \beta . \beta Q \gamma . \supset . \alpha \subset \gamma \quad (2)$

$\vdash . *24·6 . *21·33 . \supset \vdash :. \text{Hp} . \supset : \alpha Q \beta . \beta Q \gamma . \supset . \exists ! \beta - \alpha . \beta \subset \gamma .$

[*24·58] $\quad \supset . \exists ! \gamma - \alpha .$

[*24·21] $\quad \supset . \alpha \neq \gamma \quad (3)$

$\vdash . (2) . (3) . \quad \supset \vdash :. \text{Hp} . \supset : \alpha Q \beta . \beta Q \gamma . \supset . \alpha Q \gamma \quad (4)$

$\vdash . (1) . (4) . \supset \vdash . \text{Prop}$

***210·12.** $\vdash : \text{Hp} *210·1·11 . \supset . Q \in \text{Ser}$

Dem.

$\vdash . *10·1 . \quad \supset \vdash :. \text{Hp} . \alpha, \beta \in \kappa . \supset : \alpha \subset \beta . \vee . \beta \subset \alpha :$

[*5·62] $\quad \supset : \alpha \subset \beta . \alpha \neq \beta . \vee . \beta \subset \alpha . \beta \neq \alpha . \vee . \alpha = \beta \quad (1)$

$\vdash . *21·33 . \quad \supset \vdash :. \text{Hp} . \supset : \alpha Q \beta . \supset_{\alpha,\beta} . \alpha, \beta \in \kappa :$

[*33·352] $\quad \supset : C`Q \subset \kappa \quad (2)$

$\vdash . (1) . (2) . \supset \vdash :: \text{Hp} . \supset :. \alpha, \beta \in C`Q . \supset : \alpha Q \beta . \vee . \beta Q \alpha . \vee . \alpha = \beta \quad (3)$

$\vdash . *210·11 . (3) . *204·12 . \supset \vdash . \text{Prop}$

$*210{\cdot}121$. $\vdash : \text{Hp} *210{\cdot}12 . \supset . \text{D}`Q = \kappa - \iota`s`\kappa . \text{ᗡ}`Q = \kappa - \iota`p`\kappa$

Dem.

$\vdash . *21{\cdot}33 . \supset \vdash :: \text{Hp} . \supset :. \alpha \epsilon \text{D}`Q . \equiv : \alpha \epsilon \kappa : (\exists\beta) . \beta \epsilon \kappa . \alpha \subset \beta . \alpha \neq \beta :$

$[*210{\cdot}1] \quad \equiv : \alpha \epsilon \kappa : (\exists\beta) . \beta \epsilon \kappa . \exists ! \beta - \alpha :$

$[*40{\cdot}151.\text{Transp}] \quad \equiv : \alpha \epsilon \kappa . \exists ! s`\kappa - \alpha :$

$[*24{\cdot}55] \quad \equiv : \alpha \epsilon \kappa . \sim (s`\kappa \subset \alpha) :$

$[*22{\cdot}41.*40{\cdot}13] \quad \equiv : \alpha \epsilon \kappa . \alpha \neq s`\kappa \quad (1)$

$\vdash . *21{\cdot}33 . \supset \vdash :: \text{Hp} . \supset :. \alpha \epsilon \text{ᗡ}`Q . \equiv : \alpha \epsilon \kappa : (\exists\beta) . \beta \epsilon \kappa . \beta \subset \alpha . \beta \neq \alpha :$

$[*210{\cdot}1] \quad \equiv : \alpha \epsilon \kappa : (\exists\beta) . \beta \epsilon \kappa . \exists ! \alpha - \beta :$

$[*40{\cdot}15.\text{Transp}] \quad \equiv : \alpha \epsilon \kappa . \exists ! \alpha - p`\kappa :$

$[*24{\cdot}55] \quad \equiv : \alpha \epsilon \kappa . \sim (\alpha \subset p`\kappa) :$

$[*22{\cdot}41.*40{\cdot}12] \quad \equiv : \alpha \epsilon \kappa . \alpha \neq p`\kappa \quad (2)$

$\vdash . (1) . (2) . \supset \vdash . \text{Prop}$

$*210{\cdot}122$. $\vdash : \text{Hp} *210{\cdot}12 . \kappa \sim \epsilon 1 . \supset . C`Q = \kappa$

Dem.

$\vdash . *52{\cdot}181 . \supset \vdash :: \text{Hp} . \supset :. \alpha \epsilon \kappa . \supset : (\exists\beta) . \beta \epsilon \kappa . \beta \neq \alpha :$

$[\text{Hp}.*10{\cdot}1] \quad \supset : (\exists\beta) : \beta \epsilon \kappa . \beta \neq \alpha : \alpha \subset \beta . \vee . \beta \subset \alpha :$

$[*21{\cdot}33] \quad \supset : (\exists\beta) : \beta \epsilon \kappa : \alpha Q \beta . \vee . \beta Q \alpha :$

$[*33{\cdot}132] \quad \supset : \alpha \epsilon C`Q \quad (1)$

$\vdash . *21{\cdot}33 . \supset \vdash :. \text{Hp} . \supset : \alpha Q \beta . \supset_{\alpha,\beta} . \alpha, \beta \epsilon \kappa :$

$[*33{\cdot}352] \quad \supset : C`Q \subset \kappa \quad (2)$

$\vdash . (1) . (2) . \supset \vdash . \text{Prop}$

$*210{\cdot}123$. $\vdash : \text{Hp} *210{\cdot}12 . \kappa \epsilon 0 \cup 1 . \supset . Q = \dot{\Lambda}$

Dem.

$\vdash . *52{\cdot}41 . \text{Transp} . \supset \vdash : \text{Hp} . \supset . \sim (\exists\alpha, \beta) . \alpha, \beta \epsilon \kappa . \alpha \neq \beta .$

$[*21{\cdot}33] \quad \supset . \sim (\exists\alpha, \beta) . \alpha Q \beta : \supset \vdash . \text{Prop}$

$*210{\cdot}124$. $\vdash :. \text{Hp} *210{\cdot}12 . \supset : \alpha Q \beta . \equiv . \alpha, \beta \epsilon \kappa . \exists ! \beta - \alpha \quad [*210{\cdot}1]$

$*210{\cdot}13$. $\vdash : \text{Hp} *210{\cdot}12 . \kappa \subset \text{Cl}`C`P . \supset . Q = P_{\text{lc}} \upharpoonright \kappa$

Dem.

$\vdash . *170{\cdot}102 . \supset \vdash :. \text{Hp} . \supset : \alpha (P_{\text{lc}} \upharpoonright \kappa) \beta . \equiv . \alpha, \beta \epsilon \kappa . \exists ! \beta - \alpha - P``(\alpha - \beta) . \quad (1)$

$[*210{\cdot}124] \quad \supset . \alpha Q \beta \quad (2)$

$\vdash . *210{\cdot}1{\cdot}124 . \supset \vdash :. \text{Hp} . \supset : \alpha Q \beta . \supset . \alpha, \beta \epsilon \kappa . \exists ! \beta - \alpha . \alpha \subset \beta .$

$[*37{\cdot}29] \quad \supset . \alpha, \beta \epsilon \kappa . \exists ! \beta - \alpha . P``(\alpha - \beta) = \Lambda .$

$[*24{\cdot}23{\cdot}313] \quad \supset . \alpha, \beta \epsilon \kappa . \exists ! \beta - \alpha - P``(\alpha - \beta) .$

$[(1)] \quad \supset . \alpha (P_{\text{lc}} \upharpoonright \kappa) \beta \quad (3)$

$\vdash . (2) . (3) . \supset \vdash . \text{Prop}$

Thus under the hypothesis of $*210{\cdot}1$, $P_{\text{lc}} \upharpoonright \kappa$ does not depend upon P, so long as $\kappa \subset \text{Cl}'C'P$. Also we have

$*210{\cdot}14$. $\vdash : \text{Hp}\, *210{\cdot}1 . \kappa \subset \text{Cl}'C'P . \supset . P_{\text{lc}} \upharpoonright \kappa \in \text{Ser}$
$[*210{\cdot}12{\cdot}13]$

$*210{\cdot}15$. $\vdash :. \text{Hp}\, *210{\cdot}12 . \alpha, \beta \in \kappa . \supset : \sim(\alpha Q \beta) . \equiv . \beta \subset \alpha$
$[*210{\cdot}124 . *24{\cdot}55]$

$*210{\cdot}16$. $\vdash :: \text{Hp}\, *210{\cdot}1 . \supset :.$

$$\alpha \in \kappa . \lambda \subset \kappa . \supset : \alpha \subset p'\lambda . \vee . p'\lambda \subset \alpha : \alpha \subset s'\lambda . \vee . s'\lambda \subset \alpha$$

Dem.

$\vdash . *10{\cdot}1 . \supset \vdash :: \text{Hp} . \alpha \in \kappa . \lambda \subset \kappa . \supset :. \beta \in \lambda . \supset_\beta : \alpha \subset \beta . \vee . \beta \subset \alpha :.$ (1)

$[*10{\cdot}57]$ $\supset :. \beta \in \lambda . \supset_\beta . \alpha \subset \beta : \vee : (\exists \beta) . \beta \in \lambda . \beta \subset \alpha :.$

$[*40{\cdot}15{\cdot}12]$ $\supset :. \alpha \subset p'\lambda . \vee . p'\lambda \subset \alpha$ (2)

$\vdash . (1) . *10{\cdot}57 . \supset$

$\vdash :: \text{Hp} . \alpha \in \kappa . \lambda \subset \kappa . \supset :. \beta \in \lambda . \supset_\beta . \beta \subset \alpha : \vee : (\exists \beta) . \beta \in \kappa . \alpha \subset \beta :.$

$[*40{\cdot}151{\cdot}13]$ $\supset :. s'\lambda \subset \alpha . \vee . \alpha \subset s'\lambda$ (3)

$\vdash . (2) . (3) . \supset \vdash . \text{Prop}$

$*210{\cdot}17$. $\vdash : \text{Hp}\, *210{\cdot}12 . \lambda \subset \kappa . \supset .$

$$\kappa - \breve{Q}``\lambda = \kappa \cap \text{Cl}'p'\lambda . \kappa - Q``\lambda = \kappa \cap \hat{\gamma}(s'\lambda \subset \gamma)$$

Dem.

$\vdash . *37{\cdot}105 . \text{Transp} . \supset \vdash :. \alpha \in \kappa - \breve{Q}``\lambda . \equiv : \alpha \in \kappa : \beta \in \lambda . \supset_\beta . \sim(\beta Q \alpha)$ (1)

$\vdash . (1) . *210{\cdot}15 . \supset$

$\vdash :: \text{Hp} . \supset :. \alpha \in \kappa - \breve{Q}``\lambda . \equiv : \alpha \in \kappa : \beta \in \lambda . \supset_\beta . \alpha \subset \beta :$

$[*40{\cdot}15]$ $\equiv : \alpha \in \kappa \cap \text{Cl}'p'\lambda$ (2)

Similarly $\vdash :. \text{Hp} . \supset : \alpha \in \kappa - Q``\lambda . \equiv . \alpha \in \kappa \cap \hat{\gamma}(s'\lambda \subset \gamma)$ (3)

$\vdash . (2) . (3) . \supset \vdash . \text{Prop}$

$*210{\cdot}2$. $\vdash : \text{Hp}\, *210{\cdot}12 . \kappa \sim \in 1 . \supset . \overrightarrow{\text{min}}_Q{}'\lambda = \lambda \cap \kappa \cap \iota'p'(\lambda \cap \kappa)$

Dem.

$\vdash . *205{\cdot}15 . *210{\cdot}122 . \supset \vdash : \text{Hp} . \supset . \overrightarrow{\text{min}}_Q{}'\lambda = \overrightarrow{\text{min}}_Q{}'(\lambda \cap \kappa)$

$[*205{\cdot}11]$ $= \lambda \cap \kappa - \breve{Q}``(\lambda \cap \kappa)$

$[*210{\cdot}17]$. $= \lambda \cap \kappa \cap \text{Cl}'p'(\lambda \cap \kappa)$ (1)

$\vdash . *40{\cdot}12 . \supset \vdash :. \alpha \in \lambda \cap \kappa . \supset : p'(\lambda \cap \kappa) \subset \alpha :$

$[*22{\cdot}41]$ $\supset : \alpha \subset p'(\lambda \cap \kappa) . \equiv . \alpha = p'(\lambda \cap \kappa)$ (2)

$\vdash . (2) . *5{\cdot}32 . \supset \vdash . \lambda \cap \kappa \cap \text{Cl}'p'(\lambda \cap \kappa) = \lambda \cap \kappa \cap \iota'p'(\lambda \cap \kappa)$ (3)

$\vdash . (1) . (3) . \supset \vdash . \text{Prop}$

Observe that $\lambda \cap \kappa \cap \iota'p'(\lambda \cap \kappa)$ is either $\iota'p'(\lambda \cap \kappa)$ or Λ, according as $p'(\lambda \cap \kappa)$ is or is not a member of $\lambda \cap \kappa$.

∗210·201. $\vdash : \mathrm{Hp} * 210{\cdot}2 \,.\, \lambda \subset \kappa \,.\, \supset .\, \overrightarrow{\min}_Q{}^{\backprime}\lambda = \lambda \cap \iota^{\backprime} p^{\backprime}\lambda$
[∗210·2 . ∗22·621]

∗210·202. $\vdash : \mathrm{Hp} * 210{\cdot}2 \,.\, \supset .\, \overrightarrow{\max}_Q{}^{\backprime}\lambda = \lambda \cap \kappa \cap \iota^{\backprime} s^{\backprime}(\lambda \cap \kappa)$
[Proof as in ∗210·2]

∗210·203. $\vdash : \mathrm{Hp} * 210{\cdot}2 \,.\, \lambda \subset \kappa \,.\, \supset .\, \overrightarrow{\max}_Q{}^{\backprime}\lambda = \lambda \cap \iota^{\backprime} s^{\backprime}\lambda$
[∗210·202 . ∗22·621]

∗210·21. $\vdash : \mathrm{Hp} * 210{\cdot}2 \,.\, \lambda \subset \kappa \,.\, p^{\backprime}\lambda \in \lambda \,.\, \supset .\, \min_Q{}^{\backprime}\lambda = p^{\backprime}\lambda$
[∗210·201 . ∗51·31]

∗210·211. $\vdash : \mathrm{Hp} * 210{\cdot}2 \,.\, \lambda \subset \kappa \,.\, s^{\backprime}\lambda \in \lambda \,.\, \supset .\, \max_Q{}^{\backprime}\lambda = s^{\backprime}\lambda$
[∗210·203 . ∗51·31]

∗210·22. $\vdash : \mathrm{Hp} * 210{\cdot}12 \,.\, \lambda \subset \kappa \,.\, p^{\backprime}\lambda \sim\in \lambda \,.\, \supset .\, \sim \exists\, !\, \overrightarrow{\min}_Q{}^{\backprime}\lambda$
[∗210·201·123 . ∗51·211]

∗210·221. $\vdash : \mathrm{Hp} * 210{\cdot}12 \,.\, \lambda \subset \kappa \,.\, s^{\backprime}\lambda \sim\in \lambda \,.\, \supset .\, \sim \exists\, !\, \overrightarrow{\max}_Q{}^{\backprime}\lambda$
[∗210·203·123 . ∗51·211]

∗210·222. $\vdash :.\, \mathrm{Hp} * 210{\cdot}2 \,.\, \lambda \subset \kappa \,.\, \supset : p^{\backprime}\lambda \in \lambda \,.\, \equiv .\, \mathrm{E}\,!\, \min_Q{}^{\backprime}\lambda$
[∗210·21·22]

∗210·223. $\vdash :.\, \mathrm{Hp} * 210{\cdot}2 \,.\, \lambda \subset \kappa \,.\, \supset : s^{\backprime}\lambda \in \lambda \,.\, \equiv .\, \mathrm{E}\,!\, \max_Q{}^{\backprime}\lambda$
[∗210·211·221]

∗210·23. $\vdash : \mathrm{Hp} * 210{\cdot}2 \,.\, \lambda \subset \kappa \,.\, p^{\backprime}\lambda \in \kappa - \lambda \,.\, \supset .\, p^{\backprime}\lambda = \mathrm{prec}_Q{}^{\backprime}\lambda = \mathrm{tl}_Q{}^{\backprime}\lambda$

Dem.

$\vdash . * 210{\cdot}22 \,.\, \supset \vdash : \mathrm{Hp} \,.\, \supset .\, \overrightarrow{\min}_Q{}^{\backprime}\lambda = \Lambda \,.$ (1)

[∗205·122.∗210·122] $\supset .\, \lambda \subset \breve{Q}^{\backprime\backprime}\lambda$ (2)

$\vdash . (2) . * 210{\cdot}12 \,.\, * 206{\cdot}174 \,.\, \supset$

$\vdash : \mathrm{Hp} \,.\, \supset .\, \overrightarrow{\mathrm{prec}}_Q{}^{\backprime}\lambda = C^{\backprime}Q \cap \hat{\alpha}\,(\overleftarrow{Q}^{\backprime}\alpha = \breve{Q}^{\backprime\backprime}\lambda)$

[∗210·122] $= \kappa \cap \hat{\alpha}\,(\overleftarrow{Q}^{\backprime}\alpha = \breve{Q}^{\backprime\backprime}\lambda)$ (3)

$\vdash . * 37{\cdot}105 \,.\, * 210{\cdot}124 \,.\, \supset$

$\vdash :.\, \mathrm{Hp} \,.\, \supset : \beta \in \breve{Q}^{\backprime\backprime}\lambda \,.\, \equiv .\, (\exists\gamma) \,.\, \gamma \in \lambda \,.\, \exists\, !\, \beta - \gamma \,.\, \beta \in \kappa \,.$

[∗40·15.Transp] $\equiv .\, \exists\, !\, \beta - p^{\backprime}\lambda \,.\, \beta \in \kappa \,.$

[∗210·124] $\equiv .\, (p^{\backprime}\lambda)\, Q \beta$ (4)

$\vdash . (4) \,.\, \supset \vdash : \mathrm{Hp} \,.\, \supset .\, p^{\backprime}\lambda \in \kappa \,.\, \overleftarrow{Q}^{\backprime} p^{\backprime}\lambda = \breve{Q}^{\backprime\backprime}\lambda \,.$

[(3)] $\supset .\, p^{\backprime}\lambda \in \overrightarrow{\mathrm{prec}}_Q{}^{\backprime}\lambda$ (5)

$\vdash . (5) . * 210{\cdot}12 \,.\, * 206{\cdot}16 \,.\, \supset \vdash : \mathrm{Hp} \,.\, \supset .\, p^{\backprime}\lambda = \mathrm{prec}_Q{}^{\backprime}\lambda$ (6)

$\vdash . (1) . (6) . * 207{\cdot}12 \,.\, \supset \vdash : \mathrm{Hp} \,.\, \supset .\, p^{\backprime}\lambda = \mathrm{tl}_Q{}^{\backprime}\lambda$ (7)

$\vdash . (6) . (7) \,.\, \supset \vdash .$ Prop

***210·231.** $\vdash : \text{Hp} * 210{\cdot}2 \,.\, \lambda \subset \kappa \,.\, s{`}\lambda \in \kappa - \lambda \,.\, \supset \,.\, s{`}\lambda = \text{seq}_Q{`}\lambda = \text{lt}_Q{`}\lambda$

[Proof as in *210·23]

In virtue of *210·21·23, every class which is contained in κ, and whose product is a member of κ, has either a minimum or a lower limit; and in virtue of *210·211·231, every class which is contained in κ, and whose sum is a member of κ, has either a maximum or an upper limit.

***210·232.** $\vdash : \text{Hp} * 210{\cdot}2 \,.\, \lambda \subset \kappa \,.\, p{`}\lambda \in \kappa \,.\, \supset \,.\, p{`}\lambda = \text{limin}_P{`}\lambda$ [*210·21·23]

***210·233.** $\vdash : \text{Hp} * 210{\cdot}2 \,.\, \lambda \subset \kappa \,.\, s{`}\lambda \in \kappa \,.\, \supset \,.\, s{`}\lambda = \text{limax}_P{`}\lambda$ [*210·211·231]

***210·24.** $\vdash : \text{Hp} * 210{\cdot}2 \,.\, \supset \,.\, \kappa \cap \iota{`}p{`}\kappa = \overrightarrow{B}{`}Q \,.\, \kappa \cap \iota{`}s{`}\kappa = \overrightarrow{B}{`}\breve{Q}$

[*205·12·121 . *210·201·203·122]

***210·241.** $\vdash : \text{Hp} * 210{\cdot}2 \,.\, p{`}\kappa \in \kappa \,.\, \supset \,.\, p{`}\kappa = B{`}Q$ [*210·24]

***210·242.** $\vdash : \text{Hp} * 210{\cdot}2 \,.\, s{`}\kappa \in \kappa \,.\, \supset \,.\, s{`}\kappa = B{`}\breve{Q}$ [*210·24]

***210·25.** $\vdash :. \text{Hp} * 210{\cdot}2 : \lambda \subset \kappa \,.\, \supset_\lambda \,.\, p{`}\lambda \in \kappa : \supset :$

$$\lambda \subset \kappa \,.\, \supset \,.\, p{`}\lambda \in (\overrightarrow{\text{min}}_Q{`}\lambda \cup \overrightarrow{\text{prec}}_Q{`}\lambda)$$

Dem.

$\vdash \,.\, *210{\cdot}21 \,.\, \supset \vdash :. \text{Hp} \,.\, \supset : \lambda \subset \kappa \,.\, p{`}\lambda \in \lambda \,.\, \supset \,.\, p{`}\lambda \in \overrightarrow{\text{min}}_Q{`}\lambda$ (1)

$\vdash \,.\, *210{\cdot}23 \,.\, \supset \vdash :. \text{Hp} \,.\, \supset : \lambda \subset \kappa \,.\, p{`}\lambda \sim\in \lambda \,.\, \supset \,.\, p{`}\lambda \in \overrightarrow{\text{prec}}_Q{`}\lambda$ (2)

$\vdash \,.\, (1) \,.\, (2) \,.\, \supset \vdash \,.\, \text{Prop}$

***210·251.** $\vdash :. \text{Hp} * 210{\cdot}2 : \lambda \subset \kappa \,.\, \supset_\lambda \,.\, s{`}\lambda \in \kappa : \supset :$

$$\lambda \subset \kappa \,.\, \supset \,.\, s{`}\lambda \in (\overrightarrow{\text{max}}_Q{`}\lambda \cup \overrightarrow{\text{seq}}_Q{`}\lambda)$$

[Proof as in *210·25]

***210·252.** $\vdash :. \text{Hp} * 210{\cdot}2 : \lambda \subset \kappa \,.\, \supset_\lambda \,.\, p{`}\lambda \cap s{`}\kappa \in \kappa : \supset :$

$$\lambda \subset \kappa \,.\, \supset \,.\, p{`}\lambda \cap s{`}\kappa \in (\overrightarrow{\text{min}}_Q{`}\lambda \cup \overrightarrow{\text{prec}}_Q{`}\lambda) \,.\, p{`}\lambda \cap s{`}\kappa = \text{limin}_P{`}\lambda$$

Dem.

$\vdash \,.\, *40{\cdot}23{\cdot}161 \,.\, \supset \vdash : \lambda \subset \kappa \,.\, \exists ! \lambda \,.\, \supset \,.\, p{`}\lambda \subset s{`}\kappa \,.$

[*22·621] $\supset \,.\, p{`}\lambda \cap s{`}\kappa = p{`}\lambda$ (1)

$\vdash \,.\, (1) \,.\, *210{\cdot}21{\cdot}23 \,.\, \supset \vdash : \text{Hp} \,.\, \lambda \subset \kappa \,.\, \exists ! \lambda \,.\, \supset \,.\, p{`}\lambda \cap s{`}\kappa \in (\overrightarrow{\text{min}}_Q{`}\lambda \cup \overrightarrow{\text{prec}}_Q{`}\lambda)$ (2)

$\vdash \,.\, *40{\cdot}2 \,.\, \supset \vdash : \sim\exists ! \lambda \,.\, \supset \,.\, p{`}\lambda \cap s{`}\kappa = s{`}\kappa$ (3)

$\vdash \,.\, (3) \,.\, *24{\cdot}12 \,.\, \supset \vdash : \text{Hp} \,.\, \supset \,.\, s{`}\kappa \in \kappa \,.$

[*210·242] $\supset \,.\, s{`}\kappa = B{`}\breve{Q} \,.$

[*206·14] $\supset \,.\, s{`}\kappa = \text{prec}_Q{`}\Lambda$ (4)

$\vdash \,.\, (3) \,.\, (4) \,.\, \supset \vdash : \text{Hp} \,.\, \sim\exists ! \lambda \,.\, \supset \,.\, p{`}\lambda \cap s{`}\kappa \in (\overrightarrow{\text{min}}_Q{`}\lambda \cup \overrightarrow{\text{prec}}_Q{`}\lambda)$ (5)

$\vdash \,.\, (2) \,.\, (5) \,.\, \supset \vdash \,.\, \text{Prop}$

This proposition is more useful than *210·25, because its hypothesis is much oftener verified. In order that the hypothesis of *210·25 may be verified, we must have $V \in \kappa$, since $\Lambda \subset \kappa . p'\Lambda = V$; hence we must also have $s'\kappa = V$. But the hypothesis of *210·252 only requires, as far as Λ is concerned, that we should have $s'\kappa \in \kappa$.

***210·253.** $\vdash : \text{Hp} *210\cdot252 . \supset . (\lambda) . \lambda \in \text{Ɑ}'\text{min}_Q \cup \text{Ɑ}'\text{prec}_Q$
[*210·252 . *205·15 . *206·131]

***210·254.** $\vdash : \text{Hp} *210\cdot251 . \supset . (\lambda) . \lambda \in \text{Ɑ}'\text{max}_Q \cup \text{Ɑ}'\text{seq}_Q$
[Proof as in *210·253]

***210·26.** $\vdash : \text{Hp} *210\cdot2 . \lambda \subset \kappa . p'\lambda \sim\in \lambda . s'(\kappa \cap \text{Cl}'p'\lambda) \in \kappa . \supset . s'(\kappa \cap \text{Cl}'p'\lambda) = \text{prec}_Q'\lambda$

Dem.

$$
\begin{aligned}
&\vdash . *210\cdot22 . \supset \vdash : \text{Hp} . \supset . \sim \exists ! \overrightarrow{\text{min}_Q}'\lambda . \\
&[*205\cdot122] \qquad \supset . \lambda \subset \breve{Q}''\lambda & (1) \\
&\vdash . *60\cdot2 . \qquad \supset \vdash : \beta \in \kappa \cap \text{Cl}'p'\lambda . \supset . \beta \subset p'\lambda : \\
&[*40\cdot151] \qquad \supset \vdash : s'(\kappa \cap \text{Cl}'p'\lambda) \subset p'\lambda & (2) \\
&\vdash . (2) . \qquad \supset \vdash : \text{Hp} . \supset . s'(\kappa \cap \text{Cl}'p'\lambda) \in \kappa \cap \text{Cl}'p'\lambda . & (3) \\
&[*210\cdot211] \qquad \supset . s'(\kappa \cap \text{Cl}'p'\lambda) = \text{max}_Q'(\kappa \cap \text{Cl}'p'\lambda) \\
&[*210\cdot17] \qquad = \text{max}_Q'(\kappa - \breve{Q}''\lambda) \\
&[(1)] \qquad = \text{max}_Q'(\kappa - \lambda - \breve{Q}''\lambda) \\
&[*210\cdot122 . *202\cdot502 . (3)] \qquad = \text{max}_Q'p'\overrightarrow{Q}''\lambda \\
&[*206\cdot1\cdot101] \qquad = \text{prec}_Q'\lambda : \supset \vdash . \text{Prop}
\end{aligned}
$$

***210·261.** $\vdash : \text{Hp} *210\cdot2 . \lambda \subset \kappa . s'\lambda \sim\in \lambda . p'\hat{\alpha}(\alpha \in \kappa . s'\lambda \subset \alpha) \in \kappa . \supset . p'\hat{\alpha}(\alpha \in \kappa . s'\lambda \subset \alpha) = \text{seq}_Q'\lambda$ [Proof as in *210·26]

***210·262.** $\vdash : \text{Hp} *210\cdot2 . \lambda \subset \kappa . s'\lambda \sim\in \lambda . s'\kappa \cap p'\hat{\alpha}(\alpha \in \kappa . s'\lambda \subset \alpha) \in \kappa . \supset . s'\kappa \cap p'\hat{\alpha}(\alpha \in \kappa . s'\lambda \subset \alpha) = \text{seq}_Q'\lambda$

Dem.

$$
\begin{aligned}
&\vdash . *40\cdot23\cdot161 . \supset \\
&\vdash : \text{Hp} . \exists ! \hat{\alpha}(\alpha \in \kappa . s'\lambda \subset \alpha) . \supset . p'\hat{\alpha}(\alpha \in \kappa . s'\lambda \subset \alpha) \subset s'\kappa . \\
&[*22\cdot621] \qquad \supset . s'\kappa \cap p'\hat{\alpha}(\alpha \in \kappa . s'\lambda \subset \alpha) = p'\hat{\alpha}(\alpha \in \kappa . s'\lambda \subset \alpha) . \\
&[*210\cdot261] \qquad \supset . s'\kappa \cap p'\hat{\alpha}(\alpha \in \kappa . s'\lambda \subset \alpha) = \text{seq}_Q'\lambda & (1) \\
&\vdash . *10\cdot51 . \qquad \supset \vdash :. \hat{\alpha}(\alpha \in \kappa . s'\lambda \subset \alpha) = \Lambda . \supset : \alpha \in \kappa . \supset_\alpha . \sim(s'\lambda \subset \alpha) & (2) \\
&\vdash . (2) . *210\cdot16 . \supset \vdash :. \text{Hp} . \hat{\alpha}(\alpha \in \kappa . s'\lambda \subset \alpha) = \Lambda . \supset : \alpha \in \kappa . \supset_\alpha . \alpha \subset s'\lambda : \\
&[*40\cdot151] \qquad \supset : s'\kappa \subset s'\lambda : \\
&[*40\cdot161] \qquad \supset : s'\kappa = s'\lambda & (3) \\
&\vdash . *40\cdot2 . \supset \vdash : \text{Hp}(3) . \supset . s'\kappa \cap p'\hat{\alpha}(\alpha \in \kappa . s'\lambda \subset \alpha) = s'\kappa . & (4) \\
&[\text{Hp} . (3)] \qquad \supset . s'\lambda \in \kappa . \\
&[*210\cdot231] \qquad \supset . s'\lambda = \text{seq}_Q'\lambda . \\
&[(3) . (4)] \qquad \supset . s'\kappa \cap p'\hat{\alpha}(\alpha \in \kappa . s'\lambda \subset \alpha) = \text{seq}_Q'\lambda & (5) \\
&\vdash . (1) . (5) . \supset \vdash . \text{Prop}
\end{aligned}
$$

The same remark applies to this proposition as to *210·252.

***210·27.** $\vdash :. \text{Hp} *210{\cdot}2 : \lambda \subset \kappa . \supset_\lambda . s{\text{'}}\lambda \in \kappa : \supset :$

$$\lambda \subset \kappa . \supset_\lambda . \exists ! (\overrightarrow{\text{max}}_Q{\text{'}}\lambda \cup \overrightarrow{\text{seq}}_Q{\text{'}}\lambda) . \exists ! (\overrightarrow{\text{min}}_Q{\text{'}}\lambda \cup \overrightarrow{\text{prec}}_Q{\text{'}}\lambda)$$

Dem.

$\vdash . *210{\cdot}251 . \supset \vdash :. \text{Hp} . \supset : \lambda \subset \kappa . \supset_\lambda . \exists ! (\overrightarrow{\text{max}}_Q{\text{'}}\lambda \cup \overrightarrow{\text{seq}}_Q{\text{'}}\lambda)$ (1)

$\vdash . *210{\cdot}222 . \supset \vdash : \text{Hp} . \lambda \subset \kappa . p{\text{'}}\lambda \in \lambda . \supset . \exists ! \overrightarrow{\text{min}}_Q{\text{'}}\lambda$ (2)

$\vdash . *10{\cdot}1 . \quad \supset \vdash :. \text{Hp} . \supset : s{\text{'}}(\kappa \cap \text{Cl}{\text{'}}p{\text{'}}\lambda) \in \kappa :$

$[*210{\cdot}26] \quad \supset : \lambda \subset \kappa . p{\text{'}}\lambda \sim\in \lambda . \supset . \exists ! \overrightarrow{\text{prec}}_P{\text{'}}\lambda$ (3)

$\vdash . (2) . (3) . \quad \supset \vdash :. \text{Hp} . \supset : \lambda \subset \kappa . \supset . \exists ! (\overrightarrow{\text{min}}_Q{\text{'}}\lambda \cup \overrightarrow{\text{prec}}_Q{\text{'}}\lambda)$ (4)

$\vdash . (1) . (4) . \supset \vdash . \text{Prop}$

***210·271.** $\vdash :. \text{Hp} *210{\cdot}2 : \lambda \subset \kappa . \supset_\lambda . p{\text{'}}\lambda \in \kappa : \supset :$

$$\lambda \subset \kappa . \supset_\lambda . \exists ! (\overrightarrow{\text{max}}_Q{\text{'}}\lambda \cup \overrightarrow{\text{seq}}_Q{\text{'}}\lambda) . \exists ! (\overrightarrow{\text{min}}_Q{\text{'}}\lambda \cup \overrightarrow{\text{prec}}_Q{\text{'}}\lambda)$$

[Proof as in *210·27]

***210·272.** $\vdash :. \text{Hp} *210{\cdot}2 : \lambda \subset \kappa . \supset_\lambda . p{\text{'}}\lambda \cap s{\text{'}}\kappa \in \kappa : \supset :$

$$\lambda \subset \kappa . \supset_\lambda . \exists ! (\overrightarrow{\text{max}}_Q{\text{'}}\lambda \cup \overrightarrow{\text{seq}}_Q{\text{'}}\lambda) . \exists ! (\overrightarrow{\text{min}}_Q{\text{'}}\lambda \cup \overrightarrow{\text{prec}}_Q{\text{'}}\lambda)$$

[Proof as in *210·27, using *210·262]

***210·28.** $\vdash : \text{Hp} *210{\cdot}2 . s{\text{''}}\text{Cl}{\text{'}}\kappa \subset \kappa . \supset .$

$$(\lambda) . \lambda \in (\text{Cl}{\text{'}}\text{max}_Q \cup \text{Cl}{\text{'}}\text{seq}_Q) \cap (\text{Cl}{\text{'}}\text{min}_Q \cup \text{Cl}{\text{'}}\text{prec}_Q)$$

Dem.

$\vdash . *37{\cdot}61 . \supset \vdash :. \text{Hp} . \supset : \lambda \subset \kappa . \supset_\lambda . s{\text{'}}\lambda \in \kappa :$

$[*210{\cdot}27] \supset : \lambda \subset \kappa . \supset_\lambda . \exists ! (\overrightarrow{\text{max}}_Q{\text{'}}\lambda \cup \overrightarrow{\text{seq}}_Q{\text{'}}\lambda) . \exists ! (\overrightarrow{\text{min}}_Q{\text{'}}\lambda \cup \overrightarrow{\text{prec}}_Q{\text{'}}\lambda)$ (1)

$\vdash . (1) . *22{\cdot}43 . \supset$

$\vdash : \text{Hp} . \supset . (\lambda) . \exists ! \{\overrightarrow{\text{max}}_Q{\text{'}}(\lambda \cap \kappa) \cup \overrightarrow{\text{seq}}_Q{\text{'}}(\lambda \cap \kappa)\} .$

$$\exists ! \{\overrightarrow{\text{min}}_Q{\text{'}}(\lambda \cap \kappa) \cup \overrightarrow{\text{prec}}_Q{\text{'}}(\lambda \cap \kappa)\} .$$

$[*210{\cdot}122] \supset . (\lambda) . \exists ! \{\overrightarrow{\text{max}}_Q{\text{'}}(\lambda \cap C{\text{'}}Q) \cup \overrightarrow{\text{seq}}_Q{\text{'}}(\lambda \cap C{\text{'}}Q)\} .$

$$\exists ! \{\overrightarrow{\text{min}}_Q{\text{'}}(\lambda \cap C{\text{'}}Q) \cup \overrightarrow{\text{prec}}_Q{\text{'}}(\lambda \cap C{\text{'}}Q)\} .$$

$[*205{\cdot}15{\cdot}151 . *206{\cdot}131] \supset . (\lambda) . \exists ! \{\overrightarrow{\text{max}}_Q{\text{'}}\lambda \cup \overrightarrow{\text{seq}}_Q{\text{'}}\lambda\} . \exists ! \{\overrightarrow{\text{min}}_Q{\text{'}}\lambda \cup \overrightarrow{\text{prec}}_Q{\text{'}}\lambda\} .$

$[*33{\cdot}41] \quad \supset . (\lambda) . \lambda \in \text{Cl}{\text{'}}\text{max}_Q \cup \text{Cl}{\text{'}}\text{seq}_Q . \lambda \in \text{Cl}{\text{'}}\text{min}_Q \cup \text{Cl}{\text{'}}\text{prec}_Q : \supset \vdash . \text{Prop}$

***210·281.** $\vdash : \text{Hp} *210{\cdot}2 . p{\text{''}}\text{Cl}{\text{'}}\kappa \subset \kappa . \supset .$

$$(\lambda) . \lambda \in (\text{Cl}{\text{'}}\text{max}_Q \cup \text{Cl}{\text{'}}\text{seq}_Q) \cap (\text{Cl}{\text{'}}\text{min}_Q \cup \text{Cl}{\text{'}}\text{prec}_Q)$$

***210·282.** $\vdash :. \text{Hp} *210{\cdot}2 : \lambda \subset \kappa . \supset_\lambda . p{\text{'}}\lambda \cap s{\text{'}}\kappa \in \kappa : \supset .$

$$(\lambda) . \lambda \in (\text{Cl}{\text{'}}\text{max}_Q \cup \text{Cl}{\text{'}}\text{seq}_Q) \cap (\text{Cl}{\text{'}}\text{min}_Q \cup \text{Cl}{\text{'}}\text{prec}_Q)$$

Thus when either of the hypotheses of *210·281·282 is fulfilled, the series Q is Dedekindian both upwards and downwards.

***210·29.** $\vdash : \text{Hp} *210{\cdot}251 . \supset . (\lambda) . \lambda \in \text{Cl}{\text{'}}\text{limax}_P \cap \text{Cl}{\text{'}}\text{limin}_P$ [*210·28 . *207·44]

***210·291.** $\vdash : \text{Hp} *210{\cdot}252 . \supset . (\lambda) . \lambda \in \text{Cl}{\text{'}}\text{limax}_P \cap \text{Cl}{\text{'}}\text{limin}_P$
[*210·282 . *207·44]

$*211$. ON SECTIONS AND SEGMENTS.

Summary of $*211$.

The theory of the modes of separation of a series into two classes, one of which wholly precedes the other, and which together make up the whole series, is of fundamental importance. When one out of a pair of such classes is given, the other is the rest of the series; we may therefore, for most purposes, confine our attention to that one of the two classes which comes first in the serial order. Any class which can be the first of such a pair we shall call a *section* of our series. If P is the series, we shall denote the class of its sections by "$\text{sect}`P$." If $\alpha$ is a section of $P$, we shall call $C`P - \alpha$ (which is the second class of our pair) the *complement* of α. The class of complements of sections is

$$(C`P -)``\text{sect}`P,$$

which is identical with $\text{sect}`\breve{P}$ ($*211\cdot75$).

In order that a class may be a section of P, it is necessary and sufficient that it should be contained in $C`P$ and should contain all its own predecessors; thus we put

$$\text{sect}`P = \hat{\alpha}(\alpha \subset C`P \,.\, P``\alpha \subset \alpha) \quad \text{Df.}$$

We have also, by $*90\cdot23$,

$$\text{sect}`P = \hat{\alpha}(\alpha = P_*``\alpha) \quad (*211\cdot13).$$

Among sections, a specially important class consists of classes which are composed of all the predecessors of some class, *i.e.* classes of the form $P``\beta$, *i.e.* classes which are members of $\text{D}`P_\epsilon$. Whenever $P$ is transitive, $P``P``\beta \subset P``\beta$; hence $P``\beta$ is a section according to the above definition. When $P$ is a series, the complement of $P``\beta$ (when β exists and is contained in $C`P$) is

$$\overrightarrow{\max}_P`\beta \cup p`\overleftarrow{P}``\beta.$$

The members of $\text{D}`P_\epsilon$ are called *segments* of the series generated by $P$. In a series in which every sub-class has a maximum or a sequent, $\text{D}`P_\epsilon = \overrightarrow{P}``C`P$ ($*211\cdot38$), *i.e.* the predecessors of a class are always the predecessors of a single term, namely the maximum of the class if it exists, or the sequent if no maximum exists. But if there are classes which have neither a maximum nor a sequent, the predecessors of such classes are not coextensive with the predecessors of any single term. Thus in general the series of segments will be larger than the original series. For example, if our original series is of the type of the series of rationals in order of magnitude, the series of segments is of the type of the series of real numbers, *i.e.* the type of the continuum.

Among segments, a specially important class consists of those which have no maximum. In this case, if α is such a segment, we have $\alpha \subset P``\alpha$; and since (provided $P$ is transitive) we also have, for all segments, $P``\alpha \subset \alpha$, the segments having no maximum are those for which $\alpha = P``\alpha$, *i.e.* they are the class $D`(P_\epsilon \dot{\cap} I)$. In compact series, all segments belong to this latter class, but in general only those segments belong to it which correspond to a "Häufungsstelle." In all cases in which the existence of a limit is not known, the segment fulfils the functions of a limit; that is to say, in those places in the series where a limit might be expected, we have a segment having no limit or maximum, which takes the same place in the series of segments as would be taken by the limit in the original series if the limit existed. Segments having no limit or maximum are limiting points in the series of segments, and every class of segments which has no maximum in the series of segments has a limit in that series.

We have thus three classes to deal with, namely

$$(1)\quad \text{sect}`P,$$
$$(2)\quad D`P_\epsilon,$$
$$(3)\quad D`(P_\epsilon \dot{\cap} I).$$

Of these the second is contained in the first when P is transitive (∗211·15), and the third is contained in the first and second (∗211·14). The second consists of those members of the first which have either a sequent or no maximum (∗211·32); the third consists of those members of the first which have no maximum (∗211·41). If every member of the third class has a limit, *i.e.* if

$$D`(P_\epsilon \dot{\cap} I) \subset \text{Cl}`\text{seq}_P,$$

then every class has either a sequent or a maximum, *i.e.* the series is Dedekindian; and the converse also holds (∗211·47).

When P is connected, of any two sections one must be contained in the other (∗211·6). Moreover, if λ is contained in any one of the three classes $\text{sect}`P$, $D`P_\epsilon$, $D`(P_\epsilon \dot{\cap} I)$, then $s`\lambda$ is a member of that class (∗211·63·64·65). Hence the propositions of ∗210 become available. It is thus that the existence of limits in series of segments or sections is proved: the maximum or upper limit of any class λ consisting of segments or sections is $s`\lambda$, and the minimum or lower limit is the sum of the segments that are contained in every λ.

We begin, in this number, with elementary properties of sect$'P$. The sections of P are the segments of P_* (*211·13) and the sections of P_{po} (*211·17). We have

*211·26. $\vdash . C'P \epsilon \text{sect}'P . s'\text{sect}'P = C'P$

We then proceed to the elementary properties of segments, *i.e.* of $\text{D}'P_\epsilon$ (*211·3—·38). We have

*211·3. $\vdash . \overrightarrow{P}``C'P \subset \text{D}'P_\epsilon$

*211·301. $\vdash . \text{D}'P \epsilon \text{D}'P_\epsilon$

*211·302. $\vdash : P \epsilon \text{Ser} . \supset . \overrightarrow{P}``C'P = \text{sect}'P \cap \text{Ɑ}'\text{seq}_P$

*211·351. $\vdash : P \epsilon \text{Ser} . \supset . \text{sect}'P - \text{D}'P_\epsilon = \overrightarrow{P_*}``(C'P - \text{D}'P_1)$

We then proceed to elementary properties of segments having no maximum, *i.e.* of $\text{D}'(P_\epsilon \dot{\cap} I)$ (*211·4—·47). We have

*211·42. $\vdash : P \epsilon \text{trans} . \supset . \text{D}'(P_\epsilon \dot{\cap} I) = \text{D}'P_\epsilon - \text{Ɑ}'\text{max}_P$

*211·44. $\vdash . \Lambda \epsilon \text{D}'(P_\epsilon \dot{\cap} I) . \Lambda \epsilon \text{D}'P_\epsilon . \Lambda \epsilon \text{sect}'P$

*211·451. $\vdash : \overrightarrow{P}'x \epsilon \text{D}'(P_\epsilon \dot{\cap} I) . \supset . x \sim \epsilon \text{Ɑ}'(P \dot{-} P^2)$

Our next set of propositions (*211·5—·553) is concerned with compact series, *i.e.* with the hypothesis $P^2 = P$. We have

*211·51. $\vdash : P^2 = P . \supset . \text{D}'P_\epsilon = \text{D}'(P_\epsilon \dot{\cap} I)$

*211·551. $\vdash :. P \epsilon \text{Ser} . \supset : \text{Ɑ}'\text{max}_P \cap \text{Ɑ}'\text{seq}_P = \Lambda . \equiv . P = P^2$

I.e. a series is compact when, and only when, no class has both a maximum and a sequent.

We come next to the application of the propositions of *210 (*211·56—·692). These propositions proceed from

*211·56. $\vdash :. P \epsilon \text{connex} . \alpha, \beta \epsilon \text{sect}'P . \supset : \alpha \subset \beta . \text{v} . \beta \subset P``\alpha$

(Here "$P_{po} \epsilon \text{connex}$" may be substituted in the hypothesis: cf. *211·561.) The propositions of this set, which are very important, have been already mentioned.

Our next set of propositions (*211·7—·762) are concerned with the complements of sections and segments. Some of these propositions have been already mentioned; others of importance are:

*211·7. $\vdash : \alpha \epsilon \text{sect}'P . \supset . C'P - \alpha \epsilon \text{sect}'\breve{P}$

*211·703. $\vdash : P \epsilon \text{connex} . \alpha \epsilon \text{sect}'P - \iota'C'P . \supset . \exists ! p'\overleftarrow{P}``\alpha$

*211·726. $\vdash : P \epsilon \text{connex} \cap \text{Rl}'J . \alpha \epsilon \text{sect}'P . \supset .$

$$\overrightarrow{\text{max}_P}'\alpha = \overrightarrow{\text{prec}_P}'(C'P - \alpha) . \overrightarrow{\text{seq}_P}'\alpha = \overrightarrow{\text{min}_P}'(C'P - \alpha)$$

***211·727.** $\vdash :. P \epsilon \text{connex} \cap \text{Rl}`J \,.\, \alpha \epsilon \text{sect}`P \,.\, \supset :$

$$\text{E}\,!\, \text{limax}_P`\alpha \,.\, \equiv .\, \text{E}\,!\, \text{limin}_P`(C`P - \alpha)$$

***211·728.** $\vdash :. P \epsilon \text{connex} \cap \text{Rl}`J \,.\, \alpha \epsilon \text{sect}`P : \sim \text{E}\,!\, \text{max}_P`\alpha \,.\, \text{v} \,.$

$$\sim \text{E}\,!\, \text{min}_P`(C`P - \alpha) : \supset .\, \overrightarrow{\text{limax}}_P`\alpha = \overrightarrow{\text{limin}}_P`(C`P - \alpha)$$

The remaining propositions are mainly occupied with relation-arithmetic. The most important of them is

***211·82.** $\vdash :: P \epsilon \text{Ser} \,.\, Q \epsilon \text{D}`P \,\char"005B\, .\, \supset :.$

$$C`Q \epsilon \text{sect}`P \,.\, \equiv : (\exists R) \,.\, P = Q \dotplus R \,.\, \text{v} \,.\, (\exists x) \,.\, P = Q \,+\!\!\!\!\rightarrow x :$$

$$\equiv : (\exists R) \,.\, P = Q \dotplus R \,.\, \text{v} \,.\, P = Q \,+\!\!\!\!\rightarrow B`\breve{P}$$

That is, given any series contained in P, if something can be added to make it into P, its field is a section of P, and vice versa.

***211·01.** $\text{sect}`P = \hat{\alpha}(\alpha \subset C`P \,.\, P``\alpha \subset \alpha)$ Df

***211·1.** $\vdash : \alpha \epsilon \text{sect}`P \,.\, \equiv .\, \alpha \subset C`P \,.\, P``\alpha \subset \alpha$ [(*211·01)]

***211·11.** $\vdash : \alpha \epsilon \text{D}`P_\epsilon \,.\, \equiv .\, (\exists \beta) \,.\, \alpha = P``\beta$ [*37·101]

***211·12.** $\vdash : \alpha \epsilon \text{D}`(P_\epsilon \dot{\cap} I) \,.\, \equiv .\, \alpha = P``\alpha$

Dem.

$\vdash .\, *37·101 \,.\, *50·1 \,.\, \supset \vdash : \alpha \epsilon \text{D}`(P_\epsilon \dot{\cap} I) \,.\, \equiv .\, (\exists \beta) \,.\, \alpha = P``\beta \,.\, \alpha = \beta \,.$

[*13·195] $\equiv .\, \alpha = P``\alpha : \supset \vdash .\, \text{Prop}$

***211·13.** $\vdash : \alpha \epsilon \text{sect}`P \,.\, \equiv .\, \alpha = P_*``\alpha \,.\, \equiv .\, \alpha \epsilon \text{D}`\{(P_*)_\epsilon \dot{\cap} I\} \,.\, \equiv .\, \alpha \epsilon \text{D}`(P_*)_\epsilon$

Dem.

$\vdash .\, *211·1 \,.\, *90·23 \,.\, \supset \vdash : \alpha \epsilon \text{sect}`P \,.\, \equiv .\, \alpha = P_*``\alpha$ (1)

$\vdash .\, *90·17 \,.\, \supset \vdash .\, P_*``P_*``\beta = P_*``\beta \,.$

[*13·12] $\supset \vdash :. \alpha = P_*``\beta \,.\, \supset .\, P_*``\alpha = \alpha :$

[*211·11] $\supset \vdash : \alpha \epsilon \text{D}`(P_*)_\epsilon \,.\, \supset .\, \alpha = P_*``\alpha$ (2)

$\vdash .\, *10·24 \,.\, *211·11 \,.\, \supset \vdash : \alpha = P_*``\alpha \,.\, \supset .\, \alpha \epsilon \text{D}`(P_*)_\epsilon$ (3)

$\vdash .\, (1) \,.\, (2) \,.\, (3) \,.\, *211·12 \,.\, \supset \vdash .\, \text{Prop}$

In virtue of the above proposition, the properties of sect$`P$ can be deduced from those of $\text{D}`P_\epsilon$ or $\text{D}`(P_\epsilon \dot{\cap} I)$ by substituting P_* for P.

***211·131.** $\vdash : \alpha \epsilon \text{sect}`P \,.\, \supset .\, P``\alpha = P_{\text{po}}``\alpha$

Dem.

$\vdash .\, *211·13 \,.\, \supset \vdash : \text{Hp} \,.\, \supset .\, P``\alpha = P``P_*``\alpha$

[*91·52] $= P_{\text{po}}``\alpha : \supset \vdash .\, \text{Prop}$

$*211{\cdot}132$. $\vdash : \alpha \epsilon \text{sect}`P . \supset . \mathrm{D}`(P \upharpoonright \alpha) = \mathrm{D}`(P_{\text{po}} \upharpoonright \alpha) . \text{ꓷ}`(P \upharpoonright \alpha) = \text{ꓷ}`(P_{\text{po}} \upharpoonright \alpha) .$
$C`(P \upharpoonright \alpha) = C`(P_{\text{po}} \upharpoonright \alpha)$

Dem.

$\vdash . *37{\cdot}41 . *211{\cdot}131 . \supset \vdash : \text{Hp} . \supset . \mathrm{D}`(P_{\text{po}} \upharpoonright \alpha) = \alpha \cap P``\alpha$

$[*37{\cdot}41] \qquad = \mathrm{D}`(P \upharpoonright \alpha) \quad (1)$

$\vdash . *91{\cdot}502 . \qquad \supset \vdash . \text{ꓷ}`(P \upharpoonright \alpha) \subset \text{ꓷ}`(P_{\text{po}} \upharpoonright \alpha) \quad (2)$

$\vdash . *37{\cdot}41 . \qquad \supset \vdash :. y \epsilon \text{ꓷ}`(P_{\text{po}} \upharpoonright \alpha) . \equiv : y \epsilon \alpha \cap \breve{P}_{\text{po}}``\alpha :$

$[*91{\cdot}57] \qquad \equiv : y \epsilon (\alpha \cap \breve{P}``\alpha) \cup (\alpha \cap \breve{P}``\breve{P}_{\text{po}}``\alpha) \quad (3)$

$\vdash . *211{\cdot}1 . \qquad \supset \vdash : \text{Hp} . y \epsilon \alpha \cap \breve{P}``\breve{P}_{\text{po}}``\alpha . \supset . (\exists z) . zPy . z \epsilon \alpha .$

$[*37{\cdot}105] \qquad \supset . y \epsilon \breve{P}``\alpha \quad (4)$

$\vdash . (3) . (4) . \qquad \supset \vdash : \text{Hp} . \supset . \text{ꓷ}`(P_{\text{po}} \upharpoonright \alpha) \subset \alpha \cap \breve{P}``\alpha .$

$[*37{\cdot}41] \qquad \supset . \text{ꓷ}`(P_{\text{po}} \upharpoonright \alpha) \subset \text{ꓷ}`(P \upharpoonright \alpha) \quad (5)$

$\vdash . (2) . (5) . \qquad \supset \vdash : \text{Hp} . \supset . \text{ꓷ}`(P_{\text{po}} \upharpoonright \alpha) = \text{ꓷ}`(P \upharpoonright \alpha) \quad (6)$

$\vdash . (1) . (6) . \supset \vdash . \text{Prop}$

$*211{\cdot}133$. $\vdash : P_{\text{po}} \epsilon \text{connex} . \alpha \epsilon \text{sect}`P - \iota`\Lambda . \supset . C`(P \upharpoonright \alpha) = \alpha$

Dem.

$\vdash . *202{\cdot}55 . \supset \vdash : \text{Hp} . \supset . C`(P_{\text{po}} \upharpoonright \alpha) = \alpha .$

$[*211{\cdot}132] \qquad \supset . C`(P \upharpoonright \alpha) = \alpha : \supset \vdash . \text{Prop}$

$*211{\cdot}14$. $\vdash . \mathrm{D}`(P_\epsilon \dot{\cap} I) \subset \mathrm{D}`P_\epsilon . \mathrm{D}`(P_\epsilon \dot{\cap} I) \subset \text{sect}`P$

Dem.

$\vdash . *33{\cdot}263 . \qquad \supset \vdash . \mathrm{D}`(P_\epsilon \dot{\cap} I) \subset \mathrm{D}`P_\epsilon \quad (1)$

$\vdash . *211{\cdot}12 . *22{\cdot}42 . \supset \vdash : \alpha \epsilon \mathrm{D}`(P_\epsilon \dot{\cap} I) . \supset . P``\alpha \subset \alpha \quad (2)$

$\vdash . *211{\cdot}12 . *37{\cdot}15 . \supset \vdash : \alpha \epsilon \mathrm{D}`(P_\epsilon \dot{\cap} I) . \supset . \alpha \subset C`P \quad (3)$

$\vdash . (2) . (3) . *211{\cdot}1 . \supset \vdash : \alpha \epsilon \mathrm{D}`(P_\epsilon \dot{\cap} I) . \supset . \alpha \epsilon \text{sect}`P \quad (4)$

$\vdash . (1) . (4) . \supset \vdash . \text{Prop}$

$*211{\cdot}15$. $\vdash : P \epsilon \text{trans} . \supset . \mathrm{D}`P_\epsilon \subset \text{sect}`P$

Dem.

$\vdash . *211{\cdot}11 . *37{\cdot}15 . \supset \vdash : \alpha \epsilon \mathrm{D}`P_\epsilon . \supset . \alpha \subset C`P \quad (1)$

$\vdash . *211{\cdot}11 . *201{\cdot}5 . \supset \vdash : P \epsilon \text{trans} . \alpha \epsilon \mathrm{D}`P_\epsilon . \supset . P``\alpha \subset \alpha \quad (2)$

$\vdash . (1) . (2) . \supset \vdash . \text{Prop}$

$*211{\cdot}16$. $\vdash . P_{\text{po}}``\alpha \epsilon \text{sect}`P$

Dem.

$\vdash . *91{\cdot}504 . *37{\cdot}15 . \supset \vdash . P_{\text{po}}``\alpha \subset C`P \quad (1)$

$\vdash . *91{\cdot}51{\cdot}511 . \qquad \supset \vdash . P``P_{\text{po}}``\alpha \subset P_{\text{po}}``\alpha \quad (2)$

$\vdash . (1) . (2) . *211{\cdot}1 . \supset \vdash . \text{Prop}$

***211·17.** $\vdash . \text{sect}`P = \text{sect}`P_{\text{po}} = \text{sect}`P_*$

[*211·13 . *90·4 . *91·602]

The following propositions are useful in dealing with sectional relations, *i.e.* relations of the form $P \upharpoonright \alpha$, where $\alpha \in \text{sect}`P$. Unit sections often need special treatment, owing to the fact that for them we do not have $C`P \upharpoonright \alpha = \alpha$.

***211·18.** $\vdash : P_{\text{po}} \in J . \supset . \text{sect}`P \cap 1 = \iota``\overrightarrow{B}`P$

Dem.

$\vdash . *211·13 . \supset \vdash : \alpha \in \text{sect}`P \cap 1 . \equiv . \alpha = P_*``\alpha . \alpha \in 1 .$

[*52·1.*53·301] $\equiv . (\exists x) . \alpha = \iota`x . \overrightarrow{P_*}`x = \iota`x .$

[*91·54.*90·12] $\equiv . (\exists x) . \alpha = \iota`x . \overrightarrow{P_{\text{po}}}`x \subset \iota`x . x \in C`P$ (1)

$\vdash . (1) . \supset \vdash :. \text{Hp} . \supset : \alpha \in \text{sect}`P \cap 1 . \equiv . (\exists x) . \alpha = \iota`x . \overrightarrow{P_{\text{po}}}`x = \Lambda . x \in C`P .$

[*91·504] $\equiv . (\exists x) . \alpha = \iota`x . x \sim \in \text{Cl}`P . x \in C`P .$

[*93·103] $\equiv . \alpha \in \iota``\overrightarrow{B}`P :. \supset \vdash . \text{Prop}$

***211·181.** $\vdash : P_{\text{po}} \in \text{Ser} . \exists ! \overrightarrow{B}`P . \supset . \text{sect}`P \cap 1 = \iota`\iota`B`P$

Dem.

$\vdash . *202·13·523 . \supset \vdash : \text{Hp} . \supset . \overrightarrow{B}`P \in 1$ (1)

$\vdash . (1) . *211·18 . *53·3 . \supset \vdash . \text{Prop}$

***211·182.** $\vdash : P_{\text{po}} \in \text{Ser} . \overrightarrow{B}`P = \Lambda . \supset . \text{sect}`P \cap 1 = \Lambda$ [*211·18]

***211·2.** $\vdash : \alpha \in \text{sect}`P . \supset . \alpha = \alpha \cap C`P = \alpha \cup P``\alpha = (\alpha \cap C`P) \cup P``\alpha = P``\alpha \cup \overrightarrow{\max_P}`\alpha$

Dem.

$\vdash . *211·1 . *22·621·62 . \supset \vdash : \text{Hp} . \supset . \alpha = \alpha \cap C`P . \alpha = \alpha \cup P``\alpha .$ (1)

[*13·12] $\supset . \alpha = (\alpha \cap C`P) \cup P``\alpha$ (2)

[*205·131] $= P``\alpha \cup \overrightarrow{\max_P}`\alpha$ (3)

$\vdash . (1) . (2) . (3) . \supset \vdash . \text{Prop}$

***211·21.** $\vdash :. \alpha \in \text{sect}`P . \supset : \sim \exists ! \overrightarrow{\max_P}`\alpha . \equiv . \alpha \in \text{D}`(P_\epsilon \dot{\cap} I)$

Dem.

$\vdash . *211·2·12 .$ $\supset \vdash : \alpha \in \text{sect}`P . \sim \exists ! \overrightarrow{\max_P}`\alpha . \supset . \alpha \in \text{D}`(P_\epsilon \dot{\cap} I)$ (1)

$\vdash . *211·12 . *205·111 . \supset \vdash : \alpha \in \text{D}`(P_\epsilon \dot{\cap} I) . \supset . \sim \exists ! \overrightarrow{\max_P}`\alpha$ (2)

$\vdash . (1) . (2) . \supset \vdash . \text{Prop}$

***211·22.** $\vdash : P \epsilon \text{connex} . \alpha \epsilon \text{sect}‘P . \supset . \alpha \cup \overrightarrow{\text{seq}}_P‘\alpha \epsilon \text{sect}‘P$

Dem.

$\vdash . *24\cdot24 . *13\cdot12 . \quad \supset \vdash : \text{Hp} . \overrightarrow{\text{seq}}_P‘\alpha = \Lambda . \supset . \alpha \cup \overrightarrow{\text{seq}}_P‘\alpha \epsilon \text{sect}‘P \qquad (1)$

$\vdash . *206\cdot16 . *53\cdot3\cdot31 . \supset \vdash : \text{Hp} . \exists ! \overrightarrow{\text{seq}}_P‘\alpha . \supset . P``(\alpha \cup \overrightarrow{\text{seq}}_P‘\alpha) = P``\alpha \cup \overrightarrow{P}‘\text{seq}_P‘\alpha$

$[*206\cdot213] \qquad \subset P``\alpha \cup (\alpha \cap C‘P) \cup P``\alpha$

$[*211\cdot2] \qquad \subset \alpha \qquad (2)$

$\vdash . *211\cdot1 . *206\cdot18 . \quad \supset \vdash : \text{Hp} . \supset . \alpha \cup \overrightarrow{\text{seq}}_P‘\alpha \subset C‘P \qquad (3)$

$\vdash . (1) . (2) . (3) . *211\cdot1 . \supset \vdash . \text{Prop}$

***211·23.** $\vdash : P \epsilon \text{connex} . \alpha \epsilon \text{sect}‘P . \text{E} ! \text{seq}_P‘\alpha . \supset . \alpha = P``(\alpha \cup \overrightarrow{\text{seq}}_P‘\alpha) = \overrightarrow{P}‘\text{seq}_P‘\alpha$

Dem.

$\vdash . *206\cdot211 . *211\cdot2 . \supset \vdash : \text{Hp} . \supset . \alpha \subset \overrightarrow{P}‘\text{seq}_P‘\alpha \qquad (1)$

$\vdash . *206\cdot213 . *211\cdot2 . \supset \vdash : \text{Hp} . \supset . \overrightarrow{P}‘\text{seq}_P‘\alpha \subset \alpha \qquad (2)$

$\vdash . (1) . (2) . \quad \supset \vdash : \text{Hp} . \supset . \alpha = \overrightarrow{P}‘\text{seq}_P‘\alpha \qquad (3)$

$\vdash . *53\cdot3\cdot31 . \quad \supset \vdash : \text{Hp} . \supset . P``(\alpha \cup \overrightarrow{\text{seq}}_P‘\alpha) = P``\alpha \cup \overrightarrow{P}‘\text{seq}_P‘\alpha$

$[(3)] \qquad = P``\alpha \cup \alpha$

$[*211\cdot2] \qquad = \alpha \qquad (4)$

$\vdash . (3) . (4) . \supset \vdash . \text{Prop}$

***211·24.** $\vdash : P \epsilon \text{connex} . \alpha \epsilon \text{sect}‘P \cap (\text{Cl}‘\text{seq}_P \cup - \text{Cl}‘\text{max}_P) . \supset . \alpha \epsilon \text{D}‘P_\epsilon$

Dem.

$\vdash . *211\cdot23\cdot11 . \supset \vdash : P \epsilon \text{connex} . \alpha \epsilon \text{sect}‘P \cap \text{Cl}‘\text{seq}_P . \supset . \alpha \epsilon \text{D}‘P_\epsilon \qquad (1)$

$\vdash . *211\cdot21\cdot14 . \supset \vdash : \alpha \epsilon \text{sect}‘P - \text{Cl}‘\text{max}_P . \supset . \alpha \epsilon \text{D}‘P_\epsilon \qquad (2)$

$\vdash . (1) . (2) . \supset \vdash . \text{Prop}$

***211·26.** $\vdash . C‘P \epsilon \text{sect}‘P . s‘\text{sect}‘P = C‘P$

Dem.

$\vdash . *22\cdot42 . *37\cdot15 . \quad \supset \vdash . C‘P \subset C‘P . P``C‘P \subset C‘P .$

$[*211\cdot1] \qquad \supset \vdash . C‘P \epsilon \text{sect}‘P \qquad (1)$

$\vdash . (1) . *40\cdot13 . \quad \supset \vdash . C‘P \subset s‘\text{sect}‘P \qquad (2)$

$\vdash . *40\cdot151 . *211\cdot1 . \supset \vdash . s‘\text{sect}‘P \subset C‘P \qquad (3)$

$\vdash . (2) . (3) . \quad \supset \vdash . s‘\text{sect}‘P = C‘P \qquad (4)$

$\vdash . (1) . (4) . \supset \vdash . \text{Prop}$

***211·27.** $\vdash : P \epsilon \text{trans} . \supset . (\alpha \cap C‘P) \cup P``\alpha \epsilon \text{sect}‘P$

Dem.

$\vdash . *22\cdot43 . *37\cdot15 . \supset \vdash . (\alpha \cap C‘P) \cup P``\alpha \subset C‘P \qquad (1)$

$\vdash . *37\cdot22\cdot265 . \quad \supset \vdash . P``\{(\alpha \cap C‘P) \cup P``\alpha\} = P``\alpha \cup P``P``\alpha \qquad (2)$

$\vdash . (2) . *201\cdot5 . \quad \supset \vdash : \text{Hp} . \supset . P``\{(\alpha \cap C‘P) \cup P``\alpha\} = P``\alpha \qquad (3)$

$\vdash . (1) . (3) . *211\cdot1 . \supset \vdash . \text{Prop}$

***211·271.** $\vdash : P \epsilon \text{trans} . \supset . (\exists \beta) . \beta \epsilon \text{sect}'P . \overrightarrow{\max}_P{}'\alpha = \overrightarrow{\max}_P{}'\beta . \overrightarrow{\text{seq}}_P{}'\alpha = \overrightarrow{\text{seq}}_P{}'\beta$

Dem.

$\vdash . *205·15·19 . \supset$

$\vdash : \text{Hp} . \supset . \overrightarrow{\max}_P{}'\alpha = \overrightarrow{\max}_P{}'\{(\alpha \cap C'P) \cup P''\alpha\}$ (1)

$\vdash . *206·131·25 . \supset$

$\vdash : \text{Hp} . \supset . \overrightarrow{\text{seq}}_P{}'\alpha = \overrightarrow{\text{seq}}_P{}'\{(\alpha \cap C'P) \cup P''\alpha\}$ (2)

$\vdash . (1) . (2) . *211·27 . \supset \vdash . \text{Prop}$

***211·272.** $\vdash :. P \epsilon \text{trans} . \supset :$

$(\alpha) . \alpha \epsilon \text{Cl}'\max_P \cup \text{Cl}'\text{seq}_P . \equiv . \text{sect}'P \subset \text{Cl}'\max_P \cup \text{Cl}'\text{seq}_P$

Dem.

$\vdash . *24·11·14 . \supset \vdash : (\alpha) . \alpha \epsilon \text{Cl}'\max_P \cup \text{Cl}'\text{seq}_P . \supset . \text{sect}'P \subset \text{Cl}'\max_P \cup \text{Cl}'\text{seq}_P$ (1)

$\vdash . *33·41 . \quad \supset \vdash :. \text{sect}'P \subset \text{Cl}'\max_P \cup \text{Cl}'\text{seq}_P . \supset :$

$\beta \epsilon \text{sect}'P . \supset_\beta . \exists ! (\overrightarrow{\max}_P{}'\beta \cup \overrightarrow{\text{seq}}_P{}'\beta) :$

[*13·12] $\supset : \beta \epsilon \text{sect}'P . \overrightarrow{\max}_P{}'\alpha = \overrightarrow{\max}_P{}'\beta . \overrightarrow{\text{seq}}_P{}'\alpha = \overrightarrow{\text{seq}}_P{}'\beta . \supset_{\alpha,\beta} .$

$\exists ! (\overrightarrow{\max}_P{}'\alpha \cup \overrightarrow{\text{seq}}_P{}'\alpha) :$

[*10·23] $\supset : (\exists \beta) . \beta \epsilon \text{sect}'P . \overrightarrow{\max}_P{}'\alpha = \overrightarrow{\max}_P{}'\beta . \overrightarrow{\text{seq}}_P{}'\alpha = \overrightarrow{\text{seq}}_P{}'\beta . \supset_\alpha .$

$\exists ! (\overrightarrow{\max}_P{}'\alpha \cup \overrightarrow{\text{seq}}_P{}'\alpha)$ (2)

$\vdash . (2) . *211·271 . \supset$

$\vdash :. \text{Hp} . \supset : \text{sect}'P \subset \text{Cl}'\max_P \cup \text{Cl}'\text{seq}_P . \supset . (\alpha) . \exists ! (\overrightarrow{\max}_P{}'\alpha \cup \overrightarrow{\text{seq}}_P{}'\alpha) .$

[*33·41] $\supset . (\alpha) . \alpha \epsilon \text{Cl}'\max_P \cup \text{Cl}'\text{seq}_P$ (3)

$\vdash . (1) . (3) . \supset \vdash . \text{Prop}$

***211·28.** $\vdash :. P \epsilon \text{Ser} . \alpha \subset C'P . \alpha \sim\epsilon 1 . (C'P - \alpha) \sim\epsilon 1 . \supset :$

$\alpha \epsilon \text{sect}'P . \equiv . P = P \upharpoonright \alpha \mathbin{\updownarrow} P \upharpoonright (C'P - \alpha)$

Dem.

$\vdash . *204·45 . \quad \supset \vdash : \text{Hp} . \alpha \epsilon \text{sect}'P . \supset . P = P \upharpoonright \alpha \mathbin{\updownarrow} P \upharpoonright (C'P - \alpha)$ (1)

$\vdash . *160·1 . *202·55 . \supset \vdash :. \text{Hp} . P = P \upharpoonright \alpha \mathbin{\updownarrow} P \upharpoonright (C'P - \alpha) . \supset :$

$x \epsilon \alpha . y \epsilon C'P - \alpha . \supset . xPy :$

[Transp.*204·3] $\supset : x \epsilon \alpha . yPx . \supset . y \epsilon \alpha :$

[*211·1] $\supset : \alpha \epsilon \text{sect}'P$ (2)

$\vdash . (1) . (2) . \supset \vdash . \text{Prop}$

***211·281.** $\vdash : P \epsilon \text{Ser} . C'Q \cap C'R = \Lambda . P = Q \mathbin{\updownarrow} R . \supset . C'Q \epsilon \text{sect}'P$

Dem.

$\vdash . *160·1 . \supset \vdash :. \text{Hp} . \supset : x \epsilon C'Q . y \epsilon C'R . \supset . xPy :$

[Transp.*204·3] $\supset : x \epsilon C'Q . yPx . \supset . y \epsilon C'Q :$

[*211·1] $\supset : C'Q \epsilon \text{sect}'P :. \supset \vdash . \text{Prop}$

***211·282.** $\vdash :. P \in \mathrm{Ser} \,.\, Q \in \mathrm{D}`P\lceil \,.\, C`P - C`Q \sim\in 1 \,.\, \supset :$

$$C`Q \in \mathrm{sect}`P \,.\equiv.\, (\exists R) \,.\, C`Q \cap C`R = \Lambda \,.\, P = Q \mathbin{\overset{\uparrow}{+}} R$$

[*211·28·281 . *200·12]

***211·283.** $\vdash : P \subseteq J \,.\, P = Q \mathbin{\overset{\uparrow}{+}} R \,.\supset.\, C`Q \cap C`R = \Lambda$

Dem.

$\vdash .\, *160{\cdot}1 \,.\supset \vdash : \mathrm{Hp} \,.\supset.\, C`Q \uparrow C`R \subseteq J \,.$

$[*200{\cdot}32] \qquad \supset.\, C`Q \cap C`R = \Lambda : \supset \vdash .\, \mathrm{Prop}$

The following propositions are concerned with $\mathrm{D}`P_\epsilon$. This is to be compared with two other classes, namely $\mathrm{sect}`P$ and $\overrightarrow{P}``C`P$. The members of $\mathrm{sect}`P$ which do not belong to $\mathrm{D}`P_\epsilon$ are those which have a maximum but no sequent, *i.e.* (if $P$ is a series), those classes which consist of a term $x$ together with all its predecessors, where $x$ has no immediate successor. In series in which every term except the last has an immediate successor, $C`P$ will be the only member of $\mathrm{sect}`P - \mathrm{D}`P_\epsilon$, if the series has a last term; if the series has no last term, $\mathrm{sect}`P = \mathrm{D}`P_\epsilon$.

The members of $\mathrm{D}`P_\epsilon$ which are not members of $\overrightarrow{P}``C`P$ are those that have no sequent, *i.e.* those that have no upper limit (for a member of $\mathrm{D}`P_\epsilon$ which has no sequent has also no maximum). These are the members of $\mathrm{D}`P_\epsilon$ corresponding to a "gap," *i.e.* to a Dedekind section in which neither the earlier terms have a maximum nor the later terms a minimum. Hence in a Dedekindian series, $\mathrm{D}`P_\epsilon = \overrightarrow{P}``C`P$; and conversely, if $\mathrm{D}`P_\epsilon = \overrightarrow{P}``C`P$, the series is Dedekindian. These properties of $\mathrm{D}`P_\epsilon$ are proved in the following propositions.

***211·3.** $\vdash .\, \overrightarrow{P}``C`P \subset \mathrm{D}`P_\epsilon$ [*53·301 . *211·11]

***211·301.** $\vdash .\, \mathrm{D}`P \in \mathrm{D}`P_\epsilon$ [*37·25 . *211·11]

***211·302.** $\vdash : P \in \mathrm{Ser} \,.\supset.\, \overrightarrow{P}``C`P = \mathrm{sect}`P \cap \mathrm{Cl}`\mathrm{seq}_P$

Dem.

$\vdash .\, *206{\cdot}4 \,. \quad \supset \vdash : \mathrm{Hp} \,.\supset.\, \overrightarrow{P}``C`P \subset \mathrm{Cl}`\mathrm{seq}_P \qquad (1)$

$\vdash .\, *211{\cdot}3{\cdot}15 \,.\supset \vdash : \mathrm{Hp} \,.\supset.\, \overrightarrow{P}``C`P \subset \mathrm{sect}`P \qquad (2)$

$\vdash .\, *211{\cdot}23 \,. \quad \supset \vdash : \mathrm{Hp} \,.\supset.\, \mathrm{sect}`P \cap \mathrm{Cl}`\mathrm{seq}_P \subset \overrightarrow{P}``C`P \qquad (3)$

$\vdash .\, (1) \,.\, (2) \,.\, (3) \,.\supset \vdash .\, \mathrm{Prop}$

***211·31.** $\vdash :. P \in \mathrm{trans} \cap \mathrm{connex} \,.\, \alpha \in \mathrm{D}`P_\epsilon \,.\supset : \mathrm{E}\,!\, \mathrm{seq}_P`\alpha \,.\vee.\, \sim\mathrm{E}\,!\, \mathrm{max}_P`\alpha$

[*206·52 . *211·11]

***211·311.** $\vdash : P \in \mathrm{trans} \cap \mathrm{connex} \,.\, \alpha \in \mathrm{D}`P_\epsilon \,.\, \mathrm{E}\,!\, \mathrm{seq}_P`\alpha \,.\supset.\, \alpha = \overrightarrow{P}`\mathrm{seq}_P`\alpha$

[*206·31 . *211·11]

$*211\cdot312.\ \vdash : P \epsilon \text{trans} \cap \text{connex} . \alpha \epsilon D'P_\epsilon . \supset . \alpha = P``(\alpha \cup \overrightarrow{\text{seq}}_P`\alpha)$

Dem.

$\vdash . *211\cdot15\cdot23 . \supset \vdash : \text{Hp} . E ! \text{seq}_P`\alpha . \supset . \alpha = P``(\alpha \cup \overrightarrow{\text{seq}}_P`\alpha) \quad (1)$

$\vdash . *211\cdot31 . \quad \supset \vdash : \text{Hp} . \sim E ! \text{seq}_P`\alpha . \supset . \sim E ! \text{max}_P`\alpha .$

$[*211\cdot21\cdot15\cdot12] \quad \supset . \alpha = P``\alpha .$

$[*24\cdot24.\text{Hp}] \quad \supset . \alpha = P``(\alpha \cup \overrightarrow{\text{seq}}_P`\alpha) \quad (2)$

$\vdash . (1) . (2) . \supset \vdash . \text{Prop}$

$*211\cdot313.\ \vdash : \alpha \epsilon \text{sect}`P \cap D'P_\epsilon . \supset . (\exists \beta) . \beta \epsilon \text{sect}`P . \alpha = P``\beta$

Dem.

$\vdash . *211\cdot1\cdot11 . \supset \vdash :. \text{Hp} . \supset : P``\alpha \subset \alpha : (\exists \beta) . \alpha = P``\beta :$

$[*37\cdot265] \quad \supset : P``\alpha \subset \alpha : (\exists \beta) . \beta \subset C`P . \alpha = P``\beta :$

$[*22\cdot62] \quad \supset : (\exists \beta) . \beta \subset C`P . \alpha = P``(\alpha \cup \beta) :$

$[*22\cdot58] \quad \supset : (\exists \beta) . \beta \subset C`P . P``(\alpha \cup \beta) \subset \alpha \cup \beta . \alpha = P``(\alpha \cup \beta) :$

$[*37\cdot15] \quad \supset : (\exists \beta) . \alpha \cup \beta \subset C`P . P``(\alpha \cup \beta) \subset \alpha \cup \beta . \alpha = P``(\alpha \cup \beta) :$

$[*211\cdot1] \quad \supset : (\exists \beta) . \alpha \cup \beta \epsilon \text{sect}`P . \alpha = P``(\alpha \cup \beta) :. \supset \vdash . \text{Prop}$

$*211\cdot314.\ \vdash : P \epsilon \text{Rl}`J \cap \text{connex} . \alpha \epsilon \text{sect}`P \cap D'P_\epsilon . E ! \text{max}_P`\alpha . \supset . E ! \text{seq}_P`\alpha$

Dem.

$\vdash . *211\cdot313 . *205\cdot7 . \supset$

$\vdash : \text{Hp} . \supset . (\exists \beta) . \beta \epsilon \text{sect}`P . \alpha = P``\beta . E ! \text{max}_P`\beta \quad (1)$

$\vdash . *37\cdot18 . \supset$

$\vdash : \beta \epsilon \text{sect}`P . \alpha = P``\beta . E ! \text{max}_P`\beta . \supset . \overrightarrow{P}`\text{max}_P`\beta \subset \alpha \quad (2)$

$\vdash . *211\cdot1 . *205\cdot111 . \supset$

$\vdash : \text{Hp}(2) . P \epsilon \text{connex} . y \epsilon P``\beta . \supset . y \epsilon \beta - \iota`\text{max}_P`\beta .$

$[*205\cdot21] \quad \supset . y P \text{max}_P`\beta \quad (3)$

$\vdash . (2) . (3) . \supset \vdash : \text{Hp}(2) . P \epsilon \text{connex} . \supset . \alpha = \overrightarrow{P}`\text{max}_P`\beta .$

$[*206\cdot4] \quad \supset . \text{max}_P`\beta \text{ seq}_P \alpha \quad (4)$

$\vdash . (1) . (4) . \supset \vdash . \text{Prop}$

The above proposition and the two following propositions enable us in certain cases to prove propositions concerning the relations of sect$`P$ and $D`P_\epsilon$ without assuming that P is transitive. An example of the use of these propositions occurs in $*211\cdot754$, where the hypothesis assumes $P \epsilon \text{Rl}`J \cap \text{connex}$. If we used $*211\cdot31$ and its consequences instead of $*211\cdot314$ and its consequences, the hypothesis of $*211\cdot754$ would have to assume $P \epsilon \text{Ser}$.

***211·315.** $\vdash :. P \epsilon \mathrm{Rl}`J \cap \mathrm{connex} . \alpha \epsilon \mathrm{sect}`P . \supset :$

$$\alpha \epsilon \mathrm{D}`P_\epsilon . \equiv . \alpha \epsilon \mathrm{C\!\!\!I}`\mathrm{seq}_P \cup - \mathrm{C\!\!\!I}`\mathrm{max}_P$$

Dem.

$\vdash . *211·314 . \supset \vdash :. \mathrm{Hp} . \supset : \alpha \epsilon \mathrm{D}`P_\epsilon . \supset . \alpha \epsilon \mathrm{C\!\!\!I}`\mathrm{seq}_P \cup - \mathrm{C\!\!\!I}`\mathrm{max}_P \quad (1)$

$\vdash . (1) . *211·24 . \supset \vdash . \mathrm{Prop}$

***211·316.** $\vdash : P \epsilon \mathrm{Rl}`J \cap \mathrm{connex} . \supset . \mathrm{sect}`P - \mathrm{D}`P_\epsilon = \mathrm{sect}`P \cap \mathrm{C\!\!\!I}`\mathrm{max}_P - \mathrm{C\!\!\!I}`\mathrm{seq}_P$

[*211·315 . Transp]

***211·317.** $\vdash : P \epsilon \mathrm{trans} . \supset . \mathrm{D}`P_\epsilon = P_\epsilon``\mathrm{sect}`P$

Dem.

$\vdash . *211·15·313 . \supset \vdash : \mathrm{Hp} . \supset . \mathrm{D}`P_\epsilon \subset P_\epsilon``\mathrm{sect}`P \quad (1)$

$\vdash . (1) . *37·15 . \supset \vdash . \mathrm{Prop}$

***211·32.** $\vdash : P \epsilon \mathrm{trans} \cap \mathrm{connex} . \supset . \mathrm{D}`P_\epsilon = \mathrm{sect}`P \cap (\mathrm{C\!\!\!I}`\mathrm{seq}_P \cup - \mathrm{C\!\!\!I}`\mathrm{max}_P)$

[*211·24·15·31]

***211·321.** $\vdash : P \epsilon \mathrm{trans} \cap \mathrm{connex} . \supset . \mathrm{sect}`P - \mathrm{D}`P_\epsilon = \mathrm{sect}`P \cap \mathrm{C\!\!\!I}`\mathrm{max}_P - \mathrm{C\!\!\!I}`\mathrm{seq}_P$

[*211·32]

***211·33.** $\vdash :. P \epsilon \mathrm{Ser} . \alpha \epsilon \mathrm{sect}`P . \supset :$

$$\alpha \sim \epsilon \mathrm{D}`P_\epsilon . \supset . \mathrm{E}!\,\mathrm{seq}_P`P``\alpha . \sim \mathrm{E}!\,\mathrm{seq}_P`\overrightarrow{\mathrm{seq}_P}`P``\alpha$$

Dem.

$\vdash . *211·321 . \quad \supset \vdash : \mathrm{Hp} . \alpha \sim \epsilon \mathrm{D}`P_\epsilon . \supset . \mathrm{E}!\,\mathrm{max}_P`\alpha . \quad (1)$

$[*206·35] \quad \supset . \mathrm{E}!\,\mathrm{seq}_P`P``\alpha . \mathrm{seq}_P`P``\alpha = \mathrm{max}_P`\alpha \quad (2)$

$\vdash . (1) . *206·46 . \supset \vdash : \mathrm{Hp} . \alpha \sim \epsilon \mathrm{D}`P_\epsilon . \supset . \overrightarrow{\mathrm{seq}_P}`\alpha = \overrightarrow{\mathrm{seq}_P}`\overrightarrow{\mathrm{max}_P}`\alpha$

$[(2)] \quad = \overrightarrow{\mathrm{seq}_P}`\overrightarrow{\mathrm{seq}_P}`P``\alpha \quad (3)$

$\vdash . *211·321 . \quad \supset \vdash : \mathrm{Hp} . \alpha \sim \epsilon \mathrm{D}`P_\epsilon . \supset . \sim \mathrm{E}!\,\mathrm{seq}_P`\alpha .$

$[(3)] \quad \supset . \sim \mathrm{E}!\,\mathrm{seq}_P`\overrightarrow{\mathrm{seq}_P}`P``\alpha \quad (4)$

$\vdash . (2) . (4) . \quad \supset \vdash : \mathrm{Hp} . \alpha \sim \epsilon \mathrm{D}`P_\epsilon . \supset .$

$$\mathrm{E}!\,\mathrm{seq}_P`P``\alpha . \sim \mathrm{E}!\,\mathrm{seq}_P`\overrightarrow{\mathrm{seq}_P}`P``\alpha : \supset \vdash . \mathrm{Prop}$$

***211·34.** $\vdash :. P \epsilon \mathrm{Ser} . \supset : \alpha \epsilon \mathrm{sect}`P - \mathrm{D}`P_\epsilon . \equiv .$

$$\alpha = P``\alpha \cup \iota`\mathrm{seq}_P`P``\alpha . \sim \mathrm{E}!\,\mathrm{seq}_P`\overrightarrow{\mathrm{seq}_P}`P``\alpha$$

Dem.

$\vdash . *211·321 . \supset \vdash : \mathrm{Hp} . \alpha \epsilon \mathrm{sect}`P - \mathrm{D}`P_\epsilon . \supset . \mathrm{E}!\,\mathrm{max}_P`\alpha .$

$[*206·35] \quad \supset . \mathrm{max}_P`\alpha = \mathrm{seq}_P`P``\alpha . \quad (1)$

$[*211·2] \quad \supset . \alpha = P``\alpha \cup \iota`\mathrm{seq}_P`P``\alpha \quad (2)$

$\vdash . *211·321 . \supset \vdash : \mathrm{Hp} . \alpha \epsilon \mathrm{sect}`P - \mathrm{D}`P_\epsilon . \supset . \sim \mathrm{E}!\,\mathrm{seq}_P`\alpha .$

$[*206·46 . (1)] \quad \supset . \sim \mathrm{E}!\,\mathrm{seq}_P`\overrightarrow{\mathrm{seq}_P}`P``\alpha \quad (3)$

$\vdash . *206·21 . *205·111 . \supset \vdash : \mathrm{Hp} . \alpha = P``\alpha \cup \iota`\mathrm{seq}_P`P``\alpha . \supset .$

$$\mathrm{seq}_P`P``\alpha = \mathrm{max}_P`\alpha \quad (4)$$

$\vdash . (4) . *206{\cdot}46 . \supset \vdash : \text{Hp} . \alpha = P``\alpha \cup \iota`\text{seq}_P`P``\alpha . \sim \text{E} ! \text{seq}_P`\overrightarrow{\text{seq}_P}`P``\alpha . \supset .$
$\sim \text{E} ! \text{seq}_P`\alpha$ (5)

$\vdash . *206{\cdot}18 . *22{\cdot}58 . \supset \vdash : \alpha = P``\alpha \cup \iota`\text{seq}_P`P``\alpha . \supset . \alpha \subset C`P . P``\alpha \subset \alpha$ (6)

$\vdash . (4) . (5) . (6) . *211{\cdot}321 . \supset$

$\vdash : \text{Hp} . \alpha = P``\alpha \cup \iota`\text{seq}_P`P``\alpha . \sim \text{E} ! \text{seq}_P`\overrightarrow{\text{seq}_P}`P``\alpha . \supset . \alpha \in \text{sect}`P - \text{D}`P_\epsilon$ (7)

$\vdash . (2) . (3) . (7) . \supset \vdash . \text{Prop}$

***211·35.** $\vdash :. P \in \text{Ser} . \supset : \alpha \in \text{sect}`P - \text{D}`P_\epsilon . \equiv .$
$(\exists x) . x \in C`P . \alpha = \overrightarrow{P}`x \cup \iota`x . \sim \text{E} ! \breve{P}_1`x$

Dem.

$\vdash . *211{\cdot}34 . \supset \vdash :. \text{Hp} . \supset :$

$\alpha \in \text{sect}`P - \text{D}`P_\epsilon . \equiv . (\exists x) . x = \text{seq}_P`P``\alpha . \alpha = P``\alpha \cup \iota`x . \sim \text{E} ! \text{seq}_P`\iota`x$

[*206·21.*205·111] $\equiv . (\exists x) . x = \text{seq}_P`P``\alpha . x = \text{max}_P`\alpha .$
$\alpha = P``\alpha \cup \iota`x . \sim \text{E} ! \text{seq}_P`\iota`x .$

[*206·35] $\equiv . (\exists x) . x = \text{max}_P`\alpha . \alpha = P``\alpha \cup \iota`x . \sim \text{E} ! \text{seq}_P`\iota`x .$

[*205·22] $\equiv . (\exists x) . x = \text{max}_P`\alpha . \alpha = \overrightarrow{P}`x \cup \iota`x . \sim \text{E} ! \text{seq}_P`\iota`x .$

[*205·197] $\equiv . (\exists x) . x \in C`P . \alpha = \overrightarrow{P}`x \cup \iota`x . \sim \text{E} ! \text{seq}_P`\iota`x .$

[*206·44] $\equiv . (\exists x) . x \in C`P . \alpha = \overrightarrow{P}`x \cup \iota`x . \sim \text{E} ! \breve{P}_1`x :. \supset \vdash . \text{Prop}$

***211·351.** $\vdash : P \in \text{Ser} . \supset . \text{sect}`P - \text{D}`P_\epsilon = \overrightarrow{P_*}``(C`P - \text{D}`P_1)$

Dem.

$\vdash . *204{\cdot}7 . *211{\cdot}35 . \supset$

$\vdash :. \text{Hp} . \supset : \alpha \in \text{sect}`P - \text{D}`P_\epsilon . \equiv . (\exists x) . x \in C`P - \text{D}`P_1 . \alpha = \overrightarrow{P}`x \cup \iota`x .$

[*201·521] $\equiv . (\exists x) . x \in C`P - \text{D}`P_1 . \alpha = \overrightarrow{P_*}`x .$

[*37·7] $\equiv . \alpha \in \overrightarrow{P_*}``(C`P - \text{D}`P_1) :. \supset \vdash . \text{Prop}$

***211·36.** $\vdash :. P \in \text{Ser} . \text{D}`P_1 = \text{D}`P . \supset : \alpha \in \text{sect}`P - \text{D}`P_\epsilon . \equiv . \alpha = C`P . \text{E} ! B`\breve{P}$

Dem.

$\vdash . *211{\cdot}351 . \quad \supset \vdash : \text{Hp} . \supset . \text{sect}`P - \text{D}`P_\epsilon = \overrightarrow{P_*}``\overrightarrow{B}`\breve{P}$ (1)

$\vdash . (1) . *202{\cdot}52 . \supset \vdash :. \text{Hp} . \supset :$

$\alpha \in \text{sect}`P - \text{D}`P_\epsilon . \equiv . (\exists x) . x = B`\breve{P} . \alpha = \overrightarrow{P_*}`x .$

[*204·11.*201·521] $\equiv . (\exists x) . x = B`\breve{P} . \alpha = C`P .$

[*14·204] $\equiv . \alpha = C`P . \text{E} ! B`\breve{P} :. \supset \vdash . \text{Prop}$

$*211{\cdot}361$. $\vdash : P \epsilon \mathrm{Ser} . \mathrm{D}`P_1 = C`P . \supset . \mathrm{sect}`P = \mathrm{D}`P_\epsilon$

Dem.

$\vdash . *201{\cdot}63 . \quad \supset \vdash : \mathrm{Hp} . \supset . \mathrm{D}`P_1 \subset \mathrm{D}`P .$

$[*93{\cdot}103] \quad \supset . \overrightarrow{B}`\breve{P} = \Lambda \quad (1)$

$\vdash . (1) . *211{\cdot}36 . \supset \vdash : \mathrm{Hp} . \supset . \mathrm{sect}`P - \mathrm{D}`P_\epsilon = \Lambda \quad (2)$

$\vdash . (2) . *211{\cdot}15 . \supset \vdash . \mathrm{Prop}$

$*211{\cdot}371$. $\vdash :. P \epsilon \mathrm{trans} \cap \mathrm{connex} : (\alpha) . \alpha \epsilon \mathrm{C\!I}`\mathrm{max}_P \cup \mathrm{C\!I}`\mathrm{seq}_P : \supset . \mathrm{D}`P_\epsilon \subset \mathrm{C\!I}`\mathrm{seq}_P$
$[*211{\cdot}32]$

$*211{\cdot}372$. $\vdash :. P \epsilon \mathrm{trans} \cap \mathrm{connex} : (\alpha) . \alpha \epsilon \mathrm{C\!I}`\mathrm{max}_P \cup \mathrm{C\!I}`\mathrm{seq}_P : \supset . \mathrm{D}`P_\epsilon = \overrightarrow{P}``C`P$

Dem.

$\vdash . *211{\cdot}371 . \supset \vdash :. \mathrm{Hp} . \supset : \alpha \epsilon \mathrm{D}`P_\epsilon . \supset . \mathrm{E} ! \mathrm{seq}_P`\alpha .$

$[*206{\cdot}3 . *211{\cdot}1{\cdot}15] \quad \supset . \alpha = \overrightarrow{P}`\mathrm{seq}_P`\alpha .$

$[*206{\cdot}18] \quad \supset . \alpha \epsilon \overrightarrow{P}``C`P \quad (1)$

$\vdash . (1) . *211{\cdot}3 . \supset \vdash . \mathrm{Prop}$

$*211{\cdot}38$. $\vdash :. P \epsilon \mathrm{Ser} . \supset : (\alpha) . \alpha \epsilon \mathrm{C\!I}`\mathrm{max}_P \cup \mathrm{C\!I}`\mathrm{seq}_P . \equiv . \mathrm{D}`P_\epsilon = \overrightarrow{P}``C`P$

Dem.

$\vdash . *211{\cdot}11 . \supset \vdash :. \mathrm{D}`P_\epsilon = \overrightarrow{P}``C`P . \equiv : (\beta) : (\exists x) . P``\beta = \overrightarrow{P}`x . x \epsilon C`P \quad (1)$

$\vdash . *206{\cdot}174 . *205{\cdot}111 . \supset$

$\vdash : P \epsilon \mathrm{Ser} . \sim \exists ! \overrightarrow{\mathrm{max}}_P`\beta . \supset . \overrightarrow{\mathrm{seq}}_P`\beta = C`P \cap \hat{x} (P``\beta = \overrightarrow{P}`x) \quad (2)$

$\vdash . (1) . (2) . \supset \vdash :. P \epsilon \mathrm{Ser} . \mathrm{D}`P_\epsilon = \overrightarrow{P}``C`P . \supset : \sim \exists ! \overrightarrow{\mathrm{max}}_P`\beta . \supset . \exists ! \overrightarrow{\mathrm{seq}}_P`\beta :$

$[*33{\cdot}41] \quad \supset : \beta \epsilon \mathrm{C\!I}`\mathrm{max}_P \cup \mathrm{C\!I}`\mathrm{seq}_P \quad (3)$

$\vdash . (3) . *211{\cdot}372 . \supset \vdash . \mathrm{Prop}$

The following propositions are concerned with $\mathrm{D}`(P_\epsilon \dot{\cap} I)$, *i.e.* with those sections of $P$ which have no maximum. If $P$ is compact (*i.e.* if $P^2 = P$), $\mathrm{D}`(P_\epsilon \dot{\cap} I) = \mathrm{D}`P_\epsilon$. If $P$ is also a Dedekindian series, $\mathrm{D}`(P_\epsilon \dot{\cap} I) = \overrightarrow{P}``C`P$. This is the mark of Dedekindian continuity, since it states that, if $P``\alpha$ has no maximum, there is an x for which $P``\alpha = \overrightarrow{P}`x$, and this $x$ is the upper limit of $P``\alpha$; while conversely, if x is any term of $C`P$, $\overrightarrow{P}`x$ has no maximum, so that the series is compact.

$*211{\cdot}4$. $\vdash . \mathrm{D}`(P_\epsilon \dot{\cap} I) \subset - \mathrm{C\!I}`\mathrm{max}_P$

Dem.

$\vdash . *211{\cdot}12 . \supset \vdash : \alpha \epsilon \mathrm{D}`(P_\epsilon \dot{\cap} I) . \supset . \alpha - P``\alpha = \Lambda .$

$[*205{\cdot}111] \quad \supset . \overrightarrow{\mathrm{max}}_P`\alpha = \Lambda : \supset \vdash . \mathrm{Prop}$

***211·41.** $\vdash . \mathrm{D}`(P_\epsilon \dot{\cap} I) = \mathrm{sect}`P - \text{Ɑ}`\max_P$

Dem.

$\vdash . *211·1 . *205·111 . \supset$

$\vdash : \alpha \epsilon \mathrm{sect}`P - \text{Ɑ}`\max_P . \equiv . \alpha \subset C`P . P``\alpha \subset \alpha . \alpha \subset P``\alpha .$

[*22·41] $\equiv . \alpha \subset C`P . \alpha = P``\alpha .$

[*37·15.*211·12] $\equiv . \alpha \epsilon \mathrm{D}`(P_\epsilon \dot{\cap} I) : \supset \vdash . \mathrm{Prop}$

***211·411.** $\vdash : P \epsilon \mathrm{trans} . \alpha = P``\beta . \alpha \subset P``\alpha . \supset . \alpha = P``\alpha$

Dem.

$\vdash . *30·37 . \quad \supset \vdash : \mathrm{Hp} . \supset . P``\alpha = P``P``\beta$

[*201·5] $\subset P``\beta$

[Hp] $\subset \alpha$ (1)

$\vdash . (1) . *22·41 . \supset \vdash : \mathrm{Hp} . \supset . \alpha = P``\alpha : \supset \vdash . \mathrm{Prop}$

***211·42.** $\vdash : P \epsilon \mathrm{trans} . \supset . \mathrm{D}`(P_\epsilon \dot{\cap} I) = \mathrm{D}`P_\epsilon - \text{Ɑ}`\max_P$

Dem.

$\vdash . *211·14·4 . \quad \supset \vdash . \mathrm{D}`(P_\epsilon \dot{\cap} I) \subset \mathrm{D}`P_\epsilon - \text{Ɑ}`\max_P$ (1)

$\vdash . *211·411·11 . *205·111 . \supset \vdash : \mathrm{Hp} . \alpha \epsilon \mathrm{D}`P_\epsilon - \text{Ɑ}`\max_P . \supset . \alpha \epsilon \mathrm{D}`(P_\epsilon \dot{\cap} I)$ (2)

$\vdash . (1) . (2) . \supset \vdash . \mathrm{Prop}$

***211·43.** $\vdash : P \epsilon \mathrm{trans} \cap \mathrm{connex} . \supset . \mathrm{D}`P_\epsilon - \text{Ɑ}`\mathrm{seq}_P \subset \mathrm{D}`(P_\epsilon \dot{\cap} I)$

Dem.

$\vdash . *211·312 . \supset \vdash :. \mathrm{Hp} . \supset : \alpha \epsilon \mathrm{D}`P_\epsilon . \overrightarrow{\mathrm{seq}_P}`\alpha = \Lambda . \supset . \alpha = P``\alpha .$

[*211·12] $\supset . \alpha \epsilon \mathrm{D}`(P_\epsilon \dot{\cap} I) :. \supset \vdash . \mathrm{Prop}$

***211·431.** $\vdash : P \epsilon \mathrm{trans} \cap \mathrm{connex} . \supset .$

$\mathrm{D}`P_\epsilon - \mathrm{D}`(P_\epsilon \dot{\cap} I) = \mathrm{sect}`P \cap \text{Ɑ}`\max_P \cap \text{Ɑ}`\mathrm{seq}_P$

[*211·32·41]

***211·44.** $\vdash . \Lambda \epsilon \mathrm{D}`(P_\epsilon \dot{\cap} I) . \Lambda \epsilon \mathrm{D}`P_\epsilon . \Lambda \epsilon \mathrm{sect}`P$

[*37·29 . *211·12·14]

***211·45.** $\vdash : P \epsilon \mathrm{trans} . x \sim \epsilon \text{Ɑ}`(P \dot{-} P^2) . \supset . \overrightarrow{P}`x \epsilon \mathrm{D}`(P_\epsilon \dot{\cap} I)$

Dem.

$\vdash . *201·501 . \quad \supset \vdash : \mathrm{Hp} . \supset . P``\overrightarrow{P}`x \subset \overrightarrow{P}`x$ (1)

$\vdash . *33·41 . *32·3·34 . \supset \vdash : \mathrm{Hp} . \supset . \overrightarrow{P}`x - P``\overrightarrow{P}`x = \Lambda$ (2)

$\vdash . (1) . (2) . \quad \supset \vdash : \mathrm{Hp} . \supset . \overrightarrow{P}`x = P``\overrightarrow{P}`x$ (3)

$\vdash . (3) . *211·12 . \supset \vdash . \mathrm{Prop}$

***211·451.** $\vdash : \overrightarrow{P}`x \epsilon \mathrm{D}`(P_\epsilon \dot{\cap} I) . \supset . x \sim \epsilon \text{Ɑ}`(P \dot{-} P^2)$

Dem.

$\vdash . *211·12 . \supset \vdash :. \mathrm{Hp} . \supset : \overrightarrow{P}`x = P``\overrightarrow{P}`x :$

[*37·3] $\supset : yPx . \equiv_y . yP^2x :$

[*10·51] $\supset : \sim(\exists y) . yPx . \sim(yP^2x) :. \supset \vdash . \mathrm{Prop}$

***211·452.** $\vdash :. P \epsilon \text{trans} . \supset : \overrightarrow{P}\text{‘}x \,\epsilon\, \text{D}\text{‘}(P_\epsilon \dot\cap I) . \equiv . x \sim\epsilon\, \text{Ɑ}\text{‘}(P \dot- P^2)$
[*211·45·451]

***211·46.** $\vdash :. P \epsilon \text{trans} \cap \text{connex} : (\alpha) . \alpha \,\epsilon\, \text{Ɑ}\text{‘}\text{max}_P \cup \text{Ɑ}\text{‘}\text{seq}_P : \supset .$

$$\text{D}\text{‘}(P_\epsilon \dot\cap I) = \overrightarrow{P}\text{“}\{C\text{‘}P - \text{Ɑ}\text{‘}(P \dot- P^2)\}$$

Dem.

$\vdash . *211\cdot452 . \supset \vdash : \text{Hp} . \supset . \overrightarrow{P}\text{“}\{C\text{‘}P - \text{Ɑ}\text{‘}(P \dot- P^2)\} \subset \text{D}\text{‘}(P_\epsilon \dot\cap I)$ (1)

$\vdash . *211\cdot372\cdot14 . \supset$

$\vdash :. \text{Hp} . \supset : \alpha \,\epsilon\, \text{D}\text{‘}(P_\epsilon \dot\cap I) . \supset . (\exists x) . x \,\epsilon\, C\text{‘}P . \alpha = \overrightarrow{P}\text{‘}x . \overrightarrow{P}\text{‘}x \,\epsilon\, \text{D}\text{‘}(P_\epsilon \dot\cap I) .$

[*211·452] $\supset . (\exists x) . x \,\epsilon\, C\text{‘}P . \alpha = \overrightarrow{P}\text{‘}x . x \sim\epsilon\, \text{Ɑ}\text{‘}(P \dot- P^2) .$

[*37·7] $\supset . \alpha \,\epsilon\, \overrightarrow{P}\text{“}\{C\text{‘}P - \text{Ɑ}\text{‘}(P \dot- P^2)\}$ (2)

$\vdash . (1) . (2) . \supset \vdash . \text{Prop}$

***211·47.** $\vdash :. P \epsilon \text{trans} . \supset : (\alpha) . \alpha \,\epsilon\, \text{Ɑ}\text{‘}\text{max}_P \cup \text{Ɑ}\text{‘}\text{seq}_P . \equiv . \text{D}\text{‘}(P_\epsilon \dot\cap I) \subset \text{Ɑ}\text{‘}\text{seq}_P$

Dem.

$\vdash . *211\cdot272 . *24\cdot43 . \supset$

$\vdash :. \text{Hp} . \supset : (\alpha) . \alpha \,\epsilon\, \text{Ɑ}\text{‘}\text{max}_P \cup \text{Ɑ}\text{‘}\text{seq}_P . \equiv . \text{sect}\text{‘}P - \text{Ɑ}\text{‘}\text{max}_P \subset \text{Ɑ}\text{‘}\text{seq}_P .$

[*211·41] $\equiv . \text{D}\text{‘}(P_\epsilon \dot\cap I) \subset \text{Ɑ}\text{‘}\text{seq}_P :. \supset \vdash . \text{Prop}$

The following propositions are concerned with certain consequences of the hypothesis $P^2 = P$. This hypothesis is important because it is the defining characteristic of compact series.

***211·5.** $\vdash : P^2 = P . \alpha = P\text{“}\beta . \supset . \alpha = P\text{“}\alpha$

Dem.

$\vdash . *37\cdot33 . \supset \vdash : \text{Hp} . \supset . P\text{“}\beta = P\text{“}P\text{“}\beta .$

[Hp.*13·12] $\supset . \alpha = P\text{“}\alpha : \supset \vdash . \text{Prop}$

***211·51.** $\vdash : P^2 = P . \supset . \text{D}\text{‘}P_\epsilon = \text{D}\text{‘}(P_\epsilon \dot\cap I)$ [*211·5·11·12]

Thus in compact series there is no distinction between the two sorts of segments.

***211·52.** $\vdash :. P^2 = P . P \epsilon \text{connex} . \supset : \text{E} ! \,\text{max}_P\text{‘}\alpha . \supset . \sim \text{E} ! \,\text{seq}_P\text{‘}\alpha$

Dem.

$\vdash . *206\cdot5 . \supset \vdash : P^2 \in P . P \epsilon \text{connex} . \text{E} ! \,\text{max}_P\text{‘}\alpha . \text{E} ! \,\text{seq}_P\text{‘}\alpha . \supset . \dot\exists ! (P \dot- P^2)$ (1)

$\vdash . (1) . \text{Transp} . \supset \vdash : P^2 = P . P \epsilon \text{connex} . \text{E} ! \,\text{max}_P\text{‘}\alpha . \supset . \sim \text{E} ! \,\text{seq}_P\text{‘}\alpha :$
$\supset \vdash . \text{Prop}$

***211·53.** $\vdash :: P^2 = P . P \epsilon \text{connex} . \supset :. \text{E} ! \,\text{max}_P\text{‘}\alpha . \vee . \text{E} ! \,\text{seq}_P\text{‘}\alpha : \equiv :$
$\text{E} ! \,\text{max}_P\text{‘}\alpha . \equiv . \sim \text{E} ! \,\text{seq}_P\text{‘}\alpha$

Dem.

$\vdash . *4\cdot64 . \supset \vdash :. \text{E} ! \,\text{max}_P\text{‘}\alpha . \vee . \text{E} ! \,\text{seq}_P\text{‘}\alpha : \equiv : \sim \text{E} ! \,\text{seq}_P\text{‘}\alpha . \supset . \text{E} ! \,\text{max}_P\text{‘}\alpha$ (1)

$\vdash . *4\cdot73 . *211\cdot52 . \supset$

$\vdash :: \text{Hp} . \supset :. \sim \text{E} ! \,\text{seq}_P\text{‘}\alpha . \supset . \text{E} ! \,\text{max}_P\text{‘}\alpha : \equiv : \text{E} ! \,\text{max}_P\text{‘}\alpha . \equiv . \sim \text{E} ! \,\text{seq}_P\text{‘}\alpha$ (2)

$\vdash . (1) . (2) . \supset \vdash . \text{Prop}$

The condition $(\alpha):\text{E}\,!\,\text{max}_P{}^{\prime}\alpha\,.\equiv.\sim\text{E}\,!\,\text{seq}_P{}^{\prime}\alpha$ is the Dedekindian definition of continuity. In virtue of the above proposition, this is equivalent, in a series, to compactness combined with Dedekind's axiom, namely

$$(\alpha):\text{E}\,!\,\text{max}_P{}^{\prime}\alpha\,.\,\mathsf{v}\,.\,\text{E}\,!\,\text{seq}_P{}^{\prime}\alpha.$$

***211·54.** $\vdash :. P \subseteq J : \exists\,!\,\overrightarrow{\text{max}}_P{}^{\prime}\alpha\,.\supset_\alpha.\sim\exists\,!\,\overrightarrow{\text{seq}}_P{}^{\prime}\alpha : \supset . P \subseteq P^2$

Dem.

$\vdash . *10\cdot1 . \supset \vdash :. \text{Hp} . \supset : \exists\,!\,\overrightarrow{\text{max}}_P{}^{\prime}\iota^{\prime}x . \supset . \sim\exists\,!\,\text{seq}_P{}^{\prime}\iota^{\prime}x :$

$[*205\cdot18] \qquad \supset : x \in C^{\prime}P . \supset . \sim\exists\,!\,\text{seq}_P{}^{\prime}\iota^{\prime}x .$

$[*206\cdot42] \qquad \supset . \sim\exists\,!\,\overleftarrow{P \dot{-} P^2}{}^{\prime}x .$

$[*33\cdot4] \qquad \supset . x \sim\in \text{D}^{\prime}(P \dot{-} P^2) \qquad (1)$

$\vdash . *33\cdot263 . \supset \vdash : x \sim\in C^{\prime}P . \supset . x \sim\in \text{D}^{\prime}(P \dot{-} P^2) \qquad (2)$

$\vdash . (1) . (2) . \supset \vdash : \text{Hp} . \supset . \text{D}^{\prime}(P \dot{-} P^2) = \Lambda .$

$[*33\cdot241 . *25\cdot3] \qquad \supset . P \subseteq P^2 : \supset \vdash . \text{Prop}$

***211·541.** $\vdash :. P \in \text{Rl}^{\prime}J \cap \text{trans} : \exists\,!\,\overrightarrow{\text{max}}_P{}^{\prime}\alpha\,.\supset_\alpha.\sim\exists\,!\,\overrightarrow{\text{seq}}_P{}^{\prime}\alpha : \supset . P = P^2$

Dem.

$\vdash . *201\cdot1 . \supset \vdash : \text{Hp} . \supset . P^2 \subseteq P \qquad (1)$

$\vdash . *211\cdot54 . \supset \vdash : \text{Hp} . \supset . P \subseteq P^2 \qquad (2)$

$\vdash . (1) . (2) . \supset \vdash . \text{Prop}$

***211·55.** $\vdash :: P \in \text{Ser} . \supset :. \exists\,!\,\overrightarrow{\text{max}}_P{}^{\prime}\alpha\,.\supset_\alpha.\sim\exists\,!\,\overrightarrow{\text{seq}}_P{}^{\prime}\alpha : \equiv . P = P^2$
$[*211\cdot52\cdot541]$

***211·551.** $\vdash :. P \in \text{Ser} . \supset : \text{Cl}^{\prime}\text{max}_P \cap \text{Cl}^{\prime}\text{seq}_P = \Lambda . \equiv . P = P^2$
$[*211\cdot55 . *33\cdot41]$

***211·552.** $\vdash :: P \in \text{Ser} . \supset :. \text{E}\,!\,\text{max}_P{}^{\prime}\alpha\,.\equiv_\alpha.\sim\text{E}\,!\,\text{seq}_P{}^{\prime}\alpha : \equiv :$
$P = P^2 : (\alpha) : \text{E}\,!\,\text{max}_P{}^{\prime}\alpha\,.\,\mathsf{v}\,.\,\text{E}\,!\,\text{seq}_P{}^{\prime}\alpha$
$[*211\cdot55]$

***211·553.** $\vdash :: P \in \text{Ser} . \supset :. \text{Cl}^{\prime}\text{max}_P = -\text{Cl}^{\prime}\text{seq}_P . \equiv :$
$P = P^2 : (\alpha) . \alpha \in \text{Cl}^{\prime}\text{max}_P \cup \text{Cl}^{\prime}\text{seq}_P$
$[*211\cdot552 . *71\cdot163]$

The following propositions are concerned in showing that $\text{sect}^{\prime}P$, $\text{D}^{\prime}P_\epsilon$, and $\text{D}^{\prime}(P_\epsilon \dot{\cap} I)$ all verify the hypotheses of *210, if taken as the κ of that number.

***211·56.** $\vdash :. P \in \text{connex} . \alpha, \beta \in \text{sect}^{\prime}P . \supset : \alpha \subset \beta . \mathsf{v} . \beta \subset P^{\prime\prime}\alpha$

Dem.

$\vdash . *211\cdot2 . \supset \vdash :. \text{Hp} . \exists\,!\,\alpha - \beta . \supset . \exists\,!\,\alpha \cap C^{\prime}P - \beta - P^{\prime\prime}\beta .$

$[*202\cdot501] \qquad \supset . \exists\,!\,\alpha \cap p^{\prime}\overleftarrow{P}^{\prime\prime}\beta .$

$[*40\cdot682] \qquad \supset . \beta \subset P^{\prime\prime}\alpha \qquad (1)$

$\vdash . (1) . *24\cdot55 . \supset \vdash . \text{Prop}$

∗211·561. $\vdash :. P_{po} \epsilon \text{connex} . \alpha, \beta \epsilon \text{sect}‘P . \supset : \alpha \subset \beta . \vee . \beta \subset P“\alpha$ [∗211·56·17·131]

∗211·562. $\vdash :. P_{po} \epsilon \text{connex} . \alpha, \beta \epsilon \text{sect}‘P . \supset : \alpha \subset \beta . \vee . \beta \subset \alpha$ [∗211·561·1]

∗211·6. $\vdash :. P \epsilon \text{connex} . \alpha, \beta \epsilon \text{sect}‘P . \supset : \alpha \subset \beta . \vee . \beta \subset \alpha$ [∗211·56·1]

∗211·61. $\vdash :. P \epsilon \text{trans} \cap \text{connex} . \alpha, \beta \epsilon \text{D}‘P_\epsilon . \supset : \alpha \subset \beta . \vee . \beta \subset \alpha$ [∗211·15·6]

∗211·62. $\vdash :. P \epsilon \text{connex} . \alpha, \beta \epsilon \text{D}‘(P_\epsilon \dot{\cap} I) . \supset : \alpha \subset \beta . \vee . \beta \subset \alpha$ [∗211·14·6]

In the hypothesis of ∗211·61, it is necessary that P should be transitive as well as connected. Take, for example,

$$P = x \downarrow y \cup y \downarrow z \cup z \downarrow x \quad (x \neq y . x \neq z . y \neq z).$$

Then P is connected, but not transitive; also we have

$$\overrightarrow{P}‘y = \iota‘x . \overrightarrow{P}‘z = \iota‘y.$$

Hence $\iota‘x, \iota‘y \epsilon \text{D}‘P_\epsilon . \sim(\iota‘x \subset \iota‘y) . \sim(\iota‘y \subset \iota‘x)$.

Thus connection is not sufficient in the hypothesis of ∗211·61.

∗211·63. $\vdash : \lambda \subset \text{sect}‘P . \supset . s‘\lambda \epsilon \text{sect}‘P$

Dem.

$\vdash . ∗211·1 . \supset \vdash :. \text{Hp} . \supset : \alpha \epsilon \lambda . \supset_\alpha . \alpha \subset C‘P :$

[∗40·151] $\supset : s‘\lambda \subset C‘P$ (1)

$\vdash . ∗211·1 . \supset \vdash :. \text{Hp} . \supset : \alpha \epsilon \lambda . \supset_\alpha . P“\alpha \subset \alpha :$

[∗40·8] $\supset : P“s‘\lambda \subset s‘\lambda$ (2)

$\vdash . (1) . (2) . ∗211·1 . \supset \vdash . \text{Prop}$

This proposition shows that sect‘P verifies the hypothesis of ∗210·251, with the exception of sect‘$P \sim \epsilon 1$, which requires $\dot{\exists} ! P$.

∗211·631. $\vdash : \lambda \subset \text{sect}‘P . \supset . p‘\lambda \cap C‘P \epsilon \text{sect}‘P$

Dem.

$\vdash . ∗22·43 . \supset \vdash . p‘\lambda \cap C‘P \subset C‘P$ (1)

$\vdash . ∗211·1 . \supset \vdash :. \text{Hp} . \supset : \alpha \epsilon \lambda . \supset_\alpha . P“\alpha \subset \alpha :$

[∗40·81] $\supset : P“p‘\lambda \subset p‘\lambda :$

[∗37·265·15] $\supset : P“(p‘\lambda \cap C‘P) \subset p‘\lambda \cap C‘P$ (2)

$\vdash . (1) . (2) . \supset \vdash . \text{Prop}$

∗211·632. $\vdash : \lambda \subset \text{sect}‘P . \exists ! \lambda . \supset . p‘\lambda \epsilon \text{sect}‘P$

Dem.

$\vdash . ∗40·23 . \supset \vdash : \text{Hp} . \supset . p‘\lambda \subset s‘\lambda .$

[∗211·63·1] $\supset . p‘\lambda \subset C‘P$ (1)

$\vdash . (1) . ∗211·631 . \supset \vdash . \text{Prop}$

∗211·633. $\vdash : \lambda \subset \text{sect}‘P . \supset . p‘\lambda \cap s‘\text{sect}‘P \epsilon \text{sect}‘P$ [∗211·631·26]

This proposition shows that sect‘P verifies the hypothesis of ∗210·252, with the exception of sect‘$P \sim \epsilon 1$, which requires $\dot{\exists} ! P$.

***211·64.** $\vdash : \lambda \subset \mathrm{D}`P_\epsilon . \supset . s`\lambda \,\epsilon\, \mathrm{D}`P_\epsilon$

Dem.

$\vdash . *72{\cdot}504 . \supset \vdash : \mathrm{Hp} . \supset . s`\lambda = s`P_\epsilon``\breve{P}_\epsilon``\lambda$

$[*40{\cdot}38] \qquad = P``s`\breve{P}_\epsilon``\lambda \qquad (1)$

$\vdash . (1) . *211{\cdot}11 . \supset \vdash . \mathrm{Prop}$

***211·65.** $\vdash : \lambda \subset \mathrm{D}`(P_\epsilon \dot{\cap} I) . \supset . s`\lambda \,\epsilon\, \mathrm{D}`(P_\epsilon \dot{\cap} I)$

Dem.

$\vdash . *211{\cdot}12 . \supset \vdash :. \mathrm{Hp} . \supset : \alpha \,\epsilon\, \lambda . \supset_\alpha . \alpha = P_\epsilon`\alpha :$

$[*50{\cdot}17] \qquad \supset : \lambda = P_\epsilon``\lambda :$

$[*40{\cdot}38] \qquad \supset : s`\lambda = P``s`\lambda :$

$[*211{\cdot}12] \qquad \supset : s`\lambda \,\epsilon\, \mathrm{D}`(P_\epsilon \dot{\cap} I) :. \supset \vdash . \mathrm{Prop}$

***211·66.** $\vdash : \dot{\exists} ! P . \supset . \mathrm{sect}`P, \mathrm{D}`P_\epsilon \sim\epsilon\, 1$

Dem.

$\vdash . *211{\cdot}44{\cdot}26 . \qquad \supset \vdash . \Lambda, C`P \,\epsilon\, \mathrm{sect}`P \qquad (1)$

$\vdash . *33{\cdot}24 . \qquad \supset \vdash : \mathrm{Hp} . \supset . \Lambda \neq C`P \qquad (2)$

$\vdash . (1) . (2) . *52{\cdot}41 . \supset \vdash : \mathrm{Hp} . \supset . \mathrm{sect}`P \sim\epsilon\, 1 \qquad (3)$

$\vdash . *211{\cdot}44{\cdot}301 . \qquad \supset \vdash : \mathrm{Hp} . \supset . \Lambda, \mathrm{D}`P \,\epsilon\, \mathrm{D}`P_\epsilon \qquad (4)$

$\vdash . *33{\cdot}24 . \qquad \supset \vdash : \mathrm{Hp} . \supset . \Lambda \neq \mathrm{D}`P \qquad (5)$

$\vdash . (4) . (5) . *52{\cdot}41 . \supset \vdash : \mathrm{Hp} . \supset . \mathrm{D}`P_\epsilon \sim\epsilon\, 1 \qquad (6)$

$\vdash . (3) . (6) . \supset \vdash . \mathrm{Prop}$

***211·661.** $\vdash : P \,\epsilon\, \mathrm{trans} . \exists ! \mathrm{Cl\,ex}`C`P - \mathrm{Cl}`\mathrm{max}_P . \supset . \mathrm{D}`(P_\epsilon \dot{\cap} I) \sim\epsilon\, 1$

Dem.

$\vdash . *205{\cdot}111 . \supset \vdash : \alpha \,\epsilon\, \mathrm{Cl\,ex}`C`P - \mathrm{Cl}`\mathrm{max}_P . \supset . \exists ! \alpha . \alpha \subset C`P . \alpha \subset P``\alpha .$

$[*24{\cdot}58 . *37{\cdot}2] \qquad \supset . \exists ! P``\alpha . P``\alpha \subset P``P``\alpha \qquad (1)$

$\vdash . (1) . *201{\cdot}5 . \supset$

$\vdash :. P \,\epsilon\, \mathrm{trans} . \supset : \alpha \,\epsilon\, \mathrm{Cl\,ex}`C`P - \mathrm{Cl}`\mathrm{max}_P . \supset . P``\alpha = P``P``\alpha . \exists ! P``\alpha .$

$[*211{\cdot}12] \qquad \supset . P``\alpha \,\epsilon\, \mathrm{D}`(P_\epsilon \dot{\cap} I) . \exists ! P``\alpha .$

$[*10{\cdot}24] \qquad \supset . \exists ! \mathrm{D}`(P_\epsilon \dot{\cap} I) - \iota`\Lambda \qquad (2)$

$\vdash . (2) . *211{\cdot}44 . \supset \vdash . \mathrm{Prop}$

The following propositions sum up the above results in relation to the hypotheses of *210. The relation P_{lc} with its field limited to sections or segments, which occurs in the following propositions, is important, and will be considered at length in the following number.

***211·67.** $\vdash : P \,\epsilon\, \mathrm{connex} . \kappa = \mathrm{sect}`P . Q = P_{\mathrm{lc}} \upharpoonright \kappa . \supset . \mathrm{Hp}\, *210{\cdot}12$

$[*211{\cdot}6 . *210{\cdot}13]$

$*211\cdot671.\ \vdash : P \in \text{connex} \,.\, \kappa = \text{sect}`P \,.\, Q = P_{1c} \upharpoonright \kappa \,.\, \dot{\exists} ! P \,.\, \supset .$
Hp $*210\cdot251$. Hp $*210\cdot252$ [$*211\cdot67\cdot66\cdot63\cdot633$]

$*211\cdot68.\ \vdash : P \in \text{trans} \cap \text{connex} \,.\, \kappa = \text{D}`P_\epsilon \,.\, Q = P_{1c} \upharpoonright \kappa \,.\, \supset .$ Hp $*210\cdot12$
[$*211\cdot61$. $*210\cdot13$]

$*211\cdot681.\ \vdash : P \in \text{trans} \cap \text{connex} \,.\, \kappa = \text{D}`P_\epsilon \,.\, Q = P_{1c} \upharpoonright \kappa \,.\, \dot{\exists} ! P \,.\, \supset .$ Hp $*210\cdot251$
[$*211\cdot68\cdot66\cdot64$]

$*211\cdot69.\ \vdash : P \in \text{connex} \,.\, \kappa = \text{D}`(P_\epsilon \dot{\cap} I) \,.\, Q = P_{1c} \upharpoonright \kappa \,.\, \supset .$ Hp $*210\cdot12$
[$*211\cdot62$. $*210\cdot13$]

$*211\cdot691.\ \vdash : P \in \text{connex} \,.\, \kappa = \text{D}`(P_\epsilon \dot{\cap} I) \,.\, Q = P_{1c} \upharpoonright \kappa \,.\, \text{D}`(P_\epsilon \dot{\cap} I) \sim \in 1 \,.\, \supset .$
Hp $*210\cdot251$ [$*211\cdot69\cdot65$]

$*211\cdot692.\ \vdash : P \in \text{trans} \cap \text{connex} \,.\, \kappa = \text{D}`(P_\epsilon \dot{\cap} I) \,.\, Q = P_{1c} \upharpoonright \kappa \,.$
$\exists ! \text{Cl ex}`C`P - \text{Cl}`\text{max}_P \,.\, \supset .$ Hp $*210\cdot251$ [$*211\cdot691\cdot661$]

The following propositions are concerned with the relations of sections and segments of P to sections and segments of $\breve{P}$. When $\alpha \in \text{sect}`P$, $C`P - \alpha \in \text{sect}`\breve{P}$, and vice versa. Also, if $P$ is connected, the maximum of $\alpha$ (if any) is the precedent with respect to $P$ (*i.e.* the sequent with respect to $\breve{P}$) of $C`P - \alpha$, and the sequent of α (if any) is the minimum with respect to P (*i.e.* the maximum with respect to $\breve{P}$) of $C`P - \alpha$. Hence the relations to be proved follow easily.

$*211\cdot7.\ \vdash : \alpha \in \text{sect}`P \,.\, \supset .\, C`P - \alpha \in \text{sect}`\breve{P}$

Dem.

$\vdash . *22\cdot43 . \quad \supset \vdash .\, C`P - \alpha \subset C`P \quad (1)$
$\vdash . *211\cdot1 . *37\cdot1 . \supset \vdash :.\, \text{Hp} \,.\, \supset : x \in \alpha \,.\, yPx \,.\, \supset .\, y \in \alpha :$
[Transp] $\supset : x \in \alpha \,.\, y \sim \in \alpha \,.\, \supset .\, \sim (yPx) :$
[$*37\cdot1$.Transp] $\supset : x \in \alpha \,.\, \supset .\, x \sim \in \breve{P}``(-\alpha) :$
[$*37\cdot265$] $\supset : \alpha \subset -\breve{P}``(C`P - \alpha) :$
[Transp] $\supset : \breve{P}``(C`P - \alpha) \subset -\alpha :$
[$*37\cdot15$] $\supset : \breve{P}``(C`P - \alpha) \subset C`P - \alpha \quad (2)$
$\vdash . (1) . (2) . *211\cdot1 . \supset \vdash .$ Prop

$*211\cdot701.\ \vdash : \alpha \in \text{sect}`P \,.\, \text{E} ! \text{max}_P`\alpha \,.\, \supset .\, p`\overleftarrow{P}``\alpha \subset \overleftarrow{P}`\text{max}_P`\alpha \subset C`P - \alpha$

Dem.

$\vdash . *40\cdot12 . \quad \supset \vdash : \text{Hp} \,.\, \supset .\, p`\overleftarrow{P}``\alpha \subset \overleftarrow{P}`\text{max}_P`\alpha \quad (1)$
$\vdash . *205\cdot101 . \supset \vdash : \text{Hp} \,.\, \supset .\, \text{max}_P`\alpha \sim \in P``\alpha \,.$
[$*37\cdot1$.Transp.$*32\cdot181$] $\supset .\, \overleftarrow{P}`\text{max}_P`\alpha \subset -\alpha \,.$
[$*33\cdot152$] $\supset .\, \overleftarrow{P}`\text{max}_P`\alpha \subset C`P - \alpha \quad (2)$
$\vdash . (1) . (2) . \supset \vdash .$ Prop

$*211\cdot702$. $\vdash : P \in \text{connex} . \alpha \in \text{sect}\text{‘}P . \supset . C\text{‘}P - \alpha \subset p\text{‘}\overleftarrow{P}\text{‘‘}\alpha$ $[*202\cdot501 . *211\cdot1]$

$*211\cdot703$. $\vdash : P \in \text{connex} . \alpha \in \text{sect}\text{‘}P - \iota\text{‘}C\text{‘}P . \supset . \exists ! p\text{‘}\overleftarrow{P}\text{‘‘}\alpha$
$[*211\cdot702\cdot1 . *24\cdot58]$

$*211\cdot71$. $\vdash : P \in \text{connex} . \alpha \in \text{sect}\text{‘}P . E ! \max_P\text{‘}\alpha . \supset .$

$$p\text{‘}\overleftarrow{P}\text{‘‘}\alpha = \overleftarrow{P}\text{‘}\max_P\text{‘}\alpha = C\text{‘}P - \alpha$$

Dem.

$\vdash . *202\cdot501 . *211\cdot2 . \supset \vdash : \text{Hp} . \supset . C\text{‘}P - \alpha \subset p\text{‘}\overleftarrow{P}\text{‘‘}\alpha$ (1)

$\vdash . (1) . *211\cdot701 . \quad \supset \vdash : \text{Hp} . \supset . C\text{‘}P - \alpha = p\text{‘}\overleftarrow{P}\text{‘‘}\alpha$ (2)

$\vdash . (2) . *211\cdot701 . \quad \supset \vdash : \text{Hp} . \supset . \overleftarrow{P}\text{‘}\max_P\text{‘}\alpha \subset p\text{‘}\overleftarrow{P}\text{‘‘}\alpha .$

$[*211\cdot701] \quad \supset . \overleftarrow{P}\text{‘}\max_P\text{‘}\alpha = p\text{‘}\overleftarrow{P}\text{‘‘}\alpha$ (3)

$\vdash . (2) . (3) . \supset \vdash . \text{Prop}$

If α is a section of P, we shall call $C\text{‘}P - \alpha$ the *complement* of α. By the above proposition, if α is a section of P having a maximum, its complement is a section of $\breve{P}$ which is a member of $\overleftarrow{P}\text{‘‘}C\text{‘}P$.

$*211\cdot711$. $\vdash : P \in \text{connex} . P^2 \in J . \alpha \in \text{sect}\text{‘}P . \supset .$

$$\alpha = C\text{‘}P - p\text{‘}\overleftarrow{P}\text{‘‘}\alpha . C\text{‘}P \cap p\text{‘}\overleftarrow{P}\text{‘‘}\alpha = C\text{‘}P - \alpha \quad [*202\cdot503 . *211\cdot2]$$

$*211\cdot712$. $\vdash : P \in \text{connex} . \alpha \in \text{sect}\text{‘}P . E ! \min_P\text{‘}(C\text{‘}P - \alpha) . \supset . \alpha = \overrightarrow{P}\text{‘}\min_P\text{‘}(C\text{‘}P - \alpha)$

Dem.

$\vdash . *211\cdot71 \dfrac{\breve{P}}{P} . \supset$

$\vdash : P \in \text{connex} . \beta \in \text{sect}\text{‘}\breve{P} . E ! \min_P\text{‘}\beta . \supset . \overrightarrow{P}\text{‘}\min_P\text{‘}\beta = C\text{‘}P - \beta$ (1)

$\vdash . *211\cdot7 . *24\cdot492 . \supset \vdash : \alpha \in \text{sect}\text{‘}P . \beta = C\text{‘}P - \alpha . \supset . \beta \in \text{sect}\text{‘}\breve{P} . \alpha = C\text{‘}P - \beta$ (2)

$\vdash . (1) . (2) . \supset \vdash . \text{Prop}$

$*211\cdot713$. $\vdash : P \in \text{connex} . \alpha \in \text{sect}\text{‘}P - D\text{‘}P_\epsilon . \supset . E ! \max_P\text{‘}\alpha . \sim E ! \min_P\text{‘}(C\text{‘}P - \alpha)$

Dem.

$\vdash . *211\cdot24 . \text{Transp} . \supset \vdash : \text{Hp} . \supset . E ! \max_P\text{‘}\alpha$ (1)

$\vdash . *211\cdot712\cdot3 . \supset \vdash : P \in \text{connex} . \alpha \in \text{sect}\text{‘}P . E ! \min_P\text{‘}(C\text{‘}P - \alpha) . \supset . \alpha \in D\text{‘}P_\epsilon$ (2)

$\vdash . (2) . \text{Transp} . \quad \supset \vdash : \text{Hp} . \supset . \sim E ! \min_P\text{‘}(C\text{‘}P - \alpha)$ (3)

$\vdash . (1) . (3) . \supset \vdash . \text{Prop}$

***211·714.** $\vdash : P \in \text{connex} . \alpha \in \text{sect}`P . \supset . \overrightarrow{\text{seq}}_P`\alpha \subset \overrightarrow{\text{min}}_P`(C`P - \alpha)$

Dem.

$\vdash . *206·18·2 . \supset \vdash : x \in \overrightarrow{\text{seq}}_P`\alpha . \supset . x \in C`P - \alpha$ (1)

$\vdash . *206·134 . \supset \vdash : x \in \overrightarrow{\text{seq}}_P`\alpha . \supset . \overrightarrow{P}`x \subset C`P - p`\overleftarrow{P}``\alpha$ (2)

$\vdash . (2) . *202·501 . *211·2 . \supset$

$\vdash :. \text{Hp} . \supset : x \in \overrightarrow{\text{seq}}_P`\alpha . \supset . \overrightarrow{P}`x \subset \alpha .$

[*37·462] $\supset . x \sim \in \breve{P}``(C`P - \alpha)$ (3)

$\vdash . (1) . (3) . *205·11 . \supset \vdash . \text{Prop}$

The above hypothesis is not sufficient to secure $\overrightarrow{\text{seq}}_P`\alpha = \overrightarrow{\text{min}}_P`(C`P - \alpha)$, as may be seen by putting

$$P = \alpha \uparrow (\alpha \cup \iota`x), \text{ where } \exists ! \alpha . x \sim \in \alpha.$$

We then have $P \in \text{connex} . P``\alpha = \alpha . C`P - \alpha = \iota`x . p`\overleftarrow{P}``\alpha = \alpha \cup \iota`x$. Thus $\overrightarrow{\text{min}}_P`(C`P - \alpha) = \iota`x . \overrightarrow{\text{seq}}_P`\alpha = \Lambda$. It will be seen that $\alpha \uparrow (\alpha \cup \iota`x) \in \text{trans}$, so that it is useless to add $P \in \text{trans}$ to the hypothesis of *211·714. A sufficient addition is $P \Subset J$, as is proved in the following proposition.

***211·715.** $\vdash : P \in \text{connex} \cap \text{Rl}`J . \alpha \in \text{sect}`P . \supset . \overrightarrow{\text{seq}}_P`\alpha = \overrightarrow{\text{min}}_P`(C`P - \alpha)$

Dem.

$\vdash . *205·14 . \supset \vdash : x \, \text{min}_P \, (C`P - \alpha) . \supset . x \in C`P - \alpha . \overrightarrow{P}`x \cap (C`P - \alpha) = \Lambda$ (1)

$\vdash . (1) . *33·152 . *211·2 . \supset$

$\vdash :. \text{Hp} . \supset : x \, \text{min}_P \, (C`P - \alpha) . \supset . x \in C`P - \alpha - P``\alpha . \overrightarrow{P}`x \subset \alpha .$

[*202·501] $\supset . x \in C`P \cap p`\overleftarrow{P}``\alpha . \overrightarrow{P}`x \subset \alpha .$

[*200·5] $\supset . x \in C`P \cap p`\overleftarrow{P}``\alpha . \overrightarrow{P}`x \subset - p`\overleftarrow{P}``\alpha .$

[*37·1.Transp] $\supset . x \in C`P \cap p`\overleftarrow{P}``\alpha - \breve{P}``p`\overleftarrow{P}``\alpha .$

[*206·11] $\supset . x \, \text{seq}_P \, \alpha$ (2)

$\vdash . (2) . *211·714 . \supset \vdash . \text{Prop}$

***211·72.** $\vdash : P \in \text{connex} . \alpha \in \text{sect}`P - \text{D}`P_\epsilon . \supset .$

$C`P - \alpha = \breve{P}``(C`P - \alpha) . C`P - \alpha \in \text{D}`\{(\breve{P})_\epsilon \dot{\cap} I\}$ [*211·21·7·713]

***211·721.** $\vdash : P \in \text{connex} . \alpha \in \text{sect}`P \cap (\text{G}`\text{max}_P \cup \text{G}`\text{seq}_P) . \supset .$

$\overrightarrow{\text{seq}}_P`\alpha = \overrightarrow{\text{min}}_P`(C`P - \alpha)$

Dem.

$\vdash . *211·71 . \supset \vdash : P \in \text{connex} . \alpha \in \text{sect}`P \cap \text{G}`\text{max}_P . \supset . p`\overleftarrow{P}``\alpha = C`P - \alpha .$

[*206·13] $\supset . \overrightarrow{\text{seq}}_P`\alpha = \overrightarrow{\text{min}}_P`(C`P - \alpha)$ (1)

$\vdash . *211 \cdot 714 . \supset$

$\vdash :. P \epsilon \text{connex} . \alpha \epsilon \text{sect}`P . \supset : \overrightarrow{\text{seq}}_P`\alpha \subset \overrightarrow{\text{min}}_P`(C`P - \alpha) :$

$[*205 \cdot 3 . *206 \cdot 16] \quad \supset : \exists ! \overrightarrow{\text{seq}}_P`\alpha . \supset . \overrightarrow{\text{seq}}_P`\alpha = \overrightarrow{\text{min}}_P`(C`P - \alpha) \qquad (2)$

$\vdash . (2) . \supset \vdash : P \epsilon \text{connex} . \alpha \epsilon \text{sect}`P \cap \text{Ꮯ}`\text{seq}_P . \supset . \overrightarrow{\text{seq}}_P`\alpha = \overrightarrow{\text{min}}_P`(C`P - \alpha) \qquad (3)$

$\vdash . (1) . (3) . \supset \vdash . \text{Prop}$

***211·722.** $\vdash : P \epsilon \text{connex} . \alpha \epsilon \text{sect}`P . E ! \text{max}_P`\alpha . E ! \text{seq}_P`\alpha . \supset .$

$\overrightarrow{\text{max}}_P`\alpha = \overrightarrow{\text{prec}}_P`(C`P - \alpha)$

Dem.

$\vdash . *211 \cdot 721 \cdot 7 . \supset \vdash : \text{Hp} . \supset . C`P - \alpha \epsilon \text{sect}`\breve{P} . E ! \text{min}_P`(C`P - \alpha) .$

$\left[*211 \cdot 721 \frac{\breve{P}, C`P - \alpha}{P, \quad \alpha} \right] \quad \supset . \overrightarrow{\text{prec}}_P`(C`P - \alpha) = \overrightarrow{\text{max}}_P`\{C`P - (C`P - \alpha)\}$

$[*24 \cdot 492] \qquad = \overrightarrow{\text{max}}_P`\alpha$

$[\text{Hp}] \qquad = \iota`\text{max}_P`\alpha : \supset \vdash . \text{Prop}$

We have always, if $P \epsilon \text{connex} . \alpha \epsilon \text{sect}`P$,

$$\overrightarrow{\text{prec}}_P`(C`P - \alpha) \subset \overrightarrow{\text{max}}_P`\alpha.$$

The converse inclusion does not always hold, as appears (on writing $\breve{P}$ in place of P) from the note to *211·714. To secure the converse implication, it is sufficient to assume $P \in J$ or $E ! \text{seq}_P`\alpha$ or $\sim E ! \text{max}_P`\alpha$.

***211·723.** $\vdash : P \epsilon \text{connex} . \alpha \epsilon \text{sect}`P . \supset . \overrightarrow{\text{prec}}_P`(C`P - \alpha) \subset \overrightarrow{\text{max}}_P`\alpha$

Dem.

$\vdash . *202 \cdot 11 . *211 \cdot 7 . \supset \vdash : \text{Hp} . \supset . \breve{P} \epsilon \text{connex} . C`P - \alpha \epsilon \text{sect}`\breve{P} .$

$\left[*211 \cdot 714 \frac{\breve{P}}{P} . *205 \cdot 102 . *206 \cdot 101 \right] \supset . \overrightarrow{\text{prec}}_P`(C`P - \alpha) \subset \overrightarrow{\text{max}}_P`\alpha : \supset \vdash . \text{Prop}$

***211·724.** $\vdash : P \epsilon \text{connex} . \alpha \epsilon \text{sect}`P \cap (\text{Ꮯ}`\text{seq}_P \cup - \text{Ꮯ}`\text{max}_P) . \supset .$

$\overrightarrow{\text{max}}_P`\alpha = \overrightarrow{\text{prec}}_P`(C`P - \alpha)$

Dem.

$\vdash . *211 \cdot 722 . \supset \vdash : P \epsilon \text{connex} . \alpha \epsilon \text{sect}`P \cap \text{Ꮯ}`\text{seq}_P \cap \text{Ꮯ}`\text{max}_P . \supset .$

$\overrightarrow{\text{max}}_P`\alpha = \overrightarrow{\text{prec}}_P`(C`P - \alpha) \qquad (1)$

$\vdash . *211 \cdot 723 . *24 \cdot 13 . \supset \vdash : P \epsilon \text{connex} . \alpha \epsilon \text{sect}`P - \text{Ꮯ}`\text{max}_P . \supset .$

$\overrightarrow{\text{max}}_P`\alpha = \overrightarrow{\text{prec}}_P`(C`P - \alpha) \qquad (2)$

$\vdash . (1) . (2) . *22 \cdot 91 . \supset \vdash . \text{Prop}$

***211·725.** $\vdash : P \epsilon \text{connex} . \alpha \epsilon \text{sect}`P \cap \text{Ꮯ}`\text{seq}_P . \supset .$

$\overrightarrow{\text{max}}_P`\alpha = \overrightarrow{\text{prec}}_P`(C`P - \alpha) . \overrightarrow{\text{seq}}_P`\alpha = \overrightarrow{\text{min}}_P`(C`P - \alpha) \quad [*211 \cdot 721 \cdot 724]$

***211·726.** $\vdash : P \epsilon \text{connex} \cap \text{Rl}'J \,.\, \alpha \epsilon \text{sect}'P \,.\supset .$

$$\overrightarrow{\max}_P{}'\alpha = \overrightarrow{\text{prec}}_P{}'(C'P - \alpha) \,.\, \overrightarrow{\text{seq}}_P{}'\alpha = \overrightarrow{\min}_P{}'(C'P - \alpha)$$

Dem.

$\vdash . *200{\cdot}11 . *202{\cdot}11 . *211{\cdot}7 . \supset \vdash : \text{Hp} . \supset . \breve{P} \epsilon \text{connex} \cap \text{Rl}'J \,.\, C'P - \alpha \epsilon \text{sect}'\breve{P} .$

$[*211{\cdot}715 . *205{\cdot}102 . *206{\cdot}101] \qquad \supset . \overrightarrow{\text{prec}}_P{}'(C'P - \alpha) = \overrightarrow{\max}_P{}'\alpha \qquad (1)$

$\vdash . (1) . *211{\cdot}715 . \supset \vdash . \text{Prop}$

***211·727.** $\vdash :. P \epsilon \text{connex} \cap \text{Rl}'J \,.\, \alpha \epsilon \text{sect}'P \,.\supset :$

$$E\,!\,\text{limax}_P{}'\alpha \,.\equiv .\, E\,!\,\text{limin}_P{}'(C'P - \alpha) \qquad [*211{\cdot}726 . *207{\cdot}44]$$

***211·728.** $\vdash :. P \epsilon \text{connex} \cap \text{Rl}'J \,.\, \alpha \epsilon \text{sect}'P : \sim E\,!\,\max_P{}'\alpha \,.\vee .$

$$\sim E\,!\,\min_P{}'(C'P - \alpha) : \supset . \overrightarrow{\text{limax}}_P{}'\alpha = \overrightarrow{\text{limin}}_P{}'(C'P - \alpha)$$

Dem.

$\vdash . *211{\cdot}726 . *207{\cdot}43{\cdot}12 . \supset \vdash : \text{Hp} . \sim E\,!\,\max_P{}'\alpha . \supset .$

$$\overrightarrow{\text{limax}}_P{}'\alpha = \overrightarrow{\min}_P{}'(C'P - \alpha)$$

$[*207{\cdot}46 . *211{\cdot}726] \qquad = \overrightarrow{\text{limin}}_P{}'(C'P - \alpha) \qquad (1)$

Similarly $\vdash : \text{Hp} . \sim E\,!\,\min_P{}'(C'P - \alpha) . \supset . \overrightarrow{\text{limax}}_P{}'\alpha = \overrightarrow{\text{limin}}_P{}'(C'P - \alpha) \qquad (2)$

$\vdash . (1) . (2) . \supset \vdash . \text{Prop}$

***211·729.** $\vdash : P \epsilon \text{connex} \cap \text{Rl}'J \,.\, \alpha \epsilon \text{sect}'P - (\text{Ɑ}'\max_P \cap \text{Ɑ}'\text{seq}_P) . \supset .$

$$\overrightarrow{\text{limax}}_P{}'\alpha = \overrightarrow{\text{limin}}_P{}'(C'P - \alpha) \qquad [*211{\cdot}728{\cdot}726]$$

***211·73.** $\vdash : P \epsilon \text{connex} \,.\, \alpha \epsilon \text{sect}'P - \text{D}'(P_\epsilon \dot{\cap} I) . \supset .$

$$C'P - \alpha \epsilon \text{D}'\{(\breve{P})_\epsilon \dot{\cap} I\} - \text{Ɑ}'\text{prec}_P \cup \{\text{sect}'P - \text{D}'(\breve{P})_\epsilon\}$$

Dem.

$\vdash . *211{\cdot}21 . \supset \vdash : \text{Hp} . \supset . \alpha \epsilon \text{sect}'P - \text{Ɑ}'\max_P .$

$[*211{\cdot}7{\cdot}723] \qquad \supset . C'P - \alpha \epsilon \text{sect}'\breve{P} - \text{Ɑ}'\text{prec}_P .$

$[*24{\cdot}41] \qquad \supset . C'P - \alpha \epsilon (\text{sect}'\breve{P} - \text{Ɑ}'\max_P - \text{Ɑ}'\text{prec}_P) \cup$

$$(\text{sect}'\breve{P} \cap \text{Ɑ}'\max_P - \text{Ɑ}'\text{prec}_P) .$$

$[*211{\cdot}31{\cdot}21] \qquad \supset . C'P - \alpha \epsilon \{\text{D}'(\breve{P})_\epsilon \dot{\cap} I\} - \text{Ɑ}'\text{prec}_P \cup \{\text{sect}'P - \text{D}'(\breve{P})_\epsilon\} :$

$$\supset \vdash . \text{Prop}$$

***211·74.** $\vdash : P \epsilon \text{trans} \cap \text{connex} \,.\, \alpha \epsilon \text{D}'P_\epsilon - \text{D}'(P_\epsilon \dot{\cap} I) . \supset .$

$$C'P - \alpha \epsilon \text{D}'(\breve{P})_\epsilon - \text{D}'\{(\breve{P})_\epsilon \dot{\cap} I\}$$

Dem.

$\vdash . *211{\cdot}431 . \supset \vdash : \text{Hp} . \supset . \alpha \epsilon \text{sect}'P \cap \text{Ɑ}'\max_P \cap \text{Ɑ}'\text{seq}_P .$

$[*211{\cdot}7{\cdot}725] \qquad \supset . C'P - \alpha \epsilon \text{sect}'\breve{P} \cap \text{Ɑ}'\text{prec}_P \cap \text{Ɑ}'\min_P .$

$\left[*211{\cdot}431 \dfrac{\breve{P}}{P}\right] \qquad \supset . C'P - \alpha \epsilon \text{D}'(\breve{P})_\epsilon - \text{D}'\{(\breve{P})_\epsilon \dot{\cap} I\} : \supset \vdash . \text{Prop}$

The following propositions sum up our previous results.

$*211 \cdot 75$. $\vdash :. \alpha \subset C`P . Q = \breve{P} . \supset : \alpha \in \text{sect}`P . \equiv . C`P - \alpha \in \text{sect}`Q \quad [*211 \cdot 7]$

$*211 \cdot 751$. $\vdash :. P \in \text{Ser} . \alpha \subset C`P . Q = \breve{P} . \supset :$

$$\alpha \in \text{D}`P_\epsilon . \equiv . C`P - \alpha \in \text{sect}`Q \cap (\text{ᗡ}`\text{max}_Q \cup - \text{ᗡ}`\text{seq}_Q)$$

Dem.

$\vdash . *211 \cdot 32 . \supset \vdash :. \text{Hp} . \supset : \alpha \in \text{D}`P_\epsilon . \equiv . \alpha \in \text{sect}`P \cap (\text{ᗡ}`\text{seq}_P \cup - \text{ᗡ}`\text{max}_P) .$

$[*211 \cdot 75 \cdot 726] \qquad \equiv . C`P - \alpha \in \text{sect}`Q \cap (\text{ᗡ}`\text{max}_Q \cup - \text{ᗡ}`\text{seq}_Q) :. \supset \vdash . \text{Prop}$

In the above proposition, "$P \in \text{trans}$" is necessary in order that $\text{D}`P_\epsilon$ may be contained in sect$`P$, and "$P \in \text{Rl}`J$" is necessary in order that "$(C`P - \alpha) \sim \in \text{ᗡ}`\text{seq}_Q$" may imply "$\alpha \sim \in \text{ᗡ}`\text{max}_P$." Hence the full hypothesis "$P \in \text{Ser}$" becomes necessary.

$*211 \cdot 752$. $\vdash :. P \in \text{connex} . \alpha \subset C`P . Q = \breve{P} . \supset :$

$$\alpha \in \text{D}`(P_\epsilon \dot{\cap} I) . \supset . C`P - \alpha \in \text{sect}`Q - \text{ᗡ}`\text{seq}_Q$$

Dem.

$\vdash . *211 \cdot 41 . \qquad \supset \vdash : \alpha \in \text{D}`(P_\epsilon \dot{\cap} I) . \equiv . \alpha \in \text{sect}`P - \text{ᗡ}`\text{max}_P \qquad (1)$

$\vdash . (1) . *211 \cdot 7 \cdot 723 . \supset \vdash : \text{Hp} . \alpha \in \text{D}`(P_\epsilon \dot{\cap} I) . \supset .$

$$C`P - \alpha \in \text{sect}`Q - \text{ᗡ}`\text{seq}_Q : \supset \vdash . \text{Prop}$$

$*211 \cdot 753$. $\vdash :. P \in \text{Rl}`J \cap \text{connex} . \alpha \subset C`P . Q = \breve{P} . \supset :$

$$\alpha \in \text{D}`(P_\epsilon \dot{\cap} I) . \equiv . C`P - \alpha \in \text{sect}`Q - \text{ᗡ}`\text{seq}_Q \quad [*211 \cdot 41 \cdot 7 \cdot 726]$$

$*211 \cdot 754$. $\vdash :. P \in \text{Rl}`J \cap \text{connex} . \alpha \subset C`P . Q = \breve{P} . \supset :$

$$\alpha \in \text{sect}`P - \text{D}`P_\epsilon . \equiv . C`P - \alpha \in \text{D}`(Q_\epsilon \dot{\cap} I) \cap \text{ᗡ}`\text{seq}_Q$$

Dem.

$\vdash . *211 \cdot 316 . \supset$

$\vdash :. \text{Hp} . \supset : \alpha \in \text{sect}`P - \text{D}`P_\epsilon . \equiv . \alpha \in \text{sect}`P \cap (\text{ᗡ}`\text{max}_P \cap - \text{ᗡ}`\text{seq}_P) .$

$[*211 \cdot 7 \cdot 726] \qquad \equiv . C`P - \alpha \in \text{sect}`Q \cap (\text{ᗡ}`\text{seq}_Q \cap - \text{ᗡ}`\text{max}_Q) .$

$[*211 \cdot 41] \qquad \equiv . C`P - \alpha \in \text{D}`(Q_\epsilon \dot{\cap} I) \cap \text{ᗡ}`\text{seq}_Q :. \supset \vdash . \text{Prop}$

$*211 \cdot 755$. $\vdash :. P \in \text{trans} \cap \text{connex} . \alpha \subset C`P . Q = \breve{P} . \supset :$

$$\alpha \in \text{D}`P_\epsilon - \text{D}`(P_\epsilon \dot{\cap} I) . \equiv . C`P - \alpha \in \text{D}`Q_\epsilon - \text{D}`(Q_\epsilon \dot{\cap} I) \quad [*211 \cdot 74]$$

$*211 \cdot 756$. $\vdash :. P \in \text{Rl}`J \cap \text{connex} . \alpha \subset C`P . Q = \breve{P} . \supset :$

$$\alpha \in \text{sect}`P - \text{D}`(P_\epsilon \dot{\cap} I) . \equiv . C`P - \alpha \in \text{sect}`Q \cap \text{ᗡ}`\text{seq}_Q \quad [*211 \cdot 41 \cdot 7 \cdot 726]$$

$*211 \cdot 757$. $\vdash :. P \in \text{Ser} . \alpha \subset C`P . \supset :$

$$\alpha \in \text{sect}`P - \text{D}`(P_\epsilon \dot{\cap} I) . \equiv . C`P - \alpha \in \overleftarrow{P}``C`P \quad [*211 \cdot 756 \cdot 302]$$

***211·76.** $\vdash : P \in \text{Ser} . \supset . \text{D}`P_\epsilon = (C`P -)``(\text{sect}`\breve{P} - \text{Cl}`\text{tl}_P)$

Dem.

$\vdash . *207·13 . \text{Transp} . \supset \vdash . - \text{Cl}`\text{tl}_P = \text{Cl}`\text{min}_P \cup - \text{Cl}`\text{seq}_P$ (1)

$\vdash . (1) . *211·751 . \supset \vdash :. \text{Hp} . \supset : \alpha \in \text{D}`P_\epsilon . \equiv . \alpha \subset C`P . C`P - \alpha \in \text{sect}`\breve{P} - \text{Cl}`\text{tl}_P .$

[*24·492] $\equiv . (\exists \beta) . \beta \in \text{sect}`\breve{P} - \text{Cl}`\text{tl}_P . \alpha = C`P - \beta .$

[*38·13] $\equiv . \alpha \in (C`P -)``(\text{sect}`\breve{P} - \text{Cl}`\text{tl}_P) :. \supset \vdash . \text{Prop}$

***211·761.** $\vdash : P \in \text{Ser} . \supset . \text{sect}`P \cap \text{Cl}`\text{lt}_P = (C`P -)``\{\text{sect}`\breve{P} - \text{D}`(\breve{P})_\epsilon\}$

[Proof as in *211·76]

***211·762.** $\vdash : P \in \text{Ser} . \supset . \text{D}`(P_\epsilon \dot{\cap} I) = (C`P -)``(\text{sect}`\breve{P} - \overleftarrow{P}``C`P)$

Dem.

$\vdash . *211·757 . \text{Transp} . \supset$

$\vdash :. \text{Hp} . \supset : \alpha \in \text{sect}`P . C`P - \alpha \sim\in \overleftarrow{P}``C`P . \equiv . \alpha \in \text{D}`(P_\epsilon \dot{\cap} I)$ (1)

$\vdash . (1) . *24·492 . *38·13 . \supset \vdash . \text{Prop}$

***211·8.** $\vdash : P_{\text{po}} \in \text{Ser} . \alpha \in \text{sect}`P . \supset .$

$\overrightarrow{\text{max}}_P`\alpha = \overrightarrow{\text{max}}\,(P_{\text{po}})`\alpha . \overrightarrow{\text{min}}_P`(C`P - \alpha) = \overrightarrow{\text{min}}\,(P_{\text{po}})`(C`P - \alpha) = \overrightarrow{\text{seq}}\,(P_{\text{po}})`\alpha$

Dem.

$\vdash . *211·13 . *91·602 . \quad \supset \vdash : \text{Hp} . \supset . \alpha \in \text{sect}`P_{\text{po}}$ (1)

$\vdash . *211·131 . *205·111 . \supset \vdash : \text{Hp} . \supset . \overrightarrow{\text{max}}_P`\alpha = \overrightarrow{\text{max}}\,(P_{\text{po}})`\alpha$ (2)

$\vdash . (2) \frac{\breve{P}}{P} . *211·7 . (1) . \quad \supset \vdash : \text{Hp} . \supset . \overrightarrow{\text{min}}_P`(C`P - \alpha) = \overrightarrow{\text{min}}\,(P_{\text{po}})`(C`P - \alpha)$ (3)

[*211·726] $= \overrightarrow{\text{seq}}\,(P_{\text{po}})`\alpha$ (4)

$\vdash . (2) . (3) . (4) . \supset \vdash . \text{Prop}$

The above proposition is used in *232·352 and *234·242.

The following propositions lead up to *211·82, which is used in *213·4. *211·83·841·9 are also used in *213.

***211·81.** $\vdash : P \in \text{Ser} . \alpha \in \text{sect}`P . \alpha \sim\in 1 . C`P - \alpha \in 1 . \supset .$

$C`P - \alpha = \iota`B`\breve{P} . P = P \upharpoonright \alpha \rightarrow B`\breve{P} . \alpha = \text{D}`P$

Dem.

$\vdash . *211·7·181·182 \frac{\breve{P}}{P} . \supset \vdash : \text{Hp} . \supset . C`P - \alpha = \iota`B`\breve{P}$ (1)

$\vdash . *204·461 . \quad \supset \vdash : \text{Hp} . \supset . P = P \upharpoonright \text{D}`P \rightarrow B`\breve{P}$ (2)

$\vdash . (1) . *211·1 . \quad \supset \vdash : \text{Hp} . \supset . \alpha = \text{D}`P$ (3)

$\vdash . (1) . (2) . (3) . \supset \vdash . \text{Prop}$

***211·811.** $\vdash : P \epsilon \mathrm{Ser} - \iota\text{‘}\dot{\Lambda} \,.\, P = Q \nrightarrow x \,.\, \supset \,.\, C\text{‘}Q \epsilon \mathrm{sect}\text{‘}P \,.\, x = B\text{‘}\breve{P} \,.\, C\text{‘}Q = \mathrm{D}\text{‘}P$

Dem.

$\vdash . *161\cdot11 . \supset \vdash :. \mathrm{Hp} . \supset : y \epsilon C\text{‘}Q . \supset_y . yPx :$

$[*204\cdot1] \quad \supset : x \sim \epsilon C\text{‘}Q :$

$[*161\cdot15] \quad \supset : x = B\text{‘}\breve{P} \quad (1)$

$\vdash . *161\cdot13 . \supset \vdash : \mathrm{Hp} . \supset . C\text{‘}Q = \mathrm{D}\text{‘}P \quad (2)$

$\vdash . (1) . (2) . *211\cdot1 . \supset \vdash . \mathrm{Prop}$

***211·812.** $\vdash :. P \epsilon \mathrm{Ser} - \iota\text{‘}\dot{\Lambda} \,.\, Q \epsilon \mathrm{D}\text{‘}P \upharpoonleft . \supset :$

$C\text{‘}Q \epsilon \mathrm{sect}\text{‘}P \,.\, C\text{‘}P - C\text{‘}Q \epsilon 1 \,.\, \equiv . (\exists x) . P = Q \nrightarrow x \,.\, \equiv . P = Q \nrightarrow B\text{‘}\breve{P}$

Dem.

$\vdash . *204\cdot4 . *201\cdot12 . \supset \vdash : \mathrm{Hp} . \supset . C\text{‘}Q \sim \epsilon 1 \quad (1)$

$\vdash . *204\cdot41 . \supset \vdash : \mathrm{Hp} . \supset . Q = P \upharpoonleft C\text{‘}Q \quad (2)$

$\vdash . (1) . (2) . *211\cdot81 . \supset \vdash : \mathrm{Hp} . C\text{‘}Q \epsilon \mathrm{sect}\text{‘}P . C\text{‘}P - C\text{‘}Q \epsilon 1 . \supset .$

$C\text{‘}P - C\text{‘}Q = \iota\text{‘}B\text{‘}\breve{P} . P = Q \nrightarrow B\text{‘}\breve{P} \quad (3)$

$\vdash . *211\cdot811 . \supset \vdash :. \mathrm{Hp} . \supset : (\exists x) . P = Q \nrightarrow x . \equiv . P = Q \nrightarrow B\text{‘}\breve{P} \quad (4)$

$\vdash . *211\cdot811 . \supset \vdash : \mathrm{Hp} . P = Q \nrightarrow x . \supset . C\text{‘}Q \epsilon \mathrm{sect}\text{‘}P . C\text{‘}P - C\text{‘}Q \epsilon 1 \quad (5)$

$\vdash . (3) . (4) . (5) . \supset \vdash . \mathrm{Prop}$

***211·82.** $\vdash :: P \epsilon \mathrm{Ser} . Q \epsilon \mathrm{D}\text{‘}P \upharpoonleft . \supset :.$

$C\text{‘}Q \epsilon \mathrm{sect}\text{‘}P . \equiv : (\exists R) . P = Q \dagger R . \vee . (\exists x) . P = Q \nrightarrow x :$

$\equiv : (\exists R) . P = Q \dagger R . \vee . P = Q \nrightarrow B\text{‘}\breve{P}$

$[*211\cdot282\cdot283\cdot812 . *160\cdot22 . *161\cdot2]$

***211·83.** $\vdash : \dot{\exists} ! P . x \sim \epsilon C\text{‘}P . \supset . \mathrm{sect}\text{‘}(P \nrightarrow x) = \mathrm{sect}\text{‘}P \cup \iota\text{‘}(C\text{‘}P \cup \iota\text{‘}x)$

Dem.

$\vdash . *211\cdot1 . \supset$

$\vdash :. \mathrm{Hp} . \supset : \alpha \epsilon \mathrm{sect}\text{‘}(P \nrightarrow x) . \equiv . \alpha \subset C\text{‘}P \cup \iota\text{‘}x . (P \nrightarrow x)\text{“}\alpha \subset \alpha \quad (1)$

$\vdash . (1) . *161\cdot11 . \supset$

$\vdash :. \mathrm{Hp} . \supset : \alpha \epsilon \mathrm{sect}\text{‘}(P \nrightarrow x) . x \epsilon \alpha . \equiv . \alpha \subset C\text{‘}P \cup \iota\text{‘}x . P\text{“}\alpha \cup C\text{‘}P \subset \alpha . x \epsilon \alpha .$

$[*22\cdot41] \quad \equiv . \alpha = C\text{‘}P \cup \iota\text{‘}x \quad (2)$

$\vdash . (1) . *161\cdot11 . \supset$

$\vdash :. \mathrm{Hp} . \supset : \alpha \epsilon \mathrm{sect}\text{‘}(P \nrightarrow x) . x \sim \epsilon \alpha . \equiv . \alpha \subset C\text{‘}P \cup \iota\text{‘}x . P\text{“}\alpha \subset \alpha . x \sim \epsilon \alpha .$

$[*51\cdot25] \quad \equiv . \alpha \subset C\text{‘}P . P\text{“}\alpha \subset \alpha .$

$[*211\cdot1] \quad \equiv . \alpha \epsilon \mathrm{sect}\text{‘}P \quad (3)$

$\vdash . (2) . (3) . \supset \vdash . \mathrm{Prop}$

$*211 \cdot 84$. $\vdash : C`P \cap C`Q = \Lambda . \supset . \text{sect}`(P \ddagger Q) = \text{sect}`P \cup (C`P \cup)``\text{sect}`Q$

$$= \text{sect}`P \cup (C`P \cup)``(\text{sect}`Q - \iota`\Lambda)$$

Dem.

$$\vdash . *211 \cdot 1 . \supset \vdash : \alpha \epsilon \text{sect}`(P \ddagger Q) . \equiv . \alpha \subset C`P \cup C`Q . (P \ddagger Q)``\alpha \subset \alpha \quad (1)$$

$$\vdash . (1) . *160 \cdot 11 . \supset$$

$$\vdash :. \text{Hp} . \supset : \alpha \epsilon \text{sect}`(P \ddagger Q) . \alpha \subset C`P . \equiv . \alpha \subset C`P . P``\alpha \subset \alpha .$$

$$[*211 \cdot 1] \qquad \equiv . \alpha \epsilon \text{sect}`P \quad (2)$$

$$\vdash . (1) . *160 \cdot 11 . \supset$$

$$\vdash :. \text{Hp} . \supset : \alpha \epsilon \text{sect}`(P \ddagger Q) . \exists ! \alpha \cap C`Q . \equiv .$$

$$\alpha \subset C`P \cup C`Q . C`P \cup Q``\alpha \subset \alpha . \exists ! \alpha \cap C`Q .$$

$$[*24 \cdot 43 \cdot 491] \equiv . \alpha - C`P \subset C`Q . C`P \subset \alpha . Q``\alpha \subset \alpha - C`P . \exists ! \alpha - C`P .$$

$$[*24 \cdot 491 . *37 \cdot 265]$$

$$\equiv . \alpha - C`P \subset C`Q . C`P \subset \alpha . Q``(\alpha - C`P) \subset \alpha - C`P . \exists ! \alpha - C`P .$$

$$[*211 \cdot 1] \qquad \equiv . \alpha - C`P \epsilon \text{sect}`Q - \iota`\Lambda . C`P \subset \alpha .$$

$$[*22 \cdot 92] \qquad \equiv . \alpha \epsilon (C`P \cup)``(\text{sect}`Q - \iota`\Lambda) \quad (3)$$

$$\vdash . *211 \cdot 26 \cdot 44 . \supset \vdash . C`P \epsilon \text{sect}`P . C`P \epsilon (C`P \cup)``(\text{sect}`Q \cap \iota`\Lambda) \quad (4)$$

$$\vdash . (2) . (3) . \supset \vdash : \text{Hp} . \supset . \text{sect}`(P \ddagger Q) = \text{sect}`P \cup (C`P \cup)``(\text{sect}`Q - \iota`\Lambda)$$

$$[(4)] \qquad = \text{sect}`P \cup (C`P \cup)``\text{sect}`Q : \supset \vdash . \text{Prop}$$

$*211 \cdot 841$. $\vdash : C`P \cap C`Q = \Lambda . \supset .$

$$\text{sect}`(P \ddagger Q) - \iota`\Lambda = (\text{sect}`P - \iota`\Lambda) \cup (C`P \cup)``(\text{sect}`Q - \iota`\Lambda) \quad [*211 \cdot 84]$$

$*211 \cdot 9$. $\vdash . \text{sect}`(x \downarrow y) = \iota`\Lambda \cup \iota`\iota`x \cup \iota`(\iota`x \cup \iota`y)$

Dem.

$$\vdash . *211 \cdot 1 \cdot 26 . \supset \vdash . \Lambda \epsilon \text{sect}`(x \downarrow y) . \iota`x \cup \iota`y \epsilon \text{sect}`(x \downarrow y) \quad (1)$$

$$\vdash . *55 \cdot 13 . \quad \supset \vdash : x \neq y . \supset . (x \downarrow y)``\iota`x = \Lambda \quad (2)$$

$$\vdash . *55 \cdot 13 . \quad \supset \vdash : x = y . \supset . (x \downarrow y)``\iota`x = \iota`x \quad (3)$$

$$\vdash . (2) . (3) . \quad \supset \vdash . (x \downarrow y)``\iota`x \subset \iota`x .$$

$$[*211 \cdot 1] \qquad \supset \vdash . \iota`x \epsilon \text{sect}`(x \downarrow y) \quad (4)$$

$$\vdash . *211 \cdot 1 . *54 \cdot 4 . \supset$$

$$\vdash :. \beta \epsilon \text{sect}`(x \downarrow y) . \supset : \beta = \Lambda . \vee . \beta = \iota`x . \vee . \beta = \iota`y . \vee . \beta = \iota`x \cup \iota`y \quad (5)$$

$$\vdash . *55 \cdot 13 . \quad \supset \vdash : x \neq y . \supset . x \epsilon (x \downarrow y)``\iota`y - \iota`y .$$

$$[*211 \cdot 1] \qquad \supset . \iota`y \sim \epsilon \text{sect}`P \quad (6)$$

$$\vdash . *51 \cdot 23 . \quad \supset \vdash : x = y . \supset . \iota`y = \iota`x \quad (7)$$

$$\vdash . (5) . (6) . (7) . \supset \vdash :. \beta \epsilon \text{sect}`(x \downarrow y) . \supset : \beta = \Lambda . \vee . \beta = \iota`x . \vee . \beta = \iota`x \cup \iota`y \quad (8)$$

$$\vdash . (1) . (4) . (8) . \supset \vdash . \text{Prop}$$

*212. THE SERIES OF SEGMENTS.

Summary of *212.

The series of segments or sections of a series may be ordered by the relation of inclusion, after the manner considered in *210. Since, as was shown in *211, sections and segments have the properties assigned to κ in the hypothesis of *210, the resulting series are such that every class has either a maximum or a sequent, and either a minimum or a precedent; *i.e.* the series of segments or sections are Dedekindian. Most of the properties of the series of sections and of the series of segments which have no maximum, only require that the original relation should be connected. The properties of the series of segments in general ($\mathrm{D}`P_\epsilon$) require also that the original relation should be transitive.

We denote the series of segments by $\varsigma`P$, putting

$$\varsigma`P = P_{\mathrm{lc}} \upharpoonright \mathrm{D}`P_\epsilon \quad \mathrm{Df.}$$

We then have, in virtue of *210·13 and *211·61,

***212·23.** $\vdash : P \in \mathrm{trans} \cap \mathrm{connex} . \supset . \varsigma`P = \hat{\alpha}\hat{\beta} \{\alpha, \beta \in \mathrm{D}`P_\epsilon . \alpha \subset \beta . \alpha \neq \beta\}$

In like manner, for the series of segments which have no maximum, we put

$$\mathrm{sgm}`P = P_{\mathrm{lc}} \upharpoonright \mathrm{D}`(P_\epsilon \dot{\cap} I) \quad \mathrm{Df,}$$

and we have

***212·22.** $\vdash : P \in \mathrm{connex} . \supset . \mathrm{sgm}`P = \hat{\alpha}\hat{\beta} \{\alpha, \beta \in \mathrm{D}`(P_\epsilon \dot{\cap} I) . \alpha \subset \beta . \alpha \neq \beta\}$

We do not need a special notation for the series of sections, since, in virtue of *211·13, it is $\varsigma`P_*$ or $\mathrm{sgm}`P_*$. Thus, by *212·23,

***212·24.** $\vdash : P_* \in \mathrm{connex} . \supset . \varsigma`P_* = \hat{\alpha}\hat{\beta} \{\alpha, \beta \in \mathrm{sect}`P . \alpha \subset \beta . \alpha \neq \beta\}$

We begin the number with various propositions on the fields, etc. of these relations, and on the conditions for their existence. We have

***212·132.** $\vdash . \mathrm{D}`\varsigma`P = \mathrm{D}`P_\epsilon - \iota`\mathrm{D}`P . \mathrm{Cl}`\varsigma`P = \mathrm{D}`P_\epsilon - \iota`\Lambda$

***212·133.** $\vdash : \dot{\exists} ! P . \supset . C`\varsigma`P = \mathrm{D}`P_\epsilon . B`\varsigma`P = \Lambda . B`\mathrm{Cnv}`\varsigma`P = \mathrm{D}`P$

***212·14.** $\vdash : \dot{\exists} ! P . \equiv . \dot{\exists} ! \varsigma`P$

***212·152.** $\vdash . \mathrm{Cl}`\mathrm{sgm}`P = \mathrm{D}`(P_\epsilon \dot{\cap} I) - \iota`\Lambda$

$*212\cdot17.\ \vdash : \dot{\exists} ! \varsigma`P_* . \equiv . \exists ! \text{sect}`P - \iota`\Lambda . \equiv . \text{sect}`P \sim\epsilon 1 . \equiv . \dot{\exists} ! P$

$*212\cdot172.\ \vdash : \dot{\exists} ! P . \supset . C`\varsigma`P_* = \text{sect}`P . B`\varsigma`P_* = \Lambda . B`\text{Cnv}`\varsigma`P_* = C`P$

Of the next set of propositions ($*212\cdot2$—$\cdot25$), several have already been mentioned. An important proposition is

$*212\cdot25.\ \vdash : P \epsilon \text{Ser} . \supset . \overrightarrow{P} ; P = (\varsigma`P) \upharpoonright \overrightarrow{P}``C`P$

for this shows that the series of segments contains a series similar to P.

We take up next the application of the propositions of $*210$ to the series of sections and segments. We show that if $P \epsilon$ connex, sgm$`P$ and $\varsigma`P_*$ are series ($*212\cdot3$), and that if P is also transitive, $\varsigma`P$ is a series ($*212\cdot31$). We have

$*212\cdot322.\ \vdash : P \epsilon \text{connex} . \dot{\exists} ! P . \lambda \subset \text{sect}`P . \supset . s`\lambda = \text{limax}(\varsigma`P_*)`\lambda$

$*212\cdot34.\ \vdash : P \epsilon \text{connex} . \dot{\exists} ! P . \lambda \subset \text{sect}`P . \supset . p`\lambda \cap C`P = \text{limin}(\varsigma`P_*)`\lambda$

so that every class of sections has both an upper limit or maximum and a lower limit or minimum ($*212\cdot35$).

We then prove similar propositions for $\varsigma`P$ and sgm$`P$, except that in place of $*212\cdot34$ we have

$*212\cdot431.\ \vdash : P \epsilon \text{trans} \cap \text{connex} . \dot{\exists} ! P . \lambda \subset \text{D}`P_\epsilon . \supset .$
$s`(\text{D}`P_\epsilon \cap \text{Cl}`p`\lambda) = \text{limin}(\varsigma`P)`\lambda$

$*212\cdot53.\ \vdash : P \epsilon \text{connex} . \dot{\exists} ! \text{sgm}`P . \lambda \subset \text{D}`(P_\epsilon \dot{\cap} I) . \supset .$
$s`\{\text{D}`(P_\epsilon \dot{\cap} I) \cap \text{Cl}`p`\lambda\} = \text{limin}(\text{sgm}`P)`\lambda$

The reason of the difference from $*212\cdot34$ is that the product of an existent class of segments may not be a segment. Suppose, for example, the segments are all those that contain a given term x, where x has no immediate successor; then their logical product is $\overrightarrow{P}`x \cup \iota`x$, which is a section but not a segment.

We have next ($*212\cdot6$—$\cdot667$) a number of propositions on the limits and maxima of sub-classes of $\overrightarrow{P}``C`P$ in the series $\varsigma`P$. The interest of this subject lies in its relation to irrationals. If $\alpha$ is a class contained in $C`P$ and having no limit or maximum, $\overrightarrow{P}``\alpha$ is contained in $C`\varsigma`P$, and has a limit in $\varsigma`P$. We may call this limit an *irrational* segment. There is no irrational term in $C`P$, because in P there is no limit to α; but the limit, in $\varsigma`P$, of $\overrightarrow{P}``\alpha$ may be called irrational, because it corresponds to no term in $C`P$. It should be observed that (as will be proved in Section F) if $P$ is similar to the series of rationals, $\varsigma`P$ is similar to the series of real numbers.

The most useful propositions in this subject are:

$*212\cdot6.\ \vdash : P \epsilon \text{Ser} . \alpha \subset C`P . \supset .$
$\overrightarrow{\max}(\varsigma`P)`\overrightarrow{P}``\alpha = \overrightarrow{\max}(\overrightarrow{P};P)`\overrightarrow{P}``\alpha = \overrightarrow{P}``\max_P`\alpha$

***212·601.** $\vdash :. P \in \mathrm{Ser} . \alpha \subset C\text{‘}P . \supset :$

$$\mathrm{E} ! \max_P\text{‘}\alpha . \equiv . \mathrm{E} ! \max (\overrightarrow{P} | P)\text{‘}\overrightarrow{P}\text{“}\alpha . \equiv . \mathrm{E} ! \max (\varsigma\text{‘}P)\text{‘}\overrightarrow{P}\text{“}\alpha$$

***212·602.** $\vdash :. P \in \mathrm{Ser} . \dot{\exists} ! P . \alpha \subset C\text{‘}P . \supset : \mathrm{E} ! \max_P\text{‘}\alpha . \equiv . P\text{“}\alpha \in \overrightarrow{P}\text{“}\alpha$

***212·61.** $\vdash : P \in \mathrm{trans} \cap \mathrm{connex} . \dot{\exists} ! P . \supset . \mathrm{limax} (\varsigma\text{‘}P)\text{‘}\overrightarrow{P}\text{“}\alpha = P\text{“}\alpha$

***212·632.** $\vdash : P \in \mathrm{Ser} . \dot{\exists} ! P . \alpha \subset C\text{‘}P . P\text{“}\alpha \sim\in \overrightarrow{P}\text{“}C\text{‘}P . \supset . P\text{“}\alpha = \mathrm{lt} (\varsigma\text{‘}P)\text{‘}\overrightarrow{P}\text{“}\alpha$

***212·661.** $\vdash : P \in \mathrm{Ser} . \kappa \subset \mathrm{D}\text{‘}P_\epsilon . \mathrm{E} ! \mathrm{lt} (\varsigma\text{‘}P)\text{‘}\kappa . \supset .$

$$\mathrm{lt} (\varsigma\text{‘}P)\text{‘}\kappa = \mathrm{lt} (\varsigma\text{‘}P)\text{‘}\overrightarrow{P}\text{“}s\text{‘}\kappa = s\text{‘}\kappa$$

This shows that every limit in the series of segments is a limit of a class of what we may call *rational* segments (*i.e.* segments of the form $\overrightarrow{P}\text{‘}x$), namely it is the limit of $\overrightarrow{P}\text{“}s\text{‘}\kappa$.

***212·667.** $\vdash : P \in \mathrm{Ser} . \supset . \mathrm{D}\text{‘}\mathrm{lt} (\varsigma\text{‘}P) - \iota\text{‘}\Lambda = \text{Ɑ}\text{‘}\mathrm{sgm}\text{‘}P$

This shows that the segments (other than Λ) which are limits of classes of segments are the segments (other than Λ) which have no maximum in P.

The number ends with a set of propositions (*212·7—·72) on the relations of the sections and segments of two correlated series. If S is a correlator of P with Q, then S_ϵ (with its converse domain limited) is a correlator of $\varsigma\text{‘}P_*$ with $\varsigma\text{‘}Q_*$, $\varsigma\text{‘}P$ with $\varsigma\text{‘}Q$ and $\mathrm{sgm}\text{‘}P$ with $\mathrm{sgm}\text{‘}Q$ (*212·71·711·712). Hence

***212·72.** $\vdash : P \,\mathrm{smor}\, Q . \supset . \varsigma\text{‘}P_* \,\mathrm{smor}\, \varsigma\text{‘}Q_* . \varsigma\text{‘}P \,\mathrm{smor}\, \varsigma\text{‘}Q . \mathrm{sgm}\text{‘}P \,\mathrm{smor}\, \mathrm{sgm}\text{‘}Q$

This proposition is used in the next number, and also in *271.

***212·01.** $\varsigma\text{‘}P = P_{\mathrm{lc}} \upharpoonright \mathrm{D}\text{‘}P_\epsilon$ Df

***212·02.** $\mathrm{sgm}\text{‘}P = P_{\mathrm{lc}} \upharpoonright \mathrm{D}\text{‘}(P_\epsilon \dot{\cap} I)$ Df

***212·1.** $\vdash : \alpha (\varsigma\text{‘}P) \beta . \equiv . \alpha, \beta \in \mathrm{D}\text{‘}P_\epsilon . \exists ! \beta - \alpha - P\text{“}(\alpha - \beta)$
[*170·102 . *37·15]

***212·11.** $\vdash : \alpha (\mathrm{sgm}\text{‘}P) \beta . \equiv . \alpha, \beta \in \mathrm{D}\text{‘}(P_\epsilon \dot{\cap} I) . \exists ! \beta - \alpha$

Dem.

$\vdash . \text{*}170\cdot102 . \text{*}37\cdot15 . \supset$

$\vdash : \alpha (\mathrm{sgm}\text{‘}P) \beta . \equiv . \alpha, \beta \in \mathrm{D}\text{‘}(P_\epsilon \dot{\cap} I) . \exists ! \beta - \alpha - P\text{“}(\alpha - \beta)$ (1)

$\vdash . \text{*}211\cdot12 . \supset \vdash : \alpha \in \mathrm{D}\text{‘}(P_\epsilon \dot{\cap} I) . \supset . -\alpha = -P\text{“}\alpha .$

[*37·2.Transp] $\supset . -\alpha \subset -P\text{“}(\alpha - \beta) .$

[*22·621] $\supset . -\alpha - P\text{“}(\alpha - \beta) = -\alpha$ (2)

$\vdash . (1) . (2) . \supset \vdash .$ Prop

***212·12.** $\vdash : \alpha (\mathrm{sgm}\text{‘}P_*) \beta . \equiv . \alpha, \beta \in \mathrm{sect}\text{‘}P . \exists ! \beta - \alpha$ [*211·13 . *212·11]

Thus $\mathrm{sgm}\text{‘}P_*$ has the same connection with $\mathrm{sect}\text{‘}P$ as $\mathrm{sgm}\text{‘}P$ has with $\mathrm{D}\text{‘}(P_\epsilon \dot{\cap} I)$. When P is transitive, $\mathrm{sgm}\text{‘}P_*$ also has the same connection with sect'P as $\varsigma'P$ has with $\text{D}'P_\epsilon$. The following proposition makes these facts more explicit.

***212·121.** $\vdash . \text{sgm}'P_* = \varsigma'P_* = P_{1c} \upharpoonright \text{sect}'P$

Dem.

$\vdash . *211\cdot13 . \supset \vdash . \text{sgm}'P_* = P_{1c} \upharpoonright \text{sect}'P \quad (1)$

$\vdash . *212\cdot1 . *211\cdot13 . \supset \vdash : \alpha (\varsigma'P_*) \beta . \equiv . \alpha, \beta \in \text{sect}'P . \exists ! \beta - \alpha - P_*``(\alpha - \beta) \quad (2)$

$\vdash . *211\cdot13 . \supset \vdash : \alpha \in \text{sect}'P . \supset . \alpha = P_*``\alpha .$

$[*37\cdot2] \qquad \supset . P_*``(\alpha - \beta) \subset \alpha .$

$[\text{Transp}. *22\cdot621] \qquad \supset . - \alpha - P_*``(\alpha - \beta) = - \alpha \quad (3)$

$\vdash . (2) . (3) . \supset \vdash : \alpha (\varsigma'P_*) \beta . \equiv . \alpha, \beta \in \text{sect}'P . \exists ! \beta - \alpha .$

$[*211\cdot12] \qquad \equiv \alpha (\text{sgm}'P_*) \beta \quad (4)$

$\vdash . (1) . (4) . \supset \vdash . \text{Prop}$

***212·122.** $\vdash . \varsigma'P, \text{sgm}'P \in \text{Rl}'J \qquad [*170\cdot17]$

***212·123.** $\vdash . C`\varsigma'P, C`\text{sgm}'P \sim\in 1 \qquad [*200\cdot12 . *212\cdot122]$

***212·13.** $\vdash : \Lambda (\varsigma'P) \beta . \equiv . \beta \in \text{D}'P_\epsilon - \iota'\Lambda \qquad [*170\cdot6]$

***212·131.** $\vdash : \alpha (\varsigma'P) (\text{D}'P) . \equiv . \alpha \in \text{D}'P_\epsilon - \iota'\text{D}'P$

Dem.

$\vdash . *212\cdot1 . \supset \vdash : \alpha (\varsigma'P) (\text{D}'P) . \equiv . \alpha, \text{D}'P \in \text{D}'P_\epsilon . \exists ! \text{D}'P - \alpha - P``(\alpha - \text{D}'P) .$

$[*211\cdot301 . *37\cdot15] \qquad \equiv . \alpha \in \text{D}'P_\epsilon . \alpha \subset \text{D}'P . \exists ! \text{D}'P - \alpha .$

$[*24\cdot55 . *22\cdot41] \qquad \equiv . \alpha \in \text{D}'P_\epsilon . \alpha \subset \text{D}'P . \alpha \neq \text{D}'P .$

$[*37\cdot15] \qquad \equiv . \alpha \in \text{D}'P_\epsilon . \alpha \neq \text{D}'P : \supset \vdash . \text{Prop}$

***212·132.** $\vdash . \text{D}'\varsigma'P = \text{D}'P_\epsilon - \iota'\text{D}'P . \text{ᗡ}'\varsigma'P = \text{D}'P_\epsilon - \iota'\Lambda$

Dem.

$\vdash . *212\cdot13\cdot131 . \supset \vdash . \text{D}'P_\epsilon - \iota'\text{D}'P \subset \text{D}'\varsigma'P . \text{D}'P_\epsilon - \iota'\Lambda \subset \text{ᗡ}'\varsigma'P \quad (1)$

$\vdash . *212\cdot1 . \supset \vdash . \text{D}'\varsigma'P \subset \text{D}'P_\epsilon . \text{ᗡ}'\varsigma'P \subset \text{D}'P_\epsilon \quad (2)$

$\vdash . *212\cdot1 . \supset \vdash : \alpha \in \text{D}'\varsigma'P . \supset . (\exists\beta) . \beta \in \text{D}'P_\epsilon . \exists ! \beta - \alpha .$

$[*37\cdot15] \qquad \supset . \exists ! \text{D}'P - \alpha \quad (3)$

$\vdash . *212\cdot1 . \supset \vdash : \beta \in \text{ᗡ}'\varsigma'P . \supset . (\exists\alpha) . \exists ! \beta - \alpha .$

$[*24\cdot561] \qquad \supset . \exists ! \beta \quad (4)$

$\vdash . (3) . (4) . \supset \vdash . \text{D}'\varsigma'P \subset - \iota'\text{D}'P . \text{ᗡ}'\varsigma'P \subset - \iota'\Lambda \quad (5)$

$\vdash . (1) . (2) . (5) . \supset \vdash . \text{Prop}$

***212·133.** $\vdash : \dot{\exists} ! P . \supset . C`\varsigma'P = \text{D}'P_\epsilon . B`\varsigma'P = \Lambda . B`\text{Cnv}'\varsigma'P = \text{D}'P$

Dem.

$\vdash . *33\cdot24 . \supset \vdash : \text{Hp} . \supset . \Lambda \neq \text{D}'P .$

$[*212\cdot132] \qquad \supset . \Lambda \in \text{D}'\varsigma'P . \text{D}'P \in \text{ᗡ}'\varsigma'P .$

[*51·221] $\supset . \text{D}`\varsigma`P = \{(\text{D}`P_\epsilon - \iota`\text{D}`P) - \iota`\Lambda\} \cup \iota`\Lambda .$

[*212·132] $\supset . C`\varsigma`P = \{(\text{D}`P_\epsilon - \iota`\Lambda) - \iota`\text{D}`P\} \cup \iota`\Lambda \cup (\text{D}`P_\epsilon - \iota`\Lambda)$

[*22·63] $= (\text{D}`P_\epsilon - \iota`\Lambda) \cup \iota`\Lambda$

[*51·221] $= \text{D}`P_\epsilon$ (1)

$\vdash . (1) . *93·103 . *212·132 . \supset \vdash : \text{Hp} . \supset . \overrightarrow{B}`\varsigma`P = \text{D}`P_\epsilon - (\text{D}`P_\epsilon - \iota`\Lambda)$

[*211·44] $= \iota`\Lambda$ (2)

$\vdash . (1) . *93·103 . *212·132 . \supset \vdash : \text{Hp} . \supset . \overrightarrow{B}`\text{Cnv}`\varsigma`P = \text{D}`P_\epsilon - (\text{D}`P_\epsilon - \iota`\text{D}`P)$

[*211·301] $= \iota`\text{D}`P$ (3)

$\vdash . (1) . (2) . (3) . \supset \vdash . \text{Prop}$

***212·134.** $\vdash : P = \dot{\Lambda} . \supset . \varsigma`P = \dot{\Lambda}$ [*170·35]

***212·14.** $\vdash : \dot{\exists} ! P . \equiv . \dot{\exists} ! \varsigma`P$

Dem.

$\vdash . *212·133 . *211·301 . \supset \vdash : \dot{\exists} ! P . \supset . \exists ! C`\varsigma`P .$

[*33·24] $\supset . \dot{\exists} ! \varsigma`P$ (1)

$\vdash . (1) . *212·134 . \supset \vdash . \text{Prop}$

***212·141.** $\vdash : \alpha \epsilon C`\varsigma`P . \equiv . \alpha \epsilon \text{D}`P_\epsilon . \dot{\exists} ! P$

Dem.

$\vdash . *10·24 . \supset \vdash : \alpha \epsilon C`\varsigma`P . \supset . \exists ! C`\varsigma`P .$

[*33·24.*212·14] $\supset . \dot{\exists} ! P .$ (1)

[*212·133.Hp] $\supset . \alpha \epsilon \text{D}`P_\epsilon$ (2)

$\vdash . *212·133 . \supset \vdash : \alpha \epsilon \text{D}`P_\epsilon . \dot{\exists} ! P . \supset . \alpha \epsilon C`\varsigma`P$ (3)

$\vdash . (1) . (2) . (3) . \supset \vdash . \text{Prop}$

***212·142.** $\vdash : \dot{\exists} ! \varsigma`P . \equiv . \text{D}`P_\epsilon \sim \epsilon 1$

Dem.

$\vdash . *211·66 . *212·14 . \supset \vdash : \dot{\exists} ! \varsigma`P . \supset . \text{D}`P_\epsilon \sim \epsilon 1$ (1)

$\vdash . *212·132 . *211·44 . \supset \vdash : \text{D}`P_\epsilon \sim \epsilon 1 . \supset . \exists ! \text{C}\!\!\text{I}`\varsigma`P .$

[*33·24] $\supset . \dot{\exists} ! \varsigma`P$ (2)

$\vdash . (1) . (2) . \supset \vdash . \text{Prop}$

***212·15.** $\vdash : \Lambda (\text{sgm}`P) \beta . \equiv . \beta \epsilon \text{D}`(P_\epsilon \dot{\cap} I) - \iota`\Lambda$ [Proof as in *212·13]

***212·151.** $\vdash : P = \dot{\Lambda} . \supset . \text{sgm}`P = \dot{\Lambda}$ [*170·35]

The converse implication does not hold in this case. For the existence of sgm‘P, it is necessary that $C`P$ should contain classes having no maximum.

***212·152.** $\vdash . \text{C}\!\!\text{I}`\text{sgm}`P = \text{D}`(P_\epsilon \dot{\cap} I) - \iota`\Lambda$ [Proof as in *212·132]

***212·153.** $\vdash : \dot{\exists} ! \text{sgm}'P . \equiv . \exists ! D'(P_\epsilon \dot{\frown} I) - \iota'\Lambda . \equiv . D'(P_\epsilon \dot{\frown} I) \sim \epsilon 1$

Dem.

$\vdash . *212·15 . \supset \vdash : \exists ! D'(P_\epsilon \dot{\frown} I) - \iota'\Lambda . \supset . \dot{\exists} ! \text{sgm}'P$ (1)

$\vdash . *212·152 . \supset \vdash : \dot{\exists} ! \text{sgm}'P . \supset . \exists ! D'(P_\epsilon \dot{\frown} I) - \iota'\Lambda$ (2)

$\vdash . *212·11 . \supset \vdash : \dot{\exists} ! \text{sgm}'P . \supset . (\exists \alpha, \beta) . \alpha, \beta \epsilon D'(P_\epsilon \dot{\frown} I) . \alpha \neq \beta .$

[*52·16.Transp] $\supset . D'(P_\epsilon \dot{\frown} I) \sim \epsilon 1$ (3)

$\vdash . *211·44 . *52·181 . \supset$

$\vdash : D'(P_\epsilon \dot{\frown} I) \sim \epsilon 1 . \supset . (\exists \beta) . \beta \epsilon D'(P_\epsilon \dot{\frown} I) . \beta \neq \Lambda .$

[*212·15] $\supset . \dot{\exists} ! \text{sgm}'P$ (4)

$\vdash . (1) . (2) . (3) . (4) . \supset \vdash .$ Prop

***212·154.** $\vdash : \dot{\exists} ! \text{sgm}'P . \supset . C'\text{sgm}'P = D'(P_\epsilon \dot{\frown} I)$

Dem.

$\vdash . *212·153·15 . \supset \vdash : \text{Hp} . \supset . \Lambda \epsilon D'\text{sgm}'P .$

[*212·152] $\supset . D'(P_\epsilon \dot{\frown} I) \subset C'\text{sgm}'P$ (1)

$\vdash . *212·11 . \supset \vdash . C'\text{sgm}'P \subset D'(P_\epsilon \dot{\frown} I)$ (2)

$\vdash . (1) . (2) . \supset \vdash .$ Prop

***212·155.** $\vdash : \dot{\exists} ! \text{sgm}'P . \supset . \Lambda = B'\text{sgm}'P$ [*212·152·154 . *93·103]

***212·156.** $\vdash : \alpha \epsilon C'\text{sgm}'P . \equiv . \alpha \epsilon D'(P_\epsilon \dot{\frown} I) . \dot{\exists} ! \text{sgm}'P .$
$\equiv . \alpha \epsilon D'(P_\epsilon \dot{\frown} I) . D'(P_\epsilon \dot{\frown} I) \sim \epsilon 1$

Dem.

$\vdash . *212·154 . \supset \vdash : \alpha \epsilon D'(P_\epsilon \dot{\frown} I) . \dot{\exists} ! \text{sgm}'P . \supset . \alpha \epsilon C'\text{sgm}'P$ (1)

$\vdash . *10·24 . *33·24 . \supset \vdash : \alpha \epsilon C'\text{sgm}'P . \supset . \dot{\exists} ! \text{sgm}'P .$ (2)

[*212·154] $\supset . \alpha \epsilon D'(P_\epsilon \dot{\frown} I)$ (3)

$\vdash . (1) . (2) . (3) . *212·153 . \supset \vdash .$ Prop

***212·16.** $\vdash : \text{Cl}'P \subset D'P . \supset . D'P \epsilon D'(P_\epsilon \dot{\frown} I)$

Dem.

$\vdash . *37·27 . \supset \vdash : \text{Hp} . \supset . P''D'P = D'P$ (1)

$\vdash . (1) . *211·12 . \supset \vdash .$ Prop

***212·161.** $\vdash : \text{Cl}'P \subset D'P . \dot{\exists} ! P . \supset . \dot{\exists} ! \text{sgm}'P$

Dem.

$\vdash . *33·24 . *212·16 . \supset \vdash : \text{Hp} . \supset . D'P \epsilon D'(P_\epsilon \dot{\frown} I) - \iota'\Lambda .$

[*212·15] $\supset . \Lambda (\text{sgm}'P) (D'P) .$

[*11·36] $\supset . \dot{\exists} ! \text{sgm}'P : \supset \vdash .$ Prop

***212·162.** $\vdash : \text{Cl}'P \subset D'P . \dot{\exists} ! P . \supset .$
$D'P = B'\text{Cnv}'\text{sgm}'P . D'\text{sgm}'P = D'(P_\epsilon \dot{\frown} I) - \iota'D'P$

Dem.

$\vdash . *212·16·152 . *33·24 . \supset \vdash : \text{Hp} . \supset . D'P \epsilon \text{Cl}'\text{sgm}'P$ (1)

$\vdash . *212\cdot11 . *37\cdot24 . \supset \vdash : \text{Hp} . \alpha \in \text{D}`(P_\epsilon \dot{\cap} I) - \iota`\text{D}`P . \supset . \alpha\,(\text{sgm}`P)\,(\text{D}`P)$ (2)

$\vdash . *37\cdot24 . \quad \supset \vdash : \alpha \in \text{D}`(P_\epsilon \dot{\cap} I) . \supset . \sim \exists ! (\alpha - \text{D}`P) .$

$[*212\cdot11] \quad \supset . \sim \{(\text{D}`P)\,(\text{sgm}`P)\,\alpha\}$ (3)

$\vdash . (2) . (3) . \supset \vdash : \text{Hp} . \supset . \text{D}`(P_\epsilon \dot{\cap} I) - \iota`\text{D}`P \subset \text{D}`\text{sgm}`P . \text{D}`P \sim\in \text{D}`\text{sgm}`P$ (4)

$\vdash . (1) . (4) . *212\cdot154 . \supset \vdash . \text{Prop}$

***212·17.** $\vdash : \dot{\exists} ! \varsigma`P_* . \equiv . \exists ! \text{sect}`P - \iota`\Lambda . \equiv . \text{sect}`P \sim\in 1 . \equiv . \dot{\exists} ! P$

Dem.

$\vdash . *212\cdot132 . *211\cdot13 . \supset \vdash : \dot{\exists} ! \varsigma`P_* . \equiv . \exists ! \text{sect}`P - \iota`\Lambda$ (1)

$\vdash . *212\cdot142 . *211\cdot13 . \supset \vdash : \dot{\exists} ! \varsigma`P_* . \equiv . \text{sect}`P \sim\in 1$ (2)

$\vdash . *212\cdot14 . \quad \supset \vdash : \dot{\exists} ! \varsigma`P_* . \equiv . \dot{\exists} ! P_* .$

$[*90\cdot141] \quad \equiv . \dot{\exists} ! P$ (3)

$\vdash . (1) . (2) . (3) . \supset \vdash . \text{Prop}$

***212·171.** $\vdash . \text{D}`\varsigma`P_* = \text{sect}`P - \iota`C`P . \text{Cl}`\varsigma`P_* = \text{sect}`P - \iota`\Lambda$
$[*212\cdot132 . *211\cdot13 . *90\cdot14]$

***212·172.** $\vdash : \dot{\exists} ! P . \supset . C`\varsigma`P_* = \text{sect}`P . B`\varsigma`P_* = \Lambda . B`\text{Cnv}`\varsigma`P_* = C`P$
$[*212\cdot133 . *211\cdot13 . *90\cdot141]$

***212·173.** $\vdash : \alpha \in C`\varsigma`P_* . \equiv . \alpha \in \text{sect}`P . \dot{\exists} ! P . \equiv . \alpha \in \text{sect}`P . \text{sect}`P \sim\in 1$
$[*212\cdot141\cdot142\cdot14 . *211\cdot13]$

***212·18.** $\vdash . \varsigma`\breve{P}_* = (C`P -)^{\boldsymbol{;}}\text{Cnv}`\varsigma`P_*$

Dem.

$\vdash . *212\cdot12\cdot121 . \supset \vdash : \alpha\,(\varsigma`\breve{P}_*)\,\beta . \equiv . \alpha, \beta \in \text{sect}`\breve{P} . \exists ! \beta - \alpha .$

$[*211\cdot7] \quad \equiv . (\exists\gamma, \delta) . \gamma, \delta \in \text{sect}`P . \alpha = C`P - \gamma . \beta = C`P - \delta . \exists ! \beta - \alpha .$

$[*24\cdot55] \quad \equiv . (\exists\gamma, \delta) . \gamma, \delta \in \text{sect}`P . \alpha = C`P - \gamma . \beta = C`P - \delta . \sim (\beta \subset \alpha) .$

$[*211\cdot1 . *24\cdot492] \equiv . (\exists\gamma, \delta) . \gamma, \delta \in \text{sect}`P . \alpha = C`P - \gamma . \beta = C`P - \delta . \sim (\gamma \subset \delta) .$

$[*212\cdot12 . *24\cdot55] \equiv . \alpha \{(C`P -)^{\boldsymbol{;}}\text{Cnv}`\varsigma`P_*\} \beta : \supset \vdash . \text{Prop}$

***212·181.** $\vdash . (\varsigma`\breve{P}_*) \,\text{smor}\, (\text{Cnv}`\varsigma`P_*)$ $[*212\cdot18]$

The above proposition is used in *252·43.

***212·2.** $\vdash . \text{sgm}`P \in \varsigma`P . \text{sgm}`P \in \varsigma`P_*$ $[*211\cdot14 . *212\cdot1\cdot11\cdot12]$

***212·21.** $\vdash : P \in \text{trans} . \supset . \varsigma`P \in \varsigma`P_*$ $[*211\cdot15 . *212\cdot12]$

***212·22.** $\vdash : P \in \text{connex} . \supset . \text{sgm}`P = \hat{\alpha}\hat{\beta} \{\alpha, \beta \in \text{D}`(P_\epsilon \dot{\cap} I) . \alpha \subset \beta . \alpha \neq \beta\}$
$[*211\cdot62 . *210\cdot1 . *212\cdot11]$

***212·23.** $\vdash : P \in \text{trans} \cap \text{connex} . \supset . \varsigma`P = \hat{\alpha}\hat{\beta} \{\alpha, \beta \in \text{D}`P_\epsilon . \alpha \subset \beta . \alpha \neq \beta\}$
$[*210\cdot13 . *211\cdot61 . (*212\cdot01)]$

$*212\cdot24$. $\vdash : P_* \epsilon \text{connex} . \supset . \varsigma`P_* = \hat{\alpha}\hat{\beta}\{\alpha, \beta \epsilon \text{sect}`P . \alpha \subset \beta . \alpha \neq \beta\}$
[$*212\cdot121\cdot22 . *211\cdot13$]

$*212\cdot25$. $\vdash : P \epsilon \text{Ser} . \supset . \overrightarrow{P};P = (\varsigma`P) \upharpoonright \overrightarrow{P}``C`P$

Dem.

$\vdash . *204\cdot33\cdot331 . \supset$

$\vdash :. \text{Hp} . \supset : \alpha(\overrightarrow{P};P)\beta . \equiv . \alpha, \beta \epsilon \overrightarrow{P}``C`P . \alpha \subset \beta . \alpha \neq \beta .$

[$*212\cdot23.*211\cdot3$] $\equiv . \alpha, \beta \epsilon \overrightarrow{P}``C`P . \alpha(\varsigma`P)\beta :. \supset \vdash . \text{Prop}$

The following propositions, down to $*212\cdot55$, consist of applications of the propositions of $*210$, where the κ of that number is replaced by sect$`P$, D$`P_\epsilon$, or D$`(P_\epsilon \dot{\cap} I)$, and the $Q$ is replaced by $P_{1c} \upharpoonright \kappa$, *i.e.* by $\varsigma`P_*$, $\varsigma`P$, or sgm$`P$. The propositions which follow are important, since the use of segments, especially in connection with continuity, depends largely upon them.

$*212\cdot3$. $\vdash : P \epsilon \text{connex} . \supset . \text{sgm}`P, \varsigma`P_* \epsilon \text{Ser}$
[$*211\cdot67 . *210\cdot14 . *212\cdot121$]

$*212\cdot31$. $\vdash : P \epsilon \text{trans} \cap \text{connex} . \supset . \varsigma`P \epsilon \text{Ser}$
[$*211\cdot68 . *210\cdot14 . (*212\cdot01)$]

$*212\cdot32$. $\vdash : P \epsilon \text{connex} . \dot{\exists} ! P . \lambda \subset \text{sect}`P . s`\lambda \epsilon \lambda . \supset . s`\lambda = \max(\varsigma`P_*)`\lambda$
[$*210\cdot211 . *211\cdot67 . *212\cdot17$]

We write max $(\varsigma`P_*)`\lambda$, instead of putting $\varsigma`P_*$ below the line, because, when we have to deal with an expression not consisting of a single letter, it is inconvenient to write it as a suffix, especially when it contains a suffix itself, as in this case.

$*212\cdot321$. $\vdash : P \epsilon \text{connex} . \dot{\exists} ! P . \lambda \subset \text{sect}`P . s`\lambda \sim \epsilon \lambda . \supset . s`\lambda = \text{seq}(\varsigma`P_*)`\lambda$
$= \text{lt}(\varsigma`P_*)`\lambda$
[$*210\cdot231 . *211\cdot67 . *212\cdot17 . *211\cdot63$]

$*212\cdot322$. $\vdash : P \epsilon \text{connex} . \dot{\exists} ! P . \lambda \subset \text{sect}`P . \supset . s`\lambda = \text{limax}(\varsigma`P_*)`\lambda$
[$*212\cdot32\cdot321 . *207\cdot46$]

$*212\cdot33$. $\vdash : P \epsilon \text{connex} . \dot{\exists} ! P . \lambda \subset \text{sect}`P . p`\lambda \cap C`P \epsilon \lambda . \supset .$
$p`\lambda \cap C`P = \min(\varsigma`P_*)`\lambda$

Dem.

$\vdash . *211\cdot671 . *210\cdot252 . *211\cdot26 . \supset$

$\vdash : \text{Hp} . \supset . p`\lambda \cap C`P \epsilon \overrightarrow{\min}(\varsigma`P_*)`\lambda \cup \overrightarrow{\text{prec}}(\varsigma`P_*)`\lambda$ (1)

$\vdash . *206\cdot2 . \supset \vdash : p`\lambda \cap C`P \epsilon \lambda . \supset . p`\lambda \cap C`P \sim \epsilon \overrightarrow{\text{prec}}(\varsigma`P_*)`\lambda$ (2)

$\vdash . (1) . (2) . \supset \vdash : \text{Hp} . \supset . p`\lambda \cap C`P \epsilon \overrightarrow{\min}(\varsigma`P_*)`\lambda$ (3)

$\vdash . (3) . *205\cdot31 . \supset \vdash . \text{Prop}$

***212·331.** $\vdash : P \epsilon \text{connex} . \dot{\exists} ! P . \lambda \subset \text{sect}`P . p`\lambda \cap C`P \sim\epsilon \lambda . \supset .$

$p`\lambda \cap C`P = \text{prec}(\varsigma`P_*)`\lambda = \text{tl}(\varsigma`P_*)`\lambda$

Dem.

$\vdash . *211\cdot671 . *210\cdot252 . *211\cdot26 . \supset$

$\vdash : \text{Hp} . \supset . p`\lambda \cap C`P = \text{limin}(\varsigma`P_*)`\lambda$ (1)

$\vdash . *205\cdot1 . \text{Transp} . \supset \vdash : p`\lambda \cap C`P \sim\epsilon \lambda . \supset . p`\lambda \cap C`P \sim\epsilon \overrightarrow{\min}(\varsigma`P_*)`\lambda$ (2)

$\vdash . (1) . (2) . *206\cdot161 . \supset \vdash . \text{Prop}$

***212·34.** $\vdash : P \epsilon \text{connex} . \dot{\exists} ! P . \lambda \subset \text{sect}`P . \supset . p`\lambda \cap C`P = \text{limin}(\varsigma`P_*)`\lambda$

[*212·33·331 . *207·46]

***212·35.** $\vdash : P \epsilon \text{connex} . \dot{\exists} ! P . \supset .$

$(\lambda) . \lambda \epsilon \{\text{Cl}`\max(\varsigma`P_*) \cup \text{Cl}`\text{seq}(\varsigma`P_*)\} \cap \{\text{Cl}`\min(\varsigma`P_*) \cup \text{Cl}`\text{prec}(\varsigma`P_*)\}$

[*210·28 . *211·671 . *212·121]

***212·36.** $\vdash :. P \epsilon \text{connex} . \supset : \lambda \epsilon C`\text{sgm}`\varsigma`P_* . \supset . E ! \text{seq}(\varsigma`P_*)`\lambda$

Dem.

$\vdash . *211\cdot47 . *212\cdot35\cdot3 . \supset$

$\vdash :. \text{Hp} . \dot{\exists} ! P . \supset : \lambda \epsilon C`\text{sgm}`\varsigma`P_* . \supset . E ! \text{seq}(\varsigma`P_*)`\lambda$ (1)

$\vdash . *33\cdot24 . \supset \vdash : \lambda \epsilon C`\text{sgm}`\varsigma`P_* . \supset . \dot{\exists} ! \text{sgm}`\varsigma`P_* .$

[*212·151.Transp] $\supset . \dot{\exists} ! \varsigma`P_* .$

[*212·17] $\supset . \dot{\exists} ! P$ (2)

$\vdash . (1) . (2) . \supset \vdash . \text{Prop}$

***212·4.** $\vdash : P \epsilon \text{trans} \cap \text{connex} . \dot{\exists} ! P . \lambda \subset D`P_\epsilon . s`\lambda \epsilon \lambda . \supset . s`\lambda = \max(\varsigma`P)`\lambda$

[*211·68·66 . *210·211]

***212·401.** $\vdash : P \epsilon \text{trans} \cap \text{connex} . \dot{\exists} ! P . \lambda \subset D`P_\epsilon . s`\lambda \sim\epsilon \lambda . \supset .$

$s`\lambda = \text{seq}(\varsigma`P)`\lambda = \text{lt}(\varsigma`P)`\lambda$

[*211·68·66·64 . *210·231]

***212·402.** $\vdash : P \epsilon \text{trans} \cap \text{connex} . \dot{\exists} ! P . \lambda \subset D`P_\epsilon . \supset . s`\lambda = \text{limax}(\varsigma`P)`\lambda$

[*212·4·401 . *207·46]

***212·41.** $\vdash : P \epsilon \text{trans} \cap \text{connex} . \dot{\exists} ! P . \lambda \subset D`P_\epsilon . p`\lambda \epsilon \lambda . \supset . p`\lambda = \min(\varsigma`P)`\lambda$

[*211·68·66 . *210·21]

***212·411.** $\vdash : P \epsilon \text{trans} \cap \text{connex} . \dot{\exists} ! P . \lambda \subset D`P_\epsilon . p`\lambda \epsilon D`P_\epsilon - \lambda . \supset .$

$p`\lambda = \text{prec}(\varsigma`P)`\lambda = \text{tl}(\varsigma`P)`\lambda$

[*211·68·66 . *210·23]

***212·42.** $\vdash : P \epsilon \text{trans} \cap \text{connex} . \dot{\exists} ! P . \lambda \subset D`P_\epsilon . p`\lambda \sim\epsilon \lambda . \supset .$

$s`(D`P_\epsilon \cap \text{Cl}`p`\lambda) = \text{prec}(\varsigma`P)`\lambda = \text{tl}(\varsigma`P)`\lambda$

[*210·26·22 . *211·68·66·64]

The cases considered in $*212\cdot411$ and $*212\cdot42$ are not mutually exclusive, since if $p`\lambda \,\epsilon\, \text{D}`P_\epsilon$, we have $s`(\text{D}`P_\epsilon \cap \text{Cl}`p`\lambda) = p`\lambda$.

$*212\cdot421$. $\vdash : P \,\epsilon\, \text{trans} \cap \text{connex} \,.\, \dot{\exists} ! P \,.\, \lambda \subset \text{D}`P_\epsilon \,.\, p`\lambda \sim\epsilon\, \text{D}`P_\epsilon \,.\supset .$
$$s`(\text{D}`P_\epsilon \cap \text{Cl}`p`\lambda) = P``p`\lambda$$

Dem.

$\vdash . *211\cdot151 . \quad \supset \vdash :. \text{Hp} . \supset : \alpha \,\epsilon\, \lambda . \supset_\alpha . P``\alpha \subset \alpha :$
$[*40\cdot81] \quad \supset : P``p`\lambda \subset p`\lambda \quad (1)$
$\vdash . (1) . *211\cdot11 . \quad \supset \vdash : \text{Hp} . \supset . P``p`\lambda \,\epsilon\, \text{D}`P_\epsilon \cap \text{Cl}`p`\lambda .$
$[*40\cdot13] \quad \supset . P``p`\lambda \subset s`(\text{D}`P_\epsilon \cap \text{Cl}`p`\lambda) \quad (2)$
$\vdash . *13\cdot196 . *60\cdot2 . \supset \vdash :. \text{Hp} . \supset : \alpha \,\epsilon\, \text{D}`P_\epsilon \cap \text{Cl}`p`\lambda . \supset_\alpha . \alpha \subset p`\lambda . \alpha \neq p`\lambda :$
$[*211\cdot56\cdot15\cdot632] \quad \supset : \exists ! \lambda . \alpha \,\epsilon\, \text{D}`P_\epsilon \cap \text{Cl}`p`\lambda . \supset_\alpha . \alpha \subset P``p`\lambda :$
$[*40\cdot151] \quad \supset : \exists ! \lambda . \supset . s`(\text{D}`P_\epsilon \cap \text{Cl}`p`\lambda) \subset P``p`\lambda \quad (3)$
$\vdash . *40\cdot2 . *37\cdot24 . \supset \vdash : \lambda = \Lambda . \supset . s`(\text{D}`P_\epsilon \cap \text{Cl}`p`\lambda) \subset \text{D}`P . P``p`\lambda = \text{D}`P \quad (4)$
$\vdash . (3) . (4) . \quad \supset \vdash : \text{Hp} . \supset . s`(\text{D}`P_\epsilon \cap \text{Cl}`p`\lambda) \subset P``p`\lambda \quad (5)$
$\vdash . (2) . (5) . \supset \vdash . \text{Prop}$

$*212\cdot43$. $\vdash : P \,\epsilon\, \text{trans} \cap \text{connex} \,.\, \dot{\exists} ! P \,.\, \lambda \subset \text{D}`P_\epsilon \,.\, p`\lambda \sim\epsilon\, \text{D}`P_\epsilon \,.\supset .$
$$P``p`\lambda = \text{prec}\,(\varsigma`P)`\lambda = \text{tl}\,(\varsigma`P)`\lambda \quad [*212\cdot42\cdot421]$$

Thus with regard to the lower end of a class chosen out of $C`\varsigma`P$, we have three cases to distinguish: (1) if $p`\lambda \,\epsilon\, \lambda$, $p`\lambda$ is the minimum; (2) if $p`\lambda \,\epsilon\, \text{D}`P_\epsilon - \lambda$, $p`\lambda$ is the lower limit; (3) if $p`\lambda \sim\epsilon\, \text{D}`P_\epsilon$, $P``p`\lambda$ is the lower limit.

$*212\cdot431$. $\vdash : P \,\epsilon\, \text{trans} \cap \text{connex} \,.\, \dot{\exists} ! P \,.\, \lambda \subset \text{D}`P_\epsilon \,.\supset .$
$$s`(\text{D}`P_\epsilon \cap \text{Cl}`p`\lambda) = \text{limin}\,(\varsigma`P)`\lambda$$

Dem.

$\vdash . *212\cdot42 . \supset \vdash : \text{Hp} . p`\lambda \sim\epsilon\, \lambda . \supset . s`(\text{D}`P_\epsilon \cap \text{Cl}`p`\lambda) = \text{tl}\,(\varsigma`P)`\lambda \quad (1)$
$\vdash . *22\cdot441 . \supset \vdash : \text{Hp} . p`\lambda \,\epsilon\, \lambda . \supset . p`\lambda \,\epsilon\, (\text{D}`P_\epsilon \cap \text{Cl}`p`\lambda) .$
$[*40\cdot13] \quad \supset . p`\lambda \subset s`(\text{D}`P_\epsilon \cap \text{Cl}`p`\lambda) \quad (2)$
$\vdash . *60\cdot2 . \quad \supset \vdash : \alpha \,\epsilon\, \text{D}`P_\epsilon \cap \text{Cl}`p`\lambda . \supset . \alpha \subset p`\lambda :$
$[*40\cdot151] \quad \supset \vdash . s`(\text{D}`P_\epsilon \cap \text{Cl}`p`\lambda) \subset p`\lambda \quad (3)$
$\vdash . (2) . (3) . \supset \vdash : \text{Hp} . p`\lambda \,\epsilon\, \lambda . \supset . s`(\text{D}`P_\epsilon \cap \text{Cl}`p`\lambda) = p`\lambda .$
$[*212\cdot41] \quad \supset . s`(\text{D}`P_\epsilon \cap \text{Cl}`p`\lambda) = \min\,(\varsigma`P)`\lambda \quad (4)$
$\vdash . (1) . (4) . *207\cdot46 . \supset \vdash . \text{Prop}$

$*212\cdot44$. $\vdash : P \,\epsilon\, \text{trans} \cap \text{connex} \,.\, \dot{\exists} ! P \,.\supset .$
$$(\lambda) . \lambda \,\epsilon\, \{\text{Ɑ}`\max\,(\varsigma`P) \cup \text{Ɑ}`\text{seq}\,(\varsigma`P)\} \cap \{\text{Ɑ}`\min\,(\varsigma`P) \cup \text{Ɑ}`\text{prec}\,(\varsigma`P)\}$$
$[*211\cdot681 . *210\cdot28]$

***212·45.** $\vdash :. P \epsilon \text{trans} \cap \text{connex} . \supset : \lambda \epsilon C\text{‘sgm‘}\varsigma\text{‘}P . \supset . \text{E} ! \text{seq} (\varsigma\text{‘}P)\text{‘}\lambda$

Dem.

$\vdash . *211{\cdot}47 . *212{\cdot}44{\cdot}31 . \supset$

$\vdash :. \text{Hp} . \dot{\exists} ! P . \supset : \lambda \epsilon C\text{‘sgm‘}\varsigma\text{‘}P . \supset . \text{E} ! \text{seq} (\varsigma\text{‘}P)\text{‘}\lambda \quad (1)$

$\vdash . *33{\cdot}24 . \supset \vdash : \lambda \epsilon C\text{‘sgm‘}\varsigma\text{‘}P . \supset . \dot{\exists} ! \text{sgm‘}\varsigma\text{‘}P .$

[*212·151.Transp] $\supset . \dot{\exists} ! \varsigma\text{‘}P .$

[*212·14] $\supset . \dot{\exists} ! P \quad (2)$

$\vdash . (1) . (2) . \supset \vdash . \text{Prop}$

The proofs of the following propositions are exactly analogous to those of the corresponding propositions on $\varsigma\text{‘}P$.

***212·5.** $\vdash : P \epsilon \text{connex} . \dot{\exists} ! \text{sgm‘}P . \lambda \subset \text{D‘}(P_\epsilon \dot{\cap} I) . s\text{‘}\lambda \epsilon \lambda . \supset .$
$s\text{‘}\lambda = \max (\text{sgm‘}P)\text{‘}\lambda$

***212·501.** $\vdash : P \epsilon \text{connex} . \dot{\exists} ! \text{sgm‘}P . \lambda \subset \text{D‘}(P_\epsilon \dot{\cap} I) . s\text{‘}\lambda \sim \epsilon \lambda . \supset .$
$s\text{‘}\lambda = \text{seq} (\text{sgm‘}P)\text{‘}\lambda = \text{lt} (\text{sgm‘}P)\text{‘}\lambda$

***212·502.** $\vdash : P \epsilon \text{connex} . \dot{\exists} ! \text{sgm‘}P . \lambda \subset \text{D‘}(P_\epsilon \dot{\cap} I) . \supset . s\text{‘}\lambda = \text{limax} (\text{sgm‘}P)\text{‘}\lambda$
[*212·5·501]

***212·51.** $\vdash : P \epsilon \text{connex} . \dot{\exists} ! \text{sgm‘}P . \lambda \subset \text{D‘}(P_\epsilon \dot{\cap} I) . p\text{‘}\lambda \epsilon \lambda . \supset .$
$p\text{‘}\lambda = \min (\text{sgm‘}P)\text{‘}\lambda$

***212·511.** $\vdash : P \epsilon \text{connex} . \dot{\exists} ! \text{sgm‘}P . \lambda \subset \text{D‘}(P_\epsilon \dot{\cap} I) . p\text{‘}\lambda \epsilon \text{D‘}(P_\epsilon \dot{\cap} I) - \lambda . \supset .$
$p\text{‘}\lambda = \text{prec} (\text{sgm‘}P)\text{‘}\lambda = \text{tl} (\text{sgm‘}P)\text{‘}\lambda$

***212·52.** $\vdash : P \epsilon \text{connex} . \dot{\exists} ! \text{sgm‘}P . \lambda \subset \text{D‘}(P_\epsilon \dot{\cap} I) . p\text{‘}\lambda \sim \epsilon \lambda . \supset .$
$s\text{‘}\{\text{D‘}(P_\epsilon \dot{\cap} I) \cap \text{Cl‘}p\text{‘}\lambda\} = \text{prec} (\text{sgm‘}P)\text{‘}\lambda = \text{tl} (\text{sgm‘}P)\text{‘}\lambda$

This proposition includes *212·511, since, if $p\text{‘}\lambda \epsilon \text{D‘}(P_\epsilon \dot{\cap} I)$, we have
$$s\text{‘}\{\text{D‘}(P_\epsilon \dot{\cap} I) \cap \text{Cl‘}p\text{‘}\lambda\} = p\text{‘}\lambda.$$

***212·53.** $\vdash : P \epsilon \text{connex} . \dot{\exists} ! \text{sgm‘}P . \lambda \subset \text{D‘}(P_\epsilon \dot{\cap} I) . \supset .$
$s\text{‘}\{\text{D‘}(P_\epsilon \dot{\cap} I) \cap \text{Cl‘}p\text{‘}\lambda\} = \text{limin} (\text{sgm‘}P)\text{‘}\lambda$ [*212·51·52]

The proof proceeds as in *212·431.

***212·54.** $\vdash : P \epsilon \text{connex} . \dot{\exists} ! \text{sgm‘}P . \supset .$
$(\lambda) . \lambda \epsilon \{\text{Ɑ‘max} (\text{sgm‘}P) \cup \text{Ɑ‘seq} (\text{sgm‘}P)\} \cap \{\text{Ɑ‘min} (\text{sgm‘}P) \cup \text{Ɑ‘prec} (\text{sgm‘}P)\}$

***212·55.** $\vdash :. P \epsilon \text{connex} . \supset : \lambda \epsilon C\text{‘sgm‘sgm‘}P . \supset . \text{E} ! \text{seq} (\text{sgm‘}P)\text{‘}\lambda$

The following propositions are concerned with the relations of maxima, limits and sequents in P and $\varsigma\text{‘}P$ respectively. The series $\overrightarrow{P};P$, which is ordinally similar to P, is contained in $\varsigma\text{‘}P$; and if α has a maximum or limit in P, the maximum or limit of $\overrightarrow{P}\text{‘‘}\alpha$ in $\varsigma\text{‘}P$ is $\overrightarrow{P}\text{‘max}_P\text{‘}\alpha$ or $\overrightarrow{P}\text{‘lt}_P\text{‘}\alpha$. In this way, a series (namely $\overrightarrow{P};P$) which has the same ordinal properties as P can be placed in a certain Dedekindian series (namely $\varsigma\text{‘}P$) in such a way that the classes which have limits in P are those whose correlates have limits which are members of $\overrightarrow{P}``C`P$, while those whose correlates have limits which are not members of $\overrightarrow{P}``C`P$ are those which have neither a maximum nor a limit in $P$. These relations are important in many connections. For example, if $P$ is of the type of the rationals, $\varsigma`P$ is of the type of the real numbers: $C`\varsigma`P - \overrightarrow{P}``C`P$ corresponds to the irrationals, and classes contained in $\overrightarrow{P}``C`P$ but having a limit not belonging to $\overrightarrow{P}``C`P$ correspond to series of rationals having an irrational limit. In the original series $P$, there are no irrational limits; but if $\alpha$ is a class in $C`P$ and having no limit, $\overrightarrow{P}``\alpha$ has an irrational limit in $\varsigma`P$.

***212·6.** $\vdash : P \in \text{Ser} . \alpha \subset C`P . \supset .$
$$\overrightarrow{\max}(\varsigma`P)`\overrightarrow{P}``\alpha = \overrightarrow{\max}(\overrightarrow{P};P)`\overrightarrow{P}``\alpha = \overrightarrow{P}``\max_P`\alpha$$

Dem.

$\vdash . *205·9 . *200·12 . \supset$
$$\vdash : \text{Hp} . \supset . \overrightarrow{\max}(\varsigma`P)`\overrightarrow{P}``\alpha = \overrightarrow{\max}(\overrightarrow{P};P)`\overrightarrow{P}``\alpha \quad (1)$$
$\vdash . *204·35 . *205·8 . \supset$
$$\vdash : \text{Hp} . \supset . \max(\overrightarrow{P};P)`\overrightarrow{P}``\alpha = \overrightarrow{P}``\max_P`\alpha \quad (2)$$
$\vdash . (1) . (2) . \supset \vdash . \text{Prop}$

***212·601.** $\vdash :. P \in \text{Ser} . \alpha \subset C`P . \supset :$
$$E ! \max_P`\alpha . \equiv . E ! \max(\overrightarrow{P};P)`\overrightarrow{P}``\alpha . \equiv . E ! \max(\varsigma`P)`\overrightarrow{P}``\alpha$$
[*212·6]

***212·602.** $\vdash :. P \in \text{Ser} . \dot{\exists} ! P . \alpha \subset C`P . \supset : E ! \max_P`\alpha . \equiv . P``\alpha \in \overrightarrow{P}``\alpha$

Dem.

$\vdash . *212·601 . *210·223 . \supset$
$$\vdash :. \text{Hp} . \supset : E ! \max_P`\alpha . \equiv . s`\overrightarrow{P}``\alpha \in \overrightarrow{P}``\alpha .$$
$$[*40·5] \qquad \equiv . P``\alpha \in \overrightarrow{P}``\alpha :. \supset \vdash . \text{Prop}$$

***212·61.** $\vdash : P \in \text{trans} \cap \text{connex} . \dot{\exists} ! P . \supset . \text{limax}(\varsigma`P)`\overrightarrow{P}``\alpha = P``\alpha$
[*212·402 . *40·5]

***212·62.** $\vdash :. P \in \text{Ser} . \dot{\exists} ! P . \supset :$
$$E ! \text{limax}_P`\alpha . \equiv . E ! \text{limax}(\overrightarrow{P};P)`\overrightarrow{P}``\alpha .$$
$$\equiv . \text{limax}(\varsigma`P)`\overrightarrow{P}``\alpha = \overrightarrow{P}`\text{limax}_P`\alpha .$$
$$\equiv . \text{limax}(\varsigma`P)`\overrightarrow{P}``\alpha \in \overrightarrow{P}``C`P$$

Dem.

$$\vdash . *204·35 . *207·65 . \supset \vdash :. \text{Hp} . \supset : E ! \text{limax}_P`\alpha . \equiv . E ! \text{limax}(\overrightarrow{P};P)`\overrightarrow{P}``\alpha \quad (1)$$
$$\vdash . *207·51 . \supset \vdash :. \text{Hp} . \supset : \overrightarrow{P}`\text{limax}_P`\alpha = P``\alpha . \equiv . \text{limax}_P`\alpha = \text{limax}_P`\alpha .$$
$$[*14·28] \qquad \equiv . E ! \text{limax}_P`\alpha \quad (2)$$

$\vdash . (2) . *212 \cdot 61 . \supset$

$\vdash :. \text{Hp} . \supset : \text{E} ! \, \text{limax}_P{}^\backprime\alpha . \equiv . \, \text{limax} \, (\varsigma{}^\backprime P){}^\backprime \overrightarrow{P}{}^{\backprime\backprime}\alpha = \overrightarrow{P}{}^\backprime \text{limax}_P{}^\backprime\alpha$ (3)

$\vdash . *207 \cdot 51 . *14 \cdot 204 . \supset$

$\vdash :. \text{Hp} . \supset : \text{E} ! \, \text{limax}_P{}^\backprime\alpha . \equiv . (\exists x) . x \in C{}^\backprime P . \overrightarrow{P}{}^\backprime x = P{}^{\backprime\backprime}\alpha .$

[*37·7] $\equiv . P{}^{\backprime\backprime}\alpha \in \overrightarrow{P}{}^{\backprime\backprime} C{}^\backprime P$ (4)

$\vdash . (1) . (3) . (4) . \supset \vdash . \text{Prop}$

***212·621.** $\vdash :. P \in \text{Ser} . \beta \subset C{}^\backprime P . \supset : \text{limax}_P{}^\backprime\alpha \in \beta . \equiv . \, \text{limax} \, (\varsigma{}^\backprime P){}^\backprime \overrightarrow{P}{}^{\backprime\backprime}\alpha \in \overrightarrow{P}{}^{\backprime\backprime}\beta$

Dem.

$\vdash . *33 \cdot 24 . \supset \vdash : \text{Hp} . \text{limax}_P{}^\backprime\alpha \in \beta . \supset . \dot{\exists} ! P . \text{limax}_P{}^\backprime\alpha \in \beta .$

[*14·21] $\supset . \dot{\exists} ! P . \text{E} ! \, \text{limax}_P{}^\backprime\alpha . \text{limax}_P{}^\backprime\alpha \in \beta .$

[*212·62] $\supset . \, \text{limax} \, (\varsigma{}^\backprime P){}^\backprime \overrightarrow{P}{}^{\backprime\backprime}\alpha \in \overrightarrow{P}{}^{\backprime\backprime}\beta$ (1)

$\vdash . *33 \cdot 24 . *22 \cdot 621 . \supset \vdash : \text{Hp} . \text{limax} \, (\varsigma{}^\backprime P){}^\backprime \overrightarrow{P}{}^{\backprime\backprime}\alpha \in \overrightarrow{P}{}^{\backprime\backprime}\beta . \supset .$

$\dot{\exists} ! P . \text{limax} \, (\varsigma{}^\backprime P){}^\backprime \overrightarrow{P}{}^{\backprime\backprime}\alpha \in \overrightarrow{P}{}^{\backprime\backprime}\beta \cap \overrightarrow{P}{}^{\backprime\backprime} C{}^\backprime P .$

[*212·62] $\supset . \, \text{limax} \, (\varsigma{}^\backprime P){}^\backprime \overrightarrow{P}{}^{\backprime\backprime}\alpha \in \overrightarrow{P}{}^{\backprime\backprime}\beta . \text{limax} \, (\varsigma{}^\backprime P){}^\backprime \overrightarrow{P}{}^{\backprime\backprime}\beta = \overrightarrow{P}{}^\backprime \text{limax}_P{}^\backprime\alpha .$

[*72·512.*204·34] $\supset . \, \text{limax}_P{}^\backprime\alpha \in \beta$ (2)

$\vdash . (1) . (2) . \supset \vdash . \text{Prop}$

***212·63.** $\vdash : P \in \text{Ser} . \dot{\exists} ! P . \alpha \subset C{}^\backprime P . \sim \text{E} ! \, \text{max}_P{}^\backprime\alpha . \supset . \, \text{lt} \, (\varsigma{}^\backprime P){}^\backprime \overrightarrow{P}{}^{\backprime\backprime}\alpha = P{}^{\backprime\backprime}\alpha$

[*212·61·601 . *207·43]

***212·631.** $\vdash :. P \in \text{Ser} . \dot{\exists} ! P . \alpha \subset C{}^\backprime P . \supset : \text{E} ! \, \text{lt}_P{}^\backprime\alpha . \equiv . \, \text{lt} \, (\varsigma{}^\backprime P){}^\backprime \overrightarrow{P}{}^{\backprime\backprime}\alpha = \overrightarrow{P}{}^\backprime \text{lt}_P{}^\backprime\alpha .$

$\equiv . \, \text{lt} \, (\varsigma{}^\backprime P){}^\backprime \overrightarrow{P}{}^{\backprime\backprime}\alpha \in \overrightarrow{P}{}^{\backprime\backprime} C{}^\backprime P$

Dem.

$\vdash . *207 \cdot 47 . \supset \vdash : \text{E} ! \, \text{lt}_P{}^\backprime\alpha . \equiv . \text{E} ! \, \text{limax}_P{}^\backprime\alpha . \sim \text{E} ! \, \text{max}_P{}^\backprime\alpha$ (1)

$\vdash . (1) . *212 \cdot 62 \cdot 601 . \supset$

$\vdash :. \text{Hp} . \supset : \text{E} ! \, \text{lt}_P{}^\backprime\alpha . \equiv . \, \text{limax} \, (\varsigma{}^\backprime P){}^\backprime \overrightarrow{P}{}^{\backprime\backprime}\alpha = \overrightarrow{P}{}^\backprime \text{limax}_P{}^\backprime\alpha . \sim \text{E} ! \, \text{max} \, (\varsigma{}^\backprime P){}^\backprime \overrightarrow{P}{}^{\backprime\backprime}\alpha .$

[*207·43·11] $\equiv . \, \text{lt} \, (\varsigma{}^\backprime P){}^\backprime \overrightarrow{P}{}^{\backprime\backprime}\alpha = \overrightarrow{P}{}^\backprime \text{lt}_P{}^\backprime\alpha$ (2)

$\vdash . (1) . *212 \cdot 62 \cdot 601 . \supset$

$\vdash :. \text{Hp} . \supset : \text{E} ! \, \text{lt}_P{}^\backprime\alpha . \equiv . \, \text{limax} \, (\varsigma{}^\backprime P){}^\backprime \overrightarrow{P}{}^{\backprime\backprime}\alpha \in \overrightarrow{P}{}^{\backprime\backprime} C{}^\backprime P . \sim \text{E} ! \, \text{max} \, (\varsigma{}^\backprime P){}^\backprime \overrightarrow{P}{}^{\backprime\backprime}\alpha .$

[*207·43·11] $\equiv . \, \text{lt} \, (\varsigma{}^\backprime P){}^\backprime \overrightarrow{P}{}^{\backprime\backprime}\alpha \in \overrightarrow{P}{}^{\backprime\backprime} C{}^\backprime P$ (3)

$\vdash . (2) . (3) . \supset \vdash . \text{Prop}$

***212·632.** $\vdash : P \in \text{Ser} . \dot{\exists} ! P . \alpha \subset C{}^\backprime P . P{}^{\backprime\backprime}\alpha \sim \in \overrightarrow{P}{}^{\backprime\backprime} C{}^\backprime P . \supset . P{}^{\backprime\backprime}\alpha = \text{lt} \, (\varsigma{}^\backprime P){}^\backprime \overrightarrow{P}{}^{\backprime\backprime}\alpha$

Dem.

$\vdash . *212 \cdot 602 . \supset \vdash : \text{Hp} . \supset . \sim \text{E} ! \, \text{max}_P{}^\backprime\alpha .$

[*212·601] $\supset . \sim \text{E} ! \, \text{max} \, (\varsigma{}^\backprime P){}^\backprime \overrightarrow{P}{}^{\backprime\backprime}\alpha .$

[*212·61] $\supset . \, \text{lt} \, (\varsigma{}^\backprime P){}^\backprime \overrightarrow{P}{}^{\backprime\backprime}\alpha = P{}^{\backprime\backprime}\alpha : \supset \vdash . \text{Prop}$

***212·633.** $\vdash :. P \in \text{Ser} . \dot{\exists} ! P . x \in C`P . \beta \subset C`P . \supset :$

$$x = \text{lt}_P`\beta . \equiv . \overrightarrow{P}`x = \text{lt}(\varsigma`P)`\overrightarrow{P}``\beta$$

Dem.

$\vdash . *212·631 . *14·21 . \supset$

$\vdash :. \text{Hp} . \supset : x = \text{lt}_P`\beta . \supset . \overrightarrow{P}`x = \text{lt}(\varsigma`P)`\overrightarrow{P}``\beta$ (1)

$\vdash . *212·402 . \supset \vdash : \text{Hp} . \overrightarrow{P}`x = \text{lt}(\varsigma`P)`\overrightarrow{P}``\beta . \supset . \overrightarrow{P}`x = P``\beta$ (2)

$\vdash . *206·2 . \quad \supset \vdash : \text{Hp} . \overrightarrow{P}`x = \text{lt}(\varsigma`P)`\overrightarrow{P}``\beta . \supset . \overrightarrow{P}`x \sim \in \overrightarrow{P}``\beta .$

[*72·512.*204·34] $\supset . x \sim \in \beta$ (3)

$\vdash . (2) . (3) . *207·232 . \supset \vdash : \text{Hp} . \overrightarrow{P}`x = \text{lt}(\varsigma`P)`\overrightarrow{P}``\beta . \supset . x = \text{lt}_P`\beta$ (4)

$\vdash . (1) . (4) . \supset \vdash . \text{Prop}$

***212·65.** $\vdash :. P \in \text{Ser} . \alpha \subset C`P . \supset : E ! \text{seq}_P`\alpha . \equiv . \overrightarrow{P}`\text{seq}_P`\alpha = \text{seq}(\varsigma`P)`\overrightarrow{P}``\alpha$

Dem.

$\vdash . *206·17 . *210·15 . *211·3 . \supset$

$\vdash :: \text{Hp} . \supset :. \overrightarrow{P}`\text{seq}_P`\alpha = \text{seq}(\varsigma`P)`\overrightarrow{P}``\alpha . \equiv :$

$y \in \alpha \cap C`P . \supset_y . \overrightarrow{P}`y \subset \overrightarrow{P}`\text{seq}_P`\alpha . \overrightarrow{P}`y \neq \overrightarrow{P}`\text{seq}_P`\alpha :$

$\gamma \in D`P_\epsilon . \gamma \subset \overrightarrow{P}`\text{seq}_P`\alpha . \gamma \neq \overrightarrow{P}`\text{seq}_P`\alpha . \supset_\gamma . (\exists z) . z \in \alpha . \gamma \subset \overrightarrow{P}`z :$

[*204·33.*206·22] $\equiv : y \in \alpha \cap C`P . \supset_y . yP\,\text{seq}_P`\alpha :$

$\gamma \in D`P_\epsilon . \gamma \subset (\alpha \cap C`P) \cup P``\alpha . \gamma \neq (\alpha \cap C`P) \cup P``\alpha . \supset_\gamma . (\exists z) . z \in \alpha . \gamma \subset \overrightarrow{P}`z$ (1)

$\vdash . *211·56 . \supset \vdash : \text{Hp} . \gamma \in D`P_\epsilon . z \in C`P - \gamma . \supset . \gamma \subset \overrightarrow{P}`z$ (2)

$\vdash . (2) . \supset \vdash : \text{Hp} . \gamma \in D`P_\epsilon . \gamma \subset (\alpha \cap C`P) \cup P``\alpha . \gamma \neq (\alpha \cap C`P) \cup P``\alpha . \supset .$

$(\exists z) . z \in \alpha . \gamma \subset \overrightarrow{P}`z$ (3)

$\vdash . (1) . (3) . \supset$

$\vdash :: \text{Hp} . \supset :. \overrightarrow{P}`\text{seq}_P`\alpha = \text{seq}(\varsigma`P)`\overrightarrow{P}``\alpha . \equiv : y \in \alpha \cap C`P . \supset_y . yP\,\text{seq}_P`\alpha :$

[*206·211.*14·21] $\equiv : E ! \text{seq}_P`\alpha :: \supset \vdash . \text{Prop}$

***212·651.** $\vdash :. P \in \text{Ser} . \alpha \subset C`P . \supset :$

$$E ! \text{seq}_P`\alpha . \equiv . \text{seq}(\varsigma`P)`\overrightarrow{P}``\alpha \in \overrightarrow{P}``C`P . \equiv . E ! \text{seq}(\overrightarrow{P};P)`\overrightarrow{P}``\alpha$$

Dem.

$\vdash . *212·65 . \supset \vdash :. \text{Hp} . \supset : E ! \text{seq}_P`\alpha . \supset . \text{seq}(\varsigma`P)`\overrightarrow{P}``\alpha \in \overrightarrow{P}``C`P$ (1)

$\vdash . *206·17 . *210·15 . *211·3 . \supset$

$\vdash :: \text{Hp} . \supset :. \text{seq}(\varsigma`P)`\overrightarrow{P}``\alpha = \overrightarrow{P}`w . w \in C`P . \equiv :$

$y \in \alpha \cap C`P . \supset_y . \overrightarrow{P}`y \subset \overrightarrow{P}`w . \overrightarrow{P}`y \neq \overrightarrow{P}`w :$

$\gamma \in D`P_\epsilon . \gamma \subset \overrightarrow{P}`w . \gamma \neq \overrightarrow{P}`w . \supset_\gamma . (\exists z) . z \in \alpha . \gamma \subset \overrightarrow{P}`z : w \in C`P :$

[*204·33.*211·3] $\supset : \alpha \cap C`P \subset \overrightarrow{P}`w : yPw . \supset_y . (\exists z) . z \in \alpha . \overrightarrow{P}`y \subset \overrightarrow{P}`z : w \in C`P :$

[*204·32] $\supset : \alpha \cap C`P \subset \overrightarrow{P}`w . \overrightarrow{P}`w \subset \alpha \cup P``\alpha . w \in C`P :$

[*206·171.*33·15] $\supset : w = \text{seq}_P`\alpha$ (2)

$\vdash . (2) . *37\cdot7 . *14\cdot204 . \supset$

$\vdash :. \text{Hp} . \supset : \text{seq}\,(\varsigma\text{'}P)\text{'}\overrightarrow{P}\text{''}\alpha \,\epsilon\, \overrightarrow{P}\text{''}C\text{'}P . \supset . \text{E}\,!\, \text{seq}_P\text{'}\alpha$ (3)

$\vdash . (1) . (3) . *206\cdot62 . \supset \vdash . \text{Prop}$

***212·652.** $\vdash : P \,\epsilon\, \text{Ser} . \alpha \subset C\text{'}P . \text{E}\,!\, \text{max}_P\text{'}\alpha . \text{E}\,!\, \text{seq}\,(\varsigma\text{'}P)\text{'}\overrightarrow{P}\text{''}\alpha . \supset .$

$\text{seq}\,(\varsigma\text{'}P)\text{'}\overrightarrow{P}\text{''}\alpha = \alpha \cup P\text{''}\alpha$

Dem.

$\vdash . *212\cdot6\cdot601 . *206\cdot46 . \supset \vdash : \text{Hp} . \supset . \text{seq}(\varsigma\text{'}P)\text{'}\overrightarrow{P}\text{''}\alpha = \text{seq}(\varsigma\text{'}P)\text{'}\iota\text{'}\overrightarrow{P}\text{'}\text{max}_P\text{'}\alpha$ (1)

$\vdash . *206\cdot17 . *210\cdot15 . *211\cdot3 . \supset$

$\vdash :: \text{Hp} . \supset :. \beta = \text{seq}\,(\varsigma\text{'}P)\text{'}\iota\text{'}\overrightarrow{P}\text{'}\text{max}_P\text{'}\alpha . \equiv :$

$\beta \,\epsilon\, \text{D'}P_\epsilon . \overrightarrow{P}\text{'}\text{max}_P\text{'}\alpha \subset \beta . \overrightarrow{P}\text{'}\text{max}_P\text{'}\alpha \neq \beta :$

$\gamma \,\epsilon\, \text{D'}P_\epsilon . \gamma \subset \beta . \gamma \neq \beta . \supset_\gamma . \gamma \subset \overrightarrow{P}\text{'}\text{max}_P\text{'}\alpha :$

[*201·55.*210·1] $\supset : \beta \,\epsilon\, \text{D'}P_\epsilon . \exists\,!\, \beta - P\text{''}(\overrightarrow{P}\text{'}\text{max}_P\text{'}\alpha \cup \iota\text{'}\text{max}_P\text{'}\alpha) :$

$\gamma \,\epsilon\, \text{D'}P_\epsilon . \gamma \subset \beta . \gamma \neq \beta . \supset_\gamma . \gamma \subset \overrightarrow{P}\text{'}\text{max}_P\text{'}\alpha :$

[*211·56] $\supset : \beta \,\epsilon\, \text{D'}P_\epsilon . \overrightarrow{P}\text{'}\text{max}_P\text{'}\alpha \cup \iota\text{'}\text{max}_P\text{'}\alpha \subset \beta :$

$\gamma \,\epsilon\, \text{D'}P_\epsilon . \gamma \subset \beta . \gamma \neq \beta . \supset_\gamma . \gamma \subset \overrightarrow{P}\text{'}\text{max}_P\text{'}\alpha :$

[*211·3] $\supset : \beta \,\epsilon\, \text{D'}P_\epsilon . \overrightarrow{P}\text{'}\text{max}_P\text{'}\alpha \cup \iota\text{'}\text{max}_P\text{'}\alpha \subset \beta : x \,\epsilon\, \beta . \supset_x . \overrightarrow{P}\text{'}x \subset \overrightarrow{P}\text{'}\text{max}_P\text{'}\alpha :$

[*40·5] $\supset : \beta \,\epsilon\, \text{D'}P_\epsilon . \overrightarrow{P}\text{'}\text{max}_P\text{'}\alpha \cup \iota\text{'}\text{max}_P\text{'}\alpha \subset \beta . P\text{''}\beta \subset \overrightarrow{P}\text{'}\text{max}_P\text{'}\alpha :$

[*202·56] $\supset : \beta \,\epsilon\, \text{D'}P_\epsilon . \overrightarrow{P}\text{'}\text{max}_P\text{'}\alpha \cup \iota\text{'}\text{max}_P\text{'}\alpha = \beta :$

[*205·131·22] $\supset : \beta = \alpha \cup P\text{''}\alpha$ (2)

$\vdash . (1) . (2) . \supset \vdash . \text{Prop}$

***212·653.** $\vdash :. P \,\epsilon\, \text{Ser} . \text{E}\,!\, \text{max}_P\text{'}\alpha . \alpha \subset C\text{'}P . \supset : \text{E}\,!\, \text{seq}_P\text{'}\alpha . \equiv . \text{E}\,!\, \text{seq}(\varsigma\text{'}P)\text{'}\overrightarrow{P}\text{''}\alpha$

Dem.

$\vdash . *212\cdot652 . \quad \supset \vdash :. \text{Hp} . \text{E}\,!\, \text{seq}\,(\varsigma\text{'}P)\text{'}\overrightarrow{P}\text{''}\alpha . \supset . \alpha \cup P\text{''}\alpha \,\epsilon\, \text{D'}P_\epsilon$ (1)

$\vdash . *205\cdot191 . \quad \supset \vdash : \text{Hp} . \supset . \text{E}\,!\, \text{max}_P\text{'}(\alpha \cup P\text{''}\alpha)$ (2)

$\vdash . (1) . (2) . *211\cdot31 . \supset \vdash : \text{Hp}\,(1) . \supset . \text{E}\,!\, \text{seq}_P\text{'}(\alpha \cup P\text{''}\alpha) .$

[*206·25] $\supset . \text{E}\,!\, \text{seq}_P\text{'}\alpha$ (3)

$\vdash . *212\cdot65 . \quad \supset \vdash : \text{Hp} . \text{E}\,!\, \text{seq}_P\text{'}\alpha . \supset . \text{E}\,!\, \text{seq}\,(\varsigma\text{'}P)\text{'}\overrightarrow{P}\text{''}\alpha$ (4)

$\vdash . (3) . (4) . \supset \vdash . \text{Prop}$

***212·66.** $\vdash : P \,\epsilon\, \text{trans} \cap \text{connex} . \kappa \subset \text{D'}P_\epsilon . \sim\text{E}\,!\, \text{max}\,(\varsigma\text{'}P)\text{'}\kappa . \supset . \sim\text{E}\,!\, \text{max}_P\text{'}s\text{'}\kappa$

Dem.

$\vdash . *210\cdot1 . *212\cdot23 . \supset$

$\vdash :. \text{Hp} . \supset : \beta \,\epsilon\, \kappa . \supset_\beta . (\exists\gamma) . \gamma \,\epsilon\, \kappa . \beta \subset \gamma . \exists\,!\, \gamma - \beta :$

[*201·5] $\supset : \beta \,\epsilon\, \kappa . x \,\epsilon\, \beta . \supset_{\beta, x} . (\exists\gamma) . \gamma \,\epsilon\, \kappa . \exists\,!\, \gamma - \overrightarrow{P}\text{'}x - \iota\text{'}x :$

[*202·101] $\supset : x \,\epsilon\, s\text{'}\kappa . \supset_x . (\exists\gamma) . \gamma \,\epsilon\, \kappa . \exists\,!\, \gamma \cap \overleftarrow{P}\text{'}x .$

[*37·46] $\supset_x . x \,\epsilon\, P\text{''}s\text{'}\kappa :. \supset \vdash . \text{Prop}$

***212·661.** $\vdash : P \epsilon \text{Ser} . \kappa \subset \text{D}'P_\epsilon . \text{E} ! \text{lt}(\varsigma'P)'\kappa . \supset . \text{lt}(\varsigma'P)'\kappa = \text{lt}(\varsigma'P)'\overrightarrow{P}``s'\kappa = s'\kappa$

Dem.

$\vdash . *212·402 . \quad \supset \vdash : \text{Hp} . \supset . \text{lt}(\varsigma'P)'\kappa = s'\kappa \quad (1)$

$\vdash . *212·402 . \quad \supset \vdash : \text{Hp} . \supset . \text{limax}(\varsigma'P)'\overrightarrow{P}``s'\kappa = P``s'\kappa$

$[*212·66] \quad = s'\kappa \quad (2)$

$\vdash . *212·601·66 . \supset \vdash : \text{Hp} . \supset . \sim \text{E} ! \max(\varsigma'P)'\overrightarrow{P}``s'\kappa \quad (3)$

$\vdash . (2) . (3) . \quad \supset \vdash : \text{Hp} . \supset . \text{lt}(\varsigma'P)'\overrightarrow{P}``s'\kappa = s'\kappa \quad (4)$

$\vdash . (1) . (4) . \supset \vdash . \text{Prop}$

***212·662.** $\vdash : P \epsilon \text{Ser} . \kappa \subset \text{D}'P_\epsilon . \text{E} ! \text{lt}(\varsigma'P)'\kappa . \supset .$

$(\exists\lambda) . \lambda \subset \overrightarrow{P}``C'P . \text{lt}(\varsigma'P)'\kappa = \text{lt}(\varsigma'P)'\lambda$

[*212·661]

***212·663.** $\vdash : P \epsilon \text{Ser} . x \epsilon C'P . \overrightarrow{P}'x \epsilon \text{D}'\text{lt}(\varsigma'P) . \supset .$

$\overrightarrow{P}'x = \text{lt}(\varsigma'P)'\overrightarrow{P}``\overrightarrow{P}'x . x = \text{lt}_P{}'\overrightarrow{P}'x$

Dem.

$\vdash . *212·661 . \supset \vdash : P \epsilon \text{Ser} . x \epsilon C'P . \overrightarrow{P}'x = \text{lt}(\varsigma'P)'\kappa . \supset .$

$\overrightarrow{P}'x = s'\kappa . \overrightarrow{P}'x = \text{lt}(\varsigma'P)'\overrightarrow{P}``s'\kappa .$

$[*13·12 . *212·66] \quad \supset . \overrightarrow{P}'x = \text{lt}(\varsigma'P)'\overrightarrow{P}``\overrightarrow{P}'x . \sim \text{E} ! \max_P{}'\overrightarrow{P}'x .$

$[*206·4] \quad \supset . \overrightarrow{P}'x = \text{lt}(\varsigma'P)'\overrightarrow{P}``\overrightarrow{P}'x . x = \text{lt}_P{}'\overrightarrow{P}'x : \supset \vdash . \text{Prop}$

***212·664.** $\vdash :. P \epsilon \text{Ser} . x \epsilon C'P . \supset : x \epsilon \text{D}'\text{lt}_P . \equiv . \overrightarrow{P}'x \epsilon \text{D}'\text{lt}(\varsigma'P)$

Dem.

$\vdash . *212·631 . \supset \vdash : \text{Hp} . x = \text{lt}_P{}'\alpha . \supset . \overrightarrow{P}'x = \text{lt}_P{}'\overrightarrow{P}``\alpha \quad (1)$

$\vdash . (1) . *212·663 . \supset \vdash . \text{Prop}$

***212·665.** $\vdash : P \epsilon \text{Ser} . \dot{\exists} ! P . \alpha \epsilon \text{D}'(P_\epsilon \dot{\frown} I) . \supset . \text{lt}(\varsigma'P)'\overrightarrow{P}``\alpha = \alpha$

Dem.

$\vdash . *211·4 . \supset \vdash : \text{Hp} . \supset . \sim \text{E} ! \max_P{}'\alpha .$

$[*212·601·44] \quad \supset . \text{lt}(\varsigma'P)'\overrightarrow{P}``\alpha = \text{limax}(\varsigma'P)'\overrightarrow{P}``\alpha$

$[*212·402 . *40·5] \quad = P``\alpha$

$[*211·12] \quad = \alpha : \supset \vdash . \text{Prop}$

***212·666.** $\vdash : P \epsilon \text{Ser} . \dot{\exists} ! P . \supset . \text{D}'\text{lt}(\varsigma'P) = \text{D}'(P_\epsilon \dot{\frown} I)$

Dem.

$\vdash . *212·66·661 . \supset \vdash : \text{Hp} . \kappa \subset \text{D}'P_\epsilon . \gamma = \text{lt}(\varsigma'P)'\kappa . \supset . \gamma = s'\kappa . \sim \text{E} ! \max_P{}'\gamma .$

$[*211·64·42] \quad \supset . \gamma \epsilon \text{D}'(P_\epsilon \dot{\frown} I) \quad (1)$

$\vdash . (1) . *212·665 . \supset \vdash . \text{Prop}$

***212·667.** $\vdash : P \epsilon \text{Ser} . \supset . \text{D}'\text{lt}(\varsigma'P) - \iota'\Lambda = \text{Cl}'\text{sgm}'P \quad [*212·152·151·666]$

***212·7.** $\vdash : S \in P \,\overline{\text{smor}}\, Q \,.\supset .\, \text{sect}`P = S_\epsilon``\text{sect}`Q \,.\, C`\varsigma`P_* = S_\epsilon``C`\varsigma`Q_*$

Dem.

$\vdash .\, *151\cdot11\cdot131 \,.\quad \supset \vdash : \text{Hp} \,.\, \beta \subset C`Q \,.\supset .\, S``\beta \subset C`P \quad (1)$

$\vdash .\, *37\cdot2 \,.\quad \supset \vdash : Q``\beta \subset \beta \,.\supset .\, S``Q``\beta \subset S``\beta \quad (2)$

$\vdash .\, (2) \,.\, *72\cdot503 \,.\quad \supset \vdash : \text{Hp} \,.\, \beta \subset C`Q \,.\, Q``\beta \subset \beta \,.\supset .\, S``Q``\breve{S}``S``\beta \subset S``\beta \,.$

$[*151\cdot11] \quad \supset .\, P``S``\beta \subset S``\beta \quad (3)$

$\vdash .\, (1) \,.\, (3) \,.\, *211\cdot1 \,.\supset \vdash : \text{Hp} \,.\, \beta \in \text{sect}`Q \,.\supset .\, S``\beta \in \text{sect}`P \quad (4)$

$\vdash .\, (4) \,.\, *151\cdot131 \,.\quad \supset \vdash : \text{Hp} \,.\, \alpha \in \text{sect}`P \,.\supset .\, \breve{S}``\alpha \in \text{sect}`Q \,.$

$[*72\cdot502] \quad \supset .\, \alpha \in S_\epsilon``\text{sect}`Q \quad (5)$

$\vdash .\, (4) \,.\, (5) \,.\quad \supset \vdash : \text{Hp} \,.\supset .\, \text{sect}`P = S_\epsilon``\text{sect}`Q \quad (6)$

$\vdash .\, (6) \,.\, *212\cdot17\cdot172 \,.\supset \vdash .\, \text{Prop}$

***212·701.** $\vdash : S \in P \,\overline{\text{smor}}\, Q \,.\supset .\, \text{D}`P_\epsilon = S_\epsilon``\text{D}`Q_\epsilon \,.\, C`\varsigma`P = S_\epsilon``C`\varsigma`Q$

[Proof as in *212·7]

***212·702.** $\vdash : S \in P \,\overline{\text{smor}}\, Q \,.\supset .$

$\text{D}`(P_\epsilon \dot{\frown} I) = S_\epsilon``\text{D}`(Q_\epsilon \dot{\frown} I) \,.\, C`\text{sgm}`P = S_\epsilon``C`\text{sgm}`Q$

[Proof as in *212·7]

***212·71.** $\vdash : S \in P \,\overline{\text{smor}}\, Q \,.\supset .\, S_\epsilon \upharpoonright C`\varsigma`Q_* \in (\varsigma`P_*) \,\overline{\text{smor}}\, (\varsigma`Q_*)$

Dem.

$\vdash .\, *71\cdot381 \,.\supset \vdash :: \text{Hp} \,.\supset :.\, \alpha, \beta \in \text{sect}`Q \,.\supset : \exists ! \beta - \alpha \,.\equiv .\, \exists ! S``\beta - S``\alpha \quad (1)$

$\vdash .\, (1) \,.\, *212\cdot7 \,.\supset \vdash :.\, \text{Hp} \,.\, \alpha, \beta \in \text{sect}`Q \,.\supset : \alpha (\varsigma`Q_*) \beta \,.\equiv .\, S``\alpha \,(\varsigma`P_*)\, S``\beta \,.$

$[*150\cdot41] \quad \equiv .\, \alpha \,\{\breve{S}_\epsilon ; (\varsigma`P_*)\}\, \beta \quad (2)$

$\vdash .\, (2) \,.\, *212\cdot172 \,.\supset \vdash : \text{Hp} \,.\supset .\, \varsigma`Q_* \in \breve{S}_\epsilon ; (\varsigma`P_*) \quad (3)$

Similarly $\quad \vdash : \text{Hp} \,.\supset .\, \varsigma`P_* \in S_\epsilon ; (\varsigma`Q_*) \quad (4)$

$\vdash .\, *72\cdot451 \,.\quad \supset \vdash : \text{Hp} \,.\supset .\, S_\epsilon \upharpoonright C`\varsigma`Q_* \in 1 \to 1 \quad (5)$

$\vdash .\, (3) \,.\, (4) \,.\, (5) \,.\, *151\cdot27 \,.\supset \vdash .\, \text{Prop}$

***212·711.** $\vdash : S \in P \,\overline{\text{smor}}\, Q \,.\supset .\, S_\epsilon \upharpoonright C`\varsigma`Q \in (\varsigma`P) \,\overline{\text{smor}}\, (\varsigma`Q)$

[Proof as in *212·71]

***212·712.** $\vdash : S \in P \,\overline{\text{smor}}\, Q \,.\supset .\, S_\epsilon \upharpoonright C`\text{sgm}`P \in (\text{sgm}`P) \,\overline{\text{smor}}\, (\text{sgm}`Q)$

[Proof as in *212·7]

***212·72.** $\vdash : P \,\text{smor}\, Q \,.\supset .\, \varsigma`P_* \,\text{smor}\, \varsigma`Q_* \,.\, \varsigma`P \,\text{smor}\, \varsigma`Q \,.\, \text{sgm}`P \,\text{smor}\, \text{sgm}`Q$

[*212·71·711·712]

$*213$. SECTIONAL RELATIONS.

Summary of $*213$.

If α is a section of P, $P \lceil \alpha$ is called a *sectional relation* of P; and if α is a segment of P, $P \lceil \alpha$ is called a *segmental relation* of P. If P_{po} is serial, sectional relations may be arranged in a series by the relation of inclusion ($*213{\cdot}153$). That is, if we call the series of sectional relations P_ς, we shall so define P_ς as to secure that if P_{po} is serial,

$$QP_\varsigma R \,.\equiv .\, Q, R \in P \lceil ``(\text{sect}'P - \iota'\Lambda) \,.\, Q \subset R \,.\, Q \neq R \quad (*213{\cdot}21).$$

The natural definition to take would be

$$P_\varsigma = P \lceil ; \mathfrak{s}'P_*.$$

But this has the disadvantage that if xBP,

$$P \lceil \iota'x = P \lceil \Lambda \,.\, \Lambda, \iota'x \in \text{sect}'P.$$

Thus $P \lceil \alpha = P \lceil \beta$ does not imply $\alpha = \beta$; and when P is serial, $P \lceil ; \mathfrak{s}'P_*$ is not serial, because $\dot{\Lambda}\,(P \lceil ; \mathfrak{s}'P_*)\,\dot{\Lambda}$. In order to obviate this inconvenience, we confine ourselves to sections which are not null, putting

$$P_\varsigma = P \lceil ; (\mathfrak{s}'P_*) \lceil (-\iota'\Lambda) \quad \text{Df.}$$

With the above definition, we have ($*213{\cdot}151{\cdot}152$), if $P_{po} \in \text{Ser}$,

$$(P \lceil) \upharpoonright C'\{(\mathfrak{s}'P_*) \lceil -\iota'\Lambda\} \in 1 \rightarrow 1$$

and

$$P_\varsigma \text{ smor } (\mathfrak{s}'P_*) \lceil (-\iota'\Lambda).$$

The relation P_ς is very useful in dealing with well-ordered series; in this case, we have (as will be shown later)

$$P_\varsigma = P \lceil ; \overrightarrow{P} ; P \lceil \text{C}'P \mathrel{+\!\!\!\rightarrow} P.$$

It will be seen that, if $P_{po} \in \text{Ser}$, whenever P exists, $P = B'\breve{P}_\varsigma$ ($*213{\cdot}158$); and whenever $\overrightarrow{B}'P$ exists, $\dot{\Lambda} = B'P_\varsigma$ ($*213{\cdot}155$).

We have, if $P_{po} \in \text{Ser}$,

$$QP_\varsigma R \,.\equiv .\, R \in C'P_\varsigma \,.\, Q \in \text{D}'R_\varsigma \quad (*213{\cdot}245).$$

Hence $\quad R \in C'P_\varsigma \,.\supset .\, \overrightarrow{P_\varsigma}'R = \text{D}'R_\varsigma \,.\, R_\varsigma = P_\varsigma \lceil C'R_\varsigma \quad (*213{\cdot}246{\cdot}242)$.

If P is serial, the sectional relations of P are all relations such that by adding something to them they become P, *i.e.* they are

$$\hat{Q}\{(\exists R) \,.\, P = Q \mathbin{\ddagger} R \,.\vee .\, (\exists x) \,.\, P = Q \mathrel{+\!\!\!\rightarrow} x\} \quad (*213{\cdot}4).$$

Hence their relation-numbers are those that can be made equal to that of P by being added to. This fact is important in connection with the theory of greater and less among relation-numbers.

The propositions of this number are rendered complicated by the necessity of taking account of the possibility of a section being a unit class. This necessitates a good many propositions which are merely lemmas; but in the end the complications mostly disappear.

We begin with propositions on the field, etc., of P_s. We have

***213·141.** $\vdash . \text{D}`P_s = P \upharpoonright ``(\text{sect}`P - \iota`\Lambda - \iota`C`P)$

***213·142.** $\vdash : P_{\text{po}} \subseteq J . \supset . C`P_s = P \upharpoonright ``(\text{sect}`P - \iota`\Lambda)$

***213·16.** $\vdash . \text{D}`P_s = P \upharpoonright ``(\text{sect}`P - \iota`\Lambda) - \iota`P$

***213·161.** $\vdash : P_{\text{po}} \subseteq J . \exists ! \overrightarrow{B}`P . \supset . P \upharpoonright ``\text{sect}`P = P \upharpoonright ``(\text{sect}`P - \iota`\Lambda) = C`P_s$

***213·162.** $\vdash : P_{\text{po}} \in \text{Ser} . \supset . \text{Cl}`P_s = P \upharpoonright ``\text{sect}`P - \iota`\dot{\Lambda}$

We then prove:

***213·17.** $\vdash : P_{\text{po}} \in \text{Ser} . \supset . \text{Nr}`\mathfrak{s}`P_* = \dot{1} \dotplus \text{Nr}`P_s .$
$\text{Nr}`(\mathfrak{s}`P_*) \upharpoonright (\text{Cl}`\mathfrak{s}`P_*) = \text{Nr}`P_s$

If P is finite, it follows from the above that $\mathfrak{s}`P_*$ is not similar to P_s; but if P is infinite and has a beginning and is well-ordered, we find

$$\text{Nr}`\mathfrak{s}`P_* = \text{Nr}`P_s.$$

***213·172.** $\vdash : P_{\text{po}}, Q_{\text{po}} \in \text{Ser} . P \,\text{smor}\, Q . \supset . P_s \,\text{smor}\, Q_s$

We then have a set of propositions (*213·2—·251) chiefly concerned with the sections of R, where $R \in C`P_s$. Besides those already mentioned, the following are important:

***213·24.** $\vdash : \beta \in \text{sect}`P . R = P \upharpoonright \beta . \supset . \text{sect}`R = \text{sect}`P \cap \text{Cl}`C`R$

***213·243.** $\vdash . \overrightarrow{P_s}`P = \text{D}`P_s$

***213·25.** $\vdash :. P_{\text{po}} \in \text{Ser} . Q, R \in C`P_s . \supset : Q \in \text{D}`R_s . \vee . R \in \text{D}`Q_s . \vee . Q = R$

Our next set (*213·3—·32) is concerned with $\dot{\Lambda}$ and $x \downarrow y$. We have

***213·3.** $\vdash : P = \dot{\Lambda} . \supset . P_s = \dot{\Lambda}$

***213·32.** $\vdash : P \in 2_r . \supset . P_s = \dot{\Lambda} \downarrow P . P_s \in 2_r$

We then have three propositions (*213·4·41·42) showing that a sectional relation of P is one which becomes P by being added to. We proceed to a set of propositions (*213·5—·58) on $(P \dotplus\!\!\rightarrow x)_s$ and $(P \ddagger Q)_s$, leading to

***213·57.** $\vdash : P_{\text{po}} \subseteq J . \text{Nr}`Q = \text{Nr}`P \dotplus \dot{1} . \supset . \text{Nr}`Q_s = \text{Nr}`P_s \dotplus \dot{1}$

***213·58.** $\vdash : P_{\text{po}} \subseteq J . Q_{\text{po}} \in \text{Ser} . C`P \cap C`Q = \Lambda . \supset .$
$\text{Nr}`(P \ddagger Q)_s = \text{Nr}`P_s \dotplus \text{Nr}`Q_s$

***213·01.** $P_\varsigma = P \upharpoonright ; (\varsigma`P_*) \upharpoonright (-\iota`\Lambda)$ Df

***213·1.** $\vdash : QP_\varsigma R . \equiv .$

$(\exists \alpha, \beta) . \alpha, \beta \in \text{sect}`P - \iota`\Lambda . \exists ! \beta - \alpha . Q = P \upharpoonright \alpha . R = P \upharpoonright \beta$

[*212·12·121 . (*213·01)]

***213·11.** $\vdash :. P_{po} \in \text{connex} . \supset : QP_\varsigma R . \equiv .$

$(\exists \alpha, \beta) . \alpha, \beta \in \text{sect}`P - \iota`\Lambda . \alpha \subset \beta . \alpha \neq \beta . Q = P \upharpoonright \alpha . R = P \upharpoonright \beta$

[*213·1 . *211·6·17 . *210·1]

***213·12.** $\vdash . \text{D}`(\varsigma`P_*) \upharpoonright (-\iota`\Lambda) = \text{sect}`P - \iota`\Lambda - \iota`C`P$

Dem.

$\vdash . *212{\cdot}12 . \supset \vdash : \alpha \in \text{D}`(\varsigma`P_*) \upharpoonright (-\iota`\Lambda) . \equiv . (\exists \beta) . \beta \in \text{sect}`P . \exists ! \beta - \alpha . \alpha \neq \Lambda .$

$[*212{\cdot}12] \qquad \equiv . \alpha \neq \Lambda . \alpha \in \text{D}`\varsigma`P_* .$

$[*212{\cdot}171] \qquad \equiv . \alpha \in \text{sect}`P - \iota`\Lambda - \iota`C`P : \supset \vdash . \text{Prop}$

***213·121.** $\vdash : P_{po} \in \text{Ser} . \supset . \overrightarrow{B}`(\varsigma`P_*) \upharpoonright (-\iota`\Lambda) = \text{sect}`P \cap 1 = \iota``\overrightarrow{B}`P$

Dem.

$\vdash . *212{\cdot}12 . *213{\cdot}12 . \supset \vdash :. \beta \in \overrightarrow{B}`(\varsigma`P_*) \upharpoonright (-\iota`\Lambda) . \equiv :$

$\beta \in \text{sect}`P - \iota`\Lambda - \iota`C`P : \alpha \in \text{sect}`P . \exists ! \beta - \alpha . \supset_\alpha . \alpha = \Lambda \quad (1)$

$\vdash . *211{\cdot}3{\cdot}13{\cdot}1 . *37{\cdot}18 . \supset$

$\vdash : \beta \in \text{sect}`P . x \in \beta . \supset . \overrightarrow{P_*}`x \in \text{sect}`P . \overrightarrow{P_*}`x \subset \beta . \exists ! \overrightarrow{P_*}`x \quad (2)$

$\vdash . (1) . \text{Transp} . (2) . \supset \vdash :. \beta \in \overrightarrow{B}`(\varsigma`P_*) \upharpoonright (-\iota`\Lambda) . \supset :$

$\beta \in \text{sect}`P - \iota`\Lambda - \iota`C`P : x \in \beta . \supset_x . \overrightarrow{P_*}`x = \beta \quad (3)$

$\vdash . *200{\cdot}391 . \supset \vdash :. \text{Hp} . \beta \in \text{sect}`P - \iota`\Lambda : x \in \beta . \supset_x . \overrightarrow{P_*}`x = \beta . \supset :$

$\beta \in \text{sect}`P - \iota`\Lambda : x, y \in \beta . \supset_{x,y} . x = y :$

$[*52{\cdot}16] \quad \supset : \beta \in \text{sect}`P \cap 1 \quad (4)$

$\vdash . (3) . (4) . \qquad \supset \vdash : \text{Hp} . \supset . \overrightarrow{B}`(\varsigma`P_*) \upharpoonright (-\iota`\Lambda) \subset \text{sect}`P \cap 1 \quad (5)$

$\vdash . *213{\cdot}12 . *200{\cdot}12 . \supset \vdash : \text{Hp} . \supset . \text{sect}`P \cap 1 \subset \text{D}`(\varsigma`P_*) \upharpoonright (-\iota`\Lambda) \quad (6)$

$\vdash . *51{\cdot}401 . \qquad \supset \vdash :. \beta \in \text{sect}`P \cap 1 . \supset : \alpha \subset \beta . \alpha \neq \beta . \supset . \alpha = \Lambda \quad (7)$

$\vdash . (7) . *212{\cdot}22{\cdot}121 . \supset \vdash : \text{Hp} . \supset . \text{sect}`P \cap 1 \subset -\text{Cl}`(\varsigma`P_*) \upharpoonright (-\iota`\Lambda) \quad (8)$

$\vdash . (5) . (6) . (8) . *211{\cdot}18 . \supset \vdash . \text{Prop}$

***213·122.** $\vdash : P_{po} \in \text{Ser} . \exists ! \overrightarrow{B}`P . \supset . B`(\varsigma`P) \upharpoonright (-\iota`\Lambda) = \iota`B`P$

[*213·121 . *211·181]

***213·123.** $\vdash : P_{po} \in \text{Ser} . \overrightarrow{B}`P = \Lambda . \supset . \overrightarrow{B}`(\varsigma`P) \upharpoonright (-\iota`\Lambda) = \Lambda$

[*213·121]

***213·124.** $\vdash :. P_{po} \in \text{Ser} . \supset : \text{E} ! B`(\varsigma`P) \upharpoonright (-\iota`\Lambda) . \equiv . \text{E} ! B`P$

[*213·122·123]

***213·125.** $\vdash : P_{po} \subseteq J . \supset . C\text{'}\mathfrak{s}\text{'}P_* - \iota\text{'}\Lambda \sim\epsilon 1$

Dem.

$\vdash . *212·17 . \supset \vdash : P = \dot{\Lambda} . \supset . C\text{'}\mathfrak{s}\text{'}P_* = \Lambda .$

$[*52·21] \quad \supset . C\text{'}\mathfrak{s}\text{'}P_* - \iota\text{'}\Lambda \sim\epsilon 1 \quad (1)$

$\vdash . *212·172 . \supset \vdash : \dot{\exists} ! P . \supset . C\text{'}\mathfrak{s}\text{'}P_* = \text{sect}\text{'}P . C\text{'}P \epsilon C\text{'}\mathfrak{s}\text{'}P_* - \iota\text{'}\Lambda \quad (2)$

$\vdash . *211·13·3 . *200·39 . \supset \vdash : \text{Hp} . x \epsilon D\text{'}P . \supset . \overrightarrow{P_*}\text{'}x \epsilon \text{sect}\text{'}P . \exists ! C\text{'}P - \overrightarrow{P_*}\text{'}x \quad (3)$

$\vdash . (2) . (3) . \quad \supset \vdash : \text{Hp} . \dot{\exists} ! P . \supset . C\text{'}\mathfrak{s}\text{'}P_* - \iota\text{'}\Lambda \sim\epsilon 1 \quad (4)$

$\vdash . (1) . (4) . \supset \vdash . \text{Prop}$

The hypothesis $P_{po} \subseteq J$, in the above proposition, restricts P more than is necessary for the truth of the conclusion. What we really require is $P = \dot{\Lambda} . \mathsf{v} . (\exists x) . x \epsilon C\text{'}P . \overrightarrow{P_*}\text{'}x \neq C\text{'}P$, *i.e.* $\overrightarrow{P_*}\text{''}C\text{'}P \neq \iota\text{'}C\text{'}P$. This holds if either (1) the field of P does not consist of a single family, or (2) there is a member of $C\text{'}P$ which does not have the relation P_{po} to itself. Thus the only case excluded is that of a single cyclic family. The hypothesis $\overrightarrow{P_*}\text{''}C\text{'}P \neq \iota\text{'}C\text{'}P$ may be substituted for $P_{po} \subseteq J$ in most of the subsequent propositions of this number in which $P_{po} \subseteq J$ occurs in the hypothesis. We have, however, preferred the hypothesis $P_{po} \subseteq J$, as it gives a more immediate application to the case of $P \epsilon \text{Ser}$, which is the case in which the propositions of the present number are important.

***213·126.** $\vdash : P_{po} \subseteq J . \dot{\exists} ! P . \supset . \exists ! \text{sect}\text{'}P - \iota\text{'}\Lambda - \iota\text{'}C\text{'}P$

Dem.

$\vdash . *213·125 . *212·172 . \supset \vdash : \text{Hp} . \supset . \text{sect}\text{'}P - \iota\text{'}\Lambda \sim\epsilon 1 \quad (1)$

$\vdash . *211·26 . *33·24 . \quad \supset \vdash : \text{Hp} . \supset . C\text{'}P \epsilon \text{sect}\text{'}P - \iota\text{'}\Lambda \quad (2)$

$\vdash . (1) . (2) . *52·181 . \supset \vdash . \text{Prop}$

***213·13.** $\vdash : P_{po} \subseteq J . \supset . C\text{'}(\mathfrak{s}\text{'}P_*) \upharpoonright (-\iota\text{'}\Lambda) = \text{sect}\text{'}P - \iota\text{'}\Lambda$

Dem.

$\vdash . *213·125 . \supset$

$\vdash :: \text{Hp} . \supset :. \alpha \epsilon \text{sect}\text{'}P - \iota\text{'}\Lambda . \supset : (\exists \beta) : \beta \epsilon \text{sect}\text{'}P - \iota\text{'}\Lambda : \exists ! \alpha - \beta . \mathsf{v} . \exists ! \beta - \alpha :$

$[*212·12] \quad \supset : (\exists \beta) : \alpha \{(\mathfrak{s}\text{'}P_*) \upharpoonright (-\iota\text{'}\Lambda)\} \beta . \mathsf{v} . \beta \{(\mathfrak{s}\text{'}P_*) \upharpoonright (-\iota\text{'}\Lambda)\} \alpha :$

$[*33·132] \quad \supset : \alpha \epsilon C\text{'}(\mathfrak{s}\text{'}P_*) \upharpoonright (-\iota\text{'}\Lambda) \quad (1)$

$\vdash . (1) . *212·172 . \quad \supset \vdash : \text{Hp} . \dot{\exists} ! P . \supset . C\text{'}(\mathfrak{s}\text{'}P_*) \upharpoonright (-\iota\text{'}\Lambda) = \text{sect}\text{'}P - \iota\text{'}\Lambda \quad (2)$

$\vdash . *212·17 . *211·1 . \supset \vdash : P = \dot{\Lambda} . \supset . C\text{'}(\mathfrak{s}\text{'}P_*) \upharpoonright (-\iota\text{'}\Lambda) = \text{sect}\text{'}P - \iota\text{'}\Lambda \quad (3)$

$\vdash . (2) . (3) . \supset \vdash . \text{Prop}$

***213·131.** $\vdash : P_{po} \epsilon \text{Ser} . \supset . \text{G}\text{'}(\mathfrak{s}\text{'}P_*) \upharpoonright (-\iota\text{'}\Lambda) = \text{sect}\text{'}P - \iota\text{'}\Lambda - \iota\text{''}\overrightarrow{B}\text{'}P$
[*213·13·121]

***213·132.** $\vdash : P_{po} \epsilon \text{Ser} . \exists ! \overrightarrow{B}\text{'}P . \supset . \text{G}\text{'}(\mathfrak{s}\text{'}P_*) \upharpoonright (-\iota\text{'}\Lambda) = \text{sect}\text{'}P - \iota\text{'}\Lambda - \iota\text{'}\iota\text{'}B\text{'}P$
[*213·13·122]

***213·133.** $\vdash : P_{po} \in \text{Ser} . \overrightarrow{B}\text{‘}P = \Lambda . \supset . \text{G‘}(\varsigma\text{‘}P_*) \upharpoonright (-\iota\text{‘}\Lambda) = \text{sect‘}P - \iota\text{‘}\Lambda$
[*213·13·123]

***213·134.** $\vdash : P_{po} \subset J . \dot{\exists} ! P . \supset . B\text{‘Cnv‘}(\varsigma\text{‘}P_*) \upharpoonright (-\iota\text{‘}\Lambda) = C\text{‘}P$ [*213·12·13]

***213·14.** $\vdash . \text{D‘}P_s = P \upharpoonright \text{“D‘}(\varsigma\text{‘}P_*) \upharpoonright (-\iota\text{‘}\Lambda) . \text{G‘}P_s = P \upharpoonright \text{“G‘}(\varsigma\text{‘}P_*) \upharpoonright (-\iota\text{‘}\Lambda) .$
$C\text{‘}P_s = P \upharpoonright \text{“}C\text{‘}(\varsigma\text{‘}P_*) \upharpoonright (-\iota\text{‘}\Lambda)$
[*150·21·211·22]

***213·141.** $\vdash . \text{D‘}P_s = P \upharpoonright \text{“}(\text{sect‘}P - \iota\text{‘}\Lambda - \iota\text{‘}C\text{‘}P)$
[*213·12·14]

***213·142.** $\vdash : P_{po} \subset J . \supset . C\text{‘}P_s = P \upharpoonright \text{“}(\text{sect‘}P - \iota\text{‘}\Lambda)$
[*213·13·14]

***213·143.** $\vdash : P_{po} \in \text{Ser} . \supset . \text{G‘}P_s = P \upharpoonright \text{“}(\text{sect‘}P - \iota\text{‘}\Lambda - \iota\text{“}\overrightarrow{B}\text{‘}P)$
[*213·131·14]

***213·144.** $\vdash : P_{po} \in \text{Ser} . \exists ! \overrightarrow{B}\text{‘}P . \supset . \text{G‘}P_s = P \upharpoonright \text{“}(\text{sect‘}P - \iota\text{‘}\Lambda - \iota\text{‘}\iota\text{‘}B\text{‘}P)$
[*213·132·14]

***213·145.** $\vdash : P_{po} \in \text{Ser} . \overrightarrow{B}\text{‘}P = \Lambda . \supset . \text{G‘}P_s = P \upharpoonright \text{“}(\text{sect‘}P - \iota\text{‘}\Lambda)$
[*213·143]

***213·146.** $\vdash : P \subset J . \supset . P \upharpoonright \text{“sect‘}P = P \upharpoonright \text{“}(\text{sect‘}P - 1)$

Dem.

$\vdash . *37·22 . \supset \vdash . P \upharpoonright \text{“sect‘}P = P \upharpoonright \text{“}(\text{sect‘}P - 1) \cup P \upharpoonright \text{“}(\text{sect‘}P \cap 1)$ (1)
$\vdash . *200·35 . \supset \vdash : Q \in P \upharpoonright \text{“}(\text{sect‘}P \cap 1) . \supset . Q = \dot{\Lambda} .$
[*36·27] $\supset . Q = P \upharpoonright \Lambda .$
[*211·44] $\supset . Q \in P \upharpoonright \text{“}(\text{sect‘}P - 1)$ (2)
$\vdash . (1) . (2) . \supset \vdash .$ Prop

***213·15.** $\vdash :. P_{po} \in \text{Ser} . \alpha \in \text{sect‘}P - \iota\text{‘}\Lambda . \supset : P \upharpoonright \alpha = \dot{\Lambda} . \equiv . \alpha \in 1$

Dem.

$\vdash . *200·35 . \supset \vdash : \text{Hp} . \alpha \in 1 . \supset . P \upharpoonright \alpha = \dot{\Lambda}$ (1)
$\vdash . *52·41 . \supset \vdash :. \text{Hp} . \alpha \sim\in 1 . \supset : (\exists x, y) . x, y \in \alpha . x \neq y :$
[*211·1.*202·103] $\supset : (\exists x, y) : x (P_{po} \upharpoonright \alpha) y . \vee . y (P_{po} \upharpoonright \alpha) x :$
[*11·7] $\supset : \dot{\exists} ! P_{po} \upharpoonright \alpha :$
[*37·41] $\supset : \exists ! \alpha \cap P_{po}\text{“}\alpha :$
[*211·131] $\supset : \exists ! \alpha \cap P\text{“}\alpha :$
[*37·41] $\supset : \dot{\exists} ! P \upharpoonright \alpha$ (2)
$\vdash . (1) . (2) . \supset \vdash .$ Prop

***213·151.** $\vdash : P_{po} \in \mathrm{Ser} . \supset . (P \upharpoonright) \restriction (\mathrm{sect}`P - \iota`\Lambda) \in 1 \to 1$

Dem.

$\vdash . *213·15 . \supset$

$\vdash : \mathrm{Hp} . \alpha \in \mathrm{sect}`P - \iota`\Lambda - 1 . \beta \in \mathrm{sect}`P - \iota`\Lambda . P \upharpoonright \alpha = P \upharpoonright \beta . \supset . \beta \sim\in 1 .$

[*211·133] $\supset . C`P \upharpoonright \beta = \beta . C`P \upharpoonright \alpha = \alpha .$

[Hp] $\supset . \alpha = \beta$ (1)

$\vdash . *213·15 . \supset$

$\vdash : \mathrm{Hp} . \alpha \in \mathrm{sect}`P \cap 1 . \beta \in \mathrm{sect}`P - \iota`\Lambda . P \upharpoonright \alpha = P \upharpoonright \beta . \supset . \beta \in 1$ (2)

$\vdash . (2) . *211·18 . \supset \vdash : \mathrm{Hp}(2) . \supset . \alpha, \beta \in \iota``\overrightarrow{B}`P .$

[*202·523·13] $\supset . \alpha = \beta$ (3)

$\vdash . (1) . (3) . \supset \vdash :. \mathrm{Hp} . \supset : \alpha, \beta \in \mathrm{sect}`P - \iota`\Lambda . P \upharpoonright \alpha = P \upharpoonright \beta . \supset . \alpha = \beta :.$

$\supset \vdash .$ Prop

***213·152.** $\vdash : P_{po} \in \mathrm{Ser} . \supset . P_s \,\mathrm{smor}\, (\varsigma`P_*) \upharpoonright (-\iota`\Lambda)$ [*213·151·13]

***213·153.** $\vdash : P_{po} \in \mathrm{Ser} . \supset . P_s \in \mathrm{Ser}$ [*213·152 . *212·3 . *204·4·21]

***213·154.** $\vdash : P_{po} \in \mathrm{Ser} . \supset . \overrightarrow{B}`P_s = P \upharpoonright ``\iota``\overrightarrow{B}`P$ [*213·151·121 . *151·5]

***213·155.** $\vdash : P_{po} \in \mathrm{Ser} . \exists ! \overrightarrow{B}`P . \supset . B`P_s = \dot{\Lambda}$

Dem.

$\vdash . *213·151·122 . *151·5 . \supset$

$\vdash : \mathrm{Hp} . \supset . B`P_s = P \upharpoonright (\iota`B`P)$

[*200·35] $= \dot{\Lambda} : \supset \vdash .$ Prop

***213·156.** $\vdash : P_{po} \in \mathrm{Ser} . \overrightarrow{B}`P = \Lambda . \supset . \overrightarrow{B}`P_s = \Lambda$ [*213·154]

***213·157.** $\vdash :. P_{po} \in \mathrm{Ser} . \supset : E ! B`P . \equiv . E ! B`P_s$ [*213·155·156]

***213·158.** $\vdash : P_{po} \in \mathrm{Ser} . \dot{\exists} ! P . \supset . B`\breve{P}_s = P$

Dem.

$\vdash . *213·151·134 . *151·5 . \supset \vdash : \mathrm{Hp} . \supset . B`\breve{P}_s = P \upharpoonright C`P : \supset \vdash .$ Prop

***213·16.** $\vdash . \mathrm{D}`P_s = P \upharpoonright ``(\mathrm{sect}`P - \iota`\Lambda) - \iota`P$

Dem.

$\vdash . *213·141 . \supset$

$\vdash : Q \in \mathrm{D}`P_s . \equiv . (\exists \alpha) . \alpha \in \mathrm{sect}`P - \iota`\Lambda . Q = P \upharpoonright \alpha . \alpha \neq C`P$ (1)

$\vdash . *211·1 . \supset \vdash :. \alpha \in \mathrm{sect}`P . Q = P \upharpoonright \alpha . \supset : \alpha \neq C`P . \equiv . \exists ! C`P - \alpha .$

[*36·25.Transp] $\equiv . Q \neq P$ (2)

$\vdash . (1) . (2) . \supset \vdash : Q \in \mathrm{D}`P_s . \equiv . (\exists \alpha) . \alpha \in \mathrm{sect}`P - \iota`\Lambda . Q = P \upharpoonright \alpha . Q \neq P :$

$\supset \vdash .$ Prop

***213·161.** $\vdash : P_{po} \subseteq J \,.\, \exists ! \overrightarrow{B}`P \,.\, \supset .\, P \upharpoonright``\text{sect}`P = P \upharpoonright``(\text{sect}`P - \iota`\Lambda) = C`P_s$

Dem.

$\vdash .\, *211·18 \,.\, \supset \vdash : \text{Hp} \,.\, \supset .\, \iota``\overrightarrow{B}`P \subset \text{sect}`P \cap 1 \,.\, \exists ! \iota``\overrightarrow{B}`P \,.$

[*37·2·45] $\supset .\, P \upharpoonright``\iota``\overrightarrow{B}`P \subset P \upharpoonright``(\text{sect}`P - \iota`\Lambda) \,.\, \exists ! P \upharpoonright``\iota``\overrightarrow{B}`P \,.$

[*200·35] $\supset .\, \dot\Lambda \in P \upharpoonright``(\text{sect}`P - \iota`\Lambda) \,.$

[*36·27] $\supset .\, P \upharpoonright \Lambda \in P \upharpoonright``(\text{sect}`P - \iota`\Lambda) \,.$

[*37·22] $\supset .\, P \upharpoonright``\text{sect}`P = P \upharpoonright``(\text{sect}`P - \iota`\Lambda)$

[*213·142] $= C`P_s : \supset \vdash .\, \text{Prop}$

***213·162.** $\vdash : P_{po} \in \text{Ser} \,.\, \supset .\, \text{C}\!\!\text{I}`P_s = P \upharpoonright``\text{sect}`P - \iota`\dot\Lambda$

Dem.

$\vdash .\, *213·143 \,.\, \supset \vdash :.\, \text{Hp} \,.\, \supset : Q \in \text{C}\!\!\text{I}`P_s \,.\, \equiv .$

$(\exists \alpha) \,.\, \alpha \in \text{sect}`P - \iota`\Lambda - \iota``\overrightarrow{B}`P \,.\, Q = P \upharpoonright \alpha \,.$ (1)

[*213·15.*211·18] $\supset .\, Q \in P \upharpoonright``\text{sect}`P - \iota`\dot\Lambda$ (2)

$\vdash .\, *213·15 \,.\, \supset \vdash :.\, \text{Hp} \,.\, \supset : Q \in P \upharpoonright``\text{sect}`P - \iota`\dot\Lambda \,.\, \supset .$

$(\exists \alpha) \,.\, \alpha \in \text{sect}`P - \iota`\Lambda - \iota``\overrightarrow{B}`P \,.\, Q = P \upharpoonright \alpha \,.$

[(1)] $\supset .\, Q \in \text{C}\!\!\text{I}`P_s$ (3)

$\vdash .\, (2) \,.\, (3) \,.\, \supset \vdash .\, \text{Prop}$

***213·163.** $\vdash : P_{po} \in \text{Ser} \,.\, \overrightarrow{B}`P = \Lambda \,.\, \supset .\, C`P_s = P \upharpoonright``\text{sect}`P - \iota`\dot\Lambda$

Dem.

$\vdash .\, *213·156 \,.\, \supset \vdash : \text{Hp} \,.\, \supset .\, C`P_s = \text{C}\!\!\text{I}`P_s$ (1)

$\vdash .\, (1) \,.\, *213·162 \,.\, \supset \vdash .\, \text{Prop}$

***213·164.** $\vdash : P_{po} \in \text{Ser} \,.\, \overrightarrow{B}`P = \Lambda \,.\, \supset .\, \text{D}`P_s = P \upharpoonright``\text{sect}`P - \iota`\dot\Lambda - \iota`P$ [*213·142·163·16]

***213·17.** $\vdash : P_{po} \in \text{Ser} \,.\, \supset .\, \text{Nr}`\mathfrak{s}`P_* = \dot 1 \dotplus \text{Nr}`P_s \,.$
$\text{Nr}`(\mathfrak{s}`P_*) \upharpoonright (\text{C}\!\!\text{I}`\mathfrak{s}`P_*) = \text{Nr}`P_s$

Dem.

$\vdash .\, *212·171·172 \,.\, \supset \vdash : \dot\exists ! P \,.\, \supset .\, B`\mathfrak{s}`P_* = \Lambda \,.\, \text{C}\!\!\text{I}`\mathfrak{s}`P_* = \text{sect}`P - \iota`\Lambda \,.$
$(\mathfrak{s}`P_*) \upharpoonright (-\iota`\Lambda) = (\mathfrak{s}`P_*) \upharpoonright (\text{C}\!\!\text{I}`\mathfrak{s}`P_*)$ (1)

$\vdash .\, (1) \,.\, *213·125 \,.\, \supset \vdash : \text{Hp} \,.\, \dot\exists ! P \,.\, \supset .\, \text{C}\!\!\text{I}`\mathfrak{s}`P_* \sim\in 1$ (2)

$\vdash .\, *212·3 \,.\, *91·602 \,.\, \supset \vdash : \text{Hp} \,.\, \supset .\, \mathfrak{s}`P_* \in \text{connex}$ (3)

$\vdash .\, (1) \,.\, *213·152 \,.\, \supset \vdash : \text{Hp} \,.\, \dot\exists ! P \,.\, \supset .\, \text{Nr}`(\mathfrak{s}`P_*) \upharpoonright (\text{C}\!\!\text{I}`\mathfrak{s}`P_*) = \text{Nr}`P_s$ (4)

$\vdash .\, (1) \,.\, (2) \,.\, (3) \,.\, *204·46 \,.\, \supset$

$\vdash : \text{Hp} \,.\, \dot\exists ! P \,.\, \supset .\, \text{Nr}`\mathfrak{s}`P_* = \dot 1 \dotplus \text{Nr}`(\mathfrak{s}`P_*) \upharpoonright (\text{C}\!\!\text{I}`\mathfrak{s}`P_*)$

[(4)] $= \dot 1 \dotplus \text{Nr}`P_s$ (5)

$\vdash .\, *212·17 \,.\, *150·42 \,.\, \supset \vdash : P = \dot\Lambda \,.\, \supset .\, \mathfrak{s}`P_* = \dot\Lambda \,.\, P_s = \dot\Lambda \,.$

[*161·201] $\supset .\, \text{Nr}`\mathfrak{s}`P_* = \dot 1 \dotplus \text{Nr}`P_s \,.\, \text{Nr}`(\mathfrak{s}`P_*) \upharpoonright (\text{C}\!\!\text{I}`\mathfrak{s}`P_*) = \text{Nr}`P_s$ (6)

$\vdash .\, (4) \,.\, (5) \,.\, (6) \,.\, \supset \vdash .\, \text{Prop}$

***213·171.** $\vdash :. P_{\text{po}}, Q_{\text{po}} \epsilon \text{Ser} . \supset : P_\varsigma \text{ smor } Q_\varsigma . \equiv . \varsigma\text{‘}P_* \text{ smor } \varsigma\text{‘}Q_*$

Dem.

$\vdash . *212\cdot172 . \supset \vdash :. \text{Hp} . \dot{\exists} ! P . \dot{\exists} ! Q . \supset : \text{E} ! B\text{‘}\varsigma\text{‘}P_* . \text{E} ! B\text{‘}\varsigma\text{‘}Q_* :$

$[*204\cdot47 . *91\cdot602 . *212\cdot3]$

$\supset : \varsigma\text{‘}P_* \text{ smor } \varsigma\text{‘}Q_* . \equiv . (\varsigma\text{‘}P_*) \upharpoonright (\text{Ɑ}\text{‘}\varsigma\text{‘}P_*) \text{ smor } (\varsigma\text{‘}Q_*) \upharpoonright (\text{Ɑ}\text{‘}\varsigma\text{‘}Q_*) .$

$[*213\cdot17] \quad \equiv . P_\varsigma \text{ smor } Q_\varsigma \quad (1)$

$\vdash . *213\cdot158 . \quad \supset \vdash : \text{Hp} . \dot{\exists} ! P . P_\varsigma \text{ smor } Q_\varsigma . \supset . \dot{\exists} ! Q_\varsigma .$

$[*212\cdot17 . *150\cdot42] \quad \supset . \dot{\exists} ! Q \quad (2)$

$\vdash . *212\cdot17 . \quad \supset \vdash : \text{Hp} . \dot{\exists} ! P . \varsigma\text{‘}P_* \text{ smor } \varsigma\text{‘}Q_* . \supset . \dot{\exists} ! Q \quad (3)$

$\vdash . (1) . (2) . (3) . \supset \vdash :. \text{Hp} . \dot{\exists} ! P . \supset : \varsigma\text{‘}P_* \text{ smor } \varsigma\text{‘}Q_* . \equiv . P_\varsigma \text{ smor } Q_\varsigma \quad (4)$

$\vdash . *212\cdot17 . \quad \supset \vdash : \text{Hp} . P = \dot{\Lambda} . \varsigma\text{‘}P_* \text{ smor } \varsigma\text{‘}Q_* . \supset . \varsigma\text{‘}P_* = \dot{\Lambda} . \varsigma\text{‘}Q_* = \dot{\Lambda} .$

$[*150\cdot42] \quad \supset . P_\varsigma = \dot{\Lambda} . Q_\varsigma = \dot{\Lambda} \quad (5)$

$\vdash . *213\cdot17 . \quad \supset \vdash : \text{Hp} . P_\varsigma \text{ smor } Q_\varsigma . \supset . \varsigma\text{‘}P_* \text{ smor } \varsigma\text{‘}Q_* \quad (6)$

$\vdash . (5) . (6) . \quad \supset \vdash :. \text{Hp} . P = \dot{\Lambda} . \supset : \varsigma\text{‘}P_* \text{ smor } \varsigma\text{‘}Q_* . \equiv . P_\varsigma \text{ smor } Q_\varsigma \quad (7)$

$\vdash . (4) . (7) . \quad \supset \vdash . \text{Prop}$

***213·172.** $\vdash : P_{\text{po}}, Q_{\text{po}} \epsilon \text{Ser} . P \text{ smor } Q . \supset . P_\varsigma \text{ smor } Q_\varsigma \quad [*212\cdot72 . *213\cdot171]$

***213·18.** $\vdash : P \epsilon \text{connex} . R \epsilon \text{D}\text{‘}P_\varsigma . \supset . \exists ! C\text{‘}P \cap p\text{‘}\overleftarrow{P}\text{“}C\text{‘}R$

Dem.

$\vdash . *213\cdot1 . \supset \vdash : R \epsilon \text{D}\text{‘}P_\varsigma . \supset . (\exists\alpha) . \alpha \epsilon \text{sect}\text{‘}P - \iota\text{‘}C\text{‘}P . R = P \upharpoonright \alpha .$

$[*37\cdot41] \quad \supset . (\exists\alpha) . \alpha \epsilon \text{sect}\text{‘}P - \iota\text{‘}C\text{‘}P . C\text{‘}R \subset \alpha .$

$[*40\cdot16] \quad \supset . (\exists\alpha) . \alpha \epsilon \text{sect}\text{‘}P - \iota\text{‘}C\text{‘}P . p\text{‘}\overleftarrow{P}\text{“}\alpha \subset p\text{‘}\overleftarrow{P}\text{“}C\text{‘}R \quad (1)$

$\vdash . *211\cdot703 . \supset \vdash : \text{Hp} . \alpha \epsilon \text{sect}\text{‘}P - \iota\text{‘}C\text{‘}P . \supset . \exists ! p\text{‘}\overleftarrow{P}\text{“}\alpha \quad (2)$

$\vdash . *211\cdot1 . \quad \supset \vdash : \alpha \epsilon \text{sect}\text{‘}P - \iota\text{‘}C\text{‘}P . \supset . \exists ! C\text{‘}P - \alpha .$

$[*33\cdot24] \quad \supset . \dot{\exists} ! P \quad (3)$

$\vdash . (2) . (3) . *40\cdot69 . \supset \vdash : \text{Hp} . \alpha \epsilon \text{sect}\text{‘}P - \iota\text{‘}C\text{‘}P . \supset . \exists ! C\text{‘}P \cap p\text{‘}\overleftarrow{P}\text{“}\alpha \quad (4)$

$\vdash . (1) . (4) . \supset$

$\vdash : \text{Hp} . R \epsilon \text{D}\text{‘}P_\varsigma . \supset . (\exists\alpha) . \exists ! C\text{‘}P \cap p\text{‘}\overleftarrow{P}\text{“}\alpha . p\text{‘}\overleftarrow{P}\text{“}\alpha \subset p\text{‘}\overleftarrow{P}\text{“}C\text{‘}R : \supset \vdash . \text{Prop}$

***213·2.** $\vdash :. P_{\text{po}} \epsilon \text{Ser} . \alpha, \beta \epsilon \text{sect}\text{‘}P - \iota\text{‘}\Lambda . Q = P \upharpoonright \alpha . R = P \upharpoonright \beta . \supset :$

$\exists ! \beta - \alpha . \equiv . \dot{\exists} ! R \dot{-} Q . \equiv . Q \Subset R . Q \neq R . \equiv . \alpha \subset \beta . \alpha \neq \beta$

Dem.

$\vdash . *36\cdot24 . \quad \supset \vdash : \alpha \subset \beta . \supset . P \upharpoonright \alpha \Subset P \upharpoonright \beta \quad (1)$

$\vdash . *211\cdot133 . \quad \supset \vdash : \text{Hp} . \alpha, \beta \sim\epsilon 1 . P \upharpoonright \alpha \Subset P \upharpoonright \beta . \supset . \alpha \subset \beta \quad (2)$

$\vdash . *211\cdot181\cdot182 . \supset \vdash : \text{Hp} . \alpha \epsilon 1 . \supset . \alpha = \iota\text{‘}B\text{‘}P .$

$[*202\cdot521] \quad \supset . \alpha \subset \beta \quad (3)$

$\vdash . *213\cdot15 . \quad \supset \vdash : \text{Hp} . \beta \epsilon 1 . \alpha \sim\epsilon 1 . \supset . \sim(P \upharpoonright \alpha \Subset P \upharpoonright \beta) \quad (4)$

$\vdash . (2) . (3) . (4) . \quad \supset \vdash : \text{Hp} . P \upharpoonright \alpha \Subset P \upharpoonright \beta . \supset . \alpha \subset \beta \quad (5)$

$\vdash . (1) . (5) . \supset \vdash :. \text{Hp} . \supset : \alpha \subset \beta . \equiv . Q \mathrel{G} R :$ (6)

[Transp] $\supset : \exists ! \alpha - \beta . \equiv . \dot{\exists} ! Q \dot{-} R$ (7)

$\vdash . (6) . *213\cdot151 . \supset$

$\vdash :. \text{Hp} . \supset : \alpha \subset \beta . \alpha \neq \beta . \equiv . Q \mathrel{G} R . Q \neq R$ (8)

$\vdash . (7) . (8) . *210\cdot1 . *211\cdot562 . \supset \vdash . \text{Prop}$

***213·21.** $\vdash :. P_{po} \in \text{Ser} . \supset : Q P_s R . \equiv . Q, R \in P \upharpoonright \text{``}(\text{sect}\text{`}P - \iota\text{`}\Lambda) . \dot{\exists} ! R \dot{-} Q .$
$\equiv . Q, R \in P \upharpoonright \text{``}(\text{sect}\text{`}P - \iota\text{`}\Lambda) . Q \mathrel{G} R . Q \neq R$

Dem.

$\vdash . *213\cdot1\cdot2 . \supset \vdash :. \text{Hp} . \supset :$

$Q P_s R . \equiv . (\exists \alpha, \beta) . \alpha, \beta \in \text{sect}\text{`}P - \iota\text{`}\Lambda . Q = P \upharpoonright \alpha . R = P \upharpoonright \beta . \dot{\exists} ! R \dot{-} Q .$

$\equiv . (\exists \alpha, \beta) . \alpha, \beta \in \text{sect}\text{`}P - \iota\text{`}\Lambda . Q = P \upharpoonright \alpha . R = P \upharpoonright \beta . Q \mathrel{G} R . Q \neq R$ (1)

$\vdash . (1) . *37\cdot6 . \supset \vdash . \text{Prop}$

***213·22.** $\vdash :. P_{po} \in \text{Ser} . \exists ! \overrightarrow{B}\text{`}P . \supset :$

$Q P_s R . \equiv . Q, R \in P \upharpoonright \text{``sect`}P . \dot{\exists} ! R \dot{-} Q . \equiv . Q, R \in P \upharpoonright \text{``sect`}P . Q \mathrel{G} R . Q \neq R$

[*213·21·161]

***213·23.** $\vdash :. P_{po} \in \text{connex} . Q, R \in C\text{`}P_s . \supset : Q \mathrel{G} R . \vee . R \mathrel{G} Q$

[*213·1 . *211·6·17 . *36·24]

***213·24.** $\vdash : \beta \in \text{sect}\text{`}P . R = P \upharpoonright \beta . \supset . \text{sect}\text{`}R = \text{sect}\text{`}P \cap \text{Cl}\text{`}C\text{`}R$

Dem.

$\vdash . *36\cdot29 . \supset \vdash :. \text{Hp} . \supset : R \mathrel{G} P :$ (1)

[*211·1] $\supset : \alpha \in \text{sect}\text{`}P \cap \text{Cl}\text{`}C\text{`}R . \supset . \alpha \subset C\text{`}R . R\text{``}\alpha \subset \alpha .$
$\supset . \alpha \in \text{sect}\text{`}R$ (2)

$\vdash . (1) . *211\cdot1 . \supset$

$\vdash :. \text{Hp} . \supset : \alpha \in \text{sect}\text{`}R . \supset . \alpha \subset C\text{`}R . \alpha \subset C\text{`}P . (P \upharpoonright \beta)\text{``}\alpha \subset \alpha$ (3)

$\vdash . (3) . *37\cdot41\cdot413 . \supset$

$\vdash : \text{Hp} . \alpha \in \text{sect}\text{`}R . \supset . \alpha \subset \beta . \beta \cap P\text{``}(\alpha \cap \beta) \subset \alpha .$

[*22·621.*37·2] $\supset . \beta \cap P\text{``}\alpha \subset \alpha . P\text{``}\alpha \subset P\text{``}\beta .$

[*211·1] $\supset . \beta \cap P\text{``}\alpha \subset \alpha . P\text{``}\alpha \subset \beta .$

[*22·621] $\supset . P\text{``}\alpha \subset \alpha$ (4)

$\vdash . (3) . (4) . \supset \vdash :. \text{Hp} . \supset : \alpha \in \text{sect}\text{`}R . \supset . \alpha \subset C\text{`}R . \alpha \in \text{sect}\text{`}P$ (5)

$\vdash . (2) . (5) . \supset \vdash . \text{Prop}$

***213·241.** $\vdash : R \epsilon P \upharpoonright$"$\text{sect}$'$P . \supset . R_s \subseteq P_s \upharpoonright C$'$R_s$

Dem.

$\vdash . *213·1 . \supset$

$\vdash :. \text{Hp} . \supset : QR_sQ' . \equiv . (\exists \alpha, \alpha') . \alpha, \alpha' \epsilon \text{sect}$'$R - \iota$'$\Lambda .$

$Q = R \upharpoonright \alpha . Q' = R \upharpoonright \alpha' . \exists ! \alpha' - \alpha .$

[*213·24] $\equiv . (\exists \alpha, \alpha') . \alpha, \alpha' \epsilon \text{sect}$'$P \cap \text{Cl}$'$C$'$R - \iota$'$\Lambda .$

$Q = R \upharpoonright \alpha . Q' = R \upharpoonright \alpha' . \exists ! \alpha' - \alpha .$

[*213·1] $\supset . QP_sQ'$ (1)

$\vdash . (1) . *33·17 . \supset \vdash : \text{Hp} . \supset . R_s \subseteq P_s \upharpoonright C$'$R_s : \supset \vdash . \text{Prop}$

***213·242.** $\vdash : P_{po} \epsilon \text{Ser} . R \epsilon P \upharpoonright$"$\text{sect}$'$P . \supset . R_s = P_s \upharpoonright C$'$R_s$

Dem.

$\vdash . *213·1 . *211·1 . \supset \vdash :. Q (P_s \upharpoonright C$'$R_s) Q' . \supset :$

$(\exists \alpha, \alpha') . \alpha, \alpha' \epsilon \text{sect}$'$P - \iota$'$\Lambda . Q = P \upharpoonright \alpha . Q' = P \upharpoonright \alpha' . \exists ! \alpha' - \alpha :$

$(\exists \gamma, \gamma') . \gamma, \gamma' \epsilon \text{sect}$'$R - \iota$'$\Lambda . Q = R \upharpoonright \gamma . Q' = R \upharpoonright \gamma'$ (1)

$\vdash . *213·24·151 . \supset$

$\vdash :. \text{Hp} . \supset : \alpha \epsilon \text{sect}$'$P . \gamma \epsilon \text{sect}$'$R . Q = P \upharpoonright \alpha = R \upharpoonright \gamma . \supset . \alpha = \gamma$ (2)

$\vdash . (1) . (2) . \supset \vdash :. \text{Hp} . \supset : Q (P_s \upharpoonright C$'$R_s) Q' . \supset .$

$(\exists \gamma, \gamma') . \gamma, \gamma' \epsilon \text{sect}$'$R - \iota$'$\Lambda . Q = R \upharpoonright \gamma . Q' = R \upharpoonright \gamma' . \exists ! \gamma' - \gamma .$

[*213·1] $\supset . QR_sQ'$ (3)

$\vdash . (3) . *213·241 . \supset \vdash . \text{Prop}$

***213·243.** $\vdash . \overrightarrow{P_s}$'$P = \text{D}$'$P_s$

Dem.

$\vdash . *213·1 . \supset \vdash : R \epsilon \overrightarrow{P_s}$'$P . \equiv . (\exists \alpha, \beta) . \alpha, \beta \epsilon \text{sect}$'$P - \iota$'$\Lambda .$

$R = P \upharpoonright \alpha . P = P \upharpoonright \beta . \exists ! \beta - \alpha$ (1)

$\vdash . *37·41 . \supset \vdash . C$'$(P \upharpoonright \beta) \subset \beta$ (2)

$\vdash . (2) . \supset \vdash : \exists ! C$'$P - \beta . \supset . \exists ! C$'$P - C$'$(P \upharpoonright \beta) .$

[*13·14] $\supset . P \neq P \upharpoonright \beta$ (3)

$\vdash . (3) . \text{Transp} . \supset \vdash : \beta \epsilon \text{sect}$'$P . P = P \upharpoonright \beta . \supset . C$'$P = \beta$ (4)

$\vdash . (1) . (4) . \supset \vdash : R \epsilon \overrightarrow{P_s}$'$P . \equiv . (\exists \alpha) . \alpha \epsilon \text{sect}$'$P - \iota$'$\Lambda . R = P \upharpoonright \alpha . \exists ! C$'$P - \alpha .$

[*211·1] $\equiv . (\exists \alpha) . \alpha \epsilon \text{sect}$'$P - \iota$'$\Lambda - \iota$'$C$'$P . R = P \upharpoonright \alpha .$

[*213·141] $\equiv . R \epsilon \text{D}$'$P_s : \supset \vdash . \text{Prop}$

***213·244.** $\vdash : R \epsilon C$'$P_s . Q \epsilon \text{D}$'$R_s . \supset . QP_sR$

Dem.

$\vdash . *213·243 . \supset \vdash :. R \epsilon C$'$P_s . \supset : Q \epsilon \text{D}$'$R_s . \supset . QR_sR .$

[*213·241] $\supset . QP_sR : \supset \vdash . \text{Prop}$

***213·245.** $\vdash :. P_{po} \epsilon \text{Ser} . \supset : QP_sR . \equiv . R \epsilon C\text{‘}P_s . Q \epsilon \text{D}\text{‘}R_s$

Dem.

$\vdash . *213{\cdot}11 . \supset \vdash :. \text{Hp} . \supset :$

$QP_sR . \equiv . (\exists \alpha, \beta) . \alpha, \beta \epsilon \text{sect}\text{‘}P - \iota\text{‘}\Lambda . Q = P \upharpoonright \alpha . R = P \upharpoonright \beta . \alpha \subset \beta . \alpha \neq \beta .$

$[*213{\cdot}24] \equiv . (\exists \alpha, \beta) . \beta \epsilon \text{sect}\text{‘}P - \iota\text{‘}\Lambda . R = P \upharpoonright \beta . \alpha \epsilon \text{sect}\text{‘}R - \iota\text{‘}\Lambda .$
$\alpha \subset \beta . \alpha \neq \beta . Q = P \upharpoonright \alpha .$

$[*213{\cdot}142 . *211{\cdot}133 . *36{\cdot}21]$

$\equiv . (\exists \alpha) . R \epsilon C\text{‘}P_s . \alpha \epsilon \text{sect}\text{‘}R - \iota\text{‘}\Lambda . \alpha \subset C\text{‘}R . \alpha \neq C\text{‘}R . Q = R \upharpoonright \alpha .$

$[*213{\cdot}141] \equiv . R \epsilon C\text{‘}P_s . Q \epsilon \text{D}\text{‘}R_s :. \supset \vdash . \text{Prop}$

***213·246.** $\vdash : P_{po} \epsilon \text{Ser} . R \epsilon C\text{‘}P_s . \supset . \overrightarrow{P_s}\text{‘}R = \text{D}\text{‘}R_s \quad [*213{\cdot}245]$

***213·247.** $\vdash :. P_{po} \epsilon \text{Ser} . \supset : Q (P_s \upharpoonright \text{D}\text{‘}P_s) R . \equiv . R \epsilon \text{D}\text{‘}P_s . Q \epsilon \text{D}\text{‘}R_s$
$[*213{\cdot}245]$

***213·25.** $\vdash :. P_{po} \epsilon \text{Ser} . Q, R \epsilon C\text{‘}P_s . \supset : Q \epsilon \text{D}\text{‘}R_s . \vee . R \epsilon \text{D}\text{‘}Q_s . \vee . Q = R$

Dem.

$\vdash . *213{\cdot}153 . \supset \vdash :. \text{Hp} . \supset : QP_sR . \vee . RP_sQ . \vee . Q = R :$

$[*213{\cdot}245] \qquad \supset : Q \epsilon \text{D}\text{‘}R_s . \vee . R \epsilon \text{D}\text{‘}Q_s . \vee . Q = R :. \supset \vdash . \text{Prop}$

***213·251.** $\vdash :. P_{po} \epsilon \text{Ser} . Q, R \epsilon C\text{‘}P_s . \sim (Q = \dot{\Lambda} . R = \dot{\Lambda}) . \supset :$
$Q \epsilon C\text{‘}R_s . \vee . R \epsilon \text{D}\text{‘}Q_s$

Dem.

$\vdash . *213{\cdot}158 . \quad \supset \vdash : \text{Hp} . \dot{\exists} ! R . Q = R . \supset . Q \epsilon C\text{‘}R_s \quad (1)$

$\vdash . (1) . *13{\cdot}12 . \supset \vdash : \text{Hp} . \dot{\exists} ! Q . Q = R . \supset . Q \epsilon C\text{‘}R_s \quad (2)$

$\vdash . (1) . (2) . \quad \supset \vdash : \text{Hp} . Q = R . \supset . Q \epsilon C\text{‘}R_s \quad (3)$

$\vdash . (3) . *213{\cdot}25 . \supset \vdash . \text{Prop}$

***213·3.** $\vdash : P = \dot{\Lambda} . \supset . P_s = \dot{\Lambda}$

Dem.

$\vdash . *212{\cdot}17 . \supset \vdash : \text{Hp} . \supset . \varsigma\text{‘}P_* = \dot{\Lambda} .$

$[*150{\cdot}42] \qquad \supset . P_s = \dot{\Lambda} : \supset \vdash . \text{Prop}$

***213·301.** $\vdash : \exists ! \text{sect}\text{‘}P - \iota\text{‘}\Lambda - \iota\text{‘}C\text{‘}P . \supset . \dot{\exists} ! P_s \quad [*213{\cdot}141]$

***213·302.** $\vdash :. P_{po} \dot{\subset} J . \supset : \dot{\exists} ! P . \equiv . \dot{\exists} ! P_s$

Dem.

$\vdash . *213{\cdot}126{\cdot}301 . \supset \vdash : \text{Hp} . \dot{\exists} ! P . \supset . \dot{\exists} ! P_s \quad (1)$

$\vdash . (1) . *213{\cdot}3 . \supset \vdash . \text{Prop}$

***213·31.** $\vdash : x \neq y . \supset . (x \downarrow y)_s = \dot{\Lambda} \downarrow (x \downarrow y)$

Dem.

$\vdash . *211{\cdot}9 . \supset$

$\vdash : \text{Hp} . \supset . \text{sect}\text{‘}(x \downarrow y) - \iota\text{‘}\Lambda = \iota\text{‘}\iota\text{‘}x \cup \iota\text{‘}(\iota\text{‘}x \cup \iota\text{‘}y) . \exists ! (\iota\text{‘}x \cup \iota\text{‘}y) - \iota\text{‘}x .$

[$*213 \cdot 1 \cdot 141$] $\supset . \{(x \downarrow y) \upharpoonright \iota`x\} (x \downarrow y)_s \{(x \downarrow y) \upharpoonright (\iota`x \cup \iota`y)\} .$
$D`(x \downarrow y)_s = \iota`(x \downarrow y) \upharpoonright \iota`x .$

[$*200 \cdot 35 . *55 \cdot 15$] $\supset . \dot{\Lambda} (x \downarrow y)_s (x \downarrow y) . D`(x \downarrow y)_s = \iota`\dot{\Lambda}$ (1)

$\vdash . *213 \cdot 153 . *204 \cdot 25 . \supset \vdash : Hp . \supset . (x \downarrow y)_s \in Ser$ (2)

$\vdash . (1) . (2) . *204 \cdot 27 . \supset \vdash .$ Prop

$*213 \cdot 32$. $\vdash : P \in 2_r . \supset . P_s = \dot{\Lambda} \downarrow P . P_s \in 2_r$ [$*213 \cdot 31$]

$*213 \cdot 4$. $\vdash : P \in Ser . \supset .$

$$P \upharpoonright``sect`P = \hat{Q} \{(\exists R) . P = Q \ddagger R . v . (\exists x) . P = Q \nrightarrow x\}$$

Dem.

$\vdash . *211 \cdot 82 . *5 \cdot 32 . \supset$

$\vdash :: Hp . \supset :. Q \in P \upharpoonright``sect`P . \equiv :$

$Q \in D`P \upharpoonright : (\exists R) . P = Q \ddagger R . v . (\exists x) . P = Q \nrightarrow x$ (1)

$\vdash . *211 \cdot 283 . *160 \cdot 5 . \supset \vdash : Hp . P = Q \ddagger R . \supset . Q \in D`P \upharpoonright$ (2)

$\vdash . *161 \cdot 11 . \supset \vdash : Hp . P = Q \nrightarrow x . \supset . Q = P \upharpoonright C`P$ (3)

$\vdash . (2) . (3) . \supset \vdash :. Hp : (\exists R) . P = Q \ddagger R . v . (\exists x) . P = Q \nrightarrow x : \supset .$
$Q \in D`P \upharpoonright$ (4)

$\vdash . (1) . (4) . \supset \vdash .$ Prop

$*213 \cdot 41$. $\vdash : P \in Ser . \exists ! \overrightarrow{B}`P . \supset .$

$C`P_s = \hat{Q} \{(\exists R) . P = Q \ddagger R . v . (\exists x) . P = Q \nrightarrow x\}$ [$*213 \cdot 4 \cdot 161$]

$*213 \cdot 42$. $\vdash : P \in Ser . \overrightarrow{B}`P = \Lambda . \supset .$

$C`P_s = \hat{Q} \{(\exists R) . P = Q \ddagger R . v . (\exists x) . P = Q \nrightarrow x\} - \iota`\dot{\Lambda}$ [$*213 \cdot 4 \cdot 163$]

$*213 \cdot 5$. $\vdash : P_{po} \in J . x \sim \in C`P . \supset . D`(P \nrightarrow x)_s = C`P_s$

Dem.

$\vdash . *213 \cdot 141 . *211 \cdot 83 . \supset$

$\vdash : Hp . \dot{\exists} ! P . \supset . D`(P \nrightarrow x)_s = (P \nrightarrow x) \upharpoonright``(sect`P - \iota`\Lambda)$

[$*36 \cdot 4 . *161 \cdot 1$] $= P \upharpoonright``(sect`P - \iota`\Lambda)$

[$*213 \cdot 142$] $= C`P_s$ (1)

$\vdash . *213 \cdot 3 . *161 \cdot 2 . \supset \vdash : P = \dot{\Lambda} . \supset . D`(P \nrightarrow x)_s = \Lambda . C`P_s = \Lambda$ (2)

$\vdash . (1) . (2) . \supset \vdash .$ Prop

$*213 \cdot 51$. $\vdash : P_{po} \in J . x \sim \in C`P . \supset . (P \nrightarrow x)_s = P_s \nrightarrow (P \nrightarrow x)$

Dem.

$\vdash . *213 \cdot 1 . *211 \cdot 83 . \supset \vdash :: Hp . \dot{\exists} ! P . \supset :. Q (P \nrightarrow x)_s R . \equiv :$

$(\exists \alpha, \beta) . \alpha, \beta \in sect`P - \iota`\Lambda \cup \iota`(C`P \cup \iota`x) .$

$\exists ! \beta - \alpha . Q = (P \nrightarrow x) \upharpoonright \alpha . R = (P \nrightarrow x) \upharpoonright \beta .$

$[*211\cdot1.*36\cdot4] \equiv : (\exists\alpha,\beta) . \alpha,\beta \in \text{sect}`P - \iota`\Lambda . \exists ! \beta - \alpha . Q = P \upharpoonright \alpha . R = P \upharpoonright \beta . \vee .$

$(\exists\alpha) . \alpha \in \text{sect}`P - \iota`\Lambda . Q = P \upharpoonright \alpha . R = P \nrightarrow x :$

$[*213\cdot1\cdot142] \quad \equiv : QP_sR . \vee . Q \in C`P_s . R = P \nrightarrow x :$

$[*161\cdot11] \quad \equiv : Q \{P_s \nrightarrow (P \nrightarrow x)\} R \quad (1)$

$\vdash . *213\cdot3 . *161\cdot2 . \supset \vdash : P = \dot\Lambda . \supset . (P \nrightarrow x)_s = \dot\Lambda . P_s \nrightarrow (P \nrightarrow x) = \dot\Lambda \quad (2)$

$\vdash . (1) . (2) . \supset \vdash . \text{Prop}$

$*213\cdot52$. $\vdash :. Q_{po} \in \text{connex} . C`P \cap C`Q = \Lambda . \supset :$

$(\exists\beta) . \beta \cap C`Q \sim\in 1 . \beta \in (C`P \cup)``(\text{sect}`Q - \iota`\Lambda) . S = (P \ddagger Q) \upharpoonright \beta . \equiv .$

$(\exists\gamma) . \gamma \in \text{sect}`Q - \iota`\Lambda - 1 . S = P \ddagger Q \upharpoonright \gamma$

Dem.

$\vdash . *37\cdot6 . \supset \vdash : \beta \in (C`P \cup)``(\text{sect}`Q - \iota`\Lambda) . S = (P \ddagger Q) \upharpoonright \beta . \equiv .$

$(\exists\gamma) . \gamma \in \text{sect}`Q - \iota`\Lambda . \beta = C`P \cup \gamma . S = (P \ddagger Q) \upharpoonright (C`P \cup \gamma) \quad (1)$

$\vdash . *160\cdot11 . \supset \vdash :: \text{Hp} . \gamma \in \text{sect}`Q . \supset :. x \{(P \ddagger Q) \upharpoonright (C`P \cup \gamma)\} y . \equiv :$

$xPy . \vee . x \in C`P . y \in \gamma . \vee . x (Q \upharpoonright \gamma) y :.$

$[*211\cdot133.*160\cdot11] \supset :. \gamma \sim\in 1 . \supset . (P \ddagger Q) \upharpoonright (C`P \cup \gamma) = P \ddagger Q \upharpoonright \gamma \quad (2)$

$\vdash . *24\cdot24 . \supset \vdash : \text{Hp} . \beta = C`P \cup \gamma . \supset . \beta \cap C`Q = \gamma \cap C`Q \quad (3)$

$\vdash . (1) . (2) . (3) . \supset$

$\vdash :. \text{Hp} . \supset : \beta \cap C`Q \sim\in 1 . \beta \in (C`P \cup)``(\text{sect}`Q - \iota`\Lambda) . S = (P \ddagger Q) \upharpoonright \beta . \equiv .$

$(\exists\gamma) . \gamma \in \text{sect}`Q - \iota`\Lambda - 1 . S = P \ddagger Q \upharpoonright \gamma . \beta = C`P \cup \gamma \quad (4)$

$\vdash . (4) . *10\cdot281 . *13\cdot19 . \supset \vdash . \text{Prop}$

$*213\cdot53$. $\vdash : P_{po} \in J . Q_{po} \in \text{Ser} . \overrightarrow{B}`Q = \Lambda . C`P \cap C`Q = \Lambda . \supset .$

$(P \ddagger Q)_s = P_s \ddagger (P \ddagger; Q_s)$

Dem.

$\vdash . *213\cdot1 . *211\cdot841 . \supset \vdash :: \text{Hp} . \supset :. R (P \ddagger Q)_s S . \equiv :$

$(\exists\alpha,\beta) . \alpha, \beta \in \text{sect}`P - \iota`\Lambda \cup (C`P \cup)``(\text{sect}`Q - \iota`\Lambda) .$

$\exists ! \beta - \alpha . R = (P \ddagger Q) \upharpoonright \alpha . S = (P \ddagger Q) \upharpoonright \beta :$

$[*211\cdot182] \equiv : (\exists\alpha,\beta) . \alpha, \beta \in \text{sect}`P - \iota`\Lambda \cup (C`P \cup)``(\text{sect}`Q - 1 - \iota`\Lambda) .$

$\exists ! \beta - \alpha . R = (P \ddagger Q) \upharpoonright \alpha . S = (P \ddagger Q) \upharpoonright \beta :$

$[*160\cdot1.*213\cdot52]$

$\equiv : (\exists\alpha,\beta) . \alpha, \beta \in \text{sect}`P - \iota`\Lambda . \exists ! \beta - \alpha . R = P \upharpoonright \alpha . S = P \upharpoonright \beta . \vee .$

$(\exists\alpha,\gamma) . \alpha \in \text{sect}`P - \iota`\Lambda . \gamma \in \text{sect}`Q - \iota`\Lambda . R = P \upharpoonright \alpha . S = P \ddagger Q \upharpoonright \gamma . \vee .$

$(\exists\gamma,\delta) . \gamma, \delta \in \text{sect}`Q - \iota`\Lambda . \exists ! \delta - \gamma . R = P \ddagger Q \upharpoonright \gamma . S = P \ddagger Q \upharpoonright \delta :$

$[*213\cdot1\cdot142] \equiv : RP_sS . \vee . R \in C`P_s . S \in C`P \ddagger; Q_s . \vee . R (P \ddagger; Q_s) S :$

$[*160\cdot11] . \quad \equiv : R \{P_s \ddagger (P \ddagger; Q_s)\} S :: \supset \vdash . \text{Prop}$

✳213·531. $\vdash :: Q_{po} \in \mathrm{Ser} \,.\, \exists \,!\, \overrightarrow{B}\text{‘}Q \,.\, C\text{‘}P \cap C\text{‘}Q = \Lambda \,.\, \supset :.$

$(\exists\beta) \,.\, \beta \in (C\text{‘}P \cup)\text{“}(\mathrm{sect}\text{‘}Q - \iota\text{‘}\Lambda) \,.\, S = (P \ddagger Q) \upharpoonright \beta \,.\, \equiv :$

$S = P \nrightarrow B\text{‘}Q \,.\, \vee \,.\, (\exists\gamma) \,.\, \gamma \in \mathrm{sect}\text{‘}Q - \iota\text{‘}\Lambda - \iota\text{‘}\iota\text{‘}B\text{‘}Q \,.\, S = P \ddagger Q \upharpoonright \gamma$

Dem.

$\vdash \,.\, ✳213\cdot52 \,.\, \supset$

$\vdash :: \mathrm{Hp} \,.\, \supset :. (\exists\beta) \,.\, \beta \in (C\text{‘}P \cup)\text{“}(\mathrm{sect}\text{‘}Q - \iota\text{‘}\Lambda) \,.\, S = (P \ddagger Q) \upharpoonright \beta \,.\, \equiv :$

$(\exists\beta) \,.\, \beta \in (C\text{‘}P \cup)\text{“}(\mathrm{sect}\text{‘}Q \cap 1) \,.\, S = (P \ddagger Q) \upharpoonright \beta \,.\, \vee \,.$

$(\exists\gamma) \,.\, \gamma \in \mathrm{sect}\text{‘}Q - \iota\text{‘}\Lambda - 1 \,.\, S = P \ddagger Q \upharpoonright \gamma :$

$[✳211\cdot181] \equiv : (\exists\beta) \,.\, \beta = C\text{‘}P \cup \iota\text{‘}B\text{‘}Q \,.\, S = (P \ddagger Q) \upharpoonright \beta \,.\, \vee \,.$

$(\exists\gamma) \,.\, \gamma \in \mathrm{sect}\text{‘}Q - \iota\text{‘}\Lambda - \iota\text{‘}\iota\text{‘}B\text{‘}Q \,.\, S = P \ddagger Q \upharpoonright \gamma \qquad (1)$

$\vdash \,.\, ✳160\cdot11 \,.\, \supset \vdash :: \mathrm{Hp} \,.\, \supset :. x \{(P \ddagger Q) \upharpoonright (C\text{‘}P \cup \iota\text{‘}B\text{‘}Q)\} y \,.\, \equiv :$

$xPy \,.\, \vee \,.\, x \in C\text{‘}P \,.\, y = B\text{‘}Q :$

$[✳161\cdot11] \qquad \equiv : x (P \nrightarrow B\text{‘}Q) y \qquad (2)$

$\vdash \,.\, (1) \,.\, (2) \,.\, \supset \vdash \,.\, \mathrm{Prop}$

✳213·54. $\vdash : \dot{\exists} \,!\, P \,.\, P_{po} \subset J \,.\, Q_{po} \in \mathrm{Ser} \,.\, \exists \,!\, \overrightarrow{B}\text{‘}Q \,.\, C\text{‘}P \cap C\text{‘}Q = \Lambda \,.\, \text{Ɑ}\text{‘}Q_s \sim\in 1 \,.\, \supset .$

$(P \ddagger Q)_s = P_s \nrightarrow (P \nrightarrow B\text{‘}Q) \ddagger \{P \ddagger ; (Q_s \upharpoonright \text{Ɑ}\text{‘}Q_s)\}$

Dem.

$\vdash \,.\, ✳213\cdot1 \,.\, ✳211\cdot841 \,.\, \supset \vdash :: \mathrm{Hp} \,.\, \supset :. R (P \ddagger Q)_s S \,.\, \equiv :$

$(\exists\alpha, \beta) \,.\, \alpha, \beta \in \mathrm{sect}\text{‘}P - \iota\text{‘}\Lambda \cup (C\text{‘}P \cup)\text{“}(\mathrm{sect}\text{‘}Q - \iota\text{‘}\Lambda) \,.\, \exists \,!\, \beta - \alpha \,.$

$R = (P \ddagger Q) \upharpoonright \alpha \,.\, S = (P \ddagger Q) \upharpoonright \beta :$

$[✳213\cdot531]$

$\equiv : (\exists\alpha, \beta) \,.\, \alpha, \beta \in \mathrm{sect}\text{‘}P - \iota\text{‘}\Lambda \,.\, \exists \,!\, \beta - \alpha \,.\, R = P \upharpoonright \alpha \,.\, Q = P \upharpoonright \beta \,.\, \vee \,.$

$(\exists\alpha) \,.\, \alpha \in \mathrm{sect}\text{‘}P - \iota\text{‘}\Lambda \,.\, R = P \upharpoonright \alpha \,.\, S = P \nrightarrow B\text{‘}Q \,.\, \vee \,.$

$(\exists\alpha, \gamma) \,.\, \alpha \in \mathrm{sect}\text{‘}P - \iota\text{‘}\Lambda \,.\, R = P \upharpoonright \alpha \,.\, \beta \in \mathrm{sect}\text{‘}Q - \iota\text{‘}\Lambda - \iota\text{‘}\iota\text{‘}B\text{‘}Q \,.$

$S = P \ddagger Q \upharpoonright \gamma \,.\, \vee \,.$

$(\exists\gamma) \,.\, R = P \nrightarrow B\text{‘}Q \,.\, \beta \in \mathrm{sect}\text{‘}Q - \iota\text{‘}\Lambda - \iota\text{‘}\iota\text{‘}B\text{‘}Q \,.\, S = P \ddagger Q \upharpoonright \gamma \,.\, \vee \,.$

$(\exists\gamma, \delta) \,.\, \gamma, \delta \in \mathrm{sect}\text{‘}Q - \iota\text{‘}\Lambda - \iota\text{‘}\iota\text{‘}B\text{‘}Q \,.\, \exists \,!\, \delta - \gamma \,.\, R = P \ddagger Q \upharpoonright \gamma \,.$

$S = P \ddagger Q \upharpoonright \delta :$

$[✳213\cdot1\cdot142\cdot132]$

$\equiv : RP_sS \,.\, \vee \,.\, R \in C\text{‘}P_s \,.\, S = P \nrightarrow B\text{‘}Q \,.\, \vee \,.\, R = P \nrightarrow B\text{‘}Q \,.\, S \in P \ddagger\text{“}\text{Ɑ}\text{‘}Q_s \,.$

$\vee \,.\, R, S \in (P \ddagger\text{“}) \text{Ɑ}\text{‘}Q_s \,.\, R (P \ddagger ; Q) S \,.\, \vee \,.\, R \in C\text{‘}P_s \,.\, S \in P \ddagger\text{“}\text{Ɑ}\text{‘}Q_s :$

$[✳161\cdot11 \,.\, ✳211\cdot133 \,.\, ✳160\cdot11]$

$\equiv : R \{P_s \nrightarrow (P \nrightarrow B\text{‘}Q) \ddagger (P \ddagger ; Q_s \upharpoonright \text{Ɑ}\text{‘}Q_s)\} S :: \supset \vdash \,.\, \mathrm{Prop}$

***213·541.** $\vdash : P_{po} \in \text{Ser} \,.\, \exists ! \overrightarrow{B}‘P \,.\, \text{Ɑ}‘P_s \in 1 \,.\, \supset \,.\, P \in 2_r$

Dem.

$\vdash \,.\, *213·144 \,.\, *211·26 \,.\, \supset \vdash :. \text{Hp} \,.\, \supset : P \upharpoonright “(\text{sect}‘P - \iota‘\Lambda - \iota‘\iota‘B‘P) = \iota‘P :$

[*211·3·13] $\supset : x \in \text{Ɑ}‘P \,.\, \supset \,.\, P \upharpoonright \overrightarrow{P_*}‘x = P \,.$

[*202·55] $\supset \,.\, \overrightarrow{P_*}‘x = C‘P \,.$

[*200·39] $\supset \,.\, \overleftarrow{P_{po}}‘x = \Lambda \,.$

[*202·522·523] $\supset \,.\, x = B‘\breve{P} :$

[*204·271] $\supset : P_{po} \in 2_r :$

[*56·111.*91·504] $\supset : P \in 2_r :. \supset \vdash \,.\, \text{Prop}$

***213·55.** $\vdash : \dot{\exists} ! P \,.\, P_{po} \in J \,.\, Q \in 2_r \,.\, C‘P \cap C‘Q = \Lambda \,.\, \supset \,.$
$$(P \ddagger Q)_s = P_s \dotplus\!\!\rightarrow (P \dotplus\!\!\rightarrow B‘Q) \dotplus\!\!\rightarrow (P \ddagger Q)$$

Dem.

As in *213·54,

$\vdash :: \text{Hp} \,.\, \supset :. R\,(P \ddagger Q)_s S \,.$

$\equiv : R P_s S \,.\, \vee \,.\, R \in C‘P_s \,.\, S = P \dotplus\!\!\rightarrow B‘Q \,.\, \vee \,.\, R \in C‘P_s \,.\, S \in P \ddagger “\text{Ɑ}‘Q_s \,.$
$\vee \,.\, R = P \dotplus\!\!\rightarrow B‘Q \,.\, S \in P \ddagger “\text{Ɑ}‘Q_s \,.\, \vee \,.\, R, S \in P \ddagger “\text{Ɑ}‘Q_s \,.$
$R\,(P \ddagger ; Q_s)\,S :$

[*213·32] $\equiv : R P_s S \,.\, \vee \,.\, R \in C‘P_s \,.\, S = P \dotplus\!\!\rightarrow B‘Q \,.\, \vee \,.\, R \in C‘P_s \,.\, S = P \ddagger Q \,.$
$\vee \,.\, R = P \dotplus\!\!\rightarrow B‘Q \,.\, S = P \ddagger Q \,.\, \vee \,.\, R = P \ddagger Q \,.\, S = P \ddagger Q \,.$
$R\,(P \ddagger ; Q_s)\,S :$

[*213·32] $\equiv : R P_s S \,.\, \vee \,.\, R \in C‘P_s \,.\, S = P \dotplus\!\!\rightarrow B‘Q \,.\, \vee \,.\, R \in C‘P_s \,.\, S = P \ddagger Q \,.$
$\vee \,.\, R = P \dotplus\!\!\rightarrow B‘Q \,.\, S = P \ddagger Q :$

[*161·11] $\equiv : R\,\{P_s \dotplus\!\!\rightarrow (P \dotplus\!\!\rightarrow B‘Q) \dotplus\!\!\rightarrow (P \ddagger Q)\}\,S :: \supset \vdash \,.\, \text{Prop}$

***213·56.** $\vdash :. P_{po} \in J \,.\, Q_{po} \in \text{Ser} \,.\, C‘P \cap C‘Q = \Lambda \,.\, \supset :$
$\overrightarrow{B}‘Q = \Lambda \,.\, \supset \,.\, (P \ddagger Q)_s = P_s \ddagger (P \ddagger ; Q_s) :$
$\dot{\exists} ! P \,.\, \exists ! \overrightarrow{B}‘Q \,.\, Q \sim\in 2_r \,.\, \supset \,.$
$$(P \ddagger Q)_s = P_s \dotplus\!\!\rightarrow (P \dotplus\!\!\rightarrow B‘Q) \ddagger \{P \ddagger ; (Q_s \upharpoonright \text{Ɑ}‘Q_s)\} :$$
$\dot{\exists} ! P \,.\, Q \in 2_r \,.\, \supset \,.\, (P \ddagger Q)_s = P_s \dotplus\!\!\rightarrow (P \dotplus\!\!\rightarrow B‘Q) \dotplus\!\!\rightarrow (P \ddagger Q) :$
$P = \dot{\Lambda} \,.\, \supset \,.\, (P \ddagger Q)_s = Q_s$ [*213·53·54·541·55 . *160·22]

***213·561.** $\vdash : C‘P \cap C‘Q = \Lambda \,.\, \supset \,.\, (P \ddagger) \upharpoonright C‘Q_s \in 1 \to 1$

Dem.

$\vdash \,.\, *213·1 \,.\, \supset \vdash : R \in C‘Q_s \,.\, \supset \,.\, C‘R \subset C‘Q$ (1)

$\vdash \,.\, (1) \,.\, \supset \vdash :. \text{Hp} \,.\, R, S \in C‘Q_s \,.\, \supset : C‘P \cap C‘R = \Lambda \,.\, C‘P \cap C‘S = \Lambda :$

[*160·52] $\supset : P \ddagger R = P \ddagger S \,.\, \supset \,.\, R = S :. \supset \vdash \,.\, \text{Prop}$

***213·57.** $\vdash : P_{po} \subset J . \mathrm{Nr}‘Q = \mathrm{Nr}‘P \dotplus \dot{1} . \supset . \mathrm{Nr}‘Q_s = \mathrm{Nr}‘P_s \dotplus \dot{1}$

Dem.

$\vdash . *181·2·12 . (*181·01) . \supset$

$\vdash : \mathrm{Hp} . \supset . (\exists R, x) . R \,\mathrm{smor}\, P . x \sim \epsilon C‘R . Q = R \leftrightarrow x .$

[*213·51] $\supset . (\exists R, x) . R \,\mathrm{smor}\, P . x \sim \epsilon C‘R . Q_s = R_s \leftrightarrow (R \leftrightarrow x) .$

[*181·32] $\supset . (\exists R) . R \,\mathrm{smor}\, P . \mathrm{Nr}‘Q_s = \mathrm{Nr}‘R_s \dotplus \dot{1} .$

[*213·172] $\supset . \mathrm{Nr}‘Q_s = \mathrm{Nr}‘P_s \dotplus \dot{1} : \supset \vdash . \mathrm{Prop}$

***213·58.** $\vdash : P_{po} \subset J . Q_{po} \epsilon \mathrm{Ser} . C‘P \cap C‘Q = \Lambda . \supset . \mathrm{Nr}‘(P \ddagger Q)_s = \mathrm{Nr}‘P_s \dotplus \mathrm{Nr}‘Q_s$

Dem.

$\vdash . *213·53·561 . \supset$

$\vdash : \mathrm{Hp} . \overrightarrow{B}‘Q = \Lambda . \supset . (P \ddagger Q)_s = P_s \ddagger (P \ddagger ; Q_s) . \mathrm{Nr}‘P \ddagger ; Q_s = \mathrm{Nr}‘Q_s .$

[*180·32] $\supset . \mathrm{Nr}‘(P \ddagger Q)_s = \mathrm{Nr}‘P_s \dotplus \mathrm{Nr}‘Q_s$ (1)

$\vdash . *213·54·561 . *181·32 . \supset$

$\vdash : \mathrm{Hp} . \dot{\exists} ! P . \exists ! \overrightarrow{B}‘Q . \mathrm{C}‘Q_s \sim \epsilon 1 . \supset . \mathrm{Nr}‘(P \ddagger Q)_s = \mathrm{Nr}‘P_s \dotplus \dot{1} \dotplus \mathrm{Nr}‘Q_s \upharpoonright \mathrm{C}‘Q_s$

[*204·46.*213·157] $= \mathrm{Nr}‘P_s \dotplus \mathrm{Nr}‘Q_s$ (2)

$\vdash . *213·541·55 . *181·32 . \supset$

$\vdash : \mathrm{Hp} . \dot{\exists} ! P . \mathrm{C}‘Q_s \epsilon 1 . \supset . Q \epsilon 2_r . \mathrm{Nr}‘(P \ddagger Q)_s = \mathrm{Nr}‘P_s \dotplus \dot{1} \dotplus \dot{1} .$

[*181·56] $\supset . Q \epsilon 2_r . \mathrm{Nr}‘(P \ddagger Q)_s = \mathrm{Nr}‘P_s \dotplus 2_r .$

[*213·32] $\supset . \mathrm{Nr}‘(P \ddagger Q)_s = \mathrm{Nr}‘P_s \dotplus \mathrm{Nr}‘Q_s$ (3)

$\vdash . *160·22 . *213·3 . \supset \vdash : P = \dot{\Lambda} . \supset . \mathrm{Nr}‘(P \ddagger Q)_s = \mathrm{Nr}‘P_s \dotplus \mathrm{Nr}‘Q_s$ (4)

$\vdash . (1) . (2) . (3) . (4) . \supset \vdash . \mathrm{Prop}$

$*214$. DEDEKINDIAN RELATIONS.

Summary of $*214$.

We call a relation "Dedekindian" when it is such that every class has either a maximum or a sequent with respect to it. As a rule, the hypothesis that a relation is Dedekindian is only important in the case of serial relations. Dedekindian series have considerable importance, especially in connection with limits.

When P is transitive, the hypothesis that P is Dedekindian is equivalent to the hypothesis that every section of P has a maximum or a sequent ($*214 \cdot 13$); it is also equivalent to the assumption that every segment of P has a maximum or a sequent ($*214 \cdot 131$), *i.e.* to the assumption that every segment of P which has no maximum has a limit, *i.e.* to

$$\mathrm{D}‘(P_\epsilon \dot{\cap} I) \subset \text{Ɑ}‘\mathrm{lt}_P.$$

When P is a series, the hypothesis that it is Dedekindian is equivalent to the hypothesis that every segment has a sequent ($*214 \cdot 15$), *i.e.* to the hypothesis that the class of segments is the class $\overrightarrow{P}“C‘P$ ($*214 \cdot 151$). If P is a Dedekindian series, so is $\breve{P}$, and vice versa ($*214 \cdot 14$). Whenever P is connected and not null, $\mathfrak{s}‘P_*$ is a Dedekindian series ($*214 \cdot 32$), and so is $\mathrm{sgm}‘P$ if it exists ($*214 \cdot 34$); whenever P is transitive and connected and not null, $\mathfrak{s}‘P$ is a Dedekindian series ($*214 \cdot 33$). All these propositions have been virtually proved already: almost the only thing new in the present number is the definition, which is

$$\mathrm{Ded} = \hat{P}\{(\alpha) \,.\, \alpha \in \text{Ɑ}‘\mathrm{max}_P \cup \text{Ɑ}‘\mathrm{seq}_P\} \quad \mathrm{Df.}$$

$*214 \cdot 4$—$\cdot 43$ give properties of series which have Dedekindian continuity. We have

$*214 \cdot 4$. $\vdash :. P^2 = P \,.\, P \in \mathrm{connex} \,.\, \supset : P \in \mathrm{Ded} \,.\, \equiv \,.\, \text{Ɑ}‘\mathrm{max}_P = - \text{Ɑ}‘\mathrm{seq}_P$

$*214 \cdot 41$. $\vdash :. P \in \mathrm{Ser} \,.\, \supset : P^2 = P \,.\, P \in \mathrm{Ded} \,.\, \equiv \,.\, \text{Ɑ}‘\mathrm{max}_P = - \text{Ɑ}‘\mathrm{seq}_P$

I.e. in a series, Dedekindian continuity is equivalent to the assumption that the classes which have a maximum are the same as the classes which have no sequent.

***214·42.** $\vdash : P \epsilon \text{Ser} \cap \text{Ded} \,.\, P^2 = P \,.\, \alpha \epsilon \text{sect}`P \,.\supset .\, \text{limax}_P`\alpha = \text{limin}_P`(C`P - \alpha)$

This proposition is important in dealing with Dedekind "cuts."

***214·43.** $\vdash :. P \epsilon \text{Ser} \cap \text{Ded} \,.\, \alpha \epsilon \text{sect}`P \,.\supset :$

$$\text{limax}_P`\alpha = \text{limin}_P`(C`P - \alpha) \,.\vee .\, \text{max}_P`\alpha \, P_1 \, \text{min}_P`(C`P - \alpha)$$

*214·5 shows that a Dedekindian relation has a beginning and an end; the following propositions deal with $P \dot{\cap} J$ when P is Dedekindian.

*214·6 shows that a relation which is similar to a Dedekindian relation is Dedekindian.

We call a relation "semi-Dedekindian" if it becomes Dedekindian by the addition of one term at the end; the definition is

***214·02.** $\text{semi Ded} = \hat{P}(\text{sect}`P - \iota`C`P \subset \text{Cl}`\text{max}_P \cup \text{Cl}`\text{seq}_P)$ Df

---

***214·01.** $\text{Ded} = \hat{P}\{(\alpha) \,.\, \alpha \epsilon \text{Cl}`\text{max}_P \cup \text{Cl}`\text{seq}_P\}$ Df

***214·02.** $\text{semi Ded} = \hat{P}(\text{sect}`P - \iota`C`P \subset \text{Cl}`\text{max}_P \cup \text{Cl}`\text{seq}_P)$ Df

***214·1.** $\vdash : P \epsilon \text{Ded} \,.\equiv .\, (\alpha) \,.\, \alpha \epsilon \text{Cl}`\text{max}_P \cup \text{Cl}`\text{seq}_P$ [(*214·01)]

***214·101.** $\vdash : P \epsilon \text{Ded} \,.\equiv .\, - \text{Cl}`\text{max}_P \subset \text{Cl}`\text{seq}_P \,.\equiv .\, - \text{Cl}`\text{max}_P \subset \text{Cl}`\text{lt}_P$
[*214·1 . *24·312 . *207·12]

***214·11.** $\vdash : P \epsilon \text{Ded} \,.\equiv .\, (\alpha) \,.\, \alpha \epsilon \text{Cl}`\text{max}_P \cup \text{Cl}`\text{lt}_P \,.\equiv .\, (\alpha) \,.\, \alpha \epsilon \text{Cl}`\text{limax}_P$
[*214·1 . *207·14·44]

***214·12.** $\vdash :. P \epsilon \text{Ded} \,.\equiv : \alpha \subset C`P \,.\supset_\alpha .\, \alpha \epsilon \text{Cl}`\text{max}_P \cup \text{Cl}`\text{seq}_P$
[*214·1 . *205·151 . *206·131]

***214·13.** $\vdash :. P \epsilon \text{trans} \,.\supset : P \epsilon \text{Ded} \,.\equiv .\, \text{sect}`P \subset \text{Cl}`\text{max}_P \cup \text{Cl}`\text{seq}_P$
[*211·272 . *214·1]

***214·131.** $\vdash :. P \epsilon \text{trans} \,.\supset : P \epsilon \text{Ded} \,.\equiv .\, \text{D}`(P_\epsilon \dot{\cap} I) \subset \text{Cl}`\text{seq}_P$ [*211·47 . *214·1]

***214·132.** $\vdash :. P \epsilon \text{trans} \,.\supset : P \epsilon \text{Ded} \,.\equiv .\, \text{D}`P_\epsilon \subset \text{Cl}`\text{max}_P \cup \text{Cl}`\text{seq}_P$
[*214·131 . *211·42]

***214·14.** $\vdash :. P \epsilon \text{Ser} \,.\supset : P \epsilon \text{Ded} \,.\equiv .\, \breve{P} \epsilon \text{Ded}$ [*206·57 . *214·1]

***214·141.** $\vdash :. P \epsilon \text{Ser} \,.\supset : P \epsilon \text{Ded} \,.\equiv .\, (\alpha) \,.\, p`\overrightarrow{P}``(\alpha \cap C`P) \epsilon \text{Cl}`\text{max}_P \cup \text{Cl}`\text{seq}_P$
[*206·56 . *214·1]

***214·15.** $\vdash :. P \epsilon \text{Ser} \,.\supset : P \epsilon \text{Ded} \,.\equiv .\, \text{D}`P_\epsilon \subset \text{Cl}`\text{seq}_P$
[*206·36 . *214·1 . *211·11]

***214·151.** $\vdash :. P \epsilon \text{Ser} \,.\supset : P \epsilon \text{Ded} \,.\equiv .\, \text{D}`P_\epsilon = \overrightarrow{P}``C`P$ [*211·38 . *214·1]

***214·2.** $\vdash : P \epsilon \text{trans} \cap \text{connex} \cap \text{Ded} . \supset . \text{D}'P_\epsilon \subset \text{Ɑ}'\text{seq}_P$ [*211·371]

***214·21.** $\vdash : P \epsilon \text{trans} \cap \text{connex} \cap \text{Ded} . \supset . \text{D}'P_\epsilon = \overrightarrow{P}``C'P$ [*211·372]

***214·22.** $\vdash : P \epsilon \text{trans} \cap \text{connex} \cap \text{Ded} . \supset . \text{D}'(P_\epsilon \dot{\cap} I) = \overrightarrow{P}``\{C'P - \text{Ɑ}'(P \dot{-} P^2)\}$
[*211·46]

***214·23.** $\vdash : P \epsilon \text{trans} \cap \text{connex} \cap \text{Ded} . \sim \text{E} ! \max_P`\alpha . \supset .$
$\text{seq}_P`\alpha = \max_P`(\alpha \cup \iota`\text{seq}_P`\alpha) . \text{E} ! \max_P`(\alpha \cup \iota`\text{seq}_P`\alpha)$

Dem.

$\vdash . *214·101 . \supset \vdash : \text{Hp} . \supset . \text{E} ! \text{seq}_P`\alpha .$

[*206·47] $\supset . \text{seq}_P`\alpha = \max_P`(\alpha \cup \iota`\text{seq}_P`\alpha) .$ (1)

[*14·21] $\supset . \text{E} ! \max_P`(\alpha \cup \iota`\text{seq}_P`\alpha)$ (2)

$\vdash . (1) . (2) . \supset \vdash . \text{Prop}$

***214·24.** $\vdash : P \epsilon \text{connex} \cap \text{Ded} . \alpha \epsilon \text{sect}`P . \supset . \overrightarrow{\text{seq}}_P`\alpha = \overrightarrow{\min}_P`(C`P - \alpha)$
[*211·721]

***214·241.** $\vdash : P \epsilon \text{connex} . \breve{P} \epsilon \text{Ded} . \alpha \epsilon \text{sect}`P . \supset . \overrightarrow{\max}_P`\alpha = \overrightarrow{\text{prec}}_P`(C`P - \alpha)$
$\left[*214·24 \frac{\breve{P}}{P} . *211·7 \right]$

***214·3.** $\vdash :: \alpha, \beta \epsilon \kappa . \supset_{\alpha,\beta} : \alpha \subset \beta . \vee . \beta \subset \alpha :.$
$\kappa \sim \epsilon 1 . Q = \hat{\alpha}\hat{\beta}(\alpha, \beta \epsilon \kappa . \alpha \subset \beta . \alpha \neq \beta) :. \supset :.$
$\lambda \subset \kappa . \supset_\lambda . s`\lambda \epsilon \kappa : \supset . Q \epsilon \text{Ser} \cap \text{Ded}$
[*210·12·253]

***214·31.** $\vdash :. \text{Hp} *214·3 : \lambda \subset \kappa . \supset_\lambda . p`\lambda \cap s`\kappa \epsilon \kappa : \supset . Q \epsilon \text{Ser} \cap \text{Ded}$
[*210·12·254]

***214·32.** $\vdash : P \epsilon \text{connex} . \dot{\exists} ! P . \supset . \varsigma`P_* \epsilon \text{Ser} \cap \text{Ded}$ [*212·3·35]

***214·33.** $\vdash : P \epsilon \text{trans} \cap \text{connex} . \dot{\exists} ! P . \supset . \varsigma`P \epsilon \text{Ser} \cap \text{Ded}$ [*212·31·44]

***214·34.** $\vdash : P \epsilon \text{connex} . \dot{\exists} ! \text{sgm}`P . \supset . \text{sgm}`P \epsilon \text{Ser} \cap \text{Ded}$ [*212·3·54]

***214·4.** $\vdash :. P^2 = P . P \epsilon \text{connex} . \supset : P \epsilon \text{Ded} . \equiv . \text{Ɑ}`\max_P = - \text{Ɑ}`\text{seq}_P$
[*211·53]

***214·41.** $\vdash :. P \epsilon \text{Ser} . \supset : P^2 = P . P \epsilon \text{Ded} . \equiv . \text{Ɑ}`\max_P = - \text{Ɑ}`\text{seq}_P$
[*211·552]

***214·42.** $\vdash : P \epsilon \text{Ser} \cap \text{Ded} . P^2 = P . \alpha \epsilon \text{sect}`P . \supset . \text{limax}_P`\alpha = \text{limin}_P`(C`P - \alpha)$

Dem.

$\vdash . *211·721 . \supset \vdash :. \text{Hp} . \supset : \overrightarrow{\text{seq}}_P`\alpha = \overrightarrow{\min}_P`(C`P - \alpha) :$

[*214·101] $\supset : \sim \text{E} ! \max_P`\alpha . \supset . \text{lt}_P`\alpha = \min_P`(C`P - \alpha)$ (1)

$\vdash . *211·726 . \supset \vdash : \text{Hp} . \text{E} ! \max_P`\alpha . \supset . \max_P`\alpha = \text{prec}_P`(C`P - \alpha)$ (2)

$\vdash . *214{\cdot}14{\cdot}41 . \supset \vdash : \text{Hp} . \text{E} ! \text{prec}_P`(C`P - \alpha) . \supset . \sim \text{E} ! \max_P`(C`P - \alpha) .$

$[*207{\cdot}12] \quad \supset . \text{prec}_P`(C`P - \alpha) = \text{tl}_P`(C`P - \alpha) \quad (3)$

$\vdash . (2) . (3) . \quad \supset \vdash : \text{Hp} . \text{E} ! \max_P`\alpha . \supset . \max_P`\alpha = \text{tl}_P`(C`P - \alpha) \quad (4)$

$\vdash . (1) . (4) . *207{\cdot}46 . \supset$

$\vdash :. \text{Hp} . \supset : \text{limax}_P`\alpha = \min_P`(C`P - \alpha) . \vee . \text{limax}_P`\alpha = \text{tl}_P`(C`P - \alpha) :$

$[*207{\cdot}46] \supset : \text{limax}_P`\alpha = \text{limin}_P`(C`P - \alpha) :. \supset \vdash . \text{Prop}$

***214·43.** $\vdash :. P \in \text{Ser} \cap \text{Ded} . \alpha \in \text{sect}`P . \supset :$

$\text{limax}_P`\alpha = \text{limin}_P`(C`P - \alpha) . \vee . \max_P`\alpha \, P_1 \min_P`(C`P - \alpha)$

Dem.

$\vdash . *214{\cdot}11 . \supset \vdash :. \text{Hp} . \supset : \sim \text{E} ! \max_P`\alpha . \supset . \text{limax}_P`\alpha = \text{seq}_P`\alpha$

$[*211{\cdot}715] \quad = \min_P`(C`P - \alpha) \quad (1)$

$\vdash . *211{\cdot}726 . \supset \vdash : \text{E} ! \max_P`\alpha . \sim \text{E} ! \min_P`(C`P - \alpha) . \supset .$

$\text{limax}_P`\alpha = \text{tl}_P`(C`P - \alpha) \quad (2)$

$\vdash . (1) . (2) . *207{\cdot}46 . \supset \vdash :. \text{Hp} : \sim \text{E} ! \max_P`\alpha . \vee . \sim \text{E} ! \min_P`(C`P - \alpha) : \supset .$

$\text{limax}_P`\alpha = \text{limin}_P`(C`P - \alpha) \quad (3)$

$\vdash . *211{\cdot}726 . \supset \vdash : \text{Hp} . \text{E} ! \max_P`\alpha . \text{E} ! \min_P`(C`P - \alpha) . \supset .$

$\text{E} ! \max_P`\alpha . \text{E} ! \text{seq}_P`\alpha . \text{seq}_P`\alpha = \min_P`(C`P - \alpha) .$

$[*206{\cdot}5] \quad \supset . \max_P`\alpha \, P_1 \min_P`(C`P - \alpha) \quad (4)$

$\vdash . (3) . (4) . \supset \vdash . \text{Prop}$

The following propositions are no longer mere restatements of previous results.

***214·5.** $\vdash : P \in \text{Ded} . \supset . \exists ! \overrightarrow{B}`P . \exists ! \overrightarrow{B}`\breve{P} . \overrightarrow{B}`P = \overrightarrow{\text{seq}}_P`\Lambda . \overrightarrow{B}`\breve{P} = \overrightarrow{\max}_P`C`P$

Dem.

$\vdash . *205{\cdot}161 . *214{\cdot}101 . \supset \vdash : \text{Hp} . \supset . \exists ! \overrightarrow{\text{seq}}_P`\Lambda .$

$[*206{\cdot}14] \quad \supset . \exists ! \overrightarrow{B}`P . \overrightarrow{B}`P = \overrightarrow{\text{seq}}_P`\Lambda \quad (1)$

$\vdash . *206{\cdot}18{\cdot}2 . \supset \vdash . \overrightarrow{\text{seq}}_P`C`P = \Lambda \quad (2)$

$\vdash . (2) . *214{\cdot}1 . \supset \vdash : \text{Hp} . \supset . \exists ! \overrightarrow{\max}_P`C`P .$

$[*93{\cdot}117] \quad \supset . \exists ! \overrightarrow{B}`\breve{P} . \overrightarrow{B}`\breve{P} = \overrightarrow{\max}_P`C`P \quad (3)$

$\vdash . (1) . (3) . \supset \vdash . \text{Prop}$

***214·51.** $\vdash :. P \in \text{Ded} . \supset : \sim (xPx) . \vee . x \in \text{D}`(P \dot{-} P^2)$

Dem.

$\vdash . *214{\cdot}1 . \supset \vdash :. \text{Hp} . \supset : \exists ! \overrightarrow{\max}_P`\iota`x . \vee . \exists ! \overrightarrow{\text{seq}}_P`\iota`x :$

$[*53{\cdot}301 . *206{\cdot}42] \quad \supset : \exists ! \iota`x - \overrightarrow{P}`x . \vee . \exists ! \overleftarrow{P \dot{-} P^2}`x :$

$[*51{\cdot}31 . *33{\cdot}4] \quad \supset : \sim (xPx) . \vee . x \in \text{D}`(P \dot{-} P^2) :. \supset \vdash . \text{Prop}$

***214·52.** $\vdash : P \epsilon \mathrm{Ded} \,.\, P \mathrel{\dot{\subset}} P^2 \,.\, \supset \,.\, P \mathrel{\dot{\subset}} J$ [*214·51]

***214·53.** $\vdash : P \epsilon \mathrm{Ded} \,.\, \supset \,.\, \mathrm{D}\text{‘}P = \mathrm{D}\text{‘}(P \mathbin{\dot{\cap}} J)$

Dem.

$\vdash \,.\, {*}214\cdot51 \,.\, \supset \vdash : \mathrm{Hp} \,.\, xPx \,.\, \supset \,.\, x \epsilon \mathrm{D}\text{‘}(P \mathbin{\dot{-}} P^2) \,.$

[*33·13] $\supset \,.\, (\exists y) \,.\, xPy \,.\, x \mathbin{\dot{-}} P^2 y \,.$

[*34·54.Transp] $\supset \,.\, (\exists y) \,.\, xPy \,.\, x \neq y$ (1)

$\vdash \,.\, (1) \,.\, {*}13\cdot195 \,.\, \supset \vdash :. \mathrm{Hp} \,.\, \supset : (\exists y) \,.\, xPy \,.\, \supset \,.\, (\exists y) \,.\, xPy \,.\, x \neq y :$

[*33·13] $\supset : \mathrm{D}\text{‘}P \subset \mathrm{D}\text{‘}(P \mathbin{\dot{\cap}} J) :$

[*33·25] $\supset : \mathrm{D}\text{‘}P = \mathrm{D}\text{‘}(P \mathbin{\dot{\cap}} J) :. \supset \vdash \,.\, \mathrm{Prop}$

***214·531.** $\vdash : P \epsilon \mathrm{Ded} \,.\, \supset \,.\, C\text{‘}P = C\text{‘}(P \mathbin{\dot{\cap}} J)$

Dem.

$\vdash \,.\, {*}93\cdot12 \,.\, \supset \vdash :. x \epsilon \overrightarrow{B}\text{‘}\breve{P} \,.\, \supset : x \sim\epsilon \mathrm{D}\text{‘}P : (\exists y) \,.\, yPx :$

[*13·14] $\supset : (\exists y) \,.\, yPx \,.\, x \neq y :$

[*33·13] $\supset : x \epsilon \text{Ɑ}\text{‘}(P \mathbin{\dot{\cap}} J)$ (1)

$\vdash \,.\, (1) \,.\, {*}214\cdot53 \,.\, \supset \vdash : \mathrm{Hp} \,.\, \supset \,.\, \mathrm{D}\text{‘}P \cup \overrightarrow{B}\text{‘}\breve{P} \subset C\text{‘}(P \mathbin{\dot{\cap}} J) \,.$

[*93·12] $\supset \,.\, C\text{‘}P \subset C\text{‘}(P \mathbin{\dot{\cap}} J) \,.$

[*33·252] $\supset \,.\, C\text{‘}P = C\text{‘}(P \mathbin{\dot{\cap}} J) : \supset \vdash \,.\, \mathrm{Prop}$

***214·532.** $\vdash : P \epsilon \mathrm{Ded} \,.\, \supset \,.\, \text{Ɑ}\text{‘}P = \text{Ɑ}\text{‘}(P \mathbin{\dot{\cap}} J)$

Dem.

$\vdash \,.\, {*}34\cdot54 \,.\, \supset \vdash : \overrightarrow{P}\text{‘}x = \iota\text{‘}x \,.\, \supset \,.\, \overrightarrow{P}\text{‘}x = P\text{‘‘}\overrightarrow{P}\text{‘}x \,.$

[*205·123] $\supset \,.\, \overrightarrow{\mathrm{max}}_P\text{‘}\iota\text{‘}x = \Lambda$ (1)

$\vdash \,.\, {*}206\cdot134 \,.\, \supset \vdash : \overrightarrow{P}\text{‘}x = \iota\text{‘}x \,.\, \supset \,.$

$\overrightarrow{\mathrm{seq}}_P\text{‘}\overrightarrow{P}\text{‘}x = C\text{‘}P \cap \hat{y}\, \{\iota\text{‘}x \subset \overrightarrow{P}\text{‘}y \,.\, \overrightarrow{P}\text{‘}y \subset -p\text{‘}\overleftarrow{P}\text{‘‘}\iota\text{‘}x\}$

[*53·301·01] $= C\text{‘}P \cap \hat{y}\, \{\iota\text{‘}x \subset \overrightarrow{P}\text{‘}y \,.\, \overrightarrow{P}\text{‘}y \subset -\overleftarrow{P}\text{‘}x\}$

[Hp] $= C\text{‘}P \cap \hat{y}\, \{\iota\text{‘}x \subset \overrightarrow{P}\text{‘}y \,.\, \overrightarrow{P}\text{‘}y \subset -\iota\text{‘}x\}$

[*51·161] $= \Lambda$ (2)

$\vdash \,.\, (1) \,.\, (2) \,.\, \supset \vdash : \overrightarrow{P}\text{‘}x = \iota\text{‘}x \,.\, \supset \,.\, \overrightarrow{\mathrm{max}}_P\text{‘}\overrightarrow{P}\text{‘}x = \Lambda \,.\, \overrightarrow{\mathrm{seq}}_P\text{‘}\overrightarrow{P}\text{‘}x = \Lambda$ (3)

$\vdash \,.\, (3) \,.\, \mathrm{Transp} \,.\, \supset \vdash :. \mathrm{Hp} \,.\, \supset : (x) \,.\, \overrightarrow{P}\text{‘}x \neq \iota\text{‘}x :$

[*51·401.Transp] $\supset : (x) : \exists !\, \overrightarrow{P}\text{‘}x \,.\, \supset \,.\, \exists !\, \overrightarrow{P}\text{‘}x - \iota\text{‘}x :$

[*33·41] $\supset : x \epsilon \text{Ɑ}\text{‘}P \,.\, \supset \,.\, x \epsilon \text{Ɑ}\text{‘}(P \mathbin{\dot{\cap}} J)$ (4)

$\vdash \,.\, (4) \,.\, {*}33\cdot251 \,.\, \supset \vdash \,.\, \mathrm{Prop}$

***214·54.** $\vdash : P \epsilon \text{Ded} . \supset . P \dot{\frown} J \epsilon \text{Ded}$

Dem.

$\vdash . *205{\cdot}111{\cdot}195 . \supset \vdash . \overrightarrow{\max}_P{}^{\backprime}\alpha \subset \alpha \cap C^{\backprime}(P \dot{\frown} J) - P^{\backprime\backprime}\alpha$

$[*37{\cdot}201] \qquad \subset \alpha \cap C^{\backprime}(P \dot{\frown} J) - (P \dot{\frown} J)^{\backprime\backprime}\alpha$

$[*205{\cdot}111] \qquad \subset \overrightarrow{\max}(P \dot{\frown} J)^{\backprime}\alpha .$

$[*24{\cdot}59] \qquad \supset \vdash : \sim \exists ! \overrightarrow{\max}(P \dot{\frown} J)^{\backprime}\alpha . \supset . \sim \exists ! \overrightarrow{\max}_P{}^{\backprime}\alpha \qquad (1)$

$\vdash . (1) . *214{\cdot}1 . \quad \supset \vdash : \text{Hp} . \sim \exists ! \overrightarrow{\max}(P \dot{\frown} J)^{\backprime}\alpha . \supset . \exists ! \overrightarrow{\text{seq}}_P{}^{\backprime}\alpha \qquad (2)$

$\vdash . *206{\cdot}2{\cdot}17 . \supset$

$\vdash :. x \, \text{seq}_P \, \alpha . \equiv : y \epsilon \alpha \cap C^{\backprime}P . \supset_y . yPx . y \neq x : x \epsilon C^{\backprime}P :$

$yPx . \supset_y . (\exists z) . z \epsilon \alpha . \sim (zPy) :$

$[*214{\cdot}531] \quad \equiv : y \epsilon \alpha \cap C^{\backprime}(P \dot{\frown} J) . \supset_y . y (P \dot{\frown} J) x : x \epsilon C^{\backprime}(P \dot{\frown} J) :$

$yPx . \supset_y . (\exists z) . z \epsilon \alpha . \sim (zPy) :$

$[*23{\cdot}43 . *3{\cdot}14] \supset : y \epsilon \alpha \cap C^{\backprime}(P \dot{\frown} J) . \supset_y . y (P \dot{\frown} J) x : x \epsilon C^{\backprime}(P \dot{\frown} J) :$

$y (P \dot{\frown} J) x . \supset_y . (\exists z) . z \epsilon \alpha . \sim \{z (P \dot{\frown} J) y\} :$

$[*206{\cdot}17] \qquad \supset : x \, \text{seq} (P \dot{\frown} J) \, \alpha \qquad (3)$

$\vdash . (2) . (3) . \supset \vdash : \text{Hp} . \sim \exists ! \overrightarrow{\max}(P \dot{\frown} J)^{\backprime}\alpha . \supset . \exists ! \text{seq} (P \dot{\frown} J)^{\backprime}\alpha \qquad (4)$

$\vdash . (4) . *214{\cdot}1 . \supset \vdash . \text{Prop}$

***214·6.** $\vdash : P \epsilon \text{Ded} . P \, \text{smor} \, Q . \supset . Q \epsilon \text{Ded}$

Dem.

$\vdash . *207{\cdot}65 . *214{\cdot}11 . \supset$

$\vdash : P \epsilon \text{Ded} . S \epsilon P \, \overline{\text{smor}} \, Q . \supset . (\alpha) . \breve{S}^{\backprime\backprime}\alpha \epsilon \text{Cl}^{\backprime}\text{limax}_Q .$

$[*71{\cdot}481] \qquad \supset . \text{Cl}^{\backprime}\text{Cl}^{\backprime}S \subset \text{Cl}^{\backprime}\text{limax}_Q .$

$[*151{\cdot}11 . *214{\cdot}12] \qquad \supset . Q \epsilon \text{Ded} : \supset \vdash . \text{Prop}$

***214·7.** $\vdash :. P \epsilon \text{semi Ded} . \equiv : \alpha \epsilon \text{sect}^{\backprime}P . \alpha \neq C^{\backprime}P . \supset_\alpha . \exists ! (\overrightarrow{\max}_P{}^{\backprime}\alpha \cup \overrightarrow{\text{seq}}_P{}^{\backprime}\alpha)$ $[(*214{\cdot}02)]$

***214·71.** $\vdash . \text{Ded} \subset \text{semi Ded} \quad [*214{\cdot}1{\cdot}7]$

***214·72.** $\vdash :. P \epsilon \text{trans} . \supset : P \epsilon \text{Ded} . \equiv . P \epsilon \text{semi Ded} . \exists ! \overrightarrow{B}^{\backprime}\breve{P}$ $[*214{\cdot}7{\cdot}13 . *205{\cdot}121]$

***214·73.** $\vdash . \text{semi Ded} - \iota^{\backprime}\dot{\Lambda} \subset \text{Cl}^{\backprime}B \quad [*206{\cdot}14 . *211{\cdot}44 . *214{\cdot}7]$

The proof of the following proposition is given in a somewhat compressed form, since, if given with the usual fullness, it would require various lemmas not required elsewhere.

$*214 \cdot 74$. $\vdash : P \in \text{Ser} \cap \text{semi Ded} . \supset . P \upharpoonright \overleftarrow{P_*}`x \in \text{semi Ded}$

Dem.

$\vdash . *214 \cdot 7 . \supset \vdash : \text{Hp} . \alpha \in \text{sect}`P . \alpha \neq C`P . \supset . \exists ! (\overrightarrow{\text{max}}_P`\alpha \cup \overrightarrow{\text{seq}}_P`\alpha)$ (1)

$\vdash . *205 \cdot 261 . \supset$

$\vdash : \text{Hp}(1) . \overleftarrow{P_*}`x \sim \in 1 . x \in \alpha . \supset . \overrightarrow{\text{max}}(P \upharpoonright \overleftarrow{P_*}`x)`\alpha = \overrightarrow{\text{max}}_P`(\alpha \cap \overleftarrow{P_*}`x)$

[$*205 \cdot 262$] $= \overrightarrow{\text{max}}_P`\alpha$ (2)

$\vdash . *211 \cdot 75 \cdot 56 . \supset \vdash : \text{Hp}(2) . \supset . C`P - \alpha \subset \overleftarrow{P_*}`x$ (3)

$\vdash . (3) . *211 \cdot 715 . \supset \vdash : \text{Hp}(2) . Q = P \upharpoonright \overleftarrow{P_*}`x . \supset . \overrightarrow{\text{seq}}_P`\alpha = \overrightarrow{\text{min}}_P`(\overleftarrow{P_*}`x - \alpha)$

[$*205 \cdot 261$] $= \overrightarrow{\text{min}}_Q`(-\alpha)$

[$*211 \cdot 715 . *206 \cdot 25$] $= \overrightarrow{\text{seq}}_Q`(\alpha \cap \overleftarrow{P_*}`x)$ (4)

$\vdash . (2) . (4) . \supset$

$\vdash : \text{Hp}(4) . \supset . \overrightarrow{\text{max}}_P`\alpha \cup \overrightarrow{\text{seq}}_P`\alpha = \overrightarrow{\text{max}}_Q`(\alpha \cap \overleftarrow{P_*}`x) \cup \overrightarrow{\text{seq}}_Q`(\alpha \cap \overleftarrow{P_*}`x)$ (5)

$\vdash . (1) . (5) . \supset \vdash : \text{Hp} . \alpha \in \text{sect}`P . \alpha \neq C`P . \overleftarrow{P_*}`x \sim \in 1 . x \in \alpha . Q = P \upharpoonright \overleftarrow{P_*}`x . \supset .$

$\exists ! \{\overrightarrow{\text{max}}_Q`(\alpha \cap \overleftarrow{P_*}`x) \cup \overrightarrow{\text{seq}}_Q`(\alpha \cap \overleftarrow{P_*}`x)\}$ (6)

$\vdash . *211 \cdot 715 . \supset \vdash : \text{Hp} . \overleftarrow{P_*}`x \sim \in 1 . \alpha = \overrightarrow{P}`x . \supset . \overrightarrow{\text{seq}}_P`\alpha = \overrightarrow{\text{min}}_P`\overleftarrow{P_*}`x$

[$*205 \cdot 261$] $= \overrightarrow{\text{min}}(P \upharpoonright \overleftarrow{P_*}`x)`\overleftarrow{P_*}`x$

[$*206 \cdot 14$] $= \overrightarrow{\text{seq}}(P \upharpoonright \overleftarrow{P_*}`x)`\Lambda$ (7)

$\vdash . (7) . *206 \cdot 401 . \supset \vdash : \text{Hp} . \overleftarrow{P_*}`x \sim \in 1 . \supset . \exists ! \overrightarrow{\text{seq}}(P \upharpoonright \overleftarrow{P_*}`x)`\Lambda$ (8)

$\vdash . (6) . (8) . \supset \vdash :. \text{Hp} . \overleftarrow{P_*}`x \sim \in 1 . Q = P \upharpoonright \overleftarrow{P_*}`x . \supset :$

$\beta \in \text{sect}`Q - \iota`C`Q . \supset_\beta . \exists ! (\overrightarrow{\text{max}}_Q`\beta \cup \overrightarrow{\text{seq}}_Q`\beta) :$

[$*214 \cdot 7$] $\supset : Q \in \text{semi Ded}$ (9)

$\vdash . *214 \cdot 7 . *200 \cdot 35 . \supset \vdash : \text{Hp} . \overleftarrow{P_*}`x \in 1 . \supset . P \upharpoonright \overleftarrow{P_*}`x \in \text{semi Ded}$ (10)

$\vdash . (9) . (10) . \supset \vdash . \text{Prop}$

$*214 \cdot 75$. $\vdash : P \in \text{semi Ded} . P \text{ smor } Q . \supset . Q \in \text{semi Ded}$
[$*205 \cdot 8 . *206 \cdot 61 . *212 \cdot 7$]

$*215$. STRETCHES.

Summary of $*215$.

A *stretch* of a series is any piece taken out of it, and not having any gaps; that is, it is a class contained in the series, and containing all terms which come between any two of its terms. Thus it is defined as

$$\hat{\alpha}\,(\alpha \subset C\text{‘}P \,.\, P\text{“}\alpha \cap \breve{P}\text{“}\alpha \subset \alpha).$$

We denote the class of stretches by "str‘P," where "str" stands for "stretch" or "Strecke." A stretch which has no predecessors is a section of P; one which has no successors is a section of $\breve{P}$. The properties of stretches are chiefly important in connection with compact series. In discrete series, stretches are the same as intervals.

If P is transitive, stretches of P are the products of sections of P and sections of $\breve{P}$, *i.e.* of upper and lower sections of P ($*215{\cdot}16$). If P is connected, and α is a lower section, β an upper section, then if the two have a stretch $\alpha \cap \beta$ in common, we have

$$\alpha = P\text{“}(\alpha \cap \beta) \cup (\alpha \cap \beta) \,.\, \beta = \breve{P}\text{“}(\alpha \cap \beta) \cup (\alpha \cap \beta) \quad (*215{\cdot}161).$$

A slightly more general form of this proposition is

$*215{\cdot}165$. $\vdash : P_{\text{po}} \in \text{connex} \,.\, \alpha \in \text{sect}\text{‘}P \,.\, \beta \in \text{sect}\text{‘}\breve{P} \,.\, \exists \,! \, \alpha \cap \beta \,.\, \supset .$

$$\alpha = P_*\text{“}(\alpha \cap \beta) \,.\, \beta = \breve{P}_*\text{“}(\alpha \cap \beta) \,.\, P\text{“}\alpha = P_{\text{po}}\text{“}(\alpha \cap \beta) \,.\, \breve{P}\text{“}\beta = \breve{P}_{\text{po}}\text{“}(\alpha \cap \beta)$$

A specially important case is when α and β have just one term in common. In this case we have

$*215{\cdot}166$. $\vdash : P_{\text{po}} \in \text{Ser} \,.\, \alpha \in \text{sect}\text{‘}P \,.\, \beta \in \text{sect}\text{‘}\breve{P} \,.\, \alpha \cap \beta \in 1 \,.\, \supset .$

$$\alpha \cap \beta = \iota\text{‘}\text{max}_P\text{‘}\alpha = \iota\text{‘}\text{min}_P\text{‘}\beta$$

When $\alpha \cap \beta$ has more than one term, if the upper limit or maximum of α and the lower limit or minimum of β both exist, the latter precedes the former ($*215{\cdot}52$); if α and β have no common part, but together exhaust the field of P, we have either $\text{limax}_P\text{‘}\alpha = \text{limin}_P\text{‘}\beta$ or $\text{limax}_P\text{‘}\alpha \, P_1 \, \text{limin}_P\text{‘}\beta$, assuming $E\,!\,\text{limax}_P\text{‘}\alpha \,.\, E\,!\,\text{limin}_P\text{‘}\beta$ ($*215{\cdot}54$). Hence if $\text{limax}_P\text{‘}\alpha$ has no immediate successor, it must be identical with $\text{limin}_P\text{‘}\beta$. Thus we have

∗215·543. $\vdash : P \in \text{Ser} . \alpha \in \text{sect}‘P . \beta \in \text{sect}‘\breve{P} . \alpha \cup \beta = C‘P . \alpha \cap \beta \in 0 \cup 1 .$
$\text{E} ! \text{limax}_P‘\alpha . \text{limax}_P‘\alpha \sim \in \text{D}‘P_1 . \supset . \text{limax}_P‘\alpha = \text{limin}_P‘\beta$

The above propositions will be useful in Section C (∗231 and ∗233).

∗215·01. $\text{str}‘P = \hat{\alpha} (\alpha \subset C‘P . P“\alpha \cap \breve{P}“\alpha \subset \alpha)$ Df

∗215·1. $\vdash : \alpha \in \text{str}‘P . \equiv . \alpha \subset C‘P . P“\alpha \cap \breve{P}“\alpha \subset \alpha$ [(∗215·01)]

∗215·11. $\vdash . \text{str}‘P = \text{str}‘\breve{P}$ [(∗215·01) . ∗33·22]

∗215·13. $\vdash . \text{sect}‘P \subset \text{str}‘P . \text{sect}‘\breve{P} \subset \text{str}‘P$ [∗215·1 . ∗211·1]

∗215·14. $\vdash : \alpha \in \text{sect}‘P . \beta \in \text{sect}‘\breve{P} . \supset . \alpha \cap \beta \in \text{str}‘P$

Dem.

$\vdash . ∗211·1 . \supset \vdash : \text{Hp} . \supset . \alpha \subset C‘P . P“\alpha \subset \alpha . \breve{P}“\beta \subset \beta .$

[∗22·43.∗37·21] $\supset . \alpha \cap \beta \subset C‘P . P“(\alpha \cap \beta) \subset \alpha . \breve{P}“(\alpha \cap \beta) \subset \beta .$

[∗22·49] $\supset . \alpha \cap \beta \subset C‘P . P“(\alpha \cap \beta) \cap \breve{P}“(\alpha \cap \beta) \subset \alpha \cap \beta .$

[∗215·1] $\supset . \alpha \cap \beta \in \text{str}‘P : \supset \vdash . \text{Prop}$

∗215·15. $\vdash : P \in \text{trans} . \alpha \in \text{str}‘P . \supset . \alpha \cup P“\alpha \in \text{sect}‘P . \alpha \cup \breve{P}“\alpha \in \text{sect}‘\breve{P} .$
$\alpha = (\alpha \cup P“\alpha) \cap (\alpha \cup \breve{P}“\alpha)$

Dem.

$\vdash . ∗211·27 . ∗215·1 . \supset \vdash : \text{Hp} . \supset . \alpha \cup P“\alpha \in \text{sect}‘P . \alpha \cup \breve{P}“\alpha \in \text{sect}‘\breve{P}$ (1)

$\vdash . ∗215·1 . ∗22·62 . \supset \vdash : \text{Hp} . \supset . \alpha = \alpha \cup (P“\alpha \cap \breve{P}“\alpha)$

[∗22·69] $= (\alpha \cup P“\alpha) \cap (\alpha \cup \breve{P}“\alpha)$ (2)

$\vdash . (1) . (2) . \supset \vdash . \text{Prop}$

∗215·16. $\vdash : P \in \text{trans} . \supset . \text{str}‘P = \hat{\gamma} \{(\exists \alpha, \beta) . \alpha \in \text{sect}‘P . \beta \in \text{sect}‘\breve{P} . \gamma = \alpha \cap \beta\}$
$= s‘\{(\text{sect}‘P) \cap_{\shortparallel} “\text{sect}‘\breve{P}\}$

[∗215·14·15 . ∗40·7]

∗215·161. $\vdash : P \in \text{connex} . \alpha \in \text{sect}‘P . \beta \in \text{sect}‘\breve{P} . \exists ! \alpha \cap \beta . \supset .$
$\alpha = P“(\alpha \cap \beta) \cup (\alpha \cap \beta) . \beta = \breve{P}“(\alpha \cap \beta) \cup (\alpha \cap \beta)$

Dem.

$\vdash . ∗211·1 . ∗37·2 . \supset \vdash : \text{Hp} . \supset . P“(\alpha \cap \beta) \cup (\alpha \cap \beta) \subset \alpha$ (1)

$\vdash . ∗211·702 . \supset \vdash :. \text{Hp} . x \in \alpha - \beta . \supset : y \in \beta . \supset . xPy :$

[∗37·1] $\supset : \exists ! (\alpha \cap \beta) . \supset . x \in P“(\alpha \cap \beta)$ (2)

$\vdash . (2) . \supset \vdash : \text{Hp} . x \in \alpha . \supset . x \in P“(\alpha \cap \beta) \cup (\alpha \cap \beta)$ (3)

$\vdash . (1) . (3) . \supset \vdash : \text{Hp} . \supset . \alpha = P“(\alpha \cap \beta) \cup (\alpha \cap \beta)$ (4)

$\vdash . (4) \frac{\breve{P}}{P} . \supset \vdash : \text{Hp} . \supset . \beta = \breve{P}“(\alpha \cap \beta) \cup (\alpha \cap \beta)$ (5)

$\vdash . (4) . (5) . \supset \vdash . \text{Prop}$

***215·162.** $\vdash : P \epsilon \text{trans} \cap \text{connex} . \alpha \epsilon \text{sect}`P . \beta \epsilon \text{sect}`\breve{P} . \exists ! \alpha \cap \beta . \supset .$
$$P``\alpha = P``(\alpha \cap \beta) . \breve{P}``\beta = \breve{P}``(\alpha \cap \beta)$$

Dem.

$\vdash . *215·161 . \supset \vdash : \text{Hp} . \supset . P``\alpha = P``P``(\alpha \cap \beta) \cup P``(\alpha \cap \beta)$

$[*201·5] \quad = P``(\alpha \cap \beta)$ (1)

Similarly $\vdash : \text{Hp} . \supset . \breve{P}``\beta = \breve{P}``(\alpha \cap \beta)$ (2)

$\vdash . (1) . (2) . \supset \vdash . \text{Prop}$

***215·163.** $\vdash : P \epsilon \text{trans} \cap \text{connex} . \alpha \epsilon \text{sect}`P . \beta \epsilon \text{sect}`\breve{P} . \exists ! \alpha \cap \beta . \supset .$
$$p`\overleftarrow{P}``\alpha = p`\overleftarrow{P}``(\alpha \cap \beta)$$

Dem.

$\vdash . *40·16 . \quad \supset \vdash . p`\overleftarrow{P}``\alpha \subset p`\overleftarrow{P}``(\alpha \cap \beta)$ (1)

$\vdash . *10·56 . *37·1 . \supset \vdash :. \text{Hp} : y \epsilon \alpha \cap \beta . \supset_y . yPx : z \epsilon P``(\alpha \cap \beta) : \supset . zPx$ (2)

$\vdash . (2) . *215·161 . \supset \vdash :. \text{Hp} : y \epsilon \alpha \cap \beta . \supset_y . yPx : \supset : z \epsilon \alpha . \supset_z . zPx$ (3)

$\vdash . (1) . (3) . \supset \vdash . \text{Prop}$

***215·164.** $\vdash : \text{Hp} *215·162 . \supset . \overrightarrow{\text{min}}_P`\beta = \overrightarrow{\text{min}}_P`(\alpha \cap \beta) . \overrightarrow{\text{max}}_P`\alpha = \overrightarrow{\text{max}}_P`(\alpha \cap \beta) .$
$$\overrightarrow{\text{seq}}_P`\alpha = \overrightarrow{\text{seq}}_P`(\alpha \cap \beta) . \overrightarrow{\text{prec}}_P`\beta = \overrightarrow{\text{prec}}_P`(\alpha \cap \beta) .$$
$$\overrightarrow{\text{lt}}_P`\alpha = \overrightarrow{\text{lt}}_P`(\alpha \cap \beta) . \overrightarrow{\text{limax}}_P`\alpha = \overrightarrow{\text{limax}}_P`(\alpha \cap \beta)$$

Dem.

$\vdash . *215·162 . \quad \supset \vdash : \text{Hp} . \supset . \overrightarrow{\text{max}}_P`\alpha = \alpha - P``(\alpha \cap \beta)$

$[*215·161] \quad = \alpha \cap \beta - P``(\alpha \cap \beta)$

$[*205·111] \quad = \overrightarrow{\text{max}}_P`(\alpha \cup \beta)$ (1)

Similarly $\vdash : \text{Hp} . \supset . \overrightarrow{\text{min}}_P`\beta = \overrightarrow{\text{min}}_P`(\alpha \cap \beta)$ (2)

$\vdash . *215·163 . *206·13 . \supset \vdash : \text{Hp} . \supset . \overrightarrow{\text{seq}}_P`\alpha = \overrightarrow{\text{seq}}_P`(\alpha \cap \beta)$ (3)

Similarly $\vdash : \text{Hp} . \supset . \overrightarrow{\text{prec}}_P`\beta = \overrightarrow{\text{prec}}_P`(\alpha \cap \beta)$ (4)

$\vdash . (1) . (3) . *207·11·12 . \supset \vdash : \text{Hp} . \supset . \overrightarrow{\text{lt}}_P`\alpha = \overrightarrow{\text{lt}}_P`(\alpha \cap \beta)$ (5)

$\vdash . (1) . (5) . *207·45 . \quad \supset \vdash : \text{Hp} . \supset . \overrightarrow{\text{limax}}_P`\alpha = \overrightarrow{\text{limax}}_P`(\alpha \cap \beta)$ (6)

$\vdash . (1) . (2) . (3) . (4) . (5) . (6) . \supset \vdash . \text{Prop}$

***215·165.** $\vdash : P_{\text{po}} \epsilon \text{connex} . \alpha \epsilon \text{sect}`P . \beta \epsilon \text{sect}`\breve{P} . \exists ! \alpha \cap \beta . \supset .$
$$\alpha = P_*``(\alpha \cap \beta) . \beta = \breve{P}_*``(\alpha \cap \beta) . P``\alpha = P_{\text{po}}``(\alpha \cap \beta) . \breve{P}``\beta = \breve{P}_{\text{po}}``(\alpha \cap \beta)$$

Dem.

$\vdash . *211·17 . \supset \vdash : \text{Hp} . \supset . \alpha \epsilon \text{sect}`P_{\text{po}} . \beta \epsilon \text{sect}`\breve{P}_{\text{po}} . \exists ! \alpha \cap \beta .$

$[*215·161] \quad \supset . \alpha = P_*``(\alpha \cap \beta) . \beta = \breve{P}_*``(\alpha \cap \beta) .$ (1)

$[*91·52] \quad \supset . P``\alpha = P_{\text{po}}``(\alpha \cap \beta) . \breve{P}``\beta = \breve{P}_{\text{po}}``(\alpha \cap \beta)$ (2)

$\vdash . (1) . (2) . \supset \vdash . \text{Prop}$

***215·166.** $\vdash : P_{po} \in \text{Ser} . \alpha \in \text{sect}`P . \beta \in \text{sect}`\breve{P} . \alpha \cap \beta \in 1 . \supset .$
$\alpha \cap \beta = \iota`\text{max}_P`\alpha = \iota`\text{min}_P`\beta$

Dem.

$\vdash . *215·161 . *211·17 . \supset \vdash : \text{Hp} . \supset . \alpha = (\alpha \cap \beta) \cup P_{po}``(\alpha \cap \beta) .$

[*215·165] $\supset . \alpha - P``\alpha = (\alpha \cap \beta) - P_{po}``(\alpha \cap \beta) .$

[*205·11] $\supset . \overrightarrow{\text{max}}_P`\alpha = \overrightarrow{\text{max}}(P_{po})`(\alpha \cap \beta)$

[*205·17] $= \alpha \cap \beta$ (1)

Similarly $\vdash : \text{Hp} . \supset . \overrightarrow{\text{min}}_P`\beta = \alpha \cap \beta$ (2)

$\vdash . (1) . (2) . \supset \vdash . \text{Prop}$

***215·17.** $\vdash : P \in \text{trans} . \supset . \breve{P}``\alpha \cap P``\beta \in \text{str}`P$

Dem.

$\vdash . *211·15·11 . \supset \vdash : \text{Hp} . \supset . P``\beta \in \text{sect}`P . \breve{P}``\alpha \in \text{sect}`\breve{P}$ (1)

$\vdash . (1) . *215·14 . \supset \vdash . \text{Prop}$

***215·18.** $\vdash . P(x \mapsto y), P(x \vdash y), P(x \dashv y), P(x - y) \in \text{str}`P$

Dem.

$\vdash . *211·13·3 . \supset \vdash . \overrightarrow{P_*}`y \in \text{sect}`P . \overleftarrow{P_*}`x \in \text{sect}`\breve{P}$ (1)

$\vdash . *211·16 . \supset \vdash . \overrightarrow{P_{po}}`y \in \text{sect}`P . \overleftarrow{P_{po}}`x \in \text{sect}`\breve{P}$ (2)

$\vdash . (1) . (2) . *215·14 . \supset \vdash . \text{Prop}$

***215·19.** $\vdash : P^2 \subseteq J . x \in C`P . \supset . \iota`x \in \text{str}`P$

Dem.

$\vdash . *53·301 . \supset \vdash . P``\iota`x \cap \breve{P}``\iota`x = \overrightarrow{P}`x \cap \overleftarrow{P}`x$ (1)

$\vdash . (1) . *50·43 . \supset \vdash : \text{Hp} . \supset . P``\iota`x \cap \breve{P}``\iota`x = \Lambda$ (2)

$\vdash . (2) . *215·1 . \supset \vdash . \text{Prop}$

***215·2.** $\vdash : P \in \text{connex} . \alpha \in \text{str}`P . x \in \alpha . \supset . P``\alpha = \alpha - \overrightarrow{\text{max}}_P`\alpha \cup \overrightarrow{P}`x .$
$\breve{P}``\alpha = \alpha - \overrightarrow{\text{min}}_P`\alpha \cup \overleftarrow{P}`x$

Dem.

$\vdash . *205·111 . \supset \vdash . \alpha - \overrightarrow{\text{max}}_P`\alpha \subset P``\alpha$ (1)

$\vdash . *37·18 . \supset \vdash : \text{Hp} . \supset . \overrightarrow{P}`x \subset P``\alpha$ (2)

$\vdash . (1) . (2) . \supset \vdash : \text{Hp} . \supset . \alpha - \overrightarrow{\text{max}}_P`\alpha \cup \overrightarrow{P}`x \subset P``\alpha$ (3)

$\vdash . *202·103 . \supset \vdash :. \text{Hp} . y \in P``\alpha . \supset : y \in \overrightarrow{P}`x \cup \iota`x \cup \overleftarrow{P}`x :$

[*37·181] $\supset : y \in \overrightarrow{P}`x \cup \iota`x . \vee . y \in \breve{P}``\alpha :$

[*4·73] $\supset : y \in \overrightarrow{P}`x \cup \iota`x . \vee . y \in P``\alpha \cap \breve{P}``\alpha :$

[*215·1] $\supset : y \in \overrightarrow{P}`x \cup \iota`x \cup \alpha :$

[Hp] $\supset : y \in \overrightarrow{P}`x \cup \alpha$ (4)

$\vdash . *205\cdot111 . \supset \vdash . y \in P``\alpha . \supset . y \sim\in \overrightarrow{\max}_P`\alpha$ (5)

$\vdash . (4) . (5) . \quad \supset \vdash : \text{Hp} . y \in P``\alpha . \supset . y \in \alpha - \overrightarrow{\max}_P`\alpha \cup \overrightarrow{P}`x$ (6)

$\vdash . (3) . (6) . \quad \supset \vdash : \text{Hp} . \supset . P``\alpha = \alpha - \overrightarrow{\max}_P`\alpha \cup \overrightarrow{P}`x$ (7)

Similarly $\quad \vdash : \text{Hp} . \supset . \breve{P}``\alpha = \alpha - \overrightarrow{\min}_P`\alpha \cup \overleftarrow{P}`x$ (8)

$\vdash . (7) . (8) . \supset \vdash . \text{Prop}$

$*215\cdot21$. $\vdash : P \in \text{connex} . \alpha, \beta \in \text{str}`P . \exists ! \alpha \cap \beta . \supset . \alpha \cap \beta \in \text{str}`P$

Dem.

$\vdash . *215\cdot2 . \supset \vdash : \text{Hp} . \supset . (\exists x) . x \in \alpha \cap \beta . P``\alpha \subset \alpha \cup \overrightarrow{P}`x . P``\beta \subset \beta \cup \overrightarrow{P}`x .$
$\breve{P}``\alpha \subset \alpha \cup \overleftarrow{P}`x . \breve{P}``\beta \subset \beta \cup \overleftarrow{P}`x .$

$[*22\cdot68] \supset . (\exists x) . x \in \alpha \cap \beta . P``\alpha \cap P``\beta \subset (\alpha \cap \beta) \cup \overrightarrow{P}`x .$
$\breve{P}``\alpha \cap \breve{P}``\beta \subset (\alpha \cap \beta) \cup \overleftarrow{P}`x .$

$[*37\cdot21] \supset . (\exists x) . x \in \alpha \cap \beta . P``(\alpha \cap \beta) \subset (\alpha \cap \beta) \cup \overrightarrow{P}`x . \breve{P}``(\alpha \cap \beta) \subset (\alpha \cap \beta) \cup \overleftarrow{P}`x .$

$[*22\cdot69] \supset . (\exists x) . x \in \alpha \cap \beta . P``(\alpha \cap \beta) \cap \breve{P}``(\alpha \cap \beta) \subset (\alpha \cap \beta) \cup (\overrightarrow{P}`x \cap \overleftarrow{P}`x)$ (1)

$\vdash . *37\cdot18 . \supset \vdash : x \in \alpha \cap \beta . \supset . \overrightarrow{P}`x \subset P``\alpha \cap P``\beta . \overleftarrow{P}`x \subset \breve{P}``\alpha \cap \breve{P}``\beta .$

$[*22\cdot49] \quad \supset . \overrightarrow{P}`x \cap \overleftarrow{P}`x \subset P``\alpha \cap \breve{P}``\alpha \cap P``\beta \cap \breve{P}``\beta$ (2)

$\vdash . (2) . *215\cdot1 . \supset \vdash :. \text{Hp} . \supset : x \in \alpha \cap \beta . \supset . \overrightarrow{P}`x \cap \overleftarrow{P}`x \subset \alpha \cap \beta$ (3)

$\vdash . (1) . (3) . \quad \supset \vdash : \text{Hp} . \supset . (\exists x) . x \in \alpha \cap \beta . P``(\alpha \cap \beta) \cap \breve{P}``(\alpha \cap \beta) \subset \alpha \cap \beta .$

$[*215\cdot1] \quad \supset . \alpha \cap \beta \in \text{str}`P : \supset \vdash . \text{Prop}$

$*215\cdot22$. $\vdash : \alpha, \beta \in \text{str}`P . \supset . \alpha \cap \beta \in \text{str}`P$

Dem.

$\vdash . *215\cdot1 . \supset \vdash : \text{Hp} . \supset . \alpha \subset C`P . \beta \subset C`P . P``\alpha \cap \breve{P}``\alpha \subset \alpha . P``\beta \cap \breve{P}``\beta \subset \beta .$

$[*22\cdot47\cdot49] \quad \supset . \alpha \cap \beta \subset C`P . P``\alpha \cap P``\beta \cap \breve{P}``\alpha \cap \breve{P}``\beta \subset \alpha \cap \beta .$

$[*37\cdot21] \quad \supset . \alpha \cap \beta \subset C`P . P``(\alpha \cap \beta) \cap \breve{P}``(\alpha \cap \beta) \subset \alpha \cap \beta .$

$[*215\cdot1] \quad \supset . \alpha \cap \beta \in \text{str}`P : \supset \vdash . \text{Prop}$

$*215\cdot23$. $\vdash : P \in \text{connex} . \mu \subset \text{str}`P . \exists ! p`\mu . \supset . s`\mu \in \text{str}`P$

Dem.

$\vdash . *215\cdot2 . \supset \vdash :. \text{Hp} . x \in p`\mu . \supset : \alpha \in \mu . \supset_\alpha . P``\alpha \subset \alpha \cup \overrightarrow{P}`x . \breve{P}``\alpha \subset \alpha \cup \overleftarrow{P}`x :$

$[*40\cdot13] \quad \supset : \alpha \in \mu . \supset_\alpha . P``\alpha \subset s`\mu \cup \overrightarrow{P}`x . \breve{P}``\alpha \subset s`\mu \cup \overleftarrow{P}`x :$

$[*40\cdot43\cdot38] \quad \supset : P``s`\mu \subset s`\mu \cup \overrightarrow{P}`x . \breve{P}``s`\mu \subset s`\mu \cup \overleftarrow{P}`x :$

$[*22\cdot49\cdot69] \quad \supset : P``s`\mu \cap \breve{P}``s`\mu \subset s`\mu \cup (\overrightarrow{P}`x \cap \overleftarrow{P}`x)$ (1)

$\vdash . *40 \cdot 14 . \supset \vdash : \text{Hp} . x \in p`\mu . \alpha \in \mu . \supset . x \in \alpha . \alpha \in \text{str}`P .$

[*37·18] $\supset . \overrightarrow{P}`x \cap \overleftarrow{P}`x \subset P``\alpha \cap \breve{P}``\alpha . \alpha \in \text{str}`P .$

[*215·1] $\supset . \overrightarrow{P}`x \cap \overleftarrow{P}`x \subset \alpha .$

[*40·13] $\supset . \overrightarrow{P}`x \cap \overleftarrow{P}`x \subset s`\mu$ (2)

$\vdash . (1) . (2) . \supset \vdash : \text{Hp} . \exists ! \mu . \supset . P``s`\mu \cap \breve{P}``s`\mu \subset s`\mu$ (3)

$\vdash . *37 \cdot 29 . \supset \vdash : \mu = \Lambda . \supset . P``s`\mu \cap \breve{P}``s`\mu \subset s`\mu$ (4)

$\vdash . (3) . (4) . \supset \vdash .$ Prop

***215·24**. $\vdash : \mu \subset \text{str}`P . \supset . C`P \cap p`\mu \in \text{str}`P$

Dem.

$\vdash . *37 \cdot 265 . \supset \vdash . P``(p`\mu \cap C`P) \cap \breve{P}``(p`\mu \cap C`P) = P``p`\mu \cap \breve{P}``p`\mu$ (1)

$\vdash . *37 \cdot 2 . \supset \vdash : \alpha \in \mu . \supset . P``p`\mu \cap \breve{P}``p`\mu \subset P``\alpha \cap \breve{P}``\alpha$ (2)

$\vdash . (2) . *215 \cdot 1 . \supset \vdash :. \text{Hp} . \supset : \alpha \in \mu . \supset . P``p`\mu \cap \breve{P}``p`\mu \subset \alpha :$

[*40·15] $\supset : P``p`\mu \cap \breve{P}``p`\mu \subset p`\mu$ (3)

$\vdash . (1) . (3) . *215 \cdot 1 . \supset \vdash .$ Prop

***215·25**. $\vdash : \mu \subset \text{str}`P . \exists ! \mu . \supset . p`\mu \in \text{str}`P$

Dem.

$\vdash . *40 \cdot 24 . *215 \cdot 1 . \supset \vdash : \text{Hp} . \supset . p`\mu \subset C`P$ (1)

$\vdash . (1) . *215 \cdot 24 . \supset \vdash .$ Prop

***215·3**. $\vdash :. P \in \text{connex} . \alpha, \beta \in \text{str}`P - \iota`\Lambda . \alpha \cap \beta = \Lambda . \supset :$

$\alpha \subset P``\beta . \equiv . \alpha \subset p`\overrightarrow{P}``\beta . \equiv . \beta \subset p`\overleftarrow{P}``\alpha . \equiv . \beta \subset \breve{P}``\alpha$

Dem.

$\vdash . *215 \cdot 1 . \supset \vdash : \text{Hp} . \supset . \alpha \subset C`P - \beta$ (1)

$\vdash . *22 \cdot 48 . \supset \vdash : \alpha \subset P``\beta . \supset . \alpha \cap \breve{P}``\beta \subset P``\beta \cap \breve{P}``\beta :$

[*215·1] $\supset \vdash : \text{Hp} . \alpha \subset P``\beta . \supset . \alpha \cap \breve{P}``\beta \subset \beta .$

[*22·621.Hp] $\supset . \alpha \cap \breve{P}``\beta = \Lambda$ (2)

$\vdash . (1) . (2) . \supset \vdash : \text{Hp} . \alpha \subset P``\beta . \supset . \alpha \subset C`P - \beta - \breve{P}``\beta .$

[*202·501] $\supset . \alpha \subset p`\overrightarrow{P}``\beta$ (3)

$\vdash . *40 \cdot 61 . \supset \vdash :. \text{Hp} . \alpha \subset p`\overrightarrow{P}``\beta . \supset . \alpha \subset P``\beta$ (4)

$\vdash . (3) . (4) . \supset \vdash :. \text{Hp} . \supset : \alpha \subset P``\beta . \equiv . \alpha \subset p`\overrightarrow{P}``\beta .$ (5)

[*40·67] $\equiv . \beta \subset p`\overleftarrow{P}``\alpha .$ (6)

$\left[(5) \frac{\breve{P}}{P}\right]$ $\equiv . \beta \subset \breve{P}``\alpha$ (7)

$\vdash . (5) . (6) . (7) . \supset \vdash .$ Prop

***215·31.** $\vdash : P \in \text{trans} \cap \text{connex} \,.\, \alpha \in \text{str}'P \,.\, E!\,\text{min}_P{}'\alpha \,.\, E!\,\text{max}_P{}'\alpha \,.\, \supset .$
$\alpha = P\,(\text{min}_P{}'\alpha \mapsto \text{max}_P{}'\alpha)$

Dem.

$\vdash . *205{\cdot}2 . *90{\cdot}15{\cdot}151 . \supset \vdash : \text{Hp} . y \in \alpha . \supset . \text{min}_P{}'\alpha \, P_* y$ (1)

$\vdash . (1) \frac{\breve{P}}{P} . *205{\cdot}102 . \quad \supset \vdash : \text{Hp} . y \in \alpha . \supset . y P_* \, \text{max}_P{}'\alpha$ (2)

$\vdash . (1) . (2) . *121{\cdot}103 . \quad \supset \vdash : \text{Hp} . \supset . \alpha \subset P\,(\text{min}_P{}'\alpha \mapsto \text{max}_P{}'\alpha)$ (3)

$\vdash . *121{\cdot}242 . *201{\cdot}19 . *205{\cdot}2 . \supset \vdash : \text{Hp} . \supset .$

$P\,(\text{min}_P{}'\alpha \mapsto \text{max}_P{}'\alpha) = \iota'\text{min}_P{}'\alpha \cup (\overleftarrow{P}'\text{min}_P{}'\alpha \cap \overrightarrow{P}'\text{max}_P{}'\alpha) \cup \iota'\text{max}_P{}'\alpha$

[*37·18] $\subset \iota'\text{min}_P{}'\alpha \cup (\breve{P}''\alpha \cap P''\alpha) \cup \iota'\text{max}_P{}'\alpha$

[*205·11·111.*215·1] $\subset \alpha$ (4)

$\vdash . (3) . (4) . \supset \vdash . \text{Prop}$

***215·32.** $\vdash : P \in \text{trans} \cap \text{connex} \,.\, \alpha \in \text{str}'P \,.\, E!\,\text{min}_P{}'\alpha \,.\, E!\,\text{seq}_P{}'\alpha \,.\, \supset .$
$\alpha = P\,(\text{min}_P{}'\alpha \vdash \text{seq}_P{}'\alpha)$

Dem.

$\vdash . *206{\cdot}211 . *205{\cdot}2 . \supset \vdash : \text{Hp} . \supset . \alpha \subset \overleftarrow{P_*}'\text{min}_P{}'\alpha \cap \overrightarrow{P}'\text{seq}_P{}'\alpha$ (1)

$\vdash . *206{\cdot}22 . *205{\cdot}22 . \supset \vdash : \text{Hp} . \supset . \overleftarrow{P}'\text{min}_P{}'\alpha \cap \overrightarrow{P}'\text{seq}_P{}'\alpha = \breve{P}''\alpha \cap (\alpha \cup P''\alpha)$

[*215·1] $\subset \alpha .$

[*201·19.*121·241] $\supset . P\,(\text{min}_P{}'\alpha \vdash \text{seq}_P{}'\alpha) \subset \alpha$ (2)

$\vdash . (1) . (2) . \supset \vdash . \text{Prop}$

***215·33.** $\vdash : P \in \text{trans} \cap \text{connex} \,.\, \alpha \in \text{str}'P \,.\, E!\,\text{prec}_P{}'\alpha \,.\, E!\,\text{seq}_P{}'\alpha \,.\, \supset .$
$\alpha = P\,(\text{prec}_P{}'\alpha - \text{seq}_P{}'\alpha)$ [*206·22 . *215·1]

***215·4.** $\vdash : P \in \text{connex} \,.\, \mu \in \text{Cl excl}'(\text{str}'P - \iota'\Lambda) \,.\, \supset .\, P_{\text{cl}} \upharpoonright \mu = P_{\text{lc}} \upharpoonright \mu$

Dem.

$\vdash . *84{\cdot}12 . \supset \vdash :. \text{Hp} . \supset : \alpha, \beta \in \mu . \alpha \neq \beta . \supset . \alpha \cap \beta = \Lambda :$ (1)

[*170·1] $\supset : \alpha\,(P_{\text{cl}} \upharpoonright \mu)\,\beta . \equiv . \alpha, \beta \in \mu . \exists!\,\alpha - \breve{P}''\beta .$

[*215·3.Transp] $\equiv . \alpha, \beta \in \mu . \exists!\,\beta - P''\alpha .$

[(1).*170·102] $\equiv . \alpha, \beta \in \mu . \alpha \, P_{\text{lc}} \mu :. \supset \vdash . \text{Prop}$

***215·41.** $\vdash : P \in \text{trans} \cap \text{connex} \,.\, \mu \in \text{Cl excl}'(\text{str}'P - \iota'\Lambda) \,.\, \supset .\, P_{\text{lc}} \upharpoonright \mu \in \text{Ser}$

Dem.

$\vdash . *84{\cdot}12 . *170{\cdot}102 . \supset \vdash :. \text{Hp} . \supset : \alpha\,(P_{\text{lc}} \upharpoonright \mu)\,\beta . \equiv . \exists!\,\beta - P''\alpha$ (1)

$\vdash . *215{\cdot}3 . \supset$

$\vdash : \text{Hp} . \alpha, \beta \in \mu . \alpha \subset P''\beta . \beta \subset P''\alpha . \supset . \alpha \subset P''\beta . \alpha \subset \breve{P}''\beta .$

[*215·1] $\supset . \alpha \subset \beta$ (2)

Similarly $\vdash : \text{Hp}\,(2) . \supset . \beta \subset \alpha$ (3)

$\vdash . (2) . (3) . \supset \vdash :: \text{Hp} . \alpha, \beta \epsilon \mu . \supset :. \alpha \subset P``\beta . \beta \subset P``\alpha . \supset . \alpha = \beta :.$

[Transp.(1)] $\supset :. \alpha \neq \beta . \supset : \alpha (P_{\text{lc}} \upharpoonright \mu) \beta . \text{v} . \beta (P_{\text{lc}} \upharpoonright \mu) \alpha$ (4)

$\vdash . *37 \cdot 1 . \supset$

$\vdash :: \text{Hp} . \beta \cap \gamma = \Lambda . \sim (\gamma \subset P``\beta) . \supset :. (\exists z) : z \epsilon \gamma : y \epsilon \beta . \supset_y . \sim (zPy) . z \neq y :.$

[*202·103] $\supset :. (\exists z) : z \epsilon \gamma : y \epsilon \beta . \supset_y . yPz :.$

[*11·61] $\supset :. y \epsilon \beta . \supset_y . (\exists z) . z \epsilon \gamma . yPz :.$

[*37·1] $\supset :. \beta \subset P``\gamma :.$

[*201·5.*37·2] $\supset :. \gamma \subset P``\alpha . \supset . \beta \subset P``\alpha :.$

[Transp] $\supset :. \exists ! \beta - P``\alpha . \supset . \exists ! \gamma - P``\alpha$ (5)

$\vdash . (5) . (1) . \supset \vdash :. \text{Hp} . \supset : \alpha (P_{\text{lc}} \upharpoonright \mu) \beta . \beta (P_{\text{lc}} \upharpoonright \mu) \gamma . \supset . \alpha (P_{\text{lc}} \upharpoonright \mu) \gamma$ (6)

$\vdash . (4) . (6) . *170 \cdot 17 . \supset \vdash . \text{Prop}$

***215·42.** $\vdash : P \epsilon \text{trans} \cap \text{connex} . \mu \epsilon \text{Cl excl}`(\text{str}`P - \iota`\Lambda) . \mu \sim \epsilon 1 . \supset . C`P_{\text{lc}} \upharpoonright \mu = \mu$

[*202·55 . *215·41]

***215·5.** $\vdash :. P \epsilon \text{trans} \cap \text{connex} . \alpha \epsilon \text{sect}`P . \beta \epsilon \text{sect}`\breve{P} . \supset :$

$\exists ! \alpha \cap \beta . \text{limax}_P`\alpha = \text{limin}_P`\beta . \supset . \alpha \cap \beta \epsilon 1$ [*207·71 . *215·164]

***215·51.** $\vdash : P \epsilon \text{Ser} . \alpha \epsilon \text{sect}`P . \beta \epsilon \text{sect}`\breve{P} . \alpha \cap \beta \epsilon 1 . \supset .$

$\text{limax}_P`\alpha = \text{limin}_P`\beta = \breve{\iota}`(\alpha \cap \beta)$ [*207·72 . *215·164]

***215·52.** $\vdash : \text{Hp} *215 \cdot 5 . \alpha \cap \beta \sim \epsilon 0 \cup 1 . \text{E} ! \text{limax}_P`\alpha . \text{E} ! \text{limin}_P`\beta . \supset .$

$\text{limin}_P`\beta \, P \, \text{limax}_P`\alpha$

Dem.

$\vdash . *215 \cdot 164 . \supset \vdash :. \text{Hp} . \supset : \text{limax}_P`\alpha = \text{max}_P`(\alpha \cap \beta) . \text{v} . \text{limax}_P`\alpha = \text{seq}_P`(\alpha \cap \beta) :$

$\text{limin}_P`\beta = \text{min}_P`(\alpha \cap \beta) . \text{v} . \text{limin}_P`\beta = \text{prec}_P`(\alpha \cap \beta)$ (1)

$\vdash . *205 \cdot 732 . \supset \vdash : \text{Hp} . \text{limax}_P`\alpha = \text{max}_P`(\alpha \cap \beta) . \text{limin}_P`\beta = \text{min}_P`(\alpha \cap \beta) . \supset .$

$\text{limin}_P`\beta \, P \, \text{limax}_P`\alpha$ (2)

$\vdash . *206 \cdot 15 . \supset \vdash : \text{Hp} . \text{limax}_P`\alpha = \text{seq}_P`(\alpha \cap \beta) . \text{limin}_P`\beta = \text{min}_P`(\alpha \cap \beta) . \supset .$

$\text{limin}_P`\beta \, P \, \text{limax}_P`\alpha$ (3)

$\vdash . (3) \dfrac{\breve{P}, \beta, \alpha}{P, \alpha, \beta} . \supset \vdash : \text{Hp} . \text{limax}_P`\alpha = \text{max}_P`(\alpha \cap \beta) . \text{limin}_P`\beta = \text{prec}_P`(\alpha \cap \beta) . \supset .$

$\text{limin}_P`\beta \, P \, \text{limax}_P`\alpha$ (4)

$\vdash . *206 \cdot 73 . \supset \vdash : \text{Hp} . \text{limax}_P`\alpha = \text{seq}_P`(\alpha \cap \beta) . \text{limin}_P`\beta = \text{prec}_P`(\alpha \cap \beta) . \supset .$

$\text{limin}_P`\beta \, P \, \text{limax}_P`\alpha$ (5)

$\vdash . (1) . (2) . (3) . (4) . (5) . \supset \vdash . \text{Prop}$

***215·53.** $\vdash : \text{Hp} *215\cdot5 \,.\, \alpha \cap \beta = \Lambda \,.\, \text{E}\,!\,\text{limax}_P\text{‘}\alpha \,.\, \text{E}\,!\,\text{limin}_P\text{‘}\beta \,.\, \supset .$

$\text{limax}_P\text{‘}\alpha \, P_* \, \text{limin}_P\text{‘}\beta$

Dem.

$\vdash . *207\cdot2 . *205\cdot22 . \supset \vdash : \text{Hp} . \supset . \overrightarrow{P}\text{‘limax}_P\text{‘}\alpha \subset P\text{‘‘}\alpha . \overleftarrow{P}\text{‘limin}_P\text{‘}\beta \subset \breve{P}\text{‘‘}\beta .$

$[*211\cdot1] \qquad \supset . \overrightarrow{P}\text{‘limax}_P\text{‘}\alpha \subset P\text{‘‘}\alpha . \overleftarrow{P}\text{‘limin}_P\text{‘}\beta \subset \beta \qquad (1)$

$\vdash . (1) . *37\cdot1 . \supset \vdash : \text{Hp} . \text{limin}_P\text{‘}\beta \, P \, \text{limax}_P\text{‘}\alpha . \supset . (\exists x) . x \in \alpha . \text{limin}_P\text{‘}\beta \, Px .$

$[(1)] \qquad \supset . \exists \,!\, \alpha \cap \beta \qquad (2)$

$\vdash . (2) . \text{Transp} . \supset \vdash . \text{Prop}$

***215·54.** $\vdash : P \in \text{Ser} . \alpha \in \text{sect‘}P . \beta \in \text{sect‘}\breve{P} . \alpha \cap \beta = \Lambda . \alpha \cup \beta = C\text{‘}P .$

$\text{E}\,!\,\text{limax}_P\text{‘}\alpha . \text{E}\,!\,\text{limin}_P\text{‘}\beta . \supset : \text{limax}_P\text{‘}\alpha = \text{limin}_P\text{‘}\beta . \vee .$

$\text{limax}_P\text{‘}\alpha \, P_1 \, \text{limin}_P\text{‘}\beta$

Dem.

$\vdash . *211\cdot726 . \supset \vdash : \text{Hp} . \text{E}\,!\,\text{max}_P\text{‘}\alpha . \text{E}\,!\,\text{min}_P\text{‘}\beta . \supset .$

$\text{limax}_P\text{‘}\alpha = \text{max}_P\text{‘}\alpha . \text{limin}_P\text{‘}\beta = \text{seq}_P\text{‘}\alpha .$

$[*206\cdot5] \qquad \supset . \text{limax}_P\text{‘}\alpha \, P_1 \, \text{limin}_P\text{‘}\beta \qquad (1)$

$\vdash . *211\cdot726 . \supset \vdash : \text{Hp} . \sim \text{E}\,!\,\text{max}_P\text{‘}\alpha . \supset .$

$\text{limax}_P\text{‘}\alpha = \text{min}_P\text{‘}\beta . \text{limin}_P\text{‘}\beta = \text{min}_P\text{‘}\beta \qquad (2)$

$\vdash . *211\cdot726 . \supset \vdash : \text{Hp} . \sim \text{E}\,!\,\text{min}_P\text{‘}\beta . \supset .$

$\text{limax}_P\text{‘}\alpha = \text{max}_P\text{‘}\alpha . \text{limin}_P\text{‘}\beta = \text{max}_P\text{‘}\alpha \qquad (3)$

$\vdash . (1) . (2) . (3) . \supset \vdash . \text{Prop}$

***215·541.** $\vdash :: P \in \text{Ser} . \alpha \in \text{sect‘}P . \beta \in \text{sect‘}\breve{P} . \alpha \cup \beta = C\text{‘}P . \supset :.$

$\alpha \cap \beta \in 0 \cup 1 . \supset : \text{E}\,!\,\text{limax}_P\text{‘}\alpha . \equiv . \text{E}\,!\,\text{limin}_P\text{‘}\beta \quad [*211\cdot727 . *215\cdot51]$

***215·542.** $\vdash : \text{Hp} *215\cdot541 . \alpha \cap \beta = \Lambda . \text{E}\,!\,\text{limax}_P\text{‘}\alpha . \text{limax}_P\text{‘}\alpha \sim \in \text{D‘}P_1 . \supset .$

$\text{limax}_P\text{‘}\alpha = \text{limin}_P\text{‘}\beta \quad [*215\cdot54 . *211\cdot727]$

***215·543.** $\vdash : P \in \text{Ser} . \alpha \in \text{sect‘}P . \beta \in \text{sect‘}\breve{P} . \alpha \cup \beta = C\text{‘}P . \alpha \cap \beta \in 0 \cup 1 .$

$\text{E}\,!\,\text{limax}_P\text{‘}\alpha . \text{limax}_P\text{‘}\alpha \sim \in \text{D‘}P_1 . \supset . \text{limax}_P\text{‘}\alpha = \text{limin}_P\text{‘}\beta \quad [*215\cdot542\cdot51]$

*216. DERIVATIVES.

Summary of *216.

If α is any class, and P is any series, the *derivative* (or *first derivative*) of α with respect to P is the class of limits of existent sub-classes of $\alpha \cap C'P$, *i.e.* $\text{lt}_P``\text{Cl ex}`(\alpha \cap C'P)$. That is, a term $x$ belongs to the derivative of $\alpha$ if a set of terms exists which is contained both in $\alpha$ and in $C'P$, and has $x$ for its limit. The derivative of $\alpha$ with respect to $P$ will be denoted by $\delta_P`\alpha$.

In general, there will be members of α not contained in $\delta_P`\alpha$, and members of $\delta_P`\alpha$ not contained in α. α is said to be *dense* in P if all its terms except the first (if there is a first) belong to $\delta_P`\alpha$, that is, if all its terms except the first are limits of existent classes contained in α. α is said to be *closed* in P if every existent sub-class of α which has no maximum has a limit which belongs to α, *i.e.* if every existent sub-class of α has a limit or a maximum, and the derivative of α is contained in α. If α is both dense and closed, it is called *perfect*. In this case, all its terms are limits of classes chosen out of α, and every class chosen out of α has a limit or maximum in α.

The second derivative of α is $\delta_P`\delta_P`\alpha$, *i.e.* $\delta_P{}^2`\alpha$, and so on. (Derivatives of infinite order cannot be dealt with till a later stage.) If P is serial, the second derivative of α is always contained in the first (*216·14).

If P is a Dedekindian series, α is closed whenever $\delta_P`\alpha \subset \alpha$. In order to secure a Dedekindian series, it is sometimes convenient to replace $P$ by the ordinally similar series $\overrightarrow{P};P$, which is contained in the Dedekindian series $\varsigma`P$. Then α is replaced by $\overrightarrow{P}``\alpha$, and $\alpha$ is closed if the derivative of $\overrightarrow{P}``\alpha$ with respect to $\varsigma`P$ is contained in $\overrightarrow{P}``\alpha$. The relation of the derivative of $\alpha$ in $P$ to the derivative of $\overrightarrow{P}``\alpha$ in $\varsigma`P$ has been treated in *212·6 and following propositions. This subject is resumed below (*216·5 ff.).

The derivative of the series P will be defined as the series of its limit-points, and denoted by $\nabla`P$. Thus we put

$$\nabla`P = P \upharpoonright D`\text{lt}_P.$$

If P is a series, the derivative of a class α consists of those members x of $\text{C}'P$ which are such that members of α exist in every interval which ends in x, *i.e.*

***216·13.** $\vdash :: P \epsilon \text{Ser} . \supset :. x \epsilon \delta_P{}'\alpha . \equiv : x \epsilon \text{C}'P : yPx . \supset_y . \exists ! \alpha \cap \overleftarrow{P}'y \cap \overrightarrow{P}'x$

We have

***216·2.** $\vdash . \delta_P{}'C'P = \text{D}'\text{lt}_P - \overrightarrow{B}'P$

***216·3.** $\vdash : \alpha \epsilon \text{dense}'P . \equiv . \alpha - \overrightarrow{\text{min}}_P{}'\alpha \subset \delta_P{}'\alpha$

***216·32.** $\vdash : \alpha \epsilon \text{closed}'P . \equiv . \text{Cl ex}'(\alpha \cap C'P) \subset \text{C}'\text{limax}_P . \delta_P{}'\alpha \subset \alpha$

We prove (*216·4—·412) that the properties of α with respect to P, as regards being dense, closed, or perfect, belong to $\breve{S}``\alpha$ with respect to Q if S is a correlator of P with Q.

We next consider the relation of α in P to $\overrightarrow{P}``\alpha$ in $\varsigma'P$ (*216·5—·56). The point of these propositions is that $\varsigma'P$ is Dedekindian, so that a class is closed in $\varsigma'P$ if it contains its first derivative. (It is usual to *define* a class as closed whenever it contains its first derivative; but this involves the tacit assumption that the series P is Dedekindian. If P is the series of real numbers, this assumption is of course verified.) We prove (*216·52) that the derivative of $\overrightarrow{P}``\alpha$ in $\varsigma'P$ is $P```(\text{Cl ex}'\alpha - \text{C}'\text{max}_P)$, *i.e.* is the class of segments defined by such existent sub-classes of α as have no maximum; we show that α is dense, closed, or perfect in P according as $\overrightarrow{P}``\alpha$ is dense, closed, or perfect in $\varsigma'P$ (*216·53·54·56), and that $\alpha$ and $\overrightarrow{P}``\alpha$ are closed if $\overrightarrow{P}``\alpha$ contains its first derivative (*216·54).

We end with various propositions on $\nabla'P$ (*216·6—·621), of which the chief is

***216·611.** $\vdash : P \epsilon \text{Ser} . \dot{\exists} ! \nabla'P . \supset . C'\nabla'P = C'P - \text{C}'P_1 = \delta_P{}'C'P \cup \overrightarrow{B}'P$

This subject will be resumed in connection with well-ordered series in *264.

***216·01.** $\delta_P{}'\alpha = \text{lt}_P``\text{Cl ex}'(\alpha \cap C'P)$ Df

***216·02.** $\text{dense}'P = \hat{\alpha}(\alpha - \overrightarrow{\text{min}}_P{}'\alpha \subset \delta_P{}'\alpha)$ Df

***216·03.** $\text{closed}'P = \hat{\alpha}\{\text{Cl ex}'(\alpha \cap C'P) \subset \text{C}'\text{limax}_P . \delta_P{}'\alpha \subset \alpha\}$ Df

***216·04.** $\text{perf}'P = \text{dense}'P \cap \text{closed}'P$ Df

***216·05.** $\nabla'P = P \upharpoonright \text{D}'\text{lt}_P$ Df

***216·1.** $\vdash : x \epsilon \delta_P{}'\alpha . \equiv . (\exists\beta) . \beta \subset \alpha \cap C'P . \exists ! \beta . x \,\text{lt}_P\, \beta$ [(*216·01)]

$*216\cdot101$. $\vdash : x \in \delta_P`\alpha . \equiv . (\exists\beta) . \beta \subset \alpha . \exists ! \beta . \beta \subset P``\beta . x \,\mathrm{seq}_P\, \beta$

Dem.

$\vdash . *216\cdot1 . *207\cdot1 . \supset$

$\vdash : x \in \delta_P`\alpha . \equiv . (\exists\beta) . \beta \subset \alpha \cap C`P . \exists ! \beta . \beta \cap C`P \subset P``\beta . x \,\mathrm{seq}_P\, \beta .$

$[*37\cdot15] \quad \equiv . (\exists\beta) . \beta \subset \alpha . \exists ! \beta . \beta \subset P``\beta . x \,\mathrm{seq}_P\, \beta : \supset \vdash . \mathrm{Prop}$

$*216\cdot11$. $\vdash . \delta_P`\alpha \subset \breve{P}``\alpha$

Dem.

$\vdash . *216\cdot101 . *206\cdot142 . \supset \vdash : x \in \delta_P`\alpha . \supset . (\exists\beta) . \beta \subset \alpha . \exists ! \beta . x \in \breve{P}``\beta .$

$[*37\cdot2] \quad \supset . x \in \breve{P}``\alpha : \supset \vdash . \mathrm{Prop}$

$*216\cdot111$. $\vdash . \delta_P`\alpha \subset \text{Ɑ}`P \quad [*216\cdot11 . *37\cdot16]$

$*216\cdot12$. $\vdash . \delta_P`\alpha = \delta_P`(\alpha \cap C`P) \quad [*22\cdot5 . (*216\cdot01)]$

$*216\cdot13$. $\vdash :: P \in \mathrm{Ser} . \supset :. x \in \delta_P`\alpha . \equiv : x \in \text{Ɑ}`P : yPx . \supset_y . \exists ! \alpha \cap \overleftarrow{P}`y \cap \overrightarrow{P}`x$

Dem.

$\vdash . *206\cdot173 . *216\cdot101 . \supset \vdash :: P \in \mathrm{connex} . P^2 \in J . \supset :.$

$x \in \delta_P`\alpha . \equiv : (\exists\beta) . \beta \subset \alpha . \exists ! \beta . \beta \subset \overrightarrow{P}`x . \overrightarrow{P}`x \subset P``\beta :$

$[*37\cdot46] \equiv : (\exists\beta) : \beta \subset \alpha . \exists ! \beta . \beta \subset \overrightarrow{P}`x : yPx . \supset_y . \exists ! \beta \cap \overleftarrow{P}`y :$

$[*24\cdot58] \supset : yPx . \supset_y . \exists ! \alpha \cap \overrightarrow{P}`x \cap \overleftarrow{P}`y \quad (1)$

$\vdash . *33\cdot41\cdot152 . \supset$

$\vdash :. x \in \text{Ɑ}`P : yPx . \supset_y . \exists ! \alpha \cap \overrightarrow{P}`x \cap \overleftarrow{P}`y : \supset : \exists ! \alpha \cap \overrightarrow{P}`x . \alpha \cap \overrightarrow{P}`x \subset \alpha \cap C`P :$

$[*216\cdot1] \quad \supset : x \,\mathrm{lt}_P\, (\alpha \cap \overrightarrow{P}`x) . \supset . x \in \delta_P`\alpha \quad (2)$

$\vdash . *37\cdot2 . *201\cdot501 . \supset \vdash : \mathrm{Hp} . \supset . P``(\alpha \cap \overrightarrow{P}`x) \subset \overrightarrow{P}`x \quad (3)$

$\vdash . *50\cdot24 . \quad \supset \vdash : \mathrm{Hp} . \supset . x \sim\in (\alpha \cap \overrightarrow{P}`x) \quad (4)$

$\vdash . (3) . (4) . *207\cdot232 . \supset$

$\vdash :: \mathrm{Hp} . x \in \text{Ɑ}`P . \supset :. x \,\mathrm{lt}_P\, (\alpha \cap \overrightarrow{P}`x) . \equiv : \overrightarrow{P}`x \subset P``(\alpha \cap \overrightarrow{P}`x) :$

$[*37\cdot46] \quad \equiv : yPx . \supset_y . \exists ! \alpha \cap \overrightarrow{P}`x \cap \overleftarrow{P}`y \quad (5)$

$\vdash . (2) . (5) . \supset \vdash :. \mathrm{Hp} . x \in \text{Ɑ}`P : yPx . \supset_y . \exists ! \alpha \cap \overrightarrow{P}`x \cap \overleftarrow{P}`y : \supset . x \in \delta_P`\alpha \quad (6)$

$\vdash . (1) . (6) . *216\cdot111 . \supset \vdash . \mathrm{Prop}$

$*216\cdot14$. $\vdash : P \in \mathrm{Ser} . \supset . \delta_{P^2}`\alpha \subset \delta_P`\alpha$

Dem.

$\vdash . *71\cdot47 . \supset \vdash :. \mathrm{Hp} . \supset : \beta \subset \mathrm{lt}_P``\mathrm{Cl\,ex}`\alpha . \exists ! \beta . \supset .$

$(\exists\kappa) . \kappa \subset \mathrm{Cl\,ex}`\alpha . \beta = \mathrm{lt}_P``\kappa . \exists ! \beta .$

$[*37\cdot26] \quad \supset . (\exists\lambda) . \lambda \subset \mathrm{Cl\,ex}`\alpha \cap \text{Ɑ}`\mathrm{lt}_P . \beta = \mathrm{lt}_P``\lambda . \exists ! \beta .$

$[*207{\cdot}54] \supset . (\exists\lambda) . \lambda \subset \mathrm{Cl\ ex}`\alpha \cap \mathrm{Ɑ}`\mathrm{lt}_P . \beta = \mathrm{lt}_P``\lambda . \exists ! \beta . \overrightarrow{\mathrm{limax}}_P`\beta = \overrightarrow{\mathrm{lt}}_P`s`\lambda .$

$[*216{\cdot}1 . *37{\cdot}29 . *53{\cdot}24 . \mathrm{Transp}] \supset . \overrightarrow{\mathrm{limax}}_P`\beta \subset \delta_P`\alpha .$

$[*207{\cdot}45] \quad \supset . \overrightarrow{\mathrm{lt}}_P`\beta \subset \delta_P`\alpha \quad (1)$

$\vdash . (1) . (*216{\cdot}01) . \supset \vdash :. \mathrm{Hp} . \supset : \beta \,\epsilon\, \mathrm{Cl\ ex}`\delta_P`\alpha . \supset . \overrightarrow{\mathrm{lt}}_P`\beta \subset \delta_P`\alpha :$

$[*40{\cdot}43{\cdot}5] \quad \supset : \mathrm{lt}_P``\mathrm{Cl\ ex}`\delta_P`\alpha \subset \delta_P`\alpha :. \supset \vdash . \mathrm{Prop}$

***216·15.** $\vdash : \alpha \subset \beta . \supset . \delta_P`\alpha \subset \delta_P`\beta \quad [*37{\cdot}2 . (*216{\cdot}01)]$

***216·16.** $\vdash : P \,\epsilon\, \mathrm{trans} \cap \mathrm{connex} . \supset . \delta_P`\alpha = \delta_P`(\alpha - \overrightarrow{\mathrm{min}}_P`\alpha)$

Dem.

$\vdash . *24{\cdot}26{\cdot}101 . \supset \vdash : \overrightarrow{\mathrm{min}}_P`\alpha = \Lambda . \supset . \delta_P`\alpha = \delta_P`(\alpha - \overrightarrow{\mathrm{min}}_P`\alpha) \quad (1)$

$\vdash . *51{\cdot}36 . \supset \vdash : \beta \subset \alpha . \exists ! \beta . \mathrm{E} ! \mathrm{min}_P`\alpha . \mathrm{min}_P`\alpha \sim \epsilon\, \beta . \supset .$

$\beta \,\epsilon\, \mathrm{Cl\ ex}`(\alpha - \iota`\mathrm{min}_P`\alpha) .$

$[*37{\cdot}18] \quad \supset . \overrightarrow{\mathrm{lt}}_P`\beta \subset \delta_P`(\alpha - \iota`\mathrm{min}_P`\alpha) \quad (2)$

$\vdash . *205{\cdot}5 . \supset \vdash : \mathrm{Hp} . \beta \subset \alpha . \exists ! \beta . \mathrm{min}_P`\alpha \,\epsilon\, \beta . \supset . \mathrm{min}_P`\alpha = \mathrm{min}_P`\beta .$

$[*207{\cdot}262] \quad \supset . \overrightarrow{\mathrm{lt}}_P`\beta \subset \overrightarrow{\mathrm{lt}}_P`(\beta - \iota`\mathrm{min}_P`\alpha) \quad (3)$

$\vdash . (3) . *37{\cdot}18 . \supset \vdash : \mathrm{Hp}\,(3) . \beta \neq \iota`\mathrm{min}_P`\alpha . \supset . \overrightarrow{\mathrm{lt}}_P`\beta \subset \delta_P`(\alpha - \iota`\mathrm{min}_P`\alpha) \quad (4)$

$\vdash . *205{\cdot}194{\cdot}8 . \supset \vdash : \mathrm{E} ! \mathrm{min}_P`\alpha . \supset . \mathrm{min}_P`\alpha = \mathrm{max}_P`\iota`\mathrm{min}_P`\alpha .$

$[*207{\cdot}11] \quad \supset . \overrightarrow{\mathrm{lt}}_P`\iota`\mathrm{min}_P`\alpha = \Lambda \quad (5)$

$\vdash . (5) . *24{\cdot}12 . \supset \vdash : \mathrm{E} ! \mathrm{min}_P`\alpha . \beta = \iota`\mathrm{min}_P`\alpha . \supset . \overrightarrow{\mathrm{lt}}_P`\beta \subset \delta_P`(\alpha - \iota`\mathrm{min}_P`\alpha) \quad (6)$

$\vdash . (4) . (6) . \supset \vdash : \mathrm{Hp}\,(3) . \supset . \overrightarrow{\mathrm{lt}}_P`\beta \subset \delta_P`(\alpha - \iota`\mathrm{min}_P`\alpha) \quad (7)$

$\vdash . (2) . (7) . \supset \vdash : \mathrm{Hp} . \beta \subset \alpha . \exists ! \beta . \mathrm{E} ! \mathrm{min}_P`\alpha . \supset . \overrightarrow{\mathrm{lt}}_P`\beta \subset \delta_P`(\alpha - \iota`\mathrm{min}_P`\alpha) \quad (8)$

$\vdash . (8) . *40{\cdot}5{\cdot}43 . \supset \vdash : \mathrm{Hp} . \mathrm{E} ! \mathrm{min}_P`\alpha . \supset . \delta_P`\alpha \subset \delta_P`(\alpha - \iota`\mathrm{min}_P`\alpha) \quad (9)$

$\vdash . (9) . *216{\cdot}15 . \supset \vdash : \mathrm{Hp} . \mathrm{E} ! \mathrm{min}_P`\alpha . \supset . \delta_P`\alpha = \delta_P`(\alpha - \iota`\mathrm{min}_P`\alpha) \quad (10)$

$\vdash . (1) . (10) . \supset \vdash . \mathrm{Prop}$

***216·2.** $\vdash . \delta_P`C`P = \mathrm{D}`\mathrm{lt}_P - \overrightarrow{B}`P$

Dem.

$\vdash . *37{\cdot}15 . *216{\cdot}111 . \supset \vdash . \delta_P`C`P \subset \mathrm{D}`\mathrm{lt}_P - \overrightarrow{B}`P \quad (1)$

$\vdash . *216{\cdot}1 . \quad \supset \vdash : x \,\epsilon\, \mathrm{D}`\mathrm{lt}_P - \delta_P`C`P . \supset . x \,\mathrm{lt}_P\, \Lambda .$

$[*207{\cdot}3] \quad \supset . x \,\epsilon\, \overrightarrow{B}`P \quad (2)$

$\vdash . (2) . \mathrm{Transp} . \quad \supset \vdash . \mathrm{D}`\mathrm{lt}_P - \overrightarrow{B}`P \subset \delta_P`C`P \quad (3)$

$\vdash . (1) . (3) . \supset \vdash . \mathrm{Prop}$

***216·21.** $\vdash : P \,\epsilon\, \mathrm{Rl}`J \cap \mathrm{connex} . \supset . \delta_P`C`P = \mathrm{Ɑ}`P - \mathrm{Ɑ}`(P \dot{-} P^2)$
$[*207{\cdot}35 . *216{\cdot}2]$

***216·22.** $\vdash : P \,\epsilon\, \mathrm{Rl}`J \cap \mathrm{connex} . P \Subset P^2 . \supset . \delta_P`C`P = \mathrm{Ɑ}`P \quad [*216{\cdot}21]$

***216·23.** $\vdash : P \in \text{trans} . \supset . \delta_P\text{‘}C\text{‘}P = \text{seq}_P\text{“}\text{Ɑ}\text{‘sgm‘}P = \text{lt}_P\text{“}\text{Ɑ}\text{‘sgm‘}P$

Dem.

$\vdash . *206{\cdot}25 . *216{\cdot}101 . \supset$

$\vdash :. \text{Hp} . \supset : x \in \delta_P\text{‘}C\text{‘}P . \equiv . (\exists\beta) . \exists ! \beta . \beta \subset P\text{“}\beta . x \,\text{seq}_P (P\text{“}\beta) .$

$[*24{\cdot}58 . *37{\cdot}29] \quad \equiv . (\exists\beta) . \beta \subset P\text{“}\beta . \exists ! P\text{“}\beta . x \,\text{seq}_P (P\text{“}\beta) .$

$[*201{\cdot}55] \quad \supset . (\exists\beta) . P\text{“}P\text{“}\beta = P\text{“}\beta . \exists ! P\text{“}\beta . x \,\text{seq}_P (P\text{“}\beta) .$

$[*212{\cdot}152] \quad \supset . x \in \text{seq}_P\text{“}\text{Ɑ}\text{‘sgm‘}P \quad (1)$

$\vdash . *211{\cdot}4 . \supset \vdash . \text{seq}_P\text{“}\text{Ɑ}\text{‘sgm‘}P = \text{lt}_P\text{“}\text{Ɑ}\text{‘sgm‘}P \quad (2)$

$\vdash . *212{\cdot}152 . (*216{\cdot}01) . \supset \vdash . \text{lt}_P\text{“}\text{Ɑ}\text{‘sgm‘}P \subset \delta_P\text{‘}C\text{‘}P \quad (3)$

$\vdash . (1) . (2) . (3) . \supset \vdash . \text{Prop}$

***216·3.** $\vdash : \alpha \in \text{dense‘}P . \equiv . \alpha - \overrightarrow{\text{min}}_P\text{‘}\alpha \subset \delta_P\text{‘}\alpha \quad [(*216{\cdot}02)]$

***216·31.** $\vdash : \alpha \in \text{dense‘}P . \equiv . \alpha \subset C\text{‘}P . \alpha \cap \breve{P}\text{“}\alpha \subset \delta_P\text{‘}\alpha$

Dem.

$\vdash . *216{\cdot}3{\cdot}111 . \supset \vdash : \alpha \in \text{dense‘}P . \supset . \alpha - \overrightarrow{\text{min}}_P\text{‘}\alpha \subset \text{Ɑ}\text{‘}P .$

$[*205{\cdot}11] \quad \supset . \alpha \subset C\text{‘}P . \quad (1)$

$[*205{\cdot}11] \quad \supset . \alpha - \overrightarrow{\text{min}}_P\text{‘}\alpha = \alpha \cap \breve{P}\text{“}\alpha \quad (2)$

$\vdash . (1) . (2) . \supset \vdash . \text{Prop}$

***216·32.** $\vdash : \alpha \in \text{closed‘}P . \equiv . \text{Cl ex‘}(\alpha \cap C\text{‘}P) \subset \text{Ɑ}\text{‘limax}_P . \delta_P\text{‘}\alpha \subset \alpha$
$[(*216{\cdot}03)]$

***216·33.** $\vdash :. \alpha \in \text{closed‘}P . \equiv : \beta \subset \alpha . \exists ! \beta . \beta \subset P\text{“}\beta . \supset_\beta . \exists ! \overrightarrow{\text{lt}}_P\text{‘}\beta . \overrightarrow{\text{lt}}_P\text{‘}\beta \subset \alpha$

Dem.

$\vdash . *207{\cdot}45 . *205{\cdot}123 . \supset \vdash :. \text{Cl ex‘}(\alpha \cap C\text{‘}P) \subset \text{Ɑ}\text{‘limax}_P . \equiv :$

$\beta \subset \alpha . \exists ! \beta . \beta \subset C\text{‘}P . \beta \subset P\text{“}\beta . \supset_\beta . \exists ! \overrightarrow{\text{lt}}_P\text{‘}\beta :$

$[*37{\cdot}15] \quad \equiv : \beta \subset \alpha . \exists ! \beta . \beta \subset P\text{“}\beta . \supset_\beta . \exists ! \overrightarrow{\text{lt}}_P\text{‘}\beta \quad (1)$

$\vdash . *40{\cdot}43{\cdot}5 . \supset \vdash :. \delta_P\text{‘}\alpha \subset \alpha . \equiv : \beta \subset \alpha . \exists ! \beta . \beta \subset C\text{‘}P . \supset_\beta . \overrightarrow{\text{lt}}_P\text{‘}\beta \subset \alpha :$

$[*207{\cdot}11 . *24{\cdot}12] \equiv : \beta \subset \alpha . \exists ! \beta . \beta \subset C\text{‘}P . \overrightarrow{\text{max}}_P\text{‘}\beta = \Lambda . \supset_\beta . \overrightarrow{\text{lt}}_P\text{‘}\beta \subset \alpha :$

$[*205{\cdot}123 . *37{\cdot}15] \equiv : \beta \subset \alpha . \exists ! \beta . \beta \subset P\text{“}\beta . \supset_\beta . \overrightarrow{\text{lt}}_P\text{‘}\beta \subset \alpha \quad (2)$

$\vdash . (1) . (2) . *216{\cdot}32 . \supset \vdash . \text{Prop}$

***216·34.** $\vdash :: P \in \text{connex} . \supset :. \alpha \in \text{closed‘}P . \equiv :$
$\beta \subset \alpha . \exists ! \beta . \beta \subset P\text{“}\beta . \supset_\beta . \text{lt}_P\text{‘}\beta \in \alpha \quad [*216{\cdot}33 . *71{\cdot}332 . *207{\cdot}24]$

***216·35.** $\vdash : P \epsilon \text{Ser} . \text{Cl ex}`\alpha \subset \text{G}`\text{limax}_P . \supset . \text{Cl ex}`\delta_P`\alpha \subset \text{G}`\text{limax}_P$

Dem.

$\vdash . *71·47 . *37·26 . \supset$

$\vdash :. \text{Hp} . \supset : \beta \epsilon \text{Cl ex}`\delta_P`\alpha . \supset . (\exists\lambda) . \lambda \subset \text{Cl ex}`\alpha \cap \text{G}`\text{lt}_P . \beta = \text{lt}_P``\lambda . \exists ! \beta .$

[*207·54]

$\supset . (\exists\lambda) . \lambda \subset \text{Cl ex}`\alpha \cap \text{G}`\text{lt}_P . \beta = \text{lt}_P``\lambda . \exists ! \beta . \overrightarrow{\text{limax}}_P`\beta = \overrightarrow{\text{limax}}_P`s`\lambda .$

[*37·29.Transp]

$\supset . (\exists\lambda) . \lambda \subset \text{Cl ex}`\alpha \cap \text{G}`\text{lt}_P . \beta = \text{lt}_P``\lambda . \exists ! \lambda . \overrightarrow{\text{limax}}_P`\beta = \overrightarrow{\text{limax}}_P`s`\lambda .$

[*53·24.Transp] $\supset . (\exists\lambda) . s`\lambda \epsilon \text{Cl ex}`\alpha . \overrightarrow{\text{limax}}_P`\beta = \overrightarrow{\text{limax}}_P`s`\lambda .$

[Hp] $\supset . \exists ! \overrightarrow{\text{limax}}_P`\beta :. \supset \vdash . \text{Prop}$

***216·36.** $\vdash : \alpha \epsilon \text{perf}`P . \equiv . \alpha \epsilon \text{dense}`P \cap \text{closed}`P$ [(*216·04)]

***216·37.** $\vdash : \alpha \epsilon \text{perf}`P . \equiv . \text{Cl ex}`\alpha \subset \text{G}`\text{limax}_P . \delta_P`\alpha = \alpha - \overrightarrow{\text{min}}_P`\alpha$
[*216·3·32·36]

***216·371.** $\vdash : \alpha \epsilon \text{perf}`P . \equiv . \text{Cl ex}`\alpha \subset \text{G}`\text{limax}_P . \alpha \subset C`P . \delta_P`\alpha = \alpha \cap \breve{P}``\alpha$
[*216·31·32·11·36]

***216·38.** $\vdash : P \epsilon \text{trans} \cap \text{connex} . \alpha \epsilon \text{dense}`P . \supset . \delta_P`\alpha \epsilon \text{dense}`P . \delta_P`\alpha \subset \delta_P`\delta_P`\alpha$

Dem.

$\vdash . *216·3·15 . \supset \vdash : \text{Hp} . \supset . \delta_P`(\alpha - \overrightarrow{\text{min}}_P`\alpha) \subset \delta_P`\delta_P`\alpha .$

[*216·16] $\supset . \delta_P`\alpha \subset \delta_P`\delta_P`\alpha .$

[*216·3] $\supset . \delta_P`\alpha \epsilon \text{dense}`P : \supset \vdash . \text{Prop}$

***216·381.** $\vdash : P \epsilon \text{Ser} . \alpha \epsilon \text{dense}`P . \supset . \delta_P`\alpha = \delta_P`\delta_P`\alpha . \overrightarrow{\text{min}}_P`\delta_P`\alpha = \Lambda$
[*216·38·14·11]

***216·382.** $\vdash : P \epsilon \text{Ser} . \alpha \epsilon \text{dense}`P . \text{Cl ex}`\alpha \subset \text{G}`\text{limax}_P . \supset . \delta_P`\alpha \epsilon \text{perf}`P$
[*216·35·381·37]

***216·4.** $\vdash : S \epsilon P \,\overline{\text{smor}}\, Q . \supset . \delta_P`\alpha = S``\delta_Q`\breve{S}``\alpha . \breve{S}``\delta_P`\alpha = \delta_Q`\breve{S}``\alpha$

Dem.

$\vdash . *207·63 . \supset \vdash : \text{Hp} . \supset . \delta_P`\alpha = S``\text{lt}_Q``\breve{S}```\text{Cl ex}`\alpha$

[*71·491] $= S``\text{lt}_Q``\text{Cl ex}`\breve{S}``\alpha$

[(*216·01)] $= S``\delta_Q`\breve{S}``\alpha$ (1)

$\vdash . (1) . *72·52 . *216·111 . \supset \vdash . \text{Prop}$

***216·401.** $\vdash : S \epsilon P \,\overline{\text{smor}}\, Q . \supset . P \upharpoonright \delta_P`\alpha = S ; (Q \upharpoonright \delta_Q`\breve{S}``\alpha)$

Dem.

$\vdash . *150·37 . \supset \vdash : \text{Hp} . \supset . S ; (Q \upharpoonright \delta_Q`\breve{S}``\alpha) = (S ; Q) \upharpoonright S``\delta_Q`\breve{S}``\alpha$

[*216·4.*151·11] $= P \upharpoonright \delta_P`\alpha : \supset \vdash . \text{Prop}$

$*216{\cdot}41$. $\vdash :. S \in P \,\overline{\text{smor}}\, Q \,.\, \alpha \subset C\text{‘}P \,.\supset :\, \alpha \in \text{dense‘}P \,.\equiv.\, \breve{S}\text{“}\alpha \in \text{dense‘}Q$

Dem.

$\vdash . *216{\cdot}3 . *37{\cdot}2 . \supset$

$\vdash :. \text{Hp} . \supset : \alpha \in \text{dense‘}P . \supset . \breve{S}\text{“}(\alpha - \overrightarrow{\min}_P\text{‘}\alpha) \subset \breve{S}\text{“}\delta_P\text{‘}\alpha .$

$[*71{\cdot}38 . *205{\cdot}8] \quad \supset . \breve{S}\text{“}\alpha - \overrightarrow{\min}_Q\text{‘}\breve{S}\text{“}\alpha \subset \breve{S}\text{“}\delta_P\text{‘}\alpha .$

$[*216{\cdot}4] \quad \supset . \breve{S}\text{“}\alpha - \overrightarrow{\min}_Q\text{‘}\breve{S}\text{“}\alpha \subset \delta_Q\text{‘}\breve{S}\text{“}\alpha .$

$[*216{\cdot}3] \quad \supset . \breve{S}\text{“}\alpha \in \text{dense‘}Q \qquad (1)$

$\vdash . (1) \dfrac{Q, P, \breve{S}\text{“}\alpha}{P, Q, \alpha} . \supset$

$\vdash :. \text{Hp} . \supset : \breve{S}\text{“}\alpha \in \text{dense‘}Q . \supset . S\text{“}\breve{S}\text{“}\alpha \in \text{dense‘}P .$

$[*72{\cdot}502] \quad \supset . \alpha \in \text{dense‘}P \qquad (2)$

$\vdash . (1) . (2) . \supset \vdash . \text{Prop}$

$*216{\cdot}411$. $\vdash :. S \in P \,\overline{\text{smor}}\, Q \,.\, \alpha \subset C\text{‘}P \,.\supset :\, \alpha \in \text{closed‘}P \,.\equiv.\, \breve{S}\text{“}\alpha \in \text{closed‘}Q$

Dem.

$\vdash . *207{\cdot}64 . *37{\cdot}431 . \supset$

$\vdash :. \text{Hp} . \supset : \beta \subset C\text{‘}P . \exists ! \beta . \beta \in \text{Ɑ‘limax}_P . \supset .$

$\breve{S}\text{“}\beta \subset C\text{‘}Q . \exists ! \breve{S}\text{“}\beta . \breve{S}\text{“}\beta \in \text{Ɑ‘limax}_Q :$

$[*71{\cdot}49] \supset : \text{Cl ex‘}\alpha \subset \text{Ɑ‘limax}_P . \supset . \text{Cl ex‘}\breve{S}\text{“}\alpha \subset \text{Ɑ‘limax}_Q \qquad (1)$

$\vdash . *37{\cdot}2 . *216{\cdot}4 . \quad \supset \vdash :. \text{Hp} . \supset : \delta_P\text{‘}\alpha \subset \alpha . \supset . \delta_Q\text{‘}\breve{S}\text{“}\alpha \subset \breve{S}\text{“}\alpha \qquad (2)$

$\vdash . (1) . (2) . *216{\cdot}32 . \supset \vdash :. \text{Hp} . \supset : \alpha \in \text{closed‘}P . \supset . \breve{S}\text{“}\alpha \in \text{closed‘}Q \qquad (3)$

$\vdash . (3) \dfrac{Q, P, \breve{S}\text{“}\alpha}{P, Q, \alpha} . \quad \supset \vdash :. \text{Hp} . \supset : \breve{S}\text{“}\alpha \in \text{closed‘}Q . \supset . S\text{“}\breve{S}\text{“}\alpha \in \text{closed‘}P .$

$[*72{\cdot}502] \quad \supset . \alpha \in \text{closed‘}P \qquad (4)$

$\vdash . (3) . (4) . \supset \vdash . \text{Prop}$

$*216{\cdot}412$. $\vdash :. S \in P \,\overline{\text{smor}}\, Q \,.\, \alpha \subset C\text{‘}P \,.\supset :\, \alpha \in \text{perf‘}P \,.\equiv.\, \breve{S}\text{“}\alpha \in \text{perf‘}Q$

$[*216{\cdot}41{\cdot}411{\cdot}36]$

$*216{\cdot}5$. $\vdash : P \in \text{Ser} . \supset . \text{Ɑ‘}\mathfrak{s}\text{‘}P - \overrightarrow{P}\text{“}C\text{‘}P \subset \delta(\mathfrak{s}\text{‘}P)\text{‘}\overrightarrow{P}\text{“}C\text{‘}P$

Dem.

$\vdash . *212{\cdot}134 . *216{\cdot}111 . \supset$

$\vdash : \text{Hp} . P = \dot{\Lambda} . \supset . \text{Ɑ‘}\mathfrak{s}\text{‘}P - \overrightarrow{P}\text{“}C\text{‘}P = \Lambda . \delta(\mathfrak{s}\text{‘}P)\text{‘}\overrightarrow{P}\text{“}C\text{‘}P = \Lambda \qquad (1)$

$\vdash . *212{\cdot}632 . \supset$

$\vdash :. \text{Hp} . \dot{\exists} ! P . P\text{“}\alpha \sim\in \overrightarrow{P}\text{“}C\text{‘}P . \supset : P\text{“}\alpha = \text{lt}(\mathfrak{s}\text{‘}P)\text{‘}\overrightarrow{P}\text{“}\alpha :$

$[*216{\cdot}1] \supset : \exists ! \alpha . \alpha \subset C\text{‘}P . \supset . P\text{“}\alpha \in \delta(\mathfrak{s}\text{‘}P)\text{‘}\overrightarrow{P}\text{“}C\text{‘}P \qquad (2)$

$\vdash . (2) . *212\cdot132 . *37\cdot265 . \supset$

$\vdash : \text{Hp} . \dot{\exists} ! P . \beta \in \text{Ɑ}\text{‘}\varsigma\text{‘}P - \overrightarrow{P}\text{‘‘}C\text{‘}P . \supset . \beta \in \delta(\varsigma\text{‘}P)\text{‘}\overrightarrow{P}\text{‘‘}C\text{‘}P \qquad (3)$

$\vdash . (1) . (3) . \supset \vdash . \text{Prop}$

$*216\cdot51$. $\vdash : P \in \text{Ser} . \supset .$

$$\delta(\varsigma\text{‘}P)\text{‘}\overrightarrow{P}\text{‘‘}C\text{‘}P = \delta(\varsigma\text{‘}P)\text{‘}C\text{‘}\varsigma\text{‘}P = \text{D}\text{‘}\text{lt}(\varsigma\text{‘}P) - \iota\text{‘}\Lambda = \text{Ɑ}\text{‘}\text{sgm}\text{‘}P$$

Dem.

$\vdash . *212\cdot661 . \supset \vdash : \text{Hp} . \kappa \subset \text{D}\text{‘}P_\epsilon . x = \text{lt}(\varsigma\text{‘}P)\text{‘}\kappa . \supset . x = \text{lt}(\varsigma\text{‘}P)\text{‘}\overrightarrow{P}\text{‘‘}s\text{‘}\kappa \qquad (1)$

$\vdash . *207\cdot13 . *212\cdot133 . \supset$

$\vdash : \text{Hp} . \kappa \subset \text{D}\text{‘}P_\epsilon . x = \text{lt}(\varsigma\text{‘}P)\text{‘}\kappa . \supset . \kappa \neq \iota\text{‘}\Lambda \qquad (2)$

$\vdash . (1) . (2) . *40\cdot26 . \supset$

$\vdash : \text{Hp} . \kappa \subset \text{D}\text{‘}P_\epsilon . \exists ! \kappa . x = \text{lt}(\varsigma\text{‘}P)\text{‘}\kappa . \supset . \exists ! s\text{‘}\kappa . x = \text{lt}(\varsigma\text{‘}P)\text{‘}\overrightarrow{P}\text{‘‘}s\text{‘}\kappa .$

[$*216\cdot1$] $\supset . x \in \delta(\varsigma\text{‘}P)\text{‘}\overrightarrow{P}\text{‘‘}C\text{‘}P \qquad (3)$

$\vdash . (3) . *216\cdot1 . \quad \supset \vdash : \text{Hp} . \supset . \delta(\varsigma\text{‘}P)\text{‘}C\text{‘}\varsigma\text{‘}P \subset \delta(\varsigma\text{‘}P)\text{‘}\overrightarrow{P}\text{‘‘}C\text{‘}P \qquad (4)$

$\vdash . *211\cdot3 . *216\cdot15 . \supset \vdash . \delta(\varsigma\text{‘}P)\text{‘}\overrightarrow{P}\text{‘‘}C\text{‘}P \subset \delta(\varsigma\text{‘}P)\text{‘}C\text{‘}\varsigma\text{‘}P \qquad (5)$

$\vdash . (4) . (5) . \quad \supset \vdash : \text{Hp} . \supset . \delta(\varsigma\text{‘}P)\text{‘}\overrightarrow{P}\text{‘‘}C\text{‘}P = \delta(\varsigma\text{‘}P)\text{‘}C\text{‘}\varsigma\text{‘}P \qquad (6)$

[$*216\cdot2 . *212\cdot133$] $= \text{D}\text{‘}\text{lt}(\varsigma\text{‘}P) - \iota\text{‘}\Lambda \qquad (7)$

$\vdash . (6) . (7) . *212\cdot667 . \supset \vdash . \text{Prop}$

$*216\cdot52$. $\vdash : P \in \text{Ser} . \dot{\exists} ! P . \alpha \subset C\text{‘}P . \supset . \delta(\varsigma\text{‘}P)\text{‘}\overrightarrow{P}\text{‘‘}\alpha = P\text{‘‘‘}(\text{Cl ex}\text{‘}\alpha - \text{Ɑ}\text{‘}\text{max}_P)$

Dem.

$\vdash . *216\cdot1 . \supset \vdash :. \text{Hp} . \supset : \gamma \in \delta(\varsigma\text{‘}P)\text{‘}\overrightarrow{P}\text{‘‘}\alpha . \equiv . (\exists\kappa) . \kappa \subset \overrightarrow{P}\text{‘‘}\alpha . \exists ! \kappa . \gamma = \text{lt}(\varsigma\text{‘}P)\text{‘}\kappa .$

[$*212\cdot402$] $\equiv . (\exists\kappa) . \kappa \subset \overrightarrow{P}\text{‘‘}\alpha . \exists ! \kappa . \sim \text{E} ! \max(\varsigma\text{‘}P)\text{‘}\kappa . \gamma = s\text{‘}\kappa .$

[$*71\cdot47 . *37\cdot2$] $\equiv . (\exists\beta) . \beta \subset \alpha . \exists ! \beta . \sim \text{E} ! \max(\varsigma\text{‘}P)\text{‘}\overrightarrow{P}\text{‘‘}\beta . \gamma = s\text{‘}\overrightarrow{P}\text{‘‘}\beta .$

[$*40\cdot5 . *212\cdot601$] $\equiv . (\exists\beta) . \beta \subset \alpha . \exists ! \beta . \sim \text{E} ! \text{max}_P\text{‘}\beta . \gamma = P\text{‘‘}\beta .$

[$*37\cdot6$] $\equiv . \gamma \in P\text{‘‘‘}(\text{Cl ex}\text{‘}\alpha - \text{Ɑ}\text{‘}\text{max}_P) :. \supset \vdash . \text{Prop}$

$*216\cdot521$. $\vdash : P \in \text{Ser} . \alpha \subset C\text{‘}P . \supset . \overrightarrow{P}\text{‘‘}(\alpha - \overrightarrow{\text{min}}_P\text{‘}\alpha) = \overrightarrow{P}\text{‘‘}\alpha - \overrightarrow{\text{min}}(\varsigma\text{‘}P)\text{‘}\overrightarrow{P}\text{‘‘}\alpha$

Dem.

$\vdash . *71\cdot381 . *204\cdot34 . \supset \vdash : \text{Hp} . \supset . \overrightarrow{P}\text{‘‘}(\alpha - \overrightarrow{\text{min}}_P\text{‘}\alpha) = \overrightarrow{P}\text{‘‘}\alpha - \overrightarrow{P}\text{‘‘}\overrightarrow{\text{min}}_P\text{‘}\alpha$

[$*212\cdot6$] $= \overrightarrow{P}\text{‘‘}\alpha - \overrightarrow{\text{min}}(\varsigma\text{‘}P)\text{‘}\overrightarrow{P}\text{‘‘}\alpha : \supset \vdash . \text{Prop}$

∗216·53. $\vdash :. P \epsilon \text{Ser} . \dot{\exists} ! P . \alpha \subset C`P . \supset : \alpha \epsilon \text{dense}`P . \equiv . \overrightarrow{P}``\alpha \epsilon \text{dense}`\varsigma`P$

Dem.

$\vdash . *216\cdot52\cdot3 . \supset$

$\vdash :: \text{Hp} . \supset :. \overrightarrow{P}``\alpha \epsilon \text{dense}`\varsigma`P . \equiv :$

$\overrightarrow{P}``\alpha - \overrightarrow{\min}(\varsigma`P)`\overrightarrow{P}``\alpha \subset P```(\text{Cl ex}`\alpha - \text{G}`\max_P) :$

$[*216\cdot521] \equiv : \overrightarrow{P}``(\alpha - \overrightarrow{\min}_P`\alpha) \subset P```(\text{Cl ex}`\alpha - \text{G}`\max_P) :$

$[*37\cdot6] \quad \equiv : x \epsilon \alpha - \overrightarrow{\min}_P`\alpha . \supset_x . (\exists\beta) . \beta \subset \alpha . \exists ! \beta . \sim \text{E} ! \max_P`\beta . \overrightarrow{P}`x = P``\beta :$

$[*207\cdot521] \equiv : x \epsilon \alpha - \overrightarrow{\min}_P`\alpha . \supset_x . (\exists\beta) . \beta \subset \alpha . \exists ! \beta . x = \text{lt}_P`\beta :$

$[*216\cdot1] \quad \equiv : x \epsilon \alpha - \overrightarrow{\min}_P`\alpha . \supset_x . x \epsilon \delta_P`\alpha :$

$[*216\cdot3] \quad \equiv : \alpha \epsilon \text{dense}`P :: \supset \vdash . \text{Prop}$

∗216·54. $\vdash :. P \epsilon \text{Ser} . \dot{\exists} ! P . \alpha \subset C`P . \supset : \alpha \epsilon \text{closed}`P . \equiv . \delta(\varsigma`P)`\overrightarrow{P}``\alpha \subset \overrightarrow{P}``\alpha$

Dem.

$\vdash . *216\cdot52 . \supset$

$\vdash :: \text{Hp} . \supset :. \delta(\varsigma`P)`\overrightarrow{P}``\alpha \subset \overrightarrow{P}``\alpha . \equiv : P```(\text{Cl ex}`\alpha - \text{G}`\max_P) \subset \overrightarrow{P}``\alpha :$

$[*37\cdot6] \quad \equiv : \beta \subset \alpha . \exists ! \beta . \sim \text{E} ! \max_P`\beta . \supset_\beta . (\exists x) . x \epsilon \alpha . P``\beta = \overrightarrow{P}`x :$

$[*207\cdot521] \equiv : \beta \subset \alpha . \exists ! \beta . \sim \text{E} ! \max_P`\beta . \supset_\beta . \text{lt}_P`\beta \epsilon \alpha :$

$[*216\cdot34] \quad \equiv : \alpha \epsilon \text{closed}`P :: \supset \vdash . \text{Prop}$

∗216·55. $\vdash :. P \epsilon \text{Ser} . \dot{\exists} ! P . \alpha \subset C`P . \supset : \alpha \epsilon \text{closed}`P . \equiv . \overrightarrow{P}``\alpha \epsilon \text{closed}`\varsigma`P$

Dem.

$\vdash . *212\cdot44 . \supset \vdash : \text{Hp} . \supset . \overrightarrow{P}``\alpha \subset \text{G}`\text{limax}(\varsigma`P) \quad (1)$

$\vdash . (1) . *212\cdot54\cdot32 . \supset \vdash . \text{Prop}$

∗216·56. $\vdash :. P \epsilon \text{Ser} . \dot{\exists} ! P . \alpha \subset C`P . \supset : \alpha \epsilon \text{perf}`P . \equiv . \overrightarrow{P}``\alpha \epsilon \text{perf}`\varsigma`P .$

$\equiv . \delta(\varsigma`P)`\overrightarrow{P}``\alpha = \overrightarrow{P}``\alpha - \overrightarrow{\min}(\varsigma`P)`\overrightarrow{P}``\alpha$

[∗216·53·54·55·36·37 . ∗212·44]

∗216·6. $\vdash : x(\nabla`P)y . \equiv . x, y \epsilon \text{D}`\text{lt}_P . xPy$ [(∗216·05)]

∗216·601. $\vdash : x \epsilon \text{D}`\text{lt}_P \cap \text{G}`P . P \epsilon \text{connex} . \text{E} ! B`P . \supset . (B`P)(\nabla`P)x$

Dem.

$\vdash . *206\cdot14 . \supset \vdash : \text{Hp} . \supset . B`P \epsilon \text{D}`\text{lt}_P \quad (1)$

$\vdash . *202\cdot524 . \supset \vdash : \text{Hp} . \supset . (B`P)Px \quad (2)$

$\vdash . (1) . (2) . *216\cdot6 . \supset \vdash . \text{Prop}$

***216·602.** $\vdash : P \in \text{connex} \,.\, \text{E} \,!\, B\text{‘}P \,.\, \supset .\, \text{ᗡ‘}\nabla\text{‘}P = \text{D‘lt}_P - \overrightarrow{B}\text{‘}P = \delta_P\text{‘}C\text{‘}P$

Dem.

$\vdash .\, *216·601 \,.\, \supset \vdash : \text{Hp} \,.\, \supset .\, \text{D‘lt}_P - \overrightarrow{B}\text{‘}P \subset \text{ᗡ‘}\nabla\text{‘}P \quad (1)$

$\vdash .\, *216·6 \,.\, \supset \vdash .\, \text{ᗡ‘}\nabla\text{‘}P \subset \text{D‘lt}_P - \overrightarrow{B}\text{‘}P \quad (2)$

$\vdash .\, (1) \,.\, (2) \,.\, *216·2 \,.\, \supset \vdash .\, \text{Prop}$

***216·603.** $\vdash : P \in \text{connex} \,.\, \dot{\exists} \,!\, \nabla\text{‘}P \,.\, \supset .\, C\text{‘}\nabla\text{‘}P = \text{D‘lt}_P$

Dem.

$\vdash .\, *200·35 \,.\, \supset \vdash : \text{Hp} \,.\, \supset .\, \text{D‘lt}_P \sim \in 1 \,.$

$[*202·55] \qquad \supset .\, C\text{‘}\nabla\text{‘}P = \text{D‘lt}_P : \supset \vdash .\, \text{Prop}$

***216·61.** $\vdash : P \in \text{Ser} \,.\, \text{E} \,!\, B\text{‘}P \,.\, \supset .\, \text{ᗡ‘}\nabla\text{‘}P = \text{ᗡ‘}P - \text{ᗡ‘}P_1$ [*216·602·21]

***216·611.** $\vdash : P \in \text{Ser} \,.\, \dot{\exists} \,!\, \nabla\text{‘}P \,.\, \supset .\, C\text{‘}\nabla\text{‘}P = C\text{‘}P - \text{ᗡ‘}P_1 = \delta_P\text{‘}C\text{‘}P \cup \overrightarrow{B}\text{‘}P$

Dem.

$\vdash .\, *216·603 \,.\, *206·14 \,.\, \supset \vdash : \text{Hp} \,.\, \supset .\, C\text{‘}\nabla\text{‘}P = (\text{D‘lt}_P - \overrightarrow{B}\text{‘}P) \cup \overrightarrow{B}\text{‘}P$

$[*216·2] \qquad = \delta_P\text{‘}C\text{‘}P \cup \overrightarrow{B}\text{‘}P \quad (1)$

$[*216·21] \qquad = (\text{ᗡ‘}P - \text{ᗡ‘}P_1) \cup \overrightarrow{B}\text{‘}P$

$[*93·103 . *24·412] \qquad = C\text{‘}P - \text{ᗡ‘}P_1 \quad (2)$

$\vdash .\, (1) \,.\, (2) \,.\, \supset \vdash .\, \text{Prop}$

***216·612.** $\vdash : P \in \text{Ser} \,.\, \supset .\, \text{ᗡ‘}\nabla\text{‘}P \subset \text{ᗡ‘}P - \text{ᗡ‘}P_1$

Dem.

$\vdash .\, *216·6 \,.\, \supset \vdash .\, \text{ᗡ‘}\nabla\text{‘}P \subset \text{D‘lt}_P - \overrightarrow{B}\text{‘}P \quad (1)$

$\vdash .\, *216·2·21 \,.\, \supset \vdash : \text{Hp} \,.\, \supset .\, \text{D‘lt}_P - \overrightarrow{B}\text{‘}P = \text{ᗡ‘}P - \text{ᗡ‘}P_1 \quad (2)$

$\vdash .\, (1) \,.\, (2) \,.\, \supset \vdash .\, \text{Prop}$

***216·62.** $\vdash : P \in \text{Ser} \,.\, \dot{\exists} \,!\, \nabla\text{‘}P \,.\, \supset .\, C\text{‘}\nabla\text{‘}P = \text{seq}_P\text{‘‘}C\text{‘sgm‘}P = \text{lt}_P\text{‘‘}C\text{‘sgm‘}P$

Dem.

$\vdash .\, *216·611 \,.\, \supset \vdash : \text{Hp} \,.\, \supset .\, C\text{‘}\nabla\text{‘}P = \delta_P\text{‘}C\text{‘}P \cup \overrightarrow{B}\text{‘}P$

$[*216·23] \qquad = \text{seq}_P\text{‘‘ᗡ‘sgm‘}P \cup \overrightarrow{B}\text{‘}P \quad (1)$

$[*206·14] \qquad = \text{seq}_P\text{‘‘}(\text{ᗡ‘sgm‘}P \cup \iota\text{‘}\Lambda) \quad (2)$

$\vdash .\, *211·45 \,.\, \supset \vdash : \text{Hp} \,.\, \exists \,!\, \text{ᗡ‘}P - \text{ᗡ‘}P_1 \,.\, \supset .\, \exists \,!\, \text{D‘}(P_\epsilon \dot{\frown} I) - \iota\text{‘}\Lambda \,.$

$[*212·153] \qquad \supset .\, \dot{\exists} \,!\, \text{sgm‘}P \,.$

$[*212·155] \qquad \supset .\, \text{ᗡ‘sgm‘}P \cup \iota\text{‘}\Lambda = C\text{‘sgm‘}P \quad (3)$

$\vdash .\, (1) \,.\, *216·23 \,.\, *207·17 \,.\, \supset \vdash : \text{Hp} \,.\, \supset .\, C\text{‘}\nabla\text{‘}P = \text{lt}_P\text{‘‘}(\text{ᗡ‘sgm‘}P \cup \iota\text{‘}\Lambda) \quad (4)$

$\vdash .\, (2) \,.\, (3) \,.\, (4) \,.\, \supset \vdash .\, \text{Prop}$

***216·621.** $\vdash : P \in \text{Ser} \,.\, \dot{\exists} \,!\, \nabla\text{‘}P \,.\, \supset .\, \dot{\exists} \,!\, \text{sgm‘}P \,.\, \exists \,!\, \text{ᗡ‘}P - \text{ᗡ‘}P_1$ [*216·62·612]

$*217$. ON SEGMENTS OF SUMS AND CONVERSES.

Summary of $*217$.

The purpose of the present number is to prove $*217{\cdot}43$, which is required in the theory of real numbers (Part VI, Section A), where P will be the series of positive ratios including zero, Q will be the series of negative ratios in the order from zero to $-\infty$ (both excluded), α the real number zero, and Z and W two different series either of which may be taken as the series of negative and positive real numbers. In virtue of $*217{\cdot}43$, these two series are ordinally similar.

$*217{\cdot}1$. $\vdash : \alpha \cap C\text{‘}Q = \Lambda \,.\supset.\, (P \ddagger Q)\text{‘‘}\alpha = P\text{‘‘}\alpha$ $\quad$ [$*160{\cdot}1$]

$*217{\cdot}11$. $\vdash : \exists ! \, \alpha \cap C\text{‘}Q \,.\supset.\, (P \ddagger Q)\text{‘‘}\alpha = C\text{‘}P \cup Q\text{‘‘}\alpha$ $\quad$ [$*160{\cdot}1$]

$*217{\cdot}12$. $\vdash . \mathrm{D}\text{‘}(P \ddagger Q)_\epsilon \subset \mathrm{D}\text{‘}P_\epsilon \cup (C\text{‘}P \cup)\text{‘‘}\mathrm{D}\text{‘}Q_\epsilon$ $\quad$ [$*217{\cdot}1{\cdot}11 \,.\, *211{\cdot}11$]

$*217{\cdot}13$. $\vdash : C\text{‘}P \cap C\text{‘}Q = \Lambda \,.\supset.\, P\text{‘‘}\alpha = (P \ddagger Q)\text{‘‘}(\alpha - C\text{‘}Q)$ $\quad$ [$*217{\cdot}1$]

$*217{\cdot}14$. $\vdash : \exists ! \, Q\text{‘‘}\alpha \,.\supset.\, C\text{‘}P \cup Q\text{‘‘}\alpha = (P \ddagger Q)\text{‘‘}\alpha$ $\quad$ [$*217{\cdot}11$]

$*217{\cdot}15$. $\vdash : C\text{‘}P \cap C\text{‘}Q = \Lambda \,.\supset.\, \mathrm{D}\text{‘}P_\epsilon \cup (C\text{‘}P \cup)\text{‘‘}(\mathrm{D}\text{‘}Q_\epsilon - \iota\text{‘}\Lambda) \subset \mathrm{D}\text{‘}(P \ddagger Q)_\epsilon$
[$*217{\cdot}13{\cdot}14$]

$*217{\cdot}16$. $\vdash :. C\text{‘}P \cap C\text{‘}Q = \Lambda : \sim \exists ! \, \overrightarrow{B}\text{‘}\breve{P} \,.\vee.\, \exists ! \, \overrightarrow{B}\text{‘}Q : \supset .\, C\text{‘}P \in \mathrm{D}\text{‘}(P \ddagger Q)_\epsilon$

Dem.

$\vdash . *211{\cdot}301 . \quad \supset \vdash : \sim \exists ! \, \overrightarrow{B}\text{‘}\breve{P} \,.\supset.\, C\text{‘}P \in \mathrm{D}\text{‘}P_\epsilon \quad (1)$

$\vdash . (1) . *217{\cdot}15 . \supset \vdash : \mathrm{Hp} . \sim \exists ! \, \overrightarrow{B}\text{‘}\breve{P} \,.\supset.\, C\text{‘}P \in \mathrm{D}\text{‘}(P \ddagger Q)_\epsilon \quad (2)$

$\vdash . *217{\cdot}11 . \quad \supset \vdash : \exists ! \, \overrightarrow{B}\text{‘}Q \,.\supset.\, (P \ddagger Q)\text{‘‘}\overrightarrow{B}\text{‘}Q = C\text{‘}P \quad (3)$

$\vdash . (2) . (3) . \supset \vdash . \mathrm{Prop}$

$*217{\cdot}17$. $\vdash :. C\text{‘}P \cap C\text{‘}Q = \Lambda : \sim \exists ! \, \overrightarrow{B}\text{‘}\breve{P} \,.\vee.\, \exists ! \, \overrightarrow{B}\text{‘}Q : \supset .$
$\mathrm{D}\text{‘}(P \ddagger Q)_\epsilon = \mathrm{D}\text{‘}P_\epsilon \cup (C\text{‘}P \cup)\text{‘‘}\mathrm{D}\text{‘}Q_\epsilon$ $\quad$ [$*217{\cdot}12{\cdot}15{\cdot}16$]

$*217{\cdot}18$. $\vdash :. C\text{‘}P \cap C\text{‘}Q = \Lambda : \sim \exists ! \, \overrightarrow{B}\text{‘}\breve{P} \,.\vee.\, \sim \exists ! \, \overrightarrow{B}\text{‘}Q : \supset .$
$\mathrm{D}\text{‘}(P \ddagger Q)_\epsilon = \mathrm{D}\text{‘}P_\epsilon \cup (C\text{‘}P \cup)\text{‘‘}(\mathrm{D}\text{‘}Q_\epsilon - \iota\text{‘}\Lambda)$

Dem.

$\vdash . *211{\cdot}301 . \supset \vdash : \sim \exists ! \, \overrightarrow{B}\text{‘}\breve{P} \,.\supset.\, (C\text{‘}P \cup)\text{‘‘}\iota\text{‘}\Lambda \subset \mathrm{D}\text{‘}P_\epsilon \quad (1)$

$\vdash . (1) . *217{\cdot}17 . \supset \vdash : \text{Hp} . \sim \exists ! \overrightarrow{B}`\breve{P} . \supset .$

$$\text{D}`(P \dagger Q)_\epsilon = \text{D}`P_\epsilon \cup (C`P \cup)``(\text{D}`Q_\epsilon - \iota`\Lambda) \quad (2)$$

$\vdash . *217{\cdot}11 . \quad \supset \vdash : \sim \exists ! \overrightarrow{B}`Q . \exists ! \alpha \cap C`Q . \supset . \exists ! (P \dagger Q)``\alpha \cap C`Q \quad (3)$

$\vdash . *217{\cdot}1 . \quad \supset \vdash : \sim \exists ! \overrightarrow{B}`Q . \exists ! \overrightarrow{B}`\breve{P} . \alpha \cap C`Q = \Lambda . \supset . (P \dagger Q)``\alpha \neq C`P \quad (4)$

$\vdash . (3) . (4) . \quad \supset \vdash : \text{Hp} . \sim \exists ! \overrightarrow{B}`Q . \exists ! \overrightarrow{B}`\breve{P} . \supset . C`P \sim \epsilon \, \text{D}`(P \dagger Q)_\epsilon \quad (5)$

$\vdash . (5) . *217{\cdot}12{\cdot}15 . \supset$

$\vdash : \text{Hp}(5) . \supset . \text{D}`(P \dagger Q)_\epsilon = \text{D}`P_\epsilon \cup (C`P \cup)``(\text{D}`Q_\epsilon - \iota`\Lambda) \quad (6)$

$\vdash . (2) . (6) . \supset \vdash . \text{Prop}$

***217·2.** $\vdash : C`P \cap C`Q = \Lambda . \supset . \text{D}`P_\epsilon \cap (C`P \cup)``(\text{D}`Q_\epsilon - \iota`\Lambda) = \Lambda$

Dem.

$\vdash . *211{\cdot}11 . \supset$

$\vdash : \text{D}`P_\epsilon \subset \text{Cl}`C`P : \alpha \epsilon (C`P \cup)``(\text{D}`Q_\epsilon - \iota`\Lambda) . \supset . \exists ! \alpha \cap C`Q : \supset \vdash . \text{Prop}$

***217·21.** $\vdash : \exists ! \overrightarrow{B}`\breve{P} . \supset . \text{D}`P_\epsilon \cap (C`P \cup)``\text{D}`Q_\epsilon = \Lambda$

Dem.

$\vdash . *211{\cdot}11 . \supset \vdash : \text{Hp} . \alpha \epsilon \text{D}`P_\epsilon . \supset . \exists ! C`P - \alpha : \supset \vdash . \text{Prop}$

***217·22.** $\vdash : P, Q \epsilon \text{trans} \cap \text{connex} . C`P \cap C`Q = \Lambda . \exists ! \overrightarrow{B}`\breve{P} . \exists ! \overrightarrow{B}`Q . \supset .$

$$s`(P \dagger Q) = s`P \dagger (C`P \cup)\text{;}s`Q$$

Dem.

$\vdash . *201{\cdot}401 . *202{\cdot}401 . \supset \vdash : \text{Hp} . \supset . P \dagger Q \epsilon \text{trans} \cap \text{connex} \quad (1)$

$\vdash . (1) . *212{\cdot}23 . \supset$

$\vdash :: \text{Hp} . \supset :. \alpha \{s`(P \dagger Q)\} \beta . \equiv : \alpha, \beta \epsilon \text{D}`(P \dagger Q)_\epsilon . \alpha \subset \beta . \alpha \neq \beta :$

$[*217{\cdot}17{\cdot}21] \equiv : \alpha, \beta \epsilon \text{D}`P_\epsilon . \alpha \subset \beta . \alpha \neq \beta . \vee . \alpha \epsilon \text{D}`P_\epsilon . \beta \epsilon (C`P \cup)``\text{D}`Q_\epsilon .$

$$\vee . \alpha, \beta \epsilon (C`P \cup)``\text{D}`Q_\epsilon . \alpha \subset \beta . \alpha \neq \beta :$$

$[*212{\cdot}23] \quad \equiv : \alpha (s`P) \beta . \vee . \alpha \epsilon C`s`P . \beta \epsilon C`(C`P \cup)\text{;}s`Q .$

$$\vee . \alpha \{(C`P \cup)\text{;}s`Q\} \beta :$$

$[*160{\cdot}11] \quad \equiv : \alpha \{s`P \dagger (C`P \cup)\text{;}s`Q\} \beta :: \supset \vdash . \text{Prop}$

***217·23.** $\vdash :. P, Q \epsilon \text{trans} \cap \text{connex} . C`P \cap C`Q = \Lambda : \sim \exists ! \overrightarrow{B}`\breve{P} . \vee . \sim \exists ! \overrightarrow{B}`Q : \supset .$

$$s`(P \dagger Q) = s`P \dagger (C`P \cup)\text{;}(s`Q) \upharpoonright (-\iota`\Lambda)$$

Dem.

$\vdash . *201{\cdot}401 . *202{\cdot}401 . *212{\cdot}23 . \supset$

$\vdash :: \text{Hp} . \supset :. \alpha \{s`(P \dagger Q)\} \beta . \equiv : \alpha, \beta \epsilon \text{D}`(P \dagger Q)_\epsilon . \alpha \subset \beta . \alpha \neq \beta :$

$[*217{\cdot}18{\cdot}2] \equiv : \alpha, \beta \epsilon \text{D}`P_\epsilon . \alpha \subset \beta . \alpha \neq \beta . \vee . \alpha \epsilon \text{D}`P_\epsilon . \beta \epsilon (C`P \cup)``(\text{D}`Q_\epsilon - \iota`\Lambda) .$

$$\vee . \alpha, \beta \epsilon (C`P \cup)``(\text{D}`Q_\epsilon - \iota`\Lambda) . \alpha \subset \beta . \alpha \neq \beta :$$

$[*212{\cdot}23 . *160{\cdot}11] \equiv : \alpha \{s`P \dagger (C`P \cup)\text{;}(s`Q) \upharpoonright (-\iota`\Lambda)\} \beta :: \supset \vdash . \text{Prop}$

***217·24.** $\vdash : \alpha \cap \beta = \Lambda . \supset . (\alpha \cup) \upharpoonright \text{Cl}`\beta \epsilon 1 \rightarrow 1 \quad [*24{\cdot}481]$

***217·25.** $\vdash : C`P \cap C`Q = \Lambda . \supset . (C`P \cup) \upharpoonright C`s`Q \epsilon \{(C`P \cup)\text{;}s`Q\} \overline{\text{smor}} (s`Q)$

$[*217{\cdot}24]$

***217·3.** $\vdash : P \epsilon \mathrm{Ser} \,.\supset .\, \mathrm{D}`P_\epsilon = \overrightarrow{P}``C`P \cup \mathrm{D}`(P_\epsilon \dot{\cap} I) - \mathrm{G}`\mathrm{seq}_P$ [*211·32·302·41]

***217·301.** $\vdash : P \epsilon \mathrm{Ser} \,.\, \gamma \epsilon \mathrm{D}`(P_\epsilon \dot{\cap} I) - \mathrm{G}`\mathrm{seq}_P \,.\supset .\, \gamma = C`P - \breve{P}``(C`P - \gamma)$

Dem.

$\vdash .\, *211{\cdot}727 \,.\supset \vdash : \mathrm{Hp} \,.\supset .\, \sim \mathrm{E}!\, \mathrm{limin}_P`(C`P - \gamma) \,.$

$[*207{\cdot}44 . *211{\cdot}7] \quad \supset .\, C`P - \gamma \,\epsilon\, \mathrm{sect}`\breve{P} - \mathrm{G}`\mathrm{min}_P \,.$

$[*211{\cdot}41{\cdot}12] \quad \supset .\, C`P - \gamma = \breve{P}``(C`P - \gamma) : \supset \vdash .\, \mathrm{Prop}$

***217·31.** $\vdash : P \epsilon \mathrm{Ser} \,.\, \gamma \epsilon \mathrm{D}`P_\epsilon \,.\supset .\, (\exists \beta) \,.\, \gamma = P``(C`P - \breve{P}``\beta)$

Dem.

$\vdash .\, *201{\cdot}53 \,.\supset$

$\vdash : \mathrm{Hp} \,.\, \gamma = \overrightarrow{P}`x \,.\, \beta = \overleftarrow{P_*}`x \,.\supset .\, C`P - \breve{P}``\beta = \overrightarrow{P_*}`x \,.$

$[*201{\cdot}53] \quad \supset .\, P``(C`P - \breve{P}``\beta) = \gamma \qquad (1)$

$\vdash .\, (1) \,.\, *217{\cdot}3{\cdot}301 \,.\supset \vdash .\, \mathrm{Prop}$

***217·32.** $\vdash : P \epsilon \mathrm{Ser} \,.\supset .\, \mathrm{D}`(\breve{P})_\epsilon = (\breve{P})_\epsilon ``(C`P -)``\mathrm{D}`P_\epsilon$

Dem.

$\vdash .\, *217{\cdot}31 \dfrac{\breve{P}}{P} \,.\supset \vdash : \mathrm{Hp} \,.\supset .\, \mathrm{D}`(\breve{P})_\epsilon \subset (\breve{P})_\epsilon ``(C`P -)``\mathrm{D}`P_\epsilon \qquad (1)$

$\vdash .\, (1) \,.\, *37{\cdot}16 \,.\supset \vdash .\, \mathrm{Prop}$

***217·33.** $\vdash .\, (\alpha -) \upharpoonright \mathrm{Cl}`\alpha \,\epsilon\, 1 \rightarrow 1$

Dem.

$\vdash .\, *24{\cdot}492 \,.\supset \vdash : \beta \subset \alpha \,.\, \gamma \subset \alpha \,.\, \alpha - \beta = \alpha - \gamma \,.\supset .\, \beta = \gamma : \supset \vdash .\, \mathrm{Prop}$

***217·34.** $\vdash : P \epsilon \mathrm{Ser} \,.\supset .\, P_\epsilon \upharpoonright (\mathrm{sect}`P - \mathrm{G}`\mathrm{lt}_P) \,\epsilon\, 1 \rightarrow 1$

Dem.

$\vdash .\, *211{\cdot}1 \,.\supset \vdash : \alpha, \beta \,\epsilon\, \mathrm{sect}`P \,.\, P``\alpha = P``\beta \,.\, \exists !\, \beta - \alpha \,.\supset .\, \exists !\, \beta - P``\beta \qquad (1)$

$\vdash .\, (1) \,.\, *205{\cdot}111 \,. \quad \supset \vdash : \mathrm{Hp} \,.\, \mathrm{Hp}(1) \,.\supset .\, \mathrm{E}!\, \mathrm{max}_P`\beta \qquad (2)$

$\vdash .\, *211{\cdot}56 \,. \quad \supset \vdash : \mathrm{Hp}(2) \,.\supset .\, \alpha \subset P``\beta \,. \qquad (3)$

$[*205{\cdot}111 . (2)] \quad \supset .\, \mathrm{max}_P`\beta \sim\epsilon\, \alpha \qquad (4)$

$\vdash .\, (3) \,. \quad \supset \vdash : \mathrm{Hp}(2) \,.\supset .\, \alpha = P``\beta \,.$

$[*205{\cdot}22 . (2) . \mathrm{Hp}] \quad \supset .\, \alpha = \overrightarrow{P}`\mathrm{max}_P`\beta = P``\alpha \qquad (5)$

$\vdash .\, (4) \,.\, (5) \,.\, *207{\cdot}232 \,.\supset \vdash : \mathrm{Hp}(2) \,.\supset .\, \mathrm{max}_P`\beta = \mathrm{lt}_P`\alpha \qquad (6)$

$\vdash .\, (6) \,.\, \mathrm{Transp} \,.\supset \vdash : \mathrm{Hp} \,.\, \alpha, \beta \,\epsilon\, \mathrm{sect}`P \,.\, P``\alpha = P``\beta \,.\sim \mathrm{E}!\, \mathrm{lt}_P`\alpha \,.\supset .\, \beta \subset \alpha \quad (7)$

Similarly $\vdash : \mathrm{Hp} \,.\, \alpha, \beta \,\epsilon\, \mathrm{sect}`P \,.\, P``\alpha = P``\beta \,.\sim \mathrm{E}!\, \mathrm{lt}_P`\beta \,.\supset .\, \alpha \subset \beta \quad (8)$

$\vdash .\, (7) \,.\, (8) \,.\supset \vdash : \mathrm{Hp} \,.\, \alpha, \beta \,\epsilon\, \mathrm{sect}`P - \mathrm{G}`\mathrm{lt}_P \,.\, P``\alpha = P``\beta \,.\supset .\, \alpha = \beta : \supset \vdash .\, \mathrm{Prop}$

***217·35.** $\vdash : P \epsilon \mathrm{Ser} . \supset . (\breve{P})_\epsilon \mid (C`P -) \upharpoonright \mathrm{D}`P_\epsilon \epsilon 1 \rightarrow 1$

Dem.

$\vdash . *217·33 . \supset \vdash . (C`P -) \upharpoonright \mathrm{D}`P_\epsilon \epsilon 1 \rightarrow 1$ (1)

$\vdash . *211·76 . \supset \vdash : \mathrm{Hp} . \supset . (C`P -)``\mathrm{D}`P_\epsilon = \mathrm{sect}`\breve{P} - \text{Ɑ}`\mathrm{tl}_P .$

[*217·34] $\supset . (\breve{P})_\epsilon \upharpoonright (C`P -)``\mathrm{D}`P_\epsilon \epsilon 1 \rightarrow 1$ (2)

$\vdash . (1) . (2) . \supset \vdash .$ Prop

***217·36.** $\vdash : P \epsilon \mathrm{Ser} . \supset . \varsigma`\breve{P} = (\breve{P})_\epsilon ; (C`P -) ; \mathrm{Cnv}`\varsigma`P$

Dem.

$\vdash . *212·23 . \supset$

$\vdash :. \mathrm{Hp} . \supset : \beta (\varsigma`P) \alpha . \gamma = \breve{P}``(C`P - \alpha) . \delta = \breve{P}``(C`P - \beta) . \supset .$

$\beta \subset \alpha . \alpha \neq \beta . C`P - \alpha \subset C`P - \beta .$

[*37·2.*217·35] $\supset . \gamma \subset \delta . \gamma \neq \delta .$

[*212·23] $\supset . \gamma (\varsigma`\breve{P}) \delta$ (1)

$\vdash . (1) . \supset \vdash : \mathrm{Hp} . \supset . (\breve{P})_\epsilon ; (C`P -) ; \mathrm{Cnv}`\varsigma`P \in \varsigma`\breve{P}$ (2)

$\vdash . (1) . \mathrm{Transp} . \supset$

$\vdash : \mathrm{Hp} . \delta (\varsigma`\breve{P}) \gamma . \gamma = \breve{P}``(C`P - \alpha) . \delta = \breve{P}``(C`P - \beta) . \alpha, \beta \epsilon \mathrm{D}`P_\epsilon . \supset . \alpha \subset \beta$ (3)

$\vdash . *217·35 . \supset \vdash : \mathrm{Hp}(3) . \supset . \alpha \neq \beta$ (4)

$\vdash . (3) . (4) . *212·23 . \supset \vdash : \mathrm{Hp}(3) . \supset . \delta \{(\breve{P})_\epsilon ; (C`P -) ; \mathrm{Cnv}`\varsigma`P\} \gamma$ (5)

$\vdash . *217·31 . \supset \vdash : \mathrm{Hp} . \delta (\varsigma`\breve{P}) \gamma . \supset .$

$(\exists \alpha, \beta) . \gamma = \breve{P}``(C`P - \alpha) . \delta = \breve{P}``(C`P - \beta) . \alpha, \beta \epsilon \mathrm{D}`P_\epsilon$ (6)

$\vdash . (5) . (6) . \supset \vdash : \mathrm{Hp} . \supset . \varsigma`\breve{P} \in (\breve{P})_\epsilon ; (C`P -) ; \mathrm{Cnv}`\varsigma`P$ (7)

$\vdash . (2) . (7) . \supset \vdash .$ Prop

***217·37.** $\vdash : P \epsilon \mathrm{Ser} . \supset . (\breve{P})_\epsilon \mid (C`P -) \upharpoonright \mathrm{D}`P_\epsilon \epsilon (\varsigma`\breve{P}) \overline{\mathrm{smor}} (\mathrm{Cnv}`\varsigma`P)$

[*217·35·36]

***217·38.** $\vdash : P \epsilon \mathrm{Ser} . \supset . (\varsigma`\breve{P}) \mathrm{smor} (\mathrm{Cnv}`\varsigma`P)$ [*217·37]

***217·4.** $\vdash : P, Q \epsilon \mathrm{Ser} . C`P \cap C`Q = \Lambda . \mathrm{E} ! B`P . \mathrm{E} ! B`Q . \supset .$

$\varsigma`(\breve{P} \dotplus Q) = (\breve{P})_\epsilon ; (C`P -) ; \mathrm{Cnv}`\varsigma`P \dotplus (C`P \cup) ; \varsigma`Q$ [*217·22·36]

***217·41.** $\vdash :. P, Q \epsilon \mathrm{Ser} . C`P \cap C`Q = \Lambda : \sim \mathrm{E} ! B`P . \mathrm{v} . \sim \mathrm{E} ! B`Q : \supset .$

$\varsigma`(\breve{P} \dotplus Q) = (\breve{P})_\epsilon ; (C`P -) ; \mathrm{Cnv}`\varsigma`P \dotplus (C`P \cup) ; (\varsigma`Q) \upharpoonright (- \iota`\Lambda)$

[*217·23·36]

$*217{\cdot}411$. $\vdash : \mathrm{Hp}\, *217{\cdot}41 \,.\supset .\, \{\varsigma\text{‘}(\breve{P} \mathbin{\text{⤉}} Q)\} \upharpoonleft (-\iota\text{‘}\Lambda) =$

$$(\breve{P})_\epsilon \,;\, (C\text{‘}P -) \,;\, \mathrm{Cnv}\text{‘}(\varsigma\text{‘}P) \upharpoonleft (-\iota\text{‘}\mathrm{D}\text{‘}P) \mathbin{\text{⤉}} (C\text{‘}P \cup) \,;\, (\varsigma\text{‘}Q) \upharpoonleft (-\iota\text{‘}\Lambda)$$

[$*217{\cdot}41$]

$*217{\cdot}42$. $\vdash : \mathrm{Hp}\, *217{\cdot}41 \,.\supset .\, \{\varsigma\text{‘}(\breve{P} \mathbin{\text{⤉}} Q)\} \upharpoonleft \{-\iota\text{‘}\Lambda - \iota\text{‘}\mathrm{D}\text{‘}(P \mathbin{\text{⤉}} Q)\} =$

$$(\breve{P})_\epsilon \,;\, (C\text{‘}P -) \,;\, \mathrm{Cnv}\text{‘}(\varsigma\text{‘}P) \upharpoonleft (-\iota\text{‘}\Lambda - \iota\text{‘}\mathrm{D}\text{‘}P) \mathbin{\text{⥅}} \text{ᗡ}\text{‘}P$$

$$\mathbin{\text{⤉}} (C\text{‘}P \cup) \,;\, (\varsigma\text{‘}Q) \upharpoonleft (-\iota\text{‘}\Lambda - \iota\text{‘}\mathrm{D}\text{‘}Q) \quad [*217{\cdot}411]$$

$*217{\cdot}43$. $\vdash :.\, P, Q \,\epsilon\, \mathrm{Ser} \,.\, C\text{‘}P \cap C\text{‘}Q = \Lambda : \sim \mathrm{E}\,!\, B\text{‘}P \,.\mathbf{v}.\, \sim \mathrm{E}\,!\, B\text{‘}Q :$

$X = (\varsigma\text{‘}P) \upharpoonleft (-\iota\text{‘}\Lambda - \iota\text{‘}\mathrm{D}\text{‘}P) \,.\, Y = (\varsigma\text{‘}Q) \upharpoonleft (-\iota\text{‘}\Lambda - \iota\text{‘}\mathrm{D}\text{‘}Q) \,.$

$Z = \{\varsigma\text{‘}(\breve{P} \mathbin{\text{⤉}} Q)\} \upharpoonleft \{-\iota\text{‘}\Lambda - \iota\text{‘}\mathrm{D}\text{‘}(\breve{P} \mathbin{\text{⤉}} Q)\} \,.$

$W = \breve{X} \mathbin{\text{⥅}} \alpha \mathbin{\text{⤉}} Y \,.\, \alpha \sim\epsilon\, C\text{‘}X \cup C\text{‘}Y \,.\supset .$

$(\breve{P})_\epsilon \,|\, (C\text{‘}P -) \upharpoonright (\mathrm{D}\text{‘}P_\epsilon - \iota\text{‘}\Lambda - \iota\text{‘}\mathrm{D}\text{‘}P) \mathbin{\dot\cup} (\mathrm{D}\text{‘}P) \downarrow \alpha$

$\mathbin{\dot\cup} (C\text{‘}P \cup) \upharpoonright (\mathrm{D}\text{‘}Q_\epsilon - \iota\text{‘}\Lambda - \iota\text{‘}\mathrm{D}\text{‘}Q) \,\epsilon\, Z \,\overline{\mathrm{smor}}\, W \qquad [*217{\cdot}37{\cdot}25{\cdot}42]$

SECTION C.

ON CONVERGENCE, AND THE LIMITS OF FUNCTIONS.

The purpose of this section is to express in a general form the definitions of convergence, the limits of functions, the continuity of functions, and kindred notions, and to give such elementary consequences of these definitions as may seem illustrative.

In the definitions usually given in treatises on analysis, it is assumed that both the arguments and the values of the function are numbers of some kind, generally real numbers, and limits are taken with respect to the order of magnitude. There is, however, nothing essential in the definitions to demand so narrow a hypothesis. What is essential is that the arguments should be given as belonging to a series, and that the values should also be given as belonging to a series, which need not be the same series as that to which the arguments belong. In what follows, therefore, we assume that all the possible arguments to our function, or at any rate all the arguments which we consider, belong to the field of a certain relation Q, which, in cases where our definitions are useful, will be a serial relation; we assume similarly that the values of our function, at least for arguments belonging to $C\text{‘}Q$, belong to the field of a relation P, which, in all important cases, will be a serial relation. The function itself we represent by the relation of the value to the argument; that is, the relation of $f(x)$ to x is to be R, so that, if the function is one-valued, $f(x)=R\text{‘}x$. (If the function is not one-valued, $f(x)$ is any member of $\overrightarrow{R}\text{‘}x$.) Thus we may speak of R as the function, Q as the argument-series, and P as the value-series.

To take an illustration: Suppose we are given a set of real numbers $x_1, x_2, \ldots x_\nu, \ldots$, where ν may be any finite integer. Here x_ν is a function of ν; the argument-series is that of the finite integers in order of magnitude, the value-series is that of the real numbers (or any part of this series which contains all the values $x_1, x_2, \ldots x_\nu, \ldots$). The function R is the relation of x_ν to ν, so that $x_\nu = R\text{‘}\nu$. In this case, calling the argument-series Q and the value-series P (as will be done throughout this section), we have $\text{Ɑ}\text{‘}R = C\text{‘}Q=$ the finite integers, $R\text{‘‘}C\text{‘}Q = \text{D}\text{‘}R=$ the *class* $x_1, x_2, \ldots x_\nu, \ldots$, and $R;Q$ = the *series* $x_1, x_2, \ldots x_\nu, \ldots$. The series which arranges $x_1, x_2, \ldots x_\nu, \ldots$ in the order of their own magnitudes, instead of the order of magnitude of their suffixes, is $P \upharpoonright \mathrm{D}`R$ or $P \upharpoonright R``C`Q$. This will not be equal to $R;Q$ unless the function is one which continually increases, *i.e.* one for which $\mu < \nu . \supset . x_\mu < x_\nu$.

In general, the propositions of the present section are only important when P and Q are series. If our assertions are not to be trivial, we must have $\exists ! C`Q \cap \mathrm{D}`R$ and $\exists ! C`P \cap R``C`Q$, *i.e.* there must be arguments in $C`Q$ which lead to values in $C`P$. It will also generally happen that the function is one-valued, *i.e.* that $R \in 1 \rightarrow \mathrm{Cls}$. But the above conditions, though necessary to the *importance* of our propositions, are in general much narrower than the hypotheses that are necessary for the *truth* of our propositions.

The present section is wholly self-contained, that is to say, its propositions are not referred to in the sequel. We have, in this section, carried the subject as far as seemed suitable for the present work; its further development belongs to treatises on analysis.

We begin (*230) with a general conception which is involved in the notion of convergency. We shall say that the values of a function converge (or, simply, that the function itself converges) into the class α, if for late enough arguments the values always belong to the class α, *i.e.* if there is a term y such that, if $yQ_* z$, $R`z \in \alpha$, or, to avoid assuming that $R$ is one-valued, $\overrightarrow{R}`z \subset \alpha$. Thus the values of the function converge into the class α if

$$(\exists y) . y \in C`Q \cap \text{Ǝ}`R . R``\overleftarrow{Q_*}`y \subset \alpha.$$

If a term y is one such that, from y onward, all values belong to α, we write $y \in R\overline{Q}_{\mathrm{cn}}\alpha$ (where "cn" stands for "convergent"), *i.e.* we put

$$R\overline{Q}_{\mathrm{cn}}\alpha = \hat{y}\{y \in C`Q \cap \text{Ǝ}`R . R``\overleftarrow{Q_*}`y \subset \alpha\} \quad \text{Df.}$$

When there is such a y, *i.e.* when the function converges into the class α, we write "$RQ_{\mathrm{cn}}\alpha$," *i.e.* we put

$$Q_{\mathrm{cn}} = \hat{R}\hat{\alpha}(\exists ! R\overline{Q}_{\mathrm{cn}}\alpha) \quad \text{Df.}$$

"$RQ_{\mathrm{cn}}\alpha$" may be read "R is Q-convergent into α." This means that for arguments sufficiently late in the Q-series, the value of the function is always a member of α. Thus *e.g.* if $R`x = 1/x$, and $\alpha = \hat{y}(y < 1)$, $RQ_{\mathrm{cn}}\alpha$, and if $z > 1$, $z \in R\overline{Q}_{\mathrm{cn}}\alpha$.

We next consider (*231) *limiting sections* and *ultimate oscillations* of functions. For this purpose, we proceed as follows. If $RQ_{\mathrm{cn}}\alpha$, then $P_*``\alpha$ is a section of the $P$-series such that, for sufficiently late arguments, the values of the function must belong to $P_*``\alpha$. Hence if we take all possible values of α for which $RQ_{\mathrm{cn}}\alpha$, and take the logical product of all the resulting sections $P_*``\alpha$, we get a section containing all the "ultimate" values of the function; moreover this is obviously the smallest section which has this property, because, if we take any section β which contains all the "ultimate" values, we have $RQ_{cn}\beta$, and $P_*\text{‘‘}\beta = \beta$, and therefore the logical product in question is contained in β. The logical product in question is

$$p\text{‘}P_*\text{‘‘‘}\overleftarrow{Q_{cn}}\text{‘}R.$$

In order to avoid trivial exceptions which arise when $C\text{‘}Q \cap \text{Ɑ}\text{‘}R = \Lambda$, we define the "limiting section" as

$$p\text{‘}P_*\text{‘‘‘}\overleftarrow{Q_{cn}}\text{‘}R \cap C\text{‘}P.$$

This "limiting section" we denote by $P\overline{R}_{sc}Q$, where the letters "sc" stand for "section." Thus we put

$$P\overline{R}_{sc}Q = p\text{‘}P_*\text{‘‘‘}\overleftarrow{Q_{cn}}\text{‘}R \cap C\text{‘}P \quad \text{Df.}$$

$P\overline{R}_{sc}Q$ is the class of those members x of the series P which are such that, given any argument however late, there are still arguments as late or later for which the value of the function is not less than x. In like manner, $\breve{P}\overline{R}_{sc}Q$, which we will call the "limiting upper section," consists of those members x of the series P which are such that, given any argument however late, there are still arguments as late or later for which the value of the function is not greater than x. Thus the product of $P\overline{R}_{sc}Q$ and $\breve{P}\overline{R}_{sc}Q$ is the smallest stretch which contains all the "ultimate" values of the function, *i.e.* it is the stretch consisting of those terms x which are such that, however late an argument we take, there are arguments as late or later for which the value of the function is not greater than x, and also arguments for which it is not less than x. Thus the product of $P\overline{R}_{sc}Q$ and $\breve{P}\overline{R}_{sc}Q$ represents what we may call the "ultimate oscillation" of the function. We shall denote it by $P\overline{R}_{os}Q$, putting

$$P\overline{R}_{os}Q = P\overline{R}_{sc}Q \cap \breve{P}\overline{R}_{sc}Q \quad \text{Df.}$$

We may express $P\overline{R}_{sc}Q$ in a form not involving Q_{cn}, namely ($*231{\cdot}12$)

$$P\overline{R}_{sc}Q = p\text{‘}P_*\text{‘‘‘}R\text{‘‘‘}\overleftarrow{Q_*}\text{‘‘}(C\text{‘}Q \cap \text{Ɑ}\text{‘}R) \cap C\text{‘}P.$$

This formula for $P\overline{R}_{sc}Q$ may be elucidated by the following considerations. If y is any member of $C\text{‘}Q$, then $\text{Ɑ}\text{‘}R \cap \overleftarrow{Q_*}\text{‘}y$ consists of all arguments from y onwards. Hence $R\text{‘‘}(\text{Ɑ}\text{‘}R \cap \overleftarrow{Q_*}\text{‘}y)$. *i.e.* $R\text{‘‘}\overleftarrow{Q_*}\text{‘}y$, consists of all values of the function for arguments from y onwards. Hence $P_*\text{‘‘}R\text{‘‘}\overleftarrow{Q_*}\text{‘}y$ consists of all members of the P-series which are equalled or surpassed by values of the function for arguments equal to or later than y. Now if a term x belongs to the class $P_*\text{‘‘}R\text{‘‘}\overleftarrow{Q_*}\text{‘}y$ for every argument y, it is a term such that, however far up the argument-series Q we go, we shall still find values as great as or greater than x. When this is the case, we may say that x is

P-persistent. In this case, x may be regarded as not greater than the "ultimate" values of the function. Now the class of arguments concerned is $C\text{‘}Q \cap \text{Ɑ}\text{‘}R$. Hence the class of P-persistent terms is

$$p\text{‘}P_*\text{‘‘‘}R\text{‘‘‘}\overleftarrow{Q_*}\text{‘‘}(C\text{‘}Q \cap \text{Ɑ}\text{‘}R),$$

where the factor $C\text{‘}P$ may be added in order to accommodate the formula to the trivial case where $C\text{‘}Q \cap \text{Ɑ}\text{‘}R = \Lambda$ (the only case in which the factor $C\text{‘}P$ makes any difference). Thus the class of P-persistent terms is the limiting section. Similarly the $\breve{P}$-persistent terms are the limiting upper section. These are the terms which are not less than the "ultimate" values of the function. Thus the product $P\overline{R}_{os}Q$ is the terms which are neither greater than all ultimate values, nor less; hence it is the class of ultimate values, which may be appropriately called the "ultimate oscillation."

It will be seen that $P\overline{R}_{os}Q$, being the product of an upper and lower section, is itself a stretch: we may call it (alternatively) the "limiting stretch." It consists of all members x of the P-series such that the function does not, however great we make the argument, become and remain less than x, nor yet become and remain greater than x. If $P\overline{R}_{os}Q$ consists of a single term, that term is the limit of the function as the argument travels up the series Q. (This is, of course, in general different from the limit of the values of the function considered simply as a class of members of $C\text{‘}P$, *i.e.* it is different from $\text{lt}_P\text{‘}R\text{‘‘}C\text{‘}Q$.) If $P\overline{R}_{os}Q$ does not consist of a single term or none, we shall have two limits to consider, namely $\text{limax}_P\text{‘}P\overline{R}_{os}Q$ and $\text{limin}_P\text{‘}P\overline{R}_{os}Q$, which give the two boundaries of the ultimate values of the function. When the class $P\overline{R}_{os}Q$ is null, the function may be regarded as having a definite limit: in this case, $P\overline{R}_{sc}Q$ and $\breve{P}\overline{R}_{sc}Q$ are the two parts of an "irrational" Dedekind cut, *i.e.* a cut in which the first portion has no maximum and the second no minimum. Thus $P\overline{R}_{os}Q \in 0 \cup 1$ is the condition for a definite limit of the function as the argument grows indefinitely.

The above gives the generalization of the limit of a function when the argument may be any member of $C\text{‘}Q \cap \text{Ɑ}\text{‘}R$. In order to obtain limits for other classes of arguments, it is only necessary, as a rule, to limit the field of Q to the class of arguments in question, *i.e.* to replace Q by $Q \upharpoonright \alpha$ (cf. $*232$). In order, however, to avoid vexatious and trivial exceptions arising when $\alpha \in 1$, it is more convenient to replace Q by $Q_* \upharpoonright \alpha$. Thus the section of P defined by the class of arguments α is $P\overline{R}_{sc}(Q_* \upharpoonright \alpha)$. We put

$$(P\overline{R}Q)_{sc}\text{‘}\alpha = P\overline{R}_{sc}(Q_* \upharpoonright \alpha) \quad \text{Df.}$$

This definition is useful because we very often wish to be able to exhibit the limiting section defined by α as a function of α. The section $(P\overline{R}Q)_{sc}\text{‘}\alpha$ is such that, if x is any member of it, and y is any argument belonging to α, there is in α an argument equal to or later than y, for which the function has a value equal to or later than x. Thus x is such that the function does not ultimately become less than x as the argument increases in the class α. The limit or maximum of such terms as x is the limit or maximum of the ultimate values of the function as the argument approaches the top of α. The class of ultimate values is

$$(P\overline{R}Q)_{\text{sc}}\text{‘}\alpha \cap (\breve{P}\overline{R}Q)_{\text{sc}}\text{‘}\alpha, \text{ which we call } (P\overline{R}Q)_{\text{os}}\text{‘}\alpha.$$

If the function has a definite limit as the argument increases in α, the class of ultimate values must not contain more than one term.

Our next number (*233) deals with the limit of a function for a given argument. The limit or maximum of the class of ultimate values is not necessarily the value for the limit of α. It will be found, however, that, with a suitable hypothesis, the limiting section $(P\overline{R}Q)_{\text{sc}}\text{‘}\alpha$ depends only upon $Q_*\text{“}(\alpha \cap \text{Ꭰ‘}R)$, and if $\alpha \cap \text{Ꭰ‘}R$ has no maximum, it depends only upon $Q\text{“}(\alpha \cap \text{Ꭰ‘}R)$. Thus if $\alpha \cap \text{Ꭰ‘}R$ and $\beta \cap \text{Ꭰ‘}R$ both have the same limit, they define the same limiting section. Hence if a is the limit of α, the limiting section of α is $(P\overline{R}Q)_{\text{sc}}\text{‘}\overrightarrow{Q}\text{‘}a$. The upper limit of this is the upper limit of the ultimate values as the argument approaches a from below. We put

$$R\,(PQ)\text{‘}a = \text{limax}_P\text{‘}(P\overline{R}Q)_{\text{sc}}\text{‘}\overrightarrow{Q}\text{‘}a \quad \text{Df.}$$

We have thus four limits of the function as the argument approaches a, namely

$$R\,(PQ)\text{‘}a,\ \ R\,(\breve{P}Q)\text{‘}a,\ \ R\,(P\breve{Q})\text{‘}a,\ \ R\,(\breve{P}\breve{Q})\text{‘}a.$$

If R is a continuous function, these four are all equal to $R\text{‘}a$; but in general they are different from each other and from $R\text{‘}a$. The subject of the continuity of functions is dealt with in *234. When $R\,(PQ)\text{‘}a = R\,(\breve{P}Q)\text{‘}a$, each is the limit of the function for the argument a for approaches from below. It should be observed that if R is defined for a set of arguments which are dense in Q, *i.e.* if $\delta_Q\text{‘Ꭰ‘}R = C\text{‘}Q$, then $R\,(PQ)\text{‘}a$ and $R\,(\breve{P}Q)\text{‘}a$ are defined for all arguments in $C\text{‘}Q$.

$*230$. ON CONVERGENTS.

Summary of $*230$.

In the present number, we have to consider the notion of a function converging into a given class, or, as we may express it, the notion that the value of the function "ultimately" belongs to the given class. If R is the function in question, α the given class, and Q a series to which the arguments belong, we say that "R is Q-convergent into α" if there is an argument y such that, for all arguments from y onward (in the Q-order), the value of the function is an α. That is, R is Q-convergent into α if

$$(\exists y) \,.\, y \in C\text{‘}Q \cap \text{Ɑ}\text{‘}R \,.\, R\text{“}\overleftarrow{Q_*}\text{‘}y \subset \alpha.$$

A term y which is of this nature is said to belong to the class $R\overline{Q}_{\text{cn}}\alpha$. Thus R is Q-convergent into α if the class $R\overline{Q}_{\text{cn}}\alpha$ is not null. Hence we have the following pair of definitions:

$$R\overline{Q}_{\text{cn}}\alpha = C\text{‘}Q \cap \text{Ɑ}\text{‘}R \cap \hat{y}\,(R\text{“}\overleftarrow{Q_*}\text{‘}y \subset \alpha) \quad \text{Df},$$

$$Q_{\text{cn}} = \hat{R}\hat{\alpha}\,(\exists \,!\, R\overline{Q}_{\text{cn}}\alpha) \quad \text{Df}.$$

In all the cases that have any importance, R will be a one-valued function (*i.e.* a one-many relation), Q will be a series, and $C\text{‘}Q \cap \text{Ɑ}\text{‘}R$ will be a class having no maximum in Q. For, if $C\text{‘}Q \cap \text{Ɑ}\text{‘}R$ has a maximum in Q, then the classes into which R converges are simply those to which the value for this maximum belongs. The following propositions, though only *important* under the above circumstances, are in general *true* under much wider hypotheses.

It is possible to generalize still further the notion of convergence, so as to apply to any property which belongs to R when confined to sufficiently late arguments. For this purpose, we have to consider $R \upharpoonright \overleftarrow{Q_*}\text{‘}z$, where z is to be confined to terms later than or equal to some term y. If, under these circumstances, $R \upharpoonright \overleftarrow{Q_*}\text{‘}z$ always belongs to the class λ, we may say that R ultimately becomes a λ. We may put

$$R\overline{Q}_{\text{cng}}\lambda = \hat{y}\,\{y \in C\text{‘}Q \cap \text{Ɑ}\text{‘}R : yQ_*z \,.\, \supset_z .\, R \upharpoonright \overleftarrow{Q_*}\text{‘}z \in \lambda\} \quad \text{Df},$$

$$Q_{\text{cng}} = \hat{R}\hat{\lambda}\,(\exists \,!\, R\overline{Q}_{\text{cng}}\lambda) \quad \text{Df}.$$

This is the general conception of which Q_{cn} is a particular case; in fact,

$$\vdash : RQ_{cn}\alpha . \equiv . RQ_{cng}(\breve{D}``Cl`\alpha).$$

Q_{cng} will have to be used when the ultimate properties of the function with which we are concerned are not properties of its values; but when they are properties of its values, Q_{cn} enables us to deal with them more easily than Q_{cng}.

In this number, we prove the following propositions among others:

***230·171.** $\vdash : y \epsilon R\overline{Q}_{cn}(\overleftarrow{P_*}`x) . \supset . x \epsilon P_*``R``\overleftarrow{Q_*}`y$

***230·211.** $\vdash :. \alpha \subset \beta . \supset : RQ_{cn}\alpha . \supset . RQ_{cn}\beta$

***230·253.** $\vdash :. R``C`Q \subset \alpha . \supset : RQ_{cn}\alpha . \equiv . \exists ! C`Q \cap \text{C\!I}`R .$
$\equiv . \exists ! R``C`Q . \equiv . \dot{\exists} ! (R \upharpoonright C`Q)$

***230·4.** $\vdash . R\overline{Q}_{cn}\alpha = \text{C\!I}`R \cap \breve{Q}_*``(R\overline{Q}_{cn}\alpha)$

***230·42.** $\vdash :. Q_* \epsilon \text{connex} . \supset : RQ_{cn}\alpha . RQ_{cn}\beta . \equiv . RQ_{cn}(\alpha \cap \beta)$

***230·53.** $\vdash :. Q \epsilon \text{trans} \cap \text{connex} . E ! \max_Q`\text{C\!I}`R . \supset : RQ_{cn}\alpha . \equiv . \overrightarrow{R}`\max_Q`\text{C\!I}`R \subset \alpha$

In virtue of this proposition, the case when $E ! \max_Q`\text{C\!I}`R$ is uninteresting, and in order to obtain interesting interpretations of our propositions, it is necessary to suppose that $\text{C\!I}`R$ has no maximum. Similarly when, in later numbers, we consider $\text{C\!I}`R \cap \overrightarrow{Q}`x$, we shall only obtain interesting results when this has no maximum, which requires that $Q$ should be a compact series ($Q^2 = Q$) and $\text{C\!I}`R$ should be dense in Q. These assumptions are, however, not usually required for the *truth* of our propositions.

***230·01.** $R\overline{Q}_{cn}\alpha = C`Q \cap \text{C\!I}`R \cap \hat{y}(R``\overleftarrow{Q_*}`y \subset \alpha)$ Df

***230·02.** $Q_{cn} = \hat{R}\hat{\alpha}(\exists ! R\overline{Q}_{cn}\alpha)$ Df

***230·1.** $\vdash : y \epsilon R\overline{Q}_{cn}\alpha . \equiv . y \epsilon C`Q \cap \text{C\!I}`R . R``\overleftarrow{Q_*}`y \subset \alpha$ [(*230·01)]

***230·11.** $\vdash : RQ_{cn}\alpha . \equiv . \exists ! R\overline{Q}_{cn}\alpha . \equiv . (\exists y) . y \epsilon C`Q \cap \text{C\!I}`R . R``\overleftarrow{Q_*}`y \subset \alpha$
[(*230·02)]

***230·12.** $\vdash : y \epsilon R\overline{Q}_{cn}\alpha . \supset . \overleftarrow{Q_*}`y \cap \text{C\!I}`R \subset R\overline{Q}_{cn}\alpha$

Dem.

$\vdash . *230·1 . *201·14·15 . \supset$

$\vdash : y \epsilon R\overline{Q}_{cn}\alpha . yQ_*z . z \epsilon \text{C\!I}`R . \supset . R``\overleftarrow{Q_*}`y \subset \alpha . z \epsilon C`Q \cap \text{C\!I}`R . \overleftarrow{Q_*}`z \subset \overleftarrow{Q_*}`y .$

[*37·2] $\supset . z \epsilon C`Q \cap \text{C\!I}`R . R``\overleftarrow{Q_*}`z \subset \alpha .$

[*230·1] $\supset . z \epsilon R\overline{Q}_{cn}\alpha : \supset \vdash . \text{Prop}$

$*230{\cdot}13$. $\vdash . R\overline{Q}_{cn}\alpha = (R \upharpoonright C'Q)\,\overline{Q}_{cn}\alpha$

Dem.

$\vdash . *35{\cdot}64 . \supset \vdash . C'Q \cap \text{Ɑ}'R = C'Q \cap \text{Ɑ}'(R \upharpoonright C'Q)$ (1)

$\vdash . *37{\cdot}421 . \supset \vdash . R``\overleftarrow{Q_*}`y = (R \upharpoonright C'Q)``\overleftarrow{Q_*}`y$ (2)

$\vdash . (1) . (2) . *230{\cdot}1 . \supset \vdash . \mathrm{Prop}$

$*230{\cdot}131$. $\vdash : R \upharpoonright C'Q = T \upharpoonright C'Q . \supset . R\overline{Q}_{cn}\alpha = T\overline{Q}_{cn}\alpha$ [$*230{\cdot}13$]

$*230{\cdot}14$. $\vdash : y \in R\overline{Q}_{cn}\alpha . \supset . \exists ! C'Q \cap \text{Ɑ}'R . \exists ! \alpha \cap \mathrm{D}'R$

Dem.

$\vdash . *230{\cdot}1 . \supset \vdash : \mathrm{Hp} . \supset . y \in C'Q \cap \text{Ɑ}'R . \overrightarrow{R}`y \subset \alpha .$

[$*33{\cdot}41$] $\supset . y \in C'Q \cap \text{Ɑ}'R . \exists ! \overrightarrow{R}`y . \overrightarrow{R}`y \subset \alpha .$

[$*22{\cdot}621 . *33{\cdot}15$] $\supset . \exists ! C'Q \cap \text{Ɑ}'R . \exists ! \alpha \cap \mathrm{D}'R : \supset \vdash . \mathrm{Prop}$

$*230{\cdot}141$. $\vdash . R\overline{Q}_{cn}\Lambda = \Lambda$ [$*230{\cdot}14$. Transp]

$*230{\cdot}142$. $\vdash :. R = \dot{\Lambda} . \vee . Q = \dot{\Lambda} : \supset . R\overline{Q}_{cn}\alpha = \Lambda$ [$*230{\cdot}14$. Transp . $*33{\cdot}24$]

$*230{\cdot}15$. $\vdash : RQ_{cn}\alpha . \supset . \exists ! C'Q \cap \text{Ɑ}'R . \exists ! \alpha \cap \mathrm{D}'R$ [$*230{\cdot}14{\cdot}11$]

$*230{\cdot}151$. $\vdash : RQ_{cn}\alpha . \supset . \dot{\exists} ! R . \dot{\exists} ! Q . \exists ! \alpha$ [$*230{\cdot}15$]

$*230{\cdot}152$. $\vdash :. R = \dot{\Lambda} . \vee . Q = \dot{\Lambda} . \vee . \alpha = \Lambda : \supset . \sim(RQ_{cn}\alpha)$ [$*230{\cdot}151$. Transp]

$*230{\cdot}16$. $\vdash . R\overline{Q}_{cn}\alpha = R\,(\overline{Q_* \upharpoonleft \text{Ɑ}'R})_{cn}\alpha$

Dem.

$\vdash . *230{\cdot}14 . \supset \vdash : C'Q \cap \text{Ɑ}'R = \Lambda . \supset . R\overline{Q}_{cn}\alpha = \Lambda . R\,(\overline{Q_* \upharpoonleft \text{Ɑ}'R})_{cn}\alpha = \Lambda$ (1)

$\vdash . *90{\cdot}41 . \supset \vdash : \exists ! C'Q \cap \text{Ɑ}'R . \supset . C'(Q_* \upharpoonleft \text{Ɑ}'R) = C'Q \cap \text{Ɑ}'R$ (2)

$\vdash . *37{\cdot}26 . \supset \vdash . R``\overleftarrow{Q_*}`y = R``(\overleftarrow{Q_*}`y \cap \text{Ɑ}'R)$ (3)

$\vdash . (3) . *35{\cdot}102 . \supset \vdash : y \in \text{Ɑ}'R . \supset . R``\overleftarrow{Q_*}`y = R``\overleftarrow{Q_* \upharpoonleft \text{Ɑ}'R}`y$ (4)

$\vdash . (2) . (4) . *230{\cdot}1 . \supset$

$\vdash :. \exists ! C'Q \cap \text{Ɑ}'R . \supset : y \in R\overline{Q}_{cn}\alpha . \equiv . y \in C'(Q_* \upharpoonleft \text{Ɑ}'R) \cap \text{Ɑ}'R . R``\overleftarrow{Q_* \upharpoonleft \text{Ɑ}'R}`y \subset \alpha .$

[$*230{\cdot}1$] $\equiv . y \in R\,(\overline{Q_* \upharpoonleft \text{Ɑ}'R})_{cn}\alpha$ (5)

$\vdash . (1) . (5) . \supset \vdash . \mathrm{Prop}$

$*230{\cdot}161$. $\vdash : Q_* \upharpoonleft \text{Ɑ}'R = S_* \upharpoonleft \text{Ɑ}'R . \supset . R\overline{Q}_{cn}\alpha = R\overline{S}_{cn}\alpha$ [$*230{\cdot}16$]

$*230{\cdot}17$. $\vdash : y \in C'Q \cap \text{Ɑ}'R . \supset . \exists ! R``\overleftarrow{Q_*}`y$

Dem.

$\vdash . *90{\cdot}12 . *33{\cdot}41 . \supset \vdash : \mathrm{Hp} . \supset . y \in \overleftarrow{Q_*}`y . \exists ! \overrightarrow{R}`y .$

[$*37{\cdot}18$] $\supset . \exists ! R``\overleftarrow{Q_*}`y : \supset \vdash . \mathrm{Prop}$

***230·171.** $\vdash : y \in R\bar{Q}_{cn}(\overleftarrow{P_*}\text{‘}x) . \supset . x \in P_*\text{“}R\text{“}\overleftarrow{Q_*}\text{‘}y$

Dem.

$\vdash . *230\cdot1\cdot17 . \supset \vdash : \text{Hp} . \supset . \exists ! R\text{“}\overleftarrow{Q_*}\text{‘}y . R\text{“}\overleftarrow{Q_*}\text{‘}y \subset \overleftarrow{P_*}\text{‘}x .$

$[*22\cdot621] \qquad \supset . \exists ! R\text{“}\overleftarrow{Q_*}\text{‘}y \cap \overleftarrow{P_*}\text{‘}x .$

$[*37\cdot46] \qquad \supset . x \in P_*\text{“}R\text{“}\overleftarrow{Q_*}\text{‘}y : \supset \vdash . \text{Prop}$

***230·21.** $\vdash : \alpha \subset \beta . \supset . R\bar{Q}_{cn}\alpha \subset R\bar{Q}_{cn}\beta \qquad [*230\cdot1 . *22\cdot44]$

***230·211.** $\vdash :. \alpha \subset \beta . \supset : RQ_{cn}\alpha . \supset . RQ_{cn}\beta \qquad [*230\cdot21\cdot11]$

***230·22.** $\vdash . R\bar{Q}_{cn}\alpha \cup R\bar{Q}_{cn}\beta \subset R\bar{Q}_{cn}(\alpha \cup \beta) \qquad [*230\cdot21]$

***230·221.** $\vdash :. RQ_{cn}\alpha . \vee . RQ_{cn}\beta : \supset . RQ_{cn}(\alpha \cup \beta) \qquad [*230\cdot211]$

***230·23.** $\vdash . R\bar{Q}_{cn}\alpha \cap R\bar{Q}_{cn}\beta = R\bar{Q}_{cn}(\alpha \cap \beta)$

Dem.

$\vdash . *230\cdot1 . \supset \vdash : y \in R\bar{Q}_{cn}\alpha \cap R\bar{Q}_{cn}\beta . \equiv . y \in C\text{‘}Q \cap \text{Ɑ}\text{‘}R . R\text{“}\overleftarrow{Q_*}\text{‘}y \subset \alpha . R\text{“}\overleftarrow{Q_*}\text{‘}y \subset \beta .$

$[\text{Comp}.*230\cdot1] \qquad \equiv . y \in R\bar{Q}_{cn}(\alpha \cap \beta) : \supset \vdash . \text{Prop}$

***230·231.** $\vdash : RQ_{cn}(\alpha \cap \beta) . \supset . RQ_{cn}\alpha . RQ_{cn}\beta \qquad [*230\cdot211]$

***230·24.** $\vdash . R\bar{Q}_{cn}\alpha \cap R\bar{Q}_{cn}(\beta - \alpha) = \Lambda \qquad [*230\cdot23\cdot141]$

***230·25.** $\vdash . R\bar{Q}_{cn}\alpha = R\bar{Q}_{cn}(\alpha \cap \text{D}\text{‘}R) = R\bar{Q}_{cn}(\alpha \cap R\text{“}\breve{Q}_*\text{“}\text{Ɑ}\text{‘}R) = R\bar{Q}_{cn}(\alpha \cap R\text{“}C\text{‘}Q)$

Dem.

$\vdash . *37\cdot15 . \supset \vdash : R\text{“}\overleftarrow{Q_*}\text{‘}y \subset \alpha . \equiv . R\text{“}\overleftarrow{Q_*}\text{‘}y \subset \alpha \cap \text{D}\text{‘}R \qquad (1)$

$\vdash . *37\cdot18 . \supset \vdash :. y \in \text{Ɑ}\text{‘}R . \supset : R\text{“}\overleftarrow{Q_*}\text{‘}y \subset R\text{“}\breve{Q}_*\text{“}\text{Ɑ}\text{‘}R :$

$[\text{Comp}] \qquad \supset : R\text{“}\overleftarrow{Q_*}\text{‘}y \subset \alpha . \equiv . R\text{“}\overleftarrow{Q_*}\text{‘}y \subset \alpha \cap R\text{“}\breve{Q}_*\text{“}\text{Ɑ}\text{‘}R \qquad (2)$

$\vdash . *37\cdot2\cdot18 . \supset \vdash . R\text{“}\overleftarrow{Q_*}\text{‘}y \subset R\text{“}C\text{‘}Q .$

$[\text{Comp}] \qquad \supset \vdash : R\text{“}\overleftarrow{Q_*}\text{‘}y \subset \alpha . \equiv . R\text{“}\overleftarrow{Q_*}\text{‘}y \subset \alpha \cap R\text{“}C\text{‘}Q \qquad (3)$

$\vdash . (1) . (2) . (3) . *230\cdot1 . \supset \vdash . \text{Prop}$

***230·251.** $\vdash . R\bar{Q}_{cn}(R\text{“}C\text{‘}Q) = C\text{‘}Q \cap \text{Ɑ}\text{‘}R$

Dem.

$\vdash . *33\cdot15 . *37\cdot2 . \supset \vdash . (y) . R\text{“}\overleftarrow{Q_*}\text{‘}y \subset R\text{“}C\text{‘}Q \qquad (1)$

$\vdash . (1) . *230\cdot1 . \supset \vdash . \text{Prop}$

***230·252.** $\vdash : R\text{“}C\text{‘}Q \subset \alpha . \supset . R\bar{Q}_{cn}\alpha = C\text{‘}Q \cap \text{Ɑ}\text{‘}R \qquad [*230\cdot25\cdot251]$

***230·253.** $\vdash :. R\text{“}C\text{‘}Q \subset \alpha . \supset : RQ_{cn}\alpha . \equiv . \exists ! C\text{‘}Q \cap \text{Ɑ}\text{‘}R . \equiv .$

$\exists ! R\text{“}C\text{‘}Q . \equiv . \dot{\exists} ! (R \upharpoonright C\text{‘}Q) \qquad [*230\cdot11\cdot252 . *37\cdot401 . *35\cdot64]$

***230·31.** $\vdash . s`R\overline{Q}_{cn}``\kappa \subset R\overline{Q}_{cn}(s`\kappa)$

Dem.

$\vdash . *230·21 . \supset \vdash : \alpha \epsilon \kappa . \supset . R\overline{Q}_{cn}\alpha \subset R\overline{Q}_{cn}(s`\kappa) : \supset \vdash . \text{Prop}$

***230·311.** $\vdash . Q_{cn}``\kappa \subset \overrightarrow{Q_{cn}}`s`\kappa$

Dem.

$\vdash . *230·211 . \supset \vdash : \alpha \epsilon \kappa . RQ_{cn}\alpha . \supset . RQ_{cn}(s`\kappa) : \supset \vdash . \text{Prop}$

***230·32.** $\vdash . R\overline{Q}_{cn}(p`\kappa) = C`Q \cap \text{Ɑ}`R \cap p`R\overline{Q}_{cn}``\kappa$

Dem.

$\vdash . *230·1 . \supset$

$\vdash :. y \epsilon R\overline{Q}_{cn}(p`\kappa) . \equiv : y \epsilon C`Q \cap \text{Ɑ}`R . R``\overleftarrow{Q_*}`y \subset p`\kappa :$

[*40·15] $\equiv : y \epsilon C`Q \cap \text{Ɑ}`R : \alpha \epsilon \kappa . \supset_\alpha . R``\overleftarrow{Q_*}`y \subset \alpha :$

[*4·73] $\equiv : y \epsilon C`Q \cap \text{Ɑ}`R : \alpha \epsilon \kappa . \supset_\alpha . y \epsilon C`Q \cap \text{Ɑ}`R . R``\overleftarrow{Q_*}`y \subset \alpha :$

[*230·1] $\equiv : y \epsilon C`Q \cap \text{Ɑ}`R \cap p`R\overline{Q}_{cn}``\kappa :. \supset \vdash . \text{Prop}$

***230·321.** $\vdash : \exists ! \kappa . \supset . R\overline{Q}_{cn}(p`\kappa) = p`R\overline{Q}_{cn}``\kappa . p`R\overline{Q}_{cn}``\kappa \subset C`Q \cap \text{Ɑ}`R$

Dem.

$\vdash . *230·1 . \quad \supset \vdash : \alpha \epsilon \kappa . \supset_\alpha . R\overline{Q}_{cn}\alpha \subset C`Q \cap \text{Ɑ}`R \quad (1)$

$\vdash . (1) . *40·23·151 . \supset \vdash : \text{Hp} . \supset . p`R\overline{Q}_{cn}``\kappa \subset C`Q \cap \text{Ɑ}`R \quad (2)$

$\vdash . (2) . *230·32 . \supset \vdash . \text{Prop}$

***230·4.** $\vdash . R\overline{Q}_{cn}\alpha = \text{Ɑ}`R \cap \breve{Q}_*``(R\overline{Q}_{cn}\alpha)$

Dem.

$\vdash . *230·11 . *90·21 . \supset \vdash . R\overline{Q}_{cn}\alpha \subset \text{Ɑ}`R . R\overline{Q}_{cn}\alpha \subset \breve{Q}_*``(R\overline{Q}_{cn}\alpha) \quad (1)$

$\vdash . *201·14·15 . \quad \supset \vdash : R``\overleftarrow{Q_*}`y \subset \alpha . yQ_*z . \supset . R``\overleftarrow{Q_*}`z \subset \alpha \quad (2)$

$\vdash . (2) . *230·1 . \quad \supset \vdash : y \epsilon R\overline{Q}_{cn}\alpha . yQ_*z . z \epsilon \text{Ɑ}`R . \supset . z \epsilon R\overline{Q}_{cn}\alpha :$

[*37·105] $\supset \vdash . \text{Ɑ}`R \cap \breve{Q}_*``(R\overline{Q}_{cn}\alpha) \subset R\overline{Q}_{cn}\alpha \quad (3)$

$\vdash . (1) . (3) . \supset \vdash . \text{Prop}$

***230·41.** $\vdash :. Q_* \epsilon \text{connex} . \supset : R\overline{Q}_{cn}\alpha \subset R\overline{Q}_{cn}\beta . \vee . R\overline{Q}_{cn}\beta \subset R\overline{Q}_{cn}\alpha$

Dem.

$\vdash . *211·61 . *201·15 . \supset$

$\vdash :. \text{Hp} . \supset : \breve{Q}_*``(R\overline{Q}_{cn}\alpha) \subset \breve{Q}_*``(R\overline{Q}_{cn}\beta) . \vee . \breve{Q}_*``(R\overline{Q}_{cn}\beta) \subset \breve{Q}_*``(R\overline{Q}_{cn}\alpha) :$

[Fact.*230·4] $\supset : R\overline{Q}_{cn}\alpha \subset R\overline{Q}_{cn}\beta . \vee . R\overline{Q}_{cn}\beta \subset R\overline{Q}_{cn}\alpha :. \supset \vdash . \text{Prop}$

***230·42.** $\vdash :. Q_* \epsilon \text{connex} . \supset : RQ_{cn}\alpha . RQ_{cn}\beta . \equiv . RQ_{cn}(\alpha \cap \beta)$

Dem.

$\vdash . *230·41 . \supset \vdash :. \text{Hp} . \supset : R\overline{Q}_{cn}\alpha \cap R\overline{Q}_{cn}\beta = R\overline{Q}_{cn}\alpha . \vee . R\overline{Q}_{cn}\alpha \cap R\overline{Q}_{cn}\beta = R\overline{Q}_{cn}\beta :$

[*230·23] $\supset : R\overline{Q}_{cn}(\alpha \cap \beta) = R\overline{Q}_{cn}\alpha . \vee . R\overline{Q}_{cn}(\alpha \cap \beta) = R\overline{Q}_{cn}\beta :$

[*230·11] $\supset : RQ_{cn}\alpha . RQ_{cn}\beta . \supset . RQ_{cn}(\alpha \cap \beta) \quad (1)$

$\vdash . (1) . *230·231 . \supset \vdash . \text{Prop}$

***230·421.** $\vdash : Q_* \in \text{connex} \,.\, \alpha \cap \beta = \Lambda \,.\, \supset .\, \sim \{R Q_{\text{cn}} \alpha \,.\, R Q_{\text{cn}} \beta\}$ [*230·42·141]

***230·51.** $\vdash : R Q_{\text{cn}} \alpha \,.\, \supset .\, p\text{‘}\overleftarrow{Q_*}\text{“}C\text{‘}Q \cap \text{ᗡ}\text{‘}R \subset R\overline{Q}_{\text{cn}} \alpha$

Dem.

$\vdash .\, *201·14 \,.\, \supset \vdash : y \in C\text{‘}Q \cap \text{ᗡ}\text{‘}R \,.\, R\text{“}\overleftarrow{Q_*}\text{‘}y \subset \alpha \,.\, z \in p\text{‘}\overleftarrow{Q_*}\text{“}C\text{‘}Q \,.\, \supset .\, R\text{“}\overleftarrow{Q_*}\text{‘}z \subset \alpha$ (1)

$\vdash .\, *230·151 \,.\, *40·62 \,.\, \supset \vdash : \text{Hp} \,.\, \supset .\, p\text{‘}\overleftarrow{Q_*}\text{“}C\text{‘}Q \subset C\text{‘}Q$ (2)

$\vdash .\, (1) \,.\, (2) \,.\, *230·1 \,.\, \supset \vdash : \text{Hp} \,.\, y \in R\overline{Q}_{\text{cn}} \alpha \,.\, z \in p\text{‘}\overleftarrow{Q_*}\text{“}C\text{‘}Q \cap \text{ᗡ}\text{‘}R \,.\, \supset .$

$z \in C\text{‘}Q \cap \text{ᗡ}\text{‘}R \,.\, R\text{“}\overleftarrow{Q_*}\text{‘}z \subset \alpha$ (3)

$\vdash .\, (3) \,.\, *230·1·11 \,.\, \supset \vdash .$ Prop

***230·511.** $\vdash : y \in p\text{‘}\overleftarrow{Q_*}\text{“}C\text{‘}Q \,.\, \supset .\, \overleftarrow{Q_*}\text{‘}y = p\text{‘}\overleftarrow{Q_*}\text{“}C\text{‘}Q$

Dem.

$\vdash .\, *40·12 \,.\, \supset \vdash : \text{Hp} \,.\, \supset .\, p\text{‘}\overleftarrow{Q_*}\text{“}C\text{‘}Q \subset \overleftarrow{Q_*}\text{‘}y$ (1)

$\vdash .\, *40·53 \,.\, \supset \vdash :. \text{Hp} \,.\, z \in \overleftarrow{Q_*}\text{‘}y \,.\, \supset : x \in C\text{‘}Q \,.\, \supset_x .\, x Q_* y : y Q_* z :$

[*201·15] $\supset : x \in C\text{‘}Q \,.\, \supset_x .\, x Q_* z :$

[*40·53] $\supset : z \in p\text{‘}\overleftarrow{Q_*}\text{“}C\text{‘}Q$ (2)

$\vdash .\, (1) \,.\, (2) \,.\, \supset \vdash .$ Prop

***230·512.** $\vdash : \text{ᗡ}\text{‘}R \cap p\text{‘}\overleftarrow{Q_*}\text{“}C\text{‘}Q \subset R\overline{Q}_{\text{cn}} \alpha \,.\, \supset .\, R\text{“}p\text{‘}\overleftarrow{Q_*}\text{“}C\text{‘}Q \subset \alpha$

Dem.

$\vdash .\, *230·1 \,.\, \supset \vdash :. \text{Hp} \,.\, \supset : y \in \text{ᗡ}\text{‘}R \cap p\text{‘}\overleftarrow{Q_*}\text{“}C\text{‘}Q \,.\, \supset .\, R\text{“}\overleftarrow{Q_*}\text{‘}y \subset \alpha \,.$

[*230·511] $\supset .\, R\text{“}p\text{‘}\overleftarrow{Q_*}\text{“}C\text{‘}Q \subset \alpha$ (1)

$\vdash .\, (1) \,.\, *10·23 \,.\, \supset \vdash : \text{Hp} \,.\, \exists ! \, \text{ᗡ}\text{‘}R \cap p\text{‘}\overleftarrow{Q_*}\text{“}C\text{‘}Q \,.\, \supset .\, R\text{“}p\text{‘}\overleftarrow{Q_*}\text{“}C\text{‘}Q \subset \alpha$ (2)

$\vdash .\, *37·26·29 \,.\, \supset \vdash : \text{ᗡ}\text{‘}R \cap p\text{‘}\overleftarrow{Q_*}\text{“}C\text{‘}Q = \Lambda \,.\, \supset .\, R\text{“}p\text{‘}\overleftarrow{Q_*}\text{“}C\text{‘}Q = \Lambda \,.$

[*24·12] $\supset .\, R\text{“}p\text{‘}\overleftarrow{Q_*}\text{“}C\text{‘}Q \subset \alpha$ (3)

$\vdash .\, (2) \,.\, (3) \,.\, \supset \vdash .$ Prop

***230·513.** $\vdash :. \dot{\exists} ! \, Q \,.\, \supset : R\text{“}p\text{‘}\overleftarrow{Q_*}\text{“}C\text{‘}Q \subset \alpha \,.\, \equiv .\, \text{ᗡ}\text{‘}R \cap p\text{‘}\overleftarrow{Q_*}\text{“}C\text{‘}Q \subset R\overline{Q}_{\text{cn}} \alpha$

Dem.

$\vdash .\, *230·511 \,.\, \supset \vdash :. y \in \text{ᗡ}\text{‘}R \cap p\text{‘}\overleftarrow{Q_*}\text{“}C\text{‘}Q \,.\, \supset :$

$R\text{“}p\text{‘}\overleftarrow{Q_*}\text{“}C\text{‘}Q \subset \alpha \,.\, \supset .\, y \in \text{ᗡ}\text{‘}R \,.\, R\text{“}\overleftarrow{Q_*}\text{‘}y \subset \alpha$ (1)

$\vdash .\, (1) \,.\, *40·62 \,.\, *230·1 \,.\, \supset$

$\vdash : \dot{\exists} ! \, Q \,.\, y \in \text{ᗡ}\text{‘}R \cap p\text{‘}\overleftarrow{Q_*}\text{“}C\text{‘}Q \,.\, R\text{“}p\text{‘}\overleftarrow{Q_*}\text{“}C\text{‘}Q \subset \alpha \,.\, \supset .\, y \in R\overline{Q}_{\text{cn}} \alpha$ (2)

$\vdash .\, (2) \,.\, \text{Comm} \,.\, \supset \vdash :. \dot{\exists} ! \, Q \,.\, \supset :$

$R\text{“}p\text{‘}\overleftarrow{Q_*}\text{“}C\text{‘}Q \subset \alpha \,.\, \supset .\, \text{ᗡ}\text{‘}R \cap p\text{‘}\overleftarrow{Q_*}\text{“}C\text{‘}Q \subset R\overline{Q}_{\text{cn}} \alpha$ (3)

$\vdash .\, (3) \,.\, *230·512 \,.\, \supset \vdash .$ Prop

***230·514.** $\vdash : \dot{\exists} ! Q . \exists ! p'\overleftarrow{Q_*}"C'Q \cap \text{Ɑ}'R . R"p'\overleftarrow{Q_*}"C'Q \subset \alpha . \supset . RQ_{cn}\alpha$

Dem.

$\vdash . *230·513 . \supset \vdash : \text{Hp} . \supset . \exists ! p'\overleftarrow{Q_*}"C'Q \cap \text{Ɑ}'R . p'\overleftarrow{Q_*}"C'Q \cap \text{Ɑ}'R \subset R\overline{Q}_{cn}\alpha .$

$[*24·58.*230·11] \quad \supset . RQ_{cn}\alpha : \supset \vdash . \text{Prop}$

***230·52.** $\vdash : \exists ! C'Q \cap \text{Ɑ}'R . \exists ! \text{Ɑ}'R \cap p'\overleftarrow{Q_*}"s'R\overline{Q}_{cn}"\kappa . \kappa \subset \overleftarrow{Q_{cn}}'R . \supset . p'\kappa \in \overleftarrow{Q_{cn}}'R$

Dem.

$\vdash . *40·16 . \supset \vdash : \alpha \in \kappa . \supset . p'\overleftarrow{Q_*}"s'R\overline{Q}_{cn}"\kappa \subset p'\overleftarrow{Q_*}"R\overline{Q}_{cn}\alpha \quad (1)$

$\vdash . (1) . *40·61 . \supset$

$\vdash :. \text{Hp} . \supset : \alpha \in \kappa . \supset . p'\overleftarrow{Q_*}"s'R\overline{Q}_{cn}"\kappa \subset \breve{Q}_*"R\overline{Q}_{cn}\alpha .$

$[\text{Fact}.*230·4] \quad \supset . \text{Ɑ}'R \cap p'\overleftarrow{Q_*}"s'R\overline{Q}_{cn}"\kappa \subset R\overline{Q}_{cn}\alpha :$

$[*40·44] \quad \supset : \text{Ɑ}'R \cap p'\overleftarrow{Q_*}"s'R\overline{Q}_{cn}"\kappa \subset p'R\overline{Q}_{cn}"\kappa :$

$[*230·321] \quad \supset : \exists ! \kappa . \supset . \text{Ɑ}'R \cap p'\overleftarrow{Q_*}"s'R\overline{Q}_{cn}"\kappa \subset R\overline{Q}_{cn}(p'\kappa) .$

$[\text{Hp}.*24·58] \quad \supset . \exists ! R\overline{Q}_{cn}(p'\kappa) .$

$[*230·11] \quad \supset . p'\kappa \in \overleftarrow{Q_{cn}}'R \quad (2)$

$\vdash . *230·253 . *40·2 . \supset \vdash : \exists ! C'Q \cap \text{Ɑ}'R . \kappa = \Lambda . \supset . p'\kappa \in \overleftarrow{Q_{cn}}'R \quad (3)$

$\vdash . (2) . (3) . \supset \vdash . \text{Prop}$

***230·53.** $\vdash :. Q \in \text{trans} \cap \text{connex} . \text{E} ! \max_Q'\text{Ɑ}'R . \supset : RQ_{cn}\alpha . \equiv . \overrightarrow{R}'\max_Q'\text{Ɑ}'R \subset \alpha$

Dem.

$\vdash . *205·111 . \quad \supset \vdash : \text{Hp} . \supset . \max_Q'\text{Ɑ}'R \in C'Q \cap \text{Ɑ}'R \quad (1)$

$\vdash . *205·141 . *201·18 . \supset \vdash : \text{Hp} . \supset . \overleftarrow{Q_*}'\max_Q'\text{Ɑ}'R \cap \text{Ɑ}'R = \iota'\max_Q'\text{Ɑ}'R .$

$[*37·26.*53·301] \quad \supset . R"(\overleftarrow{Q_*}'\max_Q'\text{Ɑ}'R) = \overrightarrow{R}'\max_Q'\text{Ɑ}'R \quad (2)$

$\vdash . (1) . (2) . \supset \vdash :. \text{Hp} . \supset : \overrightarrow{R}'\max_Q'\text{Ɑ}'R \subset \alpha . \supset . \max_Q'\text{Ɑ}'R \in (R\overline{Q}_{cn}\alpha) .$

$[*230·11] \quad \supset . RQ_{cn}\alpha \quad (3)$

$\vdash . *205·36 . \supset \vdash :. \text{Hp} . \supset : y \in C'Q \cap \text{Ɑ}'R . R"\overleftarrow{Q_*}'y \subset \alpha . \supset . R"\overleftarrow{Q_*}'\max_Q'\text{Ɑ}'R \subset \alpha :$

$[*230·11] \quad \supset : RQ_{cn}\alpha . \supset . R"\overleftarrow{Q_*}'\max_Q'\text{Ɑ}'R \subset \alpha .$

$[(2)] \quad \supset . \overrightarrow{R}'\max_Q'\text{Ɑ}'R \subset \alpha \quad (4)$

$\vdash . (3) . (4) . \supset \vdash . \text{Prop}$

***230·54.** $\vdash :. Q \in \text{trans} \cap \text{connex} . \text{E} ! \max_Q'\text{Ɑ}'R . \supset : \kappa \subset \overleftarrow{Q_{cn}}'R . \equiv . p'\kappa \in \overleftarrow{Q_{cn}}'R$

Dem.

$\vdash . *230·53 . \supset \vdash :: \text{Hp} . \supset :. \kappa \subset \overleftarrow{Q_{cn}}'R . \equiv : \alpha \in \kappa . \supset_\alpha . \overrightarrow{R}'\max_Q'\text{Ɑ}'R \subset \alpha :$

$[*40·15] \quad \equiv : \overrightarrow{R}'\max_Q'\text{Ɑ}'R \subset p'\kappa :$

$[*230·53] \quad \equiv : p'\kappa \in \overleftarrow{Q_{cn}}'R :: \supset \vdash . \text{Prop}$

*231. LIMITING SECTIONS AND ULTIMATE OSCILLATION OF A FUNCTION.

Summary of *231.

In the present number we are concerned with the limiting section defined in a series P, to which the values of a function R belong, as the arguments to the function increase in the argument-series Q. That is, we are concerned with the section consisting of those terms x of $C\text{‘}P$ which are such that, however great the argument to R becomes, there are still values at least as great as x. Such terms as x may be said to be P-persistent; x is P-persistent if the function does not ultimately become and remain less than x. The class of persistent terms is called the *limiting section*. The limiting section may be defined as follows. If α is any class into which R is Q-convergent, then the section $P_*\text{‘‘}\alpha$ is such that the values of the function are ultimately contained in it. The product of such terms as $P_*\text{‘‘}\alpha$ is the smallest section having this property. Hence if x be any member of this section, then ultimately (*i.e.* for arguments far enough along the Q series) the values of the function R do not persistently remain less than x in the P series. Thus the product of such terms as $P_*\text{‘‘}\alpha$ is the limiting section, and we may therefore put

$$P\overline{R}_{sc}Q = p\text{‘}P_*\text{‘‘‘}\overleftarrow{Q_{cn}}\text{‘}R \cap C\text{‘}P \quad \text{Df},$$

where the letters "sc" are intended to suggest "section." (The factor $C\text{‘}P$ on the right is superfluous except when $\overleftarrow{Q_{cn}}\text{‘}R = \Lambda$, *i.e.* when $C\text{‘}Q \cap \text{Ɑ}\text{‘}R = \Lambda$.)

We will call the limiting section of $\breve{P}$, *i.e.* $\breve{P}\overline{R}_{sc}Q$, the "limiting upper section." It will be seen that if x is a member of $\breve{P}\overline{R}_{sc}Q$, then the function does not ultimately become and remain, as far as some of its arguments are concerned, greater than x, that is, however great we make the argument, we still find values not greater than x. Hence if x belongs to both $P\overline{R}_{sc}Q$ and $\breve{P}\overline{R}_{sc}Q$, we find values not less than x and values not greater than x however great we make the argument. This class, $P\overline{R}_{sc}Q \cap \breve{P}\overline{R}_{sc}Q$, may therefore be regarded as the class of ultimate values of the function. We will call it the "ultimate oscillation" of the function, since, as the argument approaches ∞, the value of the function ultimately oscillates in this stretch of P, and no smaller stretch has the same property. We will denote this class by "$P\overline{R}_{os}Q$," where "os" is intended to suggest "oscillation." $P\overline{R}_{os}Q$ is a stretch in $C\text{‘}P$, because it is the product of two sections. Hence we shall also call it the "limiting stretch." When the function has a definite limit as the argument approaches ∞, the limiting stretch must not contain more than one term.

Limits of functions for arguments x in the middle of $C'Q \cap \text{Ɑ}'R$, which will be considered later, are derived from the limits considered in the present number by limiting the field of Q to predecessors of x.

In this number we prove the following propositions among others:

$*231{\cdot}103.\ \vdash . P\overline{R}_{os}Q = P_{po}\overline{R}_{os}Q = P_*\overline{R}_{os}Q$

$*231{\cdot}12.\ \vdash . P\overline{R}_{sc}Q = p'P_*'''R'''\overleftarrow{Q}_*''(C'Q \cap \text{Ɑ}'R) \cap C'P$

$*231{\cdot}13.\ \vdash . P\overline{R}_{sc}Q \in \text{sect}'P$

$*231{\cdot}141.\ \vdash : Q_* \in \text{connex} . RQ_{cn}(\overleftarrow{P_*'x}) . \supset . x \in P\overline{R}_{sc}Q$

$*231{\cdot}191.\ \vdash : P_{po} \in \text{connex} . \exists ! P\overline{R}_{os}Q . \supset .$

$$P\overline{R}_{sc}Q = P_*''(P\overline{R}_{os}Q) . P''(P\overline{R}_{sc}Q) = P_{po}''(P\overline{R}_{os}Q)$$

$*231{\cdot}192.\ \vdash :. P_{po} \in \text{connex} . \exists ! P\overline{R}_{os}Q . \exists ! P\overline{R}_{os}Q' . \supset :$

$$P\overline{R}_{os}Q = P\overline{R}_{os}Q' . \equiv . P\overline{R}_{sc}Q = P\overline{R}_{sc}Q' . \breve{P}\overline{R}_{sc}Q = \breve{P}\overline{R}_{sc}Q'$$

$*231{\cdot}193.\ \vdash : P_{po} \in \text{Ser} . P\overline{R}_{os}Q \in 1 . \supset .$

$$P\overline{R}_{os}Q = \iota'\max_P'(P\overline{R}_{sc}Q) = \iota'\min_P'(\breve{P}\overline{R}_{sc}Q)$$

This proposition is frequently used in the present section.

In all ordinary circumstances, we shall have $C'P = P\overline{R}_{sc}Q \cup \breve{P}\overline{R}_{sc}Q$, so that if the upper and lower limiting sections do not have more than one term in common (*i.e.* if $P\overline{R}_{os}Q \in 1$), they define a Dedekind cut in P. The following propositions are concerned with this fact:

$*231{\cdot}202.\ \vdash : P_*, Q_* \in \text{connex} . \exists ! P\overline{R}_{sc}Q . \supset . C'P - (P\overline{R}_{sc}Q) \subset \breve{P}\overline{R}_{sc}Q$

$*231{\cdot}21.\ \vdash : P_*, Q_* \in \text{connex} . C'Q \cap \text{Ɑ}'R \subset Q_*''\breve{R}''C'P . \supset .$

$$C'P = P\overline{R}_{sc}Q \cup \breve{P}\overline{R}_{sc}Q$$

$*231{\cdot}22.\ \vdash : P_*, Q_* \in \text{connex} . R''C'Q \subset C'P . \supset . C'P = P\overline{R}_{sc}Q \cup \breve{P}\overline{R}_{sc}Q$

Note that "$R''C'Q \subset C'P$" is the hypothesis that for arguments belonging to $C'Q$, the values belong to $C'P$.

$*231{\cdot}24.\ \vdash : P_* \in \text{connex} . R''C'Q \subset C'P . \sim\{RQ_{cn}(\overrightarrow{P_*'x})\} . \supset . \overrightarrow{P}_*'x \subset P\overline{R}_{sc}Q$

$*231{\cdot}01.\quad P\overline{R}_{sc}Q = p'P_*'''\overleftarrow{Q}_{cn}'R \cap C'P \quad$ Df

$*231{\cdot}02.\quad P\overline{R}_{os}Q = P\overline{R}_{sc}Q \cap \breve{P}\overline{R}_{sc}Q \quad$ Df

$*231{\cdot}1.\quad \vdash . P\overline{R}_{sc}Q = p'P_*'''\overleftarrow{Q}_{cn}'R \cap C'P \quad [(*231{\cdot}01)]$

***231·101.** $\vdash . P\overline{R}_{os}Q = P\overline{R}_{sc}Q \cap \breve{P}\overline{R}_{sc}Q$ [(*231·02)]

***231·102.** $\vdash . P\overline{R}_{sc}Q = P_{po}\overline{R}_{sc}Q = P_{*}\overline{R}_{sc}Q$ [*231·1 . *91·602 . *90·4]

***231·103.** $\vdash . P\overline{R}_{os}Q = P_{po}\overline{R}_{os}Q = P_{*}\overline{R}_{os}Q$ [*231·102·101]

***231·11.** $\vdash :. x \in P\overline{R}_{sc}Q . \equiv : RQ_{cn}\alpha . \supset_{\alpha} . x \in P_{*}``\alpha : x \in C`P$ [*231·1]

***231·111.** $\vdash :. x \in P\overline{R}_{sc}Q . \equiv : y \in C`Q \cap \text{Ɑ}`R . R``\overleftarrow{Q_{*}}`y \subset \alpha . \supset_{y,\alpha} . x \in P_{*}``\alpha : x \in C`P$
[*231·11 . *230·11]

***231·112.** $\vdash :. x \in P\overline{R}_{sc}Q . \equiv : y \in C`Q \cap \text{Ɑ}`R . \supset_{y} . x \in P_{*}``R``\overleftarrow{Q_{*}}`y : x \in C`P$

Dem.

$\vdash . *231·111 . *22·42 . \supset$

$\vdash :. x \in P\overline{R}_{sc}Q . \supset : y \in C`Q \cap \text{Ɑ}`R . \supset_{y} . x \in P_{*}``R``\overleftarrow{Q_{*}}`y : x \in C`P \quad (1)$

$\vdash . *37·2 . \supset$

$\vdash :. y \in C`Q \cap \text{Ɑ}`R . \supset_{y} . x \in P_{*}``R``\overleftarrow{Q_{*}}`y : x \in C`P : \supset :$

$y \in C`Q \cap \text{Ɑ}`R . R``\overleftarrow{Q_{*}}`y \subset \alpha . \supset_{y,\alpha} . x \in P_{*}``\alpha : x \in C`P :$

[*231·111] $\supset : x \in P\overline{R}_{sc}Q \quad (2)$

$\vdash . (1) . (2) . \supset \vdash .$ Prop

***231·113.** $\vdash :. x \in P\overline{R}_{sc}Q . \equiv : y \in C`Q \cap \text{Ɑ}`R . \supset_{y} . x (P_{*} | R | \breve{Q}_{*}) y : x \in C`P$
[*231·112 . *37·3]

If R is a one-valued function (*i.e.* a one-many relation), and if we write $x \leqslant x'$ for $xP_{*}x'$, and $y \leqslant y'$ for $yQ_{*}y'$, we have

$$x \in P\overline{R}_{sc}Q . \equiv : y \in C`Q \cap \text{Ɑ}`R . \supset_{y} . (\exists y') . y \leqslant y' . x \leqslant R`y' : x \in C`P.$$

That is, x belongs to $P\overline{R}_{sc}Q$ if, for any argument y in $C`Q$, we can find an argument y', greater than or equal to y, for which the value is greater than or equal to x.

***231·12.** $\vdash . P\overline{R}_{sc}Q = p`P_{*}```R```\overleftarrow{Q_{*}}``(C`Q \cap \text{Ɑ}`R) \cap C`P$ [*231·112]

This is usually the most convenient formula for $P\overline{R}_{sc}Q$.

***231·121.** $\vdash : \exists ! C`Q \cap \text{Ɑ}`R . \supset .$

$$P\overline{R}_{sc}Q = p`P_{*}```\overleftarrow{Q_{cn}}`R = p`P_{*}```R```\overleftarrow{Q_{*}}``(C`Q \cap \text{Ɑ}`R)$$

Dem.

$\vdash . *230·253 . \supset \vdash : \text{Hp} . \supset . \exists ! \overleftarrow{Q_{cn}}`R .$

[*40·23.*37·47] $\supset . p`P_{*}```\overleftarrow{Q_{cn}}`R \subset s`P_{*}```\overleftarrow{Q_{cn}}`R .$

[*40·38.*37·16] $\supset . p`P_{*}```\overleftarrow{Q_{cn}}`R \subset C`P \quad (1)$

$\vdash . *40·23 . \supset \vdash : \text{Hp} . \supset . p`P_{*}```R```\overleftarrow{Q_{*}}``(C`Q \cap \text{Ɑ}`R) \subset s`P_{*}```R```\overleftarrow{Q_{*}}``(C`Q \cap \text{Ɑ}`R)$

[*40·38.*37·16] $\subset C`P \quad (2)$

$\vdash . (1) . (2) . *231·1·12 . \supset \vdash .$ Prop

***231·13.** $\vdash . P\overline{R}_{sc}Q \in \text{sect}\text{‘}P$ [*211·631·13 . *231·12]

***231·131.** $\vdash . P\overline{R}_{sc}Q \subset C\text{‘}P$ [*231·1]

***231·132.** $\vdash : \exists ! C\text{‘}Q \cap \text{Ɑ}\text{‘}R . \supset . P\overline{R}_{sc}Q \subset P_*\text{“}R\text{“}\breve{Q}_*\text{“}\text{Ɑ}\text{‘}R$

Dem.

$\vdash . *40{\cdot}23 . *231{\cdot}121 . \supset \vdash : \text{Hp} . \supset . P\overline{R}_{sc}Q \subset s\text{‘}P_*\text{‘‘‘}R\text{‘‘‘}\overleftarrow{Q_*}\text{“}(C\text{‘}Q \cap \text{Ɑ}\text{‘}R)$

[*40·38] $\subset P_*\text{“}R\text{“}s\text{‘}\overleftarrow{Q_*}\text{“}(C\text{‘}Q \cap \text{Ɑ}\text{‘}R)$

[*40·52.*37·265] $\subset P_*\text{“}R\text{“}\breve{Q}_*\text{“}\text{Ɑ}\text{‘}R : \supset \vdash . \text{Prop}$

***231·133.** $\vdash : C\text{‘}P \cap \text{Ɑ}\text{‘}R = \Lambda . \supset . P\overline{R}_{sc}Q = C\text{‘}P$ [*231·12 . *37·29 . *40·2]

***231·134.** $\vdash . P_{po}\text{“}(P\overline{R}_{sc}Q) = P\text{“}(P\overline{R}_{sc}Q)$ [*211·131 . *231·13]

***231·14.** $\vdash :: R \in 1 \rightarrow \text{Cls} . \supset :. x \in P\overline{R}_{sc}Q . \equiv :$
$y \in C\text{‘}Q \cap \text{Ɑ}\text{‘}R . \supset_y . (\exists z) . yQ_*z . xP_*(R\text{‘}z) : x \in C\text{‘}P$ [*71·7 . *231·113]

***231·141.** $\vdash : Q_* \in \text{connex} . RQ_{cn}(\overleftarrow{P_*}\text{‘}x) . \supset . x \in P\overline{R}_{sc}Q$

Dem.

$\vdash . *230{\cdot}4 . \supset \vdash : y \in R\overline{Q}_{cn}(\overleftarrow{P_*}\text{‘}x) . z \in \text{Ɑ}\text{‘}R . yQ_*z . \supset . z \in R\overline{Q}_{cn}(\overleftarrow{P_*}\text{‘}x) .$

[*230·171] $\supset . x \in P_*\text{“}R\text{“}\overleftarrow{Q_*}\text{‘}z$ (1)

$\vdash . *230{\cdot}171 . *96{\cdot}3 . \supset$

$\vdash : y \in R\overline{Q}_{cn}(\overleftarrow{P_*}\text{‘}x) . zQ_*y . \supset . x \in P_*\text{“}R\text{“}\overleftarrow{Q_*}\text{‘}y . \overleftarrow{Q_*}\text{‘}y \subset \overleftarrow{Q_*}\text{‘}z .$

[*37·2] $\supset . x \in P_*\text{“}R\text{“}\overleftarrow{Q_*}\text{‘}z$ (2)

$\vdash . (1) . (2) . \supset \vdash : \text{Hp} . y \in R\overline{Q}_{cn}(\overleftarrow{P_*}\text{‘}x) . z \in C\text{‘}Q \cap \text{Ɑ}\text{‘}R . \supset . x \in P_*\text{“}R\text{“}\overleftarrow{Q_*}\text{‘}z$ (3)

$\vdash . (3) . *230{\cdot}11 . \supset \vdash :. \text{Hp} . \supset : z \in C\text{‘}Q \cap \text{Ɑ}\text{‘}R . \supset_z . x \in P_*\text{“}R\text{“}\overleftarrow{Q_*}\text{‘}z$ (4)

$\vdash . *230{\cdot}151 . \quad \supset \vdash : \text{Hp} . \supset . x \in C\text{‘}P$ (5)

$\vdash . (4) . (5) . *231{\cdot}112 . \supset \vdash . \text{Prop}$

***231·142.** $\vdash : \alpha \in \text{sect}\text{‘}P . RQ_{cn}\alpha . \supset . P\overline{R}_{sc}Q \subset \alpha$

Dem.

$\vdash . *231{\cdot}1 . *40{\cdot}12 . \supset \vdash : \text{Hp} . \supset . P\overline{R}_{sc}Q \subset P_*\text{“}\alpha .$

[*211·13] $\supset . P\overline{R}_{sc}Q \subset \alpha : \supset \vdash . \text{Prop}$

***231·143.** $\vdash : RQ_{cn}(\overrightarrow{P_*}\text{‘}x) . \supset . P\overline{R}_{sc}Q \subset \overrightarrow{P_*}\text{‘}x$ [*231·142 . *211·13]

***231·144.** $\vdash : RQ_{cn}(\overrightarrow{P_{po}}\text{‘}x) . \supset . P\overline{R}_{sc}Q \subset \overrightarrow{P_{po}}\text{‘}x$ [*231·142 . *211·16]

***231·15.** $\vdash : R\text{“}C\text{‘}Q \subset C\text{‘}P . \supset . C\text{‘}P \cap p\text{‘}\overrightarrow{P_*}\text{“}C\text{‘}P \subset P\overline{R}_{sc}Q$

Dem.

$\vdash . *37{\cdot}2 . \supset \vdash :. \text{Hp} . \supset : R\text{“}\overleftarrow{Q_*}\text{‘}y \subset C\text{‘}P :$

[*40·16] $\supset : p\text{‘}\overrightarrow{P_*}\text{“}C\text{‘}P \subset p\text{‘}\overrightarrow{P_*}\text{“}R\text{“}\overleftarrow{Q_*}\text{‘}y :$

[*40·23] $\supset : y \in C\text{‘}Q \cap \text{Ɑ}\text{‘}R . \supset_y . p\text{‘}\overrightarrow{P_*}\text{“}C\text{‘}P \subset P_*\text{“}R\text{“}\overleftarrow{Q_*}\text{‘}y :$

[*231·12] $\supset : C\text{‘}P \cap p\text{‘}\overrightarrow{P_*}\text{“}C\text{‘}P \subset P\overline{R}_{sc}Q :. \supset \vdash . \text{Prop}$

***231·151.** $\vdash : P_* \in \text{connex} . C'P \cap p'\overrightarrow{P_*}\text{“}C'P \subset P\overline{R}_{sc}Q . \supset . \overrightarrow{B}'P \subset P\overline{R}_{sc}Q$
[*202·521]

***231·152.** $\vdash : P_* \in \text{connex} . C'P \cap p'\overrightarrow{P_*}\text{“}C'P \subset P\overline{R}_{sc}Q . \exists ! \overrightarrow{B}'P . \supset . B'P \in P\overline{R}_{sc}Q$
[*231·151 . *202·523]

The hypothesis $C'P \cap p'\overrightarrow{P_*}\text{“}C'P \subset P\overline{R}_{sc}Q$ is verified not only when $R\text{“}C'Q \subset C'P$, but also under certain more general hypotheses. Two such hypotheses, namely

$$C'Q \cap \text{Ɑ}'R \subset \breve{R}\text{“}C'P$$

and $$C'Q \cap \text{Ɑ}'R \subset Q_*\text{“}\breve{R}\text{“}C'P,$$

are considered in the following propositions.

***231·153.** $\vdash : C'Q \cap \text{Ɑ}'R \subset Q_*\text{“}\breve{R}\text{“}C'P . \supset . C'P \cap p'\overrightarrow{P_*}\text{“}C'P \subset P\overline{R}_{sc}Q$

Dem.

$\vdash . *37·1 . \supset \vdash :: \text{Hp} . \supset :. y \in C'Q \cap \text{Ɑ}'R . \supset_y : (\exists z) . z \in C'P . z(R \mid \breve{Q}_*)y :$

[*40·51] $\supset_y : x \in p'\overrightarrow{P_*}\text{“}C'P . \supset_x . (\exists z) . z \in C'P . z(R \mid \breve{Q}_*)y . xP_*z .$

[*34·1] $\supset_x . x(P_* \mid R \mid \breve{Q}_*)y$ (1)

$\vdash . (1) . \text{Comm} . \supset$

$\vdash :. \text{Hp} . x \in C'P \cap p'\overrightarrow{P_*}\text{“}C'P . \supset : y \in C'Q \cap \text{Ɑ}'R . \supset_y . x(P_* \mid R \mid \breve{Q}_*)y : x \in C'P :$

[*231·113] $\supset : x \in P\overline{R}_{sc}Q :. \supset \vdash . \text{Prop}$

***231·154.** $\vdash : R\text{“}C'Q \subset C'P . \supset . C'Q \cap \text{Ɑ}'R \subset \breve{R}\text{“}C'P$

Dem.

$\vdash . *37·2 . \supset \vdash : \text{Hp} . \supset . \breve{R}\text{“}R\text{“}C'Q \subset \breve{R}\text{“}C'P .$

[*37·501] $\supset . C'Q \cap \text{Ɑ}'R \subset \breve{R}\text{“}C'P : \supset \vdash . \text{Prop}$

***231·155.** $\vdash : C'Q \cap \text{Ɑ}'R \subset \breve{R}\text{“}C'P . \supset . C'Q \cap \text{Ɑ}'R \subset Q_*\text{“}\breve{R}\text{“}C'P$

Dem.

$\vdash . *22·43·45 . \supset \vdash : \text{Hp} . \supset . C'Q \cap \text{Ɑ}'R \subset C'Q \cap \breve{R}\text{“}C'P$

[*90·33] $\subset Q_*\text{“}\breve{R}\text{“}C'P : \supset \vdash . \text{Prop}$

***231·156.** $\vdash :. C'Q \cap \text{Ɑ}'R \subset Q_*\text{“}\breve{R}\text{“}C'P . \equiv : \Lambda \sim \in P_*\text{“'}R\text{“'}\overleftarrow{Q_*}\text{“}(C'Q \cap \text{Ɑ}'R) :$
$\equiv : z \in C'Q \cap \text{Ɑ}'R . \supset_z . \exists ! C'P \cap R\text{“}\overleftarrow{Q_*}'z$

Dem.

$\vdash . *37·1 . \supset \vdash :. C'Q \cap \text{Ɑ}'R \subset Q_*\text{“}\breve{R}\text{“}C'P . \equiv :$

$z \in C'Q \cap \text{Ɑ}'R . \supset_z . (\exists x) . x \in C'P . z(Q_* \mid \breve{R})x :$

[*37·3] $\equiv : z \in C'Q \cap \text{Ɑ}'R . \supset_z . (\exists x) . x \in C'P . x \in R\text{“}\overleftarrow{Q_*}'z :$

[*22·33] $\equiv : z \in C'Q \cap \text{Ɑ}'R . \supset_z . \exists ! C'P \cap R\text{“}\overleftarrow{Q_*}'z :$ (1)

[*37·265·43] $\equiv : z \in C'Q \cap \text{Ɑ}'R . \supset_z . \exists ! P_*\text{“}R\text{“}\overleftarrow{Q_*}'z$ (2)

$\vdash . (1) . (2) . \supset \vdash . \text{Prop}$

∗231·16. $\vdash : P_* \in \text{connex} . \exists ! \overrightarrow{B}\text{‘}P . C\text{‘}Q \cap \text{Ɑ}\text{‘}R \subset Q_*\text{“}\breve{R}\text{“}C\text{‘}P . \supset .$
$\exists ! P\bar{R}_{\text{sc}}Q . B\text{‘}P \in P\bar{R}_{\text{sc}}Q$ [∗231·152·153]

∗231·161. $\vdash : P_* \in \text{connex} . \exists ! \overrightarrow{B}\text{‘}P . R\text{“}C\text{‘}Q \subset C\text{‘}P . \supset . \exists ! P\bar{R}_{\text{sc}}Q . B\text{‘}P \in P\bar{R}_{\text{sc}}Q$
[∗231·154·155·16]

∗231·17. $\vdash : R\text{“}C\text{‘}Q \subset C\text{‘}P . \supset . R\bar{Q}_{\text{cn}}\alpha \subset R\bar{Q}_{\text{cn}}(P_*\text{“}\alpha)$

Dem.

$\vdash . *90\cdot13 . \supset \vdash :. \text{Hp} . \supset : y \in C\text{‘}Q . \supset . R\text{“}\overleftarrow{Q_*}\text{‘}y \subset C\text{‘}P :$

[∗230·1] $\supset : y \in R\bar{Q}_{\text{cn}}\alpha . \supset . y \in C\text{‘}Q \cap \text{Ɑ}\text{‘}R . R\text{“}\overleftarrow{Q_*}\text{‘}y \subset \alpha \cap C\text{‘}P .$

[∗90·33] $\supset . y \in C\text{‘}Q \cap \text{Ɑ}\text{‘}R . R\text{“}\overleftarrow{Q_*}\text{‘}y \subset P_*\text{“}\alpha .$

[∗230·1] $\supset . y \in R\bar{Q}_{\text{cn}}(P_*\text{“}\alpha) :. \supset \vdash . \text{Prop}$

∗231·171. $\vdash : R\text{“}C\text{‘}Q \subset C\text{‘}P . RQ_{\text{cn}}\alpha . \supset . RQ_{\text{cn}}(P_*\text{“}\alpha)$ [∗231·17 . ∗230·11]

∗231·18. $\vdash : R\text{“}C\text{‘}Q \subset C\text{‘}P . \supset . P\bar{R}_{\text{sc}}Q = p\text{‘}(\text{sect}\text{‘}P \cap \overleftarrow{Q_{\text{cn}}}\text{‘}R) \cap C\text{‘}P$

Dem.

$\vdash . *231\cdot11 . *211\cdot13 . \supset$

$\vdash :. x \in P\bar{R}_{\text{sc}}Q . \supset : \beta Q_{\text{cn}} R . \beta \in \text{sect}\text{‘}P . \supset_\beta . x \in \beta : x \in C\text{‘}P$ (1)

$\vdash . *231\cdot171 . \supset$

$\vdash :: \text{Hp} . \supset :. RQ_{\text{cn}}(P_*\text{“}\alpha) . \supset_\alpha . x \in P_*\text{“}\alpha : \supset : RQ_{\text{cn}}\alpha . \supset_\alpha . x \in P_*\text{“}\alpha :.$

[∗13·195.∗231·11]

$\supset :. (\exists\alpha) . \beta = P_*\text{“}\alpha . RQ_{\text{cn}}\beta . \supset_\beta . x \in \beta : x \in C\text{‘}P : \supset : x \in P\bar{R}_{\text{sc}}Q :.$

[∗211·13] $\supset :. \beta \in \text{sect}\text{‘}P . RQ_{\text{cn}}\beta . \supset_\beta . x \in \beta : x \in C\text{‘}P : \supset : x \in P\bar{R}_{\text{sc}}Q$ (2)

$\vdash . (1) . (2) . \supset$

$\vdash :: \text{Hp} . \supset :. x \in P\bar{R}_{\text{sc}}Q . \equiv : \beta \in \text{sect}\text{‘}P . RQ_{\text{cn}}\beta . \supset_\beta . x \in \beta : x \in C\text{‘}P :: \supset \vdash . \text{Prop}$

∗231·181. $\vdash : P \in \text{Ser} . R\text{“}C\text{‘}Q \subset C\text{‘}P . \supset . P\bar{R}_{\text{sc}}Q = C\text{‘}P \cap p\text{‘}(\overrightarrow{P}\text{“}C\text{‘}P \cap \overleftarrow{Q_{\text{cn}}}\text{‘}R)$

Dem.

$\vdash . *231\cdot18 . *211\cdot302 . *40\cdot16 . \supset$

$\vdash : \text{Hp} . \supset . P\bar{R}_{\text{sc}}Q \subset C\text{‘}P \cap p\text{‘}(\overrightarrow{P}\text{“}C\text{‘}P \cap \overleftarrow{Q_{\text{cn}}}\text{‘}R)$ (1)

$\vdash . *40\cdot55 . *230\cdot211 . \supset \vdash :. \alpha \in \text{sect}\text{‘}P \cap \overleftarrow{Q_{\text{cn}}}\text{‘}R . z \in C\text{‘}P \cap p\text{‘}\overleftarrow{P}\text{“}\alpha . \supset :$
$\overrightarrow{P}\text{‘}z \in \overleftarrow{Q_{\text{cn}}}\text{‘}R :$

[∗40·12] $\supset : x \in p\text{‘}(\overrightarrow{P}\text{“}C\text{‘}P \cap \overleftarrow{Q_{\text{cn}}}\text{‘}R) . \supset . x \in \overrightarrow{P}\text{‘}z$ (2)

$\vdash . (2) . \text{Comm} . \supset \vdash :. x \in C\text{‘}P \cap p\text{‘}(\overrightarrow{P}\text{“}C\text{‘}P \cap \overleftarrow{Q_{\text{cn}}}\text{‘}R) . \alpha \in \text{sect}\text{‘}P \cap \overleftarrow{Q_{\text{cn}}}\text{‘}R . \supset :$
$x \in C\text{‘}P : z \in C\text{‘}P \cap p\text{‘}\overleftarrow{P}\text{“}\alpha . \supset_z . x \in \overrightarrow{P}\text{‘}z :$

[∗40·41] $\supset : x \in C\text{‘}P \cap p\text{‘}\overrightarrow{P}\text{“}(C\text{‘}P \cap p\text{‘}\overleftarrow{P}\text{“}\alpha)$ (3)

$\vdash . *211{\cdot}711 . \supset$

$\vdash : \text{Hp} . \alpha \epsilon \text{sect}`P . \supset . C`P \cap p`\overrightarrow{P}``(C`P \cap p`\overleftarrow{P}``\alpha) = C`P \cap p`\overrightarrow{P}``(C`P - \alpha)$

$[*211{\cdot}7{\cdot}711] \qquad = C`P - (C`P - \alpha) \qquad (4)$

$\vdash . (3) . (4) . \supset$

$\vdash :. \text{Hp} . x \epsilon C`P \cap p`(\overrightarrow{P}``C`P \cap \overleftarrow{Q_{cn}}`R) . \supset : \alpha \epsilon \text{sect}`P \cap \overleftarrow{Q_{cn}}`R . \supset_\alpha . x \epsilon \alpha :$

$[*231{\cdot}18] \qquad \supset : x \epsilon P\overline{R}_{sc}Q \qquad (5)$

$\vdash . (1) . (5) . \supset \vdash . \text{Prop}$

$*231{\cdot}182$. $\vdash : P \epsilon \text{Ser} . R``C`Q \subset C`P . \exists ! C`P - (P\overline{R}_{sc}Q) . \supset .$

$$P\overline{R}_{sc}Q = p`(\overrightarrow{P}``C`P \cap \overleftarrow{Q_{cn}}`R) . \exists ! (\overrightarrow{P}``C`P \cap \overleftarrow{Q_{cn}}`R)$$

Dem.

$\vdash . *231{\cdot}181 . \supset$

$\vdash : \text{Hp} . \exists ! (\overrightarrow{P}``C`P \cap \overleftarrow{Q_{cn}}`R) . \supset . P\overline{R}_{sc}Q = p`(\overrightarrow{P}``C`P \cap \overleftarrow{Q_{cn}}`R) \qquad (1)$

$\vdash . *230{\cdot}11 . \supset \vdash :. \text{Hp} . \overrightarrow{P}``C`P \cap \overleftarrow{Q_{cn}}`R = \Lambda . \supset :$

$$x \epsilon C`P . y \epsilon C`Q \cap \text{Ɑ}`R . \supset_{x,y} . \sim (R``\overleftarrow{Q_*}`y \subset \overrightarrow{P}`x) .$$

$[*90{\cdot}33.\text{Hp}] \qquad \supset_{x,y} . \sim (P_*``R``\overleftarrow{Q_*}`y \subset \overrightarrow{P}`x) .$

$[*211{\cdot}56] \qquad \supset_{x,y} . \overrightarrow{P_*}`x \subset P_*``R``\overleftarrow{Q_*}`y .$

$[*90{\cdot}13] \qquad \supset_{x,y} . x \epsilon P_*``R``\overleftarrow{Q_*}`y \qquad (2)$

$\vdash . (2) . *231{\cdot}12 . \supset \vdash : \text{Hp} . \overrightarrow{P}``C`P \cap \overleftarrow{Q_{cn}}`R = \Lambda . \supset . C`P \subset P\overline{R}_{sc}Q \qquad (3)$

$\vdash . (3) . \text{Transp} . \supset \vdash . \text{Hp} . \supset . \exists ! \overrightarrow{P}``C`P \cap \overleftarrow{Q_{cn}}`R \qquad (4)$

$\vdash . (1) . (4) . \supset \vdash . \text{Prop}$

$*231{\cdot}19$. $\vdash : P \epsilon \text{trans} . Q_* \epsilon \text{connex} . R``C`Q \subset C`P . \supset .$

$$P\overline{R}_{os}Q = p`(\text{str}`P \cap \overleftarrow{Q_{cn}}`R) \cap C`P$$

Dem.

$\vdash . *231{\cdot}18{\cdot}101 . \supset \vdash :: \text{Hp} . \supset :.$

$x \epsilon P\overline{R}_{os}Q . \equiv : \alpha \epsilon \text{sect}`P . \beta \epsilon \text{sect}`\breve{P} . \alpha, \beta \epsilon \overleftarrow{Q_{cn}}`R . \supset_{\alpha,\beta} . x \epsilon \alpha \cap \beta : x \epsilon C`P :$

$[*13{\cdot}191.*11{\cdot}35] \equiv : (\exists \alpha, \beta) . \alpha \epsilon \text{sect}`P . \beta \epsilon \text{sect}`\breve{P} . \alpha, \beta \epsilon \overleftarrow{Q_{cn}}`R . \gamma = \alpha \cap \beta . \supset_\gamma .$

$$x \epsilon \gamma : x \epsilon C`P \qquad (1)$$

$\vdash . *230{\cdot}42 . \supset \vdash :. \text{Hp} . \supset : \alpha, \beta \epsilon \overleftarrow{Q_{cn}}`R . \equiv . \alpha \cap \beta \epsilon \overleftarrow{Q_{cn}}`R \qquad (2)$

$\vdash . *215{\cdot}16 . \supset$

$\vdash :. \text{Hp} . \supset : (\exists \alpha, \beta) . \alpha \epsilon \text{sect}`P . \beta \epsilon \text{sect}`\breve{P} . \gamma = \alpha \cap \beta . \equiv . \gamma \epsilon \text{str}`P \qquad (3)$

$\vdash . (1) . (2) . (3) . \supset$

$\vdash :: \text{Hp} . \supset :. x \epsilon P\overline{R}_{os}Q . \equiv : \gamma \epsilon \text{str}`P . RQ_{cn}\gamma . \supset_\gamma . x \epsilon \gamma :: \supset \vdash . \text{Prop}$

***231·191.** $\vdash : P_{po} \epsilon \text{connex} . \exists ! P\overline{R}_{os}Q . \supset .$
$$P\overline{R}_{sc}Q = P_*``(P\overline{R}_{os}Q) . P``(P\overline{R}_{sc}Q) = P_{po}``(P\overline{R}_{os}Q)$$
[*215·165 . *231·13·101]

***231·192.** $\vdash :. P_{po} \epsilon \text{connex} . \exists ! P\overline{R}_{os}Q . \exists ! P\overline{R}_{os}Q' . \supset :$
$$P\overline{R}_{os}Q = P\overline{R}_{os}Q' . \equiv . P\overline{R}_{sc}Q = P\overline{R}_{sc}Q' . \breve{P}\overline{R}_{sc}Q = \breve{P}\overline{R}_{sc}Q'$$
[*231·191·101]

***231·193.** $\vdash : P_{po} \epsilon \text{Ser} . P\overline{R}_{os}Q \epsilon 1 . \supset .$
$$P\overline{R}_{os}Q = \iota`\text{max}_P`(P\overline{R}_{sc}Q) = \iota`\text{min}_P`(\breve{P}\overline{R}_{sc}Q)$$
[*215·166 . *231·13·101]

This proposition is of fundamental importance.

***231·2.** $\vdash : P_*, Q_* \epsilon \text{connex} . C`Q \cap \text{Ꭰ}`R \subset Q_*``\breve{R}``C`P . \supset .$
$$C`P - (P\overline{R}_{sc}Q) \subset \breve{P}\overline{R}_{sc}Q$$

Dem.

$\vdash . *231·112 . \supset$

$\vdash : x \epsilon C`P - (P\overline{R}_{sc}Q) . \supset . (\exists y) . y \epsilon C`Q \cap \text{Ꭰ}`R . x \epsilon C`P - P_*``R``\overleftarrow{Q_*}`y$ (1)

$\vdash . *202·501 . *90·33 . \supset$

$\vdash :. \text{Hp} . x \epsilon C`P - P_*``R``\overleftarrow{Q_*}`y . \supset : x \epsilon p`\overleftarrow{P_*}``(R``\overleftarrow{Q_*}`y \cap C`P) :$

[*96·3] $\supset : yQ_*z . \supset_z . x \epsilon p`\overleftarrow{P_*}``(R``\overleftarrow{Q_*}`z \cap C`P) :$

[*40·61] $\supset : yQ_*z . z \epsilon \text{Ꭰ}`R . \supset_z . x \epsilon \breve{P}_*``(R``\overleftarrow{Q_*}`z \cap C`P) .$

[*37·265] $\supset_z . x \epsilon \breve{P}_*``R``\overleftarrow{Q_*}`z :$ (2)

[*90·12] $\supset : y \epsilon C`Q \cap \text{Ꭰ}`R . \supset . x \epsilon \breve{P}_*``R``\overleftarrow{Q_*}`y :$

[*96·3.*37·2] $\supset : y \epsilon C`Q \cap \text{Ꭰ}`R . zQ_*y . \supset_z . x \epsilon \breve{P}_*``R``\overleftarrow{Q_*}`z$ (3)

$\vdash . (2) . (3) . *202·137 . \supset \vdash :. \text{Hp} . y \epsilon C`Q \cap \text{Ꭰ}`R . x \epsilon C`P - P_*``R``\overleftarrow{Q_*}`y . \supset :$
$$z \epsilon C`Q \cap \text{Ꭰ}`R . \supset_z . x \epsilon \breve{P}_*``R``\overleftarrow{Q_*}`z : x \epsilon C`P :$$

[*231·112] $\supset : x \epsilon \breve{P}\overline{R}_{sc}Q$ (4)

$\vdash . (1) . (4) . \supset \vdash :. \text{Hp} . \supset : x \epsilon C`P - (P\overline{R}_{sc}Q) . \supset . x \epsilon \breve{P}\overline{R}_{sc}Q :. \supset \vdash . \text{Prop}$

This proposition is fundamental in the theory of limiting segments.

***231·201.** $\vdash : P_*, Q_* \epsilon \text{connex} . R``C`Q \subset C`P . \supset . C`P - (P\overline{R}_{sc}Q) \subset \breve{P}\overline{R}_{sc}Q$
[*231·2·154·155]

***231·202.** $\vdash : P_*, Q_* \epsilon \text{connex} . \exists ! P\overline{R}_{sc}Q . \supset . C`P - (P\overline{R}_{sc}Q) \subset \breve{P}\overline{R}_{sc}Q$

Dem.

$\vdash . *40·22 . \text{Transp} . *231·12 . \supset$

$\vdash : \text{Hp} . \supset . \Lambda \sim \epsilon P_*```R```\overleftarrow{Q_*}``(C`Q \cap \text{Ꭰ}`R) .$

[*231·156] $\supset . C`Q \cap \text{Ꭰ}`R \subset Q_*``\breve{R}``C`P .$

[*231·2] $\supset . C`P - (P\overline{R}_{sc}Q) \subset \breve{P}\overline{R}_{sc}Q : \supset \vdash . \text{Prop}$

***231·21.** $\vdash : P_*, Q_* \in \text{connex} . C'Q \cap \text{Ɑ}'R \subset Q_*``\breve{R}``C'P . \supset .$
$$C'P = P\overline{R}_{sc}Q \cup \breve{P}\overline{R}_{sc}Q$$

Dem.

$\vdash . *231·13 . \supset \vdash : P\overline{R}_{sc}Q \subset C'P . \breve{P}\overline{R}_{sc}Q \subset C'P$ (1)

$\vdash . *231·2 . \supset \vdash : \text{Hp} . \supset . C'P \subset P\overline{R}_{sc}Q \cup \breve{P}\overline{R}_{sc}Q$ (2)

$\vdash . (1) . (2) . \supset \vdash . \text{Prop}$

***231·22.** $\vdash : P_*, Q_* \in \text{connex} . R``C'Q \subset C'P . \supset . C'P = P\overline{R}_{sc}Q \cup \breve{P}\overline{R}_{sc}Q$
[*231·201·13]

***231·23.** $\vdash :. P_* \in \text{connex} . R``C'Q \subset C'P . \supset :$
$$\exists ! R``\overleftarrow{Q_*}'y - \overrightarrow{P_*}'x . \supset . \overrightarrow{P_*}'x \subset P_{po}``R``\overleftarrow{Q_*}'y$$

Dem.

$\vdash . *90·14 . \supset \vdash : \text{Hp} . \supset . R``\overleftarrow{Q_*}'y \subset C'P$ (1)

$\vdash . (1) . *90·21 . \supset \vdash : \text{Hp} . \exists ! R``\overleftarrow{Q_*}'y - \overrightarrow{P_*}'x . \supset . \exists ! P_*``R``\overleftarrow{Q_*}'y - \overrightarrow{P_*}'x .$

[*211·56.*202·13] $\supset . \overrightarrow{P_*}'x \subset P_{po}``R``\overleftarrow{Q_*}'y : \supset \vdash . \text{Prop}$

***231·24.** $\vdash : P_* \in \text{connex} . R``C'Q \subset C'P . \sim \{RQ_{cn}(\overrightarrow{P_*}'x)\} . \supset . \overrightarrow{P_*}'x \subset P\overline{R}_{sc}Q$

Dem.

$\vdash . *230·11 . \supset \vdash :. \text{Hp} . \supset : y \in C'Q \cap \text{Ɑ}'R . \supset_y . \exists ! R``\overleftarrow{Q_*}'y - \overrightarrow{P_*}'x .$

[*231·23] $\supset_y . \overrightarrow{P_*}'x \subset P_{po}``R``\overleftarrow{Q_*}'y :$

[*91·54.*40·44] $\supset : \overrightarrow{P_*}'x \subset p'P_*```R```\overleftarrow{Q_*}``(C'Q \cap \text{Ɑ}'R)$ (1)

$\vdash . (1) . *231·12 . *90·14 . \supset \vdash . \text{Prop}$

***231·25.** $\vdash :. P \in \text{Ser} . Q_* \in \text{connex} . R``C'Q \subset C'P . P\overline{R}_{os}Q = \Lambda .$
$$\text{E} ! \text{limax}_P`(P\overline{R}_{sc}Q) . \supset : \text{limax}_P`(P\overline{R}_{sc}Q = \text{limin}_P`(\breve{P}\overline{R}_{sc}Q) . \vee .$$
$$\text{limax}_P`(P\overline{R}_{sc}Q) P_1 \text{limin}_P`(\breve{P}\overline{R}_{sc}Q)$$
[*215·54·541 . *231·13·22]

***231·251.** $\vdash : \text{Hp} *231·25 . \text{limax}_P`(P\overline{R}_{sc}Q) \sim\in \text{D}'P_1 . \supset .$
$$\text{limax}_P`(P\overline{R}_{sc}Q) = \text{limin}_P`(\breve{P}\overline{R}_{sc}Q) \quad [*231·25]$$

***231·252.** $\vdash : P \in \text{Ser} . Q_* \in \text{connex} . R``C'Q \subset C'P . P\overline{R}_{os}Q \in 0 \cup 1 .$
$$\text{E} ! \text{limax}_P`(P\overline{R}_{sc}Q) . \text{limax}_P`(P\overline{R}_{sc}Q) \sim\in \text{D}'P_1 . \supset .$$
$$\text{limax}_P`(P\overline{R}_{sc}Q) = \text{limin}_P`(\breve{P}\overline{R}_{sc}Q)$$
[*215·543 . *231·13·22]

***231·4.** $\vdash : Q \in \text{trans} \cap \text{connex} \,.\, E!\, \text{max}_Q\text{‘}\text{Ɑ}\text{‘}R \,.\supset.\, P\overline{R}_{sc}Q = P_*\text{“}\overrightarrow{R}\text{‘}\text{max}_Q\text{‘}\text{Ɑ}\text{‘}R$

Dem.

$\vdash \,.\, *230·53 \,.\, *231·121 \,.\supset$

$\vdash :: \text{Hp} \,.\supset :.\, x \in P\overline{R}_{sc}Q \,.\equiv : \overrightarrow{R}\text{‘}\text{max}_Q\text{‘}\text{Ɑ}\text{‘}R \subset \alpha \,.\supset_\alpha .\, x \in P_*\text{“}\alpha :$

$[*37·2.*22·42] \quad \equiv : x \in P_*\text{“}\overrightarrow{R}\text{‘}\text{max}_Q\text{‘}\text{Ɑ}\text{‘}R :: \supset \vdash .\, \text{Prop}$

***231·41.** $\vdash : Q \in \text{trans} \cap \text{connex} \,.\, E!\, R\text{‘}\text{max}_Q\text{‘}\text{Ɑ}\text{‘}R \,.\supset.\, P\overline{R}_{sc}Q = \overrightarrow{P_*}\text{‘}R\text{‘}\text{max}_Q\text{‘}\text{Ɑ}\text{‘}R$

Dem.

$\vdash \,.\, *30·5 \,.\, *231·4 \,.\, *53·31 \,.\supset$

$\vdash : \text{Hp} \,.\supset .\, P\overline{R}_{sc}Q = P_*\text{“}\iota\text{‘}R\text{‘}\text{max}_Q\text{‘}\text{Ɑ}\text{‘}R$

$[*53·301] \quad = \overrightarrow{P_*}\text{‘}R\text{‘}\text{max}_Q\text{‘}\text{Ɑ}\text{‘}R : \supset \vdash .\, \text{Prop}$

*232. ON THE OSCILLATION OF A FUNCTION AS THE ARGUMENT APPROACHES A GIVEN LIMIT.

Summary of *232.

In the preceding number, we considered the ultimate oscillation of a function when the argument grows without limit. If, in the propositions of the last number, we confine the field of Q to $\overrightarrow{Q}$'x, where $x \in$ Ɑ'Q, the ultimate oscillation becomes the ultimate oscillation as the argument approaches x from below. If the ultimate oscillation consists of a single term, this is the limit of the function as the argument approaches x from below. If, instead of confining the argument to $\overrightarrow{Q}$'x, we confine it to any other class whose limit is x, we shall, under a very usual hypothesis, obtain the same value for the ultimate oscillation as if we confined it to $\overrightarrow{Q}$'x. And more generally, under a similar hypothesis, if α and β are two classes of arguments which define the same section (*i.e.* such that Q_*''$\alpha = Q_*$''β), then, whether or not this section has a limit, the ultimate sections and the ultimate oscillation are the same for α as they are for β. Hence we are led to consider first the result of confining the field of Q, not to $\overrightarrow{Q}$'x, but to any class α. In order not to have to exclude explicitly the case in which $\alpha \in 1$, we deal with $Q_* \upharpoonright \alpha$, not $Q \upharpoonright \alpha$. Hence we are led to the following definitions:

***232·01.** $(P\bar{R}Q)_{sc}$'$\alpha = P\bar{R}_{sc}(Q_* \upharpoonright \alpha)$ Df

***232·02.** $(P\bar{R}Q)_{os}$'$\alpha = P\bar{R}_{os}(Q_* \upharpoonright \alpha)$ Df

Most of the propositions of the present number are immediate consequences of corresponding propositions in *231. The most important application of the propositions of the present number is to the case where α is of the form $\overrightarrow{Q}$'x, x being a member of δ_Q'Ɑ'R. We may, in this case, take in place of $\overrightarrow{Q}$'x any other class of arguments (*e.g.* a progression of arguments $x_1, x_2, \ldots x_\nu, \ldots$) having x for its limit, without altering the limiting sections or the ultimate oscillation. Hence the limit of the function for a given argument (if it exists) may be determined by choosing any selection of arguments having the given argument as their limit (cf. *233·142, below).

From the definition of $(P\overline{R}Q)_{sc}\text{‘}\alpha$ we obtain immediately

***232·11.** $\vdash :. \, x \in (P\overline{R}Q)_{sc}\text{‘}\alpha \, . \equiv :$

$$y \in \alpha \cap C\text{‘}Q \cap \text{Ɑ}\text{‘}R \, . \supset_y . \, x \in P_*\text{“}R\text{“}(\alpha \cap \overleftarrow{Q_*}\text{‘}y) : x \in C\text{‘}P$$

We prove that $(P\overline{R}Q)_{sc}\text{‘}\alpha = (P\overline{R}Q)_{sc}\text{‘}(\alpha \cap C\text{‘}Q \cap \text{Ɑ}\text{‘}R)$ (*232·131), and that if $\alpha \cap C\text{‘}Q \cap \text{Ɑ}\text{‘}R = \Lambda$, the two limiting sections and the ultimate oscillation are all equal to $C\text{‘}P$ (*232·15). Also we have

***232·14.** $\vdash : Q \in \text{trans} \cap \text{connex} \, . \, \alpha \cap C\text{‘}Q \sim \in 1 \, . \supset . \, (P\overline{R}Q)_{sc}\text{‘}\alpha = P\overline{R}_{sc}(Q \upharpoonright \alpha)$

Thus the substitution of Q_* for Q in our definitions has the effect of making them applicable to unit classes, and of enabling us to substitute the hypothesis $Q_* \in \text{connex}$ for $Q \in \text{trans} \cap \text{connex}$. But when Q is transitive and connected (and therefore when Q is a series), the substitution of Q_* for Q in the definitions makes no difference unless α is a unit class. This case is trivial, since the only interest of our definitions is when α has no maximum in Q.

From *231·22 we obtain

***232·22.** $\vdash : P_*, Q_* \upharpoonright \alpha \in \text{connex} \, . \, R\text{“}(\alpha \cap C\text{‘}Q) \subset C\text{‘}P \, . \supset .$

$$C\text{‘}P = (P\overline{R}Q)_{sc}\text{‘}\alpha \cup (\breve{P}\overline{R}Q)_{sc}\text{‘}\alpha$$

We have next a set of propositions concerned in discovering circumstances under which two classes α and β which determine the same section in Q (and therefore have the same limit, if any) give the same values for the two limiting sections. For this purpose, it is only necessary to discover circumstances under which we may substitute $Q_*\text{“}(\alpha \cap \text{Ɑ}\text{‘}R)$ for α. When this can be done, the ultimate oscillation of the function as the argument approaches the limit of α can be determined by taking any set of arguments having this limit. We have

***232·301.** $\vdash . \, (P\overline{R}Q)_{sc}\text{‘}\alpha \subset (P\overline{R}Q)_{sc}\text{‘}Q_*\text{“}(\alpha \cap \text{Ɑ}\text{‘}R)$

***232·32.** $\vdash : (P\overline{R}Q)_{os}\text{‘}Q_*\text{“}(\alpha \cap \text{Ɑ}\text{‘}R) \in 0 \cup 1 \, . \supset . \, (P\overline{R}Q)_{os}\text{‘}\alpha \in 0 \cup 1$

Thus if the function has a limit as the argument approaches the limit of $Q_*\text{“}(\alpha \cap \text{Ɑ}\text{‘}R)$, it also has a limit as the argument approaches the limit of α.

***232·33.** $\vdash : P_*, Q_* \upharpoonright \alpha \in \text{connex} \, . \, R\text{“}(\alpha \cap C\text{‘}Q) \subset C\text{‘}P \, . \supset .$

$$(P\overline{R}Q)_{sc}\text{‘}\alpha \cup (\breve{P}\overline{R}Q)_{sc}\text{‘}\alpha = (P\overline{R}Q)_{sc}\text{‘}Q_*\text{“}(\alpha \cap \text{Ɑ}\text{‘}R) \cup (\breve{P}\overline{R}Q)_{sc}\text{‘}Q_*\text{“}(\alpha \cap \text{Ɑ}\text{‘}R) = C\text{‘}P$$

whence

***232·34.** $\vdash : \text{Hp} \, *232{\cdot}33 \, . \, (P\overline{R}Q)_{os}\text{‘}Q_*\text{“}(\alpha \cap \text{Ɑ}\text{‘}R) = \Lambda \, . \supset .$

$$(P\overline{R}Q)_{sc}\text{‘}\alpha = (P\overline{R}Q)_{sc}\text{‘}Q_*\text{“}(\alpha \cap \text{Ɑ}\text{‘}R) \, . \, (\breve{P}\overline{R}Q)_{sc}\text{‘}\alpha = (\breve{P}\overline{R}Q)_{sc}\text{‘}Q_*\text{“}(\alpha \cap \text{Ɑ}\text{‘}R)$$

We have also

***232·341.** $\vdash : P_* \in \text{connex} \, . \, \exists ! \, (P\overline{R}Q)_{os}\text{‘}\alpha \, . \, (P\overline{R}Q)_{os}\text{‘}Q_*\text{“}(\alpha \cap \text{Ɑ}\text{‘}R) \in 1 \, . \supset .$

$$(P\overline{R}Q)_{sc}\text{‘}\alpha = (P\overline{R}Q)_{sc}\text{‘}Q_*\text{“}(\alpha \cap \text{Ɑ}\text{‘}R) \, . \, (\breve{P}\overline{R}Q)_{sc}\text{‘}\alpha = (\breve{P}\overline{R}Q)_{sc}\text{‘}Q_*\text{“}(\alpha \cap \text{Ɑ}\text{‘}R)$$

Hence we arrive at the conclusion that, if P_{po} is a series, and x is the limit of the function for the class $Q_*``(\alpha \cap \text{Ɑ}`R)$, if $x$ is a member of $(P\bar{R}Q)_{\text{sc}}`\alpha$, it is its maximum (*232·352), while if x is not a member of $(P\bar{R}Q)_{\text{sc}}`\alpha$, it is its sequent (*232·356), assuming $(P\bar{R}Q)_{\text{sc}}`\alpha \cup (\breve{P}\bar{R}Q)_{\text{sc}}`\alpha = C`P$, which, as we saw (*233·22), is generally the case, and assuming also $P \epsilon \text{Ser}$. On the other hand, if $(P\bar{R}Q)_{\text{sc}}`\alpha$ has no maximum, $x$ is the minimum of $(\breve{P}\bar{R}Q)_{\text{sc}}`\alpha$; and if $(P\bar{R}Q)_{\text{sc}}`\alpha$ has a maximum other than $x$, this is $P_1`x$ (*232·357·358). This latter case is impossible unless x has an immediate predecessor. Hence we arrive at the following proposition :

*232·38. $\vdash : P \epsilon \text{Ser} \,.\, Q_* \upharpoonright \alpha \epsilon \text{connex} \,.\, R``(\alpha \cap C`Q) \subset C`P \,.$
$$(P\bar{R}Q)_{\text{os}}`Q_*``(\alpha \cap \text{Ɑ}`R) \epsilon 0 \cup (1 - \text{Cl}`C`P_1) \,.\supset .$$
$$\overrightarrow{\text{limax}}_P`(P\bar{R}Q)_{\text{sc}}`\alpha = \overrightarrow{\text{limax}}_P`(P\bar{R}Q)_{\text{sc}}`Q_*``(\alpha \cap \text{Ɑ}`R) \,.$$
$$\overrightarrow{\text{limin}}_P`(\breve{P}\bar{R}Q)_{\text{sc}}`\alpha = \overrightarrow{\text{limin}}_P`(\breve{P}\bar{R}Q)_{\text{sc}}`Q_*``(\alpha \cap \text{Ɑ}`R)$$

Applying this to a series having Dedekindian continuity, we know that $P_1 = \dot{\Lambda}$, and that the limax and limin always exist. Hence

*232·39. $\vdash :. P \epsilon \text{Ser} \cap \text{Ded} \,.\, P^2 = P \,.\, Q_* \epsilon \text{connex} \,.\, R``C`Q \subset C`P \,.\supset :$
$$(P\bar{R}Q)_{\text{os}}`Q_*``(\alpha \cap \text{Ɑ}`R) \epsilon 0 \cup 1 \,.\supset_\alpha .$$
$$\text{limax}_P`(P\bar{R}Q)_{\text{sc}}`\alpha = \text{limax}_P`(P\bar{R}Q)_{\text{sc}}`Q_*``(\alpha \cap \text{Ɑ}`R) =$$
$$\text{limin}_P`(\breve{P}\bar{R}Q)_{\text{sc}}`\alpha = \text{limin}_P`(\breve{P}\bar{R}Q)_{\text{sc}}`Q_*``(\alpha \cap \text{Ɑ}`R)$$

That is to say, if the value-series P has Dedekindian continuity, and contains all values for arguments in $C`Q$, then, provided the function has a definite limit for the class $Q_*``(\alpha \cap \text{Ɑ}`R)$, this is its limit also for the class α; that is to say, any collection of arguments having the same limit or maximum as a given section will give the same limit for the function.

---

*232·01. $(P\bar{R}Q)_{\text{sc}}`\alpha = P\bar{R}_{\text{sc}}(Q_* \upharpoonright \alpha)$ Df

*232·02. $(P\bar{R}Q)_{\text{os}}`\alpha = P\bar{R}_{\text{os}}(Q_* \upharpoonright \alpha)$ Df

*232·1. $\vdash . (P\bar{R}Q)_{\text{sc}}`\alpha = P\bar{R}_{\text{sc}}(Q_* \upharpoonright \alpha)$ [(*232·01)]

*232·101. $\vdash . (P\bar{R}Q)_{\text{os}}`\alpha = P\bar{R}_{\text{os}}(Q_* \upharpoonright \alpha) = (P\bar{R}Q)_{\text{sc}}`\alpha \cap (\breve{P}\bar{R}Q)_{\text{sc}}`\alpha$ [(*232·02)]

*232·11. $\vdash :. x \epsilon (P\bar{R}Q)_{\text{sc}}`\alpha \,.\equiv :$
$$y \epsilon \alpha \cap C`Q \cap \text{Ɑ}`R \,.\supset_y .\, x \epsilon P_*``R``(\alpha \cap \overleftarrow{Q_*}`y) : x \epsilon C`P$$

Dem.

$\vdash$. *90·41·42 . *231·112 . $\supset$

$\vdash :. x \epsilon (P\bar{R}Q)_{\text{sc}}`\alpha \,.\equiv : y \epsilon \alpha \cap C`Q \cap \text{Ɑ}`R \,.\supset_y .\, x \epsilon P_*``R``\overleftarrow{(Q_* \upharpoonright \alpha)}`y : x \epsilon C`P :$

[*35·102] $\equiv : y \epsilon \alpha \cap C`Q \cap \text{Ɑ}`R \,.\supset_y .\, x \epsilon P_*``R``(\alpha \cap \overleftarrow{Q_*}`y) : x \epsilon C`P :. \supset \vdash$. Prop

$*232{\cdot}12.\ \vdash . (P\bar{R}Q)_{sc}{}^{\backprime}\alpha = p^{\backprime}P_*{}^{\backprime\backprime\backprime}R^{\backprime\backprime\backprime}(\alpha \cap)^{\backprime\backprime}\overleftarrow{Q_*}{}^{\backprime\backprime}(\alpha \cap C^{\backprime}Q \cap \text{ɑ}^{\backprime}R) \cap C^{\backprime}P$
$[*232{\cdot}11]$

$*232{\cdot}121.\ \vdash : \gamma = \alpha \cap C^{\backprime}Q \cap \text{ɑ}^{\backprime}R . \supset . (P\bar{R}Q)_{sc}{}^{\backprime}\alpha = p^{\backprime}P_*{}^{\backprime\backprime\backprime}R^{\backprime\backprime\backprime}(\gamma \cap)^{\backprime\backprime}\overleftarrow{Q_*}{}^{\backprime\backprime}\gamma \cap C^{\backprime}P$

Dem.

$\vdash . *90{\cdot}13 . \supset \vdash . \alpha \cap \overleftarrow{Q_*}{}^{\backprime}y = \alpha \cap C^{\backprime}Q \cap \overleftarrow{Q_*}{}^{\backprime}y .$

$[*37{\cdot}26]\quad \supset \vdash . R^{\backprime\backprime}(\alpha \cap \overleftarrow{Q_*}{}^{\backprime}y) = R^{\backprime\backprime}(\alpha \cap C^{\backprime}Q \cap \text{ɑ}^{\backprime}R \cap \overleftarrow{Q_*}{}^{\backprime}y) \quad (1)$

$\vdash . (1) . *232{\cdot}11 . \supset \vdash .$ Prop

$*232{\cdot}13.\ \vdash : \alpha \cap C^{\backprime}Q \cap \text{ɑ}^{\backprime}R = \beta \cap C^{\backprime}Q \cap \text{ɑ}^{\backprime}R . \supset . (P\bar{R}Q)_{sc}{}^{\backprime}\alpha = (P\bar{R}Q)_{sc}{}^{\backprime}\beta$
$[*232{\cdot}121]$

$*232{\cdot}131.\ \vdash . (P\bar{R}Q)_{sc}{}^{\backprime}\alpha = (P\bar{R}Q)_{sc}{}^{\backprime}(\alpha \cap C^{\backprime}Q \cap \text{ɑ}^{\backprime}R)\quad [*232{\cdot}13]$

From the above propositions it follows that the values of $(P\bar{R}Q)_{sc}{}^{\backprime}\alpha$, $(\breve{P}\bar{R}Q)_{sc}{}^{\backprime}\alpha$, and $(P\bar{R}Q)_{os}{}^{\backprime}\alpha$ depend only upon $\alpha \cap C^{\backprime}Q \cap \text{ɑ}^{\backprime}R$; thus if α is not contained in $C^{\backprime}Q \cap \text{ɑ}^{\backprime}R$, the part not contained in $C^{\backprime}Q \cap \text{ɑ}^{\backprime}R$ is irrelevant.

$*232{\cdot}14.\ \vdash : Q \in \text{trans} \cap \text{connex} . \alpha \cap C^{\backprime}Q \sim\in 1 . \supset . (P\bar{R}Q)_{sc}{}^{\backprime}\alpha = P\bar{R}_{sc}(Q \upharpoonright \alpha)$
$[*232{\cdot}1 . *202{\cdot}54{\cdot}541]$

$*232{\cdot}15.\ \vdash : \alpha \cap C^{\backprime}Q \cap \text{ɑ}^{\backprime}R = \Lambda . \supset . (P\bar{R}Q)_{sc}{}^{\backprime}\alpha = (\breve{P}\bar{R}Q)_{sc}{}^{\backprime}\alpha = (P\bar{R}Q)_{os}{}^{\backprime}\alpha = C^{\backprime}P$
$[*232{\cdot}12{\cdot}101 . *37{\cdot}29 . *40{\cdot}2]$

$*232{\cdot}151.\ \vdash : \dot{\exists} ! P . (P\bar{R}Q)_{os}{}^{\backprime}\alpha = \Lambda . \supset . \exists ! \alpha \cap C^{\backprime}Q \cap \text{ɑ}^{\backprime}R$
$[*232{\cdot}15 . \text{Transp} . *33{\cdot}24]$

$*232{\cdot}2.\ \vdash : C^{\backprime}Q \cap \text{ɑ}^{\backprime}R \subset \alpha . \supset . (P\bar{R}Q)_{sc}{}^{\backprime}\alpha = P\bar{R}_{sc}Q$

Dem.

$\vdash . *22{\cdot}621 . *232{\cdot}11 . \supset$

$\vdash :: \text{Hp} . \supset :. x \in (P\bar{R}Q)_{sc}{}^{\backprime}\alpha . \equiv : y \in C^{\backprime}Q \cap \text{ɑ}^{\backprime}R . \supset_y . x \in P_*{}^{\backprime\backprime}R^{\backprime\backprime}\overleftarrow{Q_*}{}^{\backprime}y :$

$[*231{\cdot}112]\qquad \equiv : x \in P\bar{R}_{sc}Q :: \supset \vdash .$ Prop

$*232{\cdot}21.\ \vdash : P_*, Q_* \upharpoonright \alpha \in \text{connex} . \alpha \cap C^{\backprime}Q \cap \text{ɑ}^{\backprime}R \subset Q_*{}^{\backprime\backprime}(\alpha \cap \breve{R}^{\backprime\backprime}C^{\backprime}P) . \supset .$
$C^{\backprime}P = (P\bar{R}Q)_{sc}{}^{\backprime}\alpha \cup (\breve{P}\bar{R}Q)_{sc}{}^{\backprime}\alpha$

$\left[*231{\cdot}21 \dfrac{Q_* \upharpoonright \alpha}{Q}\right]$

$*232{\cdot}22.\ \vdash : P_*, Q_* \upharpoonright \alpha \in \text{connex} . R^{\backprime\backprime}(\alpha \cap C^{\backprime}Q) \subset C^{\backprime}P . \supset .$
$C^{\backprime}P = (P\bar{R}Q)_{sc}{}^{\backprime}\alpha \cup (\breve{P}\bar{R}Q)_{sc}{}^{\backprime}\alpha \quad [*231{\cdot}22]$

$*232{\cdot}23.\ \vdash : y \in C^{\backprime}Q \cap \text{ɑ}^{\backprime}R . \supset . (P\bar{R}Q)_{sc}{}^{\backprime}\iota^{\backprime}y = P_*{}^{\backprime\backprime}\overrightarrow{R}{}^{\backprime}y$

Dem.

$\vdash . *232{\cdot}11 . *13{\cdot}191 . \supset$

$\vdash :. \text{Hp} . \supset : x \in (P\bar{R}Q)_{sc}{}^{\backprime}\iota^{\backprime}y . \equiv . x \in P_*{}^{\backprime\backprime}R^{\backprime\backprime}(\iota^{\backprime}y \cap \overleftarrow{Q_*}{}^{\backprime}y) :. \supset \vdash .$ Prop

***232·24.** $\vdash : Q \epsilon \text{trans} \cap \text{connex} . E ! \max_Q{}'(\alpha \cap \text{Cl}'R) . \supset .$
$$(P\bar{R}Q)_{sc}{}'\alpha = P_*{}''\overrightarrow{R}{}'\max_Q{}'(\alpha \cap \text{Cl}'R)$$

Dem.

$\vdash . *232·14 . \supset \vdash : \text{Hp} . \alpha \cap C'Q \cap \text{Cl}'R \sim \epsilon 1 . \supset .$

$(P\bar{R}Q)_{sc}{}'\alpha = P\bar{R}_{sc}\{Q \upharpoonright (\alpha \cap \text{Cl}'R)\}$

$[*231·4 . *205·9] = P_*{}''\overrightarrow{R}{}'\max_Q{}'(\alpha \cap \text{Cl}'R)$ (1)

$\vdash . *205·17 . *232·23·131 . \supset$

$\vdash : \alpha \cap C'Q \cap \text{Cl}'R \epsilon 1 . \supset . (P\bar{R}Q)_{sc}{}'\alpha = P_*{}''\overrightarrow{R}{}'\max_Q{}'(\alpha \cap \text{Cl}'R)$ (2)

$\vdash . (1) . (2) . \supset \vdash . \text{Prop}$

***232·3.** $\vdash : \alpha \subset \text{Cl}'R . \supset . (P\bar{R}Q)_{sc}{}'\alpha \subset (P\bar{R}Q)_{sc}{}'Q_*{}''\alpha$

Dem.

$\vdash . *96·3 . \supset \vdash : y \epsilon Q_*{}''\alpha . \supset . (\exists z) . z \epsilon \alpha \cap C'Q . \overleftarrow{Q_*}{}'z \subset \overleftarrow{Q_*}{}'y .$

$[\text{Fact} . *37·2] \supset . (\exists z) . z \epsilon \alpha \cap C'Q . P_*{}''R''(\alpha \cap \overleftarrow{Q_*}{}'z) \subset P_*{}''R''(\alpha \cap \overleftarrow{Q_*}{}'y)$ (1)

$\vdash . (1) . *232·11 . \supset \vdash :. \text{Hp} . x \epsilon (P\bar{R}Q)_{sc}{}'\alpha . \supset : y \epsilon Q_*{}''\alpha . \supset . x \epsilon P_*{}''R''(\alpha \cap \overleftarrow{Q_*}{}'y) .$

$[*90·33] \qquad \supset . x \epsilon P_*{}''R''(Q_*{}''\alpha \cap \overleftarrow{Q_*}{}'y) :$

$[*232·11] \qquad \supset : x \epsilon (P\bar{R}Q)_{sc}{}'Q_*{}''\alpha :. \supset \vdash . \text{Prop}$

***232·301.** $\vdash . (P\bar{R}Q)_{sc}{}'\alpha \subset (P\bar{R}Q)_{sc}{}'Q_*{}''(\alpha \cap \text{Cl}'R)$

Dem.

$\vdash . *232·13 . \supset \vdash . (P\bar{R}Q)_{sc}{}'\alpha = (P\bar{R}Q)_{sc}{}'(\alpha \cap \text{Cl}'R)$

$[*232·3] \qquad \subset (P\bar{R}Q)_{sc}{}'Q_*{}''(\alpha \cap \text{Cl}'R) . \supset \vdash . \text{Prop}$

***232·31.** $\vdash . (P\bar{R}Q)_{os}{}'\alpha \subset (P\bar{R}Q)_{os}{}'Q_*{}''(\alpha \cap \text{Cl}'R)$

Dem.

$\vdash . *232·301 \dfrac{\breve{P}}{P} . \supset \vdash . (\breve{P}\bar{R}Q)_{sc}{}'\alpha \subset (\breve{P}\bar{R}Q)_{sc}{}'Q_*{}''(\alpha \cap \text{Cl}'R)$ (1)

$\vdash . *232·301 . (1) . *232·101 . \supset \vdash . \text{Prop}$

***232·32.** $\vdash : (P\bar{R}Q)_{os}{}'Q_*{}''(\alpha \cap \text{Cl}'R) \epsilon 0 \cup 1 . \supset . (P\bar{R}Q)_{os}{}'\alpha \epsilon 0 \cup 1$ [*232·31]

***232·33.** $\vdash : P_*, Q_* \upharpoonright \alpha \epsilon \text{connex} . R''(\alpha \cap C'Q) \subset C'P . \supset .$

$(P\bar{R}Q)_{sc}{}'\alpha \cup (\breve{P}\bar{R}Q)_{sc}{}'\alpha = (P\bar{R}Q)_{sc}{}'Q_*{}''(\alpha \cap \text{Cl}'R) \cup (\breve{P}\bar{R}Q)_{sc}{}'Q_*{}''(\alpha \cap \text{Cl}'R) = C'P$

Dem.

$\vdash . *232·22·301 . \supset$

$\vdash : \text{Hp} . \supset . C'P \subset (P\bar{R}Q)_{sc}{}'Q_*{}''(\alpha \cap \text{Cl}'R) \cup (\breve{P}\bar{R}Q)_{sc}{}'Q_*{}''(\alpha \cap \text{Cl}'R)$ (1)

$\vdash . (1) . *231·131 . \supset \vdash . \text{Prop}$

***232·34.** $\vdash : \text{Hp} *232·33 . (P\bar{R}Q)_{os}{}'Q_*{}''(\alpha \cap \text{Cl}'R) = \Lambda . \supset .$

$(P\bar{R}Q)_{sc}{}'\alpha = (P\bar{R}Q)_{sc}{}'Q_*{}''(\alpha \cap \text{Cl}'R) . (\breve{P}\bar{R}Q)_{sc}{}'\alpha = (\breve{P}\bar{R}Q)_{sc}{}'Q_*{}''(\alpha \cap \text{Cl}'R)$

[*232·33·301 . *24·482]

***232·341.** $\vdash : P_* \epsilon \text{connex} . \exists ! (P\bar{R}Q)_{os}`\alpha . (P\bar{R}Q)_{os}`Q_*``(\alpha \cap \text{Ɑ}`R) \epsilon 1 . \supset .$
$(P\bar{R}Q)_{sc}`\alpha = (P\bar{R}Q)_{sc}`Q_*``(\alpha \cap \text{Ɑ}`R) . (\breve{P}RQ)_{sc}`\alpha = (\breve{P}RQ)_{sc}`Q_*``(\alpha \cap \text{Ɑ}`R)$
[*231·192 . *232·31 . *60·38]

***232·35.** $\vdash : P_* \epsilon \text{connex} . (P\bar{R}Q)_{os}`Q_*``(\alpha \cap \text{Ɑ}`R) = \iota`x . \supset .$
$(P\bar{R}Q)_{sc}`\alpha \subset \overrightarrow{P_*}`x . (\breve{P}\bar{R}Q)_{sc}`\alpha \subset \overleftarrow{P_*}`x$
[*232·301 . *231·191]

***232·351.** $\vdash : \text{Hp} *232·35 . x \epsilon (P\bar{R}Q)_{sc}`\alpha . \supset . (P\bar{R}Q)_{sc}`\alpha = \overrightarrow{P_*}`x$

Dem.

$\vdash . *231·13 . \supset \vdash : \text{Hp} . \supset . \overrightarrow{P_*}`x \subset (P\bar{R}Q)_{sc}`\alpha$ (1)
$\vdash . (1) . *232·35 . \supset \vdash . \text{Prop}$

***232·352.** $\vdash : \text{Hp} *232·351 . P_{po} \in J . \supset . x = \max_P`(P\bar{R}Q)_{sc}`\alpha$
[*211·8 . *205·197 . *232·351]

***232·353.** $\vdash : \text{Hp} *232·35 . (P\bar{R}Q)_{sc}`\alpha \cup (\breve{P}\bar{R}Q)_{sc}`\alpha = C`P . x \sim\epsilon (P\bar{R}Q)_{sc}`\alpha . \supset .$
$(\breve{P}\bar{R}Q)_{sc}`\alpha = \overleftarrow{P_*}`x$

Dem.

$\vdash . *231·13 . \supset \vdash : \text{Hp} . \supset . x \epsilon (\breve{P}\bar{R}Q)_{sc}`\alpha .$
[*232·351] $\supset . (\breve{P}\bar{R}Q)_{sc}`\alpha = \overleftarrow{P_*}`x : \supset \vdash . \text{Prop}$

***232·354.** $\vdash : \text{Hp} *232·353 . P_{po} \in J . \supset . x = \min_P`(\breve{P}\bar{R}Q)_{sc}`\alpha$ $\left[*232·352 \frac{\breve{P}}{P}\right]$

***232·355.** $\vdash : \text{Hp} *232·353 . \supset . (P\bar{R}Q)_{sc}`\alpha = \overrightarrow{P_{po}}`x$

Dem.

$\vdash . *232·35 . \supset \vdash : \text{Hp} . \supset . (P\bar{R}Q)_{sc}`\alpha \subset \overrightarrow{P_*}`x - \iota`x$
[*91·542] $\subset \overrightarrow{P_{po}}`x$ (1)
$\vdash . *232·353 . \supset \vdash : \text{Hp} . \supset . C`P - \overleftarrow{P_*}`x \subset C`P - (\breve{P}\bar{R}Q)_{so}`\alpha .$
[*202·101.Hp] $\supset . \overrightarrow{P_{po}}`x \subset (P\bar{R}Q)_{sc}`\alpha$ (2)
$\vdash . (1) . (2) . \supset \vdash . \text{Prop}$

***232·356.** $\vdash :. P \epsilon \text{Ser} . (P\bar{R}Q)_{os}`Q_*``(\alpha \cap \text{Ɑ}`R) = \iota`x .$
$(P\bar{R}Q)_{sc}`\alpha \cup (\breve{P}\bar{R}Q)_{sc}`\alpha = C`P . \supset : x \sim\epsilon (P\bar{R}Q)_{sc}`\alpha . \supset . x = \text{seq}_P`(P\bar{R}Q)_{sc}`\alpha$
[*206·172 . *231·13 . *232·355]

***232·357.** $\vdash : \text{Hp} *232·35 . P_{po} \in J . \sim E ! \max_P`(P\bar{R}Q)_{sc}`\alpha . \supset . x = \min_P`(\breve{P}\bar{R}Q)_{sc}`\alpha$

Dem.

$\vdash . *232·352 . \text{Transp} . \supset \vdash : \text{Hp} . \supset . x \sim\epsilon (P\bar{R}Q)_{sc}`\alpha .$
[*232·354] $\supset . x = \min_P`(\breve{P}\bar{R}Q)_{sc}`\alpha : \supset \vdash . \text{Prop}$

***232·358.** $\vdash : \text{Hp} *232\cdot35 . P_{po} \in J . (P\bar{R}Q)_{sc}`\alpha \cup (\breve{P}\bar{R}Q)_{sc}`\alpha = C`P .$
$E ! \max_P`(P\bar{R}Q)_{sc}`\alpha . \max_P`(P\bar{R}Q)_{sc}`\alpha \neq x . \supset . \max_P`(P\bar{R}Q)_{sc}`\alpha \, P_1 x$

Dem.

$\vdash . *232\cdot352 . \text{Transp} . \supset \vdash : \text{Hp} . \supset . x \sim \epsilon (P\bar{R}Q)_{sc}`\alpha .$
$[*232\cdot356] \quad \supset . x = \text{seq}_P`(P\bar{R}Q)_{sc}`\alpha .$
$[*206\cdot5] \quad \supset . \max_P`(P\bar{R}Q)_{sc}`\alpha \, P_1 x : \supset \vdash . \text{Prop}$

***232·36.** $\vdash :. P \in \text{Ser} . (P\bar{R}Q)_{os}`Q_*``(\alpha \cap \text{Π}`R) = \iota`x .$
$(P\bar{R}Q)_{sc}`\alpha \cup (\breve{P}\bar{R}Q)_{sc}`\alpha = C`P . \supset :$
$x \in (P\bar{R}Q)_{os}`\alpha . \supset . x = \max_P`(P\bar{R}Q)_{sc}`\alpha = \min_P`(\breve{P}\bar{R}Q)_{sc}`\alpha :$
$x \in (P\bar{R}Q)_{sc}`\alpha - (\breve{P}\bar{R}Q)_{sc}`\alpha . \supset . x = \max_P`(\breve{P}\bar{R}Q)_{sc}`\alpha = \text{prec}_P`(P\bar{R}Q)_{sc}`\alpha :$
$x \in (\breve{P}\bar{R}Q)_{sc}`\alpha - (P\bar{R}Q)_{sc}`\alpha . \supset . x = \text{seq}_P`(P\bar{R}Q)_{sc}`\alpha = \min_P`(\breve{P}\bar{R}Q)_{sc}`\alpha$
[*232·352·354·356]

***232·361.** $\vdash : \text{Hp} *232\cdot36 . x \sim \epsilon \text{Π}`P_1 . \supset . x = \text{limax}_P`(P\bar{R}Q)_{sc}`\alpha$

Dem.

$\vdash . *232\cdot358 . \text{Transp} . \supset$
$\vdash :. \text{Hp} . \supset : E ! \max_P`(P\bar{R}Q)_{sc}`\alpha . \supset . \max_P`(P\bar{R}Q)_{sc}`\alpha = x \quad (1)$
$\vdash . *232\cdot352 . \text{Transp} . \supset$
$\vdash :. \text{Hp} . \supset : \sim E ! \max_P`(P\bar{R}Q)_{sc}`\alpha . \supset . x \sim \epsilon (P\bar{R}Q)_{sc}`\alpha .$
$[*232\cdot356] \quad \supset . x = \text{seq}_P`(P\bar{R}Q)`\alpha \quad (2)$
$\vdash . (1) . (2) . *207\cdot46 . \supset \vdash . \text{Prop}$

***232·37.** $\vdash : P \in \text{Ser} . (P\bar{R}Q)_{os}`Q_*``(\alpha \cap \text{Π}`R) \in 1 - \text{Cl}`\text{Π}`P_1 .$
$(P\bar{R}Q)_{sc}`\alpha \cup (\breve{P}\bar{R}Q)_{sc}`\alpha = C`P . \supset .$
$\text{limax}_P`(P\bar{R}Q)_{sc}`\alpha = \max_P`(P\bar{R}Q)_{sc}`Q_*``(\alpha \cap \text{Π}`R)$
$= \breve{\iota}`(P\bar{R}Q)_{os}`Q_*``(\alpha \cap \text{Π}`R)$
[*232·361 . *231·193]

***232·38.** $\vdash : P \in \text{Ser} . Q_* \upharpoonright \alpha \in \text{connex} . R``(\alpha \cap C`Q) \subset C`P .$
$(P\bar{R}Q)_{os}`Q_*``(\alpha \cap \text{Π}`R) \in 0 \cup (1 - \text{Cl}`C`P_1) . \supset .$
$\overrightarrow{\text{limax}}_P`(P\bar{R}Q)_{sc}`\alpha = \overrightarrow{\text{limax}}_P`(P\bar{R}Q)_{sc}`Q_*``(\alpha \cap \text{Π}`R) .$
$\overrightarrow{\text{limin}}_P`(\breve{P}\bar{R}Q)_{sc}`\alpha = \overrightarrow{\text{limin}}_P`(\breve{P}\bar{R}Q)_{sc}`Q_*``(\alpha \cap \text{Π}`R)$
[*232·33·34·37]

***232·39.** $\vdash :. P \in \mathrm{Ser} \cap \mathrm{Ded} \,.\, P^2 = P \,.\, Q_* \in \mathrm{connex} \,.\, R\lq\lq C\lq Q \subset C\lq P \,.\, \supset :$

$$(P\bar{R}Q)_{os}\lq Q_*\lq\lq(\alpha \cap \text{Ɑ}\lq R) \in 0 \cup 1 \,.\, \supset_\alpha .$$
$$\mathrm{limax}_P\lq(P\bar{R}Q)_{sc}\lq\alpha = \mathrm{limax}_P\lq(P\bar{R}Q)_{sc}\lq Q_*\lq\lq(\alpha \cap \text{Ɑ}\lq R)$$
$$= \mathrm{limin}_P\lq(\breve{P}\bar{R}Q)_{sc}\lq\alpha = \mathrm{limin}_P\lq(\breve{P}\bar{R}Q)_{sc}\lq Q_*\lq\lq(\alpha \cap \text{Ɑ}\lq R)$$

Dem.

$\vdash . *201{\cdot}63 . *232{\cdot}38 . \supset \vdash : \mathrm{Hp} \,.\, (P\bar{R}Q)_{os}\lq Q_*\lq\lq(\alpha \cap \text{Ɑ}\lq R) \in 0 \cup 1 \,.\, \supset .$

$\quad \mathrm{limax}_P\lq(P\bar{R}Q)_{sc}\lq\alpha = \mathrm{limax}_P\lq(P\bar{R}Q)_{sc}\lq Q_*\lq\lq(\alpha \cap \text{Ɑ}\lq R) \,.$

$\quad \mathrm{limin}_P\lq(\breve{P}\bar{R}Q)_{sc}\lq\alpha = \mathrm{limin}_P\lq(\breve{P}\bar{R}Q)_{sc}\lq Q_*\lq\lq(\alpha \cap \text{Ɑ}\lq R)$ (1)

$\vdash . *231{\cdot}193 . \supset \vdash : \mathrm{Hp}(1) \,.\, (P\bar{R}Q)_{os}\lq Q_*\lq\lq(\alpha \cap \text{Ɑ}\lq R) \in 1 \,.\, \supset .$

$\quad \mathrm{limax}_P\lq(P\bar{R}Q)_{sc}\lq Q_*\lq\lq(\alpha \cap \text{Ɑ}\lq R) = \mathrm{limin}_P\lq(\breve{P}\bar{R}Q)_{sc}\lq Q_*\lq\lq(\alpha \cap \text{Ɑ}\lq R)$ (2)

$\vdash . *214{\cdot}42 . *232{\cdot}33 . \supset \vdash : \mathrm{Hp}(1) \,.\, (P\bar{R}Q)_{os}\lq Q_*\lq\lq(\alpha \cap \text{Ɑ}\lq R) = \Lambda \,.\, \supset .$

$\quad \mathrm{limax}_P\lq(P\bar{R}Q)_{sc}\lq Q_*\lq\lq(\alpha \cap \text{Ɑ}\lq R) = \mathrm{limin}_P\lq(\breve{P}\bar{R}Q)_{sc}\lq Q_*\lq\lq(\alpha \cap \text{Ɑ}\lq R)$ (3)

$\vdash . (1) . (2) . (3) . \supset \vdash . \mathrm{Prop}$

***232·5.** $\vdash . (P\bar{R}Q)_{sc}\lq\overrightarrow{Q}\lq x = C\lq P \cap \hat{y}\,\{z \in \overrightarrow{Q}\lq x \cap \text{Ɑ}\lq R \,.\, \supset_z .\, y \in P_*\lq\lq R\lq\lq(\overrightarrow{Q}\lq x \cap \overleftarrow{Q_*}\lq z)\}$
[*232·11]

***232·51.** $\vdash : Q \in \mathrm{trans} \cap \mathrm{connex} \,.\, \mathrm{E}\,!\, \mathrm{max}_Q\lq(\overrightarrow{Q}\lq x \cap \text{Ɑ}\lq R) \,.\, \supset .$

$$(P\bar{R}Q)_{sc}\lq\overrightarrow{Q}\lq x = P_*\lq\lq\overrightarrow{R}\lq\mathrm{max}_Q\lq(\overrightarrow{Q}\lq x \cap \text{Ɑ}\lq R) \quad [*232{\cdot}24]$$

***232·511.** $\vdash : Q \in \mathrm{trans} \cap \mathrm{connex} \,.\, \mathrm{E}\,!\, R\lq\mathrm{max}_Q\lq(\overrightarrow{Q}\lq x \cap \text{Ɑ}\lq R) \,.\, \supset .$

$$(P\bar{R}Q)_{sc}\lq\overrightarrow{Q}\lq x = \overrightarrow{P_*}\lq R\lq\mathrm{max}_Q\lq(\overrightarrow{Q}\lq x \cap \text{Ɑ}\lq R) \quad [*232{\cdot}51]$$

***232·52.** $\vdash : Q \in \mathrm{connex} \,.\, yQx \,.\, \overrightarrow{Q}\lq x \cap (\overleftarrow{Q}\lq y \cup \iota\lq y) \cap \text{Ɑ}\lq R = \Lambda \,.\, \supset .$

$$(P\bar{R}Q)_{sc}\lq\overrightarrow{Q}\lq x = (P\bar{R}Q)_{sc}\lq\overrightarrow{Q}\lq y \quad [*232{\cdot}13]$$

***232·53.** $\vdash : Q \in \mathrm{connex} \,.\, z \in \overrightarrow{Q}\lq x \cap \text{Ɑ}\lq R \,.\, \supset . (P\bar{R}Q)_{sc}\lq\overrightarrow{Q}\lq x = (P\bar{R}Q)_{sc}\lq(\overrightarrow{Q}\lq x \cap \overleftarrow{Q_*}\lq z)$

Dem.

$\vdash . *232{\cdot}5 . *96{\cdot}3 . \supset \vdash :. \mathrm{Hp} \,.\, y \in (P\bar{R}Q)_{sc}\lq\overrightarrow{Q}\lq x \,.\, \supset :$

$\quad u \in \overrightarrow{Q}\lq x \cap \overleftarrow{Q_*}\lq z \cap \text{Ɑ}\lq R \,.\, \supset_u .\, y \in P_*\lq\lq R\lq\lq(\overrightarrow{Q}\lq x \cap \overleftarrow{Q_*}\lq u) \,.\, \overleftarrow{Q_*}\lq u \subset \overleftarrow{Q_*}\lq z :$

$[*22{\cdot}621 . *232{\cdot}11] \supset : y \in (P\bar{R}Q)_{sc}\lq(\overrightarrow{Q}\lq x \cap \overleftarrow{Q_*}\lq z)$ (1)

$\vdash . *232{\cdot}11 . *37{\cdot}2 . \supset \vdash :. \mathrm{Hp} \,.\, y \in (P\bar{R}Q)_{sc}\lq(\overrightarrow{Q}\lq x \cap \overleftarrow{Q_*}\lq z) \,.\, \supset :$

$\quad u \in \overrightarrow{Q}\lq x \cap \overleftarrow{Q_*}\lq z \cap \text{Ɑ}\lq R \,.\, \supset_u .\, y \in P_*\lq\lq R\lq\lq(\overrightarrow{Q}\lq x \cap \overleftarrow{Q_*}\lq u)$ (2)

$[*96{\cdot}3] \supset : u \in \overrightarrow{Q}\lq x \cap \overrightarrow{Q_*}\lq z \cap \text{Ɑ}\lq R \,.\, \supset_u .\, y \in P_*\lq\lq R\lq\lq(\overrightarrow{Q}\lq x \cap \overleftarrow{Q_*}\lq u)$ (3)

$\vdash . (2) . (3) . \supset \vdash :. \mathrm{Hp}(2) \,.\, \supset : u \in \overrightarrow{Q}\lq x \cap \text{Ɑ}\lq R \,.\, \supset_u .\, y \in P_*\lq\lq R\lq\lq(\overrightarrow{Q}\lq x \cap \overleftarrow{Q_*}\lq u) :$

$[*232{\cdot}5] \quad \supset : y \in (P\bar{R}Q)_{sc}\lq\overrightarrow{Q}\lq x$ (4)

$\vdash . (1) . (4) . \supset \vdash . \mathrm{Prop}$

Summary of *233.

There are four limits of a function as the argument approaches some term a in the argument-series, namely the upper and lower limits of the ultimate oscillation for approaches from below and above respectively. If the ultimate oscillation for approaches to a from below reduces to a single term, *i.e.* if $(P\overline{R}Q)_{\text{os}}\text{‘}\overrightarrow{Q}\text{‘}a \in 1$, that one term is *the* limit of the function for approaches to a from below. If this one term is also the ultimate oscillation for approaches from above, we may call it simply *the* limit of the function for the argument a. This may or may not (when it exists) be equal to the value for the argument a. It is characteristic of *continuous* functions that *the* limit exists for every argument, and is always equal to the value for that argument. Continuous functions will be considered in *234.

The upper limit or maximum of the ultimate oscillation as the argument approaches a is the upper limit or maximum of the ultimate section. Hence if we put

$$R\,(PQ)\text{‘}a = \text{limax}_P\text{‘}(P\overline{R}Q)_{\text{sc}}\text{‘}\overrightarrow{Q}\text{‘}a \quad \text{Df},$$

the four limits of the function as the argument approaches a will be

$$R\,(PQ)\text{‘}a, \quad R\,(\breve{P}Q)\text{‘}a, \quad R\,(P\breve{Q})\text{‘}a, \quad R\,(\breve{P}\breve{Q})\text{‘}a.$$

It will be seen that $R\,(PQ)\text{‘}a$ is a function of $\overrightarrow{Q}\text{‘}a$. It may happen that, if we put α in place of $\overrightarrow{Q}\text{‘}a$, the function will have a definite limit as the argument increases in α, although α has no limit or maximum. Thus if, for example, Q consists of the series of rationals, and P of the series of real numbers, if α is a class of rationals not having a rational limit, we may regard the limit of the function (if it exists), as the argument increases in α, as the value of the function for the irrational limit of α. In this way we can extend the domain of definition of a function.

In order to be able to deal with the cases in which α has no limit, we put

$$(P\overline{R}Q)_{\text{lmx}}\text{‘}\alpha = \text{limax}_P\text{‘}(P\overline{R}Q)_{\text{sc}}\text{‘}\alpha \quad \text{Df}.$$

If P is a Dedekindian series, $(P\overline{R}Q)_{\text{lmx}}\text{‘}\alpha$ always exists. If we take α to be any segment of Q, we thus get a new function, derived from R, but having segments of Q instead of members of $C\text{‘}Q$ as its arguments. Thus if R had rationals for its arguments, this new function will have real numbers for its arguments. (Real numbers may be regarded as segments of the series of rationals.)

The function $R(PQ)\text{‘}a$ is a particular case of the above; thus we take as our definition

$$R(PQ)\text{‘}a = (P\bar{R}Q)_{\text{lmx}}\text{‘}\overrightarrow{Q}\text{‘}a \quad \text{Df},$$

or, what comes to the same thing,

$$R(PQ) = (P\bar{R}Q)_{\text{lmx}} \,|\, \overrightarrow{Q} \quad \text{Df}.$$

The following propositions of this number are important:

$*233{\cdot}15.\ \vdash :.\ P \in \text{Ser} \cap \text{Ded}\,.\,(P\bar{R}Q)_{\text{sc}}\text{‘}\alpha \cup (\breve{P}\bar{R}Q)_{\text{sc}}\text{‘}\alpha = C\text{‘}P\,.\,(P\bar{R}Q)_{\text{os}}\text{‘}\alpha = \Lambda\,.\supset:$
$(P\bar{R}Q)_{\text{lmx}}\text{‘}\alpha = (\breve{P}\bar{R}Q)_{\text{lmx}}\text{‘}\alpha\,.\,\text{v}\,.\,\{(P\bar{R}Q)_{\text{lmx}}\text{‘}\alpha\}\,P_1\,\{(\breve{P}\bar{R}Q)_{\text{lmx}}\text{‘}\alpha\}$

$*233{\cdot}16.\ \vdash :.\ P \in \text{Ser} \cap \text{Ded}\,.\,P^2 = P\,.\,Q_* \in \text{connex}\,.\,R\text{‘‘}C\text{‘}Q \subset C\text{‘}P\,.\supset:$
$(P\bar{R}Q)_{\text{os}}\text{‘}\alpha \in 0 \cup 1\,.\supset_\alpha.\,(P\bar{R}Q)_{\text{lmx}}\text{‘}\alpha = (\breve{P}\bar{R}Q)_{\text{lmx}}\text{‘}\alpha$

$*233{\cdot}2$—$\cdot 25$ are applications of the more important of the propositions $*232{\cdot}34$—$\cdot 39$, showing circumstances under which the limit of the function for the class α is the same as for the class $Q_*\text{‘‘}(\alpha \cap \text{Π}\text{‘}R)$.

$*233{\cdot}4$ and following propositions apply the earlier propositions of $*233$ to the case where α is replaced by $\overrightarrow{Q}\text{‘}a$, and therefore $(P\bar{R}Q)_{\text{lmx}}\text{‘}\alpha$ is replaced by $R(PQ)\text{‘}a$. We have

$*233{\cdot}43.\ \vdash : P_{\text{po}} \in \text{Ser}\,.\,(P\bar{R}Q)_{\text{os}}\text{‘}\overrightarrow{Q}\text{‘}a \in 1\,.\supset.$
$R(PQ)\text{‘}a = R(\breve{P}Q)\text{‘}a = \iota\text{‘}(P\bar{R}Q)_{\text{os}}\text{‘}\overrightarrow{Q}\text{‘}a$

$*233{\cdot}433.\ \vdash :.\ P \in \text{Ser}\,.\,Q_* \upharpoonright \overrightarrow{Q}\text{‘}a \in \text{connex}\,.\,R\text{‘‘}\overrightarrow{Q}\text{‘}a \subset C\text{‘}P\,.\,(P\bar{R}Q)_{\text{os}}\text{‘}\overrightarrow{Q}\text{‘}a = \Lambda\,.$
$\text{E}\,!\,R(PQ)\text{‘}a\,.\,\text{E}\,!\,R(\breve{P}Q)\text{‘}a\,.\supset:$
$R(PQ)\text{‘}a = R(\breve{P}Q)\text{‘}a\,.\,\text{v}\,.\,\{R(PQ)\text{‘}a\}\,P_1\,\{R(\breve{P}Q)\text{‘}a\}$

$*233{\cdot}45.\ \vdash :.\ P \in \text{Ser} \cap \text{Ded}\,.\,P^2 = P\,.\,Q_* \in \text{connex}\,.\,R\text{‘‘}C\text{‘}Q \subset C\text{‘}P\,.\supset:$
$R(PQ)\text{‘}a = R(\breve{P}Q)\text{‘}a\,.\equiv_a.\,(P\bar{R}Q)_{\text{os}}\text{‘}\overrightarrow{Q}\text{‘}a \in 0 \cup 1$

I.e. in a series having Dedekindian continuity, the necessary and sufficient condition that the two limits of the function as the argument approaches a from below should be equal is that the ultimate oscillation should not have more than one term.

We have next a set of propositions ($*233{\cdot}5$—$\cdot 53$) on the possibility of replacing $\overrightarrow{Q}\text{‘}a$ by a class α having a for its limit, without altering the limits of the function. We have to begin with

*233·5. $\vdash : Q \in \text{Ser} . a = \text{lt}_Q`(\alpha \cap \text{Ꭰ}`R) . \supset . \overrightarrow{Q}`a = Q_*``(\alpha \cap \text{Ꭰ}`R)$

in virtue of *207·291. Thence by earlier propositions of this number,

*233·512. $\vdash :. \text{Hp} *233·5 . P \in \text{Ser} . R``(\alpha \cap C`Q) \subset C`P . (P\bar{R}Q)_{\text{os}}`\overrightarrow{Q}`a = \iota`x . \supset :$

$x = R(PQ)`a = R(\breve{P}Q)`a : x = (P\bar{R}Q)_{\text{lmx}}`\alpha . \vee . (P\bar{R}Q)_{\text{lmx}}`\alpha P_1 x$

whence we obtain

*233·514. $\vdash : \text{Hp} *233·512 . x \sim\in C`P_1 . \supset . x = (P\bar{R}Q)_{\text{lmx}}`\alpha = (\breve{P}\bar{R}Q)_{\text{lmx}}`\alpha$

Thus if P, Q are series, and x is the limit of the function for the argument a (x being a term which has no immediate successor or predecessor), x is the limit of the function for any class of arguments whose limit is a. Hence we arrive at the proposition

*233·53. $\vdash : Q \in \text{Ser} . P \in \text{Ser} \cap \text{Ded} . P^2 = P . R``C`Q \subset C`P . \alpha \subset \text{Ꭰ}`R . \text{E}! \text{lt}_Q`\alpha .$

$(P\bar{R}Q)_{\text{os}}`Q_*``\alpha \in 0 \cup 1 . \supset .$

$(P\bar{R}Q)_{\text{lmx}}`\alpha = (\breve{P}\bar{R}Q)_{\text{lmx}}`\alpha = R(PQ)`\text{lt}_Q`\alpha = R(\breve{P}Q)`\text{lt}_Q`\alpha$

Thus if P has Dedekindian continuity, and α is a class of arguments having a limit, and if the ultimate oscillation as the argument approaches this limit has not more than one term, the limit of the function for the class α exists, and is equal to the limit of the function for the argument $\text{lt}_Q`\alpha$.

---

*233·01. $(P\bar{R}Q)_{\text{lmx}} = \text{limax}_P | (P\bar{R}Q)_{\text{sc}}$ Df

*233·02. $R(PQ) = (P\bar{R}Q)_{\text{lmx}} | \overrightarrow{Q}$ Df

*233·1. $\vdash : y \{(P\bar{R}Q)_{\text{lmx}}\} \alpha . \equiv . y (\text{limax}_P) \{(P\bar{R}Q)_{\text{sc}}`\alpha\}$ [(*233·01)]

*233·101. $\vdash : y = (P\bar{R}Q)_{\text{lmx}}`\alpha . \equiv . y = \text{limax}_P`(P\bar{R}Q)_{\text{sc}}`\alpha$ [*233·1]

*233·102. $\vdash : \text{E}! \text{limax}_P`(P\bar{R}Q)_{\text{sc}}`\alpha . \equiv . (P\bar{R}Q)_{\text{lmx}}`\alpha = \text{limax}_P`(P\bar{R}Q)_{\text{sc}}`\alpha .$

$\equiv . \text{E}! (P\bar{R}Q)_{\text{lmx}}`\alpha$ [*233·101 . *14·28]

*233·103. $\vdash : P \in \text{connex} . \supset . (P\bar{R}Q)_{\text{lmx}} \in 1 \rightarrow \text{Cls}$ [*207·41 . *233·1]

*233·11. $\vdash :. P \in \text{Ser} . \supset : y = (P\bar{R}Q)_{\text{lmx}}`\alpha . \equiv . y \in C`P . \overrightarrow{P}`y = P``(P\bar{R}Q)_{\text{sc}}`\alpha$
[*207·51 . *233·101]

*233·111. $\vdash :. P \in \text{Ser} . \exists ! P``(P\bar{R}Q)_{\text{sc}}`\alpha . \supset :$

$y = (P\bar{R}Q)_{\text{lmx}}`\alpha . \equiv . \overrightarrow{P}`y = P``(P\bar{R}Q)_{\text{sc}}`\alpha$ [*207·52 . *233·101]

*233·12. $\vdash :. P \in \text{Ser} . \sim \text{E}! \max_P`(P\bar{R}Q)_{\text{sc}}`\alpha . \supset :$

$y = (P\bar{R}Q)_{\text{lmx}}`\alpha . \equiv . y \in C`P . \overrightarrow{P}`y = (P\bar{R}Q)_{\text{sc}}`\alpha$

Dem.

$\vdash . *231·13 . *211·41 . \supset \vdash : \text{Hp} . \supset . (P\bar{R}Q)_{\text{sc}}`\alpha = P``(P\bar{R}Q)_{\text{sc}}`\alpha$ (1)

$\vdash . (1) . *233·11 . \supset \vdash . \text{Prop}$

***233·13.** $\vdash : P \in \text{connex} \cap \text{Ded} \,.\, \supset .$

$$\text{E}\,!\,(P\overline{R}Q)_{\text{lmx}}\text{‘}\alpha \,.\, (P\overline{R}Q)_{\text{lmx}}\text{‘}\alpha = \text{limax}_P\text{‘}(P\overline{R}Q)_{\text{sc}}\text{‘}\alpha$$

[*233·102·103 . *214·11]

***233·14.** $\vdash : P \in \text{Ser} \,.\, (P\overline{R}Q)_{\text{os}}\text{‘}\alpha \in 1 \,.\, \supset .\, (P\overline{R}Q)_{\text{lmx}}\text{‘}\alpha = (\breve{P}\overline{R}Q)_{\text{lmx}}\text{‘}\alpha = \iota\text{‘}(P\overline{R}Q)_{\text{os}}\text{‘}\alpha$

[*231·193 . *233·102]

***233·141.** $\vdash :. P \in \text{Ser} \,.\, (P\overline{R}Q)_{\text{sc}}\text{‘}\alpha \cup (\breve{P}\overline{R}Q)_{\text{sc}}\text{‘}\alpha = C\text{‘}P \,.\, (P\overline{R}Q)_{\text{os}}\text{‘}\alpha = \Lambda \,.\, \supset :$

$$\text{E}\,!\,(P\overline{R}Q)_{\text{lmx}}\text{‘}\alpha \,.\, \equiv .\, \text{E}\,!\,(\breve{P}\overline{R}Q)_{\text{lmx}}\text{‘}\alpha$$

[*211·727 . *233·102 . *231·13]

***233·142.** $\vdash : P \in \text{Ser} \,.\, Q_* \upharpoonright \alpha \in \text{connex} \,.$

$R\text{‘‘}(\alpha \cap C\text{‘}Q) \subset C\text{‘}P \,.\, (P\overline{R}Q)_{\text{os}}\text{‘}Q_*\text{‘‘}(\alpha \cap \text{Ɑ}\text{‘}R) \in 0 \cup 1 \,.$

$\text{E}\,!\,(P\overline{R}Q)_{\text{lmx}}\text{‘}Q_*\text{‘‘}(\alpha \cap \text{Ɑ}\text{‘}R) \,.\, (P\overline{R}Q)_{\text{lmx}}\text{‘}Q_*\text{‘‘}(\alpha \cap \text{Ɑ}\text{‘}R) \sim\in C\text{‘}P_1 \,.\, \supset .$

$(P\overline{R}Q)_{\text{lmx}}\text{‘}\alpha = (\breve{P}\overline{R}Q)_{\text{lmx}}\text{‘}\alpha = (P\overline{R}Q)_{\text{lmx}}\text{‘}Q_*\text{‘‘}(\alpha \cap \text{Ɑ}\text{‘}R)$

$$= (\breve{P}\overline{R}Q)_{\text{lmx}}\text{‘}Q_*\text{‘‘}(\alpha \cap \text{Ɑ}\text{‘}R)$$

Dem.

$\vdash .\, *231\cdot252 \,.\, \supset \vdash : \text{Hp} \,.\, \supset .\, (P\overline{R}Q)_{\text{lmx}}\text{‘}Q_*\text{‘‘}(\alpha \cap \text{Ɑ}\text{‘}R) = (\breve{P}\overline{R}Q)_{\text{lmx}}\text{‘}Q_*\text{‘‘}(\alpha \cap \text{Ɑ}\text{‘}R)$ (1)

$\vdash .\, *232\cdot37 \,.\, *233\cdot14 \,.\, \supset \vdash : \text{Hp} \,.\, (P\overline{R}Q)_{\text{os}}\text{‘}Q_*\text{‘‘}(\alpha \cap \text{Ɑ}\text{‘}R) \in 1 \,.\, \supset .$

$$(P\overline{R}Q)_{\text{lmx}}\text{‘}\alpha = (\breve{P}\overline{R}Q)_{\text{lmx}}\text{‘}\alpha = (P\overline{R}Q)_{\text{lmx}}\text{‘}Q_*\text{‘‘}(\alpha \cap \text{Ɑ}\text{‘}R)$$

$$= (\breve{P}\overline{R}Q)_{\text{lmx}}\text{‘}Q_*\text{‘‘}(\alpha \cap \text{Ɑ}\text{‘}R) \quad (2)$$

$\vdash .\, (1) \,.\, *232\cdot34 \,.\, \supset \vdash : \text{Hp} \,.\, (P\overline{R}Q)_{\text{os}}\text{‘}Q_*\text{‘‘}(\alpha \cap \text{Ɑ}\text{‘}R) = \Lambda \,.\, \supset .$

$$(P\overline{R}Q)_{\text{lmx}}\text{‘}\alpha = (P\overline{R}Q)_{\text{lmx}}\text{‘}Q_*\text{‘‘}(\alpha \cap \text{Ɑ}\text{‘}R) = (\breve{P}\overline{R}Q)_{\text{lmx}}\text{‘}Q_*\text{‘‘}(\alpha \cap \text{Ɑ}\text{‘}R)$$

$$= (\breve{P}\overline{R}Q)_{\text{lmx}}\text{‘}\alpha \quad (3)$$

$\vdash .\, (2) \,.\, (3) \,.\, \supset \vdash .\, \text{Prop}$

***233·15.** $\vdash :. P \in \text{Ser} \cap \text{Ded} \,.\, (P\overline{R}Q)_{\text{sc}}\text{‘}\alpha \cup (\breve{P}\overline{R}Q)_{\text{sc}}\text{‘}\alpha = C\text{‘}P \,.\, (P\overline{R}Q)_{\text{os}}\text{‘}\alpha = \Lambda \,.\, \supset :$

$$(P\overline{R}Q)_{\text{lmx}}\text{‘}\alpha = (\breve{P}\overline{R}Q)_{\text{lmx}}\text{‘}\alpha \,.\, \mathsf{v} \,.\, \{(P\overline{R}Q)_{\text{lmx}}\text{‘}\alpha\}\, P_1 \,\{(\breve{P}\overline{R}Q)_{\text{lmx}}\text{‘}\alpha\}$$

[*214·43 . *233·13 . *231·13]

***233·16.** $\vdash :. P \in \text{Ser} \cap \text{Ded} \,.\, P^2 = P \,.\, Q_* \in \text{connex} \,.\, R\text{‘‘}C\text{‘}Q \subset C\text{‘}P \,.\, \supset :$

$$(P\overline{R}Q)_{\text{os}}\text{‘}\alpha \in 0 \cup 1 \,.\, \supset_\alpha .\, (P\overline{R}Q)_{\text{lmx}}\text{‘}\alpha = (\breve{P}\overline{R}Q)_{\text{lmx}}\text{‘}\alpha$$

Dem.

$\vdash .\, *232\cdot22 \,.\, \supset \vdash : \text{Hp} \,.\, \supset .\, C\text{‘}P = (P\overline{R}Q)_{\text{sc}}\text{‘}\alpha \cup (\breve{P}\overline{R}Q)_{\text{sc}}\text{‘}\alpha$ (1)

$\vdash .\, *201\cdot65 \,.\, \supset \vdash : \text{Hp} \,.\, \supset .\, P_1 = \dot{\Lambda}$ (2)

$\vdash .\, (1) \,.\, (2) \,.\, *233\cdot14\cdot15 \,.\, \supset \vdash .\, \text{Prop}$

***233·17.** $\vdash :. \alpha \cap C‘Q \cap ᗡ‘R = \Lambda . \supset : y = (P\bar{R}Q)_{\mathrm{lmx}}‘\alpha . \equiv . y = B‘\breve{P}$

Dem.

$\vdash . *232·15 . *233·101 . \supset$

$\vdash :. \mathrm{Hp} . \supset : y = (P\bar{R}Q)_{\mathrm{lmx}}‘\alpha . \equiv . y = \mathrm{limax}_P‘C‘P .$

$[*206·2 . *93·117] \qquad \equiv . y = B‘\breve{P} :. \supset \vdash . \mathrm{Prop}$

***233·171.** $\vdash : \alpha \cap C‘Q \cap ᗡ‘R = \Lambda . \supset . \sim \{(P\bar{R}Q)_{\mathrm{lmx}}‘\alpha = (\breve{P}\bar{R}Q)_{\mathrm{lmx}}‘\alpha\}$

Dem.

$\vdash . *93·102 . \supset \vdash . \sim (B‘P = B‘\breve{P})$ (1)

$\vdash . (1) . *233·17 . \supset \vdash . \mathrm{Prop}$

***233·172.** $\vdash : \alpha \cap C‘Q \cap ᗡ‘R = \Lambda . E ! (P\bar{R}Q)_{\mathrm{lmx}}‘\alpha . E ! (\breve{P}\bar{R}Q)_{\mathrm{lmx}}‘\alpha . \supset .$
$(P\bar{R}Q)_{\mathrm{os}}‘\alpha \sim \epsilon\, 0 \cup 1$

Dem.

$\vdash . *233·171 . *232·15 . \supset$

$\vdash : \mathrm{Hp} . \supset . (P\bar{R}Q)_{\mathrm{lmx}}‘\alpha, (\breve{P}\bar{R}Q)_{\mathrm{lmx}}‘\alpha \,\epsilon\, (P\bar{R}Q)_{\mathrm{os}}‘\alpha . (P\bar{R}Q)_{\mathrm{lmx}}‘\alpha \neq (\breve{P}\bar{R}Q)_{\mathrm{lmx}}‘\alpha .$

$[*52·41] \supset . (P\bar{R}Q)_{\mathrm{os}}‘\alpha \sim \epsilon\, 0 \cup 1 : \supset \vdash . \mathrm{Prop}$

***233·173.** $\vdash : (P\bar{R}Q)_{\mathrm{os}}‘\alpha \,\epsilon\, 0 \cup 1 . E ! (P\bar{R}Q)_{\mathrm{lmx}}‘\alpha . E ! (\breve{P}\bar{R}Q)_{\mathrm{lmx}}‘\alpha . \supset .$
$\exists ! \alpha \cap C‘Q \cap ᗡ‘R \quad [*233·172 . \mathrm{Transp}]$

***233·174.** $\vdash : P \Subset J . (P\bar{R}Q)_{\mathrm{os}}‘\alpha \,\epsilon\, 1 . \supset . \exists ! \alpha \cap C‘Q \cap ᗡ‘R$

Dem.

$\vdash . *200·12 . \supset \vdash : \mathrm{Hp} . \supset . \sim \{C‘P \subset (P\bar{R}Q)_{\mathrm{os}}‘\alpha\} .$

$[*232·15] \qquad \supset . \exists ! \alpha \cap C‘Q \cap ᗡ‘R : \supset \vdash . \mathrm{Prop}$

***233·2.** $\vdash : Q_* \upharpoonright \alpha \,\epsilon\, \mathrm{connex} . P \,\epsilon\, \mathrm{Ser} . R``(\alpha \cap C‘Q) \subset C‘P .$
$(P\bar{R}Q)_{\mathrm{os}}‘Q_*``(\alpha \cap ᗡ‘R) = \Lambda . E ! (P\bar{R}Q)_{\mathrm{lmx}}‘Q_*``(\alpha \cap ᗡ‘R) . \supset .$
$(P\bar{R}Q)_{\mathrm{lmx}}‘\alpha = (P\bar{R}Q)_{\mathrm{lmx}}‘Q_*``(\alpha \cap ᗡ‘R) . (\breve{P}\bar{R}Q)_{\mathrm{lmx}}‘\alpha = (\breve{P}RQ)_{\mathrm{lmx}}‘Q_*``(\alpha \cap ᗡ‘R)$
$[*232·34 . *211·727 . *233·102]$

***233·21.** $\vdash : P_{\mathrm{po}} \,\epsilon\, \mathrm{Ser} . \exists ! (P\bar{R}Q)_{\mathrm{os}}‘\alpha . (P\bar{R}Q)_{\mathrm{os}}‘Q_*``(\alpha \cap ᗡ‘R) \,\epsilon\, 1 . \supset .$
$(P\bar{R}Q)_{\mathrm{lmx}}‘\alpha = (\breve{P}\bar{R}Q)_{\mathrm{lmx}}‘\alpha = (P\bar{R}Q)_{\mathrm{lmx}}‘Q_*``(\alpha \cap ᗡ‘R)$
$= (\breve{P}\bar{R}Q)_{\mathrm{lmx}}‘Q_*``(\alpha \cap ᗡ‘R) = \iota‘(P\bar{R}Q)_{\mathrm{os}}‘Q_*``(\alpha \cap ᗡ‘R)$
$[*232·341 . *231·193]$

***233·22.** $\vdash :. P \,\epsilon\, \mathrm{Ser} . (P\bar{R}Q)_{\mathrm{os}}‘Q_*``(\alpha \cap ᗡ‘R) = \iota‘x .$
$(P\bar{R}Q)_{\mathrm{sc}}‘\alpha \cup (\breve{P}\bar{R}Q)_{\mathrm{sc}}‘\alpha = C‘P . \supset :$
$x = (P\bar{R}Q)_{\mathrm{lmx}}‘\alpha . \vee . (P\bar{R}Q)_{\mathrm{lmx}}‘\alpha P_1 x . (P\bar{R}Q)_{\mathrm{lmx}}‘\alpha = \max_P‘(P\bar{R}Q)_{\mathrm{sc}}‘\alpha$

Dem.

$\vdash . *232·352 . \supset \vdash : \mathrm{Hp} . x \,\epsilon\, (P\bar{R}Q)_{\mathrm{sc}}‘\alpha . \supset . x = (P\bar{R}Q)_{\mathrm{lmx}}‘\alpha$ (1)

$\vdash . *232\cdot356 . \supset \vdash : \text{Hp} . x \sim\epsilon (P\bar{R}Q)_{\text{sc}}‘\alpha . \sim \text{E} ! \max_P‘(P\bar{R}Q)_{\text{sc}}‘\alpha . \supset .$

$x = (P\bar{R}Q)_{\text{lmx}}‘\alpha \quad (2)$

$\vdash . *232\cdot358 . *207\cdot42 . \supset \vdash : \text{Hp} . x \sim\epsilon (P\bar{R}Q)_{\text{sc}}‘\alpha . \text{E} ! \max_P‘(P\bar{R}Q)_{\text{sc}}‘\alpha . \supset .$

$\max_P‘(P\bar{R}Q)‘\alpha \, P_1 x . (P\bar{R}Q)_{\text{lmx}}‘\alpha = \max_P‘(P\bar{R}Q)_{\text{sc}}‘\alpha \quad (3)$

$\vdash . (1) . (2) . (3) . \supset \vdash . \text{Prop}$

***233·23.** $\vdash : \text{Hp} *233\cdot22 . \supset .$

$x = (P\bar{R}Q)_{\text{lmx}}‘Q_*‘‘(\alpha \cap \text{ɑ}‘R) = (\breve{P}\bar{R}Q)_{\text{lmx}}‘Q_*‘‘(\alpha \cap \text{ɑ}‘R) \quad [*231\cdot193]$

***233·24.** $\vdash : \text{Hp} *233\cdot22 . x \sim\epsilon \text{ɑ}‘P_1 . \supset . x = (P\bar{R}Q)_{\text{lmx}}‘\alpha \quad [*233\cdot22]$

***233·241.** $\vdash : \text{Hp} *233\cdot22 . x \sim\epsilon C‘P_1 . \supset . x = (P\bar{R}Q)_{\text{lmx}}‘\alpha = (\breve{P}\bar{R}Q)_{\text{lmx}}‘\alpha$

$\left[*233\cdot24 \frac{\breve{P}}{P} . *233\cdot24 \right]$

***233·25.** $\vdash :. P \epsilon \text{Ser} \cap \text{Ded} . P^2 = P . Q_* \epsilon \text{connex} . R‘‘C‘Q \subset C‘P . \supset :$

$(P\bar{R}Q)_{\text{os}}‘Q_*‘‘(\alpha \cap \text{ɑ}‘R) \epsilon 0 \cup 1 . \supset .$

$(P\bar{R}Q)_{\text{lmx}}‘\alpha = (P\bar{R}Q)_{\text{lmx}}‘Q_*‘‘(\alpha \cap \text{ɑ}‘R) = (\breve{P}\bar{R}Q)_{\text{lmx}}‘\alpha = (\breve{P}\bar{R}Q)_{\text{lmx}}‘Q_*‘‘(\alpha \cap \text{ɑ}‘R)$

$[*232\cdot39]$

***233·4.** $\vdash : y \{R(PQ)\} a . \equiv . y \{(P\bar{R}Q)_{\text{lmx}}\} \overrightarrow{Q}‘a \quad [(*233\cdot02)]$

***233·401.** $\vdash : y = R(PQ)‘a . \equiv . y = (P\bar{R}Q)_{\text{lmx}}‘\overrightarrow{Q}‘a \quad [*233\cdot4]$

***233·402.** $\vdash : P \epsilon \text{connex} . \supset . R(PQ) \epsilon 1 \rightarrow \text{Cls} \quad [*207\cdot41]$

***233·41.** $\vdash : y = R(PQ)‘a . \equiv . y = (P\bar{R}Q)_{\text{lmx}}‘(\overrightarrow{Q}‘a \cap \text{ɑ}‘R)$

Dem.

$\vdash . *232\cdot13 . \supset \vdash . (P\bar{R}Q)_{\text{sc}}‘\overrightarrow{Q}‘a = (P\bar{R}Q)_{\text{sc}}‘(\overrightarrow{Q}‘a \cap \text{ɑ}‘R) \quad (1)$

$\vdash . (1) . *233\cdot401\cdot101 . \supset \vdash . \text{Prop}$

***233·42.** $\vdash :. Q \epsilon \text{trans} \cap \text{connex} . \text{E} ! \max_Q‘(\overrightarrow{Q}‘a \cap \text{ɑ}‘R) . \supset :$

$y = R(PQ)‘a . \equiv . y = \text{limax}_P‘P_*‘‘\overrightarrow{R}‘\max_Q‘(\overrightarrow{Q}‘a \cap \text{ɑ}‘R)$

$[*232\cdot24 . *233\cdot401\cdot101]$

***233·421.** $\vdash : P \epsilon \text{Rl}‘J \cap \text{trans} . Q \epsilon \text{trans} \cap \text{connex} . R‘\max_Q‘(\alpha \cap \text{ɑ}‘R) \epsilon C‘P . \supset .$

$R(PQ)‘a = R‘\max_Q‘(\overrightarrow{Q}‘a \cap \text{ɑ}‘R)$

Dem.

$\vdash . *233\cdot42 . \supset \vdash :. \text{Hp} . \supset : y = R(PQ)‘a . \equiv . y = \text{limax}_P‘\overrightarrow{P_*}‘R‘\max_Q‘(\overrightarrow{Q}‘a \cap \text{ɑ}‘R) .$

$[*205\cdot197] \quad \equiv . y = R‘\max_Q‘(\overrightarrow{Q}‘a \cap \text{ɑ}‘R) :. \supset \vdash . \text{Prop}$

***233·422.** $\vdash :. \overrightarrow{Q}‘a \cap \text{ɑ}‘R = \Lambda . \supset : y = R(PQ)‘a . \equiv . y = B‘\breve{P} \quad [*233\cdot17]$

***233·423.** $\vdash : \overrightarrow{Q}\text{‘}a \cap \text{Cl‘}R = \Lambda . \supset . \sim \{R(PQ)\text{‘}a = R(\breve{P}Q)\text{‘}a\}$ [*233·171]

***233·424.** $\vdash : \overrightarrow{Q}\text{‘}a \cap \text{Cl‘}R = \Lambda . E ! R(PQ)\text{‘}a . E ! R(\breve{P}Q)\text{‘}a . \supset . (P\bar{R}Q)_{os}\text{‘}\overrightarrow{Q}\text{‘}a \sim\epsilon 0 \cup 1$
[*233·172]

***233·425.** $\vdash : (P\bar{R}Q)_{os}\text{‘}\overrightarrow{Q}\text{‘}a \epsilon 0 \cup 1 . E ! R(PQ)\text{‘}a . E ! R(\breve{P}Q)\text{‘}a . \supset . \exists ! \overrightarrow{Q}\text{‘}a \cap \text{Cl‘}R$
[*233·424 . Transp]

***233·426.** $\vdash : P \in J . (P\bar{R}Q)_{os}\text{‘}\overrightarrow{Q}\text{‘}a \epsilon 1 . \supset . \exists ! \overrightarrow{Q}\text{‘}a \cap \text{Cl‘}R$ [*233·174]

***233·43.** $\vdash : P_{po} \epsilon \text{Ser} . (P\bar{R}Q)_{os}\text{‘}\overrightarrow{Q}\text{‘}a \epsilon 1 . \supset .$
$R(PQ)\text{‘}a = R(\breve{P}Q)\text{‘}a = \breve{\iota}\text{‘}(P\bar{R}Q)_{os}\text{‘}\overrightarrow{Q}\text{‘}\alpha$ [*231·193]

***233·431.** $\vdash : P \epsilon \text{trans} \cap \text{connex} . (P\bar{R}Q)_{os}\text{‘}\overrightarrow{Q}\text{‘}a \sim\epsilon 0 \cup 1 .$
$E ! R(PQ)\text{‘}a . E ! R(\breve{P}Q)\text{‘}a . \supset . \{R(\breve{P}Q)\text{‘}a\} P \{R(PQ)\text{‘}a\}$
[*215·52 . *231·13·101]

***233·432.** $\vdash : P \epsilon \text{trans} \cap \text{connex} . (P\bar{R}Q)_{os}\text{‘}\overrightarrow{Q}\text{‘}a = \Lambda .$
$E ! R(PQ)\text{‘}a . E ! R(\breve{P}Q)\text{‘}a . \supset . \{R(PQ)\text{‘}a\} P_* \{R(\breve{P}Q)\text{‘}a\}$ [*215·53]

***233·433.** $\vdash :. P \epsilon \text{Ser} . Q_* \upharpoonright \overrightarrow{Q}\text{‘}a \epsilon \text{connex} . R\text{‘‘}\overrightarrow{Q}\text{‘}a \subset C\text{‘}P . (P\bar{R}Q)_{os}\text{‘}\overrightarrow{Q}\text{‘}a = \Lambda .$
$E ! R(PQ)\text{‘}a . E ! R(\breve{P}Q)\text{‘}a . \supset : R(PQ)\text{‘}a = R(\breve{P}Q)\text{‘}a . \vee . \{R(PQ)\text{‘}a\} P_1 \{R(\breve{P}Q)\text{‘}a\}$
[*215·54 . *232·22]

***233·434.** $\vdash : P \epsilon \text{Ser} . Q_* \upharpoonright \overrightarrow{Q}\text{‘}a \epsilon \text{connex} . R\text{‘‘}\overrightarrow{Q}\text{‘}a \subset C\text{‘}P . E ! R(PQ)\text{‘}a .$
$E ! R(\breve{P}Q)\text{‘}a . \supset . \{R(PQ)\text{‘}a\} (P_1 \cup \breve{P}_*) \{R(\breve{P}Q)\text{‘}a\}$ [*233·43·431·433]

***233·435.** $\vdash : P \epsilon \text{Ser} . R(PQ)\text{‘}a = R(\breve{P}Q)\text{‘}a . \supset . (P\bar{R}Q)_{os}\text{‘}\overrightarrow{Q}\text{‘}a \epsilon 0 \cup 1$
[*233·431 . Transp]

***233·44.** $\vdash :. P \epsilon \text{Ser} . Q_* \upharpoonright \overrightarrow{Q}\text{‘}a \epsilon \text{connex} . R\text{‘‘}\overrightarrow{Q}\text{‘}a \subset C\text{‘}P . E ! R(PQ)\text{‘}a .$
$E ! R(\breve{P}Q)\text{‘}a . \sim \{R(PQ)\text{‘}a \epsilon \text{D‘}P_1 . R(\breve{P}Q)\text{‘}a \epsilon \text{Cl‘}P_1\} . \supset :$
$R(PQ)\text{‘}a = R(\breve{P}Q)\text{‘}a . \equiv . (P\bar{R}Q)_{os}\text{‘}\overrightarrow{Q}\text{‘}a \epsilon 0 \cup 1$ [*233·426·43·433·435]

***233·45.** $\vdash :. P \epsilon \text{Ser} \cap \text{Ded} . P^2 = P . Q_* \epsilon \text{connex} . R\text{‘‘}C\text{‘}Q \subset C\text{‘}P . \supset :$
$R(PQ)\text{‘}a = R(\breve{P}Q)\text{‘}a . \equiv_a . (P\bar{R}Q)_{os}\text{‘}\overrightarrow{Q}\text{‘}a \epsilon 0 \cup 1$
[*233·13 . *201·65 . *233·44]

***233·5.** $\vdash : Q \epsilon \text{Ser} . a = \text{lt}_Q\text{‘}(\alpha \cap \text{Cl‘}R) . \supset . \overrightarrow{Q}\text{‘}a = Q_*\text{‘‘}(\alpha \cap \text{Cl‘}R)$ [*207·291]

***233·501.** $\vdash :. Q \in \text{Ser} . a = \text{lt}_Q{}^{\text{‘}}(\alpha \cap \text{Ɑ}{}^{\text{‘}}R) . \supset : \exists ! \overrightarrow{Q}{}^{\text{‘}}a \cap \text{Ɑ}{}^{\text{‘}}R . \equiv . \exists ! \alpha \cap C{}^{\text{‘}}Q \cap \text{Ɑ}{}^{\text{‘}}R$

Dem.

$\vdash . *233{\cdot}5 . \supset \vdash :. \text{Hp} . \supset : \exists ! \overrightarrow{Q}{}^{\text{‘}}a \cap \text{Ɑ}{}^{\text{‘}}R . \equiv . \exists ! Q_*{}^{\text{‘‘}}(\alpha \cap \text{Ɑ}{}^{\text{‘}}R) \cap \text{Ɑ}{}^{\text{‘}}R .$ (1)

$[*37{\cdot}29{\cdot}265]$ $\supset . \exists ! \alpha \cap \text{Ɑ}{}^{\text{‘}}R \cap C{}^{\text{‘}}Q$ (2)

$\vdash . *90{\cdot}33 . *22{\cdot}43 . \supset \vdash : x \in \alpha \cap C{}^{\text{‘}}Q \cap \text{Ɑ}{}^{\text{‘}}R . \supset . x \in Q_*{}^{\text{‘‘}}(\alpha \cap \text{Ɑ}{}^{\text{‘}}R) . x \in \text{Ɑ}{}^{\text{‘}}R$ (3)

$\vdash . (3) . *10{\cdot}28 . \supset \vdash : \exists ! \alpha \cap C{}^{\text{‘}}Q \cap \text{Ɑ}{}^{\text{‘}}R . \supset . \exists ! Q_*{}^{\text{‘‘}}(\alpha \cap \text{Ɑ}{}^{\text{‘}}R) \cap \text{Ɑ}{}^{\text{‘}}R$ (4)

$\vdash . (1) . (2) . (4) . \supset \vdash . \text{Prop}$

***233·51.** $\vdash : \text{Hp} *233{\cdot}5 . P \in \text{Ser} . R{}^{\text{‘‘}}(\alpha \cap C{}^{\text{‘}}Q) \subset C{}^{\text{‘}}P . (P\bar{R}Q)_{\text{os}}{}^{\text{‘}}\overrightarrow{Q}{}^{\text{‘}}a = \Lambda .$
$\text{E} ! R(PQ){}^{\text{‘}}a . \supset . (P\bar{R}Q)_{\text{lmx}}{}^{\text{‘}}\alpha = R(PQ){}^{\text{‘}}a$ $[*233{\cdot}2{\cdot}5]$

***233·511.** $\vdash : \text{Hp} *233{\cdot}5 . P \in \text{Ser} . \exists ! (P\bar{R}Q)_{\text{os}}{}^{\text{‘}}\alpha . (P\bar{R}Q)_{\text{os}}{}^{\text{‘}}\overrightarrow{Q}{}^{\text{‘}}a \in 1 . \supset .$
$(P\bar{R}Q)_{\text{lmx}}{}^{\text{‘}}\alpha = (\breve{P}\bar{R}Q)_{\text{lmx}}{}^{\text{‘}}\alpha = R(PQ){}^{\text{‘}}a = R(\breve{P}Q){}^{\text{‘}}a = \breve{\iota}{}^{\text{‘}}(P\bar{R}Q)_{\text{os}}{}^{\text{‘}}\overrightarrow{Q}{}^{\text{‘}}a$
$[*233{\cdot}501{\cdot}5{\cdot}21]$

***233·512.** $\vdash :. \text{Hp} *233{\cdot}5 . P \in \text{Ser} . R{}^{\text{‘‘}}(\alpha \cap C{}^{\text{‘}}Q) \subset C{}^{\text{‘}}P . (P\bar{R}Q)_{\text{os}}{}^{\text{‘}}\overrightarrow{Q}{}^{\text{‘}}a = \iota{}^{\text{‘}}x . \supset :$
$x = R(PQ){}^{\text{‘}}a = R(\breve{P}Q){}^{\text{‘}}a : x = (P\bar{R}Q)_{\text{lmx}}{}^{\text{‘}}\alpha . \vee . (P\bar{R}Q)_{\text{lmx}}{}^{\text{‘}}\alpha \, P_1 x$
$[*233{\cdot}22{\cdot}23 . *232{\cdot}22]$

***233·513.** $\vdash : \text{Hp} *233{\cdot}512 . x \sim\in \text{Ɑ}{}^{\text{‘}}P_1 . \supset . x = (P\bar{R}Q)_{\text{lmx}}{}^{\text{‘}}\alpha$ $[*233{\cdot}512]$

***233·514.** $\vdash : \text{Hp} *233{\cdot}512 . x \sim\in C{}^{\text{‘}}P_1 . \supset . x = (P\bar{R}Q)_{\text{lmx}}{}^{\text{‘}}\alpha = (\breve{P}\bar{R}Q)_{\text{lmx}}{}^{\text{‘}}\alpha$

$\left[*233{\cdot}513 \dfrac{\breve{P}}{P} . *233{\cdot}513\right]$

***233·515.** $\vdash : P, Q \in \text{Ser} . a = \text{lt}_Q{}^{\text{‘}}(\alpha \cap \text{Ɑ}{}^{\text{‘}}R) . R{}^{\text{‘‘}}(C{}^{\text{‘}}Q \cap \alpha) \subset C{}^{\text{‘}}P .$
$(P\bar{R}Q)_{\text{os}}{}^{\text{‘}}\overrightarrow{Q}{}^{\text{‘}}a \in 0 \cup 1 . \text{E} ! R(PQ){}^{\text{‘}}a . R(PQ){}^{\text{‘}}a \sim\in C{}^{\text{‘}}P_1 . \supset .$
$(P\bar{R}Q)_{\text{lmx}}{}^{\text{‘}}\alpha = (\breve{P}\bar{R}Q)_{\text{lmx}}{}^{\text{‘}}\alpha = R(PQ){}^{\text{‘}}a = R(\breve{P}Q){}^{\text{‘}}a$
$[*233{\cdot}142{\cdot}5]$

***233·516.** $\vdash : P, Q \in \text{Ser} . \text{E} ! R(PQ){}^{\text{‘}}\text{lt}_Q{}^{\text{‘}}\alpha . R(PQ){}^{\text{‘}}\text{lt}_Q{}^{\text{‘}}\alpha \sim\in C{}^{\text{‘}}P_1 .$
$R{}^{\text{‘‘}}(C{}^{\text{‘}}Q \cap \alpha) \subset C{}^{\text{‘}}P . (P\bar{R}Q)_{\text{os}}{}^{\text{‘}}\overrightarrow{Q}{}^{\text{‘}}\text{lt}_Q{}^{\text{‘}}\alpha \in 0 \cup 1 . \supset .$
$(P\bar{R}Q)_{\text{lmx}}{}^{\text{‘}}\alpha = (\breve{P}\bar{R}Q)_{\text{lmx}}{}^{\text{‘}}\alpha = R(PQ){}^{\text{‘}}\text{lt}_Q{}^{\text{‘}}\alpha = R(\breve{P}Q){}^{\text{‘}}\text{lt}_Q{}^{\text{‘}}\alpha$
$[*233{\cdot}515]$

***233·52.** $\vdash :. \text{Hp} *233{\cdot}5 . P \in \text{Ser} \cap \text{Ded} . P^2 = P . R{}^{\text{‘‘}}C{}^{\text{‘}}Q \subset C{}^{\text{‘}}P . \supset :$
$(P\bar{R}Q)_{\text{os}}{}^{\text{‘}}\overrightarrow{Q}{}^{\text{‘}}a \in 0 \cup 1 . \supset .$
$(P\bar{R}Q)_{\text{lmx}}{}^{\text{‘}}\alpha = (\breve{P}\bar{R}Q)_{\text{lmx}}{}^{\text{‘}}\alpha = R(PQ){}^{\text{‘}}a = R(\breve{P}Q){}^{\text{‘}}a$ $[*233{\cdot}25]$

***233·53.** $\vdash : Q \in \text{Ser} . P \in \text{Ser} \cap \text{Ded} . P^2 = P . R{}^{\text{‘‘}}C{}^{\text{‘}}Q \subset C{}^{\text{‘}}P . \alpha \subset \text{Ɑ}{}^{\text{‘}}R . \text{E} ! \text{lt}_Q{}^{\text{‘}}\alpha .$
$(P\bar{R}Q)_{\text{os}}{}^{\text{‘}}Q_*{}^{\text{‘‘}}\alpha \in 0 \cup 1 . \supset .$
$(P\bar{R}Q)_{\text{lmx}}{}^{\text{‘}}\alpha = (\breve{P}\bar{R}Q)_{\text{lmx}}{}^{\text{‘}}\alpha = R(PQ){}^{\text{‘}}\text{lt}_Q{}^{\text{‘}}\alpha = R(\breve{P}Q){}^{\text{‘}}\text{lt}_Q{}^{\text{‘}}\alpha$
$[*233{\cdot}52]$

✱234. CONTINUITY OF FUNCTIONS.

Summary of ✱234.

In the present number we are concerned with the definition and analysis of the continuity of functions. The following definition of continuity is given by Dini*:

"We call it [the function] *continuous* for $x = a$, or in the point a, in which it has the value $f(a)$, if, for every positive number σ, different from 0 but as small as we please, there exists a positive number ϵ, different from 0, such that, for all values of δ which are numerically less than ϵ, the difference $f(a+\delta) - f(a)$ is numerically less than σ. In other words, $f(x)$ is continuous in the point $x = a$, where it has the value $f(a)$, if the limit of its values to the right and left of a is the same and equal to $f(a)$...."

By the second form of the above definition, the function R of previous numbers is to be called continuous at the point a if

$$R(PQ)\text{‘}a = R(\breve{P}Q)\text{‘}a = R(P\breve{Q})\text{‘}a = R(\breve{P}\breve{Q})\text{‘}a = R\text{‘}a.$$

The first form of the definition can also be so stated as to be free from any reference to number, and derivable from the ideas dealt with in the previous numbers of the present section. For this purpose, instead of "a positive number σ," we take an interval in which $R\text{‘}a$ is contained, say $P(z - w)$. Similarly the "values of δ which are numerically less than ϵ" are replaced by arguments in a certain interval containing a.

By ✱233·423, if the limits of the function as the argument approaches a are to be all equal, a must not be the maximum or minimum of $\text{Ɑ}\text{‘}R$. We therefore take the interval containing a to be an interval in which the end-points are not included, say $Q(y - y')$. Thus our definition becomes

(A) $\quad R\text{‘}a \,\epsilon\, P(z - w) \,.\supset_{z,w}.$

$$(\exists y, y') \,.\, y, y' \,\epsilon\, \text{Ɑ}\text{‘}R \,.\, a \,\epsilon\, Q(y - y') \,.\, R\text{“}Q(y \vdash\!\dashv y') \subset P(z - w)$$

We require further, what is tacitly assumed in Dini's definition, that $R\text{‘}a$ is a member of $C\text{‘}P$ which has no immediate predecessor or successor, *i.e.*

$$R\text{‘}a \,\epsilon\, C\text{‘}P - C\text{‘}P_1.$$

* *Theorie der Functionen einer veränderlichen reallen Grösse*, Chap. IV. § 30, p. 50.

In order to deal more easily with the above definition, we analyse it into the product of four factors, which concern respectively P and Q, $\breve{P}$ and Q, P and $\breve{Q}$, $\breve{P}$ and $\breve{Q}$. In the first place, it is obvious that (A) is the product of

(B) $R\text{'}a \in P(z-w) . \supset_{z,w} . (\exists y) . y \in \text{Ꭰ}\text{'}R . y \in \overrightarrow{Q}\text{'}a . R\text{''}Q(y \mapsto a) \subset P(z-w)$

and a factor obtained by substituting $\breve{Q}$ for Q in (B). If $Q_* \in$ connex, and $P_{po} \in$ Ser, (B) is the product of

(C) $R\text{'}a \in \overrightarrow{P_{po}}\text{'}w . \supset_w . (\exists y) . y \in \text{Ꭰ}\text{'}R . y \in \overrightarrow{Q}\text{'}a . R\text{''}Q(y \mapsto a) \subset \overrightarrow{P_*}\text{'}w$

and a factor obtained by writing $\breve{P}$ for P and z for w in (C); and in virtue of $R\text{'}a \sim \in C\text{'}P_1$, (C) becomes

$$R\text{'}a \in \overrightarrow{P_{po}}\text{'}w . \supset_w . (\exists y) . y \in \text{Ꭰ}\text{'}R . y \in \overrightarrow{Q}\text{'}a . R\text{''}Q(y \mapsto a) \subset \overrightarrow{P_{po}}\text{'}w,$$

i.e. if Q is transitive,

(D) $R\text{'}a \in \overrightarrow{P_{po}}\text{'}w . \supset_w . R(Q_* \upharpoonright \overrightarrow{Q}\text{'}a)_{cn}(\overrightarrow{P_{po}}\text{'}w)$

Hence the function is continuous for the argument a if a satisfies (D) and the three other hypotheses resulting from replacing P by $\breve{P}$, or Q by $\breve{Q}$, or P and Q by $\breve{P}$ and $\breve{Q}$. If we substitute x for $R\text{'}a$, and Q for $Q_* \upharpoonright \overrightarrow{Q}\text{'}a$, (D) becomes

(E) $\overrightarrow{P_{po}}\text{''}\overleftarrow{P_{po}}\text{'}x \subset \overleftarrow{Q_{cn}}\text{'}R$

Hence continuity can be studied by studying the hypothesis (E), and replacing x by $R\text{'}a$ and Q by $Q_* \upharpoonright \overrightarrow{Q}\text{'}a$.

The hypothesis (E) is an interesting one on its own account. We put

$$\text{sc}(P, Q)\text{'}R = C\text{'}P \cap \hat{x}(\overrightarrow{P_{po}}\text{''}\overleftarrow{P_{po}}\text{'}x \subset \overleftarrow{Q_{cn}}\text{'}R) \quad \text{Df.}$$

Thus "$x \in \text{sc}(P, Q)\text{'}R$" means that x is a member of the value-series such that, if y is any later member, the function ultimately becomes less than y. If we put further

$$\text{os}(P, Q)\text{'}R = \text{sc}(P, Q)\text{'}R \cap \text{sc}(\breve{P}, Q)\text{'}R \quad \text{Df,}$$

then, if x is a member of $\text{os}(P, Q)\text{'}R$, the function ultimately becomes less than any later member of $C\text{'}P$, and greater than any earlier member. Hence x is the limit of the function as the argument increases indefinitely. Hence, if we substitute $Q_* \upharpoonright \overrightarrow{Q}\text{'}a$ for Q, and if $x \in \text{os}(P, Q_* \upharpoonright \overrightarrow{Q}\text{'}a)\text{'}R$, x is the limit of the function as the argument approaches a from below, *i.e.*

$$R(PQ)\text{'}a = R(\breve{P}Q)\text{'}a = x.$$

(This is proved in $*234{\cdot}462$.) Hence, putting $R\text{'}a$ in place of x, the function is continuous from below at the point a if

$$R\text{'}a \in \text{os}(P, Q_* \upharpoonright \overrightarrow{Q}\text{'}a)\text{'}R,$$

and is continuous from above if

$$R`a \in \text{os}(P, \breve{Q}_* \upharpoonright \overleftarrow{Q}`a)`R.$$

These results, and various others connected with them, are proved below. The equivalence of Dini's two definitions is proved in ∗234·63. It will be observed that practically nothing in the theory of continuous functions requires the use of numbers.

We use the symbol "$\text{ct}(PQ)`R$" for the class of arguments $a$ for which the limit of the function for approaches to $a$ from below is $R`a$. Thus, in virtue of what was said above, we may put

$$\text{ct}(PQ)`R = \hat{a}\{R`a \in \text{os}(P, Q_* \upharpoonright \overrightarrow{Q_{\text{po}}}`a)`R - C`P_1\} \quad \text{Df.}$$

Then a function is continuous at the point a if a belongs to the two classes $\text{ct}(PQ)`R$ and $\text{ct}(P\breve{Q})`R$. Hence we put

$$\text{contin}(PQ)`R = \text{ct}(PQ)`R \cap \text{ct}(P\breve{Q})`R \quad \text{Df.}$$

The function R is continuous with respect to P and Q if it is continuous for all arguments in $C`Q$. Thus we put

$$P \,\overline{\text{contin}}\, Q = \hat{R}\{\exists ! C`Q \cap \text{ɑ}`R \,.\, C`Q \cap \text{ɑ}`R \subset \text{contin}(PQ)`R\} \quad \text{Df.}$$

Our propositions in this number begin with the properties of $\text{sc}(P, Q)`R$ and $\text{os}(P, Q)`R$. We have

∗234·103. $\vdash : P_{\text{po}} \in \text{Ser} \,.\, \exists ! \text{os}(P, Q)`R \,.\, \supset \,.\, P\overline{R}_{\text{os}}Q \in 0 \cup 1$

Thus the hypothesis $\exists ! \text{os}(P, Q)`R$ enables us to use propositions of previous numbers having the hypothesis $P\overline{R}_{\text{os}}Q \in 0 \cup 1$.

The identification of our definitions with the usual definitions of continuity of functions proceeds by means of the proposition

∗234·12. $\vdash :: Q_* \in \text{connex} \,.\, \supset :. \, x \in \text{os}(P, Q)`R \cap \text{D}`P \cap \text{ɑ}`P \,.\, \equiv :$
$$x \in \text{D}`P \cap \text{ɑ}`P : x \in P(z - w) \,.\, \supset_{z,w} .\, RQ_{\text{cn}}\{P(z - w)\}$$

We have a collection of propositions dealing with the relations of $\text{sc}(P, Q)`R$ to $P\overline{R}_{\text{sc}}Q$ and $\breve{P}\overline{R}_{\text{sc}}Q$. $\text{sc}(P, Q)`R$ is an upper section of P (∗234·131); $\text{sc}(P, Q)`R$ is the complement of $P``(P\overline{R}_{\text{sc}}Q)$, *i.e.* of $P\overline{R}_{\text{sc}}Q$ without its maximum (if any). This is expressed in the following proposition:

∗234·174. $\vdash : P_{\text{po}} \in \text{Ser} \,.\, Q_* \in \text{connex} \,.\, R``C`Q \subset C`P \,.\, \supset \,.$
$$C`P \cap p`\overrightarrow{P_{\text{po}}}``\text{sc}(P, Q)`R = P``(P\overline{R}_{\text{sc}}Q) = C`P - \text{sc}(P, Q)`R$$

We thus arrive at

∗234·182. $\vdash : P \in \text{Ser} \,.\, Q_* \in \text{connex} \,.\, R``C`Q \subset C`P \,.\, \supset \,.$
$$\overrightarrow{\text{limax}}_P`(P\overline{R}_{\text{sc}}Q) = \overrightarrow{\text{min}}_P`\text{sc}(P, Q)`R$$

Thus $\text{os}\,(P, Q)‘R$ is contained in $\overrightarrow{\text{max}}_P‘(P\overline{R}_{\text{sc}}Q) \cup \overrightarrow{\text{min}}_P‘(\breve{P}\overline{R}_{\text{sc}}Q)$ (*234·201), and therefore has not more than two terms (*234·202). If $P\overline{R}_{\text{os}}Q$ has one term, this is the only member of $\text{os}\,(P, Q)‘R$ (*234·203). If $\text{os}\,(P, Q)‘R$ has two terms, they have the relation P_1 (*234·242); hence if P is a compact series, and $\text{os}\,(P, Q)‘R$ is not null, its only member is both $\text{limax}_P‘(P\overline{R}_{\text{sc}}Q)$ and $\text{limin}_P‘(\breve{P}\overline{R}_{\text{sc}}Q)$ (*234·25), while conversely, if $\text{limax}_P‘(P\overline{R}_{\text{sc}}Q)$ and $\text{limin}_P‘(\breve{P}\overline{R}_{\text{sc}}Q)$ are equal, each is the only member of $\text{os}\,(P, Q)‘R$ (*234·251).

We now apply the above results to the limits of a function as its argument approaches the limit of a class α. This is done, as before, by substituting $Q_* \upharpoonright \alpha$ for Q. We arrive at the proposition (*234·33) that if P has Dedekindian continuity, and $\text{os}\,(P, Q_* \upharpoonright \alpha)‘R$ is not null, its only member is both $(P\overline{R}Q)_{\text{lmx}}‘\alpha$ and $(\breve{P}\overline{R}Q)_{\text{lmx}}‘\alpha$, *i.e.* is the limit of the function as the argument increases in α.

We then take for α the particular value $\overrightarrow{Q_{\text{po}}}‘a$, so that we become concerned with what happens when the argument approaches a from below. For the comparison of our definition of continuity with such definitions as the one quoted from Dini above, we have

***234·41.** $\vdash :: Q \,\epsilon\, \text{trans} \,.\, Q_* \upharpoonright \overrightarrow{Q}‘a \,\epsilon\, \text{connex} \,.\, \supset :.$

$$x \,\epsilon\, \text{os}\,(P, Q_* \upharpoonright \overrightarrow{Q}‘a)‘R \cap \text{D}‘P \cap \text{Cl}‘P \,.\, \equiv :$$
$$x \,\epsilon\, \text{D}‘P \cap \text{Cl}‘P : x \,\epsilon\, P\,(z - w) \,.\, \supset_{z,w} .$$
$$(\exists y) \,.\, y \,\epsilon\, \overrightarrow{Q}‘a \cap \text{Cl}‘R \,.\, R‘‘Q\,(y \vdash a) \subset P\,(z - w)$$

I.e. if x is neither the first nor the last member of the P-series, x belongs to $\text{os}\,(P, Q_* \upharpoonright \overrightarrow{Q}‘a)‘R$ when, and only when, given any interval $P\,(z - w)$, however small, in which x is contained, there is an argument y earlier than a, such that the value of the function for all arguments earlier than a but not earlier than y lies in the interval $P\,(z - w)$.

We deduce from previous propositions that, with the usual hypothesis as to Q, if P is a Dedekindian series,

$$R\,(PQ)‘a = \text{limin}_P‘\text{sc}\,(P, Q_* \upharpoonright \overrightarrow{Q}‘a)‘R \quad (*234·422),$$

and if P is a series and $\text{os}\,(P, Q_* \upharpoonright \overrightarrow{Q}‘a)$ is a unit class, its only member is both $R\,(PQ)‘a$ and $R\,(\breve{P}Q)‘a$, *i.e.* is the limit of the function for approaches to a from below (*234·43). The following proposition sums up our results:

***234·45.** $\vdash :. P \,\epsilon\, \text{Ser} \,.\, Q \,\epsilon\, \text{trans} \,.\, Q_* \upharpoonright \overrightarrow{Q}‘a \,\epsilon\, \text{connex} \,.\, R‘‘\overrightarrow{Q}‘a \subset C‘P \,.\, P^2 = P \,.\, \supset :$

$$\exists !\, \text{os}\,(P, Q_* \upharpoonright \overrightarrow{Q}‘a)‘R \,.\, \equiv .\, \text{os}\,(P, Q_* \upharpoonright \overrightarrow{Q}‘a)‘R = \iota‘R\,(PQ)‘a \,.$$
$$\equiv .\, \text{os}\,(P, Q_* \upharpoonright \overrightarrow{Q}‘a)‘R = \iota‘R\,(\breve{P}Q)‘a \,.$$
$$\equiv .\, R\,(PQ)‘a = R\,(\breve{P}Q)‘a$$

Thus $\exists ! \text{os}(P, Q_* \upharpoonright \overrightarrow{Q}\text{‘}a)\text{‘}R$ is, in a compact series, the necessary and sufficient condition for the existence of a definite limit of the function as the argument approaches a from below.

Without assuming $P^2 = P$, if x is a member of $\text{os}(P, Q_* \upharpoonright \overrightarrow{Q}\text{‘}a)\text{‘}R$, and if x has no immediate predecessor or successor, so that in the neighbourhood of x the series is compact, we still have $x = R(PQ)\text{‘}a = R(\breve{P}Q)\text{‘}a$ (*234·462).

We next consider $\text{ct}(PQ)\text{‘}R$. By the definition we have

*234·5. $\vdash : a \in \text{ct}(PQ)\text{‘}R . \equiv . R\text{‘}a \in \text{os}(P, Q_* \upharpoonright \overrightarrow{Q_{\text{po}}}\text{‘}a)\text{‘}R - C\text{‘}P_1$

Thus a is an argument for which the function has a single value which has no immediate predecessor or successor in P, and which, in virtue of *234·462, is the limit of the function as the argument approaches a from below (*234·52). The cases when $R\text{‘}a = B\text{‘}P$ or $R\text{‘}a = B\text{‘}\breve{P}$ require special attention; excluding these cases, we arrive at

*234·51. $\vdash :: Q \in \text{trans} . Q_* \upharpoonright \overrightarrow{Q}\text{‘}a \in \text{connex} . R\text{‘}a \in \text{D}\text{‘}P \cap \text{Ɑ}\text{‘}P . \supset :.$

$$a \in \text{ct}(PQ)\text{‘}R . \equiv : R\text{‘}a \sim\in C\text{‘}P_1 : R\text{‘}a \in P(z - w) . \supset_{z,w} .$$
$$(\exists y) . y \in \overrightarrow{Q}\text{‘}a \cap \text{Ɑ}\text{‘}R . R\text{‘‘}Q(y \mapsto a) \subset P(z - w)$$

This proposition is analogous to *234·41.

We prove (*234·562) that if P, Q are series, and α is any class of arguments for which all the values belong to $C\text{‘}P$, and if α has a limit at which the function is continuous from below, then the limit of the function, as the argument increases in α, is the value of the function at the limit of α.

We next consider $\text{contin}(PQ)\text{‘}R$, which is defined as $\text{ct}(PQ)\text{‘}R \cap \text{ct}(P\breve{Q})\text{‘}R$. We show that if P is a series whose field contains $R\text{‘‘}\overleftrightarrow{Q}\text{‘}a$, and Q is transitive, and $Q_* \upharpoonright \overleftrightarrow{Q}\text{‘}a$ is connected, and $R\text{‘}a$ is neither $B\text{‘}P$ nor $B\text{‘}\breve{P}$, then if a belongs to the class $\text{contin}(PQ)\text{‘}R$, $R\text{‘}a$ is the limit of the function for the argument a for approaches either from below or from above (*234·62). If P is compact, the converse also holds (*234·63). Our definition of a point of continuity is thus identified with the second form of Dini's definition quoted above. It is identified with the first form by the following proposition: In the circumstances of *234·62, if $R\text{‘}a \in \text{D}\text{‘}P \cap \text{Ɑ}\text{‘}P$, we have (*234·64)

$$a \in \text{contin}(PQ)\text{‘}R . \equiv : R\text{‘}a \in C\text{‘}P - C\text{‘}P_1 : R\text{‘}a \in P(z - w) . \supset_{z,w} .$$
$$(\exists y, y') . y, y' \in \text{Ɑ}\text{‘}R . a \in Q(y - y') . R\text{‘‘}Q(y \mapsto y') \subset P(z - w),$$

i.e. a is a point of continuity when, and only when, the value $R\text{‘}a$ for the argument a is a member of the P-series having no immediate predecessor or successor, and if $R\text{‘}a$ is contained in the interval $P(z - w)$, then, however small this interval may be, two arguments y, y' can be found such that a lies between them, and the values for all arguments from y to y' (both included) lie in the interval $P(z-w)$.

We end with a few propositions on continuous functions. The last of these (*234·73) states that, if P is a compact series and Q is transitive and connected, then R is continuous with respect to P and Q when, and only when, it has arguments in $C‘Q$, and for all such arguments a we have

$$R(PQ)‘a = R(\breve{P}Q)‘a = R(P\breve{Q})‘a = R(\breve{P}\breve{Q})‘a = R‘a,$$

i.e. the value for every argument is the limit for that argument for approaches either from above or from below.

***234·01.** $\mathrm{sc}(P, Q)‘R = C‘P \cap \hat{x}(\overrightarrow{P_{\mathrm{po}}}‘‘\overleftarrow{P_{\mathrm{po}}}‘x \subset \overleftarrow{Q_{\mathrm{cn}}}‘R)$ Df

***234·02.** $\mathrm{os}(P, Q)‘R = \mathrm{sc}(P, Q)‘R \cap \mathrm{sc}(\breve{P}, Q)‘R$ Df

***234·03.** $\mathrm{ct}(PQ)‘R = \hat{a}\{R‘a \in \mathrm{os}(P, Q_* \upharpoonright \overrightarrow{Q_{\mathrm{po}}}‘a)‘R - C‘P_1\}$ Df

***234·04.** $\mathrm{contin}(PQ)‘R = \mathrm{ct}(PQ)‘R \cap \mathrm{ct}(P\breve{Q})‘R$ Df

***234·05.** $P \,\overline{\mathrm{contin}}\, Q = \hat{R}\{\exists ! C‘Q \cap \text{ᗡ}‘R \,.\, C‘Q \cap \text{ᗡ}‘R \subset \mathrm{contin}(PQ)‘R\}$ Df

***234·1.** $\vdash :. x \in \mathrm{sc}(P, Q)‘R \,.\equiv: x \in C‘P : xP_{\mathrm{po}}w \,.\supset_w .\, RQ_{\mathrm{cn}}(\overrightarrow{P_{\mathrm{po}}}‘w) :$

$\equiv : x \in C‘P : xP_{\mathrm{po}}w \,.\supset_w .\, (\exists y) \,.\, y \in C‘Q \cap \text{ᗡ}‘R \,.\, R‘‘\overleftarrow{Q_*}‘y \subset \overrightarrow{P_{\mathrm{po}}}‘w$

[*230·11 . (*234·01)]

***234·101.** $\vdash : P_{\mathrm{po}} \in \mathrm{Ser} \,.\, x \in \mathrm{sc}(P, Q)‘R \,.\supset .\, P\overline{R}_{\mathrm{sc}}Q \subset \overrightarrow{P_*}‘x$

Dem.

$\vdash . *40·16 . (*234·01) . \supset$

$\vdash : \mathrm{Hp} \,.\supset .\, x \in C‘P \,.\, p‘P_*‘‘‘\overleftarrow{Q_{\mathrm{cn}}}‘R \cap C‘P \subset p‘P_*‘‘‘\overrightarrow{P_{\mathrm{po}}}‘‘\overleftarrow{P_{\mathrm{po}}}‘x \cap C‘P$

[*91·574] $\subset p‘\overrightarrow{P_{\mathrm{po}}}‘‘\overleftarrow{P_{\mathrm{po}}}‘x \cap C‘P$

[*204·65.*91·602] $\subset \overrightarrow{P_*}‘x$ (1)

$\vdash . (1) . *231·1 . \supset \vdash . \mathrm{Prop}$

***234·102.** $\vdash : P_{\mathrm{po}} \in \mathrm{Ser} \,.\, x \in \mathrm{os}(P, Q)‘R \,.\supset .\, P\overline{R}_{\mathrm{os}}Q \subset \iota‘x$

Dem.

$\vdash . *234·1·101 . (*234·02) . \supset \vdash : \mathrm{Hp} \,.\supset .\, x \in C‘P \,.\, P\overline{R}_{\mathrm{os}}Q \subset \overrightarrow{P_*}‘x \cap \overleftarrow{P_*}‘x \,.$

[*200·39] $\supset .\, P\overline{R}_{\mathrm{os}}Q \subset \iota‘x : \supset \vdash . \mathrm{Prop}$

***234·103.** $\vdash : P_{\mathrm{po}} \in \mathrm{Ser} \,.\, \exists ! \mathrm{os}(P, Q)‘R \,.\supset .\, P\overline{R}_{\mathrm{os}}Q \in 0 \cup 1$

Dem.

$\vdash . *234·102 . \supset \vdash : \mathrm{Hp} \,.\supset .\, (\exists x) \,.\, P\overline{R}_{\mathrm{os}}Q \subset \iota‘x \,.$

[*51·401] $\supset .\, P\overline{R}_{\mathrm{os}}Q \in 0 \cup 1 : \supset \vdash . \mathrm{Prop}$

***234·104.** $\vdash : RQ_{cn}(\overrightarrow{P_*}`x) . \supset . x \in sc(P, Q)`R$

Dem.

$\vdash . *91·52 . \quad \supset \vdash : xP_{po}z . \supset . \overrightarrow{P_*}`x \subset \overrightarrow{P_{po}}`z \quad (1)$

$\vdash . (1) . *230·211·151 . \supset \vdash :. Hp . \supset : xP_{po}z . \supset_z . RQ_{cn}(\overrightarrow{P_{po}}`z) : x \in C`P :$

[*234·1] $\quad \supset : x \in sc(P, Q)`R :. \supset \vdash . Prop$

***234·105.** $\vdash : P_{po} \in Ser . x \in sc(P, Q)`R \cap D`P_1 . \supset . RQ_{cn}(\overrightarrow{P_*}`x)$

Dem.

$\vdash . *201·63 . *121·254 . \supset \vdash :: Hp . xP_1z . \supset :. yP_{po}z . \supset : \sim(xP_{po}y) :$

[*202·103] $\quad \supset : yP_{po}x . \vee . y = x \quad (1)$

$\vdash . (1) . *91·54 . \quad \supset \vdash :. Hp . xP_1z . \supset : \overrightarrow{P_{po}}`z \subset \overrightarrow{P_*}`x :$

[*230·211] $\quad \supset : RQ_{cn}(\overrightarrow{P_{po}}`z) . \supset . RQ_{cn}(\overrightarrow{P_*}`x) \quad (2)$

$\vdash . *234·1 . \quad \supset \vdash : Hp . \supset . (\exists z) . xP_1z . RQ_{cn}(\overrightarrow{P_{po}}`z) \quad (3)$

$\vdash . (2) . (3) . \supset \vdash . Prop$

When $x \sim\in D`P_1$, the above proposition is not necessarily true: it may fail if $x = \min_P`sc(P, Q)`R$.

It is to be observed that $sc(P, Q)`R$ and $os(P, Q)`R$ are functions of P_{po}, so that they are unchanged when P_{po} is substituted for P. Hence the hypothesis $P_{po} \in Ser$ is as effective, with regard to them, as the hypothesis $P \in Ser$. This is stated in the following proposition.

***234·106.** $\vdash . sc(P, Q)`R = sc(P_{po}, Q)`R . os(P, Q)`R = os(P_{po}, Q)`R \quad [*234·1]$

***234·107.** $\vdash :. x \in C`P - D`P_1 . \supset : x \in sc(P, Q)`R . \equiv . \overrightarrow{P_*}``\overleftarrow{P_{po}}`x \subset \overleftarrow{Q_{cn}}`R$

Dem.

$\vdash . *121·254 . \supset \vdash :. Hp . \supset : x \sim\in D`(P_{po})_1 :$

[*201·61] $\quad \supset : x \sim\in D`\{P_{po} \dot{-} P_{po}^2\} :$

[*10·51] $\quad \supset : xP_{po}y . \supset . xP_{po}^2y .$

[*91·574] $\quad \supset . (\exists z) . xP_{po}z . \overrightarrow{P_*}`z \subset \overrightarrow{P_{po}}`y \quad (1)$

$\vdash . (1) . *230·211 . \supset$

$\vdash :. Hp : xP_{po}y . \supset_y . RQ_{cn}\overrightarrow{P_*}`y : \supset : xP_{po}y . \supset_y . RQ_{cn}\overrightarrow{P_{po}}`y \quad (2)$

$\vdash . *91·54 . *230·211 . \supset$

$\vdash :. xP_{po}y . \supset_y . RQ_{cn}\overrightarrow{P_{po}}`y : \supset : xP_{po}y . \supset_y . RQ_{cn}\overrightarrow{P_*}`y \quad (3)$

$\vdash . (2) . (3) . \supset \vdash :. Hp . \supset : \overrightarrow{P_*}``\overleftarrow{P_{po}}`x \subset \overleftarrow{Q_{cn}}`R . \equiv . \overrightarrow{P_{po}}``\overleftarrow{P_{po}}`x \subset \overleftarrow{Q_{cn}}`R \quad (4)$

$\vdash . (4) . *234·1 . \supset \vdash . Prop$

***234·11.** $\vdash :. x \in D`P \cap \text{Ꭰ}`P : x \in P(z - w) . \supset_{z,w} . RQ_{cn}\{P(z - w)\} : \equiv :$

$x \in D`P \cap \text{Ꭰ}`P : x \in P(z - w) . \supset_{z,w} .$

$(\exists y) . y \in C`Q \cap \text{Ꭰ}`R . R``\overleftarrow{Q_*}`y \subset P(z - w) \quad [*230·11]$

***234·111.** $\vdash :. x \in \mathrm{D}‘P \cap \text{Ɑ}‘P : x \in P(z-w) . \supset_{z,w} . RQ_{cn}\{P(z-w)\} : \supset .$
$x \in \mathrm{os}(P,Q)‘R$

Dem.

$\vdash . *230·211 . \supset$

$\vdash :: \mathrm{Hp} . \supset :. x \in \mathrm{D}‘P \cap \text{Ɑ}‘P :. xP_{po}w : (\exists z) . zP_{po}x : \supset_w . RQ_{cn}\overrightarrow{P_{po}}‘w :.$

$[*91·504] \quad \supset :. x \in \mathrm{D}‘P : xP_{po}w . \supset_w . RQ_{cn}\overrightarrow{P_{po}}‘w :.$

$[*234·1] \quad \supset :. x \in \mathrm{sc}(P,Q)‘R \quad (1)$

Similarly $\quad \vdash : \mathrm{Hp} . \supset . \mathrm{sc}(\breve{P},Q)‘R \quad (2)$

$\vdash . (1) . (2) . \supset \vdash . \mathrm{Prop}$

***234·12.** $\vdash :: Q_* \in \mathrm{connex} . \supset :. x \in \mathrm{os}(P,Q)‘R \cap \mathrm{D}‘P \cap \text{Ɑ}‘P . \equiv :$
$x \in \mathrm{D}‘P \cap \text{Ɑ}‘P : x \in P(z-w) . \supset_{z,w} . RQ_{cn}\{P(z-w)\}$

Dem.

$\vdash . *234·1 . \supset \vdash :. x \in \mathrm{os}(P,Q)‘R \cap \mathrm{D}‘P \cap \text{Ɑ}‘P . \equiv :$

$x \in \mathrm{D}‘P \cap \text{Ɑ}‘P : xP_{po}w . \supset_w . RQ_{cn}(\overrightarrow{P_{po}}‘w) : zP_{po}x . \supset_z . RQ_{cn}(\overleftarrow{P_{po}}‘z) :$

$[*11·71] \equiv : x \in \mathrm{D}‘P \cap \text{Ɑ}‘P : zP_{po}x . xP_{po}w . \supset_{z,w} . RQ_{cn}(\overrightarrow{P_{po}}‘w) . RQ_{cn}(\overleftarrow{P_{po}}‘z) \quad (1)$

$\vdash . *230·42 . \supset \vdash :. \mathrm{Hp} . \supset : RQ_{cn}\overrightarrow{P_{po}}‘w . RQ_{cn}(\overleftarrow{P_{po}}‘z) . \equiv .$

$RQ_{cn}(\overleftarrow{P_{po}}‘z \cap \overrightarrow{P_{po}}‘w) \quad (2)$

$\vdash . (1) . (2) . *121·1 . \supset \vdash . \mathrm{Prop}$

***234·121.** $\vdash . \overrightarrow{B}‘\breve{P} \subset \mathrm{sc}(P,Q)‘R \quad [*93·104 . (*234·01)]$

***234·122.** $\vdash :. P_{po} \in \mathrm{connex} . x = B‘P . \supset :$
$x \in \mathrm{os}(P,Q)‘R . \equiv . x \in \mathrm{sc}(P,Q)‘R . \equiv . \overrightarrow{P_{po}}“\text{Ɑ}‘P \subset \overleftarrow{Q_{cn}}‘R$
$[*234·121 . (*234·02) . *234·1 . *205·253]$

***234·13.** $\vdash : x \in \mathrm{sc}(P,Q)‘R . \supset . \overleftarrow{P_*}‘x \subset \mathrm{sc}(P,Q)‘R$

Dem.

$\vdash . *96·3 . *91·74 . *90·13 . \supset \vdash : xP_*z . \supset . \overleftarrow{P_{po}}‘z \subset \overleftarrow{P_{po}}‘x . z \in C‘P .$

$[*37·2] \quad \supset . \overrightarrow{P_{po}}“\overleftarrow{P_{po}}‘z \subset \overleftarrow{P_{po}}“\overleftarrow{P_{po}}‘x . z \in C‘P \quad (1)$

$\vdash . (1) . (*234·01) . \supset \vdash . \mathrm{Prop}$

***234·131.** $\vdash . \mathrm{sc}(P,Q)‘R = \breve{P}_*“\mathrm{sc}(P,Q)‘R . \mathrm{sc}(P,Q)‘R \in \mathrm{sect}‘\breve{P}$

Dem.

$\vdash . *90·21 . *234·1 . \supset \vdash . \mathrm{sc}(P,Q)‘R \subset \breve{P}_*“\mathrm{sc}(P,Q)‘R \quad (1)$

$\vdash . *234·13 . \quad \supset \vdash . \breve{P}_*“\mathrm{sc}(P,Q)‘R \subset \mathrm{sc}(P,Q)‘R \quad (2)$

$\vdash . (1) . (2) . *211·13 . \supset \vdash . \mathrm{Prop}$

***234·14.** $\vdash : Q_* \epsilon \text{connex} . x \epsilon \text{sc}(P, Q)`R . \supset . x \epsilon C`P . \overleftarrow{P_{po}}`x \subset \breve{P}\overline{R}_{sc}Q$

Dem.

$\vdash . *234·1 . \supset \vdash :. \text{Hp} . \supset : x \epsilon C`P : xP_{po}z . \supset_z . RQ_{cn}(\overrightarrow{P_{po}}`z) .$

[*230·211] $\supset_z . RQ_{cn}(\overrightarrow{P_*}`z) .$

[*231·141] $\supset_z . z \epsilon \breve{P}\overline{R}_{sc}Q :. \supset \vdash . \text{Prop}$

***234·141.** $\vdash : Q_* \epsilon \text{connex} . \exists ! \text{sc}(P, Q)`R . \supset . \exists ! \breve{P}\overline{R}_{sc}Q$ [*234·14]

***234·142.** $\vdash : \exists ! \text{sc}(P, Q)`R \cap D`P . \supset . \exists ! C`Q \cap \text{Cl}`R$

Dem.

$\vdash . *234·1 . \supset$

$\vdash :. x \epsilon \text{sc}(P, Q)`R \cap D`P . \supset : x \epsilon D`P : (\exists w) . xP_{po}w . \supset . \exists ! C`Q \cap \text{Cl}`R :$

[*91·504] $\supset : \exists ! C`Q \cap \text{Cl}`R :. \supset \vdash . \text{Prop}$

***234·15.** $\vdash : P_*, Q_* \epsilon \text{connex} . \exists ! \text{sc}(P, Q)`R . \supset . P\overline{R}_{sc}Q \cup \breve{P}\overline{R}_{sc}Q = C`P$

Dem.

$\vdash . *231·202 . *234·141 . \supset \vdash : \text{Hp} . \supset . C`P - P\overline{R}_{sc}Q \subset \breve{P}\overline{R}_{sc}Q \quad (1)$

$\vdash . *231·1 . \supset \vdash . P\overline{R}_{sc}Q \cup \breve{P}\overline{R}_{sc}Q \subset C`P \quad (2)$

$\vdash . (1) . (2) . \supset \vdash . \text{Prop}$

***234·16.** $\vdash : P_{po} \epsilon \text{Ser} . Q_* \epsilon \text{connex} . \supset .$

$P\overline{R}_{sc}Q \subset p`\overrightarrow{P_*}``\text{sc}(P, Q)`R . \breve{P}_{po}``\text{sc}(P, Q)`R \subset \breve{P}\overline{R}_{sc}Q$ [*234·101·14]

***234·161.** $\vdash :. P_{po} \epsilon \text{Ser} . R``C`Q \subset C`P . P\overline{R}_{sc}Q \subset \overrightarrow{P_*}`x . \supset :$

$P\overline{R}_{sc}Q = \overrightarrow{P_*}`x . \vee . RQ_{cn}(\overrightarrow{P_*}`x)$

Dem.

$\vdash . *231·24 . \supset \vdash : \text{Hp} . \sim \{RQ_{cn}(\overrightarrow{P_*}`x)\} . \supset . \overrightarrow{P_*}`x \subset P\overline{R}_{sc}Q .$

[Hp.*22·41] $\supset . \overrightarrow{P_*}`x = P\overline{R}_{sc}Q : \supset \vdash . \text{Prop}$

***234·162.** $\vdash : P_{po} \epsilon \text{Ser} . R``C`Q \subset C`P . \overrightarrow{P_*}`x = P\overline{R}_{sc}Q . x \epsilon C`P . \supset .$

$x \epsilon \text{sc}(P, Q)`R$

Dem.

$\vdash . *202·5 . \supset \vdash :. \text{Hp} . xP_{po}z . \supset : z \sim\epsilon P\overline{R}_{sc}Q :$

[*231·12] $\supset : (\exists y) . y \epsilon C`Q \cap \text{Cl}`R . z \sim\epsilon P_*``R``\overleftarrow{Q_*}`y :$

[*211·56] $\supset : (\exists y) . y \epsilon C`Q \cap \text{Cl}`R . P_*``R``\overleftarrow{Q_*}`y \subset \overrightarrow{P_{po}}`z :$

[*90·33] $\supset : (\exists y) . y \epsilon C`Q \cap \text{Cl}`R . R``\overleftarrow{Q_*}`y \subset \overrightarrow{P_{po}}`z :$

[*230·11] $\supset : RQ_{cn}(\overrightarrow{P_{po}}`z) \quad (1)$

$\vdash . (1) . *234·1 . \supset \vdash . \text{Prop}$

***234·17.** $\vdash :. P_{po} \epsilon \text{Ser} . R``C`Q \subset C`P . \supset :$

$$x \epsilon \text{sc}(P, Q)`R . \equiv . x \epsilon C`P . P\bar{R}_{sc}Q \subset \overrightarrow{P_*}`x$$

Dem.

$\vdash . *234·1·101 . \supset \vdash :. \text{Hp} . \supset : x \epsilon \text{sc}(P, Q)`R . \supset . x \epsilon C`P . P\bar{R}_{sc}Q \subset \overrightarrow{P_*}`x$ (1)

$\vdash . *234·161·162·104 . \supset \vdash :. \text{Hp} . \supset : x \epsilon C`P . P\bar{R}_{sc}Q \subset \overrightarrow{P_*}`x . \supset .$

$x \epsilon \text{sc}(P, Q)`R$ (2)

$\vdash . (1) . (2) . \supset \vdash . \text{Prop}$

***234·171.** $\vdash : P_{po} \epsilon \text{Ser} . R``C`Q \subset C`P . x \epsilon C`P - \text{sc}(P, Q)`R . \supset .$

$$\overrightarrow{P_*}`x \subset P``(P\bar{R}_{sc}Q)$$

Dem.

$\vdash . *234·17 . \supset \vdash : \text{Hp} . \supset . \exists ! P\bar{R}_{sc}Q - \overrightarrow{P_*}`x$ (1)

$\vdash . (1) . *211·56 . *231·13 . \supset \vdash : \text{Hp} . \supset . \overrightarrow{P_*}`x \subset P_{po}``(P\bar{R}_{sc}Q)$ (2)

$\vdash . (2) . *231·134 . \supset \vdash . \text{Prop}$

***234·172.** $\vdash : P_{po} \epsilon \text{Ser} . \supset . C`P - \text{sc}(P, Q)`R = C`P \cap p`\overrightarrow{P_{po}}``\text{sc}(P, Q)`R$

Dem.

$\vdash . *200·5 . \supset \vdash : \text{Hp} . \supset . C`P \cap p`\overrightarrow{P_{po}}``\text{sc}(P, Q)`R \subset C`P - \text{sc}(P, Q)`R$ (1)

$\vdash . *234·131 . \supset$

$\vdash : x \epsilon \text{sc}(P, Q)`R . y \epsilon C`P - \text{sc}(P, Q)`R . \supset . \sim(xP_*y) . x, y \epsilon C`P .$

[*202·103] $\supset . yP_{po}x$ (2)

$\vdash . (1) . (2) . \supset \vdash . \text{Prop}$

***234·173.** $\vdash : P_{po} \epsilon \text{Ser} . \exists ! \text{sc}(P,Q)`R . \supset . C`P - \text{sc}(P, Q)`R = p`\overrightarrow{P_{po}}``\text{sc}(P, Q)`R$

[*234·172 . *40·61 . *37·15]

***234·174.** $\vdash : P_{po} \epsilon \text{Ser} . Q_* \epsilon \text{connex} . R``C`Q \subset C`P . \supset .$

$$C`P \cap p`\overrightarrow{P_{po}}``\text{sc}(P, Q)`R = P``(P\bar{R}_{sc}Q) = C`P - \text{sc}(P, Q)`R$$

Dem.

$\vdash . *234·171·172 . \supset \vdash : \text{Hp} . \supset . P_*``\{C`P \cap p`\overrightarrow{P_{po}}``\text{sc}(P, Q)`R\} \subset P``(P\bar{R}_{sc}Q) .$

[*90·21] $\supset . C`P \cap p`\overrightarrow{P_{po}}``\text{sc}(P, Q)`R \subset P``(P\bar{R}_{sc}Q)$ (1)

$\vdash . *234·16 . *37·2 . \supset \vdash : \text{Hp} . \supset . P``(P\bar{R}_{sc}Q) \subset P``p`\overrightarrow{P_*}``\text{sc}(P, Q)`R$

[*40·37 . *91·52] $\subset p`\overrightarrow{P_{po}}``\text{sc}(P, Q)`R$ (2)

$\vdash . *37·15 . \supset \vdash . P``(P\bar{R}_{sc}Q) \subset D`P$ (3)

$\vdash . (1) . (2) . (3) . *234·172 . \supset \vdash . \text{Prop}$

***234·175.** $\vdash : \text{Hp } *234·174 . \exists ! \text{sc}(P,Q)`R . \supset . p`\overrightarrow{P_{po}}``\text{sc}(P, Q)`R = P``(P\bar{R}_{sc}Q)$

[*234·174 . *40·61 . *37·15]

$*234\cdot18$. $\vdash : P_{po} \in \text{Ser} . Q_* \in \text{connex} . R``C`Q \subset C`P . \supset .$
$C`P = \text{sc}(P, Q)`R \cup P``(P\bar{R}_{sc}Q) . \text{sc}(P, Q)`R \cap P``(P\bar{R}_{sc}Q) = \Lambda .$
$\text{sc}(P, Q)`R = C`P - P``(P\bar{R}_{sc}Q)$

Dem.

$\vdash . *234\cdot174 . *24\cdot411 . \supset \vdash : \text{Hp} . \supset . C`P = \text{sc}(P, Q)`R \cup P``(P\bar{R}_{sc}Q)$ (1)

$\vdash . *234\cdot174 . \quad \supset \vdash : \text{Hp} . \supset . P``(P\bar{R}_{sc}Q) \subset p`\overrightarrow{P_{po}}``\text{sc}(P, Q)`R .$

$[*200\cdot5] \quad \supset . \text{sc}(P, Q)`R \cap P``(P\bar{R}_{sc}Q) = \Lambda$ (2)

$\vdash . *24\cdot492 . *234\cdot174 . \supset \vdash : \text{Hp} . \supset . \text{sc}(P, Q)`R = C`P - P``(P\bar{R}_{sc}Q)$ (3)

$\vdash . (1) . (2) . (3) . \supset \vdash . \text{Prop}$

In virtue of this proposition, $P``(P\bar{R}_{sc}Q)$ and $\text{sc}(P, Q)`R$ are complementary sections of P, *i.e.* they constitute a Dedekind cut in P.

$*234\cdot181$. $\vdash : P_{po} \in \text{Ser} . Q_* \in \text{connex} . R``C`Q \subset C`P . \supset .$
$P\bar{R}_{sc}Q \cap \text{sc}(P, Q)`R = \overrightarrow{\max}_P`(P\bar{R}_{sc}Q) .$
$\text{sc}(P, Q)`R = (C`P - P\bar{R}_{sc}Q) \cup \overrightarrow{\max}_P`(P\bar{R}_{sc}Q)$

Dem.

$\vdash . *234\cdot18 . \supset \vdash : \text{Hp} . \supset . P\bar{R}_{sc}Q \cap \text{sc}(P, Q)`R = P\bar{R}_{sc}Q - P``(P\bar{R}_{sc}Q)$

$[*205\cdot111] \quad = \overrightarrow{\max}_P`(P\bar{R}_{sc}Q)$ (1)

$\vdash . *24\cdot412 . *231\cdot13 . \supset$

$\vdash : \text{Hp} . \supset . C`P - P``(P\bar{R}_{sc}Q) = \{C`P - (P\bar{R}_{sc}Q)\} \cup \{(P\bar{R}_{sc}Q) - P``(P\bar{R}_{sc}Q)\} .$

$[*234\cdot18 . *205\cdot111] \quad \supset . \text{sc}(P, Q)`R = (C`P - P\bar{R}_{sc}Q) \cup \overrightarrow{\max}_P`(P\bar{R}_{sc}Q)$ (2)

$\vdash . (1) . (2) . \supset \vdash . \text{Prop}$

$*234\cdot182$. $\vdash : P \in \text{Ser} . Q_* \in \text{connex} . R``C`Q \subset C`P . \supset .$
$\overrightarrow{\text{limax}}_P`(P\bar{R}_{sc}Q) = \overrightarrow{\min}_P`\text{sc}(P, Q)`R$

Dem.

$\vdash . *207\cdot51 . \supset \vdash :. \text{Hp} . \supset : x = \text{limax}_P`(P\bar{R}_{sc}Q) . \equiv . x \in C`P . \overrightarrow{P}`x = P``(P\bar{R}_{sc}Q) .$

$[*234\cdot174] \quad \equiv . x \in C`P . \overrightarrow{P}`x = C`P \cap p`\overrightarrow{P}``\text{sc}(P, Q)`R$ (1)

$\vdash . *200\cdot52 . \supset$

$\vdash : \text{Hp} . x \in C`P . \overrightarrow{P}`x = C`P \cap p`\overrightarrow{P}``\text{sc}(P,Q)`R . \supset . C`P \neq C`P \cap p`\overrightarrow{P}``\text{sc}(P,Q)`R .$

$[*40\cdot2 . \text{Transp}] \quad \supset . \exists ! \text{sc}(P, Q)`R .$

$[*40\cdot62] \quad \supset . C`P \cap p`\overrightarrow{P}``\text{sc}(P, Q)`R = p`\overrightarrow{P}``\text{sc}(P, Q)`R .$

$[*13\cdot12] \quad \supset . \overrightarrow{P}`x = p`\overrightarrow{P}``\text{sc}(P, Q)`R$ (2)

$\vdash . *22\cdot621 . \supset$

$\vdash : \overrightarrow{P}`x = p`\overrightarrow{P}``\text{sc}(P, Q)`R . \supset . \overrightarrow{P}`x = C`P \cap p`\overrightarrow{P}``\text{sc}(P, Q)`R$ (3)

$\vdash . (2) . (3) . \supset \vdash :. \text{Hp} . x \in C`P . \supset :$

$\overrightarrow{P}`x = C`P \cap p`\overrightarrow{P}``\text{sc}(P, Q)`R . \equiv . \overrightarrow{P}`x = p`\overrightarrow{P}``\text{sc}(P, Q)`R$ (4)

$\vdash . (1) . (4) . \supset$

$\vdash :. \text{Hp} . \supset : x = \text{limax}_P`(P\bar{R}_{sc}Q) . \equiv . x \in C`P . \overrightarrow{P}`x = p`\overrightarrow{P}``\text{sc}(P, Q)`R .$

$[*205\cdot67] \quad \equiv . x = \text{min}_P`\text{sc}(P, Q)`R :. \supset \vdash . \text{Prop}$

***234·183.** $\vdash : \text{Hp} * 234\cdot18 \,.\, \text{sc}(P,Q)\text{‘}R = \Lambda \,.\, \supset .\, P\overline{R}_{sc}Q = C\text{‘}P \,.\, \sim E!\, B\text{‘}\breve{P}$ [*234·181·121]

***234·2.** $\vdash : P_{po} \in \text{Ser} \,.\, R\text{‘‘}C\text{‘}Q \subset C\text{‘}P \,.\, Q_* \in \text{connex} \,.\, \supset .$

$$\text{os}(P,Q)\text{‘}R = \{\overrightarrow{\min}_P\text{‘}(\breve{P}\overline{R}_{sc}Q) - P\overline{R}_{sc}Q\} \cup \{\overrightarrow{\max}_P\text{‘}(P\overline{R}_{sc}Q) - \breve{P}\overline{R}_{sc}Q\} \cup \{\overrightarrow{\max}_P\text{‘}(P\overline{R}_{sc}Q) \cap \overrightarrow{\min}_P\text{‘}(\breve{P}\overline{R}_{sc}Q)\}$$

Dem.

$\vdash . *234\cdot181 \,.\, \supset \vdash : \text{Hp} \,.\, \supset .\, \text{os}(P,Q)\text{‘}R = \{(C\text{‘}P - P\overline{R}_{sc}Q) \cup \overrightarrow{\max}_P\text{‘}(P\overline{R}_{sc}Q)\} \cap \{(C\text{‘}P - \breve{P}\overline{R}_{sc}Q) \cap \overrightarrow{\min}_P\text{‘}(\breve{P}\overline{R}_{sc}Q)\}$ (1)

$\vdash . *231\cdot201 \,.\, \supset \vdash : \text{Hp} \,.\, \supset .\, (C\text{‘}P - P\overline{R}_{sc}Q) \cap (C\text{‘}P - \breve{P}\overline{R}_{sc}Q) = \Lambda$ (2)

$\vdash . (1) . (2) . \supset \vdash . \text{Prop}$

***234·201.** $\vdash : \text{Hp} * 234\cdot2 \,.\, \supset .\, \text{os}(P,Q)\text{‘}R \subset \overrightarrow{\max}_P\text{‘}(P\overline{R}_{sc}Q) \cup \overrightarrow{\min}_P\text{‘}(\breve{P}\overline{R}_{sc}Q)$ [*234·2]

***234·202.** $\vdash : \text{Hp} * 234\cdot2 \,.\, \supset .\, \text{os}(P,Q)\text{‘}R \in 0 \cup 1 \cup 2$ [*234·201 . *205·681 . *60·391]

***234·203.** $\vdash : \text{Hp} * 234\cdot2 \,.\, P\overline{R}_{os}Q \in 1 \,.\, \supset .$

$$\text{os}(P,Q)\text{‘}R \in 1 \,.\, \text{os}(P,Q)\text{‘}R = \iota\text{‘}\max_P\text{‘}(P\overline{R}_{sc}Q) = \iota\text{‘}\min_P\text{‘}(\breve{P}\overline{R}_{sc}Q) = P\overline{R}_{os}Q$$

[*231·193·103 . *205·68 . *234·2]

***234·204.** $\vdash : P_{po} \in \text{Ser} \,.\, P\overline{R}_{os}Q \sim\in 0 \cup 1 \,.\, \supset .\, \text{os}(P,Q)\text{‘}R = \Lambda$ [*234·103]

***234·21.** $\vdash : \text{Hp} * 234\cdot2 \,.\, P\overline{R}_{os}Q = \Lambda \,.\, \supset .$

$$\text{os}(P,Q)\text{‘}R = \overrightarrow{\max}_P\text{‘}(P\overline{R}_{sc}Q) \cup \overrightarrow{\min}_P\text{‘}(\breve{P}\overline{R}_{sc}Q)$$

Dem.

$\vdash . *205\cdot11\cdot111 \,.\, \supset$

$\vdash : \text{Hp} \,.\, \supset .\, \overrightarrow{\max}_P\text{‘}(P\overline{R}_{sc}Q) \subset -(\breve{P}\overline{R}_{sc}Q) \,.\, \overrightarrow{\min}_P\text{‘}(\breve{P}\overline{R}_{sc}Q) \subset -(P\overline{R}_{sc}Q)$ (1)

$\vdash . (1) . *234\cdot2 \,.\, \supset \vdash . \text{Prop}$

***234·23.** $\vdash :. \text{Hp} * 234\cdot2 \,.\, P\overline{R}_{os}Q \sim\in 1 \,.\, \text{os}(P,Q)\text{‘}R \in 1 \,.\, \supset :$

$$P\overline{R}_{os}Q = \Lambda : \text{os}(P,Q)\text{‘}R = \iota\text{‘}\max_P\text{‘}(P\overline{R}_{sc}Q) \,.\, \sim E!\min_P\text{‘}(\breve{P}\overline{R}_{sc}Q) \,.\, \vee .\, \text{os}(P,Q)\text{‘}R = \iota\text{‘}\min_P\text{‘}(\breve{P}\overline{R}_{sc}Q) \,.\, \sim E!\max_P\text{‘}(P\overline{R}_{sc}Q)$$

Dem.

$\vdash . *234\cdot103 \,.\, \supset \vdash : \text{Hp} \,.\, \supset .\, P\overline{R}_{os}Q = \Lambda$ (1)

[*234·21] $\supset .\, \text{os}(P,Q)\text{‘}R = \overrightarrow{\max}_P\text{‘}(P\overline{R}_{sc}Q) \cup \overrightarrow{\min}_P\text{‘}(\breve{P}\overline{R}_{sc}Q)$ (2)

$\vdash . *52\cdot41 \,.\, \supset \vdash : P\overline{R}_{os}Q = \Lambda \,.\, E!\max_P\text{‘}(P\overline{R}_{sc}Q) \,.\, E!\min_P\text{‘}(\breve{P}\overline{R}_{sc}Q) \,.\, \supset .\, \{\overrightarrow{\max}_P\text{‘}(P\overline{R}_{sc}Q) \cup \overrightarrow{\min}_P\text{‘}(\breve{P}\overline{R}_{sc}Q)\} \sim\in 1$ (3)

$\vdash . (1) . (2) . (3) . \text{Transp} \,.\, \supset$

$\vdash :. \text{Hp} \,.\, \supset : \sim E!\max_P\text{‘}(P\overline{R}_{sc}Q) \,.\, \vee .\, \sim E!\min_P\text{‘}(\breve{P}\overline{R}_{sc}Q)$ (4)

$\vdash . (2) . *205\cdot681 \,.\, \supset \vdash :. \text{Hp} \,.\, \supset : E!\max_P\text{‘}(P\overline{R}_{sc}Q) \,.\, \vee .\, E!\min_P\text{‘}(\breve{P}\overline{R}_{sc}Q)$ (5)

$\vdash . (1) . (2) . (4) . (5) . \supset \vdash . \text{Prop}$

***234·24.** $\vdash :. P \epsilon \mathrm{Ser} . Q_* \epsilon \mathrm{connex} . R``C`Q \subset C`P . \supset :$

$$\mathrm{os}(P, Q)`R \epsilon 1 . \supset . \mathrm{os}(P, Q)`R = \iota`\mathrm{limax}_P`(P\overline{R}_{sc}Q) = \iota`\mathrm{limin}_P`(\breve{P}\overline{R}_{sc}Q)$$

Dem.

$\vdash . *234·203 . *207·42 . \supset$

$\vdash : \mathrm{Hp} . P\overline{R}_{os}Q \epsilon 1 . \supset . \mathrm{os}(P, Q)`R = \iota`\mathrm{limax}_P`(P\overline{R}_{sc}Q) = \iota`\mathrm{limin}_P`(\breve{P}\overline{R}_{sc}Q)$ (1)

$\vdash . *234·23 . *211·728 . *207·42 . \supset$

$\vdash : \mathrm{Hp} . P\overline{R}_{os}Q \sim\epsilon 1 . \supset . \mathrm{os}(P, Q)`R = \iota`\mathrm{limax}_P`(P\overline{R}_{sc}Q) = \iota`\mathrm{limin}_P`(\breve{P}\overline{R}_{sc}Q)$ (2)

$\vdash . (1) . (2) . \supset \vdash . \mathrm{Prop}$

***234·241.** $\vdash : \mathrm{Hp} *234·2 . \mathrm{os}(P, Q)`R \epsilon 2 . \supset . P\overline{R}_{os}Q = \Lambda$

Dem.

$\vdash . *234·103 . \quad \supset \vdash : \mathrm{Hp} . \supset . P\overline{R}_{os}Q \epsilon 0 \cup 1$ (1)

$\vdash . *234·203 . \mathrm{Transp} . \supset \vdash : \mathrm{Hp} . \supset . P\overline{R}_{os}Q \sim\epsilon 1$ (2)

$\vdash . (1) . (2) . \supset \vdash . \mathrm{Prop}$

***234·242.** $\vdash : \mathrm{Hp} *234·2 . \mathrm{os}(P, Q)`R \epsilon 2 . \supset .$

$$\mathrm{os}(P,Q)`R = \iota`\mathrm{max}_P`(P\overline{R}_{sc}Q) \cup \iota`\mathrm{min}_P`(\breve{P}\overline{R}_{sc}Q) . \mathrm{max}_P`(P\overline{R}_{sc}Q) P_1 \mathrm{min}_P`(\breve{P}\overline{R}_{sc}Q)$$

Dem.

$\vdash . *234·201 . *205·3 . \supset \vdash : \mathrm{Hp} . \supset . \mathrm{E!}\, \mathrm{max}_P`(P\overline{R}_{sc}Q) . \mathrm{E!}\, \mathrm{min}_P`(\breve{P}\overline{R}_{sc}Q) .$

$\mathrm{max}_P`(P\overline{R}_{sc}Q) \neq \mathrm{min}_P`(\breve{P}\overline{R}_{sc}Q)$ (1)

$\vdash . *234·241·15 . \supset \vdash : \mathrm{Hp} . \supset . \breve{P}\overline{R}_{sc}Q = C`P - P\overline{R}_{sc}Q .$

[*211·8.(1)] $\supset . \mathrm{max}_P`(P\overline{R}_{sc}Q) = \mathrm{max}(P_{po})`(P\overline{R}_{sc}Q) .$

$\mathrm{min}_P`(\breve{P}\overline{R}_{sc}Q) = \mathrm{seq}(P_{po})`(P\overline{R}_{sc}Q) .$

[*206·5.*201·63] $\supset . \{\mathrm{max}_P`(P\overline{R}_{sc}Q)\} (P_{po})_1 \{\mathrm{min}_P`(\breve{P}\overline{R}_{sc}Q)\} .$

[*121·254] $\supset . \{\mathrm{max}_P`(P\overline{R}_{sc}Q)\} P_1 \{\mathrm{min}_P`(\breve{P}\overline{R}_{sc}Q)\}$ (2)

$\vdash . (1) . (2) . *234·201 . \supset \vdash . \mathrm{Prop}$

***234·243.** $\vdash : \mathrm{Hp} *234·24 . \exists ! \mathrm{os}(P, Q)`R . \supset .$

$$\mathrm{E!}\, \mathrm{limax}_P`(P\overline{R}_{sc}Q) . \mathrm{E!}\, \mathrm{limin}_P`(\breve{P}\overline{R}_{sc}Q)$$

Dem.

$\vdash . *234·202 . \supset \vdash : \mathrm{Hp} . \supset . \mathrm{os}(P, Q)`R \epsilon 1 \cup 2$ (1)

$\vdash . (1) . *234·24·242 . \supset \vdash . \mathrm{Prop}$

***234·244.** $\vdash : \mathrm{Hp} *234·2 . P^2 = P . \supset . \mathrm{os}(P, Q)`R \epsilon 0 \cup 1$

Dem.

$\vdash . *234·242·202 . \supset \vdash : \mathrm{Hp} *234·2 . \mathrm{os}(P, Q)`R \sim\epsilon 0 \cup 1 . \supset . \dot{\exists} ! P_1$ (1)

$\vdash . (1) . \mathrm{Transp} . *201·65 . \supset \vdash . \mathrm{Prop}$

***234·25.** $\vdash : \mathrm{Hp} *234·2 . P^2 = P . \exists ! \mathrm{os}(P, Q)`R . \supset .$

$$\mathrm{os}(P, Q)`R = \iota`\mathrm{limax}_P`(P\overline{R}_{sc}Q) = \iota`\mathrm{limin}_P`(\breve{P}\overline{R}_{sc}Q)$$

[*234·244·24]

***234·251.** $\vdash$: Hp *234·24 . $\text{limax}_P\text{'}(P\bar{R}_{sc}Q) = \text{limin}_P\text{'}(\breve{P}\bar{R}_{sc}Q)$. $\supset$.

$\text{os}\,(P, Q)\text{'}R = \iota\text{'}\text{limax}_P\text{'}(P\bar{R}_{sc}Q) = \iota\text{'}\text{min}_P\text{'}\text{sc}\,(P, Q)\text{'}R = \iota\text{'}\text{max}_P\text{'}\text{sc}\,(\breve{P}, Q)\text{'}R$

Dem.

$\vdash$. *234·18 . *207·51 . $\supset$

$\vdash$: Hp . $\supset$. $\text{sc}\,(P, Q)\text{'}R = C\text{'}P - \overrightarrow{P}\text{'}\text{limax}_P\text{'}(P\bar{R}_{sc}Q)$.

$\text{sc}\,(\breve{P}, Q)\text{'}R = C\text{'}P - \overleftarrow{P}\text{'}\text{limin}_P\text{'}(\breve{P}\bar{R}_{sc}Q)$.

[Hp.*202·101] $\supset$. $\text{os}\,(P, Q)\text{'}R = C\text{'}P \cap \iota\text{'}\text{limax}_P\text{'}(P\bar{R}_{sc}Q)$.

[*51·31] $= \iota\text{'}\text{limax}_P\text{'}(P\bar{R}_{sc}Q)$ (1)

[*234·182] $= \iota\text{'}\text{min}_P\text{'}\text{sc}\,(P, Q)\text{'}R$ (2)

$\left[(2)\dfrac{\breve{P}}{P}\right]$ $= \iota\text{'}\text{max}_P\text{'}\text{sc}\,(\breve{P}, Q)\text{'}R$ (3)

$\vdash$. (1) . (2) . (3) . $\supset \vdash$. Prop

***234·26.** $\vdash$:. Hp *234·2 . $P^2 = P$. $\supset$:

$\exists\,!\,\text{os}\,(P, Q)\text{'}R$. $\equiv$. $\text{os}\,(P, Q)\text{'}R = \iota\text{'}\text{limax}_P\text{'}(P\bar{R}_{sc}Q)$.

$\equiv$. $\text{os}\,(P, Q)\text{'}R = \iota\text{'}\text{limin}_P\text{'}(\breve{P}\bar{R}_{sc}Q)$.

$\equiv$. $\text{os}\,(P, Q)\text{'}R = \iota\text{'}\text{min}_P\text{'}\text{sc}\,(PQ)\text{'}R$.

$\equiv$. $\text{os}\,(P, Q)\text{'}R = \iota\text{'}\text{max}_P\text{'}\text{sc}\,(\breve{P}Q)\text{'}R$.

$\equiv$. $\text{limax}_P\text{'}(P\bar{R}_{sc}Q) = \text{limin}_P\text{'}(\breve{P}\bar{R}_{sc}Q)$

[*234·25·251·182 . *51·161]

***234·27.** $\vdash$: Hp *234·24 . $x \in \text{os}\,(P, Q)\text{'}R - \text{Ꮧ}\text{'}P_1$. $\supset$. $x = \text{limax}_P\text{'}(P\bar{R}_{sc}Q)$

Dem.

$\vdash$. *234·24 . $\supset \vdash$: Hp . $\text{os}\,(P, Q)\text{'}R \in 1$. $\supset$. $x = \text{limax}_P\text{'}(P\bar{R}_{sc}Q)$ (1)

$\vdash$. *234·242 . $\supset \vdash$: Hp . $\text{os}\,(P, Q)\text{'}R \in 2$. $\supset$. $x = \text{limax}_P\text{'}(P\bar{R}_{sc}Q)$ (2)

$\vdash$. *234·202 . $\supset \vdash$: Hp . $\supset$. $\text{os}\,(P, Q)\text{'}R \in 1 \cup 2$ (3)

$\vdash$. (1) . (2) . (3) . $\supset \vdash$. Prop

***234·271.** $\vdash$: Hp 234·24 . $x \in \text{os}\,(P, Q)\text{'}R - \text{D}\text{'}P_1$. $\supset$. $x = \text{limin}_P\text{'}(\breve{P}\bar{R}_{sc}Q)$

$\left[\text{*234·27}\,\dfrac{\breve{P}}{P}\right]$

***234·272.** $\vdash$: Hp *234·24 . $x \in \text{os}\,(P, Q)\text{'}R - C\text{'}P_1$. $\supset$.

$x = \text{limax}_P\text{'}(P\bar{R}_{sc}Q) = \text{limin}_P\text{'}(\breve{P}\bar{R}_{sc}Q)$ [*234·27·271]

The remaining propositions of the present number are for the most part immediate consequences of those already proved. In order to obtain, from propositions already proved, propositions concerning the limit of a function as the argument approaches the limit of some class of arguments α, we only have to substitute $Q_* \upharpoonright \alpha$ for Q. In order to obtain the limit of a function as the argument approaches a given term a, we take $Q_* \upharpoonright \overrightarrow{Q}\text{'}a$ in place of Q.

***234·3.** $\vdash :. x \in \text{sc}(P, Q_* \upharpoonright \alpha)`R \,.\, \equiv :$

$x \in C`P : x P_{po} w \,.\, \supset_w \,.\, (\exists y) \,.\, y \in \alpha \cap C`Q \cap \text{Ꮖ}`R \,.\, R``(\alpha \cap \overrightarrow{Q_*}`y) \subset \overrightarrow{P_{po}}`w$

[*234·1]

***234·301.** $\vdash :: Q_* \upharpoonright \alpha \in \text{connex} \,.\, \supset :. x \in \text{os}(P, Q_* \upharpoonright \alpha)`R \cap \text{D}`P \cap \text{Ꮖ}`P \,.\, \equiv :$

$x \in \text{D}`P \cap \text{Ꮖ}`P : x \in P(z - w) \,.\, \supset_{z,w} \,.$

$(\exists y) \,.\, y \in \alpha \cap C`Q \cap \text{Ꮖ}`R \,.\, R``(\alpha \cap \overleftarrow{Q_*}`y) \subset P(z - w)$

[*234·12]

***234·31.** $\vdash : P_{po} \in \text{Ser} \,.\, Q_* \upharpoonright \alpha \in \text{connex} \,.\, R``(\alpha \cap C`Q) \subset C`P \,.\, \supset \,.$

$C`P - \text{sc}(P, Q_* \upharpoonright \alpha)`R = C`P \cap p`\overrightarrow{P_{po}}``\text{sc}(P, Q_* \upharpoonright \alpha)`R = P``(P\bar{R}Q)_{sc}`\alpha$

[*234·174]

***234·311.** $\vdash : \text{Hp} *234·31 \,.\, \supset \,.\, C`P = \text{sc}(P, Q_* \upharpoonright \alpha)`R \cup P``(P\bar{R}Q)_{sc}`\alpha \,.$

$\text{sc}(P, Q_* \upharpoonright \alpha)`R \cap P``(P\bar{R}Q)_{sc}`\alpha = \Lambda \,.$

$\text{sc}(P, Q_* \upharpoonright \alpha)`R = C`P - P``(P\bar{R}Q)_{sc}`\alpha$

[*234·18]

***234·312.** $\vdash :. P \in \text{Ser} \,.\, Q_* \upharpoonright \alpha \in \text{connex} \,.\, R``(\alpha \cap C`Q) \subset C`P \,.\, \supset :$

$\text{E} ! (P\bar{R}Q)_{lmx}`\alpha \,.\, \equiv \,.\, \text{E} ! \min_P`\text{sc}(P, Q_* \upharpoonright \alpha)`R \,.$

$\equiv \,.\, (P\bar{R}Q)_{lmx}`\alpha = \min_P`\text{sc}(P, Q_* \upharpoonright \alpha)`R$

[*234·182]

***234·32.** $\vdash :. P_{po} \in \text{Ser} \,.\, Q_* \upharpoonright \alpha \in \text{connex} \,.\, R``(\alpha \cap C`Q) \subset C`P \,.\, \supset : (P\bar{R}Q)_{os}`\alpha \in 1 \,.\, \supset \,.$

$\text{os}(P, Q_* \upharpoonright \alpha)`R = (P\bar{R}Q)_{os}`\alpha = \iota`\max_P`(P\bar{R}Q)_{sc}`\alpha = \iota`\min_P`(\breve{P}\bar{R}Q)_{sc}`\alpha$

[*234·203]

***234·321.** $\vdash :: \text{Hp} *234·32 \,.\, \text{os}(P, Q_* \upharpoonright \alpha)`R \in 1 \,.\, \supset :. (P\bar{R}Q)_{os}`\alpha \sim\in 1 \,.\, \supset :$

$(P\bar{R}Q)_{os}`\alpha = \Lambda : \text{os}(P, Q_* \upharpoonright \alpha)`R = \iota`\max_P`(P\bar{R}Q)_{sc}`\alpha \,.\, \sim \text{E} ! \min_P`(\breve{P}\bar{R}Q)_{sc}`\alpha \,.$

$\vee \,.\, \text{os}(P, Q_* \upharpoonright \alpha)`R = \iota`\min_P`(\breve{P}\bar{R}Q)_{sc}`\alpha \,.\, \sim \text{E} ! \max_P`(P\bar{R}Q)_{sc}`\alpha$

[*234·23]

***234·322.** $\vdash : \text{Hp} *234·312 \,.\, \text{os}(P, Q_* \upharpoonright \alpha)`R \in 1 \,.\, \supset \,.$

$\text{os}(P, Q_* \upharpoonright \alpha)`R = \iota`(P\bar{R}Q)_{lmx}`\alpha = \iota`(\breve{P}\bar{R}Q)_{lmx}`\alpha$ [*234·24]

***234·329.** $\vdash : \text{Hp} *234·32 \,.\, \text{os}(P, Q_* \upharpoonright \alpha)`R \in 2 \,.\, \supset \,.$

$\text{os}(P, Q_* \upharpoonright \alpha)`R = \iota`\max_P`(P\bar{R}Q)_{sc}`\alpha \cup \iota`\min_P`(\breve{P}\bar{R}Q)_{sc}`\alpha \,.$

$\{\max_P`(P\bar{R}Q)_{sc}`\alpha\} \, P_1 \, \{\min_P`(\breve{P}\bar{R}Q)_{sc}`\alpha\}$

[*234·242]

***234·33.** $\vdash : \text{Hp} *234·32 \,.\, P^2 = P \,.\, \exists ! \,\text{os}(P, Q_* \upharpoonright \alpha)`R \,.\, \supset \,.$

$\text{os}(P, Q_* \upharpoonright \alpha)`R = \iota`(P\bar{R}Q)_{lmx}`\alpha = \iota`(\breve{P}\bar{R}Q)_{lmx}`\alpha$ [*234·25]

***234·331.** $\vdash : \text{Hp} *234·312 \,.\, (P\bar{R}Q)_{lmx}`\alpha = (\breve{P}\bar{R}Q)_{lmx}`\alpha \,.\, \supset \,.$

$\text{os}(P, Q_* \upharpoonright \alpha)`R = \iota`(P\bar{R}Q)_{lmx}`\alpha = \iota`(\breve{P}\bar{R}Q)_{lmx}`\alpha$

$= \iota`\min_P`\text{sc}(P, Q_* \upharpoonright \alpha)`R = \iota`\max_P`\text{sc}(\breve{P}, Q_* \upharpoonright \alpha)`R$

[*234·251]

$*234\cdot34$. $\vdash :.\ \text{Hp}\ *234\cdot32\ .\ P^2 = P\ .\ \supset :$

$\exists !\ \text{os}\,(P, Q_* \upharpoonright \alpha)'R\ .\ \equiv .\ \text{os}\,(P, Q_* \upharpoonright \alpha)'R = \iota'(P\overline{R}Q)_{\text{lmx}}'\alpha\ .$

$\equiv .\ \text{os}\,(P, Q_* \upharpoonright \alpha)'R = \iota'(\breve{P}\overline{R}Q)_{\text{lmx}}'\alpha\ .$

$\equiv .\ (P\overline{R}Q)_{\text{lmx}}'\alpha = (\breve{P}\overline{R}Q)_{\text{lmx}}'\alpha$

[$*234\cdot26$]

$*234\cdot35$. $\vdash :\ \text{Hp}\ *234\cdot312\ .\ x \in \text{os}\,(P, Q_* \upharpoonright \alpha)'R - \text{ᗡ}'P_1\ .\ \supset .\ x = (P\overline{R}Q)_{\text{lmx}}'\alpha$
[$*234\cdot27$]

$*234\cdot351$. $\vdash :\ \text{Hp}\ *234\cdot312\ .\ x \in \text{os}\,(P, Q_* \upharpoonright \alpha)'R - \text{D}'P_1\ .\ \supset .\ x = (\breve{P}\overline{R}Q)_{\text{lmx}}'\alpha$

$\left[*234\cdot35\,\dfrac{\breve{P}}{P}\right]$

$*234\cdot352$. $\vdash :\ \text{Hp}\ *234\cdot312\ .\ x \in \text{os}\,(P, Q_* \upharpoonright \alpha)'R - C'P_1\ .\ \supset .$

$x = (P\overline{R}Q)_{\text{lmx}}'\alpha = (\breve{P}\overline{R}Q)_{\text{lmx}}'\alpha$ [$*234\cdot35\cdot351$]

$*234\cdot4$. $\vdash :.\ x \in \text{sc}\,(P, Q_* \upharpoonright \overrightarrow{Q_{\text{po}}}'a)'R\ .\ \equiv :$

$x \in C'P : xP_{\text{po}}w\ .\ \supset_w .\ (\exists y)\ .\ y \in \overrightarrow{Q_{\text{po}}}'a \cap \text{ᗡ}'R\ .\ R''Q\,(y \vdash a) \subset \overrightarrow{P_{\text{po}}}'w$
[$*234\cdot3\ .\ (*121\cdot012)$]

$*234\cdot41$. $\vdash ::\ Q \in \text{trans}\ .\ Q_* \upharpoonright \overrightarrow{Q}'a \in \text{connex}\ .\ \supset :.$

$x \in \text{os}\,(P, Q_* \upharpoonright \overrightarrow{Q}'a)'R \cap \text{D}'P \cap \text{ᗡ}'P\ .\ \equiv : x \in \text{D}'P \cap \text{ᗡ}'P :$

$x \in P\,(z - w)\ .\ \supset_{z,w} .\ (\exists y)\ .\ y \in \overrightarrow{Q}'a \cap \text{ᗡ}'R\ .\ R''Q\,(y \vdash a) \subset P\,(z - w)$
[$*234\cdot301\ .\ (*121\cdot012)\ .\ *201\cdot18$]

$*234\cdot42$. $\vdash :.\ P \in \text{Ser}\ .\ Q \in \text{trans}\ .\ Q_* \upharpoonright \overrightarrow{Q}'a \in \text{connex}\ .\ R''\overrightarrow{Q}'a \subset C'P\ .\ \supset :$

$\overrightarrow{R\,(PQ)}'a = \overrightarrow{\text{min}_P}'\text{sc}\,(P, Q_* \upharpoonright \overrightarrow{Q}'a)'R$ [$*234\cdot182$]

$*234\cdot421$. $\vdash :.\ P_{\text{po}} \in \text{Ser}\ .\ Q \in \text{trans}\ .\ Q_* \upharpoonright \overrightarrow{Q}'a \in \text{connex}\ .\ R''\overrightarrow{Q}'a \subset C'P\ .\ \supset :$

$\text{sc}\,(P, Q_* \upharpoonright \overrightarrow{Q}'a)'R = \Lambda\ .\ \supset .\ \overrightarrow{R\,(PQ)}'a = \overrightarrow{B}'\breve{P}$ [$*234\cdot183$]

$*234\cdot422$. $\vdash :\ \text{Hp}\ *234\cdot42\ .\ P \in \text{Ded}\ .\ \supset .\ R\,(PQ)'a = \text{limin}_P'\text{sc}\,(P, Q_* \upharpoonright \overrightarrow{Q}'a)'R$
[$*233\cdot13\ .\ *234\cdot42$]

$*234\cdot43$. $\vdash :\ \text{Hp}\ *234\cdot42\ .\ \text{os}\,(P, Q_* \upharpoonright \overrightarrow{Q}'a)'R \in 1\ .\ \supset .$

$\text{os}\,(P, Q_* \upharpoonright \overrightarrow{Q}'a) = \iota'R\,(PQ)'a = \iota'R\,(\breve{P}Q)'a$ [$*234\cdot322$]

$*234\cdot439$. $\vdash :\ \text{Hp}\ *234\cdot421\ .\ \text{os}\,(P, Q_* \upharpoonright \overrightarrow{Q}'a)'R \in 2\ .\ \supset .$

$\text{os}\,(P, Q_* \upharpoonright \overrightarrow{Q}'a)'R = \iota'R\,(PQ)'a \cup \iota'R\,(\breve{P}Q)'a\ .$

$\{R\,(PQ)'a\}\ P_1\ \{R\,(\breve{P}Q)'a\}$ [$*234\cdot329$]

$*234\cdot44$. $\vdash :\ \text{Hp}\ *234\cdot421\ .\ P^2 = P\ .\ \exists !\ \text{os}\,(P, Q_* \upharpoonright \overrightarrow{Q}'a)'R\ .\ \supset .$

$\text{os}\,(P, Q_* \upharpoonright \overrightarrow{Q}'a)'R = \iota'R\,(PQ)'a = \iota'R\,(\breve{P}Q)'a$ [$*234\cdot33$]

***234·441.** $\vdash : \text{Hp} *234{\cdot}42 \,.\, R(PQ)\text{‘}a = R(\breve{P}Q)\text{‘}a \,.\, \supset .$

$$\text{os}(P, Q_* \upharpoonright \overrightarrow{Q}\text{‘}a) = \iota\text{‘}R(PQ)\text{‘}a = \iota\text{‘}R(\breve{P}Q)\text{‘}a \quad [*234{\cdot}331]$$

***234·45.** $\vdash :. P \in \text{Ser} \,.\, Q \in \text{trans} \,.\, Q_* \upharpoonright \overrightarrow{Q}\text{‘}a \in \text{connex} \,.\, R\text{“}\overrightarrow{Q}\text{‘}a \subset C\text{‘}P \,.\, P^2 = P \,.\, \supset :$

$$\exists ! \,\text{os}(P, Q_* \upharpoonright \overrightarrow{Q}\text{‘}a)\text{‘}R \,.\, \equiv .\, \text{os}(P, Q_* \upharpoonright \overrightarrow{Q}\text{‘}a)\text{‘}R = \iota\text{‘}R(PQ)\text{‘}a \,.$$
$$\equiv .\, \text{os}(P, Q_* \upharpoonright \overrightarrow{Q}\text{‘}a)\text{‘}R = \iota\text{‘}R(\breve{P}Q)\text{‘}a \,.$$
$$\equiv .\, R(PQ)\text{‘}a = R(\breve{P}Q)\text{‘}a \quad [*234{\cdot}34]$$

***234·46.** $\vdash : \text{Hp} *234{\cdot}42 \,.\, x \in \text{os}(P, Q_* \upharpoonright \overrightarrow{Q}\text{‘}a)\text{‘}R - \text{Ɑ}\text{‘}P_1 \,.\, \supset .\, x = R(PQ)\text{‘}a$ $[*234{\cdot}35]$

***234·461.** $\vdash : \text{Hp} *234{\cdot}42 \,.\, x \in \text{os}(P, Q_* \upharpoonright \overrightarrow{Q}\text{‘}a)\text{‘}R - \text{D}\text{‘}P_1 \,.\, \supset .\, x = R(\breve{P}Q)\text{‘}a$

$$\left[*234{\cdot}46 \frac{\breve{P}}{P}\right]$$

***234·462.** $\vdash : \text{Hp} *234{\cdot}42 \,.\, x \in \text{os}(P, Q_* \upharpoonright \overrightarrow{Q}\text{‘}a)\text{‘}R - C\text{‘}P_1 \,.\, \supset .$

$$x = R(PQ)\text{‘}a = R(\breve{P}Q)\text{‘}a \quad [*234{\cdot}46{\cdot}461]$$

***234·5.** $\vdash : a \in \text{ct}(PQ)\text{‘}R \,.\, \equiv .\, R\text{‘}a \in \text{os}(P, Q_* \upharpoonright \overrightarrow{Q}_{\text{po}}\text{‘}a)\text{‘}R - C\text{‘}P_1 \quad [(*234{\cdot}03)]$

***234·51.** $\vdash :: Q \in \text{trans} \,.\, Q_* \upharpoonright \overrightarrow{Q}\text{‘}a \in \text{connex} \,.\, R\text{‘}a \in \text{D}\text{‘}P \cap \text{Ɑ}\text{‘}P \,.\, \supset :.$

$a \in \text{ct}(PQ)\text{‘}R \,.\, \equiv : R\text{‘}a \sim\in C\text{‘}P_1 : R\text{‘}a \in P(z - w) \,.\, \supset_{z,w} .$

$(\exists y) \,.\, y \in \overrightarrow{Q}\text{‘}a \cap \text{Ɑ}\text{‘}R \,.\, R\text{“}Q(y \mapsto a) \subset P(z - w)$

Dem.

$\vdash .\, *234{\cdot}5{\cdot}4 \,.\, *53{\cdot}31 \,.\, \supset$

$\vdash :: \text{Hp} \,.\, \supset :.\, a \in \text{ct}(PQ)\text{‘}R \,.\, \equiv : R\text{‘}a \in \text{D}\text{‘}P \cap \text{Ɑ}\text{‘}P - C\text{‘}P_1 : R\text{‘}a \in P(z - w) \,.\, \supset_{z,w} .$

$(\exists y) \,.\, y \in \overrightarrow{Q}\text{‘}a \cap \text{Ɑ}\text{‘}R \,.\, R\text{“}Q(y \vdash a) \subset P(z - w) \,.\, R\text{“}\iota\text{‘}a \subset P(z - w) \quad (1)$

$\vdash .\, (1) \,.\, *121{\cdot}242 \,.\, \supset \vdash .\, \text{Prop}$

***234·52.** $\vdash :. P \in \text{Ser} \,.\, Q \in \text{trans} \,.\, Q_* \upharpoonright \overrightarrow{Q}\text{‘}a \in \text{connex} \,.\, R\text{“}\overrightarrow{Q}\text{‘}a \subset C\text{‘}P \,.\, \supset :$

$$a \in \text{ct}(PQ)\text{‘}R \,.\, \supset .\, R(PQ)\text{‘}a = R(\breve{P}Q)\text{‘}a = R\text{‘}a \quad [*234{\cdot}462{\cdot}5]$$

***234·521.** $\vdash : \text{Hp} *234{\cdot}52 \,.\, a \in \text{ct}(PQ)\text{‘}R \,.\, \supset .\, \text{os}(P, Q_* \upharpoonright \overrightarrow{Q}\text{‘}a) = \iota\text{‘}R\text{‘}a$ $[*234{\cdot}441{\cdot}52]$

***234·522.** $\vdash :. \text{Hp} *234{\cdot}52 \,.\, P^2 = P \,.\, \supset :$

$$a \in \text{ct}(PQ)\text{‘}R \,.\, \equiv .\, R(PQ)\text{‘}a = R(\breve{P}Q)\text{‘}a = R\text{‘}a$$

Dem.

$\vdash .\, *234{\cdot}45 \,.\, \supset$

$\vdash :. \text{Hp} \,.\, \supset : R(PQ)\text{‘}a = R(\breve{P}Q)\text{‘}a = R\text{‘}a \,.\, \supset .\, \text{os}(P, Q_* \upharpoonright \overrightarrow{Q}\text{‘}a)\text{‘}R = \iota\text{‘}R\text{‘}a \,.$

$[*234{\cdot}5 . *201{\cdot}65] \quad \supset .\, a \in \text{ct}(PQ)\text{‘}R \quad (1)$

$\vdash .\, (1) \,.\, *234{\cdot}52 \,.\, \supset \vdash .\, \text{Prop}$

***234·53.** $\vdash :: P_{po} \epsilon \text{connex} . Q \epsilon \text{trans} . R'a = B'P . \supset :.$

$$a \epsilon \text{ct}(PQ)'R . \equiv : B'P \sim\epsilon \text{D}'P_1 : w \epsilon \text{Ɑ}'P . \supset_w .$$
$$(\exists y) . y \epsilon \overrightarrow{Q}'a \cap \text{Ɑ}'R . R``Q(y \vdash\!\dashv a) \subset \overrightarrow{P}_{po}'w$$

Dem.

$\vdash . *234{\cdot}122 . *53{\cdot}31 . *234{\cdot}5 . \supset$

$\vdash :: \text{Hp} . \supset :. a \epsilon \text{ct}(PQ)'R . \equiv : B'P \sim\epsilon \text{D}'P_1 : (B'P) P_{po} w . \supset_w .$

$$(\exists y) . y \epsilon \overrightarrow{Q}'a \cap \text{Ɑ}'R . R``(\overleftarrow{Q_*}'y \cap \overrightarrow{Q}'a) \subset \overrightarrow{P}_{po}'w . R``\iota'a \subset \overrightarrow{P}_{po}'w :$$

$[*202{\cdot}522 . *205{\cdot}253 . *201{\cdot}18] \equiv : B'P \sim\epsilon \text{D}'P_1 :$

$$w \epsilon \text{Ɑ}'P . \supset_w . (\exists y) . y \epsilon \overrightarrow{Q}'a \cap \text{Ɑ}'R . R``Q(y \vdash\!\dashv a) \subset \overrightarrow{P}_{po}'w :: \supset \vdash . \text{Prop}$$

***234·54.** $\vdash : a \epsilon \text{ct}(PQ)'R . \supset . a \epsilon \text{Ɑ}'R \cap \breve{Q}_{po}``\text{Ɑ}'R . R'a \epsilon C'P$

Dem.

$\vdash . *234{\cdot}51 . (*234{\cdot}02) . \supset \vdash : \text{Hp} . \supset . R'a \epsilon C'P$ (1)

$\vdash . (1) . *234{\cdot}5 . (*234{\cdot}02) . \supset$

$\vdash :. \text{Hp} . \supset : \exists ! \text{sc}(P, Q_* \upharpoonright \overrightarrow{Q}_{po}'a)'R \cap \text{D}'P . \vee . \exists ! \text{sc}(\breve{P}, Q_* \upharpoonright \overrightarrow{Q}_{po}'a)'R \cap \text{Ɑ}'P :$

$[*234{\cdot}142] \supset : \exists ! \overrightarrow{Q}_{po}'a \cap \text{Ɑ}'R :$

$[*37{\cdot}46] \quad \supset : a \epsilon \breve{Q}_{po}``\text{Ɑ}'R$ (2)

$\vdash . (1) . *14{\cdot}21 . *33{\cdot}43 . \supset \vdash : \text{Hp} . \supset . a \epsilon \text{Ɑ}'R$ (3)

$\vdash . (1) . (2) . (3) . \supset \vdash . \text{Prop}$

***234·55.** $\vdash . \sim \{\min(Q_{po})'\text{Ɑ}'R \epsilon \text{ct}(PQ)'R\}$ $[*234{\cdot}54 . \text{Transp}]$

***234·56.** $\vdash : \text{Hp} *234{\cdot}52 . a \epsilon \text{ct}(PQ)'R . \supset .$

$$(P\bar{R}Q)_{os}'\overrightarrow{Q}'a \epsilon 0 \cup 1 . \text{E} ! R(PQ)'a . R(PQ)'a \sim\epsilon C'P_1 . R'a = R(PQ)'a$$

Dem.

$\vdash . *234{\cdot}5 . \supset \vdash : \text{Hp} . \supset . \exists ! \text{os}(P, Q_* \upharpoonright \overrightarrow{Q}'a)'R . R'a \sim\epsilon C'P_1 .$

$[*234{\cdot}103] \qquad \supset . (P\bar{R}Q)_{os}'\overrightarrow{Q}'a \epsilon 0 \cup 1 . R'a \sim\epsilon C'P_1$ (1)

$\vdash . *234{\cdot}52 . \supset \vdash : \text{Hp} . \supset . R'a = R(PQ)'a . \text{E} ! R(PQ)'a$ (2)

$\vdash . (1) . (2) . \supset \vdash . \text{Prop}$

***234·561.** $\vdash : P, Q \epsilon \text{Ser} . a \epsilon \text{ct}(PQ)'R . a = \text{lt}_Q'(\alpha \cap \text{Ɑ}'R) . R``\overrightarrow{Q}'a \subset C'P . \supset .$

$$(P\bar{R}Q)_{\text{lmx}}'\alpha = R'a = (\breve{P}\bar{R}Q)_{\text{lmx}}'\alpha \quad [*233{\cdot}515 . *234{\cdot}56]$$

***234·562.** $\vdash : P, Q \epsilon \text{Ser} . \text{lt}_Q'(\alpha \cap \text{Ɑ}'R) \epsilon \text{ct}(PQ)'R . R``(\alpha \cap C'Q) \subset C'P . \supset .$

$$(P\bar{R}Q)_{\text{lmx}}'\alpha = (\breve{P}\bar{R}Q)_{\text{lmx}}'\alpha = R'\text{lt}_Q'\alpha \quad [*233{\cdot}516 . *234{\cdot}56]$$

That is, if α is any class of arguments having a limit at which the function is continuous, then the limit of the function, as the argument approaches the limit of the set of arguments, is the value of the function for that limit.

***234·6.** $\vdash : a \,\epsilon\, \text{contin}\,(PQ)\text{‘}R \,.\equiv.\, a \,\epsilon\, \text{ct}\,(PQ)\text{‘}R \cap \text{ct}\,(P\breve{Q})\text{‘}R$ [(*234·04)]

***234·61.** $\vdash :: P_{\text{po}} \,\epsilon\, \text{Ser} \,.\, Q \,\epsilon\, \text{trans} \,.\, Q_* \upharpoonright \overleftrightarrow{Q}\text{‘}a \,\epsilon\, \text{connex} \,.\, R\text{‘}a \,\epsilon\, \text{D‘}P \cap \text{Ɑ‘}P \,.\supset:.$

$$a \,\epsilon\, \text{contin}\,(PQ)\text{‘}R \,.\equiv:\, R\text{‘}a \sim\epsilon\, C\text{‘}P_1 : R\text{‘}a \,\epsilon\, P(z-w) \,.\supset_{z,w}.$$

$$(\exists y, y') \,.\, a \,\epsilon\, Q(y-y') \,.\, y, y' \,\epsilon\, \text{Ɑ‘}R \,.\, R\text{‘‘}Q(y \mapsto y') \subset P(z-w)$$

Dem.

$$\vdash . *234\cdot51 \,.\supset \vdash :: \text{Hp} \,.\supset:.\, a \,\epsilon\, \text{contin}\,(PQ)\text{‘}R \,.\equiv:$$

$$R\text{‘}a \,\epsilon\, \text{D‘}P \cap \text{Ɑ‘}P - C\text{‘}P_1 : R\text{‘}a \,\epsilon\, P(z-w) \,.\supset_{z,w}.$$

$$(\exists y, y') \,.\, y \,\epsilon\, \overrightarrow{Q}\text{‘}a \cap \text{Ɑ‘}R \,.\, y' \,\epsilon\, \overleftarrow{Q}\text{‘}a \cap \text{Ɑ‘}R \,.$$

$$R\text{‘‘}Q(y \mapsto a) \cup R\text{‘‘}Q(a \mapsto y') \subset P(z-w) \quad (1)$$

$$\vdash . (1) . *201\cdot19 . *202\cdot17 \,.\supset \vdash . \text{Prop}$$

***234·62.** $\vdash :. \text{Hp}\,*234\cdot61 \,.\, P \,\epsilon\, \text{trans} \,.\, R\text{‘‘}\overleftrightarrow{Q}\text{‘}a \subset C\text{‘}P \,.\supset: a \,\epsilon\, \text{contin}\,(PQ)\text{‘}R \,.\supset.$

$$R\,(PQ)\text{‘}a = R\,(\breve{P}Q)\text{‘}a = R\,(P\breve{Q})\text{‘}a = R\,(\breve{P}\breve{Q})\text{‘}a = R\text{‘}a$$

[*234·52·6]

***234·63.** $\vdash :. \text{Hp}\,*234\cdot62 \,.\, P^2 = P \,.\supset: a \,\epsilon\, \text{contin}\,(PQ)\text{‘}R \,.\equiv.$

$$R\,(PQ)\text{‘}a = R\,(\breve{P}Q)\text{‘}a = R\,(P\breve{Q})\text{‘}a = R\,(\breve{P}\breve{Q})\text{‘}a = R\text{‘}a$$

[*234·522·6]

***234·64.** $\vdash :: \text{Hp}\,*234\cdot62 \,.\, R\text{‘}a \,\epsilon\, \text{D‘}P \cap \text{Ɑ‘}P \,.\supset:.\, a \,\epsilon\, \text{contin}\,(PQ)\text{‘}R \,.\equiv:$

$$R\text{‘}a \,\epsilon\, C\text{‘}P - C\text{‘}P_1 : R\text{‘}a \,\epsilon\, P(z-w) \,.\supset_{z,w}.$$

$$(\exists y, y') \,.\, y, y' \,\epsilon\, \text{Ɑ‘}R \,.\, a \,\epsilon\, Q(y-y') \,.\, R\text{‘‘}Q(y \mapsto y') \subset P(z-w)$$

[*234·51·6]

***234·7.** $\vdash : R \,\epsilon\, P \,\overline{\text{contin}}\, Q \,.\equiv.\, \exists\,!\, C\text{‘}Q \cap \text{Ɑ‘}R \,.\, C\text{‘}Q \cap \text{Ɑ‘}R \subset \text{contin}\,(PQ)\text{‘}R$

[(*234·04)]

***234·71.** $\vdash : R \,\epsilon\, P \,\overline{\text{contin}}\, Q \,.\supset.\, R \upharpoonright C\text{‘}Q \,\epsilon\, 1 \to \text{Cls} \,.\, R\text{‘‘}C\text{‘}Q \subset C\text{‘}P$

Dem.

$$\vdash . *234\cdot7\cdot6\cdot5 \,.\supset$$

$$\vdash :. \text{Hp} \,.\supset: a \,\epsilon\, C\text{‘}Q \cap \text{Ɑ‘}R \,.\supset.\, R\text{‘}a \,\epsilon\, \text{os}\,(P, Q_* \upharpoonright \overrightarrow{Q}_{\text{po}}\text{‘}a)\text{‘}R \,.$$

$$[*234\cdot1] \qquad \supset . R\text{‘}a \,\epsilon\, C\text{‘}P \,. \quad (1)$$

$$[*14\cdot21] \qquad \supset . E\,!\, R\text{‘}a \quad (2)$$

$$\vdash . (2) . *71\cdot572 \,. \quad \supset \vdash : \text{Hp} \,.\supset.\, R \upharpoonright C\text{‘}Q \,\epsilon\, 1 \to \text{Cls} \quad (3)$$

$$\vdash . (1) . (2) . *37\cdot61 \,.\supset \vdash : \text{Hp} \,.\supset.\, R\text{‘‘}(C\text{‘}Q \cap \text{Ɑ‘}R) \subset C\text{‘}P \quad (4)$$

$$\vdash . (3) . (4) . *37\cdot26 \,.\supset \vdash . \text{Prop}$$

***234·72.** $\vdash :. P \,\epsilon\, \text{Ser} \,.\, Q \,\epsilon\, \text{trans} \cap \text{connex} \,.\, R \,\epsilon\, P \,\overline{\text{contin}}\, Q \,.\supset:$

$$a \,\epsilon\, C\text{‘}Q \cap \text{Ɑ‘}R \,.\supset_a.\, R\,(PQ)\text{‘}a = R\,(\breve{P}Q)\text{‘}a = R\,(P\breve{Q})\text{‘}a = R\,(\breve{P}\breve{Q})\text{‘}a = R\text{‘}a$$

[*234·62·7]

***234·73.** $\vdash :: P \in \text{Ser} . P^2 = P . Q \in \text{trans} \cap \text{connex} . \supset :.$

$$R \in P \overline{\text{contin}} Q . \equiv : \exists ! C'Q \cap \text{ɑ}'R : a \in C'Q \cap \text{ɑ}'R . \supset_a . R(PQ)'a = R(\breve{P}Q)'a = R(P\breve{Q})'a = R(\breve{P}\breve{Q})'a = R'a$$

Dem.

$\vdash . *234·7·71 . \supset \vdash :: \text{Hp} . \supset :. R \in P \overline{\text{contin}} Q . \equiv : \exists ! C'Q \cap \text{ɑ}'R . R``C'Q \subset C'P :$

$a \in C'Q \cap \text{ɑ}'R . \supset_a . a \in \text{contin}(PQ)'R :$

$[*234·63] \equiv : \exists ! C'Q \cap \text{ɑ}'R . R``C'Q \subset C'P : a \in C'Q \cap \text{ɑ}'R . \supset_a .$

$$R(PQ)'a = R(\breve{P}Q)'a = R(P\breve{Q})'a = R(\breve{P}\breve{Q})'a = R'a \quad (1)$$

$\vdash . *233·401·101 . \supset$

$\vdash :. a \in C'Q \cap \text{ɑ}'R . \supset_a . R(PQ)'a = R'a : \supset : a \in C'Q \cap \text{ɑ}'R . \supset_a . R'a \in C'P :$

$[*37·61·26] \qquad \supset : R``C'Q \subset C'P \quad (2)$

$\vdash . (1) . (2) . \supset \vdash . \text{Prop}$

Lightning Source UK Ltd.
Milton Keynes UK
27 October 2010

161968UK00006B/102/P